T4-AQL-649

COMPREHENSIVE ORGANIC SYNTHESIS

IN 9 VOLUMES

COMPREHENSIVE ORGANIC SYNTHESIS

Selectivity, Strategy & Efficiency in Modern Organic Chemistry

Editor-in-Chief
BARRY M. TROST
Stanford University, CA, USA

Deputy Editor-in-Chief
IAN FLEMING
University of Cambridge, UK

Volume 1

ADDITIONS TO C—X π-BONDS, PART 1

Volume Editor
STUART L. SCHREIBER
Harvard University, Cambridge, MA, USA

PERGAMON PRESS
OXFORD • NEW YORK • SEOUL • TOKYO

ELSEVIER SCIENCE Ltd
The Boulevard, Langford Lane
Kidlington, Oxford OX5 1GB, UK

First edition 1991
Second impression 1993
Third impression 1999

Library of Congress Cataloging in Publication Data

Comprehensive organic synthesis: selectivity, strategy and efficiency in modern organic chemistry/editor[s] Barry M. Trost, Ian Fleming.
p. cm.
Includes indexes.
Contents: Vol. 1.–2. Additions to C-X[pi]-Bonds — v. 3. Carbon–carbon sigma-Bond formation — v. 4. Additions to and substitutions at C-C[pi]-Bonds — v. 5. Combining C-C[pi]-Bonds — v. 6. Heteroatom manipulation — v. 7. Oxidation — v. 8. Reduction — v. 9. Cumulative indexes.
1. Organic Compounds — Synthesis I. Trost, Barry M. 1941–
II. Fleming, Ian, 1935–
QD262.C535 1991
547.2—dc20 90-26621

British Library Cataloguing in Publication Data

Comprehensive organic synthesis
1. Organic compounds. Synthesis
I. Trost, Barry M. (Barry Martin) 1941–
547.2

ISBN 0-08-040592-4 (Vol. 1)
ISBN 0-08-035929-9 (set)

∞ ™ The paper used in this publication meets the minimum requirements of American National Standard for Information Sciences — Permanence of Paper for Printed Library Materials, ANSI Z39.48-1984.

Contents

Preface

The emergence of organic chemistry as a scientific discipline heralded a new era in human development. Applications of organic chemistry contributed significantly to satisfying the basic needs for food, clothing and shelter. While expanding our ability to cope with our basic needs remained an important goal, we could, for the first time, worry about the quality of life. Indeed, there appears to be an excellent correlation between investment in research and applications of organic chemistry and the standard of living. Such advances arise from the creation of compounds and materials. Continuation of these contributions requires a vigorous effort in research and development, for which information such as that provided by the *Comprehensive* series of Pergamon Press is a valuable resource.

Since the publication in 1979 of *Comprehensive Organic Chemistry*, it has become an important first source of information. However, considering the pace of advancements and the ever-shrinking timeframe in which initial discoveries are rapidly assimilated into the basic fabric of the science, it is clear that a new treatment is needed. It was tempting simply to update a series that had been so successful. However, this new series took a totally different approach. In deciding to embark upon *Comprehensive Organic Synthesis*, the Editors and Publisher recognized that synthesis stands at the heart of organic chemistry.

The construction of molecules and molecular systems transcends many fields of science. Needs in electronics, agriculture, medicine and textiles, to name but a few, provide a powerful driving force for more effective ways to make known materials and for routes to new materials. Physical and theoretical studies, extrapolations from current knowledge, and serendipity all help to identify the direction in which research should be moving. All of these forces help the synthetic chemist in translating vague notions to specific structures, in executing complex multistep sequences, and in seeking new knowledge to develop new reactions and reagents. The increasing degree of sophistication of the types of problems that need to be addressed require increasingly complex molecular architecture to target better the function of the resulting substances. The ability to make such substances available depends upon the sharpening of our sculptors' tools: the reactions and reagents of synthesis.

The Volume Editors have spent great time and effort in considering the format of the work. The intention is to focus on transformations in the way that synthetic chemists think about their problems. In terms of organic molecules, the work divides into the formation of carbon–carbon bonds, the introduction of heteroatoms, and heteroatom interconversions. Thus, Volumes 1–5 focus mainly on carbon–carbon bond formation, but also include many aspects of the introduction of heteroatoms. Volumes 6–8 focus on interconversion of heteroatoms, but also deal with exchange of carbon–carbon bonds for carbon–heteroatom bonds.

The Editors recognize that the assignment of subjects to any particular volume may be arbitrary in part. For example, reactions of enolates can be considered to be additions to C—C π-bonds. However, the vastness of the field leads it to be subdivided into components based upon the nature of the bond-forming process. Some subjects will undoubtedly appear in more than one place.

In attacking a synthetic target, the critical question about the suitability of any method involves selectivity: chemo-, regio-, diastereo- and enantio-selectivity. Both from an educational point-of-view for the reader who wants to learn about a new field, and an experimental viewpoint for the practitioner who seeks a reference source for practical information, an organization of the chapters along the theme of selectivity becomes most informative.

The Editors believe this organization will help emphasize the common threads that underlie many seemingly disparate areas of organic chemistry. The relationships among various transformations becomes clearer and the applicability of transformations across a large number of compound classes becomes apparent. Thus, it is intended that an integration of many specialized areas such as terpenoid, heterocyclic, carbohydrate, nucleic acid chemistry, *etc.* within the more general transformation class will provide an impetus to the consideration of methods to solve problems outside the traditional ones for any specialist.

In general, presentation of topics concentrates on work of the last decade. Reference to earlier work, as necessary and relevant, is made by citing key reviews. All topics in organic synthesis cannot be treated with equal depth within the constraints of any single series. Decisions as to which aspects of a topic require greater depth are guided by the topics covered in other recent *Comprehensive* series. This new treatise focuses on being comprehensive in the context of synthetically useful concepts.

The Editors and Publisher believe that *Comprehensive Organic Synthesis* will serve all those who must face the problem of preparing organic compounds. We intend it to be an essential reference work for the experienced practitioner who seeks information to solve a particular problem. At the same time, we must also serve the chemist whose major interest lies outside organic synthesis and therefore is only an occasional practitioner. In addition, the series has an educational role. We hope to instruct experienced investigators who want to learn the essential facts and concepts of an area new to them. We also hope to teach the novice student by providing an authoritative account of an area and by conveying the excitement of the field.

The need for this series was evident from the enthusiastic response from the scientific community in the most meaningful way — their willingness to devote their time to the task. I am deeply indebted to an exceptional board of editors, beginning with my deputy editor-in-chief Ian Fleming, and extending to the entire board — Clayton H. Heathcock, Ryoji Noyori, Steven V. Ley, Leo A. Paquette, Gerald Pattenden, Martin F. Semmelhack, Stuart L. Schreiber and Ekkehard Winterfeldt.

The substance of the work was created by over 250 authors from 15 countries, illustrating the truly international nature of the effort. I thank each and every one for the magnificent effort put forth. Finally, such a work is impossible without a publisher. The continuing commitment of Pergamon Press to serve the scientific community by providing this *Comprehensive* series is commendable. Specific credit goes to Colin Drayton for the critical role he played in allowing us to realize this work and also to Helen McPherson for guiding it through the publishing maze.

A work of this kind, which obviously summarizes accomplishments, may engender in some the feeling that there is little more to achieve. Quite the opposite is the case. In looking back and seeing how far we have come, it becomes only more obvious how very much more we have yet to achieve. The vastness of the problems and opportunities ensures that research in organic synthesis will be vibrant for a very long time to come.

BARRY M. TROST
Palo Alto, California

Contributors to Volume 1

Professor J. Aubé
Department of Medicinal Chemistry, University of Kansas, Lawrence, KS 66045-2506, USA

Dr R. Caputo
Dipartimento di Chimica Organica e Biologica, Università di Napoli, Via Mezzocannone 16, I-80134 Napoli, Italy

Dr C. Ferreri
Dipartimento di Chimica Organica e Biologica, Università di Napoli, Via Mezzocannone 16, I-80134 Napoli, Italy

Professor R. E. Gawley
Department of Chemistry, University of Miami, PO Box 249118, Coral Gables, FL 33124, USA

Professor A. Hassner
Department of Chemistry, Bar-Ilan University, Ramat-Gan 59100, Israel

Dr J. R. Hauske
Pfizer Central Research, Eastern Point Road, Groton, CT 06340, USA

Dr D. M. Huryn
Building 76, Hoffmann-La Roche Inc, 340 Kingsland Street, Nutley, NJ 07110-1199, USA

Professor T. Imamoto
Department of Chemistry, Faculty of Science, Chiba University, Yayoi-cho, Chiba 260, Japan

Dr S. E. Kelly
Pfizer Central Research, Eastern Point Road, Groton, CT 06340, USA

Professor P. Knochel
Department of Chemistry, University of Michigan, Ann Arbor, MI 48109-1055, USA

Professor A. Krief
Departement de Chemie, Facultés Universitaires Notre-Dame de la Paix, Rue de Bruxelles 61, B-5000 Namur, Belgium

Professor B. H. Lipshutz
Department of Chemistry, University of California, Santa Barbara, CA 93106, USA

Professor G. A. Molander
Department of Chemistry & Biochemistry, University of Colorado, Campus Box 215, Boulder, CO 80309-0215, USA

Professor K. Ogura
Department of Synthetic Chemistry, Chiba University, 1-33 Yayoi-cho, Chiba 260, Japan

Dr B. T. O'Neill
Pfizer Central Research, Eastern Point Road, Groton, CT 06340, USA

Dr G. Palumbo
Dipartimento di Chimica Organica e Biologica, Università di Napoli, Via Mezzocannone 16, I-80134 Napoli, Italy

Professor J. S. Panek
Department of Chemistry, Boston University, 590 Commonwealth Avenue, Boston, MA 02215, USA

Professor A. Pelter
Department of Chemistry, University College Swansea, Singleton Park, Swansea SA2 8PP, UK

Dr K. M. L. Rai
Department of Chemistry, University of Mysore, Manasa Gangotri, Mysore 570006, India

Mrs K. Rein
Department of Chemistry, University of Miami, PO Box 249118, Coral Gables, FL 33124, USA

Dr N. A. Saccomano
Pfizer Central Research, Eastern Point Road, Groton, CT 06340, USA

Professor S. L. Schreiber
Department of Chemistry, Harvard University, 12 Oxford Street, Cambridge, MA 02138, USA

Mr S. Shambayati
Department of Chemistry, Harvard University, 12 Oxford Street, Cambridge, MA 02138, USA

Professor K. Smith
Department of Chemistry, University College Swansea, Singleton Park, Swansea SA2 8PP, UK

Dr R. A. Volkmann
Pfizer Central Research, Eastern Point Road, Groton, CT 06340, USA

Professor P. G. Williard
Department of Chemistry, Brown University, Providence, RI 02912, USA

Dr P. M. Wovkulich
Building 76, Hoffmann-La Roche Inc, 340 Kingsland Street, Nutley, NJ 07110-1199, USA

Professor M. Yamaguchi
Department of Chemistry, Faculty of Science, Tohuku University, Aoba, Sendai 980, Japan

Abbreviations

The following abbreviations have been used where relevant. All other abbreviations have been defined the first time they occur in a chapter.

Techniques

CD	circular dichroism
CIDNP	chemically induced dynamic nuclear polarization
CNDO	complete neglect of differential overlap
CT	charge transfer
GLC	gas–liquid chromatography
HOMO	highest occupied molecular orbital
HPLC	high-performance liquid chromatography
ICR	ion cyclotron resonance
INDO	incomplete neglect of differential overlap
IR	infrared
LCAO	linear combination of atomic orbitals
LUMO	lowest unoccupied molecular orbital
MS	mass spectrometry
NMR	nuclear magnetic resonance
ORD	optical rotatory dispersion
PE	photoelectron
SCF	self-consistent field
TLC	thin layer chromatography
UV	ultraviolet

***Reagents, solvents,* etc.**

Ac	acetyl
acac	acetylacetonate
AIBN	2,2′-azobisisobutyronitrile
Ar	aryl
ATP	adenosine triphosphate
9-BBN	9-borabicyclo[3.3.1]nonyl
9-BBN-H	9-borabicyclo[3.3.1]nonane
BHT	2,6-di-*t*-butyl-4-methylphenol (butylated hydroxytoluene)
bipy	2,2′-bipyridyl
Bn	benzyl
t-BOC	*t*-butoxycarbonyl
BSA	*N,O*-bis(trimethylsilyl)acetamide
BSTFA	*N,O*-bis(trimethylsilyl)trifluoroacetamide
BTAF	benzyltrimethylammonium fluoride
Bz	benzoyl
CAN	ceric ammonium nitrate
COD	1,5-cyclooctadiene
COT	cyclooctatetraene
Cp	cyclopentadienyl
Cp*	pentamethylcyclopentadienyl
18-crown-6	1,4,7,10,13,16-hexaoxacyclooctadecane
CSA	camphorsulfonic acid
CSI	chlorosulfonyl isocyanate
DABCO	1,4-diazabicyclo[2.2.2]octane
DBA	dibenzylideneacetone
DBN	1,5-diazabicyclo[4.3.0]non-5-ene
DBU	1,8-diazabicyclo[5.4.0]undec-7-ene

DCC	dicyclohexylcarbodiimide
DDQ	2,3-dichloro-5,6-dicyano-1,4-benzoquinone
DEAC	diethylaluminum chloride
DEAD	diethyl azodicarboxylate
DET	diethyl tartrate (+ or –)
DHP	dihydropyran
DIBAL-H	diisobutylaluminum hydride
diglyme	diethylene glycol dimethyl ether
dimsyl Na	sodium methylsulfinylmethide
DIOP	2,3-*O*-isopropylidene-2,3-dihydroxy-1,4-bis(diphenylphosphino)butane
DIPT	diisopropyl tartrate (+ or –)
DMA	dimethylacetamide
DMAC	dimethylaluminum chloride
DMAD	dimethyl acetylenedicarboxylate
DMAP	4-dimethylaminopyridine
DME	dimethoxyethane
DMF	dimethylformamide
DMI	*N,N'*-dimethylimidazolone
DMSO	dimethyl sulfoxide
DMTSF	dimethyl(methylthio)sulfonium fluoroborate
DPPB	1,4-bis(diphenylphosphino)butane
DPPE	1,2-bis(diphenylphosphino)ethane
DPPF	1,1'-bis(diphenylphosphino)ferrocene
DPPP	1,3-bis(diphenylphosphino)propane
E^+	electrophile
EADC	ethylaluminum dichloride
EDG	electron-donating group
EDTA	ethylenediaminetetraacetic acid
EEDQ	*N*-ethoxycarbonyl-2-ethoxy-1,2-dihydroquinoline
EWG	electron-withdrawing group
HMPA	hexamethylphosphoric triamide
HOBT	hydroxybenzotriazole
$IpcBH_2$	isopinocampheylborane
Ipc_2BH	diisopinocampheylborane
KAPA	potassium 3-aminopropylamide
K-selectride	potassium tri-*s*-butylborohydride
LAH	lithium aluminum hydride
LDA	lithium diisopropylamide
LICA	lithium isopropylcyclohexylamide
LITMP	lithium tetramethylpiperidide
L-selectride	lithium tri-*s*-butylborohydride
LTA	lead tetraacetate
MCPBA	*m*-chloroperbenzoic acid
MEM	methoxyethoxymethyl
MEM-Cl	β-methoxyethoxymethyl chloride
MMA	methyl methacrylate
MMC	methylmagnesium carbonate
MOM	methoxymethyl
Ms	methanesulfonyl
MSA	methanesulfonic acid
MsCl	methanesulfonyl chloride
MVK	methyl vinyl ketone
NBS	*N*-bromosuccinimide
NCS	*N*-chlorosuccinimide

NMO	*N*-methylmorpholine *N*-oxide
NMP	*N*-methyl-2-pyrrolidone
Nu^-	nucleophile
PPA	polyphosphoric acid
PCC	pyridinium chlorochromate
PDC	pyridinium dichromate
phen	1,10-phenanthroline
Phth	phthaloyl
PPE	polyphosphate ester
PPTS	pyridinium *p*-toluenesulfonate
Red-Al	sodium bis(methoxyethoxy)aluminum dihydride
SEM	β-trimethylsilylethoxymethyl
Sia_2BH	disiamylborane
TAS	tris(diethylamino)sulfonium
TBAF	tetra-*n*-butylammonium fluoride
TBDMS	*t*-butyldimethylsilyl
TBDMS-Cl	*t*-butyldimethylsilyl chloride
TBHP	*t*-butyl hydroperoxide
TCE	2,2,2-trichloroethanol
TCNE	tetracyanoethylene
TES	triethylsilyl
Tf	triflyl (trifluoromethanesulfonyl)
TFA	trifluoroacetic acid
TFAA	trifluoroacetic anhydride
THF	tetrahydrofuran
THP	tetrahydropyranyl
TIPBS-Cl	2,4,6-triisopropylbenzenesulfonyl chloride
TIPS-Cl	1,3-dichloro-1,1,3,3-tetraisopropyldisiloxane
TMEDA	tetramethylethylenediamine [1,2-bis(dimethylamino)ethane]
TMS	trimethylsilyl
TMS-Cl	trimethylsilyl chloride
TMS-CN	trimethylsilyl cyanide
Tol	tolyl
TosMIC	tosylmethyl isocyanide
TPP	*meso*-tetraphenylporphyrin
Tr	trityl (triphenylmethyl)
Ts	tosyl (*p*-toluenesulfonyl)
TTFA	thallium trifluoroacetate
TTN	thallium(III) nitrate

Contents of All Volumes

Volume 5 Combining C—C π-Bonds

Volume 6 Heteroatom Manipulation

Displacement by Substitution Processes

Acylation-type Reactions

Protecting Groups

Functional Group Interconversion

Elimination Reactions

Volume 7 Oxidation

Oxidation of Unactivated C—H Bonds

Volume 9

1.1
Carbanions of Alkali and Alkaline Earth Cations: (i) Synthesis and Structural Characterization

PAUL G. WILLIARD
Brown University, Providence, RI, USA

1.1.1 INTRODUCTION

In this chapter the focus is primarily on the recent structural work concerning carbanions of alkali and alkaline earth cations that are widely utilized in synthetic organic chemistry. In this context the year 1981 is significant because the first detailed X-ray diffraction analyses of two lithium enolates of simple ketones, *i.e.* 3,3-dimethyl-2-butanone and cyclopentanone, were published.[1] Since 1981 a number of detailed X-ray diffraction analyses of synthetically useful enolate anions of alkali and alkaline earth cations have been described. Within this chapter, many recent structural characterizations will be examined with the overall goal of collating this new information especially as it pertains to increasing our knowledge and control over the reactivity of these most useful and important synthetic reagents. The chapter is organized by functional group because this classification is quite natural to synthetic chemists. The examples chosen have come to my attention while thinking about the role of these species in synthetic reactions. It is neither practical nor feasible to include in this chapter an exhaustive review of all structural characterizations of carbanions of alkali and alkaline earth cations.[2] Should the complete list of all such structures be required, a comprehensive search of the Cambridge Structural Database (CSD) is recommended.[3] Throughout this chapter structural references are given to six letter CSD reference codes as follows, ⟨XXXXXX⟩. These refcodes will assist in obtaining crystallographic coordinates directly from the CSD.

At the outset it is especially useful to tabulate previous review articles containing a significant body of structural information about carbanions of alkali and alkaline earth cations, since these articles supplement the work reviewed herein. The first of these articles is an excellent review entitled 'Structure and Reactivity of Alkali Metal Enolates' by Jackman and Lange published in 1977.[4] It is significant that the fundamental details of the structure and the aggregation state of alkali metal ketone enolate anions in solution were outlined by Jackman mainly from NMR experiments and that this work predates the X-ray diffraction analyses. An earlier book by Schlosser entitled 'Struktur und Reaktivität polarer Organometalle' describes alkali and alkaline earth aggregates and their reactivities.[5] Some additional relevant structural information is reviewed in previous titles in this series, *i.e.* by Wakefield in Vol. 3 of 'Comprehensive Organic Chemistry'[6] and by the same author in Vol. 7 of 'Comprehensive Organometallic Chemistry'[7] and by O'Neill, Wade, Wardell, Bell and Lindsell in Vol. 1 of 'Comprehensive Organometallic Chemistry'.[8] A short review by Fraenkel *et al.* summarizes the solution structure and dynamic behavior of some aliphatic and alkynic lithium compounds by ^{13}C, ^{6}Li and ^{7}Li NMR studies.[9] Additional comprehensive reviews regarding NMR spectroscopy of organometallic compounds contain information related to this topic.[10] A thorough listing and classification of the X-ray structural analyses of organo lithium, sodium, potassium, rubidium and cesium compounds sifted from the Cambridge Structural Database has been prepared by Schleyer and coworkers and covers published work until the latter 1980s.[11] Finally there are a few recent specialized reviews by Seebach, entitled 'Structure and Reactivity of Lithium Enolates. From Pinacolone to Selective *C*-Alkylations of Peptides. Difficulties and Opportunities Afforded by Complex Structures',[12] by Power, entitled 'Free Inorganic, Organic, and Organometallic Ions by Treatment of Their Lithium Salts with 12-Crown-4',[13] and by Boche, entitled 'Structure of Lithium Compounds of Sulfones, Sulfoximides, Sulfoxides, Thio Ethers and 1,3-Dithianes, Nitriles, Nitro Compounds and Hydrazones',[14] that mainly summarize the author's own recent contributions to the area. The reviews by Seebach and Boche are especially relevant to synthetic organic chemists and are highly recommended. Several additional articles may justifiably be included in this list; however, the reader is referred to the aforementioned publications, especially the Seebach, Boche and Schleyer reviews, for an exhaustive bibliography, since it will be unnecessary to repeat their bibliographic compilations.

Alkali and alkaline earth metal cations are associated with numerous carbanions in reactions found in nearly every contemporary total synthesis. The basis for our current mechanistic interpretation of the role of the these metal cations in synthetic reactions has been derived largely from correlating the stereochemistry of reaction products with the starting materials. These stereochemical correlations utilize as a foundation the conformational analysis of carbocyclic rings.[15] One simply notes how often chair-like or boat-like intermediates/transition states are employed to rationalize the stereochemical outcome of synthetic reactions incorporating alkali metal cations to verify the veracity of the previous statement. In almost all mechanistic pictures, one also notes that the metal cation occupies a prominent role in the purported intermediate and/or transition state. However, it has become increasingly clear that we still possess only an incomplete understanding of the aggregation state and of the structural features of many of the alkali or alkaline earth metal coordinated carbanions in solution. Presently the following conclusions about organic reactions in which carbanions of alkali and alkaline earth cations are involved will be made: (i) these carbanions are utilized almost routinely in nearly every organic synthetic endeavor; (ii) there exists a poignant lack of detailed structural information about the reactive species themselves; (iii) the development of new reactive intermediates especially those designed to enhance and to control stereoselectivity continues to grow; and (iv) the basic ideas for the design of new reagents emanates almost exclusively from detailed, but as yet largely speculative, structural postulates about these reactive organometallic species based nearly exclusively upon carbocyclic conformational analysis.

Perhaps the increasing number of intermediates and/or transition states[16,17] that have been proposed to explain the stereochemical outcome of enolate reactions can serve as a barometer of our attempts to analyze the situation. Currently we have set an all time high for the number of new mechanistic interpretations of enolate reactions. It is my feeling that this will not turn out to be as simple as an open and closed case that the present models suggest. On the contrary, there exists increasing evidence[18] for the role of highly organized, oligomeric species which play crucial roles in enolate reactions; especially in those reactions that are fast and reversible (*i.e.* thermodynamically controlled), such as the aldol reaction.

A precocious explanation of the complex role of alkali metal enolates was presented in a manuscript published in 1971.[19] A paragraph from this paper is reproduced below. It represents the manuscript's authors' explanation for the counterintuitive observation that more highly substituted (*i.e.* more sterically hindered) enolate anions undergo alkylation reactions faster than less highly substituted (*i.e.* less sterically hindered) enolates.

'The fact that less highly substituted alkali metal enolates may sometimes react more slowly with alkyl halides than their analogs having additional α-substituents has been noted in several studies.[20] These observations initially seem curious since adding α-substituents would be expected to increase the steric interference to forming a new bond at the α-carbon atom. However, there is considerable evidence that many of the metal enolates (and related metal alkoxides) exist in ethereal solvents either as tightly associated ion pairs or as aggregates (dimers, trimers, tetramers) of these ion pairs;[21] structures such as (**1**)–(**4**) (M = metal; n = 1, 2, or 3; R = alkyl or the substituted vinyl portion of an enolate) have been suggested for such material with the smaller aggregates being favored as the steric bulk of the group R increases. Thus, the bromomagnesium enolate of isopropyl mesityl ketone is suggested to have structure (**1**) (M = MgBr), whereas the enolate of the analogous methyl ketone is believed to have structure (**2**) (M = MgBr).[22] The sodium enolates of several ketones are suggested to have the trimeric structures (**3**) in various ethereal solvents. Since the reactivities of metal enolates toward alkyl halides are very dependent on the degree of association and/or aggregation,[23] we suggest that the decreased reactivity observed for less highly substituted metal enolates both in this study and elsewhere may be attributable to a greater degree of aggregation of these enolates.' (Reproduced by permission of the American Chemical Society from *J. Org. Chem.*, 1971, **36**, 2361.)

ROM(solvent)$_n$

(**1**)

R–O bridged: (solvent)$_n$M(μ-OR)$_2$M(solvent)$_n$

(**2**)

[(solvent)$_n$M–OR]$_3$ cyclic trimer

(**3**)

(MOR)$_4$ cubic tetramer

(**4**)

The above quotation aptly rationalizes a number of experimental observations having to do with alkylation reactions of enolate anions. It suggests that reactivity and, by logical extension, the stereochemical selectivity, of enolate reactions are related to the aggregation of the enolates. To me, this statement represents a general but very daring explanation. This quotation is now over 20 years old; however, the significance of the conclusions reflected here is only now becoming more widely acknowledged.[24]

Exactly 10 years after the previous statement appeared, the first lithium enolate crystal structures were published as (**5**) and (**6**).[1] Thus, structural information derived from X-ray diffraction analysis proved the tetrameric, cubic geometry for the THF-solvated, lithium enolates derived from *t*-butyl methyl ketone (pinacolone) and from cyclopentanone.[25] Hence, the tetrameric aggregate characterized previously by NMR[26] as (**7**) was now defined unambiguously. Moreover, the general tetrameric aggregate (**7**) now became embellished in (**5**) and (**6**) by the inclusion of coordinating solvent molecules, *i.e.* THF. A representative quotation from this 1981 crystal structure analysis is given below.

'There is increasing evidence that lithium enolates, the most widely used class of d^2-reagents in organic synthesis, form solvated, cubic, tetrameric aggregates of type (**8**). For the solid state this type of

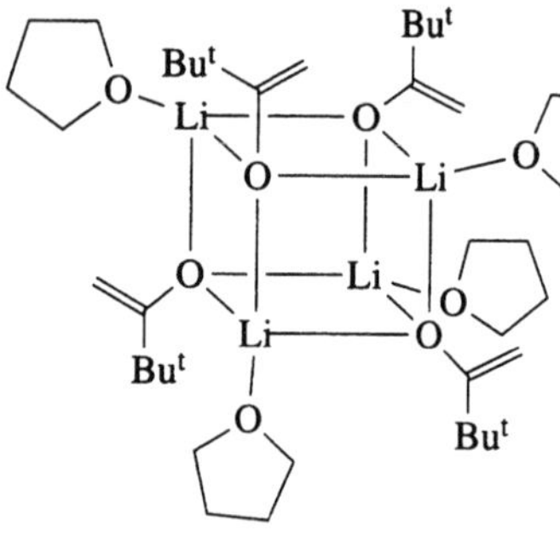

(5) 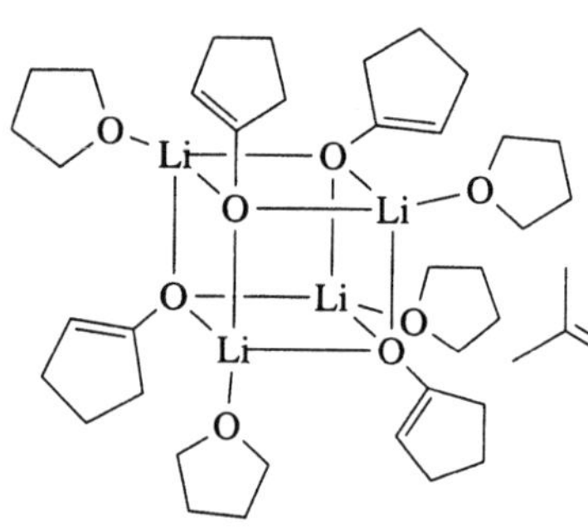 (6) 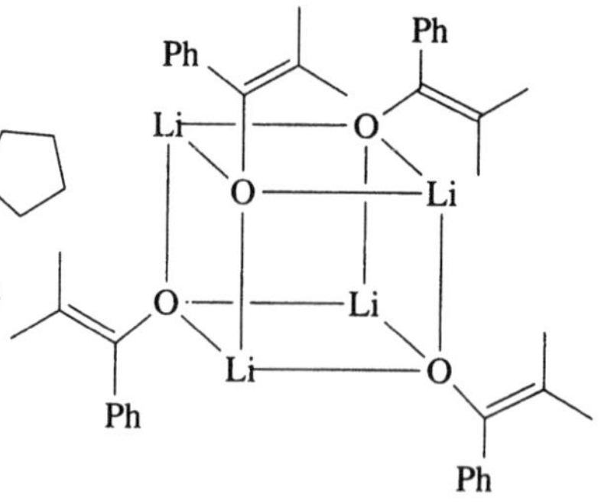 (7)

structure was definitely established for two crystalline lithium enolates and is strongly indicated for several others by their stoichiometry.... In aprotic solvents only aggregated species[27] are detected by NMR spectroscopy; even during reactions with electrophiles these aggregates are preserved and appear to be the actual reacting species, as indicated by reaction rates, which are first-order and not broken-order[28] in enolate concentration.' (Reproduced by permission of the Swiss Chemical Society from *Helv. Chim. Acta*, 1981, **64**, 2617.)

The authors of this quotation proceed to postulate the highly speculative but not unreasonable mechanism for the aldol reaction shown in Scheme 1. Justification for this mechanism appears to be based mainly upon characterization by X-ray diffraction analysis of the tetrameric cubic aggregates (**5**) and (**6**). Hence, X-ray diffraction analysis unambiguously provided the intimate structural details unobtainable by other methods.

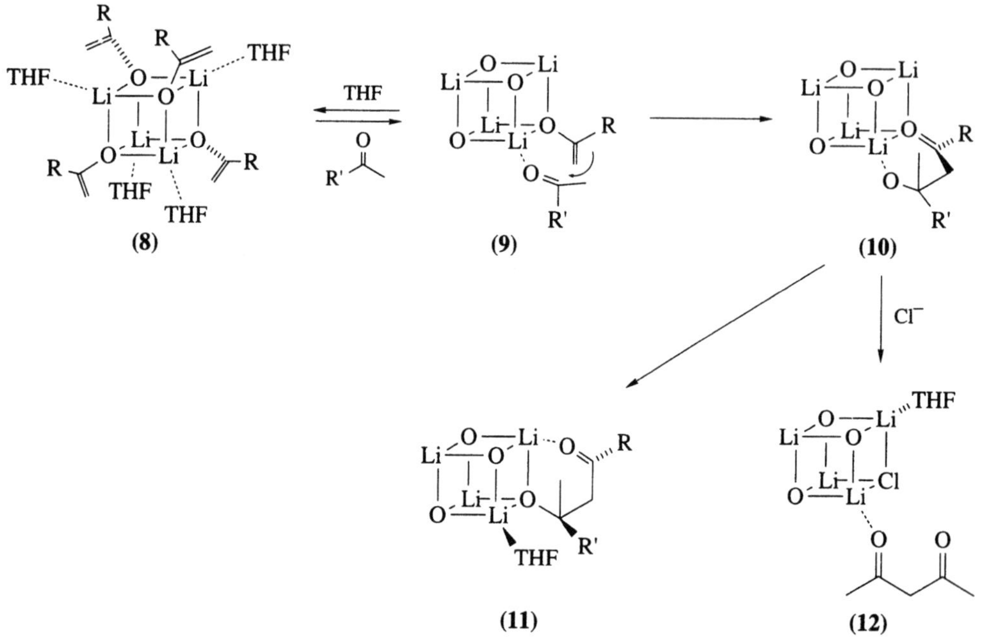

(8) (9) (10) (11) (12)

Scheme 1

Currently, the significance of the structural work in this area is aptly summarized by pointing out that it has been possible to obtain and to characterize the structure of aggregates corresponding to the intermediates (**8**) and (**9**) (M = Na), and (**11**) in the aldol reaction mechanism shown in Scheme 1.[29] At present we assume that ample evidence points to the existence of aggregated intermediates in several alkylation and aldol-like reactions.[30] Thus, the following sections of this chapter are classified roughly by functional group, and they contain structural results obtained by X-ray diffraction analysis. The examples were chosen with the thought of providing structural details about the reactive intermediates utilized in synthetic organic reactions, but it must be repeated that they do not represent a complete and comprehensive list of all such structures. As additional structural information is obtained, perhaps it will be possible to expand and to refine carbanion reaction mechanisms to include aggregated intermediates rather than simple monomers.

The results reviewed in the following pages of this chapter may provide fundamental information for the conduct, planning and strategy of organic synthesis. The origin of stereoselectivity in many organic reactions can be put on a more rational basis as more intimate structural details about the intermediates involved in these reactions are discovered. Of course, the long range goal and ultimate significance of this structural information is to provide a more thorough basis for accurate prediction and control of stereochemistry in organic reactions. Since enolate anions are universally utilized in all synthetic schemes, the successful obtention of additional structural results will have a great impact on the ability and the ease by which organic compounds will be prepared. I begin with a survey of known structural types.

1.1.2 STRUCTURAL FEATURES

1.1.2.1 Aggregation State

It is vital to recognize that metal cations impart a degree of order to the anions with which they are associated. Typically the first characteristic feature described is the stoichiometry. A simple chemical formula such as M^+A^- requires additional clarification to denote a higher degree of association such as $(M^+A^-)_x$ where the subscript x denotes the aggregation state of the species. The common descriptors of the aggregation state are monomer, dimer, trimer, tetramer, *etc*. Knowledge of the aggregation state is crucial since the reactivity of the anion is related to the aggregation state as well as to its structure.[31] The structures of the aggregates also depend critically upon the solvation of the cations. Fortunately, the majority of known structures can be built from a few simple structural patterns.

A motif found in the majority of alkali metal stabilized carbanion crystal structures is a nearly planar four-membered ring (**13**) with two metal atoms (M^+) and two anions (A^-), *i.e.* dimer. This simple pattern is rarely observed unadorned as in (**13**), yet almost every alkali metal and alkaline earth carbanion aggregate can be built up from this basic unit. The simplest possible embellishment to (**13**) is addition of two substituents (S) which produces a planar aggregate (**14**). Typically the substituents (S) in (**14**) are solvent molecules with heteroatoms that serve to donate a lone pair of electrons to the metal (M). Only slightly more complex than (**14**) is the four coordinate metal dimer (**15**). Often the substituents (S) in (**15**) are joined by a linear chain. The most common of these chains are tetramethylethylenediamine (TMEDA) or dimethoxyethane (DME) so that the spirocyclic structure (**16**) ensues. Alternatively the donors (S) in (**16**) have been observed as halide anions (X^-) when the metal (M^{2+}) is a divalent cation, *e.g.* (**17**) or (**18**). Obviously, the chelate rings found in (**16**) are entropically favorable relative to monodentate donors (S) in (**14**), (**15**), (**17**) or (**18**) (Scheme 2).

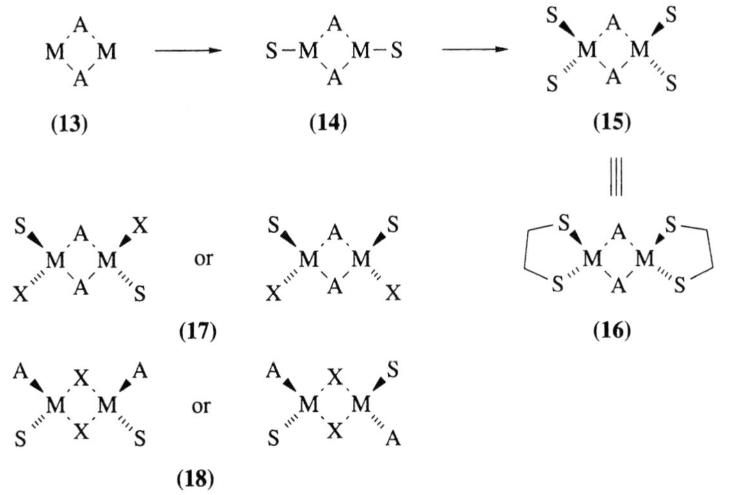

(**13**) (**14**) (**15**)

or

(**17**) (**16**)

or

(**18**)

Scheme 2

Several structural types are based on the combination of two units of (**13**). The edge-to-edge combination of (**13**) yields a 'ladder'-type structure (**19**). Of course there are various combinations of solvent donor ligands and/or chelate donors possible in (**20**) and (**21**). The face-to-face combination of (**13**) can produce a relatively cubic infrastructure (**22**) as previously seen in the enolates (**5**) and (**6**). Distorted variations of the cube (**22**) are observed, such as (**23**), or alternatively as another variation (**24**), with opposite square faces offset from one another (by nearly 90° in **24**). Such variations may be described as a 'tetrahedron within a tetrahedron'. It is noteworthy that the cube (**22**) can be derived from the ladder (**21**) simply by decreasing the appropriate internal bond angles to about 90° as indicated by the sequence of formulae (**13**) → (**19**) → (**21**) → (**22**). An advantage of the closed cubic structure (**22**) over the ladder (**19**) is the additional coordination of the terminal metal cations (M) to a third anion. The cube (**22**) is most frequently observed with four-coordinate metal cations, as in (**25**) and not in its unsolvated form (**22**) (Scheme 3).

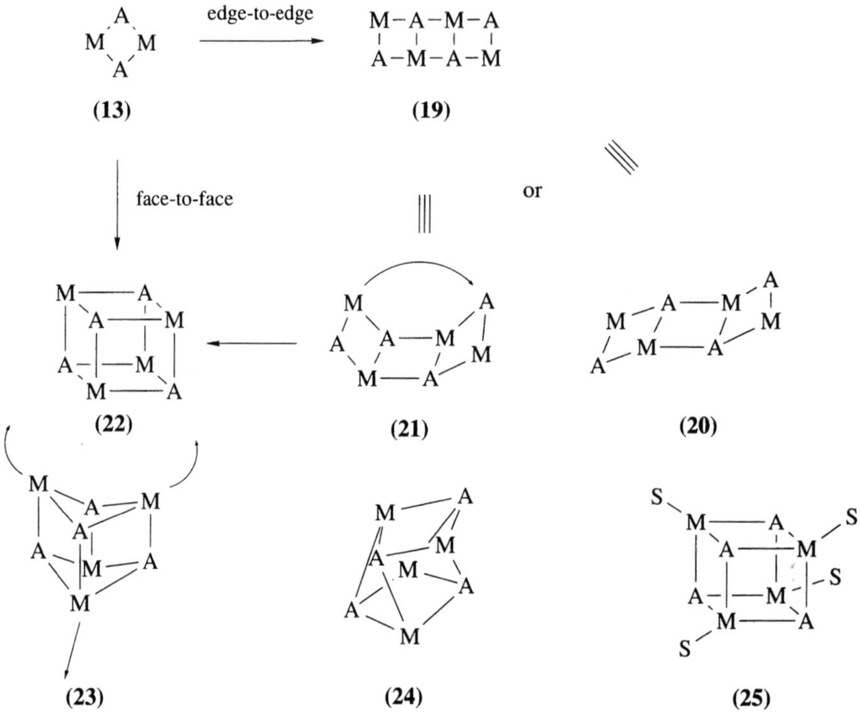

edge-to-edge

(**13**) (**19**)

face-to-face

or

(**22**) (**21**) (**20**)

(**23**) (**24**) (**25**)

Scheme 3

Edge-to-edge combination of additional units of (**13**) leads to the longer ladders (**26**) or (**27**). We have already obtained one unusual lithium enolate crystal structure corresponding to (**26**) but with additional external chelate rings. Closure of (**27**), analogous to the closure of (**21**) to (**22**), produces a hexagonal prism (**28**). Examples of structural type (**28**) are observed in addition to the solvent-coordinated hexamer (**29**). Distortion of (**29**), shown as (**30**), will lead to a somewhat less sterically hindered structure allowing for solvation of the metal cations by solvent (Scheme 4).

An alternative dissection of the hexagonal prism (**28**) is given as (**31**) (Scheme 4). Hence the hexamer (**28**) could be built up from two units of a planar trimer (**31**). This is plausible, because an example of a planar trimeric structure corresponding to (**31**) is known, *i.e.* the trimeric, unsolvated lithium hexamethyldisilazide structure.[32]

Additional structural types are known for alkali carbanions in the solid state. Examples of these are the monocyclic tetramer (**32**) or the pentacyclic tetramer (**33**), the hexamer (**34**), the dodecamer (**35**) and the infinite polymer (**36**). Undoubtedly several new structural types will be observed as mixed aggregates containing different metal cations (M^+ and M'^+) and/or different anions (A^- and A'^-) are characterized. Relatively long ladders, *i.e.* (**37**), corresponding to oligomeric chains of the dimer (**14**) combined edge-to-edge, are also likely to be characterized in the future. It is to be anticipated that the carbanions of limited solubility correspond to these extended ladders and that solubilization occurs by breaking these oligomers. A recent discussion of the propensity of lithiated amides to form either ladder structures or closed ring structures along with some *ab initio* calculations of these structural models has been presented by Snaith *et al.*[33]

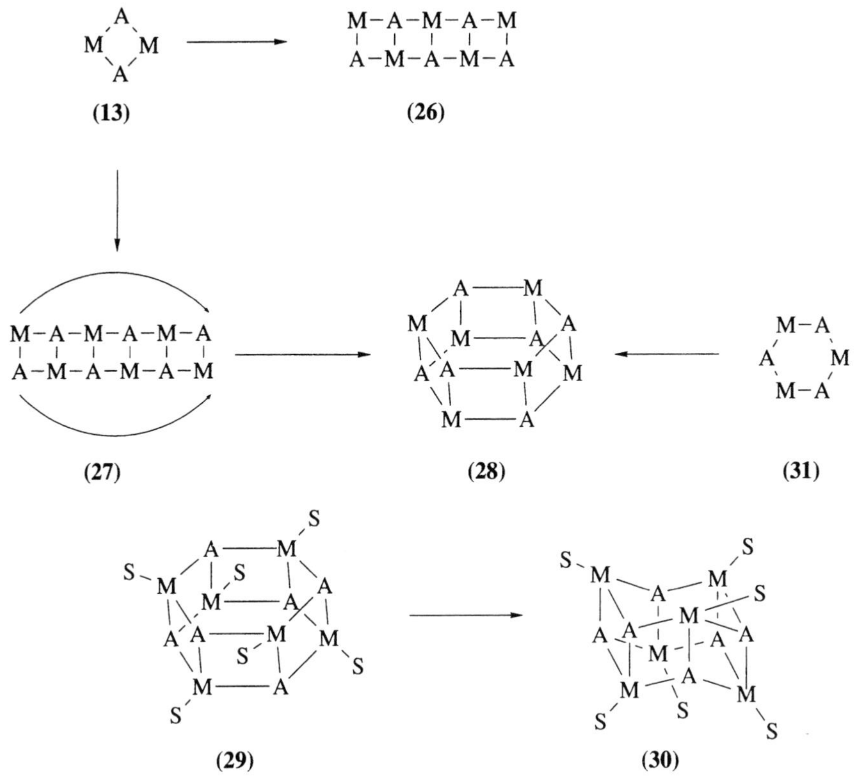

(13) (26)

(27) (28) (31)

(29) (30)

Scheme 4

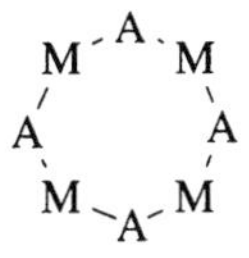

(32) 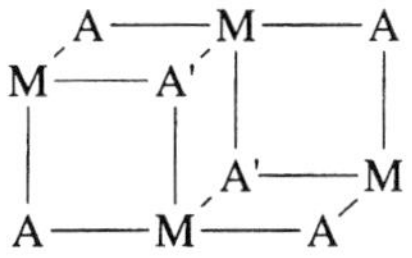 (33) 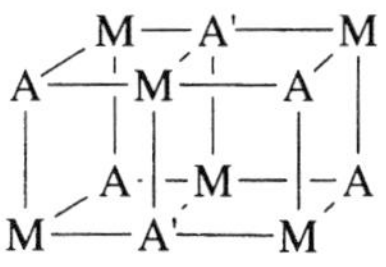 (34)

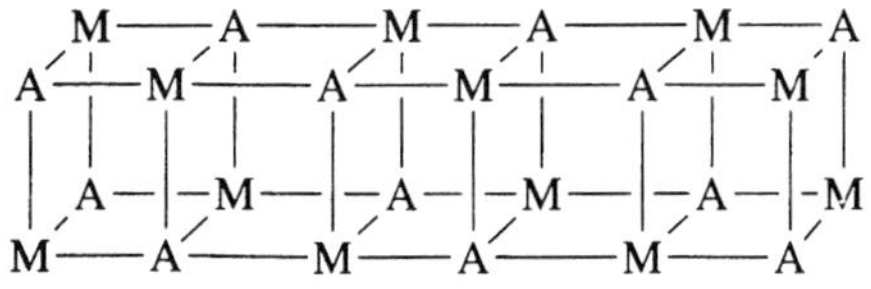

(35) 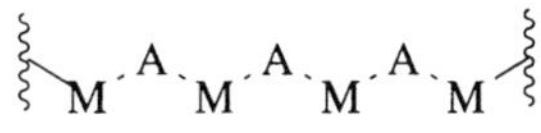 (36)

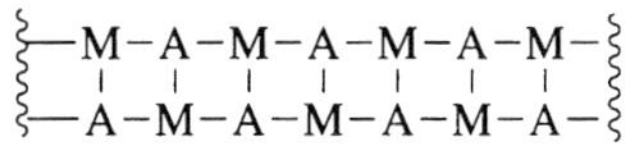

(37)

1.1.2.2 Coordination Geometry and Number

The directional preferences for coordination to the alkali metal and alkaline earth cations is obviously related to the number of substituents coordinated to the cation. As yet there is little predictability of the coordination number among these cations. For example, the first member of this series, the Li^+ cation, is the best characterized with well over 500 X-ray crystal structures containing this ion. Coordination numbers to Li^+ ranging from two through seven and all values in between can be found. The Li^+ cation is also found symmetrically π-complexed to the faces of aryl anions and to conjugated linear anions (see ref. 11). At present enough evidence exists to deduce only that the coordination number to the alkali metal and alkaline earth cations, and consequently the coordination geometry about these cations, is governed primarily by steric factors. Unfortunately the predictability of any individual unknown structure is relatively low.

In general the metal cation to substituent distances are found spanning a range of values. A working criterion for coordination to the metal cations is that the M—A distance not be greater than the sum of the van der Waals radii of M and A as listed by Pauling.[34] This criterion is particularly convenient when the anion is a typical heteroatom, such as O or N, or a halide, X. In such cases it is usually possible to derive accurate estimates of these distances from compiled sources.[35] However, the values of the M—A distance for cases where A is carbon and M is a Group Ia or IIa metal are not particularly well defined. Hence, Table 1 represents a recent search of the CSD for these values.[36]

Table 1 Carbanion–Metal Bond Lengths

Bond	*Mean*	*S.D.*	*Minimum* (Å)	*Maximum* (Å)	N_{obs}
C—Li	2.259	0.087	2.041	2.557	354
C—Na	2.646	0.060	2.566	2.756	12
C—K	No examples found				
C—Rb, C—Cs, C—Fr	No examples found				
C—Be	1.874	0.081	1.707	2.043	38
C—Mg	2.256	0.015	2.095	2.602	100
C—Ca, C—Sr, C—Ba, C—Ra	No examples found				

This table includes all examples listed in the CSD (version 4.20, 1990) located by a fragment search (*i.e.* CONNSER) for C—M bonds where M = group Ia or IIa metals irrespective of the hybridization of carbon.

Related structural aspects of metal ion coordination geometry are covered in some recent publications and are worthy of note. The directional preferences of ether oxygen atoms towards alkali and alkaline earth cations are reported by Chakrabarti and Dunitz.[37] The conclusion of this work is that the larger cations show an apparent preference to approach the ether oxygen along a tetrahedral lone pair direction, whereas Li^+ cations tend to be found along the C—O—C bisector, *i.e.* along the trigonal lone pair direction. Metal cation coordination to the *syn* and *anti* lone pair of electrons of the oxygen atoms in a carboxylate group have been reviewed by Glusker *et al.*[38] Scatter plots of M—O distances *versus* C—O—M angles for a wide variety of cation types led to the conclusion that both the coordination geometry and the distances of coordination to carboxylate lone pairs are largely governed by steric influences. Recently, the geometry of carboxyl oxygen complexation to several Lewis acids has been summarized by Schreiber *et al.*[39] Although only a few alkali metal Lewis acid–carbonyl structures are known, the general conclusion is that alkali metal cations do not show a strong directional preference for binding to carbonyls and that coordination numbers and coordination geometries vary greatly in these complexes.

1.1.3 CARBANION CRYSTAL STRUCTURES

With the general background of structural types described as above, it is now appropriate to review a number of examples of X-ray crystal structures of alkali metal and alkaline earth cations. The choice of examples is biased in favor of those species that are relevant to synthetic organic chemists. Hence, I begin with a comparison of structures of aliphatic carbanions. This group of aliphatic carbanion structures is the most widely varied and surely the least predictable. Of particular significance will be the aggregation state, the coordination number and the relative geometry about the metal cation. The figures drawn in the following sections are not computer generated plots of the actual X-ray crystal structures but are approximations of the actual structures. It is not practical to enumerate all of the specific details such as all bond lengths, bond angles and torsion angles in these structures and the reader is referred to the original publications for this specific information.

1.1.3.1 Aliphatic Carbanions

1.1.3.1.1 Unsubstituted aliphatic carbanions

Among the earliest aliphatic carbanions to be structurally characterized by X-ray diffraction analysis are the simple unsubstituted alkyllithium reagents, *i.e.* methyl-,[40] ethyl-[41] and cyclohexyl-lithium.[42] Methyl- and ethyl-lithium have also been examined in detail by quantum mechanical calculations and by electrostatic calculations.[43] The structures of methyl- and ethyl-lithium are similar. Both of these compounds crystallize as tetrameric aggregates from hydrocarbon solvents. These tetramers are generally depicted as (**38**). The aggregate (**38**) is described as a tetrahedral arrangement of lithium atoms with a single alkyl group located on each of the four faces of the Li tetrahedron. The carbanionic carbons are not necessarily equidistant from the three closest lithium atoms. However, it is clear that the three covalently bound substituents on the carbon atoms (H or alkyl) are found at the locations expected for an sp^3-hybridized atom. The carbon–lithium interactions have been referred to as two-electron four-center bonds in these structures. A low temperature crystal structure of ethyllithium[44] reveals some small changes relative to the room temperature structure, but the basic tetramer remains intact.

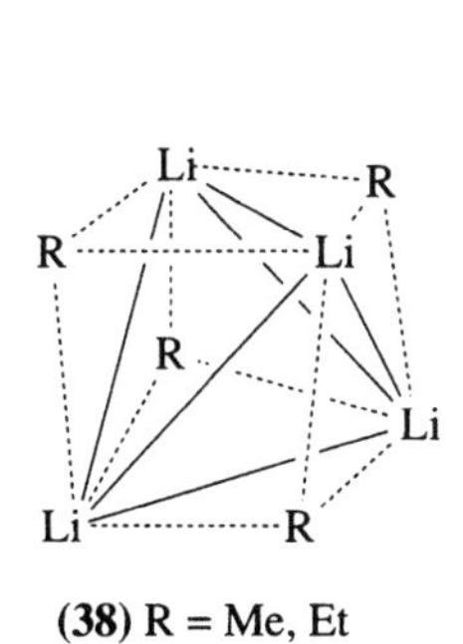

(**38**) R = Me, Et

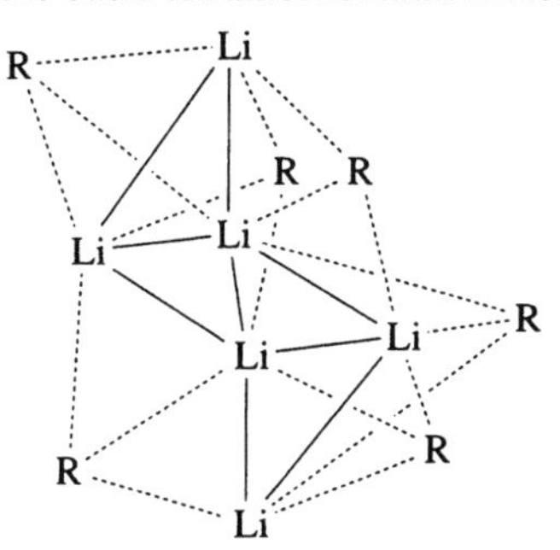

(**39**) R = cyclohexyl or
R = tetramethylcyclopropylmethyl

Cyclohexyllithium was prepared in hexane from cyclohexyl chloride and lithium sand and subsequently extracted and recrystallized from benzene solution to produce the hexameric aggregate (**39**).[42] The lithium atoms in this aggregate are nearly in an octahedral configuration, although the triangular faces of this octahedron have two short (~2.40 Å) and one long (~2.97 Å) Li—Li distance. A carbanionic carbon is found on six of the triangular faces and is most closely associated with the two lithium atoms which possess the longest Li—Li atom distance. The orientation of the cyclohexyl group is apparently determined by the interaction of α- and β-protons with the lithium atoms. The Li—C interactions in this hexamer are described as localized four-center bonds, as in the methyl- and ethyl-lithium tetramer (**38**). Two benzene molecules are occluded in the solid cyclohexyllithium hexamer, but these solvent molecules do not appear to interact with the hexamer.

Solvent-free (tetramethylcyclopropyl)methyllithium (**40**) also forms the hexameric aggregate (**39**), similar to hexameric cyclohexyllithium.[45] Both hexamers are characterized as a trigonal antiprism with triangular Li^+ faces. The (tetramethylcyclopropyl)methyllithium hexamer (**40**) was prepared in diethyl ether solution from both the Cl and the Hg compounds (**41**) as well as from the open chain compound (**42**). In contrast to the cyclohexyllithium hexamer, the hexamer (**40**) is obtained solvent free (Scheme 5).

Cl or 1/2 Hg

(**41**)

Li

Li

Cl

6

(**42**)

(**40**)

Scheme 5

An interesting cubic geometry is maintained in the mixed aggregate obtained from the reaction of cyclopropyl bromide (**43**) with lithium metal in diethyl ether solution.[46] The composition of the crystalline material is $(c\text{-}C_3H_5Li)_2{\cdot}(LiBr)_2{\cdot}4Et_2O$. The aggregate was characterized as structure (**44**). Note the similarity of (**44**) with the cubic tetramers (**38**), except for the substitution of two carbanion residues by two bromides in (**44**). Note also that each of the lithium atoms in (**44**) is coordinated to an oxygen of a diethyl ether molecule. This solvation serves to increase the coordination number of the lithiums, but does not break up the overall tetrameric nature of the aggregate. This solid loses ether at room temperature and is transformed into an ether-insoluble, tetrahydrofuran-soluble, amorphous product. The mass spectrum of the ether-free substance shows only halide-free aggregates. It is likely that differing reactivity of the salt-containing and salt-free lithium alkyls is related to the direct incorporation of lithium halide into the carbanion aggregates.

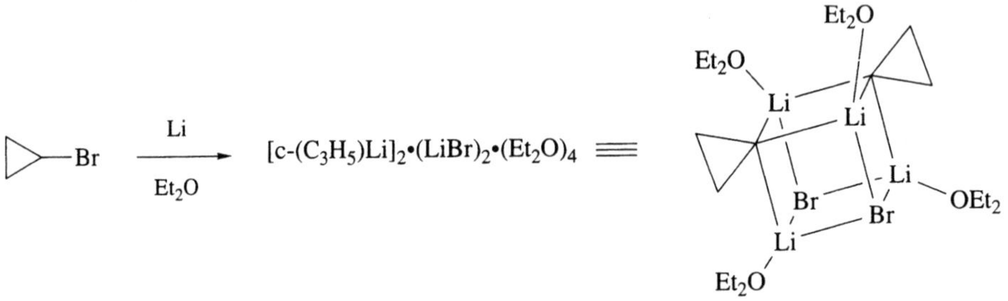

$$\text{(43)} \xrightarrow[\text{Et}_2\text{O}]{\text{Li}} [c\text{-}(C_3H_5)Li]_2{\bullet}(LiBr)_2{\bullet}(Et_2O)_4 \equiv \text{(44)}$$

(**43**) (**44**)

Another mixed aggregate complex consisting of Bu^{n}Li and *t*-butoxide was reported in 1990 as the tetramer (**45**).[47] This complex was first isolated by Lochmann[48] and has been shown to be tetrameric and dimeric in benzene and THF, respectively, by cryoscopic measurements,[49] and it has also been studied by rapid injection NMR techniques.[50] This species has received much attention because it is related to the synthetically useful 'superbasic' or 'LiKOR' reagents prepared by mixing alkali metal alkoxides with lithium alkyls or lithium amides.[51]

$$[Bu^nLi{\bullet}LiOBu^t]_4 \equiv \text{(45)}$$

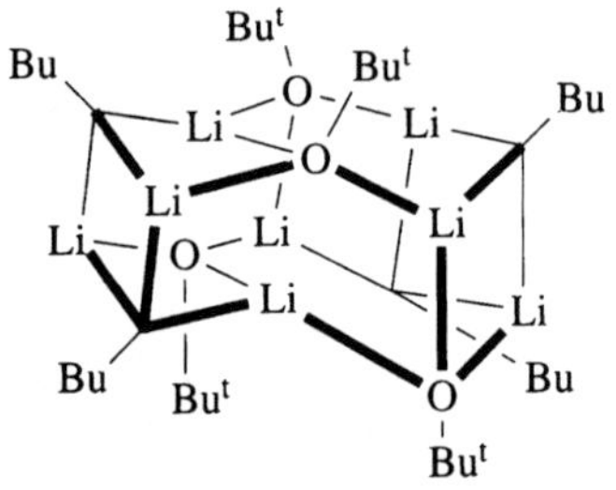

(**45**)

Another example of a solvated, cubic tetramer is methyllithium—TMEDA (**46**).[52] In this example an aggregate of composition $[(MeLi)_4{\cdot}2TMEDA]_n$ with almost ideal T_d symmetry crystallized from an ethereal solution of methyllithium and TMEDA at room temperature. This material consists of infinitely long chains of cubic tetramer linked by TMEDA molecules. Since TMEDA usually has a strong preference for formation of a chelate ring with a single lithium atom, it is somewhat unusual that such an intramolecular chelate is not observed here.

Deprotonation of bicyclobutane (**47**) by *n*-butyllithium in hexane containing a slight excess of TMEDA, followed by solvent evaporation, filtration and recrystallization from benzene yields the dimeric, bis-chelated aggregate (**48**).[53] This aggregate corresponds exactly to structural type (**16**). It is perhaps surprising that many more examples of aliphatic carbanions have not yet been characterized with this general bis-chelated dimeric structure.

Intramolecular solvated tetramers are observed for 3-lithio-1-methoxybutane (**49**)[54] and from 1-dimethylamino-3-lithiopropane (**50**)[55] in the solid state. These tetramers are shown in generalized form as (**51**) and (**52**), respectively. Note the significant difference between the aggregates (**51**) and (**52**). Variable temperature ^{7}Li NMR as well as ^{1}H NMR suggest that although the major form of 1-dimethylamino-3-lithiopropane (**50**) is the diastereomer (**51**), this structure is presumed to be in equilibrium with (**52**)

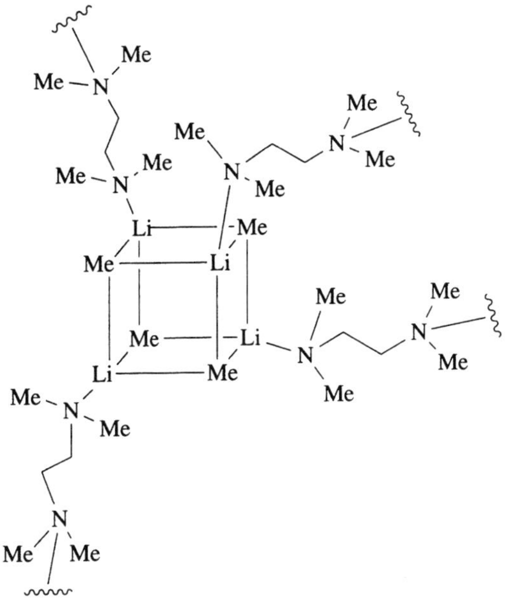

(46)

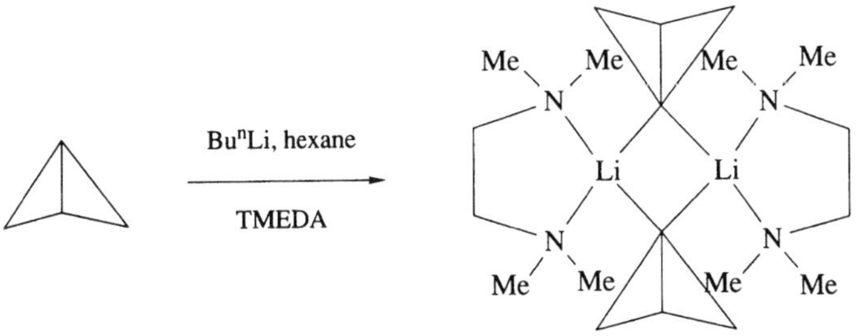

(47) **(48)**

in toluene (or cyclopentane) with activation parameters $\Delta H\ddagger = 17\ (16) \pm 2$ kcal mol^{-1}, $\Delta S\ddagger = 13\ (10) \pm 3$ cal (mol deg)$^{-1}$ (1 cal = 4.184 J).[55b]

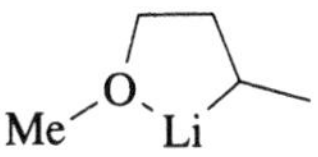

(49) 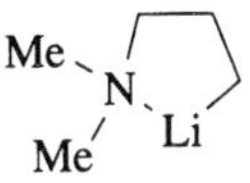 **(50)** 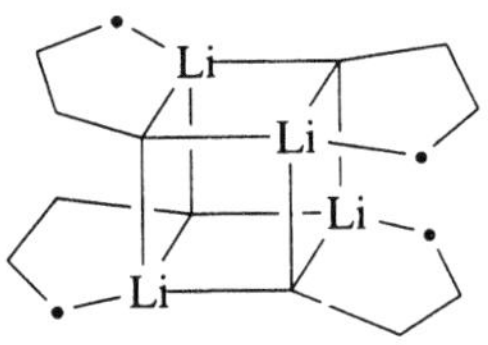 **(51)** 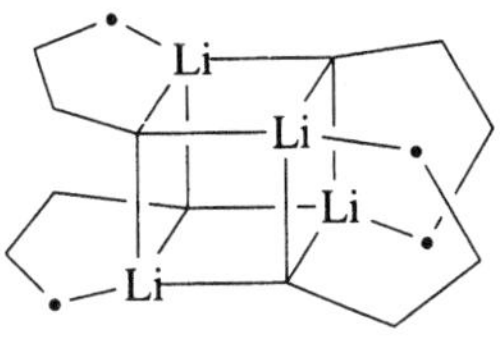 **(52)**

Benzyllithium crystallizes from hexane/toluene solution in the presence of 1,4-diazabicyclo[2.2.2]octane (DABCO) in infinite polymeric chains.[56] Inspection of the individual monomeric units of this structure reveals a unique interaction of the lithium atoms in an η^3-manner with the benzylic carbanion. This bonding is based upon the three relatively short Li—C contacts as indicated in structure (**53**). The two protons on the benzylic carbon center were located crystallographically; one of these lies in the plane of the aromatic ring and the other is significantly out of this plane. A similar η^3-Li—CCC interaction is observed in the diethyl ether solvate of triphenylmethyllithium (**54**).[57] This latter structure is depicted as (**55**).

When the lithium cation is unable to associate with the carbanion, as is the case for the Li^+ (12-crown-4) complexed lithium diphenylmethane carbanion (**56**) or Li^+ (12-crown-4) triphenylmethyl carbanion (**57**), the entire aromatic carbanions are relatively planar.[58] The planarity of (**56**) and (**57**) is indicative of

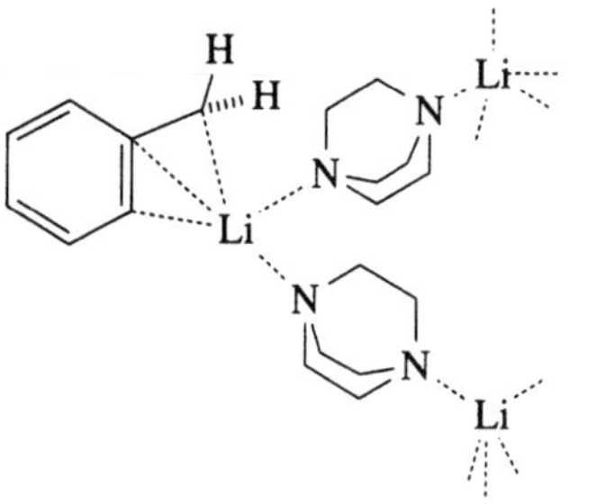

$Ph_3CLi{\bullet}(Et_2O)_2$

(53) **(54)** **(55)**

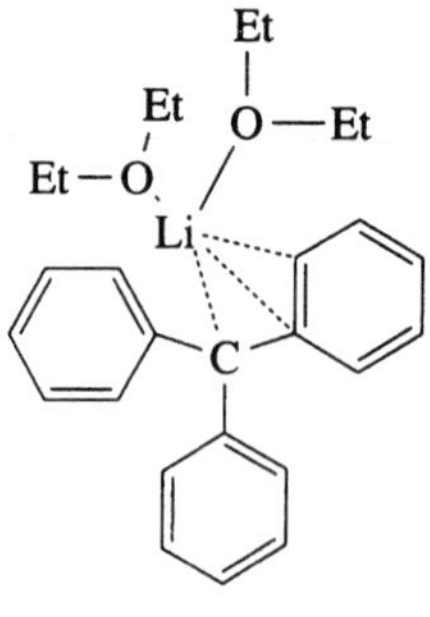

extensive delocalization in these structures. The triphenylmethyl carbanion in **(57)** can also be compared with this same species as it appears associated with Li^+·TMEDA[59] and Na^+·TMEDA[60] cations.

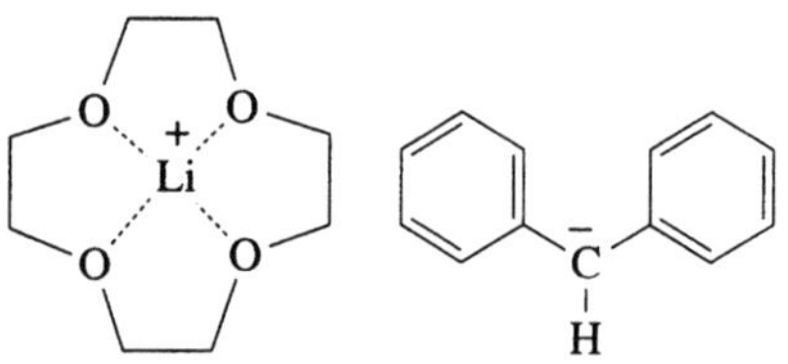

(56) 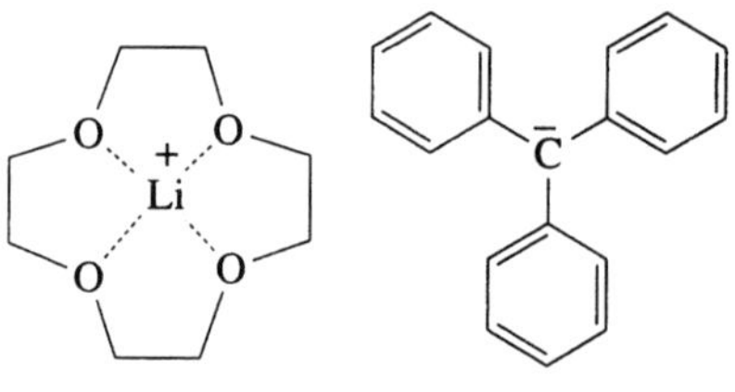 **(57)**

A dimer **(58)** of α-lithiated 2,6-dimethylpyridine crystallizes with TMEDA solvation.[61] This dimer is completely unlike the polymeric benzyllithium **(53)** in that no η^3-intramolecular bonding is observed. The central core of the dimer **(58)** consists of an eight-membered ring formed from two intermolecular chelated Li^+ atoms and nearly ideal perpendicular conformations of the α-CH_2Li^+ groups. Dimer **(58)** is a relatively rare example of a lithiated π-system where Li^+ exhibits only one carbon contact.

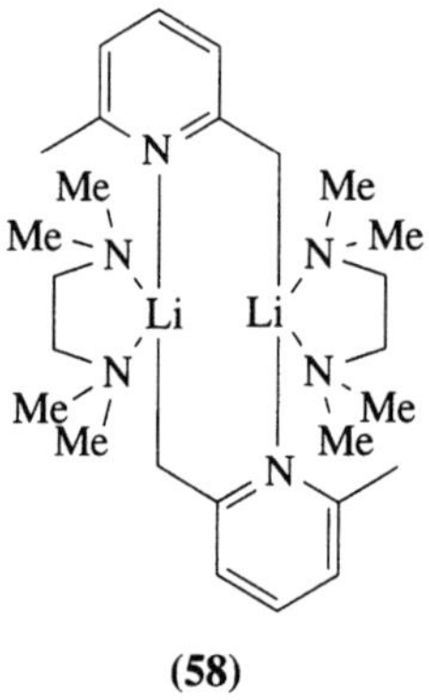

(58)

This discussion of aliphatic carbanion structures has included mainly organolithium compounds simply because the structures of most aliphatic carbanions incorporate lithium as the counterion and also because this alkali metal cation is the most widely used by synthetic organic chemists. For comparison the entire series of Group 1a methyl carbanion structures, *i.e.* MeNa, MeK, MeRb and MeCs, have been determined. Methylsodium was prepared by reaction of methyllithium with sodium *t*-butoxide.[62] Depending upon the reaction conditions, the products obtained by this procedure contain variable amounts of methyllithium and methylsodium (Na:Li atom ratios from 36:1 to 3:1). The crystal structure of these methylsodium preparations resembles the cubic tetramer **(38)** obtained for methyllithium with the Na—Na distances of 3.12 and 3.19 Å and Na—C distances of 2.58 and 2.64 Å.

Methylpotassium, prepared from MeHg and K/Na alloy or from methyllithium and potassium *t*-butoxide, has a hexagonal structure corresponding to the NiAs type **(59)**.[63] Each methyl group is considered to be coordinated to six K^+ ions in a trigonal prismatic array. Methylrubidium and methylcesium, prepared from rubidium *t*-butoxide and cesium 2-methylpentanoate respectively, also possess hexagonal structures of the same type as methylpotassium.[64]

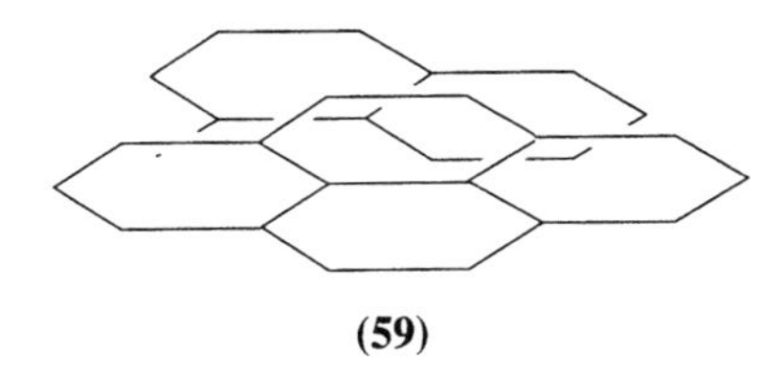

(59)

An extremely unusual pentacoordinate carbon with trigonal bipyramidal symmetry is observed in the crystalline, TMEDA solvate of benzylsodium.[65] This benzylsodium complex is best described as a tetramer with approximate D_{2d} symmetry. The four sodium atoms define a square with a benzyl carbanion bisecting each edge. The resulting eight-membered ring is slightly puckered to alleviate crowding. This structure is depicted as (**60**).

(60)

Other aliphatic carbanion structures associated with Group IIa cations are known. Some examples of these are dimethylberyllium[66] and lithium tri-*t*-butyl beryllate.[67] Since the beryllium alkyl carbanions have not yet been utilized as common synthetic reagents, these structures will not be discussed further.

Magnesium^{2+} stands out among Group IIa metal cations that are commonly utilized in synthetic organic chemistry. Indeed there have been several structural investigations of aliphatic Grignard reagents and dialkylmagnesium reagents. The simplest Grignard reagents, *i.e.* RMgX, whose structures have been determined are MeMgBr·3THF (**61**),[68] (EtMgBr·OPr^i_2)$_2$ (**62**),[69] (EtMgBr·Et_3N)$_2$ (**63**),[70] EtMgBr·$2Et_2O$ (**64**)[71] and the complex (EtMgCl·$MgCl_2$·3THF)$_2$ (**65**).[72] The crystal structures of these reagents exhibit a remarkable diversity for such seemingly similar species. As indicated in the aggregate molecular formulae above, both ethylmagnesium bromide diethyl ether solvate (**64**) and methylmagnesium bromide THF solvate (**61**) are monomeric. However, the magnesium in complex (**64**) is approximately tetrahedral and the magnesium in (**61**) is approximately trigonal bipyramidally coordinated. The general features of these latter two structures are depicted as (**66**) and (**67**). In the complex (**67**), the methyl groups and the bromine atom are disordered and the tetrahydrofuran rings are significantly distorted. The two dimeric complexes, (EtMgBr·OPr^i_2)$_2$ and (EtMgBr·Et_3N)$_2$ are similar. They both incorporate bridging bromine atoms and four-coordinate, tetrahedral Mg^{2+} ions. The general structural type of both of these compounds is given as (**68**).

The ethylmagnesium chloride complex (**65**), depicted as (**69**), is extremely complex, but can be simplified if it is seen as a dimer of EtMgCl·$MgCl_2$ containing five four-membered bridging units of magnesium and chlorine atoms. Two different types of magnesium atoms are seen in this structure. These two types of metals exhibit five and six coordination. Additionally there are two three-coordinate bridging chlorine atoms and four two-coordinate chlorine atoms in this structure.

Treatment of hexamethyldisilazane (**70**) in hexane with a slight excess of a solution of the dialkylmagnesium reagent, Bu^nBu^sMg, initially yields the dimeric complex [Bu^sMg·N(TMS)$_2$]$_2$.[73] This material is characterized as an unsolvated dimer (**71**).

Optically active diamines (–)-sparteine (**72**) and (–)-isosparteine (**73**) form complexes with ethylmagnesium bromide which crystallize in a form suitable for diffraction analysis. In both of these structures,

$EtMgBr \cdot 2Et_2O$ ≡ (64) (66)

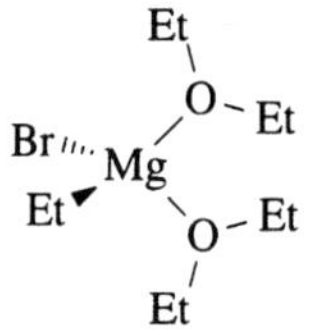

$MeMgBr \cdot 3THF$ ≡ (61) (67)

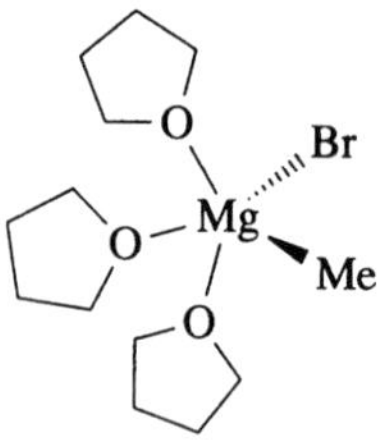

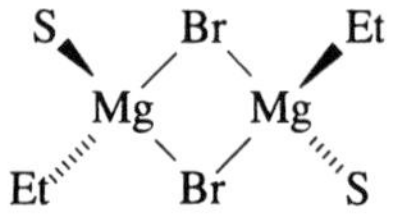

(68) S = Pr^i_2O or Et_3N

$(EtMgCl \cdot MgCl_2 \cdot STHF)_2$ ≡ (65) (69)

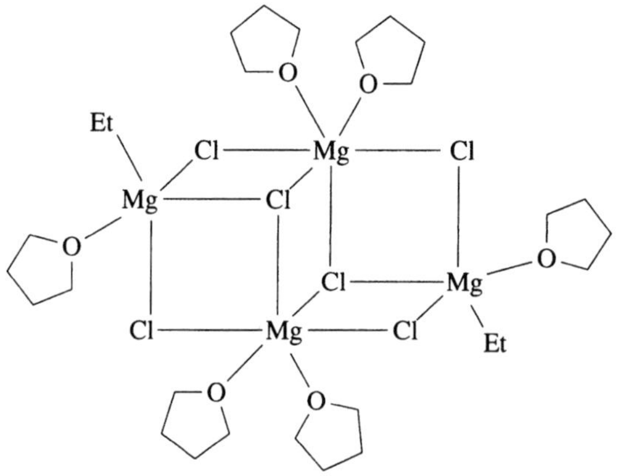

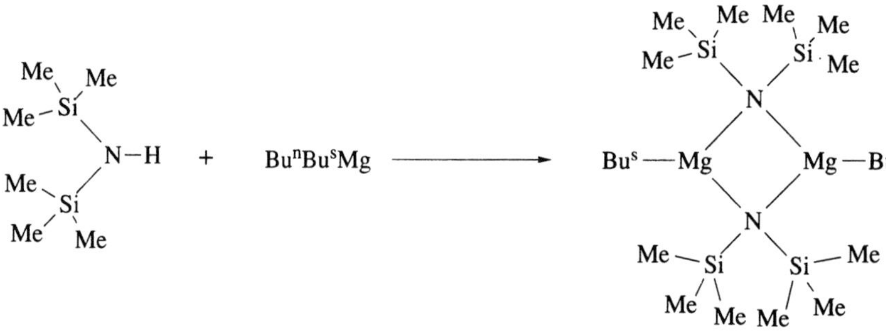

(70) + Bu^nBu^sMg → (71)

(70) **(71)**

depicted as (**74**) and (**75**), the Mg^{2+} is tetrahedrally coordinated by the carbon atom of the ethyl group, the bromine atom and two nitrogens of the (iso)sparteine residue, respectively. The complex (**74**) of ethylmagnesium bromide with the chiral bidentate ligand (–)-sparteine is catalytically active in the asymmetric, selective polymerization of racemic methacrylates.[74] Similar structures are found for the complexes of *t*-butylmagnesium chloride with (–)-sparteine[75] and for ethylmagnesium bromide with (+)-6-benzylsparteine.[76]

Reaction of MgH_2, prepared by homogeneous catalysis, with 4-methoxy-1-butene in the presence of catalytic amounts of $ZrCl_4$ yielded the monomeric magnesium inner ion complex $Mg(C_4H_8OMe)_2$.[77] This complex crystallizes with the tetrahedrally four-coordinate magnesium as shown in (**76**). In a similar reaction, treatment of bis(dialkylamino)propylmagnesium inner complexes (**77**) or (**78**) with $MgEt_2$ yielded the crystalline dimer of ethyl-3-(*N*,*N*-dimethylamino)propylmagnesium (**79**) and ethyl-3-(*N*-cyclohexyl-*N*-methylamino)propylmagnesium (**80**).[78]

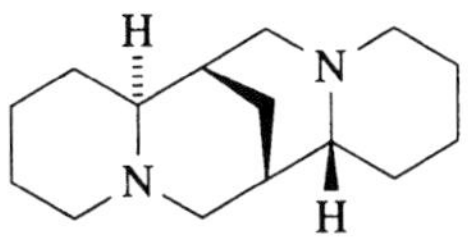

(72) 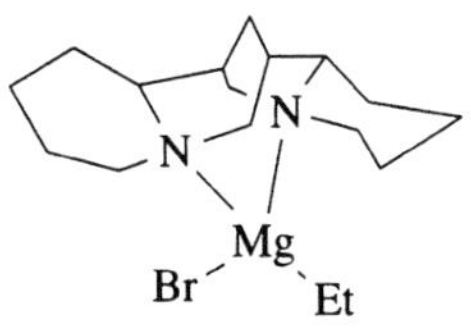 **(74)** 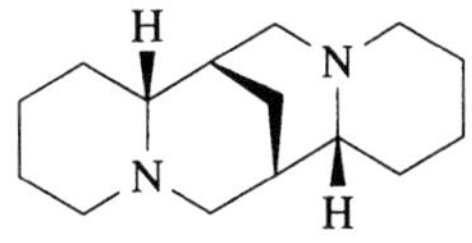 **(73)** 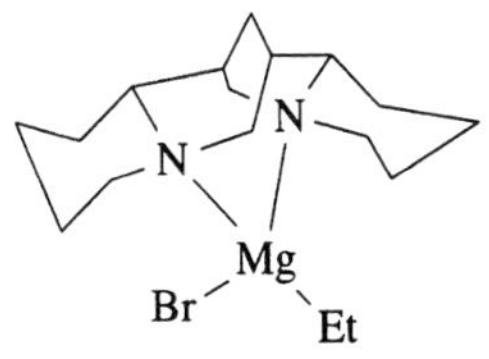 **(75)**

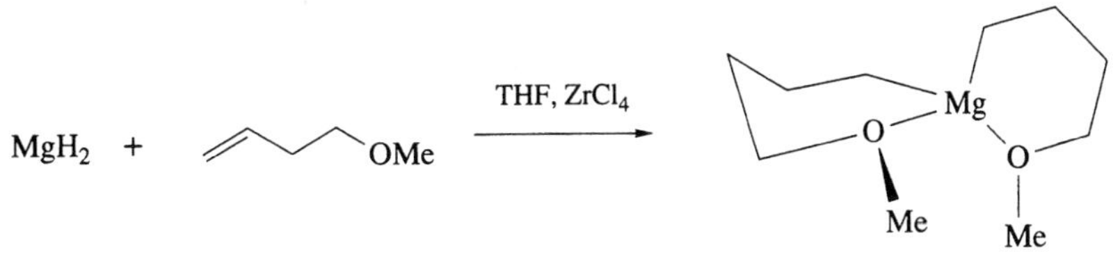

MgH_2 + OMe —(THF, $ZrCl_4$)→ **(76)**

$MgEt_2$ + $Mg[(CH_2)_3NMeR]_2$ ——→

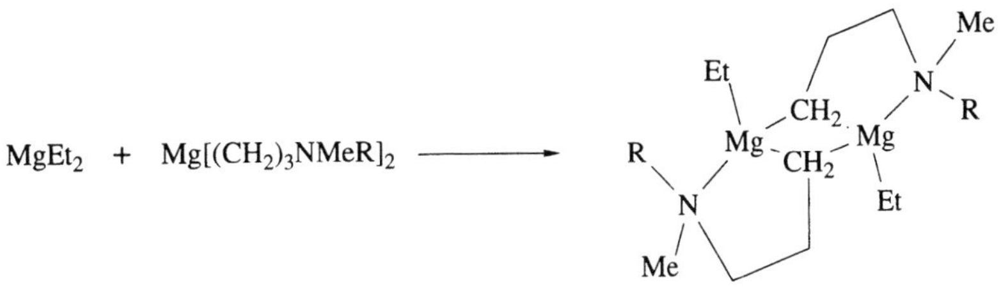

(77) R = Me
(78) R = cyclohexyl

(79) R = Me
(80) R = cyclohexyl

A triple ion was crystallized by Richey *et al.* from a solution made up by adding 2,1,1-cryptand to diethylmagnesium.[79] Diffraction analysis reveals that this triple ion consists of $[EtMg^+(2,1,1\text{-cryptand})]_2$ cations and an $(Et_6Mg_2)^{2-}$ anion. The magnesium of the cation is bound to five heteroatoms of the cryptand and to an ethyl group. The two magnesiums in the dianion exhibit identical four coordination and they form a symmetrical dimer with two bridging ethyl groups and four terminal ethyl groups. The anion is depicted as (**81**). A different structure was found for the product of the reaction of dineopentylmagnesium (Np_2Mg) with 2,1,1-cryptand.[80] In this latter reaction, crystalline $NpMg^+(2,1,1\text{-cryptand})$ cations and Np_3Mg anions are formed. The coordination geometry of magnesium in the Np cation is essentially that of a trigonal bipyramid with bonds to all six heteroatoms of the cryptand and a bond to the neopentyl group. Only the three-coordinate anion (**82**) is illustrated here. The ^1H NMR spectrum of a benzene solution of $NpMg^+(2,1,1\text{-cryptand})\cdot Np_3Mg^-$ is consistent with the presence of the same ions in solution. Diethylmagnesium cryptand complex reacts faster with pyridine than the diakylmagnesium reagent alone, and it also modifies the regioselectivity of this reaction.

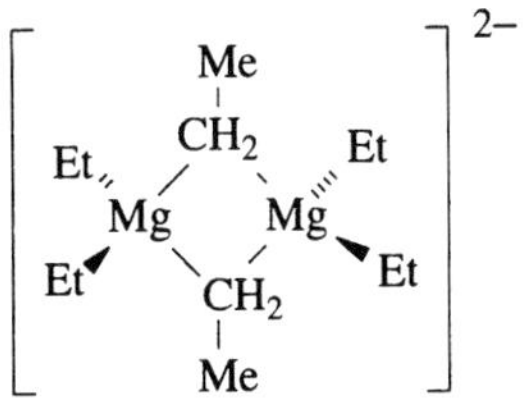

(81) 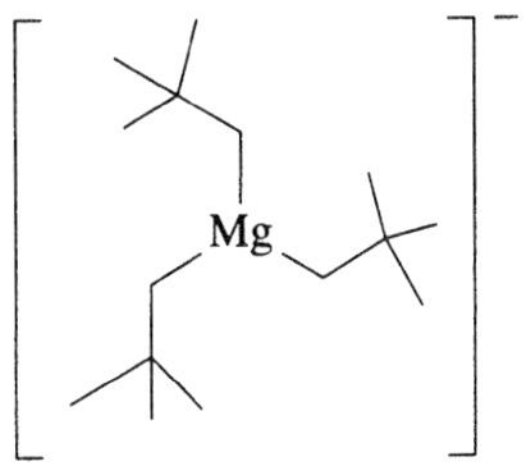 **(82)**

Diethylmagnesium and 18-crown-6 react to form a complex with six oxygens surrounding the magnesium in a quasiequatorial plane and with the ethyl groups occupying *trans* apical positions.[80] This structure is illustrated as (**83**). It has been described as a rotaxane[81] or 'threaded' structure. A related, but slightly different, structure is found for the $MeMg^+(15\text{-crown-5})\cdot Me_5Mg_2^-$ complex.[82]

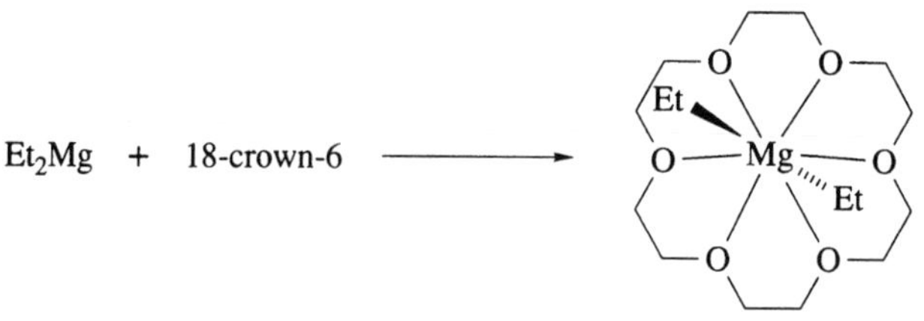

Et_2Mg + 18-crown-6 ⟶

(83)

The structures of a few dialkylmagnesium reagents have been characterized. These include $(Me_2Mg)_n$,[83] $(Et_2Mg)_n$,[84] $Me_2Mg{\cdot}TMEDA$,[85] $[(CH_2)_5Mg]_2{\cdot}4THF$,[86] $Me_2Mg{\cdot}(quinuclidine)_2$[87] and *catena*-poly-dineopentylmagnesium-μ-dioxane = $[Mg(C_5H_{11})_2{\cdot}2THF]_n$.[88] While the diakylmagnesium reagents also exhibit several different structural types, all of these complexes are related by the fact that they incorporate magnesium atoms that are four coordinate with distorted tetrahedral geometry. Unsolvated dimethylmagnesium and diethylmagnesium both form linear, polymeric chains with adjoining Mg atoms linked by two bridging alkyl groups. The solvated dimethylmagnesium complex, *i.e.* $Me_2Mg{\cdot}TMEDA$, is illustrated as (**84**) with the bond angle as shown. The pentamethylenemagnesium complex, $[(CH_2)_5Mg]_2{\cdot}4THF$, crystallizes as the dimer (**85**) with two magnesium atoms in a 12-membered ring. This tendency to form a 12-membered ring is ascribed to the large C—Mg—C valence angle of 141° which would cause severe ring strain in a monomeric magnesiocyclohexane. The polymeric dineopentylmagnesium exhibits a structure which consists of dineopentyl units linked through dioxanes forming parallel linear chains as shown in (**86**).

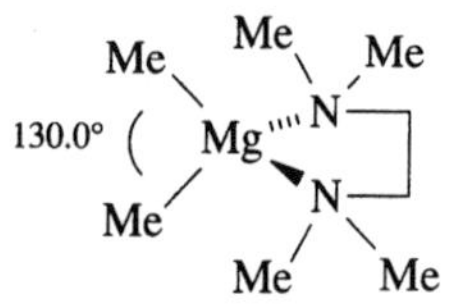

130.0° 141.5°

(84) 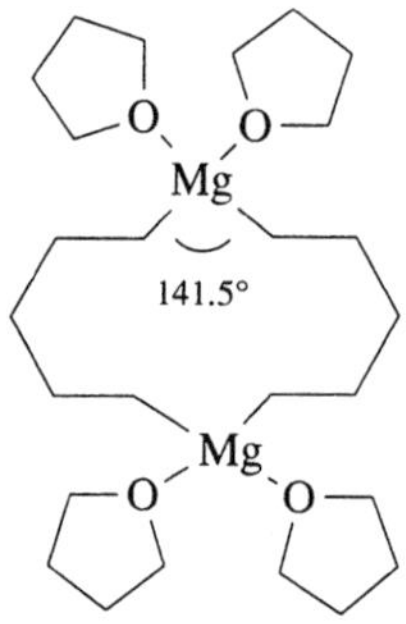 **(85)** 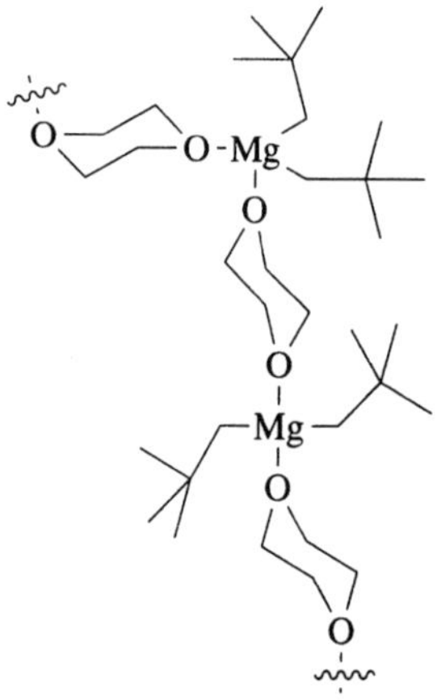 **(86)**

One final diakylmagnesium structure is illustrated as (**87**).[89] This material was obtained from the room temperature crystallization of the magnesium reagent formed from *o*-bis(chloromethyl)benzene in THF. As in all previous dialkylmagnesium reagents, the highly symmetrical trimer (**87**) includes four-coordinate magnesium atoms with distorted tetrahedral geometry.

1.1.3.1.2 α-Silyl-substituted aliphatic carbanions

There are several crystal structures of aliphatic, α-silyl-substituted carbanions. A listing of many of these structures is given in Table 2, along with some references to the original literature where these structures are described. Only a few of these compounds can be considered as generally useful synthetic reagents. The aggregates formed by α-silyl-substituted carbanions are in many cases similar to those described previously for one of the unsubstituted aliphatic alkali metal or magnesium carbanions. A few of the α-silyl-substituted carbanions display unique aggregate structures and these are described as follows.

Tris(trimethylsilyl)methyllithium (**88**), prepared from methyllithium and tris(trimethylsilyl)methane in THF, can be recrystallized from toluene to give colorless, transparent needles of the unique ate complex $[Li(THF)_4]^+[Li\{C(TMS)_3\}_2]^-$.[90] The structure of one of the anions in this complex is shown as (**89**). This

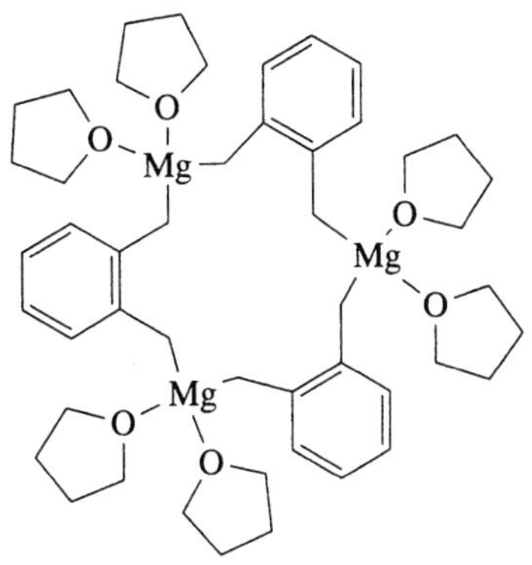

(87)

Table 2 α-Silyl-substitued carbanions

Compound	*Aggregation state*	*CSD refcode*	*Ref.*
TMS-methyllithium Me_3SiCH_2Li	Hexamer	JAFMUY	205
Bis(TMS)methyllithium $(Me_3Si)_2CHLi$	Polymer	CIMVUP	206
Bis(TMS)methyllithium·PMDETA $(Me_3Si)_2CHLi$·PMDETA	Monomer	BIYXOW	207
Tris(Me_2PhSi)methyllitium·THF	Monomer	CATZAY	208
Bis(TMS)-2-methylpyridinelithium	Dimer	CAWMUI	209
Bis(TMS)-2-methylpyridinelithium·$2Et_2O$	Dimer	COXTOY	210
Bis(TMS methyl)pyridinelithium·2TMEDA	Dimer	FERKOC	211
TMS(cyclopentadiene)·TMEDA	Monomer	CEZTIK	212
Bis(TMS)methyldiphenylphosphinelithium·TMEDA	Dimer	FIKPUK	213
TMS-methyl(PMe_2)Li·TMEDA	Dimer	GIGGOS	214
TMS-methyl(PMe_2)Li·THF	Dimer	GIKWS	215
Bis(TMS)methyl·MgCl·Et_2MgCl7Et	Monomer	MSIMMG	216
9,10-Bis(TMS)anthracene-9-diylmagnesium	Polymer	FOXBID	217
Bis(TMS)-2-methylpyridinemagnesium	Dimer	DONVUX	218
(TMS)methyl(*o*-diphenylphosphinophenyl)lithium	Dimer	VAGHUG	219

anion is linear, with the $C(TMS)_3$ groups staggered about the C—Li—C direction. The cation consists of a lithium tetrahedrally coordinated by four oxygen atoms. Treatment of this ate complex, $[Li(THF)_4]^+[Li\{C(TMS)_3\}_2]^-$, in toluene with pentamethyldiethylenetriamine (PMDETA) containing lithium chloride yields a complex whose formula is given as $[(PMDETA)\text{-}Li\text{-}(\mu\text{-}Cl)\text{-}Li\text{-}(PMDETA)]^+[Li\{C(TMS)_3\}_2]^-$.[91] The anion is the same as in the $Li(THF)_4$ complex, but the new cation contains a PMDETA-complexed, linear Li—Cl—Li geometry depicted as **(90)** (Scheme 6).

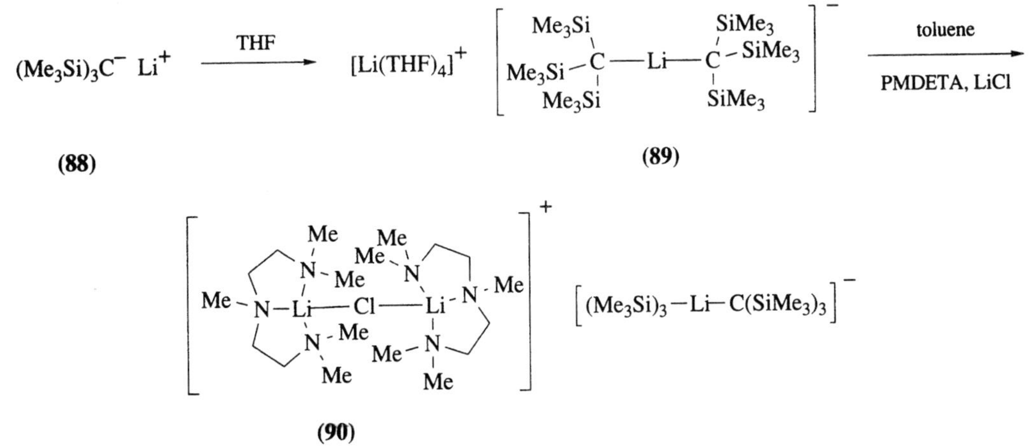

$(Me_3Si)_3C^-\ Li^+$ **(88)** $\xrightarrow{THF}$ $[Li(THF)_4]^+[(Me_3Si)_3C\text{—}Li\text{—}C(SiMe_3)_3]^-$ **(89)** $\xrightarrow[\text{PMDETA, LiCl}]{\text{toluene}}$

(90) $[(Me_3Si)_3\text{—}Li\text{—}C(SiMe_3)_3]^-$

Scheme 6

1.1.3.2 Allylic Carbanions

Several examples of crystal structures are reported for the allyl carbanion. The first to be considered is allyllithium complexed with TMEDA (**91**).[92] In this complex, the terminal carbon atoms of the allyl group are linked to different lithium atoms forming polymeric chains. Each lithium atom is also coordinated to a chelating TMEDA. There is no evidence of η^3-bonding in this allyllithium·TMEDA structure. The structure of monomeric allyllithium solvated by PMDETA subsequently revealed a structure with a single lithium atom having relatively close but unsymmetrical contacts with the terminal carbon atoms and that the C_3H_5 anion is not planar.[93] A rough outline of this structure is given in formula (**92**).

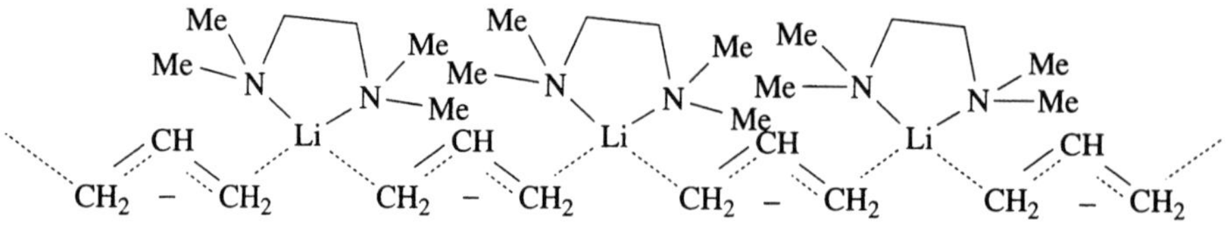

(91)

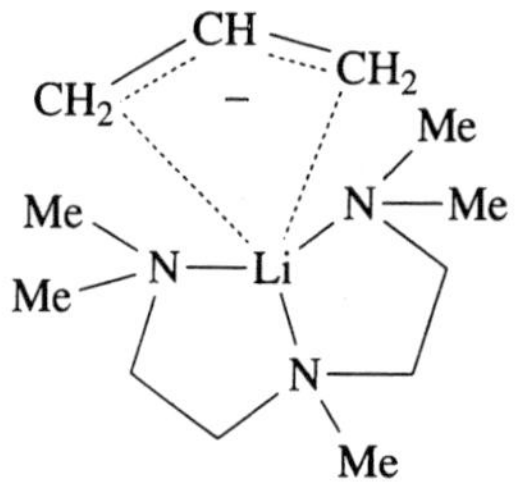

(92)

The first example of an η^3-allyllithium was reported for the polymeric 1,3-diphenylallyllithium·diethyl ether complex (**93**).[94] In this polymer, the lithium atoms lie almost symmetrically above and below the allyl group. The C(1)—C(2)—C(3) angle is quite large, *i.e.* 131°, similar to that in structures (**91**) and (**92**), but the C(2)—Li distance in (**93**) is shorter than either the C(1)—Li or C(3)—Li distances. The structural features of (**93**) correspond to those found in many allyl transition metal complexes which display η^3-bonding of the transition metal to the allyl group.[95]

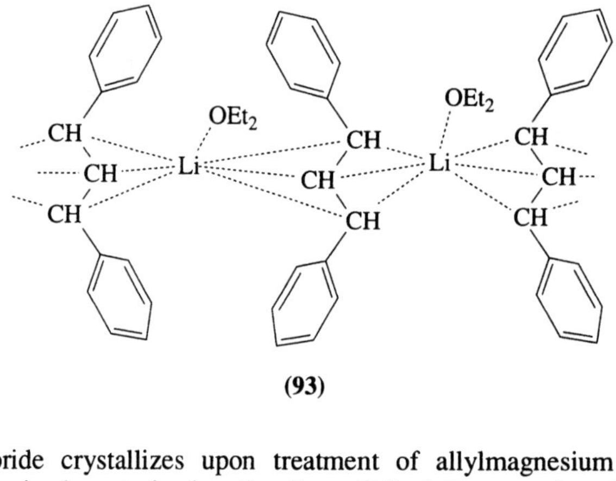

(93)

Allylmagnesium chloride crystallizes upon treatment of allylmagnesium chloride in THF with TMEDA.[96] This complex is characterized as the dimer (**94**). A four-membered Mg—Cl—Mg—Cl ring with bridging chlorine atoms forms the core of this dimer. The allyl group is clearly associated with the magnesium at C(1), *i.e.* C(1)—Mg = 2.18 Å, and the other structural features, *i.e.* the C(1)—C(2) and C(2)—C(3) bonds and the tetrahedral geometry at C(1), are commensurate with an η^1-structure. The η^1-structure of allylmagnesium compounds in solution was confirmed by NMR techniques, as well as by calculations.[97]

Bis(2,4-dimethyl-2,4-pentadienyl)magnesium (**95**), prepared by the reaction of the potassium dienide with anhydrous magnesium halide in THF and TMEDA, also exhibits a terminally bonded η^1-structure (**96**), with two dienides attached to a tetrahedral, four-coordinate Mg that is chelated by a single TMEDA

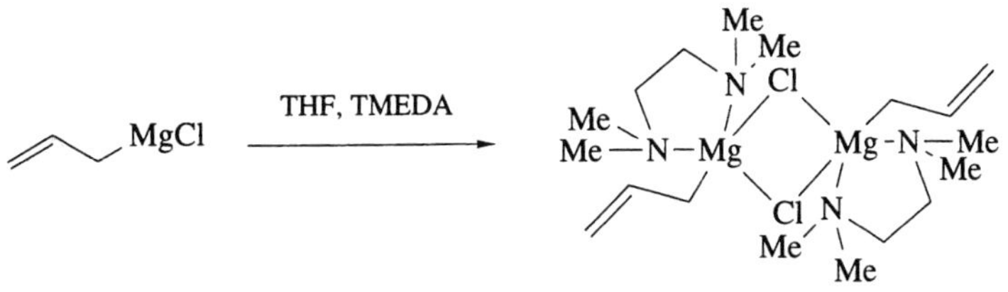

THF, TMEDA

(94)

(Scheme 7).[98] This structure is consistent with the hydrolysis product of this complex which is obtained as a 1,3-diene, although either a 1,3-diene or a 1,4-diene is obtained upon addition of (**96**) and similar homologous dienylmagnesium complexes to ketones.

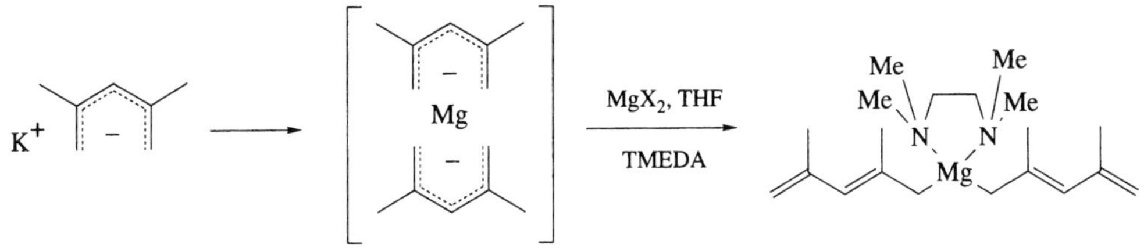

MgX_2, THF

TMEDA

(95) **(96)**

Scheme 7

The crystal structures of indenyllithium·TMEDA,[99] indenylsodium·TMEDA[100] and bis(indenyl)magnesium[101] are also known. These three structures are all different. The lithium compound is monomeric with η^5-coordination; the sodium compound is infinitely aggregated with both η^1- and η^2-bonding; and the magnesium compound appears to include all three types of interactions, *i.e.* η^1-, η^2- and η^5-bonding, of magnesium to the indenyl anion.

1.1.3.3 Vinylic Carbanions

Relatively few vinyl carbanions have been characterized by X-ray diffraction analysis. The majority of these structures were determined by Schleyer and coworkers. Representative examples of some of these compounds are given as (**97**) and (**98**).

Compound (**97**) crystallized as a dimeric, bis-THF-solvated adduct with two different types of lithium atoms, illustrated as (**99**).[102] Compound (**98**) is a doubly lithium bridged dimer chelated with one TMEDA per lithium atom and roughly depicted as (**100**).[103] In the solid state, the sodium anion of (**101**) is monomeric with some of the relevant structural parameters shown in formula (**102**).[104]

The functional equivalent of an enolate dianion (**103**) was prepared by Stork *et al.* by treatment of enamine (**104**) with *t*-butyllithium.[105] This anion crystallized as the symmetrical dimer (**105**) with the carbanionic carbon nearly symmetrically bridging two lithium atoms as shown in (**105**). Doubly bridging carbons represent a characteristic feature of these lithiated vinylic anions and this structural feature is normally expected in these compounds as well as in aryl anions (see ref. 11).

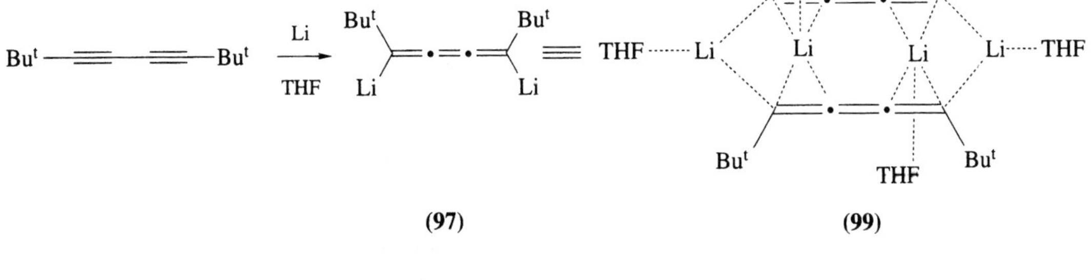

Li

THF

(97) **(99)**

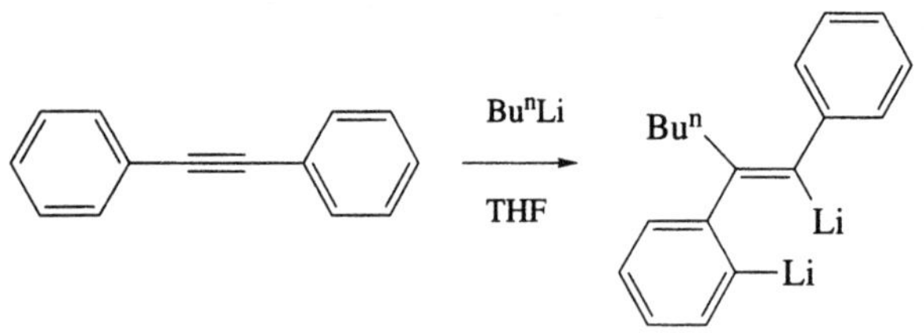

(98) 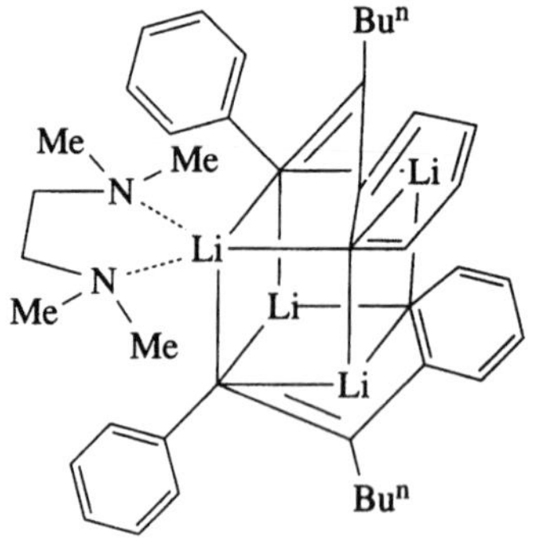 **(100)**

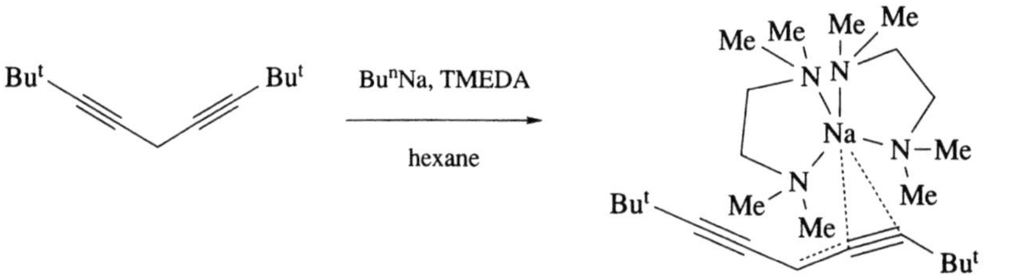

(101) **(102)**

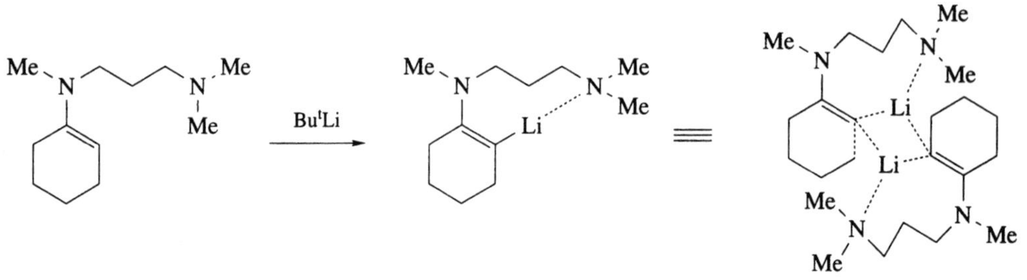

(104) **(103)** **(105)**

1.1.3.4 Alkynic Carbanions

A systematic investigation of alkynic carbanion structures has been reported by Weiss and coworkers. These structures all incorporate either *t*-butylacetylide or phenylacetylide anion. They differ by the ligand that is incorporated. Perhaps this sequence of structures most aptly demonstrates the rich variety of aggregate structural types that an acetylide anion can choose.

From the reaction of phenylacetylene with *n*-butyllithium and *N,N,N′,N′*-tetramethylpropanediamine (TMPDA) the dimeric complex (**106**) is obtained.[106] Note that the carbanionic acetylene carbon bridges only two lithium atoms in this structure. Dimeric phenylethynyllithium is also detected in THF by low temperature ^{13}C NMR investigations and by cryoscopic measurements.[107] However, *t*-butylethynyllithium crystallizes from THF-containing solutions as either the tetramer (**107**) of composition $(Bu^tC{\equiv}CLi{\cdot}THF)_4$ or as the dodecamer (**108**) of composition $[(Bu^tC{\equiv}CLi)_{12}{\cdot}4THF]$.[108] The dodecamer represents a linear combination of three tetramers of (**107**) minus the coordinating THFs. It is not unexpected that under the proper conditions of crystallization an octamer such as (**109**), representing an intermediate between (**107**) and (**108**), can be obtained. Perhaps crystallization of the octamer (**109**) can provide an interesting challenge of the experimental skills of those chemists who are fascinated by crystals and crystallizations of organic compounds.

When crystallized in the presence of of *N,N,N′,N′*-tetramethylhexanediamine (TMHDA) the cubic tetramer (**110**) of phenylethynyllithium was obtained.[109] Adjacent $(PhC{\equiv}CLi)_4$ units in (**110**) are each bound by pairs of TMHDA ligands giving rise to polymer strands with a helix-like structure.

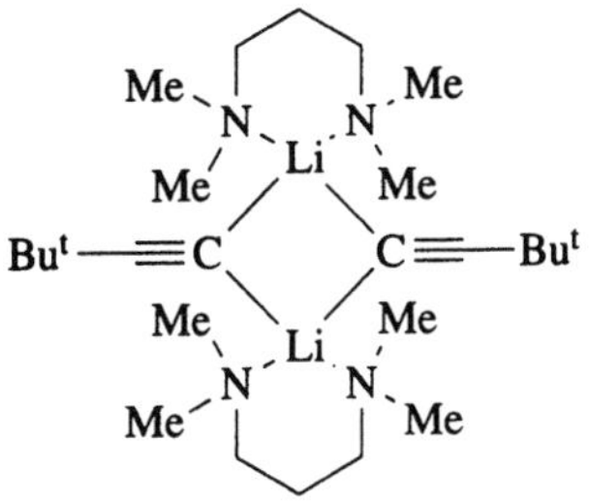

(106) 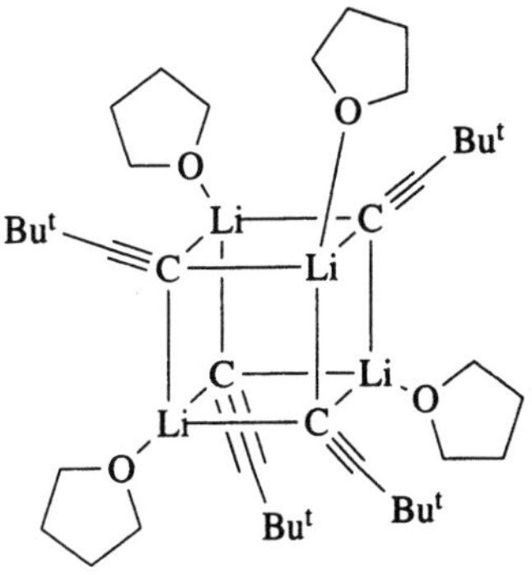 **(107)**

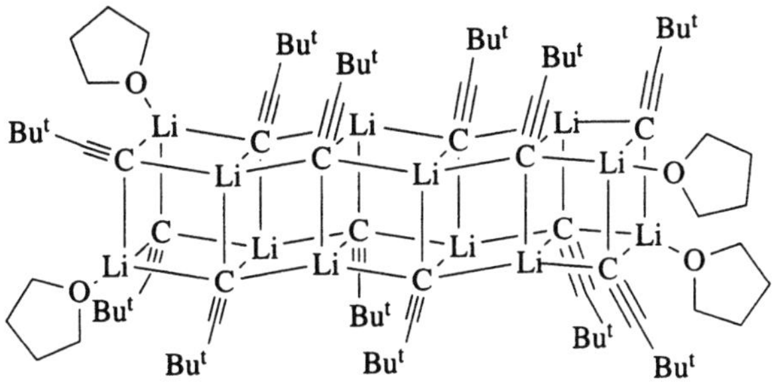

(108) 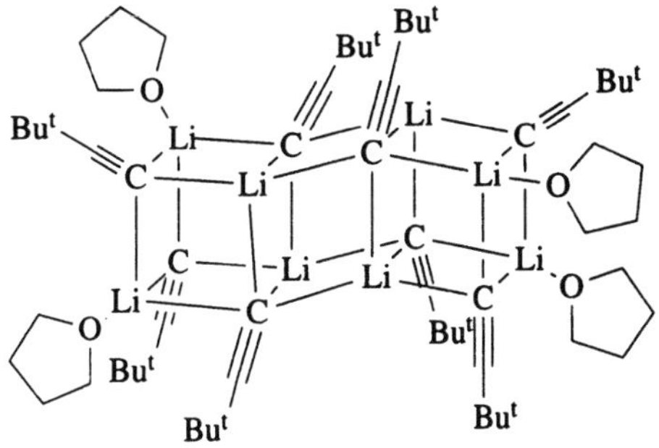 **(109)**

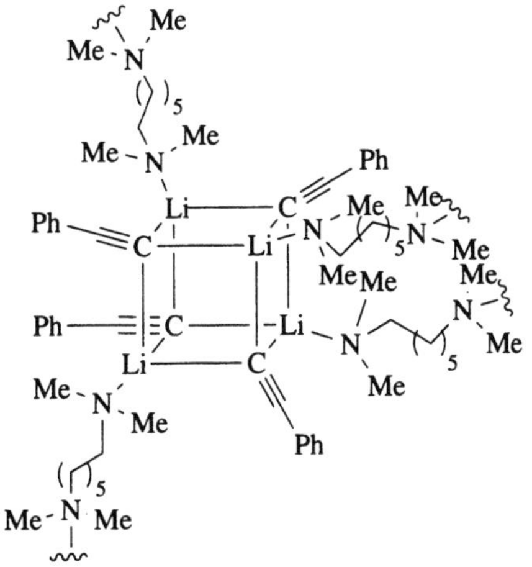

(110)

One final structure in this series is the bis-TMEDA-solvated bis(phenylethynyl)magnesium species characterized as a monomer (**111**).[110] Note the octahedral geometry of the central magnesium with two axial ethynyl ligands. This series of alkynic structures, (**106**)–(**111**), serves to underscore the unpredictability of carbanion crystal structures. The alkynic carbanions have coordination numbers of one, two or three in these complexes.

A few simple beryllium acetylide structures are known.[111] In one of these structures, illustrated as (**112**), the metal cation is symmetrically coordinated side-on to a triple bond.

1.1.3.5 Aryl Carbanions

As was the case for the alkynic carbanions, the crystalline aryl carbanions also offer the opportunity for observation and comparison of several different structural types for the same or closely related carb-

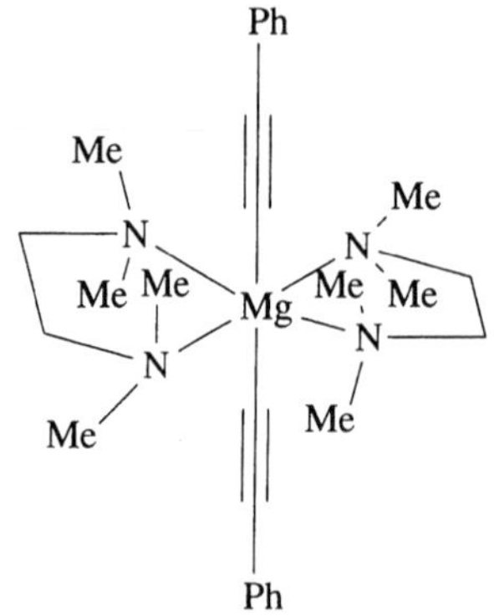

(111)

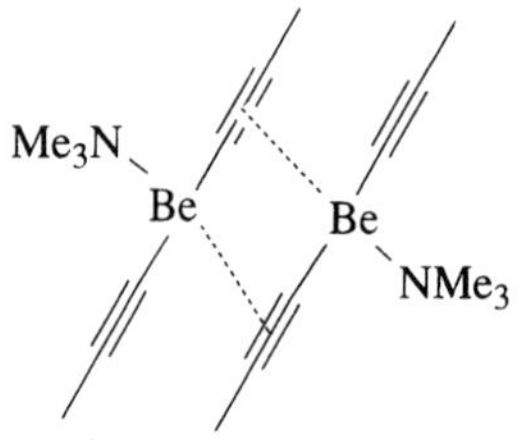

(112)

anion substrates. In many of the examples of aryl carbanion structures, the aryl carbanionic carbon is found bridging several different metal atoms. It is important to note that a large number of X-ray crystal structures of aryl carbanions have been determined, and that these have been comprehensively reviewed elsewhere.[11] Hence this discussion is necessarily limited to a subset of these structures which are representative of the whole class and which are also deemed most closely related to synthetic reagents in common usage. It is appropriate to begin this section with the structures of the simplest member of this series, *i.e.* the unsubstituted phenyl carbanion.

Phenyllithium dissolves in hexane by addition of TMEDA. The phenyllithium·TMEDA adduct subsequently crystallizes out of solution as the dimer (**113**) corresponding to general structural type (**16**).[112] With diethyl ether solvation, phenyllithium exists as a solid tetramer (**114**).[113] In ether solution PhLi is known to be either dimeric[114] or tetrameric.[115] Monomeric phenyllithium was successfully crystallized with PMDETA as the ligand.[116] This monomer is depicted as (**115**). Note the difference in the coordination number of the carbanionic center in the monomer (**115**), the dimer (**113**), and the tetramer (**114**), *i.e.* one, two and three, respectively.

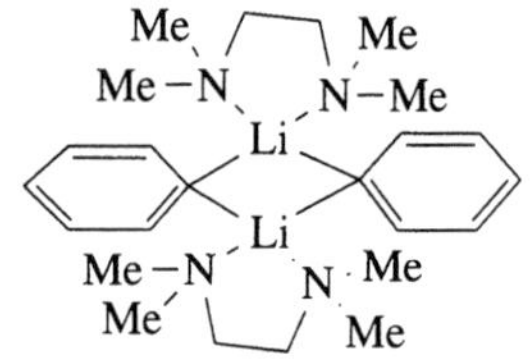

(113) 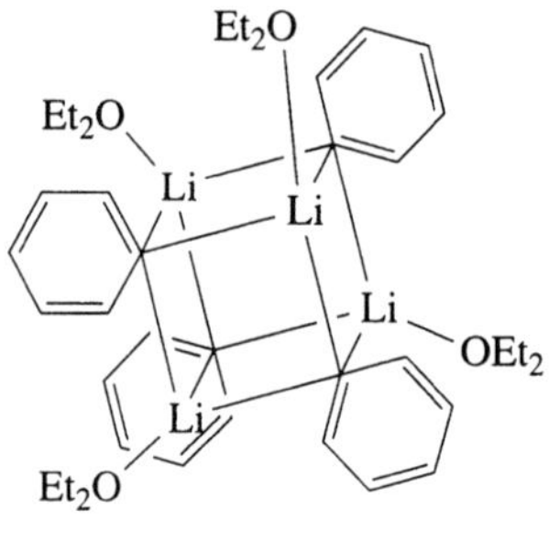 (114) 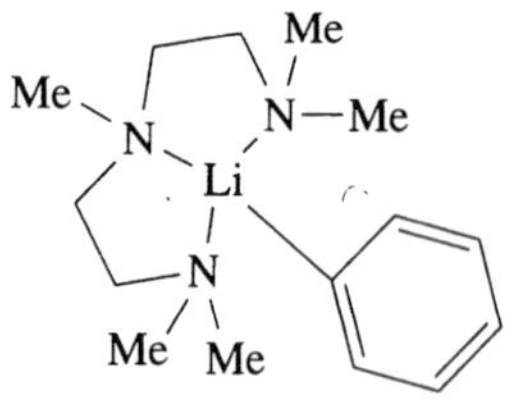 (115)

An interesting mixed tetrameric complex containing three equivalents of phenyllithium and one equivalent of lithium bromide, *i.e.* $[(PhLi)_3{\cdot}LiBr{\cdot}3Et_2O]$, depicted as (**116**), has been characterized.[117] In this mixed aggregate the lithium atom diagonally opposite the bromide in the tetramer remains unsolvated by an ether molecule. Recall that the cyclopropyllithium·lithium bromide·diethyl ether complex (**44**) with stoichiometry [$(RLi)_2\cdot(LiBr)_2\cdot4E_2O$] is structurally related to (**116**), yet in the complex (**116**) all of the lithium atoms are coordinated to a solvent molecule and this is not found in (**44**).

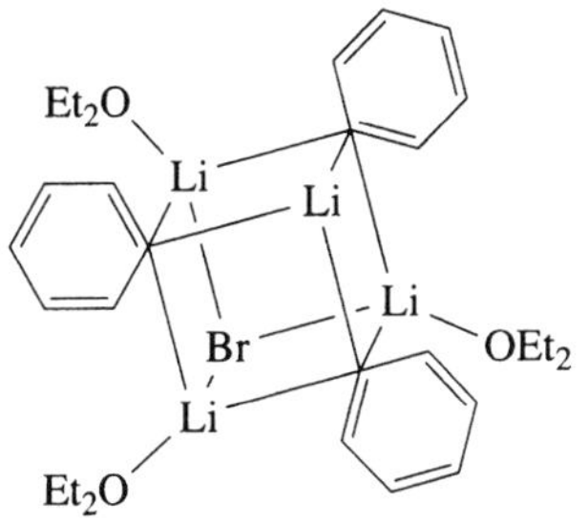

(**116**)

Alkyl substitution at the *ortho* positions of the aromatic ring influences the aggregation state and the structure by providing steric constraints. Whereas unsubstituted phenyllithium·diethyl etherate crystallized as the tetramer (**114**), mesityllithium·diethyl etherate is observed as the dimer (**117**) with stoichiometry [mesitylLi·Et_2O]$_2$.[118] Mesityllithium gains an additional solvent molecule and crystallizes as the bis-solvated dimer (**118**) with stoichiometry (mesitylLi·2THF)$_2$ from THF-containing solutions.[119] Note that the coordination number of the lithium atoms changes from three to four in these two complexes but that the coordination and the coordination geometry remain the same at the carbon center. The change in solvation of the lithium atoms in the two complexes is also reflected in the interatomic distances. Hence, the three-coordinate lithium is 1.93 Å and 2.25 Å away from the ether oxygen and the aryl carbon respectively in (**117**), whereas the corresponding distances average 2.04 Å and 2.28 Å in (**118**).

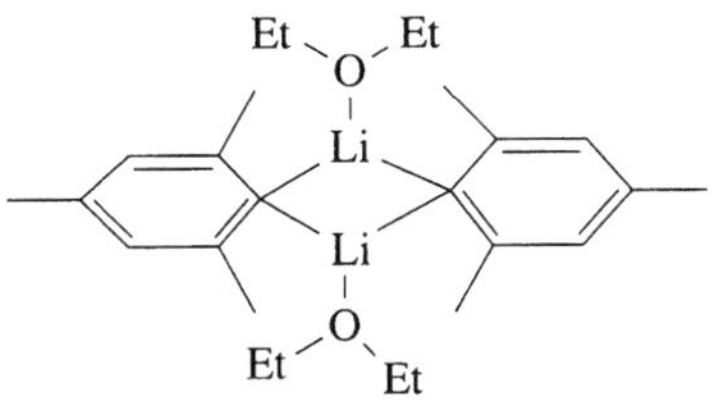

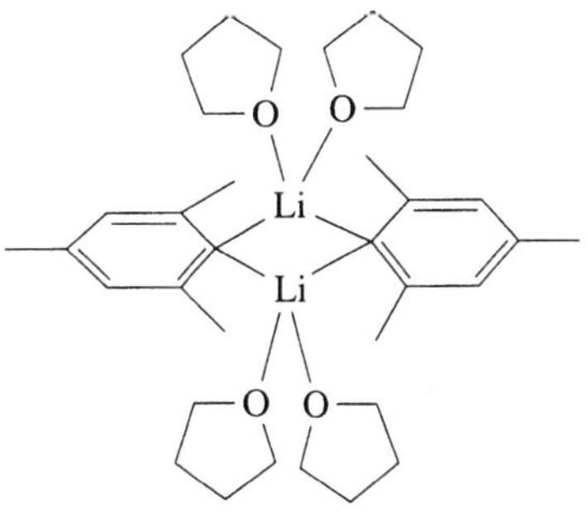

(**117**) (**118**)

Ortho heteroatom substitution provides the opportunity for internal chelation in aryl anions. This effect is successfully utilized in synthetic endeavors, is commonly referred to as '*ortho* metallation', and is reviewed elsewhere.[120] Two different but related structures of *o*-methoxyphenyllithium are known. The most symmetrical of these is the tetramer (**119**) which crystallizes from pentane solution.[121] The solution structure of this compound has been investigated by NMR spectroscopy and by cryscopic measurements and the influence of various Lewis bases on the solution structure are discussed in the same paper as the X-ray structure. A most intriguing variation of the tetramer (**119**) crystallizes from hexane solution in the presence of TMEDA (but in the absence of LiBr). The structure of *o*-methoxyphenyllithium in the presence of TMEDA consists of a pair of unsymmetrical tetramers with the general features of (**119**) but which are linked to each other by a single TMEDA unit.[122] This latter aggregate is represented by the formula (**120**). Obviously the lithium atoms in (**119**) are four coordinate, but participation of the TMEDA in (**120**) forces one of the lithium atoms to be five coordinate. This five-coordinate lithium is in contact with two oxygens of different methoxybenzene residues rather than with an oxygen and a nitrogen of the TMEDA. Hence the symmetry of a tetramer is not maintained in (**120**).

Crystal structures have been reported for 2,6-dimethoxyphenyllithium,[123] for 2,6-dimethylaminophenyllithium[124] and for *o*-*t*-butylthiophenyllithium.[125] The crystal structure of the latter compound is characterized diagrammatically as the infinite polymer (**121**) with relatively planar tetracoordination at the *ipso* carbon. In THF solution this polymer dissociates into monomers. Planar four-coordinate carbons are also observed in the 2,6-dimethoxyphenyl anion (**122**) as a dimeric unit (**123**) which forms the basic building block of the solid of this anion. In this solid two of these simple dimers (**123**) combine to form

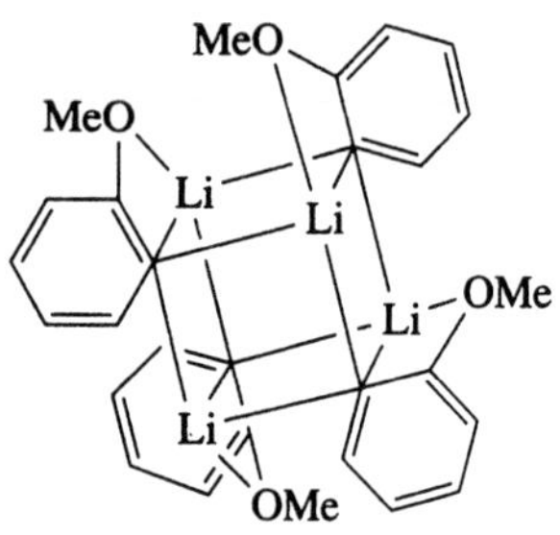

MeO MeO Li Li Li OMe Li OMe

(119)

MeO MeO Li Li Me N Me Li OMe Li OMe Me N Me OMe OMe Li Li MeO Li Li MeO

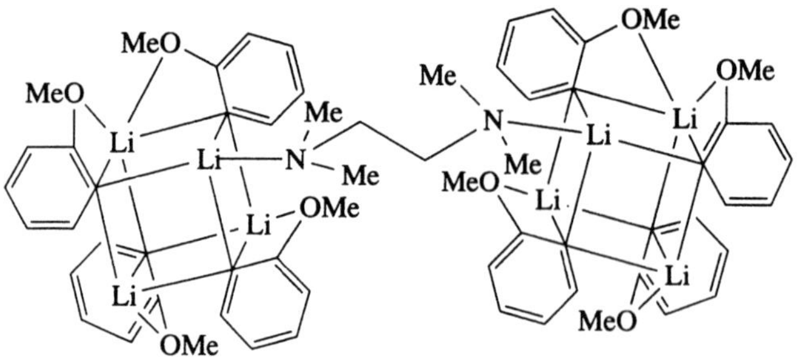

(120)

loose tetramers which are characterized as structure (**124**) in both the solid and in solution. In contrast to (**122**), recrystallization of 2,6-dimethylaminophenyllithium from hexane/ether solution yields the trimeric aggregate (**125**).

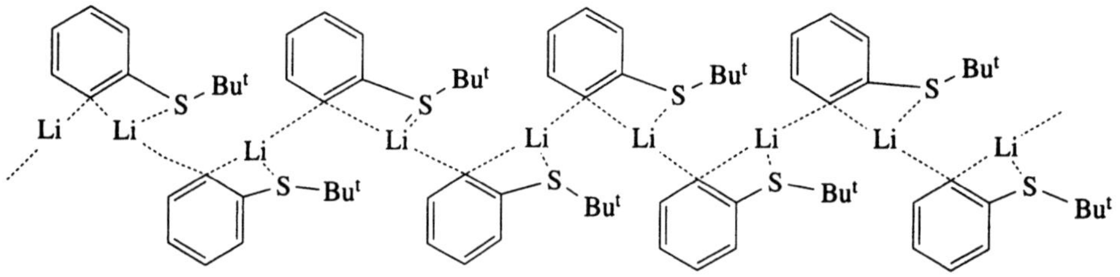

S Bu^t Li Li Li S Bu^t S Bu^t Li Li S Bu^t S Bu^t Li Li S Bu^t S Bu^t Li Li Li S Bu^t

(121)

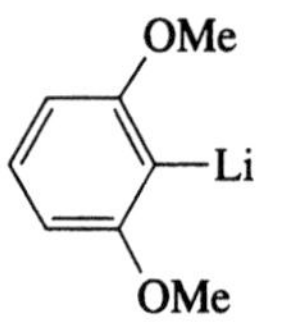

OMe Li OMe ≡ Me O O Me Li Li O O Me Me ≡ Me O O Me Li Li Me O O Me Me O O Me Li Li O O Me Me

(122) 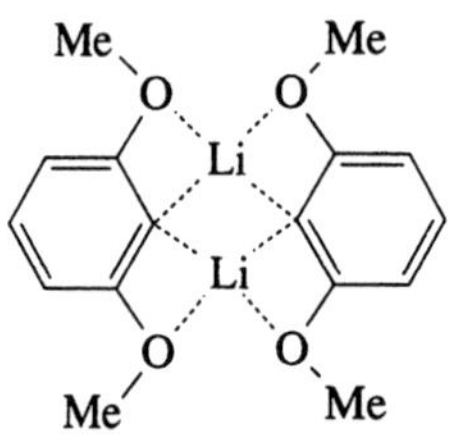 (123) 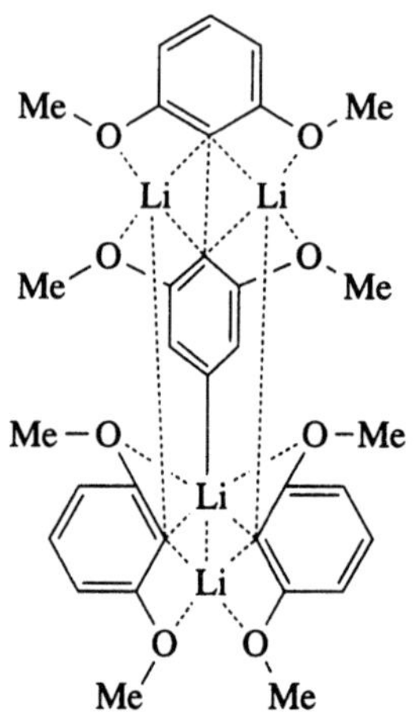 (124)

Me N Me Me N Me Li N Li Li N Me Me Me Me Me Me N N Me Me

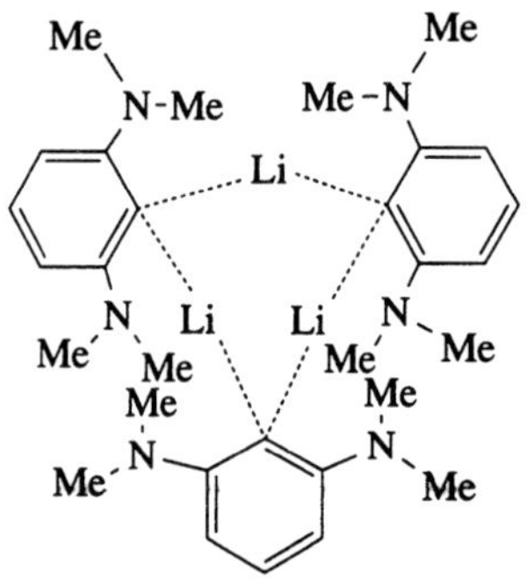

(125)

When a methylene group spacer is inserted between the *ortho* heteroatom and the carbanionic center, the coordination geometry of the anionic center is no longer restricted to be planar for intramolecular chelation to occur. Hence, *o*-(dimethylaminomethyl)phenyllithium (**126**) crystallizes from an ether/hexane solution as the internal-chelated tetramer (**127**).[126] This structure is analogous to tetrameric phenyllithium (**119**). When an additional dimethylaminomethyl group is substituted at the *ortho′* position as in 2,3,5,6-tetrakis(dimethylaminomethyl)phenyllithium, the aggregate crystallizes as the dimer (**128**).[127] The lithium atoms in both (**127**) and (**128**) are coordinated to four other nonlithium atoms; this coordination can only be achieved by dimerization and tetramerization respectively.

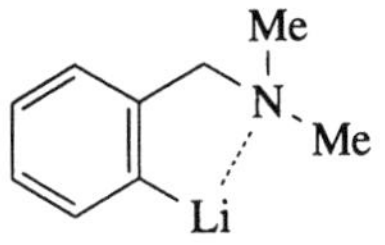

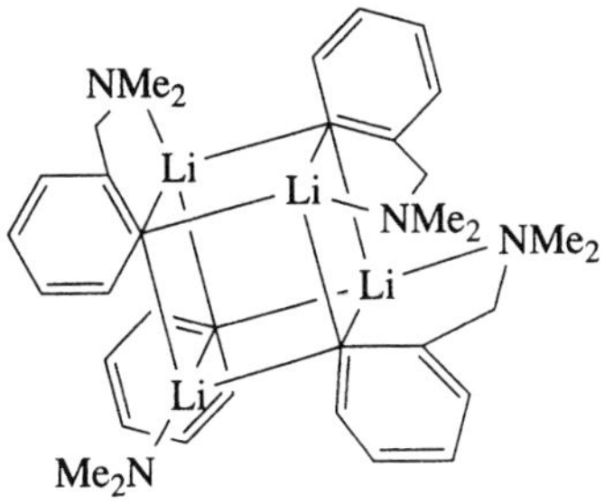

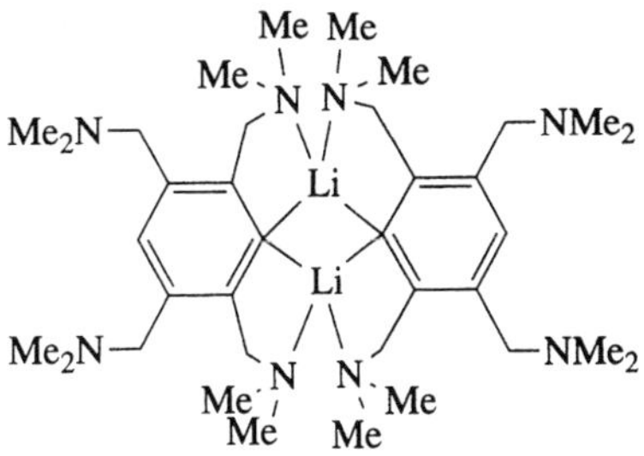

(**126**) (**127**) (**128**)

Two additional aryl crystal structures are noteworthy because they represent examples of alternative structural types of aryllithium anions. The first of these is the 2,2′-dianion of biphenyl (**129**). This material is characterized as the bis-TMEDA solvate (**130**) with two lithium atoms doubly bridging the two carbanionic centers.[128] The lithium atoms are located above and below the two aromatic rings. A completely different structure, depicted roughly as (**131**), is obtained for thc air- and moisture-sensitive, violet crystals of dilithiobenzophenonc.[129]

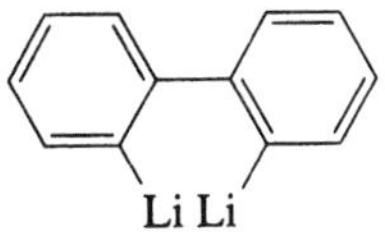

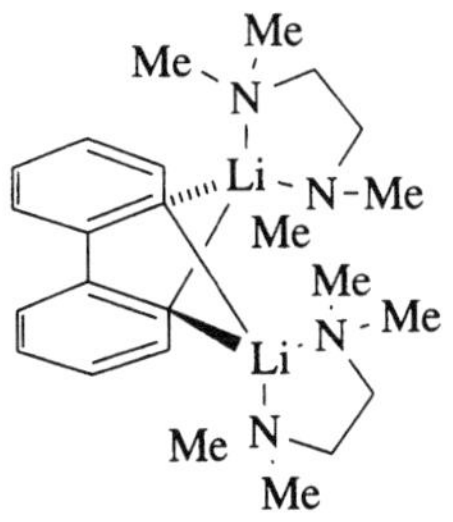

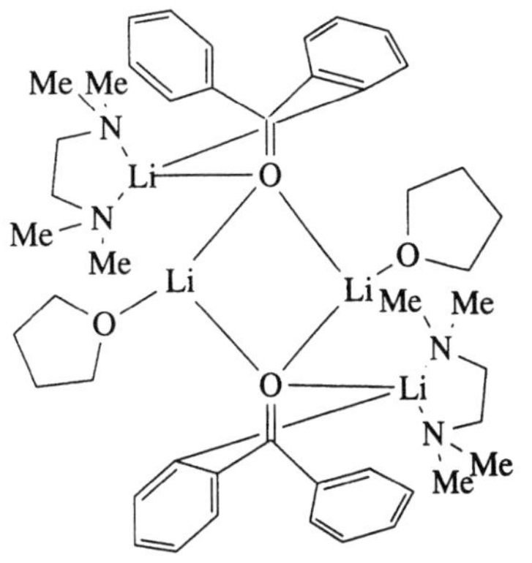

(**129**) (**130**) (**131**)

A few arylmagnesium compounds have been characterized crystallographically. As early as 1964, Stucky and Rundle determined that phenylmagnesium bromide·diethyl etherate consists of a magnesium atom tetrahedrally coordinated to two diethyl ether molecules, the phenyl group and a bromide.[130] This Grignard reagent is depicted as (**132**). Diphenylmagnesium·TMEDA also crystallizes as a monomer (**133**) with a tetrahedrally, four-coordinate magnesium atom.[131]

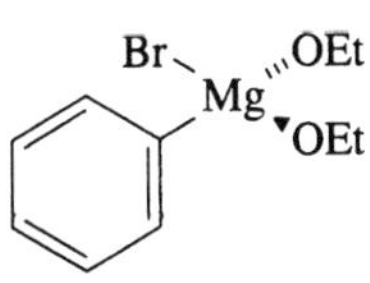

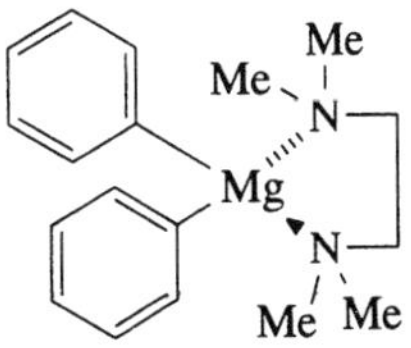

(**132**) (**133**)

Bickelhaupt and coworkers have determined the crystal structures of a series of crown ether solvated magnesium compounds. A sequence of these compounds is illustrated as the internally coordinated 15-crown-4-xylylmagnesium chloride (**134**)[132] and bromide (**135**),[133] as well as the organometallic, rotaxane (**136**).[134] Note the similarity between these structures and the corresponding aliphatic dialkylmagnesium rotaxane (**83**).

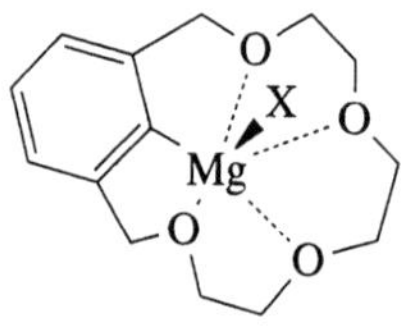

(**134**) X = Cl
(**135**) X = Br

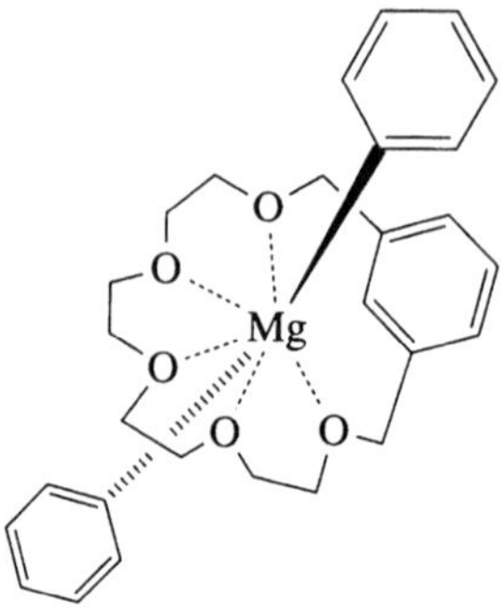

(**136**)

1.1.3.6 Enolates and Enamines and Related Species

1.1.3.6.1 Ketone enolates

One of the most relevant and fruitful areas of structural investigation for synthetic organic chemistry during the past decade has been the crystal structure determinations of a variety of enolate and closely related carbanions. Although these species have been considered only as transient reactive intermediates, a number of these enolates can be crystallized out of solution at subambient temperature and stabilized under a stream of cold, dry nitrogen gas during the 24–48 h necessary for X-ray diffraction data collection. A systematic review of these structures known to date begins with the ketone enolates.

Seebach, Dunitz and coworkers first described the THF-solvated tetrameric aggregates obtained from THF solutions of 3,3-dimethyl-2-butanone (pinacolone) and cyclopentanone lithium enolates.[1] These are represented as (**137**). The pinacolone enolate also crystallizes as the unsolvated hexamer (**138**) from hydrocarbon solution, but this hexamer rearranges instantaneously to the tetramer (**137**) in the presence of THF.[135] Williard and Carpenter completed the characterization of both the Na^+ and the K^+ pinacolone enolates.[136] Quite unexpectedly the Na^+ pinacolone enolate is obtained from hydrocarbon/THF solutions as the tetramer (**139**) with solvation of the Na^+ atoms by unenolized ketone instead of by THF. The potassium pinacolone enolate is a hexameric THF solvate depicted as (**140**) and described as a hexagonal prism. A molecular model of (**140**) reveals slight chair-like distortions of the hexagonal faces in (**140**) so that the solvating THF molecules nicely fit into the holes between the pinacolone residues.

The pinacolone enolate residue crystallizes as a dimer with solvation by *N,N,N′*-trimethylethylenediamine (TriMEDA) as indicated in formula (**141**).[137] In this structure the NH hydrogen on the secondary amine is relatively close to the terminal carbon of the enolate residue, *i.e.* NH—C═C is 2.60 Å. This

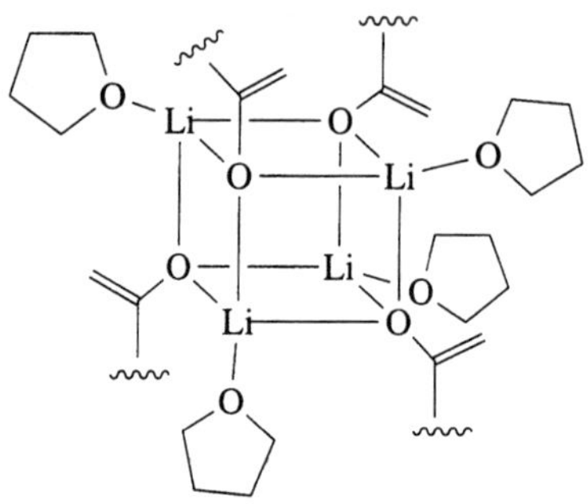

(**137**)

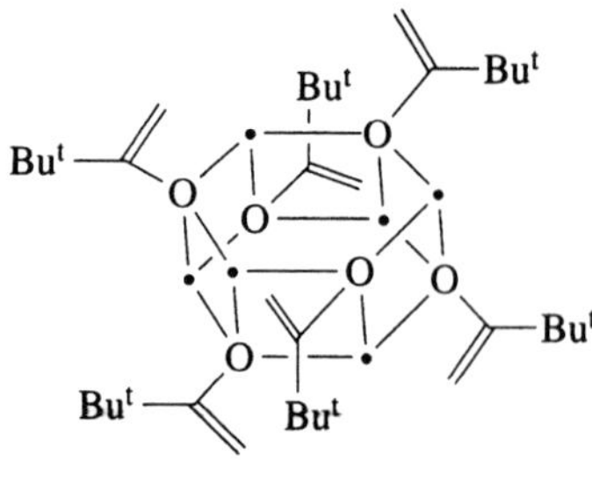

(**138**) • = Li

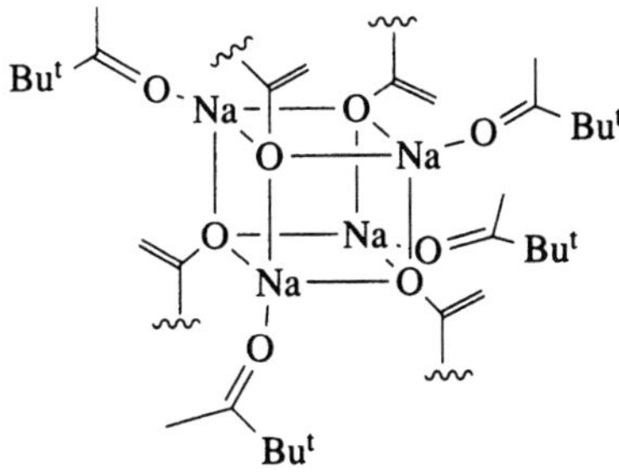

(139)

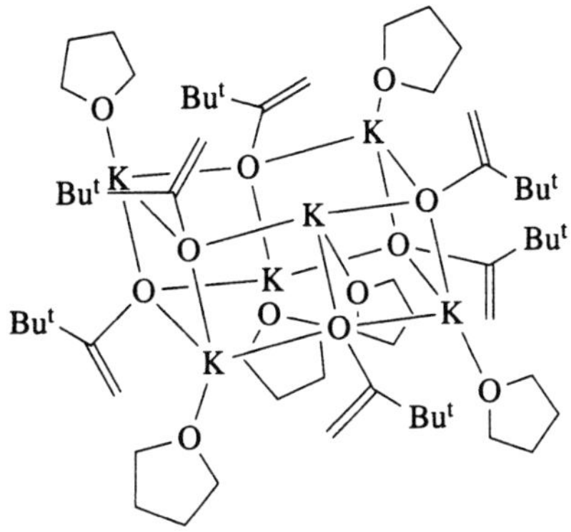

(140)

structure can prove useful in a structure–reactivity correlation, since it includes the hydrogen which is transferred in the enolization process. It is also valuable in explaining the fact that incomplete deuterium incorporation is often observed upon 'kinetic' protonation of certain enolates with deuterated acids since the NH proton can be returned directly to the enolate residue rather than a deuterium.

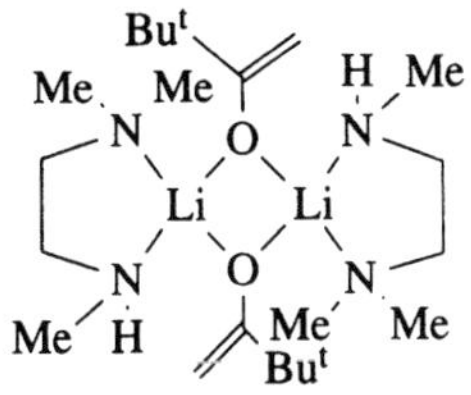

(141)

In an attempt to influence the aggregation state of a simple ketone enolate by intramolecular chelation, the homolog of pinacolone (**142**) was prepared.[138] The lithium enolate of this material cocrystallizes with lithium diisopropylamide (LDA) to produce the mixed, dimeric aggregate (**143**). A sequence of mixed enolate/amide base aggregates with five-, six- and eight-membered chelate rings similar to the aggregate (**143**) and depicted as (**144**) have been characterized.[139] It is noteworthy that the bulky silyloxy group serves as a ligand for the terminal, three-coordinate lithium atoms in (**143**).

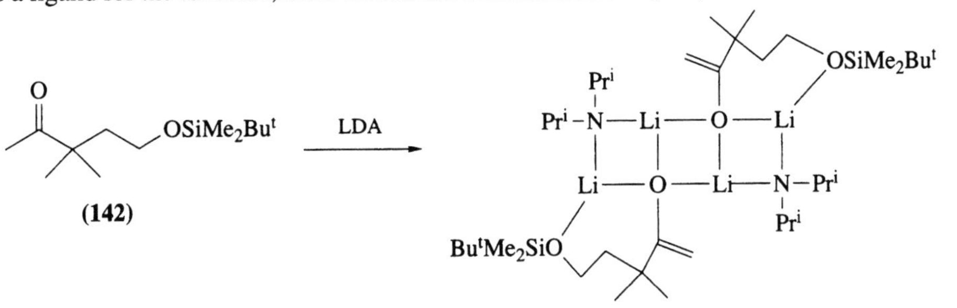

(142) LDA → (143)

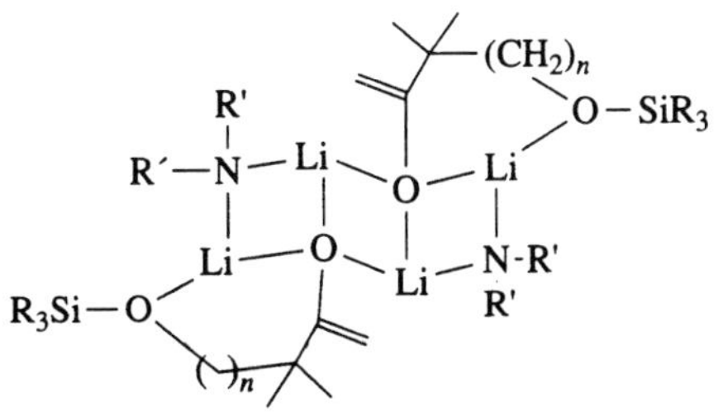

(144) n = 0, 1, 2, 3

Dimethylaminomethylacetophenone (**145**) reacts with LDA in diethyl ether to produce the tetrameric enolate aggregate (**146**).[140] This aggregate is an internally chelated variation of the tetramer (**137**). The origin of this enolate was not well defined in the original paper describing its characterization.

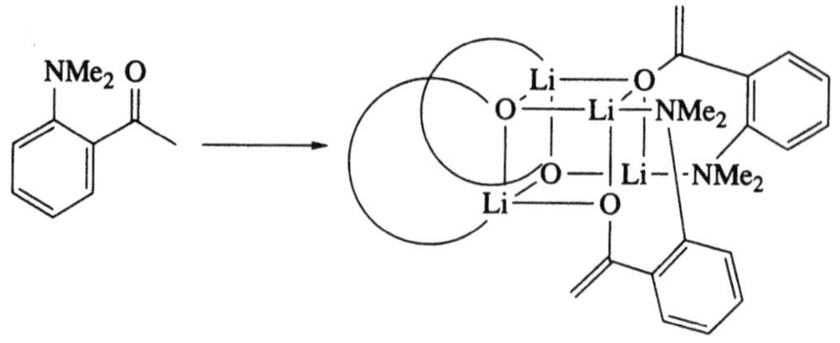

(**145**) (**146**)

The pinacolone lithium enolate condensation product with pivaldehyde (**147**) has been characterized as the tetrameric aggregate (**148**).[141] However, an attempted condensation reaction of pinacolone with itself as shown in Scheme 8 led to crystallization of a product derived from subsequent dehydration and reenolization, *i.e.* (**149**). This dienolate (**149**) was characterized as the dimer (**150**) solvated by dimethylpropyleneurea (DMPU).[142]

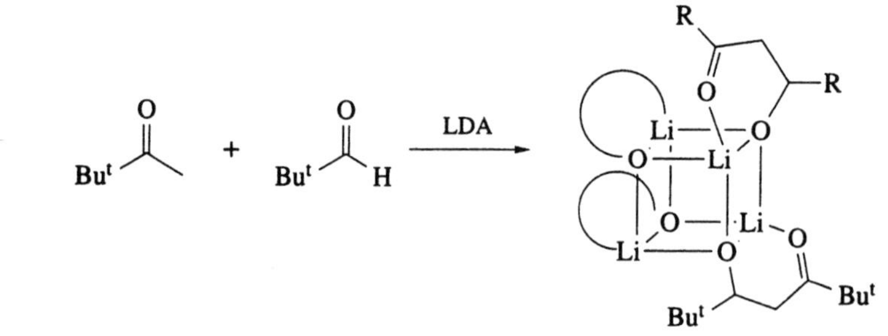

(**147**) (**148**)

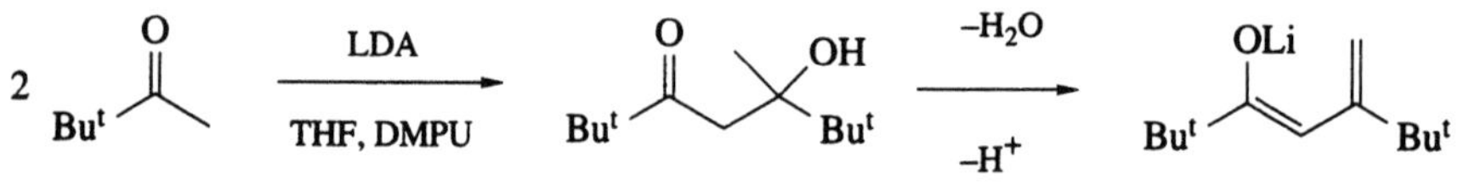

(**149**)

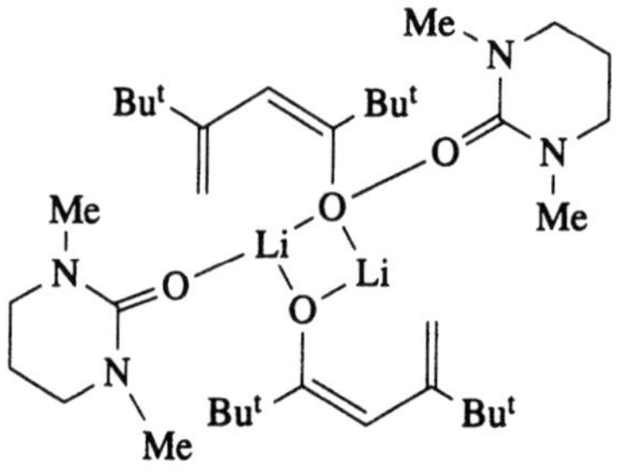

(**150**)

Scheme 8

The azaallyl enolates, *i.e.* enolates derived from ketone imines or hydrazones are synthetic equivalents of the ketone enolates and thus two examples of azaallyl enolates are included in this section. Lithiated cyclohexanonephenylimine (**151**) crystallizes out of hydrocarbon solution as the dimeric diisopropylamine solvate (**152**).[143] Significant disorder between the cyclohexyl and the phenyl moieties is observed in this crystal structure; however, it is clear that there are no η^3-azaallyl carbon contacts in this structure. This lithiated imine structure can be compared with the lithiated dimethylhydrazone of cyclohexanone

(**153**).[144] In this latter structure, roughly depicted as (**154**), there are two different lithium atoms as well as two different anion residues. In one of the residues a lithium is η^5-coordinated and in the other residue the lithium is η^1-coordinated. The possible origins of the selectivity of the alkylations of the metallated hydrazones are discussed relative to this structure. The lithiated hydrazone enolate (**155**) prepared from (*S*)-(–)-1-amino-2-(methoxymethyl)pyrrolidine (SAMP) hydrazone of 2-acetylnaphthalene (**156**) yields the monomeric bis-THF-solvated species (**157**) as ruby red crystals.[145] This is one of the few examples of the crystallization of a resolved enolate substrate.[146]

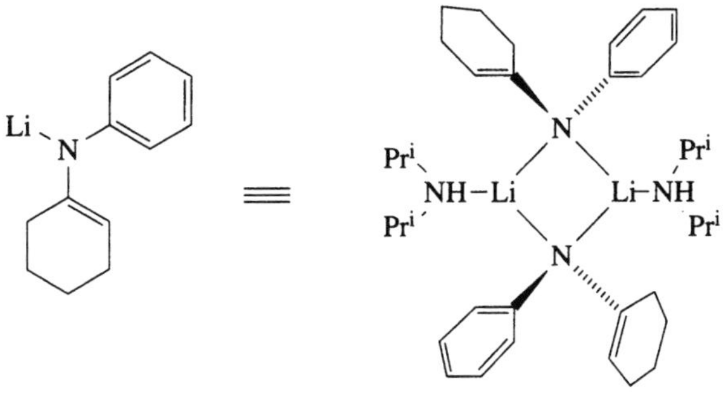

(**151**) (**152**)

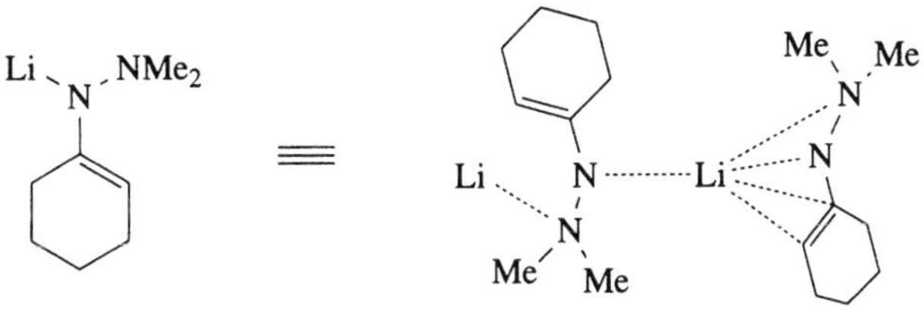

(**153**) (**154**)

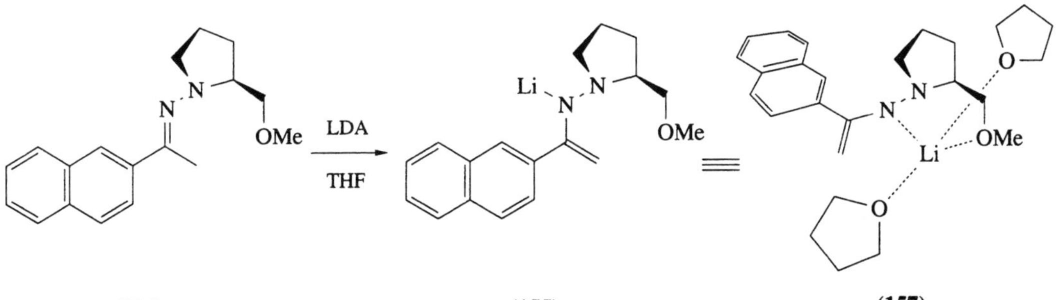

(**156**) (**155**) (**157**)

An uncharacteristic enolate coordination is observed for the α,α′-ketodianion derived from dibenzyl ketone (**158**).[147] The dianion crystallizes as a bis-TMEDA solvate with the general structure shown as (**159**). The two lithium atoms are on opposite sides of the relatively planar carbon skeleton. Each is solvated by TMEDA. Besides coordination to the oxygen, the lithium atoms are in close contact with four additional carbon atoms.

An early prediction about the structure of a magnesium ketone enolate[148] was subsequently modified when the diethyl ether solvated, magnesium bromide enolate derived from *t*-butyl ethyl ketone was characterized as the dimer (**160**) with bridging enolate residues.[149]

Recently the isolation and structure determination of the aldol product of the chiral iron enolate (**161**) with benzaldehyde was obtained as (**162**).[150] This structure is presumed to mimic closely the structure of the cyclic transition state for the aldol reaction.

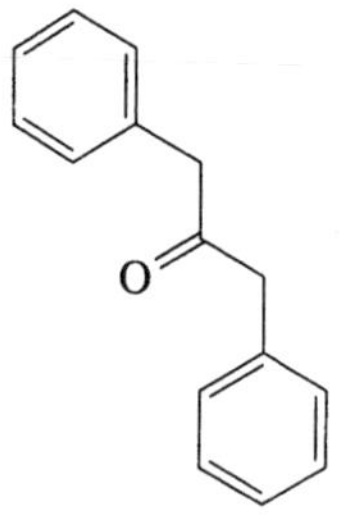

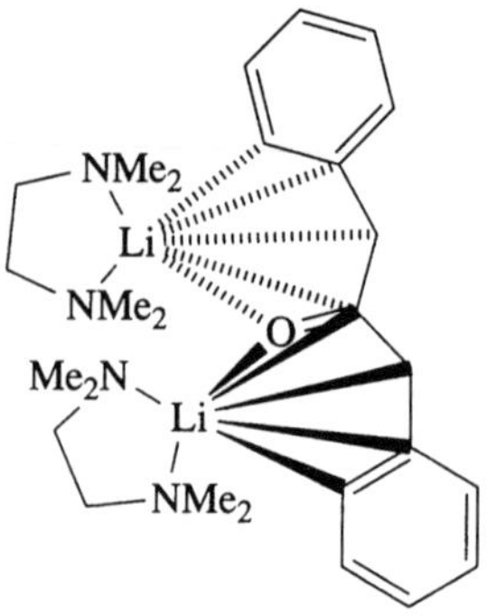

(158)

(159)

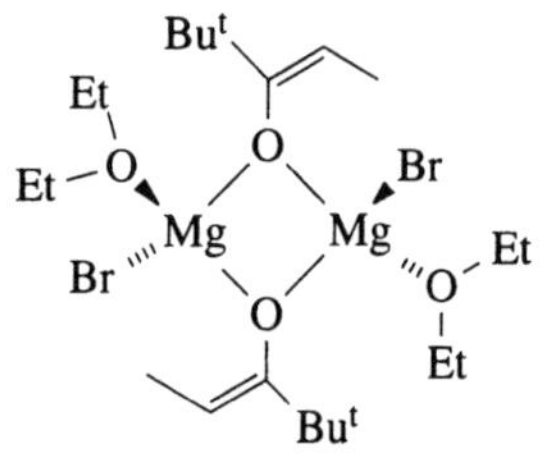

(160)

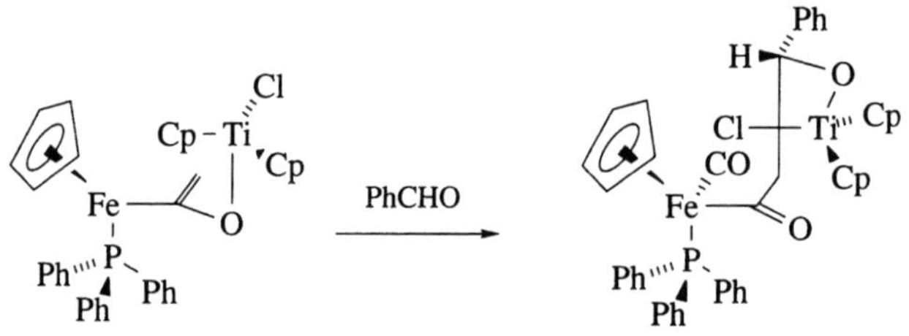

PhCHO

(161)

(162)

1.1.3.6.2 Amide and ester enolates

Few ester enolate crystal structures have been described. The lack of structural information is no doubt due to the fact that the ester enolates undergo α-elimination reactions at or below room temperature. A good discussion of the temperatures at which lithium ester enolates undergo this elimination is presented in the same paper with the crystal structures of the lithium enolates derived from *t*-butyl propionate (**163**), *t*-butyl isobutyrate (**164**) and methyl 3,3-dimethylbutanoate (**165**).[151] It is significant that two of the lithium ester enolates derived from (**163**) and (**165**) are both obtained with alkene geometry such that the alkyl group is *trans* to the enolate oxygen. It is also noteworthy that the two TMEDA-solvated enolates from (**163**) and (**164**) are dimeric, while the THF-solvated enolate from (**165**) exists as a tetramer.

Three additional ester enolates have been characterized and these can be compared to the lithium enolates (**163**)–(**165**). Recently we have obtained the mixed sodium ester/sodium hexamethyldisilazide aggregate derived from *t*-butyl isobutyrate and sodium hexamethyldisilazide as the TMEDA-solvated aggregate (**166**).[152] The second structure of interest is the zinc ester enolate, *i.e.* Reformatsky reagent, derived from *t*-butyl bromoacetate.[153] This zinc enolate (**167**) forms an eight-membered ring with the zinc atoms bonded directly to both enolate oxygens and to the α-carbon of the enolate. The observation of direct metal interaction with the enolate α-carbon of simple substrates is rare for alkali and alkaline earth metal cations. The third lithium ester enolate is derived from ethyl *N,N*-diethylglycine (**168**). It crystallizes as the internally chelated hexamer (**169**) which resembles the hexagonal prisms (**138**) and (**140**).[154]

Few amide or amide-like enolates have been characterized. This is somewhat surprising since amide enolates are expected to be less susceptible to ketene formation than the corresponding ester enolates.

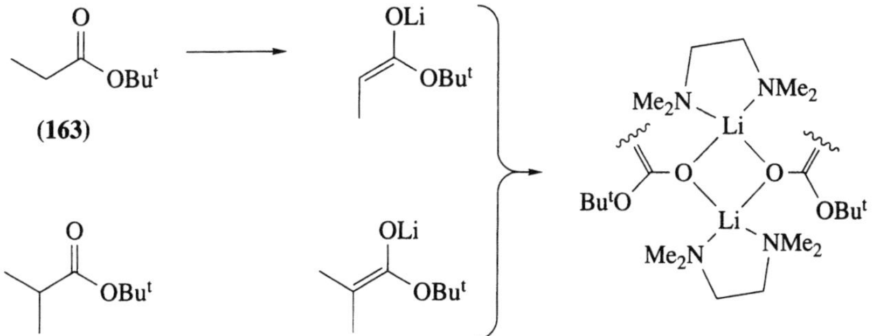

(163)

(164)

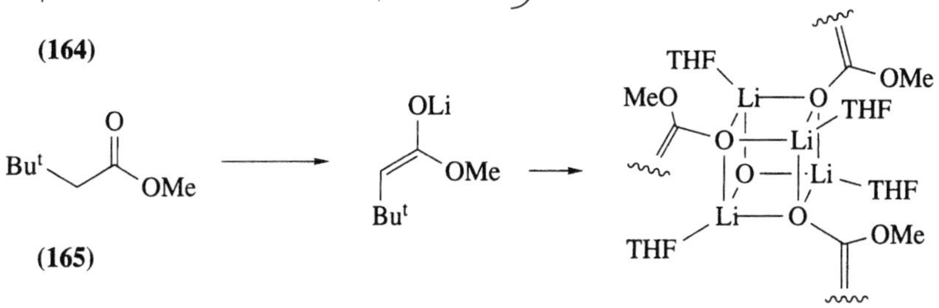

(165)

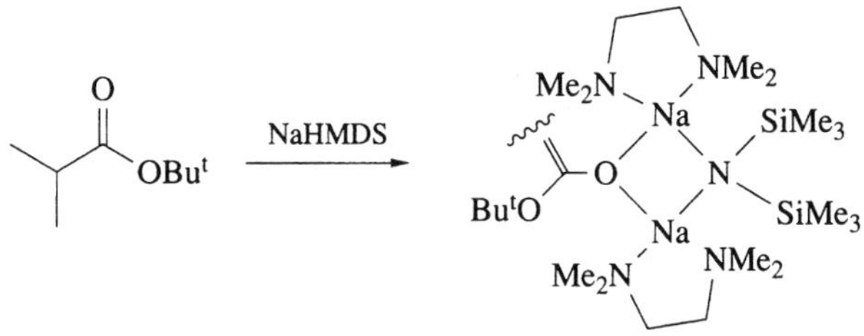

(166)

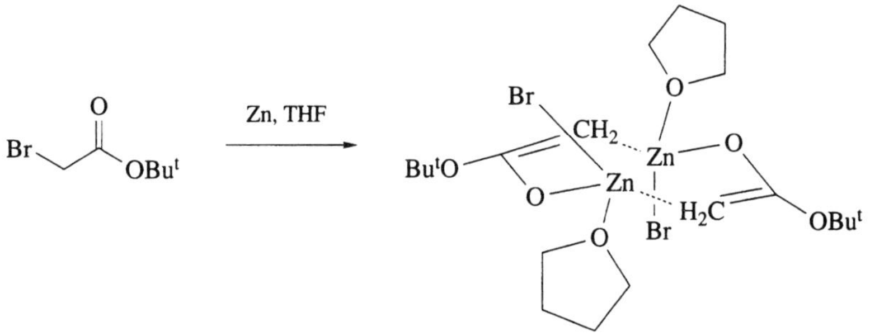

(167)

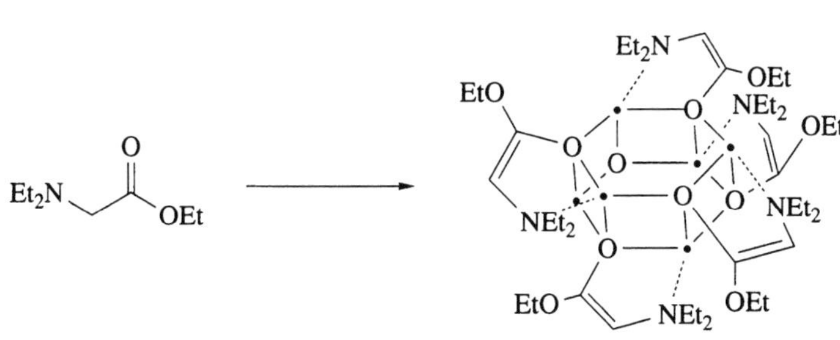

(168)

(169) • = Li

The least highly substituted amide enolate whose structure is known is the lithium enolate of *N,N*-dimethylpropionamide (**170**).[155] This enolate is obtained as a dimer solvated by TriMEDA, *i.e.* (**171**). The alkene geometry in (**171**) is opposite that found in the ester enolates from (**163**) and (**165**). Thus in the dimer (**171**), the terminal methyl group is *cis* to the enolate oxygen and the alkenic residues are on opposite sides of the Li—O—Li—O core. It was noted that the hydrogen atom of the secondary amino group in (**171**) points in the direction of the virtual lone pair of electrons on the amide nitrogen.

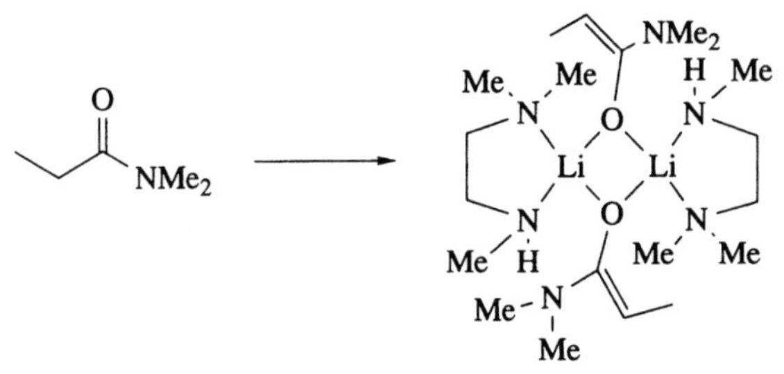

(**170**) (**171**)

The lithium enolate derived from *N,N*-dimethylcycloheptatrienecarboxamide (**172**) crystallizes as the bis-THF-solvated dimer (**173**).[156] Neither the amide nitrogens nor the extended π-system participates in complexation to the lithium atoms in this complex.

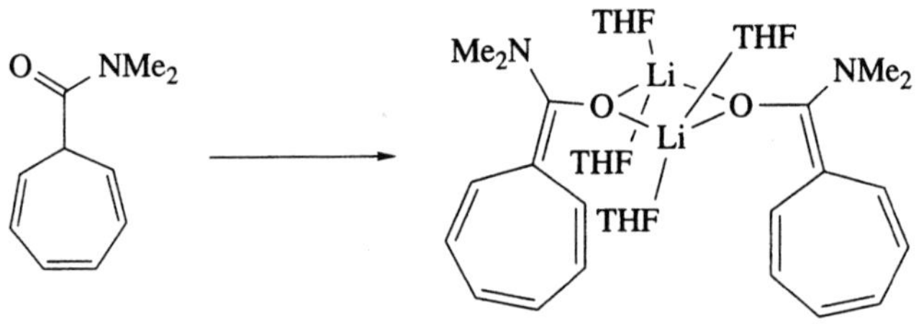

(**172**) (**173**)

Two lithium enolates (**174**) and (**175**) derived from the vinylogous urethanes (**176**) and (**177**) have been crystallized and subjected to X-ray diffraction analysis.[157] Although the individual enolate units combine to form different aggregates, they are very nearly identical in conformation, *i.e. s-trans* around the 2,3-bond; however, both the aggregation state and the diastereoselectivity of the enolates differ.[158] The enolate (**175**) is obtained from benzene solution as a tetramer and (**174**) is obtained from THF solution as a dimer. The origin of the diastereoselectivity shown by these enolates is subtle.

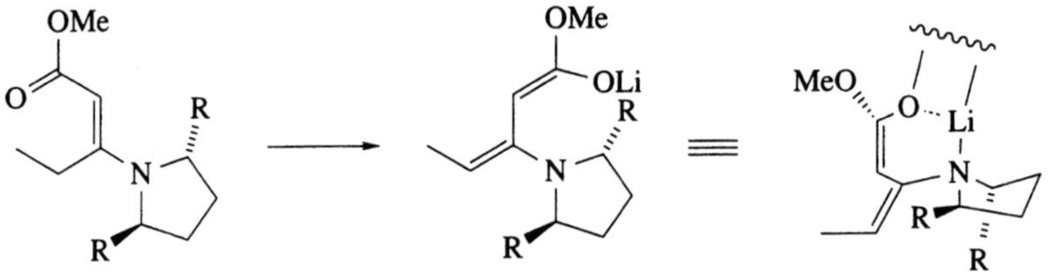

(**176**) R = H
(**177**) R = Me

(**174**) R = H
(**175**) R = Me

1.1.3.6.3 Nitrile and related enolates

Three examples of nitrile-stabilized enolates have been described by Boche *et al.* Two of these structures incorporate the anion of phenylacetonitrile. The TMEDA-solvated dimer (**178**) crystallizes out of benzene solution;[159] however, the mixed nitrile anion·LDA·$(TMEDA)_2$ complex (**179**) is obtained when excess LDA is present.[160] This latter complex has often been mistaken as a geminal dianion since it frequently gives products that appear to arise from a dianion. The crystal structure of the anion 1-cyano-2,2-dimethylcyclopropyllithium (**180**) consists of an infinite polymer (**181**) that is solvated by THF.[161] Interestingly, there are C—Li contacts in this structure and the carbanionic carbon remains tetrahedral.

The tetrahedral carbanion agrees well with the experimental results that optically active cyclopropylnitrile carbanions can undergo reprotonation with retention of configuration under certain conditions.

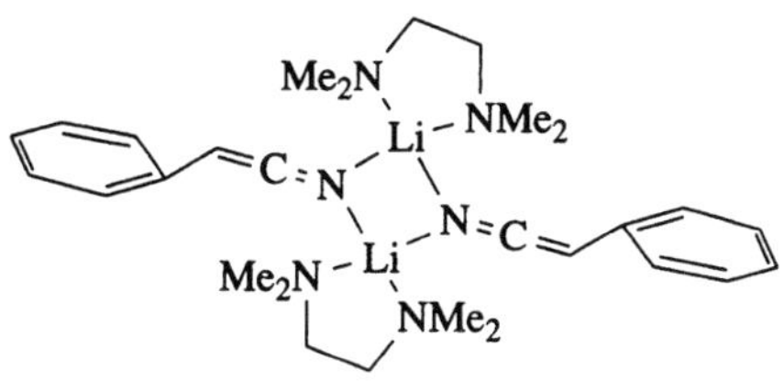

(178) 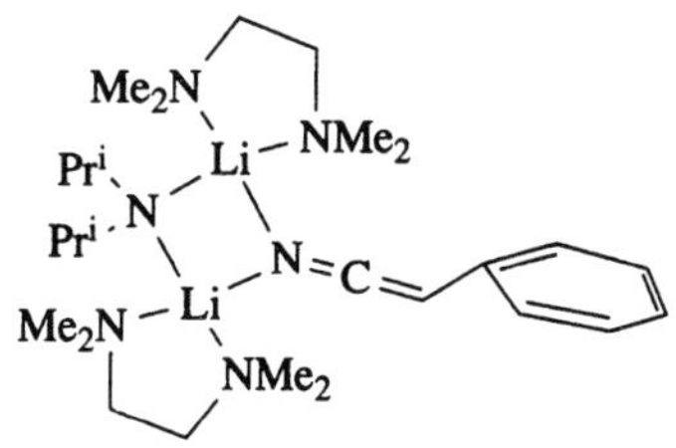 **(179)**

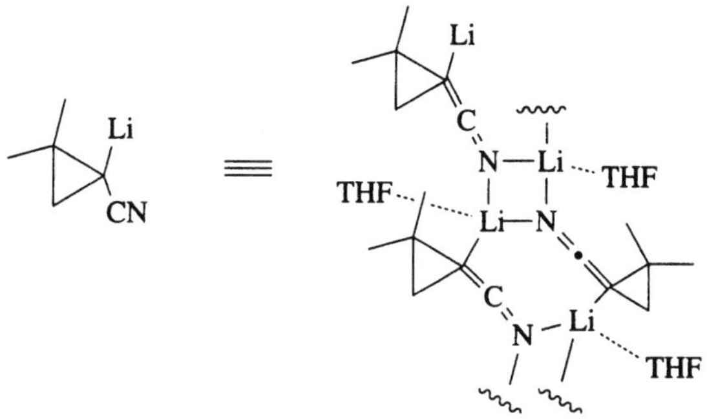

(180) **(181)**

A true dianion (**182**) is obtained in the reaction of (**183**) with base and this dianion crystallizes from ether/hexane solution with the stoichiometry $[(Me_3SiCCN)_{12}\cdot Li_{24}\cdot (Et_2O)_6\cdot (C_6H_{14})]$.[162] The crystal structure exhibits both N—Li and C—Li contacts and is best described in the original manuscript because there is no simple way to redraw this exceedingly complex aggregate structure.

(183) **(182)**

The crystal structures of both $Na^+C(CN)_3^-$ [163] and $K^+C(CN)_3^-$ [164] are known for comparison. In all examples of the nitrile-stabilized carbanions except the dianion (**182**), the metal coordination to the organic anion is through the nitrogen. No evidence of interaction between the metal and the nucleophilic carbon atom is seen. Lithiated imine (**184**) is somewhat analogous to dimer (**150**), although this species is not derived from an enolizable substrate.[165]

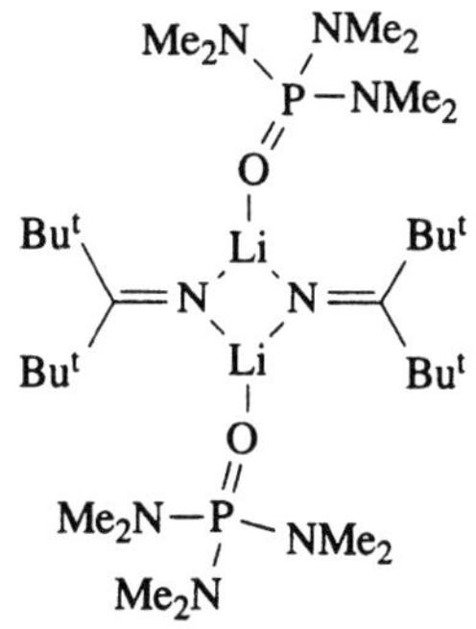

(184)

1.1.3.6.4 Other stabilized enolates

The crystal structure of a single stabilized nitronate carbanion derived from phenylnitromethane is reported as the polymeric ethanol solvate (**185**).[166] The same geometry found in the nitronate anion (**185**) was also found for the *t*-butyldimethylsilyl ester derived from quenching this nitronate.[167] References to other structures containing carbanions stabilized by a nitro group are given in Table 3; however, it is to be noted that these are derived from highly acidic carbon acids.

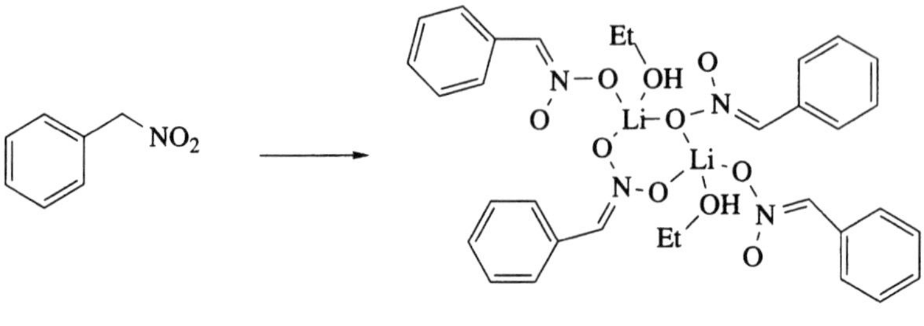

(**185**)

Table 3 Stabilized Carbanions

Compound name	*CSD refcode*	*Ref.*
$Ca(acac)_2 \cdot 2H_2O \cdot H_2O)$	BOLTIF	220
$Na(acac) \cdot H_2O$	CAFNEC	221
$Ca(acac)(CH_3CO_2\text{-}) \cdot 2H_2O)$	CUHGER	222
$Li[PhC(O)CHC(O)Ph] \cdot 2H_2O \cdot Et_3N$	DEKJUY	223
$Mg_3(acac)_6$	DENGAE	224
K-nitromalonamide	NOMLNB	225
Na(eaa)·15-crown-5	BODKUG	226
K(eaa)·18-crown-6	CREALK, CRKEAC01, CRKEAC10	227
$CsC(NO_2)_3$	CSTNME	228
K-1,1-dicyano-3-thiabut-1-en-2-olate	FAZBAJ	229
Na-2-propenal-3-olate	FUSPEO	230
K-4,4-dinitro-2-butenamide	KDNBUT	231
K-2,2-dinitroethylacetamide	NEYACM	232

The sodium salt of the stabilized enolate derived from the heteroaryl-substituted 2-oxoglutaric acid ester (**186**) is reported to have the alkene geometry as shown in formula (**187**).[168] Finally, Collum *et al.* have reported the structure of the lithiated anion derived from the *N,N*-dimethylhydrazone of 2-methoxycarbonylcyclohexanone (**188**).[169] This enolate crystallizes as the dimeric, bis-THF-solvated aggregate (**189**).

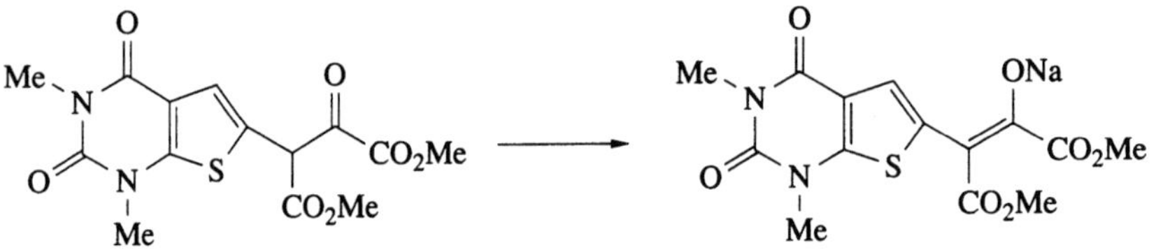

(**186**) (**187**)

1.1.3.7 Heteroatom- substituted Carbanions (α-N, α-P or α-S)

To date there are only a few synthetically relevant crystal structures with a nitrogen directly attached to the carbanionic center. X-Ray crystal structure determination of the lithium salt of bis-lactim ether (**190**), derived from alanine, has been characterized and is depicted as (**191**).[170] This structure illustrates

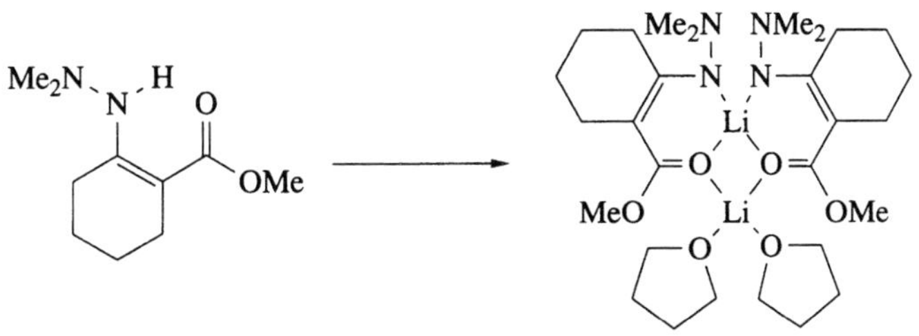

(188) **(189)**

that the lithium atoms share very different environments, *i.e.* one five coordinate and one four coordinate. The bromomagnesium derivative of *N*-pivaloyltetrahydroisoquinoline (**192**) crystallizes from THF as a monomer with octahedrally coordinated Mg atoms.[171] This Mg atom coordinates to the carbanion, the amide carbonyl oxygen, a bromine and three THF molecules. A mechanistic proposal is derived from the crystal structure (**193**) to explain the selectivity for addition of this anion to ketones.

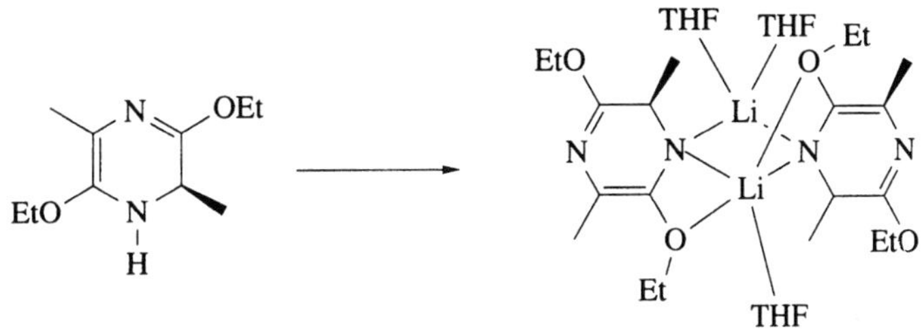

(190) **(191)**

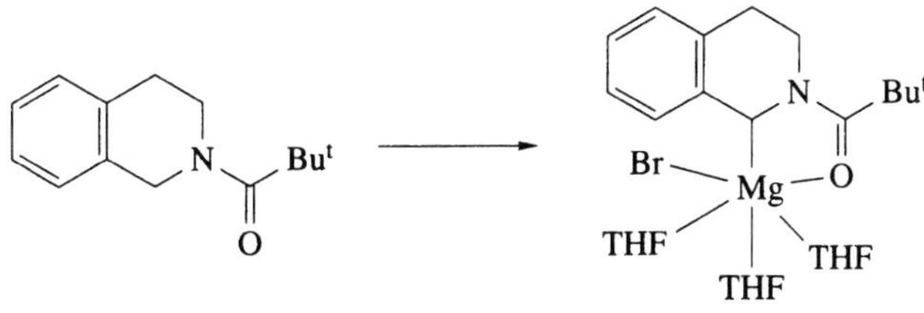

(192) **(193)**

N-Phenylpyrrole (**194**) is monolithiated at the 2-position of the heterocyclic ring. This monolithium compound crystallizes as the TMEDA-solvated dimer (**195**).[172] This structure agrees well with the ^{6}Li–^{1}H 2D heteronuclear Overhauser NMR spectroscopy (2D-HOESY). The structure serves to predict correctly that the second lithiation to a dianion occurs at the *ortho* position of the phenyl ring located closest to the lithium in the monoanion.

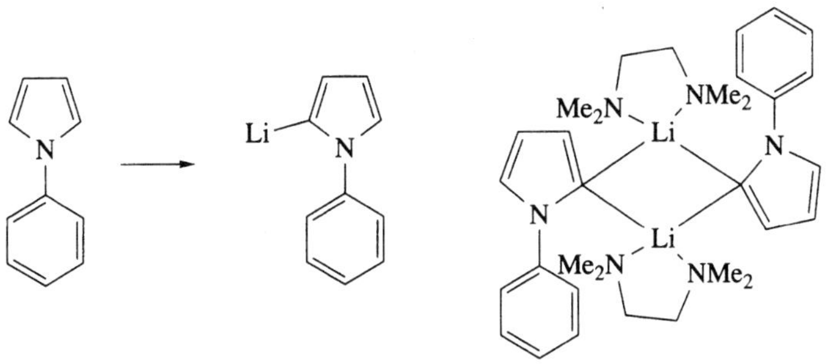

(194) **(195)**

Finally, the crystal structure of a lithiated amino nitrile (**196**) has been described in Boche's recent review article as a dimer (**197**) similar to the other nitrile anions (**179**), (**180**) and (**182**).[14] However, there is extended network throughout the solid (**197**) due to coordination of the lithium atoms with the oxygens at the *para* position of the aryl rings of adjacent molecules.

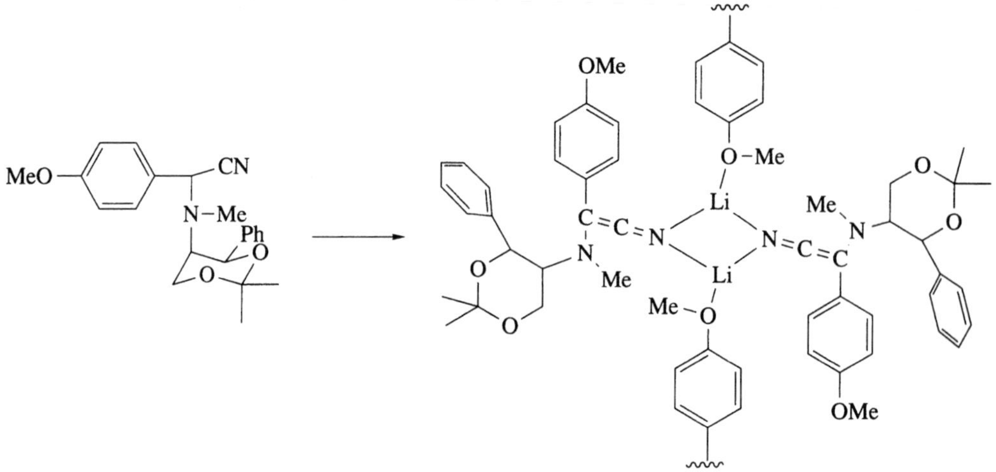

(**196**) (**197**)

Excluding the α-P-, α-Si-substituted carbanions which are listed in Table 2, there exist relatively few simple α-P-substituted carbanions whose structures are known. References to the crystal structures of some tri (alkyl or aryl) substituted phosphines are listed in Table 4. Few if any of these compounds have been utilized as synthetic reagents. Only two synthetically useful phosphorus-stabilized carbanions of Group Ia or IIa metal cations have been examined by X-ray diffraction analysis. The lithium carbanion of 2-benzyl-2-oxo-1,3,2-diazaphosphorinane (**198**) crystallizes as a monomeric bis-THF solvate (**199**) with a tricoordinate lithium atom.[173] The magnesium salt of diethoxyphosphinyl acetone (**200**) is characterized as an intramolecularly chelated trimer[174] similar in structure to $[Mg(acac)]_3$. The Cu salt of this β-keto phosphorus-stabilized anion exists only as a monomer.[175]

There has been much synthetic interest in sulfur-stabilized carbanions. These anions include sulfides, 1,3-dithianes, sulfoxides, sulfones and sulfoximides. Since the structural results in this area have been recently compiled and discussed in excellent detail by Boche,[14] it will suffice to present only the list of compounds in Table 5 whose structures have been reported. Most of these anions have varied and unique

Table 4 α-Phosphorus-stabilized Carbanions

Compound	*CSD refcode*	*Ref.*
$LiCH_2PMe_2 \cdot TMEDA$	CEDSIN, CEDSIN10	233, 234
$LiCH_2P(Me)Ph \cdot (-)$-sparteine	VAGHOA	234
$LiCH_2P(Me)Ph \cdot TMEDA$	VAGHIU	234
$[(LiCH_2PPh_2)_2(dioxane)_3] \cdot dioxane$	DUJDIV	235
$[LiC(PMe_2)_3 \cdot 2THF]_2$	CESCEI, CESCE10	236
$LiCH(PPh_2)_2 \cdot THF$	GIKXAZ	237

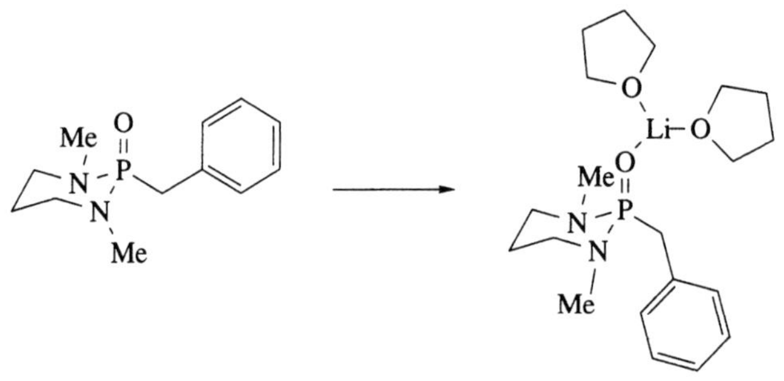

(**198**) (**199**)

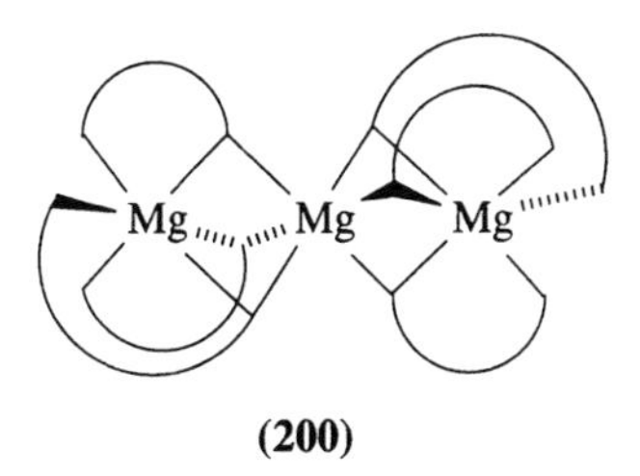

(200)

structures. These structures have proven extremely valuable for structure reactivity correlations and this is discussed in the aforementioned review.

Table 5 α-Sulfur-stabilized Carbanions

Compound type	*Compound*	*CSD refcode*	*Ref.*
Sulfides	(MeSCH$_2$Li)$_2$·(TMEDA)$_2$	CEPLEO	238
	(PhSCH$_2$Li)$_2$·(TMEDA)$_2$	CEPLAK	238
	[(*E*)-2-Butenyl-1-SBu^t]·TMEDA	GAJDAW	239
1,3-Dithianes	[Li(2-methyl-1,3-dithiane)]$_2$·(TMEDA)	LIMDTE	240
	[Li(2-phenyl-1,3-dithiane]$_2$·(TMEDA)$_2$·THF	LIPTHF	241
Sulfoxides	(α-Methylbenzyl phenyl sulfoxide·Li)$_2$·(TMEDA)$_2$	FIGHEI	242
Sulfones	[α-(Phenylsulfonyl)benzyllithium]$_2$·(TMEDA)$_2$	DIBXIV	243
	(Phenylsulfonylmethyllithium)$_2$·(TMEDA)$_2$	DOMSED	244
	2,2-Dipehnyl-1-(phenylsulfonyl)cyclopropyllithium·(DME)$_2$·(DME)	VADKIU	245
	2,0-Lithium dianion of bis(TMS)methyl phenyl sulfone	GEHZOI	246
	[α,α-lithium trimethyl(phenylsulfonylmethyl)silane]$_6$·LiO$_2$·(THF)$_{10}$	GAFXIU	247
	α-(Phenylsulfonylallyl)lithium·DME	FAGFOI	248
	α-(Methylbenzyl)phenylsulfonyldilithium·diglyme	GAVYUX	249
	Isopropylphenylsulfonyldilithium·diglyme	GAVZAE	249
	Bis(TMS)methyl phenyl sulfone potassium (Et$_2$O)(18-c-6)	SAKXOR	250
Sulfoximes	[(*S*)-(*N*-methyl-*S*-phenylsulfonimidoyl)methyllithium]$_4$·(TMEDA)$_2$	FISNOW	251
	(TMS)[*N*-(TMS)-*S*-phenylsulfonimidoyl][methyllithium]$_4$	FECRAG	252

1.1.3.8 Related Alkali Metal and Alkaline Earth Anions

1.1.3.8.1 Amides and alkoxides

Since most of the synthetically useful enolate anions described in the previous section are prepared by the reactions of enolizable substrates with alkali metal amide bases, it is appropriate to note a few structures of these amide bases. The common bases in synthetic organic chemistry include LDA and LHMDS. The structures of both of these bases are known as the THF solvates.[176,177] Both of these compounds form bis-solvated dimers corresponding to structure (**201**). The diethyl ether solvate of LHMDS also forms a bis-solvated dimer (**202**).[178] Sodium hexamethyldisilazide crystallizes as an unaggregated monomer from benzene solution.[179] Two different crystalline forms of KHMDS are known as the polymeric dioxane solvate (**203**),[180] and the unsolvated dimer (**204**).[181]

(201) R = Pr^i or SiMe$_3$ **(202)** **(203)** **(204)**

The nonalkali metal bis(trimethylsilyl)amide crystal structures are too numerous to elaborate in detail here, but a partial listing of these includes the bis(trimethylsilyl)amide anion bound to Ti, Sn, Mg, Al, *etc.* Individual references to the structures of these compounds are best found in the CSD.[3] A recent review of the various cage structures available for the main group metal amide bases and alkoxides is given.[182] Recall that a few previously described carbanion crystal structures, *i.e.* (**143**), (**166**) and (**179**), form mixed aggregates containing the lithium diisopropylamide moiety or the NHMDS moiety and a carbanion.

Recent structural investigations on lithium organo(fluorosilyl)amides have revealed that the lithium cation can form aggregates with internal lithium coordination to fluorine, and mixed aggregates of the amide and LiF.[183] Structural types such as (**205**), (**206**), (**207**) and (**208**) have been found for these compounds.

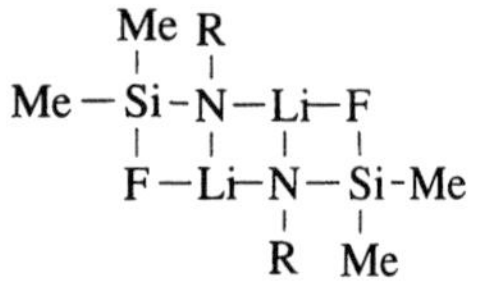

(**205**) 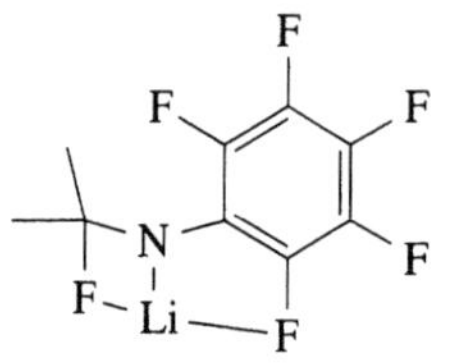 (**206**) 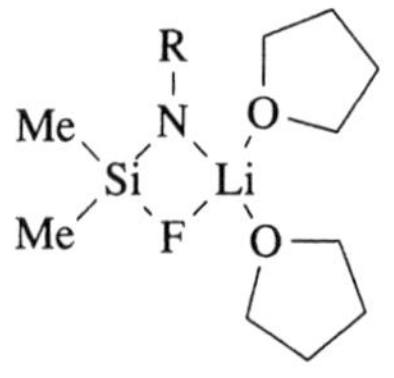 (**207**) 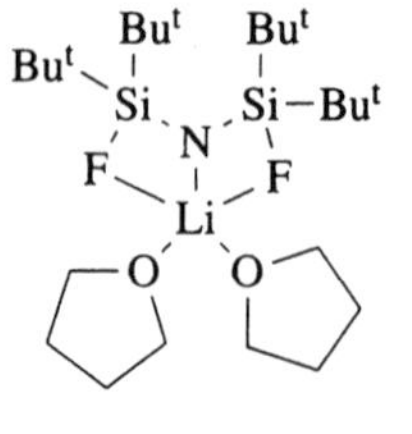 (**208**)

Many of the enolate crystal structures described in Section 1.1.3.6 are coordinated to metal cations only through the enolate oxygen atom. From a structural point of view, these aggregates might be thought of as simple alkoxide anions rather than as carbanions. Because of this structural analogy between the simple enolates and alkoxides, Table 6 is presented. References are given in this table as representative examples of aggregates of alkali metal and alkaline earth salts of relatively simple alkoxides. It is possible to compare the enolate and alkoxide anion structures and perhaps to anticipate new structural types for both groups. Recall also the interesting mixed aggregate that forms between Bu^nLi and lithium *t*-butoxide (**45**). The complex structure of a mixed metal, enolate/alkoxide aggregate has recently been described by our group as (**209**).[184]

1.1.3.8.2 Halides

The effect of added halide salts on organic reactions is recently undergoing intensive scrutiny.[185] To date only a few mixed aggregate structures containing both carbanion residues and halide anions (excluding the many MgX^- aggregates) have been described. Two of these are previously listed in this chapter as (**44**) and (**116**) plus the few mixed cuprates in the next section. References to the crystal structures of a few simple halide salts are also given in Table 6, with the expectation that these structures may provide some guidance with predicting and preparing mixed carbanion/halide aggregates whose structures remain to be determined.

Table 6 Representative Alkoxide and Halide Structures

Compound	*CSD refcode*	*Ref.*
$NaOBu^t$	NABUOX, NABUOX10	253
CsOH·MeOH	GAYCAK	254
$CsOPr^i$	IPRXCS	255
$Ba(OMe)_2$	MEOXBA	256
$Ca(OMe)_2$	MEOXCA	256
$Sr(OMe)_2$	MEOXSR	256
$(CuOBu^t)_4$	CUTBUX	257
(LiCl)·(HMPA)	CAWSIC	258
(NaBr)·$(acetamide)_2$	DIACNB, NABRAA	259
(LiCl)·(*N*-methylacetamide)	LICMAC, LICMAC10	260
(LiBr)·$(acetone)_2$	DECXEO	261
$(MgCl_2)$·(*N*-methylacetamide)	NMALIE	262
(LiI)·(Ph_3PO)	LIPPHO	263

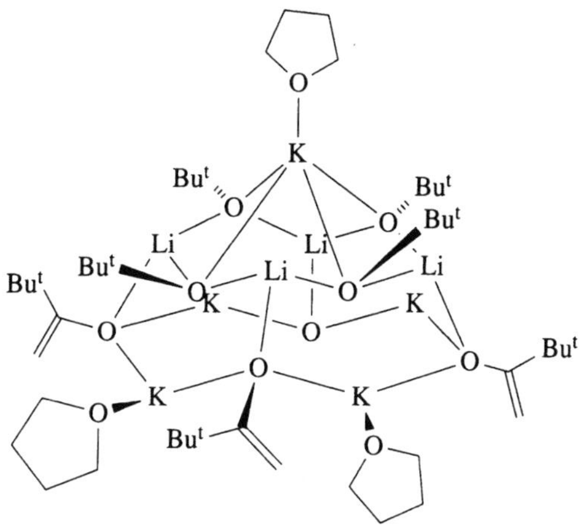

(209)

1.1.3.9 Mixed Metal Cation Structures

1.1.3.9.1 Without transition metals

There are a few carbanion aggregates formed with different metal cations in the aggregate. Three of these structures contain both Li^+ and Na^+ cations. Weiss characterized a $(MeNa)_x$:$(MeLi)_y$ preparation by powder diffraction of this material. These results suggested a geometry analogous to that of the $(MeLi)_4$ tetramer (**38**). A unique diphenyl lithium/sodium·TMEDA complex was characterized also by Weiss with stoichiometry $[Na\text{-}TMEDA]_3$ $[Li\text{-}phenyl_4]$.[186] The single lithium atom, located in the center of the ate complex, forms a pseudo-tetrahedron with four phenyl groups. The lithium atom also lies almost at the center of a triangle formed by TMEDA-coordinated Na atoms. A view from the open face of this structure is approximately as seen in (**210**). A recent characterization of a mixed Li^+/Na^+ metal amide aggregate shows a central core as drawn in (**211**).[187] In this aggregate the anion is derived from the imine of *t*-butyl phenyl ketone (**212**).

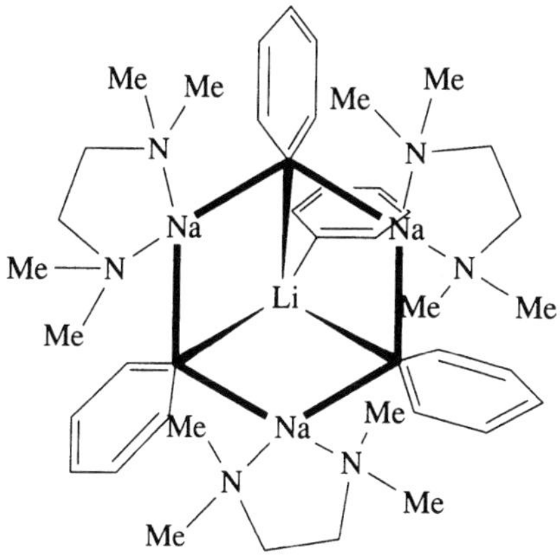

(210)

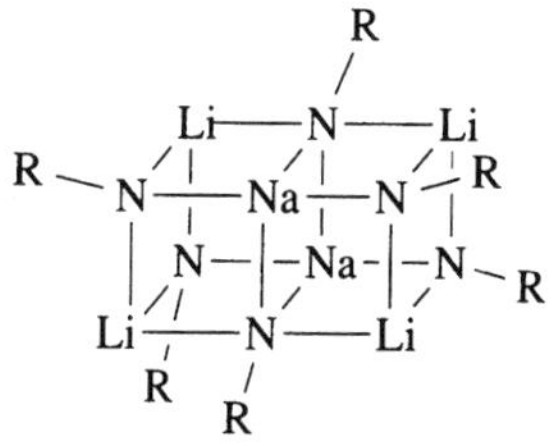

(211)

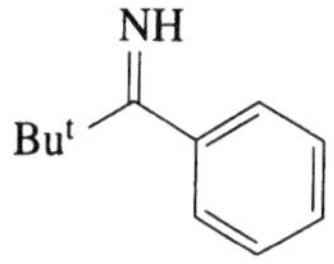

(212)

A mixed Li^+/Mg^+ aggregate corresponding to (**213**) is formed with either phenyl or methyl carbanions.[188,189] An unusual lithium/magnesium acetylide is formed with stoichiometry $Li_2[(PhC{\equiv}C)_3Mg(TMEDA)]_2$ and is depicted as (**214**). The same authors also report the ion pair characterized as the mixed benzyllithium/magnesium·TMEDA complex (**215**).[190] A different mixed lithium/magnesium aggregate depicted as (**216**) is found for the THF-solvated anion of tris(trimethylsilyl)methyl carbanion.[191]

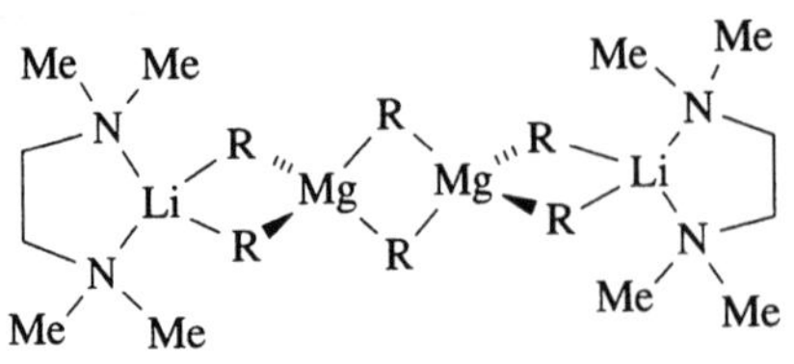

(213)

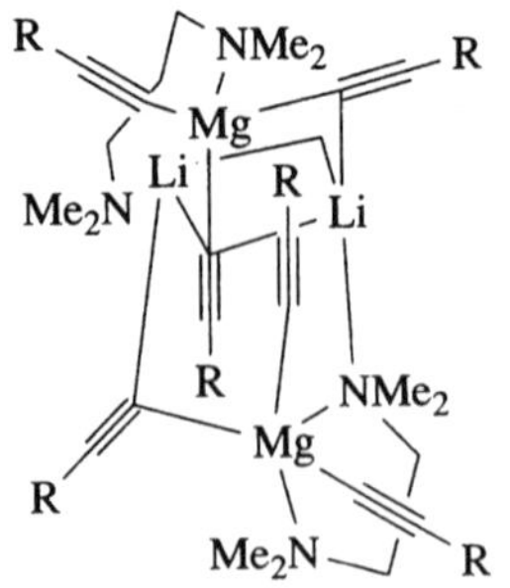

(214)

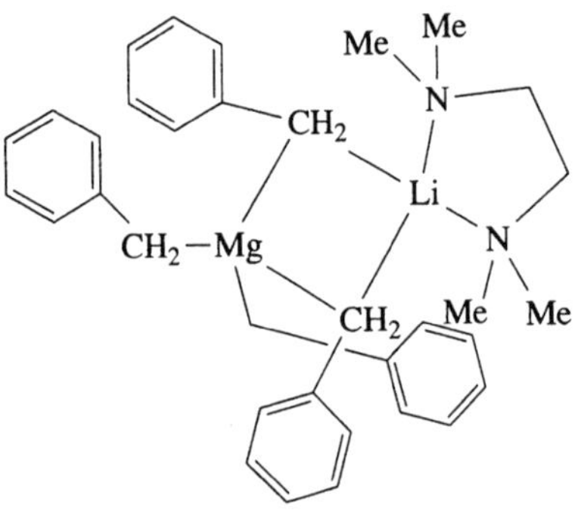

(215)

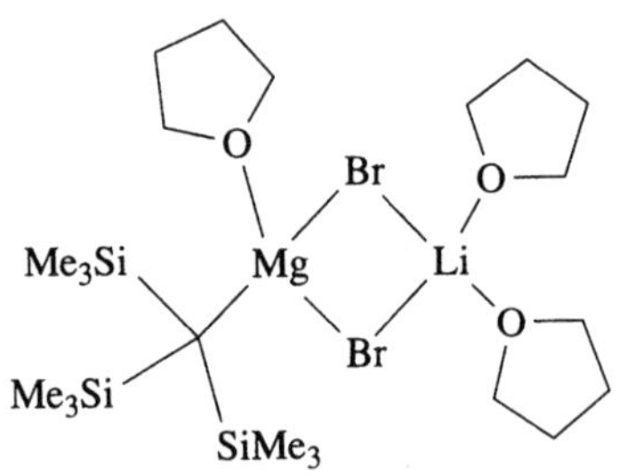

(216)

1.1.3.9.2 With transition metals (cuprates)

A dimeric complex of the type $Li_2Cu_2R_4$ where R is phenyl has recently been characterized as the diethyl ether solvate (**217**).[192] This structure is the most recent of an extremely interesting series of diorganocuprate structure determinations. These reagents find many applications in organic synthesis, and there now exist structural models for the sequence of the species as tabulated in Table 7. Almost all of these compounds incorporate aryl groups rather than alkyl groups due to the instability of the simple lithium dialkylcuprates. Many of these structures are complex and the structures are best discussed in the original manuscripts. It is noteworthy that the complexes listed in this table now provide a structural framework for the series of organocopper reagents with a number of different stoichiometries.

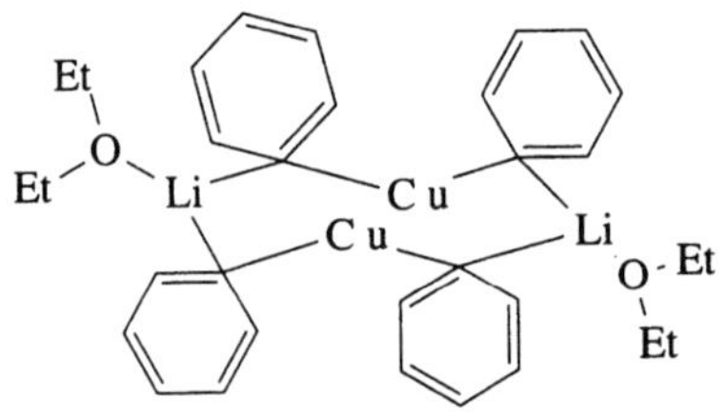

(217)

1.1.4 CRYSTAL GROWTH AND MANIPULATION

Most of the examples of X-ray crystal structure determinations cited in this review have been carried out at subambient temperature. This is necessary since many, but not all, of the carbanions are only stable as solids at low temperatures. Several special techniques exist for handling temperature- and moisture-sensitive solids. General reviews of these techniques exist.[193] On the average these diffraction analyses require a data collection time spanning the range from 12 to 50 h. The carbanions are most commonly sealed in a thin glass capillary and maintained in a stream of cold, dry nitrogen gas during the

Table 7 X-Ray Crystal Structure of Cu/Li Carbanions

Compound	*CSD refcode*	*Ref.*
$[Li(THF)_4]$	CESDAF	264
$[Li(12\text{-crown-}4)_2][CuMe_2]$	DAZWIK	265
$[Li(THF)(12\text{-crown-}4)_2][CuPh_2]$	DAZWOQ	265
$[Li(12\text{-crown-}4)_2][Cu(Br)CH(TMS)_3]\cdot(toluene)$	DAZWUW	265
$[Li(THF)_4][Cu_5Ph_6]$	BEYROM	266
$[Li(PMDETA)(THF)][Cu_5Ph_6]$	BEYRUS	266
$[Li(Et_2O)_4][LiCu_4Ph_6]\cdot(Et_2O)_2$	CUTCEZ	267
$[LiCl_2(Et_2O)_{10}][Li_2Cu_3Ph_6]_2$	CESFIP	268
$Li_2Cu_2(C_6H_4CH_2NMe_2)_4$	CUWTUJ	269
$[Li(THF)_3][CuMe(PBu^t_2)]$	VACFEK	270
$[Li_3Cu_2Ph_5(SMe_2)_4]$		271
$[\{Li(Et_2O)\}(CuPh_2)]_2$		272

period of data collection. All commercial vendors of X-ray diffraction equipment offer as an optional accessory the specialized attachment to their equipment that allows for low temperature data collection. However, the most tedious step in the entire process remains that of obtaining a single crystal suitable for diffraction analysis. Once a suitable crystal is obtained, a specialized piece of glassware for examining, selecting and manipulating the crystal has been described by Seebach *et al.*[151]

A few comments concerning the crystallization of carbanions are in order. These comments are based upon the personal experience developed in our own laboratory and also upon observations noted in the literature in the course of crystallizing enolate anions. Although alkali metal enolate anions are relatively unstable compounds, they have been prepared in the solid state, isolated, and characterized by IR and UV spectroscopy in the 1970s.[194] Thus the α-lithiated esters of a number of simple esters of isobutyric acid are prepared by metallation of the esters with lithium diisopropylamide in benzene or toluene solution. The soluble lithiated esters are quite stable at room temperature in aliphatic or aromatic hydrocarbon solvents and are crystallized out of solution at low temperature (*e.g.* –70 °C.). Alternatively the less soluble enolates tend to precipitate out of solution and are isolated by centrifugation and subsequent removal of the solvent. Recrystallization from a suitable solvent can then be attempted. The thermal stability of the lithiated ester enolates is dramatically decreased in the presence of a solvent with a donor atom such as tetrahydrofuran.

The guidelines we use for obtaining enolate anion crystals are to find a suitable solvent, concentration and temperature combination such that the crystals grow in a matter of 24 h or so. Typically this involved concentrations of 0.5 to 1.0 M in a solvent combination (hydrocarbon plus donor) that allows for complete dissolution of the anion and slow crystallization. It is preferable to crystallize the enolate anions in the range of –20 to –50 °C since the crystals obtained at this temperature can often be transferred directly to the diffractometer with a minimum effort. Exact conditions for the crystallizations of many of the compounds described in this chapter are described in the original literature.

It is noteworthy that Etter has described a recent technique of cocrystallizing stable organic compounds with triphenylphosphine oxide.[195] It is possible that additional enolate anions can be crystallized by addition of this addend to assist with the solid phase formation; however, many of the carbanions already include donors such as TMEDA, THF, *etc.* In summary the crystallization of enolate anions differs little from the crystallization of neutral organic molecules, except that it is often carried out at somewhat lower temperatures. The patience, skill and experience of the chemist often determine whether the crystallization procedure is successful.

1.1.5 THEORY, NMR AND OTHER TECHNIQUES

In concluding this review it must be noted that there are many other techniques that are being utilized to increase our understanding about the structure of synthetically important carbanions. A partial listing of these techniques would include the theoretical approaches taken by Schleyer,[196] Streitweiser,[197] Houk[198] and others[199] and classical spectroscopic techniques.[200] There exist also a number of useful NMR techniques in addition to the 2D-HOESY method previously mentioned. These NMR techniques include analysis of ^{13}C chemical shifts, ^{6}Li–^{15}N spin–spin splitting, ^{7}Li quadrupolar coupling[201] and rapid injection NMR which has proven useful as a technique for structural investigations of aliphatic carbanions.[202] Last, but certainly not least, the excellent thermochemical measurements recently reported by Arnett and coworkers serve to correlate the solid state structural studies with solution species.[203] A comprehensive list and discussion of all of the techniques that have been utilized to analyze carbanion structures is beyond the scope of this review.

In conclusion it must be stated that much work remains before it is possible to predict with confidence the structural characteristics of any new carbanion. That such structures may be utilized to rationalize unusual reactions has already been demonstrated by Grutzner.[204] Although many of the main structural types may have been uncovered already, much additional investigation is necessary before these results can be generally applied to control the stereochemical outcome of reactions of synthetically useful carbanions.

1.1.6 REFERENCES

1. R. Amstutz, W. B. Schweizer, D. Seebach and J. D. Dunitz, *Helv. Chim. Acta*, 1981, **64**, 2617 ⟨BEDYOY, BEDYUE⟩.
2. As of January 1990 there exist nearly 1500 structural characterizations of alkali metal cation balanced organic anions.
3. (a) Information about how to obtain the database itself or access thereto is available from Dr. Olga Kennard OBE, FRS, Crystallographic Data Centre, University Chemical Laboratory, Lensfield Road, Cambridge, CB2 1EW, UK; see also (b) F. H. Allen, S. Bellard, M. D. Brice, B. A. Cartwright, A. Doubleday, H. Higgs, T. Hummelink, B. G. Hummelink-Peters, O. Kennard, W. D. S. Motherwell, J. R. Rodgers and D. G. Watson, *Acta Crystallogr.*, 1979, **B35**, 2331; (c) F. H. Allen, O. Kennard and R. Taylor, *Acc. Chem. Res.*, 1983, **16**, 146.
4. L. M. Jackman and B. M. Lange, *Tetrahedron*, 1977, **33**, 2737.
5. M. Schlosser, 'Struktur und Reaktivität polarer Organometlle', Springer, Berlin, 1973.
6. B. J. Wakefield, in 'Comprehensive Organic Chemistry', ed. D. H. R. Barton and W. D. Ollis, Pergamon Press, Oxford, 1979, vol. 3, p. 943.
7. B. J. Wakefield, in 'Comprehensive Organometallic Chemistry', ed. G. Wilkinson, F. G. A. Stone and E. W. Abel, Pergamon Press, Oxford, 1982, vol. 7, p. 1.
8. M. E. O'Neill, K. Wade, J. L. Wardell, N. A. Bell and W. E. Lindsell, in 'Comprehensive Organometallic Chemistry', ed. G. Wilkinson, F. G. A. Stone and E. W. Abel, Pergamon Press, Oxford, 1982, vol. 1, p. 1.
9. G. Fraenkel, H. Hsu and B. M. Su, in 'Lithium — Current Applications in Science, Medicine and Technology', ed. R. D. Bach, Wiley, New York, 1985, p. 273.
10. (a) R. Benn and H. Rufińska, *Angew. Chem., Int. Ed. Engl.*, 1986, **25**, 861; *Angew. Chem.*, 1986, **98**, 851; (b) H. Günther, D. Mosaku, P. Bast and D. Schmalz, *Angew. Chem., Int. Ed. Engl.*, 1987, **26**, 1212; *Angew. Chem.*, 1987, **99**, 1242.
11. W. N. Setzer and P. von R. Schleyer, *Adv. Organomet. Chem.*, 1985, **24**, 353; (b) C. Schade and P. von R. Schleyer, *Adv. Organomet. Chem.*, 1987, **27**, 169.
12. D. Seebach, *Angew. Chem., Int. Ed. Engl.*, 1988, **27**, 1624.
13. P. P. Power, *Acc. Chem. Res.*, 1988, **21**, 147.
14. G. Boche, *Angew. Chem., Int. Ed. Engl.*, 1989, **28**, 277; *Angew. Chem.*, 1989, **101**, 286.
15. E. L. Eliel, N. L. Allinger, S. J. Angyal and G. A. Morrison, in 'Conformational Analysis', Wiley, New York, 1965.
16. C. H. Heathcock, in 'Asymmetric Synthesis', ed. J. D. Morrison, Academic Press, New York, 1984, vol. 1, pp. 155, 158.
17. While it is extremely unlikely that transition states can be easily characterized by physical measurements such as X-ray diffraction analyses, the Dunitz structure correlation principle allows a reasonable estimate to be made of the structural changes which occur along a reaction coordinate. Thus, structural characterizations of reaction intermediates can sometimes give rise to transition state pictures.
18. (a) H. O. House, M. Gall and H. D. Olmstead, *J. Org. Chem.*, 1971, **36**, 2361; (b) H. D. Zook, T. J. Russo, E. F. Ferrand and D. S. Stotz, *J. Org. Chem.*, 1968, **33**, 2222; (c) H. D. Zook and T. J. Russo, *J. Am. Chem. Soc.*, 1960, **82**, 1258; (d) H. D. Zook and W. L. Gumby, *J. Am. Chem. Soc.*, 1960, **82**, 1368; (e) W. L. Rellahan, W. L. Gumby and H. D. Zook, *J. Org. Chem.*, 1959, **24**, 709; (f) H. D. Zook, W. L. Kelly and I. Y. Posey, *J. Org. Chem.*, 1968, **33**, 3477.
19. H. O. House, M. Gall and H. D. Olmstead, *J. Org. Chem.*, 1971, **36**, 2361.
20. (a) D. Caine and B. Huff, *Tetrahedron Lett.*, 1966, 4695; (b) B. Huff, F. N. Tuller and D. Caine, *J. Org. Chem.*, 1969, **34**, 3070; (c) H. D. Zook, T. J. Russo, E. F. Ferrand and D. S. Stotz, *J. Org. Chem.*, 1968, **33**, 2222; (d) K. G. Hampton, T. M. Harris and C. R. Hauser, *J. Org. Chem.*, 1966, **31**, 1035.
21. (a) H. D. Zook and T. J. Russo, *J. Am. Chem. Soc.*, 1960, **82**, 1258; (b) H. D. Zook and W. L. Gumby, *J. Am. Chem. Soc.*, 1960, **82**, 1386; (c) W. L. Rellahan, W. L. Gumby and H. D. Zook, *J. Org. Chem.*, 1959, **24**, 709; (d) H. D. Zook, W. L. Kelly and I. Y. Posey, *J. Org. Chem.*, 1968, **33**, 3477; (e) A. G. Pinkus, J. G. Lindberg and A. B. Wu, *J. Chem. Soc., Chem. Commun.*, 1969, 1350; 1970, 859; (f) G. E. Coates, J. A. Heslop, M. E. Redwood and D. D. Ridley, *J. Chem. Soc. A*, 1968, 1118; (g) E. Weiss, H. Alsdorf and H. Kuhr, *Angew. Chem., Int. Ed. Engl.*, 1967, **6**, 801; (h) see also ref. 20c.
22. See ref 21e.
23. (a) H. O. House in 'Modern Synthetic Reactions', 2nd edn., Benjamin, Menlo Park, New York, 1972, p. 492; (b) W. J. le Noble and H. F. Morris, *J. Org. Chem.*, 1969, **34**, 1969; (c) R. M. Coates and R. E. Shaw, *J. Org. Chem.*, 1970, **35**, 2597, 2601; (d) A. L. Kurz, I. P. Beletskaya, A. Macias and O. A. Reutov, *Tetrahedron Lett.*, 1968, 3679; (e) S. J. Rhodes and R. W. Holden, *Tetrahedron*, 1969, **25**, 5443; (f) H. Normant, *Angew. Chem., Int. Ed. Engl.*, 1967, **6**, 1046; (g) see also ref. 20c, 21a–c.

24. (a) C. H. Heathcock and D. A. Oare, *J. Org. Chem.*, 1985, **50**, 3022; (b) R. H. Schlessinger, E. J. Iwanowicz and J. P. Springer, *J. Org. Chem.*, 1986, **51**, 3070.
25. see ref. 1.
26. L. M. Jackman and N. M. Szeverenyi, *J. Am. Chem. Soc.*, 1977, **99**, 4954.
27. (a) L. M. Jackman and B. C. Lange, *J. Am. Chem. Soc.*, 1981, **103**, 4494; (b) see also refs. 4 and 26.
28. C. A. Wilkie and R. Des, *J. Am. Chem. Soc.*, 1972, **94**, 4555.
29. For intermediate (**8**) see ref. 1; for intermediate (**9**) see ref. 136; and for an intermediate analogous to (**11**) see ref. 141.
30. See refs 12 and 14.
31. See ref. 9.
32. (a) D. Mootz, A. Zinnius and B. Bottcher, *Angew. Chem.*, 1969, **81**, 398 ⟨MSINLI⟩; (b) R. D. Rogers, J. L. Atwood and R. Gruning, *J. Organomet. Chem.*, 1978, **157**, 229 ⟨TMSIAL⟩.
33. (a) D. R. Armstrong, D. Barr, W. Clegg, R. E. Mulvey, D. Reed, R. Snaith and K. Wade, *J. Chem. Soc., Chem. Commun.*, 1986, 869 ⟨DOKBOU⟩; (b) D. R. Armstrong, D. Barr, W. Clegg, S. M. Hodgson, R. E. Mulvey, D. Reed, R. Snaith and D. S. Wright, *J. Am. Chem. Soc.*, 1989, **111**, 4719.
34. L. Pauling, 'The Nature of the Chemical Bond', 3rd edn., Cornell University Press, New York, 1960, pp. 257, 260.
35. H. Ondik and D. Smith, in 'International Tables for X-Ray Crystallography', Kynoch Press, Birmingham, 1962, vol. 3, p. 257.
36. It is to be noted that the examples in Table 1 are only those chosen by the algorithm CONNSER to have a C—M bond. It is possible that additional structures in the CSD may be considered to have an alkali metal or an alkaline earth metal to carbon contact.
37. P. Chakrabarti and J. D. Dunitz, *Helv. Chim Acta*, 1982, **65**, 1482.
38. C. J. Carrell, H. L. Carrell, J. Erlebacher and J. P. Glusker, *J. Am. Chem. Soc.*, 1988, **110**, 8651.
39. S. Shambayati, W. E. Crowe and S. L. Schreiber, *Angew. Chem., Int. Ed. Engl.*, 1990, **29**, 256.
40. E. Weiss and G. Hencken, *J. Organomet. Chem.*, 1970, **21**, 265 ⟨METHLI⟩.
41. H. Dietrich, *Acta Crystallogr.*, 1963, **16**, 681 ⟨ETHYLI⟩.
42. R. P. Zerger, W. E. Rhine and G. D. Stucky, *J. Am. Chem. Soc.*, 1974, **96**, 6048 ⟨CHXYLI⟩.
43. T. L. Brown, L. M. Seitz and B. Y. Kimura, *J. Am. Chem. Soc.*, 1968, **90**, 3245; M. F. Guest, I. H. Hillier and V. R. Saunders, *J. Organomet. Chem.*, 1972, **44**, 59.
44. H. Dietrich, *J. Organomet. Chem.*, 1981, **205**, 291 ⟨ETHYLJ⟩.
45. A. Maercker, M. Bsata, W. Buchmeier and B. Engelen, *Chem. Ber.*, 1984, **117**, 2547 ⟨COHKIT⟩.
46. H. Schmidbaur, A. Schier and U. Schubert, *Chem. Ber.*, 1983, **116**, 1938 ⟨BULNIF⟩.
47. M. Marsch, K. Harms, L. Lochmann and G. Boche, *Angew. Chem., Int. Ed. Engl.*, 1990, **29**, 308.
48. L. Lochmann, J. Pospisil, J. Vodnansky, J. Trekoval and D. Lim, *Collect. Czech. Chem. Commun.*, 1965, **30**, 2187.
49. V. Halaska and L. Lochmann, *Collect. Czech. Chem. Commun.*, 1973, **38**, 1780.
50. J. F. McGarrity and C. A. Ogle, *J. Am. Chem. Soc.*, 1985, **107**, 1805.
51. M. Schlosser, *Pure Appl. Chem.*, 1988, **60**, 1627.
52. H. Koster, D. Thoennes and E. Weiss, *J. Organomet. Chem.*, 1978, **160**, 1 ⟨MELIME⟩.
53. R. P. Zerger and G. D. Stucky, *J. Chem. Soc., Chem. Commun.*, 1973, 44 ⟨BUTLIE⟩.
54. (a) G. W. Klumpp, P. J. A. Geurink, A. L. Spek and A. J. M. Duisenberg, *J. Chem. Soc., Chem. Commun.*, 1983, 814 ⟨BUVPAJ⟩; (b) A. L. Spek, A. J. M. Duisenberg, G. W. Klumpp and P. J. A. Geurink, *Acta Crystallogr.*, 1984, **C40**, 372 ⟨BUPVAJ10⟩.
55. (a) K. S. Lee, P. G. Williard and J. W. Suggs, *J. Organomet. Chem.*, 1986, **299**, 311 ⟨DIJJEL⟩; (b) G. W. Klumpp, M. Vos, F. J. J. de Kanter, C. Slob, H. Krabbendam and A. L. Spek, *J. Am. Chem. Soc.*, 1985, **107**, 8292 ⟨DIRWIK⟩.
56. S. P. Patterman, I. L. Karle and G. D. Stucky, *J. Am. Chem. Soc.*, 1970, **92**, 1150 ⟨BZLITE⟩.
57. R. A. Bartlett, H. V. R. Dias and P. P. Power, *J. Organomet. Chem.*, 1988, **341**, 1 ⟨GABHEW⟩.
58. M. M. Olmstead and P. P. Power, *J. Am. Chem. Soc.*, 1985, **107**, 2174 ⟨CUXXIC, CUXXOI⟩.
59. J. J. Brooks and G. D. Stucky, *J. Am. Chem. Soc.*, 1972, **94**, 7333 ⟨TPMLIE⟩.
60. H. Foster and E. Weiss, *J. Organomet. Chem.*, 1979, **168**, 273 ⟨TPMNAE⟩.
61. P. von R. Schleyer, R. Hacker, H. Dietrich and W. Mahdi, *J. Chem. Soc., Chem. Commun.*, 1985, 622 ⟨CUPVAK⟩.
62. E. Weiss, G. Sauermann and G. Thirase, *Chem. Ber.*, 1983, **116**, 74 ⟨BOKZIK⟩.
63. (a) E. Weiss and G. Sauermann, *Angew. Chem., Int. Ed. Engl.*, 1968, **7**, 133; (b) E. Weiss and G. Sauermann, *Chem. Ber.*, 1970, **103**, 265 ⟨METPOT⟩.
64. E. Weiss and H. Koster, *Chem. Ber.*, 1977, **110**, 717.
65. C. Schade, P. von R. Schleyer, H. Dietrich and W. Mahdi, *J. Am. Chem. Soc.*, 1986, **108**, 2484 ⟨DOKHEQ⟩.
66. A. I. Snow and R. E. Rundle, *Acta Crystallogr.*, 1951, **4**, 348 ⟨DMETBE⟩.
67. J. R. Wermer, D. F. Gaines and H. A. Harris, *Organometallics*, 1988, **7**, 2421 ⟨GILRAU⟩.
68. M. Vallino, *J. Organomet. Chem.*, 1969, **20**, 1 ⟨MGTHFB⟩.
69. A. L. Spek, P. Voorbergen, G. Schat, C. Blomberg and F. Bickelhaupt, *J. Organomet. Chem.*, 1974, **77**, 147 ⟨EMGBIP⟩.
70. J. Toney and G. D. Stucky, *J. Chem. Soc., Chem. Commun.*, 1967, 1168 ⟨EMGTEA⟩.
71. L. J. Guggenberger and R. E. Rundle, *J. Am. Chem. Soc.*, 1968, **90**, 5375 ⟨EMGBRE10⟩.
72. J. Toney and G. D. Stucky, *J. Organomet. Chem.*, 1971, **28**, 5 ⟨MGCLTF⟩.
73. L. M. Engelhardt, B. S. Jolly, P. C. Junk, C. L. Raston, B. W. Skelton and A. H. White, *Aust. J. Chem.*, 1986, **39**, 1337 ⟨FABDIV⟩.
74. Y. Okamoto, K. Suzuki, T. Kitayama, H. Yuki, H. Kageyama, K. Miki, N. Tanaka and N. Kasai, *J. Am. Chem. Soc.*, 1982, **104**, 4618.
75. H. Kageyama, K. Miki, Y. Kai, N. Kasai, Y. Okamoto and H. Yuki, *Bull. Chem. Soc. Jpn.*, 1983, **56**, 2411 ⟨CAVCEH⟩.

76. H. Kageyama, K. Miki, Y. Kai, N. Kasai, Y. Okamoto and H. Yuki, *Bull. Chem. Soc. Jpn.*, 1984, **57**, 1189 ⟨CIPTUQ⟩.
77. K. Angermund, B. Bogdanovic, G. Koppetsch, C. Kruger, R. Mynott, M. Schwickardi and Y.-H. Tsay, *Z. Naturforsch., Teil B*, 1986, **41**, 455 ⟨DOBXIB⟩.
78. (a) B. Bogdanovic, G. Koppetsch, C. Kruger and R. Mynott, *Z. Naturforsch., Teil B*, 1986, **41**, 617 ⟨DOLGAM⟩; (b) B. Bogdanovic, G. Koppetsch, C. Kruger and R. Mynott, *Z. Naturforsch., Teil B*, 1986, **41**, 617 ⟨DOLGEQ⟩.
79. E. P. Squiller, R. R. Whittle and H. G. Richey, Jr., *J. Am. Chem. Soc.*, 1985, **107**, 432 ⟨CUZWAV, CUZVUO⟩.
80. A. D. Pajerski, G. L. BergStresser, M. Parvez and H. G. Richey, Jr., *J. Am. Chem. Soc.*, 1988, **110**, 4844 ⟨GIMNIZ⟩.
81. G. Schill, 'Catenanes, Rotanaxes and Knots', Academic Press, New York, 1971.
82. A. D. Pajerski, M. Parvez and H. G. Richey, Jr., *J. Am. Chem. Soc.*, 1988, **110**, 2660 ⟨GEGKEI⟩.
83. E. Weiss, *J. Organomet. Chem.*, 1964, **2**, 314.
84. E. Weiss, *J. Organomet. Chem.*, 1965, **4**, 101.
85. T. Greiser, J. Kopf, D. Thoennes and E. Weiss, *J. Organomet. Chem.*, 1980, **191**, 1 ⟨MEENMG⟩.
86. A. L. Spek, G. Schat, H. C. Holtkamp, C. Blomberg and F. Bickelhaupt, *J. Organomet. Chem.*, 1977, **131**, 331 ⟨MGCDOD⟩.
87. J. Toney and G. D. Stucky, *J. Organomet. Chem.*, 1970, **22**, 241 ⟨MQUNMG⟩.
88. M. Parvez, A. D. Pajerski and H. G. Richey, Jr., *Acta Crystallogr.*, 1988, **C44**, 1212 ⟨GIDBEA⟩.
89. M. F. Lappert, T. R. Martin, C. L. Raston, B. W. Skelton and A. H. White, *J. Chem. Soc., Dalton Trans.*, 1982, 1959 ⟨BOFSAQ⟩.
90. C. Eaborn, P. B. Hitchcock, J. D. Smith and A. C. Sullivan, *J. Chem. Soc., Chem. Commun.*, 1983, 827 ⟨BUTXUJ⟩.
91. N. H. Buttrus, C. Eaborn, P. B. Hitchcock, J. D. Smith, J. G. Stamper and A. C. Sullivan, *J. Chem. Soc., Chem. Commun.*, 1986, 969 ⟨DONYAG⟩.
92. H. Koster and E. Weiss, *Chem. Ber.*, 1982, **115**, 3422 ⟨BITNEX⟩.
93. U. Schumann, E. Weiss, H. Dietrich and W. Mahdi, *J. Organomet. Chem.*, 1987, **322**, 299 ⟨FOBHAF⟩.
94. G. Boche, H. Etzrodt, M. Marsch, W. Massa, G. Baum, H. Dietrich and W. Mahdi, *Angew. Chem., Int. Ed. Engl.*, 1986, **25**, 104; *Angew. Chem.*, 1986, **98**, 84 ⟨DIVLID⟩.
95. J. A. Kuduk, A. T. Poulos, and J. A. Ibers, *J. Organomet. Chem.*, 1977, **127**, 245.
96. M. Marsch, K. Harms, W. Massa and G. Boche, *Angew. Chem., Int. Ed. Engl.*, 1987, **26**, 696; *Angew. Chem.*, 1987, **99**, 706 ⟨FOGPAS⟩.
97. See footnotes 8 through 12 in ref. 96.
98. H. Yasuda, M. Yamaguchi, A. Nakamura, T. Sei, Y. Kai, N. Yasuoka and N. Kasai, *Bull. Chem. Soc. Jpn.*, 1980, **53**, 1089 ⟨MEPDMG⟩.
99. W. E. Rhine and G. D. Stucky, *J. Am. Chem. Soc.*, 1975, **97**, 737 ⟨INDYLI⟩.
100. C. Schade, P. von R. Schleyer, P. Gregory, H. Dietrich and W. Mahdi, *J. Organomet. Chem.*, 1988, **341**, 19 ⟨GABHAS⟩.
101. J. L. Atwood and K. D. Smith, *J. Am. Chem. Soc.*, 1974, **96**, 994 ⟨INDYMG⟩.
102. W. Neugebauer, G. A. P. Geiger, A. J. Kos, J. J. Stezowski and P. von R. Schleyer, *Chem. Ber.*, 1985, **118**, 1504 ⟨DANTAN⟩.
103. W. Bauer, M. Feigel, G. Muller and P. von R. Schleyer, *J. Am. Chem. Soc.*, 1988, **110**, 6033 ⟨GINVEE⟩.
104. C. Schade, P. von R. Schleyer, M. Geissler and E. Weiss, *Angew. Chem., Int. Ed. Engl.*, 1986, **25**, 902; *Angew. Chem.*, 1986, **98**, 922 ⟨FOVLOR⟩.
105. R. L. Polt, G. Stork, G. B. Carpenter and P. G. Williard, *J. Am. Chem. Soc.*, 1984, **106**, 4276 ⟨CIYDUJ⟩.
106. B. Schubert and E. Weiss, *Chem. Ber.*, 1983, **116**, 3212 ⟨CIGXUL⟩.
107. (a) R. Hässig and D. Seebach, *Helv. Chim. Acta*, 1983, **66**, 2269; (b) W. Bauer and D. Seebach, *Helv. Chim. Acta*, 1984, **67**, 1972.
108. M. Geissler, J. Kopf, B. Schubert, E. Weiss, W. Neugebauer and P. von R. Schleyer, *Angew. Chem., Int. Ed. Engl.*, 1987, **26**, 587; *Angew. Chem.*, 1987, **99**, 569 ⟨GIFHIM⟩.
109. B. Schubert and E. Weiss, *Angew. Chem., Int. Ed. Engl.*, 1983, **22**, 496; *Angew. Chem.*, 1983, **95**, 499 ⟨CAMNEJ⟩.
110. B. Schubert, U. Behrens and E. Weiss, *Chem. Ber.*, 1981, **114**, 2640 ⟨PHETMG⟩.
111. (a) N. A. Bell, I. W. Nowell and H. M. M. Shearer, *J. Chem. Soc., Chem. Commun.*, 1982, 147 ⟨BAXWAY⟩; (b) N. A. Bell, I. W. Nowell, G. E. Coates and H. M. M. Shearer, *J. Organomet. Chem.*, 1984, **273**, 179 ⟨BAXWAY10⟩; (c) B. Morosin and J. Howatson, *J. Organomet. Chem.*, 1971, **29**, 7 ⟨MPNBET10⟩.
112. D. Thoennes and E. Weiss, *Chem. Ber.*, 1978, **111**, 3157 ⟨PHENLI⟩.
113. H. Hope and P. P. Power, *J. Am. Chem. Soc.*, 1983, **105**, 5320 ⟨CALKOP⟩.
114. D. West and R. Waack, *J. Am. Chem. Soc.*, 1967, **89**, 4395.
115. L. M. Jackman and L. M. Scarmoutzos, *J. Am. Chem. Soc.*, 1984, **106**, 4627.
116. U. Schumann, J. Kopf and E. Weiss, *Angew. Chem., Int. Ed. Engl.*, 1985, **24**, 215 ⟨DANKUY⟩.
117. H. Hope and P. P. Power, *J. Am. Chem. Soc.*, 1983, **105**, 5320 ⟨CALKUV⟩.
118. R. A. Bartlett, H. V. R. Dias and P. P. Power, *J. Organomet. Chem.*, 1988, **341**, 1 ⟨GABHIA⟩.
119. M. A. Beno, H. Hope, M. M. Olmstead and P. P. Power, *Organometallics*, 1985, **4**, 2117 ⟨GAFCAR⟩.
120. (a) H. W. Gschwend and H. R. Rodriguez, *Org. React. (N. Y.)*, 1979, **26**, 1; (b) P. Beak and A. I. Meyers, *Acc. Chem. Res.*, 1986, **19**, 356.
121. S. Harder, J. Boersma, L. Brandsma, G. P. M. van Mier and J. A. Kanters, *J. Organomet. Chem.*, 1989, **364**, 1 ⟨JAJHIL⟩.
122. S. Harder, J. Boersma, L. Brandsma and J. A. Kanters, *J. Organomet. Chem.*, 1988, **339**, 7 ⟨GANBAY⟩.
123. (a) H. Dietrich, W. Mahdi and W. Storck, *J. Organomet. Chem.*, 1988, **349**, 1 ⟨GIBTOA⟩; (b) S. Harder, J. Boersma, L. Brandsma, A. van Heteren, J. A. Kanters, W. Bauer and P. von R. Schleyer, *J. Am. Chem. Soc.*, 1988, **110**, 7802 ⟨VADTUP⟩.

124. S. Harder, J. Boersma, L. Brandsma, J. A. Kanters, W. Bauer and P. von R. Schleyer, *Organometallics*, 1989, **8**, 1696.
125. (a) S. Harder, L. Brandsma, J. A. Kanters and A. J. M. Duisenberg, *Acta Crystallogr.*, 1987, **C43**, 1535 ⟨FOWDEA⟩; (b) W. Bauer, P. A. A. Klusener, S. Harder, J. A. Kanters, A. J. M. Duisenberg, L. Brandsma and P. von R. Schleyer, *Organometallics*, 1988, **7**, 552 ⟨FOWDEA10⟩.
126. J. T. B. H. Jastrzebski, G. van Koten, M. Konijn and C. H. Stam, *J. Am. Chem. Soc.*, 1982, **104**, 5490 ⟨BOFVAT⟩.
127. W. J. J. Smeets, A. L. Spek, A. A. H. van der Zeijden and G. van Koten, *Acta Crystallogr.*, 1987, **C43**, 1429 ⟨FISWEJ⟩.
128. U. Schubert, W. Neugebauer and P. von R. Schleyer, *J. Chem. Soc., Chem. Commun.*, 1982, 1184 ⟨BOBJIL⟩.
129. B. Bogdanovic, C. Kruger and B. Wermeckes, *Angew. Chem., Int. Ed. Engl.*, 1980, **19**, 817 ⟨BZPHLI⟩.
130. G. D. Stucky and R. E. Rundle, *J. Am. Chem. Soc.*, 1964, **86**, 4825 ⟨PHMGBE⟩.
131. D. Thoennes and E. Weiss, *Chem. Ber.*, 1978, **111**, 3381 ⟨DPMGEN⟩.
132. P. R. Markies, O. S. Akkerman, F. Bickelhaupt, W. J. J. Smeets and A. L. Spek, *J. Am. Chem. Soc.*, 1988, **110**, 4284 ⟨GETMVN⟩.
133. P. R. Markies, T. Nomoto, O. S. Akkerman, F. Bickelhaupt, W. J. J. Smeets and A. L. Spek, *Angew. Chem., Int. Ed. Engl.*, 1988, **27**, 1084 ⟨VACSIB⟩.
134. P. R. Markies, T. Nomoto, O. S. Akkerman, F. Bickelhaupt, W. J. J. Smeets and A. L. Spek, *J. Am. Chem. Soc.*, 1988, **110**, 4845 ⟨GEXMEB⟩.
135. P. G. Williard and G. B. Carpenter, *J. Am. Chem. Soc.*, 1985, **107**, 3345 ⟨CUYVOH⟩.
136. P. G. Williard and G. B. Carpenter, *J. Am. Chem. Soc.*, 1986, **108**, 462 ⟨DIPSAW, DIPSEA⟩.
137. T. Laube, J. D. Dunitz and D. Seebach, *Helv. Chim. Acta*, 1985, **68**, 1373 ⟨DETRAV⟩.
138. P. G. Williard and M. J. Hintze, *J. Am. Chem. Soc.*, 1987, **109**, 5339 ⟨FOGRIC⟩.
139. P. G. Williard and M. J. Hintze, unpublished results.
140. J. T. B. H. Jastrzebski, G. van Koten, M. J. N. Christophersen and C. H. Stam, *J. Organomet. Chem.*, 1985, **292**, 319 ⟨DIKRUK⟩.
141. P. G. Williard and J. M. Salvino, *Tetrahedron Lett.*, 1985, **26**, 3931 ⟨DEWBIQ⟩.
142. R. Amstutz, J. D. Dunitz, T. Laube, W. B. Schweizer and D. Seebach, *Chem. Ber.*, 1986, **119**, 434 ⟨DIXWIQ⟩.
143. R. A. Wanat, D. B. Collum, G. van Duyne, J. Clardy and R. T. DePue, *J. Am. Chem. Soc.*, 1986, **108**, 3415 ⟨DOZHEF⟩.
144. D. B. Collum, D. Kahne, S. A. Gut, R. T. DePue, F. Mohamadi, R. A. Wanat, J. Clardy and G. van Duyne, *J. Am. Chem. Soc.*, 1984, **106**, 4865 ⟨COCNOX⟩.
145. D. Enders, G. Bachstadter, K. A. M. Kremer, M. Marsch, K. Harms and G. Boche, *Angew. Chem., Int. Ed. Engl.*, 1988, **27**, 1522 ⟨SAMNEZ⟩.
146. Some enolate crystals derived from achiral substrates undergo spontaneous resolution upon crystallization. These include the aggregates (**139**) and (**150**) and perhaps others. Presumably this occurs to yield the conglomerate.
147. H. Dietrich, W. Mahdi, D. Wilhelm, T. Clark and P. von R. Schleyer, *Angew. Chem., Int. Ed. Engl.*, 1984, **23**, 621 ⟨COKSEA, COKSEA10⟩.
148. See ref 21e.
149. P. G. Williard and J. M. Salvino, *J. Chem. Soc., Chem. Commun.*, 1986, 153 ⟨DILPUJ⟩.
150. P. Berno, C. Floriani, A. Chiesi-Villa and C. Guastini, *Organometallics*, 1990, **9**, 1995.
151. D. Seebach, R. Amstutz, T. Laube, W. B. Schweizer and J. D. Dunitz, *J. Am. Chem. Soc.*, 1985, **107**, 5403 ⟨DEDXEP, DEDXIT, DEDXOZ⟩.
152. P. G. Williard and M. J. Hintze, *J. Am. Chem. Soc.*, 1990, **112**, 8602.
153. J. Dekker, J. Boersma and G. J. M. van der Kerk, *J. Chem. Soc., Chem. Commun.*, 1983, 553; J. Dekker, P. H. M. Budzelaar, J. Boersma, G. J. M. van der Kerk and A. L. Spek, *Organometallics*, 1984, **3**, 1403 ⟨BUDKAM⟩.
154. J. T. B. H. Jastrzebski, G. van Koten and W. F. van de Mieroop, *Inorg. Chim. Acta*, 1988, 142, 169 ⟨GAJCUP⟩.
155. T. Laube, J. D. Dunitz and D. Seebach, *Helv. Chim. Acta*, 1985, **68**, 1373 ⟨DETPUN⟩.
156. W. Bauer, T. Laube and D. Seebach, *Chem. Ber.*, 1985, **118**, 764 ⟨GAYYIO⟩.
157. P. G. Williard, J. R. Tata, R. H. Schlessinger, A. D. Adams and E. J. Iwanowicz, *J. Am. Chem. Soc.*, 1988, **110**, 7901 ⟨GIRKEX, GIRKIB⟩.
158. (a) R. H. Schlessinger, M. A. Poss and S. Richardson, *J. Am. Chem. Soc.*, 1986, **108**, 3112; (b) R. H. Schlessinger, E. J. Iwanowicz and J. P. Springer, *J. Org. Chem.*, 1986, **51**, 3070; (c) A. D. Adams, R. H. Schlessinger and J. R. Tata, *J. Org. Chem.*, 1986, **51**, 3068.
159. G. Boche, M. Marsch and K. Harms, *Angew. Chem., Int. Ed. Engl.*, 1986, **25**, 373 ⟨FAHBEV⟩.
160. W. Zarges, M. Marsch, K. Harms and G. Boche, *Angew. Chem., Int. Ed. Engl.*, 1989, **28**, 1392.
161. G. Boche, K. Harms and M. Marsch, *J. Am. Chem. Soc.*, 1988, **110**, 6925 ⟨GIBHOO⟩.
162. W. Zarges, M. Marsch, K. Harms and G. Boche, *Chem. Ber.*, 1989, **122**, 1307.
163. P. Andersen, B. Keewe and E. Thon, *Acta Chem. Scand.*, 1967, **21**, 1530.
164. J. R. Witt and D. Britton, *Acta Crystallogr.*, 1971, **B27**, 1835 ⟨KTCYME01⟩.
165. D. Barr, W. Clegg, R. E. Mulvey, D. Reed and R. Snaith, *Angew. Chem., Int. Ed. Engl.*, 1985, **24**, 328; *Angew. Chem.*, 1985, **97**, 322 ⟨DAPGAC⟩.
166. G. Klebe, K. H. Bohn, M. Marsch and G. Boche, *Angew. Chem., Int. Ed. Engl.*, 1987, **26**, 78; *Angew. Chem.*, 1987, **99**, 62 ⟨FECGUP⟩.
167. E. W. Colvin, A. K. Beck, B. Bastiani, D. Seebach, Y. Kai and J. D. Dunitz, *Helv. Chim. Acta*, 1980, **63**, 697 ⟨BMSINA⟩.
168. H. Kanazawa, M. Ichiba, Z. Tamura, K. Senga, K.-I. Kawai and H. Otomasu, *Chem. Pharm. Bull.*, 1987, **35**, 35 ⟨FEWBOY⟩.
169. R. A. Wanat, D. B. Collum, G. van Duyne, J. Clardy and R. T. DePue, *J. Am. Chem. Soc.*, 1986, **108**, 3415 ⟨DOZHAB⟩.

170. D. Seebach, W. Bauer, J. Hansen, T. Laube, W. B. Schweizer and J. D. Dunitz, *J. Chem. Soc., Chem. Commun.*, 1984, 853 ⟨CIFBOI⟩.
171. D. Seebach, J. Hansen, P. Seiler and J. M. Gromek, *J. Organomet. Chem.*, 1985, **285**, 1 ⟨CUXHIM⟩.
172. W. Bauer, G. Muller and P. von R. Schleyer, *Angew. Chem., Int. Ed. Engl.*, 1986, **25**, 1103 ⟨FATPEV⟩.
173. S. E. Denmark and R. L. Dorow, *J. Am. Chem. Soc.*, 1990, **112**, 864.
174. E. Weiss, J. Kopf, T. Gardein, S. Corbelin, U. Schumann, M. Kirilov and G. Petrov, *Chem. Ber.*, 1985, **118**, 3529 ⟨DENGEI⟩.
175. J. Macicek, O. Angelova, G. Petrov and M. Kirilov, *Acta Crystallogr.*, 1988, **C44**, 626 ⟨GEKYEA⟩.
176. P. G. Williard and J. M. Salvino, unpublished observation; see p. 1627 in ref. 12.
177. L. M. Englehardt, B. S. Jolly, P. C. Punk, C. L. Raston and B. W. Skelton and A. H. White, *Aust. J. Chem.*, 1986, **39**, 133 ⟨FABDER⟩.
178. M. F. Lappert, M. J. Slade, A. Singh, J. L. Atwood, R. D. Rogers and R. Shakir, *J. Am. Chem. Soc.*, 1983, **105**, 302 ⟨BUXNOX⟩; L. M. Englehardt, A. S. May, C. L. Raston and A. H. White, *J. Chem. Soc., Dalton Trans.*, 1983, 1671 ⟨BUXNOS01⟩.
179. R. Gruning and J. L. Atwood, *J. Organomet. Chem.*, 1977, **137**, 101 ⟨TMSIAS⟩.
180. A. M. Domingos and G. M. Sheldrick, *Acta Crystallogr.*, 1974, **B30**, 517 ⟨DXMSIA⟩.
181. P. G. Williard, *Acta Crystallogr.*, 1988, **C44**, 270 ⟨GASVEB⟩.
182. M. Veith, *Chem. Rev.*, 1990, **1**, 1.
183. U. Pieper, S. Walter, U. Klingebeil and D. Stalke, *Angew. Chem., Int. Ed. Engl.*, 1990, **29**, 209; *Angew. Chem.*, 1990, **102**, 218.
184. P. G. Williard and G. MacEwen, *J. Am. Chem. Soc.*, 1989, **111**, 7671.
185. See ref. 12.
186. U. Schumann and E. Weiss, *Angew. Chem., Int. Ed. Engl.*, 1988, **27**, 584; *Angew. Chem.*, 1988, **100**, 573 ⟨JAHJIL⟩.
187. D. Barr, W. Clegg, R. E. Mulvey and R. Snaith, *J. Chem. Soc., Chem. Commun.*, 1989, 57 ⟨KABYOB⟩.
188. T. Greiser, J. Kopf, D. Thoennes and E. Weiss, *Chem. Ber.*, 1981, **114**, 209 ⟨ENLIMG⟩.
189. D. Thoennes and E. Weiss, *Chem. Ber.*, 1978, **111**, 3726 ⟨PHMGLI⟩.
190. B. Schubert and E. Weiss, *Chem. Ber.*, 1984, **117**, 366 ⟨CEPTOG, CEPTUM⟩.
191. N. H. Buttrus, C. Eaborn, M. N. A. El-Kheli, P. B. Hitchcock, J. D. Smith, A. C. Sullivan and K. Tavakkoli, *J. Chem. Soc., Dalton Trans.*, 1988, 381 ⟨FUPLUX⟩.
192. N. P. Lorenzen and E. Weiss, *Angew. Chem., Int. Ed. Engl.*, 1990, **29**, 300; *Angew. Chem.*, 1990, **102**, 322.
193. (a) M. Veith and W. Frank, *Chem. Rev.*, 1988, **88**, 81; (b) D. Seebach and A. Hidber, *Chimica*, 1983, **37**, 449.
194. (a) L. Lochmann and D. Lim, *J. Organomet. Chem.*, 1973, **50**, 9; (b) L. Lochmann, R. L. De and J. Trekoval, *J. Organomet. Chem.*, 1978, **156**, 307.
195. M. C. Etter and P. W. Baures, *J. Am. Chem. Soc.*, 1988, **110**, 639.
196. A. J. Kos, T. Clark and P. von R. Schleyer, *Angew. Chem., Int. Ed. Engl.*, 1984, **23**, 620; *Angew. Chem.*, 1984, **96**, 622.
197. (a) R. Glaser and A. Streitwieser, Jr., *Pure Appl. Chem.*, 1988, **60**, 195; (b) M. J. Kaufman, S. Gronert and A. Streitwieser, Jr., *J. Am. Chem. Soc.*, 1988, **110**, 2829.
198. Y. Li, M. N. Paddon-Row and K. N. Houk, *J. Am. Chem. Soc.*, 1988, **110**, 3684.
199. T. L. Brown, *Pure Appl. Chem.*, 1970, **23**, 447.
200. J. Corset, *Pure Appl. Chem.*, 1986, **58**, 1133.
201. L. M. Jackman, L. M. Scarmoutzos, D. B. Smith and P. G. Williard, *J. Am. Chem. Soc.*, 1988, **110**, 6058.
202. (a) J. F. McGarrity and C. A. Ogle, *J. Am. Chem. Soc.*, 1985, **107**, 1805; (b) J. F. McGarrity, C. A. Ogle, Z. Birch and H. R. Loosli, *J. Am. Chem. Soc.*, 1985, **107**, 1810.
203. (a) E. M. Arnett, J. F. Franklin, M. A. Nichols and A. A. Ribeiro, *J. Am. Chem. Soc.*, 1989, **111**, 748; (b) E. M. Arnett, S. G. Maroldo, G. W. Schriver, S. L. Schilling and E. Troughton, *J. Am. Chem. Soc.*, 1985, **107**, 2091.
204. J. H. Horner, M. Vera and J. B. Grutzner, *J. Org. Chem.*, 1986, **51**, 4212.
205. B. Tecle, A. F. M. M. Rahman and J. P. Oliver, *J. Organomet. Chem.*, 1986, **317**, 267.
206. J. L. Atwood, T. Fjeldberg, M. F. Lappert, N. T. Luong-Thi, R. Shakir and A. J. Thorne, *J. Chem. Soc., Chem. Commun.*, 1984, 1163.
207. M. F. Lappert, L. M. Englehardt, C. L. Raston and A. H. White, *J. Chem. Soc., Chem. Commun.*, 1982, 1323.
208. C. Eaborn, P. B. Hitchcock, J. D. Smith and A. C. Sullivan, *J. Chem. Soc., Chem. Commun.*, 1983, 1390.
209. R. I. Papasergio, C. L. Raston and A. H. White, *J. Chem. Soc., Chem. Commun.*, 1983, 1419.
210. D. Colgan, R. I. Papasergio, C. L. Raston and A. H. White, *J. Chem. Soc., Chem. Commun.*, 1984, 1708.
211. R. Hacker, P. von R. Schleyer, G. Reber, G. Muller and L. Brandsma, *J. Organomet. Chem.*, 1986, **316**, C4.
212. M. F. Lappert, A. Singh, L. M. Engelhardt and A. H. White, *J. Organomet. Chem.*, 1984, **262**, 271.
213. H. H. Karsch, A. Appelt, B. Deubelly and G. Muller, *J. Chem. Soc. ,Chem. Commun.*, 1987, 1033.
214. H. H. Karsch, B. Deubelly, J. Riede and G. Muller, *J. Organomet. Chem.*, 1988, **342**, C29.
215. H. H. Karsch, B. Deubelly and G. Muller, *J. Organomet. Chem.*, 1988, **352**, 47.
216. R. Shakir and J. L. Atwood, *Am. Crystallogr. Assoc., Ser. 2*, 1978, **6**, 11.
217. T. Alonso, S. Harvey, P. C. Junk, C. L. Raston, B. W. Skelton and A. H. White and *Organometallics*, 1987, **6**, 2110.
218. M. J. Henderson, R. I. Papasergio, C. L. Raston, A. H. White and M. F. Lappert, *J. Chem. Soc., Chem. Commun.*, 1986, 672.
219. L. T. Byrne, L. M. Engelhardt, G. E. Jacobsen, Wing-Por Leung, R. I. Papasergio, C. L. Raston, B. W. Skelton, P. Twiss and A. H. White, *J. Chem. Soc., Dalton Trans.*, 1989, 105.
220. J. J. Sahbari and M. M. Olmstead, *Acta Crystallogr.*, 1983, **C39**, 208.
221. J. J. Sahbari and M. M. Olmstead, *Acta Crystallogr.*, 1983, **C39**, 1037.
222. J. J. Sahbari and M. M. Olmstead, *Acta Crystallogr.*, 1985, **C41**, 360.
223. F. Teixidor, A. Llobet, J. Casabo, X. Solans, M. Font-Altaba and M. Aguilo, *Inorg. Chem.*, 1985, **24**, 2315.

224. E. Weiss, J. Kopf, T. Gardein, S. Corbelin, U. Schumann, M. Kirilov and G. Petrov, *Chem. Ber.*, 1985, **118**, 3529.
225. O. Simonsen, *Acta Crystallogr.*, 1981, **B37**, 344.
226. C. Cambillau, G. Bram, J. Corset and C. Riche, *Can. J. Chem.*, 1982, **60**, 2554.
227. C. Cambillau, G. Bram, J. Corset and C. Riche, *Nouv. J. Chim.*, 1979, **3**, 9; F. Baert, J. Lamiot, L. Devos and R. Fouret, *Eur. Crystallogr. Meeting*, 1982, **7**, 197; C. Cambillau, G. Bram, J. Corset, C. Riche and C. Pascard-Billy, *Tetrahedron*, 1978, **34**, 2675.
228. N. V. Grigor'eva, N. V. Margolis, I. N. Shokhor, V. V. Mel'nikov and I. V. Tselinskii, *Zh. Strukt. Khim.*, 1966, **7**, 278.
229. H.-U. Hummel and F. Beiler, *Z. Anorg. Allg. Chem.*, 1986, **543**, 207.
230. P. Groth, *Acta Chem. Scand., Ser. A*, 1987, **41**, 117.
231. J. R. Holden and C. Dickinson, *J. Am. Chem. Soc.*, 1968, **90**, 1975.
232. I. I. Bannova, O. V. Frank-Kamenetskaya, N. V. Grigor'eva, N. V. Margolis and I. V. Tselinskii, *Zh. Strukt. Khim.*, 1974, **15**, 944.
233. L. M. Engelhardt, G. E. Jacobsen, C. L. Raston and A. H. White, *J. Chem. Soc., Chem. Commun.*, 1984, 220.
234. L. T. Byrne, L. M. Engelhardt, G. E. Jacobsen, Wing-Por Leung, R. I. Papasergio, C. L. Raston, B. W. Skelton, P. Twiss and A. H. White, *J. Chem. Soc., Dalton Trans.*, 1989, 105.
235. R. E. Cramer, M. A. Bruck and J. W. Gilje, *Organometallics*, 1986, **5**, 1496.
236. H. H. Karsch and G. Muller, *J. Chem. Soc., Chem. Commun.*, 1984, 569.
237. H. H. Karsch, L. Weber, D. Wewers, R. Boese and G. Muller, *Z. Naturforsch., Teil B*, 1984, **39**, 1518.
238. R. Amstutz, T. Laube, W. B. Schweizer, D. Seebach and J. D. Dunitz, *Helv. Chim. Acta*, 1984, **67**, 224.
239. D. Seebach, T. Maetzke, R. K. Haynes, M. N. Paddon-Row and S. S. Wong, *Helv. Chim. Acta*, 1988, **71**, 299.
240. R. Amstutz, D. Seebach, P. Seiler, B. Schweizer and J. D. Dunitz, *Angew. Chem., Int. Ed. Engl.*, 1980, **19**, 53.
241. R. Amstutz, J. D. Dunitz and D. Seebach, *Angew. Chem., Int. Ed. Engl.*, 1981, **20**, 465.
242. M. Marsch, W. Massa, K. Harms, G. Baum and G. Boche, *Angew. Chem., Int. Ed. Engl.*, 1986, **25**, 1011.
243. G. Boche, M. Marsch, K. Harms and G. M. Sheldrick, *Angew. Chem., Int. Ed. Engl.*, 1985, **24**, 573.
244. H.-J. Gais, H. J. Lindner and J. Vollhardt, *Angew. Chem., Int. Ed. Engl.*, 1985, **24**, 859.
245. W. Hollstein, K. Harms, M. Marsch and G. Boche, *Angew. Chem., Int. Ed, Engl.*, 1988, **27**, 846.
246. W. Hollstein, K. Harms, M. Marsch and G. Boche, *Angew. Chem., Int. Ed. Engl.*, 1987, **26**, 1287.
247. H.-J. Gais, J. Vollhardt, H. Gunther, D. Moskau, H. J. Lindner and S. Braun, *J. Am. Chem. Soc.*, 1988, **110**, 978.
248. H.-J. Gais, J. Vollhardt and H. J. Lindner, *Angew. Chem., Int. Ed. Engl.*, 1986, **25**, 939.
249. H.-J. Gais, J. Vollhardt, G. Hellmann, H. Paulus and H. J. Lindner, *Tetrahedron Lett.*, 1988, **29**, 1259.
250. H.-J. Gais, J. Vollhardt and C. Kruger, *Angew. Chem., Int. Ed. Engl.*, 1988, **27**, 1092.
251. H.-J. Gais, U. Dingerdissen, C. Kruger and K. Angermund, *J. Am. Chem. Soc.*, 1987, **109**, 3775.
252. H.-J. Gais, I. Erdelmeier, H. J. Lindner and J. Vollhardt, *Angew. Chem., Int. Ed. Engl.*, 1986, **25**, 938.
253. T. Greiser and E. Weiss, *Chem. Ber.*, 1977, **110**, 3388; J. E. Davies, J. Kopf and E. Weiss, *Acta Crystallogr.*, 1982, **B38**, 2251
254. R. Marx, *Z. Naturforsch., Teil B*, 1988, **43**, 521.
255. T. Greiser and E. Weiss, *Chem. Ber.*, 1979, **112**, 844.
256. H. Staeglich and E. Weiss, *Chem. Ber.*, 1978, **111**, 901.
257. T. Greiser and E. Weiss, *Chem. Ber.*, 1976, **109**, 3142.
258. D. Barr, W. Clegg, R. E. Mulvey and R. Snaith, *J. Chem. Soc., Chem. Commun.*, 1984, 79.
259. J. P. Roux and J. C. A. Boeyens, *Acta Crystallogr.*, 1969, **B25**, 1700; P. Piret, L. Rodrique, Y. Gobillon and M. van Meerssche, *Acta Crystallogr.*, 1966, **20**, 482.
260. D. J. Haas, *Nature (London)*, 1964, **201**, 64; P. Chakrabarti, K. Venkatesan and C. N. R. Rao, *Proc. R. Soc. London, Ser. A*, 1981, **375**, 127.
261. R. Amstutz, J. D. Dunitz, T. Laube, W. B. Schweizer and D. Seebach, *Chem. Ber.*, 1986, **119**, 434.
262. P. Chakrabarti, K. Venkatesan and C. N. R. Rao, *Proc. R. Soc. London, Ser. A*, 1981, **375**, 127.
263. Y. M. G. Yasin, O. J. R. Hodder and H. M. Powell, *Chem. Commun.*, 1966, 705.
264. C. Eaborn, P. B. Hitchcock, J. D. Smith and A. C. Sullivan, *J. Organomet. Chem.*, 1984, **263**, C23.
265. H. Hope, M. M. Olmstead, P. P. Power, J. Sandell and X. Xu, *J. Am. Chem. Soc.*, 1985, **107**, 4337.
266. P. G. Edwards, R. W. Gellert, M. W. Marks and R. Bau, *J. Am. Chem. Soc.*, 1982, **104**, 2072.
267. S. I. Khan, P. G. Edwards, H. S. H. Yuan and R. Bau, *J. Am. Chem. Soc.*, 1985, **107**, 1682.
268. H. Hope, D. Oram and P. P. Power, *J. Am. Chem. Soc.*, 1984, **106**, 1149.
269. G. van Koten, J. T. B. H. Jastrzebski, F. Muller and C. H. Stam, *J. Am. Chem. Soc.*, 1985, **107**, 697.
270. S. F. Martin, J. R. Fishpaugh, J. M. Power, D. M. Giolando, R. A. Jones, C. M. Nunn and A. H. Cowley, *J. Am. Chem. Soc.*, 1988, **110**, 7226.
271. M. M. Olmstead and P. P. Power, *J. Am. Chem. Soc.*, 1989, **111**, 4135.
272. N. P. Lorenzen and E. Weiss, *Angew. Chem., Int. Ed. Engl.*, 1990, **29**, 300.

1.2 Carbanions of Alkali and Alkaline Earth Cations: (ii) Selectivity of Carbonyl Addition Reactions

DONNA M. HURYN
Hoffmann-La Roche, Nutley, NJ, USA

1.2.1 ADDITIONS OF ACHIRAL REAGENTS TO CHIRAL SUBSTRATES

1.2.1.1 Additions to Acyclic Systems

1.2.1.1.1 Introduction

The selective nucleophilic addition of Grignard and organolithium reagents to carbonyl compounds has been the subject of extensive study since the early 1900s when McKenzie described the asymmetric synthesis of α-hydroxy acids from the corresponding chiral α-keto esters.[1] This report is the foundation of the seminal work of Prelog,[2] Cram,[3] Cornforth,[4] Karabatsos[5] and Felkin,[6] carried out in the 1950s and 1960s, which provides consistent models of the stereochemical outcome of nucleophilic additions to carbonyl groups. At present, successful application of these models, as well as their theoretical treatment,[7,8] continues to occupy a significant body of the chemical literature. This chapter focuses on the selective addition of carbon nucleophiles generated from organolithium and organomagnesium reagents to carbonyl compounds. The reduction of carbonyl compounds (additions of hydrogen-based reagents) is the subject of Volume 8.

While thorough reviews of this area are available in a number of sources,[9–12] a brief description of the relevant aspects of these models is described below. Cram's 'open-chain model' concerns the addition of a nucleophile to a simple α-alkyl-substituted carbonyl compound.[3] Disposition of the carbonyl oxygen and the largest α-substituent (relative sizes of the substituents are designated L, M, and S) in an *anti* relationship is expected to be the conformation from which nucleophilic addition occurs (Figure 1). In the case of α-halocarbonyls, Cornforth proposes that the carbonyl and carbon–halogen dipoles prefer an *anti* orientation; nucleophilic reaction then takes place from the less-encumbered face (Figure 2).[4] When an α-substituent capable of coordination is present, the 'cyclic chelate model' is invoked. The favored conformation results from formation of a chelate between the cationic reagent, the carbonyl oxygen and the coordinating substituent. Addition then occurs from the least-hindered face (Figure 3).[3]

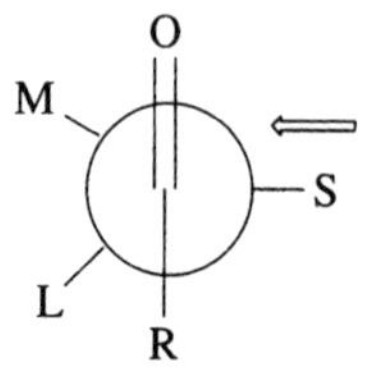

Figure 1 Open-chain model

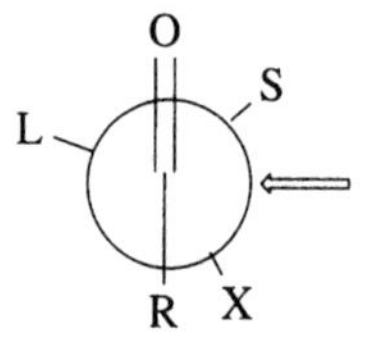

Figure 2 Dipolar model

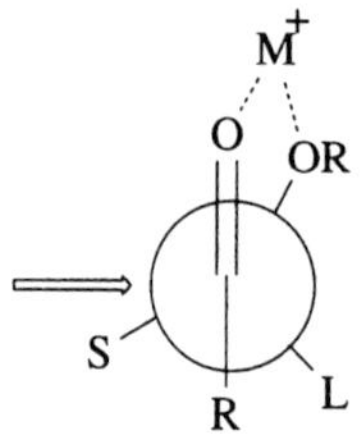

Figure 3 Cyclic chelate model

The design of reactions based on these models, in particular the cyclic chelation controlled model, often leads to high stereoselectivities. However, discrepancies between theoretical and experimental results have led to the development of alternative theories. Among these, the work of Felkin and coworkers[6] has gained the most acceptance. In an open-chain model, orientation of the largest α-substituent in a conformation perpendicular to the carbonyl group is considered most relevant (Figure 4). Because the carbonyl oxygen is deemed less sterically demanding than the substituent bonded directly to the carbonyl carbon (R), conformation A is favored over B. Calculations by Anh and Eisenstein[7] support the results derived from the Felkin model; however, they propose a different mode of nucleophilic attack. These workers suggest that with the carbonyl substrate in a conformation like that proposed by Felkin, the reacting nucleophile approaches not perpendicularly to the carbonyl bond, but tilted away from it. Preferential attack occurs on a trajectory closest to the smallest substituent (Figure 5). For the purpose of this discussion, the term 'cyclic chelation control' will be used to describe those reactions which adhere to the model depicted in Figure 3. 'Felkin–Anh' or 'nonchelation control' will refer to selective additions which can be described by models such as those shown in Figure 5.

While these models are often useful in predicting the outcome of a reaction in a qualitative sense, the degree of stereoselection can be affected by simple changes in reaction conditions. It has been observed that solvent, temperature and organometallic reagent, as well as other factors, play a vital role in determining which of the reaction modes described predominates. Conditions which enhance, diminish, or even reverse observed stereoselectivities have been reported.[11]

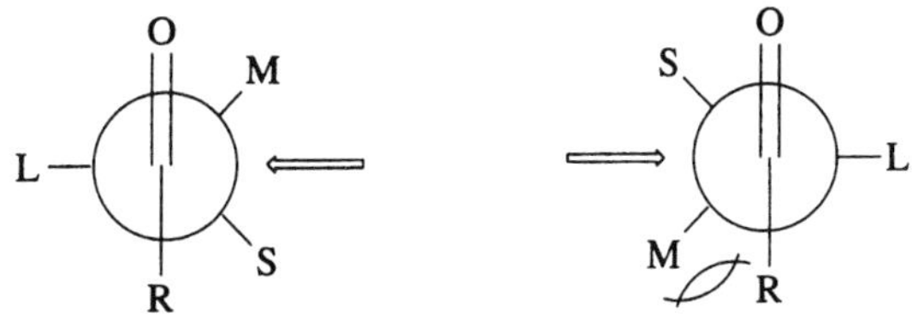

A, favored B, disfavored

Figure 4 Felkin model

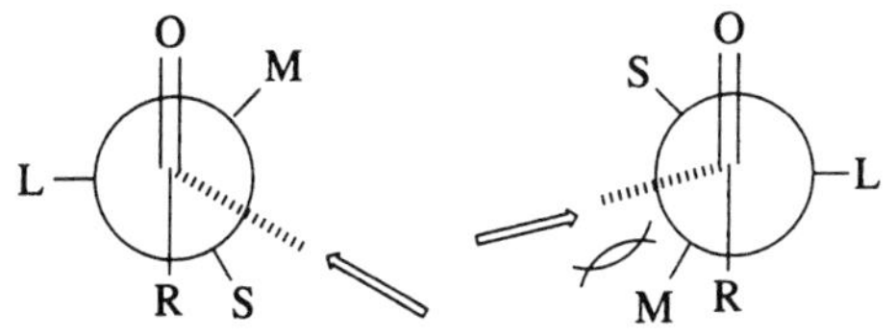

A, favored B, disfavored

Figure 5 Felkin–Anh model

1.2.1.1.2 1,2-Asymmetric induction

The addition of Grignard reagents to chiral α-alkoxy acyclic ketones is one of the most throroughly studied examples of a chelation-controlled reaction. Under certain conditions, these reactions proceed with very high (50–200:1) stereoselectivities, which can be explained by the cyclic chelate model illustrated in equation (1).[13] The nature of the solvent and organometallic reagent has a profound effect on the degree of selectivity observed (equation 2). As shown in Table 1, additions of Grignard reagents in THF are most effective, generating selectivities greater than 99:1. Organolithium nucleophiles, on the other hand, provide no useful selectivity.

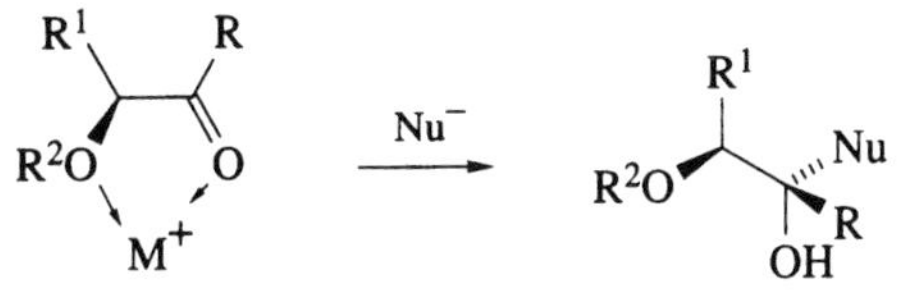

(1) $\xrightarrow{Nu^-}$ (2) (1)

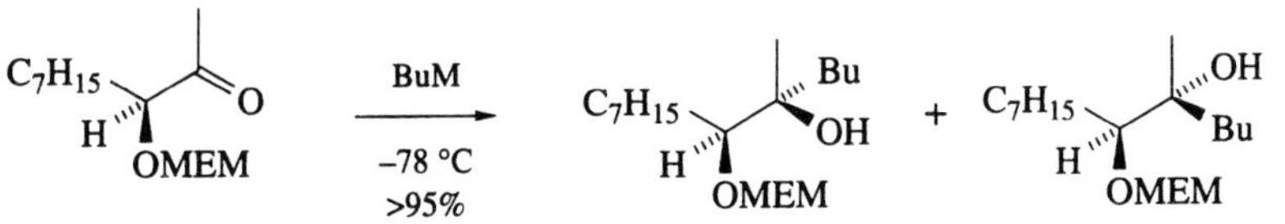

(1a) $\xrightarrow[-78\,°C,\ >95\%]{BuM}$ (2a) + (3a) (2)

Table 1 Effects of Solvent and Organometallic Reagent on the Selectivity of the Reaction of (**1a**) with BuM

Solvent	*Product ratio (**2a**):(**3a**)*	
	M = Li	*M = MgBr*
C_5H_{12}	67:33	90:10
CH_2Cl_2	75:25	93:7
Et_2O	50:50	90:10
THF	41:59	>99:1

The effects of changes in the nature of the alkoxy group (equation 3) are evident in Table 2. High levels of asymmetric induction are achieved by the use of α-substituents such as methoxymethyl ether, benzyl ether, and methylthiomethyl (MTM) ether. The sterically demanding tetrahydropyranyl ether substituent, however, interferes with chelate formation, and its use generates poor selectivities.

C_7H_{15} ... (1b) —BuMgBr, THF, –78 °C→ (2b) + (3b) (3)

Table 2 Effects of Alkoxy Group on the Selectivity of the Reaction of (**1b**) with BuMgBr

R	*Product ratio (**2b**):(**3b**)*
—MEM, —MOM, —MTM, —CH_2(2-furyl)	>99:1
—CH_2Ph	99.5:0.5
—CH_2OCH_2Ph	99:1
—THP	75:25

The high degrees of selectivity (>98:2) found in these reactions are possible not only in acyclic substrates, but also in cyclic α-alkoxy ketones and in more complex systems. The scope of this selective reaction, however, is limited to α-hydroxy ketones, as reaction with β-alkoxy aldehydes proceeds with no selectivity.[13,14]

Similar selectivities (>99:1) are observed in the cyclic chelation controlled reaction of α-benzyloxy carbonyls such as (**4a**; equation 4) with Grignard reagents.[15] However when the α-hydroxy group is protected as a silyl ether (**4b**), the selectivities observed in the addition reaction diminish (60:40), or reverse (10:90; Table 3). The nonchelating nature of a silyl group, as well as its steric bulk, are responsible for this change in selectivity. In the case of (**4b**), nucleophilic addition *via* the Felkin–Anh model effectively competes with the cyclic–chelation control mode of addition.

(**4**) —RM→ (**5**) + (**6**) (4)

(**4**) **a**: $R^1 = CH_2Ph$
b: $R^1 = Bu^tMe_2Si$

Table 3 Diastereoselective Additions to Ketones (**4a**) and (**4b**)

Ketone	*Reagent*	*Temperature (°C)*	*Time (h)*	*Solvent*	*Yield (%)*	*Ratio (**5**):(**6**)*
(**4a**)	MeMgCl	–78	2	Et_2O	85	>99:<1
(**4a**)	MeLi	–78	2	THF	90	60:40
(**4b**)	MeMgCl	–78	2	Et_2O	78	60:40
(**4b**)	(Allyl)MgCl	–78	2	THF	90	10:90

The limited conformations of an α-oxygen substituent within a ring can effectively restrict the possible modes of nucleophilic attack, and lead to highly selective nucleophilic additions. A number of studies using acrolein dimer (**7**; equation 5) illustrate how these ring-constrained systems can be manipulated by taking advantage of either the chelating ability of the α-hydroxy moiety, or the sterically demanding cyclic system.

(**7**) —RM→ (**8**) + (**9**) (5)

Simple addition of *n*-decylmagnesium bromide to (**7**) yields an approximately equal ratio of products (**8**) and (**9**).[16] The use of conditions which promote chelation (excess Grignard reagent) produces a shift in the product ratio (63:37). Addition of $ZnBr_2$ to the reaction mixture further increases the amount of chelation-controlled product formed (85:15; Table 4).

Table 4 Addition of Organometallic Reagents to Acrolein Dimer (**7**)

Reagent	*Conditions*	*Yield (%)*	*Ratio (8):(9)*
(*n*-Decyl)MgBr (1.1 equiv.)	THF, –50 °C	56	48:52
(*n*-Decyl)MgBr (3.0 equiv.)	THF, 15 °C	90	63:37
(*n*-Decyl)MgBr (6.0 equiv.)	$ZnBr_2$ (1.1 equiv.), Et_2O, –10 °C	21	85:15
(*n*-Decyl)MgBr (1.5 equiv.)	THF, HMPA (3.5 equiv.), –45 °C	71	21:79
(*n*-Decyl)MgBr (3.0 equiv.)	THF, HMPA (6.0 equiv.), –20 °C	84	15:85
EtLi	Et_2O, HMPA (4.0 equiv.), –78 °C	78	12:88
EtLi	Et_2O, TMEDA (5.0 equiv.), –78 °C	81	20:80
EtLi	Et_2O, –78 °C	—	28:72
EtMgBr	Et_2O, 0 °C → r.t.	87	70:30

Addition of HMPA to the reaction mixture to suppress chelation causes a reversal in the stereoselectivity, yielding (**9**) as the dominant product. Similar trends in the addition of ethyl metallics to (**7**) are reported.[17] Use of conditions expected to enhance chelation produces (**8**) as the major product; (**9**) is formed predominantly when chelation is inhibited (Table 4). The proposed modes of addition according to chelation control, yielding (**8**), and Felkin–Anh models, yielding (**9**), are shown in Figures 6 and 7.

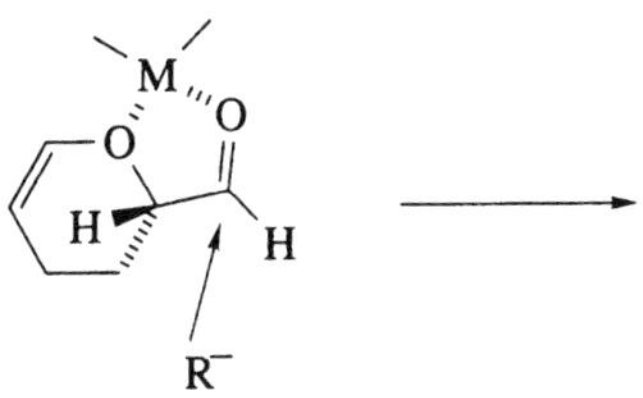

M O O H H R⁻ → (**8**)

Figure 6

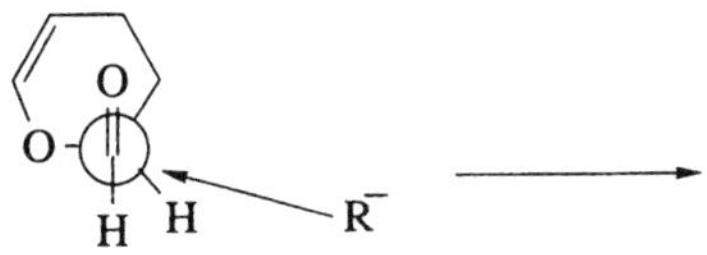

O O H H R⁻ → (**9**)

Figure 7

Nucleophilic additions to tetrahydrofurfural (**10**; equation 6) proceed under similar constraints. Grignard additions in the presence of HMPT favor formation of product (**11**), arising from Felkin–Anh addition (Table 5). In the absence of HMPT, nucleophilic addition yields the cyclic chelation control product (**12**) as the major isomer.[18]

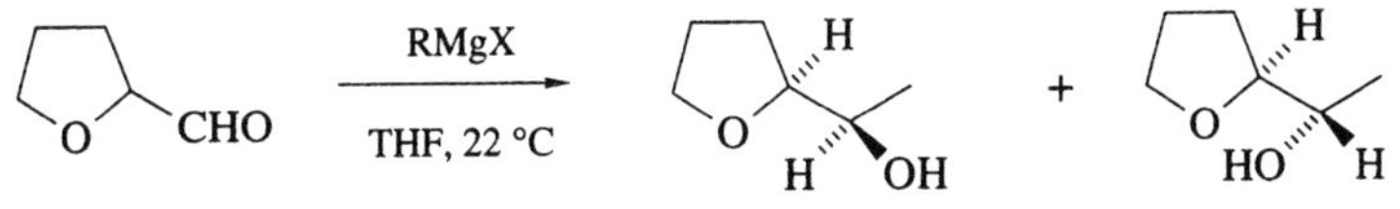

(**10**) O CHO —RMgX, THF, 22 °C→ (**11**) O H H OH + (**12**) O H HO H (6)

Table 5 Stereoselective Additions to (**10**)

RMgX	*Ratio (11):(12)*	*Yield (%)*
EtMgBr	48:52	71
EtMgBr, 2 equiv. HMPT	87:13	86
Bu^nMgBr	43:57	71
Bu^nMgBr, 2 equiv. HMPT	90:10	81
Pr^iMgBr	29:71	48
Pr^iMgBr, 2 equiv. HMPT	100:0	57
PhMgBr	30:70	73
PhMgBr, 2 equiv. HMPT	68:32	64

Considerable work has been carried out on nucleophilic additions to more complex carbonyl substrates containing both α- and β-alkoxy substituents. Optically active 2,3-isopropylideneglyceraldehyde (**13**;

equation 7) represents an ideal substrate for systematic study due to the ready availability of either enantiomer, as well as the versatile functionality present in the molecule. Factors to be considered in a stereocontrolled nucleophilic addition to (**13**) include the presence of a rigid dioxolane system adjacent to the carbonyl and the possibility of chelation to three different oxygen atoms (the carbonyl moiety and two ether oxygens).[19]

(**13**) —RM→ (**14**) + (**15**) (7)

In general, organometallic additions to (**13**), yielding alcohols (**14**) and (**15**), result in only moderate selectivities when organolithium and Grignard reagents are used.[20] The stereochemistry of the major isomer, and the degree of selectivity, depend on the nature of the organometallic reagent, as well as on the conditions under which the reaction is run. Table 6 illustrates some of these findings. Use of other organometallic reagents (Cr-, Ti- and Zn-based) yields improved selectivities.

Table 6 Stereoselectivities in the Addition of RM to (**13**)

RM	*Solvent*	*Temperature (°C)*	*(**14**):(**15**)*	*Yield (%)*
PhLi	Et_2O	−78	48:52	88
PhMgBr	Et_2O	−78	48:52	85
MeLi	Et_2O	−70	60:40	60
MeMgBr	Et_2O	−50	67:33	57
Bu^nLi	Et_2O	−78	69:31	83
Bu^nMgBr	Et_2O	−78	75:25	86
(Allyl)MgBr	Et_2O	−78	60:40	89

The presence of additives in the reaction mixture, however, can significantly enhance the selectivity of nucleophilic addition to (**13**; equation 8).[21] Very high ratios (>95:5) are observed when the addition of furyllithium to (**13**) is carried out in the presence of Zn or Sn salts (Table 7). The stereochemistry of the product is explained by a conformation in which the zinc or tin atom coordinates with the carbonyl oxygen and the 3-oxygen of the dioxolane ring (Figure 8). Nucleophilic attack from the less-hindered face selectively produces (**16**). At present, the applicability of this reaction to other substrates is not known.

(**13**) + 2-furyllithium —additive, THF→ (**16**) + (**17**) (8)

Table 7 Effect of Additive on the Stereoselective Reaction of 2-Furyllithium with (**13**)

Additive	*Temperature (°C)*	*Yield (%)*	*(**16**):(**17**)*
None	−78	68	40:60
$MgBr_2$	0	49	50:50
$SnCl_4$	0	58	95:5
$ZnCl_2$	−78	10	>95:<5
$ZnCl_2$	0	60	90:10
$ZnBr_2$	0	75	95:5
ZnI_2	0	57	>95:<5

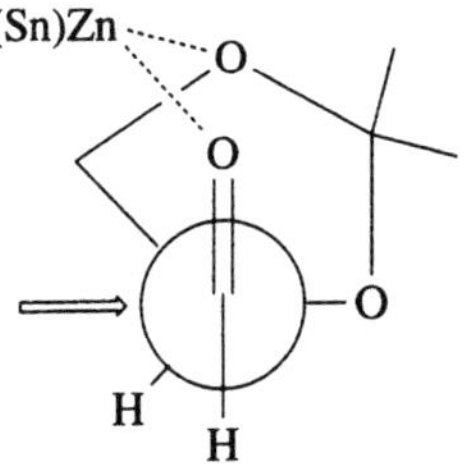

Figure 8

Variation of the diol-protecting group or of other substituents on these α,β-dihydroxycarbonyl derivatives does not lead to significant improvements in stereoselection. Moderate selectivities are reported for the Grignard and organolithium additions to the 2,3-*O,O*-dibenzylglyceraldehyde (**18**) (ranging from 45:55 to 27:73)[22] and to the homologated cyclohexylideneglyceraldehyde (**19**) (80:20 to 60:40).[23] Synthetically useful ratios can, however, be realized by the use of other organometallic reagents (Ti, Cu). The depressed selectivities seen in the cases of magnesium and lithium reagents may be due to their relatively high reactivity.

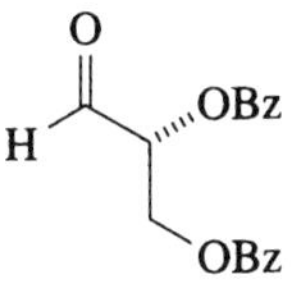

(**18**) 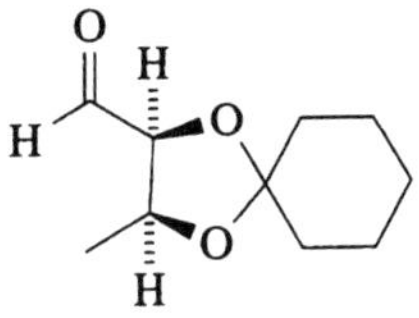 (**19**)

Several reports on selective additions to more complex carbohydrate derivatives suggest that synthetically useful selectivities are indeed obtainable. Grignard addition to aldehyde (**20**) proceeds with a moderate preference for α-chelation-controlled addition. The alcohols (**21**) and (**22**) are isolated in a ratio of about 75:25 (equation 9).[24]

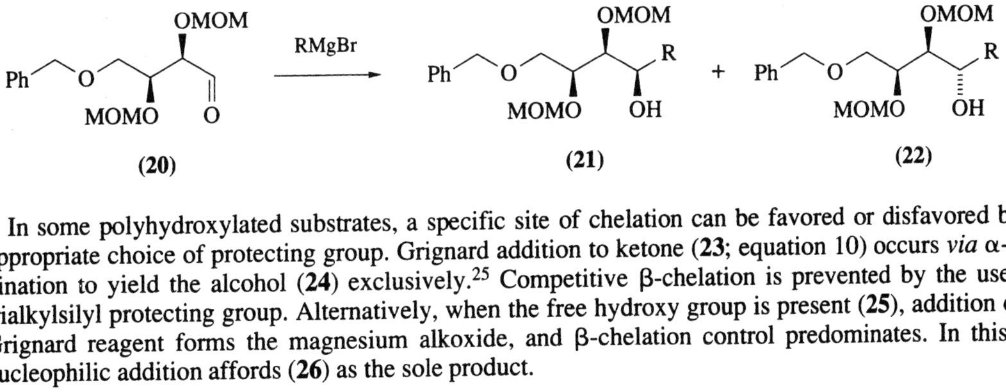

(**20**) $\xrightarrow{\text{RMgBr}}$ (**21**) + (**22**) (9)

In some polyhydroxylated substrates, a specific site of chelation can be favored or disfavored by the appropriate choice of protecting group. Grignard addition to ketone (**23**; equation 10) occurs *via* α-coordination to yield the alcohol (**24**) exclusively.[25] Competitive β-chelation is prevented by the use of a trialkylsilyl protecting group. Alternatively, when the free hydroxy group is present (**25**), addition of the Grignard reagent forms the magnesium alkoxide, and β-chelation control predominates. In this case nucleophilic addition affords (**26**) as the sole product.

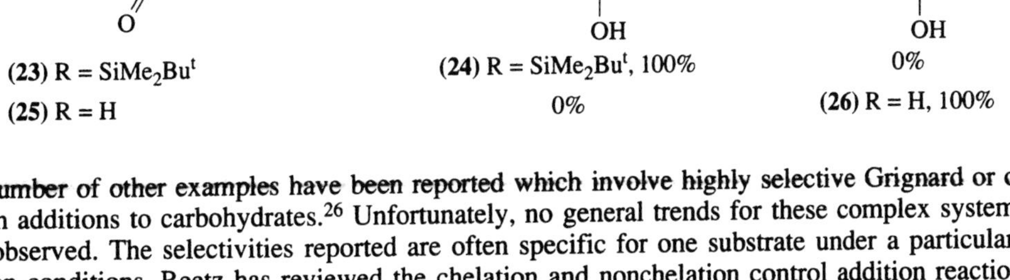

$\xrightarrow{\text{PhMgBr}}$ (10)

(**23**) R = $SiMe_2Bu^t$ (**24**) R = $SiMe_2Bu^t$, 100% 0%

(**25**) R = H 0% (**26**) R = H, 100%

A number of other examples have been reported which involve highly selective Grignard or organolithium additions to carbohydrates.[26] Unfortunately, no general trends for these complex systems have been observed. The selectivities reported are often specific for one substrate under a particular set of reaction conditions. Reetz has reviewed the chelation and nonchelation control addition reactions (not confined to organolithium or organomagnesium reagents) of α- and β-alkoxycarbonyl compounds.[27]

In simple carbonyl compounds containing an α-amino substituent, nucleophilic additions generally occur *via* the cyclic chelate model.[9,12] However, just as in the α-hydroxycarbonyl series, selectivities can be greatly influenced by substitution of the amino group. Examples of moderate to high stereoselection in both cyclic chelation and nonchelation controlled additions have been reported.

Grignard addition to the BOC-protected phenylalaninal (**27**; equation 11) occurs mainly through a cyclic chelation controlled mechanism to yield (**28**) and (**29**) in a ratio of 70:30.[28] Conditions which favor coordination of the nitrogen and magnesium atoms (high temperature) are essential for the selectivity observed.

Bu^tO–C(=O)–NH–CH(CH$_2$Ph)–CHO **(27)** —[MgBr (vinyl), THF, 25 °C]→ **(28)** + **(29)** (11)

In contrast, the use of dibenzyl-protected groups in similar systems shows opposite stereoselectivity.[29] Grignard and alkyllithium additions to dibenzyl-protected α-amino aldehydes such as (**30**; equation 12) proceed with excellent selectivities (usually >95:5) to yield the nonchelate control alcohols (**32**). The high ratios observed (Table 8) regardless of substrate or reagent suggest that the dibenzyl groups act by preventing chelation; Cornforth or Felkin–Anh modes of addition then predominate. Access to the chelation control products is not possible using Grignard or organolithium reagents. The use of titanium or tin reagents, however, provides (**31**) as the major product (Chapter 1.5, this volume).

Bn_2N–CH(R^1)–CHO **(30)** —RM→ **(31)** + **(32)** (12)

Table 8 Stereoselectivities in the Reaction of (**30**) with RM

R^1	RM	Temperature (°C)	Yield (%)	Ratio (31):(32)
Me	MeMgI	0	87	5:95
Me	MeLi	−10	91	9:91
Me	PhMgBr	0	85	3:97
Me	EtMgBr	0	85	5:95
Me	Pr^iMgBr	0	75	<3:>97
Me	Bu^tMgBr	0	72	5:95
Me	Bu^tLi	−60	88	<3:>97
CH_2Ph	MeMgI	0	85	8:92
CH_2Ph	PhMgBr	0	84	3:97
CH_2Ph	PhC≡CLi	−78	72	<4:>96
CH_2Ph	$PhCH_2CH_2MgBr$	−78	84	<4:>96
CH_2Ph	(Allyl)MgCl	−78	82	28:72
Bu^i	MeMgI	0	85	10:90
Bu^i	MeLi	−10	89	20:80
Bu^i	PhMgBr	0	84	3:97
Pr^i	MeMgI	0	87	5:95
Pr^i	MeLi	−10	81	14:86
Pr^i	PhMgBr	0	69	9:91

Earlier work further illustrates the influence of both size and basicity of the amine-protecting group on the direction and degree of stereoselectivity of nucleophilic additions.[30] While some excellent selectivities are observed in the Grignard reaction to amino aldehydes such as (**33**; equation 13), no general model adequately explains all of the results shown in Table 9.

Except for the α-dibenzylamino substrate cited, Grignard and organolithium additions to protected α-aminocarbonyls are not particularly well understood. Only modest stereoselectivities usually result, and

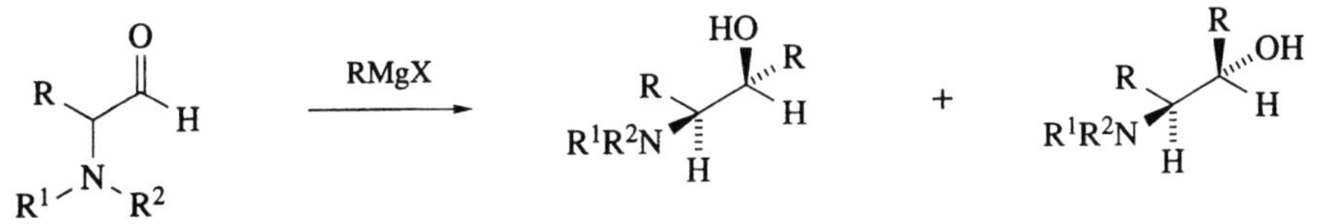

(13)

(33) (34) (35)

Table 9 Stereoselectivity in the Reaction of RMgX with (33)

R	*R*[1]	*R*[2]	*Yield (%)*	*Ratio (34):(35)*
Ph	—$CH_2CH_2CH_2CH_2$—		20	0:100
Ph	—$CH_2CH_2CH_2CH_2CH_2$—		61	16:84
Ph	—$CH_2CH_2OCH_2CH_2$—		72	45:55
Ph	Me	Me	46	0:100
Ph	Et	Et	15	12:88
Et	—$CH_2CH_2CH_2CH_2CH_2$—		45	0:100
Et	—$CH_2CH_2OCH_2CH_2$—		57	34:66
Et	Et	Et	55	52:48
Et	Pr^i	Pr^i	72	100:0

the observed selectivities may not be general. However, synthetically useful ratios can be obtained through modifications of reaction conditions, which promote one mode of addition over another.[31]

As described above, the reactions of Grignard or organolithium reagents to α-hydroxy- or α-aminocarbonyls can proceed with extremely high stereoselectivities (>99:1) when cyclic chelation control is in effect. However, attempts to generate products arising from the Cram–Felkin–Anh mode of addition exclusively have been much less successful.[11] These products are available by the use of conditions which favor nonchelation-controlled processes; however, until recently, the selectivities of these reactions (up to 80–90%) never reached those observed in cyclic chelation control additions.

This discrepancy can at least be partially explained by taking into account that in chelation-controlled reactions the acyclic substrate is essentially locked into one rigid cyclic conformation. The reactants taking part in nonchelation-controlled additions have many more degrees of freedom, and exclusive reaction with one conformer is less likely. These reactions rely on reagents which are incapable of chelation and/or substrates containing sterically or electronically differentiated substituents.[27]

Trialkylsilane substituents have been used very effectively to promote nonchelation-controlled nucleophilic reactions. Addition of Grignard and organolithium reagents to chiral acylsilane (**36**) produces the Cram–Felkin–Anh product (**37**) almost exclusively (>92:8 selective in most cases).[32] The silyl group in (**37**) can be stereospecifically replaced with hydrogen to afford the product (**40**) of formal nucleophilic addition to the parent aldehyde (Scheme 1). This sequence of reactions is one of the first examples of a general procedure for highly efficient nonchelation-controlled additions to aldehyde equivalents. Direct Grignard and organolithium additions to aldehyde (**39**) are less selective, as shown in Table 10.

Stereoselectivities observed in the reactions of the α-chiral acylsilanes are explained by consideration of the the Felkin–Anh model. The conformers depicted in Figures 9 and 10 are predicted to be those through which nucleophilic addition occurs. The sterically demanding TMS group apparently differentiates between the α-hydrogen (S) and the α-methyl (M) substituents. This preference for the conformation in Figure 9 results in a highly stereoselective reaction.

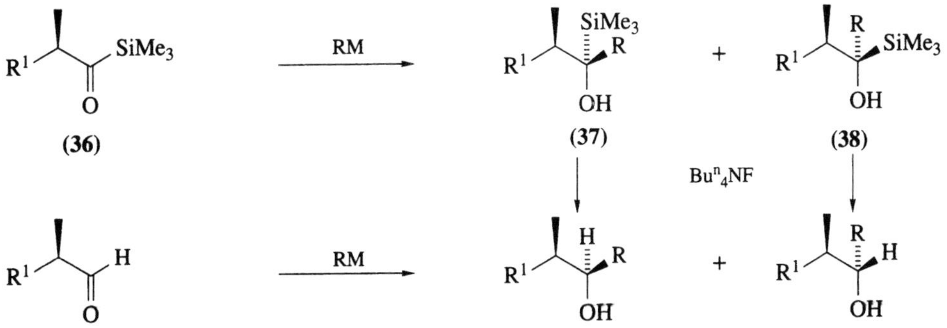

(36) (37) (38) Bu^n_4NF

(39) (40) (41)

Scheme 1

Table 10 Selectivities in the Reaction of Nucleophiles with α-Chiral Acylsilanes (**36**) and Aldehydes (**39**)

Substrate	R^1	*RM*	*Yield (%)*	*Ratio (37):(38) or (40):(41)*
(**36**)	Ph	Bu^nLi	92	>99:1
(**36**)	Ph	MeLi	96	>97.5:2.5
(**36**)	Ph	(Allyl)MgBr	96	92:8
(**36**)	1-Cyclohexenyl	Bu^nLi	56	>97:3
(**36**)	1-Cyclohexenyl	MeLi	69	>99:1
(**36**)	1-Cyclohexenyl	(Allyl)MgBr	69	92:8
(**36**)	Cyclohexyl	Bu^nLi	98	94:6
(**36**)	Cyclohexyl	MeLi	77	96:4
(**36**)	Cyclohexyl	(Allyl)MgBr	93	78:22
(**39**)	Ph	Bu^nLi	91	83:17
(**39**)	Ph	MeLi	91	80:20
(**39**)	Ph	(Allyl)MgBr	92	63:27
(**39**)	1-Cyclohexenyl	Bu^nLi	40	94:6
(**39**)	1-Cyclohexenyl	MeLi	49	66:34
(**39**)	1-Cyclohexenyl	(Allyl)MgBr	54	71:29
(**39**)	Cyclohexyl	Bu^nLi	96	78:22
(**39**)	Cyclohexyl	MeLi	75	67:33
(**39**)	Cyclohexyl	(Allyl)MgBr	59	67:33

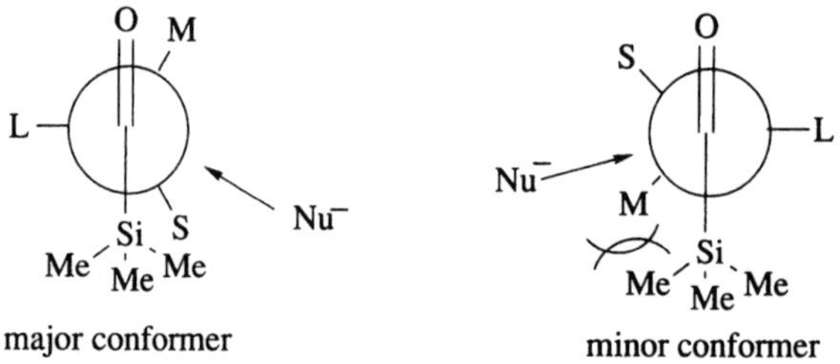

major conformer

Figure 9

minor conformer

Figure 10

Trialkylsilyl groups in other positions on the carbonyl substrate can also influence the direction of nucleophilic addition. Grignard additions to 2-alkyl- and 2-alkoxy-3-trimethylsilylalkenylcarbonyl compounds such as (**42a**), (**42b**) and (**42c**) proceed with high diastereofacial selectivity based on the Cram–Felkin–Anh model.[33] Table 11 lists results of the additions to all three derivatives. Excellent stereoselectivities (>99:1 in most cases) favoring the nonchelation-controlled product (**43**) are evident in all examples except one. Replacement of the trialkylsilyl group in the products with hydrogen, or manipulation to generate other functional groups illustrates the synthetic potential of this procedure. The key role played by the silyl group in obtaining high stereoselectivities is demonstrated by a comparison of reported nucleophilic additions to analogous substrates. In these examples, substrates which do not contain a trialkylsilyl moiety in the carbonyl substrate react with much lower (67:33) selectivities.

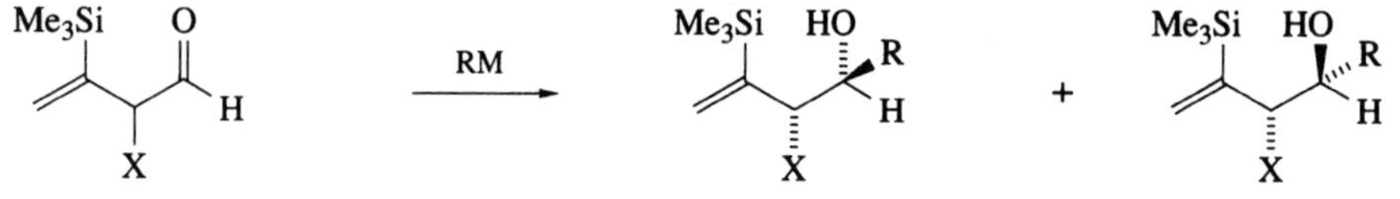

(**42**) **a**: X = Me; **b**: X = OBn; **c**: X = OMe → RM → (**43**) + (**44**) (14)

1.2.1.1.3 Remote asymmetric induction

In general, nucleophilic addition reactions using Grignard or organolithium reagents to β-chiral or other remotely chiral ketones usually yield mixtures of stereoisomers which are not synthetically useful.[9,12,27] However, some specific examples of stereoselective organolithium and Grignard additions to these remotely chiral carbonyl compounds have been reported.

Table 11 Stereoselectivities in the Reaction of (**42**) with RM

Substrate	*RM*	*Solvent*	*Yield (%)*	*Ratio (**43**):(**44**)*
(**42a**)	MeMgI	THF	84	91:9
(**42a**)	EtMgBr	THF	92	>99:<1
(**42a**)	Pr^iMgBr	THF	91	>99:<1
(**42a**)	PhMgBr	THF	94	>99:<1
(**42a**)	$CH_2{=}CH(TMS)MgBr$	THF	93	>99:<1
(**42b**)	EtMgBr	THF	63	50:50
(**42b**)	EtMgBr	Et_2O	87	93:7
(**42c**)	EtMgBr	THF	68	90:10
(**42c**)	EtMgBr	Et_2O	92	>99:<1

Diastereoselective additions of nucleophiles to the β-alkoxy-γ-hydroxy aldehyde (**45**; equation 15) are reported to generate either chelation- or nonchelation-controlled products, depending on the reaction conditions used. Chelation-controlled additions of organolithium or Grignard reagents in THF take place with reasonable selectivities to produce (**46**) as the major product.[34] These selectivities can be improved with a change of solvent (diethyl ether), and with the addition of Zn salts (Table 12). Figure 11 represents the probable mode of addition in this chelation-controlled reaction. Alternatively, the diol (**47**) results as the major product when the reaction is carried out in ether using alkylmagnesium bromides as nucleophiles. The mode of addition to generate (**47**) under these conditions is not completely understood.

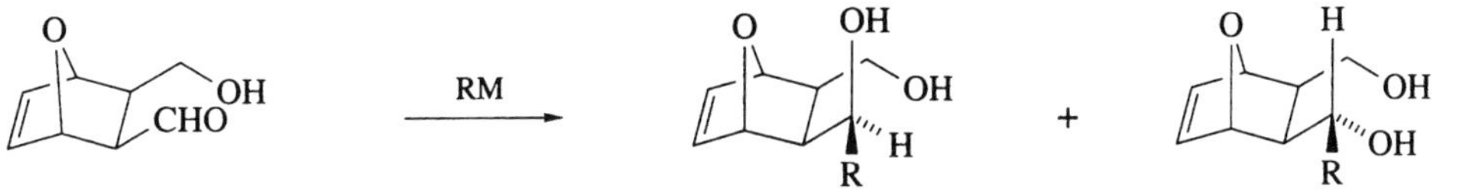

(**45**) —RM→ (**46**) + (**47**) (15)

Table 12 Diastereoselective Additions of Organometallics to (**45**)

RM	*Solvent*	*Yield (%)*	*Ratio (**46**):(**47**)*
MeMgI	THF	65	70:30
MeMgI	Ether	82	82:18
MeLi	THF/ether	80	78:22
MeLi	Ether	90	83:17
MeMgBr	THF	60	74:26
MeMgBr	Ether	77	15:85
Bu^nMgBr	THF	65	88:12
Bu^nMgBr	Ether	70	15:85
Pr^iMgBr	THF	73	91:9
Pr^iMgBr	Ether	65	15:85
$Bu^nMgBr + ZnI_2$	THF	56	94:6

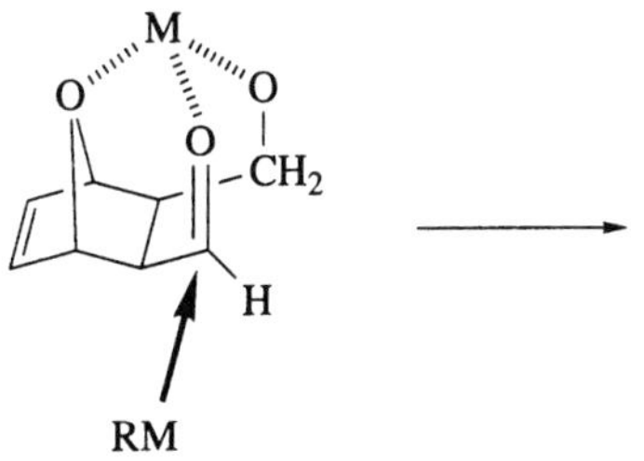

Figure 11

Systematic studies on additions to β-asymmetric amino ketones of general structure (**48**; equation 16) result in the following conclusions.[35]

(i) In reactions of (**48**) with organolithiums, isomer (**49**) predominates regardless of the nature of the reagent and the substrate. Selectivities as high as 87:13 are reported. This general trend in stereoselectivity cannot be adequately explained by one model.

(ii) Selectivities in the Grignard additions to (**48**) are highly dependent on the nature of the substrate, as well as the reagents. Neither isomer (**49**) nor (**50**) predominates, and no specific model is consistent with the experimental results. Again, the highest selectivities observed are in the 80:20 range.

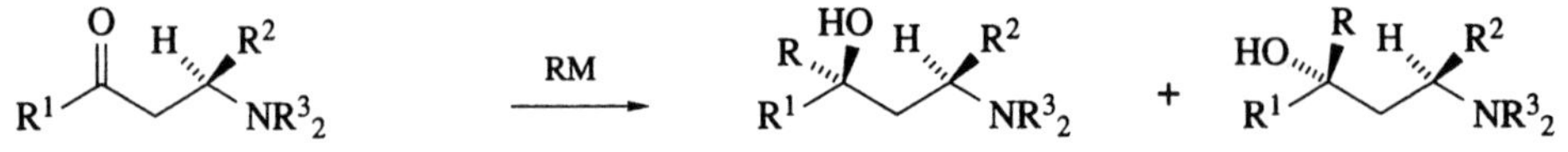

(**48**) —RM→ (**49**) + (**50**) (16)

In an example of remote asymmetric induction, chiral oxazoline (**51**) undergoes Grignard additions with moderate to good selectivity.[36] The major product formed (**52**) results from coordination of the reagent to the oxazoline moiety, followed by nucleophilic attack from the bottom, as shown in Figure 12. The phenyl substituent of the oxazoline effectively prevents addition from the top face of the molecule. Systems such as (**51**) have been used to synthesize optically active phthalides (**54**), as shown in Scheme 2. Typical optical purities of the isolated aromatic products range from 46–80%. Table 13 lists some of these results.

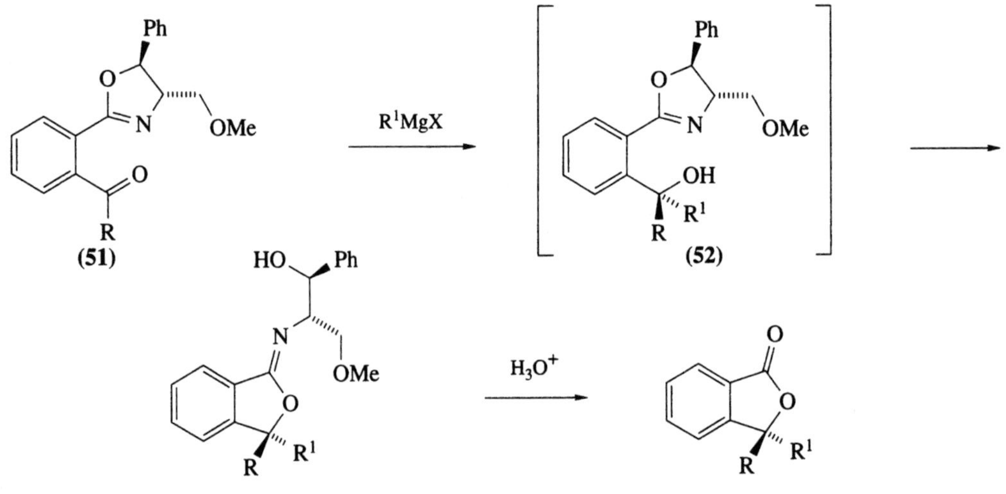

(**51**) —R^1MgX→ [(**52**)] → (**53**) —H_3O^+→ (**54**)

Scheme 2

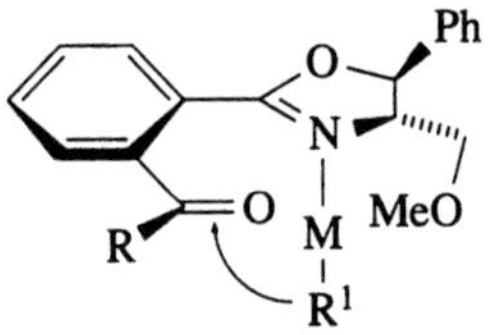

Figure 12

Table 13 Diastereoselectivity of Grignard Additions to (**51**)

R	*R¹MgX*	*Yield (%)*	*(53) Diastereomer ratios*
Me	EtMgBr	96	73:27
Me	Bu^nMgBr	92	73:27
Me	Bu^tMgBr	90	83:17
Me	PhMgBr	97	90:10
Ph	MeMgBr	99	88:12
Ph	EtMgBr	99	83:17
Et	MeMgCl	93	82:18
Bu^t	MeMgCl	66	52:48
p-BrC_6H_4	MeMgCl	95	87:13
Bu^n	MeMgCl	89	84:16

While examples of moderate to good stereoselectivities in remotely chiral molecules, such as those described above, are known, they are usually specific cases, and cannot be generalized to all systems. In general, stereoselective Grignard or organolithium additions which rely on remote or β-chelation are not efficient.

1.2.1.1.4 Chiral auxiliaries

A special class of acyclic diastereoselective reactions involves the use of chiral auxiliaries to control the absolute stereochemistry of nucleophilic additions at carbonyl centers. This process takes advantage of steric and/or electronic factors within the chiral adjuvant to promote nucleophilic addition from one face of the molecule, and thereby generate one predictable diastereomer. Removal of the auxiliary, in the best cases, generates enantiomerically pure products, as well as the recyclable chiral adjuvant. The end result of this process is the synthesis of enantiomerically pure products *via* diastereoselective reactions.

Eliel has extensively studied the 1,3-oxathiane systems (**55**) as a chiral auxiliary to control the addition of organometallics to ketones (Scheme 3).[37] Following nucleophilic addition, cleavage of the oxathiane group generates chiral α-hydroxycarbonyl compounds (**58**). The wide variety of carbonyl substituents, as well as Grignard reagents, amenable to this process is illustrated in Table 14. The enantiomeric auxiliary is available, and affords alcohols of opposite chirality in equivalent yields.

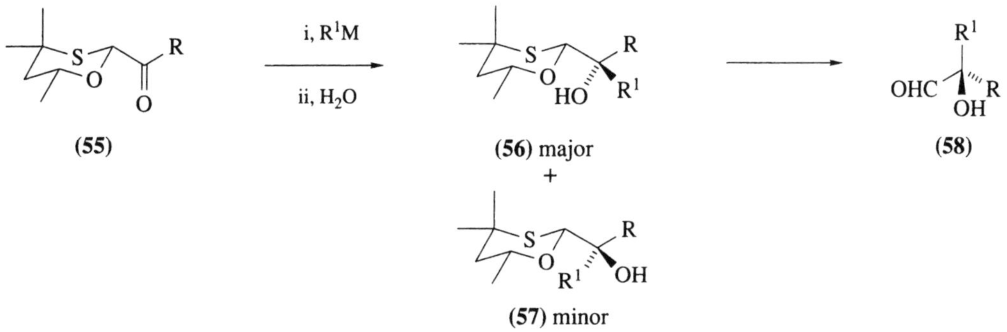

S O R O (**55**) — i, R^1M; ii, H_2O → S O R HO R^1 (**56**) major + S O R R^1 OH (**57**) minor → R^1 OHC R OH (**58**)

Scheme 3

Table 14 Stereoselective Reactions of (**55**) with Organometallics

R	*R^1M*	*Temperature (°C)*	*Solvent*	*Ratio (56):(57)*
Ph	MeMgI	Reflux	Et_2O	96:4
Ph	MeLi	Reflux	Et_2O	86:14
Ph	EtMgI	Reflux	Et_2O	>99:<1
Ph	Pr^iMgI	Reflux	Et_2O	99:1
Me	PhMgBr	Reflux	Et_2O	89:11
Me	Ph_2Mg	Reflux	Et_2O	78:22
Me	PhMgI	Reflux	Et_2O	87:13
Me	EtMgI	Reflux	Et_2O	90:10
Me	Pr^nMgI	Reflux	Et_2O	90:10
Me	Pr^nMgBr	Reflux	Et_2O	83:17
Me	Pr^nMgCl	Reflux	Et_2O	79:21
Me	$Bu^n{}_2Mg$	Reflux	Et_2O	87:13
Me	Pr^iMgI	Reflux	Et_2O	67:33
Me	(Vinyl)MgBr	Reflux	THF	91:9
Me	HC≡CMgBr	Reflux	THF	94:6
Et	PhMgBr	Reflux	Et_2O	83:17
Et	PhMgBr	−78	Et_2O	97:3
Et	MeMgI	Reflux	Et_2O	80:20
Vinyl	MeMgI	Reflux	Et_2O	87:13
Vinyl	MeMgI	−78	Et_2O	95:5
Pr^i	MeMgI	Reflux	Et_2O	68:32
Pr^i	MeMgI	−78	Et_2O	77:23
Pr^i	MeMgI	Reflux	Et_2O	69:31
Pr^i	PhMgBr	Reflux	Et_2O	85:15
Pr^i	PhMgBr	−78	Et_2O	>99:<1
Bu^t	MeMgI	Reflux	Et_2O	93:7
Bu^t	MeMgI	−78	Et_2O	96:4

Further work towards developing this process into a viable route to enantiomerically pure compounds has led to the use of the 1,3-dioxathiane (**59**; Scheme 4) as a second generation chiral auxiliary.[38] Excellent stereoselectivities (usually >90%) are found in the reaction of (**59**) with a variety of Grignard reagents. The carbinol products (**60**) are readily cleaved to the α-hydroxy aldehydes (**61**) and the sultine (**62**), which is used to regenerate (**59**). Either enantiomer of the desired product is available, by using the diastereomeric oxathiane, or by reversing the order of R-group addition. The nitrogen analog (**63**) has recently been reported to impart excellent stereoselection. Scheme 5 illustrates its utility for the synthesis of a number of α-hydroxy acids in high optical purity (see Table 15).[39]

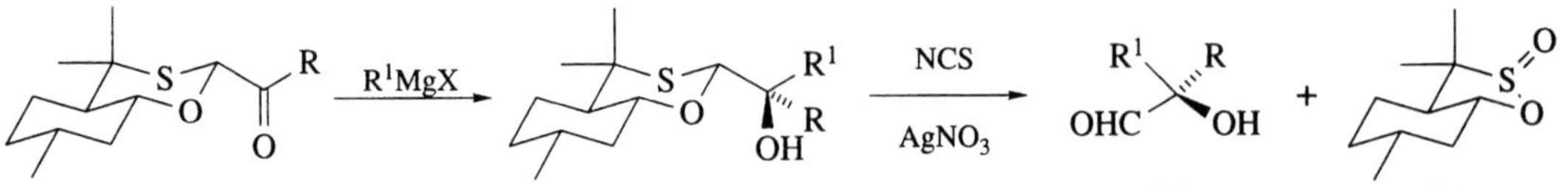

Scheme 4

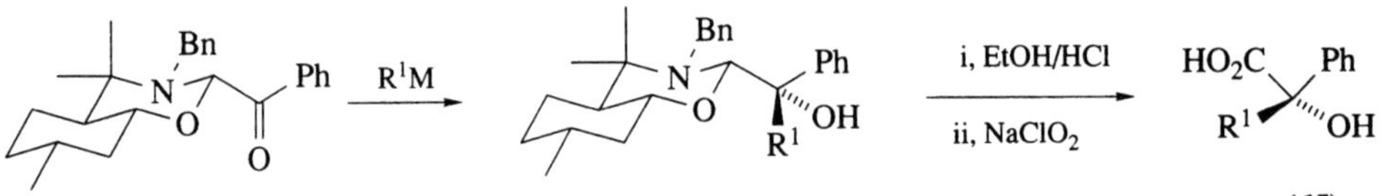

Scheme 5

Table 15 Synthesis of Optically Active α-Hydroxy Acids from (**63**)

R^1	*R^1M*	*Temperature (°C)*	*Yield (%)*	*(65) Configuration*	*ee (%)*
Me	MeMgBr	20	44	(*S*)	98
Me	MeMgBr	–70	—	(*S*)	98
Me	MeLi	–70	47	(*S*)	95
Et	EtMgBr	5	77	(*S*)	100
HC≡C	HC≡CMgBr	20	63	(*S*)	97 ± 1
α-Naphthyl	$C_{10}H_7MgBr$	20	23	(*R*)	82 ± 1

The mode of Grignard addition is through the cyclic–chelate conformation shown in Figure 13. The hard magnesium ion coordinates with the carbonyl oxygen and the hard oxygen atom of the oxathiane ring (in preference to the soft sulfur atom). Addition then occurs from the less-hindered side of the auxiliary. All reactions investigated using these ligands have led to the configuration predicted according to this model.[40]

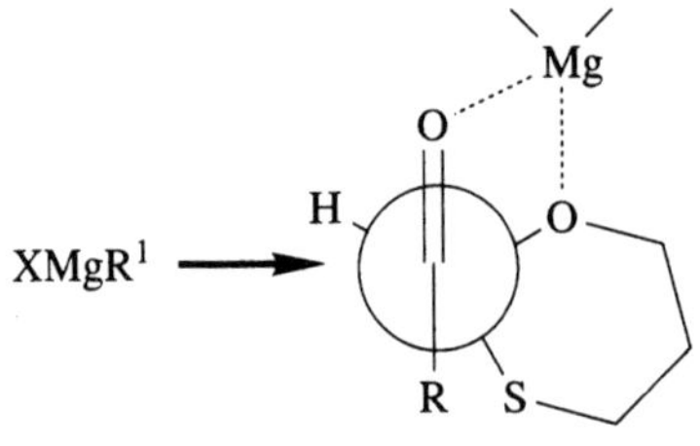

Figure 13

In a study of the factors influencing this nucleophilic addition, it was found that incorporation of an exocyclic oxygen atom into the carbonyl substituent (such as in **66**; equation 17) has a profound effect on the selectivity of the reaction.[41] Alkoxy groups capable of chelation (*e.g.* CH_2Ph) can competitively inhibit coordination of the metal to the ring oxygen, thus severely lowering, and in some cases reversing, the stereoselectivities observed. In these cases, the chelating ability of the organometallic reagent, as well as the length of the methylene linker, influence the course of the reaction. The use of a triisopropylsilyl protecting group, however, effectively prevents chelation of the exocyclic alkoxy substituent with the Grignard reagent, and yields high, predictable selectivities (see Table 16).

(66) $\xrightarrow{R^1M}$ (67) + (68) (17)

Table 16 Stereoselective Nucleophilic Additions to Oxathianyl Alkoxy Ketones (**66**)

R	*n*	R^1	*Reagent*	*(67):(68)*
CH_2Ph	1	Me	MeMgBr	33:67
$SiPr^i_3$	1	Me	MeMgBr	97:3
CH_2Ph	2	Me	MeMgBr	42:58
$SiPr^i_3$	2	Me	MeMgBr	97:3
CPh_3	2	Me	MeMgBr	86:14
CH_2Ph	3	Me	MeMgBr	81:19
$SiPr^i_3$	2	Me	$(Me)_2Mg$	98:2
TMS	2	Me	$(Me)_2Mg$	73:27

The chiral oxathianes have been broadly used for the synthesis of a number of natural products. Through these applications, these chiral auxiliaries have been shown to provide a viable procedure for the synthesis of optically pure compounds.[42]

Chiral acetals can be used as auxiliaries in the diastereoselective reactions of Grignard reagents with acyclic[43] as well as cyclic α-keto acetals.[44] Nucleophilic addition to the monoprotected diketone (**69**; equation 18) occurs with excellent stereoselectivity to generate the corresponding tertiary alcohol (**70**) as the major product, usually with greater than 95:5 selectivity. Removal of the ketal yields α-hydroxy ketones of high optical purity. In most examples, enantiomeric excesses of 95% and higher are observed in the resultant keto alcohols. Table 17 represents the results of additions to cyclic and acyclic substrates.

(69) $\xrightarrow[\text{THF, }-78\,^{\circ}\text{C}]{\text{RMgX}}$ (70) major + (71) minor (18)

Table 17 Diastereoselective Additions to α-Keto Ketals (**69**)

R^1	R^2	*RMgX*	*Yield (%)*	*Ratio (70):(71)*
—$CH_2CH_2CH_2CH_2$—		MeMgBr	93	100:0
—$CH_2CH_2CH_2$—		MeMgBr	91	98:2
—$CH_2CH_2CH_2CH_2$—		EtMgCl	95	100:0
—$CH_2CH_2CH_2$—		EtMgCl	95	100:0
—$CH_2CH_2CH_2CH_2$—		(Vinyl)MgBr	95	97:3
—$CH_2CH_2CH_2CH_2$—		PhMgBr	85	95:5
Me	Ph	EtMgCl	98	>99:1
Me	Ph	(Vinyl)MgBr	90	98:2
Me	Ph	PhMgBr	84	97:3
Me	Et	EtMgCl	92	>99:1
Me	Et	(Vinyl)MgBr	93	>99:1
Me	Et	PhMgBr	81	98:2

The stereoselectivity of these reactions can be explained by chelation of the magnesium metal with the carbonyl oxygen, the proximal methoxy oxygen atom, and one of the acetal oxygens. Migration of the alkyl group from the organometallic reagent then occurs from the least-hindered face (Figure 14). Evidence for this mechanism follows from experiments carried out using a related chiral auxiliary in which the methoxy groups were replaced with hydrogen atoms. Stereoselectivities are considerably lower (10–20% *ee*) in these cases.

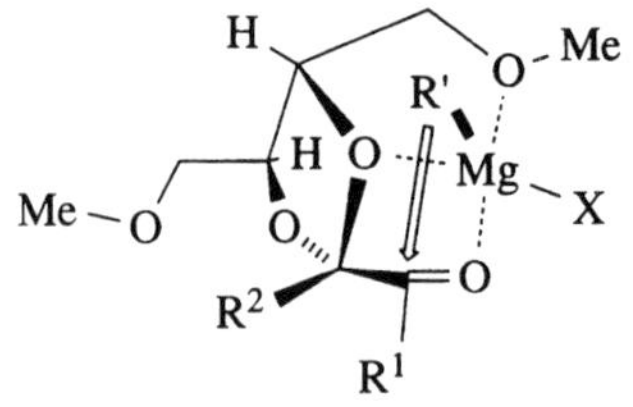

Figure 14

Extension of this methodology to the use of chiral acetals such as (**72**; equation 19) to produce optically active secondary alcohols is found to be less efficient than the ketal series. Alkyl Grignard reagents in ether (Table 18) provide the best selectivities (up to 90:10), while aryl and alkynyl organometallics show very little diastereofacial differentiation.[45]

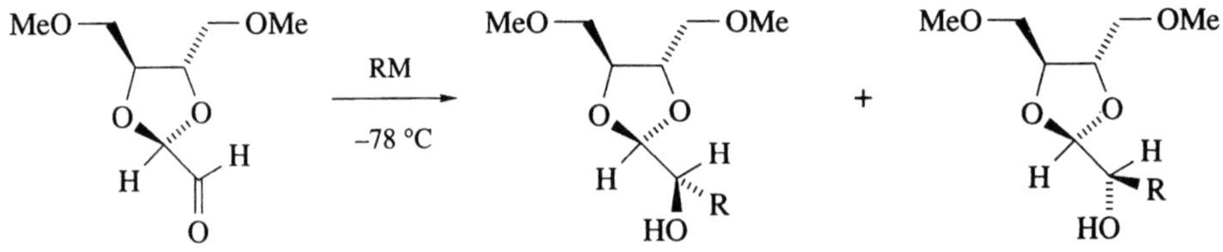

(19)

Table 18 Diastereoselective Additions to Chiral Acetal (**72**)

RM	*Solvent*	*Yield (%)*	*Ratio (73):(74)*
MeMgI	Et_2O	70	90:10
Bu^nMgBr	Et_2O	65	87:13
Bu^nLi	Et_2O	60	52:48
Pr^iMgBr	Et_2O	60	80:20
PhMgBr	Et_2O	70	34:76
PhMgBr	THF	70	42:58
PhLi	Et_2O	60	49:51
PhC≡CLi	Et_2O	65	48:52
PhC≡CMgBr	THF	65	37:63

Advantages of this chiral auxiliary include the ready availability of either enantiomer and its ease of removal, as well as the dual utility of the chiral acetal as both chiral auxiliary, and carbonyl-protecting group. A published synthesis of optically pure (–)-7-deoxydaunomycinone exemplifies the utility of this chiral adjuvant.[46]

The use of ketoaminals based on pyrrolidine as chiral auxiliaries has been demonstrated as another entry to optically active α-hydroxycarbonyls.[47] The aminals (**76**; Scheme 6) are readily obtained from a chiral diamine (**75**) and glyoxal. Addition of Grignard reagents to (**76**), followed by hydrolysis, provides a chiral α-hydroxy aldehyde (**78**) with the (*S*)-configuration. Optical purities are measured in the 94–95% range. The chiral diamine can be recovered unchanged from the reaction mixture.

The source of the asymmetry is thought to occur through two stereoselective steps. First, the preferential formation of one diastereomeric aminal with the structure shown in Figure 15 is expected due to steric arguments. Second, attack of the Grignard reagent from one face of the molecule occurs through the cyclic–chelate mode. The magnesium ion coordinates with the carbonyl oxygen, as well as with the nitrogen (N-1) of the pyrrolidine ring. Complexation with the phenyl-substituted nitrogen (N-2) is dismissed due to its electron deficiency with respect to N-1. This rigid structure leads to alkyl group attack on the carbonyl oxygen from the less-hindered side (see Figure 16).

Extension of this work to the synthesis of α-hydroxy aldehydes with substituents other than phenyl can be carried out by using the methyl ester (**79**; Scheme 7) as precursor.[48] Grignard additions afford a variety of ketoaminals (**80**) in good yield; aldehydes (R^1 = H) are available *via* diisobutylaluminum hydride reaction.[49] A second Grignard addition, followed by hydrolysis, generates α-hydroxy aldehydes (**82**) in

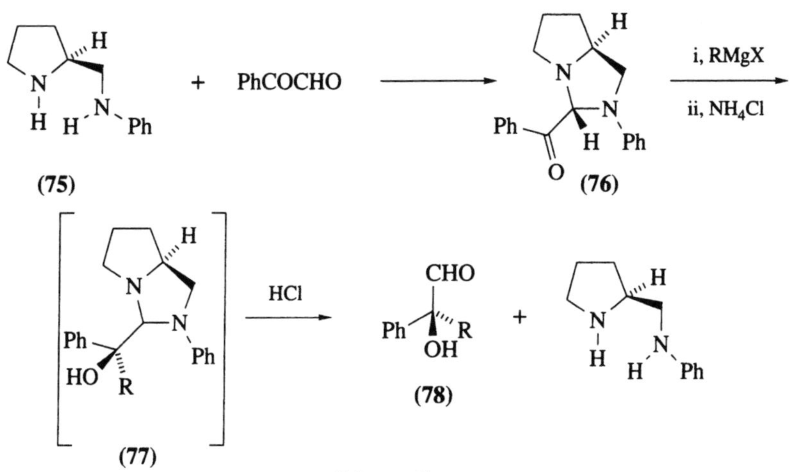

(75)

(76)

(77)

(78)

Scheme 6

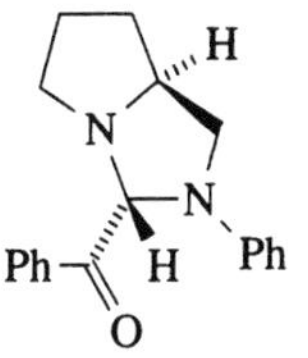

Figure 15

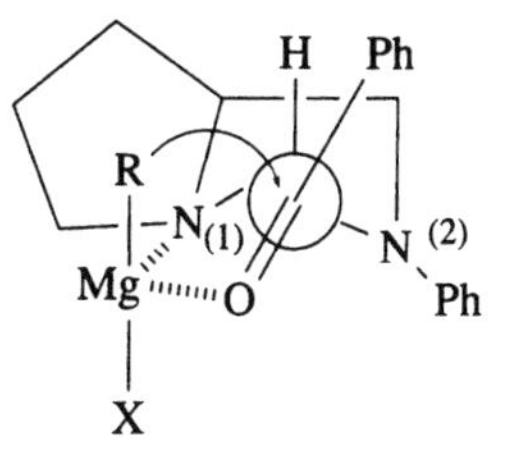

Figure 16

moderate to high optical yields (Table 19). Configurations of the products can be predicted from a model similar to that shown in Figure 16. The corresponding aldehydes (**80**; R^1 = H) react with equivalent selectivity to afford secondary alcohols.

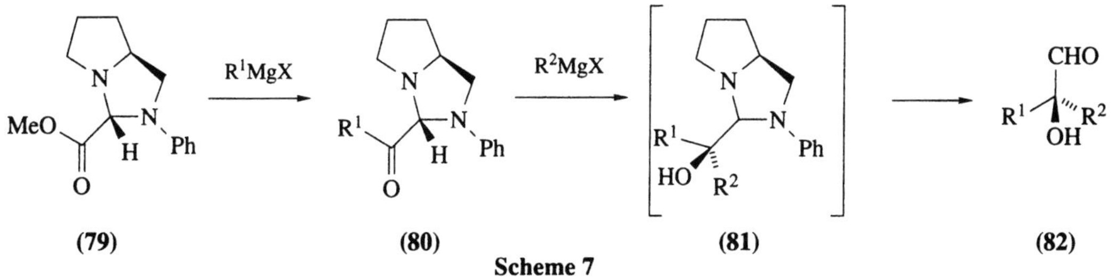

(79) (80) (81) (82)

Scheme 7

Table 19 Preparation of α-Hydroxy Aldehydes (**82**)

R^1	R^2MgX	*Yield (%)*	*(82)* *ee (%)*	*Configuration*
Me	PhMgBr	76	99	(*R*)
Me	EtMgBr	43	78	(*R*)
Me	(Vinyl)MgBr	44	93	(*R*)
Et	PhMgBr	80	100	(*R*)
Et	MeMgI	41	78	(*S*)
Pr^i	PhMgBr	75	94	(*R*)

This consecutive Grignard methodology allows the synthesis of either enantiomer of the α-hydroxy aldehyde product, since its stereochemistry is determined only by the order of Grignard reactions. The preparation of natural products such as (+)- and (–)-frontalin[50] and malyngolide[51] in high optical yield has been carried out, and demonstrates the synthetic utility of this chiral auxiliary. The use of pyrrolidine-based chiral auxiliaries is reviewed by Eliel[11] and Mukaiyama.[52]

Glyoxalate esters of phenylmenthols (**83**; equation 20) have been extensively used as chiral auxiliaries for a number of different reactions. Selective Grignard addition from the front face of the molecule affords the α-hydroxy ester (**84**) as the major product (Table 20).[53] Subsequent reduction generates the corresponding diol in optical purities ranging from 88 to 92%.[54] (This methodology is equally applicable to the synthesis of tertiary alcohols using the keto ester as starting material.) The auxiliary suffers from difficulties in its preparation in optically pure form,[55] and from the unavailability of the enantiomeric phenylmenthol. This latter limitation can be overcome by reversal of the order of nucleophilic addition.[56]

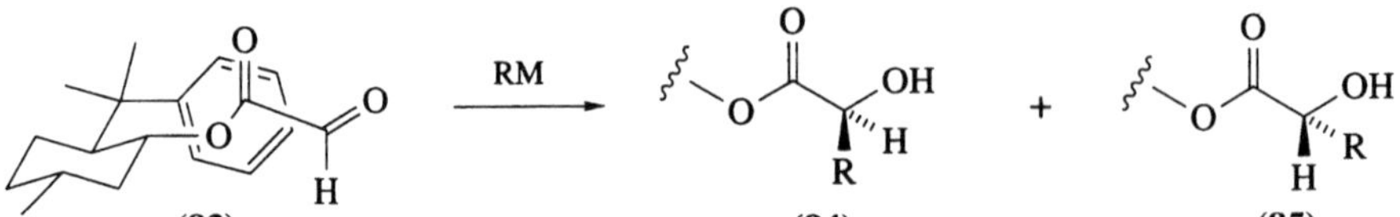

(**83**) —RM→ (**84**) + (**85**) (20)

Table 20 Selectivities in the Addition of RM to (**83**)

Reagent	*Temperature (°C)*	*Yield (%)*	*Ratio (**84**):(**85**)*
MeMgBr	0	62	95:5
MeMgBr	−78	86	>99:1
$C_6H_{13}MgBr$	−78	82	>99:1
$C_8H_{17}MgBr$	−78	80	99:1
PhMgBr	−78	90	>99:1
$C_6H_{11}MgBr$	−78	80	>99:1
MeLi	−78	74	50:50
MeLi + $LiClO_4$	−78	80	80:20

Meyers and coworkers report the synthesis of enantiomerically enriched α-hydroxy acids (enantiomeric excesses of the products generally range from 30 to 87%) from chiral ketooxazolines such as (**86**; Scheme 8).[57] In most cases, only moderate selectivities are observed. The highest ratios (62–87% *ee*) result when aryllithiums are employed as the nucleophile (Table 21). Since the course of the reaction seems to depend on a number of subtle conformational and coordinating effects, predictions of the stereochemical outcome of these processes are difficult.

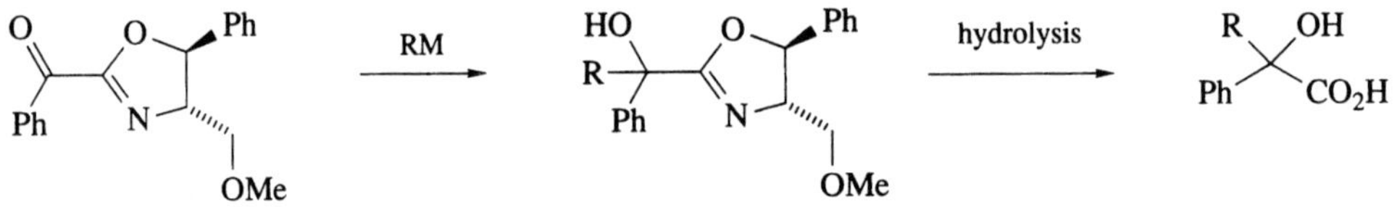

(**86**) —RM→ (**87**) —hydrolysis→ (**88**)

Scheme 8

Table 21 Synthesis of α-Hydroxy Acids (**88**)

Reagent	*(**87**) Yield (%)*	*Yield (%)*	*(**88**) ee (%)*	*Configuration*
MeMgBr/THF	95	70	9	(*S*)
MeMgBr/TMEDA/THF	99	72	32	(*R*)
MeMgBr/Et_3N/toluene	94	76	22	(*S*)
MeLi/THF	93	70	0	—
MeLi/$LiClO_4$/THF	99	73	48	(*S*)
EtMgBr/Et_3N/toluene	>99	65	33	(*S*)
Pr^nMgBr/Et_3N/toluene	>99	62	39	(*S*)
Pr^iMgBr/Et_3N/toluene	95	57	41	(*S*)
Bu^iMgBr/Et_3N/toluene	>99	55	50	(*S*)
p-Tol-Li	93	60	76	(*S*)
p-Anisyl-Li	92	55	62	(*S*)
1-Naphthyl-Li	>99	62	65	(*S*)
2-Thienyl-Li	90	73	87	(*S*)

1.2.1.2 Additions to Cyclic Systems

Nucleophilic additions to cyclic carbonyl compounds differ greatly from those of acyclic systems. In acyclic systems, only the configuration at an adjacent (1,2-asymmetric induction) or nearby center (remote asymmetric induction) is usually considered in predicting the outcome of nucleophilic attack. In cyclic systems, the conformation of the entire molecule (which is in part determined by the individual substituents) must be considered when predicting the mode of nucleophilic attack. Furthermore, a number of other factors such as torsional and electronic effects also play a role in the stereochemical course of additions to cyclic substrates. The relative importance of all of these effects (as well as others) has been the subject of considerable debate in the literature, and has not as yet been adequately resolved.[9,12,58]

Models for the nucleophilic addition of organolithium and Grignard reagents to cyclic ketones assume that the incoming group approaches the carbonyl carbon perpendicularly to the plane of the sp^2 center (Figure 17).[58] This line of approach effects maximum overlap of the orbitals in the transition state of the reaction. Whether this perpendicular nucleophilic attack then occurs from an equatorial or an axial trajectory depends on the effects mentioned above. In the case of simple cyclohexanones, if 'steric approach control' influences the reaction, the nucleophile will enter from the less-hindered equatorial position to yield the axial alcohol. Axial addition is disfavored due to steric hindrance from the axial hydrogen atoms at C-3 and C-5. When 'product development control' is in effect, formation of the more stable equatorial alcohol (from axial attack) is favored. The relative importance of all of these effects is highly dependent on the particular nature of the cyclic substrate, and has been the subject of considerable theoretical interest.[59] An excellent review of this field has been written by Ashby and Laemmle.[58]

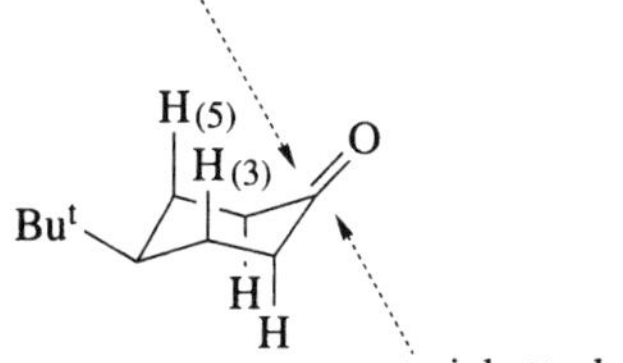

axial attack

H(5) H(3) O Bu^t H H

equatorial attack

Figure 17

The mode of addition to substituted cyclohexanones (equation 21) depends greatly on the nature and position of the substituents, as well as on the structure of the organometallic reagent. Table 22[58] lists results of nucleophilic additions to a variety of cyclohexanones. Some broad generalizations can be made.[12]

(i) With organolithiums (except acetylides) and Grignard reagents, equatorial attack is usually favored. The substitution pattern of the ketone influences the course of the reaction to a lesser extent. With acetylides, axial attack predominates due to torsional effects.

(ii) The degree of selectivity often increases with the size of the incoming nucleophile.

Predictions of the stereochemical outcome of nucleophilic additions to substituted cyclopentanones is less straightforward. The conformation of a 2-substituted five-membered ring (**92**) is such that attack from the least-hindered face (steric approach control) results in the formation of a *cis*-substituted alcohol (**93**; equation 22). Torsional strain controlled additions lead to *trans*-substituted alcohols (**94**).[58] As in the examples of cyclohexanones, steric approach control usually dominates the reaction pathway with the use of organolithiums and Grignard reagents; torsional strain control with ethynyl reagents (Table 23).[12,58] A single substituent at the 3-position of the cyclopentanone has less of an effect on the stereochemical outcome of nucleophilic addition, and product ratios in these systems are generally poor.[58]

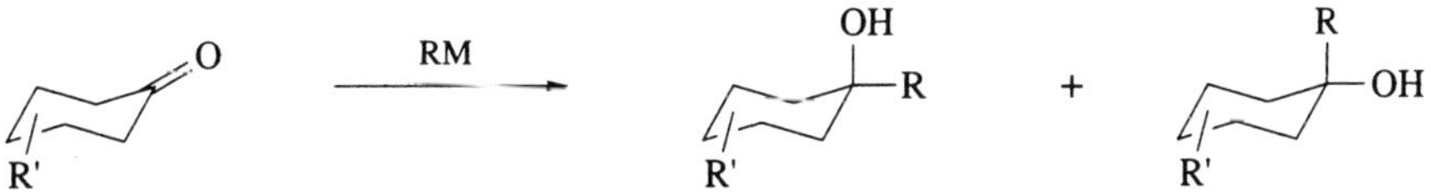

O R' (**89**) —RM→ OH R R' (**90**) equatorial attack + R OH R' (**91**) axial attack (21)

Table 22 Addition of Organometallic Reagents to Substituted Cyclohexanones (**89**)

Reagent (RM)	*Substituent (R′)*	*Ratio (**90**):(**91**)*
MeLi/Et_2O	4-Bu^t	65:35
MeLi/Et_2O	2-Me	84:16
MeLi/Et_2O	3,3,5-Me_3	100:0
PhLi	4-Bu^t	58:42
PhLi	4-Me	53:47
PhLi	3-Me	44:56
PhLi	2-Me	88:12
HC≡CNa/Et_2O + NH_3	4-Bu^t	12:88
HC≡CLi/THF + NH_3	2-Et	53:47
HC≡CLi/THF + NH_3	3-Me	18:82
HC≡CLi/THF + NH_3	2-Me	45:55
MeMgI/Et_2O	4-Bu^t	53:47
MeMgI/Et_2O	2-Me	84:16
MeMgI/Et_2O	3,3,5-Me_3	100:0
EtMgBr/Et_2O	4-Bu^t	71:29
EtMgBr/Et_2O	3-Me	68:32
EtMgBr/Et_2O	2-Me	95:5
EtMgBr/Et_2O	3,3,5-Me_3	100:0
PhMgBr/Et_2O	4-Bu^t	49:51
PhMgBr/Et_2O	4-Me	54:46
PhMgBr/Et_2O	3-Me	59:41
PhMgBr/Et_2O	2-Me	91:9
PhMgBr/Et_2O	3,3,5-Me_3	100:0
HC≡CMgBr	2-Me	45:55
HC≡CMgBr	3,3,5-Me_3	100:0

(**92**) —RM→ (**93**) *cis*-alcohol + (**94**) *trans*-alcohol (22)

Table 23 Addition of Organometallic Reagents to Substituted Cyclopentanones

Reagent	*Substituent*	*Ratio of alcohols cis:trans*
MeMgBr/Et_2O	2-Me	60:40
PhMgBr/THF	2-Me	99:1
HC≡CMgBr/THF	2-Me	50:50
(Allyl)MgCl/Et_2O	2-Me	77:23
(Vinyl)MgCl/THF	2-Me	92:8
MeLi/Et_2O	2-Me	70:30
Bu^nLi/Hexane	2-Me	86:14
PhLi/Et_2O	2-Me	95:5
HC≡CLi/THF–NH_3	2-Me	21:79
EtMgBr/Et_2O	2-Methoxy	75:25
Pr^nMgBr	2-Methoxy	87:13
HC≡CMgBr/THF	2-Methoxy	42:58
(Allyl)MgX/Et_2O	2-Methoxy	82:18
MeMgI/Et_2O	3-Me	60:40
Pr^nMgBr/Et_2O	3-Me	61:39
PhMgBr/Et_2O	3-Me	58:42
(Allyl)MgCl/Et_2O	3-Me	65:35
MeMgBr/Et_2O	3-Bu^t	54:46
Pr^iMgBr/Et_2O	3-Bu^t	61:39
HC≡CMgBr/THF	3-Bu^t	65:35
MeMgBr/Et_2O	3,4-*cis*-Me_2	92:8
PhMgBr/Et_2O	3,4-*cis*-Me_2	92:8

1.2.2 ADDITIONS OF CHIRAL REAGENTS TO ACHIRAL SUBSTRATES

1.2.2.1 Introduction

A number of reports involving the addition of chiral nucleophiles to prochiral carbonyl compounds have appeared in the literature. While many of these examples involve the use of stabilized carbanions derived from chiral sulfoxides (Chapter 2.4, this volume) or chiral ester enolates (Volume 2), several examples rely on asymmetric derivatives of organolithium and Grignard reagents. Among this class, two subsets of reactions are described. The use of organometallic nucleophiles which are chiral by virtue of covalent bonds has had limited use. A more developed field involves the use of achiral nucleophiles in a chiral environment. In these examples, close association between the nucleophile and the additive generates asymmetric reagents without virtue of a formal chemical bond.

1.2.2.2 Covalent Chiral Reagents

The aryllithium reagent (**96**; Scheme 9) can be used as a chiral organometallic reagent in nucleophilic additions to prochiral carbonyl substrates.[60] Addition to a variety of aldehydes affords, after hydrolysis of the aminal, optically active hydroxy aldehydes (**98**) in moderate to high enantiomeric excesses (Table 24). The course of the addition is proposed to occur through intramolecular coordination of the aminal auxiliary to the lithium atom, as shown in Figure 18. The rigid tricyclic structure easily differentiates the faces of the reacting aldehydes. Approach of the aldehydes from the less-hindered face affords asymmetric carbinol centers with the (*S*)-configuration according to this model.

Aromatic oxazolines such as (**99**; Scheme 10) have also been used as chiral nucleophiles.[57b,61] Additions to carbonyl compounds occur with only modest stereoselectivities. The highest ratio of diastereomers produced (64:36) occurs when acetophenone is used as substrate. A sterically undemanding transition state probably accounts for these disappointing results.

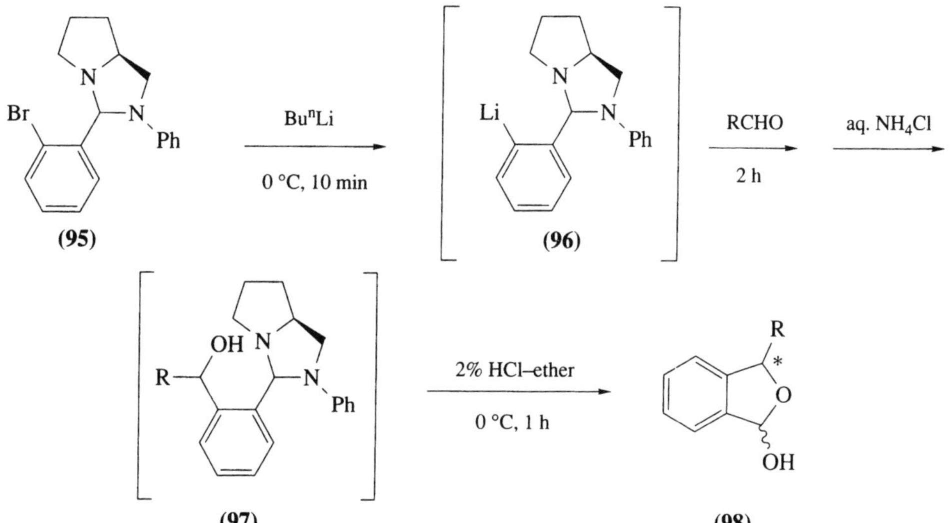

(**95**) $\xrightarrow[0\ °C,\ 10\ min]{Bu^nLi}$ [(**96**)] $\xrightarrow[2\ h]{RCHO}$ $\xrightarrow{aq.\ NH_4Cl}$ [(**97**)] $\xrightarrow[0\ °C,\ 1\ h]{2\%\ HCl\text{–}ether}$ (**98**)

Scheme 9

Table 24 Preparation of Chiral Lactols (**98**)

R	*Solvent*	*Temperature (°C)*	*Yield (%)*	*ee (%)*
Bu^n	Toluene	–78	73	65
Bu^n	Ether	–78	63	78
Bu^n	Ether	–100	62	87
Et	Ether	–100	51	88
Pr^i	Ether	–100	52	>90
n-C_8H_{17}	Ether	–100	58	90
Allyl	Toluene	–78	70	20

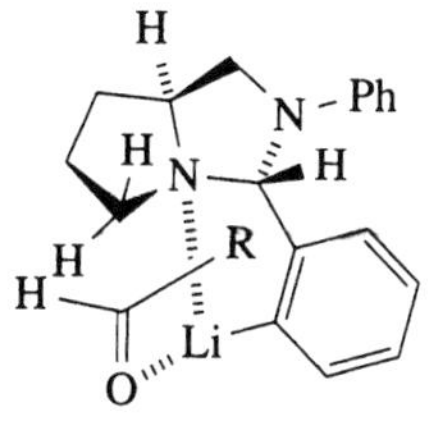

Figure 18

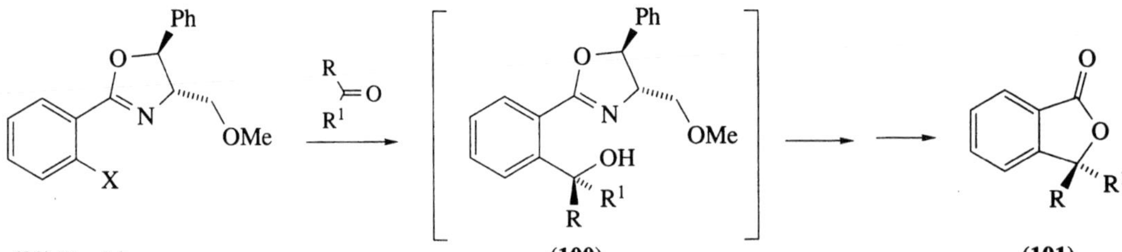

(99) X = Li

(100)

(101)

Scheme 10

1.2.2.3 Noncovalent Chiral Reagents

The stereoselective addition of achiral nucleophiles to prochiral carbonyl compounds in the presence of chiral additives is a conceptually elegant method for the synthesis of enantiomerically pure compounds from achiral starting materials. Advantages of this strategy include the ability to recover the chirality 'inducer', and the elimination of steps involving chemically introducing and removing these 'noncovalent chiral auxiliaries'. This latter point represents a significant advantage of this protocol over the use of standard chiral auxiliaries.

Several systems have been developed which exploit the well-known coordination of organolithium reagents to tertiary amines, and of Grignard reagents to ethers. This strong association generates asymmetric nucleophilic reagents; transfer of an alkyl group from this chiral organometallic results in an enantioselective process. Early work in this area led to only low asymmetric inductions. However, a number of groups have recently improved the enantioselectivity to modest, and even high levels in some cases. This area has been reviewed by Solladié.[62]

Seebach has carried out an extensive study on the use of chiral additives based on tartaric acid to induce asymmetry in the addition of *n*-butyllithium to aldehydes.[62,63] After systematically investigating a number of ligands, the tetraamine (**102**) was found to be the most effective, providing alcohols (**103**; equation 23) with optical purities in the range of 15 to 56% (Table 25).

This additive reliably causes nucleophilic attack to occur from the *si* face of the aldehyde to generate the (*S*)-alcohol. The direction of addition is irrespective of the organolithium reagent used. The authors suggest Figure 19 as the conformation responsible for the selectivities observed.

The pyrrolidine derivative (**104**; equation 24) developed by Mukaiyama and coworkers for the enantioselective addition of organolithiums to aldehydes has been one of the most studied.[52,64] Although opti-

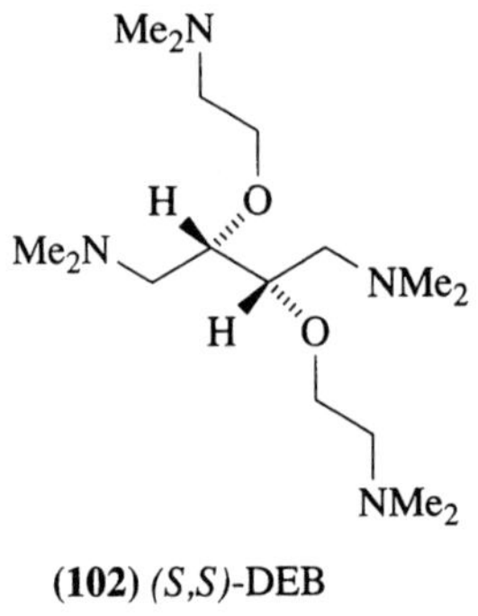

(**102**) (*S,S*)-DEB

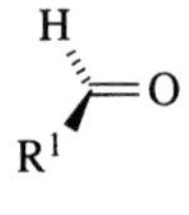

$$R^1CHO + R^2Li \xrightarrow[-78\ °C]{2\ \text{equiv. } (\mathbf{102})} R^1R^2CH(OH) \quad (23)$$

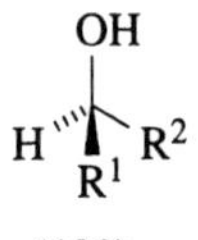

(**103**)

Table 25 Enantioselective Additions of R^2Li to Aldehydes (R^1CHO) in the Presence of (**102**)

R^1	R^2	(*103*) *ee* (%)
Me	Bu^n	46
Et	Bu^n	35
Bu^n	Me	34
Ph	Bu^n	52
Bu^n	Ph	15
o-MePh	Bu^n	56
p-MePh	Bu^n	49
Et_2CH	Bu^n	46
Pr^i	Bu^n	53
Bu^t	Bu^n	18
C_6H_{11}	Bu^n	48
Vinyl	Bu^n	24

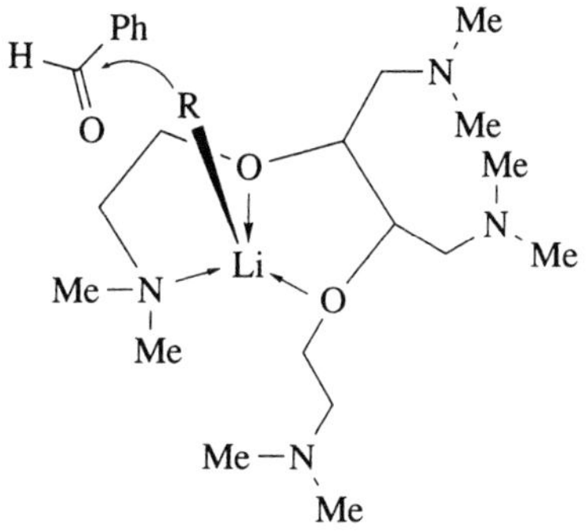

Figure 19

cal purities were generally moderate (11–72% *ee*; Table 26), reactions using alkynyllithium reagents show very good selectivities (54–92% *ee*).[65]

Extensive study of (**104**) indicates that both pyrrolidine moieties, as well as the lithiated hydroxymethyl group are crucial for high asymmetric induction. Figures 20 and 21 represent possible modes of the course of the reaction. A rigid structure containing four fused five-membered rings can be formed by coordination of the two nitrogen atoms and the oxygen atom to the lithium of the organometallic reagent.

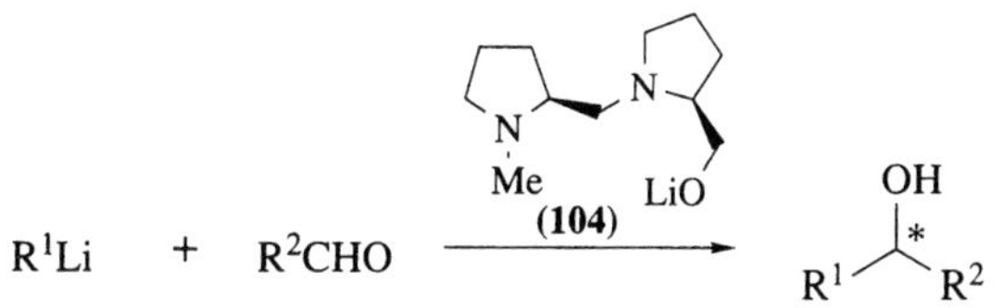

$$R^1Li \quad + \quad R^2CHO \xrightarrow{(\mathbf{104})} R^1CH^*(OH)R^2 \; (\mathbf{105}) \qquad (24)$$

Table 26 Enantioselective Additions of R^1Li to Aldehydes (R^2CHO) in the Presence of (**104**)

R^1	R^2	*Temperature (°C)*	*Yield (%)*	*Alcohol (105) Optical purity (%)*	*Configuration*
Me	Ph	–78	82	21	(*R*)
Et	Ph	–123	32	39	(*R*)
Pr^n	Ph	–123	55	39	(*S*)
Bu^n	Ph	–123	60	72	(*S*)
Bu^i	Ph	–123	59	16	(*S*)
Bu^n	Pr^i	–123	47	56	(*S*)
Ph	Bu^n	–123	46	11	(*R*)
HC≡C	Ph	–78	76	54	(*S*)
TMS—C≡C	Ph	–78	99	78	(*S*)
TMS—C≡C	Ph	–123	87	92	(*S*)
Ph_2MeSi—C≡C	Ph	–123	88	80	(*S*)

The size of the alkyllithium apparently determines the diastereofacial approach of the aldehyde, and thus determines the stereochemistry of the alcohol product.

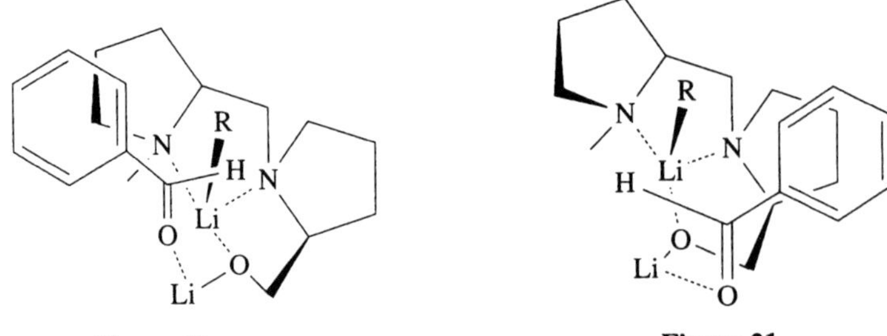

Figure 20 **Figure 21**

Conditions for optimal asymmetric induction include low temperature (–123 °C) and the use of a 1:1 mixture of dimethoxymethane and dimethyl ether as solvent.[66] Investigation of a number of different organometallic species indicated that dialkylmagnesium reagents yield the highest optical purities (equation 25). Table 27 lists representative examples of this enantioselective reaction under these optimal conditions.[67] Interestingly, all of the alcohols produced under these conditions have the (*R*)-configuration.

$$R^1CHO \quad + \quad R^2{}_2Mg \longrightarrow R^1\overset{*}{C}H(OH)R^2 \qquad (25)$$

(reagent over arrow: pyrrolidine ligand with N–Me, N and OLi; product **(105)**)

Table 27 Synthesis of Optically Active Alcohols (**105**) Using Dialkylmagnesium Reagents in the Presence of (**104**)

R[1]	*R*[2]	*Yield (%)*	*Optical purity (%)*
Ph	Me	56	34
Ph	Et	74	92
Ph	Pr^n	90	70
Ph	Pr^i	59	40
Ph	Bu^n	94	88
Ph	Bu^i	81	42
Pr^i	Bu^n	70	22

This chiral ligand can also be used as an additive in the enantioselective reaction of stabilized anions to carbonyl compounds.[68] Optical purities of these processes are moderate (up to 76% *ee*), and depend on the conditions of the reaction, as well as on the particular reactants involved.

A number of other chiral ligands are available, but have been studied much less extensively. The chiral diamines (**106**) and (**107**) are reported to mediate the reactions between aryl Grignards and aldehydes (Figure 22; equation 26).[69] The alcohols range in optical purity from 40 to 75% *ee*; selectivities increase with the bulkiness of the aldehyde substituent (see Table 28). The use of an aryloxy metal halide to complex the aldehyde moiety enhances the observed enantioselectivities.[70]

Mazaleyrat and Cram have used diamines (**108**) and (**109**) as chiral catalysts in the enantioselective addition of alkyllithiums to aldehydes (equation 27; Table 29). Optical yields of 23–95% are obtained.[71] Highest values are associated with the use of larger alkyllithium reagents, as well as with the more sterically encumbered catalyst (**108**).

A number of other chiral catalysts have been reported, among them the proline (**111**),[72] lithium amides (**112**)[73] and the tetrahydrofurylamine (**113**).[74] The optical yields of the products isolated, however, are only moderate at best. The use of optically active 2-methyltetrahydrofuran in Grignard reactions has also been reported; however, minimal induction is observed.[75]

The use of chiral ligands to enantiofacially bias a nucleophilic reaction of achiral starting materials holds great promise for asymmetric syntheses. Further development of ligands which consistently provide high enantiomeric excess, as well as generate predictable geometries at the chiral center, is necessary.

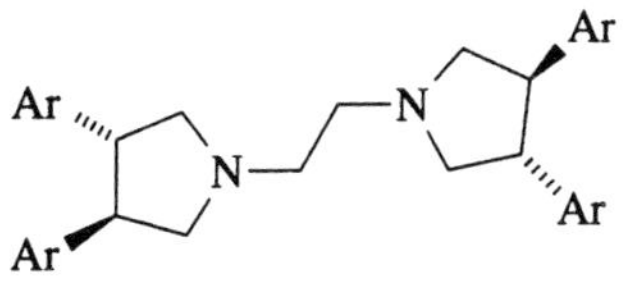

(106) Ar = Ph
(107) Ar = 3,5-xylyl

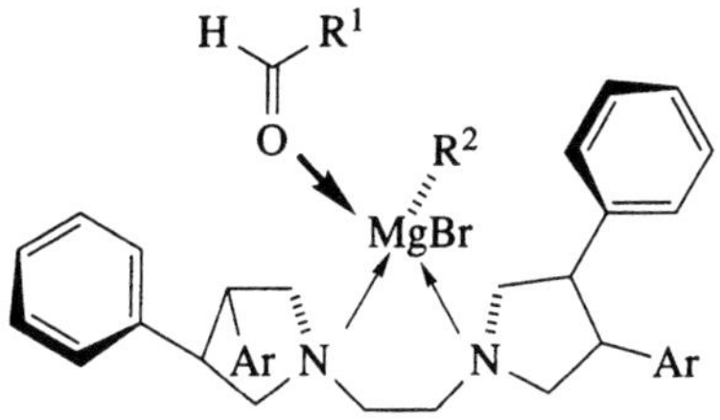

Figure 22

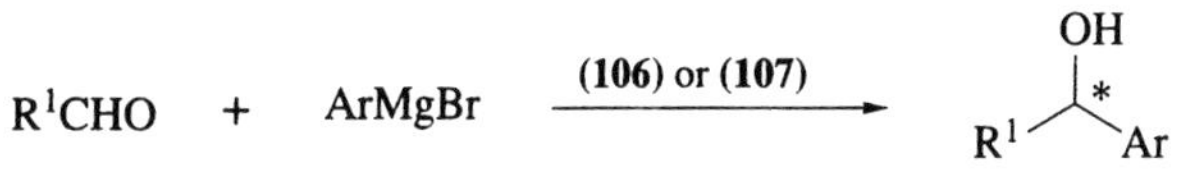

$$R^1CHO \; + \; ArMgBr \xrightarrow{\textbf{(106)} \text{ or } \textbf{(107)}} R^1CH^*(OH)Ar \qquad (26)$$

Table 28 Enantioselective Additions of Aryl Grignards with Aldehydes Mediated by (**106**) or (**107**)

R^1	*Ar*	*Ligand*	*Temperature (°C)*	*ee (%)*	*Configuration*	*Yield (%)*
Ph	α-Naphthyl	(**106**)	−100	71	(*S*)	92
Ph	α-Naphthyl	(**106**)	−78	64	(*S*)	96
Ph	α-Naphthyl	(**106**)	−45	55	(*S*)	94
Ph	α-Naphthyl	(**107**)	−100	75	(*S*)	94
Bu^t	Ph	(**106**)	−100	60	(*S*)	82
C_6H_{11}	Ph	(**107**)	−100	55	(*S*)	68
Pr^i	Ph	(**106**)	−100	47	(*S*)	68
Bu^n	Ph	(**106**)	−100	40	(*S*)	73

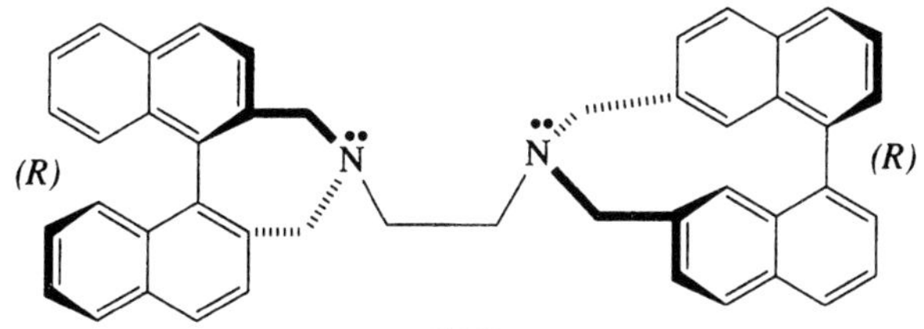

(108) 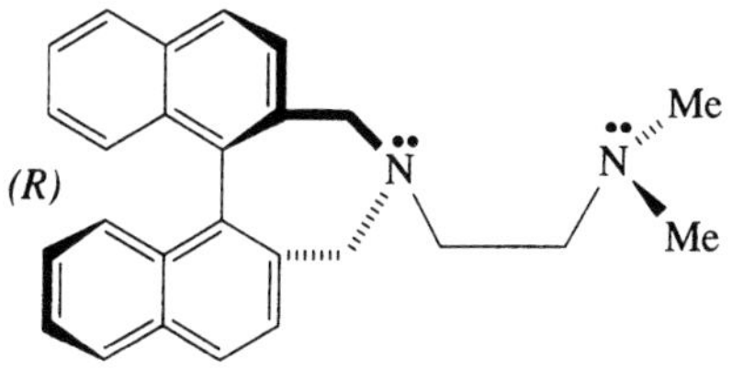 **(109)**

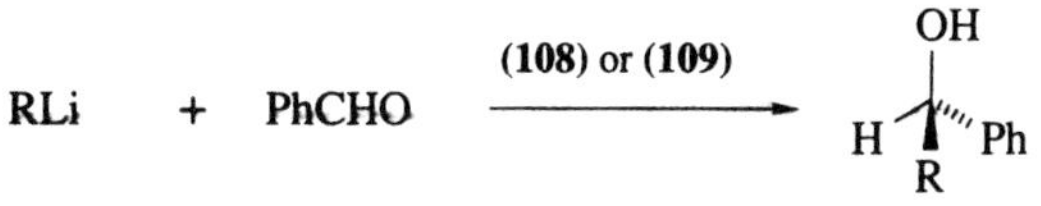

$$RLi \; + \; PhCHO \xrightarrow{\textbf{(108)} \text{ or } \textbf{(109)}} \text{H(R)C(OH)Ph} \; \textbf{(110)} \qquad (27)$$

Table 29 Enantioselective Additions of Organolithium Reagents (RLi) to Benzaldehyde in the Presence of (**108**) or (**109**)

R	*Ligand*	*Yield (%)*	*(110) ee (%)*	*Configuration*
Bu^n	108	73	95	(*R*)
Bu^n	108	63	59	(*R*)
Et	108	75	66	(*R*)
Me	108	35	36	(*R*)
Bu^n	109	71	58	(*R*)
Pr^n	109	75	53	(*R*)
Et	109	73	30	(*R*)
Me	109	65	23	(*R*)

MeO MeO N N

(**111**)

R H H Ph N Ar Li

(**112**) **a**: Ar = phenyl, R = H
b: Ar = 2-pyridyl, R = H
c: Ar = *o*-methoxyphenyl, R = H
d: Ar = phenyl, R = OMe

O N

(**113**)

1.2.3 REFERENCES

1. A. McKenzie, *J. Chem. Soc.*, 1904, **85**, 1249.
2. V. Prelog, *Helv. Chim. Acta*, 1953, **277**, 426.
3. D. J. Cram and F. A. Abd Elhafez, *J. Am. Chem. Soc.*, 1952, **74**, 5828; D. J. Cram and K. R. Kopecky, *J. Am. Chem. Soc.*, 1959, **81**, 2748; D. J. Cram and D. R. Wilson, *J. Am. Chem. Soc.*, 1963, **85**, 1245.
4. J. W. Cornforth, R. H. Cornforth and K. K. Mathew, *J. Chem. Soc.*, 1959, 112.
5. G. J. Karabatsos, *J. Am. Chem. Soc.*, 1967, **89**, 1367.
6. M. Cherest, H. Felkin and N. Prudent, *Tetrahedron Lett.*, 1968, 2199.
7. N. T. Anh and O. Eisenstein, *Nouv. J. Chim.*, 1977, **1**, 61; N. T. Anh and O. Eisenstein, *Tetrahedron Lett.*, 1976, 155.
8. For leading references involving the theoretical treatment of asymmetric nucleophilic addition reactions see: L. Salem, *J. Am. Chem. Soc.*, 1973, **95**, 94; N. T. Anh, O. Eisenstein, J.-M. Lefour and M.-E. Tran Huu Dau, *J. Am. Chem. Soc.*, 1973, **95**, 6146; H. B. Burgi, J. D. Dunitz, J.-M. Lehn and G. Wipff, *Tetrahedron*, 1974, **30**, 1563; M. N. Paddon-Row, N. G. Rondan and K. N. Houk, *J. Am. Chem. Soc.*, 1982, **104**, 7162; Y.-D. Wu and K. N. Houk, *J. Am. Chem. Soc.*, 1987, **109**, 908; E. P. Lodge and C. H. Heathcock, *J. Am. Chem. Soc.*, 1987, **109**, 2819.
9. J. D. Morrison and H. S. Mosher, 'Asymmetric Organic Reactions', Prentice-Hall, Englewood Cliffs, NJ, 1971.
10. P. A. Bartlett, *Tetrahedron*, 1980, **36**, 3.
11. E. L. Eliel, in 'Asymmetric Synthesis', ed. J. D. Morrison, Academic Press, New York, 1983, vol. 2, p. 125.
12. M. Nogradi, 'Stereoselective Synthesis', VCH, New York, 1986.
13. W. C. Still and J. H. McDonald, III, *Tetrahedron Lett.*, 1980, **21**, 1031.
14. W. C. Still and J. A. Schneider, *Tetrahedron Lett.*, 1980, **21**, 1035.
15. M. T. Reetz and M. Hüllmann, *J. Chem. Soc., Chem. Commun.*, 1986, 1600.
16. C. W. Jefford, D. Jaggi and J. Boukouvalas, *Tetrahedron Lett.*, 1986, **27**, 4011.
17. M. Bhupathy and T. Cohen, *Tetrahedron Lett.*, 1985, **26**, 2619.
18. R. Amouroux, S. Ejjiyar and M. Chastrette, *Tetrahedron Lett.*, 1986, **27**, 1035.
19. J. Jurczak, S. Pikul and T. Bauer, *Tetrahedron*, 1986, **42**, 447.
20. J. Mulzer and A. Angermann, *Tetrahedron Lett.*, 1983, **24**, 2843.
21. K. Suzuki, Y. Yuki and T. Mukaiyama, *Chem. Lett.*, 1981, 1529.
22. K. Mead and T. MacDonald, *J. Org. Chem.*, 1985, **50**, 422.
23. C. Fuganti, P. Grasselli, and G. Pedrocchi-Fantoni, *Tetrahedron Lett.*, 1981, **22**, 4017; R. Bernardi, C. Fuganti and P. Grasselli, *Tetrahedron Lett.*, 1981, **22**, 4021.
24. H. Iida, N. Yamazaki and C. Kibayashi, *Tetrahedron Lett.*, 1985, **26**, 3255; H. Iida, N. Yamazaki and C. Kibayashi, *J. Org. Chem.*, 1986, **51**, 1069, 3769.
25. J. C. Fischer, D. Horton and W. Weckerle, *Carbohydr. Res.*, 1977, **59**, 459.
26. M. L. Wolfrom and S. Hanessian, *J. Org. Chem.*, 1962, **27**, 1800; T. D. Inch, *Carbohydr. Res.*, 1967, **5**, 45; I. Höppe and U. Schöllkopf, *Liebigs Ann. Chem.*, 1980, 1474; 1983, 372.
27. M. T. Reetz, *Angew. Chem., Int. Ed. Engl.*, 1984, **23**, 556.

28. G. J. Hanson and T. Lindberg, *J. Org. Chem.*, 1985, **50**, 5399.
29. M. T. Reetz, M. W. Drewes and A. Schmitz, *Angew. Chem., Int. Ed. Engl.*, 1987, **26**, 1141.
30. P. Duhamel, L. Duhamel and J. Gralak, *Tetrahedron Lett.*, 1972, 2329.
31. M. Tramontini, *Synthesis*, 1982, 605.
32. M. Nakada, Y. Urano, S. Kobayashi and M. Ohno, *J. Am. Chem. Soc.*, 1988, **110**, 4826.
33. F. Sato, M. Kusakabe and Y. Kobayashi, *J. Chem. Soc., Chem. Commun.*, 1984, 1130; F. Sato, O. Takahashi, T. Kato and Y. Kobayashi, *J. Chem. Soc., Chem. Commun.*, 1985, 1638.
34. R. Bloch and L. Gilbert, *Tetrahedron Lett.*, 1987, **28**, 423.
35. C. Fouquey, J. Jacques, L. Angiolini and M. Tramontini, *Tetrahedron*, 1974, **30**, 2801.
36. (a) A. I. Meyers, M. A. Hanagan, L. M. Trefonas and R. J. Baker, *Tetrahedron*, 1983, **39**, 1991; (b) see also K. A. Lutomski and A. I. Meyers, in 'Asymmetric Synthesis', ed. J. D. Morrison, Academic Press, New York, 1984, vol. 3, p. 213.
37. E. L. Eliel and S. Morris-Natschke, *J. Am. Chem. Soc.*, 1984, **106**, 2937.
38. J. E. Lynch and E. L. Eliel, *J. Am. Chem. Soc.*, 1984, **106**, 2943.
39. X.-C. He and E. L. Eliel, *Tetrahedron*, 1987, **43**, 4979.
40. For a complete review of 1,3-oxathianes see: E. L. Eliel, J. K. Koskimies, B. Lohri, W. J. Frazee, S. Morris-Natschke, J. E. Lynch and K. Soai, in 'Asymmetric Reactions and Processes in Chemistry', ed. E. L. Eliel and S. Otsuka, American Chemical Society, Washington, DC, 1982, p. 37.
41. S. V. Frye and E. L. Eliel, *Tetrahedron Lett.*, 1986, **27**, 3223; S. V. Frye and E. L. Eliel, *J. Am. Chem. Soc.*, 1988, **110**, 484.
42. For examples of natural product syntheses using 1,3-oxathianes as chiral auxiliaries see: E. L. Eliel, J. K. Koskimies and B. Lohri, *J. Am. Chem. Soc.*, 1978, **100**, 1614; E. L. Eliel and W. J. Frazee, *J. Org. Chem.*, 1979, **44**, 3598; E. L. Eliel and J. E. Lynch, *Tetrahedron Lett.*, 1981, **22**, 2855; E. L. Eliel and K. Soai, *Tetrahedron Lett.*, 1981, **22**, 2859; T. Kogure and E. L. Eliel, *J. Org. Chem.*, 1984, **49**, 576; S. V. Frye and E. L. Eliel, *J. Org. Chem.*, 1985, **50**, 3402; S. V. Frye and E. L. Eliel, *Tetrahedron Lett.*, 1985, **26**, 3907; M. Ohwa, T. Kogure and E. L. Eliel, *J. Org. Chem.*, 1986, **51**, 2599; K.-Y. Ko and E. L. Eliel, *J. Org. Chem.*, 1986, **51**, 5353; M. Ohwa and E. L. Eliel, *Chem. Lett.*, 1987, 41.
43. Y. Tamura, T. Ko, H. Kondo, H. Annoura, M. Fuji, R. Takeuchi and H. Fujioka, *Tetrahedron Lett.*, 1986, **27**, 2117.
44. Y. Tamura, H. Kondo, H. Annoura, R. Takeuchi and H. Fujioka, *Tetrahedron Lett.*, 1986, **27**, 81.
45. M.-P. Heitz, F. Gellibert and C. Mioskowski, *Tetrahedron Lett.*, 1986, **27**, 3859.
46. Y. Tamura, H. Annoura, H. Yamamoto, H. Kondo, Y. Kita and H. Fujioka, *Tetrahedron Lett.*, 1987, **28**, 5709.
47. T. Mukaiyama, Y. Sakito and M. Asami, *Chem. Lett.*, 1978, 1253.
48. T. Mukaiyama, Y. Sakito and M. Asami, *Chem. Lett.*, 1979, 705.
49. M. Asami and T. Mukaiyama, *Chem. Lett.*, 1983, 93.
50. Y. Sakito and T. Mukaiyama, *Chem. Lett.*, 1979, 1027.
51. Y. Sakito, S. Tanaka, M. Asami and Y. Mukaiyama, *Chem. Lett.*, 1980, 1223.
52. T. Mukaiyama, *Tetrahedron*, 1981, **37**, 4111; T. Mukaiyama, in 'Asymmetric Reactions and Processes in Chemistry', ed. E. L. Eliel and S. Otsuka, American Chemical Society, Washington, DC, 1982, p. 21.
53. J. K. Whitesell, A. Bhattacharya and K. Henke, *J. Chem. Soc., Chem. Commun.*, 1982, 988.
54. J. K. Whitesell, D. Deyo and A. Bhattacharya, *J. Chem. Soc., Chem. Commun.*, 1983, 802.
55. J. K. Whitesell, C.-M. Liu, C. M. Buchanan, H.-H. Chen and M. A. Minton, *J. Org. Chem.*, 1986, **51**, 551.
56. For a synthesis of both enantiomers of frontalin using the same phenylmenthol auxiliary see: J. K. Whitesell and C. M. Buchanan, *J. Org. Chem.*, 1986, **51**, 5443.
57. (a) A. I. Meyers and J. Slade, *J. Org. Chem.*, 1980, **45**, 2785; (b) see also A. I. Meyers, in 'Asymmetric Reactions and Processes in Chemistry', ed. E. L. Eliel and S. Otsuka, American Chemical Society, Washington, DC, 1982, p. 83; (c) see ref. 36 for similar work using a related system.
58. E. C. Ashby and J. T. Laemmle, *Chem. Rev.*, 1975, **75**, 521.
59. For recent leading references see: E. C. Ashby and S. A. Noding, *J. Org. Chem.*, 1977, **42**, 264; A. S. Cieplak, *J. Am. Chem. Soc.*, 1981, **103**, 4540.
60. M. Asami and T. Mukaiyama, *Chem. Lett.*, 1980, 17.
61. For a related oxazoline-based chiral nucleophile see ref. 36.
62. G. Solladié, in 'Asymmetric Synthesis', ed. J. D. Morrison, Academic Press, New York, 1983, vol. 2, p. 157.
63. D. Seebach, H.-O. Kalinowski, B. Bastiani, G. Crass, H. Daum, H. Dorr, N. P. DuPreez, V. Ehrig, W. Langer, C. Nussler, H.-A. Oei and M. Schmidt, *Helv. Chim. Acta*, 1977, **60**, 301; D. Seebach and W. Langer, *Helv. Chim. Acta*, 1979, **62**, 1701; D. Seebach, G. Crass, E.-M. Wilka, D. Hilvert and E. Brunner, *Helv. Chem. Acta*, 1979, **62**, 2695.
64. T. Mukaiyama, K. Soai and S. Kobayashi, *Chem Lett.*, 1978, 219.
65. T. Mukaiyama, K. Suzuki, K. Soai and T. Sato, *Chem. Lett.*, 1979, 447; T. Mukaiyama and K. Suzuki, *Chem. Lett.*, 1980, 255.
66. K. Soai and T. Mukaiyama, *Chem. Lett.*, 1978, 491; T. Mukaiyama, K. Soai, T. Sato, H. Shimizu and K. Suzuki, *J. Am. Chem. Soc.*, 1979, **101**, 1455.
67. T. Sato, K. Soai, K. Suzuki and T. Mukaiyama, *Chem. Lett.*, 1978, 601.
68. K. Soai and T. Mukaiyama, *Bull. Chem. Soc. Jpn.*, 1979, **52**, 3371; T. Akiyama, M. Shimizu and T. Mukaiyama, *Chem. Lett.*, 1984, 611.
69. K. Tomioka, M. Nakajima and K. Koga, *Chem. Lett.*, 1987, 65.
70. K. Tomioka, M. Nakajima and K. Koga, *Tetrahedron Lett.*, 1987, **28**, 1291.
71. J. P. Mazaleyrat and D. J. Cram, *J. Am. Chem. Soc.*, 1981, **103**, 4585.
72. L. Colombo, C. Gennari, G. Poli and C. Scolastico, *Tetrahedron*, 1982, **38**, 2725.
73. M. B. Eleveld and H. Hogeveen, *Tetrahedron Lett.*, 1984, **25**, 5187.
74. J. K. Whitesell and B.-R. Jaw, *J. Org. Chem.*, 1981, **46**, 2798.
75. D. C. Iffland and J. E. Davis, *J. Org. Chem.*, 1977, **42**, 4150.

1.3
Organoaluminum Reagents

JAMES R. HAUSKE
Pfizer Central Research, Groton, CT, USA

1.3.1 INTRODUCTION

Organoaluminums have been reviewed extensively. The first thorough treatise was published in 1972 by Mole,[1] describing the preparation of organoaluminums and their reactions with various functional groups. Negishi[2,4] and Yamamoto[3] have also reviewed the general properties and reactions of organoaluminums. However, Negishi[2] focused on the 1,2-nucleophilic additions of α,β-unsaturated organoaluminums, which specifically included the reactions of alkenylalanes and alkenylalanates with carbon electrophiles, whereas the Yamamoto review[3] was of a more general nature. The preparation of alkenyl- and alkynyl organoaluminums has also been reviewed by Normant[5] and Zweifel.[6] The portion of the Normant review that focused on organoaluminum reagents described the *syn* specific carbometallation of alkynes by organoaluminums, which provides stereospecific access to alkenylaluminums. In contrast, Zweifel[6] prepared a very detailed review dealing with the preparation and reactions of alkenyl- and alkynyl-aluminums.

The stabilized anions of organoaluminums have also been discussed, inasmuch as there have been extensive reviews of allyl- and crotyl-aluminum reagents.[7–9] Yamamoto has reviewed the preparation and reactions of both allyl-[7] and crotyl-aluminum[9] reagents, while Hoffmann[8] has prepared a more general review of organocrotyl reagents, which includes some discussion of the corresponding aluminum reagents. These reviews also consider in some detail the stereochemical consequences of these processes and include some discussion of the transition state geometries. Not surprisingly, the more recent reviews[10–13] focus on the stereoselectivity and site selectivity of the organoaluminum-mediated reactions, as well as their potential applications to organic synthesis. Most notable in this regard is the review of Yamamoto,[13] which attempts to outline all of the major reactions of organoaluminum reagents not only according to the nature of the reacting substrate, but also in terms of the regioselectivity and stereoselectivity of each reaction.

Since so much has been written about the properties and the preparation of organoaluminums, this review will detail only selected reactions, with a heavy emphasis on site selectivity and stereoselectivity; moreover, the nature of the reactions that are detailed will be limited to 1,2-nucleophilic additions of nonstabilized anions to unsaturated carbon–heteroatom substrates. Obviously, this will limit the scope of the review and the very interesting regioselective and stereoselective challenges presented by stabilized anions, namely, aldol-like processes, will be omitted. Nevertheless, the scope of this review will intensify the focus on the problems of regioselectivity and stereoselectivity furnished by the 1,2-additions of organoaluminum reagents to either carbonyl systems or masked carbonyl systems. Finally, the application of organoaluminum reagents to the synthesis of natural products will be presented, again in the context of regioselective and stereoselective processes.

1.3.2 SITE SELECTIVE AND STEREOSELECTIVE ADDITIONS OF NUCLEOPHILES MEDIATED BY ORGANOALUMINUM REAGENTS

1.3.2.1 Aluminum-based Additives

In 1975 Ashby[14] extensively reviewed the stereochemical results of the 1,2-additions of a variety of organometallic reagents to prochiral carbonyl substrates. The specific section describing the addition of organoaluminum reagents is extensive, but the most interesting aspect of the review with regard to alkylaluminum-mediated additions is the observation[15,16] that the axial selectivity of alkyl addition to cyclic ketones is greatly enhanced by increasing the ratio of aluminum reagent to substrate. Thus, at ratios of 2:1 or greater, the addition results in 90% axial attack, whereas at a 1:1 ratio equatorial attack predominates. The observation was rationalized by invoking a 'compression effect', which derives from the increased steric bulk about the carbonyl as a result of complexation with the organoaluminum.[17]

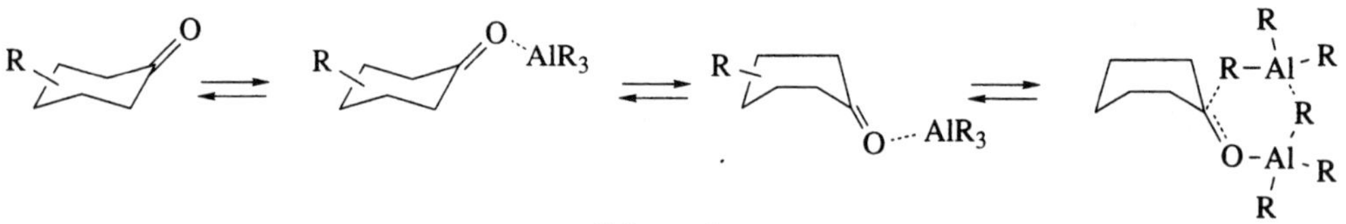

R O R O AlR$_3$ R O AlR$_3$ R R–Al–R R O–Al–R R

Scheme 1

This seminal observation has recently been utilized by Yamamoto[18] to develop an approach for the stereoselective and site selective addition of organometallics to carbonyl substrates. The approach makes use of the very bulky aluminum reagent (**1**), which is readily prepared *in situ* by exposure of trimethylaluminum to a toluene solution of 2,6-di-*t*-butyl-4-alkylphenols (molar ratio 1:2) at room temperature.

The MAD and MAT reagents are presumably monomeric in solution, since they are very bulky reagents and it is known[19] that a 1:1 mixture of benzophenone and trimethylaluminum exists as a long-lived monomeric complex at room temperature. Although no solid state structure of the complex of reagent (**1**) with carbonyl substrates is available,[20] a reasonable representation is formulated in Scheme 2.

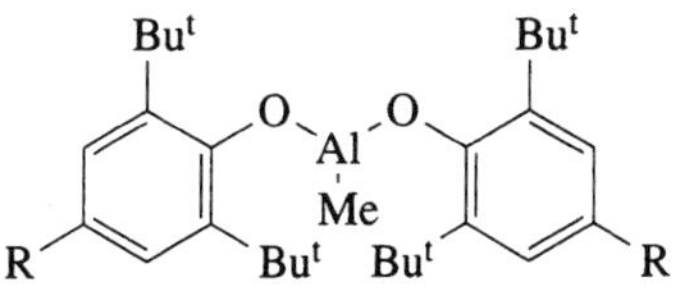

(**1**) R = Me (MAD); R = Bu^t (MAT)

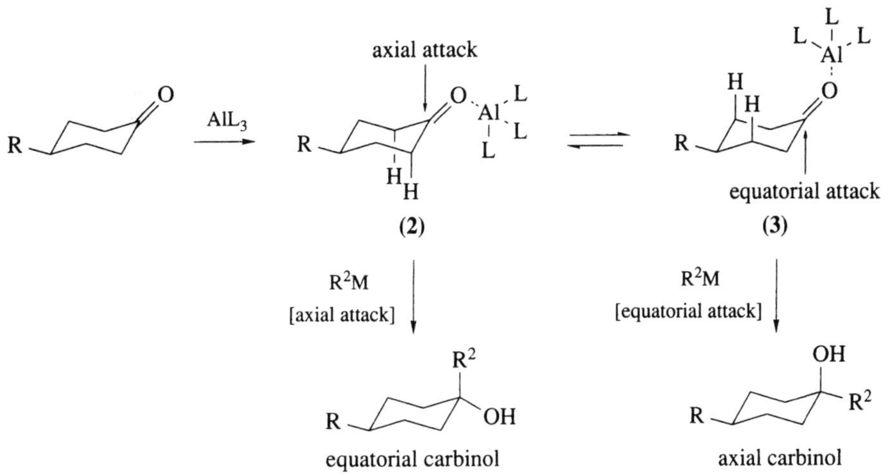

Scheme 2

The equilibrium concentration of the reactive species would greatly favor substituted cyclohexanone (**2**; Scheme 2), since it has been demonstrated[14–16] that the steric interaction of the 3,5-axial substituents with a bulky group chelating the carbonyl moiety dominates the torsional strain that develops between the chelating group and the 2,6-diaxial positions. Thus, preexposure of cyclic ketones to reagent (**1**) and subsequent addition of another organometallic would be expected to favor the formation of equatorial carbinols resulting from axial addition of the organometallic (R^1M) in Scheme 2. Table 1[8,21] summarizes the results of 1,2-additions of alkyllithium and alkyl Grignard reagents to substituted cyclohexanones in the presence or absence of reagent (**1**). Comparison of entries 4 and 5 as well as entries 6 and 7 illustrates the remarkable effect that the organoaluminum additive (**1**) has on the overall stereoselectivity of the 1,2-additions. For example, exposure of 4-*t*-butylcyclohexanone to methyllithium in the absence of (**1**) results in the preferential formation (*ca.* 4:1) of the corresponding axial alcohol, whereas in the presence of (**1**; R = *t*-butyl; MAT) the 1,2-addition proceeds with almost exclusive formation (99%) of the equatorial alcohol. Although the results outlined in Table 1 support the mechanism outlined in Scheme 2, an additional experiment was conducted attempting to further define the nature of the reacting species. When the ketone substrate is exposed to a mixture of methyllithium and reagent (**1**; R = Me; MAD) at –78 °C, the resulting ratio of axial/equatorial carbinols is similar (ax:eq = 84:16)[21] to that observed for the addition of methyllithium alone. Thus, the possibility of ate complex formation resulting from reaction of methyllithium with the aluminum additive (**1**) appears unlikely. It appears, therefore, that Scheme 2 adequately accommodates the experimental observations.

Unfortunately, this approach has limitations with regard to the nature of the ketone substrate as well as the nature of the organometallic nucleophile. For example, alkylation and reduction occurred when secondary and tertiary alkyl Grignards were utilized as nucleophiles, while attempted reactions with either vinyl or phenyl nucleophiles occurred slowly with no equatorial selectivity. Furthermore, as the steric environment about the carbonyl carbon becomes more congested, the observed stereoselectivity decreases (Table 1, entries 11–14).

The detrimental effect of the steric environment about the carbonyl moiety on the course of the reaction is very well illustrated for cyclopentanones (Table 2). Comparison of the entries in Table 2 reveals that cyclopentanone substrates are very sensitive not only to the steric bulk of the incoming nucleophile (entries 2–4), but also to the substitution pattern on the carbocycle (entries 1, 2 and 5).

Table 1 Facial Selectivities for the Addition of Organometallics to Substituted Cyclohexanones in the Presence of (**1**)

Entry	*Substrate*	*Reagent (1)*	*Organometallic*	*Yield (%)*	*Axial alcohol: equatorial alcohol*
1	4-*t*-Butylcyclohexanone	None	MeLi	—	79:21
2	4-*t*-Butylcyclohexanone	MAD	MeLi	84	1:99
3	4-*t*-Butylcyclohexanone	MAT	MeLi	92	0.5:99.5
4	4-*t*-Butylcyclohexanone	None	EtMgBr	95	48:52
5	4-*t*-Butylcyclohexanone	MAD	EtMgBr	91	0:100
6	4-*t*-Butylcyclohexanone	None	BuMgBr	58	56:44
7	4-*t*-Butylcyclohexanone	MAD	BuMgBr	67	0:100
8	4-*t*-Butylcyclohexanone	None	AllylMgBr	86	48:52
9	4-*t*-Butylcyclohexanone	MAD	AllylMgBr	90	9:91
10	2-Methylcyclohexanone	None	MeLi	—	92:8
11	2-Methylcyclohexanone	MAD	MeLi	84	14:86
12	2-Methylcyclohexanone	MAT	MeLi	80	10:90
13	2-Methylcyclohexanone	MAD	EtMgBr	0	—
14	2-Methylcyclohexanone	MAD	BuMgBr	0	—
15	3-Methylcyclohexanone	None	MeLi	—	83:17
16	3-Methylcyclohexanone	MAD	MeLi	69	9:91
17	3-Methylcyclohexanone	MAT	MeLi	95	3:97
18	3-Methylcyclohexanone	None	BuMgBr	86	79:21
19	3-Methylcyclohexanone	MAD	BuMgBr	75	1:99
20	3-Methylcyclohexanone	None	AllylMgBr	95	56:44
21	3-Methylcyclohexanone	MAD	AllylMgBr	72	24:76

Table 2 Facial Selectivities for the Addition of Organometallics to Substituted Cyclopentanones in the Presence of (**1**)

Entry	*Substrate*	*Reagent (1)*	*Organometallic*	*Yield (%)*	*Axial alcohol: equatorial alcohol*
1	2-Butylcyclopentanone	None	MeLi	—	75:25
2	2-Butylcyclopentanone	MAD	MeLi	82	1:99
3	3-Methylcyclopentanone	None	PrMgBr	54	55:45
4	3-Methylcyclopentanone	MAD	PrMgBr	79	50:50
5	2-Phenylcyclopentanone	MAD	MeMgBr	0	—

1.3.2.1.1 Cram versus *anti-Cram selectivities*

The addition of organometallics to aldehyde substrates incapable of populating intramolecularly chelated conformations preferentially results in a Cram selective process. This result is most easily rationalized by considering an aldehyde conformation that maintains the incoming organometallic reagent and the largest substituent of the aldehyde in an antiperiplanar disposition, as represented by (**5**).[22,23]

R
H O
H R^1—M Me

Cram product

(**5**)

Entries 1–4 in Table 3[21] illustrate the tendency for a Cram selective process in additions to aldehydes of type (**4**; equation 1). In contrast, when (**4**) is treated with the aluminum additive (**1**) prior to exposure to organometallics, the nucleophilic addition results in an anti-Cram product. The resulting facial selectivity may be most easily rationalized by considering transition state structure (**6**), which defines the anti-Cram face of the aldehyde to be less hindered by virtue of precoordination of the aluminum reagent (**1**) to the less sterically demanding Cram face. For example, comparison of entries 2, 6 and 9 to the corresponding entries 5, 8 and 10 in Table 3 illustrates the dramatic effect that the aluminum additive (**1**) has on the facial selectivity of the reaction. This approach to anti-Cram selectivity, however, does suffer some limitations. For example, the organometallic reagents in Table 3 are all Grignard reagents, since organolithium based nucleophiles do not work well.[21] Furthermore, additions of sp^2-hybridized nucleophiles (*e.g.* vinyl, enolate and phenyl) resulted in no stereoselectivity; also, although the *sp*-hybridized case listed in Table 3 (entry 12) proceeds in good yield, the anti-Cram facial selectivity is unimpressive (41:59; Cram:anti-Cram). Finally, the overall anti-Cram selectivity is adversely affected by increasing steric bulk of the incoming nucleophile (see entries 14, 16, 19, 24 and 28 in Table 3). Presumably, the basis of this result may be rationalized by the enhanced steric crowding of the transition state as represented in (**6**).

(**4**) —— i, additive; ii, R^1MgX ——> Cram + anti-Cram (1)

(**6**) Anti-Cram product

Table 3 Cram *versus* Anti-Cram Selectivity for the Addition of Organometallics to Aldehyde (**4**) in the Presence of (**1**; equation 1).

Entry	*R in (4)*	*Organometallic*	*Additive*	*Yield (%)*	*Cram:anti-Cram*
1	Ph	MeLi	None	—	~2:1
2	Ph	MeMgI	None	64	72:28
3	Ph	$MeTi(OPr^i)_3$	None	—	88:12
4	Ph	$MeTi(OPh)_3$	None	—	93:7
5	Ph	MeMgI	MAT	96	7:93
6	Ph	EtMgBr	None	78	84:16
7	Ph	EtMgBr	MAD	90	25:75
8	Ph	EtMgBr	MAT	98	20:80
9	Ph	BuMgBr	None	89	87:13
10	Ph	BuMgBr	MAT	98	33:67
11	Ph	BuC≡CMgBr	None	79	78:22
12	Ph	BuC≡CMgBr	MAT	96	41:59
13	1-Cyclohexenyl	MeMgI	None	64	79:21
14	1-Cyclohexenyl	MeMgI	MAT	84	2:98
15	1-Cyclohexenyl	EtMgBr	None	87	94:6
16	1-Cyclohexenyl	EtMgBr	MAD	76	34:66
17	1-Cyclohexenyl	EtMgBr	MAT	98	17:83
18	1-Cyclohexenyl	BuMgBr	None	88	94:6
19	1-Cyclohexenyl	BuMgBr	MAT	97	26:74
20	Cyclohexyl	MeMgI	None	81	82:18
21	Cyclohexyl	MeMgI	MAD	88	22:78
22	Cyclohexyl	MeMgI	MAT	75	23:77
23	Cyclohexyl	BuMgBr	None	81	89:11
24	Cyclohexyl	BuMgBr	MAT	89	77:23
25	$PhCH_2$	MeMgI	None	52	53:47
26	$PhCH_2$	MeMgI	MAT	96	45:55
27	$PhCH_2$	BuMgBr	None	68	50:50
28	$PhCH_2$	BuMgBr	MAT	94	39:61

1.3.2.1.2 Site selectivity

(i) 1,2 versus *1,4 site selectivity*

Equation (2) outlines the effect that organoaluminum additives have on the course of nucleophilic additions to α,β-unsaturated carbonyl substrates. Although organolithiums normally undergo 1,2-addition processes to α,β-unsaturated substrates (entry 8 in Table 4[21]), exposure of cyclohexenone substrates to the organoaluminum additive (**1**) prior to exposure to the nucleophile results in a total 1,4-site selective process (compare, for example, entries 1 and 8 in Table 4).

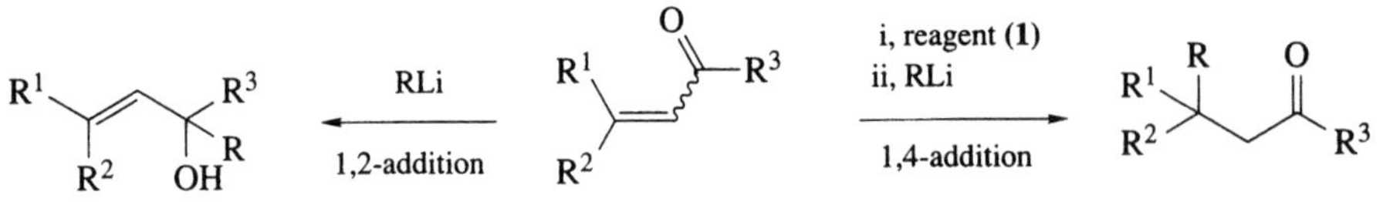

(2)

Table 4 1,2- *versus* 1,4-Regioselectivity for the Addition of Organometallics to Cyclic Enones in the Presence of (**1**; equation 2)

Entry	*Substrate*	*Organometallic*	*Additive*	*Yield (%)* *1,4-Adduct (cis/trans)*	*1,2-Adduct*
1	a	MeLi	MAD	68(29:71)	—
2	a	Bu^nLi	MAD	59(17:83)	—
3	a	Bu^tLi	MAD	73(18:82)	—
4	a	PhLi	MAD	71(33:67)	—
5	a	$CH_2{=}C(OBu^t)OLi$	MAD	87(10:90)	—
6	a	EtMgBr	MAT	35(15:85)	—
7	a	$Bu^nC{\equiv}CLi$	MAD	Recovered SM	—
8	a	MeLi	None	—	75
9	b	MeLi	MAD	70	—
10	b	PhLi	MAD	Recovered SM	—
11	b	$Me_3SiC{\equiv}CLi$	MAD	Recovered SM	—
12	c	MeLi	MAD	—	77
13	d	MeLi	MAD	26	11
14	e	MeLi	MAD	65(>95% *trans*)	16
15	e	MeLi	MAT	63(>95% *trans*)	31
16	e	Bu^sLi	MAD	65(>95% *trans*)	12
17	e	$CH_2{=}C(Me)Li$	MAD	75(>95% *trans*)	—
18	f	MeLi	MAD	Recovered SM	—
19	g	MeLi	MAD	Recovered SM	—
20	h	MeLi	MAD	74(37:63)	—
21	h	Bu^tLi	MAD	83(24:76)	—

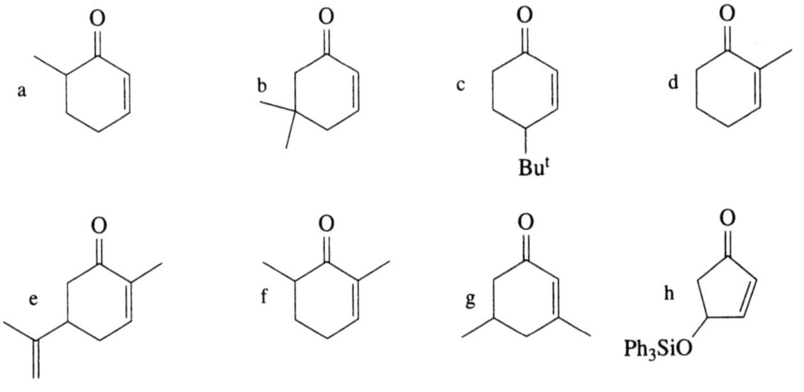

The conjugate addition processes proceed well only when the organometallic is an organolithium reagent. This is in marked contrast to the 1,2-nucleophilic additions to aldehydes detailed in the previous section, which proceeded well when the organometallic was a Grignard reagent. Furthermore, the reaction seems very sensitive to the substitution pattern of the carbonyl substrate. For example, substitution of alkyl substituents at the C-3, C-4 or C-5 position of the cyclohexenone all markedly affect the specificity (see entries 12–19, Table 4). Interestingly, the addition always proceeds with a preference for the *trans* stereoisomer for both five- and six-membered enone substrates, which complements the *cis* preference of cuprate[25] additions (see entries 1–6 and 14–17 in Table 4 for six-membered cases and entries 20 and 21 for five-membered cases). Table 5[21,24] summarizes the site selective additions of nucleophiles to acyclic α,β-unsaturated carbonyl substrates. Entries 1–4 in Table 5 demonstrate that there is a preference for the 1,2-addition products even in the presence of organoaluminum additive (**1**).

Table 5 1,2- *versus* 1,4-Regioselectivity for the Addition of Organometallics to Acyclic Enones in the Presence of (**1**; equation 2)

Entry	*Substrate*	*Organometallic*	*Additive*	*Yield (%)* *1,4-Adduct*	*1,2-Adduct*
1	(*E*)-PhCH=CHC(O)Me	MeLi	MAD	—	78
2	PhC≡CC(O)Me	MeLi	MAD	—	85
3	(*E*)-PhCH=CHC(O)Ph	MeLi	MAD	24	60
4	(*E*)-PhCH=CHC(O)Ph	MeLi	MAT	28	55
5	(*E*)-PhCH=CHC(O)Ph	Bu^tLi	MAD	77	9
6	(*E*)-PhCH=CHC(O)Bu^t	MeLi	MAD	Recovered SM	
7	(*E*)-MeCH=CHCHO	$(Me_3Si)_3Al$	None	—	—
8	$(Me)_2C$=CHC(O)Me[a]	$(Me_3Si)_3Al$	None	85	—
9	$(Me)_2C$=CHC(O)Me[b]	$(Me_3Si)_3Al$	None	—	85
10	(*Z*)-PhCH=CHCHO	$(Me_3Si)_3Al$	None	—	83
11	(*E*)-MeC[$Si(Me)_3$]=CHCHO	$(Me_3Si)_3Al$	None	—	91
12	HC≡CC(O)Me	$(Me_3Si)_3Al$	None	72 (>95% *trans*)	

[a]Conducted at –78 °C. [b]Conducted at room temperature.

The 1,4-addition product is only preferred when the steric bulk of the nucleophile is increased and the environment about the carbonyl moiety is also sterically encumbered (entry 5 in Table 5). Equation (3) outlines an approach to site selectivity that is not dependent upon additive (**1**), but does require the addition of an organoaluminum.[24]

$$\text{R(R}^1\text{)C(SiMe}_3\text{)CH}_2\text{C(O)R}^2 \xleftarrow[\text{ii, MeOH; 1,4-addition}]{\text{i, Al(SiMe}_3)_3,\ -78\ ^\circ\text{C}} \text{R(R}^1\text{)C=CHC(O)R}^2 \xrightarrow[\text{ii, MeOH; 1,2-addition}]{\text{i, Al(SiMe}_3)_3,\ 20\ ^\circ\text{C}} \text{R(R}^1\text{)C=CHC(OH)(SiMe}_3\text{)R}^2 \quad (3)$$

For example, when acyclic enones are exposed to tris(trimethylsilyl)aluminum (**7**) at room temperature, only the 1,2-addition product is observed, whereas the addition proceeds *via* a 1,4-addition process when the reaction is conducted at –78 °C (compare entries 8 and 9 in Table 5). In contrast, α,β-unsaturated aldehyde substrates only undergo 1,2-additions with reagent (**7**).

(ii) Site selective 1,2-additions—discrimination between ketones and aldehydes

Not surprisingly, there are many examples of site selective processes showing a preference for 1,2-addition to aldehydes in the presence of ketones;[26] however, in contrast, the complementary process is not readily accomplished. Thus, any site selective approach showing a preference for a ketone carbonyl must overcome the inherent reactivity difference favoring aldehydes. Equation (4) outlines an approach to site selective 1,2-nucleophilic additions that takes advantage of the more reactive nature of the aldehyde moiety.[27]

$$\underset{(\mathbf{9})}{\text{R}^1\text{CH(OH)R}^4} \xleftarrow[\text{aldehyde preference}]{\text{i, reagent (1); ii, R}^4\text{M}} \text{R}^1\text{CHO} + \text{R}^2\text{COR}^3 \xrightarrow[\text{ketone preference}]{\text{i, Me}_2\text{AlNMePh (8); ii, R}^4\text{M}} \underset{(\mathbf{10})}{\text{R}^2\text{R}^3\text{C(OH)R}^4} \quad (4)$$

For example, when a 1:1 mixture of an aldehyde and a ketone is exposed to either an equivalent of an organolithium reagent or an equivalent of a Grignard reagent, the corresponding secondary carbinol (**9**), resulting from preferential addition to the aldehyde component, is usually formed in excess (entries 2, 5, 7, 9 and 12 in Table 6[27]). However, this preference is greatly magnified when the aluminum reagent (**1**) is exposed to the carbonyl substrates prior to the addition of the organometallic (compare entries 1 and 2, 4 and 5, as well as entries 8 and 9 in Table 6).

In contrast, when the experiment is conducted with preexposure of the substrates to aluminum additive (**8**), the corresponding tertiary carbinol (**10**) resulting from preferential addition to the ketone component is produced in excess (compare entries 8 and 10, 11 and 14 in Table 6). Comparison of entries 16 and 17 demonstrates the most dramatic example of total control of this type of site selective process, since in the

Table 6 Chemospecificity for the Addition of Organometallics to Aldehydes and Ketones in the Presence of (**1**; equation 4)

Entry	*Substrates*	*Additive*	*Organometallic*	*Ratio (**9**):(**10**)*
1	$C_8H_{17}CHO$ + cyclohexanone	(**1**)	EtMgBr	36:1
2		None	EtMgBr	8:1
3		(**1**)	PhMgBr	14:1
4		(**1**)	PhLi	21:1
5		None	PhLi	1:1
6	PhCHO + PhCOMe	(**1**)	MeLi	6:1
7		None	MeLi	4:1
8		(**1**)	MeMgI	100:0
9		None	MeMgI	22:1
10		(**8**)	MeLi	1:7
11	$C_6H_{11}CHO$ + $C_6H_{11}COMe$	(**1**)	MeMgI	9:1
12		None	MeMgI	5:1
13		(**1**)	Bu^tMgCl	100:1
14		(**8**)	MeLi	1:6
15		(**8**)	PhLi	1:14
16	$OHC(CH_2)_{10}COMe$	(**1**)	MeLi	100:0
17		(**8**)	MeLi	1:62

presence of additive (**1**) site selectivity is total for the aldehyde (100:0), whereas additive (**8**) gives essentially complete reversal (1:62). The example is most interesting in that the substrate maintains both reactive centers in the same molecule and a competitive experiment of this type (an intramolecular example) is a closer analogy to the problems presented by more complicated, polyfunctional molecules.

Presumably, the selectivity of each process is a result of the generally high oxygenophilicity of aluminum reagents. Thus, aldehydes, which are inherently more reactive than ketones, are being activated by both aluminum additives. In the case of additive (**1**) this activation leads to an enhancement of the inherent reactivity difference, whereas in the case of additive (**8**) this reactivity difference results in the formation of a blocked, aminal-like intermediate of the aldehyde moiety (**11**, equation 5) preventing addition of nucleophiles to the aldehyde. Thus, the preferential addition of nucleophiles to the ketone is observed.

$$RCHO + R^1COR^2 \xrightarrow{(\mathbf{8})} RCH(NR_2)OAlL_2\ (\mathbf{11}) + R^1COR^2 \longrightarrow (\mathbf{10}) \quad (5)$$

1.3.2.2 Preparation of 1,3-Hydroxy Esters and 1,3-Hydroxy Sulfoxides *via* Organoaluminum Reagents

Although the preparation of 1,3-hydroxycarbonyl species *via* stabilized anions is beyond the scope of this review, the addition of substituted organoaluminums of type (**12**) to aldehyde or ketone substrates represents an interesting alternative to the typical aldol process.

$$CH_2=C(CO_2Me)AlBu^i_2$$

(**12**)

Reagent (**12**) is readily prepared by the addition of disobutylaluminum hydride in the presence of hexamethylphosphoric triamide to α,β-acetylenic esters.[28] When the organoaluminum (**12**) is exposed to either aldehydes or ketones, the resulting hydroxy ester (**14**) is formed in good yields (equation 6).

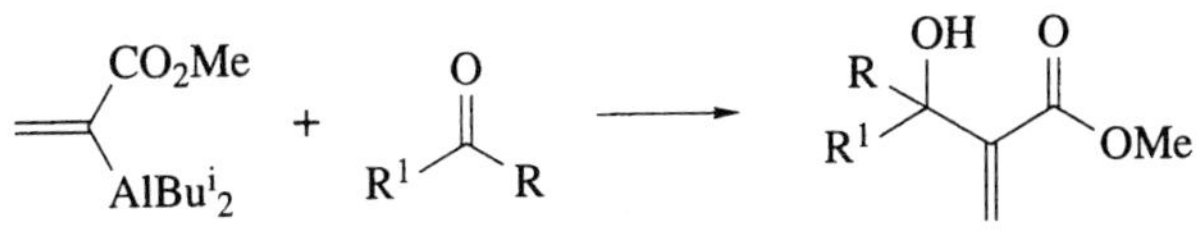

(**12**) (**13**) (**14**) (6)

Table 7[29] lists the aldehyde (entries 1–5) and ketone (entries 6 and 7) substrates that were converted to the corresponding hydroxy ester (**14**). The yields are uniformly good and there appears to be fair generality with regard to the aldehydic substrates; unfortunately, the number of ketone substrates is really not enough to assess the generality of this reaction. The reason for the paucity of ketone substrates may stem from the need to activate the ketone carbonyl with a Lewis acid (BF_3–Et_2O) in order for the reaction to proceed. The site selectivity of organoaluminum reagent (**12**) was investigated briefly. For example, the addition of (**12**) to (*E*)-2-hexenal (entry 5, Table 7), results in only the 1,2-addition product. In contrast, the addition of (**12**) to cyclohexenone results in no reaction. Since successful additions of (**12**) to the ketone substrates require Lewis acid activation, it would be interesting to expose ketones to chiral Lewis acid complexes prior to their exposure to organoaluminum (**12**). Unfortunately, the stereoselectivity of this reaction was not investigated.

Table 7 Preparation of Hydroxy Esters from the Addition of Organoaluminum (**12**) to (**13**; equation 6)

Entry	*Substrate (13)* *R*	*R*^{1}	*Yield* (%)	*Temperature* (°C)
1	H	Pr^n	87	25
2	H	Pr^i	87	25
3	H	Ph	83	25
4	H	Furanyl	80	25
5	H	(*E*)-Pr^nCH=CH	90	25
6	Pr^n	Pr^n	68	—
7	—(CH_2)$_5$—		72	—

A substrate related to the 1,3-hydroxy ester (**14**), which does permit a stereochemical investigation of an organoaluminum addition, is represented by chiral keto sulfoxide (**15**) in equation (7). Upon exposure to organometallics, the chiral keto sulfoxides afforded the corresponding diastereomeric β-hydroxy sulfoxides (**16**) and (**17**). Inspection of Table 8[30] reveals that the aluminum reagent provides diastereomer (**17**) in excess (48–92%) and in moderate yield (48–66%), whereas the titanium-based reagent provides diastereomer (**16**) in excess (60–94%) in higher overall yields (60–96%). The keto substituent (R) does have an effect on the overall facial selectivity of the reaction, since oxygen substituents in the *ortho* position of the aromatic ring do improve the relative selectivities for both aluminum- and titanium-mediated processes. The opposite selectivities generated by aluminum and titanium were rationalized on the basis of the transition state structures (**18**) and (**19**).

The titanium-chelated structure (**18**) forces the nucleophilic addition to occur from the less hindered *si*-face (*i.e. syn* to the lone pair of electrons); in contrast, the aluminum-based process proceeds *via* nonchelated transition state (**19**). In this case, the less hindered *re*-face is *syn* to the electron pair. Unfortunately, a test of this transition state hypothesis would require an organoaluminum addition under chelation con-

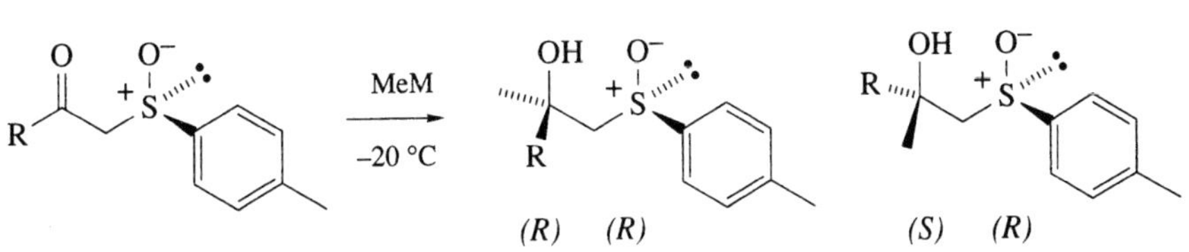

MeM, −20 °C

(*R*) (*R*) (*S*) (*R*)

(**15**) (**16**) (**17**) (7)

Table 8 Comparison of Organoaluminums *versus* Organotitaniums on the Facial Selectivity of Additions to (**15**; equation 7)

Entry	*R*	*MeM*	*Solvent*	*Yield (%)*	*Diastereomer ratio (**16**):(**17**)*
1	Ph	$MeTiCl_3$	Et_2O	79	82:18
2	Ph	Me_3Al	Toluene	66	26:74
3	*p*-MeC_6H_4	$MeTiCl_3$	Et_2O	60	80:20
4	*p*-MeC_6H_4	Me_3Al	Toluene	50	16:84
5	a	Me_3Al	Toluene	71	13:87
6	a	$MeTiCl_3$	Et_2O	96	97:3
7	b	$MeTiCl_3$	Et_2O	77	94:6
8	b	Me_3Al	Toluene	48	4:96

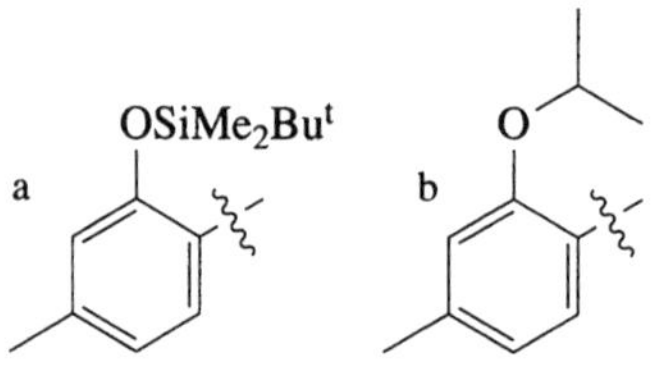

a: $OSiMe_2Bu^t$ b: O-iPr

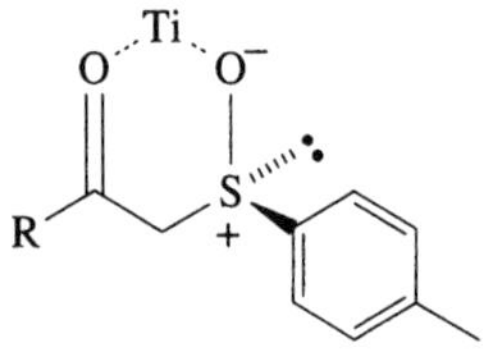

(**18**) 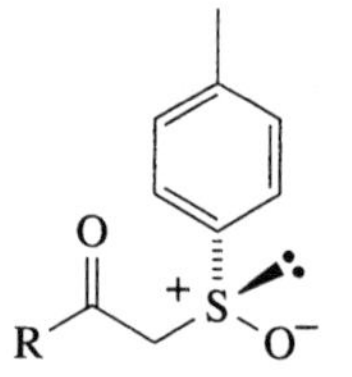 (**19**)

trol conditions. When the addition of trimethylaluminum was performed in the presence of a zinc salt, which is known to provide a chelation control element,[31] there was no reaction.

1.3.2.3 Facially Selective 1,2-Additions of Organoaluminum–Ate Complexes to Keto Ester Substrates

The 1,2-addition of chiral aluminum ate complexes to various aldehydes and ketones has been reviewed[32,33] and, generally, this protocol produces very disappointing results with respect to the overall stereospecificity of the process as compared to other organometallics;[34–36] however, the recent extension of this methodology to the preparation of chiral α-hydroxy esters and acids, *via* 1,2-additions to the prochiral ketone moieties of α-keto esters,[37–41] has renewed interest in this area.[42–44] The transformation outlined in equation (8) proceeds in excellent overall yield with exclusive selectivity for the ketone functionality; unfortunately, the stereospecificity is not as impressive.

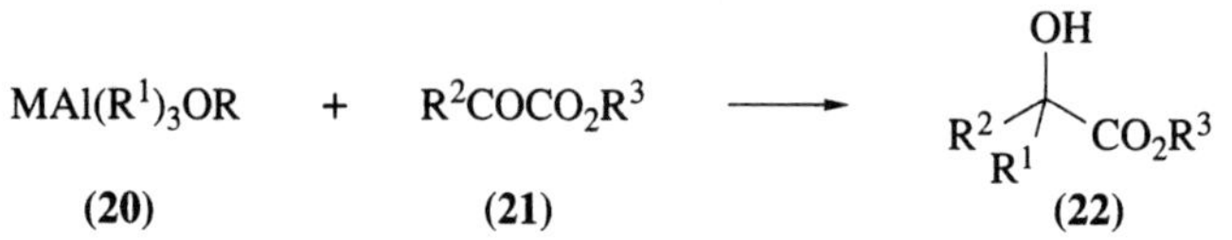

$$MAl(R^1)_3OR \ (\mathbf{20}) + R^2COCO_2R^3 \ (\mathbf{21}) \longrightarrow R^2C(OH)(R^1)CO_2R^3 \ (\mathbf{22}) \quad (8)$$

Table 9[37–41] lists the results of the additions of both chiral and achiral alkoxytrialkylaluminates (**20**) to chiral α-keto esters (**21**). Some trends are readily apparent: (i) sodium is a superior cation as compared to lithium and potassium (entries 11–14); (ii) the configuration of the chiral moiety of the ester substrate dictates the facial preference (entries 1–7) and the preference follows the dictates of Prelog's rule;[45] (iii) there is an observable solvent effect with a preference for hexane/ether mixtures; (iv) either (+)- or (±)-(2*S*,3*R*)-4-dimethylamino-1,2-diphenyl-3-methyl-2-butanol (Darvon alcohol or Chirald) is the best of the

Table 9 Results of the Addition of Chiral and Achiral Alkoxy Trialkylaluminates to Keto Esters (equation 8)

Entry	*R*	R^1	R^2	R^3	*M*	*Solvent*	*ee (%)*	*Config.*
1	(–)-*N*-Methylephedrine	Et	Me	(–)-Menthol	Li	Ether/hexane (1:1)	62	(*S*)
2	(+)-*N*-Methylephedrine	Et	Me	(–)-Menthol	Li	Hexane	15	(*S*)
3	(+)-*N*-Methylephedrine	Et	Me	(+)-Menthol	Li	Ether/hexane (1:1)	58	(*R*)
4	(–)-Menthol	Et	Me	(–)-Menthol	Li	Ether/hexane (1:1)	48	(*S*)
5	(+)-Menthol	Et	Me	(–)-Menthol	Li	Ether/hexane (1:1)	54	(*S*)
6	(–)-Menthol	Et	Me	(+)-Menthol	Li	Ether/hexane (1:1)	50	(*R*)
7	(+)-Menthol	Et	Me	(+)-Menthol	Li	Ether/hexane (1:1)	46	(*R*)
8	(–)-*N*-Methylephedrine	Et	Me	(–)-Menthol	Li	Hexane	49	(*R*)
9	(+)-*N*-Methylephedrine	Et	Ph	(–)-Menthol	Li	Hexane	0.2	(*R*)
10	(–)-*N*-Methylephedrine	Me	Ph	(–)-Menthol	Li	Hexane	38	(*R*)
11	(–)-*N*-Methylephedrine	Et	Ph	(–)-Menthol	Na	Toluene/hexane	52.5	(*R*)
12	(–)-*N*-Methylephedrine	Et	Ph	(–)-Menthol	K	Benzene/hexane	43	(*R*)
13	(–)-*N*-Methylephedrine	Et	Ph	(–)-Menthol	Na	Benzene/hexane	53	(*R*)
14	(–)-*N*-Methylephedrine	Et	Ph	(–)-Menthol	Li	Benzene/hexane	23	(*R*)
15	(+)-Darvon alcohol[a]	Et	Ph	(–)-Menthol	Li	Hexane/ether (75:100)	77	(*R*)
16	(+)-Darvon alcohol[a]	Et	Ph	(–)-Menthol	Li	Hexane/ether (30:100)	82	(*R*)
17	(+)-Darvon alcohol[a]	Et	Ph	(+)-Menthol	Li	Hexane/ether (75:100)	73	(*S*)
18	(±)-Darvon alcohol[a]	Et	Ph	(–)-Menthol	Li	Hexane/ether (30:100)	79	(*R*)
19	(±)-Darvon alcohol[a]	Et	Ph	(–)-Menthol	Li	Ether	84	(*R*)
20	*t*-Butyl alcohol	Me	Ph	(–)-Menthol	Li	Hexane	25	(*R*)
21	*t*-Butyl alcohol	Et	Ph	(–)-Menthol	Li	Hexane	27	(*R*)
22	2-Methyl-2-butanol	Et	Ph	(–)-Menthol	Li	Hexane	47	(*R*)
23	3-*t*-Butyl-3-pentanol	Et	Ph	(–)-Menthol	Li	Hexane	60	(*R*)
24	2,4-Dimethyl-3-*t*-butyl-3-pentanol	Et	Ph	(–)-Menthol	Li	Hexane/ether (75:100)	71	(*R*)

[a] Darvon alcohol = (+)- or (±)-(2*S*,3*R*)-4-dimethylamino-1,2-diphenyl-3-methyl-2-butanol.

chiral ligands (compare entries 15–19 *versus* entries 1–14); and (v) although there appears to be some double stereodifferentiation (entries 4–6 and entries 8, 9), the steric bulk of the alkoxy moiety of the organoaluminum seems to determine the overall enantioselectivity (compare entries 15–17 and 18, 19 as well as entries 20–24).

Since the facial selectivity of the process is controlled by the chiral keto ester and the observed absolute diastereomeric excess of each addition reaction is determined by steric factors within the organoaluminate, there is no apparent need to prepare chiral alkoxytrialkylaluminates. Thus, there should be very good stereodiscrimination for the additions of tetraalkylaluminates to chiral keto ester substrates (equation 9). It is advantageous to utilize tetraalkylaluminates since they are readily prepared by the titanium-catalyzed hydroalumination of alkenes (equation 10).[46,47]

$$\underset{\textbf{(23)}}{LiAl(R^1)_4} + R^2COCO_2R^3 \longrightarrow R^2C(OH)(R^1)CO_2R^3 \quad (9)$$

$$4R\text{–}CH{=}CH_2 + LiAlH_4 \xrightarrow[\text{THF}]{5\%\ TiCl_4} LiAl(CH_2CH_2R)_4 \quad (10)$$

Table 10 summarizes the data for the addition of various achiral tetraalkylaluminates (**23**) to chiral keto esters as outlined in equation (9). Presumably, the observed diastereoselectivities will reflect the inherent facial bias of the controlling chiral element, namely menthol (R^3 in Table 10). In this case the diastereoselectivities are moderate (67 to 75%), but, since Corey,[48] Oppolzer[49] and Whitesell[50] have observed superior inherent facial selectivity for the 8-substituted menthol chiral auxiliary, it would be interesting to attempt the alkyl aluminate additions on substrates incorporating this auxiliary.

Table 10 Additions of Tetraalkylaluminates to Chiral Keto Esters (equation 9)

Entry	R^1	R^2	R^3	*Solvent*	*de* (%)	*Configuration*
1	$Me(CH_2)_5$—	Ph	(–)-Menthol	THF	69	(*R*)
2	$Me_2CH(CH_2)_3$—	Ph	(–)-Menthol	THF	73	(*R*)
3	$EtMeCH(CH_2)_2$—	Ph	(–)-Menthol	THF	74	(*R*)
4	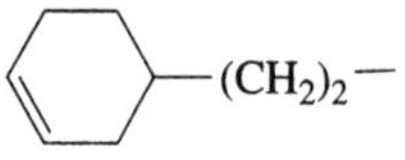 [cyclohexenyl]—$(CH_2)_2$—	Ph	(–)-Menthol	THF	72	(*R*)
5	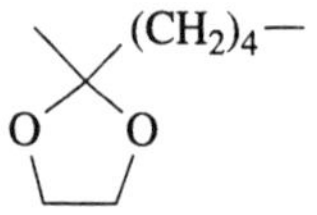 [2-methyl-1,3-dioxolan-2-yl]$(CH_2)_4$—	Ph	(–)-Menthol	THF	67	(*R*)
6	$Me(CH_2)_5$—	Me	(–)-Menthol	THF	67	—
7	$Me(CH_2)_5$—	H	(–)-Menthol	THF	75	—

1.3.2.4 Additions of Alkylaluminums to Masked Carbonyl Substrates

1.3.2.4.1 Regioselectivity and stereoselectivity of additions to α,β-unsaturated acetals and ketals

It is well known that a variety of aluminum hydride reagents cleave acetal substrates with excellent stereoselectivities.[51,53] It is not surprising, therefore, that similar experiments have been attempted with trialkylaluminum reagents. Equation (11) outlines the generalized transformation.

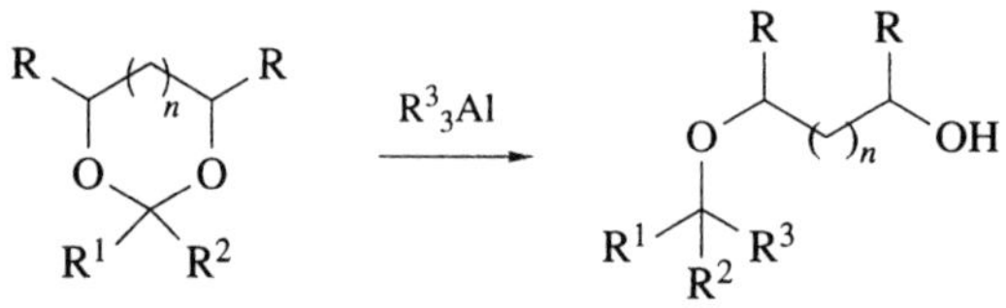

R^3_3Al (11)

n = 0, 1; R = alkyl, amide; R^1 = alkyl, alkenyl, H;
R^2 = alkyl, alkenyl, H; R^3 = alkyl, alkenyl, alkynyl

In the specific case of $n = 1$, R = Me, R^1 = cyclohexyl and R^2 = H (equation 12) exposure to trimethylaluminum resulted[52] in poor stereoselectivity and moderate yield (64%). Interestingly, although the diastereomeric ratio was defined by GC methods (2:3), the stereochemistry of the preferred diastereomer was undefined. In contrast to this result, the addition of alkylaluminums to α,β-unsaturated acetals (Scheme 3) proceeds with spectacular regioselectivity and stereoselectivity.

Allylic acetal substrates include both aspects of the selectivity problem, namely, regiochemical control and stereochemical control. Table 11[54] lists the results of alkylaluminum additions to chiral, allylic acetals. The most striking aspect of this work is the total dependence of regiochemical control on the solvent.[54,55] For example, comparison of entries 1 and 3 in Table 11 shows a complete reversal of regiochemical control in going from dichloroethane to chloroform solvent. Furthermore, the stereoselectivity of the process is dependent upon the configuration of the starting acetal. Thus, when one starts with the acetal derived from (*R*,*R*)-(4)-*N*,*N*,*N'*,*N'*-tetramethyltartaric acid diamide, the 1,2-addition product with the (*R*)-configuration (88% *ee*) is produced (entry 3, Table 11), whereas the corresponding (*S*,*S*)-isomer produced the 1,2-addition product with the (*S*)-configuration (entry 5, Table 11). It is noteworthy that, although the regioselectivity for the 1,4-process is very good, the 1,2-process is exclusively con-

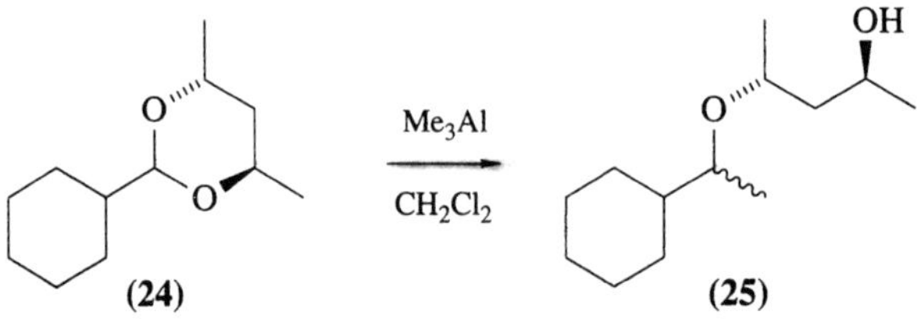

(24) $\xrightarrow[CH_2Cl_2]{Me_3Al}$ **(25)** (12)

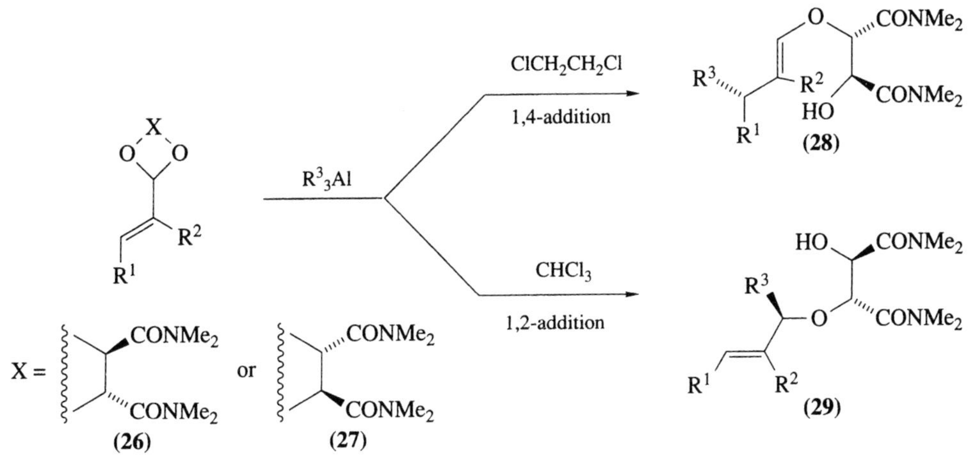

Scheme 3

trolled in chloroform solvent. On the other hand, if one wished to produce only the 1,4-addition product *via* the addition of organoaluminums to acetals, the best method appears to be that of Negishi,[56] which proceeds *via* palladium catalysis (equation 13) to produce exclusively the 1,4-addition products.

Table 11 Regioselectivity and Stereoselectivity for the Addition of Me_3Al to Chiral Allylic Acetals (Scheme 3)

(30) **(31)**

Entry	*Substrate*	*Solvent*	*Product ratio 1,2 versus 1,4*	*1,2-Product (% ee)*	*Configuration*
1	**(30)**	$ClCH_2CH_2Cl$	1:6.5	88	(*S*)
2	**(30)**	Toluene	1:1.7	96	(*S*)
3	**(30)**	$CHCl_3$	1:0	88	(*R*)
4	**(31)**	Toluene	1:2.8	96	(*R*)
5	**(31)**	$CHCl_3$	1:0	86	(*S*)

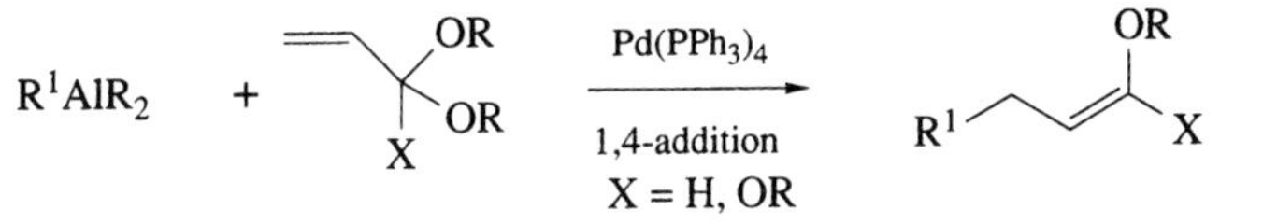

(13)

1.3.2.4.2 Chiral α-hydroxy acetals and ketals

The 1,2-addition reaction of organoaluminums may be extended to α-hydroxy ketals. An interesting example of this ketal variant appears in equation (14).

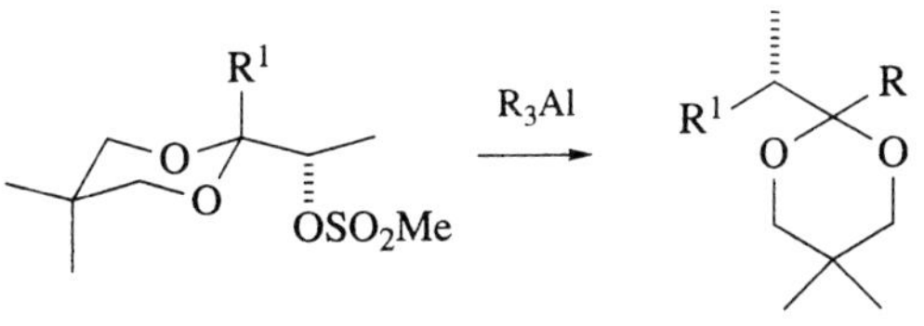

(14)

The reaction presumably takes place *via* attack of the nucleophile at the ketal carbon center with concomitant migration of the R^1 moiety to the adjacent center.[58] Organoaluminum reagents are ideally suited for such a transformation since, by virtue of their amphophilic nature, they act as nucleophiles and electrophiles. Thus, organoaluminums will not only donate the nucleophile but also activate the leaving group, which in this case is mesyl. Table 12[57,58] lists the results of alkylaluminum additions to the generalized substrate (**32**).

Table 12 The Addition of Organoaluminiums to α-Hydroxy Ketals (equation 14)

Entry	R^1	*RAl*	*Yield (%) (double bond geometry)*	*Configuration* [a]	*ee (%)*
1	*p*-$MeOC_6H_4$	Me_3Al	39	*(S)*	>95
2	*p*-$MeOC_6H_4$	Me_3Al, Bu^nLi	89	*(S)*	>95
3	*p*-$MeOC_6H_4$	Et_3Al, Bu^nLi	90	*(S)*	>95
4	$Ph(CH_2)_3$ (*cis*)	Me_3Al, Bu^nLi	80 *(cis)*	*(S)*	>95
5	$Ph(CH_2)_3$ (*trans*)	Me_3Al, Bu^nLi	94 *(trans)*	*(S)*	>95
6	*p*-$MeOC_6H_4$	$Et_2AlC{\equiv}CBu^n$, Bu^nLi	4	—	—
7	*p*-$MeOC_6H_4$	$EtAl(C{\equiv}CBu^n)_2$	81	*(S)*	>95
8	$Ph(CH_2)_3$ (*cis*)	$Et_2AlC{\equiv}CBu^n$	40 *(cis)*	*(S)*	>95
9	$Ph(CH_2)_3$ (*cis*)	$EtAl(C{\equiv}CBu^n)_2$	84 *(cis)*	*(S)*	>95

[a] The enantiomeric excess was determined by NMR methods.

The following generalizations may be drawn from the data in Table 12: (i) the 1,2-addition of alkyl groups is greatly facilitated by preparation of the ate complex (compare entries 1 and 2); (ii) the transfer of alkynyl moieties is retarded by ate complex formation (compare entries 6 and 7); (iii) the integrity of the double bond geometry of the migrating group is maintained (entries 4, 5, 8 and 9); and (iv) the stereospecificity of the process is very high (>95% *ee*). Presumably, the relative differences in nucleophilicity account for the observed reactivities, since it is known that the ate complex of alkylaluminum reagents is more nucleophilic than the corresponding alkylaluminum (as well as a weaker Lewis acid). In contrast, the electron demand of the alkynyl group bestows a reduced Lewis acid character to alkynylaluminum reagents as compared to the corresponding alkylaluminum.

The final example of 1,2-additions of organoaluminums to masked ketone substrates is outlined in equation (15).[59] The addition of a large excess (10 equiv.) of trimethylaluminum to a variety of triol acetals and ketals proceeds in relatively high overall yield. Although ketal substrates where R^1 did not equal R^2 were not attempted, a number of acetals were prepared and upon exposure to Me_3Al they afforded the corresponding diastereomeric ethers with poor diastereoselectivity (33–17% *de*).

(**34**) $\xrightarrow{Me_3Al}$ (**35**) (15)

The proposed mechanism is outlined in equation (16). A metal alkoxide (**36**) is initially formed, which undergoes a bond reorganization to betaine (**37**). The intermediate (**37**) is presumably the reacting species. Support for this process is derived from the following: (i) essentially no reaction takes place in the absence of an α-hydroxy group (see entries 1–3 in Table 13[59]); (ii) the reaction is extremely sluggish if the environment about the α-hydroxy moiety is highly congested (see entry 4 in Table 13); and (iii) the diastereoselectivity is quite poor (see entries 5 and 6 in Table 13) for the related acetal substrates.

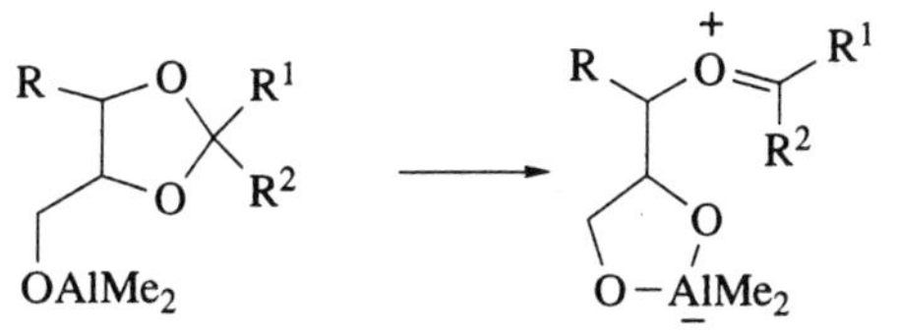

(36) → (37) (16)

Table 13 The Diastereoselective Addition of Trimethylaluminum Hydroxy Acetals and Ketals (equation 15)

Entry	*Substrate*	*Yield (%)*	*Diastereomeric ratio*
1	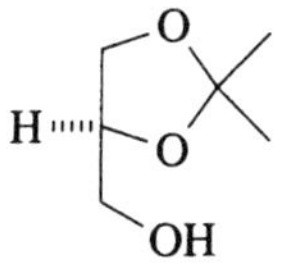	83	—
2	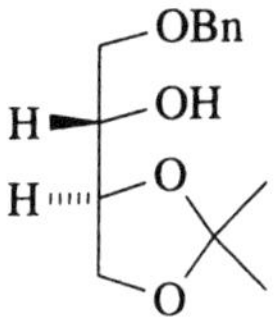	91	—
3	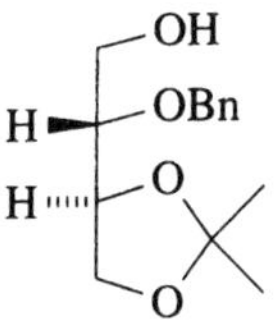	16	—
4	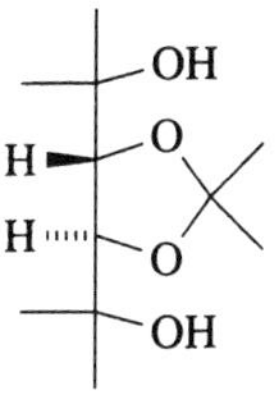	NR	—
5	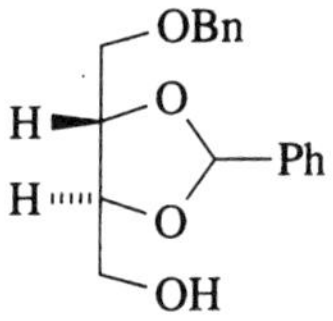	73	12:1
6	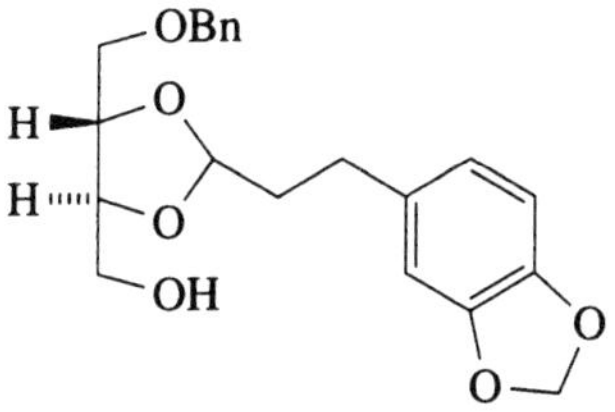	77	7.2:1

1.3.2.4.3 Application of a masked aluminum enolate in a facially selective sigmatropic protocol

3,3-Sigmatropic rearrangements (*e.g.* ene and Claisen reactions) are beyond the scope of this review; however, there are specific examples of this transformation, promoted by organoaluminum reagents, which define a 3,3-sigmatropic rearrangement protocol terminating in the 1,2-addition of an organoaluminum to an aldehyde or a ketone. Since it has been demonstrated[60,61] that sterically hindered organoaluminum reagents related to reagent (**38**) permit excellent stereochemical control of the resulting double bond geometry, it is useful to discuss a protocol that, in principle, permits not only control of the double bond geometry, but may also impart considerable bias with regard to facial selectivity.

(**38**) —$RAlX_2$ (**39**)→ [(**40**)] → (**41**) (17)

Equation (17) outlines the organoaluminum-mediated transformation. Although there is no unequivocal definition for the proposed transition state structure,[60] a chair-like version of (**40**) appears reasonable[64] in light of the resulting double bond stereochemistry.[61,62] Table 14[62,63] lists the various substrates and organoaluminum reagents that have been utilized. Although entries 1–3 (Table 14) are simple alkyl and aryl examples, it is interesting to note that the R^2 substituent may be something other than a proton (entry 3). Entries 4–8 are examples of the conversion of substituted dihydropyrans to chain-extended, unsaturated six-membered carbocycles. Obviously, in these cases the regiochemistry of the double bond is fixed by the nature of the sigmatropic process; however, an investigation of the possible facial selectivity of the 1,2-addition of the nucleophile to either the intermediate aldehyde (entries 4–7, Table 14) or the intermediate ketone (entry 8) has not been published. The protocol may also be used to prepare seven-membered unsaturated carbocycles (*e.g.* entries 9 and 10, Table 14). The reaction seems general with regard to the organoaluminum, since alkyl, aryl (entries 1–4, 7, 8 and 10), alkenyl (entry 6) and alkynyl (entries 5 and 9) substituents are readily introduced. Interestingly, if an organoaluminum containing a sulfur substituent (**39**; R = SPh, X = Et) is utilized to facilitate the transformation, the corresponding aldehyde or ketone is isolated. Presumably, a thioaryl hemiacetal or ketal is formed, which affords the corresponding aldehyde or ketone upon workup. If the organosulfur substituent is something other than *S*-phenyl (R = SEt or SBu^t), there is no rearrangement.

1.3.2.4.4 1,2-Addition to enol phosphates

Although enol phosphates are known to undergo a coupling reaction with nickel-catalyzed Grignard reagents[65] to afford substituted alkenes, the corresponding reaction with organoaluminums does not work; however, it has been demonstrated[66] that the coupling reaction of organoaluminums and enol phosphates does proceed in the presence of palladium (see equation 18 and Table 15[66]).

This reaction sequence produces alkenes in a stereospecific manner, since the double bond geometry of the enol phosphate is retained in the alkene product. Furthermore, the newly formed alkene may be introduced in a regioselective fashion *via* trapping of the regioselectively formed enolate.[67]

The success of the reaction is dependent upon the presence of $Pd(Ph_3)_4$, since the addition fails in the absence of Pd or in the presence of either nickel or $Pd(acac)_2$.[66]

1.3.3 REACTIONS OF ORGANOALUMINUM REAGENTS WITH ACID DERIVATIVES

1.3.3.1 Reactions with Ester Substrates

The conversion of esters to either amides or hydrazides may be accomplished under relatively mild conditions by the procedure outlined in equation (19).

Reagent (**42**) is readily prepared[68–70] by the treatment of primary and secondary amines and hydrazines (substituted or unsubstituted) or the corresponding hydrochlorides with trimethylaluminum. The aluminum amide reagents (**42**) readily react with a variety of ester substrates, such as conjugated esters (entries 2, 5 and 11, Table 16[68–70]), *N*-blocked amino acid esters (entry 10, Table 16), as well as alkyl esters (en-

Table 14 The Addition of Organoaluminums to **(40)** (equation 17)

Entry	R^1	R^2	*R*	*X*	*Yield* (%)	*Temperature* (° C)	*Time* (h)
1	Bu^n	H	Me	Me	91	25	0.25
2	Ph	H	Me	Me	78	25	0.25
3	H	Ph	Me	Me	71	25	0.25
4	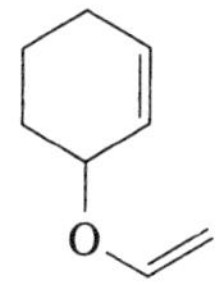		Me	Me	81	25	0.5
5	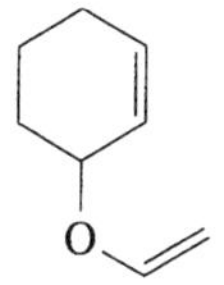		PhC≡C	Et	88	25	0.25
6	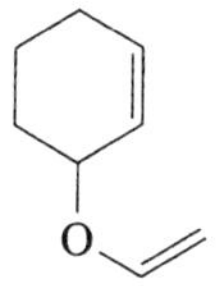			Bu^i	40	25	1.5
7	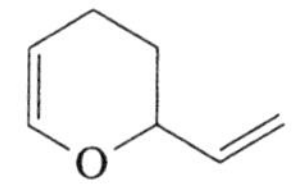		Me	Me	86	60	2
8	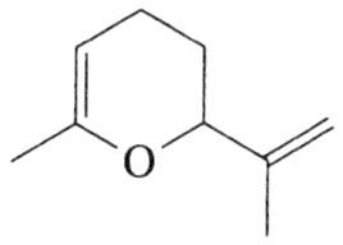		Me	Me	87	60	2
9	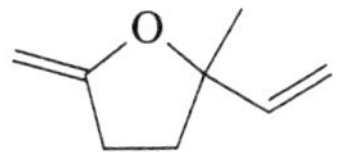		PhC≡C	Et	83	25	2
10	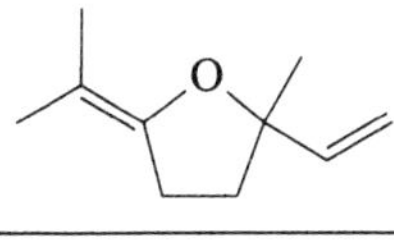		Me	Me	48	25	1

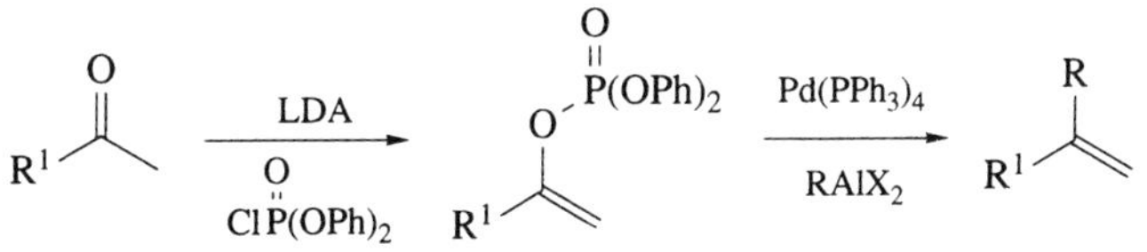

LDA, $ClP(O)(OPh)_2$; $Pd(PPh_3)_4$, $RAlX_2$ (18)

tries 3, 6–8, Table 16) and aryl esters (entries 1, 4, 9 and 12, Table 16). Dimethylaluminum hydrazides react with esters to afford the corresponding carboxylic acid hydrazides (entries 13–18, Table 16). The reaction is fairly general with regard to the nature of substituted hydrazines, since alkyl and aryl groups are tolerated on the terminal amine of the hydrazine.

Aluminum amides may also be used for the conversion of lactones to the corresponding open chain hydroxy amides (equation 20). For example, exposure of lactone **(45)** to dimethylaluminum amide **(47**; R = H or R = CH_2Ph) at 41 °C for about 24 h affords the corresponding hydroxy amide **(46)** in high yield (80%, R = CH_2Ph; 83%, R = H); in contrast, exposure of **(45)** to either benzylamine or sodium amide resulted in essentially none of the desired amide **(46)**.[68] In general the 1,2-addition of heteroatom-substituted organoaluminum reagents to carboxyl-containing substrates may be rationalized in terms of the

Table 15 The Addition of Organoaluminums to Enol Phosphates (equation 18)

Entry	R^2	*R*	*X*	*Yield (%)*
1	$Me(CH_2)_9C{=}CH_2$ (vinyl, attached at C)	Me	Me	91
2		Et	Me	71
3		PhC≡C	Et	82
4		$Me(CH_2)_3C{\equiv}C$	Et	59
5		[alkenyl structure]	Bu^i	66
6	$PhC{=}CH_2$ (vinyl, attached at C)	Me	Me	94
7		Et	Et	80
8		PhC≡C	Et	67
9	Bu^t-cyclohexenyl	PhC≡C	Et	70
10		Me	Me	72

$$RCO_2R^1 \xrightarrow[\text{25–41 °C}]{R^4R^5N{-}AlR^2R^3\ (\mathbf{42})} RC(O)NR^4R^5 \qquad (19)$$

(43) **(44)**

$R^2 = R^3 = Me$; $R^2 = Cl$, $R^3 = Me$; $R^4 = R^5 = H$, alkyl;
$R^4 = alkyl$, $R^5 = OMe$; $R^4 = H$, alkyl; $R^5 = NR_2$

Table 16 The Addition of Aluminum Amide Reagents (**42**) to Esters (equation 19)

Entry	*Substrate* *R*	R^1	R^4	*Reagent* (**42**) R^5	R^2	R^3	*No. of equivalents (42)*	*Yield (%)*	*Reaction conditions Time* (h)/*Temp.* (°C)
1	Ph	Me	H	H	Me	Me	2.2	77	17/41
2	PhCH═CH	Me	H	H	Me	Me	2.0	86	12/38
3	Cyclohexyl	Mc	H	H	Me	Me	2.2	78	17/35
4	Ph	Me	H	H	Me	Cl	3.0	83	12/50
5	PhCH═CH	Et	H	H	Me	Cl	3.0	82	12/50
6	Cyclohexyl	Me	H	H	Me	Cl	3.0	93	12/50
7	Cyclohexyl	Me	H	CH_2Ph	Me	Me	2.0	78	17/35
8	Cyclohexyl	Me	Me	Me	Me	Cl	2.0	76	12/50
9	Ph	Me	$(CH_2)_4$		Me	Me	1.1	94	5/40
10	$PhCH_2CHNHCO$ Me	Et	$(CH_2)_4$		Me	Me	1.1	77	40/40
11	[dienyl structure]	Me	$(CH_2)_4$		Me	Me	2.0	74	45/40
12	Ph	Me	$(CH_2)_5$		Me	Me	1.1	74	34/40
13	$Me(CH_2)_4$	Et	H	NH_2	Me	Me	2.2	82	16/40
14	$Me(CH_2)_4$	Et	H	PhNH	Me	Me	2.2	91	12/25
15	$Me(CH_2)_4$	Et	Me	NHMe	Me	Me	2.2	72	16/40
16	$4\text{-}MeC_6H_4$	Et	H	NMe_2	Me	Me	2.2	80	16/40
17	$4\text{-}MeC_6H_4$	Et	H	NHPh	Me	Me	2.2	87	12/25
18	$4\text{-}MeC_6H_4$	Et	Me	NHMe	Me	Me	2.2	72	12/45

hard/soft acid and base theory, first defined by Pearson.[71] The aluminum amide reagents discussed in this section all contain a weak bond between aluminum and a relatively soft substituent, when compared to the stronger (and harder) aluminum–oxygen bond which is formed upon reaction.

(45) $\xrightarrow[\text{24 h/41 °C}]{Me_2AlNHR\ \textbf{(47)}}$ (46) (20)

R = CH_2Ph, H

1.3.3.1.1 Selenol ester formation

Selenol esters are useful as active acyl equivalents, particularly with regard to macrocycle formation.[72,73] Equation (21) outlines the conversion of esters to selenol esters *via* aluminum reagent (**48**).[6,7]

$$RC(O)OR^1\ \textbf{(49)} \xrightarrow{Me_2AlSeR^2\ \textbf{(48)}} RC(O)SeR^2\ \textbf{(50)} \quad (21)$$

The reaction is fairly general with regard to the acyl substituent of ester (**49**), since aromatic and alkyl substituents are compatible (entries 1–8, Table 17[73–75]). In contrast, there are restrictions with regard to lactone substrates. For example, the five-membered lactone (entry 11, Table 17) does not undergo reaction, whereas the related six-membered lactone (entry 9, Table 17) is converted to its corresponding selenol ester in good yield (78%). Although the unsubstituted five-membered lactone did not react, the five-membered fused lactone (entry 10, Table 17) did afford the related selenol ester in good yield (80%).[73] The reaction also discriminates between axial and equatorial esters. For example, entry 14 (Table 17) presents a substrate containing an axial, as well as an equatorial, ester, which upon exposure to dimethylaluminum methaneselenolate (**48**; R^2 = Me) affords only the equatorial monoselenol ester.[75] The selectivity of this reaction is highly dependent upon the reaction solvent. When axial and equatorial methyl-4-*t*-butylcyclohexanecarboxylates (entries 12 and 13, Table 17) are exposed to organoaluminum (**48**; R^2 = Me) in diethyl ether at room temperature, the equatorial selenol ester is formed in 0.5 h, whereas the formation of the axial selenol ester requires 10 h. Equation (22) outlines the results of a site selective process.[73,76]

When cyclohexenone is exposed to either aluminum reagent (**48**) or (**51**), only the 1,4-addition product is observed in high yield (87% and 72% respectively). Thus, there appears to be excellent site selectivity, although the reasons for the selectivity are undefined.

1.3.3.2 Reactions with Acyl Chlorides

The reaction of organoaluminums with acyl chlorides typically fails to produce ketones, since the organoaluminum reacts with the desired ketone to produce carbinols.[77] However, organoaluminum reagents, in the presence of transition metal catalysts,[78–80] may be utilized to accomplish this transformation. Equation (23) outlines the overall transformation.

The copper and palladium transition metal catalysts noted in Table 18[78,79] proved to be superior to nickel, ruthenium and rhodium catalysts. The nature of the reacting species has not been unequivocally defined, but the following experimental observations may provide some insight: (i) tetrahydrofuran solvent is essential for the palladium-mediated reactions, since complex reaction mixtures (presumably containing carbinols) were observed when the reactions were performed in either benzene or methylene chloride; (ii) the reaction is truly catalytic with respect to palladium (2 mmol alkylaluminum, 0.05 mmol of $Pd(PPh_3)_4$), whereas the copper catalyst is stoichiometric; and (iii) in the case where a direct comparison may be made (entries 1–8, Table 18), the copper-based system is superior to palladium catalysis with regard to overall yield.

The palladium-catalyzed systems seem quite flexible with regard to the nature of the organoaluminum, since alkyl-, alkenyl- and alkynyl-aluminum reagents were used successfully (entries 8–14, Table 18). Furthermore, the acyl chloride substrates include alkyl, aryl and alkenyl substituents.

Table 17 The Addition of Selenoaluminum (**48**) to Esters (equation 21)

Entry	*Substrate (**49**)* *R*	*R[1]*	*Yield (**50**) (%)*	*Solvent*
1	Ph	Me	99	CH_2Cl_2
2	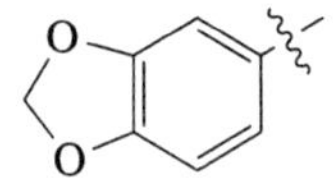	Me	Quantitative	CH_2Cl_2
3	$Ph(CH_2)_3$	Et	70	CH_2Cl_2
4	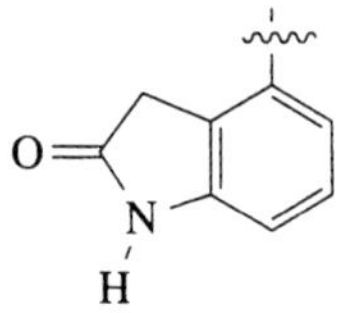	Et	80	CH_2Cl_2
5	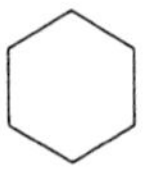	Me	93	CH_2Cl_2
6	$Me(CH_2)_5$	Me	95	CH_2Cl_2
7	$H_2C{=}CH(CH_2)_2$	Me	84	CH_2Cl_2
8	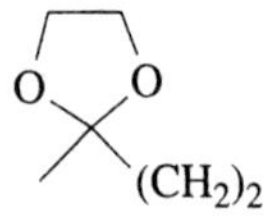 $(CH_2)_2$	Me	94	CH_2Cl_2
9	$-(CH_2)_4-$		78	CH_2Cl_2
10	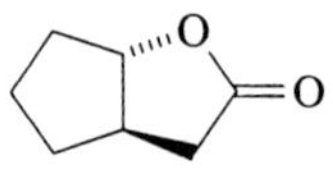		80	CH_2Cl_2
11	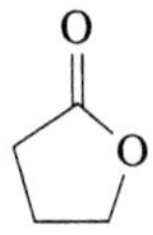		0	CH_2Cl_2
12	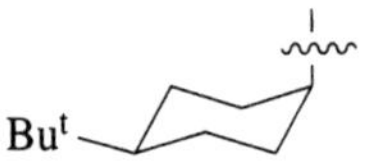	Me	92	Et_2O
13		Me	96	Et_2O
14	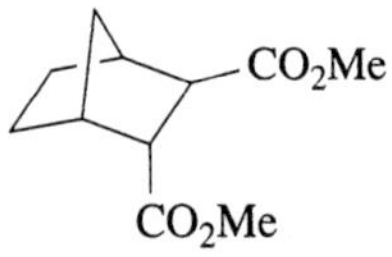		67 (equatorial monoselenol ester)	Et_2O

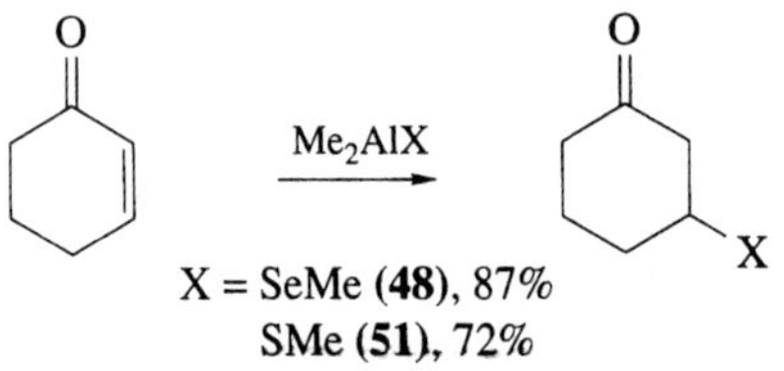

$$\text{cyclohexenone} \xrightarrow{Me_2AlX} \text{3-X-cyclohexanone} \qquad (22)$$

X = SeMe (**48**), 87%
SMe (**51**), 72%

$$R^1COCl + R_3Al \xrightarrow[\text{THF}]{\text{metal catalyst}} R^1COR \quad \text{metal = Pd, Cu} \tag{23}$$

Table 18 The Palladium-catalyzed Addition of Organoaluminums to Acyl Chlorides (equation 23)

Entry	R^1	*R*	*Catalyst*	*Yield (%)*
1	Ph	Et	$Pd(PPh_3)_4$	70
2	Ph	Et	$Cu(acac)_2PPh_3$	88
3	Ph	Me	$Cu(acac)_2PPh_3$	95
4	Ph	Me	$Pd(PPh_3)_4$	74
5	$CH_2{=}CH(CH_2)_8$	Me	$Pd(PPh_3)_4$	59
6	$CH_2{=}CH(CH_2)_8$	Me	$Cu(acac)_2PPh_3$	90
7	$CH_2{=}CH(CH_2)_8$	Et	$Cu(acac)_2PPh_3$	91
8	$CH_2{=}CH(CH_2)_8$	Et	$Pd(PPh_3)_4$	71
9	Ph	$BuC{\equiv}C$	$Pd(PPh_3)_4$	67
10	$PhCH{=}CH$	$BuC{\equiv}C$	$Pd(PPh_3)_4$	74
11	C_7H_{15}	$PhC{\equiv}C$	$Pd(PPh_3)_4$	61
12	$CH_2{=}CH(CH_2)_8$	$Me_3SiC{\equiv}C$	$Pd(PPh_3)_4$	51
13	$CH_2{=}CH(CH_2)_8$	C_6H_{13}, H / $SiMe_3$ vinyl	$Pd(PPh_3)_4$	51
14	Ph	Bu, H / $SiMe_3$ vinyl	$Pd(PPh_3)_4$	51

1.3.3.2.1 Preparation of acylsilanes via *silyl-substituted organoaluminum–ate complexes*

Acylsilanes may be prepared directly from either acyl chlorides or anhydrides by treatment with tris(trimethylsilyl)aluminum–ate complexes (equation 24).[81]

$$RCOCl \xrightarrow[\text{CuCN}]{LiAl(SiMe_3)_3\ \mathbf{(52)}} RCOSiMe_3 \tag{24}$$

The silyl organoaluminum reagent (**52**) was prepared either by the addition of 'activated' aluminum to a tetrahydrofuran solution of chlorotrimethylsilane, or by the treatment of sodium tetrakis(trimethylsilyl)aluminate with aluminum chloride. Alternatively, ate complex (**53**) may be prepared by the addition of methyllithium to tris(trimethylsilyl)aluminum.[82]

Table 19[81] lists a variety of alkyl and aryl acyl chlorides that readily undergo the transformation to the corresponding acylsilanes. The ate complex is necessary for the reaction, since tris(trimethylsilyl)aluminum afforded minute quantities of the desired acylsilane; furthermore, catalytic amounts of copper cyanide (10 mol %) are also required, although the role of the copper catalyst has not been defined. Although an excess of either reagent (**52**) or reagent (**53**) (*ca.* 2.5:1 organoaluminum to aryl substrate) is required to obtain an optimum yield of acylsilane, the ate complexes demonstrate remarkable chemoselectivity, in that they do not react with nitriles, esters, ketones, acylsilanes and carbamoyl chlorides.[82] The application of this transformation to a more highly functionalized molecule is outlined in equation (25).

$$\mathbf{(54)} \xrightarrow[\text{CuCN}]{LiMe{\bullet}Al(SiMe_3)_3\ \mathbf{(53)}} \mathbf{(55)}\ 52\% \tag{25}$$

(54): aryl COCl with CN-bearing side chain; (55): aryl $COSiMe_3$ with CN-bearing side chain

(54) **(55)** 52%

Table 19 The Copper-catalyzed Addition of Silyl Organoaluminums (**52**) to Acyl Chlorides (equation 24)

Entry	*R*	*Yield* (%)	*Reaction time* (h)
1	C_5H_{11}	95	2
2	Ph	93	1.5
3	$PhCH_2$	89	1.5
4	2-thienyl-CH_2	90	1.5
5	Bu^t	89	1.75
6	$AcOCH_2$	86	2.5

All reactions were carried out at –78 °C and with a mole ratio (substrate/reagent) of 2.5.

1.3.4 1,2-ADDITIONS OF ORGANOALUMINUMS TO CARBON–NITROGEN SYSTEMS

1.3.4.1 Stereoselective Additions

Tosylpyrazolines (**56**) stereoselectively react with trimethylaluminum to afford pyrazolols (**57**; equation 26).[83]

(**56**) $\xrightarrow{Me_3Al}$ (**57**) (26)

The reaction is highly stereoselective since the organoaluminum always approaches the face of the carbon–nitrogen double bond opposite to the hydroxy substituent. Thus, either *cis*- or *trans*-pyrazolols may be selectively prepared. The reaction does not seem to suffer any steric encumbrance since the substituent R was varied from methyl to *t*-butyl without any decrease in the overall yield. Furthermore, attempts to perform the transformation with organolithiums or Grignards failed, which presumably reflects the inherent differences in the Lewis acidity of organoaluminums and other organometallics.

1.3.4.2 Organoaluminum-promoted Beckmann Rearrangements

The Beckmann rearrangement proceeds through a transition state represented by structure (**59**) in equation (27). Thus, the organoaluminum-catalyzed protocol presents an opportunity to convert a ketone regiospecifically and stereospecifically into a chain-extended amino substrate.

(**58**) $\xrightarrow{R_2AlX}$ $[R^1\overset{+}{N}{\equiv}CR^2 \leftrightarrow R^1N{=}\overset{+}{C}R^2]$ (**59**) $\xrightarrow{X^-}$ (**60**) $\xrightarrow[R^4 = H^-]{MR^4}$ $R^1NH{-}C(R^2)(R^4)X$ (**61**) (27)

Table 20[84] summarizes the variety of substrates (R^1 and R^2) that are compatible with the reaction conditions. For example, cyclic and acyclic thioimidates (entries 1–7) are generally prepared in good to excellent yields (46–90%), with the only reported exception being entry 1. This procedure may also be utilized to stereoselectively synthesize α-monoalkylated amines as well as α,β-dialkylated amines, since the conversion of imine (**60**) to amine (**61**; equation 27) may be accomplished by exposure to either a hydride reagent (R^4 = H) or an organometallic (R^4 = alkyl). The α-monoalkylated cases are listed in entries 8–13 and the α,α-dialkylated cases are listed in entries 14–20 (Table 20). The reaction has been successfully conducted on both cyclic and acyclic substrates affording the product amines in good overall yield (51–88%). There is reasonable flexibility with regard to the alkyl group (X = methyl, propyl, substituted alkyne; Table 20), although apparently no alkenylaluminum reagents were utilized. Also, in the one case listed in Table 20 that is capable of diastereoselection (entry 10) there was no determination (other than that it was a mixture of *cis–trans* isomers) of the level of facial selectivity.[84] The dialkylated substrates (entries 14–20) are prepared by exposure of intermediate (**60**) to Grignard reagents and, although there does seem to be a fair generality (R^4 = allyl, crotyl and propargyl), it is interesting to note that apparently no alkenyl Grignards were utilized.

Table 20 Organoaluminum-mediated Beckmann Rearrangement (equation 27)

Entry	*Substrate (58)*			*Organoaluminum*		*Product*	*Yield (%)*	*Temp. (°C)*
	R^1	R^2	R^3	*R*	*X*			
1	$—(CH_2)_4—$		p-MeC_6H_4	Bu^i	SMe	(**60**)	5	40
2	$—CHMe(CH_2)_3—$		p-Me-C_6H_4	Bu^i	SMe	(**60**)	46	40
3	$—(CH_2)_5—$		Me	Me	SEt	(**60**)	62	0
4	$—CHMe(CH_2)_4—$		p-MeC_6H_4	Bu^i	SMe	(**60**)	66	0
5	$—(CH_2)_6—$		p-MeC_6H_4	Me	SEt	(**60**)	90	0
6	Ph	Me	Me	Me	SPh	(**60**)	88	–78
7	Ph	Me	Me	Bu^i	SMe	(**60**)	90	0
8	$—(CH_2)_5—$		Me	Me	Me	(**61**)R^4 = H	70	–78
9	$—(CH_2)_5—$		Me	Et	C≡CBu	(**61**)R^4 = H	67	–78
10	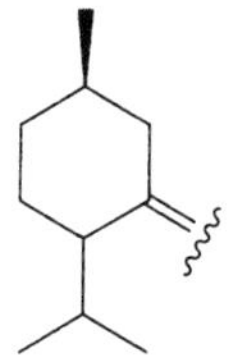		Me	Me	Me	(**61**)R^4 = H	57	–78
11	Ph	Me	Me	Me	Me	(**61**)R^4 = H	63	–78
12	Ph	Me	Me	Et	C≡CMe	(**61**)R^4 = H	60	–78
13	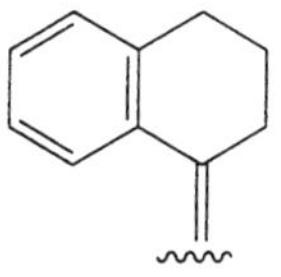		Me	Me	Pr	(**61**)R^4 = H	88	–78
14	$—(CH_2)_4—$		p-MeC_6H_4	Me	Me	(**61**)R^4 = allyl	51	–78
15	$—(CH_2)_5—$		Me	Me	Me	(**61**)R^4 = allyl	60	–78
16	$—(CH_2)_5—$		Me	Me	Me	(**61**)R^4 = propargyl	55	–78
17	Ph	Me	Me	Me	Me	(**61**)R^4 = propargyl	61	–78
18	Ph	Me	Me	Et	C≡CPh	(**61**)R^4 = allyl	88	–78
19	Ph	Me	Me	Me	Me	(**61**)R^4 = crotyl	56	–78
20	Ph	Me	Me	Et	$Me(CH_2)_3C≡C$	(**61**)R^4 = allyl	74	–78

The regioselectivity of this process is inherent to the Beckmann rearrangement, whereas the potential for useful stereoselectivities derives from the organoaluminum protocol, which permits manipulation of the prochiral imine of intermediate (**60**). Although entry 10 gave disappointing results with regard to stereoselectivity, equation (28) outlines a highly stereoselective example.

The difference in the stereoselectivity for (**64**) and (**65**) is most easily rationalized on the basis of the transition state geometry,[86,87] which is determined by the coordination of aluminum to the nitrogen lone pair of electrons. Scheme 4[85] details the effect that aluminum has on the transition state geometry and, therefore, the course of the overall reaction.

In the presence of the organoaluminum, the substituent (R) on the chiral center adjacent to nitrogen in (**67**) is forced to occupy an axial position to relieve the strain created by the steric interaction with aluminum. This would favor approach by hydride from the top face of (**67**), whereas in the absence of aluminum the favored approach of hydride is from the bottom face of transition state structure (**66**).

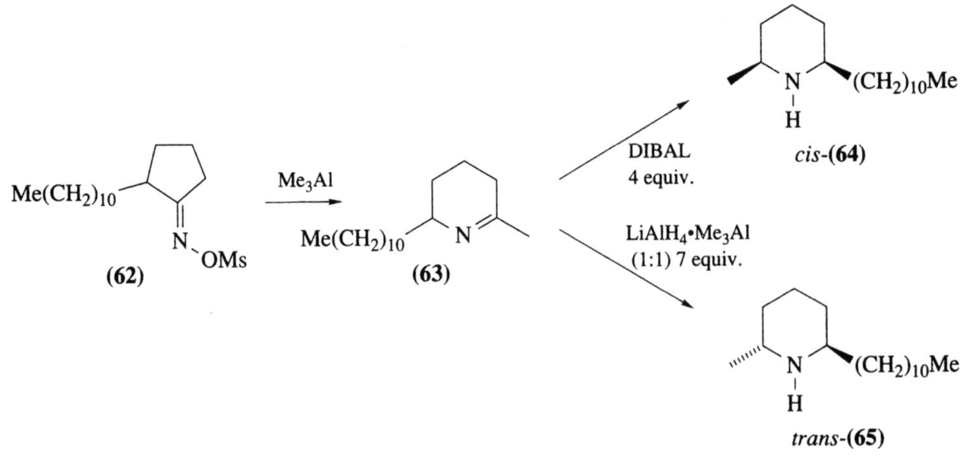

(28)

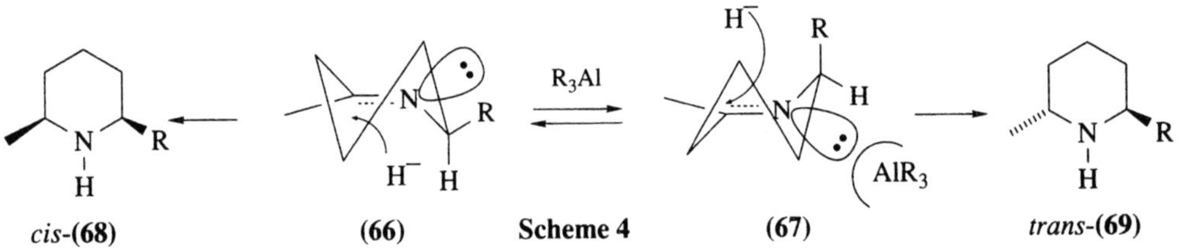

Scheme 4

1.3.5 APPLICATIONS TO NATURAL PRODUCT SYNTHESIS

1.3.5.1 Application of MAD and MAT to the Synthesis of (±)-3α-Acetoxy-15β-hydroxy-7,16-secotrinervita-7,11-diene

The synthesis of the title compound (**73**), which is the defense substance of the termite soldier,[88] is outlined in Scheme 5.[89]

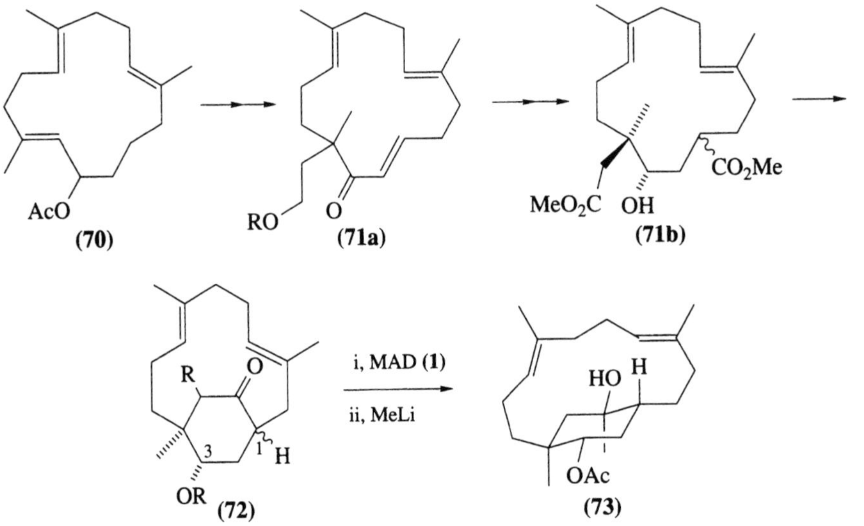

Scheme 5

The conversion of intermediate (**72**) to the natural product (**73**) was accomplished by exposure of a 1:1 complex of (**72**) and the hindered organoaluminum reagent (**1**; MAD)[18] to methyllithium (57% yield). This is a fascinating result, not only because attempted methyllithium addition failed, but also because the required configuration of the C-1 position was β with regard to the proton substituent and, therefore, approach of a nucleophile to the C-2 carbonyl moiety would also be from the less hindered β-face; unfortunately, the natural product required a β-hydroxy substituent, which requires a facial selectivity for the more hindered α-face of the carbonyl. Although the facial selectivity of the organoaluminum (**1**) mediated addition was not total, it did give a 3:1 excess of the desired α-face 1,2-addition product, which was identical to the natural product.

1.3.5.2 Application of Aminoaluminum Reagents to Natural Product Synthesis

1.3.5.2.1 Synthesis of an FK-506 fragment

FK-506 (**74**) is an interesting natural product with potent immunosuppressive properties.[90]

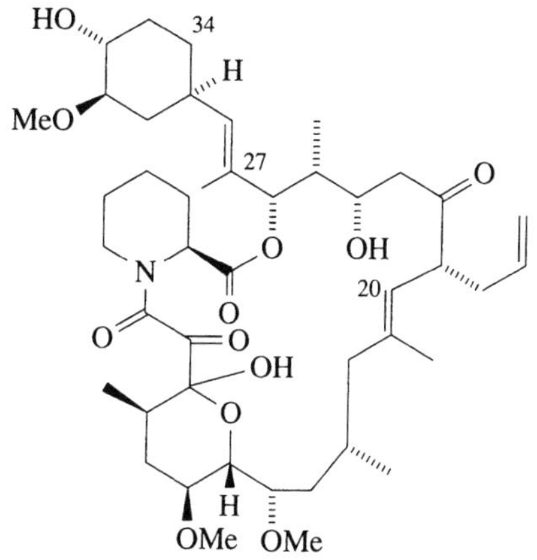

(**74**)

Recently, there have been a number of publications on the synthesis of various segments of (**74**) in the context of total synthesis and Scheme 6 outlines one approach to the synthesis of the C-20 to C-34 fragment.[91] The synthetic approach to the totally blocked C-20 to C-34 fragment relies on an aminolysis *via* aminoaluminum reagent (**82**) in three different transformations of the sequence, demonstrating the excellent selectivity of (**82**) in a very sensitive array of functionality. For example, the conversion of lactone (**77**) to hydroxyamide (**78**) selectively opens the lactone ring with no undesired epimerization; furthermore, (**82**) was also utilized in the aminolysis of the chiral auxiliary with no apparent epimerization.

1.3.5.2.2 Synthesis of granaticin

Granaticin (**83**) is a member of the pyranonaphthoquinone group of antibiotics. The portion of the total synthesis of (**83**) which requires the use of the aminoaluminum reagent (**42**) is outlined in Scheme 7.[92]

The conversion of lactone (**86**) to hydroxyamide (**87**) is complicated by the potential for epimerization, which is quite similar to the problem presented by substrate (**77**) in the previous synthesis. Once again, the transformation proceeds smoothly, with no apparent epimerization.

1.3.5.3 Application of the Nozaki Protocol

The Nozaki protocol,[93,94] which is an organoaluminum–ate mediated 1,4-addition followed by a 1,2-addition of the resulting enolate, may be considered a 1-acylethenyl anion equivalent, which does fall within the scope of this review. This process may be conducted in either an intramolecular sense or an intermolecular sense. In general, the protocol is compatible with a wide array of functionality; however, the intermolecular process requires an aldehyde, whereas the intramolecular case may be terminated by either an aldehyde or a ketone.

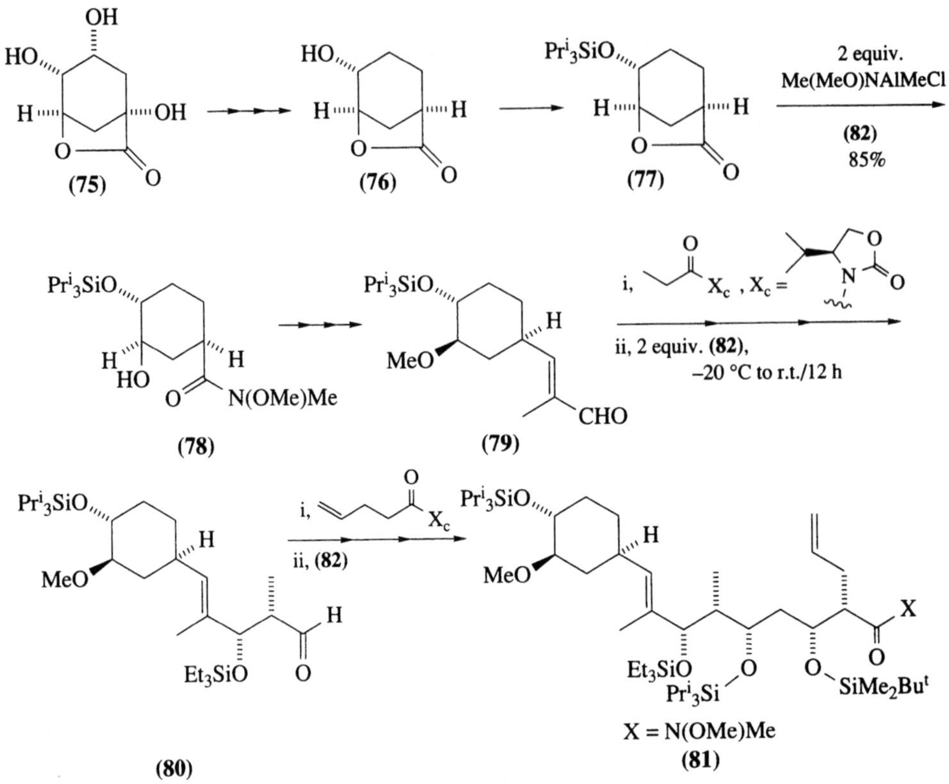

(75) → → → (76) → (77) —2 equiv. Me(MeO)NAlMeCl (82), 85%→ (78) → → → (79)

(79) —i, EtC(O)X_c, X_c = (oxazolidinone); ii, 2 equiv. (82), –20 °C to r.t./12 h→ (80)

(80) —i, $\text{CH}_2\text{=CHCH}_2\text{CH}_2\text{C(O)X}_c$; ii, (82)→ (81)

$X = N(OMe)Me$

Scheme 6

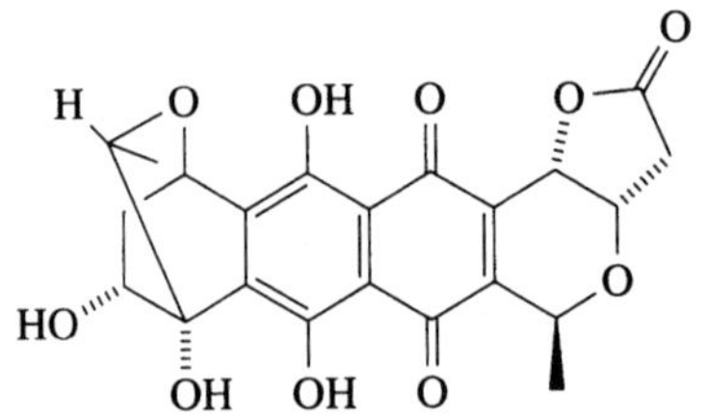

(83)

(84) → → (85) → (86) —i, 1.3 equiv. Me_2AlNMe_2, 0 °C to r.t./1.5 h; ii, [O]→ (87) → (88) 8% → → (83)

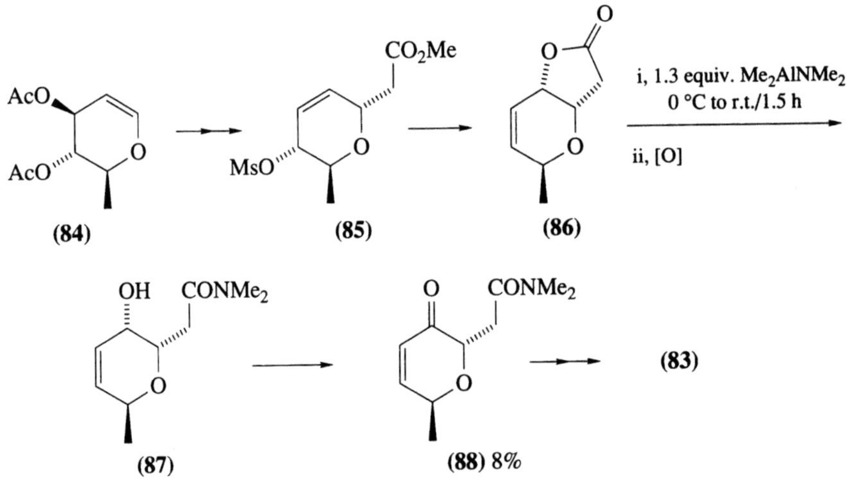

Scheme 7

1.3.5.3.1 Intramolecular protocol — application to avermectins

An example of an intramolecular Nozaki process was recently described in the total synthesis of an avermectin aglycon (**89**).[95,96] The relevant steps in the sequence are outlined in Scheme 8. The ate complex of lithium thiophenoxide and trimethylaluminum permits the addition of thiophenoxide exclusively in a 1,4-fashion to the α,β-unsaturated aldehyde (**93**), affording an aluminum enolate that is intramolecularly trapped by the electrophilic ketone moiety. The resulting hydroxy sulfide is oxidized to the corresponding hydroxy sulfoxide, which upon exposure to heat (refluxing toluene) affords the cyclized system (**95**) in 76% yield. The addition of the ate complex (**94**) to (**93**) is not only highly regioselective, but also, unlike the typical intermolecular Nozaki process, which has been reported to proceed only when the terminating carbonyl electrophile is an aldehyde,[93,94] this example is terminated by a ketone. Presumably entropic factors are driving the ring closure to the ketone in this intramolecular process.

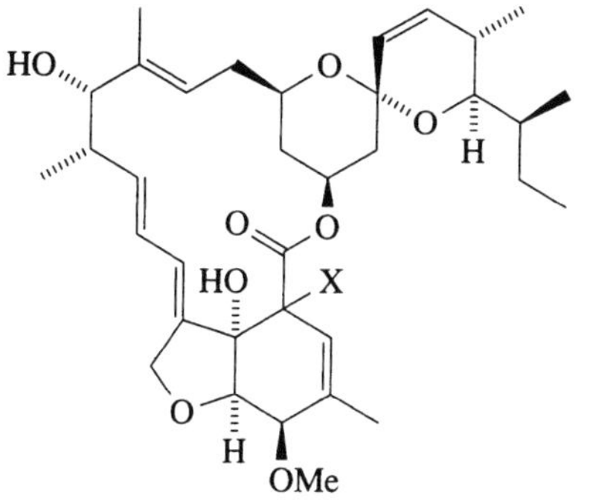

(**89**) X = β-H

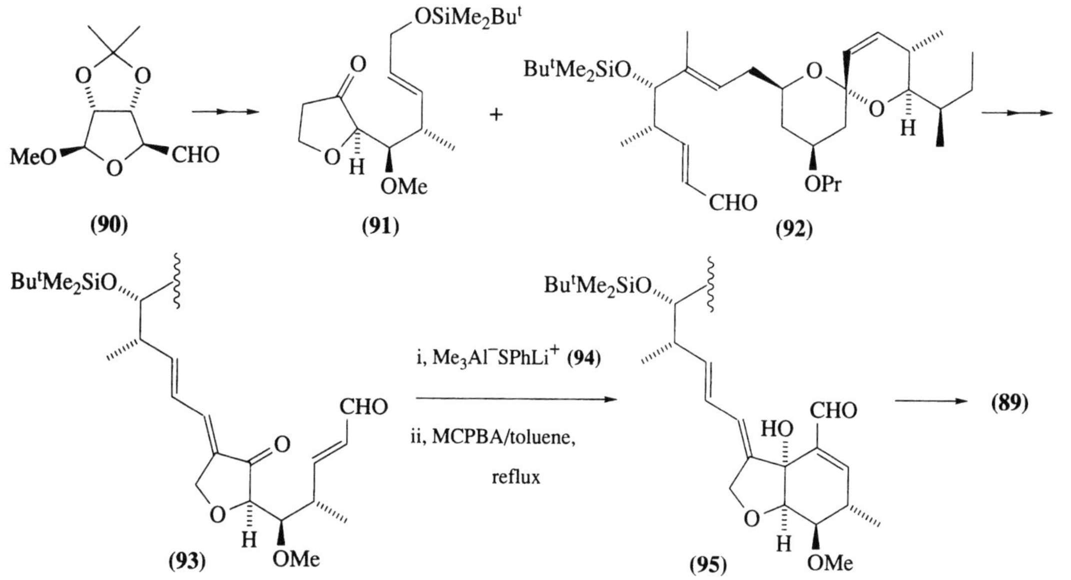

(**90**) (**91**) (**92**)

(**93**) i, $Me_3Al^-SPhLi^+$ (**94**); ii, MCPBA/toluene, reflux → (**95**) → (**89**)

Scheme 8

1.3.5.3.2 Intermolecular protocol — application to prostanoids

Scheme 9 outlines the synthesis of a prostanoid[97] intermediate (**99**) that relies on an intermolecular Nozaki process. It is important to note that unlike the intramolecular case described above, the intermolecular version of this protocol requires an aldehyde as the electrophilic trap; however, it is interesting to note that there have been no reports of the addition of Lewis acid activated ketones (presumably, as a preformed complex which would be added *via* cannula at low temperature) to the preformed aluminum enolate. Finally, in this example, the conversion of enone (**96**) to adduct (**98**) is promoted by the less reactive dimethylaluminum phenyl thiolate and not the corresponding ate complex.

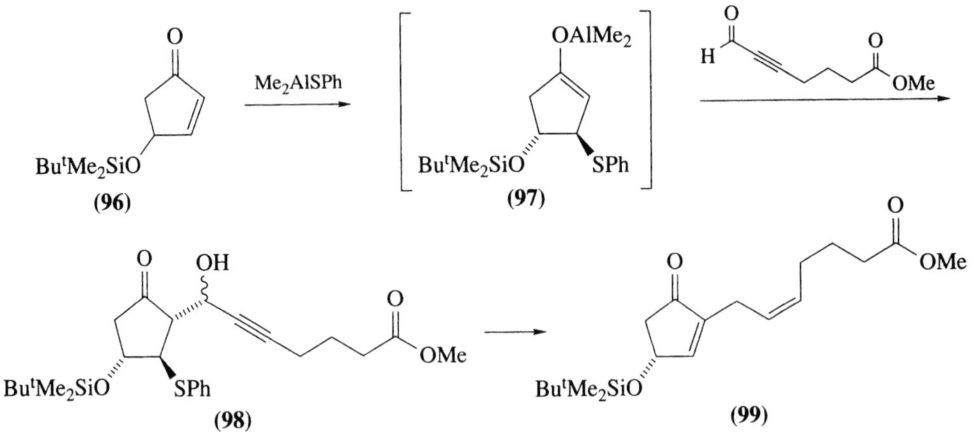

Scheme 9

1.3.5.4 Synthesis of Nonracemic Sydowic Acid

Sydowic acid (**100**) was prepared as outlined in Scheme 10.[30] The crucial transformation of ketone (**101**) to carbinol (**102**) was accomplished by stereoselective 1,2-addition of trimethylaluminum, which afforded superior facial selectivity for the *re*-face compared to Grignard reagents. In contrast, methyltrichlorotitanium addition resulted in a *si*-face stereoselectivity.

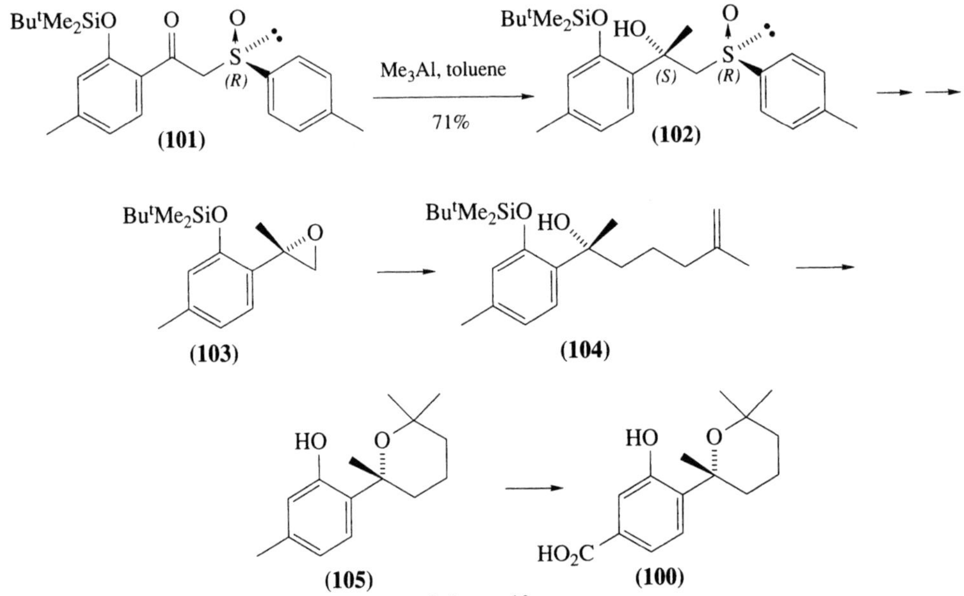

Scheme 10

1.3.5.5 Synthesis of Racemic Gephyrotoxin-223AB

Gephyrotoxin (**106**), a constituent of the skin extracts of the poison-dart frog, was synthesized as outlined in Scheme 11.[98] The organoaluminum-promoted Beckmann rearrangement produced an imine, which was stereoselectively reduced (essentially total selectivity) with DIBAL to afford piperidine (**109**; 41% yield).

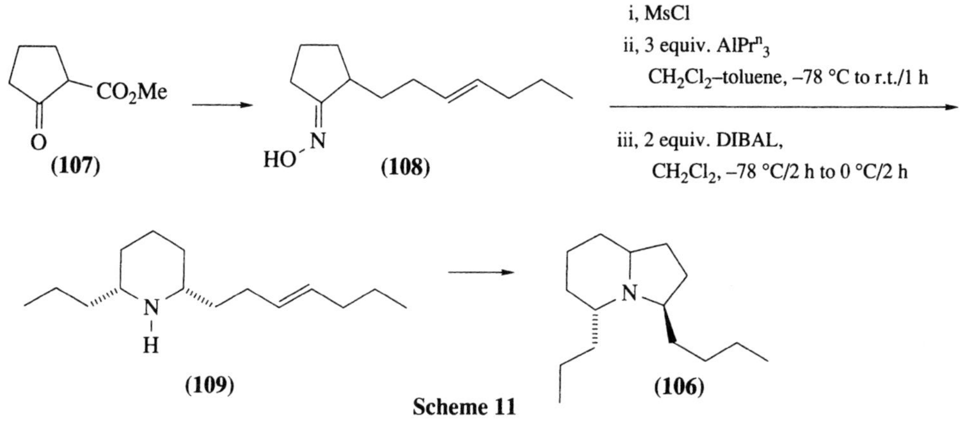

CO_2Me (107) → HO–N (108)

i, MsCl
ii, 3 equiv. $AlPr^n_3$
CH_2Cl_2–toluene, –78 °C to r.t./1 h
iii, 2 equiv. DIBAL,
CH_2Cl_2, –78 °C/2 h to 0 °C/2 h

(109) → (106)

Scheme 11

1.3.6 REFERENCES

1. T. Mole and E. Jeffery, 'Organoaluminum Compounds', Elsevier, Amsterdam, 1972.
2. E. Negishi, *J. Organomet. Chem. Libr.*, 1976, **1**, 93.
3. H. Yamamoto and H. Nozaki, *Angew. Chem., Int. Ed. Engl.*, 1978, **17**, 169.
4. E. Negishi, 'Organometallics in Organic Synthesis', Wiley, New York, vol. 1, 1980.
5. J. F. Normant and A. Alexakis, *Synthesis*, 1981, 841.
6. G. Zweifel and J. Miller, *Org. React. (N.Y.)*, 1984, **32**, 375.
7. Y. Yamamoto, *Acc. Chem. Res.*, 1987, **20**, 243.
8. R. W. Hoffmann, *Angew. Chem., Int. Ed. Engl.*, 1982, **21**, 555.
9. Y. Yamamoto and K. Maruyama, *Heterocycles*, 1982, **10**, 357.
10. K. Maruoka and H. Yamamoto, *Angew. Chem., Int. Ed. Engl.*, 1985, **24**, 668.
11. M. Nogradi, 'Stereoselective Synthesis', VCH, Weinheim, 1987.
12. H. Yamamoto, K. Maruoka and K. Furuta, in 'Stereochemistry of Organic and Bioorganic Transformations', ed. W. Bartmann and K. B. Sharpless, VCH, Weinheim, 1987.
13. K. Maruoka and H. Yamamoto, *Tetrahedron*, 1988, **44**, 5001.
14. E. C. Ashby and J. T. Laemmle, *Chem. Rev.*, 1975, **75**, 521.
15. E. C. Ashby and J. T. Laemmle, *J. Org. Chem.*, 1975, **40**, 1469.
16. J. T. Laemmle, E. C. Ashby and P. V. Roling, *J. Org. Chem.*, 1973, **38**, 2526.
17. H. M. Neumann, J. T. Laemmle and E. C. Ashby, *J. Am. Chem. Soc.*, 1973, **95**, 2597.
18. K. Maruoka, T. Itoh and H. Yamamoto, *J. Am. Chem. Soc.*, 1985, **107**, 4573.
19. T. Mole and J. R. Surtees, *Aust. J. Chem.*, 1964, **17**, 961.
20. An X-ray structure for the case of chiral Zn intermediates does exist. See, for example, R. Noyori, S. Suga, K. Kawai, S. Okada and M. Kitamura, *Pure Appl. Chem.*, 1988, **60**, 1597.
21. K. Maruoka, T. Itoh, M. Sakurai, K. Nonoshita and H. Yamamoto, *J. Am. Chem. Soc.*, 1988, **110**, 3588.
22. M. Cherest, H. Felkin and N. Prudent, *Tetrahedron Lett.*, 1968, 2199.
23. Y.-D. Wu and K. N. Houk, *J. Am. Chem. Soc.*, 1987, **109**, 908.
24. G. Altnau and L. Rosch, *Tetrahedron Lett.*, 1983, **24**, 45.
25. G. H. Posner, 'An Introduction to Synthesis Using Organocopper Reagents', Wiley-Interscience, New York, 1980.
26. M. T. Reetz, J. Westermann, R. Steinbach, B. Wenderoth, R. Peter, R. Ostarek and S. Maus, *Chem. Ber.*, 1985, **118**, 1421.
27. K. Maruoka, Y. Araki and H. Yamamoto, *Tetrahedron Lett.*, 1988, **29**, 3101.
28. T. Tsuda, T. Yoshida, T. Kawamoto and T. Saegusa, *J. Org. Chem.*, 1987, **52**, 1624.
29. T. Tsuda, T. Yoshida and T. Saegusa, *J. Org. Chem.*, 1988, **53**, 1037.
30. T. Fujisawa, A. Fujimura and Y. Ukaji, *Chem. Lett.*, 1988, 1541.
31. G. H. Posner, in 'Asymmetric Synthesis', ed. J. D. Morrison, Academic Press, New York, 1983, vol. 2, chap. 8.
32. G. Solladié, in 'Asymmetric Synthesis', ed. J. D. Morrison, Academic Press, New York, 1983, vol. 2, p. 179.
33. G. Boireau, D. Abenhaim and E. Henry-Basch, *Tetrahedron*, 1980, **36**, 3061.
34. K. Soai, A. Ookawa, K. Ogawa and T. Kaba, *J. Chem. Soc., Chem. Commun.*, 1987, 467.
35. K. Soai, S. Yokoyama, K. Ebihara and T. Hayasaka, *J. Chem. Soc., Chem. Commun.*, 1987, 1690.
36. K. Soai, A. Ookawa, T. Kaba and K. Ogawa, *J. Am. Chem. Soc.*, 1987, **109**, 7111 and refs. cited therein.
37. D. Abenhaim, G. Boireau and A. Deberly, *J. Org. Chem.*, 1985, **50**, 4045.
38. G. Boireau, A. Korenova, A. Deberly and D. Abenhaim, *Tetrahedron Lett.*, 1985, **26**, 4181.
39. A. Deberly, G. Boireau and D. Abenhaim, *Tetrahedron Lett.*, 1984, **25**, 655.
40. G. Boireau, D. Abenhaim, A. Deberly and B. Sabourault, *Tetrahedron Lett.*, 1982, **23**, 1259.
41. D. Vegh, G. Boireau and E. Henry-Basch, *J. Organomet. Chem.*, 1984, **267**, 127.
42. F. A. Davis, M. S. Haque, T. G. Ulatowski and J. C. Towson, *J. Org. Chem.*, 1986, **51**, 2402.
43. F. A. Davis and M. S. Haque, *J. Org. Chem.*, 1986, **51**, 4083.
44. F. A. Davis, T. G. Ulatowski and M. S. Haque, *J. Org. Chem.*, 1987, **52**, 5288.

45. J. D. Morrison and H. S. Mosher, 'Asymmetric Organic Reactions', Prentice-Hall, Englewood Cliffs, NJ, 1971.
46. F. Sato, Y. Mori and M. Sato, *Chem. Lett.*, 1978, 1337.
47. F. Sato, H. Kodama, Y. Tomuro and M. Sato, *Chem. Lett.*, 1979, 623.
48. E. J. Corey and H. Ensley, *J. Am. Chem. Soc.*, 1975, **97**, 6908.
49. W. Oppolzer, C. Robbiani and K. Battig, *Helv. Chim. Acta*, 1980, **63**, 2015.
50. Whitesell has observed excellent diastereoselectivities in related systems. See, for example, J. K. Whitesell, D. Deyo and A. Bhattasharya, *J. Chem. Soc., Chem. Commun.*, 1983, 802; *J. Org. Chem.*, 1986, **51**, 5443.
51. H. Yamamoto and K. Maruoka, *J. Am. Chem. Soc.*, 1981, **103**, 4186.
52. A. Mori, J. Fujiwara, K. Maruoka and H. Yamamoto, *J. Organomet. Chem.*, 1985, **285**, 83.
53. K. Weinhardt, *Tetrahedron Lett.*, 1984, **25**, 1761.
54. J. Fujiwara, Y. Fukutani, M. Hasegawa, K. Maruoka and H. Yamamoto, *J. Am. Chem. Soc.*, 1984, **106**, 5004.
55. K. Maruoka, S. Nakai, M. Sakurai and H. Yamamoto, *Synthesis*, 1986, 130.
56. S. Chatterjee and E. Negishi, *J. Org. Chem.*, 1985, **50**, 3406.
57. Y. Honda, E. Morita and G. Tsuchihashi, *Chem. Lett.*, 1986, 277.
58. Y. Honda, M. Sakai and G. Tsuchihashi, *Chem. Lett.*, 1985, 1153.
59. S. Takano, T. Ohkawa and K. Ogasawara, *Tetrahedron Lett.*, 1988, **29**, 1823.
60. K. Maruoka, K. Nonoshita, H. Banno and H. Yamamoto, *J. Am. Chem. Soc.*, 1988, **110**, 7922.
61. K. Maruoka, H. Banno, K. Nonoshita and H. Yamamoto, *Tetrahedron Lett.*, 1989, **30**, 1265.
62. K. Takai, I. Mori, K. Oshima and H. Nozaki, *Bull. Chem. Soc. Jpn.*, 1984, **57**, 446.
63. I. Mori, K. Takai, K. Oshima and H. Nozaki, *Tetrahedron*, 1984, **40**, 4013.
64. R. Vance, N. G. Rondan, K. N. Houk, F. Jensen, W. T. Borden, A. Komornicki and E. Wimmer, *J. Am. Chem. Soc.*, 1988, **110**, 2314.
65. R. Armstrong, F. Harris and L. Weiler, *Can. J. Chem.*, 1982, **60**, 673.
66. K. Takai, M. Sato, K. Oshima and H. Nozaki, *Bull. Chem. Soc. Jpn.*, 1984, **57**, 108.
67. S. J. Danishefsky and N. Mantlo, *J. Am. Chem. Soc.*, 1988, **110**, 8129.
68. A. Basha, M. Lipton and S. M. Weinreb, *Tetrahedron Lett.*, 1977, 4171.
69. J. Levin, E. Turos and S. M. Weinreb, *Synth. Commun.*, 1982, **12**, 989.
70. A. Benderly and S. Stavchansky, *Tetrahedron Lett.*, 1988, **29**, 739.
71. R. G. Pearson, *J. Am. Chem. Soc.*, 1963, **85**, 3533.
72. S. Masamune, Y. Hayase, W. Schilling, W. Chan and G. Bates, *J. Am. Chem. Soc.*, 1977, **99**, 6756.
73. A. Kozikowski and A. Ames, *Tetrahedron*, 1985, **41**, 4821.
74. A. Kozikowski and A. Ames, *J. Org. Chem.*, 1978, **43**, 2735.
75. A. Sviridov, M. Ermolenko, D. Yashunsky and N. Kochetkov, *Tetrahedron Lett.*, 1983, **24**, 4355.
76. A. Sviridov, M. Ermolenko, D. Yashunsky and N. Kochetkov, *Tetrahedron Lett.*, 1983, **24**, 4359.
77. T. Mole and E. A. Jeffery, 'Organoaluminum Compounds', Elsevier Amsterdam, 1972, p. 311.
78. K. Wakamatsu, Y. Okuda, K. Oshima and H. Nozaki, *Bull. Chem. Soc. Jpn.*, 1985, **58**, 2425.
79. K. Takai, K. Oshima and H. Nozaki, *Bull. Chem. Soc. Jpn.*, 1981, **54**, 1281.
80. E. Negishi, V. Bagheri, S. Chatterjee, F.-T Luo, J. Miller and A. Stoll, *Tetrahedron Lett.*, 1983, **24**, 5181.
81. J. Kang, J. Lee, K. Kim, J. Jeong and C. Pyun, *Tetrahedron Lett.*, 1987, **28**, 3261.
82. L. Rösch and G. Altnau, *J. Organomet. Chem.*, 1980, **195**, 47.
83. W. H. Pirkle and D. J. Hoover, *J. Org. Chem.*, 1980, **45**, 3407.
84. K. Maruoka, T. Miyazaki, M. Ando, Y. Matsumura, S. Sakane, K. Hattori and H. Yamamoto, *J. Am. Chem. Soc.*, 1983, **105**, 2831.
85. Y. Matsumura, K. Maruoka and H. Yamamoto, *Tetrahedron Lett.*, 1982, **23**, 1929.
86. A. S. Cieplak, *J. Am. Chem. Soc.*, 1981, **103**, 4540.
87. A. Narula, *Tetrahedron Lett.*, 1981, **22**, 2017.
88. J. Braekman, D. Daloze, A. Dupont, J. Pasteels and B. Tursch, *Tetrahedron Lett.*, 1980, **21**, 2761.
89. T. Kato, T. Hirukawa, T. Uyehara and Y. Yamamoto, *Tetrahedron Lett.*, 1987, **28**, 1439.
90. H. Tanaka, A. Kuroda, H. Marusawa, H. Hatanaka, T. Kino, T. Goto and M. Hashimoto, *J. Am. Chem. Soc.*, 1987, **109**, 5031.
91. S. Mills, *et al.*, *Tetrahedron Lett.*, 1988, **29**, 281.
92. K. Okazaki, K. Nomura and E. Yoshii, *J. Chem. Soc., Chem. Commun.*, 1989, 354.
93. A. Itoh, S. Ozawa, K. Oshima and H. Nozaki, *Bull. Chem. Soc. Jpn.*, 1981, **54**, 274.
94. A. Itoh, S. Itoh, S. Ozawa, K. Oshima and H. Nozaki, *Tetrahedron Lett.*, 1980, **21**, 361.
95. S. J. Danishefsky, D. M. Armistead, F. J. Wincott, H. G. Selnick and R. Hungate, *J. Am. Chem. Soc.*, 1987, **109**, 8117.
96. D. M. Armistead and S. J. Danishefsky, *Tetrahedron Lett.*, 1987, **28**, 4959.
97. J. Levin, *Tetrahedron Lett.*, 1989, **30**, 13.
98. C. Broka and K. Eng, *J. Org. Chem.*, 1986, **51**, 5043.

1.4
Organocopper Reagents

BRUCE H. LIPSHUTZ
University of California, Santa Barbara, CA, USA

1.4.1 INTRODUCTION

The direct addition of an organocopper or cuprate reagent to a carbon–heteroatom multiple bond might rightfully be considered the forgotten son of transition metal based carbon–carbon bond formation. Indeed, although organocopper reagents are potent Michael donors, their well-recognized hesitation towards competing 1,2-addition may represent their most salient feature. Still, in most circumstances, while many substitution reactions[1] (*e.g.* with a primary iodide) and especially 1,4-additions[2] (*e.g.* with α,β-unsaturated ketones) tend to be far more rapid processes, the appropriately designed substrate can often benefit considerably from a copper reagent mediated 1,2-addition, not only in terms of yield but especially with regard to diastereoselectivity.

Discussed in this chapter, for the most part, are documented cases where complexes derived from (one or more equivalents of) a Grignard or organolithium reagent, in combination with a copper(I) salt, have been used to add to an aldehyde, ketone, imine, amide or nitrile moiety. In the majority of examples, the key issue is one of stereocontrol. Hence, where available, data within the organocopper manifold *versus* those for other organometallic reagents are provided for comparison.

Several different types of copper reagents have been utilized for the 1,2-additions presented herein. Those formed from Grignard reagents (Scheme 1) may be of the type represented as (**1**) or (**2**),[3] depending upon whether ≤0.5 or 1.0 equiv. of CuX (X = I, Br) is involved. In the presence of CuCN, both 'lower order' monoanionic species (**3**)[4] and 'higher order' dianionic salts (**4**)[5] are possible. With the latter class, mixed metal clusters result from treatment of this copper(I) source with one equivalent of both RMgX and R'Li.[6] Reagents derived from organolithium precursors (Scheme 2) include mainly the Gilman cuprates R_2CuLi (**5**),[1,2,7] organocopper species (**6**) akin to (**2**) but containing lithium salts as byproducts of metathesis between the metals,[1,2] and the 'higher order' dilithiocyanocuprates (**7**).[5] Within this group, *i.e.* (**1**) to (**7**), each is distinct from the others, not only insofar as stoichiometric representations are concerned, but in the reactivity profiles displayed, as well as the regio- and stereo-chemical outcomes of their 1,2-additions (*vide infra*).

$$RMgX + \leq 0.5\ CuX \longrightarrow R_2CuMgX\ (\mathbf{1}) \equiv \text{a 'lower order' magnesium cuprate}$$

$$RMgX + CuX \longrightarrow RCu{\bullet}MgX_2\ (\mathbf{2}) \equiv \text{a Grignard-derived organocopper reagent}$$

$$RMgX + CuCN \longrightarrow RCu(CN)MgX\ (\mathbf{3}) \equiv \text{a 'lower order' Grignard-derived cyanocuprate}$$

$$RMgX + R'Li + CuCN \longrightarrow RR'Cu(CN)LiMgX\ (\mathbf{4}) \equiv \text{a 'higher order' mixed metal cyanocuprate}$$

Scheme 1

$$2\ RLi + CuX \longrightarrow R_2CuLi\ (+ LiX)\ (\mathbf{5}) \equiv \text{a 'lower order' or Gilman cuprate}$$

$$RLi + CuX \longrightarrow RCu{\bullet}LiX\ (\mathbf{6}) \equiv \text{an organolithium-derived organocopper reagent}$$

$$2\ RLi + CuCN \longrightarrow R_2Cu(CN)Li_2\ (\mathbf{7}) \equiv \text{a 'higher order' dilithium cyanocuprate}$$

Scheme 2

1.4.2 1,2-ADDITIONS TO ALDEHYDES AND KETONES

1.4.2.1 Reactions of Aldehydes

The initial observation that organocuprates react in a synthetically useful fashion with aldehydes was made by Posner in 1972 as part of a study aimed at determining functional group compatibility with R_2CuLi versus temperature.[8] Both benzaldehyde and *n*-heptanal were found to react at –90 °C in less than 10 min with lithium dimethylcuprate to afford the corresponding alcohols in good yields (equation 1).

$$RCHO + R'_2CuLi \xrightarrow[\text{-90 °C, 10 min; 80–85\%}]{Et_2O} RCH(OH)R' \quad (1)$$

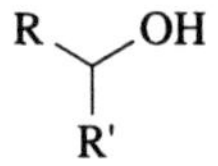

$R = Ph, n\text{-}C_6H_{13}; R' = Me, Bu^n$

The facility and efficiency with which copper reagents add to aldehydes has been examined where unsymmetrical α- or β-substitution exists, thereby raising the question of diastereoselection. Two types of situation exist in this regard: (i) where opportunities for chelation-controlled attack by the reagent can lead to a significant preference for one diastereomer; and (ii) the nonchelation-controlled delivery of an organic ligand from copper, where steric and perhaps other factors influence the directionality of addition. The former scenario was first noted by Still and Schneider,[9] concurrent with studies[10] in this group on Grignard reactions with α-alkoxy ketones. Since ketones react relatively sluggishly toward cuprates (*vide infra*), and 1,2-additions to α-alkoxy aldehydes give unimpressive results, a variety of β-alkoxy aldehydes (**8**) have been examined. For these systems, cuprates Me_2CuLi and $Bu^n{}_2CuLi$ tend to afford good ratios in favor of *anti* isomers, irrespective of the protecting group on oxygen (equation 2).[9] Organocopper species derived from vinyllithium, *i.e.* (vinyl)Cu·PBu_3 and $(vinyl)_2CuLi$, act in a similar manner, although the ratios are diminished (8:1 and 3:1, respectively) and the aldehyde is only partially consumed (50% conversion) under the standard reaction conditions in the dialkylcuprate cases. By comparison, 1,2-additions employing either the corresponding organolithium or Grignard reagents are essentially stereorandom.[9]

The initially discouraging outcome with α-alkoxy aldehydes[9] has been reexamined in detail by Mead and Macdonald.[11] Both acyclic and cyclic α,β-dialkoxy aldehydes (**9**) and (**10**) served as substrates, and reaction variables including reagents, solvents and temperatures were all considered. Excellent selectivities of the order of 94:6 to 98:2 in favor of the *syn* products (**12a**) were obtained from (**9**) using

(8) $\xrightarrow[\text{Et}_2\text{O, -78 °C}]{\text{R'}_2\text{CuLi}}$ RO–…–R' (OH) + RO–…–R' (OH) (2)

R = $PhCH_2OCH_2$	R' = Me	92%	30:1
	Bu^n	82%	17:1
R = THP, $PhCH_2$	R' = Me	90%	>20:1

$RCu·MgBr_2$ (prepared from RMgBr plus $CuBr·Me_2S$) in Et_2O at –78 °C. The α-chelate model (**11**) appears to account for the observed course of addition,[12] without interference from β-chelation or Felkin-predicted[13] modes of attack (equation 3).[14] By contrast, (**10**) did not show the same levels of stereoselectivity under identical conditions, and in fact for the one case studied ($MeCu·MgBr_2$) in this report[11] a 2:1 ratio was obtained favoring the *anti* isomer. The corresponding allylcopper·$MgBr_2$ reagent is also not efficient in terms of product ratio (68:32 *syn:anti*) with (**9**),[11] nor was it of value in additions to a trialkoxy example (**13**), where the *anti* material (**14b**) actually dominated, albeit slightly (equation 4).[15] Lithium cation containing Gilman cuprates appear to be unpredictable,[11] as reaction of Me_2CuLi with (**9**) was nonselective (47:53 *syn:anti*), although with (**10**) an improved 82:18 ratio was obtained. Subsequent use of an $MgBr_2$-modified lower order reagent on α-alkoxy aldehyde (**15**), however, did show a preference for *syn* selectivity of 10:1, (**16a**):(**16b**) (equation 5).[16] Reversal of selectivity to give high percentages of *anti* product (**12b**) can be effected using an excess (2 equiv.) of an organotitanium reagent $RTi(OPr^i)_3$ in THF at *ca.* –40 °C.[11]

(9) (10)

(9) $\xrightarrow[\text{Et}_2\text{O, Me}_2\text{S, -78 °C}]{\text{RCu•MgBr}_2}$ [(11)] ⟶ (12a) + *anti* isomer (12b) (3)

R = Me	83%	94:6
Bu^n	78%	93:7
vinyl	80%	98:2
Ph	69%	96:4

(13) $\xrightarrow[\text{Et}_2\text{O, -78 °C, 82\%}]{\text{allyl-Cu•MgBr}_2}$ (14a) + (14b) 40:60 (4)

(15) + $(\text{PhCH=CH})_2\text{CuLi}$ $\xrightarrow[\text{Et}_2\text{O, -78 °C}]{\text{MgBr}_2\text{•OEt}_2}$ (16a) + (16b) 10:1 (5)

Further work by Sato on couplings with 2,3-*O*-isopropylideneglyceraldehyde (**10**) using the same class of reagents $RCu{\cdot}MgX_2$ has established the critical role of solvent in determining the diastereomeric outcome.[17] Thus, performing the reaction in THF rather than Et_2O leads to highly selective 1,2-additions producing the *syn* products in high yields. Alkylcopper reagents give ratios ranging from 10:1 to 16:1 for (**17a**):(**17b**) (equation 6), while aryl and vinyl organocopper complexes afford virtually all of the *syn* diastereomer.[17] A cyclic mechanism[12] (*cf.* **18**) which suggests attack by RCu from the direction shown is consistent with the observed data (equation 7).[17] Whether the switch from Et_2O to THF is affecting the ease with which chelate (**18**) can form, or whether the impact is related to changes in the state of the reagent (*e.g.* aggregation), or both, is not clear at this time.

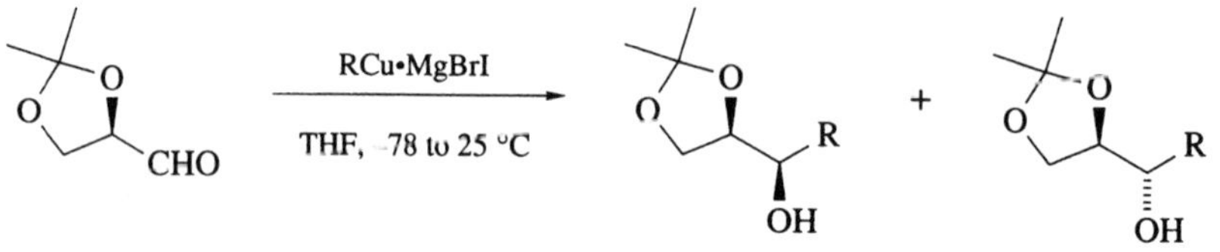

(**10**) —RCu•MgBrI, THF, –78 to 25 °C→ (**17a**) + (**17b**) (6)

R	Yield	(17a):(17b)
R = Bu^n, *n*-pentyl, n-$C_{10}H_{21}$, c-C_6H_{11}	74–89%	10:1 to 16:1
R = Ph, *p*-ClC_6H_4, *p*-$NO_2C_6H_4$	80–93%	>99:1
R = $CH_2{=}C(SiMe_3)$	86%	>98:2

(**10**) —RCu•MgBrI→ (**18**) —RCu→ (**17a**) (7)

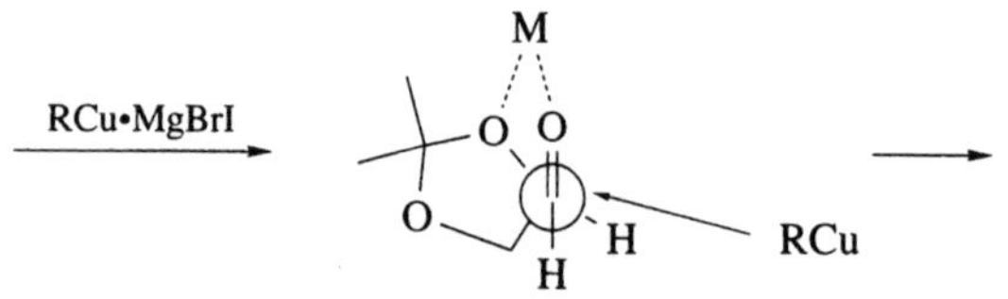

(**18**) M = MgBrI or CuR

Nonchelation-based 1,2-additions to this same aldehyde, thereby leading to high percentages of the *anti* product (**17b**), have been achieved[18] using higher order cyanocuprate technology.[5] Exposure of CuCN to an equivalent of Grignard reagent (**19a**) and methyllithium forms (stoichiometrically) the mixed metal cuprate (**20**),[6] which upon introduction of (**10**) at –78 °C gave (**21a**) and (**22a**) with a diastereomer excess (*de*) of 90% (75% chemical yield). Likewise, the corresponding reaction with (**19b**) afforded a 73% yield of (**21b**) and (**22b**), with a 96% *de* (Scheme 3). The authors point out that the lower order cuprate analog (**23**) showed considerably reduced stereoselectivity in its reaction with (**10**). Thus, by simply choosing the appropriate organocopper or organocuprate reagent, excellent *syn* or *anti* diastereoselectivity can be realized. The judicious choice of functional group manipulations in (**21a**) and (**22a**) ultimately allows for the conversion of a single starting material (*i.e.* **10**) to carbohydrate derivatives (**24**), (**25**) and (**26**).[18]

Identical treatment of aldehyde (**27**), derived from (**10**) *via* a three-step sequence (proteodesilylation, protection and ozonolysis), once again successfully produces *anti* compound (**28**), this time to the exclusion of the *syn* diastereomer (**29**) (Scheme 4).[19] Pivotal was the role of the hydroxy-protecting group, as use of benzyl in place of *t*-butyldimethylsilyl drastically lowered the selectivity. A Felkin–Anh[13,20] model is proposed to explain the stereochemical results, relying on the bulkiness of the trialkylsilyloxy moiety to assist in strongly maintaining the conformation shown in (**30**). Unfortunately, with the epimer of (**27**), compound (**31**), only a 2:1 selectivity was found (in 75% yield),[19] implying that the stereochemistry of the starting aldehyde is important in determining diastereomeric excesses from these couplings, a finding which corroborates an earlier assessment.[9]

1,2-Additions to aldehydes possessing methoxycarbonyl groups positioned three atoms removed can be an especially attractive route to 4,5-*trans*-disubstituted-γ-butyrolactones (**32**) following acid-catalyzed cyclization.[21] The two possible isomers are formed in a >95:5 ratio using Me_2CuLi or $Bu^n{}_2CuLi$, in yields between 53–82% (equation 8). The authors point to the Felkin–Anh[13,20] model (**33**), where this conformation assumes R > CH_2CO_2Me > H. Since this would not apply to the case of R = Me, a stereoelectronic effect due to the ester group may also be involved. Alternatively, the seven-membered chelate (**34**), invoked in their titanium(IV)-induced 1,2-additions,[21] may also be operative.

For aldehydes which lack α-heteroatoms and hence have no avenue for stereocontrol *via* chelation, enhanced Cram selectivity beyond that normally seen with lower order cuprates (*i.e.* 3:1[22] to 7:1[23]) can be

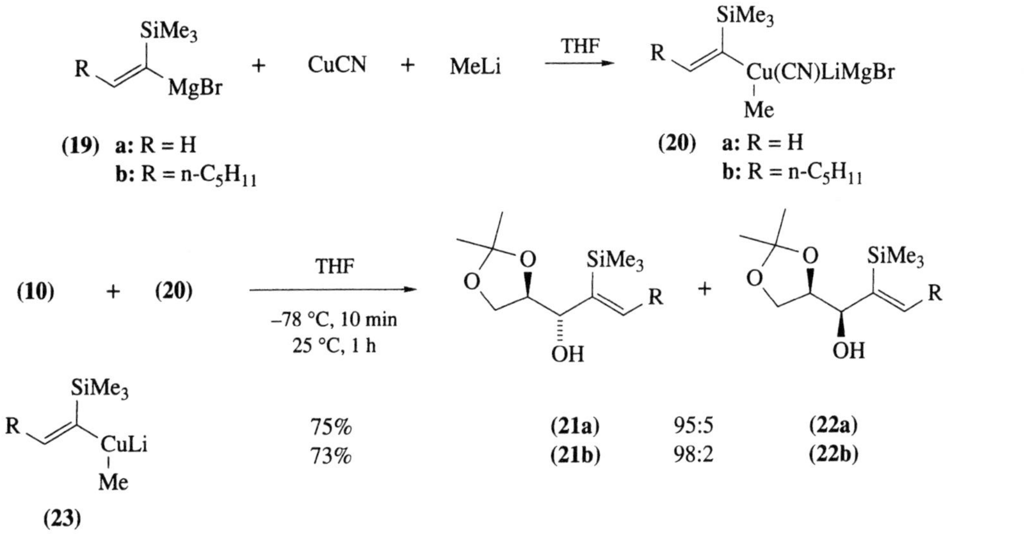

$SiMe_3$ R MgBr + CuCN + MeLi —THF→ R $SiMe_3$ Cu(CN)LiMgBr Me

(19) **a:** R = H
b: R = n-C_5H_{11}

(20) **a:** R = H
b: R = n-C_5H_{11}

(10) + **(20)** —THF, –78 °C, 10 min; 25 °C, 1 h→ O O $SiMe_3$ R OH + O O $SiMe_3$ R OH

$SiMe_3$ R CuLi Me

(23)

	Yield		Ratio	
	75%	**(21a)**	95:5	**(22a)**
	73%	**(21b)**	98:2	**(22b)**

Scheme 3

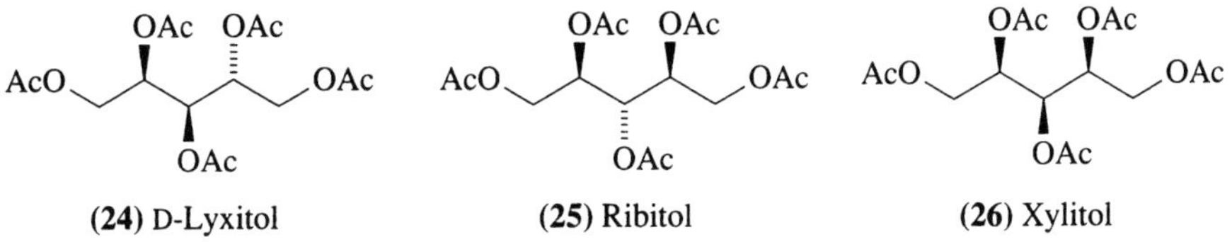

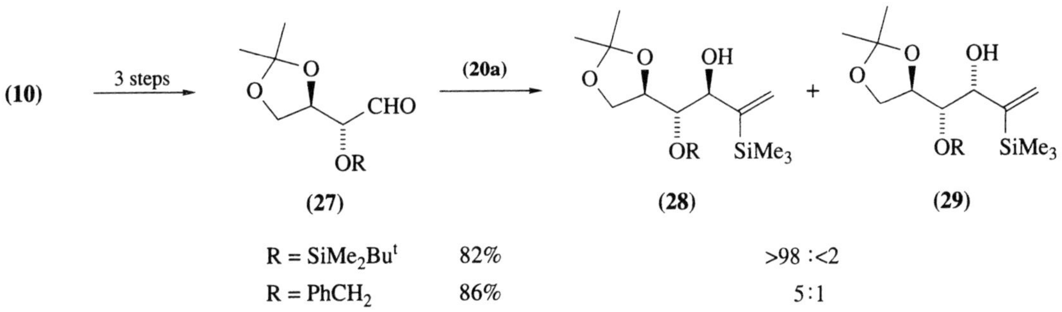

AcO OAc OAc OAc OAc — **(24)** D-Lyxitol

AcO OAc OAc OAc OAc — **(25)** Ribitol

AcO OAc OAc OAc OAc — **(26)** Xylitol

(10) —3 steps→ O O CHO OR **(27)** —**(20a)**→ O O OH OR $SiMe_3$ **(28)** + O O OH OR $SiMe_3$ **(29)**

R	Yield	(28):(29)
R = $SiMe_2Bu^t$	82%	>98 :<2
R = $PhCH_2$	86%	5:1

Scheme 4

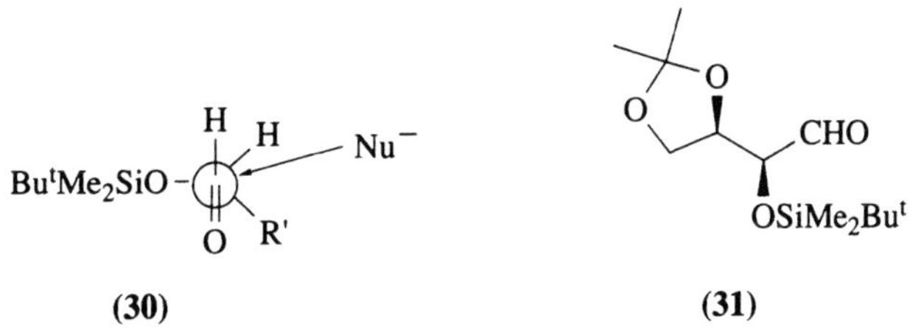

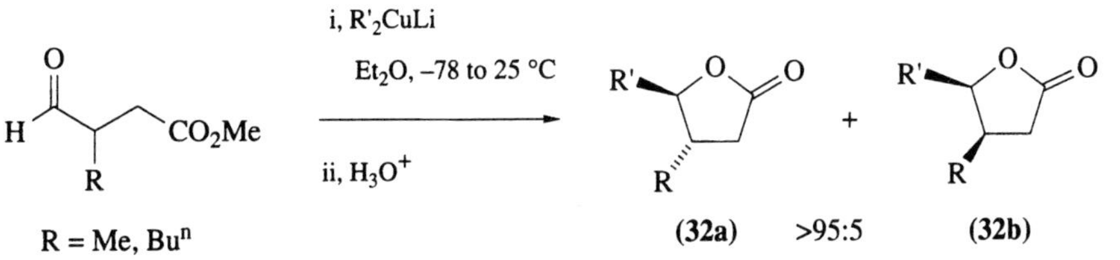

Bu^tMe_2SiO H H Nu^- O R' **(30)**

O O CHO $OSiMe_2Bu^t$ **(31)**

O H CO_2Me R — R = Me, Bu^n —i, R'_2CuLi, Et_2O, –78 to 25 °C; ii, H_3O^+→ R' O O R **(32a)** >95:5 + R' O O R **(32b)** (8)

(33) → (32a) ← (34)

induced using BF_3-modified higher order cuprates.[24] Using aldehyde (**35**) as a model with cuprate (**38**), equal amounts of the diastereomers (**36**) and (**37**) are formed at –78 °C over a 3 h period (equation 9). In the presence of both 15-crown-5 ether and $BF_3 \cdot Et_2O$, however, a quantitative conversion occurs along with an 8:1 to 10:1 ratio of *syn* and *anti* isomers. The crown ether effect could be duplicated using an ether-containing, alkynic nontransferable ligand[25] as part of the mixed cuprate (**39**). In this manner the otherwise room temperature inert reagent (**39**) reacts with $BF_3 \cdot Et_2O$ in the pot at –78 °C to return (**36**) and (**37**) to the extent of 10:1 to 12:1.[24] The importance of the crown ether effect, either internally placed or by external introduction, is clear from the case of reagent (**40**), where a 2-thienyl ligand does not possess the same virtues in this regard. The increase in diastereoselectivity has been attributed to an increase in effective size of the cuprate which bears the BF_3 on the nitrile ligand, as well as to the proximity of the crown ether–lithium complex. Hence, regardless of whether perpendicular or nonperpendicular attack prevails, an increase in *syn* selectivity is expected.

(35) —cuprate (2 equiv.), –78 °C, 3 h→ (36) + (37) (9)

$Bu^n_2Cu(CN)Li_2$ (**38**)	1:1
(**38**) + 4 BF_3 + 15-crown-5	8:1 to 10:1
Bu^n(MeO–C(Me)₂–C≡C)$Cu(CN)Li_2$ (**39**)	no reaction
(**39**) + 4 BF_3	10:1 to 12:1
Bu^n(2-Th)$Cu(CN)Li_2$ (**40**) + 4 BF_3	4:1

Still better ratios of (**36**) to (**37**) have been noted with cuprates (**38**) and (**39**) in the presence of trialkylsilyl halides.[26,27] The presence of Me_3SiCl (2 equiv.) improves the otherwise nonselective reaction of (**38**) from 1:1 to 7:1 *syn:anti*, while solutions of cuprate (**39**) containing excess Me_3SiBr lead to a 19:1 ratio in excellent yield (equation 10). The critical variable in this scheme is the Me_3SiCN produced upon sequestering of the cyano group by the silyl chloride from the higher order cuprate, producing a lower order mixed reagent (**41**; equation 11). Due to the competing reaction between cuprate (**39**) and Me_3SiCl and/or Me_3SiCN (which consumes active cuprate, forming **42**), an excess of (**39**) is needed for high conversions. The best combination, therefore, is one which utilizes the lower order cuprate (**41**; R = Bu^n) together with both Me_3SiCl and Me_3SiCN in a 1:1 ratio (equation 12). Further improvements in diastereoselectivity may be forthcoming by substituting Me_3SiBr for Me_3SiCl. The manner in which these additives act in concert (both are required for maximum yield and diastereoselectivity) to give rise to the Cram selectivity is presently unclear.[26]

(35) —cuprate + additive, THF, – 78 °C→ (36) + (37) (10)

(**38**)	+ 2 Me_3SiCl	45% + SM	7:1
(**39**)	+ 2 Me_3SiCl	88%	10:1 to 11:1
(**39**)	+ 2 Me_3SiBr	97%	17:1 to 19:1

R(MeO)—≡—Cu(CN)Li_2 **(39)** + 2 Me_3SiCl → R(MeO)—≡—CuLi **(41)** + Me_3SiCN + LiCl (major)

→ RCu(CN)Li + MeO—≡—$SiMe_3$ **(42)** + LiCl (minor) (11)

Bu^n(MeO)—≡—CuLi **(41)**

Me_3SiCl + Me_3SiCN, 96–100% → **(36)** + **(37)** 10:1 to 12:1

2 Me_3SiCl, 55–80% → **(36)** + **(37)** 8:1 to 10:1

2 Me_3SiCN, 50% → **(36)** + **(37)** 8.5:1 (12)

Me_3SiCN, 35% → **(36)** + **(37)** 7:1

syn-Products are also formed from crotylcopper additions to benzaldehyde in the presence of $BF_3{\cdot}Et_2O$.[28] Although the observed 98:2 ratio is impressive, the regiochemistry of attack is, unfortunately, relatively nonselective, the α:γ ratio being *ca.* 2:1. Without the Lewis acid in the pot, the *syn:anti* ratio is reduced to essentially 1:1 (equation 13).

PhCHO + crotyl-Cu → (Et_2O, –30 °C) → Ph–CH(OH)–CH(Me)–CH=CH₂ (*syn*) + Ph–CH(OH)–CH(Me)–CH=CH₂ (*anti*) + Ph–CH(OH)–CH₂–CH=CH–Me (13)

with BF_3	98	2
without BF_3	48	52

Interestingly, Florio has noted that by replacing the methyl group characteristic of the crotyl system with heterocyclic arrays, as in (**43**) and (**44**), these allylic copper reagents afford products of exclusive γ-attack with benzaldehyde.[29] By contrast, the counterions Li^+, $MgBr^+$, and BEt_3Li^+ strongly favor the α-regioisomers. In view of the fact that even the corresponding lithium species are found to add reversibly, it is proposed (Scheme 5) that the 1,2-addition of (**45**) is reversible, and ultimately proceeds perhaps through a four-centered transition state (**47**) to give the thermodynamically preferred products. Since only the (*E*)-stereochemistry is observed in adduct (**48**), it is suggested that complete dissociation of (**46**) does not occur, since some isomerization of the free allylcopper might be expected.

Metallated allylsilanes have also been examined in terms of the effects of various gegenions on α- *versus* γ-additions to aldehydes. Using allylaminosilane (**49**), Tamao and Ito have shown that transmetallation of the lithiated intermediate to copper using CuCN (1 equiv.) leads to a >95% preference for γ-adduct (**51**; Scheme 6).[30,31] Other metals such as magnesium, zinc and titanium show a strong preference for α-attack (>95%), which ultimately leads to dienes (**50**) following Peterson alkenation.

With silicon containing the diethylamino moiety, product (**51**) can be transformed to the isopropoxy derivative (**52**), which is susceptible to oxidation with H_2O_2, thereby converting the carbon–silicon bond to a carbon–oxygen bond.[32] In tandem with a prior MCPBA epoxidation of vinylsilane (**52**), a new entry to the 2-deoxy-*C*-nucleoside skeleton is realized (Scheme 7).

Simpler allylic copper reagents, such as lithium diallylcuprate, behave more like allyllithium in their reactions with unsaturated carbonyl systems, usually affording mainly 1,2-adducts.[33] Attempts by Normant to bypass this mode with added TMS-Cl have been completely unsuccessful.[34] High yields of homoallylic alcohols are to be expected using this reagent, even on such highly conjugated enals as cinnamaldehyde,[35,36] or with enolizable ketones (Scheme 8).[37] Recent work from our lab has shown that

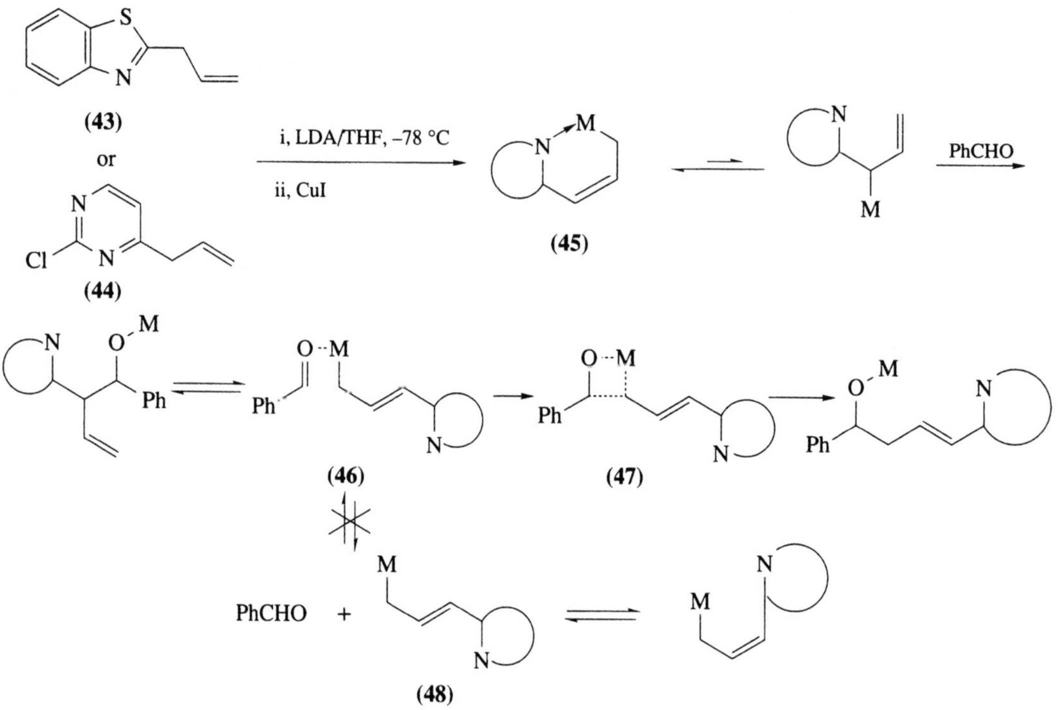

Scheme 5

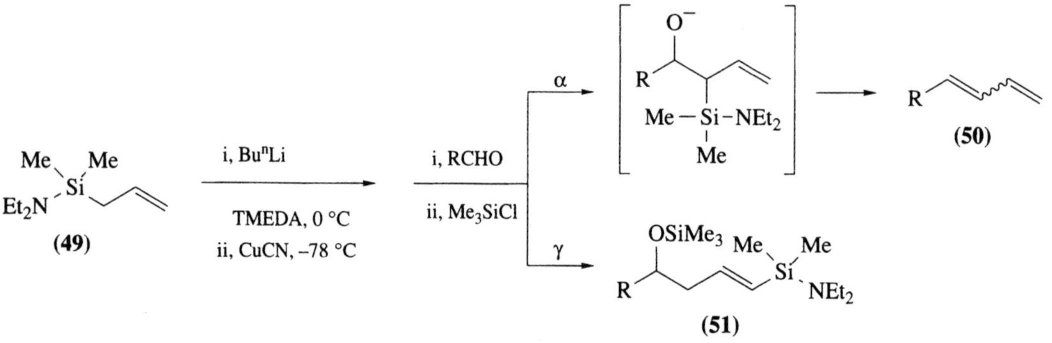

Scheme 6

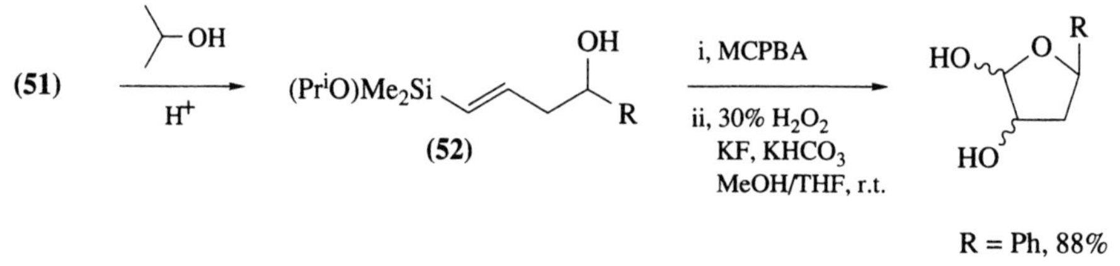

Scheme 7

allylcuprates are σ-bound (rather than π-bound) species, and that the allylic ligands are undergoing rapid α–γ exchange at –78 °C.[38] In light of these spectral studies, allylcuprates are clearly unique reagents among all those known in organocopper chemistry,[1,2,39] and may find general utility as soft, extremely reactive allylic nucleophiles toward 1,2-addition.

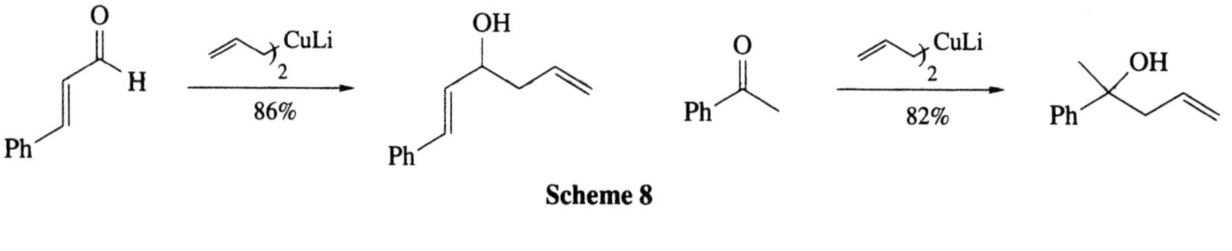

Scheme 8

Gaining preferential entry to the anti-Cram series *via* organocopper chemistry has been achieved, as described by Yamamoto, through the 'surprising'[22] effects of crown ethers on lower order cuprate 1,2-additions.[22] That is, notwithstanding an early report by Langlois,[40] treatment of an aldehyde such as (**35**) with a THF or Et_2O solution of $Bu^n{}_2CuLi$ containing one equivalent of 18-crown-6 ether gave a 1:4:2 ratio favoring the *anti* isomer (**37**; equation 14). The cuprate itself led to a 3:1 mix of *syn:anti* products (**36**) and (**37**), and hence these additives completely reverse the normal direction of diastereoselectivity. Similar, although not as pronounced, results were obtained with higher order dilithium ($Bu^n{}_2Cu(CN)Li_2$, 1:2 *syn:anti*) and lower order magnesium ($Me_2CuMgBr$, 1:2 *syn:anti*) cuprates, as well as with aggregates such as $Bu^n{}_5Cu_3Li_2$ (1:4.4 *syn:anti*).[22]

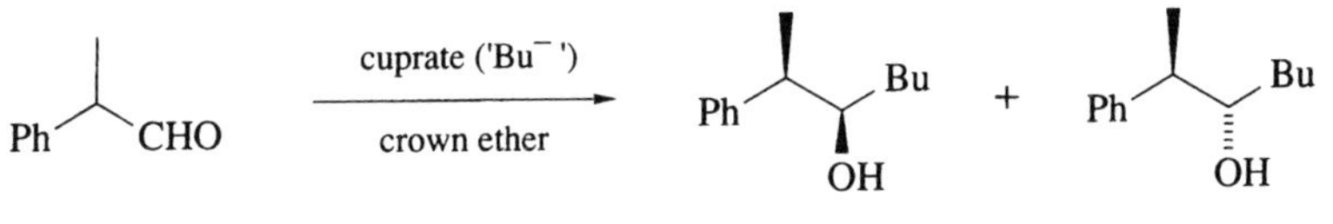

cuprate ('Bu^-') / crown ether (14)

(**35**) (**36**) (**37**)

$Bu^n{}_2CuLi$	+	18-crown-6	95%	1:4.2
$Bu^n{}_5Cu_3Li_2$	+	18-crown-6	96%	1:4.4
$Bu^n{}_2CuLi$			95%	3:1

To account for these anti-Cram selectivities, a radical mechanism is proposed based on the ability of R_2CuLi–crown ether complexes to transfer electrons more easily than R_2CuLi itself (*e.g.* to dicyclopropyl ketone).[22] If such a mechanism prevails, conformations (**53**) and (**54**) are destabilized in the transition state due to a build-up of negative charge on oxygen and the ensuing Me/O^- interaction. Moreover, the incoming nucleophile may prefer perpendicular attack, all of which taken together favors (**55**) to ultimately afford *anti* products.

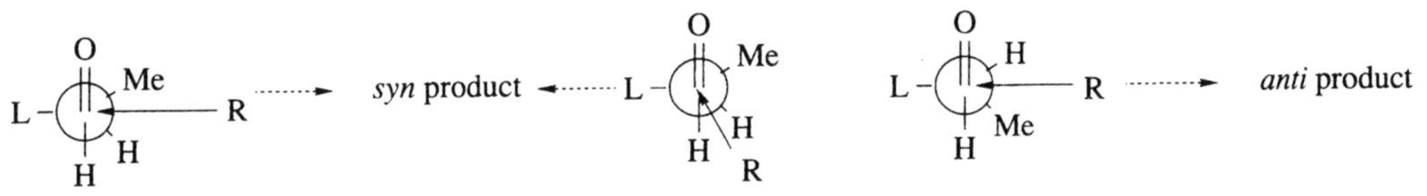

syn product ← (**53**), (**54**); (**55**) → *anti* product

(**53**) (**54**) (**55**)

The use of $BF_3{\cdot}Et_2O$ to accentuate the reactivity of otherwise sluggish cuprates toward 1,2-additions is also the subject of a recent report by Knochel.[41] Functionalized lower order cyanocuprates incorporating ZnI^+ as the gegenion in place of Li^+ or MgX^+ readily add to aldehydes at low temperatures provided excess Lewis acid is present (Scheme 9). Isolated yields of products are very good, and the observation that ketones do not react under similar conditions adds an element of chemospecificity to this method.

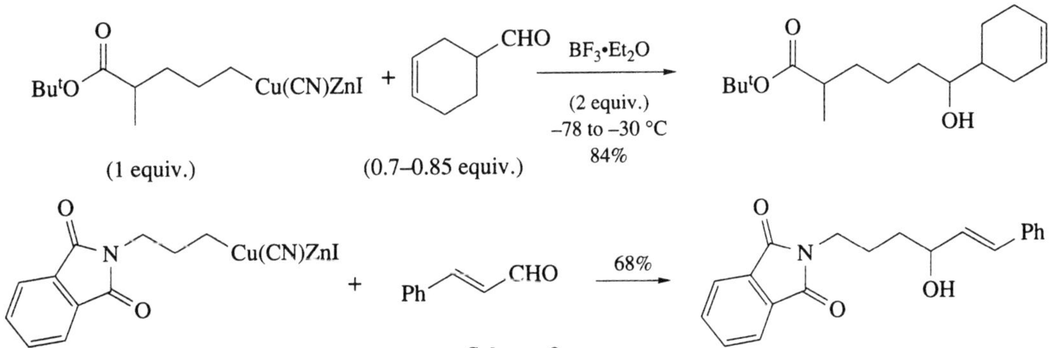

(1 equiv.) + (0.7–0.85 equiv.) → $BF_3{\cdot}Et_2O$ (2 equiv.), –78 to –30 °C, 84%

68%

Scheme 9

The iron tricarbonyl stabilized form of 2-formylbutadiene (**56**) reacts with various organometallic reagents to produce diastereomeric mixes of products (**57a**) and (**57b**).[42] While the trend is such that Grignard reagents are nonselective, organolithiums tend to add predominantly from the (*exo*) face away from the bulky iron tricarbonyl group, especially when doped with additional LiBr. Interestingly, the diastereoselectivity undergoes a complete turnabout with Gilman's reagent (equation 15). The explanation behind this reversal may involve initial attack either at the metal center or at a CO ligand, which would account for the reduced yield as several other secondary events could occur.[43] Subsequent intramolecular transfer of the methyl group to the aldehyde from the *endo* direction gives initially (**58**), as illustrated in equation (16).

(15)

(**56**) → RM → (**57a**) + (**57b**)

RM	Yield	57a:57b
PhMgBr	88%	50:50
MeLi/LiBr	85%	9:91
Me_2CuLi	40%	>90:<10

(16)

(**56**) → Me_2CuLi → [(**58**)] → H_3O^+ → (**57a**) R = Me

1.4.2.2 Reactions of Ketones

Although unhindered aldehydes are quite susceptible to attack by Gilman cuprates,[8] ketones tend to be far more resistant electrophiles. For example, Me_2CuLi requires temperatures around –10 °C before it will consume, to any significant degree, di-*n*-butyl ketone,[8] while acetophenone gives only 7% of the 1,2-adduct after 24 h at 0 °C.[44] More reactive species, *e.g.* Bu^n_2CuLi, are effective above –35 °C,[8] but still a far cry from the –90 °C conditions of aldehydic educts. Not surprisingly, esters are unresponsive at 18 °C towards Me_2CuLi, while it takes temperatures in excess of –10 °C for Bu^n_2CuLi to add in a 1,2-manner.[8] Such is not the case, however, with thioesters. As Anderson showed some years ago, thioesters react with lower order cuprates, including Me_2CuLi, even at –78 °C (equation 17).[45] Nonetheless, with respect to ketones, conditions are still sufficiently mild such that organocopper 1,2-additions can and have been used to advantage.

(17)

Me_2CuLi, Et_2O, –78 °C, 2 h, 75%

Most of the early work in this area was done by House[46] in the course of developing a model for cuprate conjugate additions to α,β-unsaturated ketones based on measured (polarographic) reduction potentials.[47] Products of 1,2-addition or reductive elimination (in the case of *e.g.* **59a, 59b**) can form upon exposure to Me_2CuLi in Et_2O depending upon the α-substitution pattern of the ketone,[48] and whether aryl alkyl[49] or diaryl[50] ketones are involved. Intermediate ketyls are usually produced *via* single-electron transfer, although with more-hindered cases (*e.g.* **60**) additional MeLi is necessary to form the 1,2-adduct.[50] This Me_2CuLi/MeLi mixture has been applied to the stereoselective synthesis of axial alcohols by Macdonald and Still.[51] Relative to several of the more common sources of methyl anion, such a cuprate/RLi combination[52] reacts predominantly to deliver the alkyl group from the equatorial face (equation 18).[51,52] The difference between results obtained with RLi or R_2CuLi alone lead to the suggestion that R_3CuLi_2 is the species responsible for the observed selectivity. Subsequent work by Ashby, however, who was the first to provide physical evidence for the existence of such higher order cuprates,[53] points to the likely complexation between R_2CuLi and the ketone (*e.g.* **61**) in a Lewis acid–Lewis base sense.[54] The newly activated substrate (**62**) now reacts rapidly with the RLi present, the stereochemistry of addition (favoring **63a** over **63b**) altered with respect to that for RLi alone since the hybridization at the complexed carbonyl carbon has changed, as has its steric environment. The effect is most noticeable in Et_2O but erodes considerably in THF as this solvent is more effective in competing for coordination sites on lithium.[54] Use of $MeLi/LiClO_4$ in Et_2O can effect the same results. Yamamoto's reagent, $RCu{\cdot}BF_3$, also adds to (**61**), although only a *ca.* 2:1 ratio of axial to equatorial alcohols is obtained.[55]

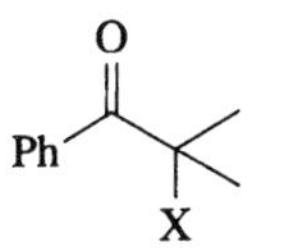

(**59**) **a:** X = OC(O)Me
b: X = Br

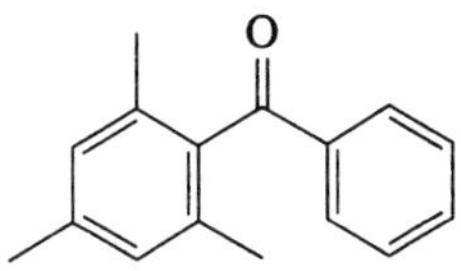

(**60**)

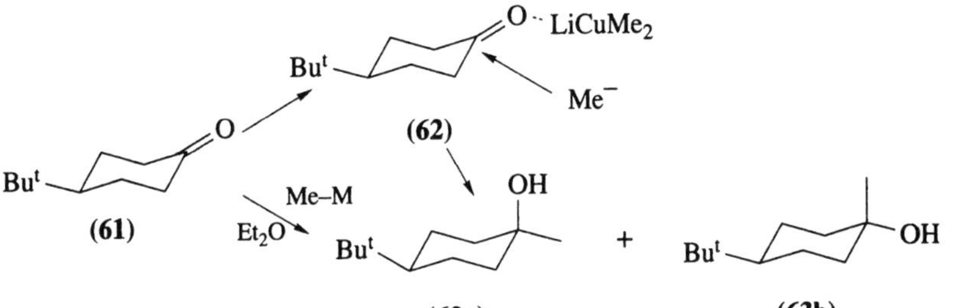

(**61**) (**62**) (**63a**) (**63b**) (18)

M = MgI, 5 °C	51:49
MgMe	70:30
Li, –78 °C	79:21
Li/Me_2CuLi, –70 °C	94:6

The (RLi + R_2CuLi) mixture was seen by Plumet as a route to the desired tertiary alcohols (**65a**) *via* 1,2-addition to 7-oxabicyclo[2.2.1]hept-5-en-2-one (**64**) from the more-hindered *endo* direction (equation 19).[56] The argument relies on complexation of the carbonyl group and/or ether oxygen with R_2CuLi, thereby blocking attack from the *exo* face. While (**65a**) was indeed the major product in all but one case (*i.e.* using $PhLi/Ph_2CuLi$), the supposedly unreactive lower order cuprates themselves gave identical results virtually upon mixing.

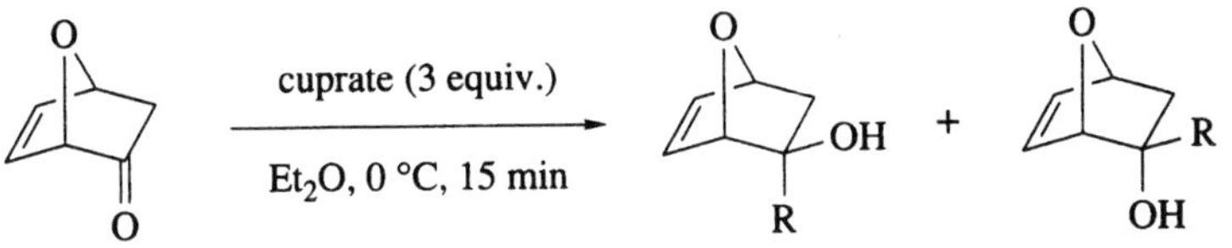

(**64**) —cuprate (3 equiv.), Et_2O, 0 °C, 15 min→ (**65a**) + (**65b**) (19)

Cuprate	*Yield* (%)	*Ratio*
MeLi + Me_2CuLi	80	6:1
Me_2CuLi	85	6:1
Ph_2CuLi	85	6:1
PhLi + Ph_2CuLi	75	1:10

Preferential 1,2-addition of Gilman's cuprate to a ketone over the usually much more facile 1,4-mode occurs in the case of keto enone (**67**).[57] When Goldsmith and Sakano treated (**67**) with 2 equiv. of Me_2CuLi at –78 °C and quenched the reaction with acetic anhydride, enone (**66**) was formed. At higher temperatures Michael addition begins to take place, although the cuprate also generates products of ketone enolization (**68**) rather than carbon–carbon bond formation (Scheme 10).

Somewhat similar trends have been found by Marino and Floyd concerning their lower order mixed acrylate cuprates (**69**) toward α,β-unsaturated ketones.[58] In Et_2O at –78 °C even simple cyclic enones, i, 2 Me_2CuLi, –78 °C; ii, Ac_2O ← (67) → i, 2 Me_2CuLi, 0 °C; ii, Ac_2O

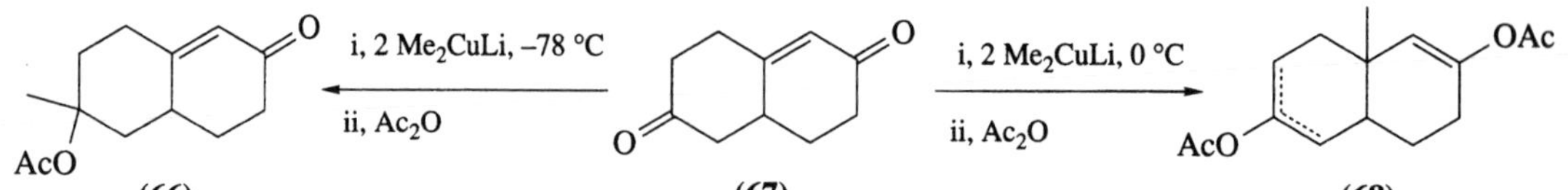

AcO (66) (67) AcO OAc (68)

Scheme 10

e.g. (**70a**) and (**70b**), give tertiary bisallylic alcohols (**71**) and (**72**), respectively, in moderate to good yields (equations 20 and 21).

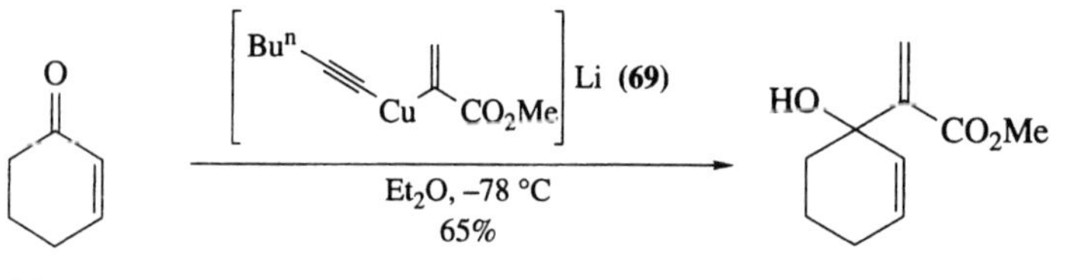

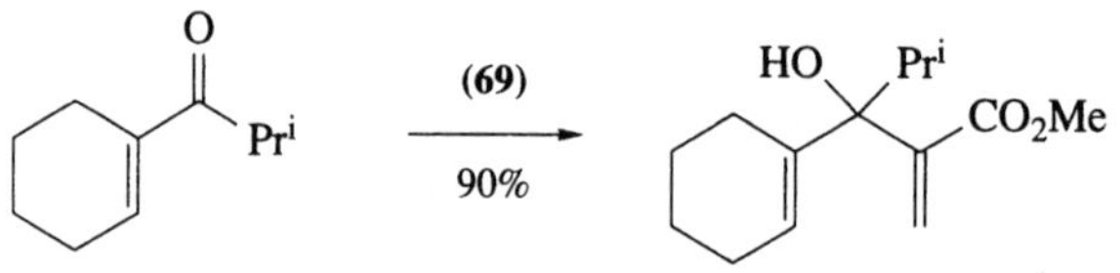

(70a) → [Bu^n—≡—Cu—C(=CH$_2$)CO_2Me] Li (**69**), Et_2O, –78 °C, 65% → HO, CO_2Me (71) (20)

(70b) Pr^i → (**69**), 90% → HO, Pr^i, CO_2Me (72) (21)

Although lithium cuprates bearing two alkynic ligands notoriously resist transfer of this group, when admixed with an equal amount of the lithium alkynide (*i.e.* 3RC≡CLi:CuI) exclusive 1,2-addition occurs with cyclic enones.[59] While prior efforts to effect this chemistry in strictly ethereal solvents (dioxane) have been unsuccessful,[60] use of HMPA (~ 20%) as cosolvent now leads to efficient couplings in these cases. The initial products may be isolated as such, or oxidatively worked up to provide, in the case of (**73**), the rearranged material (**74**; Scheme 11).

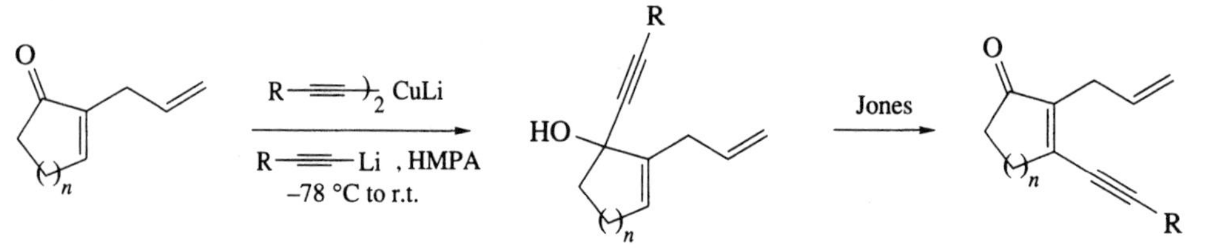

(73) → $(R—≡—)_2CuLi$, R—≡—Li, HMPA, –78 °C to r.t. → HO, R; 85–95%, n = 1, 2 → Jones → (74)

R = Pr^n, Bu^n — (74) R = Pr^n; n = 1; *ca.* 100%

Scheme 11

As part of an effort to assess the extent of neighboring group participation in solvolysis reactions of bicyclic systems of type (**76a**), use of known ketone (**75**) as starting material has been examined.[61] Both MeLi and MeMgI give excellent percentages of the undesired isomer (**76b**). Switching to Me_2CuLi is based on the notion that electron transfer[46–50] might prevail so as to provide a different stereochemical outcome. Indeed, with this reagent under otherwise identical conditions, net formal attack by methyl from the opposite (desired) direction is realized (equation 22).[61]

The direct conversion of a ketone carbonyl to a *gem*-dimethyl moiety has been accomplished, interestingly enough (*cf.* reactions of **67**[57]) on bicyclic α,β-unsaturated ketones containing a pendant COSR′ group, as in (**78**), (**79**) and (**80**) (Scheme 12).[62] In sharp contrast, the *O*-ester analog of (**77**) (*i.e.* **78**, X = OEt) leads to a 3:1 mix of tertiary alcohols (**77a**) and (**77b**), respectively. Intermediate (**82**; Scheme 13) is postulated as the precursor to (**81**), as its formation from (**78**, X = S) leads to the observed product upon treatment with Me_2CuLi.

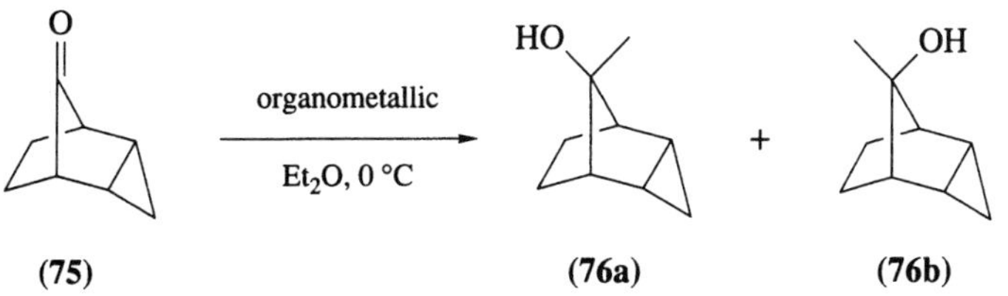

organometallic

Et_2O, 0 °C

(22)

(75) **(76a)** **(76b)**

MeLi	5:95
MeMgI	10:90
Me_2CuLi	92:8

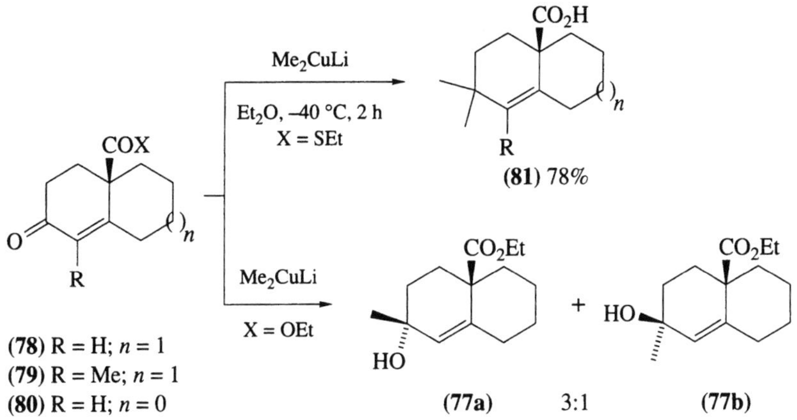

(78) R = H; *n* = 1
(79) R = Me; *n* = 1
(80) R = H; *n* = 0

Me_2CuLi, Et_2O, –40 °C, 2 h, X = SEt → **(81)** 78%

Me_2CuLi, X = OEt → **(77a)** 3:1 **(77b)**

Scheme 12

(78) i, MeLi, –78 °C; ii, NaH, r.t. → **(82)**; Me_2CuLi (3 equiv.), Et_2O, –40 °C, 2 h → **(81)** 85%

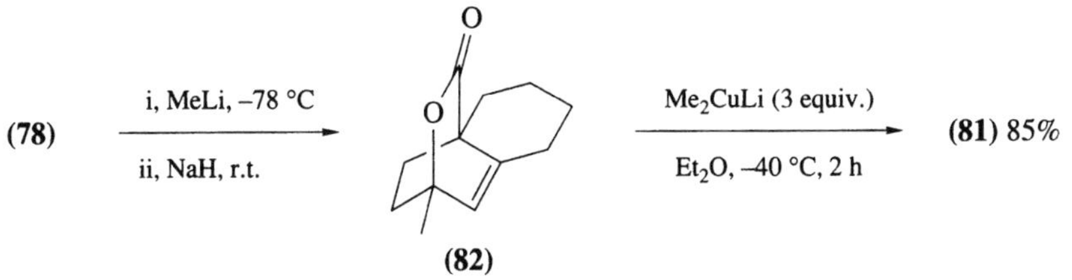

Scheme 13

Transmetallation of a lithium enolate to a copper enolate in Davies' iron carbonyl system [(η^5-C_5H_5)Fe(CO)(PPh_3)COEt] allows for highly stereoselective additions to symmetrical ketones.[63] Lithium enolates (**83**) alone give 2:1 to 6:1 ratios of (**84a**, *RR,SS*) and (**84b**, *RS,SR*), and additives (*e.g.* $SnCl_2$, BF_3, Et_2AlCl, $ZnCl_2$) either completely suppress the 1,2-addition or do little to influence the diastereoselectivity. Addition of CuCN (1 equiv.), however, gives in most cases 10:1 to 60:1 product ratios (Scheme 14).

A deacylation reaction has been effected by a reactive lower order cuprate when its reaction partner has no other pathway available. Thus, Kunieda and Hirobe have found that (–)-3-ketopinyl-2-oxazolone (**85**) undergoes smooth cleavage of the chiral auxiliary to afford the heterocycle (**86**), subsequent elaboration of which leads to either (2*S*,3*R*)-3-hydroxyglutamic acid (**87**) or (3*R*,4*R*)-4-amino-3-hydroxy-6-methylheptanoic acid (**88**) as their hydrochloride salts (Scheme 15).[64]

1.4.3 1,2-ADDITIONS TO IMINES, NITRILES AND AMIDES

1.4.3.1 Reactions of Imines

Studies on the additions of organocopper complexes across carbon–nitrogen double bonds are relatively few in number. In terms of stereochemical control in substrates bearing α- and/or β-chiral centers, the information available is even more sparse. That organocopper reagents do add to Schiff bases was shown

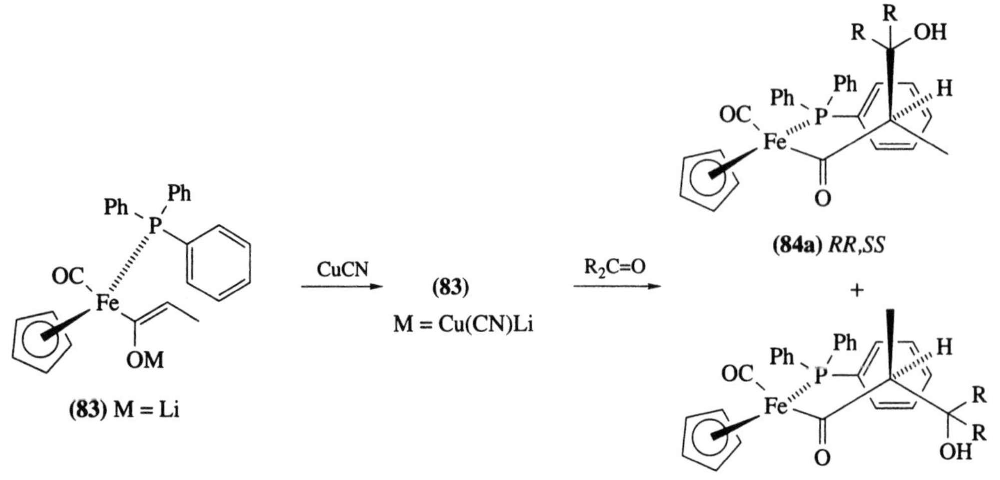

Ph Ph P OC Fe OM

(83) M = Li

CuCN → **(83)** M = Cu(CN)Li → $R_2C{=}O$

R R OH Ph Ph P OC Fe H O

(84a) *RR,SS*

+

Ph Ph P OC Fe H R R OH O

(84b) *RS,SR*

Ketone	*Ratio* ***(71a)****:****(71b)***	*Yield* (%)
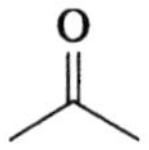 O	>40:1 (3:1)[a]	61
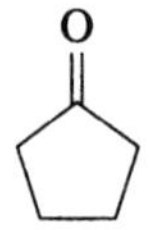 O	>60:1 (4:1)[a]	71
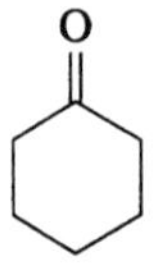 O	10:1 (2:1)[a]	68

[a]Ratio from reactions of Li enolates

Scheme 14

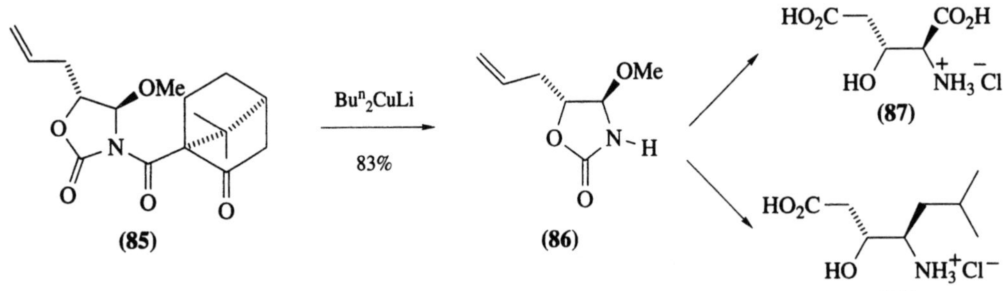

OMe O N O O O **(85)** → $Bu^n{}_2CuLi$, 83% → OMe O N-H O **(86)** →

HO_2C CO_2H HO $\overset{+}{N}H_3\,\bar{C}l$ **(87)**

HO_2C HO $NH_3^+Cl^-$ **(88)**

Scheme 15

by Akiba, most notably where acidic α-hydrogens are present in these substrates (equation 23).[65] The low basicity of $RCu{\cdot}BF_3$, prepared from CuI and RMgX in THF at –30 °C to which is added $BF_3{\cdot}Et_2O$ (1 equiv.), precludes such typical side reactions as metalloenamine formation that is observed with Grignards, the low yields of 1,2-adducts obtained from alkyl- or aryl-lithiums, and reductive dimerization. Moderate to good yields of secondary amines are normally received from unhindered aldimines, although α-branched aldimines give none of the corresponding products. This limitation can be overcome, however, using $Bu^n{}_2CuLi{\cdot}BF_3$ (1 equiv.) or the magnesium analog $Bu^n{}_2CuMgBr{\cdot}BF_3$ (equation 24). The Lewis acid is essential for success in these reactions, as no 1,2-addition is to be expected in its absence. Unfortunately, thus far there has been no success in similar reactions of ketimines.

Attempts to induce asymmetry at an imine carbon using homochiral amines to form Schiff bases followed by 1,2-additions with organocopper reagents has not produced a viable method as yet. Yamamoto

(23)

(24)

has found that imine (**89**) reacts with $Bu^n{}_2CuLi{\cdot}BF_3$ to give an 82:18 mixture of Cram (**90a**) and anti-Cram (**90b**) products (equation 25).[66] Much better results have been obtained using allylic boranes (*e.g.* allyl-9-BBN), where ratios between 94:6 and 100:0 have been achieved. Of course, these borane-mediated allylations share in the benefits associated with a highly ordered cyclic transition state (**91**) not involved with the analogous copper chemistry.

(25)

(**89**) R' = CHMePh (**90a**) (**90b**)

$Bu_2CuLi{\cdot}BF_3$	82:18
allyl-9-BBN, R' = Pr^n	96:4
allyl-9-BBN, R' = Pr^i	100:0

(**91**)

Claremon has briefly looked at the effect of catalytic CuI on the MeLi addition to glyceraldehyde acetonide *N*,*N*-dimethylhydrazone (**92**).[67] While the organolithium itself leads to a preference for the *anti* product (**93a**) on the order of 3:1 to 6:1 (depending upon the quantity of MeLi and the temperature), the cuprate reverses the ratio to favor the *syn* isomer (**93b**; equation 26). Acyclic α-alkoxyhydrazones, surprisingly, are untouched by organocopper reagents, while organolithiums give excellent (≥97:3) *anti* selectivities in high yields (85–98%).

(26)

(**92**) (**93a**) (**93b**)

MeLi, –10 to 25 °C, 95%	3:1
MeLi (5 equiv.), –55 °C, 75%	6:1
MeLi + CuI (0.1 equiv.), –20 °C, 46%	1:3

Ketoximes have been synthesized by Fujisawa using a catalytic CuI/Grignard addition to *aci*-nitroiminium chlorides, formed by *O*-acylation of a nitroalkane with *N*,*N*-dimethylchloromethyleniminium chloride.[68] Numerous organomagnesium reagents can be employed, and yields of products resulting from 1,2-addition followed by elimination of DMF range from 61–97% (Scheme 16a).

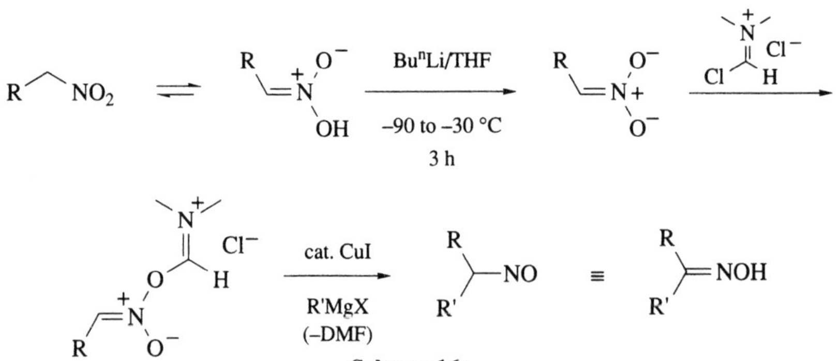

Scheme 16a

Copper(I) driven activation of the carbon–nitrogen double bond towards addition by RLi takes place in Et_2O, but not THF, where the solvent of poorer Lewis basicity does not compete for nitrogen complexation (*cf.* **95**). On this basis, Bertz converted the bis(tosylhydrazone) (**94**), *via* reaction with 'Me_3CuLi_2', to the *gem*-dimethylated product (**96**; Scheme 16b).[69] Although several other pathways associated with the various intermediates are followed to account for the five products observed in varying amounts, this study does suggest that otherwise innocuous, soluble copper(I) species may be useful additives for nucleophilic additions to imines and related functional groups.

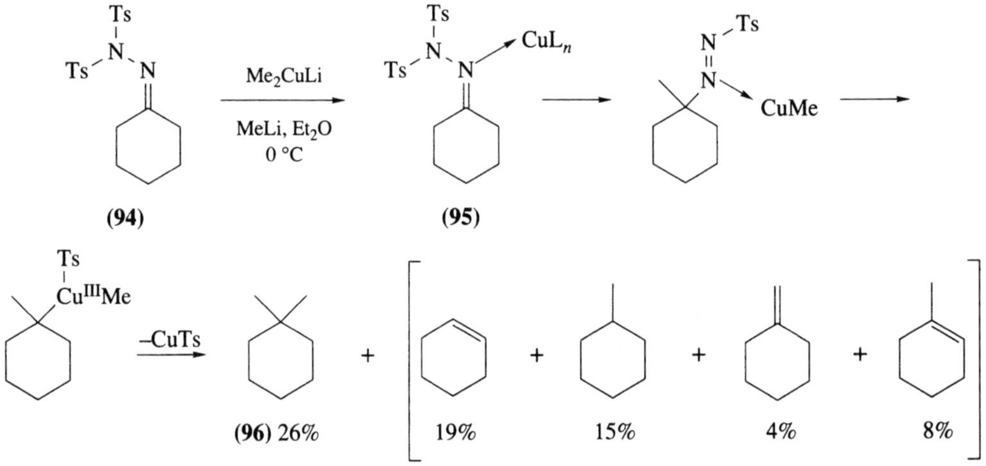

Scheme 16b

1,2-Additions of higher order cuprates to acylimines (**97**) provide a novel route to α-alkylated amino acid derivatives (**98**; equation 27).[70] Following generation of amide-stabilized systems (**97**) *via* action of singlet molecular oxygen on trisubstituted imidazoles,[71] products (**98**) can be isolated in 57–75% yields. Since Bu^nLi and Bu^sLi alone (*versus* $Bu^n_2Cu(CN)Li_2$ and $Bu^s_2Cu(CN)Li_2$) raise the yields to 86% and 61%, respectively, there seems to be no particular advantage in using a cuprate, at least for the functionality present in (**97**).

Steglich, however, has found that higher order cuprates are the reagents of choice for additions to acylimines of type (**100**), generated *in situ* from α-bromo amino acid derivatives (**99**).[72] Grignard reagents give modest yields at best, and in some cases fail altogether. Ligands transferred from copper to carbon include primary and tertiary alkyl, phenyl, 1-naphthyl and vinyl, with yields of acylated amino acid esters (**101**) ranging from 30–83%. With $(vinyl)_2Cu(CN)Li_2$, the initial adduct (**101**, R = vinyl), formed at –100 °C, can be easily isomerized to the dehydro derivative (**102**) simply upon warming the reaction mixture to room temperature (Scheme 17).

Similar studies by O'Donnell using the α-acetoxy function as the leaving group (along with the benzophenone Schiff base in **103**),[73] and by Williams on a bromide,[74] come to the same conclusion that $R_2Cu(CN)Li_2$ affords the best results (equation 28). Reactions of the former substrates are highly

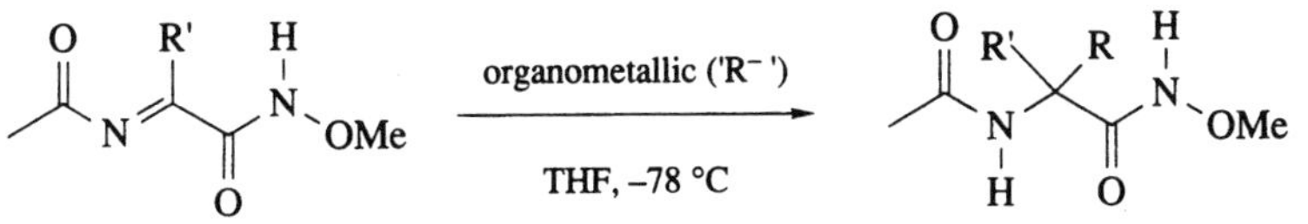

(27)

(97) (98)

$Bu^n{}_2Cu(CN)Li_2$	75%
$Bu^s{}_2Cu(CN)Li_2$	57%
Bu^nLi	86%
Bu^sLi	61%

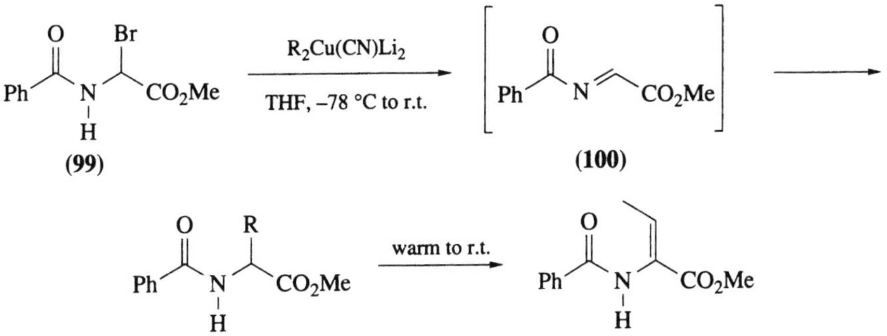

(99) (100)

(101) R = vinyl (102)

Scheme 17

sensitive to several parameters, including solvent, temperature and mode of addition. In THF at *ca* ≤5 °C using inverse addition, moderate to good yields of (**104**) are obtained. In Et_2O, only products of reduction are observed.

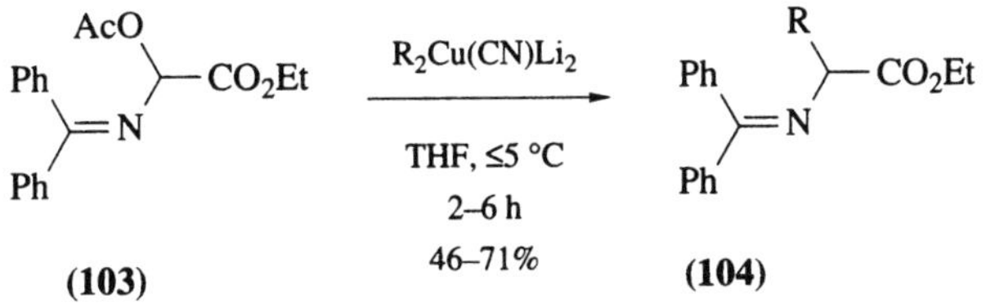

(28)

(103) (104)

R = Bu^n, Bu^t, thienyl, naphthyl

Related 1,2-additions to imines in the β-lactam area employing organocuprates have also met with considerable success.[75–79] Net substitution reactions, proceeding *via* intermediate imino derivatives, have been achieved with lithium diallyl- and dialkyl-cuprates on educts such as (**105**)[76] and (**106**).[77] When a stereochemical label is present as in (**107**)[78] and (**108**),[79] the major products reflect *trans* addition to imines (**109**) and (**110**), respectively (Scheme 18).

1.4.3.2 Reactions of Nitriles

Until recently, organocopper-mediated 1,2-additions to nitriles of any consequence were unknown. Most of those reported occurred as undesirable pathways in attempted 1,4-additions to α,β-unsaturated substrates. According to the House model, correlating reduction potentials (E_{red}) with substrate reactivity toward lithium cuprates,[47] prior complexation of the reagent with doubly bonded oxygen, rather than nitrogen (presumably of either sp^2 or sp hybridization), is an essential ingredient irrespective of E_{red}. Hence, attempted reaction of cinnamonitrile (**111**) with Me_2CuLi failed.[47] Years later, reinvestigation of this process with a higher order reagent did slowly produce a product; however, the product turned out to be methyl ketone (**112**), derived from cuprate 1,2-addition followed by hydrolysis of the intermediate imine (equation 29).[80] Nonetheless, this reaction was not synthetically useful, nor was the procedure general, as the *n*-butyl analog led to (**113**) along with several unidentified compounds with the same substrate under similar conditions.

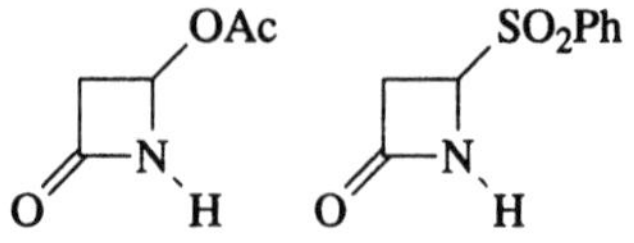

(105) (106)

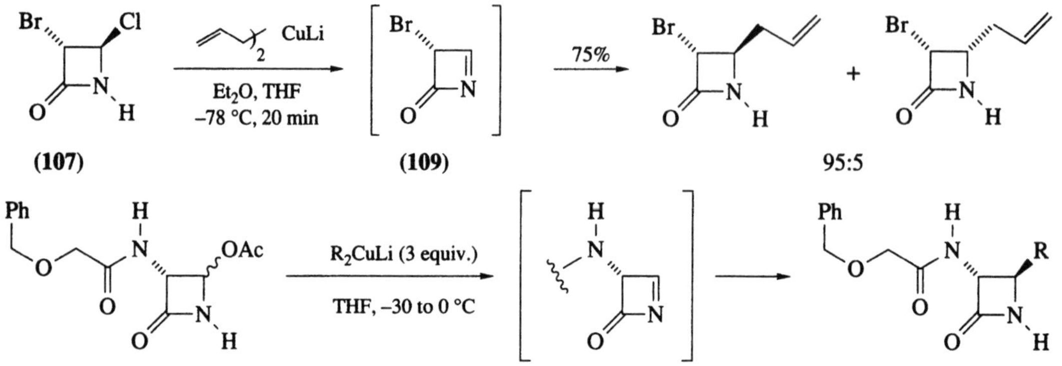

(107) $\xrightarrow[\text{Et}_2\text{O, THF, }-78\,^\circ\text{C, 20 min}]{(\text{allyl})_2\text{CuLi}}$ [(109)] $\xrightarrow{75\%}$ 95:5

(108) $\xrightarrow[\text{THF, }-30\text{ to }0\,^\circ\text{C}]{\text{R}_2\text{CuLi (3 equiv.)}}$ [(110)] $\longrightarrow$ R = Bu^n, 60%; R = allyl, 32%

Scheme 18

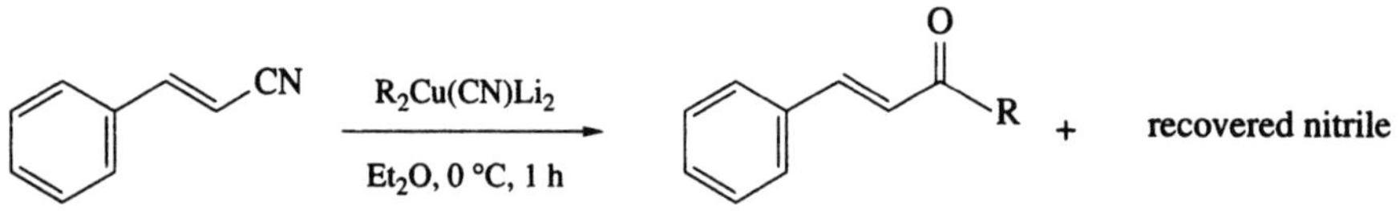

(111) $\xrightarrow[\text{Et}_2\text{O, }0\,^\circ\text{C, 1 h}]{\text{R}_2\text{Cu(CN)Li}_2}$ (112) R = Me; (113) R = Bu^n + recovered nitrile (29)

Introduction of an additive, *e.g.* TMS-Cl, did lead to good yields of an isolable product; however, the product was shown by Alexakis to be the dialkylated ketone (*e.g.* **114**) in all cases studied (equation 30).[81] Even the softer organometallic, Yamamoto's reagent '$RCu{\cdot}BF_3$', gave mixtures of mono- and dialkylated ketones with acrylonitrile and 1-cyclohexenecarbonitrile.[82]

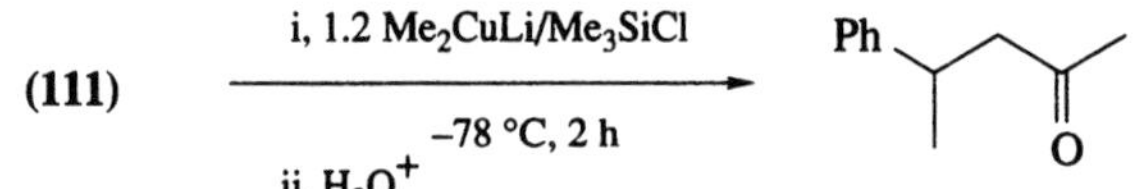

(111) $\xrightarrow[\text{ii, H}_3\text{O}^+]{\text{i, 1.2 Me}_2\text{CuLi/Me}_3\text{SiCl, }-78\,^\circ\text{C, 2 h}}$ (114) 66% (30)

A procedure which seems to skirt these problems, as described by Hall, invokes a catalytic copper salt in the presence of a Grignard reagent in refluxing THF, which affords good yields of adducts.[83] The nature of the copper(I) source is not crucial (CuCl, CuBr, CuI, CuCN were all compared), and depending upon manipulation of the initial product, ketones (from hydrolysis), sterically hindered ketimines (from protonolysis) or primary amines (from reduction) can be realized. Since reagents formed from either a 1:1 RMgX:CuX (*i.e.* $RCu{\cdot}MgX_2$) or 2:1 (*i.e.* $R_2CuMgX{\cdot}MgX_2$) ratio are clearly not responsible for the chemistry observed, higher order reagents have been suggested as the reactive species (Scheme 19a).

1.4.3.3 Reactions of Amides

Considering the proficiency associated with additions of organolithium and organomagnesium reagents to *N,N*-disubstituted amides as a means of preparing aldehydes and ketones,[84] it is not surprising that organocopper reagents are not known to function in this capacity. However, imine formation from amides is another matter, and Feringa has recently demonstrated that lower order lithium and magnesium cuprates can indeed be used with *N*-trimethylsilyl-*N*-alkylformamides to arrive at this functional group.[85]

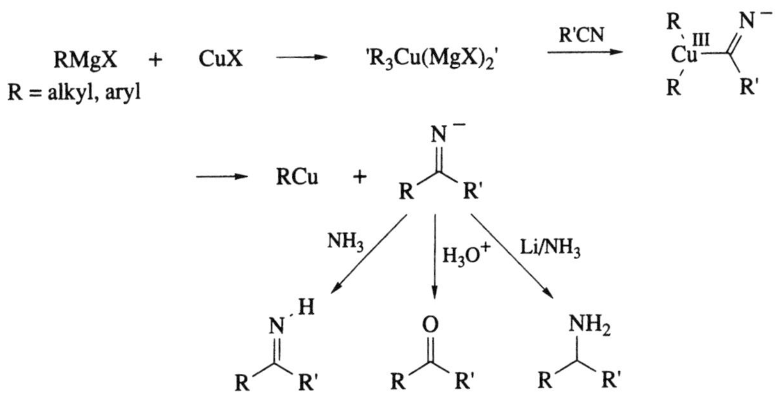

RMgX + CuX ⟶ '$R_3Cu(MgX)_2$' $\xrightarrow{R'CN}$ R(R)Cu(III)–C(=N⁻)R'

R = alkyl, aryl

⟶ RCu + RC(=N⁻)R'

NH_3 → RC(=NH)R'; H_3O^+ → RC(=O)R'; Li/NH_3 → RCH(NH_2)R'

Scheme 19a

Treatment of formamide (**115**) with either $LiCuR_2$ or catalytic CuBr/RMgBr in THF between –80 °C and –20 °C, followed by aqueous work-up, affords aldimine (**116**) exclusively of (*E*)-geometry (equation 31).

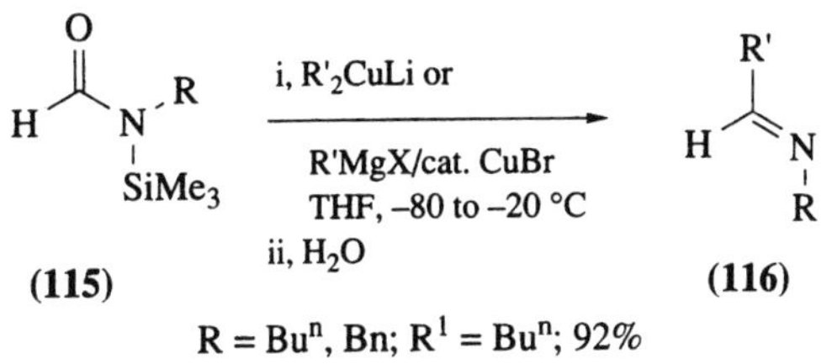

$$\text{HC(O)N(R)SiMe}_3\ \textbf{(115)} \xrightarrow[\text{ii, H}_2\text{O}]{\text{i, R'}_2\text{CuLi or R'MgX/cat. CuBr, THF, –80 to –20 °C}} \text{HC(R')=NR}\ \textbf{(116)} \quad (31)$$

R = Bu^n, Bn; R^1 = Bu^n; 92%

By switching to the corresponding *N*-acrylamide (**115**, R = Ph), secondary amines (**120**) are the major products resulting from double 1,2-additions (Scheme 19b). Presumably, the initial 1,2-adduct (**117**) undergoes a silicon migration from nitrogen to oxygen, the intermediate anion (**118**) being stabilized by the aromatic ring, which subsequently eliminates to reform imine (**119**) now susceptible to a second-stage alkylation. This secondary process does not occur with Grignard reagents, and thus arylaldimines (**119**) are formed in high yields using (**115**, R = Ph) and various RMgX in THF, without the presence of copper(I) salts.

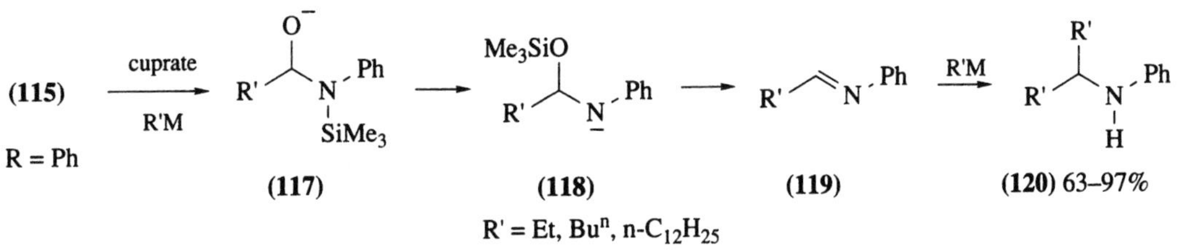

(**115**) R = Ph $\xrightarrow[\text{R'M}]{\text{cuprate}}$ R'CH(O⁻)N(Ph)SiMe$_3$ (**117**) ⟶ R'CH(OSiMe$_3$)N⁻Ph (**118**) ⟶ R'CH=NPh (**119**) $\xrightarrow{\text{R'M}}$ R'$_2$CHN(H)Ph (**120**) 63–97%

R' = Et, Bu^n, n-$C_{12}H_{25}$

Scheme 19b

1.4.4 APPLICATIONS TO NATURAL PRODUCT SYNTHESIS

As alluded to earlier, in general, 1,2-additions by organocopper reagents especially on more complex, multifunctionalized substrates are statistically few in number compared with the ever popular alkylation,[1] 1,4-addition,[2] and carbocupration[86] processes. Nonetheless, the examples which follow provide strong testimony to the value of organocopper intermediates as an alternative means of delivering a carbanion of attenuated reactivity in a controlled fashion.

Development of a novel route to the C-7 to C-13 portion (*e.g.* **122**) of erythronolides A and B (*i.e.* **121a** and **121b**) by Burke makes use of a stereoselective addition of an isopropenyllithium-derived cuprate to homochiral aldehyde (**123**) as a first, key step (Scheme 20).[87] Formation of allylic alcohol (**124**) as the major product presumably reflects a β chelation controlled pathway (*vide supra*).[9] Subsequent handling of (**124**), which included as the second critical step a dioxanone to dihydropyran enolate Claisen rearrangement,[88] produces three key subunits, including (**122**).

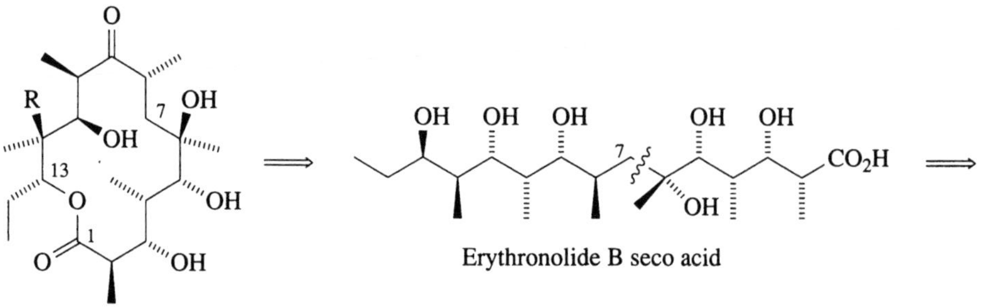

(121) **a:** R = OH; Erythronolide A
b: R = H; Erythronolide B

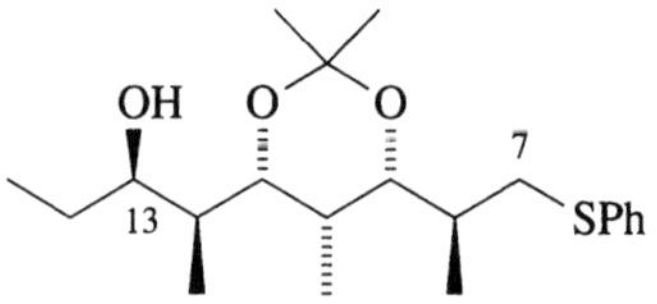

Erythronolide B seco acid

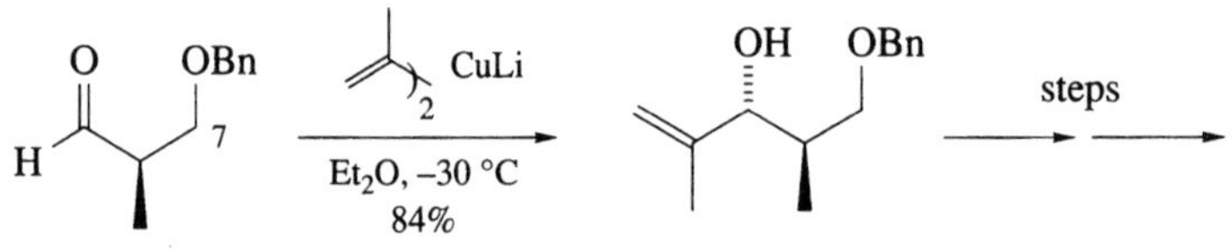

(122)

(123) $\xrightarrow[\text{Et}_2\text{O, -30 °C, 84\%}]{(\text{isopropenyl})_2\text{CuLi}}$ **(124)** $\xrightarrow{\text{steps}}$ **(122)**

Scheme 20

A similar strategy invoking these two valued processes has also been applied to the 'left wing' of the ionophoric antibiotic indanomycin (X-14547A),[89] and more generally as an entry to *C*-pyranosides (**125**; Scheme 21).[90] Treatment of aldehyde (**126**), itself prepared from elaboration of (**127**), likewise formed from (vinyl)$_2$CuLi addition to aldehyde (**128**),[89] with the Gilman reagent formed from CuI·PBu$_3$ and (*E*)-propenyllithium in Et$_2$O at low temperature affords a 24:1 ratio of alcohols (**129a**) and (**129b**) in good yield (Scheme 22).[89] Unfortunately (**129a**) is of the undesired stereochemistry and necessitates an oxidation (PCC, 93%)–reduction (Zn(BH$_4$)$_2$, 88%) sequence to arrive at (**129b**) (>100:1 **129b:129a**). Related model studies on an isopropyl analog of (**126**) (*cf.* **130**) show that the cuprate 1,2-addition approach to (**131**), with or without an intervening oxidation/reduction, could be useful for obtaining dioxanones (**132**) and (**133**) as precursors to *C*-glycosides (**134**) and (**135**), respectively. Thus, individual exposure of (**132**) and (**133**) to the Ireland ester enolate Claisen reorganization,[88] albeit at abnormally high temperatures (*ca.* 110°C for 3–4 h, affords the desired materials following aqueous acid hydrolysis and treatment with ethereal diazomethane of the initially formed silyl esters (Scheme 23).[90]

Indanomycin has also been a target of the Boeckman group, which, while employing a quite different approach to the pyran nucleus, begins the route with a cuprate 1,2-addition to the readily available homochiral aldehyde (**136**).[91] An excellent yield of two diastereomers (**137**) was obtained, epimeric at the carbon bearing the methyl group brought in as part of the cuprate (equation 32). Thus, as confirmed by ^{1}H NMR on the Mosher esters of (**137**), the stereoselectivity of the cuprate addition was >98%.

Stereochemical issues associated with polyene macrolide construction, attention being focused on the chiral centers representing C-34 to C-37 in amphotericin B in particular, have been addressed by McGarvey, wherein 1,2-additions by Me$_2$CuLi play an important role.[92] L-Aspartic acid can be parlayed into

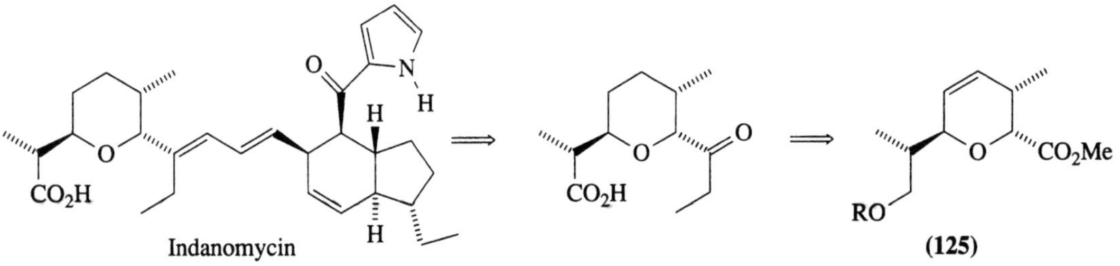

Indanomycin ⟹ ⟹ **(125)**

Scheme 21

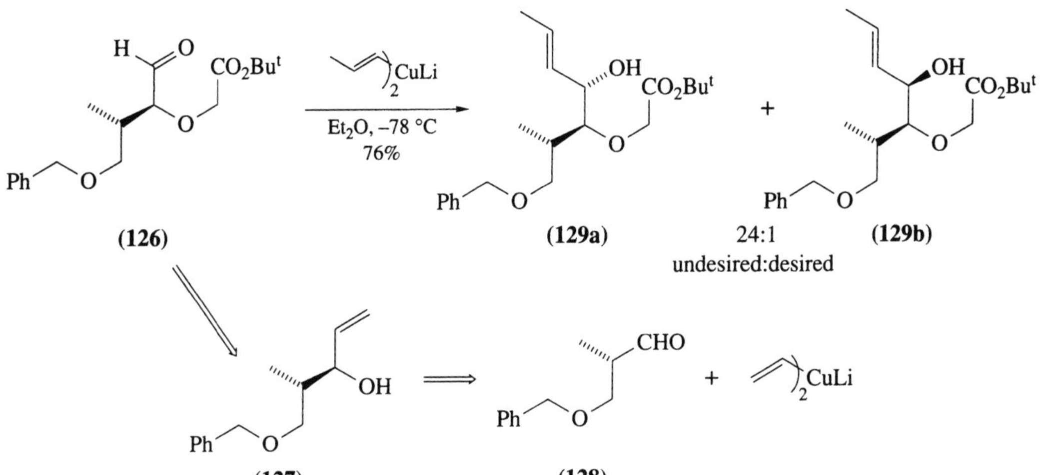

(126) (129a) (129b)

24:1 undesired:desired

(127) (128)

Scheme 22

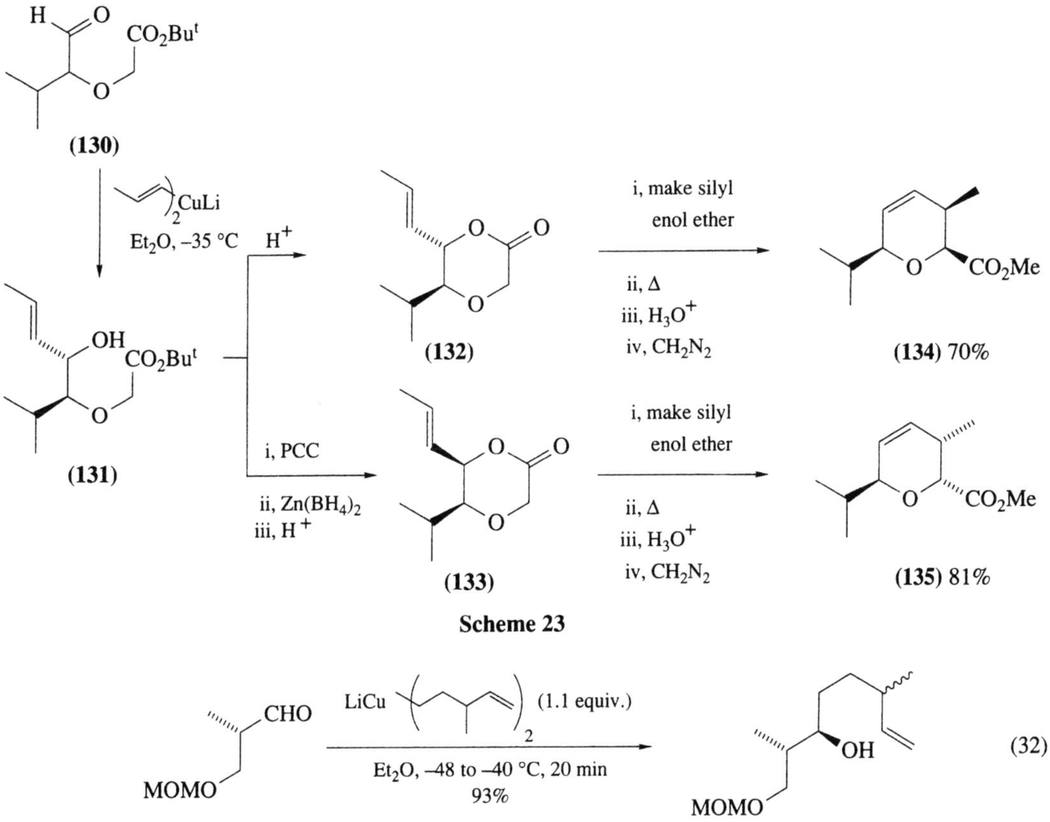

(130) (131) (132) (133) (134) 70% (135) 81%

Scheme 23

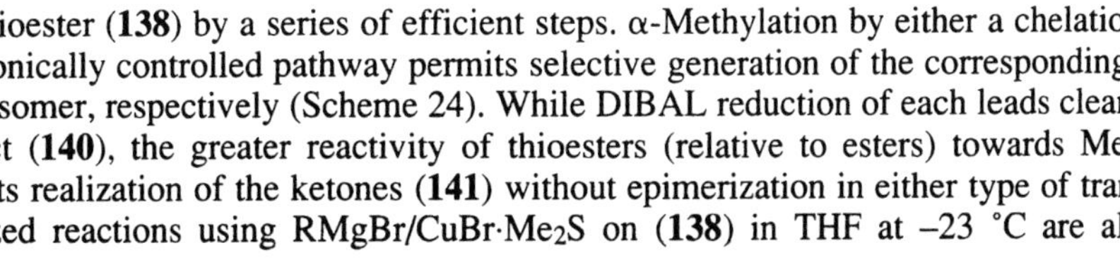

(136) (137) 1:1 mix (32)

heterocyclic thioester (**138**) by a series of efficient steps. α-Methylation by either a chelation-controlled or stereoelectronically controlled pathway permits selective generation of the corresponding *anti* (**139a**) or *syn* (**139b**) isomer, respectively (Scheme 24). While DIBAL reduction of each leads cleanly to the aldehyde product (**140**), the greater reactivity of thioesters (relative to esters) towards Me_2CuLi (*vide supra*)[45] permits realization of the ketones (**141**) without epimerization in either type of transformation. Copper-catalyzed reactions using RMgBr/CuBr·Me_2S on (**138**) in THF at −23 °C are also effective (90%).

With aldehydes (**140a**) and (**140b**) in hand, further 1,2-additions of Me_2CuLi have been examined and found to afford results (Scheme 25) opposite to those obtained with, for example, MeMgBr, in good yields (>85%). The preference is explained on the basis of attack onto the chelated bicyclic compound

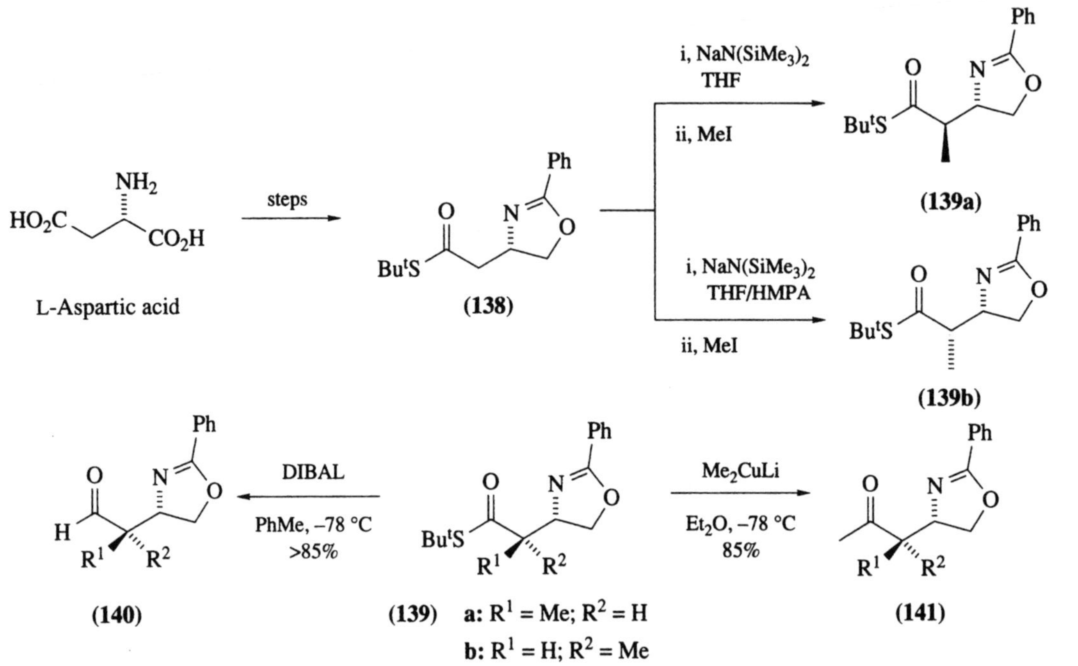

HO$_2$C, NH$_2$, CO$_2$H

L-Aspartic acid

steps

Bu^{t}S, O, N, O, Ph

(**138**)

i, NaN(SiMe$_3$)$_2$ THF

ii, MeI

(**139a**)

i, NaN(SiMe$_3$)$_2$ THF/HMPA

ii, MeI

(**139b**)

DIBAL

PhMe, –78 °C >85%

Me$_2$CuLi

Et$_2$O, –78 °C 85%

(**140**) (**139**) **a:** R^1 = Me; R^2 = H (**141**)

b: R^1 = H; R^2 = Me

Scheme 24

(**142**) from the axial direction with smaller nucleophiles, while the bulkier cuprate comes in from the bottom to avoid a 1,3-diaxial interaction.

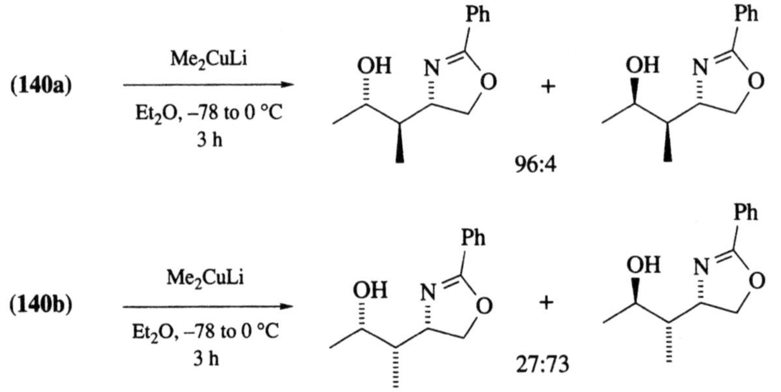

(**140a**) Me$_2$CuLi, Et$_2$O, –78 to 0 °C, 3 h → OH, N, O, Ph + OH, N, O, Ph 96:4

(**140b**) Me$_2$CuLi, Et$_2$O, –78 to 0 °C, 3 h → OH, N, O, Ph + OH, N, O, Ph 27:73

Scheme 25

M, O, N, O, Ph, H, R^1, R^2 ≡ H, MeMgBr, Ph, O, N, M, O, R^1, R^2, H, Me$_2$CuLi

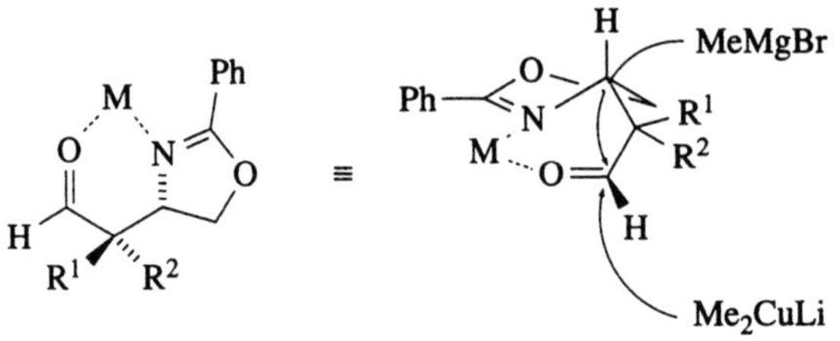

(**140**) (**142**)

As part of an extensive campaign *en route* to milbemycin β$_3$, Smith's synthesis of the 'northern hemisphere' (**143**) relied on the Ireland ester enolate Claisen rearrangements[58] of both diastereomers (**144a**) and (**144b**).[93] These esters came about from the additions of isopropenyl nucleophiles to aldehyde (**145**; equation 33). The corresponding Grignard reagent added to (**145**) to ultimately give a 2:1 mixture of (**144a**) and (**144b**), respectively. Although each product could be utilized as part of their 'divergent–

convergent maneuver',[93] a more direct route to (**144b**) was sought. While the reaction of (**145**) with isopropenyllithium occurred in a nondiscriminatory manner, the corresponding lithium cuprate gave a much-improved 7:1 (**144b**):(**144a**) ratio (equation 34). Chelation associated with the β-alkoxyspiroketal oxygen(s), as in the six-membered ring state (**146**), presumably encourages pseudoaxial attack to give mainly (**144b**).

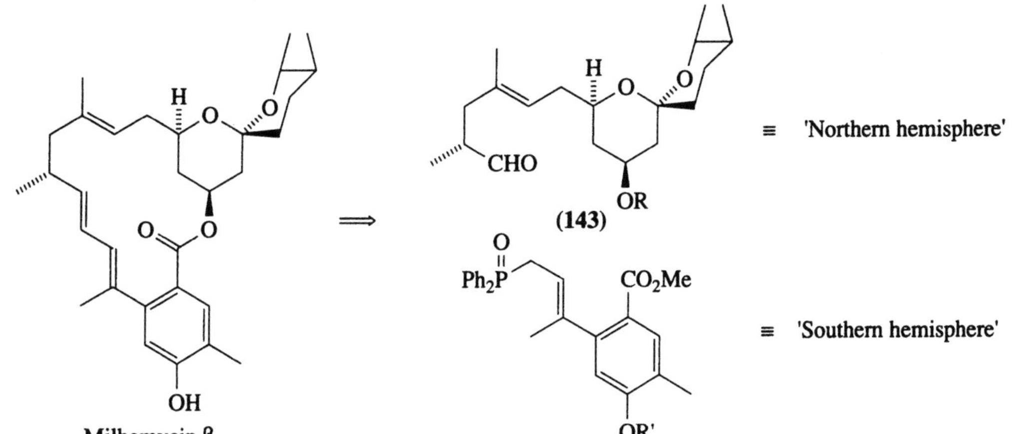
H
O
O
OH
CHO
OR
O
Ph2P
CO2Me
OR'
≡ 'Northern hemisphere'
≡ 'Southern hemisphere'

Milbemycin β_3 (**143**)

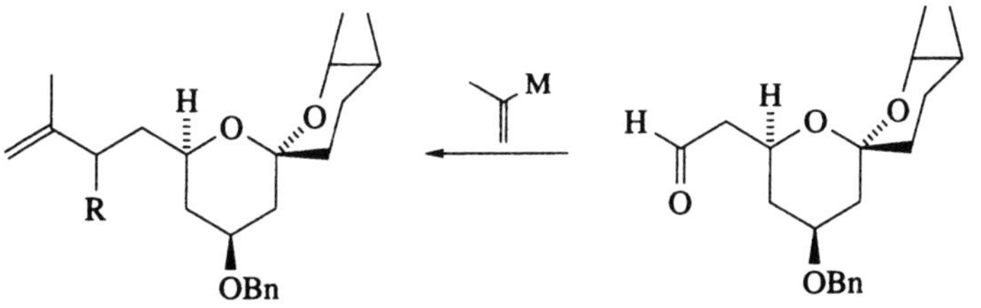
H
O
R
OBn
M
O

(**144**) **a:** R = α-$EtCO_2$ (**145**) (33)

b: R = β-$EtCO_2$

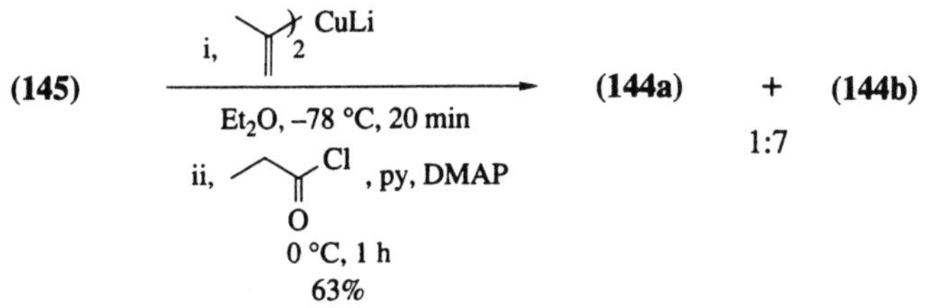
(145)
i, CuLi
2
Et2O, –78 °C, 20 min
ii, Cl , py, DMAP
O
0 °C, 1 h
63%
(144a) + (144b)
1:7

(34)

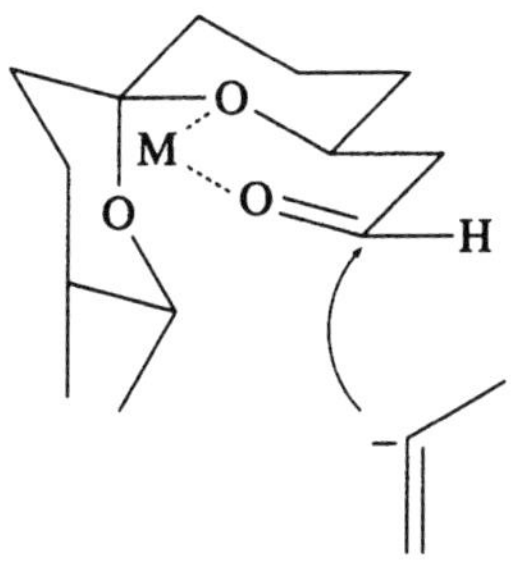
O
M
O
O
H

(**146**)

An alternative route to milbemycin β3 by Baker relies on a straightforward sequence of steps to obtain homochiral alkyne (**150**), beginning with (*S*)-methyl 3-hydroxy-2-methylpropionate (**147**).[94] A Gilman cuprate delivers the methyl group to (**148**) in a chelation-controlled manner resulting in the *threo* (*anti*) product (**149**) in high yield (Scheme 26). Coupling of (**150**) with lactone (**151**) to give (**152**), and further handling to give intermediate (**153**),[95] leads to a formal total synthesis (Scheme 27).

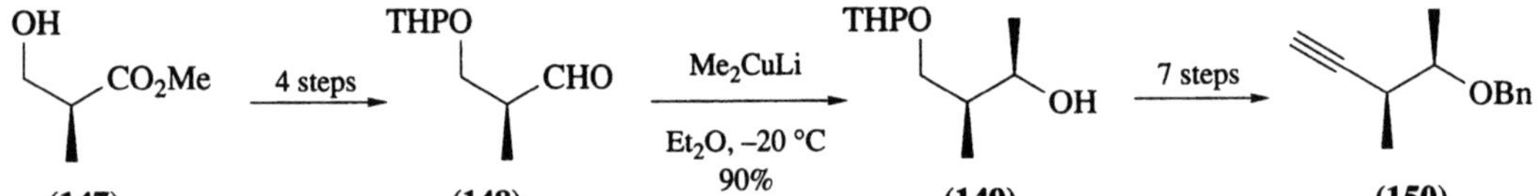

Scheme 26

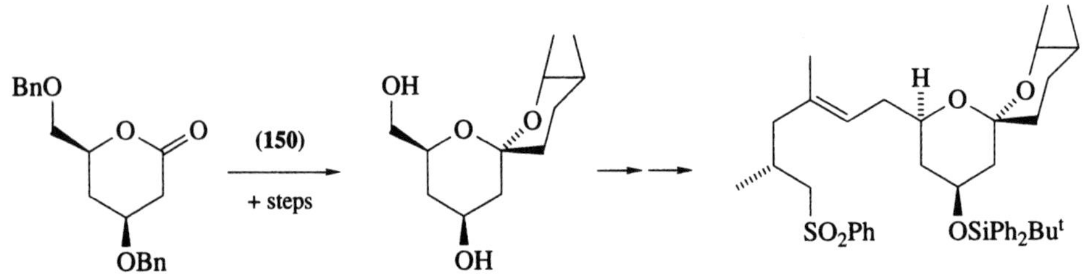

Scheme 27

Polyether antibiotic X-206, one of the first examples of this venerable class of natural products discovered close to four decades ago,[96] has recently yielded to total synthesis by Evans and coworkers.[97] Assemblage of the left half portion of the ionophore (*i.e.* **154**) requires the adjoining of two substantial homochiral pieces (**155**) and (**156**) such that the (11*R*)-configuration is generated as the C(11)—C(12) bond is made. Use of the organolithium, itself derived from (**156**), affords mainly the (11*S*)-isomer by a factor of 2:1. Incorporation of the organolithium into the higher order cuprate (**157**),[5] however, leads to clean coupling of these segments in not only excellent isolated yield, but with a 49:1 preference for the desired *anti* relationship found in (**158**; equation 35).

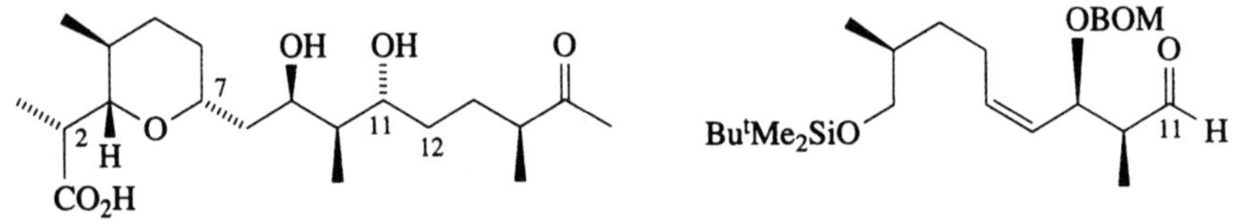

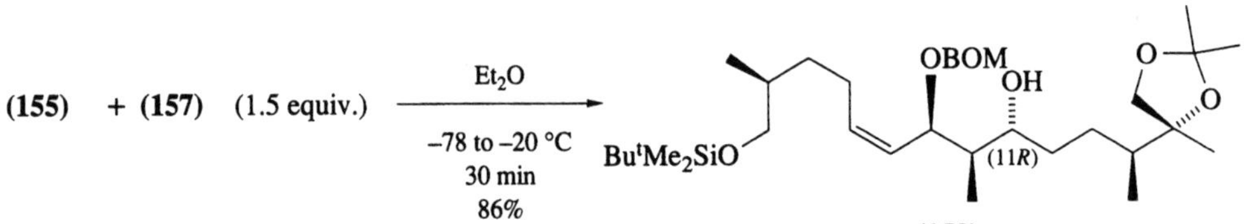

(35)

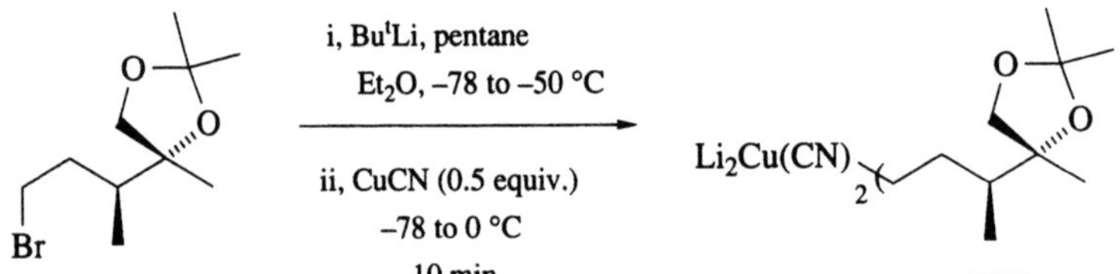

As part of an iterative sequence involving successive sigmatropic rearrangements with intervening chelation-controlled additions of vinyl nucleophiles, Kallmerten and Balestra have developed an effective combination for acyclic stereocontrol, successfully applied to the synthesis of the tocopherol side chain (**159**).[98] 1,2-Addition of a $MgBr_2 \cdot Et_2O$-modified (*E*)-1-propenyllithium-derived reagent to aldehyde (**160**) in Et_2O at –78 °C gives a (**161a**):(**161b**) ratio in excess of 50:1 (equation 36). Subsequent acylation to (**162**) sets the stage for the glycolate Claisen rearrangement[99] of the preformed silyl enol ether, the product from which is taken on to (+)-(**159**).

HO

(**159**)

OHC OBz (**160**) → (CuLi)2, $MgBr_2{\bullet}Et_2O$, Et_2O, –78 °C, 85% → OR OBz (**161a**) R = H; (**162**) R = $BnOCH_2CO$ + *anti* isomer (**161b**) (36)

In order to arrive at (±)-methyl nonactate (**164**), Baldwin envisioned the addition of a soft organometallic source of methyl anion to aldehyde (**163**), the stereochemical outcome from which would be reagent dependent (equation 37).[100] Although earlier work by Ireland had shown that using Me_2CuLi in a hydrocarbon medium on (**163**) is not selective,[101,102] the potential for controlled delivery of 'Me⁻' is predicated on related work by White,[103] which focussed on reduction of the corresponding methyl ketone. Thus, changing solvents to Et_2O does indeed alter the 1:1 ratio to 4.5:1 with Me_2CuLi, although the 8-epimethyl nonactate predominates. The desired material (**164**) is best achieved using '$MeTiCl_3$',[14b] prepared from $TiCl_4$/Me_2Zn in CH_2Cl_2.

(**163**) —'Me⁻'→ (**164**) + (**165**) (37)

	(164):(165)
Me_2CuLi, Et_2O, –78 °C, 85%	1:4.5
$TiCl_4$/Me_2Zn, CH_2Cl_2, –78 °C, 78%	24:1

Sato and coworkers have applied the organocopper methodology developed earlier by their group (*vide supra*)[17] to a multigram scale preparation of a commonly used intermediate (**166**) for the synthesis of leukotriene A_4 (LTA_4).[104] Hydromagnesiation[105] of silylalkyne (**167**), catalyzed by titanocene dichloride, is followed by transmetallation with $CuI{\cdot}Me_2S$ to give the vinylcopper species (**168**; Scheme 28). Addition of (*R*)-glyceraldehyde acetonide leads to the expected *syn* product (**169**) in 85% overall yield from (**167**). The diastereoselectivity is better than 40:1 (*syn:anti*), with allylic alcohol (**169**) well suited for Sharpless epoxidation, which establishes C-5 and C-6 in (**166**) and eventually in LTA_4. The same protocol can be used to synthesize other homochiral epoxy aldehydes in quantity.[104]

Analogs of leukotriene B_4 (LTB_4, **170**) have aroused interest based on the potent chemotactic properties of LTB_4 itself. The disconnected C-6, C-7 (*Z*)-alkene in (**170**) has been viewed by Taylor, in alkynic form, as a ready means of attaching two residues *via* a carbocupration/1,2-addition scheme (equation 38).[106] Extremely high (*Z*)-stereoselectivity was anticipated for the former step, as discussed by Normant and Alexakis,[107] although trapping of the so-formed vinylcopper complex by an aldehyde was on less secure ground. Model studies showed that lower order lithium cuprates were the most efficacious towards addition across alkynes, while reagents such as RCu·ligand (*e.g.* ligand = Me_2S, LiBr or $MgBr_2$) gave

OHC (6, 5) CO_2Me (**166**)

CO_2H LTA_4

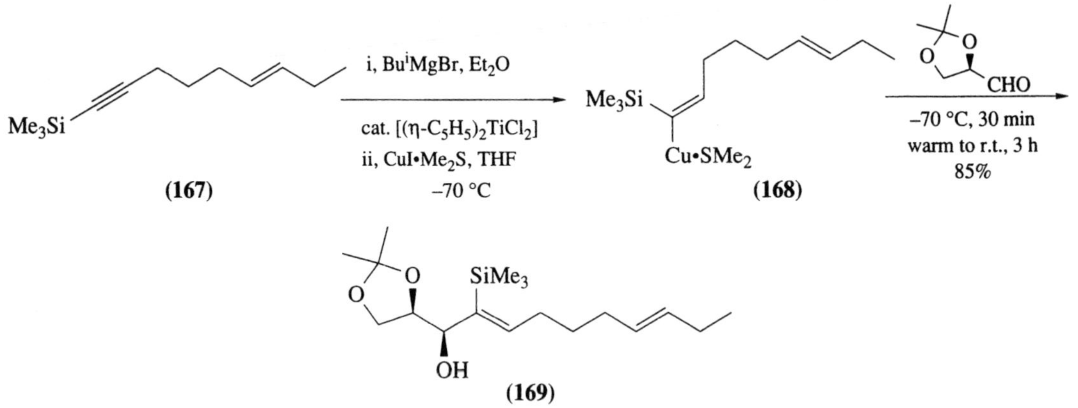

i, Bu^iMgBr, Et_2O
cat. $[(\eta\text{-}C_5H_5)_2TiCl_2]$
ii, $CuI{\cdot}Me_2S$, THF
–70 °C

–70 °C, 30 min
warm to r.t., 3 h
85%

(167) (168) (169)

Scheme 28

none of the desired product. The same negative results were found with higher order cuprates, although this is likely to be due to their greater basicity and precludes their use (as a source of carbon nucleophiles) in the presence of alkynic protons. Thus, carbocupration of acetylene with mixed cuprate **(171)** afforded vinylcuprate **(172)**, the formation of which required unusually high temperatures. The 1,2-addition to ester aldehyde **(173)** was best effected in the presence of $BF_3{\cdot}Et_2O$, thereby giving **(174)**, saponification of which gave the hexahydro LTB_4 analog (**175**; Scheme 29). A double carbocupration,[108] starting with $(C_{11}H_{23})_2CuLi$ and excess acetylene, and thence 1,2-addition to **(173)**, ultimately led to the tetrahydro LTB_4 analog (**176**; Scheme 30). Both **(175)** and **(176)** can also be converted to their corresponding lactones.

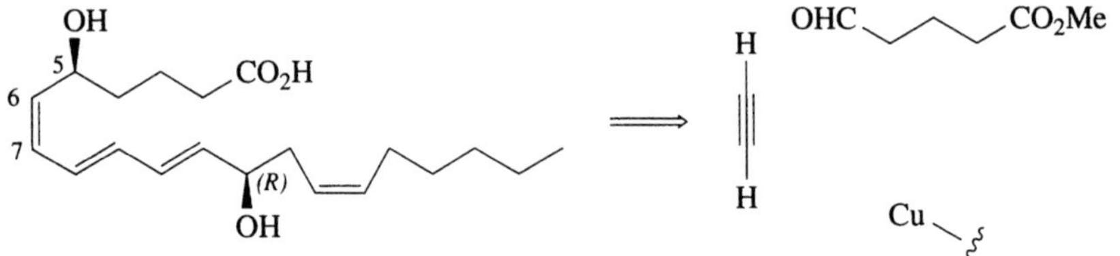

(170) (38)

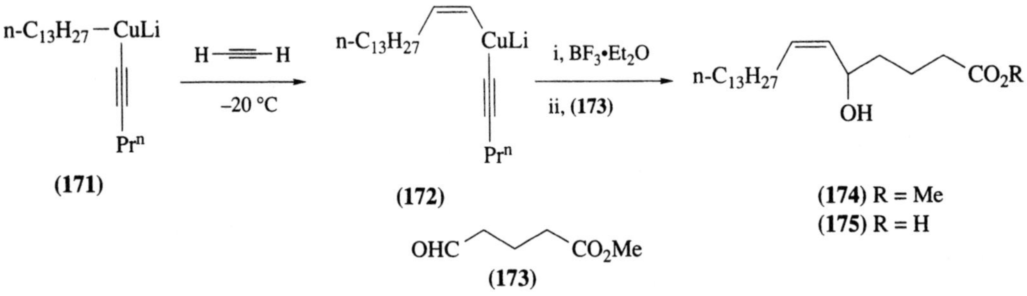

$n\text{-}C_{13}H_{27}$—CuLi; H—≡—H, –20 °C; i, $BF_3{\cdot}Et_2O$; ii, **(173)**

(171) (172) (173) (174) R = Me (175) R = H

Scheme 29

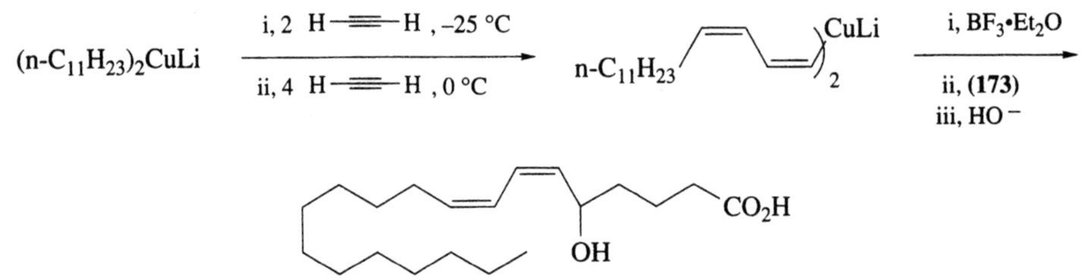

$(n\text{-}C_{11}H_{23})_2CuLi$; i, 2 H—≡—H, –25 °C; ii, 4 H—≡—H, 0 °C; i, $BF_3{\cdot}Et_2O$; ii, **(173)**; iii, HO^-

(176)

Scheme 30

Also within the domain of arachidonic acid derivatives, a 5-exo-dig cyclization[109] of an α-silyl radical onto an alkyne, followed by proteodesilylation, was applied to a synthesis of isocarbacyclin (**177**), a carbocyclic analog of the potent antihypertensive and platelet aggregation inhibitor prostacyclin (**178**). The necessary functional group array was set in place by Noyori using a 1,2-addition of a higher order silylcuprate to aldehyde (**179**; Scheme 31).[110] A mix of alcohols (**180**)[111] was obtained in good yield, the unspecified ratio being of no consequence. Radical-initiated ring closure, employing either a photochemical step or the action of hot tributyltin hydride on the derived xanthate, formed the bicyclic network (**181**). Conversion to (**177**) was subsequently achieved by cleavage of the silyl–carbon bond with TFA, followed by deblocking, isomerization of the *exo*- to the *endo*-alkene and saponification.

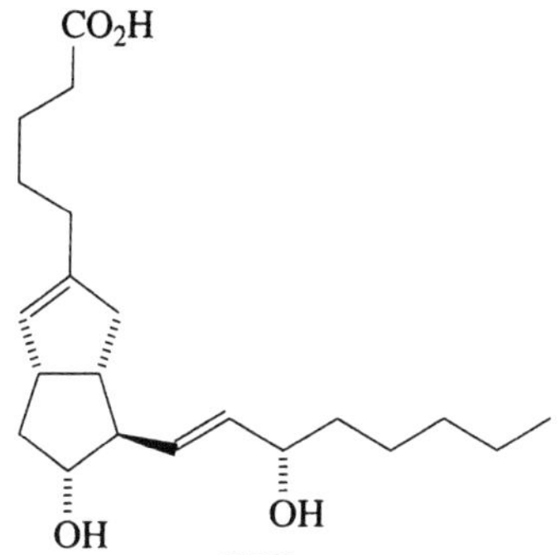

CO_2H OH OH (**177**) CO_2H O OH OH (**178**)

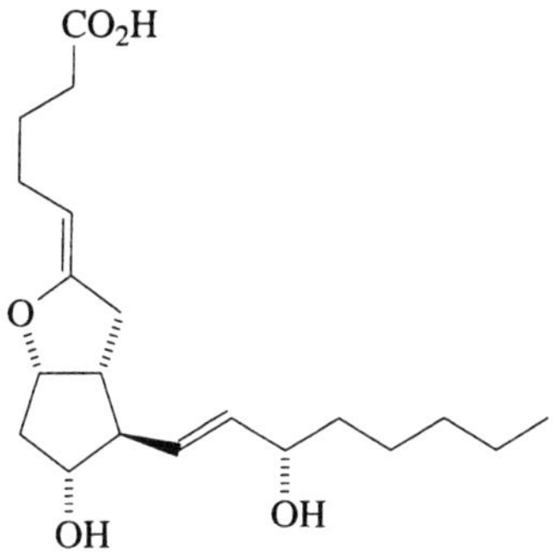

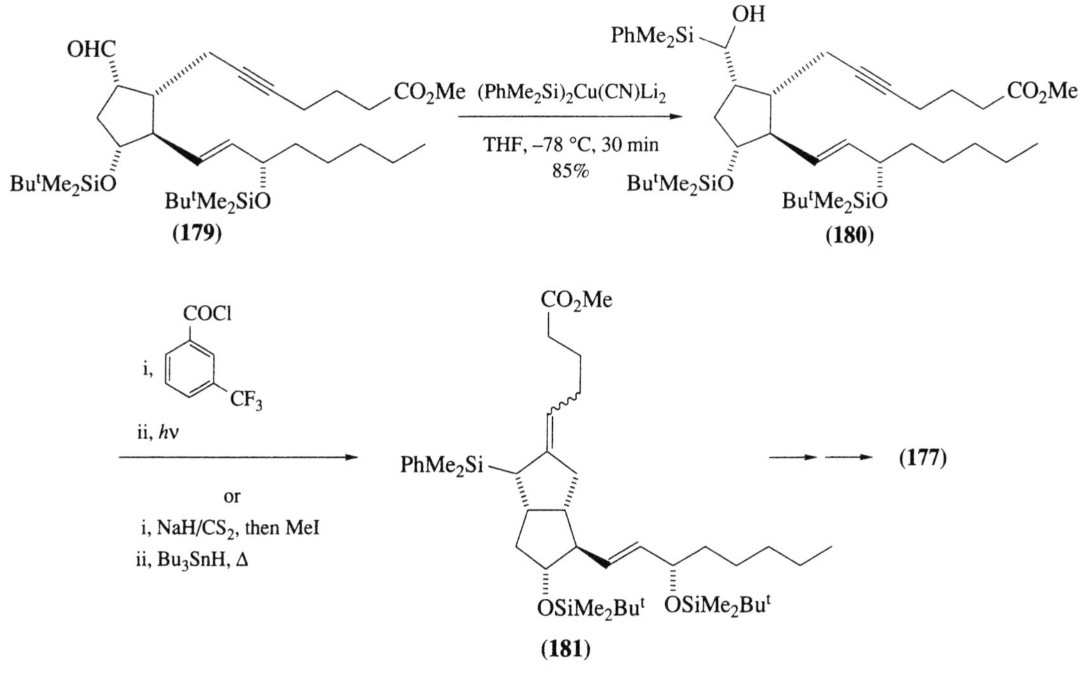

OHC CO_2Me Bu^tMe_2SiO Bu^tMe_2SiO (**179**) $(PhMe_2Si)_2Cu(CN)Li_2$ THF, –78 °C, 30 min 85% $PhMe_2Si$ OH CO_2Me Bu^tMe_2SiO Bu^tMe_2SiO (**180**)

i, COCl, CF_3 ii, *h*ν or i, NaH/CS_2, then MeI ii, Bu_3SnH, Δ CO_2Me $PhMe_2Si$ $OSiMe_2Bu^t$ $OSiMe_2Bu^t$ (**181**) → → (**177**)

Scheme 31

Tricyclic ketone (**183**), a photoproduct from the intramolecular [2 + 2] cycloaddition of silyl enol ether (**182**), was found by Pattenden and Teague to add methyl stereoselectivity through the agency of $Me_2CuLi/MeLi$.[112] The process occurs in Et_2O at –68 °C following the Macdonald and Still prescription,[51] the product (**184**) from which undergoes facile acid-induced ring expansion to cyclooctenone (**185**). Wittig methylenation, isomerization and Lewis acid promoted transannular cyclization ultimately produces (±)-pentalenene (**186**; Scheme 32).

In the β-lactam area, Grieco *et al.* have applied their development of substituted bicyclo[2.2.1]heptanes to a synthesis of the thienamycin precursor (**188a**), embedded in which are three contiguous asymmetric centers corresponding to C-5, C-6 and C-8 in the natural product (**190**).[113] Readily obtained bromo aldehyde (**187**), upon treatment with the MeLi-derived higher order cuprate $Me_2Cu(CN)Li_2$ in

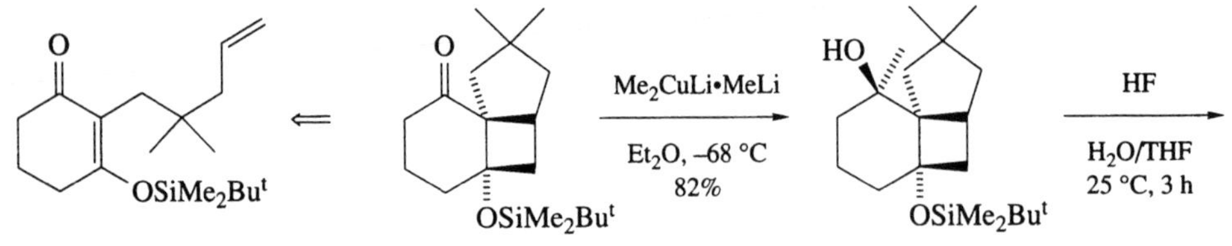

(182) (183) (184)

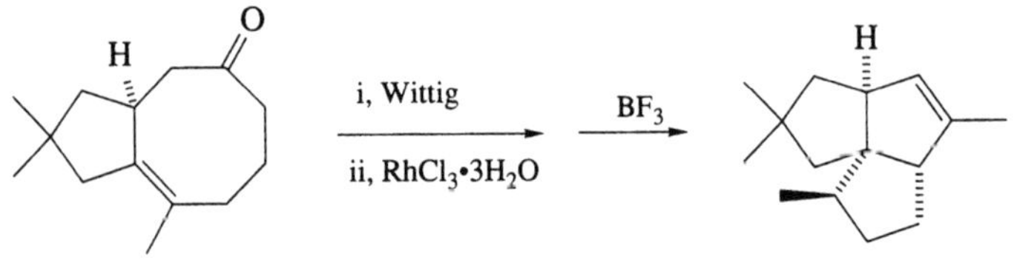

(185) (186)

Scheme 32

Et_2O at reduced temperatures, gave in good yield (79%) diastereomer (**188a**) together with 8% of the undesired epimer (**188b**). By contrast, MeLi alone showed no preference in its 1,2-addition to (**187**). Further processing of (**188a**) eventually led to β-keto ester (**189**), a known intermediate *en route* to (**190**; Scheme 33).

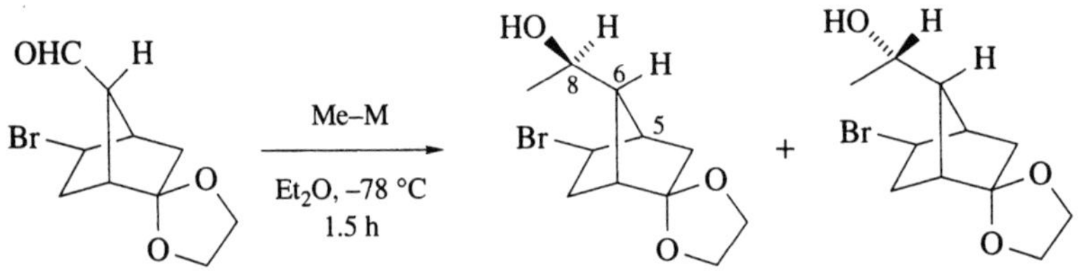

(187) (188a) (188b)

$Me_2Cu(CN)Li_2$	*ca.* 10:1
MeLi	1:1

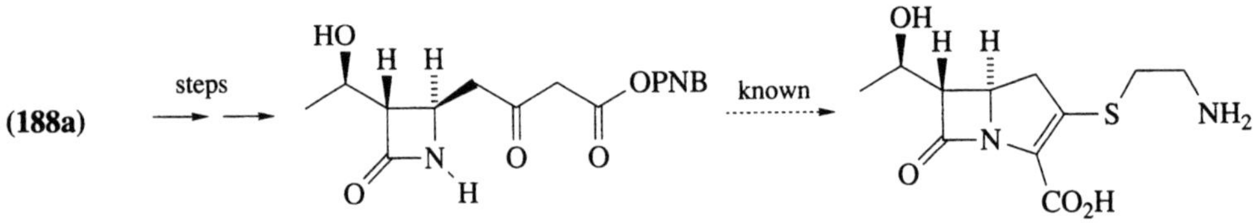

(189) (190)

Scheme 33

In search of a practical route to diastereomeric lactones (**193**) and (**194**), potential agents for mosquito control needed by entomologists for evaluation, acrolein dimer (**191**) was deemed as the logical starting point.[114] For compound (**193**), the major component of the oviposition attractant pheromone of the mosquito *Culex pipiens fatigans*, the required stereorelationship can be constructed by the initial action of a Grignard reagent on (**191**) at –20 °C in THF doped with HMPA, which minimizes chelation-enhanced *threo* (*syn*) selectivity, thereby favoring (**192a**).[9,11] TMEDA was not an acceptable substitute for HMPA. Addition of CuI to the Grignard, which according to Sato[17] forms the organocopper complex RCuMgBrI, proceeds through a Cram cyclic transition state[12] to arrive at a (**192a**):(**192b**) mix now rich in the *syn* isomer (equation 39). Separation of these 1,2-adducts and subsequent acylation–PCC oxidation culminated this brief approach by Jefford, easily amenable to scale-up (Scheme 34).

Singh and Oehlschlager have used the same starting material (*i.e.* **191**) to study the 1,2-addition of ethylmetallic reagents, in this case with an eye toward both (racemic) *exo*- and *endo*-brevicomin.[115] The *erythro* (*anti*) product (**196b**) can be obtained with the EtLi/$BF_3 \cdot Et_2O$ combination, presumably due to a strong preference for an *anti* configuration of the polar groups as in (**197**) once the Lewis acid complexation has occurred.[116] Ethylcopper reagents reversed the outcome, giving an *ca.* 89:11 split in favor of (**196a**), presumably *via* (**195**; Scheme 35), a result comparable to that obtained by Jefford (*vide supra*).[114]

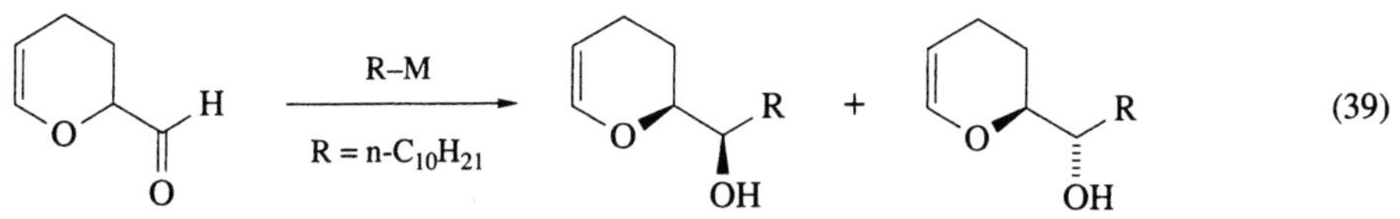

(191) → R–M, R = n-$C_{10}H_{21}$ → (192a) *anti* + (192b) *syn* (39)

RMgBr; 84% 5.67:1
THF, HMPA (6 equiv.), –20 °C
RCu•MgBrI; 84% 1:9
THF, Me_2S, –78 to 0 °C

(193) ← i, Ac_2O; ii, PCC ← (192a) (192b) → i, Ac_2O; ii, PCC → (194)

n-$C_{10}H_{21}$, OAc

Scheme 34

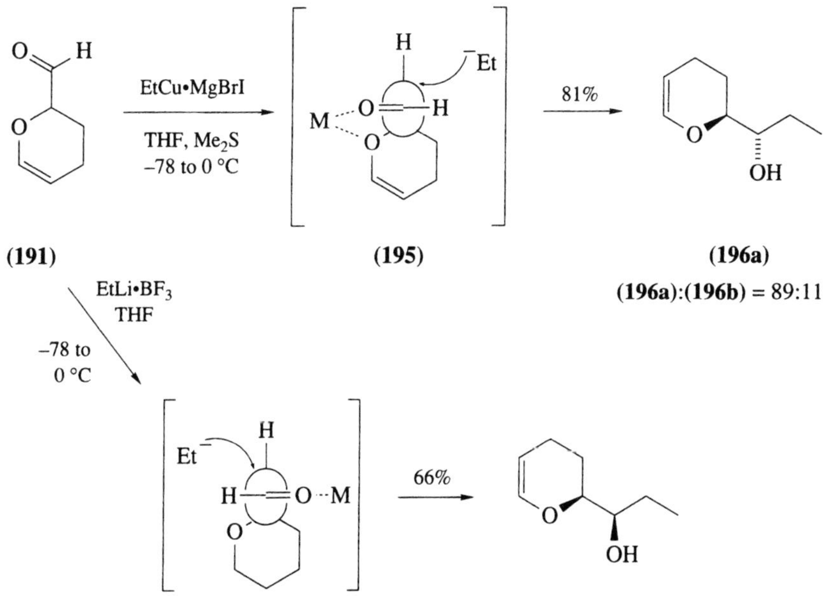

(191) → EtCu•MgBrI, THF, Me_2S, –78 to 0 °C → [(195)] → 81% → (196a)

(196a):(196b) = 89:11

(191) → $EtLi•BF_3$, THF, –78 to 0 °C → [(197)] → 66% → (196b)

(196b):(196a) = 92:8

Scheme 35

Both *exo*- and *endo*-brevicomin (*vide infra*), in optically active form, have been synthesized by elaboration of α-alkoxy aldehyde (**201**).[117] Condensation of the 1-trimethylsilylvinylcopper reagent (**198**) with the (*R*)-glyceraldehyde derivative (**199**) gave *syn* product (**200**) in >85% yield with over 98% diastereoselectivity (Scheme 36). Hydroxy protection, hydrolysis and periodate cleavage of the *vic*-diol converted (**200**) to aldehyde (**201**) in 50–65% overall yields for the two steps.

exo-Brevicomin *endo*-Brevicomin

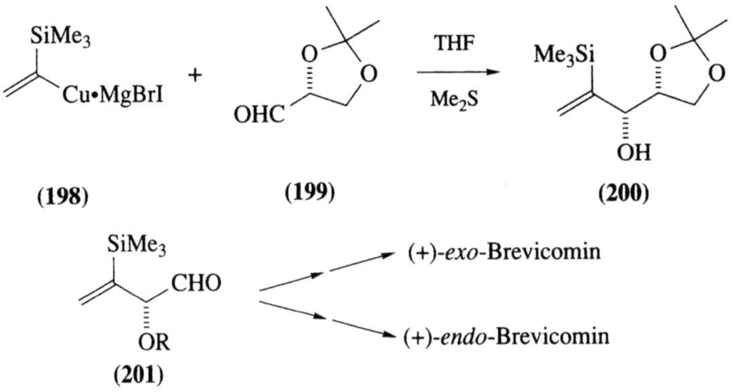

SiMe3 Cu•MgBrI + OHC O O THF Me2S Me3Si O O OH

(198) (199) (200)

SiMe3 CHO OR (201) → (+)-*exo*-Brevicomin → (+)-*endo*-Brevicomin

Scheme 36

1.4.5 REFERENCES

1. G. H. Posner, *Org. React. (N.Y.)*, 1975, **22**, 253.
2. G. H. Posner, *Org. React. (N.Y.)*, 1972, **19**, 1.
3. J. F. Normant, *Pure Appl. Chem.*, 1978, **50**, 709; A. Alexakis, C. Chuit, M. Commercon-Bourgain, J. P. Foulon, N. Jabri, P. Mangeney and J. F. Normant, *Pure Appl. Chem.*, 1984, **56**, 91; E. Erdik, *Tetrahedron*, 1984, **40**, 641.
4. J. P. Gorlier, L. Hamon, J. Levisalles and J. Wagnon, *J. Chem. Soc., Chem. Commun.*, 1973, 88; L. Hamon and J. Levisalles, *J. Organomet. Chem.*, 1983, **251**, 133; L. Hamon and J. Levisalles, *Tetrahedron*, 1989, **45**, 489.
5. B. H. Lipshutz, *Synthesis*, 1987, 325; B. H. Lipshutz, R. S. Wilhelm and J. A. Kozlowski, *Tetrahedron*, 1984, **40**, 5005; B. H. Lipshutz, *Synlett*, 1990, 119.
6. B. H. Lipshutz, D. A. Parker, S. L. Nguyen, K. E. McCarthy, J. C. Barton, S. Whitney and H. Kotsuki, *Tetrahedron*, 1986, **42**, 2873.
7. H. Gilman, R. G. Jones and L. A. Woods, *J. Org. Chem.*, 1952, **17**, 1630; G. H. Posner, 'An Introduction to Synthesis Using Organocopper Reagents', Wiley, New York, 1980.
8. G. H. Posner, C. E. Whitten and P. E. McFarland, *J. Am. Chem. Soc.*, 1972, **94**, 5106.
9. W. C. Still and J. A. Schneider, *Tetrahedron Lett.*, 1980, **21**, 1035.
10. W. C. Still and J. H. McDonald, III, *Tetrahedron Lett.*, 1980, **21**, 1031.
11. K. Mead and T. L. Macdonald, *J. Org. Chem.*, 1985, **50**, 422.
12. D. J. Cram and K. R. Kopecky, *J. Am. Chem. Soc.*, 1959, **81**, 2748; T. J. Leitereg and D. J. Cram, *J. Am. Chem. Soc.*, 1968, **90**, 4019; W. C. Still and J. A. Schneider, *J. Org. Chem.*, 1980, **45**, 3375.
13. M. Cherest, H. Felkin and N. Prudent, *Tetrahedron Lett.*, 1968, 2199; K. N. Houk, *Pure Appl. Chem.*, 1983, **55**, 277 and refs. therein.
14. For reviews on chelation- and nonchelation-controlled processes, see (a) G. J. McGarvey, M. Kimura, T. Oh and J. M. Williams, *Carbohydr. Res.*, 1984, **3**, 125; (b) M. T. Reetz, *Angew. Chem., Int. Ed. Engl.*, 1984, **23**, 556; (c) P. A. Bartlett, *Tetrahedron*, 1980, **36**, 3; (d) J. Jurczak, S. Pikul and T. Bauer, *Tetrahedron*, 1986, **42**, 447.
15. D. R. Williams and F. D. Klinger, *Tetrahedron Lett.*, 1987, **28**, 869.
16. K. Mead, *Tetrahedron Lett.*, 1987, **28**, 1019.
17. F. Sato, Y. Kobayashi, O. Takahashi, T. Chiba, Y. Takeda and M. Kusakabe, *J. Chem. Soc., Chem. Commun.*, 1985, 1636; for a review on these additions, see F. Sato and Y. Kobayashi, *Yuki Gosei Kagaku Kyokai Shi*, 1986, **44**, 558.
18. M. Kusakabe and F. Sato, *J. Chem. Soc., Chem. Commun.*, 1986, 989.
19. M. Kusakabe and F. Sato, *Chem. Lett.*, 1986, 1473.
20. N. T. Anh, *Fortschr. Chem. Forsch.*, 1980, **88**, 145.
21. T. Kunz and H.-U. Reissig, *Angew. Chem., Int. Ed. Engl.*, 1988, **27**, 268.
22. Y. Yamamoto and K. Maruyama, *J. Am. Chem. Soc.*, 1985, **107**, 6411; Y. Yamamoto, *Acc. Chem. Res.*, 1987, **20**, 243.
23. B. H. Lipshutz, E. L. Ellsworth and T. J. Siahaan, *J. Am. Chem. Soc.*, 1989, **111**, 1351.
24. B. H. Lipshutz, E. L. Ellsworth and T. J. Siahaan, *J. Am. Chem. Soc.*, 1988, **110**, 4834.
25. E. J. Corey, D. M. Floyd and B. H. Lipshutz, *J. Org. Chem.*, 1978, **43**, 3418.
26. B. H. Lipshutz, E. L. Ellsworth, T. J. Siahaan and A. Shirazi, *Tetrahedron Lett.*, 1988, **29**, 6677.
27. For other reports on the effects of trialkylsilyl halides on organocopper reactions, see (a) E. J. Corey and N. W. Boaz, *Tetrahedron Lett.*, 1985, **26**, 6015, 6019; (b) E. Nakamura, S. Matsuzawa, Y. Horiguchi and I. Kuwajima, *Tetrahedron Lett.*, 1986, **26**, 4029; (c) S. Matsuzawa, Y. Horiguchi, E. Nakamura and I. Kuwajima, *Tetrahedron*, 1989, **45**, 349; (d) A. Alexakis, J. Berlan and Y. Besace, *Tetrahedron Lett.*, 1986, **27**, 1047; (e) C. R. Johnson and T. J. Marren, *Tetrahedron Lett.*, 1987, **28**, 27; (f) E.-L. Lindstedt, M. Nilsson and T. Olsson, *J. Organomet. Chem.*, 1987, **334**, 255; (g) M. Bergdahl, E.-L. Lindstedt, M. Nilsson and T. Olsson, *Tetrahedron*, 1988, **44**, 2055; (h) M. Bergdahl, E.-L. Lindstedt, M. Nilsson and T. Olsson, *Tetrahedron*, 1989, **45**, 535.
28. Y. Yamamoto and K. Maruyama, *J. Organomet. Chem.*, 1985, **284**, C45.
29. E. Epifani, S. Florio and G. Ingrosso, *Tetrahedron*, 1988, **44**, 5869.

30. K. Tamao, E. Nakajo and Y. Ito, *Tetrahedron*, 1988, **44**, 3997.
31. Previously observed in 1,4-additions of metallated allyltrimethylsilane: *cf.* R. J. P. Corriu, C. Guerin and J. M'Boula, *Tetrahedron Lett.*, 1981, **22**, 2985.
32. K. Tamao, M. Kumada and K. Maeda, *Tetrahedron Lett.*, 1984, **25**, 321.
33. C. Chuit, J. P. Foulon and J. F. Normant, *Tetrahedron*, 1981, **37**, 1385.
34. M. Bourgain-Commercon, J. P. Foulon and J. F. Normant, *J. Organomet. Chem.*, 1982, **228**, 321.
35. G. Majetich, A. M. Casares, D. Chapman and M. Behnke, *Tetrahedron Lett.*, 1983, **24**, 1909.
36. G. Majetich, A. M. Casares, D. Chapman and M. Behnke, *J. Org. Chem.*, 1986, **51**, 1745.
37. D. K. Hutchinson and P. L. Fuchs, *Tetrahedron Lett.*, 1986, **27**, 1429.
38. B. H. Lipshutz, E. L. Ellsworth, S. H. Dimock and R. A. J. Smith, *J. Org. Chem.*, 1989, **54**, 4977.
39. B. H. Lipshutz and S. Sengupta, *Org. React. (N.Y.)*, in press.
40. C. Ouannes, G. Dressaire and Y. Langlois, *Tetrahedron Lett.*, 1977, 815.
41. M. C. P. Yeh, P. Knochel and L. E. Santa, *Tetrahedron Lett.*, 1988, **29**, 3887.
42. M. Franck-Neumann, D. Martina and M.-P. Heitz, *J. Organomet. Chem.*, 1986, **301**, 61.
43. M. Brookhart, A. R. Pinhas and A. Lukacs, *Organometallics*, 1982, **1**, 1730.
44. R. H. Schwartz and J. San Filippo, Jr., *J. Org. Chem.*, 1979, **44**, 2705.
45. R. J. Anderson, C. A. Henrick and L. D. Rosenblum, *J. Am. Chem. Soc.*, 1974, **96**, 3654.
46. H. O. House, W. L. Respess and G. M. Whitesides, *J. Org. Chem.*, 1966, **31**, 3128; H. O. House, C.-Y. Chu, J. M. Wilkins and M. J. Umen, *J. Org. Chem.*, 1975, **40**, 1460.
47. H. O. House, *Acc. Chem. Res.*, 1976, **9**, 59.
48. L. T. Scott and W. D. Cotton, *J. Chem. Soc., Chem. Commun.*, 1973, 320.
49. H. O. House, A. V. Prabhu, J. M. Wilkins and L. F. Lee, *J. Org. Chem.*, 1976, **41**, 3067.
50. H. O. House and C.-Y. Chu, *J. Org. Chem.*, 1976, **41**, 3083; see also H. Yamataka, N. Fujimura, Y. Kawafuji and T. Hanafusa, *J. Am. Chem. Soc.*, 1987, **109**, 4305.
51. T. L. Macdonald and W. C. Still, *J. Am. Chem. Soc.*, 1975, **97**, 5280.
52. W. C. Still and T. L. Macdonald, *Tetrahedron Lett.*, 1976, 2659.
53. E. C. Ashby and J. J. Watkins, *J. Chem. Soc., Chem. Commun.*, 1976, 784; *J. Am. Chem. Soc.*, 1977, **99**, 5312.
54. E. C. Ashby, J. J. Lin and J. J. Watkins, *Tetrahedron Lett.*, 1977, 1709; E. C. Ashby and S. A. Noding, *J. Org. Chem.*, 1979, **44**, 4371.
55. Y. Yamamoto, S. Yamamoto, H. Yatagai, Y. Ishihara and K. Maruyama, *J. Org. Chem.*, 1982, **47**, 119.
56. O. Arjona, R. Fernandez de la Pradilla, C. Manzano, S. Perez and J. Plumet, *Tetrahedron Lett.*, 1987, **28**, 5547.
57. D. J. Goldsmith and I. Sakano, *Tetrahedron Lett.*, 1974, 2857.
58. J. P. Marino and D. M. Floyd, *Tetrahedron Lett.*, 1975, 3897.
59. G. Palmisano and R. Pellegata, *J. Chem. Soc., Chem. Commun.*, 1975, 892.
60. H. O. House and W. F. Fischer, Jr., *J. Org. Chem.*, 1969, **34**, 3615.
61. P. G. Gassman, J. G. Schaffhausen and P. W. Raynolds, *J. Am. Chem. Soc.*, 1982, **104**, 6408.
62. H.-J. Liu, L.-K. Ho and H. K. Lai, *Can. J. Chem.*, 1981, **59**, 1685.
63. P. W. Ambler and S. G. Davies, *Tetrahedron Lett.*, 1985, **26**, 2129.
64. T. Kunieda, T. Ishizuka, T. Higuchi and M. Hirobe, *J. Org. Chem.*, 1988, **53**, 3381.
65. M. Wada, Y. Sakurai and K. Akiba, *Tetrahedron Lett.*, 1984, **25**, 1079.
66. Y. Yamamoto, S. Nishii, K. Maruyama, T. Komatsu and W. Ito, *J. Am. Chem. Soc.*, 1986, **108**, 7778; Y. Yamamoto, T. Komatsu and K. Maruyama, *J. Am. Chem. Soc.*, 1984, **106**, 5031; Y. Yamamoto and W. Ito, *Tetrahedron*, 1988, **44**, 5415.
67. D. A. Claremon, P. K. Lumma and B. T. Phillips, *J. Am. Chem. Soc.*, 1986, **108**, 8265.
68. T. Fujisawa, Y. Kurita and T. Sato, *Chem. Lett.*, 1983, 1537.
69. S. H. Bertz, *J. Org. Chem.*, 1979, **44**, 4967.
70. B. H. Lipshutz, B. Huff and W. Vaccaro, *Tetrahedron Lett.*, 1986, **27**, 4241.
71. B. H. Lipshutz and M. C. Morey, *J. Am. Chem. Soc.*, 1984, **106**, 457.
72. T. Bretschneider, W. Miltz, P. Münster and W. Steglich, *Tetrahedron*, 1988, **44**, 5403.
73. M. J. O'Donnell and J.-B. Falmagne, *Tetrahedron Lett.*, 1985, **26**, 699; *J. Chem. Soc., Chem. Commun.*, 1985, 1168.
74. P. J. Sinclair, D. Zhai, J. Reibenspies and R. M. Williams, *J. Am. Chem. Soc.*, 1986, **108**, 1103; R. M. Williams, P. J. Sinclair, D. Zhai and D. Chen, *J. Am. Chem. Soc.*, 1988, **110**, 1547.
75. H. Onoue, M. Narisada, S. Uyeo, H. Matsumura, K. Okada, T. Yano and W. Nagata, *Tetrahedron Lett.*, 1979, 3867.
76. D. H. Hua and A. Verma, *Tetrahedron Lett.*, 1985, **26**, 547.
77. T. Kobayashi, N. Ishida and T. Hiraoka, *J. Chem. Soc., Chem. Commun.*, 1980, 736.
78. A. Martel, J.-P. Daris, C. Bachand, M. Menard, T. Durst and B. Belleau, *Can. J. Chem.*, 1983, **61**, 1899.
79. W. Koller, A. Linkies, H. Pietsch, H. Rehling and D. Reuschling, *Tetrahedron Lett.*, 1982, **23**, 1545.
80. B. H. Lipshutz, R. S. Wilhelm and J. A. Kozlowski, *J. Org. Chem.*, 1984, **49**, 3938.
81. A. Alexakis, J. Berlan and Y. Besace, *Tetrahedron Lett.*, 1986, **27**, 1047.
82. Y. Yamamoto, S. Yamamoto, H. Yatagai, Y. Ishihara and K. Maruyama, *J. Org. Chem.*, 1982, **47**, 119.
83. F. J. Weiberth and S. S. Hall, *J. Org. Chem.*, 1987, **52**, 3901.
84. B. J. Wakefield, in 'Comprehensive Organometallic Chemistry', ed. G. Wilkinson, F. G. A. Stone and E. W. Abel, Pergamon Press, Oxford, 1982, chap. 44.
85. B. L. Feringa and J. F. G. A. Jansen, *Synthesis*, 1988, 184.
86. For a recent update on organocopper chemistry, including carbocupration, see ref. 39.
87. S. D. Burke, F. J. Schoenen and M. S. Nair, *Tetrahedron Lett.*, 1987, **28**, 4143.
88. R. E. Ireland, R. H. Mueller and A. K. Willard, *J. Am. Chem. Soc.*, 1976, **98**, 2868.
89. S. D. Burke, D. M. Armistead and J. M. Fevig, *Tetrahedron Lett.*, 1985, **26**, 1163.
90. S. D. Burke, D. M. Armistead, F. J. Schoenen and J. M. Fevig, *Tetrahedron*, 1986, **42**, 2787; S. D. Burke, D. M. Armistead and F. J. Schoenen, *J. Org. Chem.*, 1984, **49**, 4320.

91. R. K. Boeckman, Jr., E. J. Enholm, D. M. Demko and A. B. Charette, *J. Org. Chem.*, 1986, **51**, 4743.
92. G. J. McGarvey, J. M. Williams, R. N. Hiner, Y. Matsubara and T. Oh, *J. Am. Chem. Soc.*, 1986, **108**, 4943.
93. S. R. Schow, J. D. Bloom, A. S. Thompson, K. N. Winzenberg and A. B. Smith, III, *J. Am. Chem. Soc.*, 1986, **108**, 2662.
94. R. Baker, M. J. O'Mahony and C. J. Swain, *Tetrahedron Lett.*, 1986, **27**, 3059.
95. R. Baker, R. H. O. Boyes, D. M. P. Broom, J. A. Devlin and C. J. Swain, *J. Chem. Soc., Chem. Commun.*, 1983, 829.
96. J. Berger, A. I. Rachlin, W. E. Scott, L. H. Sternbach and M. W. Goldberg, *J. Am. Chem. Soc.*, 1951, **73**, 5295.
97. D. A. Evans, S. L. Bender and J. Morris, *J. Am. Chem. Soc.*, 1988, **110**, 2506.
98. J. Kallmerten and M. Balestra, *J. Org. Chem.*, 1986, **51**, 2855.
99. T. J. Gould, M. Balestra, M. D. Wittman, J. A. Gary, L. T. Rossano and J. Kallmerten, *J. Org. Chem.*, 1987, **52**, 3889, and refs. therein.
100. S. W. Baldwin and J. M. McIver, *J. Org. Chem.*, 1987, **52**, 320.
101. R. E. Ireland and J.-P. Vevert, *J. Org. Chem.*, 1980, **45**, 4259.
102. R. E. Ireland and J.-P. Vevert, *Can. J. Chem.*, 1981, **59**, 572.
103. M. J. Arco, M. H. Trammell and J. D. White, *J. Org. Chem.*, 1976, **41**, 2075.
104. Y. Kobayashi, Y. Kitano, T. Matsumoto and F. Sato, *Tetrahedron Lett.*, 1986, **27**, 4775.
105. F. Sato, *J. Organomet. Chem.*, 1986, **285**, 53; F. Sato, H. Ishikawa and M. Sato, *Tetrahedron Lett.*, 1981, **22**, 85.
106. M. Furber, R. J. K. Taylor and S. C. Burford, *Tetrahedron Lett.*, 1985, **26**, 2731.
107. J. F. Normant and A. Alexakis, *Synthesis*, 1981, 841.
108. A. Alexakis and J. F. Normant, *Tetrahedron Lett.*, 1982, **23**, 5151.
109. J. E. Baldwin, *J. Chem. Soc., Chem. Commun.*, 1976, 734.
110. M. Suzuki, H. Koyano and R. Noyori, *J. Org. Chem.*, 1987, **52**, 5583.
111. For an alternative route to α-silyl alcohols, see R. J. Linderman and A. Ghannam, *J. Org. Chem.*, 1988, **53**, 2878.
112. G. Pattenden and S. J. Teague, *Tetrahedron Lett.*, 1984, **25**, 3021.
113. P. A. Grieco, D. L. Flynn and R. E. Zelle, *J. Am. Chem. Soc.*, 1984, **106**, 6414.
114. C. W. Jefford, D. Jaggi and J. Boukouvalas, *Tetrahedron Lett.*, 1986, **27**, 4011.
115. S. M. Singh and A. C. Oehlschlager, *Can. J. Chem.*, 1988, **66**, 209.
116. M. Bhupathy and T. Cohen, *Tetrahedron Lett.*, 1985, **26**, 2619.
117. F. Sato, O. Takahashi, T. Kato and Y. Kobayashi, *J. Chem. Soc., Chem. Commun.*, 1985, 1638.

1.5
Organotitanium and Organozirconium Reagents

CARLA FERRERI, GIOVANNI PALUMBO and ROMUALDO CAPUTO
Università di Napoli, Italy

1.5.1 INTRODUCTION

In 1861, the first example of an organotitanium reagent was reported.[1] Nevertheless, more than ninety years were necessary before the preparation of the stable compound, phenyltitanium isopropoxide, was achieved.[2] Since then, a tremendous number of organotitanium derivatives have been synthesized,[3–5] in which the metal atom bears a broad variety of ligands sharing both σ- and π-bonds. The stable cyclopentadienyltitanium (titanocene) derivatives became popular following their introduction by Ziegler as reagents for alkene polymerization.[6]

Overcoming the theory that transition metal derivatives could be stable only with a fully coordinated metal atom (18-electron rule), finally opened the route toward the modern application of organotitanium and, later on, organozirconium reagents in synthesis.[7] This was facilitated by the work that had already been performed[8] by inorganic chemists in devising simple and high-yielding methods for the preparation of organo-titanium and -zirconium derivatives. This chapter will not deal with many of the synthetic applications of these reagents, but rather will focus on organo-titanium and -zirconium additions to C=X π-bonds and their advantages in comparison to other organometallic reagents in terms of chemo-, regio-, diastereo- and enantio-selectivity.

1.5.2 ORGANOTITANIUM AND ORGANOZIRCONIUM NUCLEOPHILES: A GENERAL VIEW

1.5.2.1 Chemical Properties and Comparison with Traditional Organometallic Reagents

The organo-titanium and -zirconium derivatives that will be considered here have the general formula R_nMX_{4-n}. The compounds consist of two distinct parts, namely the 'R' nucleophilic moiety and the 'X' ligand system. Considerable stability, selectivity and ease of handling are their most outstanding features as reagents in organic synthesis.

With regard to the 'R' nucleophilic moiety, simple linear alkyl residues should generally be avoided, due to their tendency to decompose through various processes,[9] such as reductive elimination (β-H abstraction; equation 1), and oxidative dimerization (equation 2). Attention must be paid when designing an efficient Ti or Zr nucleophile to minimize decomposition. This is achieved by using organic residues which either do not possess β-hydrogens (*e.g.* methyl, phenyl, benzyl, neopentyl, 1-norbornyl, trimethylsilylmethyl, *etc.*) or are sufficiently hindered that they do not undergo hydrogen abstraction readily.

$$[M]^n\text{CH}_2\text{CH}_2\text{R} \rightleftharpoons [M]^n(\text{H})(\text{CH}_2{=}\text{CHR}) \longrightarrow [M]^{n-2} + \text{CH}_2{=}\text{CHR} + H^+ \quad (1)$$

$$[M]^nR_2 \longrightarrow [M]^{n-2} + R{-}R \quad (2)$$

The 'X' ligands also influence the stability of the above reagents, primarily as a function of the stability of the bonds (σ or π) between the metal and the bond-forming atom. The most commonly used are alkoxy groups, *N,N*-dialkylaminyl groups, chlorine atoms or even mixed chlorine and alkoxy groups. Due to its major ability to donate electrons, nitrogen stabilizes Ti and Zr reagents more than oxygen and, consequently, much more than chlorine.

Considering the energy of the M—X bonds for various MX_4 compounds,[10] the bond strengths for Ti, Zr, and Hf decrease in the sequence M—O > M—Cl > M—N > M–C. The M–C bond strength in Ti^{IV} and Zr^{IV} compounds are comparable with that of other metal–carbon bonds,[11] whereas the rather unusual strength[10] of the M—O bond [E(Ti—O) = 115 kcal mol^{-1}; E(Zr—O) = 132 kcal mol^{-1}; 1 cal = 4.18 J] serves as a driving force for those reactions involving oxygen coordination by the metal atom. The Ti—O bond is fairly short among the metal–oxygen bonds (compare *e.g.* Li—O, Mg—O and Zr—O; Table 1),[11] a fact relevant to certain stereoselective reactions that will be discussed later (Section 1.5.3.1.4).

Table 1 Metal–Carbon and Metal–Oxygen Bond Lengths

Metal	*Bond length* Metal–carbon (Å)	Metal–oxygen (Å)
Ti	~2.10	1.70–1.90
Zr	~2.20	2.10–2.15
Li	~2.00	1.90–2.00
Mg	~2.00	2.00–2.13
B	1.5–1.6	1.36–1.48

Stability aside, the nature of the ligand system determines to some extent the Lewis acid character of Ti^{IV} and Zr^{IV} reagents. As expected, the higher the tendency of the bond-forming atom to share its electrons with the metal, the lower the acidity of the complex (*e.g.* Cl > OR > NR_2). This trend can be exploited to modulate the acidity of Ti^{IV} and Zr^{IV} reagents by replacing, for instance in the titanium series, chlorine with alkoxy groups so that their acidity decreases in the sequence $RTiCl_3$ > $RTiCl_2(OR')$ > $RTiCl(OR')_2$ > $RTi(OR')_3$.[12]

One advantage of organo-titanium and -zirconium reagents, relative to more traditional organometallic compounds in the field of nucleophilic additions, is their stability at fairly high temperature (from 0 °C to r.t.) for a reasonably long time. For example, $MeTi(OPr^i)_3$ is a yellow compound, isolated and purified by distillation at 50 °C under reduced pressure,[13,14] and the purple-red crystalline compound $MeTiCl_3$ (m.p. ~29 °C) can be stored at low temperature for several weeks.[15] They are not strongly toxic and do not afford toxic by-products.[9] Easy to handle, they require very simple work-up that includes slightly acidic conditions to avoid TiO_2-containing emulsions.[16]

Undoubtedly, the success of organo-titanium and -zirconium complexes as reagents for nucleophilic additions lies in their chemical properties, the most characteristic being chemoselectivity. Grignard reagents, organolithium compounds, and other resonance-stabilized carbanions often react more or less unselectively if two or more target functions are present in the substrate molecule. Even with other more recently reported[17] organometallic reagents, *e.g.* organo-tin, -cadmium and -chromium derivatives, such selectivity is not guaranteed. On the contrary, titanium and zirconium derivatives are capable of discriminating, for example, between an aldehyde and a ketone. More generally speaking, these reagents discriminate between similar functionalities that differ only slightly in their steric and/or electronic environment (Section 1.5.3.1.1). Furthermore, they tolerate several functional groups such as epoxides, ethers, carboxylic acid esters and thioesters, nitriles and alkyl halides.[16] Being less basic than other organometallic reagents, they can be safely utilized for nucleophilic addition to enolizable carbonyl compounds (Section 1.5.3.1.2). 1,2-Addition to α,β-unsaturated carbonyl compounds (Section 1.5.3.1.5), *gem*-dialkylation (Section 1.5.5.1) and one-step amination/alkylation of the carbonyl group (Section 1.5.5.2), represent a few examples of reactions that are more easily realized due to advances in Ti^{IV} and Zr^{IV} chemistry.

Ethyl levulinate (**1**) is reported to afford the lactone (**2**) when treated with methyltitanium triisopropoxide (equation 3).[18] In cases like this, in order to prevent lactonization, it may be expedient to use organozirconium derivatives that have a generally milder reactivity.

O CO2Et (**1**) —$MeTi(OPr^i)_3$→ $OTi(OPr^i)_3$ CO2Et ——→ O O (**2**) (3)

It is possible that the higher chemoselectivity of Ti^{IV} and Zr^{IV} reagents relative to the classical carbanions may be due to lower reaction rates. However, this cannot be the only reason, and it is likely that chemoselectivity is dependent on the nature of the C—M bond, which in the case of Ti^{IV} and Zr^{IV} is much less polarized than the analogous C—Li or C—Mg bonds.[16] Chemoselectivity is only one of the features of organo-titanium and -zirconium reagents, diastereoselectivity being a second and even more important one. The topic will be broadly covered in Sections 1.5.3.1.3 and 1.5.3.1.4 of this chapter, but as a quick illustration it is useful to compare the results of earlier experiments by Cram *et al.*[19,20] in the addition of either Grignard reagents or alkyllithium compounds to 2-phenylpropanal (**3**), and results from the addition of methyltitanium triisopropoxide to the same aldehyde (equation 4).[21] The diastereomeric products (**4**) and (**5**) are obtained, in the first instance, in about a 2:1 ratio *versus* the 9:1 ratio achieved by the titanium reagent.

H Ph O H (**3**) —$MeTi(OPr^i)_3$→ H Ph OH H (**4**) + H Ph OH H (**5**) (4)

Enantioselective addition has also been attempted with chirally modified Li or Mg reagents,[22] but only Ti^{IV} and Zr^{IV} derivatives have provided efficient tools for this purpose, due to the stable bond between chiral ligands and metal. In other cases, dissociation leads to a reduced efficacy of the chiral system.

1.5.2.2 General Preparation of Principal Organotitanium and Organozirconium Derivatives

Among the Ti^{IV} and Zr^{IV} complexes that are most commonly used for carbonyl nucleophilic additions, monoalkyl and monoaryl derivatives with alkoxylic ligand systems have the longest history of synthetic application.[9] This is likely due to the ease with which they are prepared from the readily available and inexpensive tetraalkoxy compounds (**6**). These, when treated with the tetrachloride of the same metal (equation 5), afford monochlorotrialkoxy metal derivatives (**7**), which in turn can be either alkylated or arylated by numerous lithium or magnesium derivatives[23] or by zinc, cadmium, lead and aluminum derivatives,[9] in various solvents, including THF, Et_2O, CH_2Cl_2, alkanes, *etc.* Monochlorotrialkoxytitanium compounds (**9**) are also reported to undergo ligand exchange by treatment with alcohols (equation 6),[24] a reaction which can be used to modulate the chemical properties of the reagent, as was discussed in the previous section.

$$\underset{(\mathbf{6})}{3\,M(OR)_4} \xrightarrow{MCl_4} \underset{(\mathbf{7})}{4\,ClM(OR)_3} \xrightarrow[R^1MgX]{R^1Li\ \text{or}} \underset{(\mathbf{8})}{R^1M(OR)_3} \tag{5}$$

$$\underset{(\mathbf{9})}{ClTi(OR^1)_3} \xrightarrow[-R^1OH]{R^2OH} \underset{(\mathbf{10})}{ClTi(OR^2)_3} \tag{6}$$

Monoalkyl Ti^{IV} and Zr^{IV} reagents having *N,N*-dialkylamino ligand systems (**12**) can be prepared from the corresponding tetra(*N,N*-dialkylamino) metal derivatives (**11**),[23] according to the procedure reported above for the alkoxylic ligand systems (equation 7). Recently, a large scale preparation of both tetra(*N,N*-diethylamino)- and chlorotris(*N,N*-diethylamino)-titanium(IV) compounds (**13**) and (**14**) was devised,[25] starting from lithium *N,N*-diethylamide and $TiCl_4$ (Scheme 1).

$$\underset{(\mathbf{11})}{3\,M(NR_2)_4} \xrightarrow{MX_4} 4\,XM(NR_2)_3 \xrightarrow{R^1Li} \underset{(\mathbf{12})}{4\,R^1M(NR_2)_3} \tag{7}$$

$$R = Me, Et;\ X = Br, Cl$$

$$\underset{(\mathbf{13})}{Ti(NEt_2)_4} \xleftarrow{TiCl_4\ (0.25\ \text{mol equiv.})} LiNEt_2 \xrightarrow{TiCl_4\ (0.33\ \text{mol equiv.})} \underset{(\mathbf{14})}{ClTi(NEt_2)_3}$$

Scheme 1

When the ligand system consists of chlorine atoms, monoalkyl- and monoaryl-titanium(IV) compounds are readily made from $TiCl_4$ by a chlorine atom replacement. The parent compound in this series, $MeTiCl_3$, is made conveniently from the reaction of $TiCl_4$ and dimethylzinc without the use of ethereal solvents,[26] the preparation being carried out in *n*-pentane or dichloromethane. Other methylating species can be used, such as MeMgBr, MeLi and $MeAlCl_2$. When MeLi is used in diethyl ether, an equilibrium takes place between different complexation states of Ti in the resulting $MeTiCl_3$ with the solvent (equation 8).[27] Transmetallation from organotin compounds like (**15**; equation 9) has also been utilized to make trichlorotitanium compounds.[28]

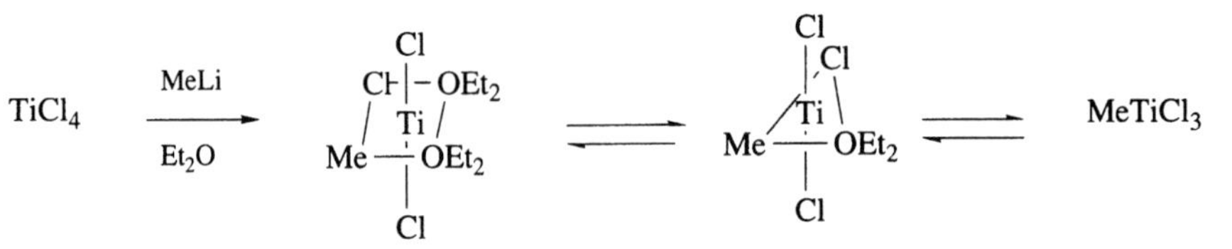

$$TiCl_4 \xrightarrow[Et_2O]{MeLi} MeTiCl_3(OEt_2)_2 \rightleftharpoons MeTiCl_3(OEt_2) \rightleftharpoons MeTiCl_3 \tag{8}$$

It is not necessary that the ligands around the metal be homogeneous, in fact mixed ligand systems can be very useful in tuning the chemical properties of the metal complexes. Their preparation is generally effected by ligand interchange, *i.e.* by mixing different Ti^{IV} species in the proper stoichiometric ratio, as is shown in equation (10) for the preparation of $MeTi(OPr^i)Cl_2$ (**16**).[13]

Ph Ph

Me_3Sn N Ph $\xrightarrow{TiCl_4}$ Cl N O Cl—Ti Cl + Me_3SnCl (9)

O Ph

(15)

$$2\,MeTiCl_3 + MeTi(OPr^i)_3 \longrightarrow 3\,MeTi(OPr^i)Cl_2 \quad (10)$$

(16)

Dialkyl and diaryl Ti^{IV} and Zr^{IV} derivatives (**18**) are somewhat less stable than their monosubstituted analogs. They are generally prepared by alkylation of the corresponding dichloro compounds (**17**; equation 11).[13,29,30] The dichloro derivatives themselves can be prepared in turn by reaction of tetrachlorides with the required amount of organometallic species. For example, zirconocene dichloride is prepared by treating zirconium(IV) chloride with cyclopentadienylsodium.[31]

$$Cl_2MCp_2 \xrightarrow{RLi} R(Cl)MCp_2 \xrightarrow{RLi} R_2MCp_2 \quad (11)$$

(17) **(18)**

Alkenyl and dienyl groups can be present in Ti^{IV} and Zr^{IV} complexes as well. Vinyltitanium compounds cannot be exploited for nucleophilic additions, however, due to their propensity to undergo oxidative coupling reactions,[32] likewise transition metal derivatives. Vinylzirconium compounds (**20**) can be prepared from alkynes (hydrozirconation; equation 12) by action of zirconium derivatives like (**19**).[33]

$R^2C{\equiv}CR^1$ $\xrightarrow{Cp_2ZrClH\ (\mathbf{19})}$ R^1 R^2 H $ZrCp_2Cl$ (12)

(20)

Allyltitanium complexes (**22**) readily add to carbonyl compounds with high regio- and stereo-selection. They are prepared by reaction of a chlorotitanium complex (**21**) with an allyl-magnesium or -lithium derivative (equation 13).[34] Some of these unsaturated Ti^{IV} complexes, like (**23**)–(**25**) in Scheme 2, obtained from allylmagnesium halides or allyllithium by reaction with titanium tetraisopropoxide or titanium tetramides,[35] are known as 'titanium ate complexes'. The structure of these ate complexes, at least from a formal point of view, can be written with a pentacoordinate Ti atom. Some ate complexes have synthetic interest, as is the case of $(allyl)Ti(OPr^i)_4MgBr$ which shows sharply enhanced selectivity towards aldehydes in comparison with the simple $(allyl)Ti(OPr^i)_3$.[16]

$ClTiX_3$ + M $\longrightarrow$ TiX_3 (13)

(21) M = MgCl or Li **(22)**

Crotyl-titanocenes and -zirconocenes like (**31**) are readily prepared from the corresponding magnesium and lithium compounds [regardless of (*E*)- or (*Z*)-configuration] by reaction with either dichlorobis(cyclopentadienyl)titanium or its zirconium analog (equation 14).[36,37] Likewise, crotylmagnesium derivatives react with other Ti^{IV} compounds [*e.g.* (**14**), (**26**) and (**27**); Scheme 3] and afford crotyltitanium complexes [*e.g.* (**28**), (**29**) and (**30**), respectively] in a stereoconvergent manner, affording solely the (*E*)-isomer. Finally, some other allylic reagents having a trivalent titanium atom are interesting for their application in synthesis.[38] The parent allyltitanium(III) derivative (**33**) is prepared by reaction of dichlorobis(cyclopentadienyl)titanium (**32**) and allylmagnesium bromide (equation 15).

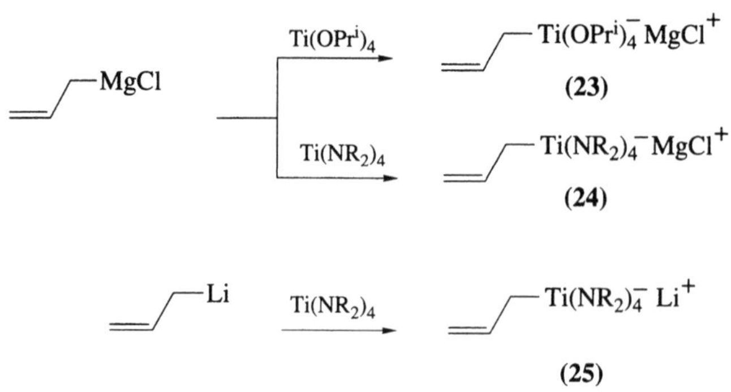

Scheme 2

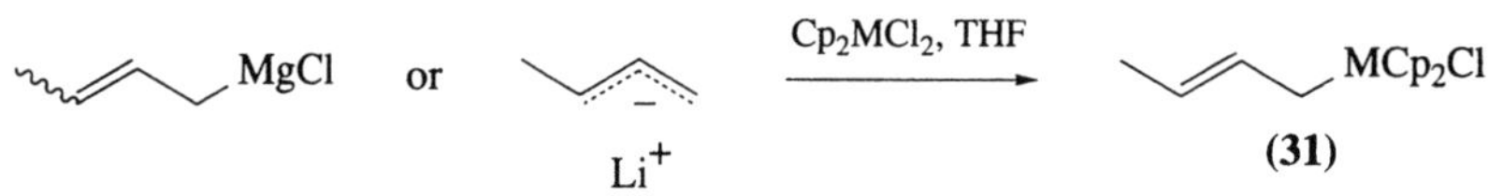

(14)

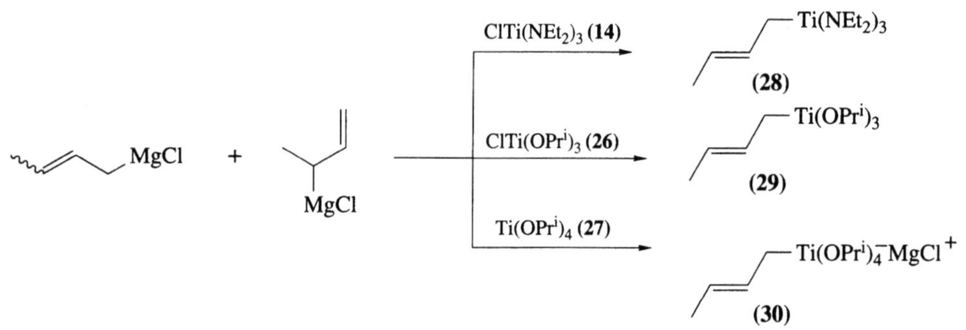

Scheme 3

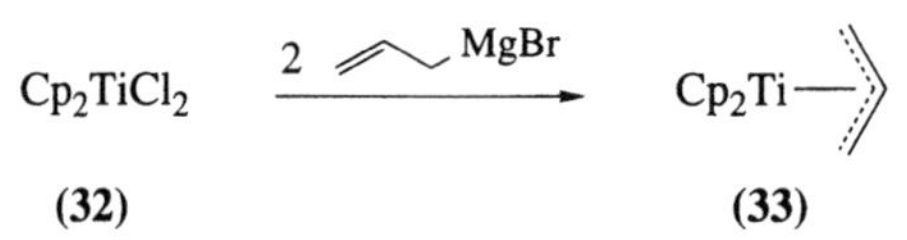

(15)

1.5.2.3 Titanium *versus* Zirconium Reactivity

Titanium and zirconium derivatives are in a certain sense complementary as reagents for carbonyl nucleophilic addition. Considering the relative position of both Ti and Zr in the periodic table, the latter is expected to share longer bonds with ligands and with the nucleophilic moiety of the reagent. As a consequence, chelation, when involved, leads to complexes of different structure and thermodynamic stability, factors which directly affect the course of the reaction. Even more important, the lower tendency towards reduction of Zr relative to Ti accounts for the stability of zirconium derivatives at room temperature. Vinylzirconium compounds can be prepared and used as such, whereas their titanium analogs readily undergo oxidative coupling with concomitant reduction of the metal.[39] Enhanced stability has also been verified for those zirconium complexes having branched alkyl groups,[9] that in the case of titanium are quite unstable.[9,11]

Generally speaking, the preparations of Ti^{IV} and Zr^{IV} reagents are quite similar, as was discussed in the previous section, but the Zr^{IV} starting materials, like $Zr(OR)_4$, are generally two or three times more expensive than the corresponding Ti^{IV} compounds and their reactions are slower. This is a characteristic of Zr^{IV} species relative to their Ti^{IV} counterparts that is verified also by the relative rates of nucleophilic addition to carbonyl compounds. The slower reaction rates can lead to side reactions: for example, alkylzirconium trialkoxides will reduce carbonyl compounds to the corresponding carbinol in a Meerwein–Ponndorf–Verley fashion.[40]

With these few exceptions, Zr^{IV} reagents perform nucleophilic additions to carbonyls with the same selectivity features as their titanium analogs,[41] under milder conditions. Moreover, due to their lower basicity,[9] they are generally more suitable to interact with easily enolizable carbonyl compounds. The reaction of **(34)** with $MeZr(OBu^n)_3$ to afford **(35)** serves to illustrate how versatile these reagents are in organic synthesis (equation 16).[39]

$MeZr(OBu^n)_3$ (16)

(34) **(35)**

1.5.3 CARBONYL ADDITION REACTIONS

1.5.3.1 Alkyl and Aryl Titanium and Zirconium Reagents

1.5.3.1.1 Addition to simple ketones and aldehydes: chemoselectivity

Methyltitanium triisopropoxide is probably the oldest and most widely studied organotitanium reagent for nucleophilic addition to carbonyls.[21] Its reaction with aldehydes and ketones (equation 17) is comparable to the reaction of more traditional organometallic reagents, like methylmagnesium halides or methyllithium, with carbonyl compounds. The already cited stability of $MeTi(OPr^i)_3$ and its solubility in various solvents would be sufficient to recommend it as a nucleophilic addition reagent but, more importantly, a wide range of functionalities (*e.g.* nitro, cyano, chloromethyl and thiol groups), as shown in Table 2, are tolerated under the reaction conditions.[16] It is also noteworthy to mention a one-pot procedure that allows the preparation of the reagent *in situ*.[42] A few undesired side reactions of methyl- and other alkyl-titanium triisopropoxides are known which are a result of transfer of the isopropoxy rather than the alkyl group. In fact, the reaction with carboxylic acid esters and with acid chlorides leads to ester interchange[43] and esterification, respectively.[9]

$MeTi(OPr^i)_3$ (17)

$R^1 = \text{or} \neq R^2 = \text{H, alkyl or aryl}$

Aldehydes react very fast and at low temperature (between –70 and 0 °C), whereas ketones require higher temperatures and longer reaction times (Tables 2 and 3). In cross experiments performed using benzaldehyde and acetophenone with $MeTi(OPr^i)_3$ in 1:1:1 molar ratios, only the aldehyde was consumed, leading almost quantitatively to its corresponding methylphenylcarbinol.[42] Such features are fully exploited in the synthesis of macrocyclic lactones like (**36**; equation 18).[45]

$MeTi(OPr^i)_3$, THF (18)

(36)

The chemoselectivity of organo-titanium and -zirconium reagents is practically unaffected by the nature of the nucleophilic moiety of their molecule. Although aldehydes always react much faster than ketones, the addition rates sometimes differ significantly, depending upon steric and electronic properties

Table 2 Carbonyl Addition Reactions of Various Alkyl- and Aryl-titanium(IV) Nucleophiles[a]

Carbonyl compound	*Nucleophile*[b]	*Solvent*	*Time (h)*	*Temperature (°C)*	*Product*	*Yield (%)*[c]
2,2-Dimethylpropanal	MeTi(OPr^i)$_3$	Et_2O	1.5	–25	3,3-Dimethyl-2-butanol	60
3-Methylbutanal	MeTi(OPr^i)$_3$	CH_2Cl_2	0.5	0	4-Methyl-2-pentanol	— (90)
Heptanal	MeTi(OPr^i)$_3$	THF	4.0	–50	2-Octanol	97
Benzaldehyde	MeTi(OPr^i)$_3$	THF	4.0	–50	1-Phenylethanol	92
Benzaldehyde	MeTi(OPr^i)$_3$	Et_2O	0.2	0	1-Phenylethanol	92
4-Nitrobenzaldehyde	MeTi(OPr^i)$_3$	THF	4.0	–50	1-(4-Nitrophenyl)ethanol	95
Cinnamaldehyde	MeTi(OPr^i)$_3$	Et_2O	4.0	–50	4-Phenyl-3-buten-2-ol[d]	91
Phenylacetaldehyde	MeTi(OPr^i)$_3$	Et_2O	2.0	–40 to 0	1-Phenyl-2-propanol	74 (90)
2-Heptanone	MeTi(OPr^i)$_3$	Et_2O	24.0	+22	2-Methyl-2-heptanol	47
2-Heptanone	MeTi(OPr^i)$_3$[e]	—	24.0	+22	2-Methyl-2-heptanol	83 (90)
Cyclohexanone	MeTi(OPr^i)$_3$[f]	THF	48.0	r.t.	1-Methylcyclohexanol	79
Acetophenone	MeTi(OPr^i)$_3$[f]	Et_2O	48.0	r.t.	2-Phenyl-2-propanol	96
(+)-1,3,3-Trimethyl-2-norbornanone	MeTi(OPr^i)$_3$[g]	Isooctane	10.0	+80	1,2,3,3-Tetramethyl-2-norbornanol[h]	84
α-Tetralone	MeTi(OPr^i)$_3$	Et_2O	24.0	r.t.	1,2,3,4-Tetrahydro-1-methylnaphthol	50
Benzaldehyde	EtTi(OPr^i)$_3$	Et_2O	6.0	–30 to +22	1-Phenylpropanol	69 (80)
4-Iodobenzaldehyde	Bu^{n}Ti(OPr^i)$_3$	Et_2O	15.0	–10 to r.t.	1-(4-Iodophenyl)pentanol	82
2-Bromobenzaldehyde	Bu^{n}Ti(OPr^i)$_3$	Et_2O	12.0	–20	1-(2-Bromophenyl)pentanol	92
4-Cyanobenzaldehyde	Bu^{n}Ti(OPr^i)$_3$	Et_2O	15.0	–10 to r.t.	1-(4-Cyanophenyl)pentanol	97
Acetaldehyde	PhTi(OPr^i)$_3$	THF	1.0	–10	1-Phenylethanol	86 (95)
Cyclohexanecarbaldehyde	PhTi(OPr^i)$_3$	THF	1.0	–10	Cyclohexylphenylmethanol	84 (95)
4-Nitrobenzaldehyde	PhTi(OPr^i)$_3$	Et_2O	0.5	–15	(4-Nitrophenyl)phenylmethanol	94
4-*t*-Butylcyclohexanone	PhTi(OPr^i)$_3$	Et_2O	15.0	r.t.	1-Phenyl-4-*t*-butylcyclohexanol[i]	95
Benzaldehyde	$CNCH_2$Ti(OPr^i)$_3$	THF	3.0	–78 to +22	2-Cyano-1-phenylethanol	81 (95)
Benzaldehyde	$CNCH_2Ti(NEt_2)_3$	THF	3.0	–78 to +22	2-Cyano-1-phenylethanol	76
Benzaldehyde	Me_3SiCH_2Ti(OPr^i)$_3$	THF	48.0	+22	2-Trimethylsilyl-1-phenylethanol	41 (60)
Benzaldehyde	(1,3-dithian-2-yl)Ti(OPr^i)$_3$	Et_2O	16.0	r.t.	(1,3-Dithian-2-yl)phenylmethanol	— (90)
Benzaldehyde	$PhSO_2CH_2$Ti(OPr^i)$_3$	THF	3.0	–78 to +22	2-Benzensulfonyl-1-phenylethanol	95 (100)
Butanal	Me_2TiCl_2	CH_2Cl_2	1.0	–40	2-Pentanol	— (90)
Cyclohexanecarbaldehyde	Me_2TiCl_2	CH_2Cl_2	1.0	–40	1-Cyclohexylethanol	78 (95)
3-Methylbutanal	$MeTiCl_3$	CH_2Cl_2	0.5	–30	4-Methyl-2-pentanol	— (90)
Cyclohexanecarbaldehyde	$MeTiCl_3$	CH_2Cl_2	0.5	–40	1-Cyclohexylethanol	— (90)
Cyclohexanecarbaldehyde	Ph_2TiCl_2	CH_2Cl_2	0.5	–78	Cyclohexylphenylmethanol	— (90)

[a]Adapted from the tables reported in refs. 21, 39 and 42–44. [b]1:1.2 molar ratio unless otherwise specified. [c]Percentages within parentheses are estimated by ^{1}H NMR. [d]No 1,4-adduct formed. [e]In the absence of solvent, using distilled MeTi(OPr^i)$_3$. [f]1:2 molar ratio. [g]1:10 molar ratio. [h]One diastereomer only. [i]1:1 *cis:trans* mixture.

Table 3 Carbonyl Addition Reactions of Various Alkyl- and Aryl-zirconium(IV) Tri-*n*-butoxides[39]

Carbonyl compound	*Nucleophile*	*Solvent*	*Time (h)*	*Temperature (°C)*	*Product*	*Yield (%)*
2-Methylcyclohexanone	MeZr(OBu^n)$_3$	CH_2Cl_2	15	–20 to +20	1,2-Dimethylcyclohexanol[a]	70
3β-Acetoxy-5-androsten-17-one	MeZr(OBu^n)$_3$	CH_2Cl_2	Overnight	–20 to +20	17α-Methyl-5-androstene-3β,17β-diol[b,c]	80
α-Tetralone	MeZr(OBu^n)$_3$	Et_2O	24	r.t.	1,2,3,4-Tetrahydro-1-methylnaphthol	90
4-Phenyl-4-ketobutyric acid	MeZr(OBu^n)$_3$	CH_2Cl_2	15	–20 to +20	4-Hydroxy-4-phenylpentanoic acid	72
5-Nitro-2-pentanone	MeZr(OBu^n)$_3$	CH_2Cl_2	2	0	2-Methyl-5-nitro-2-pentanol	72
2-Methylcyclohexanone	PhZr(OBu^n)$_3$	CH_2Cl_2	15	–20 to +20	1-Phenyl-2-methylcyclohexanol[c]	90
4-*t*-Butylcyclohexanone	Bu^{n}Zr(OBu^n)$_3$	CH_2Cl_2	8	–80 to +20	1-*n*-Butyl-4-*t*-butylcyclohexanol[d]	78
Benzaldehyde	S/S —Zr(OBu^n)$_3$	THF	Overnight	–80 to +20	(1,3-Dithian-2-yl)phenylmethanol	85

[a]Mixture of diastereomers. [b]After ester hydrolysis. [c]One diastereomer only. [d]86:14 *cis:trans* mixture.

of the substrate aldehydes as well as the structure of the reagent, including both its nucleophilic moiety and the ligand system (Table 4).[44]

Table 4 Ligand Influence on the Addition to Benzaldehyde of Various $MeTi(OR)_3$ Derivatives in Ether at –50 °C

R	*Time (min)*	*Yield (%)*
Propyl	60	20
i-Propyl	60	90
i-Propyl (distilled)	30	98
(*R*,*S*)-*s*-Butyl	30	99
(*S*)-2-Methylbutyl	5	>95

The reaction of $MeTi(OPr^i)_3$ with a 1:1 molar ratio of the isomeric hexanal (**37**) and 2-ethylbutanal (**38**) affords a mixture of the carbinols (**39**) and (**40**) in a 92:8 molar ratio (equation 19).[46] Another example of stereocontrol in the chemoselectivity of these reagents is represented by the cross reactions of the ketone pairs (**41**) and (**42**) (equation 20) or (**42**) and (**45**) (equation 21).[16] It is worthy of note that the same 1:1 molar mixture of (**42**) and (**45**) reacts more or less statistically with MeLi, affording about 50% of both (**44**) and (**46**).[16]

(**37**) + (**38**) —$MeTi(OPr^i)_3$→ (**39**) 92% + (**40**) 8% (19)

(**41**) + (**42**) → (**43**) + (**44**) (20)

	(**43**)	(**44**)
$MeTi(OPr^i)_3$	>97%	<3%
MeLi	39%	61%

(**42**) + (**45**) → (**44**) + (**46**) (21)

		(**44**)	(**46**)
$MeTi(OPr^i)_3$	(22 °C, 2 d)	15%	85%
MeLi	(0 °C, 2 min)	51%	49%

Chemoselectivity is affected by electronic factors as well. Cross experiments show that small changes in the electrophilic nature of the carbonyl group are responsible for marked differences in the nucleophilic reactivity of Ti^{IV} and Zr^{IV} reagents (equations 22 and 23).[16] The electronic nature of the nucleophilic moiety of the metal reagent also affects the chemoselectivity of the nucleophilic addition. This is the case for some special titanium(IV) triisopropoxide derivatives [*e.g.* (**53**)–(**56**)] which are selective towards aldehydes and give high addition yields, due to their resonance-stabilized residues acting as nucleophiles.[42]

(**41**) + (**47**) → (**43**) + (**48**) (22)

	(**43**)	(**48**)
$MeTi(OPr^i)_3$	94%	6%
MeLi	80%	20%

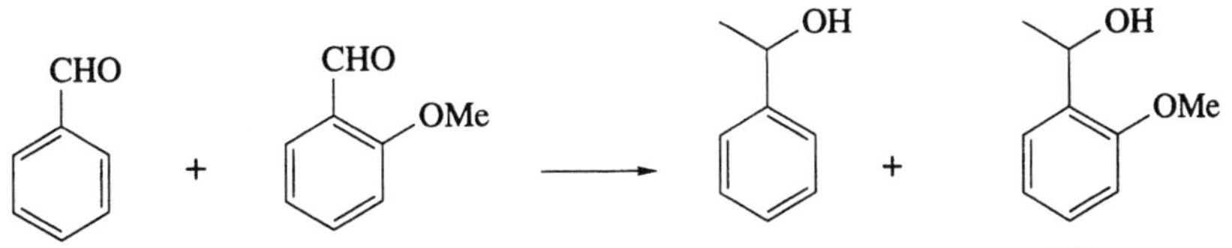

CHO (49) + CHO, OMe (50) ⟶ OH (51) + OH, OMe (52) (23)

	(51)	(52)
$MeTi(OPr^i)_3$	13%	87%
MeLi	50%	50%

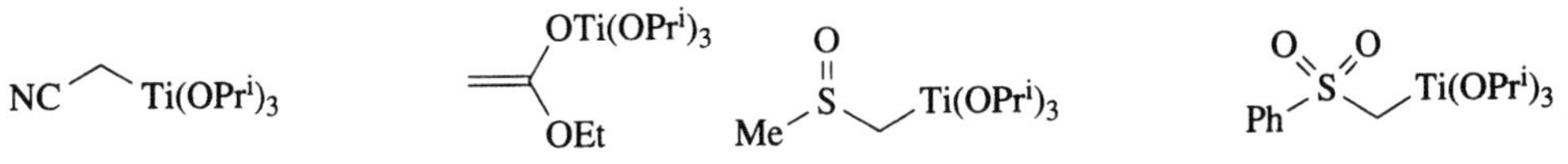

$NC\frown Ti(OPr^i)_3$ (53) $OTi(OPr^i)_3$, OEt (54) Me–S(=O)–$Ti(OPr^i)_3$ (55) Ph–SO_2–$Ti(OPr^i)_3$ (56)

The picture remains substantially unchanged moving from the trialkoxy system to other ligand systems. (Alkyl)$Ti(NMe_2)_3$, in spite of their appealing stability,[47] fail to give Grignard-type additions, and lead only to low-yielding aminoalkylation reactions (Section 1.5.5.2).[42,48] Alkyl- and aryl-chlorotitanium derivatives, like $MeTiCl_3$, Me_2TiCl_2 and Ph_2TiCl_2, characteristically exhibit good reactivity leading to nucleophilic addition in high yields (Table 2), accompanied by high chemoselectivity.[21] $MeTiCl_3$, the parent compound of the series, is a well-known and well-studied chemical. Its coordination complexes with electron-donor species, like TMEDA, glyme, THF and (–)-sparteine, are also well known.[49a] As already mentioned (equation 8), it is easily obtained from MeLi and $TiCl_4$ in ether and its Et_2O complex is very reactive towards aldehydes, acting at –30 °C, much faster and higher yielding than $MeTi(OPr^i)_3$ and having satisfactory chemoselectivity.[27] Recently, an analogous trichlorotitanium derivative (**57**) has been prepared from 1,2-diorganometallic species with $TiCl_4$.[49b] The reaction of (**57**) with carbonyl compounds is represented in Scheme 4.

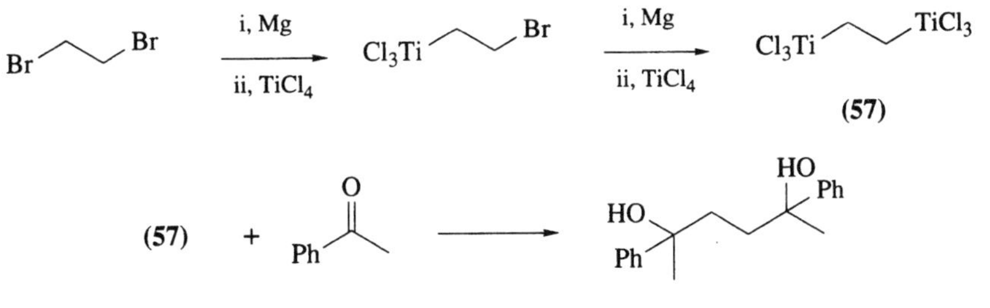

Br⌒Br —(i, Mg; ii, $TiCl_4$)→ Cl_3Ti⌒Br —(i, Mg; ii, $TiCl_4$)→ Cl_3Ti⌒$TiCl_3$ (57)

(57) + PhC(O)Me ⟶ HO, Ph … HO, Ph (diol)

Scheme 4

The instability of Ti^{IV} reagents with branched alkyl groups is due to their tendency to decompose by β-elimination.[9] For this type of compounds the corresponding zirconium reagents are more suitable, since they do not undergo decomposition. For example, *t*-butylzirconium tributoxide (**58b**) is readily prepared from *t*-butyllithium and chlorozirconium tributoxide (**58a**) (equation 24).[39] Alkyl- and aryl-zirconium(IV) tributoxides add to carbonyl compounds (Table 3) in the same fashion as Ti^{IV} derivatives, with the advantage of low basicity, as is shown by their reactions with enolizable carbonyl compounds (Section 1.5.3.1.2).

$$ClZr(OBu^n)_3 \;(\mathbf{58a}) + Bu^tLi \longrightarrow Bu^tZr(OBu^n)_3 \;(\mathbf{58b}) + LiCl \quad (24)$$

A mechanistic interpretation of all the above facts would be attractive at this point. Unfortunately, a comprehensive explanation of the experimental results is not yet available, with the exception of kinetic studies on cyclic ketones[50] and other carbonyls.[51]

1.5.3.1.2 Addition to enolizable carbonyl compounds

2-Phenylcyclopentanone (**59**) is converted to its methylated diastereomeric derivatives (**60**) and (**61**) in reasonably high yield (>90%) (equation 25), by treatment with $Me_2Ti(OPr^i)_2$.[46] When methyllithium is used, under comparable experimental conditions, only a 55% yield and lower selectivity is achieved. This can be accounted for by the relatively low basicity of titanium reagents. There is no general rule that allows one to choose the titanium reagent with the proper balance of nucleophilicity and basicity for a particular substrate. As an example, β-tetralone is methylated in 90% yield, using a $MeLi/TiCl_4$ reagent, *versus* 60% yield obtained with MeLi alone, due to competing enolization.[27]

(**59**) → (**60**) + (**61**) (25)

	(60)	(61)	
MeLi	90%	10%	(<55%)
$Me_2Ti(OPr^i)_2$	>97%	<3%	(>90%)

The above-mentioned requirements of low basicity and fast alkyl transfer are combined in alkylzirconium(IV) species,[16,41] like tetramethylzirconium which readily methylates enolizable and highly hindered carbonyl compounds such as (**62**; equation 26). The presence of several alkyl groups on the metal atom imparts enhanced reactivity to the species due to the absence of π-bond-forming electron-donor heteroatoms. In fact, tetramethylzirconium can be regarded as a supermethylating agent, evidenced by the conversion of (**65**) into (**66**; equation 27).[11]

(**62**) —(−20 °C)→ (**63**) + (**64**) (26)

	(63)	(64)
Me_4Ti	20%	35% (45% SM)
Me_4Zr	45%	0% (55% SM)

(**65**) —(Me_4Zr)→ (**66**) (27)

1.5.3.1.3 Stereocontrol in nucleophilic addition to carbonyl compounds

An attempt to prepare (+)-cuparene (**68**), although unsuccessful, illustrates diastereoselectivity of Ti^{IV} reagents.[16] Treatment of (**67**) with Me_2TiCl_2 (Scheme 5) does not give the desired dimethylated product, but rather produces the diastereomeric tertiary alcohols (**69**) and (**70**) in a 95:5 molar ratio. The diastereoselectivity, characteristic of Zr^{IV} reagents as well, is influenced by both the electronic nature and the steric hindrance of the metal ligands. Three different types of stereoselection are as follows: (i) diastereofacial selection involving carbonyl substrates bearing a chiral α-carbon atom;[52] (ii) equatorial *versus* axial selection occurring with six-membered cyclic ketones; and (iii) simple diastereoselectivity that is encountered in the reaction of prochiral crotyltitanium(IV) derivatives with prochiral carbonyl compounds. This type of selectivity will be considered in Section 1.5.3.3.

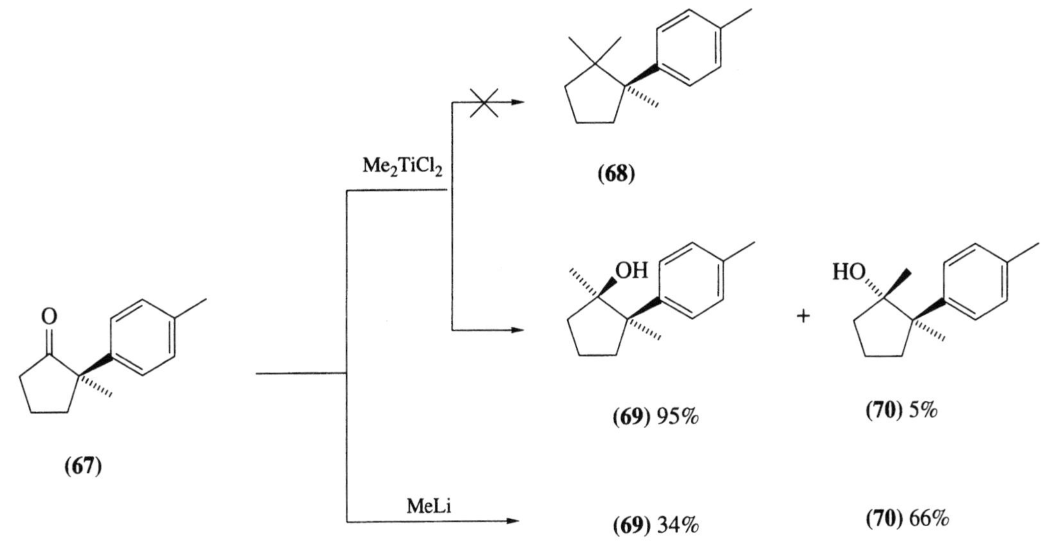

Me$_2$TiCl$_2$

(68)

(69) 95%

(70) 5%

(67)

MeLi

(69) 34%

(70) 66%

Scheme 5

(i) Diastereofacial selection

Cram's experiments using 2-phenylpropanal with various traditional organometallic reagents have already been mentioned.[19,20] The same substrate was also subjected to several Ti^{IV} and Zr^{IV} complexes (equation 28),[21] the results of which are summarized in Table 5. 2-Phenylbutanal, when treated with MeTi(OPr^i)$_3$ in Et$_2$O, affords preferentially the 'Cram' product (87:13).[53a] Zr^{IV} derivatives, like MeZr(OPr^i)$_3$, follow the same tendency although in some cases the results are somewhat less appealing.[41]

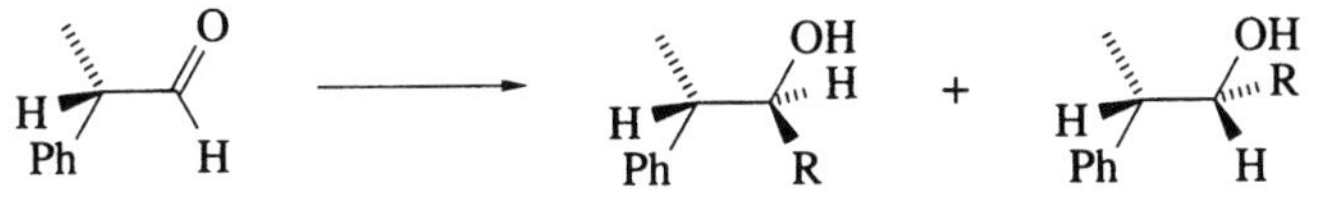

(28)

(71) 'Cram' product

(72) '*anti*-Cram' product

Table 5 Diastereoselective Addition of Ti^{IV} and Zr^{IV} Complexes to 2-Phenylpropanal (equation 28)

Reagent	*(**71**):(**72**)*	*Ref.*
MeTi(OPr^i)$_3$	87:13	21
MeTiCl$_3$	81:19	21
MeTi(OPh)$_3$	93:7	16
MeLi/TiCl$_4$/Et$_2$O	90:10	11
Ph$_2$TiCl$_2$	80:20	11
Bu^{n}Ti(OPr^i)$_3$	89:11	41
(Allyl)Ti(OPr^i)$_3$	68:32	53
[(Allyl)Ti(OPr^i)$_4$]$^-$ MgCl$^+$	75:25	11
MeZr(OPr^i)$_3$	90:10	41

Ti^{IV} complexes have been extensively used for the stereoselective introduction of side chains in steroidal molecules, *e.g.* pregnenolone acetate[53a] and the steroidal[11] C-22 aldehyde (**73**; equation 29). In these cases satisfactory results are only obtained by using Ti^{IV} reagents, like methyl, d$_3$-methyl and allyl derivatives, due to the strong steric hindrance that affects the position to be attacked. 1,3-*Anti* diastereoselection in the addition of alkyltitanium reagents to chirally β-substituted aldehydes having a dithioacetal group at the α-position (**75**) was also observed (equation 30).[53b]

$MeTi(OPr^i)_3$ (29)

(73) → **(74a)** 99% + **(74b)** 1%

$MeTiCl_2(OPr^i)$ (30)

(75) → *anti*:*syn* = 99:1

(ii) Equatorial versus *axial selection*

Nucleophilic additions to 4-*t*-butylcyclohexanone (**76**; equation 31), by various $MeTi^{IV}$ and $MeZr^{IV}$ reagents (Table 6), are reported to afford mixtures of (**77**) and its C-1 epimer (**78**) with the equatorial methyl bearing isomer (**77**) always predominating. $MeTi(OPr^i)_3$ has been used successfully in the steroid field, as in the case of cholestan-3-one[53a] or as in the case of androstane-3,17-dione (**79**; equation 32)[11] which undergoes regio- and diastereo-selective methyl addition. $MeTiCl_3$ is also reported to react with C-2- and C-3-substituted cyclohexanones, leading to significant excesses of the equatorial methyl bearing epimers (88:12 in the case of 3-methylcyclohexanone), whereas MeMgI furnishes random mixtures of both epimers.[27]

$MeTiX_3$ or $MeZrX_3$ (31)

(76) → **(77)** + **(78)**

Table 6 Equatorial *versus* Axial Addition of $MeTi^{IV}$ and $MeZr^{IV}$ Reagents to 4-*t*-Butylcyclohexanone (equation 31)

Reagent	*Solvent*	*Temperature (°C)*	*Equatorial:axial (77):(78)*	*Ref.*
$MeTi(OPr^i)_3$	CH_2Cl_2	+22	82:18	16
$MeTi(OPr^i)_3$	Et_2O	+22	86:14	16
$MeTi(OPr^i)_3$	Et_2O	0	89:11	16
$MeTi(OPr^i)_3$	*n*-Hexane	−15 to +22	94:6	53
Me_2TiCl_2	CH_2Cl_2	−78	82:18	21
$MeZr(OBu^n)_3$ + LiCl	Et_2O	+22	80:20	39

$MeTi(OPr^i)_3$ (32)

(79) → **(80)**

α-alcohol:β-alcohol = 85:15

Mechanistic interpretations of diastereoselectivity, based on the Cram's open chain model as well as more recent models, have been reported.[54,55]

1.5.3.1.4 Addition to chiral alkoxycarbonyl compounds

Chiral α-, β- or γ-alkoxycarbonyl compounds are alkylated by Ti^{IV} and Zr^{IV} reagents with high facial stereoselectivity. This is due to chelation by the metal atom which creates a bridge between the oxygen atoms belonging to the carbonyl function and the alkoxy group, respectively. The resulting cyclic intermediate is then attacked by the nucleophilic moiety of the reagent, preferentially from one side, leading to an excess of one of the potential diastereomeric products. This phenomenon has been observed with other organometallic reagents, though only asymmetric 1,2-induction is known for compounds like RMgX, RLi, R_2CuLi *etc.*,[20,56,57] whereas Ti^{IV} and Zr^{IV} reagents are capable of 1,3- and even 1,4-induction.[58] Furthermore, given that chelation is governed by the Lewis acidity of the metal atom, the chelating properties of the organometallic complexes can be tuned by varying the metal ligand system. The electron-withdrawing halogen ligands increase the acidity and enhance the chelation power. In the presence of electron-donor ligands, like alkoxy and *N,N*-dialkylamino groups, chelation will depend essentially on the nature of the substrates. For example, alkoxy ketones will undergo appreciable chelation, whereas weaker Lewis bases like alkoxy aldehydes will not be chelated, a fact that can be exploited to perform reverse stereochemical control. Examples of asymmetric 1,2- and 1,3-induction are provided by the reactions of the chiral alkoxy aldehydes (**81**) and (**82**), with $MeTiCl_3$ as shown in Scheme 6.[59–61] The induction is consistently over 90%.

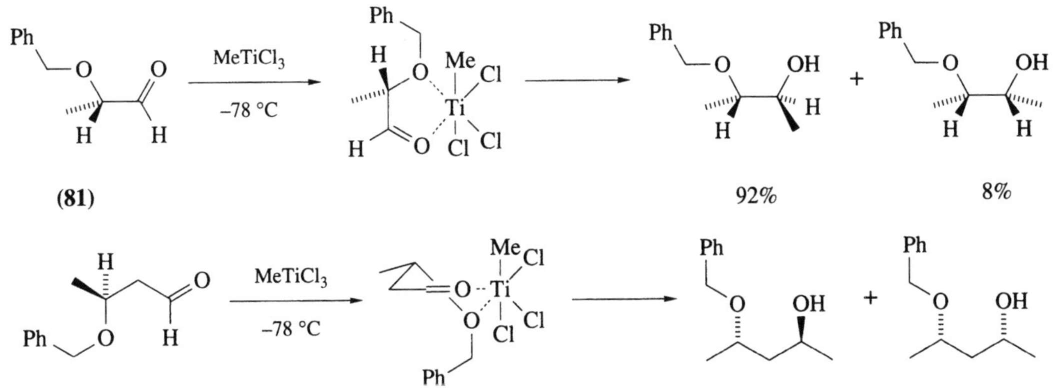

(**81**) —$MeTiCl_3$, –78 °C→ 92% + 8%

(**82**) —$MeTiCl_3$, –78 °C→ 90% + 10%

Scheme 6

Less acidic than Ti^{IV} and Zr^{IV} chloroderivatives, $MeTi(OPr^i)_3$ performs chelation-controlled addition to chiral alkoxy ketones[59] as well as or better than organomagnesium compounds,[41,58] but fails to chelate to aldehydes or hindered ketones. Should the formation of a cyclic chelation intermediate be forbidden, the reaction is subject to nonchelation control, according to the Felkin–Anh[41] (or Cornforth)[62] model. Under these circumstances, the ratio of the diastereomeric products is inverted in favor of the 'anti-Cram' product(s). In the case of benzil (**83**; Scheme 7) this can be accounted for by the unlikely formation of a cyclic intermediate such as (**85**), and thus the preferential intermediacy of the open chain intermediate (**86**) that leads to the *threo* compound (**88**).[41] This view is substantiated by the fact that replacement of titanium with zirconium, which is characterized by longer M—O bonds, restores the possibility of having a cyclic intermediate and, as a consequence, leads to the *erythro* (*meso*) compound (**87**) thus paralleling the action of Mg and Li complexes.

Nonchelation-controlled addition reactions can be very useful in organic synthesis.[58,59,63,64] An example drawn from carbohydrate chemistry is particularly significant in this regard (Scheme 8).[11] Due to chelation, the aldehyde (**89**) reacts with MeMgBr to give a mixture of the diastereomeric alcohols (**91**) and (**92**) in 88:12 molar ratio, whereas, in the reaction with $MeTi(OPr^i)_3$ under nonchelation control, only the Felkin–Anh product (**92**) is observed. 2,3-*O*-Isopropylidene-D-glyceraldehyde (**93**; equation 33) is also reported to react under nonchelation-controlled conditions with several $RTi(OPr^i)_3$ reagents.[58] Oddly, $PhTi(OPr^i)_3$ favors the formation of the *syn* product, a unique result that is difficult to rationalize.

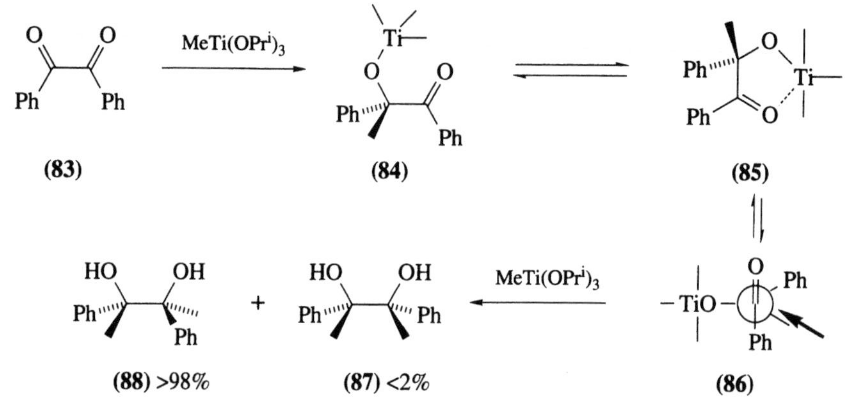

Scheme 7

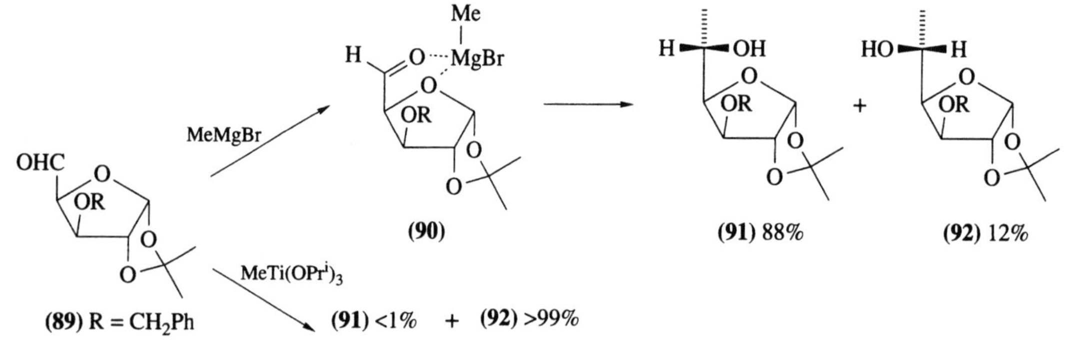

Scheme 8

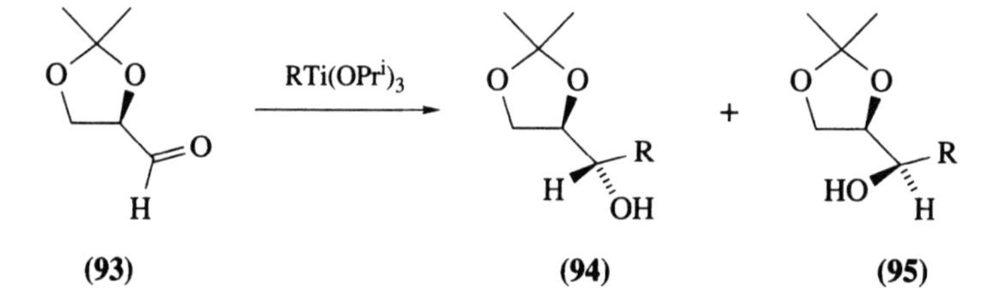

(33)

R	(94)	(95)
R = Me	75%	25%
R = Bu^n	90%	10%
R = Allyl	71%	29%
R = Ph	9%	91%

Very few examples of asymmetric 1,4-induction are reported in connection with the addition of acidic Ti^{IV} complexes to chiral γ-alkoxycarbonyl compounds. According to the Cram model, the chelation is expected to afford a flexible seven-membered ring intermediate, resulting in less efficient induction (equation 34).[59] An early example of asymmetric 1,4-induction is provided by the reaction of *o*-phthalaldehyde (**96**; equation 35)[41] with 2 equiv. of $MeTi(OPr^i)_3$ which affords an 83:17 molar mixture of *racemic*-(**97**) and *meso*-(**98**), whereas the analogous reaction with MeMgI leads to a 1:1 mixture of the above diastereomers.

The intrinsic difficulty in preparing various titanium derivatives with the desired alkyl group may be overcome by using reagent systems consisting of $TiCl_4$ as the chelating agent, and a suitable nucleophile like dialkylzinc, allylsilanes, allylstannanes, *etc.*[59] The utilization of other nucleophiles, like silyl enol ethers will be covered in Part 1 of Volume 2.

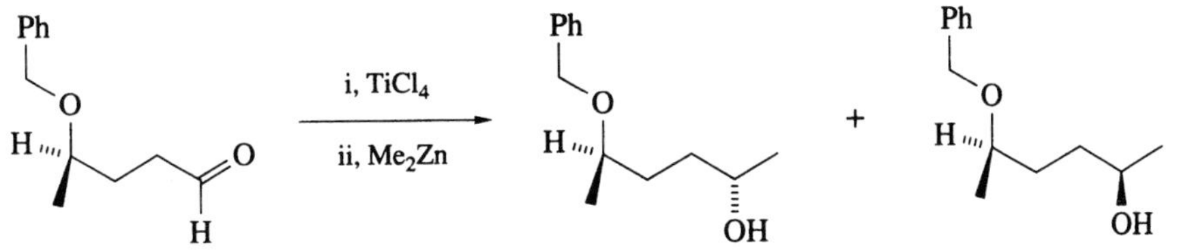

85% 15% (34)

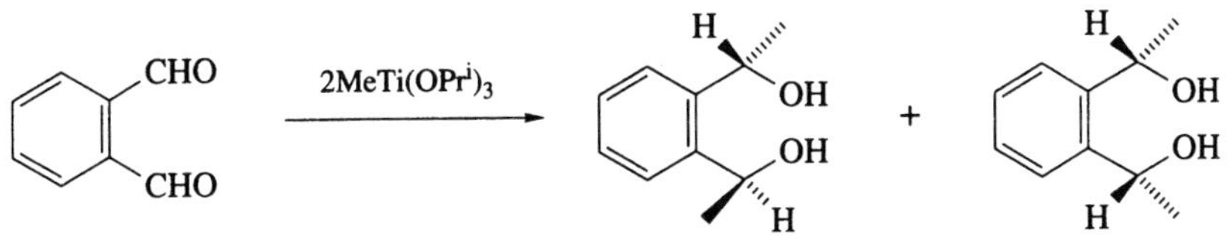

(35)

(96) **(97)** 83% **(98)** 17%

1.5.3.1.5 Addition to α,β-unsaturated carbonyl compounds

α,β-Unsaturated carbonyl compounds undergo 1,2-addition by $MeTi(OPr^i)_3$.[16,42,43] A rare case of conjugate addition is reported in the reaction of (**99**; equation 36) with $Ti(CH_2Ph)_4$.[65]

(36)

(99) 13% 87%

Ti^{IV} reagents are chemoselective and usually react with α,β-unsaturated carbonyls faster than with their saturated analogs, as is shown by several cross experiments (equation 37).[11] An exception to this generality was observed in the case of cyclohex-2-enone *versus* cyclohexanone,[16] the latter being much more reactive (equation 22, Section 1.5.3.1.1).

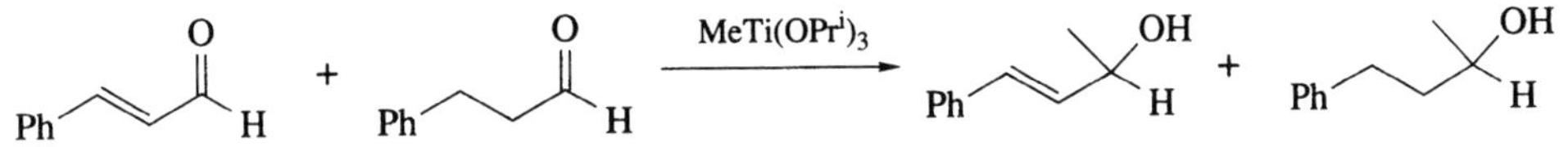

(37)

(99) 78% 22%

1.5.3.2 Vinylzirconium Reagents: Addition to α,β-Unsaturated Ketones

Vinylzirconium(IV) complexes, unlike their Ti^{IV} analogs (Section 1.5.2.2), are rather stable and can be conveniently used to perform conjugate addition to α,β-unsaturated ketones, sometimes in preference to the more popular alkenylcopper(I) reagents.[66] Vinylzirconium(IV) complexes, as such, are not very reactive and need catalytic amounts of $Ni(acac)_2$ (Scheme 9).[67,68] The reaction has been exploited in a prosta-

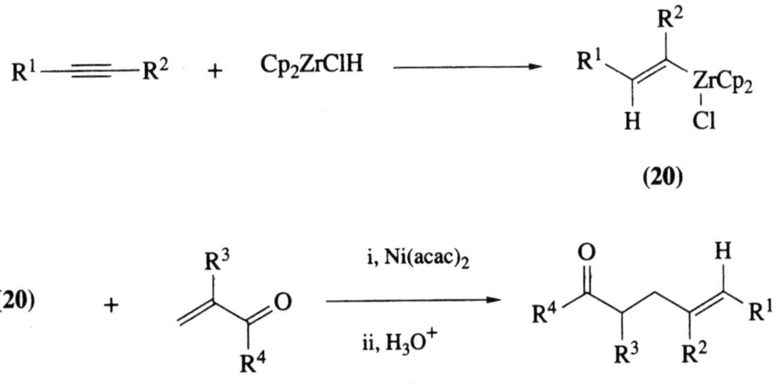

Scheme 9

glandin synthesis.[68] A mechanistic investigation of the addition reaction of vinylzirconium(IV) complexes to carbonyls has been reported.[69]

1.5.3.3 Allylic Titanium and Zirconium Reagents

As the general reactivity of allylic metal derivatives will receive proper attention in Part 1 of Volume 2, only those aspects which are very peculiar of allylic Ti^{IV} and Zr^{IV} reagents in organic synthesis will be covered in this section. From a general point of view, the allylic metal derivatives turn out to be fairly more reactive than their alkyl analogs and, as a consequence, show a diminished selectivity. As a matter of fact, (allyl)Ti(OPr^i)$_3$ reacts with a 1:1 molar mixture of heptanal and heptan-2-one (equation 38), leading to the isomeric homoallylic alcohols (**100**) and (**101**) in 86:14 ratio,[35] as well as with a 1:1 molar mixture of benzaldehyde and acetophenone giving (**102**) and (**103**) in 84:16 ratio, (equation 39),[11] whereas ketones remain substantially untouched when alkyltitanium reagents are used likewise in cross reactions (Section 1.5.3.1.1).

Ti(OPr^i)$_3$ (38)

(**100**) 86% (**101**) 14%

Ti(OPr^i)$_3$ (39)

(**102**) 84% (**103**) 16%

A few examples of chemoselective additions of allyltitanium reagents to aliphatic and aromatic carbonyl compounds are reported in Table 7. Appreciable chemoselectivity toward the aldehydic function is achieved by the titanium ate complex (**23**), whereas the reverse chemoselectivity toward ketones is realized using aminotitanium complex (**104**) and the analogous ate complexes (**24**) and (**25**), as is shown in Table 7. This is very interesting since it represents a rare case of chemoselectivity in favor of carbanion addition to ketones. A tentative explanation of this inverse chemoselection considers a fast transfer of the aminyl ligand onto the aldehyde function which becomes 'protected', as in (**105**), and thus unreactive in respect to the keto group.[35] Ketones react also selectively compared with esters, as is shown by the reaction of ethyl levulinate (**1**) with the ate complex (**23**; equation 40).[35]

Table 7 Chemoselective Addition of Allyltitanium Reagents to Equimolar Mixtures of Heptanal/Heptan-2-one (equation 38) and Benzaldehyde/Acetophenone (equation 39)[a]

Reagent		*(**100**):(**101**)*	*(**102**):(**103**)*
(Allyl)Ti(OPr^i)$_3$[b]		86:14	84:16
[(Allyl)Ti(OPr^i)$_4$]$^-$ $MgCl^+$	(**23**)	98:2	98:2
(Allyl)Ti(NMe_2)$_3$[b]	(**104**)	13:87	50:50
[(Allyl)Ti(NMe_2)$_4$]$^-$ $MgCl^+$	(**24**)	4:96	< 1:99
[(Allyl)Ti(NMe_2)$_4$]$^-$ Li^+	(**25**)	2:98	9:91

[a]Adapted from the tables reported in refs. 11, 16 and 35. [b]Prepared *in situ*.

As far as the stereoselectivity of the addition is concerned, allyltitanium reagents, unlike their alkyl analogs (Section 1.5.3.1.3), give preferentially axial addition to 4-*t*-butylcyclohexanone. The selectivity

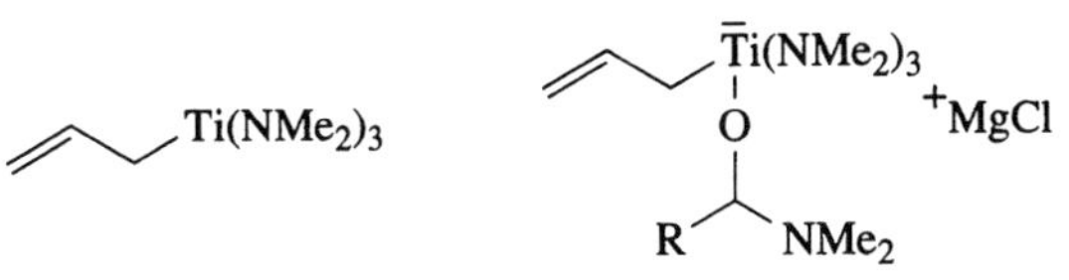

(104) (105)

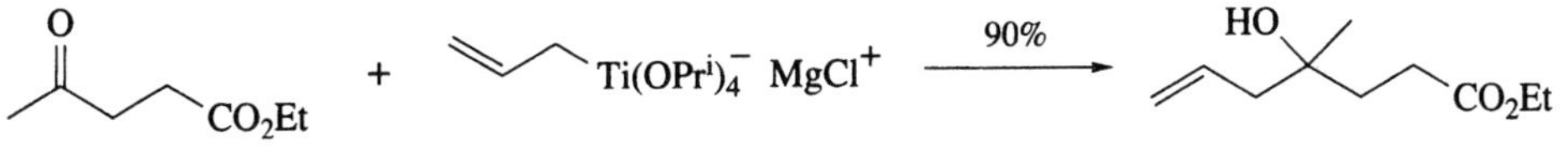

(1) (23) 90% (40)

is rather low and can be improved by using tris(dialkylamino) ligands.[53a] An example of diastereofacial selectivity of allyl derivatives is offered by tetraallylzirconium in the addition of chiral β-alkoxycarbonyl compounds;[60] 3-hydroxybutanal (**106**) reacts with tetraallylzirconium (Scheme 10) under chelation control conditions (asymmetric 1,3-induction) affording the epimeric diols (**107**) and (**108**) in 19:81 ratio. The chemistry of crotyl-Ti^{IV} and -Zr^{IV} derivatives is definitely more interesting due to their addition reactions which turn out to be sharply chemoselective, regioselective (since they attack the carbonyl substrate by their more substituted sp^2-carbon atom), and diastereoselective.

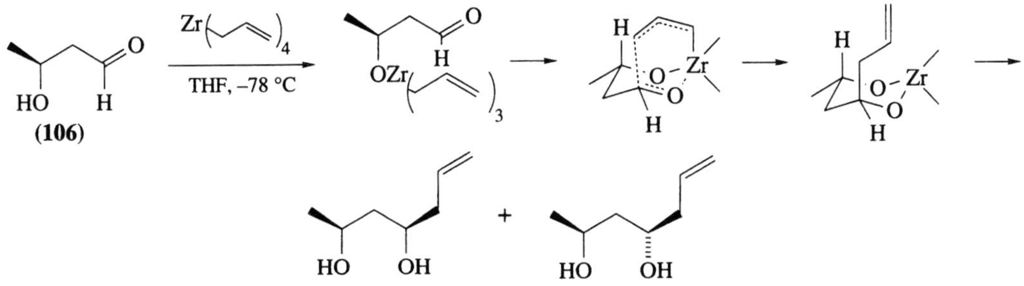

THF, –78 °C

(106)

(107) 19% (108) 81%

Scheme 10

Chemoselectivity of these reagents, tested in cross experiments (equations 41 and 42), parallels the one already discussed for the allyl derivatives.[16,35]

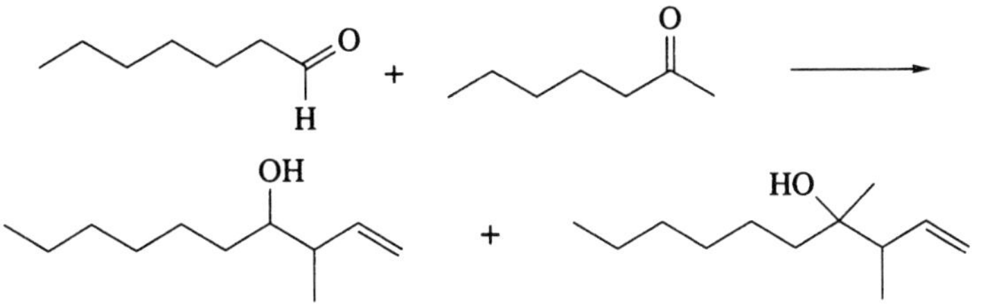

(41)

(Crotyl)$Ti(OPr^i)_4^-MgCl^+$	98%	2%
(Crotyl)$Ti(NMe_2)_4^-MgCl^+$	<2%	98%

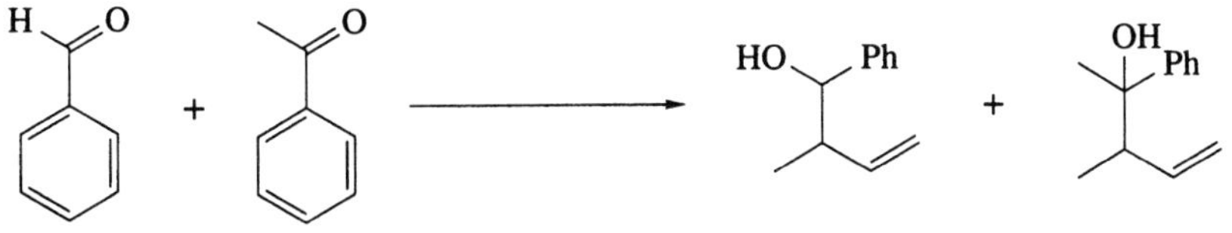

(42)

(Crotyl)$Ti(OPr^i)_4^-MgCl^+$	99%	1%
(Crotyl)$Ti(NMe_2)_4^-MgCl^+$	2%	98%

Diastereoselectivity is instead extremely important, giving rise to configurationally pure branched homoallylic alcohols. The stereochemistry of the products, which can have *threo* or *erythro* configurations, is dependent on the (*E*)- or (*Z*)-geometry of the starting crotyl metal reagent.[11] Examples of (*Z*)-crotyltitanium(IV) derivatives are not known and only ^{1}H NMR evidences allow such reagents to be assigned, not unambiguously, (*E*)-configuration.[11] The addition of several crotyl-Ti^{IV} and -Zr^{IV} derivatives to various aldehydes and ketones (equation 43) leads invariably to a mixture of *threo* (*anti*) (**109**) and *erythro* (*syn*) (**110**) homoallylic alcohols (Table 8; *cf.* also Table 9). In spite of an influence of the metal ligand system on the diastereoselectivity of the addition (Table 8), the *threo* product appears to be always predominant regardless of the Ti^{IV} reagent used. In this context, no general rule can be put forward about the choice of the suitable ligand system, although the ate complex (**30**) seems to be the most selective in the addition to aromatic aldehydes, whereas the crotyltriamide complex (**28**) should be preferred when aliphatic aldehydes are considered. The triphenoxy ligand system is also interesting, as is shown by some entries in Table 8.

R(R^1)C=O + crotyl-TiX ⟶ *threo*-(**109**) + *erythro*-(**110**) (43)

R = or ≠ H; R^1 = alkyl or aryl; X = $(NEt_2)_3$, $(OPr^i)_3$, $(OPh)_3$, $(OPr^i)_3MgCl$, $(OBu)_3$, (2,6-dimethylphenoxy)$_3$

Table 8 Diastereoselective Addition of Various Crotyltitanium(IV) Reagents to Carbonyl Compounds (equation 43)[a]

Carbonyl compound	*Reagent*	*threo:erythro*
Benzaldehyde	(Crotyl)Ti$(NEt_2)_3$ (**28**)	69:31
Hexanal	(Crotyl)Ti$(NEt_2)_3$ (**28**)	82:18
Acetaldehyde	(Crotyl)Ti$(NEt_2)_3$ (**28**)	67:33
3-Methylbutan-2-one	(Crotyl)Ti$(NEt_2)_3$ (**28**)	97:3
Heptan-2-one	(Crotyl)Ti$(NEt_2)_3$ (**28**)	72:28
3,3-Dimethylbutan-2-one	(Crotyl)Ti$(NEt_2)_3$ (**28**)	99:1
Acetophenone	(Crotyl)Ti$(NEt_2)_3$ (**28**)	85:15
Benzaldehyde	(Crotyl)Ti$(OPr^i)_3$ (**29**)	80:20
Hexanal	(Crotyl)Ti$(OPr^i)_3$ (**29**)	75:25
3-Methylbutan-2-one	(Crotyl)Ti$(OPr^i)_3$ (**29**)	88:12
Heptan-2-one	(Crotyl)Ti$(OPr^i)_3$ (**29**)	67:33
Benzaldehyde	[(Crotyl)Ti$(OPr^i)_3]^-$ $MgCl^+$ (**30**)	84:16
Hexanal	[(Crotyl)Ti$(OPr^i)_3]^-$ $MgCl^+$ (**30**)	71:29
3-Methylbutan-2-one	[(Crotyl)Ti$(OPr^i)_3]^-$ $MgCl^+$ (**30**)	78:22
Heptan-2-one	[(Crotyl)Ti$(OPr^i)_3]^-$ $MgCl^+$ (**30**)	56:44
Benzaldehyde	(Crotyl)Ti$(OPh)_3$	85:15
3,4-Dihydro-1(2*H*)-naphthalenone	(Crotyl)Ti$(OPh)_3$	65:35
Octan-2-one	(Crotyl)Ti$(OPh)_3$	70:30

[a]Adapted from the tables reported in refs. 11, 16, 53, 70 and 71.

Addition to ketones represents the real field of broad application of crotyltitanium(IV) reagents.[53a] 3-Methylbutan-2-one, by reaction with (**28**), gives a mixture of its *threo* and *erythro* addition products in a 97:3 molar ratio (>95% conversion), other ketones as reported in Table 8 also lead to their addition products in *threo:erythro* molar ratios which are of synthetic relevance. The simple aliphatic ketones react satisfactorily with (**28**). The use of (crotyl)Ti$(OPh)_3$ in this case does not give significant advantages, considering both diastereoselectivity of addition and availability of the reagent relative to (**28**), (**29**) and (**30**).[71] The enormous potential of these reagents can be fully understood considering that the corresponding Grignard reagents show no selectivity in the addition, and allylboranes either react very slowly or do not react at all with ketones.[17,73] Several mechanistic interpretations are also available, accounting for chemo- and diastereo-selectivity of crotyltitanium(IV) reagents.[11,53a,74]

Crotyl-titanocene and -zirconocene complexes, like bis(cyclopentadienyl)-Ti^{IV} and -Zr^{IV} derivatives (**111a–d**) react (equation 44) with aldehydes (Table 9) with pronounced *threo* diastereoselectivity (>90:<10 *threo:erythro* (**112**):(**113**) for Ti^{IV} reagents[37] and approximately 80:20 for their Zr^{IV} analogs).[36]

Less electronegative the halogen atom in (**111a–c**), higher the stereoselectivity of the addition.[17,37] Addition reactions of these reagents to ketones are not reported. The enhanced diastereoselectivity of titanocene-like crotyl derivatives may be ascribed to the strong electron-donating nature of the cyclopentadienyl residues which reduces the Lewis acidity of the complex and, hence, its reactivity.

$$\text{RCHO} + \text{crotyl-MCp}_2\text{X} \xrightarrow{-30\,^\circ\text{C}} (\mathbf{112}) + (\mathbf{113}) \qquad (44)$$

(**111**) **a**: X = Cl, M = Ti; **b**: X = Br, M = Ti; **c**: X = I, M = Ti; **d**: X = Cl, M = Zr

Table 9 Diasteroselective Addition of Crotyl-titanocene and -zirconocene Derivatives to Aldehydes (equation 44)[a]

Aldehyde	*Reagent*	*threo:erythro*	*Yield (%)*
Benzaldehyde	(Crotyl)TiCp$_2$Cl	60:40	96
Benzaldehyde	(Crotyl)TiCp$_2$Br	100:0	92
Benzaldehyde	(Crotyl)ZrCp$_2$Cl	81:19	90
Benzaldehyde	(Crotyl)ZrCp$_2$Cl + BF$_3$·OEt$_2$	4:96	98
Propanal	(Crotyl)TiCp$_2$Cl	66:34	92
Propanal	(Crotyl)TiCp$_2$Br	96:4	92
Propanal	(Crotyl)ZrCp$_2$Cl	86:14	88
2-Methylpropanal	(Crotyl)TiCp$_2$Br	99:1	87
2-Methylpropanal	(Crotyl)TiCp$_2$Br + BF$_3$·OEt$_2$	9:91	75
2-Methylpropanal	(Crotyl)ZrCp$_2$Cl	88:12	90

[a]Adapted from the tables reported in refs. 36, 37 and 72.

Reverse diastereoselectivity is also reported, as achieved by reaction of crotyl-Ti^{IV} and -Zr^{IV} complexes with aldehydes in the presence of BF$_3$·OEt$_2$ (Table 9).[72,75]

An interesting class of Ti^{III}-containing allylic reagents is represented by (η^5-C$_5$H$_5$)$_2$Ti(η^3-allyl) compounds (trihaptotitanium compounds), like (**114**), that are readily prepared from (η^5-C$_5$H$_5$)$_2$TiCl$_2$ (**32**) by action of allylic Grignard reagents[76] or, alternatively, alkyl Grignard species and dialkenes (equation 45).[77] When acetone is added to a purple ethereal solution of trihaptotitanium complex (**115**) under argon atmosphere,[38] a reaction takes place spontaneously at room temperature (the reaction medium becomes dark brown) and 2,3-dimethyl-4-penten-2-ol (**116**) is formed in 90% yield (Scheme 11). Benzaldehyde, under the same conditions, affords 1-phenyl-2-methyl-3-buten-1-ol in excellent yield.[38] A further advantage is the possibility of recovering the starting material (**32**), in the end, by simple air oxidation of the acidified reaction mixture (Scheme 11).[38] Due to their pronounced air sensitivity, these π-allyltitanium(III) complexes are most commonly prepared *in situ*.[38] The experiments reported in Table 10 show unambiguously the chemo- and regio-selective nature of their addition reactions.[38] In fact, other functions (*e.g.* halogens, double bonds, carboxylic acid esters) are fully tolerated; furthermore, the attack of the allyl residue occurs at the more substituted γ-carbon atom, with no exceptions. Stereoselectivity is also a relevant feature of a reaction which affords an excess of the *threo* isomer.

$$\text{Cp}_2\text{TiCl}_2\ (\mathbf{32}) \xrightarrow[\text{dialkene, 2PrMgBr}]{\text{2(allyl)MgBr or}} \text{Cp}_2\text{Ti}(\eta^3\text{-C(R}^1\text{)C(R}^2\text{)C(R}^3\text{)})\ (\mathbf{114}) \qquad (45)$$

Cyclic allyl residues enable the synthesis of cycloalkenes having a 1-hydroxyalkyl side chain.[78] Cyclic Ti^{III} complexes, like (**118**) and (**119**) are readily obtained (Scheme 12) by treatment of (η^3-C$_5$H$_5$)$_2$TiCl (**117**) with *i*-butylmagnesium chloride and cyclohexadiene or cyclopentadiene respectively, in THF at –40 °C for 10 min.[76,79] Such complexes are quite unstable and can be prepared only *in situ*. Propanal

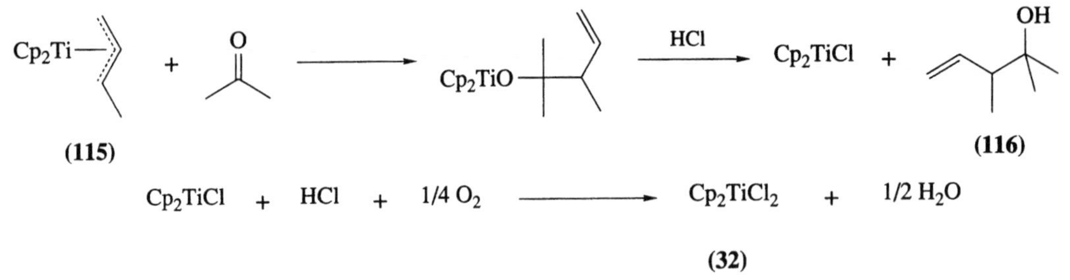

(115) (116) (32)

Cp_2TiCl + HCl + 1/4 O_2 ⟶ Cp_2TiCl_2 + 1/2 H_2O

Scheme 11

Table 10 Addition of Various Allyltitanium(III) Complexes ($RTiCp_2$) to Carbonyl Compounds

Carbonyl compound	*R*	*Product*	*Yield (%)*
Acetone	1-Methylallyl[b]	2,3-Dimethyl-4-penten-2-ol	88 (92)[a]
Methyl vinyl ketone	1-Methylallyl[b]	3,4-Dimethyl-1,5-hexadien-3-ol[c]	91
Benzaldehyde	1-Methylallyl[b]	2-Methyl-1-phenyl-3-buten-1-ol[d]	93
Chloroacetone	1,2-Dimethylallyl[b]	2,3,4-Trimethyl-1-chloro-4-penten-2-ol	95
Acrolein	1,2-Dimethylallyl[b]	4,5-Dimethyl-1,5-hexadien-3-ol[e]	90
Levulinic acid methyl ester	1,2-Dimethylallyl[b]	γ-Methyl-γ-(1,2-dimethylallyl)-γ-butyrolactone	94
Acetone	1-Ethylallyl[f]	2-Methyl-3-ethyl-4-penten-2-ol	83 (86)[a]

[a]If isolated Ti^{III} complex is utilized. [b]Prepared from (32), proper diene and PrMgBr. [c]65:35 mixture of *threo:erythro* diastereomers. [d]95:5 mixture of *threo:erythro* diastereomers. [e]90:10 mixture of *threo:erythro* diastereomers. [f]Prepared from (32) and 1-ethylallylmagnesium bromide; m.p. 90.5–91.0 °C.

reacts with (**119**) stereoselectivity,[78] affording exclusively the *erythro* compound (**120**) in 86% yield (Scheme 13); benzaldehyde also shows excellent stereoselection,[78] in spite of a somewhat lower reaction yield (Scheme 13). η^3-Cyclohexenyltitanium derivative (**118**) reacts likewise (equation 46), although showing diminished stereoselectivity.[78] A mechanistic interpretation of the stereochemical course of such addition reactions is also available.[78]

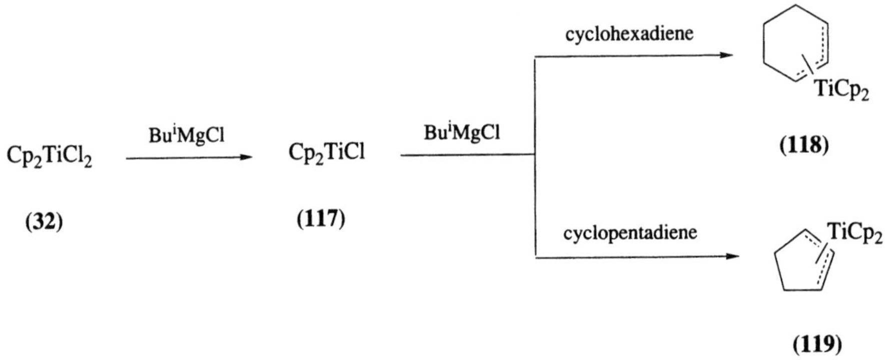

Cp_2TiCl_2 (32) —Bu^iMgCl→ Cp_2TiCl (117) —Bu^iMgCl→ cyclohexadiene → (118); cyclopentadiene → (119)

Scheme 12

(120) ← EtCHO, 86% — (119) — PhCHO, 79% → (121)

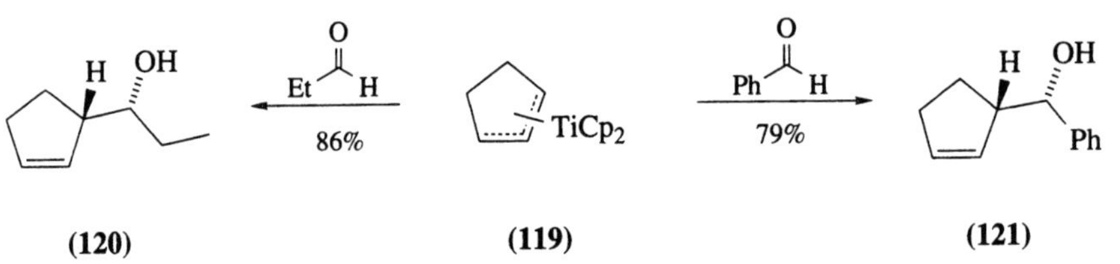

Scheme 13

Modified η^3-titanium compounds have been prepared from exocyclic dienes.[78] The compound (**123**, n = 3) is an example: it is prepared (equation 47) by addition of Bu^iMgCl to a solution of 1-vinylcyclopentene (**122**, n = 3) and $(\eta^5\text{-}C_5H_5)_2TiCl$. Its reaction with propanal takes place regiospecifically at the carbon atom belonging to the cycle, in 86% yield, affording the sole *erythro* compound (**124**, n = 3) (equation

(46)

(118) + EtCHO → *erythro* 80% + *threo* 20%

47), having (*Z*)-configuration of the double bond. The regioselectivity of the attack, however, is a function of the ring dimension and the following order is reported: 90% ($n = 5$), 75% ($n = 6$), 66% ($n = 10$).[78]

(47)

(122) —Cp_2TiCl, Bu^iMgCl→ (123) —EtCHO, 86%→ (124)

1.5.3.4 Heterosubstituted Allylic Titanium Reagents

Ti^{IV} reagents bearing heteroatom-substituted allyl residues are conveniently used in analogy with their nonsubstituted analogs to prepare homoallylic alcohols with sulfur,[80–82] silicon,[80,83–85] phosphorus,[86,87] and other β-substituents.[88] These groups can be subsequently exploited to introduce further functions. As the topic will be extensively covered in Part 1 of Volume 2, only a few salient aspects of chemo-, regio- and diastereo-selectivity of such reagents will be considered in this section.

Titanated silyl derivatives, like (**125**), are among the most interesting of these compounds. They react with carbonyl compounds[35] by addition at the γ-carbon atom and with excellent *threo* diastereoselectivity (equation 48). The reaction of (**125**) with acetophenone affords a 91:9 molar mixture of the corresponding *threo* and *erythro* products (equation 48).[35] Peterson alkenation[35,83] takes advantage of the availability of such products in a stereospecific diene synthesis. Examples of applications of these reagents in the synthesis of natural products are also reported.[89,90] The regiochemistry of the addition is often influenced by the shape and number of substituents on the allyl moiety of the reagent, as is reported for the titanated sulfur-containing compounds (**126**) which, depending upon the nature of the various substituents, perform either α- or γ-attack onto carbonyl compounds (equation 49).[81,82] In Table 11, this substituent effect is illustrated with cyclohexanecarbaldehyde.[81,82]

(48)

Me_3Si–CH=CH–CH_2–$Ti(OPr^i)_4^-\ Li^+$ (**125**) —$RCOR^1$→ *threo* + *erythro*

	threo	*erythro*
R = Ph, n-C_6H_{13}, Pr^i; R^1 = H	>99%	<1%
R = Ph; R^1 = Me	91%	9%

(49)

(**126**) + cyclohexanecarbaldehyde → (**127**) + (**128**)

Phosphorus-containing allyltitanium(IV) complexes, like (**129**), have been studied as well.[86,87] The synthetic importance of their addition products to aldehydes lies on their ready deoxygenation to afford dienes stereospecifically (equation 50).

Table 11 Influence of Substituents on the Addition of Titanated Sulfur-containing Compounds (**126**) to Cyclohexancarbaldehyde (equation 49)

R^1	R^2	R^3	R^4	(**127**):(**128**)
H	H	H	H	99:1
Me	H	H	H	99:1
H	Me	H	H	96:4
H	H	Me	H	2:98
Me	H	Me	H	85:4
H	H	Me	Me	<1:99

(50)

(**129**) (*Z*)

Metalated *N*,*N*-diisopropyl 2-alkenyl carbamates (**130**) are reported to be successfully utilized to prepare, with high γ-regioselectively (equation 51), 4-hydroxyalkenyl carbamates like (**131**), which are useful intermediates in the synthesis of substituted 4-butanolides.[91]

(51)

(**130**) (**131**)

1.5.3.5 Dienyl Titanium and Zirconium Reagents

Zirconocene complexes with conjugated dienes can be described in terms of an equilibrium between the *s-trans*-diene zirconocene (**132**) and cyclopentadienyl zirconacyclopentene (**133**) forms (equation 52).[92–94] They are readily prepared either from zirconocene dichloride,[95,96] or diphenyl zirconocene[92] in the presence of conjugated dienes. The equilibrium mixture contains (**132**) and (**133**) in approximately 45:55 molar ratio (at 25°C),[97,98] the former being the most reactive.[94] Addition of the reagent (**132** ↔ **133**) to carbonyl compounds (**134a–f**) (Scheme 14) affords oxazirconacycloheptenes (**135a–f**), which can be hydrolyzed to give alcohol mixtures (**136a–f**) and (**137a–f**).[99]

(52)

(**132**) (**133**)

Like allylzirconocene derivatives, conjugated diene zirconocene complexes are highly reactive towards various polar and nonpolar systems, and are characterized by remarkable regioselectivity. For example, isoprene zirconocene (**138**) reacts (Scheme 15) with various systems regioselectively at the sterically more congested C-1 position of isoprene.[100–102] Despite the steric congestion of this reaction site regioselectivity is higher than 95% in reactions with saturated or unsaturated aldehydes as well as with ketones, carboxylic acid esters and nitriles.[94] Furthermore, the yields of such reactions are, without exception, higher than 90%. The highly selective 1,2-addition of diene zirconocene complexes to α,β-unsaturated ketones and esters may be accounted for by the high ionic character of the C—Zr bond and the oxophilic nature of zirconium, in addition to the rigid tetrahedral geometry of the complexes.[94] Analogously, $RTiCl_3$ and alkyllanthanoids are also able to promote selective 1,2-addition to enones.[103,104] Other diene metal complexes of titanium and hafnium are known to give regioselective addition to enones,[94] although their reactivity toward such substrates is considerably lower than that of their Zr^{IV} analogs.

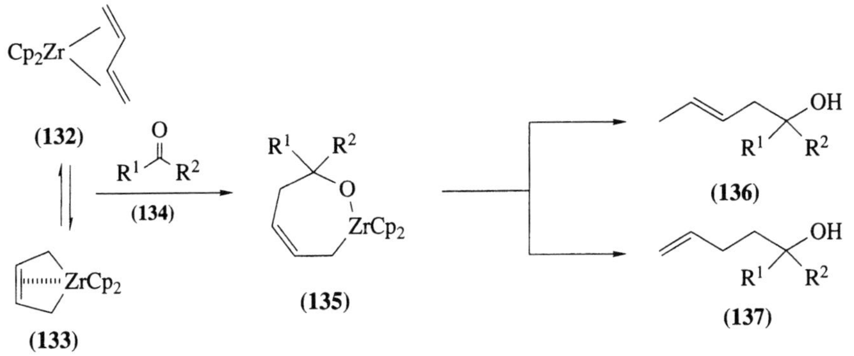

Cp_2Zr (132) ⇌ (133) $ZrCp_2$; $R^1C(O)R^2$ (134) → (135) → (136) + (137)

a: R^1 = Me, R^2 = Ph
b: R^1 = R^2 = Me
c: R^1—R^2 = —$CH_2(CH_2)_9CH_2$—
d: R^1 = R^2 = Ph
e: R^1 = Pr^i, R^2 = H
f: R^1 = Bu^t, R^2 = Me

Scheme 14

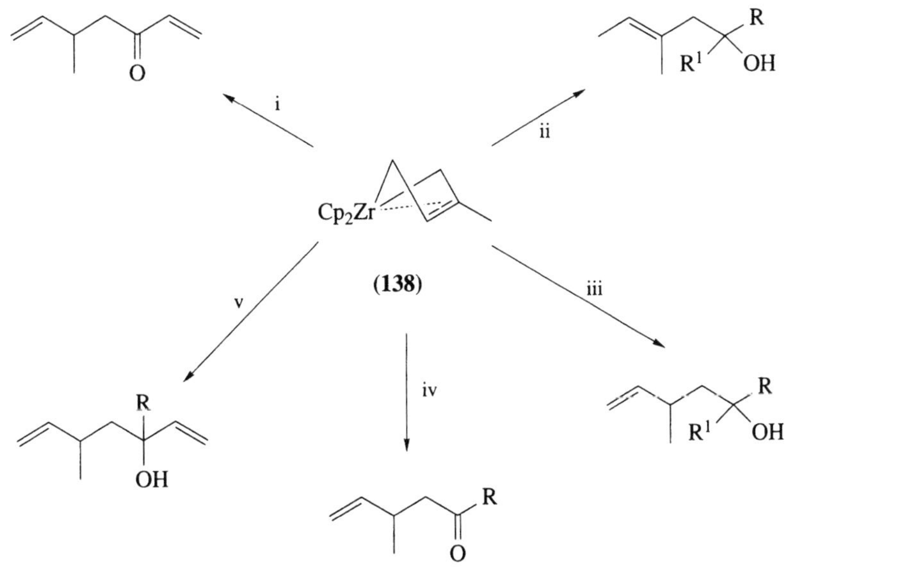

Cp_2Zr (138); i–v

i, CH_2=$CHCO_2R^1$, H^+; ii, RR^1CO; iii, RR^1CO, H^+; iv, RCO_2R^1 or RCN, H_2O/H^+; v, CH_2=CHCOR, H^+

Scheme 15

Reverse regioselectivity is observed when isoprene zirconocene is treated with carbonyl compounds under photochemical conditions, at –70 °C where the thermally induced addition is totally suppressed.[105] As a case in point, 3,3-dimethylbutan-2-one (**139**; Scheme 16) upon reaction with the zirconocene complex of isoprene in benzene at 60 °C for 2 h affords (**140a**), whereas irradiation of the reaction mixture at low temperature in toluene leads to (**140b**). Mixtures of (**140a**) and (**140b**) are obtained by irradiation of isolated (**138**) and ketone (**139**) at higher temperatures. As the reaction temperature is lowered, an increasing amount of (**140b**) is formed. Therefore, it is reasoned that the zirconocene complex reacts in the *cis* form (**138**), under thermal conditions, and in the *trans* form under photochemical conditions.

Zirconocene complexes (with *s-cis* geometry) of isoprene, 2,3-dimethylbutadiene, and 3-methyl-1,3-pentadiene are reported to give exclusively 1:1 addition with carbonyl substrates even if these are used in excess and the reaction temperature is fairly high (*ca.* 100°C).[106] On the contrary, zirconocene complexes of *s-cis*-butadiene, 1,3-pentadiene, and 2,4-hexadiene (*ca.* 1:1 mixture of the *s-cis* and *s-trans* isomers) easily accept (equation 53) 2 equiv. of either butanal or 3-pentanone, at low temperature (*ca.* 30 °C) in high yields (95%).[94,106] This can be exploited for the stepwise insertion of two different electrophiles

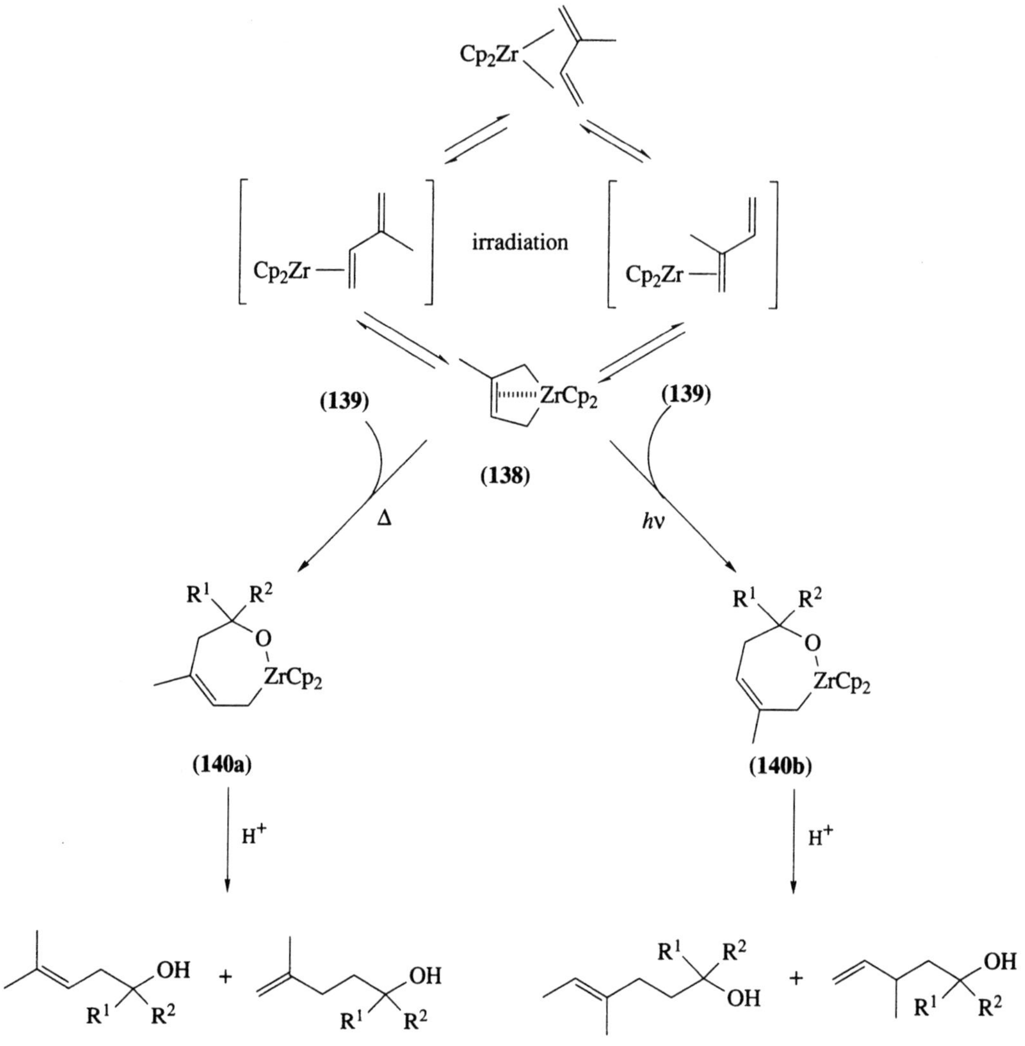

irradiation

(139) (138) (139)

Δ $h\nu$

(140a) (140b)

H^+ H^+

R^1 = Me, R^2 = Bu^t

Scheme 16

(*e.g.* isobutanal/pentan-3-one, ethyl acetate/isobutanal, pentan-3-one/acetonitrile) allowing for the preparation of variously substituted diols (Scheme 17).[94] The technique consists of the addition of the first carbonyl compound at 0 °C in hexane, followed by addition of the second electrophile at 60 °C in THF. Attempts to invert the order of addition within the pairs of electrophiles tested (*e.g.* isobutanal/ethyl acetate) are often unsuccessful.[94]

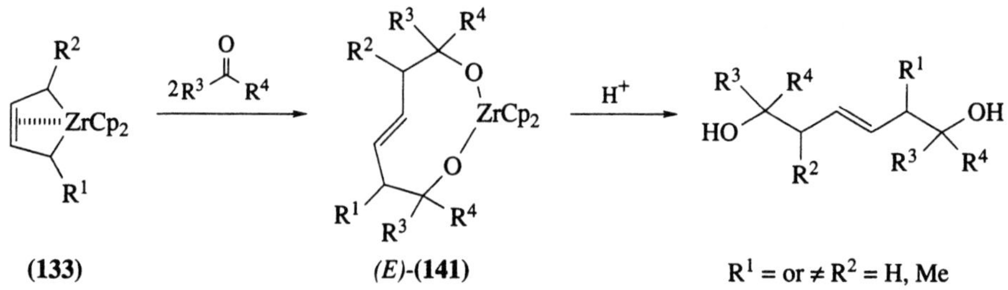

(53)

(133) (*E*)-(141) R^1 = or ≠ R^2 = H, Me

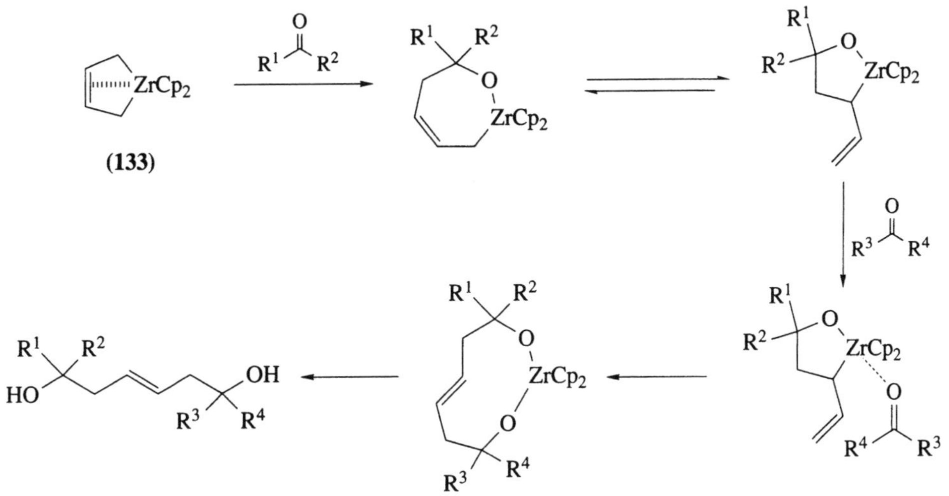

(133)

(E)-(142)

Scheme 17

1.5.3.6 Propargylic Titanium Reagents

In organic synthesis, propargyltitanium(IV) complexes are unique among the organometallic reagents in that they perform efficient addition of either alkynic or allenic residues to carbonyl compounds.[107]

Since this topic will be covered extensively in Part 1 Volume 2, only the main references concerning their preparation and their more significant synthetic applications are recalled at this point.[108,109] Hetero-substituted propargyltitanium(IV) reagents are also very useful in the synthesis of natural products.[110]

1.5.4 CHIRALLY MODIFIED TITANIUM REAGENTS: ENANTIOSELECTIVE ADDITION

Different strategies have been pursued to achieve enantioselection in the addition of specially designed chiral Ti^{IV} reagents to carbonyl compounds. They are essentially based on the use of: (i) organotitanium derivatives having chiral ligands; (ii) complexes having asymmetrically substituted chiral Ti^{IV} atoms; and (iii) $RTiX_3$-like reagents with the nucleophilic moiety bearing a removable chiral auxiliary. Also worthy of mention in this context is an indirect enantioselective addition that is achieved by the reaction of achiral $RTiX_3$ complexes with chiral acetal derivatives of the carbonyl compounds, as is shown in Scheme 18.[111]

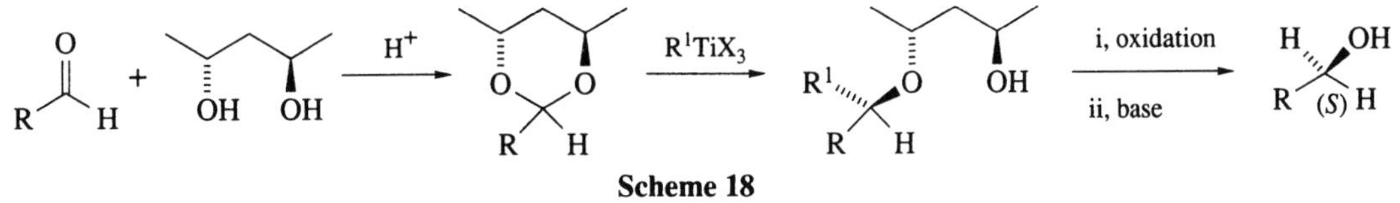

Scheme 18

1.5.4.1 Organotitanium Derivatives Having Chiral Ligands

Early experiments characterized by low enantiomeric excesses have been reported.[44,46] Ti^{IV} reagents bearing chiral alkoxy ligands are readily prepared from chlorotriisopropoxytitanium by ligand interchange (equation 54) using a chiral alcohol (or phenol), under conditions in which isopropanol can be azeotropically distilled off.[112a] The resulting chiral chlorotitanium derivative is then converted *in situ* to the desired organotitanium complex by reaction with either an RMgX or RLi species. The most commonly used chiral ligands are cinchonine, quinine, (*S*)-1,1′-binaphthol, (–)-menthol, *etc.*[112a] The enantioselectivity of the addition is influenced by the nature of both the carbonyl substrate and the ligand system, as well as by the solvent and the temperature.[112a]

Methyl- and phenyl-Ti^{IV} complexes with different chiral ligands have been tested[112a,b] in reactions with aromatic aldehydes. The enantioselectivity of the addition of chiral Ti^{IV} reagents having a bidentate 1,1′-binaphthol ligand system to aromatic aldehydes (equation 55) has been also extensively examined,

$$ClTi(OPr^i)_3 + R^*OH \xrightarrow{-3Pr^iOH} ClTi(OR^*)_3 \xrightarrow[\text{or RLi}]{RMgX} RTi(OR^*)_3 \quad (54)$$

and the results obtained are reported in Table 12.[113] The relative topicity of the addition is *lk*, leading to *re* attack of the phenyl group to the carbonyl function by the (*M*)-binaphthol Ti^{IV} reagent [hence, the (*R*)-alcohol] (equation 56) and *si* attack by its (*P*)-binaphthol analog.

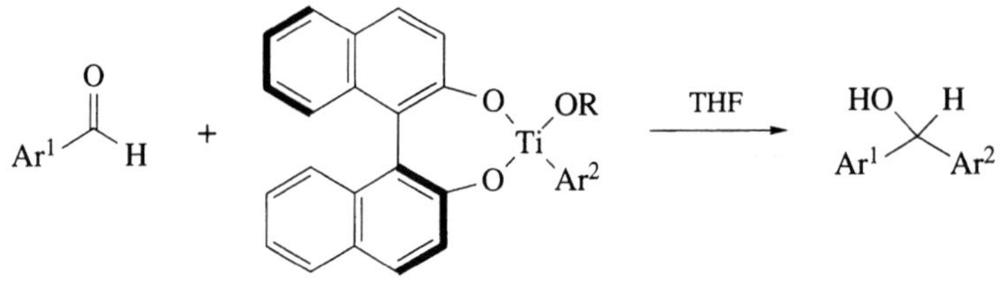

(55)

Table 12 Enantioselective Addition of (*M*)-Binaphtholphenyltitanium isopropoxide to Aromatic Aldehydes

Aldehyde	*Temperature (°C)*	*Product*	*Yield (%)*	*ee (%)*
2-Methylbenzaldehyde	–2	(2-Methylphenyl)phenylmethanol	89	85
4-Methylbenzaldehyde	–2	(4-Methylphenyl)phenylmethanol	78	86
Anisaldehyde	–69	(4-Methoxyphenyl)phenylmethanol	84	82
1-Naphthaldehyde	–15	1-Naphthylphenylmethanol	—	63
1-Naphthaldehyde	–75	1-Naphthylphenylmethanol	82	>98
2-Naphthaldehyde	–17	2-Naphthylphenylmethanol	—	57
2-Naphthaldehyde	–65	2-Naphthylphenylmethanol	85	>98

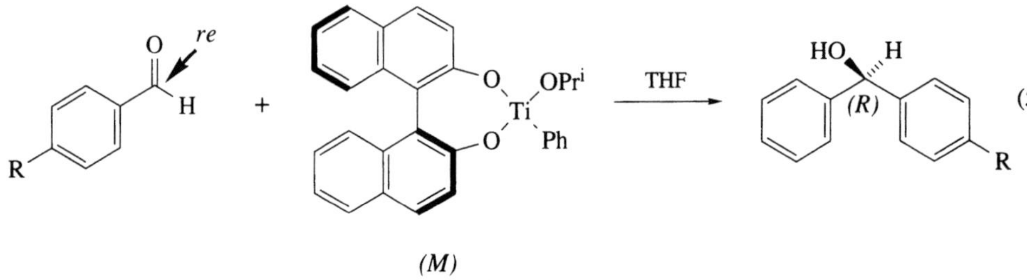

(56)

Attempts to prepare chiral Ti^{IV} reagents capable of transferring groups other than phenyl have been reported.[11] Among these reagents, the methyltitanium (*S*)-acylpyrrolidinylmethoxide diisopropoxides (**144**) are of some interest. They are prepared (Scheme 19) from (*S*)-*N*-acylpyrrolidinylmethanols (**143**) and dimethyltitanium diisopropoxide.[114] Their reactions with either benzaldehyde or 1-naphthaldehyde lead to alcohols with acceptable enantiomeric excesses, as is shown in Table 13.[114]

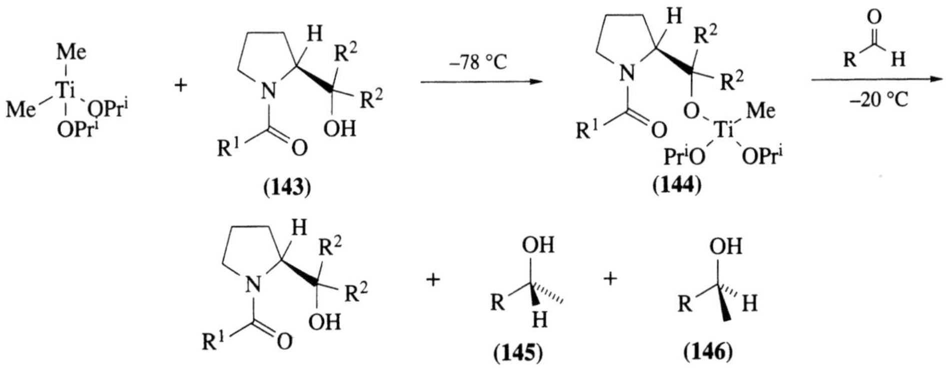

Scheme 19

Table 13 Enantioselective Addition of Methyltitanium (*S*)-*N*-Acylpyrrolidinylmethoxides (**144**) to Aromatic Aldehydes (Scheme 19)

Aldehyde	R^1	R^2	*Product*	*R*	*ee (%)*
Benzaldehyde	H	H	(**146**)	Phenyl	33.0
Benzaldehyde	H	Ph	(**146**)	Phenyl	17.6
Benzaldehyde	Me	H	(**145**)	Phenyl	23.9
1-Naphthaldehyde	Me	H	(**145**)	1-Naphthyl	27.6
Benzaldehyde	Me	Me	(**145**)	Phenyl	18.6
Benzaldehyde	Me	Ph	(**145**)	Phenyl	39.0
Benzaldehyde	Ph	H	(**145**)	Phenyl	13.9
Benzaldehyde	$PhCH_2O$	Me	(**146**)	Phenyl	23.0
1-Naphthaldehyde	$PhCH_2O$	Me	(**146**)	1-Naphthyl	27.0
Benzaldehyde	$PhCH_2O$	Ph	(**145**)	Phenyl	54.1

Further attempts to replace alkoxy groups with other ligand systems, like cyclopentadienyl, in order to reduce the Lewis acidity of the metal atom have been unsuccessful.[11]

1.5.4.2 Derivatives Having a Chiral Titanium(IV) Atom

This strategy was devised to enable the chiral center of the reagent to be closer to the carbonyl group during the addition. Ligand systems based on cyclopentadienyls lead to unsatisfactory enantiomeric excesses,[11] as is reported for the reaction of benzaldehyde with the chiral allyltitanium(IV) reagent (**148**) derived from (**147**) by *in situ* treatment with allylmagnesium chloride (Scheme 20).

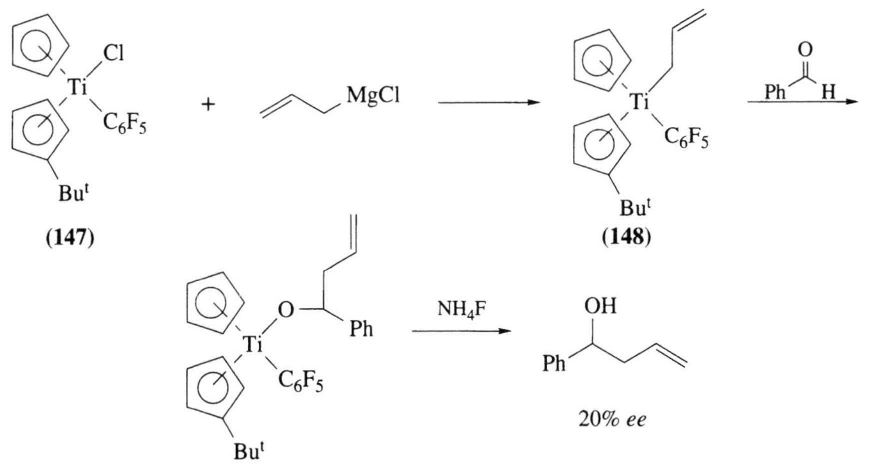

Cl, Ti, C_6F_5, Bu^t (**147**) + MgCl → Ti, C_6F_5, Bu^t (**148**) —PhCHO (Ph, H, O)→ O, Ph, Ti, C_6F_5, Bu^t —NH_4F→ OH, Ph; 20% *ee*

Scheme 20

A noteworthy improvement comes from the use of ligand systems which are themselves optically active.[11] This is the case of the complex (**150**), shown in Scheme 21, that is prepared from a sulfenylated norephedrine (**149**). The results reported for the *in situ* reactions of (**150**) with various aldehydes (Scheme 21) are summarized in Table 14.[115] Nothing is reported concerning the mechanism of the enantioselective addition, since the absolute configuration of the asymmetrically substituted Ti^{IV} atom is not known.

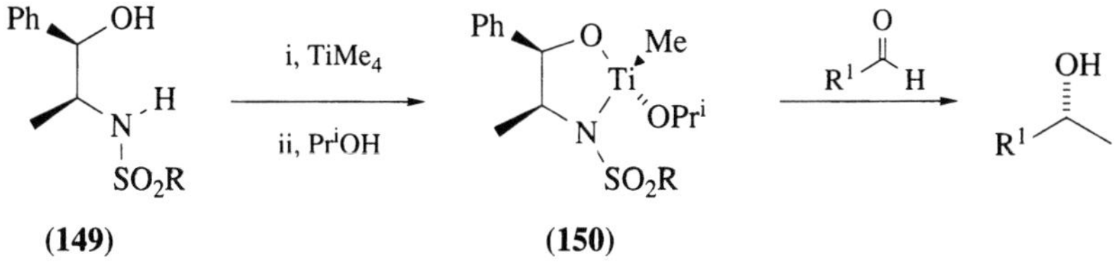

Ph, OH, N, H, SO_2R (**149**) —i, $TiMe_4$; ii, Pr^iOH→ Ph, O, Me, Ti, OPr^i, N, SO_2R (**150**) —R^1CHO (R^1, H, O)→ OH, R^1

Scheme 21

Table 14 Enantioselective Addition of Methyltitanium(IV) derivatives (**150**) of Various *N*-Sulfonylated Norephedrines to Aldehydes (Scheme 21)

*R in (**150**)*	*Aldehyde*	*Yield (%)*	*ee (%)*[a]
4-Tolyl	Benzaldehyde	78	85
4-Tolyl	2-Nitrobenzaldehyde	91	79
4-Tolyl	1-Naphthaldehyde	96	81
Methyl	Benzaldehyde	89	62
Mesityl	Benzaldehyde	93	88
Mesityl	2-Nitrobenzaldehyde	86	90
Mesityl	1-Naphthaldehyde	92	90
4-Tolyl	Octanal	81	60
Mesityl	Octanal	82	58

[a]In all cases the reaction products have the (*R*)-configuration.

1.5.4.3 Reagents with the Nucleophilic Moiety Bearing a Removable Chiral Auxiliary

These reagents can be obtained by titanation of a lithium derivative containing the chiral auxiliary (Scheme 22).[116a] Regio- and enantio-selective addition of (**151**) to aldehydes and ketones is reported to afford chiral homoaldol compounds in satisfactory yields and with appreciable enantiomeric excesses (Scheme 22). This reaction has been used in the synthesis of natural products.[116a]

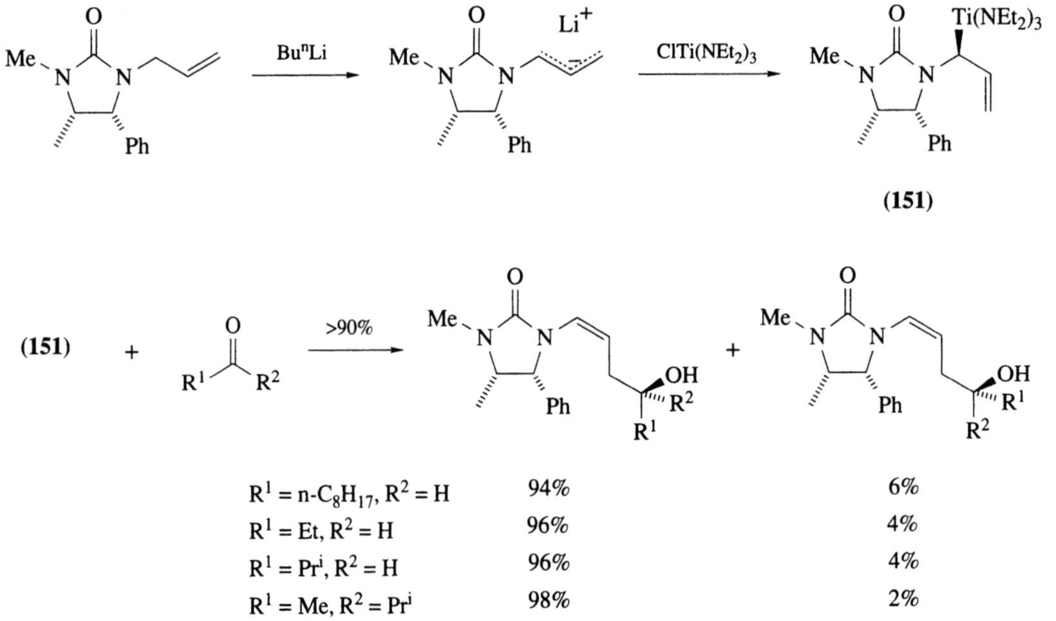

Bu^nLi ; $ClTi(NEt_2)_3$; $Ti(NEt_2)_3$; Li^+

(**151**)

(**151**) + $R^1C(O)R^2$ —>90%→

$R^1 = n\text{-}C_8H_{17}, R^2 = H$	94%	6%
$R^1 = Et, R^2 = H$	96%	4%
$R^1 = Pr^i, R^2 = H$	96%	4%
$R^1 = Me, R^2 = Pr^i$	98%	2%

Scheme 22

Lithiated chiral dihydro-1,4-dithiins (**152**) have been recently utilized[116b] in the addition to prochiral aldehydes in the presence of $Ti(OPr^i)_4$ (Scheme 23). Under such conditions (*Z*)-allylic alcohols have been obtained in good yields and with satisfactory enantioselectivity (*ee* >40%), after removal of the chiral sulfur-containing moiety of the addition product. The reaction is likely to proceed *via* the formation of a cyclic five-membered transition state (**153**) involving titanium, oxygen and one of the sulfur atoms.[53b] When $Ti(OPr^i)_4$ is replaced by $TiCl_4$, the addition becomes slower than the uncatalyzed reaction itself.

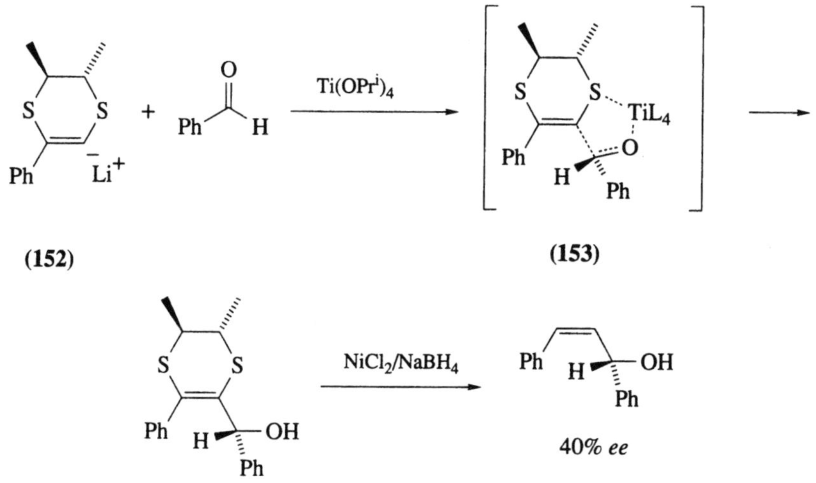

(152) (153)

Scheme 23

1.5.5 MISCELLANEOUS REACTIONS

1.5.5.1 Geminal Dialkylation of Aldehydes and Ketones

This represents probably one of the earliest interests in the chemistry of Ti^{IV} compounds.[117] Ketones, like cyclohexanone, are easily dimethylated by $MeTiCl_3$, Me_2TiCl_2 and $Me_2Zn/TiCl_4$.[118,119] Depending upon the molecular ratio of Me_2Zn and $TiCl_4$, various products **(43)**, **(154)** and **(155)** are obtained from cyclohexanone (equation 57), as is reported in Table 15.[121] The reaction proceeds through cationic intermediates;[120] various functionalities (*e.g.* carboxylic acid esters, primary and secondary alkyl halides) are well tolerated.[120] Some problems arise when α,β-unsaturated ketones are utilized, because of the formation of two distinct dimethylated products (equation 58).[120]

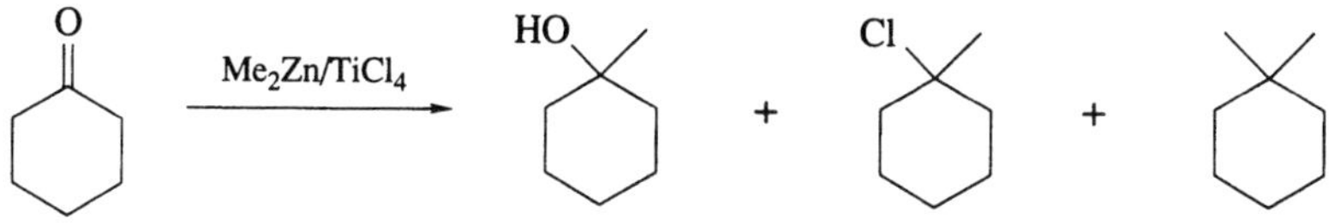

(57)

(43) (154) (155)

Table 15 Geminal Dimethylation of Cyclohexanone (equation 57)

Entry	*$TiCl_4/ZnMe_2$ (mmol)*[a]	*Temperature (°C)*	*Time (h)*	*Yield (%)* (43)	(154)	(155)
1	30/10	−30	4	—	>90	—
2	10/10	−30	4	>90	—	—
3	22/22	−30 to +22	2	—	—	>90
4	10/20	−30 to +22	4	>90	—	—

[a]In all cases 10 mmol of cyclohexanone were used in CH_2Cl_2.

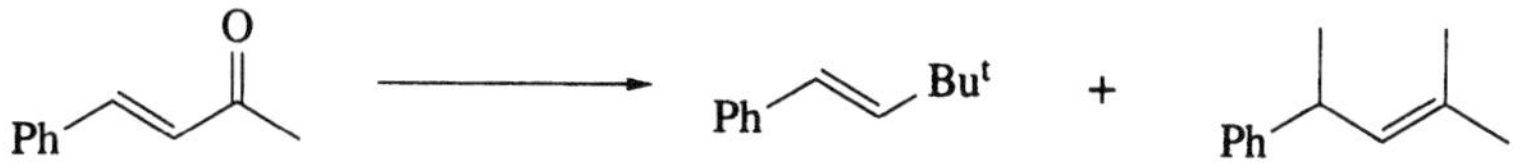

(58)

Aromatic aldehydes are also easily dimethylated by Me_2TiCl_2, and aromatic acyl chlorides form *t*-butyl derivatives directly when treated with an excess of $Me_2Zn/TiCl_4$ in methylene chloride.[121] Some examples are also known of application of such a reaction in the synthesis of natural products.[118,119]

1.5.5.2 Alkylative Amination

Alkyltitanium(IV) complexes having *N,N*-dialkylamino ligand systems, $RTi(NR^1_2)_3$, fail to give nucleophilic additions to carbonyl compounds (Section 1.5.3.1.1). Their reaction with aldehydes leads instead to tertiary amines by addition of both the alkyl moiety of the reagent and one of the *N,N*-dialkylamino ligands (equation 59).[48] The synthetic interest of the reaction is restricted to nonenolizable aldehydes, since enolizable carbonyl compounds lead to enamines.[48]

$$RTi(NR^1_2)_3 + R^2CHO \longrightarrow R^2CH(NR^1_2)R \qquad (59)$$

A few examples of alkylative amination of aldehydes are reported in Table 16.[48]

Table 16 Alkylative Amination of Various Aldehydes by $RTi(NEt_2)_3$

Aldehyde	*R*	*Product*	*Yield (%)*
Benzaldehyde	Me	*N,N*-Diethyl-1-phenylethylamine	48
4-Methylbenzaldehyde	Me	*N,N*-Diethyl-1-*p*-tolylethylamine	47
Benzaldehyde	Bu	*N,N*-Diethyl-1-phenylpentylamine	15
2-Furylaldehyde	Me	*N,N*-Diethyl-1-(2′-furyl)ethylamine	73
Cinnamaldehyde	Me	*N,N*-Diethyl-4-phenylbut-3-en-2-ylamine	44
Pivalaldehyde	Me	*N,N*-Diethyl-3,3-dimethylbut-2-ylamine	55

1.5.5.3 Protection of Aldehydes

The tendency of Ti^{IV} reagents like $Ti(NR_2)_4$ to release the *N,N*-dialkylamino group[42,48] is conveniently exploited to temporarily inactivate a carbonyl function which is fairly reactive (*e.g.* an aldehyde),[122] thus enabling a second, less reactive carbonyl function (*e.g.* a ketone), to react with another nucleophilic reagent. This leads to a sort of reverse chemoselectivity which is highly effective in discriminating between different carbonyl compounds, as well as between carbonyls having different steric environments.

1.5.6 REFERENCES

1. M. A. Cahours, *Ann. Chim. Phys.*, 1861, **62**, 257; M. A. Cahours, *Liebigs Ann. Chem.*, 1862, **122**, 48.
2. D. F. Herman and W. K. Nelson, *J. Am. Chem. Soc.*, 1952, **74**, 2693.
3. M. Bottrill, P. D. Gavens, J. W. Kelland and J. McMeeking, in 'Comprehensive Organometallic Chemistry', ed. G. Wilkinson, F. G. A. Stone and E. W. Abel, Pergamon Press, Oxford, 1982, vol. 3, chaps. 22.1–22.5; D. J. Cardin, M. F. Lappert, C. L. Raston and P. I. Riley, in 'Comprehensive Organometallic Chemistry', chaps. 23.1 and 23.2, and literature cited therein.
4. D. J. Cardin, M. F. Lappert and C. L. Raston, 'Chemistry of Organo-zirconium and -hafnium Compounds', Wiley, New York, 1986; R. S. P. Coutts, P. C. Wailes and H. Weigold, 'Organometallic Chemistry of Titanium, Zirconium, Hafnium', Academic Press, New York, 1974; G. E. Coates, M. L. H. Green and K. Wade, in 'Organometallic Compounds', Chapman and Hall, London, 1968, vol. 2.
5. U. Thewalt, 'Titan–Organische Verbindungen' in 'Gmelin Handbuch der Anorganischen Chemie', Springer-Verlag, Berlin, 1977, 1980, 1984, band 40, teil 1, 2; A. Moulik 'Zirconium–Organische Verbindungen', in 'Gmelin Handbuch der Anorganischen Chemie', 1973, band 10.
6. K. Ziegler, E. Holtzkamp, H. Breil and H. A. Martin, *Angew. Chem.*, 1955, **67**, 54.
7. G. Wilkinson, *Pure Appl. Chem.*, 1972, **30**, 627; P. J. Davidson, M. F. Lappert and R. Pearce, *Chem. Rev.*, 1976, **76**, 219; R. R. Schrock and G. W. Parshall, *Chem. Rev.*, 1976, **76**, 243.
8. C. Beermann and H. Bestian, *Angew. Chem.*, 1959, **71**, 618; D. F. Herman and W. K. Nelson, *J. Am. Chem. Soc.*, 1953, **75**, 3877; D. F. Herman and W. K. Nelson, *J. Am. Chem. Soc.*, 1953, **75**, 3882.
9. B. Weidmann and D. Seebach, *Angew. Chem., Int. Ed. Engl.*, 1983, **22**, 31; *Angew. Chem.*, 1983, **95**, 12.
10. M. F. Lappert, D. S. Patil and J. B. Pedley, *J. Chem. Soc., Chem. Commun.*, 1975, 830.
11. M. T. Reetz, 'Organotitanium Reagents in Organic Synthesis', Springer-Verlag, Berlin, 1986.
12. M. T. Reetz, *Pure Appl. Chem.*, 1985, **57**, 1781.
13. K. Clauss, *Justus Liebigs Ann. Chem.*, 1968, **711**, 19.
14. M. D. Rausch and H. B. Gordon, *J. Organomet. Chem.*, 1974, **74**, 85.
15. C. Beermann, *Angew. Chem.*, 1959, **71**, 195; H. Bestian, K. Clauss, H. Jensen and E. Prinz, *Angew. Chem., Int. Ed. Engl.*, 1963, **2**, 32; *Angew. Chem.*, 1962, **74**, 955.
16. M. T. Reetz, *Top. Curr. Chem.*, 1982, **106**, 1.
17. R. W. Hoffmann, *Angew. Chem., Int. Ed. Engl.*, 1982, **21**, 555.
18. C. T. Buse and C. H. Heathcock, *Tetrahedron Lett.*, 1978, 1685.
19. D. J. Cram and F. A. A. Elhafez, *J. Am. Chem. Soc.*, 1952, **74**, 5828; D. J. Cram and J. Allinger, *J. Am. Chem. Soc.*, 1954, **76**, 4516.

20. D. J. Cram and K. R. Kopecky, *J. Am. Chem. Soc.*, 1959, **81**, 2748.
21. M. T. Reetz, R. Steinbach, J. Westermann and R. Peter, *Angew. Chem., Int. Ed. Engl.*, 1980, **19**, 1011; *Angew. Chem.*, 1980, **92**, 1044.
22. D. Seebach, H.-O. Kalinowski, B. Bastiani, G. Crass, H. Daum, H. Dorr, N. P. DuPreez, V. Ehrig, W. Langer, C. Nussler, H.-A. Oei and M. Schmidt, *Helv. Chim. Acta*, 1977, **60**, 301; D. Seebach and W. Langer, *Helv. Chim. Acta*, 1979, **62**, 1701; W. Langer and D. Seebach, *Helv. Chim. Acta*, 1979, **62**, 1710; D. Seebach, G. Crass, E.-M. Wilka, D. Hilvert and E. Brunner, *Helv. Chim. Acta*, 1979, **62**, 2695; J. P. Mazaleyrat and D. J. Cram, *J. Am. Chem. Soc.*, 1981, **103**, 4585; T. Mukaiyama, K. Soai, T. Sato, H. Shimizu and K. Suzuki, *J. Am. Chem. Soc.*, 1979, **101**, 1455.
23. D. Seebach, B. Weidmann and L. Widler, in 'Modern Synthetic Methods', ed. R. Scheffold, Wiley, New York, 1983, vol. 3.
24. O. V. Nogina, R. K. Freidlina and A. N. Nesmeyanov, *Izv. Akad. Nauk SSSR, Ser. Khim.*, 1952, 74 (*Chem. Abstr.*, 1953, **47**, 1583).
25. M. T. Reetz, R. Urz and T. Schuster, *Synthesis*, 1983, 540.
26. A. Segnitz, *Methoden Org. Chem. (Houben–Weyl)*, 1975, **13/7**, 263.
27. M. T. Reetz, S. H. Kyung and M. Hüllmann, *Tetrahedron*, 1986, **42**, 2931.
28. W. R. Baker, *J. Org. Chem.*, 1985, **50**, 3942.
29. E. Negishi and T. Takahashi, *Synthesis*, 1988, 1.
30. K. Kuhlein and K. Clauss, *Makromol. Chem.*, 1972, **155**, 145.
31. R. K. Freidlina, E. M. Brainina, A. N. Nesmeyanov, *Dokl. Akad. Nauk SSSR*, 1961, **138**, 1369 (*Chem. Abstr.*, 1962, **56**, 11 608).
32. G. M. Whitesides, C. P. Casey and J. K. Krieger, *J. Am. Chem. Soc.*, 1971, **93**, 1379.
33. D. W. Hart, T. F. Blackburn and J. Schwartz, *J. Am. Chem. Soc.*, 1975, **97**, 679.
34. M. T. Reetz, B. Wenderoth and R. Urz, *Chem. Ber.*, 1985, **118**, 348.
35. M. T. Reetz and B. Wenderoth, *Tetrahedron Lett.*, 1982, **23**, 5259.
36. Y. Yamamoto and K. Maruyama, *Tetrahedron Lett.*, 1981, **22**, 2895.
37. F. Sato, K. Iida, S. Iijima, H. Moriya and M. Sato, *J. Chem. Soc., Chem. Commun.*, 1981, 1140.
38. F. Sato, S. Iijima and M. Sato, *Tetrahedron Lett.*, 1981, **22**, 243.
39. B. Weidmann, C. D. Maycock and D. Seebach, *Helv. Chim. Acta*, 1981, **64**, 1552.
40. H. Meerwein, B. V. Bock, B. Kirschnick, W. Lenz and A. Migge, *J. Prakt. Chem. N.F.*, 1936, **147**, 211; A. L. Wilds, *Org. React. (N. Y.)*, 1944, **2**, 178.
41. M. T. Reetz, R. Steinbach, J. Westermann, R. Urz, B. Wenderoth and R. Peter, *Angew. Chem., Int. Ed. Engl.*, 1982, **21**, 135; *Angew. Chem.*, 1982, **94**, 133; *Angew. Chem. Suppl.*, 1982, 257.
42. M. T. Reetz, J. Westermann, R. Steinbach, B. Wenderoth, R. Peter, R. Ostarek and S. Maus, *Chem. Ber.*, 1985, **118**, 1421.
43. B. Weidmann and D. Seebach, *Helv. Chim. Acta*, 1980, **63**, 2451.
44. B. Weidmann, L. Widler, A. G. Olivero, C. D. Maycock and D. Seebach, *Helv. Chim. Acta*, 1981, **64**, 357.
45. K. Kostova, A. Lorenzi-Riatsch, Y. Nakashita, M. Hesse, *Helv. Chim. Acta*, 1982, **65**, 249.
46. M. T. Reetz, R. Steinbach, B. Wenderoth and J. Westermann, *Chem. Ind. (London)*, 1981, 541.
47. H. Burger and H. J. Neese, *J. Organomet. Chem.*, 1970, **21**, 381.
48. D. Seebach and M. Schiess, *Helv. Chim. Acta*, 1982, **65**, 2598.
49. (a) M. T. Reetz and J. Westermann, *Synth. Commun.*, 1981, **11**, 647; (b) S. Achyntha Rao and M. Periasamy, *Tetrahedron Lett.*, 1988, **29**, 1583.
50. M. T. Reetz, H. Hugel and K. Dresely, *Tetrahedron*, 1987, **43**, 109.
51. M. T. Reetz and S. Maus, *Tetrahedron*, 1987, **43**, 101.
52. E. L. Eliel, *Asymm. Synth.*, 1983, **2**, part A, chap. 5.
53. (a) M. T. Reetz, R. Steinbach, J. Westermann, R. Peter and B. Wenderoth, *Chem. Ber.*, 1985, **118**, 1441; (b) Y. Honda and G. Tsuchihashi, *Chem. Lett.*, 1988, 1937.
54. M. Cherest, H. Felkin and N. Prudent, *Tetrahedron Lett.*, 1968, 2199.
55. N. T. Anh, *Top. Curr. Chem.*, 1980, **88**, 145.
56. W. C. Still and J. A. Schneider, *Tetrahedron Lett.*, 1980, **21**, 1035.
57. M. L. Wolfrom and S. Hanessian, *J. Org. Chem.*, 1962, **27**, 1800; T. Nakati and Y. Kishi, *Tetrahedron Lett.*, 1978, 2745; E. L. Eliel, J. K. Koskimies and B. Lohri, *J. Am. Chem. Soc.*, 1978, **100**, 1615; W. C. Still and J. H. McDonald, III, *Tetrahedron Lett.*, 1980, **21**, 1031; R. Bernardi, *Tetrahedron Lett.*, 1981, **22**, 4021.
58. M. T. Reetz, *Angew. Chem., Int. Ed. Engl.*, 1984, **23**, 556 and literature cited therein.
59. M. T. Reetz, K. Kesseler, S. Schmidtberger, B. Wenderoth and R. Steinbach, *Angew. Chem., Int. Ed. Engl.*, 1983, **22**, 989; *Angew. Chem.*, 1983, **95**, 1007; *Angew. Chem. Suppl.*, 1983, 1511.
60. M. T. Reetz and A. Jung, *J. Am. Chem. Soc.*, 1983, **105**, 4833.
61. M. T. Reetz, M. Hüllmann and T. Seitz, *Angew. Chem., Int. Ed. Engl.*, 1987, **26**, 477.
62. J. W. Cornforth, R. H. Cornforth and K. K. Mathew, *J. Chem. Soc.*, 1959, 112.
63. R. Block and L. Gilbert, *Tetrahedron Lett.*, 1987, **28**, 423.
64. M. T. Reetz and M. Hüllmann, *J. Chem. Soc., Chem. Commun.*, 1986, 1600.
65. D. Roulet, J. Caperos and A. Jacot-Guillarmod, *Helv. Chim. Acta*, 1984, **67**, 1475.
66. G. H. Posner, *Org. React. (N. Y.)*, 1972, **19**, 1.
67. M. J. Loots and J. Schwartz, *J. Am. Chem. Soc.*, 1977, **99**, 8045.
68. J. Schwartz, M. J. Loots and H. Kosugi, *J. Am. Chem. Soc.*, 1980, **102**, 1333.
69. F. M. Dayrit and J. Schwartz, *J. Am. Chem. Soc.*, 1981, **103**, 4466.
70. L. Widler and D. Seebach, *Helv. Chim. Acta*, 1982, **65**, 1085.
71. D. Seebach and L. Widler, *Helv. Chim. Acta*, 1982, **65**, 1972.
72. M. T. Reetz and M. Sauerwald, *J. Org. Chem.*, 1984, **49**, 2292.
73. Y. Yamamoto and K. Maruyama, *Heterocycles*, 1982, **18**, 357.
74. R. Noyori, I. Nishida and J. Sakata, *J. Am. Chem. Soc.*, 1981, **103**, 2106; Y. Yamamoto, H. Yatagai, Y. Naruta and K. Maruyama, *J. Am. Chem. Soc.*, 1980, **102**, 7107; T. Hayashi, K. Kabeta, I. Hamachi and M. Kumada,

Tetrahedron Lett., 1983, **24**, 2865; S. E. Denmark and E. Weber, *Helv. Chim. Acta*, 1983, **66**, 1655 and references cited therein.
75. Y. Yamamoto, Y. Saito and K. Maruyama, *J. Organomet. Chem.*, 1985, **292**, 311.
76. H. A. Martin and F. Jellinek, *J. Organomet. Chem.*, 1967, **8**, 115.
77. H. A. Martin and F. Jellinek, *J. Organomet. Chem.*, 1968, **12**, 149.
78. Y. Kobayashi, K. Umeyama and F. Sato, *J. Chem. Soc., Chem. Commun.*, 1984, 621.
79. F. Sato, H. Uchiyama, K. Iida, Y. Kobayashi and M. Sato, *J. Chem. Soc., Chem. Commun.*, 1983, 921.
80. Y. Ikeda, J. Ukai, N. Ikeda and H. Yamamoto, *Tetrahedron*, 1987, **43**, 731.
81. Y. Ikeda, K. Furuta, N. Meguriya, N. Ikeda and H. Yamamoto, *J. Am. Chem. Soc.*, 1982, **104**, 7663.
82. K. Furuta, Y. Ikeda, N. Meguriya, N. Ikeda and H. Yamamoto, *Bull. Chem. Soc. Jpn.*, 1984, **57**, 2781.
83. Y. Ikeda and H. Yamamoto, *Bull. Chem. Soc. Jpn.*, 1986, **59**, 657.
84. F. Sato, Y. Suzuki and M. Sato, *Tetrahedron Lett.*, 1982, **23**, 4589.
85. E. van Hulsen and D. Hoppe, *Tetrahedron Lett.*, 1985, **26**, 411.
86. J. Ukai, Y. Ikeda, N. Ikeda and H. Yamamoto, *Tetrahedron Lett.*, 1983, **24**, 4029.
87. Y. Ikeda, J. Ukai, N. Ikeda and H. Yamamoto, *Tetrahedron*, 1987, **43**, 723.
88. A. Murai, A. Abiko, N. Shimada and T. Masamune, *Tetrahedron Lett.*, 1984, **25**, 4951.
89. A. Murai, A. Abiko and T. Masamune, *Tetrahedron Lett.*, 1984, **25**, 4955.
90. Y. Ikeda, J. Ukai, N. Ikeda and H. Yamamoto, *Tetrahedron Lett.*, 1984, **25**, 5177.
91. D. Hoppe and A. Bronneke, *Tetrahedron Lett.*, 1983, **24**, 1687; T. Kramer and D. Hoppe, *Tetrahedron Lett.*, 1987, **28**, 5149.
92. G. Erker, J. Wicher, K. Engel, F. Rosenfeldt, W. Dietrich and C. Kruger, *J. Am. Chem. Soc.*, 1980, **102**, 6344.
93. Y. Kai, N. Kanehisa, K. Miki, N. Kasai, K. Mashima, K. Nagasuna, H. Yasuda and A. Nakamura, *J. Chem. Soc., Chem. Commun.*, 1982, 191.
94. H. Yasuda and A. Nakamura, *Angew. Chem., Int. Ed. Engl.*, 1987, **26**, 723.
95. H. Yasuda, Y. Kajihara, K. Mashima, K. Lee and A. Nakamura, *Chem. Lett.*, 1981, 519.
96. E. Negishi, F. E. Cederbaum and T. Takahashi, *Tetrahedron Lett.*, 1986, **27**, 2829.
97. U. Dorf, K. Engel and G. Erker, *Organometallics*, 1983, **2**, 462.
98. H. Yasuda, K. Tatsumi and A. Nakamura, *Acc. Chem. Res.*, 1985, **18**, 120.
99. G. Erker, K. Engel, J. L. Atwood and W. E. Hunter, *Angew. Chem., Int. Ed. Engl.*, 1983, **22**, 494; *Angew. Chem.*, 1983, **95**, 506; *Angew. Chem. Suppl.*, 1983, 678.
100. H. Yasuda, Y. Kajihara, K. Mashima, K. Nagasuna and A. Nakamura, *Chem. Lett.*, 1981, 671.
101. M. Akita, H. Yasuda and A. Nakamura, *Chem. Lett.*, 1983, 217.
102. Y. Kai, N. Kanehisa, K. Miki, N. Kasai, M. Akita, H. Yasuda and A. Nakamura, *Bull. Chem. Soc. Jpn.*, 1983, **56**, 3735.
103. E. C. Ashby, *Pure Appl. Chem.*, 1980, **52**, 545; M. T. Reetz, S. H. Kyung and J. Westermann, *Organometallics*, 1984, **3**, 1716; G. J. Erskine, B. H. Hunter and J. D. McCowan, *Tetrahedron Lett.*, 1985, **26**, 1371.
104. K. Yokoo, Y. Yamanaka, T. Fukagawa, H. Taniguchi and Y. Fujiwara, *Chem. Lett.*, 1983, 1301; H. Shumann, W. Genthe, E. Hahn, J. Pickardt, H. Schwarz and K. Eckert, *J. Organomet. Chem.*, 1986, **306**, 215.
105. G. Erker and U. Dorf, *Angew. Chem., Int. Ed. Engl.*, 1983, **22**, 777; *Angew. Chem.*, 1983, **95**, 506; *Angew. Chem. Suppl.*, 1983, 1120.
106. H. Yasuda, K. Nagasuna, M. Akita, K. Lee and A. Nakamura, *Organometallics*, 1984, **3**, 1470; A. Nakamura, H. Yasuda, K. Tatsumi, K. Mashima, M. Akita and K. Nagasuna, in 'Organometallic Compounds, Synthesis, Structure, and Theory', ed. B. L. Shapiro, Texas A & M University Press, Austin, 1983, vol. 1, p. 29.
107. J. Klein, in 'The Chemistry of the Carbon–Carbon Triple Bond', ed. S. Patai, Wiley, New York, 1978, p. 343.
108. K. Furuta, M. Ishiguro, R. Haruta, N. Ikeda and H. Yamamoto, *Bull. Chem. Soc. Jpn.*, 1984, **57**, 2768.
109. M. Ishiguro, N. Ikeda and H. Yamamoto, *J. Org. Chem.*, 1982, **47**, 2225.
110. H. Hiraoka, K. Furuta, N. Ikeda and H. Yamamoto, *Bull. Chem. Soc. Jpn.*, 1984, **57**, 2777.
111. A. Mori, K. Maruoka and H. Yamamoto, *Tetrahedron Lett.*, 1984, **25**, 4421; A. Mori, J. Fujiwara, K. Maruoka and H. Yamamoto, *J. Organomet. Chem.*, 1985, **285**, 83.
112. (a) A. G. Olivero, B. Weidmann and D. Seebach, *Helv. Chim. Acta*, 1981, **64**, 2485; (b) J.-T. Wang, X. Fan and Y.-M. Qian, *Synthesis*, 1989, 291.
113. D. Seebach, A. K. Beck, S. Roggo and A. Wonnacott, *Chem. Ber.*, 1985, **118**, 3673.
114. H. Takahashi, A. Kawabata, K. Higashiyama, *Chem. Pharm. Bull.*, 1987, **35**, 1604.
115. M. T. Reetz, T. Kukenhohner and P. Weinig, *Tetrahedron Lett.*, 1986, **27**, 5711.
116. (a) H. Roder, G. Helmchen, E. M. Peters, K. Peters and H. G. V. Schnering, *Angew. Chem.*, 1984, **96**, 895; *Angew. Chem., Int. Ed. Engl.*, 1984, **23**, 898; (b) G. Palumbo, C. Ferreri and R. Caputo, *Tetrahedron Asymmetry*, 1991, **2**, in press.
117. M. T. Reetz, J. Westermann and R. Steinbach, *Angew. Chem., Int. Ed. Engl.*, 1980, **19**, 900.
118. M. T. Reetz, J. Westermann and R. Steinbach, *J. Chem. Soc., Chem. Commun.*, 1981, 237.
119. M. T. Reetz, and J. Westermann, *J. Org. Chem.*, 1983, **48**, 254.
120. M. T. Reetz, J. Westermann and S. H. Kyung, *Chem. Ber.*, 1985, **118**, 1050.
121. M. T. Reetz and S. H. Kyung, *Chem. Ber.*, 1987, **120**, 123.
122. M. T. Reetz and R. Peter, *J. Chem. Soc., Chem. Commun.*, 1983, 406.

1.6
Organochromium Reagents

NICHOLAS A. SACCOMANO
Pfizer Central Research, Groton, CT, USA

1.6.1 INTRODUCTION

Organochromium(III) complexes were among the first transition metal organometallic compounds to be synthesized and have enjoyed a rich history of investigation and utility.[1] These materials can be prepared either by the addition of Grignard, organolithium or organoaluminum reagents to chromium(III) halides or by the reduction of organic halides with chromium(II) ions. Both routes provide chromium(III) complexes that are believed to possess a Cr—C σ-bond.[1] The organometallic reagents to be described herein serve as carbanion equivalents and participate in a variety of chemo- and stereo-selective carbon–carbon bond-forming reactions *via* the addition to C═X π-bonds. Such processes are the substance of this review.

1.6.2 ORGANOCHROMIUM(III) CARBANION EQUIVALENTS

1.6.2.1 Synthesis and Structure of Organochromium(III) Complexes

The synthesis of organometallic compounds of chromium was pioneered by Hein in the 1920s and 1930s, and this work has been reviewed in a monogram by Zeiss.[2] However, preparation and adequate characterization of the first simple kinetically stable chromium(III) alkyl in solution was reported by Anet and Leblanc.[3] Pentaaquabenzylchromium(III) perchlorate (**2**) was prepared from benzyl chloride (**1**) and chromium(II) perchlorate in aqueous perchloric acid (equation 1). It is known that readily available chromium(II) salts function effectively as reducing agents for a variety of compounds, including organic halides.[4] The mechanism of chromium(II) ion mediated reductions of organic halides has been thoroughly investigated. Kinetic data have verified the formation of a discrete organochromium intermediate,[5] and mechanistic studies have established that the reaction proceeds by a single-electron reduction of the C—X bond followed by reduction of the resulting radical (**3**) by a second equivalent of chromium(II) ion to provide a chromium(III) organometallic species (**4**; Scheme 1).[6–8] As the reductions are most commonly performed in aqueous media, protonolysis of the organochromium intermediate (**4**) rapidly produces the reduction products (**5**). The coexistence of water in the reaction medium represents an impediment to the mechanistic evaluation of chromium(II) ion reduction of organic halides and a limitation of organochromiums, prepared in this fashion, as potential synthetic reagents. However, the use of anhydrous chromium(II) salts (*e.g.* $CrCl_2$; see Section 1.6.3.1) has more recently allowed for the controlled generation of organochromium compounds and provides for their current and potential utility as carbanion equivalents.[9]

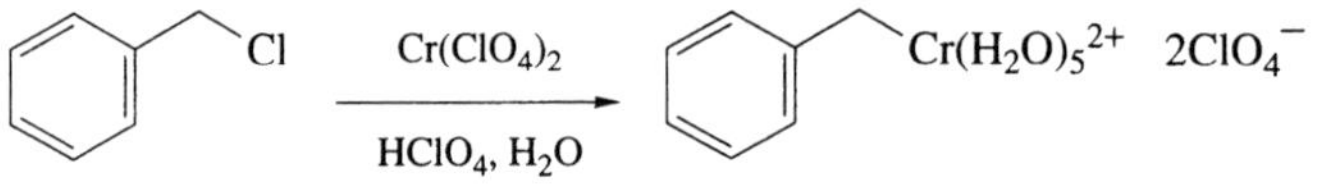

$$PhCH_2Cl \;(\mathbf{1}) \xrightarrow[HClO_4,\ H_2O]{Cr(ClO_4)_2} PhCH_2Cr(H_2O)_5^{2+}\ 2ClO_4^- \;(\mathbf{2}) \qquad (1)$$

$$-\overset{|}{\underset{|}{C}}-X \xrightarrow{Cr^{II}} -\overset{|}{\underset{|}{C}}\cdot\ (\mathbf{3}) \xrightarrow{Cr^{II}} -\overset{|}{\underset{|}{C}}-Cr^{III}\ (\mathbf{4}) \xrightarrow{H^+} -\overset{|}{\underset{|}{C}}-H\ (\mathbf{5})$$

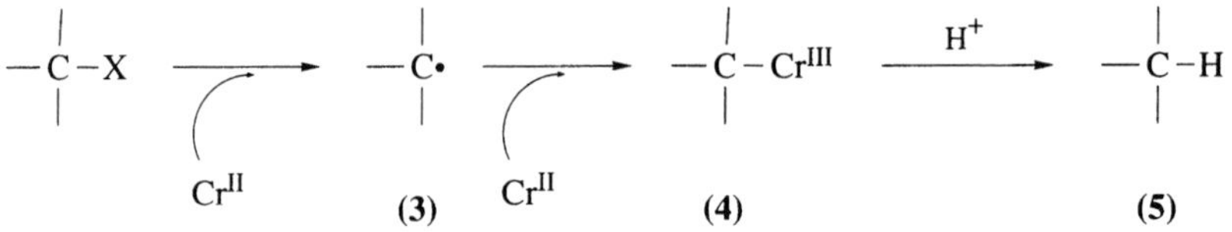

Scheme 1

An alternative synthesis of organochromium(III) compounds involves the addition of organometallic reagents to chromium(III) salts. This route provided the first example of an isolable covalent organochromium(III) compound. Herwig and Zeiss[10] reported that the addition of PhMgBr to $CrCl_3$ provided triphenyltris(tetrahydrofuran)chromium(III) (**6**; equation 2). Verification of Zeiss' structural assignment has more recently been provided in the form of X-ray data reported by Khan and Bau.[11] They have shown that (**6**) maintains three σ-bonded phenyl groups arranged in a *cis* facial fashion, as illustrated in equation (2). A related series of organochromiums was first prepared by Kurras[12] and subsequently investigated by Yamamoto and coworkers.[13] Treatment of $CrCl_3$ with organoaluminums provides organochromium(III) complexes (**7**), as shown in equation (3). The synthesis and X-ray crystal structure of dichlorotris(tetrahydrofuran)-*p*-tolylchromium(III) (**8**) was reported by Sneeden and coworkers (equation 4).[14] The crystal structure reveals a complex of the *mer*-type geometry with the Cr—O distance of the THF ligand *trans* to the σ-aryl bond appearing longer than for the other THF molecules. Other Group VI σ-alkyl and σ-aryl complexes have been studied and this area has been reviewed.[15] However, complexes (**6**), (**7**) and (**8**) are most pertinent to this review as they are representative of organochromiums that have been shown to participate in C—C bond-forming reactions.

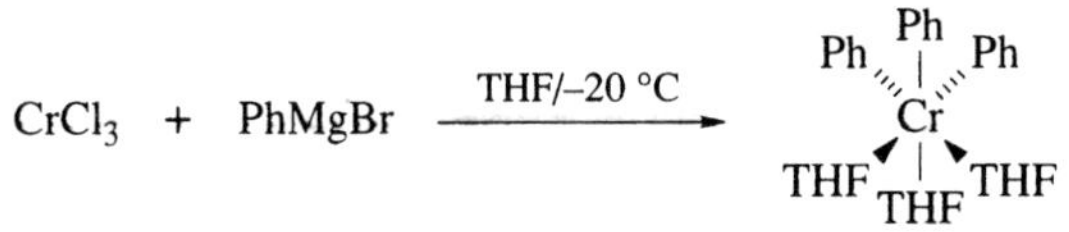

$$CrCl_3 + PhMgBr \xrightarrow{THF/-20\ ^\circ C} Ph_3Cr(THF)_3\ (\mathbf{6}) \qquad \text{Cr–C } 2.060\ \text{Å} \quad \text{Cr–O } 2.225\ \text{Å} \qquad (2)$$

$$CrCl_3 \xrightarrow[\text{THF}]{R_3Al \text{ or } R_2AlOEt} [RCrCl_2(THF)_3] \quad (7) \tag{3}$$

R = Ph, Me, Et, Pr^n, Bu^i

$$[CrCl_3(THF)_3] + p\text{-}MeC_6H_4MgCl \xrightarrow{\text{THF}} (8) \tag{4}$$

Cr–C	2.015 Å
Cr–O (THF)	2.045 Å
Cr–O (THF')	2.214 Å
Cr–Cl	2.31 Å

1.6.2.2 Formation of C—C Bonds: Background

The ability of chromium(II) ion generated organochromium(III) species to take part in C—C bond-forming reactions was discovered during mechanistic studies of the chromium(II) ion reduction of halides. Kochi and coworkers[6] have shown that the benzylchromium ion (**2**) when treated with acrylonitrile provides phenylbutyronitrile (**9**) and toluene (**10**) in 40–50% and 40–45% yield, respectively (equation 5). In addition, it was shown that (**2**) in the presence of butadiene afforded a 4–6% yield of 4-phenyl-1-butene (**11**) and 60–80% yield of toluene (equation 6), demonstrating the carbanionic nature of the organochromium species. Barton and coworkers[8] subsequently showed that 9-α-bromo-11-β-hydroxyprogesterone (**12**) when treated with chromium(II) acetate in DMSO provides 11-β-hydroxyprogesterone (**13**), the $\Delta^{9,11}$alkene (**14**) and the 5,9-cyclosteroid (**15**) with the product distribution and yield being a function of solvent and additives (equation 7). These studies[6,8] support the proposed mechanism of chromium(II) ion mediated reductions and also reveal the nucleophilic character of the *in situ* generated organochromium(III) intermediates.

$$\text{aq. PhCH}_2\text{Cr}^{2+}\ (2) \xrightarrow{\text{CH}_2=\text{CHCN}} \text{Ph(CH}_2)_3\text{CN}\ (9)\ 40–50\% + \text{PhMe}\ (10)\ 40–45\% \tag{5}$$

$$\text{aq. PhCH}_2\text{Cr}^{2+}\ (2) \xrightarrow{\text{butadiene}} \text{Ph(CH}_2)_2\text{CH=CH}_2\ (11)\ 4–6\% + \text{PhMe}\ (10)\ 60–80\% \tag{6}$$

$$(12) \xrightarrow[\text{DMSO}]{Cr(OAc)_2} (13) + (14) + (15) \tag{7}$$

Organochromium compounds prepared from chromium(III) salts and organometallic reagents have been shown to possess carbanionic character and provide the first examples of organochromium additions to carbonyl substrates. Sneeden and coworkers[16] reported that triphenyltris(tetrahydrofuran)chromium(III) (**6**) when treated with carbon monoxide provided benzpinacol (**16**; equation 8). It was also shown that treatment of 3-pentanone (**17**) with the solvated triphenylchromium complex (**6**) afforded diethylphenylcarbinol (**18**) and diol (**19**, R = Ph; equation 9). Cyclohexanone (**20**; equation 10) provided (**21**), (**22**) and (**23**) upon treatment with (**6**). These experiments are noteworthy since they demonstrate the ability of σ-organochromium complexes to add to ketones in a Grignard-like fashion. Interestingly, generation of the aldol and aldol-derived products (**19**), (**22**) and (**23**) occurred without stoichiometric consumption of organometallic reagent. The authors suggested that in these instances (**6**) does not promote aldol condensation by acting as a conventional organic base, but rather that the chromium atom is functioning as a coupling center or a reaction template at which the aldol process occurs. Monoarylchromium complex (**24**) provided addition products with aldehydes and ketones (yields for equations 11 and 12 based on **24**).[17] Reaction of (**24**) with neat acetone produced 2-phenyl-2-propanol (**25**) and mesityl oxide (**26**; equation 11). Also, (**24**) reacted with benzaldehyde to produce benzophenone (**27**) and benzyl benzoate (**28**; equation 12). It was rationalized that the benzophenone arose from a β-hydrogen elimination of the alkoxychromium intermediate (**29**; equation 13), while (**28**) was generated by a chromium-mediated Tischchenko reaction.[18]

$[Ph_3Cr(THF)_3]$ (**6**) —CO, THF; 83%→ (**16**) (8)

(**17**) —(**6**), THF, –30 °C; 70%→ (**18**) + (**19**); 81:19 (9)

(**20**) —(**6**), THF; 40%→ (**21**) + (**22**) + (**23**); 50 : 26 : 24 (10)

$[PhCrCl_2(THF)_3]$ (**24**) —acetone, r.t.→ Ph–C(Me)₂–OH (**25**) 71% + (**26**) 36% (11)

$[PhCrCl_2(THF)_3]$ (**24**) —PhCHO, Et_2O, r.t.→ PhCOPh (**27**) 84% + PhCO₂CH₂Ph (**28**) 20% (12)

$$\mathrm{H{-}C(Ph)_2{-}OCr^{III}}\ \xrightarrow{\beta\text{-hydride elimination}}\ \mathrm{HCr^{III}} + \mathbf{(27)} \qquad (13)$$

(29)

1.6.3 ALLYL-, METHALLYL-, CROTYL- AND PROPARGYL-CHROMIUM REAGENTS

1.6.3.1 Carbonyl Addition Reactions: General Features

The early studies and structural investigations of organochromium(III) compounds revealed the potential synthetic utility of these reagents as carbanion equivalents. However, it was Hiyama and coworkers[9,19] who expanded on the initial findings. These workers reported that the Barbier–Grignard-type addition of allylic halides and tosylates to carbonyl compounds was effectively mediated by anhydrous $CrCl_2$ (equation 14). Anhydrous chromium(II) chloride is commercially available[20] or can be generated *in situ* by the reduction of chromium(III) chloride[21] with either LAH[9,19] or Na(Hg)[22] in THF (equation 15). Although the exact nature of the low valent chromium reagent generated *in situ* has not been established, its functional attributes are indistinguishable from those of the commercially available material.

$$R^1R^2C{=}O + R^4R^5C{=}C(R^3)CH_2X \xrightarrow[\text{DMF or THF}]{CrCl_2} R^1R^2C(OH){-}CR^4R^5{-}C(R^3){=}CH_2 \qquad (14)$$

$$CrCl_3 \xrightarrow[\text{or Na(Hg), THF}]{LiAlH_4} CrCl_2 \qquad (15)$$

Reduction of the allylic halide (or tosylate) in this reaction is consistent with expectations.[4] Two mol equivalents of $CrCl_2$ are required for the consumption of each halide. Polar aprotic solvents (*i.e.* DMF) greatly facilitate the reduction and in some cases (*e.g.* chlorides and tosylates) are essential for the generation of the organochromium species. Under these conditions allyl chloride and tosylate add to benzaldehyde (equation 16) to provide the homoallylic alcohol (**30**). In a similar fashion an allyl group can be delivered to ketones (equations 17 and 18), but requires a larger mol ratio of allylchromium reagent to obtain comparable results. The chemoselective nature of this reagent is further revealed by its ability to

$$PhCHO \xrightarrow[\text{X = Cl; 54\%; X = OTs; 55\%}]{CrCl_2,\ DMF} PhCH(OH)CH_2CH{=}CH_2\ \mathbf{(30)} \qquad (16)$$

$$\text{cyclohexanone} \xrightarrow[\text{allyl iodide; 74\%}]{CrCl_2,\ THF} \text{1-allylcyclohexanol} \qquad (17)$$

$$\text{cyclododecanone} \xrightarrow[\text{allyl bromide; 82\%}]{CrCl_2,\ THF} \text{1-allylcyclododecanol} \qquad (18)$$

add selectively to aldehydes in the presence of ketones (equation 19), esters (equation 20) and nitriles (equation 21). In fact, the allylchromium reagent discriminates between ketones of nearly identical reactivity (equation 22). In the same study it was shown that the more-substituted γ-carbon of prenyl bromide adds to the carbonyl carbon of aldehydes (equation 23). This is a general feature of substituted allylchromiums.

$CrCl_2$, THF; allyl Br; 66% (19)

$CrCl_2$, THF; allyl Br; 75% (20)

$CrCl_2$, THF; allyl Br; 66% (21)

$CrCl_2$, THF; allyl Br; 82%; 16:84 (22)

49% at 54% conversion
(91% based on recovered ketone)

$CrCl_3$, $LiAlH_4$, DMF; 71%; **(31)** (23)

The addition to α,β-unsaturated aldehydes proceeds exclusively in a 1,2-fashion (equation 24). This regiochemical preference is general for all organochromium compounds. The addition to 4-*t*-butylcyclohexanone (**32**) occurs predominantly *via* equatorial addition (equation 25), as it provides (**33**) and (**34**) in an 88:12 ratio. Allylchromium is one of the most efficient reagents for this transformation (Table 1).

CHO + **(31)**; $CrCl_2$, THF; 82% (24)

(32); $CrCl_2$, THF; allyl Br; 85%; **(33)** + **(34)** 88:12 (25)

Table 1 Addition of Allyl Organometallics ($CH_2{=}CHCH_2M$) to 4-*t*-Butylcyclohexanone (**32**; equation 25)

M	*Conditions*	*Yield (%)*	*Ratio (33):(34)*	*Ref.*
Cr^{III}	THF/r.t.	85	88:12	9
ZnBr	THF/5 °C	79	85:15	23
$Al_{2/3}Br$	THF/5 °C	83	68:32	23
MgBr	THF/5 °C	33	45:55	23
Li	THF/–20 °C	67	35:65	23
Na	THF/–20 °C	68	35:65	23
K	THF/–20 °C	30	37:63	23
$Bu^n{}_3Sn$	$BF_3{\cdot}Et_2O$ THF/–78 °C	93	92:8	24
Sm^{III}	THF/r.t.	72	87:13	25

1.6.3.2 Allylic Chromium Reagents: 1,2-Asymmetric Induction (*Anti*/*Syn* Control)

In the original report of the $CrCl_2$-mediated carbonyl addition reaction Hiyama[9,19] reported that crotyl bromide added to benzaldehyde in the presence of chromium(II) ion and afforded a single diastereomer. Follow-up results from Heathcock and Buse,[26] and subsequent work by Hiyama and coworkers,[27] show that this reaction delivers the *anti* (*threo*) isomer (**35**) exclusively in 96% yield (equation 26). Conversion to the known β-hydroxy acid (**36**) provides proof of the stereochemistry of (**35**). The reaction of crotyl bromide with a variety of aldehydes was investigated (equation 27) and the following trends emerged (Table 2).[27] It was shown that the selective formation of the *anti* addition product occurred with unhindered aldehydes (entries 1–3), but that this preference was reversed for very large substrate aldehydes (entry 4). Moreover, solvent substitution of DMF for THF led to erosion of the *anti* preference (entries 5–8), with concomitant increase in overall chemical yield (entries 6 and 7). Hiyama also showed that the *anti* selectivity was not a function of alkene geometry in the starting bromide since both *cis*- and *trans*-1-bromo-2-butene gave exclusively the *anti* addition product with benzaldehyde. The stereochemical course of addition is rationalized in the following manner. Chromium(III) complexes prefer to exist in an octahedral configuration in which the coordination sphere is often supplemented with solvent molecules (*i.e.* THF).[11,14] A preferred transition state which is consistent with this and the stereochemical data has been proposed[26,27] and is detailed in Scheme 2. The octahedral (*E*)-crotylchromium(III) reagent (**40**) is formed from either (*E*)- or (*Z*)-crotyl bromide with two equivalents of $CrCl_2$. Such double-bond isomer-

Table 2 Addition of Crotyl Bromide (**31**) to Aldehydes (**37**) using Chromium(II) Chloride (equation 27)

Entry	*Aldehyde (37)*	*Solvent*	*Yield (%)*	*Anti:syn (38):(39)*
1	PhCHO	THF	96	100:0
2	Pr^nCHO	THF	59	93:7
3	Pr^iCHO	THF	55	95:5
4	Bu^tCHO	THF	64	35:65
5	PhCHO	DMF	92	75:25
6	n-$C_5H_{11}CHO$	DMF	77	68:32
7	Pr^iCHO	DMF	78	66:34
8	Bu^tCHO	DMF	63	37:63

PhCHO → (crotyl bromide; $CrCl_2$, THF; 96%) → (**35**) → (i, O_3, EtOAc; ii, H_2O_2, H_2O, $NaHCO_3$) → (**36**) (26)

(**37**) RCHO → (crotyl bromide; $CrCl_2$) → (**38**) *anti* + (**39**) *syn* (27)

ization of organometallic reagents to access the more stable (*E*)-double bond geometry is likely to be operative and is the situation which has been noted for crotyl-magnesium,[28] -lithium,[29] -zinc,[30] -titanium[31,41] and -zirconium compounds.[32] Ligand replacement with substrate aldehyde generates reactive complexes (**41**) and (**42**) which on transfer of the crotyl group from Cr to carbon with allylic transposition provides *anti*- and *syn*-homoallylic alcohols (**43**) and (**44**), respectively. The steric requirements of R and L should preclude transition structure (42), favor (43), and as a consequence direct the formation of *anti*-homoallylic alcohol (**43**). If the *gauche* interaction between R and the γ-methyl group becomes highly pronounced, as would be the case with very large aldehydes (*i.e.* R = Bu^t), (**41**) will be supplanted by a skew boat (**45**) as the favored transition structure. Finally, the reduction of stereoselectivity when DMF is used is interpreted as a perturbation of the reactive chromium template by the strong donor solvent.

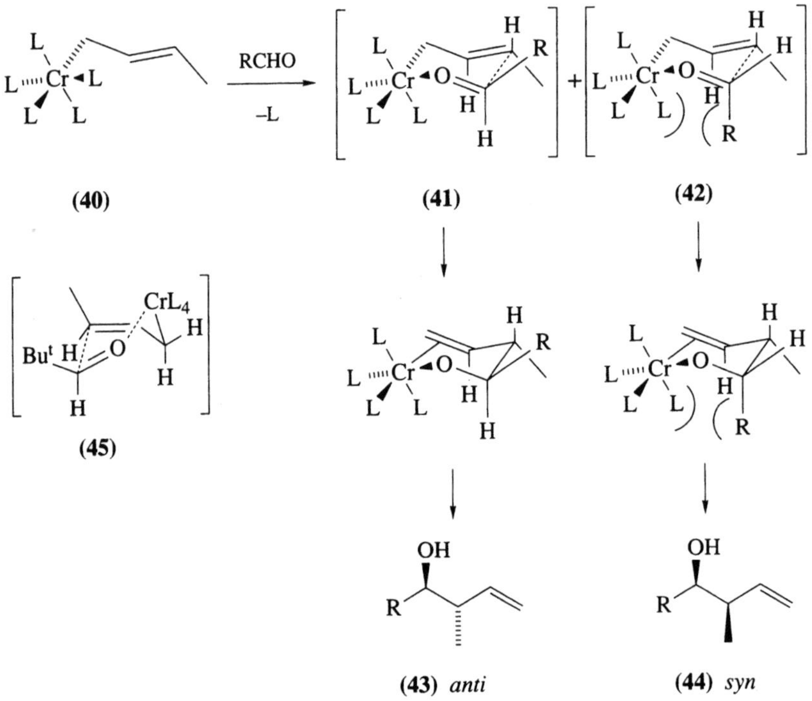

(**40**) RCHO, –L → [(**41**)] + [(**42**)]

(**45**)

(**43**) *anti* (**44**) *syn*

Scheme 2

The addition of allyl metal derivatives to aldehydes represents an important and well-developed strategy for the control of acyclic stereochemistry.[33] The transition state models which have been proposed to describe the origin of the 1,2-diastereoselection obtained with γ-substituted allyl organometallics are classified as either synclinal (cyclic) (**46**) or antiperiplanar (acyclic) (**47**; Scheme 3). Involvement of a particular transition state is a function of the metal center and the reaction conditions. Denmark and Weber[34] have suggested that substituted allyl metals are of three types with regard to alkene geometry and 1,2-diastereogenesis (*lk*/*ul*).[50] For type 1, the *anti:syn* (*lk:ul*) ratio is dependent on the (*Z*):(*E*) ratio of the alkene in the organometallic (B, Al, Si); with type 2; *syn* (*ul*) selective processes are independent of alkene geometry (Sn, Si); and for type 3; *anti* (*lk*) selective reactions are independent of alkene geometry. Examples of types 1–3 crotyl metal reagents are provided in Table 3 and equation (28). The allyl boronates (Table 3, entries 5 and 6)[37] and the pentacoordinate silicates (entries 15 and 16)[43] are crotyl metals of type 1 which react through a synclinal reaction geometry and produce *anti* products from the (*E*)-crotyl metal reagent and *syn* products from the (*Z*)-crotyl metal reagent. Trialkyl-stannanes[38] and -silanes[39] (entries 7, 8 and 9) are type 2 reagents which react under Lewis acid catalysis *via* an open antiperiplanar transition state and consequently produce *syn* products. The type 3 reagents consist of σ- and η^3-crotyltitanium,[41,42] -crotylzirconium[32,40] and organochromium(III) compounds.[27] These reagents are believed to react exclusively *via* the σ-(*E*)-crotyl metal complex through a synclinal cyclic chair transition state which leads to *anti* addition products. Set against the fabric of crotyl metal additions, the crotylchromiums represent the most convenient and perhaps the most selective reagents for the production of *anti*-homoallylic alcohols. However, with sterically demanding aldehydes, the η^3-crotyltitanocenes recently reported by Collins[42] exhibit a higher *anti* selectivity than the chromium(III) organometallics.[27]

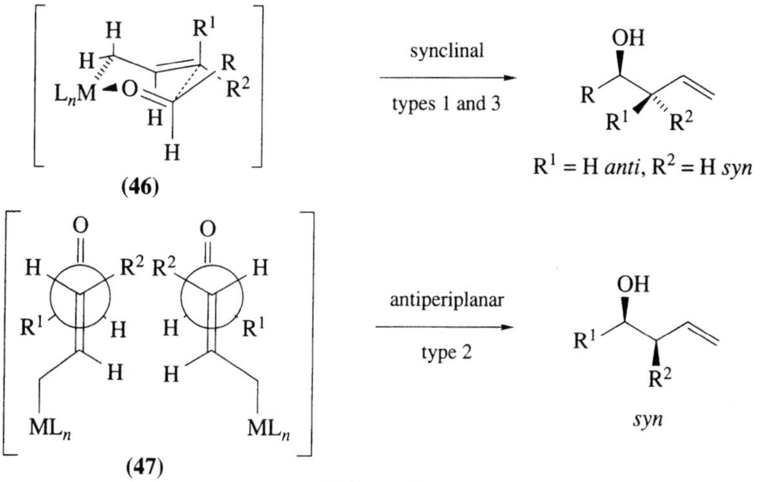

(46)

synclinal

types 1 and 3

R^1 = H *anti*, R^2 = H *syn*

(47)

antiperiplanar

type 2

syn

Scheme 3

(48)   (49) *anti*   (50) *syn*   (28)

Table 3 Addition of Crotyl Organometallics to Aldehydes (equation 28)

Entry	*M (48)*	*R*	*Conditions*	*(E):(Z) (48)*	*Anti:syn (49):(50)*	*Yield (%)*	*Ref.*
1	Li	Ph	THF/–78 °C		55:45	80	29
2	MgCl	Ph	Ether/–10 °C		38:62	93	35
3	Et_3Al^- Li^+	Ph	Ether/–70 °C		56:44	93	36
4	Et_3B^- Li^+	Ph	Ether/–70 °C		82:18	90	36
5	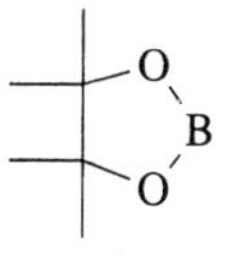	Ph	Ether/–78 °C	93:7	94:6	80	37
6		Ph	Ether/–78 °C	<5:>95	6:96	22	37
7	$Bu^n{}_3Sn$	Ph	$BF_3{\cdot}Et_2O$ CH_2Cl_2/–78 °C	100:0	6:96	90	38
8	$Bu^n{}_3Sn$	Ph	$BF_3{\cdot}Et_2O$ CH_2Cl_2/–78 °C	0:100	6:96	90	38
9	Me_3Si	Pr^i	$TiCl_4/CH_2Cl_2$/–78 °C	>99:<1	3:97	92	39
10	Me_3Si	Pr^i	$TiCl_4/CH_2Cl_2$/–78 °C	3:97	36:64	98	39
11	$CrCl_2$	Ph	THF/r.t.	100:0	100:0	96	27
12	$CrCl_2$	Ph	THF/r.t.	0:100	100:0	96	27
13	Cp_2ZrCl	Ph	THF/–78 °C		86:14	90	40
14	Cp_2TiBr	Ph	Ether/–30 °C		100:0	92	41, 42
15	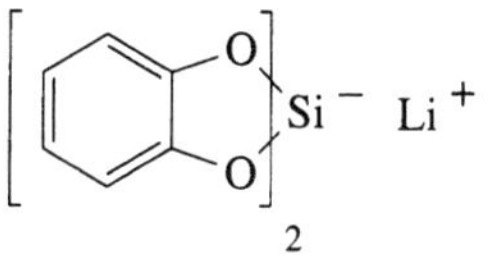 Si$^-$ Li$^+$	Ph	THF, hexane/r.t.	88:12	88:12	82	43
16	Si$^-$ Li$^+$	Ph	THF, hexane/r.t.	21:79	22:78	91	43

1.6.3.3 Crotylchromium Reagents: α- or 2,3-Asymmetric Induction

Addition of crotyl metal reagents to aldehydes bearing a stereogenic center α to the carbonyl (**51**; Scheme 4) has been used as a strategy for the controlled synthesis of the stereo triads.[49] Two of the four possible diastereomers (**52**) and (**53**) are available from the addition of an (*E*)-crotyl metal reagent to (**51**) *via* synclinal cyclic transition states. Synclinal reaction geometries (presumed to be operative for crotylchromiums, (*vide infra*) (**A**) and (**B**) (R^2 = Me) are (*lk,ul*) and (*lk,lk*) 1,2-processes, respectively.[50] Both reaction manifolds will afford the *anti*-1,2 stereochemical arrangement (a general feature of crotylchromiums), where (**A**) leads to the 2,3-*anti* product (**52**) and (**B**) leads to the 2,3-*syn* product (**53**). Transition structure (**B**) (and consequently the formation of **53**) was predicted to predominate by Cram's rule[46,48] and is consistent with the Felkin–Ahn model.[47,48] In the general model R^1 represents either the largest group or that group whose bond to C_α maintains the greatest σ^*–π^* overlap with orbitals of the carbonyl carbon. In the previous section, the ability of organochromiums to deliver a crotyl group and produce 1,2-*anti* stereochemistry selectively (**52** and **53**) was described. The potential for α- or 2,3- as well as 1,2-asymmetric induction was first addressed by Heathcock and Buse.[26] Addition of crotyl bromide/$CrCl_2$ to aldehyde (**54**; equation 29) provides two of the four possible diastereomers (**55**) and (**56**) in a 2.6:1 ratio, with complete 1,2-stereochemical control. However, the Cram:anti-Cram[46,48] ratio (**55**):(**56**) was modest. The major product (**55**) is that which is predicted by the Felkin–Anh addition model.[47] In a similar fashion (**57**) yields the two *anti* addition products (**58**) and (**59**) in nearly equal amounts (equation 30). Modest 2,3-diastereoselection has also been encountered by Hiyama and coworkers.[27] Crotylchromium addition to (**60**; equation 31) provided a 1,2-*anti*:*syn* selectivity (**61** + **62**):(**63**) of 93:7, and a disappointing Cram:anti-Cram ratio (**61** + **63**):(**62**) of 69:31. The question of 2,3-asymmetric induction has also been addressed by Kishi and coworkers.[44,45] During studies directed toward the construction of the Rifamycin ansa bridge,[44] crotylchromiums were shown to add to more complex substrate

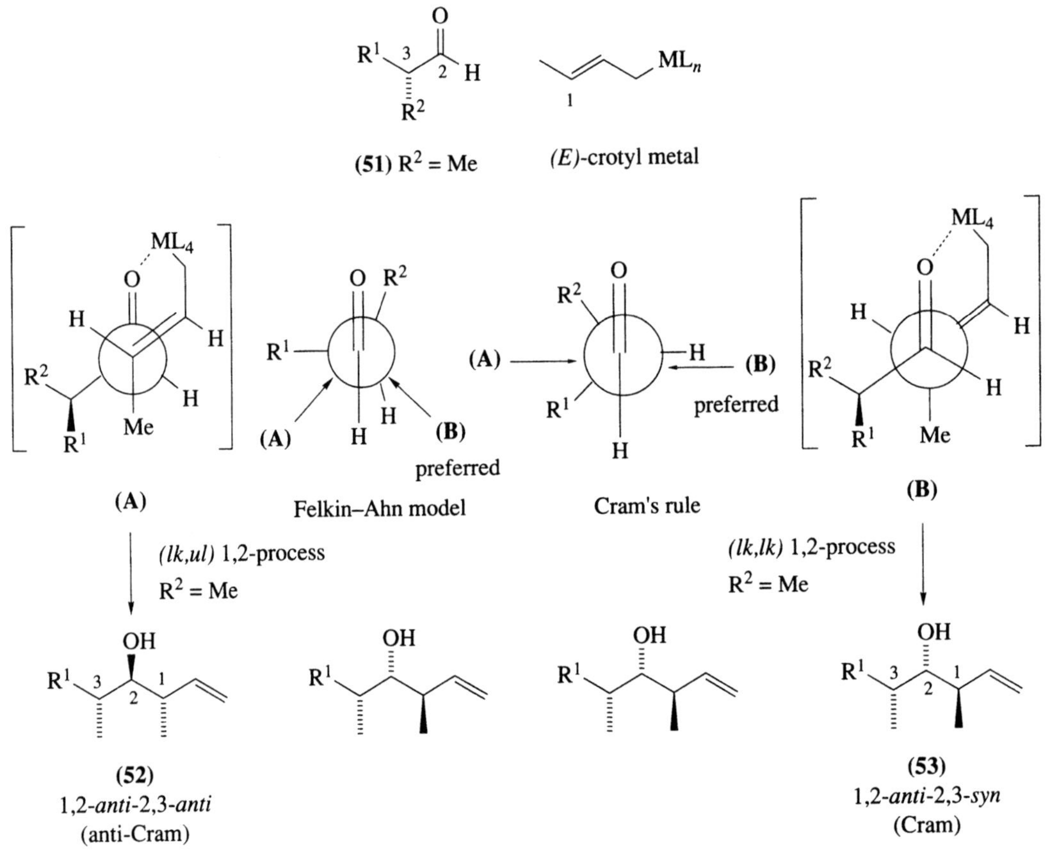

(**51**) R^2 = Me (*E*)-crotyl metal

(**A**) Felkin–Ahn model Cram's rule (**B**)

(*lk,ul*) 1,2-process R^2 = Me (*lk,lk*) 1,2-process R^2 = Me

(**52**) 1,2-*anti*-2,3-*anti* (anti-Cram) (**53**) 1,2-*anti*-2,3-*syn* (Cram)

Scheme 4

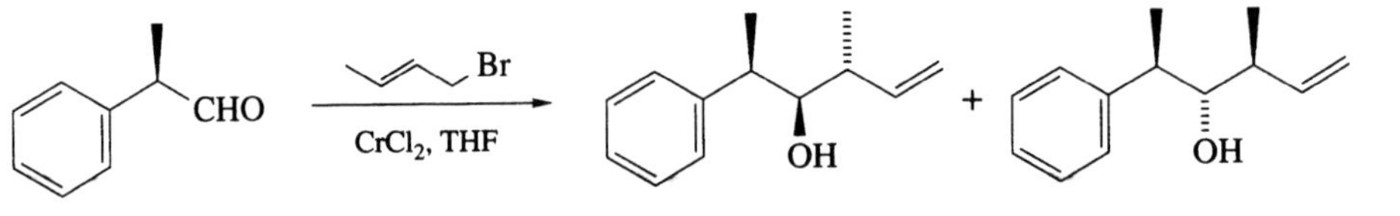

(**54**) → crotyl bromide, $CrCl_2$, THF → (**55**) + (**56**) 72:28 (29)

aldehydes with excellent 1,2- and 2,3-diastereoselectivity. Aldehyde (**64**) afforded (**65**), which maintains the 1,2-*anti* and 2,3-*syn* (or Cram) diastereochemical relationships. A similar qualitative and quantitative result was realized with aldehyde (**66**). The origin of the selectivity seen for the production of (**65**) and (**67**) was probed in a subsequent study by Kishi and Lewis.[45] The stereochemical outcome was shown not to be a consequence of chelation control[57] as is shown by the insensitivity of the Cram:anti-Cram ratio toward polarity of the reaction medium (equation 34) and toward a variety of hydroxy-protecting groups (R; equation 35). It is apparent that the origin of the 2,3-induction is a function of the steric size and chemical nature of the large α-substituent. Other examples of additions to chiral aldehydes reported by Kishi are provided in Table 4. The major homoallylic alcohols obtained all possess the 1,2-*anti*, 2,3-*syn* (Cram) relationship which arises from an (*lk,lk*) 1,2-process consistent with the expectation for a syn-clinal transition state and a Felkin–Anh-type approach. Enhanced selectivity with respect to earlier reports and differences in diastereogenesis amongst these cases is rationalized on the basis of predicted torsional preferences about the C(3)—C(4) bond in the substrate aldehydes.[45] In a related transformation, addition of allylchromium to a β-alkoxy-α-methyl aldehyde (**84**) also provided a Cram addition product (**85**) with 91% stereoselectivity (equation 41).[51]

MeO$_2$C–CH$_2$CH$_2$CH(Me)CHO (**57**) → (crotyl bromide, $CrCl_2$, THF) → (**58**) + (**59**), 50:50 (30)

(**60**) → (crotyl bromide, $CrCl_2$, THF, 90%) → (**61**) + (**62**) + (**63**), 62 : 31 : 7 (31)

(**64**) → (crotyl iodide or (Z)-crotyl iodide, $CrCl_2$, THF, 86%) → (**65**) + others, 95:5 (32)

(**66**) → (**67**) + others, >95:<5 (33)

(**68**) → (crotyl iodide, $CrCl_2$, THF) → (**69**) + (**70**) (34)

Solvent	(69):(70)
THF	61:39
Et_2O	56:44
PhMe	61:39

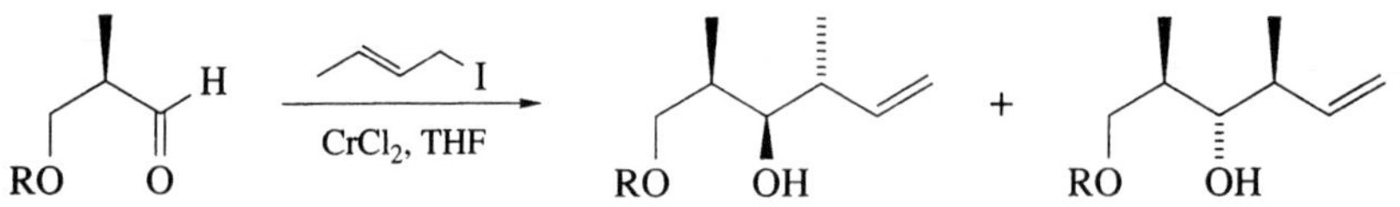

(35)

(71) (72) (73)

	(72)		(73)
R = CH_2OBn	~50	:	~50
R = THP	~50	:	~50

Table 4 Crotylchromium Addition to Chiral Aldehydes with α-Methyl and β-Alkoxy Substituents

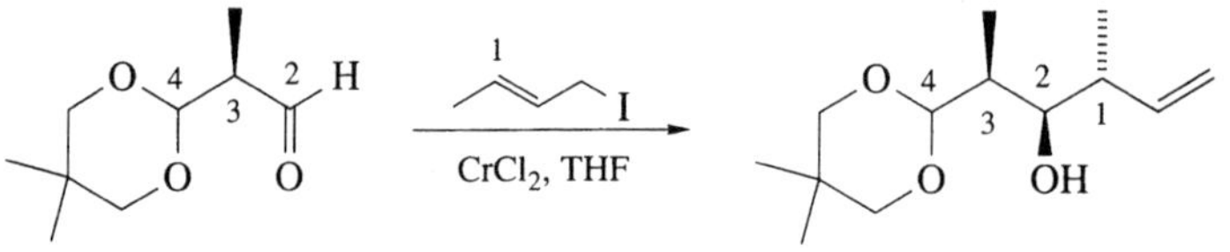

Substrate	Conditions	Product	Ratio	Eq.
(74)	$CrCl_2$, THF	(75)	92:8[a]	(36)
(76)	$CrCl_2$, THF	(77)	80:20[a]	(37)
(78)	$CrCl_2$, THF	(79)	95:5[a]	(38)
(80)	$CrCl_2$, THF	(81)	80:20[a]	(39)
(82)	$CrCl_2$, THF	(83)	91:9[a]	(40)

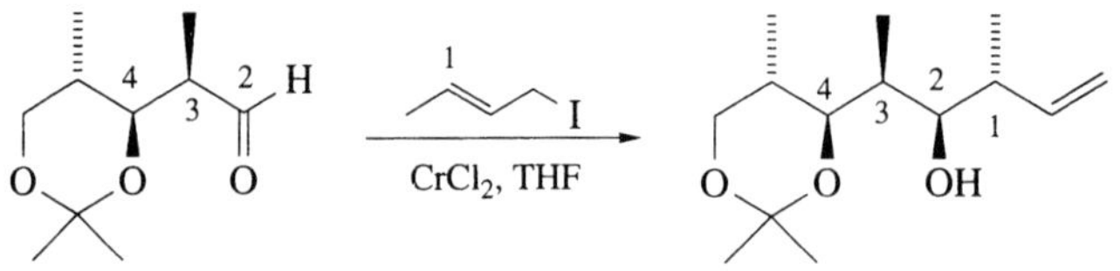

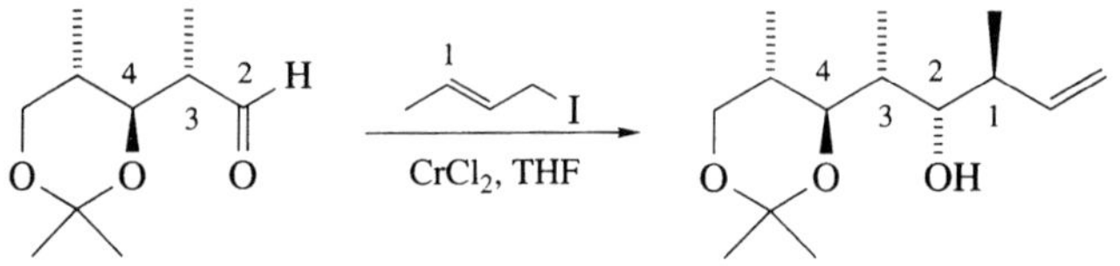

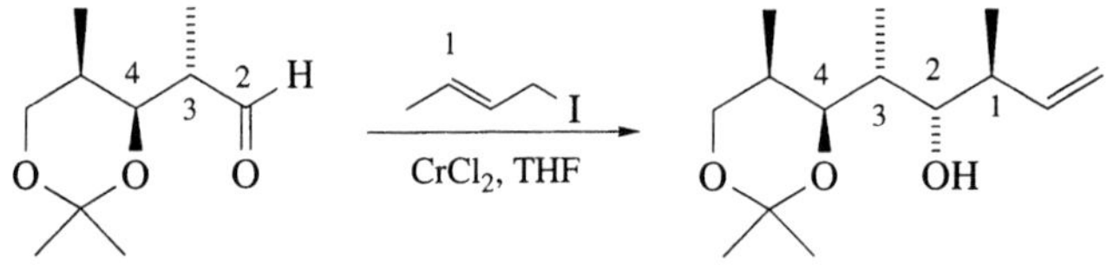

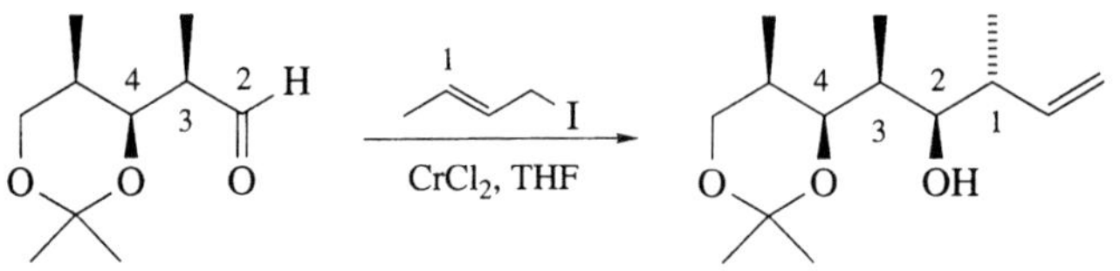

[a]Ratio of major diastereomer to next most abundant diastereomer: only structure of major diastereomer was determined.

Reaction of crotylchromium with α-methyl-β,γ-unsaturated aldehyde (**87**) afforded (**88**) as the major diastereomer.[52] The other Cram product (**89**), which is expected to arise from an antiperiplanar transition state (**46**; Scheme 3), is obtained from a BF_3-catalyzed tributylcrotylstannane addition. The remaining members of the stereo triad can be accessed by inversion of the C-2 hydroxy (*i.e.* **88** to **91** and **89** to **90**)

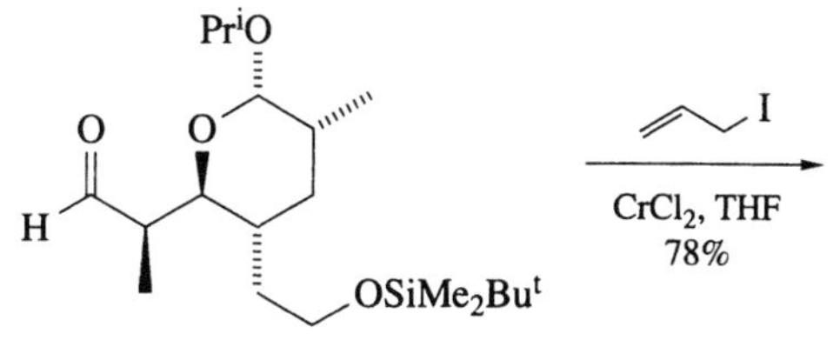

(84) → [allyl iodide, $CrCl_2$, THF, 78%] → (85) + (86) (41)

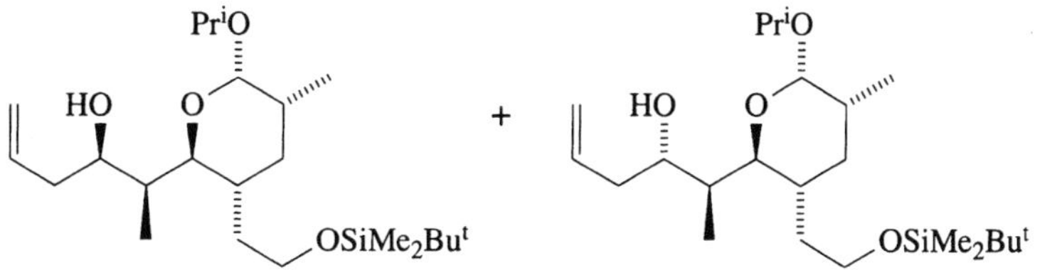

(85) 91:9 (86)

via an oxidation (PDC)–reduction ($LiBEt_3H$) sequence. In general it has been shown that (**87**) and related substrates react with nucleophiles (H^-, C^-) selectively to give products consistent with the Felkin–Anh mode of addition.[47,54]

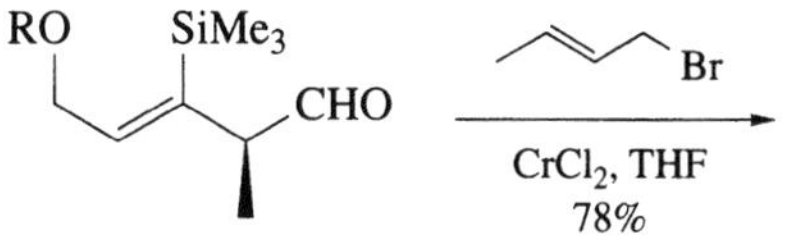

(87) → [crotyl bromide, $CrCl_2$, THF, 78%] → (88) + (89) + (90) + (91) (42)

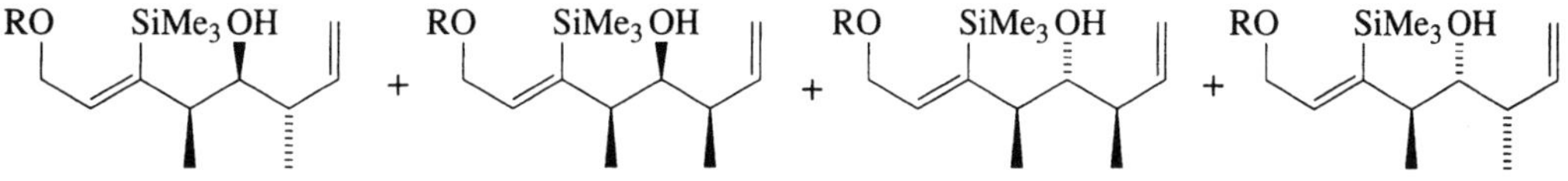

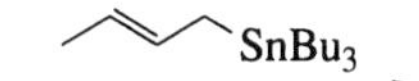

(88) >90 : (89) <1 : (90) + (91) 9

(87) → [crotyl-$SnBu_3$, $BF_3{\cdot}OEt_2$, CH_2Cl_2, 85%] → (88) + (89) + (90)/(91) (43)

3.6 : 88.4 : 8

Chromium-mediated addition of crotyl halides to α-alkoxy chiral aldehydes has been the subject of several interesting reports. Reaction of crotyl bromide/$CrCl_2$ with 2,3-*O*-isopropylideneglyceraldehyde (**92**; equation 44)[57] provides the 1,2-*anti* products with 98% stereoselectivity, given by the ratio (**93** + **95**):(**94** + **96**),[53] while 2,3-control, given by the ratio (**93** + **94**):(**95** + **96**), was nearly inoperative. However, a related substrate (**97**) produced exclusively 2,3-*syn* products with a 96:4 1,2-*anti*:*syn* ratio (equation 45).[55] Mulzer and coworkers[56] investigated the importance of the γ-substituent of the allylchromium(III) reagent with respect to 1,2- and 2,3-diastereogenesis in the addition reactions to α-alkoxy chiral aldehydes (equation 46 and Table 5). It was found in all cases that the level of 1,2-*anti* selectivity, given by the ratio (**103** + **104**):(**105** + **106**), was complete. Moreover, it was demonstrated that the 2,3-*syn* selectivity increased as a function of increasing the size of R^2. This result is consistent with both the transition state models proposed for addition of γ-substituted allylchromiums to aldehydes and with the expectation that products will be generated by Felkin–Anh-type addition (**103** and **106**).[58,60]

The addition of γ-alkoxy allylic chromium(III) reagents to aldehydes has been reported by Takai and coworkers.[76] This work is reviewed in Section 1.6.3.6.

(92)

CrCl$_2$, THF
71%

(93) : (94) : (95) : (96)
53 : 2 : 45 : 0

(44)

(97)

CrCl$_2$, THF
55%

(98) : (99) : (100) : (101)
96 : 4 : 0 : 0

(45)

(102)

CrCl$_2$, THF

(103) (104) (105) (106)

(46)

Table 5 Addition of Substituted Allylchromium Reagents to Protected α-Hydroxy Aldehydes (equation 46)[55]

Entry	R^1	R^2	*(103):(104):(105):(106)*	*Total yield (%)*
1	THP	Me	89:11:0:0	80
2	THP	Ph	91:9:0:0	71
3	THP	Bu	>99:<1:0:0	75
4	TBDMS	Bu	>99:<1:0:0	90

1.6.3.4 Other *n,m*-Asymmetric Induction

The use of allylic chromium reagents to control the relative stereochemistry of remote asymmetric centers has been described by Takeshita and coworkers.[61–64] The condensation of iridoid aldehyde synthon (**107**) and its enantiomer (**109**) with homochiral allylic chloride (**108**) provided in each case only two of the four possible diastereomers equations (47) and (48). In this reaction the γ-carbon of the organochromium offers only its *re*-face to the substrate aldehydes (**107**) and (**109**). Therefore, the two diastereomers arise from addition to either face of the aldehyde with addition to the *re*-face predominating in both cases with good to excellent selectivity. These materials were used in rather inventive syntheses of the 5-8-5 tricyclic sesterterpenes fusicocca-2,8,10-triene (**114**) and cycloaranosine (**115**). The approach to (**115**) uses a second chromium(II)-mediated aldehyde addition reaction (intramolecular) to form the eight-membered ring (see equation 52, Section 1.6.3.5).[62] A similar strategy was developed for the synthesis of ceroplastol II (**116**) and albolic acid (**117**).[63] Reaction of (**108**) with (**118**) in the presence of $CrCl_2$ provided (**120**) selectively with only 3% contamination by the hydroxy epimer (equation 49). The stereochemistry of (**120**) is consistent with the synclinal transition structure (**119**).

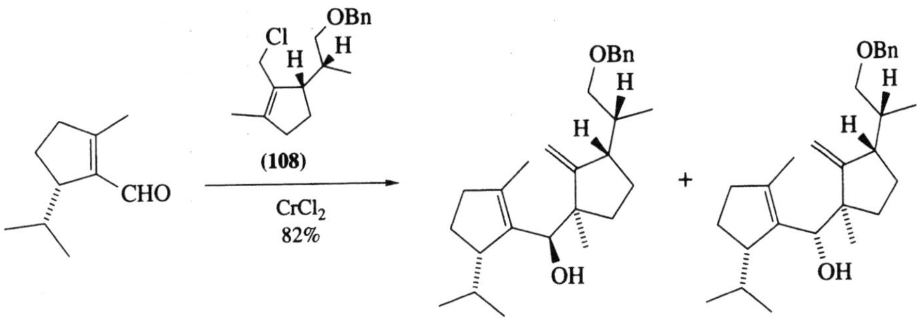

(107) —(108), $CrCl_2$, 82%→ (110) 78:22 (111) (47)

(109) —(108), $CrCl_2$, 77%→ (112) 95:5 (113) (48)

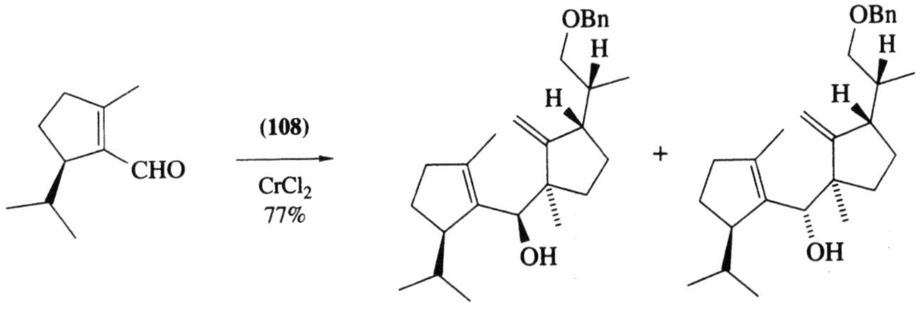

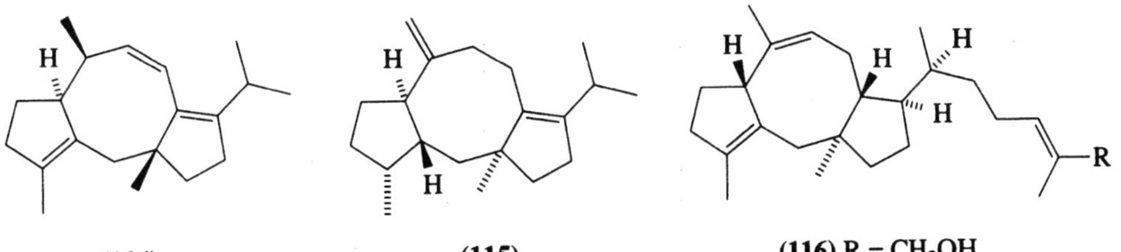

(114) (115) (116) R = CH_2OH
(117) R = CO_2H

1.6.3.5 Intramolecular Addition Reactions

The intramolecular variant of the chromium(II) ion mediated Barbier–Grignard-type addition reaction was first described by Still and Mobilio[65] in an elegant synthesis of (±)-Asperdiol (**122**; R = H). Chromium(II)-mediated cyclization of (**121**; R = Bn) provided a 4:1 mixture of (**122**) and (**123**) in 64% yield (equation 50). The relative topicity of the process is *lk*. This result is in accord with a synclinal chair transition structure (**124**) in which both hydrocarbon chains diverge from the reacting centers pseudo-

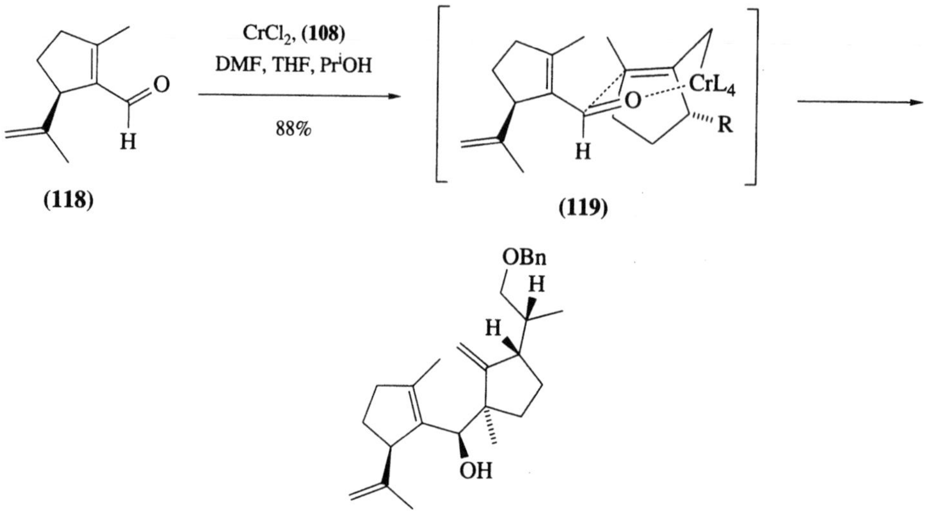

(49)

equatorially (dihedral angle C(2)—C(1)–C(14)—C(13) = *ca.* 30°; **125**). Medium[66,67] and large[68,69] ring closures employing allylstannanes,[68,69] allylzinc[66] and allylnickel[67] compounds are also described in the literature. However, the relative topicity of these intramolecular carbon–carbon bond-forming reactions is *ul* (see equations 55 and 56 in Section 1.6.3.6).[72,73]

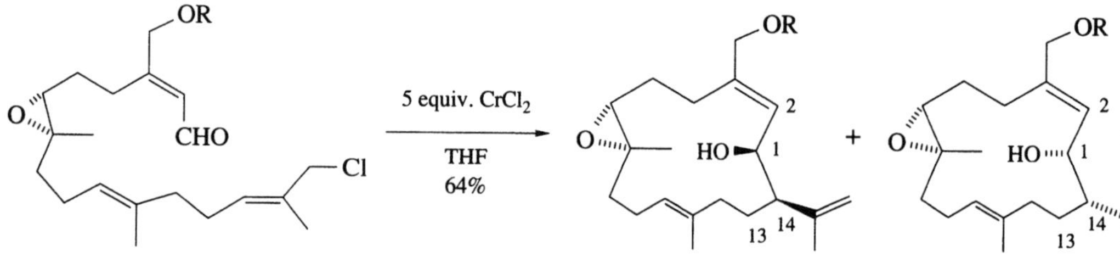

(50)

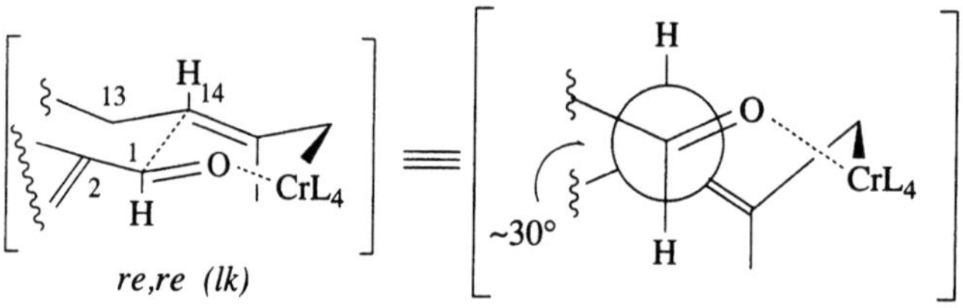

A related reaction has been described by Kitagawa and coworkers[70] in which (**126**) exclusively provides (**127**) upon treatment with $CrCl_3$–LAH (equation 51). The relative stereochemical result is qualitatively identical to that reported by Still and Mobilio.[65] Cyclodecadienol (**127**) was subsequently transformed into (±)-costunolide. A third example of a related chromium-initiated cyclization was reported by Takeshita and coworkers[62] in a synthesis of cycloaranosine (**115**). Serial treatment of (**128**) with MsCl/pyridine and $CrCl_3$–LAH provided (**129**) stereospecifically in 95% yield (equation 52). Another example of internal addition of an allylic chromium reagent to an aldehyde has been reported by Oshima and coworkers and is outlined in Section 1.6.3.6.[72]

Intramolecular alkenylchromium additions to aldehydes have been reported[94,95] and are outlined in Section 1.6.4.3. An intriguing example of an intramolecular organochromium reaction has been communicated by Gorques and coworkers.[71] Treatment of (**130**; Scheme 5) with $CrCl_2$ in HMPA elicits a reductive 1,5-acyl transposition and generates chromium phenoxide (**131**). Work-up with aqueous ammonium chloride affords phenol (**132**). However, when the reaction occurs in the presence of boron trifluoride etherate benzo[*b*]furan (**133**) is obtained directly in excellent yield.

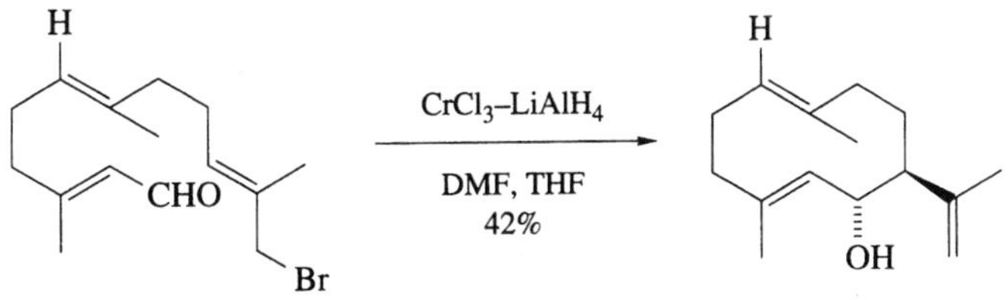

(126) $\xrightarrow[\text{DMF, THF}\ 42\%]{CrCl_3–LiAlH_4}$ (127) (51)

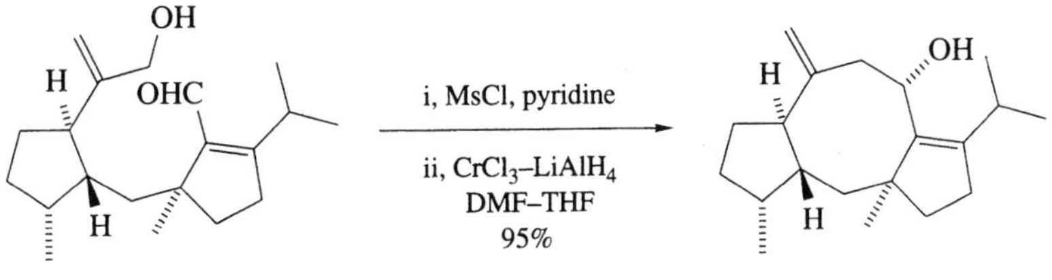

(128) $\xrightarrow[\text{ii, } CrCl_3–LiAlH_4 \text{ DMF–THF } 95\%]{\text{i, MsCl, pyridine}}$ (129) (52)

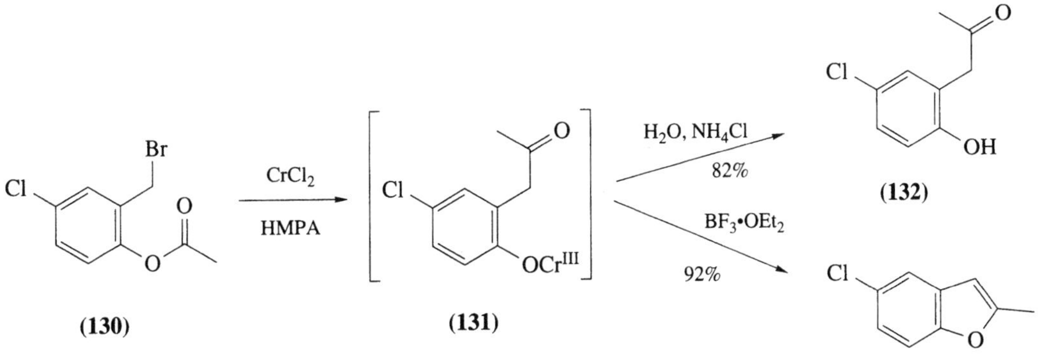

(130) $\xrightarrow[\text{HMPA}]{CrCl_2}$ [(131)] $\xrightarrow{H_2O,\ NH_4Cl\ 82\%}$ (132); $\xrightarrow{BF_3{\cdot}OEt_2\ 92\%}$ (133)

Scheme 5

1.6.3.6 Miscellaneous Substituted Substrates

Allylic chromium reagents bearing an electron-withdrawing group at the β-position of the starting halide have been the subject of several reports.[72,75] Oshima and coworkers and Drewes and Hoole have shown that $CrCl_3$–LAH facilitates the addition of ethyl α-bromomethylacrylate (**134**) to benzaldehyde to generate the α-methylenelactone (**135**; equation 53) directly.[72] It was noted that when commercially available $CrCl_2$ was used in place of Hiyama's reagent ($CrCl_3$–LAH) hydroxy ester (**138**) was obtained as the sole product. The reagent prepared from (**134**) and chromium(II) ion added specifically to the aldehyde functionality in (**135**) to provide keto lactone (**137**; equation 54). Appending a substituent at the β-position of (**134**) (*i.e.* Bu; **139**) introduces the possibility for 1,2-asymmetric induction. In sharp contrast to other allylic chromium reagents, (*E*)- or (*Z*)-(**139**) upon treatment with $CrCl_3$–LAH adds to benzaldehyde with *ul* relative topicity to provide the *syn*-α-methylenelactone (**140**) stereospecifically. An intramolecular variant of this reaction reported by Oshima and coworkers,[72] describes the transformation of (**141**) cleanly into (**142**; equation 56). Other reducing metal centers (Ni[66,74] and Zn[67,73]) have been used to initiate the construction of α-methylenelactones from allylic bromides and aldehydes. These processes also proceed with *ul* relative topicity. Another example of a trisubstituted allylic chromium reagent bearing an electron-withdrawing group at the β-position to the chromium atom was reported by Knochel and coworkers.[75] Sulfone (**143**) adds to isovaleraldehyde (**144**) in the presence of $CrCl_2$. The addition proceeds stereoselectively to provide a 96:4 ratio of *syn*-(**145**) to *anti*-(**146**) in 95% yield (equation 57). The specificity of this transformation was shown to be general (with six examples). The selective *ul* relative topicity of the addition reaction of sulfone (**143**) has been ascribed by Knochel to a synclinal transition structure (**147**). The preferred (*E*)-alkene geometry fixes the γ-substituent of the nucleophile in a pseudoaxial position. Alternatively, Oshima and coworkers explained their results by implicating an antiperiplanar reaction geometry (**148**), which is favored because coordination of the substrate aldehyde is replaced by internal ligation from an ester oxygen.

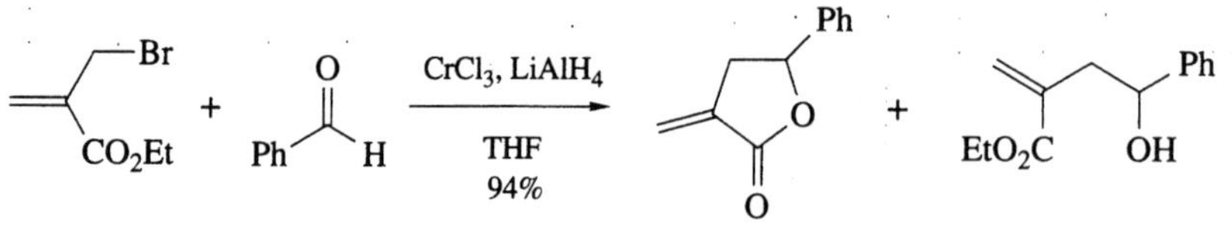

(53)

(134) **(135)** **(138)**

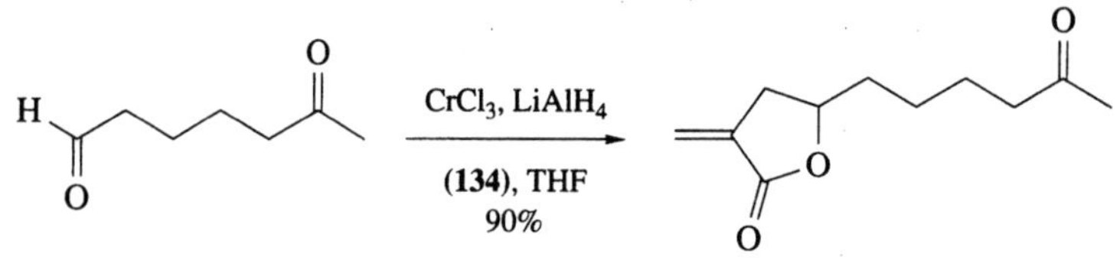

(54)

(136) **(137)**

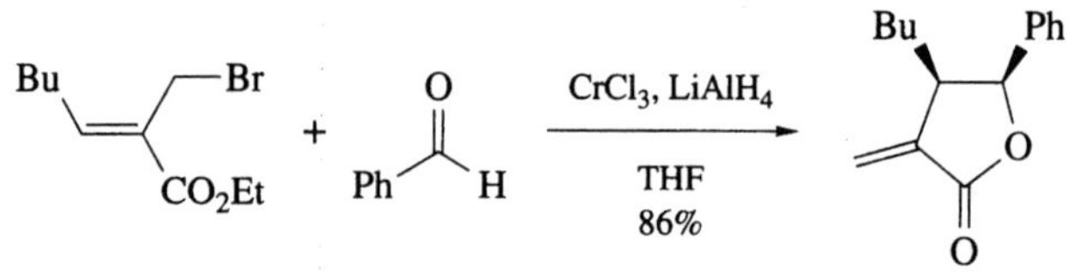

(55)

(139) **(140)**

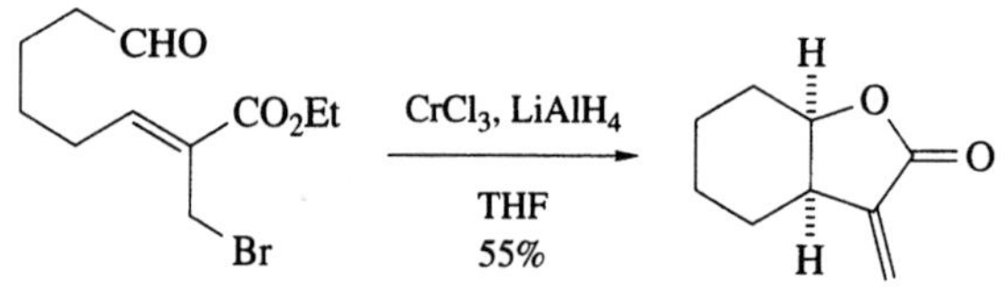

(56)

(141) **(142)**

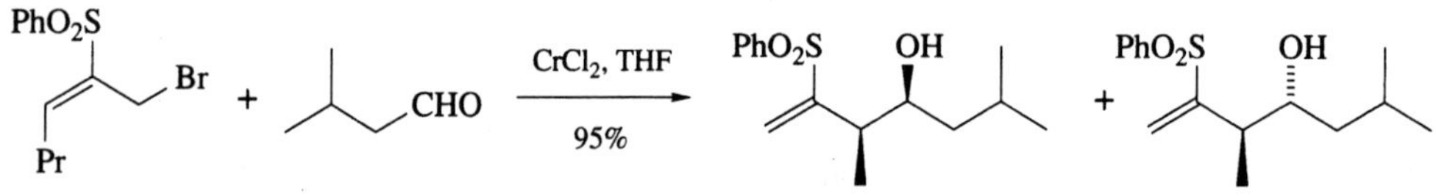

(57)

(143) **(144)** **(145)** 96:4 **(146)**

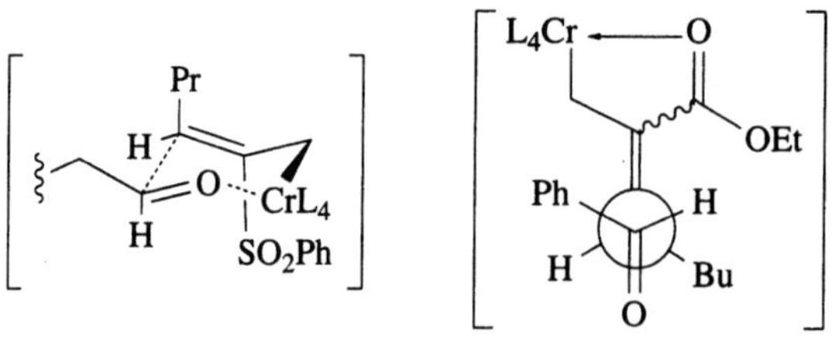

re,si (ul) *si,re (ul)*

(147) **(148)**

Takai and coworkers have recently reported their findings on the addition of γ-alkoxyallylic chromium compounds to aldehydes.[76] The reagents, which were generated by the reduction of dialkyl acetals (**149**) with $CrCl_2$ in the presence of trimethylsilyl iodide, added to aldehydes (**150**) to produce vicinal diols (equation 58 and Table 6). The addition proceeded efficiently and stereoselectively at –30 °C providing the *erythro*-1,2-diols (**151**) as the major products. The reaction tolerated substitution at the α- and β-positions (Table 6, entries 12 and 13 and entries 10 and 11, respectively) of the acetal. Suprisingly, however, the dibenzyl acetal of crotonaldehyde did not provide a useful chromium-based reagent. *Erythro*/*threo* selectivity was high in all cases except for addition to pivaldehyde (entry 8). As is the case for other organochromium(III) reagents, the oxygenated analogs generated from acrolein acetals add chemospecifically to aldehydes in the presence of ketones (equation 59). The authors suggest that the organochromium reagent was constrained to an *s-cis* configuration caused by internal ligation of the γ-oxygen atom to the metal center. As a consequence, the possible synclinal transition state geometries are boat-like arrangements (**157**) and (**158**). Preferred complexation of the aldehyde lone pair which is *syn* to hydrogen[77] and the presence of fewer eclipsing interactions make (**157**) the favored transition state and the *erythro* isomer (**151**) the predominant product.

(58)

$CrCl_2$, Me_3SiI; THF, –30 °C

(**149**) (**150**) (**151**) *erythro* (**152**) *threo*

Table 6 Reaction of Unsaturated Dialkyl Acetals with Aldehydes using $CrCl_2$/TMS-I (equation 58)[76]

Entry	R^1	R^2	R^3	R^4	*Time (h)*	*Yield (%)*	*Erythro:threo (151):(152)*
1	Me	H	H	Ph	3	99	88:12
2	Bn	H	H	Ph	1.5	97	71:29[a]
3	Bn	H	H	Ph	3	98	88:12
4	Bn	H	H	Ph	9	33	91:9[b]
5	Bn	H	H	n-C_8H_{17}	6	95	87:13
6	Bn	H	H	$PhCH_2CH_2$	2.5	99	88:12
7	Bn	H	H	Cyclohexyl	6	93	88:12
8	Bn	H	H	*t*-Butyl	7	91	33:67
9	Bn	H	H	PhCH=CH	2	97	76:24[c]
10	Bn	Me	H	Ph	3	99	85:15
11	Bn	Me	H	n-C_8H_{17}	3	99	88:12
12	Bn	H	Me	Ph	8	88	92:8
13	Bn	H	Me	n-C_8H_{17}	5	83	93:7

[a]Reaction performed at 25 °C. [b]Reaction performed at –42 °C; 42% PhCHO recovered. [c]1,2-Addition.

(59)

$CrCl_2$, Me_3SiI; THF, –30 °C

(**153**) (**154**) (**155**) (**156**)

(**151**) *erythro* ⟵ [**157**, *s-cis*] [**158**, *s-cis*] ⟶ (**152**) *threo*

(**157**) (**158**)

1.6.3.7 Propargylchromium Reagents

The chromium-mediated addition of propargyl halides to carbonyl compounds was studied by Gore and coworkers.[78–80] Unlike the crotylchromium reagents already described (*vide supra*), which react exclusively with allylic transposition, propargyl systems (**159**) react with carbonyl compounds (**160**) to provide a mixture of alkynic and allenic products (**161**) and (**162**) (equation 60). The regioselectivity was shown to be a function of the propargyl bromide and carbonyl substrate and of the presence of HMPA (hexamethylphosphoramide) in the reaction medium (Table 7). The authors suggest that the organochromium(III) reagent reacts entirely by allylic transposition (Scheme 6). Therefore, the regiochemical outcome reflects the ratio of organochromiums (**165**) and (**166**) which are formed as a mixture or equilibrate *via* mesomeric radical intermediates (**163**) and (**164**).

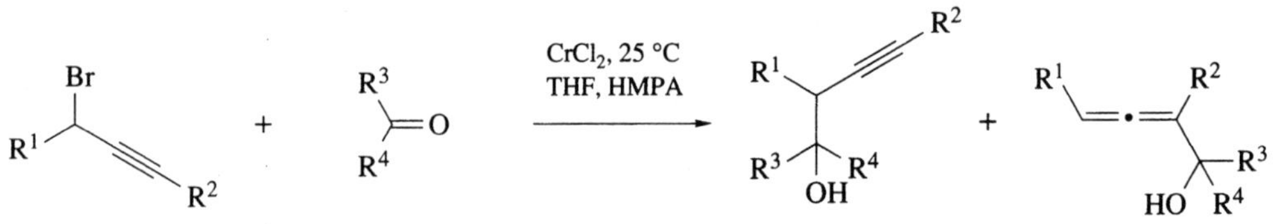

$CrCl_2$, 25 °C, THF, HMPA (60)

(**159**) (**160**) (**161**) (**162**)

Table 7 Chromium(II)-mediated Addition of Propargyl Bromides to Aldehydes and Ketones (equation 60)

Entry	R^1	R^2	R^3	R^4	*(161):(162)*	*(equiv.) HMPT*	*Yield (%)*
1	H	H	H	n-C_7H_{15}	85:15	0	72
2	H	H	—$(CH_2)_5$—		65:35	0	68
3	H	H	—$(CH_2)_5$—		21:79	5	70
4	H	C_5H_{11}	H	Pr^n	0:100	0	80
5	H	C_5H_{11}	Me	Me	0:100	0	78
6	H	C_5H_{11}	—$(CH_2)_5$—		0:100	0	78
7	n-C_7H_{15}	H	H	Pr^n	100:0	0	76
8	n-C_7H_{15}	H	Me	Me	0:100	0	66
9	n-C_7H_{15}	H	—$(CH_2)_5$—		60:40	0	75
10	n-C_7H_{15}	H	—$(CH_2)_5$—		20:80	1	68
11	Pr^n	Et	H	Pr^n	100:0	0	65
12	Pr^n	Et	Me	Me	75:25	0	60
13	Pr^n	Et	—$(CH_2)_5$—		80:20	0	50

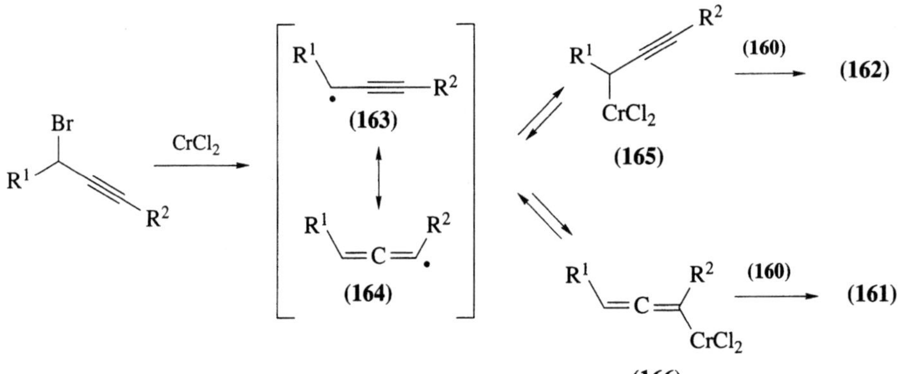

$CrCl_2$ (**163**) (**164**) (**165**) (**166**) (**160**) → (**162**); (**160**) → (**161**)

Scheme 6

1.6.3.8 Enantioselective Addition Reactions

Allylic organometallics modified at the metal center by chiral adjuvants add to aldehydes and ketones to provide optically active homoallylic alcohols. This process has been described for reagents containing boron,[81] tin[82] and chromium[83] metal centers. Gore and coworkers[83] have shown that a chromium-mediated addition reaction of allylic bromides to simple aldehydes that uses a complex of lithium *N*-methylnorephedrine and chromium(II) chloride occurs with modest (6–16% *ee*) enantioselectivity (equation 61, Table 8).

(61)

(167) (168) (170) (171)

R^{2}CHO (169), THF, 20 °C

Table 8 Enantioselective Addition of Allylchromium Reagents to Aldehydes (equation 61)

Entry	R^1	R^2	*(170):(171)*	*Yield (%)*	*ee (%)*
1	H	Pr	(SR) 58:42	49	16
2	H	CH_2CHMe_2	(SR) 58:42	58	16
3	H	c-C_7H_{15}	a	48	a
4	H	Ph	(RS) 56:44	60	11.5
5	Me	Pr	(SR) 58.5:41.5	52	17
6	Me	Ph	(RS) 53:47	53	6

[a]Not applicable.

1.6.4 ALKENYLCHROMIUM REAGENTS

1.6.4.1 General Features

The ability of the anhydrous chromium(II) ion to reduce vinyl halides and provide alkenylchromium compounds which participate in aldehyde addition reactions was first described by Takai and coworkers.[84] Treatment of 2-iodopropene (**173**) and benzaldehyde with anhydrous chromium(II) chloride in DMF afforded allylic alcohol (**174**) in quantitative yield (equation 62).[84]

General features of the process are illustrated in Table 9. The addition to aldehydes is more facile than to ketones (entries 1 and 2 *versus* 3). Vinyl bromide (**177**) adds to aldehydes in the presence of $CrCl_2$. The separate addition of (*E*)-bromostyrene (**180**) and (*Z*)-bromostyrene (**182**) to benzaldehyde occurs stereospecifically to provide (*E*)-(**181**) and (*Z*)-(**183**), respectively (entries 6 and 7). However, trisubstituted (*E*)- and (*Z*)-vinyl iodides (**184**) and (**186**) add to benzaldehyde in a stereoconvergent fashion wherein (*E*)-(**185**) is the exclusive product in both cases. Lastly, iodobenzene (**187**) adds more effectively to nonanal than does bromobenzene (**189**) The chemoselectivity of alkenylchromiums mirrors that of crotylchromium reagents (see equations 19 and 21). Selective addition to the aldehyde carbonyl of bifunctional compounds (**190**) and (**192**) yields adducts (**191**) and (**193**), respectively (equation 63 and 64). Takai and coworkers[85] have shown that alkenylchromium reagents can also be generated from enol triflates and chromium(II) chloride under nickel catalysis (equation 65). This work, as well as reports from Kishi and coworkers,[86] shows that nickel, as a trace contaminant in commercially available $CrCl_2$ is essential for most Barbier-like organochromium reactions. Consonant with this finding is the fact that high purity $CrCl_2$, free from nickel salt contamination, does not reproducibly promote organochromium formation. A catalytic cycle which is likely to be operative has been proposed (Scheme 7). Vinyl halide or triflate (**194**) undergoes oxidative addition to nickel(0) to provide a nickel(II) species (**195**)[87] which, after metal exchange with chromium(III), affords an alkenylchromium(III) reagent. The appropriate quantities of nickel(0) and chromium(III) are provided by the facile redox couple between nickel(II) and chromium(II).[88] In cases where (**194**) is an iodide, both nickel(II) and palladium(II) salts promote the catalytic cycle and the addition reaction to aldehydes. However, triflates are converted to competent alkenylchromium reagents only under nickel catalysis. Since it has been shown that palladium(0) undergoes facile oxidative addition to alkenyl and aryl triflates,[89] the alkenylpalladium triflate (**195**; M = Pd,

$CrCl_2$, DMF; 15 min, 25 °C; 100% (62)

(173) (174)

Table 9 $CrCl_2$–mediated Addition of Alkenyl Halides to Aldehydes and Ketones (equation 62)[a]

Entry	*Alkenyl halide*	*Aldehyde/ketone*	*Yield* (%)[b]	*Time* (h)	*Product*
1	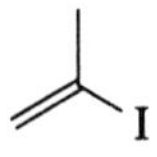 I **(173)**	PhCHO	100	0.25	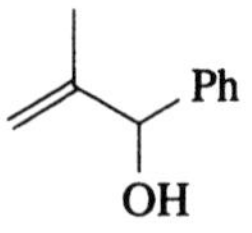 Ph OH **(174)**
2	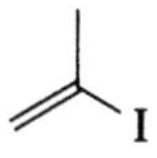 I **(173)**	n-C_8H_{17}CHO	100	0.25	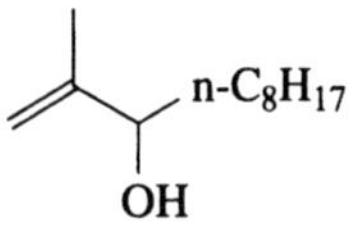 n-C_8H_{17} OH **(175)**
3	 I **(173)**	Cyclohexanone **(172)**	22[c]	3	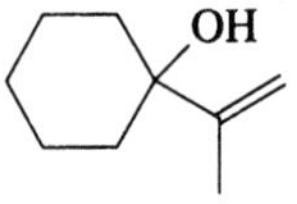 OH **(176)**
4	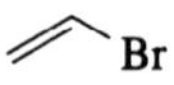 Br **(177)**	PhCHO	80	15	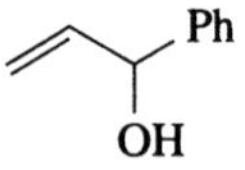 Ph OH **(178)**
5	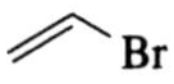 Br **(177)**	n-C_8H_{17}CHO	77	1.5	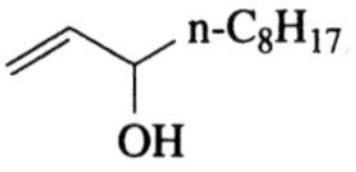 n-C_8H_{17} OH **(179)**
6	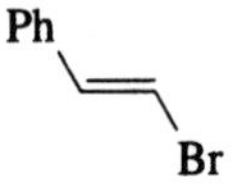 Ph Br **(180)**	PhCHO	82	1	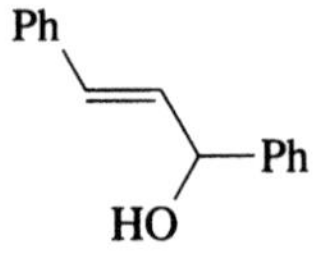 Ph Ph HO **(181)** (*E*) only
7	 Ph Br **(182)**	PhCHO	78	1	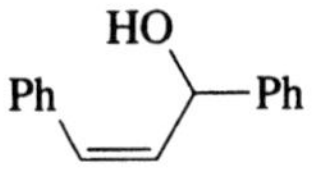 HO Ph Ph **(183)** (*Z*) only
8	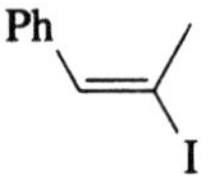 Ph I **(184)**	PhCHO	91	3	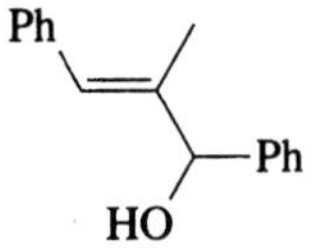 Ph Ph HO **(185)** (*E*) only

Table 9 *(continued)*

Entry	*Alkenyl halide*	*Aldehyde/ketone*	*Yield* (%)[b]	*Time* (h)	*Product*
9	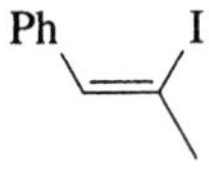 **(186)**	PhCHO	90	3	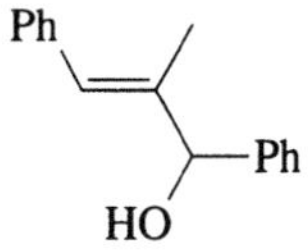 **(185)** (*E*) only
10	PhI **(187)**	n-C_8H_{17}CHO	83	3	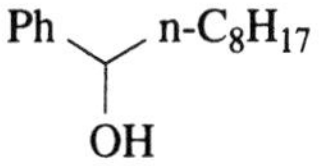 **(188)**
11	PhBr **(189)**	n-C_8H_{17}CHO	13[d]	5	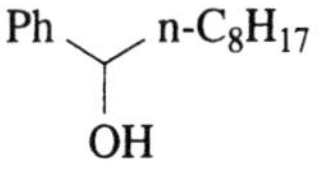 **(188)**

[a] Reaction run at 25 °C, 4 equiv. of $CrCl_2$ used. [b] Isolated yields. [c] 50 °C, 3 h. [d] 90 °C.

X = OTf) may fail to accomplish metal exchange and consequently hinder the formation of the alkenylchromium reagent. Few examples of the use of enol triflates as progenitors of vinyl carbanion equivalents have appeared.[90] Consequently, the work from Takai and coworkers is particularly useful. The alkenylchromiums generated from enol triflates are functionally indistinguishable from those generated from iodoalkenes (equation 65 and table 10).[85] This fact is disclosed by comparison of Table 9 with Table 10.

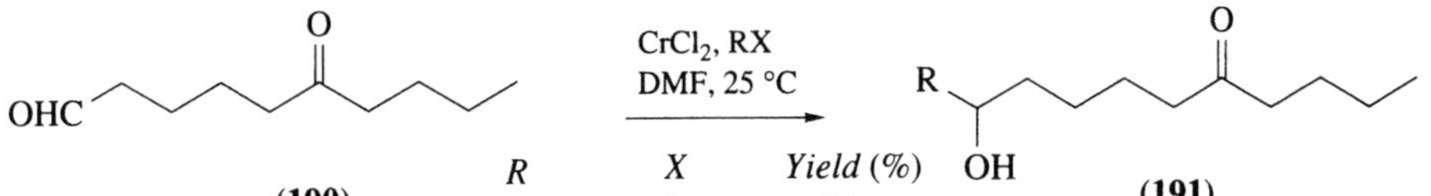

(190) → $CrCl_2$, RX, DMF, 25 °C → **(191)** (63)

R	*X*	*Yield* (%)
$H_2C{=}C(Me)$	I	94
$H_2C{=}CH$	Br	86
Ph	I	81

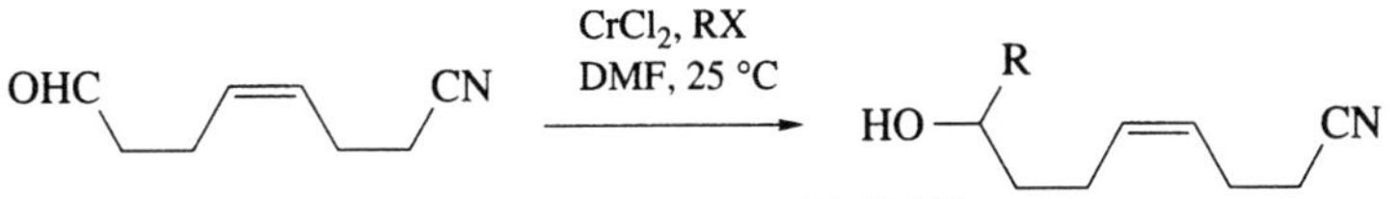

(192) → $CrCl_2$, RX, DMF, 25 °C → **(193)** (64)

R	*X*	*Yield* (%)
$H_2C{=}C(Me)$	I	96
$H_2C{=}CH$	Br	92
Ph	I	87

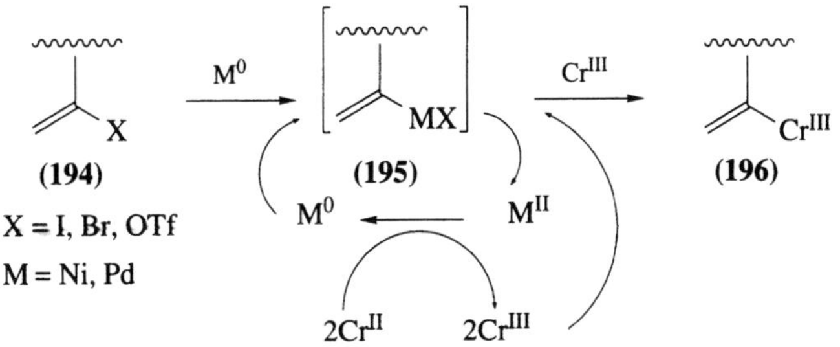

(194) X = I, Br, OTf; M = Ni, Pd → M^0 → **(195)** [MX] → Cr^{III} → **(196)** Cr^{III}; M^{II} → M^0 (2Cr^{II} → 2Cr^{III})

Scheme 7

Table 10 $CrCl_2$–mediated/$NiCl_2$–catalyzed Addition of Alkenyl Triflates to Aldehydes

Entry	*Triflate*	*Aldehyde*	*Time* (h)	*Product*	*Yield* (%)
1	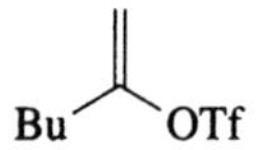 Bu OTf	PhCHO	1	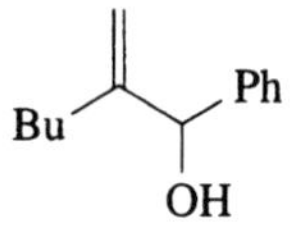 Bu Ph OH	72
2	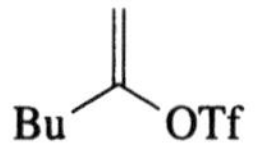 Bu OTf	n-C_8H_{17}CHO	3	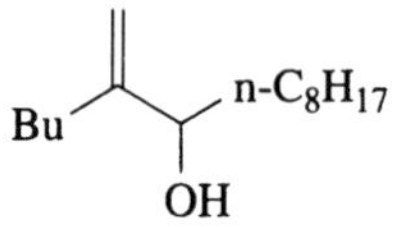 Bu n-C_8H_{17} OH	81
3	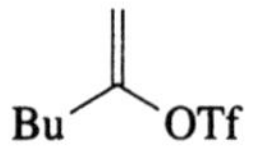 Bu OTf	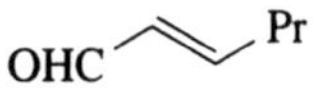 OHC Pr	4	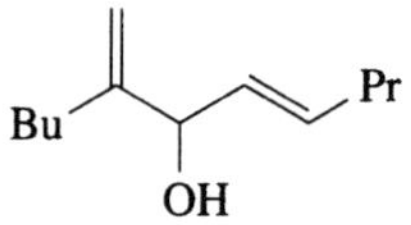 Bu Pr OH	64
4	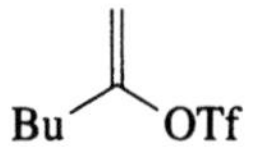 Bu OTf	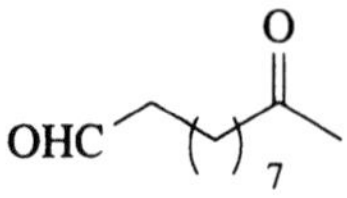 OHC ( )7 O	1	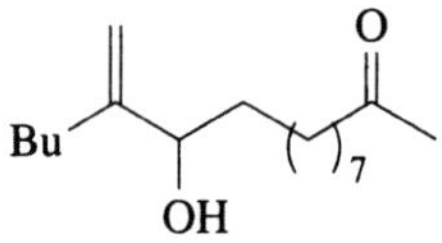 Bu ( )7 O OH	87
5	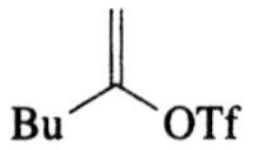 Bu OTf	 OHC CN	2	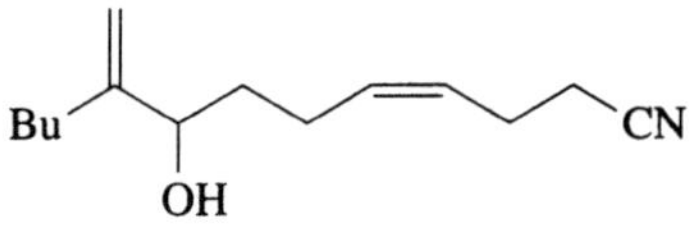 Bu CN OH	78
6	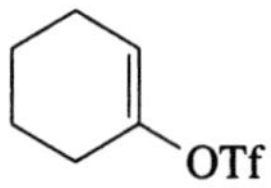 OTf	n-C_8H_{17}CHO	4	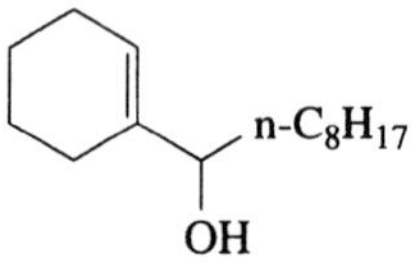 n-C_8H_{17} OH	74
7	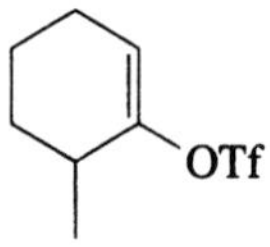 OTf	n-C_8H_{17}CHO	4	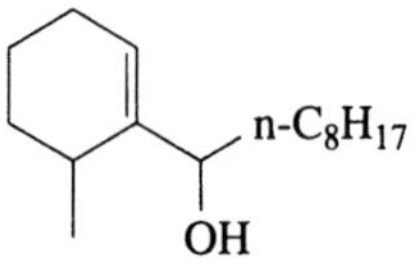 n-C_8H_{17} OH	76
8	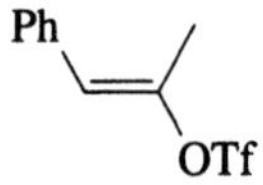 Ph OTf	PhCHO	1	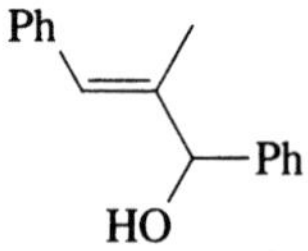 Ph Ph HO	92
9	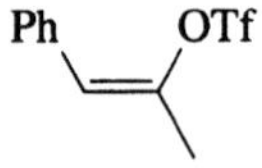 Ph OTf	PhCHO	3	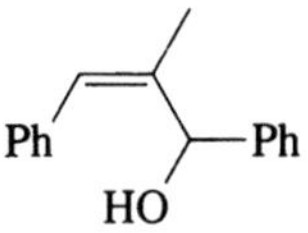 Ph Ph HO	46
10	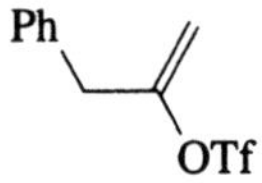 Ph OTf	PhCHO	1	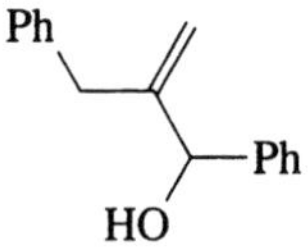 Ph Ph HO	85
11	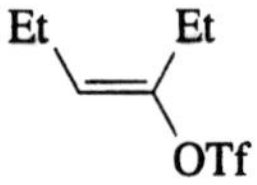 Et Et OTf	PhCHO	2	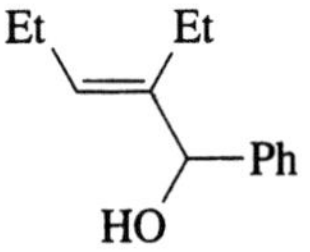 Et Et Ph HO	72

R¹ OTf (197) + R² H (198) → $CrCl_2$, cat. $NiCl_2$, DMF → R¹ R² OH (199) (65)

1.6.4.2 Synthetic Applications

A great portion of the chemical research involving alkenylchromium reagents is described by Kishi and coworkers. During synthetic studies directed towards palytoxin,[92] generation of an organocuprate derived from (**201**) proved untenable. However, the use of an *in situ* prepared organochromium reagent allowed for C(16)—C(17) bond formation (equations 66 and 67). The coupling proceeded with complete

(**200**) 1 equiv. + (**201**) 3 equiv. → $CrCl_2$, DMSO, 1% $NiCl_2$, 71% → (**202**) + (**203**) 1.3:1.0 (66)

(**200**) + (**204**) 3 equiv. → $CrCl_2$, DMSO, 1% $NiCl_2$, 58% → (**205**) 1.6 + (**206**) 1.0 (67)

retention of the C(17)—C(18) alkene geometry but with a low level of diastereogenesis at C(16). Retention of double-bond configuration is a general feature of disubstituted alkenylchromium compounds (see also entries 6 and 7, Table 9). However, alkenylchromiums derived from trisubstituted iodoalkenes and trisubstituted β-iodoenones[91] exclusively provide products with an (*E*)-alkene geometry. This stereochemical feature, also noted by Takai and coworkers (entries 8 and 9 in Table 9; entries 8 and 9 in Table 10),[84,85] is evident in equations (68) and (69). Both (*E*)- and (*Z*)-iodoalkenes provide only (*E*)-allylic alcohols as the addition product. Moreover, both reagents provide identical levels of diastereoselection, as seen from the ratios (**210**):(**211**) and (**214**):(**215**). Notwithstanding these similarities, the organometallic prepared from the (*E*)-iodoalkenes provides higher yields of the addition products. The major products obtained, (**210**) and (**215**), are consistent with the Felkin–Anh prediction for diastereoselective addition reactions to α-alkoxy aldehydes.[47] The reactions outlined in equations (68) and (69) serve as useful model studies for the coupling reaction between the C(1)—C(7) and C(8)—C(51) fragments in the actual synthesis of Palytoxin.

Kishi and coworkers[93–95] have also implemented alkenylchromium reagents for the synthesis of glycosides and *C*-methyl glycoside analogs. In the synthesis of the C-analog (**217**) of isomaltose (**216**), the C(5)—C(6) bond was established by the nickel(II) chromium(II)-mediated coupling of (*Z*)-iodoalkene

(*E*)-(**207**)
(*Z*)-(**208**)
+ (**209**)
$CrCl_2$, DMSO, 1% $NiCl_2$

(68)

		8α-(**210**)	:	8β-(**211**)
From (*E*)-(**207**)	72%	2	:	1
From (*Z*)-(**208**)	15%	2	:	1

(*E*)-(**212**)
(*Z*)-(**213**)
+ (**209**)
$CrCl_2$, 1% $NiCl_2$, DMF/DMS

(69)

		8α-(**214**)	:	8β-(**215**)
From (*E*)-(**212**)	82%	5	:	1
From (*Z*)-(**213**)	28%	5	:	1

(**218**) with alkoxy aldehyde (**219**; equation 70).[93] The reaction proceeds with excellent stereochemical control *via* a Felkin–Anh approach and provides a >15:<1 ratio of *cis*-allylic alcohols (**220**) and (**221**). Continuing studies in this area include the synthesis of *C*-sucrose (**226**).[94] The nickel(II)/chromium(II) coupling of vinyl iodide (**222**) with α-benzyloxy aldehyde (**223**) affords (**224**) and (**225**; equation 71). The major product (**225**) possesses the 3′,4′-*anti* arrangement which arises rationally from a Felkin–Anh addition pathway. This result is consistent with previously described examples (*vide supra*).

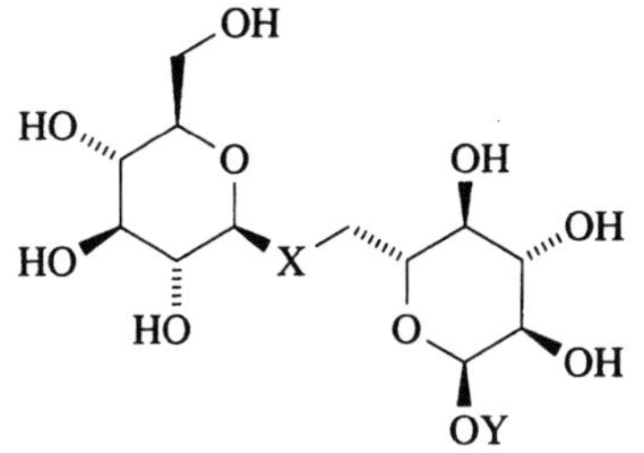

(**216**) Isomaltose: X = O, Y = H
(**217**) Methyl-*C*-glycoside: X = CH_2, Y = Me

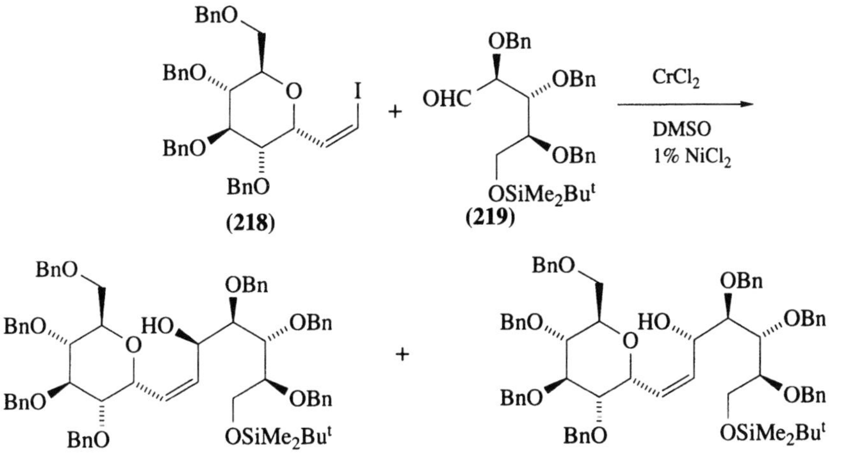

(**218**) + (**219**) $\xrightarrow[\text{DMSO, 1\% NiCl}_2]{\text{CrCl}_2}$ (**220**) >94% + (**221**) <6% (70)

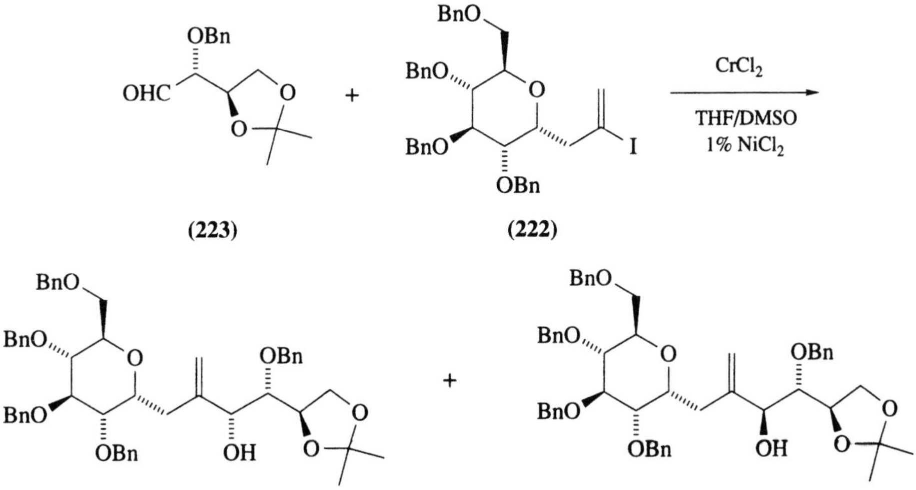

(**223**) + (**222**) $\xrightarrow[\text{THF/DMSO, 1\% NiCl}_2]{\text{CrCl}_2}$ (**224**) 9% + (**225**) 91% (71)

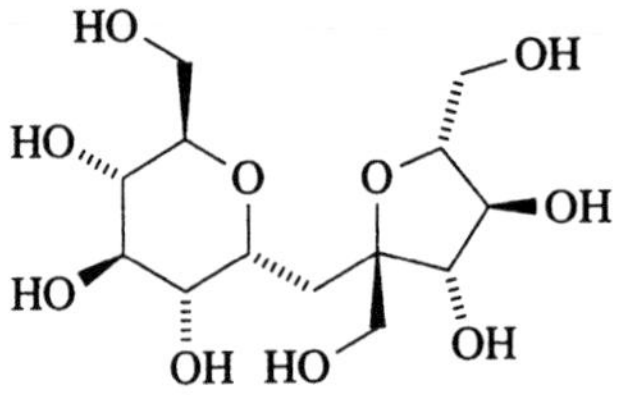

C-Sucrose (**226**)

1.6.4.3 Intramolecular Addition Reactions

Examples of intramolecular addition reactions of alkenylchromium reagents to aldehydes have appeared.[96,97] In the course of synthetic studies in the brefeldin structural series, the nickel(II)/chromium(II)-mediated intramolecular addition reactions of (*E*)-iodoalkenes (**227**) and (**230**) were studied by Schreiber and Meyers (equations 72 and 73).[96] Treatment of (**227**) with $CrCl_2$ and a catalytic portion of $[Ni(acac)_2]$ in DMF produced a 4:1 mixture of 4-epibrefeldin C (**228**) and (+)-brefeldin (**229**) in 60% yield. In a similar fashion, precursor iodide (**230**) afforded a >10:1 mixture of cyclized hydroxy lactones (**231**) and (**232**) in 70% yield. An explanation of the stereochemical preference observed has been eloquently offered in a discussion of local conformational preferences found in the starting material and the product lactone, as each is relevant to a transition structure for a 13-membered ring closure.

Another intramolecular addition of an alkenylchromium to an aldehyde was reported by Rowley and Kishi in synthetic studies toward the ophiobolins.[97] Treatment of (**223**; equation 74) with $CrCl_2/NiCl_2$ afforded the cyclized product (**234**) as a single diastereomer in 56% yield.

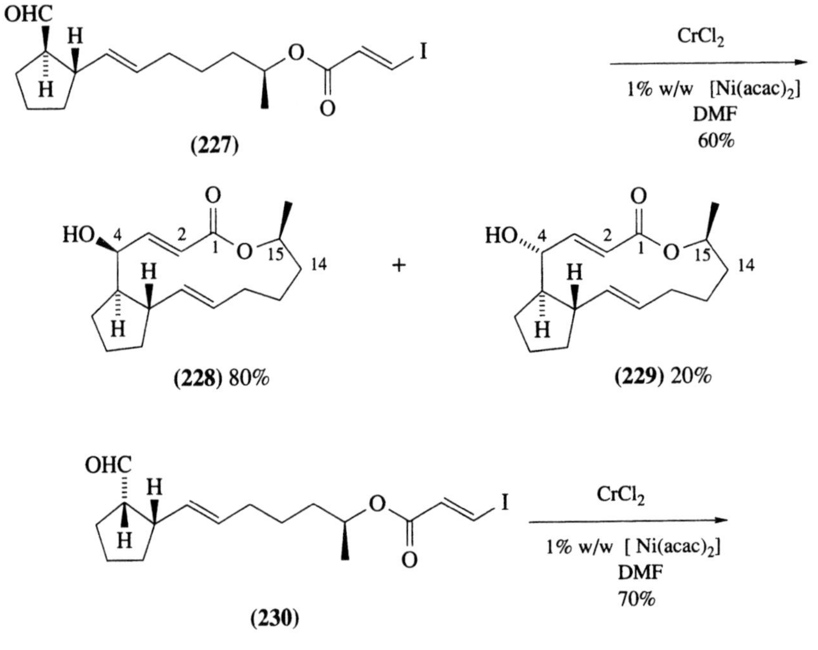

(**227**) —$CrCl_2$, 1% w/w $[Ni(acac)_2]$, DMF, 60%→ (**228**) 80% + (**229**) 20% (72)

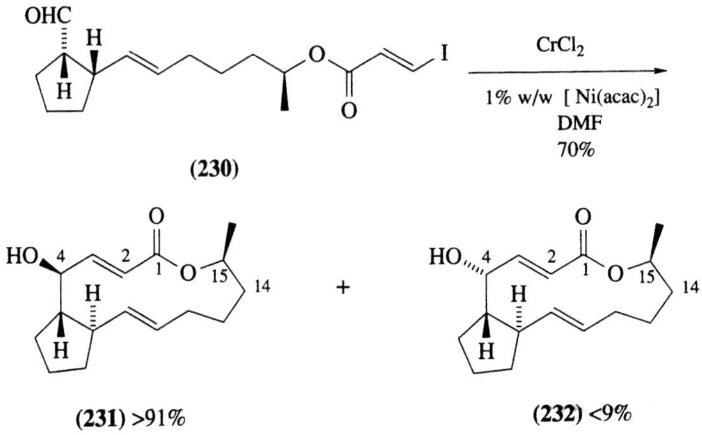

(**230**) —$CrCl_2$, 1% w/w $[Ni(acac)_2]$, DMF, 70%→ (**231**) >91% + (**232**) <9% (73)

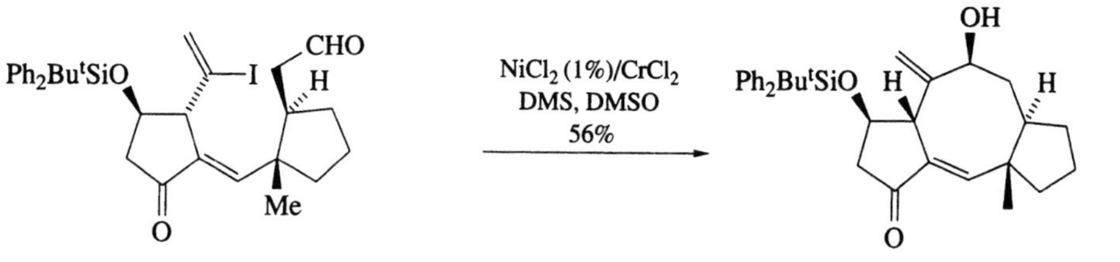

(74)

(233) **(234)**

1.6.5 ALKYNYLCHROMIUM REAGENTS

The generation of alkynylchromium reagents and their addition to carbonyl compounds has been studied by Takai and coworkers.[98] The chemoselectivity of these reagents resembles that of other organochromium(III) reagents. Preparation of alkynylchromiums can be accomplished by treatment of haloalkynes with chromium(II) chloride in DMF. Representative addition reactions of alkynylchromiums are offered in equations (75) to (79).

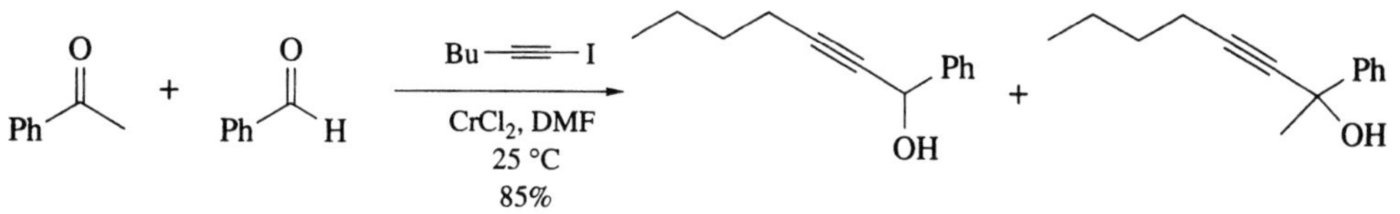

(75)

(235) 96% **(236)** 4%

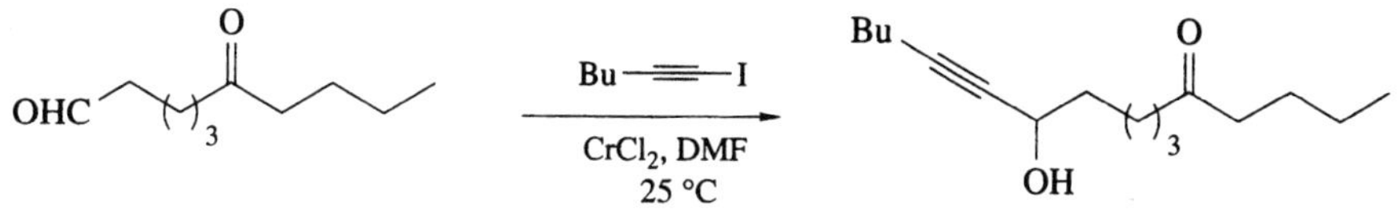

(76)

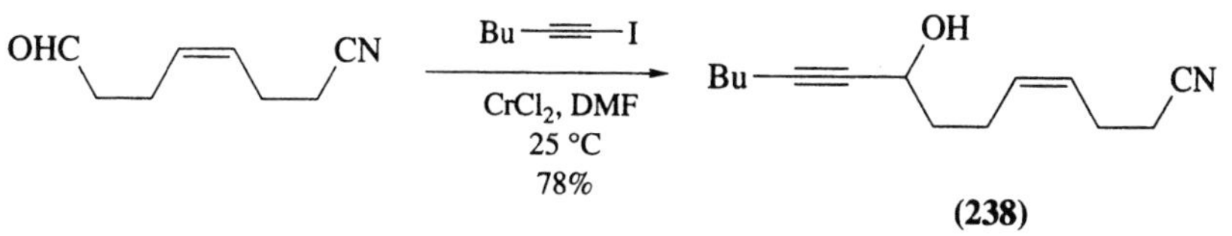

(237)

(77)

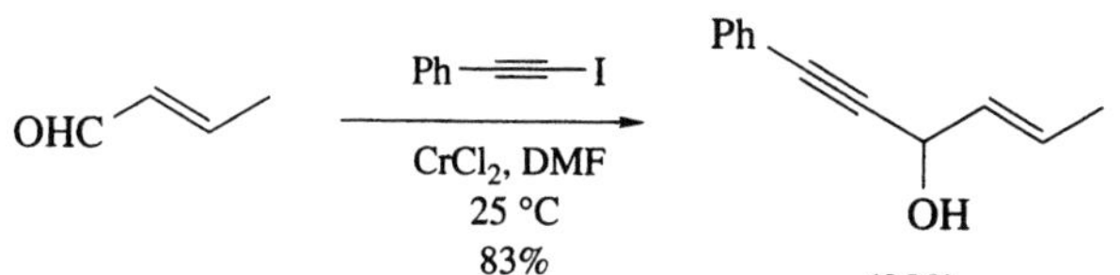

(238)

(78)

(239)

$$PhCHO \xrightarrow[\text{CrCl}_2,\ \text{DMF},\ 25\ ^\circ\text{C},\ 83\%]{\text{Bu}^t\text{Me}_2\text{Si}-\equiv-\text{Br}} \text{Bu}^t\text{Me}_2\text{Si}-\equiv-\text{CH(OH)Ph} \quad \textbf{(240)} \qquad (79)$$

1.6.6 α-ACYLCHROMIUM REAGENTS

The aldol-type reaction of α-bromo ketones with aldehydes, mediated by $CrCl_2$, has been studied by Dubois and coworkers.[99] The reaction is carried out by addition of (**241**; equation 80) to a solution of (**242**) and $CrCl_2$ in THF. The reaction proceeds with high levels of *syn* selectivity with bulky bromo ketones (**241**), independent of the substrate aldehyde used (Table 11, entries 1–6). However, the reaction is stereorandom with bromoacetophenone (entry 7) and selectively *anti* with 2-bromocyclohexanone (entry 8). No explanation of the stereoselectivity has been advanced, but the reaction is believed not to proceed *via* a simple chromium enolate since no condensation reaction is obtained by addition of (**242**) to a solution of (**241**) and $CrCl_2$. Moreover, Nozaki and coworkers[100] have demonstrated that chromium(II) chloride treatment of 1-bromocyclododecanone followed by treatment with either methyl iodide or TMS-Cl produces only cyclododecanone.

$$R^1COCH(Br)R^2\ \textbf{(241)} + R^3CHO\ \textbf{(242)} \xrightarrow[\text{THF, ~24 h}]{\text{CrCl}_2} R^1COCH(R^2)CH(OH)R^3\ \textbf{(243)}\ anti + R^1COCH(R^2)CH(OH)R^3\ \textbf{(244)}\ syn \qquad (80)$$

Table 11 Reaction of α-Bromo Ketones (**241**) with Aldehydes (**242**) using $CrCl_2$ (equation 80)[99]

Entry	*R¹*	*R²*	*R³*	*Yield (%)*	*Anti:syn (243):(244)*
1	Bu^t	Me	Me	50	0:100
2	Bu^t	Me	Et	81	0:100
3	Bu^t	Me	Pr^i	70	0:100
4	Bu^t	Me	Pr	83	0:100
5	Bu^t	Me	Ph	75	0:100
6	Bu^t	Me	Bu^t	87	0:100
7	Ph	Me	Ph	68	50:50
8	—(CH_2)$_4$—		Ph	75	100:0

1.6.7 ALKYLCHROMIUM REAGENTS

Addition of alkylchromium reagents to carbonyl compounds was studied first by Kauffmann and coworkers.[101] Organochromium reagent (**246**) can be generated *in situ* by treatment of trimethylsilylmethylmagnesium chloride (**245**) with $CrCl_3$ (Scheme 8). Condensation with aldehydes provides chromium alkoxide (**247**), which is transformed into the alkenated products (**248**) by warming with aqueous perchloric acid. Ketones are apparently inert to (**246**), providing none of the desired alkene product. A more extensive examination of alkylchromium(III) reagents has been subsequently disclosed.[102] Preparation of a variety of dichlorotris(tetrahydrofuran)alkylchromium(III) complexes (**249**) has been carried out by treatment of $CrCl_3$ with organomagnesium or organolithium reagents in tetrahydrofuran (equation 81). Complexes of this sort can also be prepared from organoaluminums and $CrCl_3$,[12,13] and X-ray data has revealed the structure of one such complex (**249**; R = *p*-tolyl). Addition of (**250**) to aldehydes (equation 82 and Table 12) proceeds smoothly (R^3 = H; Table 12, entries 2, 5 and 8), but towards ketones (**250**) is entirely unreactive ($R^3 \neq$ H; entries 3 and 6). The chemoselectivity is underscored by competition experiments (entries 4 and 7). Also, the addition reaction proceeds more efficiently with a molar excess of alkylchromium (entries 1 *versus* 2 and 8 *versus* 9). Generation of alkylchromium(III) reagents from $CrCl_2$ and organic halides has been reported by Takai and coworkers.[103] In particular, it has been demonstrated that α-thioalkylchromium(III) compounds (**254**), prepared by chromium(II) ion reduction of α-iodo sulfides (generated *in situ* from LiI and **253**),[104] add smoothly and selectively to aldehydes to afford thioether carbinols (**256**; equation 83, Table 13), (**257**; equation 84) and (**258**; equation 85). In contrast to the chemoselectivity of this organochromium compound, it has been noted that α-thiomethyllithium exhibits no chemoselectivity for aldehydes in the presence of ketones. The butylchromium reagent prepared by chromium(II) ion reduction of iodobutane provides only 11% of 1-phenyl-1-pentanol (**259**; equation 86) after 24 h. However, Kauffmann and coworkers[102] have demonstrated that the butylchromium reagent prepared from *n*-butyllithium and $CrCl_3$ yields 81% of (**259**) after 18 h. These data reveal the importance of the sulfide group for the efficient reduction of the alkyl iodide.

Diastereoselective addition of thiophenylalkylchromium(III) reagents has been studied by Takai and coworkers.[103] Treatment of aldehyde (**260**) with α-chloro sulfide (**261**), LiI and $CrCl_2$ gives β-phenylthio alcohols (**262**) and (**263**) (equation 87 and Table 14). The *syn* isomer (**262**) was generated diastereoselectively in all cases (**262**:**263** ≥ 4:1). The diastereoselectivity of the addition reactions of thiophenylalkylchromiums is especially noteworthy as the lithium reagent prepared from phenyl ethyl sulfide and Bu^tLi–HMPA adds to benzaldehyde in a stereorandom fashion (contrast entry 2, table 14).

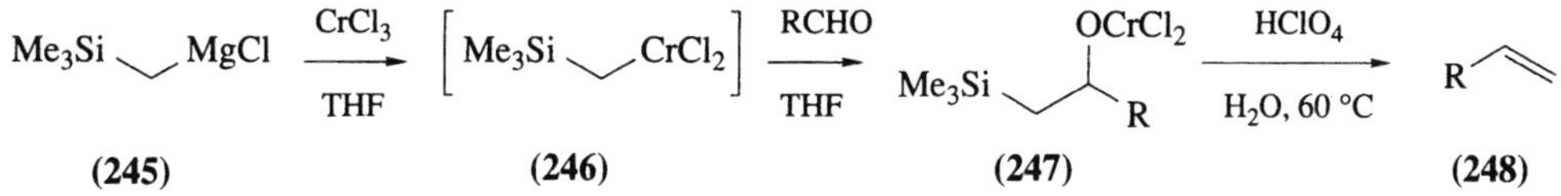

Me_3SiCH_2MgCl (**245**) $\xrightarrow[THF]{CrCl_3}$ $[Me_3SiCH_2CrCl_2]$ (**246**) $\xrightarrow[THF]{RCHO}$ $Me_3SiCH_2CH(OCrCl_2)R$ (**247**) $\xrightarrow[H_2O,\ 60\ °C]{HClO_4}$ $RCH=CH_2$ (**248**)

$R = C_8H_{17}$ 45%
$R = C_6H_{13}$ 47%

Scheme 8

$$RMgX\ \text{or}\ RLi \xrightarrow[THF]{CrCl_3} [RCr(THF)_3Cl_2]\ \mathbf{(249)} \quad (81)$$

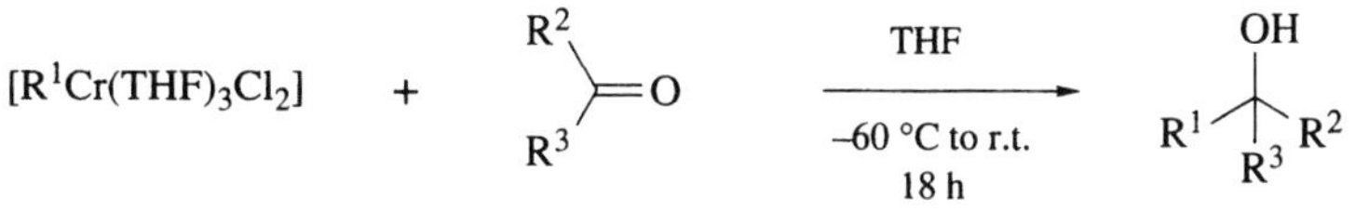

$$[R^1Cr(THF)_3Cl_2]\ \mathbf{(250)} + R^2R^3C{=}O\ \mathbf{(251)} \xrightarrow[-60\ °C\ \text{to r.t.},\ 18\ h]{THF} R^1R^2R^3C{-}OH\ \mathbf{(252)} \quad (82)$$

Table 12 Addition of Alkylchromium(III) Complexes (**250**) to Carbonyl Compounds (equation 82)

Entry	*R*1	*R*2	*R*3	*(250):(251)*	*Yield (%)*
1	Me	n-C_6H_{13}	H	1:1	36
2	Me	n-C_6H_{13}	H	3:1	85
3	Me	Me	Me	1:1	0
4	Me	n-C_6H_{13} Me	H Me	2:1:1	71 0
5	Pr^n	n-C_6H_{13}	H	3:1	73
6	Bu^n	Et	Et	1:1	0
7	Pr^n	n-C_6H_{13} Et	H Et	2:1:1	66 0
8	$PhCH_2$	n-C_6H_{13}	H	1:1	52
9	$PhCH_2$	n-C_6H_{13}	H	3:1	65
10	Bu^n	n-C_6H_{13}	H	1:1	81

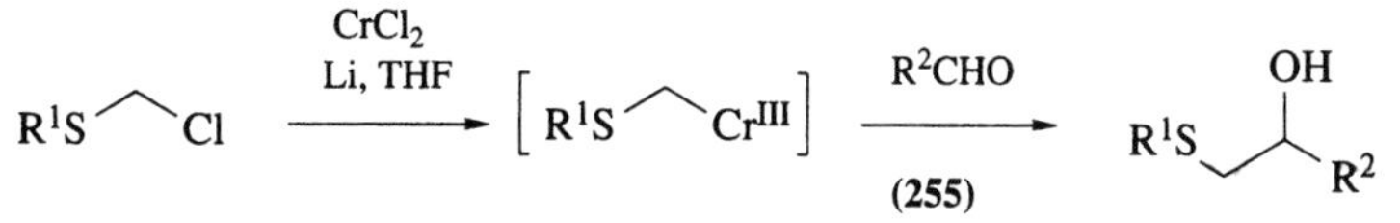

$$R^1SCH_2Cl\ \mathbf{(253)} \xrightarrow[Li,\ THF]{CrCl_2} [R^1SCH_2Cr^{III}]\ \mathbf{(254)} \xrightarrow[\mathbf{(255)}]{R^2CHO} R^1SCH_2CH(OH)R^2\ \mathbf{(256)} \quad (83)$$

Table 13 Addition of α-Halo Sulfides (**253**) to Aldehydes (**255**) Using $CrCl_2$ (equation 83)

Entry	R^1	R^2	*Temp (°C)*	*Time (h)*	*Yield (%)*
1	Me	Ph	40	5	88
2	Me	C_8H_{17}	40	9	72
3	Me	PrCH=CH	40	13	64[a]
4	Ph	Ph	60	10	63
5	Ph	C_8H_{17}	40	10	48

[a]1,2-Product only

PhCHO + PhCOMe $\xrightarrow[\text{LiI, CrCl}_2\text{, THF, 40 °C, 5 h}]{\text{MeSCH}_2\text{Cl}}$ PhCH(OH)CH₂SMe (**257**) 94% + PhCOMe 86% recovered (84)

$CH_3CO(CH_2)_7CHO \xrightarrow[\text{LiI, CrCl}_2\text{, THF, 40 °C, 3 h}]{\text{MeSCH}_2\text{Cl}}$ (**258**) (85)

PhCHO → (**259**) PhCH(OH)Bu (86)
i, Bu^nI, $CrCl_2$, DMF, r.t., 24 h, 11% or
ii, Bu^nLi, $CrCl_3$, THF, 18 h, –60 °C to r.t., 81%

The process is sensitive to solvent additives. Both yields and *syn* selectivities benefit from the addition of TMEDA (entries 1 *versus* 3 and 5 *versus* 6), HMPA (entry 5 *versus* 7), triphenylphosphine (entry 5 *versus* 8) or 1,2-bis(diphenylphosphino)ethane (entry 5 *versus* 9). The effectiveness of a given additive (*vis-a-vis* selectivity and/or % conversion) has been shown to be substrate dependent.

R^1CHO (**260**) + PhSCH(Cl)R^2 (**261**) $\xrightarrow[\text{THF, 25 °C}]{\text{CrCl}_2\text{, LiI}}$ (**262**) *syn* + (**263**) *anti* (87)

Table 14 Addition of α-Halo Sulfides (**261**) to Aldehydes (**260**) using $CrCl_2$ (equation 87)

Entry	R^1	R^2	*Ligand*	*Time (h)*	*Yield (%)*	*(262):(263)*	*Aldol*	*(261) recovered (%)*
1	Ph	Me	—	16	58	80:20	—	17
2	Ph	Me	TMEDA	6	96	88:12	—	0
3	Ph	Pr^n	TMEDA	6	95	86:14	—	0
4	Ph	Pr^i	TMEDA	25	46	80:20	—	25
5	C_8H_{17}	Me	—	17	17	86:14	11	0
6	C_8H_{17}	Me	TMEDA	1.5	25	90:10	19	0
7	C_8H_{17}	Me	HMPA	20	33	97:3	23	5
8	C_8H_{17}	Me	Ph_3P	20	23	84:16	10	24
9	C_8H_{17}	Me	DIPHOS	19	53	>98:<2	<3	32
10	C_8H_{17}	Pr^n	DIPHOS	40	8	>98:<2	<5	71
11	c-C_6H_{11}	Me	DIPHOS	18	11	90:10	<5	50
12	c-C_6H_{11}	Me	TMEDA	18	65	81:19	<5	13

1.6.8 ALKYL-*GEM*-DICHROMIUM REAGENTS: ALKENATION REACTIONS

gem-Dimetalloorganic compounds are useful reagents for the alkenation of aldehydes and ketones.[105,106] A variety of *gem*-dichromium reagents have also been used for this purpose.[107–109] Takai and coworkers[107] have reported the conversion of aldehydes (**264**) to vinyl halides (**265**) and (**266**) using haloform and chromium(II) salts in THF (equation 88 and Table 15). The (*E*)-alkenyl halide is generated selectively with the exception of the $CHI_3/CrCl_2$ reactions of α,β-unsaturated aldehydes (Table 15, entry 12), which give variable results with respect to alkene geometry. The rate of reaction has been shown to be a function of the haloform and increases in the order Cl < Br < I. In addition, the (*E*):(*Z*) ratio (**265**):(**266**) is also a function of the haloform used and increases in the order I < Br < Cl. The use of $HCBr_3/CrCl_2$ provides a mixture of alkenyl chlorides and bromides (entries 2 and 6). This problem can be alleviated by the use of $CrBr_3$/LAH as the source of $CrBr_2$, which affords the alkenyl bromides cleanly. Haloform–$CrCl_2$ reagents condense selectively with aldehydes in the presence of ketones (equation 89), and provide vinyl halides suitable for alkenylchromium preparation (see Section 1.6.4).

RCHO (**264**) —(CHX_3, Cr^{II}, THF)→ (*E*)-(**265**) + (*Z*)-(**266**) (88)

Table 15 Selective Synthesis of (*E*)-Alkenyl Halides (**265**) from Aldehydes (**264**) and CHX_3–$CrCl_2$/THF (equation 88)

Entry	*Aldehyde (264)*	*Haloform X*	*Chromium(II) source*	*Temperature (°C)*	*Time (h)*	*Yield (%)*	*(265):(266)*
1	PhCHO	I	$CrCl_2$	0	3	87	94:6
2	PhCHO	Br	$CrCl_2$	25	1.5	X = Br 32 X = Cl 43	95:5 95:5
3	PhCHO	Br	$CrBr_3/LiAlH_4$	50	1	70	95:5
4	n-C_8H_{17}CHO	Cl	$CrCl_2$	65	2	76	95:5
5	n-C_8H_{17}CHO	I	$CrCl_2$	0	2	82	83:17
6	n-C_8H_{17}CHO	Br	$CrCl_2$	25	2	X = Br 37 X = Cl 32	89:11 90:10
7	n-C_8H_{17}CHO	Br	$CrBr_3/LiAlH_4$	50	2	61	87:13
8	cyclohexyl–CHO	Cl	$CrCl_2$	65	4	76	94:6
9	cyclohexyl–CHO	I	$CrCl_2$	0	1	78	89:11
10	cyclohexyl–CHO	Br	$CrBr_3/LiAlH_4$	50	2.5	55	89:11
11	cyclohexylidene–CHO	Cl	$CrCl_2$	65	2.5	55	92:8
12	cyclohexylidene–CHO	I	$CrCl_2$	0	0.5	76	75:25 to 55:45
13	Bu^t–cyclohexanone (=O)	I	$CrCl_2$	25	4	75	—

PhCHO + PhCOMe —(CHI_3, $CrCl_2$, THF, 0 °C)→ PhCH=CHI + PhCOMe (89)

(**267**) (*E*):(*Z*) 94:6, 91%; 88% recovered

Takai and coworkers[108] have reported that the formation of alkyl-*gem*-dichromium compounds (**269**) can be achieved by the $CrCl_2$ reduction of *gem*-diiodoalkanes (**268**; equation 90). These reagents add to aldehydes and afford alkylidenation products (**271**) and (**272**) efficiently and with high levels of (*E*)-alkene selectivity (equation 90 and Table 16). The addition of 1,1-diiodoethane to aldehydes proceeds smoothly and with excellent (*E*)-selection (entries 1–5).

$$R^2CHI_2 \text{ (268)} \xrightarrow[\text{THF}]{CrCl_2\text{–DMF}} [R^2CH(Cr^{III})_2] \text{ (269)} \xrightarrow[\text{(270)}]{R^1CHO} (E)\text{-}R^2CH{=}CHR^1 \text{ (271)} + (Z)\text{-}R^2CH{=}CHR^1 \text{ (272)} \quad (90)$$

Table 16 Alkenation of Aldehydes (**270**) with *gem*-Dichromium Compounds (**269**; equation 90)

Entry	R^1	R^2	*Conditions*[a]	*Time (h)*	*Yield (%)*	*(271):(272)*
1	n-C_5H_{11}	Me	A	4.5	94	96:4
2	n-$C_{11}H_{23}$	Me	A	5.0	81	95:5
3	$Ph(CH_2)_2$	Me	A	10	85	97:3
4	Et_2CH	Me	A	2.0	99	98:2
5	4-$Pr^iC_6H_4$	Me	A	10	97	84:16
6	4$Pr^iC_6H_4$	Me	C	5.0	84	78:22
7	n-C_5H_{11}	Pr	A	24	38	96:4
8	n-C_8H_{17}	Pr	B	1.5	85	95:5
9	n-C_8H_{17}	Pr	B	1.0	96	99:1
10	Bu^t	Pr	B	1.0	87	88:12
11	Ph	Pr	C	0.5	60	51:49
12	n-C_5H_{11}	Pr^i	A	24	12	72:28
13	Ph	Pr^i	B	2.0	79	88:12
14	Ph	Bu^t	B	2.0	80	96:4
15	Ph	H	A	24	70	—
16	Ph	H	B	3.0	92	—

[a]A: (**270**) 1.0 mmol, (**268**) 2.0 mmol, $CrCl_2$ 8.0 mmol, THF; B: (**270**) 1.0 mmol, (**268**) 2.0 mmol, $CrCl_2$ 8.0 mmol, DMF 8.0 mmol, THF; C: (**270**) 1.0 mmol, (**268**) 2.0 mmol, $CrCl_3$ 8.0 mmol, Zn° 6.0 mmol, THF.

However, all other *gem*-dichromium reagents (R^2 = Pr, H, Pr^i) require DMF as a cosolvent to obtain useful yields of alkene products. The effect is attributed to the enhanced reducing ability of chromium(II) in the presence of donor ligands (*vide supra*). Use of Zn/$CrCl_3$ as a source of chromium(II) ion provides inferior stereochemical results (Table 16, entries 6 and 11). Lastly, Takai and coworkers[109] have reported the synthesis of (*E*)-alkenylsilanes. Treatment of dibromotrimethylsilane (**273**) with $CrCl_2$ yields *gem*-dichromium reagent (**274**) which reacts with aldehydes to produce (*E*)- and (*Z*)-vinylsilanes like (**276**) and (**277**). As with other *gem*-dichromiums (equation 88, Table 15; equation 90, Table 16), the alkenation process occurs efficiently with a strong preference for the generation of the (*E*)-vinylsilane (equation 91, Table 17). In corresponding carbon-substituted cases, diiodo substrates are required for useful yields of the alkene to be obtained; however, the presence of the silicon atom allows for the use of dibromo precursors. The general features of reactivity seen with other organochromiums are apparent with this reagent as well (Table 17). The authors believe that the reaction proceeds by the formation of a *gem*-dichromium species (**280**; Scheme 9).[110,111] Subsequent addition to an aldehyde provides the β-oxymetal organometallic (**281**) which then suffers elimination to afford the alkene products (**282**).[106,111] No explanation of the stereochemical outcome has been forwarded.

$$Me_3SiCHBr_2 \text{ (273)} \xrightarrow[\text{THF}]{CrCl_2,\ 25\ °C} [Me_3SiCH(Cr^{III})_2] \text{ (274)} \xrightarrow[\text{(275)}]{R^1CHO} (E)\text{-}Me_3SiCH{=}CHR^1 \text{ (276)} + (Z)\text{-}Me_3SiCH{=}CHR^1 \text{ (277)} \quad (91)$$

Table 17 Synthesis of (*E*)-Alkenylsilanes (**276**) from Aldehydes (**275**) and *gem*-Dichromium Reagents (**274**)

Entry	*Aldehyde (275)*	*Time* (h)	*Yield* (%)	*Comment*
1	PhCHO	24	82	
2	$PhCH_2CH_2CHO$	24	86	
3	n-$C_8H_{17}CHO$	24	82	
4	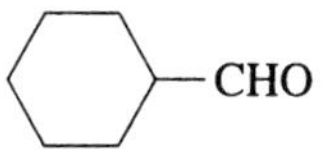 (cyclohexyl)–CHO	18	81	
5	PhCH=CHCHO	18	79	1,2-Addition
6	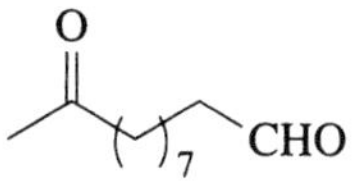 $CH_3CO(CH_2)_7$CHO	16	76	Addition to aldehyde
7	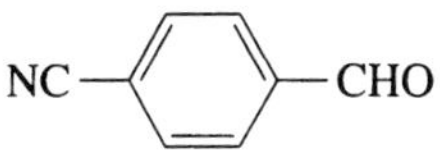 NC–C_6H_4–CHO	24	72	Addition to aldehyde
8	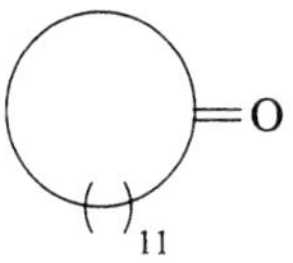 cyclododecanone ($(CH_2)_{11}$, =O)	60	0	99% recovery of starting material

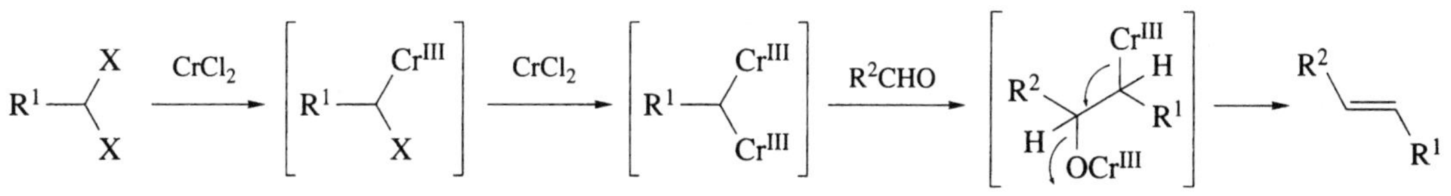

$$R^1CHX_2 \ (278) \xrightarrow{CrCl_2} [R^1CH(X)Cr^{III}] \ (279) \xrightarrow{CrCl_2} [R^1CH(Cr^{III})_2] \ (280) \xrightarrow{R^2CHO} [R^2CH(OCr^{III})CH(R^1)Cr^{III}] \ (281) \longrightarrow R^2CH{=}CHR^1 \ (282)$$

R^1 = halogen, $SiMe_3$, alkyl

Scheme 9

1.6.9 CONCLUSION

A large variety of organochromium(III) compounds has been described. The addition reactions of these materials with carbonyl substrates represent an elaborate array of chemoselective and stereoselective processes. Because of the unique reactivity and chemical properties of these reagents, organochromiums are useful reagents for organic synthesis.

1.6.10 REFERENCES

1. R. P. A. Sneeden, 'Organochromium Compounds', Academic Press, New York, 1975.
2. H. H. Zeiss, *ACS Monogr.*, 1960, **147**, 380.
3. F. A. L. Anet and E. Leblanc, *J. Am. Chem. Soc.*, 1957, **79**, 2649.
4. J. R. Hanson, *Synthesis*, 1974, 1; T.-L. Ho, *Synthesis*, 1979, 1.
5. J. K. Kochi and P. E. Mocadlo, *J. Am. Chem. Soc.*, 1966, **88**, 4094.
6. J. K. Kochi and D. D. Davis, *J. Am. Chem. Soc.*, 1964, **86**, 5264; L. H. Slaugh and J. H. Raley, *Tetrahedron*, 1964, **20**, 1005.
7. J. K. Kochi and D. Buchanan, *J. Am. Chem. Soc.*, 1965, **87**, 853; J. K. Kochi and P. E. Mocadlo, *J. Org. Chem.*, 1965, **30**, 1134.
8. D. H. R. Barton, N. K. Basu, R. H. Hesse, F. S. Morehouse and M. M. Pechet, *J. Am. Chem. Soc.*, 1966, **88**, 3016; D. H. R. Barton and N. K. Basu, *Tetrahedron Lett.*, 1964, 3151; C. H. Robinson, O. Gnoj, E. P. Oliveto and D. H. R. Barton, *J. Org. Chem.*, 1966, **31**, 2749.
9. Y. Okude, S. Hirano, T. Hiyama and H. Nozaki, *J. Am. Chem. Soc.*, 1977, **99**, 3179.
10. W. Herwig and H. H. Zeiss, *J. Am. Chem. Soc.*, 1957, **79**, 6561; W. Herwig and H. H. Zeiss, *J. Am. Chem. Soc.*, 1959, **81**, 4798.
11. S. I. Khan and R. Bau, *Organometallics*, 1983, **2**, 1896.

12. E. Kurras, *Naturwissenschaften*, 1959, **46**, 171.
13. K. Nishimura, H. Kuribayashi, A. Yamamoto and S. Ikeda, *J. Organomet. Chem.*, 1972, **37**, 317; A. Yamamoto, Y. Kano and T. Yamamoto, *J. Organomet. Chem.*, 1975, **102**, 57.
14. J. J. Daly, R. P. A. Sneeden and H. H. Zeiss, *J. Am. Chem. Soc.*, 1966, **88**, 4287; J. J. Daly and R. P. A. Sneeden, *J. Chem. Soc. A*, 1967, 736.
15. R. R. Schrock and G. W. Parshall, *Chem. Rev.*, 1976, **76**, 243.
16. R. P. A. Sneeden, T. F. Burger and H. H. Zeiss, *J. Organomet. Chem.*, 1965, **4**, 397.
17. K. Maruyama, T. Ito and A. Yamamoto, *Chem. Lett.*, 1978, 479; T. Ito, T. Ono, K. Maruyama and A. Yamamoto, *Bull. Chem. Soc. Jpn.*, 1982, **55**, 2212.
18. H. Horino, T. Ito and A. Yamamoto, *Chem. Lett.*, 1978, 17 and refs. cited therein.
19. T. Hiyama, Y. Okude, K. Kimura and H. Nozaki, *Bull. Chem. Soc. Jpn.*, 1982, **55**, 561.
20. Aldrich Chemical Co. Inc., at $83.00 for 25 g.
21. Aldrich Chemical Co. Inc., at $95.00 for 250 g.
22. P. G. M. Wuts and G. R. Callen, *Synth. Commun.*, 1986, **16**, 1833.
23. M. Guademar, *Tetrahedron*, 1976, **32**, 1689.
24. Y. Naruta, S. Ushida and K. Maruyama, *Chem. Lett.*, 1979, 919.
25. P. Girard, J.-L. Namy and H. B. Kagan, *J. Am. Chem. Soc.*, 1980, **102**, 2693.
26. C. T. Buse and C. H. Heathcock, *Tetrahedron Lett.*, 1978, 1685.
27. T. Hiyama, K. Kimura and H. Nozaki, *Tetrahedron Lett.*, 1981, **22**, 1037.
28. R. A. Benkesser, *Synthesis*, 1971, 347; M. Schlosser and M. Stahle, *Angew. Chem., Int. Ed. Engl.*, 1980, **19**, 487.
29. V. Rautenstrauch, *Helv. Chim. Acta*, 1974, **57**, 496.
30. R. Benn, H. Grondey, H. Lehmkuhl, H. Nehl, K. Angermund and C. Kruger, *Angew. Chem., Int. Ed. Engl.*, 1987, **26**, 1279.
31. D. Seebach and L. Wilder, *Helv. Chim. Acta*, 1982, **65**, 1972; R. Hanko and D. Hoppe, *Angew. Chem., Int. Ed. Engl.*, 1982, **21**, 372.
32. K. Mashima, H. Yasuda, K. Asami and A. Nakamura, *Chem. Lett.*, 1983, 219.
33. R. W. Hoffmann, *Angew. Chem., Int. Ed. Engl.*, 1982, **21**, 555; Y. Yamamoto and K. Maruyama, *Heterocycles*, 1982, **18**, 357; Y. Yamamoto, *Acc. Chem. Res.*, 1987, **20**, 243.
34. S. E. Denmark and E. Weber, *Helv. Chim. Acta*, 1983, **66**, 1655; S. E. Denmark and E. Weber, *J. Am. Chem. Soc.*, 1984, **106**, 7970 and refs. cited therein.
35. J. M. Coxon and G. S. C. Hii, *Aust. J. Chem.*, 1977, **30**, 835.
36. Y. Yamamoto, H. Yatagai and K. Maruyama, *J. Chem. Soc., Chem. Commun.*, 1980, 1072.
37. R. W. Hoffmann and H.-J. Zeiss, *J. Org. Chem.*, 1981, **46**, 1309.
38. Y. Yamamoto, H. Yatagai, Y. Naruta and K. Maruyama, *J. Am. Chem. Soc.*, 1980, **102**, 7107.
39. T. Hayashi, K. Kabeta, I. Hamachi and M. Kumada, *Tetrahedron Lett.*, 1983, **24**, 2865.
40. Y. Yamamoto and K. Maruyama, *Tetrahedron Lett.*, 1981, **22**, 2895.
41. F. Sato, K. Iida, S. Iijima, H. Moriya and M. Sato, *J. Chem. Soc., Chem. Commun.*, 1981, 1140.
42. S. Collins, W. P. Dean and D. G. Ward, *Organometallics*, 1988, **7**, 2289.
43. M. Kira, K. Sato and H. Sakurai, *J. Am. Chem. Soc.*, 1988, **110**, 4599; A. Hosomi, S. Kohra and Y. Tominaga, *J. Chem. Soc., Chem. Commun.*, 1987, 1517.
44. H. Nagaoka and Y. Kishi, *Tetrahedron*, 1981, **37**, 3873.
45. M. D. Lewis and Y. Kishi, *Tetrahedron Lett.*, 1982, **23**, 2343.
46. D. J. Cram and F. A. Abd Elhafez, *J. Am. Chem. Soc.*, 1952, **74**, 5828; D. J. Cram and K. R. Kopecky, *J. Am. Chem. Soc.*, 1959, **81**, 2748.
47. M. Cherest, H. Felkin and N. Prudent, *Tetrahedron Lett.*, 1968, 2199; N. T. Anh and O. Eisenstein, *Nouv. J. Chim.*, 1977, **1**, 61.
48. J. H. Morrison and H. S. Mosher, 'Asymmetric Organic Reactions', Prentice–Hall, New York, 1971, chap. 3; E. L. Eliel, 'Asymmetric Synthesis', Academic Press, New York, 1983, chap. 5, p. 125.
49. R. W. Hoffmann, *Angew. Chem., Int. Ed. Engl.*, 1987, **26**, 489.
50. D. Seebach and V. Prelog, *Angew. Chem., Int. Ed. Engl.*, 1982, **21**, 654; M. A. Brook, *J. Chem. Educ.*, 1987, **64**, 218.
51. T. Tanaka, Y. Oikawa, T. Hamada and O. Yonemitsu, *Chem. Pharm. Bull.*, 1987, **35**, 2209.
52. K. Suzuki, K. Tomooka, E. Katayama, T. Matsumoto and G. Tsuchihashi, *J. Am. Chem. Soc.*, 1986, **108**, 5221; K. Suzuki, E. Katayama, K. Tomooka, T. Matsumoto and G. Tsuchihashi, *Tetrahedron Lett.*, 1985, **26**, 3707.
53. J. Mulzer, P. de Lasalle and A. Freissler, *Liebigs Ann. Chem.*, 1986, 1152.
54. K. Suzuki, E. Katayama and G. Tsuchihashi, *Tetrahedron Lett.*, 1984, **25**, 2479; Y. Kobayashi, Y. Kitano and F. Sato, *J. Chem. Soc., Chem. Commun.*, 1984, 1329.
55. G. Fronza, P. Fuganti, G. Graselli, G. Petrocchi-Fantoni and C. Zirotti, *Chem. Lett.*, 1984, 335.
56. J. Mulzer, T. Schulze, A. Strecker and W. Denzer, *J. Org. Chem.*, 1988, **53**, 4098.
57. J. Jurczak, S. Pikul and T. Bauer, *Tetrahedron*, 1986, **42**, 447.
58. M. T. Reetz, *Angew. Chem., Int. Ed. Engl.*, 1984, **23**, 556.
59. N. T. Anh, *Top. Curr. Chem.*, 1980, **88**, 145.
60. Y. Yamamoto, T. Komatsu and K. Maruyama, *J. Organomet. Chem.*, 1985, **285**, 31.
61. H. Takeshita, N. Kato, K. Nakanishi, H. Tagoshi and T. Hatsui, *Chem. Lett.*, 1984, 1495; N. Kato, K. Nakanishi and H. Takeshita, *Bull. Chem. Soc. Jpn.*, 1986, **59**, 1109.
62. N. Kato, S. Tanaka and H. Takeshita, *Chem. Lett.*, 1986, 1989.
63. N. Kato, H. Kataoka, S. Ohbuchi, S. Tanaka and H. Takeshita, *J. Chem. Soc., Chem. Commun.*, 1988, 354.
64. H. Takeshita, T. Hatsui, N. Kato, T. Masuda and T. Tagoshi, *Chem. Lett.*, 1982, 1153.
65. W. C. Still and D. Mobilio, *J. Org. Chem.*, 1983, **48**, 4785.
66. M. F. Semmelhack, A. Yamashita, J. C. Tomesch and K. Hirotsu, *J. Am. Chem. Soc.*, 1978, **100**, 5565.
67. M. F. Semmelhack and E. S. C. Wu, *J. Am. Chem. Soc.*, 1976, **98**, 3384.

68. J. A. Marshall, S. L. Crooks and B. S. DeHoff, *J. Org. Chem.*, 1988, **53**, 1616; J. A. Marshall, B. S. DeHoff and S. L. Crooks, *Tetrahedron Lett.*, 1987, **28**, 527; J. A. Marshall and W. Y. Gung, *Tetrahedron Lett.*, 1988, **29**, 1657.
69. B. M. Trost and T. Sato, *J. Am. Chem. Soc.*, 1985, **107**, 719.
70. H. Shibuya, K. Ohashi, K. Kawashima, K. Hori, N. Murakami and I. Kitagawa, *Chem. Lett.*, 1986, 85.
71. B. Ledoussal, A. Gorgues and A. Le Coq, *J. Chem. Soc., Chem. Commun.*, 1986, 171; B. Ledoussal, A. Gorgues and A. Le Coq, *Tetrahedron*, 1987, **43**, 5841.
72. Y. Okuda, S. Nakatsukasa, K. Oshima and H. Nozaki, *Chem. Lett.*, 1985, 481; S. E. Drewes and R. F. A. Hoole, *Synth. Commun.*, 1985, **15**, 1067.
73. E. Ohler, K. Reininger and U. Schmidt, *Angew. Chem., Int. Ed. Engl.*, 1970, **9**, 457; G. P. Boldrini, D. Savoia, E. Tagliavini, C. Trombini and A. Umani-Ronchi, *J. Org. Chem.*, 1983, **48**, 4108.
74. L. S. Hegedus, S. D. Wagner E. L. Waterman and K. Sirala-Hansen, *J. Org. Chem.*, 1975, **40**, 593.
75. P. Auvray, P. Knochel and J. F. Normant, *Tetrahedron Lett.*, 1986, **27**, 5091.
76. K. Takai, K. Nitta and K. Utimoto, *Tetrahedron Lett.*, 1988, **29**, 5263.
77. M. T. Reetz, M. Hüllmann, W. Massa, S. Berger, P. Rademacher and P. Heymanns, *J. Am. Chem. Soc.*, 1986, **108**, 2405; S. Masamune, R. M. Kennedy, J. S. Petersen, K. N. Houk, Y. Wu, *J. Am. Chem. Soc.*, 1986, **108**, 7404.
78. P. Place, F. Delbecq and J. Gore, *Tetrahedron Lett.*, 1978, 3801.
79. C. Verniere, B. Cazes and J. Gore, *Tetrahedron Lett.*, 1981, **22**, 103.
80. P. Place, C. Verniere and J. Gore, *Tetrahedron*, 1981, **37**, 1359.
81. H. C. Brown and K. S. Bhat, *J. Am. Chem. Soc.*, 1986, **108**, 5919; R. W. Hoffmann and T. Herold, *Chem. Ber.*, 1981, **114**, 375; W. R. Roush and R. L. Halterman, *J. Am. Chem. Soc.*, 1986, **108**, 294; W. R. Roush and A. D. Palkowitz, *J. Am. Chem. Soc.*, 1987, **109**, 953; W. R. Roush, A. D. Palkowitz and M. A. J. Plamer, *J. Org. Chem.*, 1987, **52**, 316.
82. G. P. Boldrini, L. Lodi, E. Tagliavini, C. Tarasco, C. Trombini and A. Umani-Ronchi, *J. Org. Chem.*, 1987, **52**, 5447; G. P. Boldrini, E. Tagliavini, C. Trombini and A. Umani-Ronchi, *J. Chem. Soc., Chem. Commun.*, 1986, 685; J. Otera, Y. Yoshinaga, T. Yamaji, T. Yoshioka and Y. Kawasaki, *Organometallics*, 1985, **4**, 1213; T. Mukaiyama, N. Minowa, T. Oriyama and K. Narasaka, *Chem. Lett.*, 1986, 97.
83. B. Cazes, C. Verniere and J. Gore, *Synth. Commun.*, 1983, **13**, 73.
84. K. Takai, K. Kimura, T. Kuroda, T. Hiyama and H. Nozaki, *Tetrahedron Lett.*, 1983, **24**, 5281.
85. K. Takai, M. Tagashira, T. Kuroda, K. Oshima, K. Utimoto and H. Nozaki, *J. Am. Chem. Soc.*, 1986, **108**, 6048.
86. H. Jin, J. Uenishi, W. J. Christ and Y. Kishi, *J. Am. Chem. Soc.*, 1986, **108**, 5644.
87. M. F. Semmelhack, P. Helquist and J. D. Gorzynski, *J. Am. Chem. Soc.*, 1972, **94**, 9234; T. T. Tsou and J. K. Kochi, *J. Am. Chem. Soc.*, 1979, **101**, 7547.
88. F. A. Cotton and G. Wilkinson, 'Advanced Inorganic Chemistry', 5th edn., Wiley-Interscience, New York, 1988, p. 651.
89. W. J. Scott, G. T. Crisp and J. K. Stille, *J. Am. Chem. Soc.*, 1984, **106**, 4630.
90. S. Cacchi, E. Morera and G. Ortar, *Tetrahedron Lett.*, 1984, **25**, 2271.
91. S. H. Cheon, W. J. Christ, L. D. Hawkins, H. Jin, Y. Kishi and M. Taniguchi, *Tetrahedron Lett.*, 1986, **27**, 4759.
92. Y. Kishi, *Chem. Scr.*, 1987, **27**, 573 and refs. therein.
93. P. G. Geokjian, T.-C. Wu, H.-Y. Kang and Y. Kishi, *J. Org. Chem.*, 1987, **52**, 4823.
94. U. C. Dyer and Y. Kishi, *J. Org. Chem.*, 1988, **53**, 3383.
95. T.-C. Wu, P. G. Geokjian and Y. Kishi, *J. Org. Chem.*, 1987, **52**, 4819; S. A. Babirad, Y. Wang, P. G. Geokjian and Y. Kishi, *J. Org. Chem.*, 1987, **52**, 4825; Y. Wang, P. G. Geokjian, D. M. Ryckman and Y. Kishi, *J. Org. Chem.*, 1988, **53**, 4151; W. H. Miller, D. M. Ryckman, P. G. Geokjian, Y. Wang and Y. Kishi, *J. Org. Chem.*, 1988, **53**, 5580.
96. S. L. Schreiber and H. V. Meyers, *J. Am. Chem. Soc.*, 1988, **110**, 5198.
97. M. Rowley and Y. Kishi, *Tetrahedron Lett.*, 1988, **29**, 4909.
98. K. Takai, T. Kuroda, S. Nakatsukasa, K. Oshima and H. Nozaki, *Tetrahedron Lett.*, 1985, **26**, 5585.
99. J.-E. Dubois, G. P. Axiotis and E. Bertounesque, *Tetrahedron Lett.*, 1985, **26**, 4371.
100. Y. Okude, T. Hiyama and H. Nozaki, *Tetrahedron Lett.*, 1977, 3829.
101. T. Kauffmann, R. Konig, C. Pahde and A. Tannert, *Tetrahedron Lett.*, 1981, **22**, 5031.
102. T. Kauffmann, A. Hamsen and C. Beirich, *Angew. Chem., Int. Ed. Engl.*, 1982, **21**, 144.
103. S. Nakatsukasa, K. Takai and K. Utimoto, *J. Org. Chem.*, 1986, **51**, 5045.
104. R. C. Ronald, *Tetrahedron Lett.*, 1973, 3831.
105. D. R. Williams, K. Nishitani, W. Bennet and S.-Y. Sit, *Tetrahedron Lett.*, 1981, **22**, 3745; T. Okazoe, J. Hubino, K. Takai and H. Nozaki, *Tetrahedron Lett.*, 1985, **26**, 5581; J. Hubino, T. Okazoe, K. Takai and H. Nozaki, *Tetrahedron Lett.*, 1985, **26**, 5579.
106. P. Knochel and J. F. Normant, *Tetrahedron Lett.*, 1986, **27**, 1039.
107. K. Takai, K. Nitta and K. Utimoto, *J. Am. Chem. Soc.*, 1986, **108**, 7408.
108. K. Takai, Y. Kataoka, T. Okazoe and K. Utimoto, *Tetrahedron Lett.*, 1987, **28**, 1443.
109. T. Okazoe, K. Takai and K. Utimoto, *J. Am. Chem. Soc.*, 1987, **109**, 951.
110. D. Dodd and M. D. Johnson, *J. Chem. Soc. A*, 1968, 34.
111. F. Bertini, P. Grasselli, G. Zubiani and G. Cainelli, *Tetrahedron*, 1970, **26**, 1281.

1.7
Organozinc, Organocadmium and Organomercury Reagents

PAUL KNOCHEL
University of Michigan, Ann Arbor, MI, USA

1.7.1 GENERAL CONSIDERATIONS

Organozinc compounds[1,2] are readily prepared by oxidative addition of zinc to alkyl, allylic or benzylic halides, or by transmetallation reactions. Cadmium organometallics are prepared in similar ways, but show a lower thermal stability. Zinc and cadmium derivatives show a much lower reactivity and, consequently, a higher chemoselectivity than their lithium and magnesium counterparts. Organozinc halides containing various important classes of organic functional groups can be prepared in high yields and used, after transmetallation to more reactive copper or palladium organometallics, to form new carbon–carbon bonds. The addition of organozinc compounds to aldehydes and ketones is of considerable synthetic utility. Thus, alkyl- and aryl-zinc and -cadmium reagents add to aldehydes in the presence of Lewis acid catalysts with excellent chemoselectivity. If a chiral catalyst is used, very high enantioselectivities can be achieved. The addition of the more reactive allylic zinc and cadmium compounds to aldehydes and ketones does not require a catalyst and proceeds with high yields, good regioselectivity and, in some cases, excellent diastereoselectivity. The easy preparation of allylic zinc halides, combined with their high reactivity, makes them ideal nucleophilic allylating reagents.

1.7.2 PREPARATION OF ORGANOZINC REAGENTS

1.7.2.1 Introduction

The carbon–zinc bond,[1,2,3] which is a moderately strong carbon–metal bond (bond energy of Me_2Zn = 42 kcal mol^{-1}; 1 cal = 4.18 J), has a high covalent character (85%) and thus is relatively unreactive toward most organic electrophiles. However, this low reactivity allows for the preparation of a wide range of highly functionalized organozinc compounds. The presence of empty *sp*-orbitals of low energy at the zinc atom makes these organometallics very sensitive toward proton sources (such as water and alcohols) and toward oxygen. The two most important classes of zinc organometallics are organozinc halides (RZnX; **1**) and diorganozincs (R_2Zn; **2**; see Scheme 1). These are prepared either by the oxidative addition of organic halides to zinc metal or by transmetallation reactions. The organometallics (**1**) and (**2**) show different reactivities and selectivities in addition reactions to carbonyl compounds.

$$\text{Oxidative addition} \quad 2RX + 2Zn \longrightarrow \underset{(\mathbf{1})}{2RZnX} \xrightarrow[-ZnX_2]{\Delta} \underset{(\mathbf{2})}{R_2Zn}$$

$$\text{Transmetallation} \quad RM + ZnX_2 \xrightarrow[-MX]{} \underset{(\mathbf{1})}{RZnX} \xrightarrow[-MX]{RM} \underset{(\mathbf{2})}{R_2Zn}$$

Scheme 1

1.7.2.2 Preparation by Oxidative Addition

Zinc organometallics can be prepared by heating an alkyl halide/zinc mixture without solvent.[1] This procedure initially affords an alkylzinc iodide of type (**1**). Further heating and subsequent distillation converts (**1**) into the corresponding dialkylzinc (**2**; see Scheme 1). A zinc–copper couple has to be used to make the reaction reproducible, and only alkyl iodides or a mixture of alkyl iodide and the corresponding bromide are suitable substrates. This method is successful for low boiling point dialkylzinc reagents (dimethylzinc to dipentylzinc). The preparation of higher analogs requires high distillation temperatures, leading to substantial decomposition of the organometallic and to difficult separations from Wurtz coupling products and unreacted alkyl iodides.[1] The reaction can be performed under much milder conditions and with more elaborate organic halides in the presence of a solvent such as ether, ethyl acetate/toluene, 1,2-dimethoxyethane, THF, DMF, DMSO or HMPA. An alkyl halide is always less readily converted into a zinc organometallic than into the corresponding magnesium organometallic. Only alkyl iodides and activated organic bromides (benzylic, allylic or propargylic) can be used successfully. Allylic or benzylic chlorides can only be converted to the zinc derivative in very polar solvents, such as DMSO,[4] or in THF at high temperatures (50–60 °C).[5] Very mild reaction conditions can be achieved if the zinc metal has been activated. Several procedures for the activation of zinc metal have been developed.[6] The reduction of zinc chloride with various reagents such as potassium,[6b,e] potassium–graphite[6c] (C_8K) or lithium in the presence of naphthalene[6d] affords a very reactive zinc powder, which allows otherwise impossible reactions (see Scheme 2).[6d] Activation of zinc by ultrasound wave irradiation[6a] or by the use of active zinc slurries prepared by metal vapor techniques[7] has also been reported. A very convenient activation[8] is realized by the treatment of cut zinc foil or zinc dust with 4 mol % of 1,2-dibromoethane, then with 3 mol % of TMS-Cl. Under these conditions, a wide range of functionalized iodides can be converted, in THF, to the corresponding organozinc iodides in high yields (85–95%; see equation 1). Primary alkyl iodides react between 30 and 45 °C, whereas most secondary alkyl iodides are smoothly converted into the corresponding zinc derivatives between 25 and 30 °C. The tolerance of functional groups in organozinc halides is noteworthy. The zinc organometallics (**3**)–(**13**) have been prepared in THF,[5,8,11,13] benzene/*N*,*N*-dimethylacetamide[9,12,16] or benzene/HMPA[10] in good yields. In polar solvents such as DMF, zinc inserts even into C(sp^2)—I bonds of aromatic and heteroaromatic iodides affording zinc reagents such as (**11b**).[5a] Iodomethylzinc iodide (**13**) was found to undergo a new 1,2-migration with a variety of copper derivatives NuCu, affording the methylene-homologated organocopper $NuCH_2Cu{\cdot}ZnI_2$.[18] Interestingly, the presence of the THF soluble copper salt $CuI{\cdot}2LiI$ allows iodomethylzinc iodide (**13**) to convert allylic bromides directly to homoallylic iodides in high yields (see Scheme 3).[17]

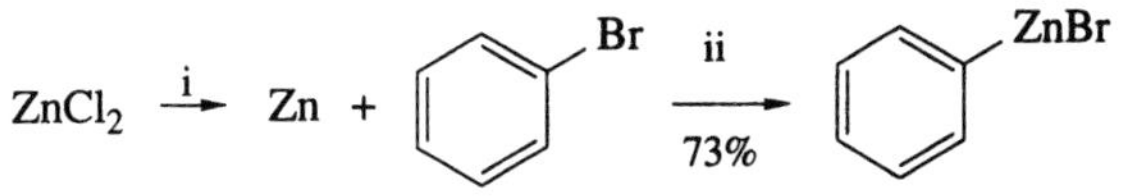

$ZnCl_2$ —i→ Zn + PhBr —ii (73%)→ PhZnBr

i, Li (2.1 equiv.), naphthalene (10 mol %), DME, 25 °C, 15 h; ii, DME, reflux, 10 h

Scheme 2

$$\text{FG—R—I} \xrightarrow[85\text{–}95\%]{\text{i}} \text{FG—R—ZnI} \qquad (1)$$

FG = OCOR, $CONR_2$, CO_2R, $N(COR)_2$, Cl, $Si(OR)_3$, $P(O)(OR)_2$, SR, S(O)R, $S(O)_2R$

i, Zn (1.5–2.5 equiv.) pretreated with 4 mol % of 1,2-dibromoethane, then 3 mol % of Me_3SiCl, addition of the alkyl iodide in THF (2–3 M solution) at 25–45 °C, then 1–12 h at 30–45 °C

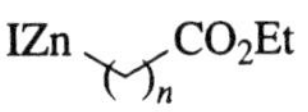

(**3**) n = 2, 3 or 4[8–12] 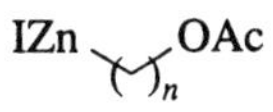 (**4**) n = 3, 4 or 5[5,11] 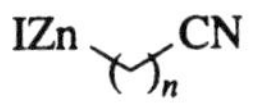 (**5**) n = 2 or 3[8,11,13] 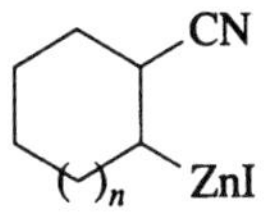 (**6**) n = 0 or 1[14]

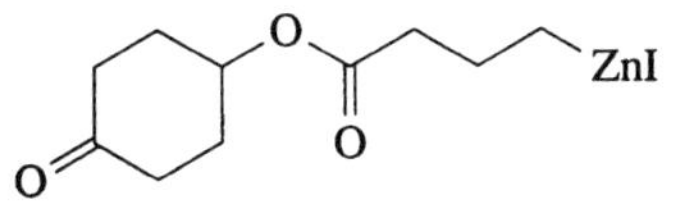

(**7**)[8] 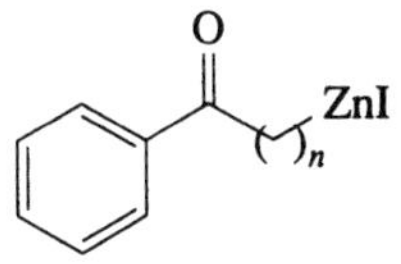 (**8**) n = 2–6[15] 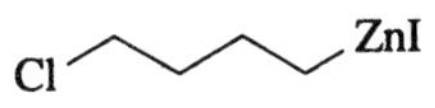 (**9**)[11,16] 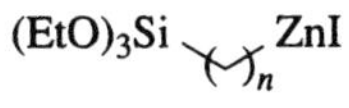 (**10**) n = 2 or 3[5]

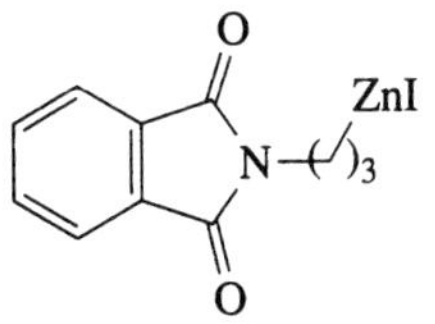

(**11a**)[10] 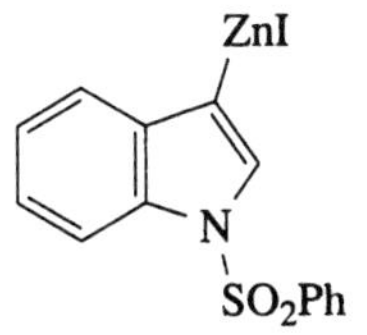 (**11b**)[5] 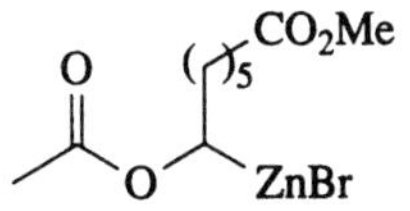 (**12**)[5,17b] 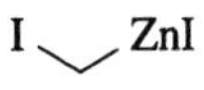 (**13**)[5,17,18]

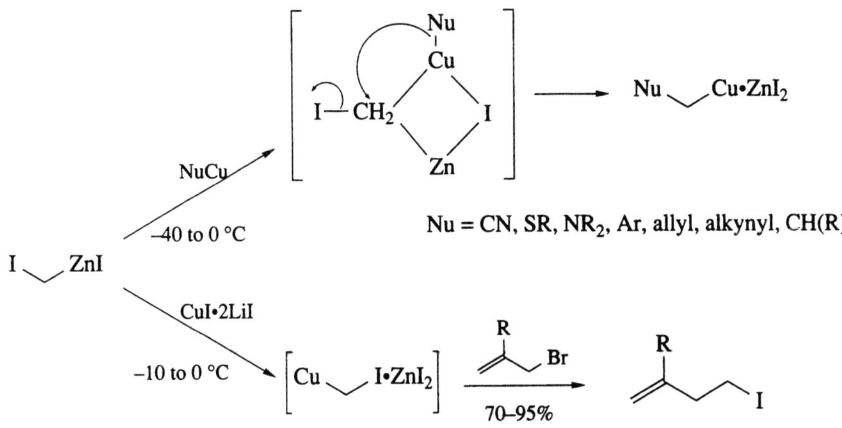

Nu = CN, SR, NR_2, Ar, allyl, alkynyl, CH(R)CN

Scheme 3

The same methodology has been used to prepare various functionalized benzylic zinc organometallics,[19] which are not available by other methods (see equation 2).[20] In the case of electron rich aromatic bromomethyl derivatives, Wurtz coupling can become a major reaction pathway. This side reaction is generally avoided by using the corresponding benzylic chloride. The formation of the zinc reagent[5b] must then be performed at 45 °C instead of 5 °C (see equation 3). *ortho*-Bis(trimethylsilyl)aminobenzylic zinc organometallics can be easily prepared and used for the synthesis of heterocycles such as indoles (see equation 4).[19b]

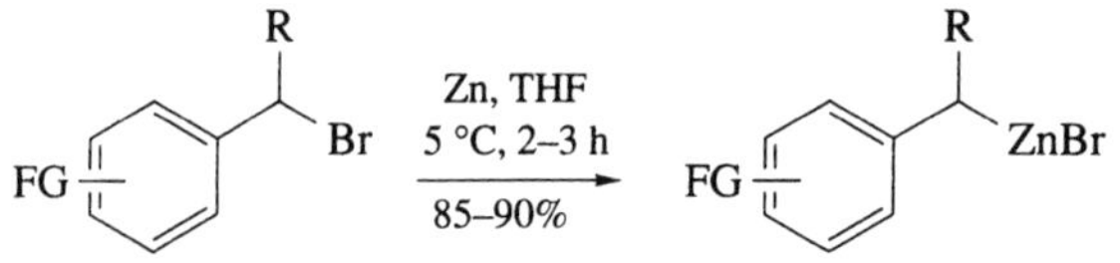

Zn, THF; 5 °C, 2–3 h; 85–90% (2)

FG = COR, OAc, OMe, CN, Cl, I, $N(SiMe_3)_2$

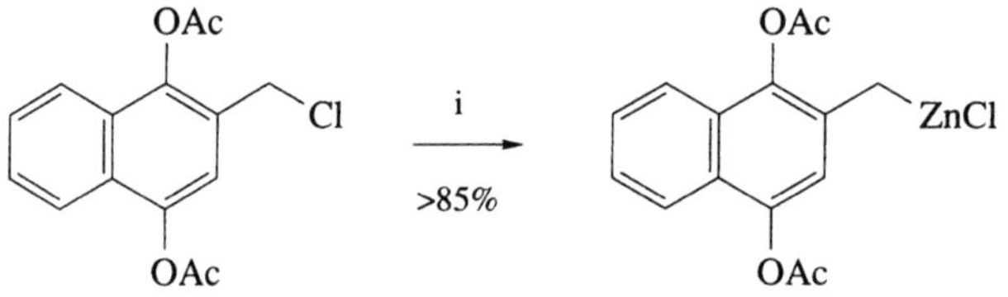

i; >85% (3)

i, Zn dust (2 equiv.), THF/DMSO (4:1), 45 °C, 4 h

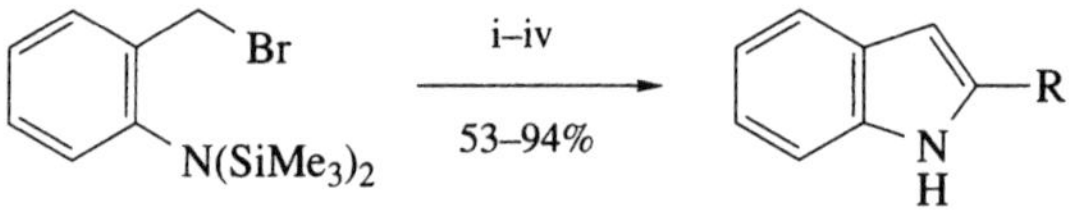

i–iv; 53–94% (4)

i, Zn, THF, 0 °C, 5 h; ii, CuCN, 2LiCl, –78 °C to –20 °C, 5 min;
iii, RCOCl, –20 °C, 14 h; iv, aqueous work-up

Various allylic zinc compounds, such as (**14**)–(**16**), can be prepared in good yields from the corresponding allylic bromide and zinc in THF. The tendency to form Wurtz coupling products increases with the number of substituents of the allylic bromide, with the presence of electron-donating groups at the double bond and with higher reaction temperatures. Thus, whereas (**15**) can be prepared[21] in high yields at 25 °C, the preparation of (**14**) has to be performed below 10 °C[22] and (**16**) can only be obtained in good yield[5c] if prepared below –5 °C. Cinnamylzinc bromide has to be prepared at –15 °C.[22a]

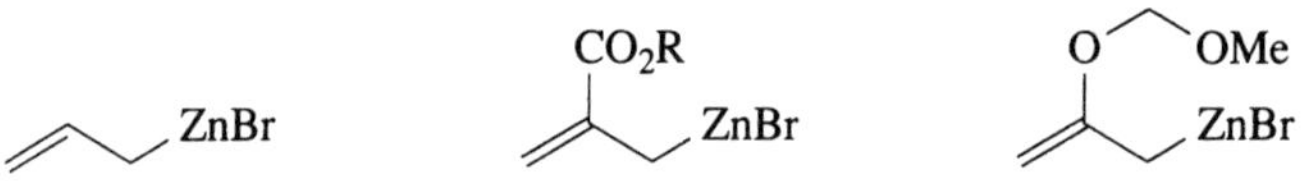

(**14**) (**15**) (**16**)

1.7.2.3 Preparation by Transmetallation Reactions

Grignard reagents[1,2,3,23] have been used extensively to prepare alkylzinc halides (**1**; see equation 5) and to some extent to prepare dialkylzinc reagents not available by oxidative addition reactions, such as di-*t*-butylzinc[24] or divinylzinc (see Scheme 4).[25] Although moderate yields are often obtained,[1] some optically active dialkylzinc reagents have been prepared in 49–81% yield.[1,26]

$$RMgX + ZnX_2 \longrightarrow \underset{(\mathbf{1})}{RZnX} + MgX_2 \qquad (5)$$

Compared to Grignard reagents, lithium organometallics[27] have been used less frequently for the preparation of organozinc compounds. Their high reactivity[28] or their easy availability[29,30] can, however, be

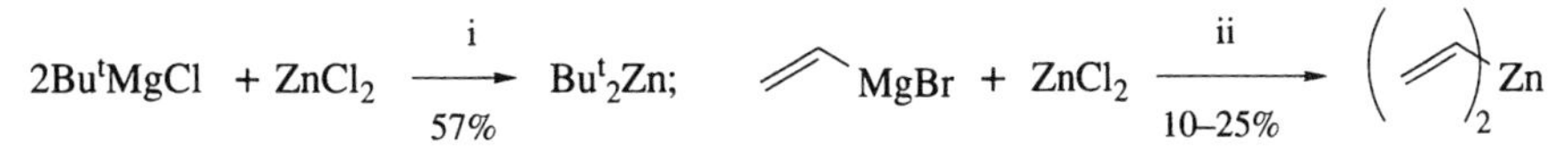

$$2Bu^tMgCl + ZnCl_2 \xrightarrow[57\%]{i} Bu^t_2Zn; \quad CH_2{=}CHMgBr + ZnCl_2 \xrightarrow[10\text{–}25\%]{ii} (CH_2{=}CH)_2Zn$$

i, ether, 25 °C, then distillation; ii, THF, 55 °C, 12 h, then distillation

Scheme 4

useful for the synthesis of several functionalized zinc organometallics, such as (**17**) and (**18**) in Scheme 5. Trialkylaluminum compounds readily react with various zinc salts affording dialkylzinc derivatives in good yields,[1,31] but only one alkyl group per molecule of R_3Al is transferred to zinc (see equation 6).[31a]

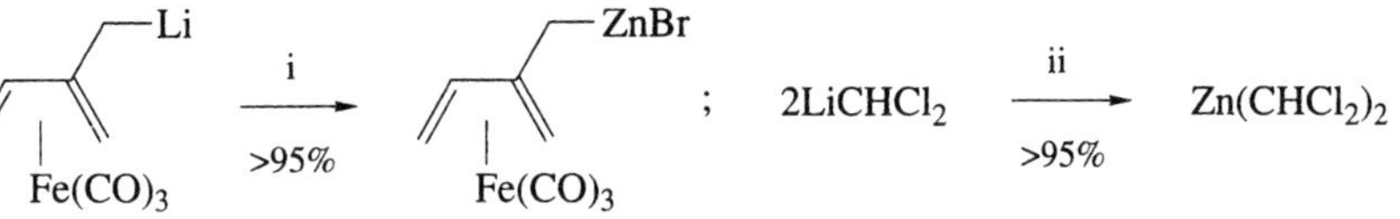

$$\text{(diene-CH}_2\text{Li)Fe(CO)}_3 \xrightarrow[>95\%]{i} \text{(diene-CH}_2\text{ZnBr)Fe(CO)}_3 \textbf{ (17)}; \quad 2LiCHCl_2 \xrightarrow[>95\%]{ii} Zn(CHCl_2)_2 \textbf{ (18)}$$

i, $ZnBr_2$ (1.1 equiv.), THF, –78 to 25 °C; ii, $ZnCl_2$ (0.5 equiv.), THF, –74 to 25 °C

Scheme 5

$$2Me_3Al + Zn(OAc)_2 \xrightarrow[88\%]{i} Me_2Zn + Me_2AlOCOMe \quad (6)$$

i, decalin, –10 to 10 °C, 2 h, then distillation

In contrast to organoaluminum compounds, the reaction with organoboranes has good synthetic potential.[32] The addition of a triallyl- or tribenzyl-borane to dimethylzinc furnishes, under mild conditions, a diallyl- or dibenzyl-zinc compound and trimethylborane (b.p. –20 °C), which escapes from the reaction mixture and rapidly drives the reaction to completion (see equation 7). Diorganomercury reagents react with zinc at higher temperatures (>100 °C), giving zinc organometallics in satisfactory to good yields.[1,33] Miscellaneous methods, such as the insertion of diazomethane into zinc halides,[34] the electrolysis of alkyl halides,[35] the opening of siloxycyclopropanes,[12b,c] a bromide–zinc exchange reaction,[36] the reduction of π-allylpalladium complexes[37] and the metallation of acidic hydrocarbons,[1,38] have been reported.

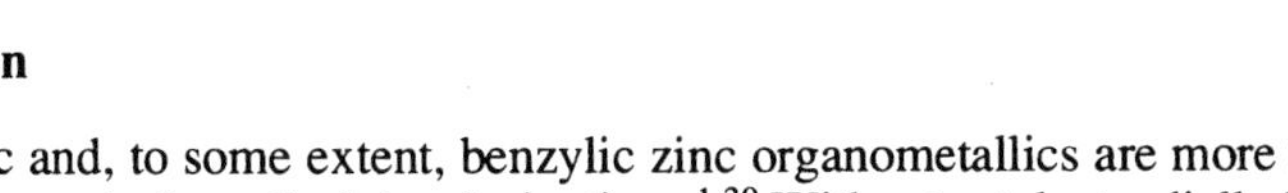

$$(CH_2{=}CHCH_2)_3B + Me_2Zn \xrightarrow[100\%]{0\ °C} (CH_2{=}CHCH_2)_3Zn + BMe_3 \quad (7)$$

1.7.3 ADDITION REACTIONS OF ORGANOZINC REAGENTS

1.7.3.1 Introduction

Allylic, propargylic and, to some extent, benzylic zinc organometallics are more reactive toward addition to carbonyl compounds than alkylzinc derivatives.[1,39] Without catalysts, dialkylzinc reagents display a very low reactivity toward aldehydes and are unreactive toward ketones. The addition of Lewis acids (electrophilic catalysis), such as magnesium or zinc bromide,[40] boron trifluoride etherate,[10] chlorotrimethylsilane,[41] and trimethylsilyl triflate,[42] or the addition of Lewis bases (nucleophilic catalysis) such as tetraalkylammonium halides[43] and amino alcohols,[44,45] can strongly enhance the rate of the addition either by activating the aldehyde, leading to an intermediate of type (**19**), or by activating the organozinc compound by the formation of a more reactive zincate of type (**20**; see Scheme 6). A common side reaction[1,40,43] with zinc organometallics having β-hydrogens is the reduction of the carbonyl compound (see Scheme 7). The carbonyl addition can be favored over the reduction by the addition of tetraalkylammonium salts,[43] as shown in Table 1 (nucleophilic catalysis).

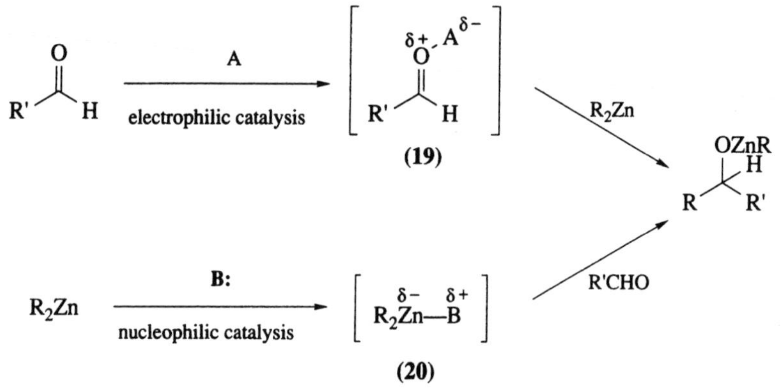

Scheme 6

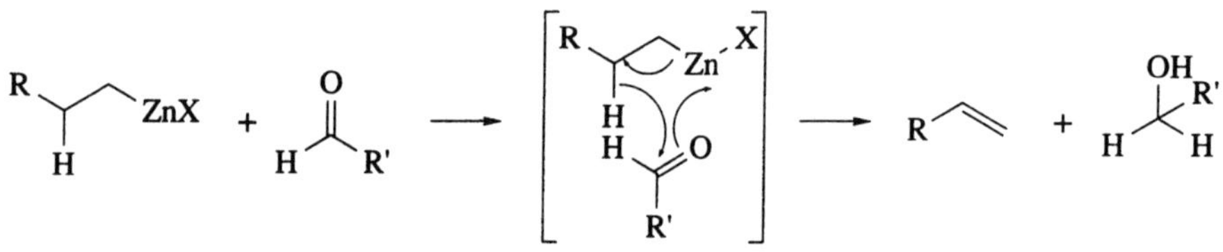

Scheme 7

Table 1 Influence of the Addition of Tetrabutylammonium Salts on the Reaction of Benzaldehyde with Diisopropylzinc in Ether[43]

Catalyst	*Reduction product (%) (benzyl alcohol)*	*Addition product (%) (1-phenylbutanol)*	*Relative rate*
None	44	56	1.0
Bu_4NI	23	77	1.1
Bu_4NBr	8	92	4.1
Bu_4NCl	6	94	8.9

1.7.3.2 Addition Reactions of Alkyl- and Aryl-zinc Reagents

Without the presence of a catalyst, the addition of alkylzinc organometallics (RZnX or R_2Zn) to aldehydes is unsatisfactory and leads to an appreciable amount of reduced product. For example, the reaction of diethylzinc and 4-chlorobenzaldehyde gives a mixture of 4-chlorobenzyl alcohol (45%) and 1-(4-chlorophenyl)propanol (38%).[44] The addition of only 5 mol % of magnesium bromide to the reaction mixture results in faster addition and higher yields.[40] Thus, the treatment of distilled Bu_2Zn with benzaldehyde affords a 15% yield of 1-phenylpentanol, whereas the same reaction in the presence of 2 equiv. of $MgBr_2$ produces 70% of the addition product.[40] Ketones usually do not react; benzophenone is reduced by diethylzinc, and only diphenylzinc gives the addition product under severe conditions (toluene, reflux, 30 h). Preparatively more useful is the reaction of ester-containing alkylzinc iodides with aldehydes in the presence of an excess of TMS-Cl and with DMA or NMP as a cosolvent (see equation 8).[41]

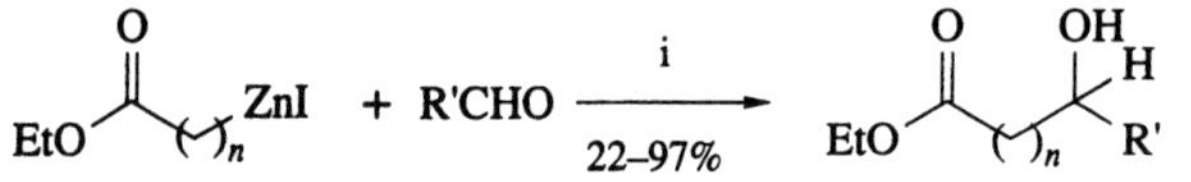

(8)

i, Me_3SiCl (3 equiv.), toluene, DMA, 25–60 °C, 3 h to 3 d

Far milder reaction conditions are possible if a transmetallation of the zinc organometallic (**21**) to the mixed copper–zinc derivative (**22**) is first performed and if the reaction is carried out in the presence of 2 equiv. of $BF_3{\cdot}OEt_2$.[10] A wide range of functional groups are tolerated in compounds (**21**) and (**22**), and high yields are usually obtained (68–93%; see Scheme 8), the reaction showing a good chemoselectivity. The treatment of a 1:1 mixture of benzaldehyde and acetophenone with the organozinc iodide (**23**; 2 h at

–30 °C) furnishes the acetoxy alcohol (**24**) in 86% isolated yield, with acetophenone recovered in 93% GLC yield (see equation 9). With an unsaturated aldehyde, such as cinnamaldehyde, a 1,2-addition reaction is observed in the presence of $BF_3·OEt_2$, but only the 1,4-addition product is obtained if the reaction is performed with TMS-Cl as an additive (see Scheme 9).[10]

FG—R—ZnI (**21**) —i→ FG—RCu(CN)ZnI (**22**) —ii (68–93%)→ FG—R–CH(OH)R'

ester, nitrile, enoate or imide

i, $CuCN•2LiCl$ (1.0 equiv.), THF, 0 °C, 10 min; ii, R'CHO (0.7–0.85 equiv.), $BF_3•OEt_2$ (2 equiv.), –78 °C, then 4–16 h at –30 °C

Scheme 8

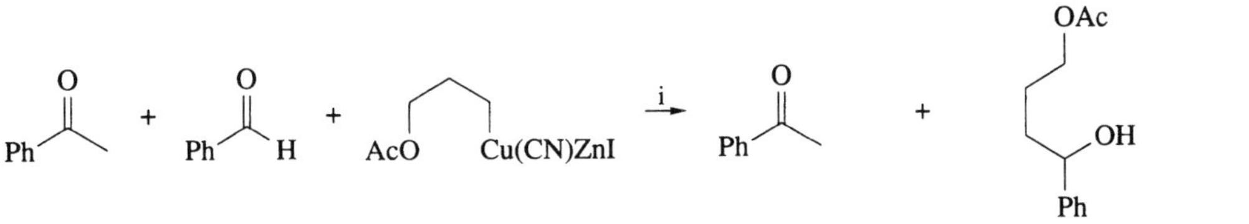

PhCOMe (1 equiv.) + PhCHO (1 equiv.) + AcO(CH₂)₃Cu(CN)ZnI (**23**) 1.4 equiv. —i→ PhCOMe 93% (GLC yield) + AcO(CH₂)₃CH(OH)Ph (**24**) 86% (isolated yield) (9)

i, $BF_3•OEt_2$ (2.8 equiv.), –78 to –30 °C, 2 h

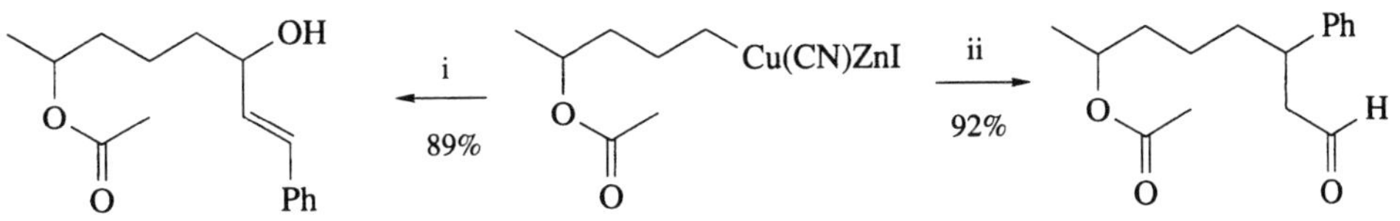

i, cinnamaldehyde, $BF_3•OEt_2$ (2 equiv.), –78 to –30 °C, then –30 °C, 4 h; (89%)
ii, cinnamaldehyde, Me_3SiCl (2 equiv.), –78 to 25 °C overnight (92%)

Scheme 9

A transmetallation to titanium derivatives[12,13,45–47] also promotes the addition of various zinc organometallics to carbonyl compounds. However, the functional group tolerance seems to be limited (see Scheme 10).[10,13,47] Interestingly, diethylzinc reacts with aromatic 1,2-diketones to give α-ethoxy ketones (see equation 10).[48]

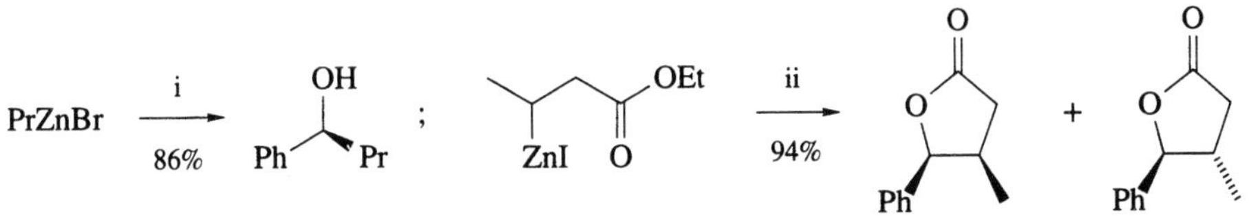

PrZnBr —i (86%)→ PhCH(OH)Pr ; EtO₂CCH₂CH(ZnI)CH₃ —ii (94%)→ lactones 99% + 1%

i, $ClTi(OPr^i)_3$ (1 equiv.), –30 °C, then PhCHO (1 equiv.), –20 to –25 °C, 2 h;[46]
ii, $ClTi(OPr^i)_3$ (1 equiv.), –30 °C, then PhCHO (0.75 equiv.), 0 °C, 2 h

Scheme 10

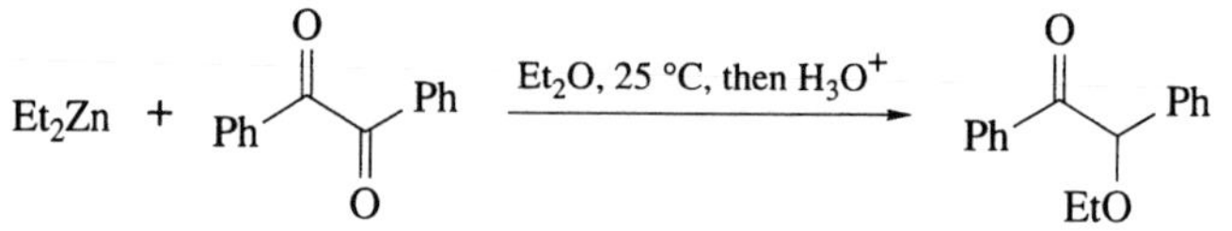

Et_2Zn + PhCOCOPh —(Et_2O, 25 °C, then H_3O^+)→ PhCOCH(OEt)Ph (10)

1.7.3.3 Addition Reactions of Allylic and Propargylic Zinc Reagents

The addition of allylic and propargylic zinc bromides to aldehydes and ketones proceeds readily[22,39,49,50] and is of high synthetic interest since the starting allylic and propargylic organometallics are easily prepared (see Scheme 11).[22] Several functionalized aldehydes or ketones can be used.[22,49,51] The reaction of allylzinc bromide with anthraquinone affords a *cis,trans* mixture of the diol (**25**) and produces only the 1,2-addition product with unsaturated ketones (see Scheme 12). Note that trialkylzincates, on the other hand, provide the 1,4-addition product in good yields.[52] Highly functionalized allylic zinc halides[29,53,54] can be prepared and afford, after addition to aldehydes or ketones, a direct approach to five-membered carbo- or hetero-cycles (see Scheme 13). A straightforward approach to the sex pheromone of the bark beetle (**26**) *via* the iron–diene complex (**27**) illustrates the synthetic utility of the Zn–Fe bimetallic isoprenoid reagent (**17**; see Scheme 14).

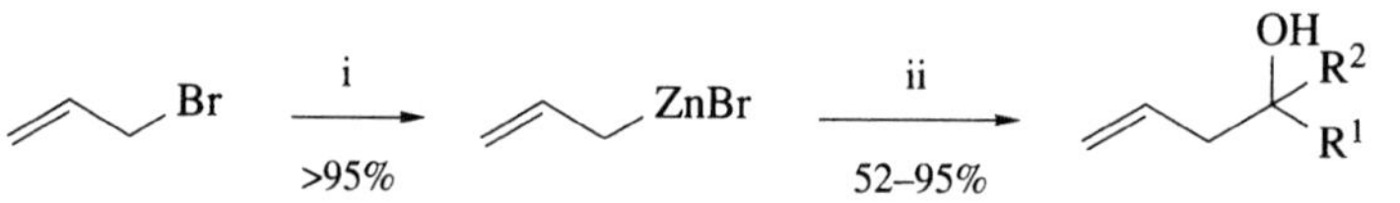

i, Zn (1 equiv.), THF, 10 °C, 3–4 h; ii, R^1COR^2 (aldehyde or ketone), 25 °C, several hours

Scheme 11

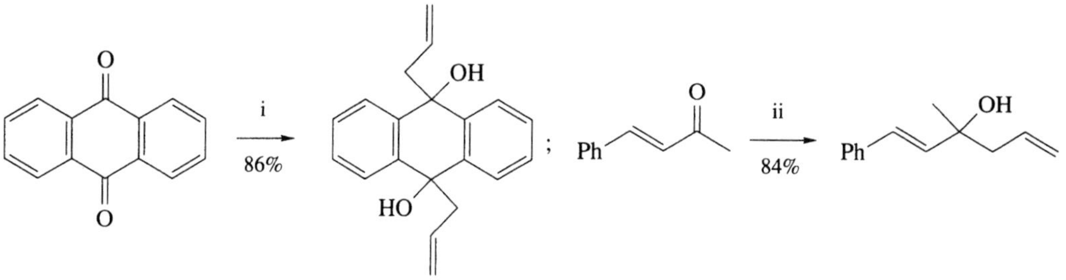

(**25**)

i, $H_2C{=}CHCH_2ZnBr$ (2.5 equiv.), THF, –5 to 0 °C, 2 h; ii, $H_2C{=}CHCH_2ZnBr$ (1.25 equiv.), 25 °C, several hours

Scheme 12

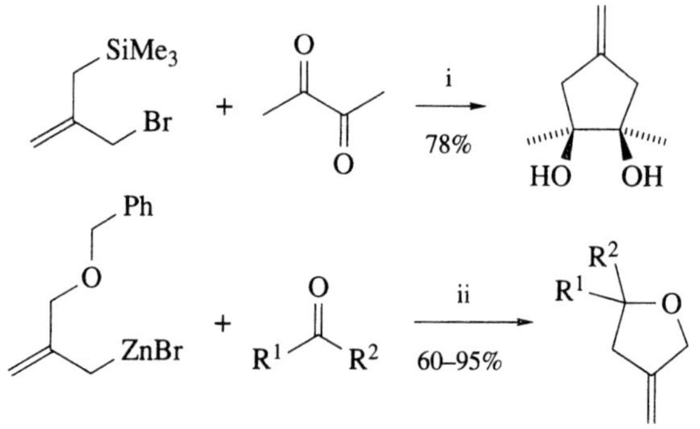

i, Zn, THF, then Bu_4NF, then H_3O^+;[53]
ii, THF, 25 °C, overnight, then $Pd(PPh_3)_4$ (5–10 mol%), 65 °C, 16–24 h

Scheme 13

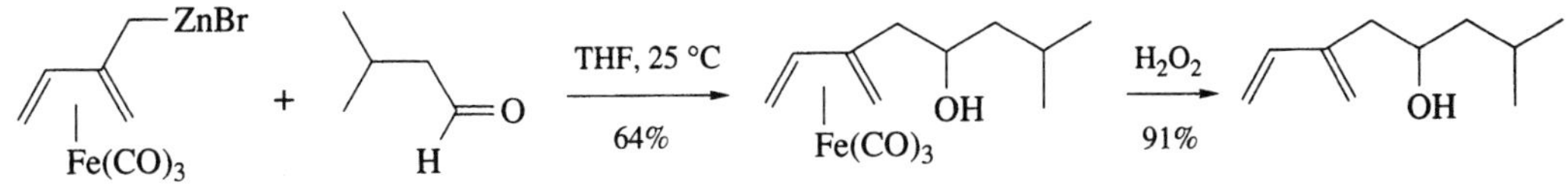

(17) (27) (26)

THF, 25 °C, 64%; H_2O_2, 91%

Scheme 14

The *in situ* generation of allylic zinc reagents in the presence of an electrophile (Barbier conditions)[55] can be advantageous compared to a conventional two-step procedure and allows several new synthetic possibilities, such as: (i) the generation of functionalized zinc organometallics of type (**28**), which add readily to various carbonyl compounds (see Scheme 15); and (ii) the generation of allylic zinc compounds in aqueous medium.[60] In this case, the reaction has a radical character[61] and does not proceed through a true zinc organometallic (see Scheme 16). The addition of substituted allylic zinc bromides proceeds with an allylic rearrangement *via* a cyclic six-membered transition state.[50,62] The new carbon–carbon bond is always formed from the most substituted end of the allylic system (see Scheme 17 and equation 11).

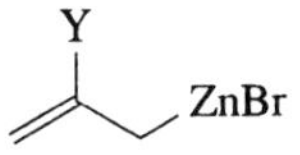

(**28**) Y = CO_2R,[56] SO_2R,[57] SOR,[42] $PO(OR)_2$,[58,56c] $SiMe_3$[59]

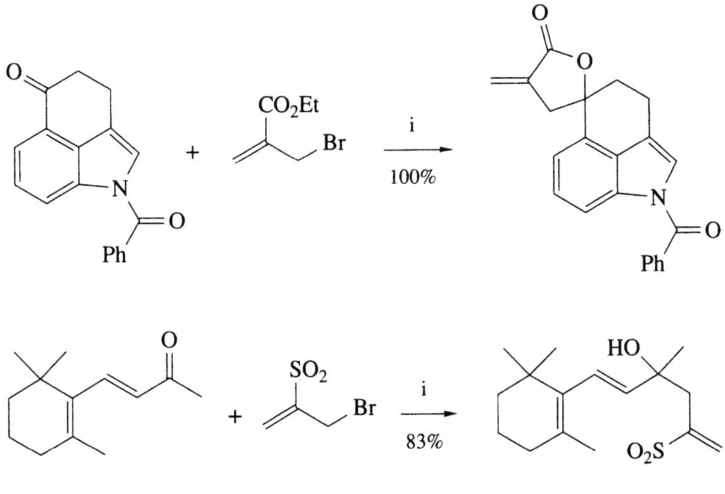

i, Zn, 40–45 °C, 1 h, then H_3O^+

Scheme 15

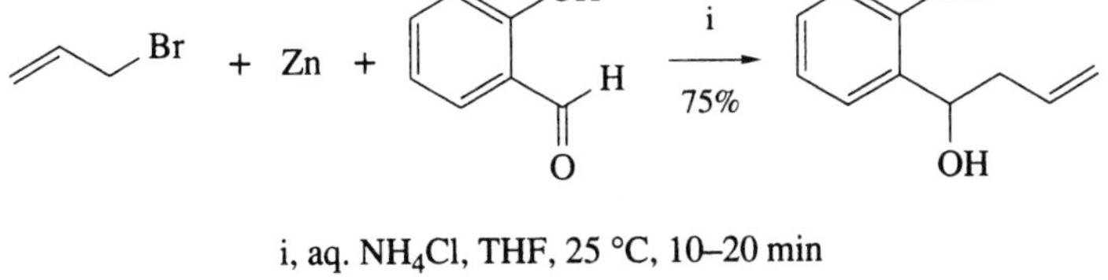

i, aq. NH_4Cl, THF, 25 °C, 10–20 min

Scheme 16

Under suitable reaction conditions,[63] the addition of allylic zinc bromides to ketones can be reversible. Thus, the reaction of diisobutyl ketone with 2-pentenylzinc bromide in THF affords, after a short reaction time, a mixture of (**29**) and (**30**) in a 44:56 ratio, whereas after 12 h, only the thermodynamically

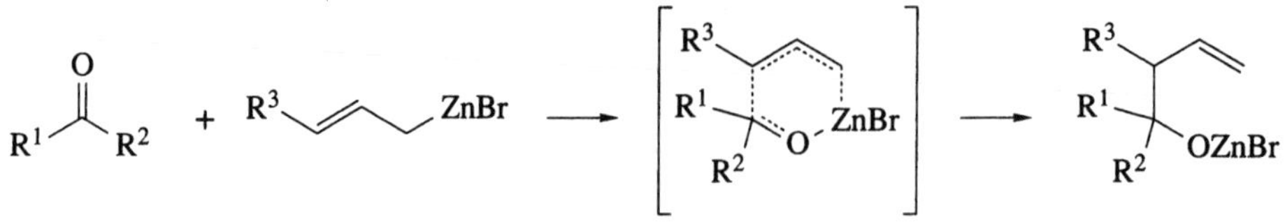

Scheme 17

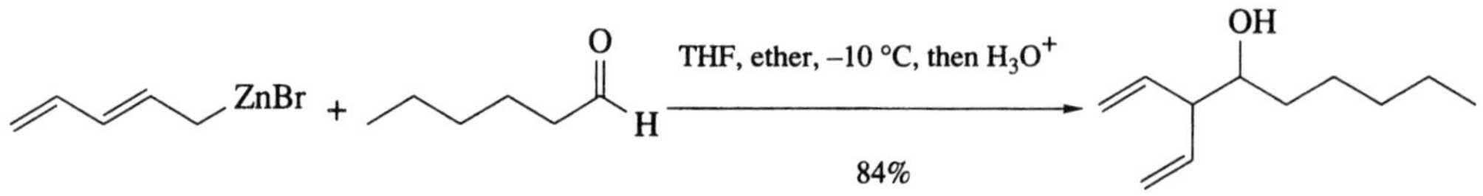

THF, ether, –10 °C, then H_3O^+

84%

(11)

more stable zinc alcoholate (**29**) is present (see equation 12). Propargylic bromides (**31**)[1,22,64,65,66] are readily converted to the corresponding zinc derivatives (Zn, THF, –10 to –5 °C), which exist as allenic organometallics of type (**32**) if R^1 = H or Ph and R^2 = alkyl or H, and as a mixture of (**32**) and (**33**) if R^1 = alkyl and R^2 = H. Their reaction with carbonyl compounds generally affords a mixture of homopropargylic and allenic alcohols of type (**34**) and (**35**) respectively in which the former predominates (see Scheme 18). The ratio between (**34**) and (**35**) strongly depends on the nature of the substituents R^1 and R^2 of (**31**), on the solvent, and/or on the carbonyl group used. With allenylzinc bromides in which R^1 = H and R^2 = alkyl, an almost exclusive formation of alkynic alcohols of type (**34**) is observed.[66] In the presence of polar cosolvents like HMPA, a reversible addition to ketones is observed.[65d]

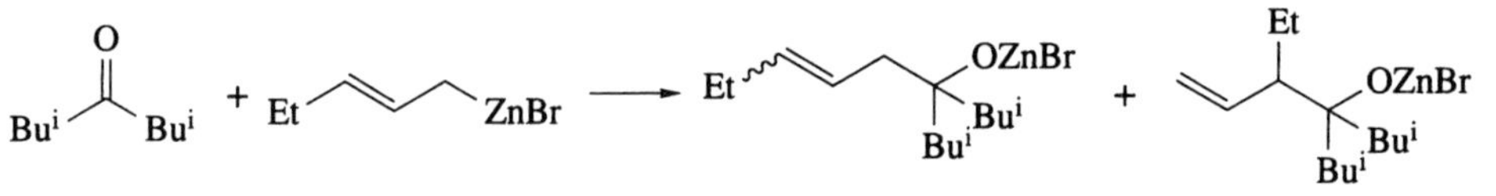

(12)

	(29)	(30)
Reaction time = 5 min	44%	56%
Reaction time = 12 h	100%	0%

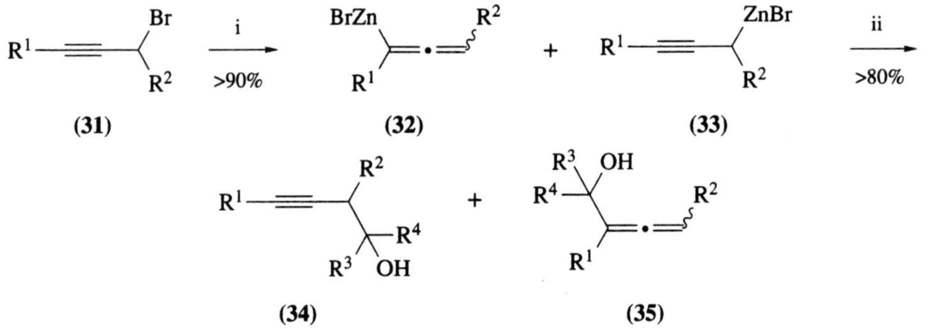

i, Zn (1 equiv.), THF, –10 to –5 °C; ii, R^3COR^4, 1 h, 0 °C

Scheme 18

1.7.3.4 Diastereoselective Addition Reactions

Zinc and cadmium reagents usually add to aldehydes and ketones with good diastereoselectivity compared to Grignard reagents. Dicrotyl-zinc and -cadmium reagents[49] add to sterically hindered aldehydes to produce mostly the *anti*-alcohols (**36**) *via* the cyclic transition state (**37**; see Scheme 19). 2-Substituted allylic zinc bromides display a *syn* diastereoselectivity. The bromo sulfones (**38**) react under Barbier conditions with aldehydes to give the *syn*-alcohols (**39**) with a high selectivity[57b] *via* a transition state such as (**40**), which minimizes the steric interactions between R^1, R^2 and the $PhSO_2$ group (see Scheme 20).

Remarkable selectivities are observed in the addition of allenic zinc chlorides to aldehydes, affording *anti*-homopropargylic alcohols *via* a transition state of type (**41**) (see Scheme 21). An extension to 1-substituted trimethylsilylallenic zinc chlorides was also possible and gave the *anti*-alcohols (**42**) in 92–99% stereoisomeric purity.

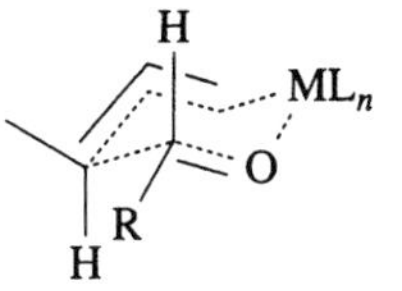

(**37**) 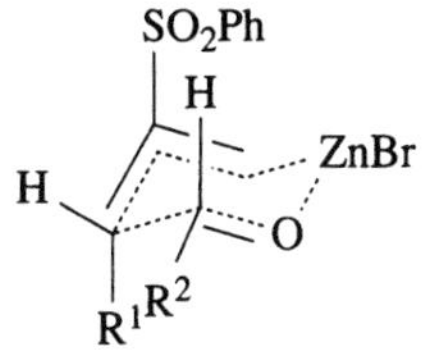 (**40**)

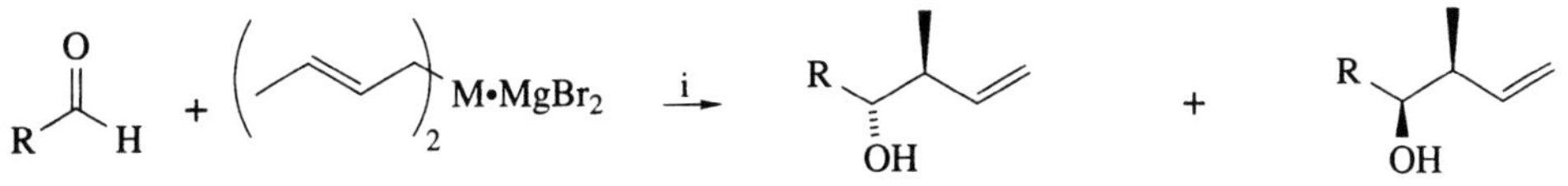

$RCHO + (CH_2{=}CHCH_2)_2M{\cdot}MgBr_2 \xrightarrow{i}$ (**36**) +

R	M	anti (%)	syn (%)
Pr^i	Mg	58	42
	Zn	70	30
	Cd	80	20
Bu^t	Mg	75	25
	Zn	84	16
	Cd	86	14

i, ether, –20 °C or –35 °C, 1 h, then H_3O^+

Scheme 19

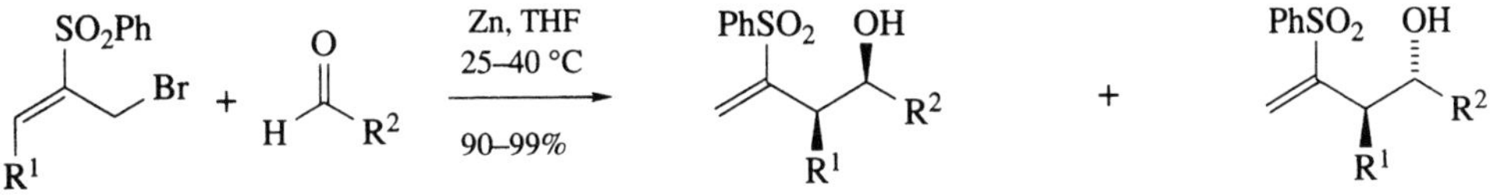

(**38**) + $HC(O)R^2$ $\xrightarrow[\text{90–99\%}]{\text{Zn, THF, 25–40 °C}}$ (**39**) +

R^1	R^2	syn (%)	anti (%)
Me	Ph	100	0
Me	c-C_6H_{11}	100	0
Me	C_5H_{11}	88	12
Pr	Ph	100	0
Pr	C_5H_{11}	67	33

Scheme 20

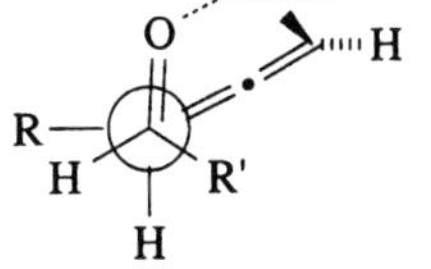

(**41**) 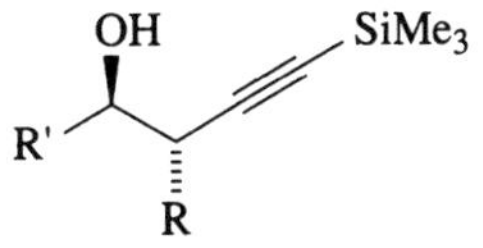 (**42**)

The addition of methylzinc (and cadmium) organometallics to chiral aldehydes[67] proceeds with low stereoselectivity and leads to a mixture of *syn*- and *anti*-alcohols (see Scheme 22). A better diastereoselectivity can be achieved by using the mixed copper–zinc reagents RCu(CN)ZnI.[10] In contrast, the more reactive diallylzinc reacts with several α-alkoxy aldehydes[68] of type (**43**) to give *anti* addition

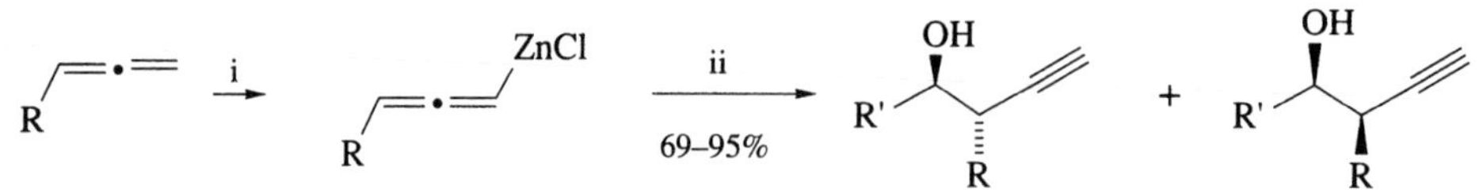

ZnCl

ii

69–95%

OH

R'

R

anti (96–99%)

+

OH

R'

R

syn (1–4%)

i, Bu^{t}Li, THF, –90 °C, 1 h, then $ZnCl_2$ in THF, –74 to –65 °C; ii, R'CHO, –74 to 25 °C, 1 h, then H_3O^+

Scheme 21

products (**44**) with over 80% *de* (diastereoisomeric excess). The addition follows Cram's rule[69] *via* a transition state such as (**45**). Zinc has a strong ability to complex with oxygen atoms[1,2,3] and this property can be used to perform several chelate-controlled additions to α-alkoxy aldehydes. Thus, diethylzinc adds in ether to the aldehyde (**46**), furnishing the *syn*-alcohol (**47**) as the major diastereoisomer (70% *de*) through a chelate-controlled transition state of type (**48**). The alcohol (**47**) could be converted into *exo*-brevicomin (**49**; see Scheme 23).[70]

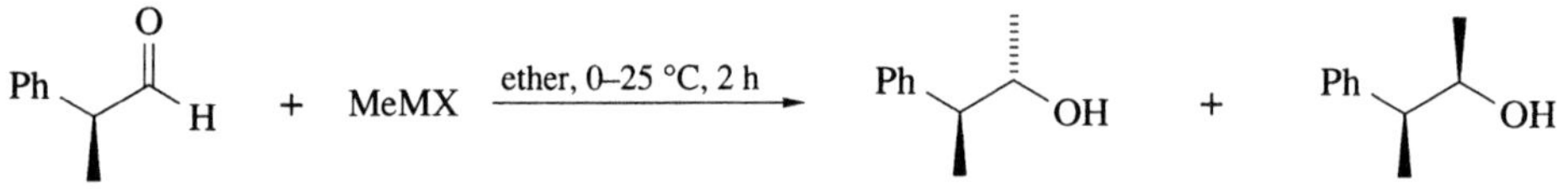

Ph + MeMX —(ether, 0–25 °C, 2 h)→ Ph ... OH + Ph ... OH

M	*X*	*anti* (%)	*syn* (%)
Mg	Br	69.5	30.5
Zn	Br	56.0	44.0
Cd	Br	62.2	37.8

Scheme 22

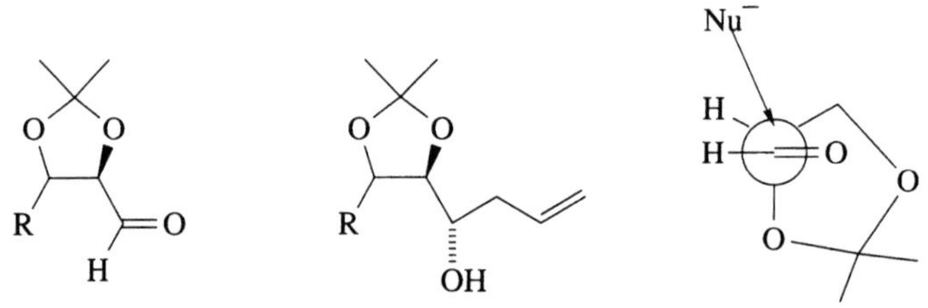

(**43**) (**44**) (**45**)

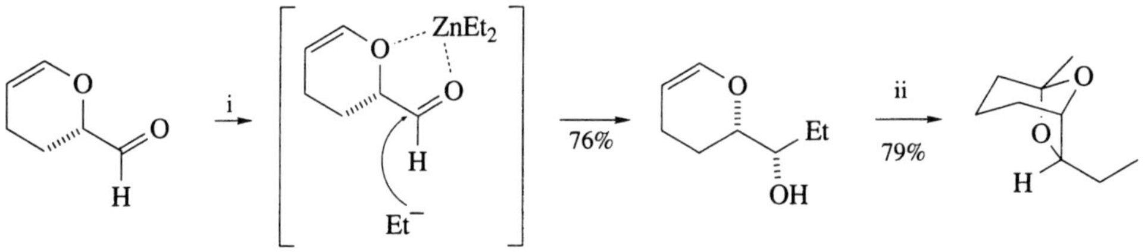

$ZnEt_2$

Et$^-$

76%

ii

79%

(**46**) (**48**) (**47**) (*syn*/*anti*:85/15) (**49**)

i, Et_2Zn, ether, 0–25 °C, 5 h; ii, TMEDA (0.25 equiv.), BuLi (2.5 equiv.), 0–25 °C, 12 h, then ether, TMEDA (2.5 equiv.), MeI (5 equiv.), 0–25 °C, 1 h, then H_3O^+

Scheme 23

In situ generated perfluoroalkylzinc iodides add to chromium tricarbonyl complexes of aromatic aldehydes with fair diastereoselectivity (44–66% *de*).[71] The low reactivity of R_2Zn toward ketones makes a stereochemical study rather difficult, since extensive reduction is observed. However, it has been found[72a] that methylzinc (and cadmium) derivatives give more axial attack with 4-*t*-butylcyclohexanone than the corresponding magnesium reagents, whereas allyl-[49] and propyl-zinc[72b] and -cadmium compounds mainly furnish the products derived from an equatorial attack (see Scheme 24). The low tendency of Pr_2Cd to reduce 4-*t*-butylcyclohexanone is noteworthy (addition/elimination ratio = 10:1 compared to 1.2–2.0:1 for propyl-zinc and –magnesium reagents). The reactive (–)-menthyl phenylglyoxalate gives α-substituted (–)-menthyl mandelates in good yields and with good stereoselectivity (71–88% *de*; see equation 13).[72c]

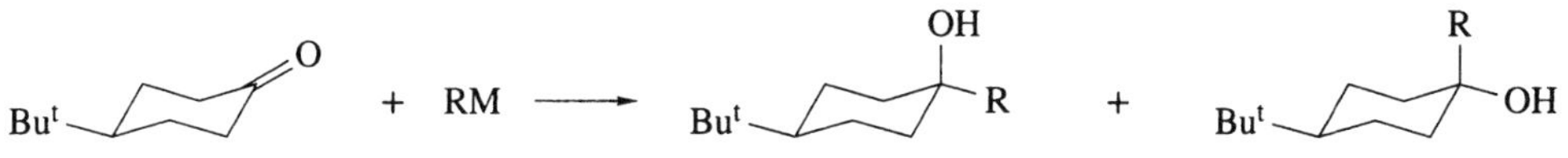

M	*Equatorial attack (%)*	*Axial attack (%)*
MeMgBr	68.4	31.6
$Me_2Zn \bullet MgBr_2$	46.5	53.5
$Me_2Cd \bullet MgI_2$	51.6	48.4
PrMgBr	69.0	31.0
$Pr_2Zn \bullet MgBr_2$	75.0	25.0
$Pr_2Cd \bullet MgBr_2$	80.0	20.0
$(Allyl)_2Mg$	44.5	55.5
$(Allyl)_2Zn$	84.0	16.0
$(Allyl)_2Cd$	77.5	22.5

Scheme 24

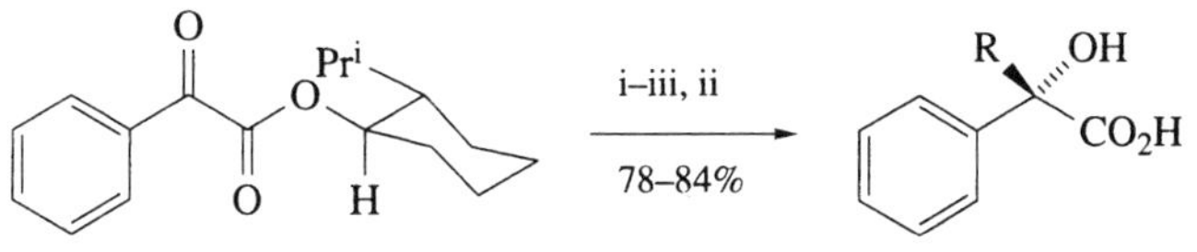

i–iii, ii; 78–84% (13)

(71–88% *de*)

i, R_2Zn, –78 °C, 3 h, then 25 °C, 1 h; ii, H_3O^+; iii, OH^-

1.7.3.5 Enantioselective Addition Reactions

The enantioselective addition of organometallics to aldehydes is a useful approach to optically active secondary alcohols.[52a,73] Diorganozinc reagents[73–84] add with excellent enantioselectivity to aldehydes in the presence of a chiral catalyst such as 1,2- or 1,3-amino alcohols (see equation 14 and Table 2). In most cases, diethylzinc has been used, but the reaction could be extended to some other dialkylzinc reagents and to divinylzinc.[25b] Alkylzinc halides afford secondary alcohols with a substantially lower enantiomeric excess.[82] Many aldehydes are good substrates,[25b,79] but the best results are usually obtained with aromatic aldehydes.[73–84]

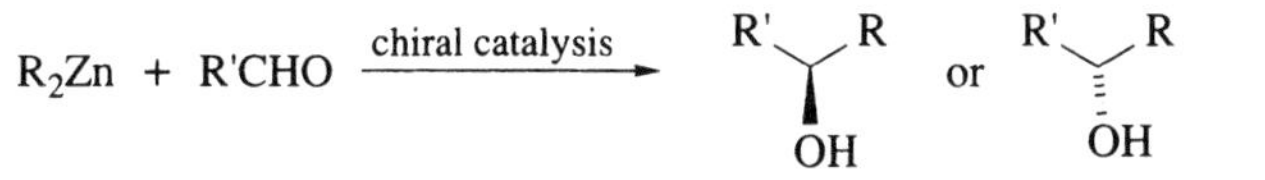

R_2Zn + R'CHO —chiral catalysis→ R'CH(OH)R or R'CH(OH)R (14)

The mechanism of the reaction has been investigated in detail and it has been established that 2 equiv. of Et_2Zn per molecule of the catalyst are required to observe an addition. If a 1:1 ratio is used, merely the reduction of the aldehyde is observed.[76] Only the ethyl groups coming from the second equivalent of Et_2Zn are transferred to the aldehyde. Thus, the sequential addition of $(C_2H_5)_2Zn$ and $(C_2D_5)_2Zn$ to the catalyst (**64**; see Table 2), followed by benzaldehyde in a 1:1:1:1 ratio affords deuterated 1-phenylpropanol, whereas the addition of $(C_2D_5)_2Zn$ first, followed by $(C_2H_5)_2Zn$ gave only nondeuterated products.[25b] A six-membered bimetallic transition state of type (**66**) has been proposed[73,76b,80,82,85] in which the first equivalent of the diethylzinc (Zn_A) activates the carbonyl group toward nucleophilic attack (electrophilic catalysis), whereas the second equivalent of Et_2Zn, which is made more nucleophilic by the coordination of a donor ligand (nucleophilic catalysis), transfers the ethyl group.[85] The catalytic cycle[73] of Scheme 25 is in agreement with most experimental results. The starting amino alcohol (**67**) is deprotonated by Et_2Zn to afford the tricoordinated ethylzinc derivative (**68**). Complexation of a second molecule of Et_2Zn at the oxygen of the zinc amino alcoholate from the less hindered side[82] affords the di-

Table 2 Catalysts (**50**)–(**65**) for the Enantioselective Addition of Diorganozinc Reagents to Aldehydes[a]

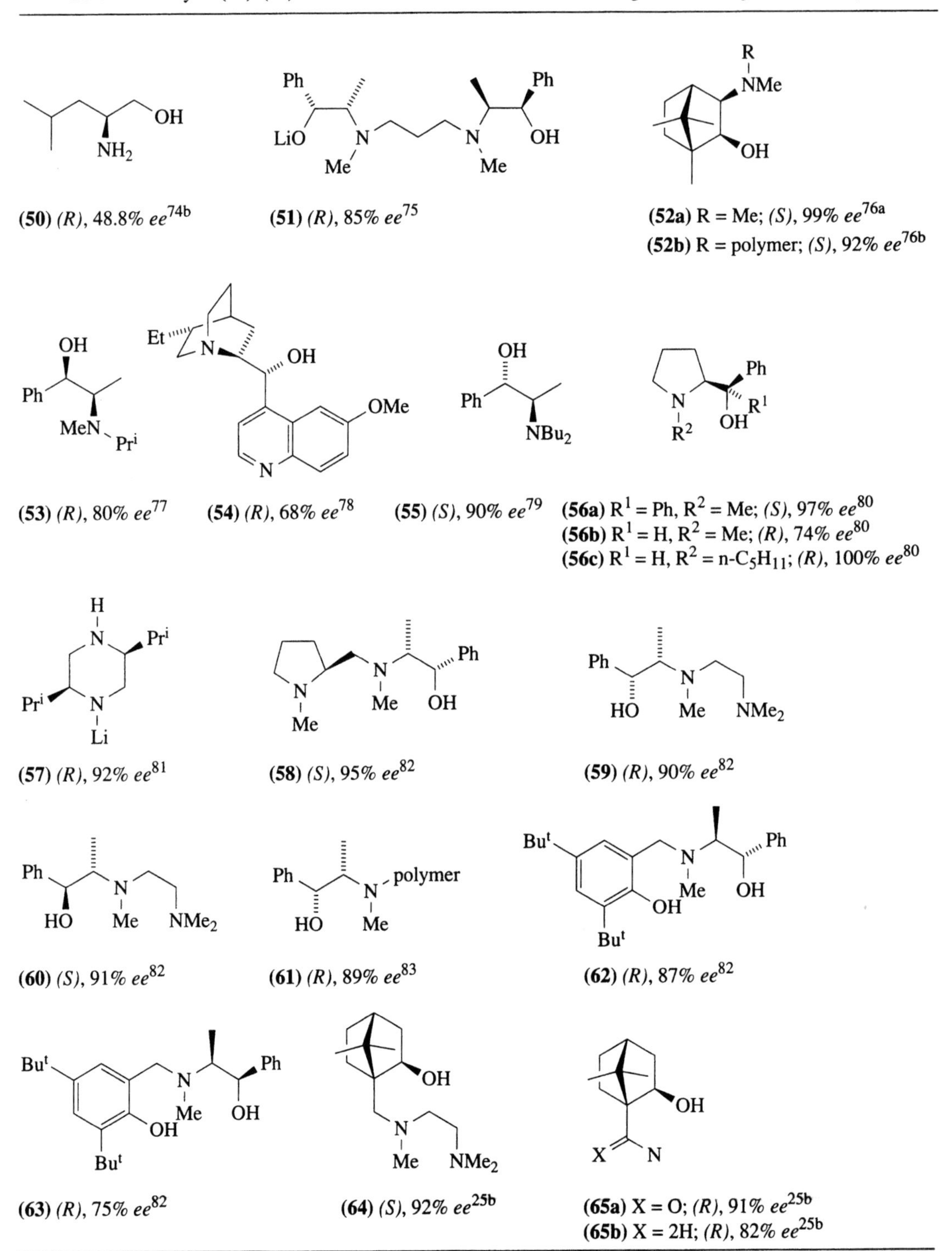

[a] The absolute configuration of 1-phenylpropanol obtained by the addition of Et_2Zn to benzaldehyde in the presence of the catalyst as well as the enantiomeric excess (% *ee*) observed are indicated.

metallic reagent (**69**), which coordinates the aldehyde leading to (**70**). This coordination occurs in the half space which contains Et_2Zn, unless a tridentate chiral inductor such as (**64**) is used.[25b] The complexation of zinc occurs at the sterically most easily available carbonyl lone pair (*cis* to hydrogen; see **71**). After the transfer of an ethyl group affording the bicyclic zinc alcoholate (**72**), the product (**73**) is liberated, regenerating the catalyst (**68**). Interestingly, the use of a configurationally impure catalyst such as 1-piperidyl-3,3-dimethyl-2-butanol (**74**; 10.7% *ee*) affords, upon addition of Et_2Zn to benzaldehyde, a product of 82% enantiomeric excess. This asymmetric amplification phenomenon[84a] can be explained by diastereomeric interactions between the enantiomers of (**74**). Furthermore, the reaction rate with a 66% *ee* catalyst is 5.5 times faster than with the racemic catalyst. These results indicate that in the case of simple bidentate amino alcohols, the actual mechanism of the enantioselective addition may be more complex than that indicated in Scheme 25. High asymmetric inductions have also been obtained recently by using dialkylzinc–orthotitanate complexes,[84b] chiral oxazaborolidines as catalysts[84c] or secondary amino alcohols derived from camphor.[84d] Applications to the preparation of optically active 2-furylcarbinols have been reported.[84e]

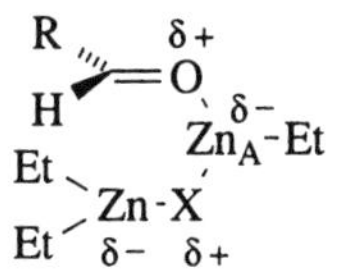

(**66**) 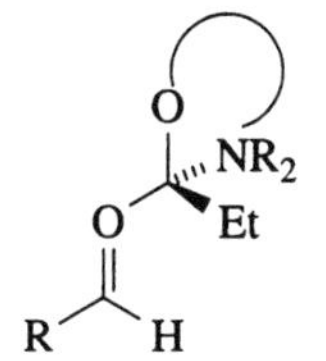 (**71**) 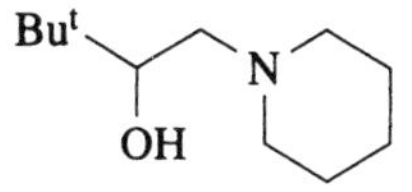 (**74**)

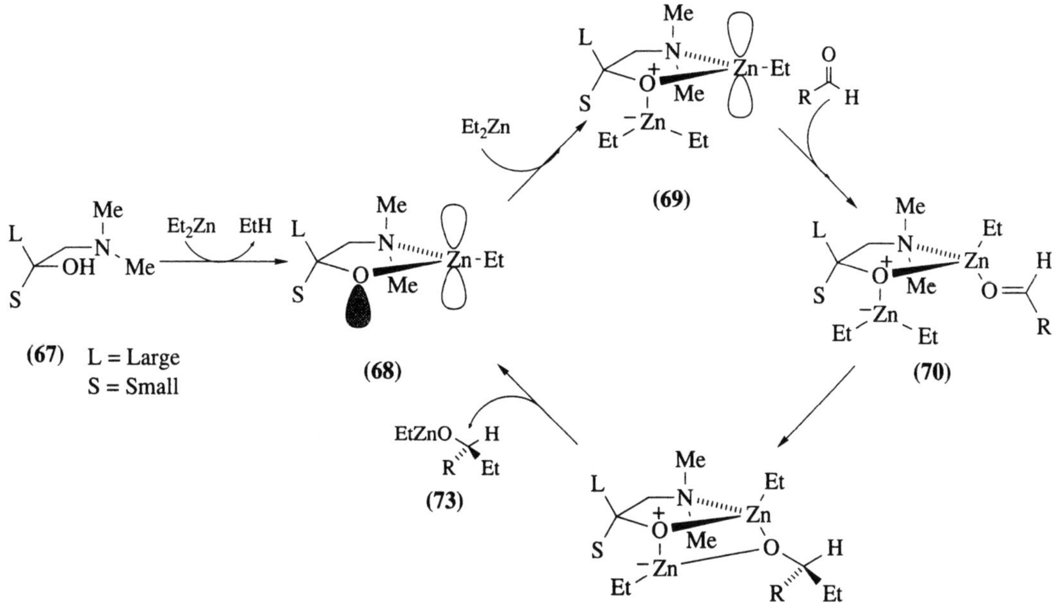

Scheme 25

1.7.4 ADDITION REACTIONS OF ORGANOCADMIUM AND ORGANOMERCURY REAGENTS

1.7.4.1 Addition Reactions of Alkyl- and Aryl-cadmium Reagents

Whereas organomercury compounds[86] do not add to aldehydes and ketones, cadmium organometallics show a useful reactivity.[2,39,87,88] As in the case of organozinc derivatives, allylic cadmium compounds are far more reactive than alkyl- or aryl-cadmium organometallics. Alkylcadmium organometallics are prepared in similar ways to the corresponding zinc compounds. However, their high thermal and photochemical instability makes their preparation and isolation more difficult.[87,88] Purified diorganocadmium compounds react very slowly with aromatic aldehydes.[89,90] However, if the reaction is conducted in the presence of magnesium, zinc, lithium or aluminum halides, a fast reaction is observed (electrophilic catalysis; see Section 1.7.3.1). Magnesium bromide and magnesium iodide are the most active promoters (see equation 15).[91] The replacement of ether by THF leads to slower addition rates.[92] Aliphatic aldehydes and ketones react less cleanly and the desired addition product is obtained in low yields (20–40%) together with reduction products.[87,93] The functionalized organocadmium compound (**75**) can be prepared and added in 50–90% yield to aldehydes (see Scheme 26).[94] Synthetically useful is the reaction of dialkylcadmium compounds with certain functionalized ketones[87,95] containing halides, an ester or a nitro group at the α-position, which furnishes polyfunctionalized molecules in fair yields (see Scheme 27). The addition of cadmium organometallics to 4-*t*-butylcyclohexanone has been reported (see Section 1.7.3.4).

$$Et_2Cd + PhCHO \xrightarrow[85\%]{i} Et{-}CH(OH)Ph \qquad (15)$$

i, $MgBr_2$ (1.2 equiv.), ether, 35 °C, 1 h

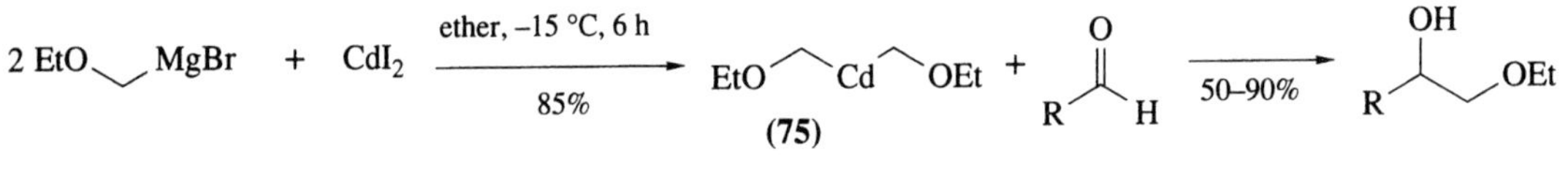

2 EtO⌒MgBr + CdI_2 —(ether, −15 °C, 6 h; 85%)→ EtO⌒Cd⌒OEt (**75**) + RCHO —(50–90%)→ RCH(OH)CH₂OEt

Scheme 26

$R_2Cd{\bullet}MgX_2$ + diketone —(ether, 35 °C, 11 h[95]; 38–70%)→ R-substituted hydroxy ketone

Bu_2Cd + nitroketone —(ref.96; 60%)→ Bu-substituted nitro alcohol (NO_2)

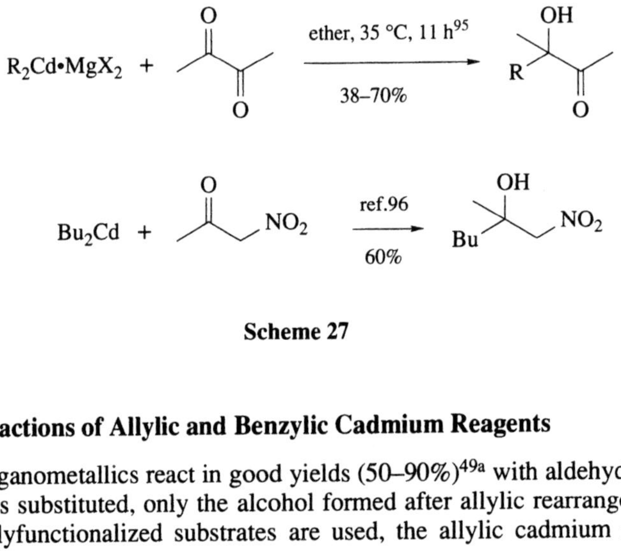

Scheme 27

1.7.4.2 Addition Reactions of Allylic and Benzylic Cadmium Reagents

Allylic cadmium organometallics react in good yields (50–90%)[49a] with aldehydes and ketones.[49a,87,88] If the allylic reagent is substituted, only the alcohol formed after allylic rearrangement is obtained (see Scheme 28).[49a] If polyfunctionalized substrates are used, the allylic cadmium reagent shows a high

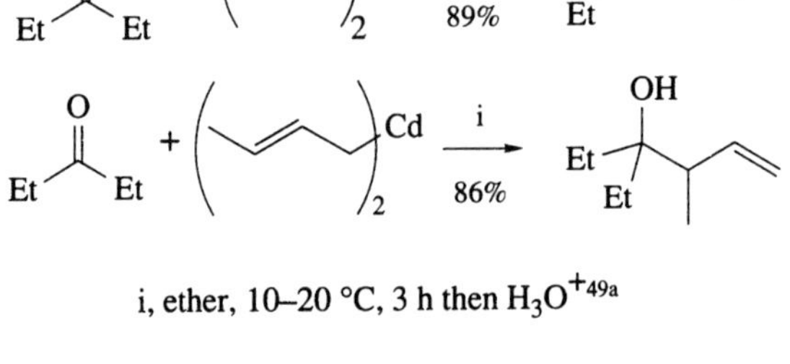

$Et_2C{=}O$ + (allyl)$_2$Cd —(i; 89%)→ $Et_2C(OH)CH_2CH{=}CH_2$

$Et_2C{=}O$ + (crotyl)$_2$Cd —(i; 86%)→ $Et_2C(OH)CH(CH_3)CH{=}CH_2$

i, ether, 10–20 °C, 3 h then H_3O^+[49a]

Scheme 28

chemoselectivity and attacks only the aldehyde function (see equation 16).[97] Enones react with allylic cadmium organometallics to give only the 1,2-addition product in high yields.[98] Benzylic cadmium[20,99] reagents display a more moderate reactivity, but add cleanly to aliphatic and aromatic aldehydes to furnish various benzylic alcohols in fair to good yields (see equation 17).[99] Sulfur-stabilized allylic cadmium derivatives react with aldehydes with high γ-selectivity.[100]

Cd (2) + MeO, OMe, MeO_2C, CHO —(i, ether, –78 °C, 1 h[97]; >70%)→ MeO, OMe, MeO_2C, OH (16)

Cd•$MgBr_2$ (FG, 2) + RCHO —(ether; 58–90%)→ R, OH, FG (17)

1.7.5 REFERENCES

1. K. Nützel, *Methoden Org. Chem. (Houben-Weyl)*, 1973, **13/2a**, 553.
2. J. Boersma, in 'Comprehensive Organometallic Chemistry', ed. G. Wilkinson, F. G. A. Stone and E. W. Abel, Pergamon Press, Oxford, 1982, vol. 2, p. 823.
3. E. Negishi, 'Organometallics in Organic Synthesis', Wiley, New York, 1980.
4. (a) L. I. Zakharkin and O. Y. Ikhlobystin, *Izv. Akad. Nauk SSSR*, 1963, 193 (*Chem. Abstr.*, 1963, **58**, 12 589a); (b) for the synthesis of organozinc derivatives in some less common solvents, see J. Grondin, M. Sebban, P. Vottero, H. Blancou and A. Commeyras, *J. Organomet. Chem.*, 1989, **362**, 237.
5. (a) T. N. Majid and P. Knochel, *Tetrahedron Lett.*, 1990, **31**, 4413; (b) S. C. Berk, M. C. P. Yeh, N. Jeong and P. Knochel, *Organometallics*, 1990, **9**, 3053; (c) P. Knochel, M. C. P. Yeh and C. Xiao, *Organometallics*, 1989, **8**, 2831.
6. For a review, see (a) E. Erdik, *Tetrahedron*, 1987, **43**, 2203; (b) R. D. Rieke and S. J. Uhm, *Synthesis*, 1975, 452; (c) R. Csuk, B. I. Glänzer and A. Fürstner, *Adv. Org. Chem.*, 1988, **28**, 85, and refs. cited therein; (d) R. D. Rieke, P. T. J. Li, T. P. Burns and S. J. Uhm, *J. Org. Chem.*, 1981, **46**, 4323; (e) R. D. Rieke, S. J. Uhm and P. M. Hudnall, *J. Chem. Soc., Chem. Commun.*, 1973, 269.
7. K. J. Klabunde and T. O. Murdock, *J. Org. Chem.*, 1979, **44**, 3901.
8. P. Knochel, M. C. P. Yeh, S. C. Berk and J. Talbert, *J. Org. Chem.*, 1988, **53**, 2390.
9. (a) Y. Tamaru, H. Ochiai, T. Nakamura, K. Tsubaki and Z. Yoshida, *Tetrahedron Lett.*, 1985, **26**, 5559; (b) Y. Tamaru, H. Ochiai, T. Nakamura and Z. Yoshida, *Org. Synth.*, 1988, **67**, 98; (c) H. Ochiai, Y. Tamaru, K. Tsubaki and Z. Yoshida, *J. Org. Chem.*, 1987, **52**, 4418, and refs. cited therein; (d) Y. Tamaru, H. Ochiai, T. Nakamura and Z. Yoshida, *Tetrahedron Lett.*, 1986, **27**, 955.
10. M. C. P. Yeh, P. Knochel and L. E. Santa, *Tetrahedron Lett.*, 1988, **29**, 3887.
11. M. C. P. Yeh, P. Knochel, W. M. Butler and S. C. Berk, *Tetrahedron Lett.*, 1988, **29**, 6693.
12. (a) E. Nakamura, K. Sekiya and I. Kuwajima, *Tetrahedron Lett.*, 1987, **28**, 337; (b) E. Nakamura, S. Aoki, K. Sekiya, H. Oshino and I. Kuwajima, *J. Am. Chem. Soc.*, 1987, **109**, 8056; (c) E. Nakamura, J. Shimada and I. Kuwajima, *Organometallics*, 1985, **4**, 641.
13. M. C. P. Yeh and P. Knochel, *Tetrahedron Lett.*, 1988, **29**, 2395.
14. T. N. Majid, M. C. P. Yeh and P. Knochel, *Tetrahedron Lett.*, 1989, **30**, 5069.
15. Y. Tamaru, H. Ochiai, T. Nakamura and Z. Yoshida, *Angew. Chem.*, 1987, **99**, 1193; *Angew. Chem., Int. Ed. Engl.*, 1987, **26**, 1157.
16. D. L. Comins and S. O'Connor, *Tetrahedron Lett.*, 1987, **28**, 1843.
17. (a) G. Wittig and M. Jautelat, *Justus Liebigs Ann. Chem.*, 1967, **702**, 24; (b) P. Knochel, T.-S. Chou, H. G. Chen, M. C. P. Yeh and M. J. Rozema, *J. Org. Chem.*, 1989, **54**, 5202.
18. (a) D. Seyferth and S. B. Andrews, *J. Organomet. Chem.*, 1971, **30**, 151; (b) P. Knochel, N. Jeong, M. J. Rozema and M. C. P. Yeh, *J. Am. Chem. Soc.*, 1989, **111**, 6474; (c) M. J. Rozema and P. Knochel, *Tetrahedron Lett.*, 1991, **32**, 1855.
19. (a) S. C. Berk, P. Knochel and M. C. P. Yeh, *J. Org. Chem.*, 1988, **53**, 5789; (b) H. G. Chen, C. Hoechstetter and P. Knochel, *Tetrahedron Lett.*, 1989, **30**, 4795.
20. For the synthesis of some functionalized benzylic organocadmium bromides by using highly reactive cadmium metal powder, see E. R. Burkhardt and R. D. Rieke, *J. Org. Chem.*, 1985, **50**, 416.
21. N. El Alami, C. Belaud and J. Villieras, *J. Organomet. Chem.*, 1987, **319**, 303; 1988, **348**, 1.
22. (a) M. Gaudemar, *Bull. Soc. Chim. Fr.*, 1962, 974; (b) M. Bellassoued, Y. Frangin and M. Gaudemar, *Synthesis*, 1977, 205; (c) M. Gaudemar, *Bull. Soc. Chim. Fr.*, 1963, 1475.
23. For some recent examples, see (a) S. Moorhouse and G. Wilkinson, *J. Organomet. Chem.*, 1973, **52**, C5; (b) E. Negishi, L. F. Valente and M. Kobayashi, *J. Am. Chem. Soc.*, 1980, **102**, 3298; (c) H. Lehmkuhl, I. Döring, R. McLane and H. Nehl, *J. Organomet. Chem.*, 1981, **221**, 1.
24. M. H. Abraham, *J. Chem. Soc.*, 1960, 4130.
25. (a) B. Bartocha, H. D. Kaesz and F. G. A. Stone, *Z. Naturforsch., Teil B*, 1959, **14**, 352; (b) W. Oppolzer and R. N. Radinov, *Tetrahedron Lett.*, 1988, **29**, 5645.
26. L. Lardici and L. Lucarini, *Ann. Chim. (Rome)*, 1964, **54**, 1233.

27. For some recent examples, see (a) E. Negishi, A. O. King and N. Okukado, *J. Org. Chem.*, 1977, **42**, 1821; (b) H. Yasuda and H. Tani, *Tetrahedron Lett.*, 1975, 11; (c) P. H. M. Budzelaar, H. J. Alberts-Jansen, K. Mollema, J. Boersma, G. J. M. van der Kerk, A. L. Spek and A. J. M. Duisenberg, *J. Organomet. Chem.*, 1983, **243**, 137.
28. C. Eaborn, N. Retta and J. D. Smith, *J. Organomet. Chem.*, 1980, **190**, 101.
29. M. F. Semmelhack and E. J. Fewkes, *Tetrahedron Lett.*, 1987, **28**, 1497.
30. G. Köbrich and H. R. Merkle, *Chem. Ber.*, 1966, **99**, 1782.
31. (a) A. L. Galyer and G. Wilkinson, *Inorg. Synth.*, 1979, **19**, 253; (b) L. Rösch and G. Altnau, *Angew. Chem.*, 1979, **91**, 62; *Angew. Chem., Int. Ed. Engl.*, 1979, **18**, 60.
32. (a) L. I. Zakharkin and O. Y. Okhlobystin, *Zh. Obshch. Khim.*, 1960, **30**, 2134; *J. Gen. Chem. USSR (Engl. Transl.)*, 1960, **30**, 2109; (b) K. H. Thiele and P. Zdunneck, *J. Organomet. Chem.*, 1965, **4**, 10; (c) K. H. Thiele, G. Engelhardt, J. Köhler and M. Arnstedt, *J. Organomet. Chem.*, 1967, **9**, 385.
33. (a) R. Taube and D. Steinborn, *J. Organomet. Chem.*, 1974, **65**, C9; (b) J. S. Denis, J. P. Oliver and J. B. Smart, *J. Organomet. Chem.*, 1972, **44**, C32; (c) J. S. Denis, J. P. Oliver, T. W. Dolzine and J. B. Smart, *J. Organomet. Chem.*, 1974, **71**, 315; (d) R. Taube, D. Steinborn and B. Alder, *J. Organomet. Chem.*, 1984, **275**, 1; (e) D. M. Heinekey and S. R. Stobart, *Inorg. Chem.*, 1978, **17**, 1463; (f) J. L. Atwood, D. E. Berry, S. R. Stobart and M. J. Zaworotko, *Inorg. Chem.*, 1983, **22**, 3480.
34. (a) G. Wittig and K. Schwarzenbach, *Justus Liebigs Ann. Chem.*, 1961, **650**, 1; (b) G. Wittig and F. Wingler, *Justus Liebigs Ann. Chem.*, 1962, **656**, 18; (c) G. Wittig and F. Wingler, *Chem. Ber.*, 1964, **97**, 2139, 2146.
35. (a) J. J. Habeeb, A. Osman and D. G. Tuck, *J. Organomet. Chem.*, 1980, **185**, 117; (b) F. F. Said and D. G. Tuck, *J. Organomet. Chem.*, 1982, **224**, 121.
36. T. Harada, D. Hara, K. Hattori and A. Oku, *Tetrahedron Lett.*, 1988, **29**, 3821.
37. Y. Masuyama, N. Kinugawa and Y. Kurusu, *J. Org. Chem.*, 1987, **52**, 3702.
38. J. Lorberth, *J. Organomet. Chem.*, 1969, **19**, 189.
39. L. Miginiac, in 'The Chemistry of the Metal–Carbon Bond', ed. F. R. Hartley and S. Patai, Wiley, New York, 1985, vol. 3, p. 99.
40. B. Marx, E. Henry-Basch and P. Fréon, *C. R. Hebd. Seances Acad. Sci., Ser. C*, 1967, **264**, 527; B. Marx, *C. R. Hebd. Seances Acad. Sci., Ser. C*, 1968, **266**, 1646.
41. Y. Tamaru, T. Nakamura, M. Sakaguchi, H. Ochiai and Z. Yoshida, *J. Chem. Soc., Chem. Commun.*, 1988, 610.
42. P. Knochel, unpublished work, 1986.
43. M. Chastrette and R. Amouroux, *Tetrahedron Lett.*, 1970, 5165.
44. H. Meerwein, *J. Prakt. Chem.*, 1936, **147**, 226.
45. M. T. Reetz, R. Steinbach and B. Wenderoth, *Synth. Commun.*, 1981, **11**, 261.
46. P. Knochel and J. F. Normant, *Tetrahedron Lett.*, 1986, **27**, 4431.
47. H. Ochiai, T. Nishihara, Y. Tamaru and Z. Yoshida, *J. Org. Chem.*, 1988, **53**, 1343.
48. B. Eistert and L. Klein, *Chem. Ber.*, 1968, **101**, 900.
49. (a) D. Abenhaïm, E. Henry-Basch and P. Fréon, *Bull. Soc. Chim. Fr.*, 1969, 4038, 4043; (b) M. Gaudemar and S. Travers, *C. R. Hebd. Seances Acad. Sci., Ser. C*, 1966, **262**, 139; (c) D. Abenhaïm and E. Henry-Basch, *C. R. Hebd. Seances Acad. Sci., Ser. C*, 1968, **267**, 87; (d) B. Maurer and A. Hauser, *Helv. Chim. Acta*, 1982, **65**, 462.
50. L. Miginiac-Groizeleau, P. Miginiac and C. Prévost, *Bull. Soc. Chim. Fr.*, 1965, 3560.
51. (a) L. Miginiac and M. Lanoiselee, *Bull. Soc. Chim. Fr.*, 1971, 2716; (b) M. Mladenova, F. Gaudemar-Bardone, S. Simova and R. Couffignal, *Bull. Soc. Chim. Fr.*, 1986, 479; (c) F. Gaudemar-Bardone, M. Mladenova and R. Couffignal, *Synthesis*, 1985, 1043.
52. (a) D. Seebach and W. Langer, *Helv. Chim. Acta*, 1979, **62**, 1701, 1710; (b) R. A. Watson and R. A. Kjonaas, *Tetrahedron Lett.*, 1986, **27**, 1437; (c) R. A. Kjonaas and E. J. Vawter, *J. Org. Chem.*, 1986, **51**, 3993; (d) W. Tückmantel, K. Oshima and H. Nozaki, *Chem. Ber.*, 1986, **119**, 1581; (e) J. F. G. A. Jansen and B. L. Feringa, *Tetrahedron Lett.*, 1988, **29**, 3593; (f) R. A. Kjonaas and R. K. Hoffer, *J. Org. Chem.*, 1988, **53**, 4133.
53. G. A. Molander and D. C. Shubert, *J. Am. Chem. Soc.*, 1986, **108**, 4683.
54. J. van der Louw, J. L. van der Baan, H. Stichter, G. J. J. Out, F. Bickelhaupt and G. W. Klumpp, *Tetrahedron Lett.*, 1988, **29**, 3579.
55. C. Blomberg and F. A. Hartog, *Synthesis*, 1977, 18.
56. (a) E. Öhler, K. Reininger and U. Schmidt, *Angew. Chem.*, 1970, **82**, 480; (b) H. Mattes and C. Benezra, *Tetrahedron Lett.*, 1985, **26**, 5697; (c) P. Knochel and J. F. Normant, *J. Organomet. Chem.*, 1986, **309**, 1.
57. (a) P. Auvray, P. Knochel and J. F. Normant, *Tetrahedron Lett.*, 1985, **26**, 2329, 4455; (b) P. Auvray, P. Knochel and J. F. Normant, *Tetrahedron Lett.*, 1986, **27**, 5091, 5095; (c) P. Auvray, P. Knochel and J. F. Normant, *Tetrahedron*, 1988, **44**, 4495, 4509.
58. J. N. Collard and C. Benezra, *Tetrahedron Lett.*, 1982, **23**, 3725.
59. P. Knochel and J. F. Normant, *Tetrahedron Lett.*, 1984, **25**, 4383.
60. (a) C. Pétrier and J. L. Luche, *J. Org. Chem.*, 1985, **50**, 910; (b) C. Einhorn and J. L. Luche, *J. Organomet. Chem.*, 1987, **322**, 177.
61. J. L. Luche, C. Allavena, C. Petrier and C. Dupuy, *Tetrahedron Lett.*, 1988, **29**, 5373.
62. B. Gross and C. Prevost, *Bull. Soc. Chim. Fr.*, 1967, 3610.
63. (a) P. Miginiac and C. Bouchoule, *Bull. Soc. Chim. Fr.*, 1968, 4675; (b) P. Miginiac, *Bull. Soc. Chim. Fr.*, 1970, 1077; (c) F. Gérard and P. Miginiac, *J. Organomet. Chem.*, 1978, **155**, 271; (d) F. Barbot and P. Miginiac, *Tetrahedron Lett.*, 1975, 3829; (e) F. Barbot and P. Miginiac, *J. Organomet. Chem.*, 1977, **132**, 445; (f) F. Gérard and P. Miginiac, *Bull. Soc. Chim. Fr.*, 1974, 1924, 2527; (g) F. Barbot and P. Miginiac, *Bull. Soc. Chim. Fr.*, 1977, 113.
64. J. L. Moreau, in 'The Chemistry of Ketenes, Allenes and Related Compounds', ed. S. Patai, Wiley, New York, 1980, p. 363.

65. (a) J. Pansard and M. Gaudemar, *Bull. Soc. Chim. Fr.*, 1968, 3332; (b) M. Gaudemar and J. L. Moreau, *Bull. Soc. Chim. Fr.*, 1968, 5037; (c) J. L. Moreau and M. Gaudemar, *Bull. Soc. Chim. Fr.*, 1970, 2171, 2175; (d) J. L. Moreau, *Bull Soc. Chim. Fr.*, 1975, 1248.
66. G. Zweifel and G. Hahn, *J. Org. Chem.*, 1984, **49**, 4565.
67. P. R. Jones, E. J. Goller and W. J. Kauffman, *J. Org. Chem.*, 1971, **36**, 3311.
68. (a) G. Fronza, C. Fuganti, P. Grasselli, G. Pedrocchi-Fantoni and C. Zirotti, *Tetrahedron Lett.*, 1982, **23**, 4143; (b) J. Mulzer, M. Kappert, G. Huttner and I. Jibril, *Angew. Chem.*, 1984, **96**, 726; *Angew. Chem., Int. Ed. Engl.*, 1984, **23**, 704; (c) T. Fujisawa, E. Kojima, T. Itoh and T. Sato, *Tetrahedron Lett.*, 1985, **26**, 6089; (d) see also Y. Yamamoto, T. Komatsu and K. Maruyama, *J. Chem. Soc., Chem. Commun.*, 1985, 814.
69. (a) D. J. Cram and F. A. Abd Elhafez, *J. Am. Chem. Soc.*, 1952, **74**, 5828; (b) M. Chérest, H. Felkin and N. Prudent, *Tetrahedron Lett.*, 1968, 2199; (c) M. Chérest and H. Felkin, *Tetrahedron Lett.*, 1968, 2205; (d) N. T. Anh and O. Eisenstein, *Nouv. J. Chim.*, 1977, **1**, 61; (e) N. T. Anh, *Top. Curr. Chem.*, 1980, **88**, 145.
70. M. Bhupathy and T. Cohen, *Tetrahedron Lett.*, 1985, **26**, 2619.
71. A. Solladié-Cavallo, D. Farkhani, S. Fritz, T. Lazrak and J. Suffert, *Tetrahedron Lett.*, 1984, **25**, 4117.
72. (a) P. R. Jones, E. J. Goller and W. J. Kauffman, *J. Org. Chem.*, 1969, **34**, 3566; (b) P. R. Jones, W. J. Kauffman and E. J. Goller, *J. Org. Chem.*, 1971, **36**, 186; (c) G. Boireau, A. Deberly and D. Abenhaïm, *Tetrahedron Lett.*, 1988, **29**, 2175.
73. For an excellent review, see D. A. Evans, *Science (Washington, D.C.)*, 1988, **240**, 420.
74. (a) N. Oguni, T. Omi, Y. Yamamoto and A. Nakamura, *Chem. Lett.*, 1983, 841; (b) N. Ogumi and T. Omi, *Tetrahedron Lett.*, 1984, **25**, 2823.
75. K. Soai, M. Nishi and Y. Ito, *Chem. Lett.*, 1987, 2405.
76. (a) M. Kitamura, S. Suga, K. Kawai and R. Noyori, *J. Am. Chem. Soc.*, 1986, **108**, 6071; (b) S. Itsuno and J. M. J. Fréchet, *J. Org. Chem.*, 1987, **52**, 4140.
77. P. A. Chaloner and S. A. R. Perera, *Tetrahedron Lett.*, 1987, **28**, 3013.
78. (a) A. A. Smaardijk and H. Wynberg, *J. Org. Chem.*, 1987, **52**, 135; (b) see also G. Muchow, Y. Vannoorenberghe and G. Buono, *Tetrahedron Lett.*, 1987, **28**, 6163.
79. K. Soai, S. Yokoyama, K. Ebihara and T. Hayasaka, *J. Chem. Soc., Chem. Commun.*, 1987, 1690.
80. (a) K. Soai, A. Ookawa, K. Ogawa and T. Kaba, *J. Chem. Soc., Chem. Commun.*, 1987, 467; (b) K. Soai, A. Ookawa, T. Kaba and K. Ogawa, *J. Am. Chem. Soc.*, 1987, **109**, 7111.
81. K. Soai, S. Niwa, Y. Yamada and H. Inoue, *Tetrahedron Lett.*, 1987, **28**, 4841.
82. E. J. Corey and F. J. Hannon, *Tetrahedron Lett.*, 1987, **28**, 5233, 5237.
83. K. Soai, S. Niwa and M. Watanabe, *J. Org. Chem.*, 1988, **53**, 927.
84. (a) N. Oguni, Y. Matsuda and T. Kaneko, *J. Am. Chem. Soc.*, 1988, **110**, 7877; (b) M. Yoshioka, T. Kawakita and M. Ohno, *Tetrahedron Lett.*, 1989, **30**, 1657; (c) N. N. Joshi, M. Srebnik and H. C. Brown, *Tetrahedron Lett.*, 1989, **30**, 5551; (d) K. Tanaka, H. Ushio and H. Suzuki, *J. Chem. Soc., Chem. Commun.*, 1989, 1700; (e) A. van Oeveren, W. Menge and B. L. Feringa, *Tetrahedron Lett.*, 1989, **30**, 6427.
85. E. C. Ashby and R. S. Smith, *J. Organomet. Chem.*, 1982, **225**, 71.
86. R. C. Larock, 'Organomercury Compounds in Organic Synthesis', Springer, New York, 1985.
87. K. Nützel, *Methoden Org. Chem. (Houben-Weyl)*, 1973, **13/2a**, 916.
88. P. R. Jones and P. J. Desio, *Chem. Rev.*, 1978, **78**, 491.
89. H. Gilman and J. F. Nelson, *Recl. Trav. Chim. Pays-Bas*, 1936, **55**, 518.
90. E. Henry-Basch, J. Michel, F. Huet, B. Marx and P. Fréon, *Bull Soc. Chim. Fr.*, 1965, 927.
91. F. Huet, E. Henry-Basch and P. Fréon, *Bull. Soc. Chim. Fr.*, 1970, 1415.
92. F. Huet, E. Henry-Basch and P. Fréon, *Bull. Soc. Chim. Fr.*, 1970, 1426.
93. G. Soussan, *C. R. Hebd. Seances Acad. Sci., Ser. C*, 1966, **263**, 954; 1969, **268**, 267.
94. C. Bernardon, E. Henry-Basch and P. Fréon, *C. R. Hebd. Seances Acad. Sci., Ser. C*, 1968, **266**, 1502.
95. G. W. Stacy, R. A. Mikulec, S. L. Razniak and L. D. Starr, *J. Am. Chem. Soc.*, 1957, **79**, 3587.
96. J. Michel, E. Henry-Basch and P. Fréon, *C. R. Hebd. Seances Acad. Sci., Ser. C*, 1964, **258**, 6171.
97. E. J. Corey, K. C. Nicolaou and T. Toru, *J. Am. Chem. Soc.*, 1975, **97**, 2287.
98. D. A. Evans, D. J. Baillargeon and J. V. Nelson, *J. Am. Chem. Soc.*, 1978, **100**, 2242.
99. C. Bernardon, *Tetrahedron Lett.*, 1979, 1581.
100. L. Bo and A. G. Fallis, *Tetrahedron Lett.*, 1986, **27**, 5193.

1.8
Organocerium Reagents

TSUNEO IMAMOTO
Chiba University, Japan

1.8.1 INTRODUCTION

Elements of the lanthanide series possess unique electronic and stereochemical properties due to their *f*-orbitals, and have great potential as reagents and catalysts in organic synthesis.[1–4] During the 1970s and 1980s many synthetic reactions and procedures using lanthanide elements were reported, in conjunction with significant development in the chemistry of organolanthanides. Several review articles covering this field of chemistry have appeared.[5–11]

Of the 15 elements from lanthanum to lutetium, cerium has the highest natural abundance, and its major inorganic salts are commercially available at moderate prices. The present author and his coworkers have utilized this relatively inexpensive element in reactions which form carbon–carbon bonds, studying the generation and reactivities of organocerium reagents. The cerium reagents, which are prepared from organolithium compounds and cerium(III) halides, have been found to be extremely useful in organic synthesis, particularly in the preparation of alcohols by carbonyl addition reactions. They react with various carbonyl compounds to afford addition products in satisfactory yields, even though the substrates are susceptible to so-called abnormal reactions when using simple organolithiums or Grignard reagents.

This chapter surveys the addition reactions of organocerium reagents to the C—X π-bond. Emphasis is placed on the utility of cerium chloride methodology, and many examples of its practical applications are given. Experimental procedures are also described in detail to enable readers to employ this method immediately in practical organic syntheses.

Other organolanthanide reagents are covered in Chapter 1.10 in this volume, while selective carbonyl addition reactions promoted by samarium and ytterbium reagents are surveyed in Chapter 1.9.

1.8.2 ORGANOCERIUM REAGENTS

1.8.2.1 Generation of Organocerium Reagents

Organocerium reagents are prepared *in situ* by the reaction of organolithium compounds with anhydrous cerium chloride or cerium iodide, as shown in equation (1).[12–14] A variety of organolithium compounds can be employed, including alkyl-, allyl-, alkenyl- and alkynyl-lithiums, which are all converted to the corresponding cerium reagents.

$$\text{RLi} + \text{CeX}_3 \xrightarrow[\text{X = Cl, I}]{\text{THF}} \text{'RCeX}_2\text{'} + \text{LiX} \qquad (1)$$

No systematic studies on the structure of organocerium reagents have been made so far. Although some experimental results indicate that no free organolithium compounds are present in the reagents, the structure of the reagents has not yet been elucidated. The cerium reagents are presumed to be σ-($RCeX_2$) or ate [$(RCeX_3)^-Li^+$] complexes, but other possibilities such as a weakly associated complex ($RLi{\cdot}CeX_3$) are not excluded. In this text, organocerium reagents are represented as '$RCeX_2$' for convenience.

1.8.2.1.1 General comments

Organocerium reagents can be generated without difficulty, but the following suggestions will help to ensure success.

Cerium chloride, rather than cerium iodide, is recommended because preparation of the iodide requires the handling of pyrophoric metallic cerium.[14] Anhydrous cerium chloride is commercially available from Aldrich, but can also be prepared in the laboratory by dehydration of cerium chloride heptahydrate using thionyl chloride,[15] or by reduction of cerium(IV) oxide using HCO_2H/HCl followed by dehydration.[16] Heating the hydrate without additive *in vacuo* is a comparatively simple method and is satisfactory for the generation of organocerium reagents.

Ethereal solvents such as tetrahydrofuran (THF) and dimethoxyethane (DME) are employed in the reactions, with THF generally being preferred. Usually, THF freshly distilled from potassium or sodium with benzophenone is used. Use of hot THF is not recommended, because on addition to cerium chloride a pebble-like material, which is not easily suspended, may be formed.

1.8.2.1.2 General procedure for the preparation of anhydrous cerium chloride

Cerium chloride heptahydrate (560 mg, 1.5 mmol) is quickly ground to a fine powder in a mortar and placed in a 30 mL two-necked flask. The flask is immersed in an oil bath and heated gradually to 135–140 °C with evacuation (*ca.* 0.1 mmHg). After 1 h, a magnetic stirrer is placed in the flask and the cerium chloride is dried completely by stirring at the same temperature *in vacuo* for a further 1 h.

The following procedure is recommended for the small-scale preparation of anhydrous cerium chloride. Cerium chloride heptahydrate (*ca.* 20 g) is placed in a round-bottomed flask connected to a dry ice trap. The flask is evacuated and heated to 100 °C for 2 h. The resulting opaque solid is quickly pulverized in a mortar and is heated again, *in vacuo* at the same temperature, for 2 h with intermittent shaking. A stirrer is then placed in the flask, which is subsequently evacuated, and the bath temperature is raised to 135–140 °C. Drying is complete after 2–3 h of stirring.

Anhydrous cerium chloride can be stored for long periods provided it is strictly protected from moisture. Cerium chloride is extremely hygroscopic; hence, it is recommended that it be dried *in vacuo* at *ca.* 140 °C for 1–2 h before use.

1.8.2.1.3 General procedure for the generation of organocerium reagents

Cerium chloride heptahydrate (560 mg, 1.5 mmol) is dried by the procedure described above. While the flask is still hot, argon gas is introduced, after which the flask is cooled in an ice bath. Tetrahydrofuran (5 mL) is added all at once with vigorous stirring. The ice bath is removed and the suspension is stirred well for 2 h or more (usually overnight) under argon at room temperature. The flask is cooled to –78 °C and an organolithium compound (1.5 mmol) is added with a syringe. Stirring for 0.5–2 h at the same temperature, or a somewhat higher temperature (–40 to –20 °C), results in the formation of a yellow or red suspension, which is ready to use for reactions.

1.8.2.1.4 Reactions with ketones and similar compounds

The addition reactions are usually carried out at –78 °C, except for reactions of the Grignard reagent/cerium chloride system (Section 1.8.4), which are conducted at 0 °C. A substrate is added to the well-stirred organocerium reagent and the mixture is stirred until the reaction is complete. Work-up is carried out in the usual manner: quenching with dilute HCl or dilute AcOH and extraction with a suitable organic solvent. When the substrates are acid sensitive, work-up using tetramethylenediamine is recommended.[17]

1.8.2.2 Scope of the Reactivity

1.8.2.2.1 Thermal stability

The reagents are generally stable at low temperature (–78 to –20 °C) and react readily with various carbonyl compounds to give the corresponding addition products in high yields. However, at temperatures above 0 °C, reagents with β-hydrogens decompose; reactions with ketones provide reduction products (secondary alcohols and pinacol coupling products), as shown in Scheme 1. Other reagents, such as methyl- and phenyl-cerium reagents, are stable at around room temperature but decompose at about 60 °C.[13,14,18]

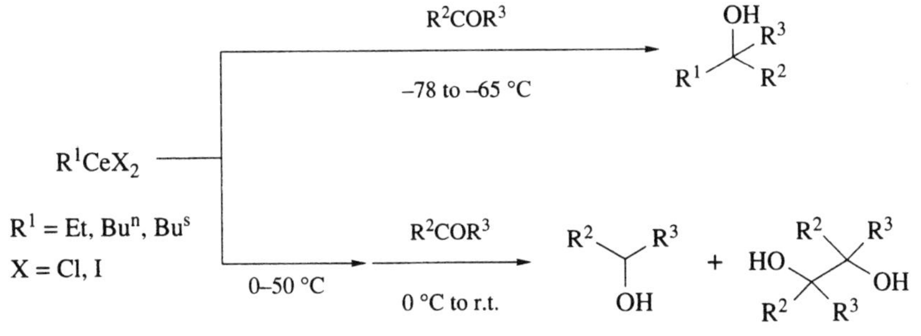

R^1CeX_2 (R^1 = Et, Bu^n, Bu^s; X = Cl, I) → (R^2COR^3, –78 to –65 °C) → $R^1R^2R^3C$–OH

R^1CeX_2 → (0–50 °C) → (R^2COR^3, 0 °C to r.t.) → R^2R^3CH–OH + HO–CR^2R^3–CR^2R^3–OH

Scheme 1

1.8.2.2.2 Reactions with organic halides, nitro compounds and epoxides

The reactivities of organocerium reagents toward organic halides are in sharp contrast to the reactivities of alkyllithiums. No metal–halogen exchange occurs at –78 °C; aryl bromides and iodides are quantitatively recovered unchanged after treatment with *n*-butyl- or *t*-butyl-cerium reagents.[18,19] Alkyl iodides are also inert to prolonged treatment with organocerium reagents at the same temperature. Benzylic halides undergo reductive coupling to give 1,2-diphenylethane derivatives upon treatment with the *n*-butylcerium reagent.[18]

Nitro compounds react immediately with organocerium reagents at –78 °C to give many products,[19] while epoxides undergo ring opening followed by deoxygenation to yield substituted alkenes.[20]

1.8.2.3 Reactions with Carbonyl Compounds

Organocerium reagents react readily with various ketones at low temperature to give the addition products in good to high yields.[12–14,21] Some representative results are shown in Table 1,[12,13] together with the results obtained on using the corresponding organolithiums alone.

Table 1 The Reaction of Organocerium Reagents with Ketones

Ketone	*Reagent*	*Product*	*Yield (%)*[a]
$(PhCH_2)_2CO$	Bu^nCeCl_2	$(PhCH_2)_2C(OH)Bu^n$	96 (33)
$(PhCH_2)_2CO$	Bu^tCeCl_2	$(PhCH_2)_2C(OH)Bu^t$	65 (trace)
$(PhCH_2)_2CO$	$HC{\equiv}CCeCl_2$	$(PhCH_2)_2C(OH)C{\equiv}CH$	95 (60)
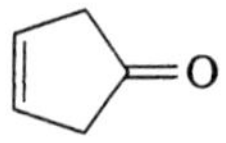 O	$PhC{\equiv}CCeCl_2$	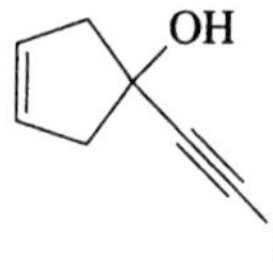 OH Ph	89 (30)
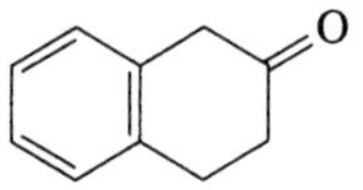 O	Bu^nCeCl_2	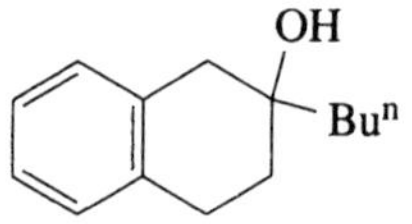 OH Bu^n	88 (trace)
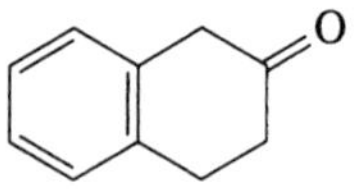 O	$H_2C{=}C(Me)CeCl_2$	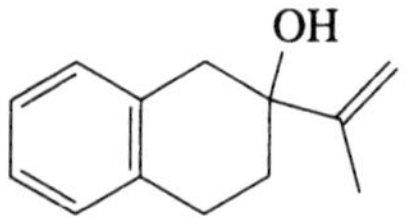 OH	88 (12)
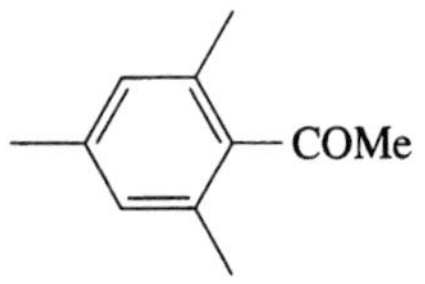 COMe	Bu^nCeCl_2	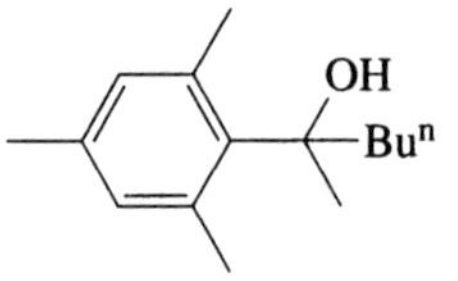 OH Bu^n	57 (<10)
p-IC_6H_4COMe	Bu^nCeCl_2	p-$IC_6H_4C(OH)(Me)Bu^n$	93 (trace)
p-BrC_6H_4COMe	Bu^nCeCl_2	p-$BrC_6H_4C(OH)(Me)Bu^n$	96 (43)
p-$BrC_6H_4COCH_2Br$	$PhC{\equiv}CCeCl_2$	p-$BrC_6H_4C(OH)(CH_2Br)C{\equiv}CPh$	95 (trace)
p-NCC_6H_4COMe	Bu^nCeCl_2	p-$NCC_6H_4C(OH)(Me)Bu^n$	48
m-$O_2NC_6H_4COMe$	Bu^nCeCl_2	Complex mixture	
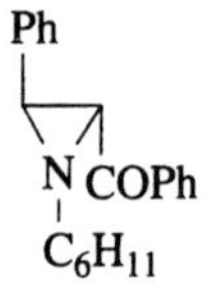 Ph N COPh C_6H_{11}	Bu^nCeCl_2	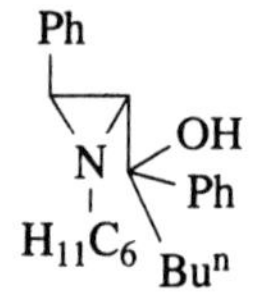 Ph N OH Ph $H_{11}C_6$ Bu^n	93
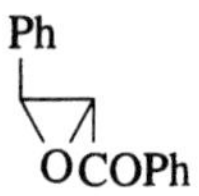 Ph O COPh	Bu^nCeCl_2	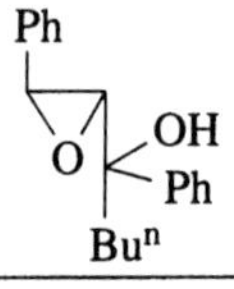 Ph O OH Ph Bu^n	52

[a] The figures in parentheses indicate the yields obtained by use of the lithium reagent alone.

It is noteworthy that the reagents are only weakly basic and react even with readily enolizable ketones in moderate to good yields. Another important fact is that selective carbonyl addition occurs in the presence of carbon–halogen bonds.

The chemoselectivity between aldehydes and ketones has been studied by Kauffmann *et al.*[22] Cerium reagents exhibit moderate aldehyde selectivities, although these are much lower than those of the organometallic reagents of titanium, zirconium and chromium (Scheme 2).[23–26]

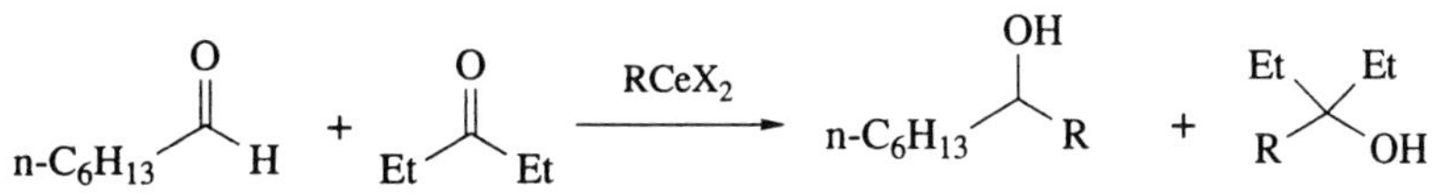

$RCeX_2$	n-$C_6H_{13}CH(OH)R$		$Et_2C(R)OH$
$MeCeCl_2$	79	:	21
$MeCeI_2$	87	:	13
Bu^nCeI_2	90	:	10

Scheme 2

Stereoselectivities have been studied by several research groups. In the case of α-heterosubstituted carbonyl compounds, both chelation control and nonchelation control have been observed, depending on the reagents and substrates (Section 1.8.2.6).

1.8.2.4 Selective Addition to α,β-Unsaturated Carbonyl Compounds

Organocerium reagents react with α,β-unsaturated carbonyl compounds to give 1,2-addition products in good to high yields (equation 2).[27,28]

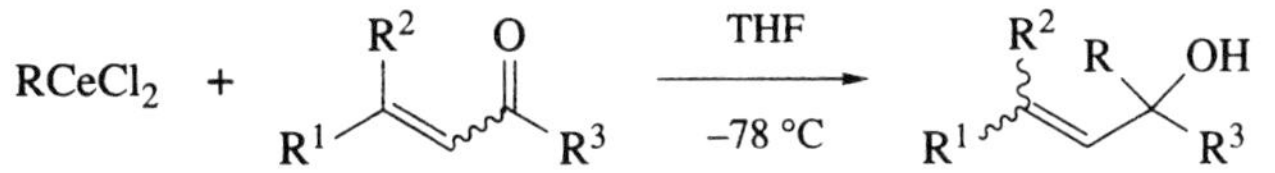

$$RCeCl_2 + R^1CH{=}C(R^2)COR^3 \xrightarrow[-78\ ^\circ C]{THF} R^1CH{=}C(R^2)C(R)(OH)R^3 \quad (2)$$

The reactions of (*E*)- and (*Z*)-1-(4′-methoxyphenyl)-3-phenyl-2-propen-1-ones (**1a** and **1b**; Scheme 3) are representative examples. The 1,2-selectivities are generally higher than those of the corresponding lithium reagents and Grignard reagents (Table 2);[28] however, the selectivity can be severely eroded by steric effects, as exemplified by the reactions of an isopropylcerium reagent.

The reactions of (*Z*)-PhCH═CHCOC$_6$H$_4$-*p*-OMe with cerium reagents provide (*Z*)-allyl alcohols in excellent yields, suggesting that the addition reactions proceed almost exclusively through a polar pathway.

Another notable difference between cerium and lithium reagents has been observed in the reaction of a stabilized carbanion with cyclohex-2-enone. The reaction with α-cyanobenzyllithium is known to be thermodynamically controlled; thus, the initially formed 1,2-adduct dissociates to the starting carbanion and α-enone, which in turn are gradually converted to the thermodynamically stable 1,4-adduct.[29] In contrast, the corresponding cerium reagent affords the 1,2-adduct exclusively in good yield, regardless of reaction time (Scheme 4 and Table 3).[28] Trivalent cerium is strongly oxophilic and intercepts the intermediate 1,2-adduct by virtue of its strong bonding to the alkoxide oxygen, thus suppressing the reverse reaction.

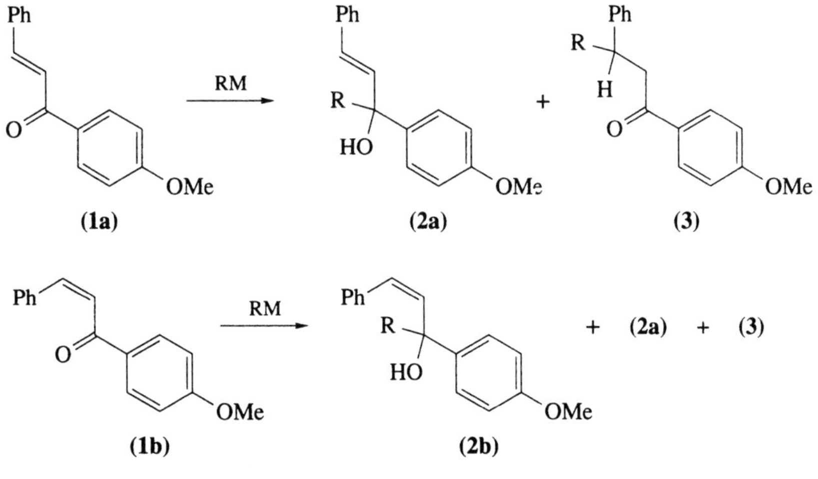

(1a) —RM→ (2a) + (3)

(1b) —RM→ (2b) + (2a) + (3)

R = Me, Bu, Ph, Pr^i; M = $CeCl_2$, Li, MgBr

Scheme 3

Table 2 Reaction of Organocerium, Organolithium or Grignard Reagents with (*E*)- or (*Z*)-PhCH=CHCOC$_6$H$_4$-OMe-*p*

Substrate	*Reagent*	*Conditions*[a] *Solvent*	*Time* (min)	*Yield of products* (%)[b] *(2a)*	*(2b)*	*(3)*
(**1a**)	MeCeCl$_2$	THF–ether (12:1)	30	98	0	Trace
	MeLi	THF–ether (5:1)	30	65	0	17
	MeMgBr	THF	30	38	0	65
	BuCeCl$_2$	THF–hexane (8:1)	30	50	0	43
	BuLi	THF–hexane (2:1)	30	36	0	50
	BuMgBr	THF	30	10	0	76
	PhCeCl$_2$	THF–ether (8:1)	30	90	0	4
	PhLi	THF–ether (3:1)	30	85	0	10
	PhMgBr	THF	30	14	0	78
	Pr^{i}CeCl$_2$	THF–hexane (4:1)	30	42	0	35
	Pr^{i}Li	THF–hexane (3:1)	30	39	0	57
	Pr^{i}MgCl	THF	30	44	0	51
(**1b**)	MeCeCl$_2$	THF–ether (12:1)	60	Trace	97	Trace
	MeLi	THF–ether (5:1)	60	Trace	96	Trace
	MeMgBr	THF	60	26	20	38
	BuCeCl$_2$	THF–hexane (8:1)	60	Trace	97	Trace
	BuLi	THF–hexane (2:1)	60	Trace	70	20
	BuMgBr	THF	60	20	25	50
	PhCeCl$_2$	THF–ether (8:1)	90	Trace	93	4
	PhLi	THF–ether (3:1)	90	Trace	90	4
	PhMgBr	THF	90	25	20	45

[a] All reactions were carried out at –78 °C. [b] Isolated yield.

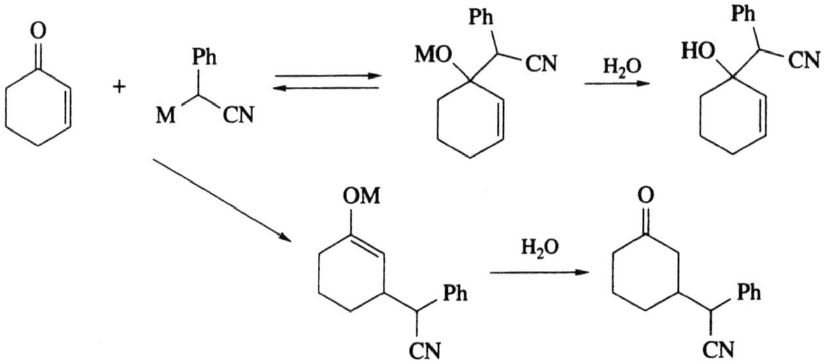

Scheme 4

Table 3

M	*Conditions*[a] *Temperature* (°C)	*Time* (min)	*Yield* (%) *1,2-Addition*	*1,4-Addition*	*Ref.*
Li	–70	1	29	36	29
Li	–70	15	21	49	29
Li	–70	180	0	90	29
CeCl$_2$	–78	4	61	0	28
CeCl$_2$	–78	15	60	0	28
CeCl$_2$	–78	240	62	0	28

[a] All reactions carried out using THF as solvent.

1.8.2.5 Addition to C—N π-Bonds

The reaction of cerium reagents with imines and nitriles which possess α-hydrogens has been studied by Wada *et al.*[30] Available data indicate that the reagents do not effectively add to these substrates, although the results are better than with alkyllithiums themselves. A modification of the procedure improves the yields of addition products. Thus, addition of the organocerium or organolithium reagent to an admixture of cerium chloride and imine or nitrile at –78 °C affords the adducts in acceptable yields, as exemplified in equation (4).[30]

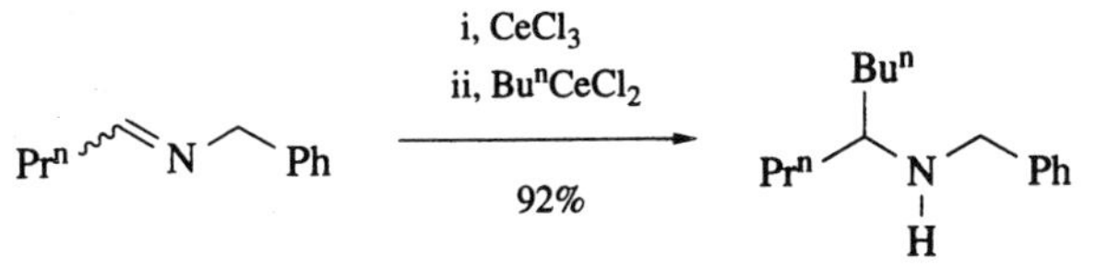

i, $CeCl_3$; ii, Bu^nCeCl_2; 92% (4)

It has been reported by Denmark *et al.* that aldehyde hydrazones react smoothly with organocerium reagents to give addition products in good to high yields (Section 1.8.2.6).[31]

1.8.2.6 Synthetic Applications

1.8.2.6.1 Alkylcerium reagents

Mash has recently employed a methylcerium reagent in the synthesis of (–)-chokol A.[32] The reagent reacts with the readily enolizable cyclopentanone derivative (**4**) to give (–)-chokol A in 80% yield, as shown in Scheme 5.

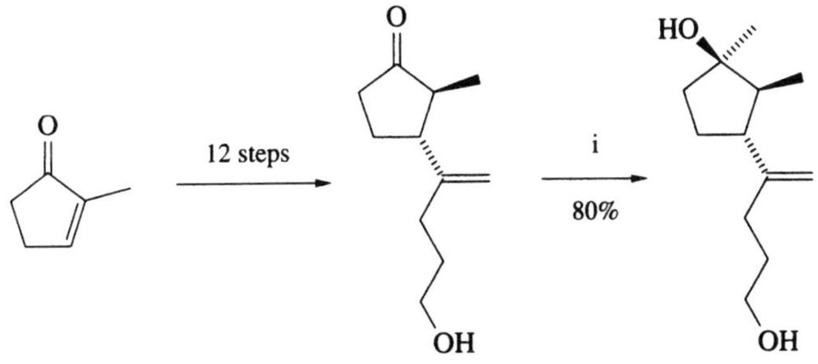

12 steps; (**4**); i, 80%; (–)-Chokol A

i, $MeCeCl_2$ (5 equiv.), THF, –78 °C, 2 h

Scheme 5

Corey and Ha have successfully employed a cerium reagent at the crucial step in the total synthesis of venustatriol, as illustrated in Scheme 6.[33]

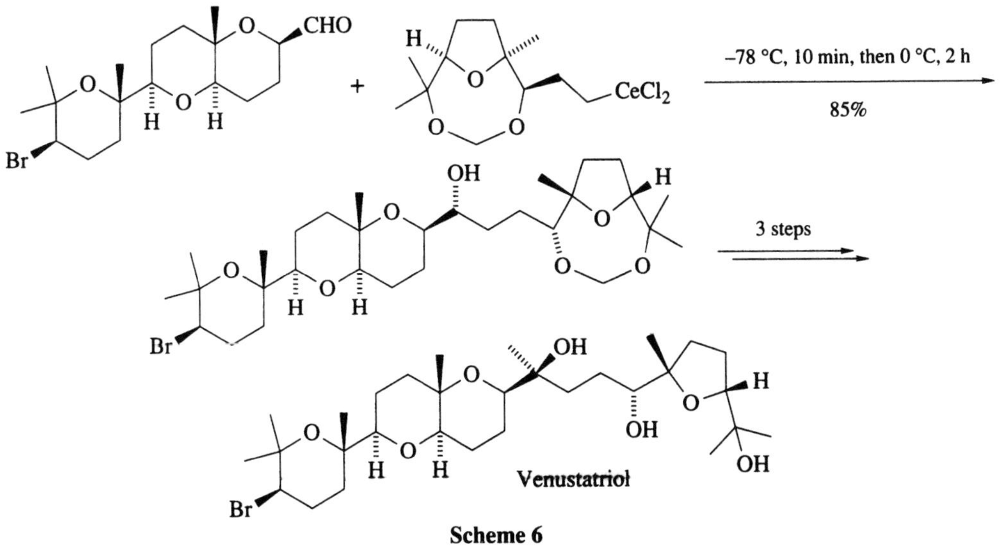

–78 °C, 10 min, then 0 °C, 2 h; 85%; 3 steps; Venustatriol

Scheme 6

In a total synthesis of (–)-bactobolin, Garigipati and Weinreb used a dichloromethylcerium reagent.[34] The intermediate product (**5**) was isolated in 54% yield as a single isomer, as shown in Scheme 7. Dichloromethyllithium alone in this reaction afforded intractable material.

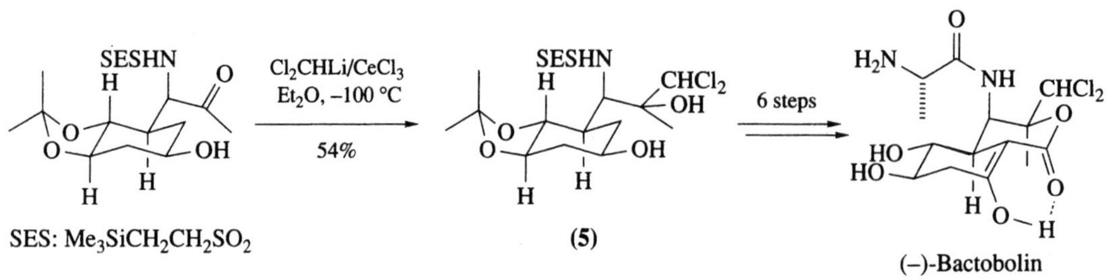

SES: $Me_3SiCH_2CH_2SO_2$

$Cl_2CHLi/CeCl_3$, Et_2O, –100 °C, 54%; 6 steps

(**5**)

(–)-Bactobolin

Scheme 7

The stereochemistry in the reactions mentioned above is consistent with a chelation-controlled addition of the organocerium reagents. On the other hand, nonchelation-controlled addition of alkylcerium reagents to carbonyl components has also been observed, as shown in Scheme 8.[35,36]

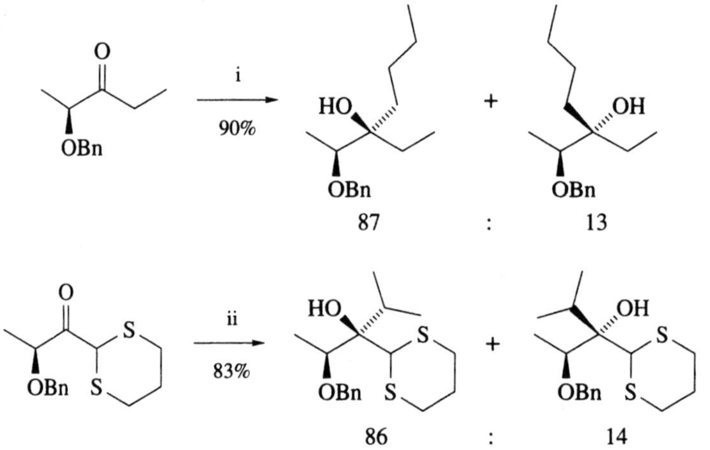

i, 90%: 87 : 13

ii, 83%: 86 : 14

i, Bu^nCeCl_2, THF, –78 °C; ii, Pr^iCeCl_2, THF, –78 to –60 °C

Scheme 8

Johnson and Tait prepared a trimethylsilylmethylcerium reagent and examined its reaction with carbonyl compounds.[17] The reagent adds to aldehydes and ketones including many readily enolizable ones to afford 2-hydroxysilanes. This modified Peterson reagent gives vastly superior yields in comparison with trimethylsilylmethyllithium. An example is shown in Scheme 9.

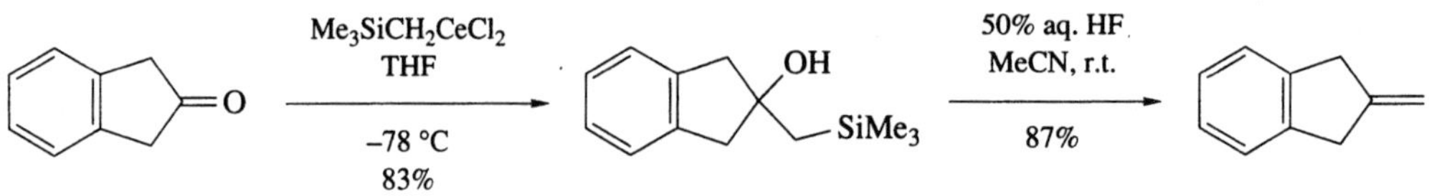

$Me_3SiCH_2CeCl_2$, THF, –78 °C, 83%; 50% aq. HF, MeCN, r.t., 87%

Scheme 9

The same reagent reacts with acyl chlorides to afford 1,3-bis(trimethylsilyl)-2-propanol derivatives, which efficiently undergo a trimethylchlorosilane-promoted Peterson reaction to afford allysilanes in high overall yields (equation 5).[37]

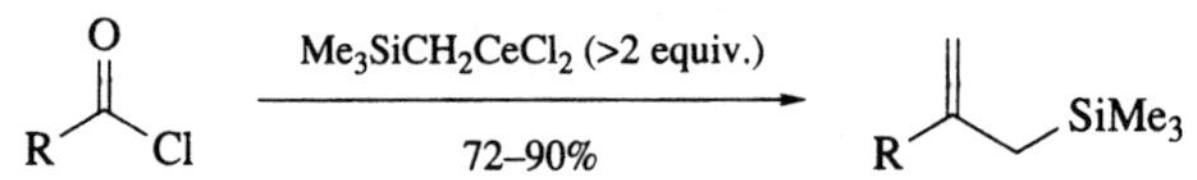

$$RCOCl \xrightarrow[72\text{–}90\%]{Me_3SiCH_2CeCl_2\ (>2\ equiv.)} RC(=CH_2)CH_2SiMe_3 \quad (5)$$

R = n-C_9H_{19}, $Ph(CH_2)_2$, PhCH=CH, PhCH=C(Me)

Recently, Mudryk and Cohen have found that reactions of lactones with organocerium reagents provide lactols in good yields, as exemplified in Scheme 10.[38] This reaction leads to an efficient one-pot synthesis of spiroketals, as illustrated in Scheme 11.[38]

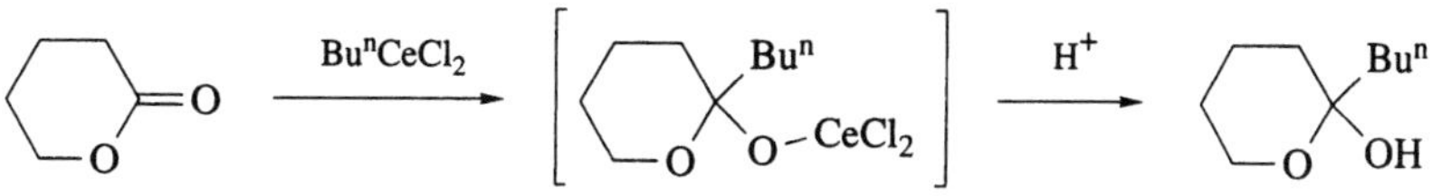

Bu^nCeCl_2 H+ Bu^n O-CeCl_2 OH

Scheme 10

radical anion Li OLi i, $CeCl_3$ ii, Et iii, H^+ 60%

Cl_2Ce $OCeCl_2$ i, ii, H^+ 65% 76% 72%

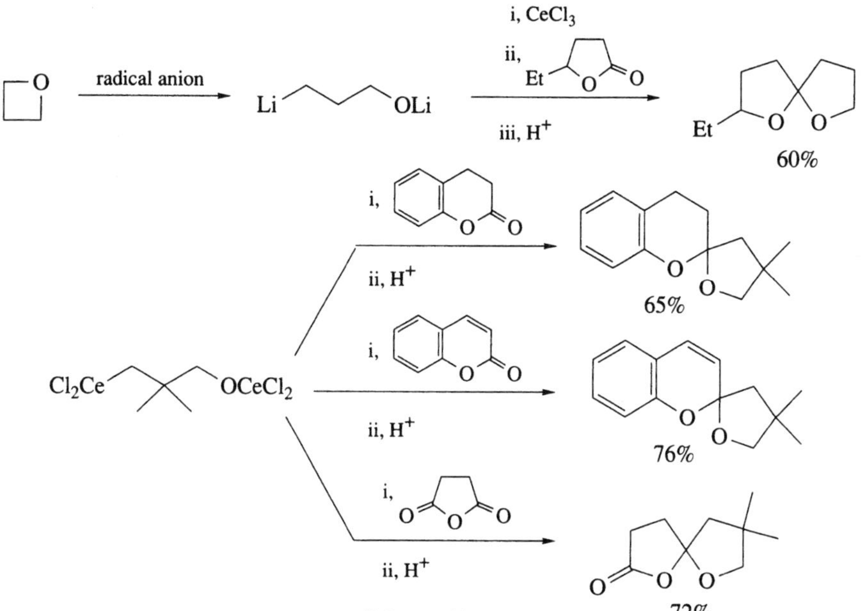

Scheme 11

Denmark *et al.* have recently reported that aldehyde SAMP-hydrazones react with various alkylcerium reagents in good yields and with high diastereoselectivities.[31] The method can be applied to the synthesis of optically active primary amines, as exemplified in Scheme 12.

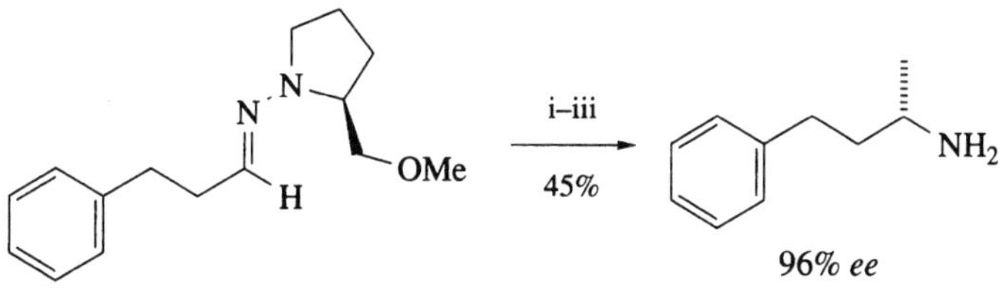

N OMe H i–iii 45% NH_2 96% *ee*

i, $MeCeCl_2$, THF; ii, MeOH; iii, H_2 (375 psi = 2.59 MPa), Raney nickel, 60 °C

Scheme 12

1.8.2.6.2 Allylcerium reagents

Cohen *et al.* have extensively studied the generation and reactivities of allylcerium reagents.[39,40] Allylcerium reagents react with α,β-unsaturated carbonyl compounds in a 1,2-selective fashion. It is particularly noteworthy that unsymmetrical allylcerium reagents react with aldehydes or enals mainly at the least-substituted terminus, as opposed to other allylorganometallics such as allyltitanium reagents. An example is shown in Scheme 13.[39]

Another prominent feature is that the reaction at –78 °C provides (*Z*)-alkenes, while at –40 °C (*E*)-alkenes are formed. Allylcerium reagents have been successfully employed in an economical synthesis of

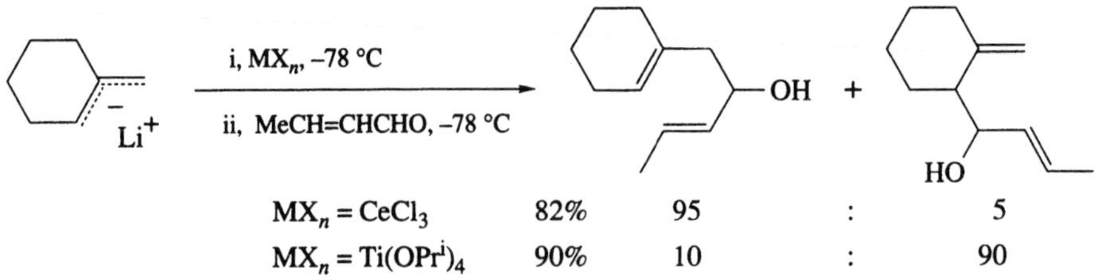

i, MX_n, –78 °C
ii, MeCH=CHCHO, –78 °C

$MX_n = CeCl_3$	82%	95	:	5
$MX_n = Ti(OPr^i)_4$	90%	10	:	90

Scheme 13

some pheromones. Typical examples, which illustrate these characteristic reactivities, are shown in Schemes 14 and 15.[40]

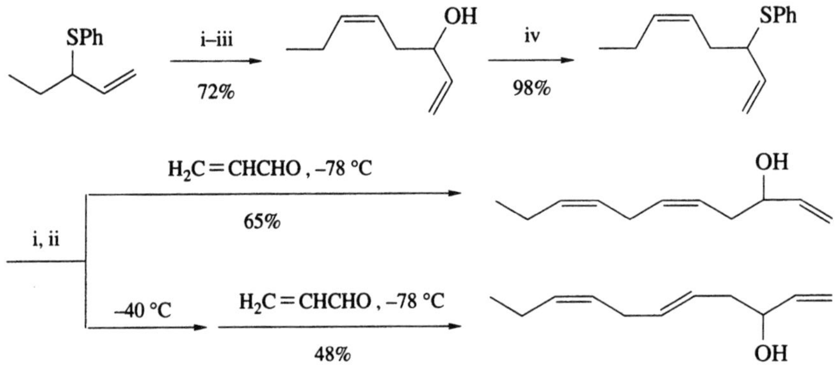

i, lithium *p,p'*-di-t-butylbiphenylide (LDBB) or lithium 1-(dimethylamino)naphthalenide (LDMAN), –60 °C; ii, $CeCl_3$, –78 °C; iii, H_2C=CHCHO; iv, $Bu^n{}_3P$/PhSSPh

Scheme 14

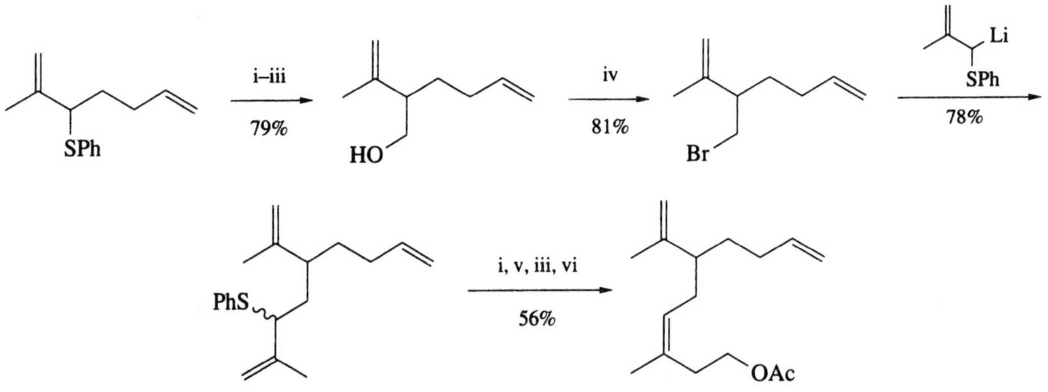

(Z):(E) = 98:2

i, LDBB; ii, $Ti(OPr^i)_4$; iii, CH_2O; iv, Ph_3P/CBr_4, MeCN; v, $CeCl_3$; vi, Ac_2O/pyridine

Scheme 15

1.8.2.6.3 Alkenyl- and aryl-cerium reagents

Suzuki *et al.* found that α-trimethylsilylvinylcerium reagents add to readily enolizable β,γ-enones. The method has been employed in the synthesis of (–)-eldanolide, as shown in Scheme 16.[41]

Paquette and his coworkers have employed alkenylcerium reagents in the efficient stereoselective synthesis of polycyclic molecules.[42,43] A typical example is illustrated in Scheme 17. β,γ-Enone (**6**) reacts with alkenylcerium reagent (**7**) with high diastereoselectivity to give adducts (**8**) and (**9**) in a ratio of 95:5. The major product (**8**) undergoes an oxy-Cope rearrangement, creating two chiral centers with high stereoselectivity to furnish (**10**).

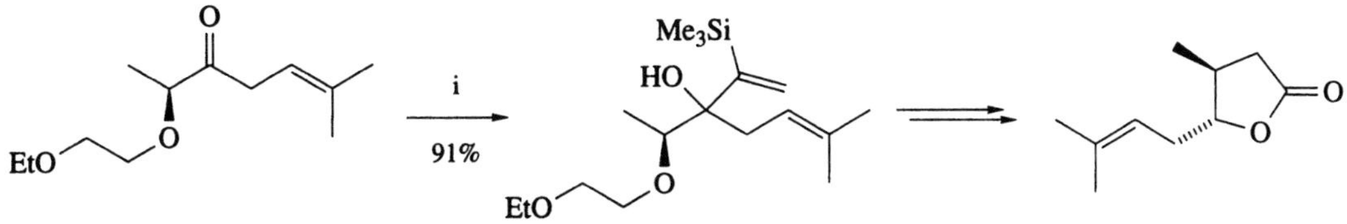

i, $H_2C{=}C(SiMe_3)CeCl_2$, THF–Et_2O–hexane (4:1:1), –78 °C, 0.5 h

Scheme 16

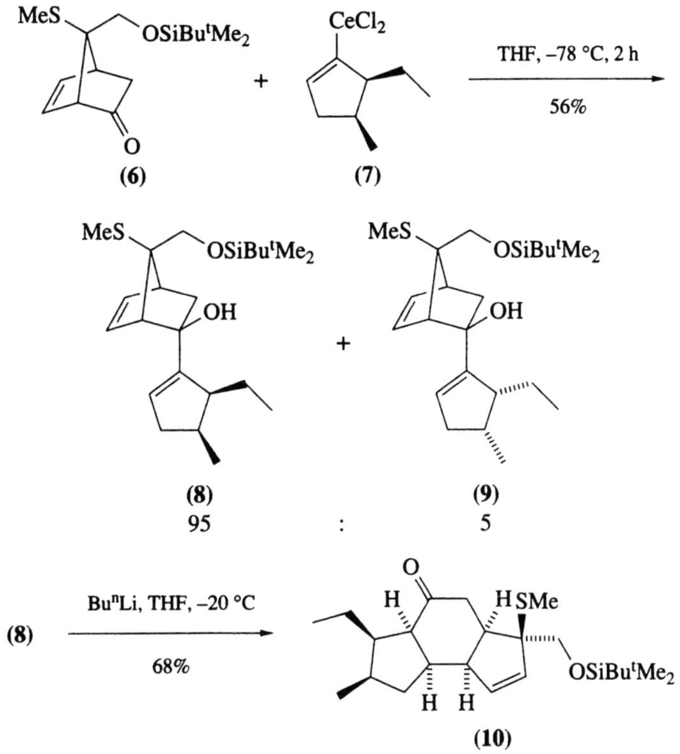

(6) **(7)**

THF, –78 °C, 2 h; 56%

(8) 95 : **(9)** 5

(8) Bu^nLi, THF, –20 °C; 68% → **(10)**

Scheme 17

A notably stereoselective reaction of an arylcerium reagent has been reported by Terashima *et al.*[44] As shown in Scheme 18, the cerium reagent provides adducts **(11)** and **(12)** in a ratio of 16:1 in 95% combined yield. In contrast, the organolithium reagent gives a lower and reversed stereoselectivity.

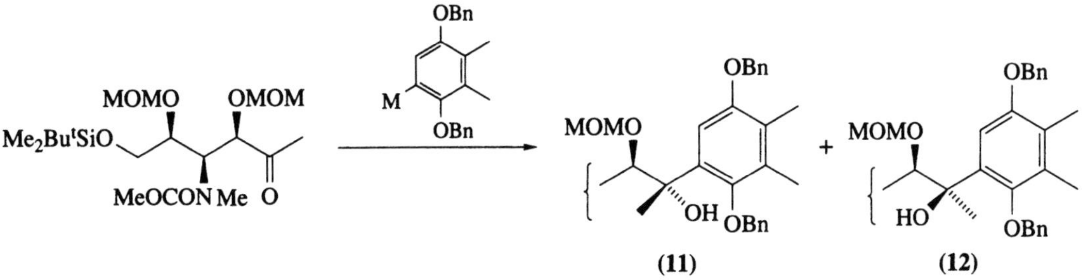

(11) **(12)**

M	*Conditions*	*Yield* (%)	*(11):(12)*
$CeCl_2$	THF, –78 °C	95	94:6
Li	THF, 0 °C	56	12:88
	Ether/THF (4:1), 0 °C	74	11:89
	Ether, 0 °C	77	34:66

Scheme 18

Recently, a new method for synthesizing coumarin derivatives has exploited the properties of arylcerium reagents, as illustrated in Scheme 19.[45] Interestingly, the bulky arylcerium reagent (**13**) adds to the easily enolizable *t*-butyl acetoacetate in satisfactory yield.

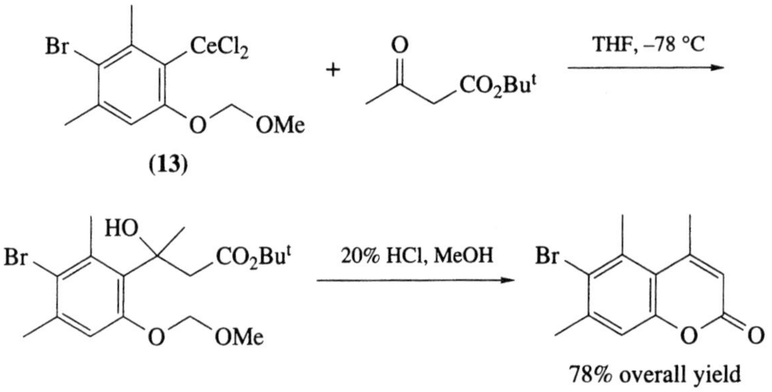

THF, –78 °C

(**13**)

20% HCl, MeOH

78% overall yield

Scheme 19

An example of the reaction of an alkenylcerium reagent with a cyclopentanone derivative has been reported. As shown in Scheme 20, the (*E*)-cerium reagent adds to the ketone to provide a single adduct in modest yield.[46] However, the (*Z*)-cerium reagent did not react, presumably due to steric effects.

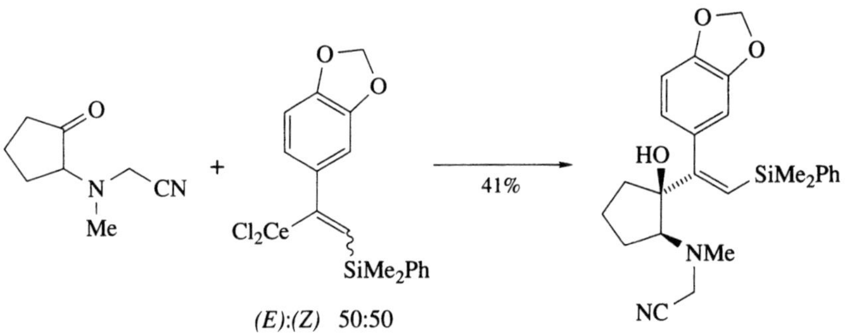

41%

(*E*):(*Z*) 50:50

Scheme 20

1.8.2.6.4 Alkynylcerium reagents

The trimethylsilylethynylcerium reagent, which was initially prepared by Terashima *et al.*, is useful for adding ethynyl groups to carbonyl moieties.[47] This method was successfully employed in the preparation of daunomycinone and related compounds.[47–51] Illustrative examples are shown in Scheme 21.[49]

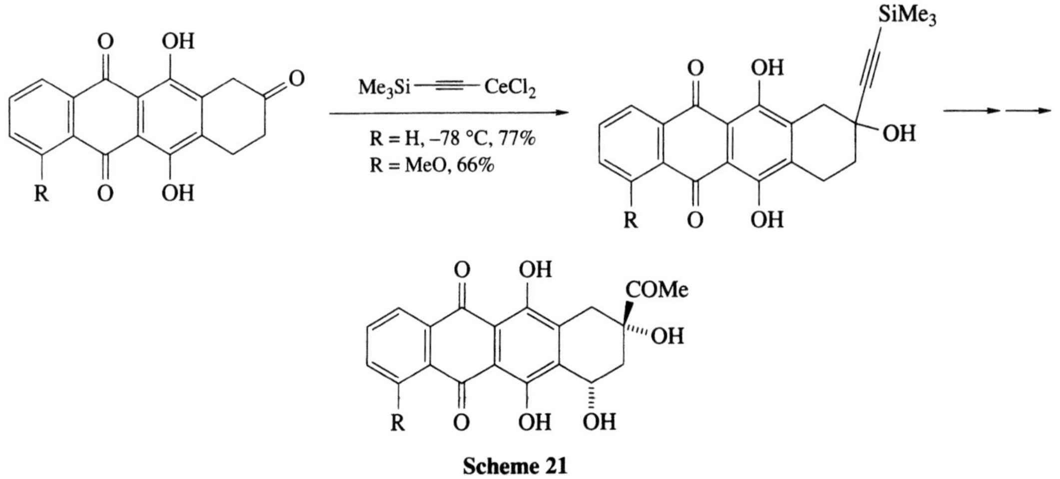

$Me_3Si—≡—CeCl_2$

R = H, –78 °C, 77%
R = MeO, 66%

Scheme 21

The utility of this reagent has been demonstrated by Tamura *et al.* in its reaction with the readily enolizable ketone (**14**).[52,53] The ethynylation of the ketone proceeds smoothly with the cerium reagent, as shown in Scheme 22; in sharp contrast, the corresponding lithium reagent provides the desired adduct (**15**) in only 11% yield.

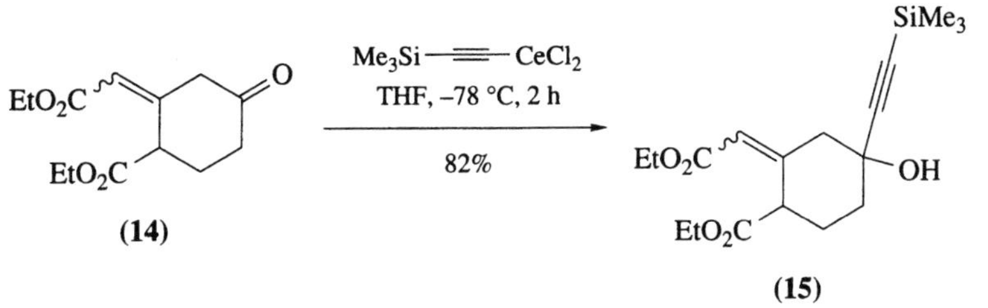

Scheme 22

Some other alkynylcerium reagents have been generated and used for the synthesis of alkynyl alcohols and related compounds in good yields.[54–56] An example is illustrated in Scheme 23.

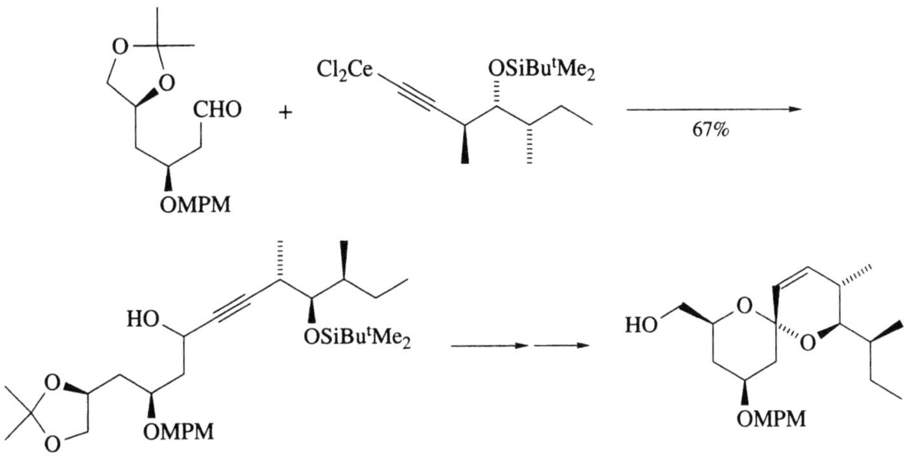

Scheme 23

1.8.3 CERIUM ENOLATES

Cerium enolates are generated by the reaction of lithium enolates with anhydrous cerium chloride in THF.[57] The cerium enolates react readily with various aldehydes and ketones at –78 °C (Scheme 24). The yields are generally higher than in reactions of lithium enolates. This is presumably due to the relative stabilities of the adducts, that of the cerium reagent being greater by virtue of coordination to the more oxophilic cerium atom. The stereochemistry of the products is almost the same as in the case of lithium enolates, as shown in Table 4. The reaction is assumed to proceed through a six-membered, chair-like transition state, as with lithium enolates.

A synthetic application of this cerium chloride methodology has been reported by Nagasawa *et al.*, as shown in Scheme 25.[58] It is noteworthy that aldol reaction of the cerium enolate proceeds in high yields, even though the acceptor carbonyl group is sterically crowded and is readily enolized by lithium enolates.

Fukuzawa *et al.* found that reduction of α-halo ketones followed by aldol reaction with aldehydes or ketones is promoted by CeI_3, $CeCl_3$/NaI or $CeCl_3$/$SnCl_2$.[59,60] These reactions are carried out at room

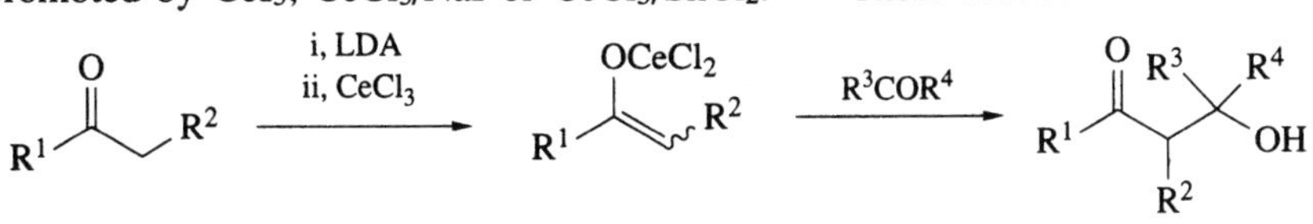

Scheme 24

Table 4

Enolized ketone	*Acceptor carbonyl compound*	*Reagent*	*Yield (%)*	*Threo:erythro*
$PhCOCH_2Me$	$MeCOCH_2Me$	$LDA–CeCl_3$	62	40:60
	$MeCOCH_2Me$	LDA	11	37:63
	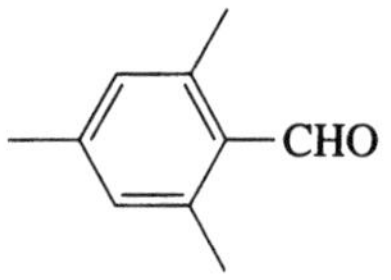 CHO	$LDA–CeCl_3$	93	20:80
		LDA	63	20:80
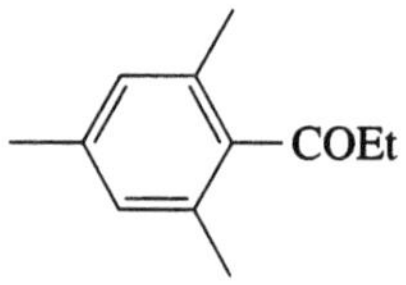 COEt	PhCHO	$LDA–CeCl_3$	94	91:9
	PhCHO	LDA	60	91:9
	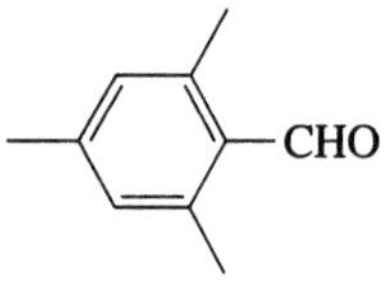 CHO	$LDA–CeCl_3$	91	93:7
		LDA	26	88:12

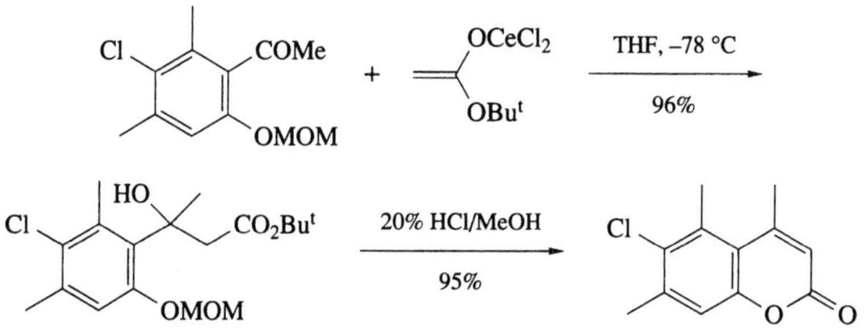

Cl COMe OMOM + $OCeCl_2$ OBu^t THF, –78 °C 96%

Cl HO CO_2Bu^t OMOM 20% HCl/MeOH 95% Cl O O

Scheme 25

temperature. Use of $CeCl_3$ provides α,β-unsaturated carbonyl compounds, while methods using $CeCl_3/NaI$ or $CeCl_3/SnCl_2$ afford exclusively β-hydroxy ketones, as shown in Scheme 26.

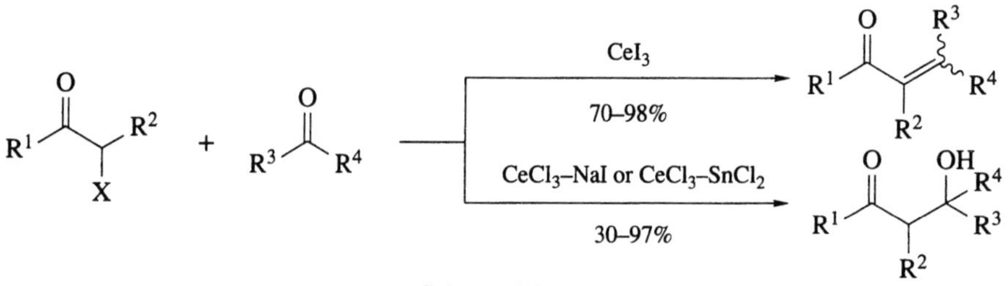

R^1 O R^2 X + O R^3 R^4 — CeI_3 70–98% → R^1 O R^3 R^4 R^2; $CeCl_3–NaI$ or $CeCl_3–SnCl_2$ 30–97% → R^1 O OH R^4 R^3 R^2

Scheme 26

1.8.4 GRIGNARD REAGENT/CERIUM CHLORIDE SYSTEMS

The addition of Grignard reagents to C—X π-bonds is undoubtedly one of the most fundamental and versatile reactions in synthetic organic chemistry. Nevertheless, it is also well recognized that these reactions are often accompanied by undesirable side reactions such as enolization, reduction, condensation, conjugate addition and pinacol coupling. In some cases, such abnormal reactions prevail over the 'normal addition reaction', resulting in poor yields of the desired products.

Table 5 Reactions of Carbonyl Compounds with Grignard Reagents in the Presence of Cerium Chloride or with Grignard Reagents Alone[a]

Carbonyl compound	*Reagent*	*Method*	*Product(s)*	*Yield (%)*
Et_3CCOMe	$MeMgBr/CeCl_3$	A	$Et_3CC(OH)Me_2$	95
Et_3CCOMe	MeMgBr		$Et_3CC(OH)Me_2$	0
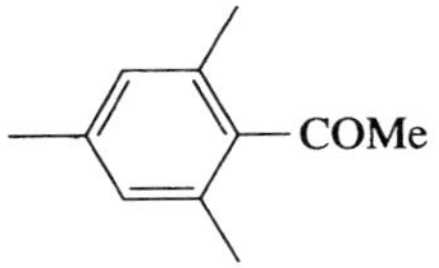 COMe	$MeMgBr/CeCl_3$	A	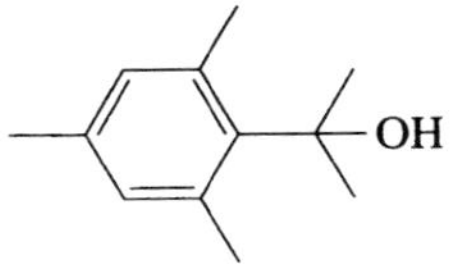 OH	47
	MeMgBr			Trace
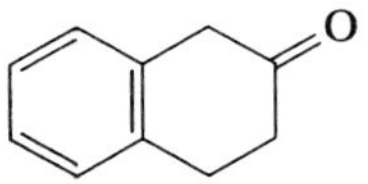 O	$EtMgCl/CeCl_3$	A	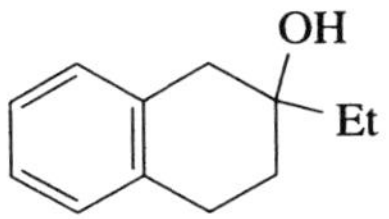 OH Et	76
	EtMgCl			8
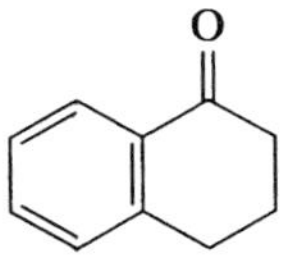 O	$Pr^iMgCl/CeCl_3$	A	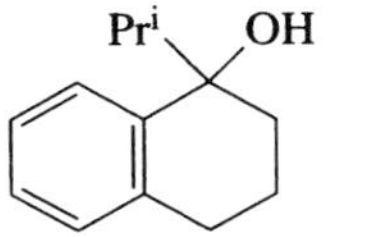 Pr^i OH	73
	Pr^iMgCl			15
PhCOMe	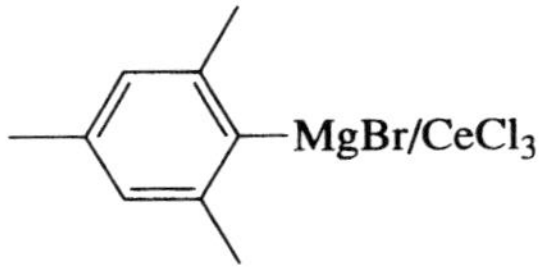 $MgBr/CeCl_3$	A	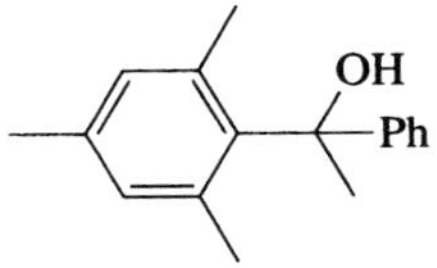 OH Ph	73
	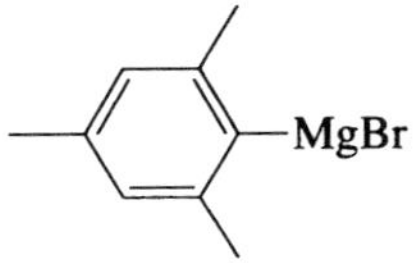 MgBr			5

Table 5 *(continued)*

Carbonyl compound	*Reagent*	*Method*	*Product(s)*	*Yield (%)*
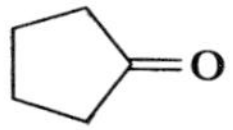	$Pr^iMgCl/CeCl_3$	B	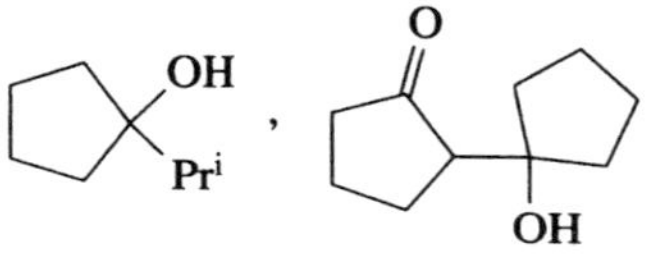	3, 80
	Pr^iMgCl			72, Trace
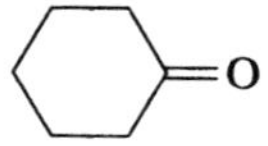	$Pr^iMgCl/CeCl_3$	B	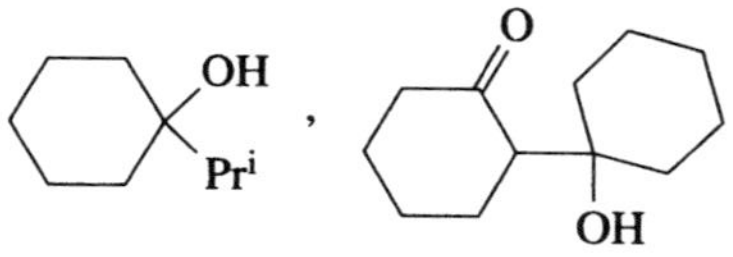	80, 0
	Pr^iMgCl			30, 35
Pr^i_2CO	$Pr^iMgCl/CeCl_3$	A	Pr^i_3COH, Pr^i_2CHOH	52, 31
Pr^i_2CO	Pr^iMgCl		Pr^i_3COH, Pr^i_2CHOH	3, 58
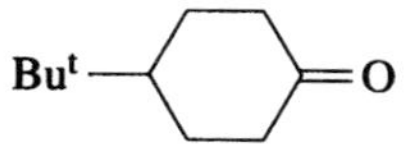	$Bu^tMgCl/CeCl_3$	A	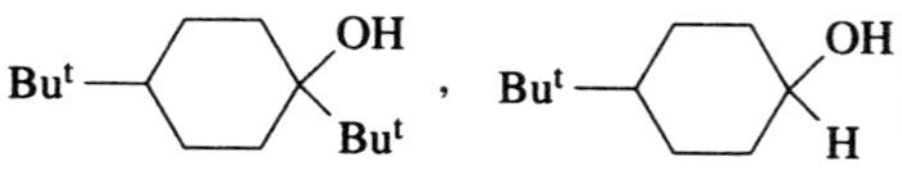	89, 10
	Bu^tMgCl			38, 40
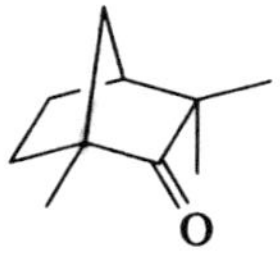	$Me_2C{=}CH(CH_2)_2MgBr/CeCl_3$	B[b]	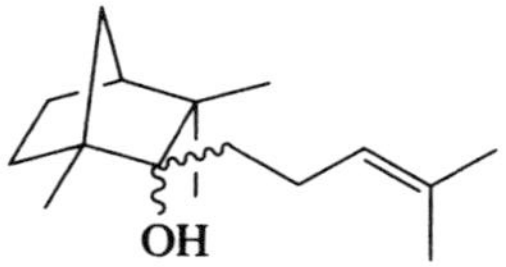	72
	$Me_2C{=}CH(CH_2)_2MgBr$			0
$PhCOCH_2Br$	$H_2C{=}CHMgBr/CeCl_3$	A	$PhC(OH)(CH_2Br)CH{=}CH_2$	95
$PhCOCH_2Br$	$H_2C{=}CHMgBr$		$PhC(OH)(CH_2Br)CH{=}CH_2$	66
PhCH=CHCOPh	$PhMgBr/CeCl_3$	A	$PhCH{=}CHC(OH)Ph_2$, Ph_2CHCH_2COPh	58, 33
PhCH=CHCOPh	$PhMgBr/CeCl_3$	A[c]	$PhCH{=}CHC(OH)Ph_2$, Ph_2CHCH_2COPh	89, 11
PhCH=CHCOPh	$PhMgBr/CeCl_3$	B[c]	$PhCH{=}CHC(OH)Ph_2$, Ph_2CHCH_2COPh	88, 9
PhCH=CHCOPh	$MeMgBr/CeCl_3$	A	PhCH=CHC(OH)MePh, $PhMeCHCH_2COPh$	92, 6
PhCH=CHCOPh	MeMgBr		PhCH=CHC(OH)MePh, $PhMeCHCH_2COPh$	52, 41
(Z)-PhCH=CHCOPh	$PhMgBr/CeCl_3$	A	$PhCH{=}CHC(OH)Ph_2$[d], Ph_2CHCH_2COPh	67, 19
(Z)-PhCH=CHCOPh	PhMgBr		$PhCH{=}CHC(OH)Ph_2$[e], Ph_2CHCH_2COPh	15, 83

Table 5 *(continued)*

Carbonyl compound	*Reagent*	*Method*	*Product(s)*	*Yield (%)*
cyclohex-2-enone	$Pr^iMgCl/CeCl_3$	A	1-isopropylcyclohex-2-en-1-ol (OH, Pr^i), 3-isopropylcyclohexanone (Pr^i, =O)	91, 5
	Pr^iMgCl			12, 53
$PhCH_2CO_2Me$	$Pr^iMgCl/CeCl_3$	A	$PhCH_2C(OH)Pr^i_2$	97
$PhCH_2CO_2Me$	Pr^iMgCl		$PhCH_2C(OH)Pr^i_2$	0
$PhCH=CHCO_2Et$	$H_2C=CHMgBr/CeCl_3$	B	$PhCH=CHC(OH)(CH=CH_2)_2$, $PhCH=CHCO(CH_2)_2CH=CH_2$	68, 9
$PhCH=CHCO_2Et$	$H_2C=CHMgBr$		$PhCH=CHC(OH)(CH=CH_2)_2$, $PhCH=CHCO(CH_2)_2CH=CH_2$	24, 39
$PhCH_2CONMe_2$	$Bu^nMgBr/CeCl_3$	B[f]	$PhCH_2COBu^n$	66
$PhCH_2CONMe_2$	Bu^nMgBr		$PhCH_2COBu^n$	8

[a] All reactions are carried out in THF at 0 °C with a molar ratio of 1:1.5:1.5 (carbonyl compound:Grignard reagent:$CeCl_3$) unless otherwise stated. [b] Molar ratio 1:2:3 (carbonyl compound:Grignard reagent:$CeCl_3$). [c] Molar ratio 1:1.5:2.5 (carbonyl compound:Grignard reagent:$CeCl_3$). [d] (*Z*):(*E*) = 91:9. [e] (*Z*):(*E*) = 60:40. [f] Molar ratio 1:3:3 (amide:Grignard reagent:$CeCl_3$).

It has been found that use of cerium chloride as an additive effectively suppresses abnormal reactions, resulting in the formation of normal addition products in significantly improved yields.[61,62] The reactions are usually carried out by one of the following two methods.

(i) The Grignard reagent is added to the suspension of cerium chloride in THF at 0 °C, the mixture is stirred well for 1–2 h at the same temperature and finally the substrate is added (method A). As vinylic Grignard reagents decompose rapidly on treatment with cerium chloride at 0 °C, reactions using these reagents should be carried out at a lower temperature.

(ii) The Grignard reagent is added at 0 °C to the mixture of substrate and cerium chloride in THF that has previously been stirred well for 1 h at room temperature (method B).

Representative examples of the reactions of various carbonyl compounds with Grignard reagents under these conditions are listed in Table 5.[62] It is emphasized that enolization, aldol reaction, ester condensation, reduction and 1,4-addition are remarkably suppressed by the use of cerium chloride. Various tertiary alcohols, which are difficult to prepare by the conventional Grignard reaction, can be synthesized by this method.

The Grignard reagent/cerium chloride system has been applied to practical organic syntheses.[63–69] A typical example is shown in Scheme 27.[63] In sharp contrast to the reaction in the presence of cerium chloride, the Grignard reagent alone affords only a 2% yield of the adduct.

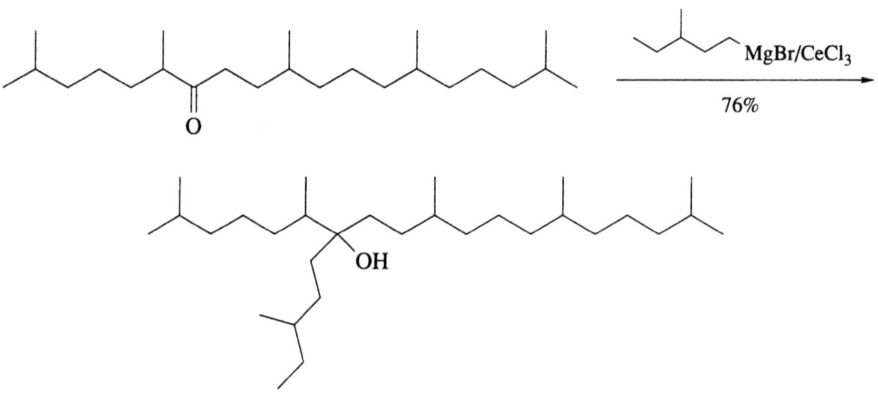

O

$MgBr/CeCl_3$

76%

OH

Scheme 27

Recently this method has been successfully applied to the preparation of substituted allylsilanes from esters. A variety of allylsilanes with other functional groups have been synthesized in good yields, as shown in Scheme 28.[70,71]

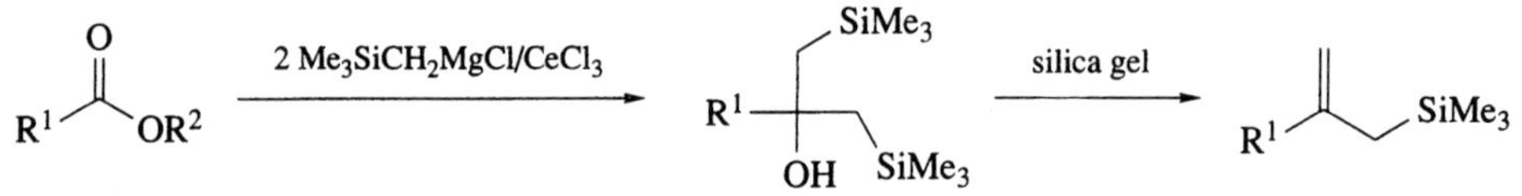

$R^1C(O)OR^2$ —— 2 $Me_3SiCH_2MgCl/CeCl_3$ ——→ $R^1C(OH)(CH_2SiMe_3)_2$ —— silica gel ——→ $R^1C(=CH_2)CH_2SiMe_3$

$R^1 = Me_3SiCH_2$, $Cl(CH_2)_n$ (n = 1, 3), $(MeO)_2CH(CH_2)_n$ (n = 0, 1, 3, 4), PhCH=CH, *etc.*; R^2 = Me or Et

Scheme 28

1.8.5 REFERENCES

1. T. J. Marks and F. D. Ernst, in 'Comprehensive Organometallic Chemistry', ed. G. Wilkinson, F. G. A. Stone and E. W. Abel, Pergamon Press, Oxford, 1982, vol. 3, p. 173.
2. T. J. Marks, *Prog. Inorg. Chem.*, 1979, **25**, 224.
3. J. H. Forsberg and T. Moeller, in 'Gmelin Handbook of Inorganic Chemistry', ed. T. Moeller, U. Kruerke and E. Schleitzer-Rust, 8th edn., Springer, Berlin, 1983, p. 137.
4. H. B. Kagan and J.-L. Namy, in 'Handbook on the Physics and Chemistry of Rare Earths', ed. K. A. Gschneidner, Jr. and L. Eyring, North-Holland, Amsterdam, 1984, vol. 6, p. 525.
5. H. B. Kagan, in 'Fundamental and Technological Aspects of Organo *f*-Elements Chemistry', ed. T. J. Marks and I. L. Fragala, Reidel, New York, 1985, p. 49.
6. H. B. Kagan and J.-L. Namy, *Tetrahedron*, 1986, **42**, 6573.
7. N. R. Natale, *Org. Prep. Proced. Int.*, 1983, **15**, 387.
8. J. R. Long, *Aldrichimica Acta*, 1985, **18**, 87.

9. H. Schumann, *Angew. Chem., Int. Ed. Engl.*, 1984, **23**, 474.
10. W. J. Evans, *Adv. Organomet. Chem.*, 1985, **24**, 131.
11. W. J. Evans, *Polyhedron*, 1987, **6**, 803.
12. T. Imamoto, Y. Sugiura and N. Takiyama, *Tetrahedron Lett.*, 1984, **25**, 4233.
13. T. Imamoto, T. Kusumoto, Y. Tawarayama, Y. Sugiura, T. Mita, Y. Hatanaka and M. Yokoyama, *J. Org. Chem.*, 1984, **49**, 3904.
14. T. Imamoto, T. Kusumoto and M. Yokoyama, *J. Chem. Soc., Chem. Commun.*, 1982, 1042.
15. A. R. Pray, *Inorg. Synth.*, 1957, **5**, 153.
16. H. J. Heeres, J. Renkema, M. Booij, A. Meetsma and J. H. Teuben, *Organometallics*, 1988, **7**, 2495.
17. C. R. Johnson and B. D. Tait, *J. Org. Chem.*, 1987, **52**, 281.
18. T. Imamoto, *Pure Appl. Chem.*, 1990, **62**, 747.
19. T. Imamoto, T. Kusumoto and T. Oshiki, unpublished results.
20. Y. Ukaji and T. Fujisawa, *Tetrahedron Lett.*, 1988, **29**, 5165; Y. Ukaji, A. Yoshida and T. Fujisawa, *Chem. Lett.*, 1990, 157.
21. T. Imamoto, T. Kusumoto, Y. Sugiura, N. Suzuki and N. Takiyama, *Nippon Kagaku Kaishi*, 1985, 445.
22. T. Kauffmann, C. Pahde, A. Tannert and D. Wingbermühle, *Tetrahedron Lett.*, 1985, **26**, 4063.
23. B. Weidmann and D. Seebach, *Angew. Chem., Int. Ed. Engl.*, 1983, **22**, 31.
24. M. T. Reetz, 'Organotitanium Reagents in Organic Synthesis', Springer, Berlin, 1986.
25. Y. Okude, S. Hirano, T. Hiyama and H. Nozaki, *J. Am. Chem. Soc.*, 1977, **99**, 3179.
26. T. Kauffmann, A. Hamsen and C. Beirich, *Angew. Chem., Int. Ed. Engl.*, 1982, **21**, 144.
27. T. Imamoto and Y. Sugiura, *J. Organomet. Chem.*, 1985, **285**, C21.
28. T. Imamoto and Y. Sugiura, *J. Phys. Org. Chem.*, 1989, **2**, 93.
29. R. Sauvetre, M.-C. Roux-Schmitt and J. Seyden-Penne, *Tetrahedron*, 1978, **34**, 2135.
30. M. Wada, K. Yabuta and K. Akiba, in '35th Symposium on Organometallic Chemistry, Japan, Osaka, 1988' Kinki Chemical Society, Osaka, 1988, p. 238.
31. S. E. Denmark, T. Weber and D. W. Piotrowski, *J. Am. Chem. Soc.*, 1987, **109**, 2224.
32. E. A. Mash, *J. Org. Chem.*, 1987, **52**, 4142.
33. E. J. Corey and D.-C. Ha, *Tetrahedron Lett.*, 1988, **29**, 3171.
34. R. S. Garigipati, D. M. Tschaen and S. M. Weinreb, *J. Am. Chem. Soc.*, 1990, **112**, 3475.
35. K. Suzuki and T. Ohkuma, private communication.
36. T. Sato, R. Kato, K. Gokyu and T. Fujisawa, *Tetrahedron Lett.*, 1988, **29**, 3955.
37. M. B. Anderson and P. L. Fuchs, *Synth. Commun.*, 1987, **17**, 621.
38. B. Mudryk, C. A. Shook and T. Cohen, *J. Am. Chem. Soc.*, 1990, **112**, 6389.
39. B.-S. Guo, W. Doubleday and T. Cohen, *J. Am. Chem. Soc.*, 1987, **109**, 4710; W. D. Abraham and T. Cohen, *J. Am. Chem. Soc.*, 1991, **113**, 2313.
40. T. Cohen and M. Bhupathy, *Acc. Chem. Res.*, 1989, **22**, 152.
41. K. Suzuki, T. Ohkuma and G. Tsuchihashi, *Tetrahedron Lett.*, 1985, **26**, 861.
42. L. A. Paquette, K. S. Learn, J. L. Romine and H.-S. Lin, *J. Am. Chem. Soc.*, 1988, **110**, 879.
43. L. A. Paquette, D. T. DeRussy and J. C. Gallucci, *J. Org. Chem.*, 1989, **54**, 2278; L. A. Paquette, N. A. Pegg, D. Toops, G. D. Maynard and R. D. Rogers, *J. Am. Chem. Soc.*, 1990, **112**, 277; L. A. Paquette, D. T. DeRussy, T. Vandenheste and R. D. Rogers, *J. Am. Chem. Soc.*, 1990, **112**, 5562.
44. M. Kawasaki, F. Matsuda and S. Terashima, *Tetrahedron*, 1988, **44**, 5695.
45. K. Nagasawa and K. Ito, *Heterocycles*, 1989, **28**, 703.
46. L. E. Overman and H. Wild, *Tetrahedron Lett.*, 1989, **30**, 647.
47. M. Suzuki, Y. Kimura and S. Terashima, *Chem. Lett.*, 1984, 1543.
48. M. Suzuki, Y. Kimura and S. Terashima, *Chem. Pharm. Bull.*, 1986, **34**, 1531.
49. Y. Tamura, M. Sasho, S. Akai, H. Kishimoto, J. Sekihachi and Y. Kita, *Tetrahedron Lett.*, 1986, **27**, 195.
50. Y. Tamura, M. Sasho, S. Akai, H. Kishimoto, J. Sekihachi and Y. Kita, *Chem. Pharm. Bull.*, 1987, **35**, 1405.
51. Y. Tamura, S. Akai, H. Kishimoto, M. Kirihara, M. Sasho and Y. Kita, *Tetrahedron Lett.*, 1987, **28**, 4583.
52. Y. Tamura, M. Sasho, H. Ohe, S. Akai and Y. Kita, *Tetrahedron Lett.*, 1985, **26**, 1549; Y. Tamura, S. Akai, H. Kishimoto, M. Sasho, M. Kirihara and Y. Kita, *Chem. Pharm. Bull.*, 1988, **36**, 3897.
53. T. Chamberlain, X. Fu, J. T. Pechacek, X. Peng, D. M. S. Wheeler and M. M. Wheeler, *Tetrahedron Lett.*, 1991, **32**, 1710.
54. C. M. J. Fox, R. N. Hiner, U. Warrier and J. D. White, *Tetrahedron Lett.*, 1988, **29**, 2923.
55. L. M. Harwood, S. A. Leeming, N. S. Isaacs, G. Jones, J. Pickardt, R. M. Thomas and D. Watkin, *Tetrahedron Lett.*, 1988, **29**, 5017.
56. K. Takeda, S. Yano and E. Yoshii, *Tetrahedron Lett.*, 1988, **29**, 6951; E. Vedejs and S. L. Dax, *Tetrahedron Lett.*, 1989, **30**, 2627; K. Narasaka, N. Saito, Y. Hayashi and H. Ichida, *Chem. Lett.*, 1990, 1411.
57. T. Imamoto, T. Kusumoto and M. Yokoyama, *Tetrahedron Lett.*, 1983, **24**, 5233.
58. K. Nagasawa, H. Kanbara, K. Matsushita and K. Ito, *Tetrahedron Lett.*, 1985, **26**, 6477.
59. S. Fukuzawa, T. Fujinami and S. Sakai, *J. Chem. Soc., Chem. Commun.*, 1985, 777.
60. S. Fukuzawa, T. Tsuruta, T. Fujinami and S. Sakai, *J. Chem. Soc., Perkin Trans. 1*, 1987, 1473.
61. T. Imamoto, N. Takiyama and K. Nakamura, *Tetrahedron Lett.*, 1985, **26**, 4763.
62. T. Imamoto, N. Takiyama, K. Nakamura, T. Hatajima and Y. Kamiya, *J. Am. Chem. Soc.*, 1989, **111**, 4392.
63. J. N. Robson and S. J. Rowland, *Tetrahedron Lett.*, 1988, **29**, 3837.
64. K. Suzuki, K. Tomooka, E. Katayama, T. Matsumoto and G. Tsuchihashi, *J. Am. Chem. Soc.*, 1986, **108**, 5221.
65. C. R. Johnson and J. F. Kadow, *J. Org. Chem.*, 1987, **52**, 1493.
66. J. H. Rigby, J. Z. Wilson and C. Senanayake, *Tetrahedron Lett.*, 1986, **27**, 3329.
67. C. P. Jasperse and D. P. Curran, *J. Am. Chem. Soc.*, 1990, **112**, 5601.
68. J. W. Herndon, L. A. McMullen and C. E. Daitch, *Tetrahedron Lett.*, 1990, **31**, 4547.
69. L. Jisheng, T. Gallardo and J. B. White, *J. Org. Chem.*, 1990, **55**, 5426.

70. T. V. Lee, J. A. Channon, C. Cregg, J. R. Porter, F. S. Roden and H. T.-L. Yeoh, *Tetrahedron* , 1989, **45**, 5877.
71. B. A. Narayanan and W. H. Bunnelle, *Tetrahedron Lett.*, 1987, **28**, 6261.

1.9
Samarium and Ytterbium Reagents

GARY A. MOLANDER
University of Colorado, Boulder, CO, USA

1.9.1 INTRODUCTION

In contrast to more traditional organometallic nucleophiles, the application of samarium and ytterbium reagents to selective organic synthesis has a relatively brief history. Early studies of lanthanide reagents simply sought to develop reactivity patterns mimicking those of more established organolithium or Grignard reagents. However, as the complexity of organic molecules requiring synthesis increased, demands for more highly selective reagents heightened accordingly. Consequently, the search for reagents which would complement those of the traditional organometallic nucleophiles brought more serious attention to the lanthanides, and an explosive expansion in the application of lanthanide reagents to organic synthesis began. This growth is perhaps best reflected by the publication of a number of excellent review articles.[1] In addition, quite thorough surveys on the synthetic and structural aspects of organolanthanide chemistry have appeared.[2] This chapter concentrates on applications of samarium and ytterbium reagents in selective organic synthesis, and specifically their employment in C—X π-bond addition reactions. Emphasis is placed on transformations that are unique to the lanthanides, and on processes that complement existing synthetic methods utilizing more traditional organometallic reagents. Since additions to C—X π-bonds comprise one of the most essential aspects of carbon–carbon bond formation, fundamental contributions made in this area are by definition of substantial importance to the art of organic synthesis.

The perception that lanthanide metals were rare and therefore inaccessible or expensive was a contributing factor to their long-lasting neglect and slow development as useful synthetic tools. In fact, 'rare earths' in general are relatively plentiful in terms of their abundance in the earth's crust. Samarium and ytterbium occur in proportions nearly equal to those of boron and tin, for example.[3] Modern separation methods have made virtually all of the lanthanides readily available in pure form at reasonable cost.

Unlike many of their main group and transition metal counterparts, inorganic lanthanide compounds are generally classified as nontoxic when introduced orally.[4] In fact, samarium chloride and ytterbium chloride exhibit similar toxicity to that of sodium chloride (LD_{50} of >2000–6700 mg kg^{-1} in mice *versus* 4000 mg kg^{-1} for NaCl). Although toxicity may obviously vary to some extent based on the ligands attached to the metal, in most cases lanthanide complexes are converted to hydroxides immediately on ingestion, and thus have limited absorption through the digestive tract. Moderate toxicity is exhibited by lanthanide salts introduced *via* the intraperitoneal route.

The +3 oxidation state is the most stable oxidation state for both samarium and ytterbium. The +2 oxidation state of ytterbium (f^{14}), and samarium (f^6) is also of great importance with regard to applications in organic synthesis. As expected on the basis of its electronic configuration, Yb^{2+} is the more stable of these two dipositive species, and Sm^{2+} is a powerful one-electron reducing agent (Sm^{2+}/Sm^{3+} = –1.55 V, Yb^{2+}/Yb^{3+} = –1.15 V). The utility of Sm^{2+} as a reductant in organic synthesis is discussed in detail below, and various aspects of its chemistry have been previously reviewed as well.[1] The type of two-electron redox chemistry on a single metal center, typical of several transition elements, is not observed in lanthanides.

It is the special combination of inherent physical and chemical properties of the lanthanides that sets them apart from all other elements, and provides a unique niche for these elements and their derivatives in selective organic synthesis. The lanthanides as a group are quite electropositive (electronegativities of samarium and ytterbium are 1.07 and 1.06, respectively, on the Allred–Rochow scale[3]) and the chemistry of these elements is predominantly ionic. This is because the 4*f*-electrons do not have significant radial extension beyond the filled $5s^25p^6$ orbitals of the xenon inert gas core.[2b] The lanthanides therefore behave as closed-shell inert gasses with a tripositive charge, and in general electrostatic and steric interactions play a greater role in the chemistry of the lanthanides than do interactions between the metal and associated ligand orbitals.[2b,5]

The *f*-orbitals do play a bonding role in complexes in which the coordination number is higher than nine. Compared with transition metals, the ionic radii of the lanthanides are large.[3] Most transition metal ionic radii lie in the range from 0.6 to 1.0 Å, whereas the lanthanides have an average ionic radius of approximately 1.2 Å. Divalent species are, of course, even larger; eight-coordinate Sm^{2+} has an ionic radius of 1.41 Å, for example. The relatively large ionic radii of the lanthanides allow the accommodation of up to 12 ligands in the coordination sphere, and coordination numbers of seven, eight and nine are common. Owing to the well-known 'lanthanide contraction', ionic radii decrease steadily across the row of lanthanides in the periodic table; eight-coordinate Sm^{3+} has an ionic radius of 1.219 Å, whereas the ionic radius of eight-coordinate Yb^{3+} is only 1.125 Å.[3] The lanthanide contraction is a consequence of poor shielding of the 4*f*-electrons, resulting in an increase in effective nuclear charge and a concomitant decrease in ionic radius. As expected, higher coordination numbers are most common in the larger, early lanthanides.

According to the concept of hard and soft acids and bases (HSAB) established by Pearson,[6] lanthanide +3 ions are considered to be hard acids, falling between Mg^{2+} and Ti^{4+} in the established scale. Lanthanides therefore complex preferentially to hard bases such as oxygen donor ligands.

The strong affinity of lanthanides for oxygen is further evidenced by the bond dissociation energies ($D°_0$) for the gas phase dissociation of diatomic lanthanide oxides (LnO).[7] Although they are among the lowest values for the lanthanides, both SmO (136 kcal mol^{-1}; 1 cal = 4.18J) and YbO (95 kcal mol^{-1}) exhibit values significantly higher than that for MgO (86.6 kcal mol^{-1}). This demonstrated oxophilicity (strong metal–oxygen bonds and hard Lewis acid character) has been used to great advantage in organic synthesis. As described below, these properties have been exploited extensively to enhance carbonyl substrate reactivity, and also to control stereochemistry in carbonyl addition reactions through chelation.

1.9.2 SAMARIUM REAGENTS

Surprisingly few organosamarium reagents have been synthesized and exploited for their utility in selective organic synthesis. However, examples of both organosamarium(III) and organosamarium(II) reagents are known, and provide some insight into potentially useful areas of further study. By far the most extensive work thus far has been carried out on use of samarium(II) species as reductants and reductive coupling agents in organic synthesis. Applications of all three of these types of reagents to C—X π-bond additions are discussed below.

1.9.2.1 Organosamarium(III) Reagents

Organosamarium(III) dihalides are typically generated by a simple transmetalation reaction involving $SmCl_3$ or SmI_3 and 1 equiv. of an organolithium or Grignard reagent. Unfortunately, little characterization of such reactions has been performed. As a consequence, limited information is available on the structure of these molecules or the exact nature of the reactive species in such mixtures. While they have been denoted as simple monomeric σ-bonded species ('$RSmX_2$'), certainly other compositions (*e.g.* 'ate' complexes or species resulting from Schlenk-type equilibria) cannot be excluded.

Reagents prepared in this fashion have been demonstrated to undergo facile carbonyl addition reactions. Two promising features of these reactions have emerged. The first is that modest selectivity in reactions of aldehydes over that of ketones can be achieved (equation 1).[8] The second, and perhaps more exciting, development is the efficient reaction of organosamarium(III) reagents with highly enolizable ketones (equation 2).[9] Good yields of 1,2-addition products can be obtained from ketone substrates that produce less than 35% of carbonyl addition product on reaction with organolithium reagents alone. Rigorously anhydrous samarium salts must be used in order to achieve high yields of the alcohols. This carbonyl addition process with enolizable ketones takes advantage of the attenuated basicity of organosamarium reagents as compared to that of their organolithium counterparts, as well as the amplified Lewis acidity of the Sm^{3+} ion. The combination of these factors apparently minimizes enolization of the carbonyl substrate, while at the same time enhancing nucleophilic addition. Although only a limited study of this phenomenon has been carried out to this point, the process rivals that of organocerium reagents (Volume 1, Chapter 1.8), and certainly the results bode well for future exploration.

Bu^nLi/SmI_3 + n-C_6H_{13}CHO/EtCOEt —(i, THF, –70 °C; ii, H_3O^+; 85%)→ n-C_6H_{13}CH(OH)Bu^n (76%) + $Et_2C(OH)Bu^n$ (24%) (1)

$Bu^nLi/SmCl_3$ + $PhCH_2COCH_2Ph$ —(i, THF, –78 °C, 3 h; ii, H_3O^+; 60%)→ $(PhCH_2)_2C(OH)Bu^n$ (2)

Reaction of benzylic halides with 2 equiv. of dicyclopentadienylsamarium reportedly generates a benzylsamarium(III) species along with dicyclopentadienylsamarium halide (equation 3).[10] Benzylsamariums undergo carbonyl addition with a variety of aldehydes and ketones, providing high yields of the corresponding alcohols (equation 4).[10,11] These complexes also react with carboxylic acid chlorides, providing modest yields of ketones and minor amounts of dibenzylated tertiary alcohol by-products (equation 5).[10,11]

$2Cp_2Sm$ + $ArCH_2X$ —(THF, r.t.)→ $ArCH_2SmCp_2$ + Cp_2SmX (3)

$PhCH_2SmCp_2$ + n-C_6H_{13}CHO —(i, THF, r.t.; ii, H_3O^+; 80%)→ n-$C_6H_{13}CH(OH)CH_2Ph$ (4)

$PhCH_2SmCp_2$ + Bu^tCOCl —(THF, –30 °C; 35%)→ Bu^tCOCH_2Ph (88%) + $Bu^tC(OH)(CH_2Ph)_2$ (12%) (5)

The preparation and reactivity of organosamarium 'ate' complexes have also been described.[12] By addition of methyllithium to samarium trichloride in Et_2O in the presence of TMEDA, $[Li(TMEDA)]_3[Sm(Me)_6]$ can be isolated in 48% yield as a crystalline solid. This complex is extremely sensitive toward air and moisture, and yet has been fully characterized. Preliminary studies indicate that it provides good selectivity for 1,2-carbonyl addition in reactions with several unsaturated aldehydes and ketones (equation 6). Unoptimized yields in all cases exceed 80%. Chemoselectivity has also been briefly studied with somewhat less impressive results. Reaction of the 'ate' complex with a 1:1 mixture of benzalacetone and cinnamaldehyde demonstrated that addition to the aldehyde was favored by a ratio of only 1.5:1 to 2:1.

$$\text{mesityl oxide} + [\text{Li(TMEDA)}]_3[\text{SmMe}_6] \xrightarrow[-78\ ^\circ\text{C, 2 h};\ >80\%]{\text{THF, Et}_2\text{O}} \text{1,2-adduct (OH)}\ (>95\%) + \text{Bu}^t\text{ ketone}\ (<5\%) \quad (6)$$

These preliminary results indicate that substantial promise holds for the further development of organosamarium(III) reagents. The synthesis and characterization of new organosamarium complexes, and application of these reagents to selective synthetic transformations awaits more extensive exploration.

1.9.2.2 Organosamarium(II) Reagents

Like the organosamarium(III) reagents discussed in Section 1.9.2.1, few organosamarium(II) reagents have been satisfactorily characterized. In many respects, however, the known chemistry of these reagents mimics that of organomagnesium halides. For example, the preparation of organosamarium(II) halides is apparently best carried out by reaction of samarium metal with organic halides (equation 7).[13] This procedure generates a mixture of divalent and trivalent organosamarium species as determined by magnetic susceptibility measurements and a variety of other analytical methods. Solutions of these reagents exhibit reactivity with ketones that is very similar to that of the corresponding Grignard reagents (equation 8).

$$\text{Sm} + \text{PhI} \xrightarrow[-30\ ^\circ\text{C}]{\text{THF}} \text{'PhSmI'} \quad (7)$$

$$\text{'PhSmI'} + \text{Ph}_2\text{C=O} \xrightarrow[41\%]{\text{THF}} \text{Ph}_3\text{COH} \quad (8)$$

Reactions of organosamarium(II) halides with aldehydes are somewhat more complicated and synthetically less useful than those with ketones. The ability of Sm^{2+} species to serve as strong reducing agents introduces a number of alternative reaction pathways.[14] For example, reaction of 'EtSmI' with benzaldehyde provides a mixture of benzyl alcohol, benzoin, hydrobenzoin, and benzyl benzoate in low yields. The first three products presumably arise from benzaldehyde ketyl, generated by single-electron transfer from the Sm^{2+} reagent to benzaldehyde. The benzyl benzoate apparently is derived from a Tischenko-type condensation reaction between a samarium alkoxide species and benzaldehyde.[14b]

The intermediate generated by the reaction of samarium metal with methyl β-bromopropionate has found a useful niche in selective synthesis. This reagent, when treated with acetophenone, generates a γ-lactone directly in yields of about 70% (equation 9).[15] Although a significant amount (20%) of pinacol product from the ketone is also generated in this process, the method does provide a useful alternative to the use of other β-metallo ester nucleophiles. Similar processes using zinc or magnesium in place of samarium provide the γ-lactones in yields of less than 30%, with unreacted starting material the predominant substance isolated. A samarium(II) ester homoenolate is postulated as the reaction intermediate in this transformation. Several other lanthanide metals (*e.g.* cerium, lanthanum and neodymium) have been found to work equally well in this transformation.

$$\text{BrCH}_2\text{CH}_2\text{CO}_2\text{Me} + \text{Sm} \xrightarrow[\text{r.t.}]{\text{THF}} [\text{BrSm}\cdots\text{O=C(OMe)CH}_2\text{CH}_2] \xrightarrow{\text{PhCOMe}} \gamma\text{-lactone (Me, Ph)}\ \ 70\% \quad (9)$$

1.9.2.3 Reactions Promoted by Samarium Diiodide and Dicyclopentadienylsamarium

Although organosamarium reagents have made little impact thus far in selective organic synthesis, the emergence of low-valent samarium reagents as reductive coupling agents has had a major influence on the field of C—X π-bond addition reactions. The disclosure of a convenient procedure for generation of samarium diiodide (SmI_2) and subsequent development of this reagent by Kagan and his coworkers precipitated explosive growth in the application of this reductant in organic synthesis. Simple functional group reductions as well as a host of reductive coupling reactions have since been investigated. In these processes, SmI_2 demonstrates reactivity and selectivity patterns which nicely complement reductants such as zinc, magnesium and other low-valent metal reductants. In addition to the advantages SmI_2 provides as a THF-soluble reductant, the Sm^{3+} ion generated as a result of electron transfer can serve as a template to control stereochemistry through chelation in C—X π-bond addition reactions. It has thus become the reagent of choice for numerous synthetic transformations.

Samarium diiodide is very conveniently prepared by oxidation of samarium metal with organic dihalides[16] or with iodine (equations 10–12).[17] Deep blue solutions of SmI_2 (0.1 M in THF) are generated in virtually quantitative yields by these processes. This salt can be stored as a solution in THF for long periods when it is kept over a small amount of samarium metal. Tetrahydrofuran solutions of SmI_2 are commercially available as well. If desired, the solvent may be removed to provide $SmI_2{\cdot}(THF)_n$ as a powder. For synthetic purposes, SmI_2 is typically generated and utilized *in situ*.

$$Sm + ICH_2CH_2I \xrightarrow[0\,^{\circ}C,\ 1\ h]{THF} SmI_2 + 0.5\ H_2C{=}CH_2 \quad (10)$$

$$Sm + ICH_2CH_2CH_2I \xrightarrow[0\,^{\circ}C,\ 1\ h]{THF} SmI_2 + H_2C{=}CH_2 \quad (11)$$

$$Sm + I_2 \xrightarrow[65\,^{\circ}C,\ 16\ h]{THF} SmI_2 \quad (12)$$

Other ether solvents (*e.g.* Et_2O, DME) are ineffective for the preparation of SmI_2, and samarium(II) salts such as $SmBr_2$ are only slowly generated by procedures analogous to those utilized for preparation of SmI_2. Furthermore, $SmBr_2$ is not nearly as soluble in THF as is SmI_2. With the exception of studies on dicyclopentadienylsamarium (*vide infra*), little effort has been made in exploring other potential samarium(II) reducing salts.

Samarium diiodide has been characterized in solution by absorption spectroscopy, magnetic susceptibility measurements, titrations of lanthanide ions with EDTA, potentiometric titrations of iodide ion, and acidometric titration and reaction of iodine, which measures the reductive capability of the solutions.[16b,c] All of these analyses are consistent with a species possessing the stoichiometry 'SmI_2'. However, little is known of the degree of aggregation or solution structure of this reagent. Crystal structure determinations of $SmI_2(NCBu^t)_2$ and $SmI_2[O(CH_2CH_2OMe)_2]_2$ have been performed.[18] The former is an infinite chain of $SmI_2(NCBu^t)_2$ with bridging iodides. The geometry about the samarium ion in this complex is a distorted octahedron. The diglyme complex is monomeric in the solid state; the geometry about the octacoordinate samarium ion is best described as a distorted hexagonal bipyramid.

Dicyclopentadienylsamarium (Cp_2Sm) is readily prepared by reaction of SmI_2 with dicyclopentadienyl sodium.[19] It is a red powder that can be stored for days at a time under an inert atmosphere without any apparent decomposition. Although it has limited solubility in most organic solvents, Cp_2Sm is emerging as a useful reductant. It appears to have reduction capabilities even greater than those of SmI_2.

Both SmI_2 and Cp_2Sm have unique characteristics which lend themselves to selective organic synthesis, and their application to a variety of problems in formal C—X π-bond addition reactions is outlined below.

1.9.2.3.1 Barbier-type reactions

As a homogeneous reductant, SmI_2 provides many advantages over more traditional reagents such as magnesium or lithium in Barbier-type syntheses. Both intermolecular and intramolecular variants of the Barbier reaction utilizing SmI_2 are finding important uses in the synthesis of complex organic molecules.

Samarium diiodide has been utilized successfully to promote intermolecular Barbier-type reactions between ketones and a variety of organic halides and other substrates.[16c] Allylic and benzylic halides (chlorides, bromides and iodides) react with ketones within a few minutes at room temperature in THF when treated with 2 equiv. of SmI_2 (equation 13). Unsymmetrical allylic halides provide mixtures of regioisomers in these coupling reactions. Diallylated (or dibenzylated) tertiary alcohols result from SmI_2-promoted reactions of allylic iodides or benzyl bromide with carboxylic acid halides (equation 14).[20] In some instances, homocoupling of the organic halide or carboxylic acid halide effectively competes with the desired cross-coupling reaction.

Ph⌒Br + n-C_6H_{13}C(O)Me —($2SmI_2$, THF; r.t., 0.5 h; 69%)→ n-C_6H_{13}C(OH)(Me)CH_2Ph (13)

n-C_7H_{15}C(O)Cl + 2 allyl iodide —($4SmI_2$, THF; r.t., 20 min; 72%)→ n-C_7H_{15}C(OH)(allyl)$_2$ (14)

Allylic halides are not as readily accessible as allylic alcohols or their ester derivatives. Thus, the requirement that allylic halides must be used as precursors for carbonyl addition reactions in conjunction with magnesium and other similar reductants is a severe restriction limiting the convenience of these routes to homoallylic alcohols. In this regard, samarium diiodide can be used to great advantage, because substrates other than allylic halides are suitable precursors for such transformations. For example, allylic phosphate esters have been reported to couple with carbonyl substrates in the presence of SmI_2 (equation 15).[21] Since esters and nitriles are unreactive under these conditions, the SmI_2-mediated process is likely to be more chemoselective than those promoted by magnesium or lithium.

n-C_6H_{13}C(O)Me + prenyl O–P(O)(OPr^i)$_2$ —($2SmI_2$, THF; r.t., 1 h; 93%)→ α-adduct n-C_6H_{13}C(OH)(Me)CH_2CH=CMe_2 (64%) + γ-adduct n-C_6H_{13}C(OH)(Me)CMe_2CH=CH_2 (36%) (15)

Curiously, the geometry about the allylic double bond is retained in these reactions (as demonstrated by stereospecific reactions with neryl phosphate and geranyl phosphate), even though the coupling lacks regioselectivity (*i.e.* homoallylic alcohols are isolated as mixtures of products in which coupling has occurred at the α- and γ-positions of the allylic phosphates). The procedure also lacks stereoselectivity; reactions of substituted allylic phosphates with prochiral aldehydes and ketones provide mixtures of diastereomers. The process has other limitations; although alkyl ketones couple nicely utilizing this procedure, aryl ketones and aldehydes provide significant amounts (50–85%) of pinacol by-products. This is a result of the reducing capabilities of SmI_2. Furthermore, α,β-unsaturated aldehydes and ketones provide complex mixtures of products.

Allylic acetates can also be utilized as substrates in SmI_2-mediated carbonyl addition reactions, but only when these reactions are performed in the presence of palladium catalysts.[22] No reaction occurs in the absence of the palladium catalyst, and a π-allylpalladium species is undoubtedly a key intermediate. Once generated, the π-allylpalladium probably undergoes oxidative–reductive transmetalation with SmI_2, generating an allylsamarium species. The latter reacts with the aldehyde or ketone, providing the observed products (Scheme 1). Significantly, palladium(II) salts can also be utilized in the reaction, indicating that SmI_2 produces a palladium(0) species *in situ* which is capable of entering the catalytic cycle. In these reactions, there is a type of synergism between the palladium(0) catalyst and SmI_2; the latter serves as the stoichiometric reductant in the process, while the palladium catalyst functions as an activator for the relatively unreactive allylic acetate. Samarium diiodide regenerates the catalyst, and brings about a charge inversion in the process by converting the electrophilic π-allylpalladium species to a nucleophilic allylsamarium.

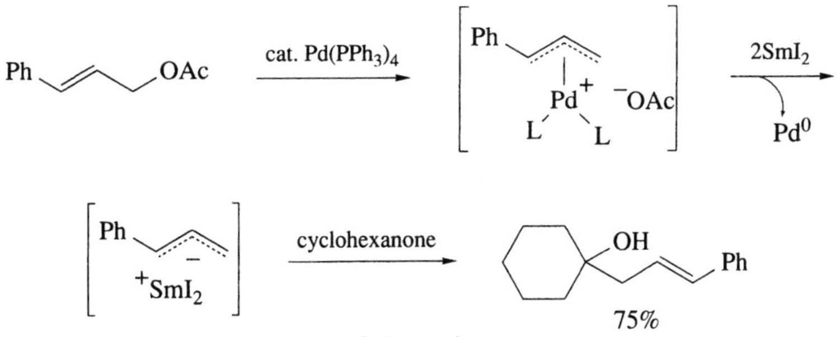

Scheme 1

In most cases, carbon–carbon bond formation occurs at the least substituted terminus of the allylic unit. A wide range of aldehydes and ketones can be utilized in the reaction, and one cyclization process has been reported (equation 16). Aromatic and α,β-unsaturated substrates cannot be used owing to competitive pinacolic coupling reactions promoted by SmI_2.

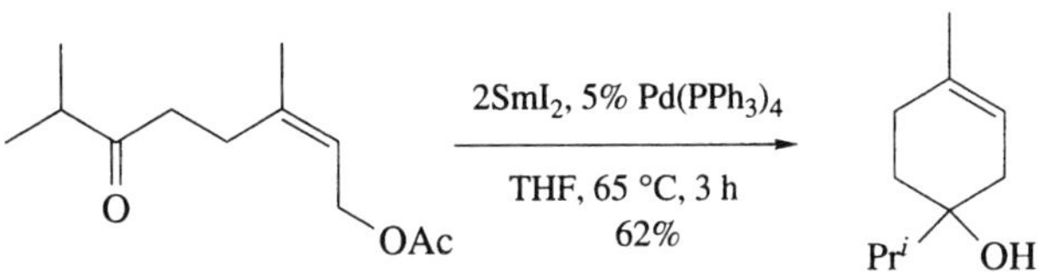

(16)

Propargylic acetates undergo analogous reactions with ketones.[23] Aldehydes can be utilized only with highly reactive propargylic acetates, again due to competitive pinacolic coupling. Primary propargylic acetates produce mixtures of allenic and homopropargylic alcohols, whereas most secondary and all tertiary propargylic carboxylates provide exclusively the allenic alcohols (equation 17). Although other transition metal salts, *e.g.* palladium(II), nickel(II) and cobalt(II), can be utilized as catalysts, lower yields are obtained.

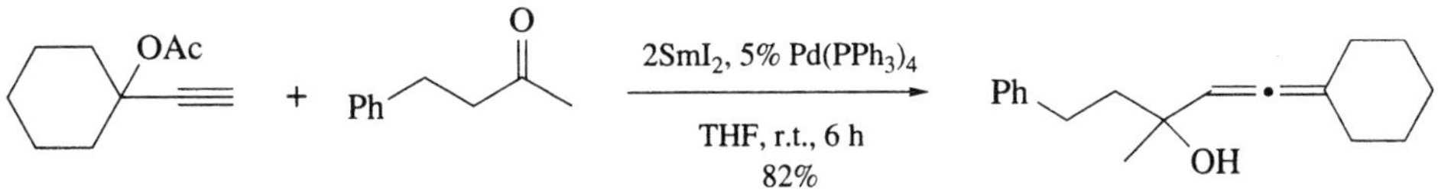

(17)

In addition to reactive allylic and benzylic substrates, other organic precursors have proven suitable for SmI_2-promoted intermolecular Barbier-type reactions. Primary organic iodides and even organic tosylates undergo carbonyl coupling, but under much harsher conditions than their allylic halide counterparts. Typically, reactions must be heated for 8–12 h in THF to accomplish complete conversion to product. Again, the ability to utilize organic tosylates in these transformations sets SmI_2 apart from more traditional reductants (equation 18). A Finkelstein-type reaction apparently converts alkyl tosylates to the corresponding iodides, which are subsequently involved in the coupling reaction. Addition of a catalytic amount of sodium iodide to the reaction mixture greatly facilitates the coupling. Alkyl bromides are less reactive than corresponding iodides and tosylates, and alkyl chlorides are virtually inert.

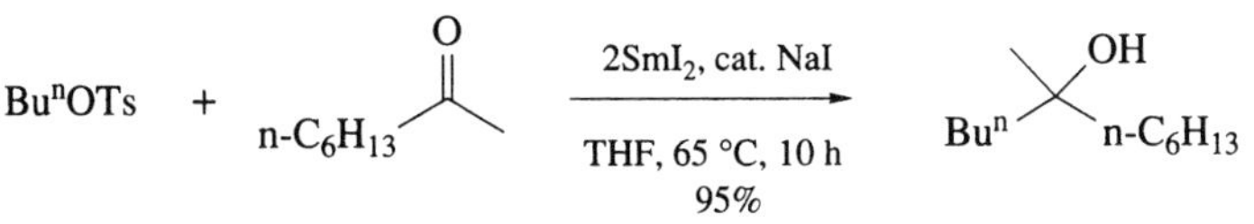

(18)

Much milder reaction conditions in the Barbier-type reaction can be employed by utilizing iron(III) salts as catalysts. For example, when 2 mol % $FeCl_3$ is added to SmI_2, the Barbier reaction between a primary organic iodide and a ketone is complete within 3 h at room temperature (equation 19). The iron(III) is probably reduced by SmI_2 to a low-valent species which serves as an efficient electron transfer catalyst, thus lowering the activation energy for the coupling process (*vide infra*).

$$Bu^nI + n\text{-}C_6H_{13}C(O)CH_3 \xrightarrow[\text{THF, r.t., 3 h; 73\%}]{2SmI_2,\ \text{cat. } FeCl_3} Bu^n(n\text{-}C_6H_{13})C(OH)CH_3 \quad (19)$$

Utilization of THF–HMPA as solvent for the reaction provides another useful technique for facilitating the SmI_2-mediated Barbier-type coupling reaction.[24] Even in the absence of a catalyst, both Bu^nBr and Bu^sBr are cleanly coupled with 2-octanone within 1 min at room temperature in this solvent system, providing greater than 90% yields of the desired tertiary alcohols.

Dicyclopentadienylsamarium (Cp_2Sm) presents a third means by which less reactive organic halides can be induced to participate in intermolecular Barbier-type processes (equation 20).[19] Experimental conditions in intermolecular Barbier reactions are much milder with Cp_2Sm (ambient temperature) than with SmI_2 (THF heated at reflux). Secondary alkyl iodides, reluctant to undergo SmI_2-mediated Barbier coupling under normal conditions, can be efficiently coupled with ketones utilizing Cp_2Sm.

$$Bu^nI + \text{4-}Bu^t\text{-cyclohexanone} \xrightarrow[\text{THF, r.t.; 65\%}]{2Cp_2Sm} \text{axial-OH, equatorial-}Bu^n\ (86\%) + \text{axial-}Bu^n\text{, equatorial-OH}\ (14\%) \quad (20)$$

Alkenyl halides and aromatic halides are unreactive with ketones in the presence of SmI_2 in THF.[16c] Pinacolic coupling products can be detected in 10–20% yield under these conditions. In THF/HMPA, iodobenzene reacts in the presence of a ketone to generate a phenyl radical, which abstracts a hydrogen from THF. Samarium diiodide induced coupling of the THF radical to the ketone (or ketyl) provides the major observed product (equation 21).[25]

$$PhI + PhCH_2CH_2C(O)CH_3 \xrightarrow[\text{THF, HMPA; 50\%}]{2SmI_2} PhCH_2CH_2C(OH)(CH_3)(\text{2-tetrahydrofuranyl}) \quad (21)$$

Aldehydes cannot be coupled to marginally reactive organic halides in SmI_2-promoted processes. A mixture of products results as a consequence of a Meerwein–Ponndorf process, initiated by reaction of the secondary samarium alkoxide intermediate with the aldehyde.[26] Highly reactive (allylic and benzylic) halides can be utilized and couple fairly efficiently with aldehydes, since they react quickly enough to suppress the undesired consecutive reaction. A two-step process (Barbier coupling followed by *in situ* oxidation) can be successfully employed with these reactive halides, providing high yields of coupled ketone (equation 22).[26b]

$$n\text{-}C_7H_{15}CHO + PhCH_2Br \xrightarrow{2SmI_2} \xrightarrow[\text{ii, } H_3O^+]{\text{i, } Bu^tCHO} n\text{-}C_7H_{15}C(O)CH_2Ph\ (74\%) + Bu^tCH_2OH \quad (22)$$

Dicyclopentadienylsamarium has been utilized to ameliorate the problem of Meerwein–Ponndorf oxidation in Barbier coupling reactions with aldehydes.[19] Dicyclopentadienylsamarium accelerates the coupling process, thereby preventing subsequent oxidation from occurring to any great extent. The enhanced reactivity of Cp_2Sm permits even secondary alkyl iodides to undergo Barbier reactions with aldehydes, providing the desired alcohols in reasonable yields (equation 23). Further studies are likely to uncover other useful reactivity patterns for Cp_2Sm that complement those of SmI_2.

$$Pr^iI + n\text{-}C_6H_{13}CHO \xrightarrow[\text{THF, r.t.; 50\%}]{3Cp_2Sm} n\text{-}C_6H_{13}CH(OH)Pr^i \quad (23)$$

The mechanism of the intermolecular SmI_2-promoted Barbier-type reaction is still open to debate. Direct S_N2-type displacement of the halide by a ketyl or a ketyl dianion is unlikely, since optically active

2-bromooctane reacts with cyclohexanone in the presence of SmI_2 to provide an optically inactive tertiary alcohol.[27] One caveat concerning this evidence is that SmI_2 itself reacts with organic halides in a Finkelstein-type reaction.[16c,27] Thus, racemization of the 2-bromooctane prior to coupling has not been ruled out in these studies.

Reduction of the organic halide to an organosamarium species might also be involved. Subsequent carbonyl addition by this organometallic reagent would provide the observed product. This mechanism finds support in studies utilizing 2-(bromomethyl)tetrahydrofuran as the alkyl halide substrate. Treatment of this halide with SmI_2 in the presence of 2-octanone produces a 60% yield of 4-penten-1-ol, and only 3% of coupled product.[27] This is indicative of the generation of a tetrahydrofurfuryl anion, which rapidly rearranges to the ring-opened alkoxide.

Another plausible mechanism for the SmI_2-mediated Barbier reaction involves coupling of ketyl and alkyl radicals in a diradical coupling mechanism.[27] Alternatively, addition of an alkyl radical to a Sm^{3+}-activated ketone carbonyl may be invoked.[28]

Highly selective synthetic transformations can be performed readily by taking advantage of the chemoselectivity of SmI_2. It has been pointed out that there is a tremendous reactivity differential in the Barbier-type reaction between primary organic iodides or tosylates on the one hand, and organic chlorides on the other. As expected, selective alkylation of ketones can be accomplished by utilizing appropriately functionalized dihalides or chlorosulfonates (equation 24).[16c] Alkenyl halides and, presumably, aryl halides can also be tolerated under these reaction conditions.

n-C_6H_{13}COMe + TsO(CH$_2$)$_4$Cl —[2SmI_2, cat. NaI; THF, 65 °C, 24 h; 62%]→ n-C_6H_{13}C(OH)(Me)(CH$_2$)$_4$Cl (24)

Nitriles and esters are also unreactive in SmI_2-promoted Barbier reactions. A very useful procedure for lactone synthesis has been developed making use of this fact. Treatment of γ-bromobutyrates or δ-bromovalerates with SmI_2 in THF/HMPA in the presence of aldehydes or ketones results in generation of lactones through a Barbier-type process (equations 25 and 26).[24] This nicely complements the β-metallo ester or 'homoenolate' chemistry of organosamarium(III) reagents described above (Section 1.9.2.1), and also the Reformatsky-type chemistry promoted by SmI_2 (Section 1.9.2.3.2). Further, it provides perhaps the most convenient route to γ- and δ-carbanionic ester equivalents yet devised.

n-$C_{11}H_{23}$CHO + Br(CH$_2$)$_4$CO$_2$Et —[2SmI_2; THF, HMPA; r.t., 1 min; 44%]→ δ-lactone bearing n-$C_{11}H_{23}$ (25)

Estrone methyl ether (MeO, H, H, H) + Br(CH$_2$)$_3$CO$_2$Me —[2SmI_2; THF, HMPA; r.t., 1 min; 41%]→ spiro γ-lactone (MeO, H, H, H) (26)

A very convenient hydroxymethylation process has been developed based on the SmI_2-mediated Barbier-type reaction.[29] Treatment of aldehydes or ketones with benzyl chloromethyl ether in the presence of SmI_2 provides the alkoxymethylated products in good to excellent yields. Subsequent reductive cleavage of the benzyl ether provides hydroxymethylated products. Even ketones with a high propensity for enolization can be alkylated by this process in reasonable yields. The method was utilized by White and Somers as a key step in the synthesis of (±)-deoxystemodinone (equation 27).[30] This particular ketone substrate resisted attack by many other nucleophilic reagents (such as methyllithium) owing to competitive enolate formation.

A unique alkoxymethylation reaction can be accomplished by treatment of α-alkoxycarboxylic acid chlorides with ketones in the presence of SmI_2 (equation 28).[31] The reaction is postulated to proceed by a reductive decarbonylation process, leading to a relatively stable α-alkoxy radical. Addition of this radical to the Sm^{3+}-activated carbonyl and further reduction and hydrolysis provides the observed product. An

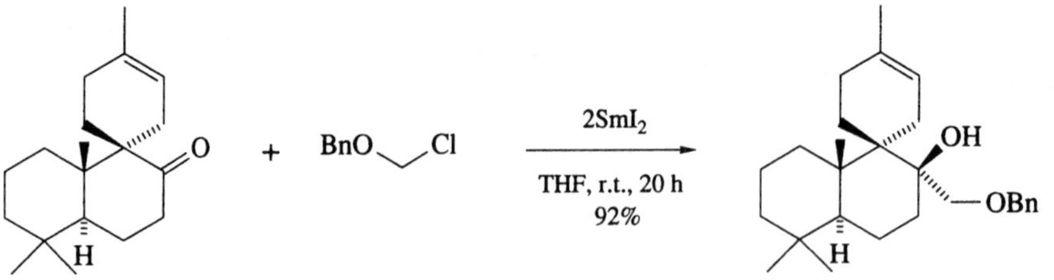

(27)

alternative mechanism involves reduction of the α-alkoxy radical to a transient anion, followed by nucleophilic addition and eventual hydrolysis. These two processes have not been distinguished at this point. With acid halides that do not afford a particularly stable radical on decarbonylation, reduction to the samarium acyl anion becomes competitive with the decarbonylation process. The chemistry of acyl radicals and samarium acyl anions is discussed separately in Section 1.9.2.3.5.

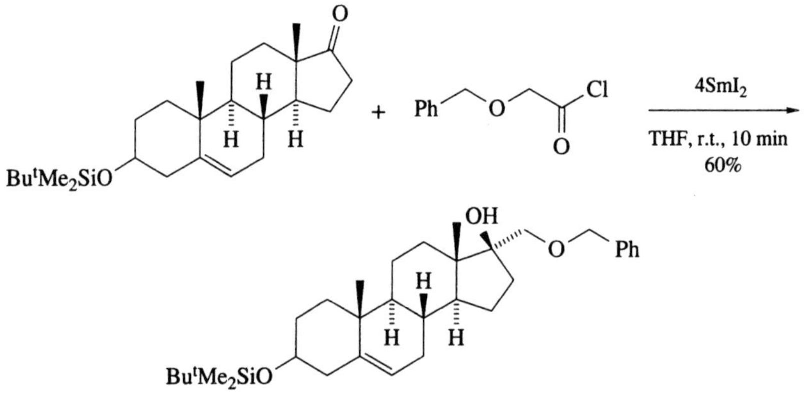

(28)

Halomethylation of aldehydes and ketones is difficult to achieve using α-halo organolithium species owing to the thermal instability of these organometallics. As an alternative, SmI_2 or samarium metal can be utilized as a reductant in conjunction with diiodomethane to induce an analogous iodomethylation reaction.[32] A wide range of aldehydes and ketones are efficiently alkylated at room temperature under these conditions. Even substrates that are susceptible to enolization react reasonably well, providing moderate yields of the iodohydrin (equation 29).[32a] Only 1,2-addition products are observed with conjugated aldehydes and ketones (equation 30).[32a] Excellent diastereoselectivity is achieved in reactions with both cyclic and acyclic ketones (equations 31 and 32).[32b]

Utilization of dibromomethane also results in the isolation of iodohydrins. Based on this observation and the fact that SmI_3 will cleave epoxides to generate iodohydrins, it has been suggested that the iodomethylsamarium alkoxide species that is initially generated cyclizes to an epoxide intermediate. The

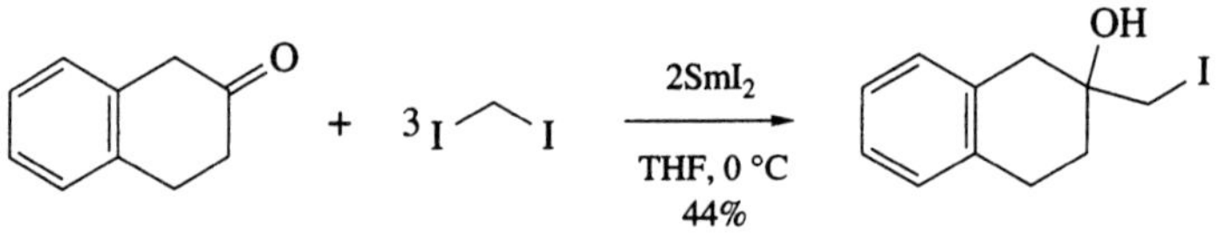

(29)

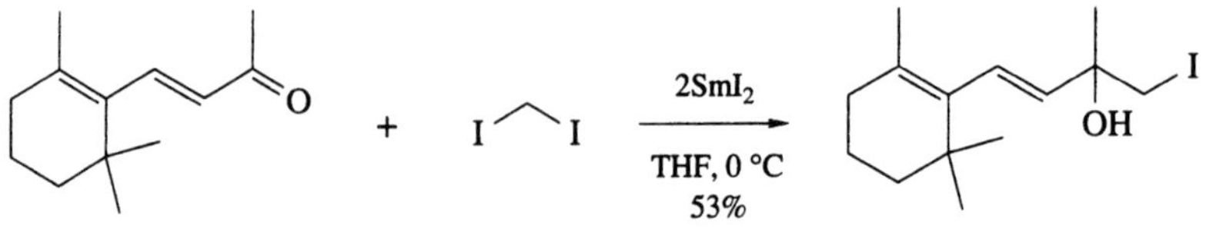

(30)

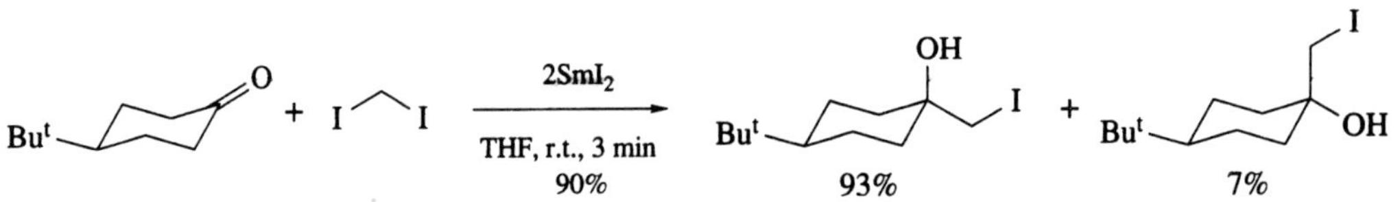

(31)

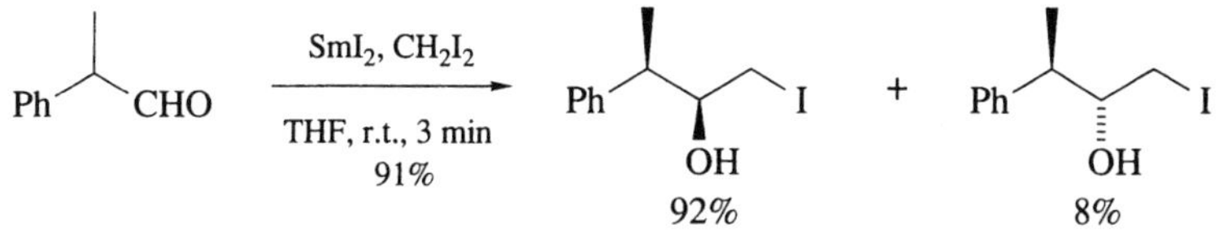

(32)

SmI_2X that is produced as a result of this process then serves to open the epoxide, generating the iodohydrin. Although this appears to be a likely scenario, a more direct route involving a Finkelstein reaction between the bromomethylsamarium alkoxide and various samarium iodide salts[16c,27] cannot be ruled out (Scheme 2).

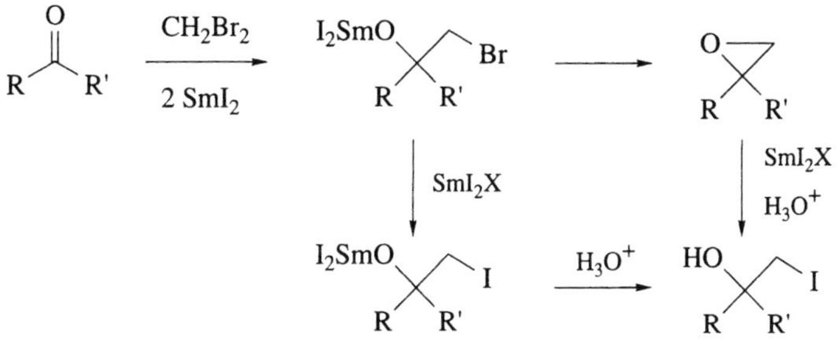

Scheme 2

A one-pot carbonyl methylenation reaction has been developed based on this iodomethylenation reaction.[33] Treatment of an iodomethylsamarium alkoxide (generated *in situ* by reaction of aldehydes or ketones with SmI_2/CH_2I_2) with SmI_2/HMPA and *N,N*-dimethylaminoethanol (DMAE) induces a reductive elimination, resulting in the generation of the corresponding methylenated material (equation 33).

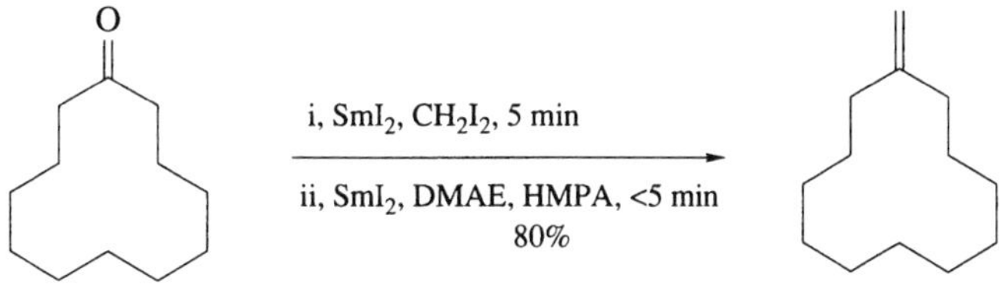

(33)

When α-halo ketones are treated with diiodomethane and samarium at 0 °C, cyclopropanols can be obtained in reasonable yields. Curiously, under the same conditions 1,2-dibenzoylethane also leads to cyclopropanol products (equations 34 and 35).[32a] Several pathways for conversion of α-halo ketones to the observed cyclopropanols can be envisioned. It has been proposed that the mechanism of this reaction involves reduction of the α-halo ketone by Sm (or SmI_2) to a samarium enolate. Cyclopropanation of this enolate with a samarium-based carbenoid subsequently provides the observed product.[32c]

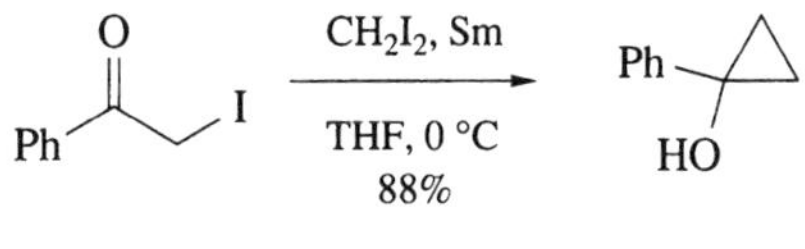

(34)

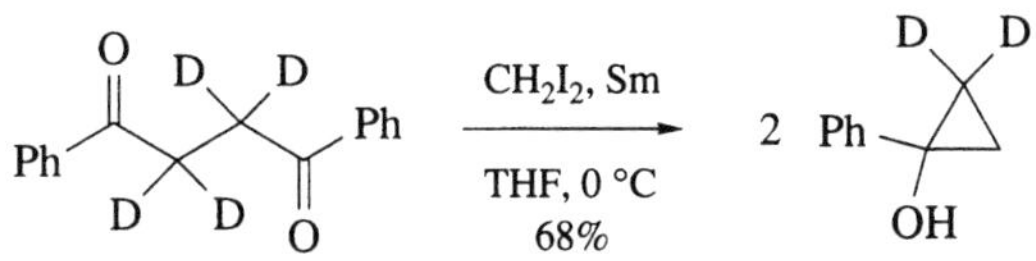

(35)

Although numerous reductants (*e.g.* magnesium, lithium, sodium, organolithiums, organocuprates and chromium(II) salts, to name only a few) have been utilized in attempts to promote intramolecular Barbier-type reactions, SmI_2 is by far the most general reductive coupling agent in terms of its utility and its scope of application. It has therefore become the reagent of choice for such processes.

Isolated cyclopentanols can be synthesized with considerable diastereoselectivity when appropriately substituted ω-iodoalkyl ketones are treated with SmI_2 in THF at –78 °C and allowed to warm to room temperature (equation 36).[34] The reaction is clearly not subject to steric inhibition about the ketone carbonyl, and provides a useful alternative to intermolecular reactions between organometallic reagents (*e.g.* RLi or RMgX) and α-substituted cyclopentanones. Intermolecular reactions between organometallic reagents and cyclopentanones often suffer from competitive enolization and/or reduction processes.

2SmI$_2$, THF, −78 °C to r.t., 77%; products 93% : 7% (36)

Perhaps more valuable is the application of the SmI_2 reductive coupling technology to the synthesis of bicyclic alcohols. Shiner and Berks have demonstrated that the procedure can be utilized to generate three-membered rings starting from α-tosyloxymethylcycloalkanones (equation 37).[35] An advantage of SmI_2 over reductants such as magnesium is that one is not restricted to organic halides in these reactions. As in this example, organic tosylates appear perfectly well suited to the Barbier process also.

2 SmI$_2$, THF, r.t., 70% (37)

Although the synthesis of four-membered rings has yet to be thoroughly explored, samarium diiodide can be utilized in the annulation of five- and six-membered rings through an intramolecular Barbier process.[36] The development of this approach to six-membered ring formation in fused bicyclic systems is particularly important. Prior to this discovery there existed no reliable and convenient method to achieve this simple annulation process. The reactions proceed with considerable diastereoselectivity when cyclopentanone substrates are utilized, or when substituents are placed at the α-position of the cycloalkanone (equations 38 and 39). Diastereoselectivity in other systems depends on whether or not an iron(III) catalyst is utilized in the reaction. In addition, in some cases higher diastereoselectivities can be obtained utilizing samarium metal, ytterbium metal, or YbI_2 as reductant (Section 1.9.3.3). Unfortunately, the sense and magnitude of stereoselectivity that can be achieved by employing these other reductants are unpredictable from substrate to substrate.

2 SmI$_2$, cat. Fe^{III}, THF, r.t., 67%; products 95% : 5% (38)

2SmI$_2$, cat. Fe^{III}, THF, −78 °C to r.t., 67% (39)

The SmI_2-mediated intramolecular Barbier procedure has been applied to several diverse systems, and in each case has been determined to be superior to other protocols. Suginome and Yamada applied the technique to syntheses of exaltone and (±)-muscone (equation 40).[37] Surprisingly, cyclization in this case generates a single diastereomer. It is claimed that the SmI_2 procedure provides better yields than procedures incorporating $Mg/HgCl_2$ or *n*-butyllithium.

Murray and coworkers have used the SmI_2-promoted intramolecular Barbier synthesis to produce 3-protoadamantanol (equation 41).[38] Although the yield in this example was not particularly high, it was the only method among several attempted that proved successful.[39]

In an elegant approach to polyquinenes, Cook and Lannoye developed a bisannulation process based on the SmI_2-mediated cyclization process (equation 42).[40] Remarkably, both of the carbon–carbon bond-

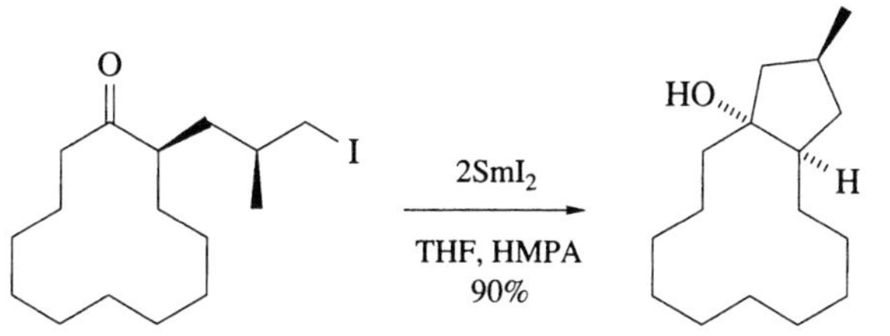

$2SmI_2$, THF, HMPA, 90% (40)

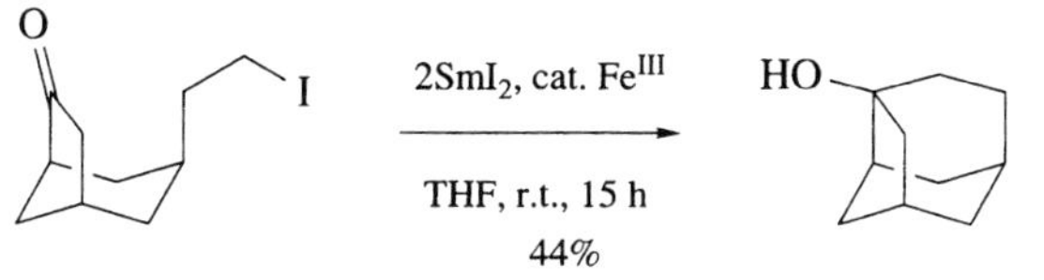

$2SmI_2$, cat. Fe^{III}, THF, r.t., 15 h, 44% (41)

forming reactions in this process proceed with approximately 90% yield, providing an incredibly efficient entry to these complex molecules.

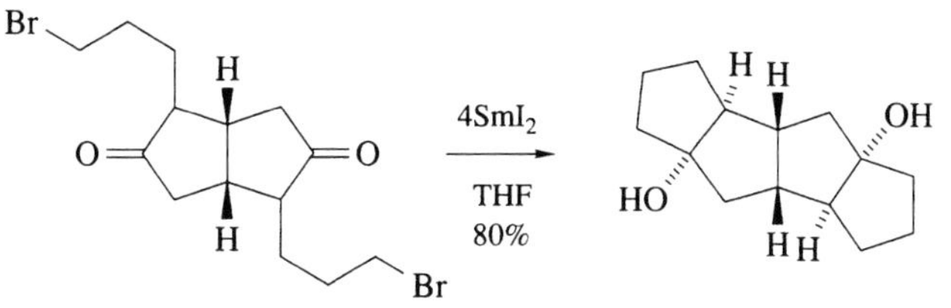

$4SmI_2$, THF, 80% (42)

Exceptionally clean cyclization can be accomplished by utilizing conjugated enones as precursors for the Barbier reaction (equation 43).[41] High diastereoselectivity is achieved in these reactions, and under the mild conditions required for cyclization the TMS ether protecting group remains intact. It is also interesting that a neopentyl halide is effective in the cyclization. This result would appear to exclude an *S*N2-type displacement of an organic halide by a samarium ketyl as a possible mechanism for the SmI_2-promoted intramolecular Barbier reaction.

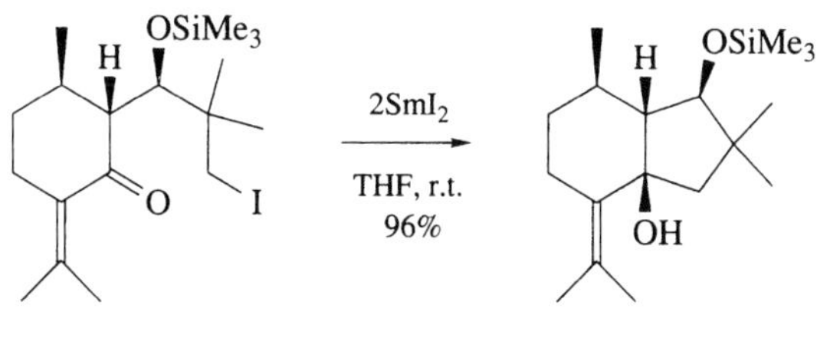

$2SmI_2$, THF, r.t., 96% (43)

Ketyls appear to be important intermediates in Barbier-type coupling reactions promoted by SmI_2. This provided the very real possibility that the Sm^{3+} ion generated on electron transfer could be utilized as an effective Lewis acid template to control stereochemistry *via* chelation in suitably functionalized substrates. Indeed, a number of systems have been designed with this idea in mind. In β-ketoamide systems, the samarium(III) ion can participate in a rigid, chelated intermediate which serves to control stereochemistry in the cyclization process (equation 44).[42] These particular cyclization reactions appear to proceed under kinetic control; there is no evidence to suggest that any equilibration takes place under the reaction conditions, and a single diastereomer is generated in each example. Six-membered rings can also be constructed by this process, although the yields are lower. By-products derived from simple reduction of the ketone to an alcohol are also isolated in these cases.

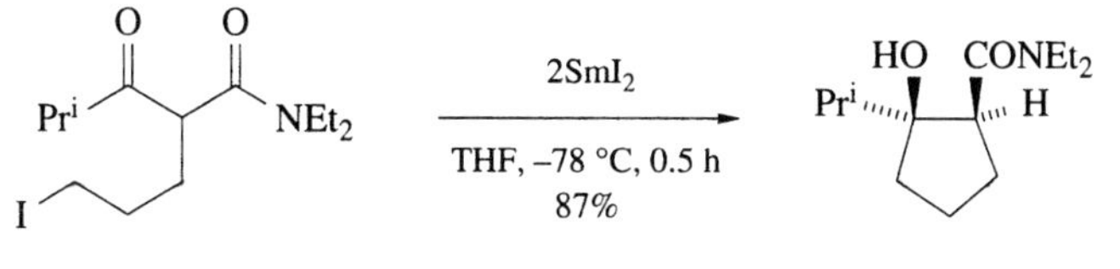

$2SmI_2$, THF, –78 °C, 0.5 h, 87% (44)

Allylic halide precursors provide exceptional yields of cyclic products, and both five- and six-membered rings comprising several different substitution patterns can be accessed by the same technology (equations 45 and 46).[42] In some cases, excellent stereochemical control at three contiguous stereocenters is established.

$2SmI_2$, THF, –78 °C, 0.5 h, 80% (45)

$2SmI_2$, THF, –78 °C, 0.5 h, 74%; 92% + 8% (46)

A number of analogous β-keto esters have also been explored as substrates for intramolecular Barbier cyclization.[42] In the alkyl halide series, a convenient route to 2-hydroxycyclopentanecarboxylates results (equation 47). However, six-membered rings are inaccessible utilizing this procedure. In contrast to β-ketoamide substrates, the β-keto ester series provide products which are clearly under thermodynamic control; that is, the observed diastereoselectivity is the result of a retroaldol–aldol equilibration, which serves to equilibrate the initially formed samarium aldolates. In most cases, diastereoselectivity is high, and the sense of relative asymmetric induction is predictable, based on a simple model for the reaction. However, the degree of diastereoselectivity is highly dependent on substituent and solvent effects. In particular, the use of coordinating solvents or additives (such as tetraglyme, 18-crown-6, or *N,N*-dimethylacetoacetamide) that serve to strip the samarium(III) ion away from the chelating center, radically diminish the diastereoselectivity observed in these reactions. It should be pointed out that these cyclizations cannot be carried out by treating the substrates with activated magnesium. Unreacted starting material is recovered under these conditions.[42]

$2SmI_2$, THF, –78 °C, 0.5 h, 83% (47)

Evidence for a radical coupling mechanism (as opposed to a carbanionic carbonyl addition mechanism) in the intramolecular SmI_2-promoted Barbier reactions has come from studies on appropriately functionalized substrates in the β-keto ester series. It is well known that heterosubstituents are rapidly eliminated when they are adjacent to a carbanionic center. Indeed, treatment of a β-methoxy organic halide (suitably functionalized for cyclization[34,43]) with an organolithium reagent leads only to alkene (equation 48). No cyclized material can be detected. On the other hand, treatment of the same substrate with SmI_2 provides cyclized product and a small amount of reduced alcohol, with none of the alkene detected by gas chromatographic analysis (equation 49).[44]

Bu^nLi, THF, –100 °C, 70% (48)

$2SmI_2$, THF, –78 °C to r.t.; 45% + 17% (49)

These results, together with the mechanistic studies by Kagan *et al.*[27] lend strong support for a radical cyclization process. Two general mechanisms are suggested (Scheme 3). In both, initial electron transfer from SmI_2 to the ketone carbonyl occurs, generating a ketyl. This chelated intermediate might suffer one of two fates. Dissociative electron transfer from the second equivalent of SmI_2 to the halide could occur (pathway A), providing a diradical species. Closure to the samarium aldolate and hydrolysis would result in formation of the observed product. Alternatively, the initially generated ketyl could undergo a dissociative intramolecular electron transfer to the halide (pathway B). Addition of the alkyl radical to the Sm^{3+}-activated ketone carbonyl,[28] subsequent reduction of that intermediate with the second equivalent of SmI_2 and hydrolysis would again complete the process. Experiments have yet to be designed and carried out to distinguish between a process involving cyclization after single-electron transfer and two-electron cyclization processes. However, it is clear that samarium carbanions are not involved in these intramolecular processes.

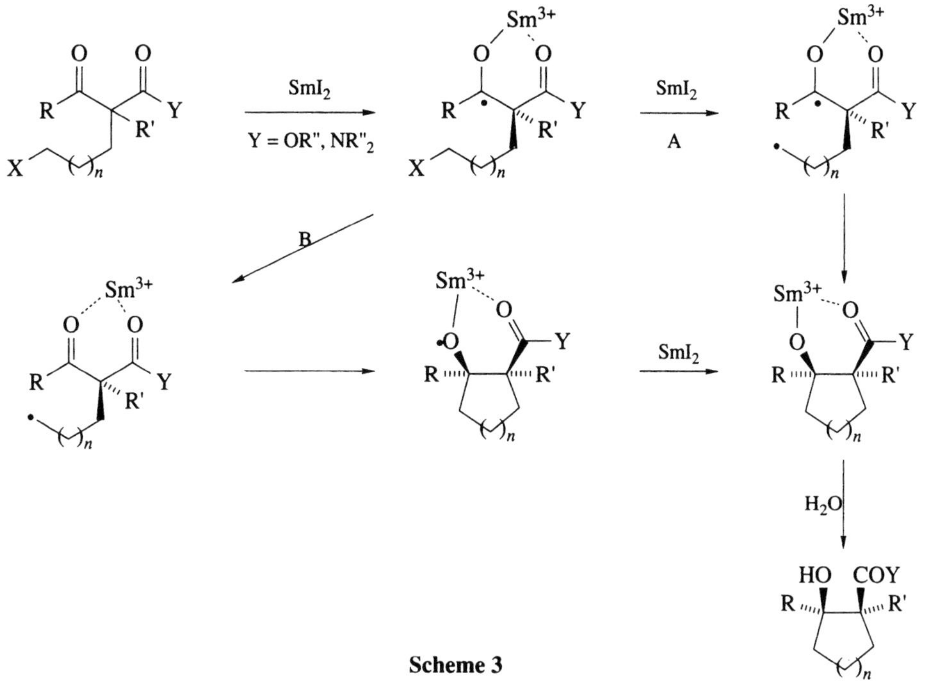

SmI_2, Y = OR", NR"$_2$; SmI_2, A; B; SmI_2; H_2O; HO COY, R, R'

Scheme 3

Allylic halide substrates in the β-keto ester series cyclize well, and convenient routes to five-, six- and seven-membered rings have been described (equations 50 and 51).[42] Unfortunately, the diastereoselectivity in these examples again is highly dependent on the substitution patterns about the dicarbonyl substrate.

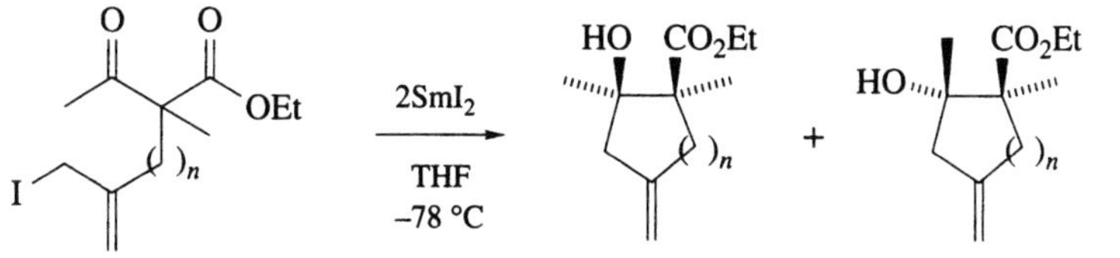

2SmI_2, THF, –78 °C → (1) + (2) (50)

n	*Yield (%)*	*Diastereoselectivity (1) : (2)*
1	84	86 : 14
2	73	64 : 36
3	64	50 : 50

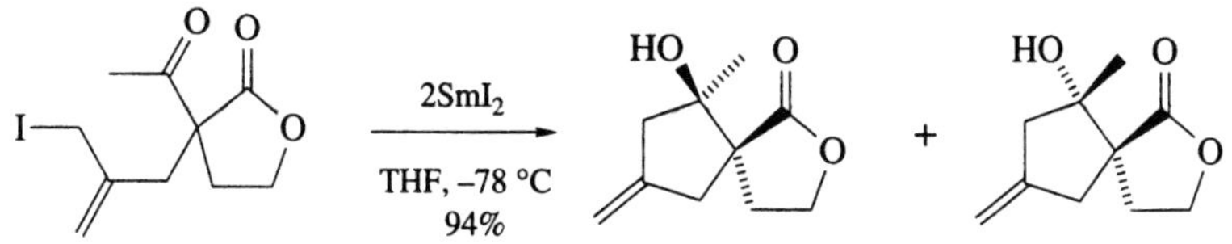

2SmI_2, THF, –78 °C, 94% → 86% + 14% (51)

Attempts to cyclize ethyl (*E*)-2-acetyl-2-methyl-6-bromo-4-hexenoate have been unsuccessful, with ethyl 2-methyl-3-oxobutanoate isolated as the major product of the reaction (equation 52).[42] Loss of butadiene, as required for this transformation, is clearly facilitated by the ability of a β-keto ester stabilized (radical or anion) intermediate to serve as an effective leaving group in the reaction. Thus, cyclization of (*E*)-8-bromo-4-methyl-6-octen-3-one proceeds smoothly to provide the expected carbocycle in 91% isolated yield (equation 53).

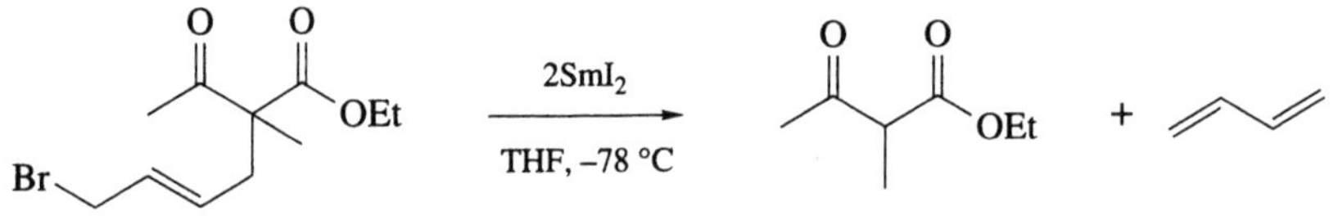

$2SmI_2$, THF, –78 °C (52)

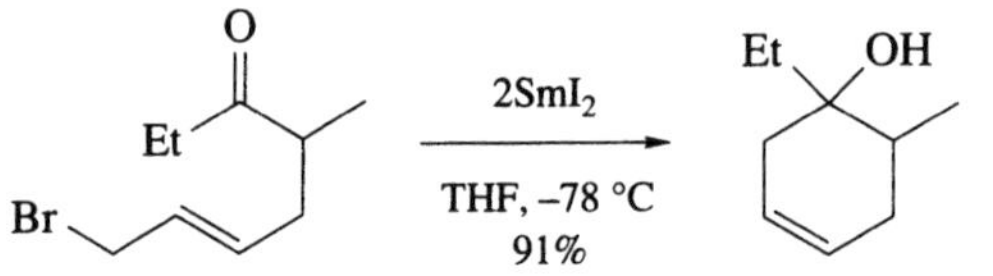

$2SmI_2$, THF, –78 °C, 91% (53)

These examples again have some mechanistic implications in that they appear to rule out cyclization *via* S_N2 displacement of the halide by a samarium ketyl. However, one cannot distinguish between a mechanism based on allylsamarium addition to the carbonyl *versus* an electron transfer mechanism as outlined for the alkyl halide substrates above. Both mechanisms allow for isomerization of the double bond (*via* 1,3-allylic transposition in the case of an allylmetallic,[45] or configurational instability in an allylic radical[46] in a diradical coupling mechanism) and also provide reasonable routes for generation of butadiene. Further mechanistic work is clearly required in order to provide a more detailed understanding of all of these intramolecular Barbier-type reactions.

1.9.2.3.2 Reformatsky-type reactions

In addition to serving as a useful replacement for lithium or magnesium in Barbier-type coupling reactions, SmI_2 also provides advantages over zinc as a reductant in Reformatsky-type coupling reactions (equation 54).[16c,27] The latter only performs well when an activated form of zinc is utilized, and thus the homogeneous conditions afforded by SmI_2 provide the advantage of enhanced reactivity under milder conditions.

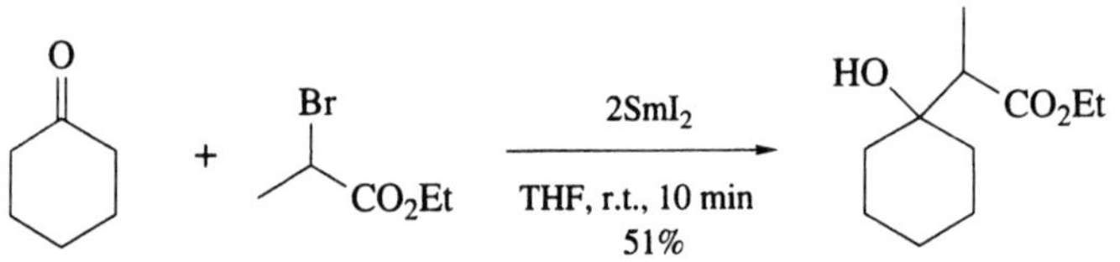

$2SmI_2$, THF, r.t., 10 min, 51% (54)

The procedure has been adapted to permit construction of medium- and large-ring lactones through an intramolecular process (equation 55).[47] Eight- to fourteen-membered ring lactones can be synthesized under high dilution conditions in 75–92% yields, and the process appears much better than procedures involving use of Zn–Ag/Et_2AlCl.[48] Diastereoselectivity in the SmI_2-mediated cyclizations utilizing α-bromopropionate ester precursors was less than 2.5:1.

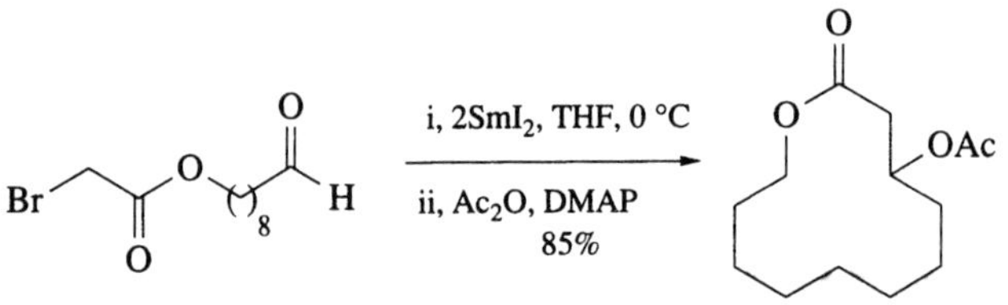

i, $2SmI_2$, THF, 0 °C; ii, Ac_2O, DMAP, 85% (55)

Vedejs and Ahmad have used this SmI_2-promoted macrocyclization technique as a key step in the total synthesis of a cytochalasin (equation 56).[49] In this reaction, the 11-membered ring product is isolated in

46% yield as a single diastereomer. Curiously, the zinc-promoted process provides a 1:1 mixture of diastereomers in 75% yield.

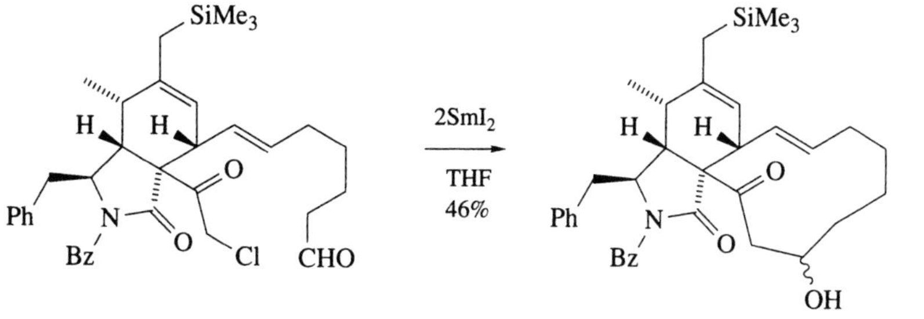

(56)

Reductive cyclizations of β-bromoacetoxy aldehydes and ketones promoted by SmI_2 afford β-hydroxyvalerolactones with unprecedented degrees of 1,3-asymmetric induction (equation 57).[50] Numerous attempts at utilizing zinc-mediated intramolecular Reformatsky reactions to access these lactones have failed. The successful development of the SmI_2-based methodology therefore provides perhaps the most convenient entry to this important class of molecules.[50]

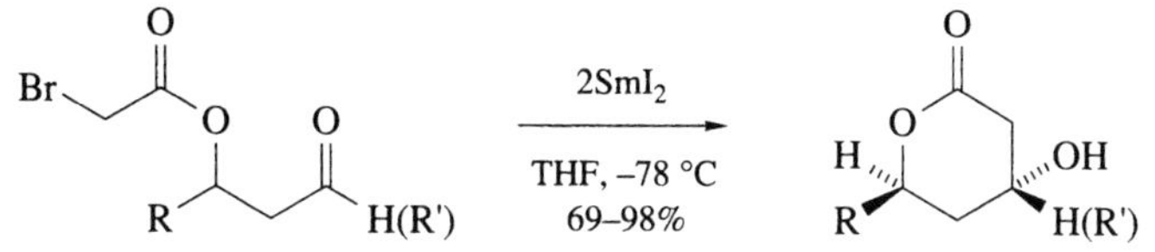

(57)

Yields in the SmI_2-promoted intramolecular Reformatsky reaction are typically higher for ketones than for aldehyde substrates, but in both series diastereoselectivity is virtually complete. It has been suggested that reaction of SmI_2 with the β-bromoacetoxy initially generates a Sm^{3+} ester enolate, with cyclization taking place through a rigid cyclic transition structure enforced by chelation (Scheme 4).[50]

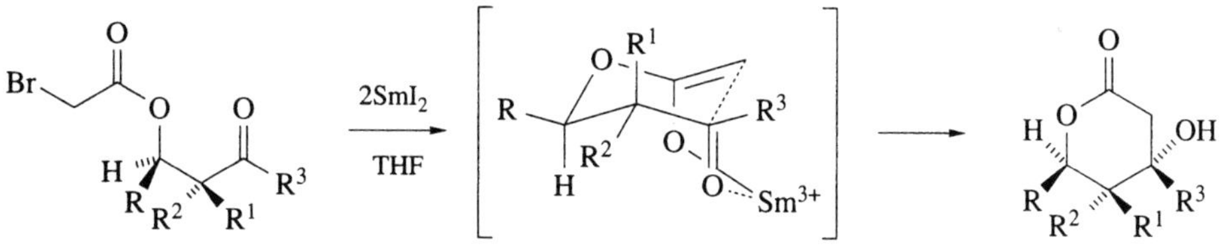

Scheme 4

In contrast to other reported methods of 1,3-asymmetric induction, the SmI_2-mediated intramolecular Reformatsky procedure permits strict control of stereochemistry even in diastereomeric pairs of substrates bearing α-substituents (equations 58 and 59).[50] Although the diastereoselectivity is somewhat lower for the *syn* diastereomeric substrate, where the α-substituent would be axially disposed in the proposed transition structure leading to the product, 1,3-asymmetric induction is still predominant and overwhelms other effects to an impressive extent.

Only a few exceptions to this general pattern of diastereoselection have been observed.[50,51] Some *syn* diastereomeric α-substituted β-bromoacetoxy aldehydes and ketones provide diastereomeric mixtures of products or the opposite diastereomeric product than is anticipated on the basis of the transition structure proposed in Scheme 4 (equation 60). Steric factors which preclude access to chair transition structures may be responsible for the change in the sense of diastereoselectivity in these examples.

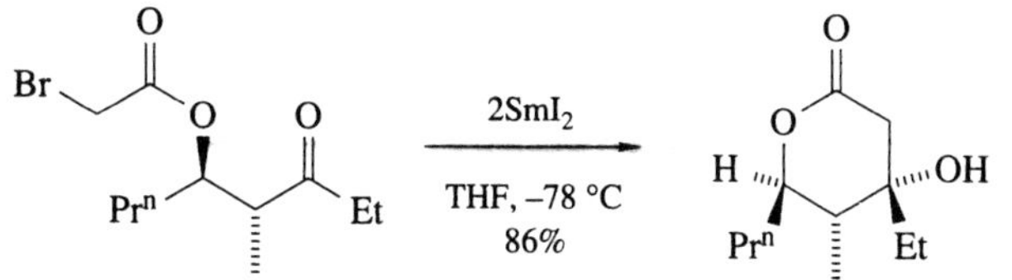

(58)

$2SmI_2$, THF, −78 °C, 65%: 93% + 7% (59)

$2SmI_2$, THF, −78 °C, 88% (60)

Surprisingly, 1,3-asymmetric induction can be relayed from a tertiary acetoxy stereocenter (equation 61). The unprecedented degree of stereochemical control exhibited by this process appears to be general for aldehydes and ketones, although the scope of the reaction with regard to substituents at the β-position is limited.[51]

$2SmI_2$, THF, −78 °C, 75%: > 97% + < 3% (61)

Seven-membered ring lactones can be accessed in excellent yields by the SmI_2-mediated intramolecular Reformatsky reaction as well. Although several substitution patterns provide exceptional relative asymmetric induction in this process (equation 62), it is clear that high diastereoselectivity cannot be achieved for all substitution patterns in the formation of seven-membered ring lactones.[52]

$2SmI_2$, THF, −78 °C, 68% (62)

1.9.2.3.3 Ketyl–alkene coupling reactions

The ability of SmI_2 to generate ketyls prompted its use for the reductive cross-coupling of ketones with alkenes. Both intermolecular and intramolecular processes of this type have been described.

Conjugated esters react with aldehydes and ketones in the presence of 2 equiv. of SmI_2 and a proton source, affording reasonable yields of butyrolactones (equation 63).[53] The presence of HMPA dramatically enhances reactivity (and yields), permitting reactions to run to completion in 1 min as opposed to 3–6 h without this additive. The method complements electroreductive,[54] photoreductive,[55] and other metal-induced ketone–alkene cyclizations[56] that have been developed. Use of unsaturated esters such as ethyl methacrylate and ethyl crotonate leads to diastereomeric mixtures of products in reactions with prochiral aldehydes and ketones.[53c] Conjugated nitriles do not fare as well as their unsaturated ester counterparts in these reactions. Yields of 17–20% are reported for the nitrile substrates.[53a] In terms of the ketyl precursor, both aliphatic and aromatic ketones and aldehydes can be utilized,[53a] and even formaldehyde is effective to some degree.[53b]

The reaction is considered to proceed by a radical process.[53c] When the reaction is carried out with MeOD as the proton source, α-monodeuterolactone is generated. Two mechanisms can be envisioned which are consistent with this observation. The first involves coupling of a samarium ketyl with an allylic radical derived from single-electron reduction of the unsaturated carbonyl substrate. Protonation

$2SmI_2$, THF, r.t., 2 h, 76%; 90% : 10% (63)

and cyclization to the lactone completes the process. A more likely mechanism involves simple ketyl addition to the conjugated ester. Subsequent reduction of the radical intermediate, protonation and cyclization would again provide the observed lactones. A third mechanism initially suggested involved reduction of the unsaturated ester by SmI_2, generating a stable samarium β-metallo ester intermediate. Direct addition of this intermediate to a ketone or aldehyde would also provide an entry to the lactone (see Section 1.9.2.2). This latter mechanism seems unlikely, since addition of an aldehyde or ketone to a mixture of ethyl acrylate and SmI_2 failed to produce a reasonable yield of lactone.

Bicyclic butyrolactones can be generated when intramolecular versions of the reaction are carried out (equation 64).[57] The yields are improved by addition of HMPA, and reactions can be carried out under milder conditions. Addition of a catalytic amount of $FeCl_3$ has little effect on the yields. In most cases, diastereoselectivities range from 2.5:1 to 4:1.

$2SmI_2$, THF, 65 °C, 4 h, 92%; 90% : 10% (64)

A much more highly diastereoselective process results when alkenic β-keto ester and β-ketoamide substrates can be utilized in the ketone–alkene reductive coupling process. Both electron deficient and unactivated alkenes can be utilized in the reaction (equations 65 and 66).[58] In such examples, one can take advantage of chelation to control the relative stereochemistry about the developing hydroxy and carboxylate stereocenters. Favorable secondary orbital interactions between the developing methylene radical center and the alkyl group of the ketyl,[54c,56a,59] and/or electrostatic interactions in the transition state[54a,55,59] account for stereochemical control at the third stereocenter.

$2SmI_2$, THF, Bu^{t}OH, –78 °C, 87% (65)

$2SmI_2$, THF, MeOH, –78 °C, 88% (66)

Since 2 equiv. of SmI_2 are required for the reaction, the reductive coupling process must be a two-electron process overall (Scheme 5).[58] Cyclization appears to occur after transfer of a single electron, with Sm^{3+} controlling the stereochemistry at this stage by chelation with the Lewis basic ester carbonyl. Subsequent reduction to a transient carbanion, followed by immediate protonation, accounts for the observed products. Only if a transient anion is generated can one account for >90% deuterium incorporation at the methyl group when the reaction is performed in MeOD (equation 67).[44]

There is an inherent competition between simple reduction of the ketone and the reductive cyclization process with unsaturated carbonyl substrates. Cyclization processes that are slower than that of the ketyl–alkene cyclization forming a five-membered ring, suffer from lower yields owing to this competition. For example, ketyl–alkyne coupling can also be achieved when mediated by SmI_2, but yields are lower than those achieved with analogous keto–alkenes (equation 68). This might have been expected on the basis that radical additions to alkynes are slower than corresponding additions to alkenes.[60] Similarly, the rate

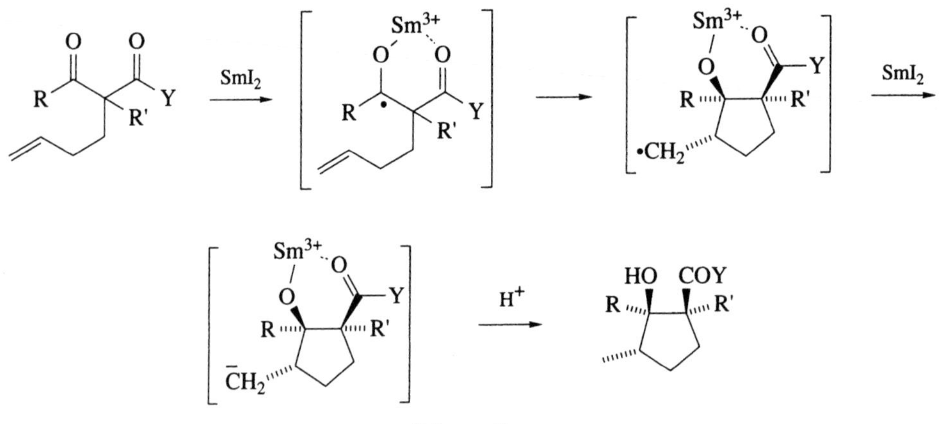

Scheme 5

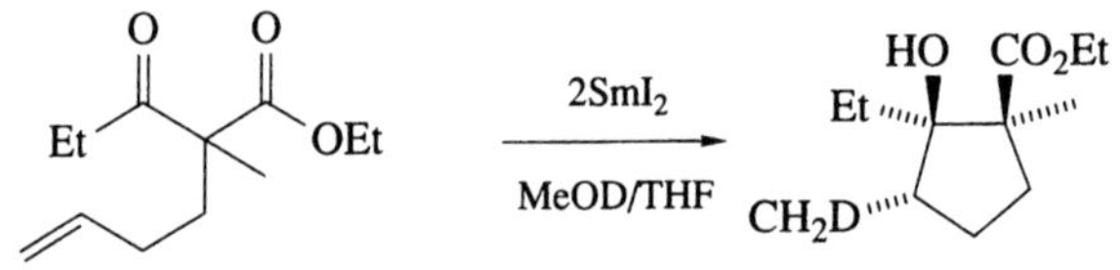

(67)

retardation encountered in formation of six-membered rings by radical processes prevents the construction of 2-hydroxycyclohexanecarboxylates by the SmI_2-promoted ketyl–alkene cyclization process.

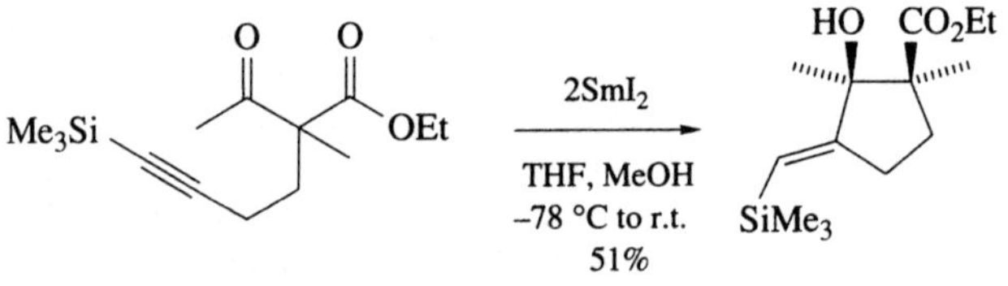

(68)

An elegant tandem radical cyclization process promoted by SmI_2 has been developed as a key step in the synthesis of (±)-hypnophilin and the formal total synthesis of (±)-coriolin (equation 69).[61] Cyclization in this case again occurs after transfer of a single electron, and in fact the entire process requires less than 2 equiv. of SmI_2. When cyclizations were quenched with D_2O, no deuterium was incorporated at the newly formed vinyl carbon. This implies that the alkenyl radical produced after tandem cyclization abstracts a hydrogen from the solvent faster than it is reduced to the anion by SmI_2. This and the work by Molander and Kenny described above[58] are completely in line with observations of Inanaga *et al.* in work on the reduction of organic halides with SmI_2.[25] Thus, alkyl halides are reduced to hydrocarbons by means of a transient anion (which can be trapped by D_2O) with SmI_2, whereas aryl (and presumably alkenyl) halides show no deuterium incorporation on reduction. With sp^2-hybridized radicals, hydrogen abstraction from the THF solvent is thus faster than reduction by SmI_2 to the anion. Further studies in ketone–alkene reductive cyclization reactions are bound to lead to exciting new entries to highly complex carbocyclic ring systems.

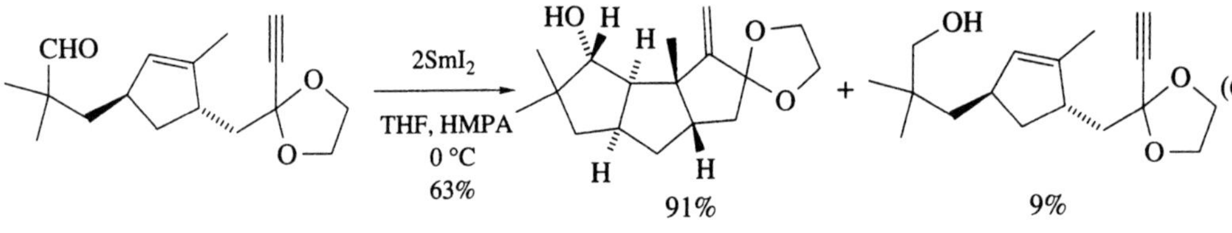

(69)

1.9.2.3.4 Pinacolic coupling reactions

As expected with a reagent that is capable of generating ketyls, intermolecular pinacolic coupling reactions can also be carried out with considerable efficiency using SmI_2. Treatment of aldehydes or ketones with SmI_2 in the presence of a proton source such as methanol results in selective reduction to the corresponding alcohols, and the formation of pinacols is negligible. However, in the absence of a proton source, both aldehydes and ketones can be cleanly coupled in the presence of SmI_2 to generate pinacols (equation 70).[62] The yields are excellent in nearly every case, and the method therefore competes effectively with other established procedures for this process. Unfortunately, roughly equimolar ratios of *threo* and *erythro* isomers are generated in these reactions. Aromatic aldehydes and ketones couple within a few seconds at room temperature in THF. Aliphatic aldehydes require a few hours under these conditions, and a day is needed for complete reaction of aliphatic ketones. Amines, nitriles, aryl halides and nitro groups are tolerated under these conditions. Samarium diiodide is thus superior to other reductants in terms of its functional group compatibility. Surprisingly, even carboxylic acids can also be incorporated into substrates with little decrease in the yields of pinacolic products. It is not clear why competitive reduction to the alcohols is not observed in this instance, since a proton source is provided by the acid.

2 O_2N–C_6H_4–CHO → ($2SmI_2$; THF, r.t., 0.5 min; 95%) (70)

Dicyclopentadienylsamarium also promotes intermolecular pinacolic coupling reactions with exceptional efficiency.[19] Both benzaldehyde and acetophenone are reported to undergo coupling very rapidly at room temperature in the presence of this reductant. After hydrolysis, pinacols are isolated in virtually quantitative yields.

Samarium diiodide has also been utilized as a reductant to promote pinacolic coupling reactions mediated by low-valent titanium species (equation 71).[63] Utilizing this protocol, high diastereoselectivity can be achieved, although the yields for this particular process were not reported.

2 PhCHO → (SmI_2, Cp_2TiCl_2; THF, –78 °C to r.t.) 92% + 8% (71)

Intramolecular pinacolic coupling reactions have also proven successful with SmI_2. Yields with simple diketones are relatively low.[62b] However, excellent yields and diastereoselectivities are achieved in intramolecular pinacolic coupling reactions of β-keto ester and β-ketoamide substrates (equation 72).[64] A variety of substitution patterns can be tolerated in these reactions to generate five-membered carbocycles. Six-membered rings can also be generated by this process, but substantially lower yields and diastereoselectivities are observed (equation 73).[44] Yields obtained for β-ketoamide substrates are also lower than those observed in the β-keto ester series.

(3) → ($2SmI_2$; THF, MeOH, –78 °C; 82%) (72)

→ ($2SmI_2$; THF, MeOH, –78 °C; 47%) 75% + 25% (73)

Curiously, the relative stereochemistry between the carboxylate and the adjacent hydroxy group in the SmI_2-mediated intramolecular pinacolic coupling reaction is opposite to that observed in the intramolecular Barbier reactions and ketone–alkene reductive coupling reactions discussed previously (compare equation 72 with equations 47 and 66, for example). From a synthetic point of view, this result is highly advantageous because it provides entry to the manifold of diastereomeric products. The results also have mechanistic implications. Unlike potential substrates for ketone–alkene reductive coupling reactions, precursors for the pinacolic coupling reaction contain two nearly equally reducible functional groups. This complicates any rational assessment of the stereochemical outcome of these reactions. Furthermore, several different mechanisms can be proposed for the intramolecular pinacolic coupling reaction. One scenario involves two-electron reduction followed by cyclization. After initial reduction of one of the carbonyl substituents to a ketyl, intermolecular reduction to generate a dianion could ensue. Subsequent nucleophilic attack by this dianion at the unreduced carbonyl and hydrolysis would provide the observed product. A mechanism of this type can be ruled out. Ketyl dianions are generally inaccessible, even under the most brutal reducing conditions. Reduction of a ketyl is highly endothermic,[65] and certainly SmI_2 is not a strong enough reducing agent to generate such a species. Furthermore, a dianion intermediate would quickly become protonated under reaction conditions utilized for these reactions (pK_a Bu^{t}OH = 17, pK_a MeOH = 16, pK_a carbonyl dianion ~ 49–51), resulting in large amounts of uncyclized reduction products.[65a]

The most feasible pathway to coupled products is intramolecular ketyl addition to the Sm^{3+}-coordinated ketone (see Scheme 6). Several examples of ketyl addition to Lewis acid activated carbonyls have been reported in the literature.[66] Clerici and Porta have demonstrated in detailed experiments that intermolecular addition of ketyls to carbonyls can be a rapid process.[66a–c] Generally, ketyl addition to carbonyls is a reversible reaction. However, reversibility can be greatly affected by Lewis acid chelation of the complex, and further reduction of the radical intermediate (**8**) by the second equivalent of SmI_2 would serve to make the process irreversible.[65a,66b]

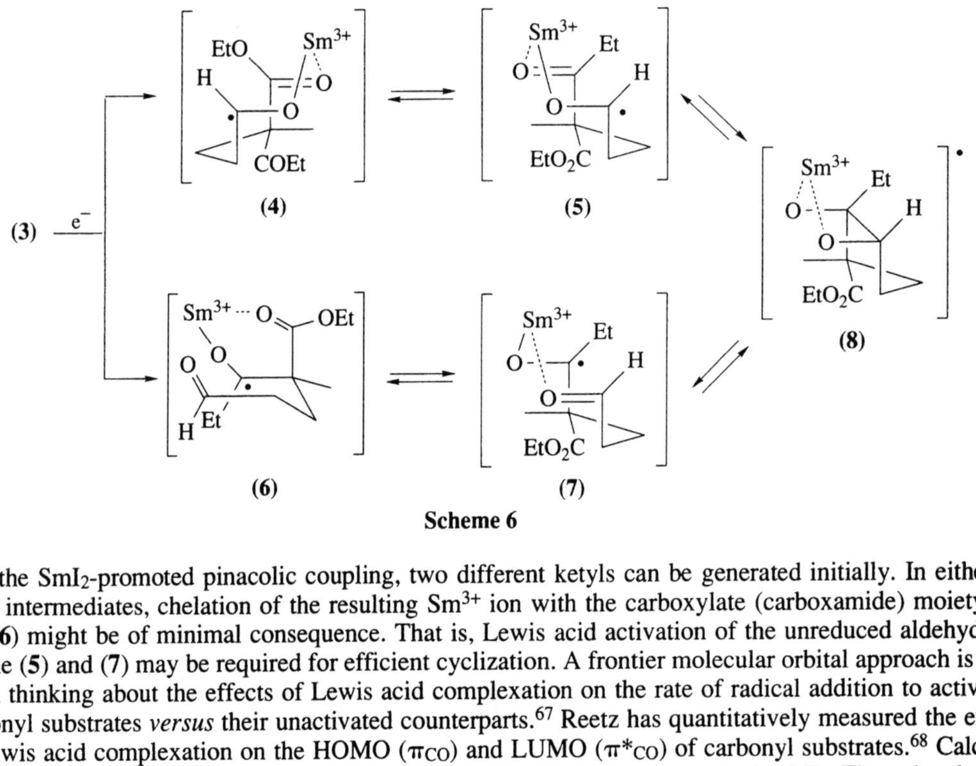

Scheme 6

In the SmI_2-promoted pinacolic coupling, two different ketyls can be generated initially. In either of these intermediates, chelation of the resulting Sm^{3+} ion with the carboxylate (carboxamide) moiety (**4**) and (**6**) might be of minimal consequence. That is, Lewis acid activation of the unreduced aldehyde or ketone (**5**) and (**7**) may be required for efficient cyclization. A frontier molecular orbital approach is useful in thinking about the effects of Lewis acid complexation on the rate of radical addition to activated carbonyl substrates *versus* their unactivated counterparts.[67] Reetz has quantitatively measured the effect of Lewis acid complexation on the HOMO (π_{CO}) and LUMO (π^*_{CO}) of carbonyl substrates.[68] Calculations indicate that the LUMO energy decreases by ~50 kcal on coordination with BF_3. Thus, the electrophilicity of the carbonyl is greatly enhanced, making it more susceptible to nucleophilic radical addition. In the SmI_2-promoted pinacolic coupling reaction, the rate of ketyl addition may be substantially increased by complexation of Sm^{3+} to the ketone as in intermediates (**5**) and (**7**). If complexation is required for efficient cyclization, this would explain the *cis*-diol stereochemistry observed for these substrates, regardless of which carbonyl is first reduced to initiate the reductive cyclization process.

Dipolar repulsion between the carboxylate moiety and the developing diol centers in intermediates (**5**) and (**7**) would account for the (*trans*) relative stereochemistry between these stereocenters. Following cyclization, intermolecular reduction of the Sm^{3+}–chelated complex (**8**) by the second equivalent of SmI_2 and protonolysis by alcohol irreversibly drives the reaction to completion, generating the observed products.

Certainly another plausible mechanism must also be considered. After initial ketyl formation, a second intermolecular reduction could follow, generating a diketyl intermediate. Subsequent carbon–carbon bond formation and protonolysis would again provide the observed products. One cannot unambiguously distinguish between this mechanism and the ketyl addition mechanism. However, both *cis*- and *trans*-diols might be expected from a diketyl coupling reaction. Corey has investigated intramolecular pinacolic coupling reactions promoted by Ti^{2+} (which also lead to generation of *cis*-diols), and argues that a diketyl coupling mechanism is unlikely.[66d] Strong dipolar repulsion between the Ti^{3+}-complexed ketyls would appear to favor generation of *trans*-diols. The same argument may apply in the SmI_2-mediated process; exclusive formation of *cis*-diols would not seem likely from coupling of a di-Sm^{3+}-complexed diketyl. Furthermore, one might speculate that Lewis acid catalyzed intramolecular carbonyl addition (by the ketyl) may be faster than intermolecular reduction of a ketone to a ketyl by SmI_2.

In a useful extension of the methodology, highly functionalized nonracemic carbocycles can be synthesized by intramolecular pinacolic coupling reactions utilizing readily available oxazolidinone precursors (equation 74).[44]

$2SmI_2$; THF, Bu^tOH; 52% (74)

Related to the intramolecular pinacolic coupling reactions in some respects is a ketone–nitrile reductive coupling process. This process also permits the construction of highly functionalized carbocycles,[44] although the yields are somewhat reduced owing to the reluctance of nitriles to undergo such radical addition reactions (equation 75). Presumably, simple reduction of the ketone to the alcohol competes with the desired process.

$2SmI_2$; THF, Bu^tOH; 45% (75)

1.9.2.3.5 *Acyl anion and acyl radical chemistry*

Lithium acyl anions, long sought as unique intermediates, have only recently been synthesized and utilized effectively in synthetic organic chemistry.[69] These reactive organometallics are generated by reaction of organolithiums with carbon monoxide at extremely low temperatures. Samarium acyl anions can be prepared in a somewhat analogous fashion. Thus, when Bu^tBr is added to Cp_2Sm while under an atmosphere of CO in THF at low temperature, a samarium acyl anionic complex is apparently generated (equation 76).[70] Addition of an aldehyde to the reaction mixture at –20 °C, followed by hydrolysis, results in the formation of an α-ketol in modest yields.

$$Bu^tBr + Cp_2Sm + CO \xrightarrow[-20\,^\circ C]{THF} [Bu^tC(O)SmCp_2] \xrightarrow[ii,\ H_3O^+]{i,\ heptanal} Bu^tC(O)CH(OH)\text{-}n\text{-}C_6H_{13}\ (42\%) \quad (76)$$

In the absence of the aldehyde, homocoupled pinacol and a compound resulting from a double carbonylation are isolated in low yields.[70] Both products are consistent with initial formation of a samarium acyl anion.

In addition to the carbon monoxide insertion route, samarium acyl anions can also be prepared under reductive conditions by reaction of SmI_2 with acyl halides.[1c,71] In the absence of any other electrophiles, the acyl halides provide moderate yields of α-diketones (equation 77). The main by-product generated in these reactions is the α-ketol.

n-C_8H_{17}COCl —(2SmI_2, THF, r.t., 10 min, 50%)→ n-C_8H_{17}COCO-n-C_8H_{17} (77)

Mechanistic studies strongly suggest the intermediacy of a samarium acyl anion. For example, the phenylacetyl radical ($PhCH_2CO\cdot$) is known to rapidly decarbonylate ($k = 5.2 \times 10^7$ s^{-1}), providing a benzyl radical which dimerizes to bibenzyl.[72] However, addition of phenylacetyl chloride to a solution of SmI_2 in THF leads to a 75% yield of the expected diketone, and neither toluene nor bibenzyl is detected. Apparently, intermolecular reduction of this acyl radical to the corresponding anion proceeds at a rate which is greater than 5.2×10^7 s^{-1}.

The acylsamarium species has not been isolated or characterized spectroscopically. Its structure has tentatively been assigned as (**9**) or (**10**), analogous to that of $Cp_2LuCOBu^t$. The latter compound has been prepared from Cp_2LuBu^t and CO.[1c]

R–CO–SmI_2 (**9**) ⟷ R–CO–SmI_2 (**10**)

Samarium acyl anions can be trapped by electrophiles other than acid halides. For example, addition of a mixture of a carboxylic acid chloride and an aldehyde or ketone to a solution of SmI_2 in THF results in the synthesis of α-hydroxy ketones (equations 78 and 79).[73] Intramolecular versions of the reaction have also been performed, although the scope of the reaction is limited owing to the difficulty in obtaining suitable substrates for the reaction (equation 80).[74]

n-C_8H_{17}COCl + EtCHO —(i, 2SmI_2; ii, H_3O^+; 63%)→ n-C_8H_{17}COCH(OH)Et (78)

Ph_2NCOCl + n-C_7H_{15}CHO —(i, 2SmI_2; ii, H_3O^+; 67%)→ Ph_2NCOCH(OH)-n-C_7H_{15} (79)

Cl —(2SmI_2, THF, –78 °C, 82%)→ HO, H (80)

Intramolecular trapping studies have verified the intermediacy of acyl radicals in the conversion of carboxylic acid chlorides to samarium acyl anions by SmI_2.[75] Treatment of 2-allyloxybenzoyl chlorides with SmI_2 resulted in a very rapid reaction, from which cyclopropanol products could be isolated in yields of up to 60% (equation 81). Apparently, initial formation of the acyl radical was followed by rapid radical cyclization. The β-keto radical generated by this process undergoes cyclization by a radical or anionic process, affording the observed cyclopropanols (Section 1.9.2.3.1).

1.9.2.3.6 Miscellaneous

A new method for the masked formylation of aldehydes and ketones has been developed which relies on the ability of SmI_2 to generate phenyl radicals from iodobenzene. As pointed out previously, aryl

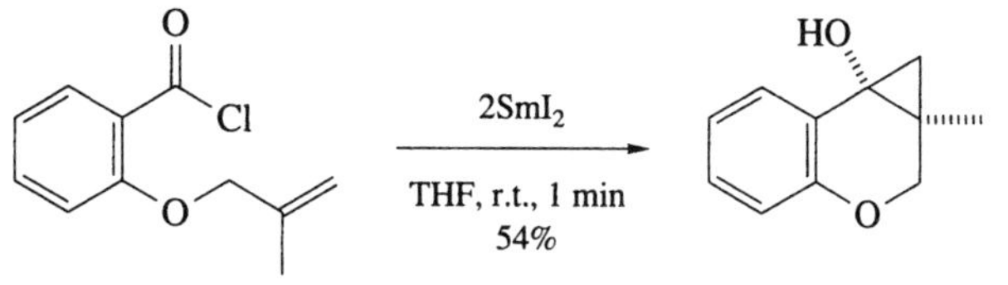

$$\xrightarrow[\text{THF, r.t., 1 min; 54\%}]{2SmI_2} \quad (81)$$

halides do not undergo Barbier-type coupling reactions with ketones in the presence of SmI_2. Instead, THF adducts of the carbonyl compounds are obtained (equation 21).[25] When 1,3-dioxolane is utilized in place of THF, the initially formed phenyl radical can abstract a hydrogen from the dioxolane. The resulting dioxolanyl radical can couple to the carbonyl, generating the observed products (Scheme 7).[76] Both aldehydes and ketones can be utilized in the reaction, with yields ranging from 73–77% for five different substrates.

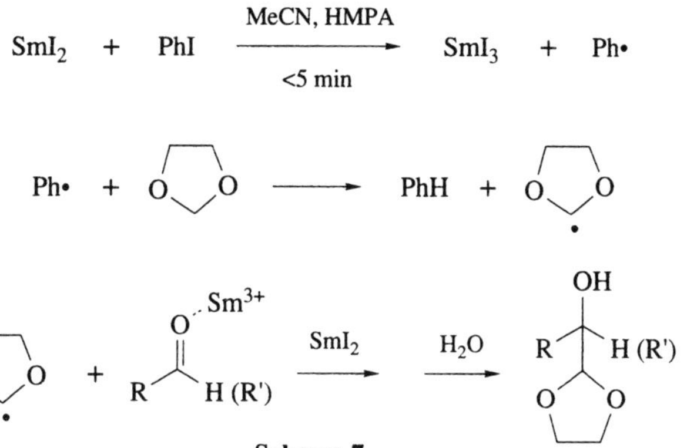

$$SmI_2 + PhI \xrightarrow[<5\ \text{min}]{\text{MeCN, HMPA}} SmI_3 + Ph\bullet$$

$$Ph\bullet + \text{1,3-dioxolane} \longrightarrow PhH + \text{dioxolanyl radical}$$

$$\text{dioxolanyl radical} + RC(O)H(R') \cdots Sm^{3+} \xrightarrow{SmI_2} \xrightarrow{H_2O} RC(OH)H(R')\text{-dioxolanyl}$$

Scheme 7

Martin *et al.* have described the reductive cyclization of ω-unsaturated α-amino radicals mediated by SmI_2.[77] Reduction of ω-unsaturated iminium salts by SmI_2 in the presence of camphorsulfonic acid generates the ω-unsaturated α-amino radicals, which cyclize to provide good yields of nitrogen heterocycles (equation 82). The process is restricted to the formation of five-membered nitrogen heterocycles, and increased steric bulk adjacent to the radical center was also found to inhibit cyclization. As expected, the presence of an activating group on the acceptor double bond (*e.g.* an aryl substituent) increases the yield of the cyclization. The process could be carried out electrochemically as well as by utilizing cobalt(I) reductants, but the relative strengths and weaknesses of these various approaches has yet to be fully assessed.

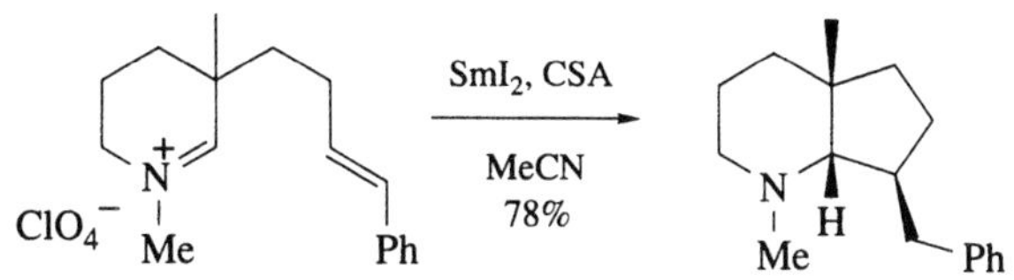

$$ClO_4^- \xrightarrow[\text{MeCN; 78\%}]{SmI_2,\ CSA} \quad (82)$$

1.9.3 YTTERBIUM REAGENTS

Ytterbium reagents have certainly not attained the status achieved by the corresponding samarium reagents in terms of their utility in selective organic synthesis. Nevertheless, there are indications that ytterbium reagents, too, have the potential to serve as selective nucleophiles in C—X π-bond addition reactions, and eventually will take their place among the other lanthanide reagents with a unique role in organic synthesis. The similarity between samarium and ytterbium reagents in many cases is quite striking. As a consequence, the organization of this section mirrors that of the samarium reagents above, and many resemblances between the two classes of reagents will become apparent.

1.9.3.1 Organoytterbium(III) Reagents

Remarkably little chemistry of organoytterbium(III) reagents has been explored as it pertains to selective organic synthesis. These reagents do show promise as useful organic nucleophiles, however. Reaction of 4-*t*-butylcyclohexanone with $Bu^nLi–YbCl_3$ provides a nearly quantitative yield of the expected carbonyl addition product (equation 83).[9] Unfortunately, no mention is made of the diastereoselectivity of this process. Although it might be anticipated that organoytterbium(III) reagents, like their cerium and samarium counterparts, will undergo clean carbonyl addition to highly enolizable ketones, this point has apparently not been addressed.

$$Bu^nLi\text{–}YbCl_3 + \text{4-}Bu^t\text{-cyclohexanone} \xrightarrow[-78\ ^\circ C,\ 3\ h;\ 97\%]{THF} \text{1-}Bu^n\text{-4-}Bu^t\text{-cyclohexan-1-ol} \quad (83)$$

1.9.3.2 Organoytterbium(II) Reagents

The accessibility of a stable +2 oxidation state for ytterbium leads to the possibility of Grignard-type reagents and chemistry. Indeed, both the methods of preparation and reactions of organoytterbiums reported to date closely mimic those of the corresponding Grignard reagents. In spite of rather significant study, these organoytterbium reagents have yet to really assume a special role in organic synthesis. Nevertheless, some unique reactivity patterns have been observed, and with further systematic study one can expect more original reaction manifolds to emerge.

Organoytterbium(II) halides are most conveniently prepared by oxidative metalation of organic iodides with ytterbium metal (equation 84).[13] Since an induction period is often noticed in such reactions, activation of the metal with a trace amount of CH_2I_2 can be utilized to facilitate this process.[78]

$$EtI + Yb \xrightarrow[-20\ ^\circ C;\ 83\%]{THF} EtYbI \quad (84)$$

Compounds prepared in this fashion have been determined to consist largely of 'RYbI' stoichiometries, although the possible existence of Schlenk-type equilibria has never been examined. Ytterbium to iodine ratios determined by elemental analysis, the measured magnetic susceptibilities, and reactivity patterns of these reagents are all consistent with this formulation.[13] From magnetic susceptibilities, the calculated percentages of 'RYbI' generated in solution by this procedure were determined to range from 83–93%, depending on the structure of the organic iodide substrate. This is drastically different from the situation involving samarium reagents discussed in Section 1.9.2.2, in which significant amounts of Sm^{3+} species were also generated. The attenuated reductive capabilities of Yb^{2+} species accounts for the increased selectivity in generation of the organoytterbium(II) reagents.

Oxidative–reductive transmetalation of ytterbium metal with diorganomercury compounds has been utilized as an entry to dialkynyl- and polyfluorinated diaryl-ytterbiums (equations 85 and 86).[79] The dialkynylytterbiums are indefinitely stable in an inert atmosphere at room temperature. On the other hand, the polyfluorinated diarylytterbiums exhibit variable stability. Isolated yields are often low owing to thermal decomposition of these organometallics. However, most can be generated in nearly quantitative yields by this procedure and simply characterized *in situ*.

$$Ph\text{—}\equiv\text{—}Hg\text{—}\equiv\text{—}Ph + Yb \xrightarrow[19\ ^\circ C,\ 4\ h;\ 98\%]{THF} Ph\text{—}\equiv\text{—}Yb\text{—}\equiv\text{—}Ph + Hg \quad (85)$$

$$(o\text{-}HC_6F_4)_2Hg + Yb \xrightarrow[0\ ^\circ C,\ 3\ h;\ 84\%]{THF} (o\text{-}HC_6F_4)_2Yb + Hg \quad (86)$$

Metal–hydrogen exchange processes have also been exploited to generate dialkynylytterbiums (equation 87).[79b,d] Clearly, this procedure is of much less synthetic value than the oxidative–reductive transmetalation method described above. Of perhaps greater synthetic utility is the metal–hydrogen exchange reaction of MeYbI with other carbon acids. For example, phenylacetylene and fluorene both react readily to generate reasonable yields of the corresponding organoytterbium iodides (equations 88 and 89).[80] Triphenylmethane and diphenylmethane do not react under these conditions. Incidentally, methyl Grignard reagents provide far lower yields of metalated products than organoytterbiums under comparable reaction conditions.

$$(C_6F_5)_2Yb \;+\; 2\,Ph{-}{\equiv} \xrightarrow[41\%]{THF} Ph{-}{\equiv}{-}Yb{-}{\equiv}{-}Ph \;+\; 2C_6F_5H \qquad (87)$$

$$MeYbI \;+\; Ph{-}{\equiv} \xrightarrow[-20\ °C]{THF} Ph{-}{\equiv}{-}YbI \;+\; CH_4 \qquad (88)$$

$$MeYbI \;+\; \text{fluorene} \xrightarrow[-20\ °C]{THF} \text{9-fluorenyl-YbI} \;+\; CH_4 \qquad (89)$$

Useful applications of organoytterbium reagents to organic synthesis pale in comparison to those of organocerium and organosamarium reagents. Most reactivity patterns of organoytterbiums closely mimic those of organomagnesium and organolithium reagents. Carbonation reactions can be carried out on organoytterbiums, but yields are modest. Alkynes can be converted to a one-carbon homologated carboxylic acid in about 50% overall yield (equation 90).[80] Carbonation of bis(pentafluorophenyl)ytterbium generates the expected carboxylic acid in 50% yield, along with nearly 20% of 2,3,4,5-tetrafluorobenzoic acid. It is proposed that the latter is generated by an *ortho* oxidative metalation reaction which is triggered by the initially formed ytterbium(II) carboxylate (equation 91).[81]

$$Ph{-}{\equiv} \xrightarrow[\text{iii, } H_3O^+ \;\; 50\%]{\text{i, MeYbI; ii, } CO_2} Ph{-}{\equiv}{-}CO_2H \qquad (90)$$

$$(C_6F_5)_2Yb \xrightarrow[\text{ii, } H_3O^+]{\text{i, } CO_2} \underset{50\%}{C_6F_5CO_2H} \;+\; \underset{16\%}{o\text{-}HC_6F_4CO_2H} \qquad (91)$$

Organoytterbium(II) reagents react with aldehydes and ketones to provide modest yields of the corresponding alcohols (equation 92).[82] Significant quantities of pinacol products are generated when diorganoytterbiums are reacted with aromatic ketones, presumably as a result of electron transfer from the ytterbium(II) organometallic. Although principally carbanion transfer reagents, it is clear that organoytterbium(II) reagents can also serve as effective reductants.

$$PhYbI \;+\; PhC(O)CH_3 \xrightarrow[78\%]{THF} Ph_2C(OH)CH_3 \qquad (92)$$

Organoytterbium(II) halides provide higher 1,2-selectivity in reactions with α,β-unsaturated aldehydes and ketones than their Grignard counterparts, although yields are sometimes low (equation 93).[78,83]

More surprising is the attenuated reactivity of organoytterbium reagents for ketones, especially when compared to carboxylic acid esters. Competitive reaction of phenylytterbium iodide with a 1:1 mixture of

$$\text{CH}_2\text{=CHC(O)CH}_3 + \text{PhYbI} \xrightarrow[\text{r.t., 18 h; 37\%}]{\text{THF}} \text{CH}_2\text{=CHC(OH)(CH}_3\text{)Ph} \quad (93)$$

methyl benzoate and acetophenone results in the formation of 34% benzophenone and only 17% 1,1-diphenylethanol.[9] Unfortunately, no account was made of the remainder of the material. However, these results imply that organoytterbium reagents are more reactive towards esters than ketones. The attenuated reactivity towards ketones has been exploited in the development of a selective ketone synthesis from carboxylic acid derivatives (equation 94).[84] Iron trichloride proved to be an effective catalyst for this reaction, providing higher selectivity than reactions utilizing copper(I) salts or with the organoytterbium reagent alone. Unfortunately, yields reported are too low to be of much value in synthesis.

$$\text{PhYbI} + \text{PhC(O)Cl} \xrightarrow[\text{THF, -20 °C; 24\%}]{\text{cat. FeCl}_3} \underset{99\%}{\text{PhC(O)Ph}} + \underset{1\%}{\text{Ph}_3\text{COH}} \quad (94)$$

Phenylytterbium(II) iodide has also been demonstrated to react selectively with *N*,*N*-dimethylbenzamide, affording a 60% yield of benzophenone (equation 95).[14a] Under comparable conditions, the corresponding Grignard reagent provided benzophenone in only 20% yield.

$$\text{PhYbI} + \text{PhC(O)NMe}_2 \xrightarrow[\text{65 °C, 2 h; 60\%}]{\text{THF}} \text{PhC(O)Ph} \quad (95)$$

Nitriles do not undergo efficient reactions with organoytterbium reagents.[9] However, isocyanates are reported to provide good yields of the corresponding amides (equation 96).[13]

$$\text{PhYbI} + \text{PhNCO} \xrightarrow[52\%]{} \text{PhHNC(O)Ph} \quad (96)$$

1.9.3.3 Barbier-type Reactions Promoted by Ytterbium Diiodide

Although the application of SmI_2 as a reductant and reductive coupling reagent has already had a major impact on selective organic synthesis, utilization of the corresponding ytterbium reagents has lagged behind. There are several important reasons for this. First, although several rapid and convenient syntheses of SmI_2 have been reported, preparation of $YbBr_2$ by reaction of ytterbium with 1,2-dibromoethane requires a reaction time of over 2 d. In addition, although SmI_2 is relatively soluble in solvents like THF (0.1 M), both YbI_2 and $YbBr_2$ have limiting solubilities of <0.04 M in the same solvent.[16b,85] Finally, the redox potential of ytterbium(II) species is borderline for the types of transformations that are of interest to synthetic organic chemists. Nevertheless, some transformations have been reported which nicely complement those accomplished by SmI_2.

Intermolecular Barbier-type reactions are reportedly not possible when YbI_2 is utilized as the reductant.[16c] However, intramolecular versions of the reaction proceed smoothly.[36] Both five- and six-membered rings can be generated in this process, and in some cases the observed diastereoselectivities exceed those achieved with SmI_2 (equation 97). Unfortunately, diastereoselectivity is not always high nor predictable, and thus mixtures of *cis* and *trans* bicyclic alcohols are usually generated.

$$\text{2-(4-iodobutyl)cyclohexanone} \xrightarrow[\text{THF; 68\%}]{2\text{YbI}_2} \underset{85\%}{\text{HO, H-decalinol (cis)}} + \underset{15\%}{\text{HO, H-decalinol (trans)}} \quad (97)$$

Other types of reactions that so successfully employ SmI_2 as a reductant have yet to be attempted utilizing YbI_2. As a milder reductant that might also provide some advantages in terms of diastereoselectivity over SmI_2, YbI_2 and other ytterbium(II) reagents may have a bright future in synthetic organic chemistry.

1.9.3.4 Miscellaneous

Ytterbium metal has been found to promote cross-coupling reactions between diaryl ketones and a variety of C—X π-bond electrophiles.[86] The reactions reportedly occur by nucleophilic addition of an ytterbium diaryl ketone dianion to the electrophiles. The net result of these transformations is that the diaryl ketones have been converted by the ytterbium from an electrophilic species to a nucleophilic diarylcarbinol anion equivalent. Although this methodology probably will not be a general one from the point of view of the ketone (alkyl ketone dianions are, in general, energetically inaccessible), the procedure does have synthetic utility when nucleophilic incorporation of diarylcarbinols is desired.

The earliest studies on this reaction began with an attempt to generate simple symmetrical pinacols.[86] Reaction of 1 equiv. of ytterbium metal with 2 equiv. of a diaryl ketone in THF/HMPA provided excellent yields of the corresponding symmetrical pinacols (equation 98). Interestingly, when equimolar quantities of ytterbium metal and benzophenone were employed, the sole product isolated after aqueous work-up was benzhydrol. When D_2O was utilized to quench this reaction mixture, *C*-deuterated benzhydrol was formed (equation 99). These latter results indicated that a discrete ketone dianionic intermediate was generated in the reaction between ytterbium metal and diaryl ketones.

$$Ph_2C{=}O + Yb \xrightarrow[\text{r.t., 10 min}]{\text{THF, HMPA}} Ph_2C(OH){-}C(OH)Ph_2 \quad 97\% \qquad (98)$$

$$Ph_2C{=}O + Yb \xrightarrow[\text{r.t., 10 min}]{\text{THF, HMPA}} \xrightarrow{D_2O} Ph_2C(D)OH \quad 98\% \qquad (99)$$

Spectral studies lend support to the existence of a discrete organoytterbium species. IR spectra taken of reaction mixtures provide evidence for a three-membered oxametallacyclic structure (**11**) incorporating a divalent ytterbium.[86] Of course, the precise nature of this intermediate is still unknown; nevertheless, available evidence does point to a unique type of intermediate which may prove useful in further synthetic transformations.

$$Ph_2C(O)Yb \text{ (three-membered ring)}$$

(11)

Indeed, it has been found that unsymmetrical pinacols can be generated in surprisingly high yields by treating 1 equiv. of a diaryl ketone with 1 equiv. of ytterbium metal, and subsequently quenching the resultant reaction mixture with a variety of aldehydes and ketones (equation 100).[86] Yields in most cases are high, and this particular transformation represents one of the very few ways in which such a process can be accomplished efficiently. Reaction of benzophenone/ytterbium with 2-cyclohexen-1-one provides mixtures of 1,2- and 1,4-addition products, together with some benzhydrol.

$$Ph_2C{=}O + Yb \xrightarrow{\text{THF}} \mathbf{(11)} \xrightarrow[\text{r.t., 15 min}]{\text{MeCHO}} Ph_2C(OH){-}CH(OH)Me \quad 82\% \qquad (100)$$

Recognition that a discrete intermediate could be trapped by aldehydes and ketones led to further studies involving a variety of electrophiles.[86] For example, CO_2 was found to react effectively with the ytterbium intermediate, providing good yields of α-hydroxycarboxylic acids (equation 101). Carbonylation did not take place at room temperature under ambient pressures of CO, and carbon disulfide provided a mixture of unidentified products.

$$\mathrm{Ph_2C{=}O} + \mathrm{Yb} \xrightarrow{\mathrm{THF}} \textbf{(11)} \xrightarrow[\text{1 atm}]{\mathrm{CO_2}} \mathrm{Ph_2C(OH)CO_2H}\ \ 61\% \qquad (101)$$

Carboxylic acid derivatives such as esters and amides undergo nucleophilic acyl substitution reactions with the ketone dianion derived from benzophenone, providing modest yields of the corresponding carbonyl products (equations 102 and 103).[86] Benzhydrol is a significant by-product in these reactions.

$$\mathrm{Ph_2C{=}O} + \mathrm{Yb} \xrightarrow{\mathrm{THF}} \textbf{(11)} \xrightarrow[\text{r.t., 2 h}]{\mathrm{MeCO_2Et}} \mathrm{Ph_2C(OH)COMe}\ \ 43\% \qquad (102)$$

$$\mathrm{Ph_2C{=}O} + \mathrm{Yb} \xrightarrow{\mathrm{THF}} \textbf{(11)} \xrightarrow[\text{r.t., 2 h}]{\mathrm{HCONMe_2}} \mathrm{Ph_2C(OH)CHO}\ \ 31\% \qquad (103)$$

Isocyanates also react with the ytterbium ketone dianion, providing the expected amides in good yields (equation 104).[86]

$$\mathrm{Ph_2C{=}O} + \mathrm{Yb} \xrightarrow{\mathrm{THF}} \textbf{(11)} \xrightarrow[\text{r.t., 30 min}]{\mathrm{PhNCO}} \mathrm{Ph_2C(OH)CONHPh}\ \ 58\% \qquad (104)$$

Finally, nitriles undergo addition with several different diaryl ketone dianions, affording α-hydroxy ketones after hydrolysis of the reaction mixture (equation 105).[86] In some cases, deoxygenated ketones appear as minor by-products of the reaction.

$$\mathrm{Ph_2C{=}O} + \mathrm{Yb} \xrightarrow{\mathrm{THF}} \textbf{(11)} \xrightarrow[\text{ii, } \mathrm{H_3O^+}]{\text{i, MeCN, r.t., 4 h}} \mathrm{Ph_2C(OH)COMe}\ \ 72\% \qquad (105)$$

1.9.4 REFERENCES

1. (a) N. R. Natale, *Org. Prep. Proced. Int.*, 1983, **15**, 387; (b) H. B. Kagan and J.-L. Namy, in 'Handbook on the Physics and Chemistry of the Rare Earths', ed. K. A. Gschneidner, Jr. and L. Eyring, Elsevier, Amsterdam, 1984, p. 525; (c) H. B. Kagan, in 'Fundamental and Technological Aspects of Organo-*f*-Element Chemistry', ed. T. J. Marks and I. L. Fragalà, Reidel, Dordrecht, 1985, p. 49; (d) H. B. Kagan and J.-L. Namy, *Tetrahedron*, 1986, **42**, 6573; (e) J. R. Long, in 'Handbook on the Physics and Chemistry of Rare Earths', ed. K. A. Gschneidner, Jr. and L. Eyring, Elsevier, Amsterdam, 1986, p. 335.
2. (a) H. Schumann and W. Genthe, in 'Handbook on the Physics and Chemistry of Rare Earths', ed. K. A. Gschneidner, Jr. and L. Eyring, Elsevier, Amsterdam, 1984, p. 445; (b) W. J. Evans, *Adv. Organomet. Chem.*, 1985, **24**, 131; (c) W. J. Evans, *Polyhedron*, 1987, **6**, 803.
3. J. E. Huheey, 'Inorganic Chemistry: Principles of Structure and Reactivity', Harper and Row, New York, 1983.
4. (a) T. J. Haley, *J. Pharm. Sci.*, 1965, **54**, 663; (b) D. W. Bruce, B. E. Hietbrink and K. P. DuBois, *Toxicol. Appl. Pharmacol.*, 1963, **5**, 750; (c) Rare Earths Reminder, Rhône-Poulenc, Paris, 1986.
5. K. M. Mackay and R. A. Mackay, 'Introduction to Modern Inorganic Chemistry', Intertext Books, London, 1969.
6. R. G. Pearson (ed.), 'Hard and Soft Acids and Bases', Dowden, Hutchinson & Ross, Stroudsburg, PA, 1973.

7. E. Murad and D. L. Hildenbrand, *J. Chem. Phys.*, 1980, **73**, 4005.
8. T. Kauffmann, C. Pahde, A. Tannert and D. Wingbermühle, *Tetrahedron Lett.*, 1985, **26**, 4063.
9. T. Imamoto, T. Kusumoto, Y. Tawarayama, Y. Sugiura, T. Mita Y. Hatanaka and M. Yokoyama, *J. Org. Chem.*, 1984, **49**, 3904.
10. J. Collin, J.-L. Namy, C. Bied and H. B. Kagan, *Inorg. Chim. Acta*, 1987, **140**, 29.
11. H. B. Kagan, M. Sasaki and J. Collin, *Pure Appl. Chem.*, 1988, **60**, 1725.
12. H. Schumann, J. Müller, N. Bruncks, H. Lauke and J. Pickardt, *Organometallics*, 1984, **3**, 69.
13. D. F. Evans, G. V. Fazakerley and R. F. Philips, *J. Chem. Soc. (A)*, 1971, 1931.
14. (a) K. Yokoo, Y. Fujiwara, T. Fukagawa and H. Taniguchi, *Polyhedron*, 1983, **2**, 1101; (b) K. Yokoo, N. Mine, H. Taniguchi and Y. Fujiwara, *J. Organomet. Chem.*, 1985, **279**, C19.
15. S. Fukuzawa, T. Fujinami and S. Sakai, *J. Chem. Soc., Chem. Commun.*, 1986, 475.
16. (a) J.-L. Namy, P. Girard and H. B. Kagan, *Nouv. J. Chim.*, 1977, **1**, 5; (b) J.-L. Namy, P. Girard and H. B. Kagan, *Nouv. J. Chim.*, 1981, **5**, 479; (c) P. Girard, J.-L. Namy and H. B. Kagan, *J. Am. Chem. Soc.*, 1980, **102**, 2693.
17. T. Imamoto and M. Ono, *Chem. Lett.*, 1987, 501.
18. V. Chebolu, R. R. Whittle and A. Sen, *Inorg. Chem.*, 1985, **24**, 3082.
19. J.-L. Namy, J. Collin, J. Zhang and H. B. Kagan, *J. Organomet. Chem.*, 1987, **328**, 81.
20. S. Araki, M. Hatano, H. Ito and Y. Butsugan, *Appl. Organomet. Chem.*, 1988, **2**, 79.
21. S. Araki, M. Hatano, H. Ito and Y. Butsugan, *J. Organomet. Chem.*, 1987, **333**, 329.
22. T. Tabuchi, J. Inanaga and M. Yamaguchi, *Tetrahedron Lett.*, 1986, **27**, 1195.
23. T. Tabuchi, J. Inanaga and M. Yamaguchi, *Chem. Lett.*, 1987, 2275.
24. K. Otsubo, K. Kawamura, J. Inanaga and M. Yamaguchi, *Chem. Lett.*, 1987, 1487.
25. J. Inanaga, M. Ishikawa and M. Yamaguchi, *Chem. Lett.*, 1987, 1485.
26. (a) J. Souppe, J.-L. Namy and H. B. Kagan, *Tetrahedron Lett.*, 1982, **23**, 3497; (b) J.-L. Namy, J. Souppe, J. Collin and H. B. Kagan, *J. Org. Chem.*, 1984, **49**, 2045; (c) J. Souppe, L. Danon, J.-L. Namy and H. B. Kagan, *J. Organomet. Chem.*, 1983, **250**, 227.
27. H. B. Kagan, J.-L. Namy and P. Girard, *Tetrahedron*, 1981, **37**, suppl. **9**, 175.
28. P. Dowd and S. Choi, *J. Am. Chem. Soc.*, 1987, **109**, 6548, and refs. cited therein.
29. T. Imamoto, T. Takeyama and M. Yokoyama, *Tetrahedron Lett.*, 1984, **25**, 3225.
30. J. D. White and T. C. Somers, *J. Am. Chem. Soc.*, 1987, **109**, 4424.
31. M. Sasaki, J. Collin and H. B. Kagan, *Tetrahedron Lett.*, 1988, **29**, 4847.
32. (a) T. Imamoto, T. Takeyama and H. Koto, *Tetrahedron Lett.*, 1986, **27**, 3243; (b) T. Tabuchi, J. Inanaga and M. Yamaguchi, *Tetrahedron Lett.*, 1986, **27**, 3891; (c) T. Imamoto and N. Takiyama, *Tetrahedron Lett.*, 1987, **28**, 1307.
33. M. Matsukawa, T. Tabuchi, J. Inanaga and M. Yamaguchi, *Chem. Lett.*, 1987, 2101.
34. G. A. Molander and J. B. Etter, *Synth. Commun.*, 1987, **17**, 901.
35. A. H. Berks, Ph.D. Thesis, University of Colorado, Boulder, 1988.
36. G. A. Molander and J. B. Etter, *J. Org. Chem.*, 1986, **51**, 1778.
37. H. Suginome and S. Yamada, *Tetrahedron Lett.*, 1987, **28**, 3963.
38. J. J. Sosnowski, E. B. Danaher and R. K. Murray, Jr., *J. Org. Chem.*, 1985, **50**, 2759.
39. R. K. Murray, Jr., private communication.
40. G. Lannoye and J. M. Cook, *Tetrahedron Lett.*, 1988, **29**, 171.
41. B. A. Barner and M. A. Rahman, 'Abstracts of the Third Chemical Congress of North America', Toronto, Canada, June 5–10, 1988, Abstract ORGN 419.
42. G. A. Molander, J. B. Etter and P. W. Zinke, *J. Am. Chem. Soc.*, 1987, **109**, 453.
43. M. P. Cooke, Jr. and I. N. Houpis, *Tetrahedron Lett.*, 1985, **26**, 4987.
44. C. Kenny, Ph.D. Thesis, University of Colorado, Boulder, 1989.
45. L. A. Fedorov, *Russ. Chem. Rev. (Engl. Transl.)*, 1970, **39**, 655.
46. H.-G. Korth, P. Lommes and R. Sustmann, *J. Am. Chem. Soc.*, 1984, **106**, 663.
47. T. Tabuchi, K. Kawamura, J. Inanaga and M. Yamaguchi, *Tetrahedron Lett.*, 1986, **27**, 3889.
48. K. Maruoka, S. Hashimoto, Y. Kitagawa, H. Yamamoto and H. Nozaki, *J. Am. Chem. Soc.*, 1977, **99**, 7705.
49. E. Vedejs and S. Ahmad, *Tetrahedron Lett.*, 1988, **29**, 2291.
50. G. A. Molander and J. B. Etter, *J. Am. Chem. Soc.*, 1987, **109**, 6556.
51. P.-J. Thorel, unpublished research.
52. L. Harring, Ph.D. Thesis, University of Colorado, Boulder, 1991.
53. (a) S. Fukuzawa, A. Nakanishi, T. Fujinami and S. Sakai, *J. Chem. Soc., Chem. Commun.*, 1986, 624; (b) K. Otsubo, J. Inanaga and M. Yamaguchi, *Tetrahedron Lett.*, 1986, **27**, 5763; (c) S. Fukuzawa, A. Nakanishi, T. Fujinami and S. Sakai, *J. Chem. Soc., Perkin Trans. 1*, 1988, 1669.
54. (a) T. Shono, I. Nishiguchi, H. Ohmizu and M. Mitani, *J. Am. Chem. Soc.*, 1978, **100**, 545; (b) D. P. Fox, R. D. Little and M. M. Baizer, *J. Org. Chem.*, 1985, **50**, 2202; (c) E. Kariv-Miller and T. J. Mahachi, *J. Org. Chem.*, 1986, **51**, 1041.
55. D. Belotti, J. Cossy, J. P. Pete and C. Portella, *J. Org. Chem.*, 1986, **51**, 4196.
56. (a) S. K. Pradhan, S. R. Kadam, J. N. Kolhe, T. V. Radhakrishnan, S. V. Sohani and V. B. Thaker, *J. Org. Chem.*, 1981, **46**, 2622; (b) E. J. Corey and S. G. Pyne, *Tetrahedron Lett.*, 1983, **24**, 2821; (c) T. Ikeda, S. Yue and C. R. Hutchinson, *J. Org. Chem.*, 1985, **50**, 5193.
57. S. Fukuzawa, M. Iida, A. Nakanishi, T. Fujinami and S. Sakai, *J. Chem. Soc., Chem. Commun.*, 1987, 920.
58. G. A. Molander and C. Kenny, *Tetrahedron Lett.*, 1987, **28**, 4367.
59. A. L. J. Beckwith, *Tetrahedron*, 1981, **37**, 3073.
60. B. Giese, 'Radicals in Organic Synthesis: Formation of Carbon–Carbon Bonds', Pergamon Press, Oxford, 1986.
61. T. L. Fevig, R. L. Elliott and D. P. Curran, *J. Am. Chem. Soc.*, 1988, **110**, 5064.
62. (a) J.-L. Namy, J. Souppe and H. B. Kagan, *Tetrahedron Lett.*, 1983, **24**, 765; (b) A. Fürstner, R. Csuk, C. Rohrer and H. Weidmann, *J. Chem. Soc., Perkin Trans. 1*, 1988, 1729.

63. Y. Handa and J. Inanaga, *Tetrahedron Lett.*, 1987, **28**, 5717.
64. G. A. Molander and C. Kenny, *J. Org. Chem.*, 1988, **53**, 2132.
65. (a) V. Rautenstrauch, *Tetrahedron*, 1988, **44**, 1613; (b) J. W. Huffman, R. H. Wallace and W. T. Pennington, *Tetrahedron Lett.*, 1988, **29**, 2527; (c) J. W. Huffman, W.-P. Liao and R. H. Wallace, *Tetrahedron Lett.*, 1987, **28**, 3315.
66. (a) A. Clerici and O. Porta, *Tetrahedron*, 1983, **39**, 1239; (b) A. Clerici and O. Porta, *J. Org. Chem.*, 1983, **48**, 1690; (c) A. Clerici and O. Porta, *J. Org. Chem.*, 1987, **52**, 5099; (d) E. J. Corey, R. L. Danheiser and S. Chandrasekaran, *J. Org. Chem.*, 1976, **41**, 260.
67. I. Fleming, 'Frontier Orbitals and Organic Chemical Reactions', Wiley-Interscience, New York, 1976.
68. M. T. Reetz, M. Hüllmann, W. Massa, S. Berger, P. Rademacher and P. Heymanns, *J. Am. Chem. Soc.*, 1986, **108**, 2405.
69. (a) D. Seyferth and R. M. Weinstein, *J. Am. Chem. Soc.*, 1982, **104**, 5534; (b) D. Seyferth, R. M. Weinstein and W.-L. Wang, *J. Org. Chem.*, 1983, **48**, 1144; (c) R. M. Weinstein, W.-L. Wang and D. Seyferth, *J. Org. Chem.*, 1983, **48**, 3367; (d) D. Seyferth, W.-L. Wang and R. C. Hui, *Tetrahedron Lett.*, 1984, **25**, 1651.
70. J. Collin and H. B. Kagan, *Tetrahedron Lett.*, 1988, **29**, 6097.
71. P. Girard, R. Couffignal and H. B. Kagan, *Tetrahedron Lett.*, 1981, **22**, 3959.
72. D. Griller and K. U. Ingold, *Acc. Chem. Res.*, 1980, **13**, 317.
73. J. Souppe, J.-L. Namy and H. B. Kagan, *Tetrahedron Lett.*, 1984, **25**, 2869.
74. P. W. Zinke, Ph. D. Thesis, University of Colorado, Boulder, 1987.
75. M. Sasaki, J. Collin and H. B. Kagan, *Tetrahedron Lett.*, 1988, **29**, 6105.
76. M. Matsukawa, J. Inanaga and M. Yamaguchi, *Tetrahedron Lett.*, 1987, **28**, 5877.
77. S. F. Martin, C.-P. Yang, W. L. Laswell and H. Rüeger, *Tetrahedron Lett.*, 1988, **29**, 6685.
78. A. B. Sigalov, E. S. Petrov, L. F. Rybakova and I. P. Beletskaya, *Izv. Akad. Nauk SSSR, Ser. Khim.*, 1983, 2615.
79. (a) G. B. Deacon, W. D. Raverty and D. G. Vince, *J. Organomet. Chem.*, 1977, **135**, 103; (b) G. B. Deacon and A. J. Koplick, *J. Organomet. Chem.*, 1978, **146**, C43; (c) G. B. Deacon, A. J. Koplick, W. D. Raverty and D. G. Vince, *J. Organomet. Chem.*, 1979, **182**, 121; (d) G. B. Deacon, A. J. Koplick and T. D. Tuong, *Aust. J. Chem.*, 1982, **35**, 941.
80. K. Yokoo, Y. Kijima, Y. Fujiwara and H. Taniguchi, *Chem. Lett.*, 1984, 1321.
81. G. B. Deacon, P. I. Mackinnon and T. D. Tuong, *Aust. J. Chem.*, 1983, **36**, 43.
82. (a) G. B. Deacon and T. D. Tuong, *J. Organomet. Chem.*, 1981, **205**, C4; (b) T. Fukagawa, Y. Fujiwara, K. Yokoo and H. Taniguchi, *Chem. Lett.*, 1981, 1771.
83. (a) K. Yokoo, Y. Yamanaka, T. Fukagawa, H. Taniguchi and Y. Fujiwara, *Chem. Lett.*, 1983, 1301; (b) Z. Hou, Y. Fujiwara, T. Jintoku, N. Mine, K. Yokoo and H. Taniguchi, *J. Org. Chem.*, 1987, **52**, 3524.
84. T. Fukagawa, Y. Fujiwara and H. Taniguchi, *Chem. Lett.*, 1982, 601.
85. P. Watson, *J. Chem. Soc., Chem. Commun.*, 1980, 652.
86. (a) Z. Hou, K. Takamine, Y. Fujiwara and H. Taniguchi, *Chem. Lett.*, 1987, 2061; (b) Z. Hou, K. Takamine, O. Aoki, H. Shiraishi, Y. Fujiwara and H. Taniguchi, *J. Chem. Soc., Chem. Commun.*, 1988, 668; (c) Z. Hou, K. Takamine, O. Aoki, H. Shiraishi, Y. Fujiwara and H. Taniguchi, *J. Org. Chem.*, 1988, **53**, 6077.

1.10
Lewis Acid Carbonyl Complexation

SOROOSH SHAMBAYATI and STUART L. SCHREIBER
Harvard University, Cambridge, MA, USA

1.10.1 INTRODUCTION

Over 100 years have passed since the appearance of the first report of the formation of crystalline complexes between boron trifluoride and aromatic aldehydes.[1] Since then, Lewis acids have found a prominent role in organic synthesis through their action on carbonyl-containing compounds. They have been used as additives or catalysts in carbonyl addition processes, and in many cases they constitute the essential component of basic carbon–carbon bond forming reactions. Most recently there has been impressive development in the area of asymmetric catalysis by Lewis acids. New chiral reagents possessing near complete asymmetric control are being reported at an increasingly rapid rate. The true origins of the 'Lewis acid effect', however, are still poorly understood. Three examples from simple nucleophilic additions to 4-*t*-butylcyclohexanone (**1**) are illustrative.

Addition of MeLi to (**1**) at –78 °C in diethyl ether requires 60 min to reach completion and affords a 65:35 mixture of axial:equatorial alcohols (Figure 1).[2,3] In the presence of $LiClO_4$, however, the same addition is complete within 5 s and proceeds with higher stereoselectivity to give predominantly the axial alcohol in a 92:8 ratio. Precomplexation of the ketone with the bulky Lewis acid, methylaluminum(2,6-di-*t*-butyl-4-methylphenoxide) (MAD), on the other hand, affords the equatorial alcohol almost exclusively.[3]

Rationalization, let alone *a priori* prediction, of these results based on common structural models is not a trivial issue. In order to address such problems one needs a better understanding of the energetics and conformational properties of Lewis acid carbonyl complexes. The aim of this chapter is to survey the lit-

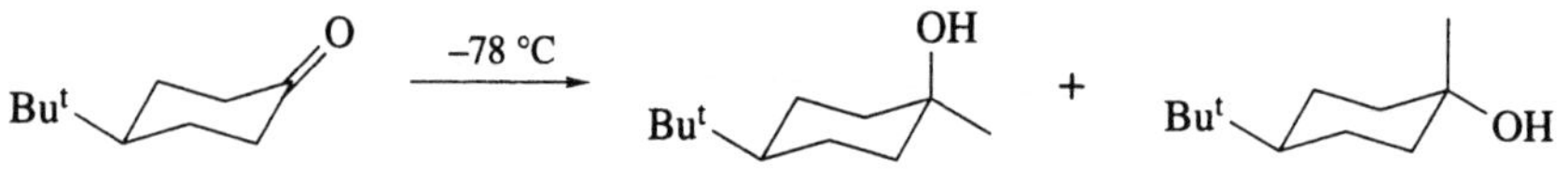

(1) (2) (3)

MeLi, 60 min	65:35
MeLi/LiClO$_4$, 5 s	92:8
MeLi/MAD, 2–3 h	1:99

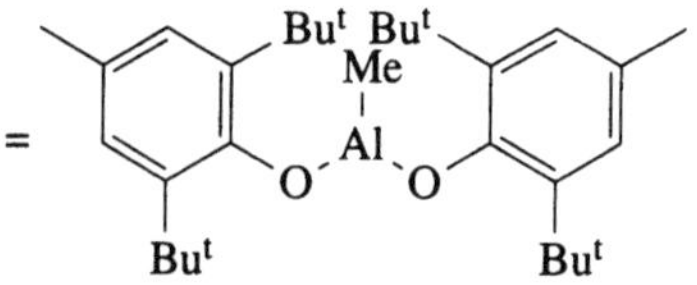

Figure 1 Effect of additives on the reaction of MeLi and *t*-butylcyclohexanone

erature that has helped to elucidate the structural consequences of complexation between Lewis acids and a carbonyl moiety. To facilitate the focus and analysis of the large body of available data on this subject, the structural questions of prime interest will be summarized in the following section. The ensuing sections will then draw upon pertinent data in an attempt to explain each particular issue.

1.10.2 STRUCTURAL ISSUES

1.10.2.1 Effects on Rate and Reactivity

Perhaps the best known effect of Lewis acids on the reaction of carbonyl-containing compounds concerns their ability to enhance reactivity. For example, the $AlCl_3$-catalyzed Diels–Alder reaction of anthracene and maleic anhydride in dichloromethane at room temperature is complete in 90 s, while the uncatalyzed reaction has been estimated to require 4800 h for 95% completion.[4] Such dramatic rate enhancements in Lewis acid promoted reactions are generally rationalized in terms of an increase in the polar character of the carbonyl group. This suggestion has direct implications for the structure of Lewis acid carbonyl complexes. Possible consequences include a longer C—O bond, increased dipole moment[5] and an influence on the conformational preferences of groups proximal to the carbonyl ligand. An examination of the theoretical and experimental results that pertain to these issues may, in turn, shed some light on the origins of rate enhancement.

1.10.2.2 σ- *versus* π-(η^2)-Bonding

In principle, a carbonyl ligand can coordinate to a Lewis acidic metal center either through its lone pair electrons to form a σ-bond or through the carbonyl π-system to afford η^2-metallooxirane complexes (Figure 2). A preference for one mode of bonding over the other will not only depend on the nature of the Lewis acid but also on the steric and electronic requirements of the carbonyl ligand. Moreover, different modes of bonding impose different constraints upon the conformation and reactivity of the ligand. For example, η^2-bonding necessarily blocks one face of the carbonyl ligand, thereby directing the approach of a reactive reagent to the opposite face. However, a simple rule of face selectivity is not available for the σ-bound carbonyl complex. On the other hand, the conformational biases of substituents *syn* and *anti* to the metal (Figure 2; R^1 and R^2) in these systems would be quite different than for η^2-complexes, where R^1 and R^2 occupy symmetric volumes of space with respect to an achiral Lewis acid. Furthermore, one might expect a more polar and perhaps more reactive C—O bond for σ-complexes. Therefore, it would be helpful to know if any preference for Lewis acid σ- or π-(η^2)-coordination and *syn* or *anti* to particular substituents, exists.

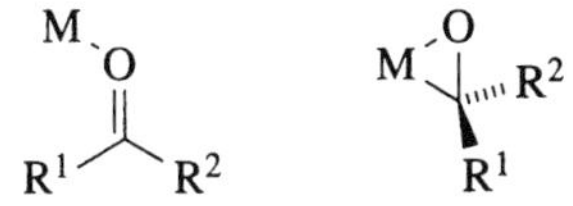

Figure 2 σ-Bonding *versus* π-bonding

1.10.2.3 Conformational Issues

1.10.2.3.1 Conjugated carbonyls: s-cis versus s-trans

Although there are many conformational issues which can influence the stereochemical outcome of the reaction of carbonyls, the central question for most simple α,β-unsaturated carbonyl systems is whether the reactive intermediate adopts an *s-cis* or *s-trans* conformation (Figure 3). Since the *s-cis* and *s-trans* arrangements expose different faces of an α,β-unsaturated moiety, this issue becomes particularly important in enantioselective additions to these systems. The question can now be posed as to how coordination to a Lewis acid may affect the equilibrium between the *s-cis* and *s-trans* arrangements.

s-trans *s-cis*

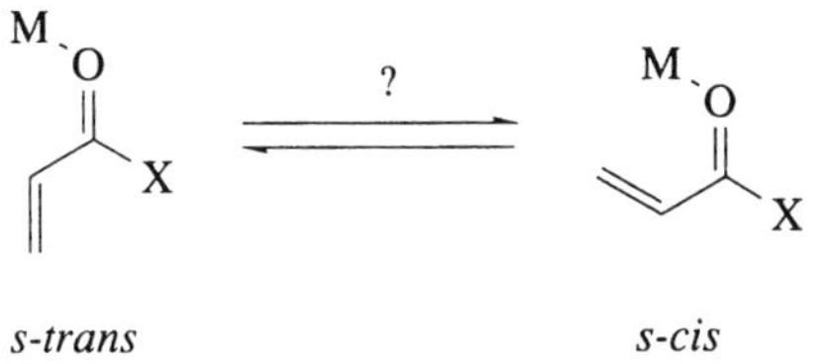

Figure 3 *s-cis versus s-trans*

1.10.2.3.2 Nonconjugated carbonyls

Despite the great deal of attention devoted to nucleophilic additions to α-chiral carbonyls, the source of stereoselectivity in these reactions (predicted by Cram's rules of asymmetric induction[6]) remains largely unresolved. Neither direct structural studies nor correlation of reactant and product stereochemistries have yielded any conclusive support for a single comprehensive model. Similarly, the effect of Lewis acids on these systems is only understood at the level of chelation-controlled additions (*vide infra*).

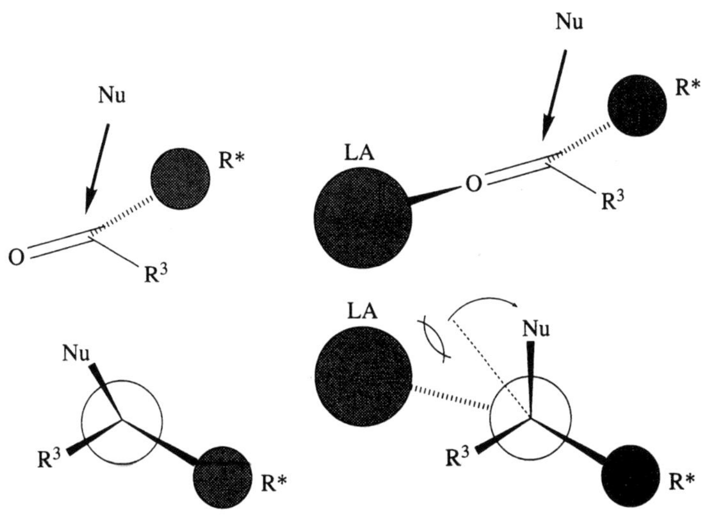

Figure 4 Postulated effect of Lewis acids (LA) on the trajectory of nucleophilic attack

It has been suggested that Lewis acids could affect the trajectory of nucleophilic additions to carbonyl electrophiles (Figure 4).[7] For α-chiral carbonyls this change may translate to a more acute angle of attack, and thereby closer proximity of the nucleophile and the center of chirality. This hypothesis is supported by MNDO calculations of the H^-/pivalic aldehyde system,[8] and also by an investigation of nucleophilic additions to chiral thionium ions.[9] Higher levels of diastereoselectivity are observed with larger aryl substituents on the thionium species (Figure 5).[9] Bulky aryl groups can thus be envisioned as playing a role similar to that of a coordinating Lewis acid and increasing the stereoselectivity by reorientation of the trajectory of attack. Still, this proposal does not address the question of the influence of Lewis acids on the inherent conformational preferences α to a carbonyl. A structural model of this effect would undoubtedly be of great value for the design of stereospecific reactions of this type, and it may also provide some insight into the origins of Cram selectivity itself.

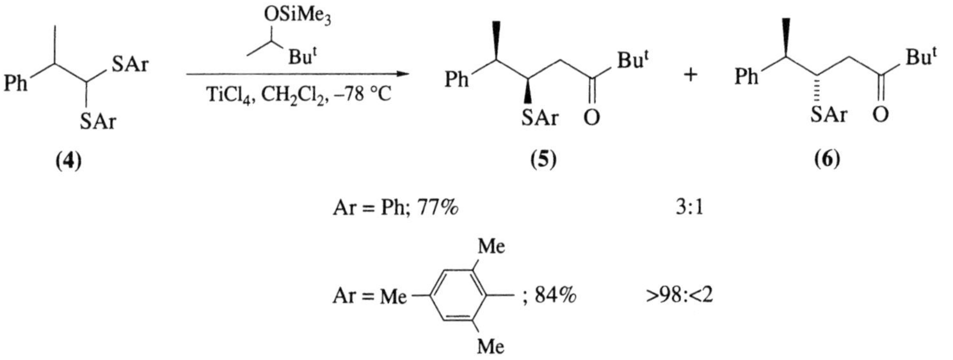

(4) → (5) + (6)

OSiMe$_3$, Bu^t; $TiCl_4$, CH_2Cl_2, −78 °C

Ar = Ph; 77% 3:1

Ar = 2,4,6-trimethylphenyl (Me, Me, Me); 84% >98:<2

Figure 5 Diastereoselectivity in nucleophilic additions to chiral thionium ions

1.10.3 THEORETICAL STUDIES

Most of the early theoretical studies of Lewis acid carbonyl interactions focused on alkali metal cations such as Li^+ and Na^+, partly due to the significance of such interactions in biological systems and partly due to computational limitations. In 1973 some of the first calculations on the formaldehyde/Li^+ system were performed.[10] Using both *ab initio* (SCF–LCAO–MO) and semiempirical (CNDO/2) methods the researchers found that in the ground state the complex possesses C_{2v} symmetry in which Li^+ lies on the axis of the carbonyl C—O bond with an O—Li distance of 1.77 Å.

Interestingly, the energy of interaction showed a strong preference for the linear geometry of C—O—Li, whereas little out of plane angular dependence was observed (Figure 6).[10] A second study of the same complex, using larger basis sets and a full configuration interaction (CI) analysis, later confirmed these results (linear C—O—Li, C_{2v} symmetry, d[O—Li] = 1.70 Å) and also indicated that the C—O bond length of the complex (1.23 Å) is close to that of isolated formaldehyde at equilibrium.[11] Thus it was argued that structural relaxation of the carbonyl group was not important. However, a Mulliken population analysis revealed a pronounced polarization of the formaldehyde electron density towards oxygen, indicating that complex stabilization is achieved mainly through electrostatic ion–dipole interactions with little covalent character for the O—Li bond.

difficult facile

Figure 6 Relative ease of distortions for the Li^+/formaldehyde complex

(based on *ab initio* results of ref. 10)

In 1979 the results of *ab initio* calculations at the 4-31G and 6-31G level on the same complex as well as a formaldehyde/H^+ complex were reported.[12,13] Structural and energetic comparisons of the two complexes showed that while Li^+ prefers a linear geometry for electrostatic ion–dipole bonding, the proton coordinates to the carbonyl through a largely covalent bond, resulting in a bent structure (C_s symmetry) for the H_2CO—H^+ complex with a CO—H angle of 124.5°. The energy of the Li^+/formaldehyde complex was then compared to the energy of Li^+ complexes of urea, acetone, carbonyl fluoride and carbonic acid. Based on these calculations the order of Li^+ affinities of Lewis bases with the general formula R_2C=O could be assigned as R = NH_2 > Me > OH > H > F, consistent with intuitive notions of electron-donating or electron-withdrawing abilities of the substituents R.

Complexation of acetone with a sodium cation was investigated through a combination of *ab initio* and semiempirical calculations.[14] The investigators concluded that, similar to the lithium complexes, Na^+/acetone adopts a linear geometry, with Na 1.22 Å from the carbonyl oxygen. These results are in full accord with an extensive *ab initio* study of proton, lithium and sodium affinities of first and second row bases.[15] Both groups, however, also pointed out deficiencies in their theoretical methods which could account for some of the discrepancies with experimentally determined Lewis acid–base bonding energies.[16]

The geometries found for the complexes of formaldehyde with first and second row cations in theoretical studies were analyzed in terms of molecular orbitals.[17] Based on the results of photoelectron spectroscopy,[18] it was argued that the carbonyl group contains two nonequivalent lone pairs; an *sp*-hybridized orbital contains one pair of electrons along the C=O axis and a second, higher energy lone pair in a *p*-like orbital lies perpendicular to the C=O axis (Figure 7).[17]

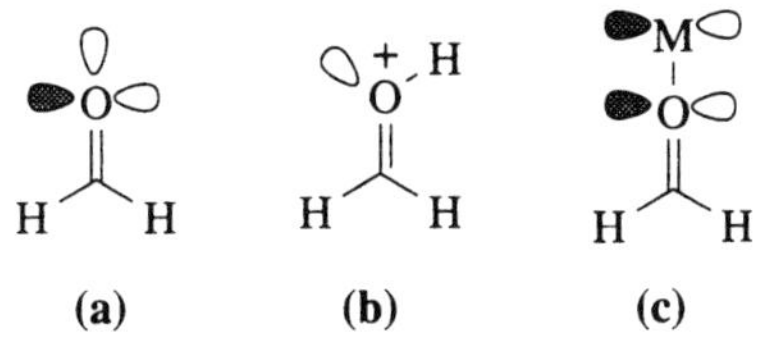

(a) (b) (c)

Figure 7 Proposed molecular orbital scheme for uncomplexed and complexed formaldehyde

When the Lewis acid interacts with only one lone pair of the carbonyl a bent geometry is observed, since the second lone pair can be stabilized by acquiring more *s*-character in an sp^2-like orbital (Figure 7b). The linear geometry is favored for Lewis acids with a second vacant orbital that can accommodate additional electron donation from the second *p* lone pair (Figure 7c). The structures of a number of Lewis acid/formaldehyde complexes were thus rationalized and it was also noted that these geometric preferences may be weak, since in some cases strong Lewis acid carbonyl interactions were found in structures distorted by as much as 30° from their optimum (bent or linear) geometries.

It has been noted that although the linear geometry is consistently predicted for cationic Lewis acid carbonyl complexes in *ab initio* calculations, extrapolation of these results to neutral Lewis acid complexes may not be justified.[19] Semiempirical MNDO calculations predicted the bent conformation as the lowest energy structure for neutral Lewis acidic derivatives of beryllium, boron and aluminum complexed with *trans*-2,3-dimethylcyclopropanone, whereas linear structures were predicted for the cationic complexes of beryllium and aluminum (Table 1).[19]

Table 1 Geometric Preferences for the Complexes of Dimethylcyclopropanone with Lewis Acids; MNDO-calculated Heats of Formation for the Bent and Linear Geometries

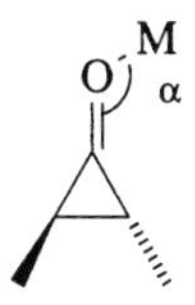

M	*Bent*		*Linear*		
	Angle (α)	ΔH (kcal mol^{-1})	*Angle* (α)	ΔH (kcal mol^{-1})	$\Delta\Delta H$
BeH^+	150.0°	138.7	180.0°	137.1	1.6
BeH_2	149.8°	−27.04	179.9°	−26.63	0.41
$AlMe_2^+$	136.0°	111.7	179.5°	108.8	2.9
$AlMe_3$	138.8°	−52.24	180.0°	−49.49	2.75
BF_2^+	133.5°	−11.27	180.0°	−6.85	4.42
BF_3	133.4°	−270.8	180.0°	−265.6	5.2

In a recent article the question of the effect of Lewis acids on the conformational preferences of α,β-unsaturated carbonyl systems was addressed.[20] Calculations were first conducted on three uncomplexed α,β-unsaturated carbonyl compounds and the relative energies of the *s-cis* and *s-trans* conformations were compared. As can be seen from Table 2,[20] while the *s-trans* conformation is favored for uncomplexed acrolein, both acrylic acid and methyl acrylate prefer the *s-cis* conformation, although this preference is not strong.

Table 2 Relative Ground State Energies of α,β-Unsaturated Carbonyls

Entry	*X*	*Relative energies* (kcal mol^{-1}) *s-trans*	*s-cis*	E_a[b,c]	*Basis set*
1	—H	0.0	1.8	8.9 (5.0)	6-31G*//3-21G*
2	—OH[a]	0.6	0.0	7.5 (3.8)	6-31G*//3-21G*
3	—OMe[a]	0.7	0.0	7.4	6-31G*//3-21G*

[a]Values are for the (*Z*)-conformations of acrylic acid and methyl acrylate. [b]Barriers to rotation: calculated relative energies for an optimized structure with the τ(C═C—C═O) angle constrained to 90°. [c]Experimentally determined values given in parentheses when available.

Next, complexation with various Lewis acids was studied extensively. Protonation of acrolein, for example, affords a bent complex in which both the C—O and the C—C double bonds are longer by 0.03–0.07 Å. The C—C single bond is comparably (0.06 Å) shorter than in uncomplexed acrolein. Although protonation *anti* to the C═C bond seems to have little effect on the equilibrium between *s-cis* and *s-trans* (compare entries 1 from Tables 2 and 3), *syn* protonation seems to strongly disfavor the *s-cis* conformation, presumably due to steric interactions (Table 3; entry 2). *Anti* coordination is slightly favored for the *s-trans* conformation but strongly preferred for the *s-cis* complex. Protonation of acrylic acid and methyl acrylate is more complicated since in addition to the *s-cis/s-trans* and *syn/anti* issues, (*E*)/(*Z*) conformational preferences of the acid or ester should also be considered. It was found that the (*E*)/(*Z*) equilibrium is not strongly influenced by Lewis acid coordination. Here the more stable (*Z*)-conformer will be assumed in order to focus on the two other issues. Thus, for both protonated (*Z*)-acrylic acid and protonated (*Z*)-methyl acrylate the *syn*, *s-trans* structure has the lowest energy (Table 3).[20] As compared to acrolein, the *syn–anti* preference is now reversed but the greater *s-trans* stability is maintained.

Table 3 Conformational Preferences of Protonated α,β-Unsaturated Carbonyls

Entry	*X*		*Relative energies* (kcal mol^{-1}) *s-trans*	*α*	*s-cis*	*β*	*Basis set*
1	—H	(*Anti*)[a]	0.0	121°	1.8	120°	6-31G*//3-21G*
2	—H	(*Syn*)[a]	0.4	239°	6.0	236°	6-31G*//3-21G*
3	—OH[b]	(*Syn*)[a]	0.0	239°	4.2	123°	6-31G*//3-21G*
4	—OH[b]	(*Anti*)[a]	5.6	123°			6-31G*//3-21G*
5	—OMe[b]	(*Syn*)[a]	0.0	240°	3.6	238°	6-31G*//3-21G*

[a]The terms *syn* and *anti* refer to the position of the Lewis acid relative to the C═C double bond. [b]Calculated relative energies for the (*Z*) conformers.

It was noted that a fourth low energy complex for methyl acrylate is structure (**7**), which is conceptually related to (*Z*)-acrylic acid by coordination of the Lewis acid Me^+ *syn* to the double bond.

At this point, a simple steric argument seems capable of rationalizing these results. Thus, the problem can be viewed in the following manner: protonation occurs on the least-hindered lone pair regardless of

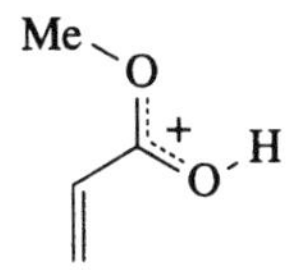

(7)

the inherent *s-cis*/*s-trans* preference of the substrate but always with retention of a (*Z*)-acid or -ester conformer; *syn* to H for acrolein and *syn* to the double bond for (*Z*)-methyl acrylate. In the latter case (*syn* to C═C) steric interactions between the Lewis acid and the double bond locks the *s-trans* conformation, yielding the (*Z*)-*syn-s-trans* structure. All other planar conformations suffer from one or more severe steric interactions, which lead to higher energy structures. Such a 'gearing effect' has been proposed before for other Lewis acid carbonyl complexes[110] and will be discussed further in light of some experimental results in the ensuing sections.

Two other Lewis acids were also included in these studies. Lithium cation coordination with acrolein gave a linear, *s-trans* complex as the most stable structure (Table 4).[20] An increase of the *s-cis–s-trans* gap by 1.4 kcal mol^{-1} is now difficult to rationalize on steric grounds since the C═O-Li bond is nearly linear. The researchers have suggested that this effect is due to an increase in the electron density on the carbonyl oxygen upon complexation, which in turn would increase 'nonbonded electron pair–hydrogen repulsion of the alkene termini in the *s-cis* conformer'. A similar effect is observed with (*Z*)-acrylic acid, where the linear *s-trans* structure is once again more stable than the *s-cis* complex by 1.5 kcal mol^{-1}.

Table 4 Conformational Preferences of α,β-Unsaturated Carbonyls Coordinated to Lithium Cation

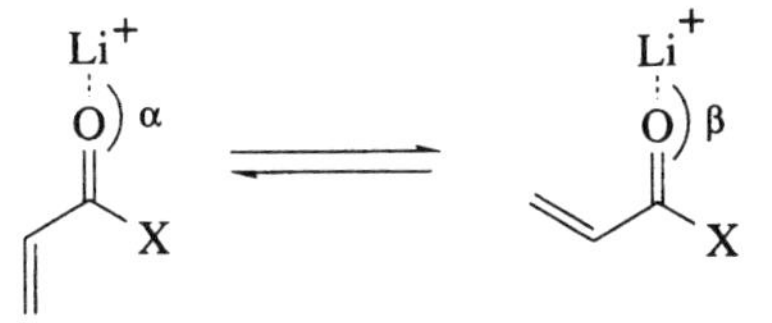

Entry	X	Relative energies (kcal mol^{-1}) *s-trans*	*α*	*s-cis*	*β*	E_a[a]	Basis set
1	—H	0.0	178°	3.2	169°	12.0	6-31G*//3-21G
2	—OH	0.0	189°	1.5	180°	8.5	6-31G*//3-21G

[a]Rotational barrier for isomerization calculated for an optimized structure with τ(C═C—C═O) angle constrained to 90°.

The borane complexes of acrylic acid and methyl acrylate were studied next (Table 5).[20] The *syn*, *s-trans* conformers are again preferred. Notably, the steric arguments discussed previously can nicely rationalize these results. Single point calculations on a linearly restricted BH_3 complex (α = β = 180°) revealed only a 0.1 kcal mol^{-1} preference for the *s-trans* structure, thereby supporting the notion that, in contrast to the Li^+ case, the steric gearing effect is dominant in determining the structure here.

Table 5 Conformational Preferences of α,β-Unsaturated Carbonyls Coordinated to Borane

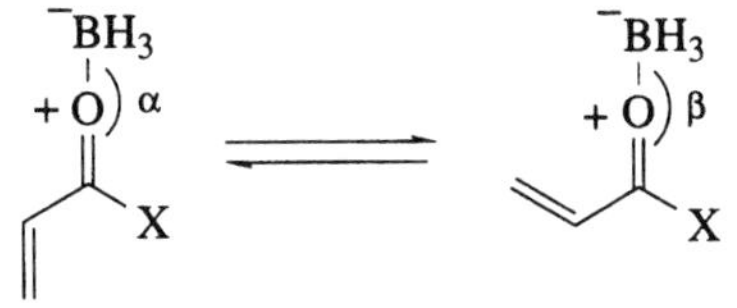

Entry	X		Relative energies (kcal mol^{-1}) *s-trans*	*α*	*s-cis*	*β*	Basis set
1	—OMe	(*Syn*)	0.0	230°	1.4	221°	3-21G//3-21G
2	—OMe	(*Anti*)	5.4	154°	4.4	154°	3-21G//3-21G
3	—OH	(*Syn*)	0.0	230°	1.3	221°	3-21G//3-21G

LePage and Wiberg have analyzed the rotational barriers of Lewis acid coordinated aldehydes and ketones.[21] Calculations on formaldehyde complexes of BH_3, BF_3, AlH_3 and $AlCl_3$ revealed a number of interesting effects. A planar [τ(R—C—O—M) = 0°] bent geometry with a C—O—M angle of 120° seems to be optimal for these neutral Lewis acids (Table 6; Figure 8). Small distortions (15°) in the nodal plane, however, can be accommodated fairly readily, with aluminum complexes being more flexible than the complexes of boron (Figure 8).[21]

Table 6 Calculated Relative Energies for the Bent and Linearly Coordinated Complexes of Formaldehyde

Entry	*M*	*Relative energies* (kcal mol^{-1}) *bent*	*α*	*linear*	*β*	*Basis set*
1	H^+	0.0	117.4°	24.2	180°	MP3/6-31G*//6-31G*
2	BH_3	0.0	122.5°	10.4	180°	MP3/6-31G*//6-21G*
3	BF_3	0.0	122.1°	4.00	180°	MP2/6-31G*//6-21G*
4	AlH_3	0.0	125.1°	5.95	180°	MP2/6-31G*//6-21G*
5	$AlCl_3$	0.0	141.2°	4.63	180°	6-31G*//6-21G*

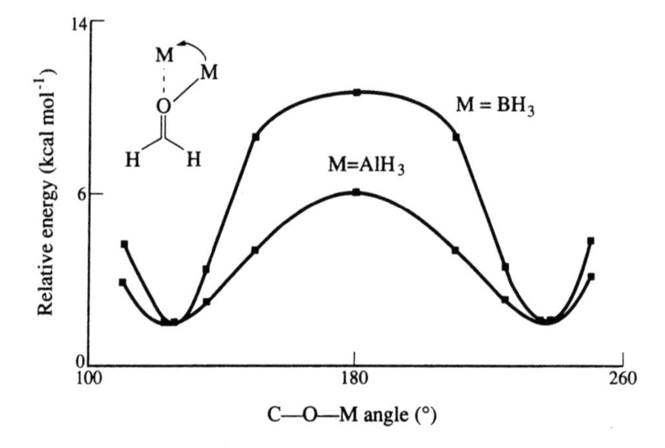

Figure 8 Rotational profile for formaldehyde complexes of BH_3 and AlH_3 [τ(H—C—O—M) = 0°]

Perpendicular π-bonding is predicted to be energetically even less favorable than the linear mode of coordination in the nodal plane (Figure 9).[21] Once again small distortions are facile, although aluminum seems to be less tolerant of π-bonding than boron.

With propanal, borane (BH_3) coordinates through the less-hindered, *anti* (*syn* to H) lone pair to yield a complex which is 2.8 kcal mol^{-1} more stable than the *syn* complex. Surprisingly, the rotational restrictions about the C—C—C—O bond of propanal are somewhat relaxed upon coordination of a Lewis acid (Figure 10).[21]

This phenomenon was rationalized by arguing that the C═O dipole moment is decreased upon Lewis acid complexation, thereby reducing dipole-induced dipole stabilization of the lowest energy (methyl) eclipsed conformer.

In addition to these provocative results several important shortcomings of theoretical methods, even at this level of sophistication, were emphasized. Among other things, the calculated coordination energies and rotational barriers are both strongly basis set and electron correlation dependent. Moreover, there are also many discrepancies with empirical results and no consistent trends in the theoretical disagreements with experiment can be found. The rotational barrier for the BH_3/formaldehyde complex was calculated to be 4–6 kcal mol^{-1} lower than the spectroscopically determined value (8–10 kcal mol^{-1}). In contrast, the coordination energy of BF_3 with dimethyl ether was found to be much larger than the experimental value.

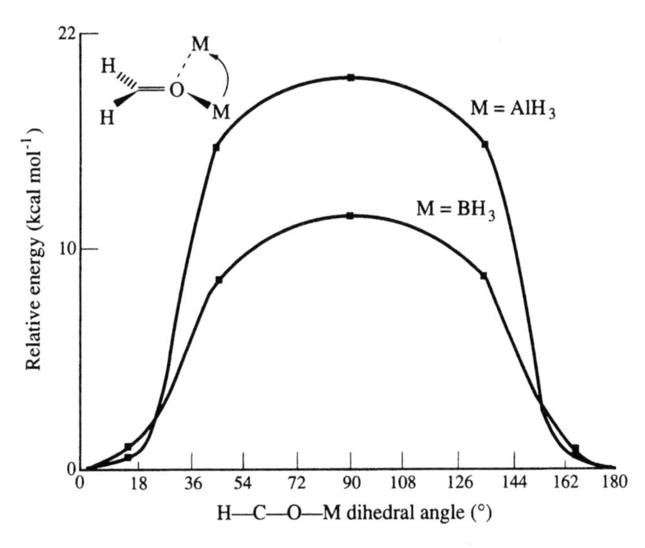

Figure 9 Energy profile for out of plane bending for formaldehyde complexes of BH_3 and AlH_3 (C—O—M = 120°C)

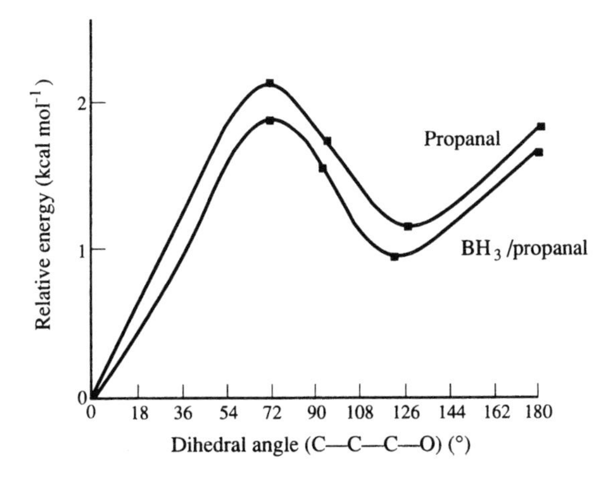

Figure 10 Energy profile for rotation about the C_α bond of propanal and the effect of BH_3 complexation

These points prompt a few words of caution regarding the results discussed in this section and some comments on theoretical results in general. In particular, besides considerations of the size and effect of basis sets, one needs to examine the energy dependence on electron correlation. For donor–acceptor complexes, such as Lewis acid–carbonyls, electron correlation seems to be crucial. It is also instructive to consider the dipole moments of various structures in theoretical studies. Assumption of a gas phase environment may overestimate destabilization due to a large dipole, as dipoles can often be stabilized quite effectively in solution.[22] At times the interpretation of theoretical results are the source of trouble and confusion. Thus two different groups propound apparently conflicting and opposite views on the effect of Lewis acids on the carbonyl dipole.

In recent years theoreticians have exhibited a serious interest in some of the central questions of experimental organic chemistry. Unfortunately, at present, due to computational limitations, theory is unable to reliably address systems of great complexity. Larger basis sets and more efficient computing machines would naturally facilitate the examination of larger systems with increasing reliability. Another essential component, however, appears to be sensitivity towards the true origins of the observed effects or the source of disagreement among different theoretical studies or with experimental values. Difficult as they may be to achieve, coherent conceptual frameworks with strong powers of prediction are central to any theory that hopes to explain a complex problem such as Lewis acid carbonyl complexation. Thus from the experimentalist's viewpoint, a predictive tool that can be readily tested and utilized is invariably more useful than the actual numerical results which may or may not be an accurate reflection of reality.

As a final note to this discussion, one may recognize that, despite their discrepancies and disagreements, the *ab initio* calculations point to a few general observations. To begin with, ionic and covalently bound Lewis acid complexes can be regarded as two distinct categories. Ionic complexes (except for that of H^+) have a primarily ion–dipole mode of coordination with a preference for a linear or near linear geometry. Slight (up to 15°) geometric distortions of these complexes seem to occur with relative ease and are independent of the direction of bending.

Covalently bound Lewis acids, on the other hand, prefer a planar, bent conformation in the direction of one of the sp^2-hybridized carbonyl lone pairs. When the two sites of coordination on the carbonyl are nonequivalent, steric effects appear to determine the coordination geometry. The Lewis acid generally lies *syn* to the small group. For α,β-unsaturated carbonyls Lewis acid coordination increases the *s-trans* preference through steric and/or electronic influences. For nonconjugated carbonyls conformational effects on the substituents of the carbonyl group are either negligible or undetectable by theory. As will be shown in the following sections many of these generalizations are manifest in experimental observations.

1.10.4 NMR SPECTROSCOPY

The nature of donor–acceptor complexes has been the subject of various NMR studies conducted as early as the 1960s. Early calorimetric studies[23–25] showed that boron trihalides are capable of forming donor–acceptor complexes with a number of Lewis bases and the heats of adduct formation for some of these complexes were determined. Gaseous boron trifluoride, for example, was shown to form a complex with ethyl acetate in a highly exothermic reaction ($-\Delta H = 32.9 \pm 0.2$ kcal mol^{-1}).[25] IR and UV analysis of BF_3 complexes of aromatic aldehydes indicated a σ-complex with a lengthened C=O bond and a highly delocalized π-system.[26] More detailed structural information, however, was acquired only after closer inspection by low temperature ^1H, ^{11}B, ^{13}C and ^{19}F NMR studies.[27]

Initially, it was established that boron trihalides form complexes of 1:1 stoichiometry with ethereal and carbonyl-containing Lewis bases. Alkyl ethers appeared to form stronger complexes than acetone and dimethyl formamide. In addition, dimethyl ether was found to complex more strongly than dimethyl sulfide, a trend which is the opposite to that observed with gallium and tin Lewis acids.[28] Low temperature ^1H NMR spectra of BF_3/methyl ketone complexes further revealed that, while BF_3 exchange is rapid at room temperature, separate signals due to free and bound species can be observed at −90 °C.[29] The Lewis acidity of various boron derivatives was correlated with the chemical shift differences between the free and complexed ligand, and the order of decreasing Lewis acidity was established as $BBr_3 > BCl_3 > BF_3 > BH_3$.

Syn–anti isomerization about the C=O bond of BF_3/diethyl ketone adduct at low temperatures gave the first indication of an asymmetrical, bent structure.[30] At −125 °C two separate signals could be distinguished for the methylene protons α to the carbonyl. Since intermolecular exchange is slow even at −90 °C it was argued that the two sets of signals are due to protons *syn* and *anti* to the Lewis acid. A coalescence temperature of −111 °C ($\Delta\upsilon = 30$ Hz) suggested a barrier of *ca.* 8 kcal mol^{-1} for the *syn–anti* isomerization.

Low temperature (−50 °C) ^1H and ^{19}F NMR spectra of α- or β-substituted ketones, esters and nitriles complexed to boron trifluoride showed that boron coordinates preferentially at the most basic and least-hindered base when more than one coordination site is present in the ligand.[31] Ordinarily a relative measure of chemical shifts for each type of complex would be derived from the spectra of a number of model complexes. For example, ^{19}F chemical shifts for BF_3 complexes of ketones (acetone), esters (methyl acetate), nitriles (acetonitrile) and ethers (diethyl ether) were measured as 149 p.p.m., 150 p.p.m., 144 p.p.m. and 156 p.p.m., respectively. Inspection of the resonance frequency and relative intensities of each peak would then reveal the types and ratios of the complexes present in solution.

Methoxyacetonitrile showed exclusive coordination at the methoxy oxygen while methoxymethyl acetate indicated interactions with BF_3 at both the ether and the carbonyl oxygens. Interestingly, a large difference was observed in the preferred site of coordination between methyl methoxyacetate and methoxymethyl acetate (equations 1 and 2). The apparent lower basicity of the ether oxygen in the latter was attributed to the electron-withdrawing effects of the proximal acetate group.

F_3B–O(Me)–CH_2–C(=O)–OMe $\xrightarrow{82:12}$ MeO–CH_2–C(=O→BF_3)–OMe (1)

$T = -105$ °C

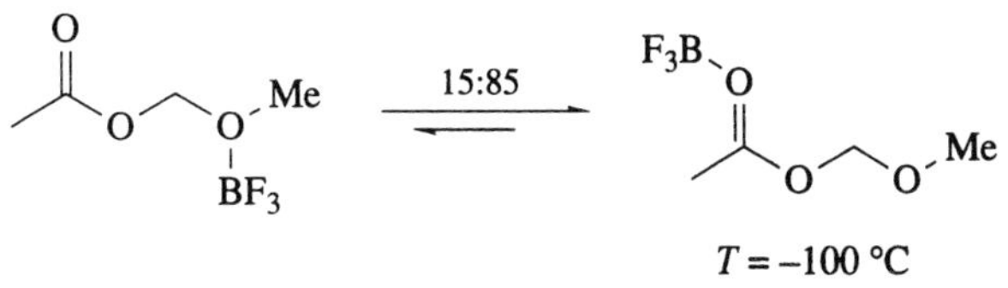

15:85 (2)

$T = -100$ °C

The BF_3/2-methoxyethyl acetate system showed a marked temperature dependence. Complexation at the carbonyl oxygen is dominant at low temperature (63% carbonyl complex at –105 °C) but a shift to the methoxy base is observed upon warming to –60 °C (61% ether complex).

A systematic study of cycloalkanones established the order of their basicity towards boron trifluoride.[32] It was concluded that steric effects control the extent of complexation. For simple cycloalkanones C_n (where n depicts the size of the ring) the basicities were shown to decrease in the order $C_8 > C_5 > C_7 > C_6 = C_9 > C_{10} > C_4$.

NMR spectra of various carbonyl-containing steroids were investigated.[33,34] Here, intermolecular exchange was slow even at 0 °C. In the presence of hydroxy or ether functionality little or no complexation at the carbonyl sites was observed.[34] For BF_3/androstene-3,17-dione (Figure 11a) complexation occurred solely on the enone carbonyl of the A-ring, in contrast to 5β-androstane-3,17-dione (Figure 11b), where the carbonyls of the A- and the D-ring were complexed in a 64:36 ratio.[33]

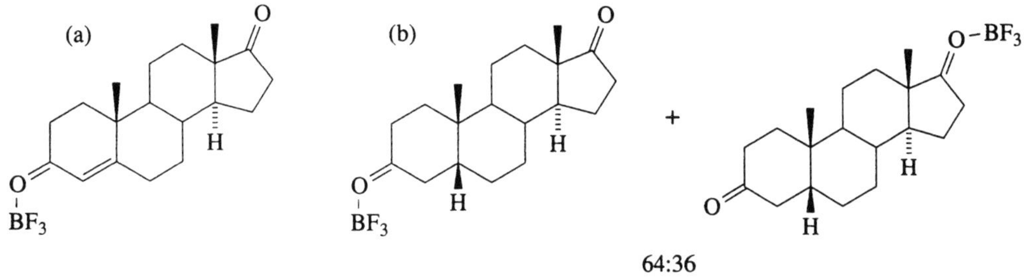

(a) (b) 64:36

Figure 11 Complexation of steroids to boron trifluoride

In the same study, 2-cyclohexen-1-one was studied as a model enone. The ^{13}C NMR spectrum at –85 °C showed two sets of signals for both the C-α and C-β carbons of the complexed ligand, presumably due to slow *syn–anti* isomerization of BF_3. However, it was not determined which conformer predominates or what the barrier to isomerization might be.

^1H and ^{13}C NMR studies of cyclic and acyclic BF_3/ketone complexes clearly demonstrated the unsymmetrical, bent nature of their structures and suggested a barrier of 8–10 kcal mol^{-1} for *syn–anti* isomerization about the carbonyl group of simple alkyl ketones.[35–37] Cyclopentanone, for example, displayed two separate sets of signals for the methylene groups *syn* and *anti* to BF_3 at –100 °C with a coalescence temperature of –64 °C ($\Delta G = 10.3$ kcal mol^{-1}). An interesting comparison can be made between the complexes of acetone (**8**) and ethyl methyl ketone (**9**).[35] Complexation of acetone at –30 °C resulted in a slight shift (0.1 p.p.m.) of the α-carbon resonance, while for ethyl methyl ketone the two α-carbons showed a much larger shift (*ca.* 2 p.p.m.), one upfield and the other downfield from their original frequencies (Figure 12).[35]

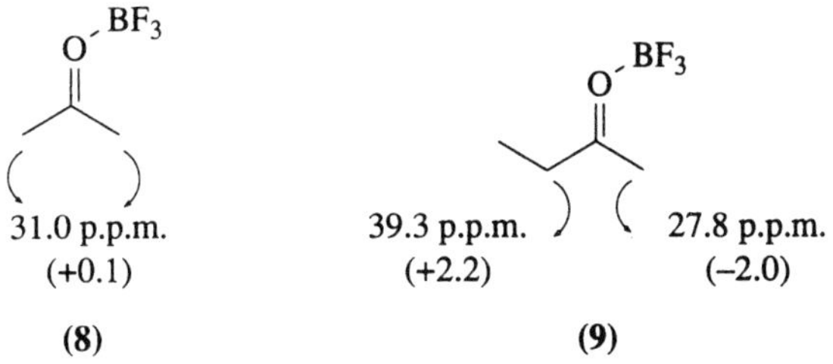

31.0 p.p.m. (+0.1) (**8**)

39.3 p.p.m. (+2.2) 27.8 p.p.m. (–2.0) (**9**)

Figure 12 Selected ^{13}C chemical shifts of BF_3-coordinated ketones. Numbers in parentheses depict the change in chemical shifts from those of the free ligand in p.p.m.

These results are consistent with a single preferred conformation for (**9**) with slow *syn–anti* exchange and probably an averaging effect for the rapidly isomerizing acetone complex. It is noteworthy that the relatively small difference in the steric bulk of the methyl and ethyl substituents can affect the *syn–anti* equilibrium.

For simple cyclic ketones (C_n) the barrier for *syn–anti* isomerization of BF_3 was shown to decrease in the order $C_5 > C_7 = C_6 > C_4$, a trend which closely follows the order of their basicities towards BF_3 (*vide supra*).

In contrast to boron-centered Lewis acids, tin and titanium derivatives prefer a 1:2 (acid:base) stoichiometry in complexation with carbonyls. ^{13}C and ^{1}H NMR spectra of an equimolar solution of $SnCl_4$ and 4-*t*-butylbenzaldehyde showed signals only for free ligand and a 1:2 (acid:base) complex over a wide temperature range.[38] Only with excesses of $SnCl_4$ was any 1:1 adduct observed and even in the presence of 10 equiv. of Lewis acid, only 25% of the 1:1 complex could be detected. Intermolecular exchange in this case was shown to be slow below –40 °C and no *syn–anti* isomerization could be detected.

Similar observations were made in a study of $TiCl_4$/carbonyl complexes but over a smaller temperature range and to a more limited extent.[39–41] Thus, ^{1}H NMR studies showed that $TiCl_4$ forms 1:2 adducts with ketones, esters, ethers and nitriles. Ethers and esters were once again stronger ligands than ketones.[39] $TiCl_4$ has also been used as a diagnostic shift reagent for NMR studies of β-lactams and several α,β-unsaturated carbonyls.[40,41] Both functionalities form strong complexes (as deduced from the changes in ^{1}H and ^{13}C chemical shifts) but more detailed structural information has not been obtained.

The structure of carbonyls complexed to lanthanide shift reagents [LSR, *e.g.* $Eu(fod)_3$] has been the subject of some debate in the literature.[42–45] The two prevailing models for the structure of these complexes are known as the one-site and the two-site models, simply referring to the linear and bent geometries of coordination, respectively. The position of the lanthanide Lewis acid is generally determined by comparing the observed induced shifts with calculated values for randomly generated structures followed by a search for the best fit. The choice between a linear geometry and a 'time-averaged geometry' of two bent structures may be a difficult one, since the corresponding calculated values are probably quite similar. Although crystallographic studies of LSR complexes show a bent geometry,[47] strong spectroscopic evidence for the linear, one-site model has also been obtained.[44–46] Both models, however, agree that the location of the Lewis acid is strongly affected by the steric environment of the carbonyl group, and also that in line with *ab initio* predictions, distortions away from the lowest energy geometry appear facile.

NMR spectroscopy of Lewis acids complexed with α,β-unsaturated carbonyls has shed light on their structure and conformational preferences. Low temperature ^{1}H and ^{13}C NMR spectra of BF_3/2-cyclohexenone indicated slow exchange at –80 °C.[48] A mixture of *syn* and *anti* complexes was detected but the *anti* preference seemed to increase with increasing bulk at the α-carbon.

^{1}H and ^{13}C NMR of a wide range of Lewis acids complexed with crotonaldehyde, tiglic aldehyde, 3-pentenone and methyl crotonate were investigated.[49] Correlation of ^{1}H and ^{13}C shifts at –20 °C consistently showed an *anti, s-trans* conformation for crotonaldehyde complexed with 12 different Lewis acids. Notably, ^{13}C NMR shifts suggested significant decrease of electron density on C-1 and C-3 but an increase of charge on C-2. This observation may be the result of a greater contribution from the enol resonance structure of the complex (Figure 13).

Figure 13 Enol resonance structure of BF_3/enal complexes

The linear correlation between the changes in chemical shifts (Δδ) of different protons for complexed and uncomplexed crotonaldehyde has a somewhat different character for α,β-unsaturated ketones and esters. Conformational isomerization, for example, was found to be rapid for enones (2-cyclohexenone, 2-pentenone) even at low temperatures, although it was evident that the *syn* conformation, in which the Lewis acid and H-2 are in close proximity, is one of the low energy conformers. The change in the resonance frequency of H-2 was attributed to 'through-space deshielding' due to the anisotropy of the Lewis acid. Complexes of methyl crotonate provided further evidence for the predominance of the *syn* conformation. Correlation of the changes in chemical shifts of the H-2 resonances of this system with those of the crotonaldehyde complexes suggested through-space interactions of the Lewis acid with H-2, consistent with an abundance of conformer (**10**).

Finally, a scale of Lewis acidity based on the induced changes in chemical shifts was established (Table 7). Of special relevance for synthetic applications may be the relative powers of BF_3 and $TiCl_4$.

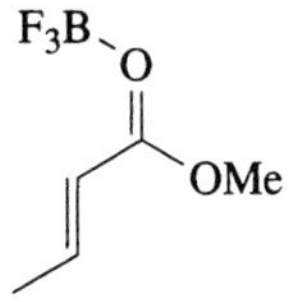

(10)

These results suggest that the mildness of BF_3/Et_2O (relative to $TiCl_4$) may simply be due to the low effective concentrations of free Lewis acid, since uncomplexed BF_3 seems to be a much stronger Lewis acid. Furthermore, in cases where a second potential site of coordination (*e.g.* an ether functionality) is present in the ligand, the effective concentration of BF_3 may be even lower.

Table 7 Relative Lewis Acidities as determined by ^{1}H NMR Spectroscopy.[a]

Lewis acid	*Relative power*	*Lewis acid*	*Relative power*
BBr_3	1.00 ± 0.005	$EtAlCl_2$	0.77
BCl_3	0.93 ± 0.020	$TiCl_4$	0.66 ± 0.030
$SbCl_5$	0.85 ± 0.030	Et_2AlCl	0.59 ± 0.030
$AlCl_3$	0.82	$SnCl_4$	0.52 ± 0.040
BF_3	0.77 ± 0.020		

[a]Relative power refers to the induced Δδ-values of the H-3 resonances of various α,β-unsaturated carbonyl bases.

Lewis acids that possess two empty sites of coordination are capable of imposing conformational constraints on a carbonyl ligand either by forming 2:1 complexes or *via* chelation when a second basic site is present in the ligand. Despite the popularity of the 'chelation-control' model since its original proposal in 1952 by Cram and coworkers,[50] rigorous and direct experimental evidence for chelation in α- and β-alkoxycarbonyls was only obtained recently from variable temperature ^{1}H and ^{13}C NMR.[51–54]

NMR titration experiments with 2-methoxycyclohexanone (**11**; equation 3) and $TiCl_4$ at 23 °C indicated the formation of a conformationally rigid 1:1 adduct, in which both the carbonyl carbon and the carbon bearing the methoxy group had lower field chemical shifts with respect to the free ligand.[51,52] The reaction of $MeTiCl_3$ and 2-benzyloxy-3-pentanone was monitored by ^{13}C NMR spectroscopy (equation 4). At –45 °C two new sets of carbonyl signals (downfield from those of the free ligand) appeared immediately upon addition of $MeTiCl_3$. Additionally, broadening and an upfield shift of the titanium-bound

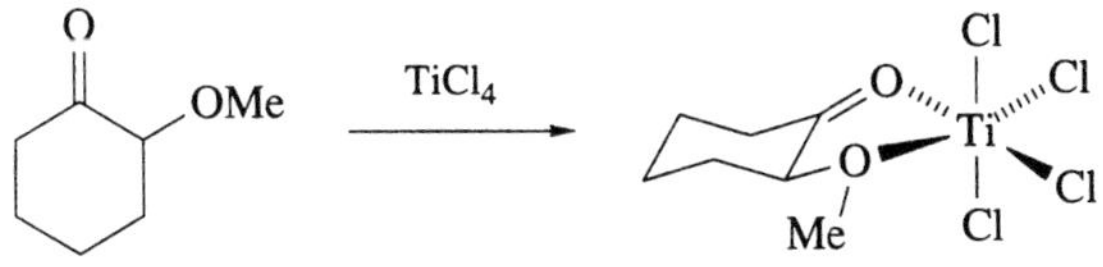

(11) $\xrightarrow{TiCl_4}$ **(12)** (3)

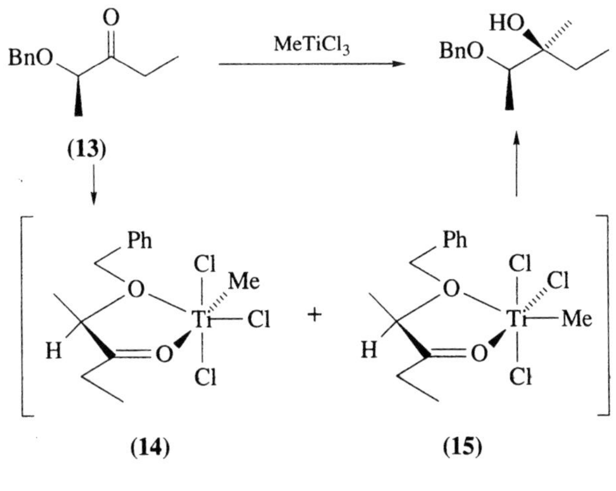

(13) $\xrightarrow{MeTiCl_3}$ (4)

(14) + **(15)**

methyl group indicated formation of an octahedral complex. These signals were assigned to the two diastereomeric structures (**14**) and (**15**), in which the methyl group occupies a meridional position *cis* or *trans* to the carbonyl ligand. Structures with axially disposed methyl substituents were discarded, based on precedence from previous NMR and crystallographic studies. Kinetics and crossover experiments also showed that product formation is due to bimolecular rather than intramolecular methyl transfer.[53]

β-Alkoxycarbonyls can also form stable and conformationally rigid chelates. Spectroscopic evidence for these structures was provided by a study of titanium-, tin- and magnesium-derived Lewis acids and β-alkoxy aldehydes (**16**) and (**17**).[54] Variable temperature (–80 °C to –20 °C) ^{13}C and ^{1}H NMR spectra showed that (**16**) can form conformationally rigid 1:1 adducts with $TiCl_4$, $SnCl_4$ and $MgBr_2/Et_2O$.

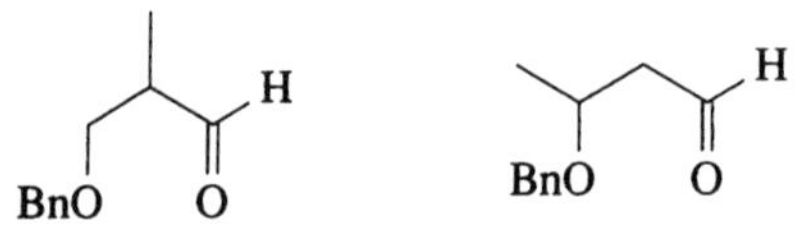

(**16**) (**17**)

The splitting pattern and coupling constants for the C-2 methine hydrogen (J_{ax-ax} = 9.7 Hz, J_{ax-eq} = 3.5 Hz) further indicated its pseudo-axial disposition, and structure (**18**) was proposed for the $TiCl_4$ complex (equation 5).

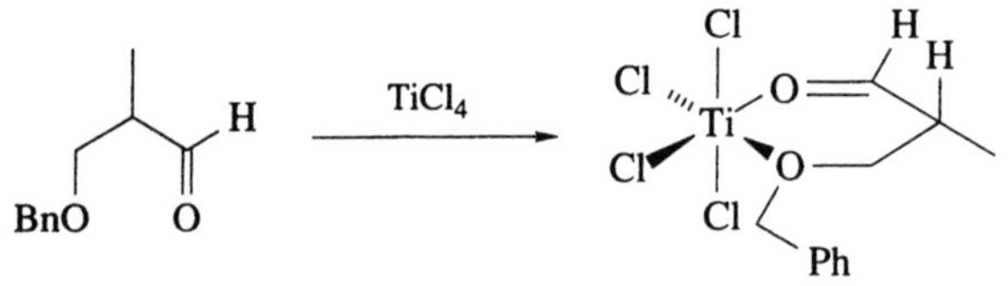

(**16**) → (**18**) (5)

Aldehyde (**17**) behaved somewhat differently (equation 6). In this case, the splitting pattern for the C-3 methine hydrogen suggested a pseudo-equatorial location in the $TiCl_4$ complex, but no discreet, stable chelate could be detected with $SnCl_4$ even at –93 °C. The preference for structure (**19**) with an axial methyl group was rationalized based on the observation that the alternative conformation (**20**) would be disfavored due to an $A^{1,3}$-like interaction between the equatorial methyl group and the benzyl protecting group. The benzyl group, in turn, should be locked in its conformation due to steric interactions with the chlorine ligands of the Lewis acid.

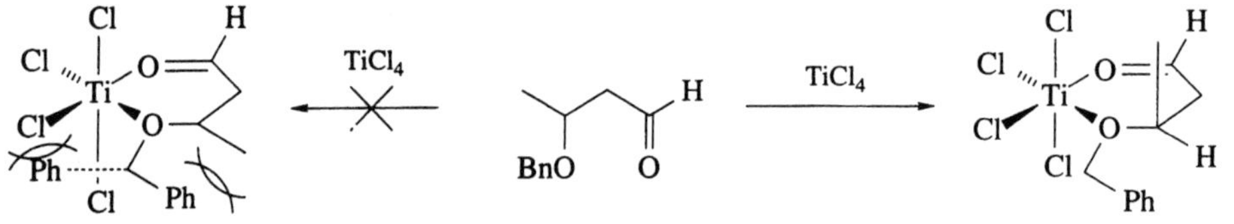

(**20**) (**17**) (**19**) (6)

Further support for this hypothesis was obtained by studying systematic changes in the steric and electronic requirements of the oxygen-protecting group.[55] The *O*-methyl derivative (**21a**), for example, was shown to form a rigid chelate with $TiCl_4$, in which the C-3 methyl substituent is equatorially (J_{ax-ax} = 10 Hz, J_{ax-eq} = 1 Hz) disposed (equation 7). Apparently, the size of the oxygen-protecting group in (**22**) is

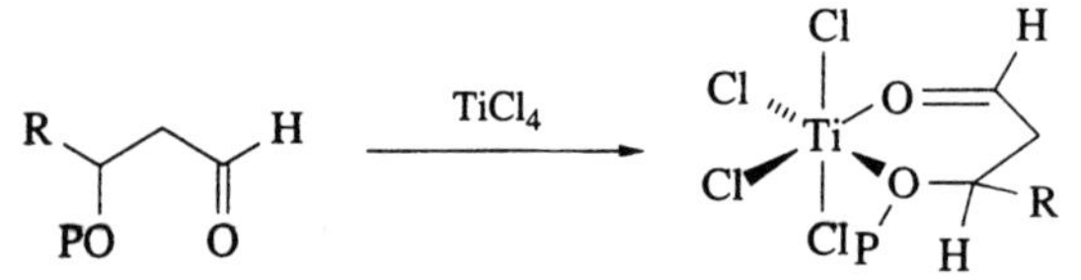

(7)

(**21a**) R = *n*-hexyl, P = Me
(**21b**) R = *n*-hexyl, P = Et

(**22**)

reduced to the extent that an equatorial substituent can now be accommodated. To appreciate the delicate balance of various influential factors here, it can be noted that in the *O*-ethylated complex (**21b**) the C-3 alkyl group is once again forced to adopt an axial disposition (*cf.* **19**).

Yet another interesting observation was made in the study of β-siloxy aldehydes.[56] ^{1}H and ^{13}C NMR spectra of silyl ether (**23**; equation 8) showed only 1:2 (acid:base) adduct formation and no evidence for chelation over a wide range of concentration and temperature (–80 °C to –20 °C). The researchers proposed that the reason for lack of chelation with silicon protecting groups may be electronic in nature: delocalization of the oxygen lone pair electrons into empty silicon *d*-orbitals could reduce oxygen basicity. *Ab initio* calculations on a number of alcohols, ethers and silyl ethers supported this proposal by indicating a large Si—O—C bond angle (131°), which may result from Si—O π-interactions, thereby ascribing double bond character to the Si—O bond.[57] A search of the Cambridge Structural Database (CSD) for the silyl ether [Si—O—C (sp^3)] substructure indicated a mean value of 126 ± 9 °C for the Si—O—C bond angle and 1.65 ± 0.05 Å for the Si—O bond length in the solid state.[58] Moreover, the Si—O bond length shows a linear decrease with increasing Si—O—C bond angle (Figure 14), suggesting that the Si—O—C bond angle may be a valid measure of the Si—O double bond character. No such correlation was found between the Si—O—C angle and the C—O bond length (Figure 15).

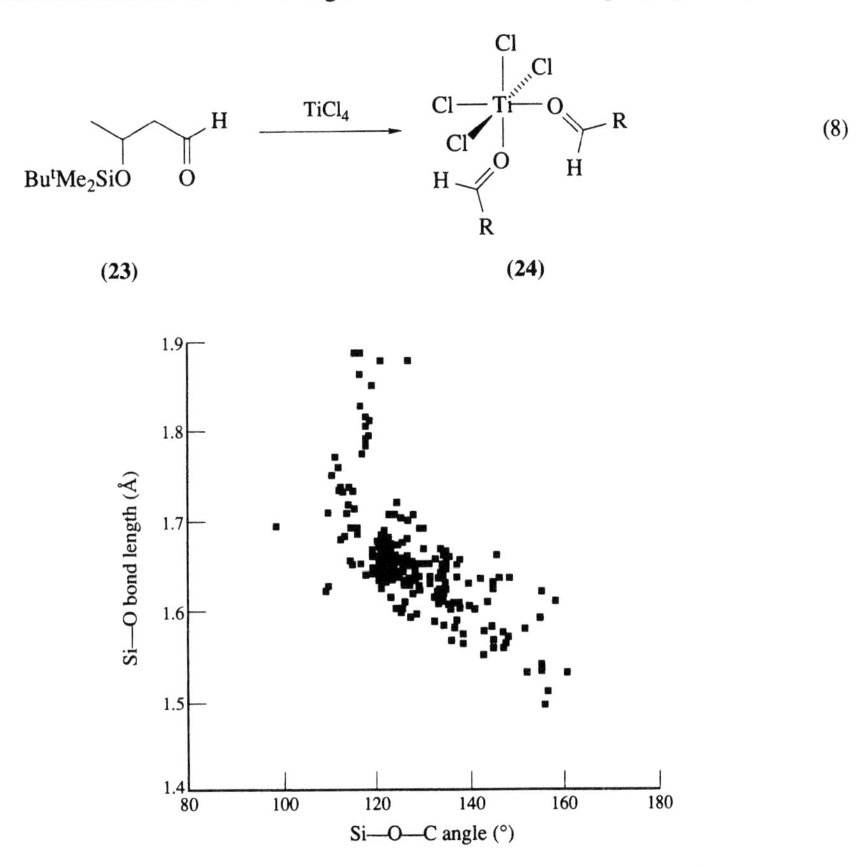

Figure 14 Correlation of the Si—O bond lengths and Si—O—C (sp^3) bond angles (data from Cambridge Crystallographic Data Base)

From this discussion it is quite clear that NMR spectroscopy is one of today's most powerful tools in the pursuit of structural information. Not only can NMR spectroscopy provide accurate and easily accessible structural data regarding a large variety of chemical species, but its ability to reveal energetic and solution dynamics data is unmatched even by X-ray crystallography.[59] Detection of Lewis acid carbonyl complexes by these methods has produced several noteworthy results. It is quite evident that Lewis acids form strong and stable complexes with a variety of oxygen bases. The order of Lewis basicity of some of the most common oxygen-containing functional groups towards Lewis acids has been established. Thermodynamically, ethers seem to be stronger bases than carbonyls, although kinetic basicities of these groups have not been measured. Carbonyl-containing compounds listed in the order of reducing basicity towards common Lewis acids are amides > esters > enones > ketones > aldehydes. Effective basicity seems to be dependent on both electronic and steric effects. Thus a sterically congested ester may

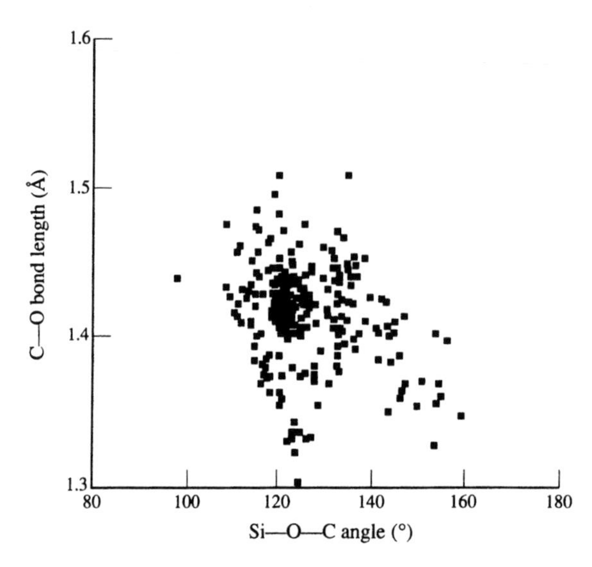

Figure 15 Correlation of the C—O bond lengths and Si—O—C (sp^3) bond angles (data from Cambridge Crystallographic Data Base)

coordinate more weakly than an unhindered aldehyde. An order of Lewis acidity towards carbonyl bases has also been deduced from NMR spectroscopy (Table 7), which may prove helpful in fine-tuning the reactivity of Lewis acidic catalysts.

Most of the available data point to a bent, planar conformation for Lewis acid carbonyl complexes. *Syn–anti* isomerization seems to be fast at room temperature on the NMR timescale but at low temperature individual isomers can be observed. The barrier to isomerization is estimated as 8–10 kcal mol^{-1}, but the mechanism of this process (*i.e.* through a parallel or perpendicular arrangement) is unresolved. The *syn–anti* equilibrium is quite sensitive towards the relative size of the carbonyl substituents and the Lewis acid prefers to lie *syn* to the smaller substituent.

In agreement with *ab initio* predictions, Lewis acid complexation with α,β-unsaturated carbonyls seems to encourage adoption of the *s-trans* conformation. In the complexes of conjugated esters and ketones the gearing effect discussed previously may be responsible for the abundance of the (*Z*)-*syn*-*s-trans* conformation.

Strong evidence for Lewis acid chelation with bidentate bases has been gathered. Subtle and intricate factors seem to contribute to the overall conformation of the chelated complex but the importance of protecting groups for the alkoxy ligands should be emphasized. Silicon protecting groups reduce oxygen basicity probably through a combination of steric and electronic effects and thereby disfavor chelation. Alkyl ethers, on the other hand, are strong chelators, but small differences in the size of the protecting group may result in large differences in the overall conformation of the Lewis acid chelate.

There are also a few interesting topics which have not yet been addressed. The question of the conformational preferences of α-chiral aldehydes upon Lewis acid complexation is among the most important issues. Theoretical predictions imply that the small magnitude of the differences in energy may prevent detection of individual conformers even at very low temperatures. However, it is not inconceivable that through the proper choice of Lewis acid and design of an appropriate carbonyl ligand one may exaggerate the energetic differences such that detection becomes possible.

Stereoelectronic effects of various remote substituents on the basicity of the carbonyl is yet another intriguing question. Electronic effects of remote substituents have been postulated to be responsible for stereoselective additions to carbonyls in rigid adamantanone systems.[60] If present, such stereoelectronic effects would similarly be expected to bias the energetics and structural features of Lewis acid complexation.

Finally, the mechanism of *syn–anti* isomerization remains elusive. Is π-bonding truly higher in energy than the linear, planar arrangement as predicted by theory? The answers to these questions will undoubtedly help to clarify our emerging view of the interactions of the carbonyl group with Lewis acids.

1.10.5 X-RAY CRYSTALLOGRAPHY

There are two distinct approaches to the analysis of X-ray structural data. In one approach a detailed consideration of individual angles and distances in single crystal structures is used to attain insights into the properties and reactivity patterns of a particular compound or class of compounds. On the other hand, a statistical analysis of a large number of crystal structures that share a common structural feature can be used to extract generalizations regarding particular types of interactions or molecular properties. Clearly, the latter approach is more useful and more reliable in deriving a general sense of a bond length or bond angle, while a detailed analysis is preferred when various factors which may contribute to the final structure of a complex (especially those with unusual structures) are to be determined. Since both methods have their merits, a combination of the two has been adopted throughout this section.

Such a combination was used in connection with a study of the crystal structure of the LiBr/acetone complex.[61] In this dimeric complex (Figure 16) each lithium ion is coordinated by two molecules of acetone through the carbonyl lone pair electrons with a bent, planar geometry in a tetrahedral environment. The Li—O bond length is measured as 2.0 Å with an Li—O—C angle of 145°. Although the observed Li—O distance is very close to various theoretical predictions (*vide supra*) the nonlinear coordination geometry disagrees with *ab initio* calculations. To probe the source of this discrepancy and to determine the true low energy conformation of alkali metal/carbonyl complexes the researchers conducted a general search of the Cambridge Structural Database (CSD). Of the nine crystal structures that were analyzed in this study, six contained Li^+, two Na^+ and one K^+ as the Lewis acid bound to the carbonyl ligands. The coordination angle for Li^+ was consistently found to be bent (120°–156.3°) but large distortions from planarity (up to 55°) seemed to occur randomly. The Li—O bond distance showed no linear dependence on the Li—O—C—X dihedral angle.

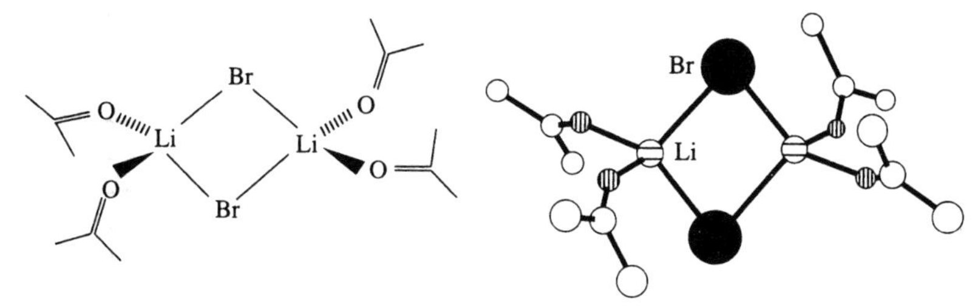

Figure 16 Crystal structure of LiBr•$(acetone)_2$ complex

The crystal structures of Na^+ complexes, however, showed that linear and bent geometries are equally accessible and bending out of the plane of the carbonyl was once again observed (Figure 17).[62]

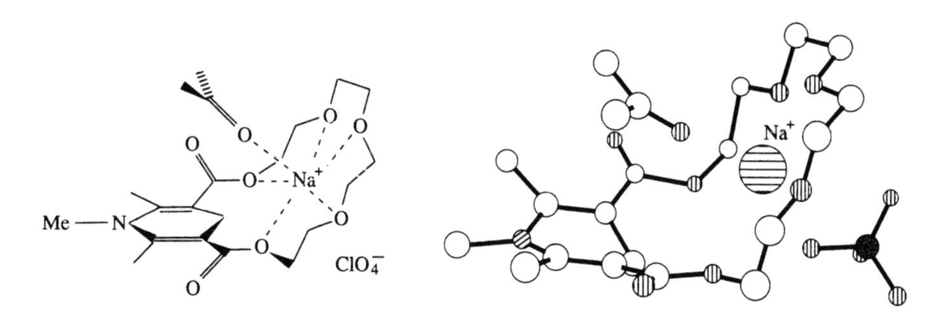

Figure 17 Linear complexation of a Na^+/acetone complex

The workers concluded that the potential energy surface for carbonyl coordination to alkali metals is likely to be fairly flat with respect to changes in the M—O═C angle and the M—O═C—X dihedral angle. This proposal would also account for the discrepancy between theory and the experimental results regarding the exact structure of the lowest energy species.

A more recent search of the CSD for alkali metal/carbonyl crystal structures also confirmed these results.[63] The average Li—O bond length and the Li—O—C bond angle were found to be 1.99 ± 0.07 Å and 139 ± 3°, respectively. Of the 23 examples found, 9 exhibited coordination by more than 10° out of the plane of the carbonyl group.

Thus, cationic alkali metal Lewis acids do not exhibit any strong directional preferences in coordination to carbonyls.[64] In particular, bending towards the π-cloud of the carbonyl with no significant lengthening of the M—O bond is commonly observed. Coordination seems to occur from the direction that best satisfies the cation's electron demand and minimizes steric interactions. These, in turn, depend on the coordination number of the cation, the relative size and electronic properties of the ligands, and, in the solid state, on crystal-packing forces. It is, therefore, not at all surprising that gas phase calculations, which assume a 1:1 stoichiometry and necessarily ignore packing forces, predict different low energy structures, since under those conditions the linear geometry might very well be favored. What theory would predict as the lowest energy conformation of a multicoordinated, alkali metal/carbonyl complex is an interesting question which, at present, remains unanswered.

Finally, the biological relevance and importance of metal cation/carbonyl interactions, although beyond the scope of the present discussion, should be noted. Such interactions may play a significant role in determining the structures and conformations of metal-binding peptides or metalloproteins. Two provocative crystal structures of Li^+ bound to the cyclic decapeptides antamanide and perhydroantamanide have been reported.[65,66] In the former, lithium was found to be pentacoordinated to four carbonyl oxygens and the nitrogen atom of an acetonitrile solvent molecule in a pseudo square pyramidal arrangement (Figure 18).[65] In this structure, the Li—O bond lengths and Li—O—C bond angles are similar to the values discussed previously and are consistent with the principles of the foregoing discussion.

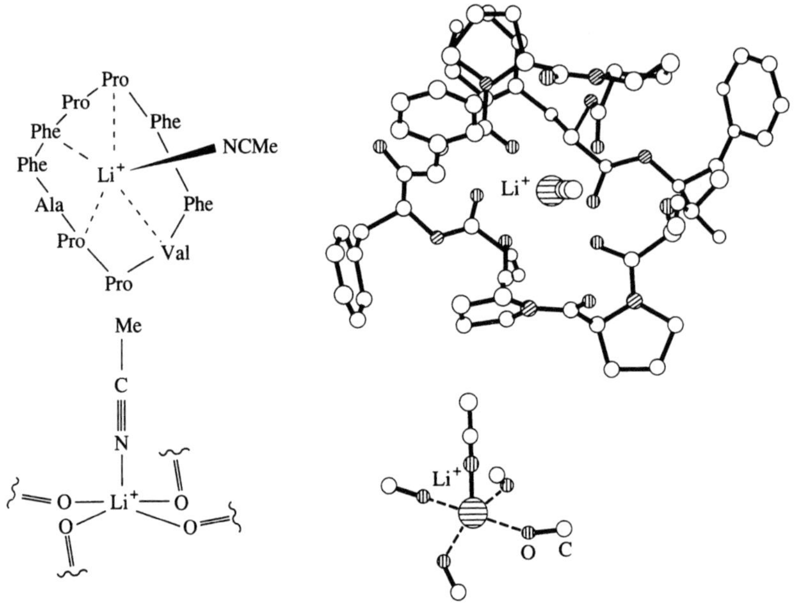

Figure 18 Crystal structure of Li^+/antamanide complex

Relative to alkali metals, neutral Lewis acids seem to perform much more consistently. Crystal structure of the BF_3/benzaldehyde complex may serve as a representative example (Figure 19).[67] Here the Lewis acid is placed 1.59 Å from the carbonyl oxygen, along the direction of the oxygen lone pair and *anti* to the larger phenyl substituent.[68]

Although crystal structures of other bimolecular complexes of carbonyls with boronic Lewis acids have not been reported, a number of intramolecular chelates have been detected in the solid state.[69–71] In all these cases boron is found to lie in the direction of the carbonyl lone pair with no more than 11° distortion away from the best plane of the carbonyl group. The average B—O bond length is 1.581 ± 0.019 Å and the B—O—C angle lies between 112° and 119°.

A search of the CSD files for the Al—O═C substructure revealed 30 such interactions with Lewis acidic aluminum atoms from 23 crystal structures.[72] Mean values for the Al—O distances and Al—O—C bond angles were found to be 1.88 ± 0.09 Å and 136 ± 4°, respectively. Aluminum was found to lie within ±8° of the carbonyl plane where θ in Figure 20 had a mean value of 83 ± 1°.

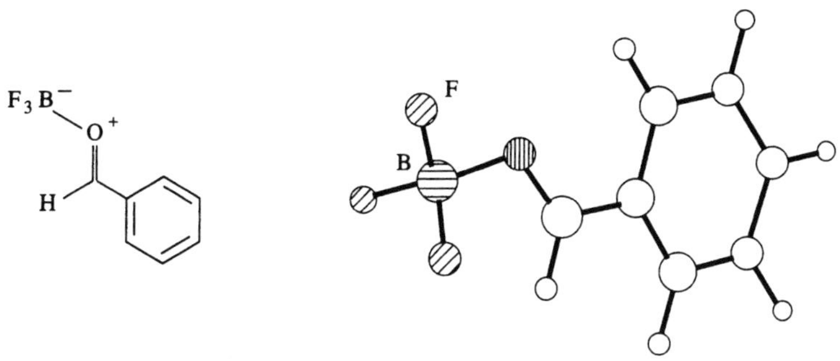

Figure 19 Crystal structure of BF_3/benzaldehyde complex

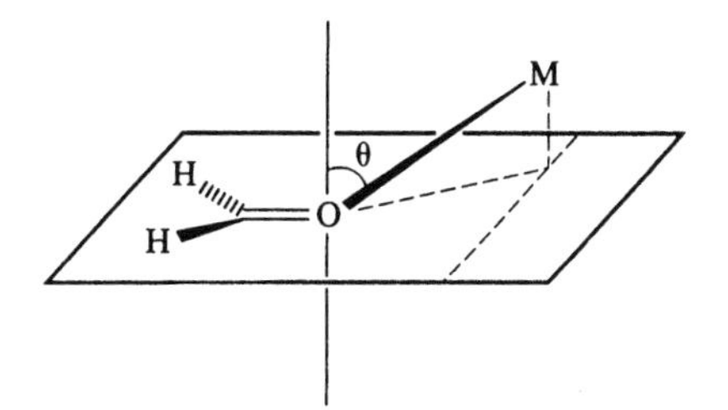

Figure 20

The crystal structure of the $AlCl_3$/tetramethylurea ($AlCl_3$/TMU) complex shows a 1:1 adduct, in which aluminum binds to the carbonyl lone pair electrons at an angle of 132.5° and at a distance of 1.78 Å (Figure 21).[73] The C—O bond length (1.239 Å) is approximately 0.06 Å longer than that of unbound TMU and 0.03 Å longer than in the corresponding Me_2SnCl_2/TMU complex. The Lewis acid is once again coplanar with the carbonyl, although the dimethylamino group *syn* to aluminum is twisted out of conjugation to avoid steric interactions.

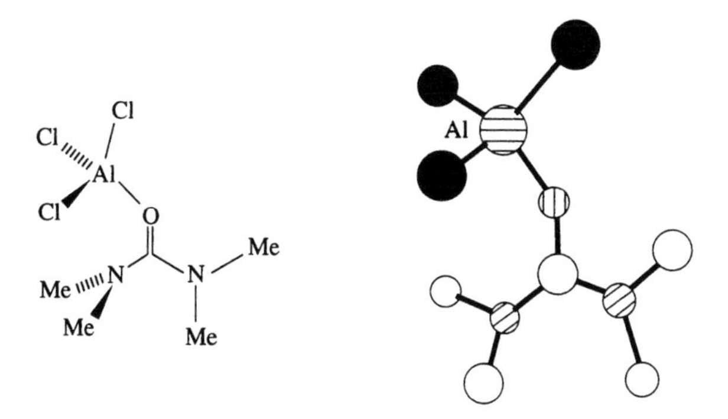

Figure 21 Crystal structure of $AlCl_3$/TMU complex

Based on statistical analyses it may be reasonable to propose that this structure reflects the idealized or low energy mode of coordination to carbonyls for aluminum-centered Lewis acids. It would be interesting to see how distortions away from such an arrangement may come about. Imposition of steric congestion through bulky substituents is one way to address this question. The crystal structure of (2,6-di-*t*-butyl-4-methyl)phenoxydiethylaluminum/methyl toluate was recently reported (Figure 22).[74]

In this structure a pseudo-tetrahedral aluminum is positioned 1.89 Å from the carbonyl ligand. The Al—O—C angle of 145.6° is on the more linear end of the Al—O—C angle spectrum, but, more interestingly, the Al—O═C—C torsion angle (49.9°) shows considerable bending towards the π-plane. Clearly, this distortion from the 'idealized' structure is a result of the unusually bulky phenoxide ligand on the Lewis acid, coupled with the fact that in (*Z*)-aromatic esters both sp^2-lone pairs are fairly hindered. Notice that other available mechanisms for relief of steric strain were not employed. Either twisting of the

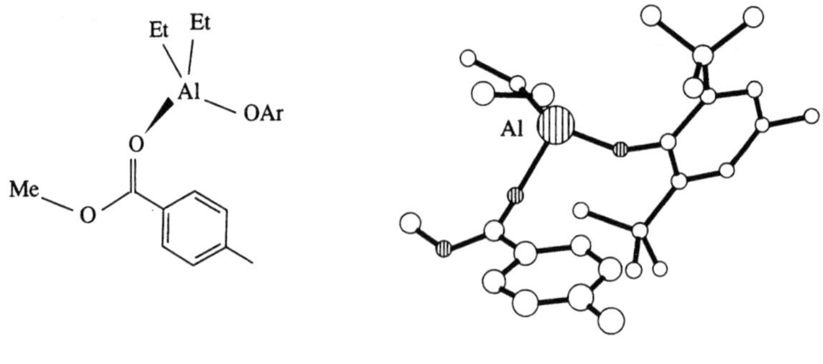

Figure 22 Crystal structure of $Et_2Al(OAr)$/methyl toluate complex

phenyl substituent or alteration of an idealized (*Z*)-ester conformation could have accommodated a more planar Lewis acid complex, but neither occurs to any considerable extent [τ(O═C—O—O) = 4.1°, τ(O═C—C═C) = 12.1°]. Furthermore, a linear Lewis acid complex would also avoid steric interactions with the carbonyl substituents but this mode of coordination seems to be excluded as well. These results suggest that the energetic barrier for out-of-plane bending of the Lewis acid is less steep and probably lower than the barrier for either (*E*)/(*Z*) ester isomerization, rotation about the phenyl–carbonyl C—C bond or in-plane distortion to a linear structure. The latter proposition is in sharp contrast to theoretical predictions.

When a Lewis acid possesses two empty coordination sites, another degree of complexity is added to the structural issues discussed above. Titanium(IV) Lewis acids, for example, show a strong preference for a six-coordinate, octahedral arrangement. Thus, the acid:carbonyl stoichiometry for these complexes is often 1:2 or, when only a single equivalent of the carbonyl base is present, dimeric structures with bridging ligands are observed. The crystalline 1:1 adduct of $TiCl_4$ and ethyl acetate, for example, contains dimeric units of two octahedral titaniums with bridging chlorine atoms (Figure 23).[75] The 2.03 Å long Ti—O bond is almost coplanar with the carbonyl group [τ(Ti—O═C—C) = 2.86°] and the Ti—O—C angle is measured as 152°.

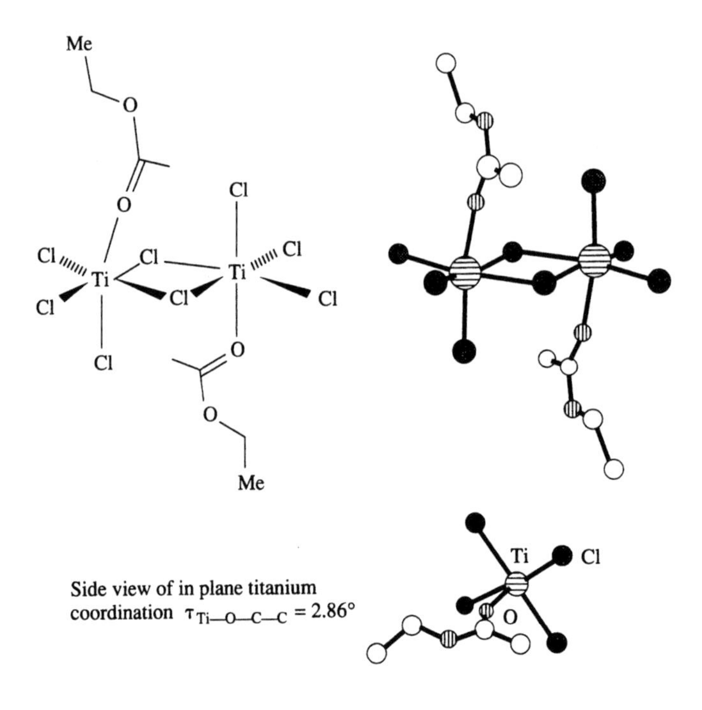

Figure 23 Crystal structure of $TiCl_4$/ethyl acetate complex

These values agree well with the average Ti—O═C bond lengths and bond angles derived from searches of the CSD files ($Ti—O_{mean} = 2.14 \pm 0.07$ Å), although the Ti—O—C angle is on the wider side of the observed range ($Ti—O—C_{mean} = 125 \pm 12°$).[76] Meanwhile, the (*Z*)-conformation of the ester is retained [τ(O═C—O—C) = 2.6°] by coordinating the Lewis acid *syn* to the methyl substituent.

In line with the behavior of aluminum-centered Lewis acids, deviations from the 'idealized' conformation are observed when steric congestion is imposed on the structure. This point is illustrated in the crystal structure of the $TiCl_4$/ethyl anisate complex (Figure 24).[77] Although at first glance the main features of this complex (dimeric complex, octahedral titanium, *etc.*) are very similar to those of the $TiCl_4$/ethyl acetate structure, out-of-plane bonding to the carbonyl clearly distinguishes the two structures. The Ti—O—C—C dihedral angle is 45.7° and the Ti—O—C angle has now opened up to 168.7°. The (*Z*)-conformation of the ester is still intact [τ(O═C—O—C) = 1.4°] and the Lewis acid is bound more or less *syn* to the phenyl substituent, which, in turn, is twisted out of conjugation and away from the Lewis acid by 13.0°. These structural features are very similar to those of the bulky aluminum Lewis acid complex discussed previously and one may expect them to be quite general in similar systems. Interestingly, distortions from the 'idealized' structures are more prominent for the titanium complex, which may suggest that, in the dimeric form, $TiCl_4$ is in fact even larger than the aryloxyaluminum reagent, or that π-distortion occurs more readily for Ti than for Al.

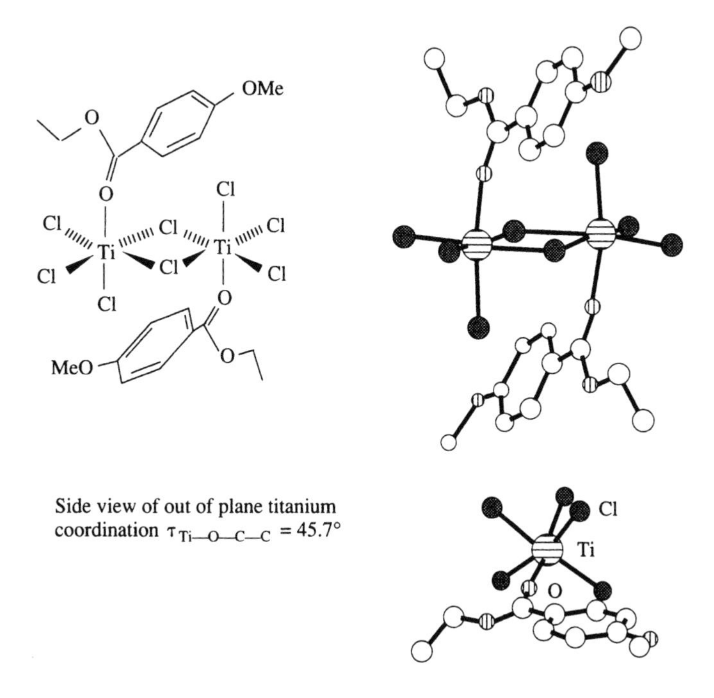

Figure 24 Crystal structure of $TiCl_4$/ethyl anisate complex

Extreme out-of-plane bonding is observed in the crystalline complex of $TiCl_4$ and acryloylmethyl lactate (Figure 25).[78] In this structure Ti is bound by two ester carbonyls in a seven-membered ring chelate. Both carbonyls are π-bonded with τ(Ti—O(1)═C—C) = 63.6° and τ(Ti—O(2)═C—C) = 48.1°. Both esters, however, remain very close to planarity ($\tau_1 = 8.8°$, $\tau_2 = 4.1°$) and in the preferred (*Z*)-conformation. Furthermore, due to the presence of a chelate ring the Lewis acid is *anti* to the acrylate double bond and the enoate adopts an *s-cis* geometry. It is difficult to rationalize π-bonding in this case since direct steric interactions appear to be unimportant. Moreover, π-bonding cannot be the result of chelation since crystal structures of $TiCl_4$ chelated by acetic anhydride[79] or 3,3-dimethyl-2,4-pentanedione[80] show no evidence for π-bonding. Even the seven-membered ring diester chelate of $TiCl_4$ with diethyl phthalate has the Lewis acid coplanar with the carbonyl ligands (Figure 26).[79,81] Another noticeable feature of the chelated structures is that they are all monomeric, probably due to the ability of the bidentate ligands to satisfy titanium's desire for hexacoordination and octahedral geometry.

In the previous section it was noted that in solution tin-derived Lewis acids, like titanium, prefer to form 1:2 acid:carbonyl or 1:1 chelated adducts. Consistent with these findings, the crystal structure of

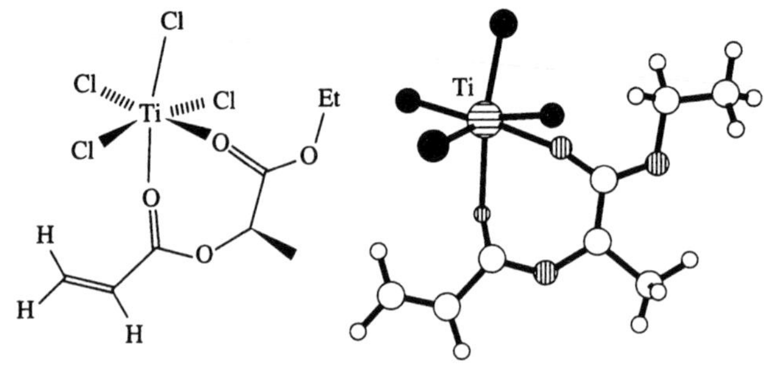

Figure 25 Crystal structure of $TiCl_4$/acryloylmethyl lactate complex

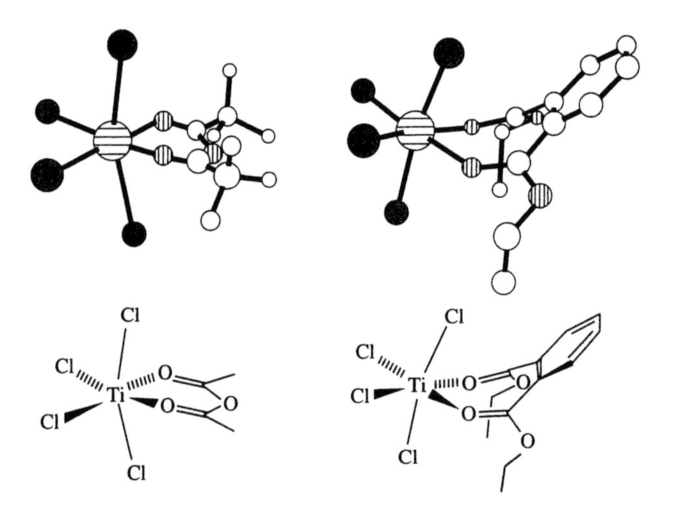

Figure 26 Crystal structures of two $TiCl_4$ chelates

$SnCl_4$/4-*t*-butylbenzaldehyde shows a 1:2 stoichiometry with two nonequivalent aromatic aldehydes, *cis* to one another around the octahedral tin atom (Figure 27).[38]

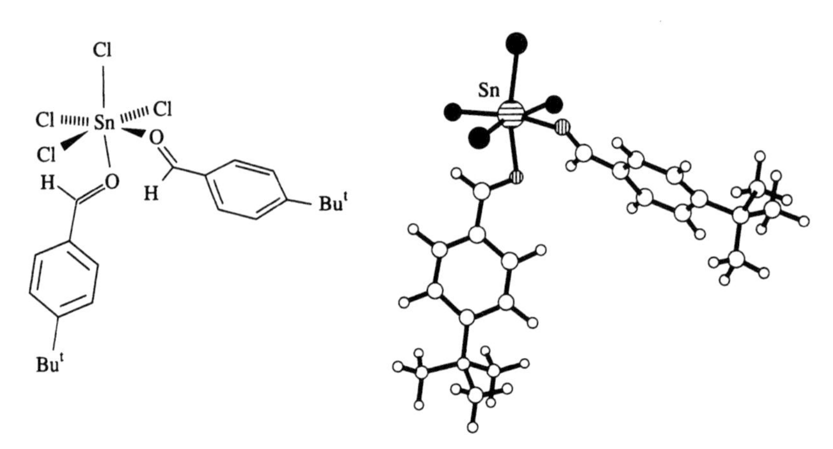

Figure 27 Crystal structure of the $SnCl_4$ /4-*t*-butylbenzaldehyde complex

Similarly to the BF_3/benzaldehyde complex, tin lies within the carbonyl plane with respect to both aldehydes (τ_1 = 2°, τ_2 = 4°), in the direction of sp^2-hybridized lone pairs (Sn—O(1)—C(1) = 128°, Sn—O(2)—C(2) = 126.2°) and *anti* to the aromatic ring.

These measurements and the Sn—O bond lengths (2.23 Å) are close to the average values of tin–carbonyl bond angles and bond lengths found in the CSD files (Sn—O = 2.3 ± 0.1 Å; Sn—O—C = 127 ± 10°).[82] However, the 1:2 stoichiometry and the *cis* relationship of the carbonyl ligands are not universal. Only slight alterations in the nature of the Lewis acid or its ligands are necessary to alter this arrangement. For example, an X-ray analysis of Ph_2SnCl_2/*p*-dimethylaminobenzaldehyde shows a monomeric 1:1 complex with a pentacoordinated, trigonal bipyramidal tin atom (Figure 28).[83]

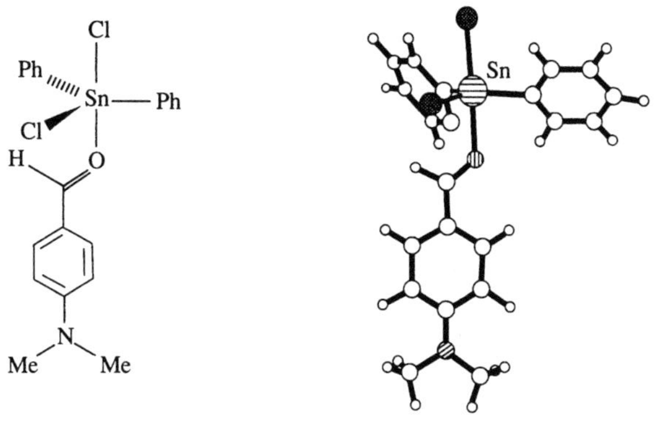

Figure 28 Crystal structure of $Ph_2Sn_2/p\text{-}NMe_2C_6H_4CHO$ complex

Although neither the Sn—O bond length (2.3 Å) nor the bond angles [Sn—O—C = 121°, τ(Sn—O═C—C) = 4°] nor the *anti* geometry of coordination have changed significantly compared to the $SnCl_4$ complex, the environment about the Sn atom is clearly different. It is likely that these changes are caused by the substitution of chlorines for aryl groups on the Lewis acid. Thus, crystal structures of Ph_3SnCl/tetramethylurea (TMU),[84] Me_2SnCl_2/TMU,[85] Me_3SnCl/triphenylphosphoranyldiacetone[86] and Me_2SnCl_2/salicylaldehyde[87] all feature pentacoordinated, trigonal bipyramidal 1:1 adducts, while $SnX_4\cdot 2L$ (X = Cl, Br, I; L = urea or thiourea) complexes contain *cis*-octahedral tin centers.[88] The *cis* and *trans* stereochemistry of the octahedral complexes seems to depend on the size and the basicity of the base, as well as the nature of the Lewis acid. Me_2SnCl_2/DMU[85] and $SnCl_4$/ethyl cinnamate[89] are both *trans*-octahedral 1:2 complexes. The latter complex is particularly relevant to the discussion of the conformational preferences of α,β-unsaturated carbonyl complexes (Figure 29).[89] Here the ligand adopts a *(Z)-s-trans* conformation with tin coordinated *syn* to the double bond. The Sn—O—C—C dihedral angle of 21° indicates some out-of-plane bonding, although not as much as in the titanium complexes.

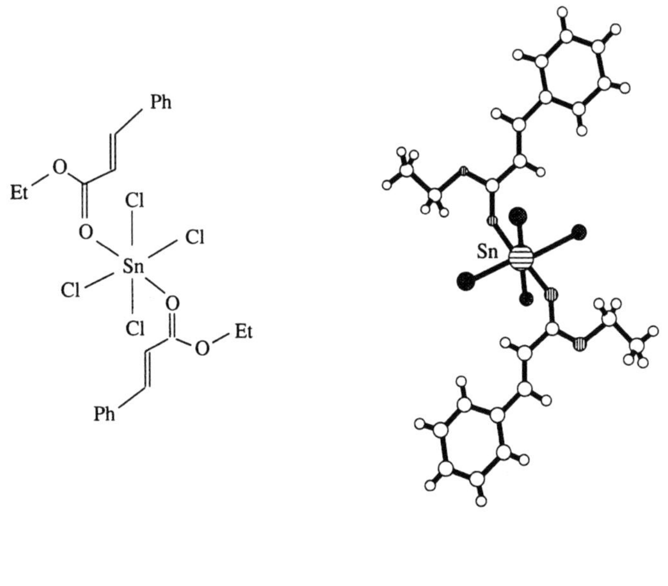

Figure 29 Crystal structure of $SnCl_4$/ethyl cinnamate complex

Crystal structures of chelated tin Lewis acids have also been obtained. Recently, X-ray analysis of two five-membered chelates of $SnCl_4$ were reported.[90] $SnCl_4$ complexes of 2-benzyloxy-3-pentanone and methoxyacetophenone are both 1:1 monomeric chelates with distorted octahedral geometry (Figure 30).[90] In the former complex, tin is apparently coplanar with the carbonyl plane (10° puckering of the stannacycle) at a distance of 2.184(3) Å away from the carbonyl oxygen. More importantly, the ether oxygen is planar and seems to coordinate to the Lewis acid through an sp^2-like lone pair. This point lends credence to the assumption of $A^{1,3}$-like interactions between the oxygen-protecting group and an α-substituent, as discussed in the last section.[54] It should also be noted that coordination along the direction of a trigonal lone pair of ether oxygen atoms is not altogether unexpected and is, in fact, well documented for other Lewis acids such as lithium cation.[91]

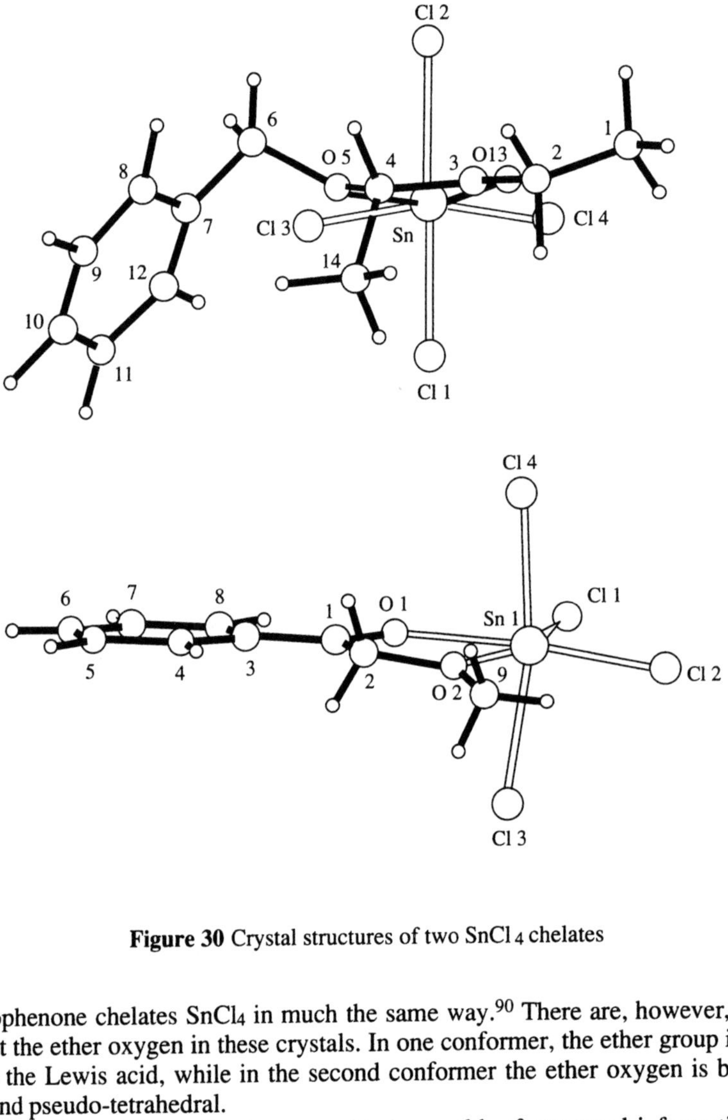

Figure 30 Crystal structures of two $SnCl_4$ chelates

Methoxyacetophenone chelates $SnCl_4$ in much the same way.[90] There are, however, two distinct conformations about the ether oxygen in these crystals. In one conformer, the ether group is planar with sp^2-like donation to the Lewis acid, while in the second conformer the ether oxygen is better described as sp^3-hybridized and pseudo-tetrahedral.

In summary, crystallographic studies have provided a wealth of structural information regarding carbonyl complexation of various Lewis acids. More specifically, it has been shown that alkali metal cations do not show a strong directional preference for binding to carbonyls and for these complexes coordination numbers and coordination geometries vary greatly. Boron, aluminum, titanium and tin Lewis acids all accept electron density from sp^2-lone pairs of a carbonyl ligand at *ca.* 130–140°, but in the presence of steric interactions they easily distort from their optimal geometry. For aluminum, titanium and tin complexes there is fairly strong evidence that, contrary to theoretical predictions, these distortions are invariably out of plane, towards the π-cloud of the carbonyl group rather than the linear, in-plane geometry. This may also suggest a mechanism for *syn–anti* isomerization of the Lewis acid, although this point is less clear.

The stoichiometry of complexation is ordinarily 1:1 (acid:carbonyl) for boron- and aluminum-centered Lewis acids, which give pseudo-tetrahedral complexes, and 1:2 for octahedral Ti^{IV} complexes. Tin(IV)-derived Lewis acids can form either 1:2 octahedral or 1:1 trigonal bipyramidal complexes, depending on the nature of their ligands and on the carbonyl base.

Lewis acids prefer to lie *syn* to the smaller substituent of the carbonyl, *e.g. syn* to H for aldehydes, *anti* to —OR for simple alkyl esters. In α,β-unsaturated systems, Lewis acid coordination *syn* to the double bond favors the *s-trans* conformation, but in two crystal structures, where coordination *anti* to the alkene occurs, *s-cis* complexes are observed.[78,111] Finally, chelation with titanium and tin occurs readily and yields stable, crystalline complexes.

Some of the shortcomings and difficulties encountered in the attainment of structural data from crystalline compounds should also be noted here. The single most important difficulty in X-ray crystallographic analyses is the task of obtaining suitable, X-ray diffracting crystals. This task is made even more formidable for the highly reactive and often unstable Lewis acid carbonyl complexes. Most of the crystal structures discussed here were obtained by crystal growth and data collection at low temperatures, under inert atmospheres. Once the X-ray diffraction data are available, care should be taken to avoid overinterpretation of single structures and invalid generalizations. In this section, statistical analyses of a large number of crystal structures have been used to distinguish the ordinary from the unusual. In unusual systems, crystal-packing forces may be the cause of deviations from the norm, and it is always important to take this factor into consideration. Lastly, it is worth noting that crystal structures represent static single point conformations and their relevance to dynamic chemical reactions is not clear. Nevertheless, in the past chemical dynamics have been cleverly inferred from crystallographic data,[59] although performing such a task for the case at hand would require a very serious and systematic effort.

1.10.6 TRANSITION METALS AS LEWIS ACIDS

Despite the focus of this chapter on the most commonly utilized Lewis acids in organic synthesis, a much larger body of data regarding the structure of donor/acceptor complexes of transition metals with carbonyls exists. Although a comprehensive treatment of this subject is beyond the scope of the present discussion, it is nonetheless worthwhile to consider the structural features of some of these complexes briefly, since many demonstrate novel and unusual ways of interacting with the carbonyl group.[92]

To begin with, one can consider those high valent transition metals that seem to behave analogously to the Lewis acids discussed thus far. The crystal structure of a cationic iridium complex $[IrH_2(Me_2CO)_2PPh_3]BF_4$ shows *cis* coordination of two acetone ligands to an octahedral iridium atom (Figure 31).[93] Iridium is coordinated to the lone pairs of the acetone ligands, coplanar with the carbonyl at an Ir—O═C angle of 133.1° for one acetone ligand and 134.9° for the other. Aside from the interest in this complex due to its role as an active dehydrogenation catalyst,[94] one can also regard it as an effective Lewis acid towards acetone, since it seems to possess all the necessary structural characteristics.

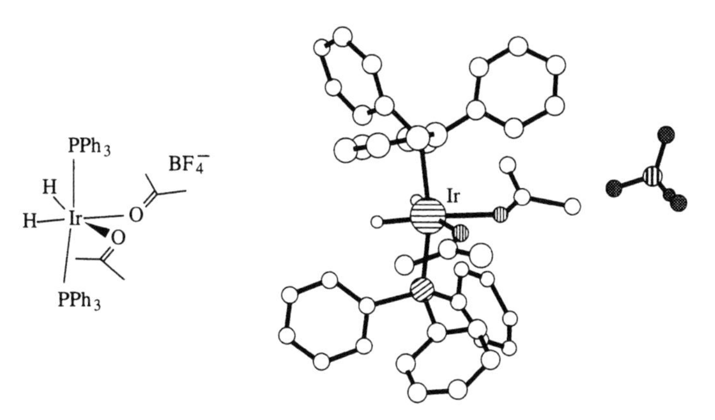

Figure 31 Crystal structure of $[IrH_2(Me_2CO)_2PPh_3]BF_4$

Dicarbonylcyclopentadienyliron cation (Fp^+) is another reactive Lewis acid. This 16-electron complex is known to coordinate ethers, nitriles and various carbonyl-containing compounds quite effectively and with 1:1 stoichiometry. The resultant 18-electron complexes are often stable, crystalline compounds and the crystal structures of a few of them have been solved.[95–97] X-ray analysis of Fp(cyclohexenone)BF_4, for instance, shows iron σ-bonded in the plane of the carbonyl [τ(Fe—O═C—C) = 3.7(9)°] at 132.81(4)° (Figure 32).[95,97] Interestingly, iron is coordinated *syn* to the double bond, which could be explained by steric or electronic arguments. Sterically, coordination next to the smaller methine, as opposed to the methylene, α to the carbonyl should be favored. A similar preference may be expected if the enone moiety is envisioned in an enol ether resonance structure (*cf.* Figure 33). Thus, the known preference of enol ethers to adopt a (*Z*)-conformation may be extended to this case to explain the observed result. The latter, electronic argument, however, does not appear to override steric factors in an aldehyde complex. Preliminary results from a crystal structure of Fp(cinnamaldehyde)PF_6 indicate metal coordination *cis* to the aldehydic hydrogen and *trans* to the enal double bond.[97]

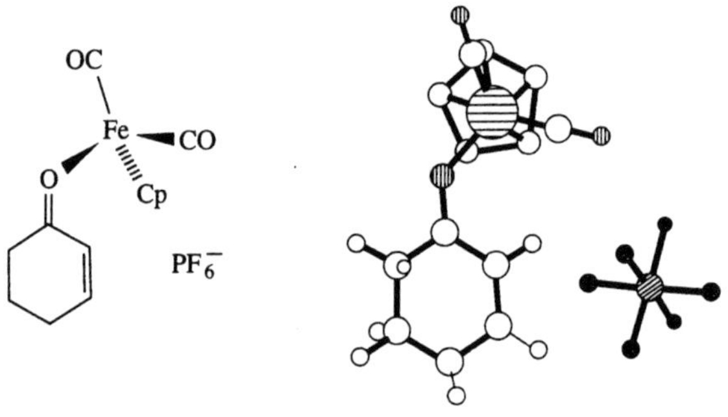

Figure 32 Crystal structure of Fp(cyclohexenone)PF_6

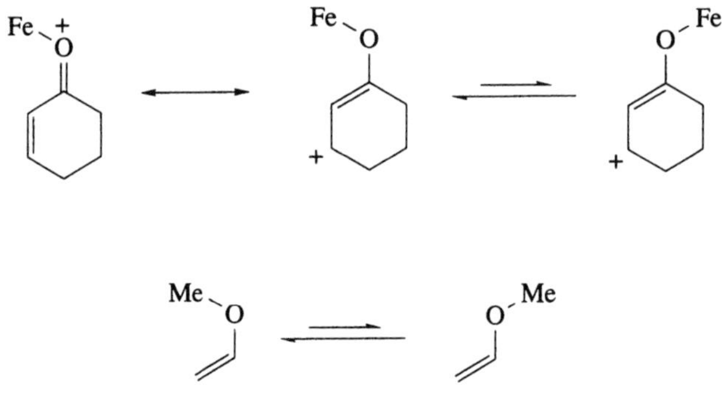

Figure 33 Analogy between *syn*/*anti* Lewis acid complexes and *(E)*/*(Z)* enol ethers

Fp(4-methoxy-3-butenone)BF_4 also shows coordination *syn* to the double bond, *s-trans* geometry and no evidence for π-bonding [Figure 34; τ(Fe—O═C—C—O) = 0(1)°].[97] Fp(tropone)BF_4 is yet another complex whose crystal structure has been solved.[96] In this case, the complex is particularly stable due to

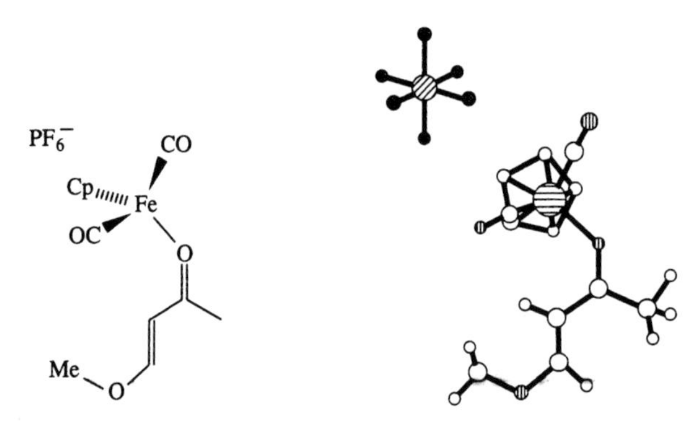

Figure 34 Crystal structure of Fp(4-methoxy-3-butenone)BF_4

the high basicity of the tropone ligand and the crystal shows very similar structural features to other Fp(carbonyl) complexes.

A tungsten-derived Lewis acid, $[(Me_3P)(CO)_3(NO)W]^+SbF_6^-$, was recently reported to catalyze diene polymerization and Diels–Alder reactions.[98] A crystal structure of the acrolein-bound complex (Figure 35) shows tungsten σ-coordinated [τ(W—O═C—C) = 180°], *syn* to hydrogen, at an angle of 137.1°. In line with *ab initio* predictions, acrolein adopts an *s-trans* conformation, despite the absence of any obvious steric interactions in the *s-cis* conformer. It is apparent from this crystal structure that the tungsten complex behaves in a similar fashion to classical Lewis acids, and its structure can be predicted based on the same principles.

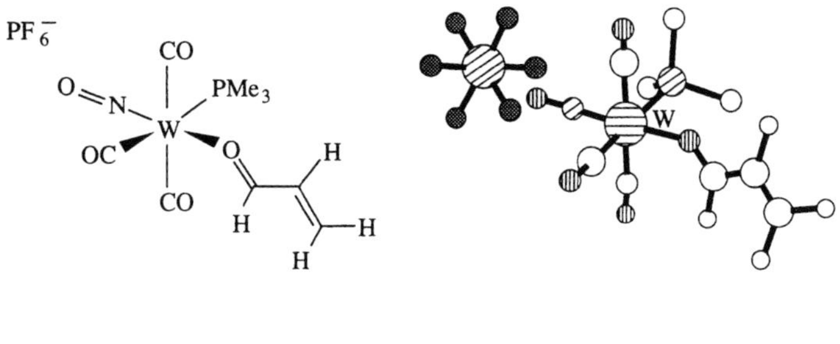

Figure 35 Crystal structure of $(Me_3P)(CO)_3(NO)WFSbF_5$

The cationic complex $[(\eta^5\text{-}C_5H_5)Re(NO)(PPh_3)]^+$ (Z) is capable of binding carbonyls either η^2, through the π-system, or by σ-bonding through the lone pair electrons.[99,100] Crystal structures of Z(phenylacetaldehyde) and Z(acetophenone) clearly show the two different modes of complexation (Figure 36).

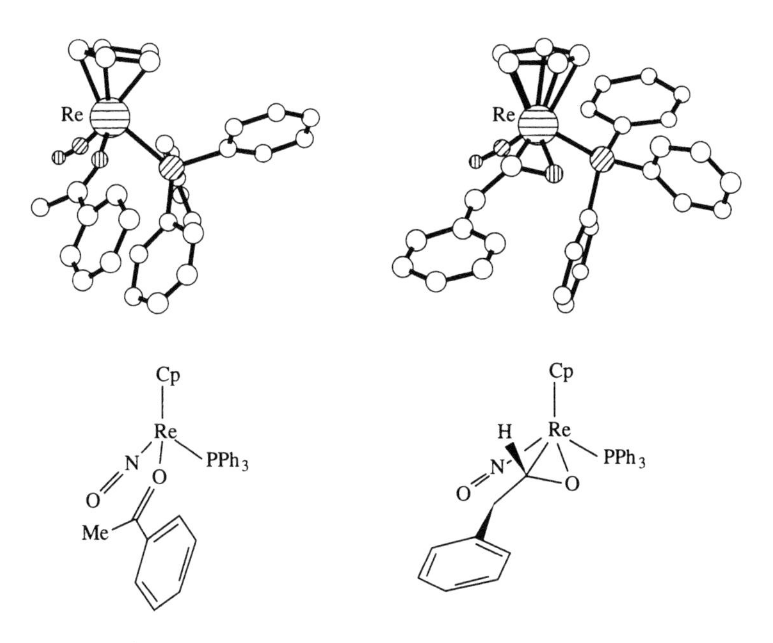

Figure 36 σ- and η^2-bonding for $Cp(Ph_3P)(CO)Re^+$

The authors attributed the difference in binding to increased steric bulk and lower π-acidity of ketones as compared to aldehydes. No crystal structures of Fp^+/aldehyde complexes are available in order to determine whether Fp^+ has a similar 'amphichelic' binding property but $(Ph_3P)(CO)_2Fe^0$ binds cinnamaldehyde in an η^4 fashion.[101] Additionally, it is worth noting that both of the rhenium complexes shown here are chiral and it has been shown that in the enantiomerically pure form, they undergo nucleophilic additions to the carbonyls with high enantioselectivities.[99,100] Finally, the significance of the phenylacetaldehyde crystal structure should not escape attention. This is the first crystal structure of a nonchelated, α-substituted Lewis acid/aldehyde complex. Despite the unusual η^2-bonding in this complex there may be direct implications for the conformational preferences of Lewis acid/α-chiral aldehyde complexes. Remarkably, the phenyl ring lies nearly perpendicular to the plane of the carbonyl at a dihedral angle of τ(C—C—C=O) = 94°, which is reminiscent of the Felkin–Anh proposal for the reactive conformation of α-substituted carbonyl systems.[102] However, it also seems reasonable to argue that the phenyl group occupies the sterically least congested area of space, and that the observed conformation is merely an artifact of the steric requirements of this particular complex, rather than a general, electronic phenomenon.

η^2-Bonding appears to occur with electron rich metals and electron deficient carbonyls. This combination allows for better back-bonding from the metal to the carbonyl π*-orbital and disfavors competitive σ-coordination by lowering the carbonyl lone pair basicity. Hexafluoroacetone, for example, is η^2-bound to $Ir(Ph_3P)_2(CO)Cl$[103] (*cf.* ref. 93), and electron rich Ni^0 complexes show π-bonding to both aldehydes[104] and ketones.[105] The crystal structure of $(TMEDA)Ni(C_2H_4)(H_2CO)$, for example, shows formaldehyde bound in a metallooxirane structure in which the C—O bond has lengthened to 1.311 Å (Figure 37).[106] Similarly, a molybdenum Lewis acid bound to benzaldehyde shows η^2-π-bonding and a C—O bond length of 1.333 Å.[107] Acetone bound to pentamineosmium(II) (C—O = 1.322 Å) has also been detected in the solid state and π-bonding is observed in this case as well.[108]

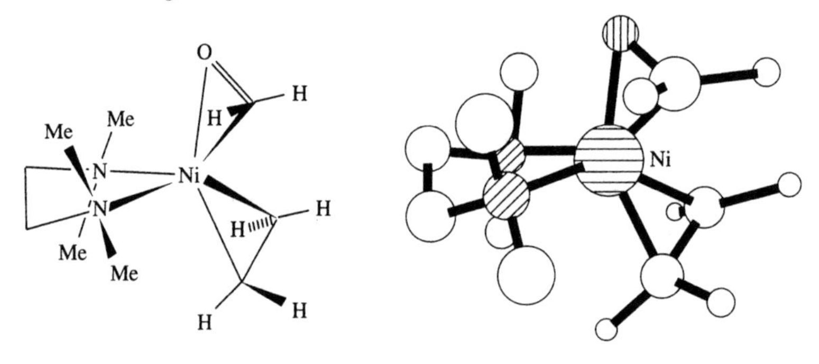

Figure 37 Crystal structure of $(TMEDA)Ni(C_2H_4)$/formaldehyde complex

Finally, carbonyls can bind two metals at once. The crystal structure of a bridging (μ), η^2-bound acetaldehyde complex, for example, shows the carbonyl coordinated to two molybdenum atoms (Figure 38).[109] It appears that in this structure the carbonyl utilizes its π as well as its lone pair electrons to bond to the two metal centers. Conceptually one can think of this molybdenum complex as a bidentate Lewis acid that chelates the carbonyl group.

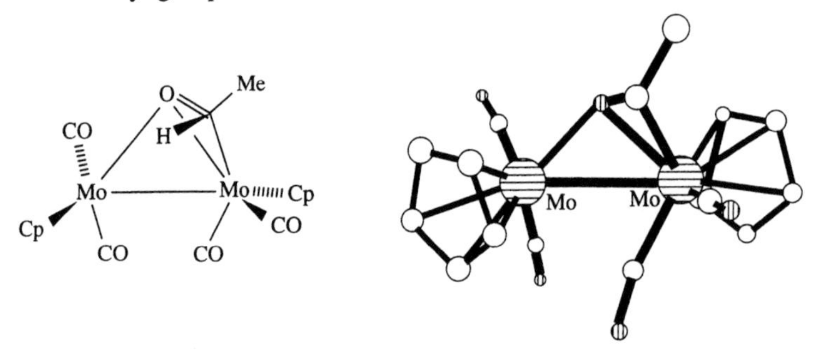

Figure 38 Crystal structure of $(\eta^2\text{-}\mu\text{-MeCHO})Cp_2(CO)_4Mo_2$

In conclusion, it can be noted that high valent transition metals seem perfectly capable of serving as effective Lewis acids. Many of the systems discussed here exhibit exceptional robustness, stability and a propensity to form crystalline complexes. This would facilitate the task of crystallization and structural analysis, and one can imagine that transition metal complexes can be used as structural probes of Lewis acid–carbonyl interactions. In this vein, the first glimpse of the origins of Cram selectivity in α-chiral aldehydes may have been obtained from the crystal structure of a rhenium aldehyde complex. Lastly, the variety of ligands and metals that can be exploited in this field provide a useful handle for the custom design of Lewis acidic reagents. These constructs may also benefit from the large body of structural (especially crystallographic) data that is available for transition metal complexes.

1.10.7 STRUCTURAL MODELS FOR LEWIS ACID MEDIATED REACTIONS

A significant aim of the structural theory of organic chemistry is to acquire the ability to predict the behavior of molecules of known chemical structure. Ideally, one would be able to predict such 'structure–reactivity' relationships based on the principles of quantum mechanics, but such an accomplishment is, at present, not possible. The shortcomings of modern theoretical chemistry in performing a reliable analysis of complex organic molecules have already been mentioned. An even greater hurdle, however, may lie still further ahead in answering the question of chemical selectivity. Thus, even if one were to predict the structure and energy of a molecular system correctly, to choose between a large number of available reaction paths and to predict the lowest lying transition state remain very difficult tasks indeed.

As a corollary, reliable prediction of the structure of Lewis acid carbonyl complexes is by no means a solution to the problem of predicting their reactivity. To address the latter problem one needs to correlate the structural information with the observed patterns of reactivity and *vice versa*.

In this section the stereochemical implications of the structural knowledge gained in the previous sections will be discussed briefly. In some cases, this knowledge has resulted in the confirmation and/or development of various transition state models. In other instances, where compounds of known structure behave contrary to the predictions of structural theory, it has revealed inadequacies in, and limitations of, a model or collection of models.

1.10.7.1 Additions to α,β-Unsaturated Carbonyls

Asymmetric Diels–Alder reactions have been the subject of some of the more thorough mechanistic studies. Fairly reliable structural models for predicting the outcome of these reactions exist. In a review of the subject, it has been suggested that the stereochemical course of the reaction of a variety of chiral acrylates could be consistently predicted based on two models (Figures 39 and 40).[110] Model A positions the complex in a *(Z)-syn-s-trans* conformation and presumes attack from the least-hindered face of the double bond. This model is consistent with almost all of the structural data for systems of this type (*e.g.* $SnCl_4$/ethyl cinnamate X-ray diffraction study). Contrapuntally, the large number of experimental observations that can be explained by this model support the assumption that the crystal structure conformation (**26**) is relevant to the course of these reactions.

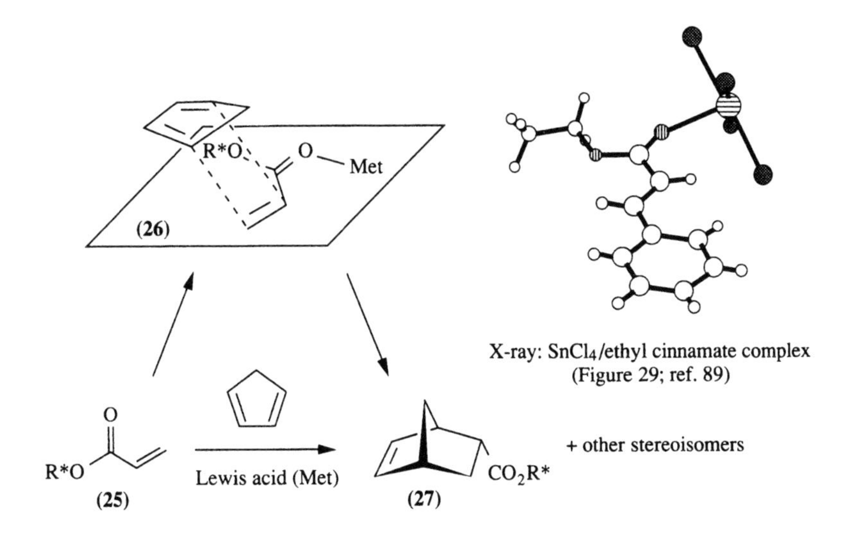

Figure 39 Model A for Lewis acid mediated Diels–Alder reactions

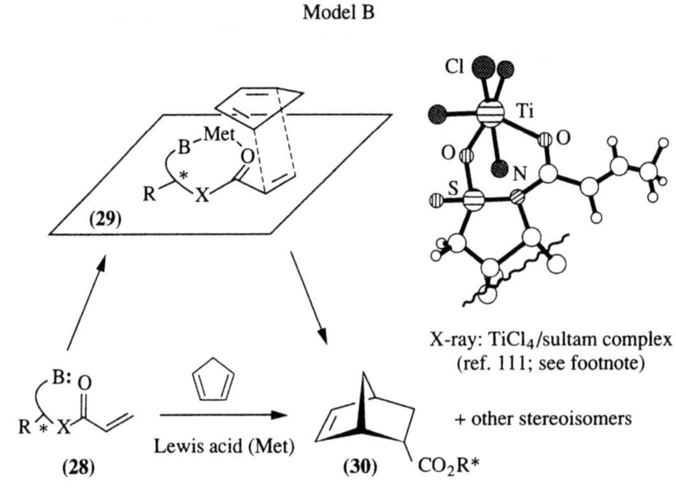

Model B

X-ray: $TiCl_4$/sultam complex (ref. 111; see footnote)

+ other stereoisomers

Figure 40 Model B for Lewis acid mediated Diels–Alder reactions

Model B describes the case in which a chelating group is present in the dienophile, as shown in Figure 40. In contrast to model A, the Lewis acid complex is now *anti-s-cis*, although the (*Z*)-ester conformation is still intact. The *s-cis* conformation is also observed in two crystal structures of chelated complexes, although one of these (*cf.* Figure 25) is somewhat unusual.[111]

A similar model was proposed in describing the asymmetric Diels–Alder reaction of chiral α,β-unsaturated *N*-acyloxazolidinones (**31**).[112–114] The researchers noted that the reaction stoichiometry and the nature of the Lewis acid are crucial in obtaining high levels of diastereoselection (Table 8).[112–114] In particular, stoichiometric amounts of chelating Lewis acids (*i.e.* $SnCl_4$, $TiCl_4$, $ZrCl_4$) resulted in increased stereoselectivities relative to Lewis acids with only a single coordination site (*i.e.* $AlCl_3$, $EtAlCl_3$, Et_2AlCl). The most diastereoselective reactions, however, were observed when a slight excess (1.4 equiv.) of Et_2AlCl was used. These results were interpreted in terms of the action of the ionic species $Et_2Al^+(Et_2AlCl_2^-)$ as the effective Lewis acid. Faster reaction rates, improved *endo* selectivity and Lewis acid concentration studies were all consistent with this interpretation and structure (**34**) was proposed as a transition state model.[112] In support of this hypothesis, the researchers have pointed to

Table 8 Lewis Acid Promoted Diels–Alder Reaction of Chiral α,β-Unsaturated Acyloxazolidinones

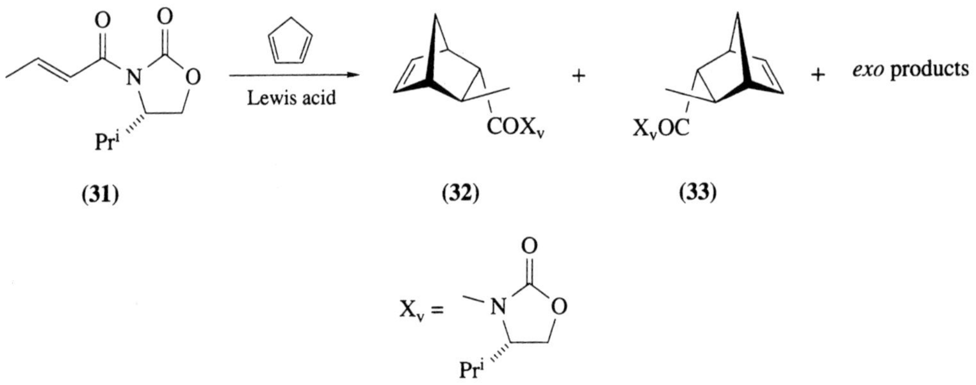

(31) → Lewis acid → (32) + (33) + *exo* products

Lewis acid (equiv.)	*Temperature* (°C)	*Time* (h)	*Conversion* (%)	*(32):(33)*	*Σ(endo):Σ(exo)*
$SnCl_4$ (1.1)	−78	2	70	3.1:1	14.9:1
$TiCl_4$ (1.1)	−78	3	100	2.7:1	9.9:1
$ZrCl_4$ (1.4)	−78	3	100	7.2:1	99:1
$AlCl_3$ (1.0)	−78	3	60	1.5:1	4.2:1
$EtAlCl_2$ (1.1)	−78	3	50	1.7:1	11:1
Et_2AlCl (0.8)	0	6	100	6.5:1	15:1
Et_2AlCl (1.4)	−78	2.5	100	17:1	50:1

work on the complexation of dialkylaluminum chlorides and amine bases, where similiar phenomena were observed.[115]

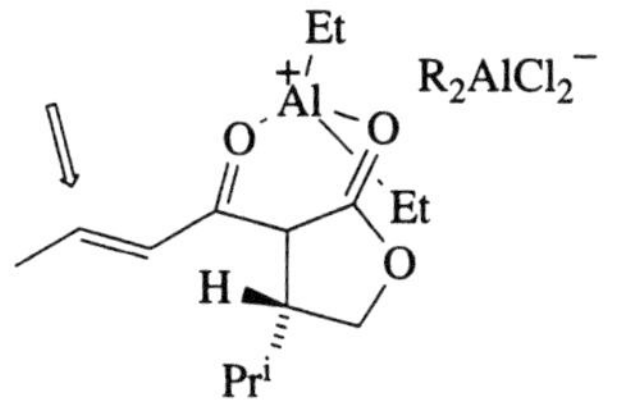

Et $R_2AlCl_2^-$ Et Pr^i

(34)

Some of the Lewis acid catalyzed Michael additions to α,β-unsaturated carbonyls can also be rationalized based on these models. For example, BF_3-mediated additions of organocopper reagents to chiral α,β-unsaturated esters such as (–)-8-phenylmenthyl crotonate (**35**) occur with high levels of diastereoselectivity.[116–118] The product stereochemistries for these reactions could be predicted by assuming the reactive conformation (**36**), which follows the basic structural tenets of model A (Figure 41).[117]

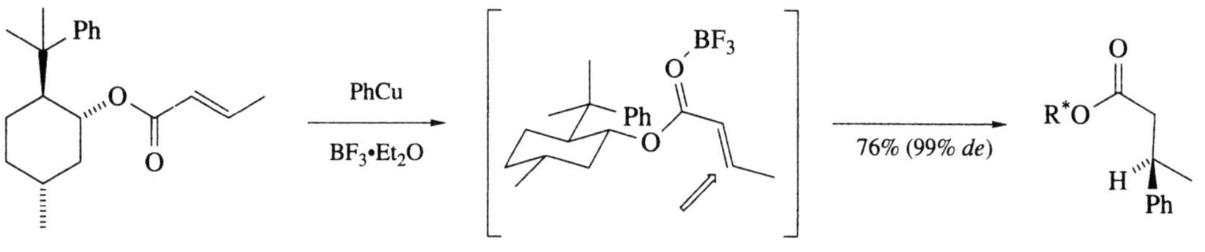

PhCu, $BF_3 \cdot Et_2O$ → [**(36)**] → 76% (99% *de*)

(35) **(36)** **(37)**

Figure 41 BF_3-mediated asymmetric Michael additions to chiral crotonates

The inventors of the chiral catalyst (**38**) for asymmetric 1,4-additions to α,β-unsaturated ketones[119] have proposed the transition structure depicted in (**40**) as a possible model for the Lewis acid catalyzed 1,4-addition to cyclohexenone (Figure 42).[120]

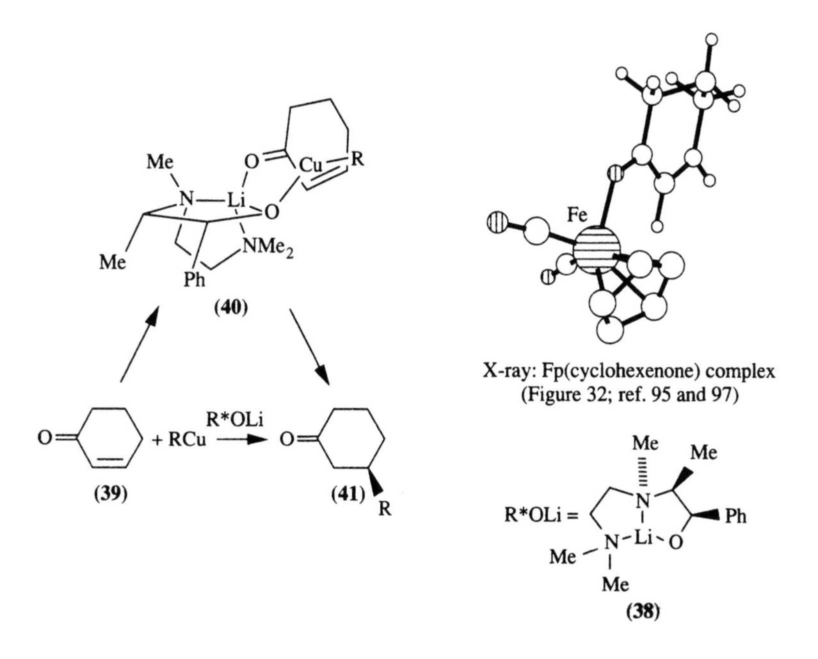

Figure 42 A model for catalytic asymmetric 1, 4-addition to cyclohexenone

This model is consistent with the conformation found in the structure of Fp(cyclohexenone)BF_4[95,97] and the similarities can be seen in Figure 42.[119,120] Notice that the assumption that Lewis acid (Li^+) coordination occurs *syn* to the double bond is essential for mediation of attack on the *re face* of the double bond. *Anti* coordination (if all else is retained) would result in *si face* addition.

Catalytic asymmetric induction in Diels–Alder reactions is somewhat more difficult to analyze based on these models. The chiral Lewis acids shown in Figure 43 all promote asymmetric Diels–Alder cycloaddition with variable degrees of enantioselectivity.[121–128]

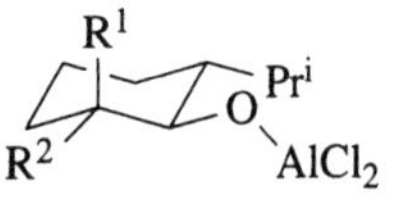

X = Cl, OPr^i

(42) R^1, R^2 = H
(43) R^1 = H, R^2 = Me

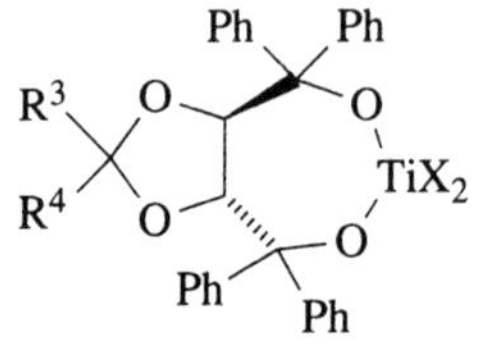

(44) R^3 = Me, R^4 = Bu^t
(45) R^3 = H, R^4 = Ph

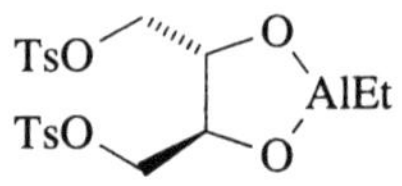

(46)

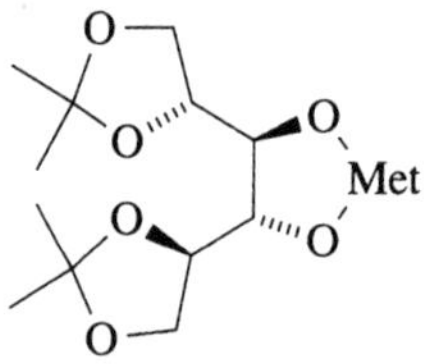

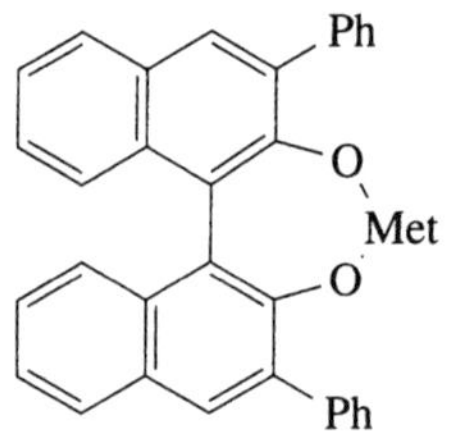

(47) Met = TiX_2 **(49)**
(48) Met = AlX **(50)**

Figure 43 Chiral catalysts for asymmetric Diels–Alder reactions

Construction of useful structural models for these reactions, however, would require not only a full structural characterization of the Lewis acid, but also knowledge of the preferred conformation of the chiral ligands. In many cases the catalysts are generated *in situ* and the stoichiometry or the aggregation state of the Lewis acid are not well defined. The latter point is particularly important since both titanium and aluminum alkoxides are known to form dimeric or polymeric species.[129]

The well-characterized aluminum Lewis acid (**52**) has been reported to catalyze hetero Diels–Alder reactions of aldehydes with high enantioselectivity (Figure 44).[130]

The model proposed for the reaction of simple aromatic aldehydes (**54**) places the Lewis acid *syn* to the aldehydic hydrogen, consistent with the frequent observation of this preference in crystal structures of Lewis acid–aldehyde complexes (*cf.* Figures 19, 27 and 28).

Chiral lanthanide shift reagents (LSR) also promote asymmetric hetero Diels–Alder reactions (Figure 45).[131–134] Lack of rigorous structural information regarding LSR/carbonyl interactions, however, does not allow meaningful speculation on the source of chirality transfer.

In addition, chiral dienes seem to exhibit a curious behavior in the presence of chiral LSRs (Table 9).[131–134] Diastereoselection is highest when a 'mismatched' diene/catalyst pair is used. The researchers have proposed a novel 'interactivity' between the catalyst and the chiral auxiliary, the nature of which is at this point unclear.

1.10.7.2 Additions to Nonconjugated Carbonyls

The design of novel Lewis acidic reagents for additions to nonconjugated carbonyls, based on structural information, has attracted much attention in recent years. An excellent overview of this subject has recently appeared, in which the mechanistic aspects of and various models for carbonyl addition processes are highlighted.[135] In particular, attention was directed towards the *ab initio* treatment of solvation effects in additions to the carbonyl group.[136,137] These calculations indicated that in the addition of water

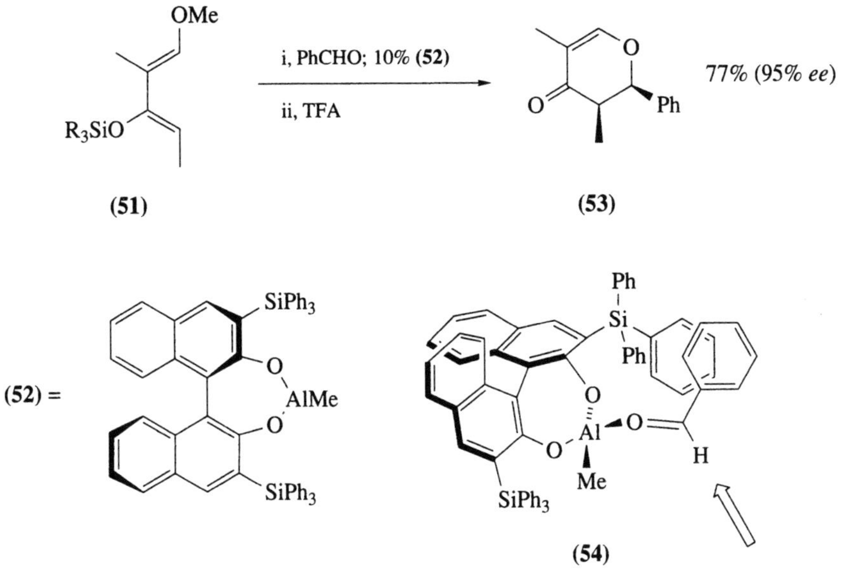

(51) (53)

i, PhCHO; 10% (52)
ii, TFA

77% (95% *ee*)

(52) =

(54)

Figure 44 Catalytic asymmetric hetero-Diels–Alder reaction

i, PhCHO; 1% (+)-Eu(hfc)$_3$
ii, TFA

42% *ee*

(55) (56)

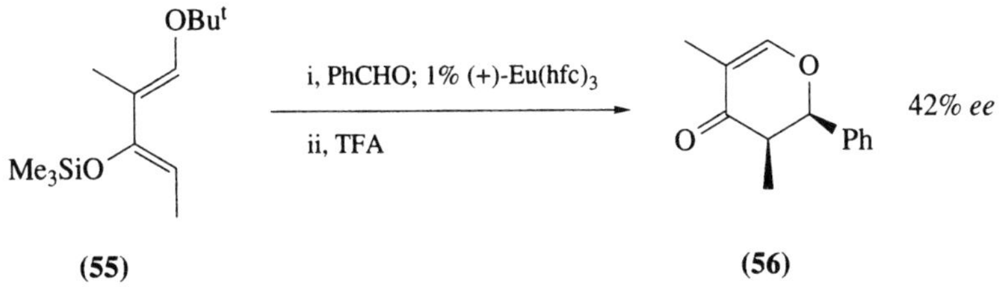

Figure 45 Asymmetric hetero-Diels–Alder reactions catalyzed by chiral lanthanide shift reagents

Table 9 Interactivity of Lanthanide Shift Reagents and Chiral Dienes in Hetero-Diels–Alder Reactions

PhCHO
Lewis acid

(57) (58) L-pyranose (59) D-pyranose

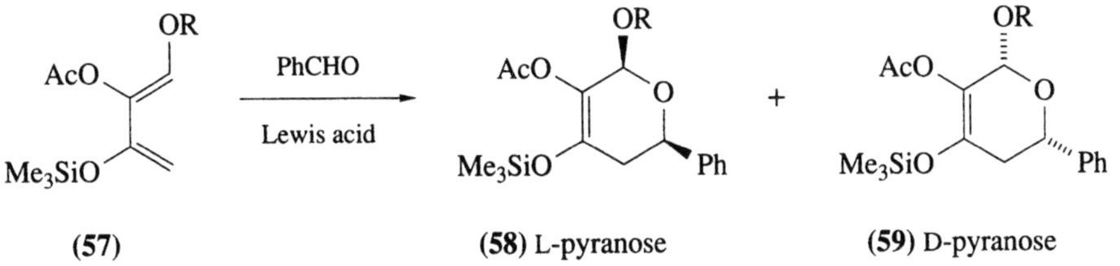

Lewis acid	*R*	*Product*	*ee (%)*
Eu(fod)$_3$	(–)-Menthyl	D-Pyranose	10
(+)Eu(hfc)$_3$	Bu^t	L-Pyranose	42
(+)Eu(hfc)$_3$	(+)-Menthyl	L-Pyranose	18
(+)Eu(hfc)$_3$	(–)-Menthyl	L-Pyranose	86

or ammonia to formaldehyde, the intermediacy of a molecule of water in a six-membered transition state can considerably lower the transition energy, relative to a four-membered transition state (Figure 46).[135–137]

This seemingly simple result may have far reaching consequences. For example, it may help to explain the effect of added lithium salts in nucleophilic additions to cyclohexanones as discussed earlier in this chapter. Thus, model (**63**) shown in Figure 47[2,135–137] can explain the enhancement of rate and may also be relevant to the origins of stereoselectivity in this reaction. Of course, the exact location of the lithium atom and the aggregation state of the adding nucleophile are subject to speculation, since for lithium these parameters seem to be highly variable.

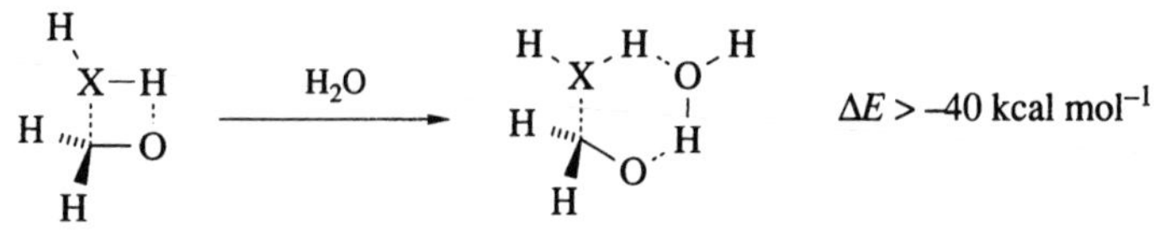

H X–H H O H (60) $\xrightarrow{H_2O}$ H H H X O H O H H (61) $\Delta E > -40$ kcal mol^{-1}

Figure 46 Four-membered *versus* six-membered transition states; role of a bridging catalyst

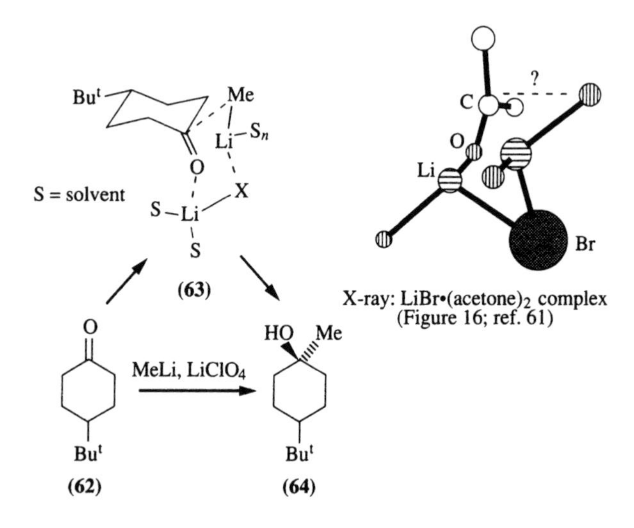

Figure 47 A model for $LiClO_4$-mediated addition of MeLi to 4-*t*-butylcyclohexanone

It may be recalled that an opposite stereochemical result is obtained by employing the bulky aluminum reagents MAD and MAT.[3] This observation has been explained by invoking out-of-plane complexation of the Lewis acid in a direction which would prevent equatorial attack (Figure 48).[3,74] The X-ray crystal structure of methyl toluate complexed with a bulky aluminum Lewis acid is fully consistent with this model.[74] However, it is worth mentioning that a six-membered transition state, perhaps involving $[Me_2(ArO)Al]^-Li^+$, has not been considered as an alternative mechanism.

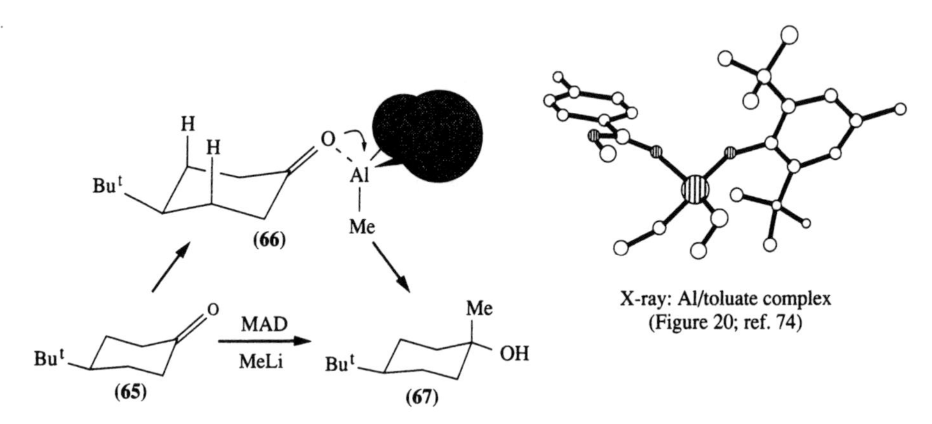

Figure 48 A model for MAD-mediated addition of MeLi to 4-*t*-butylcyclohexanone

It is more difficult to account for the remarkable anti-Cram selectivity observed in the MAT-mediated nucleophilic additions to α-chiral aldehydes, although out of plane coordination may play an important role (Figure 49).[3]

Chiral Lewis acidic catalysts derived from β-amino alcohols constitute a major field of recent development. These reagents have been used for enantioselective reduction of ketones[138–145] and for dialkylzinc additions to aldehydes.[146–155]

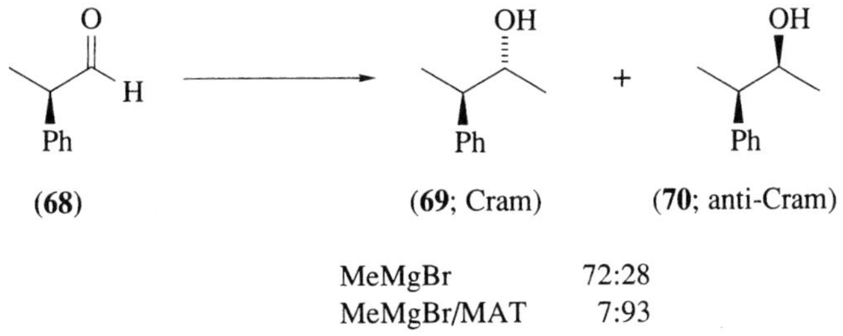

(68) **(69**; Cram) **(70**; anti-Cram)

MeMgBr	72:28
MeMgBr/MAT	7:93

Figure 49 Anti-Cram addition of MeMgBr to α-phenylpropanal mediated by the bulky aluminum Lewis acid MAT

The isolation and characterization of the reagent derived from the reaction of β-amino alcohols and borane accompanied the first report of a truly catalytic procedure for the enantioselective reduction of ketones.[143,144] A representative example is shown in Figure 50.[135,143,144] Based on ^{1}H and ^{11}B NMR spectroscopy, a six-membered, boat-like transition state model (**73**) was postulated.[135,143]

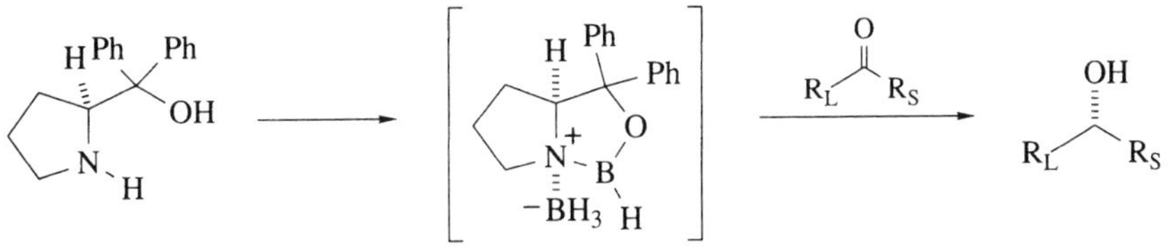

(71) **(72)**

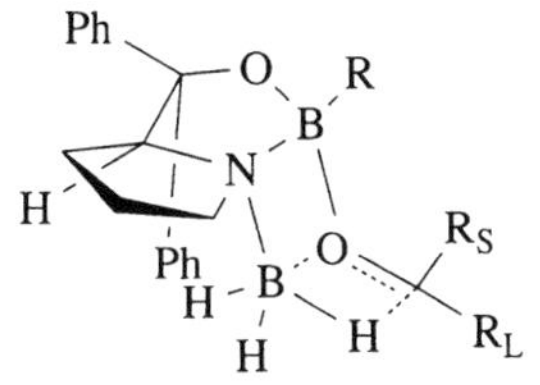

(73)

Figure 50 Catalytic asymmetric reduction of ketones

In this model the sense of asymmetric induction is controlled by two principle factors: (i) coordination of borane on the least-hindered face of the bicyclic ring system; and (ii) coordination of the Lewis acid *syn* to the small group (R_S). The latter point is in good agreement with the structural data that has been presented in this chapter, and is further supported by results from the asymmetric reduction of oxime ethers (Figure 51).[145] As predicted by the model (**75** and **78**), (*E*)- and (*Z*)-oxime ethers afford enantiomeric amines upon reduction by the reagent derived from (–)-norephedrine and borane (2 equiv.). Here, Lewis acid coordination is dictated by the (*E*)/(*Z*) stereochemistry of the oxime ether rather than by the rule of coordination *syn* to the small group.

A very similar model can be invoked to explain the results of catalytic enantioselective dialkylzinc additions to aldehydes.[135] The catalysts used in these reactions are invariably lithium- or zinc-centered Lewis acids. The transition structure shown in Figure 52[135,146,150] has been put forward by several groups and predicts the stereochemical outcome of many systems of this general type (**83–89**; Figure 53) quite nicely. In this model a dialkylzinc molecule is coordinated to a basic site (X) on the least-hindered face of the catalyst. The Lewis acidic metal can then deliver a molecule of aldehyde by coordination *syn* to the aldehydic hydrogen. The assumption of a boat transition state can be justified by invoking Zn–O interactions along the reaction path and comparison with various zinc–alkoxide crystal structures (*e.g.* MeZnOMe).

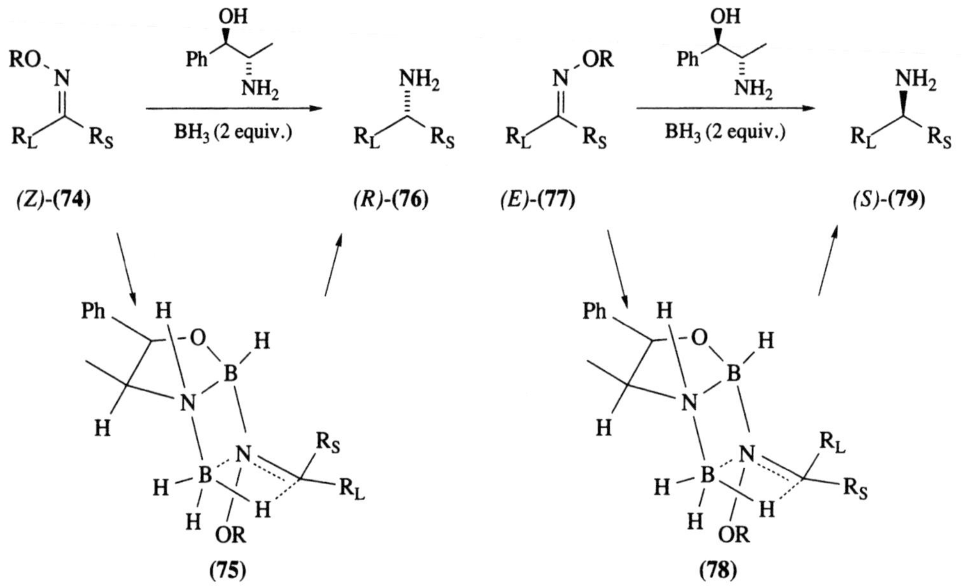

Figure 51 Asymmetric reduction of oxime ethers

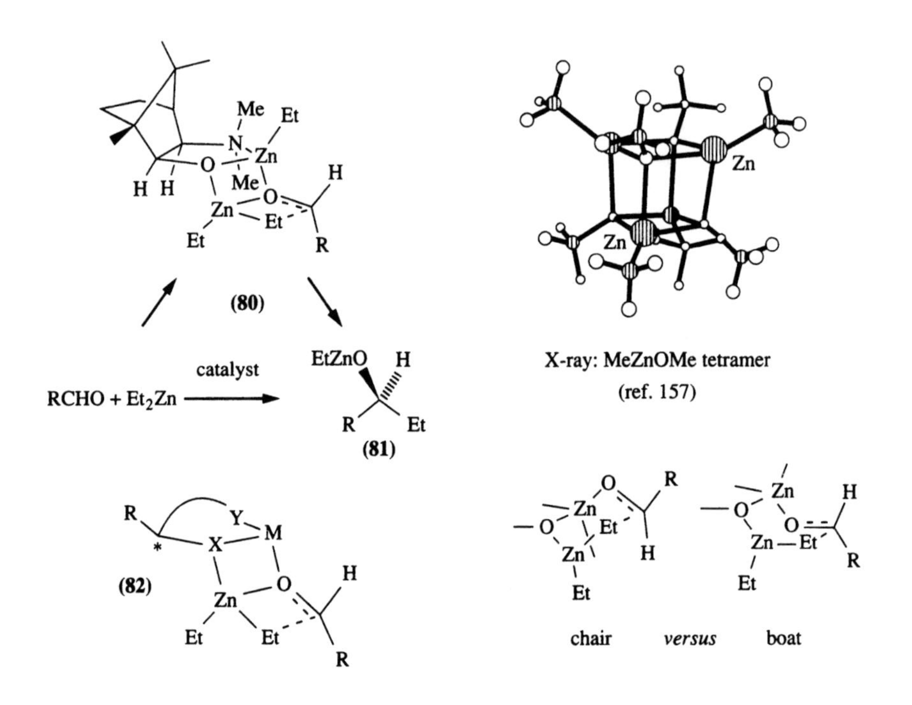

Figure 52 A model for catalytic asymmetric dialkylzinc additions to aldehydes

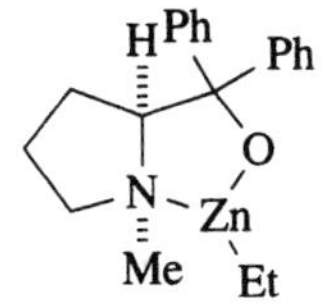

(S)-(83) (ref. 147) 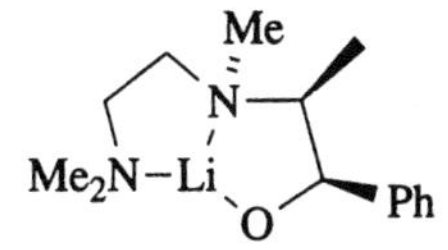 (R)-(84) (ref. 151, 152) 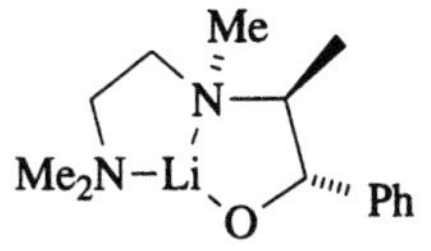 (S)-(85) (ref. 151, 152)

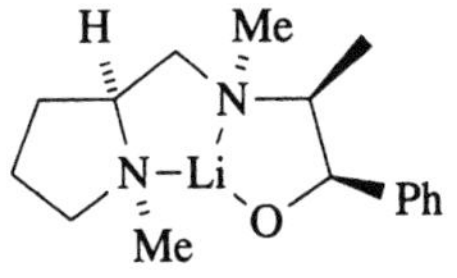

(R)-(86) (ref. 151, 152) 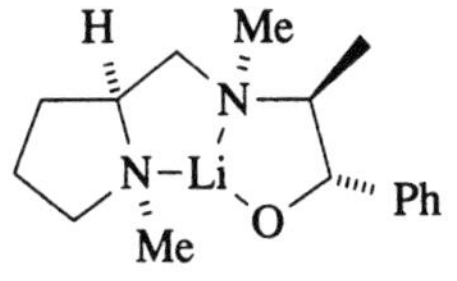 (S)-(87) (ref. 151, 152)

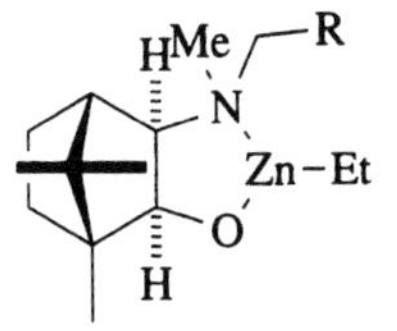

(S)-(88) (ref. 145, 149) 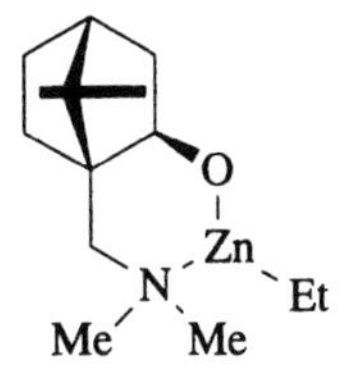 (R)-(89) (ref. 159)

Figure 53 Chiral catalysts for asymmetric dialkylzinc additions to carbonyls

Two interesting features of these models include the presence of an acid/base pair for organization of a well-defined six-membered transition state, and the role of the asymmetry adjacent to the basic group (X), which determines the face selectivity of the reaction. Note that for some catalysts (*i.e.* **85** and **87**) the 'least-hindered' site of coordination is not necessarily the convex face of a polycyclic system.

Lastly, the similarities between (**73**) and (**82**) suggest that these catalysts may be used interchangeably for reductions and dialkylzinc additions. Furthermore, the choice of small and large groups may not always be obvious. For example, in simple α,β-unsaturated ketones or in 1-chloroacetophenone the large group is not well defined.[144] Enone (**90**) is reduced to the (*R*)-allylic alcohol upon treatment with one equivalent of borane and a catalytic amount of (**91**; Figure 54).[144] Assuming that (**73**) is a valid transition structure, this result would suggest that the Lewis acid coordinates *anti* to the enone double bond, which, in light of the structural data, is somewhat surprising.

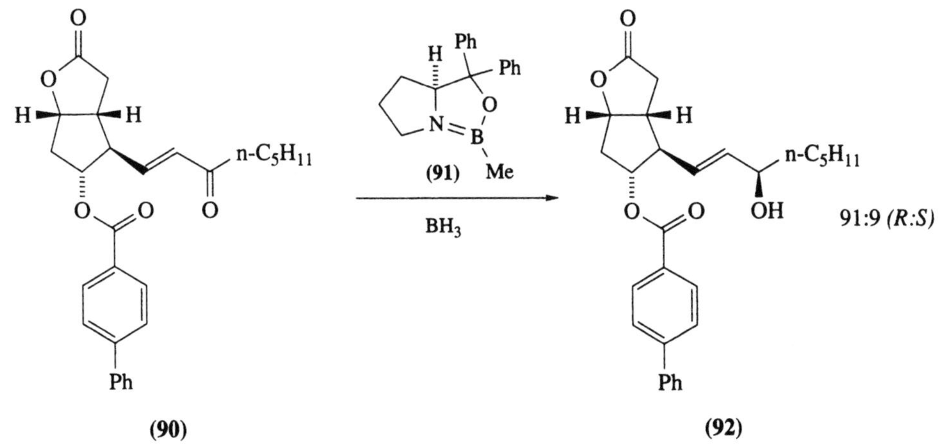

(**90**) —(**91**), BH_3→ (**92**) 91:9 (*R*:*S*)

Figure 54 Catalytic asymmetric reduction of enones

In this particular substrate one should also consider other potential sites of coordination for boron, such as the ester or the lactone carbonyl, which may be considered more basic than the carbonyl of the ketone.

Recent reports by two groups propose two new and entirely different models for similar catalytic dialkylzinc additions.[154,155] One of these[154] assumes a pentavalent, pseudotrigonal bipyramidal zinc(II) catalyst (**93**)[157] and the other[155] purports transfer of the alkyl group from the carbonyl-bearing metal (**94**; Figure 55).

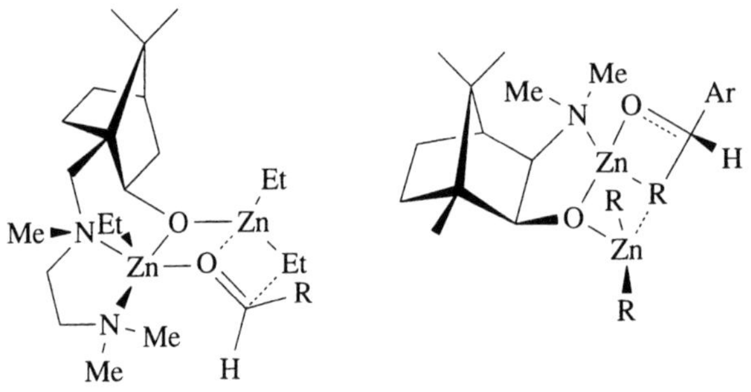

Figure 55 Two models for catalytic dialkylzinc additions to carbonyls

Structural evidence for these models has been gathered from X-ray crystal structures and, in the case of (**94**), from crossover experiments. Use of two different dialkylzinc reagents for the generation of the catalyst and the nucleophile gave a statistical mixture of possible products.[155] This result is in contrast to the findings of another group,[149] who found no crossover in a very similar system. It is interesting to note that model (**94**) essentially represents metal coordination to the π-face of the carbonyl as opposed to σ-bonding in all other models. Clearly, more mechanistic details are needed in order to identify the true mechanism of these reactions. This problem, however, is a good demonstration of a case where the validity of several plausible models cannot be discerned despite the availability of crystallographic and spectroscopic data.

The aldol reactions of isocyanoacetates and a variety of aldehydes proceeds with very high levels of asymmetric induction under the influence of catalytic amounts of a chiral Au^I complex (Figure 56).[158,159] This remarkable transformation provokes speculation regarding the possible mechanism of catalysis and provides a challenge to model-building exercises. Although a crystal structure of an Au^I/carbonyl complex is not presently available, one may draw analogies with the isoelectronic, square planar Ni^0/η^2-benzaldehyde complex (Figure 56).[160] Note that both the nickel and the gold compounds can be formally viewed as d^8-transition metal complexes with a preference for the square planar geometry.[161] Accordingly, one may expect Au^I Lewis acids to form η^2-square planar complexes with carbonyls, although the presence of other bases (*i.e.* isocyanate, tertiary amines) may completely alter this bias.[162] Thus, it should be emphasized that the proposed transition structure (**95**) is highly speculative and neither the orientation of the aldehyde nor the role and structure of the diamine linker can be predicted reliably.

1.10.8 CONCLUSIONS

Interactions of the carbonyl group with Lewis acids are colorful and varied, but by no means unpredictable. The structural information gathered in this chapter points, fairly consistently, to the same set of principles. 'Rules of complexation' such as coordination *syn* to the small substituent, *s-trans* preference due to gearing effects and distortions towards the π-cloud in response to steric strain are among these principles. Section 1.10.7 illustrated a few examples of how the predictive power of such structural rules have been applied to the design of novel and highly selective Lewis acidic catalysts.

In the context of future prospects in catalysis, transition metal complexes seem to provide a particularly diverse source of useful Lewis acids. Although the structural features of these complexes were not discussed extensively in this chapter, η^2-bonding seems unique to these complexes and warrants further exploration.

For the design of new reagents one would also like to understand the conformational changes of the carbonyl group upon complexation. The *s-trans* effect on unsaturated carbonyls is a clear example of such effects. More subtle, and perhaps more interesting, are the conformational consequences of Lewis

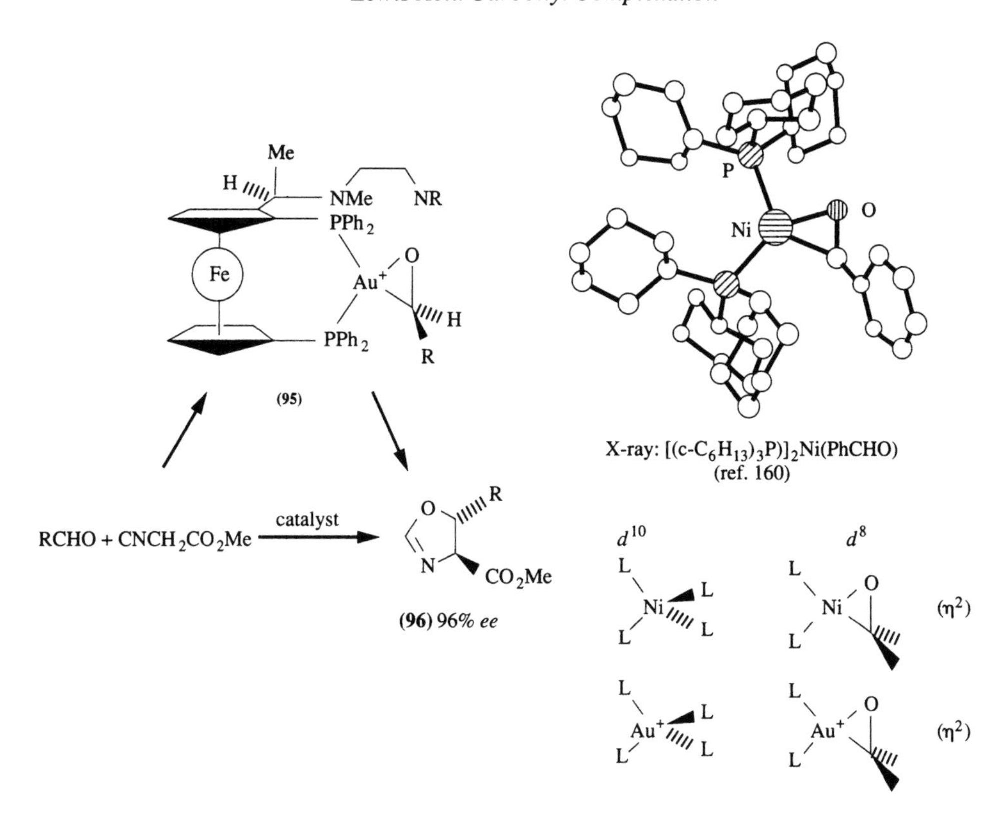

Figure 56 A possible model for the catalytic asymmetric aldol reaction of isocyano acetates and aldehydes

acid complexation for nonconjugated carbonyls. The relevance of these issues to the topic of chelation-controlled additions was discussed at length. Similar effects for nonchelating carbonyls, however, are completely unresolved, but one may hope to generate a coherent set of rules in these systems as well. Impressive achievements of the past and worthy rewards of the future provide a strong incentive to pursue this and other structural problems of Lewis acid–carbonyl complexation with vigor and optimism.

ACKNOWLEDGMENT

We thank Ms. Sally Jalbert for invaluable help with the preparation of this manuscript. Many helpful suggestions from Dr. Neville Anthony and Mr. Kurtis Macferrin are gratefully acknowledged.

This chapter was taken in part from S. Shambayati, W. E. Crowe and S. L. Schreiber, *Angew. Chem., Int. Ed. Engl.*, 1990, **29**, 256.

1.10.9 REFERENCES

1. M. Landolf, *C. R. Hebd. Seances Acad. Sci.*, 1878, **86**, 671.
2. (a) E. C. Ashby and S. A. Noding, *J. Org. Chem.*, 1979, **44**, 4371; (b) see also T. L. Macdonald and W. C. Still, *J. Am. Chem. Soc.*, 1975, **97**, 5280.
3. K. Maruoka, T. Itoh and H. Yamamoto, *J. Am. Chem. Soc.*, 1985, **107**, 4573.
4. P. Yates and P. E. Eaton, *J. Am. Chem. Soc.*, 1960, **82**, 4436.
5. There are two different views on the effect of Lewis acids on the dipole moment of the carbonyl. For an interesting discussion see ref. 21.
6. For a review and discussion see: E. L. Eliel, in 'Asymmetric Synthesis', ed. J. D. Morrison, Academic Press, New York, 1983, vol. 2, p. 125.
7. C. H. Heathcock and L. A. Flippin, *J. Am. Chem. Soc.*, 1983, **105**, 1667.
8. E. P. Lodge and C. H. Heathcock, *J. Am. Chem. Soc.*, 1987, **109**, 2819.
9. I. Mori, P. A. Bartlett and C. H. Heathcock, *J. Am. Chem. Soc.*, 1987, **109**, 7199.
10. P. Russegger and P. Schuster, *Chem. Phys. Lett.*, 1973, **19**, 254.

11. T. K. Ha, U. P. Wild, R. O. Kühne, C. Loesch, T. Schaffhauser, J. Stachel and A. Wokaun, *Helv. Chim. Acta*, 1978, **61**, 1193.
12. J. E. Del Bene, *Chem. Phys. Lett.*, 1979, **64**, 227.
13. J. E. Del Bene, *Chem. Phys.*, 1979, **40**, 329.
14. T. Weller, R. Lochman, W. Meiler and H.-J. Köhler, *J. Mol. Struct.*, 1982, **90**, 81.
15. S. F. Smith, J. Chandrasekhar and W. L. Jorgensen, *J. Phys. Chem.*, 1982, **86**, 3308.
16. For examples see: (a) R. H. Staley and J. L. Beauchamp, *J. Am. Chem. Soc.*, 1975, **97**, 5920; (b) R. L. Woodin and J. L. Beauchamp, *J. Am. Chem. Soc.*, 1978, **100**, 501; (c) I. Dzidic and P. Kebarle, *J. Phys. Chem.*, 1970, **74**, 1466.
17. D. J. Raber, N. K. Raber, J. Chandrasekhar and P. von R. Schleyer, *Inorg. Chem.*, 1984, **23**, 4076.
18. K. Kimura, S. Katsumata, Y. Achiba, T. Yamazaki and S. Iwata, 'Handbook of He(I) Photoelectron Spectra of Fundamental Organic Molecules', Halsted Press, New York, 1981, p. 140.
19. D. J. Nelson, *J. Org. Chem.*, 1986, **51**, 3185.
20. R. J. Loncharich, T. R. Schwartz and K. N. Houk, *J. Am. Chem. Soc.*, 1987, **109**, 14.
21. T. J. LePage and K. B. Wiberg, *J. Am. Chem. Soc.*, 1988, **110**, 6642.
22. See refs. 13 and 21.
23. H. C. Brown and R. R. Holm, *J. Am. Chem. Soc.*, 1956, **78**, 2173.
24. T. D. Coyle and F. G. A. Stone, *J. Am. Chem. Soc.*, 1961, **83**, 4138.
25. M. F. Lappert and J. K. Smith, *J. Chem. Soc.*, 1965, 7102.
26. (a) M. Rabinovitz and A. Grinvald, *J. Am. Chem. Soc.*, 1972, **94**, 2724; (b) P. C. Myhre, C. D. Fisher, T. Nielsen and W. M. Schubert, *J. Am. Chem. Soc.*, 1965, **87**, 29.
27. A. Fratiello, T. P. Onak and R. E. Schuster, *J. Am. Chem. Soc.*, 1968, **90**, 1194.
28. N. N. Greenwood and T. S. Srivastava, *J. Chem. Soc. (A)*, 1966, 270.
29. R. J. Gillespie and J. S. Hartman, *Can. J. Chem.*, 1968, **46**, 2147.
30. U. Henriksson and S. Forsén, *J. Chem. Soc. (D)*, 1970, 1229.
31. A. Fratiello and R. E. Schuster, *J. Org. Chem.*, 1972, **37**, 2237.
32. A. Fratiello, G. A. Vidulich and Y. Chow, *J. Org. Chem.*, 1973, **38**, 2309.
33. R. E. Schuster and R. D. Bennett, *J. Org. Chem.*, 1973, **38**, 2904.
34. A. Fratiello and C. S. Stover, *J. Org. Chem.*, 1974, **40**, 1244.
35. J. S. Hartman, P. Stilbs and S. Forsén, *Tetrahedron Lett.*, 1975, 3497.
36. P. Stilbs and S. Forsén, *Tetrahedron Lett.*, 1974, 3185.
37. A. Fratiello, R. Kubo and S. Chow, *J. Chem. Soc., Perkin Trans. 2*, 1975, 1205.
38. S. E. Denmark, B. R. Henke and E. Weber, *J. Am. Chem. Soc.*, 1987, **109**, 2512.
39. A. K. Bose, P. R. Srinivasan and G. Trainor, *J. Am. Chem. Soc.*, 1974, **96**, 3670.
40. A. K. Bose, P. R. Srinivasan and G. Trainor, *J. Magn. Reson.*, 1974, **15**, 592.
41. A. K. Bose, P. R. Srinivasan and G. Trainor, *Tetrahedron Lett.*, 1975, 1571.
42. R. E. Lenkinski and J. Reuben, *J. Am. Chem. Soc.*, 1976, **98**, 4065.
43. P. Finocchiaro, A. Recca, P. Maravigna and G. Montaudo, *Tetrahedron*, 1974, **30**, 4159.
44. D. J. Raber, C. M. Janks, M. D. Johnston, Jr. and N. K. Raber, *J. Am. Chem. Soc.*, 1980, **102**, 6591.
45. D. J. Raber and C. M. Janks, *Tetrahedron*, 1986, **42**, 4347.
46. D. J. Raber, J. A. Peters and M. S. Nieuwenhuizen, *J. Chem. Soc., Perkin Trans. 2*, 1986, 853.
47. (a) B. C. Mayo, *Chem. Soc. Rev.*, 1973, **1**, 49; (b) H. L. Ammon, P. H. Mazzocchi, W. J. Kopecky, Jr., H. J. Tamburini and P. H. Watts, *J. Am. Chem. Soc.*, 1973, **95**, 1968.
48. J. Torri and M. Azzaro, *Bull. Soc. Chim. Fr., Part 2*, 1978, 286.
49. R. F. Childs, D. L. Mulholland and A. Nixon, *Can. J. Chem.*, 1982, **60**, 801.
50. D. J. Cram and F. A. Abd Elhafez, *J. Am. Chem. Soc.*, 1952, **74**, 5828.
51. M. T. Reetz, K. Kesseler, S. Schmidtberger, B. Wenderoth and R. Steinbach, *Angew. Chem., Int. Ed. Engl.* 1983, **22**, 989.
52. M. T. Reetz, K. Kesseler, S. Schmidtberger, B. Wenderoth and R. Steinbach, *Angew. Chem. Suppl.*, 1983, 1511.
53. M. T. Reetz, M. Hüllmann and T. Seitz, *Angew. Chem., Int. Ed. Engl.*, 1987, **26**, 477.
54. G. E. Keck and S. Castellino, *J. Am. Chem. Soc.*, 1986, **108**, 3847.
55. G. E. Keck, S. Castellino and M. R. Wiley, *J. Org. Chem.*, 1986, **51**, 5478.
56. G. E. Keck and S. Castellino, *Tetrahedron Lett.*, 1987, **28**, 281.
57. S. D. Kahn, G. E. Keck and W. J. Hehre, *Tetrahedron Lett.*, 1987, **28**, 279.
58. S. Shambayati, J. F. Blake, S. G. Wierschke, W. L. Jorgensen and S. L. Schreiber, *J. Am. Chem. Soc.*, 1990, **112**, 697.
59. H. B. Burgi and J. D. Dunitz, *Acc. Chem. Res.*, 1983, **16**, 153.
60. A. S. Cieplak, *J. Am. Chem. Soc.*, 1981, **103**, 4540.
61. R. Amstutz, J. D. Dunitz, T. Laube, W. B. Scheweizer and D. Seebach, *Chem. Ber.*, 1986, **119**, 434.
62. R. H. van der Veen, R. M. Kellogg, A. Vos and T. J. van Bergen, *J. Chem. Soc., Chem. Commun.*, 1978, 923.
63. S. Shambayati, unpublished results.
64. B. Pullman and N. Goldblum (eds.), 'Metal–Ligand Interactions in Organic Chemistry and Biochemistry', Reidel, Boston, 1976.
65. I. L. Karle, *J. Am. Chem. Soc.*, 1974, **96**, 4000.
66. I. L. Karle, *Proc. Natl. Acad. Sci. USA*, 1985, **82**, 7155.
67. M. T. Reetz, M. Hüllmann, W. Massa, S. Berger, P. Rademacher and P. Heymanns, *J. Am. Chem. Soc.*, 1986, **108**, 2405.
68. Heteronuclear NOE experiments provided evidence for the *anti* geometry in solution. Enhancement at the aldehydic (but not the phenyl) proton absorption was observed upon irradiation of the fluorine atoms, see ref. 67.
69. A. J. Baskar and C. M. Lukehart, *J. Organomet. Chem.*, 1983, **254**, 149.
70. R. C. F. Jones and G. E. Peterson, *Tetrahedron Lett.*, 1983, 4757.

71. S. J. Rettig and J. Trotter, *Can. J. Chem.*, 1976, **54**, 1168.
72. S. Shambayati, unpublished results.
73. A. Bittner, D. Männig and H. Nöth, *Z. Naturforsch. Teil B*, 1986, **41**, 587.
74. A. P. Shreve, R. Mulhaupt, W. Fultz, J. Calabrese, W. Robbins and S. Ittel, *Organometallics*, 1988, **7**, 409.
75. L. Brun, *Acta Crystallogr.*, 1966, **20**, 739.
76. S. Shambayati, unpublished results.
77. I. W. Bassi, M. Calcaterra and R. Intrito, *J. Organomet. Chem.*, 1977, **127**, 305.
78. T. Poll, J. O. Metter and G. Helmchen, *Angew. Chem., Int. Ed. Engl.*, 1985, **24**, 112.
79. B. Viard, M. Poulain, D. Grandjean and J. Anandrut, *J. Chem. Res. (S)*, 1983, 850.
80. G. Maier and U. Seipp, *Tetrahedron Lett.*, 1987, **28**, 4515.
81. J. Utka, P. Sobota and T. Lis, *J. Organomet. Chem.*, 1987, **334**, 341.
82. S. Shambayati, unpublished results.
83. C. Mahadevan, M. Seshasayee and A. S. Kothiwal, *Cryst. Struct. Commun.*, 1982, **11**, 1725.
84. S. Calogero, G. Valle and U. Russo, *Organometallics*, 1984, **3**, 1205.
85. G. Valle, S. Calogero and U. Russo, *J. Organomet. Chem.*, 1982, **228**, C79.
86. J. Buckle, P. G. Harrison, T. J. King and J. A. Richards, *J. Chem. Soc., Dalton Trans*, 1975, 1552.
87. D. Cunningham, T. Donek, M. J. Frazer, M. McPartlin and J. D. Matthews, *J. Organomet. Chem.*, 1975, **90**, C23.
88. For an extensive discussion of this subject *cf.* ref. 84.
89. F. D. Lewis, J. D. Oxman and J. C. Huffman, *J. Am. Chem. Soc.*, 1984, **106**, 466.
90. M. T. Reetz, K. Harms and W. Reif, *Tetrahedron Lett.*, 1988, **29**, 5881.
91. P. Chakrabarti and J. D. Dunitz, *Helv. Chim. Acta*, 1982, **65**, 1482.
92. For a recent review of transition metal carbonyl complexes, see Y.-H. Huang and J. A. Gladysz, *J. Chem. Educ.*, 1988, **65**, 299.
93. R. H. Crabtree, G. G. Hatky, C. P. Parnell, B. E. Segmüller and P. J. Uriarte, *Inorg. Chem.*, 1984, **23**, 354.
94. R. H. Crabtree, M. F. Mellea, J. M. Mihelcic and J. Quirk, *J. Am. Chem. Soc.*, 1982, **104**, 107.
95. B. M. Foxman, P. T. Klemarczyk, R. E. Liptrot and M. Rosenblum, *J. Organomet. Chem.*, 1980, **187**, 253.
96. P. Boudjouk, J. B. Woell, L. J. Radonovich and M. W. Eyring, *Organometallics*, 1982, **1**, 582.
97. S. Shambayati, unpublished results.
98. R. V. Honeychuck, P. V. Bonnesen, J. Farahi and W. H. Hersh, *J. Org. Chem.*, 1987, **52**, 5293.
99. J. M. Fernandez, K. Emerson, R. D. Carsen and J. A. Gladysz, *J. Am. Chem. Soc.*, 1986, **108**, 8268.
100. J. M. Fernandez, K. Emerson, R. D. Carsen and J. A. Gladysz, *J. Chem. Soc., Chem. Commun.*, 1988, 37.
101. M. Sacerdoti, V. Bertolasi and G. Gilli, *Acta Crystallogr., Sect. B*, 1980, **36**, 1061.
102. (a) N. T. Anh and O. Eisenstein, *Nouv. J. Chim.*, 1977, **1**, 61; (b) M. Cherest, H. Felkin and N. Prudent, *Tetrahedron Lett.*, 1968, 2199.
103. B. Clark, M. Green, R. B. L. Osborn and F. G. A. Stone, *J. Chem. Soc. (A)*, 1968, 167.
104. See ref. 101.
105. T. T. Tsou, J. C. Huffman and J. K. Kochi, *Inorg. Chem.*, 1979, **18**, 2311.
106. W. Schröder, K. R. Pörschke, Y.-H. Tsay and K. Krüger, *Angew. Chem., Int. Ed. Engl.*, 1987, **26**, 919.
107. H. Brunner, J. Wachter, I. Bernal and M. Breswich, *Angew. Chem., Int. Ed. Engl.*, 1979, **18**, 861.
108. W. D. Harman, D. P. Fairlie and H. Taube, *J. Am. Chem. Soc.*, 1986, **108**, 8223.
109. H. Adams, N. A. Bailey, J. T. Gauntlett and M. J. Winter, *J. Chem. Soc., Chem. Commun.*, 1984, 1360.
110. W. Oppolzer, *Angew. Chem., Int. Ed. Engl.*, 1984, **23**, 876.
111. W. Oppolzer, I. Rodriquez, J. Blagg and G. Bernardinelli, *Helv. Chim. Acta*, 1989, **72**, 123. The structure shown in Figure 40 has been reproduced from this reference, based on partial bond length and bond angle data. The details of this representation may be inaccurate.
112. D. A. Evans, K. T. Chapman and J. Biasha, *J. Am. Chem. Soc.*, 1984, **106**, 4261.
113. D. A. Evans, K. T. Chapman and J. Biasha, *Tetrahedron Lett.*, 1984, **25**, 4071.
114. D. A. Evans, K. T. Chapman, D. T. Hung and A. T. Kawaguchi, *Angew. Chem., Int. Ed. Engl.*, 1987, **26**, 1184.
115. H. Lehmkuhl and H.-D. Kobs, *Justus Liebigs Ann. Chem.*, 1968, **719**, 11.
116. Y. Yamamoto, *Angew. Chem., Int. Ed. Engl.*, 1986, **25**, 947.
117. W. Oppolzer and H. J. Löher, *Helv. Chim. Acta*, 1981, **64**, 2808.
118. W. Oppolzer and T. Stevenson, *Tetrahedron Lett.*, 1986, **27**, 1139.
119. E. J. Corey, R. Naef and F. J. Hannon, *J. Am. Chem. Soc.*, 1986, **108**, 7114.
120. E. J. Corey and F. J. Hannon, *Tetrahedron Lett.*, 1987, **28**, 5233.
121. S. Hashimoto, M. Komeshima and K. Koga, *J. Chem. Soc., Chem. Commun.*, 1979, 437.
122. H. Takemura, M. Komeshima and I. Takahashi, *Tetrahedron Lett.*, 1987, **28**, 5687.
123. M. Quimpére and K. Jankowsky, *J. Chem. Soc., Chem. Commun.*, 1987, 676.
124. K. Narasaka, M. Inoue and T. Yamada, *Chem. Lett.*, 1986, 1109.
125. K. Narasaka, M. Inoue and T. Yamada, *Chem. Lett.*, 1986, 1967.
126. K. Narasaka, M. Inoue and T. Yamada, *Chem. Lett.*, 1987, 2409.
127. C. Chapius and J. Jurczak, *Helv. Chim. Acta*, 1987, **70**, 436.
128. D. Seebach, A. K. Beck, R. Imwinkelreid, S. Roggo and A. Wonnacott, *Helv. Chim. Acta*, 1987, **70**, 954.
129. For a discussion of this point see: (a) S. F. Pedersen, J. C. Dewan, R. R. Eckman and K. B. Sharpless, *J. Am. Chem. Soc.*, 1987, **109**, 1279; (b) I. D. Williams, S. F. Pedersen, K. B. Sharpless and S. J. Lippard, *J. Am. Chem. Soc.*, 1984, **106**, 6430; (c) ref. 128.
130. Y. Yamamoto, *J. Am. Chem. Soc.*, 1988, **110**, 310.
131. M. D. Bednarski, C. Maring and S. J. Danishefsky, *Tetrahedron Lett.*, 1983, **24**, 3451.
132. M. D. Bednarski and S. J. Danishefsky, *J. Am. Chem. Soc.*, 1983, **105**, 6968.
133. S. J. Danishefsky, *Aldrichimica Acta*, 1986, **19**, 59.
134. M. D. Bednarski and S. J. Danishefsky, *J. Am. Chem. Soc.*, 1986, **108**, 7060.
135. D. A. Evans, *Science (Washington, D.C.)*, 1988, **240**, 420. Many of the models presented in this chapter were based on insights provided by this article.

136. I. H. Williams, *J. Am. Chem. Soc.*, 1987, **109**, 6299.
137. I. H. Williams, D. Spangler, D. A. Fence, G. M. Maggiora and R. L. Schowen, *J. Am. Chem. Soc.*, 1983, **105**, 31.
138. S. Itsuno, K. Ito, A. Hirao and S. Nakahama, *J. Chem. Soc., Chem. Commun.*, 1983, 469.
139. S. Itsuno, K. Ito, A. Hirao and S. Nakahama, *J. Org. Chem.*, 1984, **49**, 555.
140. S. Itsuno, K. Ito, A. Hirao and S. Nakahama, *J. Chem. Soc., Perkin Trans. 1*, 1984, 2887.
141. S. Itsuno, M. Nakano, K. Miyazaki, H. Masuda, K. Ito, A. Hirao and S. Nakahama, *J. Chem. Soc., Perkin Trans. 1*, 1985, 2039.
142. S. Itsuno, M. Nakano, K. Ito, A. Hirao, M. Owa, N. Kanda and S. Nakahama, *J. Chem. Soc., Perkin Trans. 1*, 1985, 2615.
143. E. J. Corey, R. K. Bakshi and S. Shibata, *J. Am. Chem. Soc.*, 1987, **109**, 5551.
144. E. J. Corey, R. K. Bakshi, C.-P. Chen and V. K. Singh, *J. Am. Chem. Soc.*, 1987, **109**, 7925.
145. M. Kitamura, S. Suga, K. Kawai and R. Noyori, *J. Am. Chem. Soc.*, 1986, **108**, 6071.
146. K. Soai, A. Ookawa, K. Ogawa and T. Kaba, *J. Chem. Soc., Chem. Commun.*, 1987, 467.
147. K. Soai, A. Ookawa, K. Ogawa and T. Kaba, *J. Am. Chem. Soc.*, 1987, **109**, 7111.
148. A. A. Smaardijk and H. Wynberg, *J. Org. Chem.*, 1987, **52**, 135.
149. S. Itsuno and J. M. J. Fréchet, *J. Org. Chem.*, 1987, **52**, 4140.
150. P. A. Chaloner and S. A. R. Perera, *Tetrahedron Lett.*, 1987, **28**, 3013.
151. E. J. Corey and F. J. Hannon, *Tetrahedron Lett.*, 1987, **28**, 5233.
152. E. J. Corey and F. J. Hannon, *Tetrahedron Lett.*, 1987, **28**, 5237.
153. Y. Sakito, Y. Yoneyoshi and G. Suzukamo, *Tetrahedron Lett.*, 1988, **29**, 223.
154. W. Oppolzer and R. N. Radinov, *Tetrahedron Lett.*, 1988, **29**, 5645.
155. R. Noyori, S. Suga, K. Kawai, S. Okada and M. Kitamura, *Pure Appl. Chem.*, 1988, **60**, 1597.
156. N. W. Alcock, K. P. Balakrishnan, A. Berry, P. Moore and C. J. Reader, *J. Chem. Soc., Dalton Trans.*, 1973, 1537.
157. H. M. M. Shearer and C. B. Spencer, *Acta Crystallogr., Sect. B*, 1980, **36**, 2046.
158. Y. Ito, T. Sawamura and T. Hayashi, *J. Am. Chem. Soc.*, 1986, **108**, 6405.
159. Y. Ito, T. Sawamura and T. Hayashi, *Tetrahedron Lett.*, 1987, **28**, 6215.
160. J. Kaiser, J. Sieler, D. Walther, E. Dinjus and L. Golic, *Acta Crystallogr., Sect. B*, 1982, **38**, 1584.
161. R. H. Crabtree, 'The Organometallic Chemistry of Transition Metals', Wiley, New York, 1988, p. 20.
162. For recent studies on the role of central chirality in these reactions, see S. D. Pastor and A. Togni, *J. Am. Chem. Soc.*, 1989, **111**, 2333.

1.11
Lewis Acid Promoted Addition Reactions of Organometallic Compounds

MASAHIKO YAMAGUCHI
Tohoku University, Sendai, Japan

1.11.1 INTRODUCTION

The importance of the Lewis acidic nature of organometals in addition reactions to C═X bonds (X = heteroatom) has been well documented.[1] The complexation of organometallic compounds with C═X bonds is considered to be the origin of the promotion of the addition reactions and the generation of chemo-, regio- or stereo-selectivities. The complexation phenomena are discussed in detail in Chapter 1.10, of this volume. In reactions of simple organometals such as alkyl-magnesiums or -aluminums, a single organometallic species undertakes both of the tasks involved in the C—C bond formation process: complexation and nucleophilic attack. For example, reaction of benzophenone with Me_3Al in a 1:1 ratio is reported to give a monomeric 1:1 complex at room temperature, which decomposes to dimethylaluminum 1,1-diphenylethoxide at 80 °C (equation 1).[2]

$$Ph_2C{=}O + Me_3Al \longrightarrow Ph_2C{=}O{\cdot}Me_3Al \longrightarrow Ph_2C(Me){-}OAlMe_2 \quad (1)$$

In recent years, a new type of addition reaction to C═X bonds is emerging in organic synthesis which utilizes a *binary reagent system* composed of a nucleophilic organometal and a Lewis acid. In contrast to the addition of simple organometals mentioned above, this new methodology assigns the task of the nucleophilic attack and the complexation to the two separate reagents. Consequently, the selection of an appropriate Lewis acid allows a modification of the nature of the C═X bonds through complexation, which dramatically enhances the rate of organometal addition reactions, increases the yields, changes the regio- or stereo-selectivities and allows chemoselective reactions to be conducted in complex molecules possessing other sensitive functional groups. It should also be noted that the complexation of a Lewis acid with an organometal can modify the nature of the nucleophiles, for example by the ate complex formation, which offers another advantage of these reagent systems. The development of the binary reagents, especially that of effective Lewis acids, is one of the hot topics in contemporary organic synthesis. The synthetic aspects of these useful reagent systems will be reviewed here.

The addition reactions can be divided into two categories depending on the nature of the nucleophilic organometals employed: (i) the Lewis acid promoted addition reactions of relatively unreactive organometals such as alkyl-silanes, -stannanes, *etc.*, where, although the reagents are stable and storable, they turn to highly reactive species in the presence of appropriate Lewis acids; and (ii) the Lewis acid promoted addition reactions of reactive organometals such as alkyl-lithiums, -magnesiums, *etc.*, where, although the reagents themselves are generally capable of reaction with the electrophiles, the presence of the Lewis acids results in the variation of the reaction course. The present review deals with both of these categories with some emphasis on the latter.

Although these methodologies have proved to be quite useful in organic synthesis, the precise mechanisms are still not known in many cases because of the lack of mechanistic studies. The following can be presented as the candidates in the reaction of an $R—ML_n/M'L_m$ system, in which $R—ML_n$ represents the nucleophilic organometals and $M'L_m$ the Lewis acids: (i) the original organometallic compound $R—ML_n$ attacks a C═X bond activated by a Lewis acid $M'L_m$; (ii) the ate complex $ML_n^+[RM'L_m]^-$, generated from $R—ML_n$ and $M'L_m$ (equation 2), adds to a C═X bond; or (iii) a new organometallic species $R—M'L_{m-1}$ formed by the transmetalation, which produces another Lewis acid ML_{n+1} (equation 3), reacts with a C═X bond.

$$RML_n + M'L_m \longrightarrow ML_n^+ \left[RM'L_m\right]^- \quad (2)$$

$$RML_n + M'L_m \longrightarrow RM'L_{m-1} + ML_{n+1} \quad (3)$$

This chapter focuses the attention on the reactions of nonstabilized carbanionic compounds such as alkyl, vinyl, aryl, alkynyl metals, *etc.*, and the chemistry of the stabilized system, *i.e.* allylic, propargylic or oxaallylic carbanions is presented in Volume 2 of this series. Electrophiles with C═X bonds which are discussed include aldehydes, ketones, epoxides, aziridines, acetals, orthoesters and imines, all of which turn into highly reactive electrophiles in the presence of Lewis acids.

1.11.2 LEWIS ACID PROMOTED REACTIONS OF ALDEHYDES AND KETONES

1.11.2.1 Control of the Reactivity

A classic example of a Lewis acid promoted addition reaction is that of organocadmiums to aldehydes or ketones.[3,4] Despite the conventional use of the alkylcadmiums in the ketone synthesis from acid halides, organocadmium compounds add rapidly and efficiently to simple carbonyl compounds, provided that *in situ* reagents prepared from Grignard reagents and cadmium halides are employed.[5,6] Pure dialkylcadmiums show almost no reactivity towards ketones. However, addition of magnesium halides greatly increases the reactivity. These phenomena have been interpreted in terms of a prior complexation of the carbonyl group with MgX_2, followed by an attack of R_2Cd on the resultant complex. Magnesium halides promote addition reactions far more effectively than zinc, cadmium, lithium or aluminum salts (equation 4).[7]

$$RCHO \xrightarrow{R'_2Cd \cdot MgX_2} RCH(OH)R' \quad (4)$$

R	R'	Yield
R = Me	R' = n-C_6H_{13}	60%
R = Me	R' = Ph	60%
R = Ph	R' = Me	55%
R = Ph	R' = Bu^n	65%

Organozinc reagents behave similarly. Reagents prepared *in situ* from Grignard reagents and $ZnBr_2$, or reconstituted reagents from distilled R_2Zn and $MgBr_2$ react smoothly with aldehydes and ketones (equation 5).[8,9] $TiCl_4$-promoted addition reaction of alkylzincs has also been developed. It is carried out by adding R_2Zn to a solution of the carbonyl component and $TiCl_4$. Another procedure, the addition of carbonyl compounds to a solution of R_2Zn and $TiCl_4$, is reported to be less effective (equation 6).[10]

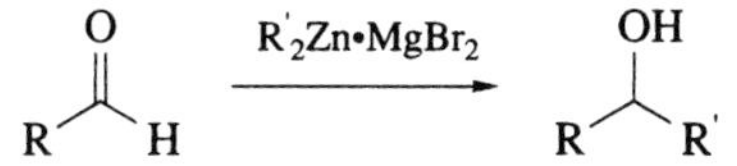

R = Et, R' = Et, 70%; R = Et, R' = Bu^n, 70%; R = Ph, R' = Et, 70%; R = Ph, R' = Bu^n, 50% (5)

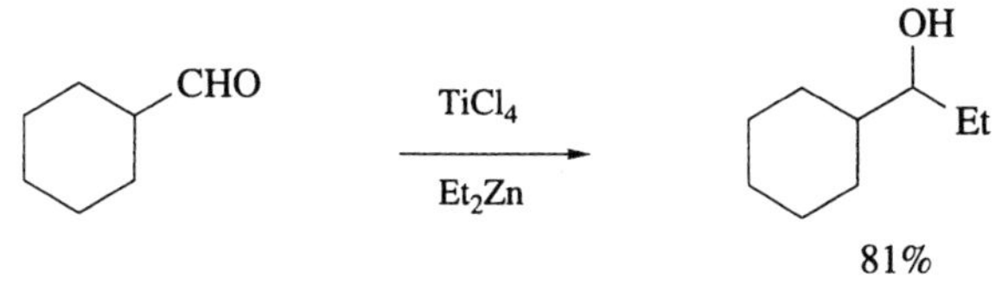

81% (6)

TMS-Cl or TMS-I, formed *in situ*, works as the promoter in addition reactions of zinc homoenolates (β-metallocarbonyl compounds), generated from 1-alkoxy-1-siloxycyclopropanes and ZnX_2 (equation 7). No reaction takes place with the purified zinc homoenolates. In contrast, titanium homoenolates are reactive enough to add to aldehydes in the absence of the Lewis acid promoter.[11] Related reactions of zinc esters with aldehydes in the presence of $(Pr^iO)_3TiCl$ have been reported (equation 8).[12]

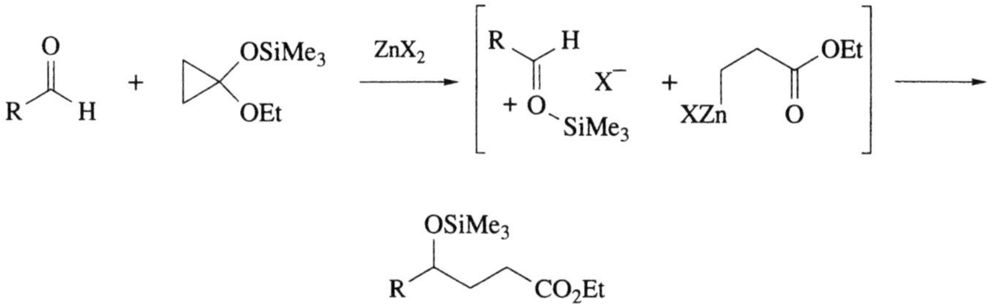

R = Ph, 89%; R = MeCH=CH, 72%; R = n-C_6H_{13}, 44% (7)

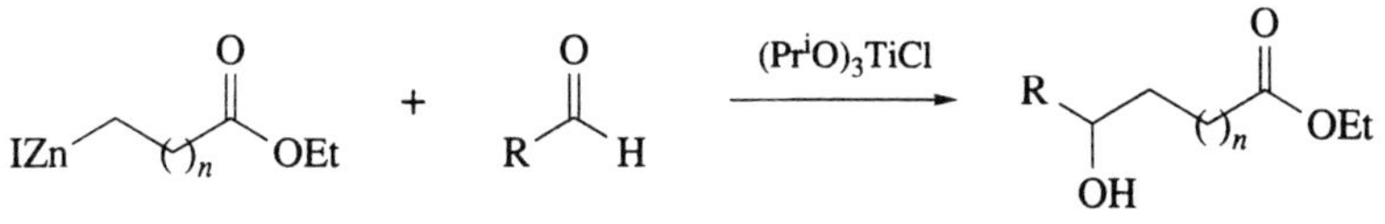

(8)

Organic synthesis utilizing the Group IVB organometals has been a growing field in the past decade.[13] Various organo-silanes or -stannanes with activated carbon–metal bonds can be used in addition reactions to ketones and aldehydes. Although certain organostannanes add to carbonyls in the absence of the Lewis acid, the presence of the promoters dramatically enhances the rate and allows the reaction to be conducted under mild conditions. In the presence of $TiCl_4$, $SnCl_4$, $AlCl_3$, $BF_3 \cdot OEt_2$, TMSOTf, Me_3O^+ BF_4^-, $TrClO_4$, *etc*., allylation of aldehydes with allyl-silanes or -stannanes proceeds smoothly, and homoallylic alcohols are obtained in high yields.[14,15] The Lewis acid promoted aldol reactions of silyl enol ethers are reviewed in Scheme 1.[16]

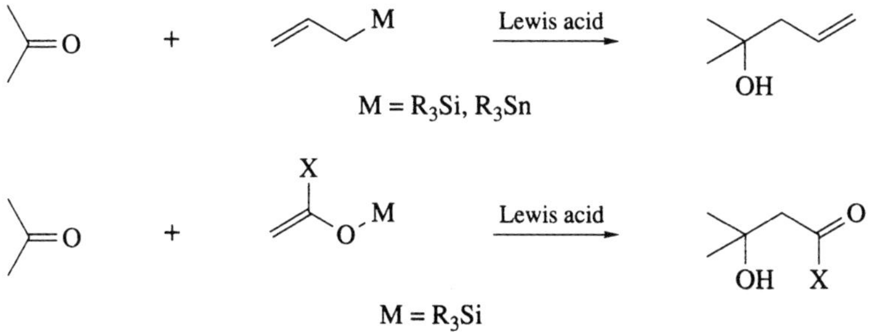

M = R_3Si, R_3Sn

M = R_3Si

Scheme 1

Organo-silanes and -stannanes possessing *sp* carbon–metal bonds also add to carbonyls in the presence of Lewis acids. Alkynylation with silyl- and stannyl-alkynes is promoted by $AlCl_3$ or $ZnCl_2$ (equations 9 and 10).[17–19] Notably, the reaction of 1,3-bis(trimethylsilyl)-1-propyne with chloral affords alkynic alcohol instead of allenylic alcohol, showing the preferential cleavage of *sp* C—Si bonds (equation 11).[17]

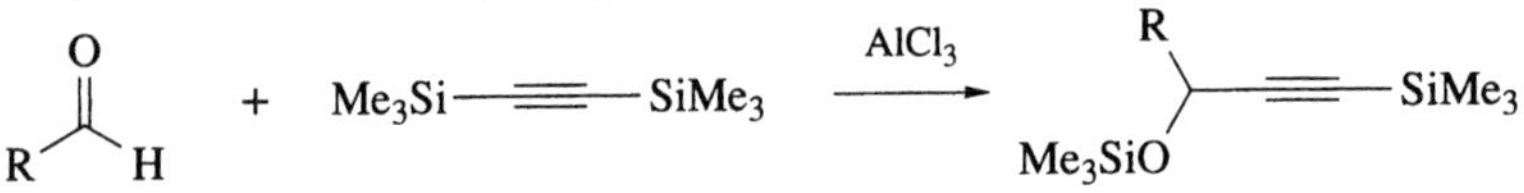

$$RCHO + Me_3Si{-}C{\equiv}C{-}SiMe_3 \xrightarrow{AlCl_3} RCH(OSiMe_3){-}C{\equiv}C{-}SiMe_3 \quad (9)$$

R = Et, 67%; R = Pr^i, 51%; R = n-C_8H_{17}, 46%

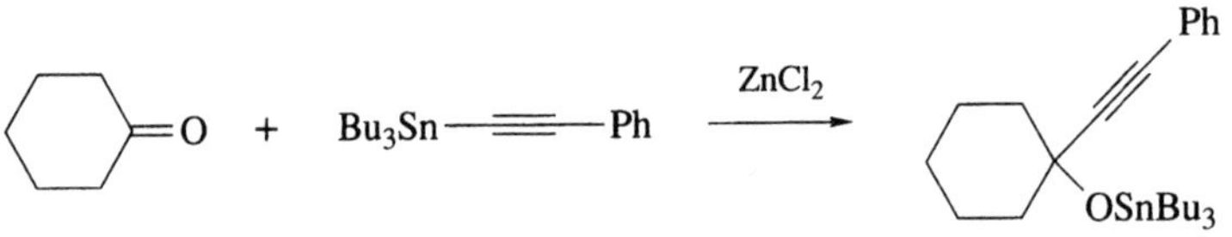

$$\text{cyclohexanone} + Bu_3Sn{-}C{\equiv}C{-}Ph \xrightarrow{ZnCl_2} \text{1-(phenylethynyl)-1-(OSnBu}_3\text{)cyclohexane} \quad (10)$$

$$Cl_3CCHO + Me_3Si{-}C{\equiv}C{-}CH_2SiMe_3 \xrightarrow[50\%]{AlCl_3} Cl_3CCH(OSiMe_3){-}C{\equiv}C{-}CH_2SiMe_3 \quad (11)$$

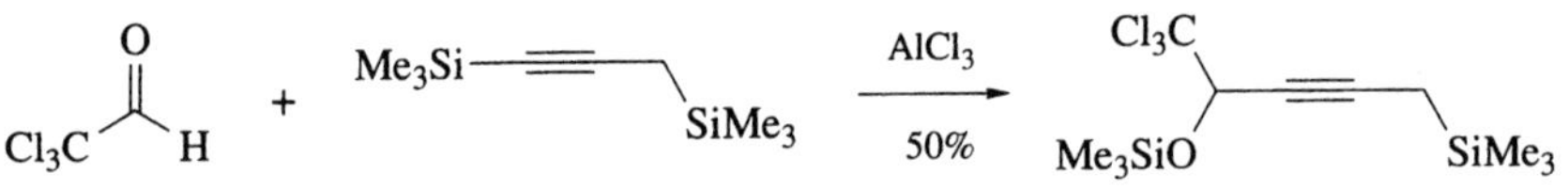

Trialkylsilyl cyanides, which also possess *sp* C—Si bonds, react with carbonyls.[20] ZnI_2,[21–24] $AlCl_3$,[25] TMSOTf,[26] $LnCl_3$ (Ln = La, Ce, Sm),[27] *etc.*, are employed as the promoter, and cyanohydrin silyl ethers are obtained in high yields even from hindered ketones (equation 12). The products are converted to various synthetically important intermediates such as cyanohydrins, α,β-unsaturated nitriles or amino alcohols.

$$R_2C{=}O + R_3SiCN \xrightarrow{\text{Lewis acid}} R_2C(OSiR_3)CN \quad (12)$$

Other organo-silanes and -stannanes are relatively inert, and addition reactions to aldehydes or ketones have been quite limited. However, there seems to be no reason why these organometals should not find applications in the future through the development of appropriate promoters. Actually, several examples appear in the literature. Aryl- or vinyl-silanes, which possess sp^2 C—Si bonds, add to chloral in the presence of $AlCl_3$ (Scheme 2).[17] The intramolecular addition reactions of alkylstannanes to ketones proceed with $TiCl_4$ (equations 13 and 14).[28,29]

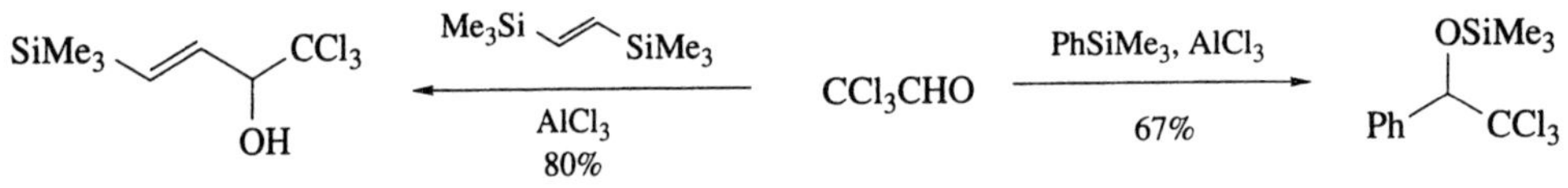

$$Me_3SiCH{=}CHCH(OH)CCl_3 \xleftarrow[AlCl_3,\ 80\%]{Me_3SiCH{=}CHSiMe_3} CCl_3CHO \xrightarrow[67\%]{PhSiMe_3,\ AlCl_3} PhCH(OSiMe_3)... $$

Correction of last arrow product: Ph–C(OSiMe$_3$)–CCl$_3$

Scheme 2

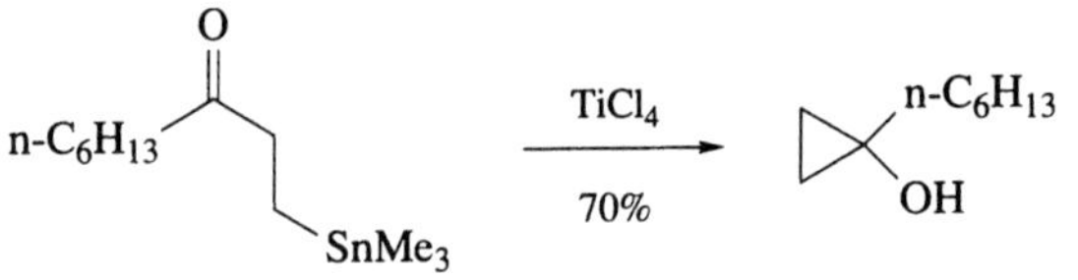

$$n\text{-}C_6H_{13}COCH_2CH_2CH_2SnMe_3 \xrightarrow[70\%]{TiCl_4} \text{1-}(n\text{-}C_6H_{13})\text{cyclopropanol} \quad (13)$$

$$\text{2-(4-trimethylstannylbutyl)cyclohex-2-enone} \xrightarrow{TiCl_4} [\text{OR, H bicyclic intermediate}] \xrightarrow{63\%} \text{chloro-octahydronaphthalene} \quad (14)$$

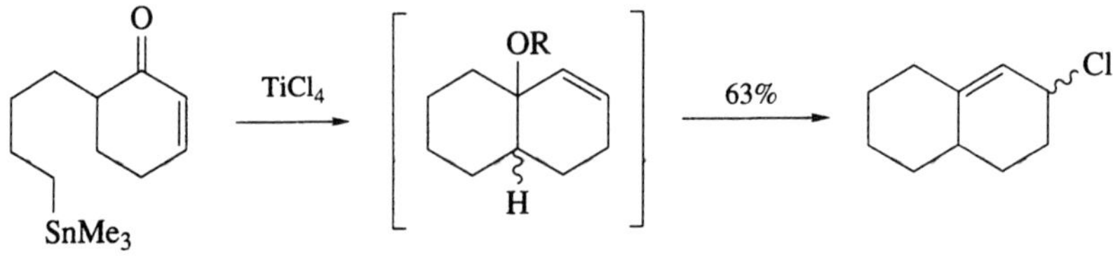

In contrast to the inertness of alkyl-silanes and -stannanes, alkylplumbanes in the presence of $TiCl_4$ add to aldehydes. The effectiveness of BF_3 as the promoter eliminates the possibility of organotitanium species occurring as intermediates. Addition of R_4Pb to a mixture of an aldehyde and $TiCl_4$ is important, as clean reaction does not occur when the order of addition is reversed, *i.e.* addition of the aldehyde to a mixture of R_4Pb and $TiCl_4$. As the reagent reacts only with aldehydes and not with ketones, octanal can be butylated selectively in the presence of 2-octanone (equation 15).[30]

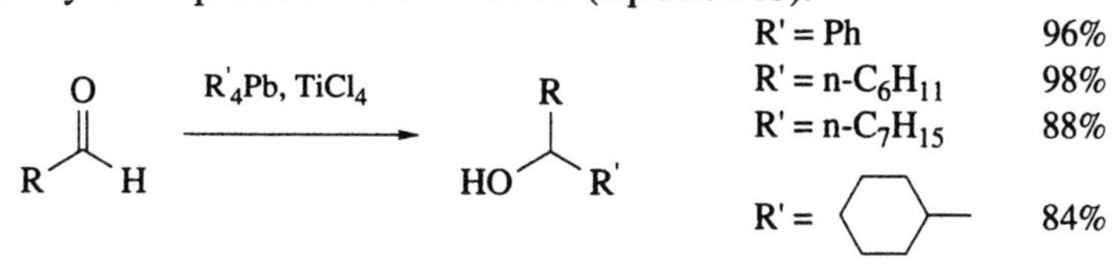

RCHO → (R'_4Pb, $TiCl_4$) → RCH(OH)R' (15)

R'	Yield
R' = Ph	96%
R' = n-C_6H_{11}	98%
R' = n-C_7H_{15}	88%
R' = cyclohexyl	84%

Alkylcoppers (RCu) were shown to add to aldehydes in the presence of $BF_3{\cdot}OEt_2$ (equation 16).[31,32]

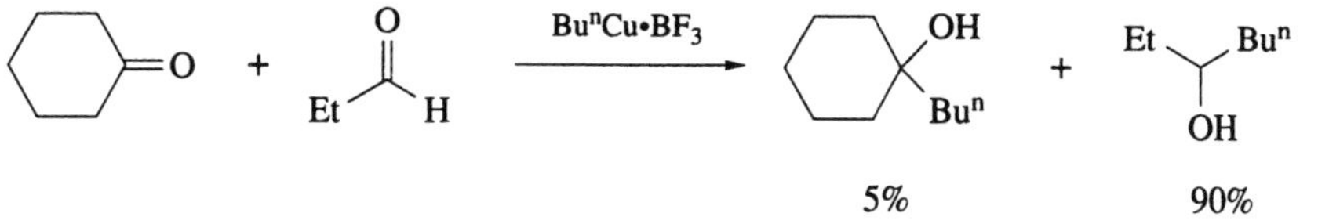

Cyclohexanone + EtCHO → ($Bu^nCu{\cdot}BF_3$) → 1-Bu^n-cyclohexanol (5%) + Et–CH(OH)–Bu^n (90%) (16)

Organolithiums are widely used nucleophiles in organic synthesis, and normally add to aldehydes and ketones smoothly. However, they sometimes fail to react for such reasons as the presence of sensitive functionalities, the deprotonation side reactions, steric hinderance, *etc.* One of the popular modifications of the alkyllithium reactions is the addition of magnesium salts. The coupling reaction of lithiated sulfones with enolizable aldehydes or ketones gives a low yield of β-hydroxy sulfones because of the deprotonation. The process is greatly improved by the presence of $MgBr_2$ (Scheme 3).[33,34] The reaction is successfully employed in the total synthesis of zincophorin (equation 17).[35] A vinyllithium–$MgBr_2$ complex is used instead of vinyllithium in the synthesis of (–)-vertinolide (equation 18).[36]

Lithium salts are also effective for promoting the organolithium or magnesium addition reactions. $LiClO_4$ was found to enhance the rate of MeLi or Me_2Mg addition to 4-*t*-butylcyclohexanone.[37] Another example is shown in the reaction of 2-acetylpyridine with alkyllithiums, where added LiBr raises the yields of the adducts.[38] This phenomenon is explained by the coordination of LiX to the carbonyl oxygen prior to the C—C bond formation.

Recently, BF_3 has also been found to be an effective promoter of organolithium reactions. At low temperatures, various organolithiums are able to coexist with the Lewis acid in solution without provoking transmetalation, which would give unreactive alkylboranes. Addition of lithiated sulfones to the enolizable aldehydes is carried out in the presence of $BF_3{\cdot}OEt_2$ at –78 °C (Scheme 4).[39,40] $BF_3{\cdot}OEt_2$ promotes the reaction of the sterically demanding vinyllithium reagents with aldehydes (equation 19).[41] $RLi{\cdot}BF_3$ reagents are finding wide use in organic synthesis, and further examples are shown in the following sections.

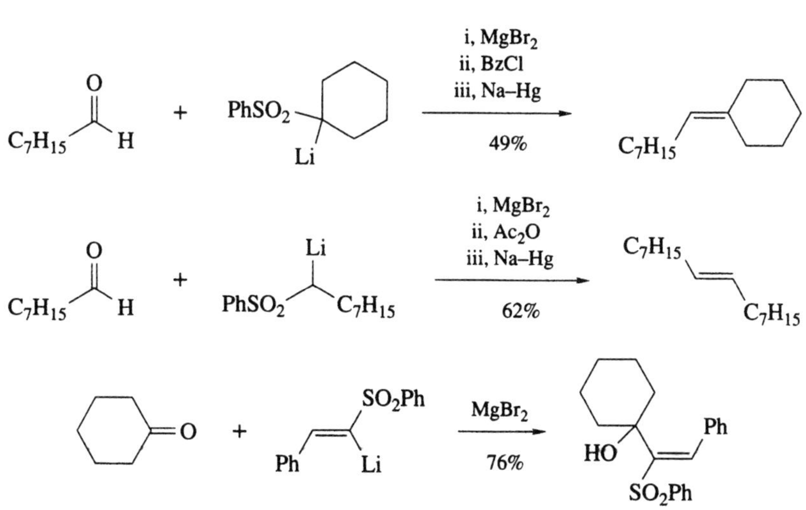

$C_7H_{15}CHO$ + 1-lithio-1-($PhSO_2$)cyclohexane → (i, $MgBr_2$; ii, BzCl; iii, Na–Hg; 49%) → $C_7H_{15}CH$=cyclohexylidene

$C_7H_{15}CHO$ + $PhSO_2CH(Li)C_7H_{15}$ → (i, $MgBr_2$; ii, Ac_2O; iii, Na–Hg; 62%) → $C_7H_{15}CH=CHC_7H_{15}$

Cyclohexanone + PhCH=C(Li)SO_2Ph → ($MgBr_2$; 76%) → 1-[PhCH=C(SO_2Ph)]-cyclohexanol

Scheme 3

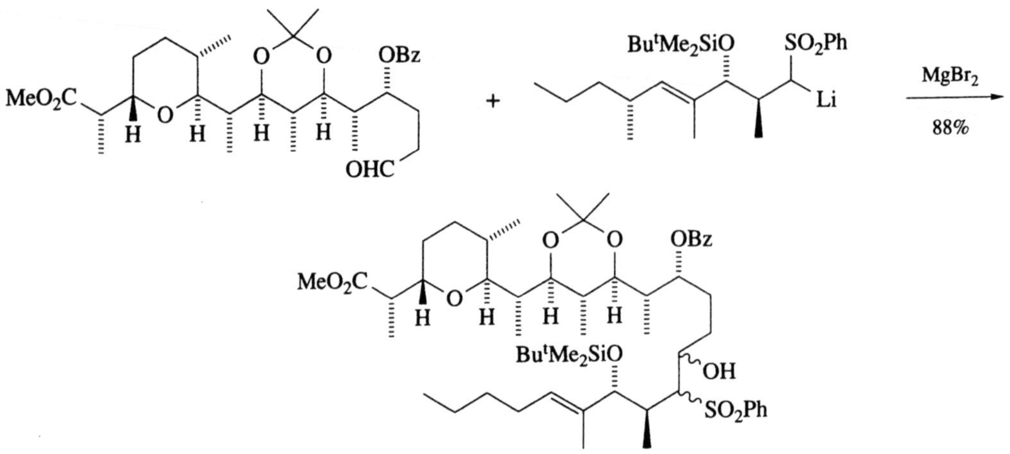

MeO$_2$C, O, O, O, OBz, H, H, H, H, OHC + Bu^{t}Me$_2$SiO, SO$_2$Ph, Li $\xrightarrow[88\%]{MgBr_2}$

MeO$_2$C, O, O, O, OBz, H, H, H, H, Bu^{t}Me$_2$SiO, OH, SO$_2$Ph (17)

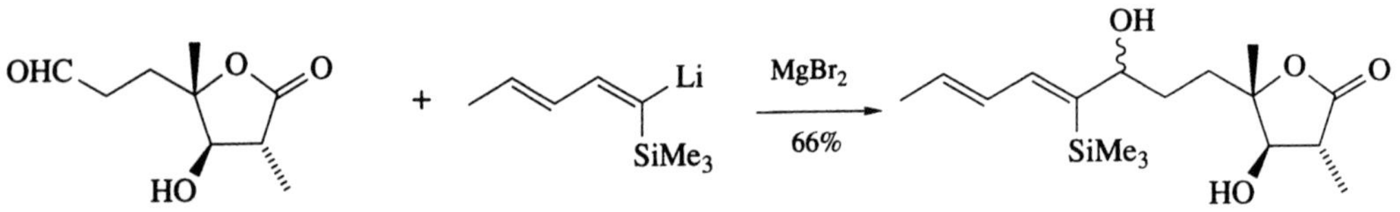

OHC, O, O, HO + Li, SiMe$_3$ $\xrightarrow[66\%]{MgBr_2}$ OH, SiMe$_3$, O, O, HO (18)

The complex reagents of organo-lithiums and -magnesiums with transition metal halides ($TiCl_4$, $CrCl_3$, UCl_4, MnI_2, $CeCl_3$, VCl_3, *etc.*) have been developed. Although the actual reacting species or the effect of the Lewis acids is not fully clear, they provide unique and useful methodologies in organic synthesis.

The chemistry of organotitanium reagents has been explored by Reetz[42–44], Weidmann and Seebach.[45] The MeLi–$TiCl_4$ system provides a nonbasic reagent which reacts chemo- and stereo-selectively. Nitro,

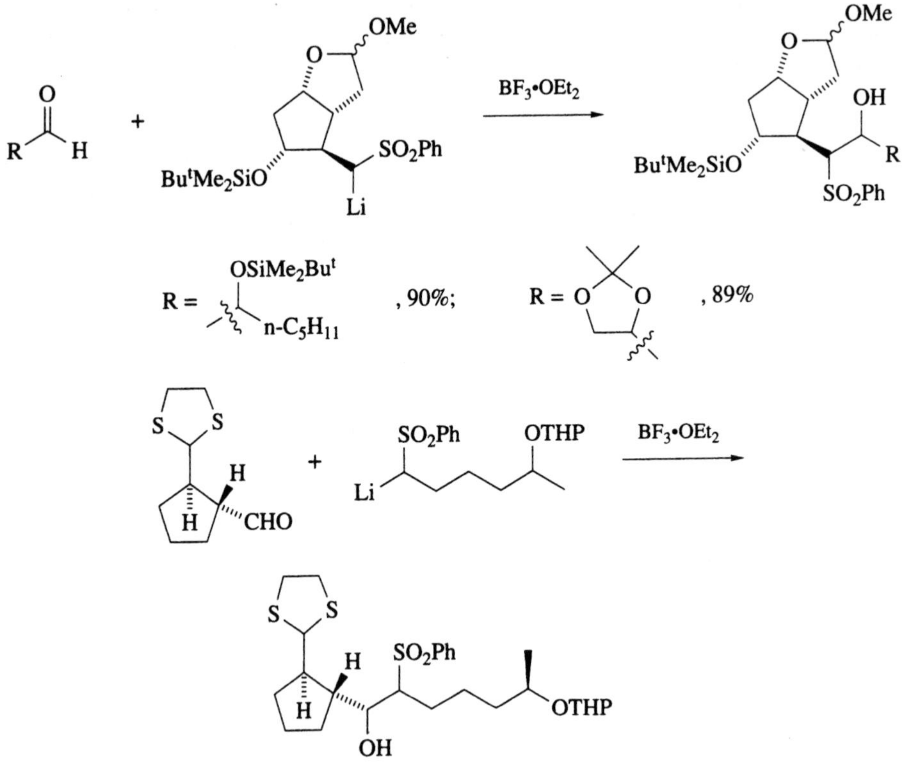

R, O, H + O, OMe, Bu^{t}Me$_2$SiO, SO$_2$Ph, Li $\xrightarrow{BF_3 \cdot OEt_2}$ O, OMe, OH, Bu^{t}Me$_2$SiO, R, SO$_2$Ph

R = OSiMe$_2$Bu^t, n-C$_5$H$_{11}$, 90%; R = O, O , 89%

S, S, H, H, CHO + SO$_2$Ph, OTHP, Li $\xrightarrow{BF_3 \cdot OEt_2}$

S, S, H, SO$_2$Ph, H, OTHP, OH

Scheme 4

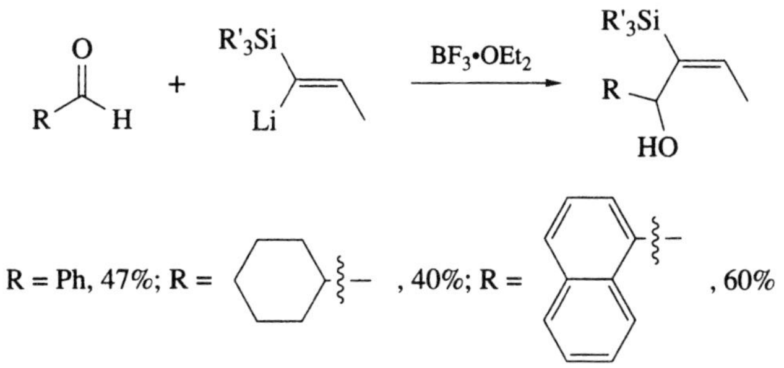

(19)

R = Ph, 47%; R = cyclohexyl, 40%; R = 1-naphthyl, 60%

cyano or ester groups do not interfere with the carbonyl addition. The reaction proceeds smoothly with enolizable ketones without proton abstraction (Scheme 5).[46]

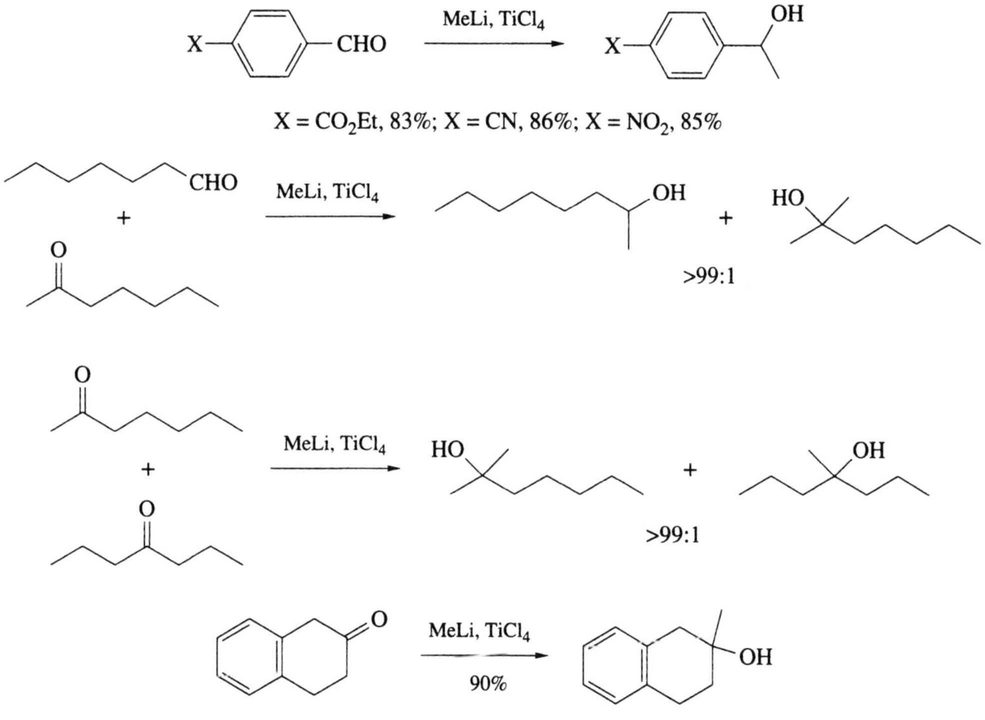

MeLi, $TiCl_4$

X = CO_2Et, 83%; X = CN, 86%; X = NO_2, 85%

MeLi, $TiCl_4$

>99:1

MeLi, $TiCl_4$

>99:1

MeLi, $TiCl_4$

90%

Scheme 5

Organochromium reagents generated from alkyl-lithiums or -magnesiums and $CrCl_3$ react with aldehydes selectively in the presence of ketones (equation 20).[47,48] RLi–UCl_4 reagents, developed recently, are also aldehyde selective.[49]

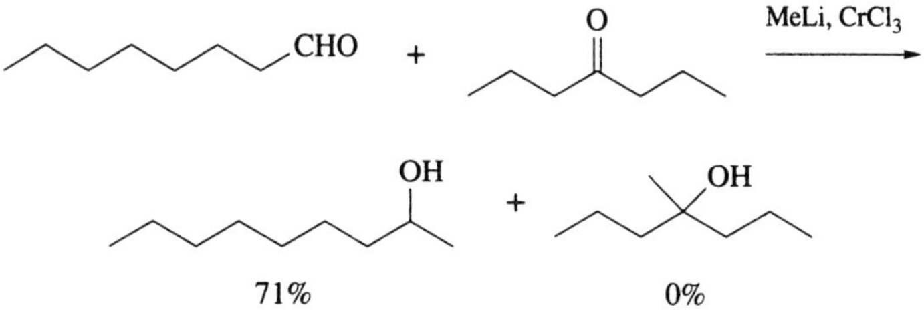

MeLi, $CrCl_3$

71% 0%

(20)

A similar chemoselectivity is observed with alkylmanganese compounds prepared from alkyl-lithiums or -magnesiums and MnI_2. The organometallic compounds react with aldehydes at –50 °C, while ketone addition occurs only at higher temperatures (Scheme 6).[50,51]

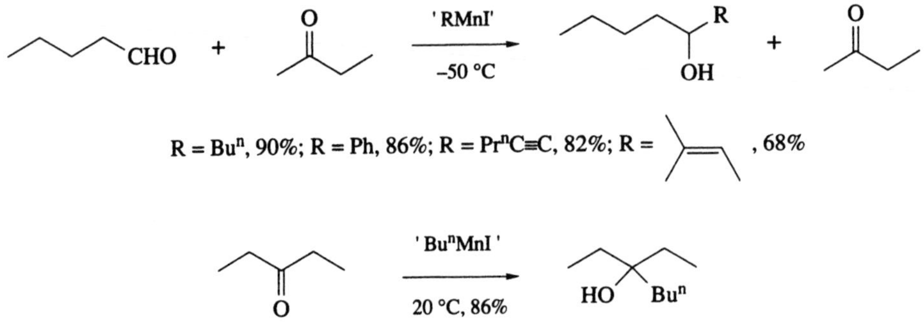
CHO
O
'RMnI'
-50 °C
R
OH
O
R = Bu^n, 90%; R = Ph, 86%; R = Pr^nC≡C, 82%; R =
, 68%
'Bu^nMnI'
20 °C, 86%
O
HO
Bu^n

Scheme 6

RLi–$CeCl_3$ is an effective alkylating reagent for easily enolizable ketones, which give very low yields of the adducts with alkyl-lithiums or -magnesiums.[52,53] Selective 1,2-addition reactions with α,β-unsaturated ketones can be conducted with the cerium reagents (Scheme 7).[54]

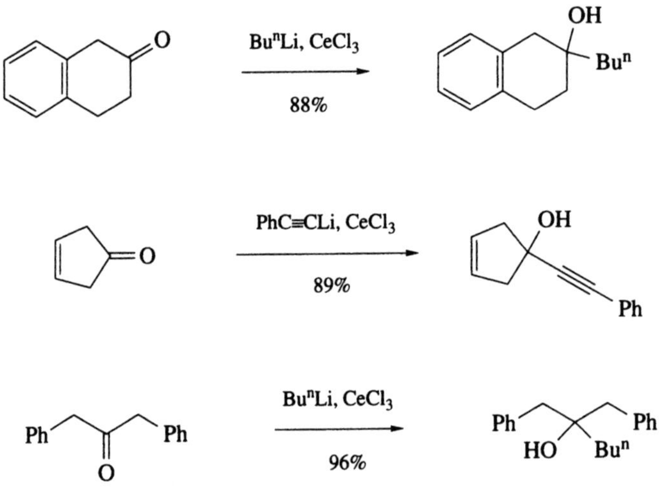
O
Bu^nLi, CeCl3
88%
OH
Bu^n
O
PhC≡CLi, CeCl3
89%
OH
Ph
Ph
Ph
O
Bu^nLi, CeCl3
96%
Ph
Ph
HO
Bu^n

Scheme 7

The reactions of RLi–VCl_3 or RMgX–VCl_3 with aldehydes result in oxidative carbon–carbon bond formation, affording ketones. Since a lithium alkoxide is converted to a ketone in the presence of VCl_3, the vanadium(III) species is considered to work as an oxidant (Scheme 8).[55,56]

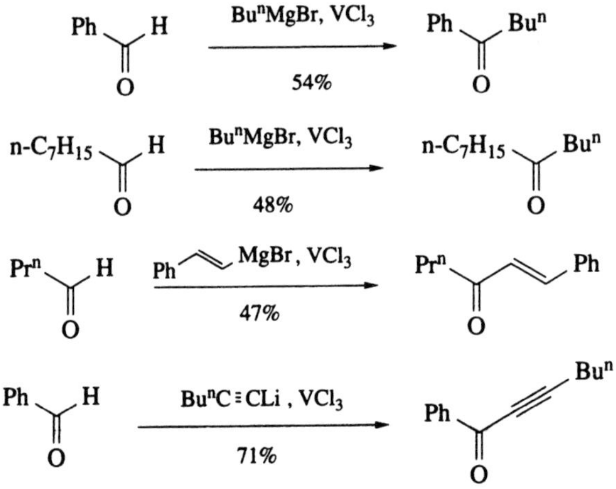
Ph
H
O
Bu^nMgBr, VCl3
54%
Ph
Bu^n
O
n-C7H15
H
O
Bu^nMgBr, VCl3
48%
n-C7H15
Bu^n
O
Pr^n
H
O
Ph
MgBr, VCl3
47%
Pr^n
Ph
O
Ph
H
O
Bu^nC≡CLi, VCl3
71%
Bu^n
Ph
O

Scheme 8

1.11.2.2 Control of the Stereoselectivity

The stereochemistry of organometal additions to cyclic ketones has been extensively studied using various alkyl metals. The results are used to gain knowledge concerning the solution state of the particular organometals and to explore the driving force behind the steric course of the addition.

Addition to cyclohexanones is considered to be influenced by two factors: (i) the steric interaction of the incoming groups with 3,5-axial substituents; and (ii) the torsional strain of the incoming groups with 2,6-axial substituents. Steric strain hinders axial attack, whereas torsional strain hinders equatorial attack. The actual stereochemistry of the addition depends upon which factor is greater in a particular case.[1] The production of the desired isomer in high stereoselectivity is required from the synthetic point of view.

In a large number of studies on the addition reactions to 4-*t*-butylcyclohexanone, acceptably high selectivities have been attained with only a limited number of methods. Selective equatorial attack is obtained with bulky nucleophiles, since steric interactions of the incoming bulky reagents with 3,5-axial hydrogens outweigh torsional effects. For example, Bu^tMgBr gives the axial alcohol exclusively.[1] Similarly, $MeTi(OPr^i)_3$, which possesses bulky ligands, is superior to MeLi for equatorial attack.[57]

The use of organometal–Lewis acid complex reagents is quite effective in controlling the axial–equatorial problem, since the reaction course can be modified by the Lewis acid complexation to carbonyl oxygen. Although MeLi showed a low selectivity in the addition to 4-*t*-butylcyclohexanone, the presence of an appropriate Lewis acid dramatically enhances either equatorial or axial attack. The MeLi–Me_2CuLi reagent shows a high tendency to deliver a methyl moiety from an equatorial site.[58] A high equatorial selectivity as well as a considerable rate enhancement is also observed with the MeLi–$LiClO_4$ reagent, and the results are interpreted as the complexation of Li^+ to carbonyl followed by the addition of MeLi (Figure 1).[37,59]

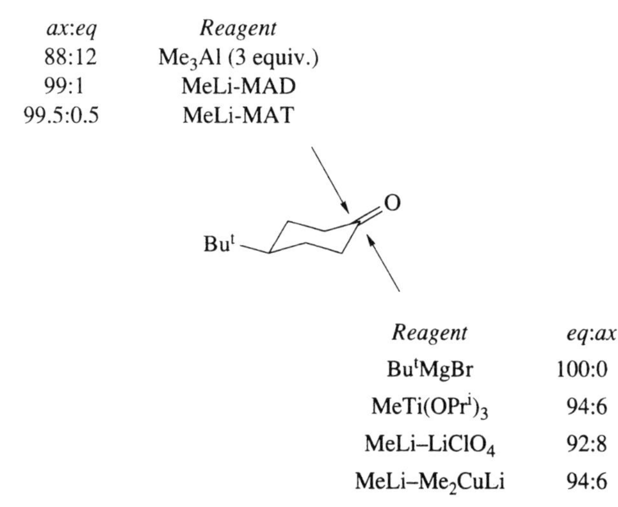

ax:eq Reagent
88:12 Me3Al (3 equiv.)
99:1 MeLi-MAD
99.5:0.5 MeLi-MAT
O
Bu^t
Reagent eq:ax
Bu^tMgBr 100:0
MeTi(OPr^i)3 94:6
MeLi–LiClO4 92:8
MeLi–Me2CuLi 94:6

Figure 1

Axial attack has been rather difficult to attain. The treatment of the ketone with 3 equiv. of Me_3Al has been known to give equatorial alcohol predominantly, although low selectivity was obtained using 1 equiv. of reagent. Methylation of the Me_3Al-complexed carbonyl was suggested. This methodology, however, was not applicable to ethyl- or butyl-aluminum because of the competitive carbonyl reduction.[1]

A selective axial attack is achieved by employing bulky organoaluminum ligands; methylaluminum bis(2,6-di-*t*-butyl-4-methylphenoxide) (MAD) or methylaluminum bis(2,4,6-tri-*t*-butylphenoxide) (MAT). Treatment of carbonyl compounds with the aluminum reagents prior to alkyllithium or Grignard addition gives equatorial alcohols in high selectivities. Various alkyl groups can be introduced by this reaction (Figure 1). Since addition of ketones to a mixture of MeLi and MAD resulted in low selectivity, the possibility of the ate complex Li^+ $[Me{\cdot}MAD]^-$ as the reactive intermediate was excluded. Preferential formation of the sterically favored isomer (**1b**) rather than the alternative (**1a**) was suggested for the transition state (Figure 2).[60,61]

Similar tendencies are observed with other cyclic ketones. With 2- or 3-methylcyclohexanone, MeLi–LiX (X = $CuMe_2$, ClO_4) or $MeTi(OPr^i)_3$ favors equatorial attack, whereas Me_3Al (3 equiv.), MeLi–MAT, or MeLi–MAD attacks from the axial site (Figure 3).[58–61]

In relation to the synthesis of chain molecules with a series of asymmetric centers, macrolides or polyether antibiotics, asymmetric induction in the addition reactions of organometals to chiral aldehydes

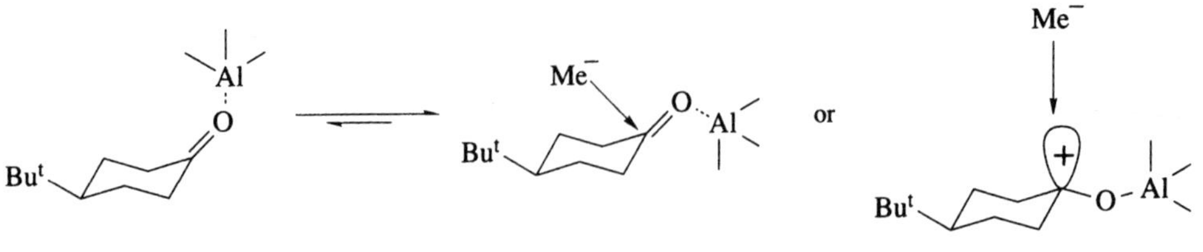

(1a) or **(1b)**

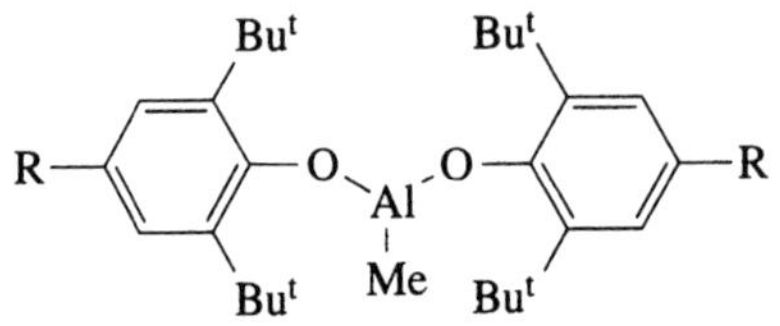

MAD: R = Me

MAT: R = Bu^t

Figure 2

MeLi–MAD *ax.:eq.* 93:7

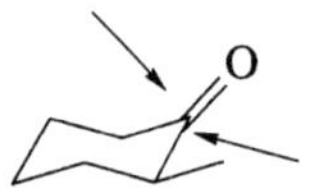

MeLi–MAT *ax.:eq.* 97:3

	eq.:ax.
MeLi	92:8
$MeTi(OPr^i)_3$	96:4
MeLi–$LiClO_4$	96:4
MeLi–Me_2CuLi	97:3

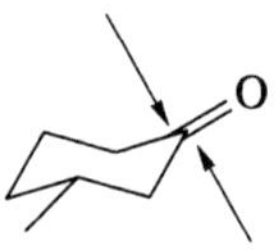

	eq.:ax.
MeLi	83:17
$MeTi(OPr^i)_3$	89:11

Figure 3

has been extensively studied.[42,62] In the following part, Cram/anti-Cram selectivity, and the chelation/nonchelation problem is discussed.

Addition to chiral aldehydes which have no additional functional group capable of interacting with metal species is governed by electronic and/or steric factors. Of several mechanistic rationalizations provided since Cram, the Felkin model best agrees with the prediction based on *ab initio* calculations. In this model, M and S represent medium and small groups, respectively, attached to the chiral α-carbon, and L represents either the largest group or the group whose bond to the α-carbon provides the greatest $\sigma^*-\pi^*$ overlap with the carbonyl π^* orbital. A nucleophile approaches from the opposite side of the L group (Figure 4). Although the selectivity of this type of reaction had not previously been very high, the Lewis acid promoted addition methodologies succeeded in attaining high Cram selectivities, or even anti-Cram selectivities (equations 21 and 22).

MeLi–$TiCl_4$[46] or $PbEt_4$–$TiCl_4$[30] reagents show high Cram selectivities in addition reactions to 2-phenylpropanal. An explanation was presented by Heathcock for the selectivities achieved by the Lewis acid promoted additions compared to simple organometal additions. In the latter cases, a trajectory is followed that brings the nucleophiles closer to H rather than to R^*, and asymmetry in R^* is transferred

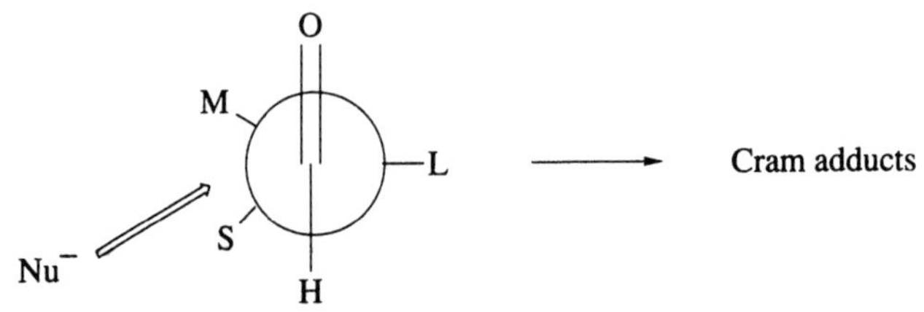

Cram adducts

Figure 4

(21)

$MeTi(OPr^i)_3$	93:7
$MeLi–TiCl_4$	90:10
$PbEt_4–TiCl_4$	93:7
MeLi–K[2.2.1]	9:1
$Bu_2CuCNLi_2$–15-crown-5–BF_3	8–10:1

(22)

MeMgI–MAT	93: 7
EtMgBr–MAT	87:13
Bu_2CuLi–K[2.1]	5:1

imperfectly. When a Lewis acid coordinates with an aldehyde occupying the position *syn* to H,[63] the nucleophile may be forced to approach the carbonyl plane in a perpendicular fashion, resulting in greater stereoselectivity (Figure 5).[64]

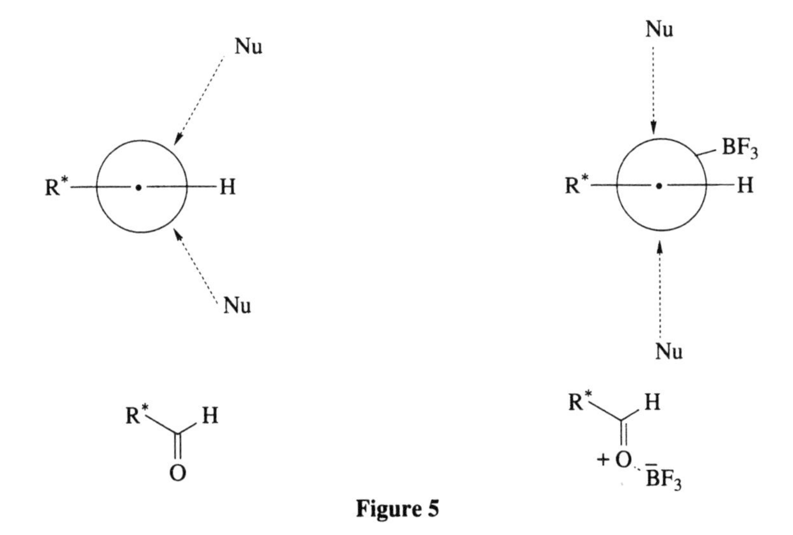

Figure 5

The presence of crown ethers was found to enhance the Cram selectivity. The MeLi/18-crown-6 reagent system shows a similar level of selectivity to the above reagents, and BuLi/15-crown-5 and allyllithium/18-crown-6 add to 2-phenylpropanal in Cram fashion almost exclusively.[65] Since a bulky reagent $MeTi(OPr^i)_3$ also exibits a high Cram selectivity,[57] the bulkiness of the RLi–crown ether reagent might be one of the important factors which control the selectivity. Interestingly, $R_2CuCNLi_2$–crown ether–BF_3 reagent also shows Cram selectivity.[66]

Stannylalkynes in the presence of $TiCl_4$ react with a steroidal aldehyde in a highly Cram selective manner. The use of lithium derivatives gives 1:1 mixtures of two diastereomers (equation 23).[67]

Recently it has been reported that the addition of Grignard reagents to MAT or MAD complexed aldehydes results in anti-Cram products in high stereoselectivities. Formation of a sterically least hindered complex and subsequent attack of the organometal from the opposite site to the bulky ligand is suggested by Yamamoto (Figure 6).[60,61] The R_2CuLi–crown ether system is also reported to show anti-Cram selectivities. An electron transfer mechanism has been suggested for this reaction.[65]

Although the stereochemistry of the organometal addition reactions to carbonyls with α-alkyl (typically α-methyl) substituents is explained mostly by the Felkin–Ahn model, different phenomena are observed with α-alkoxy- or α-hydroxy-carbonyls, which make the opposite π-face sterically more

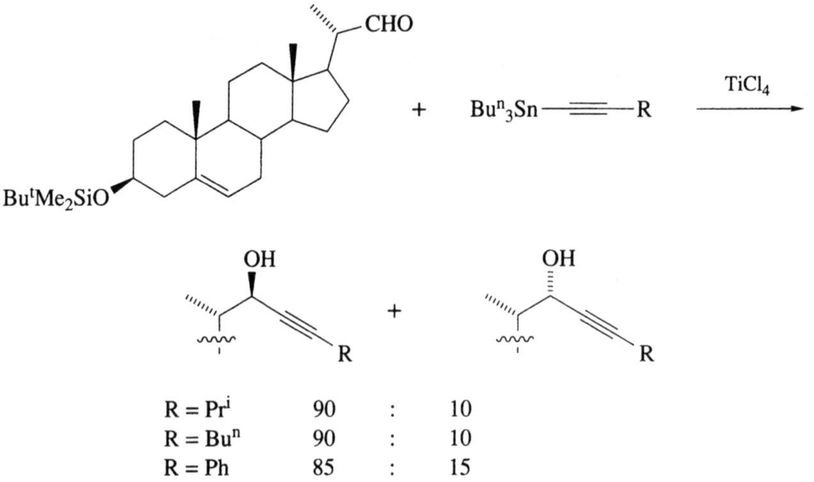

(23)

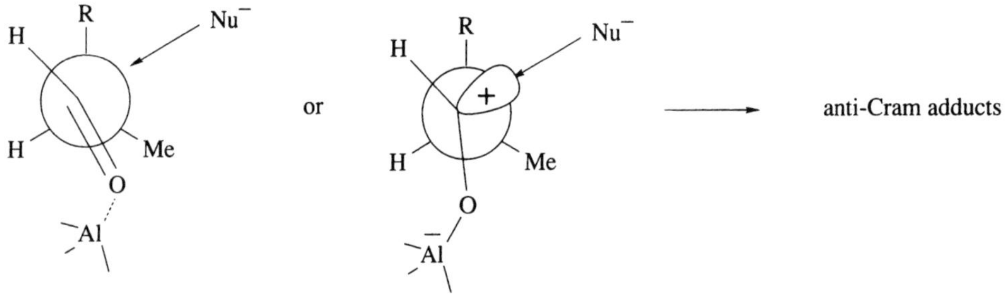

R = Pr[i]	90	:	10
R = Bu[n]	90	:	10
R = Ph	85	:	15

anti-Cram adducts

Figure 6

accessible. In such cases, Cram's cyclic model pertains, which assumes a chelation of the metal species between the carbonyl oxygen and the α-oxygen atom (chelation control).

Systematic studies on the chelation-controlled additions were carried out, varying the type of alkoxy group, the carbon nucleophile, the solvent and the temperature. It was found that α-alkoxy ketones react highly stereoselectively with Grignard reagents in THF (equation 24). Alkyllithiums were not effective.[68,69] The generalization was made use of in the synthesis of the polyether antibiotic monensin.[70]

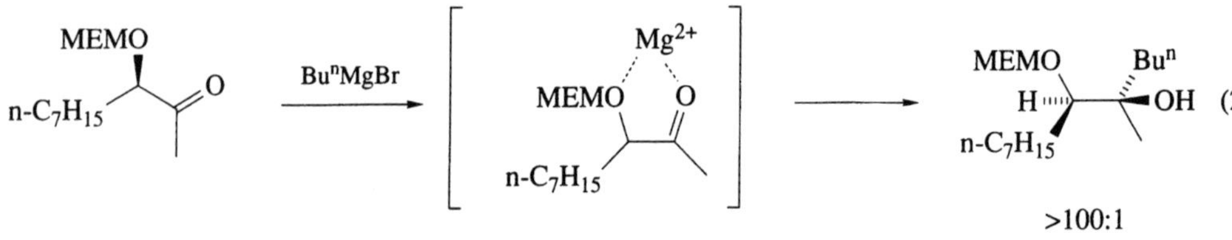

(24)

>100:1

The high levels of 1,2-asymmetric induction observed with ketones did not extend to aldehydes. However, Asami and coworkers achieved it by pretreating α-alkoxy aldehydes with $ZnBr_2$.[71,72] The reaction is used in the synthesis of *exo*-(+)-brevicomin[71] and L-rhodinose (Scheme 9).[73] Chelation-controlled addition of a Lewis acidic reagent $MeTiCl_3$ is reported,[74] and a chelation intermediate is actually detected by low temperature NMR techniques (equation 25).[75] A 'tied-up' method, which involves the precomplexation of α-alkoxy aldehydes with a Lewis acid and addition of soft C-nucleophiles, shows a high level of asymmetric induction. $SnCl_4$ and $TiCl_4$, capable of forming six-coordinate octahedral complexes, are well suited. R_2Zn, TMS-CN, allylsilanes or silyl enol ethers are employed as the nucleophiles (equation 26).[76,77]

In order to obtain the other isomer in addition reactions to α-alkoxy aldehydes, reagents incapable of chelation must be used, and electronic and/or steric factors relied upon (nonchelation control). Treatment of α-alkoxy aldehydes with an excess of gaseous BF_3 followed by silyl enol ethers results in the nonchelation-controlled adducts. Formation of a rigid conformation due to electronic repulsion is expected (equation 27). $BF_3{\cdot}OEt_2$ is reported to be considerably less efficient.[78,79] The use of $RTi(OPr^i)_3$, which

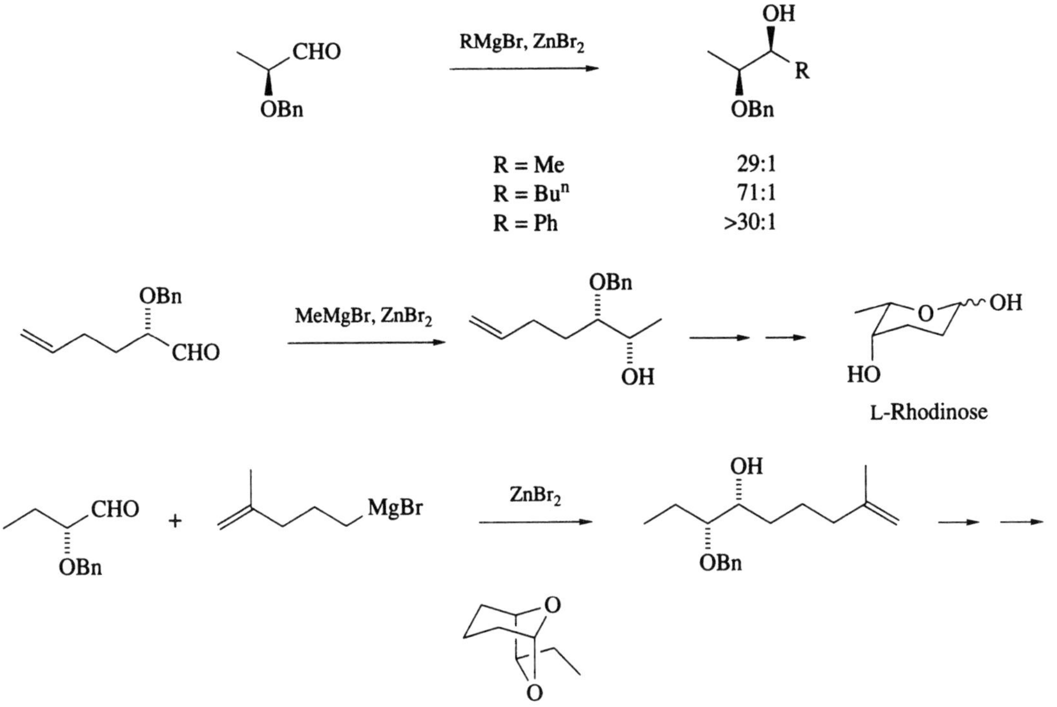

RMgBr, $ZnBr_2$

R = Me	29:1
R = Bu^n	71:1
R = Ph	>30:1

MeMgBr, $ZnBr_2$

L-Rhodinose

$ZnBr_2$

exo-(+)-Brevicomin

Scheme 9

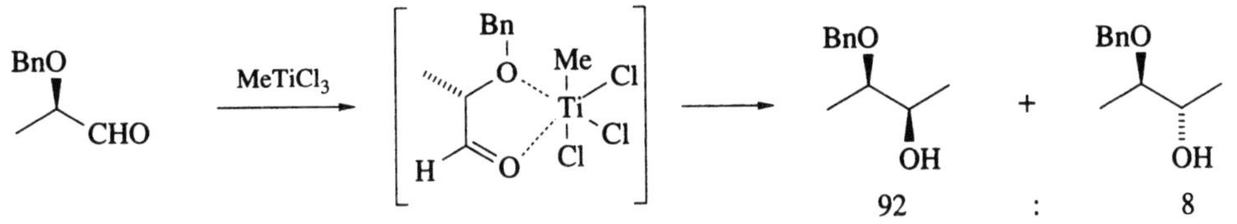

$MeTiCl_3$ (25)

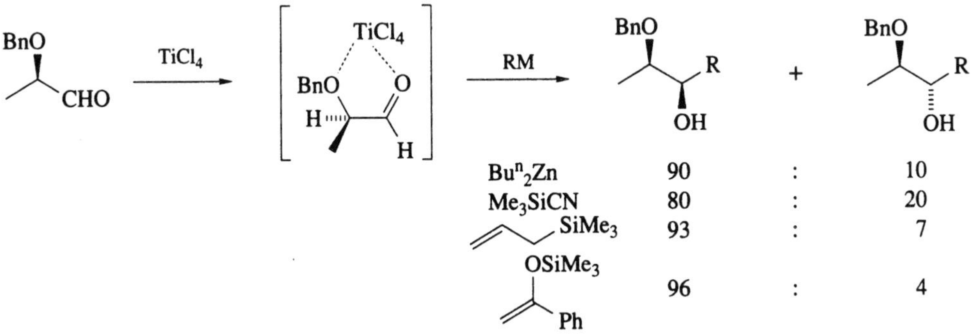

92 : 8

$TiCl_4$ RM (26)

RM		
$Bu^n{}_2Zn$	90 :	10
Me_3SiCN	80 :	20
allyl-$SiMe_3$	93 :	7
CH_2=C($OSiMe_3$)Ph	96 :	4

are weakly Lewis acidic and incapable of chelation, provides another method for nonchelation-controlled addition (equation 28).[57]

Novel nonchelation phenomena are observed with a steroidal α-hydroxy aldehyde. The reaction of a lithium or magnesium alkynide with the aldehyde gives the (20*R*,22*R*)-diastereomer predominantly, the formation of which was explained by Cram's cyclic model. When $BF_3{\cdot}OEt_2$ is added to the lithium alkynide prior to the addition of the aldehyde, the stereoselectivity is inverted, and the (20*R*,22*S*)-isomer is obtained as the principal product. Transformation of α-alkoxy aldehyde to the boron 'ate' complex is suggested. Other Lewis acids, such as $B(OMe)_3$, $AlCl_3$, *etc*., are less effective (equation 29).[80]

Organometal addition reactions to α,β-dihydroxy aldehyde derivatives have been extensively studied in relation to the stereoselective synthesis of sugar derivatives. Protecting groups play an important role in the chelation/nonchelation problem. With a benzyl protecting group at the α-oxygen, similar behavior to simple α-alkoxy aldehydes is observed. $TiCl_4$- or $SnCl_4$-promoted reactions of organo-silanes, -plum-

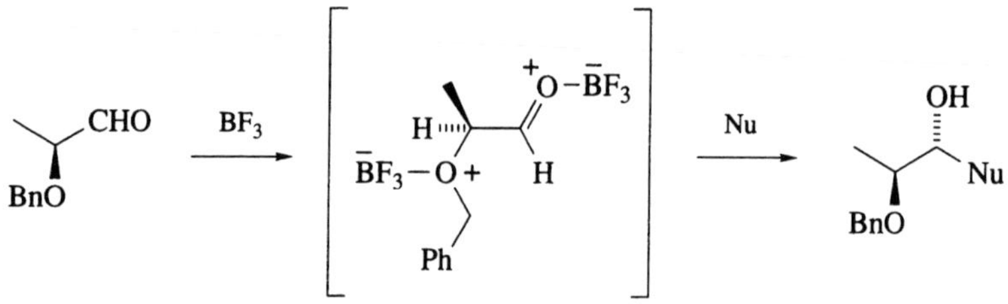

CHO BnO $\xrightarrow{BF_3}$ [H, $\bar{B}F_3$–O+, Ph, ^+O–$\bar{B}F_3$, H] $\xrightarrow{Nu}$ OH Nu BnO (27)

Nu = CH_2COBu^t, 90:10; Nu = CMe_2CO_2Me, 84:16

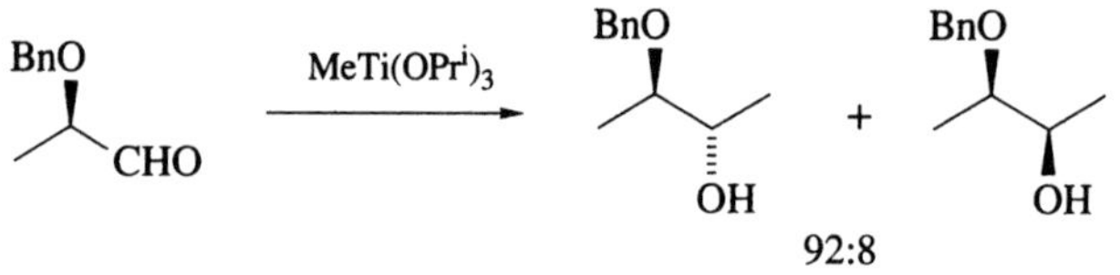

BnO CHO $\xrightarrow{MeTi(OPr^i)_3}$ BnO OH + BnO OH (28)

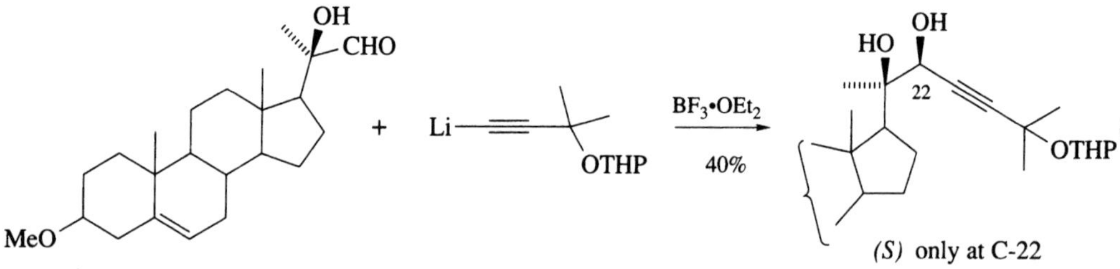

92:8

MeO OH CHO + Li OTHP $\xrightarrow[40\%]{BF_3•OEt_2}$ HO OH 22 OTHP (29)

(*S*) only at C-22

banes or -zincs give chelation-controlled addition products, while BF_3 reverses the selectivity. Bulky $RTi(OPr^i)_3$ also shows nonchelation selectivities (Scheme 10).[81–83] $RCu–MgBr_2$ reagents are reported to be superior to RLi or RMgX in conducting chelation-controlled addition to an α,β-dibenzyloxy aldehyde (equation 30).[83]

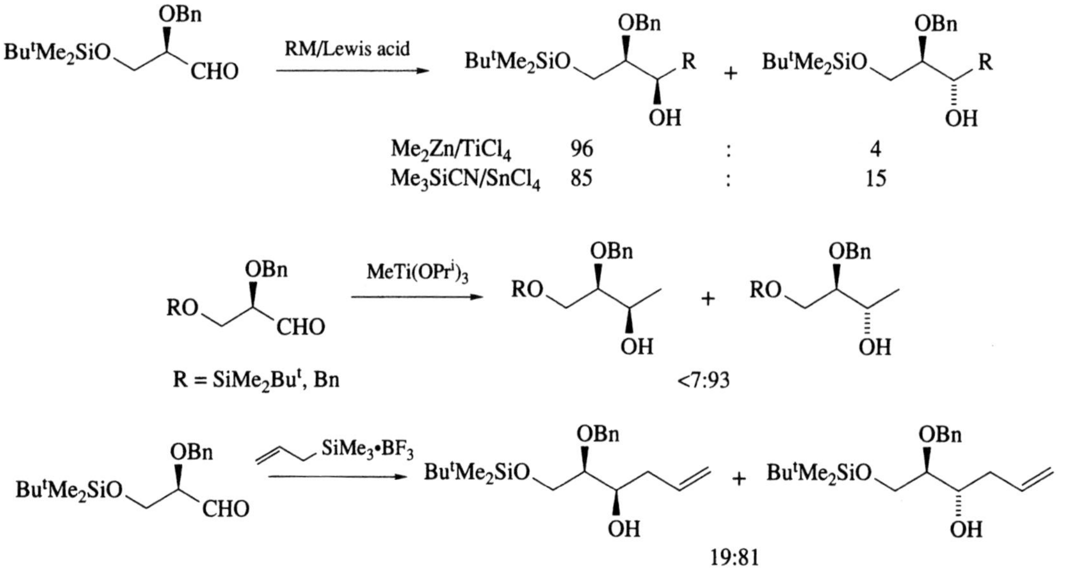

Bu^tMe_2SiO OBn CHO $\xrightarrow{RM/Lewis\ acid}$ Bu^tMe_2SiO OBn R OH + Bu^tMe_2SiO OBn R OH

$Me_2Zn/TiCl_4$	96	:	4
$Me_3SiCN/SnCl_4$	85	:	15

RO OBn CHO $\xrightarrow{MeTi(OPr^i)_3}$ RO OBn OH + RO OBn OH

R = $SiMe_2Bu^t$, Bn

<7:93

Bu^tMe_2SiO OBn CHO $\xrightarrow{SiMe_3•BF_3}$ Bu^tMe_2SiO OBn OH + Bu^tMe_2SiO OBn OH

19:81

Scheme 10

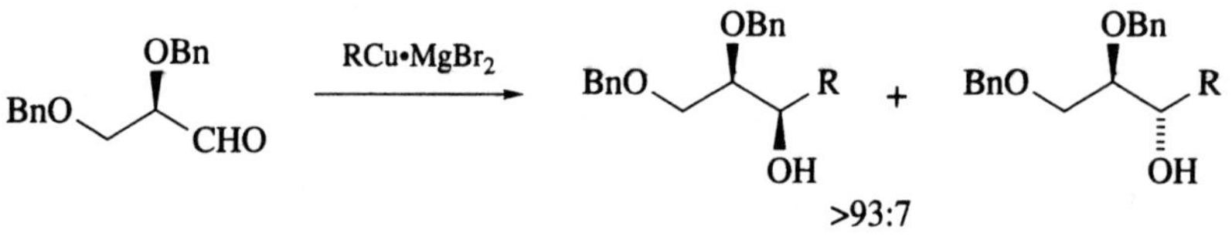

BnO OBn CHO $\xrightarrow{RCu•MgBr_2}$ BnO OBn R OH + BnO OBn R OH (30)

>93:7

In contrast to the α-alkoxy aldehydes, acetonide-protected α,β-dihydroxy aldehydes generally exhibit nonchelation selectivities in simple organometal addition reactions and the Lewis acid promoted reactions.[84] The presence of ZnX_2 was found to enhance the selectivity of furyllithium addition to acetonide aldehydes (Scheme 11). ZnX_2 is more efficient than $SnCl_4$ or $MgBr_2$. Notably, ZnX_2 enhances both nonchelation and chelation selectivity, depending on the protecting groups (*cf.* Scheme 9). The reaction is employed in the synthesis of the rare sugars L-tagatose and D-ribulose.[85] These peculiarities of acetonide aldehydes are explained by the inhibition of chelation formation, since a Felkin–Ahn type transition state predominates as a consequence of: (i) the significant ring strain which develops in the chelate structure; (ii) the depressed donor abilities of the acetonide oxygen due to inductive electron withdrawal; and (iii) the steric inhibition to chelate formation due to nonbonded interaction between the metal ligands and the acetonide methyl group.[82] Another explanation has been made which attributes the selectivity to a chelation between the β-oxygen and the carbonyl oxygen (Scheme 11).[85,86]

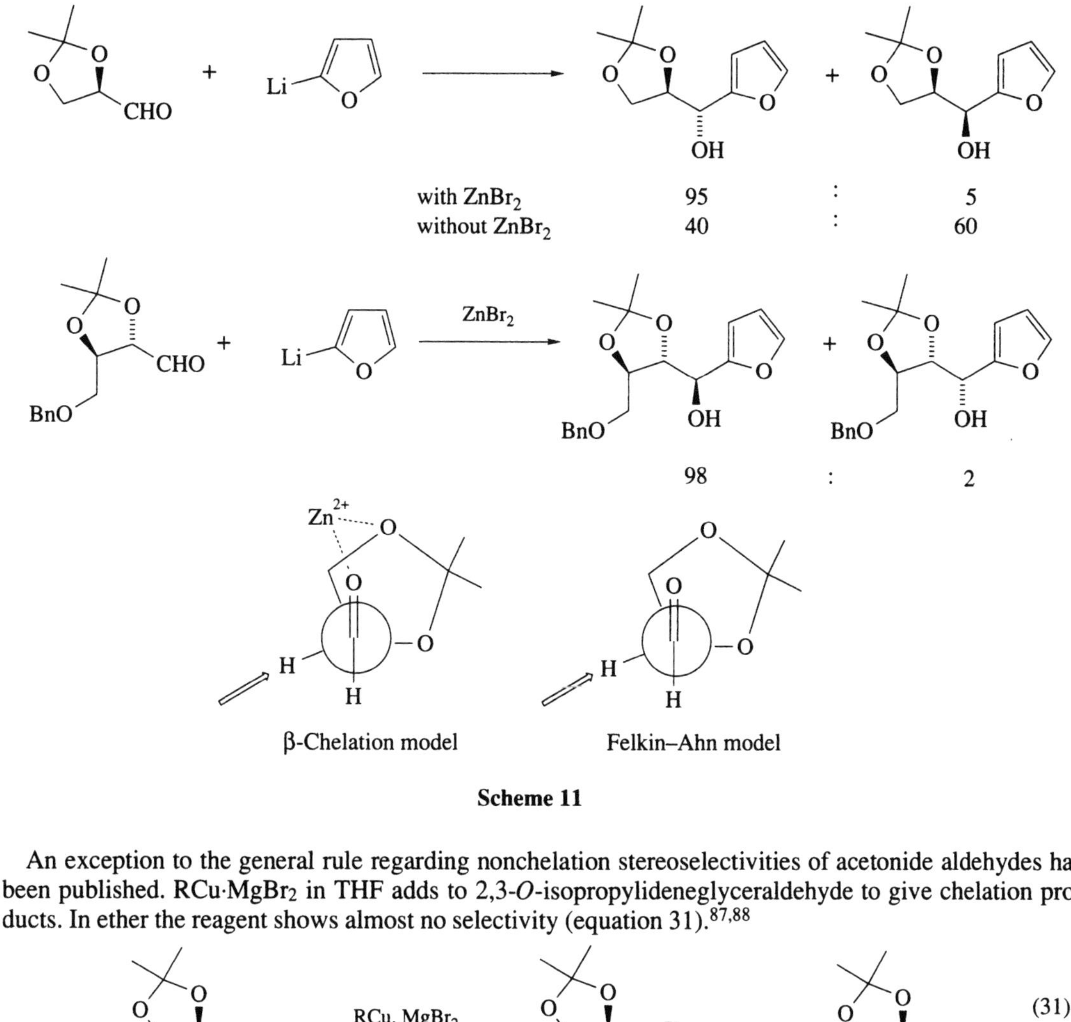

with $ZnBr_2$	95 :	5
without $ZnBr_2$	40 :	60

$ZnBr_2$: 98 : 2

β-Chelation model   Felkin–Ahn model

Scheme 11

An exception to the general rule regarding nonchelation stereoselectivities of acetonide aldehydes has been published. $RCu{\cdot}MgBr_2$ in THF adds to 2,3-*O*-isopropylideneglyceraldehyde to give chelation products. In ether the reagent shows almost no selectivity (equation 31).[87,88]

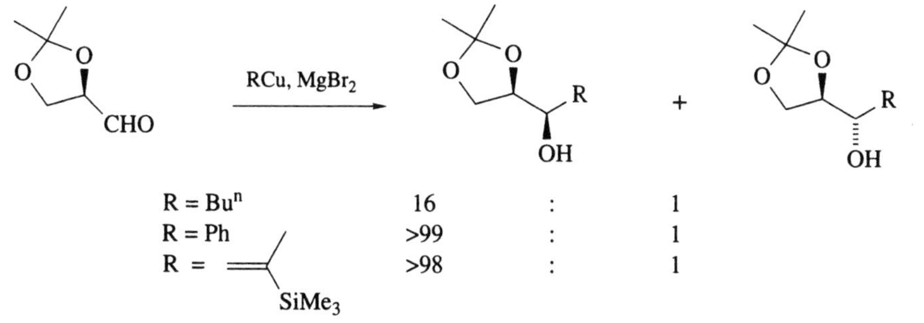

RCu, $MgBr_2$ (31)

R		
R = Bu^n	16 :	1
R = Ph	>99 :	1
R = $=C(SiMe_3)-$	>98 :	1

Addition of organometallic compounds to α,β-epoxy aldehydes gives predominantly nonchelation-controlled adducts. Grignard reagents or allylstannanes in the presence of $BF_3{\cdot}OEt_2$ give good results (equation 32).[89]

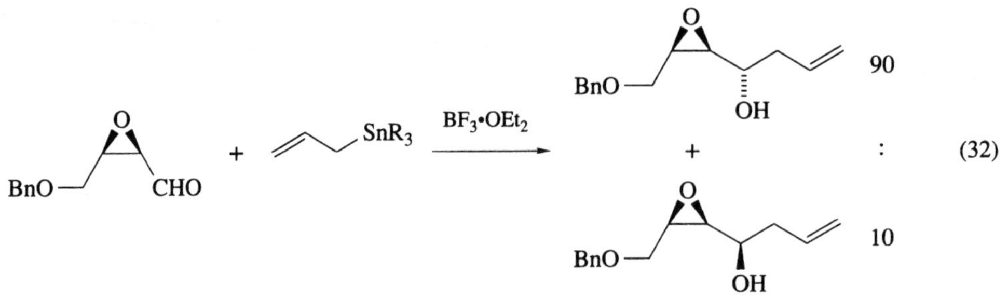

BnO, O, CHO + SnR₃ —BF₃•OEt₂→ BnO, O, OH (90) + BnO, O, OH (10); 90 : 10 (32)

A high level of 1,3-asymmetric induction was achieved by the assistance of a Lewis acid. Complexation of a β-alkoxy aldehyde with $TiCl_4$ followed by addition of Bu_2Zn, allylsilane or silyl enol ethers at –78 °C results in chelation-controlled products in >85% selectivities (Scheme 12).[77,79] Even a considerable level of 1,4-asymmetric induction is observed with the Me_2Zn–$TiCl_4$ system (equation 33).[74]

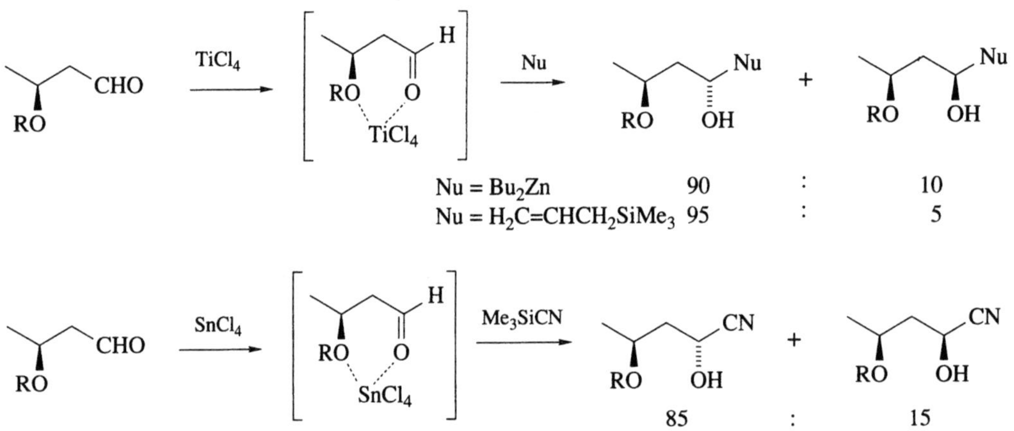

RO, CHO —$TiCl_4$→ [RO, O, H, $TiCl_4$] —Nu→ RO, Nu, OH + RO, Nu, OH

$Nu = Bu_2Zn$ 90 : 10

$Nu = H_2C{=}CHCH_2SiMe_3$ 95 : 5

RO, CHO —$SnCl_4$→ [RO, O, H, $SnCl_4$] —Me_3SiCN→ RO, CN, OH + RO, CN, OH

85 : 15

Scheme 12

BnO, CHO —$TiCl_4$→ [BnO, Cl_4, Ti, O, H, H] —Me_2Zn→

BnO, OH + BnO, OH (33)

85 : 15

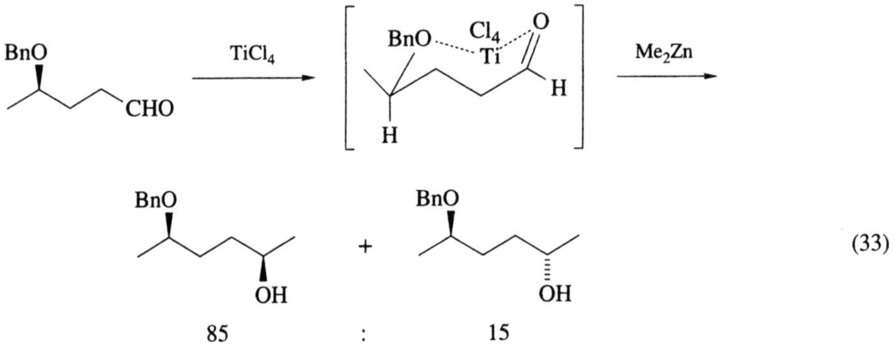

Erythro/*threo* selectivity in the reactions of crotyl-type organometals with aldehydes (simple diastereoselectivity) is markedly influenced by the presence of a Lewis acid. Crotyltitanium reagents react with aldehydes to afford *threo* adducts preferentially,[90] while *erythro* isomers are obtained in high selectivity by the addition of the organometal to a mixture of an aldehyde and $BF_3{\cdot}OEt_2$.[91] A pericyclic transition state is assumed for the former, and an open-chain transition state, in which the nucleophile attacks the BF_3-complexed aldehyde, is suggested for the latter (Scheme 13). A similar reversal of the diastereoselectivity was reported in the reactions of crotyl Cu^I, Cd, Hg^{II}, Tl^I, $ZrCp_2Cl$ and VCp_2Cl.[92] The mechanistic aspects of the Lewis acid promoted addition reactions of allyl-stannanes or -silanes have been studied in detail.[93]. Configurationally defined α-alkoxylithium is readily accessible from the corresponding organostannanes, and the addition to aldehydes has been examined. *syn*-1,2-Diol derivatives are obtained predominantly in the presence of $MgBr_2$. Enantiofacial discrimination is considered to result

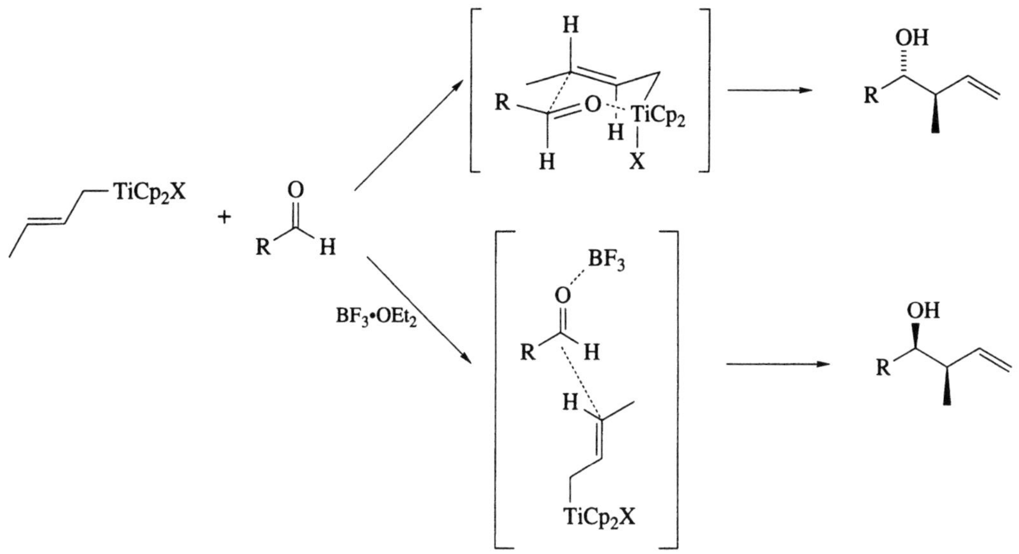

Scheme 13

from the unfavorable interaction which is absent in the transition state (**2a**), although the stereoregulating effect of the metal salt remains unclear (Scheme 14).[94]

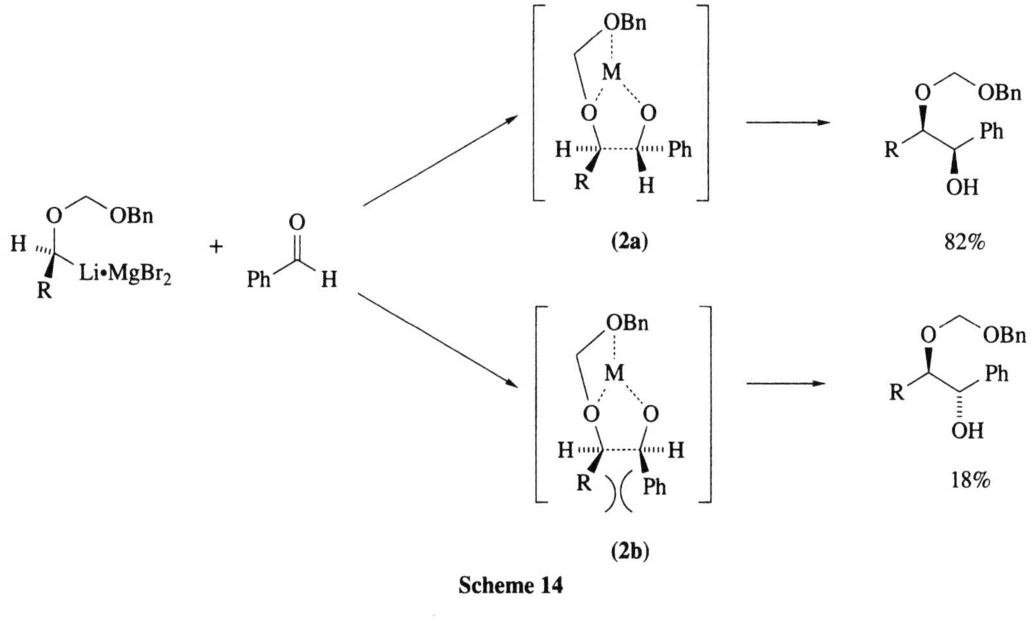

(**2a**) 82%

(**2b**) 18%

Scheme 14

Although lithiated 1,2,3,4-tetrahydroisoquinoline reacts with aldehydes with little diastereoselectivity, the presence of $MgBr_2$ dramatically enhances the selectivity. *l*-Isomers are obtained from 2-pivaloyl-1,2,3,4-tetrahydroisoquinoline in >97% *de*, and a *u*-isomer predominantly from a formamidine.[95] In the latter case, an organomagnesium reagent (**3**) formed by transmetalation was shown to be the reactive intermediate (Scheme 15).[96]

Stereoselectivity in the addition of α-lithiated (*R*)-methyl-*p*-tolyl sulfoxide to aromatic aldehydes is enhanced from 1:1 to 4:1 by the presence of $ZnCl_2$ (equation 34).[97]

Chiral organotitanium reagents generated from aryl Grignards transfer aryl groups to aromatic aldehydes with high enantiofacial selectivity. The use of aryllithium resulted in lower selectivity indicating the important role of MgX_2 in the asymmetric induction (equation 35).[98]

Tomioka found that, by the pretreatment of benzaldehyde with 2,4,6-$Me_3C_6H_2OAlCl_2$, the enantiofacial addition reaction of an RMgBr–chiral diamine complex led to carbinols with considerably higher optical purities. Steric modification of the benzaldehyde carbonyl by the complexation of the aluminum

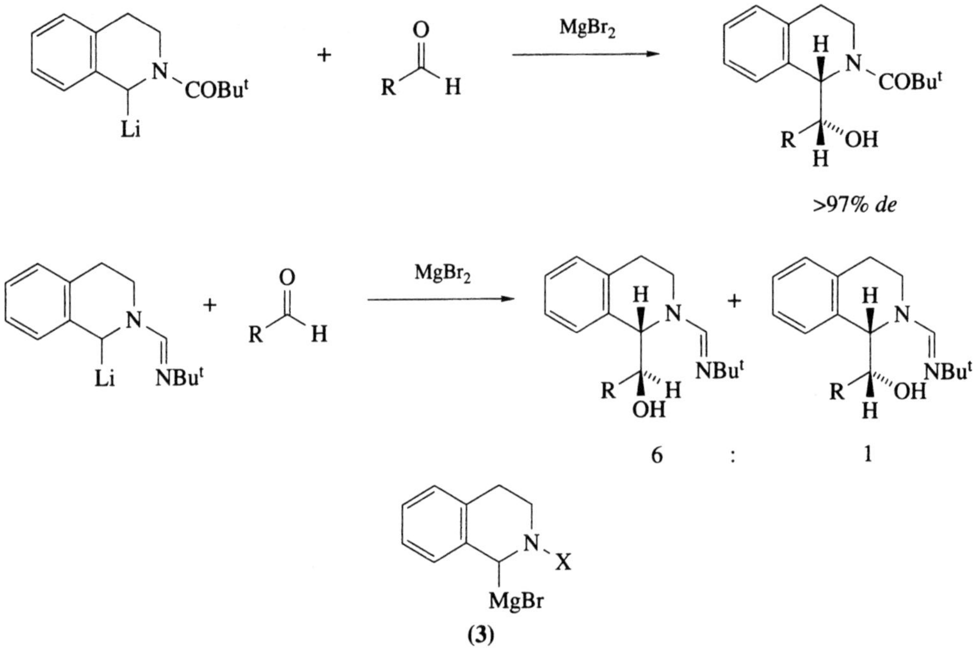

MgBr$_2$; COBu^t ; Li ; >97% *de*

MgBr$_2$; NBu^t ; 6 : 1

MgBr ; X

(3)

Scheme 15

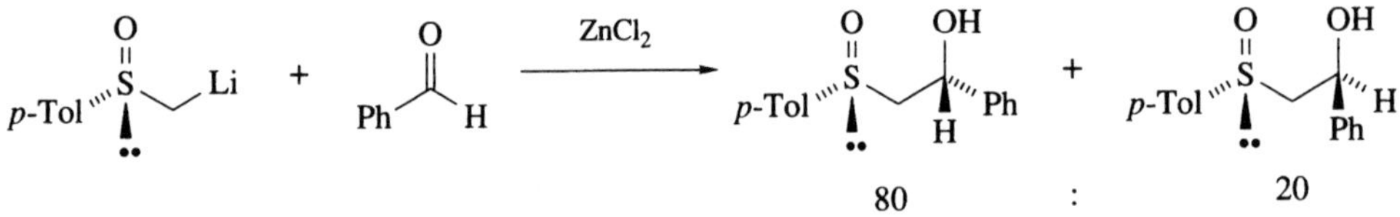

$$p\text{-TolS(O)CH}_2\text{Li} + \text{PhCHO} \xrightarrow{\text{ZnCl}_2} \quad 80 : 20 \tag{34}$$

$$\xrightarrow[\text{ii, PhMgX; iii, ArCHO}]{\text{i, ClTi(OPr}^i)_3} \text{ArCH(OH)Ph} \quad 56\text{–}>98\%\ ee \tag{35}$$

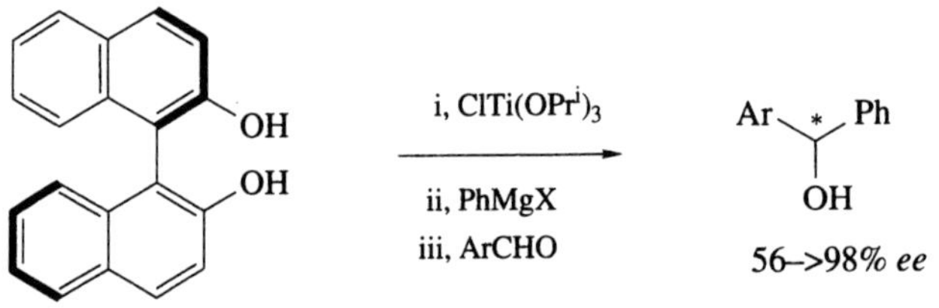

reagent was presumed (equation 36).[99] This type of approach including chiral modifications of carbonyls with chiral Lewis acids will undoubtedly find wide use in asymmetric synthesis.

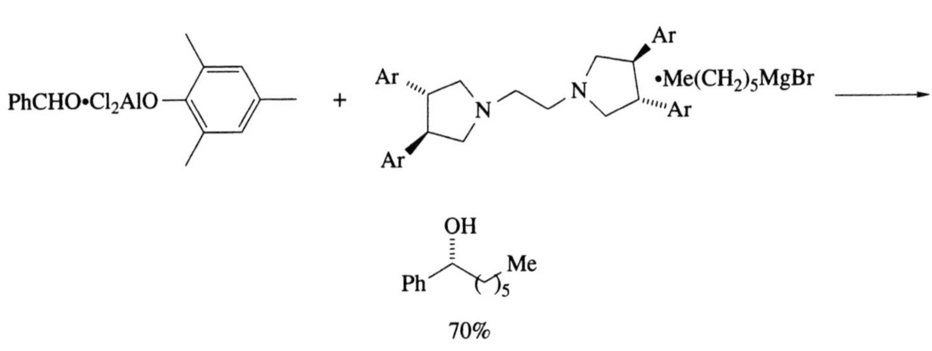

PhCHO•Cl$_2$AlO– + •Me(CH$_2$)$_5$MgBr → 70% (36)

1.11.3 LEWIS ACID PROMOTED REACTIONS OF EPOXIDES

Although nucleophilic ring opening of epoxides with organometals is a well-documented technology in organic synthesis, the reaction sometimes fails to occur due to the unreactiveness and the facile isomerization of the electrophile. The presence of an appropriate Lewis acid, notably BF_3, greatly promotes the C—C bond cleavage.

Alkynylation of oxiranes or oxetanes with lithium alkynides is effectively carried out in the presence of $BF_3{\cdot}OEt_2$ at –78 °C. The use of BF_3 gives better results than $TiCl_4$, $SnCl_4$ or $AlCl_4$. The reaction takes place stereospecifically with *anti* opening, and the attack generally occurs at the less hindered site. Several functional groups such as halogens, acetals or certain esters survive the reaction conditions (Scheme 16).[100–102]

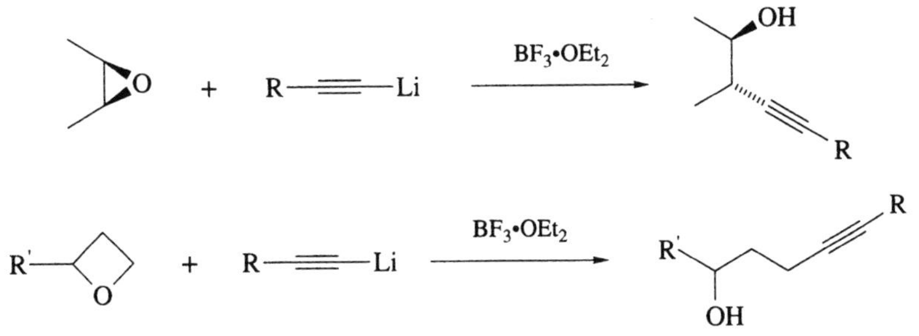

Scheme 16

A novel regioselectivity is observed with *trans*-2,3-epoxy alcohols. The C-1 attack of 1,2-epoxy alcohol formed by the Payne rearrangement proceeds predominantly, and *anti*-β,γ-dihydroxyalkynes are obtained stereospecifically (equation 37).[103] The reaction is employed in the synthesis of a pheromone, *erythro*-6-acetoxyhexadecan-5-olide.

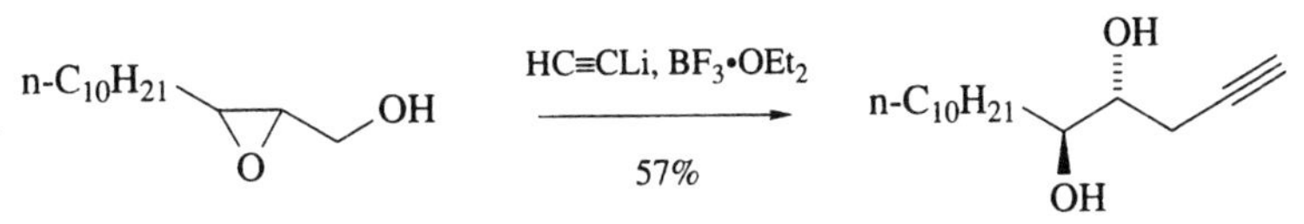

(37)

The mechanistic aspects of this Lewis acid promoted reaction have been examined by low temperature NMR studies, and reaction of the lithium alkynides with the Lewis acid activated epoxides is indicated.[104,105] The order of the addition of the reagents does not affect the product yields provided that the reaction is carried out at –78 °C; addition of $BF_3{\cdot}OEt_2$ to a mixture of an epoxide and an alkynide or addition of an epoxide to a mixture of an alkynide and $BF_3{\cdot}OEt_2$ are both possible. The transmetalation between RLi and BF_3 which produces unreactive organoboron compounds is shown to be very slow at this temperature.[102,104,105]

The alkynylation reactions have been applied to polyfunctionalized molecules (Scheme 17),[106–110] and have proved to be quite useful in natural product synthesis.

The use of Me_3Ga allows a catalytic mode of the reaction to be carried out (equation 38).[111]

Related syntheses using a variety of organolithium compounds have also been developed. Alkyllithiums, vinyllithiums or phenyllithiums in the presence of $BF_3{\cdot}OEt_2$ give the oxirane and oxetane opening products in high yields (equation 39).[105] Enolate-type nucleophiles can also be employed for this purpose (Scheme 18).[112–114]

The reaction rates of organocopper and cuprate reagents with less reactive epoxides are enhanced by the presence of $BF_3{\cdot}OEt_2$.[115,116] $R_2CuCNLi_2$–BF_3 reagents are especially effective, and even mesitylation or *t*-butylation of cyclohexene oxide can be carried out in high yields. Similar stereo- and regio-chemical features with RLi–BF_3 reagents — *anti* opening and attack at the less hindered site — are observed. The order of the addition of reagents again does not affect the yields of the products. The cleavage of sterically hindered epoxysilanes was carried out with the cuprate–BF_3 reagent, and applied to the synthesis of pheromones (Scheme 19).[117] Aziridines are cleaved effectively with R_2CuLi in the presence of $BF_3{\cdot}OEt_2$ (equation 40).[118]

A novel stereoselectivity is observed in the Lewis acid mediated S_N2'-type reactions of an epoxycycloalkene and methylating reagents. *Syn* attack occurs with MeLi–$LiClO_4$ in high selectivity, and exclusive *anti* attack with MeCu–Me_3Al. Chelation of MeLi to the epoxide or to the epoxide–$LiClO_4$ complex is suggested for the former, while attack of MeCu on the Me_3Al-coordinated epoxide from the less hindered site is presumed for the latter. The use of MeLi gives a mixture of several compounds (Scheme 20).[119]

Lewis acid promoted ring opening of oxiranes with organo-silanes or -stannanes is reported. TMS-CN cleaves the C—O bond in the presence of $AlCl_3$,[120] Et_2AlCl,[121] $Ti(OPr^i)_4$[122] or LnX_3,[27] and β-hydroxy-

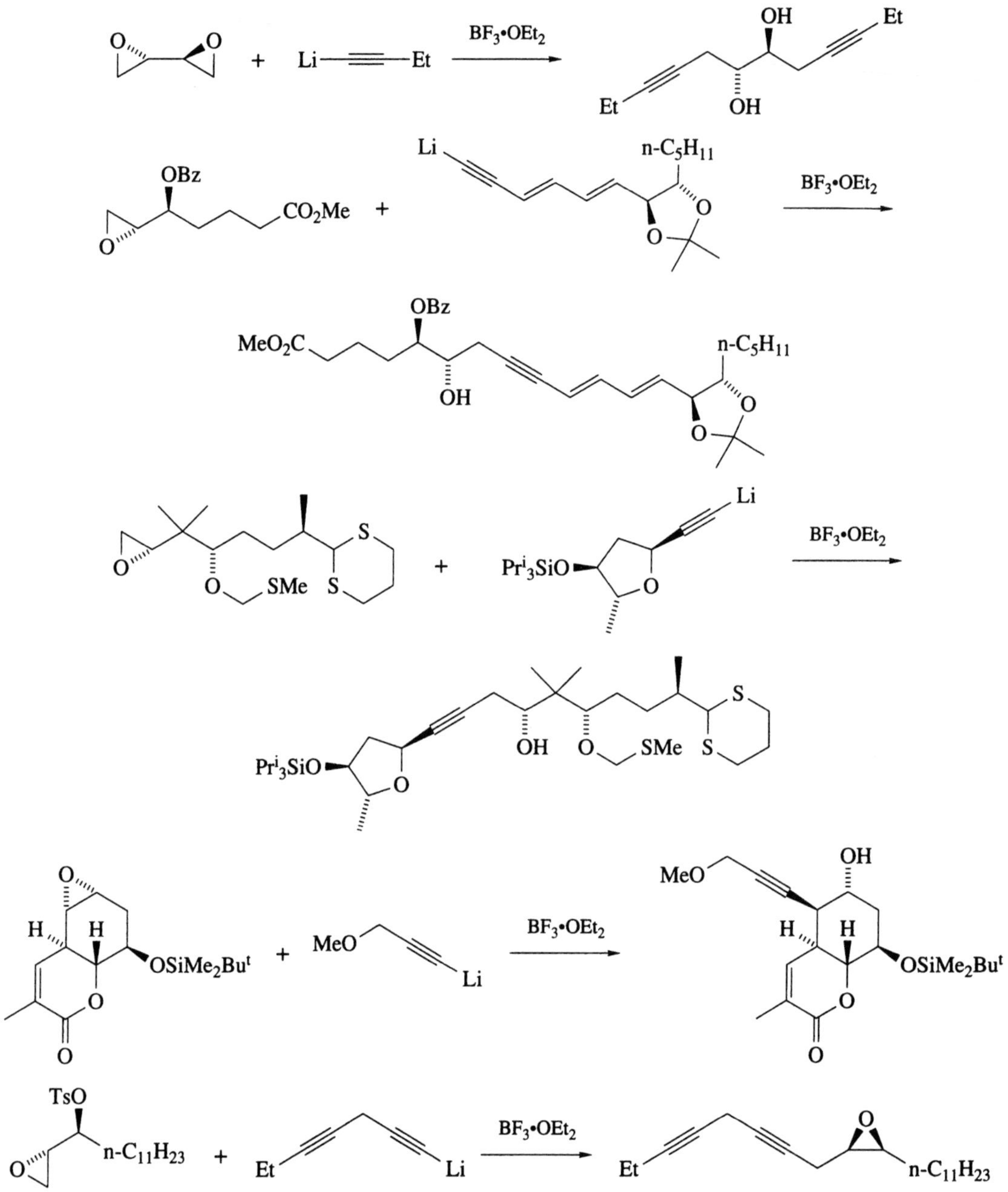

$BF_3 \cdot OEt_2$

OBz CO_2Me + Li n-C_5H_{11} O O $BF_3 \cdot OEt_2$

OBz MeO_2C OH n-C_5H_{11} O O

O SMe S S + Li Pr^i_3SiO O $BF_3 \cdot OEt_2$

Pr^i_3SiO O OH O SMe S S

O H H $OSiMe_2Bu^t$ O O + MeO Li $BF_3 \cdot OEt_2$ MeO OH H H $OSiMe_2Bu^t$ O O

TsO O n-$C_{11}H_{23}$ + Et Li $BF_3 \cdot OEt_2$ Et O n-$C_{11}H_{23}$

Scheme 17

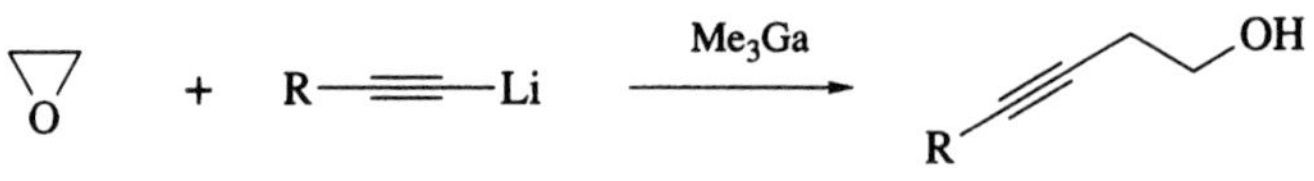

O + R—≡—Li $\xrightarrow{Me_3Ga}$ R OH (38)

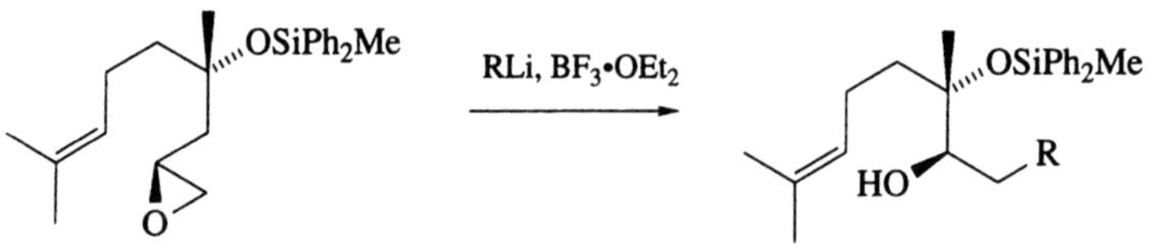

$OSiPh_2Me$ O $\xrightarrow{RLi,\ BF_3 \cdot OEt_2}$ $OSiPh_2Me$ HO R (39)

R = Bu^t, 95%; R = Ph, 85%; R = CH_2=C(Me)OEt, 99%

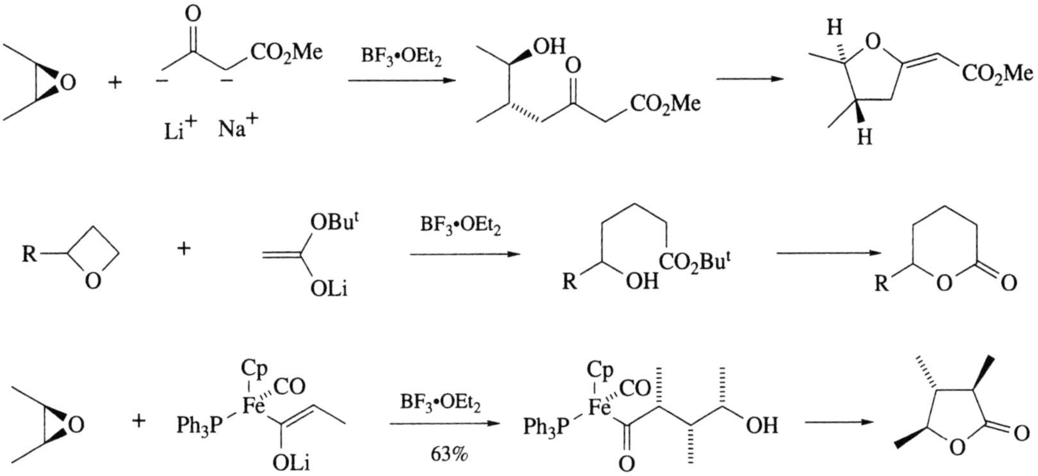

Scheme 18

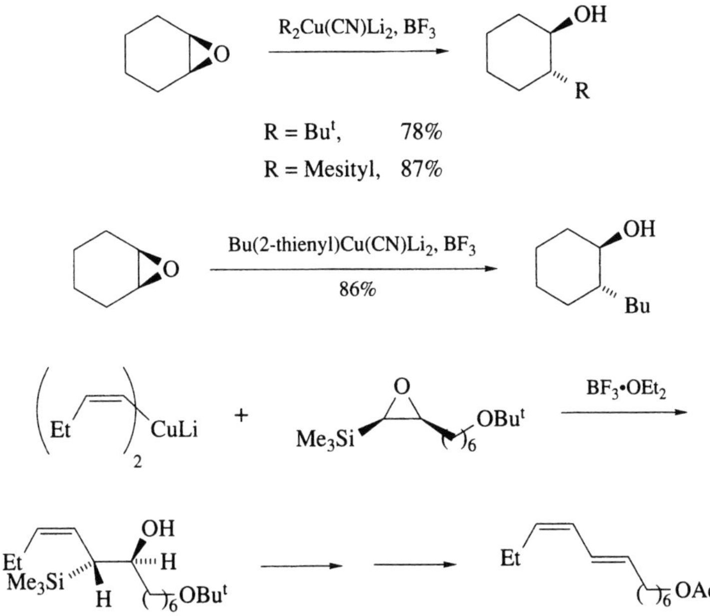

Scheme 19

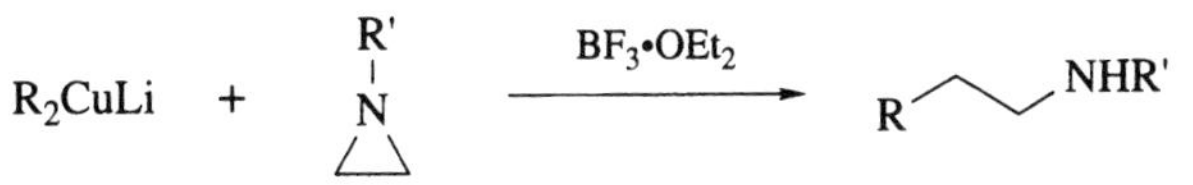

(40)

nitriles are obtained. With aluminum catalysts, propylene oxide is cyanated at the less hindered site, while attack at the more hindered site proceeds with isobutene oxide. Apparently, the cationic nature of the Lewis acid coordinated epoxide plays an important role in the latter case (Scheme 21).[120,121] $Ti(OPr^i)_4$ promotes the reaction of TMS-CN or KCN with 2,3-epoxy alcohols, C-3 attack predominating over C-2 attack (equation 41).[122] Organostannanes with unactivated C—Sn bonds alkylate epoxides intramolecularly in the presence of $TiCl_4$, and cyclopropyl alcohols are obtained (Scheme 22).[123,124]

1.11.4 LEWIS ACID PROMOTED REACTIONS OF ACETALS

Acetals are quite unreactive towards simple organometallic compounds and have been employed as a convenient protecting group. However, they turn into highly reactive species in the presence of Lewis

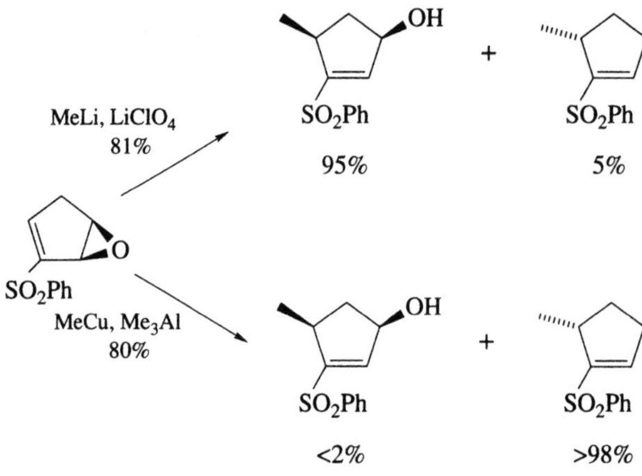

MeLi, $LiClO_4$ 81%

MeCu, Me_3Al 80%

SO_2Ph OH 95% + 5%

<2% + >98%

Scheme 20

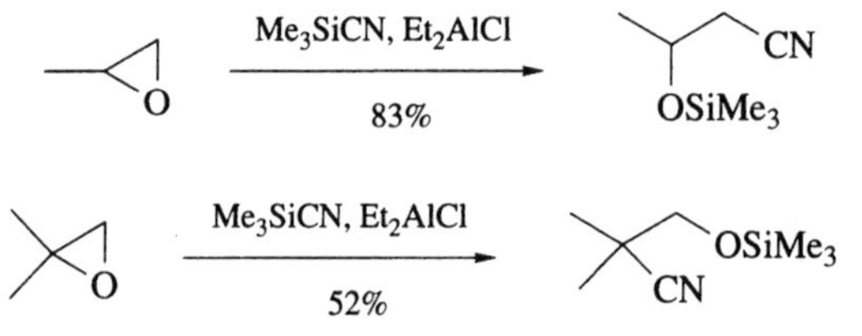

Me_3SiCN, Et_2AlCl 83% → $OSiMe_3$ CN

Me_3SiCN, Et_2AlCl 52% → $OSiMe_3$ CN

Scheme 21

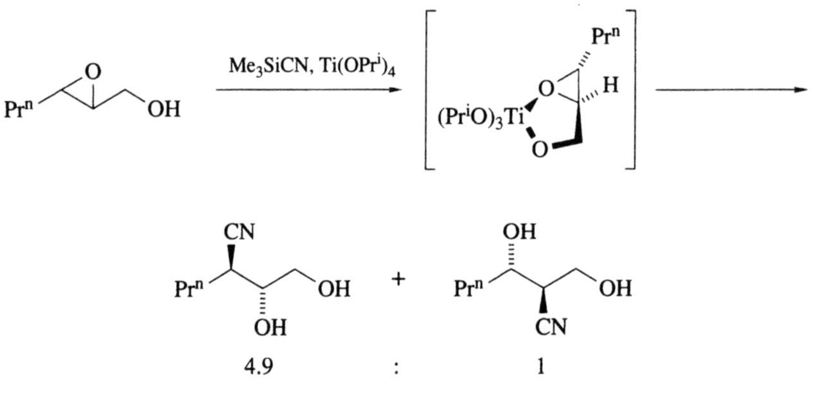

Pr^n OH —Me_3SiCN, $Ti(OPr^i)_4$→ [$(Pr^iO)_3Ti$ … Pr^n H] →

Pr^n CN OH OH + Pr^n OH OH CN (41)

4.9 : 1

Me_3Sn —$BF_3{\bullet}OEt_2$→ cyclopropyl–OH

Bu_3Sn —$BF_3{\bullet}OEt_2$, 90%→ cyclopropyl–CH_2OH

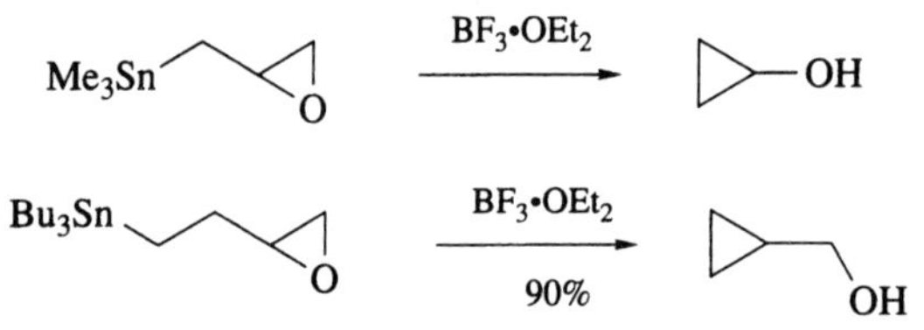

Scheme 22

acids. Aldol reactions with silyl enol ethers[16] and allylations with allylsilanes,[14] which utilize stabilized carbanionic species, have been extensively investigated in the past decade.

In recent years, nonstabilized carbanions have also been used in this type of reaction. The earliest example is the combination of an organomagnesium compound and $TiCl_4$ developed by Mukaiyama. Alkyl Grignard reagents react with α,β-unsaturated acetals activated by $TiCl_4$ at the α-position to give allylic ethers (equation 42). PhMgBr, on the other hand, gives γ-addition products.[125] Although normal alkyl acetals of aromatic and aliphatic aldehydes are unreactive towards RMgX–$TiCl_4$, reactive acetals give the corresponding products. For example, 2-alkyltetrahydropyrans are synthesized from 2-(2,4-dichlorophenoxy)tetrahydropyran (equation 43).[126,127] Reactions of RMgBr–BF_3 with *N,O*-acetals have also been reported, which give amines by the preferential cleavage of the C—O bond (equation 44).[128]

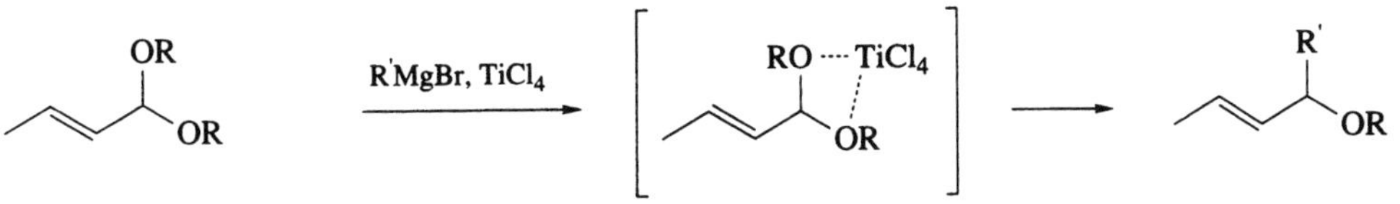

(42)

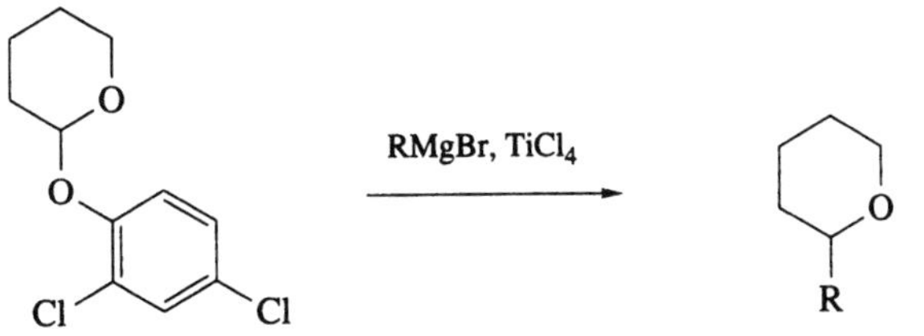

(43)

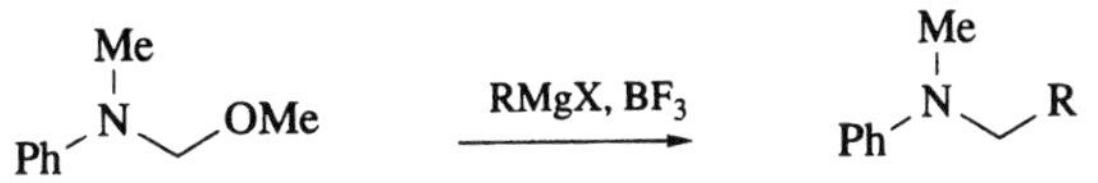

(44)

Organolithium compounds also react with acetals or orthoesters in the presence of $BF_3{\cdot}OEt_2$. Dialkoxymethylations of lithium enolates with triethyl orthoformate are carried out by adding $BF_3{\cdot}OEt_2$ to the mixture (equation 45). Prior mixture of an enolate and the Lewis acid results in a drastic decrease of the product yield. Lithium enolates are generated from silyl enol ethers and MeLi, and C—C bond formation proceeds regiospecifically with respect to the enolates. The condensation is applicable to a fully substituted enolate.[129] Butenolide anions add to acetals or orthoesters pretreated with $BF_3{\cdot}OEt_2$ at the C-5 position (equation 46).[130]

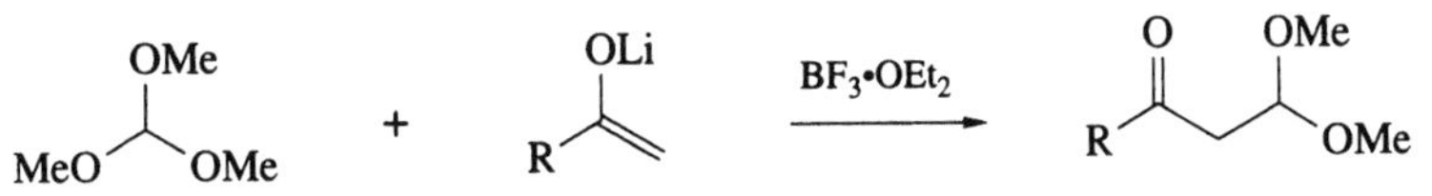

(45)

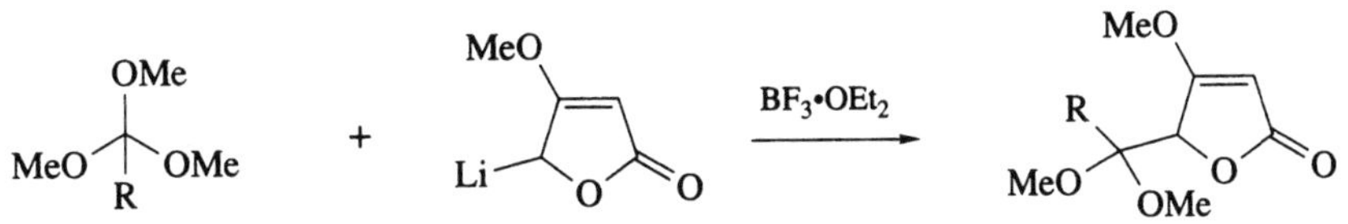

(46)

The association of BF_3 with organocopper and cuprate reagents greatly increases the reactivity towards acetals (equation 47). The reaction can be carried out by either of the following procedures with the same experimental results: (i) copper reagents are premixed with $BF_3{\cdot}OEt_2$ at –78 °C and the electrophile is added, or (ii) copper reagents are premixed with the electrophile and $BF_3{\cdot}OEt_2$ is added. These procedures can also be applied to α,β-unsaturated acetals, giving allylic ethers. In the case of tetrahydropyranyl acetal, ring-opening products are obtained exclusively, in contrast to the reaction of $RMgX–TiCl_4$, which gives 2-alkyltetrahydropyrans (*cf.* equation 43). Orthoesters are more susceptible to the Lewis acid promoted reactions.[131]

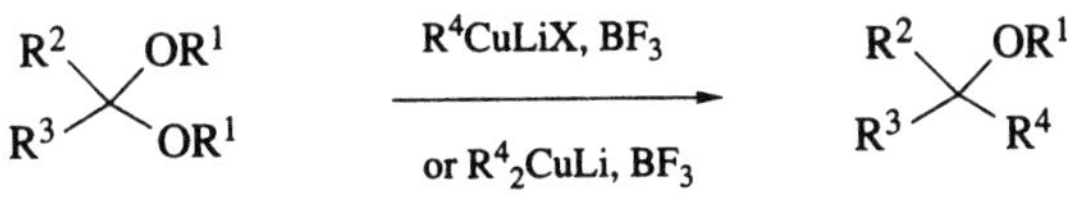

(47)

In the presence of a Lewis acid such as $SnCl_2$, $BF_3{\cdot}OEt_2$,[132–134] or $TiClO_4$,[135] TMS-CN reacts with acetals to give cyanohydrin ethers. D-Ribofuranosyl cyanide, an important intermediate of C-nucleoside synthesis, is prepared from a furanosyl acetate (Scheme 23).[133]

Acetals prepared from chiral diols and carbonyl compounds serve as a chiral synthetic equivalent of aldehydes or ketones. 1,3-Dioxanes synthesized from chiral 2,4-pentanediols are especially useful, and high asymmetric inductions are observed in the Lewis acid promoted reactions of a variety of organometallic compounds. After the removal of the chiral auxiliary by the oxidation and β-elimination procedures, optically active alcohols are obtained. Optically active propargylic alcohols and cyanohydrins are synthesized from organosilane compounds, TMS-C≡CR or TMS-CN in the presence of $TiCl_4$ (Scheme 24).[136–138] Reactive organometals such as alkyl-lithiums, -magnesiums or -coppers also react with chiral

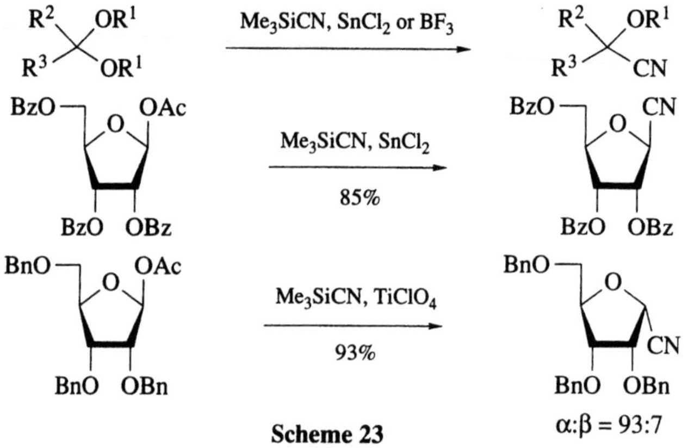

Scheme 23

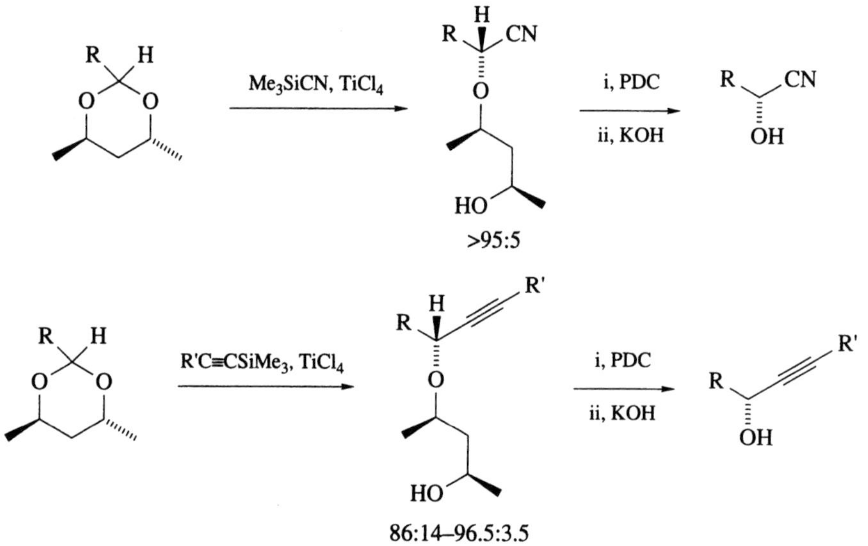

Scheme 24

acetals activated by $TiCl_4$ or BF_3 (Scheme 25).[139–141] An S_N2-like transition state, which relieves 1,3-diaxial interaction between hydrogen and the axial methyl group by a lengthening of a C—O bond, has been proposed by Johnson for the interpretation of the highly asymmetric induction (Figure 7).[138,142]

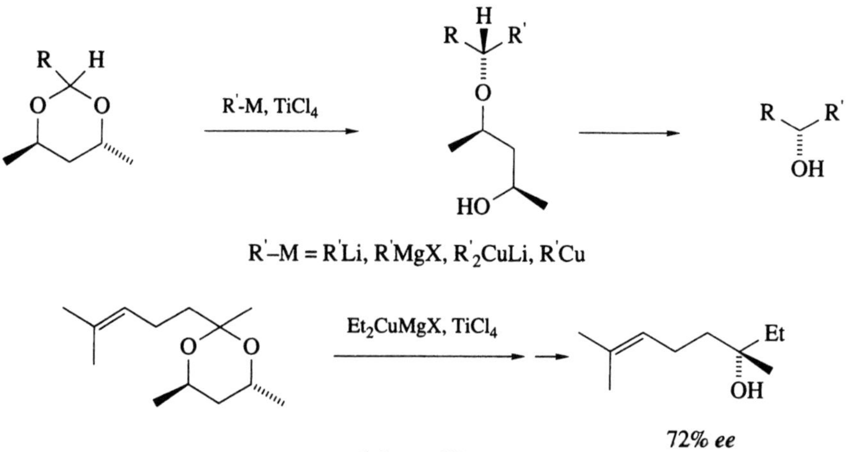

Scheme 25

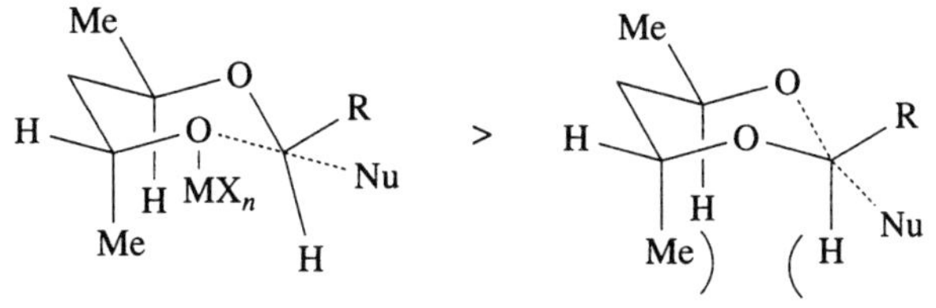

Figure 7

1.11.5 LEWIS ACID PROMOTED REACTIONS OF IMINES

Compared with aldehydes and ketones, aldimines and ketimines are less reactive towards nucleophilic addition. Furthermore, imine additions are subject to steric constraints, and rapid deprotonation proceeds with imines bearing α-hydrogen atoms. The Lewis acid promoted addition methodology has provided a solution to these problems.

In the presence of MgX_2, organo-cadmium and -zinc compounds add smoothly to imines derived from aromatic aldehydes and arylamines. Yields are very low with isolated alkyl-cadmiums or -zincs which lack a Lewis acid promoter (equation 48).[143–145]

A catalytic amount of Lewis acid, such as ZnI_2, $AlCl_3$, $TiCl_4$, $Al(OPr^i)_3$, $Al(acac)_3$, *etc.*, promotes the addition reactions of TMS-CN to imines and oximes, giving *N*-TMS-α-aminonitriles.[146,147] The product is a useful precursor of α-aminonitriles, α-aminoamides or α-amino acids. Cyanosilylation of (–)-*N*-alkylidene-(l-methylbenzyl)amines catalyzed by $ZnCl_2$ affords α-aminonitriles in 57–69% *de*. The use of $ZnCl_2$ gives better results than $AlCl_3$ or $Al(OPr^i)_3$ (Scheme 26).[148] It is also noted that the optical purities of the adducts obtained by the Lewis acid promoted reaction are much higher than those attained by the simple addition of hydrogen cyanide to imines.

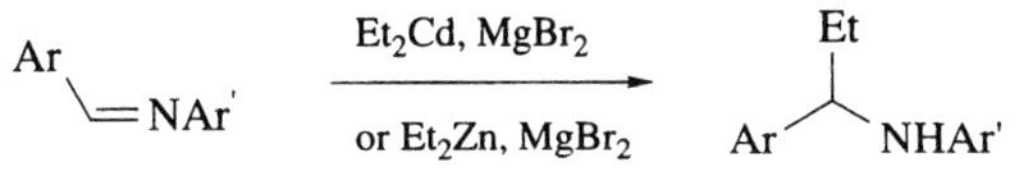

Ar–CH=NAr' —(Et_2Cd, $MgBr_2$ or Et_2Zn, $MgBr_2$)→ Ar–CH(Et)–NHAr' (48)

$R^1R^2C{=}NR^3$ —(Me_3SiCN, $AlCl_3$)→ $R^1R^2C(NC)–N(R^3)SiMe_3$

$R^1R^2C{=}NOH$ —(Me_3SiCN, ZnI_2)→ $R^1R^2C(NC)–N(H)OSiMe_3$

$R^1R^2C{=}N–CH^*(Me)Ph$ —(Me_3SiCN, ZnI_2)→ $R^1C^*H(NC)–N(SiMe_3)–C^*H(Me)Ph$

57–70% *de*

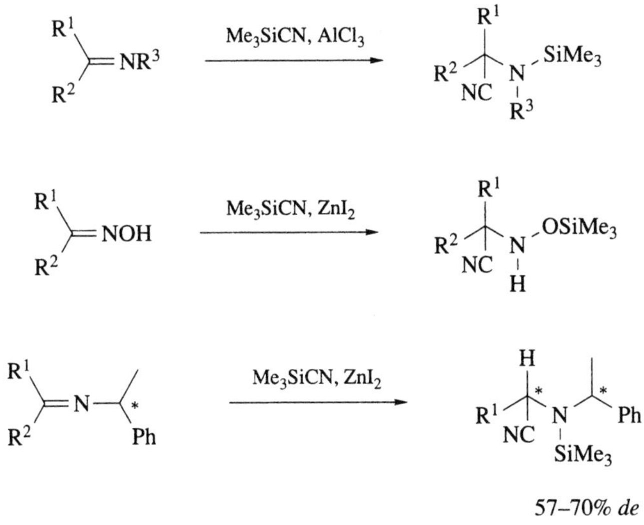

Scheme 26

In the presence of $ZnBr_2$, nonchelation controlled addition of lithiated *N*,*N*-dimethylacetamide to 2,3-*O*-cyclohexylidene-4-deoxy-L-threose benzylimine proceeds in high stereoselectivity. The absence of the Lewis acid results in a slight preference for the other isomer. The product is used in the synthesis of L-daunosamine (equation 49).[149]

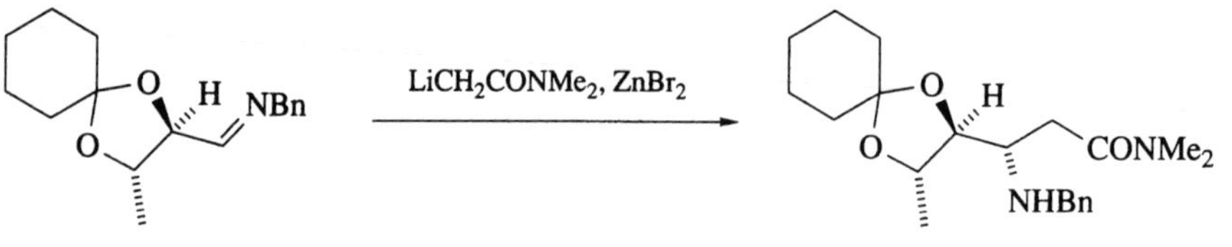

LiCH$_2$CONMe$_2$, ZnBr$_2$
NBn
CONMe$_2$
NHBn

(49)

3-Thiazolines activated with an equivalent of BF_3 readily react with a wide range of organometals, giving *trans*-4,5-disubstituted thiazoles stereoselectively. Alkyllithiums, Grignard reagents, lithium alkynides, nitronates, ester and ketone enolates have been employed as the nucleophile. Stereocontrolled construction of three contiguous asymmetric centers is performed with a lithiated isothiocyanatoacetate, and the product is successfully transformed to (+)-biotin (Scheme 27).[150,151]

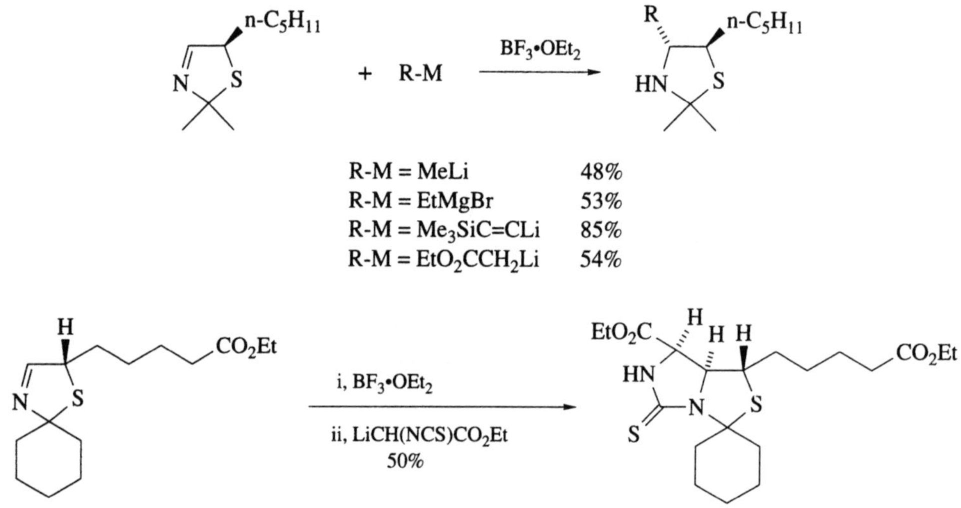

n-C$_5$H$_{11}$
+ R-M
BF$_3$•OEt$_2$
R-M = MeLi 48%
R-M = EtMgBr 53%
R-M = Me$_3$SiC=CLi 85%
R-M = EtO$_2$CCH$_2$Li 54%
CO$_2$Et
i, BF$_3$•OEt$_2$
ii, LiCH(NCS)CO$_2$Et
50%
EtO$_2$C

Scheme 27

RCu, prepared from alkylmagnesiums and CuI, reacts with aldimines in the presence of $BF_3{\cdot}OEt_2$. Grignard or copper(I) reagents do not give the addition product at all, and the starting imines are recovered. Interestingly, preparation of an RCu–BF_3 complex prior to the C—C bond formation reaction is necessary, and the addition of RCu to a mixture of an imine and $BF_3{\cdot}OEt_2$ results in low yield. R_2CuM–BF_3 (M = Li or Mg) are effective for more hindered aldimines.[152] Lithium alkynide–BF_3 reagents also add to aldimines at –78 °C, and aminoalkynes are obtained in good yields. Again the presence of the Lewis acid is reported to be essential (Scheme 28).[153]

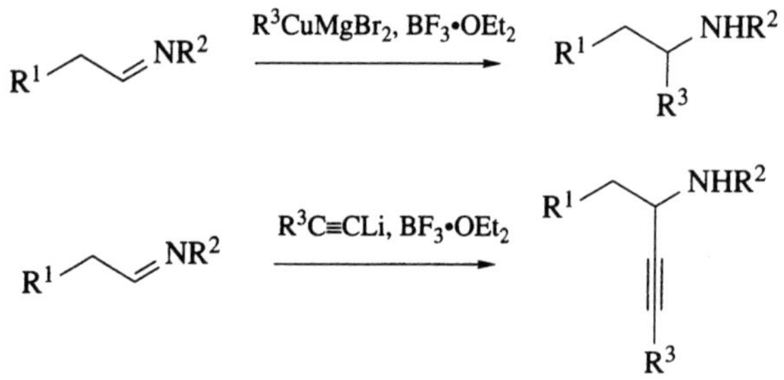

R^{3}CuMgBr$_2$, BF$_3$•OEt$_2$
R^{3}C≡CLi, BF$_3$•OEt$_2$
NR^2
NHR^2

Scheme 28

Perfluoroalkyllithiums, generated from perfluoroalkyl iodide and MeLi, add to imines pretreated with $BF_3{\cdot}OEt_2$. Since the addition of an imine to a mixture of $C_6F_{13}Li$ and $BF_3{\cdot}OEt_2$ results in the recovery of the starting material, activation of the imine, rather than an ate complex formation, is suggested as the role of the Lewis acid. In the presence of $BF_3{\cdot}OEt_2$, an imine is more reactive than the carbonyl of methyl benzoate towards the addition of $C_6F_{13}Li$. A high Cram-type asymmetric induction is observed in the addition of $C_6F_{13}Li$ to 2-phenylpropanal imine (Scheme 29).[154] The reaction of the dianion of (4-phenylsulfonyl)butanoic acid with imines activated by $BF_3{\cdot}OEt_2$ has also been reported (equation 50).[155]

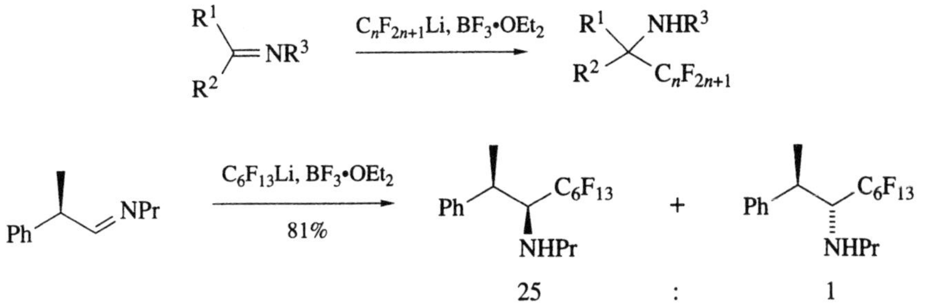

Scheme 29

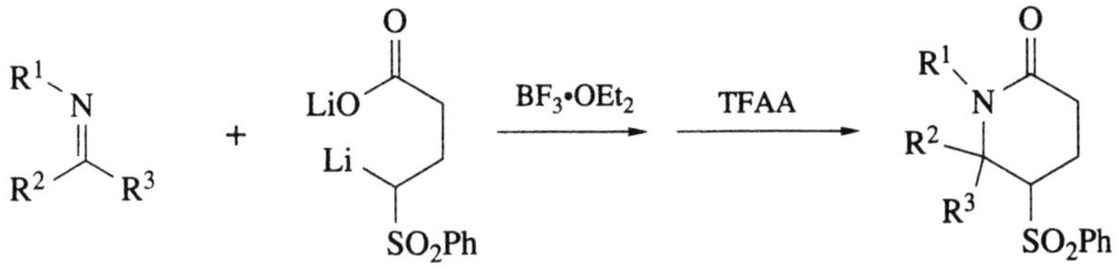

(50)

1.11.6 REFERENCES

1. E. C. Ashby and J. T. Laemmle, *Chem. Rev.*, 1975, **75**, 521, and refs. cited therein.
2. T. Mole and J. R. Surtees, *Aust. J. Chem.*, 1964, **17**, 1961.
3. P. R. Jones and P. J. Desio, *Chem. Rev.*, 1978, **78**, 491.
4. L. Miginiac, in 'The Chemistry of the Metal–Carbon Bond', ed. F. R. Hartley and S. Patai, Wiley, New York, 1985, vol. 3, p. 99.
5. J. Kollonitsch, *J. Chem. Soc. A*, 1966, 453.
6. E. Henry-Basch, F. Huet, B. Marx and P. Fréon, *C. R. Hebd. Seances Acad. Sci., Ser. C*, 1965, **260**, 3694.
7. F. Huet, E. Henry-Basch and P. Fréon, *Bull. Soc. Chim. Fr.*, 1970, 1415.
8. B. Marx, E. Henry-Basch and P. Fréon, *C. R. Hebd. Seances Acad. Sci., Ser. C*, 1967, **264**, 527.
9. B. Marx, *C. R. Hebd. Seances Acad. Sci., Ser. C*, 1968, **266**, 1646.
10. M. T. Reetz, R. Steinbach and B. Wenderoth, *Synth. Commun.*, 1981, **11**, 261.
11. H. Oshino, E. Nakamura and I. Kuwajima, *J. Org. Chem.*, 1985, **50**, 2802.
12. H. Ochiai, T. Nishihara, Y. Tamaru and Z. Yoshida, *J. Org. Chem.*, 1988, **53**, 1343.
13. V. G. Kumar Das and C.-K. Chu, in 'The Chemistry of the Metal–Carbon Bond', ed. F. R. Hartley and S. Patai, Wiley, New York, 1985, vol. 3, p. 1.
14. A. Hosomi, *Acc. Chem. Res.*, 1988, **21**, 200.
15. Y. Yamamoto, *Acc. Chem. Res.*, 1987, **20**, 243.
16. T. Mukaiyama, *Org. React. (N.Y.)*, 1982, **28**, 203.
17. G. Deleris, J. Dunogues and R. Calas, *J. Organomet. Chem.*, 1975, **93**, 43.
18. G. Deleris, J. Dunogues and R. Calas, *Tetrahedron Lett.*, 1976, 2449.
19. K. König and W. P. Neumann, *Tetrahedron Lett.*, 1967, 495.
20. W. C. Groutas and D. Felker, *Synthesis*, 1980, 861.
21. D. A. Evans, L. K. Truesdale and G. L. Carroll, *J. Chem. Soc., Chem. Commun.*, 1973, 55.
22. D. A. Evans and L. K. Truesdale, *Tetrahedron Lett.*, 1973, 4929.
23. D. A. Evans, G. L. Carroll and L. K. Truesdale, *J. Org. Chem.*, 1974, **39**, 914.
24. V. H. Rawal, J. A. Rao and M. P. Cava, *Tetrahedron Lett.*, 1985, **26**, 4275.
25. W. Lidy and W. Sundermeyer, *Chem. Ber.*, 1973, **106**, 587.
26. R. Noyori, S. Murata and M. Suzuki, *Tetrahedron*, 1981, **37**, 3899.
27. A. E. Vougioukas and H. B. Kagan, *Tetrahedron Lett.*, 1987, **28**, 5513.
28. T. L. Macdonald and S. Mahalingam, *J. Am. Chem. Soc.*, 1980, **102**, 2113.
29. T. Sato, M. Watanabe, T. Watanabe, Y. Onoda and E. Murayama, *J. Org. Chem.*, 1988, **53**, 1894.
30. Y. Yamamoto and J. Yamada, *J. Am. Chem. Soc.*, 1987, **109**, 4395.
31. Y. Yamamoto, *Angew. Chem., Int. Ed. Engl.*, 1986, **25**, 947.
32. Y. Yamamoto, S. Yamamoto, H. Yatagai, Y. Ishihara and K. Maruyama, *J. Org. Chem.*, 1982, **47**, 119.
33. P. J. Kocienski, B. Lythgoe and S. Ruston, *J. Chem. Soc., Perkin Trans. 1*, 1978, 829.
34. J. J. Eisch and J. E. Galle, *J. Org. Chem.*, 1979, **44**, 3279.
35. S. J. Danishefsky, H. G. Selnick, M. P. DeNinno and R. E. Zelle, *J. Am. Chem. Soc.*, 1987, **109**, 1572.
36. J. E. Wrobel and B. Ganem, *J. Org. Chem.*, 1983, **48**, 3761.
37. E. C. Ashby and S. A. Noding, *J. Org. Chem.*, 1979, **44**, 4371.
38. J. Epsztajn and A. Bieniek, *J. Chem. Soc., Perkin Trans. 1*, 1985, 213.
39. B. Achmatowicz, E. Baranowska, A. R. Daniewski, J. Pankowski and J. Wicha, *Tetrahedron Lett.*, 1985, **26**, 5597.
40. S. L. Schreiber and H. V. Meyers, *J. Am. Chem. Soc.*, 1988, **110**, 5198.
41. E. Torres, G. L. Larson and G. J. McGarvey, *Tetrahedron Lett.*, 1988, **29**, 1355.
42. M. T. Reetz, *Angew. Chem., Int. Ed. Engl.*, 1984, **23**, 556.

43. M. T. Reetz, *Top. Curr. Chem.*, 1982, **106**, 1.
44. M. T. Reetz, *Pure Appl. Chem.*, 1985, **57**, 1781.
45. B. Weidmann and D. Seebach, *Angew. Chem., Int. Ed. Engl.*, 1983, **22**, 31.
46. M. T. Reetz, S. H. Kyung and M. Hüllmann, *Tetrahedron*, 1986, **42**, 2931.
47. T. Kauffmann, A. Hamsen and C. Beirich, *Angew. Chem., Int. Ed. Engl.*, 1982, **21**, 144.
48. T. Kauffmann, R. Konig, C. Pahde and A. Tannert, *Tetrahedron Lett.*, 1981, **22**, 5031.
49. A. Dormond, A. Aaliti and C. Moïse, *J. Org. Chem.*, 1988, **53**, 1034.
50. G. Cahiez, D. Bernard and J. F. Normant, *Synthesis*, 1977, 130.
51. G. Cahiez and J. F. Normant, *Tetrahedron Lett.*, 1977, 3383.
52. T. Imamoto, T. Kusumoto and M. Yokoyama, *J. Chem. Soc., Chem. Commun.*, 1982, 1042.
53. T. Imamoto, Y. Sugiura and N. Takiyama, *Tetrahedron Lett.*, 1984, **25**, 4233.
54. T. Imamoto and Y. Sugiura, *J. Organomet. Chem.*, 1985, **285**, C21.
55. T. Hirao, D. Misu and T. Agawa, *J. Am. Chem. Soc.*, 1985, **107**, 7179.
56. T. Hirao, D. Misu and T. Agawa, *Tetrahedron Lett.*, 1986, **27**, 933.
57. M. T. Reetz, R. Steinbach, J. Westermann, R. Peter and B. Wenderoth, *Chem. Ber.*, 1985, **118**, 1441.
58. T. L. Macdonald and W. C. Still, *J. Am. Chem. Soc.*, 1975, **97**, 5280.
59. E. C. Ashby, J. J. Lin and J. J. Watkins, *Tetrahedron Lett.*, 1977, 1709.
60. K. Maruoka, T. Itoh and H. Yamamoto, *J. Am. Chem. Soc.*, 1985, **107**, 4573.
61. K. Maruoka, T. Itoh, M. Sakurai, K. Nonoshita and H. Yamamoto, *J. Am. Chem. Soc.*, 1988, **110**, 3588.
62. J. D. Morrison (ed.) 'Asymmetric Synthesis', Academic Press, New York, 1983, vol. 2; 1984, vol. 3; and refs. cited therein.
63. M. T. Reetz, M. Hüllmann, W. Massa, S. Berger, P. Rademacher and P. Heymanns, *J. Am. Chem. Soc.*, 1986, **108**, 2405.
64. C. H. Heathcock and L. A. Flippin, *J. Am. Chem. Soc.*, 1983, **105**, 1667.
65. Y. Yamamoto and K. Maruyama, *J. Am. Chem. Soc.*, 1985, **107**, 6411.
66. B. H. Lipshutz, E. L. Ellsworth and T. J. Siahaan, *J. Am. Chem. Soc.*, 1988, **110**, 4834.
67. Y. Yamamoto, S. Nishii and K. Maruyama, *J. Chem. Soc., Chem. Commun.*, 1986, 102.
68. W. C. Still and J. H. McDonald, III, *Tetrahedron Lett.*, 1980, **21**, 1031.
69. W. C. Still and J. A. Schneider, *Tetrahedron Lett.*, 1980, **21**, 1035.
70. D. B. Collum, J. H. McDonald, III, and W. C. Still, *J. Am. Chem. Soc.*, 1980, **102**, 2117, 2118, 2120.
71. M. Asami and T. Mukaiyama, *Chem. Lett.*, 1983, 93.
72. M. Asami and R. Kimura, *Chem. Lett.*, 1985, 1221.
73. T. Itoh, A. Yoshinaka, T. Sato and T. Fujisawa, *Chem. Lett.*, 1985, 1679.
74. M. T. Reetz, K. Kesseler, S. Schmidtberger, B. Wenderoth and R. Steinbach, *Angew. Chem., Int. Ed. Engl.*, 1983, **22**, 989; *Angew. Chem. Suppl.*, 1983, 1511.
75. M. T. Reetz, M. Hüllmann and T. Seitz, *Angew. Chem., Int. Ed. Engl.*, 1987, **26**, 477.
76. M. T. Reetz, K. Kesseler and A. Jung, *Angew. Chem., Int. Ed. Engl.*, 1985, **24**, 989.
77. M. T. Reetz and A. Jung, *J. Am. Chem. Soc.*, 1983, **105**, 4833.
78. M. T. Reetz and K. Kesseler, *J. Chem. Soc., Chem. Commun*, 1984, 1079.
79. M. T. Reetz, K. Kesseler and A. Jung, *Tetrahedron Lett.*, 1984, **25**, 729.
80. E. K. Dolence, M. Adamczyk, D. S. Watt, G. B. Russell and D. H. S. Horn, *Tetrahedron Lett.*, 1985, **26**, 1189.
81. M. T. Reetz and K. Kesseler, *J. Org. Chem.*, 1985, **50**, 5434.
82. K. Mead and T. L. Macdonald, *J. Org. Chem.*, 1985, **50**, 422.
83. J. Mulzer and A. Angermann, *Tetrahedron Lett.*, 1983, **24**, 2843.
84. J. Jurczak, S. Pikul and T. Bauer, *Tetrahedron*, 1986, **42**, 447.
85. K. Suzuki, Y. Yuki and T. Mukaiyama, *Chem. Lett.*, 1981, 1529.
86. S. Pikul and J. Jurczak, *Tetrahedron Lett.*, 1985, **26**, 4145.
87. F. Sato, Y. Kobayashi, O. Takahashi, T. Chiba, Y. Takeda and M. Kusakabe, *J. Chem. Soc., Chem. Commun*, 1985, 1636.
88. Y. Takeda, T. Matsumoto and F. Sato, *J. Org. Chem.*, 1986, **51**, 4728.
89. G. P. Howe, S. Wang and G. Procter, *Tetrahedron Lett.*, 1987, **28**, 2629.
90. F. Sato, K. Iida, S. Iijima, H. Moriya and M. Sato, *J. Chem. Soc., Chem. Commun.*, 1981, 1140.
91. M. T. Reetz and M. Sauerwald, *J. Org. Chem.*, 1984, **49**, 2292.
92. Y. Yamamoto and K. Maruyama, *J. Organomet. Chem.*, 1985, **284**, C45.
93. S. E. Denmark, T. Wilson and T. M. Willson, *J. Am. Chem. Soc.*, 1988, **110**, 984, and refs. cited therein.
94. G. J. McGarvey and M. Kimura, *J. Org. Chem.*, 1982, **47**, 5420.
95. D. Seebach and M. A. Syfrig, *Angew. Chem., Int. Ed. Engl.*, 1984, **23**, 248.
96. D. Seebach, J. Hansen, P. Seiler and J. M. Gromek, *J. Organomet. Chem.*, 1985, **285**, 1.
97. M. Braun and W. Hild, *Chem. Ber.*, 1984, **117**, 413.
98. D. Seebach, A. K. Beck, S. Roggo and A. Wonnacott, *Chem. Ber.*, 1985, **118**, 3673.
99. K. Tomioka, M. Nakajima and K. Koga, *Tetrahedron Lett.*, 1987, **28**, 1291.
100. M. Yamaguchi and I. Hirao, *Tetrahedron Lett.*, 1983, **24**, 391.
101. M. Yamaguchi, Y. Nobayashi and I. Hirao, *Tetrahedron Lett.*, 1983, **24**, 5121.
102. M. Yamaguchi, Y. Nobayashi and I. Hirao, *Tetrahedron*, 1984, **40**, 4261.
103. M. Yamaguchi and I. Hirao, *J. Chem. Soc., Chem. Commun.*, 1984, 202.
104. H. C. Brown, U. S. Racherla and S. M. Singh, *Tetrahedron Lett.*, 1984, **25**, 2411.
105. M. J. Eis, J. E. Wrobel and B. Ganem, *J. Am. Chem. Soc.*, 1984, **106**, 3693.
106. S. Hatakeyama, K. Sakurai, K. Saijo and S. Takano, *Tetrahedron Lett.*, 1985, **26**, 1333.
107. D. Askin, C. Angst and S. J. Danishefsky, *J. Org. Chem.*, 1985, **50**, 5005.
108. J. Morris and D. G. Wishka, *Tetrahedron Lett.*, 1986, **27**, 803.
109. R. M. Soll and S. P. Seitz, *Tetrahedron Lett.*, 1987, **28**, 5457.
110. T. W. Bell and J. A. Ciaccio, *Tetrahedron Lett.*, 1988, **29**, 865.
111. K. Utimoto, C. Lambert, Y. Fukuda, H. Shiragami and H. Nozaki, *Tetrahedron Lett.*, 1984, **25**, 5423.

112. M. Yamaguchi, K. Shibato and I. Hirao, *Tetrahedron Lett.*, 1984, **25**, 1159.
113. M. Yamaguchi and I. Hirao, *Chem. Lett.*, 1985, 337.
114. S. G. Davies and P. Warner, *Tetrahedron Lett.*, 1985, **26**, 4815.
115. A. Alexakis, D. Jachiet and J. F. Normant, *Tetrahedron*, 1986, **42**, 5607.
116. B. H. Lipshutz, D. A. Parker, J. A. Kozlowski and S. L. Nguyen, *Tetrahedron Lett.*, 1984, **25**, 5959.
117. A. Alexakis and D. Jachiet, *Tetrahedron Lett.*, 1988, **29**, 217.
118. M. J. Eis and B. Ganem, *Tetrahedron Lett.*, 1985, **26**, 1153.
119. J. C. Saddler and P. L. Fuchs, *J. Am. Chem. Soc.*, 1981, **103**, 2112.
120. W. Lidy and W. Sundermeyer, *Tetrahedron Lett.*, 1973, 1449.
121. J. C. Mullis and W. P. Weber, *J. Org. Chem.*, 1982, **47**, 2873.
122. J. M. Chong and K. B. Sharpless, *J. Org. Chem.*, 1985, **50**, 1560.
123. H. G. Kuilvila and N. M. Scarpa, *J. Am. Chem. Soc.*, 1970, **92**, 6990.
124. D. J. Peterson, M. D. Robbins and J. Hansen, *J. Organomet. Chem.*, 1974, **73**, 237.
125. T. Mukaiyama and H. Ishikawa, *Chem. Lett.*, 1974, 1077.
126. H. Ishikawa and T. Mukaiyama, *Chem. Lett.*, 1975, 305.
127. H. Ishikawa, T. Mukaiyama and S. Ikeda, *Bull. Chem. Soc. Jpn.*, 1981, **54**, 776.
128. T. Shono, Y. Matsumura, K. Inoue, H. Ohmizu and S. Kashimura, *J. Am. Chem. Soc.*, 1982, **104**, 5753.
129. M. Suzuki, A. Yanagisawa and R. Noyori, *Tetrahedron Lett.*, 1982, **23**, 3595.
130. A. Pelter and R. Al-Bayati, *Tetrahedron Lett.*, 1982, **23**, 5229.
131. A. Ghribi, A. Alexakis and J. F. Normant, *Tetrahedron Lett.*, 1984, **25**, 3075, 3079.
132. K. Utimoto, Y. Wakabayashi, Y. Shishiyama, M. Inoue and H. Nozaki, *Tetrahedron Lett.*, 1981, **22**, 4279.
133. K. Utimoto and T. Horiie, *Tetrahedron Lett.*, 1982, **23**, 237.
134. K. Utimoto, Y. Wakabayashi, T. Horiie, M. Inoue, Y. Shishiyama, M. Obayashi and H. Nozaki, *Tetrahedron*, 1983, **39**, 967.
135. T. Mukaiyama, S. Kobayashi and S. Shoda, *Chem. Lett.*, 1984, 1529.
136. W. S. Johnson, R. Elliot and J. D. Elliot, *J. Am. Chem. Soc.*, 1983, **105**, 2904.
137. J. D. Elliot, V. M. F. Choi and W. S. Johnson, *J. Org. Chem.*, 1983, **48**, 2294.
138. V. M. F. Choi, J. D. Elliot and W. S. Johnson, *Tetrahedron Lett.*, 1984, **25**, 591.
139. S. D. Lindell, J. D. Elliot and W. S. Johnson, *Tetrahedron Lett.*, 1984, **25**, 3947.
140. A. Ghribi, A. Alexakis and J. F. Normant, *Tetrahedron Lett.*, 1984, **25**, 3083.
141. A. Mori, K. Maruoka and H. Yamamoto, *Tetrahedron Lett.*, 1984, **25**, 4421.
142. P. A. Bartlett, W. S. Johnson and J. D. Elliot, *J. Am. Chem. Soc.*, 1983, **105**, 2088.
143. J. Thomas, E. Henry-Basch and P. Fréon, *C. R. Hebd. Seances Acad. Sci., Ser. C*, 1968, **267**, 176.
144. J. Thomas, E. Henry-Basch and P. Fréon, *Bull. Soc. Chim. Fr.*, 1969, 109.
145. J. Thomas, *Bull. Soc. Chim. Fr.*, 1973, 1300.
146. I. Ojima, S. Inaba and K. Nakatsugawa, *Chem. Lett.*, 1975, 331.
147. Y. Nakajima, T. Makino, J. Oda and Y. Inoue, *Agric. Biol. Chem.*, 1975, **39**, 571.
148. I. Ojima, S. Inaba and Y. Nagai, *Chem. Lett.*, 1975, 737.
149. T. Mukaiyama, Y. Goto and S. Shoda, *Chem. Lett.*, 1983, 671.
150. R. A. Volkmann, J. T. Davis and C. N. Meltz, *J. Am. Chem. Soc.*, 1983, **105**, 5946.
151. C. N. Meltz and R. A. Volkmann, *Tetrahedron Lett.*, 1983, **24**, 4503, 4507.
152. M. Wada, Y. Sakurai and K. Akiba, *Tetrahedron Lett.*, 1984, **25**, 1079.
153. M. Wada, Y. Sakurai and K. Akiba, *Tetrahedron Lett.*, 1984, **25**, 1083.
154. H. Uno, Y. Shiraishi, K. Shimokawa and H. Suzuki, *Chem. Lett.*, 1988, 729.
155. C. M. Thompson, D. L. C. Green and R. Kubas, *J. Org. Chem.*, 1988, **53**, 5389.

1.12
Nucleophilic Addition to Imines and Imine Derivatives

ROBERT A. VOLKMANN
Pfizer Central Research, Groton, CT, USA

1.12.1 SCOPE

Synthetic and biological interest in highly functionalized acyclic and cyclic amines has contributed to the wealth of experimental methodology developed for the addition of carbanions to the carbon–nitrogen double bond of imines/imine derivatives (azomethines). While a variety of practical methods exist for the enantio- and stereo-selective syntheses of substituted alcohols from aldehyde and ketone precursors, related imine additions have inherent structural limitations. Nonetheless imines, by virtue of nitrogen substitution, add a synthetic dimension not available to ketones. In addition, improved procedures for the preparation and activation of imines/imine derivatives have increased the scope of the imine addition reaction.

This chapter will attempt to provide a comprehensive picture of nonstabilized carbanion additions to the carbon–nitrogen double bond of imines/imine derivatives. Included are organometallic condensations with cyclic and acyclic azomethines [including *N*-(trimethylsilyl)imines, sulfenimines and sulfonimines], iminium salts, *N*-acylimines (*N*-acyliminium salts), hydrazones, oximes and nitrones (Scheme 1). While 1,4-additions to azadienes (including α,β-unsaturated and aromatic aldimines) are presented, the addition of organometallic reagents to the carbon–nitrogen bond of aromatic compounds (*i.e.* pyridines and quinolines) is not. Because the structural features of each imine derivative (class) are uniquely responsible for the chemistry of azomethines, this chapter is organized by imine structure. In concert with the theme of 'Comprehensive Organic Synthesis,' recent scientific contributions, particularly those involved in the control of stereochemistry, are highlighted. Essential background for each class of azomethines is provided. For an extensive evaluation of areas, the reader is urged to also consult appropriate review articles.

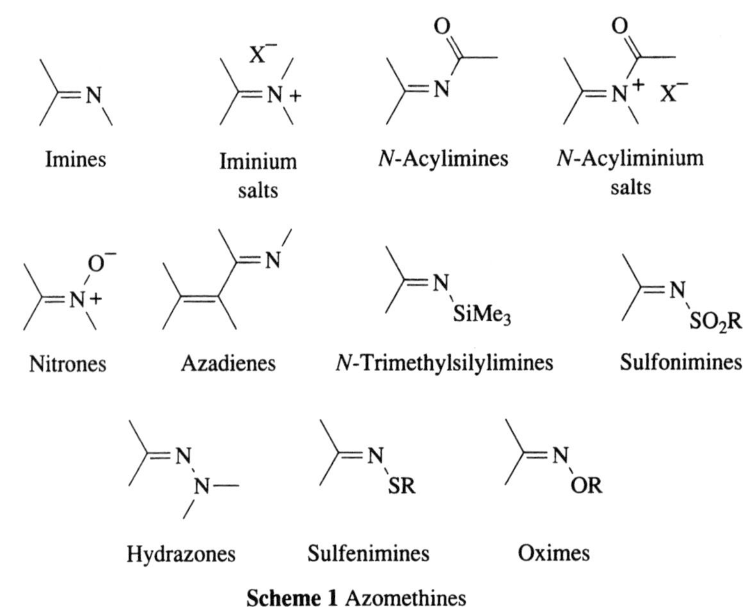

Scheme 1 Azomethines

1.12.2 INTRODUCTION

The inability of certain nucleophiles to add to the carbon–nitrogen double bond of imines/imine derivatives, coupled with the propensity of basic reagents to preferentially abstract protons α to the imine double bond, has limited the utility of the group in synthetic organic chemistry. While unique solutions exist for individual reactions, they are at best only applicable to a particular structural class of azomethines. Since structurally diverse imine derivatives have been utilized for the preparation of highly functionalized amines, an overview of some of the structural features of azomethines and nonstabilized carbanions responsible for (i) azomethine reactivity, (ii) proton abstraction, and (iii) stereochemical control in azomethine additions is included.

1.12.2.1 Azomethine Reactivity

A systematic evaluation of the relationship between azomethine structure and reactivity with nonstabilized carbanions has not been reported. Some imines/imine derivatives are inert to nucleophilic addition. The electrophilicity of imines can, however, be increased by *N*-alkylation to form highly reactive iminium salts, by *N*-oxidation to form reactive nitrones, or by *N*-acylation or *N*-sulfonylation to form reactive acylimines and sulfonimines. Sulfonimines and nitrones condense with organometallic reagents to generate sulfonamide and hydroxylamine products. 'Activated' imines need not be isolated, as the *in situ* preparation of acylimines, iminium and acyliminium salts can be employed for the generation of substituted amines (amine derivatives). These activating groups, however, do not provide a general solution to this fundamental problem of imine reactivity, since they are in certain cases not easily removed. For this reason, Lewis acid activation with $BF_3{\cdot}OEt_2$ and the *in situ* formation of iminium salts with TMS-OTf have been recently employed in amine synthesis.

1.12.2.2 Deprotonation of Azomethines

While the abstraction of protons adjacent to the carbon–nitrogen double bond of imines/imine derivatives has been utilized for the regioselective generation of azaallyl anions (which are useful in asymmetric ketone synthesis), it competes with and often prevents the addition of nucleophiles to imines. For this reason, imine additions often involve azomethines (*e.g.* benzylidineanilines) which are not capable of enolization. Many potentially useful additions, however, involve substrates capable of proton abstraction. By avoiding in certain instances some of the structural features of imines/imine derivatives and the reaction conditions responsible for proton abstraction, products resulting from this serious side reaction can be minimized.

The regioselectivity of imine deprotonations[1] is considerably more complicated than similar ketone deprotonations,[2] and can be influenced by imine geometry, nitrogen substitution, steric accessibility of α-hydrogen atoms and the deprotonation conditions (base, solvent) employed. Chelation-directed proton abstraction is implicated for metallated oximes and tosylhydrazones. For example, organometallic-mediated deprotonation of oximes (**1**) occurs with *syn* selectivity[3] (*syn* to the nitrogen alkoxide group) and provides a regioselective route to reactive azirines.[4] Likewise, the failure of aliphatic aldehyde tosylhydrazones (**2**; (*E*)-configuration) to form dianions at low temperatures is consistent with the strong preference for proton abstraction *syn* to the NSO_2Ar moeity and is presumably responsible for the susceptibility of (**2**) to nucleophilic addition.[5] Proton abstraction of ketone dimethylhydrazones and ketimines can lead to isomeric azaallyl anions. Ketone dimethylhydrazones generally deprotonate on the less-substituted α-carbon, regardless of the C=N geometry.[6] Alkyllithium-mediated deprotonation of ketimines, however, generally occurs *anti* to the nitrogen substituent.[7] For example, Bu^sLi (THF) treatment of 2-methylcyclohexanone-derived *N,N*-dimethylhydrazone (**3**) and *N*-cyclohexylimine (**4**) followed by the addition of benzyl chloride results in *syn* (100:0) and *anti* (87:13) α-alkylation, respectively.

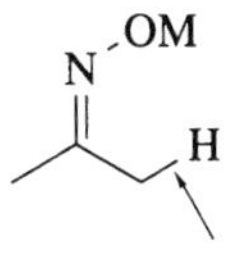

OM N H (1) M N N Ts H (2) NMe$_2$ N H (3) N H (4)

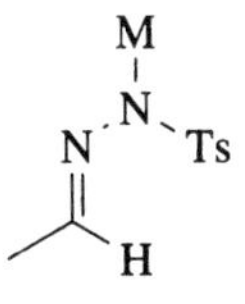

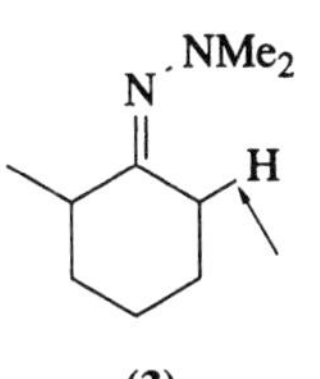

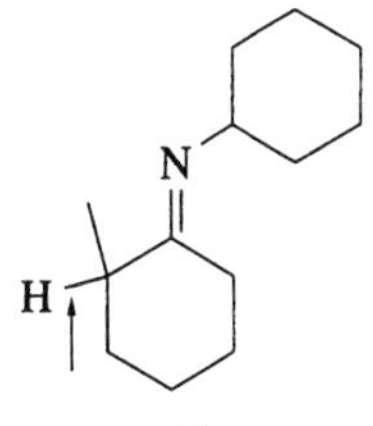

Regardless of the kinetic regioselectivity, Fraser, Houk and coworkers[8] have demonstrated that the nitrogen substituents of most metallated (lithio) aldehyde- and ketone-derived azomethines prefer the (*Z*)-orientation (Scheme 2). This thermodynamic preference is postulated to result from a minimization of dipole–dipole (electrostatic) interaction between nitrogen lone pair electrons and the carbanion.

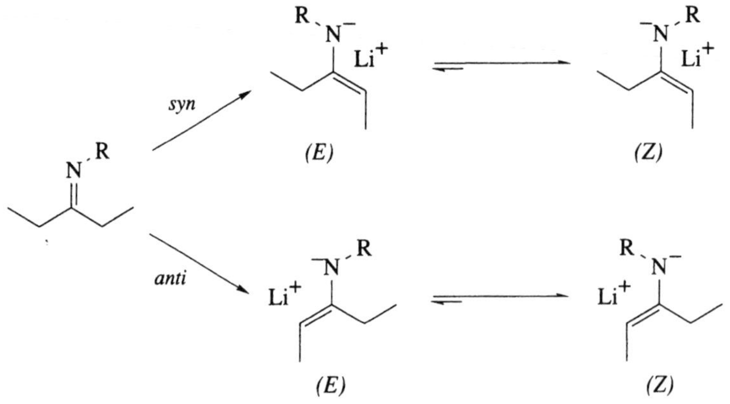

Scheme 2 Metallation of imines/imine derivatives

On the other hand, endocyclic ketimines such as 2-alkyl-1-pyrroline (**5**) have a thermodynamic bias for the (*E*)-azaallyl anion (which is postulated to result from excessive angle strain of the (*Z*)-azaallyl anion). This result suggests that by judicious selection of nucleophiles, proton abstraction might be minimal in additions to endocyclic aldimines which are devoid of 2-alkyl substitution, such as 1-pyrroline (**6**).

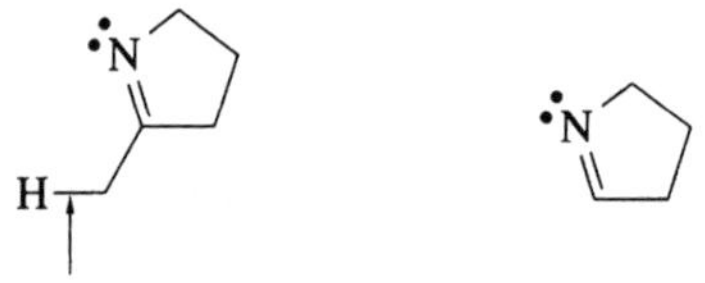

(**5**) (**6**)

Highly basic (organolithium) reagents are employed for the abstractions of protons adjacent to the carbon–nitrogen double bond of imines/imine derivatives. To minimize proton abstractions in nucleophilic additions, less basic reagents (many of which are outside the scope of this chapter)[9] such as allylboranes, allylboronates, allyl Grignards, allylzincs, allylstannanes, alkylcoppers, alkylcuprates, organocerium reagents and metal enolates (Li, B, Al, Zr, Sn, Zn) are used.

1.12.2.3 Stereochemical Control in Azomethine Additions

The design of practical methods for effectively controlling product stereochemistry in nucleophilic additions remains an important challenge in synthetic organic chemistry. The addition of carbon nucleophiles to aldehydes and ketones possessing an adjacent asymmetric center has been extensively investigated.[10] Two strategies for controlling product stereochemistry, which generally lead to opposite diasteriofacial selectivity, have emerged: (i) chelation control, in which Lewis acid reagents form intermediate chelates, rendering one face of the carbonyl moeity more accessible for nucleophilic addition (Cram's cyclic model, Figure 1a); and (ii) nonchelation control, in which the addition of reagents to substrates incapable of internal chelation is governed by steric and/or electronic factors. The bias for addition from a diastereotopic face in nonchelation-controlled additions was predicted empirically by Cram and is supported by the Cornforth and Felkin–Ahn models shown in Figures 1b and 1c.

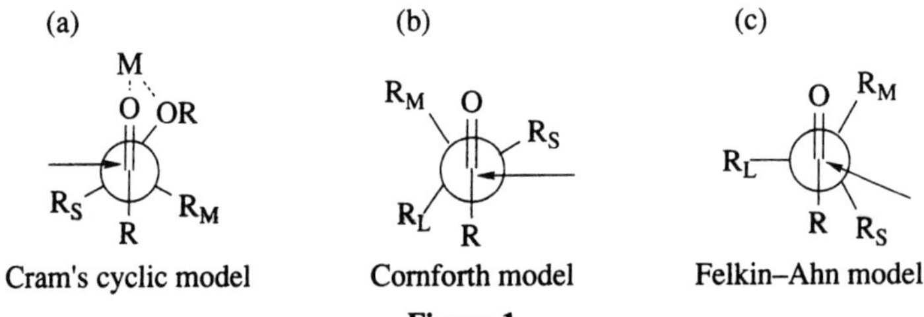

Figure 1

The relevance of these models for the addition of carbon nucleophiles to structurally similar acyclic imine/imine derivatives has not been adequately tested. Products which are consistent with chelation control are obtained in the addition of organolithium reagents to dimethylhydrazones (**7**) of α-alkoxyacetaldehydes.[11] In contrast to allylorganometallic and ester enolate additions,[9] only a few examples suggestive of nonchelation (Felkin–Ahn) control in the additions of nonstabilized organometallic reagents to chiral aldimines have been reported.[12,13] The paucity of any in depth systematic investigations examining the additions of nonstabilized carbanions to aldimines/aldimine derivatives containing an adjacent asymmetric center is not surprising. The propensity of acyclic imines to isomerize, the presence in these derivatives of other heteroatoms which are capable of chelation (*e.g.* hydrazones), and poor imine reactivity all contribute to the complexity of the problem.

(**7**) (**8**) (**9**) (**10**)

A major synthetic effort, however, has been focused on asymmetric induction resulting from the addition of organometallic reagents to aldimines/aldimine derivatives derived from chiral amines/hydrazines. A variety of chiral aldimines (**8**),[14] nitrones (**9**)[15] and hydrazones (**10**)[16–18] (all of which contain as part of the chiral auxilary a terminal alcohol, ether or ester moiety capable of chelation) have been examined. High diastereoselectivity can be obtained in these additions and, while the intimate mechanistic details which account for the stereoselectivity are not known, mechanisms involving: (i) chelation control (Figure 2a); and (ii) chelation-mediated delivery of the organometallic reagent (Figure 2b) have been suggested. The susceptibility of nitrogen–oxygen, nitrogen–nitrogen and nitrogen–aryl alkyl bonds within the addition products to reductive cleavage contributes to their value for the enantioselective synthesis of amines.

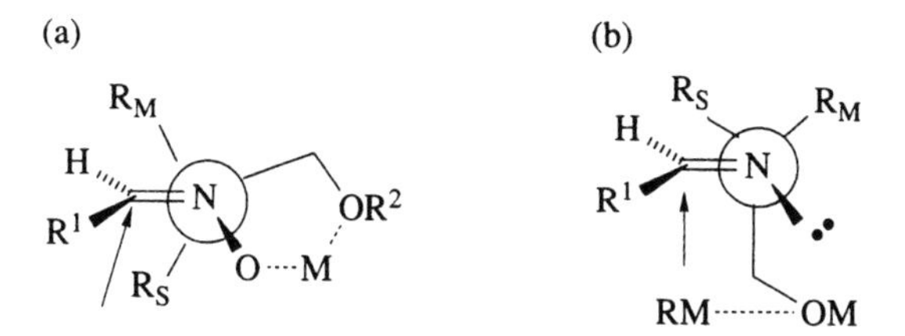

Figure 2 (a) Chelation control, *e.g.* nitrones; (b) Organometallic delivery, *e.g.* aldimines

The 1,4-addition of organometallic reagents to α,β-unsaturated aldimines (**11**)[19] and aromatic aldimines (**12**)[20] resembles additions to oxazolines (**13**)[21] and (**14**),[22] and provides a valuable method for generating a remote stereocenter (1,5-asymmetric induction). Other strategies have been employed for controlling azomethine diastereofacial selectivity. For example, organometallic treatment of chiral chromium complexes (**15**)[23] of *N*-arylaldimines leads to high stereoselectivity (presumably a result of steric control). Further investigations are required to assess the value of the chromium tricarbonyl complexes in asymmetric amine synthesis.

By constructing rigid endocyclic imines/imine derivatives (either isolated or generated *in situ*) some of the problems associated with acyclic stereocontrol can be avoided. The addition of organometallic reagents to 3-thiazolines (**16**)[24] or the putative piperidine intermediate (**17**)[25] provides excellent stereocontrol.

In summary, notable advances have been made in controlling stereoselectivity resulting from the addition of nonstabilized carbanions to chiral imine/imine derivatives. Unfortunately our level of mechanistic understanding in these additions is unsatisfactory. While additions involving chiral nitrones, hydrazones and some cyclic imines have been evaluated in reasonable detail, few systematic studies of other aldimine/aldimine derivatives are available.

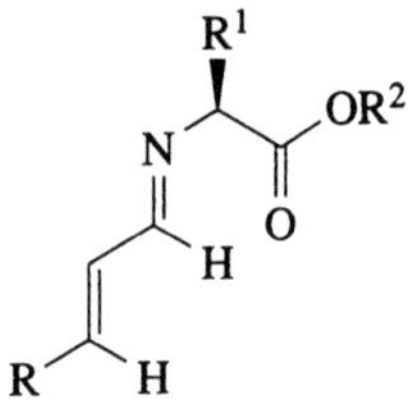

(11) 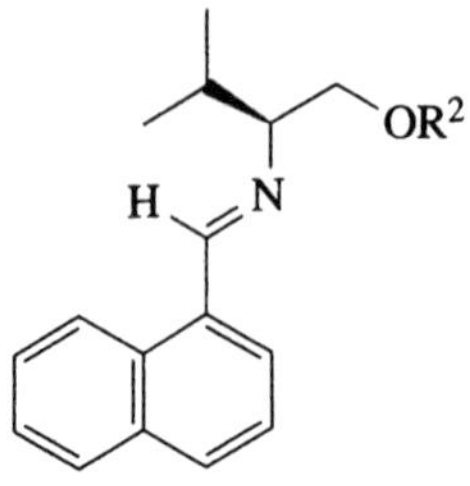 **(12)** 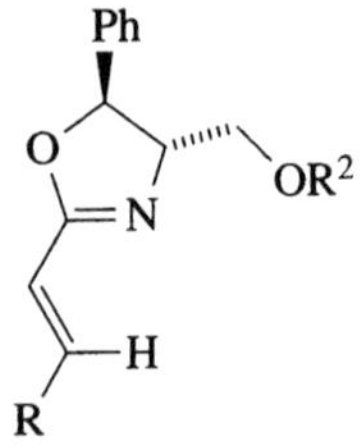 **(13)**

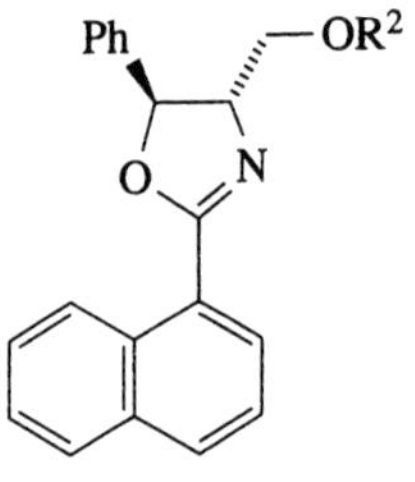

(14) 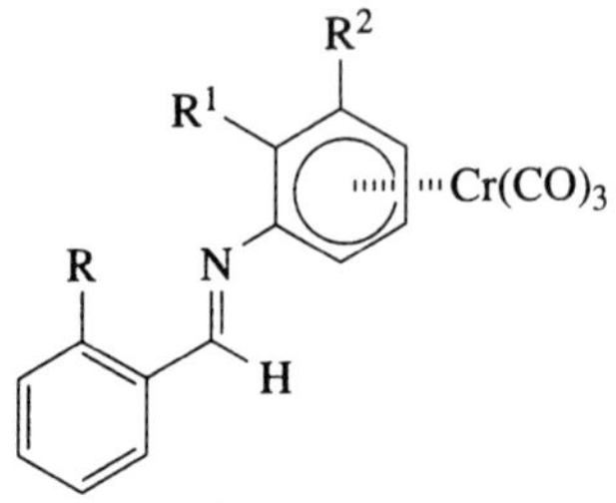 **(15)**

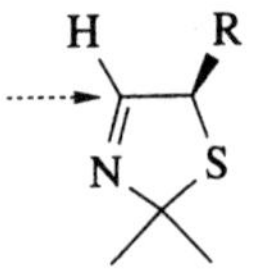

(16) 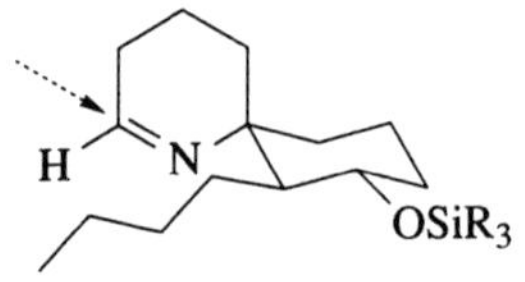 **(17)**

1.12.3 ORGANOMETALLIC ADDITIONS TO IMINES

The addition of organometallic agents to aldimines and ketimines provides a useful route to substituted amines, although this reaction is sensitive to imine/organometallic substitution. Along with addition, competitive enolization, reduction and bimolecular reduction (coupling) reactions are also possible.

1.12.3.1 Nonenolizable Imines

Addition to imines derived from aryl aldehydes has been investigated extensively and has been reviewed.[26] The addition of Schiff base (**18**) to excess Grignard reagent (2 equiv.) provides a route to secondary amines (**19**; equation 1). As the size of R and branching of R^1 increases, addition yields decrease (entries 1–3, Table 1),[12,27,28] and reduction products such as (**20**) are often generated.

Table 1 The Reaction of Organometallic Reagents with Azomethine (**18**)

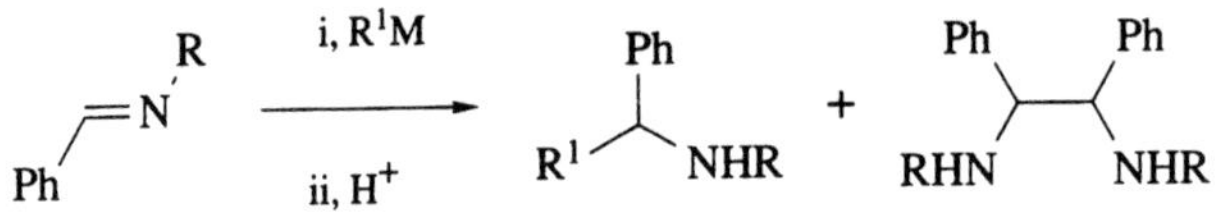

$$\text{(18)} \xrightarrow[\text{ii, H}^+]{\text{i, R}^1\text{M}} \text{(19)} + \text{(20)} \qquad (1)$$

Entry	*R*	*R^1M (Et_2O)*	*Yield* (**19**)	*Yield* (**20**)	*Ref.*
1	Me	MeMgI, Δ	72%	—	27
2	Bu^t	MeMgI, Δ	—	—	28
3	Bu	Pr^iMgI, Δ	—	56%	27
4	Bu^t	$2C_6F_{13}Li(BF_3)$, −78 °C	68%	—	12

In reactions with *N*-benzylidineaniline (PhCH=NPh), the addition of $MgBr_2$ (2 equiv.) to solutions of EtMgBr, Et_2Cd and Et_2Zn, or, in the case of Et_2Mg, lowering the solvent basicity (DME and Et_2O preferred over THF) results in dramatic yield improvements.[29] In the absence of other metal salts, Grignard addition yields increase if a 1:2 ratio of Schiff base to organometallic agent is employed.[30]

The effect of phenyl substitution on the rate of Grignard addition to *N*-benzylidineaniline has been examined. The reaction rate in ether for ethylmagnesium bromide conforms to $r = k[R^1{}_2Mg{\cdot}MgX_2][\text{Schiff base}]$. A four-centered reaction mechanism (Scheme 3) has been suggested.[30]

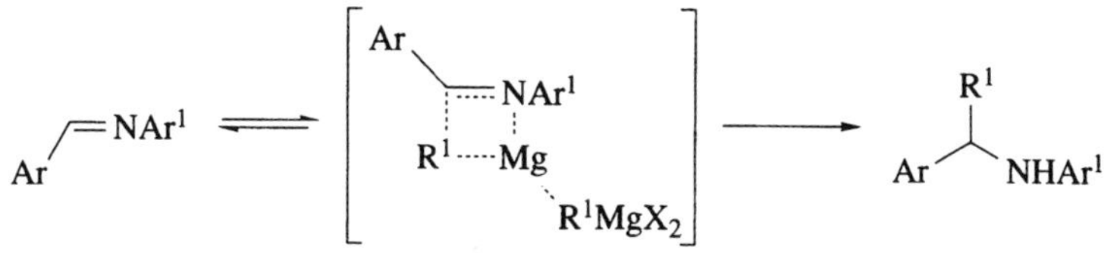

Scheme 3

In certain cases, Grignard reagents do not add to azomethines, but organolithium compounds (alone[28,31] or in the presence of $BF_3{\cdot}OEt_2$)[12] have been reported to add to many of these Schiff bases (entry 4, Table 1).

The addition of organometallic agents to *N*-methylenamines (**22**) has been utilized for the generation of unsymmetrical secondary amines. Even though these formaldehyde imines rapidly trimerize to yield 1,3,5-trialkylhexahydro-1,3,5-triazines (**24**), they can be generated *in situ* and treated with organolithium and Grignard reagents to provide secondary amines (**23**; Scheme 4). *N*-(Alkoxymethyl)aryl-,[32a] *N*-alkyl-*N*-(alkylthiomethyl)-,[32b] *N*-(cyanomethyl)-,[33] and *N*-(aminomethyl)-amines[33] (**21**) have been utilized for the preparation of formaldehyde imines (**22**). Primary, secondary, tertiary, aryl and alkenyl organometallic agents (a minimum of 2 equivalents is required) have been condensed with *N*-methylenamine precursors.

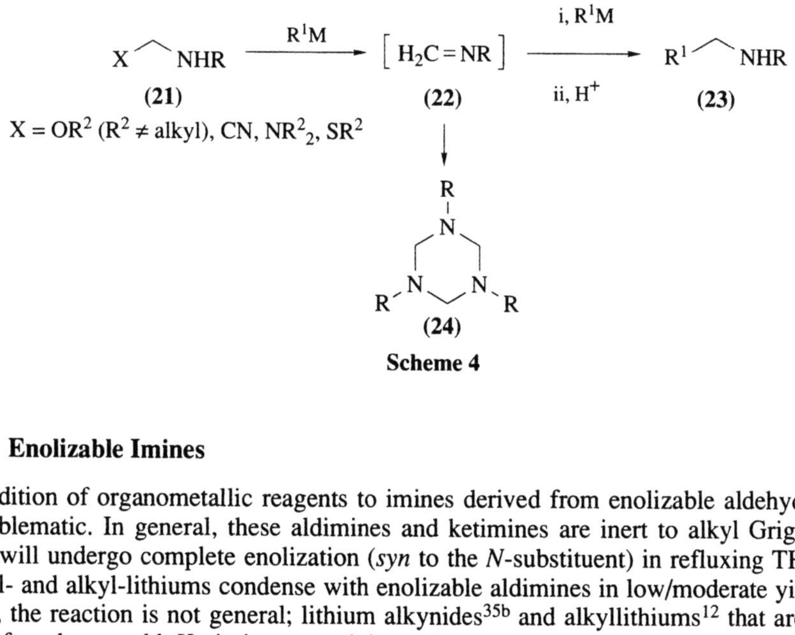

$X = OR^2$ ($R^2 \neq$ alkyl), CN, $NR^2{}_2$, SR^2

Scheme 4

1.12.3.2 Enolizable Imines

The addition of organometallic reagents to imines derived from enolizable aldehydes and ketones is more problematic. In general, these aldimines and ketimines are inert to alkyl Grignards and in their presence will undergo complete enolization (*syn* to the *N*-substituent) in refluxing THF.[34] On the other hand, aryl- and alkyl-lithiums condense with enolizable aldimines in low/moderate yields (~30–60%).[31] However, the reaction is not general; lithium alkynides[35b] and alkyllithiums[12] that are not stable above – 78 °C often do not add. Ketimines containing α-hydrogen atoms are resistant to organometallic addition. Sakurai and coworkers[7] have treated these ketimines with organolithium reagents to regioselectively generate substituted α-lithiated imines (deprotonation occurs preferentially *anti* to the *N*-substituent).

Activation of the C=N moiety by the addition of $BF_3{\cdot}OEt_2$[35,36] has increased the scope of organometallic additions (Table 2). Akiba and coworkers[35b] have shown that lithium alkynides treated with $BF_3{\cdot}OEt_2$ add to substituted aldimines (entry 1, Table 2).[12,13,35] BF_3 complexes of organocopper and dialkylcuprate reagents provide good yields of addition products. Dialkylcuprates are preferred in the condensation of branched aldimines (entry 2, Table 2).[35a]

Table 2 The Addition of Organometallic Reagents to Activated Imines Containing α-Hydrogen Atoms

Entry	*Organometallic*	*Imine*	*Product*	*Yield (%)*	*Ref.*
1	2 $LiC{\equiv}C(CH_2)_4Me$/ $BF_3{\bullet}OEt_2$/THF	$Ph(CH_2)_2CH{=}NPr^i$	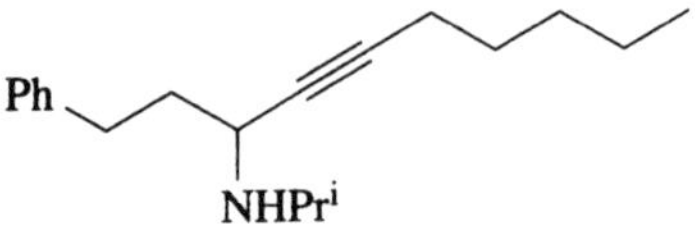 Ph, $NHPr^i$	82	35b
2	$Bu_2CuLi{\bullet}BF_3$/THF	$Me_2CHCH{=}N(CH_2)_2Ph$	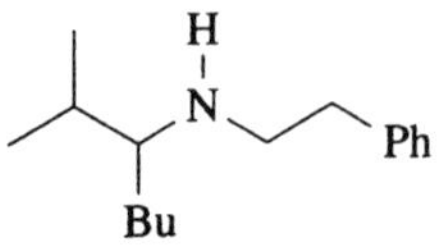 H, N, Ph, Bu	63	35a
3	$C_6F_{13}I$/MeLi/BF_3/Et_2O	$Me(Ph)C{=}NPr^n$	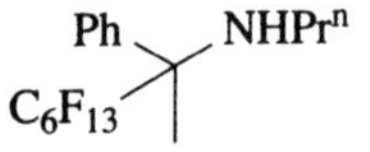 Ph, $NHPr^n$, C_6F_{13}	84	12
4	$C_6F_{13}I$/MeLi/BF_3/Et_2O	$Me(Ph)CHC{=}NPr^n$	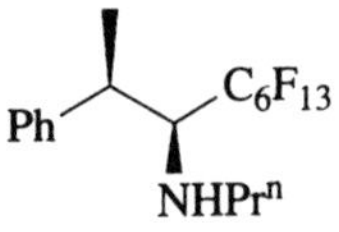 Ph, C_6F_{13}, $NHPr^n$ (25:1)	81	12
5	$Bu_2CuLi{\bullet}BF_3$	$Me(Ph)CHC{=}NPr^n$	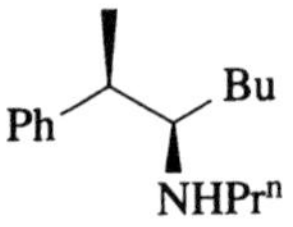 Ph, Bu, $NHPr^n$ (4:1)	NA	13

While Lewis acid activation has not provided a general solution for ketimine additions, isolated examples such as the condensation of perfluoroalkyllithiums with imines derived from acetophenone (entry 3, Table 2) have been reported.[12]

1.12.3.3 Stereochemical Control

Asymmetric induction is not high in the addition of organometallic agents to chiral aldehydes whose substituents have no chelating ability (such as 2-phenylpropanal). Although few examples involving nonstabilized carbanions to chiral aldimines have been reported, the presence of a stereogenic center adjacent to the aldimine can provide good diastereoselectivity; *e.g.* 60:40 *versus* 84:16, Cram/anti-Cram selectivity for 2-phenylpropanal *versus* its corresponding *N*-propylaldimine (**25**; equation 2) in the addition of allyl Grignard.[13] In the case of dialkylcuprate and perfluoroalkyllithium additions, BF_3 is required. Again good Cram/anti-Cram (Felkin–Ahn) selectivity is obtained (entries 4 and 5, Table 2).[12,13] Chelation/nonchelation control in the addition of stabilized allyl (crotyl) organometallics to chiral aldimines has been reviewed.[9b]

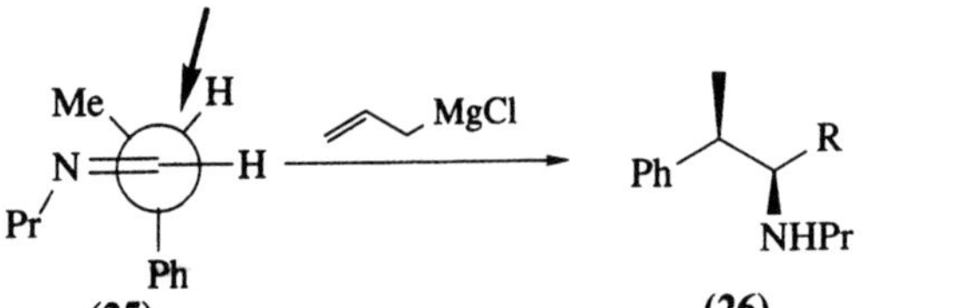

Me, H, N, H, Pr, Ph — MgCl → Ph, R, NHPr (2)

(25) **(26)** major (84:16, Cram:anti-Cram)

Fiaud and Kagan[37] examined the addition of organometallic agents to *N*-substituted imines of (–)-menthyl glyoxylates with the intention of producing substituted amino acids. Grignard condensations with *N*-[(*S*)-α-methylbenzyl]iminoacetate (**27**) produced, in modest yields, varying amounts of secondary and tertiary amines (**28**) and (**29**), demonstrating the molecules ambident electrophilicity (equation 3). While the problem of regiochemical control (**29** *versus* **28**) can be solved by the use of organocadmium agents, stereochemical control in the production of (**29**) is modest (*de* ≈ 40–50%). The menthyl ester moiety is the major stereochemical determinant, as replacement of the *N*-[(*S*)-α-methylbenzyl] substituent with the corresponding (*R*)-isomer had little effect on reaction diastereoselectivity (for R^1 = Pr^n, 40 *versus* 41% *ee*).

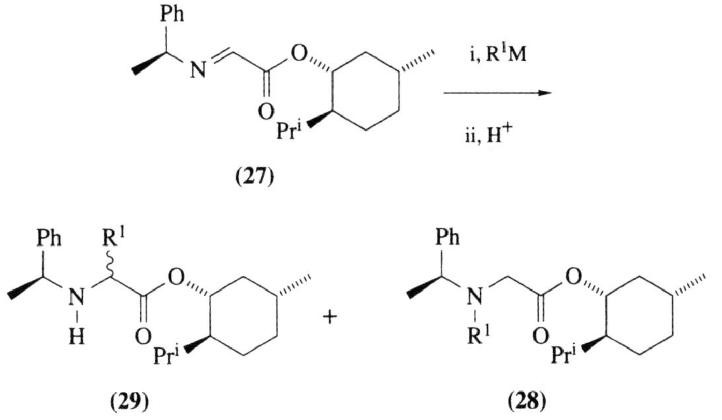

i, R^1M; ii, H^+

(**27**) → (**29**) + (**28**) (3)

Diastereoselectivity produced by 1,3-asymmetric induction in the reaction of (*S*)-valinol-derived imines with organometallic agents can be high. Takahashi, Suzuki and coworkers (equation 4) have examined the condensation of chiral azomethines (**8**), which are reported to exist exclusively as the (*E*)-isomer, with organolithium and Grignard reagents (≥4 equiv.).[14] Addition occurs predominantly from the *si*-face of Figure 3 and is consistent with an alkoxy-mediated delivery of the organometallic agent. The size of substituent R^1 at the resident stereogenic center of Figure 3 is a key stereochemical determinant of the diastereoselectivity (R^1 = Pr^i, Bu^i and Bu^s are superior to R^1 = Me; entries 1 and 2, Table 3).[14] Conversion of the alcohol funtionality to the corresponding methyl ether has little effect on the stereochemical course of the addition (entries 3, 4, 6 and 7, Table 3). By appropriate selection of azomethine and organometallic agent, both diastereomeric amines can be obtained (entries 1 and 5, Table 3).

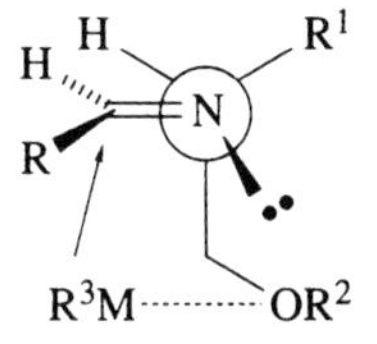

Figure 3

Table 3 The Reaction of Organometallic Reagents with Chiral Aldimines (**8**)

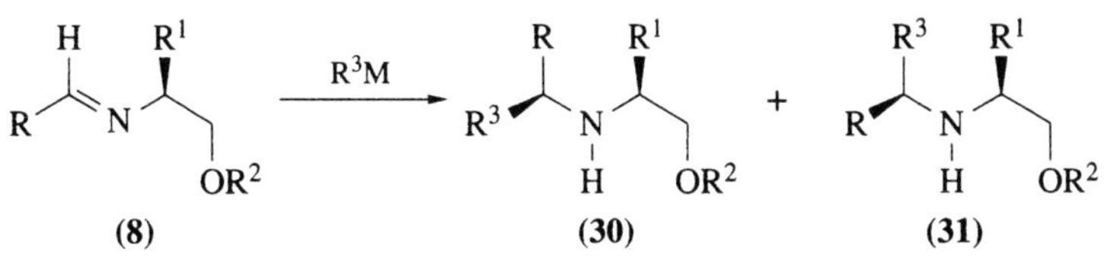

(**8**) —R^3M→ (**30**) + (**31**) (4)

Entry	*R*	*R^1*	*R^2*	*R^3M*	*Overall yield (%)*	*(30) (%)*	*(31) (%)*	*Ref.*
1	Ph	Pr^i	H	$PhCH_2MgCl/THF$	73	>98	—	14a
2	Ph	Me	H	$PhCH_2MgCl/THF$	53	78	22	14b
3	Ph	Pr^i	H	$EtMgBr/Et_2O$	63	78	22	14e
4	Ph	Pr^i	Me	$EtMgBr/Et_2O$	81	91	9	14e
5	CH_2Ph	Pr^i	H	$PhLi/Et_2O$	35	>98	—	14a
6	Et	Pr^i	H	$PhLi/Et_2O$	68	>97	—	14e
7	Et	Pr^i	Me	$PhLi/Et_2O$	60	95	5	14e

1.12.3.4 Cyclic Imines

Cyclic imines, such as 1-pyrroline (**6**) and Δ^1-piperidine (**32**), which are prone to trimerization, can be prepared from *N*-halopyrrolidine and *N*-halopiperidine respectively and treated *in situ* with organolithium reagents to provide 2-alkylated and 2-arylated pyrrolidines and piperidines in modest yield.[38] Corey *et al.*[25] treated substituted piperidine (**17**) generated in this fashion with *N*-pentyllithium to establish the stereochemistry of the remote pentyl side chain of perhydrohistrionicotoxin (**33**), as condensation occurred from the more accessible *re*-face of the imine π-system of (**17**).

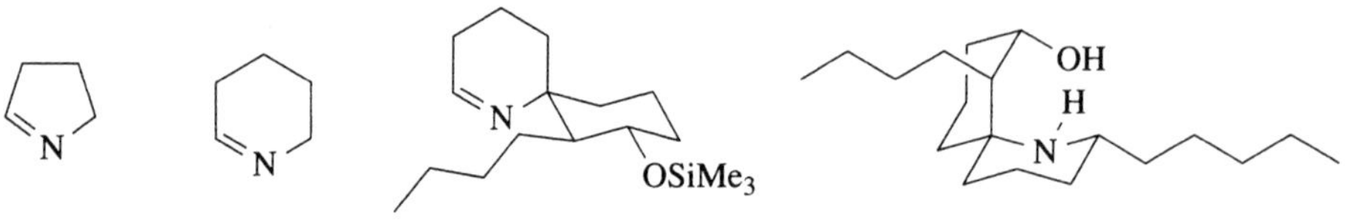

(**6**) (**32**) (**17**) (**33**)

Difficulties in generating highly functionalized alicyclic imines have limited their synthetic utility. *In situ* generation of piperidines from highly substituted α-aminonitriles has been employed in the Grignard-mediated conversion of the 1-α-cyano-1-deoxynojirimycin derivative (**34**) to the 1-α-substituted amine (**35**; equation 5).[39]

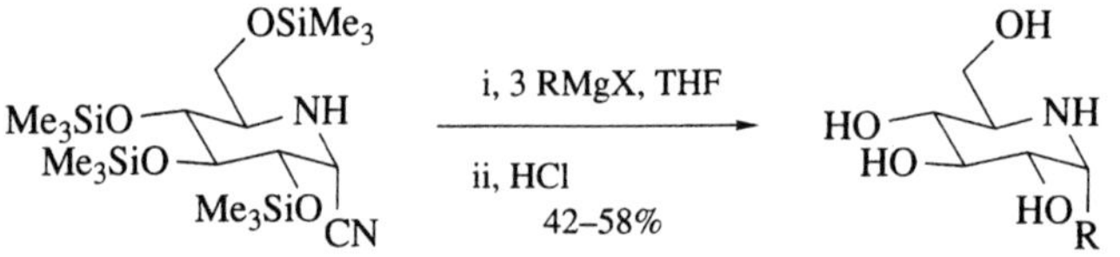

(**34**) i, 3 RMgX, THF; ii, HCl; 42–58% → (**35**) (5)

Substituted 3-thiazolines (**16**) are stable and easily prepared, but are inert to organometallic addition. Activation of (**16**) with $BF_3 \cdot OEt_2$ followed by organometallic addition (R^1MgX, R^1Li) provides *trans*-4,5-disubstituted thiazolidines (**36**) (masked 2-aminothiols) in which R^1 = aryl, alkyl or alkynyl (equation 6).[24]

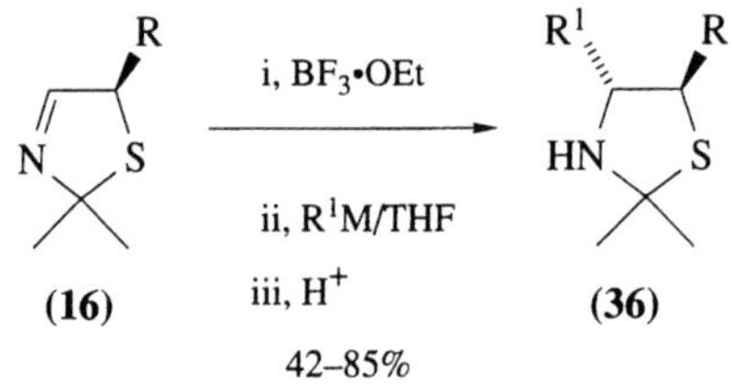

(**16**) i, $BF_3 \cdot OEt$; ii, R^1M/THF; iii, H^+; 42–85% → (**36**) (6)

The addition of organolithium agents to bicyclic imines such as (**37**) provides modest yields of alkylated product (**38**; equation 7).[40] These condensations have been reviewed.[26]

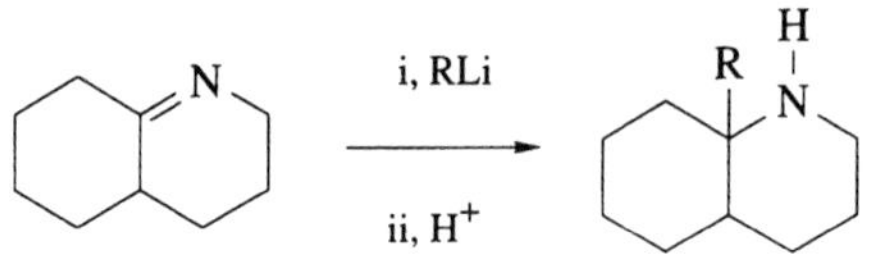

(**37**) i, RLi; ii, H^+ → (**38**) (7)

1.12.3.5 Chiral Chromium Complexes

Solladie-Cavallo and Tsamo (equation 8) have investigated the addition of Grignard reagents to chiral chromium tricarbonyl complexes of diarylimines (**15**).[23] Results of these additions are shown in Table 4. While the absolute stereochemistry of the product(s) (**39**) formed has not been determined, the results demonstrate that high diastereoselectivity can be obtained in the addition of benzyl Grignards to azomethine (**15**), in which ring A is *ortho* substituted (entries 4–6, Table 4).

Table 4 Addition of Grignard Reagents to Chiral Chromium Tricarbonyl Complexes of Diarylimines (**15**)

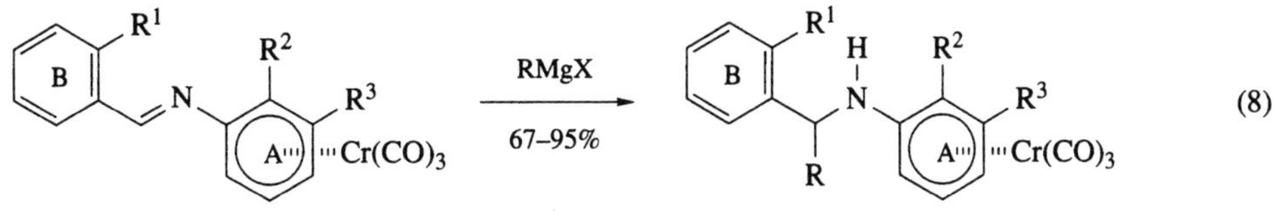

RMgX, 67–95%: (**15**) → (**39**) (8)

Entry	*RMgX*	*R^1*	*R^2*	*R^3*	*Diastereomeric ratio*[a]
1	MeMgI	H	Me	H	67:33
2	MeMgI	Me	Me	H	67:33
3	MeMgI	Me	Ph	H	66:34
4	$PhCH_2MgCl$	Me	Me	H	100:0
5	$PhCH_2MgCl$	OMe	Me	H	100:0
6	$PhCH_2MgCl$	Cl	Me	H	100:0
7	$PhCH_2MgCl$	Me	H	Me	57:43

[a]Determined by ^1H NMR (250 MHz).

The diastereoselectivity has been rationalized by the authors with the transition state model shown in Figure 4a, in which the $Cr(CO)_3$ moiety is situated in the plane of the azomethine group (σ_1) and approach of the Grignard occurs nearly orthogonal to this plane. The facial selectivity is postulated to arise from interactions of the incoming Grignard reagent and the substituents on ring A. Determination of the absolute configuration of the product(s) (**39**) should probe the relevance of this model since organometallic addition to (**15**) in which $Cr(CO)_3$ is orthogonal to the plane of the azomethine group (Figure 4b) would generate the opposite diastereomer.

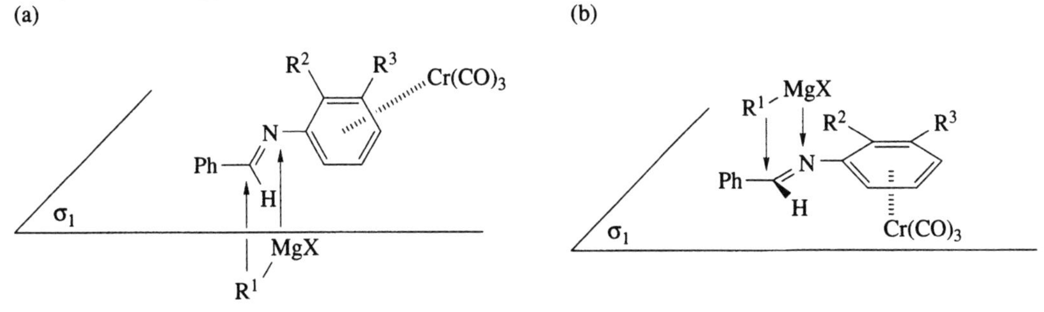

Figure 4

Alkyllithium addition to the (η^6-arene) dicarbonylchromium imine chelates has been examined.[41] Treatment of optically pure chelate (**41**) with methyllithium provides amine (**42**) with an enantiomeric excess of 94% (Scheme 5). No diastereoselectivity was reported for alkyllithium additions to (arene)tricarbonylchromium complex (**40**). In the absence of additional examples, the generality of this chromium chelate methodology for asymmetric amine synthesis cannot be assessed.

1.12.4 ORGANOMETALLIC ADDITIONS TO IMINIUM SALTS

1.12.4.1 Background

While the addition of organometallic reagents to acyclic or cyclic imines (**43**) is often compromised by poor imine reactivity, these reagents readily condense with structurally diverse, more electrophilic iminium salts (**44**), bearing a positive charge on the nitrogen atom, to provide substituted tertiary amines (**45**; Scheme 6). A review of the literature describing additions to cyclic iminium salts prior to 1966 is avail-

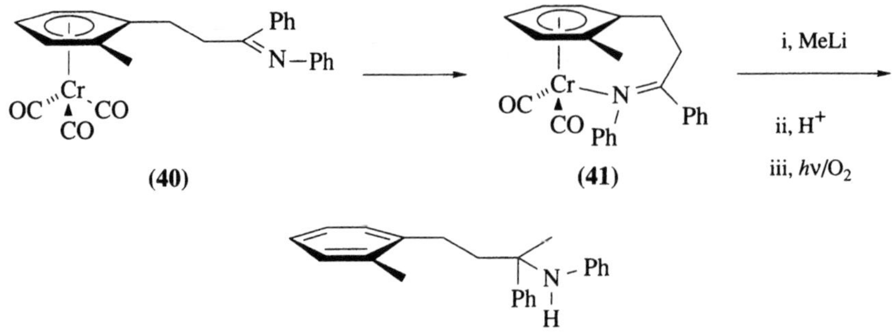

(40) → (41) → [i, MeLi; ii, H^+; iii, $h\nu/O_2$] → (42) 72% (*ee* 94%)

Scheme 5

able.[42] More recent reviews (up to 1976) summarize the addition of organometallic reagents to iminium salts derived from aliphatic, aromatic or heteroaromatic aldehydes as well as from aromatic or aliphatic ketones.[43]

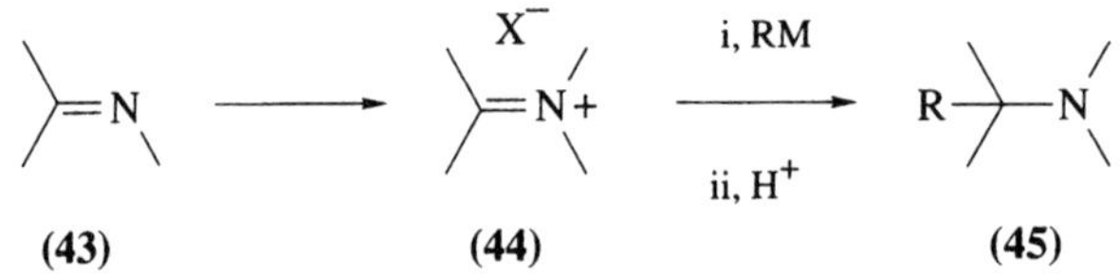

(43) → (44) → [i, RM; ii, H^+] → (45)

Scheme 6

1.12.4.2 Preformed Iminium Salts

Preformed iminium salts have been used extensively in organic synthesis. The facility of the condensation is a function of iminium salt substitution. Treatment of formaldehyde-derived *N,N*-dimethyl(methylene)ammonium halides (or trifluoroacetates) (**46**) with Grignard and lithium reagents results in the high yield formation of dimethylaminomethyl-containing compounds (**47**).[44] Subsequent oxidation[45] or alkylation[46] of these products has been employed to generate terminal alkenes (**48**; Scheme 7). As expected, addition yields are modest for the more-hindered iminium salts derived from other aldehydes and are somewhat lower for those derived from cyclic ketones.[47]

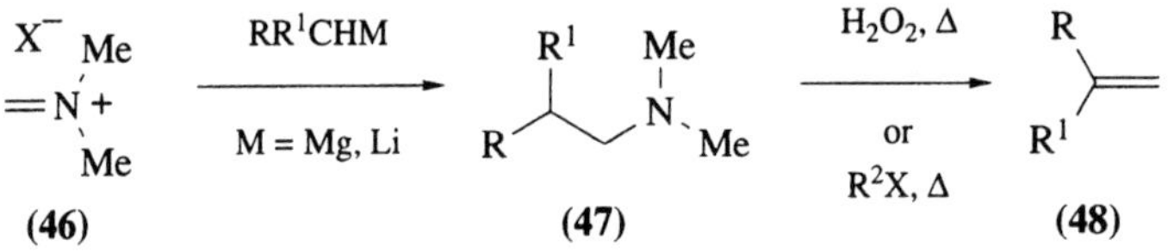

(46) → [RR^1CHM; M = Mg, Li] → (47) → [H_2O_2, Δ or R^2X, Δ] → (48)

Scheme 7

Addition to cyclic iminium salts has been utilized in alkaloid synthesis. A zinc-promoted reductive coupling reaction of iminium salts and alkyl halides has been reported by Shono *et al.* (Scheme 8).[48] Evidence delineating the mechanistic course (organozinc addition or electron transfer reaction) of the addition has not been established. In contrast to organolithium or Grignard additions, aromatic halogen and alkoxycarbonyl substituents are compatible with this methodology. The intramolecular version of this reaction has been employed for the synthesis of tricyclic amines (**53**; equation 9).

The quaternization of imines to form more reactive iminium salts has had limited synthetic utility since the activating substituents are often not easily removed. For this reason Lewis acids have been utilized to activate imines. MacLean and coworkers[49] have also addressed this problem and found that silylation of 3,4-dihydroisoquinolines and 3,4-dihydro-β-carbolines with trimethylsilyl triflate (TMS-OTf) provides a reactive, yet labile, silyl iminium salt which undergoes nucleophilic addition (Scheme 9). With this procedure, 3,4-dihydro-6,7-dialkoxyisoquinoline (**54**) was converted to amidine (**56**) in 98% yield. In the absence of TMS-OTf no addition occurred.

While 2-substituted-4,4,6-trimethyl-5,6-dihydro-1,3-oxazines[50] 2-substituted-4,4-dimethyl-2-oxazolines[51] and 1-benzyl-2-alkyl-4,5-dihydroimidazoles[52] are inert to organometallic addition, their corresponding methiodide salts (**57**)–(**59**) are reactive and have been utilized for the synthesis of ketones (**60**;

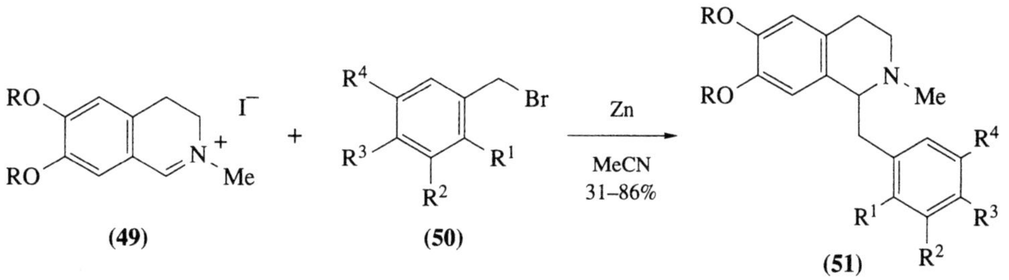

(49) (50) Zn, MeCN, 31–86% (51)

Scheme 8

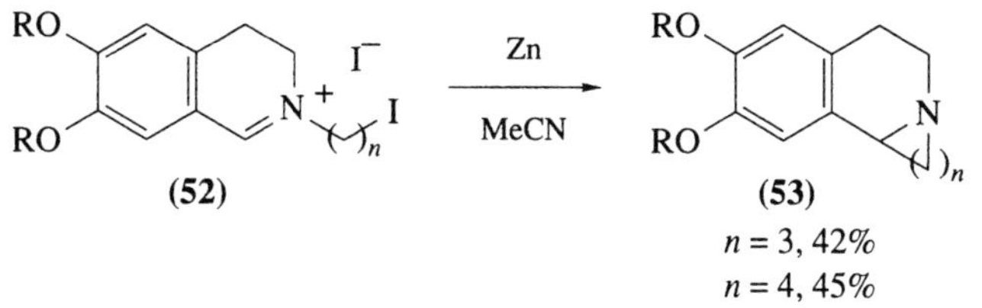

(52) Zn, MeCN (53) $n = 3$, 42%; $n = 4$, 45% (9)

(54) Me_3SiOTf (55) (56)

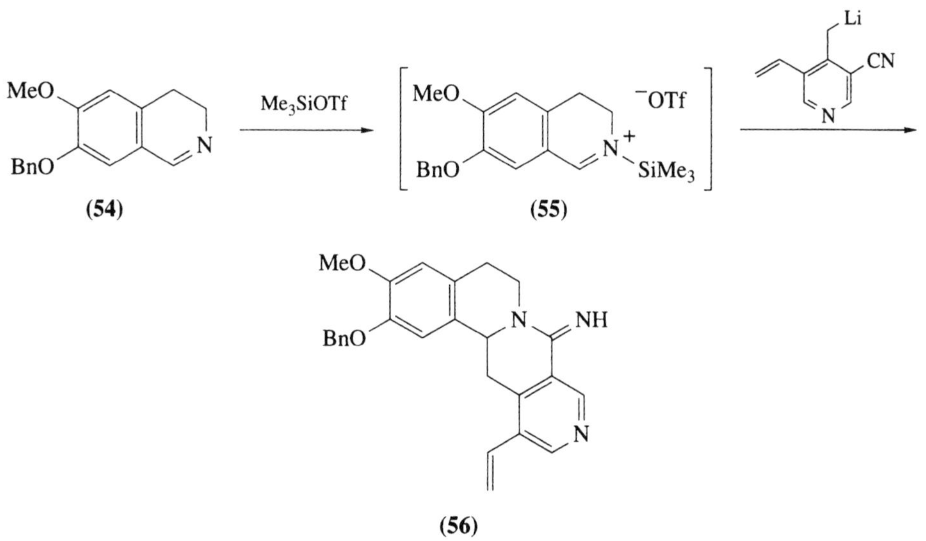

Scheme 9

Scheme 10). The use of this methodology in ketone synthesis is not widespread, given the lack of availability of these salts and the sensitivity of this reaction to substrate/organometallic substitutions.

Addition to vinylogous iminium salts has also been explored. Moriya *et al.*[53] have examined the addition of nucleophiles to 3-(1-pyrrolidinylmethylene)-3*H*-indolium salt (**61**) in the synthesis of 3-(1-dialkylamino)alkyl-1*H*-indoles (**62**; equation 10).

Similarly, Gupton *et al.*[54] have shown that acyclic amidinium salts (**63**) and vinylogous amidines (**65**) are reactive and can be utilized for the synthesis of structurally diverse aldehydes (Scheme 11).

In addition, functionalized piperidines (**68**) have been generated by the addition of organometallic reagents to 5,6-dihydropyridinium salts (**67**; equation 11).[55]

1.12.4.3 Iminium Salts Generated *In Situ*

Difficulties associated with the preparation and purification of anhydrous, soluble iminium salts (**44**) suitable for organometallic additions can in certain instances be avoided by an *in situ* generation of these reactive molecules.[56] *N,N*-(Disubstituted)aminomethyl ethers (or acetates) (**69**), sulfides (**70**) and nitriles (**71**) are frequently utilized in the generation of these salts. In addition, *N,N*-(disubstituted)aminomethylamides (**72**), -sulfonates (**73**) and halides (**74**) have been employed (Scheme 12).

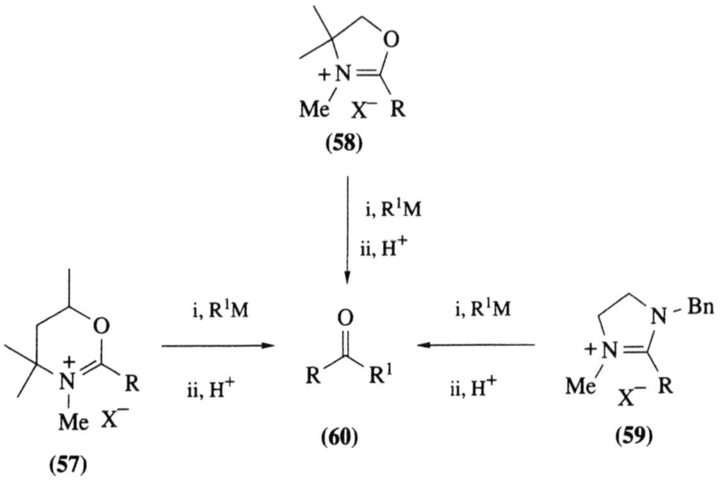

(58) (57) (60) (59)

i, R^1M ii, H^+

Scheme 10

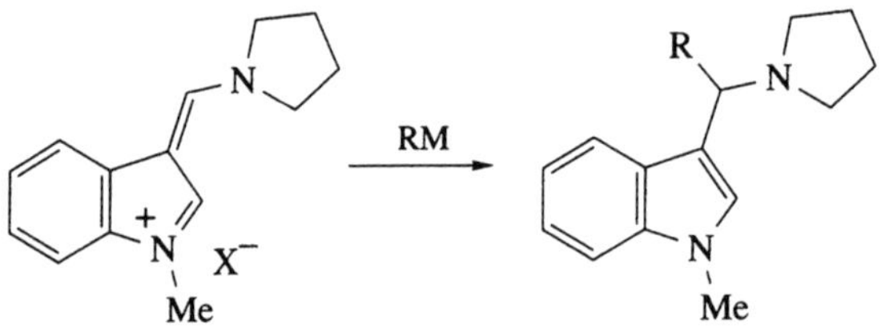

(61) (62) 76–100%

RM (10)

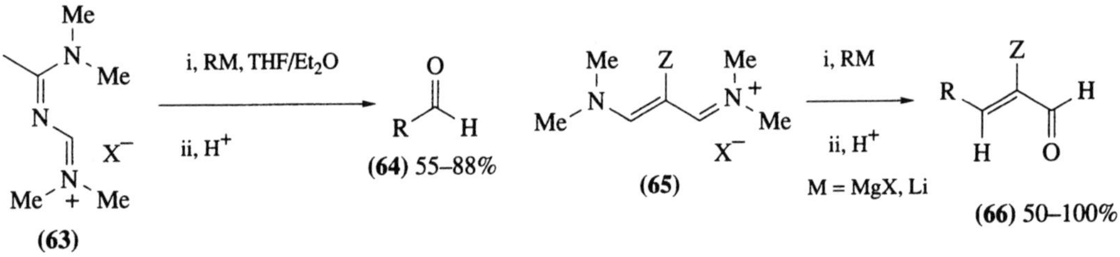

(63) i, RM, THF/Et_2O ii, H^+ (64) 55–88%

(65) i, RM ii, H^+ M = MgX, Li (66) 50–100%

Scheme 11

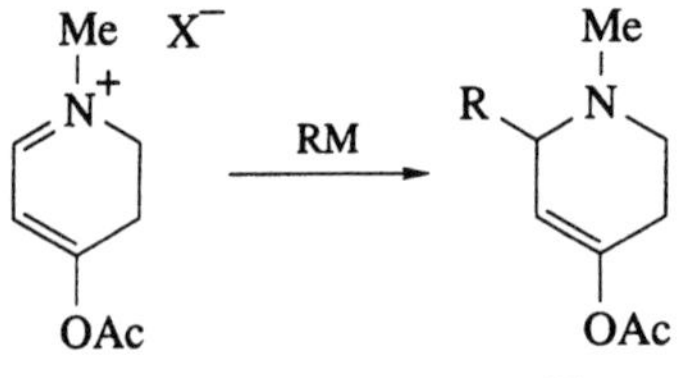

(67) RM (68) (11)

The reaction of *N,N*-(disubstituted)aminomethyl ethers (**69**) with Grignard reagents has been extensively investigated[56] and provides good yields of substituted tertiary amines. Representative examples are shown in Table 5. Acyclic[57] as well as cyclic[58] ethers (entries 1 and 2, Table 5) have been studied. Grignard treatment of bis(alkoxymethyl)amines (entry 3, Table 5)[59] and dialkylformamide acetals (entry 4, Table 5)[60] results in the displacement of both alkoxy substituents. In addition, β,γ-unsaturated-α-amino esters can be easily accessed by organozinc (R_2Zn) addition to alkoxy(dialkylamino)acetates (entry 5, Table 5).[61]

The propensity of formaldehyde imines (**22**) to trimerize to give hexahydro-1,3,5-triazines (**24**) has limited their synthetic utility. Sekiya and coworkers[62] and Bestmann *et al.*[63] have shown that *N,N*-bis(trimethylsilyl)methoxymethylamine (**75**) is a useful synthetic equivalent of formaldehyde imine. Treatment

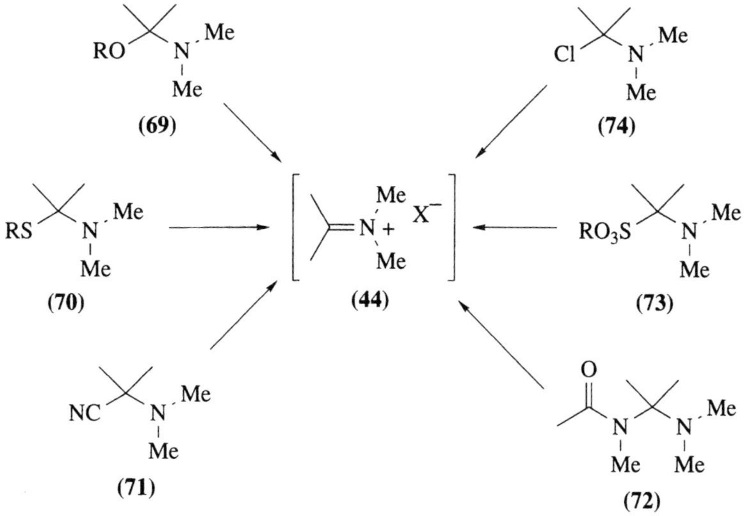

RO–C(Me₂)–NMe₂ (69); RS–C(Me₂)–NMe₂ (70); NC–C(Me₂)–NMe₂ (71); MeC(O)N(Me)–C(Me₂)–NMe₂ (72); RO_3S–C(Me₂)–NMe₂ (73); Cl–C(Me₂)–NMe₂ (74) → [$Me_2C{=}N^+Me_2$ X^-] (44)

Scheme 12

Table 5 Addition of Organometallic Reagents to *N,N*-Disubstituted Aminoalkyl Ethers

Entry	*Substrate*	*Organometallic reagent/conditions*	*Product*	*Yield (%)*	*Ref.*
1	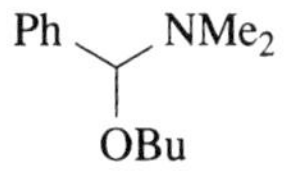 PhCH(OBu)NMe_2	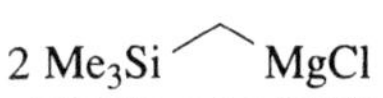 2 Me_3SiCH=CHCH₂MgCl THF/0–20 °C/15 h	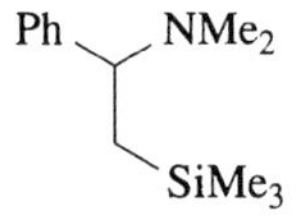 PhCH(NMe_2)CH₂CH=CH$SiMe_3$	84	57
2	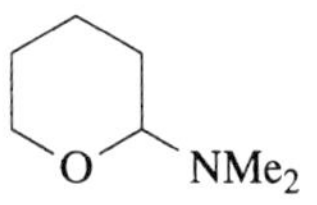 2-(NMe_2)tetrahydropyran	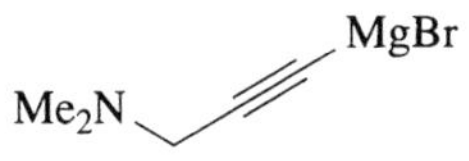 Me_2NCH₂C≡CMgBr	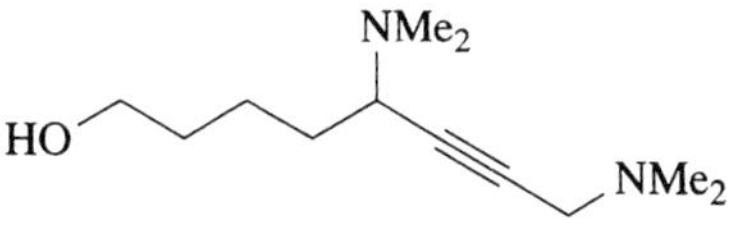 HO(CH₂)₄CH(NMe_2)C≡CCH₂NMe_2	89	58
3	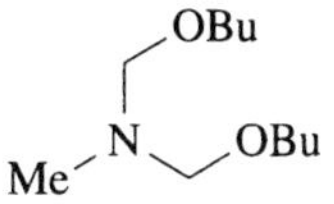 MeN(CH₂OBu)₂	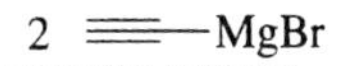 2 HC≡C–MgBr THF/50 °C/15 h	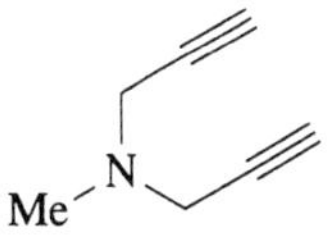 MeN(CH₂C≡CH)₂	44	59
4	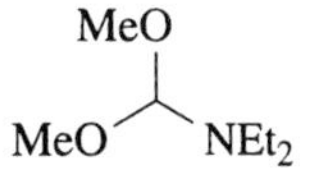 (MeO)₂CHNEt_2	2.5 PhMgBr Et_2O/0 °C to r.t./30 min	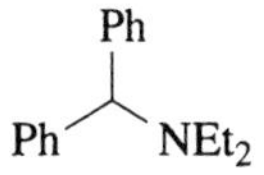 Ph₂CHNEt_2	93	60
5	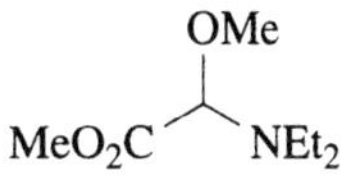 MeO_2CCH(OMe)NEt_2	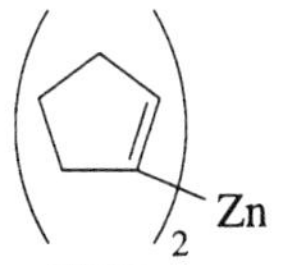 (cyclopentenyl)₂Zn THF	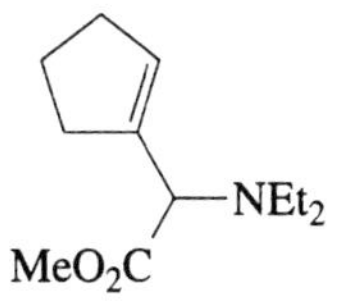 MeO_2CCH(cyclopentenyl)NEt_2	63	61

of aryl, heteroaryl, primary and secondary alkyl Grignards (organolithium reagents can also be used but require the addition of $MgBr_2$) with (**75**) generates in good yield *N,N*-bis(trimethylsilyl)amines (**76**), which can be readily desilylated to generate primary amines (**77**; Scheme 13).

α-Amino ethers need not be isolated in the synthesis of tertiary amines. α-Substituted phenethylamines (**81**), for example, have been generated in good yield from aldehydes and benzylmagnesium chloride using titanium amide complexes.[64a] The intermediacy of a titanium amide tetraisopropoxide ate complex (**79**) which can condense with aldehydes to form the reactive titanium complex (**80**), has been proposed (Scheme 14).

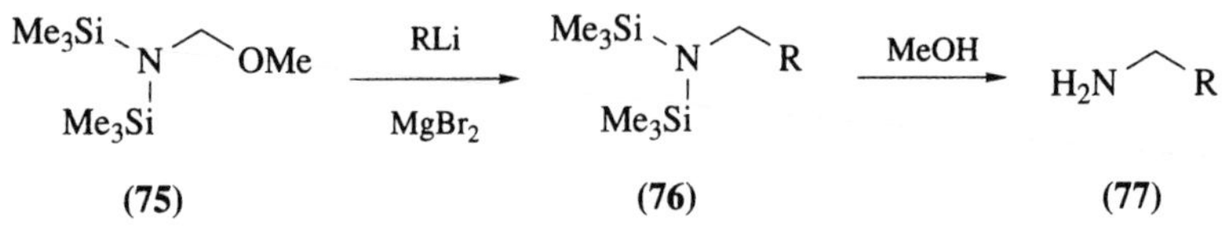

(75) **(76)** **(77)**

Scheme 13

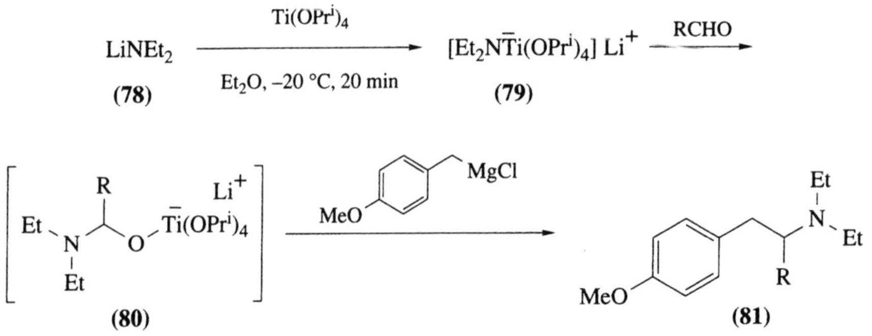

(78) **(79)** **(80)** **(81)**

Scheme 14

In addition, amine *N*-oxides (**82**) can be treated with trialkylsilyl triflates to generate, after methyllithium treatment, α-siloxyamines (**83**) which, when allowed to react with Grignard reagents (or trialkylaluminum), generate tertiary amines (**84**) in modest yields (Scheme 15).[65]

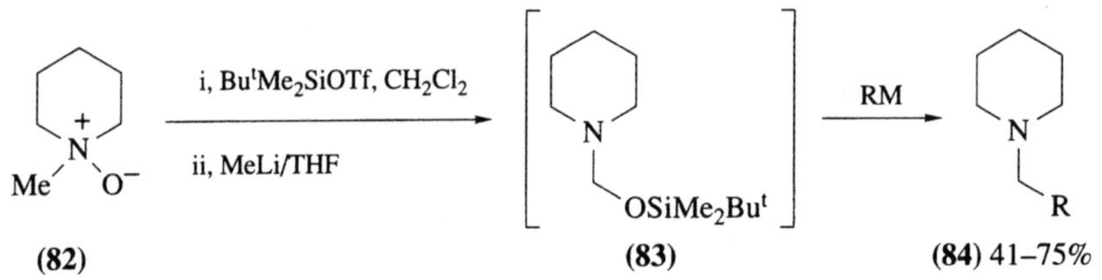

(82) **(83)** **(84)** 41–75%

Scheme 15

O'Donnell *et al.*[66] have examined benzophenone-derived glycine acetate (**85**) as a glycine cation equivalent. Higher order mixed cuprates [$R_2Cu(CN)Li_2$] react with (**85**) to afford alkylated Schiff base (**86**) in moderate yields. The reaction is sensitive to reaction conditions and reduction competes with organometallic addition. Hydrolysis of (**86**) provides access to aryl- or alkyl-substituted amino acids (**87**; Scheme 16).

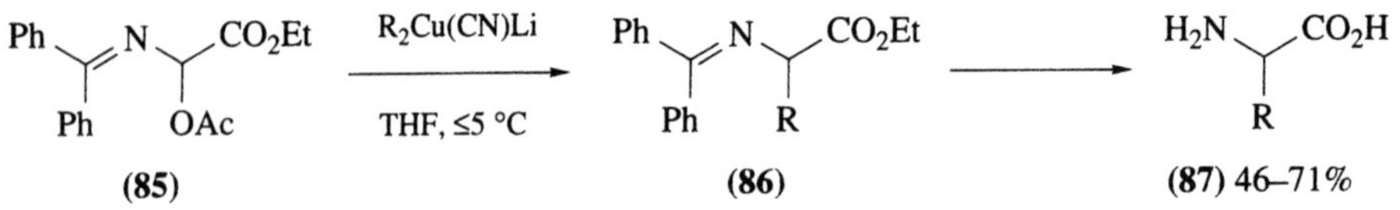

(85) **(86)** **(87)** 46–71%

Scheme 16

N-(Arylthiomethyl)amines, which are easily prepared and reportedly have better shelf stability[67] than the corresponding *N*-(alkoxymethyl)amines,[56] have also been utilized to prepare tertiary amines. Alkenyl cuprate addition to *N,N*-diethylphenylthiomethylamine produces allylic amines in high yield (entry 1, Table 6).[68] Both alkenyl groups of the cuprate react. *S*-(Dialkylaminomethyl) dithiocarbamates (entry 2, Table 6),[69] sulfonates (entry 3, Table 6)[70] and amides[71] also generate iminium salts *in situ*. Of a variety of amides examined, imides such as dialkylaminomethyl-succinimides (entry 4, Table 6)[71] and -phthalimides provide the best overall yields. Organolithium reagents are rarely utilized in aminomethylation reactions. However, *N*-chloromethylamines condense readily with lithiated anisole derivatives (entry 5, Table 6).[72]

The condensation of Grignard reagents with *N,N*-(disubstituted)aminomethylnitriles has been widely studied.[56] The reaction can accommodate hindered α-aminonitriles as exemplified by the condensation of phenyl Grignard with 1-piperidinocyclohexane carbonitrile to afford phencyclidine (entry 6, Table

Table 6 Addition of Organometallic Reagents to Iminium Salts Generated *In Situ*

Entry	*Substrate*	*Organometallic reagent/conditions*	*Product*	*Yield (%)*	*Ref.*
1	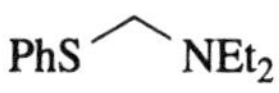 PhS–CH₂–NEt₂	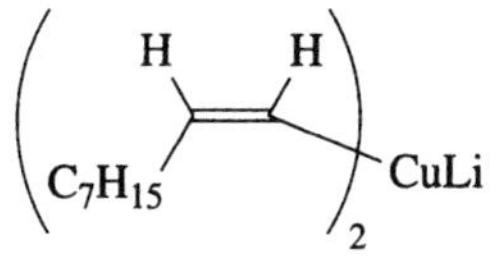 ((Z)-C_7H_{15}CH=CH)$_2$CuLi, THF/–40 °C to r.t./3 h	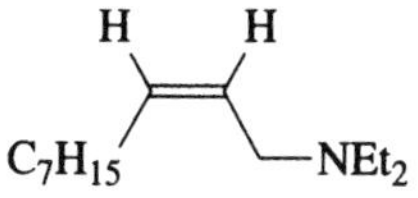 (Z)-C_7H_{15}CH=CHCH₂NEt₂	95	68
2	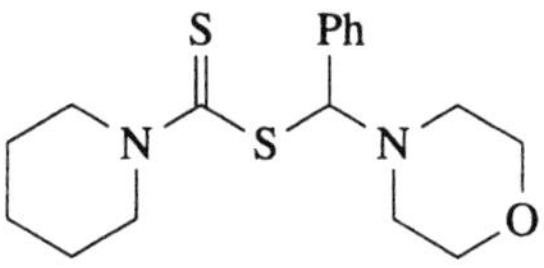 [structure: piperidine dithiocarbamate / morpholine, S, Ph]	PhMgBr, Et_2O	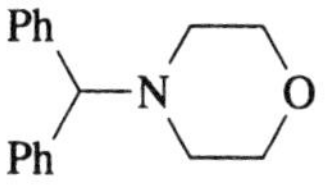 Ph_2CH–N(morpholine)	81	69
3	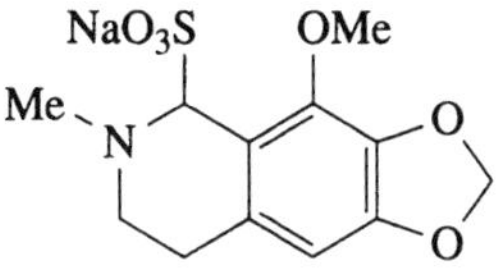 [structure: NaO_3S, Me–N, OMe, methylenedioxy tetrahydroisoquinoline]	EtMgBr, Et_2O	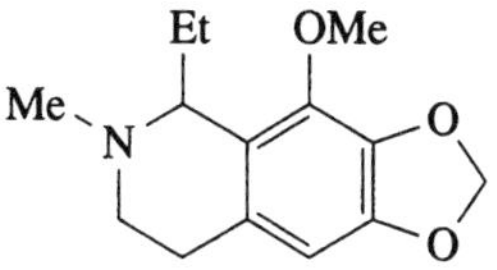 [structure: Et, Me–N, OMe, methylenedioxy tetrahydroisoquinoline]	94	70
4	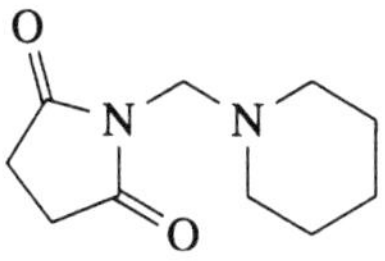 [structure: succinimide-N–CH₂–N-piperidine]	$PhCH_2CH_2MgBr$, Et_2O	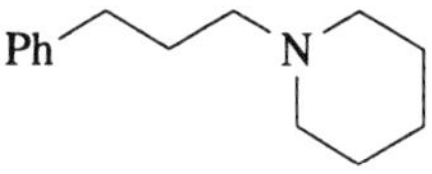 Ph(CH₂)₃–N-piperidine	80	71
5	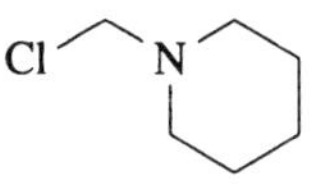 Cl–CH₂–N-piperidine	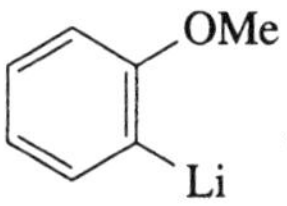 2-MeO-C_6H_4Li, Et_2O	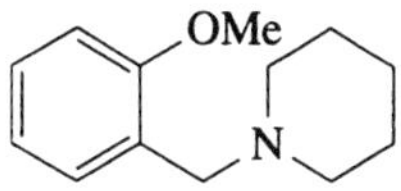 2-MeO-C_6H_4CH₂–N-piperidine	82	72
6	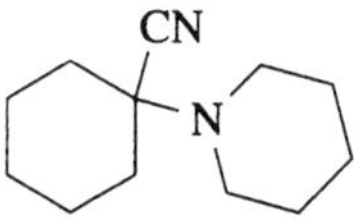 1-CN-cyclohexyl–N-azepane	PhMgBr, Et_2O, Isooctane Δ, 1 h	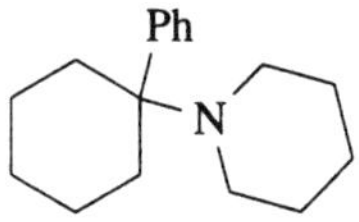 1-Ph-cyclohexyl–N-azepane	54	73

6).[73] In general, Grignards add to the α-carbon atom, while alkyllithiums add to the nitrile carbon atom of *N,N*-(disubstituted)aminomethylnitriles. If the α-aminonitrile has an active hydrogen atom on nitrogen, both Grignards and organolithium reagents add to the α-carbon atom. On the other hand, if the aminonitriles have an acetyl group on the amine nitrogen atom, both reagents condense with the nitrile moiety.[74]

1.12.5 ORGANOMETALLIC ADDITIONS TO *N*-ACYLIMINES AND *N*-ACYLIMINIUM SALTS

1.12.5.1 Background

The α-amidoalkylation reaction, which involves the addition of carbon nucleophiles (primarily aromatic rings, alkenes, cyanides, isocyanides, alkynes, organometallic and active methylene-containing compounds) to substituted amides (**89**) where X is a leaving group (Scheme 17), has been extensively reviewed.[75] A variety of organometallic condensations have since been reported which extend the scope of this reaction.[76] The reactive intermediates in these reactions are considered to be *N*-acylimines (**88**) or *N*-acyliminium salts (**90**). Direct displacement of the α-haloalkylamide precursors (**89**; X = halogen) cannot in certain cases be discounted.

A comprehensive kinetic study examining the reactivity of *N*-acylimines/*N*-acyliminium salts *versus* imines/iminium salts has not, to date, been published. Spectral data (^{13}C NMR), however, would suggest

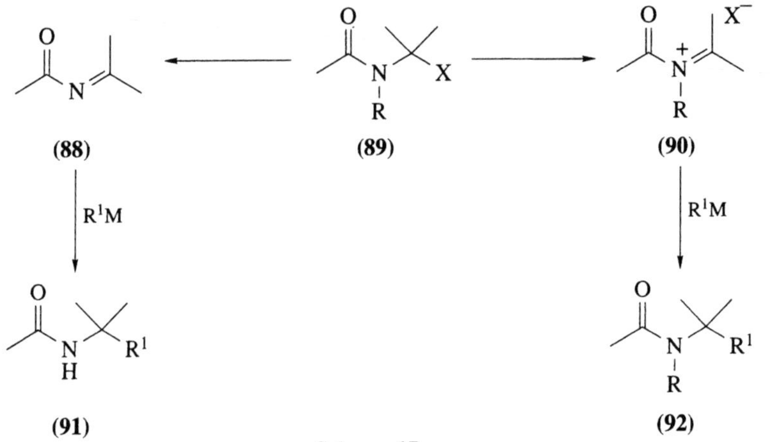

(88) (89) (90)

R^1M R^1M

(91) (92)

Scheme 17

that *N*-acyliminium salts could be even more reactive than iminium salts, since the imine carbon atom in (**93**) and (**94**) is shifted about 5 p.p.m. downfield from that in iminium salt (**95**).[75e,77]

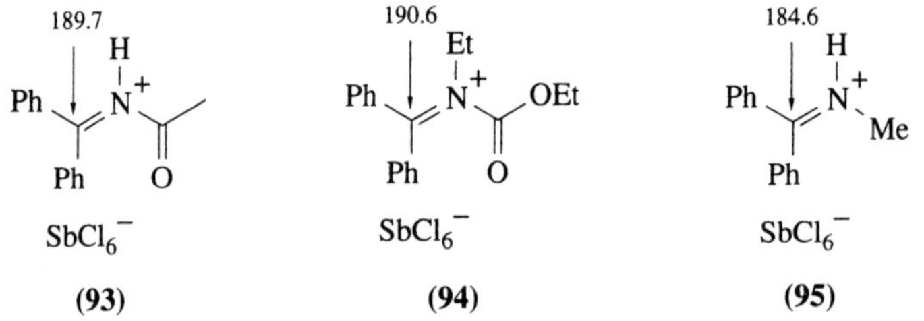

189.7 190.6 184.6

$SbCl_6^-$ $SbCl_6^-$ $SbCl_6^-$

(93) (94) (95)

1.12.5.2 α-Amidoalkylation Reactions Involving Substituted Azetidin-2-ones

No α-amidoalkylation reaction has been studied in recent years in greater detail than the addition of nucleophiles to substituted azetidin-2-ones such as (**96**), bearing a leaving group (X) in the 4-position. The addition of oxygen and sulfur nucleophiles has been employed for the synthesis of oxapenam and penem antibiotics, while the addition of stabilized and nonstabilized carbanions has been utilized for the synthesis of carbapenems (**97**; Scheme 18).[78] While a thorough comparative study examining the effect of leaving groups (X) on the reactivity of these 4-substituted azetidin-2-ones is not available, cross comparisons[79] (many of which are outside the scope of this chapter) suggest an approximate order of reactivity of Cl > SO_2Ph > OAc > SCSOEt, SCS_2Et > N_3 > OEt > SEt. High yields of 4-alkyl-, 4-allyl-, 4-vinyl- and 4-ethynyl-azetidin-2-ones have been obtained by treatment of 4-phenylsulfonylazetidin-2-ones with either lithium organocuprates or Grignard reagents.[80] Yields for the organocuprate additions are, in general, superior to those of the Grignard condensations.

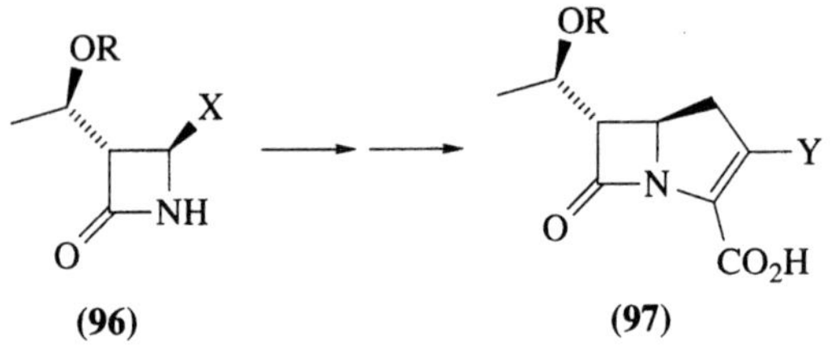

(96) (97)

Scheme 18

In the addition of phenylthioethynyl Grignard to 3-substituted azetidinone (**98**), the stereochemistry of addition is exclusively *trans*, consistent with the intermediacy of azetinone intermediate (**99**; Scheme 19).[81]

N-Substituted azetidin-2-ones are more stable than their unsubstituted counterparts and as a result are more resistant to nucleophilic additions. However, treatment of either 4β- or 4α-chloroazetidinone (**101**)

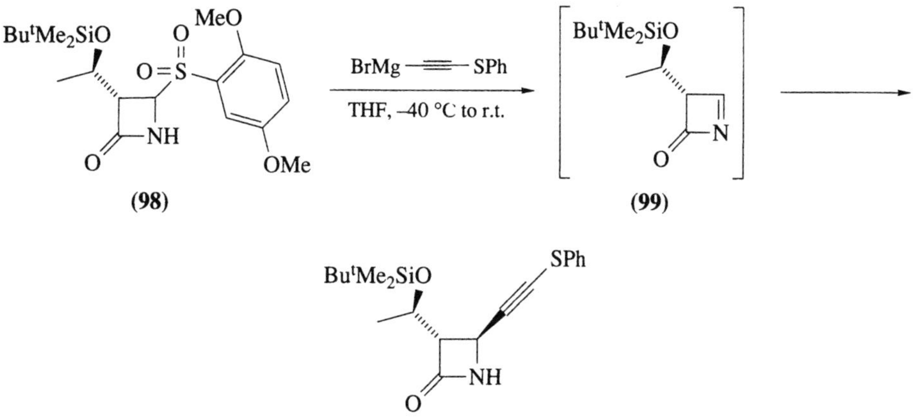

(98) (99)

(100) 80%

Scheme 19

with allylcopper reagents provides the same product ratio of substituted azetidinones (**102**), suggesting the intermediacy of an acyliminium ion (equation 12).[82]

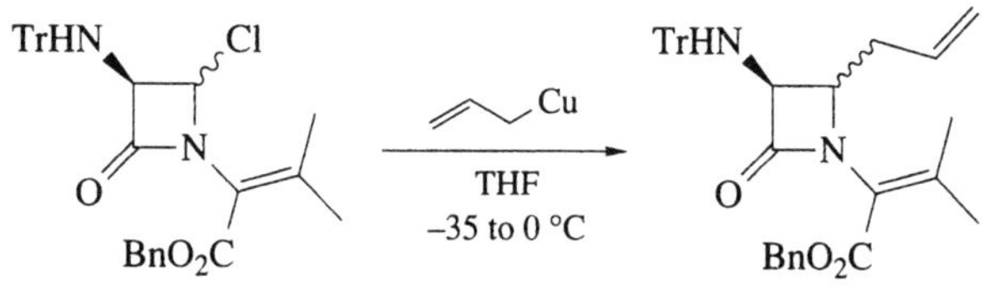

(101) (102) 4β:4α = 33:67 (12)

1.12.5.3 *N*-Acylimines/*N*-Acyliminium Salts

The addition of organometallic reagents to substituted *N*-halomethylamides (**103**) has been utilized to generate substituted amides (**104**; equation 13). Selected examples are shown in Table 7.[84–92]

N-Acylimine intermediates stabilized by electron-withdrawing groups can be isolated (entries 3–6, Table 7) and readily undergo nucleophilic addition. In most cases, however, *N*-acylimines need not be isolated since the addition of organometallic reagents (2 equiv.) to *N*-halomethylamide precursors provides the products in comparable yields. Almost all substituted *N*-halomethylamides lack β-hydrogen atoms due to the propensity of *N*-acylimines to tautomerize to acyl enamides.[83]

Complex alkenyl-copper and -cuprate derivatives, and also alanates, have been generated with high stereochemical purity, and upon treatment with *N*-chloromethyl-*N*-methylformamide or *N*-chloromethylphthalimide provide access to highly functionalized allylic amides (entries 1 and 2, Table 7).[84] Addition to acyliminomalonic esters[85] or acyliminophosphonates[86] provides acylaminomalonates and acylaminoalkylphosphonates in modest yield (entries 3 and 4, Table 7).

Carbanion amidoalkylations provide a versatile means of generating α-amino acids. The ability to oxidize alkenes[87] or hydrolyze trichloromethyl groups[88] of amidoalkylation products (entries 5 and 6, Table 7) to acids provides access to these compounds. α-Halo-*N*-(*t*-butoxycarbonyl)glycine esters have been used as 'electrophilic glycinates' for the preparation of broadly substituted amino acids (entry 7, Table 7).[89] Several approaches have been examined as possible enantioselective routes to relatively inaccessible, synthetic amino acids. The reactions of vinylmagnesium bromide with *N*-Cbz-L-Phe-α-chloro-Gly-OMe resulted in no asymmetric induction as a 1:1 mixture of diastereomeric β,γ-unsaturated

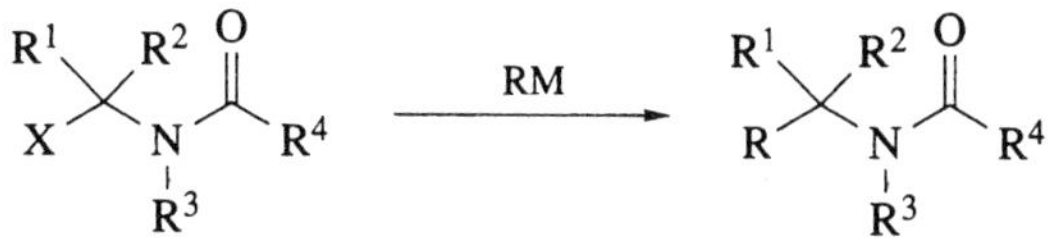

(103) (104) (13)

Table 7 The Addition of Organometallic Reagents to *N*-Acylimines

Entry	*Electrophiles*	*Organometallic Reagent*	*Conditions*	*Product*	*Yield (%)*	*Ref.*
1	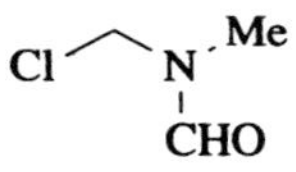 Cl, N, Me, CHO	0.62 equiv. 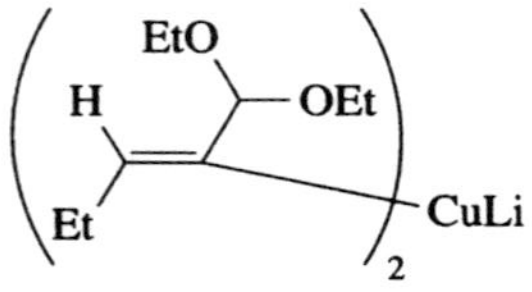 (H, Et, EtO, OEt)$_2$CuLi	THF, –20 °C to r.t., 3 h	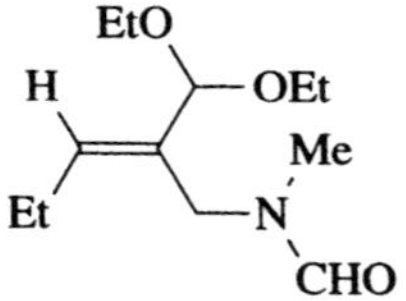 H, Et, EtO, OEt, Me, N, CHO	89	84
2	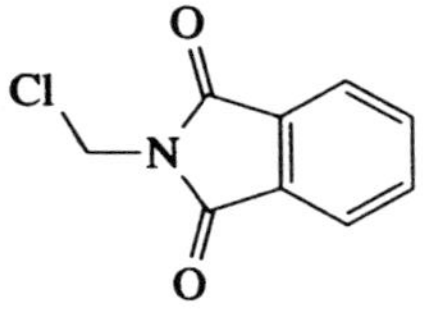 Cl, N, O, O	1.2 equiv. 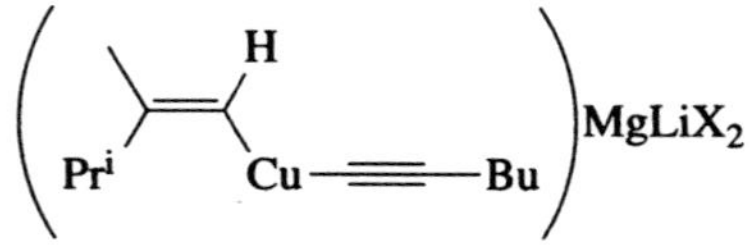 (Pr^i, H, Cu—≡—Bu)MgLiX$_2$	THF, –20 °C to r.t., 3 h	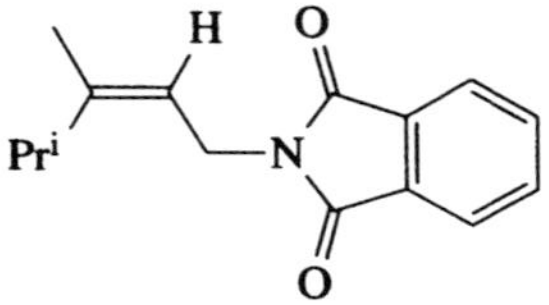 Pr^i, H, N, O, O	47	84
3	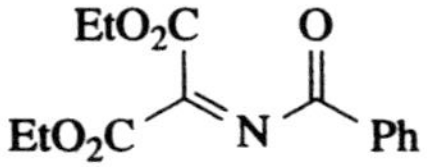 EtO_2C, EtO_2C, N, O, Ph	1 equiv. N, Li 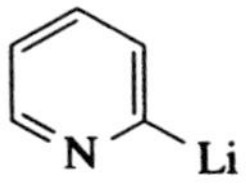	THF, –100 to –78 °C	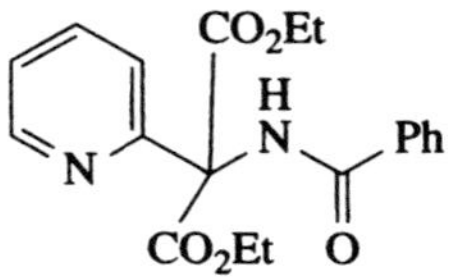 N, CO_2Et, CO_2Et, H, N, O, Ph	50	85
4	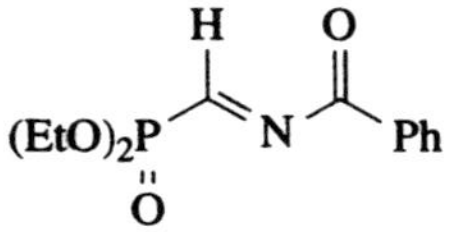 $(EtO)_2P$, O, H, N, O, Ph	1.1 equiv.  MgBr	THF, –78 °C	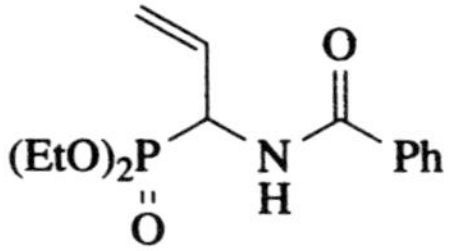 $(EtO)_2P$, O, N, H, O, Ph	45	86
5	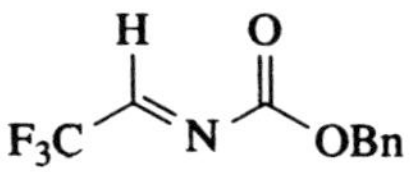 F_3C, H, N, O, OBn	 MgBr	THF, 0 °C	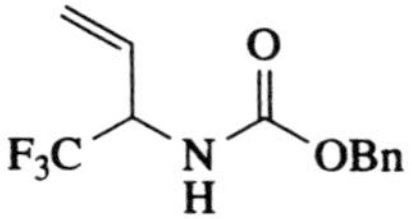 F_3C, N, H, O, OBn	81	87
6	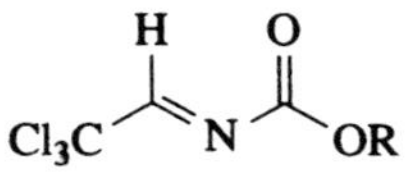 Cl_3C, H, N, O, OR	1 equiv. MeMgBr	Et_2O, 0 °C, 12 h	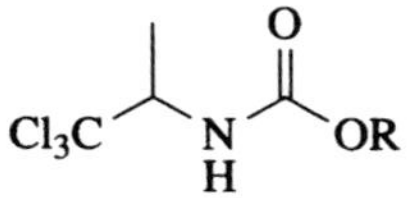 Cl_3C, N, H, O, OR	56	88

Table 7 *(continued)*

Entry	*Electrophiles*	*Organometallic Reagent*	*Conditions*	*Product*	*Yield (%)*	*Ref.*
7	Bu^tO_2C–CH(Br)–NH–C(=O)–OBu^t	2.2 equiv. 4-MeC$_6$H$_4$MgBr	THF, –78°C to r.t., 12 h	4-MeC$_6$H$_4$–CH(CO_2Bu^t)–NH–C(=O)–OBu^t	91	89
8	MeO_2C–CH(Cl)–NH–C(=O)–CH(CH_2Ph)–NH–CBz	Excess CH_2=CHMgBr	THF, –70 °C, 2 h	MeO_2C–CH(CH=CH_2)–NH–C(=O)–CH(CH_2Ph)–NH–CBz	60	90
9	BOC–NH–CH(Br)–C(=O)–O–(8-phenylmenthyl)	Bu^iMgBr	Et_2O, 0 °C	BOC–NH–CH(Bu^i)–C(=O)–O–(8-phenylmenthyl)	65 (95% *ee*)[a]	91
10	BOC–N, Ph, Ph, O, Br, =O (oxazinone)	$Bu_2CuCNLi_2$	THF, –78 °C, 30 min	BOC–N, Ph, Ph, O, Bu, =O (oxazinone)	48 (99.5% *ee*)[a]	92

[a] Optical purity of the L-norleucine produced using the chiral auxiliary.

α-amino acid esters was produced (entry 8, Table 7).[90] Obrecht and coworkers[91] have examined the addition of Grignard reagents to α-bromo-*N*-Boc-glycine-(–)-phenylmenthyl ester (entry 9, Table 7). The high enantioselectivity (82–95% *ee*, 4 examples) can be rationalized by the shielding of the *re*-face of the imine double bond by the aromatic ring of the (–)-8-phenylmenthyl moiety, as illustrated in Figure 5. A two-step conversion of these products to amino acids was developed as a result of the resistance of phenylmenthyl esters to hydrolysis.

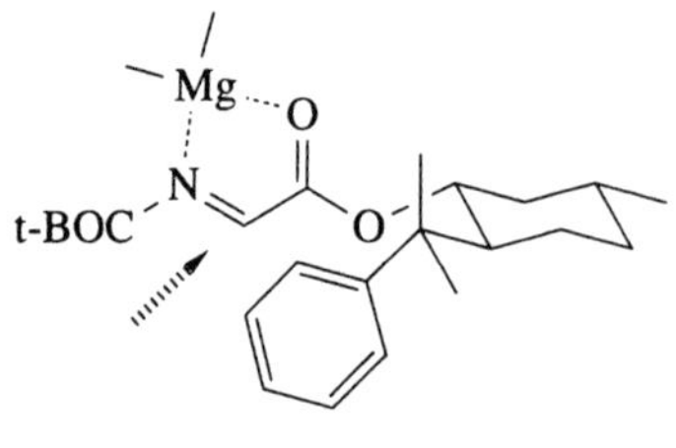

Figure 5

Williams and coworkers have constructed a rigid 3-bromotetrahydrooxazin-2-one ring from D- or L-erythro-α,β-diphenyl-β-hydroxyethylamine as a chiral template for electrophilic glycinate condensations (entry 10, Table 7).[92] Upon nucleophilic addition, very high diastereoselectivity is obtained (>91%, 11 examples) and is rationalized by a sterically controlled nucleophilic addition to a putative iminium species (Figure 6) or its bromoamide precursor. Hydrogenation or dissolving metal reduction of these products provides amino acids directly in excellent yield. The methodology is best suited for 'neutral' nucleophiles (*e.g.* silyl enol ethers, allyltrimethylsilanes) since the more basic organometallic agents competitively reduce the α-bromo ketones, resulting in lower overall yields.

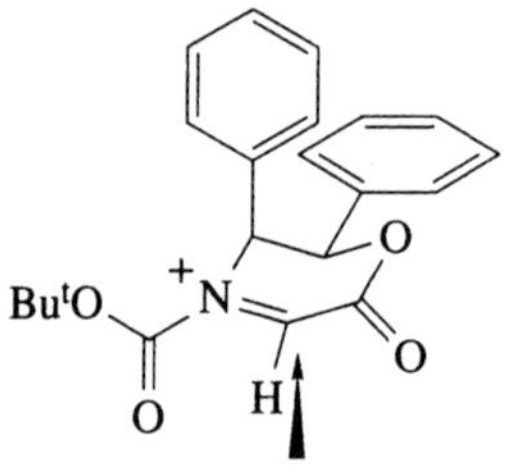

Figure 6

Lipshutz *et al.*[93] have described a singlet oxygen based approach for the conversion of 1,2,4-trisubstituted imidazoles (**105**) to *N*-acylimines (**106**) containing either alkyl or aryl substituents (Scheme 20). *N*-Acylimines (**106**; $R^1 = Bu^n, Bu^i$, Ph; R = CH_2Ph, CH_2CH_2Ph) are inert to organometallic reagents. Remarkably, hydroxamic *N*-acylimines (**106**; $R^1 = Bu^n, Bu^i$, Ph; R = OMe) react with a variety of aryl, vinyl, ethynyl, primary and secondary organometallic reagents (2.5 equiv.) to afford α,α-disubstituted amino acid bis-amides (**107**) in good yield (52–94%). The fact that hydroxamic acid acylimine derivatives react and amide acylimines are inert to nucleophilic attack is unexpected. The authors speculate that organometallic complexation with the hydroxamic moiety might be responsible for organometallic delivery. A chelation-mediated acylimine activation of (**106**) cannot be discounted.

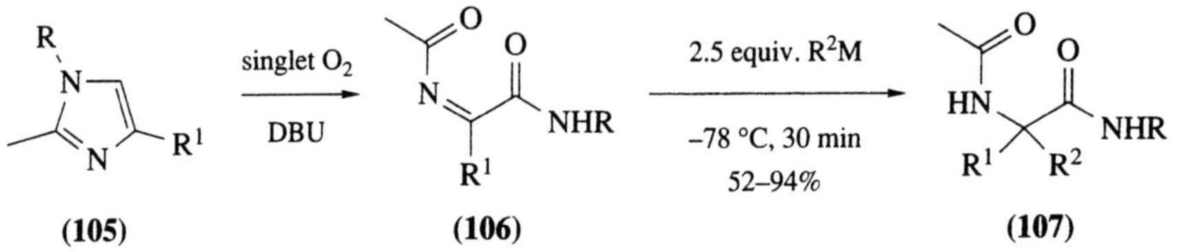

(105) —singlet O_2, DBU→ (106) —2.5 equiv. R^2M, –78 °C, 30 min, 52–94%→ (107)

Scheme 20

Weinreb *et al.*[94] have described a procedure for the preparation of acyclic *N*-alkylamides (**110**) which involves the addition of trialkylalanes (4 equiv.) to amide methylols (**108**; Scheme 21). While the methodology has also been extended to ureas and carbamates, all efforts to couple hydroxylactams such as (**111**; R = H) with trialkylalanes failed.

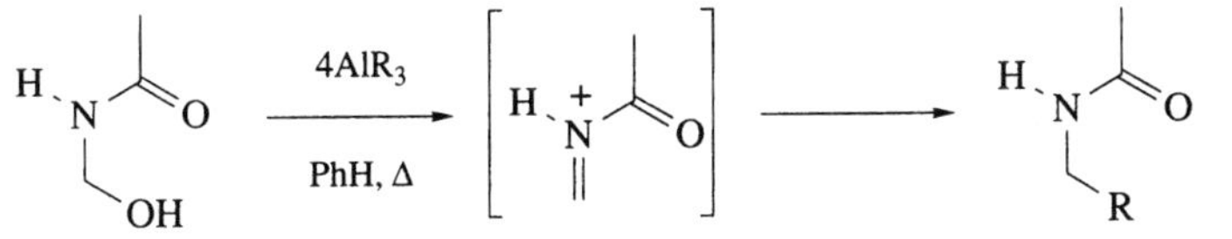

(**108**) $\xrightarrow[\text{PhH, }\Delta]{4AlR_3}$ [(**109**)] ⟶ (**110**) 19–96%

Scheme 21

Treatment of lactam methyl ether (**111**; R = Me) with trialkylalanes, however, generated desired product (**112**) in modest yield (equation 14).[95] Similarly, Shono *et al.*[96] have treated α-methoxy carbamates (**113**) with Grignard reagents in the presence of boron trifluoride etherate to afford substituted pyrrolidines (**114**; equation 15).

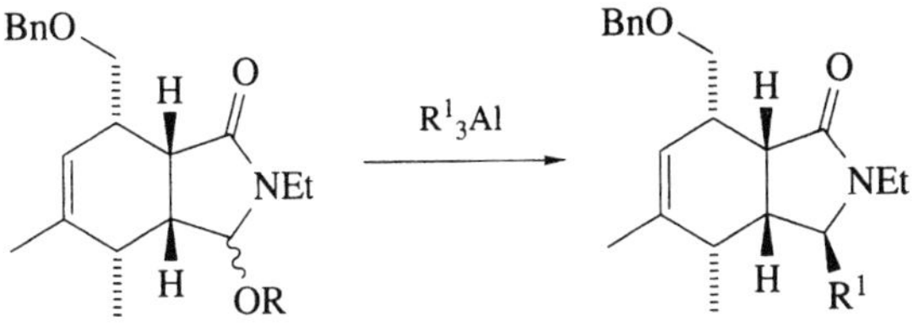

(**111**) $\xrightarrow{R^1{}_3Al}$ (**112**) 22–66% (14)

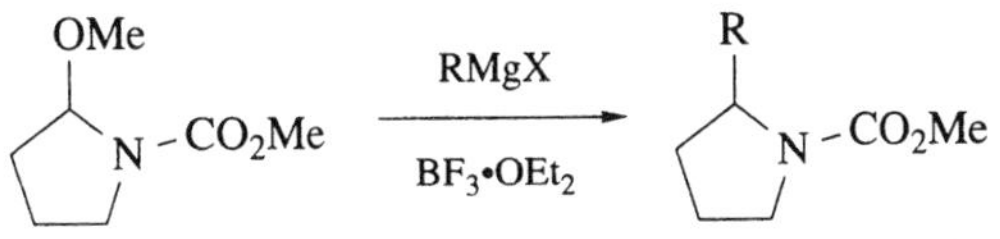

(**113**) $\xrightarrow[BF_3 \cdot OEt_2]{RMgX}$ (**114**) 76–96% (15)

1.12.6 ORGANOMETALLIC ADDITIONS TO HYDRAZONES

1.12.6.1 Tosylhydrazones

The reaction of alkyllithium reagents with acyclic and cyclic tosylhydrazones can lead to mixtures of elimination (route A) and addition (route B) products (Scheme 22). The predominant formation of the less-substituted alkene product in the former reaction (Shapiro Reaction) is a result of the strong preference for deprotonation *syn* to the *N*-tosyl group.[97] Nucleophilic addition to the carbon–nitrogen tosylhydrazone double bond competes effectively with α-deprotonation (and alkene formation) if abstraction of the α-hydrogens is slow and excess organolithium reagent is employed.[97a,98] Nucleophilic substitution is consistent with an S_N2' addition of alkyllithium followed by electrophilic capture of the resultant carbanion.

Treatment of 5α-cholestan-3-one tosylhydrazone (**115**) with excess (8 equiv.) alkyllithium reagent (primary, secondary and tertiary carbanions) provides, after axial protonation, 3β-alkylcholestane products (**116**) in addition to minor amounts of the anticipated alkene by-products (equation 16).[99]

In the case of verbenone tosylhydrazone (**117**), δ-pinene (**118**) is produced along with verbenene (**119**) following methyllithium treatment (equation 17).[100]

Good yields of substitution products can be obtained if elimination is precluded. Alkyllithium addition to fluorenone tosylhydrazone (**120**) provides access to 9,9-disubstituted fluorenes (**121**; equation 18).[101]

In a number of cases, the addition of lithium reagents to ketone mono- and di-tosylhydrazones can be improved by the inclusion of Cu^I, which activates the carbon–nitrogen bond toward nucleophilic addition.[102] However, the formation of complex product mixtures with many ketone tosylhydrazones limits its synthetic utility.[103]

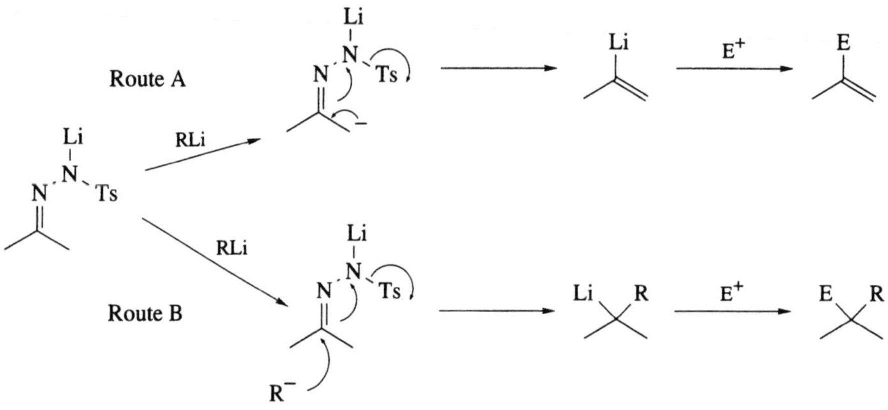

Scheme 22

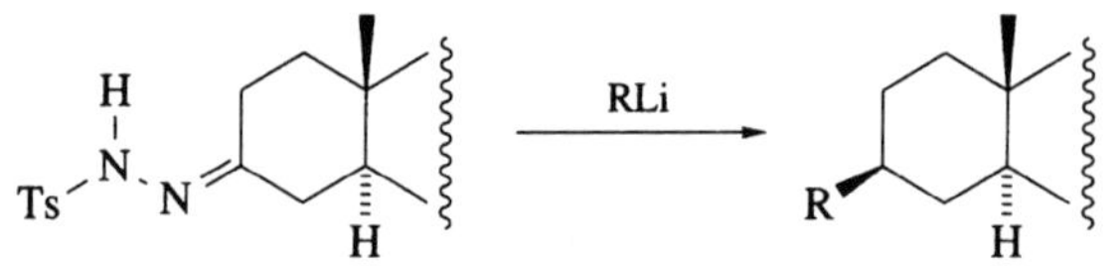

(115) (116) (16)

R = Bu^n, 55%; R = Bu^s, 50%; R = Bu^t, 48%

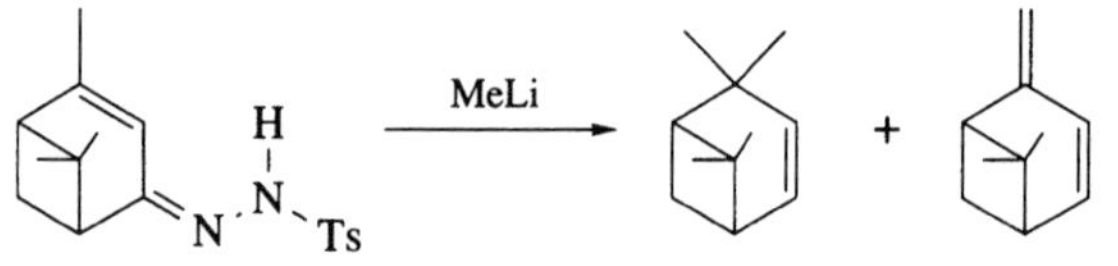

(117) (118) (119) (17)

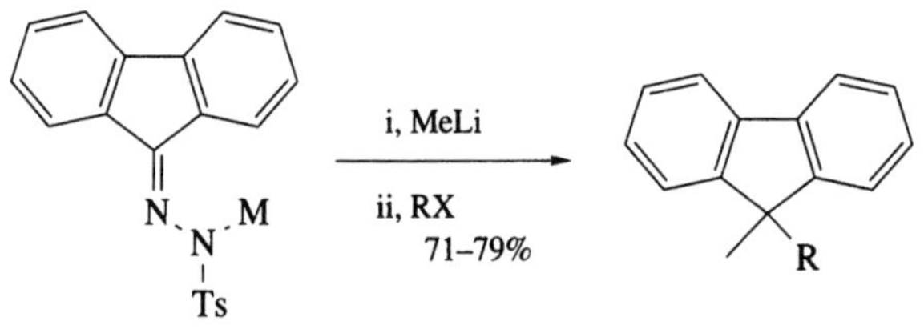

(120) (121) (18)

The failure of aldehyde tosylhydrazones (**122**), which exist predominantly in the (**E**)-configuration,[6,104] to form stable dianions at low temperature, as opposed to many ketone tosylhydrazones which do, is consistent with the strong preference for deprotonation *syn* to the *N*-tosyl group. As a result, alkyllithium reagents add rapidly to the aldehyde tosylhydrazone C=N bond. The addition of primary, secondary and tertiary alkyllithiums to aldehyde tosylhydrazones provides modest yields of the expected coupled hydrocarbons (**127**), accompanied by carbanion-derived side products (**124–126**) (Scheme 23).[5]

Vedejs *et. al.*[105] have shown that carbanions bearing leaving groups condense with aldehyde tosylhydrazones to generate disubstituted (**128**) and trisubstituted alkenes (**129**) (Scheme 24). While stereochemical control (*cis:trans*) is poor in the generation of disubstituted alkenes (compared, for example, to the Wittig reaction), the methodology is well suited for the synthesis of structurally diverse terminal alkenes.

As expected, the presence of Cu^I promotes addition to the lithium and magnesium salts of secondary and tertiary aldehyde tosylhydrazones.[106] Dilithiotrialkylcuprates react with tosylhydrazones (**130**) to generate in high yield hindered cuprates (**131**), which undergo typical cuprate reactions (Scheme 25).

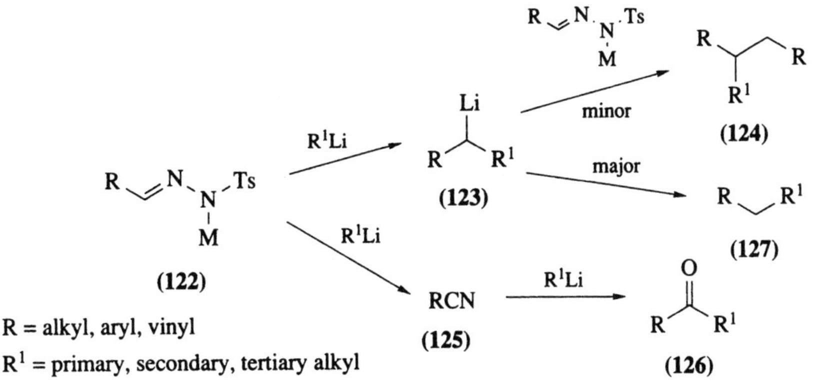

R = alkyl, aryl, vinyl
R^1 = primary, secondary, tertiary alkyl

Scheme 23

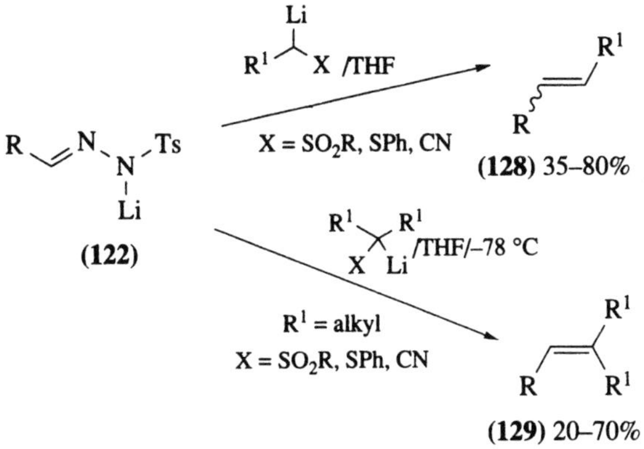

R^1 = alkyl
X = SO_2R, SPh, CN

Scheme 24

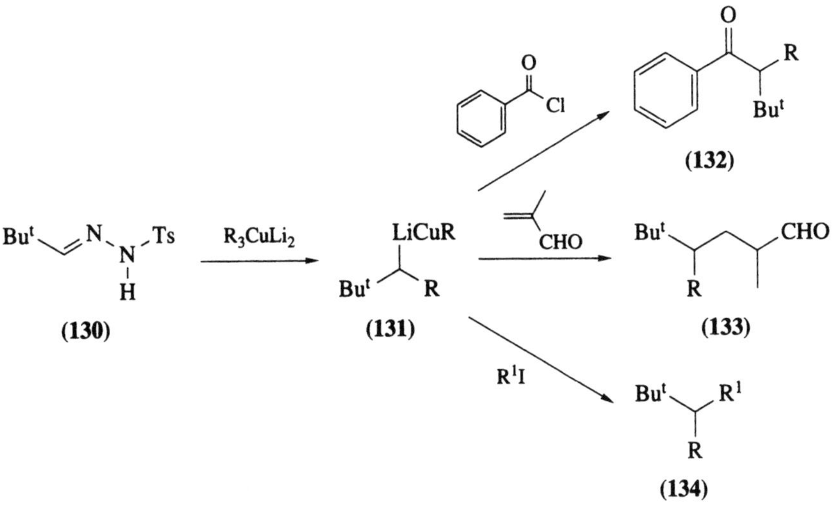

Scheme 25

1.12.6.2 Chiral *N,N*-Dialkylhydrazones

Organometallic reagents condense with aldehyde-derived *N,N*-dialkylhydrazones.[107] This methodology has since been utilized for enantioselective syntheses of amines. The addition of Grignard reagents to *N*-aminoephedrine and L-valinol derived hydrazones **(135)**[16a–c] and **(138)**[16d] to provide, after hydrogenolysis, access to (*R*)- and (*S*)-1-phenylalkylamines **(137)** and **(140)** has been studied by Takahashi *et al.* (Scheme 26). Diastereoselectivity is >98% for alkyl Grignards (Me, Et, Pr^i) but erodes (<50%) with the addition of arylalkyl(benzyl) Grignards. The *re*-facial selectivity of **(135)** and *si*-facial selectivity of

(**138**) can be rationalized by the formation of the six-membered magnesium chelates shown in Figures 7a and 7b, with the diastereoselectivity resulting from a sterically controlled Grignard addition.

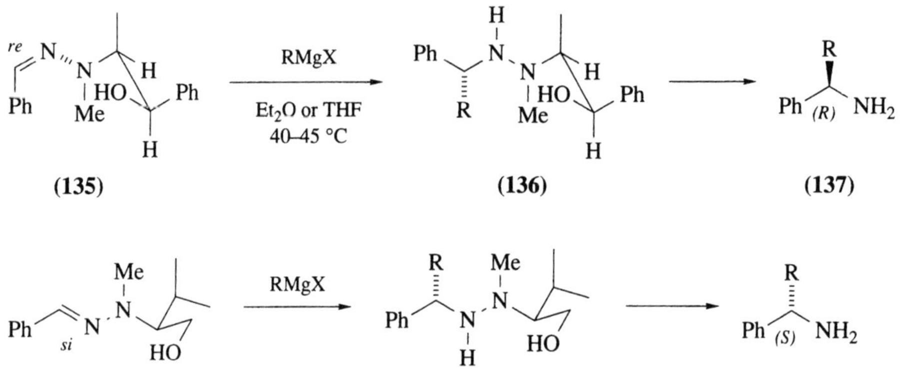

Scheme 26

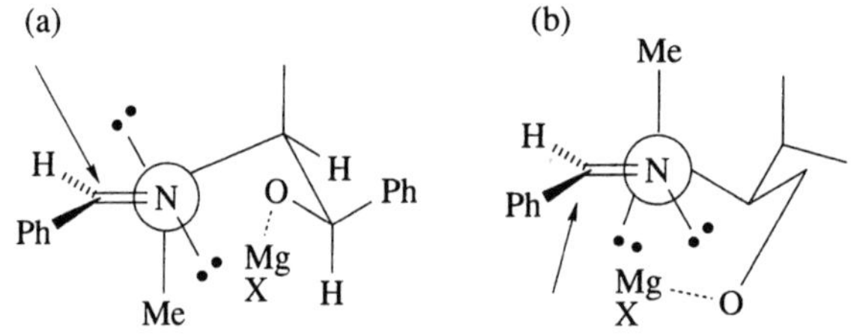

Figure 7

1.12.6.3 SAMP (RAMP) Hydrazones

Enders and Denmark have shown that SAMP (RAMP) hydrazones apparently provide a more versatile route to optically pure amines. Organolithium[17] and organocerium[18] reagents add to structurally diverse SAMP aldehyde hydrazones (**141**; Scheme 27). Representative examples are shown in Table 8. While addition diastereomeric selectivity is similar for both the organo-lithium and -cerium substrates, the overall yields in the organocerium additions are superior to those of various organometallic reagents (R^1Li, R^1MgX, $R^1{}_2CuLi$) alone or in conjunction with additives ($BF_3 \cdot Et_2O$, TMEDA). The preferred reagent stoichiometry is R^1M:$CeCl_3$:hydrazone = 2:2:1. Vinyl, primary, secondary and tertiary organocerium reagents give addition products with high diastereomeric selectivity. Aliphatic, aromatic, α,β-unsaturated and highly enolizable aldehyde hydrazones are suitable for nucleophilic addition (entries 1–4, Table 8). Only 1,2-addition is observed with α,β-unsaturated SAMP hydrazones (entry 4, Table 8). Although the intimate mechanistic details of this addition are not known, these results suggest the organolithium or organocerium ($RCeCL_2$) reagent possibly coordinates to the methoxymethyl group and delivers R^1 to the *re*-face of (**141**) to generate preominantly (**142**).

1.12.6.4 α-Alkoxyaldehyde Dimethylhydrazones

Organolithium reagents (primary, secondary, tertiary, aryl and vinyl) also add in excellent yield to α-alkoxyaldehyde dimethylhydrazones (**146**); equation 19) with high *threo* diastereoselectivity (Table 9).[11] Hydrogenolytic cleavage of the resultant hydrazines provide an attractive route to *threo*-2-amino alcohols.

Threo diastereoselectivity is consistent with a chelation-controlled (Cram cyclic model) organolithium addition (Figure 8a). Since five-membered chelation of lithium is tenuous, an alternative six-membered chelate involving the dimethylamino nitrogen atom of the thermodynamically less stable (*Z*)-hydrazone (in equilibrium with the (*E*)-isomer) cannot be discounted. The trityl ether (entry 4, Table 9) eliminates the chelation effect of the oxygen atom such that the *erythro* diastereomer predominates (*via* normal Felkin–Ahn addition) (Figure 8b).

Table 8 Addition of Organolithium and Organocerium Reagents to SAMP Hydrazones (**141**)

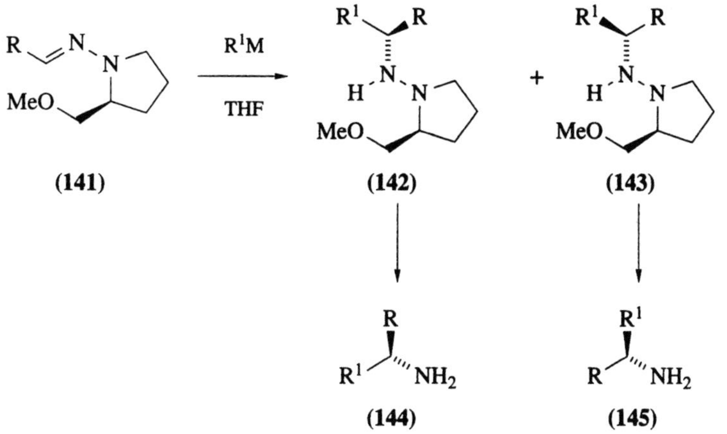

Scheme 27

Entry	*R*	*R^1M (2 equiv.)*	*Yield (%)*	*Diastereomeric ratio (**142**):(**143**)*	*Ref.*
1	$PhCH_2CH_2$	MeLi/$CeCl_3$	81[a] (59)[b]	98:2 (98:2)[c]	18
2	$PhCH_2$	MeLi/$CeCl_3$	66 (0)	96:4 (—)	18
3	Ph	MeLi/$CeCl_3$	59 (47)	91:9 (90:10)	17, 18
4	(*E*)-MeCH=CH_2	MeLi/$CeCl_3$	82 (52)	96:4 (95:5)	18

[a]Yield after chromatography of the corresponding methyl carbamates. [b]Yield for the addition without $CeCl_3$ (2 equiv. MeLi). [c]Diastereomeric ratio for the addition without $CeCl_3$ (2 equiv. MeLi).

Table 9 Addition of Organolithium Reagents to α-Alkoxyaldehyde Dimethylhydrazones (**146**)

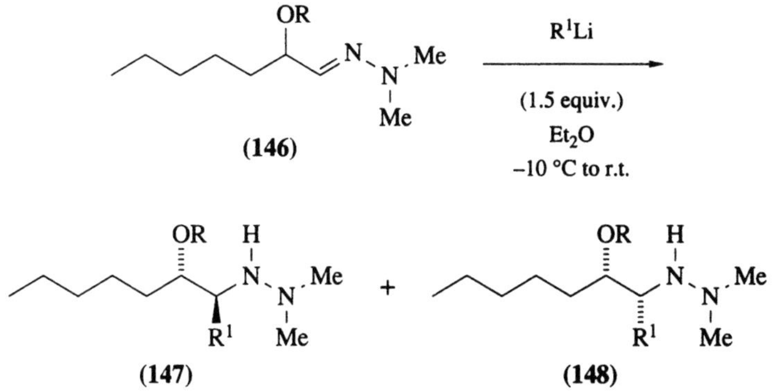

(19)

Entry	*R*	*R^1Li*	*Yield (%)*	*Diastereomeric ratio (**147**):(**148**)*
1	Bn	MeLi	98	97:3
2	Bn	Bu^tLi	98	>98:2
3	CMe_2OMe	PhLi	85	>98:2
4	Tr	MeLi	85	1:10

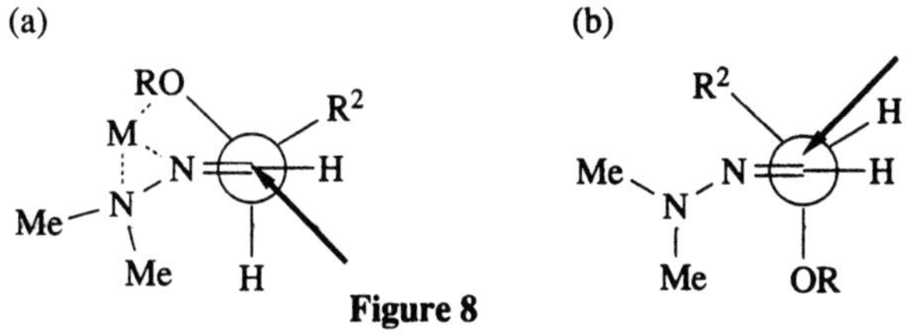

Figure 8

Only fair selectivity was obtained using the dimethylhydrazone of (*R*)-glyceraldehyde acetonides (**149**; equation 20). The addition of catalytic CuI reversed the diastereoselectivity (Table 10).[11]

Table 10 Addition of Organolithium Reagents to (*R*)-Glyceraldehyde Acetonide Dimethylhydrazones (**149**)

(**149**) —MeLi→ (**150**) + (**151**) (20)

Entry	*Organolithium reagent*	*Threo (150)*	*Erythro (151)*
1	MeLi	3	1
2	MeLi (0.1 equiv. CuI)	1	3

1.12.7 ORGANOMETALLIC ADDITIONS TO AZADIENES

1.12.7.1 Chiral α,β-Unsaturated Aldimines

Koga and coworkers[19] have examined the addition of organometallic reagents to chiral α,β-unsaturated aldimines (**11**) derived from amino acid esters (equation 21). While *n*-butyllithium and lithium di-*n*-butylcuprate add to the aldimine moiety (1,2-addition), 1,4-addition occurs preferentially with Grignard reagents (Table 11).[19a,b]

Table 11 Addition of Organometallic Reagents to Chiral α,β-Unsaturated Aldimines (**11**)

(**11**) —i, 2 equiv. R^2MgX, 5:1 Et_2O:THF, −55 °C; ii, HCl→ (**152**) + (**153**) (21)

Entry	*R*	R^1	R^2	*Isolated yield*[a] *(%)*	*(152) (%)*	*(153) (%)*
1	Me	Pr^i	Ph	42	65	35
2	Me	Bu^t	Ph	52	91	9
3	Ph	Bu^t	Et	56	95	5
4	Me	Bu^t	$(CH_2)_2CH{=}CMe_2$	48	98	2
5	Me	Bu^t	Bu^n	40	98	2

[a]Isolated as the corresponding alcohol.

Incorporation of a bulky substituent (R^1 = Bu^t) at the stereogenic center of the aldimine prevents deprotonation at this center and in addition is responsible for the high diastereoselectivity (entries 1 and 2, Table 11). The absolute configuration of the major β,β-disubstituted aldehydes produced after hydrolysis of the aldimine addition products is postulated to result from a chelation-mediated Grignard addition from the less-hindered *re*-face of the α,β-unsaturated aldimine existing in the *s-cis* conformation (Figure 9). Both diastereomeric products are accessible in theory, since R- and R^2-substituents can be reversed (entries 2 and 3, Table 11).

Similarly, Grignard addition to cycloalkenecarbaldehyde-derived aldimines (**154**) affords, upon hydrolytic work-up, optically enriched *trans*-2-substituted cyclohexenecarbaldehydes (**156**; Scheme 28).[19c–e] The magnesioenamine intermediate (**155**) generated by the Grignard additions (only phenyl and vinyl reported) can be alkylated with methyl iodide to give, in most instances, the aldehyde (**158**).[19d,e] The amino ester bidentate ligand dictates the stereochemical control in the alkylation. The formation of the other

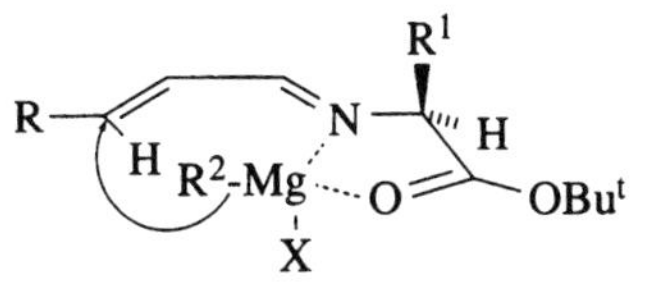

Figure 9

diasteriomeric aldehyde (**157**) in refluxing THF presumably results from a thermal isomerization of the (*Z*)-magnesioenamine (**155**), followed by methyl iodide alkylation.

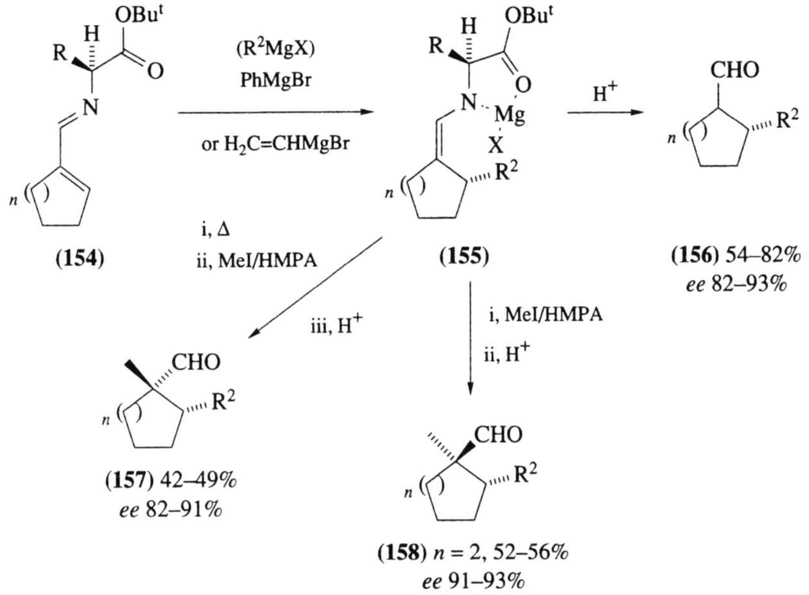

Scheme 28

1.12.7.2 2-Azadienes

Metalloenamines produced in the addition of organometallic reagents to 2-azadienes have synthetic advantages over related enolates and enamines since they are formed regioselectively, exhibit a low tendency to suffer equilibration by proton transfer processes and are very nucleophilic. Wender and Eissenstat[108] have shown that *N*-allylic and α,β-unsaturated imines undergo facile prototropic isomerization to *N*-alkenylimines (2-azadienes) in the presence of potassium *t*-butoxide. Addition of organolithium reagents to *N*-alkenylimines (**160**) provides in high yield a regiospecific generation of lithioenamine (**161**), which can be smoothly alkylated or condensed with aldehydes (directed aldol condensation) (Scheme 29).[109] A potential limitation of this methodology, involving the existence of unfavorable α,β-unsaturated-*N*-alkenylimine equilibriums (*e.g.* **159** *versus* **160**, can be circumvented by manipulating aromatic ring substitution (*p*-OMe substituent favors azadiene formation).

The one-pot procedure involving the *in situ* generation of metalloenamines has been utilized by Martin *et al.* in the synthesis of the cyclohexenone (**164**); Scheme 30).[110]

1.12.7.3 Aromatic Aldimines

Gilman *et al.*[111] first investigated the 1,4 addition of Grignard reagents to benzophenone anil under forcing conditions. Meyers *et al.*[20] have shown that conjugative addition of organometallic reagents to 1-naphthylimines provides an alternative to naphthyloxazolines[21,22] for the preparation of substituted dihydronaphthalenes. Treatment of organolithium reagents (R = Pr^i, Bu^n, Bu^t) with *O*-*t*-butylvalinol-derived imine (**12**),[20] followed by a methyl iodide quench, generates (**165**) in >95% *ee* (Scheme 31). Addition to (**12**; Figure 10) is consistent with a chelation-mediated (butoxy moiety) delivery of the organolithium reagent, paralleling the stereochemical course of the valinol-derived aldimine additions described by

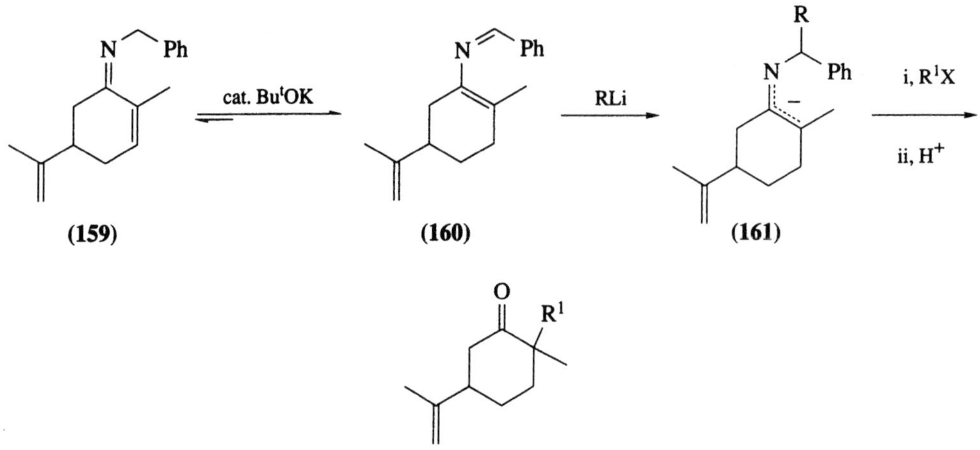

(159) (160) (161) (162)

Scheme 29

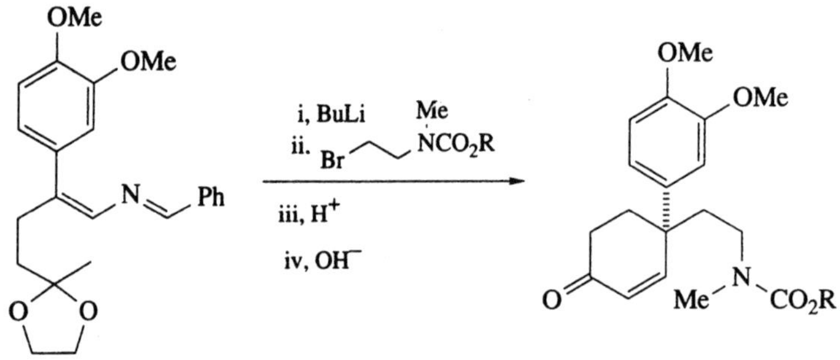

(163) (164)

Scheme 30

Takahashi and Suzuki (equation 4).[14] The use of methyl-, benzyl- and 2-propenyl-lithium gave exclusive 1,2-addition. Demonstration that this methodology can be extended to 2-naphthyl, 3-pyridyl and 3-quinolyl aldimines (paralleling oxazoline chemistry)[21,22] has not been reported.

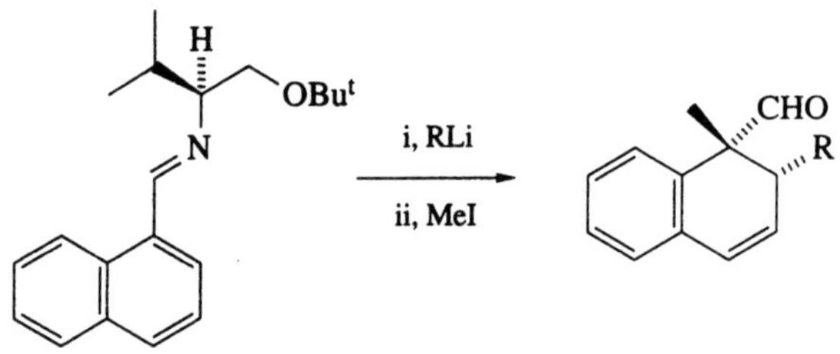

(12) (165)

Scheme 31

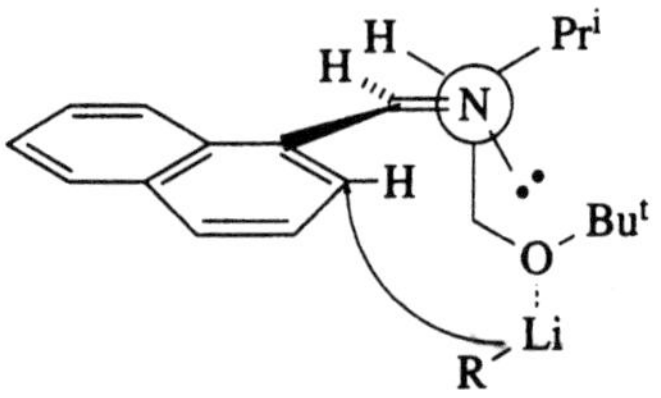

Figure 10

1.12.8 ORGANOMETALLIC ADDITIONS TO OXIMES AND OXIME ETHERS

In addition to providing hydroxy (alkoxy) amines, the reaction of oximes[112] or oxime ethers[107] with organometallic reagents can generate additional products. The propensity for proton abstraction α to the carbon–nitrogen double bond, the existence of mixtures of (*E*)- and (*Z*)-oxime isomers, the lability of the nitrogen–oxygen bond coupled with the poor oxime reactivity all contribute to the variability of this reaction.

1.12.8.1 Aldoxime Ethers/BF_3-Activated Oxime Ethers

Treatment of benzaldehyde oxime ether (**166**) with butyllithium (pentane/–10 °C) demonstrates the complexity of the reaction (Scheme 32) as the desired alkoxyamine (**167**; R = Bu) is accompanied by other oxime-derived side products[113] (entry 1, Table 12). Selectivity is reagent/solvent dependent as allyl Grignard (ether),[113] allylzinc bromide (THF),[113] and butyllithium (THF)[114] treatment produce predominantly amine (**171**; R = allyl) (the Beckmann rearrangement derived product), alkoxyamine (**167**; R = allyl) (the oxime addition product) and ketone (**169**; R = Bu) (the nitrile-derived product), respectively (entries 2–4, Table 12).

Table 12 Addition of Organometallic Reagents to Benzaldehyde Oxime

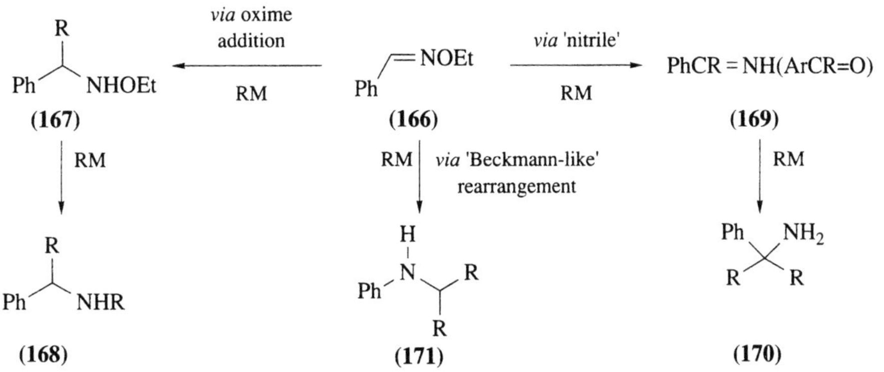

via oxime addition; RM; *via* 'nitrile'; RM; PhCR=NH(ArCR=O); RM; *via* 'Beckmann-like' rearrangement; RM

Scheme 32

Entry	*Organometallic Reagent (RM)*	*R*	*Products (%)* (**167**)	(**168**)	(**169**)	(**170**)	(**171**)	*Ref.*
1	3BuLi (pentane/–10 °C/1 h)	Butyl	57	10	20	1	3	113
2	3Allylmagnesium bromide (ether/20 °C/16 h)	Allyl					96	113
3	3Allylzinc bromide (THF/20 °C/16 h)	Allyl	60				7	113
4	2BuLi (THF/0 °C/1 h)	Butyl			80			114

The addition of organometallic reagents to formaldoxime ethers[115] has been employed to incorporate aminomethyl substituents. Lithium carbanions add rapidly (–40 °C) to formaldoxime ether (**172**) to generate lithium alkoxyamide (**173**); Scheme 33).[116] Lithium alkoxyamides ($LiRNOR^1$), in contrast to alkoxyamines ($RNHOR^1$), react under mild conditions with organolithium reagents (R^2Li) to provide amines ($RNHR^2$).[117,118] Displacement of the benzyloxy moiety of (**173**) with a second equiv. of organometallic reagent requires somewhat higher temperatures (0–40 °C) and thereby permits sequential addition of two different organolithium reagents. Benzyloxyamide (**173**) and secondary amide (**174**) can be protonated or quenched with typical acylating or alkylating agents.

Reaction with (**172**) and other aldoximes may require oxime activation, which can be achieved with the addition of 1 equiv. of $BF_3{\cdot}OEt_2$.[119–121] Yields in the addition of organometallic reagents to substituted aldoximes are modest and are a function of the isomeric composition of the oxime ethers, as the (*Z*)-oxime isomers are reported to preferentially react with organolithium reagents (entries 1 and 2, Table 13).[120] The reaction has been employed for the preparation of 6-aminoalkyl-substituted pencillins (entry 3, Table 13).[119] Cyclic oxime ether additions have also been evaluated (entries 4 and 5, Table 13).[120,121] With the lability of the nitrogen–oxygen bond, addition to 5-substituted isoxazolines provides a potential avenue for stereospecific synthesis of substituted 3-aminoalcohols (entry 5, Table 13).

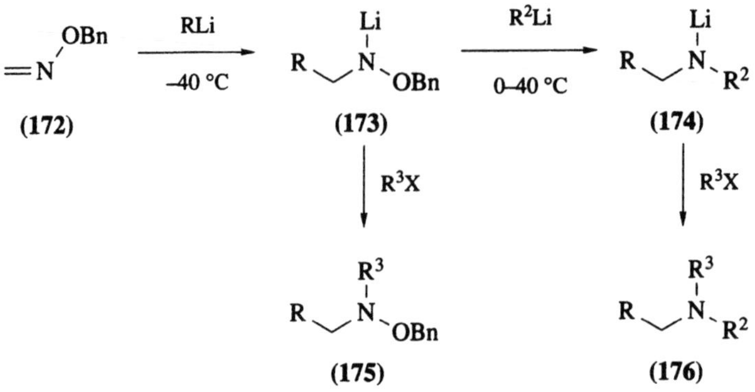

Scheme 33

Table 13 The Addition of Organometallic Reagents to Oxime Ethers in the Presence of $BF_3\bullet OEt_2$ [a]

Entry	Organometallic reagent	Oxime	Product	Yield (%)	Ref.
1	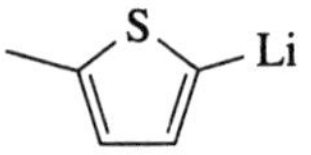	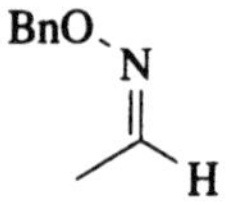	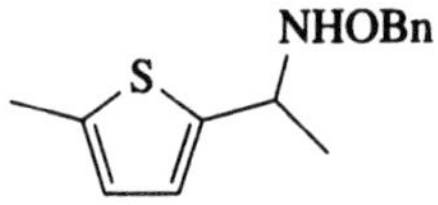	64	120
2	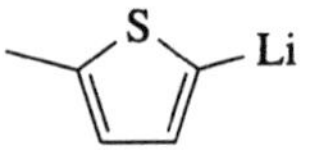	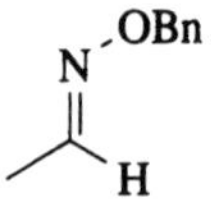	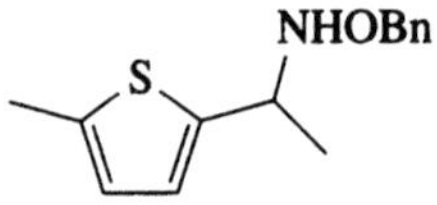	7	120
3	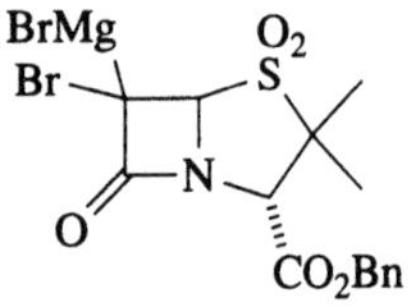	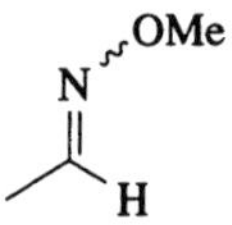	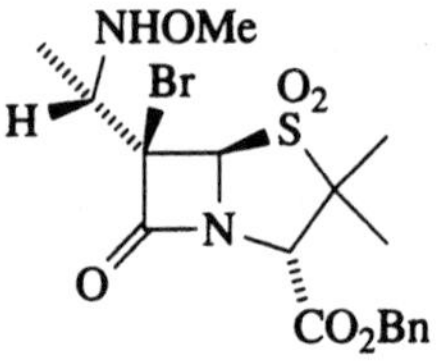	40	119
4	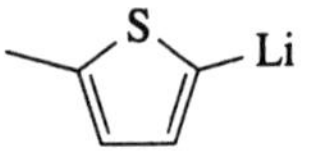	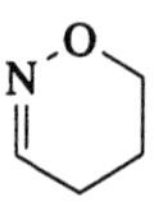	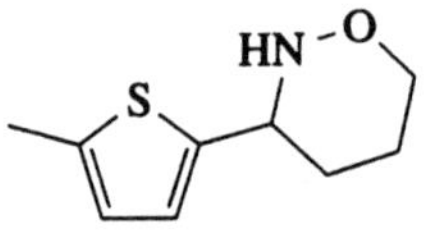	61	120
5	Me_3Si—≡—Li	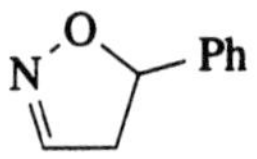	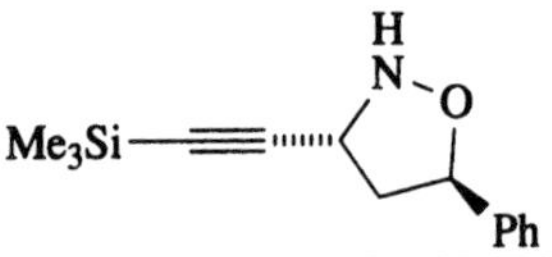	50	121

[a] THF/–78 °C.

Corey *et al.*[122] have elegantly exploited the ability of oxime ethers to stabilize carbanions *syn* to the *N*-alkoxy substituent in the boron trifluoride mediated addition of the mixed cuprate reagent derived from (**177**) to α,β-unsaturated oxime ether (**178**; Scheme 34). Stereochemistry is established by the ring *t*-butyldimethylsiloxy group. The addition does not occur in the absence of boron trifluoride. The alkylation of the cuprate adduct of (**178**) by iodoalkyne (**179**) provides the PGE_2 skeleton.

The condensation of organolithium reagents (2 equiv.) with glyoxylate-derived oxime ethers (**182**) provides a direct method for the synthesis of α-*N*-hydroxy amino acids (equation 22).[123] Both glyoxylic acid and glyoxylamide oxime ethers are compatible with this process. *N*-hydroxyaminoacetamides are also produced in low/moderate yields by the addition of isonitriles to oximes (oxime ethers) analogous to the four component condensation described by Ugi.[124]

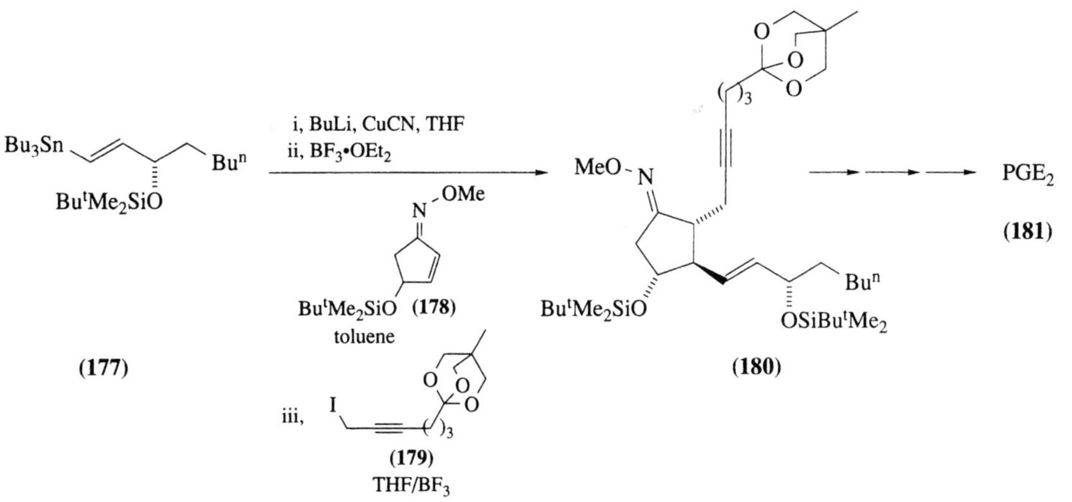

Bu₃Sn (177) i, BuLi, CuCN, THF; ii, $BF_3•OEt_2$; (178) toluene; iii, (179) THF/BF_3 → (180) → → → PGE_2 (181)

Scheme 34

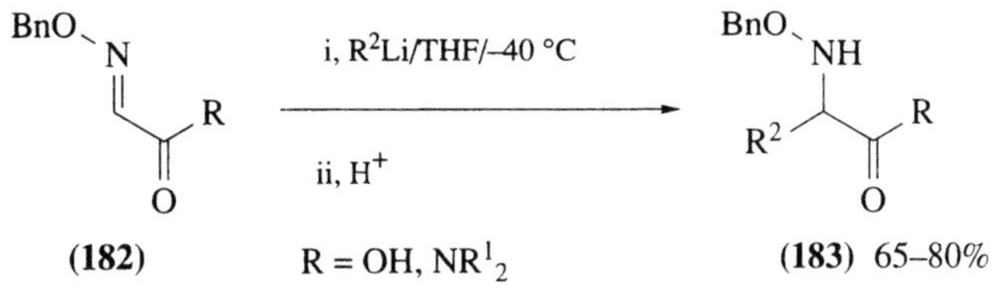

(182) i, R^2Li/THF/–40 °C; ii, H^+; R = OH, NR^1_2 → (183) 65–80% (22)

1.12.8.2 Ketoximes/Ketoxime Derivatives

The reaction of ketoximes with Grignard reagents has been extensively investigated and provides aziridines in low/moderate yields.[125–133] Examples are shown in Table 14. The stereochemical preference for substituted aziridines is consistent with the formation of an azirine intermediate (**185**; Scheme 35) followed by Grignard addition from the less-hindered side of the ring (entries 5 and 6, Table 14).[129] Azirine ring formation in turn occurs predominantly *syn* to the oxime hydroxy group (entries 1, 2, 3, 5 and 6, Table 14).[126,129] The regiospecificity of azirine formation is solvent dependent (entries 1 and 2, Table 14). Replacement of the oxime functionality with dimethylhydrazone methiodides[134] provides in many instances superior yields to substituted aziridines (entries 4 and 7, Table 14).

The acid-catalyzed Beckmann rearrangement of ketoximes provides a reliable synthetic tool for the synthesis of amides and lactams. A methyllithium-promoted Beckmann rearrangement of oxime tosylates to give substituted amines was reported by Gabel.[135] Yamamoto and coworkers[136] have in recent years increased the synthetic scope and utility of this rearrangement with the discovery of an organoaluminum-mediated reaction. Successive treatment of oxime sulfonates (**187**) with trialkylaluminum, which induces the Beckmann rearrangement and captures the intermediary iminocarbocation, followed by a progargylic or allylic Grignard reagent generates acyclic and cyclic amines (**190**) in synthetically useful yields (Scheme 36).

For example, treatment of oxime mesylate (**191**) with Me_3Al (2 equiv.) in CH_2Cl_2 (–78 °C) followed by allylmagnesium bromide (2 equiv., –78 to 0 °C) yields amine (**192**; equation 23).[136]

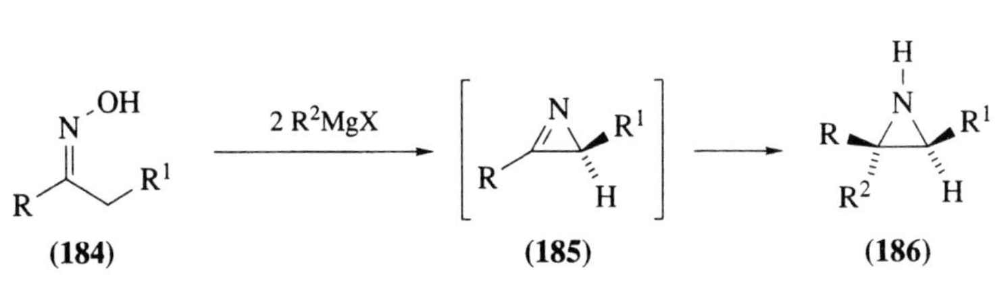

(184) 2 R^2MgX → [(185)] → (186)

Scheme 35

Table 14 Preparation of Aziridines from Oximes (Quaternary Hydrazones) and Grignard Reagents

Entry	*Substrate*	*Grignard*	*Product(s)*	*Yield (%)*	*Ref.*
1	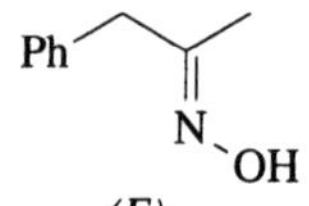 Ph, N–OH (*E*)	5 EtMgBr, toluene	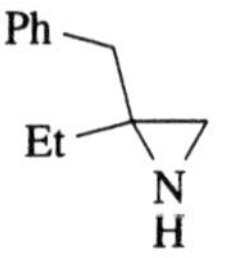 Ph, Et, N, H	33	126
2	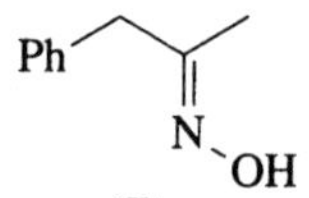 Ph, N–OH (*E*)	5 EtMgBr, THF	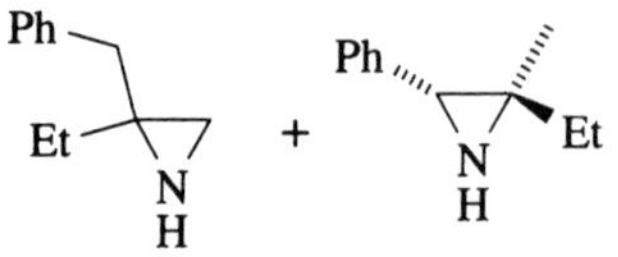 Ph, Et, N, H 85% + Ph, Et, N, H 15%	—	126
3	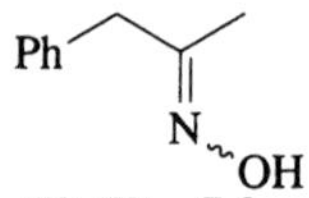 Ph, N–OH (*E*):(*Z*) = 7:3	5 EtMgBr, toluene	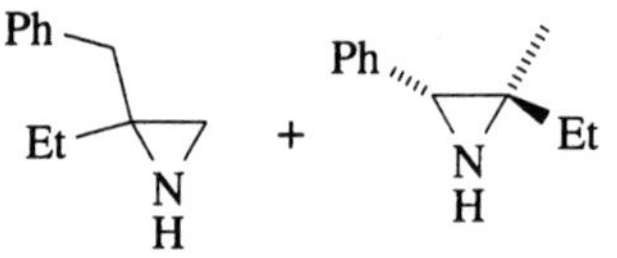 Ph, Et, N, H 52% + Ph, Et, N, H 48%	46	126
4	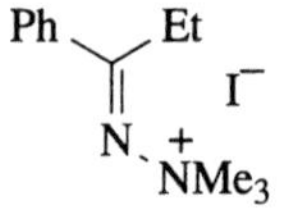 Ph, Et, I^-, N^+, NMe_3	PhMgBr	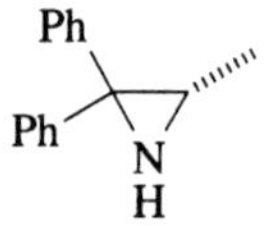 Ph, Ph, N, H	80	134
5	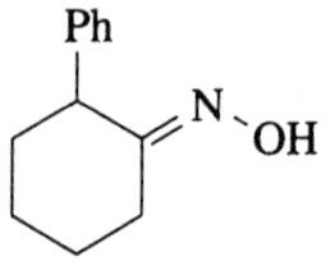 Ph, N–OH	2 EtMgBr, toluene	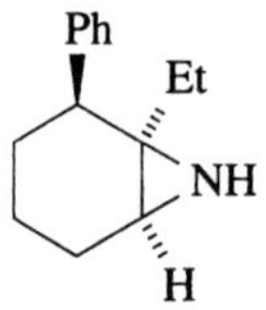 Ph, Et, NH, H	35	129
6	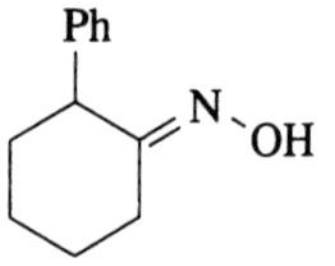 Ph, N–OH	2 MeMgBr, toluene	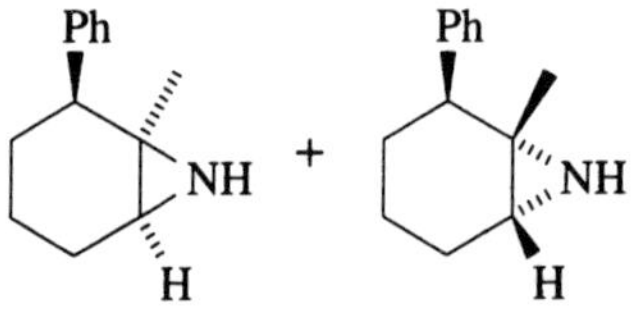 Ph, NH, H 75% + Ph, NH, H 25%	53	129
7	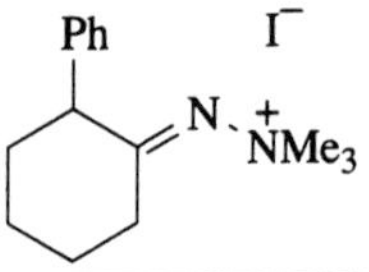 Ph, I^-, N^+, NMe_3	MeMgBr	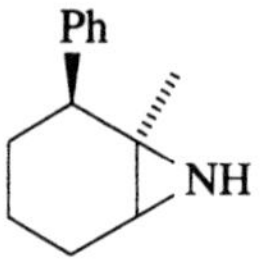 Ph, NH	93	134

$R^1R^2C{=}N{-}OSO_2R$ **(187)** $\xrightarrow{Me_2AlR}$ $[R^2{-}{\equiv}\overset{+}{N}{-}R^1]$ **(188)** ⟶ $R^2(R)C{=}N{-}R^1$ **(189)** $\xrightarrow[\text{ii, } H^+]{\text{i, } R^3MgBr}$ $R^2R^3(R)C{-}N(H)R^1$ **(190)**

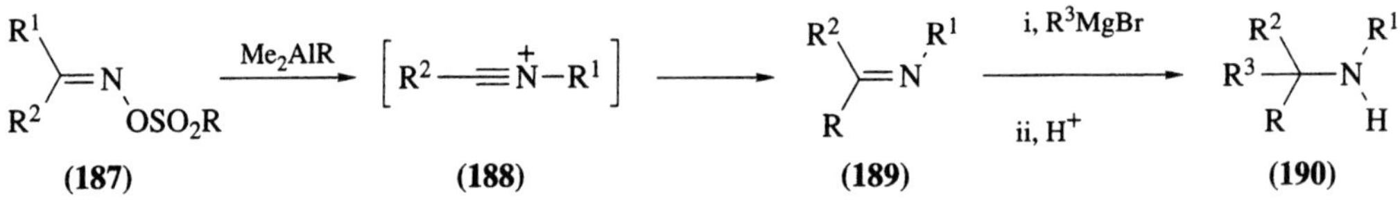

Scheme 36

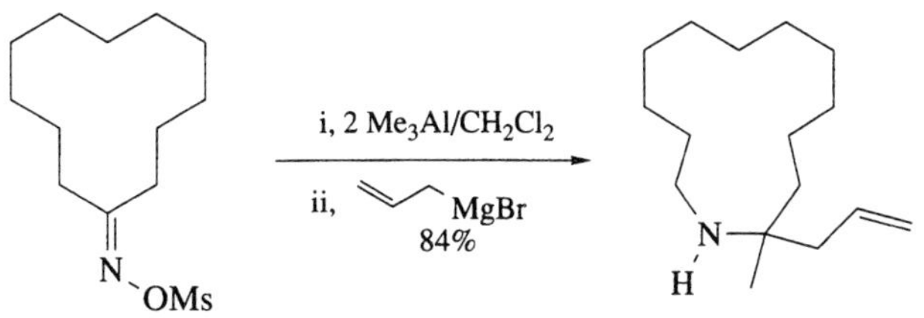

i, 2 Me_3Al/CH_2Cl_2
ii, MgBr
84%

(191) **(192)** (23)

1.12.9 ORGANOMETALLIC ADDITIONS TO *S*-ARYL SULFENIMINES

The addition of organometallic reagents to *S*-aryl sulfenimines (**193**) yields, upon aqueous work-up, substituted primary amines (**195**). The scope of these additions has not been extensively explored, largely because of the difficulties encountered in sulfenimine preparation. Recently, the generation of sulfenimines from *N,N*-bis(trimethylsilyl)sulfenamides and aldehydes or ketones has provided a more convenient access to these molecules.[137]

Davis and Mancinelli[138] examined the addition of aryl- and alkyl-lithium reagents to nonenolizable and enolizable *S*-aryl sulfenimines (Table 15). The intermediate sulfenamides (**194**) contain a relatively weak sulfur–nitrogen bond, which is cleaved on aqueous work-up to generate amines (**195**) directly. The ability to add organometallic reagents to enolizable sulfenimines (in addition to providing good diastereoselectivity in several diallylzinc[139] and enolate condensations)[140] demonstrates synthetic advantages over the corresponding oximes or trimethylsilyl imines. On the other hand, sulfenimine condensations are plagued by the functionality's ambient electrophilicity, as demonstrated in the attempted condensation of sulfenimine (**196**) with methyllithium.[141] In this case only products indicative of nitrogen–sulfur bond cleavage (imine **197** and thioanisole) were formed. Hart and coworkers[140] have found that nucleophilic attack on the sulfenimine sulfur atom can be minimized by replacement of the *S*-aryl moiety with the bulky *S*-trityl substituent.

Table 15 The Addition of Organometallic Reagents to *S*-Aryl Sulfenimines (**193**)

2 RM, Et_2O; H_2O

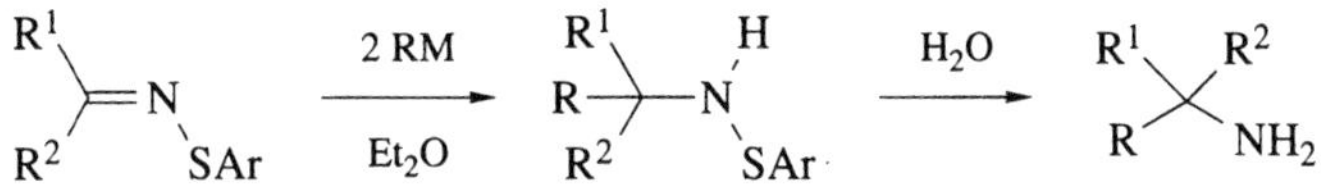

(193) **(194)** **(195)**

Scheme 37

Entry	R^1	R^2	*RM (excess)*	*Product (195)*	*Yield (%)*
1	Ph	H	MeLi	PhCH(Me)NH_2	79
2	Me	H	PhLi	PhCH(Me)NH_2	61
3	Me	Me	Bu^tLi	$Bu^tCMe_2NH_2$	43
4	—$(CH_2)_5$—		MeLi	 1-methylcyclohexyl-NH_2	68

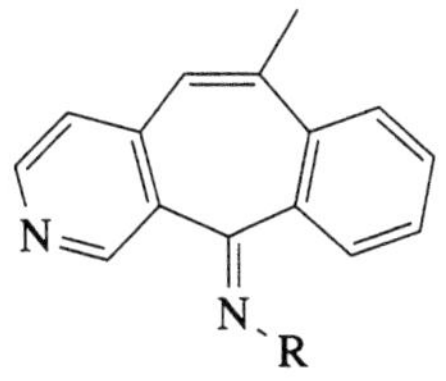

(196) R = SPh
(197) R = H

1.12.10 ORGANOMETALLIC ADDITIONS TO SULFONIMINES

Nucleophiles readily add to *N*-(arylsulfonyl)imines (**198**; Equation 24).[142] The reaction has limited synthetic utility for amine syntheses due to the harsh conditions required for the removal of the product's *N*-(arylsulfonyl) protecting group.

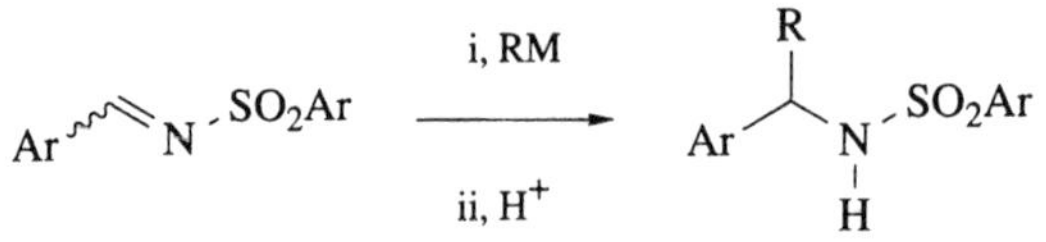

$$\text{ArCH=N-SO}_2\text{Ar} \xrightarrow[\text{ii, H}^+]{\text{i, RM}} \text{ArCH(R)-NH-SO}_2\text{Ar} \quad (24)$$

(198) **(199)**

Diaryl sulfamides (**201**) produced in the condensation of Grignard (primary, tertiary or aryl) and organolithium (primary or aryl) reagents with sulfamylimines (**200**) can, however, be hydrolyzed in refluxing aqueous pyridine to afford arylamines (**202**) in good (75–95%) overall yield (Scheme 38).[143]

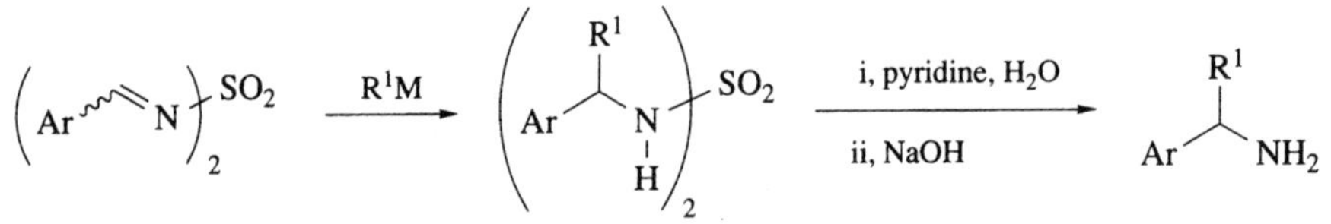

$$(\text{ArCH=N})_2\text{SO}_2 \xrightarrow{\text{R}^1\text{M}} (\text{ArCH(R}^1\text{)NH})_2\text{SO}_2 \xrightarrow[\text{ii, NaOH}]{\text{i, pyridine, H}_2\text{O}} \text{ArCH(R}^1\text{)NH}_2$$

(200) **(201)** **(202)**

Scheme 38

1.12.11 ORGANOMETALLIC ADDITIONS TO *N*-TRIMETHYLSILYLIMINES

While organometallic reagents condense with *N*-substituted imines (Schiff bases) to afford, after hydrolysis, good yields of substituted amines, the reaction with *N*-unsubstituted imines (**203**)[26a] derived from ammonia (which are easily hydrolyzed and self condense) is not synthetically useful. As a result, the use of masked imines containing labile silicon– or sulfur–nitrogen bonds, such as *N*-trimethylsilylimines (**204**) or *N*-sulfenimines (**205**), has been explored.

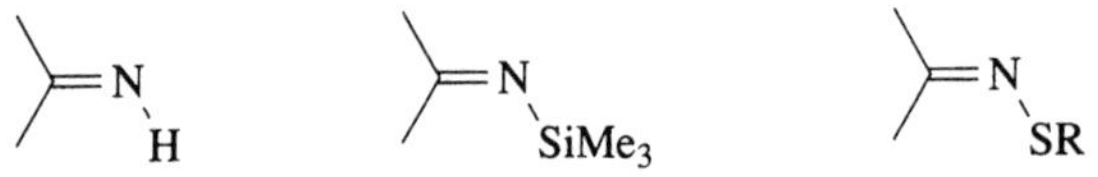

$$\text{R}_2\text{C=N-H} \qquad \text{R}_2\text{C=N-SiMe}_3 \qquad \text{R}_2\text{C=N-SR}$$

(203) **(204)** **(205)**

Demonstration that *N*-trimethylsilylimines (which are sometimes too unstable to isolate) can be generated and treated *in situ* with organometallic reagents has increased the scope and utility of these reactions for the preparation of substituted primary amines.[144] Nonenolizable aldehydes, for instance, condense with lithium bis(trimethylsilyl)amide to afford solutions of *N*-trimethylsilylaldimines. Hart[144] *et al.* and Nakahama and coworkers[145] have independently shown that trimethylsilylimines (**206**) react with organolithium and Grignard reagents to give, after aqueous work-up, primary amines (**207**) in moderate/excellent yields (equation 25). The isolation of phenyltrimethylsilane, albeit in low yield, in the addition of phenylmagnesium bromide to (**206**; R^1 = H) suggests that silicon attack may, in certain instances, compete with azomethine addition.

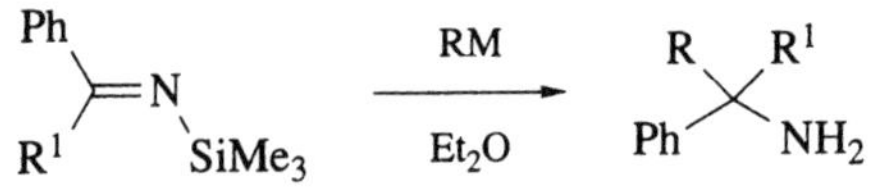

$$\text{Ph(R}^1\text{)C=N-SiMe}_3 \xrightarrow[\text{Et}_2\text{O}]{\text{RM}} \text{Ph(R)(R}^1\text{)C-NH}_2 \quad (25)$$

(206) **(207)**

R^1 = H; R = Me, Bu^n, Ph, Bu^t (87–100%); R^1 = Ph; R = Bu^t (68%)

Several other preparative methods, involving the condensation of (i) *N*-(trimethylsilyl)phosphinimines and carbonyl compounds;[146] and (ii) organometallic reagents with nitriles followed by quenching with chlorotrimethylsilane,[147] have been used for the preparation of other aldehyde- and ketone-derived silylimines. Organolithium reagents add to nonenolizable ketone-derived silylimines such as (**206**; R^1 = Ph) to generate tertiary carbinamines (**207**).[145]

Attempts to extend the organometallic addition reaction to *N*-trialkylsilylimines derived from enolizable ketones have been frustrated by difficulties encountered in the preparation of these silylimines (due to competitive enolization), in addition to the existence of a tautomeric equilibrium between desired silylimines and the corresponding enamines. As a result, addition products (formed in low yield) are accompanied by significant amounts of starting materials (presumably generated *via* enamine hydrolysis).[145] However, silylimines derived from enolizable aldehydes reportedly can be generated and trapped *in situ* with ester enolates to form β-lactams (18–60% yield).[148]

Iminium salts bearing a labile trimethylsilyl group can be generated *in situ* and undergo nucleophilic addition (see Sections 1.12.4.2 and 1.12.7.3). Bis(trimethylsilyl)methoxymethylamine (**75**), for example, has been used as a formaldehyde equivalent for the preparation of primary amines.[62,63] Cyclic imines, such as 3,4-dihydroquinolines, react with trimethylsilyl triflate (TMS-OTf) to provide reactive labile iminium salts (**55**), which condense with picoline anions.[49] The addition of nonstabilized Grignard and organolithium reagents to acyclic aromatic ketimines and aldimines, however, is often not facilitated by the presence of TMS-OTf.[149]

1.12.12 ORGANOMETALLIC ADDITIONS TO NITRONES

The highly polarized imine double bond of nitrones is responsible for the group's high electrophilic activity. The susceptibility of nitrones to nucleophilic addition has been exploited particularly in dipolar [3 + 2] cycloaddition reactions. The addition of organometallic reagents to acyclic and cyclic nitrones has been reviewed.[150] Grignards add to acyclic aldonitrones (**208**) bearing alkyl and aryl substituents to generate after work-up *N,N*-disubstituted hydroxylamines (**209**). *N,N*-disubstituted hydroxylamines can be further elaborated *via* oxidation[151,154] to substituted nitrones (**210**) or by reduction[152] to secondary amines (**211**). Acyclic ketonitrones are resistant to organometallic addition and have been reduced by Grignard reagents to the corresponding Schiff bases.[153]

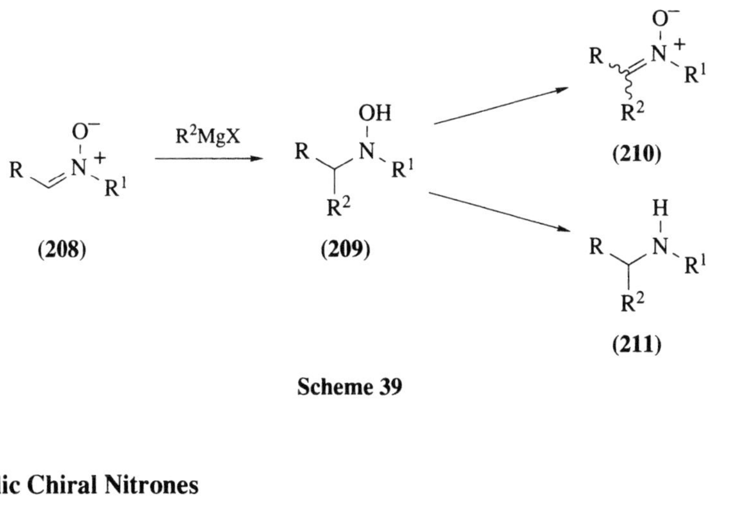

R^2MgX

(208) **(209)** **(210)** **(211)**

Scheme 39

1.12.12.1 Acyclic Chiral Nitrones

Chang and Coates[15] have examined the addition of organometallic reagents to nitrones (**212**; equation 26) containing a chiral nitrogen auxiliary (Table 16). These nitrones are prepared by the condensation of chiral *N*-alkylhydroxylamines with alkyl and aryl aldehydes and are assumed to exist in the (*Z*)-configuration. High diastereoselectivity (determined by ^1H NMR) is obtained in the addition of organolithium and Grignard reagents to nitrones bearing a β-alkoxy group on nitrogen (entries 1, 2, 4, 5 and 6, Table 16).

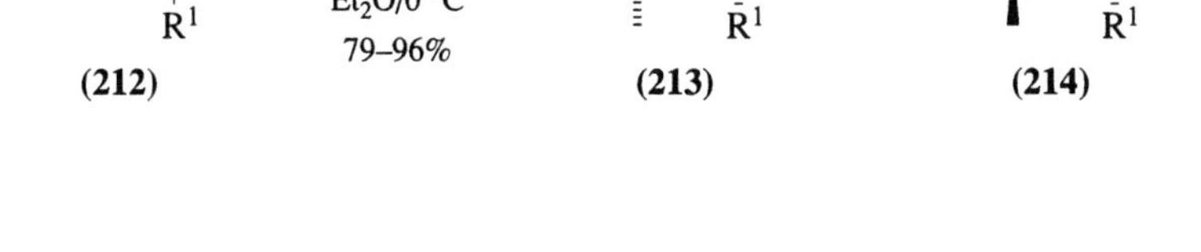

(212) $\xrightarrow[\text{Et}_2\text{O/0 °C, 79–96\%}]{R^3M}$ **(213)** + **(214)** (26)

Table 16 Additions of Organometallic Reagents to Racemic Nitrones (**212**; equation 26)

Entry	*Nitrone (212)* *R*	*R*[1]	*R*[2]	*R*^{3}*M*	*(213) (%)*	*(214) (%)*
1	Me	Ph	OMe	PhMgBr	97	3
2	Me	Ph	OBn	PhMgBr	90	10
3	Me	Ph	H	PhMgBr	54	46
4	Me	Pr^i	OMe	PhMgBr	95	5
5	Me	Pr^i	OMe	PhLi	94	6
6	Ph	Ph	OMe	MeMgBr	5	95
7	Ph	Pr^i	OTBDMS	MeMgBr	92	8

The absolute configuration of the major products (**213**) or (**214**) can be predicted by the chelation model shown in Scheme 40, as the resident stereogenic substituent controls the facial selectivity in the organometallic additions. The diastereoselectivity is affected by the β-alkoxy functionality. The methoxy substituent is preferred, as the selectivity erodes with benzyloxy substitution (entries 1 and 2, Table 16) and is reversed with the bulky *t*-butyldimethylsiloxy group (entries 6 and 7, Table 16). As expected, selectivity is poor for nitrones devoid of β-alkoxy substituents (entry 3, Table 16). By appropriate nitrone/organometallic substitution, both diastereomeric products can be preferentially generated (entries 1 and 6, Table 16).

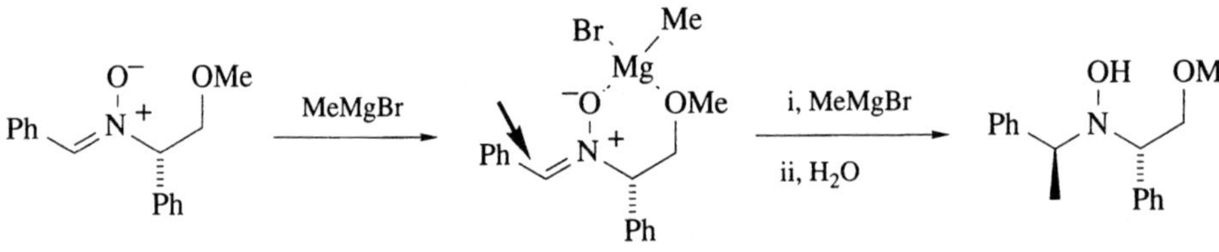

MeMgBr; i, MeMgBr; ii, H_2O

Scheme 40

Stereochemical control in the addition of Grignard reagents to chiral racemic *N*-(2-phenylpropylidene)alkylamine *N*-oxides (**215**) has been observed (equation 27).[154] Diastereoselectivity is modest (**216**:**217** = 2:1–5:1). The formation of hydroxylamines (**216**) as the major products is consistent with Felkin–Ahn Grignard addition (Figure 11).

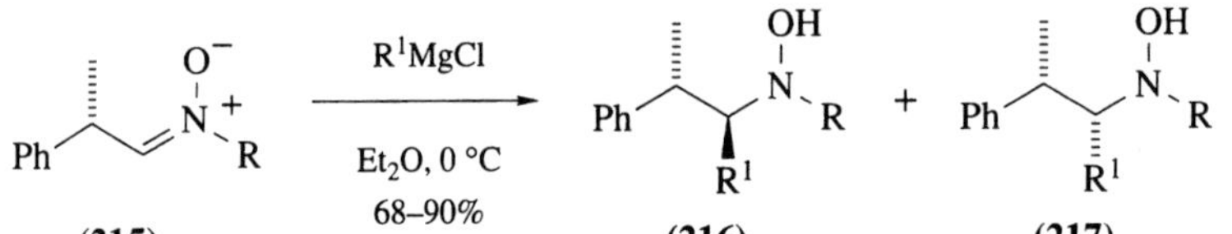

R^1MgCl, Et_2O, 0 °C, 68–90%; (**215**) → (**216**) + (**217**) (27)

R = C_6H_{11}, Me; R^1 = Me, Et, Pr^i

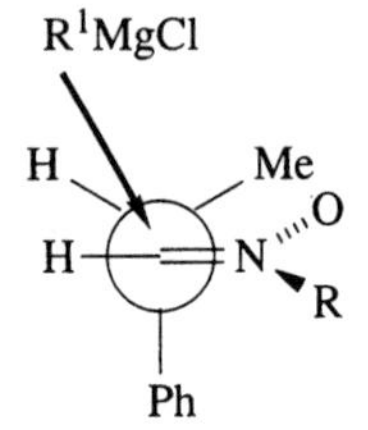

Figure 11

1.12.12.2 Cyclic Nitrones

Additions have been reported for both cyclic aldo- and keto-nitrones and have been utilized in the synthesis of structurally diverse heterocyclic systems (see Table 17).[155–157] The condensation of 3,4-dimethoxybenzyl Grignard with 3,4-dihydroisoquinoline *N*-oxide (entry 1, Table 17) provides an access to pellefierine alkaloids.[155] A synthetic use of cyclic ketonitrone condensations is demonstrated in the addition of allylmagnesium bromide to 6-methyl-2,3,4,5-tetrahydropyridine *N*-oxides, as the resultant hydroxylamine product can be readily oxidized and trapped internally to generate useful bridged isoxazolidines (entry 2, Table 17). Nitrones need not be isolated for organometallic addition as Grignard reagents readily condense with dimeric nitrones (entry 3, Table 17).[156] In addition, the condensation of alkyllithiums and Grignards with heterocyclic *N*-oxides such as 2,4,4-trialkyloxazoline *N*-oxides followed by $Cu(OAc)_2$ exposure provides a flexible entry into biologically useful Doxyl (4,4-dimethyloxazolidine-*N*-oxyl) nitroxide spin labels (entry 4, Table 17).[157]

Table 17 Additions of Organometallic Reagents to Cyclic Nitrones

Entry	*Nitrone*	*Organometallic*	*Product*	*Ref.*
1	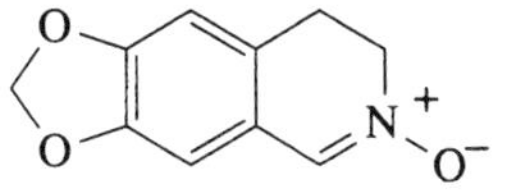	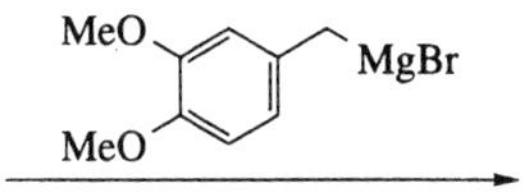 3,4-(MeO)2C6H3CH2MgBr, 65%	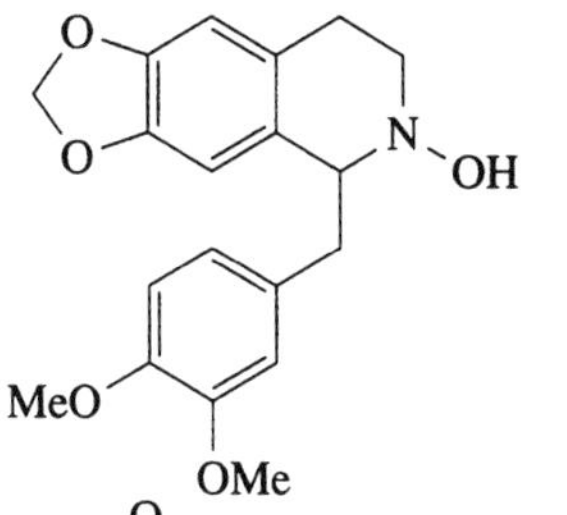	155
2	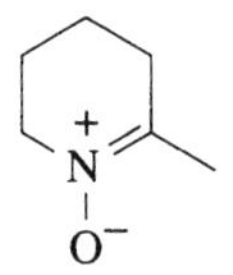	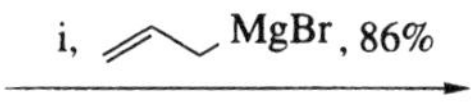 i, allyl-MgBr, 86%; ii, Pd	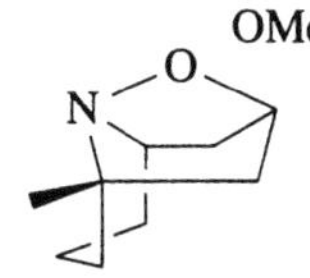	155
3	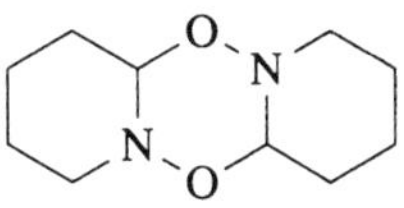	2PhMgBr, 93%	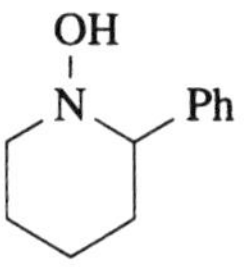	156
4	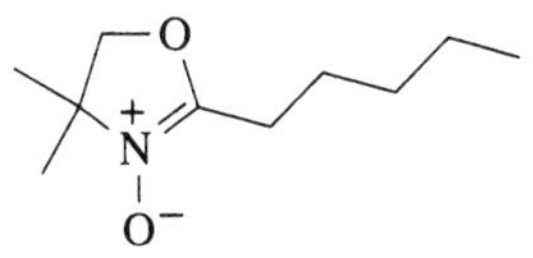	i, MeLi; ii, Cu^{2+}	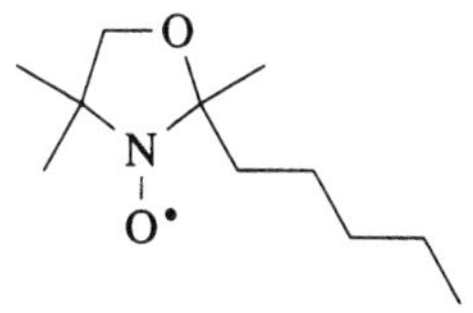	157

1.12.13 REFERENCES

1. J. K. Smith, D. E. Bergbreiter and M. Newcomb, *J. Am. Chem. Soc.*, 1983, **105**, 4396 and references therein.
2. C. H. Heathcock, C. T. Buse, W. A. Kleschick, M. C. Pirrung, J. E. Sohn and J. Lampe, *J. Org. Chem.*, 1980, **45**, 1066.
3. R. E. Lyle, H. M. Fribush, G. G. Lyle and J. E. Saavedra, *J. Org. Chem.*, 1978, **43**, 1275 and references therein.
4. A. Padwa and A. D. Woolhouse, in 'Comprehensive Heterocyclic Chemistry', ed. A. R. Katritzky and C. W. Rees, Pergamon Press, Oxford, 1984, vol. 7, p. 80.
5. E. Vedejs and W. T. Stolle, *Tetrahedron Lett.*, 1977, 135.
6. E. J. Corey and D. Enders, *Chem. Ber.*, 1978, **111**, 1337, 1362.
7. A. Hosomi, Y. Araki and H. Sakurai, *J. Am. Chem. Soc.*, 1982, **104**, 2081.
8. K. N. Houk, R. W. Strozier, N. G. Rondan, R. R. Fraser and N. Chauqui-Offermanns, *J. Am. Chem. Soc.*, 1980, **102**, 1426 and references therein.
9. (a) E. Kleinman, in 'Comprehensive Organic Synthesis', ed. B. M. Trost and I. Fleming, Pergamon Press, Oxford, 1991, vol. 2, chap. 4.1; (b) E. Kleinman and R. A. Volkmann, in 'Comprehensive Organic Synthesis', ed. B. M. Trost and I. Fleming, Pergamon Press, Oxford, 1991, vol. 2, chap. 4.3.
10. D. A. Evans, J. V. Nelson and T. R. Taber, *Top. Stereochem.*, 1982, **13**, 1.
11. D. A. Claremon, P. K. Lumma and B. T. Phillips, *J. Am. Chem. Soc.*, 1986, **108**, 8265.

12. H. Uno, Y. Shiraishi, K. Shimokawa and H. Suzuki, *Chem. Lett.*, 1988, 729.
13. Y. Yamamoto, T. Komatsu and K. Maruyama, *J. Am. Chem. Soc.*, 1984, **106**, 5031.
14. (a) H. Takahashi, Y. Suzuki and H. Inagaki, *Chem. Pharm. Bull.*, 1982, **30**, 3160; (b) Y. Suzuki and H. Takahashi, *Chem. Pharm. Bull.*, 1983, **31**, 31; (c) H. Takahashi, Y. Chida, T. Suzuki, S. Yanaura, Y. Suzuki and C. Masuda, *Chem. Pharm. Bull.*, 1983, **31**, 1659; (d) H. Takahashi, Y. Suzuki and T. Hori, *Chem. Pharm. Bull.*, 1983, **31**, 2183; (e) Y. Suzuki and H. Takahashi, *Chem. Pharm. Bull.*, 1983, **31**, 2895.
15. Z.-Y. Chang and R. M. Coates, *J. Org. Chem.*, 1990, **55**, 3464, 3475.
16. (a) H. Takahashi, K. Tomita and H. Otomasu, *J. Chem. Soc., Chem. Commun.*, 1979, 668; (b) H. Takahashi, K. Tomita and H. Noguchi, *Chem. Pharm. Bull.*, 1981, **29**, 3387; (c) H. Takahashi and H. Inagaki, *Chem. Pharm. Bull.*, 1982, **30**, 922; (d) H. Takahashi and Y. Suzuki, *Chem. Pharm. Bull.*, 1983, **31**, 4295.
17. D. Enders, H. Schubert and C. Nübling, *Angew. Chem., Int. Ed. Engl.*, 1986, **25**, 1109.
18. S. E. Denmark, T. Weber and D. W. Piotrowski, *J. Am. Chem. Soc.*, 1987, **109**, 2224.
19. (a) S. Hashimoto, S. Yamada and K. Koga, *J. Am. Chem. Soc.*, 1976, 98, 7450; (b) S. Hashimoto, S. Yamada and K. Koga, *Chem. Pharm. Bull.*, 1979, **27**, 771; (c) S. Hashimoto, H. Kogen, K. Tomioka and K. Koga, *Tetrahedron Lett.*, 1979, 3009; (d) H. Kogen, K. Tomioka, S. Hashimoto and K. Koga, *Tetrahedron Lett.*, 1980, **21**, 4005; (e) H. Kogen, K. Tomioka, S. Hashimoto and K. Koga, *Tetrahedron*, 1981, **37**, 3951; (f) K. Tomioka and K. Koga, in 'Asymmetric Synthesis', ed. J. D. Morrison, Academic Press, New York, 1983, vol. 2, p. 201.
20. (a) A. I. Meyers, J. D. Brown and D. Laucher, *Tetrahedron Lett.*, 1987, **28**, 5283; (b) A. I. Meyers, J. D. Brown and D. Laucher, *Tetrahedron Lett.*, 1987, **28**, 5279.
21. (a) A. I. Meyers and C. E. Whitten, *J. Am. Chem. Soc.*, 1975, **97**, 6266; (b) A. I. Meyers and C. E. Whitten, *Heterocycles*, 1976, **4**, 1687; (c) A. I. Meyers and C. E. Whitten, *Tetrahedron Lett.*, 1976, 1947; (d) A. I. Meyers, J. Slade, R. K. Smith, E. D. Mihelich, F. M. Herschenson and C. D. Liang, *J. Org. Chem.*, 1979, **44**, 2247; (e) A. I. Meyers, R. K. Smith and C. E. Whitten, *J. Org. Chem.*, 1979, **44**, 2250.
22. (a) B. A. Barner and A. I. Meyers, *J. Am. Chem. Soc.*, 1984, **106**, 1865; (b) A. I. Meyers and B. A. Barner, *J. Org. Chem.*, 1986, **51**, 120; (c) A. I. Meyers, G. P. Roth, D. Hoyer, B. A. Barner and D. Laucher, *J. Am. Chem. Soc.*, 1988, **110**, 4611; (d) A. I. Meyers and D. Hoyer, *Tetrahedron Lett.*, 1984, **25**, 3667; (e) A. I. Meyers, N. R. Natale, D. G. Wettlaufer, S. Rafii and J. Clardy, *Tetrahedron Lett.*, 1981, **22**, 5123; (f) A. I. Meyers and D. G. Wettlaufer, *J. Am. Chem. Soc.*, 1984, **106**, 1135.
23. A. Solladié-Cavallo and E. Tsamo, *J. Organomet. Chem.*, 1979, **172**, 165.
24. C. N. Meltz and R. A. Volkmann, *Tetrahedron Lett.*, 1983, **24**, 4503.
25. E. J. Corey, J. F. Arnett and G. N. Widiger, *J. Am. Chem. Soc.*, 1975, **97**, 430.
26. (a) R. W. Layer, *Chem. Rev.*, 1963, **63**, 499; (b) P. A. S. Smith, 'Open Chain Nitrogen Compounds', Benjamin, New York, 1965; (c) K. Harada, in 'The Chemistry of the Carbon–Nitrogen Double Bond', ed. S. Patai, Interscience, New York, 1970, p. 266.
27. H. Thies and H. Schoenenberger, *Arch. Pharm. (Weinheim, Ger.)*, 1956, **289**, 408.
28. B. L. Emling, R. J. Horvath, A. J. Saraceno, E. F. Ellermeyer, L. Haile and L. D. Hudac, *J. Org. Chem.*, 1959, **24**, 657.
29. (a) J. Thomas, *Bull. Soc. Chim. Fr.*, 1973, 1296; (b) J. Thomas, *Bull. Soc. Chim. Fr.*, 1973, 1300; (c) J. Thomas, *Bull. Soc. Chim. Fr.*, 1975, 209.
30. R. E. Dessy and R. M. Salinger, *J. Am. Chem. Soc.*, 1961, **83**, 3530.
31. (a) J. Huet, *Bull. Soc. Chim. Fr.*, 1964, 952; (b) J. Huet, *Bull. Soc. Chem. Fr.*, 1964, 960; (c) J. Huet, *Bull. Soc. Chim. Fr.*, 1964, 967.
32. (a) J. Barluenga, A. M. Bayón and G. Asensio, *J. Chem. Soc., Chem. Commun.*, 1983, 1109; (b) J. Barluenga, A. M. Bayón and G. Asensio, *J. Chem. Soc., Chem. Commun.*, 1984, 427.
33. L. E. Overman and R. M. Burk., *Tetrahedron Lett.*, 1984, **25**, 1635.
34. G. Stork and S. R. Dowd, *J. Am. Chem. Soc.*, 1963, **85**, 2178.
35. (a) M. Wada, Y. Sakurai and K. Akiba, *Tetrahedron Lett.*, 1984, **25**, 1079; (b) M. Wada, Y. Sakurai and K. Akiba., *Tetrahedron Lett.*, 1984, **25**, 1083.
36. Y. Yamamoto, *Angew. Chem., Int. Ed. Engl.*, 1986, **25**, 947.
37. (a) J.-C. Fiaud and H. B. Kagan, *Tetrahedron Lett.*, 1970, 1813; (b) J.-C. Fiaud and H. B. Kagan, *Tetrahedron Lett.*, 1971, 1019.
38. F. E. Scully, Jr., *J. Org. Chem.*, 1980, **45**, 1515.
39. H. Böshagen, W. Geiger and B. Junge, *Angew. Chem., Int. Ed. Engl.*, 1981, **20**, 806.
40. E. F. Godefroi and L. H. Simanyi, *J. Org. Chem.*, 1962, **27**, 3882.
41. (a) A. Solladié-Cavallo, J. Suffert and J.-L. Haesslein, *Angew. Chem., Int. Ed. Engl.*, 1980, **19**, 1005; (b) A. Solladié-Cavallo, J. Suffert and A. DeCian. *J. Organomet. Chem.*, 1982, **236**, 83.
42. K. Bláha and O. Cervinka, *Adv. Heterocycl. Chem.* 1966, **6**, 147.
43. (a) J. V. Paukstelis, in 'Enamines: Synthesis, Structure and Reactions', ed. A. G. Cook, Decker, New York, 1969, chap 5, p. 169; (b) H. Böhme and H. G. Viehe, in 'Advances in Organic Chemistry; Methods and Results', ed. H. Böhme and H. G. Viehe, Wiley, New York, 1976, vol. 9, p. 183.
44. (a) N. L. Holy, *Synth. Commun.*, 1976, **6**, 539; (b) T. A. Bryson, G. H. Bonitz, C. J. Reichel and R. E. Dardis, *J. Org. Chem.* 1980, **45**, 524.
45. J. L. Roberts, P. S. Borromeo and C. D. Poulter, *Tetrahedron Lett.*, 1977, 1299.
46. H. Saimoto, K. Nishio, H. Yamamoto, M. Shinoda, T. Hiyama and H. Nozaki, *Bull. Chem. Soc. Jpn.*, 1983, **56**, 3093.
47. (a) H. Böhme and P. Plappert, *Chem. Ber.*, 1975, **108**, 2827; (b) L. Duhamel, P. Duhamel and N. Mancelle, *Bull. Soc. Chim. Fr.*, 1974, 331.
48. (a) T. Shono, H. Hamaguchi, M. Sasaki, S. Fujita and K. Nagami, *J. Org. Chem.*, 1983, **48**, 1621; (b) T. Shono, M. Sasaki, K. Nagami and H. Hamaguchi, *Tetrahedron Lett.*, 1982, **23**, 97.
49. (a) Jahangir, D. B. MacLean, M. A. Brook and H. L. Holland, *J. Chem. Soc., Chem. Commun.*, 1986, 1608; (b) Jahangir, D. B. MacLean and H. L. Holland, *Can. J. Chem.*, 1987, **65**, 727; (c) Jahangir, M. A. Brook, D.

B. MacLean and H. L. Holland, *Can. J. Chem.*, 1987, **65**, 2362; (d) D. B. Repke, Jahangir, R. D. Clark and D. B. MacLean, *J. Chem. Soc., Chem. Commun.*, 1988, 439.
50. A. I. Meyers and E. M. Smith, *J. Org. Chem.*, 1972, **37**, 4289.
51. A. I. Meyers and E. W. Collington, *J. Am. Chem. Soc.*, 1970, **92**, 6676.
52. (a) M. W. Anderson, R. C. F. Jones and J. Saunders, *J. Chem. Soc., Chem. Commun.*, 1982, 282; (b) M. W. Anderson, R. C. F. Jones and J. Saunders, *J. Chem. Soc., Perkin Trans. 1*, 1986, 1995.
53. T. Moriya, K. Hagio and N. Yoneda, *Synthesis*, 1980, 728.
54. (a) J. T. Gupton and C. M. Polaski, *Synth. Commun.*, 1981, **11**, 561; (b) J. T. Gupton and D. E. Polk, *Synth. Commun.*, 1981, **11**, 571; (c) J. T. Gupton and C. Colon, *Synth. Commun.*, 1984, **14**, 271; (d) J. T. Gupton, B. Norman and E. Wysong, *Synth. Commun.*, 1985, **15**, 1305; (e) J. T. Gupton, J. Coury, M. Moebus and S. Fitzwater, *Synth. Commun.*, 1986, **16**, 1575.
55. (a) A. I. Meyers and S. Singh, *Tetrahedron Lett.*, 1967, 5319; (b) F. Tubery, D. S. Grierson and H. P. Husson, *Tetrahedron Lett.*, 1987, **28**, 6457.
56. (a) H. Hellmann and G. Opitz, 'α-Aminoalkylierung', VCH, Weinheim, 1960, p. 223; (b) J. Mathieu and J. Weill-Raynal, 'Formation of C–C Bonds', Thieme, Stuttgart, 1973, vol. 1.
57. G. Courtois and P. Miginiac, *Bull. Soc. Chim. Fr.*, 1982, **11–12**, 395.
58. D. Couturier, *Ann. Chim. (Paris)*, 1972, **7**, 19.
59. I. Iwai and Y. Yura, *Chem. Pharm. Bull.*, 1963, **11**, 1049.
60. G. Eisele and G. Simchen, *Synthesis*, 1978, 757.
61. R. Golse, M. Bourhis and J.-J. Bosc, *C. R. Hebd. Seances Acad. Sci., Ser. C*, 1978, **287**, 585.
62. T. Morimoto, T. Takahashi and M. Sekiya, *J. Chem. Soc., Chem. Commun.*, 1984, 794.
63. H. J. Bestmann and C. Wölfel, *Angew. Chem., Int. Ed. Engl.*, 1984, **23**, 53.
64. (a) H. Takahashi, T. Tsubuki and K. Higashiyama, *Synthesis*, 1988, 238; (b) B. Weidmann and D. Seebach, *Angew. Chem., Int. Ed. Engl.*, 1983, **22**, 31.
65. N. Tokitoh and R. Okazaki, *Tetrahedron Lett.*, 1984, **25**, 4677.
66. M. J. O'Donnell and J.-B. Falmagne, *Tetrahedron Lett.*, 1985, **26**, 699.
67. I. E. Pollak, A. D. Trifunac and G. F. Grillot, *J. Org. Chem.*, 1967, **32**, 272.
68. C. Germon, A. Alexakis and J. F. Normant, *Tetrahedron Lett.*, 1980, **21**, 3763.
69. N. Kreutzkamp and H.-Y. Oei, *Chem. Ber.*, 1968, **101**, 2459.
70. D. Beke and M. Martos-Bartsai, *Acta Chim. Acad. Sci. Hung.*, 1957, **11**, 295.
71. M. Sekiya and Y. Terao, *Chem. Pharm. Bull.*, 1970, **18**, 947.
72. H. Böhme and U. Bomke, *Arch. Pharm. (Weinheim, Ger.)*, 1970, **303**, 779.
73. V. H. Maddox, E. F. Godefroi and R. F. Parcell, *J. Med. Chem.*, 1965, **8**, 230.
74. (a) J. Yoshimura, Y. Ohgo and T. Sato, *Bull. Chem. Soc. Jpn.*, 1965, **38**, 1809; (b) J. Yoshimura, Y. Ohgo and T. Sato, *J. Am. Chem. Soc.*, 1964, **86**, 3858.
75. (a) H. Hellmann, *Angew. Chem.*, 1957, **69**, 463; (b) H. E. Zaugg and W. B. Martin, *Org. React. (N. Y.)*, 1965, **14**, 52; (c) H. E. Zaugg, *Synthesis*, 1970, 49; (d) H. E. Zaugg, *Synthesis*, 1984, 85; (e) W. N. Speckamp and H. Hiemstra, *Tetrahedron*, 1985, **41**, 4367.
76. H. E. Zaugg, *Synthesis*, 1984, 207.
77. E. U. Würthwein, R. Kupfer and C. Kaliba, *Angew. Chem. Suppl.*, 1983, 264.
78. T. Kametani, *Heterocycles*, 1982, **17**, 463.
79. (a) K. Clauss, D. Grimm and G. Prossel, *Justus Liebigs Ann. Chem.*, 1974, 539; (b) H. Fliri and C.-P. Mak, *J. Org. Chem.*, 1985, **50**, 3438.
80. (a) T. Kobayashi, N. Ishida and T. Hiraoka, *J. Chem. Soc., Chem. Commun.*, 1980, 736; (b) W. Koller, A. Linkies, H. Pietsch, H. Rehling and D. Reuschling, *Tetrahedron Lett.*, 1982, **23**, 1545.
81. K. Maruyama and T. Hiraoka, *J. Org. Chem.*, 1986, **51**, 399.
82. H. Onoue, M. Narisada, S. Uyeo, H. Matsumura, K. Okada, T. Yano and W. Nagata, *Tetrahedron Lett.*, 1979, 3867.
83. S. Jendrzejewski and W. Steglich, *Chem. Ber.*, 1981, **114**, 1337.
84. C. Germon, A. Alexakis and J. F. Normant, *Synthesis*, 1984, 40.
85. R. Kober, W. Hammes and W. Steglich, *Angew. Chem., Int. Ed. Engl.*, 1982, **21**, 203.
86. T. Schrader, R. Kober and W. Steglich, *Synthesis*, 1986, 372.
87. (a) F. Weygand, W. Steglich, W. Oettmeier, A. Maierhofer and R. S. Loy, *Angew. Chem., Int. Ed. Engl.*, 1966, **5**, 600; (b) F. Weygand, W. Steglich and W. Oettmeier, *Chem. Ber.*, 1970, **103**, 818.
88. C. Kashima, Y. Aoki and Y. Omote, *J. Chem. Soc., Perkin Trans. 1*, 1975, 2511.
89. P. Münster and W. Steglich, *Synthesis*, 1987, 223.
90. A. L. Castelhano, S. Horne, R. Billedeau and A. Krantz, *Tetrahedron Lett.*, 1986, **27**, 2435.
91. P. Ermert, J. Meyer, G. Stucky, J. Schneebeli and J.-P. Obrecht, *Tetrahedron Lett.*, 1988, **29**, 1265.
92. (a) P. J. Sinclair, D. Zhai, J. Reibenspies and R. M. Williams, *J. Am. Chem. Soc.*, 1986, **108**, 1103; (b) R. B. Williams, P. J. Sinclair, D. Zhai and D. Chen, *J. Am. Chem. Soc.*, 1988, **110**, 1547.
93. B. H. Lipshutz, B. Huff and W. Vaccaro, *Tetrahedron Lett.*, 1986, **27**, 4241.
94. A. Basha and S. M. Weinreb, *Tetrahedron Lett.*, 1977, 1465.
95. M. Y. Kim, J. E. Starrett, Jr. and S. M. Weinreb, *J. Org. Chem.*, 1981, **46**, 5383.
96. T. Shono, *Tetrahedron*, 1984, **40**, 811.
97. (a) R. H. Shapiro and M. J. Heath, *J. Am. Chem. Soc.*, 1967, **89**, 5734; (b) G. Kaufman, F. Cook, H. Schechter, J. Bayless and L. Friedman, *J. Am. Chem. Soc.*, 1967, **89**, 5736; (c) W. G. Dauben, M. E. Lorber, N. D. Vietmeyer, R. H. Shapiro, J. H. Duncan and K. Tomer, *J. Am. Chem. Soc.*, 1968, **90**, 4762; (d) J. E. Stemke and F. T. Bond, *Tetrahedron Lett.*, 1975, 1815; (e) R. H. Shapiro, M. Lipton, K. J. Kolonko, R. L. Buswell and L. A. Capuano, *Tetrahedron Lett.*, 1975, 1811; (f) R. H. Shapiro, *Org. React. (N. Y.)*, 1976, **23**, 405.
98. J. Meinwald and F. Uno, *J. Am. Chem. Soc.*, 1968, **90**, 800.
99. (a) J. E. Herz and E. González, *J. Chem. Soc., Chem. Commun.*, 1969, 1395; (b) J. E. Herz and C. V. Ortiz, *J. Chem. Soc. C*, 1971, 2294.
100. Y. Bessiere-Chretien and J.-P. Bras, *C. R. Hebd. Seances Acad. Sci., Ser. C*, 1969, **268**, 2221.

101. R. H. Shapiro and T. Gadek, *J. Org. Chem.*, 1974, **39**, 3418.
102. S. H. Bertz, *J. Org. Chem.*, 1979, **44**, 4967.
103. J. F. W. Keana, D. P. Dolata and J. Ollerenshaw, *J. Org. Chem.*, 1973, **38**, 3815.
104. Y. P. Kitaev, B. I. Buzykin and T. V. Troepol'skaya, *Russ. Chem. Rev. (Engl. Transl.)*, 1970, **39**, 441.
105. E. Vedejs, J. M. Dolphin and W. T. Stolle, *J. Am. Chem. Soc.*, 1979, **101**, 249.
106. S. H. Bertz, *Tetrahedron Lett.*, 1980, **21**, 3151.
107. A. Marxer and M. Horvath, *Helv. Chim. Acta*, 1964, **47**, 1101.
108. P. A. Wender and M. A. Eissenstat, *J. Am. Chem. Soc.*, 1978, **100**, 292.
109. P. A. Wender and J. M. Schaus, *J. Org. Chem.*, 1978, **43**, 782.
110. S. F. Martin, G. W. Phillips, T. A. Puckette and J. A. Colapret, *J. Am. Chem. Soc.*, 1980 **102**, 5866.
111. (a) H. Gilman, J. E. Kirby and C. R. Kinney, *J. Am. Chem. Soc.*, 1929, **51**, 2252; (b) H. Gilman and J. Morton, *J. Am. Chem. Soc.*, 1948, **70**, 2514.
112. H. G. Richey, Jr., R. McLane and C. J. Phillips, *Tetrahedron Lett.*, 1976, 233.
113. J. Pornet and L. Miginiac, *Bull. Soc. Chim. Fr.*, 1975, 841.
114. S. Itsuno, K. Miyazaki and K. Ito, *Tetrahedron Lett.*, 1986, **27**, 3033.
115. (a) K. Ikeda, Y. Yoshinaga, K. Achiwa and M. Sekiya, *Chem. Lett.*, 1984, 369; (b) K. Ikeda, K. Achiwa and M. Sekiya, *Tetrahedron Lett.*, 1983, **24**, 4707.
116. A. Basha and D. W. Brooks, *J. Chem. Soc., Chem. Commun.*, 1987, 305.
117. P. Beak, A. Basha and B. Kokko, *J. Am. Chem. Soc.*, 1984, **106**, 1511.
118. G. Boche and H.-U. Wagner, *J. Chem. Soc., Chem. Commun.*, 1984, 1591.
119. D. K. Pirie, W. W. Welch, P. D. Weeks and R. A. Volkmann, *Tetrahedron Lett.*, 1986, **27**, 1549.
120. K. E. Rodriques, A. Basha, J. B. Summers and D. W. Brooks, *Tetrahedron Lett.*, 1988, **29**, 3455.
121. R. A. Volkmann, unpublished results.
122. E. J. Corey, K. Niimura, Y. Konishi, S. Hashimoto and Y. Hamada, *Tetrahedron Lett.*, 1986, **27**, 2199.
123. T. Kolasa, S. Sharma and M. J. Miller, *Tetrahedron Lett.*, 1987, **28**, 4973.
124. G. Zinner, D. Moderhack, W. Kliegel, *Chem. Ber.*, 1969, **102**, 2536.
125. For a review see J. P. Freeman, *Chem. Rev.*, 1973, **73**, 283.
126. G. Alvernhe and A. Laurent, *Bull. Soc. Chim. Fr.*, 1970, **8–9**, 3003.
127. K. Miyano and T. Taguchi, *Chem. Pharm. Bull.*, 1970, **18**, 1806.
128. G. Alvernhe and A. Laurent, *Tetrahedron Lett.*, 1972, 1007.
129. R. Chaabouni and A. Laurent, *Bull. Soc. Chim. Fr.*, 1973, **9–10**, 2680.
130. Y. Diab, A. Laurent and P. Mison, *Tetrahedron Lett.*, 1974, 1605.
131. Y. Diab, A. Laurent and P. Mison, *Bull. Soc. Chim. Fr.*, 1974, **9–10**, 2202.
132. R. Bartnik and A. Laurent, *C. R. Hebd. Seances Acad. Sci.*, 1974, **279**, 289.
133. K. Imai, Y. Kawazoe and T. Taguchi, *Chem. Pharm. Bull.*, 1976, **5**, 1083.
134. G. Alvernhe, S. Arseniyadis, R. Chaabouni and A. Laurent, *Tetrahedron Lett.*, 1975, 355.
135. N. W. Gabel, *J. Org. Chem.*, 1964, **29**, 3129.
136. K. Maruoka, T. Miyazaki, M. Ando, Y. Matsumura, S. Sakane, K. Hattori and H. Yamamoto, *J. Am. Chem. Soc.*, 1983, **105**, 2831.
137. T. Morimoto, Y. Nezu, K. Achiwa and M. Sekiya, *J. Chem. Soc., Chem. Commun.*, 1985, 1584.
138. F. A. Davis and P. A. Mancinelli, *J. Org. Chem.*, 1977, **42**, 398.
139. C. Fuganti, P. Grasselli, G. Pedrocchi-Fantoni, *J. Org. Chem.*, 1983, **48**, 909.
140. D. A. Burnett, D. J. Hart and J. Liu, *J. Org. Chem.*, 1986, **51**, 1929.
141. D. G. Brenner, W. Halczenko and K. L. Shepard, *J. Heterocycl. Chem.*, 1986, **23**, 145.
142. (a) U. Nadir and V. K. Koul, *Synthesis*, 1983, 554; (b) P. Perlmutter and C. C. Teo, *Tetrahedron Lett.*, 1984, **25**, 5951; (c) F. A. Davis, J. Wei, A. C. Sheppard and S. Gubernick, *Tetrahedron Lett.*, 1987, **28**, 5115.
143. F. A. Davis, M. A. Giangiordano and W. E. Starner, *Tetrahedron Lett.*, 1986, **27**, 3957.
144. D. J. Hart, K. Kanai, D. G. Thomas and T.-K. Yang, *J. Org. Chem.*, 1983, **48**, 289.
145. A. Hirao, I. Hattori, K. Yamaguchi, S. Nakahama and N. Yamazaki, *Synthesis*, 1982, 461.
146. W. Sundermeyer and W. Lidy, *Chem. Ber.*, 1976, **109**, 1491.
147. L.-H. Chan and E. G. Rochow, *J. Organomet. Chem.*, 1967, **9**, 231.
148. G. Cainelli, D. Giacomini, M. Panunzio, G. Martelli and G. Spunta, *Tetrahedron Lett.*, 1987, **28**, 5369.
149. M. A. Brook and Jahangir, *Synth. Commun.*, 1988, **18**, 893.
150. (a) J. Hamer and A. Macaluso, *Chem. Rev.*, 1964, 473; (b) W. Rundel, *Methoden Org. Chem. (Houben-Weyl)*, 1968, **10/4**, 421; (c) G. Tennant, in 'Comprehensive Organic Chemistry', ed. D. H. R. Barton and W. D. Ollis, Pergamon Press, Oxford, 1979, vol. 2, p. 504; (d) G. R. Delpierre and M. Lamchen, *Q. Rev. Chem. Soc.*, 1965, **19**, 329.
151. S.-I. Murahashi, H. Mitsui, T. Watanabe and S. Zenki, *Tetrahedron Lett.*, 1983, **24**, 1049 and references cited therein.
152. Y. Kita, F. Itoh, O. Tamura, Y. Y. Ke and Y. Tamura, *Tetrahedron Lett.*, 1987, **28**, 1431.
153. A. Dornow, H. Gehrt and F. Ische, *Justus Liebigs Ann. Chem.*, 1954, **585**, 220.
154. M. P. Cowling, P. R. Jenkins and K. Cooper, *J. Chem. Soc., Chem. Commun.*, 1988, 1503.
155. H. Mitsui, S. Zenki, T. Shiota and S.-I. Murahashi, *J. Chem. Soc., Chem. Commun.*, 1984, 874.
156. J. Thesing and H. Mayer, *Chem. Ber.*, 1956, **89**, 2159.
157. (a) J. F. W. Keana and T. D. Lee, *J. Am. Chem. Soc.*, 1975, **97**, 1273; (b) T. D. Lee and J. F. W. Keana, *J. Org. Chem.*, 1976, **41**, 3237.

1.13 Nucleophilic Addition to Carboxylic Acid Derivatives

BRIAN T. O'NEILL
Pfizer Central Research, Groton, CT, USA

1.13.1 INTRODUCTION

The reaction of carbon-based nucleophiles and carboxylic acid derivatives often presents a sophisticated problem in reaction chemoselectivity, especially when selective acylation of organometallics to form a ketone is required (equation 1). Historically, this transformation has been plagued by the formation of by-products due to subsequent nucleophilic addition to the desired product. A great deal of effort has been directed toward developing gentler techniques which avoid overaddition. Alternatively, the preparation of a ketone from a carboxylic acid equivalent often relies on a three-step approach, as shown in equation (2).

$$RC(=O)OR^* \xrightarrow{R^1M} RC(=O)R^1 \qquad (1)$$

$$RC(=O)OR^* \xrightarrow{\text{reductant}} RC(=O)H \xrightarrow{R^1M} RCH(OH)R^1 \xrightarrow{\text{oxidation}} RC(=O)R^1 \qquad (2)$$

This familiar protocol begins with the selective reduction of an ester to an aldehyde. The aldehyde is then subjected to a nucleophilic addition with an organometallic to form a secondary alcohol (see Chapters 1.1 and 1.2 of this volume and Chapters 1.11 and 1.12, Volume 8). The required ketone is obtained through oxidation by one of several known reagents for the task (see Chapters 2.7 and 2.8, Volume 7). Although made simple by the ready availability of reducing reagents for the seemingly nontrivial first step and the facile oxidation depicted in the third step, it seems apparent that, even with high yields in each step, the overall transformation suffers from extended linearity. Considering methodology currently available, the development of a single-step general ketone synthesis through direct transformation of the acid derivative would be more expeditious. The goal of this chapter is to document the success of the latter approach, and to influence current practitioners to consider this strategy when designing synthetic approaches to target molecules.

The extent of the challenge can be rationalized as follows: reaction of a weakly electrophilic species such as an ester with a powerful nucleophile such as a Grignard reagent affords initially a ketone. By comparison with the ester, this highly electrophilic species cannot shield itself from the sea of excess nucleophile present in the reaction mixture and thus undergoes a second nucleophilic addition. For example, reaction of methylmagnesium bromide with ethyl acetate affords the intermediate 2-propanone only momentarily before succumbing to further nucleophilic attack with formation of *t*-butyl alcohol. It is often not possible to control a reaction of this type simply by the choice of solvents, temperature, order of addition or by the amount of the carbanionic reagent used. Although in most cases the desired product is a more reactive moiety than the starting carboxylic acid derivative, new reagents and substrates have been devised that allow for sophisticated manipulation of these reactivities. As we shall see, solutions to this problem are now abundant and our discussion will concentrate on methodology which has been developed recently for promoting ketone formation.

While most of the chemistry discussed in this chapter has been developed in the past decade, several important methods have withstood the test of time and have made important contributions in areas such as natural product synthesis. Methods such as cuprate acylation[1] and the addition of organolithiums[2] to carboxylic acids have continued to enjoy widespread use in organic synthesis, whereas older methods including the reaction of organocadmium reagents with acid halides, once virtually the only method available for acylation, has not seen extensive utilization recently.[3] In the following discussion, we shall be interested in cases where selective monoacylation of nonstabilized carbanion equivalents has been achieved. Especially of concern here are carbanion equivalents or more properly organometallics which possess no source of resonance stabilization other than the covalent carbon–metal bond. Other sources of carbanions that are intrinsically stabilized, such as enolates, will be covered in Chapter 3.6, Volume 2.

Related chemistry not covered in this chapter includes the acylation of stabilized organometallics and the broad category of enolate or metalloenamine *C*-acylation (Volume 2), the Claisen or Dieckman condensations (Chapter 3.6, Volume 2), the acylation of heteroatom-stabilized carbanions (Part 2, Volume 1), the Friedel–Crafts acylation of alkenes, aromatics or vinyl silanes (see Chapters 3.1 and 2.2, Volume 2), nor will methods that involve radical-mediated acylation be discussed in this section.

Finally, ketone formation by acylation of carbanions can proceed in several ways, including masking of the ketonic product until isolation, the use of carbanion equivalent reagents which do not readily react with the product or the use of substrates which are more reactive than the ketone towards acylation. An alternative approach involves substrates which selectively coordinate the organometallic and thus activate the substrate towards acylation. As we shall see, examples of successful acylation can occur even when the ketone is present in the reaction mixture. Additionally, transition metal mediated acylation provides a mechanistic alternative where carbon–carbon bond formation is the result of an electronic reorganization of the metal, driven by the stability of the particular oxidation state. This chapter will discuss these approaches in detail, as well as some other types of reactions that selectively result in the formation of ketones, by the presentation of examples from the natural products literature, and simultaneously review the scope and limitations of many of these methods.

1.13.2 ACYLATION OF ORGANOLITHIUM AND GRIGNARD REAGENTS

1.13.2.1 Acylation with *N*-Methoxy-*N*-methylamides

In 1981, Nahm and Weinreb reported an effective and versatile method for the direct acylation of unstabilized organometallics.[4] Organolithium and Grignard reagents, the most readily available nucleophilic agents in this class, have been used interchangeably in this ketone synthesis (equation 3). The approach has been successful in large part because of the exceptional stability of tetrahedral intermediate (**1**), the primary adduct from organometallic addition to *N*-methoxy-*N*-methylamides. The inertness of (**1**) prevents premature release of the ketone functionality and thus avoids products from secondary addition of the nucleophile. Subsequently, we shall discuss methods in which the ketone was released into the reaction mixture during acylation without seriously affecting the selectivity of acylation; however, the Weinreb approach has become one of the most generally effective for preventing overaddition. Furthermore, it is likely that prior coordination of the metal alkyl to the carbonyl and methoxyamide groups can serve to accelerate addition and limit enolization. Prior to this disclosure, few methods provided selective monoaddition of a nucleophile to carboxylic acid derivatives with stoichiometric quantities of both the acylating agent and organometallic.[4] The present method avoids the preparation of specialized organo-cadmiums[3] or -zincs for acylation, and complements the use of *S*-(2-pyridyl) thioates as acylating agents as demonstrated by Mukaiyama (see Section 1.13.2.2).

O
R N OMe
Me
R^1M, THF
M
O OMe
R N
R^1 Me
(**1**)
H^+, H_2O
O
R R^1
(3)

While nucleophilic addition to an *N*-methoxy-*N*-methylamide with an organolithium or Grignard reagent affords a ketone, the reduction of the same substrate may also be preformed with excess lithium aluminum hydride, or (more conveniently) DIBAL-H, thus providing an aldehyde directly. Table 1 shows the scope of both of these transformations.[4]

Table 1 Addition of Organometallics to *N*-Methoxy-*N*-methylamides

R	*R^1M*	*Equiv.*	*Reaction time*	*Temperature (°C)*	*Product*	*Yield (%)*
Ph	MeMgBr	1.1	1.0 h	0	PhCOMe	93
Ph	MeMgBr	75	1.0 h	0	PhCOMe	96
Ph	Bu^nLi	2.0	1.0 h	0	$PhCOBu^n$	84
Ph	PhC≡CLi	1.1	1.0 h	20	PhC≡CCOPh	90
Ph	PhC≡CMgBr	1.5	1.5 h	65	PhC≡CCOPh	92
Ph	DIBAL-H	Excess	1.0 h	–78	PhCHO	71
Ph	$LiAlH_4$	Excess	8 min	–78	PhCHO	67

The requisite *N*-methoxy-*N*-methylamides may be prepared from acid chlorides by employing a slight excess of the commercially available *N,O*-dimethylhydroxylamine hydrochloride[5] in the presence of pyridine. They have also been prepared from acylimidazoles[6] and from mixed anhydrides of carboxylic acids.[7] Once prepared, these systems possess stability equivalent to that of most tertiary amides and, in this section, we shall discuss several cases where extensive manipulation of ancillary functionality was accomplished without interference from the amide.

Recently, Hlasta has shown that *N*-methoxy-*N*-methylamides may be prepared through the addition of organolithiums to *N*-methoxy-*N,N′,N′*-trimethylurea (**2**) or the symmetrical reagent (**3**) (Scheme 1).[8] Interestingly, it is possible in some cases to introduce a second alkyllithium in order to produce a ketone directly without isolation of the intermediate amide.

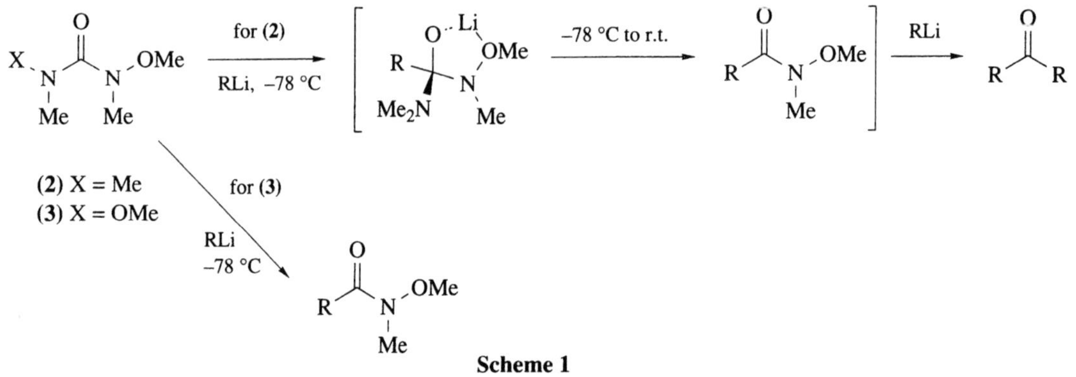

(**2**) X = Me
(**3**) X = OMe

for (**2**) RLi, –78 °C; –78 °C to r.t.; RLi; for (**3**) RLi –78 °C

Scheme 1

As an extension of their chiral aldol and alkylation technology, Evans and coworkers have reported a variety of methods for cleaving and replacing the chiral oxazolidinone auxiliary once chain construction has been completed.[9] Included in this methodology was the direct transformation of a chiral imide to an *N*-methoxy-*N*-methylamide through the use of aluminum amides (prepared *in situ*).[10] This reaction has been shown to be rather general for complex substrates (Scheme 2).[11]

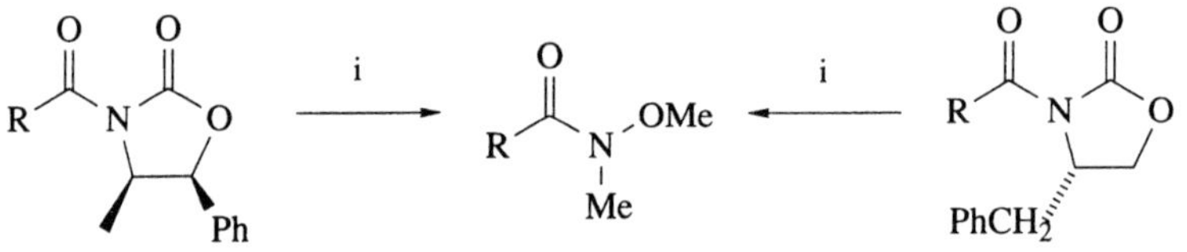

i, MeONHMe•HCl, $AlMe_3$, CH_2Cl_2, –15 to 0 °C, >90%

Scheme 2

Since it had been determined that ketone or aldehyde functionality was not directly accessible from chiral *N*-acyloxazolidinones, the transamination–metal alkyl addition procedure provided a conveniently expeditious alternative. The first step, transamination, proceeded in high yield by introduction of the *N*-acyloxazolidinone into a solution of the aluminum amide in dichloromethane at –15 °C. The reaction is favored by the presence of α-heteroatom substituents and by β-alcohol functionality (aldol adducts). Acceleration of the transamination in the latter case is most likely due to formation of a chelated intermediate (**5**) which serves to activate only the exocyclic carbonyl towards attack (equation 4). Because of the indicated activation, these aldol adducts are often the best substrates for this permutation. The effectiveness of the transamination in the case of (**4**) is noteworthy, as retroaldol fragmentation of this substrate usually occurs under mild base catalysis.[11a]

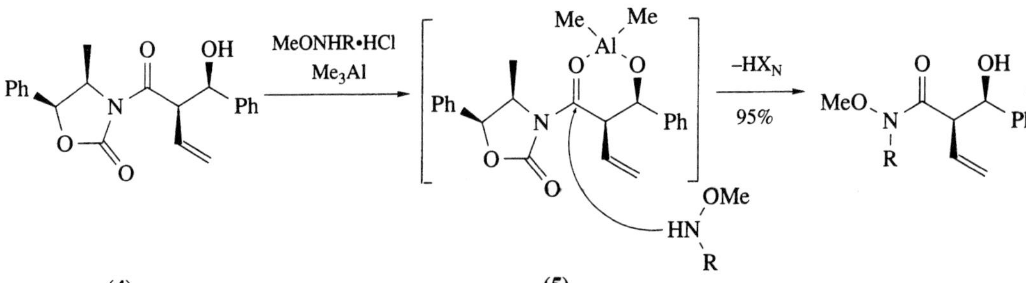

MeONHR•HCl, Me_3Al; $-HX_N$, 95% (4)

(**4**) (**5**)

In the case of a simple α-alkylated adduct such as (**6**), which cannot form a chelated intermediate of the type (**5**) or which does not contain α-heteroatom activation, and for cases that are especially hindered, attempted transamination often leads to competing attack upon the oxazolidinone carbonyl (equation 5).[15]

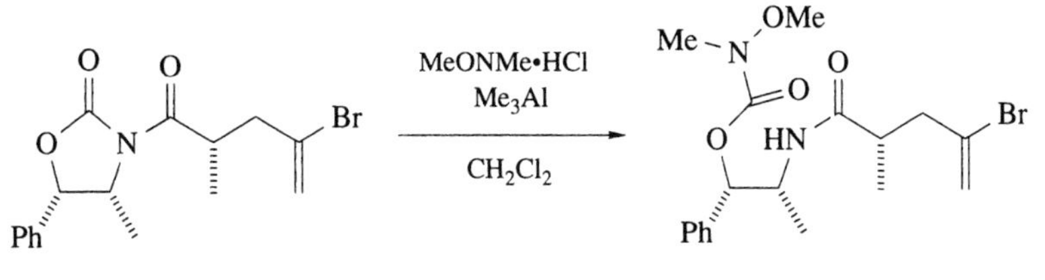

MeONMe•HCl, Me_3Al, CH_2Cl_2

(**6**) → (**7**) (5)

The preparation of fully elaborated ketones and aldehydes was demonstrated by Evans in the context of the total syntheses of the antibiotic X-206[12] and the antitumor cytovaricin.[14] The exceptional stability of the *N*-methoxy-*N*-methylamide allows application of an array of functional group transformations on distant portions of the molecule before organometallic addition is carried out. Early intermediates towards X-206, compounds (**8**) and (**9**), readily suffer ozonolytic cleavage to the aldehyde amide (**10**). (Notice that either terminus of (**9**) may later be converted in a single operation to an aldehyde.) When imide or ester functionality is substituted for the *N*-methoxy-*N*-methylamide of (**8**), attempted protection of the secondary alcohol results in retroaldolization. However, *O*-benzylation was accomplished with intermediate (**8**) without competing retroaldol cleavage, even when the sodium aldolate was used! Intermediate (**10**) was subjected to the Wittig reaction in toluene and the resulting (**11**) was then reduced by DIBAL-H to aldehyde (**12**). No epimerization was observed at either stereogenic center during these modifications (Scheme 3).

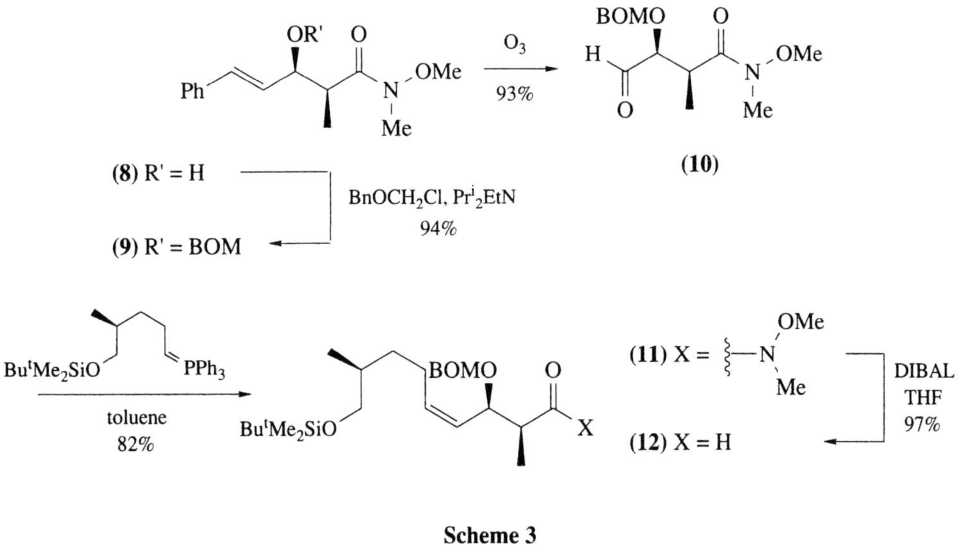

(**8**) R' = H; (**9**) R' = BOM ($BnOCH_2Cl$, $Pr^i{}_2EtN$, 94%); O_3, 93% → (**10**)

Bu^tMe_2SiO…PPh_3, toluene, 82% → (**11**) X = N(OMe)Me; (**12**) X = H (DIBAL, THF, 97%)

Scheme 3

The synthesis of cytovaricin passes through several highly functionalized intermediates including compound (**13**).[13] Glycosidation of the alcohol at C-3 was carried out with the *N*-methoxy-*N*-methylamide functionality in place. The addition of methyllithium and deprotection of the alcohol completes the formation of ketone (**14**) without competing epimerization or elimination (equation 6).

The acylation of vinylorganometallics was also possible with this technology. In equation (7), taken from the cytovaricin study, the acylation was completed directly after transamination of the starting imide (not shown), without racemization or retroaldolization and apparently without the necessity of protection of the β-hydroxy alcohol.[14]

A model study, directed at the total synthesis of X-206, showed that more functionalized vinyllithiums could be acylated in very high yield.[15] The product, an α-methylene ketone, does not suffer further attack by the vinyllithium reagent because the enone is concealed until after quenching of the reaction mixture.

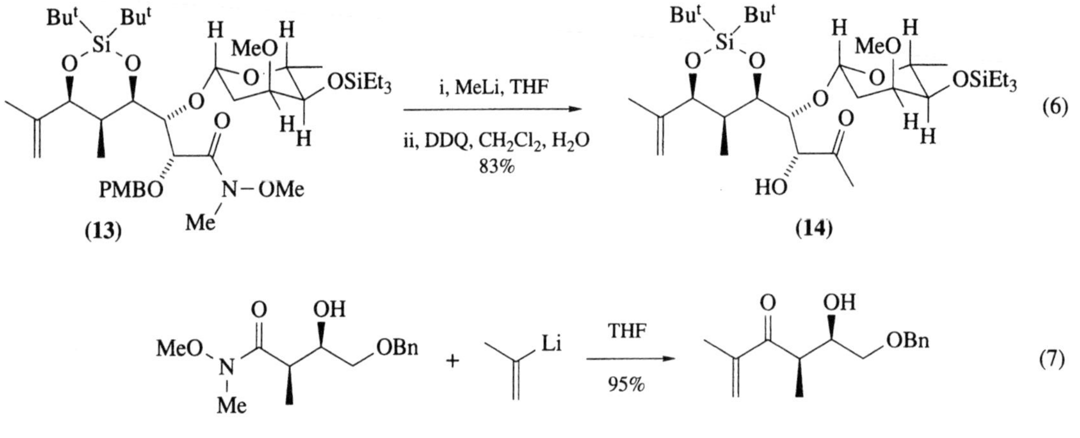

(13) → (14): i, MeLi, THF; ii, DDQ, CH_2Cl_2, H_2O; 83% (6)

THF, 95% (7)

The tetrahedral intermediate generated by organometallic addition thus provides '*in situ*' protection of the ketone (equation 8).

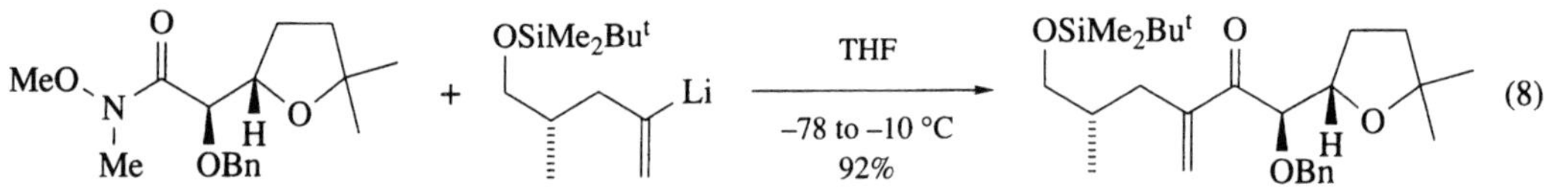

THF, −78 to −10 °C, 92% (8)

Organometallic addition to the *N*-methoxy-*N*-methylamide (**15**) also affords an exceptionally stable tetrahedral intermediate (**16**) and carbonyl-protecting group, first used in the synthesis of X-206.[12] Deprotonation of the hydrazone in intermediate (**16**) was subsequently carried out with lithium diisopropylamide. The resulting dianion initiated a novel attack upon epoxide (**17**) and in the ensuing transformation was followed by tetrahydrofuran ring formation as depicted, in 71% yield, all in one pot (Scheme 4).

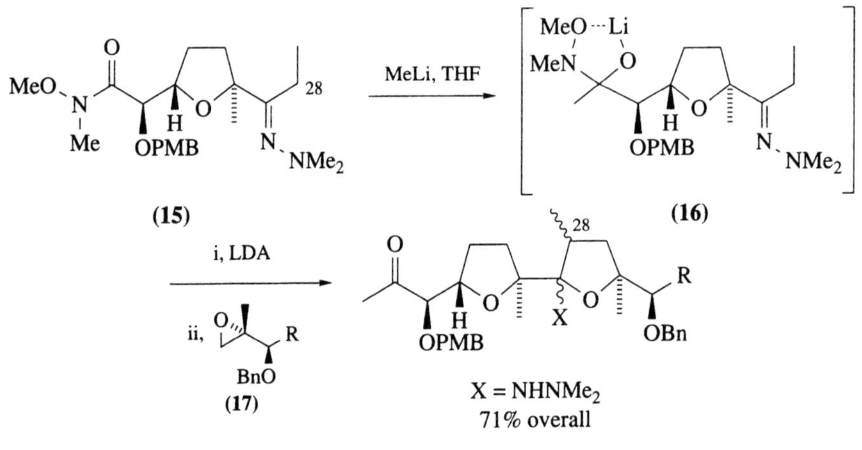

(15) → MeLi, THF → (16) → i, LDA; ii, (17)

X = $NHNMe_2$
71% overall

Scheme 4

A similar transformation was carried out on the less-oxygenated substrate (**18**) during the synthesis of a portion (C-10 to C-20) of the antibiotic ferensimycin (Scheme 5).[16]

The most complex example of this type of consecutive organometallic acylation, subsequent deprotonation, and tetrahydrofuran ring formation was recorded during the synthesis of the right hand portion of X-206 (Scheme 6).[12] The high overall yield obtained in this process is a testament to the method's generality.

This method has been applied by other researchers as well. The recent total synthesis of (−)-FK 506 by the Merck Process Group employed *N*-methoxy-*N*-methylamide functionality as a mild route to complex aldehydes (equation 9).[17]

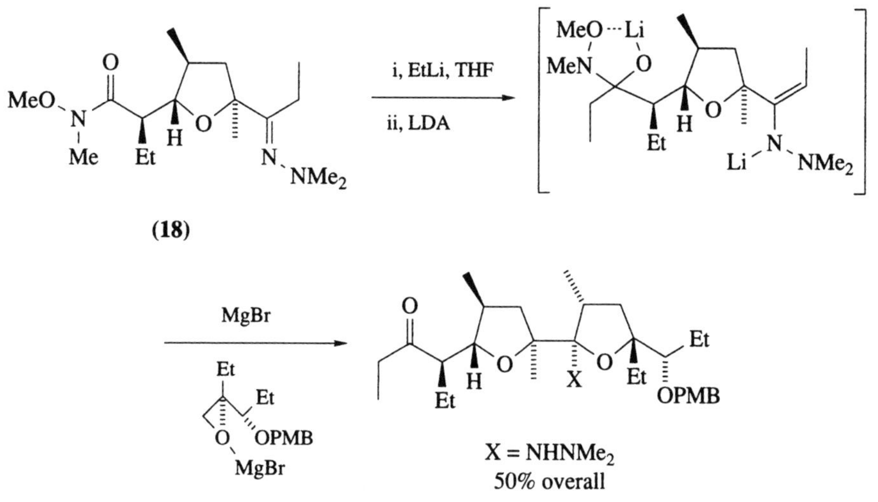

(18)

i, EtLi, THF

ii, LDA

X = NHNMe$_2$
50% overall

Scheme 5

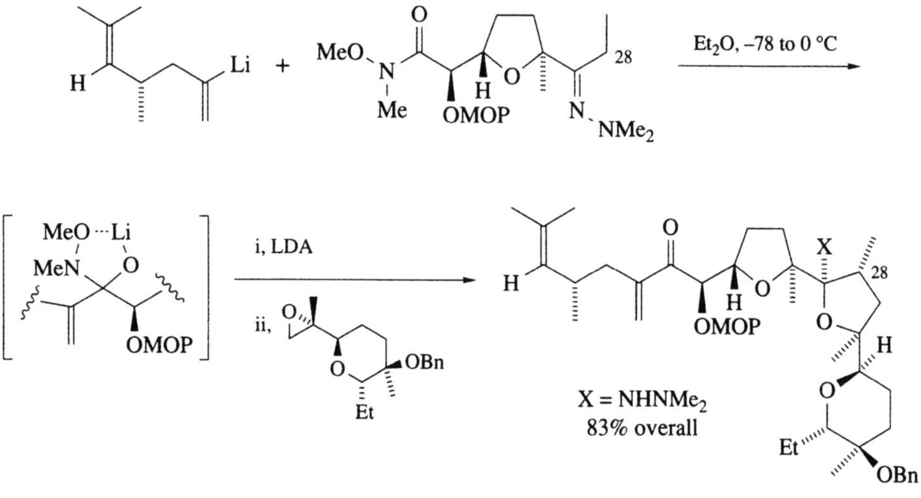

Et_2O, –78 to 0 °C

i, LDA

X = NHNMe$_2$
83% overall

Scheme 6

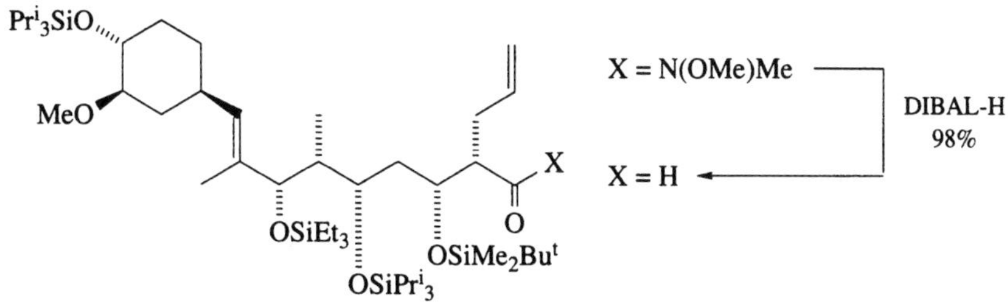

X = N(OMe)Me

DIBAL-H
98%

X = H

(9)

Patterson has prepared ethyl ketone (**20**) without racemization and in good overall yield by the sequence shown in Scheme 7. The preparation of (**20**) by standard organometallic addition to the corresponding aldehyde followed by Swern oxidation resulted in substantial racemization.[18]

By way of comparison, a very similar transformation has been accomplished without the use of the *N*-methoxy-*N*-methylamide. Hydrolysis of methyl ester (**19**) affords the corresponding carboxylic acid which was treated directly with methyllithium to afford a 70% overall yield of methyl ketone.[19]

The acetylation of a precursor to the natural respiratory toxin (±)-anatoxin-a was best achieved under the Weinreb protocol as depicted in equation (10). Other methods which relied on organocuprate or Grignard addition to thioacetates were totally unsuccessful. A proton-transfer side reaction consumed up

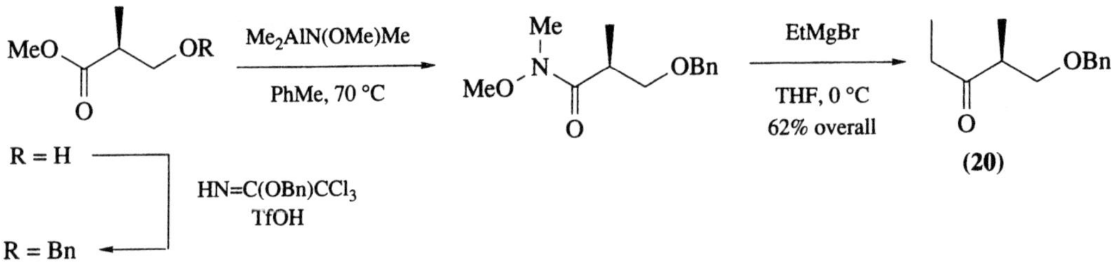

MeO–C(=O)–CH(Me)–CH₂OR → (Me₂AlN(OMe)Me, PhMe, 70 °C) → MeO–N(Me)–C(=O)–CH(Me)–CH₂OBn → (EtMgBr, THF, 0 °C, 62% overall) → (20)

R = H → (HN=C(OBn)CCl₃, TfOH) → R = Bn (19)

Scheme 7

to 10% of the vinyllithium, formed by lithium–halogen exchange under standard conditions (>2 equiv. Bu^{t}Li), but it was not clear whether this was due to deprotonation of the acetylating agent or *t*-butyl bromide.[20]

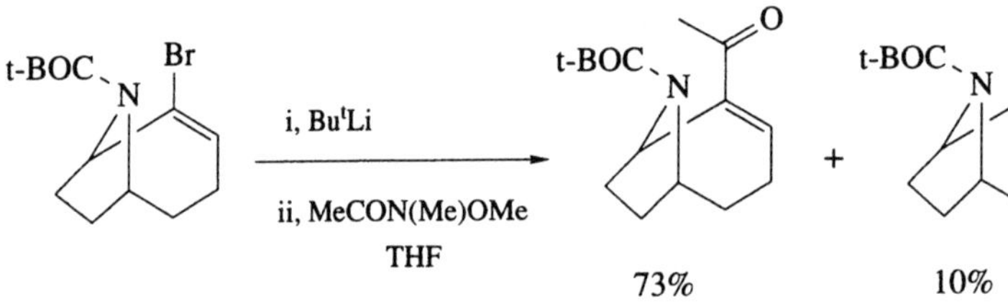

t-BOC (Br) → (i, Bu^{t}Li; ii, MeCON(Me)OMe, THF) → t-BOC (acetyl) 73% + t-BOC 10% (10)

Weinreb has reported the total syntheses of (+)-actinobolin and (–)-bactobolin from common intermediate (**21**). Both the reductive cleavage and Grignard addition proceeded in high overall yield (equation 11).[21]

(21) → (11)

(–)-Bactobolin intermediate; MeMgBr, THF, 88% X = Me
(+)-Actinobolin intermediate; $LiAlH_4$, THF, 89% X = H

A highly selective acylation of assorted organometallics by β-lactam (**22**) was achieved through the use of the *N*-methoxy-*N*-methylamide. There was no attack at the β-lactam carbonyl and the overall yields were quite good (equation 12).[22]

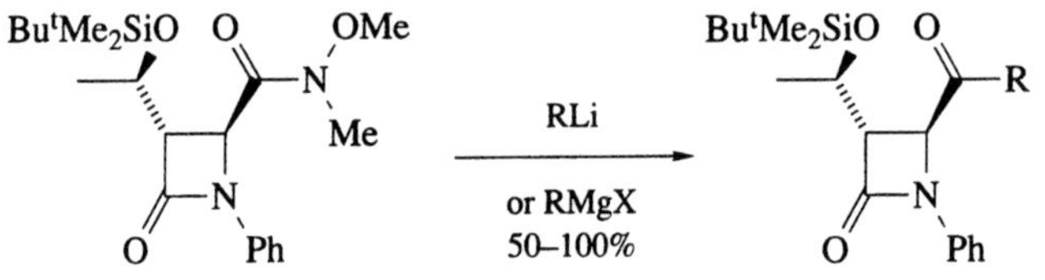

(22) → (RLi or RMgX, 50–100%) (12)

Williams has reported that *ortho*-lithiated anisole may be acylated with the *N*-methoxy-*N*-methylamide of 2-(benzyloxy)acetic acid (equation 13) in higher yield than previously demonstrated with organocadmium reagents (90% *versus* 40%). The ketone (**23**) is an early intermediate in an approach to quinocarcin.[23] Acylation of lithium 3-lithiopropoxide, generated from the stannane, afforded a hydroxy ketone without racemization at the α-position (equation 14).[24]

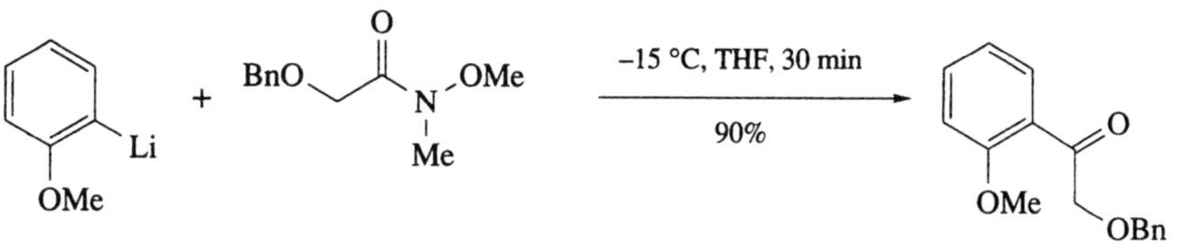

(13)

(23)

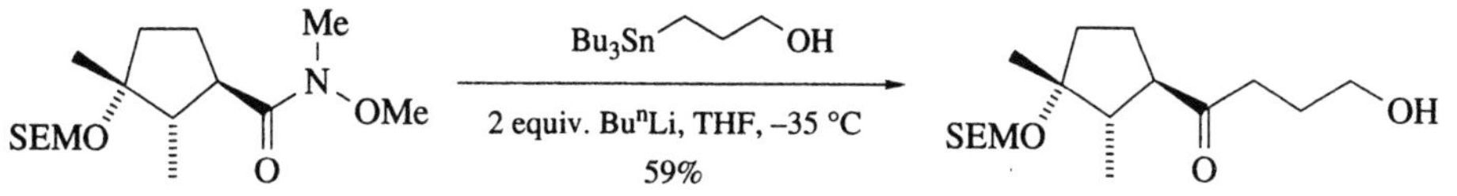

(14)

1.13.2.1.1 Formation of alkynic ketones

Alkynic ketones have been used extensively in natural product synthesis, due in large part to the contributions of Midland and coworkers and the development of general methods for enantioselective reduction of this moiety to afford optically active propargyl alcohols using chiral trialkylboranes.[25] Furthermore, the derived alkynic alcohol is a versatile system which can be manipulated directly into *cis*- or *trans*-allylic alcohols and as a precursor for vinylorganometallic species. This section will briefly cover progress made in the direct acylation of alkynic organolithiums with the acylation protocol developed by Weinreb (see also Section 1.13.2.7).

Several examples of alkynic ketone formation have been recorded since Weinreb's first examples.[4] A Diels–Alder strategy for the synthesis of mevinolin required the preparation of alkynic ketone (**24**). Standard methods, calling for the addition of the alkynide anion to an aldehyde followed by oxidation, lead to extensive degradation and by-product formation. The Weinreb methodology was clearly more effective (Scheme 8).[26]

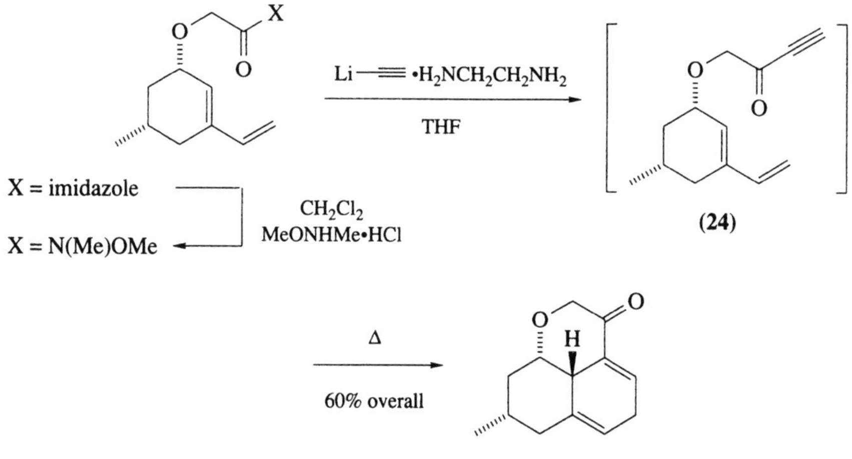

(24)

Scheme 8

Trost has prepared several alkynic ketones by direct acylation of *N*-methoxy-*N*-methylamides. Because of subsequent transformations on the product, the addition served as a method for introducing a dienylanion. Interestingly, the methoxyamide moiety is specifically activated towards nucleophilic addition, and the neighboring ester of the lactate system (**25**) was not disturbed during this transformation (equation 15). Prior coordination of the organolithium within the amide subunit and subsequent activated addition of the nucleophile serves as a reasonable explanation for the unusual selectivity of this process.[27]

Rapoport has compared the leaving-group ability of a variety of activated acyl groups towards metal alkynides and has demonstrated the usefulness of acylisoxazolidides such as (**27**) for the preparation of α′-amino-α,β-ynones (equation 16). Qualitatively, these acylating agents bear a strong resemblance to the simpler *N*-methoxy-*N*-methylamides (**26**) and appear to be as effective in acylation. The advantage of

(15)

(25)

the former substrate appears to stem from the crystalline nature of many of the resulting adducts. The ynones prepared by this route retain the optical purity of the *N*-carboethoxyamino acids. Interestingly, the study also found that the *N*-protected amino acids themselves were unreactive towards lithium alkynides, whereas alkyl- or alkenyl-lithiums added rapidly to these substrates (*vide infra*).[28]

(16)

X	M	% ynone	ynone:*t*-alcohol
(26)	M = Li (250 mol %)	84	>100:1
(27) m.p. = 95–96 °C	M = Li (250 mol %)	88	>100:1
(28)	M = MgBr (200 mol %)	<1	1:20
(29)	M = Li (100 mol %)	<5	1:10

Rapoport found that *S*-(2-pyridyl) thioates such as (**28**) did not function as selective acylating agents, and substantial amounts of tertiary alcohol were formed through overaddition (equation 16). Presumably, the tetrahedral intermediate, derived from nucleophilic addition to the *S*-(2-pyridyl) thioate, was not a stable entity in the reaction mixture. As will be discussed shortly, the lability of these intermediates had been recognized previously.[30] The novel dimethylpyrazolide moiety of substrate (**29**) also did not confer any additional stability to the tetrahedral intermediate and tertiary alcohol was the major product (equation 16). Tertiary amides, such as those derived from pyrrolidine or dimethylamine, were reactive towards lithium alkynides in the presence of BF_3, but analysis of the product indicated that it had undergone substantial racemization.[28]

Jacobi's synthesis of (±)-paniculide-A involves an intramolecular Diels–Alder reaction of the alkynic ketone (**31**). This compound was prepared in >90% yield (Scheme 9) by acylation of a lithium alkynide with the *N*-methoxy-*N*-methylamide (**30**). Addition of the anion to other derivatives related to (**30**) such as an acid chloride, a trifluoroacetic mixed anhydride, an acyl imidazole, *S*-(2-pyridyl) thiolates and a mixed carbonic anhydride (from ethyl chloroformate) led to either bis-addition or to proton abstraction. Notice should be made of the stability exhibited by the *N*-methoxy-*N*-methylamide group while the oxazole moiety was being introduced.[29]

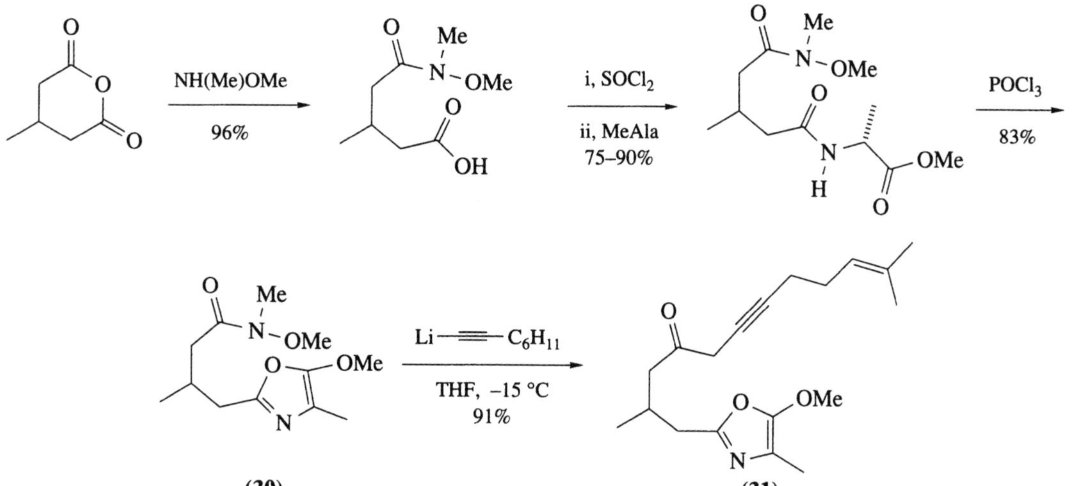

(30) (31)

Scheme 9

1.13.2.2 Acylation with *S*-(2-Pyridyl) Thioates

During the previous discussion, mention was made of the acylation of Grignard reagents with *S*-(2-pyridyl) thioates as developed by Mukaiyama. In principle, this method also avoids the preparation of cuprates and organocadmiums in favor of the more readily available Grignard reagents; however, the approach often needs modification, usually through the use of copper(I) salts, in cases where overaddition is encountered (*vide infra*). Some other limitations of the Grignard addition have surfaced during the synthesis of complex natural products; however, the original method is successful in many cases and is exemplified by the preparation of an intermediate for the synthesis of *cis*-jasmone (equation 17).[30]

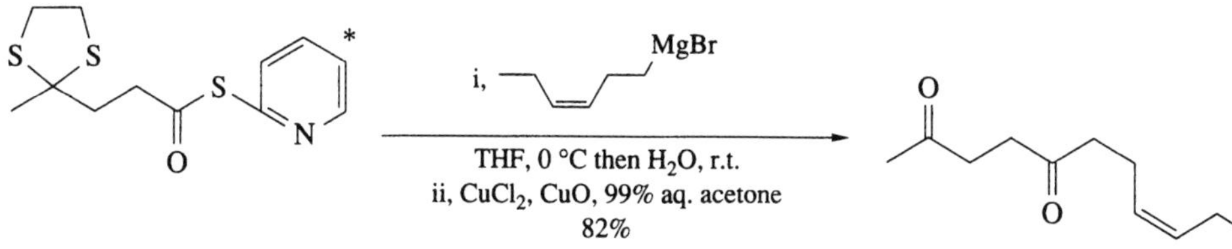

(17)

* from levulinic acid in 2 steps and 97% overall yield: i, $HS(CH_2)_2SH$; ii, 2,2′-pyr disulfide, PPh_3

The reaction depicted was run in THF at 0 °C, other solvents having been found to be inferior. The *S*-(2-pyridyl) thioates may be prepared through reaction of the corresponding acid chloride and 2-pyridinethiol in the presence of a tertiary amine. They are also available directly from carboxylic acids by reaction with 2,2′-dipyridyl disulfide (Aldrithiol-2) and triphenylphosphine.[31] In the case illustrated above, protection of the ketone would seem unnecessary if Grignard addition was selective for the thiol ester; however, the starting material, *S*-(2-pyridyl) γ-oxopentanethioate, is not stable to the lactonization shown in equation (18).

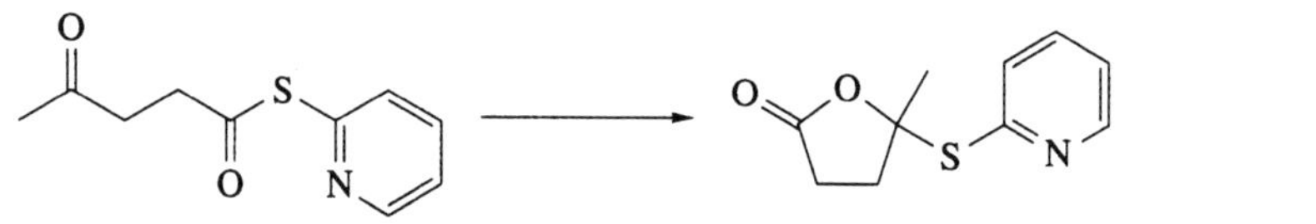

(18)

Mukaiyama has shown that in cases where lactonization was not possible, oxothioates can be used in the acylation process with selective addition only to the thiol ester (equations 19 and 20).[32]

Mechanistic studies have shown that the tetrahedral intermediate (**33**) is not stable during the course of the reaction (Scheme 10). By monitoring the reaction mixture with infrared spectroscopy, Mukaiyama demonstrated that the ketone is released before the quench and therefore could compete with the more reactive thiol ester in nucleophilic addition.[33] The authors postulate that selective addition occurs to the

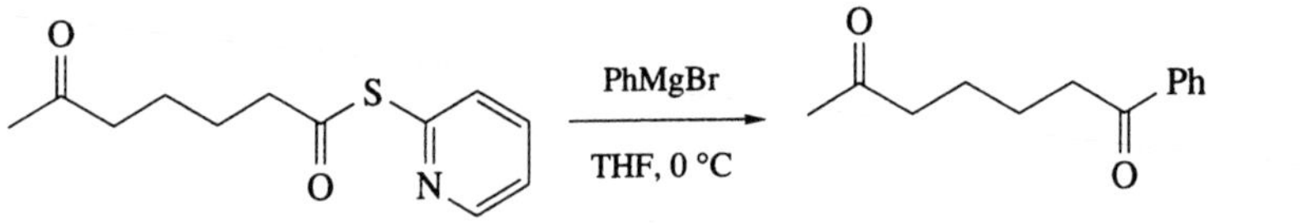

(19)

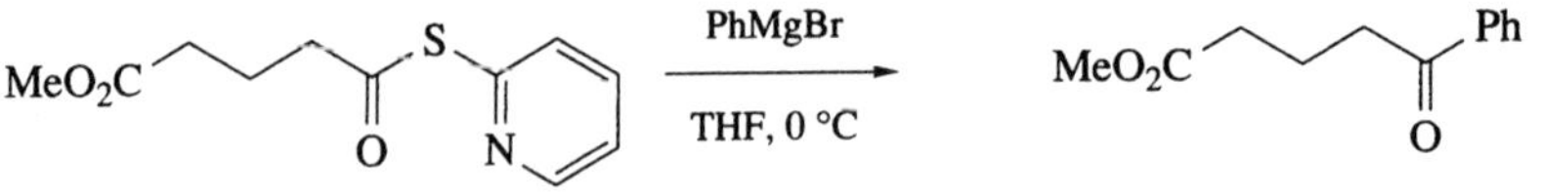

(20)

thiol ester because of the formation of the activated chelate (**32**); unfortunately, the apparently strong chelation provided by tetrahedral intermediate (**33**) is not sufficient to maintain the integrity of this species before quenching.[34] This is a conceptually different situation than was observed for Weinreb's method wherein the tetrahedral intermediate is exceptionally stable. In terms of the chemoselectivity of Mukaiyama's approach, the reactivity differences between the pyridyl thiol ester and other ketones have been shown to be substrate dependent and may not be general.

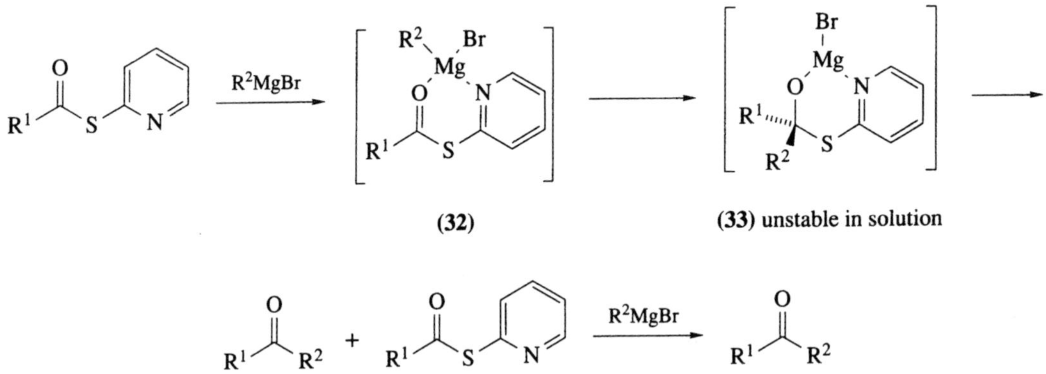

(**32**) (**33**) unstable in solution

Scheme 10

The formation of a late-stage intermediate for the total synthesis of erythronolide B by Corey and co-workers takes advantage of this methodology. The coupling of pyridyl thiol ester (**36**) with Grignard reagent (**35**), prepared indirectly from iodide (**34**), affords the erythronolide B intermediate (**37**) in high yield without racemization or attack at the lactone (Scheme 11).[35] The corresponding reaction in the erythronolide A series which contains additional oxidation at pro-C-12, indicated on (**37**), results in overaddition (see Section 1.13.3.2). Another case of overaddition by functionalized Grignard reagents has been recorded by Still (see Section 1.13.3.1).

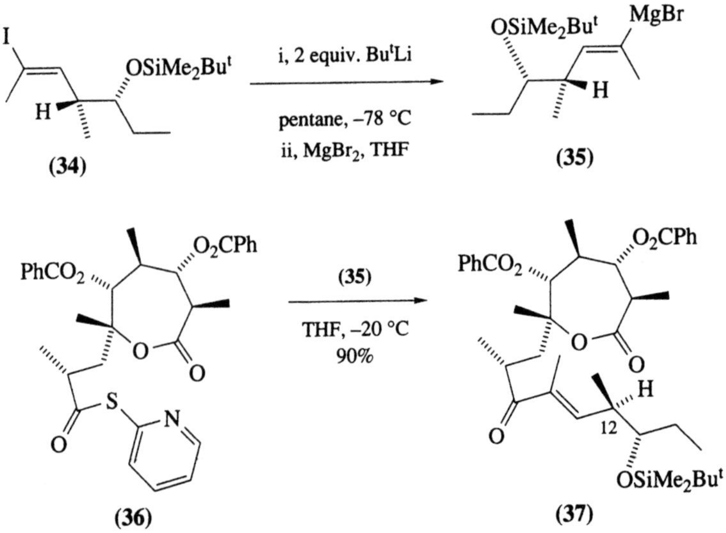

(**34**) (**35**)

(**36**) (**37**)

Scheme 11

The interesting antibiotics X-14547A and A-23187 (calcimycin) contain a novel 2-ketopyrrole moiety, which is apparently important for biological activity. Nicolaou has developed a simple method for incorporation of this unit into the sensitive intermediates required for preparation of the natural products.[36] Through use of the Mukaiyama protocol, formation of the 2-pyridyl thiol ester was effected with 2,2′-dipyridyl disulfide and triphenylphosphine. The reaction mixture was then cooled to –78 °C and was treated with a solution of pyrroylmagnesium chloride in THF (equation 21).

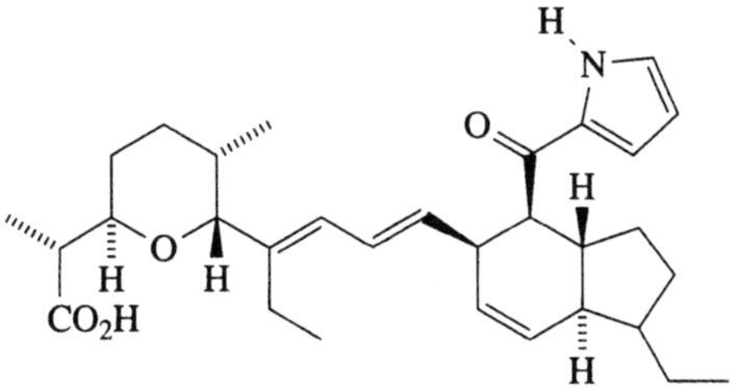

X-14547A 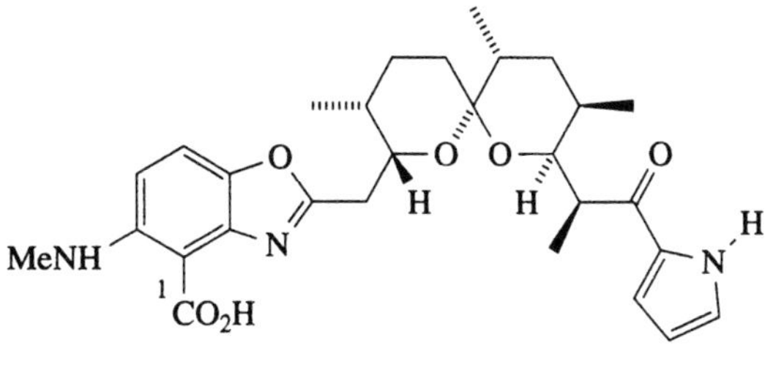 A-23187

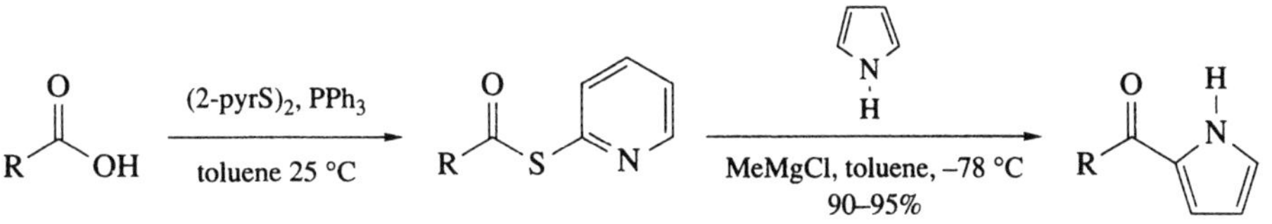

$(2\text{-pyrS})_2$, PPh_3, toluene 25 °C; pyrrole, MeMgCl, toluene, –78 °C, 90–95% (21)

The use of the acid chloride instead of the 2-pyridyl thiol ester also results in formation of the 2-ketopyrrole functionality, but significant amounts of the 3-ketopyrrole isomer were also formed. Several rather complex substrates were used as acylating agents in this process with a high degree of success. Some examples of the 2-ketopyrroles that have been prepared are shown in Scheme 12.

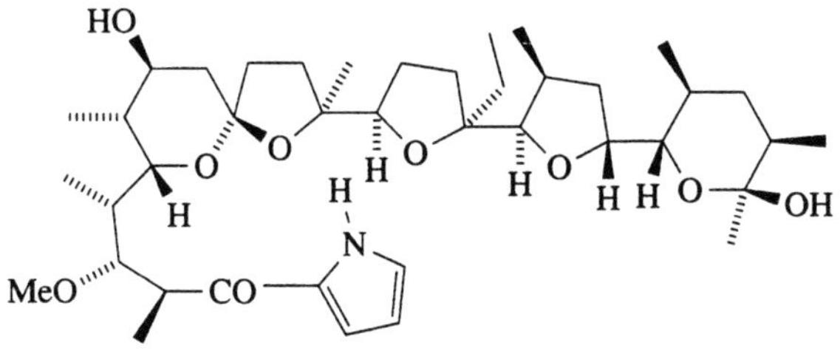

90% from monensic acid and pyrroylmagnesium bromide

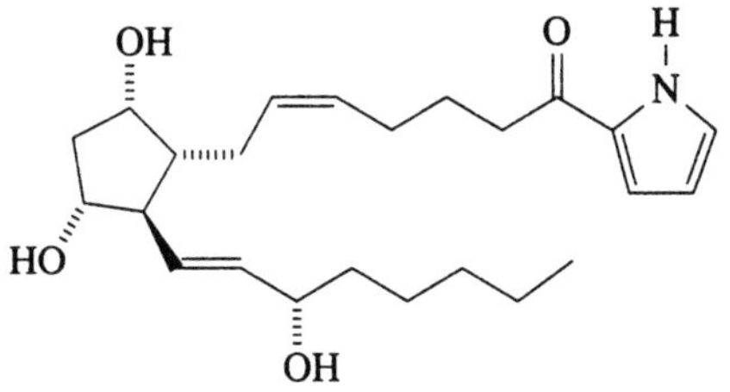

95% from $PGF_{2\alpha}$

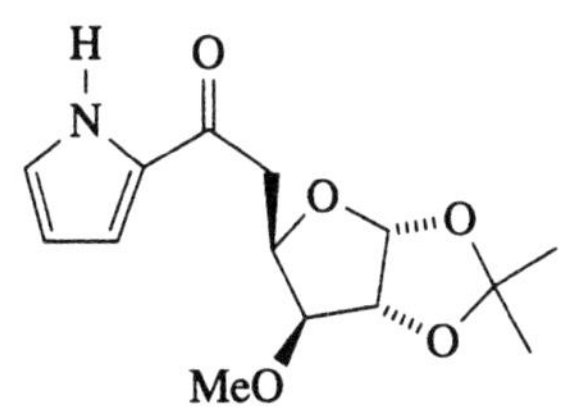

89% from carboxylic acid

Scheme 12

The synthesis of X-14547 A was completed by the transformation shown in equation (22).[37] Interestingly, the reaction proceeded without interference from the carboxylic ester group even though excess pyrrolylmagnesium chloride was used. At higher temperatures, it was possible to observe addition at both the desired position and at the distant carboxylic ester in 95% yield.

The antibiotic (–)-A-23187 (calcimycin) has been prepared by Boeckman using similar methodology (equation 23).[38] No racemization was observed at the position α to the thiol ester.

X = OH

X = S-2-pyridyl

pyrrolyl-MgCl, toluene–THF, –20 °C, 90% (22)

pyrrolyl-MgCl, toluene, –78 °C, 80% (23)

$OSiPh_2Bu^t$

A similar transformation has been carried out by Nakahara and Ogawa; however, these workers reported no success in adding the pyrrole functionality to the penultimate precursor containing the benzoxazole ring.[39] Most recently, French workers have demonstrated the late stage introduction of the pyrrole unit by prior formation of the methyl ester at position 1 and use of a catalytic amount of copper(I) iodide in the acylation step.[40] Finally, for substrates that require protection at the ring nitrogen, acylation of 2-lithio-*N*-(*N'*,*N'*-dimethylamino)pyrrole has been demonstrated by Grieco. The reaction is exemplified by the synthesis of a model system for calcimycin (equation 24).[41] Although extra steps are involved in this transformation the overall yield is slightly higher than the previous case.

i, 2-lithio-N-(NMe_2)pyrrole, $MgBr_2$, THF, –78 °C; ii, $Cr_2(OAc)_4$•H_2O, EtOH, 12 h; 86% overall (24)

1.13.2.3 Acylation by Carboxylic Acids

Methyl ketones are often directly prepared from carboxylic acids by reaction with methyllithium. Other simple alkyl ketones may also be prepared in the same fashion, making this a method that should be considered whenever these substrates are required.[42] An important demonstration of this protocol was reported by Masamune and coworkers in their synthesis of chiral propionate surrogates (Scheme 13).[43] The ethyl and cyclopropyl[44] ketones are important starting materials for macrolide total synthesis and have been prepared on a large scale. The overall yield for the ethyl ketone is 65% using 3.5 equiv. of ethyllithium without protection of the hydroxy group.

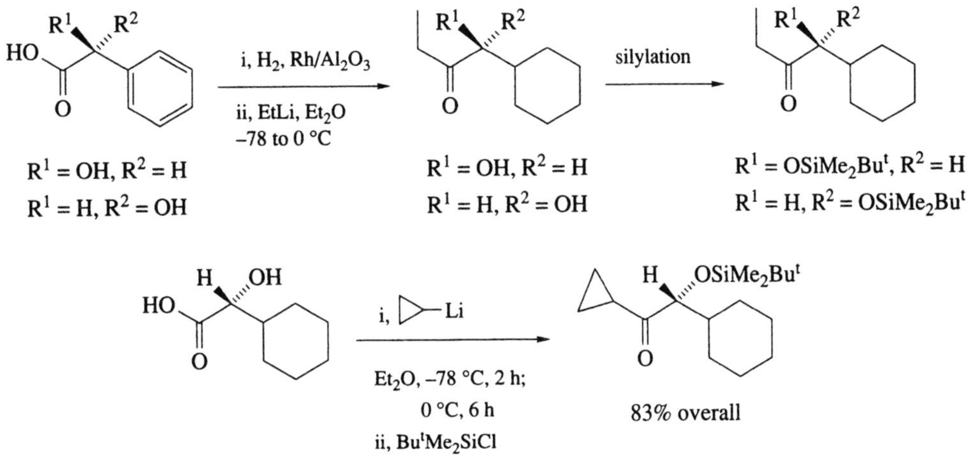

i, H_2, Rh/Al_2O_3
ii, EtLi, Et_2O
–78 to 0 °C

silylation

$R^1 = OH, R^2 = H$
$R^1 = H, R^2 = OH$

$R^1 = OH, R^2 = H$
$R^1 = H, R^2 = OH$

$R^1 = OSiMe_2Bu^t, R^2 = H$
$R^1 = H, R^2 = OSiMe_2Bu^t$

i, cyclopropyl–Li
Et_2O, –78 °C, 2 h;
0 °C, 6 h
ii, Bu^tMe_2SiCl

83% overall

Scheme 13

Many experimental procedures have been discussed for effectively converting a carboxylic acid to a ketone through the use of organolithium reagents. House has advocated a procedure which begins with formation of the lithium carboxylate in dimethoxyethane. A slight excess of methyllithium is then added at 0 °C to form a 1,1-dilithium dialkoxide. Subsequently, the manner in which the geminal dialkoxide was quenched had a dramatic effect on the yield. Breakdown of the intermediate dialkoxide can in some instances precede protonation of the alkyllithium reagent. Rapid addition of the organometallic to the exposed ketone can then take place and result in the formation of a tertiary alcohol. For this reason, House favored inverse addition of the reaction mixture to a vigorously stirred solution of dilute hydrochloric acid at 0 °C. In this way high yields of the ketone may be obtained. This procedure may be even more important for reactions which are conducted with large excesses of alkyllithiums.[45] Excess methyllithium can also be quenched by addition of ethyl formate to the reaction mixture followed by aqueous acid, as was demonstrated by Mander.[46]

Other improvements in the standard methodology for addition of alkyllithiums to carboxylic acids and esters have been reported by Rubottom[47] and Cooke.[48] Both methods utilize trimethylsilyl chloride (TMS-Cl) at some point during ketone preparation; the former method of Rubottom introduces TMS-Cl once addition of the alkyllithium to the carboxylic acid is complete. This was done in an effort to trap residual methyllithium and allow the use of large excesses of reagent. However, it is known that most simple alkyllithiums react only sluggishly with trialkylsilyl chlorides,[49] thus the actual mechanism may involve silylation of the geminal dialkoxide. The method of Cooke uses a preformed solution of TMS-Cl and carboxylic ester. The reaction mixture was then treated with an alkyllithium whilst maintaining the reaction mixture at –78 °C. Previous use of carboxylic esters in this reaction often led to the formation of tertiary alcohols. The success of this procedure has been attributed to the formation of the *O*-silylated tetrahedral intermediate (**38**) which was apparently stable to the reaction conditions (Table 2).[47,48]

Table 2 Use of Trimethylsilyl Chloride (TMS-Cl) in Acylation of Carboxylic Acids and Esters

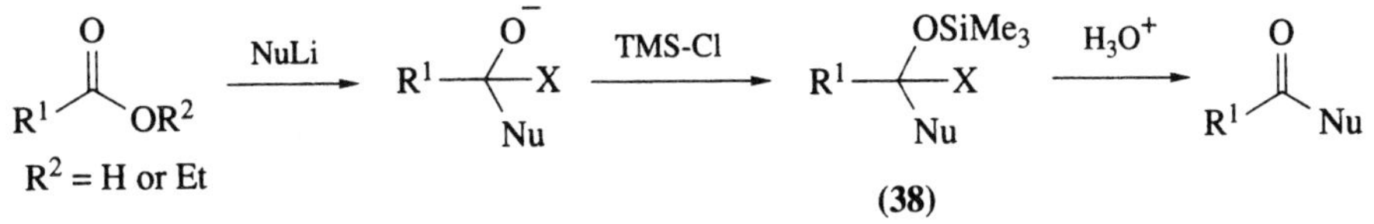

$R^1CO_2R^2 \xrightarrow{NuLi} R^1C(O^-)(X)Nu \xrightarrow{TMS\text{-}Cl} R^1C(OSiMe_3)(X)Nu \xrightarrow{H_3O^+} R^1CONu$

R^2 = H or Et; (**38**) = $R^1C(OSiMe_3)(X)Nu$

R^1	X	Nu	Equiv. TMS-Cl	Temperature (°C)	Yield (%) R^1CONu	Yield (%) $R^1(Nu)_2OH$	Ref.
Bu^n	OEt	Bu^n	25	–100	90	8	48
Et	OEt	Bu^n	5	–100	87	15	48
Ph	OEt	Bu^n	5	–100	82	21	48
C_6H_{11}	OTMS	Me	20	0	92	2	47
Ph	OTMS	Me	20	0	97	0	47
p-OHC_6H_4	OTMS	Me	20	0	87	<5	47

The acylation of organolithiums with carboxylic acids has often been used in natural product synthesis. Certainly, the method is one of the easiest approaches available when analysis of a target structure indicates that organometallic acylation may be more efficient than the two-step nucleophilic addition to an aldehyde followed by oxidation. Many simple organolithiums are commercially available or may be prepared easily by standard methodology.[50] A case in which this plan was effective can be seen in Stork's synthesis of the corticosteroid ring system (C-11 oxygenated steroid) which embodied the transformations shown in equation (25).[51] Addition of the readily available lithium reagent proceeded in greater than 90% yield, reacting with both the acid and ketone functional groups. After an *in situ* dehydration, Diels–Alder ring closure afforded the penultimate precursor to the C-11 steroid system (equation 25). The addition of vinyllithium reagents to carboxylic acids was first reported by Floyd.[52] Interestingly, the latter investigation demonstrated that the reaction of vinyllithium with stearic acid actually had to be heated to 40 °C to achieve a 70% yield of product.

OSiMe$_2$Bu^t; O; HO$_2$C; H; i, Li; ether, –78 °C; >90%; ii, CH_2Cl_2, TFA, –78 °C; OSiMe$_2$Bu^t; H; O; 70%; OSiMe$_2$Bu^t; H; H; H; O

(25)

The intermediate 1,1-dialkoxide which was formed from the addition of ethyllithium to carboxylic acid (**39**) functioned as a masked form of a ketone, and facilitated a subsequent Wittig reaction on aldehyde (**40**; equation 26).[53]

$Ph_3\overset{+}{P}(CH_2)_5CO_2H$ Br^- (**39**) —xs EtLi→ $Ph_3P=CH(CH_2)_4C(Et)(O^-Li^+)_2$ —$R(CH_2)_3CHO$ (**40**)→ HO; N; cbz; 3; 4 COEt

(26)

The well-known Parham cyclization of bromoaromatic carboxylic acid derivatives can be exemplified by the examples in equations (27)–(29). The reaction is usually run at very low temperature in ether or THF and provides surprisingly good yields of aromatic ketones. The method has been reviewed recently.[54]

Br; CO_2H —2 equiv. Bu^{n}Li, –100 °C; 77%→ O

(27)

NC; Br; CO_2H —2 equiv. Bu^{n}Li, –100 °C; 69%→ NC; O

(28)

The simple homochiral adduct from Diels–Alder cycloaddition of (*R*)-pantolactone acrylate ester with cyclopentadiene was hydrolyzed and converted to the methyl ketone in 89% yield through addition of 3

2 equiv. Bu^nLi, –100 °C

50%

(29)

equiv. of methyllithium (equation 30). The reaction mixture was treated with TMS-Cl before addition of aqueous ammonium chloride under the conditions of Rubottom.[55]

i, 3 equiv. MeLi, –25 to 0 °C, THF

ii, Me_3SiCl, then aq. NH_4Cl

89%

(30)

Paquette's synthesis of africanol advanced through the α,β-unsaturated acid (**41**). Addition of 2.2 equiv of methyllithium followed by standard work-up gave a 71% yield of the methyl ketone (**42**; equation 31).[56]

2.2 equiv. MeLi, Et_2O

–78 to 0 °C

71%

(31)

(41) **(42)**

There has been an isolated report dealing with the direct acylation of Grignard reagents by carboxylic acids mediated by nickel(II) salts. Large excesses of Grignard reagents must be used (6 equiv.) but the reaction can be run at room temperature with only 7 mol % $Ni(DPPE)Cl_2$. Normally, Grignard reagents are not useful in carboxylic acid acylation because the tetrahedral intermediate breaks down rapidly and the ketone is attacked by the organometallic. The role and generality of the nickel-mediated process has not been elucidated as yet.[57]

1.13.2.4 α-Amino Acids as Acylating Agents

α-Amino acids can be directly converted to ketones through the addition of alkyl- or vinyl-organometallics. This approach complements the addition of lithium alkynides to activated amino acid derivatives (Section 1.13.2.1.1) but is much simpler to carry out. Rapoport has shown that 3 equiv. of the organolithium are necessary to transform a protected amino acid such as *N*-(ethoxycarbonyl)alanine to an aryl or alkyl ketone. The first 2 equiv. of the organometallic function only as a base, resulting in dianion formation. This species is crucial to the process because further deprotonation is not possible and the intermediate is thus protected from racemization. *N,N*-Disubstituted amino acids were found to be racemized under the same conditions presumably because enolization was no longer impeded by nitrogen deprotonation. In the interest of economy, carboxylate anion formation may be carried out with *n*-butyllithium prior to addition of the desired nucleophile, but the acylation of a Grignard reagent requires prior formation of the lithium carboxylate to avoid breakdown of the tetrahedral intermediate in solution. The choice for nitrogen protection does not include the popular t-BOC or cbz groups but ethoxycarbonyl, benzoyl, acetyl and phenylsulfonyl may be used interchangeably (Table 3).[58]

The addition of allyllithium provides access to β,γ-unsaturated amino ketones or α,β-unsaturated amino ketones, depending upon the conditions chosen for work-up. In Scheme 14, simple interchange of the acidic reagent in the quench affected the positioning of the double bond.[58]

A stereospecific total synthesis of the antibiotic sibirosamine was carried out using this methodology. In the first step, addition of methylmagnesium iodide to *N*-(phenylsulfonyl)-L-allothreonine afforded the α-amino methyl ketone in modest yield.[59] This reaction afforded a diminished yield compared with *N*-(phenylsulfonyl)-L-threonine for some unexplained reason. A clever use of the L-amino acid serine allows for a straightforward preparation of D-amino acids such as D-dopa as is shown in Scheme 15.[60]

Table 3 Formation of α-Amino Ketones from Various α-*N*-Acylated Amino Acids and Organometallic Reagents

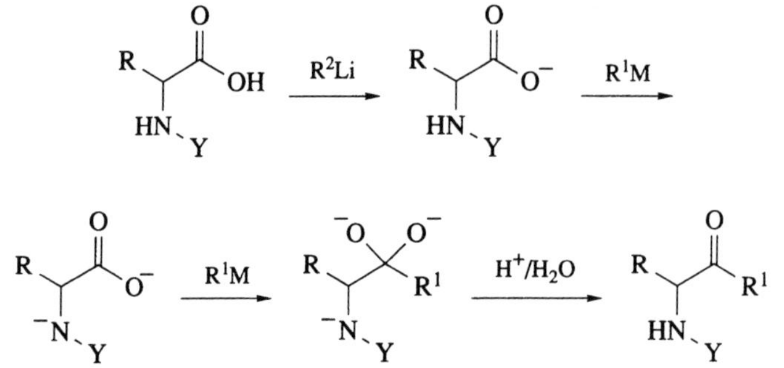

Compound	*R*	*R*[1]	*Y*	*R*[2]	*M*	*Yield (%)*
a	Me	$CH_2CH{=}CH_2$	COMe	Bu^n	MgBr	40
b	Me	$CH_2CH{=}CH_2$	COPh	Bu^n	MgBr	87
c	Me	$CH_2CH{=}CH_2$	CO_2Et	$CH_2CH{=}CH_2$	Li	74
d	Pr^i	$CH_2CH{=}CH_2$	SO_2Ph	Bu^n	MgBr	62
e	$MeSCH_2CH_2$	$CH_2(CH_2)_2CH_3$	SO_2Ph	Bu^n	Li	54
f	$MeSCH_2CH_2$	$CH{=}CH_2$	SO_2Ph	Bu^n	MgBr	35
g	o-$BnOC_6H_4CH_2$	$CH_2CH{=}CH_2$	CO_2Et	Bu^n	MgBr	71
h	$EtO_2CNH(CH_2)_3CH_2$	$CH_2CH{=}CH_2$	SO_2Ph	Bu^n	MgBr	52
i	$HOCH_2$	$CH_2CH{=}CH_2$	SO_2Ph	Bu^n	MgBr	53
j	$HOCH_2$	$CH{=}CH_2$	SO_2Ph	Bu^n	MgBr	48

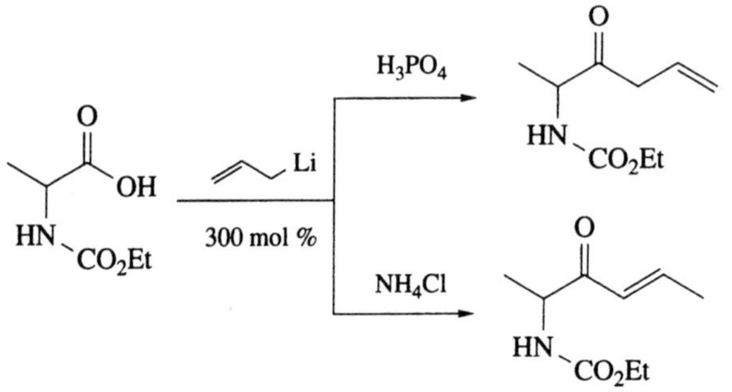

Scheme 14

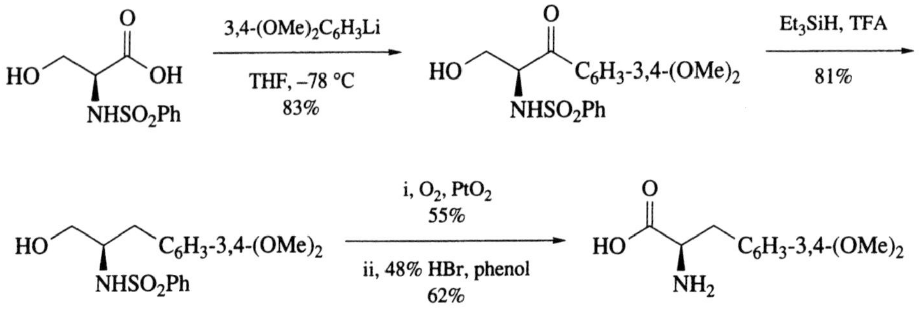

Scheme 15

1.13.2.5 Acylation with Acid Chlorides

The synthesis of ketones from acid chlorides has been demonstrated with several organometallic reagents (see Sections 1.13.3, 1.13.4 and 1.13.5); however, the acylation of organolithiums often leads to substantial amounts of overaddition product. Initially, acid chlorides appear to be a poor choice for ketone synthesis considering highly electrophilic nature and the poor stability of the tetrahedral intermediate produced by nucleophilic addition. For these reasons, substantial amounts of ketone may reside in solution during addition of the organometallic. In addition, acid chlorides offer no site for prior coordination of the organometallic, as was the case with the *S*-(2-pyridyl) thioate of Mukaiyama (Section 1.13.2.2).[30]

The fact that organomagnesium reagents can often be acylated with acid chlorides to provide high yields of ketones as the exclusive product is somewhat surprising. Although the reaction is not well understood in its present form, the scope of the process can be illustrated by the following examples. Sato and coworkers have found the choice of solvent and temperature are important for the acylation of simple Grignard reagents.[61] A partial list of examples is given in Table 4.

Table 4 Acylation of Grignard Reagents by Acid Chlorides

$$R^1MgX + R^2COCl \xrightarrow{\text{THF, −78 °C}} R^1COR^2$$

*R*¹	*Equiv.*	*X*	*R*²	*Product*	*Yield (%)*
Ph	1.0	Br	Ph	PhCOPh	89
p-MeC_6H_4	1.0	Br	Ph	*p*-MeC_6H_4COPh	84
Pr^n	2.0	Cl	Bu^t	Pr^nCOBu^t	88
n-C_6H_{13}	2.0	Br	Me	n-$C_6H_{13}COMe$	93

To avoid overaddition, the reaction was conducted by slowly introducing the Grignard reagent to a solution of the acid chloride at –78 °C. These workers reported that replacing tetrahydrofuran with diethyl ether caused a dramatic increase in the amount of tertiary alcohol formed, indicating that Grignard reagents reacted more rapidly with acid chlorides in THF rather than in ether, and that the reaction must be executed at low temperature. Many of the aromatic cases use a molar equivalent amount of R^1MgX; however, the best yields were obtained in the nonaromatic series by using 0.5 equiv. of Grignard reagent. Similar observations were reported by Eberle[62] and they also showed that functional groups such as carboxylic esters, halides and alkenes could be tolerated in the reaction. 1 equiv. of Grignard reagent was utilized and unlike the previous case the acylating agent was added directly to a cold solution of the organometallic. This should maximize the chances for tertiary alcohol formation; however, none was obtained. A characteristic example is shown in equation (32).

$$CH_2{=}CH(CH_2)_3MgBr + ClCO(CH_2)_3CO_2Me \xrightarrow[\text{81\%}]{\text{THF, −78 °C}} CH_2{=}CH(CH_2)_3CO(CH_2)_3CO_2Me \quad (32)$$

An isolated example of Grignard acylation by proline acid chloride was achieved by the use of a large excess of acid chloride (**43**). The keto aldehyde (**45**) was obtained in 71% yield (equation 33).[63] Unfortunately, the reaction with stoichiometric amounts of the acid halide were not discussed. In light of the results of Sato[61] and Eberle,[62] the real capabilities of this reaction were apparently not explored.

$$\textbf{(43)} \xrightarrow[\text{THF, −78 °C to r.t., 71\%; ii, } H_3O^+\text{, 83\%}]{\text{i, MgBr}(CH_2)_2\text{-(1,3-dioxan-2-yl)} \textbf{ (44)}} \textbf{(45)} \quad (33)$$

(43): N-cbz-prolyl chloride; (45): N-cbz-2-pyrrolidinyl–CO(CH₂)₂CHO

To show the potential of the method, the related Grignard reagent (**46**) has been acylated in good yield through use of a stoichiometric mixture of organomagnesium and acid halide (equation 34). The use of low temperatures and THF was critical to the process.[64]

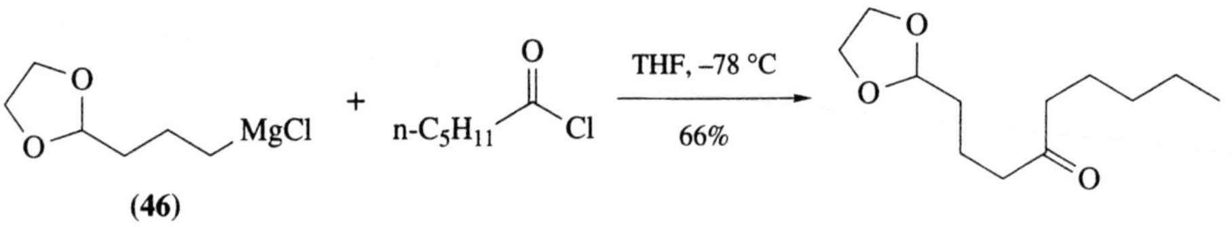

MgCl + n-C_5H_{11}COCl —(THF, –78 °C; 66%)→ (34)

(46)

Sworin[65] has studied several interesting reactions of the Grignard reagent (**47**) which may indicate it reacts as the dialkylmagnesium (**48**) in THF at –78 °C (Schlenck equilibrium; equation 35).[66] The proposal was advanced that internal coordination of magnesium by the flanking oxygen moieties resulted in a softer nucleophile, 'similar in reactivity to an organocuprate'. The dimeric material may be responsible for the somewhat 'anomalous behavior' observed in reactions of this Grignard reagent and may point to the importance of temperature and solvent for successful acylations of other Grignard reagents. These workers also recommend that at least 2 equiv. of (**47**) be employed per mole of substrate to insure that (**48**) is the sole reactive intermediate. At room temperature, the equilibrium is believed to shift to the left with formation of the monomeric and harder nucleophile (**47**). Reactions at this temperature take a different course and favor 1,2-addition to the ketone.

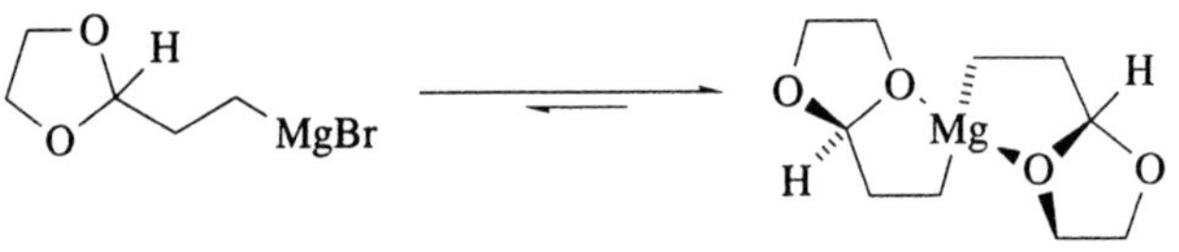

MgBr ⇌ Mg (35)

(47) (48)

Grignard reagents can also be monoacylated in the presence of iron(III) salts. The reaction uses catalytic amounts of iron and reproducible yields of ketones have been obtained.[67] The addition can be conducted at room temperature with stoichiometric amounts of the acid chloride and Grignard reagent. Only very simple substrates were considered initially; however, a synthesis of the methyl ketone derived from Mosher's acid was reported in very high yield (equation 36), as well as a synthesis of a heptatrienyl ketone (equation 37).[68] The reaction may be mediated by an acyl iron complex, although at room temperature one would have expected overaddition products from the Grignard reagent. No mechanistic studies have been carried out as yet.

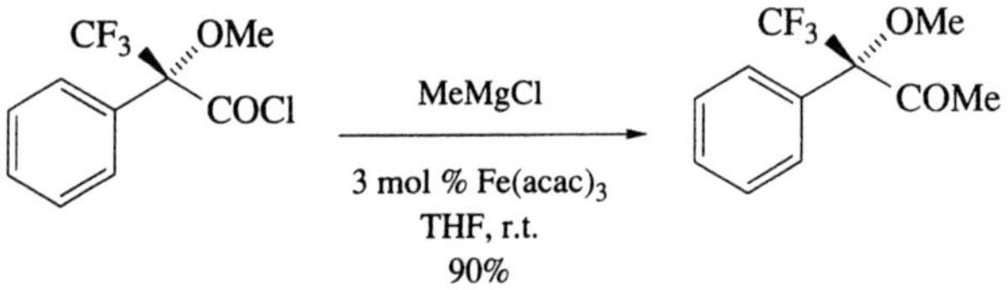

CF_3, OMe, COCl —(MeMgCl; 3 mol % $Fe(acac)_3$, THF, r.t., 90%)→ CF_3, OMe, COMe (36)

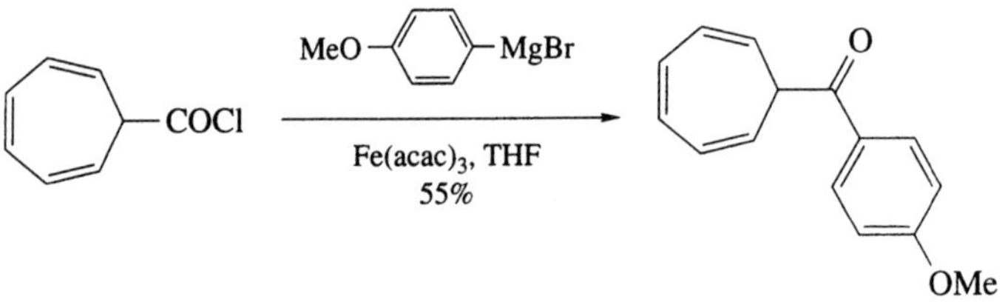

COCl —(MeO–C_6H_4–MgBr; $Fe(acac)_3$, THF, 55%)→ OMe (37)

1.13.2.6 Acylation with Carboxylic Esters

Carboxylic esters have also been used to acylate Grignard reagents with some success. Workers at Shionogi research laboratories demonstrated that the combination of methylmagnesium bromide and triethylamine could be acylated in good yield by substrates such as (**49**; equation 38). The reaction did not require THF as a solvent, but an excess of triethylamine was needed to achieve the best results. Presumably the amine alters the usual aggregation of the organomagnesium reagent and/or influences the Schlenck equilibrium between MeMgBr and Me_2Mg, as discussed above. This combined reagent may selectively add to the ester functionality in preference to the newly formed ketone. On the other hand, the authors postulate that triethylamine aids in the deprotonation of the resulting methyl ketone, which is then protected from further attack. To provide evidence for this they quenched a reaction mixture with D_2O and obtained deuterium incorporation at the newly introduced methyl group. Therefore if the ketone was converted to the enolate, then the combination of the Grignard reagent and triethylamine must be capable of serving as a base as well as a nucleophile (2 equiv. of reagent were used). Further investigations of this result would be of interest.[69]

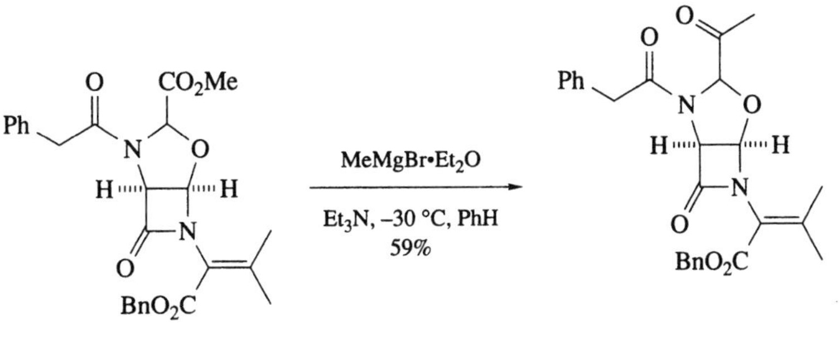

(49) — $MeMgBr{\bullet}Et_2O$; Et_3N, –30 °C, PhH, 59% — (38)

The use of other mixed reagents to promote acylation and subsequent enolization of the ketone during its formation have been reported by Fehr.[70] The success of the method depends on the ease of ketone deprotonation and thus was limited to substituted allylic nucleophiles. The final product was obtained entirely in the form of the α,β-unsaturated ketone. A combination of the nucleophilic Grignard reagent and the nonnucleophilic base lithium diisopropylamide converts sterically hindered ester (**50**) into α-damascone (**52**) *via* (**51**) (Scheme 16). The ratio of ketone to tertiary alcohol was 98:2 (many cases gave selectivity greater than 9:1); however, a few examples showed a substantial amount of tertiary alcohol formation.

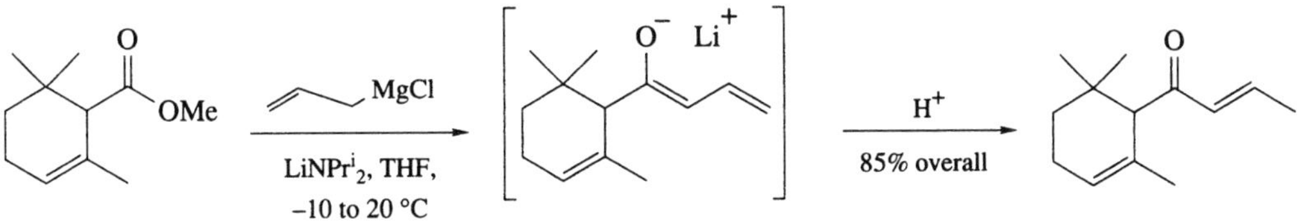

(50) — allyl-MgCl, $LiNPr^i_2$, THF, –10 to 20 °C — **(51)** — H^+, 85% overall — **(52)**

Scheme 16

The proposed enolate has been trapped in the case of amide (**53**) but there was no report if this could be accomplished on the ester substrates (equation 39). In general the amides did give higher yields of ketone and less tertiary alcohol, but this is expected based purely on the enhanced stability of the tetrahedral intermediate. It is not clear to what extent the amide base may influence the degree of aggregation of the Grignard reagent and thus alter its reactivity. All of the cases studied were extremely sterically congested, which could have influenced the rate of tetrahedral intermediate collapse and enolate formation. It should be noted that the previously discussed combination of Grignard reagent and triethylamine failed to give ketones selectively in these cases.[69]

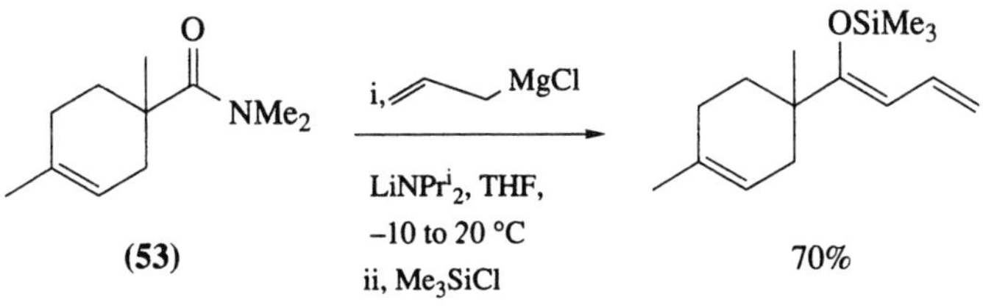

(53) — i, allyl-MgCl, $LiNPr^i_2$, THF, –10 to 20 °C; ii, Me_3SiCl — $OSiMe_3$ product, 70% — (39)

Comins has reported that simple esters can be converted to secondary alcohols in one step with a mixture of Grignard reagent and lithium borohydride ($LiBH_4$) in THF. The reaction is conducted at 0 to –10 °C to preclude undesired reduction of the ester by $LiBH_4$. Once acylation has occurred, reduction of the intermediate ketone occurs more rapidly than does addition of a second equivalent of the Grignard reagent.[71] Burke has modified the procedure slightly so as to include DIBAL-H as the reducing agent. In this fashion, either isomer of alcohol (**54**) may be obtained (equation 40).[72]

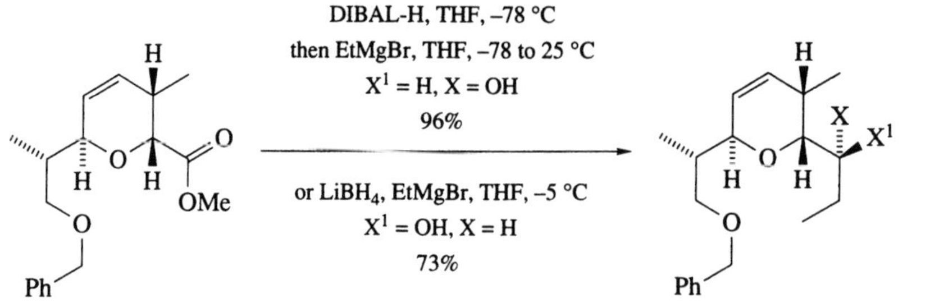

DIBAL-H, THF, –78 °C then EtMgBr, THF, –78 to 25 °C X^1 = H, X = OH 96%

or $LiBH_4$, EtMgBr, THF, –5 °C X^1 = OH, X = H 73%

(**54**) (40)

Seebach has also studied the utility of esters in organometallic acylation. In this case, preformed ester enolates of 2,6-di(*t*-butyl)-4-methylphenyl esters (BHT esters) were slowly warmed above –20 °C to form the corresponding ketene. If this was done in the presence of an additional equivalent of alkyllithium the ketene was trapped to give a ketone enolate in high yield. The same reaction failed to give any product when simple esters such as methyl, ethyl or *t*-butyl were used.[73] Scheme 17 is illustrative of the method.

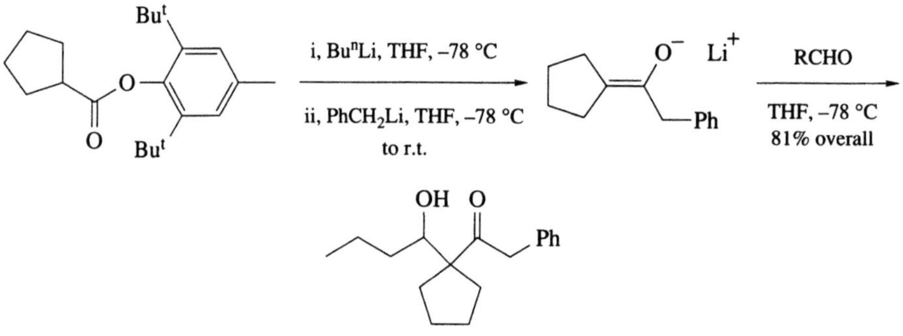

i, Bu^nLi, THF, –78 °C; ii, $PhCH_2Li$, THF, –78 °C to r.t.; RCHO, THF, –78 °C, 81% overall

Scheme 17

A variety of alkyllithiums could be acylated by this method. The enolate was then quenched with electrophiles such as aldehydes and chlorosilanes. Unfortunately only branched chain carboxylic acids could be used in this process, as monosubstituted ketenes were unstable in the presence of butyllithium.[74]

1.13.2.7 Preparation of Alkynic Ketones from Lactones

Lactones are convenient acylating agents for lithium alkynides and have been used extensively for this purpose. Lactones are stable, readily available substances with a reactivity substantially different from simple esters. The first report discussing this strategy was by Ogura in 1972.[75] This work demonstrated that γ-valerolactone undergoes monoaddition with simple lithium alkynides, but suffers multiple addition with the corresponding magnesium alkynide. Chabala discussed the effect of lactone ring size on acylation as well as proposing the mechanism by which they undergo addition.[76] In this study, simple esters were found to undergo bis-addition when treated with lithium alkynides, due to the poor stability of the tetrahedral intermediate. However, in the case of δ-valerolactone, the intermediate ketal alkoxide was found to be stable in the reaction mixture and thus serves to protect the carbonyl function from further attack (equation 41). Other ring sizes do not seem to work as well, although acceptable yields of alkynic ketones can be obtained from γ-butyrolactone (50–60%).

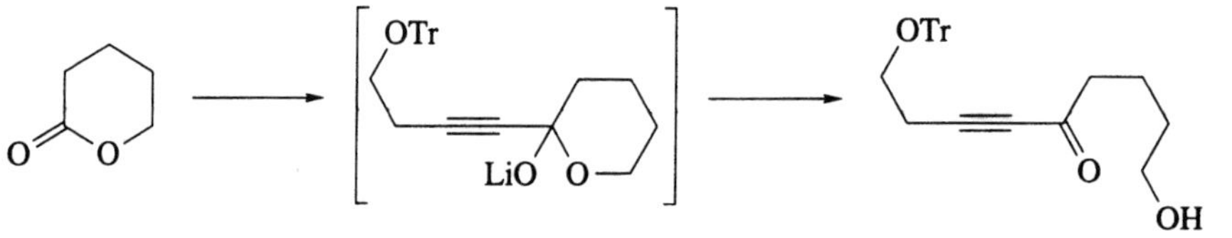

(41)

Recent evidence for the formation of hemiketal intermediates upon acylation of alkynides has been obtained from glucopyranolactones. Treatment of tetrabenzyl (**55**) with the anion from 1-benzyloxy-3-butyne gave a quantitative yield of hemiketal (**56**) which showed IR absorptions for the OH and alkyne portions of the molecule ($\lambda = 3350\ cm^{-1}$ and $2250\ cm^{-1}$, respectively). This compound was stereospecifically reduced to the *C*-glycoside (**57**) with triethylsilane/BF_3·etherate (overall yield 72%; Scheme 18). None of the other stereoisomer or ring-opened product was obtained.[77]

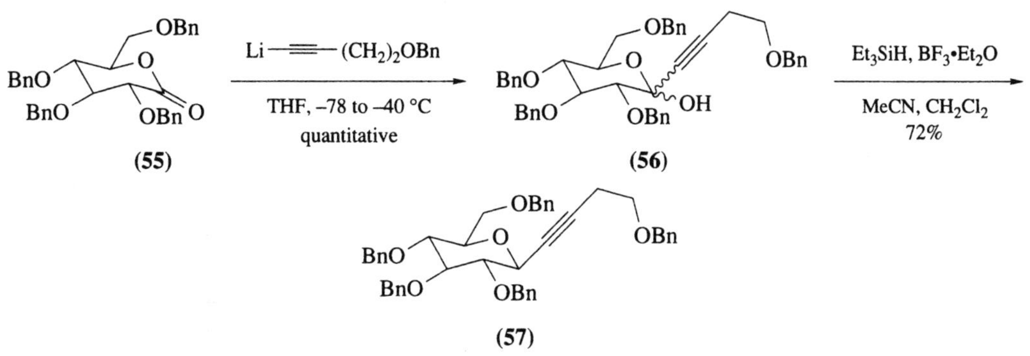

Scheme 18

The first important test of this methodology came in Hanessian's investigation of the spiroketal portion of avermectin B_{1a}. This highly convergent approach incorporates all the oxidation levels and functionality required for carbons C(15)–C(28), except for the necessity of alkyne to alkene conversion. The lithium alkynide was prepared at –78 °C and then mixed with boron trifluoride etherate under the conditions of Yamaguchi (Scheme 19).[78] (Direct condensation of the lithium salt and lactone lead to substantial amounts of α,β-unsaturated lactone.) Addition of the lactone in stoichiometric amounts to the solution of the modified alkynide led to the formation of the desired hemiketal in acceptable yield. Further improvements could be obtained by the recycling of starting material.[79]

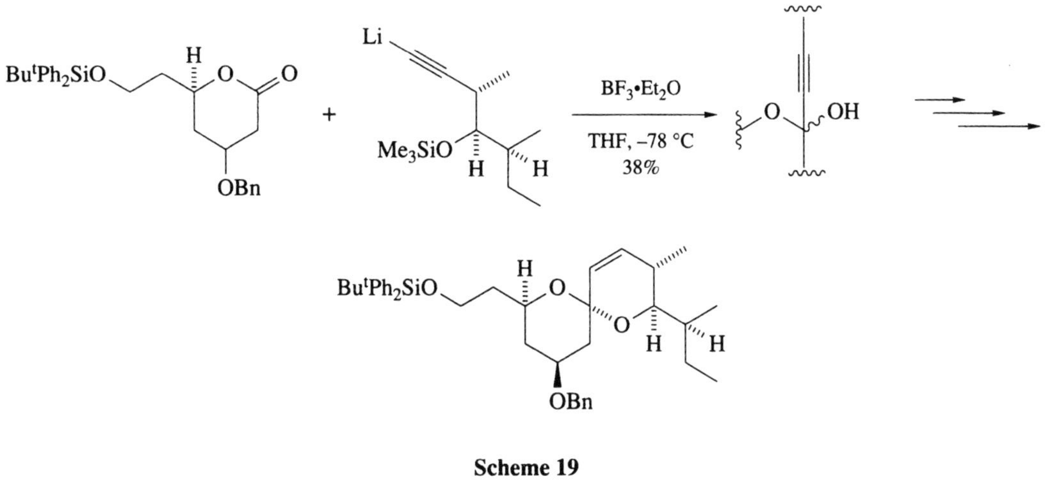

Scheme 19

A more recent example of a functionalized alkyne addition can be seen in Crimmin's synthesis of talaromycin A (equation 42).[80] This particular alkynide is an equivalent of the formyl acetone dianion and its use has been generalized as an entry into the spiroketal portion of the milbemycins (Scheme 20).[81] This approach differs from the Hanessian strategy in that formation of the C(17)–C(21) pyran ring is constructed last, through use of the alkynic unit.

An alternative construction which incorporates much of the alkyl functionality on to the alkyne portion and which carries the required methoxycarbonylalkyl substituent in the correct configuration at C-17 was reported by Langlois (equation 43).[82] This process and recently modified versions of the Hanessian spiroketal synthesis seem to correlate well.[79]

In the lactone acylations discussed above, there was never any evidence for a competing side reaction due to break down of the hemiketal moiety and Michael addition of the alcohol to the newly formed propargylic ketone. This may be taken as further evidence for the stability of the ketal-alkoxide intermediate; however, hemiketal ring opening and intramolecular Michael addition would provide an

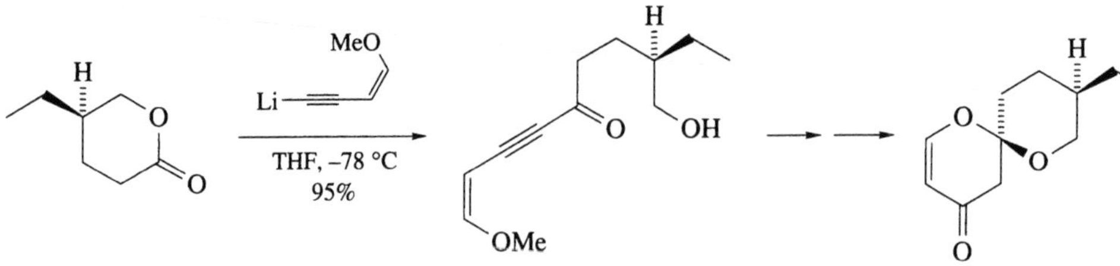

(42)

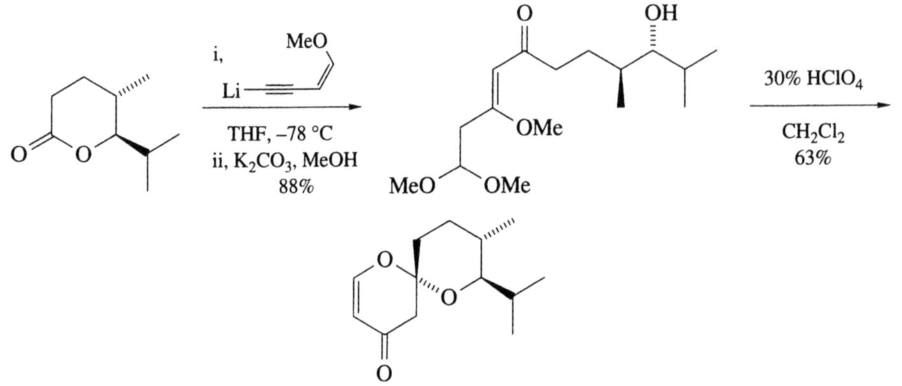

Scheme 20

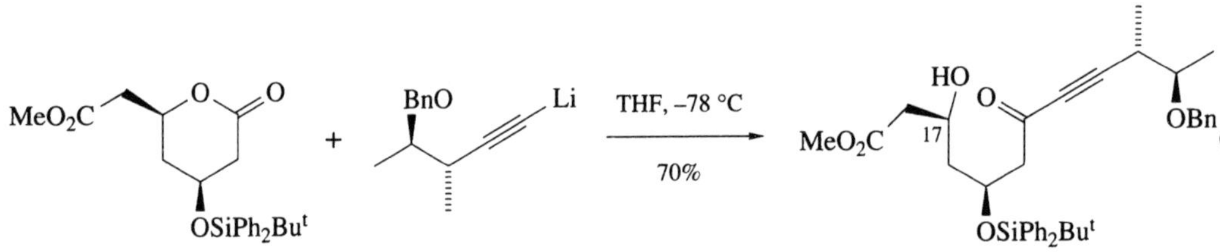

(43)

eight-membered ring vinylogous ester, seemingly a difficult process. Schreiber has discussed a method for achieving this nontrivial ring expansion in the context of the total synthesis of gloeosporone (equation 44).[83]

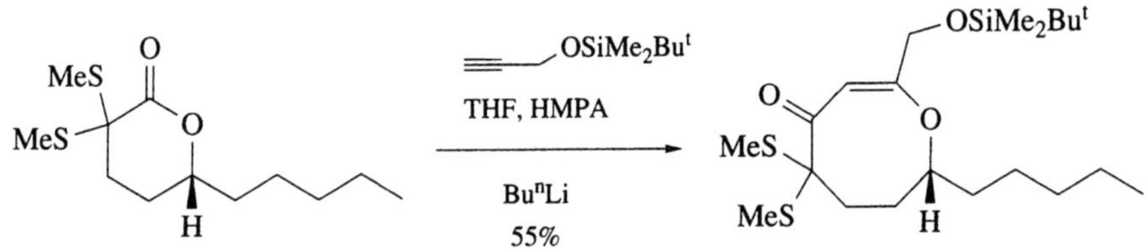

(44)

The initial hemiketal addition product is most likely induced to undergo ring opening as the reaction is warmed to room temperature and while in the presence of HMPA. Conjugate addition of the alkoxide to the alkynic ketone completes the ring formation. The reaction is promoted by α,α-disubstitution of the δ-lactone, as well as by the presence of electron-withdrawing groups on the alkynic fragment.

In a related fashion, certain types of vinylogous amides can be prepared by a modification of this same method, as reported by Suzuki (equations 45 and 46).[84] The triphenylsilyl moiety is essential for this transformation. The corresponding *t*-butyldimethylsilylacetylene gave the simple acylation product without subsequent Michael addition of the amine. In the ring expansion process the use of unmodified lithium acetylide (LiC≡CH) lead to the formation of by-products. A similar ring expansion process was reported for α-lactams with formation of five-membered ring vinylogous amides.[85]

An elegant 'reconstruction' of monensin from two chromic acid degradation products serves as an excellent example of the generality of alkynide addition to activated carboxylic acid derivatives. The alkynide (**59**) was prepared in eight steps from one degradation product and was then sequentially deprotonated with *n*-butyllithium and treated with magnesium bromide. The resultant magnesium salt

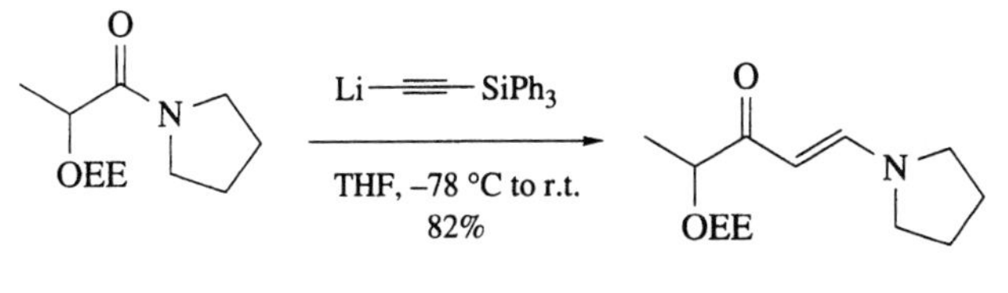

(45)

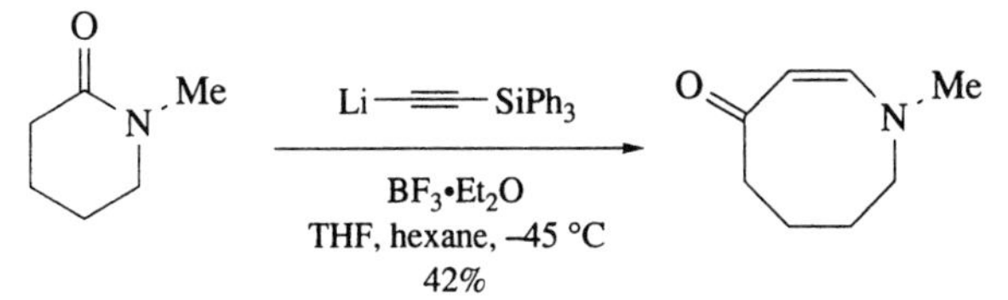

(46)

was then acylated with mixed anhydride (**58**), derived from another degradation fragment. This α,β-ynone was obtained in 87% yield (equation 47).[86] Lewis acid mediation of alkynide addition as provided by magnesium bromide or boron trifluoride[87] often improves the efficiency of the process and is useful for reactions which give substantial quantities of by-products.

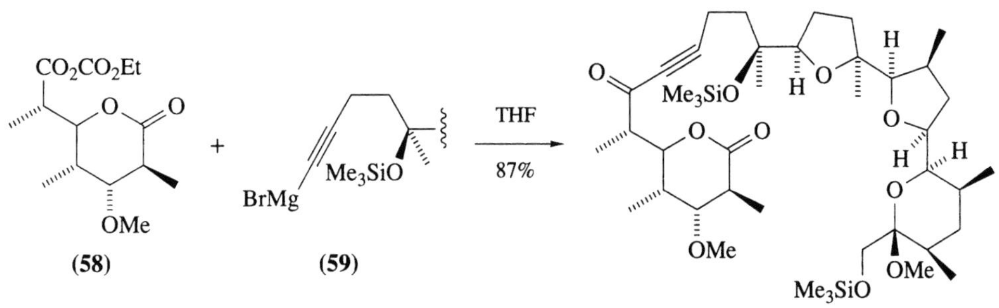

(47)

The highly reactive carbonyl of lactone (**60**), an intermediate in the synthesis of forskolin, was easily converted to propargyl ketone (**61**) by addition of the lithium alkynide as shown in equation (48).[88] It is possible that the intermediate ketal alkoxide was not stable in solution because of ring strain; however, no multiple addition products were reported, nor was there any Michael addition of the alkoxide to the resulting ynone.

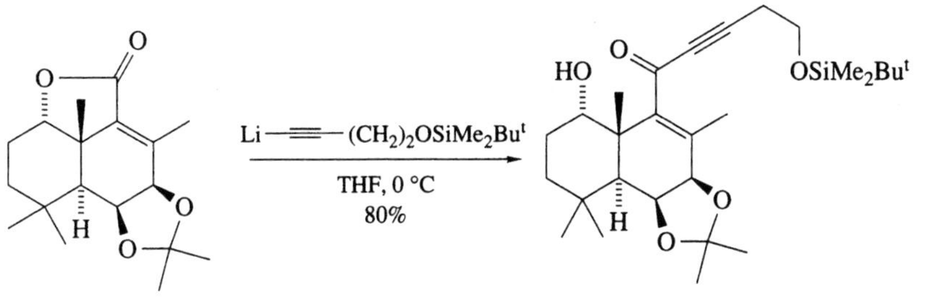

(48)

One final example of acylation of alkynes by lactones forms part of the synthesis of neomethynolide by Yamaguchi. The Prelog–Djerassi lactone serves as the acylating agent (equation 49). The functionalized alkynide undergoes addition very selectively at the lactonic carbonyl group, despite the presence of a relatively unhindered primary ester.[89]

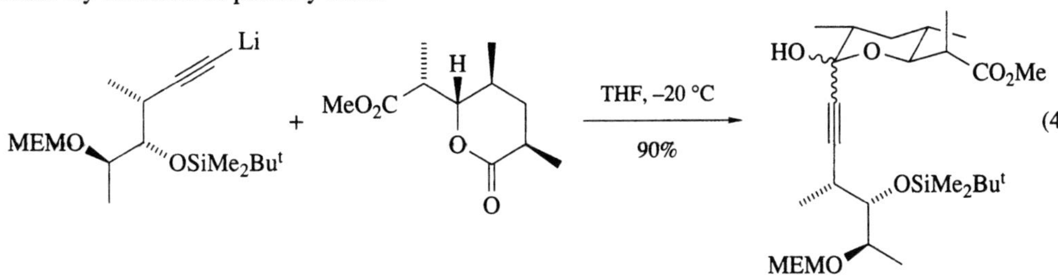

(49)

1.13.2.8 Other Activated Acylating Agents for Ketone Synthesis

In this section, several additional types of acylating agents for ketone synthesis will be discussed. Nearly all of these reagents are designed to preassociate with the organometallic reagent prior to acylation. At the present time, very infrequent use has been made of these new acylating agents, and their particular advantage remains to be demonstrated.

1.13.2.8.1 Acylating agents derived from pyridine or quinoline

Sakan and Mori have described the reaction of Grignard reagents with carboxylic esters formed from 8-hydroxyquinoline.[90,91] This acylating agent was designed to complex the organomagnesium with the substrate prior to formation of the tetrahedral intermediate, as indicated in Scheme 21. The actual tetrahedral species is not stable to the reaction conditions and a strong insoluble complex is formed between magnesium ion and 8-hydroxyquinoline. Although it is apparent that free ketone exists in solution simultaneously with complex (**62**), little if any tertiary alcohol formation was reported in the few simple cases investigated.[91] This may be due to the higher reactivity of the complexed oxoquinoline ester relative to the ketone.

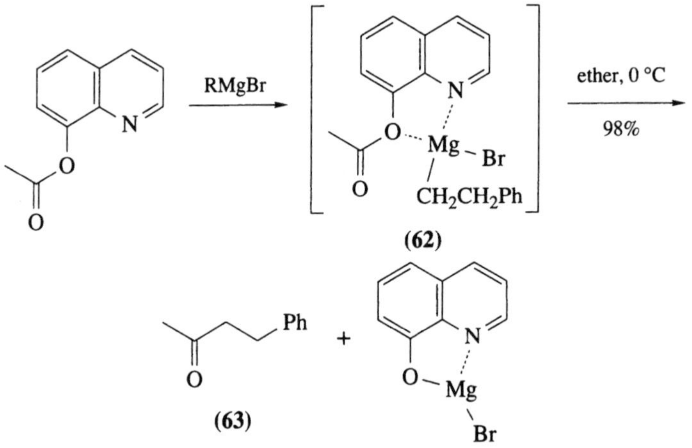

RMgBr

ether, 0 °C

98%

CH_2CH_2Ph

(**62**)

(**63**)

Scheme 21

A competition experiment (Scheme 22) was conducted to test the acylation abilities of the 8-acyloxyquinoline system and another metal-induced acylating agent, the 2-acyloxypyrazines (**65**) and (**67**). A mixture of quinoline (**64**) and pyrazine (**65**) was treated with phenethylmagnesium bromide under conditions which had previously afforded a 98% yield of the ketone (**63**) with the 8-acyloxyquinoline system. Surprisingly, the only product was that derived from addition to the acyloxypyrazine reagent, *i.e.* ketone (**68**). When substrates (**66**) and (**67**) were treated in the same fashion again only the product derived from addition to the acyloxypyrazine system was observed.[92] These results support the notion that the effect of prior complexation of the organometallic with the acylating agent may be as important to the rate of acylation as an incremental amplification in electrophilicity.

Similar results were obtained through the acylation of Grignard reagents with *O*-acyloximes (Scheme 23), although a higher degree of metal to substrate interaction is possible in this case.[93] The extent to which the species drawn in Scheme 23 actually participates in the reaction is unknown.[94]

Meyers and Comins have reported acylation of Grignard reagents with *N*-methylaminopyridylamides as a chemoselective method of ketone formation (equation 50). Once again chelation is expected to play a role in activating the substrate towards nucleophilic attack. However, as in the previously described cases, the resulting tetrahedral intermediate is not stable in the reaction mixture and a second equivalent of either a Grignard reagent or an alkyllithium can be introduced to form unsymmetrical tertiary alcohols. Alkyllithium reagents did not function as well in the formation of ketones; tertiary alcohols were the major product.[95]

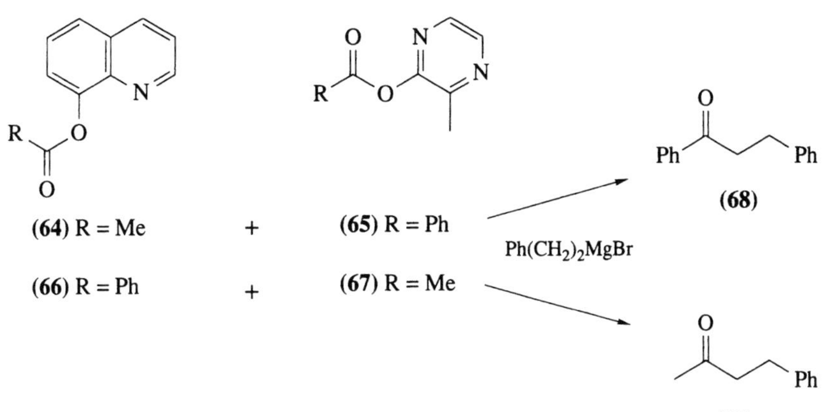

(64) R = Me + (65) R = Ph → (68)

(66) R = Ph + (67) R = Me → (63)

$Ph(CH_2)_2MgBr$

Scheme 22

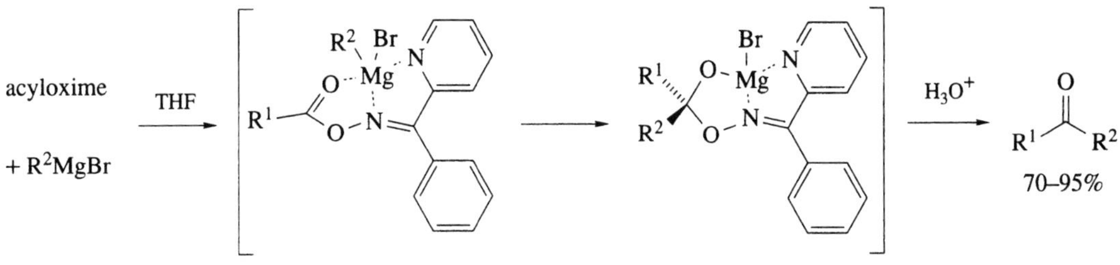

acyloxime + R^2MgBr → THF → H_3O^+ → 70–95%

Scheme 23

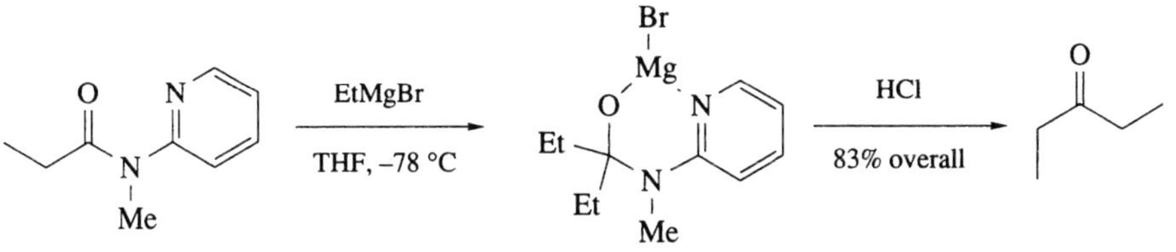

EtMGBr, THF, –78 °C; HCl, 83% overall (50)

1.13.2.8.2 *Carboxymethyleniminium salts*

The generation of active acylating agents *in situ* has certain advantages over methods which require a separate step for their preparation. Acylating agents that have previously been prepared *in situ* and used for ketone synthesis have included anhydrides and acyl imidazoles.[96] A frequent side reaction with these rather reactive agents is the formation of tertiary alcohols. Fujisawa and coworkers advocated the use of carboxymethyleniminium salts, prepared directly from carboxylic acid salts, and then subjected to reaction with Grignard reagents (Scheme 24). Good yields of ketones were observed in most cases.[97] The

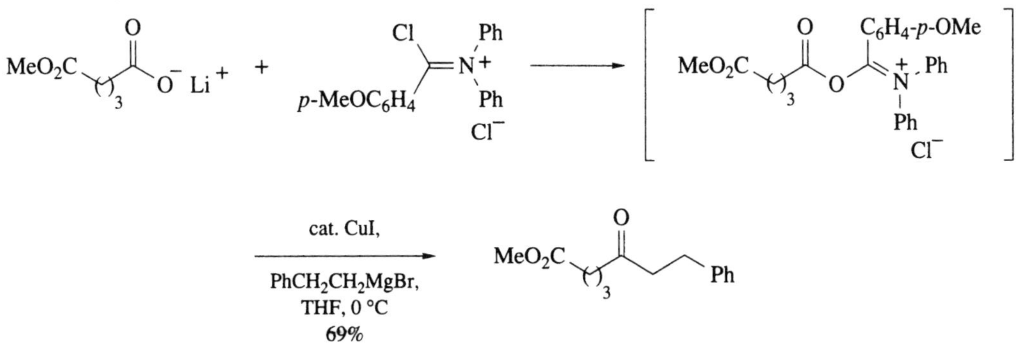

cat. CuI, $PhCH_2CH_2MgBr$, THF, 0 °C, 69%

Scheme 24

particular example shown in Scheme 24 indicates the selectivity obtained in competition with a carboxylic ester. No other acylated products or tertiary alcohols were present.

Bearing in mind that the exact nature of the intermediate in this and the following case has not been elucidated, together with the often surprising ability of acid chlorides to provide ketones when treated with Grignard reagents or organocuprates, the same authors have discussed the reaction of carboxylic acids with α-chloroenamines as a way to prepare acylating agents for ketone synthesis (Scheme 25).[98] In this case, the overall yield for the addition of organometallics may be slightly higher than the previously described method. The possible advantage is the direct transformation of a free acid to an acylating agent without the need for an extra equivalent of base.

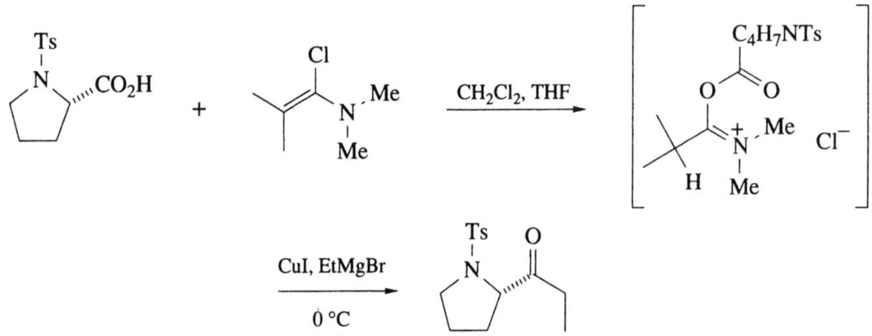

Ts N CO_2H + Cl N Me Me $\xrightarrow{CH_2Cl_2,\ THF}$ [C_4H_7NTs O O H N⁺ Me Me Cl^-]

$\xrightarrow[0\ °C,\ 80\%]{CuI,\ EtMgBr}$ Ts N O **(69)**

$[\alpha]_D = -156°$
(c 0.0218, $CHCl_3$)

Scheme 25

A comparison of the example in Scheme 25 with the results of earlier investigations with *N*-tosylproline demonstrates that the use of α-chloroenamines has some advantage. *N*-Tosylproline when treated with methyllithium in ether gave a 50% yield of methyl ketone of ≥95% *ee*. Similarly, preparation of the mixed anhydride with pivaloyl chloride followed by addition of ethylmagnesium bromide affords (**69**) in 53% yield[99] utilizing the conditions of Mukaiyama (equation 51).[100] No racemization was observed in any of these cases. Surprisingly in equation (51), the acylation of the mixed anhydride proceeds with only 1 equiv. of Grignard reagent, although there is an equimolar amount of triethylamine hydrochloride present.

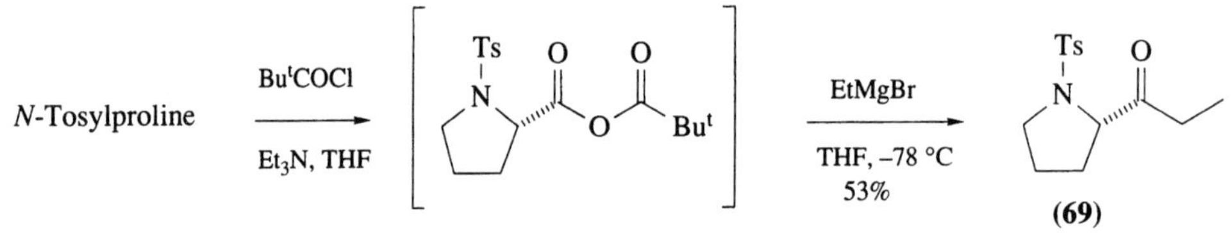

N-Tosylproline $\xrightarrow[Et_3N,\ THF]{Bu^tCOCl}$ [Ts N O O O Bu^t] $\xrightarrow[THF,\ -78\ °C,\ 53\%]{EtMgBr}$ Ts N O **(69)** (51)

$[\alpha]_D = -157.8°$ (c 0.0203, $CHCl_3$)

1.13.2.8.3 Acylation with the mixed anhydrides of phosphorus

Kende has demonstrated that the mixed anhydride from carboxylic acids and diphenylphosphinic chloride would acylate Grignard reagents to afford ketones in moderate to good yield. Tertiary carbinols were not observed unless excess Grignard reagent was added. The intermediate anhydrides were generally isolated and made free of triethylamine hydrochloride before addition of the nucleophile. The reaction shown in equation (52) gave improved yields over the simple addition of methyllithium to the carboxylic acid.[101]

Another procedure which was thought to pass through a mixed phosphorus anhydride was the acylation of Grignard reagents with an adduct formed between lithium carboxylates and triphenylphosphine dichloride (Scheme 26). The betaine (**70**) was the proposed tetrahedral intermediate; however, since no evidence is provided, the reaction may have also proceeded by way of the acid chloride. Surprisingly, good yields of ketone are preserved even in the presence of excess nucleophile and no tertiary alcohol formation was observed. Triethylamine can be used for prior deprotonation of the carboxylic acid; how-

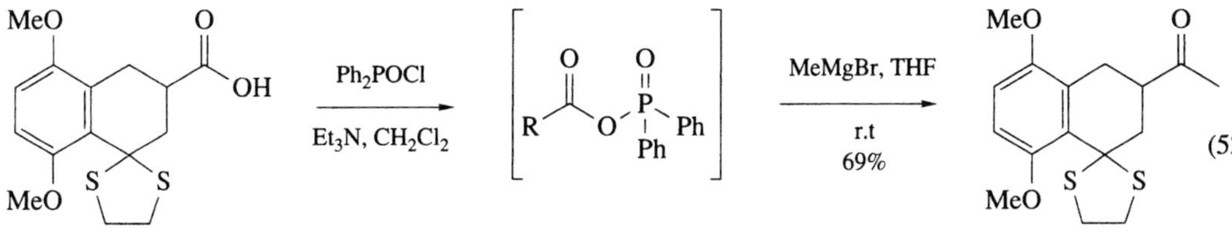

$\xrightarrow[\text{Et}_3\text{N, CH}_2\text{Cl}_2]{\text{Ph}_2\text{POCl}}$ $\xrightarrow[\text{r.t, 69\%}]{\text{MeMgBr, THF}}$ (52)

ever, unlike the case of mixed carboxylic anhydrides, 2 equiv. of Grignard reagent were required to complete the addition.[102]

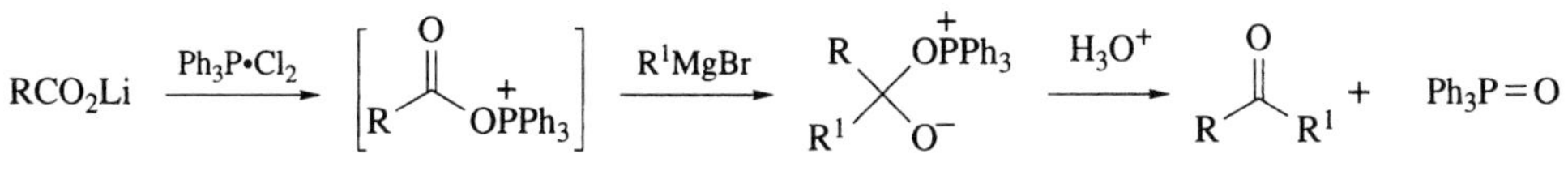

RCO_2Li $\xrightarrow{Ph_3P{\bullet}Cl_2}$ $[RC(O)\overset{+}{O}PPh_3]\ Cl^-$ $\xrightarrow{R^1MgBr}$ **(70)** $\xrightarrow{H_3O^+}$ $RC(O)R^1$ (60–95%) + $Ph_3P{=}O$

Scheme 26

1.13.2.9 Addition to Oxalic Acid Derivatives

General methods for α-keto ester synthesis are important for the preparation of a wide variety of natural substances of current interest. The direct addition of alkyllithiums or Grignard reagents to oxalate esters has been reported to give good yields of α-keto esters.[103] Presumably the tetrahedral intermediate formed from such an addition is stabilized by the strong electron withdrawal of the adjacent ester carbonyl, thus preventing or retarding reformation of the sp^2 center. Trapping of this intermediate is possible by the addition of acetyl chloride (equation 53).[104]

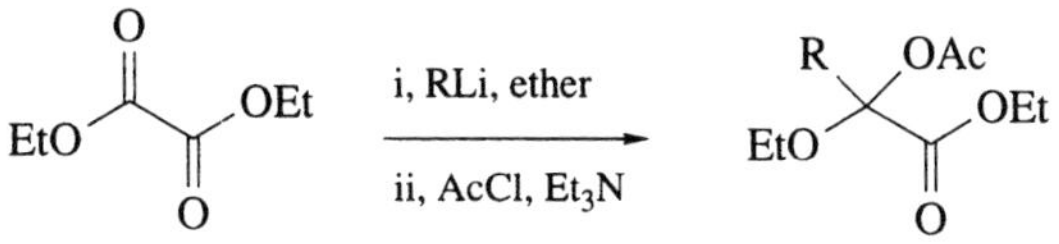

$\xrightarrow[\text{ii, AcCl, Et}_3\text{N}]{\text{i, RLi, ether}}$ (53)

R = Me, Ph

A second example of this methodology comes from the Merck synthesis of homo-tyrosine, an intermediate for a potent dopamine agonist (Scheme 27).[105] The Grignard reagent is added to a solution of 2 equiv. of diethyl oxalate at –20 °C.

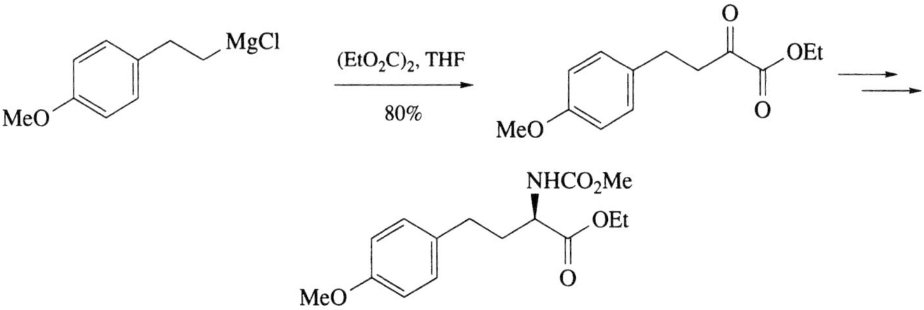

$\xrightarrow[\text{80\%}]{(\text{EtO}_2\text{C})_2\text{, THF}}$

Scheme 27

Preparation of the tricarbonyl region of the immunosuppressive FK-506 can be accomplished through acylation of a heavily functionalized dithiane as shown in equation (54).[106]

Previously, Corey had established a similar type of oxalate acylation using functionalized dithianes in the synthesis of aplasmomycin.[107] One additional entry into the α-keto ester system was through the reaction of triethoxyacetonitrile and alkyllithium reagents.[108] Only simple alkyl- and aryl-lithium reagents have been applied thus far, with yields in the range of 80%. Interestingly, the reaction took a different course with Grignard reagents; simple esters were the exclusive products.

X = OMe 30%, R = Ph

X = (piperidinyl, CO_2Bu^t) 60%, R = Ph

R = Ph

(54)

1.13.3 ACYLATION BY ORGANOCOPPER REAGENTS

1.13.3.1 Stoichiometric Organocopper Reagents

A great deal of study has surrounded the usefulness of preformed organocopper reagents (both stoichiometric and dialkylcuprates) and the utility of copper-catalyzed organometallics in ketone synthesis. When this methodology first became useful there were few other reliable methods for the direct synthesis of ketones from carboxylic acid derivatives.[109] The addition of organolithiums to carboxylic acids, or the reaction of organocadmium or organozincs with acid chlorides was often accompanied by side reactions or suffered from low yields. It was apparent that ketones did not generally react with organocuprates at low temperature, unlike the situation with the corresponding aldehydes. The conditions for acylation were therefore mild enough to prevent side reactions. Stoichiometric alkylcopper species (RCu) were found to afford lower yields of the ketone than were obtained from lithium dialkylcuprates and acid chlorides.[110] These reagents are much less reactive than dialkylcuprates and often require catalysis for addition to take place (*vide infra*). The work of Rieke represents the most recent study of the acylation of simple organocopper reagents.[111] Treatment of a functionalized alkyl bromide with a 'highly reactive copper solution' (formed *in situ* by reduction of copper(I) iodide triphenylphosphine complex with lithium naphthalide) results in the formation of an alkylcopper. This reagent adds rapidly to acid chlorides at –35 °C, affording ketones in good yield. The approach allows the preparation of organocopper reagents unavailable by methods which use organo-lithium or -magnesium reagents and copper(I) iodide. A drawback of the chemistry is that an excess of the acid chloride is required due to a serious side reaction which takes place between any excess of soluble copper metal and the acylating agent. Benzoyl chloride reacts rapidly with this form of copper(0) to form *cis*-stilbene diol dibenzoate thus consuming 4 equiv. of the reagent. Some examples of the method are included in Table 5.[111]

Table 5 Formation and Acylation of Stoichiometric Organocuprate Reagents prepared from Rieke Copper

$$\mathrm{CuIPPh_3} + \text{Li-naphthalide} \longrightarrow \mathrm{Cu^0} \xrightarrow{\mathrm{R^1X}} \mathrm{R^1Cu} \xrightarrow{\mathrm{RCOCl}} \mathrm{RCOR^1}$$

Alkyl halide	*Acid chloride*	*Equiv.*	*Temperature (°C)*	*Product*	*Yield (%)*
$Br(CH_2)_3CO_2Et$	PhCOCl	2.75	–35	$PhCO(CH_2)_3CO_2Et$	81
$Br(CH_2)_6Cl$	PhCOCl	2.80	–35	$PhCO(CH_2)_6Cl$	77
$Br(CH_2)_6$-epoxide	PhCOCl	2.80	–35	$PhCO(CH_2)_6$-epoxide	58
$Br(CH_2)_3CN$	$Me(CH_2)_2COCl$	2.85	–35	$Me(CH_2)_2CO(CH_2)_3CN$	61
o-NCC_6H_4Br	PhCOCl		0	o-NCC_6H_4COPh	71

One of the advantages of using stoichiometric organocopper reagents (RCu) instead of the more common dialkylcuprates (R_2CuLi) is that only 1 equiv. of RCu is necessary for good yields of ketones, whereas 3 equiv. of the dialkylcuprate are usually required in acylations. In many cases it is possible to substitute the more readily available alkyl or aryl Grignard reagent in the presence of stoichiometric, or in some cases catalytic, amounts of copper iodide.[112] These modified Grignard reagents are then used in equal molar amounts relative to the acid chloride. An important example of this methodology appeared in the synthesis of monensin by Still.[113] The central fragment (**71**) comprising carbons C(8)–C(15) was prepared through copper(I)-catalyzed Grignard addition to the *S*-pyridyl thiol ester (**72**), comprising carbons C(16)–C(25) (equation 55). The direct reaction of the Grignard reagent with the *S*-pyridyl thiol ester under the conditions of Mukaiyama resulted in multiple addition (Section 1.13.2.2).[30]

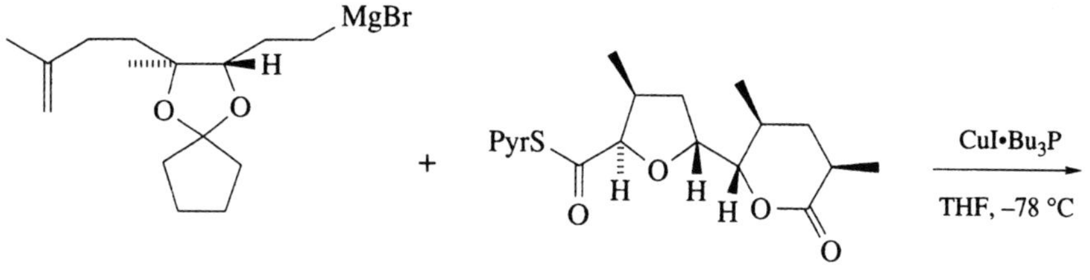

MgBr; PyrS; CuI•Bu_3P, THF, –78 °C

(**71**) (**72**) (55)

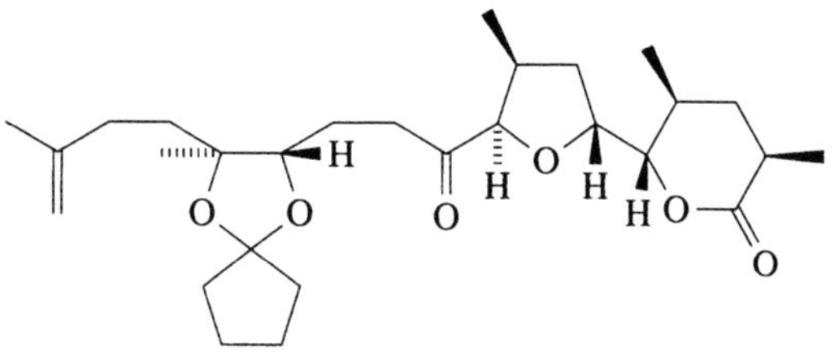

A recent example by McGarvey demonstrates the formation of chiral derivatives (**73**)–(**76**) which are useful for elaboration into the ubiquitous propionate unit through stereoselective enolate-based alkylation (equation 56).[114] The thiol ester substrates were derived from aspartic acid in seven steps. The acylation of either dimethylcuprate or the Grignard-derived organocopper reagent was extremely clean when these organometallics were used in excess. No epimerization of the adjacent center was observed during the addition.

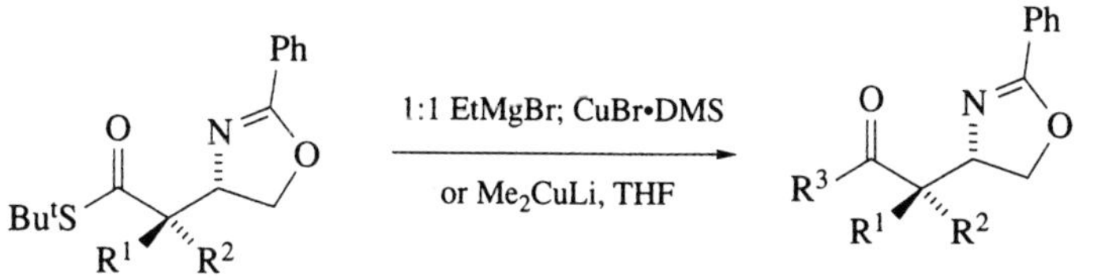

Bu^tS; 1:1 EtMgBr; CuBr•DMS or Me_2CuLi, THF (56)

$R^1 = R^2 = H$
$R^1 = Me, R^2 = H$
$R^1 = H, R^2 = Me$

(**73**) $R^3 = Et, R^1 = R^2 = H$; 90%
$R^3 = Me$ approx. 85% yield for all below
(**74**) $R^1 = R^2 = H$
(**75**) $R^1 = Me, R^2 = H$
(**76**) $R^1 = H, R^2 = Me$

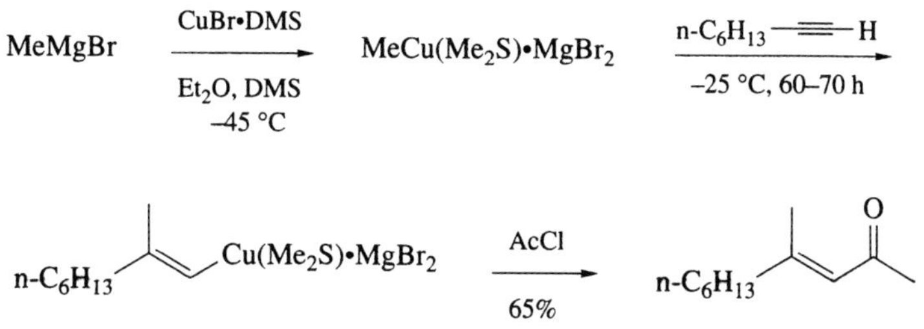

MeMgBr —CuBr•DMS, Et_2O, DMS, –45 °C→ $MeCu(Me_2S)\cdot MgBr_2$ —n-C_6H_{13}–≡–H, –25 °C, 60–70 h→

n-C_6H_{13}–C(Me)=CH–$Cu(Me_2S)\cdot MgBr_2$ —AcCl, 65%→ n-C_6H_{13}–C(Me)=CH–C(O)Me

Scheme 28

1.13.3.1.1 Acylation of vinylcuprates

The extension of organocopper-based technology to the synthesis of α,β-unsaturated ketones was accomplished by acylation of vinylcopper reagents (Scheme 28). This represented something of an unexpected advance due to the potential for subsequent conjugate addition to the product.[115] The approach has not seen extensive utility, compared to approaches using organotransition metal and organolithium methodology which have become more general. The first specific example was generated through adaptation of Normant's method for the formation of 1,1-disubstituted alkenylcuprates.[116] This trisubstituted alkene synthesis proceeds with retention of the stereochemistry generated initially by organocopper addition to an alkyne. The vinylcopper reagent must be β,β-disubstituted to avoid subsequent conjugate addition to the product. This restriction is removed when acylation is carried out with the more reactive acid bromide.[117]

The method was made considerably more general by inclusion of a catalytic amount of a palladium(0) complex during addition of the organometallic. Alkenylcopper reagents actually react relatively slowly with acid halides but, in a fashion analogous to other alkenyl metal species (see Section 1.13.4), they may be readily transmetallated to form an acylpalladium(II) complex which then undergoes reductive elimination to the product (Scheme 29).[118] A further discussion of acylation mediated by palladium complexes is included in Section 1.13.4. Interestingly, α,β-unsaturated acid chlorides react under these conditions to form divinyl ketones.

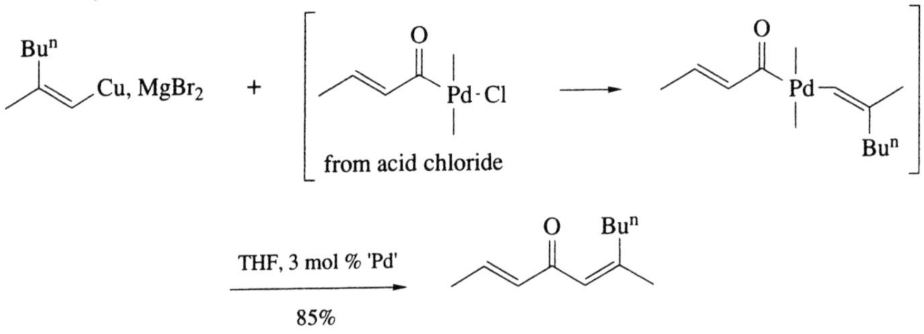

Bu^n ... Cu, $MgBr_2$ + [Pd·Cl, from acid chloride] → [Pd, Bu^n]

THF, 3 mol % 'Pd', 85%

Scheme 29

γ-Silylated vinylcopper reagents are acylated with acid chlorides followed by palladium-catalyzed 1,5-sigmatropic rearrangement to form silyloxy dienes in moderate yield (equation 57).[119]

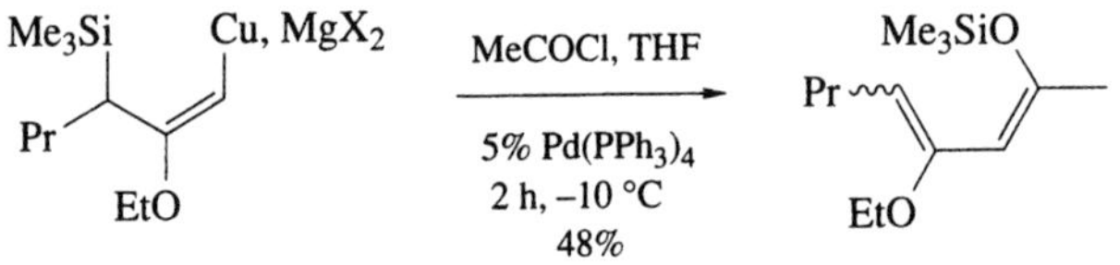

Me_3Si, Cu, MgX_2, Pr, EtO → MeCOCl, THF; 5% $Pd(PPh_3)_4$; 2 h, −10 °C; 48% → Me_3SiO, Pr, EtO (57)

The more reactive lithium dialkenylcuprates (R_2CuLi) cannot be directly acylated even at low temperature, because further addition to the product takes place. These reagents can be modified *in situ* so that the actual species undergoing reaction is an organozinc (requires palladium catalysis); however, other methods for generating organozincs are available that do not involve prior organocopper formation (Section 1.13.4.4). Stoichiometric vinylcopper and other copper-mediated acylations were not directly applicable to the case shown in equation (58)[120] (however, see Section 1.13.2.1).

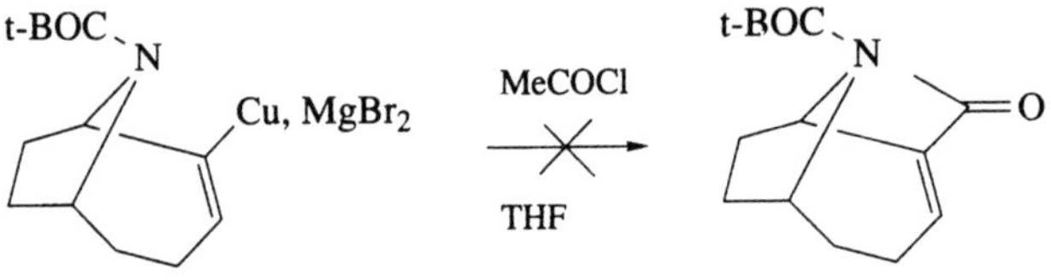

t-BOC, N, Cu, $MgBr_2$ → MeCOCl, THF (no reaction) → t-BOC, N, O (58)

1.13.3.2 Acylation of Lithium Dialkylcuprates with Acid Chlorides or Thiol Esters

The use of the more reactive lithium dialkylcuprate (R_2CuLi) species in acylation requires the complete exclusion of moisture and air.[121] This fact makes application of organocuprate acylation early in a synthetic route rather difficult if large amounts of material are to be handled or complex transferrable ligands are to be prepared. Nevertheless, the reaction has been widely used at mM scale with simple alkyllithiums. Posner has shown that lithium dimethyl- and other di-*n*-alkyl-copper reagents react with primary, secondary and tertiary acid chlorides at low temperature to provide good to excellent yields of simple ketones (equations 59 and 60).[121,122]

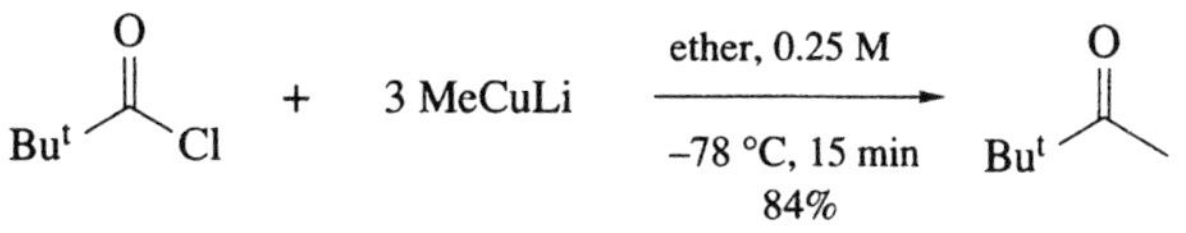

Bu^tCOCl + 3 MeCuLi → (ether, 0.25 M; –78 °C, 15 min; 84%) → Bu^tCOMe (59)

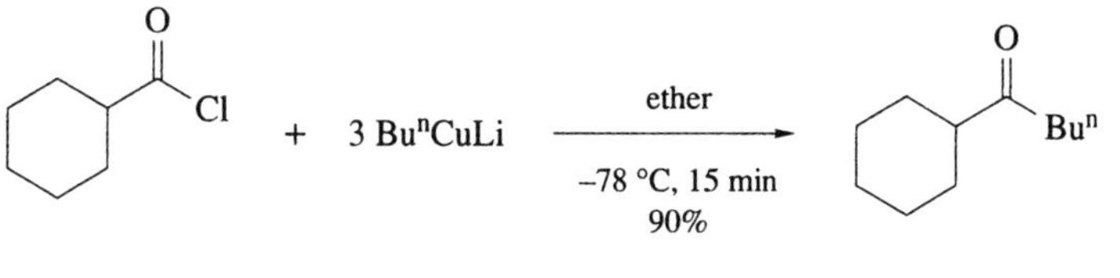

$C_6H_{11}COCl$ + 3 Bu^nCuLi → (ether; –78 °C, 15 min; 90%) → $C_6H_{11}COBu^n$ (60)

An interesting transformation, carried out by Paquette, demonstrates the selectivity that can be achieved with these reagents. The reaction of the acid chloride (**77**) and cuprates (**78**) or (**79**) takes place selectively in the presence of the lactone. The cuprates must be added to a solution of the acid chloride to obtain high yields. A later transformation demonstrated that a lactone can be converted easily to the ketone with phenyllithium (Scheme 30).[123] No loss of stereochemistry occurs in either permutation.

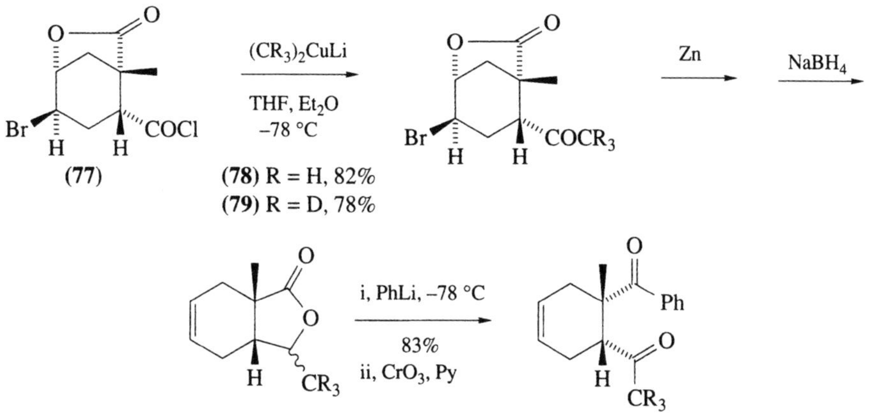

(**77**) → $(CR_3)_2CuLi$; THF, Et_2O; –78 °C → COCR$_3$ product → Zn → $NaBH_4$

(**78**) R = H, 82%
(**79**) R = D, 78%

i, PhLi, –78 °C; 83%; ii, CrO_3, Py

Scheme 30

The diacid chloride (**80**) can be easily converted to a diketone without epimerization or aldol condensation (equation 61).[124] In this case, inverse addition was not required for high yields.

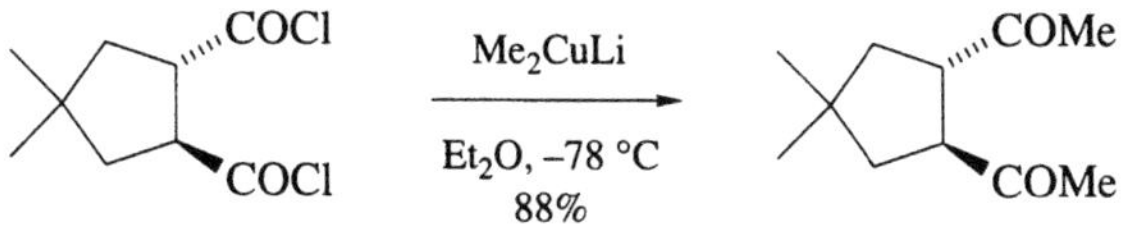

(**80**) → Me_2CuLi; Et_2O, –78 °C; 88% → bis(COMe) product (61)

Walba has used cuprate technology for the synthesis of the septamycin A-ring fragment shown in Scheme 31.[125] The use of the well-known anhydride (**81**) in this transformation was unsatisfactory.

The same acid chloride formed part of the premonensin synthesis of Sih.[126] Somewhat later in this synthesis, one of the final transformations before union of two large fragments was acylation of dimethylcuprate by acid chloride (**82**). The reaction proceeded in good yield but the requisite acid chloride had to be prepared from a methyl ester in two steps. A very similar dissection was used by Evans except that the methyl ketone (**84**) was prepared directly from the corresponding dimethylamide (**83**) using methyl-

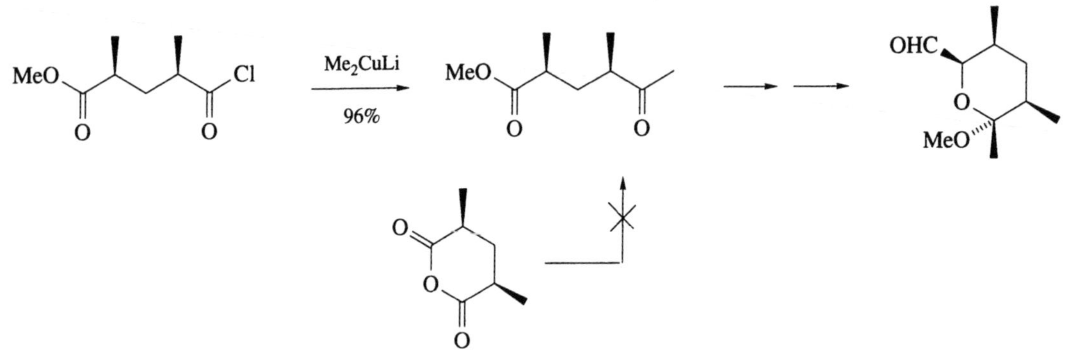

MeO, Cl; Me_2CuLi, 96%; MeO; OHC; MeO; (81)

Scheme 31

lithium (equation 62).[127] Although both procedures gave high yields, the amide was perhaps the more useful functional group, since it was carried intact through several prior synthetic permutations before acylation.

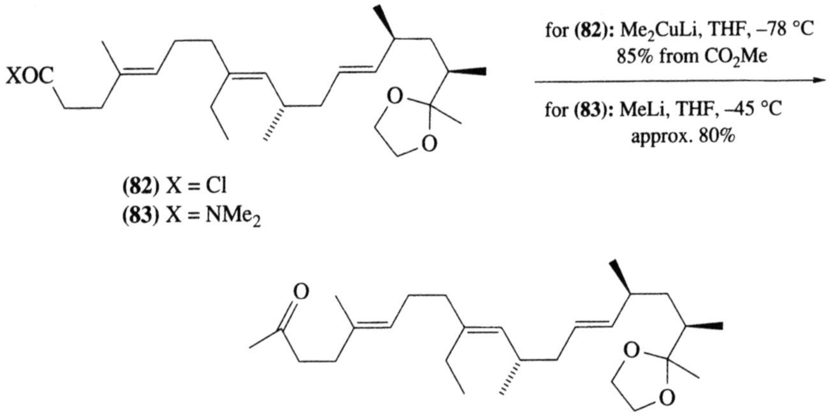

for **(82)**: Me_2CuLi, THF, –78 °C, 85% from CO_2Me

for **(83)**: MeLi, THF, –45 °C, approx. 80%

(82) X = Cl
(83) X = NMe_2

(62)

(84)

Because 3 equiv. of the organocuprate are usually required for the synthesis of ketones, the method is less effective for substrates of limited availability. Corey has described a class of mixed homocuprates R_rR_tCuLi in which one of the groups attached to copper is a nontransferable ligand (R_r) while the other group (R_t) is designed to be selectively introduced into the substrate.[128] Although this reagent was first used in the conjugate addition of valuable prostaglandin side chains to cyclopentenones, it has also found utility in acylation chemistry. Most often the residual groups are copper(I) alkynides which become strongly complexed to copper. Copper alkynides are somewhat more susceptible to acylation[129] than they are towards conjugate addition to enones, and the choice of acylating agent is more consequential. The most useful nontransferable ligand in this class is that derived from 3-methyl-3-methoxy-1-butyne.[130] In Corey's erythronolide A synthesis, the coupling of vinyl iodide (**85**) and the *S*-pyridyl thiol ester (**36**) could not be accomplished through Mukaiyama's direct acylation of Grignard reagents,[131] despite close analogy between related transformations in the erythronolide B series (see Section 1.13.2.2). The obstacle was overcome by the formation of mixed homocuprate (**86**) in which a nontransferrable ligand was used. The cuprate was selectively acylated by the *S*-pyridyl thiol ester with formation of ketone (**87**; Scheme 32).[132] The reaction was also quite solvent dependent and only a minimum amount of THF can be used in relation to nonpolar cosolvents (hexane/pentane:THF 1.2:1).

Attention should be drawn to the subtle differences between the reaction shown in Scheme 32 and the formation of the related erythronolide B intermediate (Scheme 11) by Corey and coworkers.[35] As mentioned earlier (Section 1.13.2.2), the addition of the Grignard reagent derived from iodide (**34**) was carried out easily on the identical substrate, *S*-(pyridyl) thiol ester (**36**) used in the A series, providing erythronolide B intermediate (**37**). The reaction in the erythronolide A series appears to be sensitive to the presence of the MTM-protected alcohol.

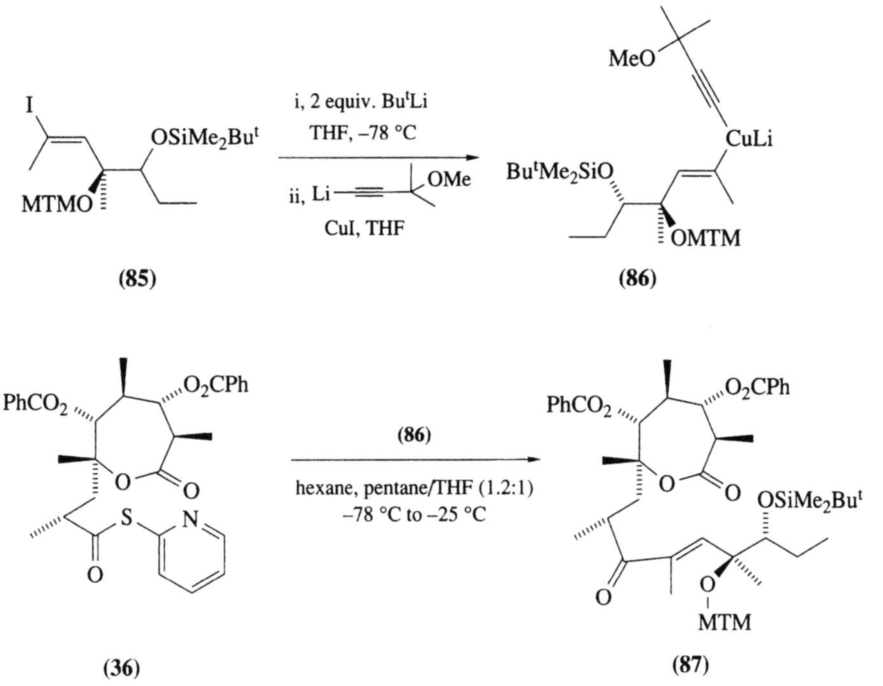

Scheme 32

There have been few transformations of vinylcuprate reagents with acid chlorides (Section 1.13.3.1.1). Marino and Linderman have reported a general preparation of divinyl ketones useful in a Nazarov sense for the formation of cyclopentenones (Scheme 33). Addition of various cuprate species to ethyl propiolate formed a mixed cuprate which is perhaps best represented as the allene (**91**). In the case of heterocuprates (**89**) and (**90**), acylation proceeded in good yield to form the divinyl ketone. Dimethylcuprate afforded none of the desired product but instead produced 1-acetylcyclohexene. The method was generalized for several different acid chlorides.[133]

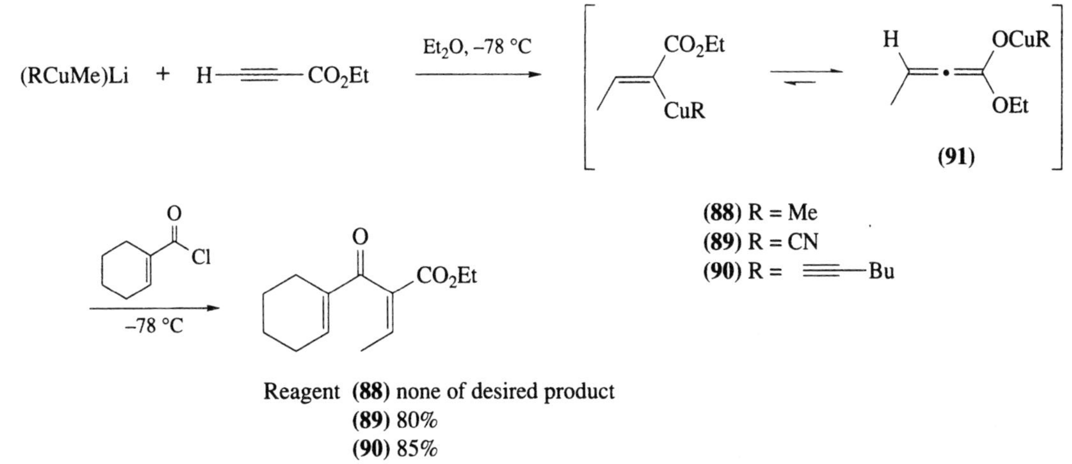

Scheme 33

1.13.3.3 Acylation of Heterocuprates

The addition of secondary and tertiary dialkylcuprates is often impractical due to the thermal instability of these reagents. For instance it is known that solutions of *s*- and *t*-alkylcopper(I) reagents cannot be cleanly prepared from 2 equiv. of the organolithium and copper(I) iodide even at –78 °C.[134] Posner has developed a method which makes preparation and acylation of these species more practical.[135] The method uses heterocuprates [R(R^{1}X)CuLi] in which R^{1}X is an alkoxide ion, thiolate or lithium amide ion and R is the transferrable ligand. The most satisfactory results have been obtained when R^{1}X was derived from *t*-butoxide. The reagents were prepared by addition of lithium *t*-butoxide to copper(I) iodide with formation of copper(I) *t*-butoxide. Addition of *s*-butyl- or *t*-butyl-lithium resulted in formation of the heterocuprate which has been shown to add efficiently to acid chlorides (Table 6).[135] As with many heterocuprates, these reagents are generally stable only to –50 °C, but acylation can proceed in high yield utilizing only 1.2–1.3 equiv. of reagent and product isolation was relatively facile. Similar organocuprates can be prepared from primary alkyllithium reagents and acylation is also facile. Unfortunately, these reagents have not seen a great deal of use in natural product synthesis as yet.

Table 6 Acylation of Heterocuprate Reagents Derived from Organolithiums and Cuprous *t*-Butoxide

$$\text{Bu}^t\text{OH} \xrightarrow[\text{ii, CuI}]{\text{i, Bu}^n\text{Li}} \text{CuOBu}^t \xrightarrow{\text{RLi}} \text{R(OBu}^t\text{)CuLi} \xrightarrow[-78\ ^\circ\text{C, 20 min}]{\text{R}^1\text{COCl, THF}} \text{R}^1\text{C(O)R}$$

Acid halide	*Cuprate (equiv.)*	*Product*	*Isolated yield (%)* $R = Bu^t$	$R = Bu^s$	$R = Bu^n$
$Br(CH_2)_{10}COCl$	1.2–1.3 (1.5%)	$Br(CH_2)_{10}COR$	78 (83)	83	
PhCOCl	1.2–1.3	PhCOR	82	87	70
$MeO_2C(CH_2)_2COCl$	1.2–1.3	$MeO_2C(CH_2)_2COR$	61	66	
$Bu^nCO(CH_2)_4COCl$	1.2–1.3	$Bu^nCO(CH_2)_4COR$	73	86	

Posner has also studied the effect of various other nontransferable ligands in the process. It was concluded that phenylthio was equivalent to *t*-butoxide as a nontransferable ligand, but phenoxy, dimethylamino and *t*-butylthio were inferior (equations 63 and 64).[136]

$$\text{EtO}_2\text{C(CH}_2)_2\text{COCl} \xrightarrow[\text{THF, }-78\ ^\circ\text{C, 15 min}]{\text{1.2 equiv. of PhS(Bu}^t\text{)CuLi}} \text{EtO}_2\text{C(CH}_2)_2\text{COBu}^t \quad 65\% \tag{63}$$

$$\text{PhCOCl} \xrightarrow[\text{THF, }-78\ ^\circ\text{C, 20 min}]{\text{1.1 equiv. of PhS(Bu}^t\text{)CuLi}} \text{PhCOBu}^t \quad 84\text{–}87\% \tag{64}$$

The importance of achieving the combined goals of high thermal stability for the organometallic and for improving the efficiency of alkyl group transfer has led Bertz and Dabbagh to continue the study of heterocuprates in a variety of transformations including acylation. Although these workers were primarily interested in assaying the thermal stability of new types of heterocuprates, they showed that $Bu^nCu(PPh_2)Li$ or $Bu^nCu(PCy_2)Li$ can be quantitatively acylated by excess PhCOCl after the cuprates had been aged for 30 min at 0 °C and 25 °C (Table 7). They provided a comparison of the reactivity of these reagents with other heterocuprates developed by other workers.[137]

An important development in cuprate acylation methodology that addresses several of the most objectionable properties of the reagents themselves has come from Knochel and his associates.[138] New highly functionalized copper reagents, represented by the formula RCu(CN)ZnI, can be prepared from readily available primary and secondary alkylzinc iodides by transmetallation with the soluble complex CuCN·2LiX in THF (equation 65).

These reagents have several advantages in acylation chemistry. Once prepared, these lower order cuprates may be used at temperatures near 0 °C rather that the usual –78 °C needed for alkylheterocuprates. Only one transferable ligand is involved per mole of reagent, which avoids the loss of precious substrates. Reaction work-up and product isolation is simplified relative to heterocuprates containing sulfur or phosphorus ligands. The organozinc and the corresponding organocuprate tolerate a wide variety of

Table 7 Comparison of Heterocuprate Reagents for Acylation with Benzoyl Chloride

$PhCOCl \xrightarrow{\text{aged heterocuprate}} PhCOR$

Entry	*Reagent*	*Solvent*	*Temperature (°C)*	*Yield (%)*	*Ref.*
1	$Bu^nCu(PPh_2)Li$	Ether	0 (25)	99 (95)	
2	$Bu^nCu(PCy_2)Li$	Ether	0 (25)	97 (89)	
3	$Bu^nCu(NCy_2)Li$	Ether	0 (25)	98 (89)	
4	$Bu^nCu(NPr^i_2)Li$	Ether	0 (25)	100 (87)	
5	$Bu^nCu(NEt_2)Li$	Ether	0 (25)	98 (73)	224
6	$Bu^nCu(SPh)Li$	Ether	0 (25)	19 (0)[a]	
7	$Bu^nCu(C{\equiv}CBu^t)Li$	Ether	0 (25)	92 (89)	225
8	$Bu^nCu(CH_2SO_2Ph)Li$	THF	0 (25)	99 (91)	226
9	$Bu^n{}_2Cu(CN)Li_2$	Ether	0 (25)	95 (84)	227

[a] The yields after 30 min at –50 °C and –25 °C were 100% and 97%.

$$RI \xrightarrow[\text{THF, 25–40 °C, 4–12 h}]{\text{activated zinc}} RZnI \xrightarrow[\text{0 °C, 10 min}]{\text{CuCN–2LiX}} RCu(CN)ZnI \xrightarrow[\text{0 °C, 3 h}]{R^1COCl} RCOR^1 \quad (65)$$

functionality contained within the original iodide substrate. Ketone, ester and nitrile functionality do not interfere with the generation of the organometallic. In addition, α,β-unsaturated esters are not attacked by these reagents. The yields of acylated organometallic are quite high (Table 8).[138] Further developments of this class of reagent should prove rewarding.

Table 8 Acylation of RCu(CN)ZnI with Acid Chlorides

Reagent	*Acyl halide*	*Product*	*Yield (%)*
$NC(CH_2)_2Cu(CN)ZnI$	PhCOCl	$NC(CH_2)_2COPh$	83
$NC(CH_2)_2Cu(CN)ZnI$	$Cl(CH_2)_3COCl$	$NC(CH_2)_2CO(CH_2)_3Cl$	77
$NC(CH_2)_2Cu(CN)ZnI$	$C_6H_{11}COCl$	$NC(CH_2)_2COC_6H_{11}$	79
$Pr^iCu(CN)ZnI$	Ph(OAc)CHCOCl	$Ph(OAc)CHCOPr^i$	82
$EtO_2C(CH_2)_3Cu(CN)ZnI$	PhCOCl	$EtO_2C(CH_2)_3COPh$	87
$C_6H_{11}Cu(CN)ZnI$	PhCOCl	$C_6H_{11}COPh$	84
$Bu^sCu(CN)ZnI$	$Cl(CH_2)_3COCl$	$Cl(CH_2)_3COBu^s$	94

1.13.3.4 Acylation with Thiol Esters

Because of the requirement for a large excess of dialkylcuprate in the reaction with acid chlorides, other acylating agents have been studied. Anderson has shown that *S*-ethyl thiol esters or *S*-phenyl thiol esters react in stoichiometric fashion with homocuprates to provide good yields of ketones (Table 9).[139]

Table 9 Acylation of Dialkylcuprates with *S*-Ethyl Thiol Esters

$$C_9H_{19}COSEt \xrightarrow{R_2CuLi} C_9H_{19}COR$$

Homocuprate	*Equiv.*	*Solvent*	*Temperature (°C)*	*Time (h)*	*R*	*Yield (%)*
Me_2CuLi	1.2	Et_2O	–78	2	Me	75
$Bu^n{}_2CuLi$	0.55	Et_2O	–40	2.5	Bu^n	89
$Bu^n{}_2CuLi$	1.0	Et_2O	–40	2.5	Bu^n	87
$Pr^i{}_2CuLi$	1.1	THF	–40	2	Pr^i	66

In contrast to the results obtained during the acylation of acid chlorides, nearly quantitative substitution of the thiol ester by each alkyl group of the cuprate takes place. Presumably, the intermediary heterocuprate R(EtS)CuLi, related to those prepared by Posner (*vide supra*), was also sufficiently reactive to provide the ketone from the thiol ester. The yields obtained depend on the reaction temperature and solvent. The transformation was performed by the rapid introduction of the thiol ester to a solution of copper(I) reagent in ether at –40 °C. Reaction times varied from one to several hours and quenching was carried out at low temperature with ammonium chloride. Methyl ketones must be prepared in ether at –78 °C to avoid selfcondensation. These thiol esters undergo bis-addition with standard Grignard or organolithium reagents to form tertiary alcohols unlike the *S*-(2-pyridyl) thioates of Mukaiyama (see Section 1.13.2.2). However, reaction of the thiol esters with 1.5 equiv. of $Bu^nMgBr \cdot CuI$ in THF gave *n*-butyl ketones in >80% yield.

The selective addition of diethyl cuprate to the dodecadienethioate shown in equation (66) demonstrates the utility of this methodology. Notice that there is only a small amount of conjugate addition product.[139] In diethyl ether, a large percentage of conjugate addition product is formed, even at –78 °C (equation 66).

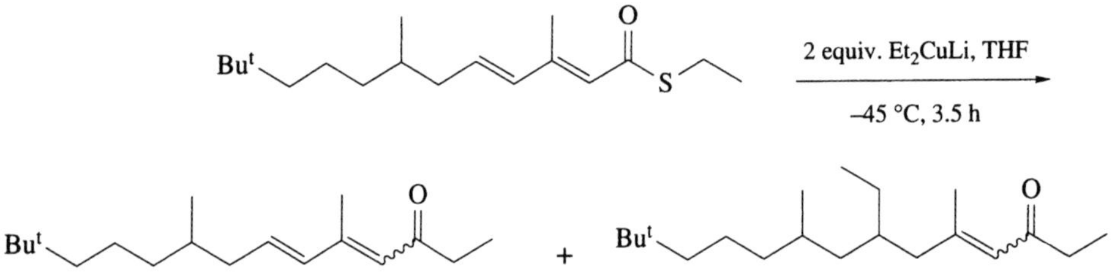

2 equiv. Et_2CuLi, THF; –45 °C, 3.5 h (66)

72% 3:1 (4*E*):(4*Z*) + <5%

Another example of this methodology has appeared recently from Masamune and coworkers in connection with a total synthesis of bryostatin (equation 67).[140] The salient point here is the demonstrated utility of the thiol ester, prepared directly through stereoselective boron enolate aldol condensation. Notice that no further activation or removal of a chiral auxiliary is necessary for this transformation, unlike other related aldol methodology.

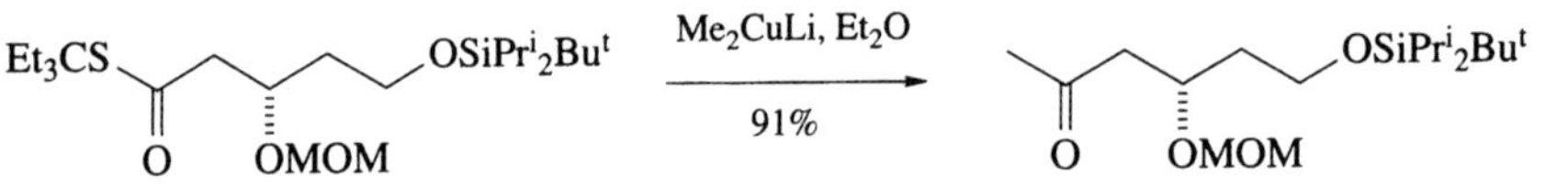

Et_3CS … $OSiPr^i_2Bu^t$ (O, OMOM) → Me_2CuLi, Et_2O; 91% → … $OSiPr^i_2Bu^t$ (O, OMOM) (67)

Researchers at Merck have studied the synthesis of the simple carbapenem system (**94**) through Wittig cyclization of the keto ylide (**93**). This interesting precursor was prepared directly from the thiol ester (**92**) by reaction with lithium dimethylcuprate or magnesium diphenylcuprate (Scheme 34).[141]

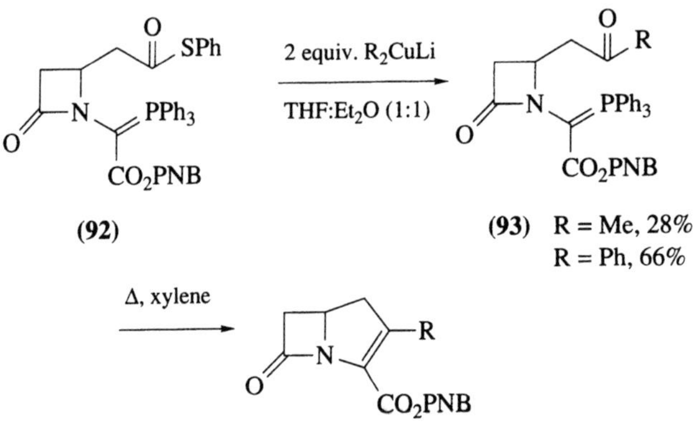

(**92**) (SPh, PPh_3, CO_2PNB) → 2 equiv. R_2CuLi; THF:Et_2O (1:1) → (**93**) R = Me, 28%; R = Ph, 66%

Δ, xylene → (**94**) (R, CO_2PNB)

Scheme 34

Other acylating agents are similarly effective in the formation of ketones by reaction with homocuprates. Kim has demonstrated that activated esters such as the 2-pyridyl carboxylates are satisfactory cuprate traps.[142] These esters can be prepared from carboxylic acids and 2-pyridyl chloroformate, provided a catalytic amount of DMAP is utilized (Scheme 35).

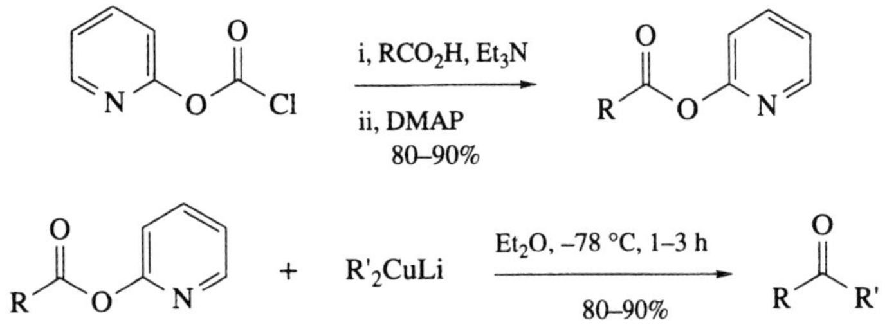

i, RCO_2H, Et_3N; ii, DMAP; 80–90%

+ R'_2CuLi — Et_2O, –78 °C, 1–3 h; 80–90%

Scheme 35

Once again these acylating agents are useful for the stoichiometric introduction of simple nucleophilic groups (methyl and *n*-butyl) into multifunctionalized substrates because the intermediate heterocuprate is also acylated (equations 68–70).

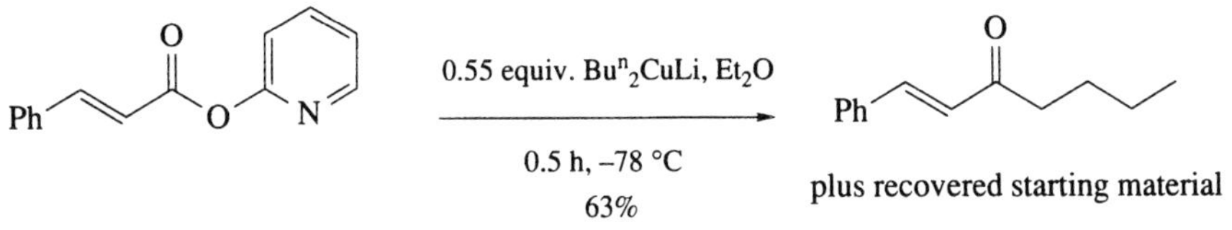

0.55 equiv. Bu^n_2CuLi, Et_2O; 0.5 h, –78 °C; 63% — plus recovered starting material (68)

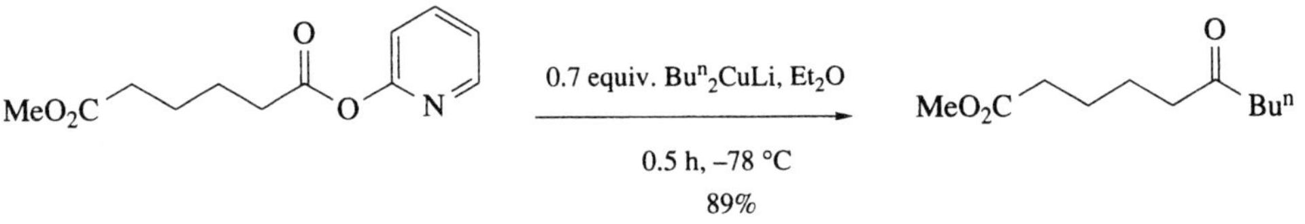

MeO_2C ... 0.7 equiv. Bu^n_2CuLi, Et_2O; 0.5 h, –78 °C; 89% → MeO_2C ... Bu^n (69)

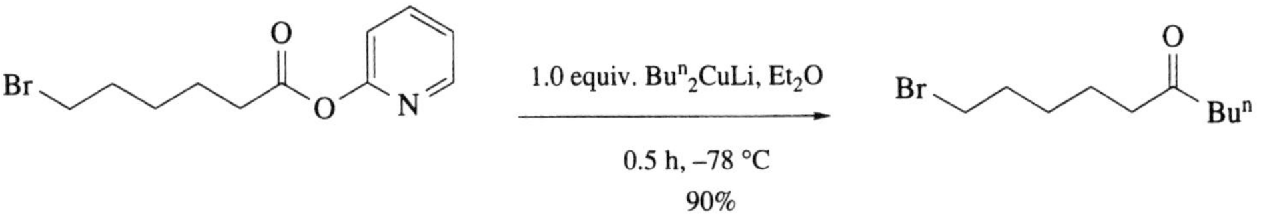

Br ... 1.0 equiv. Bu^n_2CuLi, Et_2O; 0.5 h, –78 °C; 90% → Br ... Bu^n (70)

Kim[143] has also studied the corresponding acylation of homocuprates by *S*-(2-pyridyl) thioates, discussed earlier in the context of total synthesis of monensin and erythronolide A (Sections 1.13.2.2 and 1.13.3.2). Under the standard anaerobic conditions necessary for cuprate formation, good yields of ketones could be derived from acylation of lithium dimethylcuprate (or lithium dibutylcuprate) by *S*-(2-pyridyl) thiobenzoate and other simple *S*-(pyridyl) thiol esters (equation 71). Interestingly, if the homocuprate is intentionally placed under an oxygen atmosphere before acylation and then reacted with the *S*-(2-pyridyl) thioate in oxygen at –78 °C, one obtains good yields of the corresponding ester (equation 72).

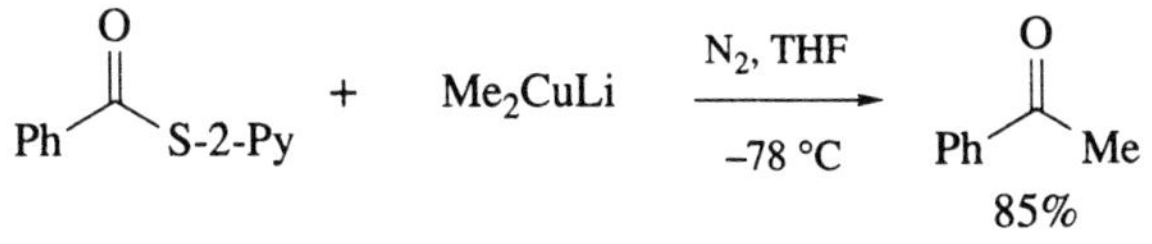

PhCO(S-2-Py) + Me_2CuLi —(N_2, THF, –78 °C)→ PhCOMe 85% (71)

PhCO(S-2-Py) + Me_2CuLi —(O_2, THF, –78 °C)→ PhCOOMe 71% 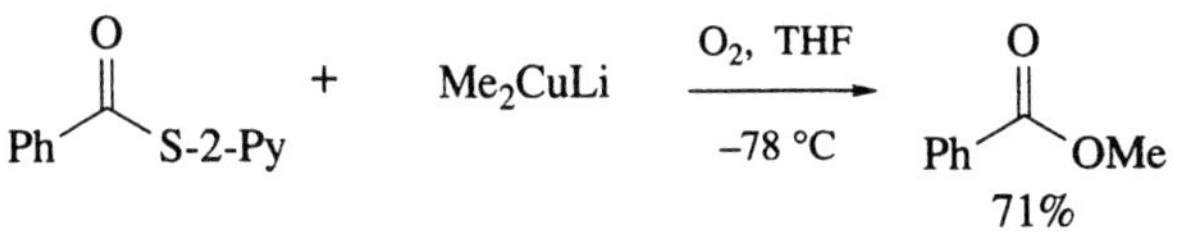 (72)

1.13.3.5 Acylation of α-Trimethylsilylmethylcopper

Potentially useful β-silyl ketones[144] may be obtained through acylation of the trimethylsilylmethylcopper reagent generated from 1 equiv. of copper(I) iodide and trimethylsilylmagnesium bromide. The results obtained with this reagent have varied depending upon the exact conditions used. Normant has found that acylation of trimethylsilylmethylcopper magnesium bromide under palladium catalysis results in the direct formation of the silyl enol ether without formation of β-silyl ketone.[145] It appears that under the latter conditions, the initial product is a β-silyl ketone which then undergoes a palladium- or acid-catalyzed C to O isomerization. Several previous investigations including one by Akiba indicate this is the case. Akiba has shown that β-silyl ketones can be obtained from the stoichiometric copper reagent in good yield, but when treated with a catalytic amount of triflic acid, they underwent immediate isomerization to silyl enol ethers.[146,147] Kishi had earlier prepared 1-(trimethylsilyl)-2-butanone by addition of $Me_3SiCH_2MgBr{\cdot}CuI$ to acetyl chloride affording an 80–90% yield.[148] The report indicated that this transformation is superior to methods which use the corresponding trimethylsilylmethyllithium.[149,150]

The construction of the naturally derived narbomycin[151] and tylosin-aglycones[152] by Masamune and coworkers employ identical methodology for seco-acid formation. In each case, Peterson alkenation of a functionalized aldehyde (not shown) and the silyl ketones (**96**; R = $SiMe_3$; Scheme 36) or (**99**; Scheme 37) efficiently introduced the required (*E*)-α,β-unsaturation. Silyl ketone formation is accomplished in each case through cuprate acylation by an activated carboxylic acid derivative. Formation of an acid chloride was not possible in the sensitive tylosin-aglycone intermediate; however, selective acylation of the silylcuprate proceeded at the pyridyl thiol ester moiety of (**98**) and not with the *t*-butyl thiol ester. In a related investigation,[153] (**97**), an advanced intermediate for 6-deoxyerythronolide B, was obtained from (**95**) *via* addition of lithium diethylcuprate to the acid chloride (84% yield). In all the above cases, no addition was observed at the *t*-butyl thiol ester.

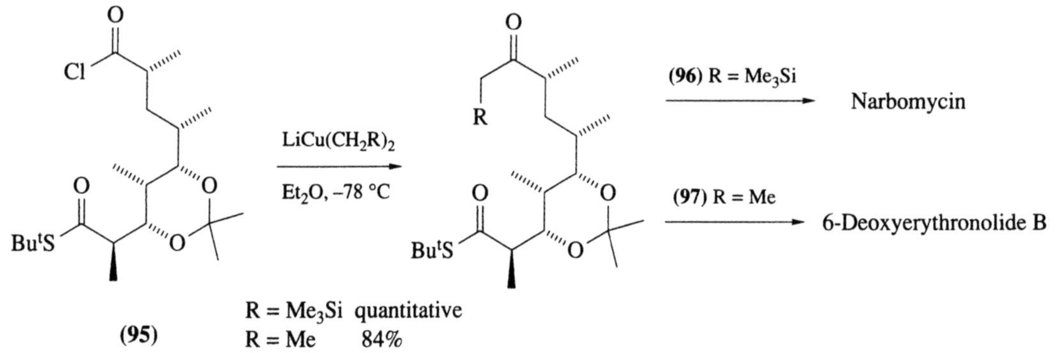

(**95**) $\xrightarrow[\text{Et}_2\text{O, –78 °C}]{\text{LiCu(CH}_2\text{R)}_2}$ R = Me_3Si quantitative; R = Me 84%

(**96**) R = Me_3Si → Narbomycin

(**97**) R = Me → 6-Deoxyerythronolide B

Scheme 36

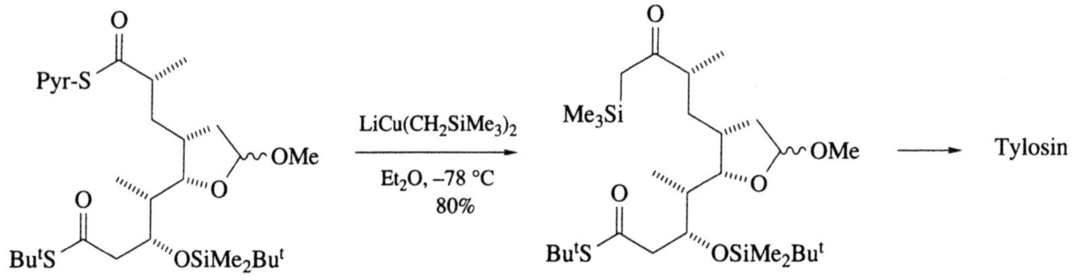

(**98**) $\xrightarrow[\text{Et}_2\text{O, –78 °C, 80\%}]{\text{LiCu(CH}_2\text{SiMe}_3\text{)}_2}$ (**99**) → Tylosin

Scheme 37

1.13.4 ACYLATION MEDIATED BY LOW-VALENT PALLADIUM COMPLEXES

1.13.4.1 Acid Chlorides and Organostannanes

The acylation of the mildly nucleophilic organostannanes was first reported by Migita in 1977[154] through the use of palladium catalysis. However, the reported conditions were harsh and of limited scope. Nearly simultaneously, Stille and Milstein reported a more comprehensive study of this method,[155] including mechanistic and synthetic studies. Their work also formed the basis for the discovery of improved reaction conditions for acylation.[156] Equimolar amounts of an acid chloride and tetrasubstituted organotin undergo acylation with palladium catalysis in HMPA to afford high yields of ketones wherein one of the tin substituents has been acylated (equation 73). Transfer of a second substituent from R_3SnCl is 100 times slower than transfer from R_4Sn but can still be useful. R_2SnCl_2 and $RSnCl_3$ do not under go acylation. The reaction can be carried out in the presence of the following functional groups without interference: NO_2, RC≡N, aryl halide (≠ Br), vinyl, methoxy, carboxylic esters and aldehydes. The latter functional group is not tolerated by any other method of ketone formation that involves organometallic addition to a carboxylic acid derivative. An especially interesting case is that of *p*-acetylbenzaldehyde (equation 74) which serves to introduce the method. The product, a keto aldehyde, is generally difficult to prepare by other methods since it undergoes rapid selfcondensation.

$$X{-}R{-}C(O)Cl + R^1{}_3SnR^2 \xrightarrow[n = 0\text{–}4]{[L_nPd^0]} X{-}R{-}C(O)R^2 + R^1{}_3SnCl \quad (73)$$

$$p\text{-OHC-}C_6H_4\text{-COCl} \xrightarrow[\text{Me}_4\text{Sn, HMPA, 65 °C, air, 86\%}]{\text{BnPd(PPh}_3)_2\text{Cl}} p\text{-OHC-}C_6H_4\text{-COMe} \quad (74)$$

Other aromatic and aliphatic acid chlorides give good to excellent yields of the desired ketones using this procedure. Hindered or α,β-unsaturated acid chlorides also function effectively, the latter forms α,β-unsaturated ketones without competing conjugate addition (equations 75 and 76).

$$PhCH{=}CHCOCl \xrightarrow[91.3\%]{\text{Me}_4\text{Sn, BnPd(PPh}_3)_2\text{Cl}} PhCH{=}CHCOMe \quad (75)$$

$$CH_2{=}CHCOCl \xrightarrow[93.3\%]{\text{Me}_4\text{Sn, BnPd(PPh}_3)_2\text{Cl}} CH_2{=}CHCOMe \quad (76)$$

There is no need for an inert atmosphere as both the catalyst and organostannane are air stable; in fact the reaction is accelerated by the presence of oxygen. Depending on the solvent, reaction times are very short, an hour or less in HMPA and under 24 h for solvents like chloroform, THF and dichloroethane. The end of the reaction is dramatically signaled by the precipitation of metallic palladium from the clear reaction solution.

The acylation is limited to the use of acid chlorides due to their unique ability to oxidatively add palladium(0); other acylating agents are not generally useful in this context. It was also known that acid chlorides do not react with organotins without Lewis acid catalysis and more importantly organotins do not generally react with the expected product, the ketone, except under very strong Lewis acidic conditions.[157] Even diacid chlorides may be utilized in this process (equation 77); however, oxalyl chloride cannot be used due to the indicated decarbonylation of the intermediate acid chloride (equation 78).[155]

$$ClC(O)(CH_2)_6C(O)Cl \xrightarrow[90.5\%]{\text{Me}_4\text{Sn, BnPd(PPh}_3)_2\text{Cl}} MeC(O)(CH_2)_6C(O)Me \quad (77)$$

$$ClC(O)C(O)Cl \xrightarrow[\text{BnPd(PPh}_3)_2\text{Cl}]{\text{Me}_4\text{Sn}} MeC(O)C(O)Cl \xrightarrow[-\text{CO}]{\text{Pd}^0} MeC(O)Cl \xrightarrow[\text{Pd}^0]{\text{Me}_4\text{Sn}} MeC(O)Me \quad (78)$$

10% overall yield

α-Diketones have been prepared by the reaction of acylstannanes and acid chlorides.[158] Few cases have been studied thus far and the yields for unsymmetrical α-diketones were moderate at best. Two by-products have been observed in the reaction. Decarbonylation made up 4–20% of the material balance and degradation of acyl-tri-*n*-butyltin resulted in *n*-butyl ketone, although the latter mode of decomposition only accounted for a few percent of the reaction product (Scheme 38).

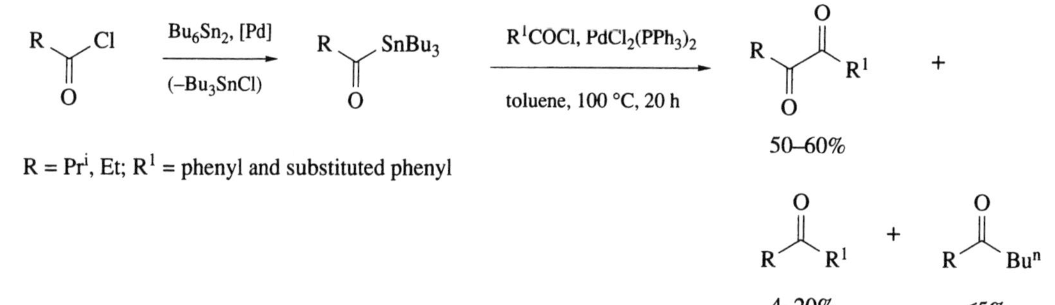

RCOCl —[Bu_6Sn_2, [Pd]; ($-Bu_3SnCl$)]→ $RCOSnBu_3$ —[R^1COCl, $PdCl_2(PPh_3)_2$; toluene, 100 °C, 20 h]→ $RCOCOR^1$ (50–60%) + $RCOR^1$ (4–20%) + $RCOBu^n$ (≤5%)

R = Pr^i, Et; R^1 = phenyl and substituted phenyl

Scheme 38

Symmetrical diketones have been prepared in a similar fashion from 2 mol of the acid chloride and 1 mol of hexabutylditin with palladium catalysis (equation 79). The yields were still moderate.[159]

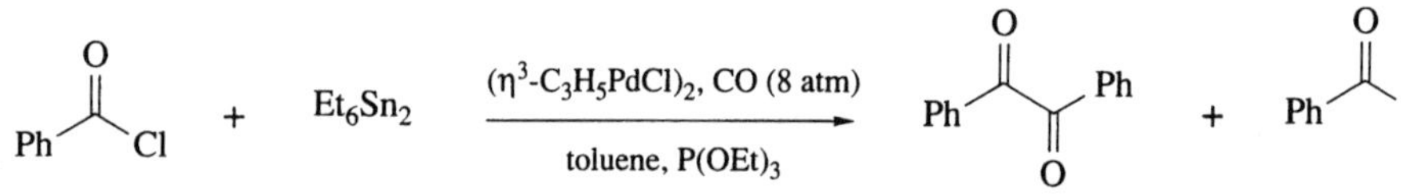

PhCOCl + Et_6Sn_2 —[$(\eta^3-C_3H_5PdCl)_2$, CO (8 atm); toluene, $P(OEt)_3$]→ PhCOCOPh + PhCOPh (79)

Soderquist has reported a slightly more effective method not subject to the losses due to decarbonylation of intermediates.[160] Acylation of α-methoxyvinyltin (**100**) under palladium(0) catalysis afforded a good yield of the α-methoxyenone (**101**). Hydrolysis in acetone/aqueous acid releases the diketo functionality (equation 80). Only the unsubstituted vinyl system has been employed thus far.

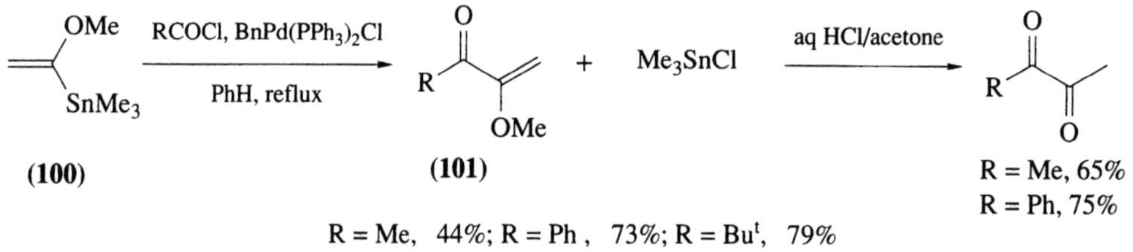

$CH_2=C(OMe)SnMe_3$ (**100**) —[RCOCl, $BnPd(PPh_3)_2Cl$; PhH, reflux]→ $RCOC(OMe)=CH_2$ (**101**) + Me_3SnCl —[aq HCl/acetone]→ $RCOCOCH_3$ (80)

R = Me, 44%; R = Ph , 73%; R = Bu^t, 79%

R = Me, 65%
R = Ph, 75%

Prior to the studies which uncovered the utility of organotin acylation as described in the preceding paragraphs, acid chlorides were also found to undergo a similar rhodium(I)-mediated acylation with allyltins to form β,γ-unsaturated ketones.[161] The palladium(0)-catalyzed coupling has been found to be more general with respect to varied substitution patterns of both reagents and could be conducted under essentially neutral conditions (see Section 1.13.5.1).

1.13.4.1.1 Mechanistic studies of palladium-catalyzed acylation

The mechanism of palladium-catalyzed coupling of organic halides with tetrasubstituted organotins has been widely studied by Stille[162] and a general understanding of the mechanism is critical to further developments in this area. The use of other organometallics in palladium-catalyzed acylation will follow this discussion, as well as further examples where this method has been used.

Thus far we have discussed numerous examples whereby selective ketone formation has been achieved through organometallic acylation. The problem was approached by choosing a less nucleophilic organometallic which can be acylated but does not interact with the desired product. Thus far, few reagents with this type of selectivity have been found (organocuprates). Most often, the strategy was to either preserve the tetrahedral intermediate formed upon nucleophilic addition or to activate the substrate towards nucleophilic attack by prior coordination of the organometallic. A conceptually different approach, achieved through the use of transition metals, alters the mechanism of acylation such that carbon–carbon bond formation is not part of the rate-determining step. Many examples of the formation of acylmetal complexes have been reported in the literature.[163] Depending upon the nature of the metal and the degree of coordination, these complexes can function as nucleophiles or as electrophiles in reactions with organic substrates. An example of a nucleophilic acylmetal complex used in ketone synthesis was reported by Collman. Nucleophilic acyliron(0) complexes such as (**102**) readily undergo oxidative addition (akin to nucleophilic displacement) with alkyl halides to form the hexacoordinate acylalkyliron(II) species (**104**) or (**105**). These are unstable with respect to reductive elimination of the alkyl and acyl ligands and the complex degrades to form a ketone and complexed iron(0).[164] The carbon–carbon bond formation is therefore a result of electronic reorganization of the metal driven by the stability (or lack thereof) of the particular oxidation state with the given ligands.

The acyliron(0) complex (**102**) has been isolated and subjected to the same nucleophilic displacement (or equivalently oxidative addition) with excellent correlation (Scheme 39). The same species is also readily available from acid chlorides (*i.e.* formation of **103**), but the overall process has not been widely used in the synthesis of ketones (Scheme 40).[165] The final step of the process, a reductive elimination of acyliron(II) complex (**104**) or (**105**), is quite rapid and it has not been possible to isolate and identify the presumed intermediates in this case (Scheme 41). Since the oxidative addition of the acyliron complex with the alkyl halide is extremely mild, the corresponding ketone formed in the reaction is not subject to attack by organometallic reagents and no tertiary alcohol is formed.

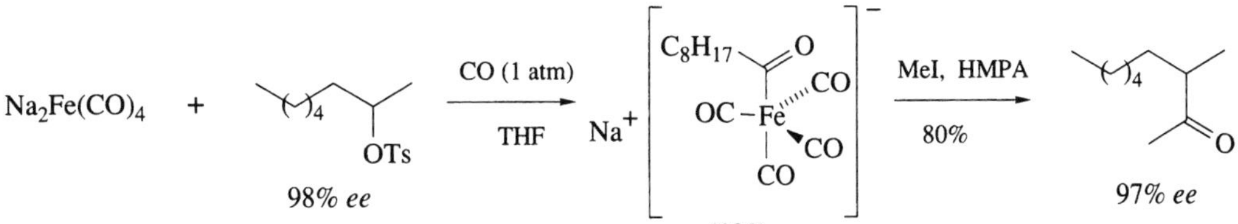

$Na_2Fe(CO)_4$ + OTs (98% *ee*) —CO (1 atm), THF→ Na^+ [$C_8H_{17}C(O)Fe(CO)_4$]$^-$ (**102**) —MeI, HMPA, 80%→ ketone (97% *ee*)

Scheme 39

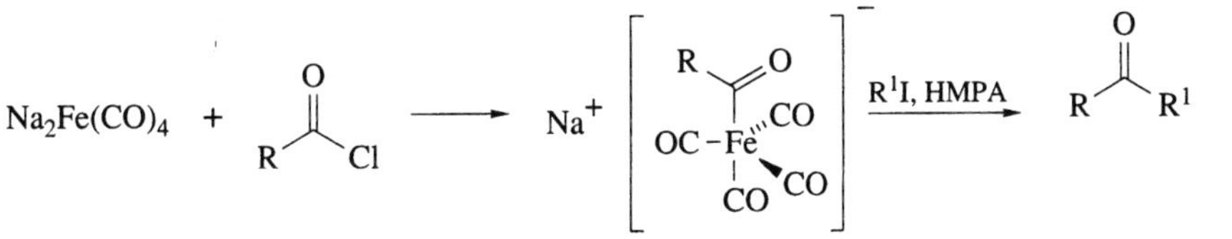

$Na_2Fe(CO)_4$ + RCOCl → Na^+ [RC(O)Fe(CO)$_4$]$^-$ (**103**) —R^1I, HMPA→ $RCOR^1$

Scheme 40

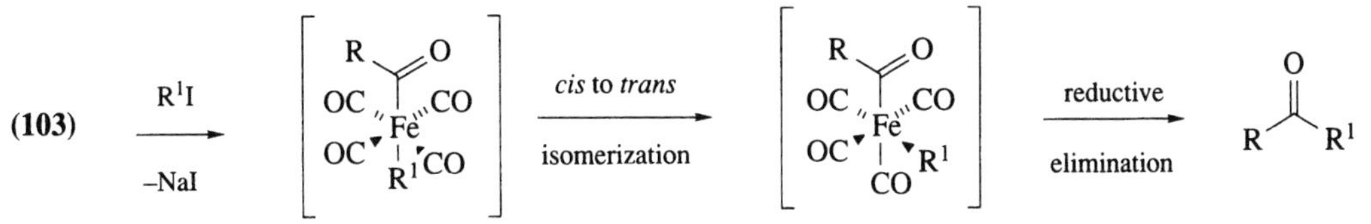

(**103**) —R^1I, –NaI→ (**104**) —*cis* to *trans* isomerization→ (**105**) —reductive elimination→ $RCOR^1$

Scheme 41

Palladium(0) complexes are well known to suffer oxidative addition with acid chlorides.[166] The resulting acylpalladium(II) complex (**106**) is, in contrast to the acyliron(II) complex discussed above, an electrophilic species which is subject to nucleophilic attack by various organometallics. Stille has studied the addition of organotins because they undergo rapid nucleophilic addition to the acylpalladium(II) complex, but do not add to either the acid chloride or react with the ketone (Scheme 42). There are also several other organometallics useful in this sense (*vide infra*).

Both of the described processes, nucleophilic acyliron or electrophilic acylpalladium, rely upon the rapid degradation of organoacyl metal complexes by a net reduction of the metal to a low-valent form with creation of a carbon–carbon bond. The precise mechanism involved in the formation of each high-valent complex is complementary and interesting but irrelevant to the resulting ketone formation, due to

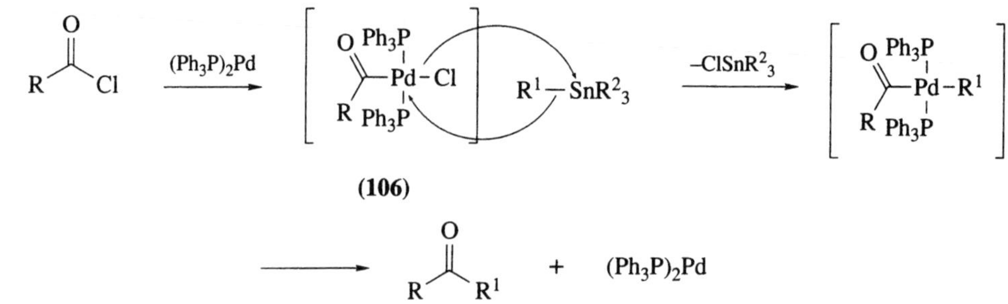

$$\mathrm{RCOCl} \xrightarrow{(Ph_3P)_2Pd} \left[\mathrm{RC(O)Pd(PPh_3)_2Cl}\right] \textbf{(106)} + \mathrm{R^1{-}SnR^2_3} \xrightarrow{-ClSnR^2_3} \left[\mathrm{RC(O)Pd(PPh_3)_2R^1}\right] \longrightarrow \mathrm{RCOR^1} + (Ph_3P)_2Pd$$

Scheme 42

reductive elimination. Many examples of these processes have been described with other metals, but are beyond the scope of this discussion.[167] The particular case of organocuprates may also fit into the same model. One can envision an oxidative addition of a dialkylcuprate (a copper(I) complex) to the acid chloride forming an acylcopper(III) species. Reductive elimination returns copper(I) and generates the ketone. At present, however, there is no evidence for this latter supposition, none of the intermediates in the organocuprate acylation have been characterized and researchers are only now beginning to understand the complex chemistry of organocuprates.[168]

1.13.4.1.2 Effect of the catalyst

Returning to the palladium-catalyzed process, we begin with a discussion of the catalytic cycle and particularly the function of the metal, the effect of phosphine ligands and the identity of the organopalladium intermediates along the route. What will become important is the rate at which these intermediates are transformed along the reaction pathway, and whether or not a step in the catalytic cycle is made rate limiting through the addition or omission of a particular reagent, solvent or ligand.

The rapid rate of oxidative addition of palladium(0) to an acid chloride has been attributed to the extended bond length and highly polarized nature of the carbon–chlorine bond.[169] The half life for oxidative addition of tetrakis(triphenylphosphine)palladium(0) to benzoyl chloride has been measured at approximately 10 min at –40 °C.[170] Palladium(0) is required for oxidative addition but large effects on the reaction rate may be observed depending on the degree and type of ligation around this metal. Electron-donating ligands, such as triarylphosphines, tend to make palladium more nucleophilic, but they also increase steric interaction with the substrate.[172] Palladium(0) catalysts are also air sensitive making their handling a problem. It is often beneficial to introduce the catalyst in the form of a palladium(II) species that is more stable to atmospheric conditions. In a separate step, reduction of the palladium(II) complex occurs by one of several mechanisms to a suitable form of solvated or ligated palladium(0), which undergoes reaction. Stille has found that benzyl(chloro)bis(triphenylphosphine)palladium(II), formed as shown in equation (81), is an excellent catalyst precursor for acylation of organotins by acid chlorides.[155]

$$\mathrm{Pd^0(PPh_3)_4 + PhCH_2Cl \longrightarrow PhCH_2Pd^{II}(PPh_3)_2Cl + 2\,PPh_3} \quad (81)$$

The actual acylation catalyst for alkyltins and acid chlorides is generated through sacrificial transmetallation of the organotin to form a new palladium(II) complex with release of trialkyltin halide (see Scheme 43). Reductive elimination then forms a small amount of benzylated adduct and the coordinatively unsaturated and highly nucleophilic catalyst bis(triphenylphosphine)palladium(0) [$Pd^0(PPh_3)_2$]. The generalized mechanism for acid chloride insertion and coupling is depicted in Scheme 43. Oxidative addition of $Pd^0(PPh_3)_2$ to the acid chloride forms an observable and isolable palladium(II) complex (**107**).[155] Transmetallation of a single tin ligand affords *trans*-acylalkylpalladium(II) complex (**108**) which is not observable but then undergoes a rapid *trans* to *cis* migration, possibly involving dissociation of one of the ligands.[190] The *cis* complex (**109**) immediately undergoes reductive elimination to the product and returns the palladium(0) complex.

Commercially available $Pd^0(PPh_3)_4$ also catalyzes the process depicted above, but observations indicate that the reaction rate is significantly depressed over those seen with $PhCH_2Pd^{II}(PPh_3)_2Cl$.[155] This is true even though the same coordinately unsaturated complex, $Pd^0(PPh_3)_2$, is responsible for catalysis in both cases through the well-known dissociative process shown in equation (82).[171]

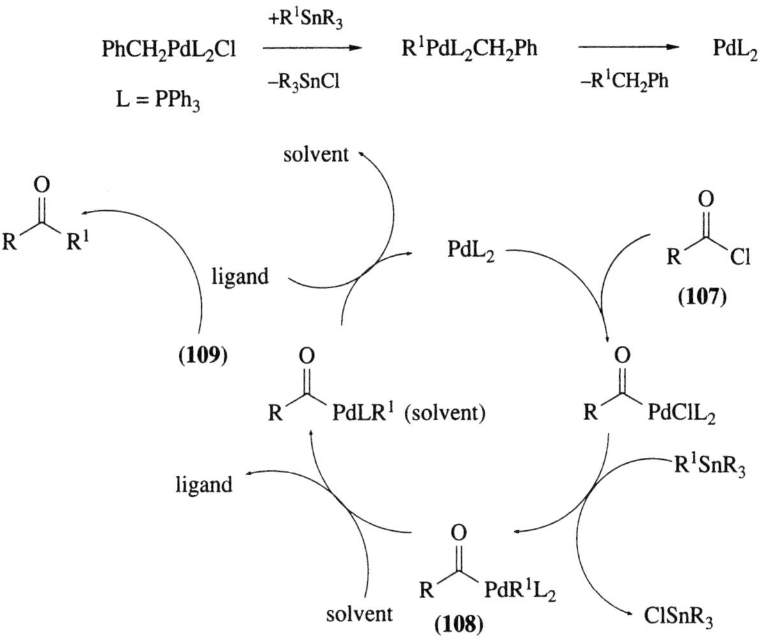

$$PhCH_2PdL_2Cl \xrightarrow[-R_3SnCl]{+R^1SnR_3} R^1PdL_2CH_2Ph \xrightarrow[-R^1CH_2Ph]{} PdL_2$$

$L = PPh_3$

RC(O)R¹ (109) ← ligand; solvent; PdL_2; RC(O)Cl (107); RC(O)PdClL₂; R^1SnR_3; ClSnR₃; RC(O)PdR¹L₂ (108); ligand; solvent; RC(O)PdLR¹ (solvent)

Scheme 43

$$Pd^0(PPh_3)_4 \rightleftharpoons Pd^0(PPh_3)_2 + 2PPh_3 \quad (82)$$

The inclusion of additional triphenylphosphine in reactions catalyzed by $PhCH_2Pd(PPh_3)_2Cl$ results in a deceleration of the oxidative addition process. The highly reactive coordinatively unsaturated catalyst $Pd^0(PPh_3)_2$ now becomes part of the equilibrium shown above. Since in this case coordinatively saturated $Pd^0(PPh_3)_4$ first undergoes dissociation, the added ligand shifts the equilibrium towards the left and decreases the rate of oxidative addition.[172] In mechanistic studies to be discussed later, the effect of added triphenylphosphine was found to be the greatest during oxidative addition. Smaller effects of this ligand were observed in the ensuing transmetallation and reductive elimination.

Other potential sources of catalyst for the acylation include the air- and moisture-stable dibenzylideneacetone complex of palladium(0); $Pd^0_2(DBA)_3{\cdot}CHCl_3$.[173] This complex reacts with triphenylphosphine to form coordinatively unsaturated complexes *in situ* but has been used infrequently for acylation.[174] So-called 'ligandless catalysts' have been exploited with promising results. Ligandless palladium complexes such as RPd^0XL_2 (L = solvent) contain no phosphine-derived ligands but instead are merely solvated by the reaction medium. These complexes are among the most active catalysts for cross-coupling and can be formed *in situ* from either $(\eta^3\text{-}C_3H_5Pd^{II}Cl)_2$, $(MeCN)_2Pd^{II}Cl_2$ or $Pd^{II}Cl_2{\cdot}LiCl$ through metathesis with the organotin. Acid chlorides will suffer oxidative addition by these complexes and good yields of ketones are obtained (Scheme 44).[175]

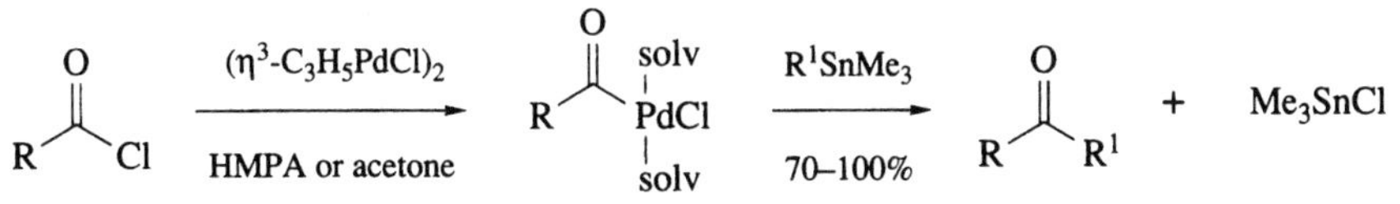

$$RC(O)Cl \xrightarrow[\text{HMPA or acetone}]{(\eta^3\text{-}C_3H_5PdCl)_2} RC(O)PdCl(solv)_2 \xrightarrow[70\text{–}100\%]{R^1SnMe_3} RC(O)R^1 + Me_3SnCl$$

R = Ph; R¹ = Me, Bn, vinyl, Ph, p-ClC_6H_4, p-CNC_6H_4, p-$NO_2C_6H_4$, 2-thenoyl,
R = Me, o-FC_6H_4, Bn, PhCH=CH, p-ClC_6H_4; R¹ = Ph, Me

Scheme 44

Aromatic or α,β-unsaturated acid chlorides were found to undergo rapid oxidative addition with these palladium complexes, and the subsequent acylation of an organotin proceeded smoothly in HMPA (1–60 min) or in acetone. Formation of simple aliphatic acylpalladium(II) complexes proceeds at markedly reduced rates and the acylation of organotins is slower but still useful (10–24 h).

The formation of the acylated palladium(II) species has been formulated as shown in Scheme 45. The sequence relies on strong solvent participation to prevent precipitation of palladium(0) during reductive elimination of intermediate complex (**110**). When tetramethyltin or benzyltrimethyltin is to be used in this process it is necessary to introduce a small amount of Me_6Sn_2 to aid in the formation of the active catalyst. These organotins are slow to react with $(\eta^3\text{-}C_3H_5PdCl)_2$.

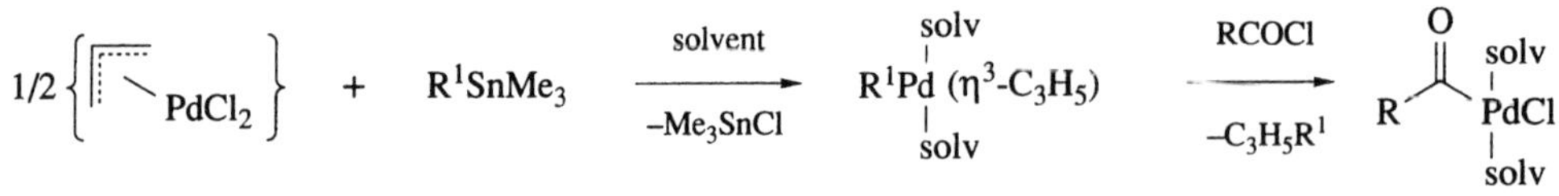

1/2 { $PdCl_2$ } + R^1SnMe_3 —solvent, $-Me_3SnCl$→ $R^1Pd(\eta^3\text{-}C_3H_5)$ (solv, solv) —RCOCl, $-C_3H_5R^1$→ RC(O)PdCl (solv, solv)

(**110**)

Scheme 45

1.13.4.1.3 Effect of solvent, catalyst concentration and oxygen

Several factors were found to influence the rate of the Stille acylation process, including an acceleration effect by oxygen and also acceleration by the use of polar solvents such as HMPA. Oxidative addition to the acid chloride forms a *trans*-acylpalladium(II) complex which can in principle degrade by the loss of carbon monoxide. In some cases it is advantageous to use an atmosphere of CO to reverse this process.[176] The initial rate was also found to be critically dependent on catalyst concentration. Surprisingly, high catalyst concentrations do not accelerate the reaction and are in fact only slightly more effective than low levels of palladium (see Table 10). The initial rate and $t_{1/2}$ were found to be sensitive to the concentration of palladium complex between 0 and 3 × 10^3 M, but became constant above this concentration. The optimal catalyst concentration for the reaction shown in Table 10 is about 7 × 10^{-4} M in $PhCH_2Pd(PPh_3)_2Cl$.[155]

Table 10 Dependence of Reaction on Catalyst Concentration

PhCOCl + Me_4Sn —$BnPd(PPh_3)_2Cl$, HMPA→ PhCOMe + Me_3SnCl

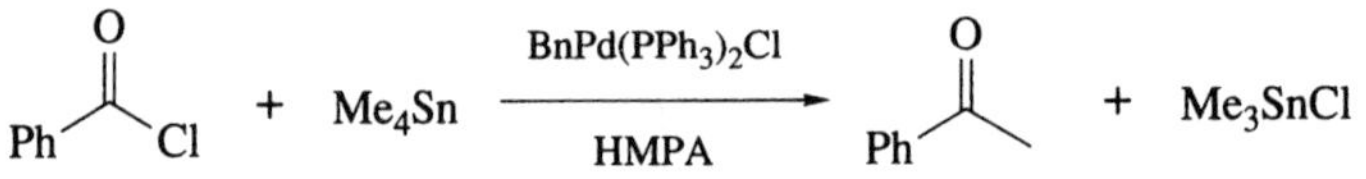

[Catalyst] (mol L^{-1} × 10^{-4})	*Turnover number*	*Initial rate (mol L^{-1} min^{-1} × 10^{-4})*	*$t_{1/2}$ (min)*
0.866	14925	30.72	125
1.73	7692	76.80	60
6.94	1852	172.80	38
8.30	1563	70.40	50
30.4	422	46.08	128
75.5	170	40.96	115
149.4	86	42.24	120

[PhCOCl] = 1.28 M, [Me_4Sn] = 1.40 M.

The exact reasons for this behavior are not clear; however, deactivation at high concentrations may be due to dimerization of the catalyst which liberates triphenylphosphine as shown in equation (83). As discussed, the reaction is significantly impeded by added phosphine.

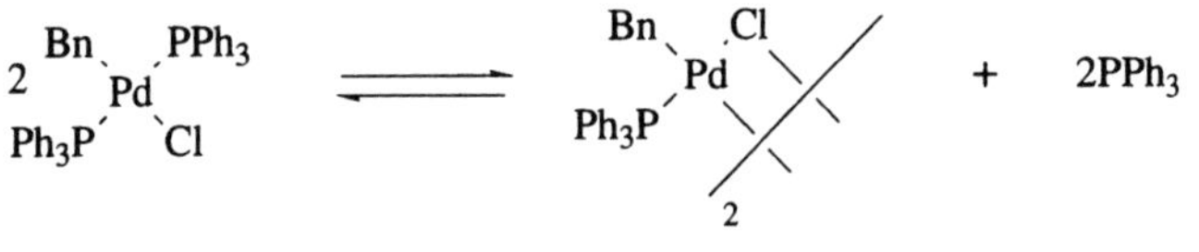

$$2\ Pd(Bn)(PPh_3)(Cl)(PPh_3) \rightleftharpoons [Pd(Bn)(Cl)(PPh_3)]_2 + 2PPh_3 \quad (83)$$

Although it is not certain by what mode added oxygen accelerates the catalytic process, it has been proposed that oxygen initiates a radical-based oxidative addition to palladium which is several orders of magnitude faster than the alternative nucleophilic displacement occurring largely under inert atmosphere.[177] Radical scavengers predictibly override the rate enhancement due to oxygen. It is possible that at least two mechanisms are active to some extent during oxidative addition.[178] One path involves nucleophilic addition, displacement of chloride with possible formation of a cationic metal complex, and collapse to $RCOPd(L_2)X$.[179] The other process may involve halide abstraction, electron transfer to form a radical pair (L_2Pd^IX + RCO·) which can collapse to $RCOPd(L_2)X$.[179] In this case oxygen can act as the electron acceptor to initiate the process. The latter proposition is based upon CIDNP effects observed during oxidative addition of platinum and palladium complexes to alkyl halides.[180] Latter stages of the process, including transmetallation and reductive elimination, are not enhanced. The stoichiometric reaction of isolated acylpalladium complex (**107**) (Scheme 43; R = Ph) with tetraalkyltins (transmetallation, *trans* to *cis* isomerization and reductive elimination) was actually shown to be retarded by the presence of oxygen.[181]

Originally, it was felt that a polar solvent such as HMPA was a requirement for the reaction; however, subsequent work has shown that even nonpolar solvents such as chloroform can be used with only a small decrease in yield and in reaction rate.[182] The work-up was made considerably simpler by this change and in some cases the reaction is made more selective over similar reactions using HMPA. Bromoaromatic compounds such as 4-bromobenzoyl chloride are converted to the acetophenone without replacement of the halogen. In HMPA, up to 26% of the 4-methylacetophenone is produced. No bromination or elimination was observed from substrates such as 6-bromohexanoyl chloride. It is known that cyclic ethers such as THF may be cleaved with acylpalladium(II) halide complexes in the absence of tetraalkyltins, but this solvent can still be used effectively for ketone synthesis since the rate of organotin acylation exceeds that of the degradation reaction.[183] A strong coordination of the acylpalladium complex by THF is indicated by these observations. Interestingly, solvent effects play a role in the selectivity of acylation when using allylic stannanes. In equations (84)–(86), the use of $CHCl_3$ rather than THF led to multiple addition products.[182]

O, Cl, O_2N + $Bu^n{}_3Sn$ — $BnPd(PPh_3)_2Cl$, $CHCl_3$, 65 °C → OH, O_2N (84)

O, Cl, O_2N + $Bu^n{}_3Sn$ — $Pd(PPh_3)_4$, THF, 65 °C, 83% → O, O_2N (85)

90:10 β,γ:α,β

O, Cl, O_2N + $Bu^n{}_3Sn$ — $BnPd(PPh_3)_2Cl$, THF, 50 °C → O, O_2N (86)

83:17 mixture with γ-isomer

The palladium(II) catalyst, because of its Lewis acidity, may play a role in the addition of allylic tin to the ketone; however, acylation of crotyltin was not reported to form a tertiary alcohol using palladium(II). It appears that solvent effects dominate in these cases. As part of the same study, substituted vinylstannanes were shown to undergo acylation with retention of configuration; however, the resulting α,β-unsaturated ketones were not configurationally stable to the reaction conditions. Isomerically pure (*Z*)-1-propenylstannane was acylated to afford a 50:50 mixture of alkenes (equation 87). The (*Z*)-α,β-unsaturated ketone was shown to isomerize to a mixture of (*Z*)- and (*E*)-isomers under the reaction conditions. Mixtures of (*Z*)- and (*E*)-2-substituted vinylstannanes were acylated to afford mainly the (*E*)-α,β-unsaturated ketone (equation 88).[184]

$$PhCOCl + Bu^n_3SnCH_2CH{=}CH_2 \xrightarrow[\text{THF, 65 °C, 74\%}]{BnPd(PPh_3)_2Cl} PhCOCH{=}CHCH_3\ (Z) + PhCOCH{=}CHCH_3\ (E) \quad (87)$$

$$PhCOCl + Bu^n_3SnCH{=}CHCH_2OSiMe_2Bu^t \xrightarrow[\text{THF, 65 °C, 78\%}]{BnPd(PPh_3)_2Cl} PhCOCH{=}CHCH_2OSiMe_2Bu^t \quad (88)$$

1.13.4.1.4 Transmetallation and reductive elimination

The acceleration of palladium-catalyzed acylation of organostannanes with polar solvents is thought to originate in the transmetallation and reductive elimination sequence. Kinetic studies have shown that the transmetallation of the tin substituent is rate limiting for reactions catalyzed by $BnPd^{II}(PPh_3)_2Cl$ and in the absence of excess triphenylphosphine. Support for transmetallation of the acylpalladium(II) chloride by the stannane, and direct evidence for the existence of the acylpalladium complex in solution was obtained from ^{31}P NMR using the reaction shown in equation (89).[185] After the reaction had proceeded for 1.5 h at 65 °C, the ^{31}P resonance due to $BnPd^{II}(PPh_3)_2Cl$ at δ 28.7 disappeared and a new resonance at δ 19.8 was observed. The latter correlated exactly with spectra of independently prepared PhCO-$Pd^{II}(PPh_3)_2Cl$.[186] The build up of this intermediate in the catalytic process suggests the transmetallation was rate limiting. Acylarylpalladium(II) complexes (**108**) and (**109**) (Scheme 43) were not observed spectroscopically, intimating that the reductive elimination was not a rate-limiting process. Also, it was not possible to observe the presence of the presumed palladium(0) catalyst $(PPh_3)_2Pd^0$ which demonstrates that oxidative addition was not rate limiting.

$$\underset{\text{2 equiv.}}{Bu^n_3SnPh} + \underset{\text{1 equiv.}}{PhCOCl} \xrightarrow[\text{65 °C, CDCl}_3]{\text{17 mol \% }BnPd(PPh_3)_2Cl} PhCOPh + Bu^n_3SnCl \quad (89)$$

The order of reactivity in the transmetallation of organostannanes to the acylpalladium(II) complex (**107**) has been found to be PhC≡C > PrC≡CH > PhCH=CH, CH_2=CH > Ar > allyl, benzyl > $MeOCH_2$ > Me > Bu.[185] The order of reactivity indicates that substantial amount of charge is borne by the migrating group and is consistent with electrophilic attack by the acylpalladium(II) complex. In most cases with diverse substituents attached to tin, alkyl groups will not be transferred and the better migrating ligand is always the one to be consumed. Alkyl groups can only be transferred with tetraalkylstannanes. Tetramethyltin is commercially available and has been used for the purpose of preparing methyl ketones. The tetraorganotin reagents transfer the first group rapidly but the second leaves about 100 times slower. Usually stoichiometric amounts of acid chloride and stannane are used.

As discussed earlier, the transmetallation segment occurs with retention of stereochemistry in the case of vinylstannanes, although the thermodynamic isomer usually predominates. Stille has provided evidence that transmetallation of alkyl substituents occurs with inversion of configuration based upon the acylation of (*S*)-(–)-(α-deuteriobenzyl)tributyltin ($[\alpha]^{20}_D$ –0.328° (neat) approx. 75% *ee*). Although the final product was sensitive to racemization, the authors could conclude that at least 65% stereospecificity was realized in this process. The last step, reductive elimination has been shown to proceed with retention of configuration (equation 90).[187]

$$PhCH(D)SnBu_3 \xrightarrow[\text{PhCOCl, HMPA, 65 °C}]{\text{4 mol \% }BnPd(PPh_3)_2Cl} PhCH(D)COPh \quad (90)$$

Reductive elimination from the acylorganopalladium(II) complex is generally a facile process. The rate is influenced by the solvent polarity and added triphenylphosphine.[188] Reductive elimination has been shown to be faster than the elimination of palladium hydride from intermediate (**111**) in both chloroform and HMPA (Scheme 46).[189] This makes the process useful even for alkyl group acylation.

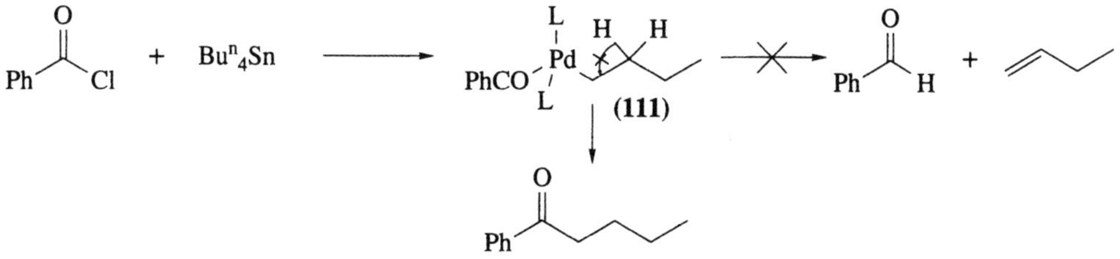

Ph–C(=O)–Cl + $Bu^n{}_4Sn$ ⟶ (111) ⟶ PhCOCH₂CH₂CH₂CH₃ (not PhCHO + butene)

Scheme 46

Kinetic studies indicate that reductive elimination is preceded by a dissociation step in which a phosphine ligand is removed. This places the migrating groups on adjacent sites for elimination from a tricoordinate species. Recall the rate retardation caused by addition of triphenylphosphine to a stoichiometric reaction of $RCOPd^{II}L_2Cl$ and R'_4Sn. This would be expected if ligand dissociation is a necessary step in *trans* to *cis* migration prior to reductive elimination (Scheme 47).[190]

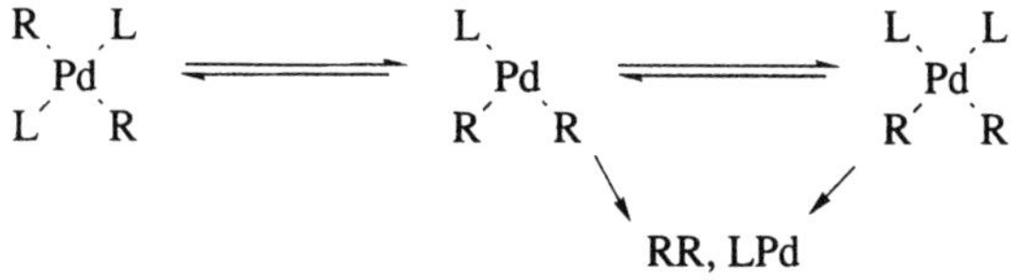

trans-PdR_2L_2 ⇌ PdR_2L ⇌ cis-PdR_2L_2 ⟶ RR, LPd

Scheme 47

Polar solvents may lower the activation energy needed for removal of a phosphine ligand by substitution with a weaker solvent ligand at the site of unsaturation. During reductive elimination, palladium obtains two electrons from the displaced groups, therefore strong σ-donation from these groups is more important than continued electron donation from a phosphine. Strongly donating phosphine ligands tend to reduce the rate of reductive elimination by keeping the electron density on the metal relatively high.[190] Although additional phosphine ligand was observed to slow the rate of reductive elimination by displacing the equilibrium shown in Scheme 47, the overall rate of the catalytic reaction is effected to a much greater extent, indicating a more substantial hindering effect during oxidative addition.[155] Beletskaya has shown that oxidative addition of palladium to the acid chloride is not dependent upon the σ-donation of phosphine ligands and advocates the use of ligandless palladium in these reactions.[175]

The end of the reaction is signified by the precipitation of metallic palladium as bis(triphenylphosphine)palladium(0) undergoes disproportionation in the absence of the acid chloride (equation 91).

$$2Pd(PPh_3)_2 \longrightarrow Pd(PPh_3)_4 + Pd \qquad (91)$$

1.13.4.2 Sources of Tetrasubstituted Stannanes

Aromatic, heterocyclic, alkynyl, alkenyl and alkyl stannanes have all been shown to be useful as the nucleophilic partner in palladium-catalyzed acylation.[191] The organostannanes are not water or air sensitive and may be prepared easily by one of several general strategies;[192] a few simple organotins are commercially available. The application of trimethyl- or tri-*n*-butyl-stannyl anions with organic electrophiles provides the most versatile approach to these derivatives; a wide range of functionality may be tolerated as part of the electrophilic substrate. Alternatively, trialkyltin halides or sulfonates may be reacted with common organometallic agents, including Grignard, organolithium and organoaluminum reagents. The functionality which is to be introduced into the organostannane by the latter method is limited by the choice of the organometallic; however, both approaches have seen a great deal of use. Once tin is incorporated into an organic substrate, selective manipulation of other functional groups on the molecule may subsequently be carried out. Selective reactions such as lithium aluminum hydride reduction or permanganate- and chromium-mediated oxidation have been demonstrated, as well as a variety of nucleophilic addition and acid–base chemistry. The organotins may be purified by distillation or silica gel chromatography. Other methods of organotin construction, which have been employed recently, include the free radical addition of organotin hydrides to substituted alkenes or alkynes, the palladium-catalyzed coupling of hexaalkyldistannanes with aryl or benzylic and allylic systems and the formation of α-stannylated ketones through reaction of trialkyltin amides with simple ketones.

1.13.4.3 Acylation of Organostannanes with Acid Chlorides

Several interesting uses of palladium-mediated acylation in organic synthesis have appeared recently. In the synthesis of the marine natural product diisocyanoadociane completed by Corey, formation of optically active enone ester (**112**) was accomplished under the conditions shown in equation (92).[193] Note that tetrakis(triphenylphosphine)palladium(0) was used as the catalyst and the reaction was conducted under an inert atmosphere without severely impeding its completion (2 h).

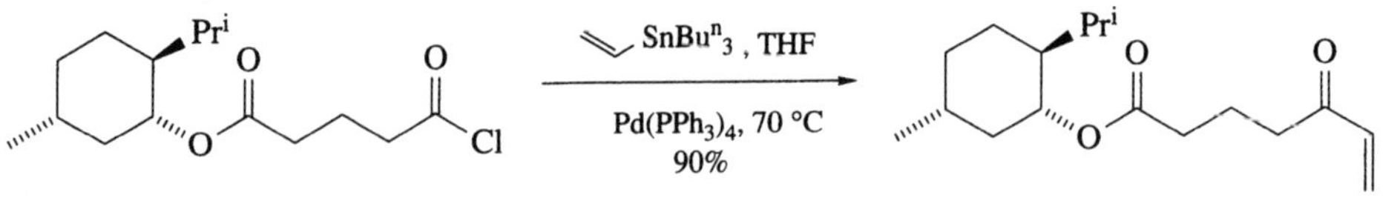

(92)

(112)

The conditions used by Stille provided an increased rate of conversion in a related synthesis of a steroid precursor (equation 93); however, HMPA was the solvent.[194] Enone (**113**) was previously synthesized in 40–44% yield by aluminum chloride mediated acylation of ethylene.[195]

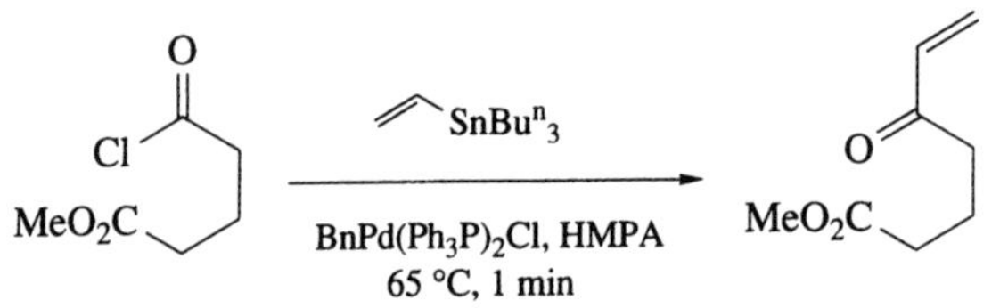

(93)

(113)

Preparation of the macrolide antibiotic pyrenophorin may be accomplished through Mitsunobu coupling of two identical fragments derived from (**114**; equation 94). Stille has completed a formal synthesis of this molecule with minimal protection of the precursor, through palladium-mediated acylation of organotins.[196,197] Several modifications of the original procedure[155] were incorporated as part of this synthesis to facilitate isolation of the product and to avoid side reactions caused by decomposition of the intermediate acylpalladium species. Replacement of the preferred solvent HMPA with chloroform caused a reduction in the reaction rate but the work-up was considerably easier. Reaction times in chloroform vary depending on the degree of substitution on the organotin. Unsubstituted vinyltins react in less than an hour; however, di- and tri-substituted vinylstannanes transfer the alkene group at progressively slower rates, ranging from 20 to 72 h, respectively. In the synthesis of enone ester (**114**) shown in equation (94), two charges of the stannane were required during the 30 h reaction period before complete consumption of the acid chloride was noted. A total of 1.6 equiv. of stannane were added. Additionally, carbon monoxide was necessary to reverse the observed decomposition of the intermediate acylpalladium complex; the latter modifications served to improve the yield by 40%.

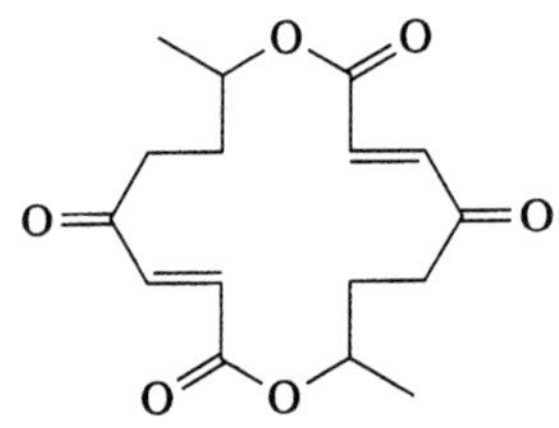

Pyrenophorin

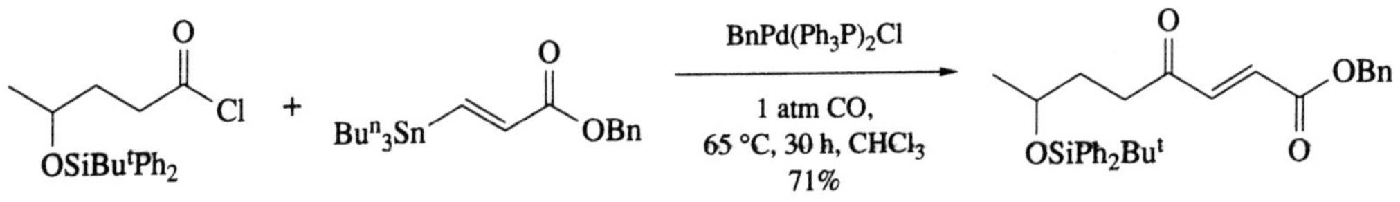

(94)

(114)

In Kende's formal total synthesis of the antitumor agent quadrone[198], a late stage intermediate, prior to closure of the third carbocyclic ring, was prepared using the method of Stille. The reaction appears to be exceedingly slow, even when HMPA is used as the solvent (equation 95). This points to the fact that transfer of simple alkyl groups from tin to the acylpalladium complex is not facile. No mention was made of alternative nucleophiles for this transformation such as dimethylcuprate or methylmagnesium chloride.

COCl; O — $BnPd(Ph_3P)_2Cl$ / 2 equiv. Me_4Sn, HMPA, 65 °C, 3 d, 82% → O; O (95)

In contrast, early in the synthesis of the hexahydrobenzofuran portion of the avermectins, Ireland reported that palladium-catalyzed acylation of tetramethyltin was the most effective method for preparing the required methyl ketone as shown in equation (96).[199] The sensitive 3,4-*O*-isopropylidene-L-threonyl chloride was converted in high yield to the corresponding methyl ketone without epimerization at C-3. To avoid decarbonylation, the reaction was run under a carbon monoxide atmosphere until completion (4 h).

O; COCl; O; $OCOBu^t$ — Me_4Sn, HMPA / $Pd(Ph_3P)_2(PhCH_2)Cl$, 25 °C, 87% → O; O; O; $OCOBu^t$ (96)

The acylation of organocadmium reagents with acid chlorides such as (**115**) formed an early method for synthesis of progesterone derivatives such as 21-methyl progesterone (equation 97). The same transformation may be accomplished more easily with palladium-catalyzed organotin acylation in 45% yield.[200]

COCl; O; (**115**) — Et_4Sn, HMPA / $Bn(Ph_3P)_2PdCl$, 80 °C, 45 min, 45% → O; O (97)

Holton has demonstrated that certain palladium(II) complexes can function as nucleophiles towards powerful acylating agents such as acetyl chloride.[201] These reactions proceed through the intermediacy of the palladium(IV) species (**116**) formed through oxidative addition as depicted in equation (98). Palladium(IV) intermediates had been proposed earlier to explain rate acceleration by reactive alkylating agents such as methyl iodide or benzyl bromide in metal-catalyzed carbon–carbon bond formation.[202] Logue has studied a substantial number of palladium-catalyzed acylations of (1-alkynyl)tributylstannanes.[203] Alkynylstannanes were acylated under very mild conditions which allowed a variety of functional groups to be present; however, in many cases the alkynylstannanes themselves had to be prepared from alkynyllithiums. Recall the earlier discussion of the acylation of alkynyllithiums by *N*-methoxy-*N*-methylamides (Section 1.13.2.1.1), and lactones (Section 1.13.2.7). In addition, it is also possible to acylate terminal alkynes directly with copper(I) iodide and palladium dichloride bistriphenylphosphine complex without the necessity of tin mediation.[204,214]

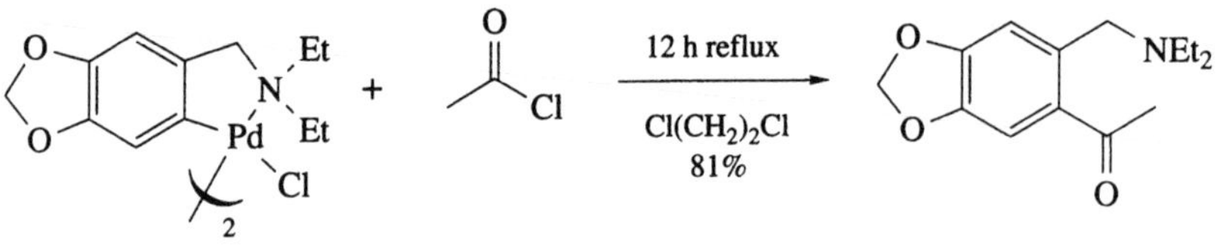

(98)

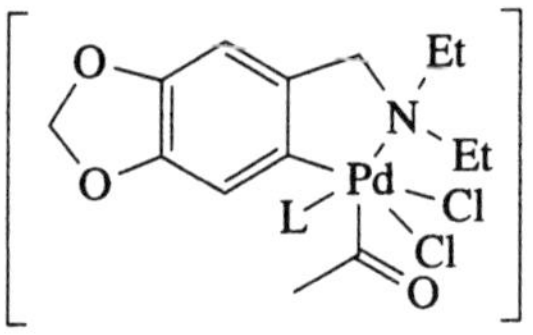

(116)

1.13.4.4 Palladium-catalyzed Acylation of Organozincs

Organometallics derived from the reaction of zinc chloride and Grignard reagents or organolithiums can be efficiently acylated by acid chlorides in the presence of palladium(0). The reaction is quite similar to the acylation of organostannanes but depending upon the case, may be easier to carry out for relatively simple alkyl organometallics which do not transfer well from tin. On the other hand, the need for an organometallic limits the functionality that can reside on the intended nucleophile. Unlike the case of organotins and acid chlorides there is some uncatalyzed acylation of zinc organometallics by the substrate; however, the latter process in and of itself is not useful. Organozinc reagents do not appear to add to ketones without Lewis acid activation.

Although not intensely studied, the mechanism of acylation with zinc reagents is expected to take the same course as that proposed for organotins. The most reactive catalyst, $(Ph_3P)_2Pd$ can be generated *in situ* from $(Ph_3P)_4Pd$ as discussed previously (Section 1.13.4.1.2) or by reduction of $Cl_2Pd^{II}(PPh_3)_2$ and $Cl_2Pd(DPPF)$ with diisobutylaluminum hydride.[206] Also Stille's catalyst, $BnPd^{II}(PPh_3)_2Cl$, is reported to be as effective for acylation under these conditions as with the organotins.[208] As with stannane acylation triphenylphosphine has a rate-retarding effect on oxidative addition of the acid chloride and would be expected to have a somewhat smaller rate-retarding effect on reductive elimination. Reactive halides such as benzylic bromides may be directly acylated with acid chlorides in the presence of zinc powder and palladium(0).[205] The most effective catalyst precursor was $Cl_2Pd^{II}(PPh_3)_2$ and the rate was decreased by suboptimal catalyst concentrations as was observed for organotins. Many of the examples have been substituted alkenyl or alkynyl zincs (equations 99–101). The acylation is >98% stereospecific for retention of configuration with disubstituted alkenes.[206]

The reaction was recently applied to the synthesis of 1,4- and 1,5-diketones (equation 102).[207] Perhaps not surprisingly, the β- and γ-ketozincs are stable to selfcondensation and proton abstraction. HMPA or some polar aprotic solvent must be used for high yields.

The corresponding dialkyl zinc reagents have also been used in acylation (equation 103). Under the reaction conditions with aromatic acid chlorides, substituted benzaldehydes are often a by-product. The

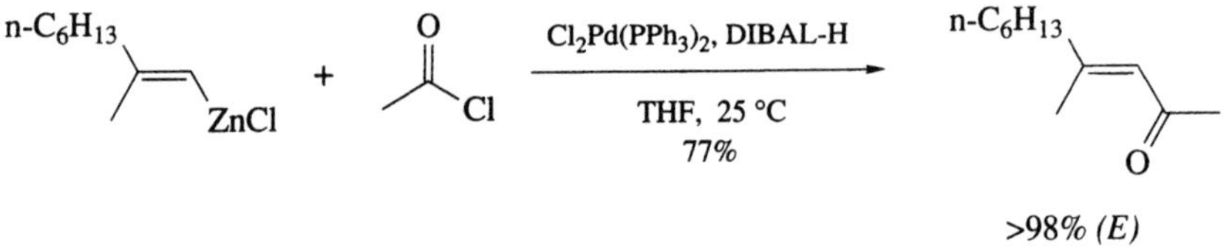

(99)

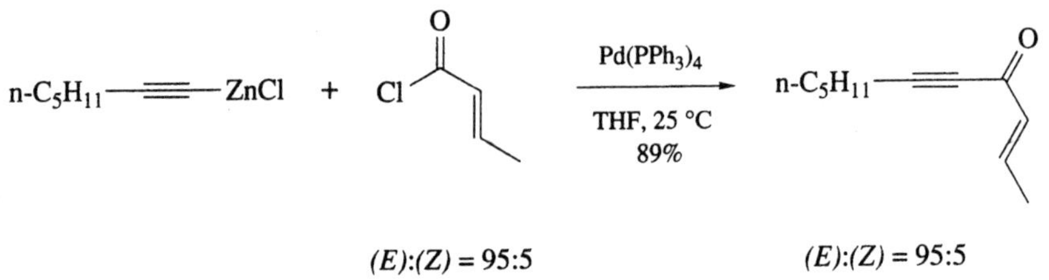

>98% *(E)*

(100)

(E):*(Z)* = 95:5 *(E)*:*(Z)* = 95:5

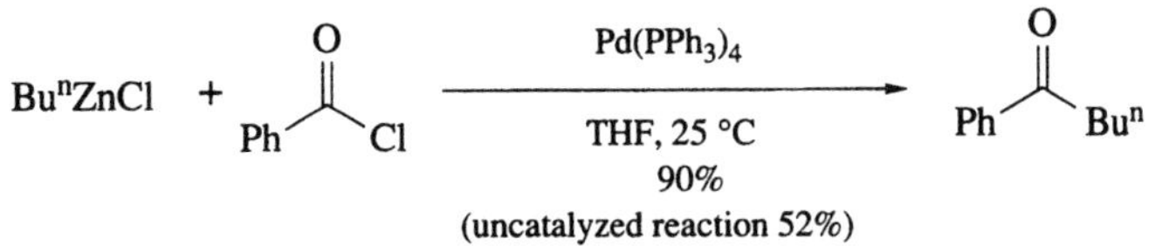

Bu^nZnCl + PhCOCl —[$Pd(PPh_3)_4$, THF, 25 °C; 90% (uncatalyzed reaction 52%)]→ PhCOBu^n (101)

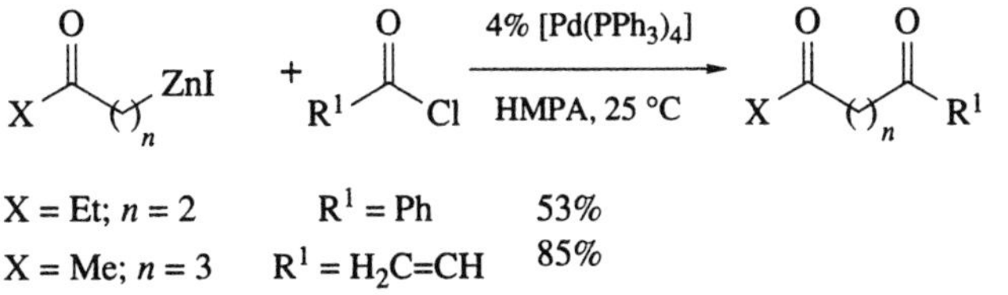

X–CO–$(CH_2)_n$–ZnI + R^1COCl —[4% [$Pd(PPh_3)_4$], HMPA, 25 °C]→ X–CO–$(CH_2)_n$–CO–R^1 (102)

X = Et; n = 2 R^1 = Ph 53%
X = Me; n = 3 R^1 = H_2C=CH 85%

yield of desired product can be maximized by using ether as the solvent and the bidentate DPPF ligand which prevents the elimination of palladium hydride. Alkyl acid chlorides do not require these special conditions in acylation.[208]

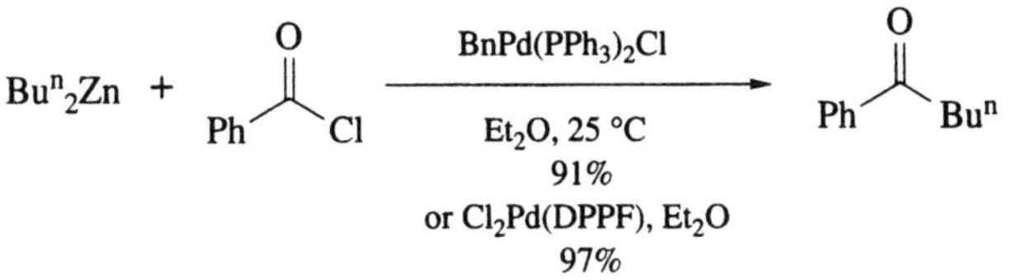

$Bu^n{}_2Zn$ + PhCOCl —[$BnPd(PPh_3)_2Cl$, Et_2O, 25 °C; 91%; or $Cl_2Pd(DPPF)$, Et_2O; 97%]→ PhCOBu^n (103)

Amino acid synthons can be prepared from iodoalanine with no loss of optical integrity (Scheme 48). The amino acid was transformed into a novel zinc reagent through reductive metallation with a zinc–copper couple in benzene/dimethyl acetamide. This organometallic was acylated under palladium catalysis in good overall yield.[209]

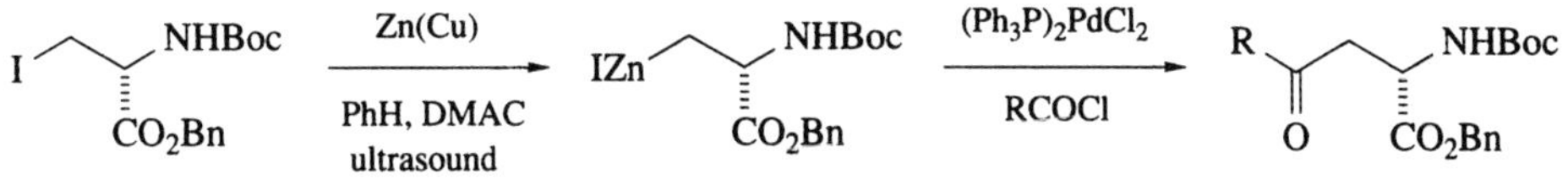

I–CH_2–CH(NHBoc)–CO_2Bn —[Zn(Cu), PhH, DMAC, ultrasound]→ IZn–CH_2–CH(NHBoc)–CO_2Bn —[$(Ph_3P)_2PdCl_2$, RCOCl]→ R–CO–CH_2–CH(NHBoc)–CO_2Bn

R = Ph, 70%; R = 2-furyl, 90%; R = Me, 80%; R = Bu^tCH_2, 84%

Scheme 48

In a related procedure, the Diels–Alder substrate (**118**) was prepared from the iodide (**117**) through reductive metallation with a zinc–copper couple followed by palladium-catalyzed acylation (Scheme 49).[210] This was a very rapid acylation in contrast to the related organotin-mediated coupling.

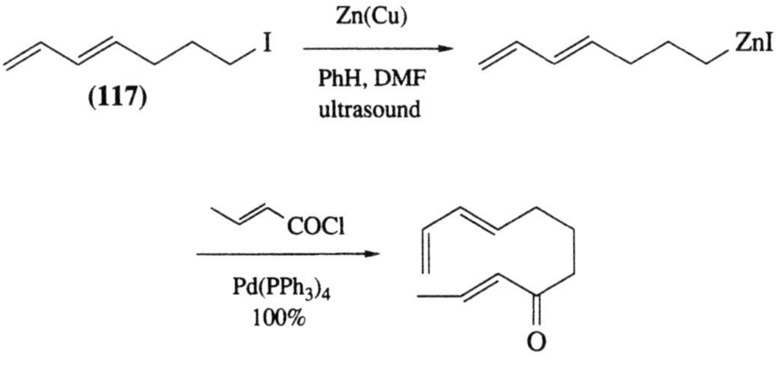

(**117**) —[Zn(Cu), PhH, DMF, ultrasound]→ dienyl–ZnI —[crotonyl COCl, $Pd(PPh_3)_4$; 100%]→ (**118**)

Scheme 49

1.13.4.5 Palladium-catalyzed Acylation of Organomercurials and other Organometallics

Many other types of organometallics which are not acylated directly by acid chlorides and which do not undergo addition to ketones may still transmetallate into the acylpalladium(II) complex. Simple alkyl organomercurials have been acylated in this fashion to give moderate to good yields of ketones.[211] Larock has studied the palladium-catalyzed acylation of vinylmercury(II) compounds with acyl halides (equation 104).[212] The reaction was only modestly productive and could not compare to the yield provided by aluminum chloride catalysis.

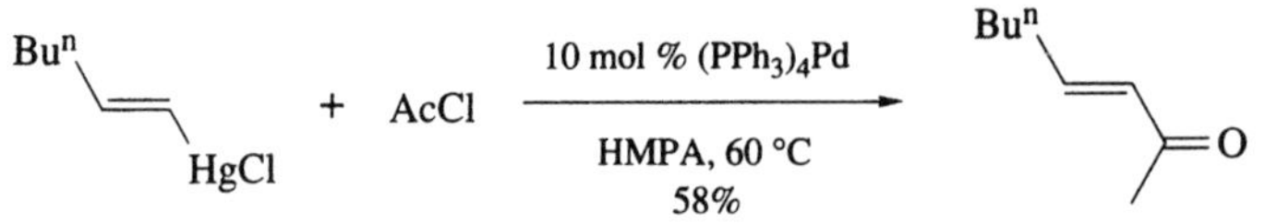

Bu^n HgCl + AcCl — 10 mol % $(PPh_3)_4Pd$, HMPA, 60 °C, 58% → Bu^n ... =O (104)

Organoaluminum reagents are known to react with ketones to form geminal dialkyl compounds, thus their use in acylation chemistry has been limited. For instance treatment of benzoyl chloride with triethylaluminum in THF gave less than 5% of propiophenone. Surprisingly, the corresponding palladium(0)-catalyzed process afforded this ketone in 70% yield with 5 mol % $Pd^0(PPh_3)_4$ (equation 105). The rate of acylation must be substantially faster than addition of the trialkylaluminum reagent to the ketone. Transmetallation of the organoaluminum with the acylpalladium(II) intermediate must also be facile. Palladium catalysts generated *in situ* from $Pd^{II}(OAc)_2$ and 2 equiv. PPh_3 were effective, as were other sources including $Cl_2Pd^{II}(PPh_3)_2$.[213] Presumably, the actual catalyst, $Pd^0(PPh_3)_2$, was generated by sacrificial transmetallation of a portion of the organoaluminum as in the analogous case of organotins (Section 1.13.4.1.2).

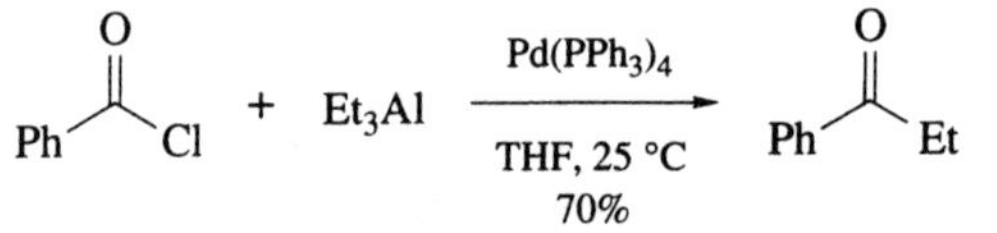

PhCOCl + Et_3Al — $Pd(PPh_3)_4$, THF, 25 °C, 70% → PhCOEt (105)

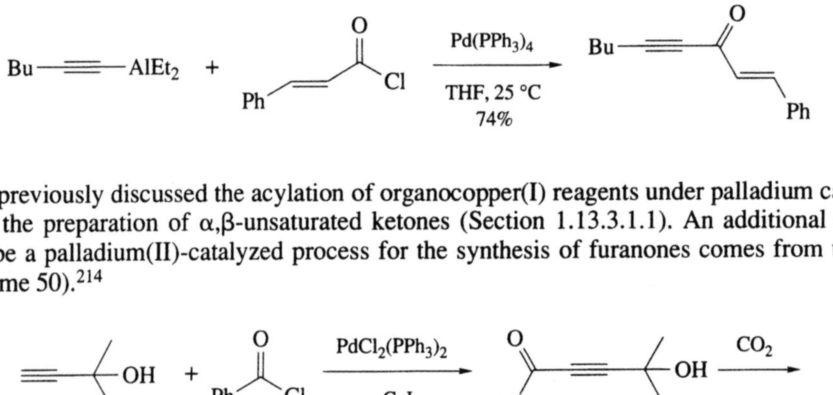

Bu—≡—$AlEt_2$ + PhCH=CHCOCl — $Pd(PPh_3)_4$, THF, 25 °C, 74% → Bu—≡—COCH=CHPh (106)

We have previously discussed the acylation of organocopper(I) reagents under palladium catalysts as a method for the preparation of α,β-unsaturated ketones (Section 1.13.3.1.1). An additional example of what must be a palladium(II)-catalyzed process for the synthesis of furanones comes from the work of Inoue (Scheme 50).[214]

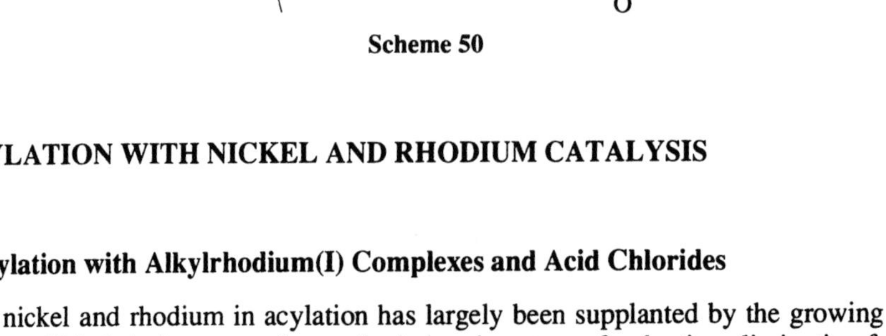

OH + PhCOCl — $PdCl_2(PPh_3)_2$, CuI → PhCO—≡—C(Me)$_2$OH — CO_2 →

Ph, O, O, =O, O — $-CO_2$, Δ, 58% overall → O, Ph, O

Scheme 50

1.13.5 ACYLATION WITH NICKEL AND RHODIUM CATALYSIS

1.13.5.1 Acylation with Alkylrhodium(I) Complexes and Acid Chlorides

The use of nickel and rhodium in acylation has largely been supplanted by the growing use of palladium complexes. This is in large part due to the lethargic nature of reductive elimination from nickel(II) complexes and the need for stoichiometric amounts of rhodium in the original protocol. Nevertheless, an exceedingly mild method for the acylation of simple primary alkyl-, aryl- or allyl-lithiums (or Grignard reagents) by acid halides is achieved through the prior transformation of these organometallics into substituted rhodium(I) complexes.[217] Although this method has not found widespread acceptance in the literature, it was one of the first examples of transition metal mediated organic synthesis applied to the preparation of ketones. Substrates complicated with extensive functionality and which are sensitive to strongly basic conditions can be acylated in much the same way that palladium mediates acylation of a variety of nucleophiles. For this reason, a short discussion of the applications of this technology are included.

Although the scope of nucleophilic reagents that can be used in this approach is not exceedingly broad, many different functional groups such as aldehydes, esters and nitriles as well as α-chloro ketones are tolerated as part of the substrate molecule. The conditions are mild and the rhodium(I) complex is returned unchanged from the reaction mixture. The addition of allyllithiums or Grignard reagents offers a selective method for the preparation of β,γ-unsaturated ketones. As mentioned, it is necessary to employ stoichiometric amounts of a rhodium complex, usually in the form of chloro(carbonyl)bis(triphenylphosphine)rhodium(I).[215] This is because of the high nucleophilicity of the organometallics chosen. The reaction is thought to proceed by addition of the nucleophile to form rhodium(I) complex (**119**) which then oxidatively adds to the acid chloride forming rhodium(III) complex (**120**). Reductive elimination releases the ketone and returns the original form of the rhodium complex, ready for reuse (Scheme 51). Attempts to isolate the intermediate rhodium(I) or rhodium(III) complex have thus far failed; however, the reaction may be monitored by infrared spectroscopy. A clear transformation of the initial rhodium complex (**121**) to the alkylrhodium(I) complex (**119**) was observed by a shift of the carbonyl absorption to lower energy, expected of the more electron-rich complex. It was not possible to observe the rhodium(III) complex (**120**) by this method. After addition of the acid chloride, infrared bands for the expected carbonyl product and for the starting rhodium(I) complex become evident. The expectation of a rhodium(III) intermediate was based upon other work in this area, but still remains a supposition. Oxidative addition to the acid chloride was not observed to take place with complex (**121**) and formation of the alkyl complex (**119**) increases the ability of the metal to add oxidatively by electron donation. Other more nucleophilic forms of rhodium such as $Rh(PMe_2Ph)_3Cl$ are known to oxidatively add to acid chlorides but have not been used in this acylation process.[216]

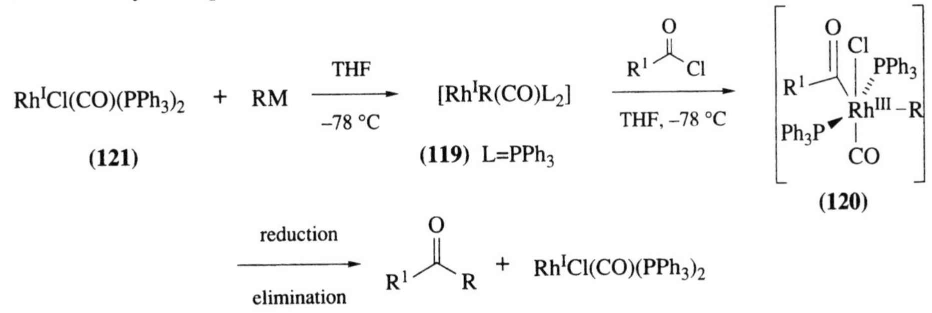

$Rh^{I}Cl(CO)(PPh_3)_2$ + RM —(THF, –78 °C)→ $[Rh^{I}R(CO)L_2]$ —(R^1COCl, THF, –78 °C)→ [$R^1C(O)Rh^{III}(Cl)(PPh_3)_2(CO)R$]

(**121**) (**119**) L=PPh_3 (**120**)

—(reduction elimination)→ R^1COR + $Rh^{I}Cl(CO)(PPh_3)_2$

Scheme 51

Some examples of this method are given in Table 11.[217] Most of the entries proceed in about 60–80% yield, with the exception of secondary and tertiary alkyllithiums. In these cases, the facility of β-hydride elimination in the intermediate alkylrhodium complex predominates over oxidative addition to the acid halide. One example of formation of an optically active ketone proceeded without racemization at the α-center.

Table 11 Acylation of Alkylrhodium Complexes with Acid Chlorides

Acid chloride	*RM*	*Product*	*Yield (%)*
n-$C_{11}H_{23}COCl$	MeLi	n-$C_{11}H_{23}CO$	77
n-$C_{11}H_{23}COCl$	MeMgBr	n-$C_{11}H_{23}COMe$	58
PhCOCl	PhLi	PhCOPh	94
PhCOCl	CH_2=$CHCH_2MgBr$	$PhCOCH_2CH$=CH_2	71
trans-PhCH=CHCOCl	MeLi	*trans*-PhCH=CHCOMe	68
$ClCH_2COCl$	Bu^nLi	$ClCH_2COBu^n$	55
(*S*)-(+)-PrCH(Me)COCl	EtLi	(*S*)-(+)-PrCH(Me)COEt	83
PhCOCl	EtCH(Me)Li	EtCH(Me)COPh	3

In an extension of this work, Pittman has found that the rhodium(I) complex may be anchored to polystyrene resin thus enabling facile catalyst regeneration (Scheme 52). Unfortunately, certain acid chlorides such as *p*-nitrobenzoyl do not oxidatively add to the rhodium(I) complex at an appreciable rate; other unreactive substrates include *p*-methoxybenzoyl and some hindered acid chlorides.[218]

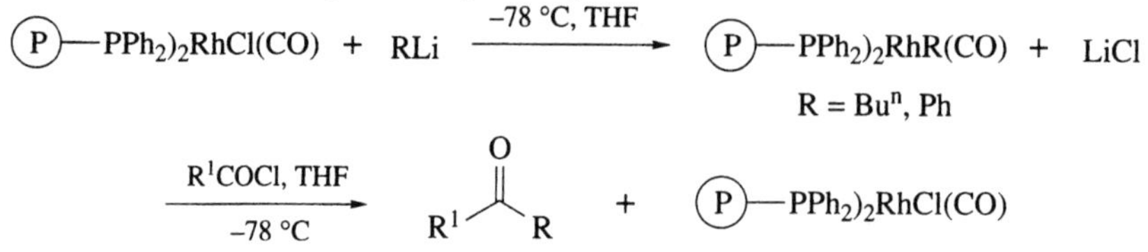

$$\text{(P)}\!-\!PPh_2)_2RhCl(CO) + RLi \xrightarrow{-78\ °C,\ THF} \text{(P)}\!-\!PPh_2)_2RhR(CO) + LiCl$$

R = Bu^n, Ph

$$\xrightarrow[-78\ °C]{R^1COCl,\ THF} R^1C(O)R + \text{(P)}\!-\!PPh_2)_2RhCl(CO)$$

$R^1 = MeO_2C(CH_2)_4COCl$ R = Ph, 56%

$R^1 = m\text{-}NCC_6H_4COCl$ R = Bu^n, 60%

Scheme 52

The advantage of the previously described palladium-mediated acylation of organotins actually lies in the fact that the tin reagents do not directly react with the acid chlorides in the same way as organolithiums and Grignard reagents. The substituent on tin readily transmetallates or acts as the nucleophile towards the acylated palladium(II) complex which then undergoes rapid reductive elimination to form the product and reform the palladium(0) catalyst. Since organolithiums and magnesiums are not compatible with acid halides, stoichiometric quantities of the rhodium complex are necessary to preform the alkylrhodium(I) complex. To make the rhodium-mediated process catalytic in metal, Migita and coworkers demonstrated that the combination of allyltins and acid chlorides will afford ketones in good yield using 2 mol % $ClRh^I(PPh_3)_3$ (equations 107 and 108).[219] They propose the mechanism of this process to involve addition of the allyltin to the rhodium(I) complex, which then suffers oxidative addition with the acid chloride. An alternative possibility begins with oxidative addition of the acid chloride to the rhodium(I) complex, followed by transmetallation of the allyltin, and reductive elimination. In equation (108) introduction of the tin substituent occurs without allylic rearrangement, as is also the case with palladium catalysis; however, compared to the methodology developed for the corresponding palladium-catalyzed process, the use of rhodium in acylation has not been shown to be advantageous.

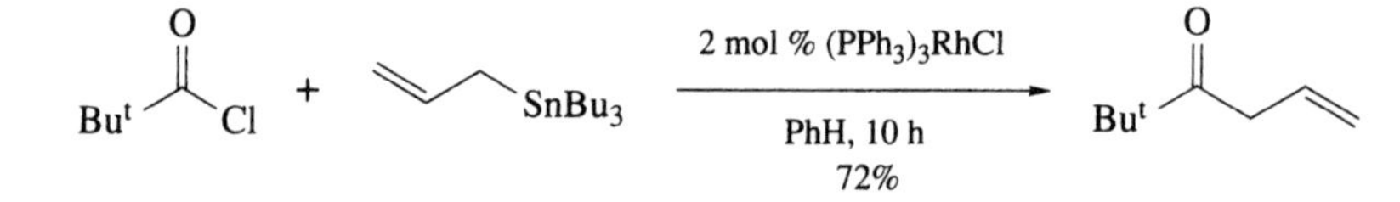

$$Bu^tCOCl + CH_2{=}CHCH_2SnBu_3 \xrightarrow[PhH,\ 10\ h,\ 72\%]{2\ mol\ \%\ (PPh_3)_3RhCl} Bu^tCOCH_2CH{=}CH_2 \quad (107)$$

$$EtCOCl + CH_3CH{=}CHCH_2SnBu_3 \xrightarrow[PhH,\ 12\ h,\ 64\%]{2\ mol\ \%\ (PPh_3)_3RhCl} EtCOCH_2CH{=}CHCH_3 \quad (108)$$

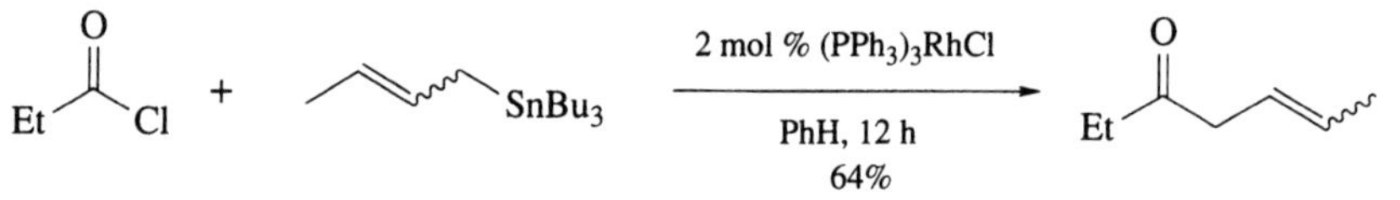

1.13.5.2 Acylation by Organonickel Complexes

Only scattered examples of the use of nickel salts in acylation have been reported in the past few years. Marchese and coworkers in a series of papers have discussed their discovery that Ni^{II} complexes such as $Ni(DPPE)Cl_2$ will moderate the acylation of Grignard reagents to afford ketones (see also Sections 1.13.2.5 and 1.13.2.6). In a particularly interesting synthesis of 1,4-diketones or 1,4-keto aldehydes, these workers selectively monoacylated the Grignard reagent derived from 2-(2-bromoethyl)-1,3-dioxolane with *S*-phenyl carbonochloridothioate using catalytic amounts of nickel(II) (Scheme 53). Subsequently, a second equivalent of another Grignard reagent was added to the resulting product, this time with mediation from iron(III) acetylacetonate.

Each step of the process proceeds in high yield without formation of tertiary alcohol by-products. The reaction can be run at 0 °C in THF, which is somewhat of an advantage compared to the low temperatures normally required in Grignard acylations. Both steps can also be conducted in one pot, although yields are higher if the intermediate thiol ester is isolated before the next step.[220]

Mukaiyama has described the use of nickel(II) salts as catalysts for the acylation of weakly nucleophilic organozincs. The advantage of this methodology is that the zinc reagent is prepared *in situ* from

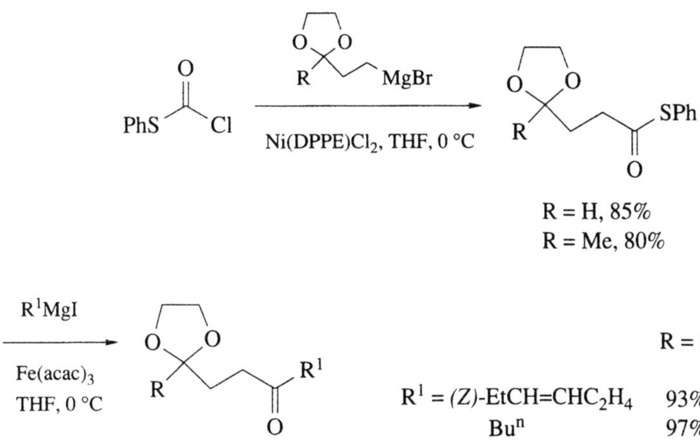

PhS(C=O)Cl → [2-(2-R-1,3-dioxolan-2-yl)ethyl]MgBr, $Ni(DPPE)Cl_2$, THF, 0 °C → dioxolane–CH_2CH_2–C(=O)SPh

R = H, 85%
R = Me, 80%

→ R^1MgI, $Fe(acac)_3$, THF, 0 °C → dioxolane–CH_2CH_2–C(=O)R^1

	R = Me	R = H
R^1 = *(Z)*-EtCH=CHC_2H_4	93%	
Bu^n	97%	
Bu^s	93%	90%
Ph	75%	73%

Scheme 53

the corresponding alkyl iodide (equation 109). Many types of functional groups are tolerated, including ketones, esters, chlorides and α,β-unsaturated carboxylic acid derivatives. Only 10 mol % of nickel catalyst is needed for this acylation.[221]

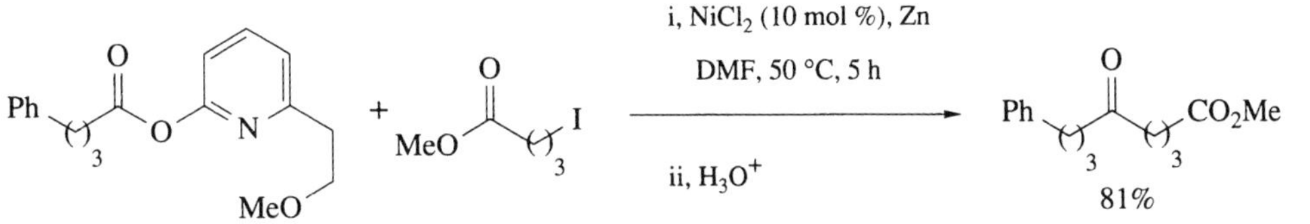

$$\mathrm{Ph(CH_2)_3CO_2\text{-}[6\text{-}(2\text{-methoxyethyl})\text{-}2\text{-pyridyl}]} + \mathrm{MeO_2C(CH_2)_3I} \xrightarrow[\text{ii, } H_3O^+]{\text{i, } NiCl_2 \text{ (10 mol \%), Zn, DMF, 50 °C, 5 h}} \mathrm{Ph(CH_2)_3CO(CH_2)_3CO_2Me} \quad 81\% \qquad (109)$$

Generally only primary and some secondary iodides have been useful in this process; tertiary iodides do not react. It was also observed that no reaction occurred between the alkyl iodide and zinc unless the nickel catalyst was present. Several carboxylic acid derivatives were tested as acylating agents; however, the best yields were afforded by the 2-[6-(2-methoxyethyl)pyridyl] carboxylate shown in equation (109). Presumably, the higher degree of coordination attainable between this functionality and the organometallic, activates the acylating agent towards nucleophilic addition. This important effect has been seen in several other acylating agents including the *S*-(2-pyridyl) thiol ester. Both the *S*-(2-pyridyl) thiol ester and the 2-pyridyl carboxylic ester gave lower yields in this reaction.[221]

β,γ-Unsaturated ketones have been prepared in moderate yield through the acylation of π-allylnickel complexes with activated 2-pyridyl carboxylates. But, isomerization of the initially formed unconjugated alkene resulted in mixtures of products and limited the value of the method. Substituted π-allylnickel complexes derived from crotyl bromide or cinnamyl bromide were acylated in 79% and 50% yields, respectively, without formation of the α,β-unsaturated ketone.[222]

Rieke and coworkers have found that a special type of activated metallic nickel, available through reduction of nickel(II) iodide with lithium metal, suffers oxidative addition of benzylic and allylic halides. The resulting nickel(II) complexes readily undergo cross-coupling with acid chlorides to form ketones. Once again it was difficult to obtain β,γ-unsaturated ketones from this method. Moderate to good yields of simple ketones may be prepared by this method.[223]

1.13.6 REFERENCES

1. G. H. Posner, 'An Introduction to Synthesis Using Organocopper Reagents', Wiley, New York, 1980.
2. M. J. Jorgenson, *Org. React. (N. Y.)*, 1970, **18**, 1.
3. For one recent application toward the total synthesis of Juvabione see D. A. Evans and J. V. Nelson, *J. Am. Chem. Soc.*, 1980, **102**, 774; for a review see D. A. Shirley, *Org. React. (N. Y.)*, 1954, **8**, 28.
4. S. Nahm and S. M. Weinreb, *Tetrahedron Lett.*, 1981, **22**, 3815.
5. Commercially available from Aldrich Chemical Co., Milwaukee, WI, or can be prepared by the procedure of O. P. Goel and U. Krolls, *Org. Prep. Proced. Int.*, 1987, **19**, 75.

6. P. D. Theisen and C. H. Heathcock, *J. Org. Chem.*, 1988, **53**, 2374.
7. O. P. Goel, U. Krolls, M. Stier and S. Kesten, *Org. Synth.*, 1988, **67**, 69.
8. D. J. Hlasta and J. J. Court, *Tetrahedron Lett.*, 1989, **30**, 1773.
9. D. A. Evans, T. C. Britton and J. A. Ellman, *Tetrahedron Lett.*, 1987, **28**, 6141 and references cited therein.
10. A. Basha, M. Lipton and S. M. Weinreb, *Tetrahedron Lett.*, 1977, 4171; J. Levin, E. Turos and S. M. Weinreb, *Synth. Commun.*, 1982, **12**, 989; A. Basha, M. Lipton and S. M. Weinreb, *Org. Synth.*, 1979, **59**, 49.
11. For recent work see (a) D. A. Evans, E. B. Sjogren, J. Bartroli and R. L. Dow, *Tetrahedron Lett.*, 1986, **27**, 4957; (b) D. A. Evans, E. B. Sjogren, A. E. Weber and R. E. Conn, *Tetrahedron Lett.*, 1987, **28**, 39 and references cited therein.
12. D. A. Evans, S. L. Bender and J. Morris, *J. Am. Chem. Soc.*, 1988, **110**, 2506.
13. S. Kaldor, Ph.D. Thesis, Harvard University, 1989.
14. D. A. Evans, S. Kaldor, T. K. Jones, J. Clardy and T. J. Stout, *J. Am. Chem. Soc.*, 1990, **112**, 70.
15. D. A. Evans and S. L. Bender, *Tetrahedron Lett.*, 1986, **27**, 799.
16. D. A. Evans, K. DeVries and D. Ginn, unpublished results, Harvard University.
17. S. Mills, R. Desmond, R. A. Reamer, R. P. Volante and I. Shinkai, *Tetrahedron Lett.*, 1988, **29**, 281; T. K. Jones, S. Mills, R. A. Reamer, D. Askin, R. Desmond, R. P. Volante and I. Shinkai, *J. Am. Chem. Soc.*, 1989, **111**, 1157.
18. I. Patterson and M. A. Lister, *Tetrahedron Lett.*, 1988, **29**, 585.
19. J. D. White, G. Nagabhushana-Reddy and G. O. Spessard, *J. Am. Chem. Soc.*, 1988, **110**, 1624.
20. R. L. Danheiser, J. M. Morin, Jr. and E. J. Salaski, *J. Am. Chem. Soc.*, 1985, **107**, 8066.
21. (+)-Actinobolin: R. S. Garigipati, D. M. Tschaen and S. M. Weinreb, *J. Am. Chem. Soc.*, 1985, **107**, 7790; (–)-Bactobolin: R. S. Garigipati and S. M. Weinreb, *J. Org. Chem.*, 1988, **53**, 4143.
22. J. S. Prasad and L. S. Liebeskind, *Tetrahedron Lett.*, 1987, **28**, 1857.
23. R. M. Williams, P. P. Ehrlich, W. Zhai and J. Hendrix, *J. Org. Chem.*, 1987, **52**, 2615.
24. W. Oppolzer and A. F. Cunningham, Jr., *Tetrahedron Lett.*, 1986, **27**, 5467.
25. M. M. Midland, A. Tramontano, A. Kazubski, R. S. Graham, D. J. S. Tsai and D. B. Cardin, *Tetrahedron*, 1984, **40**, 1371 and references cited therein.
26. S. J. Hecker and C. H. Heathcock, *J. Org. Chem.*, 1985, **50**, 5159.
27. B. M. Trost and T. Schmidt, *J. Am. Chem. Soc.*, 1988, **110**, 2301.
28. T. L. Cupps, R. H. Boutin and H. Rapoport, *J. Org. Chem.*, 1985, **50**, 3972.
29. P. A. Jacobi, C. S. R. Kaczmarek and U. E. Udodong, *Tetrahedron*, 1987, **43**, 5475 and references cited therein.
30. T. Mukaiyama, M. Araki and H. Takei, *J. Am. Chem. Soc.*, 1973, **95**, 4763.
31. T. Endo, S. Ikenaga and T. Mukaiyama, *Bull. Chem. Soc. Jpn.*, 1970, **43**, 2632.
32. M. Araki, S. Sakata, H. Takei and T. Mukaiyama, *Bull. Chem. Soc. Jpn.*, 1974, **47**, 1777.
33. M. Araki, S. Sakata, H. Takei and T. Mukaiyama, ref. 32.
34. For a similar rationalization see T. Sakan and Y. Mori, *Chem. Lett.*, 1972, 793.
35. E. J. Corey, S. Kim, S.-e. Yoo, K. C. Nicolaou, L. S. Melvin, Jr., D. J. Brunelle, J. R. Falck, E. J. Trybulski, R. Lett and P. W. Sheldrake, *J. Am. Chem. Soc.*, 1978, **100**, 4620.
36. K. C. Nicolaou, D. A. Claremon and D. P. Papahatjis, *Tetrahedron Lett.*, 1981, **22**, 4647.
37. K. C. Nicolaou, D. A. Claremon, D. P. Papahatjis and R. L. Magolda, *J. Am. Chem. Soc.*, 1981, **103**, 6969.
38. R. K. Boeckman, Jr., A. B. Charette, T. Asberom and B. H. Johnston, *J. Am. Chem. Soc.*, 1987, **109**, 7553.
39. Y. Nakahara, A. Fujita, K. Beppu and T. Ogawa, *Tetrahedron*, 1986, **42**, 6465.
40. J.-G. Gourcy, M. Prudhomme, G. Dauphin and G. Jeminet, *Tetrahedron Lett.*, 1989, **30**, 351.
41. G. R. Martinez, P. A. Grieco and C. V. Srinivasan, *J. Org. Chem.*, 1981, **46**, 3760.
42. For a comprehensive review see M. J. Jorgenson, *Org. React. (N. Y.)*, 1970, **18**, 1.
43. S. Masamune, W. Choy, F. A. J. Kerdesky and B. Imperiali, *J. Am. Chem. Soc.*, 1981, **103**, 1566; S. Masamune, S. Asrof-Ali, D. L. Snitman and D. S. Garvey, *Angew. Chem., Int. Ed. Engl.*, 1980, **19**, 557.
44. S. Masamune, T. Kaiho and D. S. Garvey, *J. Am. Chem. Soc.*, 1982, **104**, 5521.
45. H. O. House and T. M. Bare, *Org. Synth., Coll. Vol.*, 1973, **5**, 775; H. O. House and T. M. Bare, *J. Org. Chem.*, 1968, **33**, 943; R. Levine, M. J. Karten and W. M. Kadunce, *J. Org. Chem.*, 1975, **40**, 1770; R. Levine and M. J. Karten, *J. Org. Chem.*, 1976, **41**, 1176; D. E. Nicodem and M. L. P. F. C. Marchiori, *J. Org. Chem.*, 1981, **46**, 3928; R. G. Riley and R. M. Silverstein, *Tetrahedron*, 1974, **30**, 1171.
46. I. A. Blair, L. N. Mander, P. H. C. Mundill and S. G. Pyne, *Aust. J. Chem.*, 1981, **34**, 1887.
47. G. M. Rubottom and C. Kim, *J. Org. Chem.*, 1983, **48**, 1550.
48. M. P. Cooke, Jr., *J. Org. Chem.*, 1986, **51**, 951.
49. See, for example, D. Seyferth and R. M. Weinstein, *J. Am. Chem. Soc.*, 1982, **104**, 5534; D. Seyferth and K. R. Wursthorn, *J. Organomet. Chem.*, 1979, **182**, 455; A. E. Bey and D. R. Weyenberg, *J. Org. Chem.*, 1966, **31**, 2036.
50. B. J. Wakefield, 'Organolithium Methods', Academic Press, New York, 1988 and references cited therein.
51. G. Stork, G. Clark and C. S. Shiner, *J. Am. Chem. Soc.*, 1981, **103**, 4948.
52. J. C. Floyd, *Tetrahedron Lett.*, 1974, 2877.
53. M. Natsume and M. Ogawa, *Heterocycles*, 1980, **14**, 615.
54. W. E. Parham and C. K. Bradsher, *Acc. Chem. Res.*, 1982, **15**, 300.
55. E. J. Corey and Y. B. Xiang, *Tetrahedron Lett.*, 1988, **29**, 995.
56. L. A. Paquette and W. H. Ham, *J. Am. Chem. Soc.*, 1987, **109**, 3025.
57. V. Fiandanese, G. Marchese and L. Ronzioni, *Tetrahedron Lett.*, 1983, **24**, 3677.
58. C. G. Knudsen and H. Rapoport, *J. Org. Chem.*, 1983, **48**, 2260; T. F. Buckley and H. Rapoport, *J. Am. Chem. Soc.*, 1981, **103**, 6157.
59. P. J. Maurer, C. G. Knudsen, A. D. Palkowitz and H. Rapoport, *J. Org. Chem.*, 1985, **50**, 325.
60. P. J. Maurer, H. Takahata and H. Rapoport, *J. Am. Chem. Soc.*, 1984, **106**, 1095.
61. F. Sato, M. Inoue, K. Ogura and M. Sato, *Tetrahedron Lett.*, 1979, 4303.
62. M. K. Eberle and G. G. Kahle, *Tetrahedron Lett.*, 1980, **21**, 2303.

63. S.-i. Kiyooka, Y. Sekimura and K. Kawaguchi, *Synthesis*, 1988, 745.
64. S. A. Bal, A. Marfat and P. Helquist, *J. Org. Chem.*, 1982, **47**, 5045.
65. M. Sworin and W. L. Neumann, *Tetrahedron Lett.*, 1987, **28**, 3217.
66. F. A. Cotton and G. Wilkinson, 'Advanced Inorganic Chemistry', 5th edn., Wiley-Interscience, New York, 1988, p. 158.
67. (a) C. Cardellicchio, V. Fiandanese, G. Marchese and L. Ronzioni, *Tetrahedron Lett.*, 1987, **28**, 2053; (b) V. Fiandanese, G. Marchese, V. Martina and L. Ronzioni, *Tetrahedron Lett.*, 1984, **25**, 4805.
68. K. Ritter and M. Hanack, *Tetrahedron Lett.*, 1985, **26**, 1285.
69. I. Kikkawa and T. Yorifuji, *Synthesis*, 1980, 877; as an alternative mechanism, the ester may be deprotonated, followed by elimination to form the *ketene*, addition of the second equivalent of reagent results in the formation of the ketone enolate.
70. (a) C. Fehr and J. Galindo, *Helv. Chim. Acta*, 1986, **69**, 228; (b) C. Fehr, J. Galindo and R. Perret, *Helv. Chim. Acta*, 1987, **70**, 1745; (c) C. Fehr and J. Galindo, *J. Org. Chem.*, 1988, **53**, 1828.
71. D. L. Comins and J. J. Herrick, *Tetrahedron Lett.*, 1984, **25**, 1321.
72. S. D. Burke, D. N. Deaton, R. J. Olsen, D. M. Armistead and B. E. Blough, *Tetrahedron Lett.*, 1987, **28**, 3905.
73. R. Haner, T. Laube and D. Seebach, *J. Am. Chem. Soc.*, 1985, **107**, 5396; ketenes can also be prepared by reduction of α-bromoacyl bromides and treated with various alkyllithiums with similar results, L. M. Baigrie, H. R. Seiklay and T. T. Tidwell, *J. Am. Chem. Soc.*, 1985, **107**, 5391.
74. For a similar strategy see L. M. Bairgrie, R. Leung-Toung and T. T. Tidwell, *Tetrahedron Lett.*, 1988, **29**, 1673; C. Fehr and J. Galindo, *J. Am. Chem. Soc.*, 1988, **110**, 6909.
75. H. Ogura, H. Takahashi and T. Itoh, *J. Org. Chem.*, 1972, **37**, 72.
76. J. C. Chabala and J. E. Vincent, *Tetrahedron Lett.*, 1978, **19**, 937.
77. J.-M. Lancelin, P. H. A. Zollo and P. Sinay, *Tetrahedron Lett.*, 1983, **24**, 4833.
78. M. Yamaguchi and I. Hirao, *Tetrahedron Lett.*, 1983, **24**, 391.
79. S. Hanessian, A. Ugolini and M. Therien, *J. Org. Chem.*, 1983, **48**, 4427. Later reports on the total synthesis of avermectin B_{1a} indicated that the yield of this transformation had been improved to 82% by using the lithium salt of the alkyne. No further comment was made by the authors. S. Hanessian, A. Ugolini, P. J. Hodges, P. Beaulieu, D. Dube and C. Andre, *Pure Appl. Chem.*, 1987, **59**, 299; S. Hanessian, A. Ugolini, D. Dube, P. J. Hodges and C. Andre, *J. Am. Chem. Soc.*, 1986, **108**, 2776 and references cited therein.
80. M. T. Crimmins and R. O'Mahoney, *J. Org. Chem.*, 1989, **54**, 1157; (b) M. T. Crimmins and D. M. Bankaitis, *Tetrahedron Lett.*, 1983, **24**, 4551.
81. (a) M. T. Crimmins, D. M. Bankaitis-Davis and W. G. Hollis, Jr., *J. Org. Chem.*, 1988, **53**, 652; (b) M. T. Crimmins, W. G. Hollis, Jr. and D. M. Bankaitis-Davis, *Tetrahedron Lett.*, 1987, **28**, 3651.
82. N. V. Bac and Y. Langlois, *Tetrahedron Lett.*, 1988, **29**, 2819.
83. (a) S. L. Schreiber and S. E. Kelly, *Tetrahedron Lett.*, 1984, **25**, 1757; (b) S. L. Schreiber, S. E. Kelly, J. A. Porco, Jr., T. Sammakia and E. M. Suh, *J. Am. Chem. Soc.*, 1988, **110**, 6210.
84. K. Suzuki, T. Ohkuma and G. Tsuchihashi, *J. Org. Chem.*, 1987, **52**, 2929.
85. E. R. Talaty, A. R. Clague, M. O. Agho, M. N. Deshpande, P. M. Courtney, D. H. Burger and E. F. Roberts, *J. Chem. Soc., Chem. Commun.*, 1980, 889.
86. D. Cai and W. C. Still, *J. Org. Chem.*, 1988, **53**, 4641.
87. For a review see Y. Yamamoto, *Angew. Chem., Int. Ed. Engl.*, 1986, **25**, 947.
88. E. J. Corey, P. Da Silva Jardine and J. C. Rohloff, *J. Am. Chem. Soc.*, 1988, **110**, 3672.
89. J. Inanaga, Y. Kawanami and M. Yamaguchi, *Chem. Lett.*, 1981, 1415.
90. E. J. Corey and R. L. Dawson, *J. Am. Chem. Soc.*, 1962, **84**, 4899.
91. T. Sakan and Y. Mori, *Chem. Lett.*, 1972, 793.
92. K. Abe, T. Sato, N. Nakamura and T. Sakan, *Chem. Lett.*, 1977, 645.
93. T. Miyasaka, H. Monobe and S. Noguchi, *Chem. Lett.*, 1986, 449.
94. For further discussion of 5-coordinate magnesium see F. A. Cotton and G. Wilkinson, 'Advanced Inorganic Chemistry', 5th edn., Wiley-Interscience, New York, 1988, p. 159.
95. A. I. Meyers and D. L. Comins, *Tetrahedron Lett.*, 1978, 5179; D. L. Comins and A. I. Meyers, *Synthesis*, 1978, 403.
96. M. S. Newman and A. S. Smith, *J. Org. Chem.*, 1948, **13**, 592; H. A. Staab, *Angew. Chem., Int. Ed. Engl.*, 1962, **1**, 351.
97. T. Fujisawa, T. Mori and T. Sato, *Tetrahedron Lett.*, 1982, **23**, 5059.
98. T. Fujisawa, T. Mori, K. Higuchi and T. Sato, *Chem. Lett.*, 1983, 1791.
99. D. A. Evans, J. V. Nelson, E. Vogel and T. R. Taber, *J. Am. Chem. Soc.*, 1981, **103**, 3099.
100. M. Araki and T. Mukaiyama, *Chem. Lett.*, 1974, 663; M. Araki, S. Sakata, H. Takei and T. Mukaiyama, *Chem. Lett.*, 1974, 687.
101. A. S. Kende, D. Scholz and J. Schneider, *Synth. Commun.*, 1978, **8**, 59.
102. T. Fujisawa, S. Iida, H. Uehara and T. Sato, *Chem. Lett.*, 1983, 1267.
103. J. S. Nimitz and H. S. Mosher, *J. Org. Chem.*, 1981, **46**, 211; L. M. Weinstock, R. B. Currie, and A. V. Lovell, *Synth. Commun.*, 1981, **11**, 943; W. J. Middleton and E. M. Bingham, *J. Org. Chem.*, 1980, **45**, 2883.
104. X. Creary, *J. Org. Chem.*, 1987, **52**, 5026.
105. D. G. Melillo, R. D. Larsen, D. J. Mathre, W. F. Shukis, A. Wood and J. R. Colleluori, *J. Org. Chem.*, 1987, **52**, 5143.
106. M. Egbertson and S. J. Danishefsky, *J. Org. Chem.*, 1989, **54**, 11.
107. E. J. Corey, D. H. Hua, B.-C. Pan and S. P. Seitz, *J. Am. Chem. Soc.*, 1982, **104**, 6818.
108. G. P. Axiotis, *Tetrahedron Lett.*, 1981, **22**, 1509.
109. G. M. Whitesides, C. P. Casey, J. San Filippo, Jr. and E. J. Panek, *Trans. N. Y. Acad. Sci.*, 1967, **29**, 572.
110. G. H. Posner and C. E. Whitten, *Tetrahedron Lett.*, 1970, 4647.
111. R. M. Wehmeyer and R. D. Rieke, *Tetrahedron Lett.*, 1988, **29**, 4513; G. W. Ebert and R. D. Rieke, *J. Org. Chem.*, 1988, **53**, 4482.

112. Stoichiometric copper, for a review see J. F. Normant, *Pure Appl. Chem.*, 1978, **50**, 709; J. A. MacPhee, M. Boussu and J.-E. Dubois, *J. Chem. Soc., Perkin Trans. 2*, 1974, 1525; J.-E. Dubois, M. Boussu and C. Lion, *Tetrahedron Lett.*, 1971, 829; catalytic copper, for a review see E. Erdik, *Tetrahedron*, 1984, **40**, 641; T. Fujisawa and T. Sato, *Org. Synth.*, 1987, **66**, 116.
113. D. B. Collum, J. H. McDonald, III and W. C. Still, *J. Am. Chem. Soc.*, 1980, **102**, 2120.
114. G. J. McGarvey, J. M. Williams, R. N. Hiner, Y. Matsubara and T. Oh, *J. Am. Chem. Soc.*, 1986, **108**, 4943.
115. A. Marfat, P. R. McGuirk and P. Helquist, *Tetrahedron Lett.*, 1978, 1363.
116. (a) A. Marfat, P. R. McGuirk, R. Kramer and P. Helquist, *J. Am. Chem. Soc.*, 1977, **99**, 253 and references cited therein; (b) J. F. Normant, G. Cahiez, M. Bourgain, C. Chuit and J. Villieras, *Bull. Soc. Chim. Fr.*, 1974, 1656; (c) A. Alexakis, J. F. Normant and J. Villieras, *Tetrahedron Lett.*, 1976, 3461; (d) H. Westmijze, H. Kleijn and P. Vermeer, *Tetrahedron Lett.*, 1977, 2023.
117. J. F. Normant and A. Alexakis, *Synthesis*, 1981, 841.
118. N. Jabri, A. Alexakis and J. F. Normant, *Tetrahedron Lett.*, 1983, **24**, 5081.
119. J. P. Foulon, M. Bourgain-Commercon and J. F. Normant, *Tetrahedron*, 1986, **42**, 1399.
120. R. L. Danheiser, J. M. Morin, Jr. and E. J. Salaski, *J. Am. Chem. Soc.*, 1985, **107**, 8066.
121. For recent discussion of organocuprate preparation see (a) G. H. Posner, 'An Introduction to Synthesis Using Organocopper Reagents', Wiley, New York, 1980; (b) B. H. Lipshutz, *Synthesis*, 1987, 325; (c) B. H. Lipshutz, R. S. Wilhelm and J. A. Kozlowski, *Tetrahedron*, 1984, **40**, 5005; (d) J. G. Noltes and G. van Koten, in 'Comprehensive Organometallic Chemistry', ed G. Wilkinson, F. G. A. Stone and E. W. Abel, Pergamon Press, Oxford, 1982, vol. 2, p. 709; and W. Carruthers, vol. 7, p. 685; (e) S. H. Bertz, C. P. Gibson and G. Dabbagh, *Tetrahedron Lett.*, 1987, 4251; (f) B. H. Lipshutz, S. Whitney, J. A. Kozlowski and C. M. Breneman, *Tetrahedron Lett.*, 1986, 4273 and references cited therein.
122. G. H. Posner, C. E. Whitten and P. E. McFarland, *J. Am. Chem. Soc.*, 1972, **94**, 5106.
123. L. A. Paquette, J. M. Gardlik, K. J. McCullough and Y. Hanzawa, *J. Am. Chem. Soc.*, 1983, **105**, 7644.
124. T. Fex, J. Froborg, G. Magnusson and S. Thoren, *J. Org. Chem.*, 1976, **41**, 3518. Treatment of the diketone with base readily effected epimerization, which was followed by condensation to the α,β-unsaturated bicyclic ketone.
125. D. M. Walba and M. D. Wand, *Tetrahedron Lett.*, 1982, **23**, 4995.
126. D. V. Patel, F. VanMiddlesworth, J. Donaubauer, P. Gannett and C. J. Sih, *J. Am. Chem. Soc.*, 1986, **108**, 4603.
127. D. A. Evans and M. DiMare, *J. Am. Chem. Soc.*, 1986, **108**, 2476.
128. E. J. Corey and D. J. Beames, *J. Am. Chem. Soc.*, 1972, **94**, 7210.
129. For reviews see J. F. Normant, *Synthesis*, 1972, 63; Y. Yamamoto, *Angew. Chem., Int. Ed. Engl.*, 1986, 25, 947; J. F. Normant, *Tetrahedron Lett.*, 1970, 2659.
130. E. J. Corey, D. Floyd and B. H. Lipshutz, *J. Org. Chem.*, 1978, **43**, 3418.
131. T. Mukaiyama, M. Araki and H. Takei, *J. Am. Chem. Soc.*, 1973, **95**, 4763.
132. E. J. Corey, P. B. Hopkins, S. Kim, S.-e. Yoo, K. P. Nambiar and J. R. Falck, *J. Am. Chem. Soc.*, 1979, **101**, 7131.
133. J. P. Marino and R. J. Linderman, *J. Org. Chem.*, 1981, **46**, 3696.
134. G. M. Whitesides, W. F. Fischer, Jr., J. San Filippo, Jr., R. W. Bashe and H. O. House, *J. Am. Chem. Soc.*, 1969, **91**, 4871. This procedure uses copper(I) iodide·tri-*n*-butylphosphine complex; the resulting cuprates are significantly more stable at –78 °C.
135. G. H. Posner, C. E. Whitten and J. J. Sterling, *J. Am. Chem. Soc.*, 1973, **95**, 7788; G. H. Posner and C. E. Whitten, *Tetrahedron Lett.*, 1973, 1815.
136. See ref. 121a; and G. H. Posner and C. E. Whitten *Org. Synth.*, 1976, **55**, 122; see also ref. 137 for some conflicting results in the case of dialkylamino ligands.
137. S. H. Bertz and G. Dabbagh, *J. Org. Chem.*, 1984, **49**, 1119; S. H. Bertz, G. Dabbagh and G. M. Villacorta, *J. Am. Chem. Soc.*, 1982, **104**, 5824; recently Martin *et al.* have prepared a related heterocuprate ($MeCuPBu^t_2Li$) which possesses increased thermal stability, S. F. Martin, J. R. Fishpaugh, J. R. Power, D. M. Giolando, R. A. Jones, C. M. Nunn and A. H. Cowley, *J. Am. Chem. Soc.*, 1988, **110**, 7226.
138. P. Knochel, M. C. P. Yeh, S. C. Berk and J. Talbert, *J. Org. Chem.*, 1988, **53**, 2392; M. C. P. Yeh and P. Knochel, *Tetrahedron Lett.*, 1988, **29**, 2395; M. C. P. Yeh, P. Knochel and L. E. Santa, *Tetrahedron Lett.*, 1988, **29**, 3887.
139. R. J. Anderson, C. A. Henrick and L. D. Rosenblum, *J. Am. Chem. Soc.*, 1974, **96**, 3654.
140. S. Masamune, *Pure Appl. Chem.*, 1988, **60**, 1587; M. A. Blanchette, M. S. Malamas, M. H. Nantz, J. C. Roberts, P. Somfai, D. C. Whritenour and S. Masamune, *J. Org. Chem.*, 1989, **54**, 2817.
141. L. Cama and B. G. Christensen, *Tetrahedron Lett.*, 1980, **21**, 2013.
142. S. Kim and J. I. Lee, *J. Org. Chem.*, 1983, **48**, 2608; S. Kim, *Org. Prep. Proced. Int.*, 1988, **20**, 145.
143. S. Kim, J. I. Lee and B. Y. Chung, *J. Chem. Soc., Chem. Commun.*, 1981, 1231.
144. Although somewhat misleading, the term β-silyl ketone refers to the isomer in which the silyl group is attached to the carbon adjacent to the carbonyl. The term α-silyl ketone and acylsilane have been used interchangeably to indicate attachment of the silane and carbonyl carbon.
145. J. P. Foulon, M. Bourgain-Commercon and J. F. Normant, *Tetrahedron*, 1986, **42**, 1399.
146. Y. Yamamoto, K. Ohdoi, M. Nakatani and K. Akiba, *Chem. Lett.*, 1984, 1967.
147. This is the well known 'Brook rearrangement', A. G. Brook, D. M. MacRae and W. W. Limburg, *J. Am. Chem. Soc.*, 1967, **89**, 5493.
148. B. A. Pearlman, J. M. McNamara, I. Hasan, S. Hatakeyama, H. Sekizaki and Y. Kishi, *J. Am. Chem. Soc.*, 1981, **103**, 4248.
149. M. Demuth, *Helv. Chim. Acta*, 1978, **61**, 3136.
150. Recently an alternative method has appeared which involves the addition of Grignard reagents to α-silyl esters, G. L. Larson, I. M. de Lopez-Cepero and L. E. Torres, *Tetrahedron Lett.*, 1984, **25**, 1673; P. F. Hudrlik and D. J. Peterson, *Tetrahedron Lett.*, 1974, 1133.
151. T. Kaiho, S. Masamune and T. Toyoda, *J. Org. Chem.*, 1982, **47**, 1612.

152. S. Masamune, L. D.-L. Lu, W. P. Jackson, T. Kaiho and T. Toyoda, *J. Am. Chem. Soc.*, 1982, **104**, 5523.
153. S. Masamune, M. Hirama, S. Mori, S. A. Ali and D. S. Garvey, *J. Am. Chem. Soc.*, 1981, **103**, 1568.
154. M. Kosugi, Y. Shimizu and T. Migita, *Chem. Lett.*, 1977, 1423.
155. (a) D. Milstein and J. K. Stille, *J. Am. Chem. Soc.*, 1978, **100**, 3636; (b) D. Milstein and J. K. Stille, *J. Org. Chem.*, 1979, **44**, 1613; (c) J. K. Stille, *Angew. Chem., Int. Ed. Engl.*, 1986, **25**, 508.
156. M. Pereyre, J.-P. Quintard and A. Rahm, 'Tin in Organic Synthesis', Butterworths, London, 1987, p. 198.
157. A. N. Kashin, N. A. Bumagin, I. O. Kalinovskii, I. P. Beletskaya and O. A. Reutov, *Zh. Org. Khim.*, 1980, **16**, 1569.
158. J.-B. Verlhac, E. Chanson, B. Jousseaume and J.-P. Quintard, *Tetrahedron Lett.*, 1985, **26**, 6075.
159. See ref. 158 and N. A. Bumagin, Yu. V. Gulevich and I. P. Beletskaya, *J. Organomet. Chem.*, 1985, **282**, 421.
160. J. A. Soderquist and W. W.-H. Leong, *Tetrahedron Lett.*, 1983, **24**, 2361; J. A. Soderquist and G. J.-H. Hsu, *Organometallics*, 1982, **1**, 830; see also M. Kosugi, T. Sumiya, Y. Obara, M. Suzuki, H. Sano and T. Migita, *Bull. Chem. Soc. Jpn.*, 1987, **60**, 767.
161. M. Kosugi, Y. Shimizu and T. Migita, *J. Organomet. Chem.*, 1977, **129**, C36.
162. J. K. Stille, *Angew. Chem., Int. Ed. Engl.*, 1986, **25**, 508.
163. For a recent review see J. P. Collman, L. S. Hegedus, J. R. Norton and R. G. Finke, 'Principles and Applications of Organotransition Metal Chemistry', University Science Books, California, 1987.
164. J. P. Collman, *Acc. Chem. Res.*, 1975, **8**, 342; J. P. Collman, L. S. Hegedus, J. R. Norton and R. G. Finke, in ref. 163, p. 755.
165. Disodium tetracarbonylferrate is commercially available at reasonable cost; however, it is highly basic (pK_b = OH^-), extremely oxygen sensitive and spontaneously inflammable in air.
166. J. K. Stille and K. S. Y. Lau, *Acc. Chem. Res.*, 1977, **10**, 434.
167. J. P. Collman, L. S. Hegedus, J. R. Norton and R. G. Finke, in ref. 163.
168. See ref. 121.
169. M. C. Baird, J. T. Mague, J. A. Osborn and G. Wilkinson, *J. Chem. Soc. A*, 1967, 1347.
170. P. Four and F. Guibe, *J. Org. Chem.*, 1981, **46**, 4439.
171. (a) B. E. Mann and A. Musco, *J. Chem. Soc., Dalton Trans.*, 1975, 1673; (b) A. Musco, W. Kuran, A. Silvani and M. Anker, *J. Chem. Soc., Chem. Commun.*, 1973, 938; (c) J. P. Collman, L. S. Hegedus, J. R. Norton and R. G. Finke, in ref. 163, p. 67.
172. C. A. Tolman, *Chem. Rev.*, 1977, **77**, 313.
173. T. Ukai, H. Kawazura, Y. Ishii, J. J. Bonnet and J. A. Ibers, *J. Organomet. Chem.*, 1974, **65**, 253.
174. F. G. Salituro and I. A. McDonald, *J. Org. Chem.*, 1988, **53**, 6138.
175. I. P. Beletskaya, *J. Organomet. Chem.*, 1983, **250**, 551.
176. A. F. Renaldo, J. W. Labadie and J. K. Stille, *Org. Synth.*, 1988, **67**, 86.
177. D. Milstein and J. K. Stille, *J. Am. Chem. Soc.*, 1979, **101**, 4992.
178. (a) K. S. Y. Lau, R. W. Fries and J. K. Stille, *J. Am. Chem. Soc.*, 1974, **96**, 4983; (b) P. K. Wong, K. S. Y. Lau and J. K. Stille, *J. Am. Chem. Soc.*, 1974, **96**, 5956; (c) Y. Becker and J. K. Stille, *J. Am. Chem. Soc.*, 1978, **100**, 838; (d) K. S. Y. Lau, P. K. Wong and J. K. Stille, *J. Am. Chem. Soc.*, 1976, **98**, 5832; (e) J. K. Stille and K. S. Y. Lau, *J. Am. Chem. Soc.*, 1976, **98**, 5841.
179. A. V. Kramer, J. A. Labinger, J. S. Bradley and J. A. Osborn, *J. Am. Chem. Soc.*, 1974, **96**, 7145.
180. A. V. Kramer and J. A. Osborn, *J. Am. Chem. Soc.*, 1974, **96**, 7832.
181. D. Milstein and J. K. Stille, *J. Org. Chem.*, 1979, **44**, 1613.
182. J. W. Labadie, D. Tueting and J. K. Stille, *J. Org. Chem.*, 1983, **48**, 4634.
183. I. Pri-Bar and J. K. Stille, *J. Org. Chem.*, 1982, **47**, 1215.
184. J. W. Labadie and J. K. Stille, *J. Am. Chem. Soc.*, 1983, **105**, 6129.
185. J. W. Labadie and J. K. Stille, ref. 184.
186. J. W. Labadie and J. K. Stille, *J. Am. Chem. Soc.*, 1983, **105**, 669.
187. J. W. Labadie and J. K. Stille, ref. 186.
188. A. Gillie and J. K. Stille, *J. Am. Chem. Soc.*, 1980, **102**, 4933.
189. J. W. Labadie and J. K. Stille, ref. 184.
190. (a) K. Tatsumi, R. W. Hoffmann, A. Yamamoto and J. K. Stille, *Bull. Chem. Soc. Jpn.*, 1981, **54**, 1857; (b) M. K. Loar and J. K. Stille, *J. Am. Chem. Soc.*, 1981, **103**, 4174; (c) A. Moravskiy and J. K. Stille, *J. Am. Chem. Soc.*, 1981, **103**, 4182.
191. J. K. Stille, *Angew. Chem., Int. Ed. Engl.*, 1986, **25**, 508.
192. M. Pereyre, J.-P. Quintard and A. Rahm, 'Tin in Organic Synthesis', Butterworths, London, 1987, p. 8.
193. E. J. Corey and P. A. Magriotis, *J. Am. Chem. Soc.*, 1987, **109**, 287.
194. D. Milstein and J. K. Stille, ref. 155b.
195. L. B. Barkley, W. S. Knowles, H. Raffelson and Q. E. Thompson, *J. Am. Chem. Soc.*, 1956, **78**, 4111.
196. J. W. Labadie and J. K. Stille, *Tetrahedron Lett.*, 1983, **24**, 4283.
197. J. W. Labadie, D. Tueting and J. K. Stille, ref. 182.
198. A. S. Kende, B. Roth, P. Sanfilippo and T. J. Blacklock, *J. Am. Chem. Soc.*, 1982, **104**, 5808.
199. R. E. Ireland and D. M. Obrecht, *Helv. Chim. Acta*, 1986, **69**, 1273.
200. H. L. Holland and E. M. Thomas, *Can. J. Chem.*, 1982, **60**, 160.
201. R. A. Holton and K. J. Natalie, Jr., *Tetrahedron Lett.*, 1981, **22**, 267.
202. D. Milstein and J. K. Stille, *J. Am. Chem. Soc.*, 1979, 101, 4981; see also refs. 188 and 190c.
203. M. W. Logue and K. Teng, *J. Org. Chem.*, 1982, **47**, 2549.
204. Y. Tohda, K. Sonogashita and N. Hagihara, *Synthesis*, 1977, 777.
205. T. Sato, K. Naruse, M. Enokiya and T. Fujisawa, *Chem. Lett.*, 1981, 1135.
206. E. Negishi, V. Bagheri, S. Chatterjee, F.-T. Luo, J. A. Miller and A. T. Stoll, *Tetrahedron Lett.*, 1983, **24**, 5181.
207. Y. Tamura, H. Ochiai, T. Nakamura and Z. Yoshida, *Angew. Chem., Int. Ed. Engl.*, 1987, **26**, 1157; Y. Tamura, H. Ochiai, T. Nakamura and Z. Yoshida, *Org. Synth.*, 1988, **67**, 98; E. Nakamura, S. Aoki, K. Sekiya, H. Oshino and I. Kuwajima, *J. Am. Chem. Soc.*, 1987, **109**, 8056.

208. R. A. Grey, *J. Org. Chem.*, 1984, **49**, 2288.
209. R. F. W. Jackson, K. James, M. J. Wythes and A. Wood, *J. Chem. Soc., Chem. Commun.*, 1989, 644.
210. Y. Tamaru, H. Ochiai, F. Sanda and Z. Yoshida, *Tetrahedron Lett.*, 1985, **26**, 5529.
211. K. Takagi, T. Okamoto, Y. Sakakibara, A. Ohno, S. Oka and N. Hayama, *Chem. Lett.*, 1975, 951.
212. R. C. Larock and J. C. Bernhardt, *J. Org. Chem.*, 1978, **43**, 710.
213. K. Wakamatsu, Y. Okuda, K. Oshima and H. Nozaki, *Bull. Chem. Soc. Jpn.*, 1985, **58**, 2425.
214. Y. Inoue, K. Ohuchi and S. Imaizumi, *Tetrahedron Lett.*, 1988, **29**, 5941; see also refs. 203 and 204.
215. Commercially available from Lancaster Synthesis, Windham, NH, or may be prepared by the procedure described in *Inorg. Synth.*, 1968, **11**, 99.
216. M. C. Baird, J. T. Mague, J. A. Osborn and G. Wilkinson, *J. Chem. Soc. A*, 1967, 1347; J. Chatt and B. L. Shaw, *J. Chem. Soc. A*, 1966, 1437; A. J. Deeming and B. L. Shaw, *J. Chem. Soc. A*, 1969, 597.
217. (a) L. S. Hegedus, P. M. Kendall, S. M. Lo and J. R. Sheats, *J. Am. Chem. Soc.*, 1975, **97**, 5448; (b) L. S. Hegedus, S. M. Lo and D. E. Bloss, *J. Am. Chem. Soc.*, 1973, **95**, 3040.
218. C. U. Pittman, Jr. and R. M. Hanes, *J. Org. Chem.*, 1977, **42**, 1194.
219. M. Kosugi, Y. Shimizu and T. Migita, *J. Organomet. Chem.*, 1977, **129**, C36.
220. (a) V. Fiandanese, G. Marchese and F. Naso, *Tetrahedron Lett.*, 1988, **29**, 3587; (b) C. Cardellicchio, V. Fiandanese, G. Marchese and L. Ronzioni, *Tetrahedron Lett.*, 1985, **26**, 3595.
221. M. Onaka, Y. Matsuoka and T. Mukaiyama, *Chem. Lett.*, 1981, 531.
222. M. Onaka, T. Goto and T. Mukaiyama, *Chem. Lett.*, 1979, 1483.
223. (a) S. Inaba and R. D. Rieke, *Tetrahedron Lett.*, 1983, **24**, 2451; (b) S. Inaba and R. D. Rieke, *J. Org. Chem.*, 1985, **50**, 1373.
224. G. H. Posner, *J. Am. Chem. Soc.*, 1973, **95**, 7788.
225. H. O. House and M. J. Umen, *J. Org. Chem.*, 1973, **38**, 3893.
226. C. R. Johnson and D. S. Dhanoa, *J. Chem. Soc., Chem. Commun.*, 1982, 358.
227. B. H. Lipshutz, *Synthesis*, 1987, 325.

2.1
Nitrogen Stabilization

ROBERT E. GAWLEY and KATHLEEN REIN
University of Miami, Coral Gables, FL, USA

2.1.1 INTRODUCTION

This chapter covers the carbonyl addition chemistry of carbanions stabilized by a nitrogen atom or a nitrogen-containing functional group in which the nitrogen is responsible for the stabilization. In most cases, the carbanions are formed by deprotonation, but metal–halogen exchange is occasionally important. A carbanion that is stabilized by a nitrogen may exist in three oxidation states: *sp*, sp^2 or sp^3. The simplest nitrogen-stabilized carbanion is cyanide, the *sp*-hybridized case. In recent years, most efforts in this area have been expended on developing the chemistry of sp^2- and sp^3-carbanions. This chapter deals with the addition of these types of anions to carbonyl compounds. Alkylation reactions of sp^3-hybridized species are covered in Volume 3, Chapter 1.2, and alkylation of sp^2-hybridized carbanions is covered in Volume 3, Chapter 1.4. Specifically excluded from this chapter are additions of carbanions stabilized by a nitro group (the Henry nitroaldol reaction) and azaenolates, which are covered in Volume 2, Chapters 1.10, 1.16 and 1.17.

Each section of this chapter is subdivided according to the type of species being metalated: acyclic, carbocyclic or heterocyclic. The site of metalation is the criterion for classifying each species. Thus, the *ortho* metalation of anisole is classified as a carbocyclic system, whereas metalation of the *N*-methyl group of a heterocycle is an acyclic system and the α-metalation of a piperidinecarboxamide is a heterocyclic system.

2.1.2 *sp*-HYBRIDIZED CARBANIONS (CYANIDE)

The addition of HCN to aldehydes has been a well-known reaction since the 19th century, especially in the context of the Kiliani–Fischer synthesis of sugars. Even older is the Strecker synthesis of amino acids by simultaneous reaction of aldehydes with ammonia and HCN followed by hydrolysis. The challenge in recent years has been to achieve face-selectivity in the addition to chiral aldehydes. These face-selective additions, known as 'nonchelation-controlled' processes, refer to the original formulation of Cram's for the reaction of nucleophiles with acyclic chiral carbonyl compounds.[1] The 'chelation-controlled' reactions refer also to a formulation of Cram's, but whose stereochemical consequences sometimes differ.[2]

An example of this type of effort is as follows. Complexation of the carbonyl oxygen of *N,N*-dibenzyl-α-amino aldehydes with Lewis acids such as BF_3, $ZnBr_2$ or $SnCl_4$, followed by addition of trimethylsilyl cyanide leads to adducts formed by a nonchelation-controlled process.[3] Complexation with $TiCl_4$ or $MgBr_2$ affords the opposite stereochemistry preferentially, through a chelation-controlled process (Scheme 1).[3]

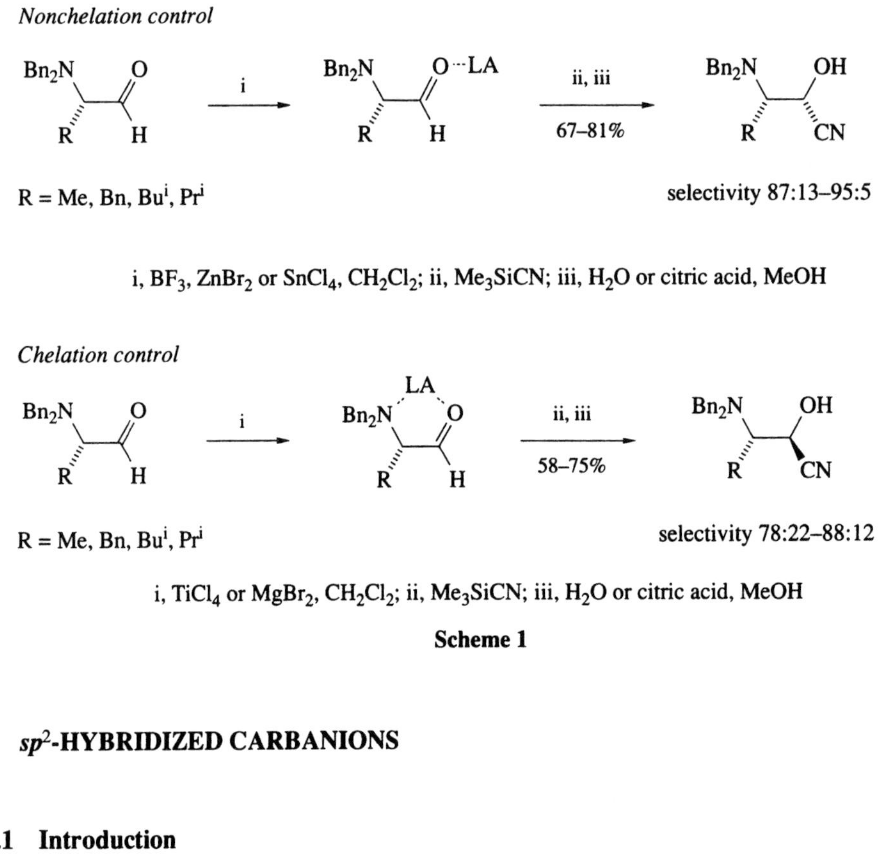

Nonchelation control

Bn_2N, O, R, H → (i) → Bn_2N, O···LA, R, H → (ii, iii; 67–81%) → Bn_2N, OH, R, CN

R = Me, Bn, Bu^i, Pr^i — selectivity 87:13–95:5

i, BF_3, $ZnBr_2$ or $SnCl_4$, CH_2Cl_2; ii, Me_3SiCN; iii, H_2O or citric acid, MeOH

Chelation control

Bn_2N, O, R, H → (i) → LA, Bn_2N, O, R, H → (ii, iii; 58–75%) → Bn_2N, OH, R, CN

R = Me, Bn, Bu^i, Pr^i — selectivity 78:22–88:12

i, $TiCl_4$ or $MgBr_2$, CH_2Cl_2; ii, Me_3SiCN; iii, H_2O or citric acid, MeOH

Scheme 1

2.1.3 *sp*^{2}-HYBRIDIZED CARBANIONS

2.1.3.1 Introduction

There are two common methods for forming *sp*^{2}-hybridized carbanions: deprotonation and metal-halogen exchange. In their 1979 review of heteroatom-facilitated lithiations, Gschwend and Rodriguez contend that there are two mechanistic extremes for a heteroatom-facilitated lithiation: a 'coordination only' mechanism and an 'acid–base mechanism'.[4] They further state that: 'between these extremes there is a continuous spectrum of cases in which both effects simultaneously contribute in varying degrees to the observed phenomena'.[4] Because the subject of this chapter is nitrogen-stabilized carbanions, and because this stabilization is most often exerted by coordination to the cation (usually lithium), it turns out that most of the deprotonations are best rationalized by prior coordination of the base to a heteroatom (the coordination only mechanism). Because these effects often render a deprotonation under conditions of kinetic control, this phenomenon has been termed a 'complex-induced proximity effect'.[5] The most important manifestation of this coordination in the examples presented below is a high degree of regioselectivity in proton removal.

We have not restricted our coverage to instances of cation coordination to a nitrogen atom, since there are a number of important functional groups (such as secondary and tertiary amides) where coordination occurs at oxygen but stabilization is also provided by nitrogen. Thus, most of the developments of the last 10 years have been in the area commonly known as 'directed metalations'. Several reviews have appeared on various aspects of these subjects.[5–9]

2.1.3.2 Additions *via* Metalation of Acyclic Systems

A novel method has been reported for the elaboration of carbonyl compounds, which is discussed in Section 2.1.4.2.[10,11] However, one of the examples falls into the present category: the transformation of cyclohexanone to keto alcohol (**4**), *via* enamidine (**1**; equation 1). Treatment of (**1**) with *t*-butyllithium effects regioselective deprotonation of the vinylic hydrogen to give (**2**), which adds to propanal to give (**3**). Hydrolysis then provides keto alcohol (**4**).[10,11]

O → Bu–N(H)–N–Bu^t (**1**) —Bu^{t}Li, THF, –78 °C→ Bu–N(Li)–N–Bu^t (**2**) —EtCHO→

Bu–N(OH)–N–Bu^t (**3**) —i, N_2H_4, H^+; ii, H_2O→ O, OH (**4**) (1)

A synthesis of pyrroles and pyridines is possible by addition of dilithium species (**5**) to carbonyls (Scheme 2). The *syn* lithiation to give (**5**) was established by quenching (**5**) with trimethylsilyl chloride.[12]

2.1.3.3 Additions *via* Metalation of Carbocyclic Systems

2.1.3.3.1 Ortho *metalation*

The vast majority of the species discussed in this and the following sections are formed by *ortho* lithiation (equation 2). In that the regioselective metalation of a given substrate at a given position is predicated on the 'directing ability' of the directing functional group, it is important to know what functional groups might take precedence over others. This ordering may be the result of either kinetics or thermodynamics. At one extreme for example, a strongly basic atom in a functional group might coordinate a lithium ion so strongly that coordination (and therefore *ortho* lithiation) is precluded at any other site (a kinetic effect). At the other extreme, metalation might occur to produce the most stable anion (a thermodynamic effect).

The relative directing abilities of several functional groups have been evaluated by both inter- and intra-molecular competition experiments.[9,13–15] The strongest directing group is a tertiary amide, but groups such as 3,3-dimethyloxazolinyl and secondary amides are also effective. The pK_a values of a

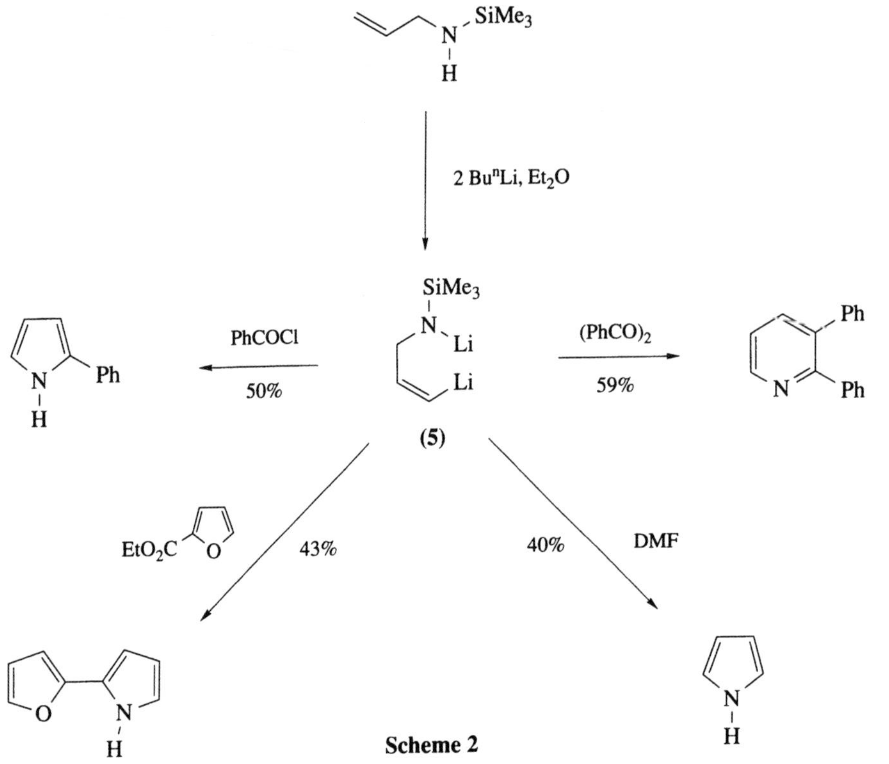

Scheme 2

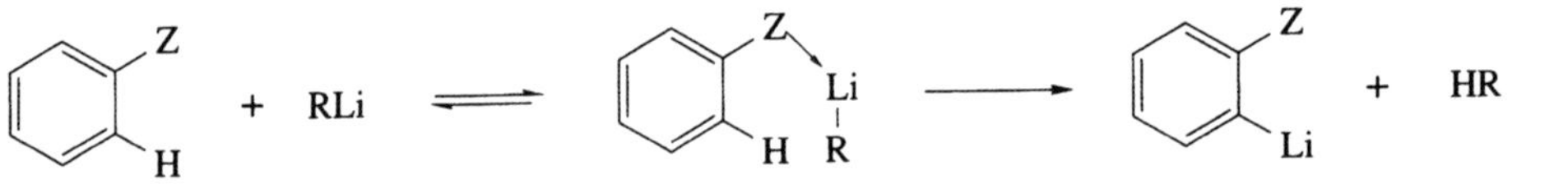

(2)

number of monosubstituted benzenes (*ortho* lithiation) were measured against tetramethylpiperidine (pK_a = 37.8) by NMR spectroscopy.[16] The results, shown in Table 1,[16] are accurate to ±0.2 pK_a units. By and large, the thermodynamic acidities parallel the directing ability of the substituent. Thus, the acidity imparted to the *ortho* position is a useful guideline for determining relative directing ability. Also pertinent to the mechanism of nitrogen-directed lithiations is a recent theoretical study on the lithiation of enamines, which concludes that 'a favorable transition state involves the achievement of both the stereoelectronic requirement for deprotonation and stabilizing coordination of the lithium cation with the base, the nitrogen of the enamine and the developing anionic center'.[17]

Table 1 pK_a Values for *Ortho* Lithiation of Substituted Benzenes in THF at 27 °C
PhR + LITMP ⇌ *o*-RC_6H_4Li + HTMP

R	pK_a (±0.2)	*R*	pK_a (±0.2)
—NMe_2	≥40.3[a]	—OPh	38.5
—CH_2NMe_2	≥40.3[a]	—SO_2NEt_2	38.2[b]
—C≡CPh	≥40.3[a]	3,3-Dimethyloxazolinyl	38.1[b]
—$NHCOBu^t$	≥40.5[a]	—CN	38.1
—OLi	≥40.5[a]	—$CONPr^i_2$	37.8[b]
—OTHP	40.0	—$OCONEt_2$	37.2[c]
—OMe	39.0		

[a]No metalation observed. [b]–40 °C. [c]–70 °C.

From a synthetic standpoint, a highly useful consequence of the directed metalation strategy for aromatic substitution is the cooperativity exerted by two directing groups that are *meta*: metalation and substitution occur at the hindered position between the two directing groups (equation 3). However, exceptions have been noted in certain instances.[18] Metalation may be directed elsewhere by the use of a trimethylsilyl blocking group at the preferred position.[19]

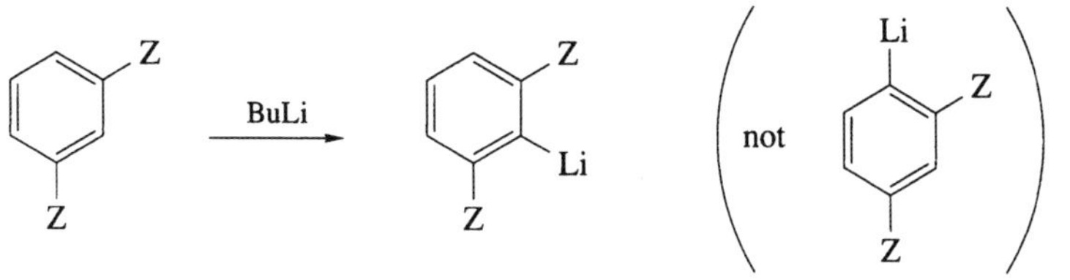

(3)

2.1.3.3.2 *Meta metalation*

In some cases, aniline or indole derivatives can be substituted *meta* to the nitrogen by lithiation of the appropriate chromium tricarbonyl complexes.[20–22] Examples are given in Scheme 3. One of the Cr—C═O bonds eclipses the C—N bond, and therefore the other carbonyls eclipse the *meta* C—H bonds. Two suggestions have been offered for the *meta* selectivity: (i) the butyllithium coordinates the chromium carbonyl oxygen and then removes the proximate proton (a kinetic effect);[20] (ii) the eclipsed conformation produces a lower electron density at the *meta* position, which in turn renders the *meta* protons more acidic (a thermodynamic effect).[21]

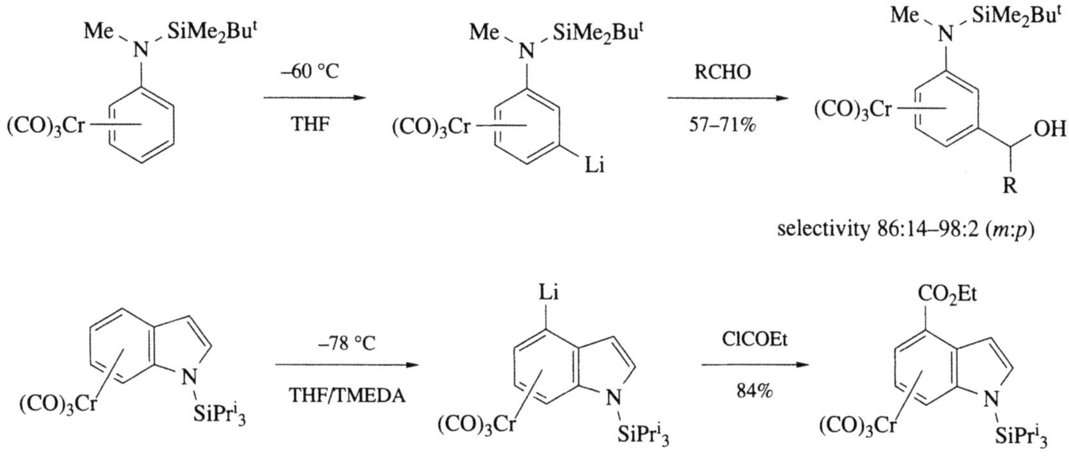

selectivity 86:14–98:2 (*m*:*p*)

Scheme 3

2.1.3.3.3 *Amines and anilides*

The classic example of amines as directing groups, and the reaction cited by Gschwend and Rodriguez as the best example of a 'coordination only' mechanism,[4] is the *ortho* metalation of *N,N*-dialkylbenzylamines, reported by Hauser in 1963.[23] Recent applications of the same directing group, working co-operatively with a *meta* methoxy to produce substitution between the two, are shown in Scheme 4.[24–26] In the illustrated examples, paraformaldehyde is the electrophile, and the benzylic alcohol (**6**) is obtained in 92% yield. The conversion of (**6**) to isochromanones such as (**7**) and berberines such as (**8**) illustrates the advantages of directed metalations over more traditional aromatic substitution methods such as Pictet–Spengler cyclizations.

Imidazolidines may also act as *ortho* directing groups: the lithiation of 1,3-dimethyl-2-phenylimidazolidine followed by addition to benzophenone proceeds in 63% yield.[27] Carbazole aminals can be metalated *ortho* to the nitrogen, while benzo[*a*]carbazole may be dilithiated at nitrogen and the 1-position.[28] A 2-amino group of a biphenyl directs lithiation to the 2′-position of the other ring in a novel synthesis of a phenanthride.[29]

Lithium amides add to benzaldehydes to form α-aminoalkoxides that direct *ortho* metalation, as shown in Scheme 5.[30,31] In a related process, the α-aminoalkoxides may be metalated by lithium–halogen exchange.[32]

In an aliphatic system, Stork has reported the regioselective lithiation of chelating enamines such as (**9**), to give vinyl carbanions such as (**10**; Scheme 6). Work-up often results in hydrolysis of the enamine, and in the case of addition to aldehydes, dehydration.[17,33]

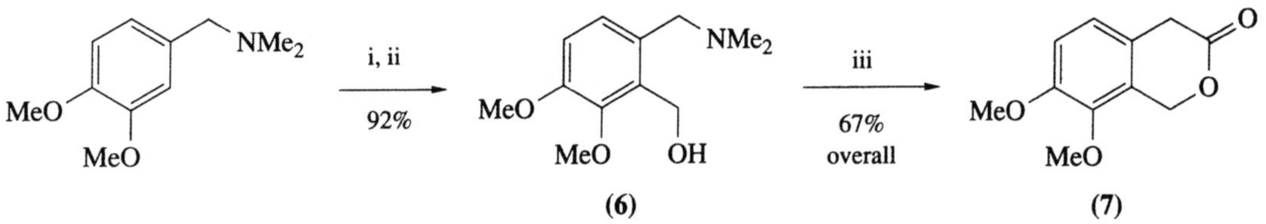

i, Bu^nLi, Et_2O, 0 °C; ii, $(CH_2O)_n$, 15 h; iii, 4 steps

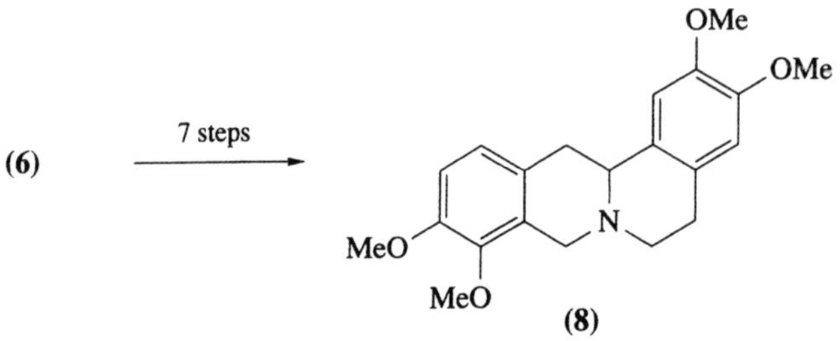

Scheme 4

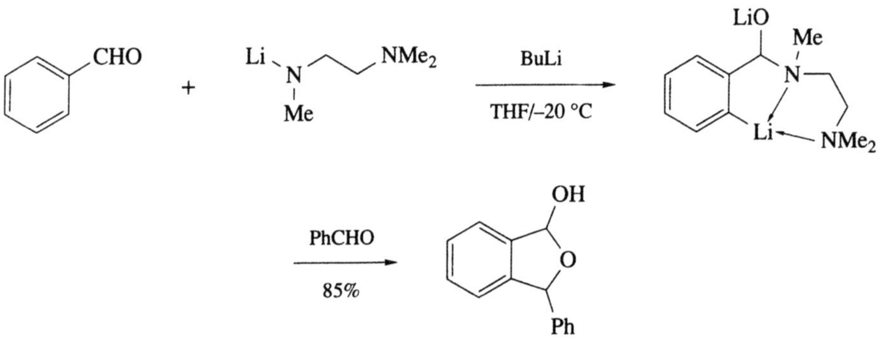

Scheme 5

Gschwend reported the *ortho* lithiation of aniline pivalamides and subsequent addition to nitriles and carbonyls in 1979.[34] A few years later, Wender used a similar aryllithium (**11**), obtained by metal–halogen exchange, in a new synthesis of indoles (Scheme 7).[35] An analogous metalation occurs when *N*-phenylimidazol-2-ones are treated with LDA in THF at –78 °C.[36]

2.1.3.3.4 Amides

Secondary amides may direct *ortho* lithiation, but they must first be deprotonated. This makes them somewhat weaker *ortho* directors than tertiary amides, but they may still serve the purpose quite well. For example, the lithiation and addition of benzamide (**12**) to aldehyde (**13**) was used in a synthesis of 11-deoxycarminomycinone (Scheme 8).[37] More recently, it has been shown that the amide monoanion may be obtained by addition of a phenylsodium to an isocyanate.[38]

The pK_a data listed in Table 1 note that tertiary amides are more acidic than several other functional groups, suggesting thermodynamic acidity as an important component of the mechanistic rationale for their lithiation. Indeed tertiary amides are the strongest *ortho* directing group, taking preference over all other functional groups tested in both intra- and inter-molecular competition experiments.[9,13–15] *N,N,N′,N′*-Tetramethylphosphonic diamides also promote efficient *ortho* metalation.[39] The use of tertiary amides as *ortho* directors was reviewed in 1982,[9] so this discussion will focus only on developments since then.

Beak has reported an aromatic ring annelation using *ortho* lithiation to regioselectively introduce an aldehyde, which is then converted to a carbene. Subsequent Diels–Alder cycloaddition of the resultant isobenzofurans result in adducts that may be oxidized to naphthalenes or reduced to tetralins.[40,41] A typical example is shown in Scheme 9.

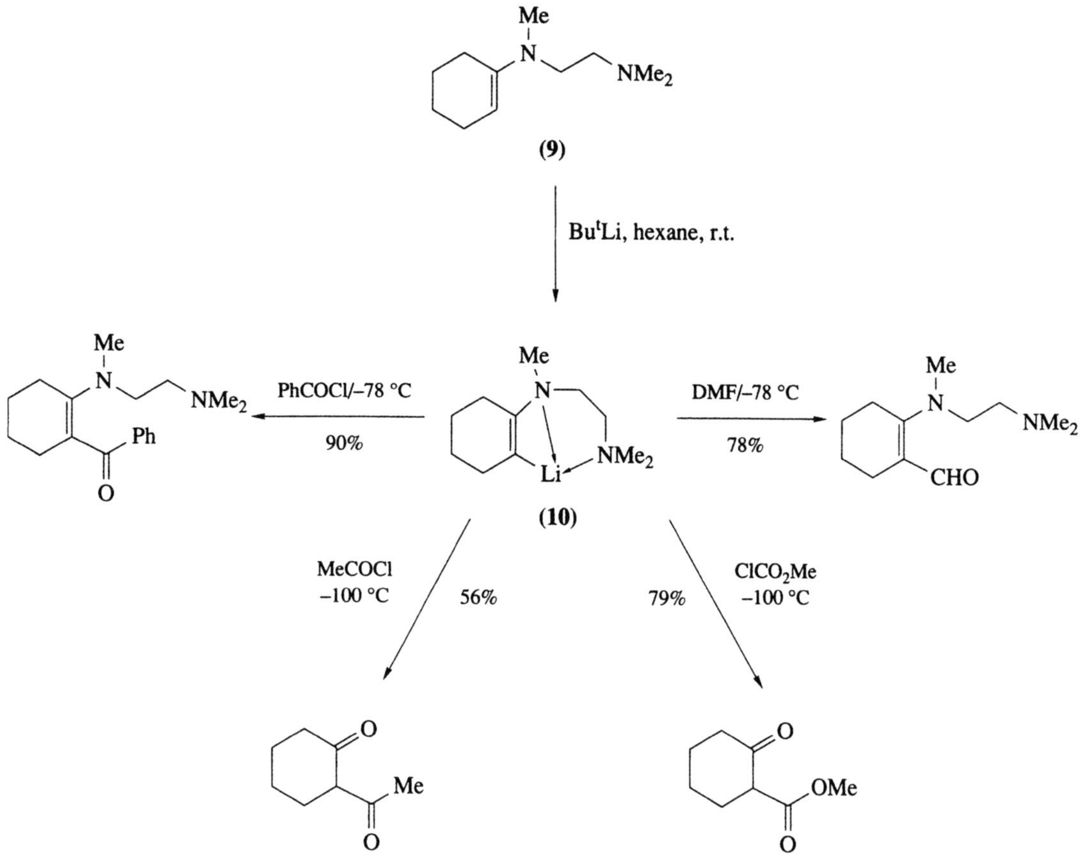

Me
N
NMe$_2$
(9)
Bu^{t}Li, hexane, r.t.
Me
N
NMe$_2$
Ph
O
PhCOCl/–78 °C
90%
Me
N
Li
NMe$_2$
(10)
DMF/–78 °C
78%
Me
N
NMe$_2$
CHO
MeCOCl
–100 °C
56%
79%
ClCO$_2$Me
–100 °C
O
Me
O
O
OMe
O

Scheme 6

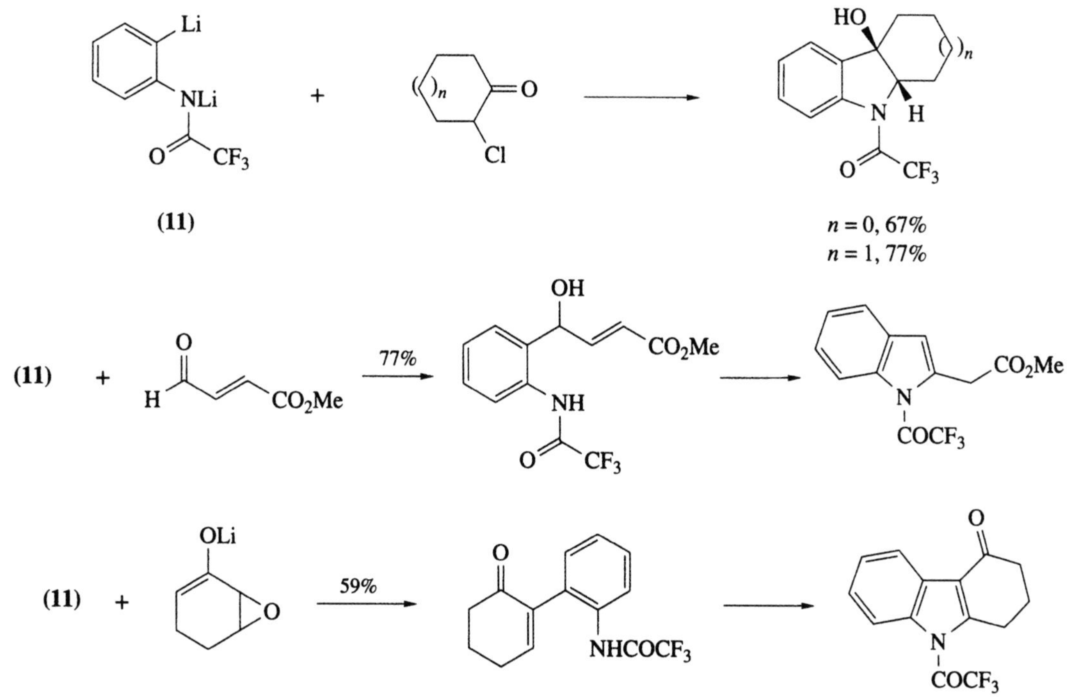

Li
NLi
O
CF$_3$
(11)
+
( )$_n$
O
Cl
HO
( )$_n$
N H
O
CF$_3$
$n = 0$, 67%
$n = 1$, 77%

(11) + O H CO$_2$Me 77% OH CO$_2$Me NH O CF$_3$ CO$_2$Me N COCF$_3$

(11) + OLi O 59% O NHCOCF$_3$ O N COCF$_3$

Scheme 7

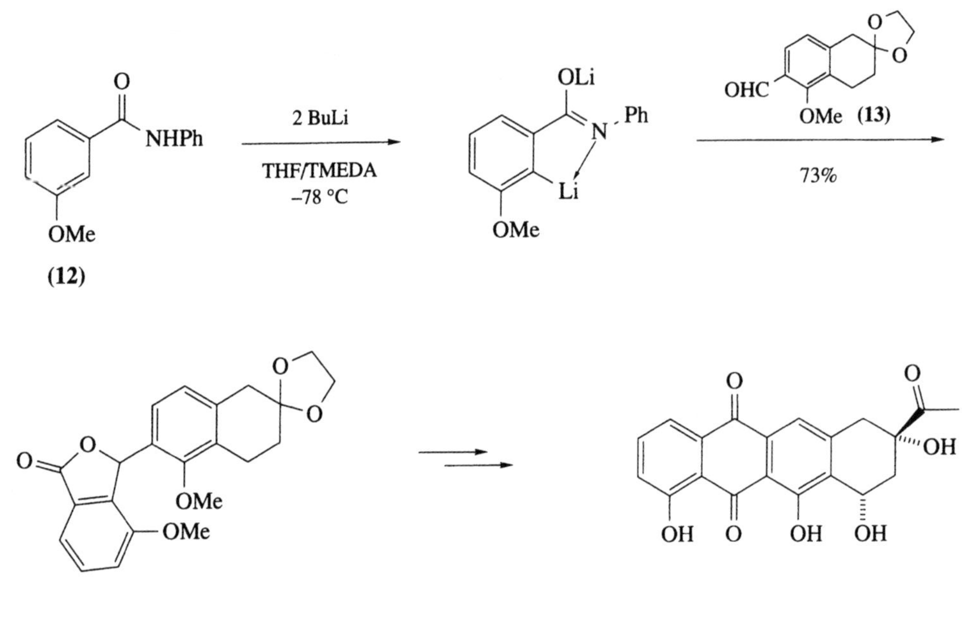

Scheme 8

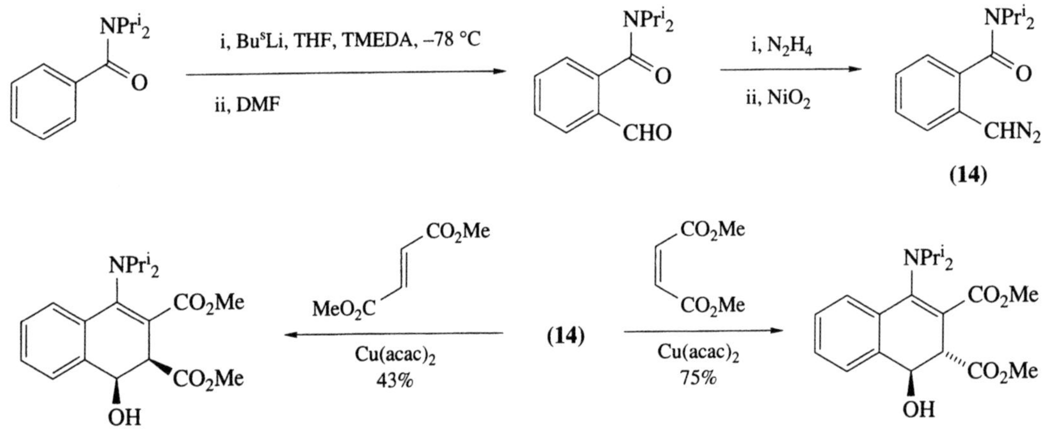

Scheme 9

Ortho lithiation of a tertiary amide and addition to 3-(phenylthio)acrolein, followed by a second lithiation *in situ*, provides a convenient 'one-pot' synthesis of naphthoquinones.[42] One of the several examples reported is shown in Scheme 10.

Two groups report the addition of metalated benzamides to aldehyde carbonyls.[43,44] The methods differ in the metal. Lithiated diethylamides must be transmetalated with $MgBr_2$ before addition to the aldehyde, but β-aminoamides react similarly as the lithium derivative. Following addition, the hydroxyamide is hydrolyzed to afford phthalides in moderate overall yield (Scheme 11).

The lithiation of a 1,6-methano[10]annulenamide occurs selectively at the '*peri*' position,[45] but the lithiation of fused ring aromatics takes place preferentially at the *ortho* (rather than *peri*) position,[46,47] as shown by the examples in Scheme 12. Subsequent transformations of the phthalides obtained in the naphthalene example also illustrate the usefulness of this method for the annelation of aromatic rings. The preference for *ortho* over *peri* lithiation holds true for phenanthrenes as well. The selective metalation of trimethoxyphenanthrenamide (**15**) followed by phthalide synthesis as above constitute the key steps in the synthesis of the phenanthroquinolizidine alkaloid cryptopleurine and the phenanthroindolizidine alkaloid antofine (Scheme 13).[48,49]

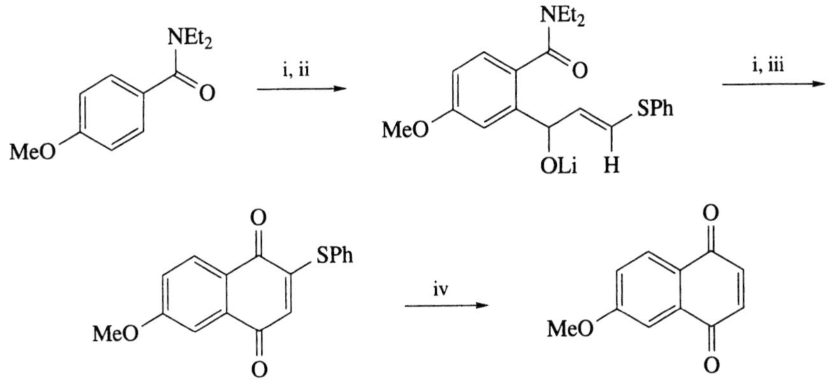

i, Bu^sLi, THF, TMEDA, –78 °C; ii, PhSCH=CHCHO; iii, air, r.t.; iv, Bu_3SnH

Scheme 10

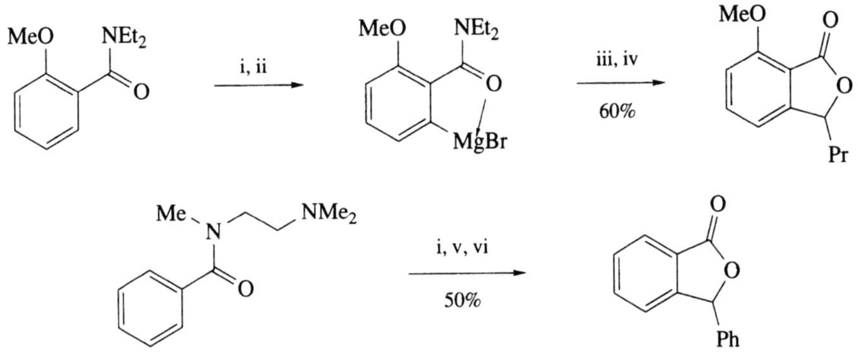

i, Bu^sLi, THF, TMEDA, –78 °C; ii, $MgBr_2$-Et_2O, to r.t.; iii, PrCHO, –78 °C;
iv, TsOH, PhH, reflux; v, PhCHO, to r.t.; vi, 6M HCl, reflux

Scheme 11

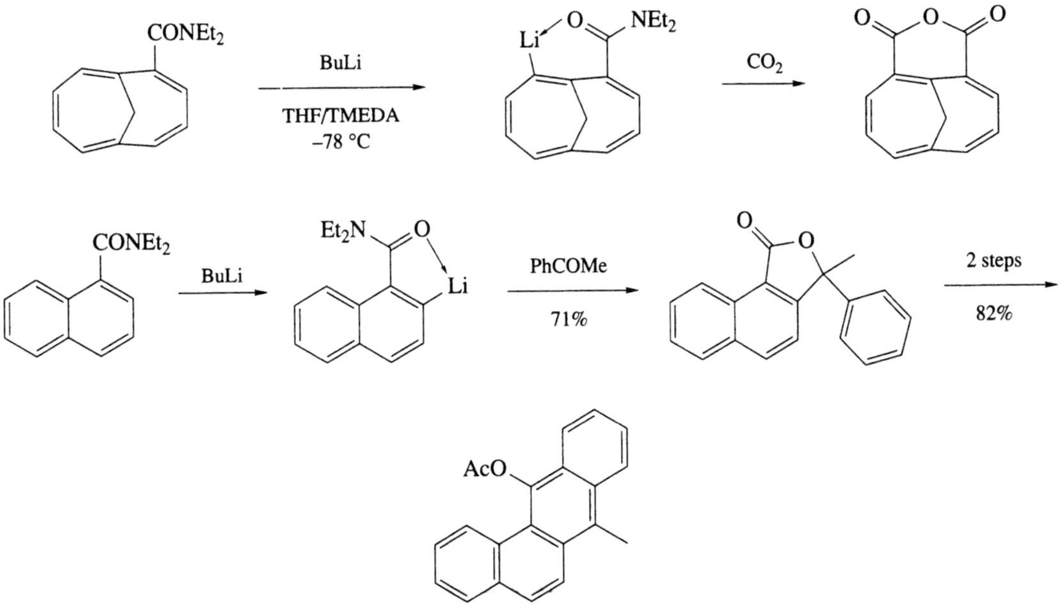

Scheme 12

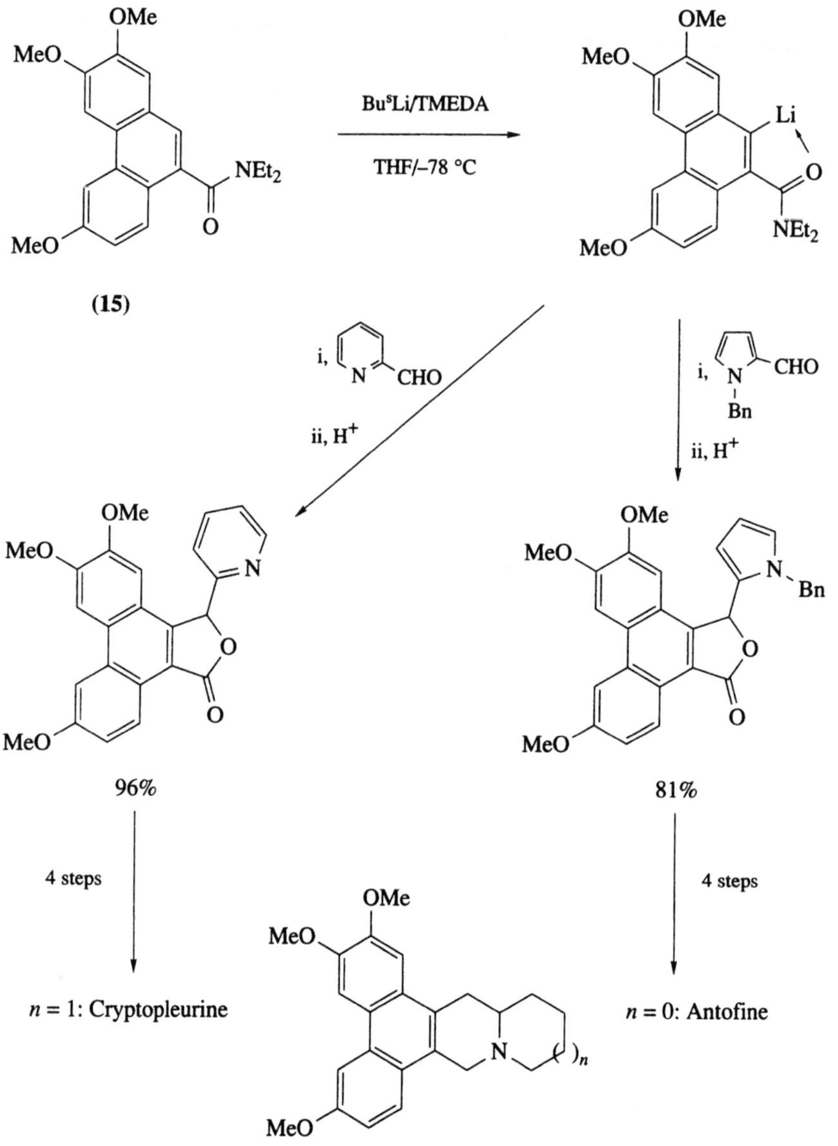

(15)

$Bu^sLi/TMEDA$

THF/–78 °C

i, 2-pyridinecarboxaldehyde (CHO)

ii, H^+

i, N-benzylpyrrole-2-carboxaldehyde (CHO, N–Bn)

ii, H^+

96%

81%

4 steps

4 steps

n = 1: Cryptopleurine

n = 0: Antofine

Scheme 13

2.1.3.3.5 *Nitriles*

In spite of the acidity imparted to the *ortho* position by a cyano group (Table 1), little use has been made of it as a directing group. Two examples are shown in Scheme 14. Because of the susceptibility of the nitrile function to addition by organometallics, the bases used are lithium amides.[50,51]

2.1.3.3.6 *Oxazolines*

Oxazolines may be used as *ortho* directors in phthalide syntheses analogous to those shown in Schemes 11–14. The phthalides derived from oxazolines have been further transformed to polycyclic aromatics by a route that is analogous to, and perhaps more general than, those shown in Scheme 12.[52,53] An efficient synthesis of the lignin lactones chinensin and justicidin along such lines was reported by Meyers (Scheme 15).[54]

The use of a chiral oxazoline to achieve an enantiofacial-selective addition to aldehydes was also reported by Meyers.[55] Although the selectivities were not high (51:49–64:36), the diastereomeric products could be separated by crystallization. A typical example is shown in Scheme 16. In this[55] and another

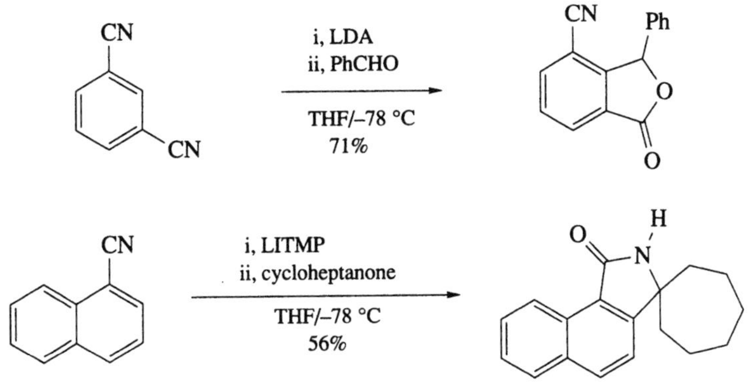

i, LDA
ii, PhCHO
THF/–78 °C
71%

i, LITMP
ii, cycloheptanone
THF/–78 °C
56%

Scheme 14

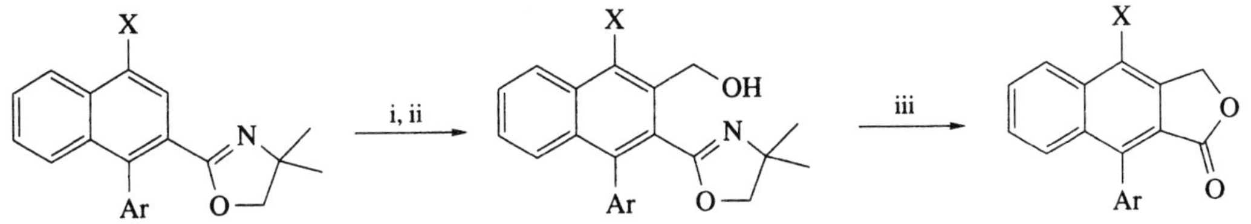

46–50%

Chinensin: X = H; Ar = 3,4-dimethoxyphenyl
Justicidin: X = OMe; Ar = 3,4-methylenedioxyphenyl

i, Bu^sLi, TMEDA, THF, –78 °C; ii, $(CH_2O)_n$ or DMF then $NaBH_4$; iii, HCl

Scheme 15

study,[56] it is shown that the alcohol addition product may subsequently attack the oxazoline causing ring opening.

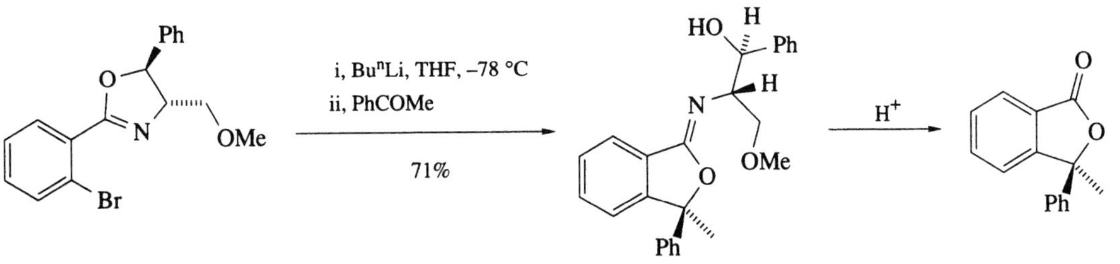

i, Bu^nLi, THF, –78 °C
ii, PhCOMe
71%
H^+

selectivity 64:36; enriched to 100% by crystallization

Scheme 16

2.1.3.3.7 Urethanes

Lithiation of both *N*-phenyl- and *O*-phenyl-urethanes has been reported. The *ortho* lithiation of *N*-*t*-butoxycarbonylaniline and subsequent addition to carbonyls, nitriles and several other electrophiles was first reported by Muchowski in 1980.[57] In some cases the adduct cyclized by attacking the urethane carbonyl. Typical examples are shown in Scheme 17. Lithiation of an *N*-*t*-butoxycarbonylaniline derivative served as one of two directed lithiation steps in Snieckus' synthesis of anthramycin (**17**; Scheme 18).[58]

Treatment of phenothiazines with 2 equiv. of butyllithium affords an *N*,*o*-dilithium species, but reaction with electrophiles occurs at both sites. Katritzky has shown that the sequence of *N*-lithiation, carbonation and *o*-lithiation protects the nitrogen from alkylation (Scheme 19).[59]

The *ortho* lithiation of phenolic urethanes was reported by Snieckus in 1983.[60] In addition to being an efficient *ortho* director (as expected considering the pK_a data in Table 1), the *O*-phenylurethane also is capable of an 'anionic Fries rearrangement'. This rearrangement allows substitution *ortho* to the urethane, then lithiation and rearrangement at the *ortho'* position, resulting in introduction of a tertiary

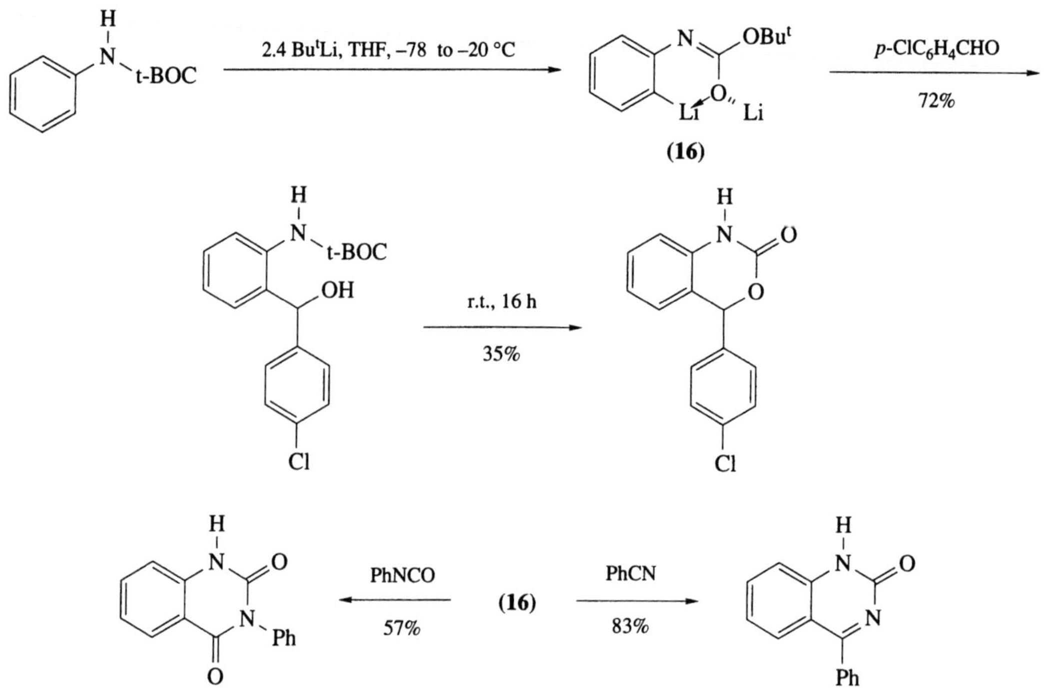

Scheme 17

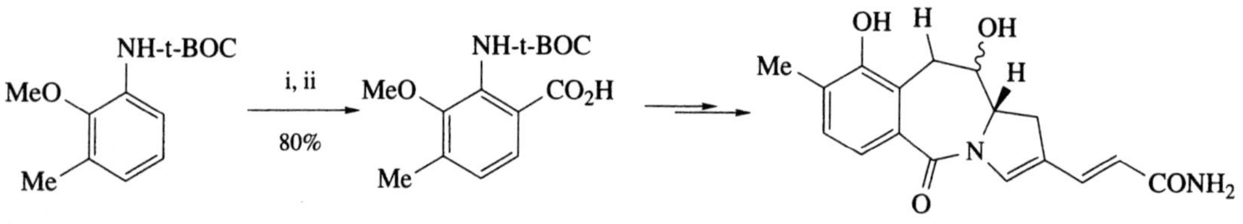

i, 2.5 Bu^tLi, THF, –78 to –20 °C; ii, CO_2 (17)

Scheme 18

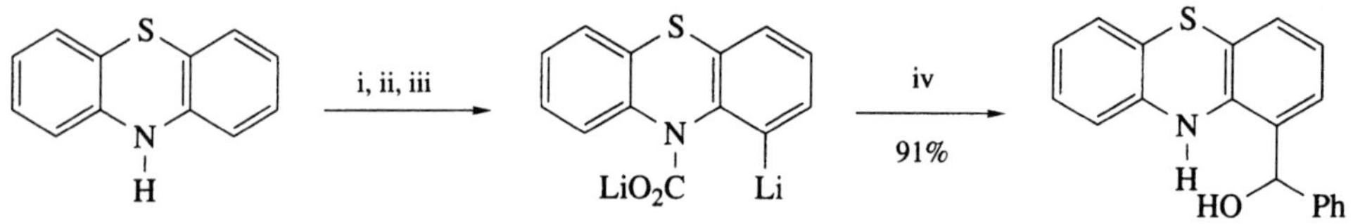

i, Bu^nLi, THF, –78 °C; ii, CO_2; iii, 2 Bu^tLi, –78 to –20 °C; iv, PhCHO, –78 °C, then H^+

Scheme 19

amide. An *ortho* lithiation, carbonyl addition and anionic Fries rearrangement (**18** → **19**) are illustrated by the sequence shown in Scheme 20, which is part of a formal synthesis of ochratoxins A and B.[61] Noteworthy in this scheme is the selective lithiation of (**18**) *ortho* to the urethane, in preference to the tertiary amide.

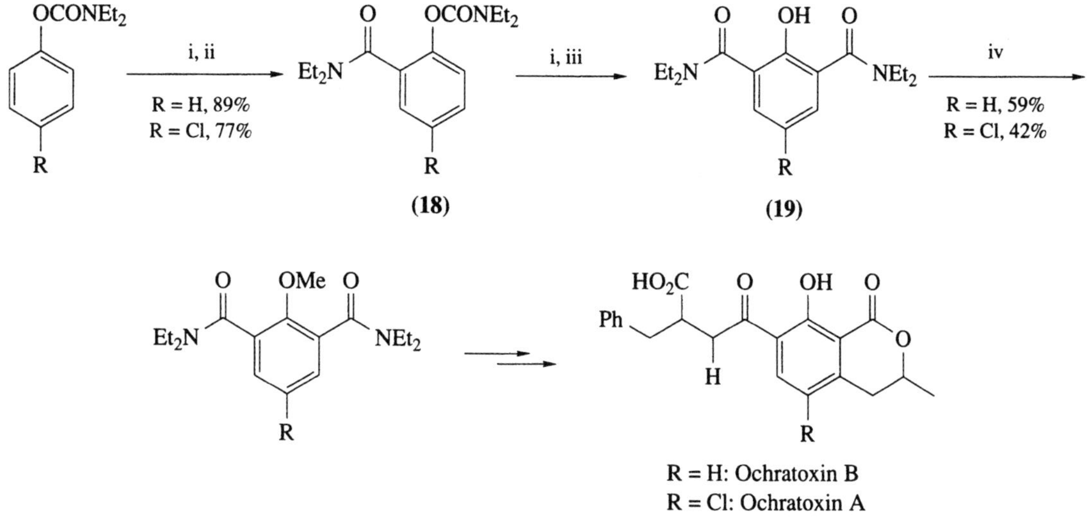

OCONEt₂ / R — i, ii (R = H, 89%; R = Cl, 77%) → (18) — i, iii → (19) — iv (R = H, 59%; R = Cl, 42%) → ⟶ Ochratoxins

R = H: Ochratoxin B
R = Cl: Ochratoxin A

i, Bu^sLi, THF, TMEDA, –78 °C; ii, $ClCONEt_2$; iii, –78 °C to r.t.; iv, MeI, K_2CO_3, acetone

Scheme 20

2.1.3.4 Additions *via* Metalation of Heterocyclic Systems

2.1.3.4.1 α-Metalation

The metalation of heterocycles is possible without the aid of a directing group. This type of reaction is most common in the π-excessive heterocycles, and is most important for thiophenes.[4] For nitrogen heterocycles, examples of unactivated lithiation of π-excessive azoles have been reported, and are summarized below. π-Deficient heterocycles such as pyridine are resistant to unactivated lithiation,[4] although pyridine can be metalated with low regioselectivity using butylsodium.[62] Pyridines also form weak complexes with fluoro ketones; the complex of 4-*t*-butylpyridine and hexafluoroacetone can be lithiated and added to benzaldehyde in 60% yield.[63]

A review on the metalation and metal–halogen exchange reactions of imidazole appeared in 1985.[64] Generally, *N*-protected imidazoles metalate at the 2-position;[65] 1,2-disubstituted imidazoles usually metalate at the 5-position, unless sterically hindered.[64] Even 2,5-dilithiation of imidazoles has been achieved.[66] 1-Substituted 1,3,4-triazoles can be metalated at the 5-position and added to carbonyls in good yield.[67] Oxazoles are easily lithiated at the 2-position, but the resultant anion readily fragments.[68] 1-(Phenylthiomethyl)benzimidazole can be lithiated at the 2-position at low temperature (Scheme 21), but higher temperatures afford rearrangement products.[69]

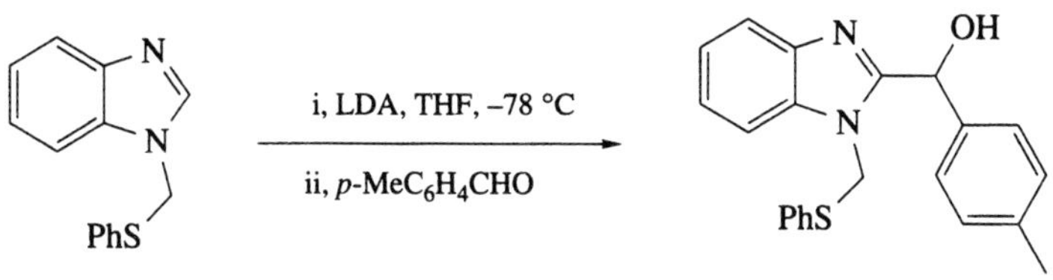

i, LDA, THF, –78 °C; ii, *p*-MeC_6H_4CHO

Scheme 21

Pyrrolopyridines can be lithiated at the 2-position,[70] in direct analogy to pyrrole itself. The reaction is sufficiently mild that it has been applied to the functionalization of purine nucleosides, as illustrated by the examples in Scheme 22.[71]

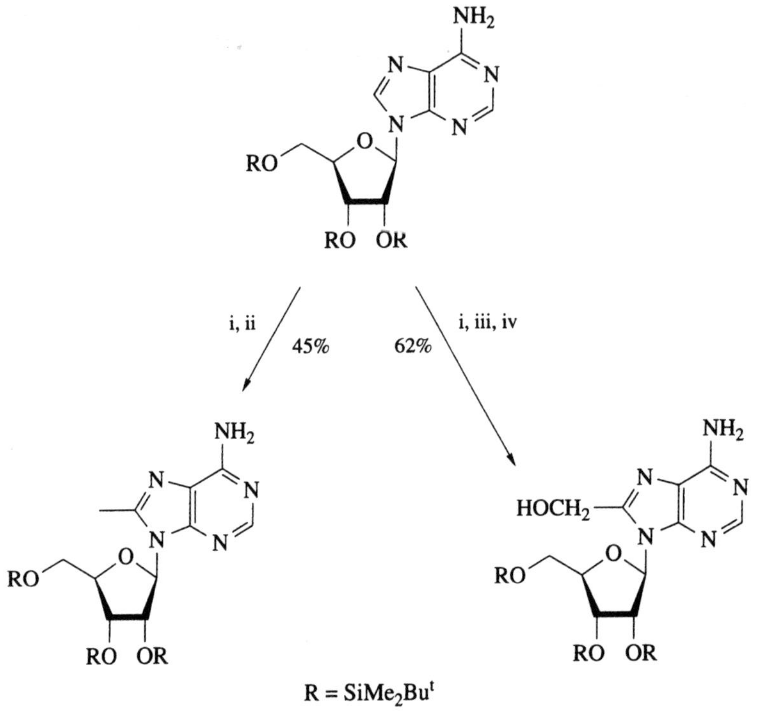

i, ii 45%

62% i, iii, iv

R = $SiMe_2Bu^t$

i, 5 LDA, THF, –78 °C; ii, MeI; iii, HCO_2Me; iv, $NaBH_4$

Scheme 22

2.1.3.4.2 **Ortho** ***metalation***

As will be seen in the examples below, the use of an activating group usually determines the site of lithiation for π-excessive heterocycles, and facilitates the lithiation of π-deficient ones.

Furans and thiophenes normally undergo α-lithiation,[4] but when substituted at the 2-position by an activating group, a competition arises between metalation at the 3-position (*ortho* lithiation) and the 5-position (α-lithiation).[4,72–74] 2-Oxazolinylthiophenes may be lithiated selectively at either the 3- or 5-position by adjusting the reaction conditions;[73] tertiary amides give little or no *ortho* selectivity,[74] but secondary amides direct *ortho* lithiation reasonably well, as seen in Scheme 23.[74] Both thiophenes and furans that are substituted with an oxazoline or tertiary amide at the 2-position may be dilithiated at the 3- and 5-positions.[75,76] Although secondary amides are less successful at directing *ortho* lithiation of furans than thiophenes,[74] *N,N,N′,N′*-tetramethyldiamido phosphates work quite well. Subsequent hydrolysis affords access to butenolides.[77] A typical example is shown in Scheme 24.

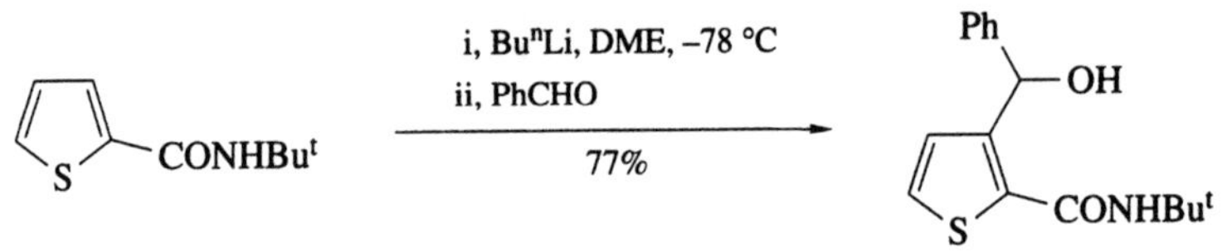

$CONHBu^t$ → i, Bu^nLi, DME, –78 °C; ii, PhCHO; 77% → $CONHBu^t$

Scheme 23

$P(NMe_2)_2$ → i, Bu^nLi, THF, –75 °C; ii, PhCOMe; 65% → $P(NMe_2)_2$ → HCO_2H; 100%

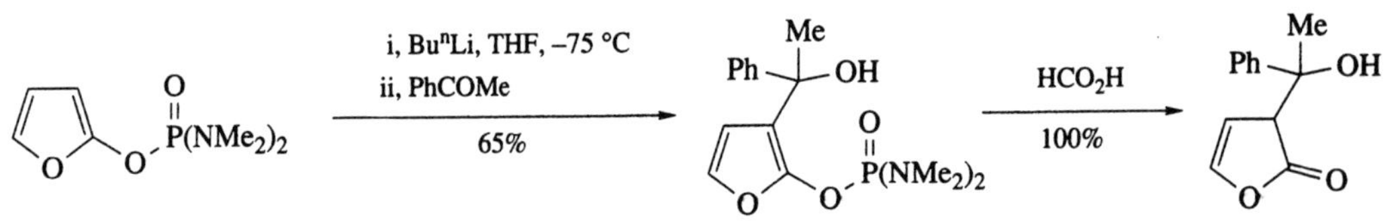

Scheme 24

N-Substituted pyrroles and indoles normally undergo lithiation at the 2-position (α-metalation),[4,78] so when there is an *ortho* director on the nitrogen, metalation is facilitated. For example, the lithiation of *N*-t-BOC pyrrole and its addition to benzaldehyde occurs in 75% yield.[79] Similar lithiations of *N*-t-BOC indole[79] and *N*-benzenesulfonylindole[80,81] have also been reported. Examples of these reactions are illustrated in Scheme 25.

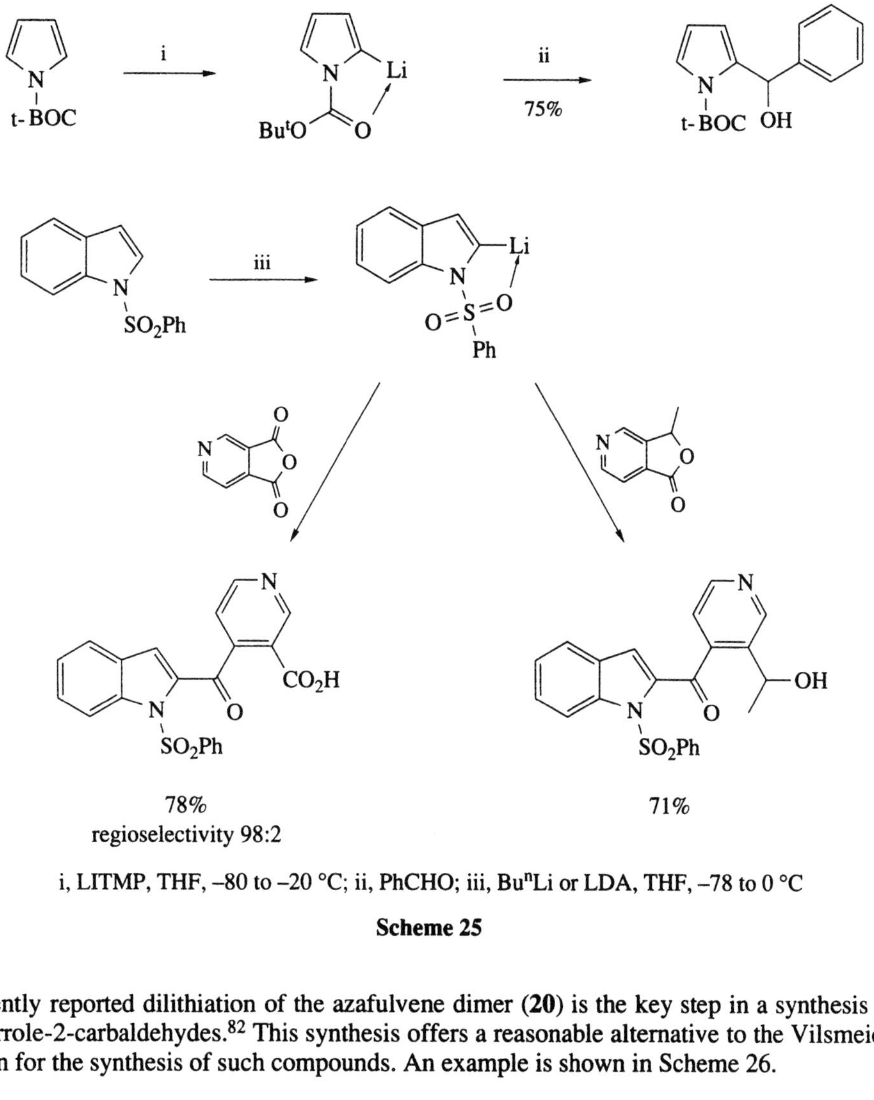

t-BOC N → (i) Li, N, Bu^tO, O → (ii, 75%) N, t-BOC, OH

N, SO_2Ph → (iii) Li, N, O=S=O, Ph

N, O, O, O → N, CO_2H, N, O, SO_2Ph — 78%, regioselectivity 98:2

N, O, O → N, OH, N, O, SO_2Ph — 71%

i, LITMP, THF, –80 to –20 °C; ii, PhCHO; iii, Bu^nLi or LDA, THF, –78 to 0 °C

Scheme 25

The recently reported dilithiation of the azafulvene dimer (**20**) is the key step in a synthesis of 5-substituted pyrrole-2-carbaldehydes.[82] This synthesis offers a reasonable alternative to the Vilsmeier–Haack formylation for the synthesis of such compounds. An example is shown in Scheme 26.

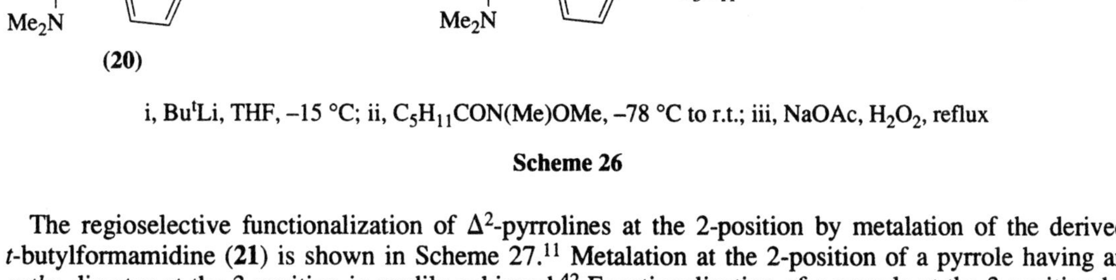

NMe_2, N, N, Me_2N (**20**) → (i, ii) C_5H_{11}, O, NMe_2, N, N, O, C_5H_{11}, Me_2N → (iii, 54%) C_5H_{11}, O, N, H, CHO

i, Bu^tLi, THF, –15 °C; ii, $C_5H_{11}CON(Me)OMe$, –78 °C to r.t.; iii, NaOAc, H_2O_2, reflux

Scheme 26

The regioselective functionalization of Δ^2-pyrrolines at the 2-position by metalation of the derived *t*-butylformamidine (**21**) is shown in Scheme 27.[11] Metalation at the 2-position of a pyrrole having an *ortho* director at the 3-position is readily achieved.[42] Functionalization of a pyrrole at the 3-position by *ortho* metalation of a 2-substituted derivative is more problematic, with a mixture of 3- and 5-substituted products usually resulting,[74,83] as summarized in a recent review.[84] An intriguing possibility is *ortho* palladation, as shown in Scheme 28.[85]

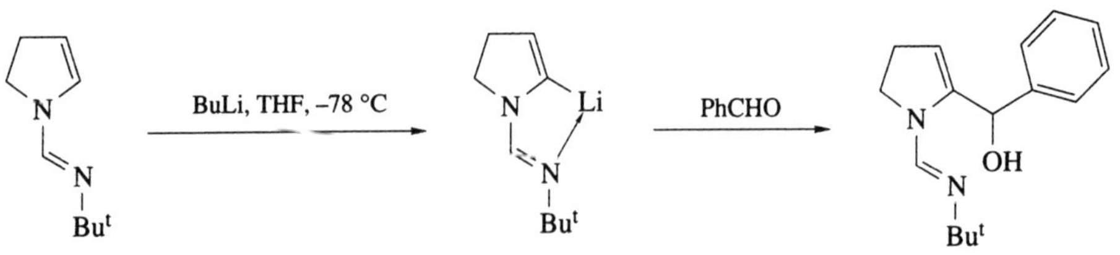

BuLi, THF, –78 °C; PhCHO

Scheme 27

Li_2PdCl_4, MeOH, r.t.; CO, MeOH

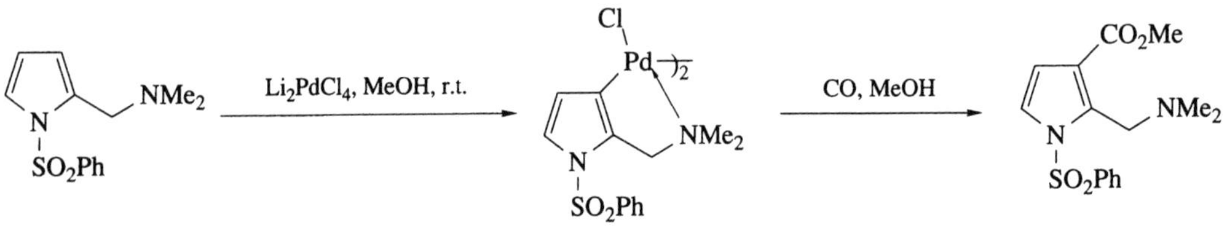

Scheme 28

Ortho lithiation of 2-substituted indoles occurs readily, but fragmentation to an alkynylanilide may occur in some instances.[86] The use of a 2-pyridyl group to facilitate the 3-lithiation of an indole was recently used in a synthesis of some indolo[2,3-*a*]quinolizine alkaloids;[87] an example is the synthesis of flavopereirine (**22**; Scheme 29).

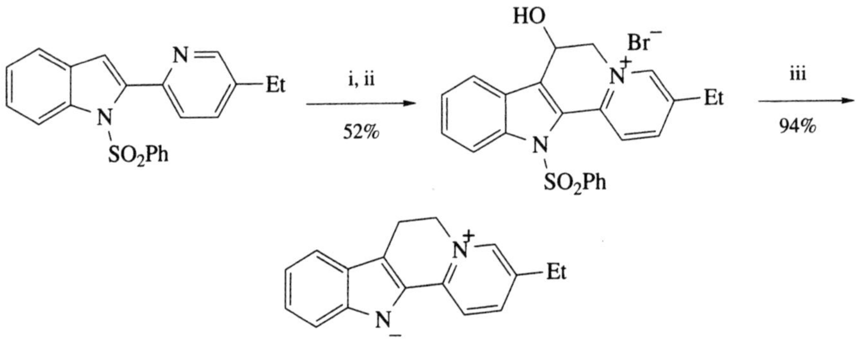

(22)

i, Bu^nLi, THF, –78 °C; ii, $BrCH_2CHO$, then AcOH; iii, NaOH, H_2O, MeOH, reflux

Scheme 29

In pyridines,[88] the *ortho* lithiation of 2-substituted secondary[89–91] and tertiary[90,92,93] amides and sulfonamides[94] has been reported. All three afford regioselective metalation at the 3-position, as illustrated by Kelly's synthesis of berninamycinic acid (**23**; Scheme 30).[95] A recent development is the use of catalytic amounts of diisopropylamine for the *ortho* metalation of 2-methoxypyridine.[96]

Pyridines substituted at the 3-position by a urethane,[97] halogen,[98,99] secondary amide[90,100] or tertiary amide[90,92] metalate regioselectively at the 4-position. An exception appears to be 3-alkoxypyridines, which metalate selectively at the 2-position.[101] The lithiation of halopyridines may be accomplished by metal–halogen exchange, or by *ortho* lithiation. In the latter instance, lithium amides are used as the base, and the temperature must be kept low to prevent pyridyne formation.[98,102,103] For 3-halo-4-lithiopyridines, the order of stability is F >> Cl > Br >> I.[98] Selective lithiation at the 4-position, directed by a tertiary amide, has been used in the synthesis of bostrycoidin (**24**)[104] and sesbanine (**25**; Scheme 31).[105]

Pyridines having a directing group at the 4-position undergo *ortho* metalation. Groups reported in this category include sulfonamides,[94] secondary amides,[90,91] tertiary amides,[106,107] halogens[98] and oxazolines.[56,108,109] As was the case in the carbocyclic series, a pyridine having directing groups in

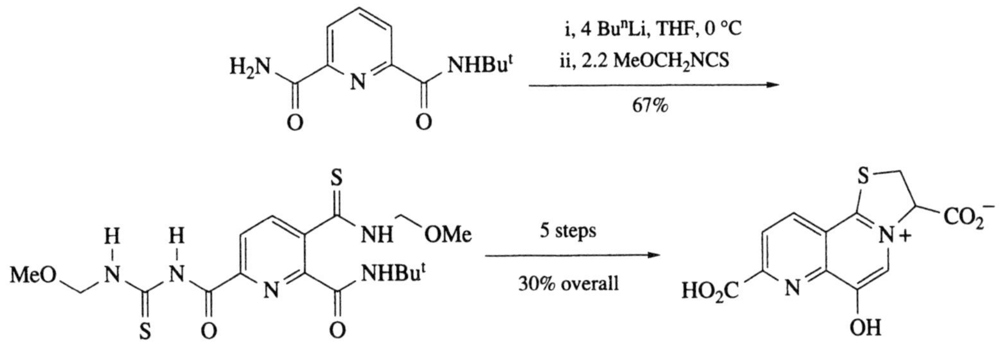

i, 4 Bu^nLi, THF, 0 °C
ii, 2.2 $MeOCH_2NCS$
67%

5 steps
30% overall

(23)

Scheme 30

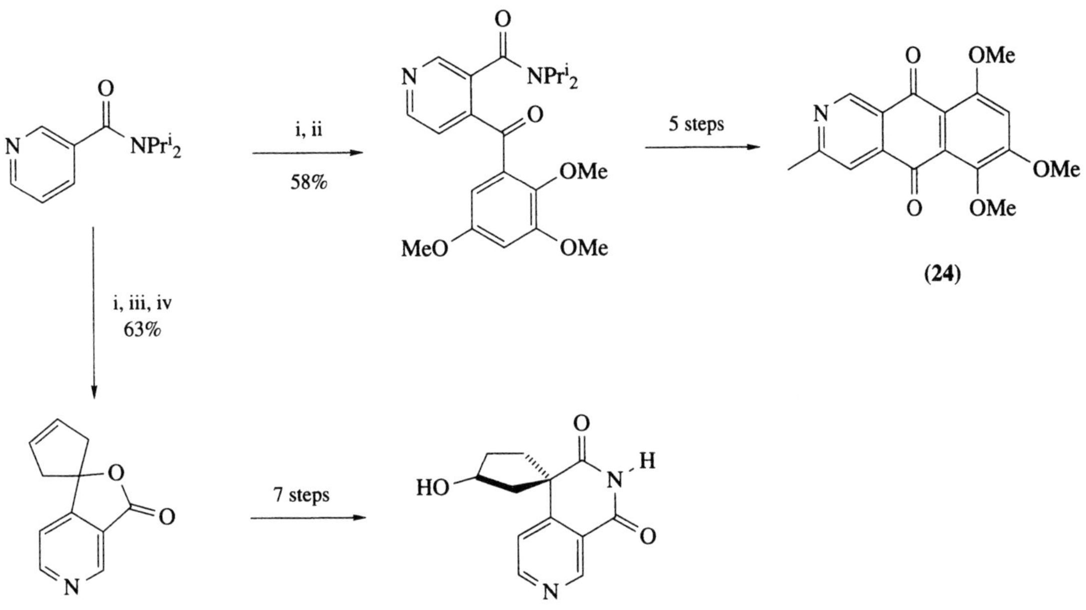

i, ii
58%

5 steps

(24)

i, iii, iv
63%

7 steps

(25)

i, LITMP, DME, –78 °C; ii, 2,3,5-$(MeO)_3C_6H_2CONMe_2$; iii, cyclopent-3-en-1-one; iv, TFA, CH_2Cl_2

Scheme 31

positions 2 and 4 is metalated in between, at the 3-position. This has been used in a synthesis of (*g*)-fused isoquinolines, whose key step is shown in Scheme 32.[109]

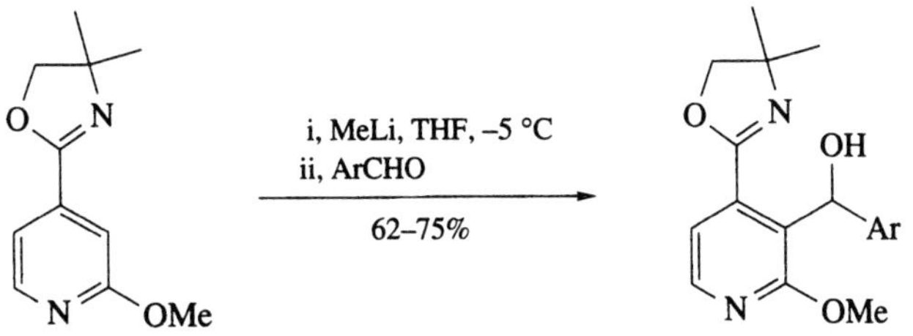

i, MeLi, THF, –5 °C
ii, ArCHO
62–75%

Ar = 1-naphthyl, 2-naphthyl, 3-methoxyphenyl, 3-thienyl

Scheme 32

2.1.4 sp^3-HYBRIDIZED CARBANIONS

2.1.4.1 Introduction

The deprotonation of an sp^3-hydrogen α to a nitrogen atom, although a relatively recent development synthetically, is now a common phenomenon. It has been stated that the 'normal' reactivity of carbon atoms attached to nitrogen or oxygen is a^1, meaning it is an acceptor site in polar reactions.[110] The deprotonation of such a site has therefore been called a charge affinity inversion or reactivity umpolung.[111] There is ample precedent in the literature to support the notion of a^1 reactivity α to nitrogen, of course, but we contend that this classification is inappropriate. The α-deprotonation of dimethyldodecylamine[112] and of triethylamine[113] were reported over 20 years ago, although the former was in low yield and no products of reaction of the latter with electrophiles were found. However in 1984, Albrecht reported that *s*-butylpotassium readily deprotonates *N*-methylpiperidine, *N*-methylpyrrolidine and trimethylamine, and that the derived organometallics add readily to aldehydes, ketones and alkyl halides.[114] In 1987, it was found that *t*-butyllithium deprotonates a methyl group of *N,N,N',N'*-tetramethylethylenediamine (TMEDA), whereas *n*-butylpotassium ($Bu^nLi/KOBu^t$) deprotonates a methylene.[115]

Furthermore, it was first shown nearly 20 years ago that α-lithioamines could be produced by the transmetalation of α-aminostannanes.[116] Thus, the thermodynamic stability of α-amino anions is reasonably good, even if the kinetic acidity of the conjugate acids is low. Although some authors have suggested that α-amino carbanions constitute a reversal of 'normal' reactivity,[110,111] we suggest that this notion is inappropriate and should be discontinued.

Most of the chemistry described in the following sections involves two types of anion stabilization by nitrogen: resonance and dipole stabilization. Since these topics were reviewed in 1984,[111] this discussion is restricted to recent developments. By and large, the chemistry of resonance-stabilized species (azaenolates) is covered in Volume 2 of this series; however, there are a few species whose inclusion here seems appropriate and consistent with the present discussion. One such example is removal of a benzylic proton by a base that is coordinated to a nitrogen or nitrogen-containing functional group.

Another resonance-stabilized species discussed here is nitrosamine anions, whose chemistry has been reviewed several times.[111,117–119] Cyclic nitrosamines normally lose the axial proton *syn* to the nitrosamine oxygen,[120] and alkylate by axial approach of the electrophile.[121] In these respects, nitrosamine anions are similar to their isoelectronic counterparts, oxime dianions, as shown in equation (4).[122,123]

Ph, N, N, O → (i, BuLi; ii, MeCOMe) → Ph, N, N, O, H, OH (4)

Dipole-stabilization is a term coined by Beak to describe the situation that results when a carbanion is stabilized by an adjacent dipole.[117] Such a situation arises when, for example, an amide is deprotonated α to nitrogen. The chemistry of these systems has been reviewed,[111,117] so only a few pertinent points will be made here. Firstly, metalation occurs *syn* to the carbonyl oxygen, and when the system is cyclic, the equatorial proton is removed selectively, and the electrophile attacks equatorially, as shown in equation (5).[124,125] Thus, in contrast to nitrosamines, amide anions give the less stable equatorial product.[124,125]

Ph, N, Ar, O → (i, BuLi; ii, MeCOMe) → Ph, N, Ar, O, H, OH (5)

Considerable theoretical work has been done to explain equatorial alkylations such as the one illustrated in equation (5).[125–127] The simplest model of a dipole-stabilized anion is *N*-methylformamide anion, $HCONHCH_2^-$. The geometric requirements for this system are strict: the lone pair on carbon is 16–18 kcal mol^{-1} more stable when oriented in the nodal plane of the amide π-system than when rotated 90° into conjugation.[126,127] To explain the removal of an equatorial hydrogen, it has been suggested that 'the electronic effects of extended amide conjugation should be felt relatively early along the reaction coordinate for proton removal'.[126] Thus equatorial protons are removed for stereoelectronic reasons, and the anions of piperidinecarboxamides[124,125] and amidines[128,129] are configurationally stable and do not invert. Note, however, that if the carbanion is also benzylic, pyramidal inversion is possible.[130,131] In spite of the intervention of pyramidal inversion in benzylic systems, the anion is still in the nodal plane of the amide π-system. Semiempirical calculations on lithiated oxazolines[129,131] and an X-ray crystal structure of a pivaloylisoquinoline Grignard[132] confirm the location of the carbon–metal bond in or near the nodal plane of the amide or amidine. At the same time, the theoretical and crystal structures show considerable overlap with the benzene *p*-orbitals, thus providing a possible explanation for the inversion process.

The mechanism of the deprotonation of dipole-stabilized anions has been studied in detail. It has been shown by IR spectroscopy that a preequilibrium exists between the butyllithium base and the amide[133,134] or amidine,[135] forming a coordination complex prior to deprotonation. A recent mechanistic study has shown that, in cyclohexane solvent, this prior coordination is between the amide (or added TMEDA) and aggregated *s*-butyllithium, and that the effect of the coordination is to increase the reactivity of the complex.[134] The diastereoselectivity of proton removal in chiral benzylic systems has also been examined,[130,131,136] but since the anions invert, this selectivity is of little consequence in the alkylation step.

2.1.4.2 Additions *via* Metalation of Acyclic Systems

The directed metalation of aromatic systems that was discussed in Section 2.1.3.3 has one ramification that was not mentioned there: the directed lithiation of an *o*-methyl group. Although the resultant species is formally a resonance-stabilized anion, and therefore covered in Volume 2 of this series, we mention it here for consistency with the other topics covered. In particular, the examples that have appeared in recent years involve substrates having a methyl *ortho* to a tertiary amide. Intentional use of such a directed lithiation has been used in the synthesis of the isocoumarin natural products hydrangenol and pyllodulcin.[137,138] Interestingly, the directed metalation of 5-methyl-oxazoles and -thiazoles occurs in preference to deprotonation at a 2-methyl group (azaenolate) (Scheme 33).[139]

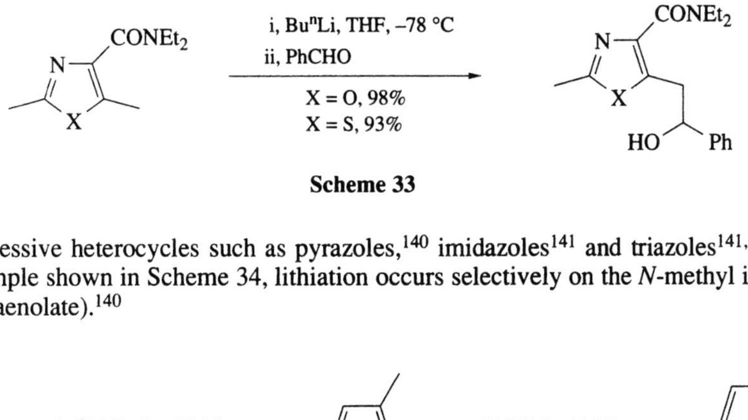

i, Bu^nLi, THF, –78 °C; ii, PhCHO; X = O, 98%; X = S, 93%

Scheme 33

N-Alkyl π-excessive heterocycles such as pyrazoles,[140] imidazoles[141] and triazoles[141,142] can be lithiated. In the example shown in Scheme 34, lithiation occurs selectively on the *N*-methyl in preference to the *C*-methyl (azaenolate).[140]

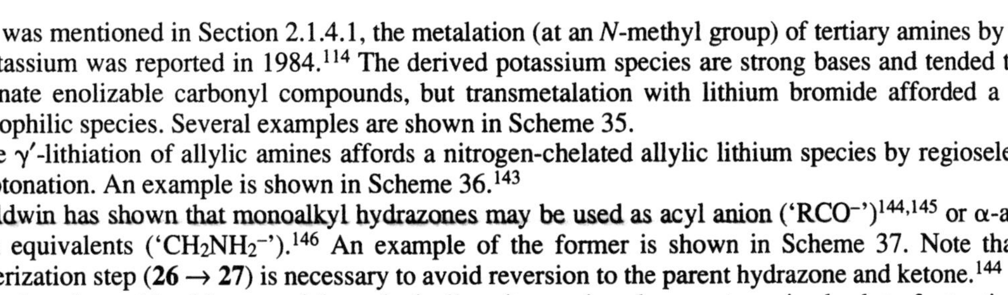

Bu^nLi, Et_2O, –78 °C; PhCOMe, 23 °C, 78%

Scheme 34

As was mentioned in Section 2.1.4.1, the metalation (at an *N*-methyl group) of tertiary amines by *s*-butylpotassium was reported in 1984.[114] The derived potassium species are strong bases and tended to deprotonate enolizable carbonyl compounds, but transmetalation with lithium bromide afforded a more nucleophilic species. Several examples are shown in Scheme 35.

The γ′-lithiation of allylic amines affords a nitrogen-chelated allylic lithium species by regioselective deprotonation. An example is shown in Scheme 36.[143]

Baldwin has shown that monoalkyl hydrazones may be used as acyl anion ('RCO^-')[144,145] or α-amino anion equivalents ('$CH_2NH_2^-$').[146] An example of the former is shown in Scheme 37. Note that the isomerization step (**26** → **27**) is necessary to avoid reversion to the parent hydrazone and ketone.[144]

In spite of considerable potential synthetically, nitrosamines have not received a lot of attention because of their high toxicity.[111,117–119] One potentially important development, reported by Seebach, is a

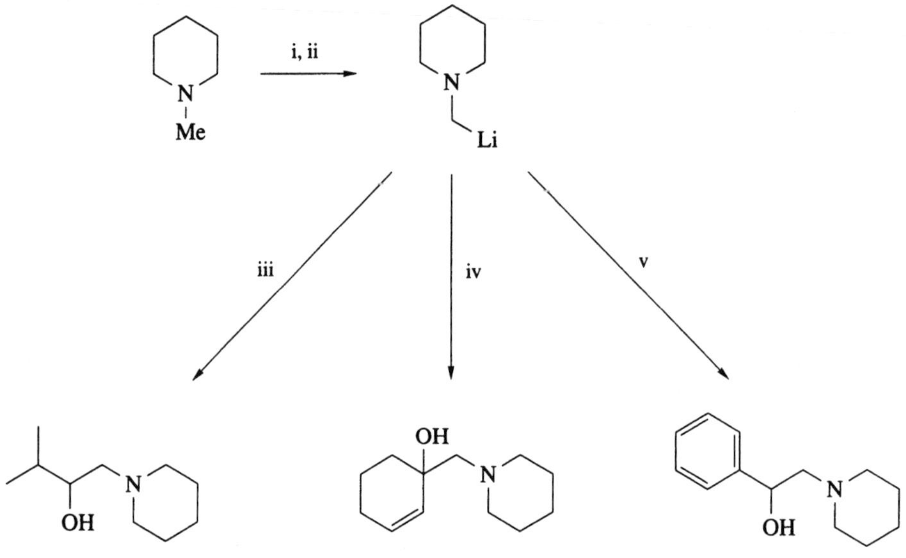

i, ii

iii iv v

62% 50% 73%

i Bu^sLi, $KOBu^t$, isopentane, –78 °C to 0 °C; ii, LiBr, Et_2O, –78 °C to 0 °C; iii, Pr^iCHO; iv, cyclohexenone, HMPA; v, PhCHO

Scheme 35

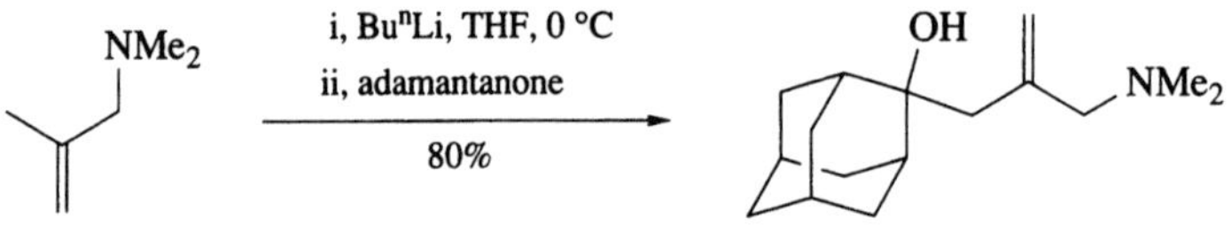

i, Bu^nLi, THF, 0 °C
ii, adamantanone
80%

Scheme 36

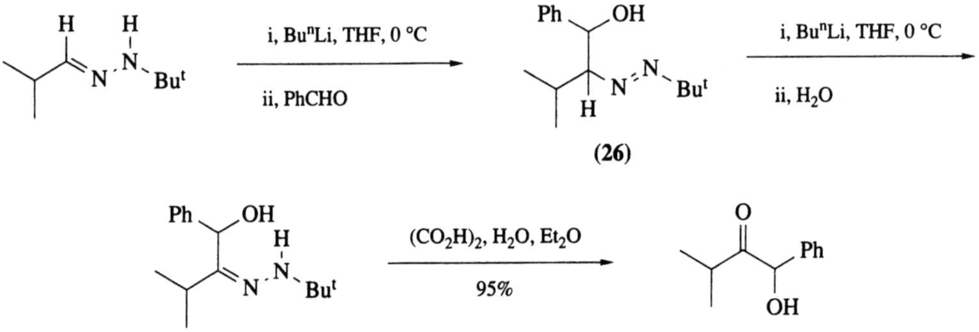

i, Bu^nLi, THF, 0 °C
ii, PhCHO

(26)

i, Bu^nLi, THF, 0 °C
ii, H_2O

(27)

$(CO_2H)_2$, H_2O, Et_2O
95%

Scheme 37

one-pot alkylation and reduction protocol for the synthesis of secondary amines.[147] An example is shown in Scheme 38.

An interesting method for the synthesis of amines and the homologation of carbonyl compounds has been reported that utilizes the condensation of a lithiated formamidine with a carbonyl compound.[10,11] Typical examples are shown in Scheme 39.

A number of heterocyclic *N*-alkyllactams have been metalated to dipole-stabilized anions, and the yields of addition to carbonyls are reasonably good.[36] An experimental and theoretical study of the competitive metalation to form enolates or dipole-stabilized anions of a series of alicyclic *N*-benzyllactams has been reported by Meyers and Still.[148] Experimentally, the regioselectivity of the deprotonation varies inconsistently with ring size. Specifically: 5-, 6- and 11-membered rings are deprotonated in the ring to form enolates (**29**), whereas 7- and 8-membered rings are deprotonated at the *N*-benzyl to form dipole-

i, Bu^{n}Li; ii, piperonal; iii, $LiAlH_4$; iv, Raney Ni, H_2

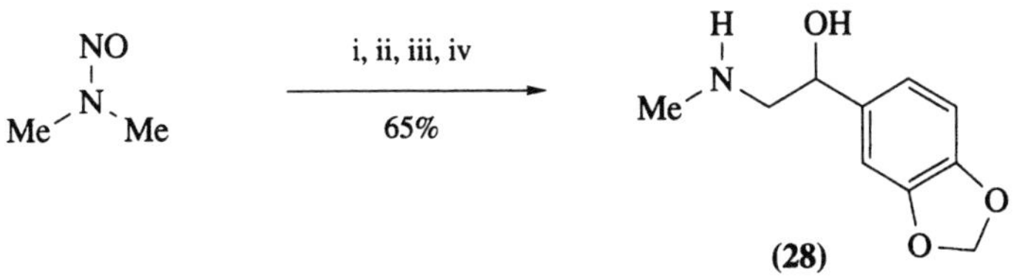

Scheme 38

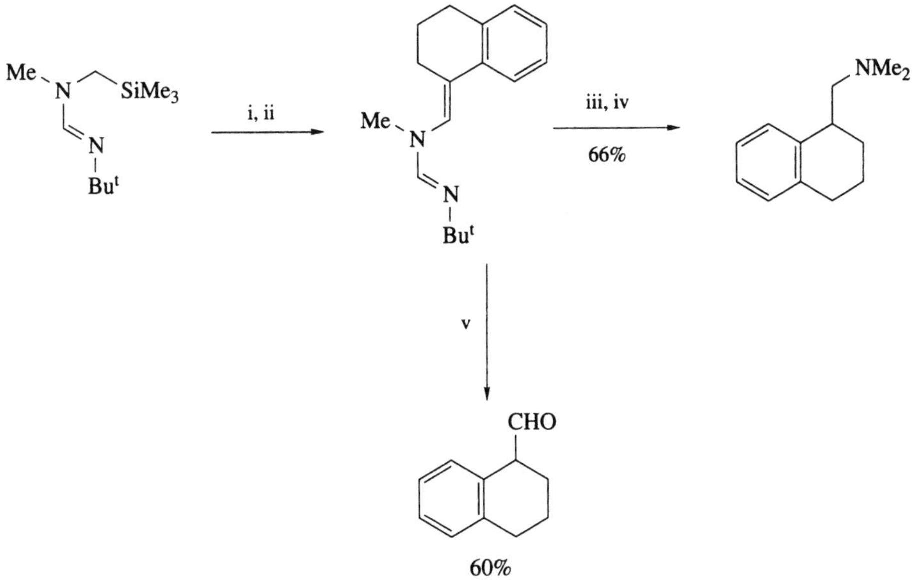

i, Bu^{s}Li, THF, –78 to –20 °C; ii, α-tetralone, –78 °C to r.t.; iii, $NaBH_4$; iv, dilute HCl; v, N_2H_4, H^+

Scheme 39

stabilized anions (**30**; Scheme 40). In contrast, 9-, 10- and 13-membered ring lactams give mixtures of enolization and benzylic metalation. A simplistic molecular mechanics model was found to predict the regioselectivity that evaluates the strain energy necessary to achieve the optimal geometry for enolate formation. The model restrains the lactam α-hydrogen in a stereoelectronically preferred 90° O═C—C—H alignment, then compares the resulting minimized energy calculated using the MM2 force field with the global minimum for each lactam. When the differences in strain energies are small (<0.1 kcal mol^{-1}), enolization is the preferred course of deprotonation. When the differences are between 1.3 and 2.2 kcal mol^{-1}, mixtures of enolization and benzylic metalation are found, and when the differences are large, 2.25 and ≥ 3.5 kcal mol^{-1}, exclusive benzylic metalation is found.

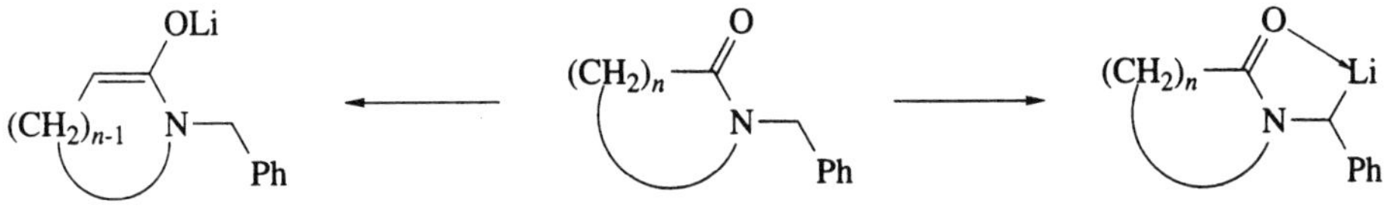

Scheme 40

The formation of α-aminolithium reagents by transmetalation of α-aminostannanes was first reported in the early 1970s.[116] However, the exploitation of this protocol was delayed until better methods were developed for the synthesis of the requisite stannanes. Quintard has shown that tributyltin Grignard reagents afford α-aminostannanes when reacted with amino acetals[149] or iminium ions,[150] as shown in Scheme 41. Transmetalation of the α-aminostannanes and addition to aldehydes and ketones has been reported for simple tertiary amines[149,150] as well as carbamates.[151] Representative examples are shown in Scheme 42.

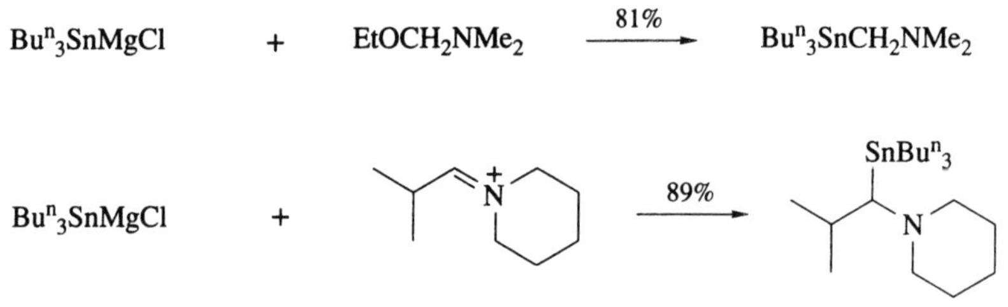

$Bu^n{}_3SnMgCl$ + $EtOCH_2NMe_2$ —81%→ $Bu^n{}_3SnCH_2NMe_2$

$Bu^n{}_3SnMgCl$ + (iminium) —89%→ ($SnBu^n{}_3$ product)

Scheme 41

$Me_2NCH_2SnBu_3$ —i, ii, iii; 95%→ Macromerine

i, ii, iv; 69%

i, ii; 82%

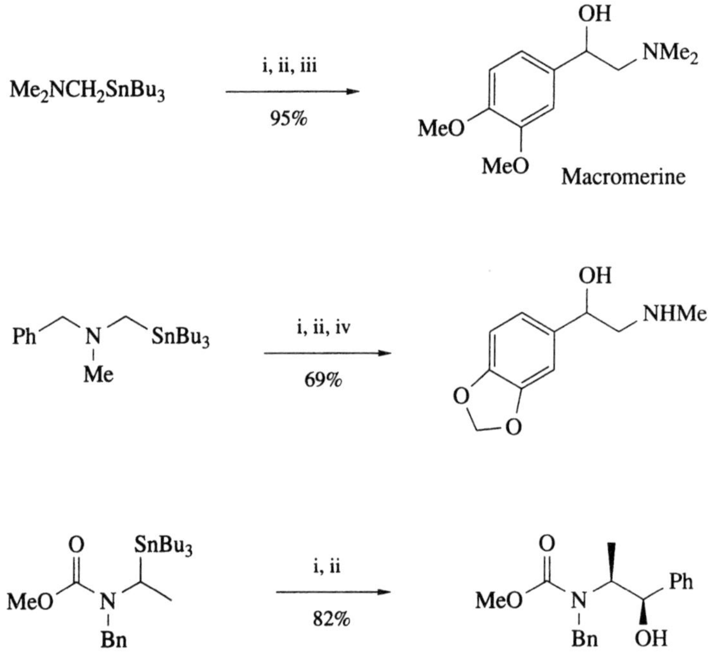

i, BuLi; ii, ArCHO; iii, H^+; iv, H_2, Pd/C

Scheme 42

2.1.4.3 Additions *via* Metalation of Carbocyclic Systems

The regioselective *syn*, vicinal lithiation of cyclopropane and cubane amides has been reported.[152–154] Transmetalation to organomercury[153,154] or zinc[152] compounds facilitates functionalization, as shown in Scheme 43.

Directed lithiations of α,β- and γ,δ-unsaturated amides[155–157] have been extensively studied.[158,159] Illustrative examples are shown in Scheme 44. Prior complexation of the alkyllithium base with the amide carbonyl oxygen directs the base to the thermodynamically less acidic β′-position in α,β-unsaturated amide (**31**), which adds to benzophenone and subsequently lactonizes. Analysis of the NMR spectrum reveals that the organolithium added the benzophenone in the equatorial position.[156,158] A different kinetic deprotonation is seen in γ,δ-unsaturated amide (**32**), where β-lithiation to form an allylic anion predominates over α-lithiation to form an enolate.[157,159] Addition of the lithium anion to acetone affords poor regioselectivity, but transmetalation to magnesium before carbonyl addition yields a species which adds exclusively at the δ-position.[157,159]

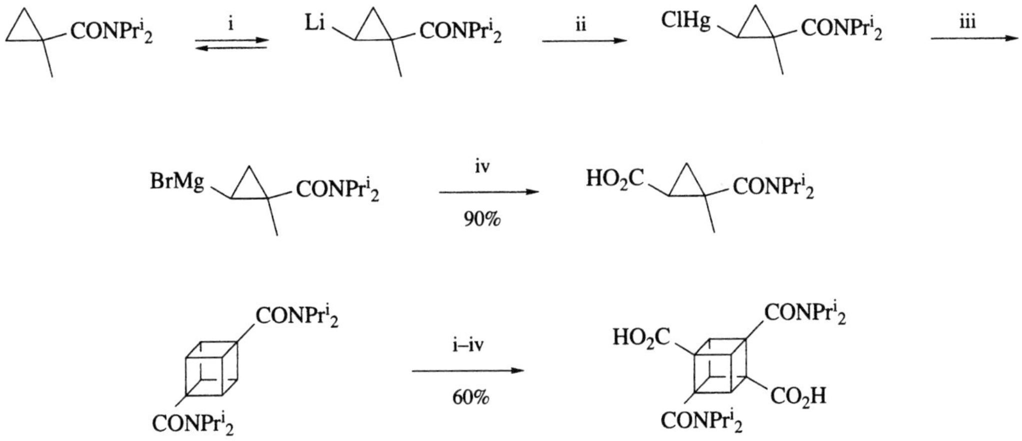

i, LITMP, THF, 0 °C; ii, $HgCl_2$; iii, MeMgBr, –20 °C; iv, CO_2

Scheme 43

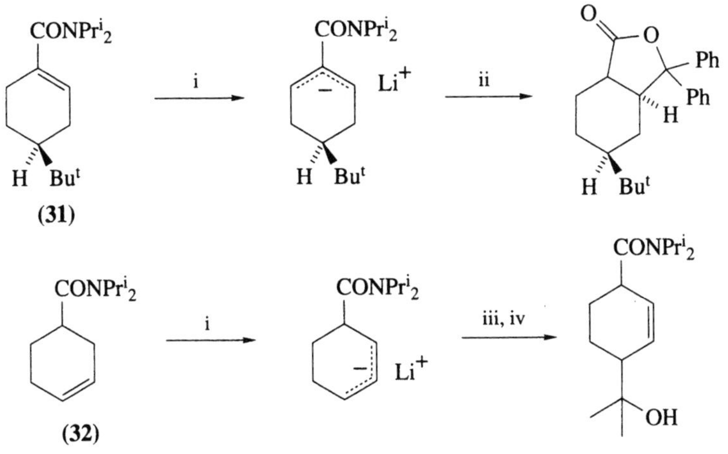

i, Bu^sLi, TMEDA, THF, –78 °C; ii, Ph_2CO; iii, $MgBr_2{\bullet}Et_2O$; iv, Me_2CO

Scheme 44

2.1.4.4 Additions *via* Metalation of Heterocyclic Systems

A one-pot procedure for the activation and metalation of tetrahydroisoquinoline involves the carbonation of the lithium amide anion and then further metalation. As is illustrated in Scheme 45, the dipole-stabilized anion species may be added to carbonyl compounds in good yield.[160] For the activation of tetrahydroisoquinoline Grignards, Seebach examined benzamides, pivalamides and phosphoramides, and found that the benzamides would not metalate, and that although the phosphoramides were most easily removed, the pivalamides were the most nucleophilic species. As is shown in Scheme 46, the lithiated pivaloylisoquinoline adds to cyclohexanone in good yield.[161]

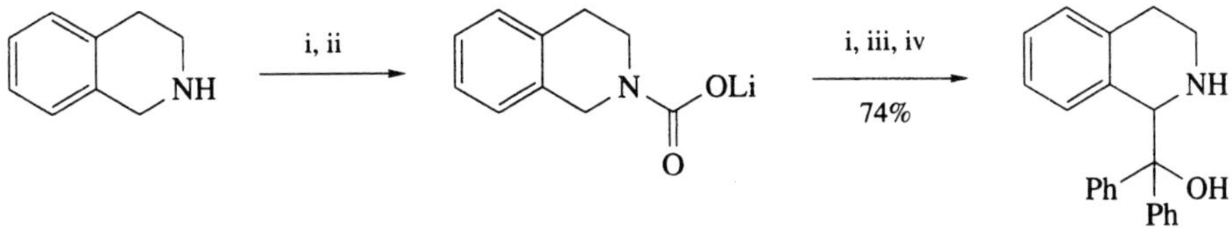

i, Bu^nLi, THF, –20 °C; ii, CO_2; iii, Ph_2CO; iv, 2M HCl

Scheme 45

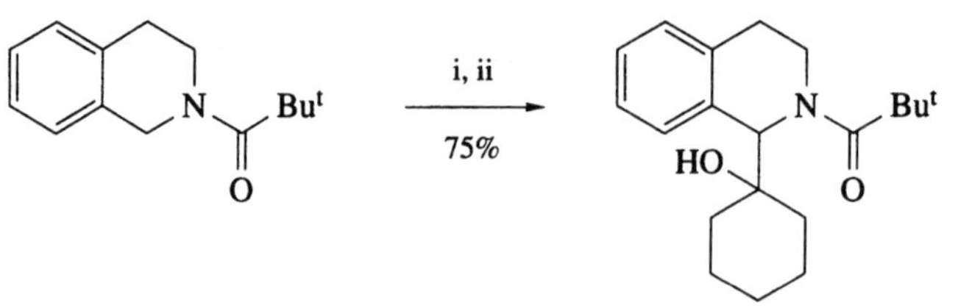

i, Bu^{t}Li, TMEDA, THF, –78 °C; ii, cyclohexanone

Scheme 46

Seebach also compared the same pivaloylisoquinoline to a tetrahydroisoquinoline formamidine to evaluate the face-selectivity in the addition of the metalated derivatives to aldehydes.[132,162,163] In both cases, the organolithium showed significantly lower diastereoselectivity than the Grignard obtained by transmetalation with $MgBr_2{\cdot}Et_2O$, as shown by the examples in Scheme 47. The transmetalation protocol was used to prepare a number of racemic isoquinoline alkaloids.[162]

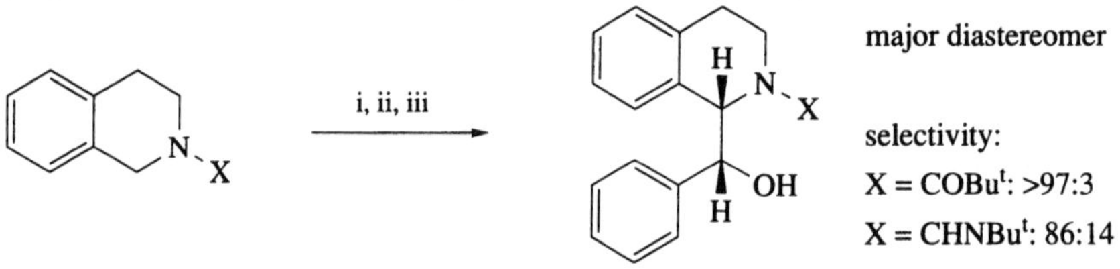

major diastereomer

selectivity:
X = COBu^t: >97:3
X = CHNBu^t: 86:14

i, Bu^{t}Li, THF, –78 °C; ii, $MgBr_2{\bullet}Et_2O$; iii, PhCHO

Scheme 47

Similarly low face-selectivity was found in the addition of lithiated formamidines of tetrahydroquinoline,[164] dihydroindole[164] and β-carboline[165] to benzaldehyde, although addition of the lithiated β-carboline to methyl chloropropyl ketone afforded an 8.5:1 selectivity (Scheme 48).[166] Lithiated formamidines of pyrrolidine and piperidine also add to benzaldehyde in excellent yield, but the diastereoselectivity was not reported.[128]

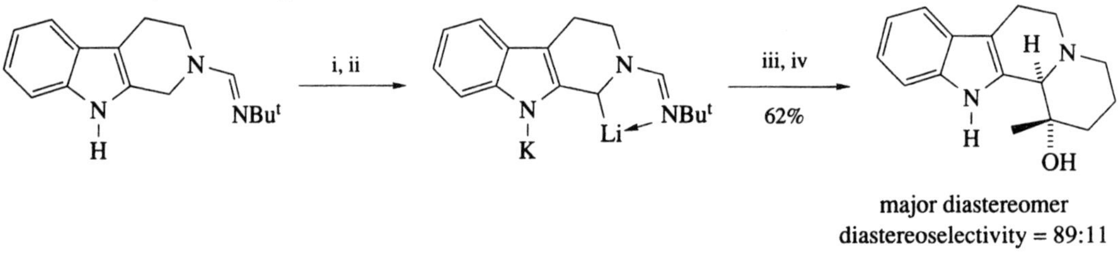

major diastereomer
diastereoselectivity = 89:11

i, KOBu^t or KH; ii, Bu^{n}Li, THF, –78 °C; iii, $MeCO(CH_2)_3Cl$; iv, N_2H_4, H^+

Scheme 48

Formamidines whose α-protons are allylic are easily metalated, but the predominant site of electrophilic attack is the γ-position, as shown by the example in Scheme 49.[128]

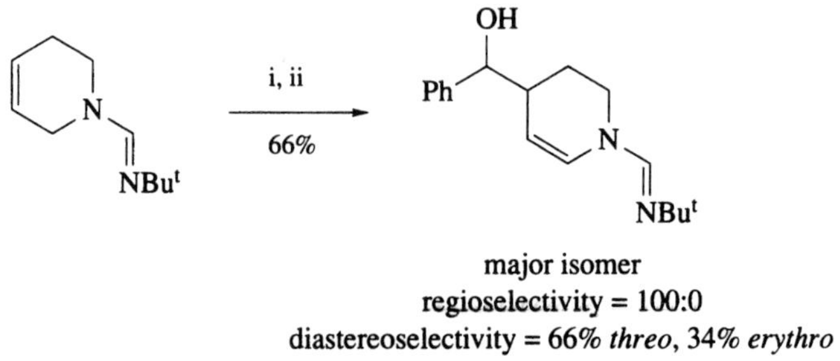

major isomer
regioselectivity = 100:0
diastereoselectivity = 66% *threo*, 34% *erythro*

i, Bu^{n}Li, THF, –78 °C; ii, PhCHO, –78 °C to r.t.

Scheme 49

The lithiation and carbonyl additions of piperidinecarboxamides has been studied by both Beak[167] and Seebach.[168] An example of the addition of a lithiated derivative to propionaldehyde is shown in Scheme 50.[167] Beak found that although the face-selectivity of the addition shown in Scheme 50 is not high, acid hydrolysis affords a single diastereomer of the product of *N*- to *O*-acyl migration. The stereospecificity must be obtained before the acyl migration; the suggested mechanism is illustrated in Scheme 51.[167]

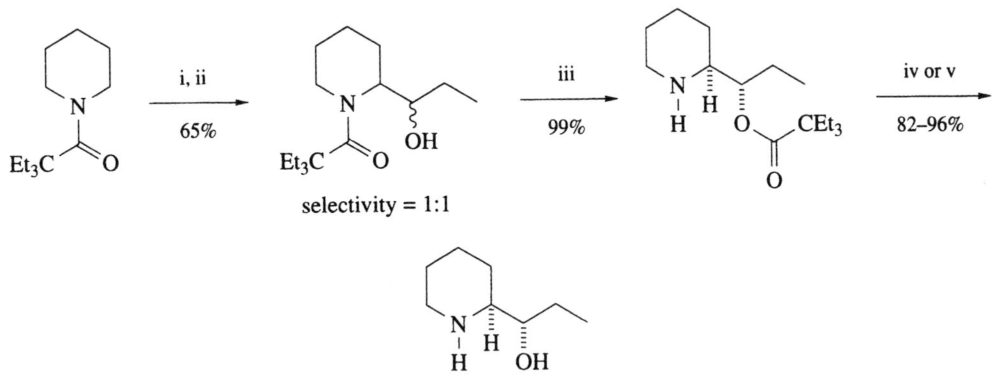

selectivity = 1:1

i, Bu^sLi, TMEDA, Et_2O; ii, EtCHO; iii, conc. HCl, MeOH; iv, $KOBu^t$, H_2O, diglyme; v, $LiAlH_4$

Scheme 50

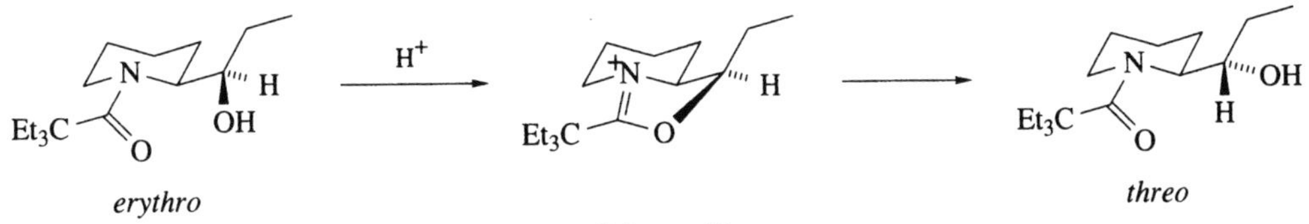

erythro *threo*

Scheme 51

2.1.5 REFERENCES

1. D. J. Cram and F. A. Abd Elhafez, *J. Am. Chem. Soc.*, 1952, **74**, 3210; leading references to more recent work: E. P. Lodge and C. H. Heathcock, *J. Am. Chem. Soc.*, 1987, **109**, 2819, 3353.
2. D. J. Cram and K. R. Kopecky, *J. Am. Chem. Soc.*, 1959, **81**, 2748; recent work: W. C. Still and J. H. McDonald, III, *Tetrahedron Lett.*, 1980, **21**, 1031; W. C. Still and J. A. Schneider, *Tetrahedron Lett.*, 1980, **21**, 1035; reviews: M. T. Reetz, *Angew. Chem., Int. Ed. Engl.*, 1984, **23**, 556; E. L. Eliel, in 'Asymmetric Synthesis', ed. J. D. Morrison, Academic Press, New York, 1983, vol. 2, p. 125.
3. M. T. Reetz, M. W. Drewes, K. Harms and W. Reif, *Tetrahedron Lett.*, 1988, **29**, 3295.
4. H. W. Gschwend and H. R. Rodriguez, *Org. React. (N.Y.)*, 1979, **26**, 1.
5. P. Beak and A. I. Meyers, *Acc. Chem. Res.*, 1986, **19**, 356.
6. A. Krief, *Tetrahedron*, 1980, **36**, 2531.
7. N. S. Narashimhan and R. S. Mali, *Synthesis*, 1983, 957.
8. A. I. Meyers and M. Reuman, *Tetrahedron*, 1985, **41**, 837.
9. P. Beak and V. Snieckus, *Acc. Chem. Res.*, 1982, **15**, 306.
10. A. I. Meyers and G. E. Jagdmann, Jr., *J. Am. Chem. Soc.*, 1982, **104**, 877.
11. A. I. Meyers, P. D. Edwards, T. R. Bailey and G. E. Jagdmann, Jr., *J. Org. Chem.*, 1985, **50**, 1019.
12. S. A. Burns, R. J. P. Corriu, V. Huynh and J. J. E. Moreau, *J. Organomet. Chem.*, 1987, **333**, 281.
13. P. Beak and R. A. Brown, *J. Org. Chem.*, 1982, **47**, 34.
14. P. Beak, A. Tse, J. Hawkins, C.-W. Chen and S. Mills, *Tetrahedron*, 1983, **39**, 1983.
15. M. Skowrońska-Ptasińska, W. Verboom and D. N. Reinhoudt, *J. Org. Chem.*, 1985, **50**, 2690.
16. R. R. Fraser, M. Bresse and T. S. Mansour, *J. Am. Chem. Soc.*, 1983, **105**, 7790.
17. G. Stork, R. L. Polt, Y. Li and K. N. Houk, *J. Am. Chem. Soc.*, 1988, **110**, 8360.
18. J. E. Macdonald and G. S. Poindexter, *Tetrahedron Lett.*, 1987, **28**, 1851.
19. R. J. Mills and V. Snieckus, *J. Org. Chem.*, 1983, **48**, 1565.
20. M. Fukui, T. Ikeda and T. Oishi, *Tetrahedron Lett.*, 1982, **23**, 1605.
21. N. F. Masters and D. A. Widdowson, *J. Chem. Soc., Chem. Commun.*, 1983, 955.
22. G. Nechvatal and D. A. Widdowson, *J. Chem. Soc., Chem. Commun.*, 1982, 467.
23. F. N. Jones, M. F. Zinn and C. R. Hauser, *J. Org. Chem.*, 1963, **28**, 663.
24. N. S. Narashimhan, R. S. Mali and B. K. Kulkarni, *Tetrahedron Lett.*, 1981, **22**, 2797.
25. R. S. Mali, P. D. Sharadbala and S. L. Patil, *Tetrahedron*, 1986, **42**, 2075.
26. N. S. Narashimhan, R. S. Mali and B. K. Kulkarni, *Tetrahedron*, 1983, **39**, 1975.
27. T. D. Harris and G. P. Roth, *J. Org. Chem.*, 1979, **44**, 2004.
28. A. R. Katritzky, G. W. Rewcastle and L. M. Vasquez de Miguel, *J. Org. Chem.*, 1988, **53**, 794.

29. P. S. Chandrachood and N. S. Narashimhan, *Tetrahedron*, 1981, **37**, 825.
30. D. L. Comins, J. D. Brown and N. Mantlo, *Tetrahedron Lett.*, 1982, **23**, 3979.
31. D. L. Comins and J. D. Brown, *J. Org. Chem.*, 1984, **49**, 1078.
32. A. Sinhababa and R. T. Borchardt, *J. Org. Chem.*, 1983, **48**, 2356.
33. G. Stork, C. S. Shiner, C.-W. Cheng and R. L. Polt, *J. Am. Chem. Soc.*, 1986, **108**, 304.
34. W. Fuhrer and H. W. Gschwend, *J. Org. Chem.*, 1979, **44**, 1133.
35. P. A. Wender and A. A. White, *Tetrahedron*, 1983, **39**, 3767.
36. A. R. Katritzky, N. E. Grzeskowaik, T. Siddiqui, C. Jayaram and S. N. Vassilatos, *J. Chem. Res. (S)*, 1986, 12.
37. A. S. Kende and S. D. Boettger, *J. Org. Chem.*, 1981, **46**, 2799.
38. J. Einhorn and J. L. Luche, *Tetrahedron Lett.*, 1986, **27**, 501.
39. L. Dashan and S. Trippett, *Tetrahedron Lett.*, 1983, **24**, 2039.
40. P. Beak and C.-W. Chen, *Tetrahedron Lett.*, 1983, **24**, 2945.
41. C.-W. Chen and P. Beak, *J. Org. Chem.*, 1986, **51**, 3325.
42. I. Mastatomo and K. Tsukasa, *Tetrahedron Lett.*, 1985, **26**, 6213.
43. M. P. Sibi, M. A. J. Miah and V. Snieckus, *J. Org. Chem.*, 1984, **49**, 737.
44. D. L. Comins and J. D. Brown, *J. Org. Chem.*, 1986, **51**, 3566.
45. R. Neidlein and W. Wirth, *Helv. Chim. Acta*, 1986, **69**, 1263.
46. S. A. Jacobs and R. G. Harvey, *Tetrahedron Lett.*, 1981, **22**, 1093.
47. R. G. Harvey, C. Cortez and S. A. Jacobs, *J. Org. Chem.*, 1982, **47**, 2120.
48. M. Iwao, M. Watanabe, S. O. de Silva and V. Snieckus, *Tetrahedron Lett.*, 1981, **22**, 2349.
49. M. Iwao, K. K. Mahalanabis, M. Watanabe, S. O. de Silva and V. Snieckus, *Tetrahedron*, 1983, **39**, 1955.
50. J. C. Martin and T. D. Krizan, *J. Org. Chem.*, 1982, **47**, 2681.
51. R. R. Fraser and S. Savard, *Can. J. Chem.*, 1986, **64**, 621.
52. M. S. Newman and R. Kannan, *J. Org. Chem.*, 1979, **44**, 3388.
53. M. S. Newman and S. Veeraraghavan, *J. Org. Chem.*, 1983, **48**, 3246.
54. A. I. Meyers and W. B. Avila, *J. Org. Chem.*, 1981, **46**, 3881.
55. A. I. Meyers, M. A. Hanagan, L. M. Trefonas and R. J. Baker, *Tetrahedron*, 1983, **39**, 1991.
56. I. M. Dorder, J. M. Mellor and P. D. Kennewell, *J. Chem. Soc., Perkin Trans. 1*, 1984, 1247.
57. J. M. Muchowski and M. C. Venuti, *J. Org. Chem.*, 1980, **45**, 4798.
58. J. N. Reed and V. Snieckus, *Tetrahedron Lett.*, 1984, **25**, 5505.
59. A. R. Katritzky, L. M. Vazquez de Miguel and G. W. Rewcastle, *Synthesis*, 1988, 215.
60. M. P. Sibi and V. Snieckus, *J. Org. Chem.*, 1983, **48**, 1935.
61. M. P. Sibi, S. Chattopadhyay, J. W. Dankwardt and V. Snieckus, *J. Am. Chem. Soc.*, 1985, **107**, 6312.
62. J. Verbeek, A. V. E. George, R. L. P. de Jong and L. Brandsma, *J. Chem. Soc., Chem. Commun.*, 1984, 257.
63. S. L. Taylor, D. Y. Lee and J. C. Martin, *J. Org. Chem.*, 1983, **48**, 4156.
64. B. Iddon, *Heterocycles*, 1985, **23**, 417.
65. D. J. Chadwick and R. I. Ngochindo, *J. Chem. Soc., Perkin Trans. 1*, 1984, 481.
66. A. J. Carpenter, D. J. Chadwick and R. I. Ngochindo, *J. Chem. Res. (M)*, 1983, 1913.
67. K. D. Anderson, J. A. Sikorski, D. B. Reitz and L. T. Pilla, *J. Heterocycl. Chem.*, 1986, **23**, 1257.
68. A. Dondoni, T. Dall' Occo, G. Fantin, M. Iogagnolo, A. Medici and P. Dedrini, *J. Chem. Soc., Chem. Commun.*, 1984, 258.
69. A. R. Katritzky, W. H. Ramer and J. N. Lam, *J. Chem. Soc., Perkin Trans. 1*, 1987, 775.
70. E. Bisagni, N. C. Hung and J. M. Lhoste, *Tetrahedron*, 1983, **39**, 1777.
71. H. Hayakawa, K. Haraguchi, H. Tanaka and T. Miyasaka, *Chem. Pharm. Bull.*, 1987, **35**, 72.
72. P. Ribéreau and G. Queguiner, *Tetrahedron*, 1983, **39**, 3593.
73. A. J. Carpenter and D. J. Chadwick, *J. Chem. Soc., Perkin Trans. 1*, 1985, 173.
74. A. J. Carpenter and D. J. Chadwick, *J. Org. Chem.*, 1985, **50**, 4362.
75. E. G. Doat and V. Snieckus, *Tetrahedron Lett.*, 1985, **26**, 1149.
76. A. J. Carpenter and D. J. Chadwick, *Tetrahedron Lett.*, 1985, **26**, 5335.
77. J. H. Nasman, N. Kopola and P. Göran, *Tetrahedron Lett.*, 1986, **27**, 1391.
78. G. R. Martinez, P. A. Grieco and C. V. Srinivasan, *J. Org. Chem.*, 1981, **46**, 3760.
79. I. Hasan, E. R. Marinelli, L.-C. C. Lin, F. W. Fowler and A. B. Levy, *J. Org. Chem.*, 1981, **46**, 157.
80. M. M. Cooper, G. J. Hignett and J. A. Joule, *J. Chem. Soc., Perkin Trans. 1*, 1981, 3008.
81. M. G. Saulnier and G. W. Gribble, *J. Org. Chem.*, 1982, **47**, 2810.
82. J. M. Muchowski and P. Hess, *Tetrahedron Lett.*, 1988, **29**, 777.
83. D. J. Chadwick, M. V. McKnight and R. I. Ngochindo, *J. Chem. Soc., Perkin Trans. 1*, 1982, 1343.
84. H. J. Anderson and C. E. Loader, *Synthesis*, 1985, 353.
85. M. E. K. Cartoon and G. W. H. Cheeseman, *J. Organomet. Chem.*, 1982, **234**, 123.
86. G. W. Gribble and D. A. Johnson, *Heterocycles*, 1986, **24**, 2127.
87. G. W. Gribble and D. A. Johnson, *Tetrahedron Lett.*, 1987, **28**, 5259.
88. F. Marsais and G. Queguiner, *Tetrahedron*, 1983, **39**, 2009.
89. A. R. Katritzky, S. Rahimi-Rastgoo and N. K. Ponkshe, *Synthesis*, 1981, 127.
90. J. Epsztajn, A. Bieniek, J. Z. Brzezinski and A. Józwiak, *Tetrahedron Lett.*, 1983, **24**, 4735.
91. J. Epsztajn, A. Bieniek and M. W. Plotka, *J. Chem. Res.(S)*, 1986, 20.
92. J. Epsztajn, Z. Berski, J. Z. Brzezinski and A. Józwiak, *Tetrahedron Lett.*, 1980, **21**, 4739.
93. J. Epsztajn, J. Z. Brzezinski and A. Józwiak, *J. Chem. Res.(S)*, 1986, 18.
94. F. Marsais, A. Cronnier, F. Trecourt and G. Queguiner, *J. Org. Chem.*, 1987, **52**, 1133.
95. T. R. Kelly, A. Echavarren, N. S. Chandrakumar and Y. Köksal, *Tetrahedron Lett.*, 1984, **25**, 2127.
96. F. Trécourt, M. Mallet, F. Marsais and G. Quéguiner, *J. Org. Chem.*, 1988, **53**, 1367.
97. T. Güngör, F. Marsais and G. Queguiner, *Synthesis*, 1982, 499.
98. G. W. Gribble and M. G. Saulnier, *Tetrahedron Lett.*, 1980, **21**, 4137.
99. F. Marsais, P. Breant, A. Guinguene and G. Quenguiner, *J. Organomet. Chem.*, 1981, **216**, 139.

100. Y. Tamura, M. Fujita, L.-C. Chen, M. Inoue and Y. Kita, *J. Org. Chem.*, 1981, **46**, 3564.
101. F. Marsais, G. LeNard and G. Queguiner, *Synthesis*, 1982, 235.
102. T. Gungor, F. Marsais and G. Queguiner, *J. Organomet. Chem.*, 1981, **215**, 139.
103. M. Mallet and G. Queguiner, *Tetrahedron*, 1985, **41**, 3433.
104. M. Watanabe, E. Shinoda, Y. Shimizu and S. Furukawa, *Tetrahedron*, 1987, **43**, 5281.
105. M. Iwao and T. Kuraishi, *Tetrahedron Lett.*, 1983, **24**, 2649.
106. W. Lubosch and D. Seebach, *Helv. Chim. Acta*, 1980, **63**, 102.
107. R. Schlecker and D. Seebach, *Helv. Chim. Acta*, 1978, **61**, 512.
108. A. I. Meyers and R. A. Gable, *J. Org. Chem.*, 1982, **47**, 2633.
109. E. Bisagni and M. Rautureau, *Synthesis*, 1987, 142.
110. D. Seebach, *Angew. Chem., Int. Ed. Engl.*, 1979, **18**, 239.
111. P. Beak, W. J. Zajdel and D. B. Reitz, *Chem. Rev.*, 1984, **84**, 471.
112. D. J. Peterson and H. R. Hays, *J. Org. Chem.*, 1965, **30**, 1939.
113. A. R. Lepley and W. A. Khan, *J. Org. Chem.*, 1966, **31**, 2061.
114. H. Albrect and H. Dollinger, *Tetrahedron Lett.*, 1984, **25**, 1353.
115. F. H. Köhler, N. Herktorn and J. Blümel, *Chem. Ber.*, 1987, **120**, 2081.
116. D. J. Peterson, *J. Am. Chem. Soc.*, 1971, **93**, 4027.
117. P. Beak and D. B. Reitz, *Chem. Rev.*, 1978, **78**, 275.
118. D. Seebach and D. Enders, *Angew. Chem., Int. Ed. Engl.*, 1978, **14**, 15.
119. J. E. Saavedra, *Org. Prep. Proced. Int.*, 1987, **19**, 83.
120. R. R. Fraser and L. K. Ng, *J. Am. Chem. Soc.*, 1976, **98**, 5895.
121. R. R. Fraser, T. B. Grindley and S. Passannanti, *Can. J. Chem.*, 1975, **53**, 2473.
122. W. G. Kofron and M. K. Yeh, *J. Org. Chem.*, 1976, **41**, 439.
123. M. E. Jung, P. A. Blair and A. Lowe, *Tetrahedron Lett.*, 1976, 1439.
124. D. Seebach and W. Wykpiel, *Helv. Chim. Acta*, 1978, **61**, 3100.
125. N. G. Rondan, K. N. Houk, P. Beak, W. J. Zajdel, J. Chandrasekhar and P. von R. Schleyer, *J. Org. Chem.*, 1981, **46**, 4108.
126. R. D. Bach, M. L. Braden and G. J. Wolber, *J. Org. Chem.*, 1983, **48**, 1509.
127. L. J. Bartolotti and R. E. Gawley, *J. Org. Chem.*, 1989, **54**, 2980.
128. A. I. Meyers, P. D. Edwards, W. F. Rieker and T. R. Bailey, *J. Am. Chem. Soc.*, 1984, **106**, 3270.
129. R. E. Gawley, G. C. Hart and L. J. Bartolotti, *J. Org. Chem.*, 1989, **54**, 175.
130. R. E. Gawley, *J. Am. Chem. Soc.*, 1987, **109**, 1265.
131. K. Rein, M. Goicoechea-Pappas, T. V. Anklekar, G. C. Hart, G. A. Smith and R. E. Gawley, *J. Am. Chem. Soc.*, 1989, **111**, 2211.
132. D. Seebach, J. Hansen, P. Seiler and J. M. Gromek, *J. Organomet. Chem.*, 1985, **285**, 1.
133. M. Al-Aseer, P. Beak, D. R. Hay, D. J. Kempf, S. Mills and S. G. Smith, *J. Am. Chem. Soc.*, 1983, **105**, 2080.
134. D. R. Hay, Z. Song, S. G. Smith and P. Beak, *J. Am. Chem. Soc.*, 1988, **110**, 8145.
135. A. I. Meyers, W. F. Rieker and L. M. Fuentes, *J. Am. Chem. Soc.*, 1983, **105**, 2082.
136. A. I. Meyers and D. A. Dickman, *J. Am. Chem. Soc.*, 1987, **109**, 1263.
137. M. Watanabe, M. Sahara, S. Furukawa, R. Billedeau and V. Snieckus, *Tetrahedron Lett.*, 1982, **23**, 1647.
138. M. Watanabe, M. Sahara, M. Kubo, S. Furukawa, R. Billedeau and V. Snieckus, *J. Org. Chem.*, 1984, **49**, 742.
139. P. Cornwall, C. P. Dell and D. W. Knight, *Tetrahedron Lett.*, 1987, **28**, 3585.
140. A. R. Katritzky, C. Jayaram and S. N. Vassilatos, *Tetrahedron*, 1983, **39**, 2023.
141. M. R. Cuberes, M. Moreno-Mañas and A. Trius, *Synthesis*, 1985, 302.
142. S. Shimizu and M. Ogata, *J. Org. Chem.*, 1986, **51**, 3897.
143. L. R. Hillis and R. C. Ronald, *J. Org. Chem.*, 1981, **46**, 3349.
144. R. M. Adlington, J. E. Baldwin, J. C. Bottaro and M. W. D. Perry, *J. Chem. Soc., Chem. Commun.*, 1983, 1040.
145. J. E. Baldwin, J. C. Bottaro, J. N. Kolhe and R. M. Adlington, *J. Chem. Soc., Chem. Commun.*, 1984, 22.
146. J. E. Baldwin, R. M. Adlington and I. M. Newington, *J. Chem. Soc., Chem. Commun.*, 1986, 176.
147. D. Seebach and W. Wykpiel, *Synthesis*, 1979, 423.
148. A. I. Meyers, K. B. Kunnen and W. C. Still, *J. Am. Chem. Soc.*, 1987, **109**, 4405.
149. J.-P. Quintard, B. Elissondo and B. Jousseaume, *Synthesis*, 1984, 495.
150. B. Elissondo, J.-B. Verlhac, J.-P. Quintard and M. Pereyre, *J. Organomet. Chem.*, 1988, **339**, 267.
151. W. H. Pearson and A. C. Lindbeck, *J. Org. Chem.*, 1989, **54**, 5651.
152. P. E. Eaton, H. Higuchi and R. Millikan, *Tetrahedron Lett.*, 1987, **28**, 1055.
153. P. E. Eaton, G. T. Cunkle, G. Marchioro and R. M. Martin, *J. Am. Chem. Soc.*, 1987, **109**, 948.
154. P. E. Eaton, R. G. Daniels, D. Casucci and G. T. Cunkle, *J. Org. Chem.*, 1987, **52**, 2100.
155. J. J. Fitt and H. W. Gschwend, *J. Org. Chem.*, 1980, **45**, 4257.
156. P. Beak and D. J. Kempf, *J. Am. Chem. Soc.*, 1980, **102**, 4550.
157. P. Beak, J. E. Hunter and Y. M. Jun, *J. Am. Chem. Soc.*, 1983, **105**, 6350.
158. P. Beak, D. J. Kempf and K. D. Wilson, *J. Am. Chem. Soc.*, 1985, **107**, 4745.
159. P. Beak, J. E. Hunter, Y. M. Jun and A. P. Wallin, *J. Am. Chem. Soc.*, 1987, **109**, 5403.
160. A. R. Katritzky and K. Akutagawa, *Tetrahedron*, 1986, **42**, 2571.
161. D. Seebach, J.-J. Lohmann, M. A. Syfrig and M. Yoshifuji, *Tetrahedron*, 1983, **39**, 1963.
162. D. Seebach, I. M. P. Huber and M. A. Syfrig, *Helv. Chim. Acta*, 1987, **70**, 1357.
163. D. Seebach and M. A. Syfrig, *Angew. Chem., Int. Ed. Engl.*, 1984, **23**, 248.
164. A. I. Meyers and S. Hellring, *Tetrahedron Lett.*, 1981, **22**, 5119.
165. A. I. Meyers and S. Hellring, *J. Org. Chem.*, 1982, **47**, 2229.
166. A. I. Meyers and M. F. Loewe, *Tetrahedron Lett.*, 1984, **25**, 2641.
167. P. Beak and W. J. Zajdel, *J. Am. Chem. Soc.*, 1984, **106**, 1010.
168. W. Wykpiel, J.-J. Lohmann and D. Seebach, *Helv. Chim. Acta*, 1981, **64**, 1337.

2.2
Boron Stabilization

ANDREW PELTER and KEITH SMITH
University College Swansea, UK

2.2.1 INTRODUCTION

Based upon their knowledge of the chemistry of carbonyl, nitro, sulfonyl, cyano compounds *etc.*, organic chemists have long believed that carbanions XCH_2^- are stabilized if the group X is electron withdrawing.

Recently, calculations have been carried out on anions XCH^-_2 in which X is a member of the first row element set: Li, BeH, BH_2, CH_3, NH_2, OH and F.[1–4] Schleyer's calculations (Table 1)[1] suggest that the methyl group is destabilizing, the amino group is borderline, while all other groups are stabilizing. These results are qualitatively similar to those obtained from other calculations. In particular, very large π-effects are exhibited by BeH and planar BH_2 groups, while inductive stabilization by the electronegative F and OH groups is less effective.

In Table 2 are given similar calculations for stabilization by first row elements, together with calculated and experimental values for a variety of stabilizing organic groups.[4]

Clearly, whichever method of calculation is used, there is a remarkable stabilization by a boron atom, comparable with that of a carbonyl group. When diffuse function-augmented basis sets are used, the stabilization energies calculated are generally lower,[1] and in the case of organic stabilizing groups this brings them closer to the experimental values (Table 2).

There is an appreciable C—B bond shortening of 0.13 Å in planar $H_2B—CH_2^-$ when compared to the perpendicular form, and at the 4-31G level the planar form is 57.2 kcal mol^{-1} (1 cal = 4.18 J) lower in energy than the perpendicular form.[4]

Table 1 Calculated[1] Stabilization Energies (kcal mol^{-1}) for Carbanions XCH_2^-

X	*STO-3G//STO-3G*	*4-31G//4-31G*[2,3]	*6-31G*//4-31G*[3]	*MP2/6-31G*//4-31G*[3]	*4-31+G//4-31+G*	*MP2/4-31+G//4-31+G*	*6-31+G*//4-31+G*	*MP2/6-31+G*//4-31+G*
Li	−54.9	−17.5	−13.7	−25.8	−2.4	−6.6	−4.8	−8.1
BeH	−65.8	−40.4	−38.1	−46.9	−31.8	−33.3	−31.3	−32.9
BH_2	−81.6	−67.7	−61.4	−71.8	−54.7	−58.0	−53.2	−57.4
CH_3	−9.0	−2.1	−1.4	−3.3	+5.7	+4.8	+4.0	+3.0
NH_2	−16.6	−5.2	−3.3	−5.6	+1.9	+0.6	+1.5	−0.6
OH	−20.0	−15.8	−7.9	−10.9	−7.7	−8.1	−3.7	−5.6
F	−21.7	−24.6	−14.6	−16.1	−15.6	−13.3	−9.3	−9.0

Table 2[4] Calculated and Experimental Stabilization Energies (kcal mol^{-1}) for Carbanions XCH_2^-

X	*4-31G*	*6-31G**	*MP2/6-31G**	*Experimental*
Li	−17.4	−13.7	−25.8	—
BeH	−40.4	−38.1	−46.9	—
BH_2	−67.6	−61.4	−71.8	—
CH_3	−2.1	−1.4	−3.3	—
NH_2	−5.2	−3.3	−5.6	—
OH	−15.7	−7.9	−10.9	—
F	−24.6	−14.6	−16.1	—
CN	−61.1	−55.0	−57.9	−44.4
NO_2	−98.1	−75.9	—	−57.9
CH_3CH_2	−5.6	−4.4	—	—
$CH_2{=}CH$	−37.5	−31.8	−37.1	−25.8
$HC{\equiv}C$	−44.1	−42.9	−46.5	−35.2
CF_3	−57.0	−37.0	—	—
CHO	−71.5	−60.5	−66.2	−50.2
Ph	−44.4	—	—	−37.6

Since the degree of stabilization of boron-stabilized carbanions is similar to that for anions stabilized by a formyl or a cyano group, the comparative lack of knowledge of boron-stabilized carbanions must be due either to kinetic factors associated with their production or to the availability and nature of their precursors.

Boron-stabilized carbanions are expected to assume a geometry which allows maximum overlap between the lone pair on the α-carbon atom and the vacant boron orbital. Dynamic NMR studies[5] on Mes_2BC^-HPh (Mes = 2,4,6-trimethylphenyl) demonstrate this by showing that the anion decomposes at 140 °C, before observation of rotation effects about the B—C$^-$HPh bond. This gives a $\Delta G^{\ddagger}$ rotation of >22 kcal mol^{-1}, which is similar to that of the isoelectronic Mes_2B—NHR[6,10] and greater than that found[7,10] for rotation about the B—S bond in Mes_2B—SR (19.25 kcal mol^{-1}; calc.[8] 22 kcal mol^{-1}) and for rotation about the B—O bond in Mes_2B—OMe (13.2 kcal mol^{-1}).[9,10] The B—C rotational barrier[11] of $(Mes_2B)_2CHLi$ is 17 kcal mol^{-1}, very similar to that found for a series of isoelectronic substituted allyl cations.[12] It was concluded[11] that structure (**1**) makes a significant contribution to the overall electronic structure of $Mes_2BCH_2^-$.

The crystal structure of (**2**), as its 12-crown-4 derivative, has been determined.[13] The C_2BCH_2 core of the molecule is essentially planar and the length of the B—CH_2 bond is 1.444 Å. This is within the 1.42–1.45 Å predicted for B$=CH_2$ and quite distinct from the B—Me distance of the parent compound (**3**).[13] The distances and angles found are in accord with formulation (**1**), as is the upfield shift of 43.2 p.p.m. in the ^{11}B NMR spectrum on passing from (**3**) to (**2**).[13] Thus there is a high degree of orbital overlap in species of type (**4**), which might be better represented in general as the highly stabilized forms (**5**).

$\bar{B}{=}CH_2$ Mes_2B Li Mes_2BMe

(**1**) (**2**) (**3**)

$$R^1{}_2B-\overset{R^2}{\underset{R^2}{C^-}} \qquad R^1{}_2\bar{B}=C(R^2)_2$$

(4) **(5)**

2.2.2 PREPARATION OF BORON-STABILIZED CARBANIONS

2.2.2.1 Introduction

This topic has been summarized in references 14 and 15, of which the former is the most recent and wide ranging of the reviews available over the whole field of boron chemistry.

There are currently three general methods for the production of boron-stabilized carbanions, each of which has analogies in carbonyl chemistry. The cleavage of a 1,1-diborylalkane by base readily yields the desired anion in a fashion similar to the base cleavage of a β-dicarbonyl compound. Deprotonation α to a boron atom can be accomplished directly in special circumstances, as can the addition of an organometallic compound to a vinylborane, which is similar to conjugate addition to an α,β-unsaturated carbonyl compound. Each of these methods is treated in detail in the next three sections.

2.2.2.2 Preparation and Cleavage of 1,1-Diboryl Compounds

1,1-Diboryl compounds (**6**) are made by reaction of certain dialkylboranes, generally dicyclohexylborane [$(C\text{–}C_6H_{11})_2BH$], disiamylborane (Sia_2BH) or 9-borabicyclo[3.3.1]nonane (9-BBN-H), with 1-alkynes (equation 1).[14,16–18]

$$R^1-\equiv \quad + \quad 2R^2{}_2BH \longrightarrow R^1CH_2CH(BR^2{}_2)_2 \;\textbf{(6)} \qquad (1)$$

The products (**6**) are very prone to hydrolytic cleavage by base; this is presumably due to the ready production of an intermediate carbanion followed by its rapid protonation. This possibility was tested[19] by treating a series of compounds (**6**; R^2 = cyclohexyl) with Bu^nLi at –78 °C in THF. The products had properties consistent with the production of (**7**; equation 2). The by-product, a simple trialkylborane, also reacts with butyllithium, and so 2 equiv. of the latter must be added. Thus, the process is wasteful in that one atom of boron and 1 equiv. of base are consumed in an unproductive fashion (equation 3).

$$R^1CH_2CH(BR^2{}_2)_2 \;\textbf{(6)} + BuLi \longrightarrow R^1CH_2CH(BR^2{}_2)Li \;\textbf{(7)} + BuBR^2{}_2 \qquad (2)$$

$$BuBR^2{}_2 + BuLi \longrightarrow Bu_2\bar{B}R^2{}_2Li^+ \qquad (3)$$

The reaction of pent-1-yne with borane, followed by base cleavage, has been studied,[20] yields having been estimated by alkylation with ethyl bromide. Although the reaction occurred with sodium and lithium methoxides, it went better with butyllithium and best with 2 equiv. of methyllithium. It was later shown[17] that diborane gave *ca.* 10% of 1,2-addition, which did not occur with dicyclohexyl- and disiamyl-borane. Hydroboration with 9-BBN-H[21] readily gave (**6**) ($BR^2{}_2$ = 9-BBN), which on cleavage with 2.5 mol equiv. of MeLi gave high yields of the corresponding product (**7**).

Tris(dialkoxyboryl)alkanes such as (**8**), produced as in equation (4), may also be used as precursors of boron-stabilized carbanions.[22] Reaction of (**8**) with base gives (**9**),[22,23] which is stabilized by two boron atoms (equation 5). It is not clear whether a similar reaction succeeds with $CH_2[B(OR)_2]_2$.

Although cleavage of *gem*-diboryl compounds is a mild and general process for the production of boron-stabilized carbanions, it nevertheless suffers from two disadvantages. Firstly, although

$$3(RO)_2BCl + CHCl_3 + 6Li \longrightarrow HC[B(OR)_2]_3 \; (\mathbf{8}) + 6LiCl \qquad (4)$$

$$(\mathbf{8}) + MeLi \longrightarrow LiCH[B(OR)_2]_2 \; (\mathbf{9}) + MeB(OR)_2 \qquad (5)$$

1,1-diborylalkanes are available from terminal alkynes, *gem*-diboryl compounds within a carbon chain are not readily available. Compounds such as (**8**) are very interesting as one-carbon synthons, but are confined to this function as yields drop when CH is replaced by CMe or CPh.[24] Secondly, there is the wasteful use of base and boron referred to previously.

2.2.2.3 Cleavage of α-Substituted Organoboranes

Boron-stabilized carbanions may also be produced by selective cleavage of a heteroatom group from an α-substituted organoborane such as a borylstannylmethane (equations 6–8).[25,26]

$$\text{(pinacolato)B–CH(SnMe}_3\text{)–Me} + MeLi \xrightarrow{-100\,^\circ C} \text{(pinacolato)B–CH(Li)–Me} \qquad (6)$$

$$Mes_2B\text{–}CH_2\text{–}SnR_3 \xrightarrow{\text{MesLi or PhSLi}} Mes_2B\text{–}CH_2\text{–}Li + XSnR_3 \qquad (7)$$

$$Mes_2B\text{–}CH(R)\text{–}SiMe_3 \xrightarrow{LiF} Mes_2B\text{–}CH(R)\text{–}Li + FSiMe_3 \qquad (8)$$

The reactions of bases with borylstannylalkanes generally lead to cleavage of the tin moiety,[25,26] and lithium thiophenoxide is especially specific.[25] It is of interest that the reaction of fluoride anion with dimesitylboryl(trimethylsilyl)alkanes also gives α-boryl carbanions, thus obviating the need to use organometallics.[26,27]

2.2.2.4 Deprotonation of Organoboranes

It appears that the first example of deprotonation to give a boron-stabilized carbanion was that shown in equation (9).[28] In general, however, early attempts to deprotonate organoboranes (equation 10) foundered because borate formation (equation 11) was favored.

$$\text{1-phenyl-1,4-dihydroborinine} \xrightarrow{Bu^tLi} \text{boratabenzene anion (resonance forms, B–Ph)} \qquad (9)$$

Successful deprotonations must discourage equation (11) so as to allow equation (10) to proceed. This may be achieved in several ways as follows: (a) the reagent can be a very hindered, non-nucleophilic base; (b) the groups around boron can be large so that attack on boron is inhibited on steric grounds; or (c) the electrophilicity of the boron atom may be lowered by the use of heteroatom substituents (*e.g.* equation 10; X = OR). All three of these approaches have been used, either separately or in conjunction with one another.

$$R_2CH{-}BX_2 + MY \longrightarrow R_2C(M){-}BX_2 + YH \qquad (10)$$

$$R_2CH{-}BX_2 + MY \longrightarrow R_2CH{-}\bar{B}(Y)X_2\ M^+ \qquad (11)$$

The first successful attempt to deprotonate a simple alkylborane is shown in equation (12).[29] *B*-Methyl-9-BBN (**10**) was reacted with a variety of lithium amides and the yield of (**12**) estimated by deuterium incorporation on quenching with D_2O. Neither lithium diethylamide nor lithium diisopropylamide gave any of (**12**) at all, but the hindered piperidide (**11**) (LITMP) produced up to 75% of (**12**) using 100% excess of base.

(12)

(**10**) (**11**) (**12**)

Even using LITMP it was not possible to convert (**13**) into (**14**) when X = H (equation 13). It appears that stabilization by one dialkoxyboryl group is not sufficient and that at least one other stabilizing group is required for anion formation.[30] Deprotonation with LITMP was successful for (**13**) with X = Ph,[30] SPh,[31] TMS,[32,33] Ph_3P^+ [33] and $CH{=}CH_2$,[33] but failed for X = Me_2S^+ [33] and R_3N^+.[34] Compound (**15**; X = H; equation 14) is successfully deprotonated by LITMP in the presence of TMEDA to give (**16**; X = H). Unfortunately, with (**15**; X = Ph), cleavage takes precedence over deprotonation and (**14**; X = Ph) is produced rather than (**16**; X = Ph).[30]

(13)

(**13**) (**14**)

(14)

(**15**) (**16**)

Compound (**17**; equation 15) is deprotonated by LDA to give (**19**), but treatment with butyllithium leads to cleavage of the B—C bond (equation 16).[35]

(15)

(**17**) (**18**)

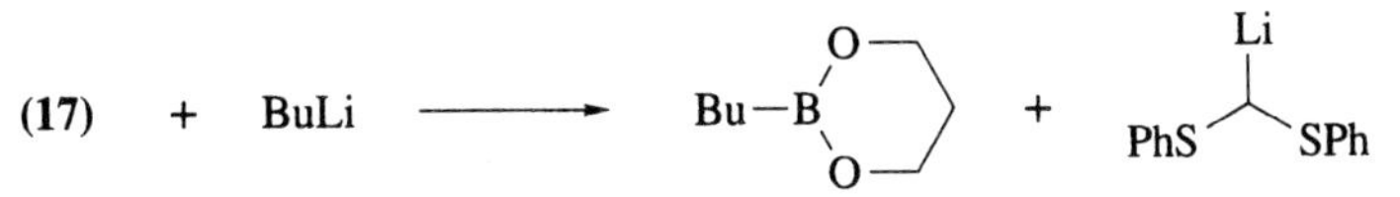

$$(\mathbf{17}) + BuLi \longrightarrow Bu{-}B(O_2C_3H_6) + (PhS)_2CHLi \qquad (16)$$

The production of boron-stabilized carbanions using steric hindrance on the borane to inhibit borate formation was first demonstrated[36] as part of a study of the properties of dimesitylboryl derivatives. Unlike the situation with dialkoxyboryl derivatives, it was possible to carry out the deprotonation with only one boron atom present and with no extra stabilizing groups. Either LDA or lithium dicyclohexylamide may be used as base, the latter being rather more efficient (equation 17).[36] The initial study showed that the reaction with $\mathrm{Mes_2BCHR^1R^2}$ was successful with: R^1, R^2 = H; R^1 = H, R^2 = Me; R^1 = H, R^2 = Ph; and R^1, R^2 = Me.

$$\mathrm{Mes_2B{-}CHR^1R^2} \longrightarrow \mathrm{Mes_2B{-}CLiR^1R^2} \qquad (17)$$

A study of the reactions of various bases with dimesitylmethylborane (Scheme 1) showed that the outcome critically depends on the nature of the base used.[5] Both mesityllithium and lithium dicyclohexylamide give high yields of the required anion, mesityllithium proving to be a superior base for a wide variety of alkyldimesitylboranes. *n*-Butyllithium attacks at boron to give the borate and *t*-butyllithium gives the hydroborate by β-hydrogen transfer. Sodium hydride also gives the hydroborate, while potassium metal gives the radical anion.

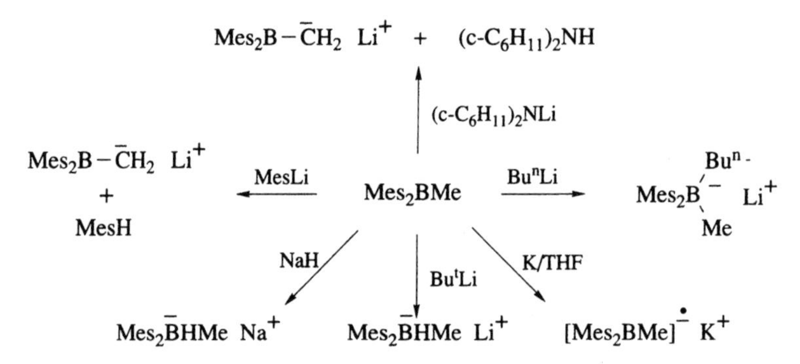

Scheme 1 Reactions of various bases with dimesitylmethylborane

Further studies showed that a variety of substituted anions, $\mathrm{Mes_2BCHLiX}$, in which X = SPh,[37] TMS,[37] $\mathrm{SnPh_3}$[26] amd $\mathrm{BMes_2}$,[37] are available by deprotonation of the corresponding boranes. It is noteworthy that the triphenyltin derivative gives the anion without cleavage, while both the trimethyltin and tri-*n*-butyltin derivatives are cleaved at the carbon–tin bond by treatment with mesityllithium.

Vinylboranes, readily available by the hydroboration of 1-alkynes with dialkylboranes such as $\mathrm{Sia_2BH}$,[38] react readily with LITMP to give high yields of the corresponding allyl carbanions (equation 18).[38]

$$\mathrm{Sia_2BH} + \mathrm{RCH_2C{\equiv}CH} \longrightarrow \mathrm{RCH_2CH{=}CH{-}BSia_2} \longrightarrow [\mathrm{RCH{\cdots}CH{\cdots}CH{-}BSia_2}]^- \mathrm{Li^+} \qquad (18)$$

Mesityllithium or lithium dicyclohexylamide react similarly with allyldimesitylborane (equation 19).[39] Thus, boron-stabilized allyl units are available from either allyl- or vinyl-boranes.

$$\mathrm{Mes_2B{-}CH_2CH{=}CH_2} \longrightarrow [\mathrm{Mes_2B{-}CH{\cdots}CH{\cdots}CH_2}]^- \mathrm{Li^+} \qquad (19)$$

2.2.2.5 Addition to Vinylboranes

As vinylboranes are electronically similar to α,β-unsaturated ketones, reactions analogous to conjugate addition might be expected. Unfortunately, many attempts to add organometallic compounds to simple alkenyldimesitylboranes gave only very low yields (*ca.* 5%) of the required carbanions.[40] However, additional steric hindrance and/or carbanion-stabilizing ability at the α-position, caused by the introduction of a TMS group, allows the process to be realized and so a wide variety of organometallics have been added to 1-dimesitylboryl-1-trimethylsilylethylene (**19**; R^1 = H; equation 20).[41]

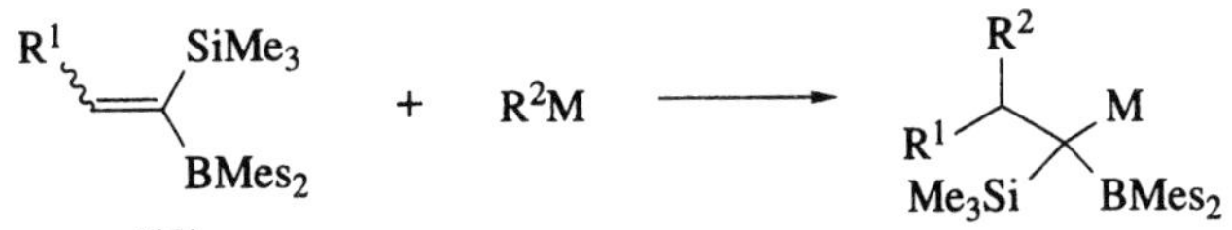

(20)

For the case of R^1 = H, R^2M can be Bu^nLi, Bu^tLi, PhLi, $(RS)_2CHLi$, $Bu^tO_2CCH_2Li$, $CH_2{=}CH(CH_2)_4Li$ and $Bu_2Cu(CN)Li$. However, there was no addition of BuMgCl or PhC≡CLi, nor could an enolate anion be added. For (**19**; R^1 = Ph or $CH{=}CH_2$) addition of butyllithium was successful and in the latter case the addition was entirely to the terminal position (equation 21).[41] Somewhat strangely, the addition of BuLi to (**19**; R^1 = Bu^n) was unsuccessful.[41]

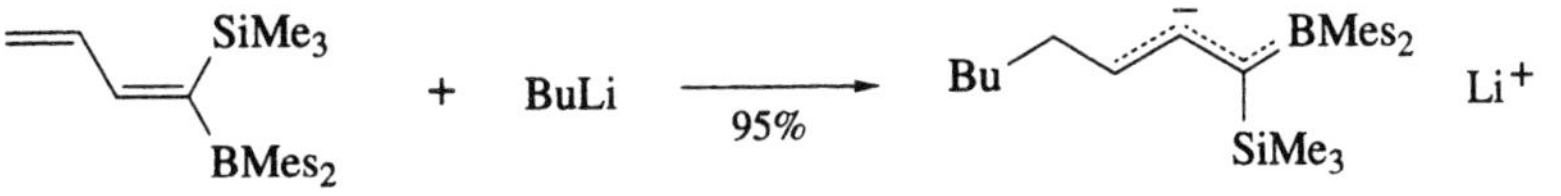

(21)

Of particular interest is the process shown in equation (22), in which the (*E*)-alkene geometry of (**20**) precludes intramolecular ate complex formation and hence an addition–cyclization reaction occurs to give the α-unsubstituted carbanion (**21**), which may be trapped by protonation, deuteration or alkylation.[41]

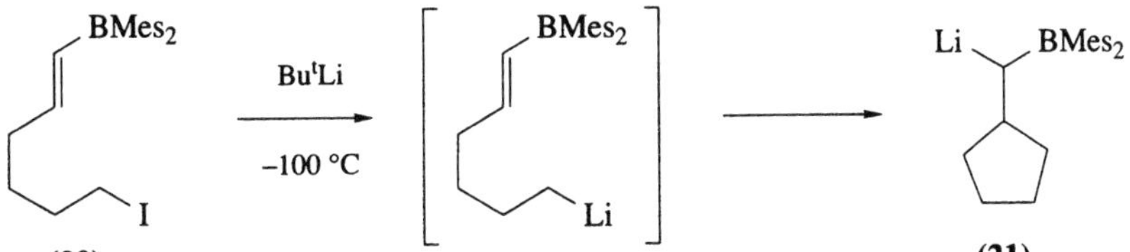

(22)

As much recent work has concerned dimesitylboryl species, it is worth remarking that the related compounds (**22**)–(**24**) also readily yield carbanions.[42] Unlike mesityl compounds, (**22**) cannot be deprotonated at the 4′-methyl group, while compounds (**23**) have some special properties due to the great increase in steric hindrance around boron.[42a] Compounds (**24**) promise to be of utility due to their very ready solvolysis with water or alcohols in the presence of catalytic quantities of mineral acid. They are even selectively hydrolyzed in preference to vinyl as well as alkyl groups (equation 23).[42b]

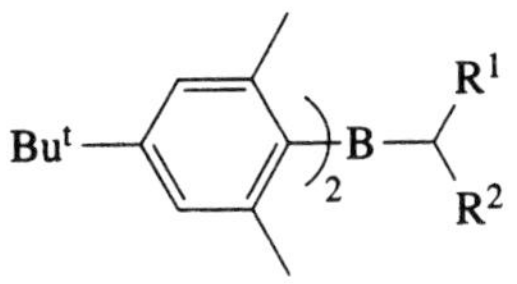

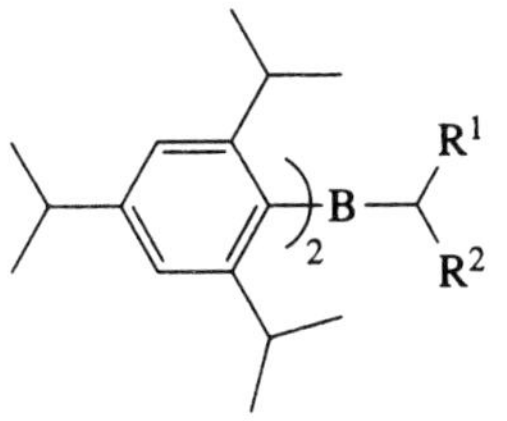

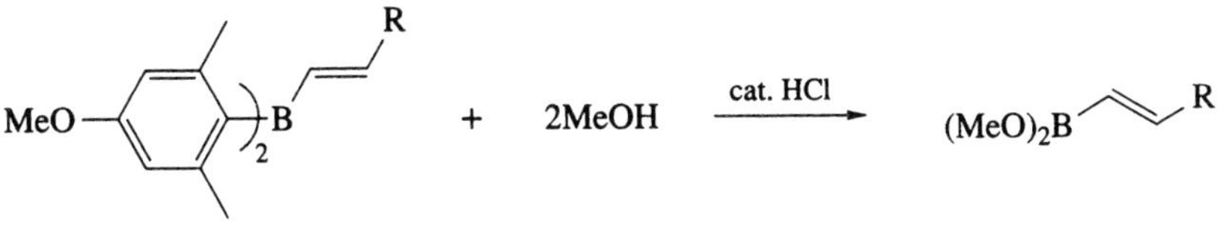

(23)

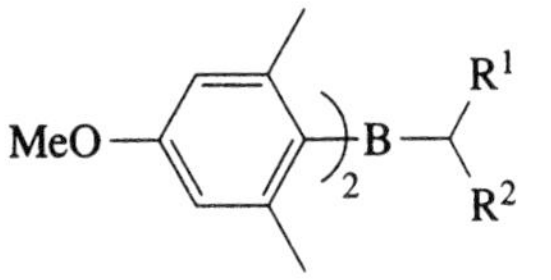

2.2.3 REACTIONS OF NONALLYLIC BORON-STABILIZED CARBANIONS

2.2.3.1 Reactions with Metal Halides

The reactions shown in equation (24) proved relatively facile for M = Ge, Sn or Pb. Although chlorotriphenylsilane failed to react, the corresponding TMS derivative was readily produced.[22]

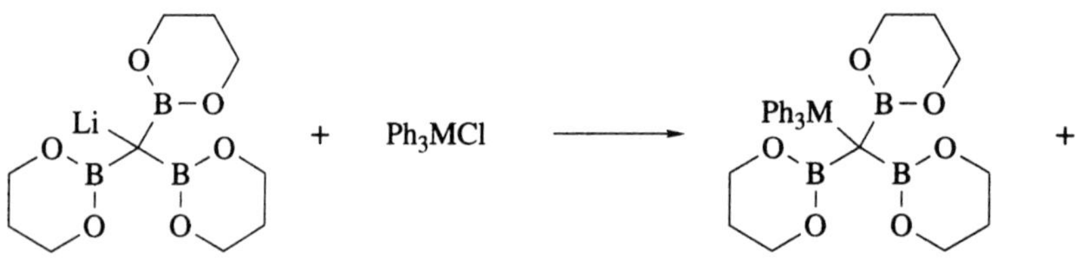

(24)

(25)

The products (**25**; M = Ge, Sn, Pb) react with butyllithium to give (**26**) by C—B cleavage and further reactions of these anions with $Ph_3M'Cl$ (M′ = Sn, Pb) proceed to give compounds (**27**; equation 25) with no evidence of disproportionation.[22] If (**28**; equation 26) is treated in a similar fashion, the very stable tristannylmethylborane (**29**) results.[22]

(25)

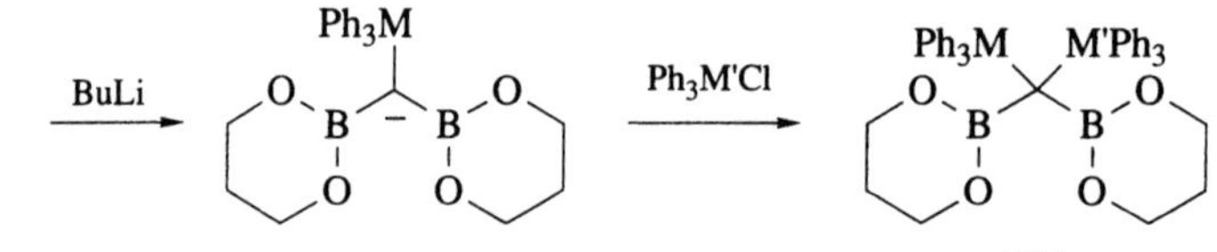

(26)

To obtain analogous organometallic compounds containing α-protons, it is better to use ethylenedioxyboryl derivatives (equation 27).[43]

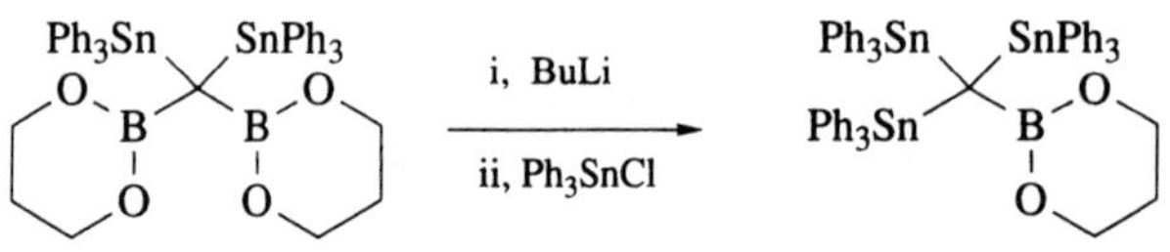

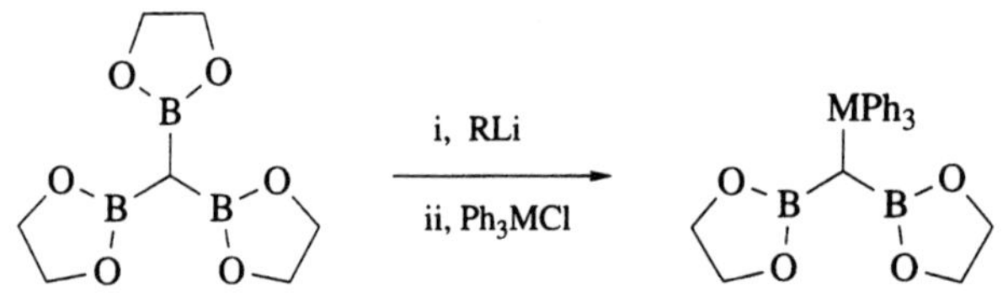

(27)

The reactions shown in equation (28) have been used[37] to make a set of dimesitylboryl compounds (**30**) in which M = Si, R = Me, n = 3; M = Sn, R = Me, Bu, Ph, n = 3; M = S, R = Ph, n = 1; M = Hg, R = CH_2BMes_2, n = 1. In most cases R_nMCl was used as the reactant, but for introducing sulfur to give (**31**), $PhSSO_2Ph$ was advantageous in yielding pure product.[26] Otherwise the reaction shown in equation (29) was utilized.[37] Products (**30**) are stable, crystalline compounds that may act as precursors of boron-stabilized carbanions.[37]

$$Mes_2B\!\diagup\!\!\diagdown Li + R_nMX \longrightarrow Mes_2B\!\diagup\!\!\diagdown MR_n + LiX \quad (28)$$

(30)

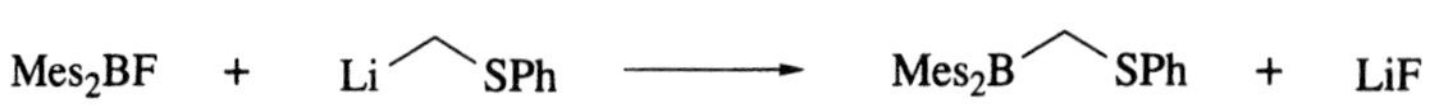

$$Mes_2BF + Li\!\diagup\!\!\diagdown SPh \longrightarrow Mes_2B\!\diagup\!\!\diagdown SPh + LiF \quad (29)$$

(31)

2.2.3.2 Alkylation Reactions

In general, all types of boron-stabilized carbanions are readily alkylated by primary alkyl halides. The situation for the alkylation of dimesitylboryl compounds is summarized in Scheme 2.[44] These reactions represent highly efficient homologation processes. Oxidation of the tertiary alkyl organoboranes is slow, but use of 4-methoxy-2,6-dimethylphenyl groups instead of mesityl groups renders the products sensitive to solvolysis.[42]

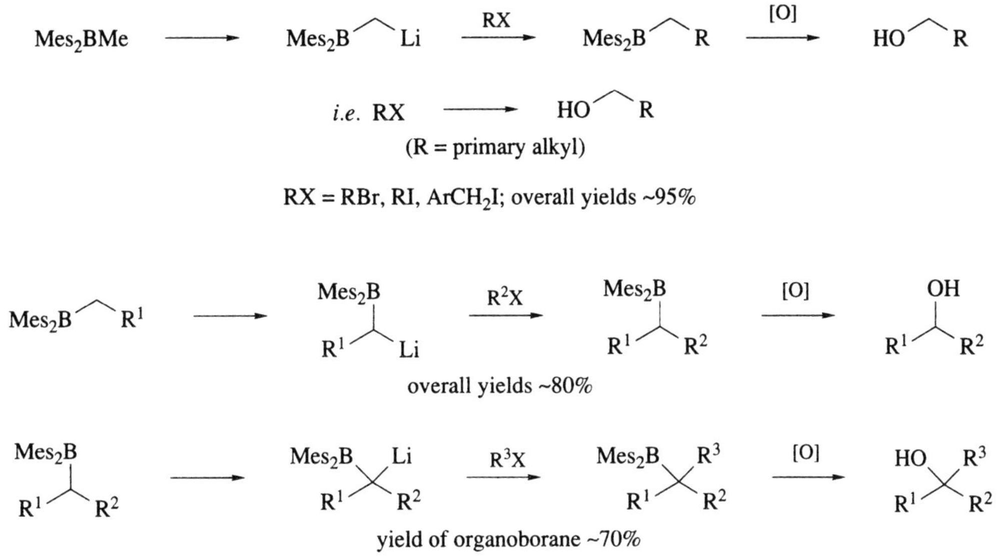

Mes_2BMe ⟶ Mes_2B⌒Li —RX→ Mes_2B⌒R —[O]→ HO⌒R

i.e. RX ⟶ HO⌒R
(R = primary alkyl)

RX = RBr, RI, $ArCH_2I$; overall yields ~95%

Mes_2B⌒R^1 ⟶ Mes_2B–CH(R^1)Li —R^2X→ Mes_2B–CH(R^1)R^2 —[O]→ R^1CH(OH)R^2
overall yields ~80%

Mes_2B–CH(R^1)R^2 ⟶ Mes_2BC(Li)(R^1)R^2 —R^3X→ Mes_2BC(R^3)(R^1)R^2 —[O]→ HOC(R^3)(R^1)R^2
yield of organoborane ~70%

Scheme 2 Alkylation of alkyldimesitylboranes

Alkylations with *s*-alkyl halides give lower yields due to competitive elimination reactions.[44]

Alkylations of boron-stabilized carbanions have been carried out with primary alkyl halides containing acetal,[22,45] alkene,[41] alkyne,[22] chloride,[45] cyano,[45] ester[41] and tosylate[45] groups, though a ketone group was not tolerated.[22]

The anion derived from (**31**) reacted in a remarkable fashion with primary alkyl halides so that alkylation occurred only on sulfur to give the corresponding ylide (equation 30).[46]

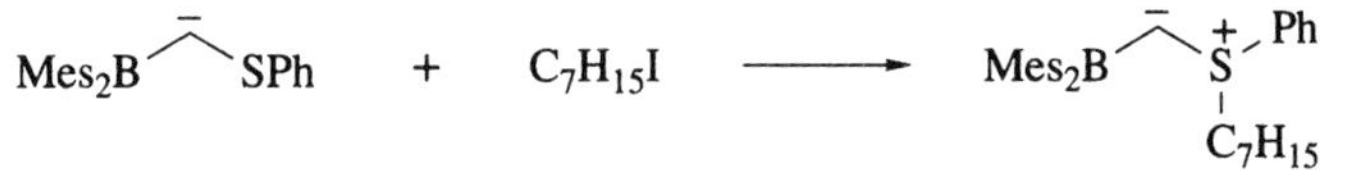

$$Mes_2B\overset{-}{C}H SPh + C_7H_{15}I \longrightarrow Mes_2B\overset{-}{C}H\overset{+}{S}(Ph)C_7H_{15} \quad (30)$$

Alkylations of bis(dialkoxyboryl)methyl anions give products that may be oxidized to aldehydes or ketones (Scheme 3).[31]

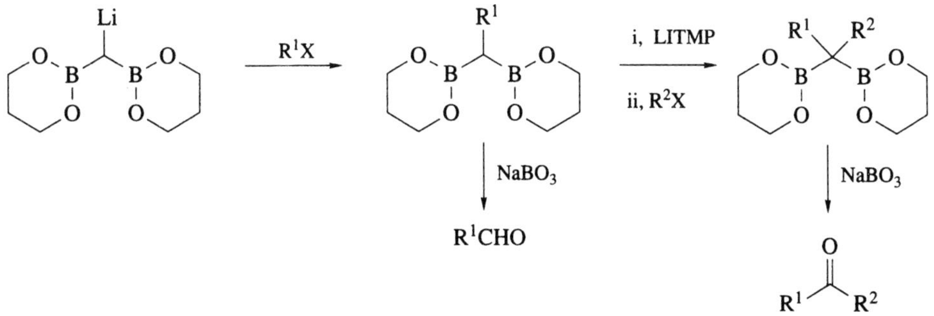

(bis(1,3,2-dioxaborinan-2-yl)methyl)Li —R^1X→ (bis(dioxaborinanyl))CHR^1 —i, LITMP; ii, R^2X→ (bis(dioxaborinanyl))CR^1R^2

(bis(dioxaborinanyl))CHR^1 —$NaBO_3$→ R^1CHO

(bis(dioxaborinanyl))CR^1R^2 —$NaBO_3$→ R^1C(=O)R^2

Scheme 3 Alkylation of bis(dialkoxyboryl)methyl anions

Monoalkylation of (**32**)[31] gives products that are readily converted by NCS into monothioacetals (equation 31).[47]

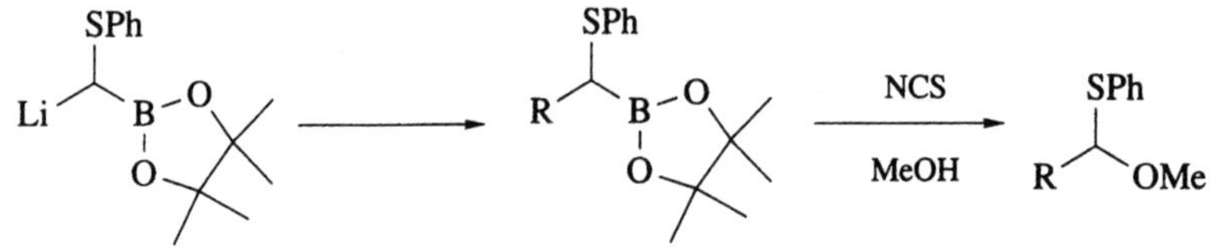

SPh, Li, B, O, O (32) —→ R, SPh, B, O, O —(NCS / MeOH)→ R, SPh, OMe (31)

2.2.3.3 Reactions with Epoxides

The reactions of epoxides with dialkoxyboryl-stabilized carbanions are complicated by interaction of the oxyanion produced with the dialkoxyboryl grouping. When such an interaction is sterically inhibited, as with cyclohexene oxide, a single product results (equation 32).[31]

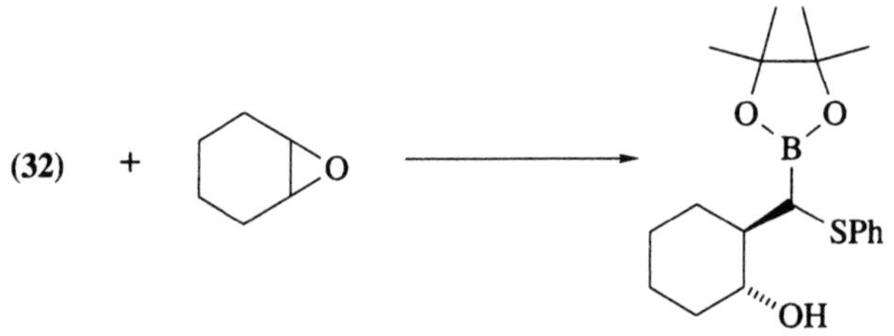

(32) + cyclohexene oxide —→ O, B, O, SPh, OH (32)

Dimesitylboryl-stabilized anions generally react readily with epoxides to give products that can be oxidized to 1,3-diols (Scheme 4).[48] The regioselectivity is high and is dominated by the bulk of the dimesitylboron group, so that even (**2**), the anion derived from methyldimesitylborane, attacks styrene oxide regiospecifically. While (**2**) shows little regioselectivity in its reaction with *trans*-1-methyl-2-pentyl-

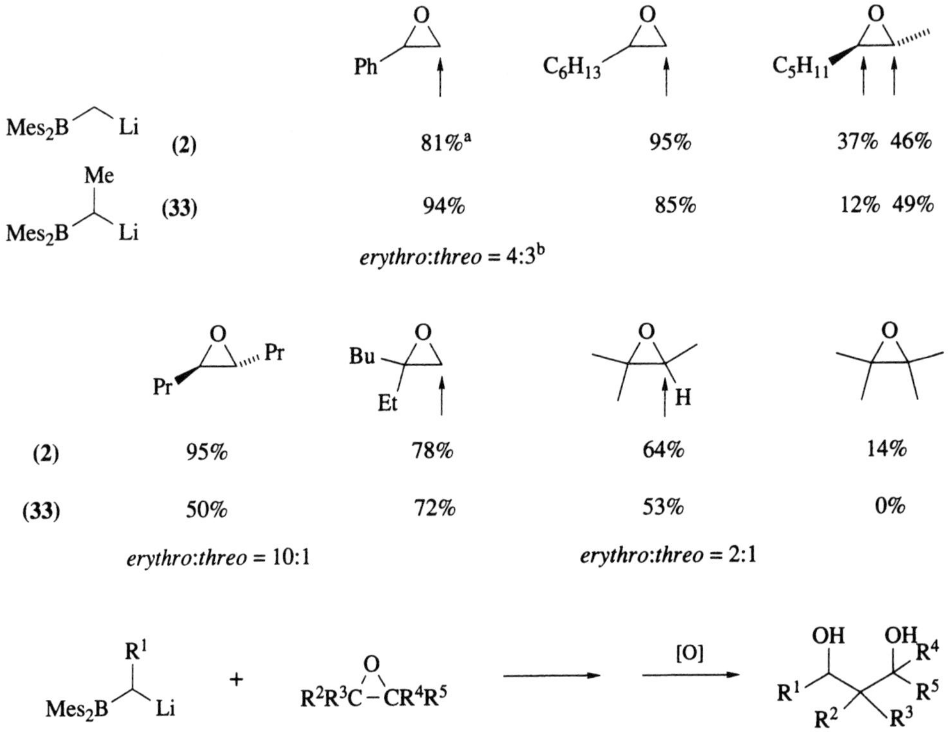

	Ph-epoxide	C_6H_{13}-epoxide	C_5H_{11}-epoxide (trans-Me)
Mes_2B–CH2–Li (**2**)	81%[a]	95%	37% 46%
Mes_2B–CH(Me)–Li (**33**)	94%	85%	12% 49%
	erythro:threo = 4:3[b]		

	Pr/Pr-epoxide (trans)	Bu, Et-epoxide	trisubstituted epoxide (H)	tetramethyl epoxide
(**2**)	95%	78%	64%	14%
(**33**)	50%	72%	53%	0%
	erythro:threo = 10:1		*erythro:threo* = 2:1	

Mes_2B–CH(R^1)–Li + R^2R^3C–O–CR^4R^5 —→ —[O]→ R^1, OH, R^2, R^3, OH, R^4, R^5

[a] Yields are of isolated 1,3-diol and are based on starting organoboranes. The arrows indicate the position(s) of attack on the epoxide based on the isolated diol. For symmetrical epoxides no arrows are shown.

[b] The *erythro:threo* ratio is defined by the relationship of the 1,3-alcohol units.

Scheme 4 Reactions of dimesitylboryl-stabilized carbanions with epoxides

oxirane, anion (**33**), derived from ethyldimesitylborane, reacts with unusual selectivity with the same epoxide. In certain cases, stereo- as well as regio-selectivity is obtained. The *erythro:threo* ratio of 10:1 in the 1,3-diols derived by reaction of (**33**) with *trans*-1,2-di-*n*-propyloxirane is extremely unusual.[48]

With more hindered epoxides, yields are lower but only become negligible with tetrasubstituted epoxides. The method therefore provides a widely applicable synthesis of 1,3-diols.

2.2.3.4 Acylation Reactions

The overall process shown in equation (33) has been fairly well investigated. The presumed initial acylation products (**34**) undergo rearrangement to give alkenyloxyboranes (enol borinates) (**35**), the isolable products of the reaction. There has been no recorded instance of polyacylation in this process.

(33)

(**34**) (**35**)

A specific example of the process, useful for the synthesis of phenyl ketones, is shown in equation (34).[21] Unfortunately, the further reactions of the intermediate enol borinates (**36**) with aldehydes show little diastereoselectivity (equation 35).[21]

(34)

(**36**)

(35)

The reactions of (**37**) with esters give poor or moderate yields of ketones (equation 36).[30]

(36)

(**37**)

The reactions with esters of stabilized anions (**32**) and (**38**) give α-phenylthio ketones in yields of 75–82% (equation 37).[31] Acylations with succinic anhydride and butyrolactone to give γ-keto acids and γ-hydroxy ketones respectively, also proceed in good yields.[31]

(37)

(**38**)

The acylation of the pinacol derivative of dihydroxy[lithio(trimethylsilyl)methyl]borane with methyl benzoate has been recorded.[33] No details were given due to difficulties in the isolation of the products.

Acylation of Mes_2BCH_2Li (**2**) with methyl benzoate for 5 min gave a 72% yield of acetophenone, marginally better than for use of benzoyl chloride or benzoic anhydride.[49] However, acylation of other dimesitylboryl-stabilized anions does not appear to be of general use, often giving only modest yields.[49]

The ethoxycarbonylation shown in equation (38) leads to ethyl octanoate in 50% yield.[49] When dimethyl carbonate was used, α-methylation occurred in preference to ester homologation.[49]

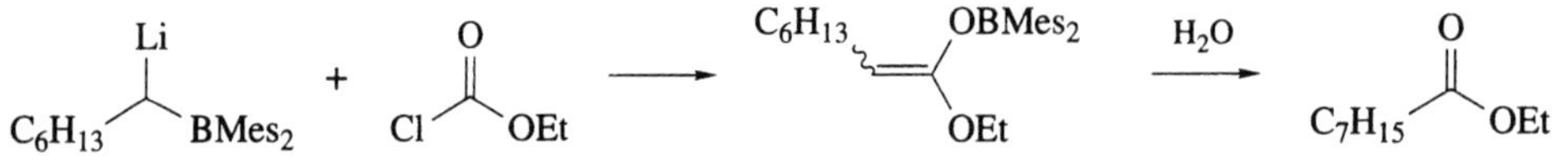

(38)

Acylation of (**2**) with benzonitrile gives an intermediate which yields acetophenone (52%) on hydrolysis with 3 M HCl.[49] This seems to be the only recorded reaction of a boron-stabilized carbanion with a nitrile.[49]

The carboxylation of boron-stabilized carbanions followed by acidification has been reported to give malonic acids in yields of 65–70% (equation 39).[19] The carboxylation of (**39**), however, did not yield any of the corresponding malonic acids.[26]

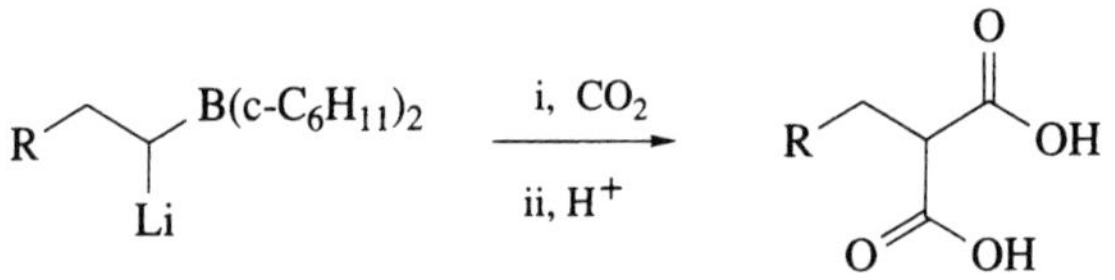

(39)

2.2.3.5 Reactions with Aldehydes and Ketones

It was early reported[50] that boron-stabilized carbanions reacted with benzaldehyde and ketones to yield alkenes (equation 40). The reaction gave a 45–50% yield in the one case quantified, with (*E*):(*Z*) ratios varying over a wide range with temperature. Benzophenone gave a similar yield but the yields with aliphatic ketones were considerably less.[50]

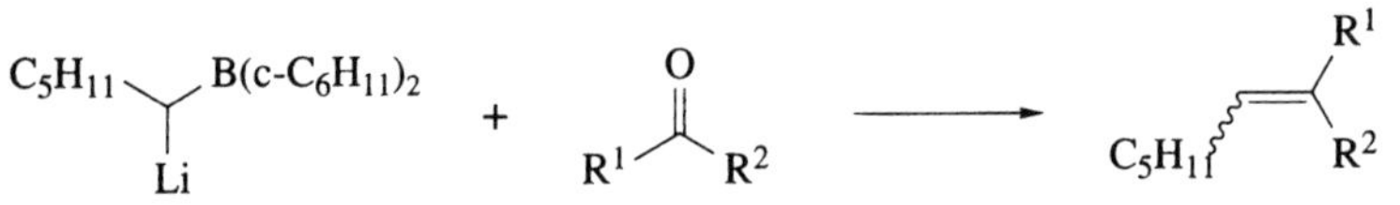

(40)

An analogous reaction with cyclohexanone gave methylenecyclohexane in 55–60% yield (equation 41).[29]

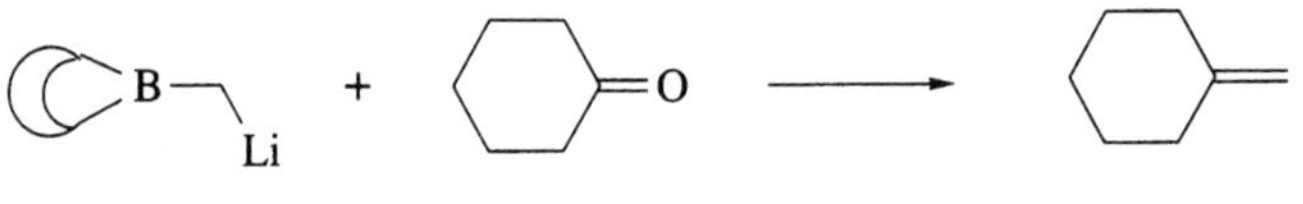

(41)

The reactions of dimesitylboryl-stabilized carbanions (**39**; R = H, Me, C_7H_{15}) with aldehydes and ketones are extremely interesting as they can lead to different products, depending on the nature of R, the carbonyl compounds and the reagents used in the work-up.

Aromatic ketones react to give excellent yields (70–90%) of the corresponding alkenes directly (equation 42), a reaction that was postulated to proceed by *syn* elimination from oxaboretane intermediates by analogy with the Wittig and Peterson reactions.[51] In one case an intermediate β-hydroxyborane has been isolated by chromatography, and in many cases intermediate salts separate, but dissolution of the salts in chloroform gives alkenes upon warming.[51]

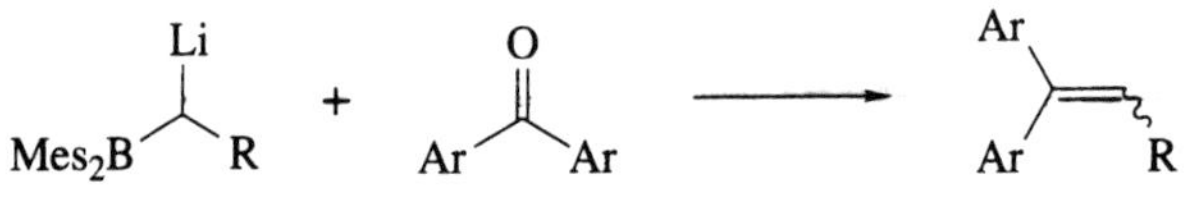

(42)

The direct reactions of (**39**) with benzaldehyde are complex, giving products as shown in equation (43) in proportions varying with the conditions. The alkene, however, was overwhelmingly (*E*)-alkene, in contrast to the similar reactions of 'unstabilized' Wittig reagents.

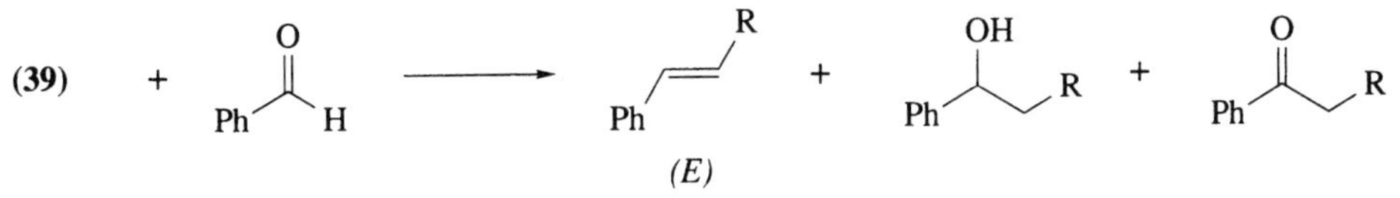

(43)

If the reactions are carried out at low temperature and the reaction mixtures then oxidized also at low temperature, the products are *erythro*-1,2-diols (**41**) obtained in good yields. This is a synthetically useful process, unique among Wittig-type reactions.[52] On the assumption that the oxidation proceeds with retention of configuration at carbon, as for all other C—B bond cleavages by alkaline hydrogen peroxide,[18] then the intermediate has the stereochemistry (**40**; Scheme 5). This can exist as the acyclic form (**40a**) or the cyclic form (**40b**). The latter, however, would have to give (*Z*)-alkene, whereas (*E*)-alkene is actually observed. This must arise therefore by *anti* elimination from (**40a**), which is a most unusual pathway in Wittig-type reactions.

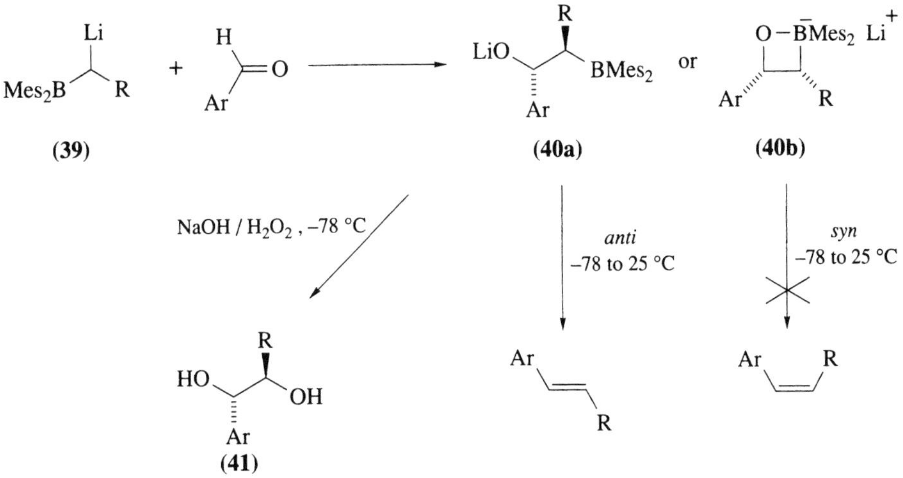

Scheme 5 Reactions of aromatic aldehydes with dimesitylboryl-stabilized carbanions

Intermediates (**40**) react at –110 °C with TMS-Cl to give compounds (**42**), which are stable, readily purified products. The ^{1}H NMR of (**42**) is in accord with the assigned stereochemistry and shows that the oxidation of (**40**) had indeed proceeded with retention of configuration. Reaction of (**42**) with HF/MeCN gives (*E*)-alkenes in excellent yields (Scheme 6) with none of the many by-products seen in the original reaction. The (*E*):(*Z*) ratios range from 100:0 to 95:5 and the reaction tolerates NO_2, Cl, OMe and alkyl groups in the aromatic aldehyde.[53]

By contrast if (**40a**) is reacted with trifluoroacetic anhydride (TFAA) at low temperatures and the reaction then allowed to warm, the (*Z*)-alkene is obtained, presumably by a cyclic ester type elimination (Scheme 6).[53] The isolated yields (72–77%) of (*Z*)-alkene are rather lower than the yields for the (*E*)-alkene process, as are the stereoselectivities. In the case of 4-nitrobenzaldehyde the process shows little stereoselectivity.[53]

The reactions of (**39**; R = H) with aliphatic aldehydes give alkenes in yields that are strongly temperature dependent.[51] If the reaction is carried out in the presence of TFAA, then excellent yields of the methylene compounds result.[54] However, reaction of aliphatic aldehydes with (**39**; R = C_7H_{15}) in the same conditions proceeds by a most unexpected redox process to yield ketones in good yields after aqueous work-up.[54] The appearance of ketones as products is unique in any Wittig-type of reaction at this oxidation level. Alkenyloxyboranes appear to be the intermediates in the reactions, and these react with excess TFAA to give enol trifluoroacetates, which can be isolated and characterized.[54] A possible sequence is shown in Scheme 7.[54]

If the added aldehyde is premixed with a protic acid, the intermediates (**43**) are still formed, but are immediately discharged to give (**44**), which then decompose to give alkenes in good yields. Acetic acid shows little stereochemical discrimination in the production of alkenes except when R = Bu^t (100% of *Z*)

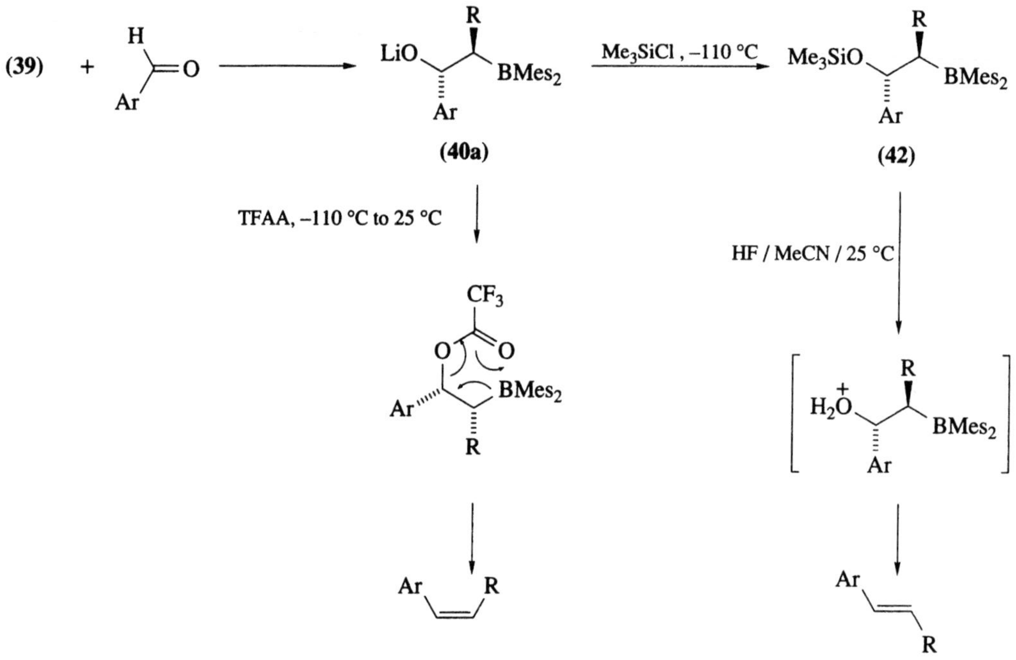

Scheme 6 Stereoselective alkene formation using the boron Wittig reaction

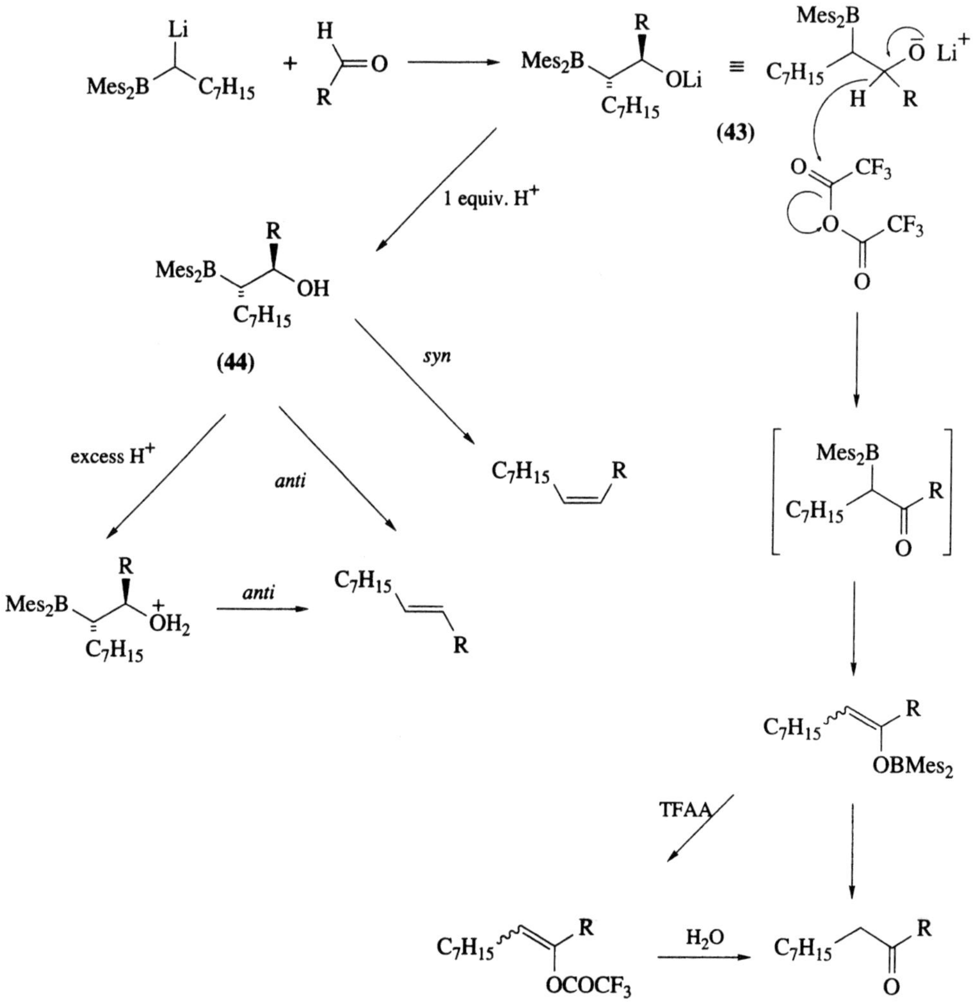

Scheme 7 Reactions of 1-dimesitylboryl-1-lithiooctane with aliphatic aldehydes

and cyclohexyl (90% of *Z*), but excess of stronger acids (HCl, CF_3SO_3H) gives >90% of (*E*)-alkene for less hindered aldehydes.[54] The overall situation for (**39**; R = C_7H_{15}) is summarized in Scheme 7.

Bis(dialkoxyboryl)-stabilized anions react with aldehydes to give alkenylboranes which may be oxidized to the homologated aldehydes (equation 44).[29] Addition of TMEDA and careful control of conditions are required to give good yields.

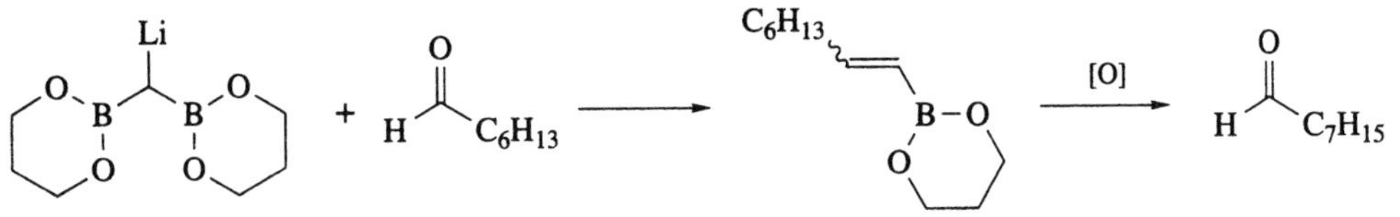

(44)

The reactions of α-TMS-substituted boranes with aldehydes proceed with elimination of the TMS group to give alkenylboranes as in equations (45) and (46). Such products can be hydrolyzed to alkenes or oxidized to aldehydes.[33] The intermediate in the case of the dimesitylborane reaction was directly oxidized to aldehyde in 95% overall yield, a good formyl homologation process.[37]

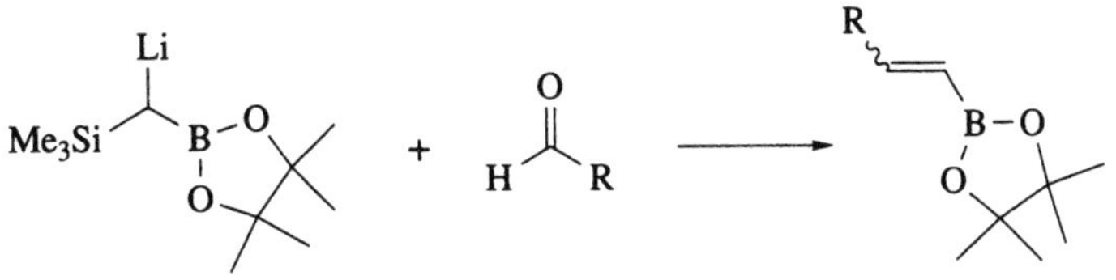

(45)

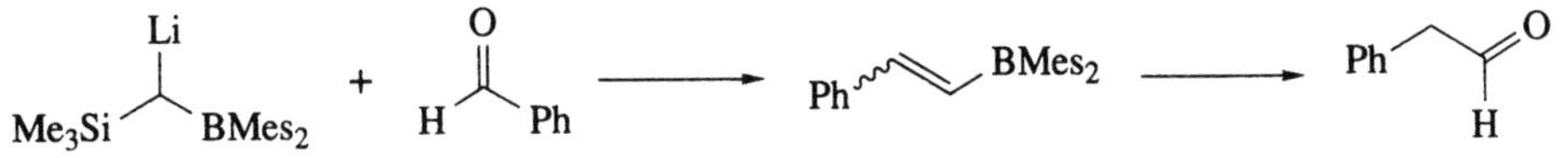

(46)

In reactions with benzophenone and cyclohexanone the dialkoxyboryl-stabilized carbanions still eliminate silicon[33] but the dimesitylboryl derivatives eliminate boron and silicon competitively so that a mixture of products results.[36]

1-Lithio-1-phenylthioalkane-1-boronates react with aldehydes or ketones to give phenylthioalkenes as mixtures of geometric isomers, when this is possible, in yields of 61–86% (equation 47).[31]

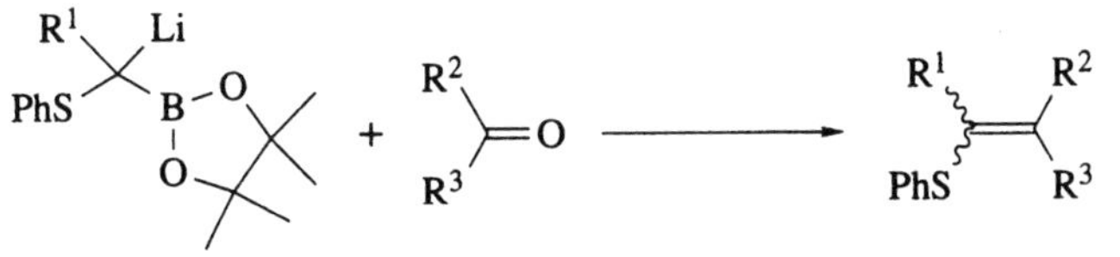

(47)

The dialkoxyboryl-substituted Wittig reagent (**45**) also loses boron upon reaction with benzophenone to yield another Wittig reagent that has been used in the synthesis of allenes (equation 48).[33]

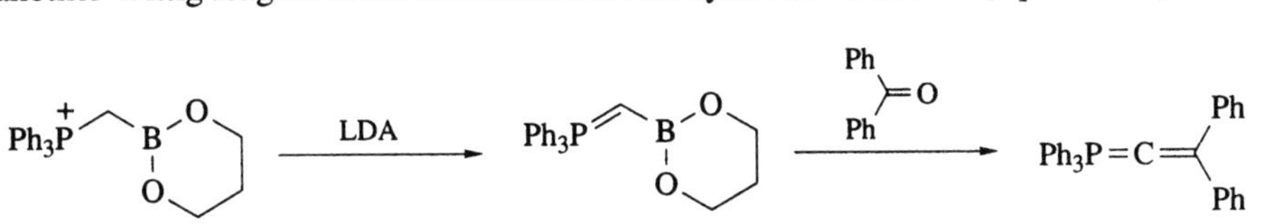

(48)

(**45**)

2.2.3.6 Halogenation Reactions

The halogenation of boryl-substituted carbanions has received little attention. However, bromination of the appropriate carbanions has been accomplished to give compounds (**46**),[22] (**47**)[22] and (**48**).[31] The latter undergoes reaction with phenylmagnesium bromide to give (**49**), though only 57% of impure material has been isolated. With aliphatic Grignard reagents only cleavage products were isolated.[31]

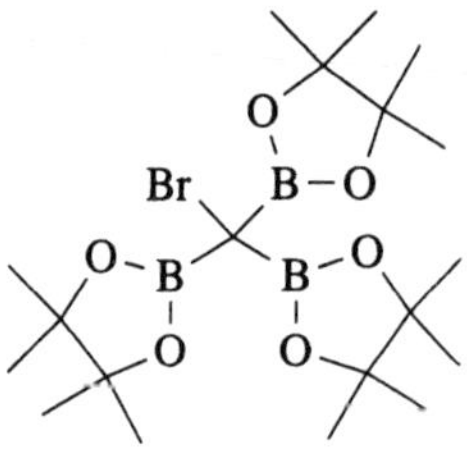

(46) 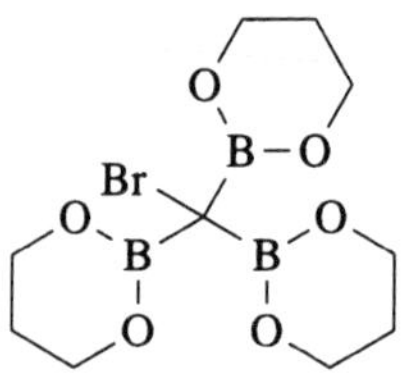 (47) 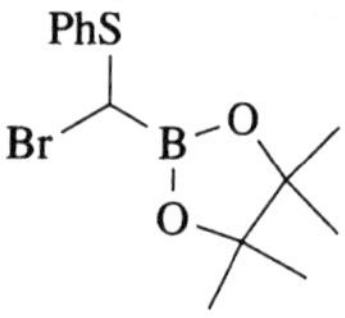 (48)

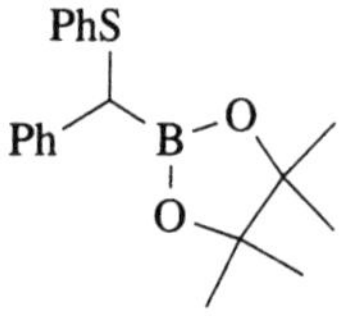

(49) 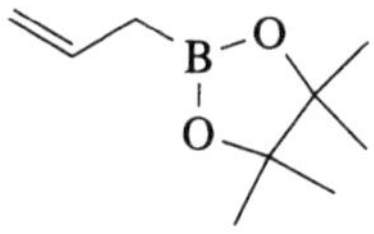 (50)

2.2.4 REACTIONS OF ALLYLIC BORON-STABILIZED CARBANIONS

The anion derived from (**50**) gives mixtures of products from reactions with D_2O, MeI, BuI and TMS-Cl and it was concluded that it cannot function as a useful synthetic reagent.[33]

Anion (**51**) protonates and methylates at the α-position but reacts with TMS-Cl and acetone (equation 49) specifically at the γ-position. No product derived from a boron-Wittig reaction was noted in the reaction with acetone, which upon work-up with propionic acid gave the corresponding alkene (**52**).[38]

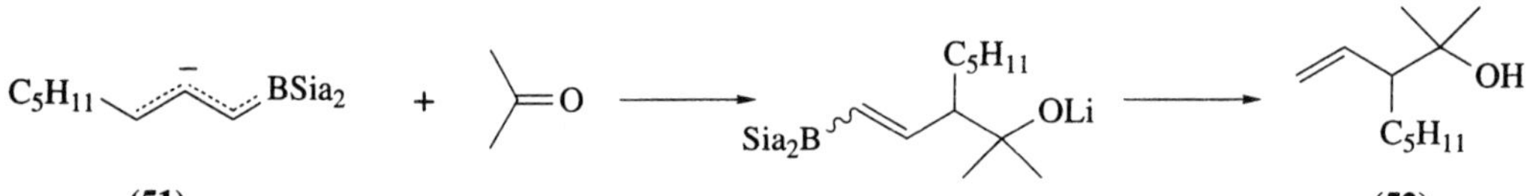

(49)

(51) (52)

Anions (**53**) undergo only α-attack with TMS-Cl or Bu_3SnCl to give (**54**; equation 50).[55] The stereochemistry of compounds (**54**) has not been defined, but upon reaction with water they give only the (*Z*)-allyl-silicon or -tin derivatives (**55**).[55]

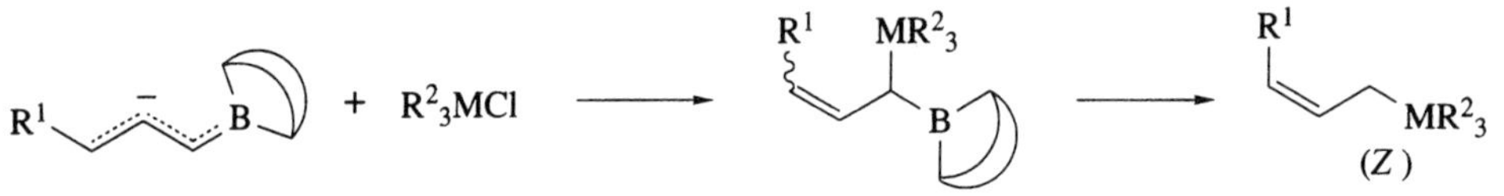

(50)

(53) (54) (55)

The reactions of anion (**56**) proceed entirely at the γ-position with D_2O, Me_2SO_4, EtI, PrI, $C_6H_{13}I$, $C_7H_{15}I$, $C_8H_{17}I$, $PhCH_2I$ and TMS-Cl, and in every case (*E*)-alkenylboranes result.[39] Alkylations proceed in high yields (90–96%) and buffered oxidation of the products gives aldehydes (90–95% overall), giving an efficient three-carbon homologation of alkyl iodides (equation 51).[39]

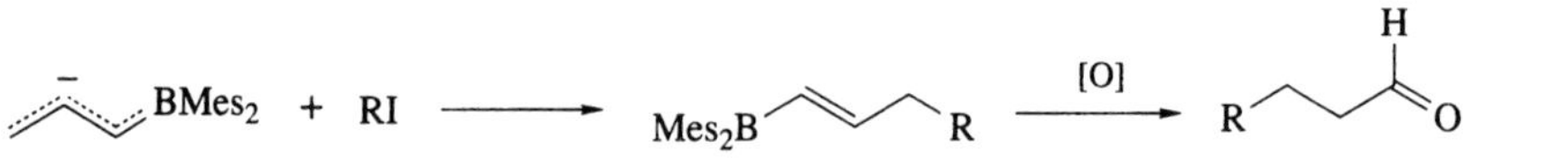

(51)

(56) (*E*)

Benzaldehyde also reacts with (**56**) at the γ-position and oxidation of the intermediate yields γ-lactols (equation 52).[39]

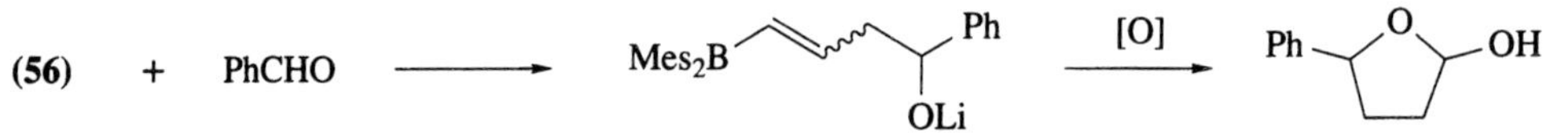

(**56**) + PhCHO ⟶ Mes_2B–CH=CH–CH$_2$–CH(OLi)Ph —[O]⟶ (52)

2.2.5 CONCLUSION

It is clear that the chemistry of carbanions stabilized by boron is so far undeveloped compared, say, with the chemistry of carbanions stabilized by sulfur. Despite this, a variety of unique reactions have been produced and it is certain that more await discovery and development.

2.2.6 REFERENCES

1. G. W. Spitznagel, T. Clark, J. Chandrasekhar and P. von R. Schleyer, *J. Comput. Chem.*, 1982, **3**, 363.
2. T. Clark, H. Korner and P. von R. Schleyer, *Tetrahedron Lett.*, 1980, **21**, 743.
3. A. C. Hopkinson and M. H. Lien, *Int. J. Quantum Chem.*, 1980, **18**, 1371.
4. A. Pross, D. J. De Frees, B. A. Levi, S. K. Pollack, L. Radom and W. J. Hehre, *J. Org. Chem.*, 1981, **46**, 1693.
5. A. Pelter, B. Singaram, L. Williams and J. W. Wilson, *Tetrahedron Lett.*, 1983, **24**, 623.
6. N. M. D. Brown, F. Davidson and J. W. Wilson, *J. Organomet. Chem.*, 1980, **192**, 133.
7. F. Davidson and J. W. Wilson, *J. Organomet. Chem.*, 1981, **204**, 147.
8. O. Gropen, E. Wisloff Nilssen and H. M. Seip, *J. Mol. Struct.*, 1974, **23**, 289.
9. P. Finocchiaro, D. Gust and K. Mislow, *J. Am. Chem. Soc.*, 1973, **95**, 7029.
10. N. M. D. Brown, F. Davidson and J. W. Wilson, *J. Organomet. Chem.*, 1981, **210**, 1.
11. M. V. Garad and J. W. Wilson, *J. Chem. Res. (S)*, 1982, 132.
12. J. M. Bollinger, J. M. Brinich and G. A. Olah, *J. Am. Chem. Soc.*, 1970, **92**, 4025.
13. M. M. Olmstead, P. P. Power, K. J. Weese and R. J. Doedens, *J. Am. Chem. Soc.*, 1987, **109**, 2541.
14. A. Pelter, K. Smith and H. C. Brown, 'Boron Reagents', Academic Press, New York, 1988.
15. A. Pelter, in 'Boron Chemistry', ed. S. Hermanek, World Scientific Publishing, Singapore, 1987, p. 416.
16. H. C. Brown and G. Zweifel, *J. Am. Chem. Soc.*, 1961, **83**, 3834.
17. G. Zweifel and H. Arzoumanian, *J. Am. Chem. Soc.*, 1967, **89**, 291.
18. A. Pelter and K. Smith, in 'Comprehensive Organic Chemistry', ed. D. H. R. Barton and W. D. Ollis, Pergamon Press, Oxford, 1979, vol. 3, p. 689.
19. G. Cainelli, G. Dal Bello and G. Zubiani, *Tetrahedron Lett.*, 1965, 3429.
20. G. Zweifel and H. Arzoumanian, *Tetrahedron Lett.*, 1966, 2535.
21. T. Mukaiyama, M. Murakami, T. Oriyama and M. Yamaguchi, *Chem. Lett.*, 1981, 1193.
22. D. S. Matteson, *Synthesis*, 1975, 147.
23. D. S. Matteson, R. J. Moody and P. K. Jesthi, *J. Am. Chem. Soc.*, 1975, **97**, 5608.
24. R. B. Castle and D. S. Matteson, *J. Am. Chem. Soc.*, 1968, **90**, 2194; R. B. Castle and D. S. Matteson, *J. Organomet. Chem.*, 1969, **20**, 19.
25. D. S. Matteson and J. W. Wilson, *Organometallics*, 1985, **4**, 1690.
26. A. Pelter and R. Pardasani, unpublished results.
27. D. J. S. Tsai and D. S. Matteson, *Organometallics*, 1983, **2**, 236.
28. A. J. Ashe, III and P. Shu, *J. Am. Chem. Soc.*, 1971, **93**, 1804.
29. M. W. Rathke and R. Kow, *J. Am. Chem. Soc.*, 1972, **94**, 6854.
30. D. S. Matteson and R. J. Moody, *Organometallics*, 1982, **1**, 20.
31. D. S. Matteson and K. H. Arne, *J. Am. Chem. Soc.*, 1978, **100**, 1325; *Organometallics*, 1982, **1**, 280.
32. D. S. Matteson and D. J. Majumdar, *J. Chem. Soc., Chem. Commun.*, 1980, 39.
33. D. S. Matteson and D. J. Majumdar, *Organometallics*, 1983, **2**, 230.
34. D. S. Matteson and D. J. Majumdar, *J. Organomet. Chem.*, 1979, **170**, 259.
35. A. Mendoza and D. S. Matteson, *J. Org. Chem.*, 1979, **44**, 1352.
36. J. W. Wilson, *J. Organomet. Chem.*, 1980, **186**, 297.
37. M. V. Garad, A. Pelter, B. Singaram and J. W. Wilson, *Tetrahedron Lett.*, 1983, **24**, 637.
38. R. Kow and M. W. Rathke, *J. Am. Chem. Soc.*, 1973, **95**, 2715.
39. A. Pelter, B. Singaram and J. W. Wilson, *Tetrahedron Lett.*, 1983, **24**, 631.
40. A. Pelter, unpublished results.
41. M. P. Cooke, Jr. and R. K. Widener, *J. Am. Chem. Soc.*, 1987, **109**, 931.
42. (a) A. Norbury, Ph.D. Thesis, University College Swansea, 1990; (b) A. Pelter, R. Drake and M. Stewart, *Tetrahedron Lett.*, 1989, **30**, 3086.
43. D. S. Matteson and P. K. Jesthi, *J. Organomet. Chem.*, 1976, **110**, 25.
44. A. Pelter, L. Williams and J. W. Wilson, *Tetrahedron Lett.*, 1983, **24**, 627.
45. D. S. Matteson and R. J. Moody, *J. Am. Chem. Soc.*, 1977, **99**, 3196.
46. A. Pelter, G. Bugden, R. Pardasani and J. W. Wilson, *Tetrahedron Lett.*, 1986, **27**, 5033.
47. A. Mendoza and D. S. Matteson, *J. Organomet. Chem.*, 1978, **156**, 149.
48. A. Pelter, G. Bugden and R. Rosser, *Tetrahedron Lett.*, 1985, **26**, 5097.
49. L. Williams, Ph.D. Thesis, University College of Swansea, 1983.
50. G. Cainelli, G. Dal Bello and G. Zubiani, *Tetrahedron Lett.*, 1966, 4315.
51. A. Pelter, B. Singaram and J. W. Wilson, *Tetrahedron Lett.*, 1983, **24**, 635.
52. A. Pelter, D. Buss and A. Pitchford, *Tetrahedron Lett.*, 1985, **26**, 5093.
53. A. Pelter, D. Buss and E. Colclough, *J. Chem. Soc., Chem. Commun.*, 1987, 297.
54. A. Pelter, K. Smith, M. Rowlands and S. Elgendy, *Tetrahedron Lett.*, 1989, **30**, 5643, 5647.
55. H. Yatagai, Y. Yamamoto and K. Maruyama, *J. Am. Chem. Soc.*, 1980, **102**, 4548.

2.3
Sulfur Stabilization

KATSUYUKI OGURA
Chiba University, Japan

2.3.1 INTRODUCTION

Sulfur is one of the most frequently employed elements in organic synthesis. Typical functionalities containing sulfur atoms are illustrated below (**1a–1d**); they all stabilize the adjacent carbanion, which serves as a nucleophile. The carbanion reacts with alkyl halides or adds to C=X π-bonds (X = C, O, N, *etc.*) to form a C—C bond, which is an essential process in organic synthesis.

This section deals with addition reactions of sulfur-stabilized carbanions to C=X π-bonds from the standpoint of stereo- and regio-selectivity.

(**1a**) Sulfide (**1b**) Sulfoxide (**1c**) Sulfone (**1d**) Sulfoximine

2.3.2 SULFENYL-STABILIZED CARBANIONS

2.3.2.1 Configuration of the Carbanion

The action of butyllithium on thioanisole in THF generates (phenylthio)methyllithium in a low yield of 35%.[1,2] Corey and Seebach found that reaction of equimolar amounts of butyllithium, DABCO and thioanisole in THF at 0 °C produces (phenylthio)methyllithium in *ca.* 97% yield.[3] Dimethyl sulfide can be metalated with a butyllithium–TMEDA complex at room temperature (equation 1).[4] Treatment of chloromethyl *p*-tolyl sulfide with magnesium produces the corresponding Grignard reagent; a reaction temperature between 10 and 20 °C is crucial for its efficient generation (equation 2).[5]

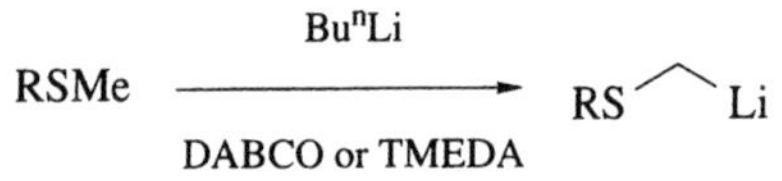

$$\mathrm{RSMe} \xrightarrow[\text{DABCO or TMEDA}]{\mathrm{Bu^nLi}} \mathrm{RSCH_2Li} \quad (1)$$

$$\mathrm{RSCH_2Cl} \xrightarrow[10\text{–}20\ ^\circ\mathrm{C}]{\mathrm{Mg}} \mathrm{RSCH_2MgCl} \quad (2)$$

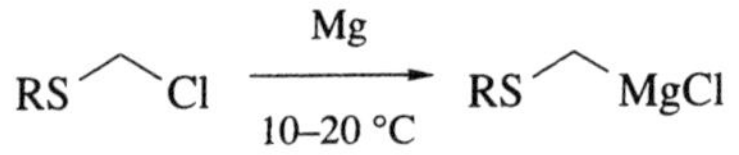

The pK_a value for 2-CH_2 of 1,3-dithiane is 31.1,[6] indicating that the sulfenyl group stabilizes the adjacent carbanion. Theoretical studies present convincing evidence for the unimportance of *d*-orbital participation in the acidification of C—H bonds α to sulfur atoms.[6,7] Whether or not 3*d*-orbitals are included in the basis set, the preferred conformation of $^-CH_2SH$ is predicted to be that of an sp^3 carbanion (**2**).[8] The calculation, NMR studies and crystallography all support a tetrahedral structure for the α-lithiated sulfide.[9]

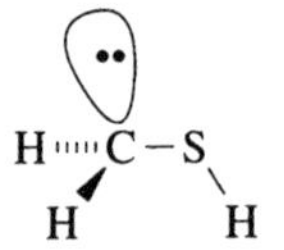

(**2**)

Anderson and coworkers emphasized the polarizability of sulfur, which accounts for the regiochemistry of C—H acidification by sulfur.[10] Epiotis *et al.* developed a hyperconjugative model involving delocalization of the unshared pair on carbon into the low-lying adjacent S—R antibonding orbital,[11] which accounts for the stereochemical aspects of C—H bond acidification by sulfur.

2.3.2.2 Addition to C═O Bonds

(Phenylthio)methyl metal (metal = Li or MgCl) adds to the carbonyl group of ketones and aldehydes [3–5,12,13] Benzoyl derivatives of these adducts are converted to alkenes by reductive elimination with Li–NH_3,[14] $TiCl_4$–Zn[15] or Ti.[16] Transformation of (**3**) into an alkene *via* its phosphoric ester has also been reported (equation 3).[17]

The above ketone methylenation is applicable to highly hindered ketones, which are usually inert to the Wittig method. For example, a highly hindered tricyclic ketone (norzizanone, **4a**) undergoes the methylenylation to give zizaene (**4b**; equation 4). Δ^5-Cholesten-3-one (**5a**) was also converted to the nonconjugated diene 3-methylene-Δ^5-cholestene (**5b**; equation 5).[14]

Condensation of the lithio derivative of benzyl phenyl sulfide (**6**) with benzaldehyde gave two diastereomers of 2-phenylthio-1,2-diphenyl-1-ethanol (**7**) in a ratio of 60:40 (equation 6).[18]

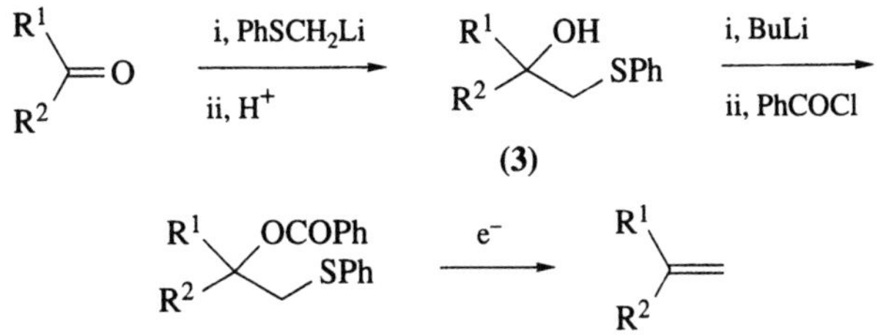

(3)

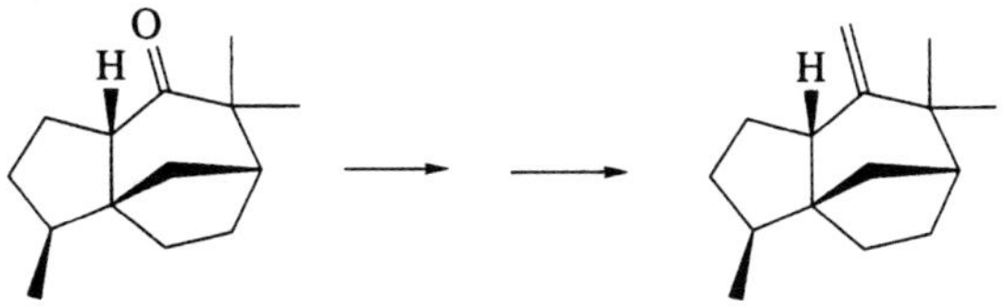

(4)

(4a) (4b)

(5)

(5a) (5b)

(6)

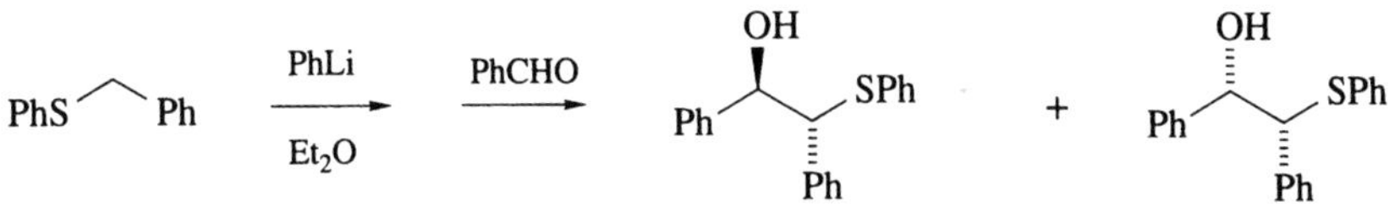

(6) (7a) (7b)

In the addition of the sulfide (**8**), which has a chiral center at the β-position, 1,2-induction was completely stereoselective, while 1,3-asymmetric induction occurred with 80% efficiency (Scheme 1). It has been shown that the critical factor for obtaining high stereoselectivity is a thermodynamic preference in the lithio derivative (**9**) and not a diastereoselective deprotonation. In essence, C—Li bonds to sulfur are similar to their oxygen and nitrogen counterparts and differ only in the fact that the epimerization rate of (**9**) is seemingly faster.[19]

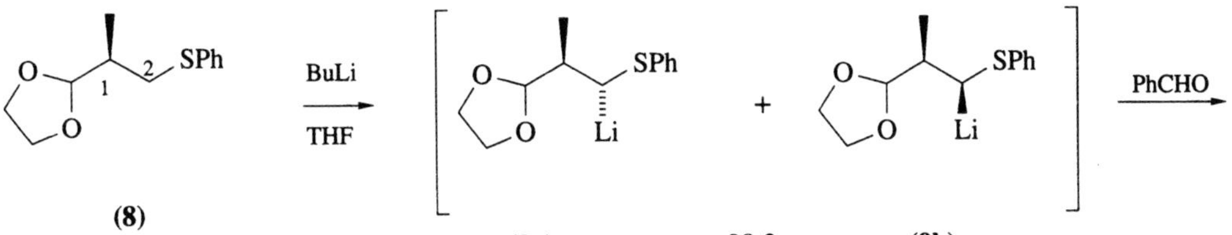

(8) (9a) 98:2 (9b)

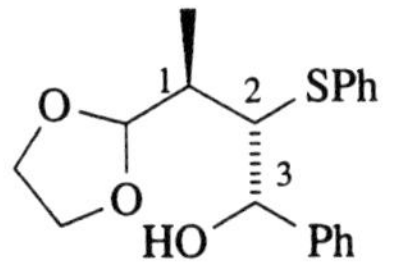

α-OH:β-OH = 4:1

Scheme 1

2.3.2.3 α-Sulfenylated Allylic Carbanions

The reaction of α-sulfenylated allylic carbanions with electrophiles may give both the α- and γ-products (equation 7). Regiochemical control of this ambident anion is dependent upon many factors, including substituents, counterions, the solvent system, the type of electrophile and steric effects.

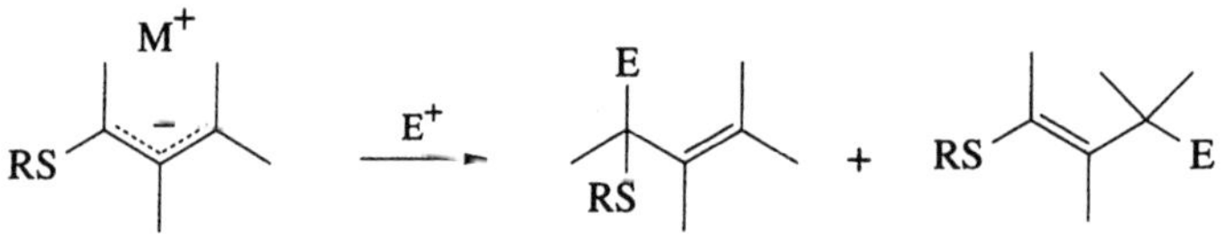

M^+ RS ⟶ (E^+) E, RS + RS, E (7)

When the counterion is lithium, the α:γ regioselectivity depends in part upon the electrophile employed (Scheme 2). 3-Methyl-2-butenyl phenyl sulfide generates an allylic carbanion by the action of *n*-butyllithium–THF. Methyl iodide reacts with α-selectivity (exclusively α at –78 °C), whereas acetone predominantly affords the γ-adduct (α:γ = 25:75). The regioselectivity of alkylation and of reaction with acetone also depends on the presence of solvating species. The results are best explained by assuming that in THF without a complexing agent, the carbanion exists as an ion pair, the lithium ion being closely associated with the α-carbon. In contrast, the cryptate [2.2.2] accommodates the lithium in its cavity so that solvent-separated ion pairs or free anions are present. The reaction now occurs at both α- and γ-carbons with methyl iodide (α:γ = 60:40) and only at the α-carbon with acetone.[20] Addition of the lithio derivative of an aryl allyl sulfide to benzaldehyde (equation 8) was also reported to afford predominantly the corresponding γ-adduct (α:γ = 15:85 or 28:72 in the cases of aryl = phenyl or *p*-methoxyphenyl, respectively). The diastereomeric ratio of the product resulting from α-attack ranges from 34:66 (aryl = mesityl) to 25:75 (aryl = *p*-methoxyphenyl).[21]

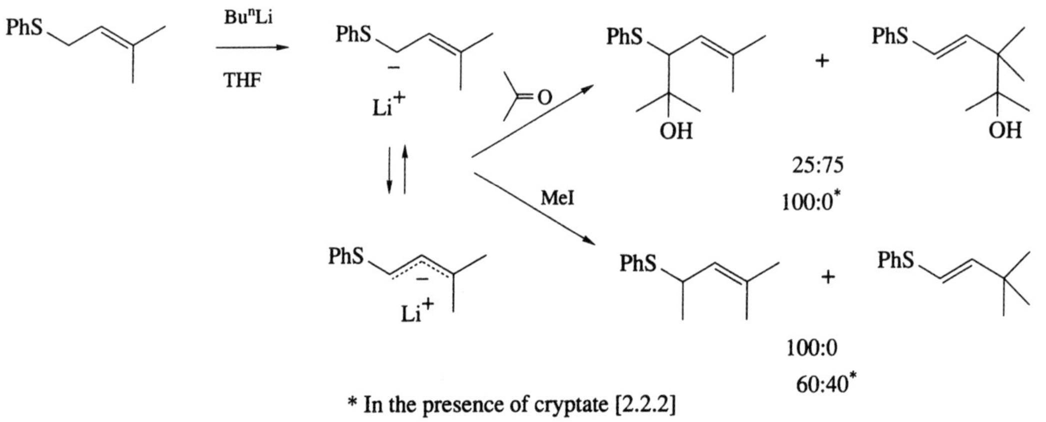

PhS ⟶ (Bu^nLi, THF) PhS, Li^+ ⇅ PhS, Li^+; acetone ⟶ PhS, OH + PhS, OH (25:75; 100:0*); MeI ⟶ PhS + PhS (100:0; 60:40*)

* In the presence of cryptate [2.2.2]

Scheme 2

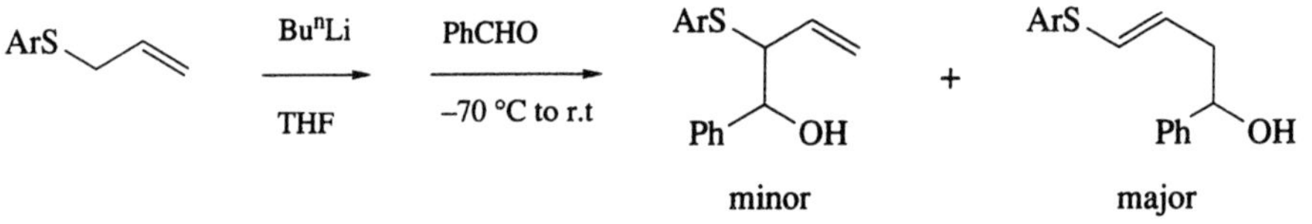

ArS ⟶ (Bu^nLi, THF) ⟶ (PhCHO, –70 °C to r.t) ArS, Ph, OH (minor) + ArS, Ph, OH (major) (8)

In contrast to lithio derivatives, (isopropylthio)allyl copper reacts with acetone in ether at –78 °C to yield an α-adduct as a major product (Scheme 3).[22]

The reaction between a wide variety of (alkylthio)allyltitanium reagents of type (**11**) and carbonyl compounds has also been reported (equation 9). Table 1 illustrates how the relative proportions of α- and γ-adduct formed vary according to the reagent.

The substitution pattern of the starting sulfide (**10**) can have a pronounced effect on the α:γ ratio in the final condensation products. With unsubstituted sulfides ($R^1 = R^2 = R^3 = R^4 = H$) or α- and β-mono- and di-substituted sulfides (R^1 and/or R^2 = Me), the α-selectivities are about 97–99%. On the other hand, a dramatic alteration in product distribution occurs when the condensation with aldehydes is carried out using γ-substituted sulfides (R^3 and/or R^4 = Me); excellent γ-selectivity was observed. It is highly interesting that α- and γ-disubstituted sulfide ($R^1 = R^3$ = Me) gives the α-adduct almost exclusively. The

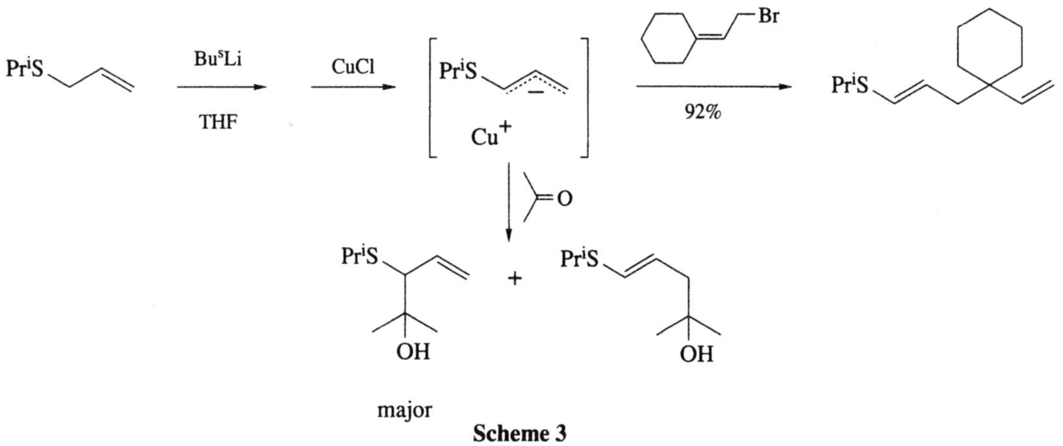

Scheme 3

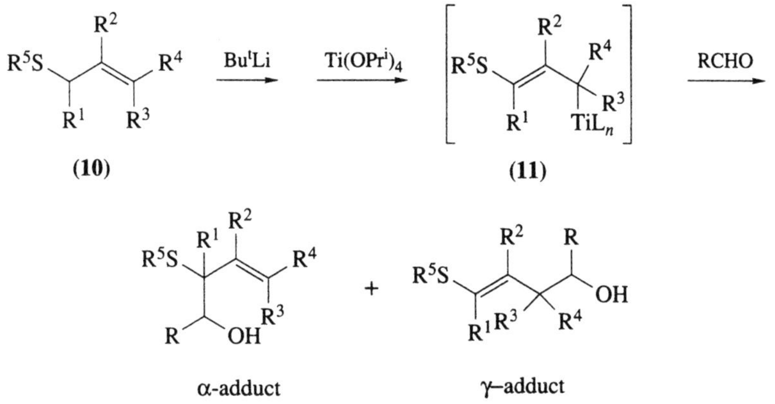

(9)

Table 1 Reaction of the Anion (**11**) with Aldehydes

R^1	R^2	R^3	R^4	R^5	RCHO	*Yield (%)* α-Adduct[a]	β-Adduct
H	H	H	H	Ph	c-$C_6H_{11}CHO$	99 (>30:1)	<1
H	H	H	H	Et	n-$C_5H_{11}CHO$	87 (>30:1)	5
H	H	H	H	Et	PhCHO	94 (6:1)	2
Me	H	H	H	Ph	c-$C_6H_{11}CHO$	99 (>30:1)	<1
H	Me	H	H	Et	c-$C_6H_{11}CHO$	98 (>30:1)	2
H	H	Me	H	Et	c-$C_6H_{11}CHO$	>1	93
Me	Me	H	H	Ph	c-$C_6H_{11}CHO$	83 (>30:1)	<1
Me	H	Me	H	Ph	c-$C_6H_{11}CHO$	71 (>30:1)	4
H	H	Me	Me	Ph	c-$C_6H_{11}CHO$	>1	99

[a]The value in parenthesis is the ratio of *erythro:threo*.

second important trend is the exceedingly high *erythro* selectivity of the reaction. Even an aromatic aldehyde gives the *erythro*-alcohol selectively. Furthermore, the *erythro* stereoselectivity is not affected by the α-branches of the starting sulfides.[23]

The well-known procedure for desulfurization *via* sulfonium salt formation followed by base-catalyzed cyclization converts this reaction into a stereoselective alkenyloxirane synthesis. Some examples are illustrated in Scheme 4.

The triethylaluminum or triethylborane ate complexes (**12**) of the (isopropylthio)allyl carbanion react with carbonyl compounds at the α-position (equation 10). In the reactions with carbonyl compounds, very high regioselectivity (for example, butanal 95:5, 3-methylbutanal 99:1, cyclohexanone 92:8 and acetophenone 95:5) was achieved by using the aluminum ate complex. On the other hand, the α-regioselectivity with ketones decreases if the boron ate complex is used (cyclohexanone 72:28, acetophenone 45:55). It is noteworthy that the stereoselectivity of the α-adduct from an aldehyde is low. Presumably the geometry of the double bond in the ate complex (**12**) is not homogeneous.[24]

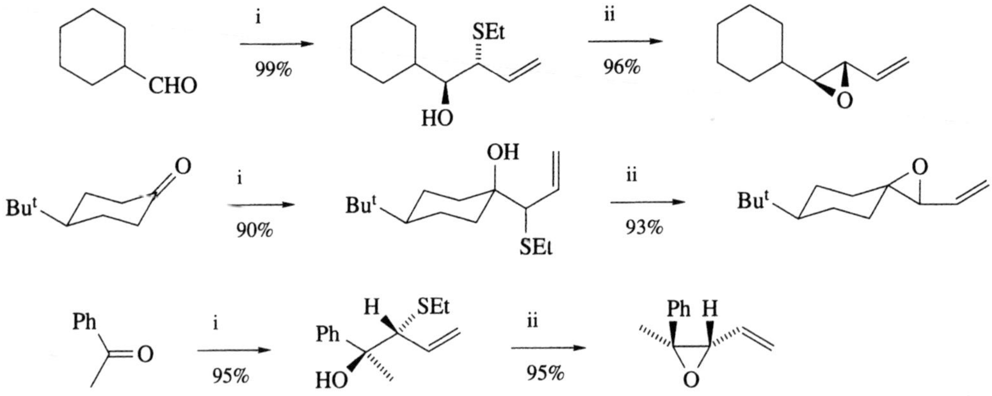

i, CH_2=$CHCH_2SEt/Bu^tLi/Ti(OPr^i)_4$, THF, –78 °C; ii, Me_3OBF_4, CH_2Cl_2, 0 °C, then aqueous NaOH, 25 °C

Scheme 4

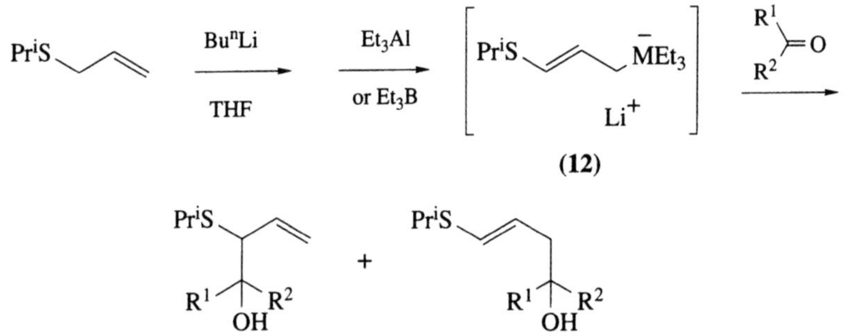

(12)

(10)

Seebach and coworkers reported that the doubly lithiated 2-propenethiol (**13**) reacts preferentially in the γ-position with a variety of electrophiles, including aldehydes and ketones. The selectivity is 67–75%.[25] The magnesium derivative (**14**), readily accessible by addition of $MgBr_2$ (1 equiv.), reacts with aldehydes and ketones to give α-adducts almost exclusively (more than 90%). Successive addition of a saturated or α,β-unsaturated aldehyde or ketone followed by methyl iodide to a solution of the magnesium derivative at –80 °C affords very high yields of the α-adducts, which are synthetic precursors for vinyloxiranes (Scheme 5).[26]

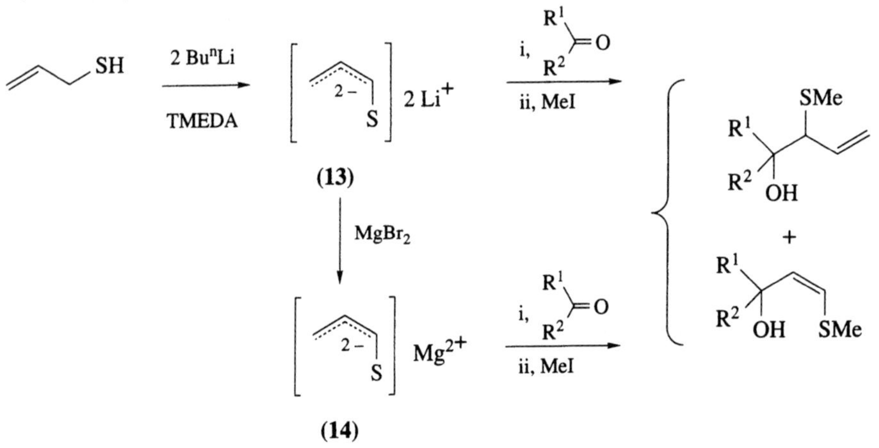

(13)

(14)

Scheme 5

2.3.2.4 Miscellaneous

Formaldehyde dithioacetals are often utilized as synthetic equivalents of formyl anion ($^-$CHO) and carbonyl dianion ($^{2-}$C=O). 1,3-Dithiane is commonly used; its 2-H is quite acidic (pK_a = 31.1)[6] due to stabilization by its two adjacent sulfenyl groups. 2-Lithio-1,3-dithiane adds to aldehydes and ketones. Its addition to 4-*t*-butylcyclohexanone proceeds under kinetic control, and the rates of equatorial and axial attack are comparable.[27] 2-Phenyl-1,3-dithiane generates a more stable carbanion at the 2-position (pK_a = 28.5 in cyclohexylamine), thus enabling thermodynamic control, under which attack is exclusively equatorial. The dianion (**15**) of 2-hydroxymethyl-1,3-dithiane adds to an epoxycyclohexanone (**16**) in a highly stereoselective manner, which is the key step in a novel approach to the avermectin southern hexahydrobenzofuran unit (Scheme 6).[28]

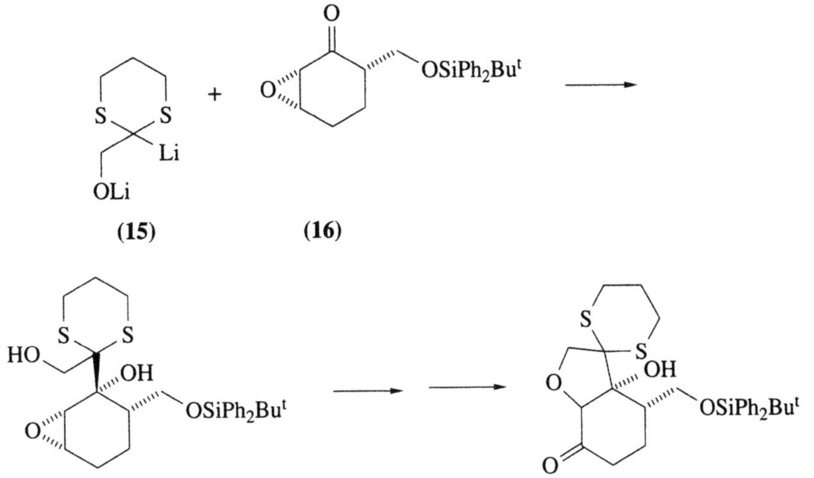

Scheme 6

In the reaction with α,β-unsaturated ketones (**18**), 2-lithio-1,3-dithiane (**17**; R = H) produces 1,2-adduct (**20**) in THF. By addition of HMPA (1–2 equiv.) to the reaction mixture, 1,4-addition is successfully attained (equation 11).[29] It is noteworthy that the lithio derivative of 2-aryl-1,3-dithiane regiospecifically adds to 2-cyclohexenone in a 1,4-manner (25 °C in THF).[30]

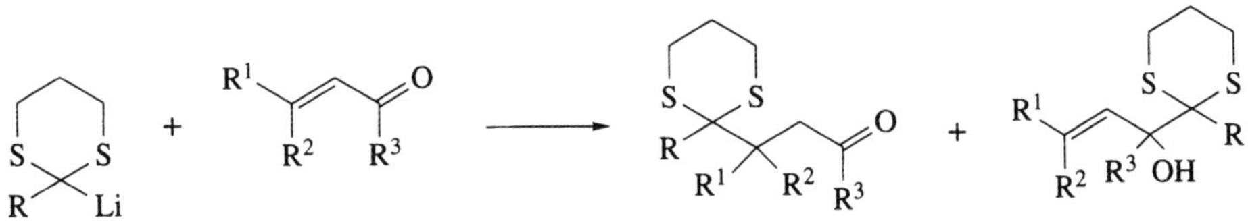

(11)

The reactions of lithiated tris(phenylthio)-,[31] methoxyphenylthiotrimethylsilyl-,[32] bis(methylthio)trimethylsilyl- and bis(methylthio)stannylmethanes[33] with conjugated ketones give the corresponding 1,4-adducts exclusively.

The allyllithium derived from 2-ethylidene-1,3-dithiane (**22**) and LDA reacts with an aldehyde (**21**) at predominantly the α-position (α:γ = 90:10). When the counterion is changed from lithium to cadmium, the regioselectivity is reversed (α:γ = 10:90), as shown in Scheme 7.[34]

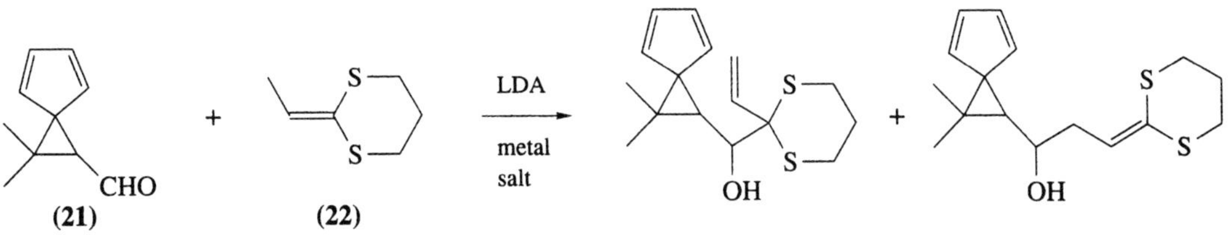

LDA	96% (90:10)	LDA/CuI•P(OMe)$_3$	61% (35:65)
LDA/ZnCl$_2$	70% (60:40)	LDA/CdCl$_2$	90% (10:90)

Scheme 7

The crotyllithium compound (**23**) generated from (*E*)-2-(1-propen-1-yl)-1,3-dithiane reacts with aldehydes to give exclusively γ-adducts, which favor the *anti* isomer (Scheme 8).[35] However, crotyllithium (**23**) reacts with ketones at either the α- or the γ-site, depending on the nature of the ketone (for 2-butanone, α:γ = 84:16, *anti*:*syn* = 77:23). To account for the high regioselectivity and *anti* selectivity of the aldehyde addition, a chelation model with a chair-like transition state (**25**) was proposed. This regio- and diastereo-selective reaction is applicable to syntheses of *trans*-β,γ-disubstituted γ-lactone (**27**). The *anti* γ-adducts are prone to cyclization to afford (**26**), which undergoes hydrolysis with $HgCl_2$, leading to (**27**; equation 12).

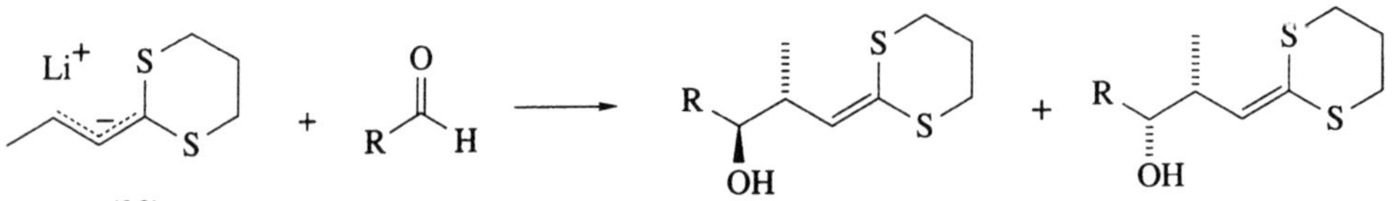

(**23**)   *anti*-(**24**)   *syn*-(**24**)

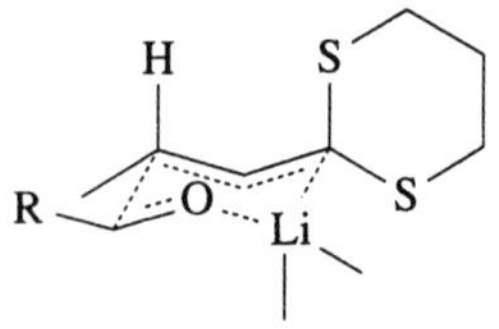

R = Me,	86%	(80:20)	R = Me_2C=CH,	88% (84:16)
R = Et,	91%	(~100:0)	R = PhCH=CH,	86% (85:15)
R = Pr^i,	93%	(~100:0)		

(**25**)

Scheme 8

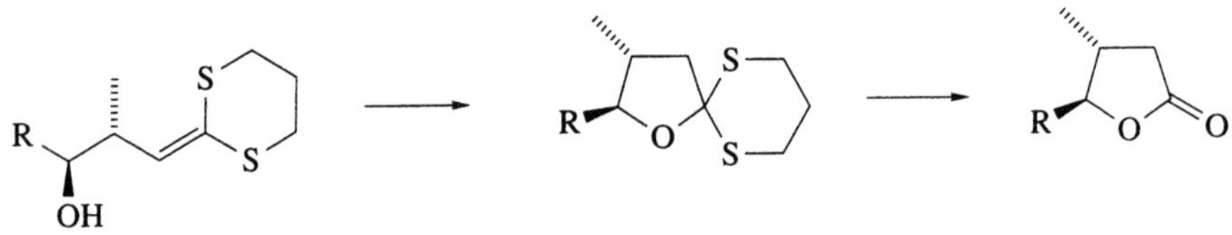

anti-(**24**) → (**26**) → (**27**)   (12)

2.3.3 SULFINYL-STABILIZED CARBANIONS

2.3.3.1 Configuration of the Carbanion

Sulfoxides bearing two different substituents are dissymmetric and resolvable into optically active enantiomeric forms. Since the sulfinyl group is pyramidal, a diastereotopic relationship exists for the methylene protons (H_R and H_S) and differential acidity of these protons is observed.

It has been demonstrated that *pro-(R)* hydrogen (H_R) of (*S*)-benzyl methyl sulfoxide undergoes exchange in alkaline D_2O or in MeOD containing NaOMe at a rate approximately 16 times faster than the *pro-(S)* hydrogen (H_S). The results hitherto reported are summarized in Figure 1.[36–40] Deuteration and methylation of benzyl methyl sulfoxide occur with opposite stereospecificity, whereas the corresponding actions of benzyl *t*-butyl sulfoxide in THF proceed with the same stereospecificity. The lithiated α-sulfinyl carbanions produced from (S_S) and (R_S)-benzyl methyl sulfoxides may have (S_C,S_S) and (S_C,R_S) configurations, respectively. Water comes from the lithiated side of the carbanions because its polarization causes it to interact initially with the countercation. However, methyl iodide is a nonpolar substrate and prefers to react on the more nucleophilic side, which is *anti* to the sulfur lone pair. Thus, the sulfur lone pair can exert an α-effect to make the *anti* lone pair more polarizable.[41]

Ab initio MO calculations on the hypothetical anion $^-CH_2S(O)H$ suggest a nonplanar geometry to be preferable at the anionic center, and the order of carbanion stability is shown in Figure 2.[42,43]

This order is not in agreement with most of the results from experiments conducted to determine the conformational preferences of α-sulfinyl carbanions in solution, where solvation and ion-pairing effects are important. For example, Biellmann found that the product stereochemistry varies with solvent and temperature, and with the presence or absence of lithium chelating agents such as cryptate.[44] Lithium cation–oxygen chelation may control the conformation in THF (Figure 3), while a free anion is predominant in the presence of the chelating agent.

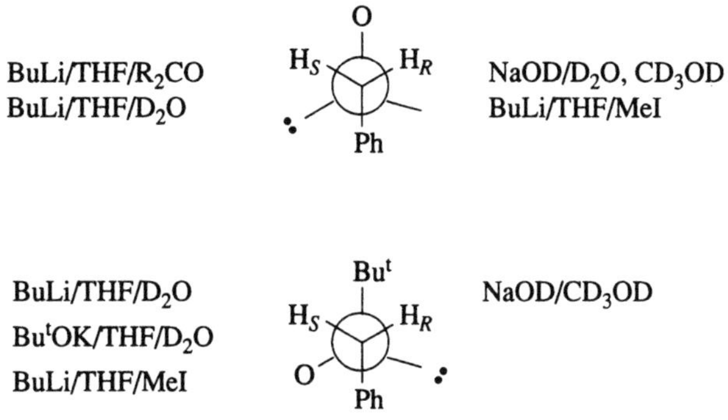

BuLi/THF/R_2CO
BuLi/THF/D_2O
H_S H_R O Ph
NaOD/D_2O, CD_3OD
BuLi/THF/MeI

BuLi/THF/D_2O
$Bu^tOK/THF/D_2O$
BuLi/THF/MeI
Bu^t H_S H_R O Ph
NaOD/CD_3OD

Figure 1

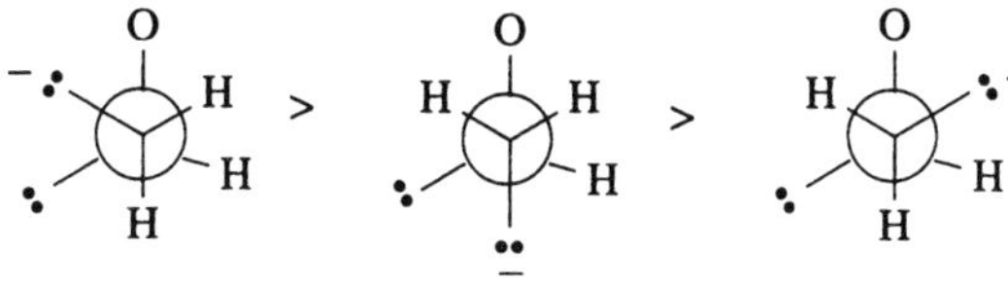

Figure 2

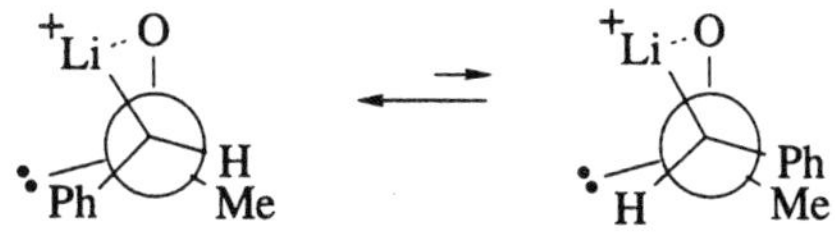

Figure 3

Even the widely accepted assumption of pyramidal sp^3-hybridized α-sulfinyl carbanions has been challenged. A ^{13}C NMR study provides evidence for a chelated α-lithio thiane *S*-oxide with a planar metalated carbon.[45] The stereoselectivity observed in methylation with MeI is completely reversed with $(MeO)_3PO$. These results are explained in terms of *trans* approach of MeI to the chelated face and *cis* approach of $(MeO)_3PO$, which requires coordination to Li^+. It was also reported that the carbanion carbon of PhCHLiS(O)Me or $Bu^tS(O)CH_2Li$ is primarily sp^2-hybridized or intermediate between sp^2- and sp^3-hybridized, respectively.[46]

2.3.3.2 Addition to C=O Bonds

In 1972, Tsuchihashi disclosed that the carbanion (**28**; Ar = *p*-tolyl), generated from (*R*)-methyl *p*-tolyl sulfoxide with lithium diethylamide, adds to benzaldehyde or α-tetralone to give an adduct (**29**) in a diastereomeric ratio of 50:50 or 64:36, respectively.[47] Additions of this carbanion to various unsymmetrical ketones are also reported to be poorly diastereoselective (for example, EtCOMe 50:50, Bu^tCOMe 55:45, Bu^tCOPh 70:30).[48] Note that in the case of Ar = 2-pyridyl a chiral sulfinyl group increases the asymmetric induction observed in the addition of the corresponding carbanion to carbonyl compounds (PhCHO 80:20, *n*-$C_9H_{19}CHO$ 70:30).[49] Since diastereomer pairs of (**29**) are separable, chromatographic separation followed by reductive desulfurization with Raney Ni provides a method for obtaining optically active alcohols (**30**; Scheme 9).

Reaction of the α-carbanion of an alkyl aryl sulfoxide (RCH_2SOAr) with aldehydes may give four diastereomers. In general, the reaction is highly diastereoselective with respect to the α-sulfinyl carbon, but poorly diastereofacially selective with respect to attack on the carbonyl component. In fact, the α-carbanion (**31**) of benzyl *t*-butyl sulfoxide adds to an aldehyde to produce only two diastereomers (**32a**) and (**32b**). As shown in Scheme 10, the selectivity increases when the counterion is Zn^{2+}. A transition state structure (**33**) is proposed to account for the *anti* stereoselection.[50] Addition of the dianion of (*R*)-3-(*p*-tolylsulfinyl)propionic acid (**34**) to aldehydes affords two main diastereoisomeric β-(*p*-tolylsulfinyl)-γ-lactones (**35**; R = Ph and Bu^t, *ca.* 60:40). These isomers (**35**) were separated by chromatography, and their

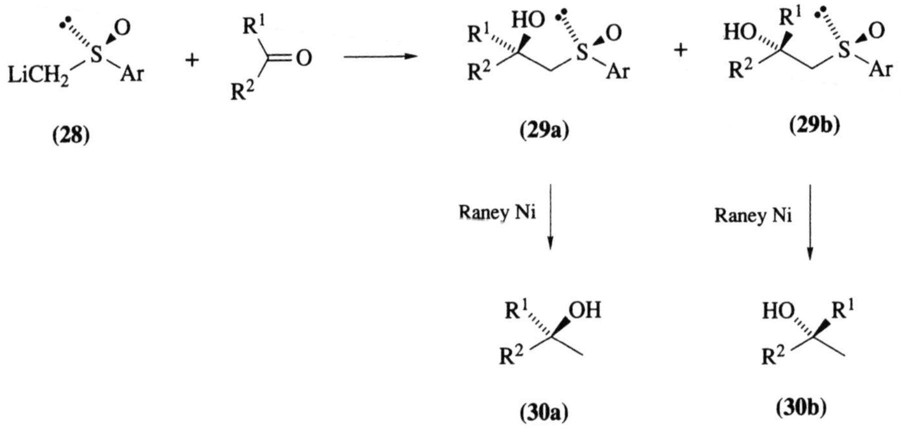

(28) (29a) (29b)

Raney Ni Raney Ni

(30a) (30b)

Scheme 9

pyrolysis gave optically pure 5-substituted furan-2(5*H*)-ones (**36**; Scheme 11).[51] Analogously, optically active *cis*-7,8-epoxy-2-methyloctadecane (**39**), the sex attractant of the gypsy moth, was synthesized by utilizing the addition of the α-carbanion of optically active *t*-butyl 5-methylhexyl sulfoxide (**37**) to undecanal (**38a:38b** = *ca*. 67:33) as shown in Scheme 12.[52]

Condensation of α-sulfinyl carbanions with aldehydes provides a useful method for generating 1,2-asymmetry as well as construction of 1,3-asymmetric relationships in acyclic systems. α-Sulfinyl carbanions such as (**40**) undergo condensation with aldehydes, yielding the β-hydroxy sulfoxides (**41**), thus generating two asymmetric centers at C-4 and C-5 (equation 13). Successive treatment of the sulfoxide (**42a**) with LDA and then with benzaldehyde at –78 °C gave adducts (**43**) and (**44**) in 85% yield in a 91:9 ratio, whereas the diastereomer (**42b**) gave four adducts (**45, 46, 47** and **48**) in 94% yield of 67:17:13:3 composition, respectively (Scheme 13). In analogous fashion, condensation of the racemic sulfoxides which have the opposite configuration of the β-methyl substituent at C-3 afforded 85–90% yields with benzaldehyde.[53–55] It is noteworthy that major products (**43**) and (**45**) share the same relative stereochemistry along the carbon backbone, in spite of the inversions of sulfinyl configuration. The general mode of the addition demonstrates the same preference for '*erythro*' orientations at C-4 and C-5 as observed in the addition of an anion to a carbonyl compound, which presumably does not involve a cyclic, chair-like, transition state. An indolizidinyl sulfoxide (**49**) also adds to butanal *via* its α-sulfinyl carbanion to afford two diastereomers (**50a**) and (**50b**) in a ratio of 2:1. Since the configuration of the newly created

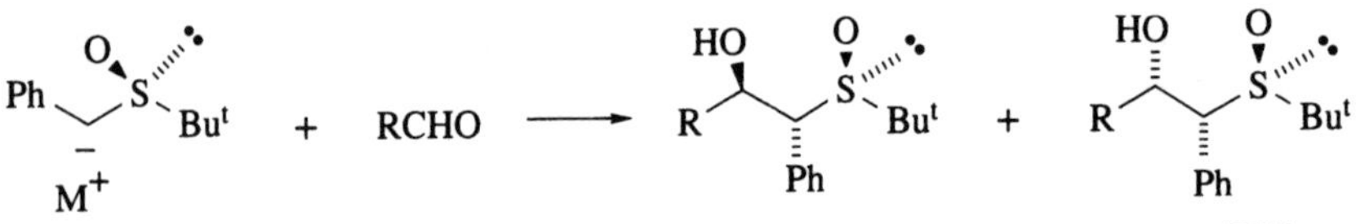

anti-(**32a**) *syn*-(**32b**)

R	Countercation	*anti*:*syn*	Yield
R = Ph, countercation =	Li^+	63 : 37	(81%)
	Zn^{2+}	84 : 16	(98%)
R = Pr^i, countercation =	Li^+	78 : 22	(90%)
	Zn^{2+}	91 : 9	(72%)
R 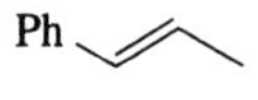 = PhCH=CH, countercation =	Li^+	49 : 51	(100%)
	Zn^{2+}	57 : 43	(82%)

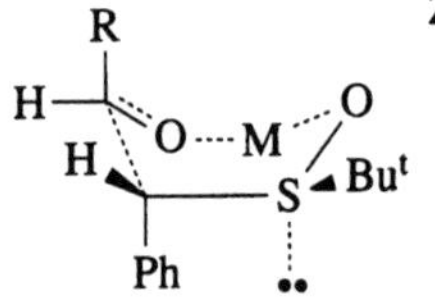

(**33**)

Scheme 10

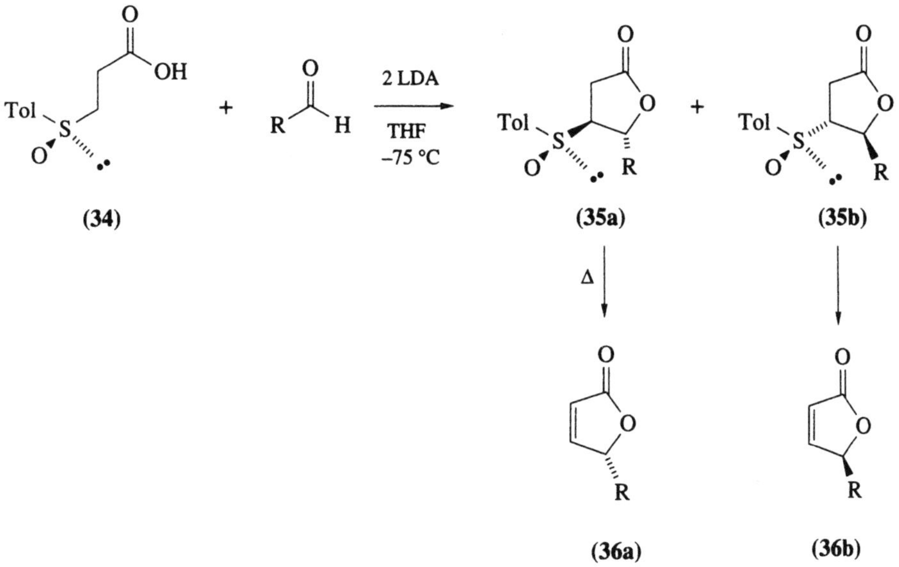

(34) (35a) (35b) (36a) (36b)

2 LDA, THF, –75 °C; Δ

Scheme 11

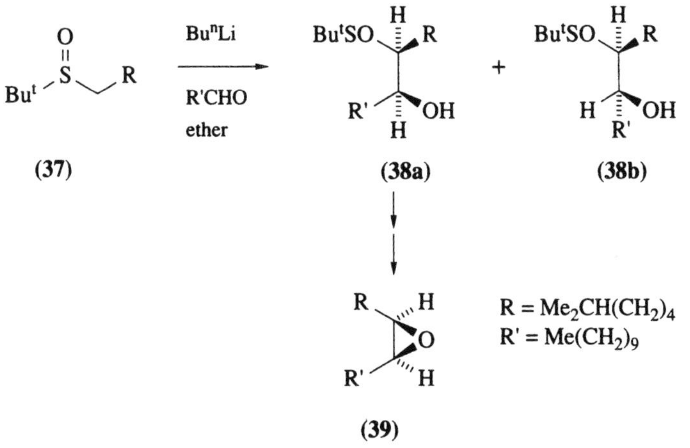

(37) (38a) (38b) (39)

Bu^nLi, R'CHO, ether

R = $Me_2CH(CH_2)_4$
R' = $Me(CH_2)_9$

Scheme 12

(40) (41)

RCHO; [H]

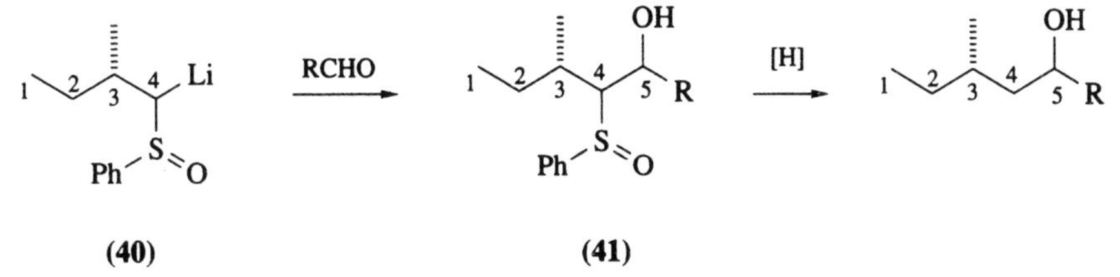

(13)

hydroxycarbon is always (*S*), pyrolytic dehydrosulfinylation of **(50a)** and **(50b)** furnishes the same product, (–)-elaeokanine B.[56]

2.3.3.3 Addition to C═N Bonds

In 1973, addition of a carbanion **(52**; Ar = *p*-tolyl) to benzylideneaniline **(51a**; Scheme 15) at –10 to –20 °C was found to be a highly diastereoselective process.[57] Later, this and its related reactions were re-examined in detail.[58] An anion **(52**; Ar = Ph) adds smoothly to **(51a)** at –78 °C over a period of 5 h to give a mixture of the diastereoisomeric adducts with modest diastereoselection (18:82). Longer reaction

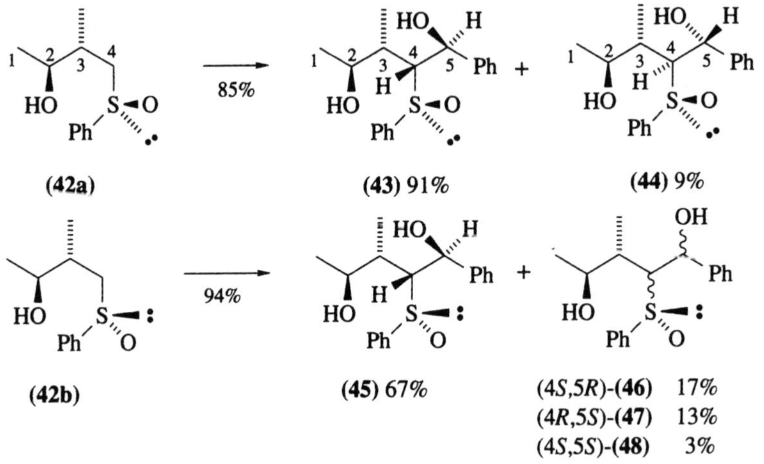

(42a) → (85%) **(43)** 91% + **(44)** 9%

(42b) → (94%) **(45)** 67% + (4*S*,5*R*)-**(46)** 17%; (4*R*,5*S*)-**(47)** 13%; (4*S*,5*S*)-**(48)** 3%

Scheme 13

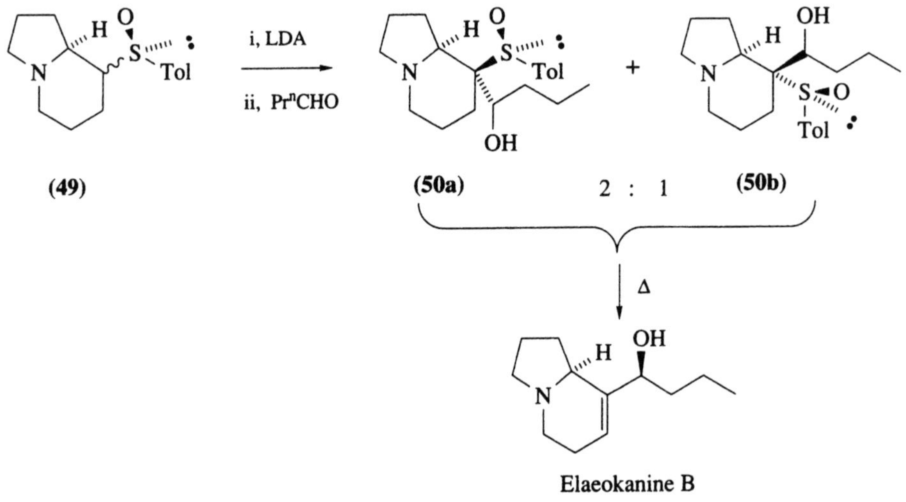

(49) → (i, LDA; ii, Pr^nCHO) **(50a)** : **(50b)** 2 : 1 → (Δ) Elaeokanine B

Scheme 14

times (0.5–20 h) at 0 °C result in very poor diastereoselection (30:70–43:57), suggesting an equilibrium between the anions (**53**) of the diastereoisomeric adducts. Under kinetically controlled conditions (0 °C/10 min), the reaction of (**52**; Ar = *p*-tolyl) with (**51b**) or (**51c**) is highly diastereoselective (9:91). The chair transition state (**55**; equation 14) can account for the preference of the diastereoisomeric adduct (**54a**) over (**54b**).

Addition of (**52**; Ar = *p*-tolyl) to 3,4-dihydro-6,7-dimethoxyisoquinoline (**56**) gives a diastereomeric mixture of the adduct (**57a**:**57b** = 77:23) under kinetically controlled conditions (–78 °C ~ –40 °C/2 h), whereas the diastereoselection was reversed to a high degree (8:92) at 0 °C (equation 15).

Since these adducts undergo reductive desulfurization with Raney Ni, the optically active aryl methyl sulfoxide is a versatile reagent for the synthesis of optically active amines from imines.

2.3.3.4 Addition to Nonactivated C═C Bonds

Intramolecular addition of an α-sulfinyl carbanion to an isolated double bond occurs on treatment of medium-ring *E*-homoallylic sulfoxides (**58**), (**59**), (**60**) and (**61**) with butyllithium (0.5 equiv.; equations 16–19)[59,60] Kinetic data suggest that this process is a nucleophilic addition of a carbanion (an α-lithiated sulfoxide) to a nonactivated double bond and takes place readily, provided a suitable proton source is available. Normally the free sulfoxide is the proton donor species.[60]

Under the same conditions, the corresponding (*Z*)-isomers are inert, even though the *E–Z* differential strain is negligible in 9- and 10-membered ring systems. The presence of an (*E*) double bond induces a

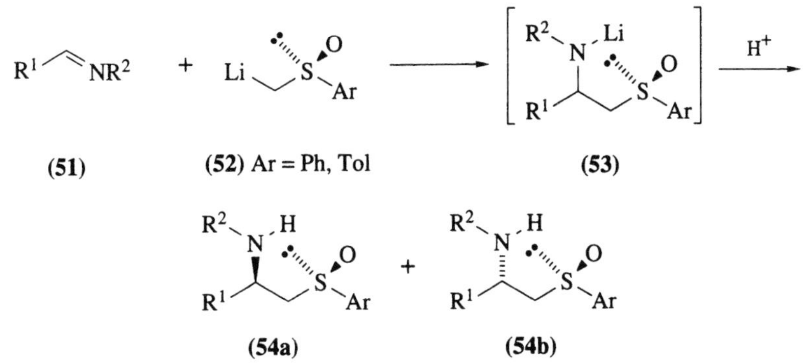

(51) (52) Ar = Ph, Tol (53)

(54a) (54b)

$R^1 = R^2 = Ph$ 18:82

$R^1 = Ph$, $R^2 = Me$; or R^1 = 2-furyl, $R^2 = Ph$ 9:91

Scheme 15

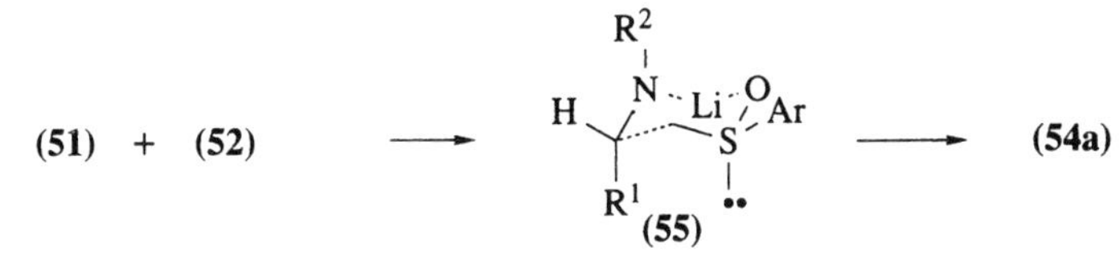

(51) + (52) → (55) → (54a) (14)

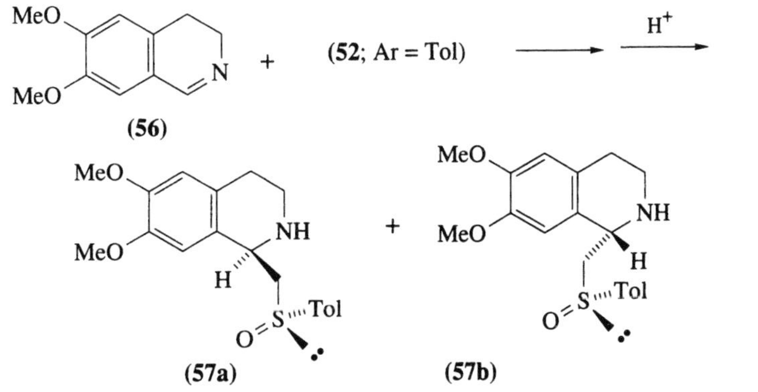

(56) + (52; Ar = Tol) → H^+ → (57a) + (57b) (15)

conformation in which the double bond lies in a plane almost perpendicular to the mean ring plane and is faced transannularly by the anionic orbital α to the sulfinyl group. Although the cyclization of (**58**) and (**59**) is stereospecific, treatment of (**60**) or (**61**) with butyllithium (0.5 equiv.) at –30 °C or –20 °C for 2 h produces two isomeric products in about equal amounts. This result is due to kinetic control. When the resulting mixture is allowed to stand at room temperature in the presence of butyllithium, an equilibrium is established between these products to afford the *trans*-junction isomer (**64b**) or (**65b**) exclusively or predominantly.[61]

This transannular addition of α-sulfinyl carbanions to nonactivated double bonds is utilized as the key step in a synthesis of *trans*-1-thiadecalin (**70**) in enantiomerically pure form.[62] The required (*E*)-thiacyclodec-4-ene *S*-oxide (**66a,b**) was prepared *via* several steps from (*R*,*R*)-1,6-dibromo-3,4-hexanediol. Upon treatment with butyllithium, a 4:1 mixture of isomeric sulfoxides (**66a** and **66b**) undergoes smooth cyclization to give a mixture of isomeric bicyclic sulfoxides (**67a** and **67b**) in the same 4:1 ratio as the starting material, suggesting that the cyclization is essentially stereospecific. The major isomer (**67a**) is reduced with PCl_3 to a sulfide (**68**) from which the desired (**70**) is derived *via* a thiaoctaline (**69**; Scheme 16).

2.3.3.5 Addition of Allylic Sulfinyl Carbanions to C=O Bonds

Deprotonation of allyl sulfoxides with LDA generates the corresponding allylic carbanions, which are able to add to aldehydes. Addition of the anion of aryl allyl sulfoxides to aromatic aldehydes proceeds

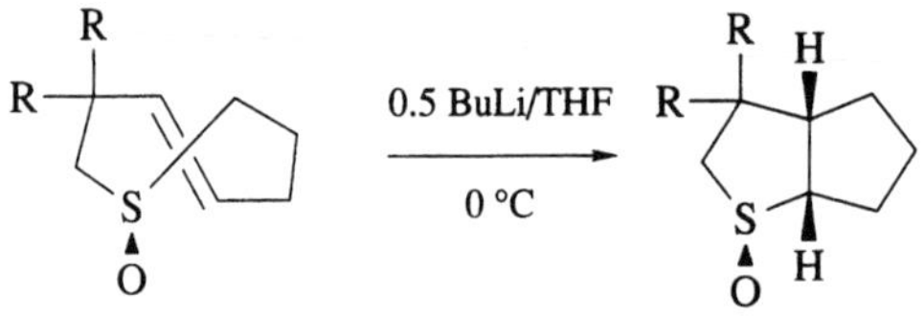

(58) → (62): 0.5 BuLi/THF, 0 °C; R = H, Me (16)

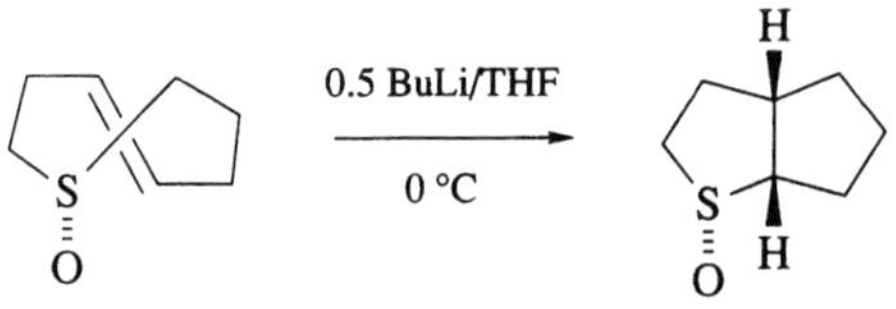

(59) → (63): 0.5 BuLi/THF, 0 °C (17)

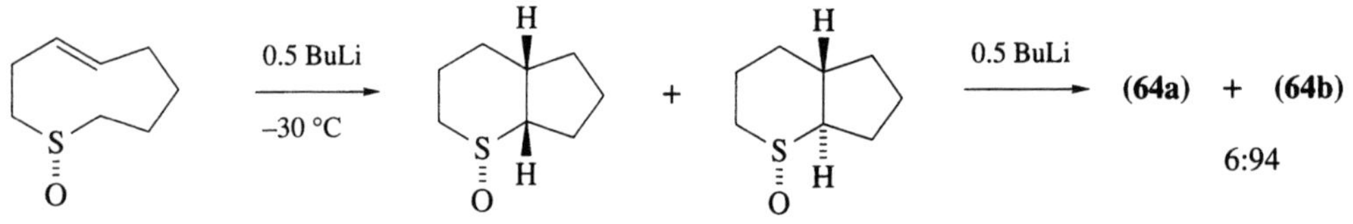

(60) → (64a) + (64b) 59:41: 0.5 BuLi, –30 °C; 0.5 BuLi → (64a) + (64b) 6:94 (18)

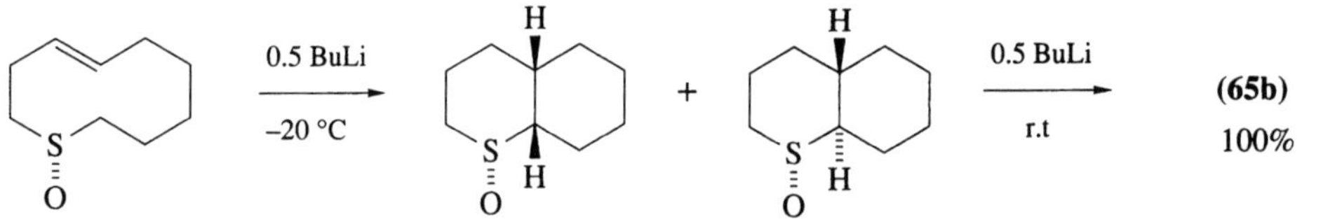

(61) → (65a) + (65b) 50:50: 0.5 BuLi, –20 °C; 0.5 BuLi, r.t → (65b) 100% (19)

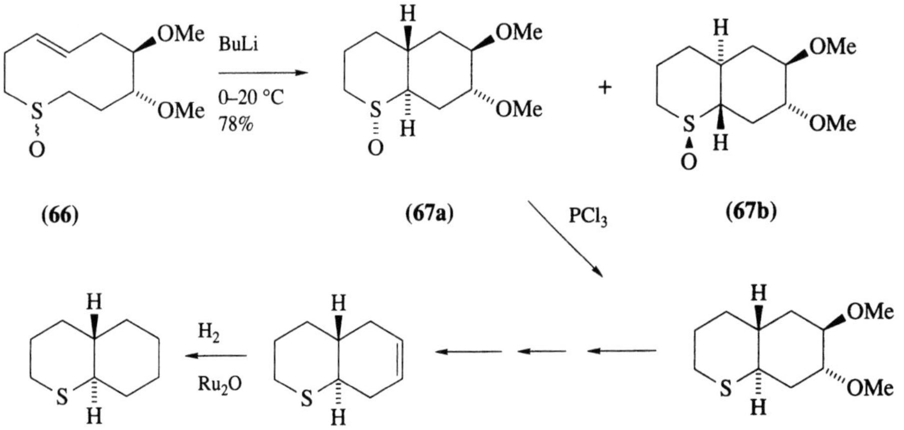

(66) → (67a) + (67b): BuLi, 0–20 °C, 78%; (67a) → PCl_3 → (68) → → → (69) → H_2, Ru_2O → (70)

Scheme 16

readily in moderate yields to afford a mixture of products resulting from α- and γ-attack on the allyl anion (equation 20). All four possible diastereomers are formed from α-attack. Rapid epimerization at two chiral centers (sulfinyl sulfur and α-carbon) occurs *via* allylic sulfoxide–sulfenate 2,3-sigmatropic rearrangement, causing the low stereoselection summarized in Table 2. The γ-product possesses (*E*)-geometry at the double bond, and 1,5-asymmetric induction is observed to the extent that the major/minor diastereomeric ratio exceeds 2:1. The metalated carbon in the α-lithiated sulfoxide should be planar and the lithium cation should chelate with the sulfinyl oxygen. By analogy with the model proposed to explain diastereoselection in the reaction of aldehydes with enolates, the transition states **(71)** and **(72)**, involving coordination in a six-membered ring, were proposed (Scheme 17).[63,64]

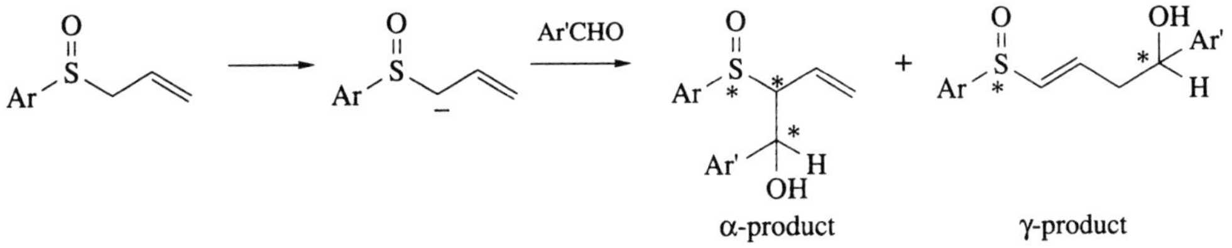

Ar'CHO (20)

α-product γ-product

Table 2 Results of the Addition Reaction in Equation (20)

Ar	*Ar′*	*α/γ*	*Diastereomer ratio* *a-Product*	*γ-Product*
Ph	Ph	1/2.1	3/6/7/2	3.0/1
p-Tol	Ph	1/1.7	3/3/6/2	3.0/1
p-$NO_2C_6H_4$	Ph	1/2.2	5/5/5/1	2.0/1
p-Tol	*p*-$MeOC_6H_4$	1/1.0	4/3/1/1	2.0/1
p-Tol	*o*-$MeOC_6H_4$	1/1.1	8/5/2/1	3.4/1
p-Tol	*p*-$NO_2C_6H_4$	1/1.5	6/3/2/1	2.5/1
p-Tol	2-naphthyl	1/1.2	5/4/2/1	2.4/1

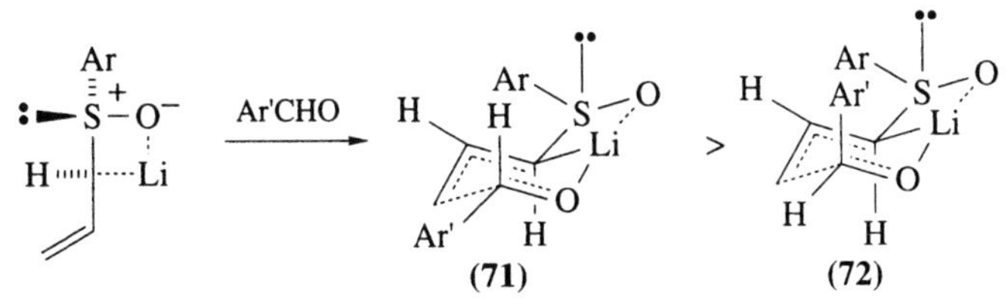

Ar'CHO

(71) > **(72)**

Scheme 17

Treatment of allyl *p*-tolyl sulfoxide with LDA followed by addition of HMPA (4.4 mol equiv.) and chiral 2-methylalkanal (3 mol equiv.) at –78 °C gave a mixture of readily separable α- and γ-adducts, where the α:γ regioisomer ratios are markedly higher than those obtained in the reaction with aromatic aldehydes. Exposure of the α-adduct to an excess of a thiophile (trimethyl phosphite or dimethylamine in MeOH) results in quenching of the allylic sulfoxide–sulfenate equilibrium and affords diastereoisomeric mixtures of the *syn*- and *anti*-diols (Scheme 18),[65,66] for which some yields and product ratios are shown in Table 3.

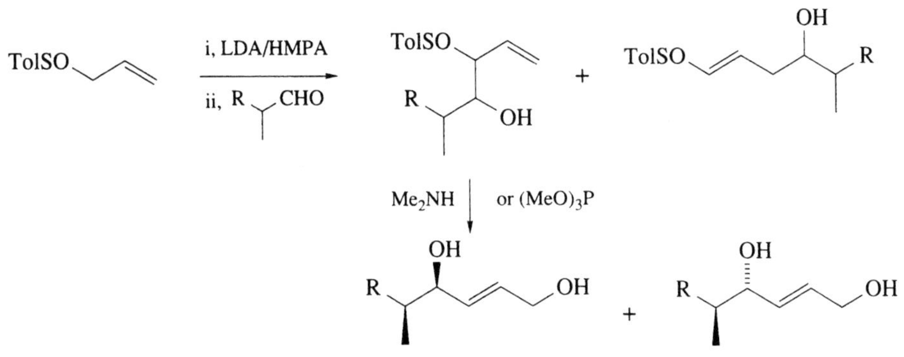

i, LDA/HMPA; ii, R–CHO

Me_2NH or $(MeO)_3P$

(73a) **(73b)**

Scheme 18

Table 3 Synthesis of Diols (**73**)

R	*α/γ*	*Yield (%)*	*Syn/anti*
Et	6.1/1	40	2.1/1
Pr^i	10/1	40	5.1/1
c-C_6H_{11}	5.6/1	57	6.4/1
Ph	9/1	67	28/1

Tandem condensation of chiral α-alkoxy aldehydes with the anion of allyl 2-pyridyl sulfoxide and $(MeO)_3P$-promoted desulfurization of the resulting α-adduct provide (*E*)-alkoxydiols. The major stereochemical path in their preparation may be accounted for by the Felkin–Anh (nonchelation control) model (Scheme 19).[67]

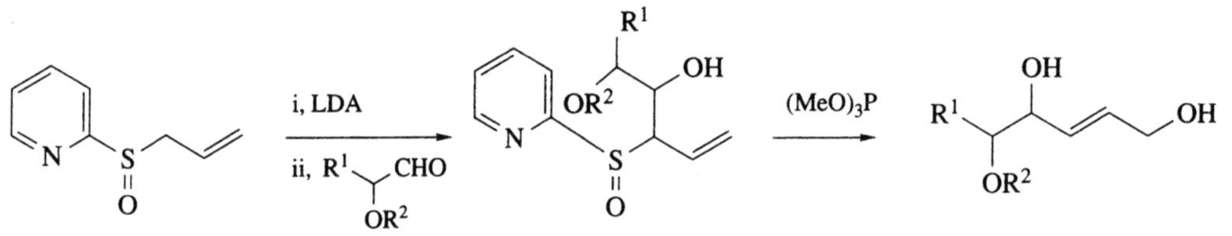

i, LDA; ii, $R^1CH(OR^2)CHO$; $(MeO)_3P$

R¹	*R²*	*Yield (%)*	*syn/anti*
Me	$PhCH_2$	51	1 / 2.5
Me	Bu^tMe_2Si	40	1 / 5.5
$PhCH_2OCH_2$	$PhCH_2$	68	1 / 3.0
$C_6H_{10}(Me)CH_2O$		60	1 / 6.5

Scheme 19

2.3.3.6 Addition of Allylic Sulfinyl Carbanions to C=C—C=O Bonds

In preliminary reports, the γ-carbon of the carbanion of allyl phenyl sulfoxide has been shown to attack cyclopentenone and cyclohexenone by 1,4-addition to deliver vinyl sulfoxides.[68–71] The lithiated carbanion (**75**) of 1-(phenylsulfinyl)-2-octene (*E*:*Z* = 85:15) adds to 4-*t*-butoxycyclopent-2-en-1-one (**74**) to give *syn-(E)*-vinylic sulfoxide (**76**) and *anti-(E)*-vinylic sulfoxide (**77**) in the ratio of 79:21. It has been suggested that (**76**) arises almost exclusively from the (*E*)-(**75**), and (**77**) derives from the (*Z*)-(**75**). Both the products have the same geometry about the double bond, but differ in configuration at the allylic carbon atom (equation 21).[72,73]

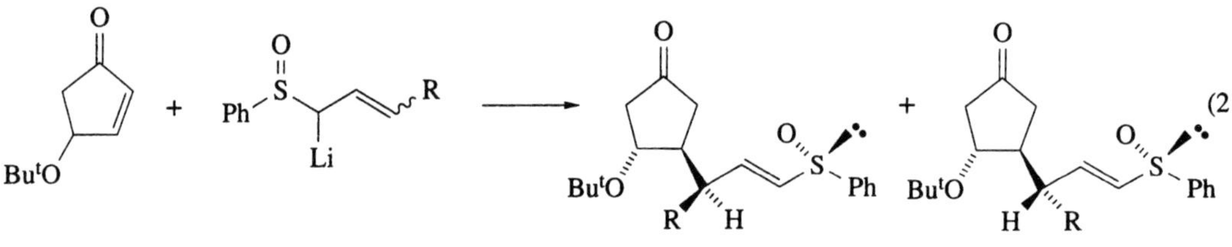

(21)

(**74**) (**75**) R = n-C_5H_{11} (**76**) (**77**)

Since the 1,4-γ-adducts can be converted to various types of bicyclic compounds,[74] 1,4-addition of allylic sulfoxides to enones provides a useful tool for stereoselective synthesis of cyclic compounds. Further, the 1,4-addition can be utilized in a three-component coupling process for synthesis of prostanoic acid derivatives (Scheme 20).[75] Typical examples of addition of the lithio derivative (**78**) of (*R*)-allyl *p*-tolyl sulfoxide to cyclic enones are summarized in Scheme 21.[76]

Only 1,4-γ-adducts are formed in the five-membered ring system and the carbanion, which has *trans* geometry, attacks from the *si* face. The '*trans*-fused 10-membered' cyclic transition state offers an explanation for this highly stereoselective reaction. Six- and seven-membered cyclic enones provide both 1,4-γ- and 1,2-γ-adducts, with the larger ring providing a greater percentage of the 1,2-γ-adduct. With the seven-membered ring, a 58% yield of the 1,2-γ-adduct is isolated. These 1,4- and 1,2-addition re-

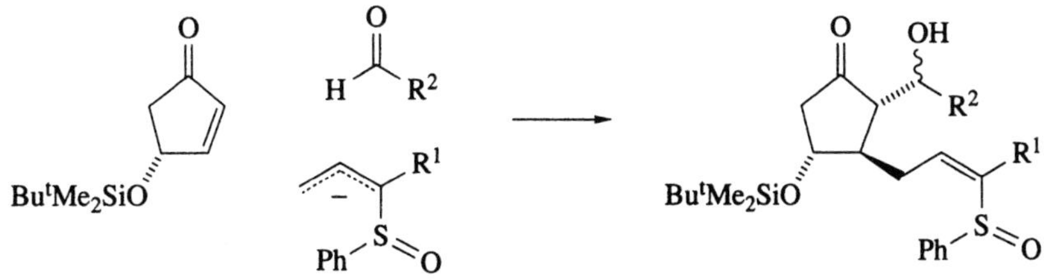

$R^1 = (CH_2)_4Me$, $R^2 = (CH_2)_5CO_2Me$ 68%

$R^1 = (CH_2)_3CO_2Bu^t$, $R^2 = CH{=}CHCH(OSiMe_2Bu^t)(CH_2)_3Me$ 62%

Scheme 20

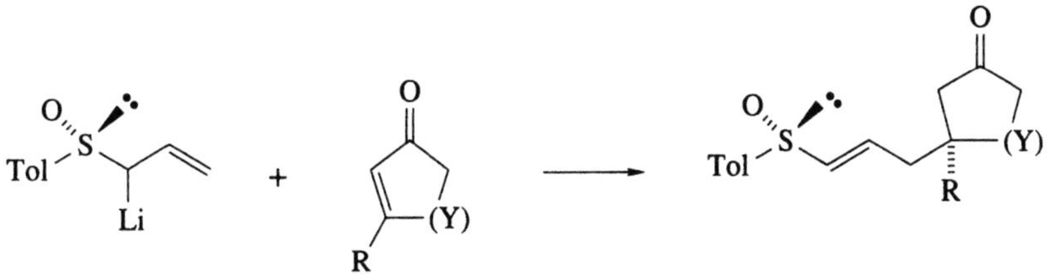

(78)

$Y = CH_2$,	R = H	91% (96% *ee*)
$Y = CH_2$,	R = Me	80% (95% *ee*)
Y = O,	R = H	70% (95% *ee*)
$Y = CH_2CH_2$,	R = H	79% (50% *ee*)[a]
$Y = CH_2CH_2CH_2$,	R = H	25% (59% *ee*)[b]

[a] 1,2-adduct 14% (50% *ee*)

[b] 1,2-adduct 58% (50% *ee*)

Scheme 21

actions appear to be kinetically controlled. The same product or product ratio is observed whether the addition reactions are performed at –100 °C or at –50 °C, and the 1,2-γ- and 1,4-γ-adducts do not interconvert under these reaction conditions, or even at 25 °C.

Kinetic resolution is operative in the reaction with 4-substituted cyclopent-2-en-1-ones.[76] For instance, when 2 equiv. of racemic 4-(1-methyl-1-phenylethoxy)-2-cyclopentenone (**79**) is reacted with the carbanion (**78**), (*S*)-(**79**) reacts to yield predominantly the product derived by approach of the carbanion from the *si* face of (**79**). The formation of the *cis* adduct probably results from the chelation of lithium counterion with the C-4 oxygen (equation 22). Kinetically selective resolution of a bicyclic enone (**80**) is also effected. On treatment of the carbanion of (*R*)-(**78**) with racemic (**80**) (2 equiv.), an enantiomer of (**82**) and (*S*)-(**80**) are obtained in 80% and 45% yields respectively. Addition of the lithio derivative (2 equiv.) of

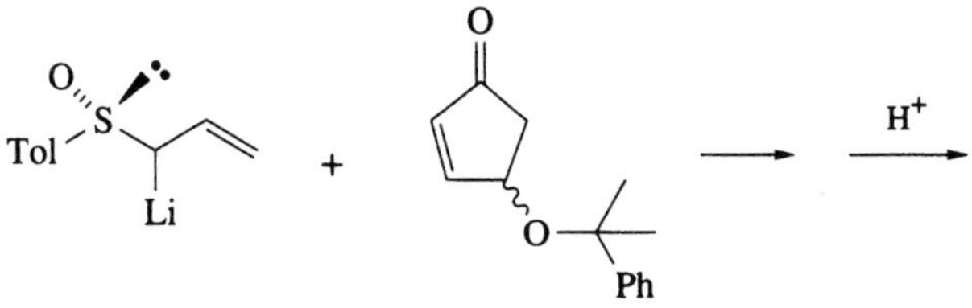

(78) **(79)** 2 equiv.

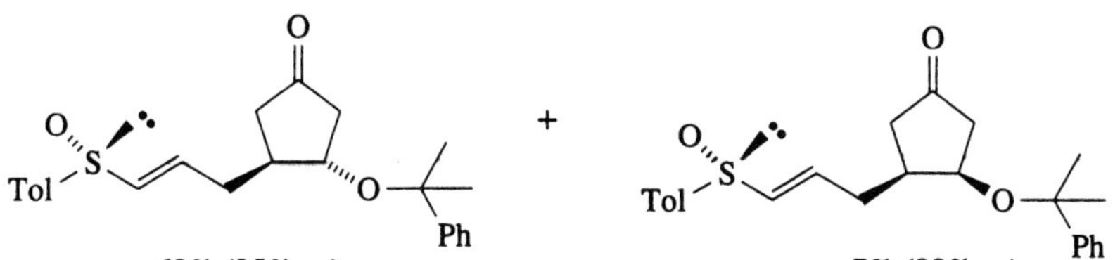

68% (95% *ee*) + 7% (90% *ee*) (22)

racemic *cis*-crotyl phenyl sulfoxide (**81**) to (*S*)-(**80**) affords a 1,4-γ-adduct (**83**; 91% yield/82% *ee*), which is a synthetic precursor of (+)-pentalenene (Scheme 22).[76,77]

Stereospecific 1,4-addition of the carbanion (**78**) to (siloxymethyl)bicyclooctenone (**84**) yields mainly the corresponding 1,4-γ-adduct. In seven steps, this adduct is converted into the known δ-lactol (**85**), from which pentalenolactone (*E*)-methyl ester (**86**) may then be derived (Scheme 23).[78]

In addition, the 1,4-addition reaction of the anion (**87**) derived from various cyclic allylic sulfoxides with 2-cyclopentenones (**88**) were investigated (Scheme 24).[79] Methyl substitution at C-3 of (**88**) hinders the 1,4-addition. An activated enone (**88e**), however, affords excellent chemical and optical yields of the 1,4-adducts (**89**). (+)-12,13-Epoxytrichothec-9-ene (**91**) and its antipode have been enantioselectively synthesized from (*S*)-4-methyl-2-cyclohexenone (**90**) in 11 steps (Scheme 25).[79]

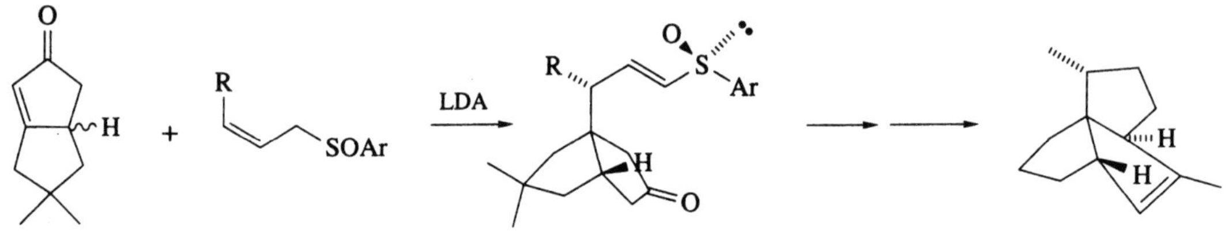

(**80**) (**78**) Ar = Tol, R = H; (**81**) Ar = Ph, R = Me; LDA; (**82**) Ar = Tol, R = H; (**83**) Ar = Ph, R = Me; (+)-Pentalenene

Scheme 22

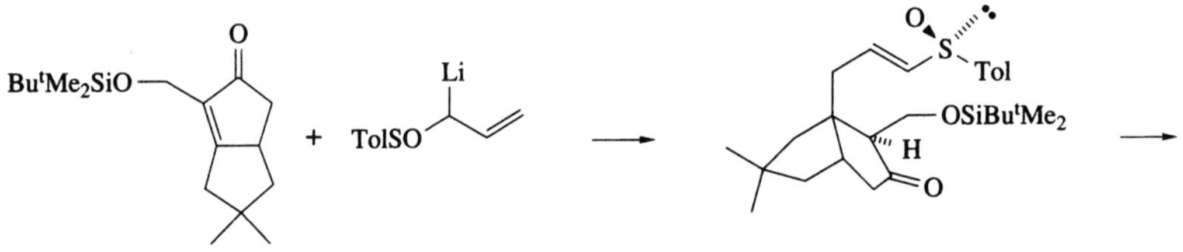

(**84**) (**78**) 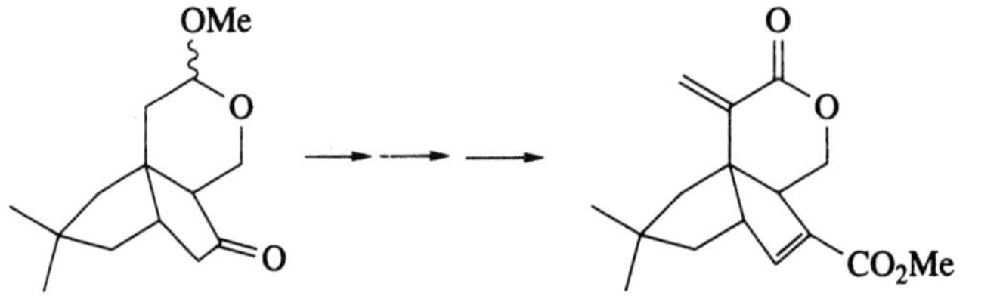 (**85**) (**86**)

Scheme 23

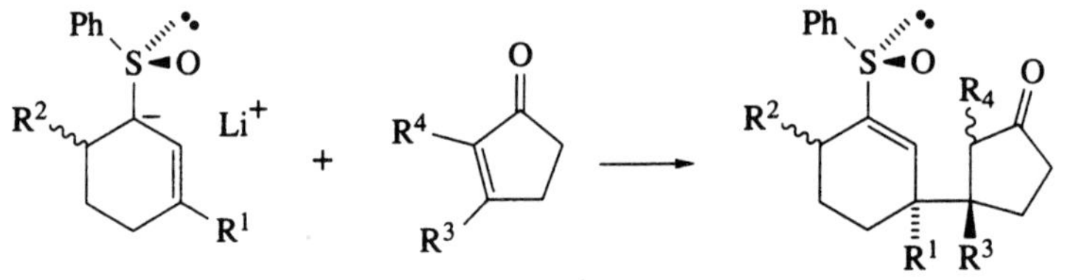

(**87**) (**88**) (**89**)

a: $R^1 = R^2 = R^3 = R^4 = H$ 50%

b: $R^1 = Me$, $R^2 = R^3 = R^4 = H$ 60%

c: $R^1 = R^2 = Me$, $R^3 = R^4 = H$ 61%

d: $R^1 = R^2 = R^3 = Me$, $R^4 = H$ 5%

e: $R^1 = R^2 = R^3 = Me$, $R^4 = CO_2Me$ 87%

Scheme 24

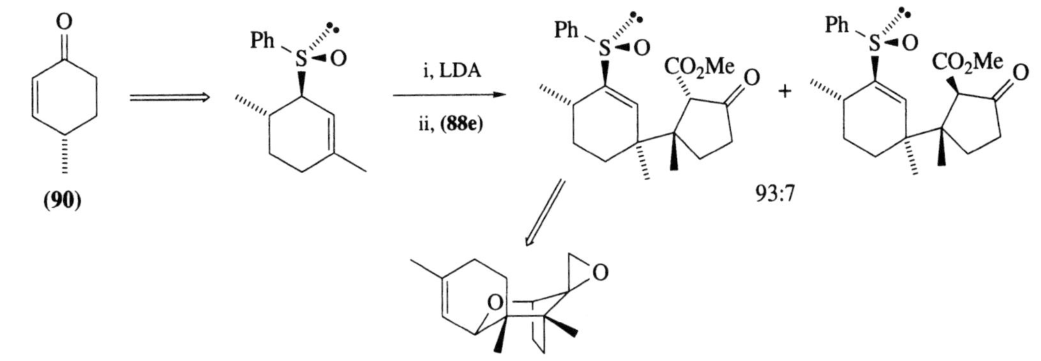

(90)

i, LDA

ii, (88e)

93:7

(91)

Scheme 25

2.3.3.7 Addition of α-Sulfinyl Carbonyl Compounds to C═O Bonds

The metal enolate of *(R)-N,N*-dimethyl-α-(*p*-tolylsulfinyl)acetamide (**92**) reacts with an aldehyde to afford a diastereomeric mixture of adduct (**93**), which undergoes desulfurization with 10% Na(Hg) in MeOH (NaH_2PO_4) to give an optically active β-hydroxyamide (**94**). By the use of Li^+ as a countercation, low to medium levels of enantioselection are achieved. However, the magnesium enolate of (**92**) realizes a high enantioselectivity, as summarized in Table 4. A rigid chelation model, shown in Scheme 26, is suggested for the transition state of this addition.[80]

Similarly, the magnesium enolate of *t*-butyl (*R*)-(*p*-tolylsulfinyl)acetate adds to aldehydes and ketones. The subsequent reductive desulfurization yields the corresponding optically active β-hydroxy esters (Scheme 27).[81,82] This process can be utilized for preparation of a synthetic intermediate (**95**) of maytansine with a selectivity of 93:7 (equation 23).[83] Michael-type addition of *t*-butyl (*R*)-(*p*-tolylsulfinyl)acetate to crotonic ester with NaH in DMF has also been reported, but the enantiomeric excess of the desulfurized product is very low (12%).[84]

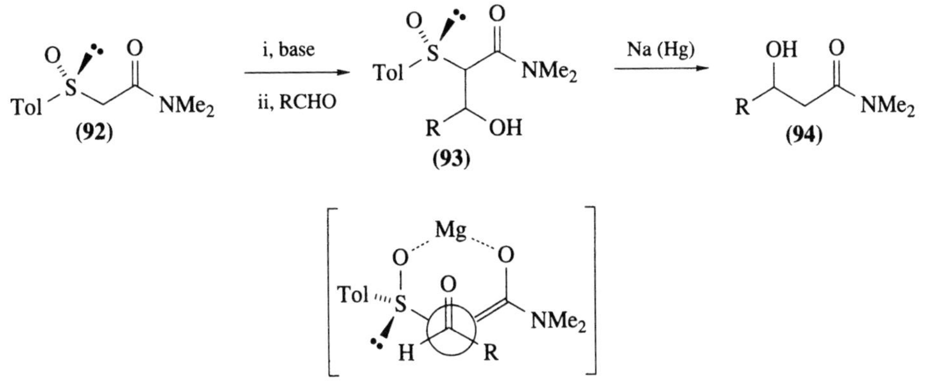

(92) i, base ii, RCHO (93) Na (Hg) (94)

Scheme 26

Table 4 Addition of (**92**) to RCHO

R	*BuLi as base* *Yield (%)*	*ee (%)*	*α*	*Bu^tMgBr as base* *Yield (%)*	*ee (%)*	*α*
Me	65	47	+	68	>>99	–
Bu^i	77	45	+	71	98	–
Pr^i	78	31	+	66	95	–
Bu^t	20	8	+	56	90	–

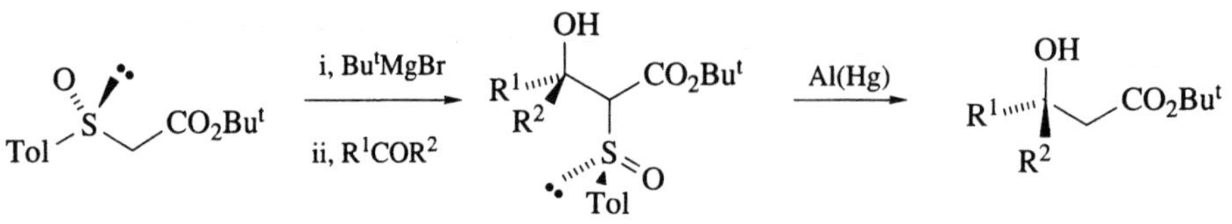

i, Bu^tMgBr
ii, R^1COR^2
Al(Hg)

R^1	R^2	Yield (%)	ee (%)
H	Ph	85	91 (S)
Me	Ph	75	68 (S)
H	n-C_7H_{15}	80	20 (R)
Me	 (cyclohexyl)	88	95 (S)
Me	CO_2Et	80	85 (S)

Scheme 27

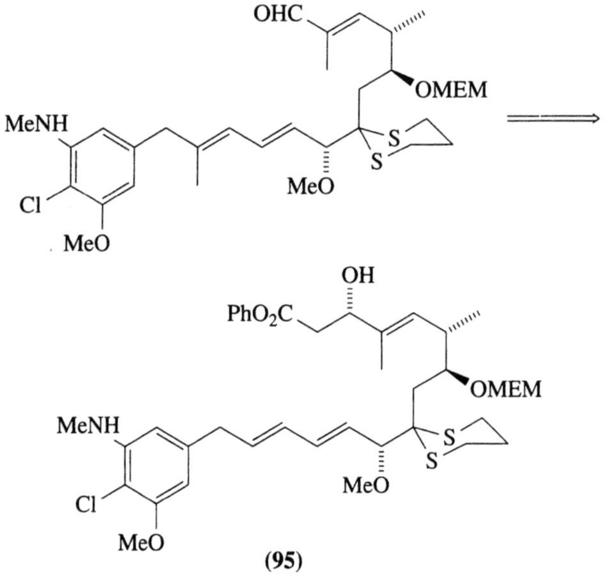

(23)

(95)

Direct condensation of (*p*-tolylsulfinyl)methyl ketones (**96**) with aldehydes followed by reductive desulfurization is an alternative method for obtaining optically active β-hydroxy ketones (**97**), but this reaction occurs with only moderate enantiomeric excess (Scheme 28).[85]

Aldol-type condensation of chiral α-sulfinyl hydrazones with aldehydes also provides a route leading to chiral β-hydroxy ketones.[86,87] This is exemplified by the synthesis of (–)-(*R*)-[6]-gingerol (**98**; equation 24).[88]

2.3.3.8 Addition of α-Halo Sulfoxides to C=X Bonds

Durst reported that lithiation of chloromethyl phenyl sulfoxide (**99**) with butyllithium followed by treatment of a symmetrical ketone (–78 °C to –20 °C) affords an adduct (**100**). This reaction is so highly stereoselective that only one diastereomer is produced. Reaction of (**100**) with dilute methanolic KOH gives an epoxy sulfoxide (**101**; equation 25).[89,90]

When chloromethyl *p*-tolyl sulfoxide (**102**; R = *p*-tolyl) is treated with carbonyl compounds (**103**) and potassium *t*-butoxide in *t*-butyl alcohol ether, (*p*-tolylsulfinyl)oxiranes (**104**) are directly formed.[91] Chloromethyl methyl sulfoxide (**102**; R = Me) exhibits the same behavior (Scheme 29).[92] These reactions proceed with high stereoselection at the position α to the sulfinyl group. The stereochemical course proposed for the attack of the carbanion of an α-halo sulfoxide on a carbonyl compound is shown in Figure 4.[93]

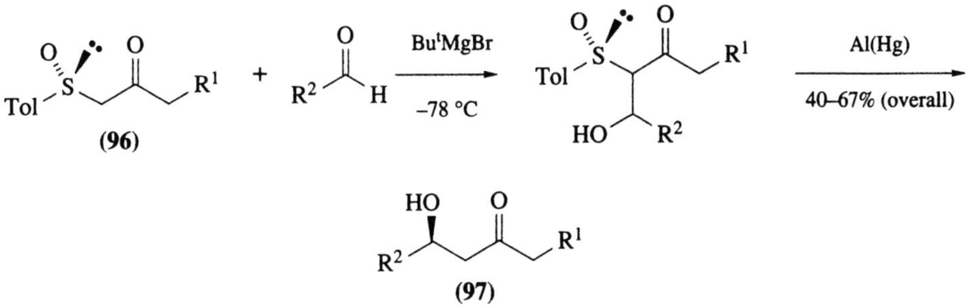

(96) + R^2CHO —(Bu^tMgBr, –78 °C)→ —(Al(Hg), 40–67% (overall))→ (97)

R^1	R^2	ee (%)	R^1	R^2	ee (%)
H	Et	64	Et	Et	78
H	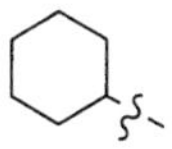 cyclohexyl	54	Et	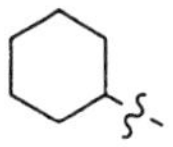 cyclohexyl	74
H	Ph	72			

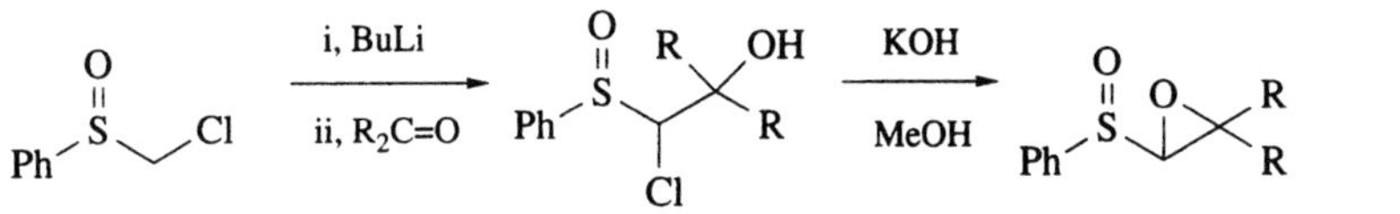

Scheme 28

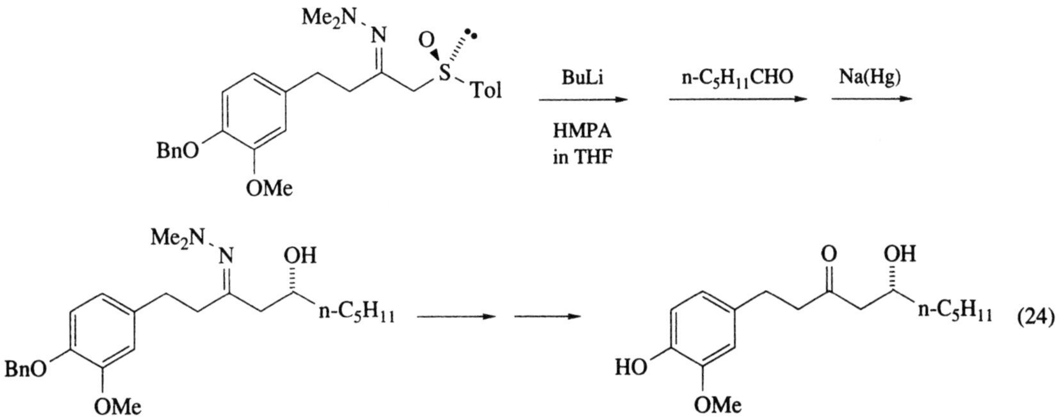

—(BuLi, HMPA in THF)→ —($n\text{-}C_5H_{11}CHO$)→ —(Na(Hg))→ —→ —→ (98) optical purity of 60% (24)

(99) —(i, BuLi; ii, $R_2C{=}O$)→ (100) **a**: R = Me; **b**: 2R = $-(CH_2)_5-$; **c**: R = Ph —(KOH, MeOH)→ (101) one diastereomer (25)

It is noteworthy that the sterically hindered epoxide is preferentially formed in this Darzens-type condensation, which is in sharp contrast with that of α-halo sulfones (see Section 2.3.4.4). This intriguing

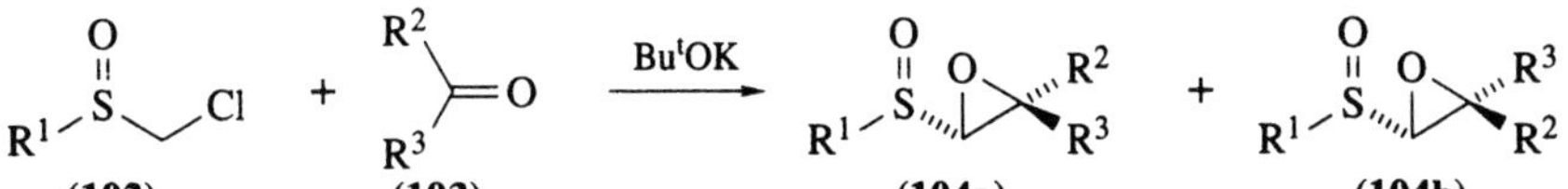

(102) + (103) —(Bu^tOK)→ (104a) + (104b)

R^1 = Tol, R^2 = Ph, R^3 = H, 100:0

R^1 = Me, R^2 = Bu^t, R^3 = Me, 75:25

Scheme 29

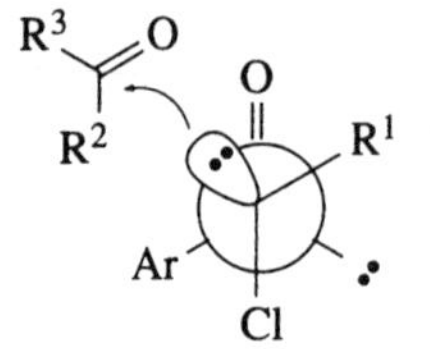

Figure 4

phenomenon may be rationalized in terms of a kinetically controlled approach of the anion of α-halo sulfoxide to aldehydes or ketones.[92]

Lithio-fluoro (or -chloro) methyl phenyl sulfoxide reacts with *N*-(benzylidene)aniline derivatives (**105**) to afford the corresponding aziridines (**106**) in good yields. Two diastereomeric isomers are produced, but the *trans* isomer is the major product except in the case of *N*-benzylidene-*p*-nitroaniline (Scheme 30).[94]

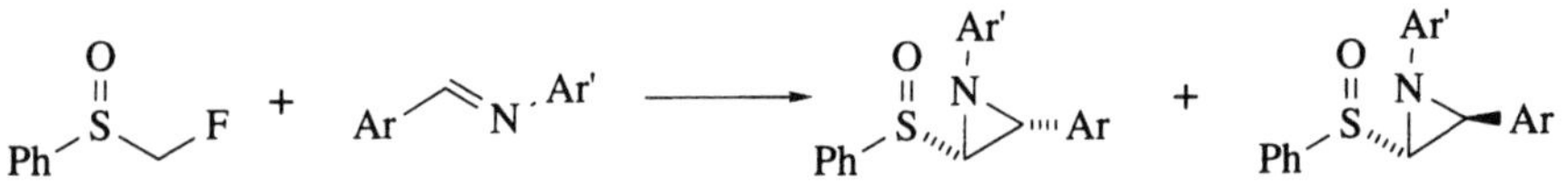

(**105**) (**106a**) (**106b**)

Ar = Ph,	Ar' = Ph,	38:72
Ar = *p*-ClC_6H_4,	Ar' = Ph,	45:55
Ar = *p*-$NO_2C_6H_4$,	Ar' = Ph,	100:0
Ar = Ph,	Ar' = *p*-BrC_6H_4,	29:71

Scheme 30

The Darzens-type products from α-halo sulfoxides and carbonyl compounds undergo reductive desulfurization by the action of butyllithium. This is a useful synthetic route to various epoxides. Yamakawa and coworkers disclosed a novel asymmetric synthesis of epoxides (**109**) using optically active 1-chloroalkyl *p*-tolyl sulfoxide (**107**).[93,95] This method is most effective when the starting ketones are symmetrical, because only one diastereomer is then formed in the Darzens-type condensation (equation 26).

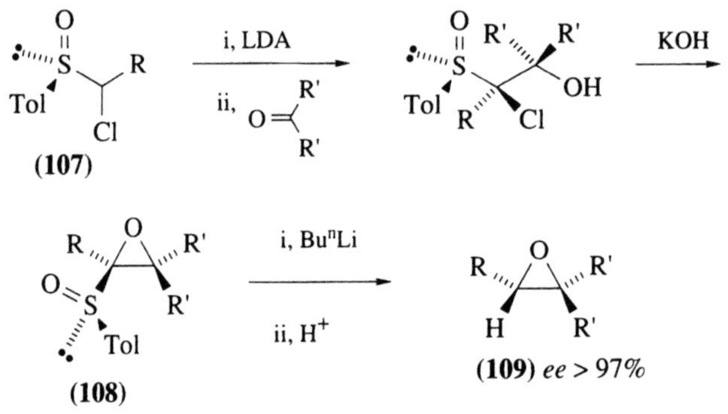

(**107**) i, LDA; ii, $O{=}CR'_2$; KOH

(**108**) i, Bu^nLi; ii, H^+ (**109**) *ee* > 97% (26)

R = n-$C_{10}H_{23}$, R' = Me or 2R' = $(CH_2)_5$ or $(CH_2)_4$

2.3.3.9 Addition of Dithioacetal *S*-Oxides and *S*,*S'*-Dioxides to C=X Bonds

Dropwise addition of butyllithium to a solution of (±)-1,3-dithiane *S*,*S'*-dioxide (**110**) in pyridine–THF (1.5:1) generates an anion (**111**), which reacts with an aldehyde to give an adduct (**112**) as a 1:1 diastereomeric mixture. The reaction is extremely rapid at –78 °C, but the kinetic selectivity is moderate. In the reaction with benzaldehyde or pivalaldehyde, equilibration is attained at 0 °C to give predominantly a single diastereomer in good yield (Scheme 31).[96]

Formaldehyde dimethyl dithioacetal *S*-oxide (FAMSO; **113**) and its ethyl analog are widely used as synthetic carbonyl equivalents.[97,98] Addition of the lithio derivative of (**113**) to aldehydes and ketones followed by acidic hydrolysis is a preparative method for α-hydroxy aldehyde derivatives (equation 27).[99] A chiral analog of (**113**), (*S*)-formaldehyde di-*p*-tolyl dithioacetal *S*-oxide (**114**), can be synthesized from (–)-mentyl *(R)-p*-toluenesulfinate.[5,100] The reaction of the lithio derivative of (*S*)-(**114**)

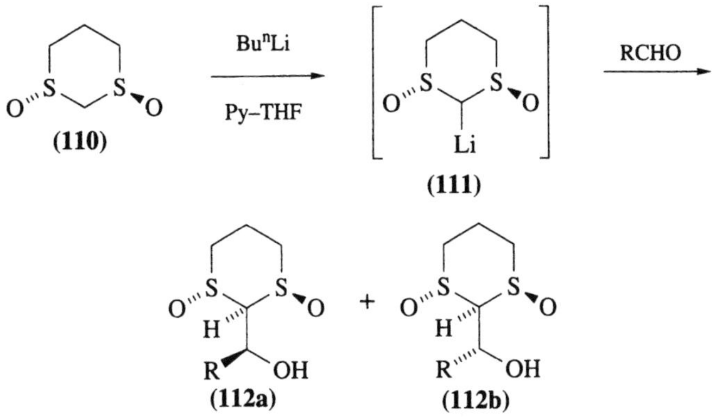

(110) $\xrightarrow[\text{Py–THF}]{\text{Bu}^n\text{Li}}$ [(111)] $\xrightarrow{\text{RCHO}}$ (112a) + (112b)

		(112a):(112b)
Bu^nCHO,	–78 °C	63:37
	0 °C/60 min	49:51
Bu^tCHO,	–78 °C	64:36
	0 °C/60 min	14:86
PhCHO,	–78 °C	66:34
	0 °C/60 min	36:64

Scheme 31

with benzaldehyde in THF at –78 °C affords an adduct (**115**), which is transformed into (*R*)-α-methoxyphenylacetaldehyde (**116**) with an enantiomeric excess of more than 70%.[101,102] The intermediary adduct (**115**) should consist of four stereoisomers, but the ratio is reported to be 55:30:15:0 [tentatively assigned (1*S*,2*S*,3*R*):(1*S*,2*R*,3*R*):(1*S*,2*S*,3*S*):(1*S*,2*R*,3*R*)]. It is noteworthy that a highly stereoselective formation of the adduct is achieved by $NaBH_4$ reduction of the benzoyl derivative (**117**) in $MeOH–NH_3$ (Scheme 32).[103]

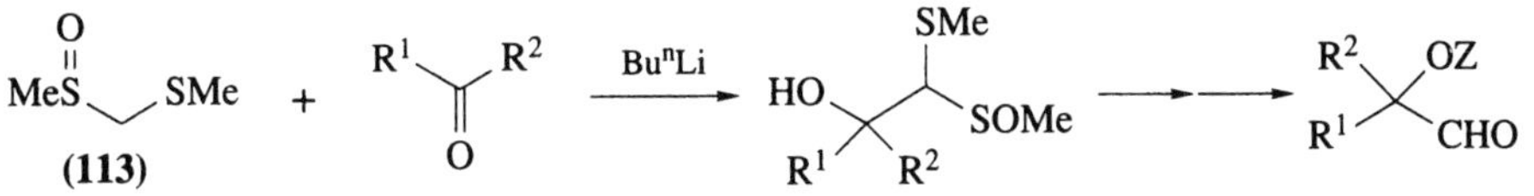

(113) + R^1COR^2 $\xrightarrow{\text{Bu}^n\text{Li}}$ $HOC(R^1)(R^2)CH(SMe)(SOMe)$ ⟶ ⟶ $R^1R^2C(OZ)CHO$ (27)

PhCHO + (114) $\xrightarrow[\text{THF, –78 °C}]{\text{Bu}^n\text{Li}}$ (115) ⟶ ⟶ (116)

(117) $\xrightarrow[\text{MeOH–NH}_3]{\text{NaBH}_4}$ (115)

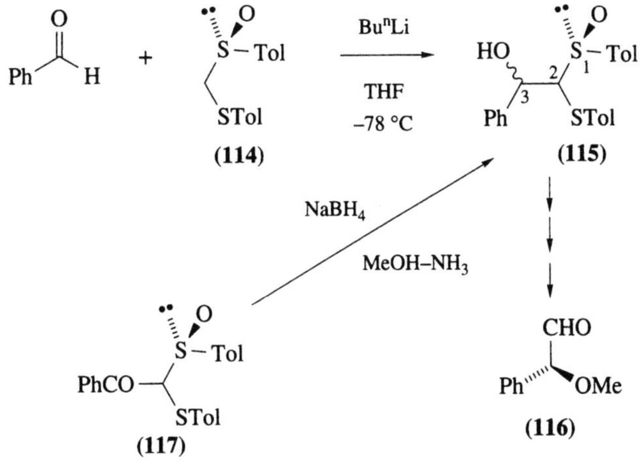

Scheme 32

The anion of an aldehyde dithioacetal *S*-oxide is well known to add to α,β-unsaturated carbonyl compounds.[104–106] Conjugate addition of formaldehyde di-*p*-tolyl dithioacetal *S*-oxide (**114**) to open-chain and cyclic enones is achieved by using HMPA as a polar cosolvent in THF (–78 °C).[107] The lithio derivative of (*S*)-(**114**) was found to add to 2-cyclopentenone with asymmetric induction. Transformation of the dithioacetal part into a formyl group gives 3-formylcyclopentanone in 39% enantiomeric excess (equation 28).[107] Interestingly, highly asymmetric induction is observed in the conjugate addition of the lithiated (*S*)-(**114**) to 2-[6-(methoxycarbonyl)-1-hexyl]-2-cyclopentenone (**118**).[108] The enone (**118**) was added at –78 °C to the anion of (*S*)-(**114**) in THF containing HMPA (10 mol equiv.). After 30 min at –78 °C, usual work-up gave the 1,4-addition product (**119**) in 45% yield. The ratio of the four *trans* diastereomers was 48:44:5:3 [(1*S*,2*R* or *S*,3*R*,4*S*):(1*S*,2*S* or *R*,3*R*,4*S*):(1*S*,2*R* or *S*,3*S*,4*R*):(1*S*,2*S* or *R*,3*S*,4*R*)]. Formation of significant amounts (>90%) of only two diastereomers should be noted. The key step of this asymmetric synthesis is the addition of a chiral formyl anion equivalent to the enone (**118**). The reaction proceeds with high β- and γ-asymmetric induction of 92% and with poor α-stereoselection (52:48), probably due to the presence of α- and β-substituents of (**118**), which create a more demanding steric array in the transition state.

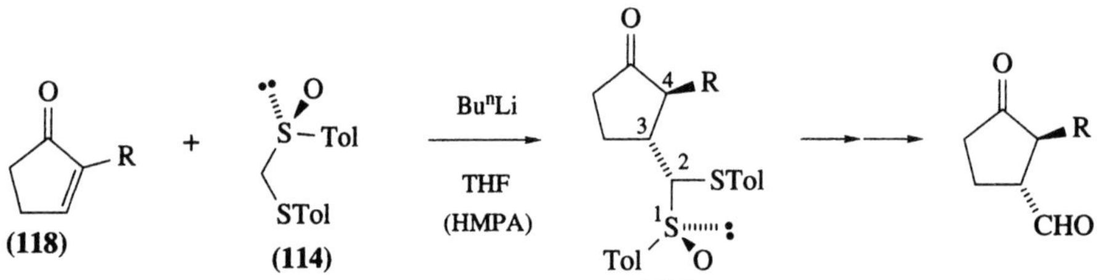

(**118**) + (**114**) —Bu^nLi, THF (HMPA)→ (**119**) ⟶ ⟶ (28)

R = $(CH_2)_6CO_2Me$

2.3.4 SULFONYL-STABILIZED CARBANIONS

2.3.4.1 Configuration of the Carbanion

Configuration of an α-sulfonyl carbanion has been much investigated both theoretically and experimentally. Several experiments were designed to distinguish between planar and pyramidal structures for α-sulfonyl carbanions, which investigated: (i) the rate of H/D exchange of 1-phenylethyl phenyl sulfone;[109,110] (ii) the intramolecular H/D exchange of cyclopropyl alkyl sulfones;[111–114] and (iii) the base-catalyzed decarboxylation of optically active 2-methyltetrahydrothiophene-2-carboxylic acid 1,1-dioxide derivatives. The results suggest that α-sulfonyl carbanions are symmetrical (planar) as discrete reaction intermediates in asymmetric environments.[115]

Ab initio calculations on the hypothetical α-sulfonyl carbanion $HSO_2CH^-_2$ suggested that the geometry of the anionic carbon is intermediate between sp^2 and sp^3 structures.[116] Further, HSO_2CH_2Li was calculated to have only an unsymmetrically chelated structure (**120**; Figure 5).[117] Bors and Streitwieser, Jr. also reported theoretical (*ab initio*) studies of the anion of dimethyl sulfone, showing that the pyramidal anion is 0.57 kcal mol^{-1} (1 cal = 4.184 J) higher in energy over the planar form. The most stable geometry of the lithium salt of this anion is a structure (**121**), which is similar to (**120**). Another minimum with an unusual structure (**122**), is 1.1 kcal mol^{-1} higher in energy than (**121**; see Figure 6).[118]

It should be noted that the crystal structure of the Li/TMEDA salt of benzyl phenyl sulfone reported by Boche[119] is conceptually similar to the theoretical structure (**122**). X-Ray structure determination of α-(phenylsulfonyl)benzyllithium–TMEDA, α-(phenylsulfonyl)allyllithium and the potassium salt of bis(methylsulfonyl)-3-(2,6-dimethyoxypyridyl)sulfonylmethane shows that the coordination about the anionic carbon atom is nearly planar, and that the *p*-orbital on the anionic carbon is approximately *gauche* to the two oxygen atoms on the sulfur.[119–122] The anionic carbon atom can be described as interacting with the sulfur atoms in an ylide-like manner with a barrier to rotation about the $^-C—SO_2$ bond.

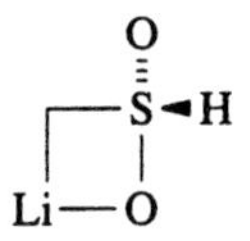

(**120**)

Figure 5

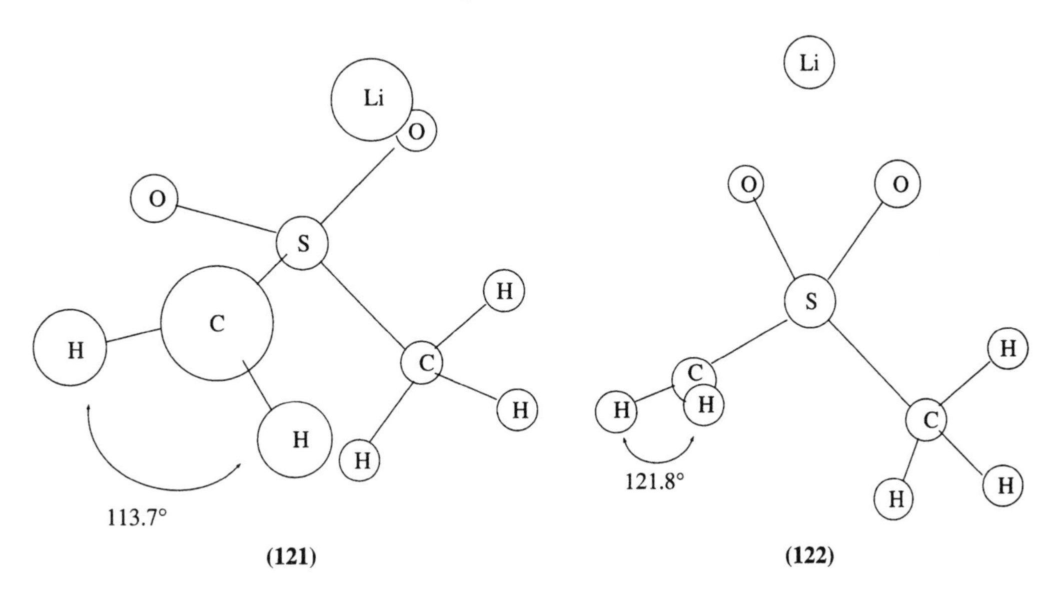

Figure 6

2.3.4.2 Addition to C=O Bonds

Sulfonyl groups are useful for organic synthesis, and several methods have been elucidated for generation of their α-anions, for example: (i) action of butyllithium; (ii) treatment of α-silyl sulfone with fluoride anion; (iii) addition of an anion to an α,β-unsaturated sulfone, and so on. Although the carbanion α to the sulfonyl group has the ability to add inter- or intra-molecularly to C=O bonds, the stereoselectivity of this addition is relatively low. Fluoride-induced reaction of (**123**) gave a cyclization product (**124**) as a 75:25 mixture of two diastereomers (equation 29).[123] Therefore, addition of the α-sulfonyl carbanion to a formyl group is usually utilized in combination with a subsequent reaction to delete the newly created asymmetry. In the above case, reductive cleavage of (**124**) gave the corresponding alkene (**125**). Intermolecular addition of an α-sulfonyl carbanion to an aldehyde was utilized as an efficient method for coupling of subunits in a synthesis of FK506. The lithium salt of (**126**) reacted with an aldehyde (**127**) in THF at –78 °C. Oxidation of the resultant product mixture (**128**) with Dess–Martin periodinane afforded keto sulfone diastereomers which, upon reduction with lithium naphthalenide, were converted into the homogeneous ketone (**129**) in 60% overall yield (equation 30).[124]

Me_3Si, $PhSO_2$, OHC — TBAF, Et_2O; 25 °C/0.75 h; 78% → HO, SO_2Ph → (29)

(**123**) (**124**) 75:25 (**125**)

2.3.4.3 Addition of Allylic Sulfonyl Carbanions to C=X Bonds

The butyllithium-generated anion of an allylic sulfone was reported to add to conjugate enones, but different regioselectivity was observed for 2-cyclohexenone (1,4-γ) and 3-penten-2-one (1,4-α), as shown in equation (31).[125] Clean 1,4-α additions to both acyclic and cyclic enones can be realized when the lithio carbanions are allowed to react in the presence of HMPA (2 mol equiv.) at –78 °C.[126] Under these conditions, allyl phenyl sulfone reacts with 2-cyclohexenone, 2-methyl-2-cyclopentenone and 3-methyl-2-cyclohexenone, giving the corresponding 1,4-α adducts as a 75:25 mixture of two diastereoisomers. In the reaction of allylic phenyl sulfone with an acyclic enone, 4-methyl-3-penten-2-one,

(126) (127) (128) (129) (30)

a 1,4-α adduct is exclusively produced. From several experiments, it has been concluded that the rapid and highly regioselective formation of 1,4-α adduct in the THF–HMPA (2 mol equiv.) system results from a direct conjugate addition rather than a kinetically controlled 1,2-α-addition followed by a rearrangement.

(31)

2.3.4.4 Addition of α-Halo Sulfones to C=O Bonds

When chloromethyl phenyl sulfone (**130**; Ar = phenyl) was subjected to the Darzens-type reaction with an aldehyde, a thermodynamically stable *trans* isomer (**133**) was produced exclusively (equation 32). This is in sharp contrast with the corresponding reaction of chloromethyl phenyl sulfoxide. Tavares proposed that the initial nucleophilic attack of the α-sulfonyl carbanion upon a carbonyl compound is rapidly reversible due to its stability, and that the product-determining step is the ring closure. Thermodynamic equilibrium between the two diastereomers of (**132**) allows predominant formation of the thermodynamically stable isomer (**133**) from the preferred transition state.[127]

(130) (131) (132) (133) (32)

Treatment of the epoxy sulfones obtained thus with $MgBr_2$ produces α-bromo ketones or α-bromo aldehyde.[128] This conversion can be applied to stereoselective preparations of α-bromo methyl ketones from cyclohexanone derivatives. Thus, treatment of 17-β-hydroxy-5-α-androstan-3-one (**134**) with 1-chloroethyl phenyl sulfone provides an epoxy sulfone (**135**), which is cleaved by $MgBr_2$ to give a 99:1 mixture of 3-acetyl-3-bromoandrostanes (**136** and **137**; equation 33), whereas similar treatment of 17-β-hydroxy-5-β-androstan-3-one (**138**) gives a 4:96 mixture of 3-acetyl-3-bromoandrostanes (**139** and **140**; equation 34). These facts may reflect the steric course of the initial attack of the α-sulfonyl carbanion on the carbonyl face.

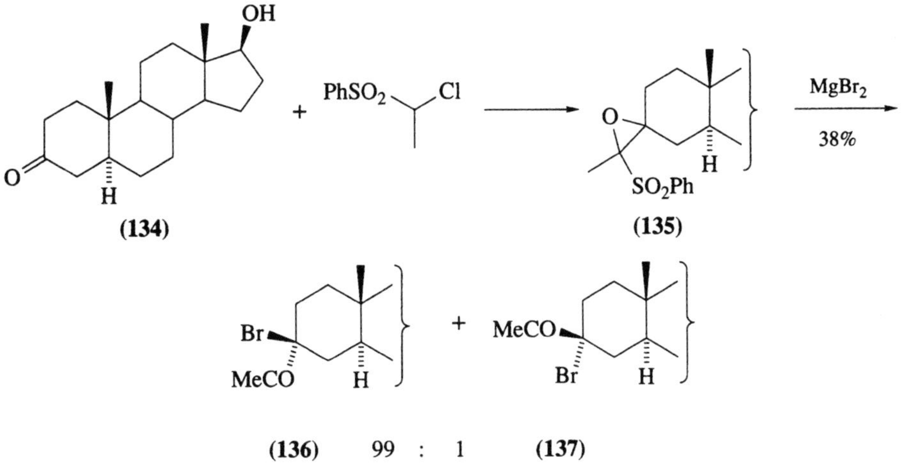

(134) (135) MgBr$_2$ 38%

(136) 99 : 1 (137)

(33)

(138) MgBr$_2$ 51%

(139) 96:4 (140)

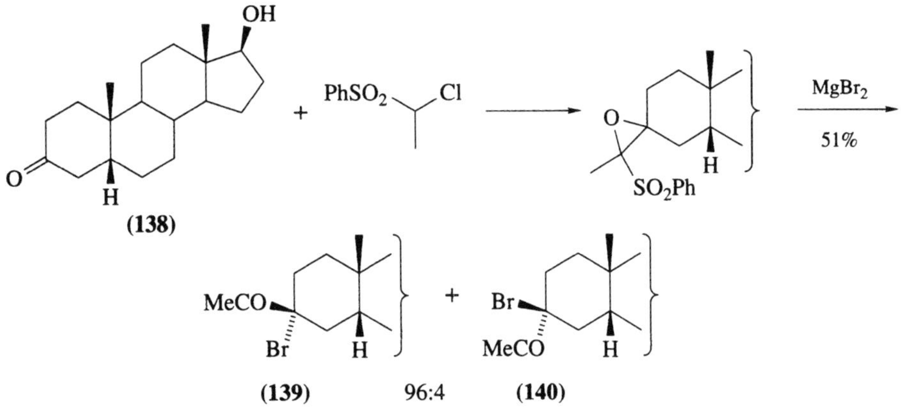

(34)

2.3.5 SULFONIMIDOYL-STABILIZED CARBANIONS

2.3.5.1 Configuration of the Carbanion

Bordwell estimated the pK_a of *N,S*-dimethyl-*S*-phenylsulfoximine (**141**) in DMSO to be about 33, less acidic than methyl phenyl sulfone by about 4 p*K* units. From a cyclopropyl effect on the equilibrium acidity of *N*-phenylsulfonyl-*S*-methyl-*S*-phenylsulfoximine (**142**), it was concluded that the carbanion is planar or nearly planar to the sulfonimidoyl group.[129] However, on the basis of the $^{1}J_{CH}$ coupling constant in a ^{13}C NMR spectroscopic study, it was concluded by Marquet that the configuration of the lithiated α-carbon (**142**) is pyramidal.[130]

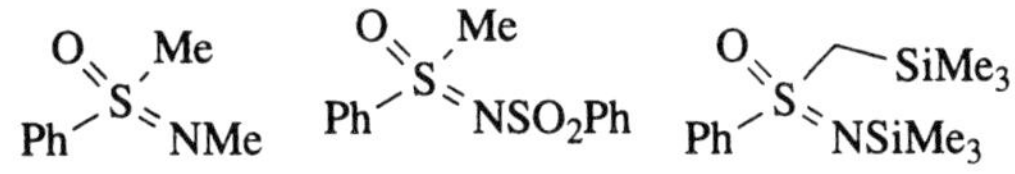

(141) (142) (143)

X-Ray structure analyses of the α-lithiated (**141**) and *N*-trimethylsilyl-*S*-trimethylsilylmethyl-*S*-phenylsulfoximine (**143**) were reported.[131,132] The latter compound can be crystallized as a tetramer including two molecules of cyclohexane, and its α-carbon is shown to be markedly pyramidalized. The $^{1}J_{CH}$ value of its α-carbon varies depending on the solvent employed, suggesting a hybrization between sp^2 and sp^3 in cyclohexane and a nearly sp^2 hybrization in THF.[133]

2.3.5.2 Addition to C=O Bonds

In 1971, Johnson and a coworker reported that the lithio derivative of (*S*)-*N*,*S*-dimethyl-*S*-phenylsulfoximine (**141**) added to benzaldehyde in THF at room temperature to give an adduct as a 75:25 mixture of two diastereomers. Fractional crystallization gave one of the diastereomers in a pure form. Treatment of this material with aluminum amalgam in aqueous THF yielded optically pure (*R*)-1-phenylethanol.[134] This preliminary result was extended to a general method for syntheses of optically active alcohols. By addition of the lithium derivative of (*S*)-(**141**) to prochiral ketones and aldehydes, optically active β-hydroxysulfoximines were prepared as diastereomeric pairs. The ketone adducts (**144**) showed moderate diastereoselectivities (for example, 60:40 for propiophenone). After separation by medium pressure chromatography on silica gel, each diastereomer was desulfurized with Raney nickel to yield an optically active tertiary alcohol (**145**; Scheme 33). When the aldehyde adducts (**146**) are treated with Raney nickel without separation into each diastereomer, the corresponding secondary alcohols (**147**) are yielded in optical purities of 25–46% (equation 35).[135]

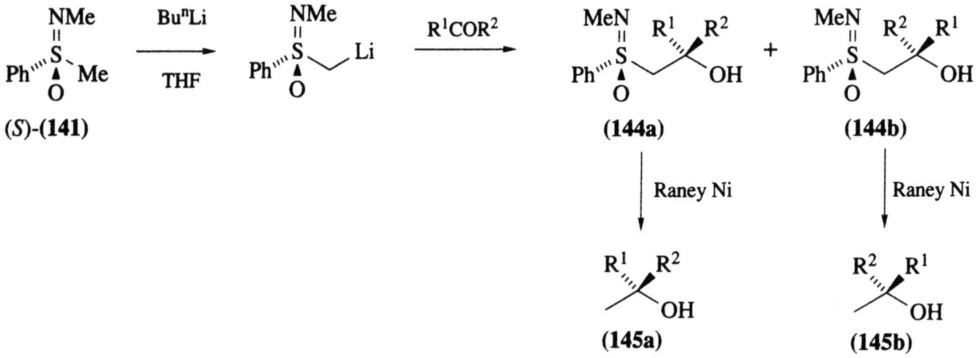

(*S*)-(**141**) — Bu^nLi, THF → lithio derivative — R^1COR^2 → (**144a**) + (**144b**); Raney Ni → (**145a**), (**145b**)

Scheme 33

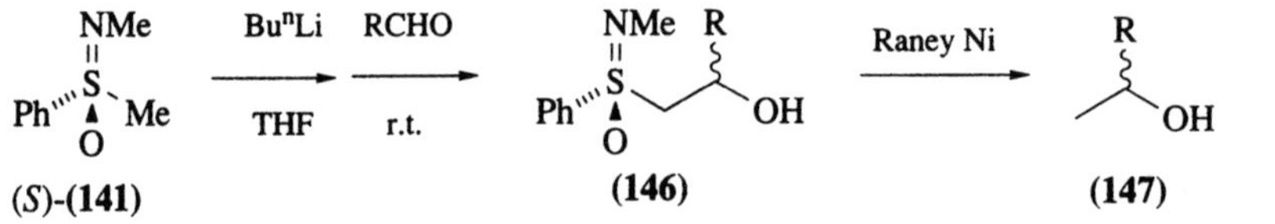

(*S*)-(**141**) — Bu^nLi, THF; RCHO, r.t. → (**146**) — Raney Ni → (**147**) (35)

N-Silyl-*S*-methyl-*S*-phenylsulfoximine (**148**) adds to aldehydes in ether at –78 °C. Desilylation of the resulting adducts with hydrochloric acid affords *N*-unsubstituted β-hydroxyalkylsulfoximines (**149**; Scheme 34).[136] The product ratios are insensitive to the steric bulk of the reacting aldehydes and to reaction temperature (–78 to –10 °C). The use of a larger *N*-silyl group enhances the diastereoselectivity of 1,2-addition, particularly when a sterically bulky aldehyde is used. The oxygen-coordinated model (**150**) provides for interaction of the silylated sulfoximine nitrogen with the R group of the incoming aldehyde.

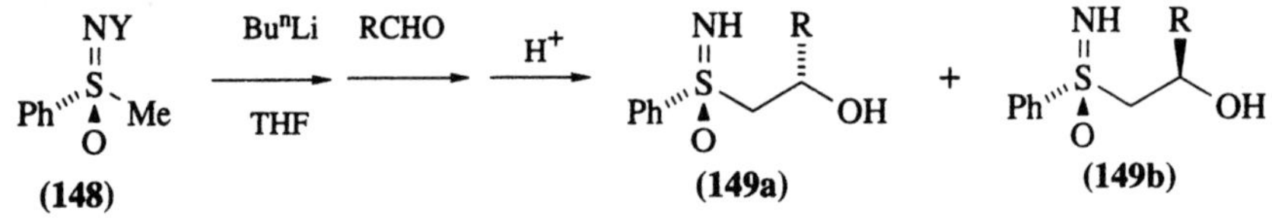

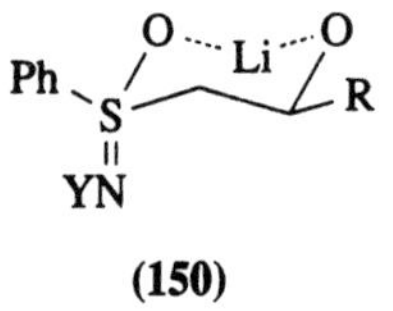

(**148**) — Bu^nLi, THF; RCHO; H^+ → (**149a**) + (**149b**)

(**150**)

Y	*R = Me*	*R = Bu^t*
$SiMe_3$	2:1	2.2:1
$SiMe_2Bu^t$	4:1	8:1
$SiMePh_2$	6:1	8:1

Scheme 34

(*S*)-(**141**) is also known to be a reagent for ketone methylenation with optical resolution.[137] Thus, addition of (*S*)-(**141**) to a racemic tricyclic ketone (**151**) resulted in the production of two diastereomers, (+)-(**152**) and (+)-(**153**), which were readily separated by flash chromatography. Treatment of (+)-(**152**) with aluminum amalgam and acetic acid in aqueous THF gave (–)-β-panasinsene (**154**) in 96% yield. Similar treatment of (+)-(**153**) gave the unnatural enantiomeric (+)-β-panasinsene (**155**; Scheme 35).

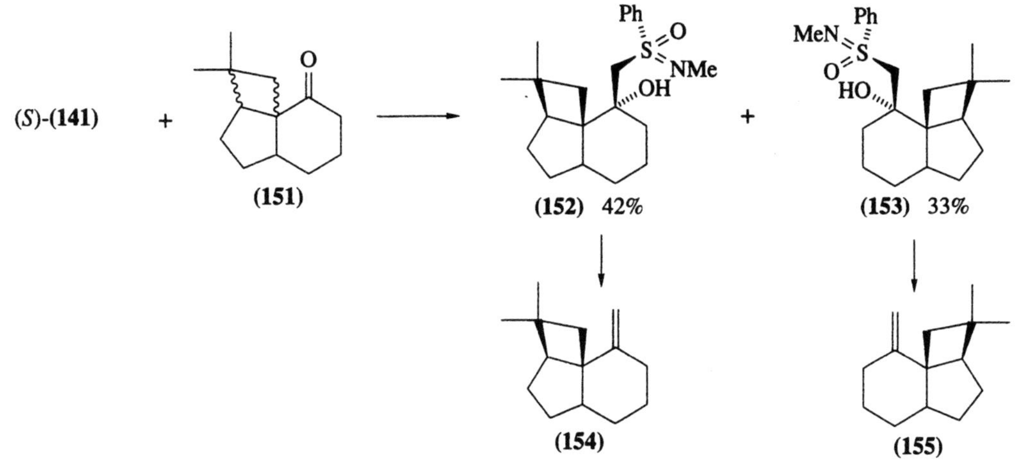

Scheme 35

The adduct (**157**) derived from (*S*)-(**141**) and an α,β-unsaturated ketone (**156**) undergoes Simmons–Smith cyclopropanation which is directed by the coexisting β-hydroxysulfoximine chiral center. As outlined in Scheme 36, this is a new methodology for optical activation of cyclopropyl ketones (**159**).[138] Indeed, optically active tricyclic ketones (–)-(**161**) and (+)-(**161**) were synthesized from a bicyclic ketone (**160**) in optical purities of 96% and 94%, respectively (equation 36).

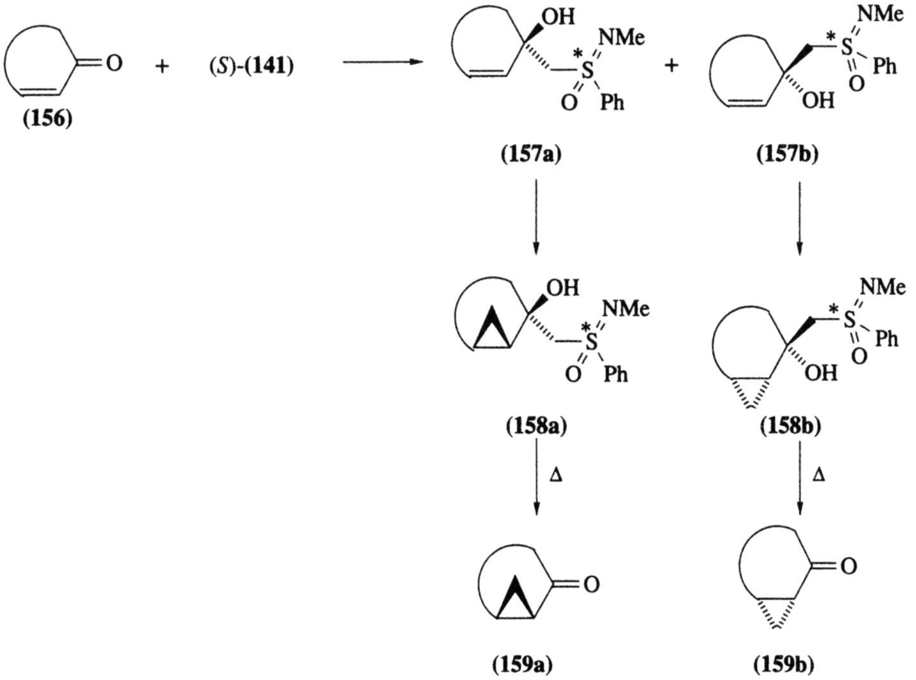

Scheme 36

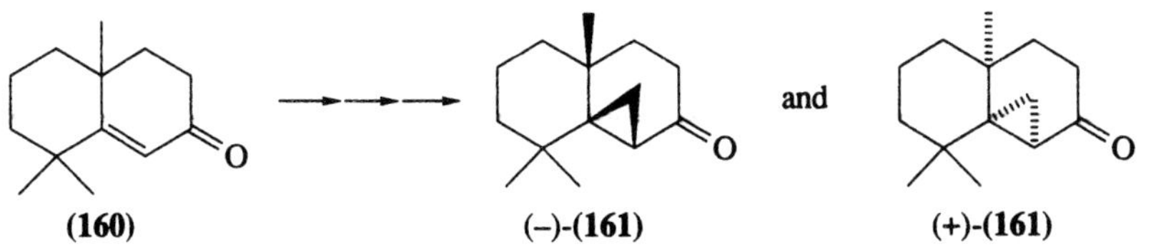

(36)

The same type of process using osmylation instead of the Simmons–Smith reaction leads to a synthesis of enantiomerically pure 2,3-dihydroxycycloalkanones from 2-cycloalkenones.[139] Addition of (*S*)-(**141**) to 3,5,5-trimethyl-2-cyclohexenone (**162**) produces a mixture of diastereomers (**163a**) and (**163b**), which are easily separable by silica gel chromatography. Treatment of the individual diastereomers with triethylamine *N*-oxide and OsO_4 (5 mol %) affords crude triols. In each case, a single diastereomeric product is produced. The subsequent thermolysis of the diastereomeric triols gives enantiomeric 2,3-dihydroxy-3,5,5-trimethyl-2-cyclohexanones (**164**; Scheme 37).

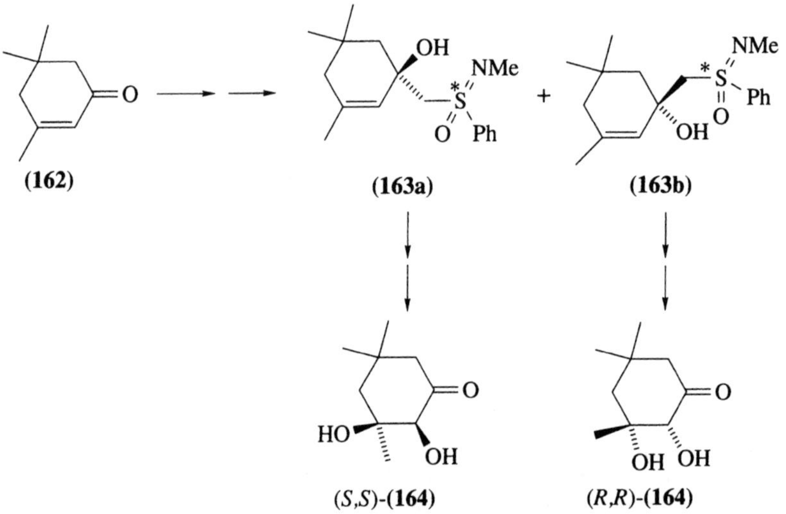

(**162**) (**163a**) (**163b**)

(*S,S*)-(**164**) (*R,R*)-(**164**)

Scheme 37

Optical resolution of ketones using (*S*)-(**141**) has also been reported. As mentioned above, the addition of (*S*)-(**141**) as its α-lithio derivative to ketones occurs readily and irreversibly at –78 °C to give an adduct in excellent yield. The adduct usually consists of two or three diastereomers, which are separated by chromatography. Envelope-shaped bicyclic ketones such as (**165**) and (**166**) give diastereofacial specificity on addition of the lithio derivative of (*S*)-(**141**) to form two diastereomers in a 50:50 ratio. Simple 2-substituted cyclohexanones such as (**167**) and (**168**) generally give rise to three diastereomeric adducts (61:22:17 and 57:25:18, respectively): two major adducts resulting from equatorial addition and a single minor product resulting from axial addition. Steric complications arise during axial attack, but with only one of the two diastereomeric transition states. The diastereofacial selectivity of chiral acyclic ketones (or aldehydes) is usually low, and four diastereomers result from the addition of (*S*)-(**141**). Thermolysis of the individual diastereomers occurs smoothly at 80–120 °C to regenerate the optically active ketone and the starting (*S*)-(**141**).[140,141] This technique is applicable to resolving various cyclic ketones (**169**, **170** and **171**), which are useful intermediates for the synthesis of biologically active compounds.[140,142,143]

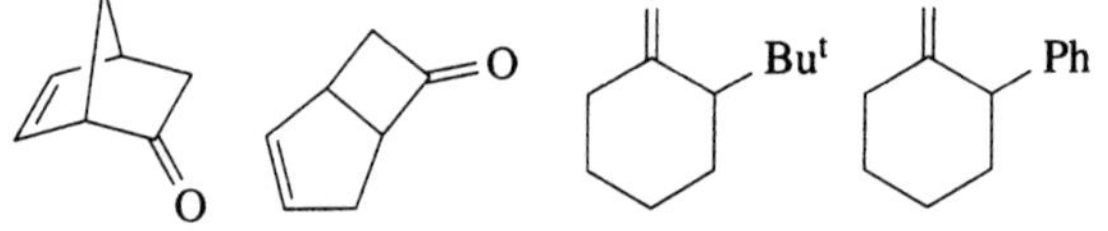

(**165**) (**166**) (**167**) (**168**)

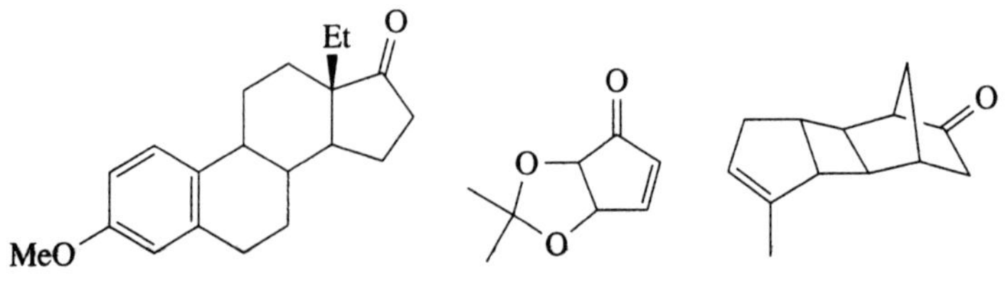

(**169**) (**170**) (**171**)

2.3.5.3 Others

The α-carbanion of an *N*-substituted sulfoximine adds to cycloalkadiene–molybdenum complexes (**173**) with asymmetric induction.[144] Although the reaction of *N*-substituted *S*-methyl-*S*-phenylsulfoximine is problematic, giving multiple products, the enolate ion of its *S*-(methoxycarbonyl)methyl analog (**172**) satisfactorily reacts with these complexes. It was established that the enolates derived from (*R*)-(**172**) preferentially add to the *pro*-(*R*) terminus of the complexes. Desulfonylation of the adducts (**174**) gave enantiomerically enriched monoester derivatives (**175**; equation 37), which could be further functionalized by hydride abstraction and a second nucleophilic addition. Typical results are summarized in Table 5. The reactions show a marked dependence on the sulfoximine *N*-substituent. The corresponding sodium and potassium enolates gave quite similar enantiomeric excesses, higher in all cases than those observed for the lithium enolates. The asymmetric inductions observed for reactions of complexes (**173**) are quite respectable and synthetically useful. To explain this stereoselectivity, open transition state models are proposed. The similar reaction of (**172**) to give dienyliron complexes (**176**) and (**177**) was also reported. Since the observed asymmetric inductions are lower, these reactions seem to be less useful.

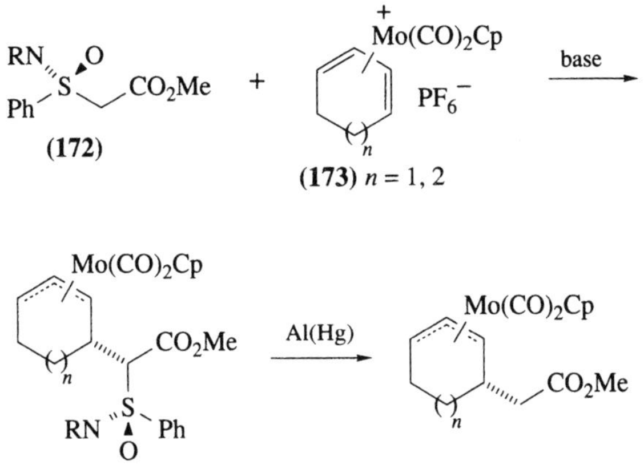

(**172**) + (**173**) n = 1, 2 —base→ (**174**) n = 1, 2 —Al(Hg)→ (**175**) n = 1, 2 (37)

Table 5 Asymmetric Induction in Equation (37)

(173) n	R^a	Base	(175) Yield (%)[b]	ee (%)[a]
1	Ts (+)	Bu^{t}OK	79	16 (+)
1	Me (+)	NaH	45	35 (+)
1	TBDMS (+)	Bu^{t}OK	80	78 (–)
1	DMTS (–)	LDA	77	55 (–)
1	DMTS (–)	NaH	83	75 (–)
1	DMTS (–)	Bu^{t}OK	80	80 (–)
2	DMTS (–)	Bu^{t}OK	80	85 (–)

[a](+) or (–) indicates enantiomer. [b]Overall yield after desulfonylation.

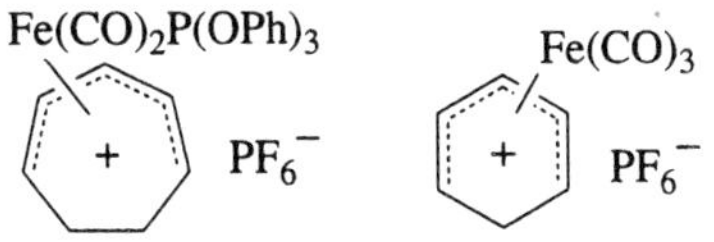

(**176**) (**177**)

Recently Gais and coworkers disclosed an asymmetric synthesis of alkyl-substituted exocyclic alkenes from ketones by the use of optically active *N*,*S*-dimethyl-*S*-phenylsulfoximine (**141**).[145] Scheme 38 shows the conversion of 4-*t*-butylcyclohexanone (**178**) into optically active 4-*t*-butyl-1-ethylidenecyclohexane (**182**). Addition of the lithiated (*S*)-(**141**) to (**178**) affords two diastereomeric adducts (**179**) and (**180**), which are easily separable.[146] Trimethylsilylation followed by action of butyllithium brings about asymmetric elimination of $LiOSiMe_3$, leading to an alkenylsulfoximine (**181**).[147] Treatment of (**181**) with dimethylzinc in the presence of $MgBr_2$ (2 equiv.) gives the enantiomerically pure alkene (**182**) in 74% yield. This method is effectively applicable to stereoselective synthesis of a precursor (**186**) for carbacyclins. Highly selective *exo* addition of the lithiated (*R*)-(**141**) to bicyclic ketone (**183**) yields the β-hydroxysulfoximine (**184**). Silylation and then subsequent asymmetric elimination of $LiOSiMe_3$ affords the (*Z*)-alkenylsulfoximine (**185**) with more than 98% diastereoselectivity.[147] In the reaction of (**185**) with a dialkylzinc in ether in the presence of $NiCl_2$(DPPP), a carbacyclin precursor (**186**) is obtained in 70% yield with 99:1 diastereoselectivity (Scheme 39).[145]

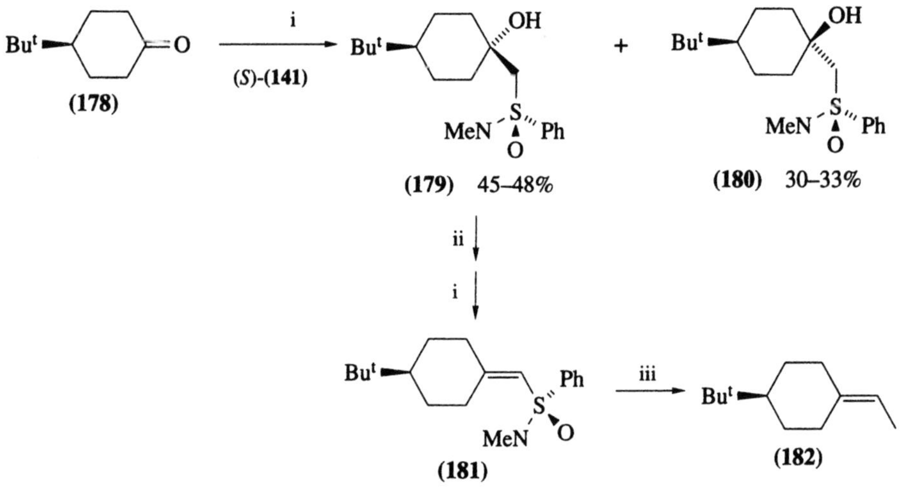

(**178**) (*S*)-(**141**) (**179**) 45–48% (**180**) 30–33% (**181**) (**182**)

i, Bu^nLi, THF, –78 °C; ii, Bu^nLi, THF, –78 °C, then Me_3SiCl, –78 to 25 °C;
iii, $Me_2Zn/MgBr_2/NiCl_2$(DPPP), ether

Scheme 38

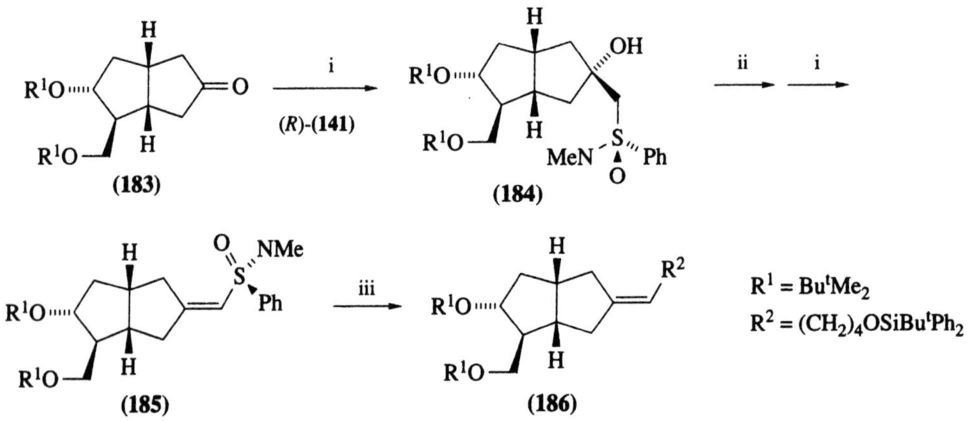

(**183**) (*R*)-(**141**) (**184**) (**185**) (**186**)

$R^1 = Bu^tMe_2$
$R^2 = (CH_2)_4OSiBu^tPh_2$

i, Bu^nLi, THF, –78 °C; ii, Bu^nLi, THF, –78 °C, then Me_3SiCl, –78 to 25 °C;
iii, $R^2{}_2Zn/MgBr_2/NiCl_2$(DPPP), ether

Scheme 39

2.3.6 REFERENCES

1. H. Gilman and F. J. Webb, *J. Am. Chem. Soc.*, 1940, **62**, 987.
2. H. Gilman and F. J. Webb, *J. Am. Chem. Soc.*, 1949, **71**, 4062.
3. E. J. Corey and D. Seebach, *J. Org. Chem.*, 1966, **31**, 4097.
4. D. J. Peterson, *J. Org. Chem.*, 1967, **32**, 1717.
5. K. Orgura, M. Fujita, K. Takahashi and H. Iida, *Chem. Lett.*, 1982, 1697.
6. A. Streitwieser, Jr. and S. P. Ewing, *J. Am. Chem. Soc.*, 1975, **97**, 190.
7. A. Streitwieser, Jr. and J. E. Williams, Jr., *J. Am. Chem. Soc.*, 1975, **97**, 191.
8. F. Bernardi, I. G. Csizmadia, A. Mangini, H. B. Schlegel, M.-H. Whangbo and S. Wolfe, *J. Am. Chem. Soc.*, 1975, **97**, 2209.
9. D. Seebach, J. Gabriel and R. Hässig, *Helv. Chim. Acta*, 1984, **67**, 1083.

10. W. T. Borden, E. R. Davidson, N. H. Andersen, A. D. Denniston and N. D. Epiotis, *J. Am. Chem. Soc.*, 1978, **100**, 1604.
11. N. D. Epiotis, R. L. Yates, F. Bernardi and S. Wolfe, *J. Am. Chem. Soc.*, 1976, **98**, 5435.
12. A. G. Schultz, W. Y. Fu, R. D. Lucci, B. G. Kurr, K. M. Lo and M. Boxer, *J. Am. Chem. Soc.*, 1978, **100**, 2140.
13. A. P. Kozikowski and E. M. Huie, *J. Am. Chem. Soc.*, 1982, **104**, 2923.
14. R. L. Sowerby and R. M. Coates, *J. Am. Chem. Soc.*, 1972, **94**, 4758.
15. T. Mukaiyama, Y. Watanabe and M. Shiono, *Chem. Lett.*, 1974, 1523.
16. S. C. Welch and J.-P. Loh, *J. Org. Chem.*, 1981, **46**, 4072.
17. I. Kuwajima, S. Sato and Y. Kurata, *Tetrahedron Lett.*, 1972, 737.
18. C. A. Kingsbury, *J. Org. Chem.*, 1972, **37**, 102.
19. P. G. McDougal, B. D. Condon, M. D. Laffosse, Jr., A. M. Lauro and D. Van Derveer, *Tetrahedron Lett.*, 1988, **29**, 2547.
20. P. M. Atlani, J. F. Biellmann, S. Dube and J. J. Vicens, *Tetrahedron Lett.*, 1974, 2665.
21. D. D. Ridley and M. A. Smal, *Aust. J. Chem.*, 1980, **33**, 1345.
22. K. Oshima, H. Yamamoto and H. Nozaki, *J. Am. Chem. Soc.*, 1973, **95**, 7926.
23. Y. Ikeda, K. Furuta, N. Meguriya, N. Ikeda and H. Yamamoto, *J. Am. Chem. Soc.*, 1982, **104**, 7663.
24. Y. Yamamoto, H. Yatagai, Y. Saito and K. Maruyama, *J. Org. Chem.*, 1984, **49**, 1096.
25. K.-H. Geiss, B. Seuring, R. Pieter and D. Seebach, *Angew. Chem., Int. Ed. Engl.*, 1974, **13**, 479.
26. D. Seebach, K.-H. Geiss and M. Pohmakotr, *Angew. Chem., Int. Ed. Engl.*, 1976, **15**, 437.
27. E. Juaristi, J. S. Cruz-Sanchez and F. R. Ramos-Morales, *J. Org. Chem.*, 1984, **49**, 4912.
28. A. G. M. Barrett and N. K. Capps, *Tetrahedron Lett.*, 1986, **27**, 5571.
29. C. A. Brown and A. Yamaichi, *J. Chem. Soc., Chem. Commun.*, 1979, 100.
30. P. C. Ostrowski and V. V. Kane, *Tetrahedron Lett.*, 1977, 3549.
31. A.-R. B. Manas and R. A. J. Smith, *J. Chem. Soc., Chem. Commun.*, 1975, 216.
32. J. Otera, Y. Niibo and H. Aikawa, *Tetrahedron Lett.*, 1987, **28**, 2147.
33. D. Seebach and R. Bürstinghaus, *Angew. Chem., Int. Ed. Engl.*, 1975, **14**, 57.
34. L. Bo and A. G. Fallis, *Tetrahedron Lett.*, 1986, **27**, 5193.
35. J.-M. Fang, B.-C. Hong and L.-F. Liao, *J. Org. Chem.*, 1987, **52**, 855.
36. J. E. Baldwin, R. E. Hackler and R. M. Scott, *Chem. Commun.*, 1969, 1415.
37. T. Durst, R. R. Fraser, M. R. McClory, R. B. Swingle, R. Viau and Y. Y. Wigfield, *Can. J. Chem.*, 1970, **48**, 2148.
38. T. Durst, R. Viau and M. R. McClory, *J. Am. Chem. Soc.*, 1971, **93**, 3077.
39. Y. Iitaka, A. Itai, N. Tomioka, Y. Kodama, K. Ichikawa, K. Nishihata, M. Nishio, M. Izumi and K. Doi, *Bull. Chem. Soc. Jpn.*, 1986, **59**, 2801.
40. K. Nakamura, M. Higaki, S. Adachi, S. Oka and A. Ohno, *J. Org. Chem.*, 1987, **52**, 1414.
41. J. O. Edwards and R. G. Pearson, *J. Am. Chem. Soc.*, 1962, **84**, 16.
42. S. Wolfe, A. Rauk and I. G. Csizmadia, *J. Am. Chem. Soc.*, 1967, **89**, 5710.
43. A. Rauk, S. Wolfe and I. G. Csizmadia, *Can. J. Chem.*, 1969, **47**, 113.
44. J. F. Biellmann and J. J. Vicens, *Tetrahedron Lett.*, 1974, 2915.
45. G. Chassaing, R. Lett and A. Marquet, *Tetrahedron Lett.*, 1978, 471.
46. R. Lett, G. Chassaing and A. Marquet, *J. Organomet. Chem.*, 1976, **111**, C17.
47. G. Tsuchihashi, S. Iriuchijima and M. Ishibashi, *Tetrahedron Lett.*, 1972, 4605.
48. N. Kunieda, M. Kinoshita and J. Nokami, *Chem. Lett.*, 1977, 289.
49. G. Demailly, C. Greck and G. Solladié, *Tetrahedron Lett.*, 1984, **25**, 4113.
50. S. G. Pyne and G. Boche, *J. Org. Chem.*, 1989, **54**, 2663.
51. P. Bravo, P. Carrera, G. Resnati and C. Ticozzi, *J. Chem. Soc., Chem. Commun.*, 1984, 19.
52. D. G. Farnum, T. Veysoglu, A. M. Carde, B. A. Duhl-Emswiler, T. A. Pancoast, T. J. Reitz and R. T. Cardé, *Tetrahedron Lett.*, 1977, 4009.
53. D. R. Williams, J. G. Phillips and J. C. Huffman, *J. Org. Chem.*, 1981, **46**, 4101.
54. D. R. Williams, J. G. Phillips, F. H. White and J. C. Huffman, *Tetrahedron*, 1986, **42**, 3003.
55. D. R. Williams and J. G. Phillips, *Tetrahedron*, 1986, **42**, 3013.
56. D. H. Hua, S. N. Bharathi, P. D. Robinson and A. Tsujimoto, *J. Org. Chem.*, 1990, **55**, 2128.
57. G. Tsuchihashi, S. Iriuchijima and K. Maniwa, *Tetrahedron Lett.*, 1973, 3389.
58. S. G. Pyne and B. Dikic, *J. Chem. Soc., Chem. Commun.*, 1989, 826.
59. V. Cerè, C. Paolucci, S. Pollicino, E. Sandri and A. Fava, *J. Chem. Soc., Chem. Commun.*, 1981, 764.
60. V. Cerè, C. Paolucci, S. Pollicino, E. Sandri and A. Fava, *J. Org. Chem.*, 1986, **51**, 4880.
61. V. Cerè, C. Paolucci, S. Pollicino, E. Sandri and A. Fava, *J. Chem. Soc., Chem. Commun.*, 1986, 223.
62. V. Cerè, C. Paolucci, S. Pollicino, E. Sandri and A. Fava, *J. Org. Chem.*, 1988, **53**, 5689.
63. D. J. Antonjuk, D. D. Ridley and M. A. Smal, *Aust. J. Chem.*, 1980, **36**, 2635.
64. D. D. Ridley and M. A. Smal, *Aust. J. Chem.*, 1983, **36**, 1049.
65. R. Annunziata, M. Cinquini, F. Cozzi and L. Raimondi, *J. Chem. Soc., Chem. Commun.*, 1986, 366.
66. R. Annunziata, M. Cinquini, F. Cozzi, L. Raimondi and S. Stefanelli, *Tetrahedron*, 1986, **42**, 5443.
67. R. Annunziata, M. Cinquini, F. Cozzi, L. Raimondi and S. Stefanelli, *Tetrahedron*, 1986, **42**, 5451.
68. L. L. Vasil'eva, V. I. Mel'nikova, E. T. Gainullina and K. K. Pivnitskii, *Zh. Org. Khim.*, 1980, **16**, 2618; *cf. Chem. Abstr.*, 1981, **94**, 191 837b.
69. M. R. Binns, R. K. Haynes, T. L. Houston and W. R. Jackson, *Aust. J. Chem.*, 1981, **34**, 2465.
70. M. R. Binns, R. K. Haynes, T. L. Houston and W. R. Jackson, *Tetrahedron Lett.*, 1980, **21**, 573.
71. J. Nokami, T. Ono, A. Iwao and S. Wakabayashi, *Bull. Chem. Soc. Jpn.*, 1982, **55**, 3043.
72. M. R. Binns, R. K. Haynes, A. G. Katsifis, P. A. Schober and S. C. Vonwiller, *Tetrahedron Lett.*, 1985, **26**, 1565.
73. M. R. Binns, R. K. Haynes, A. G. Katsifis and A. H. White, *Aust. J. Chem.*, 1987, **40**, 291.
74. R. K. Haynes and A. G. Katsifis, *J. Chem. Soc., Chem. Commun.*, 1987, 340.

75. J. Nokami, T. Ono and S. Wakabayashi, *Tetrahedron Lett.*, 1985, **26**, 1985.
76. D. H. Hua, S. Venkataraman, M. J. Coulter and G. Sinai-Zingde, *J. Org. Chem.*, 1987, **52**, 719.
77. D. H. Hua, *J. Am. Chem. Soc.*, 1986, **108**, 3835.
78. D. H. Hua, M. J. Coulter and I. Badejo, *Tetrahedron Lett.*, 1987, **28**, 5465.
79. D. H. Hua, S. Venkataraman, R. Chan-Yu-King and J. V. Paukstelis, *J. Am. Chem. Soc.*, 1988, **110**, 4741.
80. R. Annunziata, M. Cinquini, F. Cozzi, F. Montanari and A. Restelli, *J. Chem. Soc., Chem. Commun.*, 1983, 1138.
81. C. Mioskowski and G. Solladié, *J. Chem. Soc., Chem. Commun.*, 1977, 162.
82. C. Mioskowski and G. Solladié, *Tetrahedron*, 1980, **36**, 227.
83. E. J. Corey, L. O. Weigel, A. R. Chamberlin, H. Cho and D. H. Hua, *J. Am. Chem. Soc.*, 1980, **102**, 6613.
84. F. Matloubi and G. Solladié, *Tetrahedron Lett.*, 1979, 2141.
85. F. Schneider and R. Simon, *Synthesis*, 1986, 582.
86. L. Colombo, C. Gennari, G. Poli, C. Scolastico, R. Annunziata, M. Cinquini and F. Cozzi, *J. Chem. Soc., Chem. Commun.*, 1983, 403.
87. R. Annunziata, F. Cozzi, M. Cinquini, L. Colombo, C. Gennari, G. Poli and C. Scolastico, *J. Chem. Soc., Perkin Trans. 1*, 1985, 251.
88. R. Annunziata, S. Cardani, C. Gennari and G. Poli, *Synthesis*, 1984, 702.
89. T. Durst, *J. Am. Chem. Soc.*, 1969, **91**, 1034.
90. T. Durst and K.-C. Tin, *Tetrahedron Lett.*, 1970, 2369.
91. D. F. Tavares, R. E. Estep and M. Blezard, *Tetrahedron Lett.*, 1970, 2373.
92. G. Tsuchihashi and K. Ogura, *Bull. Chem. Soc. Jpn.*, 1972, **45**, 2023.
93. T. Satoh, T. Oohara and K. Yamakawa, *Tetrahedron Lett.*, 1988, **29**, 2851.
94. C. Mahidol, V. Reutrakul, V. Prapansiri and C. Panyachotipun, *Chem. Lett.*, 1984, 969.
95. T. Satoh, T. Oohara, Y. Ueda and K. Yamakawa, *Tetrahedron Lett.*, 1988, **29**, 313.
96. V. K. Aggarwal, I. W. Davies, J. Maddock, M. F. Mahon and K. C. Molloy, *Tetrahedron Lett.*, 1990, **31**, 135.
97. K. Ogura, *Pure Appl. Chem.*, 1987, **59**, 1033.
98. K. Ogura, in 'Studies in Natural Products Chemistry', ed. Atta-ur-Rahman, Elsevier, Amsterdam, 1990, vol. 6, p. 307.
99. K. Ogura and G. Tsuchihashi, *Tetrahedron Lett.*, 1972, 2681.
100. L. Colombo, C. Gennari and E. Narisano, *Tetrahedron Lett.*, 1978, 3861.
101. L. Colombo, C. Gennari, C. Scolastico, G. Guanti and E. Narisano, *J. Chem. Soc., Chem. Commun.*, 1979, 591.
102. L. Colombo, C. Gennari, C. Scolastico, G. Guanti and E. Narisano, *J. Chem. Soc., Perkin Trans. 1*, 1981, 1278.
103. K. Ogura, M. Fujita, T. Inaba, K. Takahashi and H. Iida, *Tetrahedron Lett.*, 1983, **24**, 503.
104. K. Ogura, M. Yamashita and G. Tsuchihashi, *Tetrahedron Lett.*, 1978, 1303.
105. J. L. Herrmann, J. E. Richman and R. H. Schlessinger, *Tetrahedron Lett.*, 1973, 3271.
106. J. L. Herrmann, J. E. Richman and R. H. Schlessinger, *Tetrahedron Lett.*, 1973, 3275.
107. L. Colombo, C. Gennari, G. Resnati and C. Scolastico, *Synthesis*, 1981, 74.
108. L. Colombo, C. Gennari, G. Resnati and C. Scolastico, *J. Chem. Soc., Perkin Trans. 1*, 1981, 1284.
109. E. J. Corey and T. H. Lowry, *Tetrahedron Lett.*, 1965, 793.
110. E. J. Corey and T. H. Lowry, *Tetrahedron Lett.*, 1965, 803.
111. W. Kirmse and U. Mrotzeck, *J. Chem. Soc., Chem. Commun.*, 1987, 709.
112. H. E. Zimmerman and B. S. Thyagarajan, *J. Am. Chem. Soc.*, 1960, **82**, 2505.
113. A. Ratajczak, F. A. L. Anet and D. J. Cram, *J. Am. Chem. Soc.*, 1967, **89**, 2072.
114. W. Th. van Wijnen, H. Steinberg and T. J. de Boer, *Tetrahedron*, 1972, **86**, 5423.
115. D. J. Cram and T. A. Whitney, *J. Am. Chem. Soc.*, 1967, **89**, 4651.
116. S. Wolfe, A. Rauk and I. G. Csizmadia, *J. Am. Chem. Soc.*, 1969, **91**, 1567.
117. S. Wolfe, L. A. LaJohn and D. F. Weaver, *Tetrahedron Lett.*, 1984, **25**, 2863.
118. D. A. Bors and A. Streitwieser, Jr., *J. Am. Chem. Soc.*, 1986, **108**, 1397.
119. G. Boche, M. Marsch, K. Harms and G. M. Sheldrick, *Angew. Chem., Int. Ed. Engl.*, 1985, **24**, 573.
120. H.-J. Gais, H. J. Lindner and J. Vollhardt, *Angew. Chem., Int. Ed. Engl.*, 1985, **24**, 859.
121. H.-J. Gais, J. Vollhardt and H. J. Lindner, *Angew. Chem., Int. Ed. Engl.*, 1986, **25**, 939.
122. J. S. Grossert, J. Hoyle, T. S. Cameron, S. P. Roe and B. R. Vincent, *Can. J. Chem.*, 1987, **65**, 1407.
123. M. B. Anderson and P. L. Fuchs, *J. Org. Chem.*, 1990, **55**, 337.
124. A. B. Jones, A. Villalobos, R. G. Linde, II and S. J. Danishefsky, *J. Org. Chem.*, 1990, **55**, 2786.
125. G. A. Krauss and K. Frazier, *Synth. Commun.*, 1978, **8**, 483.
126. M. Hirama, *Tetrahedron Lett.*, 1981, **22**, 1905.
127. P. F. Vogt and D. F. Tavares, *Can. J. Chem.*, 1969, **47**, 2875.
128. J. P. Bégué, D. Bonnet-Delpon, M. Charpentier-Morize and J. Sansoulet, *Can. J. Chem.*, 1982, **60**, 2087.
129. F. G. Bordwell, J. C. Branca, C. R. Johnson and N. R. Vanier, *J. Org. Chem.*, 1980, **45**, 3884.
130. G. Chassaing and A. Marquet, *Tetrahedron*, 1978, **34**, 1399.
131. H.-J. Gais, I. Erdelmeier, H. J. Lindner and J. Vollhardt, *Angew. Chem., Int. Ed. Engl.*, 1986, **25**, 938.
132. H.-J. Gais, U. Dingerdissen, C. Kruger and K. Angermund, *J. Am. Chem. Soc.*, 1987, **109**, 3775.
133. G. Boche, *Angew. Chem., Int. Ed. Engl.*, 1989, **28**, 277.
134. C. W. Schroeck and C. R. Johnson, *J. Am. Chem. Soc.*, 1971, **93**, 5305.
135. C. R. Johnson and C. J. Stark, Jr., *J. Org. Chem.*, 1982, **47**, 1193.
136. K.-J. Hwang, E. W. Logusch, L. H. Brannigan and M. R. Thompson, *J. Org. Chem.*, 1987, **52**, 3435.
137. C. R. Johnson and N. A. Meanwell, *J. Am. Chem. Soc.*, 1981, **103**, 7667.
138. C. R. Johnson and M. R. Barbachyn, *J. Am. Chem. Soc.*, 1982, **104**, 4290.
139. C. R. Johnson and M. R. Barbachyn, *J. Am. Chem. Soc.*, 1984, **106**, 2459.
140. C. R. Johnson and J. R. Zeller, *J. Am. Chem. Soc.*, 1982, **104**, 4021.
141. C. R. Johnson and J. R. Zeller, *Tetrahedron*, 1984, **40**, 1225.

142. C. R. Johnson and T. D. Penning, *J. Am. Chem. Soc.*, 1988, **110**, 4726.
143. R. G. Salomon, B. Basu, S. Roy and R. B. Sharma, *Tetrahedron Lett.*, 1989, **30**, 4621.
144. A. J. Pearson, S. L. Blystone, H. Nar, A. A. Pinkerton, B. A. Roden and J. Yoon, *J. Am. Chem. Soc.*, 1989, **111**, 134.
145. I. Erdelmeier and H.-J. Gais, *J. Am. Chem. Soc.*, 1989, **111**, 1125.
146. C. R. Johnson, C. W. Schroeck and J. R. Shanklin, *J. Am. Chem. Soc.*, 1973, **95**, 7424.
147. I. Erdelmeier, H.-J. Gais and H. J. Lindner, *Angew. Chem., Int. Ed. Engl.*, 1986, **25**, 935.

2.4 The Benzoin and Related Acyl Anion Equivalent Reactions

ALFRED HASSNER
Bar-Ilan University, Ramat-Gan, Israel

and

K. M. LOKANATHA RAI
University of Mysore, India

2.4.1 INTRODUCTION

C—C bond formation is a pivotal process in organic synthesis.[1,2] An effective pathway for C—C bond formation involves the reaction of an enolate at the α-position of a carbonyl compound with a carbon electrophile (alkyl halide, epoxide, carbonyl or unsaturated carbonyl compound). An alternative pathway would involve reaction of an acyl anion with an electrophile. Since acyl anions are in general unattainable, a great deal of effort has been expended to find masked acyl anions (acyl anion equivalents),[3] usually compounds of type $^{-}$RCXY, in which CXY can be reconverted into a carbonyl group. The benzoin condensation (Scheme 1), one of the oldest reactions in organic chemistry, represents a special case of such masked acyl anions. In this chapter we do not aim at exhaustive coverage of the subject but rather to give examples that highlight the utility of acyl anion equivalents, especially those related to cyanohydrins, in synthetic manipulations.

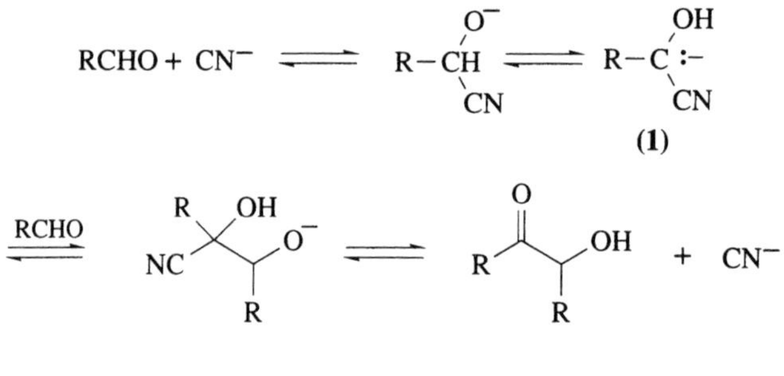

RCHO + CN⁻ ⇌ R–CH(O⁻)(CN) ⇌ R–C(OH)(CN):⁻ **(1)**

⇌(RCHO) NC–C(R)(OH)–CH(R)–O⁻ ⇌ R–CO–CH(R)–OH + CN⁻

Scheme 1

In the benzoin condensation, both aromatic and heterocyclic aldehydes are transformed into α-hydroxy ketones of the general formula ArCHOHCOAr, often called benzoins.[4–6] This class of compounds is frequently encountered in natural products, hence the benzoin and related reactions have received much attention.[7–13] The reaction employs a cyanide ion as the catalyst and the mechanism, proposed by Lapworth,[4] involves formation of carbanions stabilized by the nitrile group (Scheme 1).

ArCOCH(OH)Ar + Ar¹CHO →(KCN) Ar¹COCH(OH)Ar + ArCHO (1)

The fact that treatment of a benzoin with potassium cyanide in the presence of a second aldehyde leads to mixed benzoins indicates that all the steps are reversible (equation 1).

The carbanion (**1**) can be generated independently from cyanohydrins and can be added to the double bond of α,β-unsaturated ketones, esters and nitriles by an irreversible reaction which ultimately leads to γ-diketones, 4-oxocarboxylic esters and 4-oxonitriles.[7]

Studies on thiamine (vitamin B_1) catalyzed formation of acyloins from aliphatic aldehydes and on thiamine or thiamine diphosphate catalyzed decarboxylation of pyruvate[7,8] have established the mechanism for the catalytic activity of 1,3-thiazolium salts in carbonyl condensation reactions. In the presence of bases, quaternary thiazolium salts are transformed into the ylide structure (**2**), the ylide being able to exert a catalytic effect resembling that of the cyanide ion in the benzoin condensation (Scheme 2). Like cyanide, the zwitterion (**2**), formed by the reaction of thiazolium salts with base, is nucleophilic and reacts at the carbonyl group of aldehydes. The resultant intermediate can undergo base-catalyzed proton transfer to give a carbanion (**3**), which is stabilized by the thiazolium ring. The carbanion (**3**), like (**1**), undergoes 1,4-addition to α,β-unsaturated esters, ketones or nitriles.

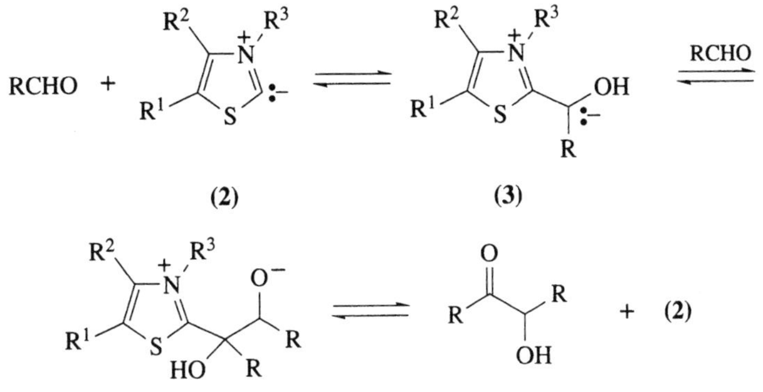

RCHO + (**2**) ⇌ (**3**) ⇌(RCHO) ... ⇌ RCOCH(OH)R + (**2**)

Scheme 2

2.4.1.1 Catalysts for Benzoin Formation

Generally benzoins are generated by the action of sodium cyanide or potassium cyanide on aromatic aldehydes in aqueous ethanol *via* cyanohydrin intermediates.[4] Benzoins may also be prepared in good yields by treating aromatic aldehydes with potassium cyanide in the presence of crown ethers in water or aprotic solvents.[14] Other sources for cyanide in this type of condensation are tetrabutylammonium cyanide,[5] polymer-supported cyanide[15] and acetone cyanohydrin with K_2CO_3.[16] Similarly, addition of aromatic aldehydes to α,β-unsaturated ketones can be accomplished by means of cyanide catalysis in DMF.[7]

The use of thiazolium salts enables the benzoin condensation to proceed at room temperature.[5] It can also be performed in dipolar aprotic solvents or under phase transfer conditions.[5] Thiazolium salts such as vitamin B_1,[17] thiazolium salts attached to γ-cyclodextrin,[18] macrobicyclic thiazolium salts,[19] thiazolium carboxylate,[20] naphtho[2,1-*d*]thiazolium and benzothiazolium salts catalyze the benzoin condensation[8] and quaternary salts of 1-methylbenzimidazole and 4-(4-chlorophenyl)-4*H*-1,2,4-triazole are reported to have similar catalytic activity.[7,8] Alkylation of 2-hydroxyethyl-4-methyl-1,3-thiazole with benzyl chloride, methyl iodide, ethyl bromide and 2-ethoxyethyl bromide yields useful salts for catalyzing 1,4-addition of aldehydes to activated double bonds.[8] Insoluble polymer-supported thiazolium salts are catalysts for the benzoin condensation and for Michael addition of aldehydes.[21–24] Electron rich alkenes such as bis(1,3-dialkylimidazolidin-2-ylidenes) bearing primary alkyl substituents at the nitrogen atoms[25] or bis(thiazolin-2-ylidene) bearing benzyl groups at the nitrogen atoms[26] are examples of a new class of catalyst for the conversion of ArCHO into ArCHOHCOAr.

The yeast-mediated condensation of benzaldehyde with acetaldehyde is of particular interest since it represents one of the first industrially useful microbial transformations, with the acyloin produced being subsequently converted chemically to D-ephidrine.[27] Other illustrations of synthetic value are the yeast-induced condensation of aldehyde (**4a,b**) with fermentatively generated acetaldehyde. The initially formed acyloins (**5a,b**) are not isolated but are further reduced, again with enantiotopic specificity, to give the pheromone synthon (**6a**; R = β-styryl) and the α-tocopherol chromanyl moiety precursor (**6b**; R = 2-propenylfuran) respectively (Scheme 3).[28]

α-Hydroxy ketones bearing a trifluoromethyl group were prepared from an aromatic aldehyde in high yield *via* a hydrazone.[29]

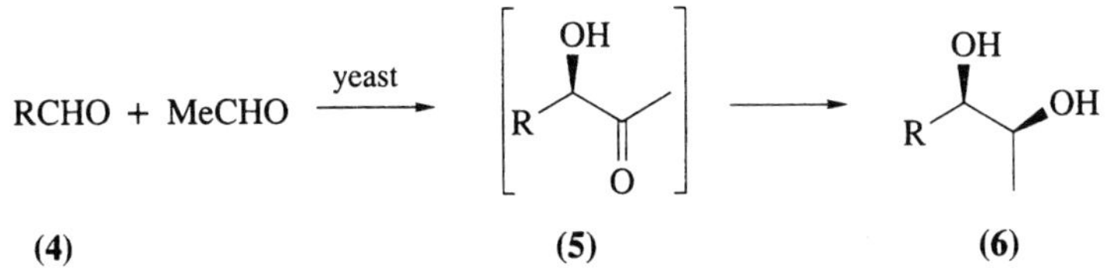

RCHO + MeCHO (**4**) —yeast→ [(**5**)] → (**6**)

Scheme 3

2.4.1.2 Masked Acyl Anion Equivalents

The benzoin condensation has recently been recognized as belonging to the general class of reactions that involve masked acyl anions as intermediates.[5,8,11,12] For example, an aldehyde is converted into an addition product RCH(OX)Y, which renders the C—H acidic. Then under basic conditions, a masked acyl anion (see **1**) can be formed and may react with an electrophilic component E^+. Decomposition of the product RCE(OX)Y should regenerate the carbonyl group with formation of RC(O)E. Intermediates such as (**1**) are used in the conversion of aldehydes into α-hydroxy ketones, α-diketones and 1,4-dicarbonyl compounds, proving to be a powerful strategy in the development of new synthetic methods.[3,8–12]

The most systematically investigated acyl anion equivalents have been the TMS ethers of aromatic and heteroaromatic aldehyde cyanohydrins,[30] TBDMS-protected cyanohydrins,[31,32] benzoyl-protected cyanohydrins,[33] alkoxycarbonyl-protected cyanohydrins,[34] THP-protected cyanohydrins,[35] ethoxyethyl-protected cyanohydrins,[36] α-(dialkylamino)nitriles,[8] cyanophosphates,[37] diethyl 1-(trimethylsiloxy)-phenylmethyl phosphonate[38] and dithioacetals.[3] Deprotonation of these masked acyl anions under the action of strong base, usually LDA, followed by treatment with a wide variety of electrophiles is of great synthetic value. If the electrophile is another aldehyde, α-hydroxy ketones or benzoins are formed. More recently, the acyl carbanion equivalents formed by electroreduction of oxazolium salts[39] were found to be useful for the formation of ketones, aldehydes or α-hydroxy ketones (Scheme 4). α-Methoxyvinyllithium also can act as an acyl anion equivalent and can be used for the formation of α-hydroxy ketones, α-diketones, ketones, γ-diketones[40] and silyl ketones.[13,41,42]

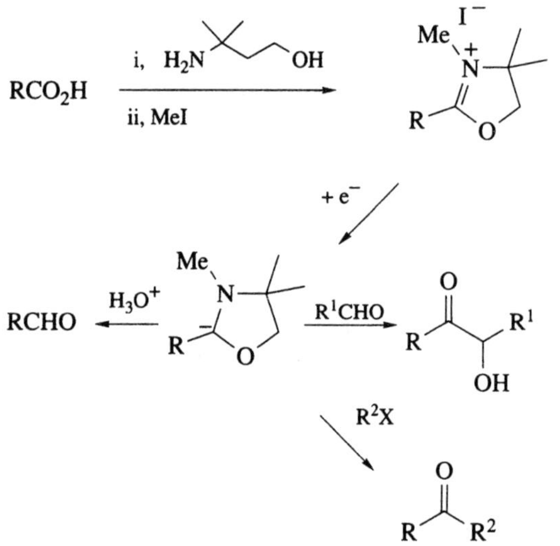

RCO_2H → (i, H_2N–C(Me)$_2$–CH_2OH; ii, MeI) → N-methyloxazolinium iodide; $+e^-$ → anion; R^1CHO → $RC(O)CH(OH)R^1$; R^2X → $RC(O)R^2$; H_3O^+ → RCHO

Scheme 4

2.4.1.3 Electrophiles

In the benzoin condensation, one molecule of aldehyde serves as an electrophile. If a carbanion is generated from protected cyanohydrins, α-aminonitriles or dithioacetals, it can react with electrophiles such as alkyl halides, strongly activated aryl halides or alkyl tosylates to form ketones. Amongst other electrophiles which are attacked by the above carbanions are heterocyclic *N*-oxides, carbonyl compounds, α,β-unsaturated carbonyl compounds, α,β-unsaturated nitriles, acyl halides, Mannich bases, epoxides and chlorotrimethyl derivatives of silicon, germanium and tin.

2.4.1.4 Miscellaneous

Benzoins can also be synthesized by the photolysis of aldehydes at high concentration.[43] The formation of benzoin can be rationalized as arising from intermolecular hydrogen abstraction followed by collapse of the radicals.

The addition of lithium and Grignard reagents to isocyanides which do not contain α-hydrogens proceeds by an α-addition to produce a metalloaldimine (**7**, an acyl anion equivalent). The lithium aldimines are versatile reagents which can be used as precursors for the preparation of aldehydes, ketones, α-hydroxy ketones, α-keto acids, α- and β-hydroxy acids, silyl ketones and α-amino acids (Scheme 5).[44,45]

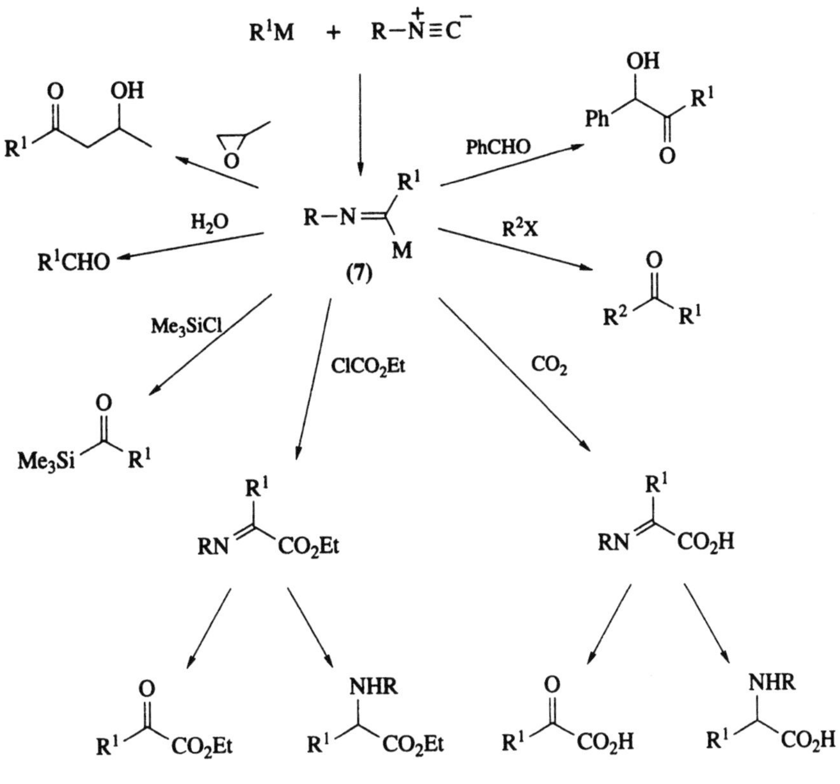

Scheme 5

The lithium salts of aldehyde *t*-butylhydrazones react with electrophiles (aldehydes, ketones, alkyl halides) to form C-trapped *t*-butylazo compounds; isomerization and hydrolysis give α-hydroxy ketones or ketones in good yields, thereby providing a convenient path *via* a new acyl anion equivalent (Scheme 6).[46] Reaction of these lithium salts with aldehydes and ketones, followed by elimination, provides a new route to azaalkenes,[47] whereas homolytic decomposition of C-trapped azo compounds of trityl and diphenyl-4-pyridylmethylhydrazones lead to the formation of alkanes, alkenes, alcohols or saturated esters.[48]

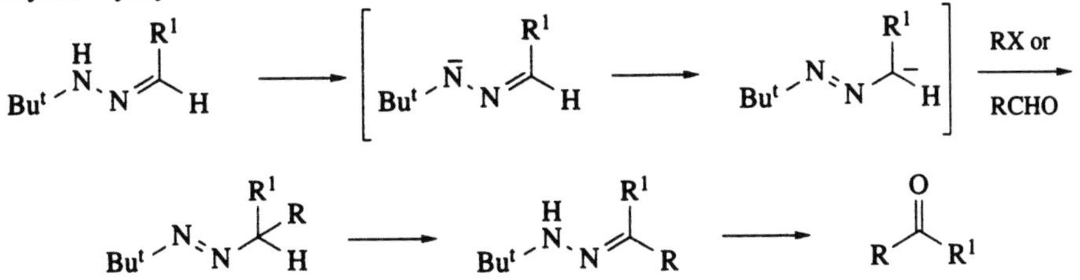

Scheme 6

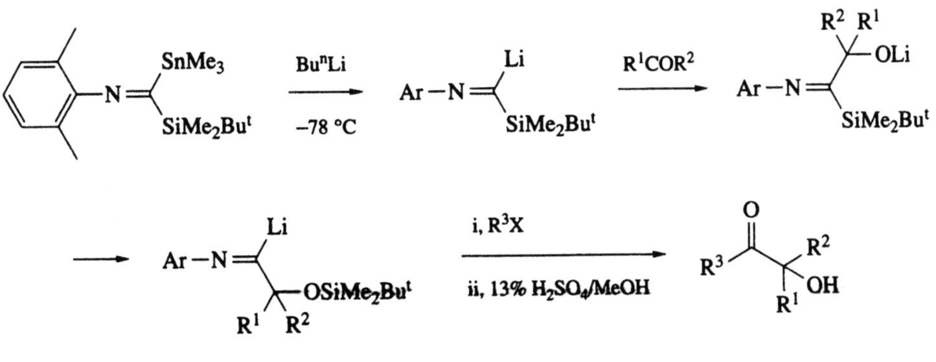

Scheme 7

When the 2,6-xylylimine of trialkylsilyl trimethylstannyl ketone is selectively transmetallated at –78 °C with Bu^nLi to generate a derivative, the latter serves as a versatile masked acyl anion that can react with various electrophiles. For example, α-hydroxy ketones were prepared using this new reagent (Scheme 7).[49]

2.4.2 CHIRAL CYANOHYDRINS

Asymmetric synthesis by means of a cyanohydrin is an important process in organic synthesis, because the cyanohydrin can be easily converted into a variety of valuable synthetic intermediates, such as α-hydroxy ketones, α-hydroxy acids, γ-diketones, β-amino alcohols, 4-oxocarboxylic esters, 4-oxonitriles, α-amino acids and acyl cyanides. More specifically, the (*S*)-cyanohydrin of *m*-phenoxybenzaldehyde is a building block for the synthesis of the insecticide deltamethrin,[50] or (1*R*)-*cis*-pyrethroids.[51]

In catalytic processes with enzymes such as D-oxynitrilase[52] and (*R*)-oxynitrilase (mandelonitrilase)[53] or synthetic peptides such as cyclo[(*S*)-phenylalanyl-(*S*)-histidyl],[54] or in reaction with TMS-CN promoted by chiral titanium(IV) reagents[55,56] or with lanthanide trichlorides,[57] hydrogen cyanide adds to numerous aldehydes to form optically active cyanohydrins. The optically active Lewis acids (**8**) can also be used as a catalyst.[58] Cyanation of chiral cyclic acetals with TMS-CN in the presence of titanium(IV) chloride gives cyanohydrin ethers, which on hydrolysis lead to optically active cyanohydrins.[59] An optically active cyanohydrin can also be prepared from racemic RR′C(OH)CN by complexation with brucine.[60]

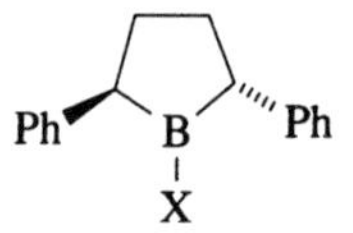

Ph B Ph
X

(**8**) X = OMe or Cl

2.4.2.1 1,2-Addition to Aldehydes and Ketones

The oldest known reaction involving masked acyl anions of the type (**1**) is the preparation of benzoin from aromatic aldehydes as illustrated in Scheme 1 (R = Ph). The reaction is applicable only to aldehydes with no α-hydrogen atoms (otherwise the aldol reaction competes); attempts to prepare a mixed acyloin yield a mixture of products. It is essential that the reaction be carried out in aprotic solvents, general preference being given to DMF. Aldehydes which normally give poor yields of benzoin in aqueous alcohol react in DMF–DMSO to give good yields of benzoins, especially if tetrabutylammonium cyanide is used as the base.[5,7] Under these conditions, the cyanide ion becomes more nucleophilic and basic; the reaction is usually complete in a few hours at room temperature.

Quite often the products of the benzoin condensation are unstable to oxygen and lead to α-diketones.[5] This is the case in intramolecular condensations when the initial product is an *o*-quinol, which is easily oxidized to the quinone.[5] 4-Substituted tetramides can be prepared by the reaction of an aromatic aldehyde with oxalaldehyde in the presence of cyanide ion, *via* the benzoin condensation (Scheme 8).[6]

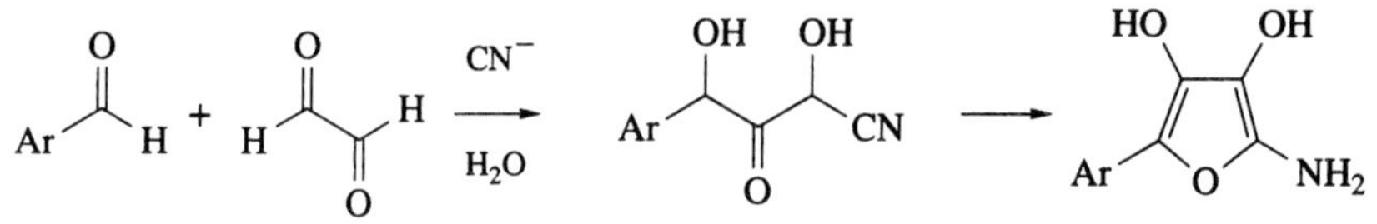

ArCHO + OHC–CHO $\xrightarrow[H_2O]{CN^-}$ Ar–CH(OH)–CO–CH(OH)–CN ⟶ 2-amino-3,4-dihydroxy-5-arylfuran (Ar, HO, OH, O, NH_2)

Scheme 8

2.4.2.2 1,4-Addition to α,β-Unsaturated Carbonyl Compounds

The most suitable technique for addition of aromatic aldehydes to α,β-unsaturated ketones utilizes cyanide catalysis in DMF (equation 2).[7] Unsubstituted aromatic aldehydes and those bearing substituents in the ring, except for *ortho*-substituted benzaldehydes, can add to an aliphatic, aromatic or heterocyclic α,β-unsaturated ketone. Addition of heterocyclic aldehydes can be catalyzed by cyanide ion as well. The products are γ-diketones, which can be converted to cyclopentenones, furans, pyrroles, thiophenes and pyrazines, as illustrated for the synthesis of *N*-methyl-2,3,5-tris(3-pyridyl)pyrrole, quinolizidines and 1-(3-phenyl-1-indolizidinyl)-1-propanol.[7]

(1) + $R^1CH{=}CHC(O)R^2$ ⟶ $RC(O)CH(R^1)CH_2C(O)R^2$ + CN^- (2)

Aromatic or heterocyclic aldehydes also add smoothly to α,β-unsaturated esters and nitriles.[7]

Dialdehydes such as 2,5-thiophenedicarbaldehyde can be added to acrylonitrile in the presence of cyanide ion and the product can be converted to 4,9-dioxododecanedioic acid.[7]

2.4.2.3 Addition to Mannich Bases

Aromatic and heterocyclic aldehydes add readily to Mannich bases in a cyanide-catalyzed reaction in DMF to produce γ-diketones (equation 3).

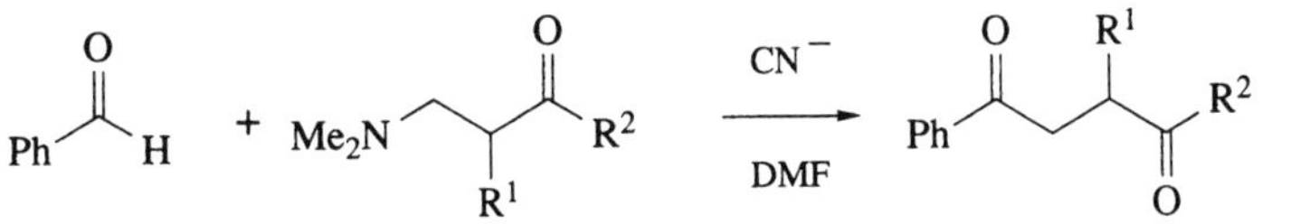

$PhCHO$ + $Me_2NCH_2CH(R^1)C(O)R^2$ $\xrightarrow[\text{DMF}]{CN^-}$ $PhC(O)CH_2CH(R^1)C(O)R^2$ (3)

R^1 = H, Me, Ph; R^2 = Me, Pr^i, Ph; $R^1R^2 = -(CH_2)_n-$

Examples include the reactions of 3-pyridinecarbaldehyde with 3-dimethylamino-1-phenyl-1-propanone, of furfural with 3-dimethylamino-1-(2-furyl)-1-propanone and of 2-thiophenecarbaldehyde with 3-dimethylamino-1-(2-thienyl)-1-propanone.[7]

2.4.2.4 Miscellaneous Reactions of Cyanohydrins

Direct conversion of aldehydes to esters was carried out *via* oxidative benzoin reactions, using an aromatic nitro compound as an oxidizing agent under the catalytic action of cyanide ion or of a conjugate base of a thiazolium ion (Scheme 9).[61]

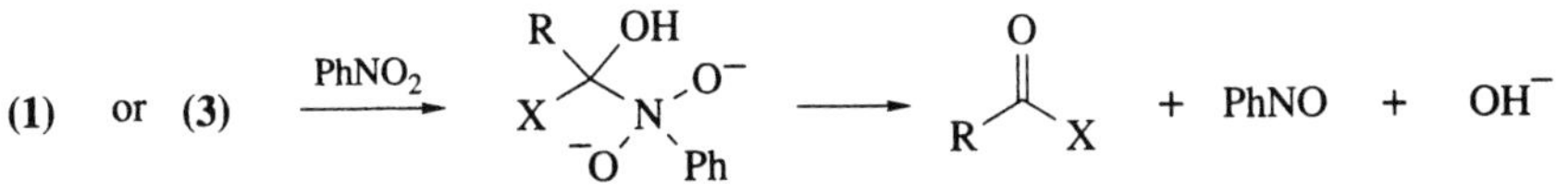

(1) or (3) $\xrightarrow{PhNO_2}$ $R(OH)C(X)-O-N(O^-)Ph$ ⟶ $RC(O)X$ + PhNO + OH^-

X = CN, Tz

Scheme 9

2.4.3 *O*-PROTECTED CYANOHYDRINS

The mechanism of the benzoin condensation,[4] as depicted in Scheme 1, suggested that anions derived from a protected aldehyde cyanohydrin should function as nucleophilic acylating reagents. The use of protected cyanohydrins as carbanion equivalents has been studied by Stork[62] and by Hunig[63] and has found wide applicability in chemical synthesis. Such species may serve as either acyl anion equivalents[12] or homoenolate anions.[25,36]

Anions of protected cyanohydrins of aliphatic, aromatic or α,β-unsaturated aldehydes undergo 1,4-addition to cyclic and acyclic enones. The synthetic utility of protected cyanohydrins in 1,4-addition depends on regioselectivity, since a competing reaction is 1,2-addition to the carbonyl group. The regioselectivity of these reactions (1,4- *versus* 1,2-addition) is dependent on the structure of the protected cyanohydrin, the enone and the reaction solvent.[8] Some general principles which influence the regioselectivity can be defined.

Conjugate additions predominate with bulky anions or with an enone containing a hindered carbonyl function. Anions derived from protected cyanohydrins of α,β-unsaturated aldehydes favor 1,4-additions. Anions derived from aryl aldehydes, especially if substituted with electron-withdrawing substituents, give predominantly conjugate addition. Increased bulk at the β-position of the enone, such as in β,β-disubstituted enones, leads to increased amounts of 1,2-addition.[8]

The anions of protected cyanohydrins are excellent nucleophiles in reaction with both primary and secondary alkyl halides.[8]

2.4.3.1 *O*-Silyl-protected Cyanohydrins

O-Silylated cyanohydrins have found considerable utility in the regioselective protection of *p*-quinones,[64] as intermediates for the preparation of β-amino alcohols[65] and as precursors to acyl anion equivalents.[36] Such compounds are typically prepared in high yield by either thermal or Lewis acid catalyzed addition of TMS-CN across the carbonyl group.[64] This cyanosilylation has a variety of disadvantages and modified one-pot cyanosilylation procedures have been reported.[31,66] The carbonyl group can be regenerated by treatment with acid, silver fluoride[64] or triethylaluminum hydrofluoride followed by base.[63]

2.4.3.1.1 1,2-Addition to aldehydes and ketones

The anion of the adduct of TMS-CN with benzaldehyde reacts with aldehydes and ketones to form acyloin silyl ethers, by way of a 1,4-*O*-silyl rearrangement (Scheme 10). This method gives α-hydroxy ketones in excellent yield (80–90%) and allows the selective synthesis of unsymmetrical benzoins.[67]

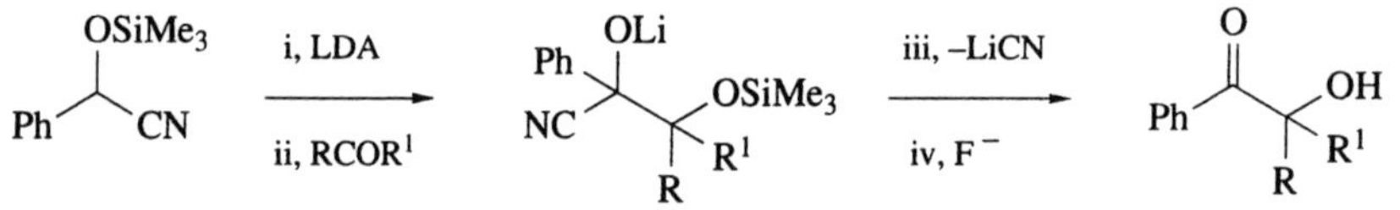

$PhCH(OSiMe_3)CN$ —(i, LDA; ii, $RCOR^1$)→ $Ph(NC)C(OLi)–C(OSiMe_3)RR^1$ —(iii, –LiCN; iv, F^-)→ $PhCO–C(OH)RR^1$

Scheme 10

Trimethylsiloxy cyanohydrins (**9**) derived from an α,β-unsaturated aldehyde form ambident anions (**9a**) on deprotonation. The latter can react with electrophiles at the α-position as an acyl anion equivalent (at –78 °C)[7] or at the γ-position as a homoenolate equivalent (at 0 °C).[36] The lithium salt of (**9**) reacts exclusively at the α-position with aldehydes and ketones.[36] The initial kinetic product (**10**) formed at –78 °C undergoes an intramolecular 1,4-silyl rearrangement at higher temperature to give (**11**). Thus the initial kinetic product is trapped and only products resulting from α-attack are observed (see Scheme 11). The α-hydroxyenones (**12**), γ-lactones (**13**) and α-trimethylsiloxyenones (**11**) formed are useful precursors to cyclopentenones and the overall reaction sequence constitutes a three-carbon annelation procedure.

2.4.3.1.2 1,4-Addition to α,β-unsaturated carbonyl compounds

The anion (**9a**) derived from crotonaldehyde reacts with 2-cyclohexen-1-one (**14**) to give exclusively a 1,4-addition product (equation 4).[8] Both the solvent and the nature of the protecting group affect the regioselectivity in the reactions of anions (**9**) with 4-methyl-3-penten-2-one (**15**). For instance, in THF the 1,2-adduct (**16**) is formed in 74% yield (equation 5), while in ether only the 1,4-addition product (**17**) is isolated.

The introduction of electron-donating or electron-withdrawing substituents in the *para* position of anion (**18**) strongly influences the regioselectivity in additions to (**15**).[8] In DME the amount of 1,4-addition increases from 0% for the *p*-dimethylamino to 100% for the *p*-cyano derivative. In ether only the 1,4-product (**19**) is observed when the *para* substituent is hydrogen, chloro, trifluoromethyl or cyano. The *p*-OMe derivative in ether gives 70% of (**19**), while the *p*-dimethylamino derivative gives mainly 1,2-product (**20**; equation 6; 86%).

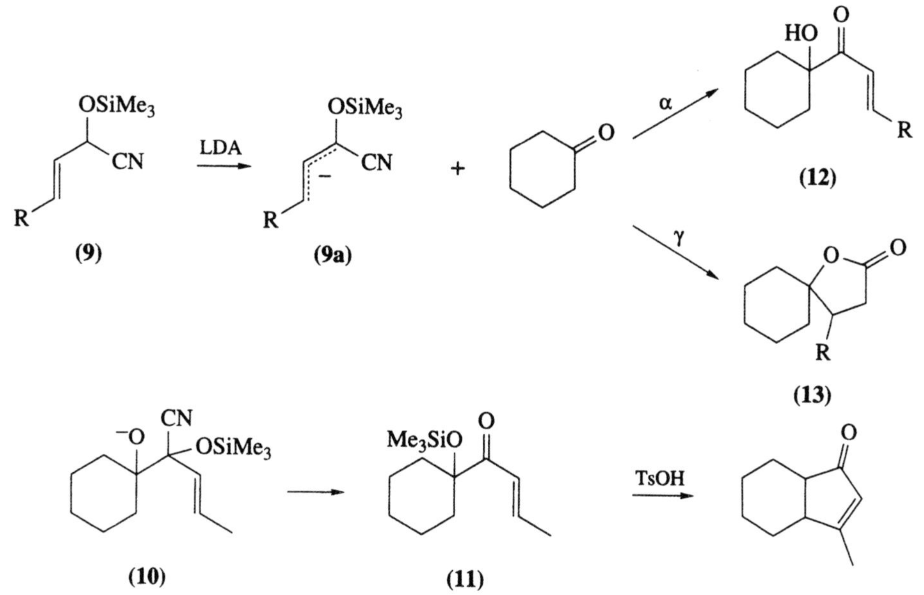

OSiMe3 CN R (9) LDA OSiMe3 CN R (9a) + O α HO O R (12) γ O O R (13)

−O CN OSiMe3 (10) Me3SiO O (11) TsOH O

Scheme 11

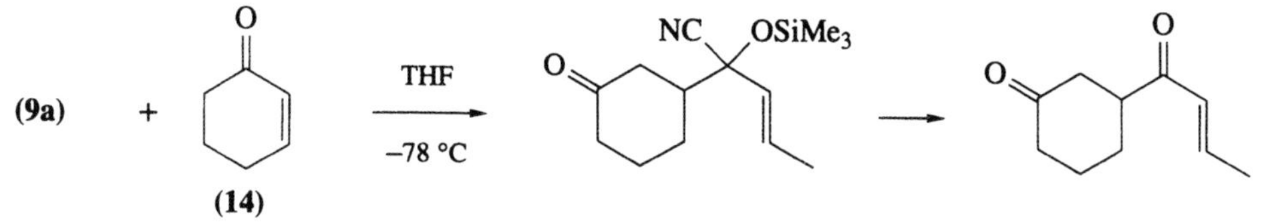

(9a) + O (14) THF −78 °C O NC OSiMe3 O O (4)

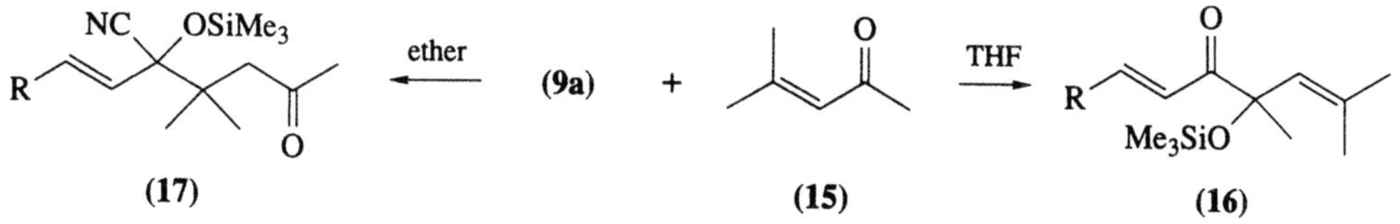

NC OSiMe3 R O (17) ether (9a) + O (15) THF R O Me3SiO (16) (5)

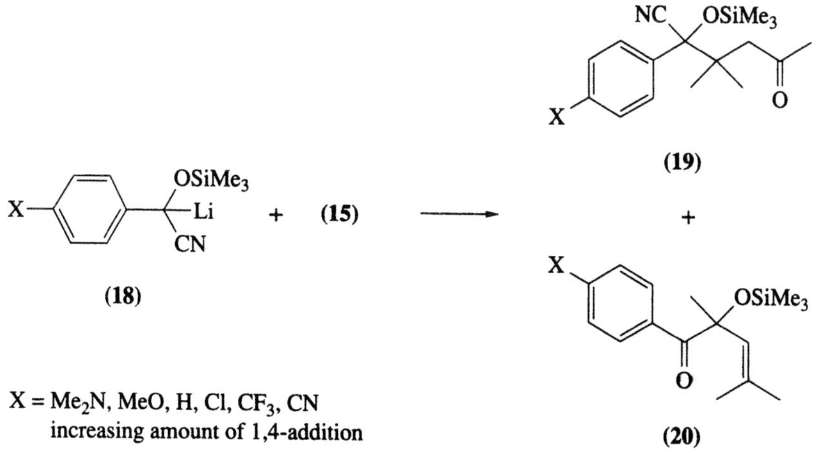

X OSiMe3 Li CN (18) + (15) NC OSiMe3 X O (19) + X OSiMe3 O (20) (6)

X = Me_2N, MeO, H, Cl, CF_3, CN
increasing amount of 1,4-addition

Addition of the α,β-unsaturated anion (**21**) to the Michael acceptor (**22**), in which either alkylation or 1,4-addition is possible, affords only the Michael product. Internal alkylation of the intermediate ester enolates leads to cyclopropyl derivatives (equation 7).[8] Terpenoid polyenes are prepared through conjugate addition of the lithiated protected cyanohydrins (**23**) to dienyl sulfoxide (**24**; equation 8).[8]

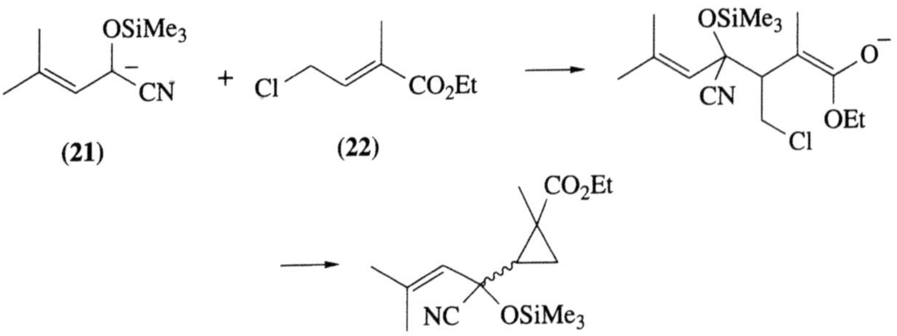

(**21**) (**22**) (7)

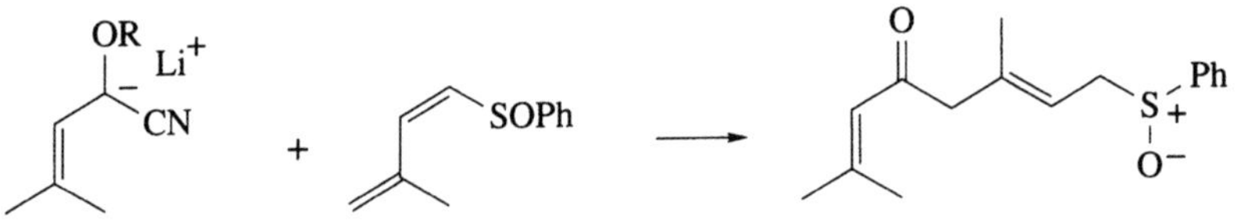

(**23**) (**24**) (8)

R = TMS,

2.4.3.1.3 *Alkylation with R—X*

Anions (**25**) derived from aryl and heterocyclic aldehydes have been used by Hunig *et al.* for the preparation of a large number of ketones in which one residue is aromatic or heterocyclic (equation 9).[8] In view of the competition between displacement and elimination so frequently encountered in reactions of nucleophiles with secondary halides, it is striking that alkylation of (**25**; R = Ph) proceeds well even with a tertiary iodide. 1-Bromo-4-*t*-butylcyclohexane reacts with (**25**) to give the *trans* product (**26**) with inversion of configuration (equation 10).

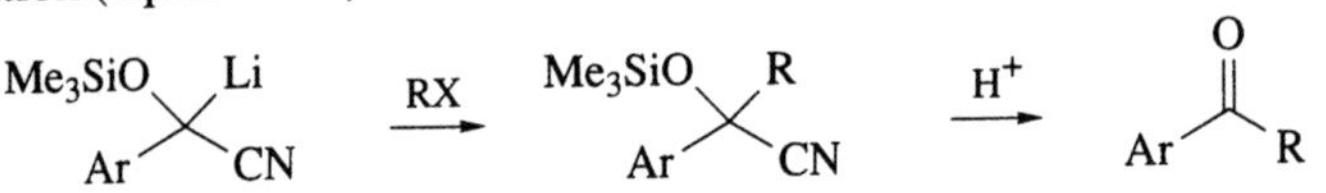

(**25**) (9)

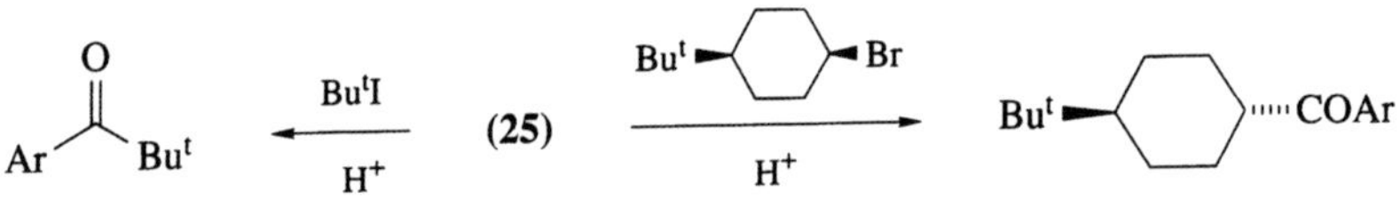

(**26**) (10)

An intramolecular S_N2 displacement at a neopentyl center in (**27**) gave (**28**), a precursor for the total synthesis of steroids (equation 11).[8]

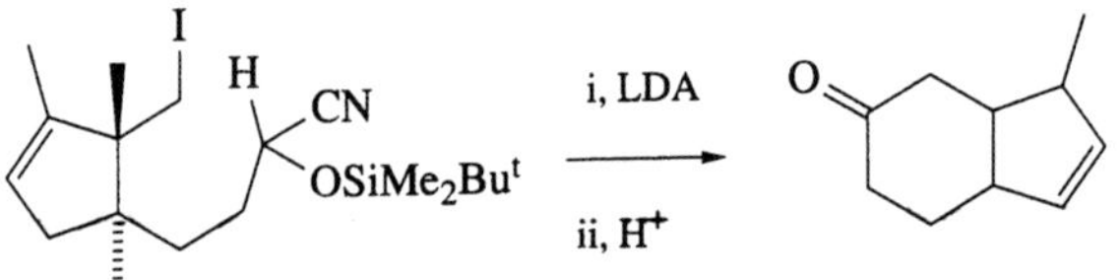

(**27**) (**28**) (11)

2.4.3.1.4 Miscellaneous reactions of O*-silylated cyanohydrins*

Oxidation of a cyanohydrin derived from a conjugated aldehyde (as the O-TMS derivative) using pyridinium dichromate (PDC) in DMF gave an α,β-unsaturated lactone (δ^2-butenolide) as the major product (equation 12).[68] Simple nonconjugated cyanohydrins are not satisfactory substrates for the synthesis of acyl cyanides using PDC, because they seem to add to the initially formed acyl cyanides, leading ultimately to cyanohydrin esters. Oxidation of cyanohydrin to acyl cyanides can be carried out either by means of manganese dioxide,[69] ruthenium-catalyzed oxidation with *t*-butyl hydroperoxide[70] or NBS.[71]

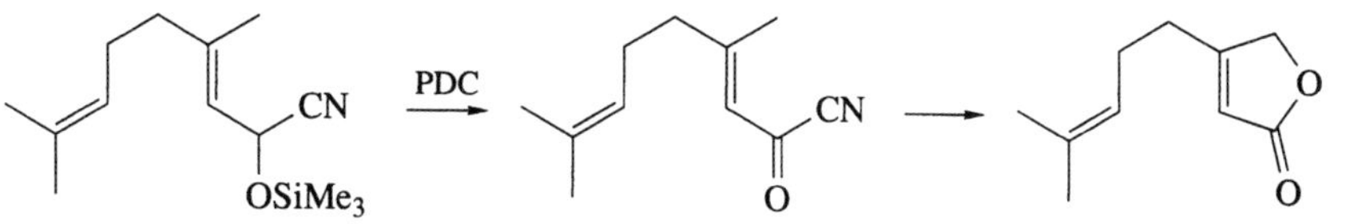

CN OSiMe$_3$ —PDC→ CN O → O O (12)

Tetronic acids and β-keto-γ-butyrolactones are easily prepared by reaction of an *O*-trimethylsilylated cyanohydrin with α-bromo esters in the presence of a Zn–Cu couple in a Reformatsky-type reaction (Scheme 12).[72,73]

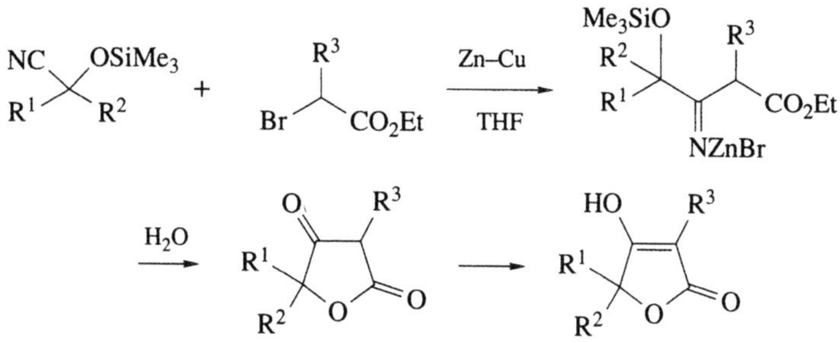

NC OSiMe$_3$ R^1 R^2 + Br CO$_2$Et R^3 —Zn–Cu, THF→ Me$_3$SiO R^2 R^1 R^3 CO$_2$Et NZnBr

—H$_2$O→ O R^3 R^1 R^2 O O → HO R^3 R^1 R^2 O O

Scheme 12

Acyloins are prepared in high yields by attack of a Grignard reagent on the cyano group of *O*-trimethylsilylated cyanohydrins. The method is particularly useful for the preparation of unsymmetrical acyloins.[74]

2.4.3.2 Cyanohydrin Ethers and Esters

Acid-catalyzed addition of aliphatic, aromatic or heteroaromatic cyanohydrins to ethyl vinyl ether, *n*-butyl vinyl ether or dihydro-4*H*-pyran provides base stable, protected cyanohydrin derivatives.[9] Phase transfer catalyzed alkylation of aliphatic cyanohydrins with allylic bromides gave α-substituted α-allyloxyacetonitrile.[75] Carbonyl compounds react with cyanide under phase transfer catalysis to give cyanohydrin anions, which are trapped by an acyl chloride or ethyl chloroformate to give acyl- or alkoxycarbonyl-protected cyanohydrins respectively.[34] The reduction of the carbonyl group of an acyl cyanide by $NaBH_4$ under phase transfer conditions followed by esterification serves as an alternative route to aldehyde-derived cyanohydrin esters.[76]

Cyanation of acetals was achieved either by means of *t*-butyl isocyanide or β-trimethylsilylethyl isocyanide in the presence of titanium(IV) chloride (equation 13)[77] or by TMS-CN in the presence of electrogenerated acid.[78]

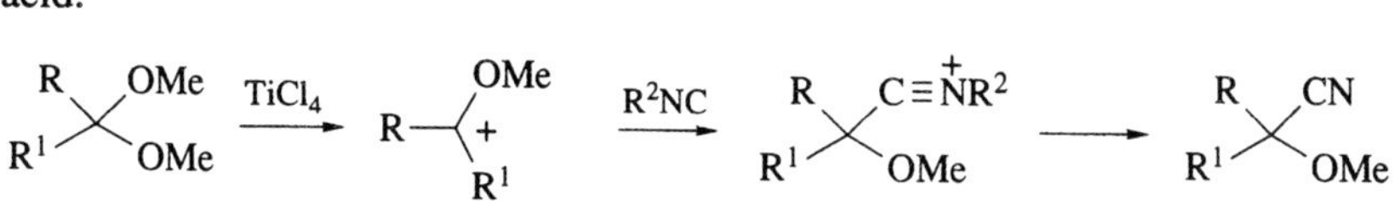

R OMe R^1 OMe —TiCl$_4$→ R–C$^+$(OMe)R^1 —R^{2}NC→ R C≡$\overset{+}{N}$R^2 R^1 OMe → R CN R^1 OMe (13)

2.4.3.2.1 1,2-Addition to aldehydes and ketones

The condensation of cyanohydrin ethers with aldehydes or ketones provides α-hydroxy ketones.[9] *O*-Benzoyl-protected cyanohydrins react with aldehydes to give α-hydroxy ketones *via* intramolecular deprotective benzoylation analogous to TMS-protected cyanohydrins (Scheme 10).[33]

A comparative study was reported using 1,3-dithianes and *O*-benzoylated cyanohydrins respectively for the synthesis of several analogs of secoisoquinoline alkaloids bearing a dimethylamino side chain and a benzilic or reduced benzilic group.[79]

2.4.3.2.2 1,4-Addition to α,β-unsaturated carbonyl compounds

The anions of protected cyanohydrins derived from saturated aliphatic aldehydes undergo competitive 1,2- and 1,4-addition to unsaturated carbonyl electrophiles. The proportion of the adducts appears to vary as a function of both structure and solvent. Steric interactions that favor dissociation of the reversibly formed 1,2-addition product increase the proportion of the 1,4-addition product. For example, increasing the size of the substituent R in a protected cyanohydrin (**29**) from methyl to *n*-pentyl increases the ratio of 1,4-addition product (**30**) to 1,2-addition product (**31**) from 1.5:1 to 2.7:1 (equation 14).[62]

LDA

(**29**) (**30**) (**31**) (14)

Conjugate addition of an aryl cyanohydrin has been used in the synthesis of tetracyclines, 13-oxyprostanoids and β-cuparenone.[8]

2.4.3.2.3 Alkylation with R—X

The alkylation of protected cyanohydrin anions constitutes an excellent method for ketone synthesis.[9,62] Generally the anions are generated from aliphatic or aromatic aldehyde protected cyanohydrins with LDA under nitrogen at –78 °C. The addition of an alkyl halide produces the protected ketone cyanohydrin. The carbonyl group is then liberated by successive treatment with dilute acid and dilute aqueous base. This method is applicable for the synthesis of buflomedil.[80]

Regiocontrolled synthesis of 2,4-disubstituted pyrroles is achieved using the alkylation of a protected cyanohydrin with an alkynyl bromide (equation 15).[81]

i, LDA; ii, Br–CH$_2$–C≡C–R^1 → i, $LiAlH_4$; ii, Hg^{2+}/H^+ (15)

EE = ethoxyethyl

The side chain of brassinolide, (20*S*,22*R*), can be stereoselectively introduced by alkylation (S_N2) of the (20*R*)-tosyloxy steroid with a protected cyanohydrin followed by the stereoselective reduction of the resulting 23-en-22-one.[82]

α-Hydroxy-γ-butyrolactones are prepared by the alkylation of a protected cyanohydrin with an epoxide.[83] Maldonado *et al.* applied this method to the synthesis of α-bisbololone (equation 16).[84]

i, LDA; ii, (epoxide) (16)

Of special interest is the use of protected cyanohydrins in the formation of carbocyclic rings. Ring closure of an acyclic intermediate to form a five-membered ring (75–85%) has been described in the synthesis of prostaglandins (equation 17).[85,86] In addition this method is applicable to the formation of cyclopropyl,[87] cyclobutyl[85] and cyclohexyl rings (60–70%).[88]

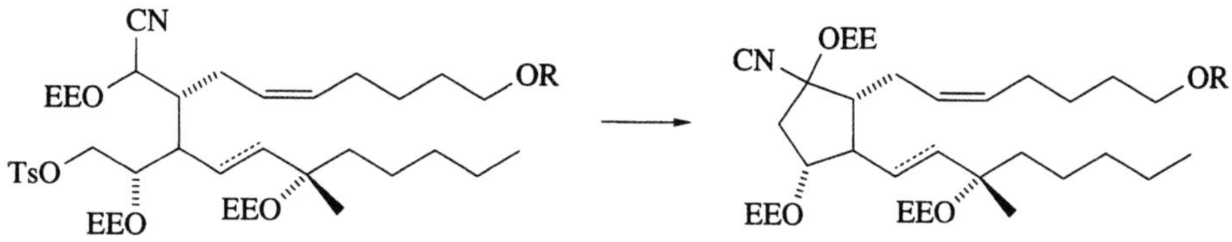

(17)

Macrocyclic ketones are prepared by intramolecular alkylation without the use of high dilution conditions.[89] These cyclizations have the following characteristic features:[85,87] (i) the alkylation is irreversible and very rapid, and the cyclized product is stable under basic conditions. Therefore, this method of cyclization requires short reaction times and no high dilution conditions and provides the macrocycle in a satisfactory yield; (ii) the carbanion derived from an α,β-unsaturated cyanohydrin acts *via* α-attack, so that γ-attack and isomerization of double bonds are not observed; and (iii) the cyclized products are easily converted by mild acid and base treatment to the corresponding enones in high yields with excellent stereoselectivity. This cyclization method can be applied to the synthesis of a wide variety of naturally occurring compounds such as the pheromone periplanone B,[90] the sesquiterpenes acoragermacrone (equation 18),[91] germacrone,[92] humulene,[93] mukulol,[91] germacrone lactones such as costunolide and haagenolide,[94] (*E,E*)- and (*E,Z*)-2,6-cyclodecadienones,[95] *trans*-2-cyclopentadecanone, a precursor of the natural products (+)-muscone and exaltone,[89] zearalenone,[96] dihydroxy-*trans*-resorcylide[97] and the 14-membered (*E,E,E*)-macrocyclic triene, a precursor for the Diels–Alder synthesis of a steroid skeleton (equation 19).[98] Such cyclizations have been carried out on a 60 mmol scale.[90]

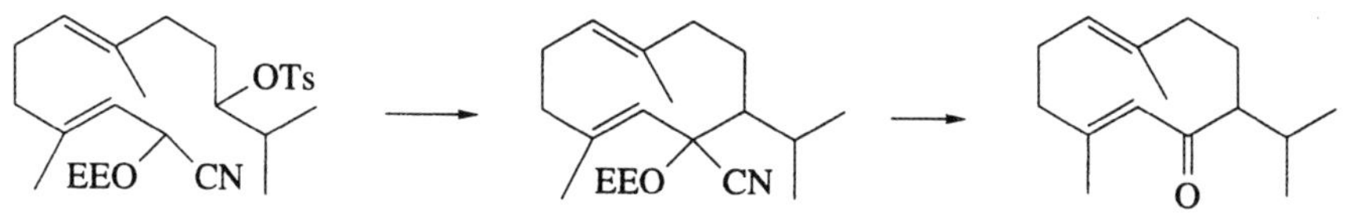

(18)

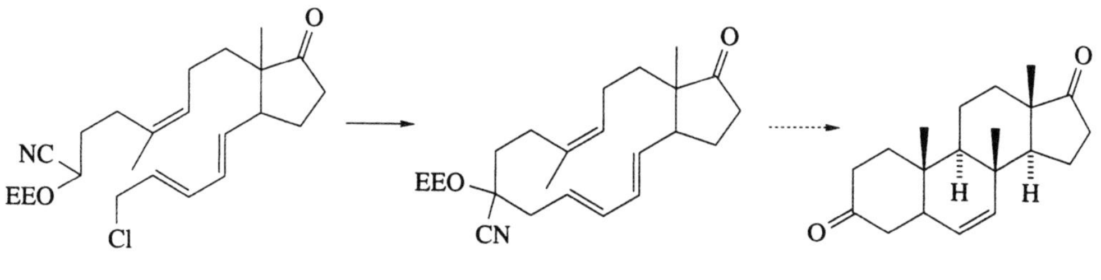

(19)

2.4.3.2.4 Miscellaneous reactions

The classic S_NAr substitution of activated aryl halides by protected cyanohydrin anions provides substituted benzophenones.[8] Another procedure for the arylation of protected cyanohydrin anions involves the use of aromatic substrates activated as their π-chromium tricarbonyl complexes.[9,99] Addition of the anion of (**32**) to the 1,3-dimethoxybenzene complex, for example, leads principally to the *meta*-substituted isomer (**33**; equation 20). Preferential *meta* regioselectivity is also noted with other π-chromium tricarbonyl complexes of arenes. Other arylations of cyanohydrin anions include interesting but synthetically limited additions at the α-position of quinoline *N*-oxides.[8] In a similar manner, cyanohydrin carbonates of aromatic aldehydes react with *N*-oxides of quinoline and isoquinoline.[34]

β,γ-Unsaturated ethers of cyanohydrins, on formation of lithio derivatives, undergo a 2,3-sigmatropic rearrangement[8] to form β,γ-unsaturated ketones (equation 21), whereas benzylic ethers of aliphatic cyanohydrins gave *o*-methylaryl ketones.[8] The method has been used to prepare 3-methyl-1-(3-methyl-2-furyl)-1-butanones, a naturally occurring C_{10} terpene,[8] α-allenic ketones[8] and enolic monoethers of γ-keto aldehydes *via* 2,3-sigmatropic rearrangement of their respective carbanions.[8]

An annelation reaction of α,β-unsaturated carbonyl compounds with cyanophthalides (**34**) can be used for the construction of naphthoquinone and anthraquinone systems in biologically active natural

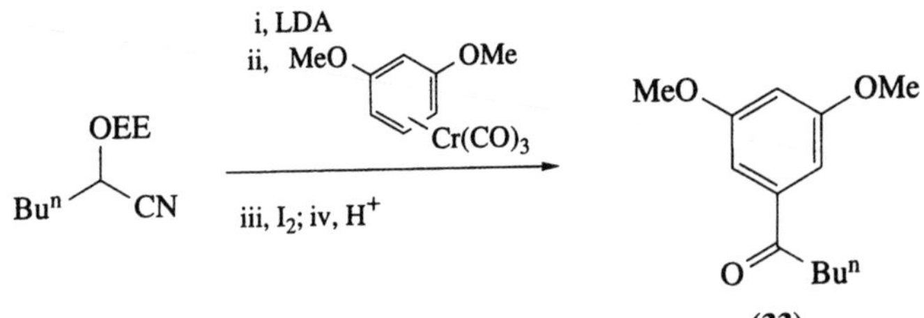

(20)

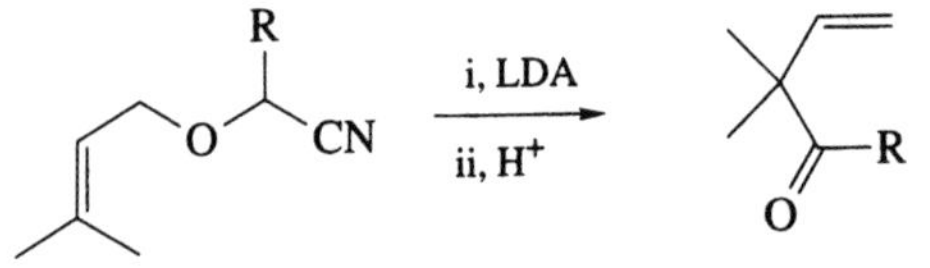

(21)

products.[100] For example, synthesis of racemic 1-fluoro-, 4-fluoro-, 2,3-difluoro- and 1,4-difluoro-4-demethoxy-daunomycinones (**35**) is achieved by annelation of highly functionalized quinone monoketal with the appropriate cyanophthalide anion followed by deprotection (equation 22).[101,102] Annelative reactions of arynes generated *in situ* from haloarenes and LDA in THF with lithiated cyanophthalide provide a convenient way of preparing a wide variety of anthraquinones and anthracyclinones.[103] More recently, Yoshii *et al.* utilized a tetracyclic cyanophthalide anion in an annelative reaction with 5-*t*-butoxy-2-furfurylideneacetone in their synthesis of (±)-granticin.[104]

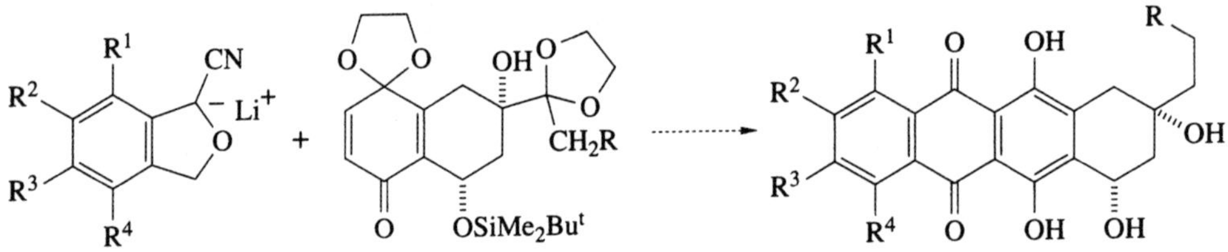

(22)

2-Substituted cyclohexenones are prepared by the alkylation of 2-cyano-6-methoxytetrahydropyran, a cyclic protected cyanohydrin (Scheme 13).[105]

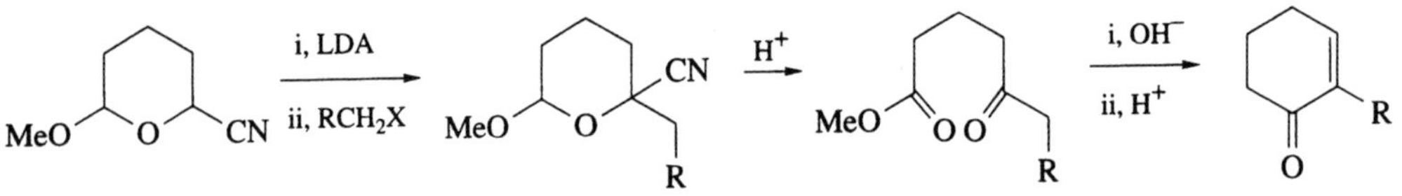

Scheme 13

An interesting variant of these Michael-type additions is the 1,6-addition of the anion of type (**29**) to a dienyl sulfone as a route to tagetones[106] and 1,4-addition to nitrostyrene to form β-nitro ketones.[107] A key step in a synthesis of 11-deoxyanthracyclinone involves the regioselective reaction of complexed styrene with lithiated protected acetaldehyde cyanohydrin.[108]

2.4.4 α-(DIALKYLAMINO)NITRILES

From a historical perspective, the α-(dialkylamino)nitrile anions were the first acyl anion equivalents to undergo systematic investigation.[9] More recent studies indicate that anions of α-(dialkylamino)nitriles derived from aliphatic, aromatic or heteroaromatic aldehydes intercept an array of electrophiles[9] including alkyl halides, alkyl sulfonates, epoxides, aldehydes, ketones, acyl chlorides, chloroformates, unsaturated ketones, unsaturated esters and unsaturated nitriles. Aminonitriles are readily prepared and their anions are formed with a variety of bases[9] such as sodium methoxide, KOH in alcohol, NaH, LDA, PhLi, sodium amide, 70% NaOH and potassium amide. Regeneration of the carbonyl group can be achieved either by the usual acid hydrolysis or by other mild hydrolytic agents such as copper sulfate,[109] iron(II) sulfate,[109] copper acetate,[110] silver nitrate in D_2O–THF–diethyl ether[111] and silica gel.[111]

Aromatic and aliphatic aldehydes in the presence of dialkylamines and an equivalent of acid such as hydrochloric, perchloric or *p*-toluenesulfonic acid give iminium salts, which add cyanide ion to form α-(dialkylamino)nitriles.[9] An alternative preparation involves the reaction of the aldehyde with dialkylamines in the presence of acetone–cyanohydrin, α-(*N,N*-dialkylamino)isobutyronitriles,[112] diethyl phosphorocyanidate[113] or TMS-CN.[114] Another route to α-aminonitrile starts with an aldehyde, the salt of an amine and KCN in organic solvents under solid–liquid two-phase conditions by combined use of alumina and ultrasound.[115] Chiral α-aminonitriles were prepared by Strecker-type reactions,[116] cyanosilylation of Schiff's bases,[117] amination of α-siloxynitriles[118] or from an *N*-cyanomethyl-1,3-oxazolidine synthon.[119] Reaction of tertiary amines with ClO_2 in the presence of 5.7 mol equiv. of aqueous NaCN as an external nucleophile affords α-aminonitrile.[120]

2.4.4.1 1,2-Addition to Aldehydes and Ketones

α-Hydroxy ketones can be prepared by the addition of carbanions derived from easily accessible α-(dialkylamino)nitriles to carbonyl compounds followed by hydrolysis.[121] Optically active α-hydroxy ketones are obtained by the nucleophilic acylation of chiral α-aminonitriles.[122] 1,2-Addition to cyclic ketones, aryl aldehydes and aliphatic aldehydes leads to ethanolamines.[8a] Stereoselectivity in favor of the *threo* isomer occurs with α-(dimethylamino)propionitrile, while reversed stereoselectivity is observed with the open chain Reissert compound (**36**; Scheme 14).[8b] A high degree of stereoselectivity is also observed in the condensation of *N*-benzoyl-2-cyanopiperidine with propanal, a key step in the synthesis of (+)-conhydrine (*erythro*-α-ethyl-2-piperidinemethanol) (**37**; Scheme 15).[8]

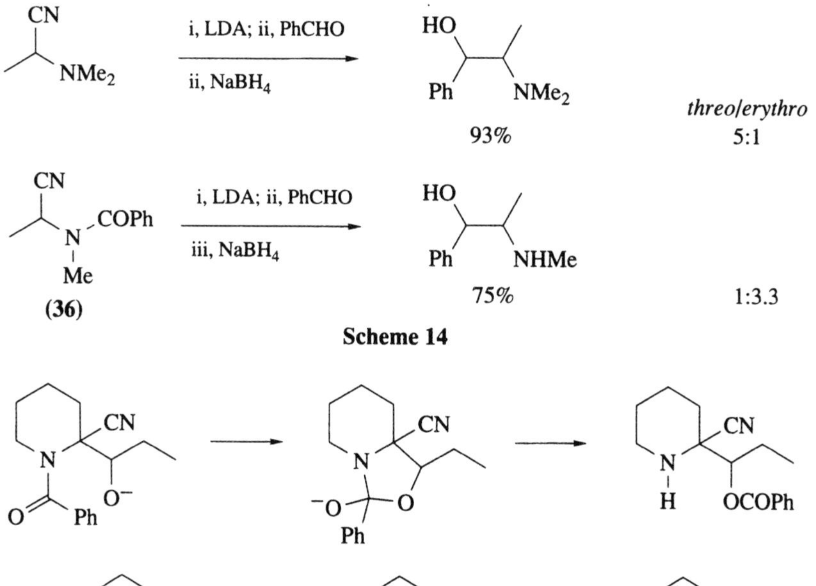

Scheme 14

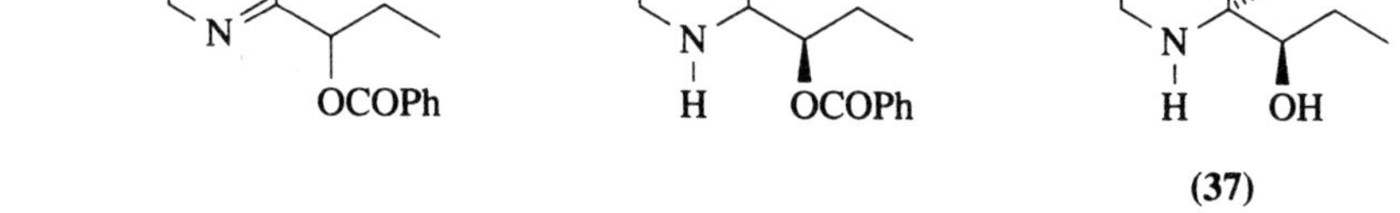

Scheme 15

Unsymmetrical amino ketones were prepared by 1,2-addition of an α-aminonitrile to an aldehyde (Scheme 16).[123]

α-(Dialkylamino)nitriles (**38**) can react as either acyl anion or β-homoenolate ion equivalents on varying the reaction conditions.[8] This constitutes a general entry to the formation of cyclopentenone derivatives (Scheme 17).

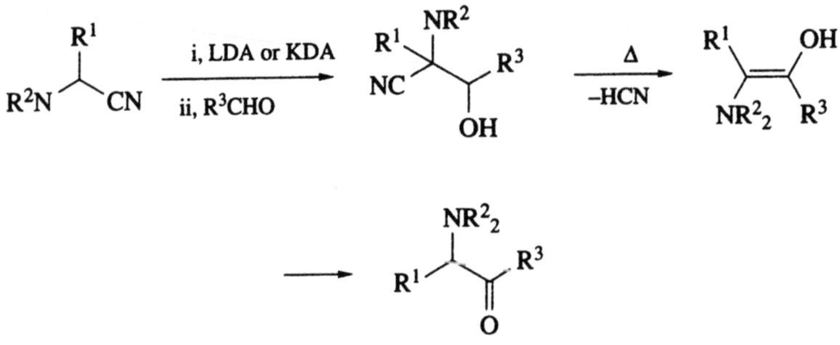

Scheme 16

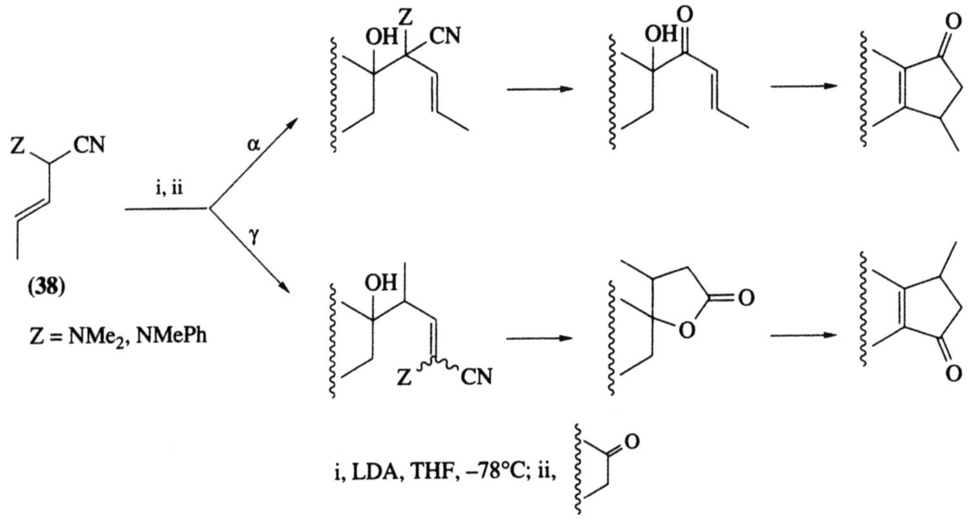

i, LDA, THF, –78°C; ii,

Scheme 17

The lithium salt of the unsaturated α-aminonitrile (**39**) reacts as a homoenolate (γ-attack) with aldehydes or ketones to give 1,2-addition products (**40**). Deprotonation of the masked carbonyl affords substituted lactones (equation 23).[8]

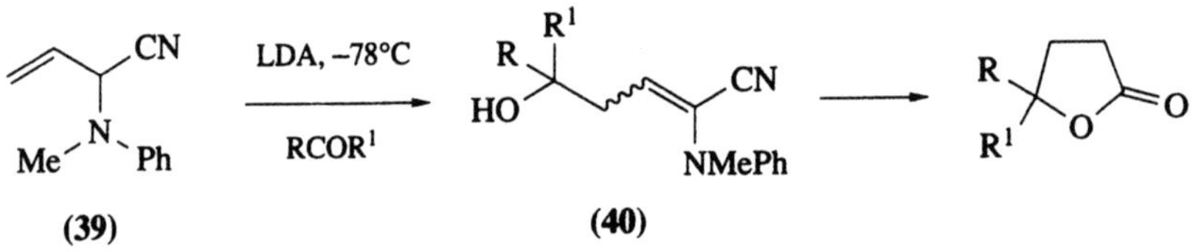

(23)

2.4.4.2 1,4-Addition to α,β-Unsaturated Carbonyl Compounds

Addition of α-aminonitriles to an α,β-unsaturated ketone affords 1,4-diketones. High yields are also observed in 1,4-additions to ethyl acrylate or acrylonitrile.[7,109] α-Phenyl-α-(*N*-morpholino)acetonitrile gives good yields of 1,4-adducts with methyl crotonate, methyl methacrylate, diethyl maleate and ethyl propiolate under catalysis of sodium methoxide in THF. Reaction of the lithium salt of (**41**) with methyl acrylate is reported to give pyrrole derivatives (Scheme 18).

Conjugate addition of the benzoyl anion equivalent (**42**) to unsubstituted and 2-substituted cyclic enones can be highly stereoselective (equation 24).[124] Lewis acids such as $BF_3{\cdot}Et_2O$, $Ti(OPr^i)_4$ or $ZnCl_2$ make possible the Michael addition of (**42**) even to β,β-disubstituted α-enones.[125]

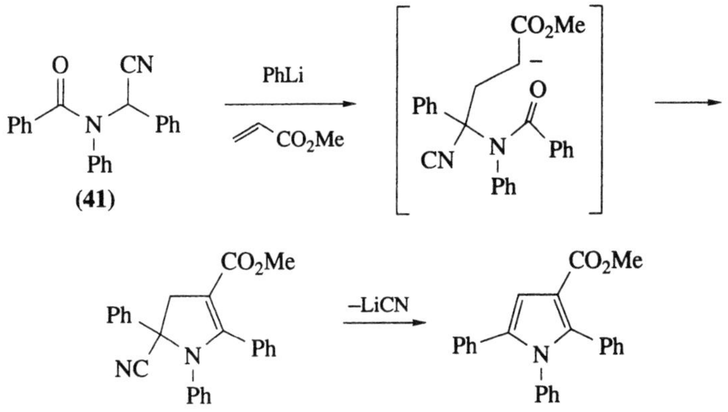

Scheme 18

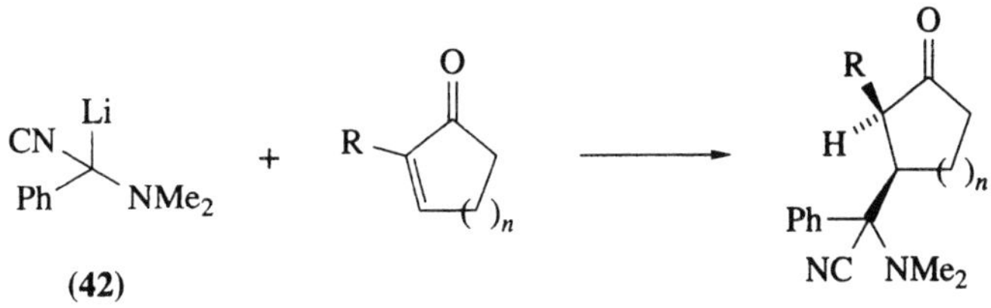

(24)

2.4.4.3 Alkylation with R—X

Alkylation of metallated α-(dialkylamino)nitriles with alkyl halides, epichlorohydrin, allyl halides or ethyl bromoacetate followed by hydrolysis furnishes an array of ketones in good yield.[8,9] For example, reaction of (**42**) with benzyl chloride leads to the formation of deoxybenzoins.[8] Alkylation of α-(dialkylamino)nitriles is also possible if the latter are derived from aliphatic aldehydes.[8,9]

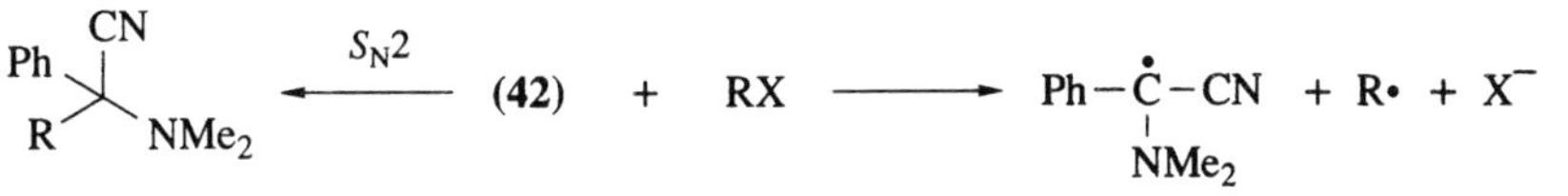

Scheme 19

The alkylation of (**42**) with chiral 1-methylheptyl halides in liquid ammonia proceeds with partial inversion of configuration. The accompanying racemization is dependent on the basic reagent, on the configuration of the alkylating agent and on the extent of participation of the electron transfer process in the alkylation (Scheme 19).[111]

The key step for the conversion of open-chained α,β-unsaturated aldehydes to their corresponding cyclized carbonyl compounds involves the regioselective alkylation of (**39**) with dihalides.[126]

2-Alkyl- and 2,6-dialkyl-piperidine alkaloids have been synthesized by the alkylation of cyclic α-aminonitriles such as *N*-benzyl-2-cyano-6-methylpiperidine[127,128] or 1-benzyl-2,6-dicyanopiperidine.[129]

Unsaturated α-(dialkylamino)nitriles (**43**) form ambident anions which are alkylated at either the α- or γ-position.[8] Alkylation of (**43a**) with methyl iodide is completely selective to give only the γ-alkylated regioisomer, while alkylation of (**43b**) with ethyl bromide gave exclusively the α-alkylated product (equation 25). The structure of the secondary α-amino group as well as the steric bulk of the alkylating reagent influence the regioselectivity. With dimethylamine or piperidine as the amine component, the alkylation with methyl iodide gives approximately equal amounts of α- and γ-alkylation product, while with the morpholino derivative (**43b**) α-alkylation is the major product. Increased amounts of γ-alkylation are observed with isopropyl bromide.

Ring closures through intramolecular alkylation of Reissert compounds have been described.[8] The sequential dialkylation of dimethylaminoacetonitrile (**44**) first with methallyl chloride and then isopentenyl bromide, followed by hydrolysis and isomerization, affords artemesia ketone in 77% overall yield (Scheme 20).[130]

3-Substituted pyrazoles can be prepared by the alkylation of α-(dimethylamino)nitriles with bromodiethyl acetal followed by heating with hydrazine dihydrochloride in ethanol.[131] 3,5-Dialkylpyrazoles were

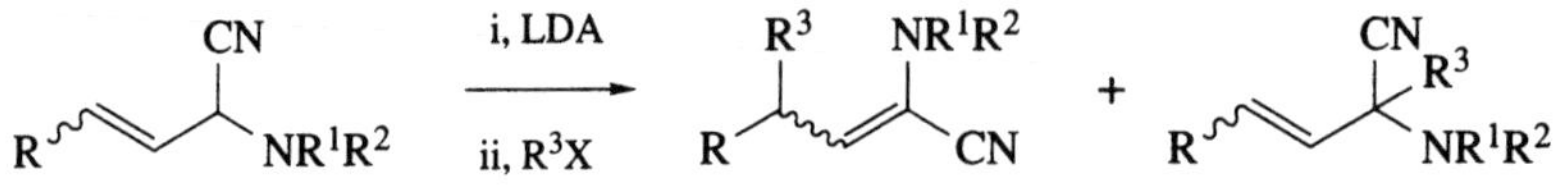

CN
R NR[1]R[2]
i, LDA
ii, R[3]X
R[3] NR[1]R[2]
R CN
+
CN R[3]
R NR[1]R[2]
(25)

(**43a**) R = H; R^1 = Me; R^2 = Ph

(**43b**) R = Me; R^1R^2 = $-(CH_2)_2O(CH_2)_2-$

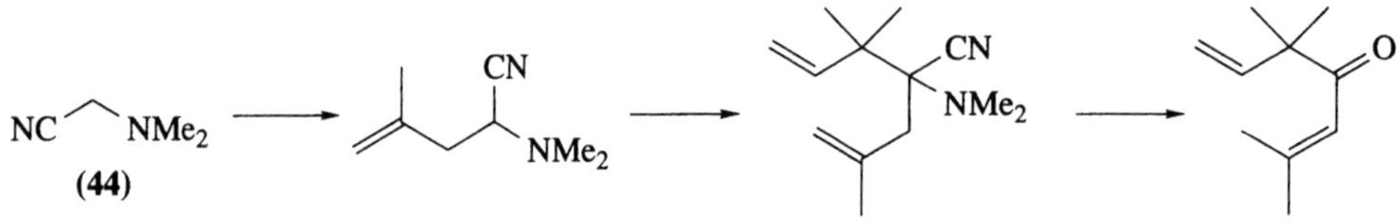

NC NMe$_2$
(**44**)
CN
NMe$_2$
CN
NMe$_2$
O

Scheme 20

synthesized by reaction of α,γ-dialkyl-α,γ-dipyrrolidinylglutaronitriles with hydrazine.[132] Vinyl ketones in general and the himachalene skeleton specifically can be synthesized by the alkylation of (**45**) with an alkyl halide, *e.g.* (**46**; Scheme 21).[133]

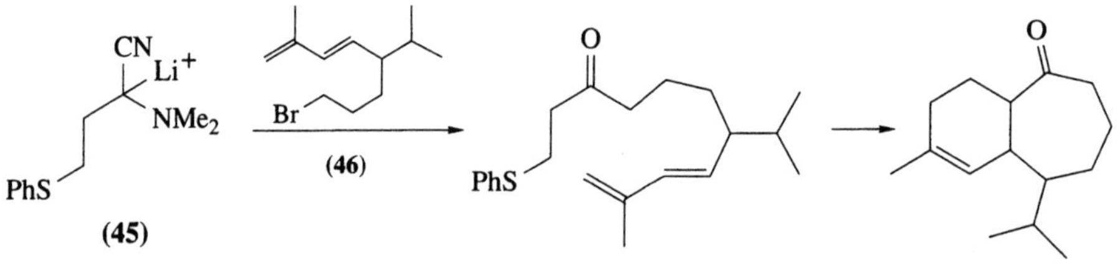

CN
Li$^+$
NMe$_2$
PhS
(**45**)
Br
(**46**)
O
PhS
O

Scheme 21

Reaction of 1-benzyl-2,5-dicyanopyrrolidine with an alkyl halide gives unsymmetrical 2,5-dialkylated products (**47**) in high yield. Hydrolysis of (**47**) provides a γ-diketone, which serves as a precursor for jasmone analogs (Scheme 22).[134]

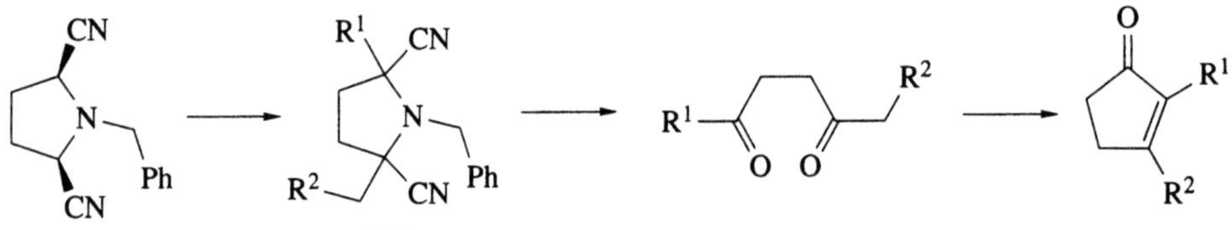

CN
N
Ph
CN
R[1] CN
N
R[2] CN Ph
(**47**)
R[2]
R[1]
O O
O
R[1]
R[2]

Scheme 22

δ-Diketones and their aldol-derived cyclohexenones are obtained from alkylation of 1-benzyl-2,6-dicyanopiperidine followed by hydrolysis.[135,136] This dicyano analog of (**47**) provides a new method for the synthesis of *trans*-2-methyl-6-undecylpiperidine (solenopsin), α-propylpiperidine (coniine) and other 2,6-dialkylpiperidine alkaloids.[129]

The pyrrolidine synthon (**48**)[137] is useful for the chiral synthesis of pyrrolizidine,[138] pyrrolidine,[139,140] indolizidine[138] and azabicyclic alkaloids[141] (see Scheme 23).

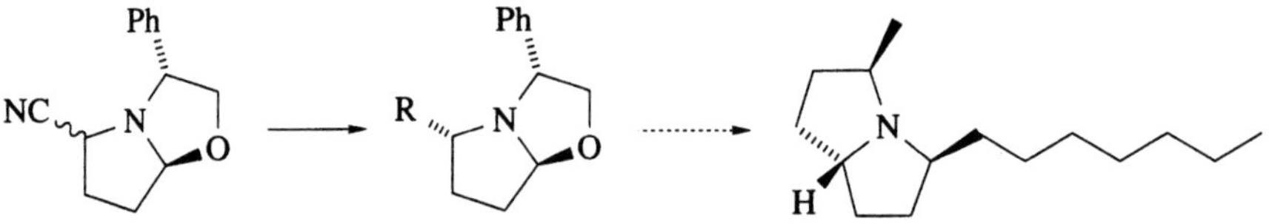

Ph
NC N
O
(**48**)
Ph
R N
O
N
H

Scheme 23

In a similar manner, the 2-cyano-6-oxazolopiperidine synthon is useful for the chiral synthesis of indolizidine (monomerine I),[142] piperidine [(+)- and (–)-coniine and dihydropinidine][143] and quinolizidine alkaloids.[144,145] 2-Hydroxymethyl-1-amino-1-cyclopropanecarboxylic acid[146] (–)-(2*R*)-hydroxy-(3*S*)-nonylamine[147] and α-substituted phenylethylamines[148] are obtained in optically active form from (–)-*N*-cyanomethyl-4-phenyloxazolidine.

Alkylation of the α-aminonitrile (**49**) with propyl bromide followed by reductive decyanation gives the key intermediate (**50**), which on treatment with diethyl cyanophosphonate leads to the formation of a new α-aminonitrile (**51**). Compound (**51**) reacts with K and 18-crown-6 in THF to form gephyrotoxin-223AB, an indolizidine alkaloid (Scheme 24).[149]

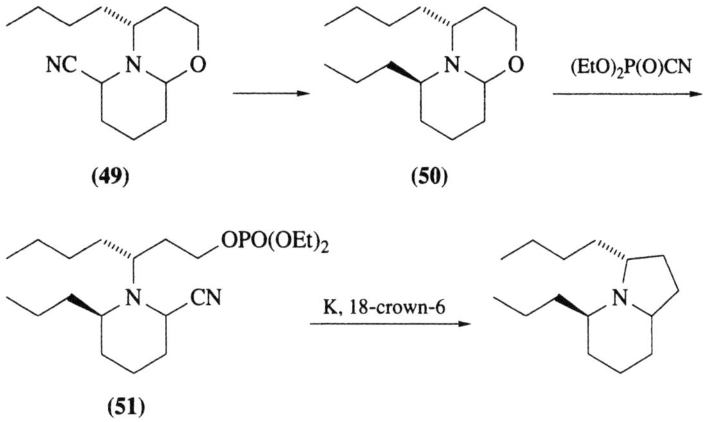

NC N O (**49**) → N O (**50**) $(EtO)_2P(O)CN$ →

$OPO(OEt)_2$ N CN (**51**) K, 18-crown-6 → N

Scheme 24

α-Ethoxyurethanes react with TMS-CN in the presence of Lewis acid to afford the corresponding α-cyanourethanes, analogous to α-aminonitrile, which *via* the carbanion are transformed to α-alkylated products in moderate yield.[150] This synthon is useful for the synthesis of coniine and *trans*-quinolizidines.[151]

2.4.4.4 Miscellaneous Reactions of α-Aminonitriles

1,2,5-Trisubstituted pyrroles are obtained in 50–100% yields by the addition of open chain analogs of a Reissert compound to the vinyltriphenylphosphonium cation, with subsequent cyclization by an intramolecular Wittig reaction and base-catalyzed elimination of HCN (Scheme 25).[152]

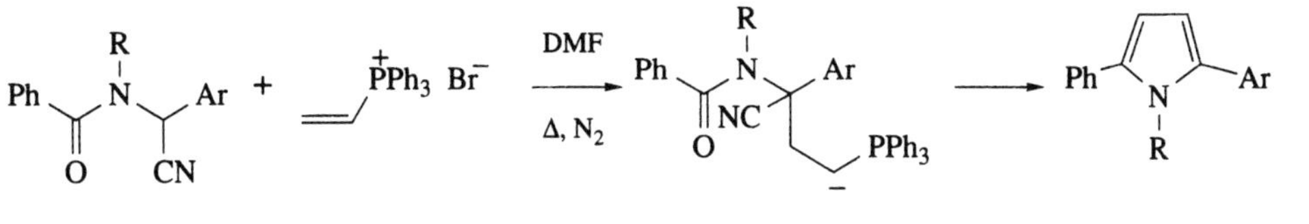

Ph(C=O)N(R)CH(CN)Ar + $CH_2=CH\overset{+}{P}Ph_3$ Br^- —DMF, Δ, N_2→ Ph(C=O)N(R)C(NC)(Ar)$CH_2\overset{-}{C}HPPh_3$ → Ph, Ar, N, R (pyrrole)

Scheme 25

Allylamines are prepared regio- and stereo-selectively by reaction of Grignard reagents with an α-aminonitrile.[153] α-Phenyl-4-morpholinoacetonitrile (**52**) reacts with 4-chlorobenzoyl chloride to give diketone (**53**; equation 26).[109]

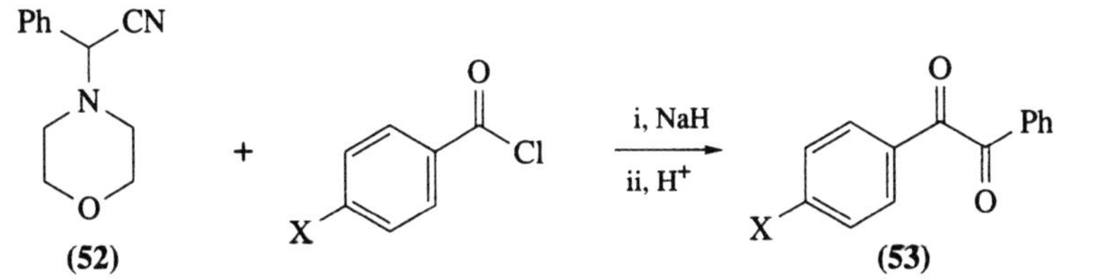

Ph CN N O (**52**) + X–C_6H_4–COCl —i, NaH; ii, H^+→ X–C_6H_4–CO–CO–Ph (**53**) (26)

The phenylcyanoamine derivative (**54**) has been used as a benzoyl anion equivalent in an intramolecular process to effect the replacement of the hydroxy group of enols or phenols by the benzoyl group (Scheme 26).[154]

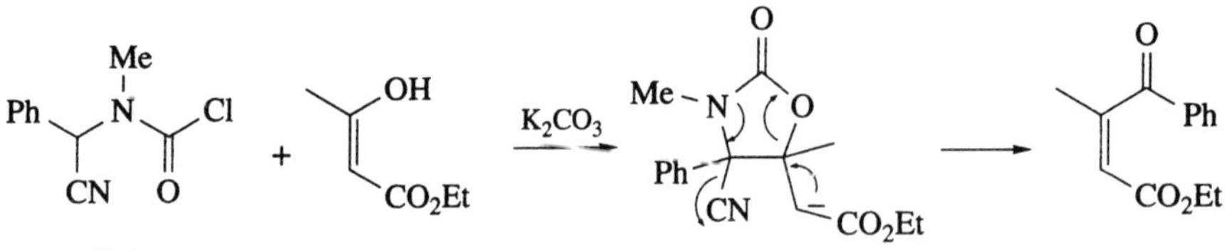

(**54**)

Scheme 26

A particularly interesting development involves the alkylation of the nitrogen of an α-aminonitrile with allylic halides and subsequent 2,3-sigmatropic rearrangement of the allylic ammonium ylides.[9] This method provides a convenient route to 2-methyl-3-formylpyridines or 2-methyl-3-acylpyridines. With the use of 1-(cyanomethyl)pyrrolidine, the reaction with various allylic halides proceeds with a stereochemical bias that varies with the substrate to give β,γ-unsaturated aldehydes, as illustrated in the preparation of (**55**; equation 27). The procedure also succeeds with benzylic halides to furnish *ortho*-oriented carbonyl and methyl groups on an aromatic ring.

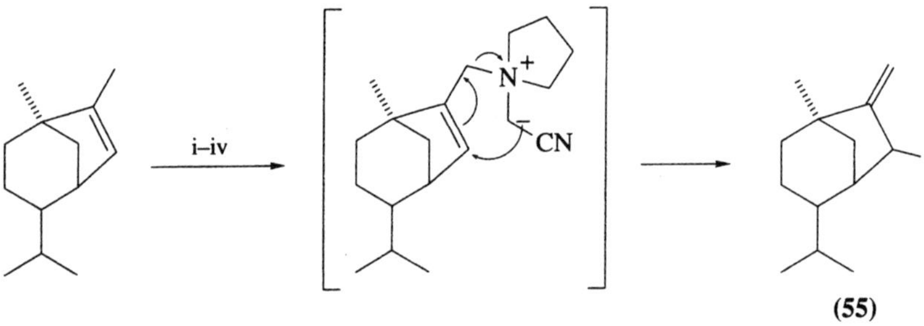

(**55**) (27)

i, NBS^-; ii, 1-(cyanomethyl)pyrrolidine; iii, Bu^tOK; iv, H^+

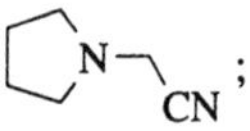

Ester-stabilized ammonium ylides of α-(dialkylamino)nitriles, formed by treatment of ammonium salts with DBU (1,5-diazabicyclo[5.4.0]undec-5-ene), undergo spontaneous fragmentation to give α,β-unsaturated nitriles.[8]

Asymmetric induction is observed in 2,3-sigmatropic rearrangements *via* chiral ammonium chlorides such as (**56**), which was obtained from (*S*)-proline ethyl ester.[8] Ylide formation with potassium *t*-butoxide, rearrangement and acid hydrolysis of the aminonitrile affords (*R*)-(+)-methyl-2-phenyl-3-butenal (90% optical purity) (equation 28). Application of this sequence furnishes intermediates in a new approach to steroids, to various sesquiterpenes such as α-sinsenal and to a highly substituted pyridine, as part of an impressive synthesis of streptonegrin.[9]

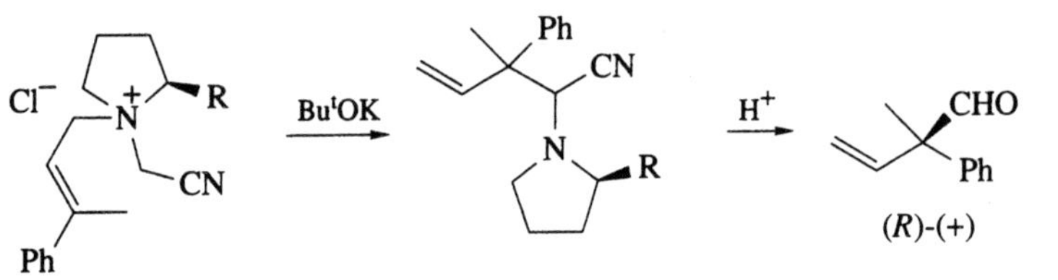

(**56**) $R = CH_2OCH_2Ph$ (*R*)-(+) (28)

2.4.5 CYANOPHOSPHATES

Cyanophosphates can be prepared from aromatic or α,β-unsaturated aldehydes with diethyl phosphorocyanidate (DEPC, $(EtO)_2P(O)CN$) in THF using LiCN[37] or LDA as a base.[155] Deprotonation of a cyanophosphate with Bu^nLi in the presence of tetramethylethylenediamine in THF at –78 °C, followed by reaction with alkyl (or acyl) halides, carbonyl compounds and α,β-unsaturated nitriles or esters, as well as cyanophosphates themselves, affords alkylated (or acylated) products, mixed benzoin and acyloin phosphates, polysubstituted cyclopropanes and diarylfumaronitriles in moderate yields respectively (Scheme 27).[37,156]

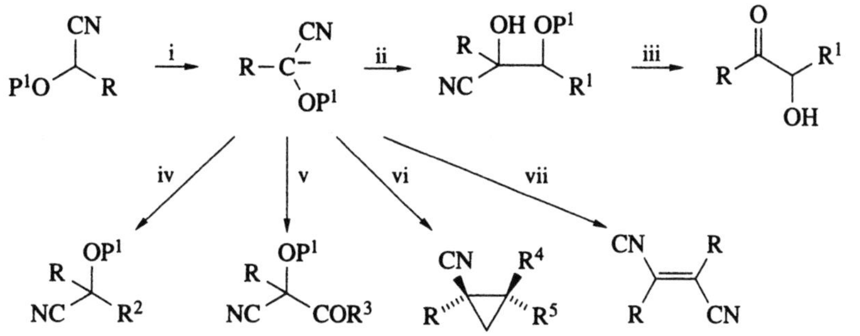

i, $Bu^nLi/TMEDA$; ii, R^1CHO; iii, NaOH; iv, R^2X; v, R^3COX;
vi, $CH_2{=}CR^4R^5$, R^4 = H, Me, R^5 = CN, CO_2Et; vii, $RCH(OP^1)CN$

Scheme 27

2.4.6 α-THIONITRILES

α-Thionitriles also serve as acyl anion equivalents. The parent compound $RSCH_2CN$ is accessible by the reaction of $RSCH_2Cl$ with $Hg(CN)_2$, or by an S_N2 process involving $ClCH_2CN$ and RSNa. Aliphatic nitriles on treatment with two equivalents of LDA followed by RSSR afford α-thionitriles (**57**), as does the reaction of thioacetals (alkyl or aryl) with TMS-CN in the presence of $SnCl_4$ (Scheme 28).[157] They can also be prepared by the reaction of thioacetals with $Hg(CN)_2$ and iodine[158] or from a 1-(phenylthio)vinylsilane with TMS-CN *via* a thionium ion intermediate.[159] Alternatively such compounds are prepared by the reaction of an aldehyde with phenyl thiocyanate and tributylphosphine.[160]

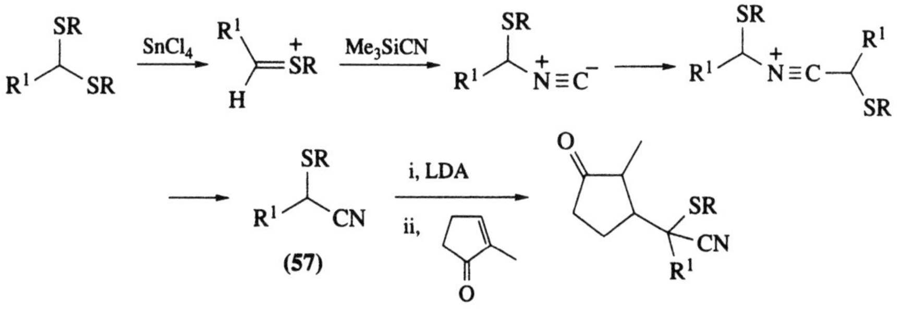

Scheme 28

Reaction of an α-(phenylthio)nitrile with base, such as LDA, followed by an aldehyde or a ketone gives a 1,2-addition product in good yield, while the use of α,β-unsaturated aldehydes and ketones usually leads to the 1,4-addition product.[161] Conjugate addition of the α-(phenylthio)acetonitrile anions to 2-methyl- or 2-phenyl-2-cyclohexenones or 2-methyl-2-cyclopentenones, followed by acid quenching under kinetic control, leads to different ratios of *cis*- and *trans*-2,3-disubstituted cyclanones depending on ring size.[162]

An intramolecular alkylation of α-thionitrile (**58**) leads to a macrocycle, an intermediate for the synthesis of the sesquiterpene (–)-dihydrogermacrene D (Scheme 29).[163] Alkylation of an α-cyanosulfone followed by reduction gave α-substituted acetonitriles.[164] Cyclization of a 1,3-dibromoalkane with (methylthio)acetonitrile yields a 1-cyano-1-(methylthio)cyclobutane, a precursor for the synthesis of cyclopentanones.[165] Reaction of α-(arylthio)nitriles with aryl halides in the presence of LDA and Bu_3P at –78 °C affords cyanostilbenes.[160]

A chiral cyclopentenoid building block (**60**) has been reported to result in a one-pot cyclization process from epoxide (**59**) (readily accessible from (*R*,*R*)-(+)-tartaric acid) with the carbanion derived from (phenylthio)acetonitrile (equation 29).[166]

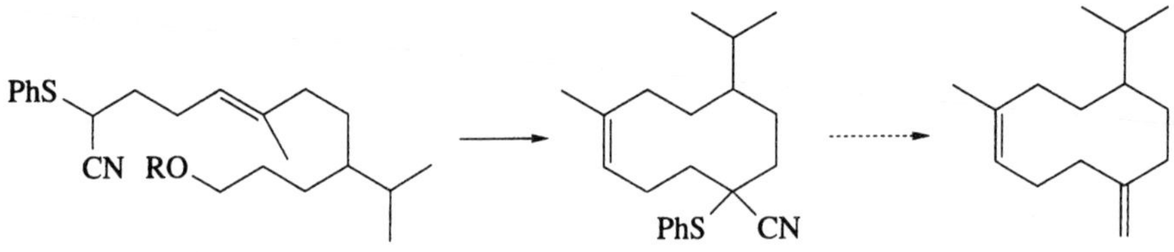

(58)

Scheme 29

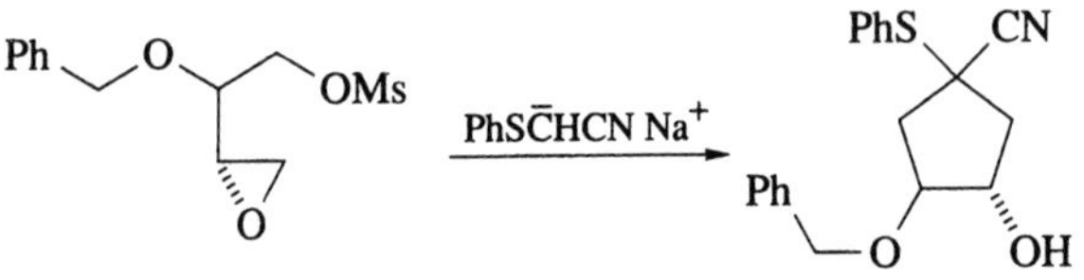

(59) (60) (29)

2.4.7 α-(ARYLSELENO)NITRILES

Cyanoselenation of aldehydes with phenyl selenocyanate in the presence of Bu_3P gave α-(arylseleno)nitriles.[167] Selenylation of nitriles after treatment with 2 equiv. of base also led to the formation of α-selenonitriles.[168] Their lithiated derivatives undergo smooth alkylation with methyl iodide (80%) and Michael addition to cyclohexenone in 90% yield (Scheme 30).[167]

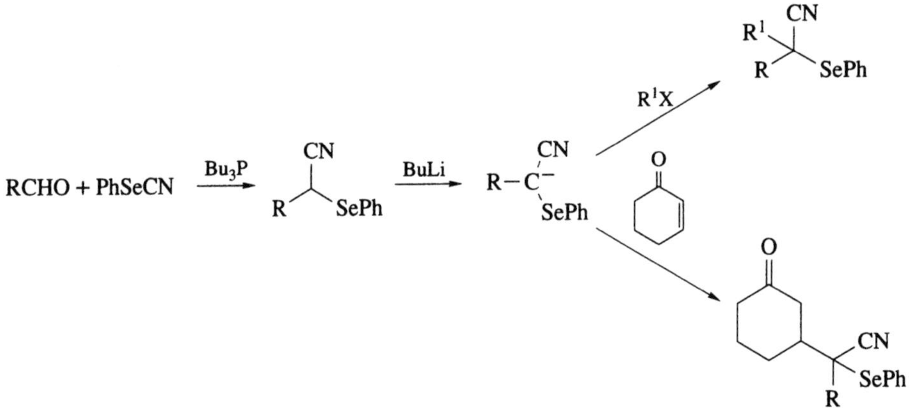

Scheme 30

2.4.8 α-HETEROSUBSTITUTED PHOSPHONATE CARBANIONS AS ACYL ANION EQUIVALENTS

A system analogous to a cyanohydrin is a siloxyalkylphosphonate. Thus, the diethyl 1-phenyl-1-trimethylsiloxymethylphosphonate carbanion derived from **(61)** on treatment with LDA[169] acts as an effective acyl anion equivalent in the preparation of α-hydroxy ketones and ketones.[38] The phosphonate **(61)** is formed from triethyl phosphite, benzaldehyde and TMS-Cl in excellent yield.[170]

2.4.8.1 1,2-Addition to Aldehydes and Ketones

When the carbanion of **(61)** reacts with aldehydes or ketones, a 1,4-silicon migration from O to O with subsequent loss of diethyl phosphite is observed (Scheme 31).[38] Hydrolysis of the resulting rearranged product leads to α-hydroxy ketones in good yield.

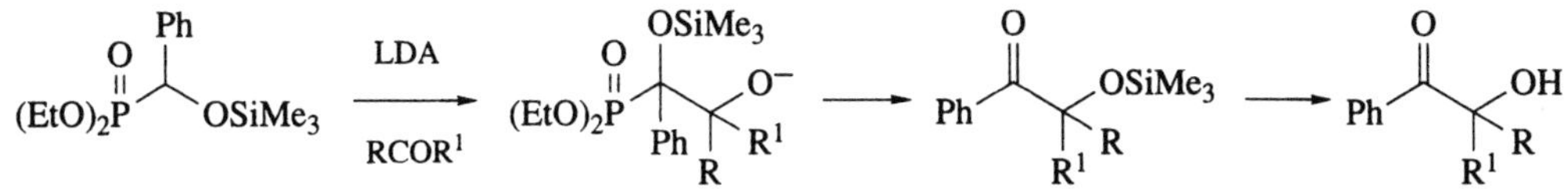

Scheme 31

2.4.8.2 Alkylation with R—X

Unsymmetrical ketones result from the alkylation of the carbanion generated *in situ* from phosphonate (**61**) with LDA, as shown in equation (30).[171]

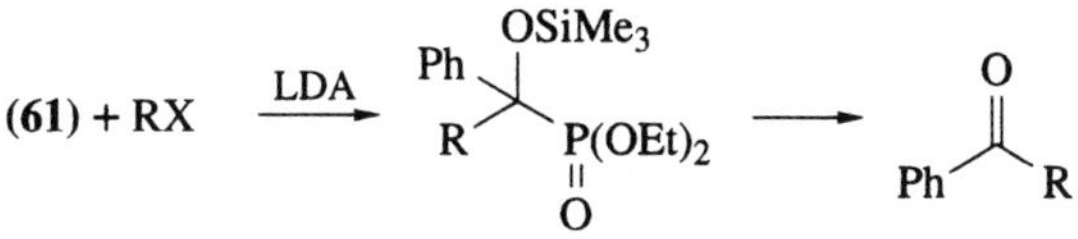

(30)

A new acyl anion equivalent derived from an aldehyde, a secondary amine and diphenylalkylphosphine oxide, *e.g.* (*N*-morpholinomethyl)diphenylphosphine oxide, reacts with aldehydes or ketones to form enamines, which on acid hydrolysis afford aldehydes or ketones (Scheme 32).[172,173] The usefulness of this acyl anion equivalent is demonstrated by the synthesis of dihydrojasmone and (*Z*)-6-henicosen-11-one.[174]

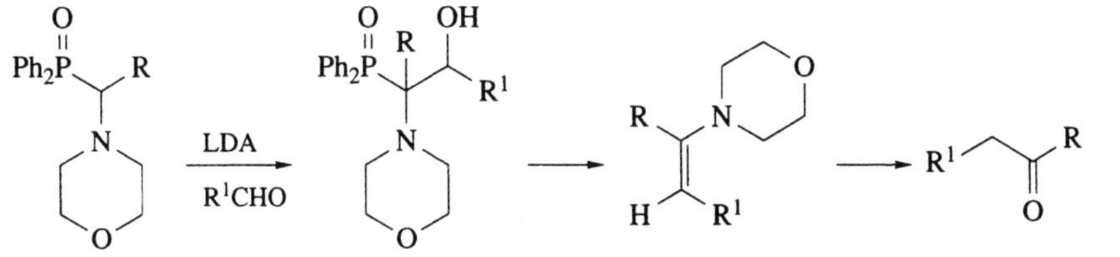

Scheme 32

2.4.9 DITHIOACETALS AND DITHIANES

In dithioacetals the proton geminal to the sulfur atoms can be abstracted at low temperature with bases such as Bu^nLi. Lithium ion complexing bases such as DABCO, HMPA and TMEDA enhance the process. The resulting anion is a masked acyl carbanion, which enables an assortment of synthetic sequences to be realized *via* reaction with electrophiles.[3,12,13,175,176] Thus, a dithioacetal derived from an aldehyde can be further functionalized at the aldehyde carbon with an alkyl halide, followed by thioacetal cleavage to produce a ketone. Dithiane carbanions allow the assemblage of polyfunctional systems in ways complementary to traditional synthetic routes. For instance, the β-hydroxy ketone systems, conventionally obtained by an aldol process, can now be constructed from different sets of carbon groups.[176]

Dithioacetals, 1,3-dithianes or 1,3-dithiolanes are prepared by reaction of the corresponding carbonyl compound in the presence of an acid catalyst (conc. HCl, Lewis acids such as ZnI_2, $BF_3{\cdot}Et_2O$, TMS-Cl, *etc.*) with a thiol or dithiol.[177] Silica gel treated with thionyl chloride was found to be an effective as well as selective catalyst for thioacetalization of aldehydes.[178] Thioacetalization can also be achieved using a (polystyryl)diphenylphosphine–iodine complex as a catalyst.[179] Conversion of aldehydes or acetals into 1,3-dithianes is achieved with the aid of organotin thioalkoxides and organotin triflates[180] or with 2,2-dimethyl-2-sila-1,3-dithiane.[181] Direct conversion of carboxylic acids to 1,3-dithianes can be carried out by reaction with 1,3,2-dithiabornenane–dimethyl sulfide and tin(II) chloride[182] or 1,3,2-dithiaborolene with trichloromethyllithium followed by basic hydrolysis.[183]

The dithiane-derived anion can be generated by the action of Bu^nLi in THF at –78 °C or with complex bases $NaNH_2$–RONa at room temperature.[184] Lithiated dithiane can also be prepared *in situ* by sonication of *n*-butyl chloride with lithium in the presence of dithiane.[185] Dithioacetals or ketals are resistant to acidic or basic hydrolysis. Regeneration of the carbonyl group from the dithioketal sometimes presents difficulties but can be carried out by hydrolysis in polar solvents (acetone, alcohols, acetonitrile) in the presence of metallic ions such as Hg^{II}, Cu^{II}, Ag^{I}, Ti^{IV}[177] or Tl^{III}.[186] Alternatively, alkylative hydrolysis using methyl iodide, methyl fluorosulfonate or Meerwein salts, or oxidative cleavage using chloramine-T, *N*-halosuccinimides, peroxy acids, ceric ammonium nitrate (CAN) or 1-chlorobenzotriazoles leads to the carbonyl compound.[177] In other cases one can use *n*-tributyltin hydride, trialkoxonium tetrafluoroborate, isopentyl nitrite, thionyl chloride–silica gel–water, $Fe(NO_3)_2$ or $Cu(NO_3)_2$ supported on clay,[187] nitrosonium ion (NO^+) sources,[188] methyltriphenylphosphonium tribromide, trimethylphenylammonium tribromide,[189] phenyldichlorophosphonate–DMF–NaI[190] or photochemistry[3,177] to achieve the hydrolytic cleavage.

It has been shown that thioacetal monosulfoxides undergo hydrolysis with greater ease than thioacetals or dithianes, *i.e.* the hydrolysis can be carried out with dilute sulfuric or perchloric acid.[175] Not only is the hydrolysis of the thioacetal monosulfoxides more facile but the addition of the lithium salt proceeds 1,4 rather than 1,2 to α,β-unsaturated carbonyl compounds.[175,191]

2.4.9.1 1,2-Addition to Aldehydes and Ketones

1,2-Addition of lithiodithianes to carbonyl compounds followed by hydrolysis gives α-hydroxy ketones in good yield (equation 31).[3,175,192] The reaction of sugar lactones with lithiodithiane provides a preparative route to higher sugars.[193] The reaction of 2-phenyl-1,3-dithianyllithium with 4-*t*-butylcyclohexanone in cyclohexane or THF proceeds with thermodynamic control, involving exclusively equatorial attack on the carbonyl group. In contrast only kinetic control (axial attack) is seen in the reaction of the same ketone with 1,3-dithianyllithium.[194]

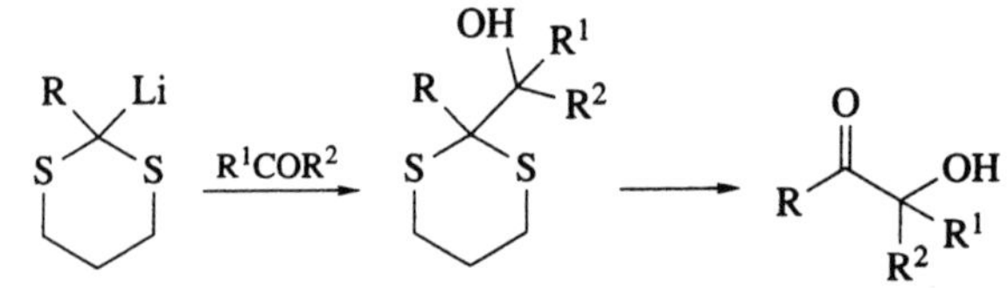

R^1COR^2 (31)

Stereoselective addition of a dithiane anion to chiral 2-methyl-3-trimethylsilyl-3-butenal combined with the stereoselective addition of a Grignard reagent to the chiral α-alkoxy ketone affords a practical method for the construction of α,γ-dimethyl-α,β-dihydroxy compounds, useful intermediates for the synthesis of erythronolides (Scheme 33).[195] β-Hydroxy carboxylic esters were synthesized by the addition of ethyl 1,3-dithiolanyl-2-carboxylate enolate to a chiral aldehyde, followed by desulfurization.[196]

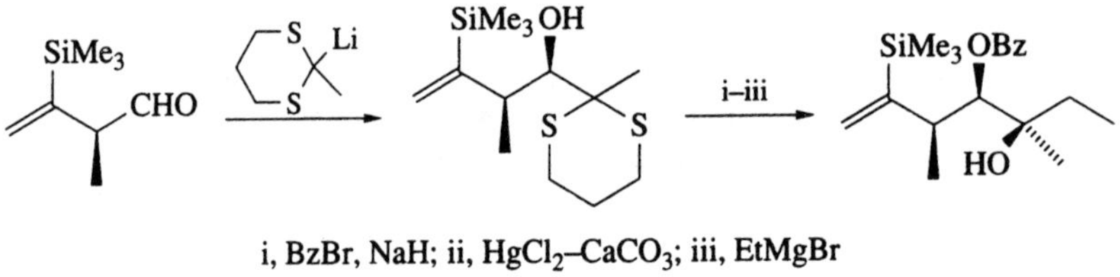

i, BzBr, NaH; ii, $HgCl_2$–$CaCO_3$; iii, EtMgBr

Scheme 33

The intramolecular carbonyl addition of a lithiated dithiane to a ketone was used in the synthesis of rocaglamide, an anticancer compound, in 64% yield.[197]

α-Hydroxy ketones having fungicidal activity were prepared from the reaction of lithiated 2-alkyldithianes with 3-pyridyl ketones.[198]

The reaction of 2-cyclohexenone with 2-lithiodithiane occurs in a 1,2-fashion to give the unsaturated hydroxy derivative, which, after rearrangement followed by deprotection, gives 3-hydroxy-1-cyclohexenecarbaldehyde.[199]

1,2-Addition of dithianyl anion to the appropriate aldehydes is a key step in the multistep synthesis of (i) amphotericin B (Scheme 34);[200,201] (ii) the dioxabicyclononane unit of tirandamycin (Scheme 35);[202] (iii) secoisoquinoline alkaloids such as peshawarine, cryptopleurospermine, (±)-corydalisol, (±)-aobamine and hypercornine[203] and baldulin; and (iv) sesquiterpene lactones[204] and chiral lycoserone.[205]

Reaction of dianion (**62**) with nicotinaldehyde or *N*-methylpiperidone gave the corresponding alcohols in good yield (equation 32),[206] whereas the *N*-protected anion led to the formation of the ring-opened alkyne derivative (equation 33).

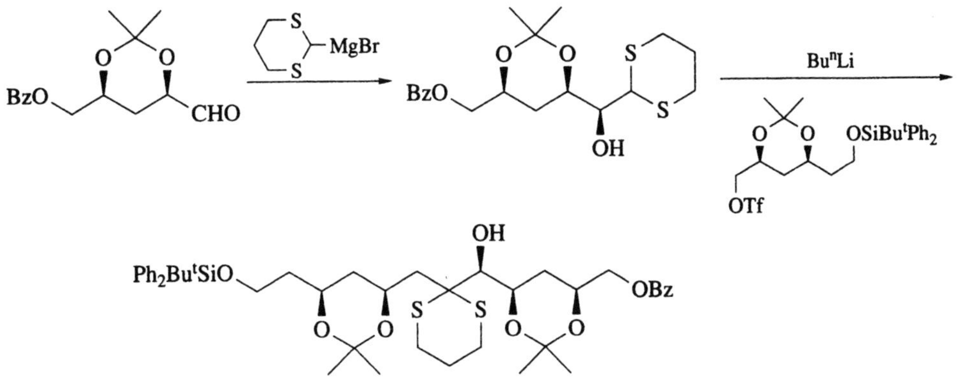

Scheme 34

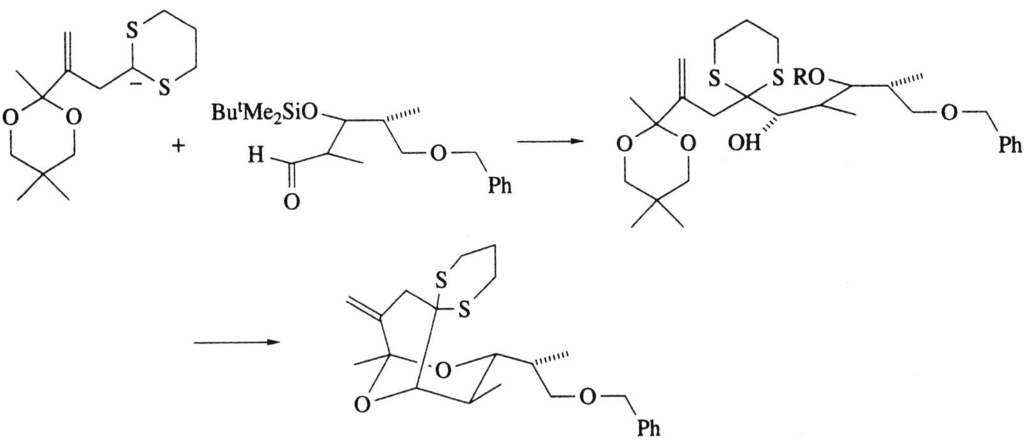

Scheme 35

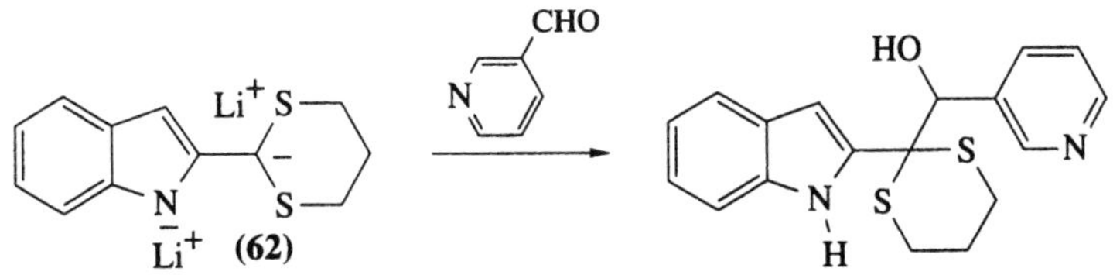

(32)

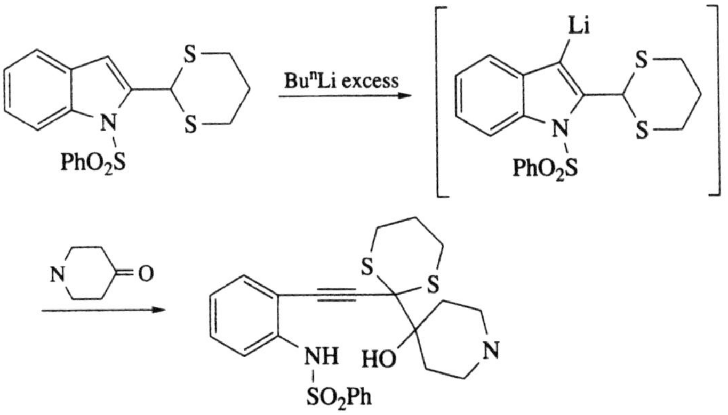

(33)

The regio- and diastereo-selective reaction of the anion generated from 2-(1-propen-1-yl)-1,3-dithiane with an aldehyde is applicable to the synthesis of *trans*-β,γ-disubstituted γ-lactones, including the natural products (±)-eldnolide and (±)-*trans*-quercus lactone.[207]

2-Lithio-2-(1-methyl-2-alkenyl)-1,3-dithianes showed 1,3-*syn* selectivity on addition to aldehydes (equation 34). The presence of a bulky substituent R^3 on the aldehyde gave increased *syn* selectivity in the following order: alkyl > phenyl > 1-silylalkenyl > alkenyl > alkynyl. Activation of an aldehyde by $BF_3 \cdot Et_2O$ reduced the ratio.[208]

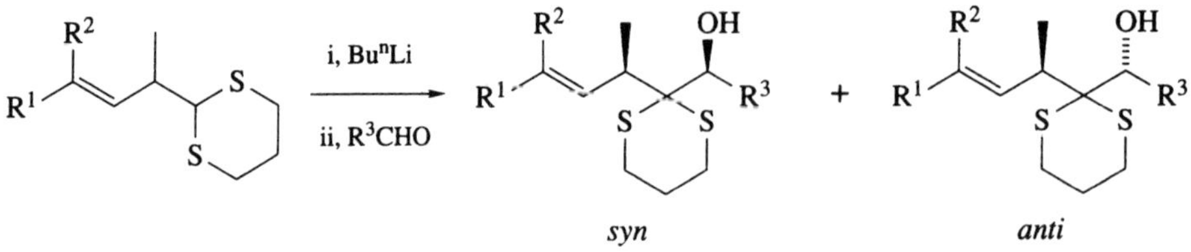

i, Bu^nLi; ii, R^3CHO — *syn* + *anti* (34)

Intramolecular addition of lithiodithiane to an acetal leads to the formation of cyclic derivatives (equation 35).[209] Carbonyl additions have been extended to the analogous imonium salts ($R_2C{=}^+NR_2$), which furnish α-amino ketones.[210]

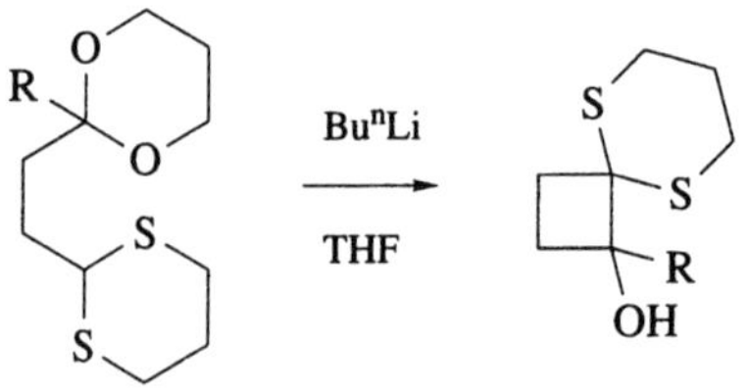

Bu^nLi, THF (35)

The reaction of 6-deoxy-6-iodo-1,2;3,4-di-*O*-isopropylidene-α-D-galactopyranose with 2-lithiodithiane yields an unsaturated open chain heptose derivative (equation 36).[211]

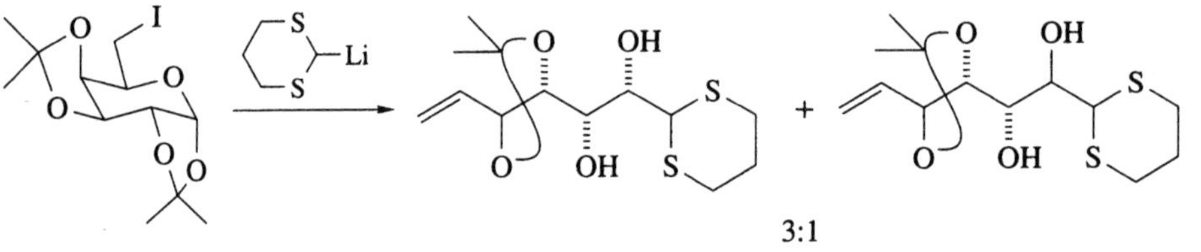

3:1 (36)

2.4.9.2 1,4-Addition to α,β-Unsaturated Carbonyl Compounds

This reaction is of importance since it enables the conversion of α,β-unsaturated carbonyl systems into 1,4-dicarbonyl compounds, which are precursors of cyclopentenones (*e.g.* jasmonoids, rethronoids and prostanoids).[212] The 1,4-addition proceeds in the presence of 1–2 equiv. of HMPT[213] and additional anion-stabilizing groups on the alkene enhance the tendency of metallated *S,S*-acetals to undergo the conjugate addition.

Chirally mediated conjugate addition reactions of lithiated dithioacetals to a prochiral α,β-unsaturated ester proceed enantioselectively to the corresponding adduct.[214] Alkylation of an acyclic α,β-unsaturated ester was achieved with high diastereoselectivity using a dithioacetal as the stereocontrolling unit (Scheme 36).[215] The conjugate addition of 2-lithiodithiane to 2-cyclohexenones, followed by quenching of the intermediate enolate anion with 2-chloroacyl chloride, affords 2-substituted 4-(1,3-dithian-2-yl)-4,5,6,7-tetrahydro-3(2*H*)-benzofuranones.[216]

The benzylic alcohols (**63**) are readily prepared by the conjugate addition of appropriate sulfur-stabilized carbanions, either as aryldithianes or as arylbis(phenylthiomethanes), to butenolides, followed by trapping of the generated enolate with an aromatic aldehyde.[217] Cyclization of the derived ketone provides a short efficient synthesis of clinically important podophyllotoxin derivatives (Scheme 37),[218] whereas Raney nickel desulfurization, followed by hydrolysis under various conditions, gave lignan lactones like (–)-burseran, (–)-cubebin, hinokinin[219] and podorhizol.[217] More significantly, the use of a chiral butenolide leads to an asymmetric synthesis of the lignans (+)-burseran, isostegane and (+)-steganacin.[217]

A total synthesis of (±)-aromatin has utilized the lithium anion of the dithiane of (*E*)-2-methyl-2-butenal as a functional equivalent of the thermodynamic enolate of methyl ethyl ketone in an aprotic

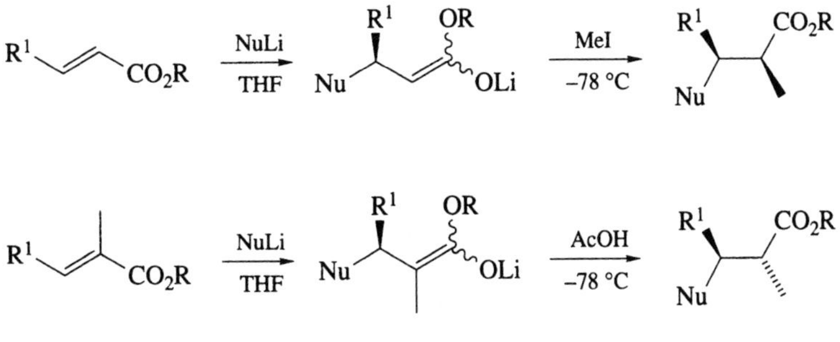

R1 CO2R —NuLi, THF→ R1 OR Nu OLi —MeI, −78 °C→ R1 CO2R Nu

R1 CO2R —NuLi, THF→ R1 OR Nu OLi —AcOH, −78 °C→ R1 CO2R Nu

Nu = dithiane

Scheme 36

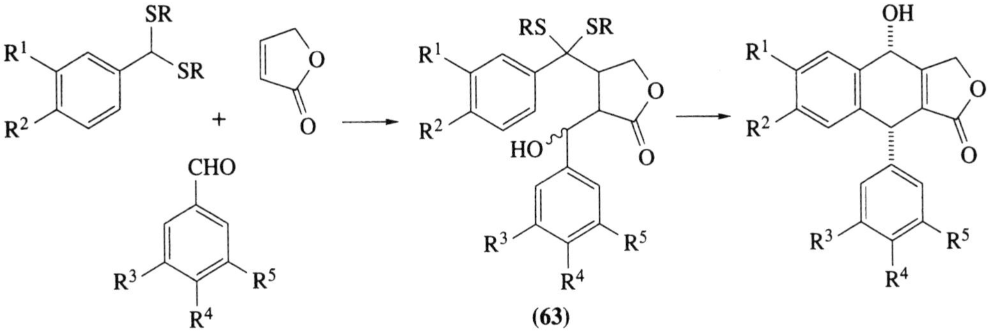

R1 R2 SR SR + O O CHO R3 R4 R5 → R1 R2 RS SR HO O O R3 R4 R5 **(63)** → R1 R2 OH O O R3 R4 R5

Scheme 37

Michael addition.[220] The synthesis of the naturally occurring benz[*a*]anthraquinones, X-14881 (**64**; R = Me) and ochromycinone (R = H) utilizes 2-phenyldithiane anion (Scheme 38),[221] while 11-deoxydaunomycinone was prepared using 2-methyldithiane.[222] The formation of bicyclo[3.3.0]oct-1(5)-ene-2,6-dione utilizes 2-(2,2-diethoxyethyl)-1,3-dithiane anion.[223] Conjugate addition of the highly stabilized anion (**65**) occurs readily with enamide (**66**), whereas the related acyclic anion (**67**) appears to be less reactive than the cyclic analog (Scheme 39).[12] Conjugate addition of lithiated α-alkylthio monosulfoxide provides a new route for the synthesis of dihydrojasmone.[224]

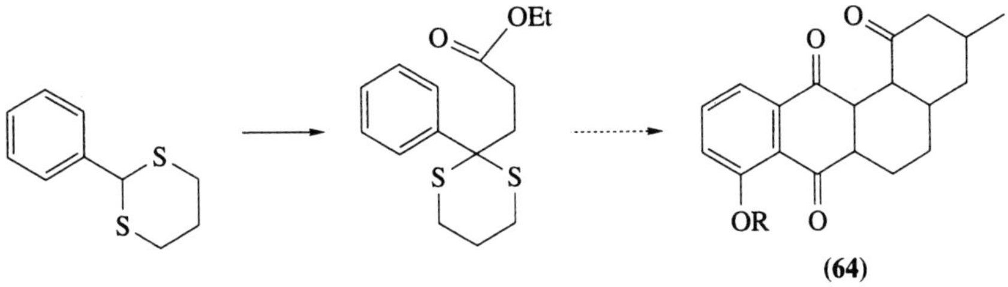

S S → O OEt S S ····→ O O O OR O **(64)**

Scheme 38

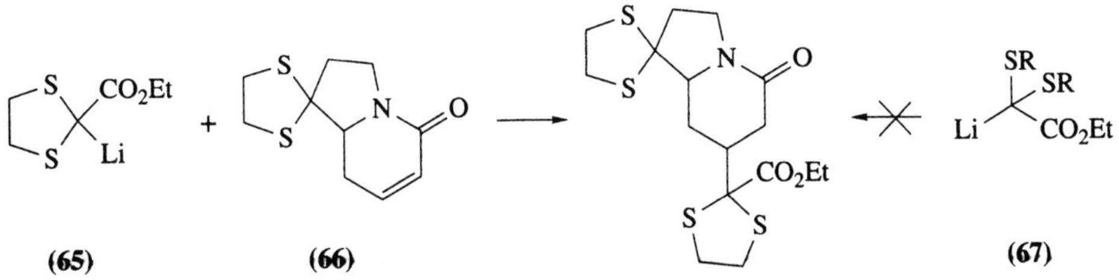

S S CO2Et Li **(65)** + S S N O **(66)** → S S N O CO2Et S S ←✕ SR SR Li CO2Et **(67)**

Scheme 39

2.4.9.3 Alkylation with R—X

The lithio derivatives of 1,3-dithiane react readily with alkyl halides to afford 2-alkylated dithianes in excellent yields.[3,225] Dithiane anions are useful for the formation of small- and medium-sized rings *via* alkylation.[226]

The alkylation method has been used for the multistep synthesis of vermiculine,[3] homofernascene, the trail pheromone components of the fire ant *Solinopis invicta*,[227] the spiroacetal fragment of milbemycin E,[228] chiral segments of roflamycin, the macrolide gloeosporone,[229] an autoinhibitor of spore germination,[230] antibiotic A23187, calcimycin,[231] a fragment containing the backbone of the antibiotic boromycin,[232] taloromycin B, an acute avian toxin,[233] and baiyunol, an aglycon of a sweat substance.[234]

Alkylation of 2-lithio-2-ethoxycarbonyl-1,3-dithiane with 5-bromo-1,2-pentadiene gave a precursor for the total synthesis of lycorenine alkaloids.[235]

The aliphatic portion of (±)-zearalenone has been synthesized starting from 2,3-dihydropyran and iodopentanone ketal by employing dithiane for C—C bond formation *via* alkylation.[236] The method also proved to be useful for the preparation of long chain alcohols starting from 5-alkyl-2-(1,3-dithianyl)thiophene.[237] Similarly two rigid crown ethers (**68**), which possess a tripyridine subunit, are prepared by the construction of the middle pyridine moiety *via* a dithiane precursor (Scheme 40).[238]

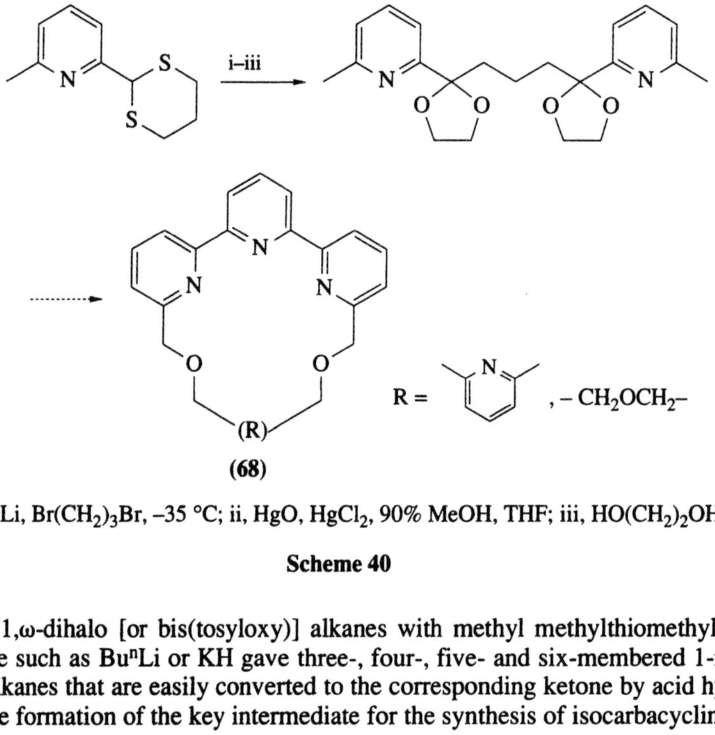

i–iii

R = , – CH_2OCH_2–

(**68**)

i, Bu^nLi, $Br(CH_2)_3Br$, –35 °C; ii, HgO, $HgCl_2$, 90% MeOH, THF; iii, $HO(CH_2)_2OH$, TsOH

Scheme 40

Cyclization of 1,ω-dihalo [or bis(tosyloxy)] alkanes with methyl methylthiomethyl sulfoxide in the presence of a base such as Bu^nLi or KH gave three-, four-, five- and six-membered 1-methylsulfinyl-1-methylthiocycloalkanes that are easily converted to the corresponding ketone by acid hydrolysis.[239] This is applicable to the formation of the key intermediate for the synthesis of isocarbacyclin, a potent prostacyclin analog.[240]

Bis(benzenesulfonyl)methane[241] and 1,3-benzodithiole tetroxide[242] are useful formyl anion equivalents for alkylation as well as cycloalkylation reactions.

2.4.9.4 Acylation with Lithiodithianes

Reaction of lithiodithianes with acyl chlorides, esters or nitriles leads to the formation of 1,2-dicarbonyl compounds in which one of the carbonyl groups is protected as the thioacetal.[175,176,243,244] Optically active amino ketones of type (**69**) are prepared *via* acylation of dithiane with an oxazoline-protected (*S*)-serine methyl ester (Scheme 41).[245] Optically active (*S*)-2-alkoxy-1-(1,3-dithian-2-yl)-1-propanones were prepared by the reaction of the corresponding methyl (*S*)-lactate with 2-lithio-1,3-dithiane, which can be used for the enantioselective synthesis of (–)-trachelanthic acid.[246] Enantioselective synthesis of L-glyceraldehyde involves the acylation of a dithiane glycolic acid derivative followed by bakers' yeast mediated reduction.[247]

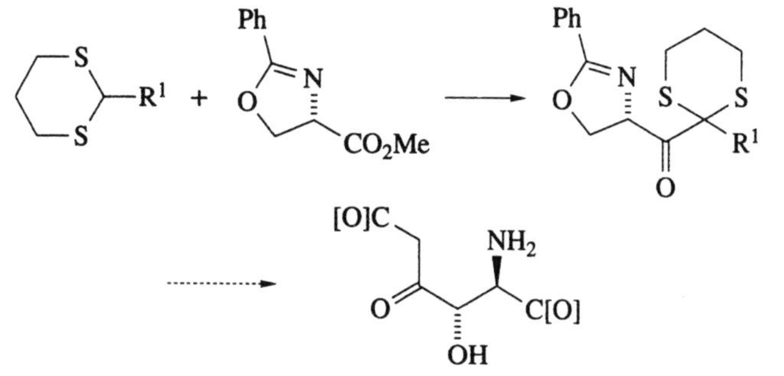

(69)

Scheme 41

2.4.9.5 Miscellaneous Reactions of Dithioacetals

Arylation of dithiane can be achieved by the reaction of the anion with activated haloarenes such as halopyridines or haloarylchromium(tricarbonyls).[3]

α-Diketones are prepared either by the reaction of 2-lithio-1,3-dithianes with a nitrile oxide (Scheme 42)[248] or by the reaction of 2-lithiodithiolane with iron(pentacarbonyl) followed by alkylation with an alkyl halide.[249]

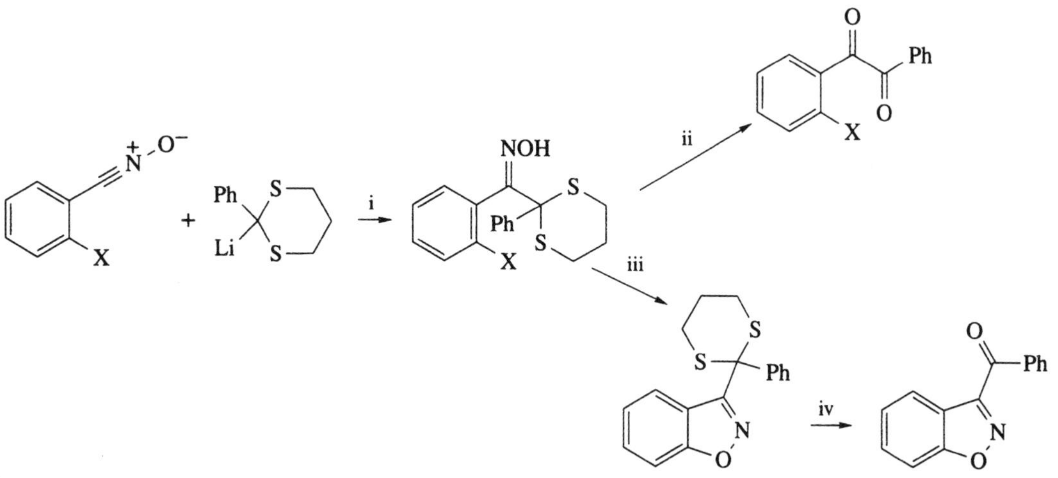

i, THF; ii, CF_3CO_2H, 35%; iii, KOH, EtOH; iv, HgO

Scheme 42

Epoxides and oxetanes react readily with lithiodithiane to give derivatives of β- or γ-hydroxy aldehydes or ketones.[176] The method has been used for the multistep synthesis of an inomycin fragment,[250] thietane prostaglandin analogs,[251] branched-chain nucleoside sugars,[252] the southern hexahydrobenzofuran unit avermectin,[253] the anti-Gram-positive bacterial elaiphylin[254] and *syn*-1,3-polyols.[255]

Allylic anions generated from 2-propenyl-1,3-dithiane and 2-styryl-1,3-dithiane react exclusively at α-carbon atoms either with carbonyl compounds[256] or with three- to six-membered cyclic ethers in the presence of $BF_3{\cdot}Et_2O$.[257]

Reaction of dithiolane (**70**) with BuLi followed by a hindered organoborane gives (**71**), which after oxidative work-up affords an unsymmetrical ketone in good yield (Scheme 43).[258]

Reaction of dithiane with DMF leads to the formation of an aldehyde, a key intermediate for the synthesis of the macrocyclic antibiotic (±)-pyrenophorin.[259,260]

2-Diethoxyphosphinyl-2-buten-4-olide reacts with lithiated dithianes followed by an intermolecular Wittig–Horner reaction to produce fused γ-lactones (equation 37).[261] The reaction of 2-lithio-1,3-dithianes with nitroarenes gives 2- or 4-[(1,3-dithian)-2′-yl]cyclohexa-3,5(or 2,5-)-diene-1-nitronate compounds (conjugate addition products), free nitroarene radical anions (redox products), 1,3-dithianes and 2,2′-bis(1,3-dithianes).[262]

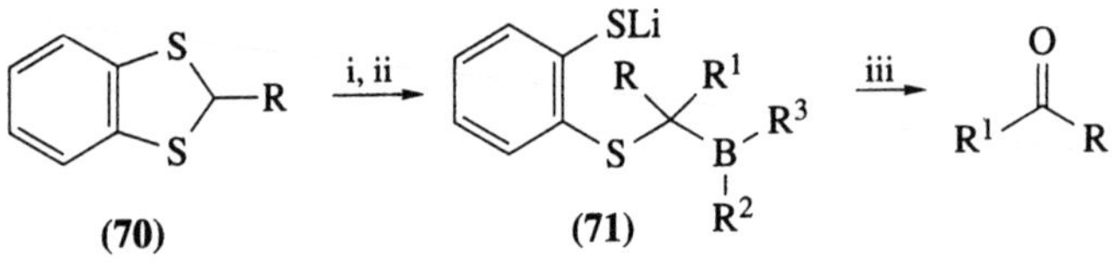

(70) (71)

i, Bu^nLi, THF, 30 °C; ii, $BR^1R^2R^3$; iii, $NaOH/H_2O_2$, r.t.

Scheme 43

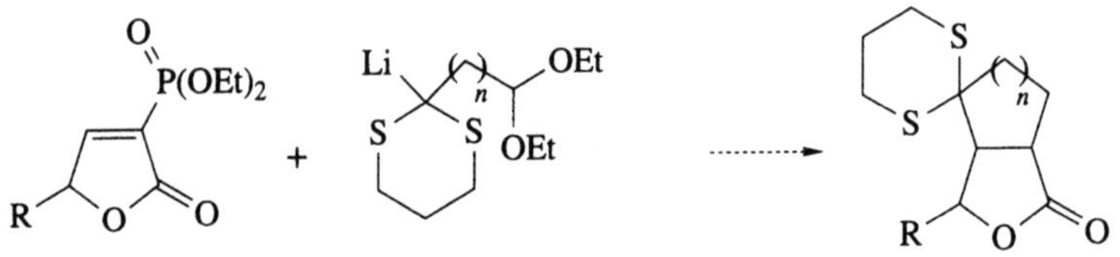

(37)

A key step for the synthesis of prostaglandin D_1 methyl ester involves the reaction of dithiane **(72)** with (1-methylthiovinyl)triphenylphosphonium chloride (**73**; Scheme 44).[263]

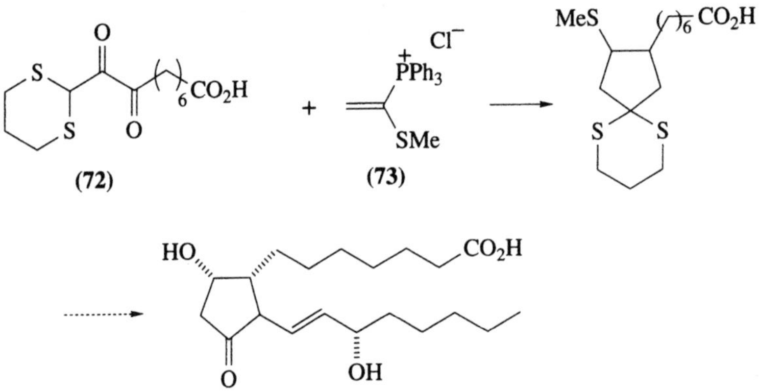

(72) (73)

Scheme 44

Lithiodithiane is aminated by reaction with a vinyl azide, as shown in equation (38).[264]

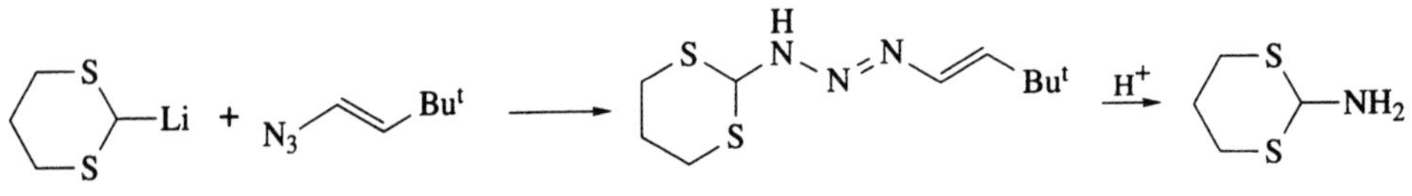

(38)

Methoxy(phenylthio)methane and methoxy(phenylsulfonyl)methane are useful formyl anion equivalents for one-carbon homologation.[265] For instance, addition of methoxy(phenylthio)methyllithium to ketones, followed by rearrangement of the adduct, provides a new method for the preparation of α-(phenylthio) aldehydes (equation 39). The rearrangement is stereospecific.[266] This method has been used for the total synthesis of annelated furans such as (+)-euryfuran,[267] butenolides such as isoderminin and confertifolin[268] and spirocyclic tetrahydrofurans and butenolides.[269]

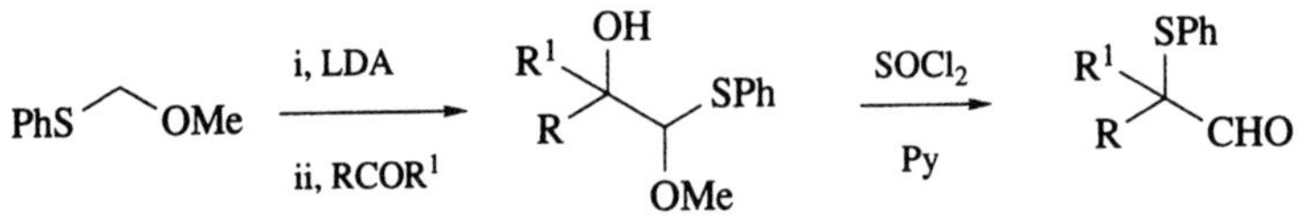

(39)

Alkylation of phenylthiomethyl(trimethyl)silane or phenylthiophenyl(trimethylsilyl)methane with alkyl halides followed by deprotection gave aldehydes[270] or ketones[271] respectively.

2.4.10 DISELENOACETALS

Diselenoacetals are also useful for the protection or umpolung of the carbonyl group.[2,272,273] Bis(alkylseleno)alkanes and arylseleno analogs are prepared by the reaction of carbonyl compounds with selenols and trimethylsilyl selenides in the presence of acid catalysts (conc. HCl, conc. H_2SO_4, Lewis acids such as $BF_3{\cdot}Et_2O$, $TiCl_4$, $ZnCl_2$ or $AlCl_3$) or with tris(methylseleno)borane in the presence of LAH and $BF_3{\cdot}Et_2O$.[274]

The selenium-stabilized carbanions derived by deprotonation of selenoacetals by strong bases, such as a mixture of KDA–lithium *t*-butoxide, LiTMP in HMPT/THF or LBDA in THF at –78 °C, react readily with a variety of electrophiles including primary or secondary halides, epoxides, ketones, aldehydes and enones, followed by deprotection, to give ketones, β-hydroxy ketones, α-hydroxy ketones and 1,4-dicarbonyl compounds respectively.[2,275,276]

Regeneration of the carbonyl group from a diseleno-ketal or -acetal can be carried out by clay-supported iron(III) or copper(II) nitrate,[277] $HgCl_2/CaCO_3$/MeCN, $(PhSeO)_2O$/THF, $NaIO_4$/EtOH–H_2O, HgOAc/MeCN or $CuCl_2$/CuO.[275]

The presence of DME or HMPT allows the C-3 regio- and stereo-selective 1,4-addition of 1,1-bis(methylseleno)alkanes to methylcyclohexenones (see Scheme 45). The resulting adducts have been stereoselectively transformed into *cis*- or *trans*-3-ethyl-2-methylcyclohexanone using Bu_3SnH and AIBN.[278]

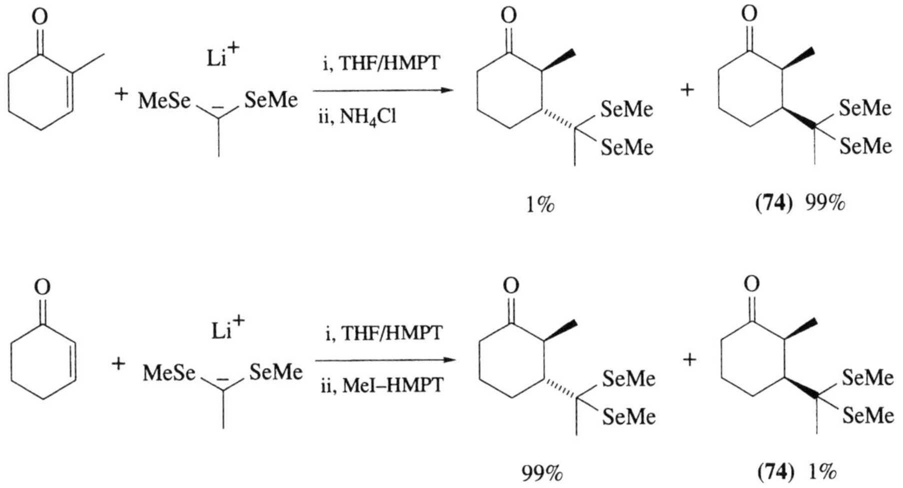

Scheme 45

Selective transformation of the diseleno-substituted cyclohexanone to the corresponding diketone can be carried out from the *cis* product (**74**), whereas the *trans* isomer does not produce the desired diketone but instead leads to a ketovinyl selenide.[275]

α-Phenylselenyl cyclic ethers prepared by the reactions of either lactols or lactol acetate with benzeneselenol in the presence of a Lewis acid act as acyl anion equivalents, which on deprotonation with BuLi followed by reaction with alkyl halides lead to the formation of alkylated products.[279]

The lithio reagents generated from silylselenylmethanes (Me_3SiCH_2SePh) and $Me_3SiCH(SePh)CH_2R$ are useful synthetic equivalents of a protected formyl anion and an acyl anion respectively.[2] Silylselenylmethanes are valuable reagents for the transformation of halides to aldehydes.[2]

2.4.11 TOSYLMETHYL ISOCYANIDE

Tosylmethyl isocyanide (TosMIC) (**75**; R = H), a versatile reagent in synthesis, can also be used as an acyl anion equivalent. For instance symmetrical and unsymmetrical diketones were prepared by using this TosMIC synthon (equation 40).[280] Ketones are homologated to enones by alkylating the condensation product derived from TosMIC, followed by acid hydrolysis (Scheme 46).[281]

1-Isocyano-1-tosyl-1-alkenes (**76**), formed by the reaction of TosMIC with an aldehyde or ketone, react with a primary amine or ammonia to give 1,5-disubstituted (or 5-monosubstituted) imidazoles in high yield (Scheme 47).[282]

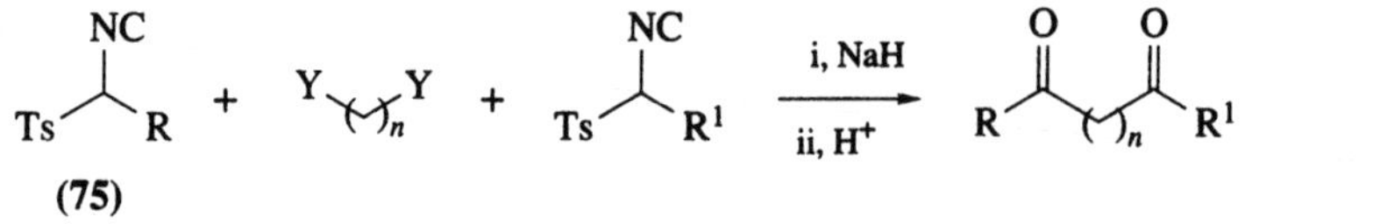

(40)

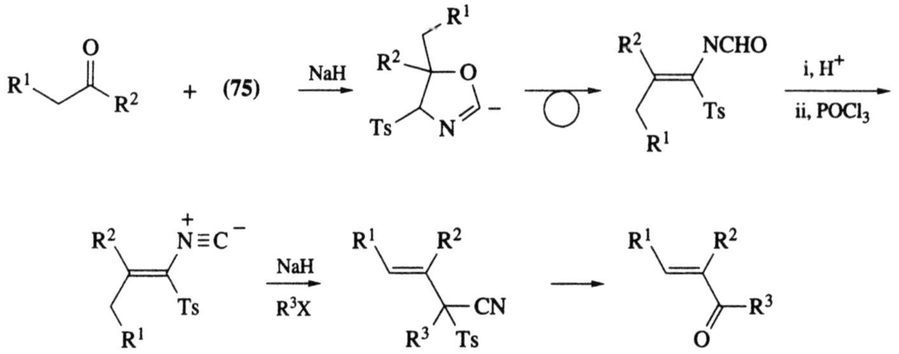

Scheme 46

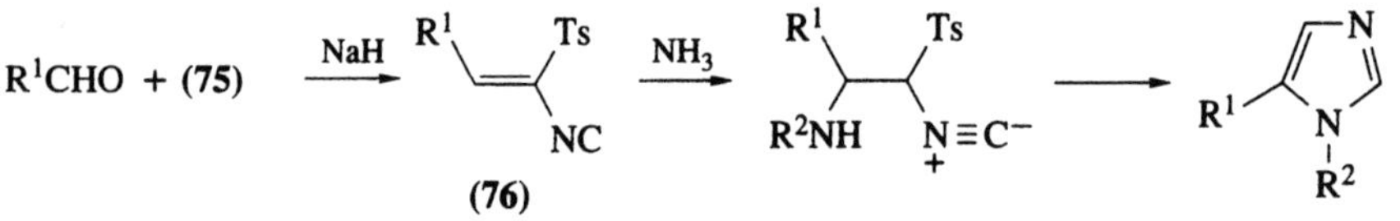

Scheme 47

2.4.12 REFERENCES

1. S. F. Martin, *Synthesis*, 1979, 633.
2. A. Krief, D. Seebach, *Tetrahedron*, 1980, **36**, 2531.
3. B. T. Grobel and D. Seebach, *Synthesis*, 1977, 357.
4. W. S. Ide and J. S. Buck, *Org. React. (N.Y.)*, 1957, **4**, 269.
5. T. Laird, in 'Comprehensive Organic Chemistry', ed. D. H. R. Barton and W. D. Ollis, Pergamon Press, Oxford, 1979, vol. 1, p. 1142.
6. H. Herlinger, *Methoden Org. Chem. (Houben-Weyl)*, 1973, **VII/2a**, 653.
7. H. Stetter, *Angew. Chem., Int. Ed. Engl.*, 1976, **15**, 639.
8. (a) J. D. Albright, *Tetrahedron*, 1983, **39**, 3207; (b) G. Stork, R. M. Jacobson and R. Levitz, *Tetrahedron Lett.*, 1979, 771.
9. S. Arseniyadis, K. S. Kyler and D. S. Watt, *Org. React. (N.Y.)*, 1984, **31**, 47.
10. D. Seebach, *Angew. Chem., Int. Ed. Engl.*, 1969, **8**, 639.
11. D. Seebach, *Angew. Chem., Int. Ed. Engl.*, 1979, **18**, 239.
12. O. W. Lever, Jr., *Tetrahedron*, 1976, **32**, 1943.
13. T. Hase and J. K. Koskimies, *Aldrichimica Acta*, 1981, **14**, 73; 1982, **15**, 35.
14. S. Akabori, M. Ohtomi and K. Arai, *Bull. Chem. Soc. Jpn.*, 1976, **49**, 746.
15. X. Songkai, T. Jingren, J. Fengying, Z. Zingqing and H. Wenhong, *Zhongshan Daxue Xuebao, Ziran Kexueban*, 1984, 11 (*Chem. Abstr.*, 1985, **102**, 131 297z); J. Castells and E. Dunach, *Chem. Lett.*, 1984, 1859.
16. Kh. M. Shakhidayatov, N. F. Abdulaev and Ch. Sh. Kadyrov, *USSR Pat.* 436 814 (1974) (*Chem. Abstr.*, 1974, **81**, 135 733j).
17. K. Jianshe, *Daxue Huaxue*, 1987, **2**, 34 (*Chem. Abstr.*, 1988, **108**, 74 387b).
18. R. Breslow and E. Kool, *Tetrahedron Lett.*, 1988, **29**, 1635.
19. L. H. Dieter and D. Francois, *Angew. Chem.*, 1986, **98**, 1125.
20. J. Castells, F. Lopez-Calahorra and L. Domingo, *Tetrahedron Lett.*, 1985, **26**, 5457.
21. J. Castells and F. Dunach, *Chem. Lett.*, 1984, 1859.
22. C. S. Sell and L. A. Dorman, *J. Chem. Soc., Chem. Commun.*, 1982, 629.
23. I. Zilniece and A. Zicmanis, *Latv. PSR Zinat. Akad. Vestis, Kim. Ser.*, 1986, 732 (*Chem. Abstr.*, 1987, **107**, 175 606g).
24. M. Bassedas, M. Carreras, J. Castells and F. Lopez-Calahorra, *React. Polym. Ion Exch. Sorbents*, 1987, **6**, 109 (*Chem. Abstr.*, 1988, **108**, 6570f).
25. M. F. Lappert and R. K. Maskell, *J. Chem. Soc., Chem. Commun.*, 1982, 580.

26. J. Castells, F. Lopez-Calahorra, M. Bassedas and P. Urrios, *Synthesis*, 1988, 314; J. Castells, F. Lopez-Calahorra, F. Geijo, R. Perez-Dolz and M. Bassedas, *J. Heterocycl. Chem.*, 1986, **23**, 715.
27. A. H. Rose, 'Industrial Microbiology', Butterworths, London, 1961, p. 264.
28. C. Fuganti and P. Grasselli, *J. Chem. Soc., Chem. Commun.*, 1982, 205; R. Bernardi, C. Fuganti, P. Grasselli and G. Marinoni, *Synthesis*, 1980, 50.
29. Y. Kamitori, M. Hojo, R. Musada, T. Fujitani, S. Ohara and T. Yokoyama, *Synthesis*, 1988, 208.
30. W. P. Weber, 'Silicon Reagents for Organic Synthesis', Springer, New York, 1983, p. 6; E. W. Colvin, 'Silicon in Organic Synthesis', Butterworths, Boston, 1981, pp. 296, 310.
31. V. H. Rawal, J. Appa Rao and M. P. Cava, *Tetrahedron Lett.*, 1985, **26**, 4275.
32. R. Yoneda, H. Hisakawa, S. Harusawa and T. Kurihara, *Chem. Pharm. Bull.*, 1987, **35**, 3850.
33. M. D. Rozwadowska, *Tetrahedron*, 1985, **41**, 3135.
34. B. C. Uff, A. Al Kolla, K. E. Adamali and V. Harutunian, *Synth. Commun.*, 1978, **8**, 163; A. T. Au, *Synth. Commun.*, 1984, **14**, 743, 749.
35. H. Barrios, C. Sandoval, B. Ortiz, R. Sanchez-Obregon and F. Yuste, *Org. Prep. Proced. Int.*, 1987, **19**, 427.
36. R. M. Jacobson, G. P. Lahm and J. W. Clader, *J. Org. Chem.*, 1980, **45**, 395 and refs. therein.
37. T. Kurihara, K. Santo, S. Harusawa and R. Yoneda, *Chem. Pharm. Bull.*, 1987, **35**, 4777.
38. R. E. Koenigkramer and H. Zimmer, *J. Org. Chem.*, 1980, **45**, 3994.
39. T. Shono, S. Kashimura, Y. Yamaguchi and F. Kuwata, *Tetrahedron Lett.*, 1987, **28**, 4411; T. Shono, S. Kashimura, Y. Yamaguchi, O. Ishige, H. Uyama and F. Kuwata, *Chem. Lett.*, 1987, 1511.
40. C. G. Chavdarian and C. H. Heathcock, *J. Am. Chem. Soc.*, 1975, **97**, 3822.
41. A. Hassner and J. A. Soderquist, *J. Organomet. Chem.*, 1977, **131**, C1.
42. J. A. Soderquist and A. Hassner, *J. Am. Chem. Soc.*, 1980, **102**, 1577; *J. Org. Chem.*, 1980, **45**, 541.
43. D. M. Findlay and M. F. Tchir, *J. Chem. Soc., Chem. Commun.*, 1974, 514.
44. Y. Ito, H. Imai, T. Matsuura and T. Saegusa, *Tetrahedron Lett.*, 1984, **25**, 3091 and refs. therein.
45. D. Seyferth and R. C. Hui, *J. Am. Chem. Soc.*, 1985, **107**, 4551.
46. R. M. Adlington, J. E. Baldwin, J. C. Bottaro and M. W. D. Perry, *J. Chem. Soc., Chem. Commun.*, 1983, 1040.
47. J. E. Baldwin, R. M. Adlington, J. C. Bottaro, J. N. Kolhe, M. W. D. Perry and A. V. Jain, *Tetrahedron*, 1986, **42**, 4223.
48. J. E. Baldwin, R. M. Adlington, J. C. Bottaro, J. N. Kolhe, I. M. Newington and M. W. D. Perry, *Tetrahedron*, 1986, **42**, 4235.
49. Y. Ito, T. Matsuura and M. Murakami, *J. Am. Chem. Soc.*, 1987, **109**, 7888.
50. M. Elliot, N. F. Janes, B. P. S. Khambay and D. A. Pulman, *Pestic. Sci.*, 1983, **14**, 182.
51. R. B. Mitra, G. H. Kulkarni, Z. Muljiani and P. N. Khanna, *Synth. Commun.*, 1988, **18**, 1139.
52. J. Brussee, E. C. Roos and A. van der Gen, *Tetrahedron Lett.*, 1988, **29**, 4485.
53. F. Effenberger, T. Ziegler and S. Forster, *Angew. Chem., Int. Ed. Engl.*, 1987, **26**, 458; *Angew. Chem.*, 1987, **99**, 491.
54. Y. Kobayashi, H. Hayashi, K. Miyaji and S. Inoue, *Chem. Lett.*, 1986, 931 and refs. therein.
55. K. Narasaka, T. Yamada and H. Minamikawa, *Chem. Lett.*, 1987, 2073.
56. K. Inoue, M. Malsumoto, S. Takahashi, T. Ohashi and K. Watanabe, *Eur. Pat.* 271 868 (1988) (*Chem. Abstr.*, 1988, **109**, 210 734n).
57. A. E. Vougioukas and M. B. Kagan, *Tetrahedron Lett.*, 1987, **28**, 5513.
58. M. T. Reetz, F. Kunisch and P. Heitmann, *Tetrahedron Lett.*, 1986, **27**, 4721.
59. V. M. F. Choi, J. D. Elliot and W. S. Johnson, *Tetrahedron Lett.*, 1984, **25**, 591.
60. K. Tanaka and F. Toda, *Nippon Kagaku Kaishi*, 1987, 456 (*Chem. Abstr.* 1987, **107**, 197 525g).
61. J. Castells, F. Pujol, H. Llitjos and M. Moreno-Manas, *Tetrahedron*, 1982, **38**, 337.
62. G. Stork and L. A. Maldonado, *J. Am. Chem. Soc.*, 1971, **93**, 5286; 1974, **96**, 5272.
63. U. Hertenstein, S. Hunig and M. Oller, *Synthesis*, 1976, 416.
64. T. W. Green, 'Protective Groups in Organic Synthesis', Wiley, New York, 1981, p. 141.
65. L. R. Krepski, K. M. Jensen, S. M. Heilmann and J. K. Rasmussen, *Synthesis*, 1986, 301.
66. R. Yoneda, K. Santo, S. Harusawa and T. Kurihara, *Synthesis*, 1986, 1054.
67. S. Hunig and G. Wehner, *Chem. Ber.*, 1979, **112**, 2062.
68. E. J. Corey and G. Schimdt, *Tetrahedron Lett.*, 1980, 731.
69. B. S. Bal, W. E. Childers, Jr. and H. W. Pinnick, *Tetrahedron*, 1981, **37**, 2091.
70. S.-I. Murahashi, T. Naota and N. Nakajima, *Tetrahedron Lett.*, 1985, **26**, 925.
71. H. Helmut and J. Johannes, *Chem. Ber.*, 1986, **119**, 1400.
72. L. R. Krepski, L. E. Lynch, S. M. Heilmann and J. K. Rasmussen, *Tetrahedron Lett.*, 1985, **26**, 981.
73. T. Kitazume, *Synthesis*, 1986, 855.
74. M. Gill, M. J. Kiefel and D. A. Lally, *Tetrahedron Lett.*, 1986, **27**, 1933; L. R. Krepski, S. M. Heilmann and J. K. Rasmussen, *Tetrahedron Lett.*, 1983, **24**, 4075.
75. B. Cazes and S. Julia, *Bull. Soc. Chim. Fr.*, 1977, 925, 931; J. M. McIntosh, *Can. J. Chem.*, 1977, **55**, 4200.
76. J. M. Photis, *J. Org. Chem.*, 1981, **46**, 182.
77. Y. Ito, H. Imai, K. Segoe and T. Saegusa, *Chem. Lett.*, 1984, 937.
78. S. Torii, T. Inokuchi and T. Kobayashi, *Chem. Lett.*, 1984, 897.
79. M. D. Rozwadowska and M. Chrzanowska, *Tetrahedron*, 1985, **41**, 2885.
80. R. Sanchez-Obregon, E. Marquez, H. Barrios, B. Ortez, F. Yuste and F. Walls, in 'Proceedings of the 3rd Chemical Congress of North America on Organic Chemistry, Toronto, Canada, 1988', American Chemical Society, 1988, p. 185.
81. G. A. Garcia, G. R. Yvonne and G. Q. Yolanda, in 'Proceedings of the 3rd Chemical Congress of North America on Organic Chemistry, Toronto, Canada, 1988', American Chemical Society, 1988, p. 472.
82. T. Takahashi, A. Ootake, H. Yamada and J. Tsuji, *Tetrahedron Lett.*, 1985, **26**, 69.
83. G. A. Garcia, H. Munoz and J. Tamariz, *Synth. Commun.*, 1983, **13**, 569.
84. F. L. Malanco and L. A. Maldonado, *Synth. Commun.*, 1976, **6**, 515.

85. G. Stork and T. Takahashi, *J. Am. Chem. Soc.*, 1977, **99**, 1275.
86. G. Stork, T. Takahashi, I. Kawamoto and T. Suzuki, *J. Am. Chem. Soc.*, 1978, **100**, 8272.
87. G. Stork, J.-C. Depezay and J. d'Angelo, *Tetrahedron Lett.*, 1975, 389.
88. T. Takahashi, K. Hori and J. Tsuji, *Tetrahedron Lett.*, 1981, **22**, 119.
89. T. Takahashi, T. Nagashima and J. Tsuji, *Tetrahedron Lett.*, 1981, **22**, 1359.
90. T. Takahashi, Y. Kanda, H. Nemoto, K. Kitamura, J. Tsuji and Y. Fukazawa, *J. Org. Chem.*, 1986, **51**, 3393.
91. T. Takahashi, H. Nemoto, J. Tsuji and I. Miura, *Tetrahedron Lett.*, 1983, **24**, 3485.
92. T. Takahashi, K. Kitamura, H. Nemoto, J. Tsuji and I. Miura, *Tetrahedron Lett.*, 1983, **24**, 3489.
93. T. Takahashi, K. Kitamura and J. Tsuji, *Tetrahedron Lett.*, 1983, **24**, 4695.
94. T. Takahashi, H. Nemoto, Y. Kanda, J. Tsuji, Y. Fukazawa, T. Okajima and Y. Fujise, *Tetrahedron*, 1987, **43**, 5499; *Heterocycles*, 1987, **25**, 139.
95. T. Takahashi, H. Nemoto and J. Tsuji, *Tetrahedron Lett.*, 1983, **24**, 2005.
96. T. Takahashi, T. Nagashima, H. Ikeda and J. Tsuji, *Tetrahedron Lett.*, 1982, **23**, 4361.
97. T. Takahashi, I. Minami and J. Tsuji, *Tetrahedron Lett.*, 1981, **22**, 2651.
98. T. Takahashi, K. Shimizu, T. Doi, J. Tsuji and Y. Fukazawa, *J. Am. Chem. Soc.*, 1988, **110**, 2674.
99. M. F. Semmelhack, in 'New Application of Organometallic Reagents in Organic Synthesis', ed. D. Seyferth, Elsevier, New York, 1976, p. 361.
100. K. Okazaki, K. Nomura and E. Yoshii, *Synth. Commun.*, 1987, **17**, 1021 and refs. therein.
101. G. W. Morrow, J. S. Swenton, J. A. Filppi and R. L. Wolgemuth, *J. Org. Chem.*, 1987, **52**, 713.
102. R. W. Irvine, S. A. Kinloch, A. S. McCormick, R. A. Russell and R. N. Warrener, *Tetrahedron*, 1988, **44**, 4591.
103. S. R. Khanapure, R. T. Reddy and E. R. Biehl, *J. Org. Chem.*, 1987, **52**, 5685.
104. K. Nomura, K. Okazaki, K. Hori and E. Yoshii, *J. Am. Chem. Soc.*, 1987, **109**, 3402.
105. G. C. Laredo and L. A. Maldonado, *Heterocycles*, 1987, **25**, 179.
106. E. Guittet and S. Julia, *Tetrahedron Lett.*, 1978, 1155.
107. S. A. Ferrino and L. A. Maldonado, *Synth. Commun.*, 1980, **10**, 717.
108. M. Uemura, T. Minami and Y. Hayashi, *J. Chem. Soc., Chem. Commun.*, 1984, 1193.
109. J. D. Albright, F. J. McEvoy and D. B. Moran, *J. Heterocycl. Chem.*, 1978, **15**, 881.
110. K. Takahashi, A. Honma, K. Ogura and H. Iida, *Chem. Lett.*, 1982, 1263.
111. J. Chauffaille, E. Hebert and Z. Welvart, *J. Chem. Soc., Perkin Trans. 2*, 1982, 1645.
112. L. A. Yanovskaya, Kh. Shakhidayatov, E. P. Prokofiev, G. M. Andrianova and V. F. Kucherov, *Tetrahedron*, 1968, **24**, 4677.
113. S. Harusawa, Y. Hamada and T. Shioiri, *Tetrahedron Lett.*, 1979, 4663.
114. K. Mai and G. Patil, *Tetrahedron Lett.*, 1984, **25**, 4583; K. Mai and G. Patil, *Synth. Commun.*, 1985, **15**, 157; W. J. Greenle, *J. Org. Chem.*, 1984, **49**, 2632.
115. T. Hanafusa, J. Ichihara and T. Ashida, *Chem. Lett.*, 1987, 687.
116. D. S. Stout, L. A. Black and W. L. Matier, *J. Org. Chem.*, 1983, **48**, 5369 and refs. therein.
117. I. Ojima and S. Inaba, *Chem. Lett.*, 1975, 737.
118. K. Mai and G. Patil, *Synth. Commun.*, 1984, **14**, 1299.
119. J. L. Marco, J. Royer and H. P. Husson, *Tetrahedron Lett.*, 1985, **26**, 3567.
120. C. K. Chen, A. G. Hortmann and M. R. Marzabadi, *J. Am. Chem. Soc.*, 1988, **110**, 4829.
121. V. Reutrakul, P. Ratananukul and S. Nimirawath, *Chem. Lett.*, 1980, 71.
122. N. Maigrot, J. P. Mazaleyrat and Z. Welvart, *J. Org. Chem.*, 1985, **50**, 3916.
123. D. Enders, in 'Current Trends in Organic Synthesis', ed. H. Nozaki, Pergamon Press, Oxford, 1983, p. 151.
124. M. Zervos and L. Wartski, *Tetrahedron Lett.*, 1984, **25**, 4641.
125. M. Zervos and L. Wartski, *Tetrahedron Lett.*, 1986, **27**, 2985.
126. J.-M. Fang, H. T. Chang and C. C. Lin, *J. Chem. Soc., Chem. Commun.*, 1988, 1385.
127. M. Bonin, J. R. Romero, D. S. Grierson and H. P. Husson, *Tetrahedron Lett.*, 1982, **23**, 3369.
128. M. Harris, D. S. Grierson and H. P. Husson, *Tetrahedron Lett.*, 1981, **22**, 1511.
129. K. Takahashi, H. Kurita, K. Ogura and H. Iida, *J. Org. Chem.*, 1985, **50**, 4368.
130. L. Stella and A. Amrollah-Modjdabadi, *Synth. Commun.*, 1984, **14**, 1141.
131. K. Takahashi, Y. Urano, K. Ogura and H. Iida, *Synthesis*, 1985, 690.
132. K. Takahashi, A. Suzuki, K. Ogura and H. Iida, *Heterocycles*, 1986, **24**, 1075.
133. P. Tuchinda, V. Prapansiri, W. Naengchomnong and V. Reutrakul, *Chem. Lett.*, 1984, 1427.
134. K. Takahashi, T. Aihara and K. Ogura, *Chem. Lett.*, 1987, 2359.
135. K. Takahashi, T. Mikajiri, H. Kurita, K. Ogura and H. Iida, *J. Org. Chem.*, 1985, **50**, 4372.
136. K. Takahashi, M. Asakawa and K. Ogura, *Chem. Lett.*, 1988, 1109.
137. J. Royer and H. P. Husson, *Tetrahedron Lett.*, 1987, **28**, 6175.
138. S. Arseniyadis, P. Q. Huang and H. P. Husson, *Tetrahedron Lett.*, 1988, **29**, 1391.
139. S. Arseniyadis, P. Q. Huang, D. Piveteau and H. P. Husson, *Tetrahedron*, 1988, **44**, 2457.
140. P. Q. Huang, S. Arseniyadis and H. P. Husson, *Tetrahedron Lett.*, 1987, **28**, 547.
141. S. Arseniyadis, P. Q. Huang and H. P. Husson, *Tetrahedron Lett.*, 1988, **29**, 631.
142. J. Royer and H. P. Husson, *J. Org. Chem.*, 1985, **50**, 670.
143. L. Guerrier, J. Royer, D. S. Grierson and H. P. Husson, *J. Am. Chem. Soc.*, 1983, **105**, 7754; H. P. Husson, *Youji Huaxue*, 1987, 314 (*Chem. Abstr.*, 1988, **109**, 73 693g).
144. M. Bonin, J. Royer, D. S. Grierson and H. P. Husson, *Tetrahedron Lett.*, 1986, **27**, 1569.
145. J. M. McIntosh and L. C. Matassa, *J. Org. Chem.*, 1988, **53**, 4452.
146. D. J. Aitken, J. Royer and H. P. Husson, *Tetrahedron Lett.*, 1988, **29**, 3315.
147. J. L. Marco, J. Royer and H. P. Husson, *Tetrahedron Lett.*, 1985, **26**, 6345.
148. J. L. Marco, J. Royer and H. P. Husson, *Synth. Commun.*, 1987, **17**, 669.
149. E. Zeller and D. S. Grierson, *Heterocycles*, 1988, **27**, 1575.
150. T. Nagasaka, H. Tamano, T. Maekawa and F. Hamaguchi, *Heterocycles*, 1987, **26**, 617.
151. T. Nagasaka, H. Hayashi and F. Hamaguchi, *Heterocycles*, 1988, **27**, 1685.

152. J. V. Cooney and W. E. McEwen, *J. Org. Chem.*, 1981, **46**, 2570.
153. H. Ahlbrecht and H. Dollinger, *Synthesis*, 1985, 743.
154. I. T. Kay and S. E. J. Glue, *Tetrahedron Lett.*, 1986, **27**, 113.
155. S. Harusawa, R. Yoneda, T. Kurihara, Y. Hamada and T. Shioiri, *Chem. Pharm. Bull.*, 1983, **31**, 2932.
156. R. Yoneda, K. Santo, S. Harusawa and T. Kurihara, *Synth. Commun.*, 1987, **17**, 921.
157. M. T. Reetz and H. M. Starke, *Tetrahedron Lett.*, 1984, **25**, 3301.
158. F. Pochat and E. Levas, *Tetrahedron Lett.*, 1976, 1491.
159. T. Takeda, Y. Kaneko, H. Nakagawa and T. Fujiwara, *Chem. Lett.*, 1987, 1963.
160. M. Yokoyama, H. Ohteki, M. Kurauchi, K. Hoshi, E. Yanagisawa, A. Suzuki and T. Imamoto, *J. Chem. Soc., Perkin Trans. 1*, 1984, 2635.
161. D. Morgans, Jr and G. B. Feigelson, *J. Org. Chem.*, 1982, **47**, 1131.
162. E. Hatzgrigoriou, L. Wartski, J. Seyden-Penne and E. Toromanoff, *Tetrahedron*, 1985, **41**, 5045.
163. T. Kitahara and K. Mori, *J. Org. Chem.*, 1984, **49**, 3281.
164. T. Sakamoto, E. Katoh, Y. Kondo and H. Yamanaka, *Heterocycles*, 1988, **27**, 1353; H. Suzuki, Q. Yi, J. Inoue, K. Kusume and T. Ogawa, *Chem. Lett.*, 1987, 887.
165. M. Yamashita, J. Onozuka, G. Tsuchihashi and K. Ogura, *Tetrahedron Lett.*, 1983, **24**, 79.
166. F. Barrière, J. C. Barrière, D. H. R. Barton, J. Cleophax, A. Gateau-Oleskar, S. D. Gero and F. Tadj, *Tetrahedron Lett.*, 1985, **26**, 3119, 3121.
167. P. A. Grieco and Y. Yokoyama, *J. Am. Chem. Soc.*, 1977, **99**, 5210.
168. D. N. Brattesani and C. H. Heathcock, *Tetrahedron Lett.*, 1974, 2279.
169. R. E. Koenigkramer and H. Zimmer, *Tetrahedron Lett.*, 1980, **21**, 1017.
170. G. H. Birum and G. Richardson (Monsanto Inc.), *US Pat.* 3 113 139 (1963) (*Chem. Abstr.*, 1961, **60**, 5551d).
171. M. Sekine, M. Nakajima, A. Kume and T. Hata, *Tetrahedron Lett.*, 1979, 4475; T. Hata, A. Hashizume, M. Nakajima and M. Sekine, *Tetrahedron Lett.*, 1978, 363.
172. N. C. J. M. Broekhof and A. van der Gen, *Recl. Trav. Chim. Pays-Bas*, 1984, **103**, 305, 312.
173. E. Enders and T. Gassner, *Chem. Ber.*, 1984, **117**, 1034.
174. N. C. J. M. Broekhof, P. van Elburg, D. J. Hoff and A. van der Gen, *Recl. Trav. Chim. Pays-Bas*, 1984, **103**, 317.
175. M. Venet, *Janssen Chim. Acta*, 1985, **3**, 18.
176. D. Seebach and E. J. Corey, *J. Org. Chem.*, 1975, **40**, 231.
177. T. W. Green, 'Protective Groups in Organic Synthesis', Wiley, New York, 1981, p. 129.
178. Y. Kamitori, M. Hojo, R. Masuda, T. Kimura and T. Yoshida, *J. Org. Chem.*, 1986, **51**, 1427.
179. R. Caputo, C. Ferreri and G. Palumbo, *Synthesis*, 1987, 386.
180. T. Sato, E. Yoshida, T. Kobayashi, J. Otera and H. Nozaki, *Tetrahedron Lett.*, 1988, **29**, 3971.
181. J. A. Soderquist and E. I. Miranda, *Tetrahedron Lett.*, 1986, **27**, 6305.
182. S. Kim, S. S. Kim, S. T. Lim and S. C. Shim, *J. Org. Chem.*, 1987, **52**, 2114.
183. H. C. Brown and T. Imai, *J. Org. Chem.*, 1984, **49**, 892.
184. M. C. Carre, G. Ndebeka, A. Riondel, P. Bourgasser and P. Caubère, *Tetrahedron Lett.*, 1984, **25**, 1551.
185. J. Einhorn and J. L. Luche, *J. Org. Chem.*, 1987, **52**, 4124.
186. T.-L. Ho, R. J. Hill and C. M. Wong, *Heterocycles*, 1988, **27**, 1718.
187. M. Balogh, A. Cornelis and P. Laszlo, *Tetrahedron Lett.*, 1984, **25**, 3313.
188. A. Cornelis and P. Laszlo, *Synthesis*, 1985, 909.
189. H. J. Cristau, A. Bazbouz, P. Morand and E. Torrielles, *Tetrahedron Lett.*, 1986, **27**, 2965.
190. H.-J. Liu and V. Wiszniewski, *Tetrahedron Lett.*, 1988, **29**, 5471.
191. J. E. Richman, J. L. Herrmann and R. H. Schlessinger, *Tetrahedron Lett.*, 1973, 3267.
192. B. C. Ranu and D. C. Sarkar, *J. Chem. Soc., Chem. Commun.*, 1988, 245.
193. D. Horton and W. Priebe, *Carbohydr. Res.*, 1981, **94**, 27.
194. E. Juraristi, J. S. Cruz-Sanchez and F. R. Ramos-Morales, *J. Org. Chem.*, 1984, **49**, 4912.
195. A. K. Samaddar, T. Chiba, Y. Kobayashi and F. Sato, *J. Chem. Soc., Chem. Commun.*, 1985, 329.
196. L. A. Flippin and M. A. Dombroski, *Tetrahedron Lett.*, 1985, **26**, 2977.
197. A. E. Davey and R. J. K. Taylor, *J. Chem. Soc., Chem. Commun.*, 1987, 25.
198. B. Zehnder (Hoffmann-La Roche), *Eur. Pat.* 209 854 (*Chem. Abstr.*, 1987, **106**, 213 769b).
199. B. C. Ranu and D. C. Sarkar, *Synth. Commun.*, 1987, **17**, 155.
200. G. Solladié and J. Hutt, *Tetrahedron Lett.*, 1987, **28**, 797.
201. M. Kinoshita, M. Taniguchi, M. Morioka, H. Takami and Y. Mizusawa, *Bull. Chem. Soc. Jpn.*, 1988, **61**, 2147; M. Kinoshita, M. Morioka, M. Taniguchi and T. Shimizu, *Bull. Chem. Soc. Jpn.*, 1987, **60**, 4005.
202. T. R. Kelly, N. S. Chandrakumar, J. D. Cutting, R. R. Goehring and F. R. Weibel, *Tetrahedron Lett.*, 1985, **26**, 2173.
203. M. D. Rozwadowska and D. Matecka, *Tetrahedron*, 1988, **44**, 1221; M. Chrzanowska and M. D. Rozwadowska, *Tetrahedron*, 1986, **42**, 6021; M. D. Rozwadowska and M. Chrzanowska, *Tetrahedron*, 1986, **42**, 6669.
204. P. T. Lansbury, D. J. Mazur and J. P. Springer, *J. Org. Chem.*, 1985, **50**, 1632.
205. J. Kuhnke and F. Bohlmann, *Liebigs Ann. Chem.*, 1988, 743.
206. M. Rubiralta, N. Casamitjana, D. S. Grierson and H. P. Husson, *Tetrahedron*, 1988, **44**, 443.
207. J.-M. Fang, B. C. Hong and L.-F. Liao, *J. Org. Chem.*, 1987, **52**, 855; *J. Org. Chem.*, 1986, **51**, 2828; J.-M. Fang and B.-C. Hong, *J. Org. Chem.*, 1987, **52**, 3162.
208. Y. Honda, E. Morita, K. Ohshiro and G. Tsuchihashi, *Chem. Lett.*, 1988, 21.
209. A. Murai, M. Ono and T. Masamune, *J. Chem. Soc., Chem. Commun.*, 1977, 573.
210. L. Duhamel, P. Duhamel and N. Mancelle, *Bull. Soc. Chim. Fr.*, 1974, 331.
211. H. Redlich and J. B. Lenfers, *Liebigs Ann. Chem.*, 1988, 597.
212. R. A. Ellison, *Synthesis*, 1973, 397; T.-L. Ho, *Synth. Commun.*, 1974, **4**, 265.
213. C. A. Brown and A. Yamaichi, *J. Chem. Soc., Chem. Commun.*, 1979, 101.
214. K. Tomioka, M. Sudani, Y. Shinmi and K. Koga, *Chem. Lett.*, 1985, 329.

215. H. Kawasaki, K. Tomioka and K. Koga, *Tetrahedron Lett.*, 1985, **26**, 3031.
216. S. Bernasconi, G. Jommi, S. Montanari and M. Sisti, *Synthesis*, 1987, 1126.
217. R. S. Ward, *Chem. Soc. Rev.*, 1982, **11**, 102.
218. A. Pelter, R. S. Ward, M. C. Pritchard and I. T. Kay, *Tetrahedron Lett.*, 1985, **26**, 6377; *J. Chem. Soc., Perkin Trans. 1*, 1988, 1603.
219. N. Rehnberg and G. Magnusson, *Tetrahedron Lett.*, 1988, **29**, 3599.
220. F. E. Ziegler and J.-M. Fang, *J. Org. Chem.*, 1981, **46**, 825.
221. K. Katsuura and V. Snieckus, *Tetrahedron Lett.*, 1985, **26**, 9.
222. A. A. Abdallah, J. P. Gesson, J. C. Jacquasy and M. Mandon, *Bull. Soc. Chim. Fr.*, 1986, 93.
223. D. R. S. Laurent and L. A. Paquette, *J. Org. Chem.*, 1986, **51**, 3861.
224. J. L. Herrmann, J. E. Richman and R. H. Schlessinger, *Tetrahedron Lett.*, 1973, 3275.
225. J. F. Biellmann and J. B. Ducep, *Org. React. (N.Y.)*, 1982, **27**, 103.
226. D. Seebach, N. R. Jones and E. J. Corey, *J. Org. Chem.*, 1968, **33**, 300.
227. F. M. Alvarez, R. K. V. Meer and C. S. Lofgren, *Tetrahedron*, 1987, **43**, 2897.
228. G. Khandekar, G. C. Robinson, N. A. Stacey, P. G. Steel, E. J. Thomas and S. Vather, *J. Chem. Soc., Chem. Commun.*, 1987, 877.
229. B. H. Lipshutz, H. Kotsuki and W. Lew, *Tetrahedron Lett.*, 1986, **27**, 4825.
230. S. L. Schreiber, S. E. Kelly, J. A. Porco, Jr., T. Sammakia and E. M. Suh, *J. Am. Chem. Soc.*, 1988, **110**, 6210.
231. Y. Nakahara, A. Fujita, K. Beppu and T. Ogawa, *Tetrahedron*, 1986, **42**, 6465.
232. R. M. Soll and S. P. Seitz, *Tetrahedron Lett.*, 1987, **28**, 5457.
233. S. L. Schreiber and J. J. Sommer, *Tetrahedron Lett.*, 1983, **24**, 4781.
234. M. Nishizawa, H. Yamada and H. Hayashi, *Tetrahedron Lett.*, 1986, **27**, 3255.
235. P. N. Confalone and R. A. Earl, *Tetrahedron Lett.*, 1986, **27**, 2695.
236. A. V. Rama Rao, M. W. Deshmukh and G. V. M. Sharma, *Tetrahedron*, 1987, **43**, 779.
237. S. J. Rao, U. T. Bhalerao and B. D. Tilak, *Indian J. Chem., Sect. B*, 1987, **26**, 208.
238. G. R. Newkome, H. W. Lee and F. P. Fronczek, *Isr. J. Chem.*, 1986, 87.
239. K. Ogura, M. Yamashita, M. Suzuki, S. Furukawa and G. Tsuchihashi, *Bull. Chem. Soc. Jpn.*, 1984, **57**, 1637.
240. Y. Torisawa, H. Okabe and S. Ikegami, *J. Chem. Soc., Chem. Commun.*, 1984, 1602.
241. A. F. Cunningham, Jr. and E. P. Kundig, *J. Org. Chem.*, 1988, **53**, 1823.
242. E. P. Kundig and A. F. Cunningham, Jr., *Tetrahedron*, 1988, **44**, 6855.
243. P. C. B. Page, M. B. van Niel and D. Westwood, *J. Chem. Soc., Perkin Trans. 1*, 1988, 269; *J. Chem. Soc., Chem. Commun.*, 1987, 775.
244. G. Guanti, L. Banfi, A. Guaragna and E. Narisano, *J. Chem. Soc., Perkin Trans. 1*, 1988, 2369.
245. S. J. Blarer, *Tetrahedron Lett.*, 1985, **26**, 4055.
246. T. Sato, R. Kato, K. Gokyu and T. Fujisawa, *Tetrahedron Lett.*, 1988, **29**, 3955.
247. G. Guanti, L. Banfi and E. Narisano, *Tetrahedron Lett.*, 1986, **27**, 3547.
248. T. Yamamori and I. Adachi, *Tetrahedron Lett.*, 1980, **21**, 1747.
249. M. Yamashita and R. Suemitsu, *J. Chem. Soc., Chem. Commun.*, 1977, 691.
250. K. P. Shelly and L. Weiler, *Can. J. Chem.*, 1988, **66**, 1359.
251. G. Roger, C. A. Keal and M. A. Sackville, *J. Chem. Res. (S)*, 1986, 282 (*Chem. Abstr.*, 1987, **106**, 84 228y).
252. M. Ashwell, A. S. Jones and R. T. Walker, *Nucleic Acids Res.*, 1987, **15**, 2157 (*Chem. Abstr.*, 1987, **107**, 154 665).
253. A. G. M. Barrett and N. K. Capps, *Tetrahedron Lett.*, 1986, **27**, 5571.
254. K. Toshima, K. Tatsuta and M. Kinoshita, *Bull. Chem. Soc. Jpn.*, 1988, **61**, 2369.
255. Y. Mori, A. Takeuchi, H. Kageyama and M. Suzuki, *Tetrahedron Lett.*, 1988, **29**, 5423.
256. J.-M. Fang, M. Y. Chen and W. J. Yang, *Tetrahedron Lett.*, 1988, **29**, 5937.
257. J.-M. Fang and M. Y. Chen, *Tetrahedron Lett.*, 1988, **29**, 5939.
258. S. Ncube, A. Pelter and K. Smith, *Tetrahedron Lett.*, 1979, 1893, 1895.
259. E. W. Colvin, T. A. Purcell and R. A. Raphael, *J. Chem. Soc., Perkin Trans. 1*, 1976, 1718.
260. S. Hatakeyama, K. Satoh, K. Sakurai and S. Takano, *Tetrahedron Lett.*, 1987, **28**, 2717.
261. T. Minami, K. Watanabe, T. Chikugo and Y. Kitajima, *Chem. Lett.*, 1987, 2369.
262. G. Bartoli, R. Dalpozzo, L. Grossi and P. E. Todesco, *Tetrahedron*, 1986, **42**, 2563.
263. A. G. Cameron, A. T. Hewson and A. H. Wadsworth, *Tetrahedron Lett.*, 1982, **23**, 561.
264. A. Hassner, P. Munger and B. A. Belinka, Jr., *Tetrahedron Lett.*, 1982, **23**, 699.
265. J. Otera, *Synthesis*, 1988, 95; T. Sato, H. Okazaki, J. Otera and H. Nozaki, *J. Am. Chem. Soc.*, 1988, **110**, 5209.
266. B. J. M. Jansen, R. M. Peperzak and A. de Groot, *Recl. Trav. Chim. Pays-Bas*, 1987, **106**, 489.
267. B. J. M. Jansen, R. M. Peperzak and A. de Groot, *Recl. Trav. Chim. Pays-Bas*, 1987, **106**, 549.
268. B. J. M. Jansen, R. M. Peperzak and A. de Groot, *Recl. Trav. Chim. Pays-Bas*, 1987, **106**, 505.
269. I. Coldham, E. W. Collington, P. Hallett and S. Warren, *Tetrahedron Lett.*, 1988, **29**, 5321.
270. P. J. Kocienski, *Tetrahedron Lett.*, 1980, 1559.
271. D. J. Ager, *Tetrahedron Lett.*, 1980, **21**, 4759; 1981, **22**, 587; 1983, **24**, 95.
272. A. Krief, in 'The Chemistry of Organic Selenium and Tellurium Compounds', ed. S. Patai, Wiley, New York, 1987, vol. 2, p. 675.
273. D. L. J. Clive, *Tetrahedron*, 1978, **39**, 1049.
274. M. Clarembeau, A. Cravador, W. Dumont, L. Hevesi, A. Krief, J. Lucchetti and D. van Ende, *Tetrahedron*, 1985, **41**, 4793.
275. J. Lucchetti and A. Krief, *Synth. Commun.*, 1983, **13**, 1153.
276. M. Clarembeau and A. Krief, *Tetrahedron Lett.*, 1986, **27**, 1723.
277. P. Laszlo, P. Pennetreau and A. Krief, *Tetrahedron Lett.*, 1986, **27**, 3153.
278. C. Lucchetti and A. Krief, *Tetrahedron Lett.*, 1981, **22**, 1623.
279. D. J. Goldsmith, D. Liotta, M. Volmer, W. Hoekstra and L. Waykole, *Tetrahedron*, 1985, **41**, 4873.

280. A. M. van Leusen, R. Oosterwijk, E. van Echten and D. van Leusen, *Recl. Trav. Chim. Pays-Bas*, 1985, **104**, 50.
281. J. Moskal and A. M. van Leusen, *Tetrahedron Lett.*, 1984, **25**, 2581, 2585.
282. A. M. van Leusen, F. J. Schaart and D. van Leusen, *Recl. Trav. Chim. Pays-Bas*, 1979, **98**, 258.

2.5
Silicon Stabilization

JAMES S. PANEK
Boston University, MA, USA

2.5.1 INTRODUCTION

The chemistry of organic silicon compounds is one of the fastest growing fields in synthetic organic chemistry. Consequently, it has become one of the most important areas of research and the development of new methodology which provides general stereoselective reaction processes continues to be actively investigated.[1] This chapter will review the synthetic utility of silicon-stabilized carbanion equivalents and their use in addition reactions to C═X π-bonds. No attempt has been made to provide an exhaustive literature search, rather the discussion focuses on more recent developments in stereoselective bond-forming processes and their practical usage. The chapter has been organized by the various types of stabilized carbanion equivalents, with issues of selectivity being treated in the context of each specific type. Particular attention has been given to the treatment of σ–π-bonded silicon-stabilized carbanions, *e.g.* vinylsilanes, allenylsilanes, alkynylsilanes, propargylsilanes, allylsilanes and α-metallated organosilanes. Primary emphasis has been placed on practical enantio- and stereo-selective synthetic methods and yields; reaction conditions and proposed mechanisms, however, are frequently discussed. Reactions which have not been judged to be synthetically useful, due to isolation of the desired addition product in low yield, have been neglected.

2.5.2 CHARACTERISTICS OF ORGANOSILICON COMPOUNDS

Organic silicon compounds display a multitude of functions in organic synthesis.[1] The wide applicability of organosilicon reagents in stereoselective bond-forming reactions can be attributed to the large number of functional groups and the variety of reaction conditions that can be accommodated by silicon, and to its ability to function as an electron donor and acceptor. The reactivity and selectivity of reactions involving organosilanes is dependent upon both steric requirements and electronic contributions. The electronic effects associated with silicon include: (i) inductive effects; (ii) field effects; (iii) *p–d* π-bonding; and (iv) hyperconjugative effects. A brief summary of the physical properties of organosilicon compounds is helpful.

2.5.2.1 Inductive Effects

Inductive effects are generally considered to be transmitted through the σ-framework of a molecule. The electronegativity of an element is usually a measure of its ability to attract σ-electrons.[2] In many synthetic operations selectivity reflects the energy differences between reagents, activated intermediates and transition states. Thus, caution must be exercised when considering the influence of electronic effects on selectivities of reactions involving organosilanes. Through purely inductive effects trialkylsilyl groups are electron donating; however, the inductive effects of silicon are weak and in general only influence atoms directly bonded to it.

2.5.2.2 Field Effects

Field effects describe the polarization of an adjacent π-system to the σ-dipole moment of the entire R_3Si group.[2,3] A π-inductive effect alters a nearby π-system without charge transfer to or from that system. Two types of π-inductive effects have been described.[4] The first is termed π_σ and is the result of charge differences in the σ-bonding system, as a consequence of inductive effects. The effect is illustrated in Figure 1, where an electropositive substituent Y induces a partial positive charge at the C-1 position of the benzene ring. This induces redistribution of the π-electron density in a fashion that can be predicted by consideration of the contributing resonance forms in the illustrated structures. In the case of an electron-withdrawing substituent Y, the π_σ-effect predicts diminished electronegativity at C-2 and C-3 and increased electron density at C-1 and C-4.

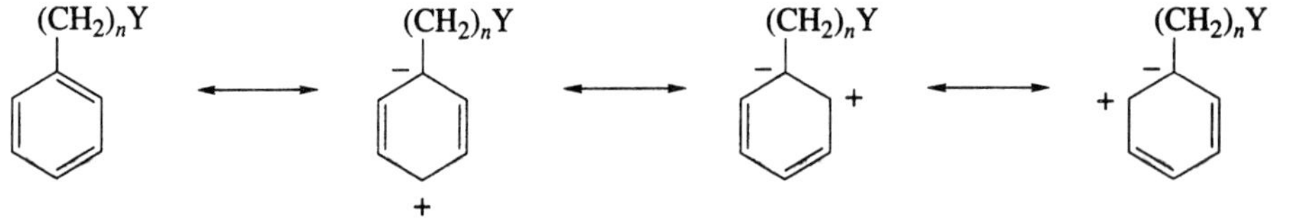

$(CH_2)_nY$ $(CH_2)_nY$ $(CH_2)_nY$ $(CH_2)_nY$

Figure 1

The second π-inductive effect, π_F, is the field effect, which arises when the electric dipole of $(CH_2)_nY$ affects the entire π-system through polarization. This process is illustrated in Figure 2.[2] Both π_σ and π_F contribute to the overall π-electron density, but it is difficult to separate these effects when they operate mainly at the C-1 position. For trialkylsilyl groups it is not easy to predict the magnitude of the field effect because although the Si—C bond is polarized such that silicon bears a partial positive charge, the trialkylsilyl group can be electron withdrawing, depending on the alkyl substituents.

$(CH_2)_nY$
δ–
δδ– δδ–
δδ+ δδ+
δ+

Figure 2

2.5.2.3 *p–d* π-Bonding

The mechanism by which the trialkylsilyl group can function as a π-electron-withdrawing group is influenced by the physical properties of silicon. The most widely recognized explanation is that the low-lying, unoccupied silicon *d*-orbitals can participate in *p–d* π-bonding, as illustrated in Figure 3.[2,5] In this illustration the electron density from the *p*-orbital on X can be delocalized onto silicon through a donor–acceptor interaction with the vacant Si 3*d*-orbital. Pauling was the first to introduce this concept to provide an explanation for the short lengths of silicon–oxygen and silicon–halogen bonds.[6] The *p–d* π-bonding model is most easily applied to systems in which electron density in a *p*-type orbital adjacent to silicon is transferred onto the silicon atom.

Figure 3

2.5.2.4 Hyperconjugation

When two adjacent molecular orbitals are relatively close in energy and have appropriate symmetry, they can undergo perturbation resulting in the lowering of the energy of one orbital and an increase in the other. This phenomenon is shown in Figure 4 for the interaction of the Si—C σ^*- orbital with a π-orbital.[1f,2] In this example the *p*-orbital is lowered in energy through hyperconjugation, as reflected in the ionization potential. The magnitude of the hyperconjugative interaction is directly proportional to the orbital energy difference and the orbital coefficients.

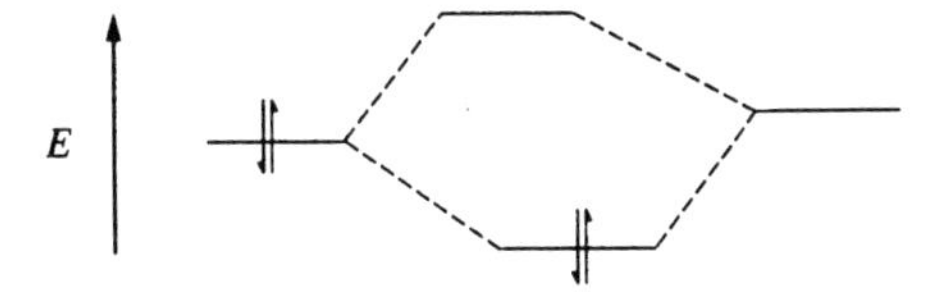

Figure 4 The hyperconjugative interaction of π- and σ^*- orbitals

The hyperconjugation model when applied to carbocation stabilization in organosilane systems is illustrated in Figure 5.

δ+
H δ– H
δ+SiR₃ ⟷ H H +SiR₃

Figure 5

For organosilicon compounds the conformational requirements are well defined; for maximum stabilization of the carbocation β to the silicon group it is necessary for the C—Si bond to be orientated antiperiplanar to the empty *p*-orbital. Another system that is amenable to the simple hyperconjugation model is the α-silyl carbanion, $R_3SiCH_2^-$, as illustrated in Figure 6. Since the C—Si bonding orbital is higher in energy than C—C or C—H bonding orbitals and has a large coefficient on the adjacent carbon atom, through-space hyperconjugative stabilization by silicon is more influential in stabilizing an electron deficient center than that of an alkyl substituent or hydrogen.[1a,f,2]

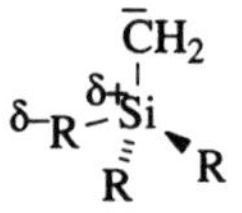

$\bar{C}H_2$
δ+ Si, δ− R, R, R

Figure 6

2.5.2.5 Nucleophilic Substitution Reactions

In the periodic table a column of elements will exhibit a trend of electronegativity that decreases in magnitude from top to bottom. The same trend is observed for silicon and carbon; the Pauling electronegativity of carbon is 2.5 eV, while that of silicon is 1.8,[1b,6] indicating that the C–Si bond is polarized such that silicon is the electron deficient atom and is responsible for accelerating nucleophilic attack at silicon (Figure 7). The reaction rate is in fact accelerated if the group undergoing displacement is a good leaving group.[1b] As a result, organosilanes are relatively nonpolar compounds compared to other organometallic reagents, so they can tolerate the presence of a wide variety of functional groups and in most cases can be handled without special precaution.

$$\overset{\delta+}{Si}-\overset{\delta-}{C}$$

Figure 7

The σ-bonds between silicon and electronegative elements such as the halogens are quite strong. A comparison of the bond dissociation energies between carbon/halogen and silicon/halogen is listed in Table 1.[1b]

Table 1 Bond Energies (kJ mol^{-1})

C—F	552	Si—F	552.7 ± 2.1	C—Br	280 ± 21	Si—Br	343 ± 50
C—Cl	397 ± 29	Si—Cl	456 ± 42	C—I	209 ± 21	Si—I	293

The driving force for many addition reactions involves the formation of a new silicon compound with a stronger bond to the silicon atom. Treatment of the trimethylsilyl ether (**1**) with fluoride ion (F^-) results in the formation of fluorotrimethylsilane (**2**) because the derived Si—F bond has a higher bond dissociation energy (BDE) than the Si—O bond undergoing heterolytic bond cleavage (Scheme 1). The facile removal of the silyl group by use of a fluoride ion is commonly referred to as desilylation and is responsible for the extensive use of trialkylsilanes as protecting groups in synthesis.[7]

$$\underset{(\mathbf{1})}{RO-SiMe_3} \xrightarrow{F^-} \underset{(\mathbf{2})}{FSiMe_3} + RO^-$$

Scheme 1

The result of carbon–silicon σ-bond stabilization of an adjacent carbocation in electrophilic additions to C—X π-bonds is that the reactions normally occur with high levels of stereo- and regio-control. For instance, Lewis acid catalyzed addition reactions of allylsilanes to C—X π-systems proceed with C-3-regiospecificity and electrophilic additions of vinylsilanes occur predominantly at the silicon-bearing carbon. Trapping of the β-carbocation through the loss of the silyl group (or silicon substituent) then gives the addition product.

2.5.3 SILICON ATTACHED TO PARTICIPATING π-BONDED CARBON NUCLEOPHILES

The spectrum of silicon-stabilized carbanions containing σ-bonded silyl groups attached to a participating π-bond (π-nucleophile) and their use in organic synthesis has grown considerably over the last few years. In this context vinylsilanes, alkynylsilanes and allenylsilanes have emerged as remarkably useful and versatile carbon nucleophiles for the regio- and stereo-selective formation of carbon–carbon bonds. This section will discuss the stereochemical aspects of σ–π-bonded silicon-stabilized carbanions in addition reactions to C—X π-systems. Excellent reviews have appeared covering vinylsilane-[1a,8] and alkynylsilane-terminated cyclization reactions.[8]

2.5.3.1 Vinylsilanes

Vinylsilanes function as vinyl anion equivalents in addition reactions to C—X π-systems. The reaction has been successfully conducted under a variety of conditions. The strong σ-donating capability of the silyl group is primarily responsible for directing the initially formed cation along a single reaction pathway. The adjacent carbanion center is stabilized by overlap with the low energy, unoccupied 3*d*-orbital of silicon or by σ → σ* orbital overlap (Figure 8). As a consequence, characteristic regio- and stereoselectivity are observed in vinylsilane addition reactions; as shown in Figure 9, electrophilic substitution takes place at the silicon-bearing carbon and occurs stereospecifically with retention of configuration.[8]

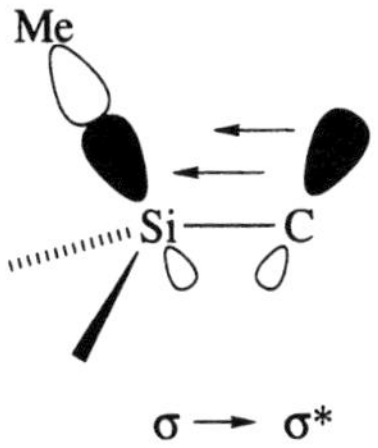

Me Si C σ → σ*

Figure 8

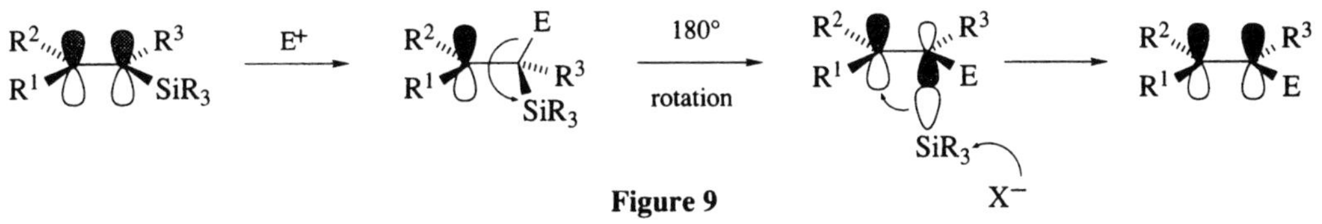

R^2 R^3 R^1 SiR_3 E^+ → R^2 E R^1 R^3 SiR_3 180° rotation → R^2 R^3 R^1 E SiR_3 X^- → R^2 R^3 R^1 E

Figure 9

The intermolecular addition of vinylsilanes to C—X π-systems is a well-studied reaction;[1j,8a,b] however, not until 1981 had the intramolecular variant been demonstrated. Thus, coverage of vinylsilane additions in this chapter will focus on the regio- and stereo-selectivity in intramolecular (cyclization) reactions. Two modes of cyclization are possible, as determined by the position of the alkene with respect to the ring system: cyclization can occur in an endocyclic or an exocyclic mode with respect to the vinylsilane group (Figure 10).

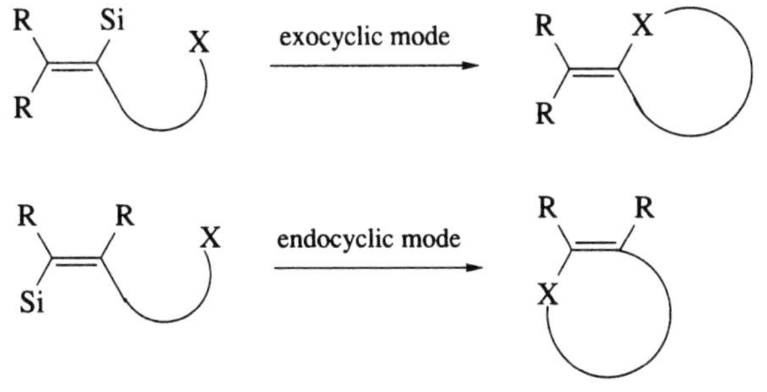

R Si X R exocyclic mode → R X R

R R X Si endocyclic mode → R R X

X = electrophilic initiating center; Si = nucleophilic terminating center

Figure 10

The first report of an intramolecular addition of a vinylsilane came from the Burke laboratory.[9] The reaction was used for the preparation of spiro[4.5]decadienones. As shown in Scheme 2, the addition of the vinylsilane to the intermediate acylium ion generated using titanium tetrachloride resulted in the formation of the enone system **(3)** → **(4)**.

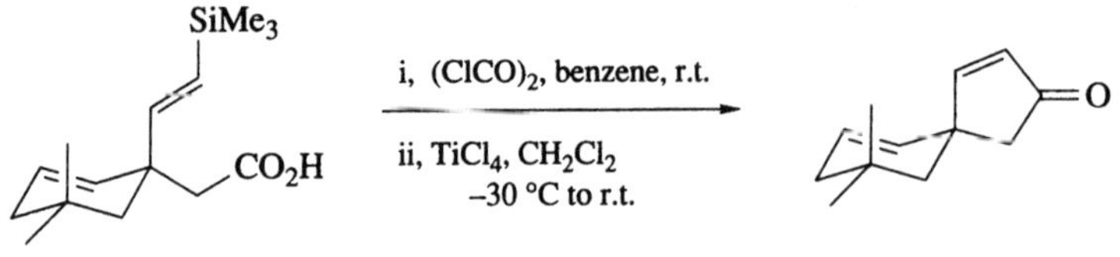

$SiMe_3$ CO_2H i, $(ClCO)_2$, benzene, r.t. ii, $TiCl_4$, CH_2Cl_2 –30 °C to r.t.

(3) **(4)**

Scheme 2

Kuwajima has reported similar internal addition reactions resulting in the formation of spirocyclic enones. Treatment of the acyl chlorides (**5a,b**; Scheme 3) with $TiCl_4$ resulted in the formation of enone systems (**6a**) and (**6b**) in good yields.[10]

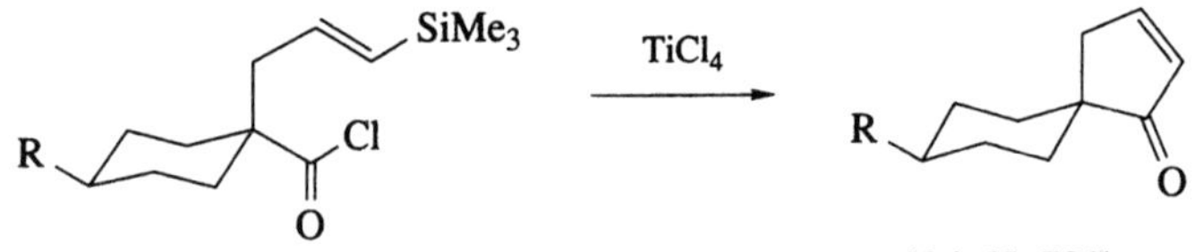

$SiMe_3$ $TiCl_4$

(**5a**) R = H
(**5b**) R = Bu^t

(**6a**) 65–70%
(**6b**) 61%

Scheme 3

Competition studies reported by Kuwajima,[10] which also complement the results of Nakai,[11] illustrate the limitations of the β-effect as a tool for predicting the outcome of vinylsilane-terminated cyclizations (Scheme 4). Acylium ion initiated cyclizations of (**7a**) and (**7b**) gave the expected cyclopentenones (**8a**) and (**8b**). However, compound (**7c**), upon treatment with titanium tetrachloride, gave exclusively the cyclopentenone product (**8c**) arising from the chemoselective addition on the 1,1-disubstituted alkene followed by protodesilylation of the vinylsilane. The reversal observed in the mode of addition may be a reflection of the relative stabilities of the carbocation intermediates. The internal competition experiments of Kuwajima indicate that secondary β-silyl cations are generated in preference to secondary carbocations (compare Schemes 3 and 4), while tertiary carbocations appear to be more stable than secondary β-silyl carbocations, as judged by the formation of compound (**8c**).

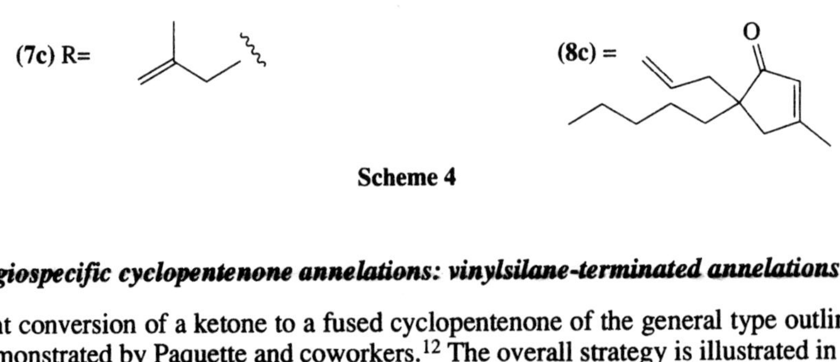

R COCl $SiMe_3$ $TiCl_4$

(**7a**) R=
(**7b**) R=
(**7c**) R=

(**8a**) 69%
(**8b**) 70%
(**8c**) =

Scheme 4

2.5.3.1.1 *Regiospecific cyclopentenone annelations: vinylsilane-terminated annelations*

The efficient conversion of a ketone to a fused cyclopentenone of the general type outlined in Scheme 5 has been demonstrated by Paquette and coworkers.[12] The overall strategy is illustrated in Scheme 6 and begins with Freidel–Crafts acylation of the cyclic vinylsilane (**8**) to give the enone (**9**), which undergoes tin tetrachloride mediated cyclization to give the fused 5,5-pentenones (**10a**) and (**10b**), as a 10:1 mixture of double bond isomers.

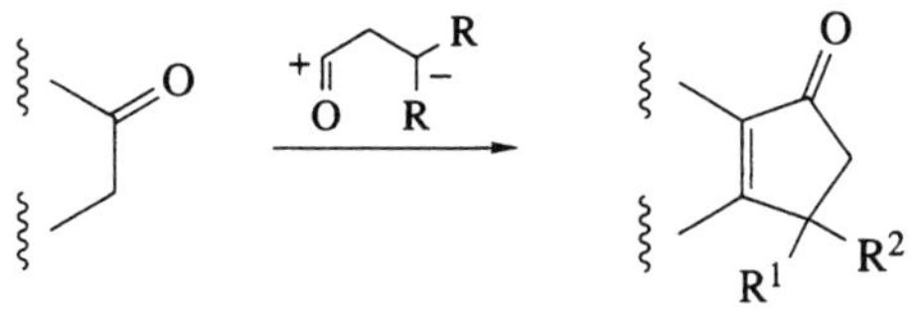

Scheme 5

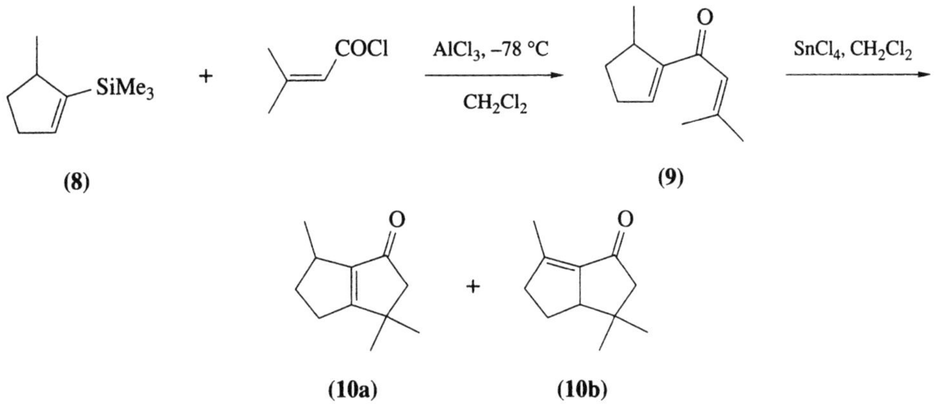

(8) + COCl; $AlCl_3$, –78 °C, CH_2Cl_2 → (9); $SnCl_4$, CH_2Cl_2 → (10a) + (10b)

Scheme 6

Subsequent investigations by Denmark have shown that vinylsilanes can direct the course of Nazarov-type cyclization reactions.[13] Once again the regiochemical outcome is controlled by the β-effect as substitution takes place at the carbon bearing the silyl group (Scheme 7). This clever modification provided a solution to the serious problem of double bond isomerization that is frequently observed in the classical Nazarov reaction. Treatment of the enone (**11**) with commercially available 'anhydrous' iron(III) chloride ($FeCl_3$) promoted cyclization to give one double bond isomer of the *cis*-hydrindinone (**12**) in excellent yield. Table 2 summarizes a representative collection of cyclization reactions catalyzed by $FeCl_3$.

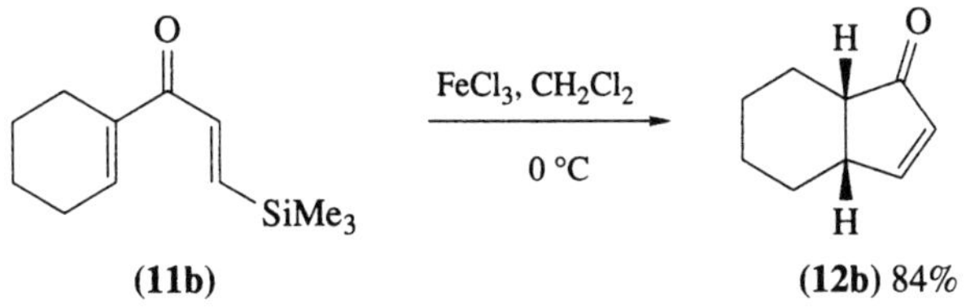

(11b) $FeCl_3$, CH_2Cl_2, 0 °C → (12b) 84%

Scheme 7

2.5.3.1.2 *Acetal- and carbonyl-initiated cyclizations*

In general, carbonyl groups do not function as good initiators in vinylsilane-mediated cyclization reactions and relatively few examples exist. The cyclization of vinylsilanes with ketones or aldehydes as initiators is a highly underdeveloped reaction that holds considerable potential. As reported by Tius and coworkers,[14] treatment of aldehyde (**13**) with a catalytic amount of *p*-toluenesulfonic acid gave a mixture of the tetralins (**14**) and (**15**) in a combined yield of 53%, with lesser amounts of the enone (**16**; Scheme 8). The enone system presumably arises from the intramolecular 1,3-hydride transfer of the intermediate α-silyl carbocation (**17**). Similarly, the vinylsilane (**18**) undergoes cyclization to produce the disubstituted benzene derivative (**19**), although yields are low.

Activated carbonyl groups serve as excellent substrates in Lewis acid catalyzed cyclization reactions, with faster reaction rates and higher yields being obtained in related cyclization processes involving acetals and ketals.[15] Thus, treatment of the dimethyl acetals (**20a–c**) with $TiCl_4$ at –78 °C afforded the biaryl

Table 2 Iron(III) Chloride Induced Cyclizations of Vinylsilanes

Substrate	*Solvent*	*Temperature* (° C)	*Time* (h)	*Product*	*Yield* (%)[a]
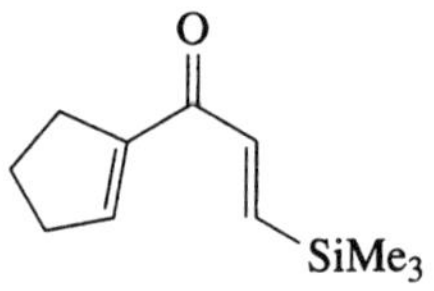 (11a)	$(CH_2Cl)_2$[b]	20	2.5	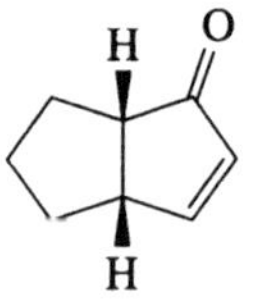 (12a)	55
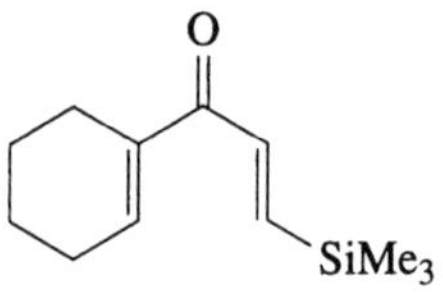 (11b)	CH_2Cl_2	0	4.0	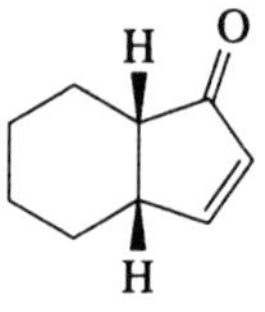 (12b)	84
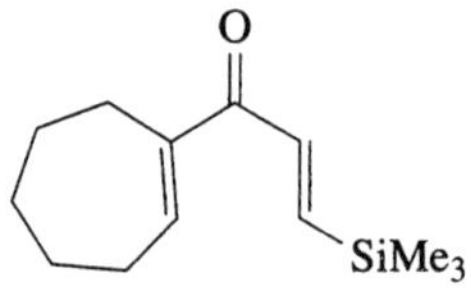 (11c)	CH_2Cl_2	0	1.0	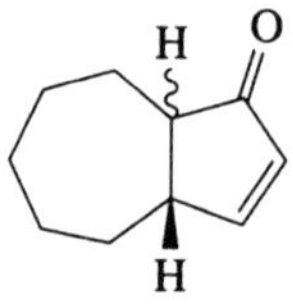 (12c)	74
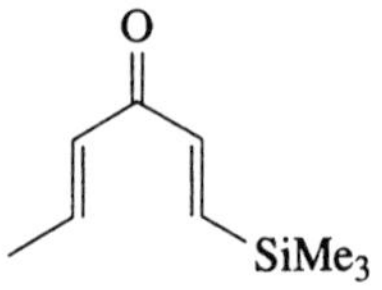 (11d)	CH_2Cl_2	20	12.0	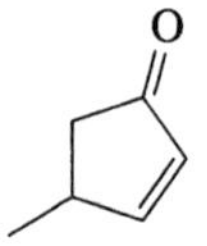 (12d)	54
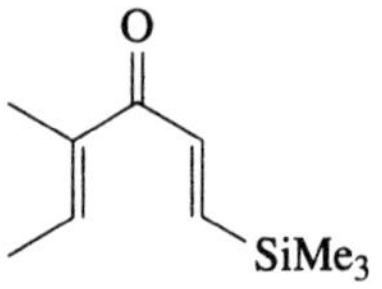 (11e)	CH_2Cl_2	20	12.0	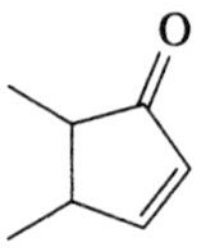 (12e)	95

[a] All reactions were run with 1.05 equiv. of $FeCl_3$ (0.08 M) in vinylsilane. [b] CH_2Cl_2 gave poor yields.

ring systems (**23a–c**). It was presumed by the authors that the products arose through the intermediates (**21**) and (**22**) (Scheme 9).

2.5.3.1.3 Effect of vinylsilane double bond configuration

Both vinylsilane dimethyl acetals (*E*)-(**24**) and (*Z*)-(**24**) undergo cyclization to form the substituted cyclohexene (**25**; Scheme 10).[1a,16] The vinylsilane (*Z*)-(**24**) cyclizes nearly twice as fast as the (*E*)-stereoisomer and in higher yield. This greater efficiency may be due to the lability of the product allyl ether to the Lewis acid catalyst.

Trost has reported the use of dithioacetals as initiators in vinylsilane-terminated cyclizations (Scheme 11).[17] Reaction of the dimethyl thioketal (**26**) with dimethyl(methylthio)sulfonium tetrafluoroborate (DMTSF)[18] led to the formation of the allylic sulfide (**28**) *via* a 2,3-sigmatropic rearrangement of the initially formed sulfonium salt (**27a**), which may have been catalyzed by complexation with excess DMTSF.

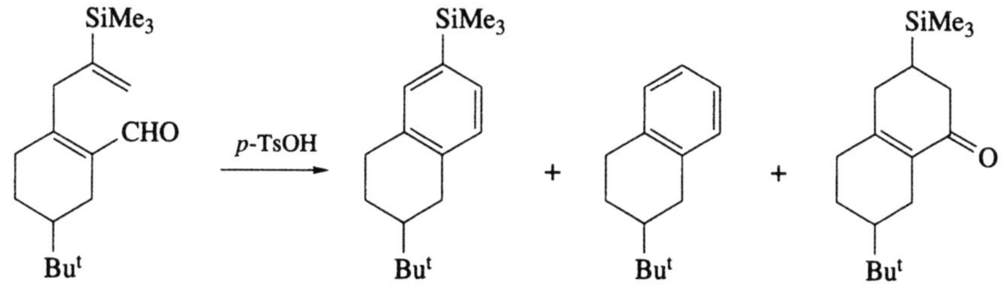

(13) (14) (15) (16)

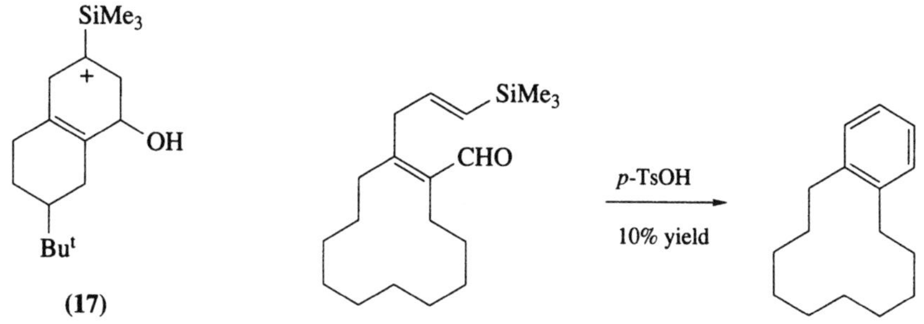

(17) (18) (19)

Scheme 8

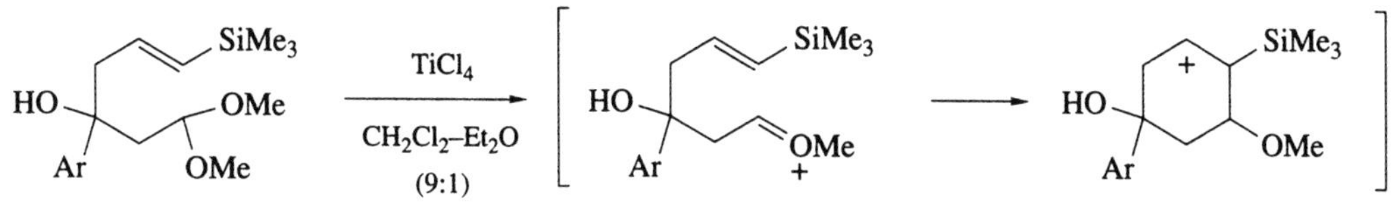

(20a) Ar = *p*-tolyl
(20b) Ar = β-naphthyl
(20c) Ar = 4-ethoxyphenyl

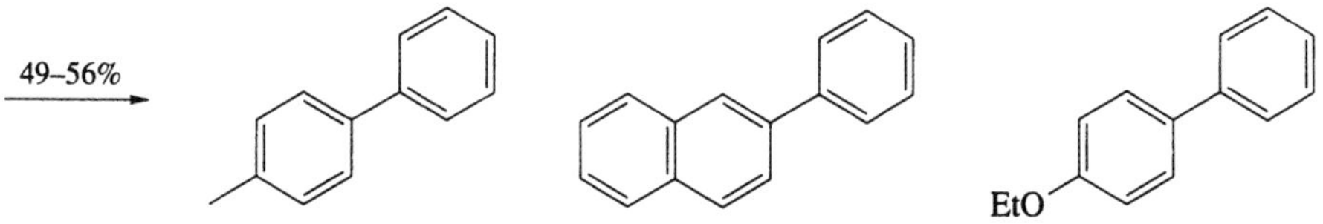

(21) (22)

(23a) (23b) (23c)

Scheme 9

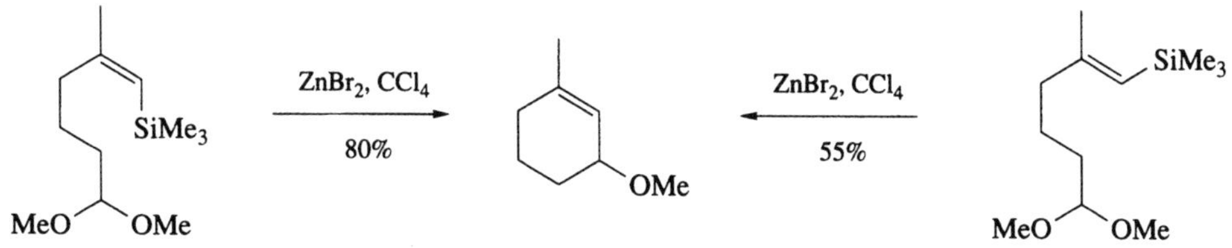

(E)-(24) (25) *(Z)*-(24)

Scheme 10

Competition experiments between two π-bonded nucleophiles within the same molecule were studied in an attempt to identify reaction parameters and factors responsible for regioselectivity.[17] These experiments, summarized in Scheme 12, demonstrated that the dithioacetal (initiating carbocation) is in competition with two nucleophilic functional groups within the same molecule, a silyl enol ether and a vinylsilane. In this instance, when the thioacetal (**29**) was treated with DMTSF, complete chemoselectiv-

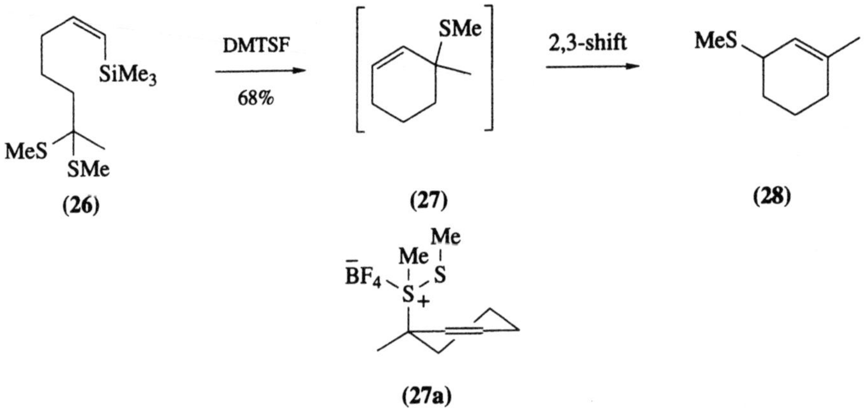

SiMe$_3$ MeS SMe (26) DMTSF 68% SMe (27) 2,3-shift MeS (28)

Me Me $\bar{B}F_4$ S S+ (27a)

Scheme 11

ity was observed for the addition of the more nucleophilic enol ether to the intermediate thionium ion to give the cyclohexanone derivative (**30b**). Cyclization product (**30a**), arising from reaction with the vinylsilane functional group, was not detected. However, cyclization of the corresponding ketone (**31**) gave the spirocyclic ketone (**32**). In this case, the first of the two consecutive cyclizations must occur *via* addition of the vinylsilane moiety, as neither ketone (**30b**) nor enone (**30a**) yielded the spirocyclic ketone (**32**) upon treatment with DMTSF (Scheme 13).

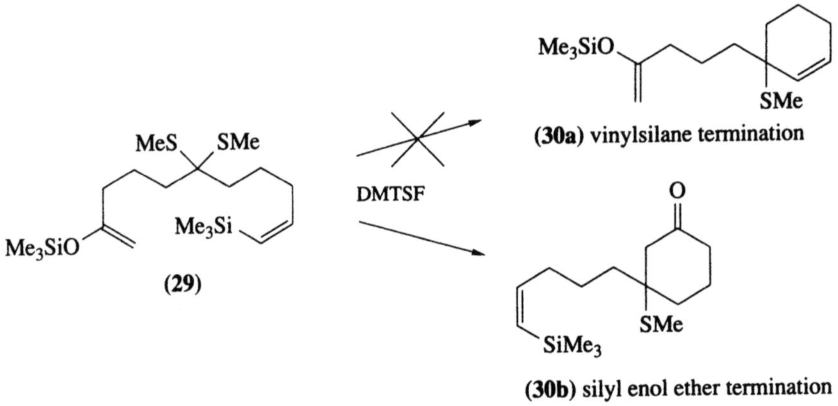

MeS SMe Me$_3$Si Me$_3$SiO (29) DMTSF

Me$_3$SiO SMe (**30a**) vinylsilane termination

O SMe SiMe$_3$ (**30b**) silyl enol ether termination

Scheme 12

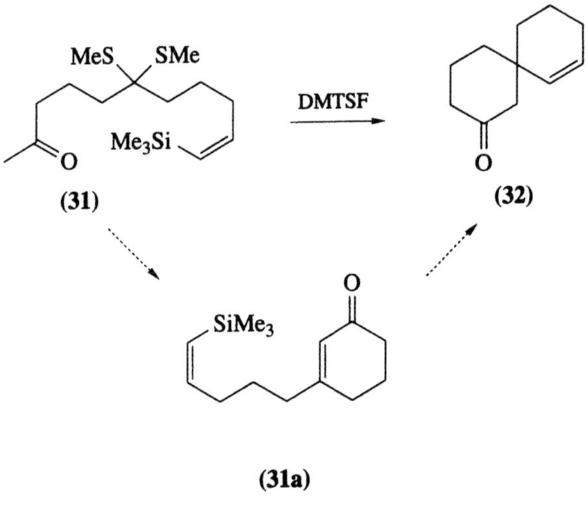

MeS SMe Me$_3$Si O (31) DMTSF O (32)

SiMe$_3$ O (31a)

Scheme 13

Related cyclizations that occur in an endocyclic mode with respect to the initiating acetal function lead to a variety of different sized oxygen heterocycles, including eight-membered cyclic ethers. As summarized earlier by Blumenkopf and Overman, four modes of oxonium ion promoted cyclizations are possible.[8a] The sense of cyclization is dependent on the orientation of the initiating and terminating functional groups (Figure 11; equations 1–4).

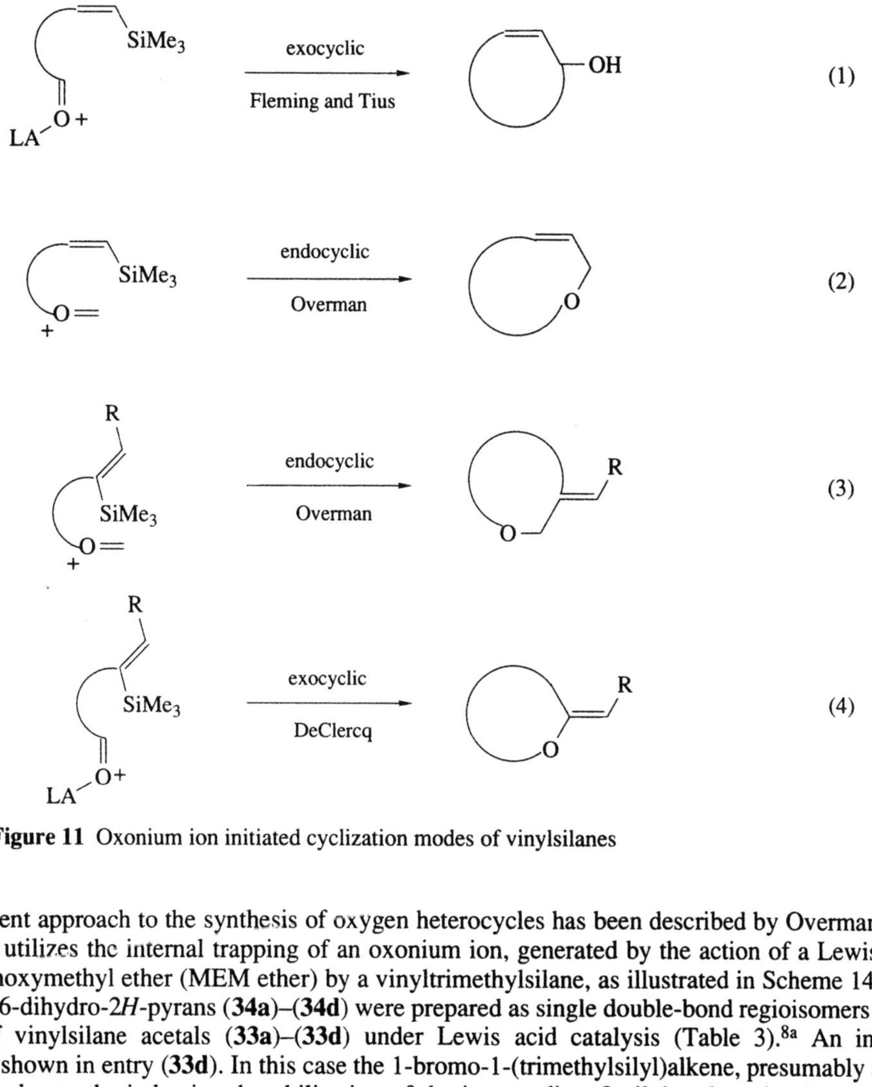

SiMe3 exocyclic → Fleming and Tius OH (1)
LA O+

SiMe3 endocyclic → Overman O (2)

R SiMe3 endocyclic → Overman R O (3)

R SiMe3 exocyclic → DeClercq R O (4)
LA O+

Figure 11 Oxonium ion initiated cyclization modes of vinylsilanes

An efficient approach to the synthesis of oxygen heterocycles has been described by Overman and coworkers. It utilizes the internal trapping of an oxonium ion, generated by the action of a Lewis acid on methoxyethoxymethyl ether (MEM ether) by a vinyltrimethylsilane, as illustrated in Scheme 14.[8a,19] For instance, 5,6-dihydro-2*H*-pyrans (**34a**)–(**34d**) were prepared as single double-bond regioisomers from the reaction of vinylsilane acetals (**33a**)–(**33d**) under Lewis acid catalysis (Table 3).[8a] An interesting example is shown in entry (**33d**). In this case the 1-bromo-1-(trimethylsilyl)alkene, presumably a weaker nucleophile due to the inductive destabilization of the intermediate β-silyl carbocation by the bromine atom, still underwent cyclization in good yield. More importantly, from a synthetic viewpoint the reaction resulted in the formation of a regiospecifically functionalized pyran system in good yield.

The construction of medium-sized ring systems has been a persistent and difficult problem in organic synthesis. The cyclizations that occur in an exocyclic mode with respect to the vinylsilane terminator have been used to prepare five-, six- and even seven- and eight-membered oxygen heterocycles.[19,20a,b] As illustrated in Scheme 15, for the construction of pyran systems, acetals (**35a**) and (**35b**) underwent cyclization to yield the tetrahydropyrans (**36a**) and (**36b**) in excellent yields with high levels of stereoselection (>99.5% retention of stereochemistry about the double bond).

The first stereocontrolled method for the preparation of 3-alkylideneoxepanes was demonstrated using this cyclization strategy.[20] Thus, Lewis acid promoted cyclizations of vinylsilane acetals (**37a**) and (**37b**) yielded the seven-membered ring alkylideneoxacycles (**38a**) and (**38b**), as illustrated in Scheme 16.

The vinylsilane-terminated cyclization strategy has been extended to the preparation of eight-membered cyclic ethers.[20a,b] Oxocenes (Scheme 17) with Δ^4-unsaturation (3,6,7,8-tetrahydro-2*H*-oxocins) were prepared efficiently by the $SnCl_4$-catalyzed cyclization of the mixed acetals (**39**) with complete regiochemical control. Electrophilic addition on the 2-(trimethylsilyl)-1-alkene occurs predominantly at the terminal position of the alkene to form a tertiary α-silyl, rather than a primary β-silyl, carbocation.[8a] The

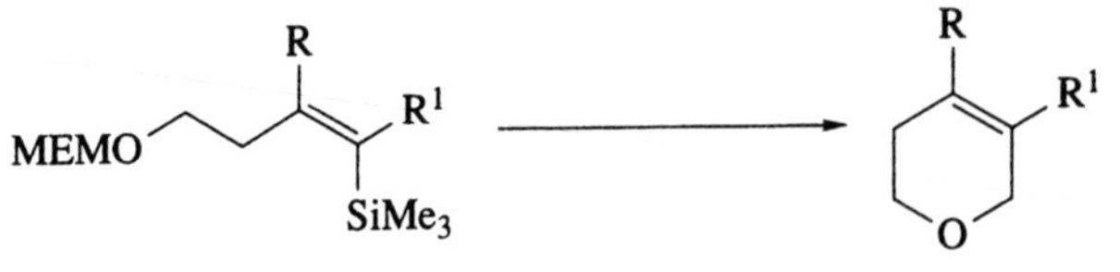

(33) (34)

Scheme 14

Table 3 Preparation of Dihydro-2*H*-pyrans (**34**)

Entry	*R*	R^1	*Lewis acid*	*Temperature* (°C)	*Time* (h)	*Yield* (%)
(**33a**)	H	H	$SnCl_4$	–20	1.0	71
(**33b**)	H	$(CH_2)_3Ph$	$SnCl_4$	–20	1.0	65
(**33c**)	Me	H	$TiCl_3(OPr^i)$	–20	2.0	83
(**33d**)	H	Br	$TiCl_4$	–60	2.0	78

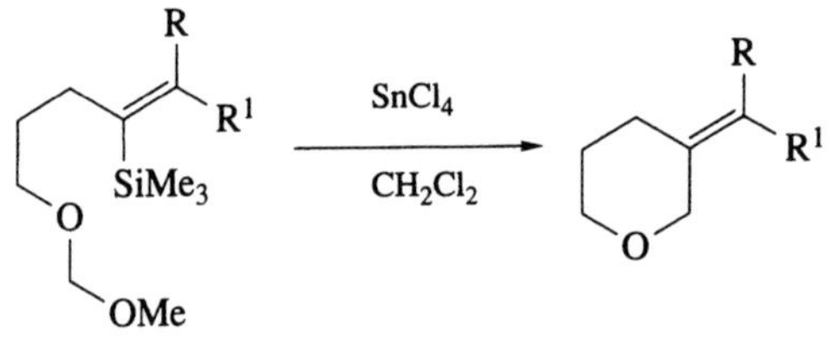

(**35a**) R = H, R^1 = Bu^n (**36a**) 89%

(**35b**) R = Bu^n, R^1 = H (**36b**) 92%

Scheme 15

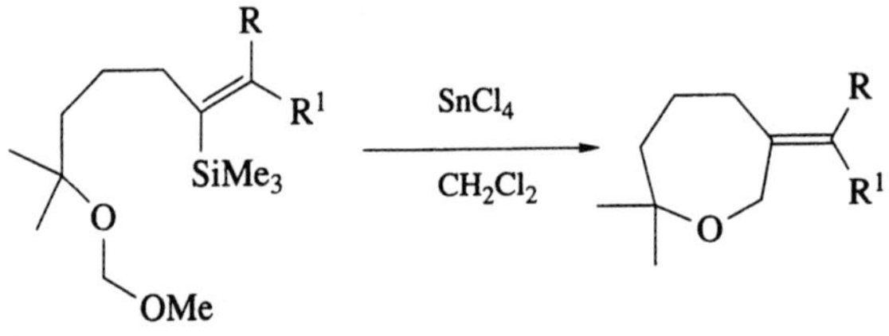

(**37a**) R = H, R^1 = Bu^n (**38a**) 71%

(**37b**) R = Bu^n, R^1 = H (**38b**) 57%

Scheme 16

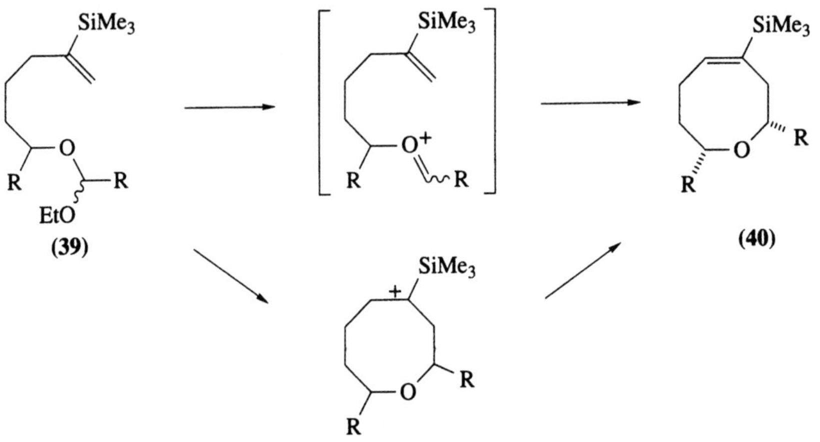

(**39**) (**40**)

Scheme 17

cis stereochemistry of cyclizations that form 2,8-disubstituted-Δ^4-oxocenes of structural type (**40**) can be rationalized on the basis that cyclization of the more stable (*E*)-oxonium ion (**41**) $\rightarrow$ (**42**) should occur preferentially in the conformation having the hydrogen (smallest substituent) at the α-carbon in the plane of the partial C—O π-bond (Scheme 18). Alternatively, Overman has suggested that the *cis* stereochemistry could arise from trapping of the (*E*)-oxonium ion in an intramolecular ene reaction, which proceeds through a boat–chair bicyclo[3.3.1]nonane transition state, as illustrated in Scheme 19.[20b]

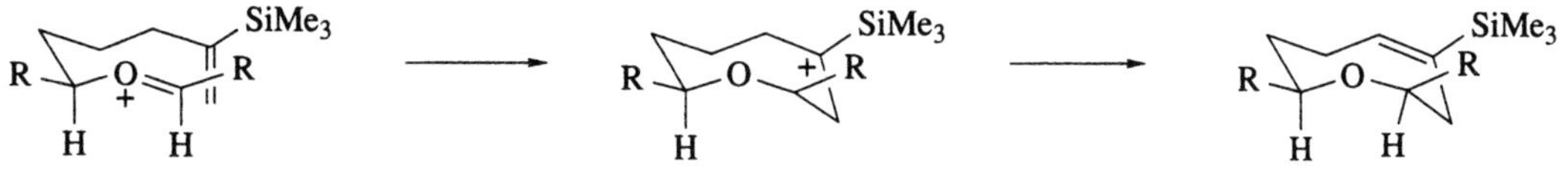

(*E*)-Oxonium ion (**41**) (**42**)

Scheme 18

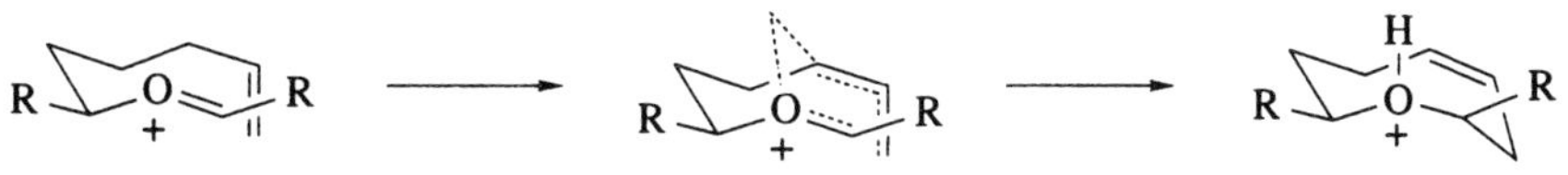

Scheme 19 Oxonium ion trapping *via* an intramolecular ene reaction

In contrast, vinylsilane acetals (**43a**)–(**43c**), with a two-carbon tether between the vinylsilane and the MEM group, cyclize in the presence of tin tetrachloride to form tetrasubstituted alkylidenetetrahydrofurans (**44a**)–(**44c**). However, little or no stereochemical preference for either (*E*)- or (*Z*)-alkene stereoisomers was observed in these reactions (Scheme 20 and Table 4).[19]

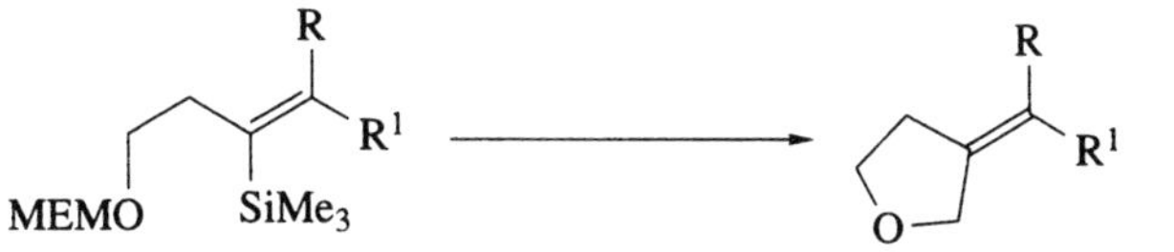

(**43**)

(**44a**) R = Et, R^1 = Bu^n
(**44b**) R = Bu^n, R^1 = Et
(**44c**) R = Bu^n, R^1 = H

Scheme 20

Table 4 $SnCl_4$-catalyzed Cyclizations of MEM Ethers

Entry	*R*	*R^1*	*Lewis acid*	*(E):(Z) ratio*	*Yield (%)*
(**43a**)	Et	Bu^n	$SnCl_4$	40:60	81
(**43b**)	Bu^n	Et	$SnCl_4$	40:60	86
(**43c**)	Bu^n	H	$SnCl_4$	97:3	81

Through the efforts of the Overman research group, the acetal–vinylsilane cyclization reaction has been shown to be a useful strategy for the stereoselective preparation of a variety of medium-sized oxygen heterocycles. This cyclization strategy has been used successfully in the asymmetric synthesis of the marine natural product (–)-laurenyne (**40a**), an eight-membered cyclic ether.[21] These molecules are members of an unusual class of C-15 nonisoprenoid metabolites.

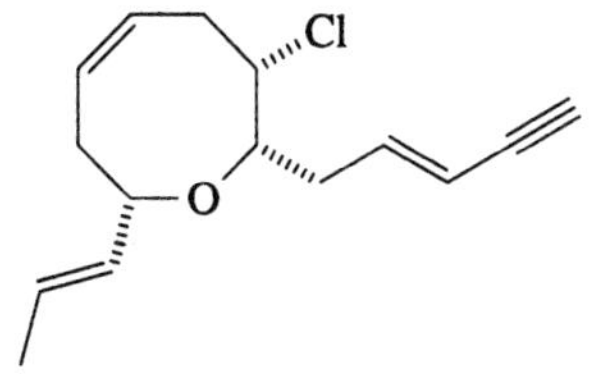

(**40a**) (–)-Laurenyne

In contrast, the related cyclizations used to form carbocyclic systems have been of limited synthetic value due to the acid sensitive nature of the derived allylic ether product (equation 1; Figure 11). The sensitivity of these carbocycles may be rationalized by considering the poor orbital overlap between the allylic C—O bond with the adjacent π-bond in a five-, six- or seven-membered ring.[8a,20a]

2.5.3.1.4 Acyliminium and iminium ion initiated cyclizations

Activated C=N bonds have been used extensively as initiators in cyclization reactions. In particular, acyliminium and iminium ions serve as reliable initiators in vinylsilane-terminated cyclizations and thus represent an exceptionally useful method for the construction of nitrogen heterocycles.[8a] As with oxonium ions, iminium ions may be generated using similar conditions that are compatible with the terminating vinyltrimethylsilanes. Acyliminium and iminium ions have been applied in key ring-forming reactions in a number of alkaloid syntheses. This section will examine the use of this methodology in stereoselective bond-forming reactions relevant to the synthesis of nitrogen heterocycles (Scheme 21).

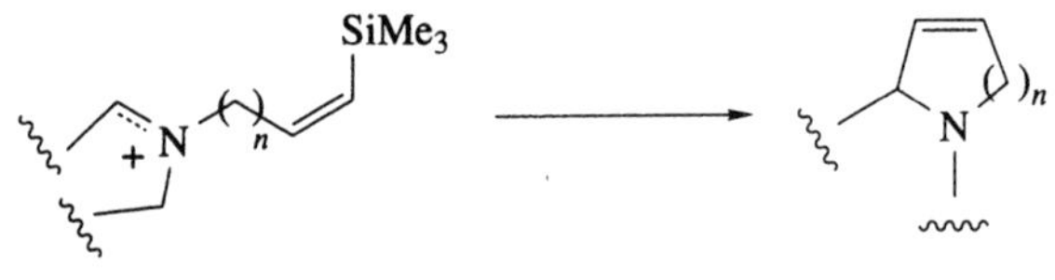

Scheme 21

An acyliminium ion terminated cyclization served as the key step in a short synthesis of the indolizinone alkaloids elaeocanines B (**47a**) and A (**47b**).[22] The key intermediates and cyclizations are summarized in Scheme 22.

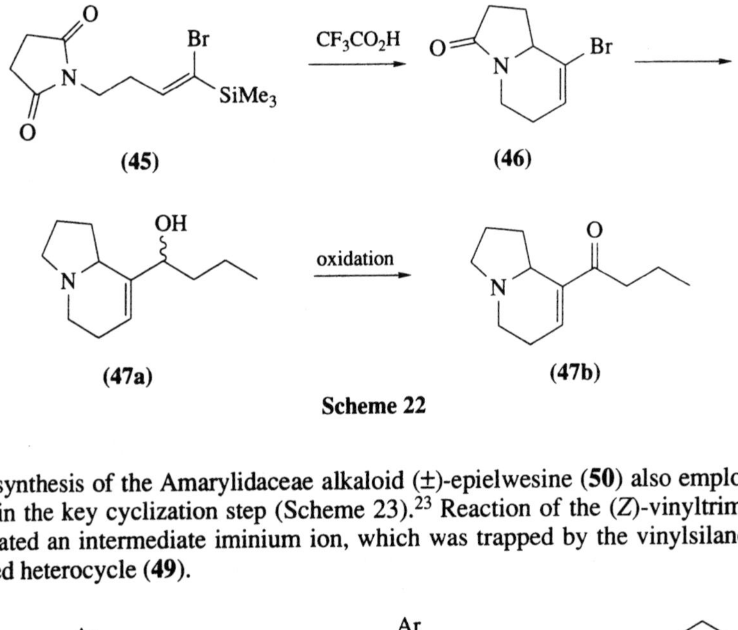

Scheme 22

The reported synthesis of the Amarylidaceae alkaloid (±)-epielwesine (**50**) also employed an iminium ion–vinylsilane in the key cyclization step (Scheme 23).[23] Reaction of the (*Z*)-vinyltrimethylsilane (**48**) with TFA generated an intermediate iminium ion, which was trapped by the vinylsilane nucleophile to yield the *cis* fused heterocycle (**49**).

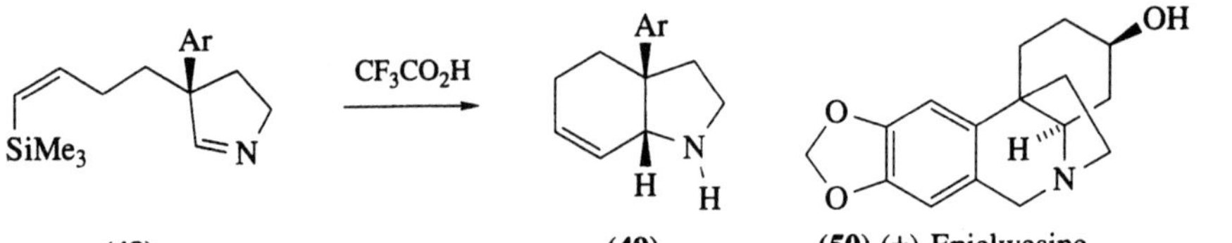

Scheme 23

The synthesis of the indoloquinolizadine alkaloid deplancheine (**52**; Scheme 24), using an exocyclic variant of an iminium ion initiated vinylsilane cyclization, has been reported by Malone and Overman.[24] Thus, treatment of vinylsilane (**51**) with formaldehyde resulted in the generation of an intermediate iminium ion that underwent smooth cyclization, forming the carbon framework of deplancheine. The (*Z*)-vinylsilane (**51b**) also cyclized to the (*Z*)-stereoisomer of deplancheine (**52b**) *via* the corresponding iminium ion to give the tetracyclic ring system in greater than 98% isomeric purity. This series of experiments provides an excellent demonstration of the high levels of stereoselection that can be achieved in vinylsilane cyclization reactions that occur in the exocyclic mode with respect to the attacking nucleophile.

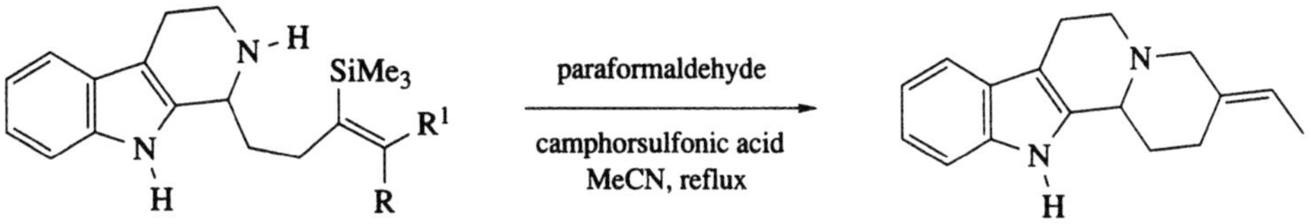

paraformaldehyde; camphorsulfonic acid, MeCN, reflux

(**51a**) R = Me, R^1 = H

(**51b**) R = H, R^1 = Me

(**52**) (±)-Deplancheine

Scheme 24

A successful enantioselective approach to the pumiliotoxin class of alkaloids has been developed using vinylsilane cyclization strategies.[25] Three members of this class, pumiliotoxin A (**55**), pumiliotoxin B (**56**) and pumiliotoxin 251D (**57**) have been synthesized in optically active forms through the use of iminium ion–vinylsilane cyclization reactions. The strategy employed in these syntheses relied on the chiral pool to provide optical activity. L-Proline was used as the optically active starting material; its (*S*)-stereocenter became the C-8a stereocenter of the indolidizidine ring system in all three alkaloids. The iminium ion–vinylsilane strategy was employed in constructing the piperidine ring and establishing the (*Z*)-stereochemistry of the alkylidene side chain. The key bond-forming reaction in each synthesis is detailed in Scheme 25 and involved the generation and subsequent trapping of an iminium ion by the vinylsilane. The total syntheses of (+)-geissoschizine (**58**) and (±)-(*Z*)-isositsirkine (**59**) have also been documented. The general strategy is outlined in Scheme 26 below and utilized, as a key step, stereocontrolled formation of the ethylidene side chain by stereospecific cyclization of an (*E*)- or (*Z*)-vinylsilane iminium (**A**) → (**B**).

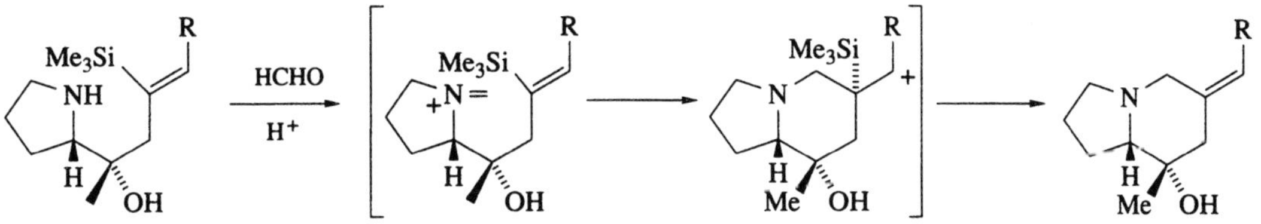

HCHO, H^+

(**53**) (**54**)

Scheme 25

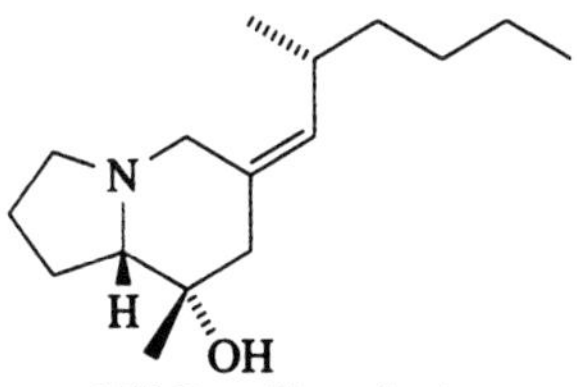

(**55**) Pumiliotoxin A

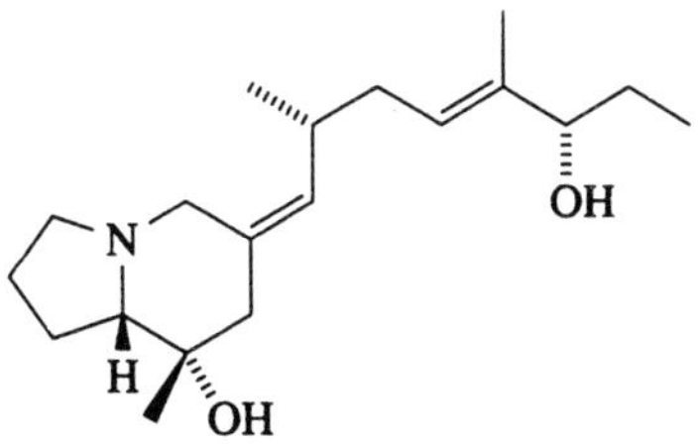

(**56**) Pumiliotoxin B

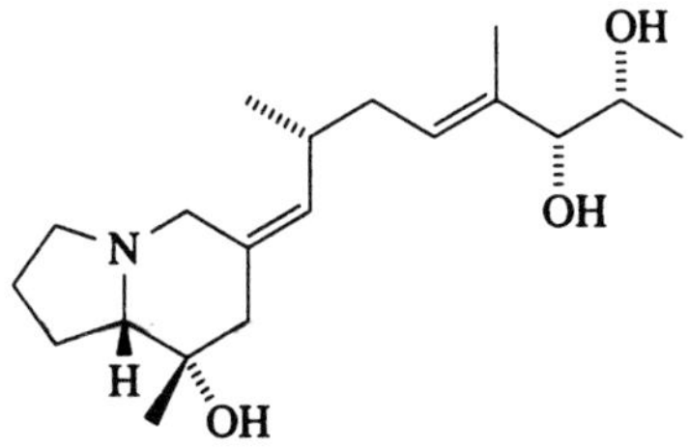

(**57**) Pumiliotoxin 251D

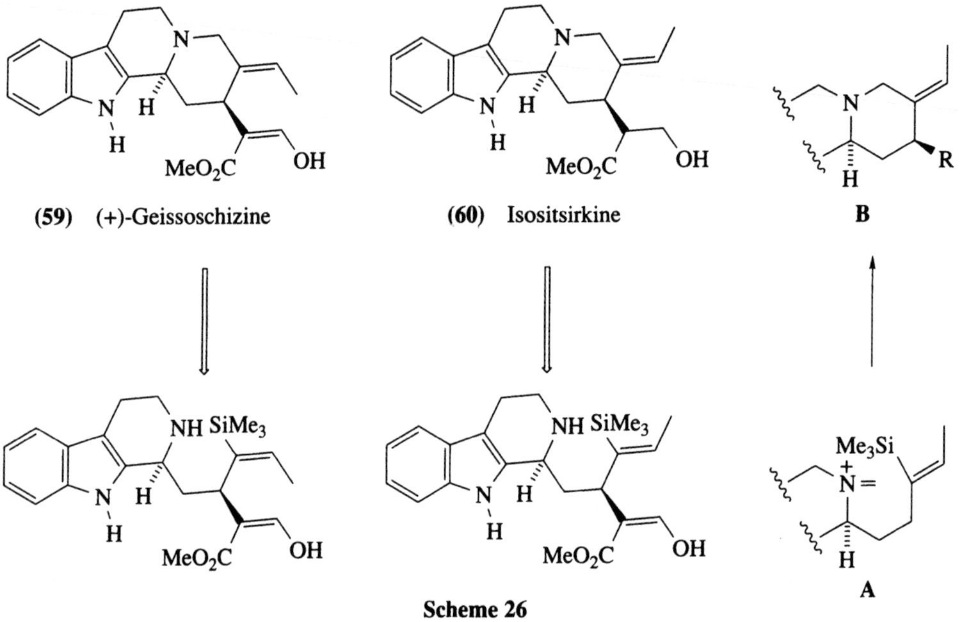

(59) (+)-Geissoschizine

(60) Isositsirkine

B

A

Scheme 26

An efficient enantiodivergent synthesis of levorotatory (1*S*,2*R*,8a*S*)-1,2-dihydroxyindolizidine (**62a**) and its antipode (**62b**) has been developed starting from commercially available D-isoascorbic acid (**61**; Scheme 27).[27] This strategy illustrates the complementary use of iminium and *N*-acyliminium ion inter-

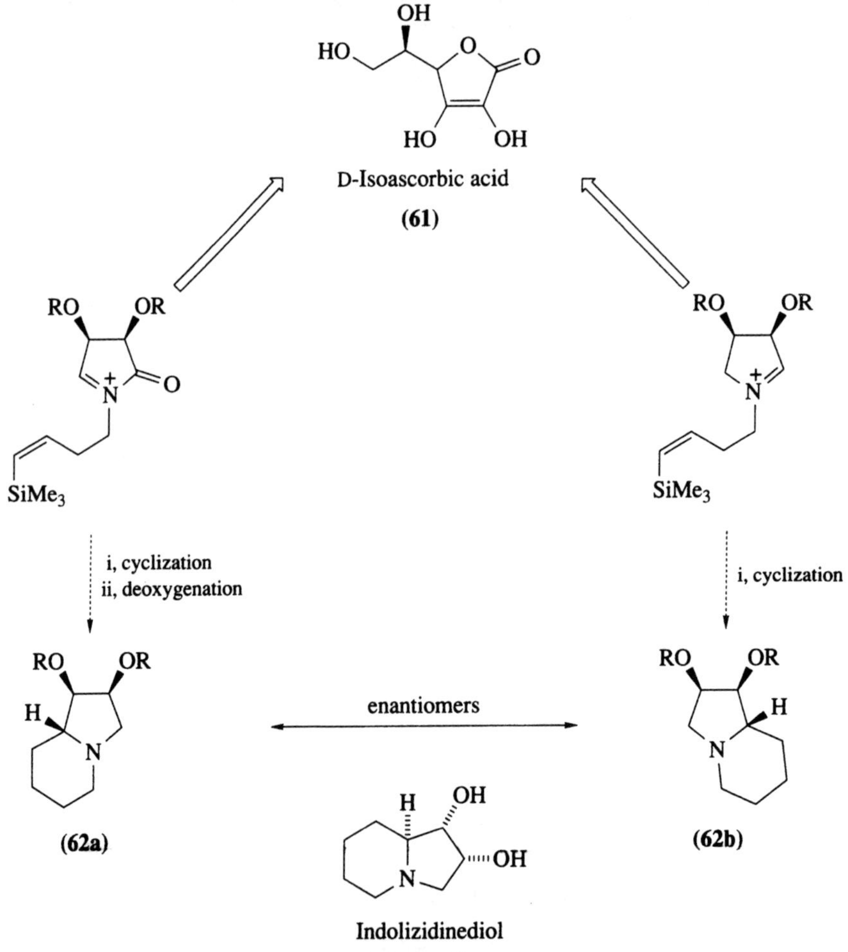

D-Isoascorbic acid

(61)

i, cyclization
ii, deoxygenation

i, cyclization

enantiomers

(62a)

(62b)

Indolizidinediol

Scheme 27

mediates as initiators for the enantiodivergent synthesis of the indolizidinediol alkaloids. The different types of iminium ion initiated cyclizations, and their related electron deficient intermediates, have become some of the most useful methods for the preparation of nitrogen heterocycles. These studies have illustrated the compatibility of iminium ion–vinylsilane cyclizations with complex alkaloids containing sensitive functional groups.[8a,25–27]

2.5.3.2 Allenylsilanes: Multifunctional Propargylic Anion Equivalents

Alkynic intermediates serve as important functional groups in organic synthesis. Many important reactions exploiting the unique and versatile chemistry of the carbon–carbon triple bond have been devised over the last few years.[28] A general strategy for the synthesis of substituted alkynes involves substitution and addition reactions of propargylic anion equivalents; this approach is particularly well suited for the preparation of homopropargylic alcohols (Scheme 28).

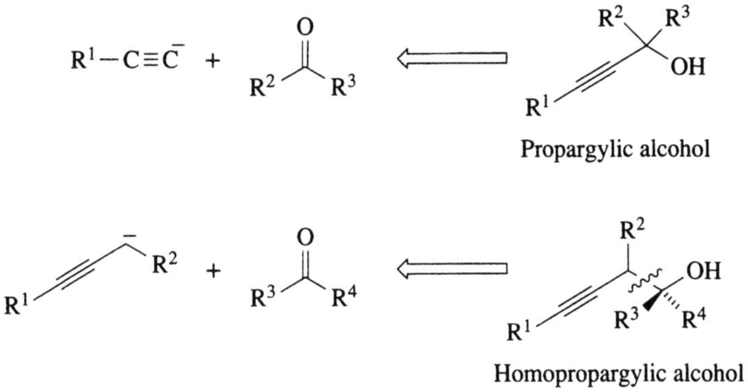

Propargylic alcohol

Homopropargylic alcohol

Scheme 28

This methodology has been limited by the tendency of metallated propargylic anions to add to electrophiles to produce a mixture of regioisomers. This lack of regiochemical control results from the fact that these anions exist as an equilibrating mixture of allenic and propargylic anions, which can both add to electrophiles to produce allenic and propargylic alcohols (Scheme 29). Thus, product ratios are determined by the position of the equilibrium between the two organometallic species. Pioneering work by Danheiser and coworkers has resulted in the development of allenylsilanes as useful propargylic anion equivalents.[29] Importantly, these readily available organometallic compounds do not participate in the dynamic equilibrium described above and, as a result, undergo additions with virtual regiospecificity. As described in Sections 2.5.4.1 to 2.5.4.8, allenylsilanes have proven to be useful carbon nucleophiles and are known to participate in regio- and stereo-controlled addition reactions to a variety of electrophiles.

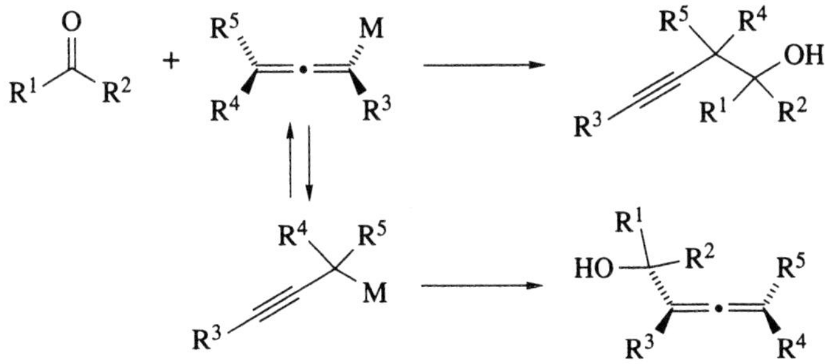

Scheme 29

2.5.3.2.1 Titanium tetrachloride promoted additions of allenylsilanes to carbonyl compounds

The $TiCl_4$-mediated addition of allenylsilanes (**66**; Scheme 30) to aldehydes and ketones provides a general regiocontrolled route to a wide variety of substituted homopropargylic alcohols.[29] Importantly, all three types of homopropargylic alcohols (**63, 64** and **65**) are accessible through allenylsilane methodology.

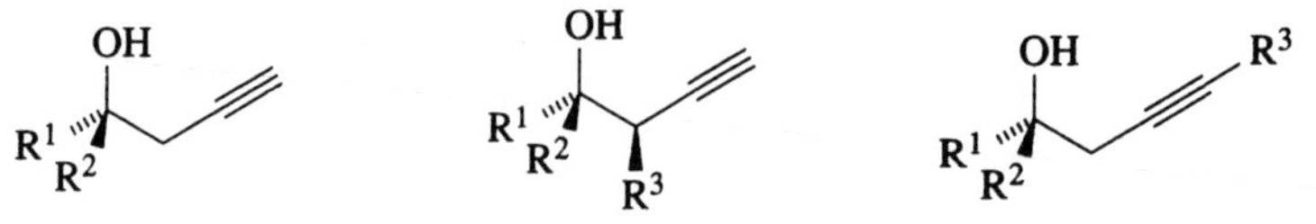

(63) (64) (65)

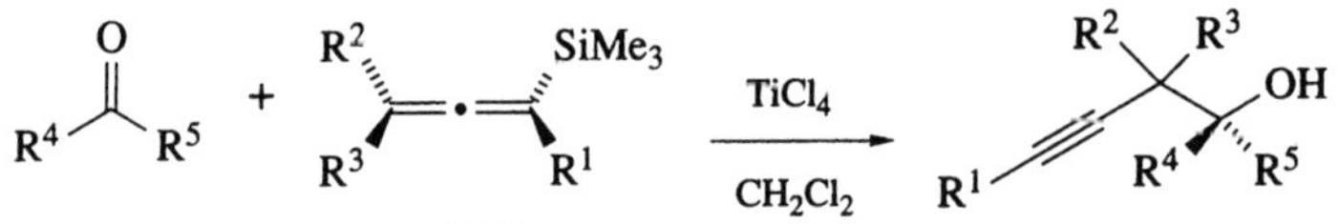

(66)

$TiCl_4$, CH_2Cl_2

Homopropargylic alcohol

a: R^1 = Me, R^2 = R^3 = H **d**: R^1 = R^3 = H, R^2 = –$(CH_2)_2Ph$

b: R^1 = Pr^i, R^2 = R^3 = H **e**: R^1 = R^2 = H, R^3 = Me

c: R^1 = R^3 = H, R^2 = Me **f**: R^1 = Me, R^3 = Et, R^3 = H

Scheme 30

Reactions of chiral allenes proceed with a preference for the formation of the *syn* diastereomer. The stereochemical outcome of these reactions can be rationalized by invoking an open transition state model for the addition reactions (Figure 12), which depicts an antiperiplanar orientation of the chiral allenylsilane to the aldehyde carbonyl. In this model, steric repulsion between the allenyl methyl and the aldehyde substituent is most likely responsible for the destabilization of transition state (**B**), which leads to the *anti* (minor) stereoisomer. This destabilizing interaction is minimized in transition state (**A**). Table 5[29] illustrates representative examples and summarizes the scope of the regiocontrolled synthesis of homopropargylic alcohols using allenylsilanes.

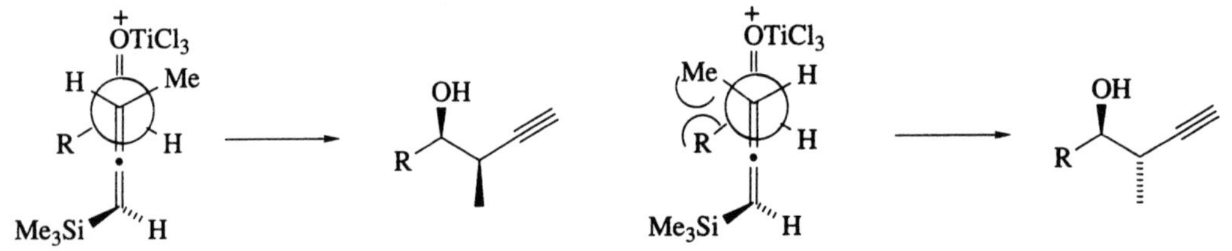

Transition state A *syn* Transition state B *anti*

Figure 12

2.5.3.2.2 *[3 + 2] annelation strategies involving allenylsilanes*

[3 + 2] annelation tactics involving allenylsilanes as the three-carbon nucleophile have been extensively investigated by Danheiser and coworkers.[30] Their efforts in this area have resulted in the generation of an effective strategy for the preparation of a variety of both carbocyclic and heterocyclic compounds. These processes are summarized in Figure 13 with the illustrated transformations. The allenylsilane (**66**) reacts with α,β-unsaturated carbonyl compounds to give carbocyclic derivatives (**67a**)–(**69a**). Allenylsilanes react with heteroallenophiles, *e.g.* aldehydes, acylium, acyliminium, iminium and nitronium ions to yield the corresponding heterocyclic products (**70**)–(**72**). Sections 2.5.4.2–2.5.4.8 offer a detailed discussion of these reactions.

2.5.3.2.3 *Additions to α,β-unsaturated carbonyl compounds*

The reaction of allenylsilanes with electron deficient π-systems constitutes a powerful and general method for the regio- and stereo-selective preparation of substituted cyclopentenes.[30] As first reported, (trimethylsilyl)allenes function as three-carbon nucleophiles in $TiCl_4$-promoted (trimethylsilyl)cyclopentene annelations (Scheme 31). The annelation process involves the polarization of an α,β-unsaturated carbonyl compound by $TiCl_4$ to generate an alkoxyallylic cation. Regiospecific electrophilic substitution of this cation at C-3 of the allenylsilane produces a silicon-stablized vinyl cation. A 1,2-shift of the trimethylsilyl group then yields an isomeric vinyl cation, which is trapped by the titanium enolate to generate the annelated (trimethylsilyl)cyclopentene products (**67b**).

Table 5 Regiocontrolled Synthesis of Homopropargylic Alcohols (see Scheme 30)

Entry	*Carbonyl compound*	*Allenylsilane*	*Method* [a]	*Product(s)*	*Yield (%)*
1	Cyclohexanecarbaldehyde	**66a**	A	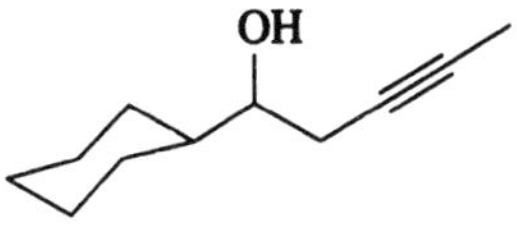 OH	68
2	$PhCH_2CH_2CHO$	**66a**	A	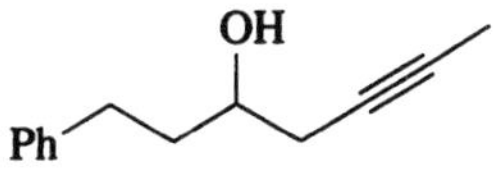 OH, Ph	85
3	Pr^iCOMe	**66a**	A	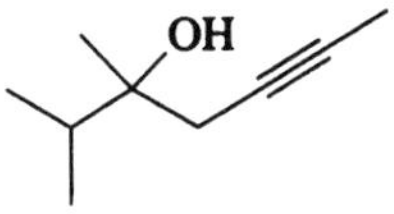 OH	86
4	$PhCH_2CH_2CHO$	**66b**	A	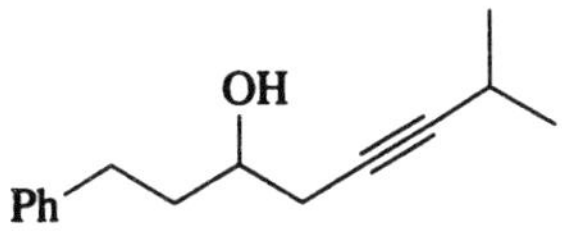 OH, Ph	89
5	Cyclohexanone	**66b**	A	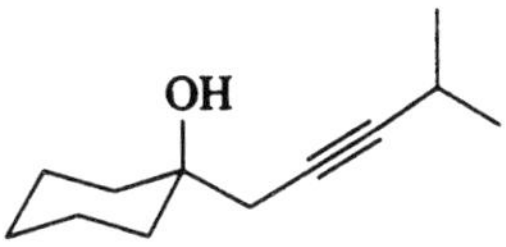 OH	84
6	Pr^iCOMe	**66b**	A	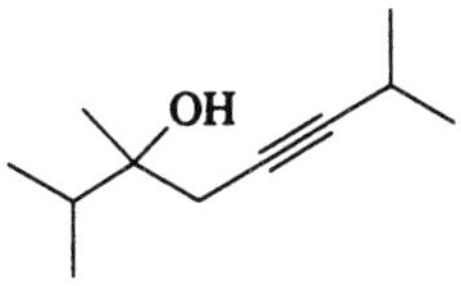 OH	51
7	$PhCH_2CH_2CHO$	**66c**	B	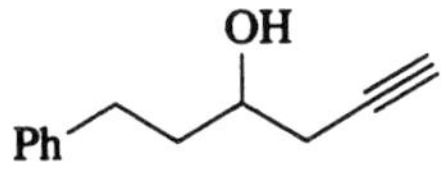 OH, Ph	84
8	$PhCH_2COMe$	**66c**	B	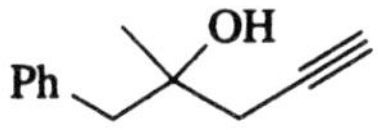 Ph, OH	72
9	Cyclohexanone	**66c**	B	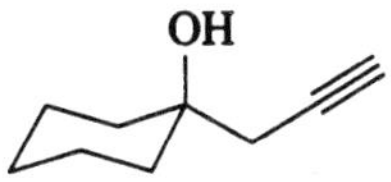 OH	89
10	MeCOMe	**66d**	B	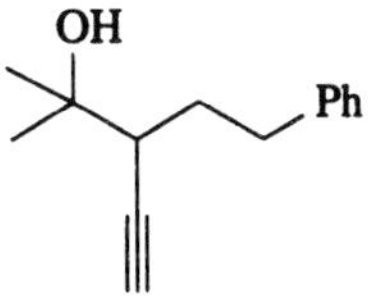 OH, Ph	38
11	Cyclohexanone	**66d**	B	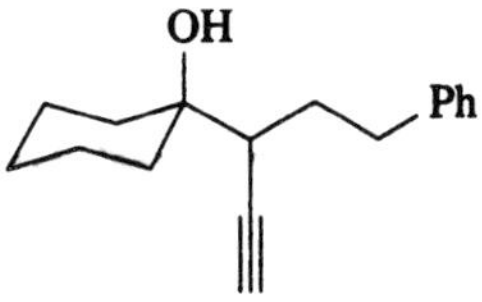 OH, Ph	49

Table 5 *(continued)*

Entry	*Carbonyl compound*	*Allenylsilane*	*Method* [a]	*Product(s)*	*Yield (%)*
12	$PhCH_2CH_2CHO$	**66e**	B	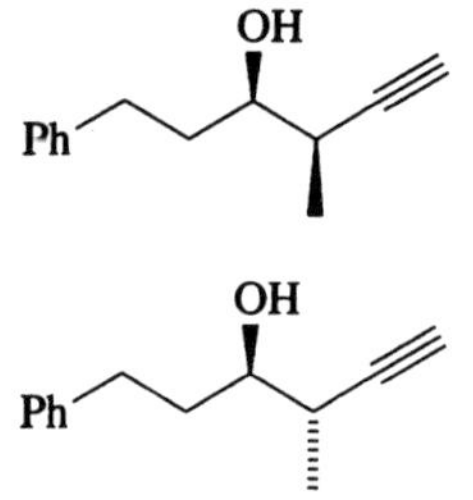 OH, Ph; OH, Ph	89 (3.1:1)
13	Cyclohexanecarbaldehyde	**66e**	B	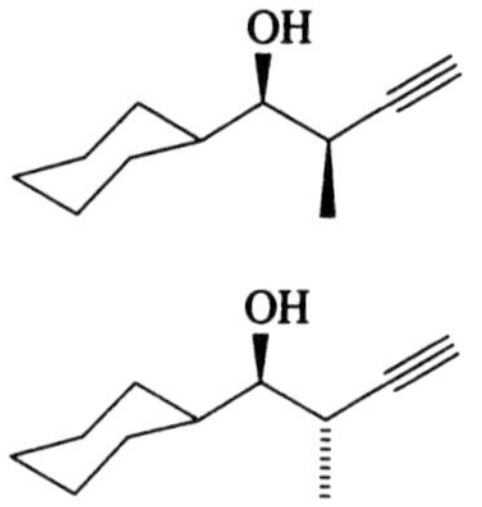 OH; OH	81 (4.1:1)
14	Cyclohexanone	**66e**	B	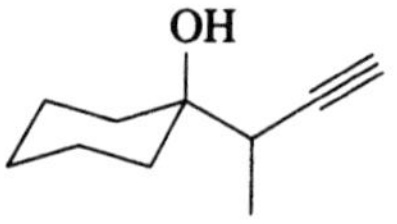 OH	77

[a] Method A: $TiCl_4$ (1.15 equiv.), –78 to 25 °C. Method B: $TiCl_4$ (1.15 equiv.), –78 to 25 °C then 2.0–2.5 equiv. of KF in DMSO.

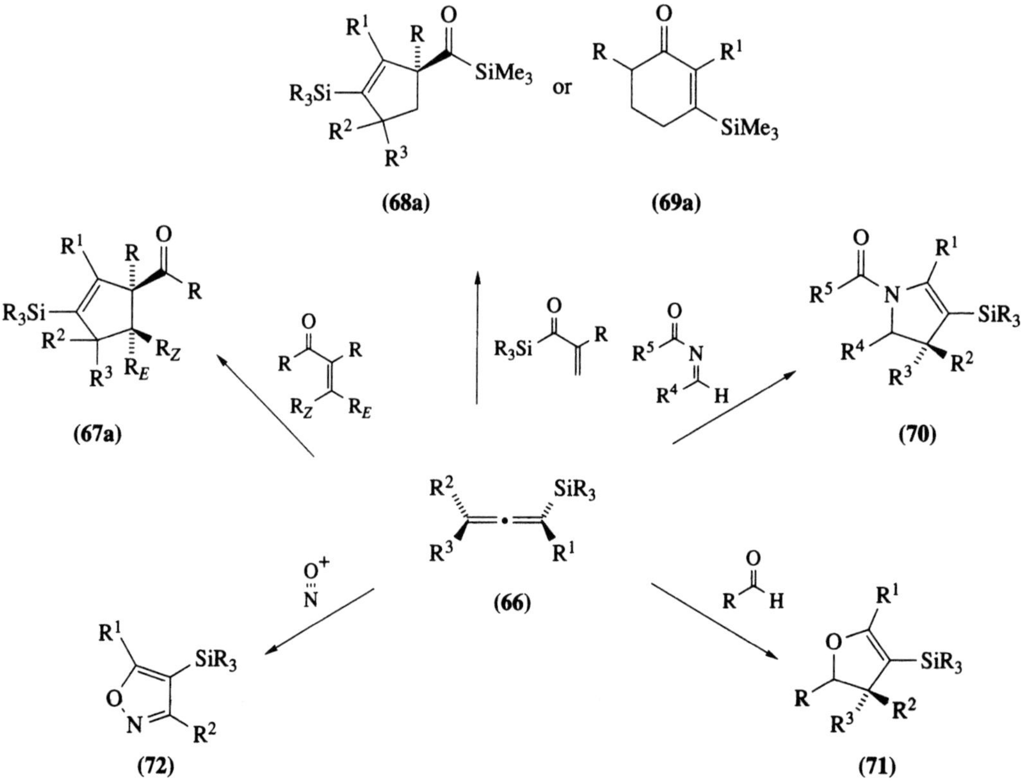

Figure 13

Table 6 (Trimethylsilyl)cyclopentene Annelations (see Scheme 30)

Entry	*α,β-Unsaturated ketone*	*Allenylsilane*	*(Trimethylsilyl)cyclopentene product*	*Yield (%)*
1	Methyl vinyl ketone	**66a**	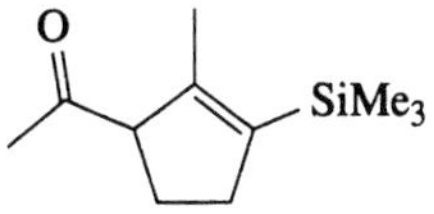	68–75
2	Cyclohexenone	**66a**	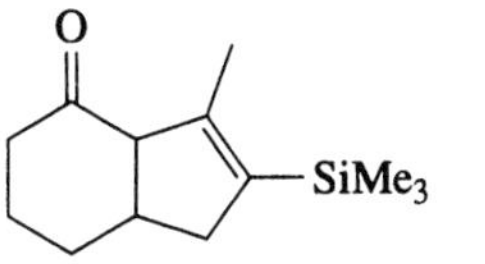	85
3	Cycloheptenone	**66a**	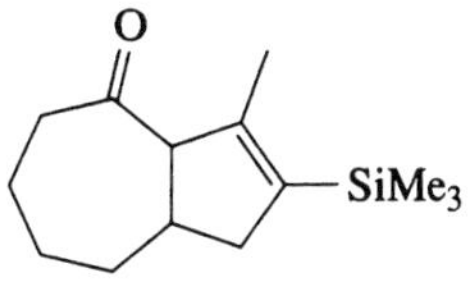 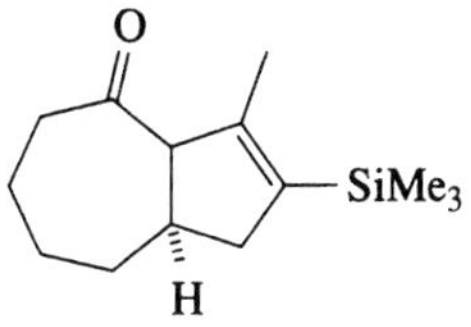	90–94 (*cis*:*trans* = 83:17)
4	Cyclopentenone	**66a**	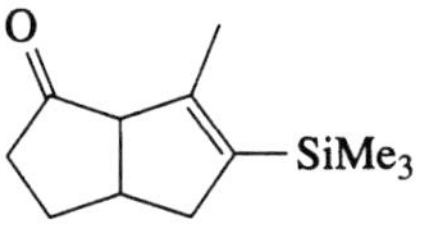	48
5	*trans*-3-Penten-2-one	**66a**	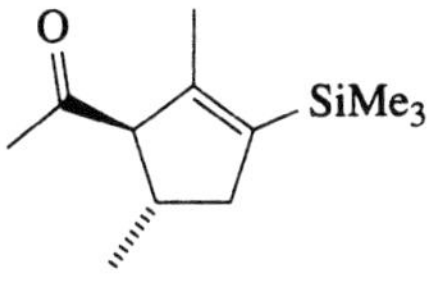	79
6	Carvone	**66a**	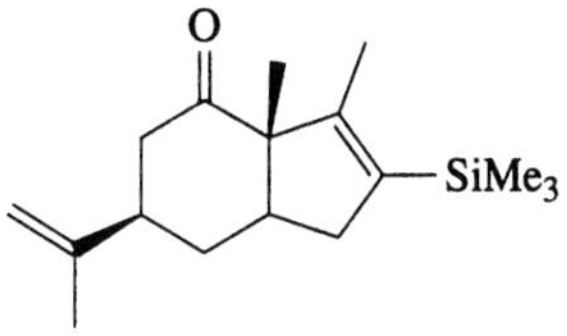	87
7	1-Acetylcyclohexene	**66a**	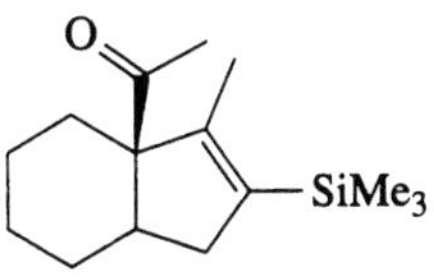	91
8	2-Methylene-α-tetralone	**66a**	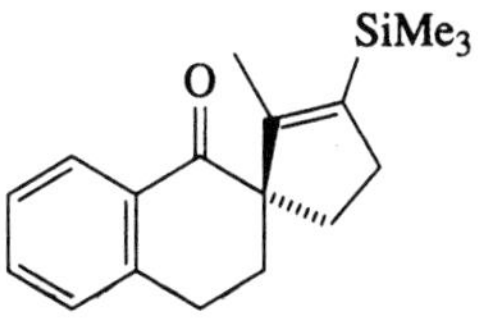	80–84

Table 6 *(continued)*

Entry	*α,β-Unsaturated keton*	*Allenylsilane*	*(Trimethylsilyl)cyclopentene product*	*Yield (%)*
9	Phenyl vinyl ketone	**66b**	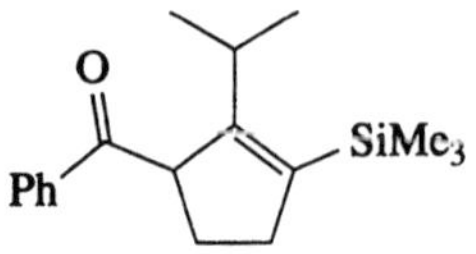	69–73
10	Cyclohexenone	**66b**	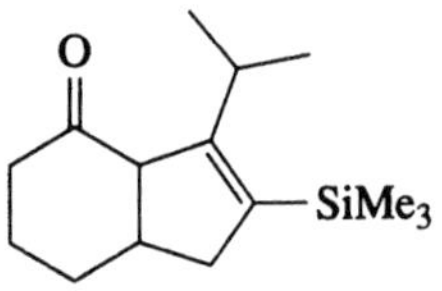	81–85
11	Cyclohexenone	**66b**	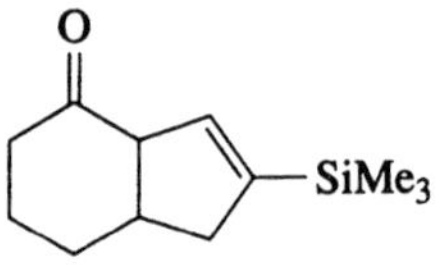	17–19
12	Methyl vinyl ketone	**66e**	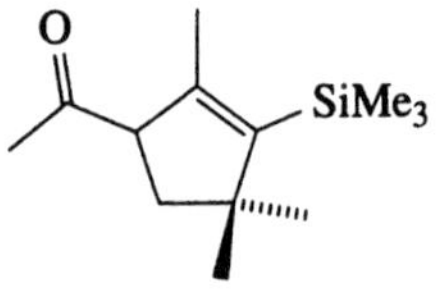	80
13	Cyclohexenone	**66e**	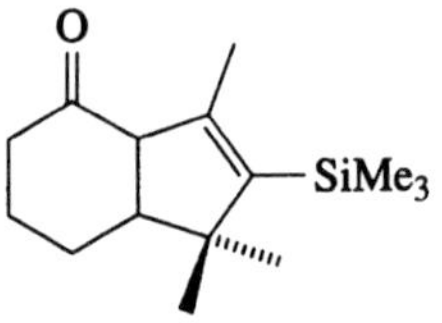	61–63
14	Cyclohexenone	**66f**	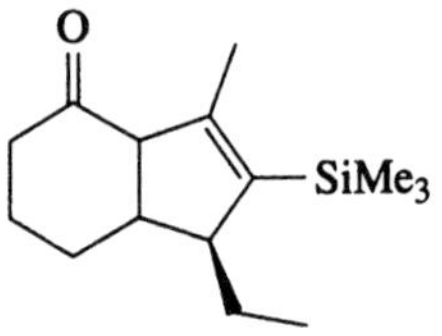 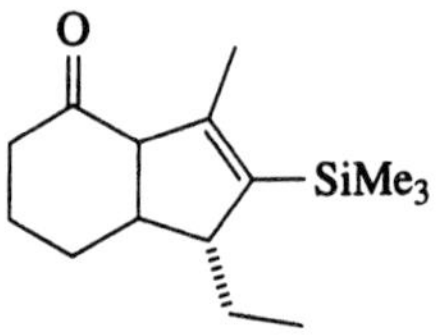	79 (*cis*:*trans* = 95:5)
15	Cyclopentenone	**66c**	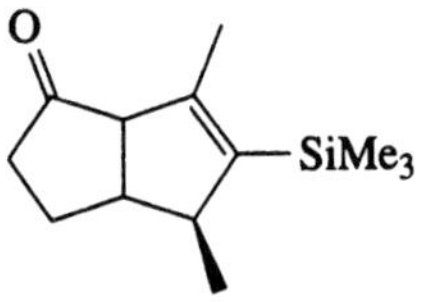 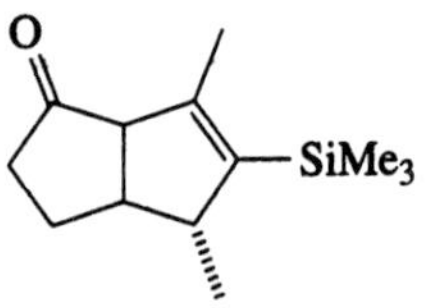	68 (*cis*:*trans* = 75:25)

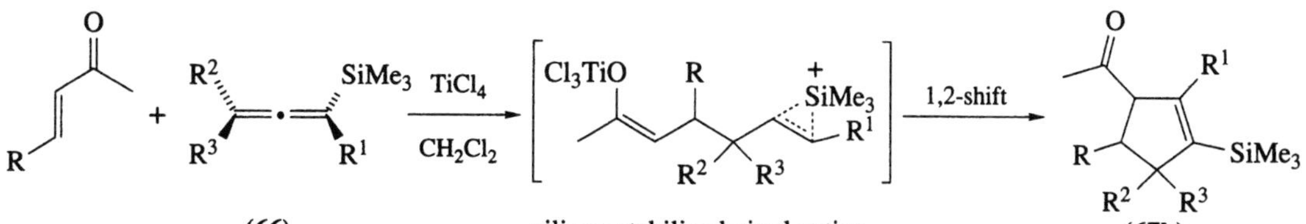

O; R; + ; R^2; $SiMe_3$; R^3; R^1; $TiCl_4$; CH_2Cl_2; Cl_3TiO; R; +; $SiMe_3$; R^1; R^2; R^3; 1,2-shift; O; R^1; R; $SiMe_3$; R^2; R^3

(66)

silicon-stabilized vinyl cation

(67b)

Scheme 31

Both cyclic and acyclic enone systems participate in the (trimethylsilyl)cyclopentene annelation (Table 6).[40] α-Methylene ketones react to form spiro-fused systems and the intermediates derived from acetylcyclohexanone, cyclohexenone and cyclopentanone cyclize to yield 5,5- and 6,5-fused ring systems.[30]

2.5.3.2.4 Additions to α,β-unsaturated acylsilanes

α,β-Unsaturated acylsilanes serve as highly reactive carboxylic acid equivalents in conjugate addition reactions with allenylsilanes.[31] The trimethylsilyl acylsilanes provide the basis for a [3 + 3] annelation approach to six-membered carbocycles. By manipulating the trialkylsilyl group of the acylsilane, the course of the annelation reaction can be controlled to produce either five- or six-membered rings (**68** and **69**; Figure 13 and Scheme 32). α,β-Unsaturated acylsilanes combine with allenylsilanes at –78 °C in the presence of $TiCl_4$ to produce trimethylsilyl–cyclopentene annelation products in good yield (Table 7).[31]

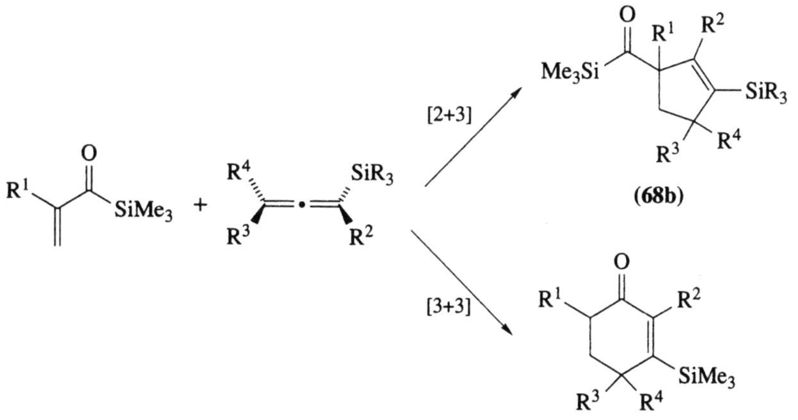

O; R^1; $SiMe_3$; +; R^4; SiR_3; R^3; R^2; [2+3]; [3+3]; O; R^1; R^2; Me_3Si; SiR_3; R^3; R^4

(68b)

O; R^1; R^2; $SiMe_3$; R^3; R^4

(69b)

Scheme 32

O; R^1; SiR_3; +; $SiMe_3$; R^2; $OTiCl_3$; R_3Si; R^1; +; R^2; $SiMe_3$; 1,2-shift; $OTiCl_3$; R_3Si; R^1; R^2; +; Me_3Si

+; Cl_3TiO; R^1; R^2; R_3Si; $SiMe_3$; R_3Si; $OTiCl_3$; R^1; R^2; +; $SiMe_3$; R^1; O; R_3Si; R^2; $SiMe_3$

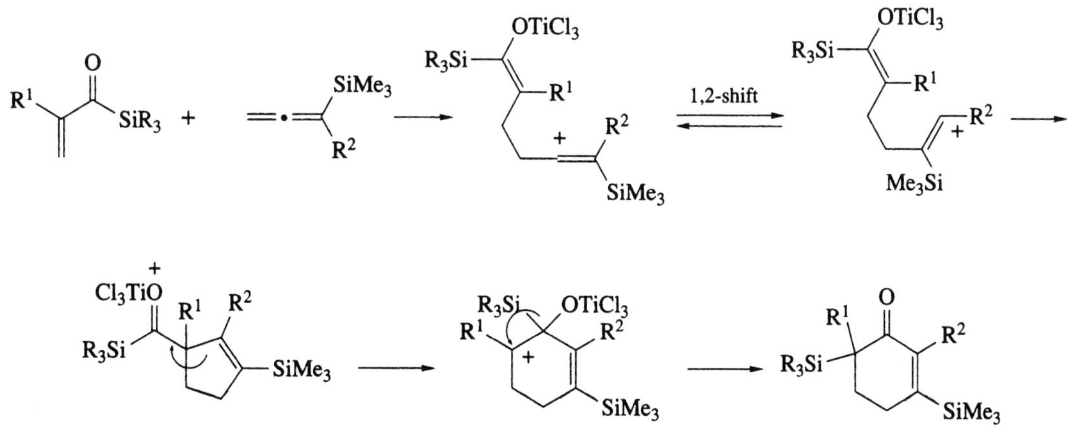

(69c)

Scheme 33

The annelation products derived from 2-alkyl-substituted α,β-unsaturated acylsilanes undergo a rearrangement to β-silylcyclohexanone derivatives when treated with $TiCl_4$. The proposed mechanism is illustrated in Scheme 33.[31] The annelation process commences with the regiospecific electrophilic substitution at the C(3)-position of the allenylsilane, producing a vinyl carbocation which undergoes a 1,2-cationic trimethylsilyl shift to yield an isomeric vinyl carbocation. Cyclization then gives the [3 + 2] annelation product. Ring expansion of the cyclopentene generates a tertiary carbocation, which undergoes a second 1,2-anionic trimethylsilyl shift to produce the cyclohexanone (**69c**).

Table 7 Annelations of α,β-Unsaturated Acylsilanes

Acylsilane	*Allenylsilane*	*Product*	*Yield (%)*[a]
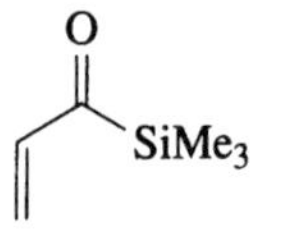 O, $SiMe_3$	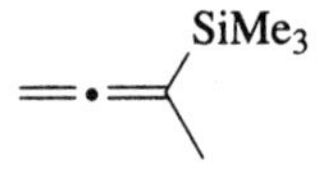 $SiMe_3$	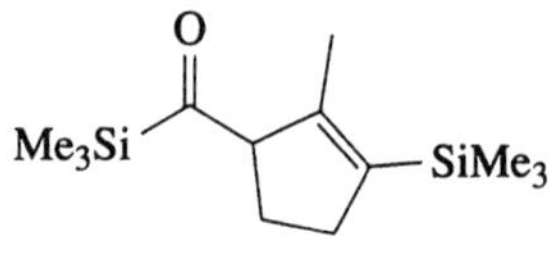 O, Me_3Si, $SiMe_3$ (**68c**)	72
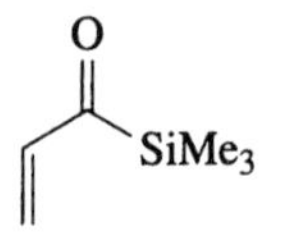 O, $SiMe_3$	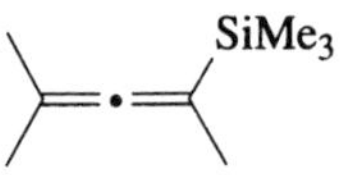 $SiMe_3$	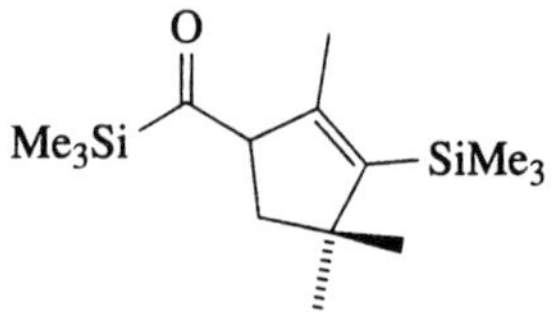 O, Me_3Si, $SiMe_3$ (**68d**)	55
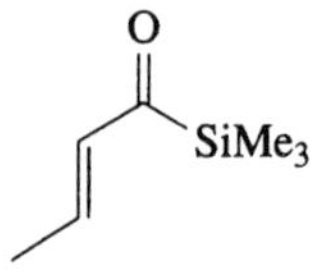 O, $SiMe_3$	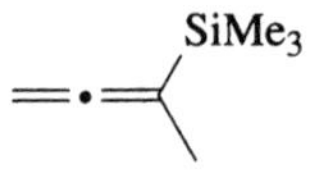 $SiMe_3$	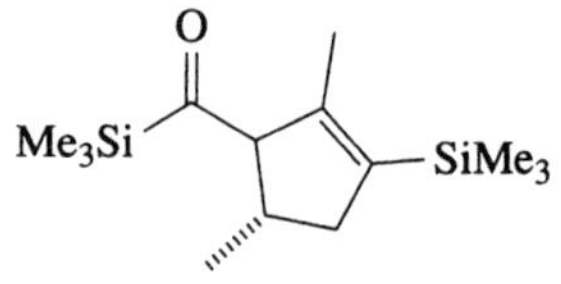 O, Me_3Si, $SiMe_3$ (**68e**)	66
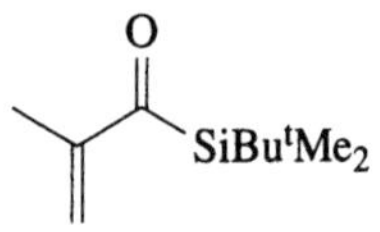 O, $SiBu^tMe_2$	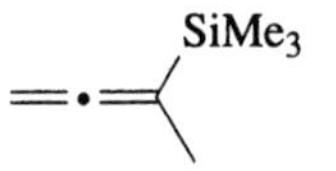 $SiMe_3$	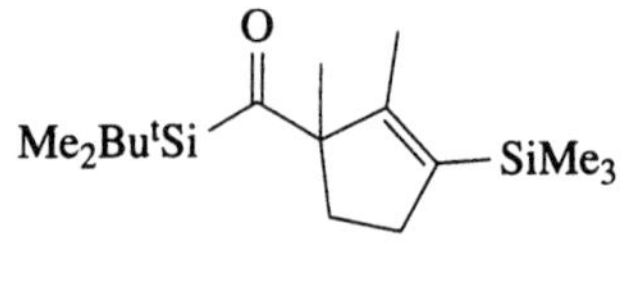 O, Me_2Bu^tSi, $SiMe_3$ (**68f**)	78
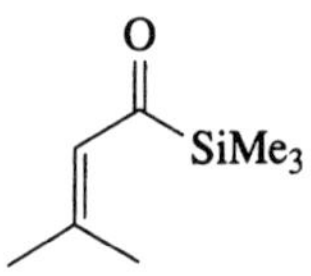 O, $SiMe_3$	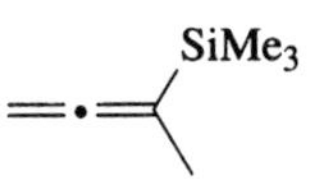 $SiMe_3$	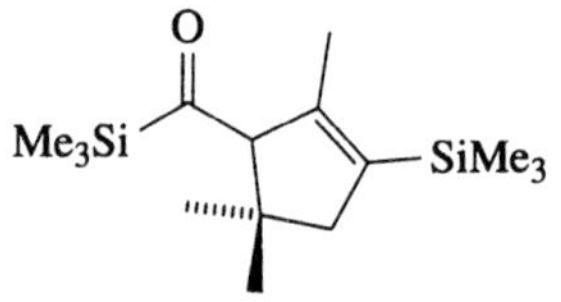 O, Me_3Si, $SiMe_3$ (**68g**)	74

[a] Reactions were conducted using 1.0–1.5 equiv. of allenylsilane and $TiCl_4$ (1.5 equiv.) in methylene chloride at –78 °C for 0.5–1.0 h or 30 s in the case of (**68e**).

2.5.3.2.5 Additions to acyliminium ions

The reactions of sterically encumbered allenylsilanes of structure type (**66**) with *N*-acyliminium ions generated by the titanium tetrachloride promoted reaction of ethoxypyrrolidinone produced the nitrogen heterocycle (**70**), as depicted in Figure 13. In this reaction it is likely the lactam product is generated by a pathway similar to that involved in the [3 + 2] cyclopentene annelation.[32] Thus, a regiospecific electrophilic substitution of the *N*-acyliminium ion derived from substitution at the C-3 position of the allenylsilane produces a vinyl cation stabilized by the hyperconjugative interaction of the adjacent σ-donor silicon group. A 1,2-cationic trialkylsilyl shift then occurs, affording an isomeric vinyl cation, which is intercepted by the nucleophilic nitrogen atom to form the new heterocycle (**70**) as summarized in Table 8.[32]

Table 8 Preparation of Nitrogen Heterocycles *via* [3 + 2] Annelation

Allenophile	*Allene*	*Annelation product(s)*	*Yield (%)*
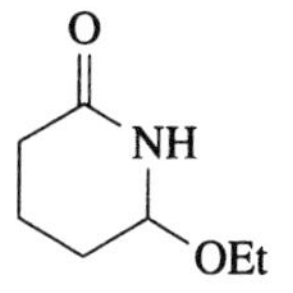 O, NH, OEt	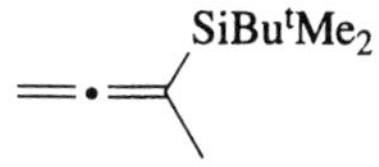 $SiBu^tMe_2$	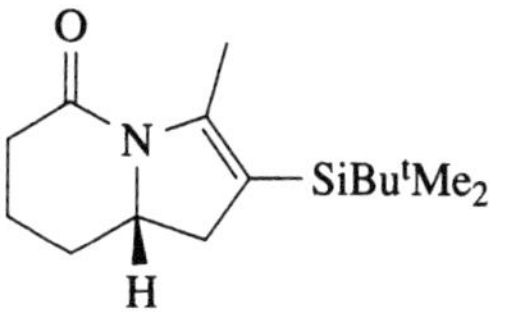 O, N, $SiBu^tMe_2$, H (**70a**)	67
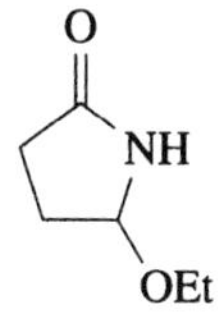 O, NH, OEt	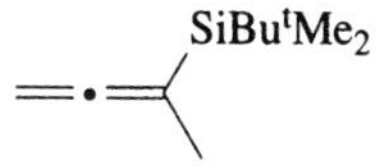 $SiBu^tMe_2$	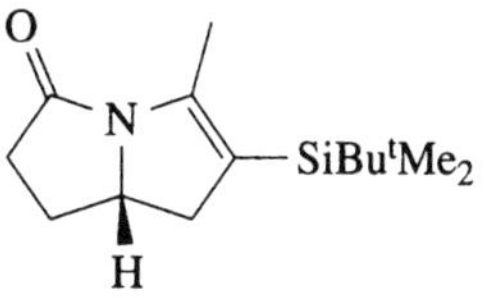 O, N, $SiBu^tMe_2$, H (**70b**)	63
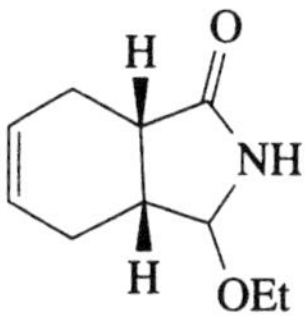 H, O, NH, H, OEt	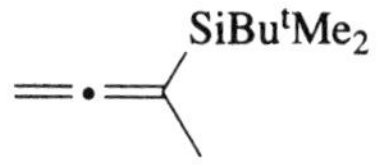 $SiBu^tMe_2$	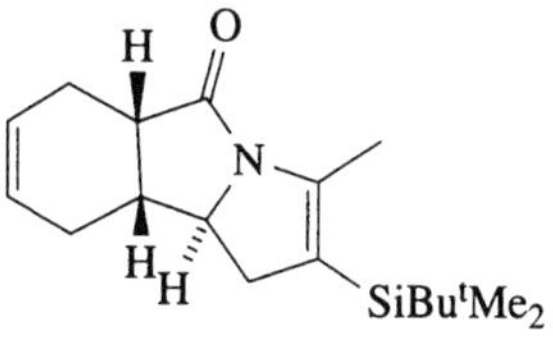 H, O, N, H, H, $SiBu^tMe_2$ (**70c**)	76
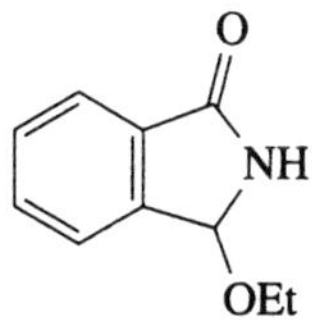 O, NH, OEt	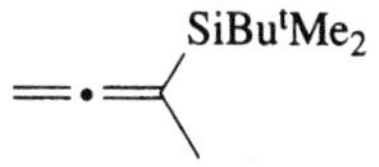 $SiBu^tMe_2$	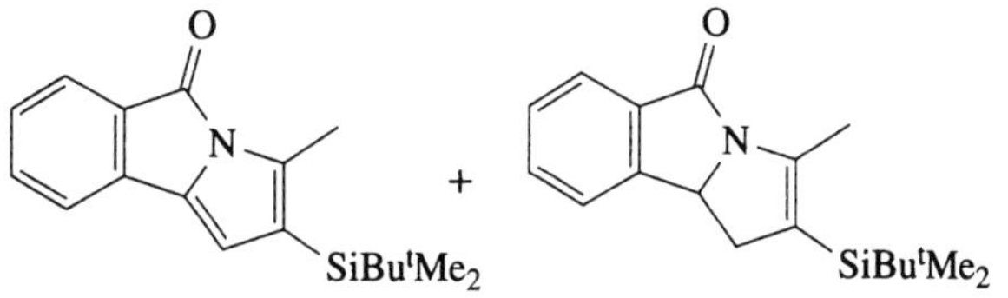 O, N, $SiBu^tMe_2$ + O, N, $SiBu^tMe_2$ (**70d**) 2.8:1	76
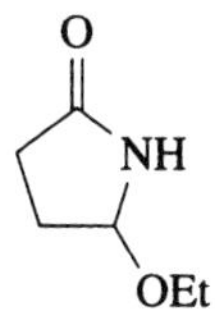 O, NH, OEt	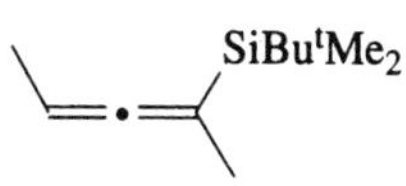 $SiBu^tMe_2$	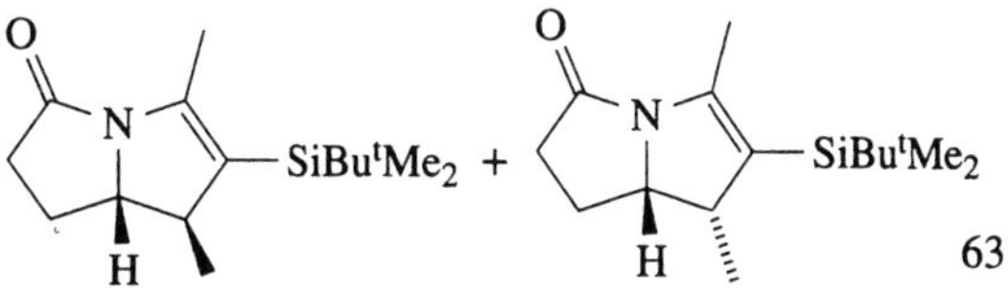 O, N, $SiBu^tMe_2$, H + O, N, $SiBu^tMe_2$, H (**70e**) 1.2:1	63

2.5.3.2.6 Additions to aldehydes

A variety of aldehydes can function as heteroallenophiles in this [3 + 2] annelation (Figure 13 and Table 9).[32] Reactions of the C(3)-substituted allenylsilanes (**66**) gave predominantly the *cis*-substituted dihydrofurans (**71**). The *cis* stereochemistry was anticipated, based on the well-documented stereochemical course of Lewis acid catalyzed additions of 3-substituted allylsilanes to aldehydes.[32]

Table 9 Preparation of Oxygen Heterocycles *via* [3 + 2] Annelation

Allenophile	*Allene*	*Annelation product(s)*	*Yield (%)*
 CHO	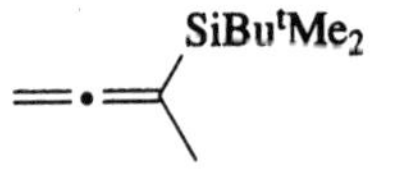 $SiBu^tMe_2$	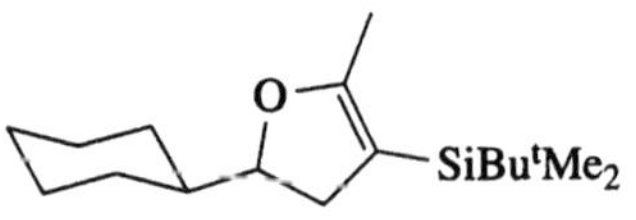 O, $SiBu^tMe_2$ **(71a)**	76
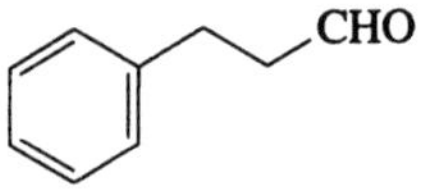 CHO	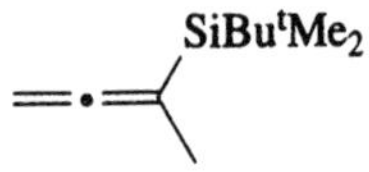 $SiBu^tMe_2$	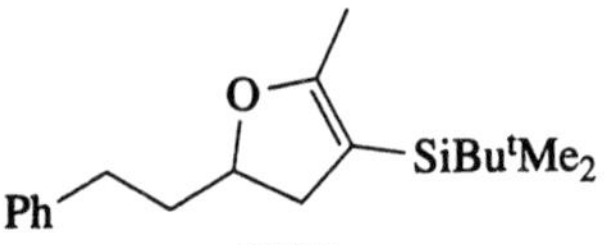 Ph, O, $SiBu^tMe_2$ **(71b)**	70
MeCHO	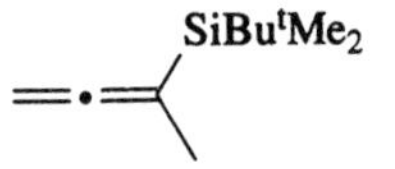 $SiBu^tMe_2$	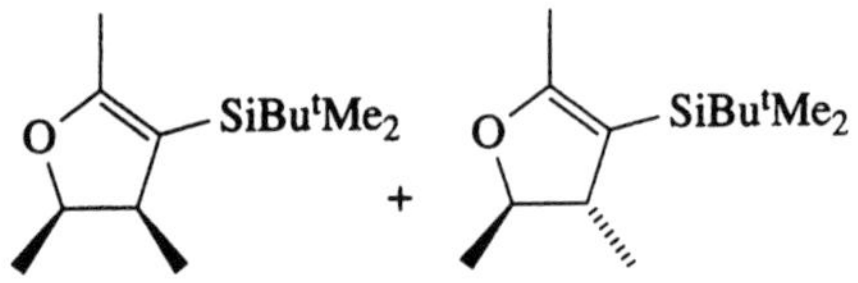 O, $SiBu^tMe_2$ + O, $SiBu^tMe_2$ **(71c)** 1.4:1	78
 CHO	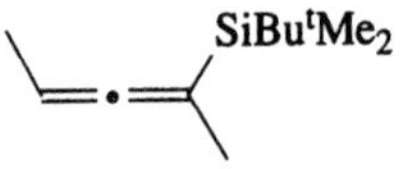 $SiBu^tMe_2$	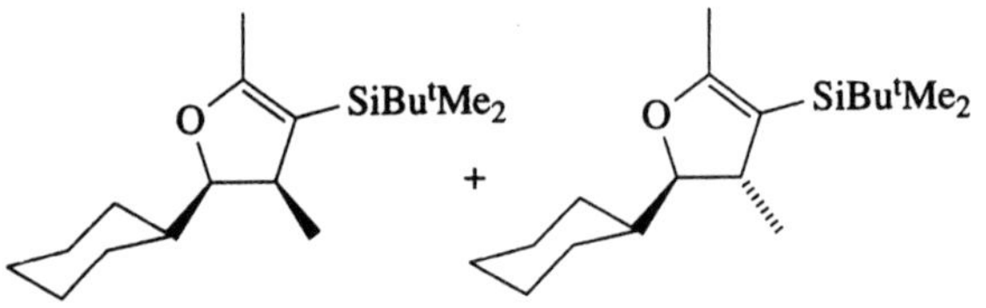 O, $SiBu^tMe_2$ + O, $SiBu^tMe_2$ **(71d)** 7:1	97
Bu^tCHO	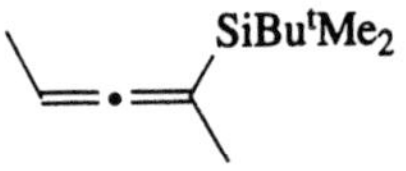 $SiBu^tMe_2$	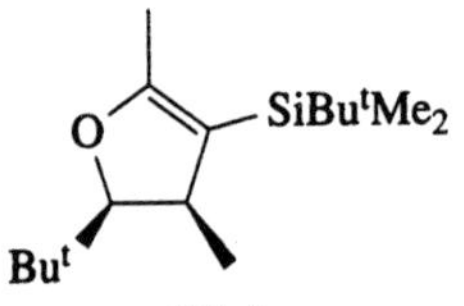 O, $SiBu^tMe_2$, Bu^t **(71e)**	92
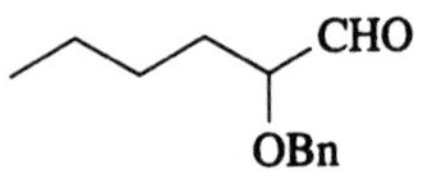 CHO, OBn	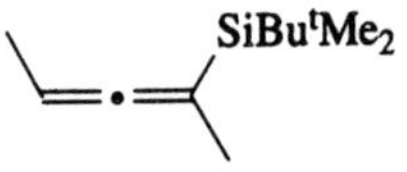 $SiBu^tMe_2$	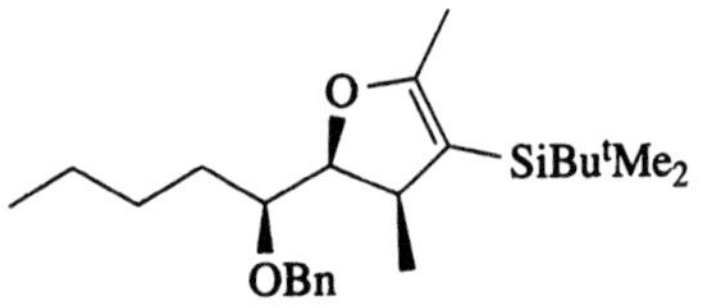 O, $SiBu^tMe_2$, OBn **(71f)**	88

2.5.3.2.7 Additions to activated C═N π-bonds

A general strategy for the synthesis of five-membered heteroaromatic compounds has been developed which involves the reaction of allenylsilanes with π-bonded electrophilic species of the general type $X{\equiv}Y^+$ (heteroallenophile).[33] Scheme 34 details a stepwise mechanism to interpret the annelation. Addition of the heteroallenophile at the C-3 carbon of the allenylsilane produces a vinyl carbocation, which rapidly rearranges to the isomeric vinyl cation. The trialkylsilyl group serves two roles in this initial step: (i) it directs the regiochemical course of the electrophilic addition reaction (*i.e.* β-silicon effect); and (ii) it promotes the transfer of electrophilic character from the central carbon to the terminus of the three-carbon annelation unit, indicating the high migratory aptitude of silyl groups in cationic rearrangements. Cyclization furnishes the intermediate, which subsequently undergoes proton elimination to generate the aromatic heterocycle.

This [3 + 2] annelation strategy has been employed in a regiocontrolled synthesis of 4-silylisoxazoles. In this instance, nitrosonium tetrafluoroborate functions as a heteroallenophile in the isoxazole annelation (Scheme 35). The electrophilic addition of nitrosonium tetrafluoroborate ($NOBF_4$) to the C-3 position of the allenylsilane produces the silicon-stabilized vinyl cation. A 1,2-trialkylsilyl shift then occurs affording an isomeric vinyl cation, which is trapped by the nucleophilic oxygen of the nitroso group to form the intermediate (**73**). Deprotonation yields the aromatized 4-silylisoxazole system (**72**). This regiocontrolled, one-step annelation method allows access to a wide range of substituted 4-silylisoxazoles.

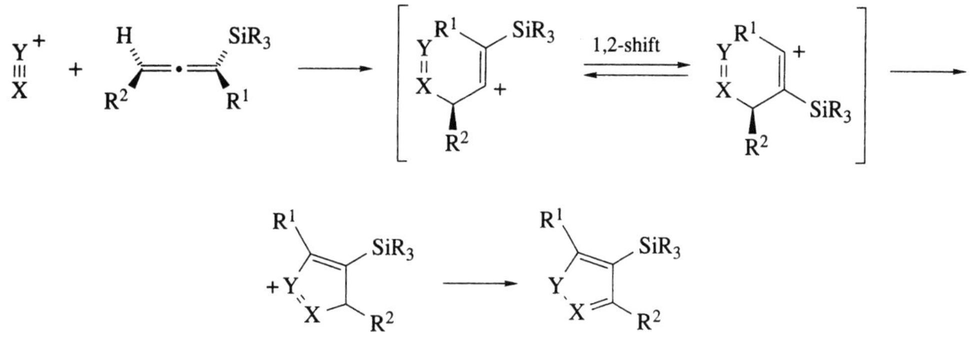

Scheme 34

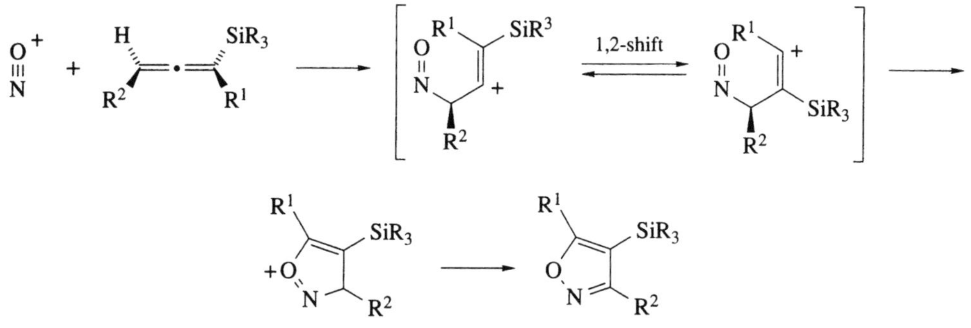

(**73**) (**72**)

Scheme 35

2.5.3.2.8 Additions to tropylium ions

A new [3 + 2] annelation route to azulenes has been reported and involves the addition of a tropylium cation to allenylsilanes. The approach is based on the general [3 + 2] annelation strategy used in the preparation of five-membered carbocycles and heterocycles and uses tropylium tetrafluoroborate ($TpBF_4$) as the tropylium cation source.[34] The proposed mechanism is depicted in Scheme 36 and begins with addition of the tropylium ion to the C(3)-position of the allenylsilane, generating a vinyl cation which rapidly rearranges by a 1,2-shift to the isomeric vinyl cation. Cyclization of the latter cation furnishes the cycloheptyldienyl cation, which undergoes deprotonation to generate the dihydroazulene. Dehydrogenation occurs if a second equivalent of tropylium ion is added to the reaction: hydride abstraction by the tropylium ion generates a more stable (delocalized) tropylium ion derivative, which finally aromatizes by the elimination of the C(3)-proton to yield the product azulene (**74**). As indicated in Table 10,[34] azulene annelations which employ TBDMS alkenes are more efficient than those involving TMS derivatives, since TMS allenes undergo desilylation of the intermediate vinyl cations to produce propargyl-substituted cycloheptatrienes as by-products.

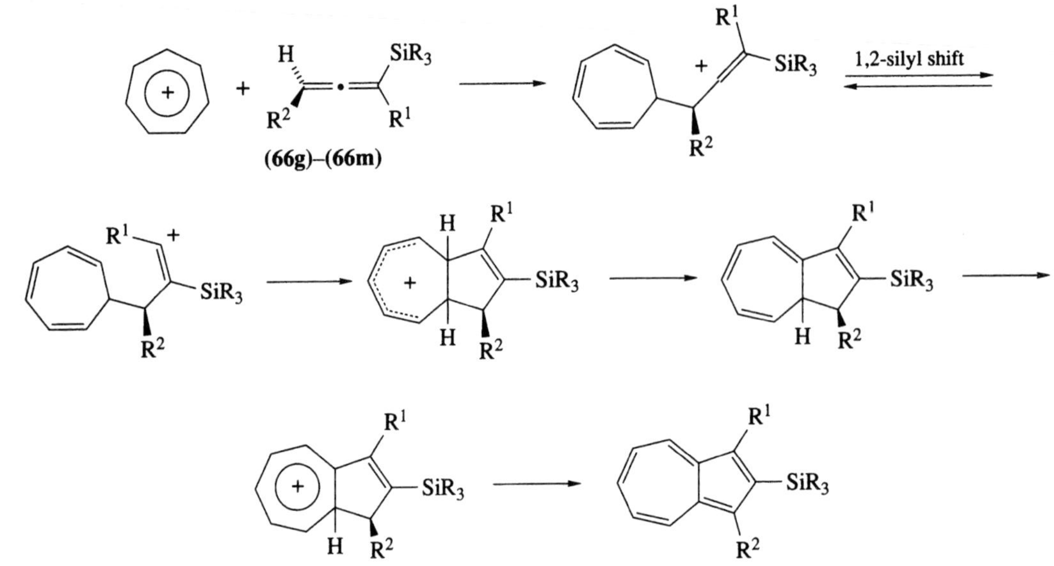

(66g)–(66m)

1,2-silyl shift

(74)

Scheme 36

Table 10 Synthesis of Substituted Azulenes (see Scheme 36)

Entry	*Allenylsilane*	*Method* [a]	*Product*	*Yield (%)*
1	**(66g)** $R^1 = R^2 = Me$	A	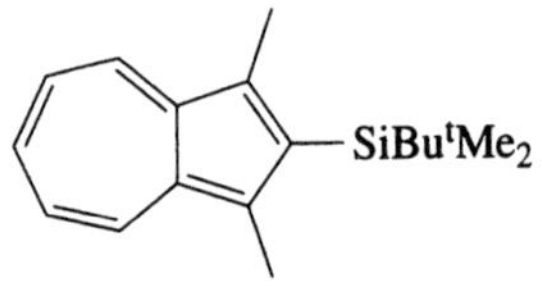 **(74a)**	52–59
2	**(66h)** $R^1 = Me$; $R^2 = (CH_2)_3Ph$	B	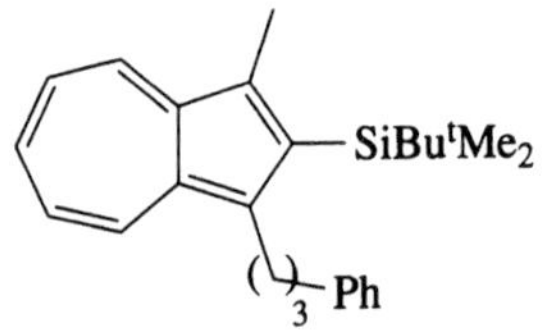 **(74b)**	57
3	**(66i)** $R^1 = Me$; $R^2 = (CH_2)_3CH{=}CH_2$	B	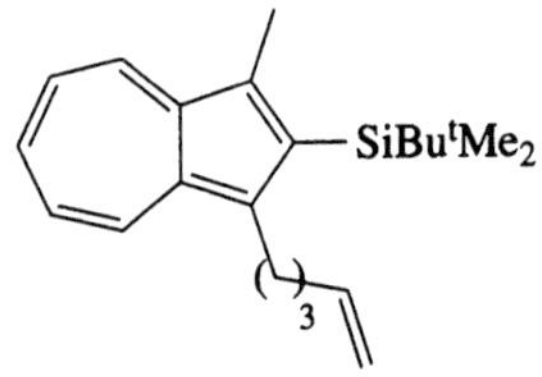 **(74c)**	43

Table 10 *(continued)*

Entry	*Allenylsilane*	*Method* [a]	*Product*	*Yield (%)*
4	**(66j)** R^1 = cyclohexyl; R^2 = CH_2Me	B	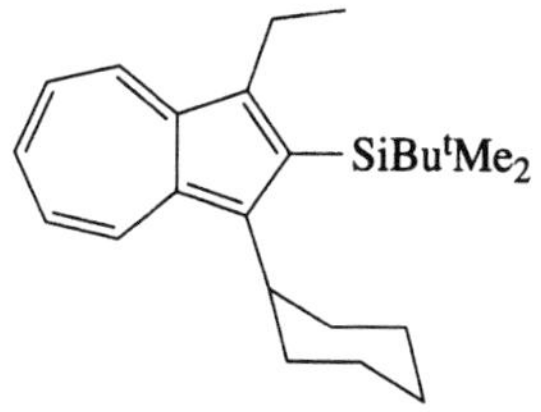 **(74d)**	44 (58)
5	**(66k)** R^1 = Me; R^2 = $(CH_2)_4Br$	B	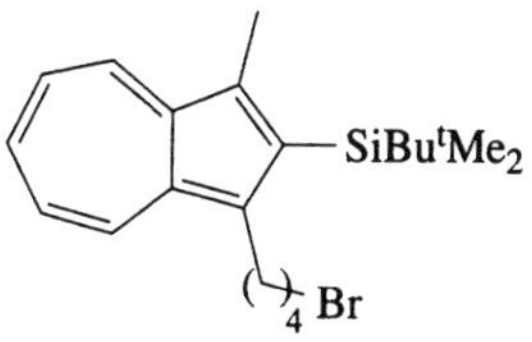 **(74e)**	55
6	**(66l)** R^1 = Me; R^2 = $(CH_2)_2(o\text{-}BrC_6H_4)$	B	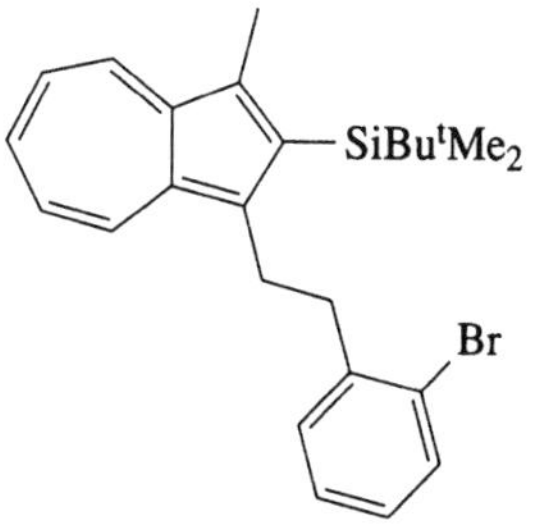 **(74f)**	63
7	**(66m)** R^1 = Me; R^2 = H	C	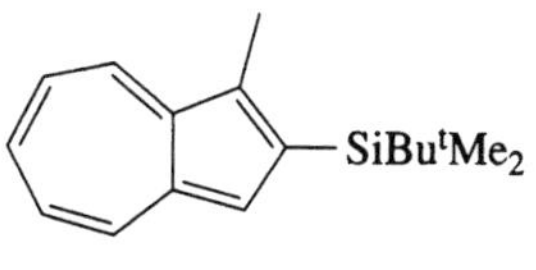 **(74g)**	22

[a] Method A: $TpBF_4$ (2.0 equiv.) and 158 mg poly(4-vinylpyridine) $mmol^{-1}$ of $TpBF_4$ in MeCN at 25 °C for 24 h. Method B: $TpBF_4$ (4.0 equiv.) $MeSi(OMe)_3$ in MeCN at 25 °C for 48 h. Method C: $TpBF_4$ (3.0 equiv.) and 123 mg poly(4-vinylpyridine) $mmol^{-1}$ of $TpBF_4$ in MeCN/heptane (1:1) at 65 °C for 2.0 h.

2.5.3.3 Alkynylsilanes: Additions to C—X π-Bonds

Alkynylsilanes can function as carbon nucleophiles in addition reactions to electrophilic π-systems. In principle electrophilic addition reactions to alkynylsilanes can occur to produce α- or β-silyl-substituted vinyl cations, as illustrated in Scheme 37. The α-silyl carbocation is not the most stabilized cation, the reason being that the carbon–silicon bond can achieve coplanarity with the vacant orbital on the β-carbonium ion, making possible β-stabilization through hyperconjugation. Depending on the configuration of the carbocation, the developing vacant orbital can exist as a *p*-orbital, as in structure **(75a)**, or an sp^2-hybrid, as in structure **(75b)**.

Alkynylsilanes have been used as terminators in allylic alcohol and ketene thioacetal initiated cyclizations. In the late 1970s Heathcock[35] and Johnson[36] independently established the utility of alkynylsilanes in cyclization reactions. The example reported by Heathcock illustrated the electronic effect of silicon in the formic acid promoted transformation of the alkynylsilane (**76**) to the bicyclic ketone (**78**) in 76% yield. In contrast cyclization of the analogous alkyne (**77**) produced the bicyclo[2.2.2]octene (**79**), as illustrated in Scheme 38.

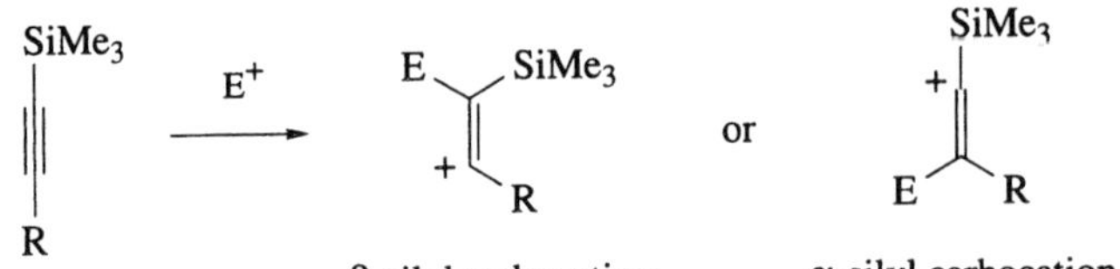

$SiMe_3$ E^+ E $SiMe_3$ + R or $SiMe_3$ + E R

β-silyl carbocation α-silyl carbocation

Scheme 37

R $SiMe_3$ E $SiMe_3$ R E

empty *p*-orbital (**75a**) *sp*^{2}-hybridized (**75b**)

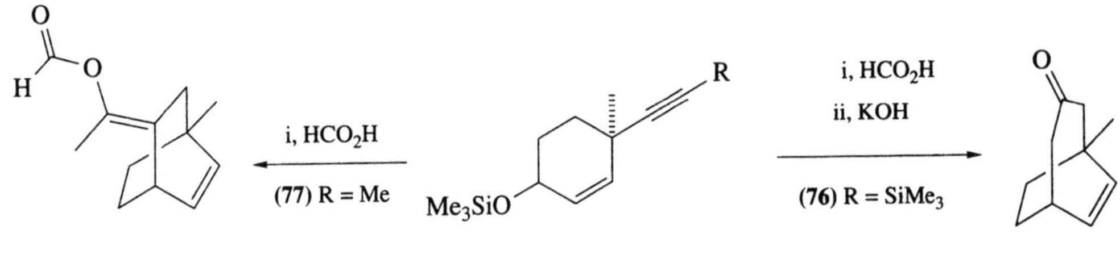

O H O   i, HCO_2H (**77**) R = Me Me_3SiO R   i, HCO_2H ii, KOH (**76**) R = $SiMe_3$ O

(**79**) (**78**)

Scheme 38

Johnson *et al.* have investigated the use of alkynylsilanes as terminators in the synthesis of steroids and triterpenes through biomimetic polyene cyclizations.[37] This strategy was used in the stereospecific synthesis of the tetracyclic ketone (**81**) using an alkynylsilane as a terminator. Thus, treatment of (**80**) with trifluoroacetic acid under carefully optimized reaction conditions yielded, after hydrolysis of the ortho ester, the tetracyclic ketone (**81**; Scheme 39).

Clearly the course of these cyclization reactions is dependent upon the silicon group. In this regard, cyclization of (**82**) affords the steroid nucleus (**83**; Scheme 40). The formation of (**83**) was attributed, in part, to a transition state preference for the formation of the linear vinyl carbocation (**84b**) rather than the bent vinyl cation (**84a**), which would be produced in an endocyclic cyclization.[38,39] The formation of (**81**) was controlled by the generation of the β-silyl carbocation (**85a**), which may be a precursor to an α-silyl ketone, which undergoes protodesilylation. It is not known whether the formation of (**81**) as the major cyclization product occurs through a kinetic pathway or by Wagner–Meerwein rearrangement[40] of the kinetically preferred linear carbocation (**85b**).

Ketene dithioacetals have also been used as initiators for alkynylsilane-terminated cyclization reactions.[41] Treatment of the thioacetal (**86**) with trifluoroacetic acid gave the bicyclic ketodithioketal (**87**) in 76% yield. In this reaction the protonated ketene dithioacetal functions as an acylium ion equivalent, which is trapped by the nucleophilic alkynylsilane.

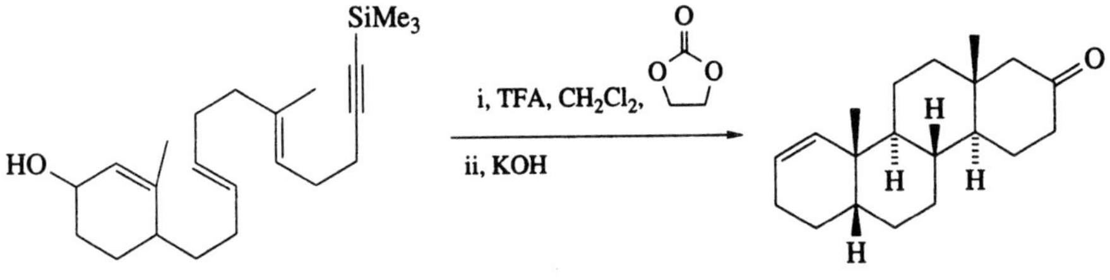

SiMe3
i, TFA, CH_2Cl_2,
ii, KOH
HO
O
H
H H
H
(80) (81)

Scheme 39

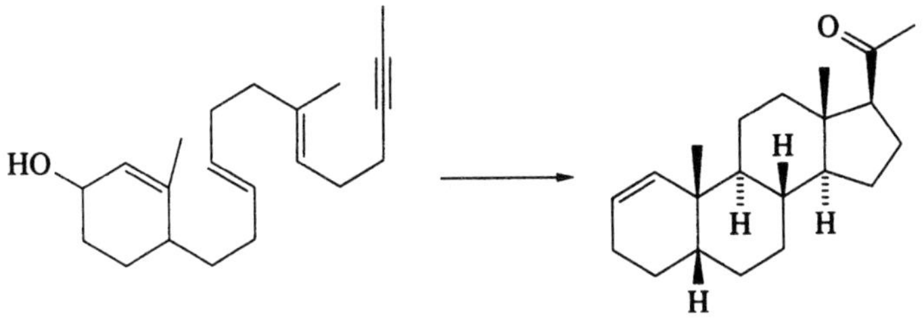

HO
O
H
H H
H
(82) (83)

Scheme 40

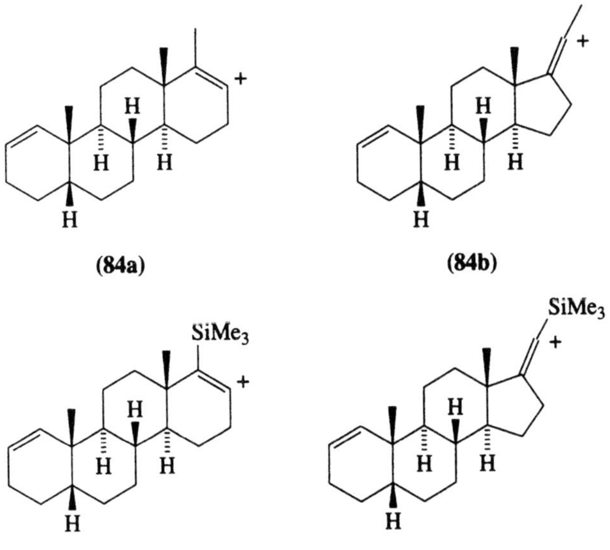

(84a) (84b)

SiMe3
(85a) (85b)

SiMe3
TFA/trifluoroethanol
S S
(86) (87)

Scheme 41

2.5.4 SILICON ATTACHED TO ADJACENT PARTICIPATING π-BONDED CARBON SYSTEMS

2.5.4.1 Allylsilanes

Fleming has published a comprehensive review on the chemistry of allylsilanes.[1a] The following sections will provide an overview of the mechanistic proposals currently used to rationalize the regio- and stereo-selectivities in addition reactions involving allylsilanes, and focus on recent contributions to this important area of research. Allylsilanes have proven to be an exceedingly useful class of organometallic reagents and continue to show enormous potential in stereoselective bond formation reactions. They participate in addition reactions with a wide variety of C—X π-systems. Their use as carbon nucleophiles has received considerable attention due in part to the large number of chemical transformations that can be achieved and the wide range of reaction conditions that can be tolerated by these reagents.

2.5.4.1.1 Stereochemistry of addition reactions involving allylsilanes

Generally, electrophilic addition reactions to acyclic allylsilanes proceed with *anti* stereochemistry.[42–44] The stereochemistry of these reactions has been interpreted on the basis of ground state conformational energies of the allylsilane. As illustrated in Figure 14, the preferred ground state conformer **(88a)** *versus* **(88b)** orients the smallest substituent, usually hydrogen, in a position eclipsing (inside) the adjacent double bond.[45] The bulky silicon group may direct attack of the electrophile to the opposite face of the π-system, generating a secondary carbocation. Subsequent bond rotation orients the C—Si bond periplanar to the empty *p*-orbital, providing stabilization through hyperconjugation to the electron deficient center. Apparently this additional stabilization is responsible for the high levels of selectivity that are observed in the elimination reaction (second step) to form the (*E*)-double bond, as illustrated for structure **(89a)**.

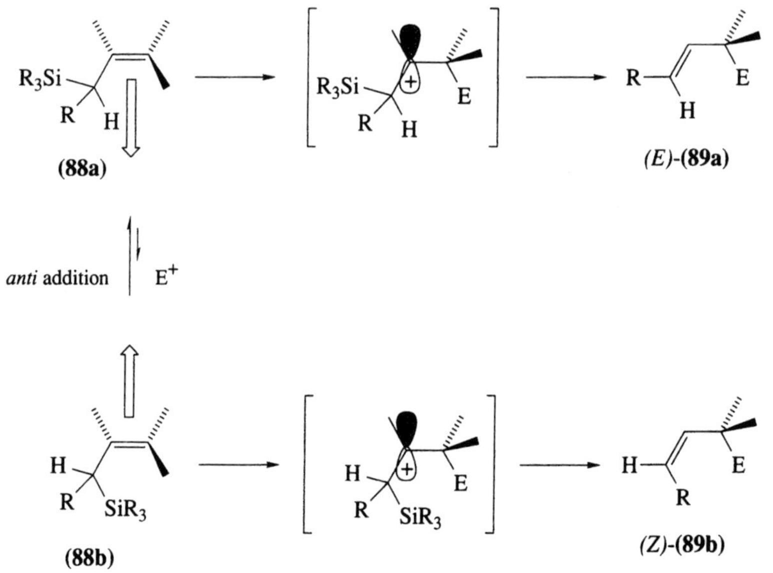

Figure 14 Stereochemistry of electrophilic additions to chiral allylic silanes

2.5.4.1.2 Intermolecular additions to aldehydes, ketones and acetals

C(3)-substituted allylsilanes (crotylsilanes) participate in chelation-controlled addition reactions with aldehydes to give *syn* addition products as the major stereoisomers.[1a,2a] Generally, the (*E*)-crotylsilanes are highly selective in the *syn* sense (>95:5). In contrast, the (*Z*)-crotylsilanes are much less selective (60–70:40–30 *syn*:*anti*; Scheme 42). Hayashi and Kumada have reported a successful approach to optically active homoallyl alcohols using this strategy.[46] They have reported that useful levels of asymmetric induction can be achieved in Lewis acid catalyzed addition reactions of optically active (*R*)-(*E*)- and (*R*)-(*Z*)-crotylsilanes (**90**) and (**92**) to achiral aldehydes (Schemes 43 and 44).

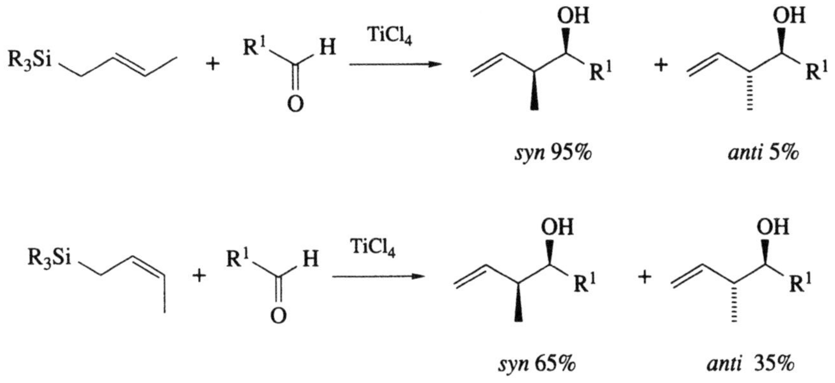

syn 95% *anti* 5%

syn 65% *anti* 35%

Scheme 42

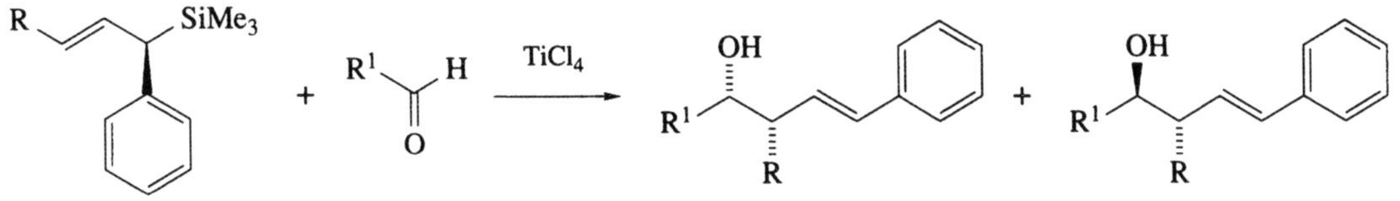

(*R*)-(*E*)-(**90**) *syn*-(**91**) *anti*-(**91**)

Scheme 43

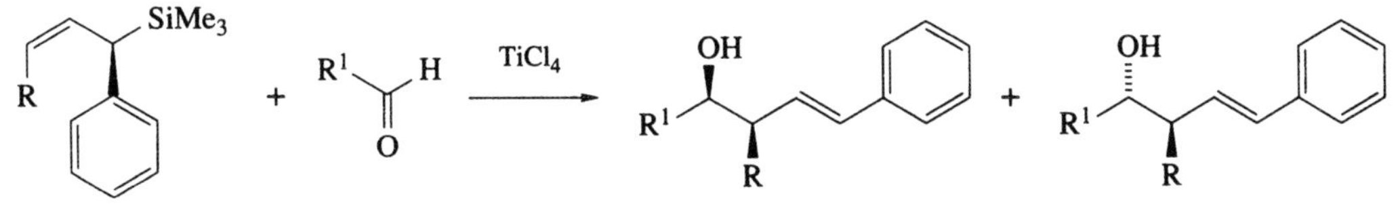

(*R*)-(*Z*)-(**92**) *syn*-(**93**) *anti*-(**93**)

Scheme 44

Table 11[46] summarizes the important results of asymmetric induction in addition reactions of optically active allylsilanes (**90**) and (**92**). The origin of the π-facial selectivity in these reactions can be traced to the *anti* selectivity that is commonly observed for S_E2' reactions of allylsilanes (Figure 14).[1a,2a] The *syn* stereochemistry with respect to the double bond configuration (*E* or *Z*) is also very high even with aldehydes bearing sterically bulky substituent groups, as illustrated by entry (*R*)-(*Z*)-(**92b**).

Table 11 Asymmetric Induction in Additions of Optically Active Allylsilanes (*R*)-(*E*)-(**90**) and (*R*)-(*Z*)-(**92**) to Achiral Aldehydes

Allylsilane	*R*	*R^{1}CHO*	(**91**) *syn:anti*	(**93**) *syn:anti*	*Yield (%)*
(*R*)-(*E*)-(**90a**)	Me	Bu^{t}CHO	>99:1		47
(*R*)-(*Z*)-(**92a**)	Me	Bu^{t}CHO		>99:1	27
(*R*)-(*E*)-(**90a**)	Me	Pr^{i}CHO	>95:5		67
(*R*)-(*Z*)-(**92a**)	Me	Pr^{i}CHO		>65:35	61
(*R*)-(*E*)-(**90b**)	Ph	Bu^{t}CHO	>99:1		44
(*R*)-(*Z*)-(**92b**)	Ph	Bu^{t}CHO		>99:1	10

The stereochemical outcome of these reactions has been interpreted on the basis of two types of transition state models. These models use very different orientations of the reacting double bonds to explain the stereoselectivities. The first is referred to as an 'open' or 'extended' antiperiplanar transition state model, illustrated in Figure 15 for both (*E*)- and (*Z*)-crotylsilanes. In this case the carbon nucleophile and the activated carbonyl (aldehyde) are *anti* to each other and also coplanar. Using this model transition states (**I**) and (**II**) are favored due to diminished steric interactions between the aldehyde substituent and the vinylmethyl group. The destabilizing interactions created by these substituents are greatest for transition states (**III**) and (**IV**), which place them in a *gauche* orientation.[1a]

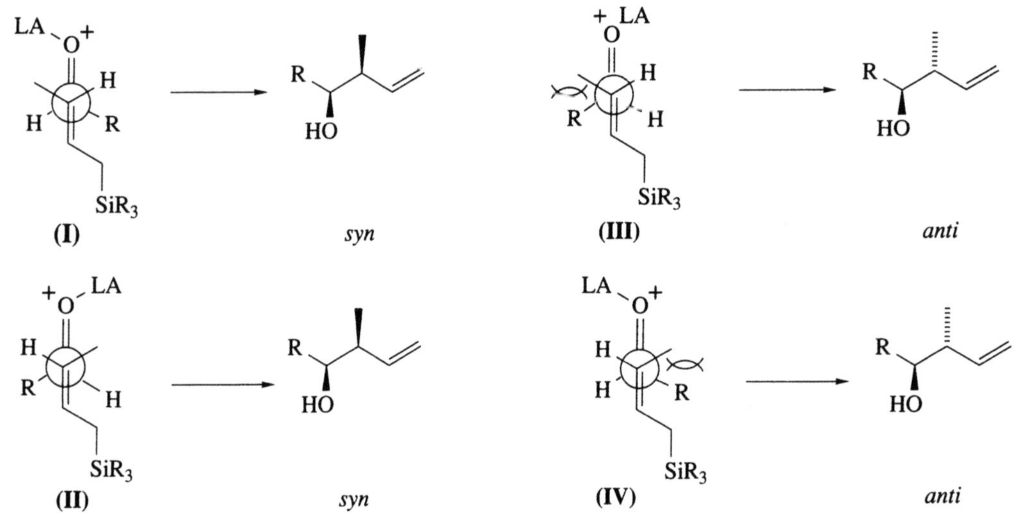

Figure 15 Open transition state models for intermolecular additions of C(3)-substituted allylsilanes

An alternate model using the transition states (**V**)–(**XII**) can be used to rationalize the selectivities in these reactions (Figure 16). This model requires synclinal orientation of the reacting double bonds. The transition states (**V**) or (**VI**) and (**IX**) or (**X**) are favored over (**VII**), (**VIII**), (**XI**) and (**XII**) because steric interactions between the Lewis acid bonded to the oxygen lone pair *anti* to the R substituent of the aldehyde and the vinyl methyl group are minimized.

Useful levels of stereoselectivity were obtained in intermolecular addition reactions of C(3)-substituted allylsilanes to chiral aldehydes. Lewis acids that are capable of chelating to heteroatoms have been used to direct the stereochemical course of allylsilane additions to α-alkoxy and α,β-dialkoxy carbonyl compounds. The allylation of α-benzyloxy aldehyde (**94**) in the presence of $TiCl_4$ and $SnCl_4$ furnished products with high levels of *syn* stereoselection (*syn*-**95**).[47] In contrast, under nonchelation-controlled reaction conditions ($BF_3 \cdot OEt_2$) allyltrimethylsilane reacted to form predominantly the *anti*-1,2-diol product (*anti*-**95**), as shown in Scheme 45.

Taddei and coworkers reported that chiral allyltrimethylsilanes containing an optically active ligand derived from (–)-myrtenal attached to silicon (**96**) underwent enantioselective addition reactions with achiral aldehydes (Scheme 46) to give, after acid hydrolysis, optically active homoallyl alcohols (**98**).[48] A variety of Lewis acids were examined to optimize enantiomeric excess, and $TiCl_4$ was found to be the most effective catalyst. The results of the $TiCl_4$-promoted additions are reported in Table 12.[48]

2.5.4.1.3 Intramolecular additions of allyltrimethylsilanes to aldehydes and ketones

Intramolecular addition of allylsilanes performed under chelation-controlled conditions represents an efficient method in which useful levels of stereoselectivity can be achieved. This process has been demonstrated for a variety of α-alkoxy, α,β-dialkoxy and β-dicarbonyl compounds, including β-keto esters, β-keto amides and β-keto lactones.

The use of chelation to direct the stereochemical outcome of intramolecular additions of allylsilanes to β-dicarbonyl compounds can provide excellent levels of diastereoselectivity.[49] Cyclizations of this type proceed at low temperature under mild reaction conditions and are highly chemoselective, providing routes to highly functionalized five-, six- and seven-membered rings (**100a, 100b** and **100c**; Scheme 47). For the cases examined, $SnCl_4$ and $FeCl_3$ proved to be less effective than $TiCl_4$ in the cyclization of ethyl 2-alkyl-2-alkanoyl-4-(trimethylsilyl)methyl-4-pentenoate (**99**; $n = 1$; Table 13).[49]

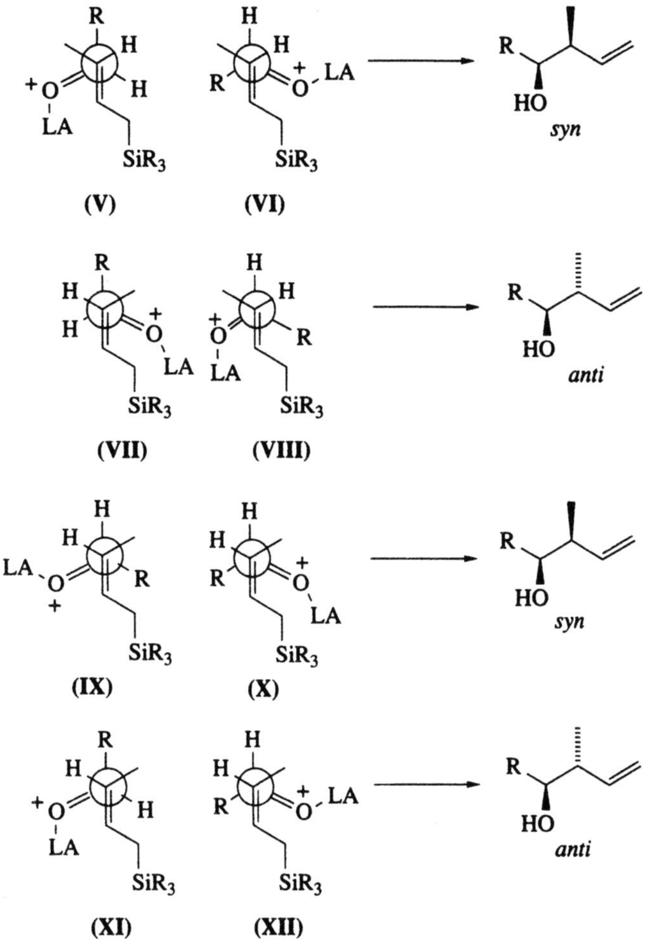

(V) (VI) → *syn*

(VII) (VIII) → *anti*

(IX) (X) → *syn*

(XI) (XII) → *anti*

Figure 16 Synclinal transition state model for the intermolecular addition of C(3)-substituted allylsilanes to aldehydes

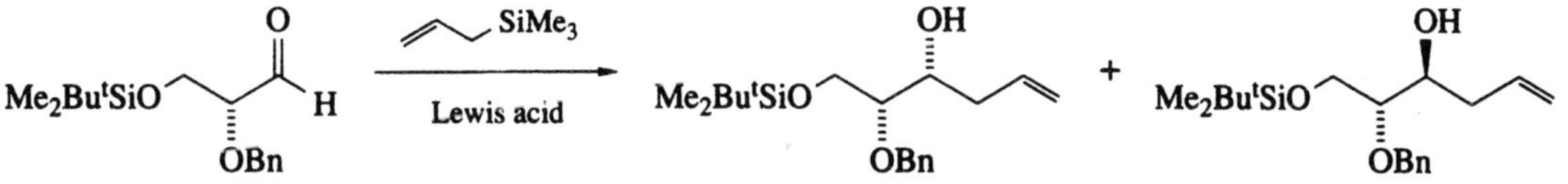

(94) → *syn*-(95) + *anti*-(95)

Lewis acid	*syn*-(95)	*anti*-(95)
$TiCl_4$	>98%	<2%
$SnCl_4$	>98%	<2%
$BF_3 \cdot OEt_2$	19%	81%

Scheme 45

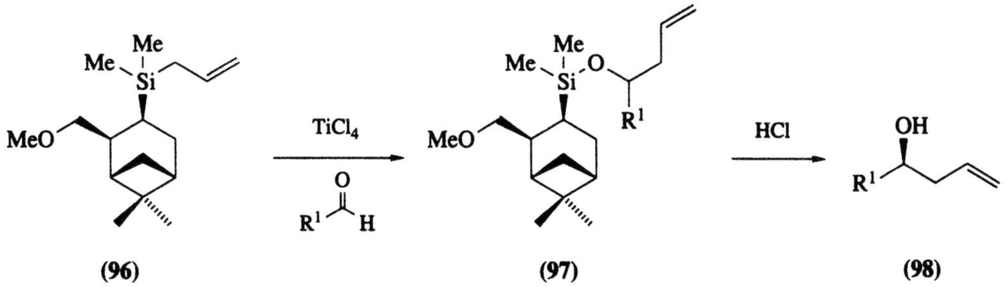

(96) → (97) → (98)

Scheme 46

Table 12 Enantioselective Allylation of Achiral Aldehydes Catalyzed by $TiCl_4$

Aldehyde	*Product*	*Yield (%)*	*ee (%)*[a]
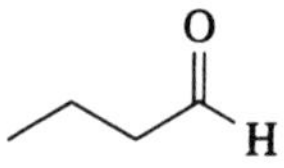	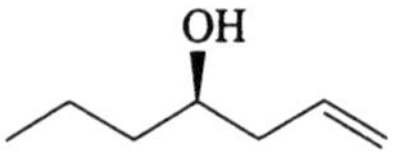 **(98a)**	58	46
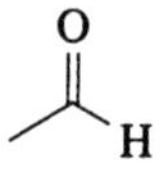	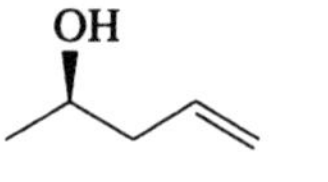 **(98b)**	61	40
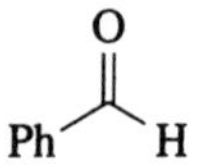	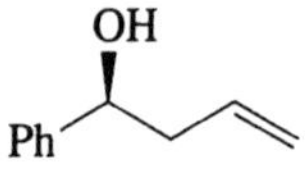 **(98c)**	45	21
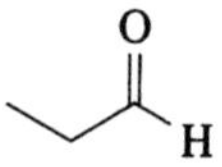	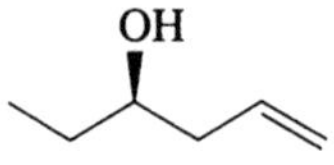 **(98d)**	39	40

[a] Enantiomeric excess determination performed by Mesher analysis and literature comparison.

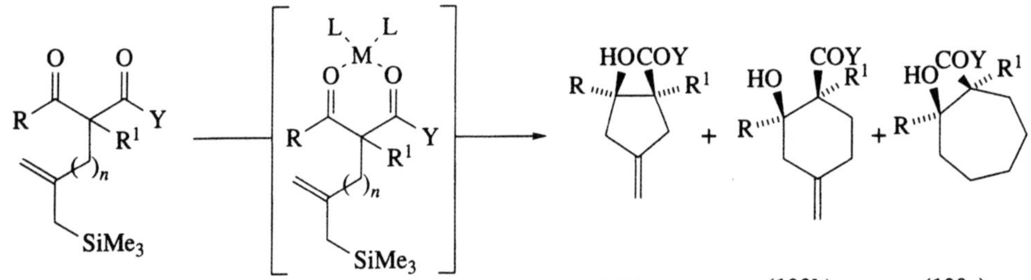

(99) $n = 0, 1, 2$ **(100a)** **(100b)** **(100c)**

Scheme 47

Table 13 Cyclizations of Ethyl 2-Alkyl-2-alkanoyl-4-(trimethylsilyl)methyl-4-pentenoate (**99**) Catalyzed by Titanium Tetrachloride (Scheme 47)

Substrate	*R*	*R*[1]	*Isolated yield of (**100**) (%)*[a]
(99a)	Me	Me	88
(99b)	Me	Et	78
(99c)	Me	Pr^i	65
(99d)	Me	Ph	72
(99e)	Et	Me	78
(99f)	Pr^i	Me	66
(99g)	Bu^t	Me	74
(99h)	Ph	Me	74
(99i)	Me	H	57

[a] Diastereomer ratio was >200:1.

The Lewis acid promoted intramolecular additions of β-allylsiloxy aldehydes represent an efficient method for the construction of 1,3-stereocenters, as illustrated in Scheme 48.[50] The sense of 1,3-asymmetric induction in $TiCl_4$-catalyzed allylsilane addition to *O*-allylsilyl-protected aldehydes is opposite (>90% *syn* selection) to that reported for the intermolecular addition to the *O*-benzyl derivative.

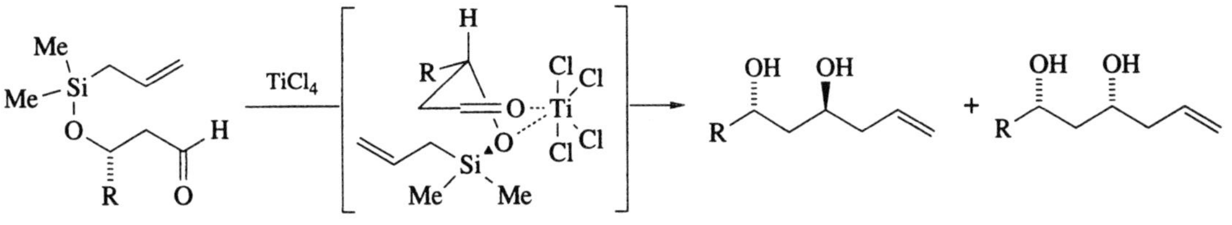

(101a) R = Me
(101b) R = Bu^n

anti-**(102a)** *syn*-**(102a)**
anti-**(102b)** *syn*-**(102b)**

Scheme 48

Reaction of aldehydes (**101a**) and (**101b**) containing equivalent amounts of the Lewis acids $TiCl_4$, $SnCl_4$ and $BF_3{\cdot}OEt_2$ are summarized in Table 14. Interestingly, $TiCl_4$ and $SnCl_4$ lead to opposite diastereomers (*syn versus anti*). The results of $TiCl_4$-mediated reactions (Table 14)[50] support the original hypothesis of an internal allyl transfer in which silicon and titanium act as templates. However, as pointed out by Reetz,[50] the levels of stereoselection do not unequivocally demonstrate an intramolecular pathway.

Table 14 Lewis Acid Catalyzed Intramolecular Additions of β-Allylsiloxy Aldehydes

β-Allylsiloxy aldehyde	*R*	*Lewis acid*	*Yield (%)*	*(**102**) anti:syn*
(**101a**)	Me	$TiCl_4$	70	8:92
(**101a**)	Me	$SnCl_4$	70	92:8
(**101a**)	Me	$BF_3{\cdot}OEt_2$	60	70:30
(**101b**)	Bu^n	$TiCl_4$	80	10:90

2.5.4.1.4 Stereochemistry of internal allylsilane additions to aldehydes and acetals

In view of the importance of stereoselective addition reactions of allylsilanes to C—X π-bonds, the stereochemical course of intramolecular addition reactions involving aldehydes has been investigated by Denmark and coworkers.[51] The results are summarized below in Tables 15[51] and 16.[52] In the cases involving additions to aldehydes, the authors have made the assumption that the coordination of the Lewis acid to the carbonyl oxygen generates the '(*E*)-complex' (**103**; Scheme 49), which after cyclization generates the *syn* and *anti* bicyclic alcohols (**104**). The major steric contribution arises from interaction between the Lewis acid and the trimethylsilylmethyl group. This interpretation suggests that there may

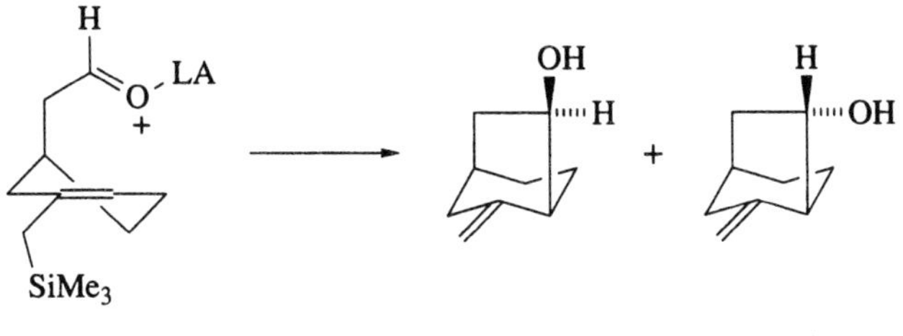

(103) *syn*-**(104)** *anti*-**(104)**

Scheme 49

Table 15 Lewis Acid Catalyzed Cyclization of Aldehyde (**103**)

Lewis acid	*Solvent*	*Temperature (°C)*	*(**104**) syn:anti*
$SnCl_4$	CH_2Cl_2	–70	49:51
Et_2AlCl	CH_2Cl_2	–70	66:34
$FeCl_3$	CH_2Cl_2	–70	70:30
$AlCl_3$	CH_2Cl_2	–70	79:21
$BF_3{\cdot}OEt_2$	CH_2Cl_2	–70	80:20
Bu^n_4NF	THF	67	30:70

be a stereoelectronic advantage for the synclinal orientation of reactants under electrophilic conditions. Reactions catalyzed by fluoride ions show a reversal of stereoselectivity and most likely involve a different mechanism.

Analogous systems related to the models for allylsilane–acetal cyclizations have been studied.[52] In these cases, however, cyclization of the acetals (**105**; Scheme 50) forms the corresponding *syn* and *anti* bicyclic ethers (**106**). The authors concluded that the stereochemistry of cyclization of acetal models (**105**) is dependent upon the mechanism of activation. In the presence of TMS-OTf the acetals (**105a–c**) reacted through an S_N2-type process, while the isopropyl acetal (**105d**) reacted through prior ionization to oxonium ion (**107**).

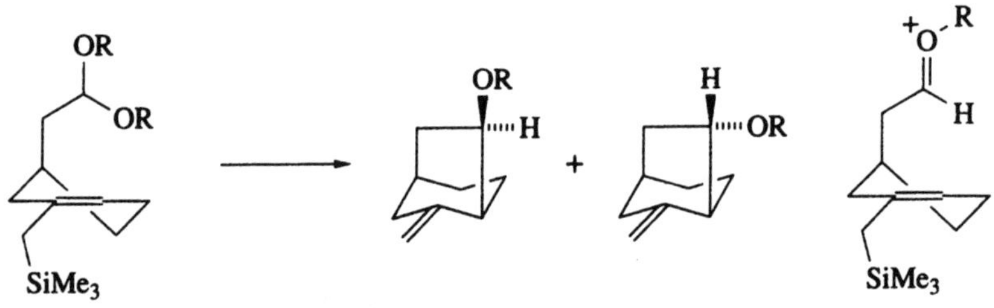

OR OR $SiMe_3$ OR H H OR ^{+}O R H $SiMe_3$

(**105a**) = Me
(**105b**) = Et
(**105c**) = Bu^i
(**105d**) = Pr^i

syn-(**106**) *anti*-(**106**) (**107**)

Scheme 50

Table 16 Effect of Lewis Acid in the Cyclization of Dimethyl Acetal (**105a**)

Reagent	*Temperature* (°C)	*(**106**) Syn:anti*	*Yield* (%)
TMS-OTf	–70	96:4	100
TfOH	–70	96:4	62
$Ti(OPr^i)_2Cl_2$	–20	87:13	21
$AlCl_3$	–20	86:14	33
BCl_3	–70	82:18	57
$BF_3·OEt_2$	–20	77:23	95
$TiCl_4$	–90	47:53	55
$SnCl_4$	–70	45:55	35
$SnCl_4$	–60	71:29	81

For the cases examined, the structure of the acetal had a dramatic effect on the stereochemical outcome of the reaction. Thus, allylsilanes (**105a–c**) all showed *syn* selectivity. However, the isopropyl case (**105d**) showed a slight *anti* preference. This reversal of selectivity has been interpreted as a change in mechanism rather than a steric phenomenon related to the added steric bulk of the diisopropyl acetal. The results of these experiments are summarized in Table 17.[52]

Table 17 Effect of Acetal Structure on the Stereochemical Outcome of the Cyclization of (**105a**)–(**105d**) with TMS-OTf[a]

Substrate	*R*	*syn* (%)	*anti* (%)
(**105a**)	Me	96	4
(**105b**)	Et	92	8
(**105c**)	Bu^i	90	10
(**105d**)	Pr^i	38	62

[a]All reactions were run in CH_2Cl_2 (0.05 M) with 1.0 equiv. of TMS-OTf.

2.5.4.2 Propargylsilanes

Propargylsilanes undergo electrophilic addition reactions to generate β-silyl carbocations (**109**) that can, in principle, react further to give either addition (**110**) or substitution (**111**) products, as illustrated in Scheme 51. As in the case of allylsilanes, however, substitution predominates.[2a]

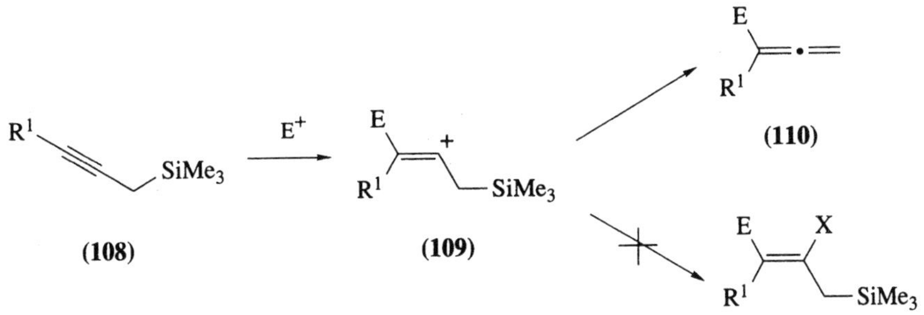

Scheme 51

The stereochemical course of the addition reactions to propargylsilanes can be rationalized using a mechanism similar to that for allylsilanes (Scheme 52).[53] The propargylsilane would be expected to exist in conformation (**113**; Scheme 53) with the carbon–silicon bond overlapping with the π-lobes of one of the carbon–carbon π-bonds. The electrophile attacks the π-bond conjugated with the carbon–silicon bond *anti* to the trialkylsilane group, generating the stabilized vinyl cation (**114**). Loss of the trialkylsilicon group then affords the expected allene.

Scheme 52

Scheme 53

Propargylsilanes show predictable behavior as cyclization terminators. Internal additions of propynylsilanes to acyclic *N*-acyliminium ions promoted by Lewis acids or protic acids yield α-allenic amides or carbamates (Scheme 54).[54]

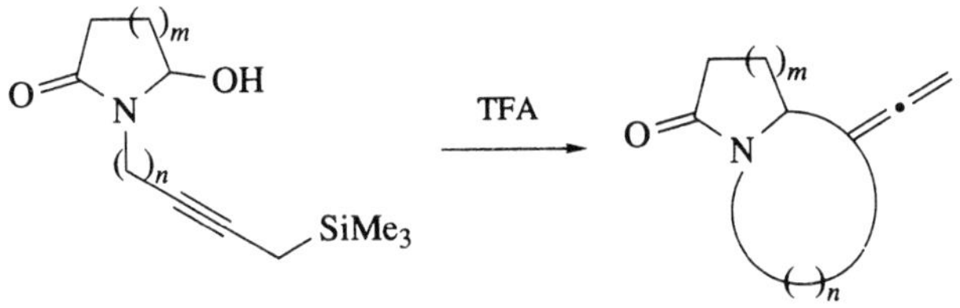

Scheme 54

The intramolecular acid-catalyzed reaction of 2-propargylsilanes of structural type (**117**) with *N*-acyliumion ion precursors leads to bridged azabicyclic ring systems (**118**; Scheme 55) which contain an α-allenic amide function. These bicyclic products serve as intermediates to highly functionalized *trans*-fused ring systems possessing a 1,3-diene moiety.

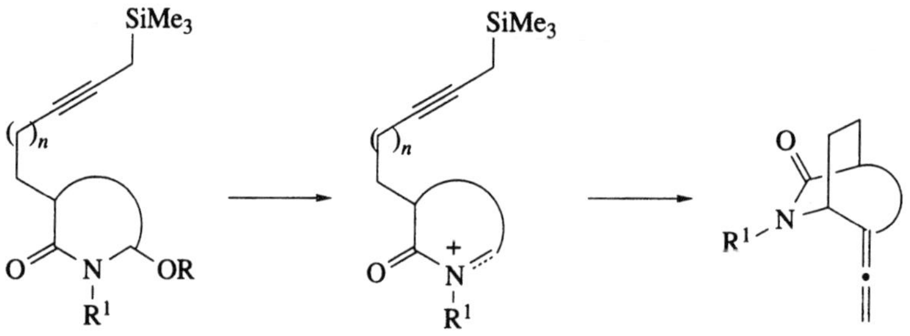

(117) **(118)** **(119)**

Scheme 55

2.5.5 α-SILYLORGANOMETALLIC REAGENTS

Silicon is capable of stabilizing an adjacent carbon–metal bond although it is more electropositive than hydrogen or carbon. The origins of this stabilization are discussed in Sections 2.5.2.1–2.5.2.5.

2.5.5.1 General Methods for the Formation of α-Silyl Carbanions

The variety of methods available for the formation of α-silyl carbanions can be divided into four categories.

(i) Deprotonation: alkyllithium reagents (*e.g.* Bu^nLi) can easily abstract protons adjacent to trialkylsilyl groups (**120** → **121**; Scheme 56) generating the α-silyl carbanion.[55]

Me_3Si ⁀R —BuLi/TMEDA→ Me_3Si ⁀R Li^+

(120) **(121)**

Scheme 56

(ii) Metal–halogen exchange: α-halosilanes undergo metal–halogen exchange upon treatment with alkyllithium reagents (**122** → **123**; Scheme 57).[56]

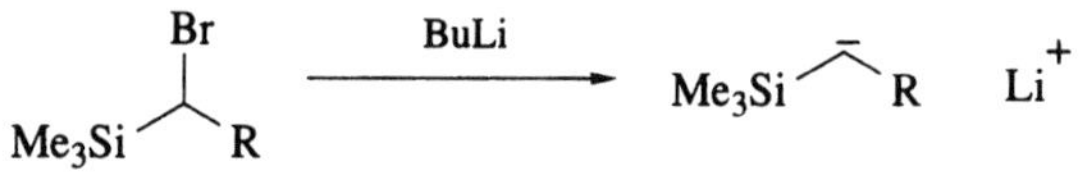

(122) —BuLi→ **(123)**

Scheme 57

(iii) Transmetallation: α-selenosilanes are transmetallated in the presence of alkyllithium reagents (**124** → **125**; Scheme 58).[57] Included in this category are desilylation of bis(silyl) compounds (**126**) by fluoride and alkoxide anions.[58] This procedure has been shown to be a particularly clean method for the generation of α-silyl carbanions (**127**), as illustrated in Scheme 59.

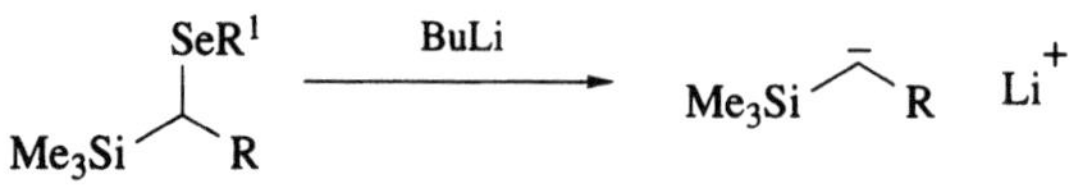

(124) —BuLi→ **(125)**

Scheme 58

(iv) Addition of an organometallic reagent to vinylsilanes: The example provided in Scheme 60 illustrates the ability of silicon to activate and direct the formation of carbon nucleophiles. These addition

ROLi + (126) → $ROSiMe_3$ + (127)

Scheme 59

reactions proceed with excellent levels of selectivity, as only α-silyl carbanions are observed; such levels of selectivity are not reached with unsubstituted or unsymmetrical alkenes.[59]

$R^1{}_3Si$–CH=CH$_2$ + R^2Li → $R^1{}_3Si$–CH(Li)–CH$_2R^2$

Scheme 60

The directing ability of silicon is further illustrated by the observation that isopropylmagnesium chloride adds to the triethoxyvinylsilane (**128**) to generate the addition product (**129**; Scheme 61); however, the triethoxyallylsilane (**130**) undergoes clean substitution at silicon when treated with isopropylmagnesium chloride to yield the substitution product (**131**; Scheme 62).

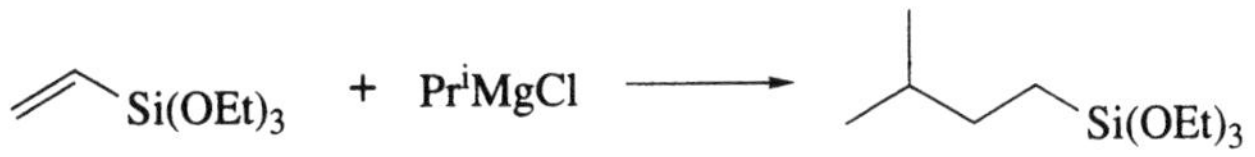

(128) + Pr^iMgCl → (129)

Scheme 61

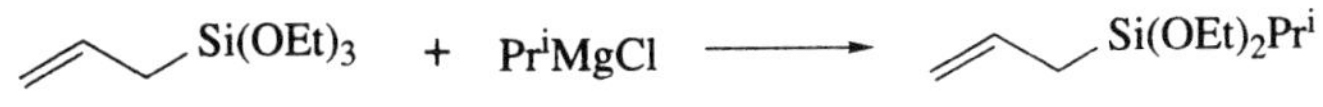

(130) + Pr^iMgCl → (131)

Scheme 62

Both organolithium[60] and Grignard reagents[61] undergo additions to vinylsilanes to yield α-silyl carbanions. This directing effect can be overcome by the influence of a stronger electron-withdrawing group, such as a nitro group, adjacent[62] to the silicon group, as illustrated in Scheme 63 with the transformation of (**132**) to (**133**).

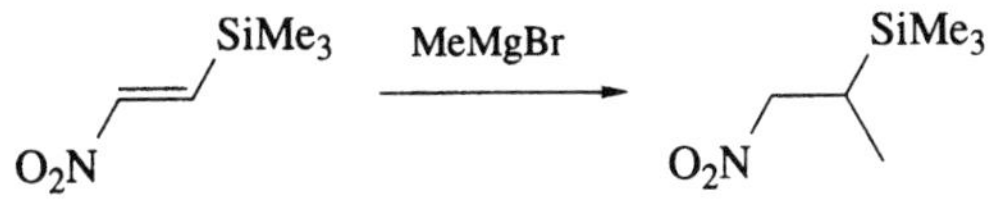

(132) —MeMgBr→ (133)

Scheme 63

One of the most significant demonstrations of the ability of silicon to promote nucleophilic attack at carbon–carbon bonds was reported by Stork within the context of the development of an improved variant of the Robinson annelation.[63,64] Existing annelation methods which depend on conjugate addition of an enolate to an α,β-unsaturated ketone suffer from a variety of drawbacks, which are consequences of the reversible nature of the Michael addition reaction and the ease with which most electrophilic alkenes undergo multiple additions and oligomerization. α-Silylvinyl ketones of structure type (**134**) are not prone to these problems and thus serve as excellent four-carbon components (Michael acceptors) in the Robinson annelation. The ability of silicon to stabilize adjacent carbanions is shown in Scheme 64 with the α-silyl carbanion and its resonance form (**135** and **136**). The additional stabilization of the intermediate enolate ion halts the reversibility of the process, ultimately generating the octalone system in good yield with minimum side reactions. The added steric bulk created by the trialkylsilane group diminishes the reactivity of the intermediate enolate, thus providing an increase in chemoselectivity.

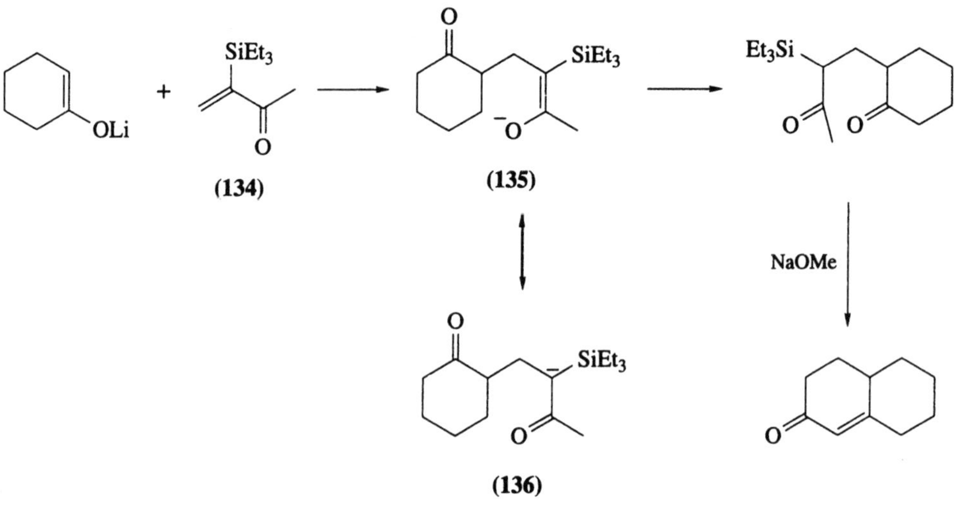

(134) (135) (136)

Scheme 64

2.5.5.2 The Peterson Reaction: Introduction

The use of α-silyl organometallic reagents, principally lithium and Grignard reagents, to convert aldehydes and ketones to alkenes has been applied extensively in organic synthesis. Peterson was the first to investigate this type of reaction sequence and many useful applications were derived from these initial observations and experiments. The reaction has been reviewed by Ager.[65] This section is concerned with the mechanism of the Peterson reaction and some recent applications of the Peterson-type alkenation of carbonyl compounds.

As illustrated in Scheme 65, the intermediate β-silyl alkoxide (**138**) is subject to either protonation to give the corresponding β-silyl alcohol (**139**), or elimination of R_3SiO to generate the alkene (**140**). The pathway leading to the alkene is known as the carbonyl alkenation reaction, or the Peterson reaction.[55]

(137) (138) (139) (140)

Scheme 65

2.5.5.2.1 *The mechanism of the Peterson reaction*

As a consequence of the presumed irreversibility of the first step of the Peterson reaction, the stereochemistry of the elimination was determined solely by the relative rates (k_1 and k_1') of formation of *threo*- and *erythro*-β-silyl alkoxides (**141**) and (**142**), as indicated in Scheme 66. Support for the irreversibility of this step comes from the experiments on the nucleophilic opening of the diastereomeric epoxides (**143**) and (**144**). Thus, the *syn*- and *anti*-epoxides, when treated with lithium dipropyl cuprate, yielded the *threo*- and *erythro*-β-silyl alcohols (**145**) and (**146**), respectively.[66] Treatment of the *threo* isomer with base gave the (*Z*)-alkene exclusively and the *erythro* isomer gave the (*E*)-alkene. The high levels of selectivity for this elimination indicated that the initial addition is not reversible. The elimination was also shown to be stereospecific and only the *syn* stereoisomer was generated under basic conditions. Under acidic conditions the stereochemistry of the elimination product is inverted. Thus, β-silyl alcohols (**143**) and (**144**) yield the (*E*)-alkene (**147**) and the (*Z*)-alkene (**148**) respectively, resulting from *anti* elimination (Scheme 66).

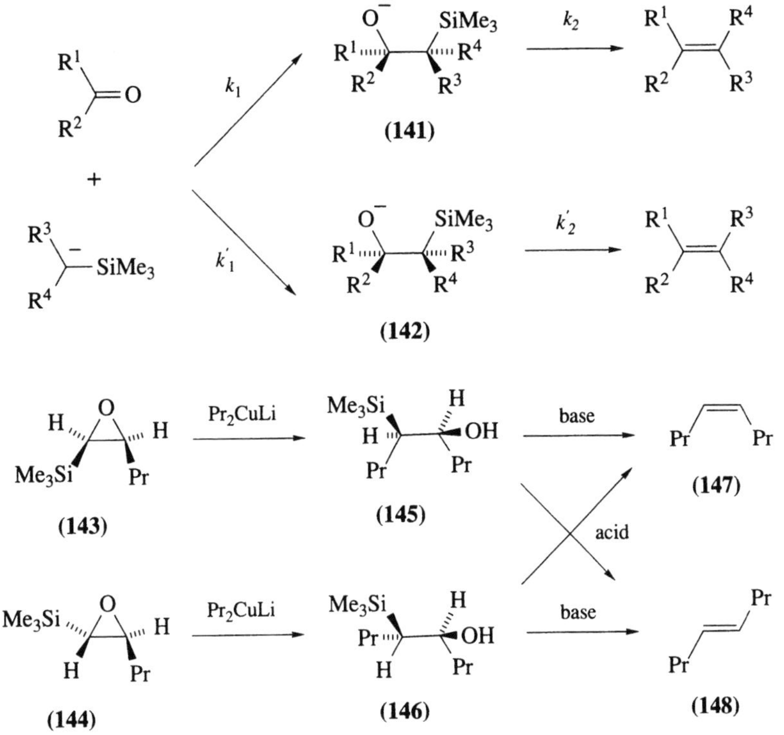

Scheme 66 Mechanism of the Peterson reaction

2.5.5.2.2 *Stereochemistry of the Peterson reaction*

The *(E):(Z)* ratio of the derived alkene varied considerably with the size of the silyl group. A transition state model has been developed to rationalize the relative transition state energies.[67] Nucleophilic attack on carbonyl compounds occurs in the plane of the C—O π-bond with a Nu—C—O angle of approximately 109°.[68,69] The smallest group R, usually a hydrogen, is placed in the least sterically demanding position and attack occurs between the hydrogen and the R substituent of the aldehyde in order to minimize steric repulsion. The remaining substituents are arranged such that the largest group is positioned *anti* to the R substituent of the aldehyde. Thus, as the steric size of the silyl group increases, transition state (**149**) becomes more favorable than transition state (**150**), resulting in the preferential formation of the *erythro* isomer leading to formation of the (*Z*)-alkene.

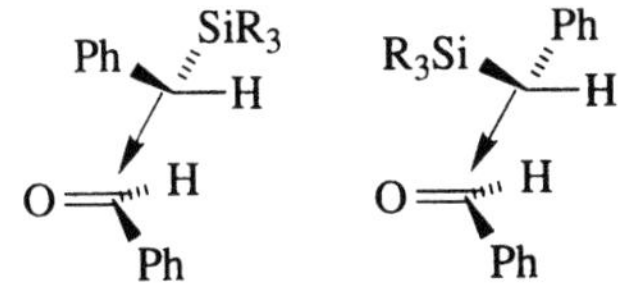

2.5.5.3 Functionalized α-Silyl Carbanions

The addition of trimethylsilylmethyllithium to enolizable ketones generally gives significant amounts of the enolized carbonyl compound. Johnson and Tait[70] have shown that when the lithium anion is converted to the cerium reagent it gives high yields of the addition product. For example, α-tetralone (**151**) reacts with trimethylsilylmethyllithium to give after protonation an ~1:1 mixture of the desired exocyclic alkene (**152**) and recovered ketone. In sharp contrast, the corresponding cerium reagent gave a 95:5 ratio of alkene to ketone (see Scheme 67).

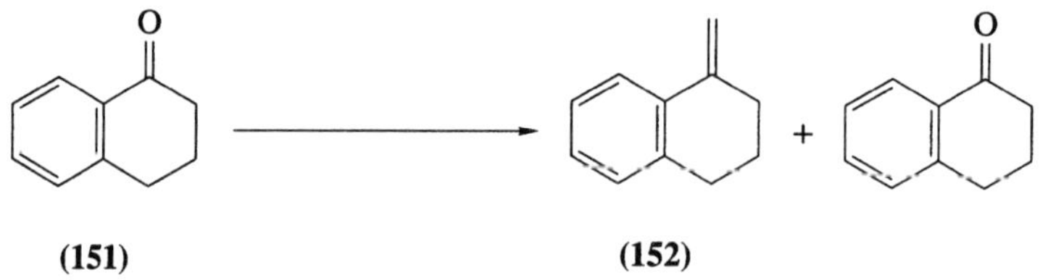

(**151**) → (**152**) +

	(152)	
Me_3SiCH_2Li	50	50
$Me_3SiCH_2Li/CeCl_3$	>95	5

Scheme 67

Anions of α-silyl phosphonates of type (**153**) also undergo additions to carbonyl compounds. The corresponding addition products, β-silyl alkoxides, can react with ketones to yield the product of the Peterson alkenation or the Wittig reaction. In practice only the Peterson product (**154**) is obtained, indicating that loss of $OSiMe_3$ is faster than elimination of $O{=}PPh_3$ (Scheme 68).[71,72] If the α-silyl carbanion is adjacent to a chlorine atom (**155**), an internal displacement reaction follows the initial formation of the β-silyl alkoxide, and epoxides (**156**) are formed (Scheme 69).[73,74]

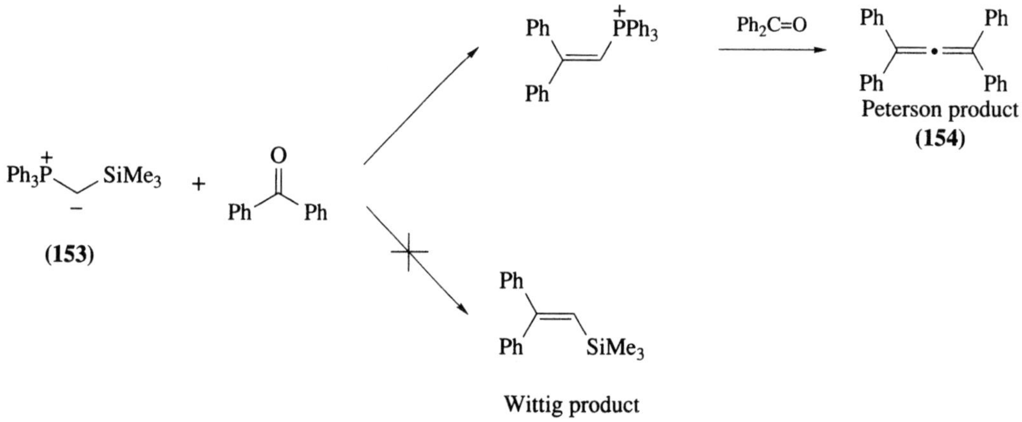

$Ph_3\overset{+}{P}$ SiMe3 (**153**) + Ph(C=O)Ph

Ph, Ph, $\overset{+}{P}Ph_3$ —$Ph_2C{=}O$→ Ph, Ph, Ph, Ph
Peterson product
(**154**)

Ph, Ph, SiMe3
Wittig product

Scheme 68

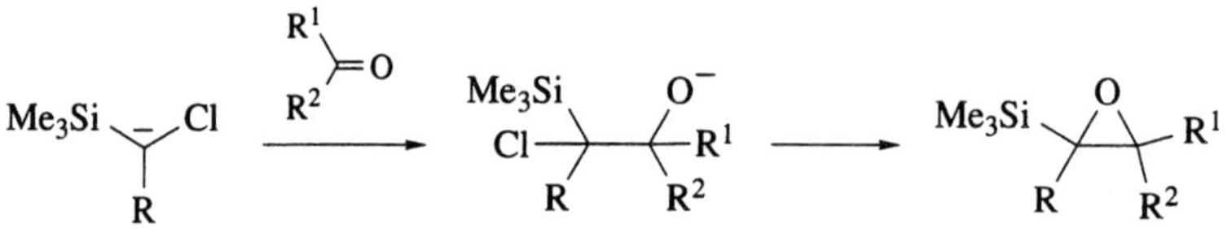

Me3Si, Cl, R (**155**) —$R^1R^2C{=}O$→ Me3Si, Cl, R, O^-, R^1, R^2 (**156**) → Me3Si, O, R, R^1, R^2

Scheme 69

Mixed ketene *O,S*-acetal derivatives (**157**) can be readily prepared from the reaction of [methoxy(phenylthio)(trimethylsilyl)methyl]lithium,[75] generated from the mixed acetal (**158**), with aldehydes and ketones (Scheme 70).

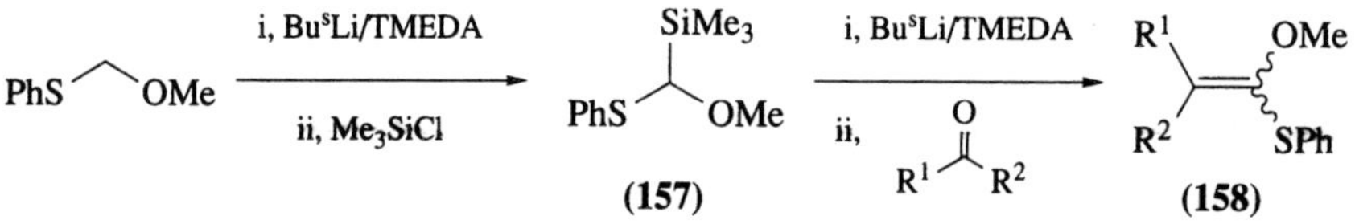

PhS OMe —i, Bu^sLi/TMEDA; ii, Me_3SiCl→ PhS, SiMe3, OMe (**157**) —i, Bu^sLi/TMEDA; ii, R^1COR^2→ R^1, R^2, OMe, SPh (**158**)

Scheme 70

The lithium reagent of phosphonate (**159**) reacts with aldehydes and ketones to produce vinylphosphonate (**160**). These compounds were oxidized (OsO_4/NMO) and deprotected to provide a convenient and high yielding route to α-hydroxy acids[76] of structure type (**161**; Scheme 71).

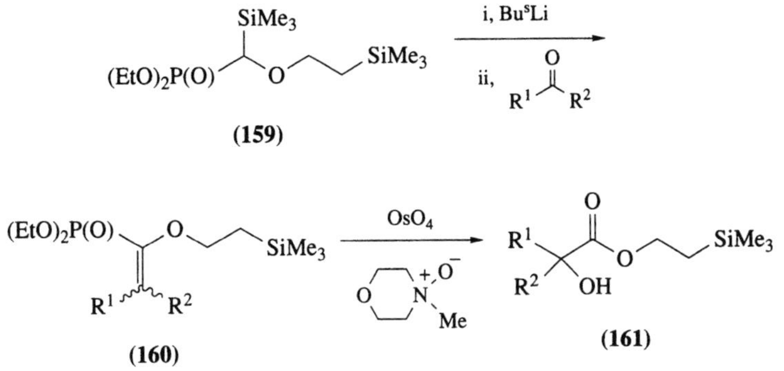

Scheme 71

2.5.5.3.1 *Reactions involving ambient silyl-substituted carbanions*

The Peterson reaction has been successfully employed using α-silylallyl anions (**162a** and **162b**; Figure 17).

Figure 17

In addition reactions to carbonyl compounds C-3 attack generally predominates,[77–79] as illustrated in Scheme 72 with the formation of vinylsilane (**163**). C-1 addition also predominates in alkylation and acylation reactions involving α-silylallyl anions when alkyl halides[80] and acyl halides[81] are used as electrophiles.

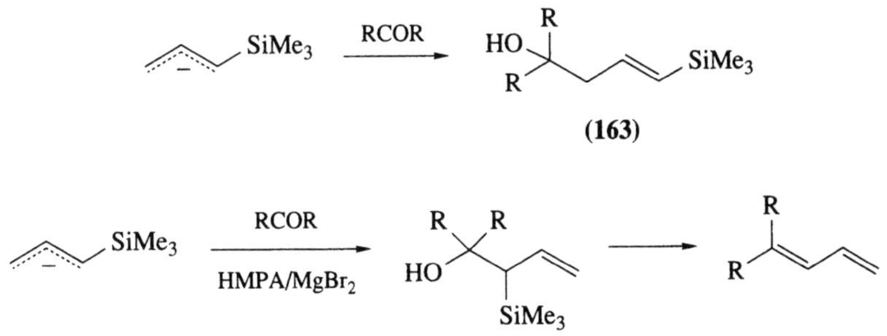

Scheme 72

Interestingly, in the presence of HMPT and magnesium bromide, C-1 attack predominates and is followed by elimination, yielding the 1,3-diene system[82] as shown in Scheme 72. Similarly, allylboronates of structural type (**164**) bearing a vinylsilane at the C-3 position gave predominantly C-1 addition products to form the *syn (erythro)* isomer (**165**) as the major diastereomer (Scheme 73).[83]

Metallation of the bis(silyl) derivative (**166**) generates a symmetrical carbanion (**167**), which also reacts stereoselectively with aldehydes to produce *anti* β-silyl alcohol derivatives (**168**), which undergo base-catalyzed elimination reactions to generate the *(E)/(Z)* diene system (**169**; Scheme 74).[84]

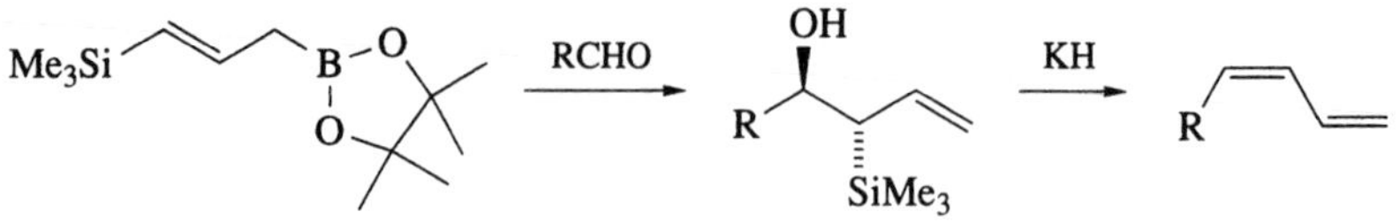

(164) **(165)**

Scheme 73

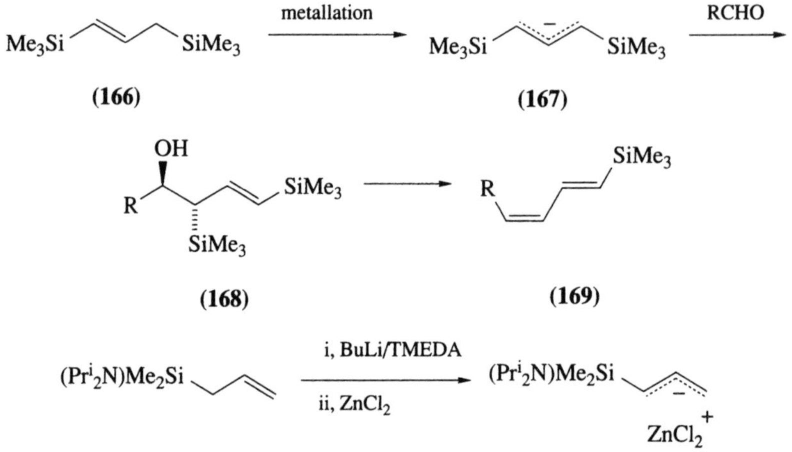

(166) **(167)**

(168) **(169)**

(170)

Scheme 74

2.5.5.3.2 *Metallated allylaminosilanes*

The organozinc reagent (**170**) derived from allyl(diisopropylamino)dimethylsilane reacts with aldehydes to form *anti*-3-silyl-2-alken-4-ols (**171**), which are further oxidized ($KHCO_3/H_2O_2$) to *anti*-1-alkene-3,4-diols (**172**).[85] The use of a sterically bulky $Pr^i_2NMe_2Si$ group in the zinc reagent was required to suppress the elimination pathway and the formation of conjugated 1,3-diene systems. These allylamino anions thus function as useful α-alkoxyallyl anion equivalents, as illustrated in Scheme 75. The important results of these experiments are summarized in Table 18.[85]

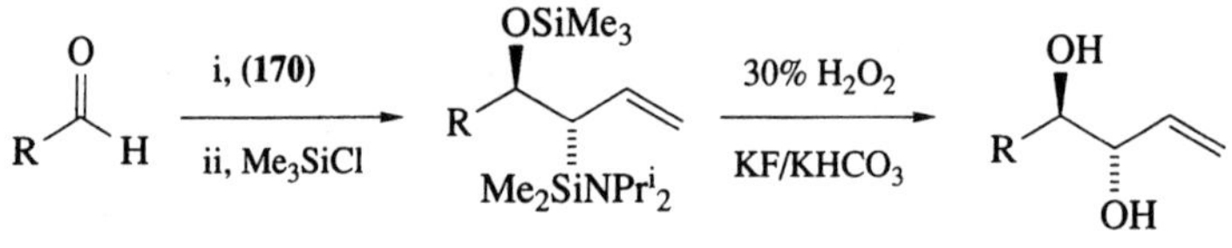

(171) **(172)**

Scheme 75

2.5.5.3.3 *Addition reactions of α-silyl anions to C═N π-bonds*

An interesting variant of the Peterson reaction involves the addition of an α-silylbenzylic anion (**173**) to a carbon–nitrogen double bond such as an imine. In this case the loss of RN—$SiMe_3$ occurs less readily than the loss of alkoxytrialkylsilane (O—SiR_3).[86] The initial addition reaction is reversible and the stereochemistry of the elimination reaction is predominantly *trans* (Scheme 76). Both acid- and base-catalyzed elimination reactions lead to the *trans* product (**174**).

Table 18 Stereoselective Additions of Metallated Aminosilanes to Aldehydes

Aldehyde	*α-Silyloxysilane*	*Yield (%)*	*Diol*	*Yield (%)*
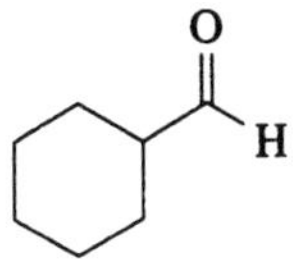 cyclohexanecarbaldehyde	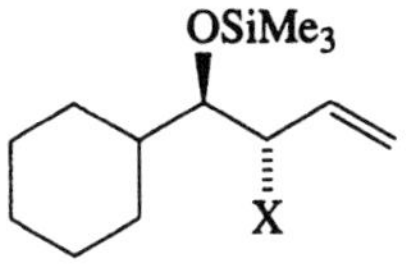 **(171a)**	54	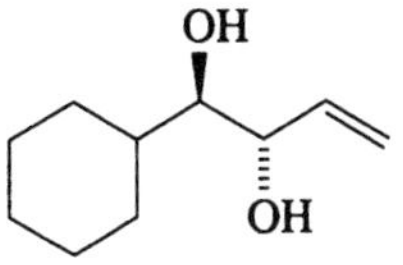 **(172a)**	78
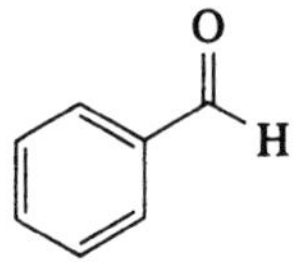 benzaldehyde	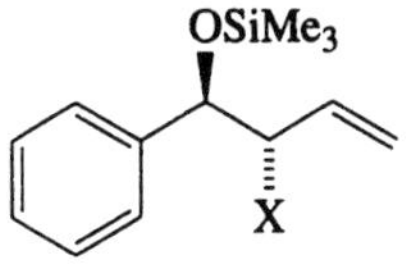 **(171b)**	97	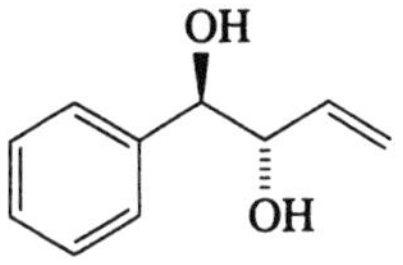 **(172b)**	80
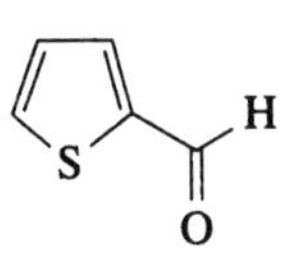 thiophene-2-carbaldehyde	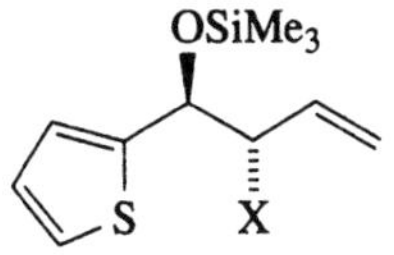 **(171c)**	87	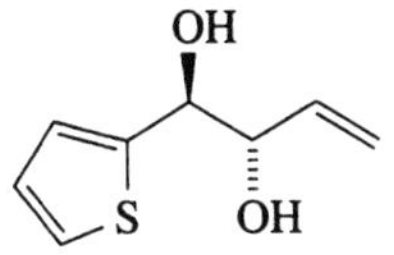 **(172c)**	72
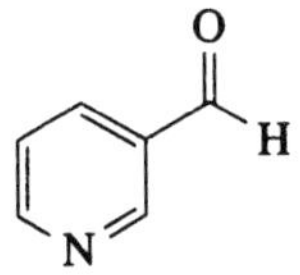 pyridine-3-carbaldehyde	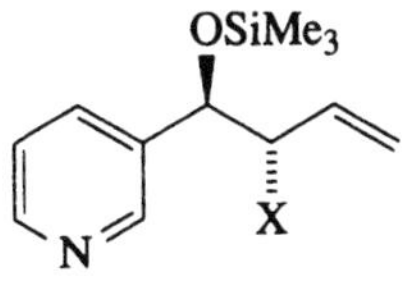 **(171d)**	75	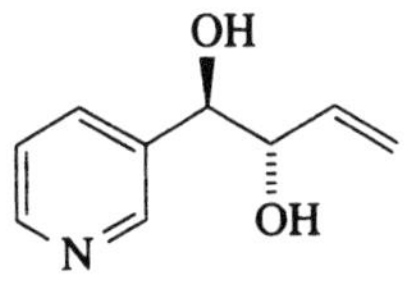 **(172d)**	93

X = $SiMe_2(NPr^i{}_2)$.

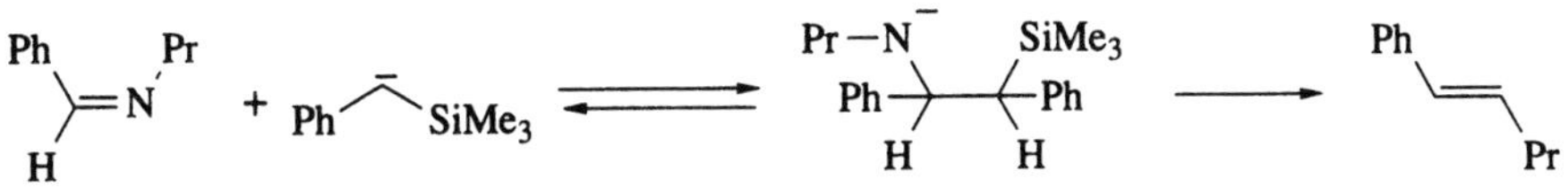

Ph(H)C=NPr + $Ph\overset{-}{C}HSiMe_3$ **(173)** ⇌ $Pr{-}\overset{-}{N}$–CH(Ph)–CH(Ph)–$SiMe_3$ → PhCH=CHPr **(174)**

Scheme 76

2.5.6 REFERENCES

1. For general reviews of organosilicon chemistry, see (a) I. Fleming, *Org. React. (N.Y.)*, 1989, **37**, 57; (b) G. Majetich, *Org. Synth. Theory Appl.*, 1989, **1**, 173; (c) H. Sakurai (ed.), 'Organosilicon and Bioorganosilicon Chemistry: Structure, Bonding, Reactivity and Synthetic Application', Halsted, New York, 1985; (d) W. P. Weber, 'Silicon Reagents for Organic Synthesis', Springer; Berlin, 1983; (e) E. W. Colvin, 'Silicon in Organic Synthesis', Butterworths, London, 1983; (f) I. Fleming, in 'Comprehensive Organic Chemistry', ed. D. H. R. Barton and W. D. Ollis, Pergamon Press, Oxford, 1979, vol. 3, p. 539; (g) P. Magnus, T. K. Sarkar and S. Djuric, in 'Comprehensive Organometallic Chemistry', ed. G. W. Wilkinson, F. G. A. Stone and E. W. Abel, Pergamon Press, Oxford, 1982, vol. 7, p. 515; (h) L. A. Paquette, *Science (Washington, D.C.)*, 1982, **217**, 793; (i) I. Fleming, *Chem. Soc. Rev.*, 1981, **10**, 83; (j) P. Magnus, *Aldrichimica Acta*, 1980, **13**, 43; (k) R. Calas, *J. Organomet. Chem.*, 1980, **200**, 11; (l) T. H. Chan and I. Fleming, *Synthesis*, 1979, 761; (m) E. W. Colvin, *Chem. Soc. Rev.*, 1978, **7**, 15; (n) P. F. Hudrlik, *J. Organomet. Chem. Libr.*, 1976, **1**, 127.
2. (a) For a review on the activating and directing effects of silicon, see A. R. Bassindale and P. G. Taylor, in 'The Chemistry of Organic Silicon Compounds', ed. S. Patai and Z. Rappoport, Wiley, New York, 1989, part 2, p. 893; (b) for a recent theoretical study of the β-effect and leading references, see S. G. Wierschke, J. Chandrasekhar and W. L. Jorgensen, *J. Am. Chem. Soc.*, 1985, **107**, 1496.

3. R. K. Topsom, *Prog. Phys. Org. Chem.*, 1976, **12**, 1.
4. (a) M. J. S. Dewar and P. J. Grisdale, *J. Am. Chem. Soc.*, 1962, **84**, 3539; (b) M. Charton, in 'Correlation Analysis in Chemistry', ed. N. B. Chapman and J. Shorter, Plenum Press, New York, 1978, chap. 5.
5. C. G. Pitt, *J. Organomet. Chem.*, 1973, **61**, 49.
6. L. Pauling, 'The Nature of the Chemical Bond', 3rd edn., Cornell University Press, Ithaca, NY, 1960.
7. For a review on the use of the trimethylsilyl group as a protecting group in organic chemistry, see N. H. Andersen, D. A. McCrae, D. B. Grotjahn, S. Y. Gabhe, L. J. Theodore, R. M. Ippolito and T. K. Sarkar, *Tetrahedron*, 1981, **37**, 4069.
8. (a) L. E. Overman and T. A. Blumenkopf, *Chem. Rev.*, 1986, **86**, 857; (b) L. Birkofer and O. Stuhl, in 'The Chemistry of Organic Silicon Compounds', ed. S. Patai and Z. Rappoport, Wiley, New York, 1989, chap. 10.
9. S. D. Burke, C. W. Murtiashaw, M. S. Dike, S. M. Smith-Strickland and J. O. Saunders, *J. Org. Chem.*, 1981, **46**, 2400.
10. E. Nakamura, K. Fuzuzaki and I. Kuwajima, *J. Chem. Soc., Chem. Commun.*, 1975, 633.
11. K. Mikami, N. Kishi and T. Nakai, *Tetrahedron Lett.*, 1983, **24**, 795.
12. W. F. Fristad, D. S. Dime, T. R. Bailey and L. A. Paquette, *Tetrahedron Lett.*, 1979, 1999.
13. (a) S. E. Denmark and T. K. Jones, *J. Am. Chem. Soc.*, 1982, **104**, 2642; (b) T. K. Jones and S. E. Denmark, *Helv. Chim. Acta*, 1983, **66**, 2377; (c) T. K. Jones and S. E. Denmark, *Helv. Chim. Acta*, 1983, **66**, 2397; (d) S. E. Denmark, K. Habermas, G. A. Hite and T. K. Jones, *Tetrahedron*, 1986, **42**, 2821.
14. M. A. Tius and S. Ali, *J. Org. Chem.*, 1982, **47**, 3163.
15. M. A. Tius, *Tetrahedron Lett.*, 1981, **22**, 3335.
16. H.-F. Chow and I. Fleming, *J. Chem. Soc., Perkin Trans. 1*, 1984, 1815.
17. B. M. Trost and E. Murayama, *J. Am. Chem. Soc.*, 1981, **103**, 6529.
18. J. K. Kim, M. L. Kline and M. C. Caserio, *J. Am. Chem. Soc.*, 1978, **100**, 6243.
19. L. E. Overman, A. Castañeda and T. A. Blumenkopf, *J. Am. Chem. Soc.*, 1986, **108**, 1303.
20. (a) T. A. Blumenkopf, M. Bratz, A. Castañeda, G. C. Look, L. E. Overman, D. Rodriguez and A. S. Thompson, *J. Am. Chem. Soc.*, 1990, **112**, 4386; (b) T. A. Blumenkopf, G. C. Look and L. E. Overman, *J. Am. Chem. Soc.*, 1990, **112**, 4399; (c) A. Castañeda, D. J. Kucera and L. E. Overman, *J. Org. Chem.*, 1989, **54**, 5695.
21. L. E. Overman and A. S. Thompson, *J. Am. Chem. Soc.*, 1988, **110**, 2248.
22. L. E. Overman, T. C. Malone and G. P. Meier, *J. Am. Chem. Soc.*, 1983, **105**, 6993.
23. L. E. Overman and R. M. Burk, *Tetrahedron Lett.*, 1984, **25**, 5739.
24. L. E. Overman and T. C. Malone, *J. Org. Chem.*, 1982, **47**, 5297.
25. (a) L. E. Overman and K. L. Bell, *J. Am. Chem. Soc.*, 1981, **103**, 1851; (b) L. E. Overman, K. L. Bell and F. Ito, *J. Am. Chem. Soc.*, 1984, **106**, 4192.
26. L. E. Overman and A. J. Robichaud, *J. Am. Chem. Soc.*, 1989, **111**, 300.
27. M.-P. Heitz and L. E. Overman, *J. Org. Chem.*, 1989, **54**, 2591.
28. For a review concerning the chemistry of propargylic anion equivalents, see R. Epsztein, in 'Comprehensive Carbanion Chemistry', ed. E. Buncel and T. Durst, Elsevier, Amsterdam, 1984, part B, p. 107.
29. (a) R. L. Danheiser, D. J. Carini and C. A. Kwasigroch, *J. Org. Chem.*, 1986, **51**, 3870; (b) R. L. Danheiser and D. J. Carini, *J. Org. Chem.*, 1980, **45**, 3925.
30. (a) R. L. Danheiser, D. J. Carini and A. Basak, *J. Am. Chem. Soc.*, 1981, **103**, 1604; (b) R. L. Danheiser, D. J. Carini, D. M. Fink and A. Basak, *Tetrahedron*, 1983, **39**, 935.
31. R. L. Danheiser and D. M. Fink, *Tetrahedron Lett.*, 1985, **26**, 2513.
32. R. L. Danheiser, C. A. Kwasigroch and Y.-M. Tasi, *J. Am. Chem. Soc.*, 1985, **107**, 7233.
33. R. L. Danheiser and D. A. Becker, *Heterocycles*, 1987, **25**, 277.
34. D. A. Becker and R. L. Danheiser, *J. Am. Chem. Soc.*, 1988, **111**, 389.
35. L. D. Lozar, R. D. Clark and C. H. Heathcock, *J. Org. Chem.*, 1977, **42**, 1386.
36. W. S. Johnson, T. M. Yarnell, R. F. Myers and D. R. Morton, Jr., *Tetrahedron Lett.*, 1978, **26**, 3201.
37. W. S. Johnson, T. M. Yarnell, R. F. Myers, D. R. Morton, Jr. and D. R. Boots, *J. Org. Chem.*, 1980, **45**, 1254.
38. For a detailed discussion of the stereochemical course of biomimetic polyene cyclizations, see P. A. Bartlett, in 'Asymmetric Synthesis', ed. D. J. Morrison, Academic; New York, 1984, vol 3, chaps. 5 and 6.
39. Y. L. Baukov and L. F. Lutsenko *Organomet. Chem. Rev., Sect. A*, 1970, **6**, 355; A. G. Brook, *J. Am. Chem. Soc.*, 1957, 4373; A. G. Brook and N. V. Schwartz, *J. Org. Chem.*, 1962, 2311.
40. W. S. Johnson, M. B. Gravestock, R. J. Parry and A. Okorie, *J. Am. Chem. Soc.*, 1972, **94**, 8604.
41. R. S. Brinkmeyer, *Tetrahedron Lett.*, 1979, 207.
42. (a) G. Wickham and W. Kitching, *Organometallics*, 1983, **2**, 541; (b) G. Wickham and W. Kitching, *J. Org. Chem.*, 1983, **48**, 612.
43. H. Wetter and P. Scherer, *Helv. Chim. Acta*, 1983, **66**, 118.
44. I. Fleming and N. K. Terrett, *J. Organomet. Chem.*, 1984, **264**, 99.
45. (a) M. N. Paddon-Row, N. G. Rondan and K. N. Houk, *J. Am. Chem. Soc.*, 1982, **104**, 7162; (b) S. D. Khan, C. F. Pau, A. R. Chamberlin and W. J. Hehre, *J. Am. Chem. Soc.*, 1987, **109**, 650.
46. (a) T. Hayashi, M. Konishi and M. Kumada, *J. Am. Chem. Soc.*, 1982, **104**, 4963; (b) T. Hayashi, Y. Okamoto and M. Kumada, *Tetrahedron Lett.*, 1983, **24**, 807.
47. M. T. Reetz and K. Kesseler, *J. Org. Chem.*, 1985, **50**, 5434.
48. L. Coppi, A. Mordini and M. Taddei, *Tetrahedron Lett.*, 1987, **28**, 969; for the preparation of carbon-centered optically active allylsilanes, see L. Coppi, A. Ricci and M. Taddei, *Tetrahedron Lett.*, 1987, **28**, 965.
49. For a compilation of these internal addition reactions, see G. A. Molander and S. W. Andrews, *Tetrahedron*, 1988, **44**, 3869.
50. For stereoselective intramolecular allylsilane additions, see M. T. Reetz, A. Jung and C. Bolm, *Tetrahedron*, 1988, **44**, 3889.
51. S. E. Denmark and E. Weber, *Helv. Chim. Acta*, 1983, **66**, 1655.
52. S. E. Denmark and T. M. Willson, *J. Am. Chem. Soc.*, 1989, **111**, 3475.
53. T. Hayashi, Y. Okamoto and M. Kumada, *Tetrahedron Lett.*, 1983, **24**, 807.

54. For a recent review, see (a) W. N. Speckamp and H. Hiemstra, *Tetrahedron*, 1985, **41**, 4367; (b) W. J. Klaver, M. J. Moolenaar, H. Hiemstra and W. N. Speckamp, *Tetrahedron*, 1988, **44**, 3805.
55. D. J. Peterson, *J. Org. Chem.*, 1968, **33**, 780.
56. A. G. Brook, J. M. Duff and D. G. Anderson, *Can. J. Chem.*, 1970, **48**, 561.
57. W. Dumont and A. Krief, *Angew. Chem., Int. Ed. Engl.*, 1976, **15**, 161.
58. A. R. Bassindale, R. J. Ellis and P. G. Taylor, *Tetrahedron Lett.*, 1984, **25**, 2705.
59. G. R. Buell, R. J. P. Corriu, C. Guerin and L. Spialter, *J. Am. Chem. Soc.*, 1970, **92**, 7424.
60. (a) D. Seyferth, T. Wada and G. Raab, *Tetrahedron Lett.*, 1960, 20; (b) M. R. Stober, K. W. Michael and J. L. Speier, *J. Org. Chem.*, 1967, **32**, 2740; (c) D. Seyferth and T. Wada, *Inorg. Chem.*, 1962, **1**, 78; (d) L. F. Cason and H. G. Brooks, *J. Am. Chem. Soc.*, 1952, **74**, 4582; (e) L. F. Cason and H. G. Brooks, *J. Org. Chem.*, 1954, **19**, 1278.
61. K. Tamao, R. Kanatani and M. Kumada, *Tetrahedron Lett.*, 1984, **25**, 1905.
62. T. Hayama, S. Tomoda, Y. Takeuchi and Y. Nomura, *Tetrahedron Lett.*, 1983, **24**, 2795.
63. (a) G. Stork and B. Ganem, *J. Am. Chem. Soc.*, 1973, **95**, 6152; G. Stork and J. Singh, *J. Am. Chem. Soc.*, 1974, **96**, 6181; (b) R. K. Boeckman, Jr., *J. Am. Chem. Soc.*, 1974, **96**, 6179.
64. R. K. Boeckman, Jr., *Tetrahedron*, 1983, **39**, 925.
65. D. J. Ager, *Synthesis*, 1984, 384.
66. P. F. Hudrlik, D. J. Peterson and R. J. Rona, *J. Org. Chem.*, 1975, **40**, 2263.
67. A. R. Bassindale, R. J. Ellis, J. C. Y. Lau and P. G. Taylor, *J. Chem. Soc., Chem. Commun.*, 1986, 98.
68. H. B. Burgi and J. D. Dunitz, *Acc. Chem. Res.*, 1983, **16**, 153.
69. J. E. Baldwin, *J. Chem. Soc., Chem. Commun.*, 1976, 734.
70. C. R. Johnson and B. D. Tait, *J. Org. Chem.*, 1987, **52**, 281.
71. F. A. Carey and A. S. Court, *J. Org. Chem.*, 1972, **37**, 939.
72. H. Gilman and R. A. Tomasi, *J. Org. Chem.*, 1962, **27**, 3647.
73. C. Burford, F. Cooke, E. Ehlinger and P. Magnus, *J. Am. Chem. Soc.*, 1977, **99**, 4536.
74. F. Cooke and P. Magnus, *J. Chem. Soc., Chem. Commun.*, 1977, 513.
75. S. Hackett and T. Livinghouse, *J. Org. Chem.*, 1986, **51**, 879.
76. J. Binder and E. Zbiral, *Tetrahedron Lett.*, 1986, **27**, 5829.
77. E. Ehlinger and P. Magnus, *J. Chem. Soc., Chem. Commun.*, 1979, 421.
78. M. J. Carter and I. Fleming, *J. Chem. Soc., Chem. Commun.*, 1976, 679.
79. E. Ehlinger and P. Magnus, *J. Am. Chem. Soc.*, 1980, **102**, 5004.
80. T. H. Chan and K. Koumaglo, *J. Organomet. Chem.*, 1985, **285**, 109.
81. R. J. P. Corriu, C. Guerin and J. M'Boula, *Tetrahedron Lett.*, 1981, **22**, 2985.
82. P. W. K. Lau and T. H. Chan, *Tetrahedron Lett.*, 1978, 2383.
83. D. J. S. Tsai and D. S. Matteson, *Tetrahedron Lett.*, 1981, **22**, 2751.
84. T. H. Chan and J.-S. Li, *J. Chem. Soc., Chem. Commun.*, 1982, 969.
85. K. Tamao, E. Nakajo and Y. Ito, *J. Org. Chem.*, 1987, **52**, 957.
86. (a) T. Konakahara and Y. Takagi, *Synthesis*, 1979, 192; (b) T. Konakahara and Y. Takagi, *Tetrahedron Lett.*, 1980, **21**, 2073.

2.6
Selenium Stabilization

ALAIN KRIEF
Facultés Universitaire Notre Dame de la Paix, Namur, Belgium

2.6.1 INTRODUCTION

Over the past 15 years, organoselenium reagents have been increasingly used in organic synthesis, especially for the construction of complex molecules.[1] This is due to the ease with which selenium can be introduced into organic molecules,[2–7] and the large variety of selective reactions which can be performed on organoselenium compounds. These allow the synthesis of different selenium-free functional groups or molecules.[1–15] Organometallics bearing a selenium-stabilized carbanion have played an important role in such developments.[7,9,12,16,17] They are usually readily available and can be particularly good nucleophiles, especially towards compounds bearing an electrophilic carbon atom. In this respect they have been widely used, since they allow the synthesis of more complex organoselenium compounds with the concomitant formation of a new carbon–carbon bond attached at one end to the selenium-containing moiety. This review article deals with the synthesis of selenium-stabilized carbanions and only discloses their reactivity towards carbonyl compounds, especially aldehydes and ketones. Some further transformations of functionalized selenides to selenium-free molecules will also be discussed. We have in most cases chosen examples involving the trapping of organometallics, with aldehydes and ketones, as a definite proof of their existence. In some cases, however, such information was not available and we have therefore described trapping experiments with alkylating agents instead.

2.6.2 SYNTHESIS OF ORGANOMETALLICS BEARING A SELENIUM-STABILIZED CARBANION

2.6.2.1 Generalities on Selenium-stabilized Carbanions

2.6.2.1.1 Reaction of alkyl metals with selenides and functionalized selenides

Organometallics bearing a selenium-stabilized carbanion[7,9,12,16,17] belong to the well-known family of α-heterosubstituted organometallics which have proved particularly useful in organic synthesis over the past 30 years.[7,18–33] Although they share similar features with other members of this family, they possess exceptional properties due to the special behavior of the selenium atom.[1–4,6–9,12,14–16,34]

Several organometallics bearing a seleno moiety and belonging to the aryl- (especially phenyl-) and methyl-seleno series, having their selenium atom in different oxidation states (Scheme 1), such as α-metalloalkyl selenides, α-metalloalkyl selenoxides, α-metalloalkyl selenones and α-metalloalkyl selenonium salts, as well as α-metallovinyl selenides and α-metallovinyl selenoxides, have been prepared, usually by metallation of the corresponding carbon acid with lithium or potassium amide or with potassium *t*-butoxide.

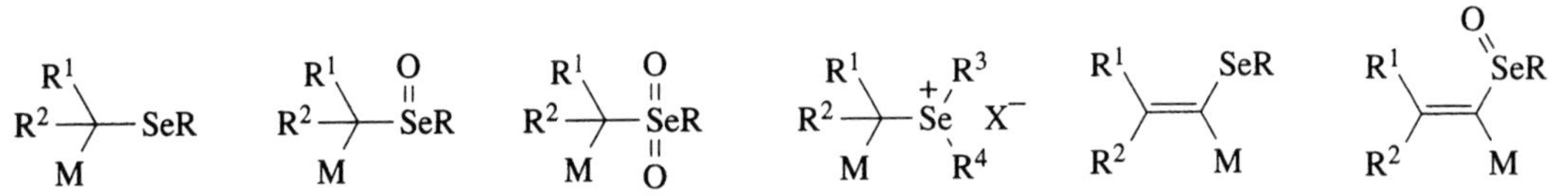

Scheme 1

These bases, however, are not strong enough to metallate[35,36] dialkyl and alkyl aryl selenides. Amylsodium[37,38] and alkyllithiums,[7,37–39] which have proved particularly useful for the metallation of the corresponding sulfides and phosphines, are not useful for the metallation of dialkyl or alkyl aryl selenides[7,17] owing to their tendency to react on the selenium atom and to produce a novel selenide and a novel organometallic (Scheme 2).[7,37–45] This latter reaction is easier when carried out in THF than in ether (Scheme

2, a),[16,17,43] is faster with *s*- or *t*-butyllithium[16,17,43,44] than with *n*-butyllithium[7,16,35,36,38–40,43–50] or methyllithium (Scheme 2, a)[45] and delivers the organometallic with the more stable carbanionic center of the three possible carbanions (Scheme 2, compare a with b). Thus butyllithiums react with methyl phenyl selenide and produce butyl methyl selenides and phenyllithium.[16,38–40,44] Substitution of the methyl group of methyl phenyl selenide by a group able to stabilize a carbanionic center, such as a vinyl,[45] aryl,[43,44,46] thio[47–49] or seleno[7,16,35,36] moiety, not only increases the reactivity of the selenide (the cleavage reaction takes place faster and at lower temperature: Scheme 2; compare a with b) but also dramatically changes the regiochemistry of the reaction. Phenyllithium is no longer produced but instead allyl-,[45] benzyl-,[43,44,46] thiomethyl-[47–49] and selenomethyl-lithiums[7,16,35,36] are obtained, respectively, in almost quantitative yield (Scheme 2, b).

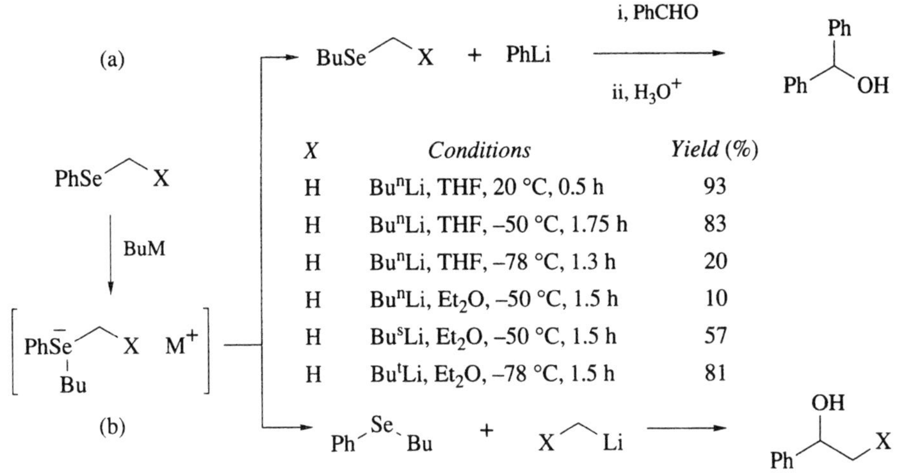

(a) $PhSeCH_2X$ —BuM→ [$PhSe^{-}(Bu)CH_2X$ M^+] → $BuSeCH_2X$ + PhLi —(i, PhCHO; ii, H_3O^+)→ Ph_2CHOH

X	*Conditions*	*Yield (%)*	*Ref.*
H	Bu^nLi, THF, 20 °C, 0.5 h	93	40
H	Bu^nLi, THF, –50 °C, 1.75 h	83	40
H	Bu^nLi, THF, –78 °C, 1.3 h	20	16, 44
H	Bu^nLi, Et_2O, –50 °C, 1.5 h	10	16, 44
H	Bu^sLi, Et_2O, –50 °C, 1.5 h	57	16, 44
H	Bu^tLi, Et_2O, –78 °C, 1.5 h	81	16, 44

(b) → PhSeBu + XCH_2Li → $PhCH(OH)CH_2X$

X	*Conditions*	*Yield (%)*	*Ref.*
$CH{=}CH_2$	Bu^nLi, THF, –78 °C, 0.1 h	86	45
Ph	Bu^nLi, THF, –78 °C, 0.3 h	86	46
SPh	Bu^nLi, THF, –78 °C, 0.2 h	94	47, 48, 49
SePh	Bu^nLi, THF, –78 °C, 0.1 h	82	35, 36

Scheme 2

This reaction has proven to be of wide applicability: the selenium–metal exchange is a valuable alternative to the hydrogen–metal, halogen–metal or tin–metal exchange. Although less general than the two former methods, it takes advantage of the easy synthesis of selenides and functionalized selenides and of their thermal stability, even in the case of allyl or benzyl derivatives. Moreover, the butyl selenides produced concomitantly with the organometallic and resulting from the C—Se bond cleavage are generally inert in the reaction medium.[7,12,16] This differs markedly from the halogen–metal exchange by butyllithiums, which instead leads to butyl halides that react further with the organometallics present in the reaction medium.[51,52]

For the synthesis of α-selenoalkyllithiums, the selenium–lithium exchange reaction is a good alternative to the almost impossible metallation of unactivated selenides.[7,16,17] Thus it has been found that a large variety of selenoacetals, often readily available from carbonyl compounds and selenols,[35,53] react with butyllithiums to provide α-selenoalkyllithiums[4,7,9,12,16,17,35,36,40,48,49,54–56] in very high yields (Scheme 2; see also Section 2.6.2.3).

2.6.2.1.2 Stabilization of carbanionic centers by selenium-containing moieties

The seleno moiety, when linked to a carbanionic center, provides[17,56–59] an extra stabilization of about 10 pK_a unit. This is of the same order of magnitude as that provided by the corresponding thio moiety,[17] which stabilizes an sp^3 carbanion slightly better than does a seleno moiety.[7,17,57,60] These results differ greatly from those obtained by calculation,[59] which suggest a much higher difference (4 kcal; 1 cal = 4.18 J) and in the reverse order. Recent results from our laboratory[7,16,60] seem to show that heteroatom substituents affect the stabilization of an sp^3-hybridized carbanionic center much more than the heteroatom itself (S, Se). Thus a phenylseleno group stabilizes a carbanionic center to a higher extent than a methylseleno or (probably) a methylthio moiety.[60] Results which support the above mentioned tendencies are collected in Schemes 3, 4, 5 and 6.[60]

Compound	*Conditions*	*Measured*	*Sulfur/selenium*	*Ref.*
$PhXCD_3$	KNH_2, NH_3, –33 °C	k_{isotop}	10	64
m-$CF_3C_6H_4XMe$	LiTMP, THF, –78 °C	k_{dep}	3.8	56
$PhXCH_2CH{=}CH_2$	LDA, THF, –78 °C	k_{dep}	7.5	65
$PhXCH_2C{\equiv}CH$	NaOEt, EtOH, 25 °C	$k_{isom}{}^{a}$	12.6	62, 63
$PhXCH{=}C{=}CH_2$	NaOEt, EtOH, 25 °C	$k_{isom}{}^{b}$	7	62, 63
$PhCOCH_2XPh$	$MeSOCH_2Na$, DMSO	K_{eq}	32	66
Me_3XI	NaOD, D_2O, 62 °C	K_{isotop}	45	67
$(PhX)_2CH_2$	0 °C	$pK_S/pK_{Se}{}^{c}$	32/35	36

[a] Isomerization of propargyl to allenyl selenide. [b] Isomerization of allenyl to 1-propynyl selenide.
[c] $pK_{aS} = 17.1$, $pK_{bSe} = 18.6$

Scheme 3 Relative kinetic and thermodynamic acidity of sulfides and selenides

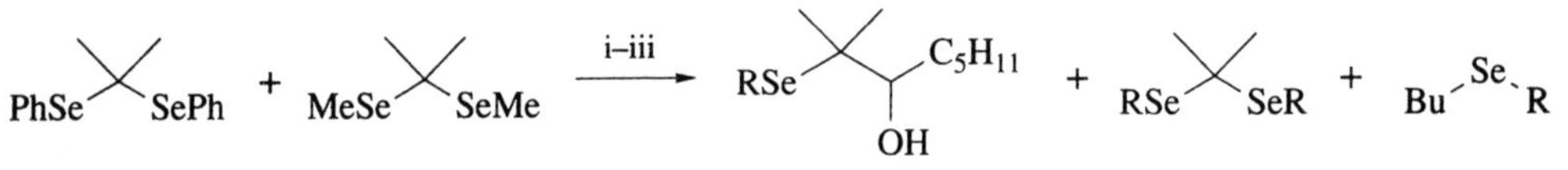

$Me_2C(SePh)_2$ + $Me_2C(SeMe)_2$ —(i–iii)→ $RSeC(Me)_2CH(OH)C_5H_{11}$ + $Me_2C(SeR)_2$ + BuSeR

R = Ph	62%	65%	77%
R = Me	10%	10%	23%

i, Bu^nLi, THF–hexane, –78 °C, 0.1 h; ii, $C_5H_{11}CHO$; iii, H_3O^+

Scheme 4

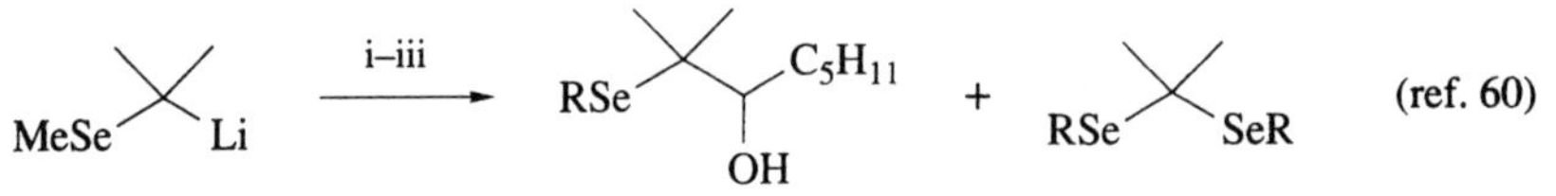

$Me_2C(SeMe)Li$ —(i–iii)→ $RSeC(Me)_2CH(OH)C_5H_{11}$ + $Me_2C(SeR)_2$ (ref. 60)

Solvent	*Temperature* (°C)	*Time* (h)	*R*	*Yield* (%)	
THF–hexane	–78	0.1	Me	44	12
			Ph	30	50
THF–hexane	–78	0.3	Me	30	20
			Ph	40	40

i, $Me_2C(SePh)_2$; ii, $C_5H_{11}CHO$; iii, H_3O^+

Scheme 5

The results in Schemes 3 to 6 are derived[17,36,56,61–67] from kinetic or thermodynamic acidity measurements on functionalized selenides and on the corresponding sulfides (Scheme 3), as well as from equilibration reactions[7,60] involving 2-phenylseleno-2-propyllithium and 2-phenylseleno-2-phenylthiopropane, and those involving 2-phenylthio-2-propyllithium and 2,2-bis(phenylseleno)propane (Scheme 6).

Surprisingly, for sp^2-hybridized carbanionic centers the reverse order of stabilization is found.[17,65,68] The seleno moiety is now more stabilizing than a thio moiety (Scheme 7, *cf.* Scheme 3).[17,69]

Several factors[59,66,69] might govern the relative stabilization of a negative charge by sulfur and selenium. The selenium atom is larger and more polarizable than sulfur, and is therefore expected to disperse the charge more efficiently. Sulfur, however, is slightly more electronegative and better stabilizes the charge by an inductive effect.[69] Conjugative interactions *via d*-orbitals should be more favorable for sulfur; however, conjugation is less favorable for sp^2-hybridized carbanions for which the greater polarizability of selenium dominates.

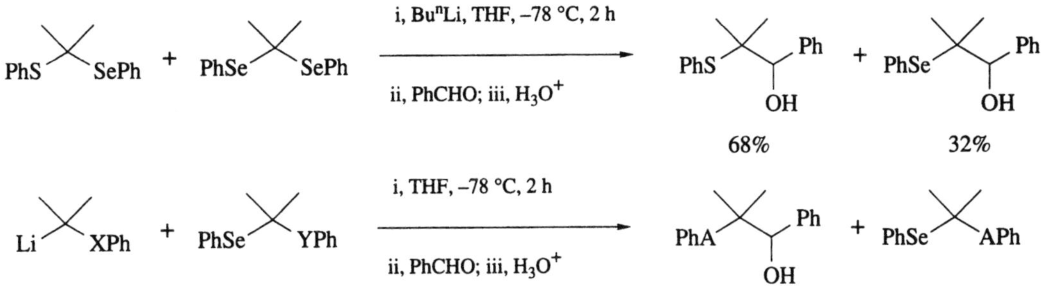

PhS–C(Me₂)–SePh + PhSe–C(Me₂)–SePh —(i, Bu^nLi, THF, –78 °C, 2 h; ii, PhCHO; iii, H_3O^+)→ PhS–C(Me₂)–CH(OH)Ph + PhSe–C(Me₂)–CH(OH)Ph

68% 32%

Li–C(Me₂)–XPh + PhSe–C(Me₂)–YPh —(i, THF, –78 °C, 2 h; ii, PhCHO; iii, H_3O^+)→ PhA–C(Me₂)–CH(OH)Ph + PhSe–C(Me₂)–APh

(ref. 60)

	A	*Yield* (%)	
X = S, Y = Se	S	66	19
	Se	33	66
X = Se, Y = S	S	66	25
	Se	33	63

Scheme 6

Compound	*Conditions*	*Measured*	*Sulfur/selenium*
 5-membered ring (X)	$KOBu^t$, DMSO, 25 °C	K_{isotop}	0.67
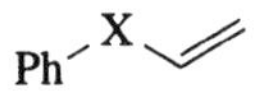 PhXCH=CH₂	LDA, THF, –78 °C	K_{eq}	0.21
	LiTMP, THF, –78 °C	k_{dep}	0.37
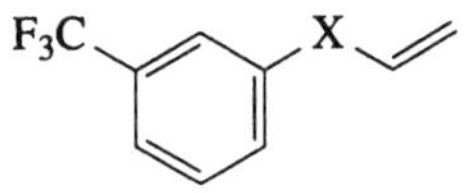 m-F_3C-C₆H₄XCH=CH₂	LDA, THF, –78 °C	K_{eq}	0.37
	LiTMP, THF, –78 °C	k_{dep}	0.42

Scheme 7 Relative kinetic and thermodynamic acidity of vinyl sulfides and selenides (refs. 17 and 69)

The group attached to the seleno moiety modulates its ability to stabilize a carbanionic center (this effect is often higher than the one produced by the heteroatom itself). Thus a m-CF_3PhSe group enhances[56,69,70] the kinetic acidity of a methylene group by approximately a factor of 20 over a PhSe group (Scheme 8),[7,71,72] which is in turn a much better stabilizing group than MeSe.

m-R-C₆H₄SeCH=CH₂ → m-R-C₆H₄SeC(Li)=CH₂

LiTMP $k(CF_3/H) = 14.9^a$

LDA $k(CF_3/H) = 64^b$

m-R-C₆H₄SeCH₂CH=CH₂ → m-R-C₆H₄SeCH(Li)CH=CH₂

LDA $k(CF_3/H) = 11.7^a$

[a] $k(CF_3/H)$ is the relative rate of deprotonation, $k(CF_3/k_H) = \log M(CF_3)/\log M(H)$, where $M(CF_3)$ and M(H) are the mole fractions of components m-CF_3RH and RH remaining when the reaction is quenched.

[b] Equilibrium constant $K(CF_3/H)$ where $K = [CF_3RH][CF_3RLi]/RH/RLi$ (ref. 69)

Scheme 8

Thus, whereas 1,1-bis[(m-trifluoromethyl)phenylseleno]alkanes[65] and 1,1-bis(phenylseleno)-alkanes[35,36,71,73] are successfully metallated under various conditions (see below), their methylseleno analogs are not.[71] 1-Phenylseleno- and 1-methylseleno-1-arylethanes exhibit[72] a similar tendency, since the former is efficiently metallated with KDA in THF but not the latter (Scheme 9, compare c to f).

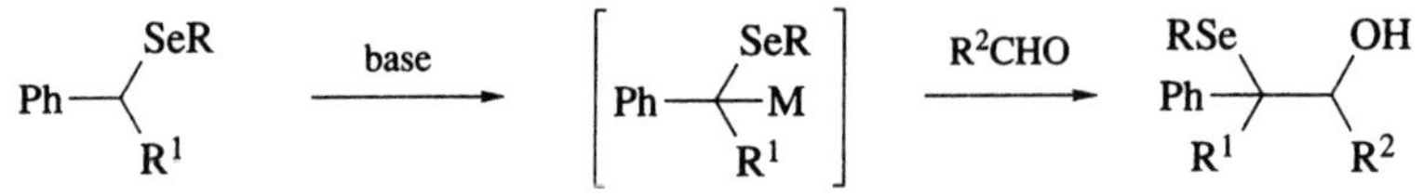

Ph–CH(SeR)(R¹) —base→ [Ph–C(SeR)(M)(R¹)] —R^2CHO→ Ph–C(RSe)(R¹)–CH(OH)(R²)

Entry	*R*	*R¹*	*Conditions*	*R²*	*Yield (%)*	*Ref.*
a	Ph	H	LDA, THF–hexane, –78 °C, 0.1 h	Et	70	56
b	Ph	Me	LDA, THF–hexane, –78 °C, 0.1 h	Ph	0	72
c	Ph	Me	KDA, THF–hexane, –50 °C, 0.3 h	Ph	87	72
d	Ph	Bu	KDA, THF–hexane, –50 °C, 0.3 h	Ph	70	72
e	Me	H	KDA, THF–hexane, –50 °C, 0.3 h	Ph	79	72
f	Me	Me	KDA, THF–hexane, –50 °C, 0.3 h	Ph	0	72

Scheme 9

It also has been found that butyllithiums react faster with 2,2-bis(phenylseleno)propane than with its methylseleno analog.[60] Thus, addition of *n*-butyllithium to a 1:1 mixture of 2,2-bis(phenylseleno)propane and 2,2-bis(methylseleno)propane leads selectively to the formation of 2-lithio-2-phenylselenopropane under kinetically controlled conditions, providing the α-selenoalkyllithium with the more stable carbanionic center (Scheme 4). Furthermore, 2-methylselenopropyllithium reacts[60] with 2,2-bis(phenylseleno)propane and leads, under thermodynamically controlled conditions, to 2-phenylseleno-2-propyllithium, thus demonstrating the better propensity of a phenylseleno moiety over a methylseleno moiety to stabilize a carbanionic center (Scheme 5).

The presence of a group able to delocalize the charge from the carbanionic center favors the synthesis of the carbanion whatever the method used, whereas substitution of the carbanionic center by alkyl groups disfavors it. Thus, whereas methyl phenyl and dimethyl selenide cannot be metallated with LDA at –78 °C or 0 °C, phenylseleno and methylseleno acetates and propionates lead, on reaction with the same base at –78 °C, to the corresponding organometallics (Scheme 10).[74,75]

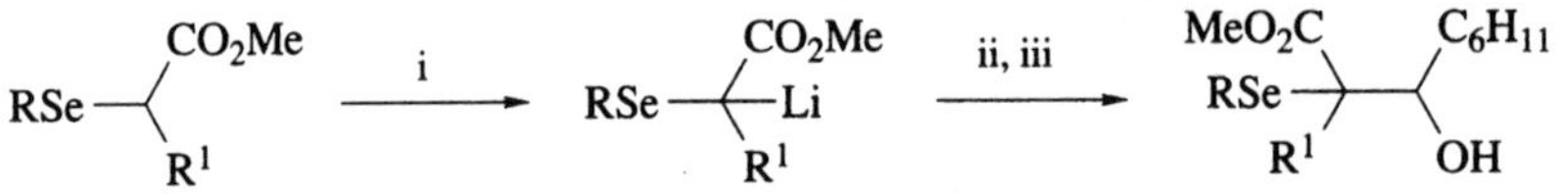

RSe–CH(CO_2Me)(R¹) —i→ RSe–C(CO_2Me)(Li)(R¹) —ii, iii→ MeO_2C(RSe)(R¹)C–CH(OH)(C_6H_{11}) (ref. 74)

R	*R¹*	*Yield (%)*
Ph	H	93
Ph	Me	60
Me	H	80
p-ClC_6H_4	H	96

i, LDA, THF–hexane. –78 °C, 0.5 h; ii, $C_6H_{13}CHO$; iii, H_3O^+

Scheme 10

Yet another important difference of reactivity towards *n*-butyllithium has been found between phenyl selenoacetals derived from aldehydes and ketones.[60] This distinction allowed,[60] when the reaction was performed in ether–hexane, the selective synthesis of an α-selenoalkyllithium derived from the phenyl selenoacetal of an aldehyde in the presence of the phenyl selenoacetal of a methyl ketone, which remained untouched (Scheme 11).[60]

α-Selenoalkyl metals have been prepared[7,9,12,16,17] by a large array of methods such as metallation of selenides, selenoxides, selenones and selenonium salts; selenium–metal exchange between selenoacetals and alkyllithiums or metals; halogen–lithium exchange between α-haloalkyl selenides and alkyllithiums; addition of organolithiums to vinyl selenides; and selenophilic addition of organometallics to selones (Scheme 12).

Among these methods, the metallation reaction is the method of choice for selenides bearing electron-withdrawing groups able to stabilize the carbanionic center. This also applies to selenoxides, selenones and selenonium salts.

The reaction of selenoacetals with butyllithiums allows the synthesis of a large variety of α-selenoalkyllithiums whose carbanionic centers are unsubstituted, monoalkyl substituted or dialkyl substituted.

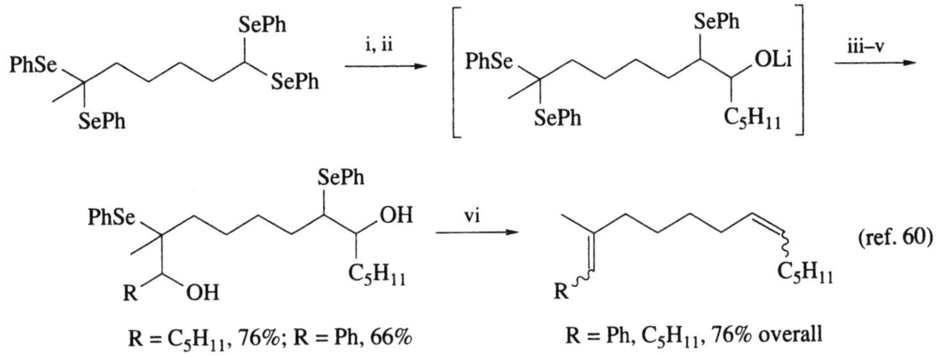

(ref. 60)

R = C_5H_{11}, 76%; R = Ph, 66%

R = Ph, C_5H_{11}, 76% overall

i, Bu^nLi (1 equiv.), ether–hexane, 45 °C, 0.5 h; ii, $C_5H_{11}CHO$; iii, Bu^sLi (1 equiv.), ether–hexane, –78 °C, 0.5 h; iv, RCHO; v, H_3O^+; vi, P_2I_4

Scheme 11

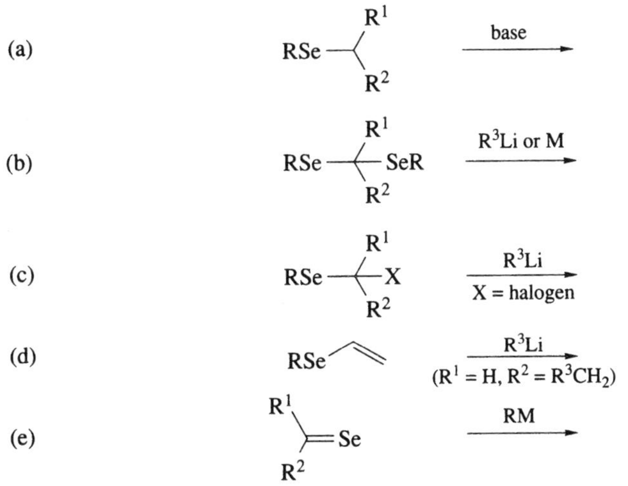

(a) base

(b) R^3Li or M

(c) R^3Li, X = halogen

(d) R^3Li (R^1 = H, R^2 = R^3CH_2)

(e) RM

Scheme 12

The other methods cited above are synthetically less useful, especially the selenophilic addition to selones.[76]

All the different types of selenium-substituted organometallics with almost all the possible substitution patterns around the carbanionic center are now available, and therefore the intermediates are among the few α-heterosubstituted organometallics to be so widely available.[7]

2.6.2.2 Synthesis by Metallation of Organometallics Bearing a Selenium-stabilized Carbanion

2.6.2.2.1 Synthesis of α-selenoalkyl metals by metallation of selenides

Alkyl aryl and dialkyl selenides, aside from the parent compounds methyl phenyl selenide (Scheme 13, a)[36] and trifluoromethylphenyl methyl selenide (Scheme 13, b),[56] cannot be easily metallated. It was expected that the extra stabilization provided by the cyclopropyl group would be sufficient to permit the metallation of cyclopropyl selenides, but that proved not to be the case.[56,77]

Cyclopropyl phenyl selenides are inert towards lithium diisopropylamide (LDA) and lithium tetramethylpiperidide (LiTMP) (Scheme 14, a). Butyllithiums instead react on the selenium atom and produce[77] butyl cyclopropyl selenides and phenyllithium rather than cyclopropyllithiums and butyl phenyl selenides (Scheme 14, b).

Aryl methyl selenides with the methyl group substituted by an aryl (Scheme 9 and 15),[56,70,72,78,79] a vinyl (Schemes 16–18),[56,69,80–90] a 1-butadienyl (Scheme 17, d)[91] or an ethynyl (Schemes 19 and

ArSe–CH(X)(R) ⟶ ArSe–C(X)(R)–Li

	Ar	X	R	Conditions	Yield (%)	Ref.
(a)	Ph	H	H	Bu^nLi–TMEDA, THF, 0 °C	38	38
(b)	4-$CF_3C_6H_4$	H	H	LiTMP, THF–hexane, –55 °C, 2 h	100	56
(c)	4-$CF_3C_6H_4$	H	Me	LiTMP, THF–hexane, –55 °C, 2 h	0	56

Scheme 13

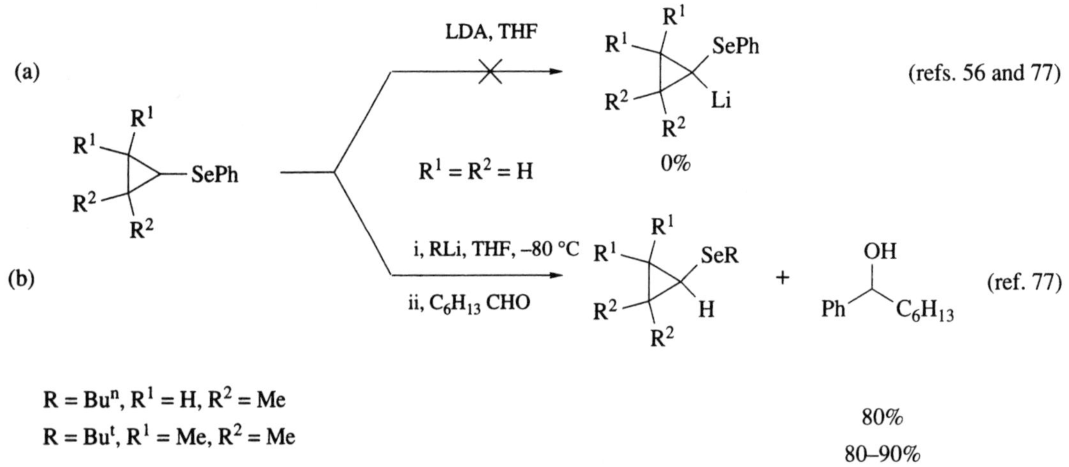

(a) LDA, THF (reaction does not occur); $R^1 = R^2 = H$; lithiated cyclopropyl phenyl selenide, 0% (refs. 56 and 77)

(b) i, RLi, THF, –80 °C; ii, C_6H_{13} CHO; cyclopropyl selenide (SeR, H) + PhCH(OH)C_6H_{13} (ref. 77)

$R = Bu^n, R^1 = H, R^2 = Me$ — 80%

$R = Bu^t, R^1 = Me, R^2 = Me$ — 80–90%

Scheme 14

20)[56,92,93] group, or a methoxy (Scheme 21, a),[56] a phenylthio (Scheme 22),[94] an arylseleno (Scheme 21, b; Scheme 23, a),[35,36,49,56,65,71,73,95–97] a silyl (Scheme 24, a–d)[56,70,98–100] or a telluryl (Scheme 25, a)[101] moiety, some of which provide a further stabilization of the carbanionic center, have been successfully metallated by metalloamides.

As a general trend in this series of compounds, *m*-trifluoromethylphenyl selenides are more efficiently metallated than phenyl selenides, methylseleno derivatives are scarcely reactive, and the presence of an alkyl substituent on the carbon to be metallated often inhibits the reaction (Scheme 9 and Schemes 13–25).

As far as the basic system is concerned, KDA in THF–hexane (a 1:1 mixture of lithium diisopropylamide (LDA) and potassium *t*-butoxide)[45,72,73] is by far more efficient than lithium tetramethylpiperidide (LiTMP) (in THF–HMPA[71] or THF–hexane[45,56,71,72]) which is in fact at least a power of 10 more reactive[17,56] than lithium amides (in HMPA[41] or THF–hexane[35,36,41,56,65,70,78–82,84–86,90,92,93,95,97,99,101–103]). Thus, although bis(phenylseleno)methane[35,36] and its bis(*m*-trifluoromethyl) analogs[65] are almost quantitatively metallated with lithium diisobutyramide or LDA in THF, respectively (Scheme 21, b and c;

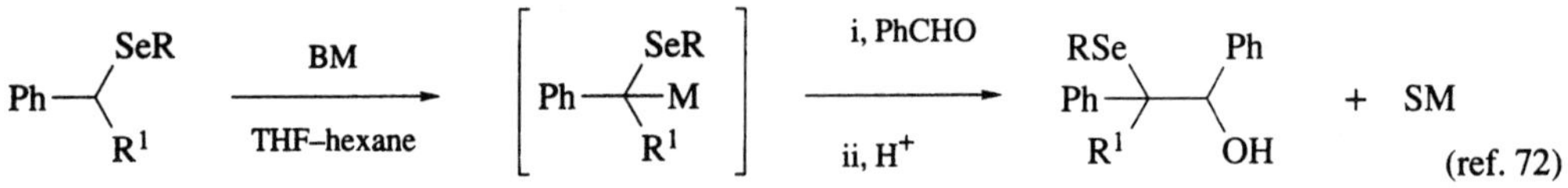

Ph–CH(SeR)(R¹) —BM, THF–hexane→ [Ph–C(SeR)(M)(R¹)] —i, PhCHO; ii, H^+→ RSe(Ph)(R¹)C–CH(Ph)OH + SM (ref. 72)

	R	R¹	Conditions	Yield (%)	
(a)	Ph	Me	KDA, –50 °C, 0.3 h	87	0
(b)	Ph	Me	LDA, 0 °C, 0.5 h	15	85
(c)	Ph	Me	LiTMP, 20 °C, 0.5 h	62	0
(d)	Me	H	KDA, –78 °C, 0.3 h	79	—
(e)	Me	H	LDA, –50 °C, 0.5 h	61	16
(f)	Me	H	LiTMP, –78 °C, 0.3 h	67	17

Scheme 15

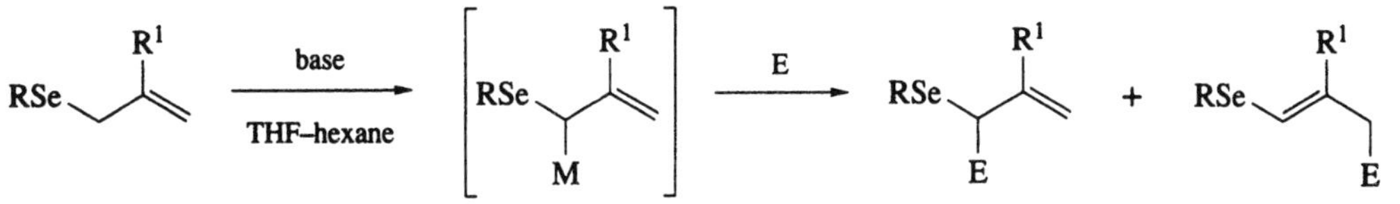

	R	*R¹*	*Conditions*	*E*	*Yield (%)*	*Ratio*	*Ref.*
(a)	Ph	H	LDA, –78 °C, 0.2 h	$Ph(CH_2)_3Br$	68	100:0	80
(b)	Ph	H	LDA, –78 °C, 0.2 h	PhCOMe	70–85	15:85	80
(c)	Ph	H	LDA, –78 °C, 0.2 h, then Et_3Al	PhCOMe	70–85	64:36 (64:36)[a]	81, 82
(d)	Ph	H	LDA, –78 °C, 0.2 h	EtCHO	70–85	12:88	81, 82
(e)	Ph	H	LDA, –78 °C, then Et_3Al	EtCHO	70–85	100:0 (81:19)[a]	81, 82
(f)	Ph	Me	LDA, –78 °C, 0.5 h	Bu_3SnCl	100	0:100	84
(g)	Me	H	LiTMP	PhCHO	76	18:82	45

[a] *Threo:erythro* ratio

Scheme 16

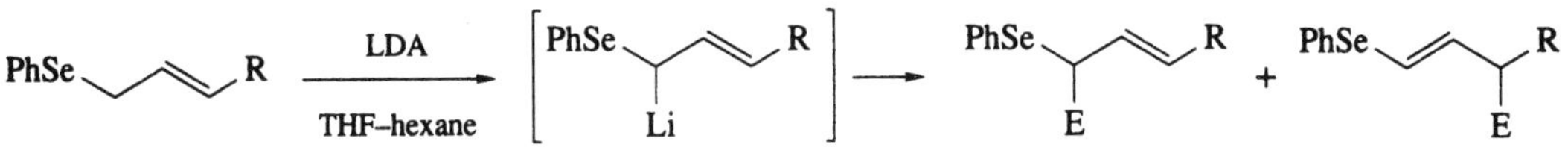

	R	*Conditions*	*E*	*Yield (%)*	*Ratio*	*Ref.*
(a)	Me	–78 °C, 0.6 h	$Ph(CH_2)_2Br$	80	100:0	80
(b)	Ph	–78 °C, 0.1 h	Me_2CO	55	100:0	80
(c)	SePh	–78 °C, 0.1 h	Pr^iCHO	88	—	85

(d) i, LDA, THF; ii, Me_2CO; iii, H_3O^+; 90 °C — 71% overall (ref. 91)

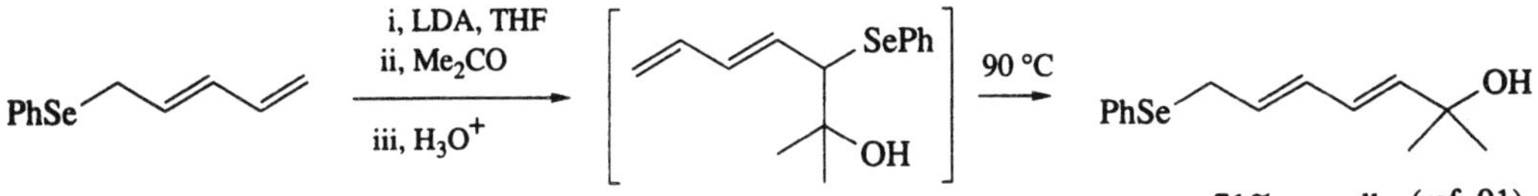

Scheme 17

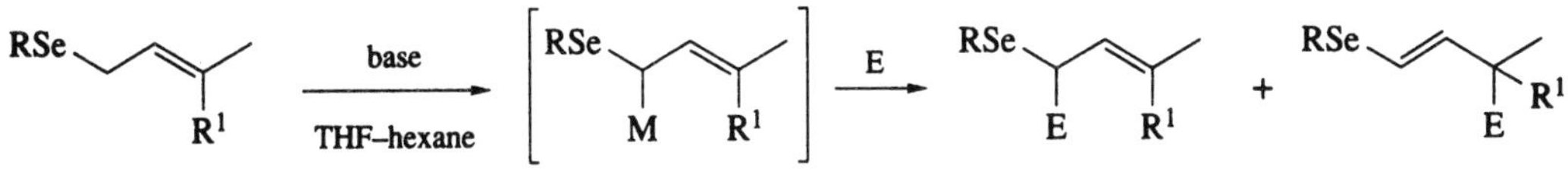

	R	*R¹*	*Conditions*	*E*	*Yield (%)*	*Ratio*	*Ref.*
(a)	Ph	Cl	LDA, –78 °C, <0.1 h	$Ph(CH_2)_2Br$	85	100:0	80
(b)	Ph	Me	LDA, 0 °C, 0.5 h	$Me_2PhSiCl$	71	100:0	80
(c)	Me	Me	LiTMP, –78 °C, 0.5 h	PhCHO	11	0:100	45
(d)	Me	Me	LiTMP, –25 °C, 0.5 h	PhCHO	47	0:100	45
(e)	Me	Me	KDA, –78 °C, 0.3 h	PhCHO	78	35:65	45

Scheme 18

Scheme 23, a), bis(methylseleno)methane remains untouched[71] under these or even under stronger conditions. Higher homologs in the arylseleno series require the concomitant use of LiTMP and HMPT,[71] or better, KDA in THF (Scheme 23, b–d).[73]

In fact, the metallation of some of these compounds can be achieved in minute amounts using even LDA or Bu^tOK, but the equilibrium between the different partners is not in favor of the formation of the

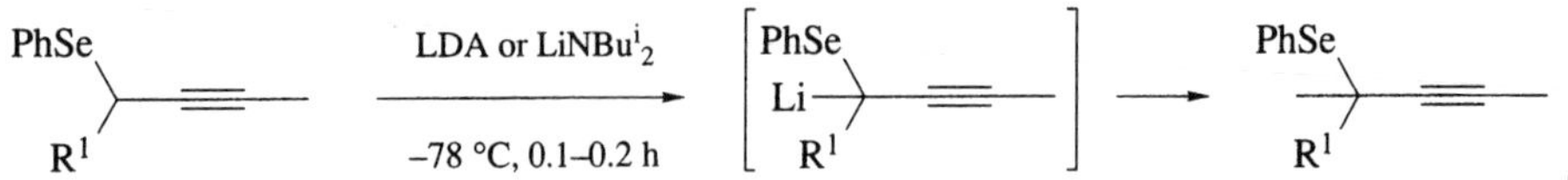

PhSe, R^1 — LDA or $LiNBu^i_2$, −78 °C, 0.1–0.2 h → [PhSe, Li, R^1] → PhSe, R^1 (ref. 93)

(a) R^1 = H, exclusive

(b) R^1 = Me, predominant

Scheme 19

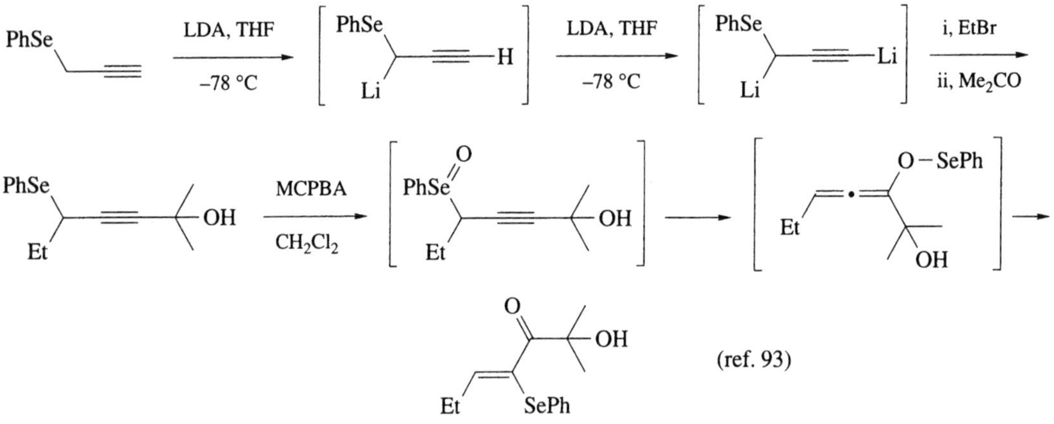

PhSe — LDA, THF, −78 °C → [PhSe, Li, H] — LDA, THF, −78 °C → [PhSe, Li, Li] — i, EtBr; ii, Me_2CO →

PhSe, Et, OH — MCPBA, CH_2Cl_2 → [PhSe(O), Et, OH] → [O−SePh, Et, OH] →

O, OH, Et, SePh (ref. 93)

59% overall

Scheme 20

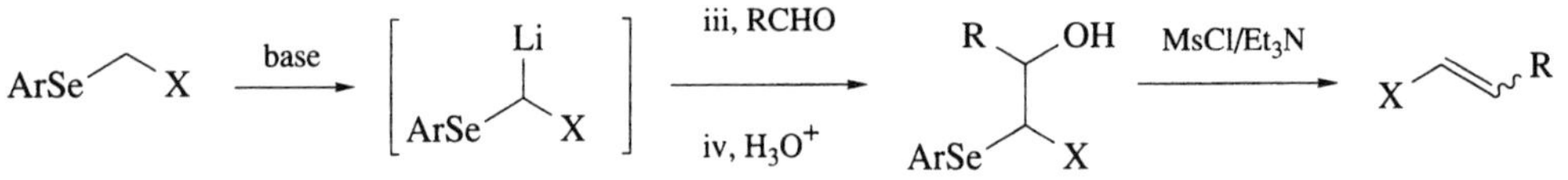

ArSe, X — base → [Li, ArSe, X] — iii, RCHO; iv, H_3O^+ → R, OH, ArSe, X — MsCl/Et_3N → X, R

	Base/conditions	*Ar*	*X*	*R*	*Yield (%)*	*Temperature/time*	*Yield (%)*	*(Z):(E)*
(a)	i	*m*-$CF_3C_6H_4$	OMe	Ph	76		72	46:54
(b)	ii	Ph	SePh	Ph	95	0 °C, 0.2 h	74	0:100
	ii	Ph	SePh	Me	92	0 °C, 0.2 h	84	50:50
	ii	Ph	SePh	Et	98	0 °C, 0.2 h	74	50:50
	ii	Ph	SePh	Pr^i	71	0 °C, 0.2 h	81	25:75
(c)	ii	*m*-$CF_3C_6H_4$	*m*-$CF_3C_6H_4Se$	Ph	–	20 °C, 3 h	85	0:100
	ii	*m*-$CF_3C_6H_4$	*m*-$CF_3C_6H_4Se$	Me	91	20 °C, 3 h	63	50:50
	ii	*m*-$CF_3C_6H_4$	*m*-$CF_3C_6H_4Se$	Et	98	20 °C, 3 h	72	50:50
	ii	*m*-$CF_3C_6H_4$	*m*-$CF_3C_6H_4Se$	Pr^i	98	20 °C, 3 h	79	30:70

i, LiTMP, THF, −78 °C, 0.1 h; ii, LDA, THF, −78 °C, 0.5 h (refs. 56 and 65)

Scheme 21

α-selenoalkyl metal. In some cases, however, the equilibrium can be driven to the other side by, for example, trapping the organometallic, once it is formed, with an alkylating agent. This is effectively the case for 3-chloro-1,1-bis(phenylseleno)propane and its bis(methylseleno) analogs, as well as 3-chloro-1,1-bis(phenylseleno)hexane, which produce[104] the corresponding selenoacetals of cyclopropanone on reaction with LDA in THF (Scheme 26, a and b). 3-Chloro-1,1-bis(methylseleno)hexane, however, does not react under similar conditions (Scheme 26, d). Such difference of reactivity between phenylseleno and methylseleno derivatives is also observed when potassium *t*-butoxide is used as the base. Thus, whereas 3-chloro-1,1-bis(phenylseleno)propane is almost quantitatively cyclized (Scheme 26, b),[12,77] its methylseleno analogs unexpectedly produce[12,77] 1,1-bis(methylseleno)-1-propene (Scheme 26, c). 1,1-Bis(phenylseleno)cyclopropane has also been prepared by Reich[97] from 3-tosyloxy-1-phenylseleno-

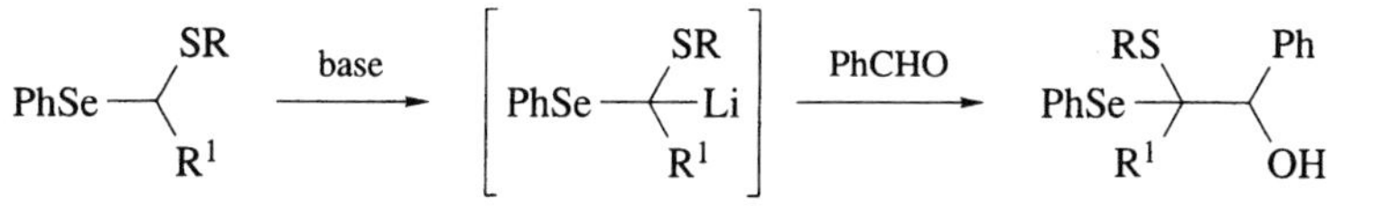

PhSe–CH(SR)(R[1]) →(base) [PhSe–C(SR)(Li)(R[1])] →(PhCHO) PhSe–C(RS)(R[1])–CH(Ph)(OH) (ref. 94)

R	*R[1]*	*Base/conditions*	*Yield (%)*	*Recovered SM (%)*
Ph	H	LiTMP, THF, –24 °C, 1.5 h	91	–
Ph	Me	LiTMP, THF, –24 °C, 1.5 h	68	10
Ph	C_5H_{11}	LiTMP, THF, –24 °C, 1.5 h	0	85
Ph	C_5H_{11}	LiTMP, diglyme, –24 °C, 0.5 h	65	17
Me	C_5H_{11}	LiTMP, diglyme, –78 °C, 1 h	75	0

Scheme 22

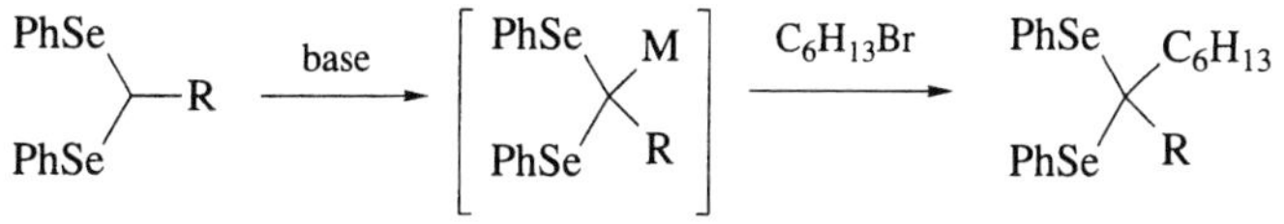

$(PhSe)_2CH$–R →(base) [$(PhSe)_2C(M)R$] →($C_6H_{13}Br$) $(PhSe)_2C(C_6H_{13})R$

	Conditions	*R*	*Yield (%)*	*Ref.*
(a)	Bu^i_2NLi, THF, –78 °C, then –30 °C, 0.2 h	H	80	35, 36
(b)	LDA, THF–HMPA, –30 °C, then 20 °C, 0.2 h	Me	74	71
	LDA, THF–HMPA, –30 °C, then 20 °C, 0.2 h	C_6H_{13}	44	71
(c)	LiTMP, THF–HMPA, –30 °C, then 20 °C, 0.2 h	Me	83	71
	LiTMP, THF–HMPA, –30 °C, then 20 °C, 0.2 h	C_6H_{13}	86	71
(d)	KDA, THF, –78 °C	Me	95	73

$(PhSe)_2CH$–R →(i, KDA, THF, –78 °C) [$(PhSe)_2C(K)R$]

(e) →(ii, Pr^iCHO; iii, H_3O^+) $(PhSe)_2C(R)$–CH(OH)Pr^i (ref. 73)
R = Me, 90%
R = Pr^i, 63%
R = $C_{10}H_{21}$, 84%

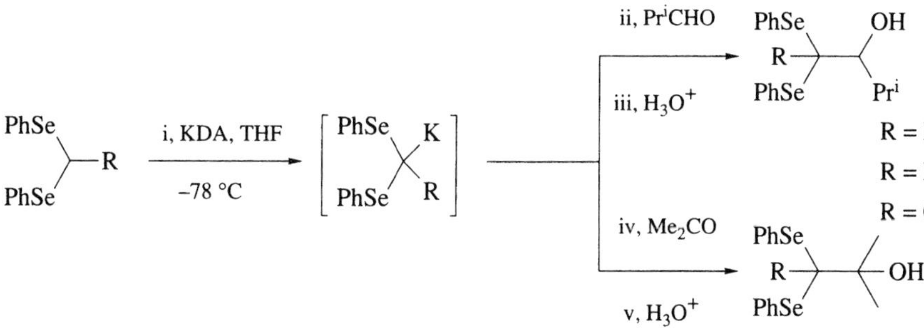

(f) →(iv, Me_2CO; v, H_3O^+) $(PhSe)_2C(R)$–C(OH)Me_2 (ref. 73)
R = Me, 77%

Scheme 23

Me_3Si–CH(SeR)(R[1]) →(i) [Me_3Si–C(SeR)(M)(R[1])] →(ii, iii) Me_3Si–C(RSe)(R[1])–CH(OH)C_6H_{13} + RSe–CH=CH–C_6H_{13}

	R	*R[1]*	*Conditions*	*Yield (%)*	*Ratio*	*Ref.*
(a)	Me	H	–30 °C, 0.5 h	0	–	100
(b)	Ph	H	–78 °C, 0.5 h	5	100:0	100
(c)	Ph	H	–30 °C, 0.5 h	30	57:43	100
(d)	m-$CF_3C_6H_4$	H	–78 °C, 0.5 h	83	–	70
(e)	Ph	$C_{10}H_{21}$	–30 °C, 0.5 h	0	–	100
(f)	Ph	Ph	0 °C, 0.5 h	77	–	70
(g)	m-$CF_3C_6H_4$	Ph	–78 °C, 0.8 h	81	–	70

i, LDA, THF–hexane; ii, $C_6H_{13}CHO$; iii, H_3O^+

Scheme 24

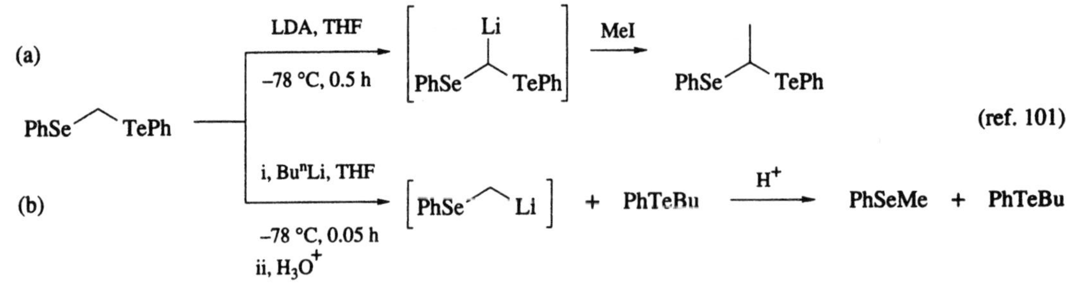

(a) PhSe⌒TePh → LDA, THF, –78 °C, 0.5 h → [PhSe–CH(Li)–TePh] → MeI → PhSe–CH(Me)–TePh (ref. 101)

(b) PhSe⌒TePh → i, Bu^nLi, THF, –78 °C, 0.05 h; ii, H_3O^+ → [PhSe⌒Li] + PhTeBu → H^+ → PhSeMe + PhTeBu

Scheme 25

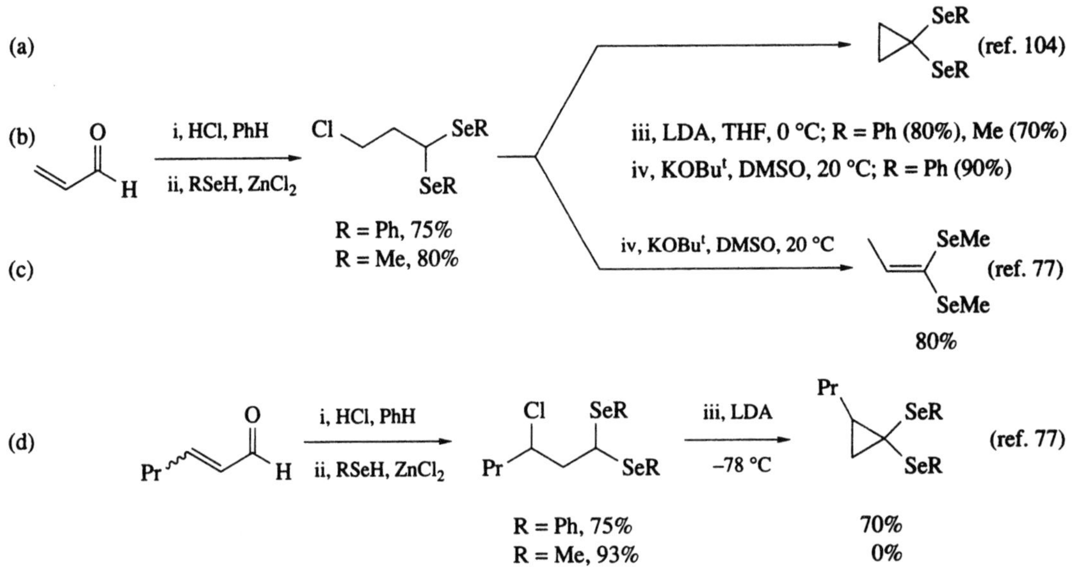

(a) → 1,1-bis(SeR)cyclopropane (ref. 104)

(b) acrolein (O, H) → i, HCl, PhH; ii, RSeH, $ZnCl_2$ → Cl⌒⌒CH(SeR)SeR (R = Ph, 75%; R = Me, 80%) → iii, LDA, THF, 0 °C; R = Ph (80%), Me (70%); iv, $KOBu^t$, DMSO, 20 °C; R = Ph (90%)

(c) iv, $KOBu^t$, DMSO, 20 °C → CH₃CH=C(SeMe)SeMe, 80% (ref. 77)

(d) Pr-CH=CH-CHO (O, H) → i, HCl, PhH; ii, RSeH, $ZnCl_2$ → Pr-CH(Cl)-CH₂-CH(SeR)SeR (R = Ph, 75%; R = Me, 93%) → iii, LDA, –78 °C → 2-Pr-1,1-bis(SeR)cyclopropane (70%; 0%) (ref. 77)

Scheme 26

TsO(CH₂)₂CH(SePh)SePh → i, MCPBA; ii, LDA → [TsO(CH₂)₂C(Li)(SePh)Se(=O)Ph] → cyclopropyl(SePh)Se(=O)Ph → i, H^+; ii, I^- or SO_3^{2-} → 1,1-bis(SePh)cyclopropane (ref. 97), 57% overall

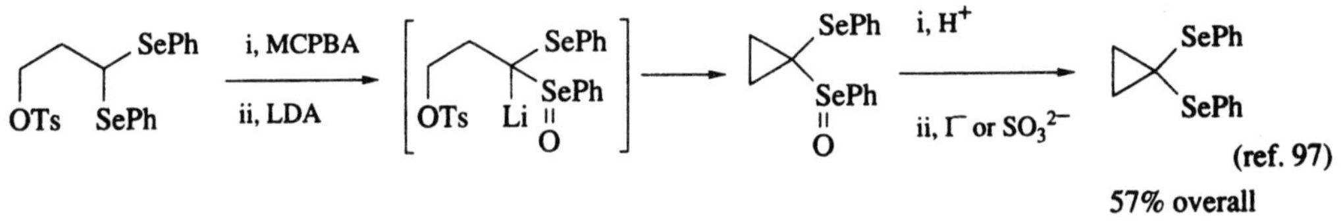

Scheme 27

1-phenyl-selenoxypropane and LDA in THF followed by reduction of the resulting cyclopropyl phenyl selenoxide with, for example, iodide anion (Scheme 27).[97]

Benzyl selenides exhibit a similar pattern of reactivity. Thus, although the parent aryl benzyl selenides are successfully metallated with LDA at –78 °C,[56,72,78,79] their higher homologs, as well as benzyl methyl selenides,[72] are not. Use of stronger bases such as LiTMP, or better, KDA, allows the metallation of various aryl benzyl selenides, even those bearing an alkyl substituent at the benzylic site (Scheme 13, c; Scheme 15 a and c; Scheme 23, b–d). They are, however, of limited use in the methylseleno series since the parent compound is the only one to react (Scheme 15, d–f).[72]

Allylic[45,80–82,84,85] and propargylic[93] selenides seem to be more acidic. In the phenylseleno series, for example,[80–82,84,85] the synthesis of those compounds bearing an alkyl substituent on the carbanionic center has been already achieved with LDA and does not require the use of stronger bases (Schemes 16–20). In the methylseleno series,[45] however, the use of LiTMP (Scheme 16, h; Scheme 18, c and d) or KDA (Scheme 18, e compared with c and d) is often needed.

Finally, the metallation is particularly difficult with selenides bearing a methoxymethyl substituent,[56] and is only successful[56] when a *m*-(trifluoromethyl)phenyl moiety is attached to the selenium atom (Scheme 21, a).

As already mentioned, LiTMP acts usually as a stronger base than LDA or lithium diethylamide; in some rare cases the reverse order of reactivity has been observed. Thus it has been found[56] that the trimethylsilyl substituent reduces the kinetic acidity of $CF_3PhSeCH_2SiMe_3$ as compared with $CF_3PhSeMe$ towards LiTMP by a factor of 2.8. With LDA as base, however, the silyl substituent has the opposite effect,[56] since *m*-$CF_3PhSeCH_2SiMe_3$ is deprotonated at least 25 times as fast as *m*-$CF_3PhSeMe$ (Scheme 24, d). Presumably, steric effects cancel the normal acidifying effect of the trimethylsilyl group when a hindered base such as LiTMP is used.

Phenylselenomethyl phosphonates[102] and (phenylseleno)chloromethane[105] have been metallated by NaH and Bu^tOK, respectively (Scheme 28, a and b; Scheme 29). The former organometallics react with carbonyl compounds and directly produce vinyl selenides (Scheme 28, a and b),[102] whereas the second decomposes to phenylselenomethylene[105] which adds stereospecifically to alkenes and leads to phenylselenocyclopropanes in reasonably good yields (Scheme 29).[105]

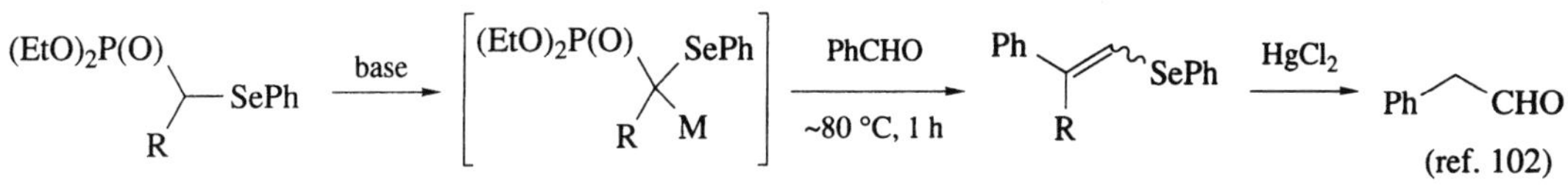

$(EtO)_2P(O)CH(R)SePh$ —base→ $[(EtO)_2P(O)C(R)(M)SePh]$ —PhCHO, ~80 °C, 1 h→ PhCH=C(R)SePh —$HgCl_2$→ $PhCH_2CHO$ (ref. 102)

	Conditions	R	Yield (%)	(Z):(E)	Yield (%)
(a)	NaH, THF–HMPA (10:1), 80 °C, 1 h	H	80	6:94	90
(b)	NaH, THF–HMPA (10:1), 80 °C, 6 h	Me	86	47:53	–
	NaH, THF–HMPA (10:1), 80 °C, 6 h	Et	70	8:92	–
(c)	Bu^nLi, THF–hexane, –78 °C, 4 h	H	70	16:84	–

Scheme 28

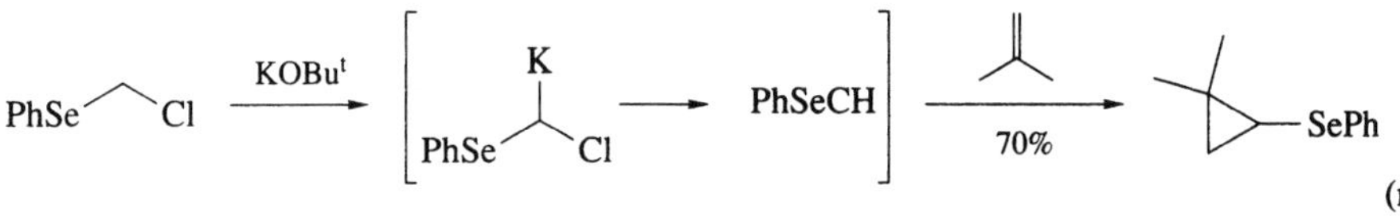

$PhSeCH_2Cl$ —$KOBu^t$→ [PhSeCH(K)Cl → PhSeCH] —(alkene), 70%→ phenylselenocyclopropane (SePh) (ref. 105)

Scheme 29

Bu^nLi, which usually performs the Se–Li exchange[7,9,12,16,17] in selenides, effects the metallation of some functionalized selenides. This is particularly the case of phenylselenomethyl phosphonates[102] and phosphonium salts (Scheme 28, c).[106,107] This is also the case of methyl phenyl selenide (Scheme 13, a compared with Scheme 2, a),[36] phenyl trimethylsilylmethyl selenide (Scheme 30, a)[98] and bis(phenylseleno)methane (Scheme 30, c and d),[36,71] which are metallated with butyllithium but only in coordinating solvents such as TMEDA,[36,98] dimethoxyethane (DME)[71] or HMPA.[71]

PhSeCH(X)R ——→ [PhSeC(X)(R)Li] + [LiCH(X)R]

	X	R	Conditions	Yield (%) PhSeC(X)(R)Li	Yield (%) LiCH(X)R	Ref.
(a)	Me_3Si	H	Bu^sLi–TMEDA, 25 °C, 1.5 h	83	15	98
(b)	PhSe	H	Bu^nLi, DME–hexane, –78 °C, 1 h	50	40	71
(c)	PhSe	Me	Bu^nLi, DME–hexane, –78 °C, 1 h	25	35	71

Scheme 30

Selenides bearing two groups capable of stabilizing a carbanionic center on the α-carbon are much more acidic than the above mentioned derivatives and both the phenylseleno and the methylseleno derivatives can be readily metallated. Thus, tris(phenylseleno)methane,[35,36] tris(methylseleno)methane (Scheme 31),[71,108] arylbis(phenylseleno)methane and arylbis(methylseleno)methane (Scheme 32)[72] lead to the corresponding methyllithiums with LiDBA,[35,36] LDA,[71,72,100] LiTMP[72] or KDA[72] in THF. Surprisingly, with arylbis(seleno)methanes selenophilic attack of LiTMP[72] and LDA,[72] which produces α-selenobenzyllithiums, competes with the metallation reaction (Scheme 32, d and e). This reaction can be prevented if KDA is used instead of LDA (Scheme 32, a and b).[72]

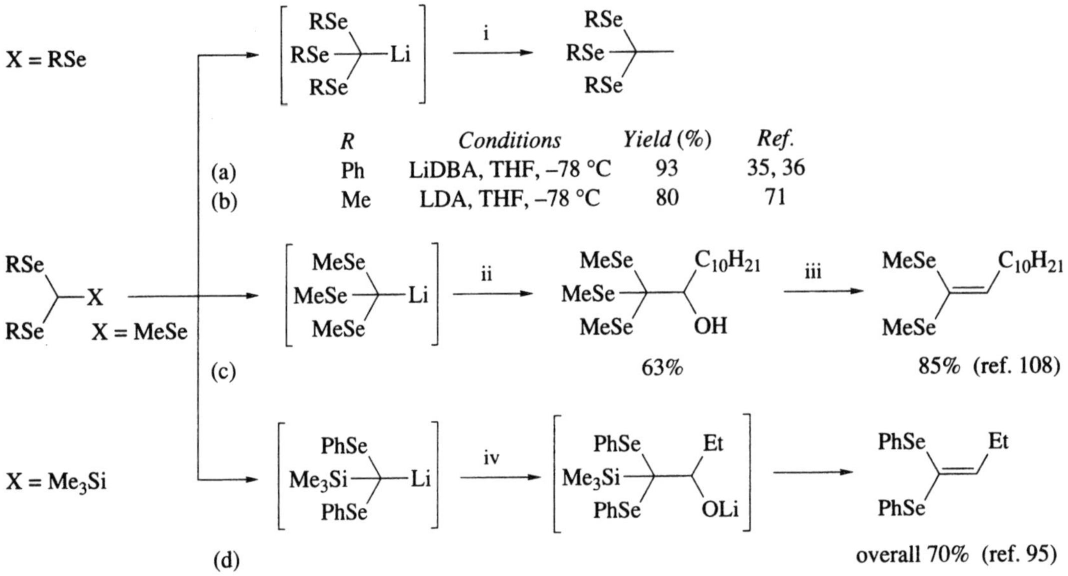

	R	Conditions	Yield (%)	Ref.
(a)	Ph	LiDBA, THF, –78 °C	93	35, 36
(b)	Me	LDA, THF, –78 °C	80	71

(c) 63% ; 85% (ref. 108)

(d) overall 70% (ref. 95)

i, MeI; ii, $C_{10}H_{21}CHO$; iii, PI_3, Et_3N, CH_2Cl_2, 20 °C, 1 h; iv, EtCHO

Scheme 31

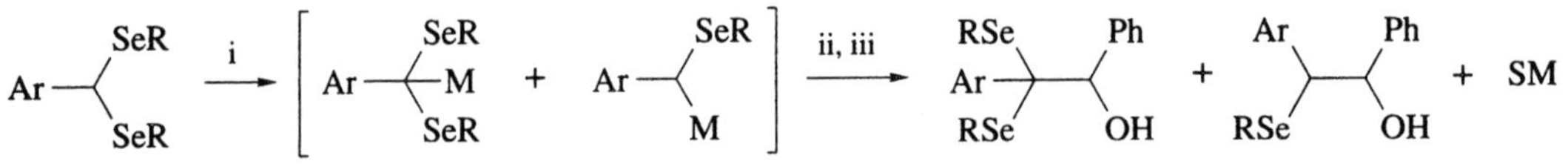

	R	Ar	Base	Yield (%)	Ratio
(a)	Me	Ph or 4-$MeOC_6H_4$	KDA	79	100:0:0
(b)	Ph	Ph or 4-$MeOC_6H_4$	KDA	81	100:0:0
(c)	Me	Ph	LDA	–	50:0:50
(d)	Me	Ph	LiTMP	–	16:30:54
(e)	Ph	Ph	LDA	–	26:11:50[a]

(ref. 72)

[a] Other products are formed.

i, base, THF–hexane, –78 °C, 0.3 h; ii, PhCHO; iii, H^+

Scheme 32

It has been noticed that lithium diethylamide is far superior to LDA for the metallation of vinyl-[85] and phenyl-(trimethylsilyl)phenylselenomethane (Scheme 33).[70] Presumably, the steric hindrance due to the trimethylsilyl group prevents the attack of the bulky LDA.

The situation is clearly different for α-seleno ketones (Schemes 34–36),[96,109–111] α-seleno esters (Schemes 10 and 37),[56,74,75,112,113] selenolactones (Scheme 38),[74,114] α-seleno acids (Scheme 39)[56,115] and α-selenonitriles (Scheme 40),[116,117] which have been easily metallated according to Rathke's procedure[118,119] using LDA as a base. The reaction is reasonably general and takes place on a large variety of arylseleno and methylseleno derivatives including those that bear an alkyl group in the α-position. In the latter case, however, some competitive C—Se bond cleavage has been observed from time to time.[96,111]

α-Phenylselenonitroalkanes (Scheme 41)[120] and α-phenylseleno-β-lactams (Scheme 42, a) derived from penicillin[121] have been transformed to their enolates on reaction with calcium hydroxide and butyllithium, respectively. In the latter case, however, successful metallation has been achieved exclusively on the stereoisomer bearing an *endo*-phenylseleno moiety (Scheme 42, compare a with b).[121]

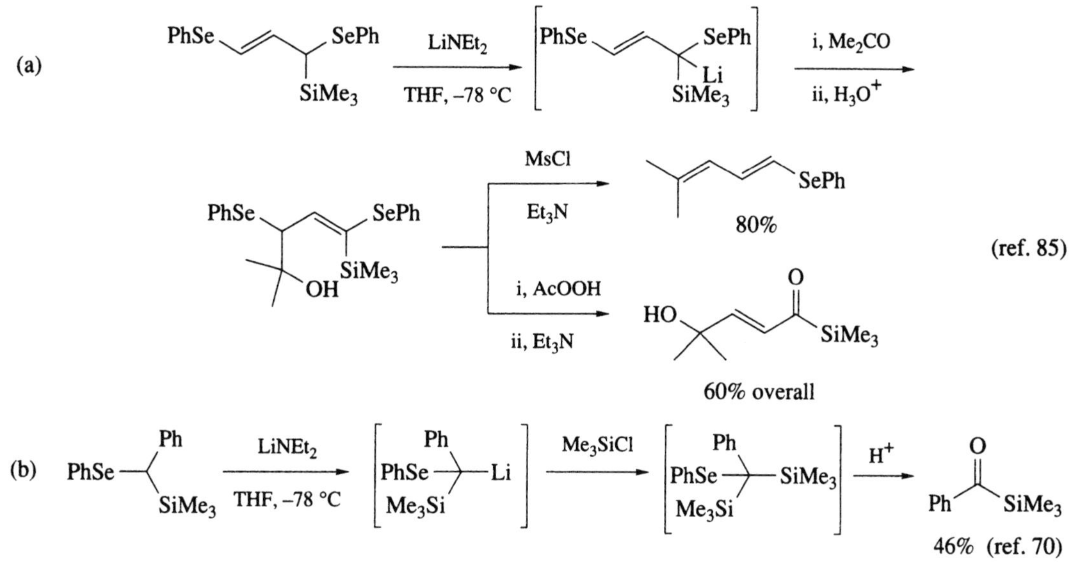

(a) PhSe–CH=CH–CH(SePh)(SiMe$_3$) $\xrightarrow[\text{THF, –78 °C}]{\text{LiNEt}_2}$ [PhSe–CH=CH–C(Li)(SePh)(SiMe$_3$)] $\xrightarrow[\text{ii, H}_3\text{O}^+]{\text{i, Me}_2\text{CO}}$

PhSe, SePh, SiMe$_3$, OH — MsCl / Et$_3$N → SePh 80%; i, AcOOH / ii, Et$_3$N → HO, O, SiMe$_3$ 60% overall (ref. 85)

(b) PhSe–CH(Ph)(SiMe$_3$) $\xrightarrow[\text{THF, –78 °C}]{\text{LiNEt}_2}$ [PhSe–C(Ph)(Li)(Me$_3$Si)] $\xrightarrow{\text{Me}_3\text{SiCl}}$ [PhSe–C(Ph)(SiMe$_3$)(Me$_3$Si)] $\xrightarrow{\text{H}^+}$ PhC(O)SiMe$_3$ 46% (ref. 70)

Scheme 33

OMe, O, O, SePh, O $\xrightarrow{\text{i, ii}}$ OMe, O, O, SePh, O 61% $\xrightarrow{\text{iii}}$ OMe, O, O, O 65% (ref. 110) → → Vernolepin

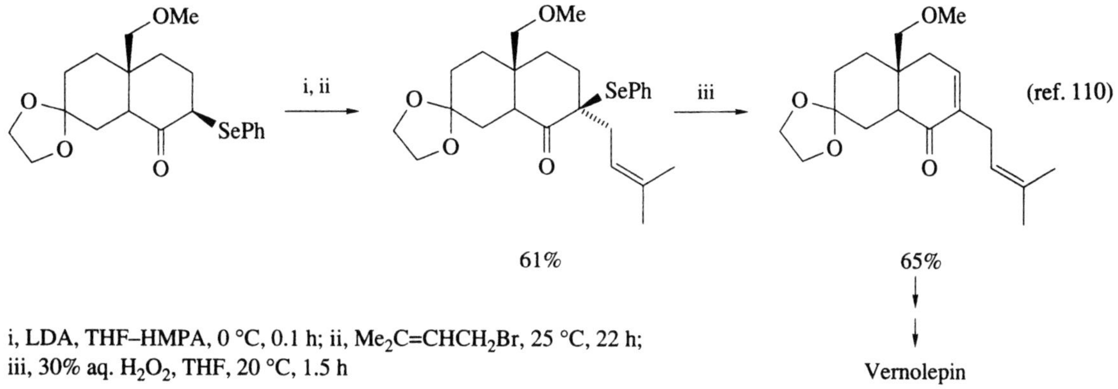

i, LDA, THF–HMPA, 0 °C, 0.1 h; ii, Me$_2$C=CHCH$_2$Br, 25 °C, 22 h;
iii, 30% aq. H$_2$O$_2$, THF, 20 °C, 1.5 h

Scheme 34

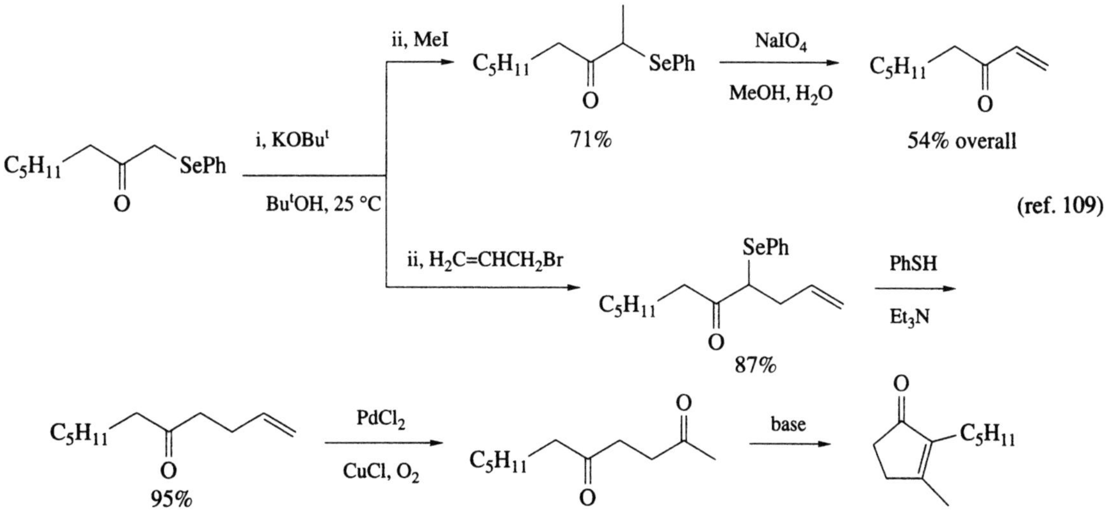

C$_5$H$_{11}$CH$_2$C(O)CH$_2$SePh $\xrightarrow[\text{Bu}^t\text{OH, 25 °C}]{\text{i, KOBu}^t}$

ii, MeI → C$_5$H$_{11}$CH$_2$C(O)CH(CH$_3$)SePh 71% $\xrightarrow[\text{MeOH, H}_2\text{O}]{\text{NaIO}_4}$ C$_5$H$_{11}$CH$_2$C(O)CH=CH$_2$ 54% overall (ref. 109)

ii, H$_2$C=CHCH$_2$Br → C$_5$H$_{11}$CH$_2$C(O)CH(SePh)CH$_2$CH=CH$_2$ 87% $\xrightarrow[\text{Et}_3\text{N}]{\text{PhSH}}$

C$_5$H$_{11}$CH$_2$C(O)CH$_2$CH$_2$CH=CH$_2$ 95% $\xrightarrow[\text{CuCl, O}_2]{\text{PdCl}_2}$ C$_5$H$_{11}$CH$_2$C(O)CH$_2$CH$_2$C(O)CH$_3$ $\xrightarrow{\text{base}}$ O, C$_5$H$_{11}$

Scheme 35

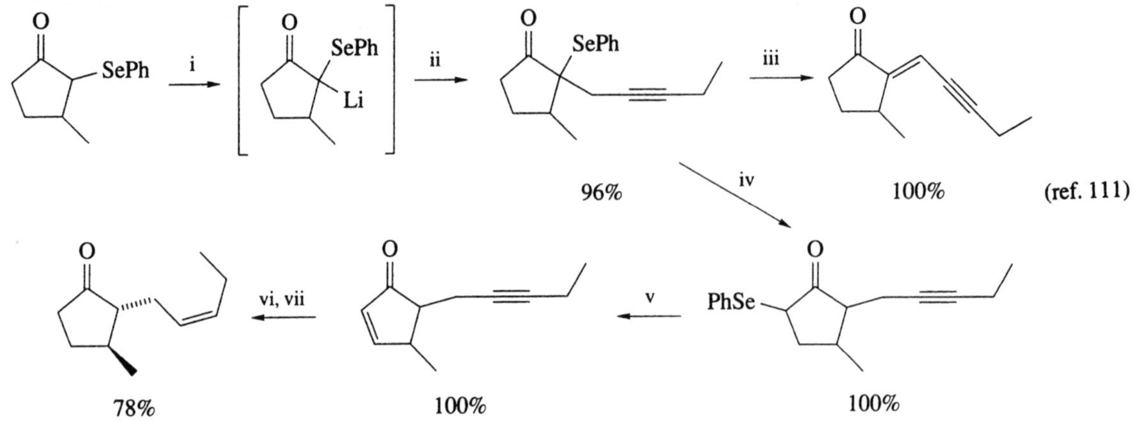

(ref. 111)

96% 100%

78% 100% 100%

i, LDA, THF, –78 °C, 0.2 h; ii, EtC≡CCH_2Br, THF–HMPA, –78 °C, 2 h, then 20 °C, 10 h;
iii, 30% aq. H_2O_2, CH_2Cl_2; iv, LDA, THF–HMPA, –78 °C, 0.5 h, then 20 °C, 1 h;
v, 30% aq. H_2O_2, 20 °C, 1 h; vi, NaOMe, 20 °C, 5 h; vii, H_2, $BaSO_4$, EtOAc

Scheme 36

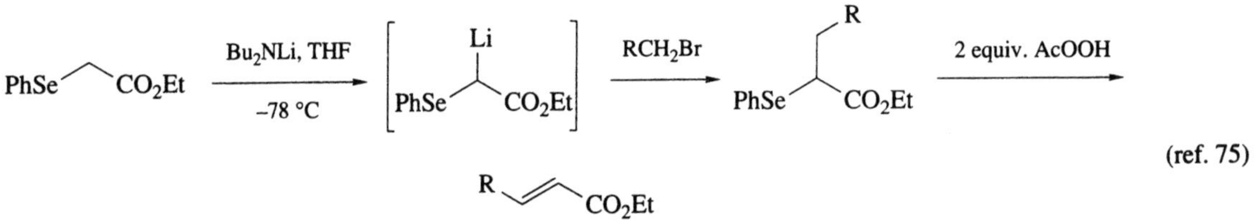

Bu_2NLi, THF, –78 °C; RCH_2Br; 2 equiv. AcOOH (ref. 75)

Overall yields: R = Me, 66%; R = Ph, 65%

Scheme 37

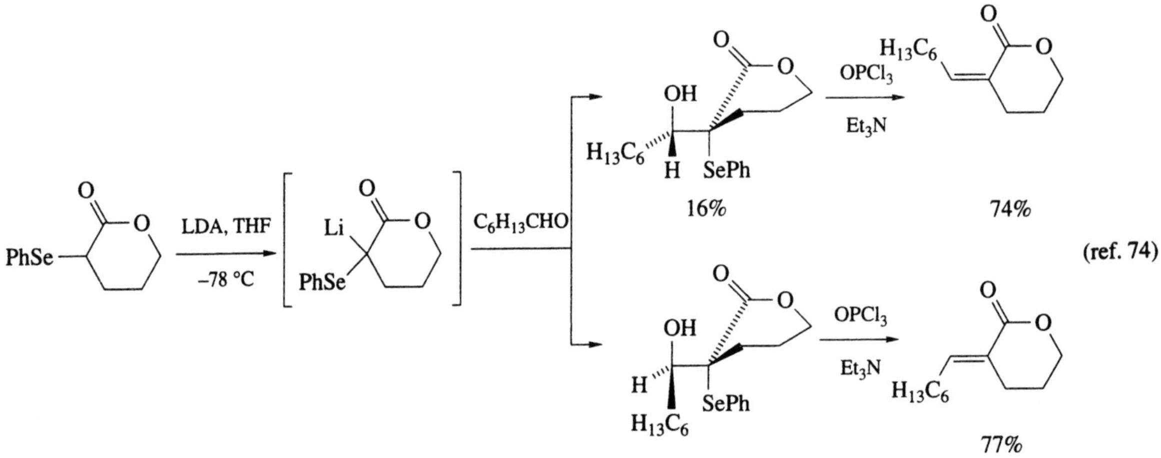

LDA, THF, –78 °C; $C_6H_{13}CHO$; $OPCl_3$, Et_3N (ref. 74)

16% 74%

77%

Scheme 38

2.6.2.2.2 *Synthesis of α-selenovinyl metals (1-seleno-1-alkenyl metals) by metallation of vinyl selenides*

Metallation of vinyl selenides can be more readily achieved than that of alkyl selenides since two different effects now work in the same direction. Hydrogens linked to an sp^2 carbon are more acidic than those linked to an sp^3 carbon, and the selenium atom stabilizes the resulting sp^2 carbanion to a larger extent than it does in the sp^3 case (see Section 2.6.2.1.2).[17,65,68] Except in the special cases of selenophene[122] and benzoselenophene (Scheme 43),[123] which also belong to that series of compounds, butyllithiums are not suitable metallating agents. Depending upon the solvent and the nature of the vinyl

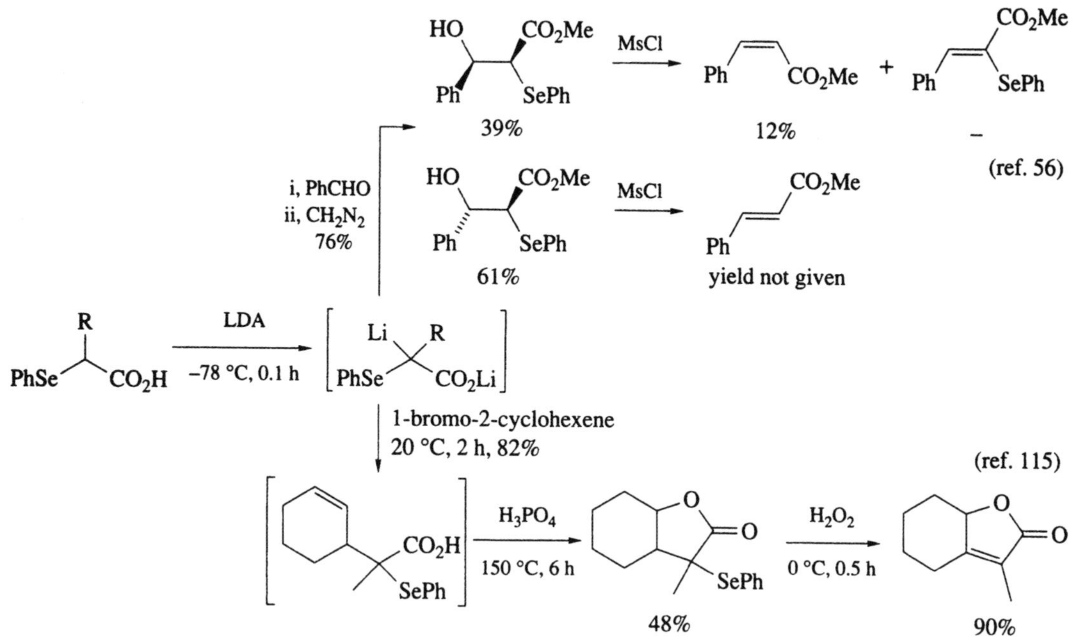

HO CO$_2$Me Ph SePh 39% —MsCl→ Ph CO$_2$Me 12% + Ph CO$_2$Me SePh –

(ref. 56)

i, PhCHO ii, CH$_2$N$_2$ 76%

HO CO$_2$Me Ph SePh 61% —MsCl→ Ph CO$_2$Me yield not given

R PhSe CO$_2$H —LDA, –78 °C, 0.1 h→ [Li R PhSe CO$_2$Li]

1-bromo-2-cyclohexene 20 °C, 2 h, 82%

(ref. 115)

[CO$_2$H SePh] —H$_3$PO$_4$, 150 °C, 6 h→ O O SePh 48% —H$_2$O$_2$, 0 °C, 0.5 h→ O O 90%

Scheme 39

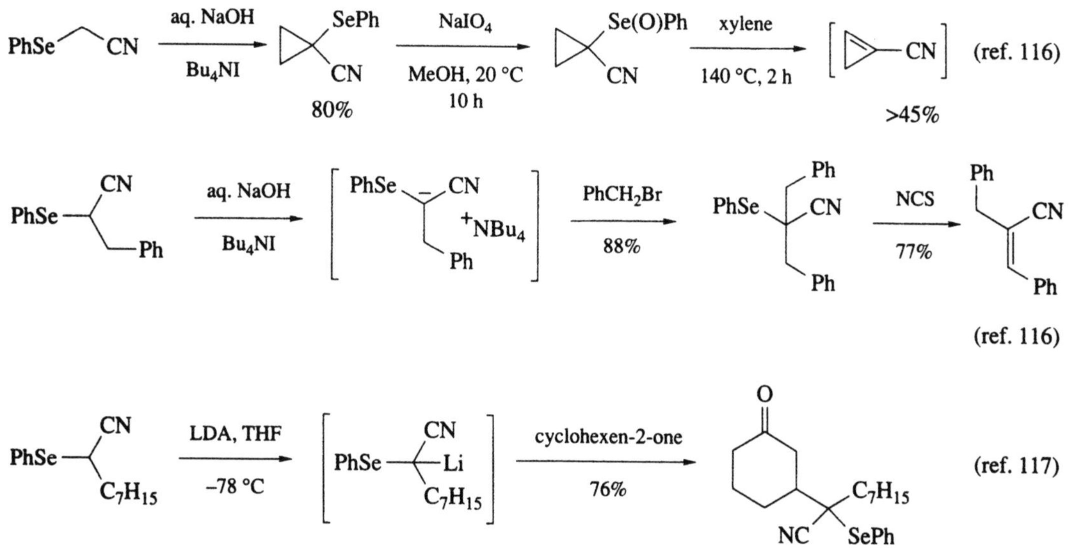

PhSe CN —aq. NaOH, Bu$_4$NI→ SePh CN 80% —NaIO$_4$, MeOH, 20 °C 10 h→ Se(O)Ph CN —xylene, 140 °C, 2 h→ [CN] >45% (ref. 116)

PhSe CN Ph —aq. NaOH, Bu$_4$NI→ [PhSe – CN $^+$NBu$_4$ Ph] —PhCH$_2$Br, 88%→ PhSe Ph CN Ph —NCS, 77%→ Ph CN Ph

(ref. 116)

PhSe CN C$_7$H$_{15}$ —LDA, THF, –78 °C→ [PhSe CN Li C$_7$H$_{15}$] —cyclohexen-2-one, 76%→ O NC SePh C$_7$H$_{15}$ (ref. 117)

Scheme 40

selenide different reactions take place. Thus, alkyllithiums in ether[41,65,124] or DME[42] add across the carbon–carbon double bond of phenyl vinyl selenide (see Section 2.6.2.5), whereas the C—Se bond cleavage leading to phenyllithium and butyl vinyl selenide occurs[41,65,125] when the reaction is performed in THF at

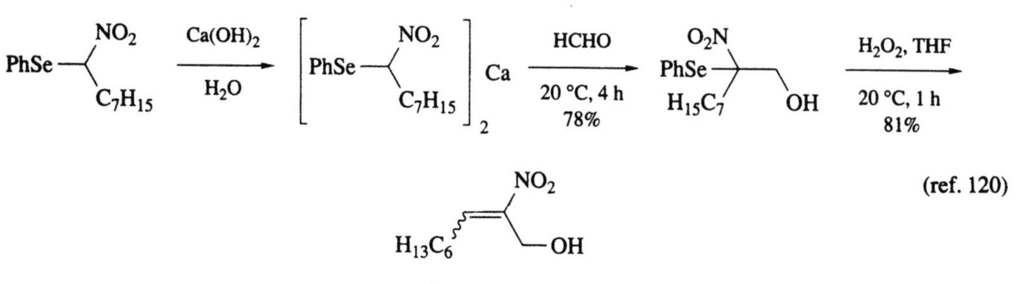

PhSe NO$_2$ C$_7$H$_{15}$ —Ca(OH)$_2$, H$_2$O→ [PhSe NO$_2$ C$_7$H$_{15}$]$_2$ Ca —HCHO, 20 °C, 4 h, 78%→ O$_2$N PhSe H$_{15}$C$_7$ OH —H$_2$O$_2$, THF, 20 °C, 1 h, 81%→

(ref. 120)

NO$_2$ H$_{13}$C$_6$ OH

Scheme 41

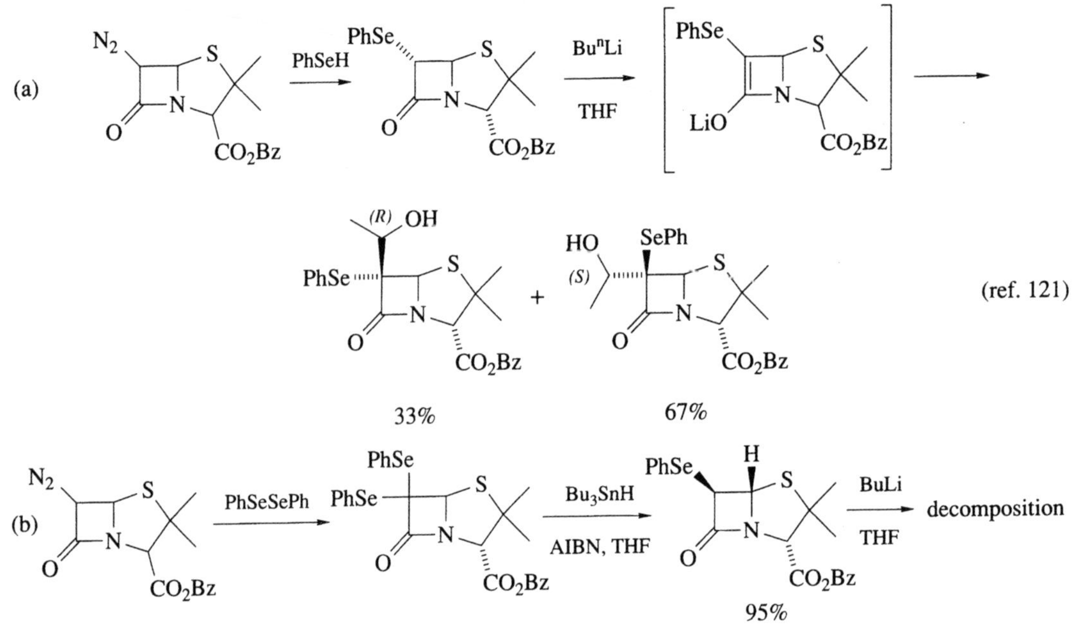

(ref. 121)

33% 67%

95%

decomposition

Scheme 42

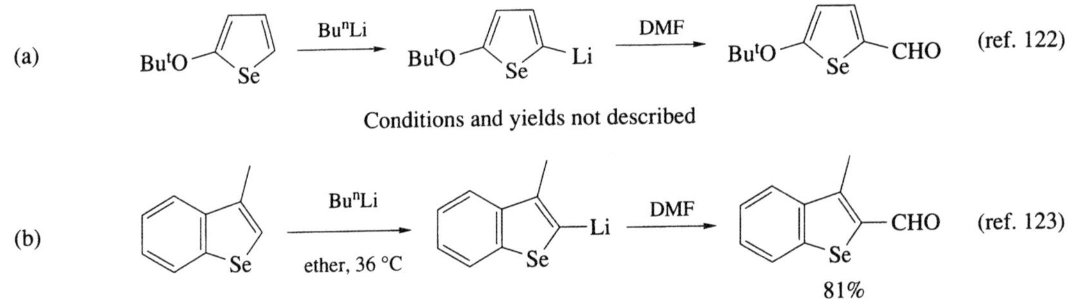

(ref. 122)

Conditions and yields not described

(ref. 123)

81%

Scheme 43

low temperature. Under the latter conditions methyl vinyl selenides are metallated instead on their methyl group.[125]

Metalloamides are suitable for the synthesis of 1-metallovinyl aryl selenides (Scheme 44)[41,65,73] and 1-metalloallenyl phenyl selenides (Scheme 45)[93] from the corresponding vinyl or allenyl selenides. The synthesis of the latter organometallics is even more conveniently achieved[93] by metallation of 1-phenylseleno-1-propyne or by a two-step, one-pot sequence of reactions from 2-chloroallyl phenyl selenide (Scheme 45).

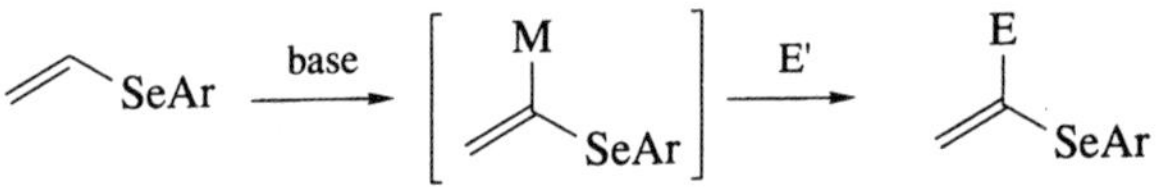

	Ar	*Conditions*	*E'*	*E*	*Yield (%)*	*Ref.*
(a)	Ph	LDA (1.5 equiv.), THF, –78 °C, 1 h	D_2O	D	80	65
(b)	Ph	LDA (1.5 equiv.), THF, –78 °C, 1 h	Me_2CO	Me_2CHOH	80	65
(c)	Ph	LDA (1.5 equiv.), THF, –78 °C, 1 h	$C_6H_{13}CHO$	$C_6H_{13}CHOH$	40	41
(d)	Ph	LDA (1.5 equiv.), THF, –78 °C, 1 h	CO_2	CO_2H	77	65
(e)	Ph	LDA (1 equiv.), THF–HMPA (20:1), –78 °C	$C_{10}H_{21}Br$	$C_{10}H_{21}$	80	41
(f)	Ph	KDA (1 equiv.), THF, –78 °C	$C_{10}H_{21}Br$	$C_{10}H_{21}$	94	73
(g)	$4-CF_3C_6H_4$	LDA (1 equiv.), THF, –78 °C	CO_2	CO_2H	80	65
(h)	$4-CF_3C_6H_4$	LDA (1 equiv.), THF–HMPA, –78 °C	BuI	Bu	85	65

Scheme 44

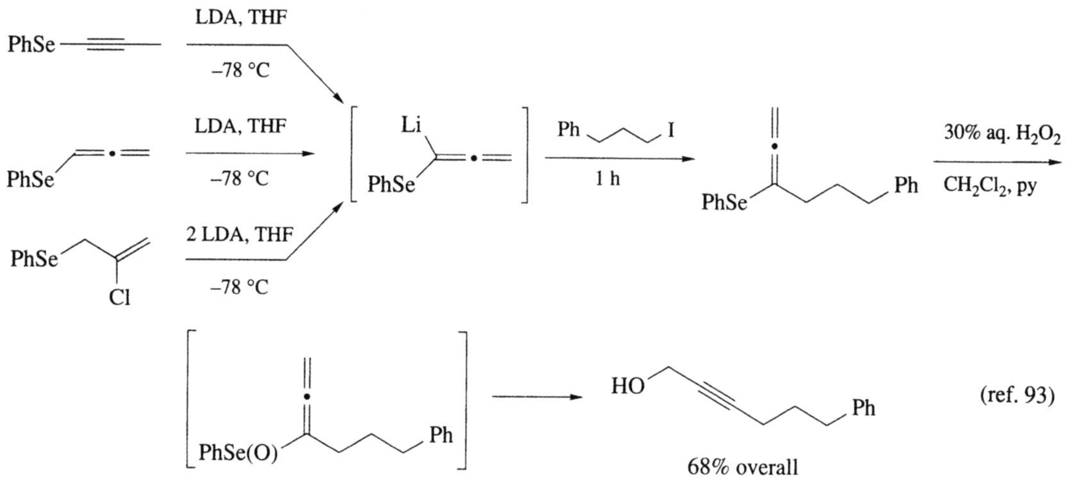

(ref. 93)

68% overall

Scheme 45

LDA in THF[41,65] or in THF–HMPT,[41] LiTMP in THF[65] and KDA in THF[73] have all been successfully used for the metallation of phenyl[41,65,73] or *m*-(trifluoromethyl)phenyl[65] vinyl selenide (Scheme 44). The deprotonation was shown to be reversible[65] with LDA in THF and irreversible with LiTMP[65] when performed in the same solvent. Extension of this reaction to homologous derivatives proved difficult[65,73] since metallation at the allylic sites often competes. Allylic metallation is particularly favorable with aryl 1-propenyl selenides (Scheme 46)[65,73] and, whatever the conditions used (LiTMP[65] or KDA[73]), with aryl 1-(2-methyl-1-propenyl) selenides (Scheme 47). In the latter case both the (*Z*)- and the (*E*)-methyl groups have been metallated leading to the corresponding α-metalloallyl selenides (Scheme 47).[65,73]

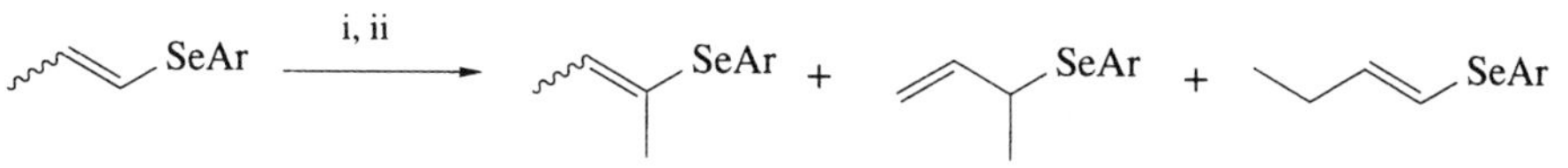

(ref. 65)

Ar	*Time* (h)	*Overall yield* (%)^a	*Ratio*
Ph	2	42	9:73:12
Ph	9	88	4:74:18
m-$CF_3C_6H_4$	1	66	30:51:7
m-$CF_3C_6H_4$	7	93	12:68:11

^{a} Products resulting from decomposition of the carbanion are also formed.

i, LDA (1.25 equiv.), THF, –78 °C, times as above; ii, MeI

Scheme 46

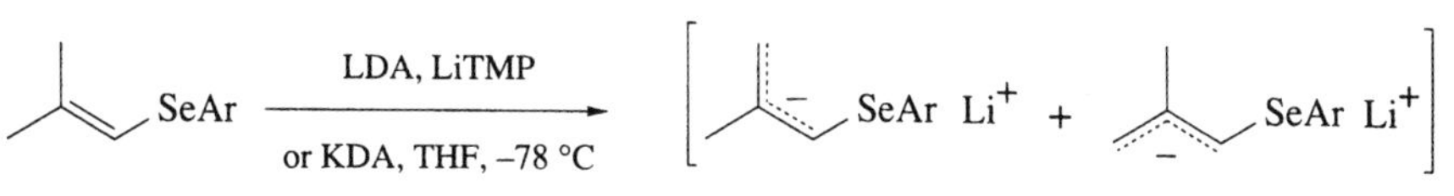

Ar = Ph or *m*-$CF_3C_6H_4$

(refs. 55 and 73)

Scheme 47

Increasing the substitution of the alkyl group on the β-carbon of 1-alkenyl phenyl selenides again favors metallation at the vinylic site (Scheme 48).[65,73] KDA in THF has proved to be, without contest, the most successful reagent for this purpose.[73] The reaction takes place at low temperature (–78 °C) and the 1-phenylselenoalkenyl metal, produced in almost quantitative yield,[73] retains its stereochemistry (Scheme 48, b–d).

Satisfactory deprotonations of 1-(*E*)-butenyl and 1-(3-methyl-1-(*E*)-butenyl) selenides in either the phenyl or *m*-(trifluoromethyl)phenyl series cannot be achieved[65] with LDA in THF (Scheme 48, compare a, e and g). Even LiTMP does not deprotonate the phenyl vinyl selenides (Scheme 48, a), and complete

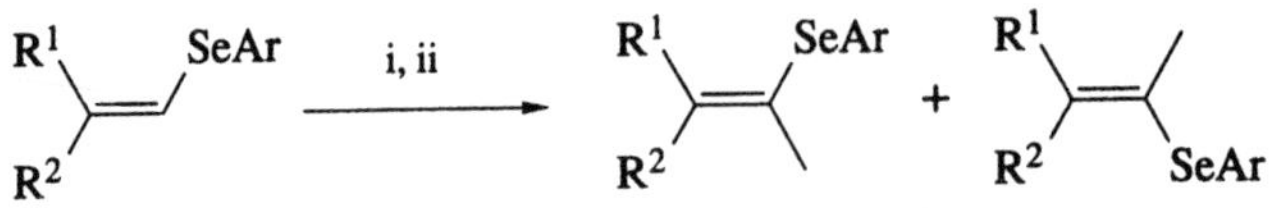

	Ar	*R¹*	*R²*	*Reagent*	*Yield (%)*		*Ref.*
(a)	Ph	H	Et or Pr^i	LDA or LiTMP	Unsatisfactory metallation		65
(b)	Ph	H	Bu	KDA	85	–	73
(c)	Ph	Bu	H	KDA	–	85	73
(d)	Ph	H	Pr^i	KDA	80	–	73
(e)	m-$CF_3C_6H_4$	H	Et	LDA	Unsatisfactory metallation		65
(f)	m-$CF_3C_6H_4$	H	Et	LiTMP	32	32	65
(g)	m-$CF_3C_6H_4$	H	Pr^i	LDA	Unsatisfactory metallation		65
(h)	m-$CF_3C_6H_4$	H	Pr^i	LiTMP	43	19	65
(i)	Py	H	C_6H_{13}	LDA	94	–	126
(j)	Py	C_6H_{13}	H	LDA	100	–	126

i, LDA (1.5 equiv.), THF, –78 °C or LiTMP (1.5 equiv.), THF, –50 °C or KDA (1 equiv.), THF, –78 °C; ii, MeI

Scheme 48

deprotonation of the m-CF_3 aryl substituted selenides requires[65] the use of an excess of LiTMP (1.5 equiv.) and a higher temperature (–50 °C instead of –78 °C). Methylation of the resulting organometallics, furthermore, leads to the corresponding vinyl selenides in modest yield and as a mixture of (*Z*)- and (*E*)-stereoisomers. These results differ substantially[65] from those obtained with KDA (see above).

Related results have been described[126] from pyridyl alkenyl selenides which, contrary to other aryl alkenyl selenides (see below),[65] are easily metallated at the vinyl site with LDA (Scheme 48, compare i and j with f and h). The ready deprotonation and stereoselective methylation of these selenides have been ascribed[126] to the presence of the nitrogen atom which can reduce the electron density of the double bond and also chelate with the lithium counterion.[65,103]

Phenyl vinyl selenoxide was expected to be considerably more acidic than the corresponding selenide and it was thought that it might be deprotonated more easily and with greater selectivity at the vinylic carbon. Deprotonation with LDA occurs[65] at –78 °C as well as at –90 °C (Scheme 49), but the decomposition of the resulting α-lithiovinyl selenoxide is so rapid ($t_{1/2}$ = 0.5 h at –78 °C in THF) that the methylation product is formed in less than 50% yield (Scheme 49).[65]

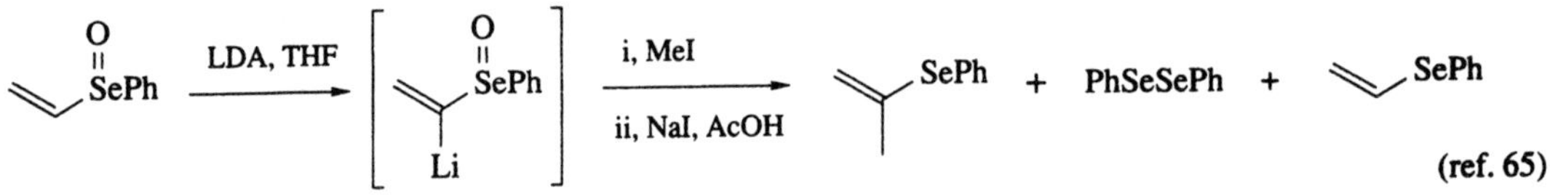

LDA, THF; i, MeI; ii, NaI, AcOH; SePh + PhSeSePh + SePh (ref. 65)

(a)	0.2 h, –78 °C	35%	44%	22%
(b)	0.5 h, –78 °C	19%	67%	14%

Scheme 49

2.6.2.2.3 *Synthesis of α-metalloalkyl selenoxides, selenones and selenonium salts*

Alkyl selenoxides,[97,98,127,128] selenones[129–131] and selenonium salts[132–137] are far more acidic than the corresponding selenides and have been successfully deprotonated by LDA,[98,127,128,132] KDA[130,131] or Bu^tOK[129–132,135,136] in THF or DMSO, KOH in DMSO[134,135] or $NaNH_2$ in liquid ammonia.[137]

Alkyl metals are not suitable for the metallation of these compounds. For example, Bu^nLi reacts at –78 °C on the selenium atom of selenonium salts and produces a novel organolithium in which the carbanionic center is more stabilized (Scheme 50, a and b), whereas alkylmagnesium bromides and chlorides unexpectedly lead[129] to decyl bromide and chloride respectively on reaction with decyl phenyl selenone (Scheme 50, c).

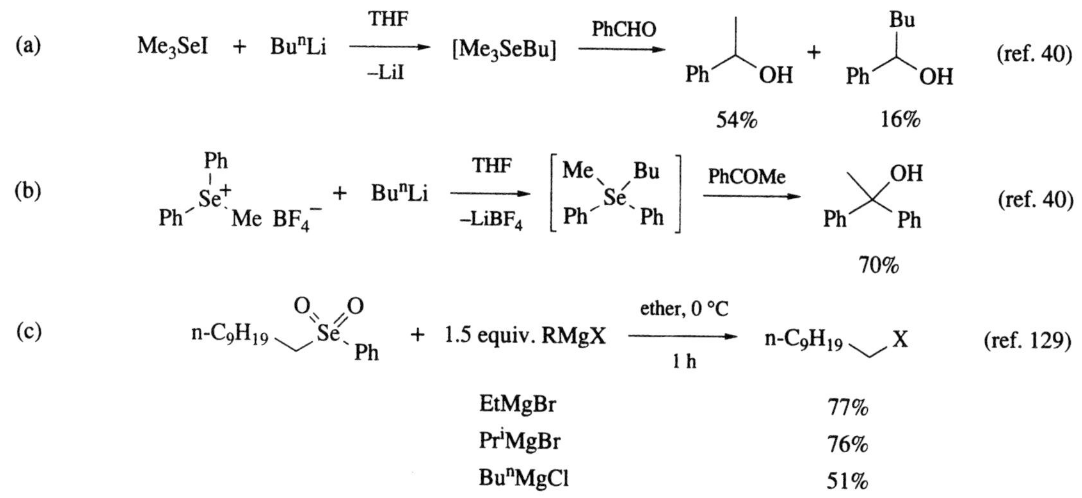

(a) Me_3SeI + Bu^nLi —THF, –LiI→ $[Me_3SeBu]$ —PhCHO→ Ph–CH(Me)–OH (54%) + Ph–CH(Bu)–OH (16%) (ref. 40)

(b) $Ph_2Se^+Me\ BF_4^-$ + Bu^nLi —THF, $-LiBF_4$→ [Me(Bu)Se(Ph)Ph] —PhCOMe→ $Ph_2C(Me)OH$ (70%) (ref. 40)

(c) $n\text{-}C_9H_{19}CH_2SeO_2Ph$ + 1.5 equiv. RMgX —ether, 0 °C, 1 h→ $n\text{-}C_9H_{19}CH_2X$ (ref. 129)

RMgX	Yield
EtMgBr	77%
Pr^iMgBr	76%
Bu^nMgCl	51%

Scheme 50

(i) Metallation of selenoxides

Due to the thermal instability and hygroscopicity of alkyl phenyl selenoxides their metallation is best achieved in the same pot in which they are prepared. Thus, differently substituted selenides have been oxidized[97,98] at low temperature (–78 °C) with ozone or a stoichiometric amount of *m*-chloroperoxybenzoic acid and the resulting mixture has been directly subjected[97,98] to the reaction of 2 equiv. of LDA, which neutralizes the benzoic acid present in the medium and at the same time achieves the desired metallation (Schemes 51–54). This reaction has been used for the synthesis of a large variety of organolithium reagents including those bearing two hydrogens, one alkyl or two alkyl groups on the carbanionic center,[97,98,127,128] as well as those bearing functionalized alkyl substituents there.[97,128]

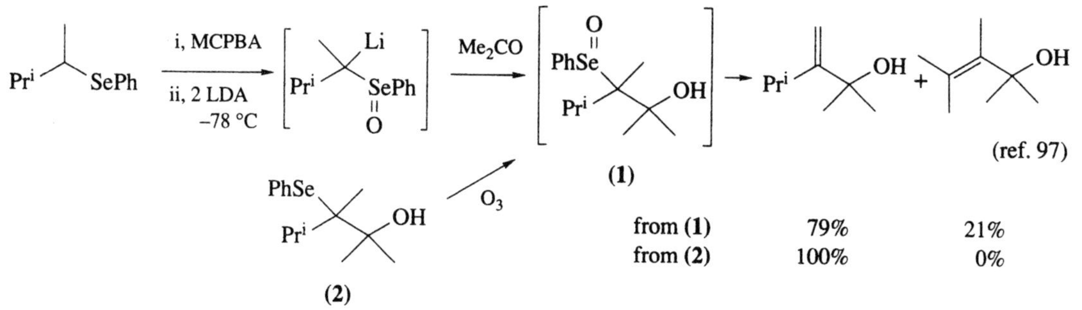

Pr^i–CH(Me)–SePh —i, MCPBA; ii, 2 LDA, –78 °C→ [Pr^iC(Me)(Li)Se(O)Ph] —Me_2CO→ [PhSe(O)C(Me)(Pr^i)C(Me)_2OH] **(1)** → allylic alcohol + allylic alcohol (ref. 97)

PhSeC(Me)(Pr^i)C(Me)_2OH **(2)** —O_3→ **(1)**

from **(1)**	79%	21%
from **(2)**	100%	0%

Scheme 51

PhSe–CH_2CH_2–R —i, MCPBA, THF, –78 °C; ii, 2 LDA, –78 °C→ [PhSe(O)CH(Li)CH_2R] —iii, cyclohexanone; iv, H_3O^+→ [1-(PhSe(O)CH(CH_2R))cyclohexanol] → 1-(RCH=CH)cyclohexanol, (*E*)-isomer (ref. 97)

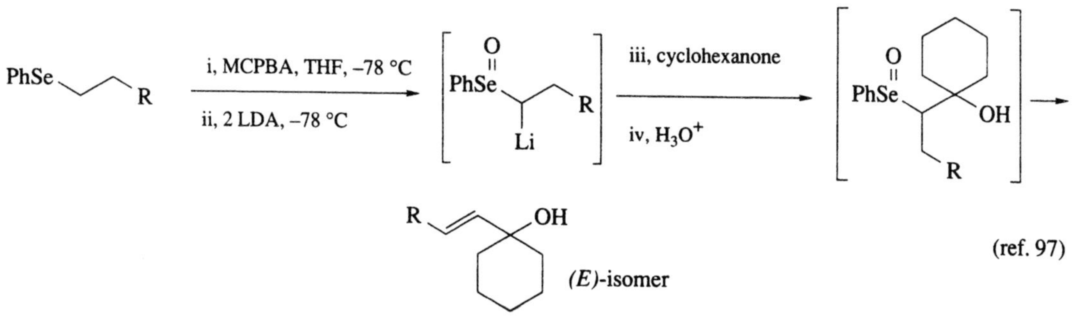

R = H (66%), R = Et (78%)

Scheme 52

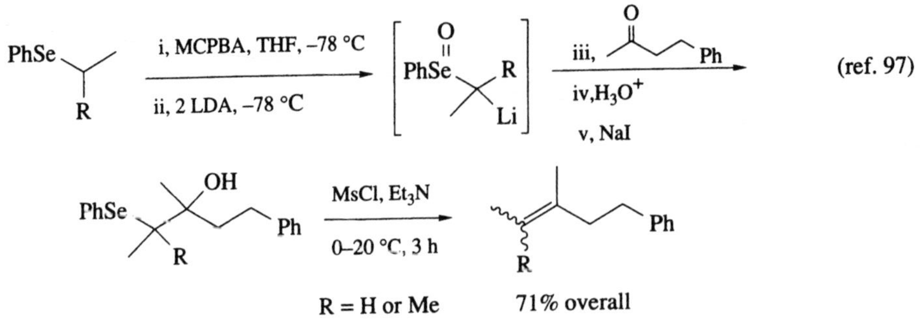

i, MCPBA, THF, –78 °C; ii, 2 LDA, –78 °C; iii, $PhCH_2CH_2COCH_3$; iv, H_3O^+; v, NaI (ref. 97)

MsCl, Et_3N, 0–20 °C, 3 h

R = H or Me — 71% overall

Scheme 53

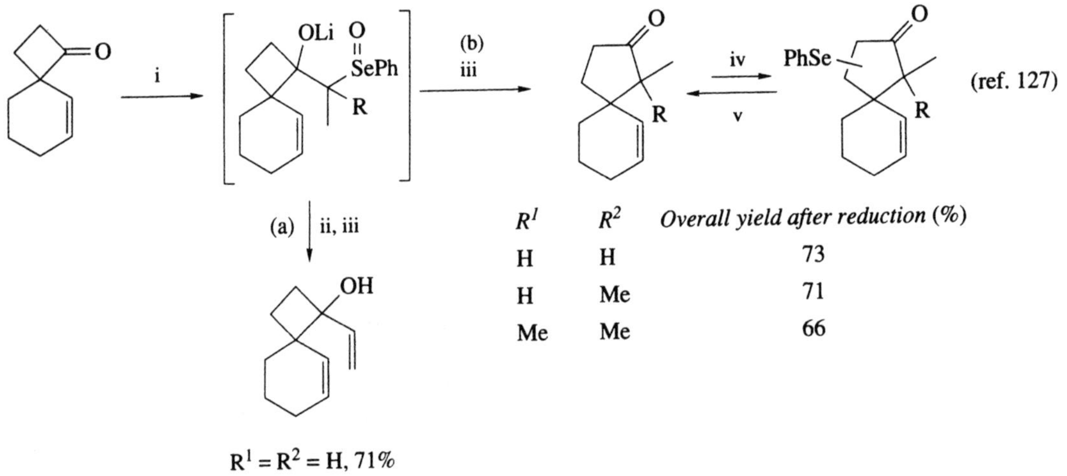

(ref. 127)

(a) ii, iii — $R^1 = R^2 = H$, 71%

(b) iii; iv; v

R^1	R^2	*Overall yield after reduction (%)*
H	H	73
H	Me	71
Me	Me	66

i, $PhSe(O)CR^1R^2Li$, THF, –78 °C; ii, AcOH, THF; iii, 55 °C, 0.1 h, –PhSeOH; iv, PhSeOH; v, Al–Hg

Scheme 54

Further reaction of these species with carbonyl compounds and hydrolysis of the resulting alkoxide leads to β-oxidoalkyl selenoxides which have been transformed[97,127] into allyl alcohols on thermal decomposition (Schemes 51, 52 and 54, entry a; see Section 2.6.4.4) or reduced to β-hydroxyalkyl selenides[97] or to alkenes (Scheme 53).[97] β-Oxidoalkyl selenoxides derived from cyclobutanones react in a different way since they rearrange[127,128] to cyclopentanones upon heating (Scheme 54, b, Schemes 120 and 121 and Section 2.6.4.5.3).

(ii) Metallation of selenones

Alkyl phenyl selenones can be metallated with LDA in THF but the reaction is slow and suffers from side reactions attributed to the substitution of the starting material with LDA or with the α-lithioalkyl selenones produced. KDA[131] and Bu^tOK[129–131,138] are much more efficient. The former base allows for the deprotonation of methyl,[131,138] *n*-alkyl[129] and cyclopropyl[130,131] phenyl selenones at low temperature (–78 °C), but α-potassioalkyl selenones have to be used immediately upon generation since they are particularly unstable. Otherwise, metallation of phenyl selenones is even more conveniently achieved with Bu^tOK in THF, and it even occurs in the presence of carbonyl compounds (Schemes 55 and 56)[129,130,138] or α,β-unsaturated esters (Schemes 57 and 58)[131] to provide epoxides and cyclopropyl esters, respectively, in reasonably good yields.

One of the major drawbacks of these reactions is clearly the unavailability of the whole set of alkyl phenyl selenones due to side reactions which occur during their preparations. Methyl, *n*-alkyl and some cyclopropyl phenyl selenones are available from the corresponding selenides and potassium permanganate,[130,139] peroxycarboxylic acids[130,139] or hydrogen peroxide/seleninic acids mixtures,[131] whereas *s*-alkyl phenyl selenones,[139] 2,2,3,3-tetramethylcyclopropyl selenone[130] and benzyl phenyl selenone[139] are not available in these ways.

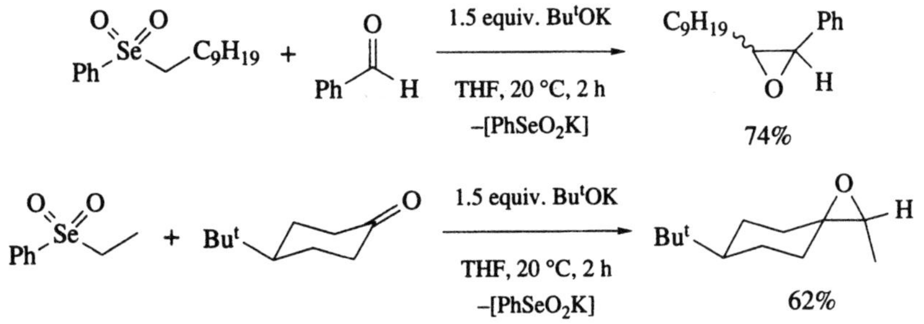

(ref. 129)

1.5 equiv. Bu^tOK; THF, 20 °C, 2 h; $-[PhSeO_2K]$; 74%

(ref. 138)

1.5 equiv. Bu^tOK; THF, 20 °C, 2 h; $-[PhSeO_2K]$; 62%

Scheme 55

aq. $KMnO_4$, CH_2Cl_2, 20 °C; R = H, 5 h, 82%; R = Me, 24 h, 70%

Bu^tOK, THF, 20 °C, 3 h; $R^1C(O)R^2$

(a) R = H, R^2 = Me: $R^1 = C_6H_{13}$, 92%; R^1 = Ph, 98%

(b) R = H: R^3 = H, 98%; $R^3 = Bu^t$, 94%

(c) R^2 = H

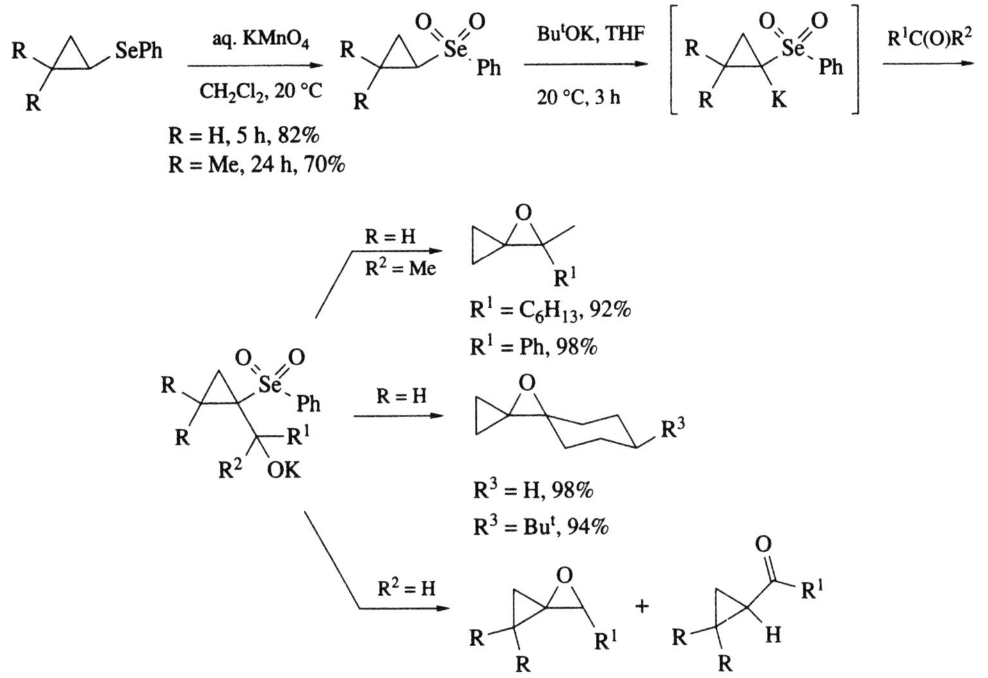

R	*R¹*	*Yield (%)*	*Ratio*		
H	Ph	70	50	:	50
H	C_6H_{13}	74	75	:	25
Me	Ph	45	30	:	70

(ref. 130)

Scheme 56

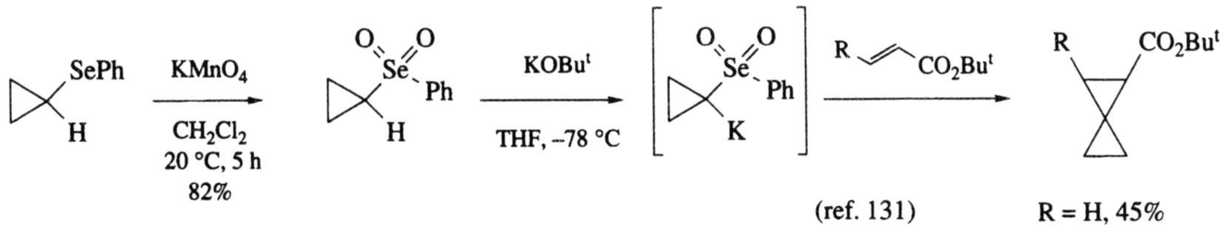

$KMnO_4$, CH_2Cl_2, 20 °C, 5 h, 82%; $KOBu^t$, THF, −78 °C; $R\text{-}CH{=}CH\text{-}CO_2Bu^t$

(ref. 131)

R = H, 45%; $R = C_9H_{19}$, 70%

Scheme 57

(iii) Metallation of selenonium salts

Methyl- and ethyl-selenuranes, prepared from trimethyl-, methyldiphenyl- and ethyldiphenyl-selenonium salts, respectively, and Bu^tOK in DMSO, react with aromatic aldehydes and ketones, including those that are α,β-unsaturated, to produce the corresponding epoxides in good yields (Scheme 59).[132] Metal-

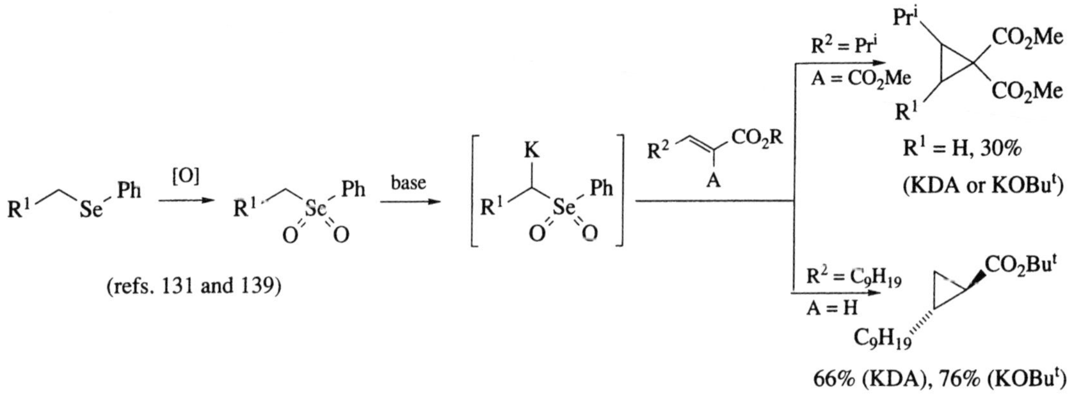

(refs. 131 and 139)

$R^2 = Pr^i$, $A = CO_2Me$: $R^1 = H$, 30% (KDA or $KOBu^t$)

$R^2 = C_9H_{19}$, $A = H$: 66% (KDA), 76% ($KOBu^t$)

Scheme 58

lation of trimethylselenonium iodide has been performed at –30 °C using dichloromethyllithium in DME (Scheme 60), but further reaction with acetophenone leads to oxidomethylstyrene in low yield.

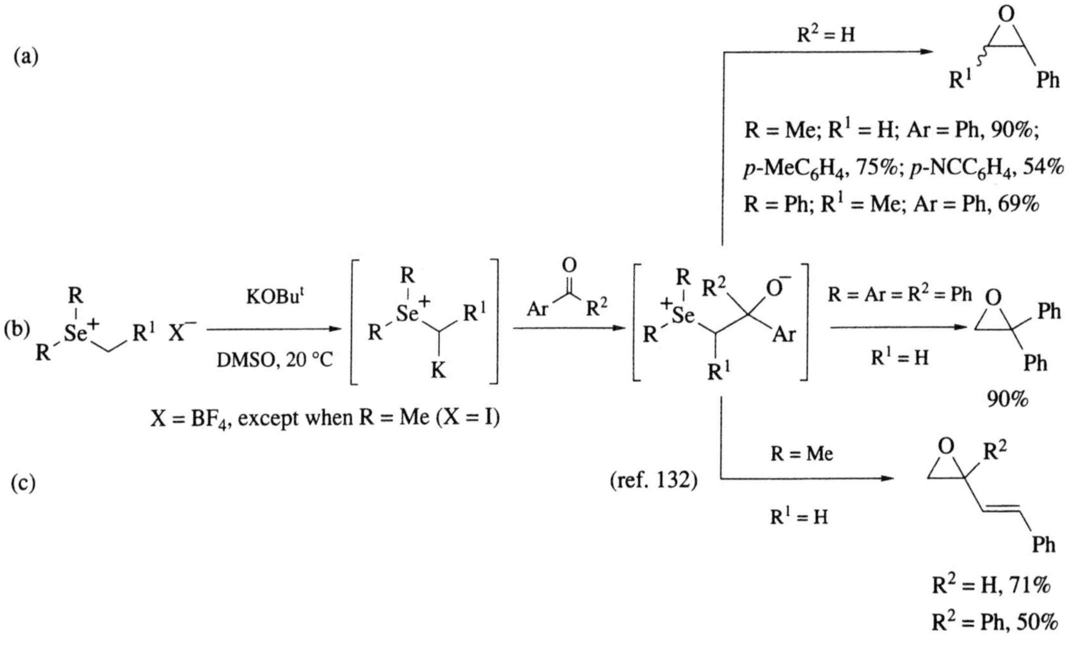

(a) $R^2 = H$

R = Me; R^1 = H; Ar = Ph, 90%; p-MeC_6H_4, 75%; p-NCC_6H_4, 54%
R = Ph; R^1 = Me; Ar = Ph, 69%

(b) $KOBu^t$, DMSO, 20 °C

$X = BF_4$, except when R = Me (X = I)

R = Ar = R^2 = Ph, R^1 = H: 90%

(c) (ref. 132) R = Me, R^1 = H

R^2 = H, 71%
R^2 = Ph, 50%

Scheme 59

$Me_3Se^+ I^-$ → Cl_2CHLi, DME, –30 °C → PhCOMe →

(ref. 132) 30%

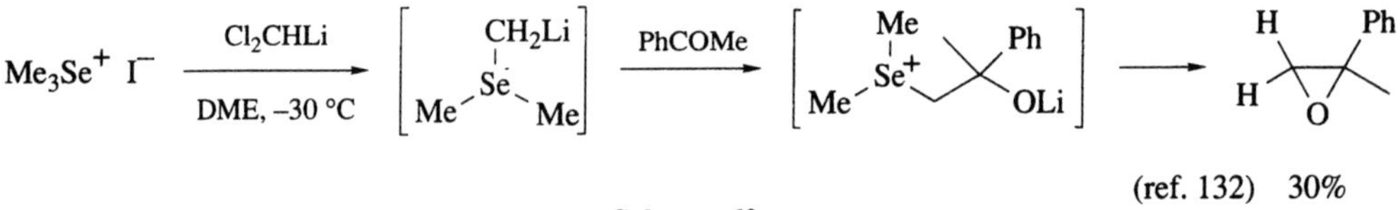

Scheme 60

Interestingly, the same reaction performed on decyldimethylselenonium fluoroborate and benzaldehyde exclusively leads[77] to styrene oxide and decyl methyl selenide (Scheme 61, a). If benzaldehyde is omitted, 1-decene is formed[133] in very good yield (Scheme 61, b). The above mentioned observations, as well as the fact that decyldiphenylselenonium fluoroborate does not produce 1-decene when reacted with potassium *t*-butoxide, suggests that a *syn* elimination formally related to the selenoxide[140] *syn* elimination reaction is taking place on the decylmethylmethyleneselenurane intermediate. Although the selenoxide elimination reaction offers the advantage of taking place in a neutral medium, the ylide elimination reaction offers the advantage of permitting the synthesis of alkylidenecyclopropanes[133] in very high yields from the corresponding selenides (Scheme 62, a),[141] and the synthesis of allylidenecyclopropane from 1-vinylcyclopropyl methyl selenide (Scheme 62, b).[7,135] These are unavailable *via* the selenoxide elimination reaction on related derivatives.[141]

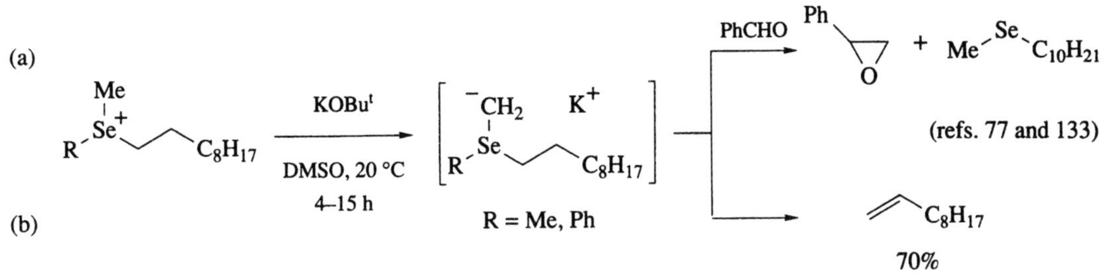

Scheme 61

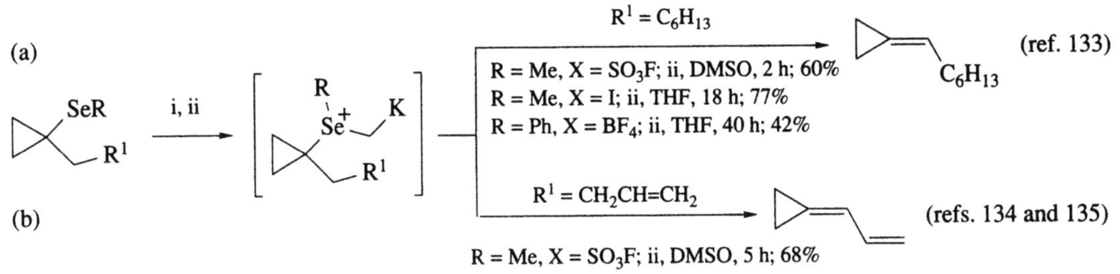

i, MeX, 20 °C; ii, KOBu^t, solvent, 20 °C

Scheme 62

The regiochemistry of this reaction is reminiscent of the related selenoxide elimination since it mainly leads to the less-substituted alkene (Scheme 63).[133] However, striking differences exist in the case of 1-octylcyclobutyl methyl selenide, where the reaction regioselectively produces octylidenecyclobutane *via* the ylide route (Scheme 64, a),[136] whereas a 1:1 mixture of octylidenecyclobutane and 1-octylcyclobutene is formed *via* the selenoxide route (Scheme 64, b).[136]

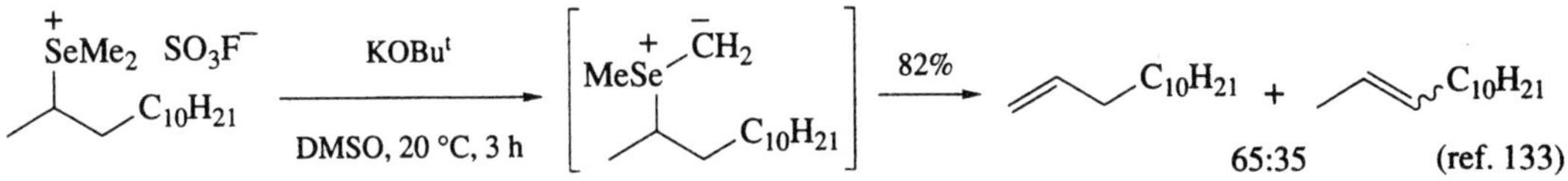

Scheme 63

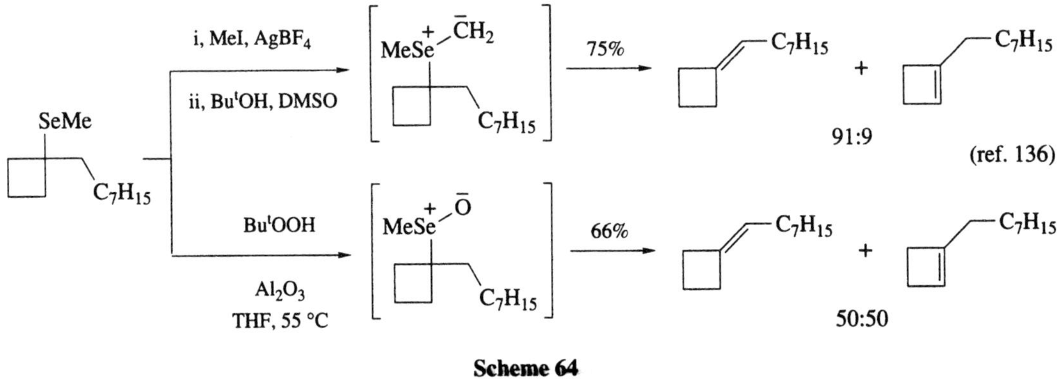

Scheme 64

Allyldimethyl- and allylmethylphenyl-selenonium salts react under similar conditions with potassium *t*-butoxide (THF, 20 °C, 20 h) and lead to homoallyl selenides in very high yields, including the particularly strained alkylidenecyclopropane derivatives (Scheme 65, b).[141] These result from the metallation of the methyl group of the selenonium salts, leading to the corresponding ylides which then suffer a [2,3] sigmatropic rearrangement.

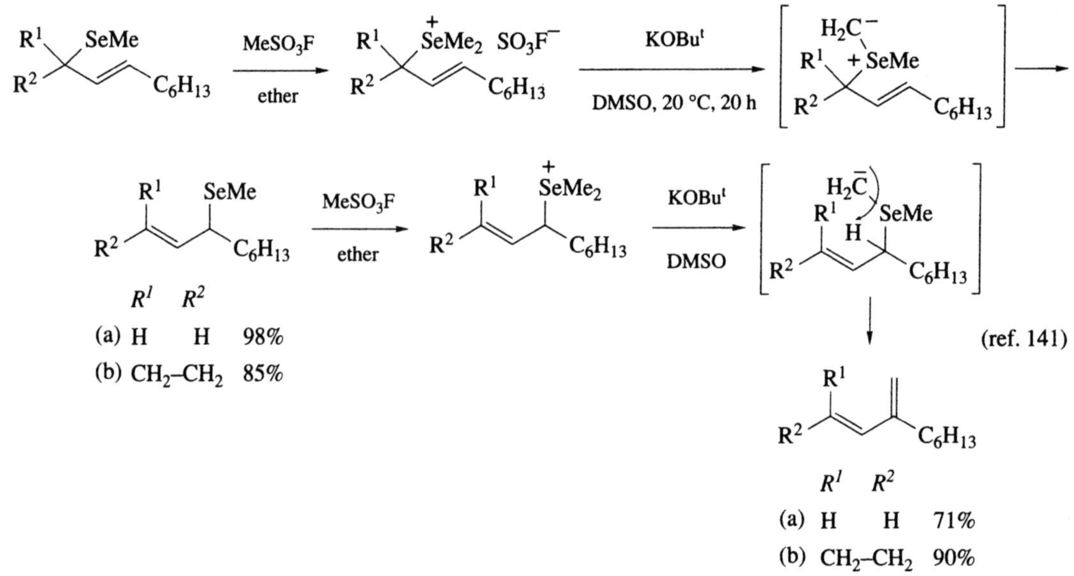

(ref. 141)

Scheme 65

Homoallyl selenides are valuable precursors of dienes and allylidenecyclopropanes (Scheme 65).[141] The dienes have been produced using the same sequence of reactions outlined above (Schemes 63 and 64), which involves the alkylation of the homoallyl selenide followed by treatment of the resulting salt with Bu^tOK in DMSO (Scheme 65).

Treatment of benzylmethylphenylselenonium tetrafluoroborate with potassium *t*-butoxide gave[137] a 9:1 mixture of methyl phenyl and benzyl phenyl selenides but no products which would have resulted had the appropriate ylide been formed (Scheme 66, a). This demonstrated that at least this salt is extremely susceptible to nucleophilic attack, even by nucleophiles as hindered as the *t*-butoxide anion (Scheme 66, a).[137]

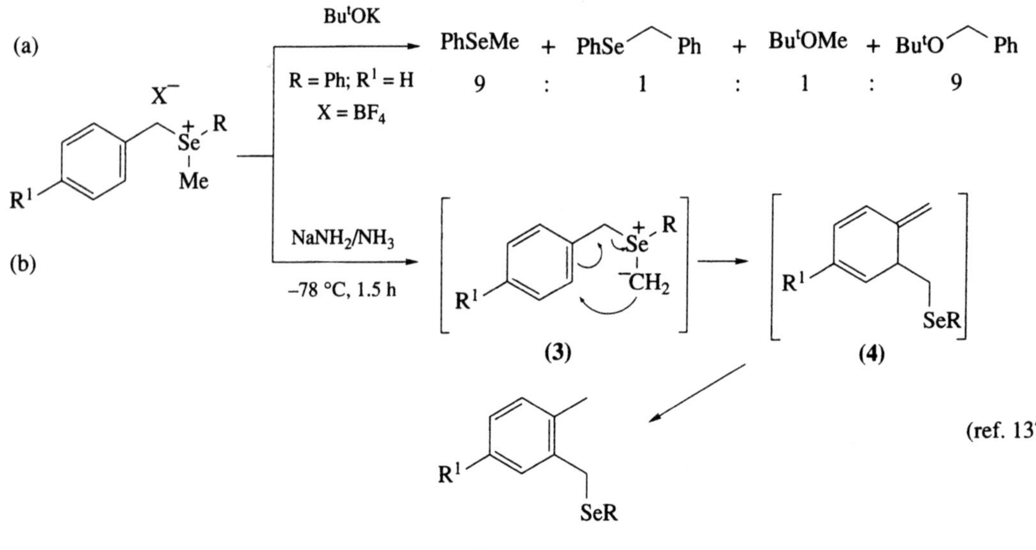

(ref. 137)

R = Ph; X = BF_4; R^1 = H, 17%; Me, 38%; Cl, 37%

R = Me; X = SO_3F; R^1 = H, 43%; Cl, 50%

Scheme 66

Ylide formation has been accomplished, however, through the use of sodium amide in liquid ammonia.[137] This reaction has been extended to analogous salts (Scheme 66, b and Scheme 67). *o*-Methylbenzyl phenyl and methyl selenides are produced in modest to good yields, respectively, by metallation on the methyl group of the corresponding selenonium salts followed by spontaneous [2,3] sigmatropic rearrangement and rearomatization of the methylidenecyclohexadiene intermediate (**4**; Scheme 66).[137] Stilbenes, probably resulting from metallation at the benzylic site, are also produced in this reaction (Scheme 67, for example).[137]

Of special interest were the results obtained from the *m*-chlorobenzyldimethylselenonium fluoroborate since the chlorine appears to direct the attack of the ylide predominantly to the more-hindered position *ortho* to itself (Scheme 67).

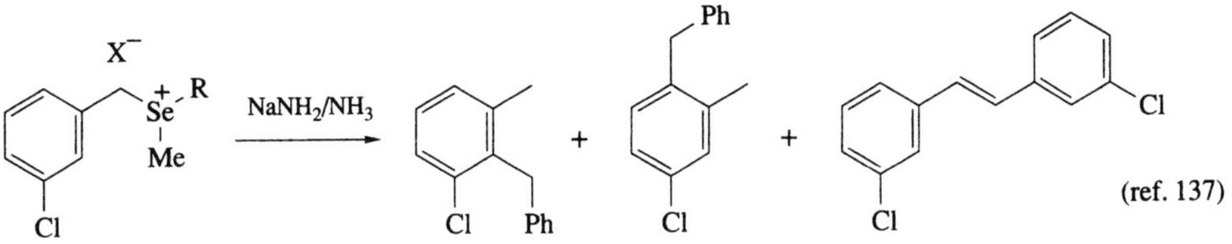

X⁻
Se
R
Me
Cl
NaNH2/NH3
Cl
Ph
Ph
Cl
Cl
Cl
(ref. 137)

R = Ph	23%	–	28%
R = Me	56%	14%	–

Scheme 67

Phenacylmethyl(dimethyl)selenonium bromide is far more acidic than the selenonium salts presented above since deprotonation can be achieved with aqueous potassium hydroxide (Scheme 68).[142] The resulting ylide is stable at room temperature and decomposes to 1,2,3-triphenacylcyclopropane on irradiation or heating (Scheme 68, a).[142] It reacts with benzalacetophenone in a different manner from methylene(dimethyl)selenurane (Scheme 59, c)[132] and provides, rather than the oxirane, 1-phenyl-2,3-phenacylcyclopropane in very high yield (Scheme 68, b).[142]

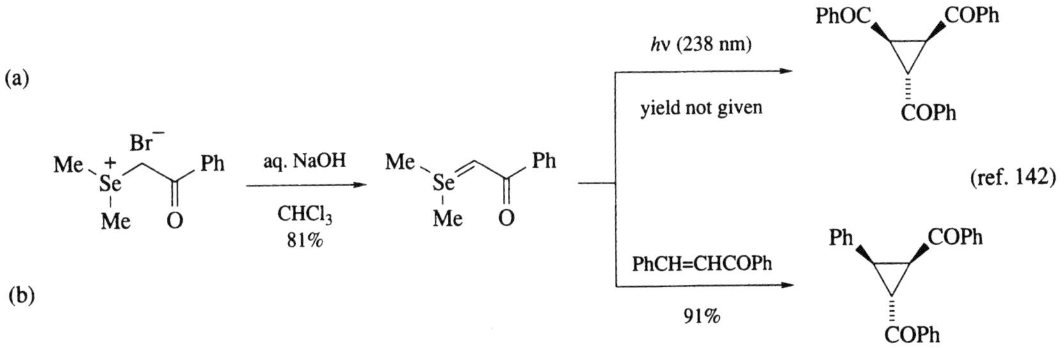

(a)
(b)
Br⁻
Me
Se
Ph
Me
O
aq. NaOH
CHCl3
81%
Me
Se
Ph
Me
O
hν (238 nm)
yield not given
PhCH=CHCOPh
91%
PhOC
COPh
COPh
Ph
COPh
COPh
(ref. 142)

Scheme 68

2.6.2.3 Synthesis of Organometallics Bearing an α-Seleno Carbanion by Selenium–Metal Exchange

2.6.2.3.1 Synthesis of α-selenoalkyllithiums by selenium–lithium exchange

(i) Generalities

α-Selenoalkyllithiums with carbanionic centers bearing hydrogens or alkyl groups and which cannot be synthesized by metallation of selenides (see Section 2.6.2.2.1) have been conveniently prepared from selenoacetals and butyllithiums. This approach relies not only on the great availability of selenoacetals, which are accessible from a wide range of commercially available starting materials, but also on the efficiency of the Se–Li exchange. It finally allows wider structural variations on the α-selenoalkyllithium than those possible for related α-heterosubstituted organometallics[7] usually prepared by metallation.

Although the work has been routinely carried out with both phenyl and methyl selenoacetals with equal success, the compounds belonging to the last series are, in some specific cases, much more accessible and the resulting organometallic more reactive towards electrophiles in general and towards carbonyl compounds in particular.[7,9,12,16,17]

(ii) Synthesis of selenoacetals

Selenoacetals are readily available compounds which have been prepared: (i) by selenoacetalization[53] of carbonyl compounds with selenols[36,40,49,55,143–145] in acidic media, with tris(seleno)boranes[53,144,146–148] in neutral or acidic media or with trimethylsilyl selenides and aluminum trichloride (Schemes 69 and 70);[53,144] (ii) by trans-selenoacetalization of *O,O*-acetals with selenols[77] or triselenoboranes (Scheme 71);[149] (iii) by substitution of dihaloalkanes with selenolates (Scheme 72);[36,55,56] (iv) by insertion of carbenes into diselenides (Scheme 73);[53,121,150–153] and (v) by addition of selenols to vinyl selenides (Scheme 74),[154] by selenenylation of α-selenoalkyl anions (Scheme 75)[155,156] and by alkylation of 1,1-bis(seleno)alkyl metals (Schemes 26, 76 and 77).[12,35,36,71,73,100,104]

The direct selenoacetalization of carbonyl compounds by selenols is by far the shortest and most convenient route to selenoacetals.[53] The reaction is usually carried out at 20 °C with zinc chloride (0.5 equiv. *versus* the carbonyl compound) and delivers rapidly (<3 h) and in reasonably good yields methyl and phenyl selenoacetals derived from aliphatic aldehydes and ketones and cyclic ketones (Scheme 69). Selenoacetalization is more difficult to achieve with hindered ketones, such as adamantanone and diisopropyl ketone, and with hindered aromatic carbonyl compounds.[53] In these cases the reaction is best achieved[53] with titanium tetrachloride instead of zinc chloride and is often limited to the methylseleno derivatives (Scheme 78). Tris(methylseleno)borane offers the advantage of not requiring an acid catalyst and is particularly useful for the selenoacetalization of acid labile aldehydes such as citronellal (Scheme 70, e).

Transacetalization in our opinion does not offer a definite advantage over the direct method except for those cases, such as that of cyclopropanone,[77] where the *O,O*-acetal is more accessible than the parent carbonyl compound (Scheme 71). The reaction, however, does not seem to be general (Scheme 71, b).

(a) $R^1C(O)CH_3 \xrightarrow{i} R^1C(SeR)_2CH_3$

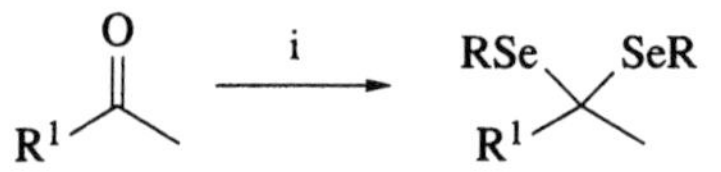

R^1	H	H	Me	Me	n-C_6H_{13}	n-C_6H_{13}	Bu^t	Bu^t
R	Ph	Me	Ph	Me	Ph	Me	Ph	Me
Yield (%)	90	91	85	90	80	77	32	81

(b) cycloalkanone $\xrightarrow{i}$ 1,1-bis(seleno)cycloalkane

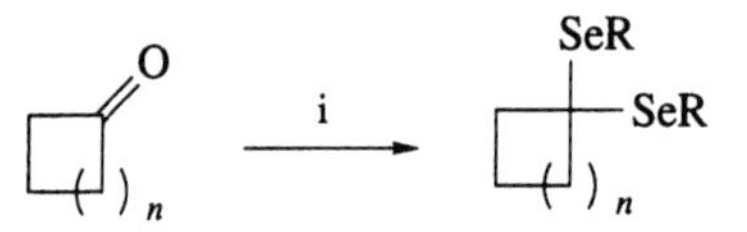

R	Ph	Ph	Ph	Me	Me	Me
n	1	2	3	1	2	3
Yield (%)	84	81	81	89	87	91

(c) adamantanone $\xrightarrow{i}$ 2,2-bis(seleno)adamantane

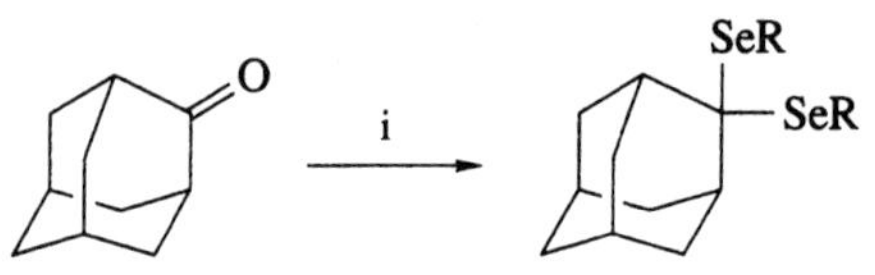

R = Ph, 27%
R = Me, 88%

i, RSeH, $ZnCl_2$, 20 °C, 3 h (ref. 53)

Scheme 69

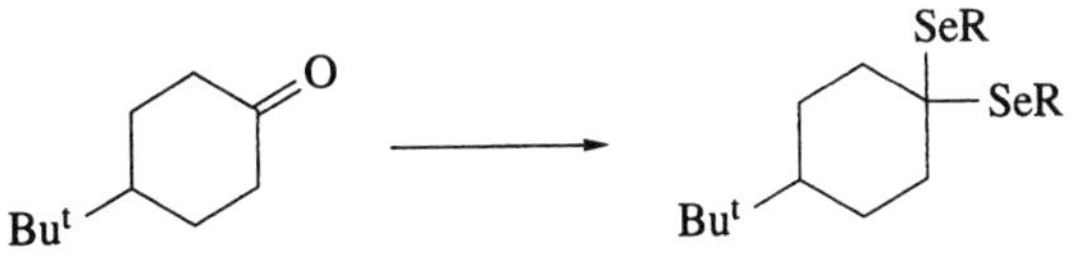

	Conditions	*R*	*Yield (%)*	*Ref.*
(a)	RSeH, $ZnCl_2$, 20 °C, 6 h	Ph	70	53
(a)	RSeH, $ZnCl_2$, 20 °C, 6 h	Me	85	53
(b)	$B(SeMe)_3$, $CHCl_3$, 20 °C, 31 h	Me	84	53
(c)	Me_3SiSeR, $AlCl_3$, 3 h	Ph	82	53
(c)	Me_3SiSeR, $AlCl_3$, 3 h	Me	90	53

	Conditions	*Yield (%)*	*Ref.*
(d)	MeSeH, $ZnCl_2$, 20 °C	0	144
(e)	$B(SeMe)_3$, $CHCl_3$, 20 °C, 5 h	60	144
(f)	$Me_3SiSeMe$, $AlCl_3$, $CHCl_3$, 20 °C	8	144

Scheme 70

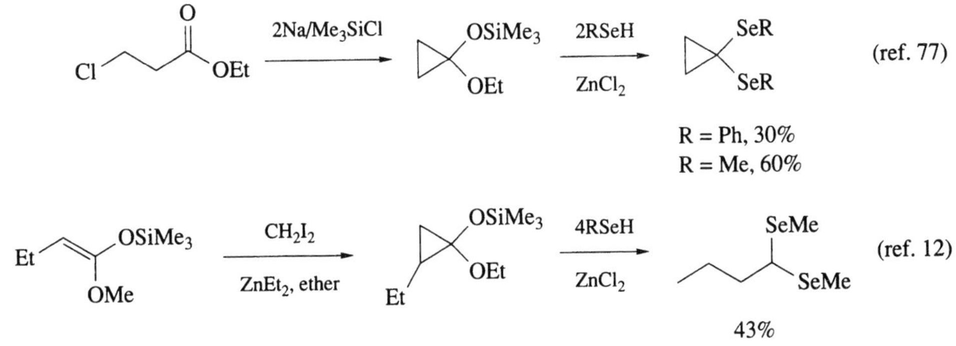

Cl–CH₂CH₂–C(O)OEt → (2Na/Me_3SiCl) → 1-OSiMe₃-1-OEt-cyclopropane → (2RSeH, $ZnCl_2$) → 1,1-bis(SeR)cyclopropane (ref. 77)

R = Ph, 30%
R = Me, 60%

Et–CH=C(OSiMe₃)(OMe) → (CH_2I_2, $ZnEt_2$, ether) → cyclopropane (OSiMe₃, OEt, Et) → (4RSeH, $ZnCl_2$) → 1,1-bis(SeMe)butane, 43% (ref. 12)

Scheme 71

(a) X–CH₂–X + RSeNa → RSe–CH₂–SeR (refs. 35, 36, 55)
X = I; R = Me, Ph

(b) 1,1-dibromo-2-ethylcyclopropane → (MeSeK, DMF) → 1,1-bis(SeMe)-2-ethylcyclopropane (55%) + 1-SeMe-2-ethylcyclopropane (5%) + SM (13%) (ref. 12)

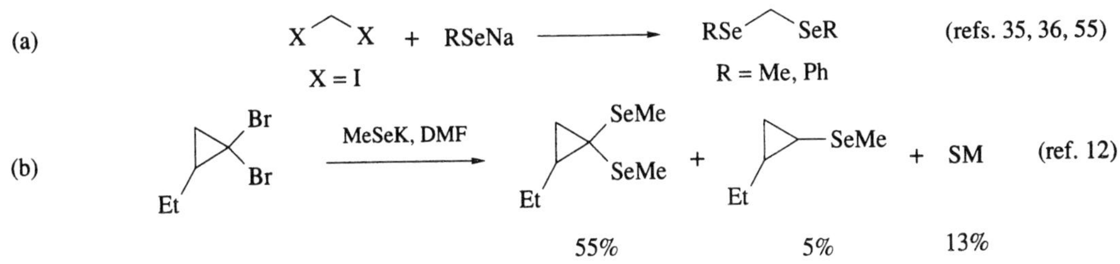

Scheme 72

The reaction of 1-metallo-1,1-bis(seleno)alkanes with electrophiles is another interesting approach to selenoacetals (Schemes 76 and 77),[36,71,73] which *inter alia* allows the synthesis of 1,1-bis(seleno)cyclopropanes (Scheme 26, a, b and d; Scheme 77, c) and also permits the synthesis of functionalized selenoacetals by concomitant C—C bond formation (Scheme 79).[157] Other methods listed above are of limited use. The substitution of dihaloalkanes by selenolates is the method of choice for the synthesis of bis(seleno)methanes (Scheme 72, a),[36,55,56] and can be applied to the synthesis of the relatively unhindered selenoacetal of cyclopropanone (Scheme 72, b).

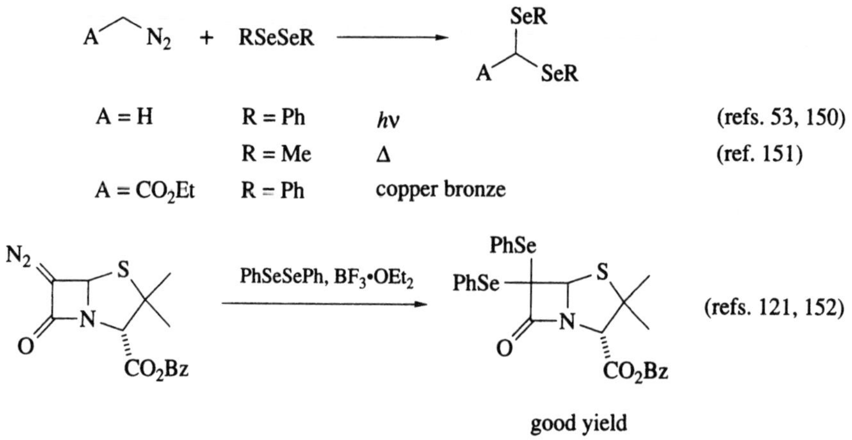

$A\!\frown N_2$ + RSeSeR ⟶ $A\!\frown CH(SeR)_2$

A	R		
A = H	R = Ph	$h\nu$	(refs. 53, 150)
	R = Me	Δ	(ref. 151)
A = CO_2Et	R = Ph	copper bronze	

PhSeSePh, BF_3•OEt_2 (refs. 121, 152)

good yield

Scheme 73

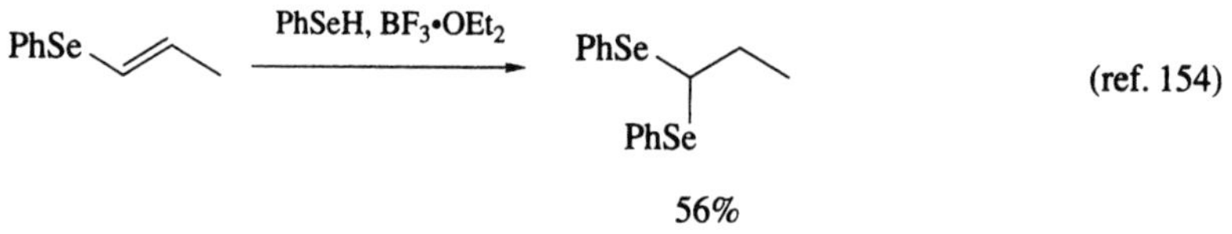

PhSeH, BF_3•OEt_2 (ref. 154)

56%

Scheme 74

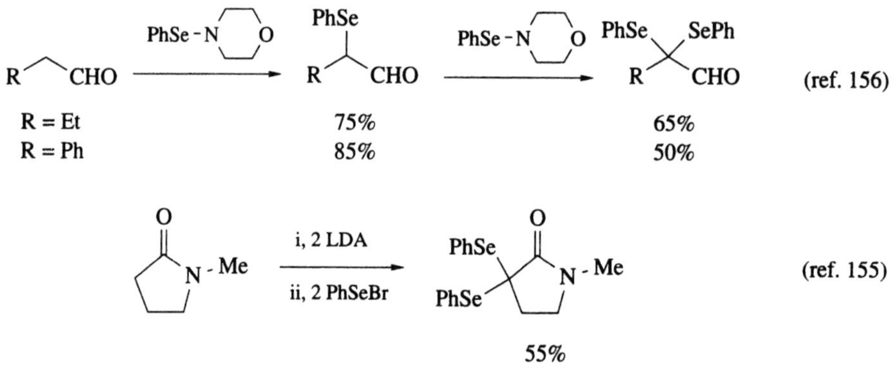

(ref. 156)

R = Et	75%	65%
R = Ph	85%	50%

i, 2 LDA; ii, 2 PhSeBr (ref. 155)

55%

Scheme 75

base; R^2X

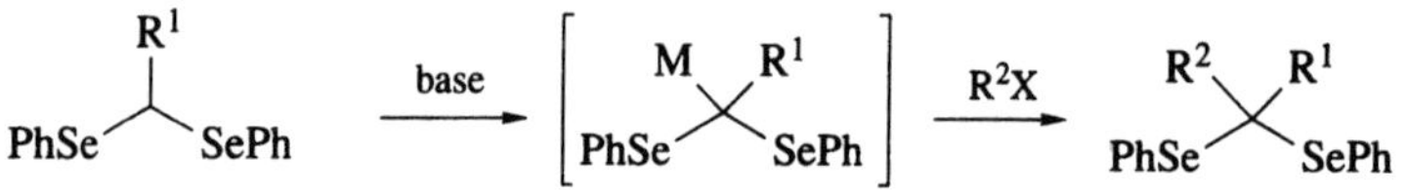

R^1	Conditions	R^2X	Yield (%)	Ref.
H	$LiNBu_2$, THF, –78 °C	MeI	91	35, 36
C_6H_{13}	LiTMP, THF–HMPA, –30 °C	$C_6H_{13}Br$	86	71
$C_{10}H_{21}$	KDA, THF, –78 °C	$PhCH_2Br$	98	73

Scheme 76

(iii) Synthesis of α-selenoalkyl metals by selenium–metal exchange

Phenyl and methyl selenoacetals usually react with butyllithiums and produce the corresponding α-selenoalkyllithiums and butyl selenides.[4,7,9,12,13,16,17,35,36,48–50,54,55,60,158] The reaction is usually carried out at –78 °C (a temperature at which these organometallics are stable for a long period), with *n*-butyllithium in THF or *s*-butyllithium in ether. It provides a large variety of α-selenoalkyllithiums including

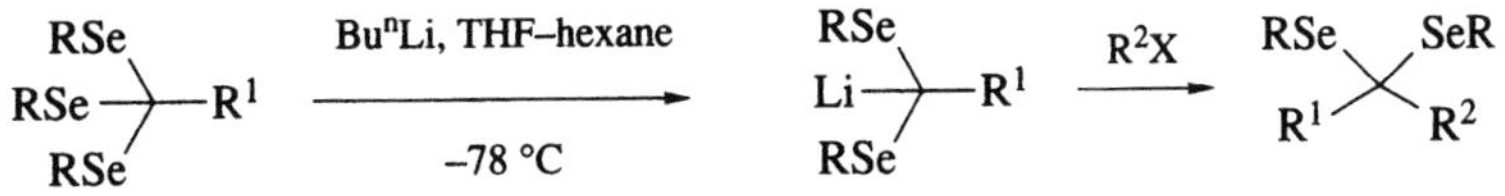

RSe / RSe / RSe —R[1] → (Bu^nLi, THF–hexane, –78 °C) → RSe / Li / RSe —R[1] → (R^2X) → RSe, SeR, R^1, R^2

	R	*R[1]*	*R^2X*	*Yield (%)*	*Ref.*
(a)	Ph	Me	MeI	72	35, 36
(b)	Me	C_6H_{13}	BuBr	80	100

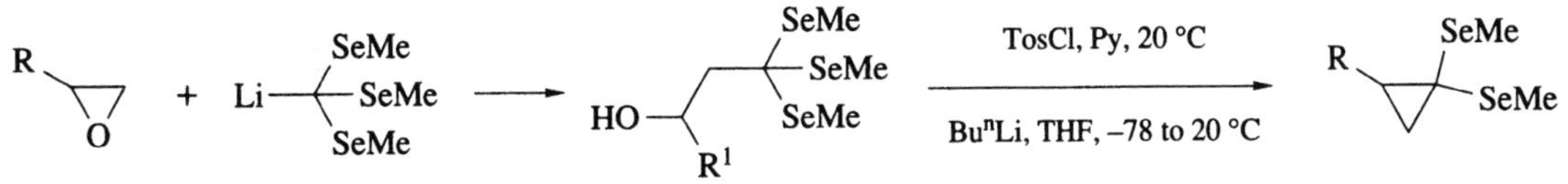

R-epoxide + Li–C(SeMe)$_3$ → HO–CH(R^1)–CH_2–C(SeMe)$_3$ → (TosCl, Py, 20 °C; Bu^nLi, THF, –78 to 20 °C) → R-cyclopropane(SeMe)$_2$

(c)

R[1]	*Yield (%)*	*Yield (%)*	*Ref.*
H	71	60	104
Me	60	60	104
C_6H_{13}	61	47	104

Scheme 77

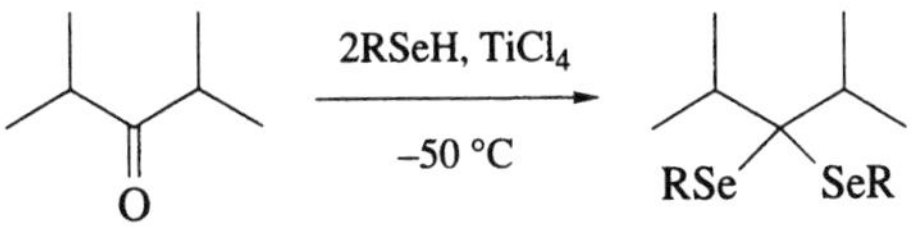

Diisopropyl ketone → (2RSeH, $TiCl_4$, –50 °C) → RSe, SeR (ref. 53)

R = Ph, 0%
R = Me, 69%

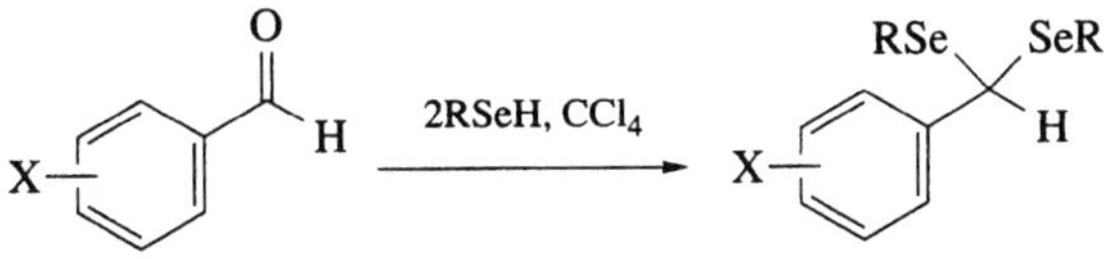

X–C_6H_4–CHO → (2RSeH, CCl_4) → X–C_6H_4–CH(SeR)(SeR) (ref. 53)

R = Ph 0.35 equiv. $TiCl_4$, –50 to 20 °C X = H, 64%; 2-Cl, 55%
R = Me 0.5 equiv. $ZnCl_2$, 20 °C, 2 h X = H, 92%; 4-EtO, 93%; 4-Cl, 78%

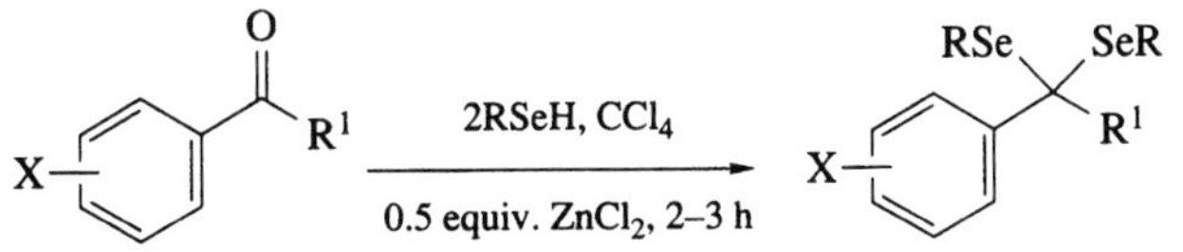

X–C_6H_4–CO–R^1 → (2RSeH, CCl_4, 0.5 equiv. $ZnCl_2$, 2–3 h) → X–C_6H_4–C(SeR)(SeR)R^1

R = Ph; R^1 = Me X = H, 55%; 4-Me, 39%; 4-MeO, 10% (ref. 53)
R = Me; R^1 = Me X = H, 99%; 4-MeO, 78%; 4-Cl, 35% (ref. 72)

Scheme 78

those where the carbanionic center is unsubstituted,[35,36,54,55,160] monoalkyl[40,48–50,54,55] or dialkyl substituted,[40,48–50,54,55,104,135,136,159–163] or that bear an aryl[54,164] or a heteroaryl[158] moiety (Scheme 80).[54]

In general, the Se–Li exchange is easier: (i) when carried out in THF rather than in ether; (ii) when *s*- or *t*-butyllithium is used instead of *n*-butyllithium. Methyllithium is almost unreactive in most cases; and (iii) if a better stabilization of the carbanionic center can be achieved. Thus phenyl selenoacetals react more rapidly than their methyl analogs (see Section 2.6.2.1.2), and α-selenoalkyllithiums where the carbanionic center is part of a three-membered cycle or is substituted by an aryl group are more readily obtained than those which are monoalkyl or dialkyl substituted.

The presence of bulky groups around the reactive site does not have a marked effect[54] on the reactivity of selenoacetals derived from aldehydes but it dramatically lowers[54] the reaction rates of those derived from ketones such as adamantanone or diisopropyl ketone. In these cases, the reaction rate is greatly enhanced when *s*- or *t*-butyllithium is used in place of *n*-butyllithium.[54]

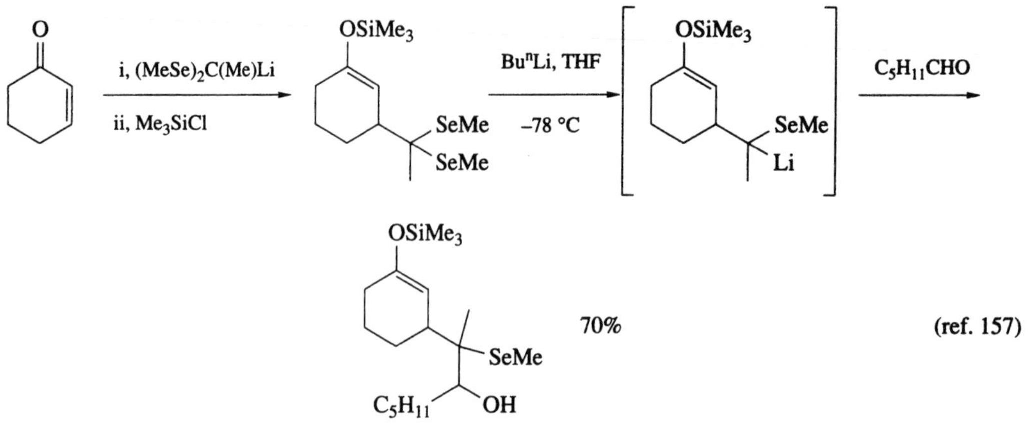

O — i, $(MeSe)_2C(Me)Li$; ii, Me_3SiCl → $OSiMe_3$, SeMe, SeMe — Bu^nLi, THF, –78 °C → [$OSiMe_3$, SeMe, Li] — $C_5H_{11}CHO$ → $OSiMe_3$, SeMe, C_5H_{11}, OH 70% (ref. 157)

Scheme 79

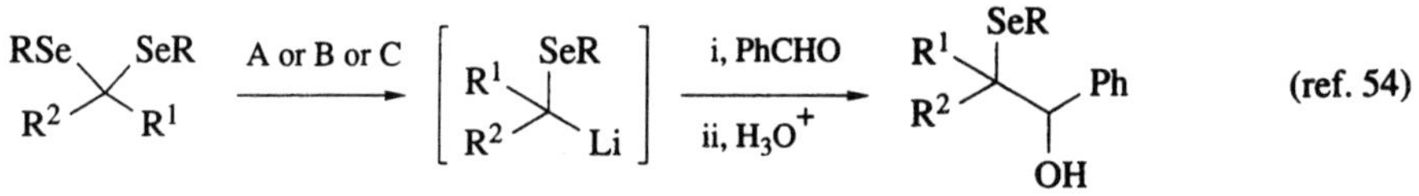

RSe, SeR, R^2, R^1 — A or B or C → [R^1, SeR, R^2, Li] — i, PhCHO; ii, H_3O^+ → R^1, SeR, R^2, Ph, OH (ref. 54)

Intermediate	*R*	*Yield (%)*	*Method*	*Intermediate*	*R*	*Yield (%)*	*Method*
$RSeCH_2Li$	Me	80	A	$RSeC(Me)(n\text{-}C_6H_{13})Li$	Me	90	A, 2 h
	Me	82	B		Ph	78	A
	Ph	82	A	$RSeC(Me)(Bu^t)Li$	Me	32	A
$RSeCH(Me)Li$	Me	80	A		Me	77	A, 2 h
	Ph	70	A		Ph	59	A
$RSeCH(n\text{-}C_6H_{13})Li$	Me	84	A	$RSeC(n\text{-}C_6H_{13})_2Li$	Me	17	A
	Me	82	B		Me	93	A, 4 h
	Ph	89	A		Ph	47	A
	Ph	74	B	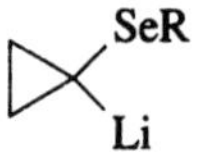 SeR, Li	Me	91	A
$RSeCH(Bu^t)Li$	Me	82	A		Me	68	B
	Ph	92	A		Ph	80	A
$RSeCMe_2Li$	Me	75	A	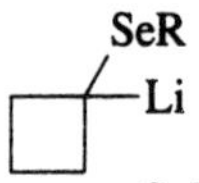 SeR, Li	Me	87	A
	Me	80	B		Ph	90	A
	Ph	94	A	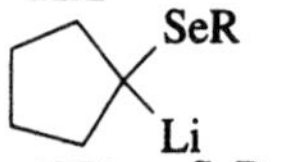 SeR, Li	Me	82	A
	Ph	83	B		Ph	91	A
$RSeCPr^i_2Li$	Me	77	A, 2 h	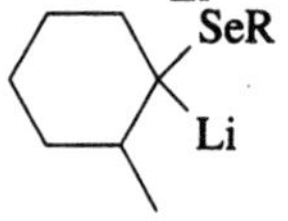 SeR, Li	Me	25	A
 SeR, Li	Me	87	A, 2 h		Me	78	B
	Ph	88	A		Me	87	C
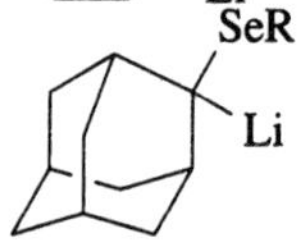 SeR, Li	Me	0	A		Ph	55	A, 4 h
	Me	81	C		Ph	79	B

A = Bu^nLi, THF–hexane, –78 °C, 0.9 h; B = Bu^sLi, ether–hexane, –78 °C, 0.5 h; C = Bu^sLi, THF–hexane, –78 °C, 0.5 h

Scheme 80

Metallation is never a competing reaction even for those selenoacetals derived from aromatic aldehydes (Scheme 81).[54] Only in the case of phenyl selenoacetals derived from formaldehyde and acetaldehyde are the corresponding 1-metallo-1,1-bis(seleno)alkanes produced (Scheme 82).[54] In fact these compounds arise from the metallation of the selenoacetals not by the butyllithium but by the α-selenoalkyllithium just produced.[54] This side reaction, which does not occur with methylseleno analogs and with higher homologs in the phenylseleno series, can be prevented by slow reverse addition of the reactants (Scheme 82).[54]

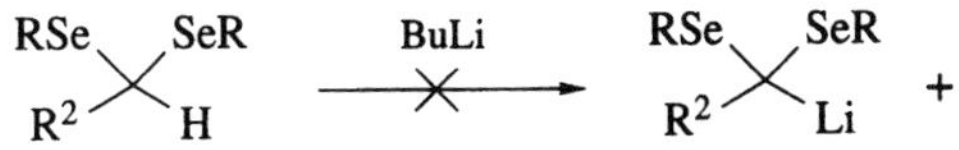

(ref. 54)

Scheme 81

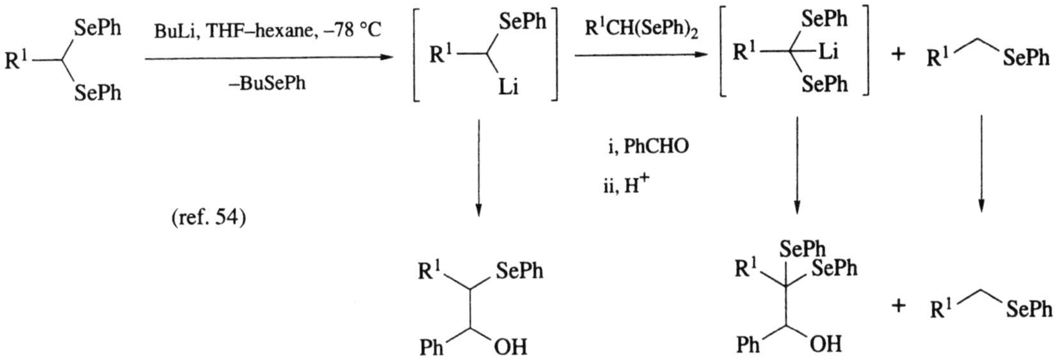

(ref. 54)

	Reaction	R^1		Yield (%)	
(a)	'Normal addition'	H	61	23	16
(b)	'Normal addition'	Me	70	8	8
(c)	'Reverse addition'	H	82	6	5

Scheme 82

Except for the case shown in Scheme 83,[165,166] α-selenoalkyllithiums are stable intermediates at or below −78 °C.[7,16,54] They do not have a high tendency to decompose to carbenes (Scheme 84, a), nor are the phenyl selenoacetals transformed to their ring-metallated isomers (Scheme 84, b) or to the more stable α-alkylselenomethyllithiums (methylseleno derivatives) (Scheme 84, c).[54] Interestingly, they are not alkylated by the concomitantly produced butyl selenide (Scheme 85, a).[54]

Reaction of *n*-butyllithium with 1,1-bis(seleno)-4-*t*-butylcyclohexane,[54,167a–c] 1,1-bis(methylseleno)-2-methylcyclohexane[54] or 1,1-bis(phenylseleno)-3-silyloxybutane[167d] leads to, in each case, one of the stereoisomers of the α-selenoalkyllithium, as observed by ^{77}Se NMR and trapping experiments (Scheme 86). In the former case, the reaction exclusively produces the axially oriented lithio derivative and in the latter case it leads almost exclusively to the stereoisomer shown in Scheme 86. It has been secured in the case of 1,1-bis(seleno)-4-*t*-butylcyclohexanes, that the C—Se bond cleavage is operating stereoselectively on the axial seleno group.

Only a few functionalized α-selenoalkyllithiums have been synthesized apart from the ones shown in Schemes 11,[60] 79[157] and 86.[167]

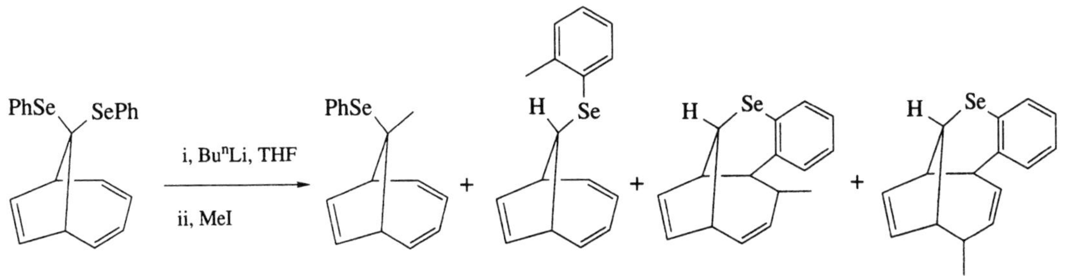

(ref. 165)

	Conditions	Yield (%)			Ratio	
(a)	−78 °C, 0.1 h	96	15	85	–	–
(b)	20 °C, 1 h	84	–	–	73	27

Scheme 83

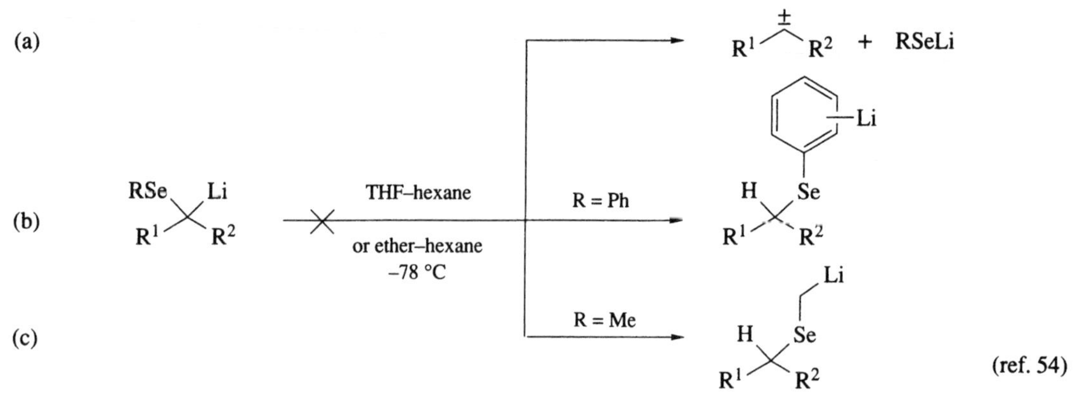

(a) R^1 R^2 (±) + RSeLi

(b) RSe Li / R^1 R^2 — THF–hexane or ether–hexane, –78 °C — R = Ph → H Se (C₆H₄Li) / R^1 R^2

(c) R = Me → H Se(CH_2Li) / R^1 R^2 (ref. 54)

Scheme 84

RSe SeR / R^1 R^2 —BuLi→ RSe Li / R^1 R^2 + BuSeR ↛ RSe Bu / R^1 R^2 (ref. 54)

R^1, R^2 = H, alkyl, allyl, aryl

Scheme 85

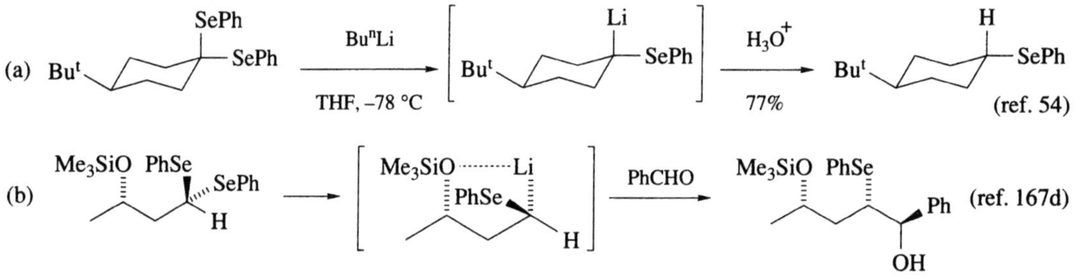

(a) Bu^t (cyclohexane) SePh SePh —Bu^nLi, THF, –78 °C→ [Bu^t (cyclohexane) Li SePh] —H_3O^+, 77%→ Bu^t (cyclohexane) H SePh (ref. 54)

(b) Me_3SiO PhSe ... SePh H → [Me_3SiO ····· Li, PhSe, H] —PhCHO→ Me_3SiO PhSe ... Ph OH (ref. 167d)

Scheme 86

It has been shown that the methyl selenoacetal resulting from the reaction of 1-lithio-1,1-bis(methylseleno)methane with 5-methyl-2-cyclohexen-1-one and trimethylsilyl chloride is efficiently cleaved by *s*-butyllithium in THF to afford[157] the corresponding α-selenoalkyllithium in very high yield (Scheme 79). Chemoselective cleavage of a selenoacetal derived from an aldehyde has been achieved[60] in the presence of a selenoacetal derived from a ketone (Scheme 11). The best results have been observed[60] in the phenylseleno series when the reaction is carried out at –42 °C in ether–hexane. Under these conditions the difference of reactivity between the two selenoacetals is so high that almost complete discrimination can be achieved (Scheme 87).[60]

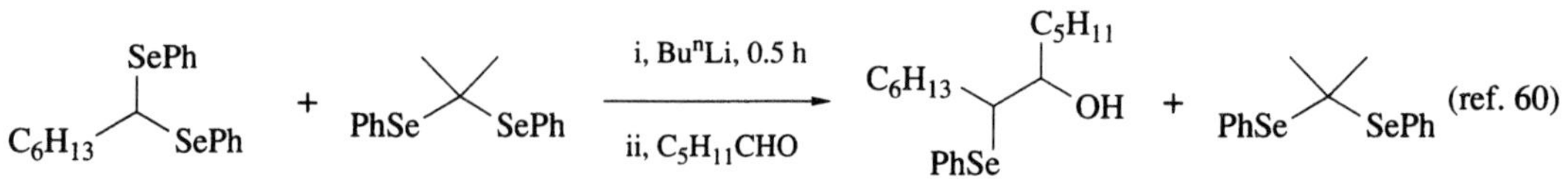

C_6H_{13} SePh SePh + PhSe SePh —i, Bu^nLi, 0.5 h; ii, $C_5H_{11}CHO$→ C_6H_{13} PhSe C_5H_{11} OH + PhSe SePh (ref. 60)

Conditions	*Yield (%)*	
Ether, –42 °C	82	94
THF, –78 °C	70	75

Scheme 87

The same selectivity cannot be achieved under similar conditions in the methylseleno series due to the inertness of these selenoacetals. Performing the reaction with *s*-butyllithium in THF–hexane (at –110 °C to –78 °C) leads[60] to rather poor selectivity (Scheme 88). A somewhat better chemoselection is achieved if the reaction is performed in THF–hexane at –78 °C and then the medium warmed to –40 °C prior to the addition of the electrophile (Scheme 88).[60] We believe that under these conditions a thermodynamic control is operative. Application of these observations has allowed the selective synthesis of 1,7-dienes from 1,1,6,6-tetrakis(phenylseleno)heptane (Scheme 11).[60]

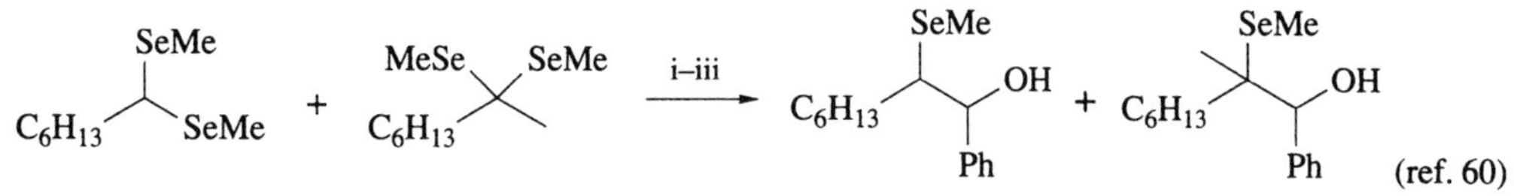

$C_6H_{13}CH(SeMe)_2$ + $C_6H_{13}C(Me)(SeMe)_2$ —i–iii→ $C_6H_{13}CH(SeMe)CH(OH)Ph$ + $C_6H_{13}C(Me)(SeMe)CH(OH)Ph$ (ref. 60)

	Conditions	Yield (%)	Yield (%)	Chemoselectivity (%)
(a)	Bu^sLi, ether/hexane, –78 °C, 0.5 h	55	25	73
(b)	Bu^nLi, THF/hexane, –100 °C, 0.5 h	15	5	75
(c)	Bu^nLi, THF/hexane, –78 °C, 0.05 h	37	11	77
(d)	Bu^nLi, THF/hexane, –78 °C, 1 h	63	23	74
(e)	Bu^nLi, THF/hexane, –78 °C, 0.1 h, then –40 °C, 1 h	78	8	90

i, RLi; ii, PhCHO; iii, H_3O^+

Scheme 88

As already mentioned, α-selenoalkyllithiums are stable for several hours at –78 °C. However, they start to decompose when the temperature is raised to around –50 °C or –40 °C. The exact temperature at which this process begins, the rate of decomposition and the nature of the products formed clearly depends upon the structure of the selenoacetal. Those bearing two alkyl groups on the carbanionic center often decompose below –40 °C. The resulting products are not always identical and in most cases their structures have not been identified. A more detailed work has been carried out on 1-seleno-1-heptyllithiums. In the methylseleno series, whatever the solvent used, only a trace amount of the expected 1-methylseleno-1-heptyllithium is present after standing for 3.5 h at 0 °C; 1-methylselenoheptane, which probably results from the protonation of the organometallic by the solvent, and 1-methylseleno-1-heptene, resulting from a hydride elimination, are produced instead (Scheme 89).[54]

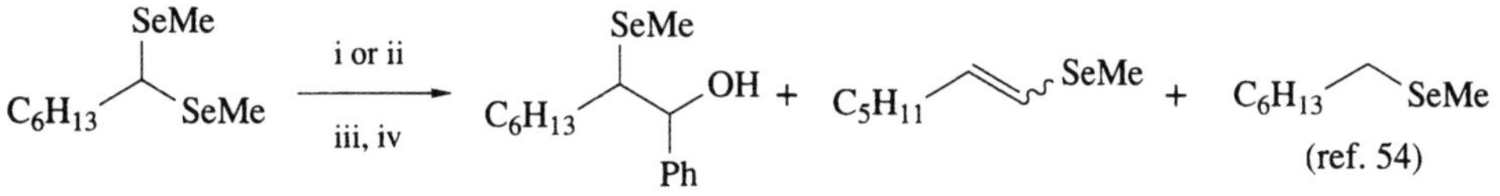

$C_6H_{13}CH(SeMe)_2$ —i or ii; iii, iv→ $C_6H_{13}CH(SeMe)CH(OH)Ph$ + $C_5H_{11}CH=CHSeMe$ + $C_6H_{13}CH_2SeMe$ (ref. 54)

	Conditions		Yield (%)	
(a)	i	25	27	small amount
(b)	ii	trace	36	35

i, Bu^nLi, THF–hexane, –78 °C, 0.2 h, then –78 to 0 °C, then 0 °C, 1 h; ii, Bu^nLi, THF–hexane, –78 °C, 0.2 h, then –78 to 0 °C, then 0 °C, 3.5 h; iii, PhCHO; iv, H_3O^+

Scheme 89

The decomposition takes another course with 1-phenylseleno-1-heptyllithium (Scheme 90).[54] In ether and THF the reagent is still present (±20%) after 3.5 h at 0 °C, along with heptylselenophenyllithiums resulting from the isomerization of the original organometallic to the more stable aryllithiums (Scheme 90). Heptyl phenyl selenide (15%) arising from the protonation of the organometallic by the solvent and substantial amounts of phenyllithium (trapped as diphenyl carbinol, ±21%) from unknown origin are also found as by-products when the reaction is carried out in THF–hexane or in ether–hexane, respectively.

The same type of exchange occurs,[165,166] but unexpectedly more rapidly and at –78 °C, with the α-selenoalkyllithium shown in Scheme 83. At higher temperature (20 °C)[165,166] the addition of the β-selenoaryllithium across the carbon–carbon double bond present in the molecule also occurs (Scheme 83, b).

α-Selenoalkyllithiums usually do not have a high tendency to add across carbon–carbon double bonds even intramolecularly. However, it has been found that 1-phenyl-2-methylseleno-2-oct-5-enyllithium cyclizes at 0 °C and leads to the benzylidenecyclopentane shown in Scheme 91.[168] The mechanism of this reaction is being investigated.[168]

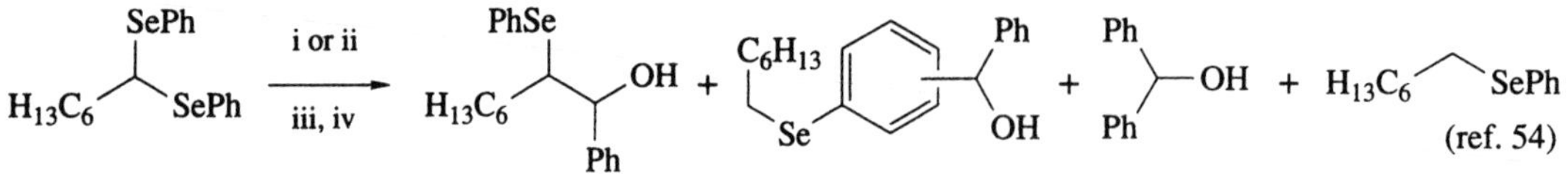

(ref. 54)

	Conditions	*Yield (%)*			
(a)	i	25	11	trace	15
(b)	ii	21	25	21	0

i, Bu^nLi, THF–hexane, –78 °C, 0.2 h, then –78 to 0 °C, then 0 °C, 3.5 h; ii, Bu^nLi, ether–hexane, –78 °C, 0.2 h, then –78 to 0 °C, then 0 °C, 3.5 h; iii, PhCHO; iv, H_3O^+

Scheme 90

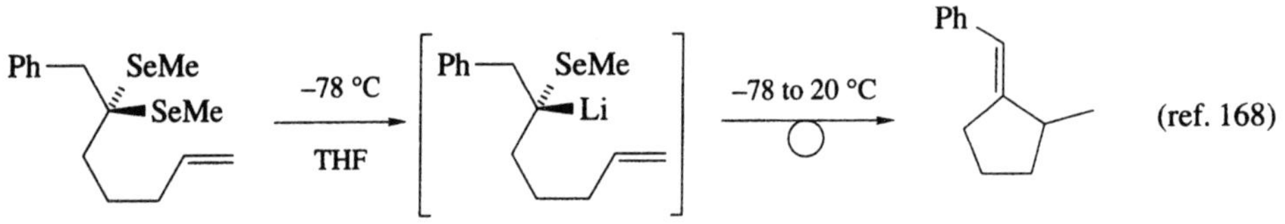

–78 °C, THF; –78 to 20 °C (ref. 168)

Scheme 91

α-Selenobenzyllithiums are much more prone to add to alkenes.[169,170] Although they are stable around –78 °C (Scheme 92, a), the addition reaction usually takes place around 0 °C and produces [3.1.0]bicyclohexanes and [4.1.0]bicycloheptanes in reasonably good yields when a carbon–carbon double bond is present in a suitable position with respect to the organometallic group (Scheme 92, b).[169,170]

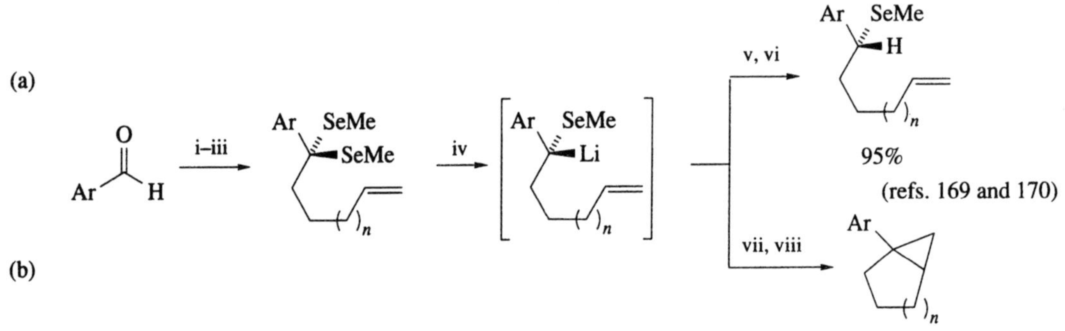

(a) 95%

(refs. 169 and 170)

(b) Ar = Ph; n = 1, 80%; n = 2, 80%
Ar = 4-$MeOC_6H_4$; n = 1, 45%

i, MeSeH (2 equiv.), $ZnCl_2$; ii, KDA, THF, –78 °C; iii, $H_2C{=}CH(CH_2)_3Br$, iv, Bu^nLi, THF, –78 °C; vi, –78 °C, 0.5 h; vii, –78 °C, 0.5 h; viii, 0 °C, 1 h

Scheme 92

Addition also occurs[169,170] with ethylene and with activated alkenes, such as styrene, butadiene, trimethylsilylethylene and phenylthioethylene, and provides in all cases the corresponding cyclopropanes in good yield (Scheme 93).[200,201] Interestingly, the α-selenoalkyllithiums do not react with the C—C bond at lower temperatures so that the anion is still available for reaction with alkyl halides or carbonyl compounds (Scheme 92, compare a with b).

The synthesis of α-selenoalkyl metals reported above is not limited to those compounds which possess alkyl or aryl groups on an sp^3 carbanionic center. For example, α-selenoalkyllithiums bearing a vinyl (Scheme 94),[87] furfuryl (Scheme 95),[158] silyl (Scheme 96),[100,171] methoxy (Scheme 97)[56] or seleno moiety (Schemes 77 and 98)[35,36,71] have been obtained in almost quantitative yield from the corresponding functionalized selenoacetals[56,87,100,171] or orthoesters[36,71] and Bu^nLi. The Se–Li exchange is in all these cases, except that of tris(phenylseleno)methane,[36] exclusively observed even with those compounds which possess a hydrogen susceptible to metallation. The 6,6-bis(phenylseleno)-β-lactam shown

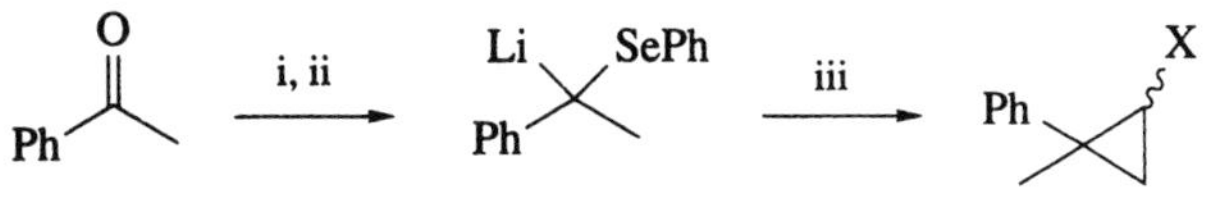

(refs. 169 and 170)

	Conditions	X	Yield (%)
(a)	0 °C, 1–2 h	H	50
	0 °C, 1–2 h	Me	0
	0 °C, 1–2 h	Ph	60
(b)	–30 °C, 0.2–3 h	$CH=CH_2$	70
	–30 °C, 0.2–3 h	$SiMe_3$	81
	–30 °C, 0.2–3 h	SPh	59

i, MeSeH, $TiCl_4$, 0 °C; ii, Bu^nLi, THF; iii, $H_2C=CHX$

Scheme 93

in Scheme 99 also reacts[121] with *n*-butyllithium and provides the corresponding 6-lithio-6-phenylseleno derivative (compare to Scheme 42, a) resulting from an attack on the selenium atom rather than on the carbonyl group. Interestingly, methylmagnesium bromide, which is usually inert towards selenoacetals, reacts efficiently with the above mentioned selenoacetal to produce the 6-phenylselenoenolate derivative (Scheme 99, b). This is probably due to efficient stabilization of the carbanionic center by delocalization on the carbonyl group.

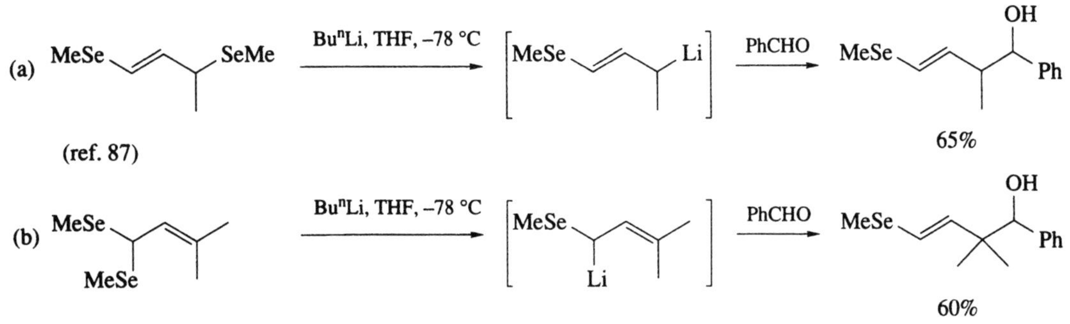
(a)
MeSe
SeMe
Bu^nLi, THF, –78 °C
MeSe
Li
PhCHO
MeSe
OH
Ph
(ref. 87)
65%
(b)
MeSe
MeSe
Bu^nLi, THF, –78 °C
MeSe
Li
PhCHO
MeSe
OH
Ph
60%

Scheme 94

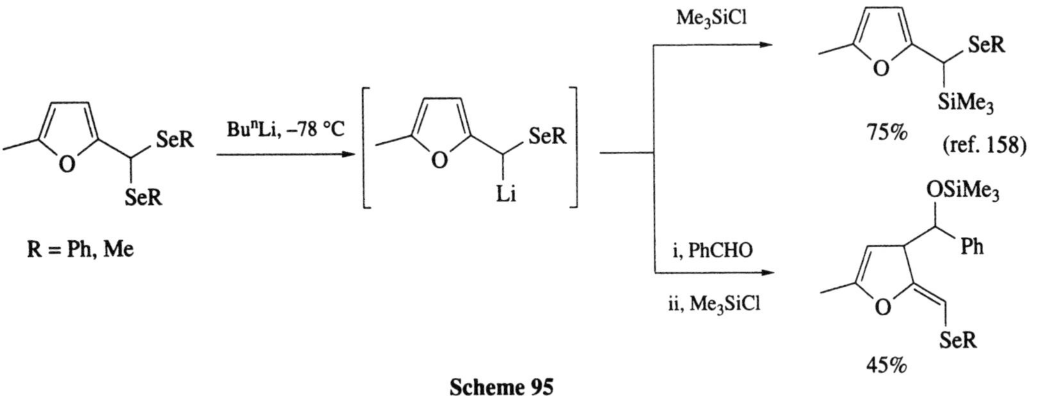
SeR
SeR
R = Ph, Me
Bu^nLi, –78 °C
SeR
Li
Me3SiCl
SeR
SiMe3
75%
(ref. 158)
i, PhCHO
ii, Me3SiCl
OSiMe3
Ph
SeR
45%

Scheme 95

2.6.2.3.2 Synthesis of α-selenovinyl metals by selenium–metal exchange

1,1-Bis(phenylseleno)alkenes, readily prepared from vinylcarbenes and diselenides (Scheme 100),[172] trimethylsilylbis(phenylseleno)methyllithium and carbonyl compounds (Scheme 31, d),[95] orthoselenoesters[35,36] and diphosphorus tetraiodide (P_2I_4)[174] or tin tetrachloride ($SnCl_4$),[175] and from β-hydroxyalkyl orthoesters and P_2I_4 (Scheme 101, b),[108] efficiently react with *n*-butyllithium in THF and provide, even at

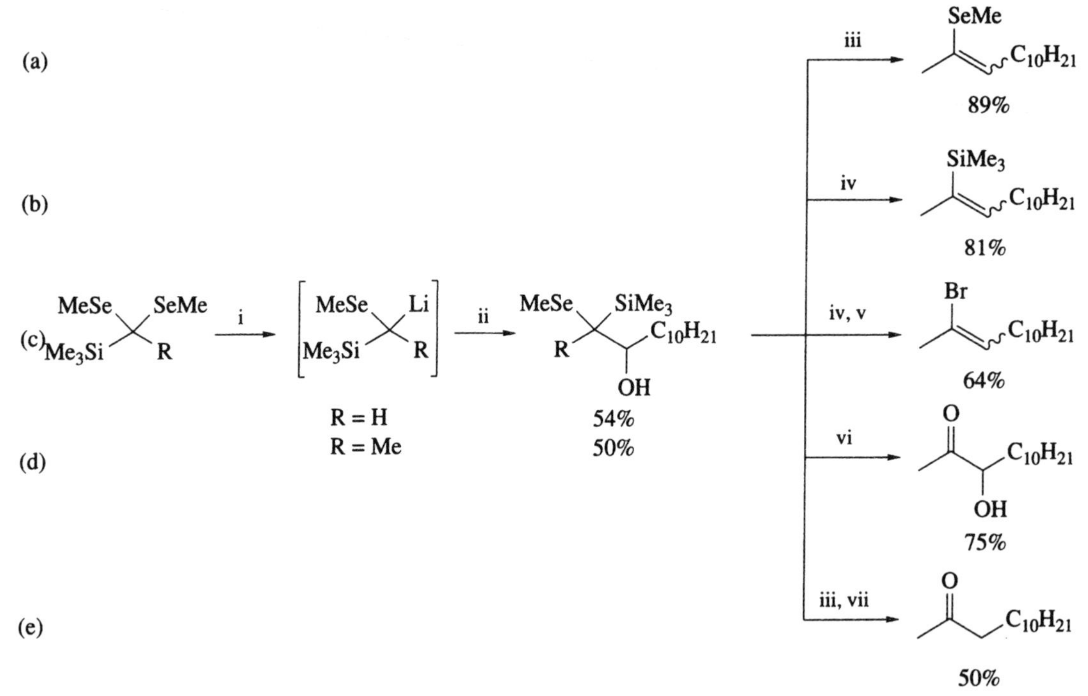

(a) iii → SeMe, $C_{10}H_{21}$ 89%

(b) iv → $SiMe_3$, $C_{10}H_{21}$ 81%

(c) MeSe SeMe / Me_3Si R —i→ [MeSe Li / Me_3Si R] —ii→ MeSe $SiMe_3$ / R $C_{10}H_{21}$ OH —iv, v→ Br, $C_{10}H_{21}$ 64%

R = H 54%

R = Me 50%

(d) vi → O, $C_{10}H_{21}$, OH 75%

(e) iii, vii → O, $C_{10}H_{21}$ 50%

i, Bu^nLi, THF, –78 °C, 0.1 h; ii, $C_{10}H_{21}CHO$; iii, Bu^tOK, THF, 55 °C; iv, $POCl_3$, Et_3N, 20 °C; v, Br_2, CCl_4, 20 °C; vi, H_2O_2, THF, 20 °C, vii, $HgCl_2$, aq. MeCN, 20 °C, 12 h (ref. 171)

Scheme 96

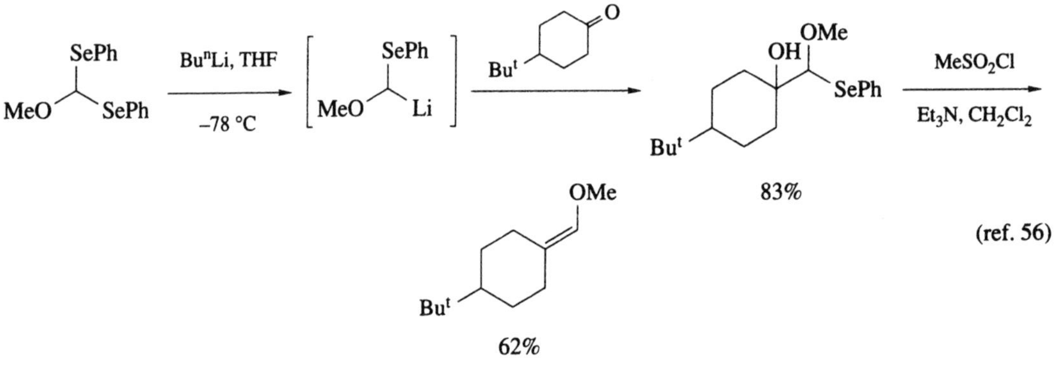

MeO(SePh)SePh —Bu^nLi, THF, –78 °C→ [MeO(SePh)Li] —Bu^t-cyclohexanone→ Bu^t-cyclohexyl(OH)CH(OMe)SePh 83% —$MeSO_2Cl$, Et_3N, CH_2Cl_2→ Bu^t-cyclohexylidene=CHOMe 62%

(ref. 56)

Scheme 97

–78 °C, 1-phenylseleno-[173] and 1-methylseleno-1-alkenyllithiums[176] in almost quantitative yields (Scheme 101). The stereochemistry of the reaction is not well defined and the only available results are those concerning the stereochemistry of the compounds resulting from further reaction of these organometallics with electrophiles.[173,176] Both stereoisomers are formed if the methylselenoalkenyllithium is hydrolyzed[176] or reacted[176] with an aldehyde or a ketone, whereas only one stereoisomer of an α,β-unsaturated carbonyl compound is found from 1-methylseleno-1-alkenyllithiums and DMF[176] and from the 1-phenylseleno-1-alkenyllithiums and phenacyl bromide.[173]

2.6.2.4 Miscellaneous Syntheses of α-Selenoalkyl Metals

More recently, lithium naphthalenide, lithium dimethylaminonaphthalenide and lithium 4,4′-dimethoxybiphenylide have been successfully reacted with selenoacetals to produce the corresponding α-selenoalkyllithiums is moderate yields.[168] Phenylselenomethyllithium has been prepared[101] from (phenylseleno)(phenyltelluro)methane and butyllithium (Scheme 25, b), and α-phenylselenoalkyllithiums have also been obtained[177] from α-bromoalkyl phenyl selenides and *n*-butyllithium in THF. The

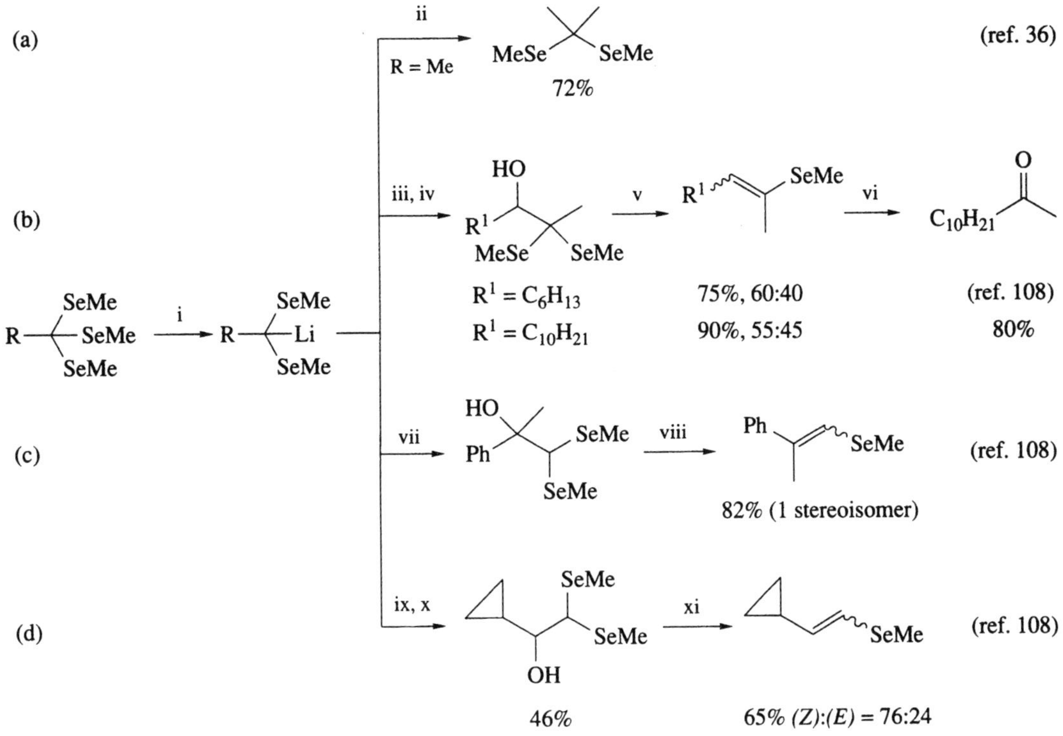

(a) ii, R = Me → MeSe–C(Me)2–SeMe, 72% (ref. 36)

(b) iii, iv → R^1CH(OH)C(Me)(SeMe)2 → v → R^1CH=C(Me)SeMe → vi → $C_{10}H_{21}$COMe

$R^1 = C_6H_{13}$ 75%, 60:40

$R^1 = C_{10}H_{21}$ 90%, 55:45 (ref. 108) 80%

R–C(SeMe)3 → i → R–C(Li)(SeMe)2

(c) vii → Ph(Me)C(OH)–C(Me)(SeMe)2 → viii → Ph(Me)C=C(Me)SeMe, 82% (1 stereoisomer) (ref. 108)

(d) ix, x → cyclopropyl–CH(OH)–C(Me)(SeMe)2, 46% → xi → cyclopropyl–CH=C(Me)SeMe, 65% *(Z)*:*(E)* = 76:24 (ref. 108)

i, Bu^{n}Li, THF, –78 °C; ii, MeI; iii, R^1CHO; iv, H_3O^+; v, P_2I_4, Et_3N, CH_2Cl_2, 20 °C; vi, $HgCl_2$, H_2O, MeCN, 20 °C; vii, PhCOMe; viii, P_2I_4, Et_3N, CH_2Cl_2, 20 °C; ix, cyclopropylaldehyde; x, H_3O^+; xi, P_2I_4, Et_3N, CH_2Cl_2, 20 °C

Scheme 98

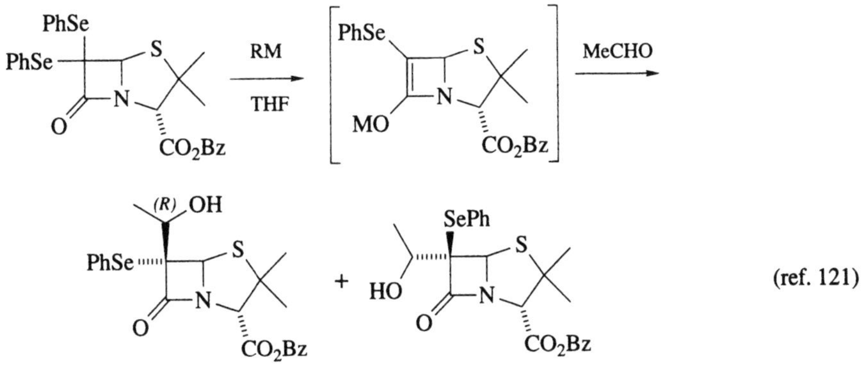

6,6-bis(PhSe)penam–CO_2Bz → RM, THF → [enolate: PhSe, MO, CO_2Bz] → MeCHO → 6-PhSe-6-[(R)-CH(OH)Me]penam–CO_2Bz + 6-SePh-6-[CH(OH)Me]penam–CO_2Bz (ref. 121)

	RM	*Yield* (%)	*Ratio*
(a)	BuLi, –78 °C	–	40:60
(b)	MeMgBr, –60 °C	78	97:3

Scheme 99

latter reaction is highly selective, readily occurs at –78 °C and provides a wide range of α-phenylselenoalkyllithiums in about 60% yield (Scheme 102).

The halogen–lithium exchange is, however, far less general than the selenium–metal exchange from selenoacetals for the synthesis of α-selenoalkyllithiums. It does not provide α-phenylselenoalkyllithiums from α-chloroalkyl phenyl selenides[177] and does not allow the synthesis of methylselenoalkyllithiums from α-haloalkyl methyl selenides.[177] It also suffers from the instability of the α-bromoalkyl phenyl selenides, especially those which produce the α-selenoalkyllithiums where the carbanionic center is dialkyl substituted.[177] Furthermore the butyl bromide produced concomitantly with the α-selenoalkyllithium

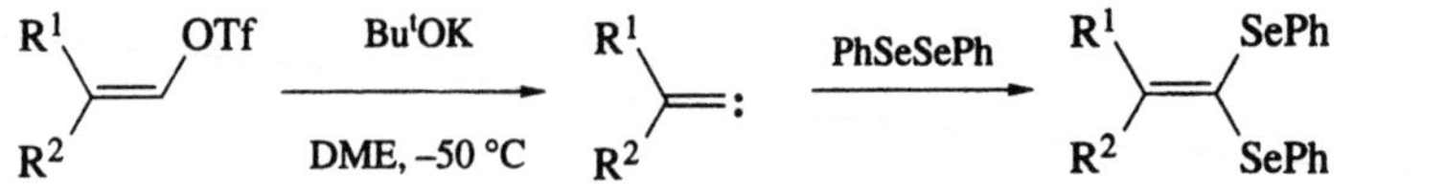

$R^1R^2C{=}CH$–OTf → (Bu^tOK, DME, –50 °C) → $R^1R^2C{=}C{:}$ → (PhSeSePh) → $R^1R^2C{=}C(SePh)_2$ (ref. 172)

R^1	R^2	Yield (%)
Me	Me	47
$CH_2(CH_2)_3CH_2$		61

Scheme 100

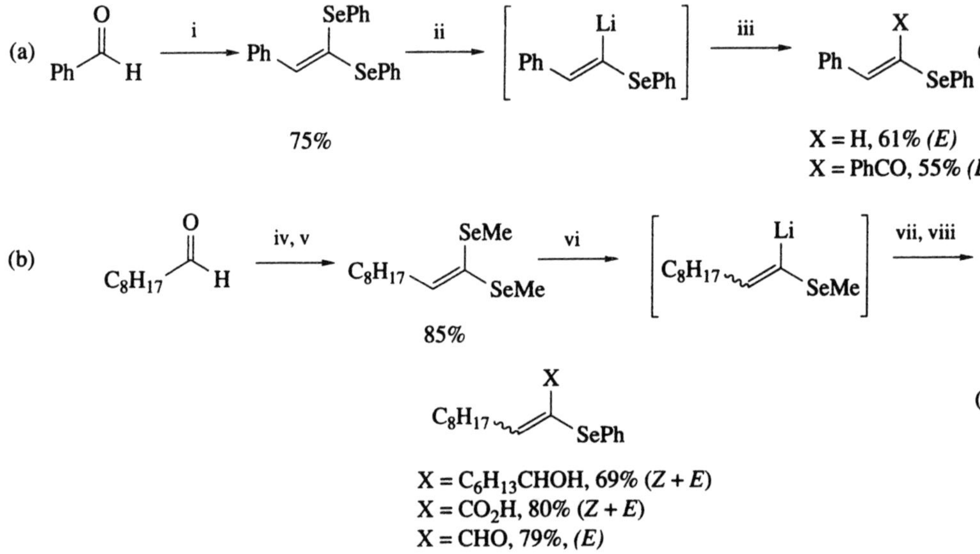

(a) PhCHO → (i) → PhCH=C(SePh)SePh (75%) → (ii) → [PhCH=C(Li)SePh] → (iii) → PhCH=C(X)SePh (ref. 173)

X = H, 61% (*E*)
X = PhCO, 55% (*E*)

(b) $C_8H_{17}CHO$ → (iv, v) → $C_8H_{17}CH{=}C(SeMe)SeMe$ (85%) → (vi) → [$C_8H_{17}CH{=}C(Li)SeMe$] → (vii, viii) → $C_8H_{17}CH{=}C(X)SePh$ (ref. 176)

X = C_6H_{13}CHOH, 69% (*Z* + *E*)
X = CO_2H, 80% (*Z* + *E*)
X = CHO, 79%, (*E*)

i, $Me_3SiCLi(SePh)_2$, THF, –78 °C, 1 h, then 30 °C, 14 h; ii, Bu^nLi, THF, –78 °C; iii, H_2O or PhCOBr; iv, $LiC(SeMe)_3$; v, P_2I_4, Et_3N, CH_2Cl_2, 20 °C, 1 h; vi, Bu^nLi, THF, –78 °C; vii, $C_6H_{13}CHO$ or CO_2 or Me_2NCHO; viii, H_3O^+

Scheme 101

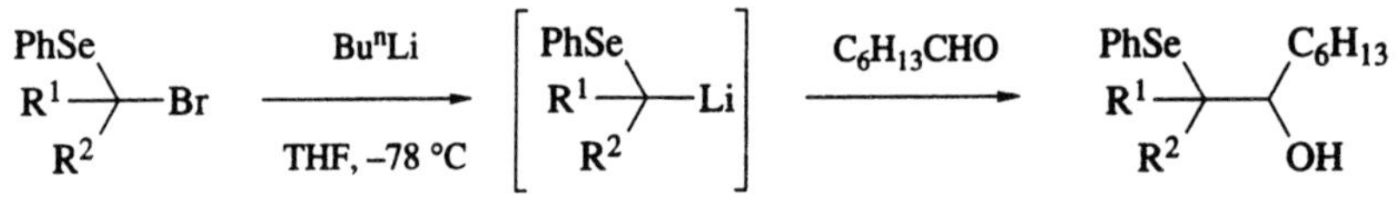

PhSe(R^1)(R^2)C–Br → (Bu^nLi, THF, –78 °C) → [PhSe(R^1)(R^2)C–Li] → ($C_6H_{13}CHO$) → PhSe(R^1)(R^2)C–CH(OH)C_6H_{13} (ref. 177)

R^1	R^2	Yield (%)
H	H	69
H	Me	63
Me	Me	62

Scheme 102

competes with other electrophiles in further reactions.[177] The Br–Li exchange has been successfully used for the synthesis of 2-lithioselenophene[122] and 2-lithiobenzoselenophene (Scheme 103).[123]

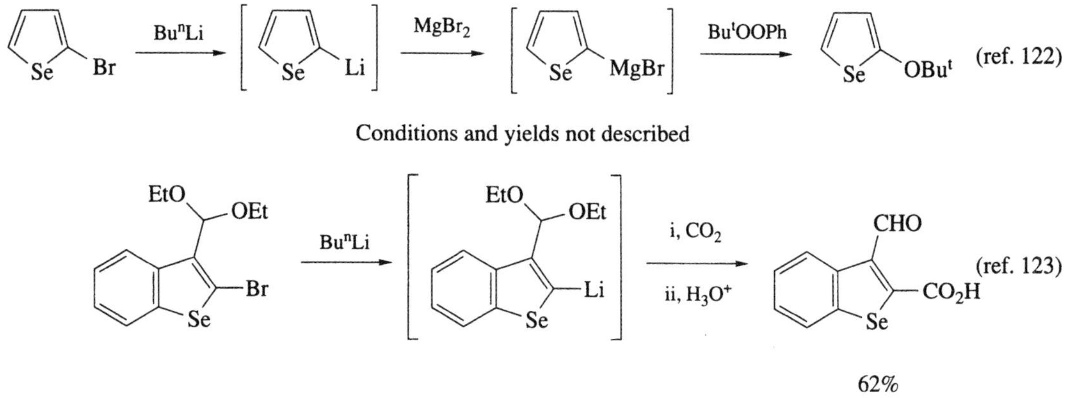

Se Br —Bu^nLi→ [Se Li] —$MgBr_2$→ [Se MgBr] —Bu^tOOPh→ Se OBu^t (ref. 122)

Conditions and yields not described

EtO OEt, Br, Se —Bu^nLi→ [EtO OEt, Li, Se] —i, CO_2; ii, H_3O^+→ CHO, CO_2H, Se (ref. 123)

62%

Scheme 103

2.6.2.5 Synthesis of α-Selenoalkyl Metals by Addition of Organometallics to Vinyl Selenides, Vinyl Selenoxides and Vinyl Selenones

The reaction of organolithium compounds with phenyl vinyl selenides[11,41,42,124,174,178,179] proceeds by at least three different routes which involve: (i) metallation, which leads to 1-phenylseleno-1-vinyllithium (Section 2.6.2.2.2);[41,65,125] (ii) Se–Li exchange, which produces phenyllithium and a novel vinyl selenide (Scheme 104, a);[41,42] or (iii) addition of the organolithium across the carbon–carbon double bond leading to α-selenoalkyllithiums (Scheme 104, b).[41,42,124] The choice of the experimental conditions, especially the solvent, is crucial for chemoselective reactions.

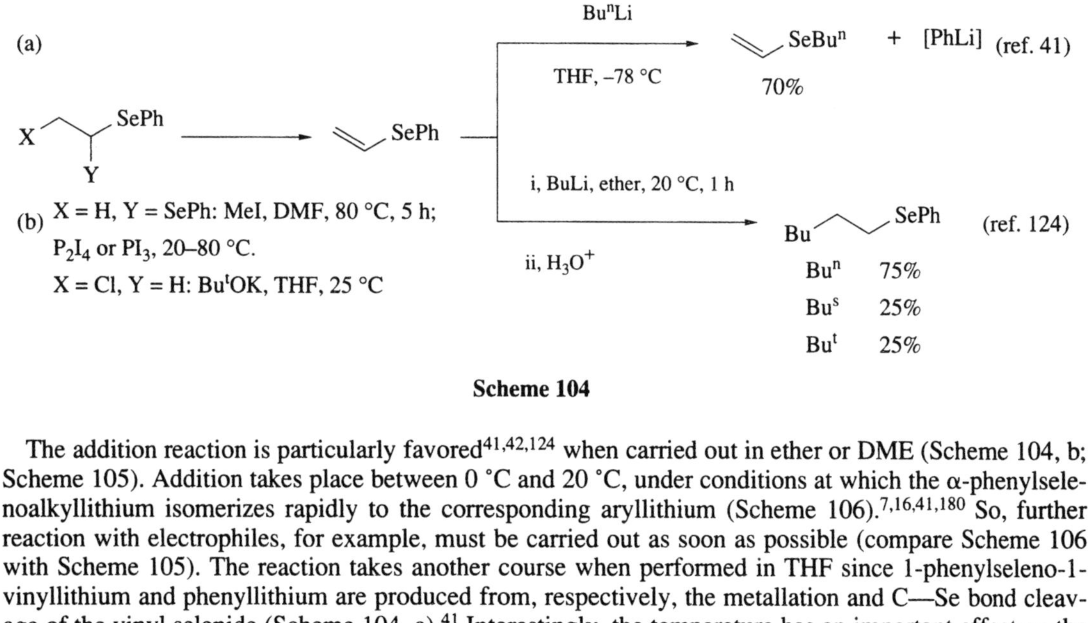

(a) —Bu^nLi, THF, –78 °C→ $SeBu^n$ + [PhLi] (ref. 41)
70%

X(Y)SePh ——→ SePh

(b) X = H, Y = SePh: MeI, DMF, 80 °C, 5 h; P_2I_4 or PI_3, 20–80 °C.
X = Cl, Y = H: Bu^tOK, THF, 25 °C

—i, BuLi, ether, 20 °C, 1 h; ii, H_3O^+→ Bu SePh (ref. 124)

Bu^n 75%
Bu^s 25%
Bu^t 25%

Scheme 104

The addition reaction is particularly favored[41,42,124] when carried out in ether or DME (Scheme 104, b; Scheme 105). Addition takes place between 0 °C and 20 °C, under conditions at which the α-phenylselenoalkyllithium isomerizes rapidly to the corresponding aryllithium (Scheme 106).[7,16,41,180] So, further reaction with electrophiles, for example, must be carried out as soon as possible (compare Scheme 106 with Scheme 105). The reaction takes another course when performed in THF since 1-phenylseleno-1-vinyllithium and phenyllithium are produced from, respectively, the metallation and C—Se bond cleavage of the vinyl selenide (Scheme 104, a).[41] Interestingly, the temperature has an important effect on the course of this reaction, since in THF the Se–Li exchange becomes predominant if the reaction is carried out at low temperature (–78 °C instead of 0 °C).[41]

α-(Methylseleno)benzyllithiums also react with phenyl vinyl selenide but surprisingly give 1-phenyl-1-methylselenocyclopropanes. This result has been tentatively rationalized in Scheme 107.[169,170]

A reaction related to those mentioned above involves an addition–cyclization sequence of 2-azaallyllithiums on phenyl vinyl selenide, which leads to seleno-substituted pyrrolidines (Scheme 108).[181,182]

γ-Selenoalkylmetals have been suggested as intermediates in a large array of reactions. Thus organocuprates efficiently add at the C-3 site of α-seleno-α,β-unsaturated ketones to produce enolates which can be further trapped with various alkyl or propargyl halides to give α,β-dialkyl-α-seleno ketones (Scheme 109). These, after proper control of the relative stereochemistry at the α- and β-carbons to the carbonyl group, have been selectively transformed to cycloalkenones (Schemes 36 and 109, a)[111,183] or

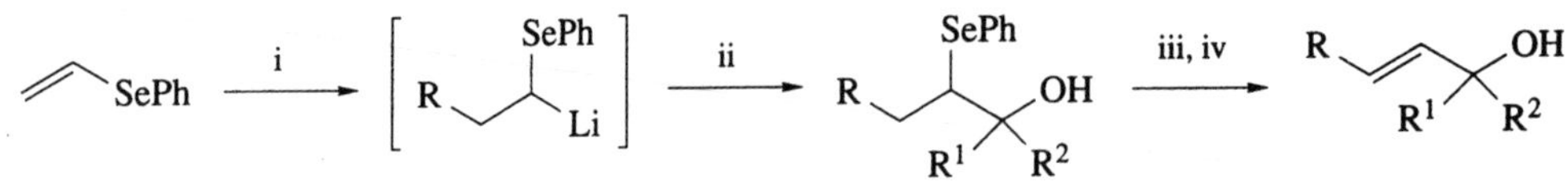

	RLi	Conditions	Carbonyl compound	Yield (%)		Ref.
(a)	Bu^nLi	Ether, 20 °C, 0.2 h	$C_6H_{13}CHO$	40		41
(b)	Bu^nLi	DME, 0 °C	PhCHO	71		42
(c)	Pr^iLi	Ether, 0 °C	Me_2CO	72	81	42

i, RLi; ii, R^1R^2CO; iii, O_3, –78 °C, 0.1 h; iv, $Pr^i{}_2NH$, CCl_4, 80 °C, 0.5 h

Scheme 105

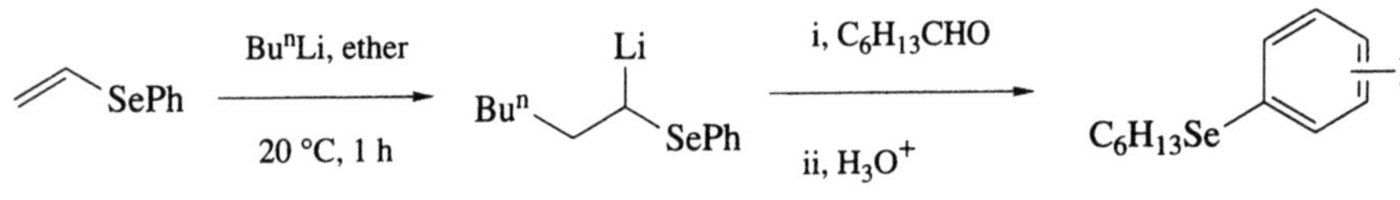

Bu^nLi, ether, 20 °C, 1 h; i, $C_6H_{13}CHO$; ii, H_3O^+ (ref. 41)

E = H, 30%; E = $C_6H_{13}CHOH$, 18%

Scheme 106

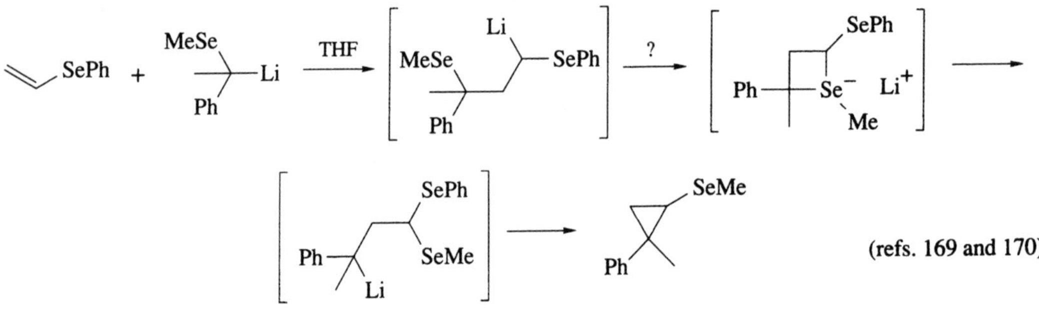

(refs. 169 and 170)

Scheme 107

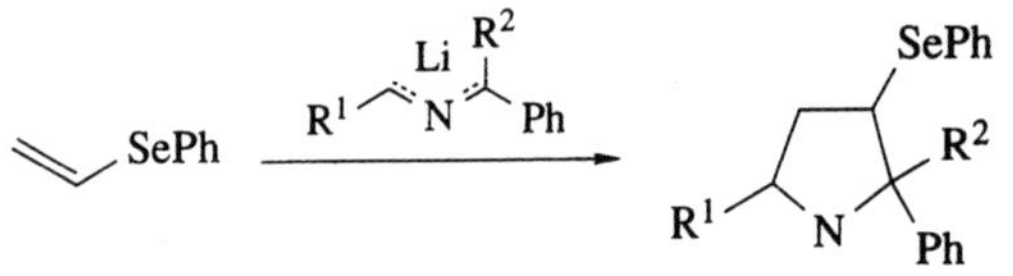

(refs. 181 and 182)

R^1	R^2	Conditions	Yield (%)
Ph	H	–40 °C, 22 h	77
H	Ph	–50 to –10 °C, 6 h	42

Scheme 108

alkylidenecycloalkanes (Scheme 109, b)[183] using the well-established selenoxide *syn* elimination reaction.

Lithium enolates derived from methyl ketones and acetates successfully add to vinyl selenoxides[184,185] and to vinyl selenones to give cyclopropyl ketones and esters (Scheme 110).[185,186] The α-lithioalkyl selenoxide or selenone intermediates presumably exchange to the lithium enolates which, after displacement of the seleno or selenoxy group, lead to the observed products.

Additions of various nucleophiles to vinyl selenoxides[184,185,187] and vinyl selenones[17,185–187] have been described. Thus potassium hydroxide[187] and amines[17] add to vinyl selenones and directly produce epoxides and aziridines, respectively, whereas 3-methoxyoxetanes have been obtained from γ-hydroxy-α-alkenyl selenones and sodium methoxide (Scheme 111).[185]

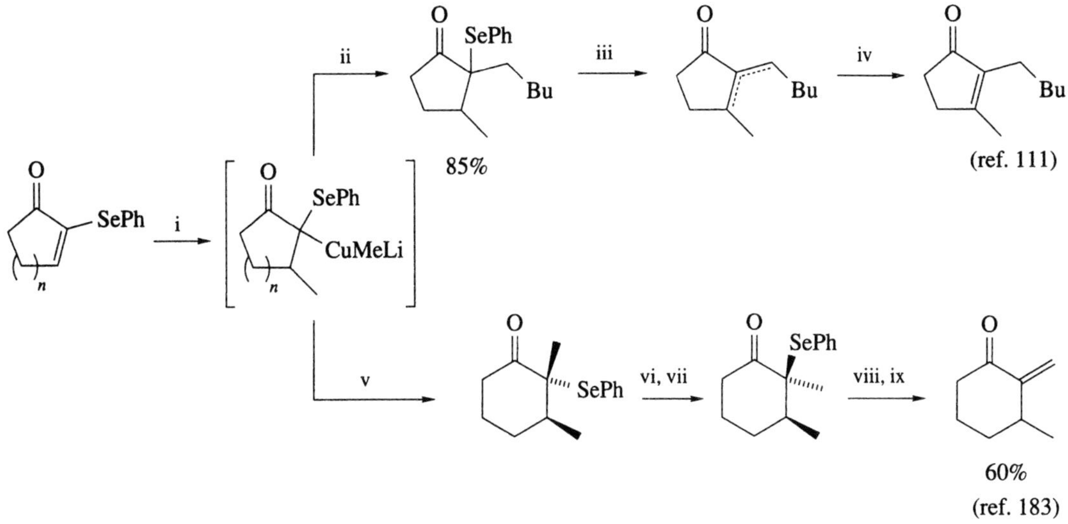

O SePh ii O SePh Bu iii O Bu iv O Bu

O SePh i [O SePh CuMeLi n] 85% (ref. 111)

v O SePh vi, vii O SePh viii, ix O

60%
(ref. 183)

i, Me_2CuLi, Et_2O, 20 °C, 0.3 h; ii, Bu^nLi, THF–HMPA; iii, aq. H_2O_2, CH_2Cl_2; iv, HCl, Bu^nOH, 90 °C, 1 h; v, MeI, THF–HMPA; vi, PhSeLi; vii, PhSeCl; viii, O_3, CH_2Cl_2; ix, Et_2NH, 50 °C

Scheme 109

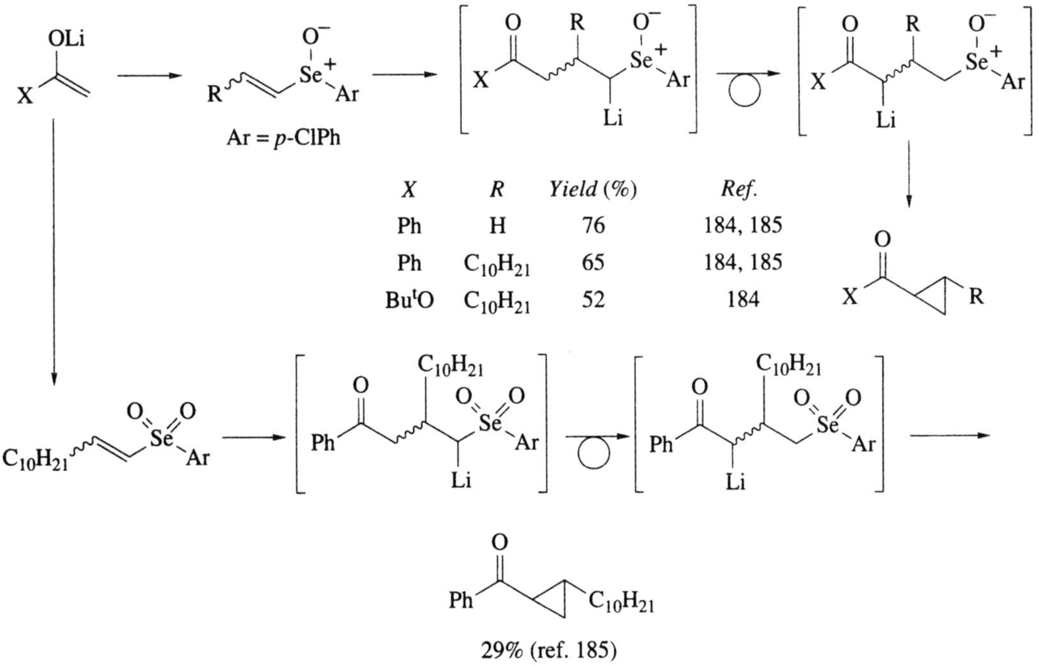

OLi X → R Se+ O− Ar → [X O R O− Se+ Ar Li] → [X O R O− Se+ Ar Li]

Ar = p-ClPh

X	R	Yield (%)	Ref.
Ph	H	76	184, 185
Ph	$C_{10}H_{21}$	65	184, 185
Bu^tO	$C_{10}H_{21}$	52	184

X O R

$C_{10}H_{21}$ O O Se Ar → [Ph O $C_{10}H_{21}$ O O Se Ar Li] → [Ph O $C_{10}H_{21}$ O O Se Ar Li] →

Ph O $C_{10}H_{21}$

29% (ref. 185)

Scheme 110

O O Se Ar R OH NaOMe [O O Se Ar OMe R Na OH] → [O O Se Ar OMe R ONa] → O OMe R

Ar = p-ClC_6H_4

(ref. 185)

R = $C_{10}H_{21}$, 78%
R = $C_5H_{11}CHCO_2Bu^t$, 81%

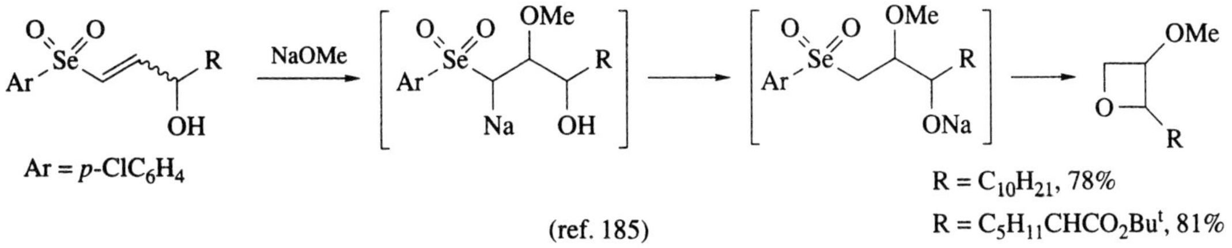

Scheme 111

2.6.3 REACTIVITY OF CARBONYL COMPOUNDS WITH ORGANOMETALLICS BEARING A SELENIUM-STABILIZED CARBANION

2.6.3.1 Generalities

The organometallics bearing a selenium atom directly attached to the carbanionic center, the synthesis of which has been discussed above, have been reacted with a large variety of electrophiles.[1,4,5,7–9,12,13,16,17] Among these electrophiles, aldehydes and ketones occupy a place of choice, usually leading, after hydrolysis, to β-hydroxyalkyl selenides. Some examples are gathered in Schemes 112 to 118.

Examples of organometallics bearing a selenium-stabilized carbanion include: (i) α-selenoalkylhydroxylithiums bearing hydrogens or alkyl groups on the carbanionic[4,7–9,13,16,17,35,36,40,48–50,54–56,77,159,160,162,163,166,177,188–206] center; (ii) α-selenocyclopropyllithiums[7,12,16,77,104,159–161] and α-selenocyclobutyllithiums;[7,12,16,136,162,163] (iii) α-selenoallyllithiums (Schemes 16–18, 94 and 119);[16,17,45,80–82,84–87,91] (iv) α-selenobenzyllithiums (Schemes 9, 15 and 118, b)[54,56,72,164,203] and (phenylseleno)(furfuryl)methyllithium (Scheme 95);[158] (v) 1-phenylselenovinyllithiums (Scheme 44);[41,65,73,176] (vi) 1-methoxy-1-phenylselenomethyllithium (Scheme 21, a; Scheme 97);[56] (vii) 1-phenylseleno-1-thioalkyllithiums (Scheme 22);[94] (viii) 1-phosphonato-1-phenylselenoalkyllithiums (Scheme 28);[102] (ix) 1-seleno-1-silylalkyllithiums (Scheme 24, b and c; Scheme 96),[100,171] 1-seleno-1-silylallyllithiums (Scheme 33, a)[85] and 1-seleno-1-silylbenzyllithiums (Scheme 33; b);[70] (x) 1,1-bis(seleno)alkyllithiums (Scheme 23, e and f),[35,36,56,65,71,73,98,108,192] 1,1-bis(seleno)benzyllithiums (Scheme 32),[72] 1,1-bis(seleno)(trimethylsilyl)methyllithiums (Scheme 31, d)[95] and tris(seleno)methyllithiums (Scheme 31, a–c);[36,71,108] (xi) selenonium ylides (Schemes 59 and 60; Scheme 61, a);[77,132,206] (xii) α-metalloalkyl selenoxides (Schemes 51–54);[97,127,128] (xiii) α-metallovinyl selenoxides (Scheme 179);[103] (xiv) α-metalloalkyl selenones (Schemes 55 and 56);[129,130,138,206] and (xv) α-selenometalloenolates derived from esters (Scheme 10),[74] lactones (Scheme 38),[74] lactams[121,207] including β-lactams[121] (Scheme 42, a; Scheme 99), and nitroalkanes (Scheme 41).[120] Most of the organometallics listed produce the corresponding alcohols by reaction with aldehydes and ketones and further hydrolysis of the resulting alcoholates.

β-Oxidoalkylselenonium salts (Schemes 59–61)[77,132,206] and β-oxidoalkyl selenones (Schemes 55 and 56)[129,130,138,206] are unstable too, and directly collapse to epoxides even at –78 °C. The former reaction is limited to nonenolizable carbonyl compounds.[132] In the case of acetophenone and trimethylselenonium iodide, for example, polymethylation of the carbonyl compound occurs (Scheme 122).[132] The results can be rationalized by considering enolate formation and alkylation of the enolate by the selenonium salt.

β-Oxidoalkyl selenides bearing a phosphonato (Scheme 28),[102] phosphonio[106,107] or silyl group (Scheme 24, b and c; Scheme 31, d)[95,100,171] are unstable too, and have a high tendency to produce vinyl selenides *via* the well-known Wittig–Horner or Peterson elimination reaction. Lithium alkoxides derived from 1-trimethylsilyl-1-selenoalkyllithiums can be trapped with water and lead to the corresponding alcohols in modest yields (Scheme 96). Their potassium analogs, as well as lithium alkoxides derived from bis(phenylseleno)trimethylsilylmethane, however, directly collapse to the corresponding alkenes (Scheme 31, d).

2.6.3.2 Nucleophilicity of α-Selenoalkyllithiums Towards Aldehydes and Ketones

α-Selenoalkyllithiums, even those where the carbanionic center is monoalkyl, dialkyl[55,188,189,193,194] or benzyl substituted,[208] are particularly nucleophilic towards carbonyl compounds (Schemes 112–116). There is some evidence that they are far more reactive with carbonyl compounds than with alkyl halides (Scheme 123).[12,77]

Phenylseleno- and methylseleno-alkyllithiums usually exhibit a closely related reactivity towards carbonyl compounds whether the reaction is performed in THF or in ether. As expected, the nucleophilicity of such species often decreases[189] by increasing the substitution around the carbanionic center (**6** and **8**; Scheme 112), but interestingly they are often far more nucleophilic than the corresponding alkyllithiums.[51] However, in some rare cases employing particularly hindered reaction partners there is a significant difference of reactivity between phenyl- and methyl-selenoalkyllithiums when the reactions are performed in THF, since the former reagents are much less nucleophilic than the latter (compare **11** and **12** in Scheme 113 and **17a**–**17f** in Scheme 114). As general trends, α-methylselenoalkyllithiums are more nucleophilic in ether than in THF (compare **15** in Scheme 113, **17d** and **17e** in Scheme 114 and **24** in Scheme 116).

β-Hydroxyalkyl selenides have been transformed to epoxides (see Section 2.6.4.5.2)[7,16,206] and to alkenes (see Section 2.6.4.3).[7,12,16] These are therefore produced from two carbonyl compounds, one of them being activated as a selenoacetal and then as an α-selenoalkyllithium. Best results are obtained

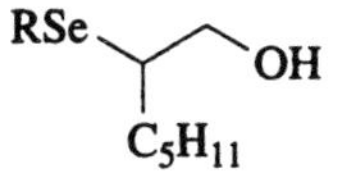

(5)

R = Me, 58%, method A (ref. 195)
R = Ph, 78%, method A

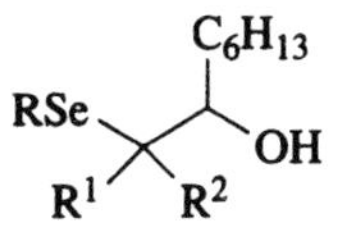

(6)

	R	*R¹*	*R²*	*Yield (%)*	*Method*	*Ref.*
(a)	Ph	H	n-C_6H_{13}	75	A	193
(b)	Me	H	n-C_6H_{13}	82	A	193
(c)	Ph	H	Et	50	A	196
(d)	Me	Me	Et	82	A	55
(e)	Me	$(CH_2)_5$		76	A	55
(f)	Me	H	Bu^t	79	A	189
(g)	Me	Me	Bu^t	57	A or B	189

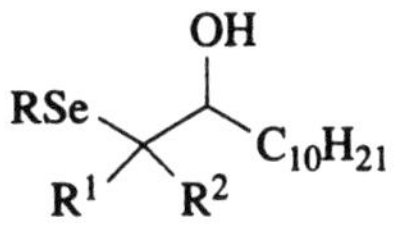

(7)

	R	*R¹*	*R²*	*Yield (%)*	*Method*	*Ref.*
(a)	Ph	H	H	71	A	193
(b)	Me	H	H	71	A	193
(c)	Ph	H	Me	80	A	193
(d)	Me	$(CH_2)_2CH(n\text{-}C_5H_{11})$		67	A	159

(8)

	R	*R¹*	*R²*	*R³*	*Yield (%)*	*Method*	*Ref.*
(a)	Ph	H	Bu^t	C_6H_{13}	75	A	193
(b)	Me	H	Bu^t	C_6H_{13}	44	B	189
(c)	Me	Me	Bu^t	C_6H_{13}	9	B	180
(d)	Ph	Me	Me	C_6H_{13}	79	B	—
(e)	Me	Me	Me	C_6H_{13}	73	B	—
(f)	Me	Me	Me	C_9H_{19}	63	A	194

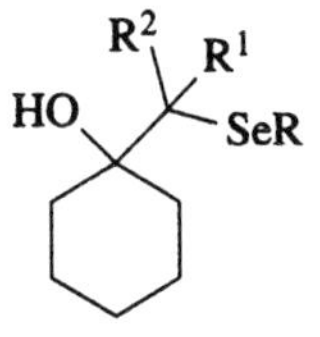

(9)

	R	*R¹*	*R²*	*Yield (%)*	*Method*	*Ref.*
(a)	Me	H	Me	94	A	55
(b)	Me	H	C_6H_{13}	60	A	55
(c)	Me	Me	Et	60	A	55
(d)	Ph	$(CH_2)_5$		70	B	223
(e)	Ph	Me	Me	78	A	132

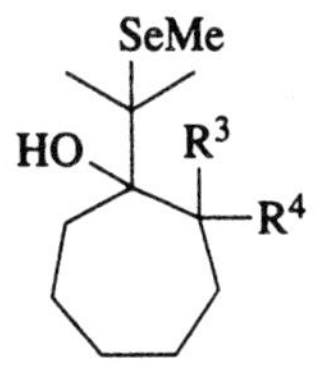

(10)

	R³	*R⁴*	*Yield (%)*	*Method*	*Ref.*
(a)	H	H	64	B	223
(b)	H	Me	64	B	223
(c)	Me	Me	64	B	223
(d)	$(CH_2)_4$		88	B	223

A = Bu^nLi, THF–hexane, –78 °C; B = Bu^sLi, ether–hexane, –78 °C; C = Bu^sLi, THF–hexane, –78 °C; D = Bu^tLi, ether–hexane, –78 °C

Scheme 112

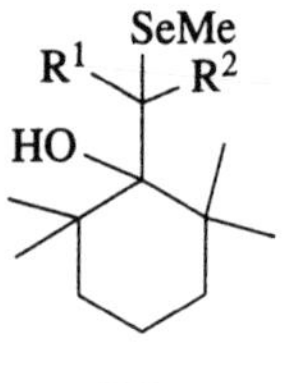

(11)

	R^1	R^2	Method A Yield (%)	Method B Yield (%)	Ref.
(a)	H	H	83	87	188, 189
(b)	H	Me	83	85	188, 189
(c)	Me	Me	77	85	188, 189

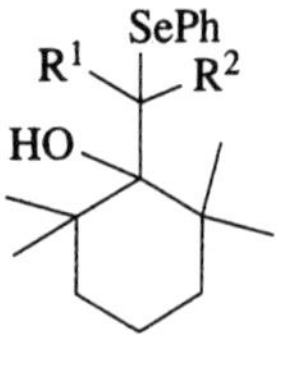

(12)

	R^1	R^2	Method A Yield (%)	Method B Yield (%)	Ref.
(a)	H	H	22	85	188, 189
(b)	H	Me	5	63	188, 189
(c)	Me	Me	0	60	188, 189

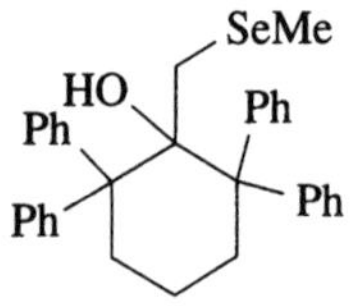

(13)

0% by method A, C or D (ref. 188)

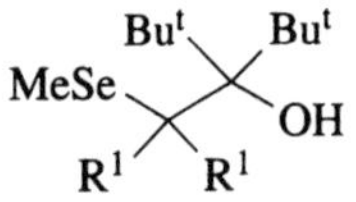

(14)

	R^1	Yield (%)	Method	Ref.
(a)	H	81	C	189, 206
(b)	Me	76	D	189, 206

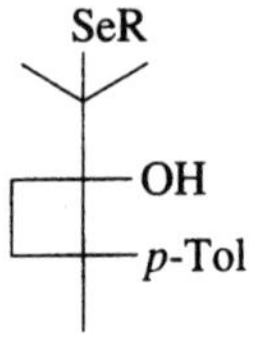

(15)

	R	Yield (%)	Method	Ref.
(a)	Me	0	A	160
	Me	66	B	160
(b)	Ph	0	A	160
	Ph	49	B	160

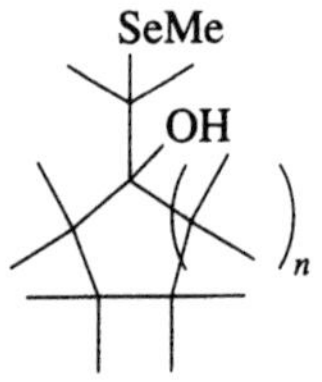

(16)

	n	Yield (%)	Method	Ref.
(a)	0	73	B	190, 191
(b)	1	73	B	190, 191

A = Bu^nLi, THF–hexane, –78 °C; B = Bu^sLi, ether–hexane, –78 °C; C = Bu^sLi, THF–hexane, –78 °C; D = Bu^tLi, ether–hexane, –45 °C

Scheme 113

when the less hindered or the less enolizable of the two carbonyl compounds is transformed to the α-selenoalkyllithium (compare **8b** and **8c** in Scheme 112 to **14** in Scheme 113).

α-Selenoalkyllithiums react efficiently with highly hindered carbonyl compounds such as 2,2,6-trimethylcyclohexanone (see **17**; Scheme 114),[12,77,188,189] 2,2,6,6-tetramethylcyclohexanone (see **11** and **12**; Scheme 113),[12,77,188,189] permethylcyclobutanone,[190,191] permethylcyclopentanone (see **16**; Scheme 113)[190,191] and di-*t*-butyl ketone (see **14**; Scheme 113). However, they do not add to 2,2,6,6-tetraphenylcyclohexanone (see **13**; Scheme 113).[189] Their reactivity towards highly enolizable carbonyl com-

		R	*R¹*	*R²*	*Yield (%)*	*Ax:eq* [a]	*Method*	*Ref.*
(17)	(a)	Me	H	H	92	18:82	A	188, 189
	(b)	Me	H	H	85	31:69	D	188, 189
	(c)	Me	H	Me	85	95:5	A or C	188, 189
	(d)	Me	Me	Me	48	33:67	A	188, 189
	(e)	Me	Me	Me	85	95:5	D	188, 189
	(f)	Ph	Me	Me	0	–	A	188, 189
	(g)	Ph	Me	Me	84	95:5	D	188, 189
(18)	(a)		H	H	85	75:25	A	189, 210
	(b)		H	H	75	69:31	D	189, 210
	(c)		H	Me	70	90:10	A	189, 210
	(d)		H	Me	79	90:10	D	189, 210
	(e)		Me	Me	75	100:0	A or D	189, 210

A = Bu^nLi, THF–hexane, –78 °C; B = Bu^sLi, ether–hexane, –78 °C; C = Bu^sLi, THF–hexane, –78 °C; D = Bu^tLi, ether–hexane, –78 °C.

[a] Axial alcohol:equatorial alcohol.

Scheme 114

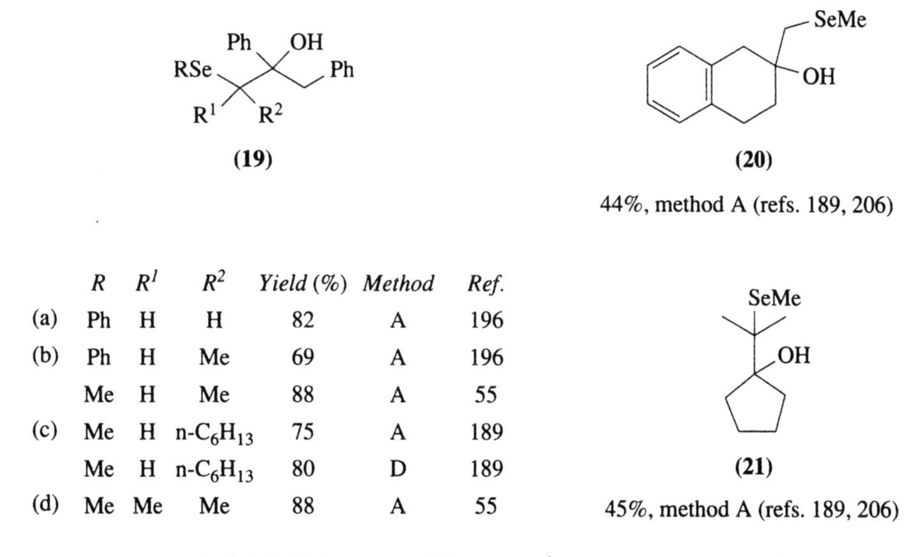

	R	*R¹*	*R²*	*Yield (%)*	*Method*	*Ref.*
(a)	Ph	H	H	82	A	196
(b)	Ph	H	Me	69	A	196
	Me	H	Me	88	A	55
(c)	Me	H	n-C_6H_{13}	75	A	189
	Me	H	n-C_6H_{13}	80	D	189
(d)	Me	Me	Me	88	A	55

(19)

(20) 44%, method A (refs. 189, 206)

(21) 45%, method A (refs. 189, 206)

A = Bu^nLi, THF–hexane, –78 °C; D = Bu^tLi, ether–hexane, –78 °C

Scheme 115

pounds is somewhat capricious. Deoxybenzoin, a particularly enolizable derivative, is efficiently transformed to β-hydroxyalkyl selenides (**19** in Scheme 115; **24** in Scheme 116),[55,193,194] whereas β-tetralone[189] and cyclopentanone[189] are partially enolized (**20** and **21**; Scheme 115).

α-Selenoalkyllithiums exhibit towards carbonyl compounds a nucleophilicity almost identical to that of the corresponding α-thioalkyllithiums[12,77,209] but slightly superior to that of α-metalloalkyl selenoxides[98,127,128,192] and several orders of magnitude superior to that of selenium[132,206] and sulfur ylides.[25,32,33] The reasons for such exceptional nucleophilicity are not yet understood. Reactions involving single-electron transfer could take place but experimental support is not available.

α-Selenoalkyllithiums which bear another heteroatomic moiety on the carbanionic center are still valuable nucleophiles towards carbonyl compounds but they are not as efficient as the α-selenoalkyllithiums reported above.[71,100] Even so, the presence of a trimethylsilyl substituent on the carbanionic center substantially decreases[100,171] the reactivity of these organometallics (Scheme 96), whereas a phospho-

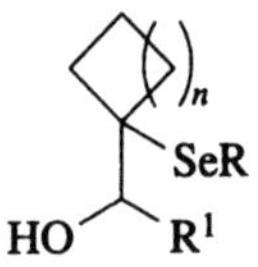

(22)

n	R	R^1	Yield (%)	Method	Ref.
0	Me	$C_{10}H_{21}$	72	A	12, 77
0	Ph	c-C_6H_{11}	63	A	12, 77
0	Me	c-C_6H_{11}	73	A	12, 77
1	Ph	$C_{10}H_{21}$	71	D	136, 156
1	Me	$C_{10}H_{21}$	90	D	136, 156

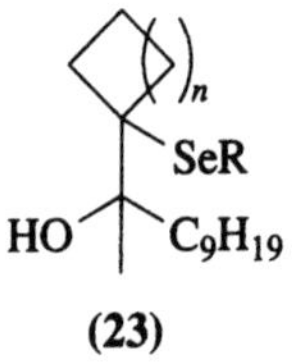

(23)

n	R	Yield (%)	Method	Ref.
0	Ph	54	A	12, 77
0	Me	69	A	136
1	Ph	70	A	136
1	Me	78	A	136

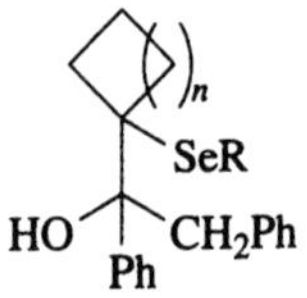

(24)

n	R	Yield (%)	Method	Ref.
0	Ph	61	A	12, 77
0	Me	56	A	12, 77
0	Me	70	B	12, 159
1	Me	68	A	12, 77

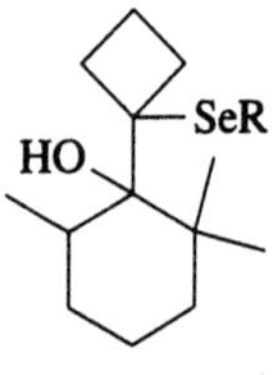

(25)

n	R	Yield (%)	Method	Ref.
0	Ph	47	A	12, 77
0	Me	64	C	12, 159
0	Ph	35	A	12, 159
1	Me	54	A	12, 159
1	Me	68	C	12, 159

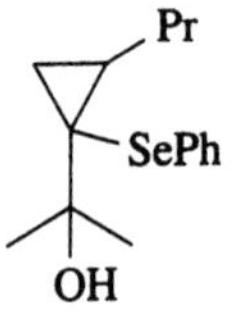

(26)

58%, method A (refs. 12 and 77)
60%, method B (refs. 12 and 77)

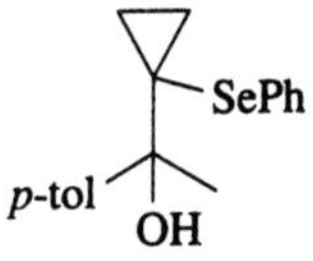

(27)

R = Ph, 81%, method C (ref. 160)
R = Me, 85%, method C (ref. 160)

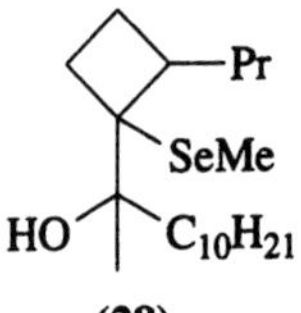

(28)

71%, method A (refs. 12 and 77)

A = Bu^{n}Li, THF–hexane, –78 °C; B = Bu^{s}Li, ether–hexane, –78 °C; C = Bu^{s}Li, THF–hexane, –78 °C; D = Bu^{t}Li, ether–hexane, –78 °C

Scheme 116

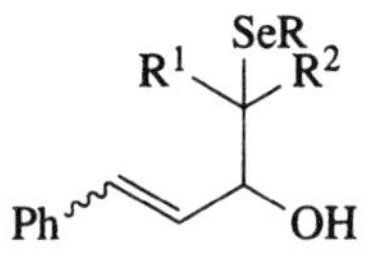

(29)

	R	*R¹*	*R²*	*Yield (%)*	*Method*	*Ref.*
(a)	Me	H	Me	90	A	222
(b)	Ph	H	Me	66	A	223
(c)	Ph	H	n-C_6H_{13}	68	A	223
(d)	Me	Me	Me	85	A	222
(e)	Ph	Me	Me	78	A	221

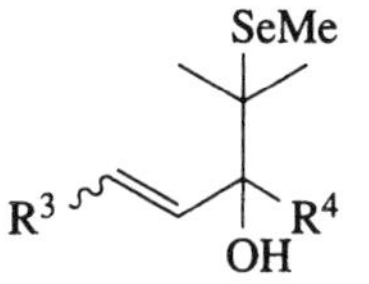

(30)

	R³	*R⁴*	*Yield (%)*	*Method*	*Ref.*
(a)	Pr^n	Me	74	B	223
(b)	Pr^n	Et	74	B	223
(c)	H	Me	55	A	221
(d)	Me	Me	45	A	221

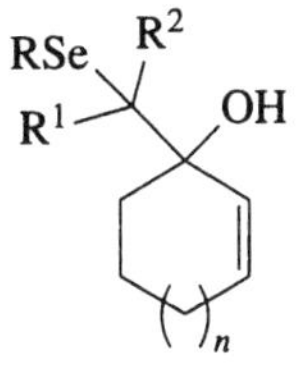

(31)

	n	*R*	*R¹*	*R²*	*Yield (%)*	*Method*	*Ref.*
(a)	1	Me	H	n-C_6H_{13}	64	A	222
(b)	1	Me	Me	Me	55	A	221
(c)	1	Ph	Me	Me	83	B	225
(d)	2	Me	Me	Me	57	B	225
(e)	2	Ph	Me	Me	51	B	225
(f)	7	Me	Me	Me	54	B	225
(g)	1	Me	CH_2-CH_2		86	A	12, 77
(h)	1	Ph	CH_2-CH_2		88	A	12, 77

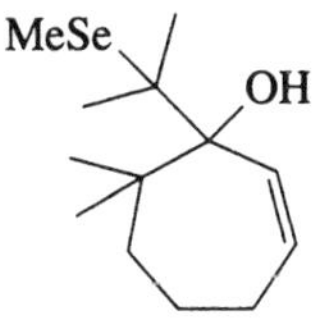

(32)

14%, method A
46%, method B (ref. 225)

A = Bu^nLi, THF–hexane, –78 °C; B = Bu^sLi, ether–hexane, –78 °C; C = Bu^sLi, THF–hexane, –78 °C; D = Bu^tLi, ether–hexane, –78 °C

Scheme 117

nio[106,107] or a phosphonato[102] group dramatically lowers its reactivity. For example, α-phosphonio- and α-phosphonato-α-selenoalkyl metals[102] efficiently react with aldehydes[102,106] but not with ketones.[102,106,107]

2.6.3.3 Stereochemistry of the Addition of α-Selenoalkyllithiums to Aldehydes and Ketones

The reactions of α-selenoalkyllithiums with aliphatic and aromatic aldehydes and ketones are not usually stereoselective regardless of the solvent used (ether or THF). Even in the most favorable cases, such as that of 1-methylseleno-2,2-dimethylpropyllithium and heptanal, in which well-differentiated bulky groups are involved, the stereoisomeric ratio ranges from 1:1 to 3:2 (Scheme 124).[189,206] However, in the case shown in Scheme 86, b,[167d] in which the lithium can coordinate to the silyloxy group, only one of the two stereoisomeric β-hydroxyalkyl selenides is formed.

α-Selenoalkyllithiums react stereoselectively with rigid cyclohexanones such as 4-*t*-butylcyclohexanone (**17**; Scheme 114)[210] and 2,2,6-trimethylcyclohexanone (**18**; Scheme 114).[188,189] In most cases the axial alcohols resulting from the equatorial attack of the organometallic are produced (either in THF or

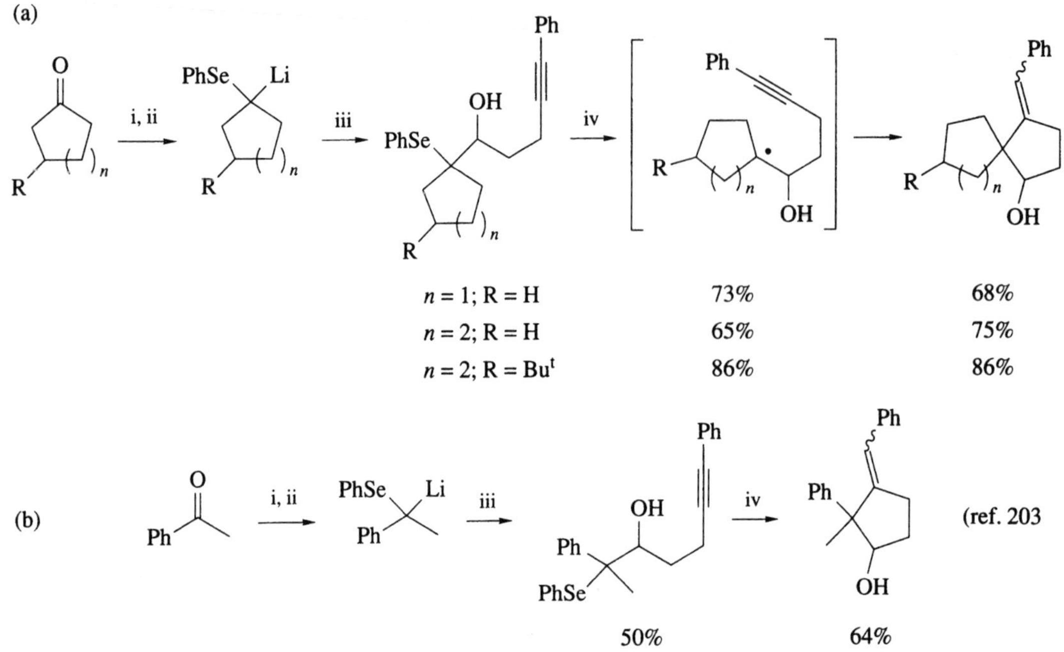

(a)

$n = 1$; R = H	73%	68%
$n = 2$; R = H	65%	75%
$n = 2$; R = Bu^t	86%	86%

(b) (ref. 203)

50% 64%

i, PhSeH, $ZnCl_2$; ii, Bu^{n}Li; iii, PhC≡C$(CH_2)_2$CHO; iv, Ph_3SnH, AIBN, benzene, 90 °C, 14 h

Scheme 118

ether). In rare cases, such as that of 2-seleno-2-propyllithium and 2,2,6-trimethylcyclohexanone (**17d** and **17e**; Scheme 114),[188] however, the axial: equatorial ratio is highly dependent upon the solvent used (see also **17a**, **17b**, **18a** and **18b** in Scheme 114). Otherwise, 1-selenocyclopropyllithiums react stereoselectively with α-methylseleno aldehydes and lead to β-hydroxyalkyl methyl selenides with stereochemistry that can be accounted for by the Cram[211,212] and Felkin rules (Scheme 125).[213,214] Finally, methylselenomethyllithium adds on the less-hindered face of protected pregnenolone and related steroidal ketones to produce the (20*R*)-derivatives in good yield (Scheme 126).[206,215,216]

2.6.3.4 The Ambident Reactivity of α-Selenoallyllithiums

α-Metalloallyl selenides are ambident nucleophiles which can react[30] at their α- and γ-sites with electrophiles. The regiochemistry of the reaction not only depends upon the nature and the number of substituents present on the allylic or vinylic carbons but also upon the nature of the metallic counterion. Thus aldehydes (Scheme 16, e)[45,81,82] and ketones (Scheme 16, c),[80] except cyclopentenone (Scheme 127, a),[83,90] exhibit a high propensity to react at the γ-site of the parent phenylseleno-[80–82,90] and methylseleno-allyllithiums.[45]

The presence of one alkyl[80] or vinyl[91] group at the γ-site usually inverts the regiochemistry of the reaction which now takes place almost exclusively at the α-site (Scheme 17, b–d), except in the case

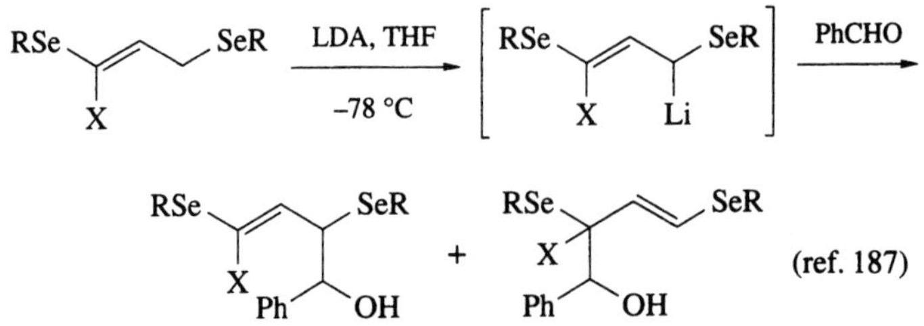

LDA, THF, –78 °C; PhCHO (ref. 187)

R = Ph; X = H; 71%; 50:50
R = Me; X = SeMe; 81%; 61:39

Scheme 119

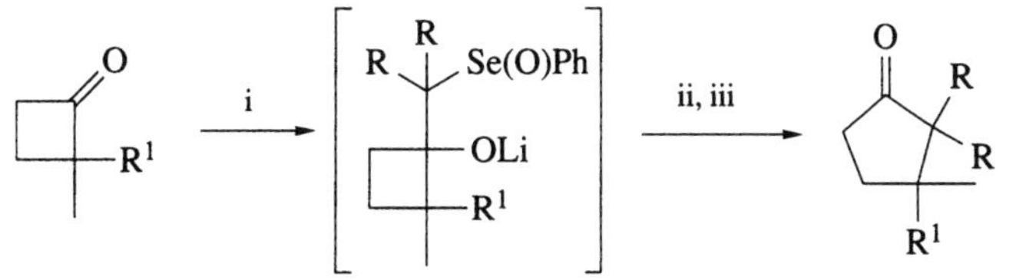

	R	*R¹*	*Yield (%)*	*Ref.*
(a)	H	n-C_6H_{13}	78	128
	H	Bu^t	9	128
	H	$CH_2CH=CMe_2$	56	128
(b)	Me	n-C_6H_{13}	36	128
(c)	Me	*p*-tol	39	127

i, $PhSe(O)CR_2Li$, THF, –78 °C; ii, 55 °C, 1 h; iii, Al–Hg

Scheme 120

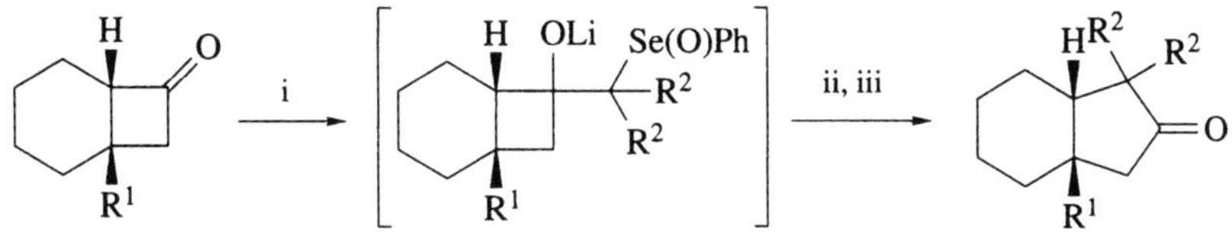

R¹	*R²*	*Yield (%)*
H	H	48
Ph	H	34
Me	Ph	46

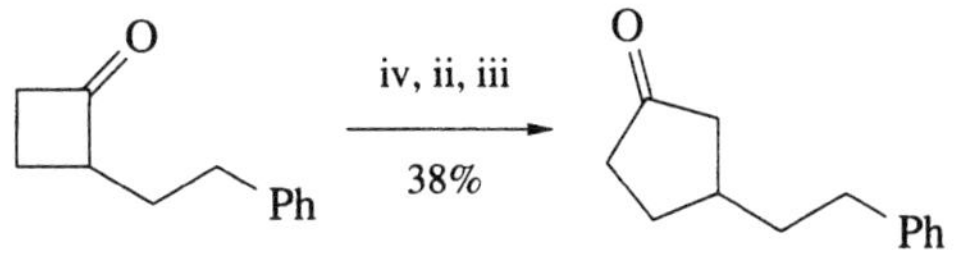

iv, ii, iii 38% (ref. 128)

i, $PhSe(O)CR^2_2Li$, THF, –78 °C; ii, 55 °C, 2–3 h; iii, Al–Hg; iv, $PhSe(O)CH_2Li$, THF, –78 °C

Scheme 121

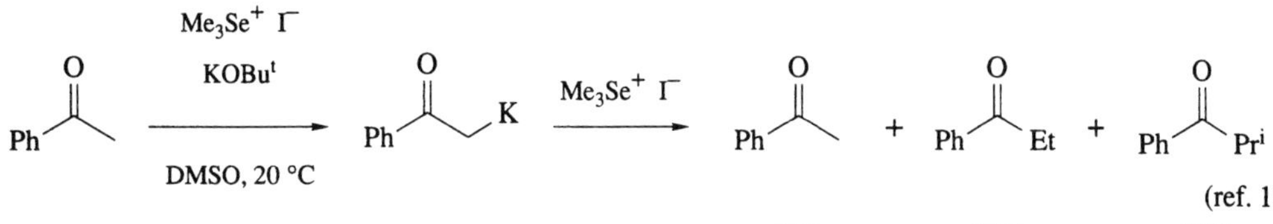

$Me_3Se^+ I^-$, $KOBu^t$, DMSO, 20 °C; $Me_3Se^+ I^-$

40% 13% 28% (ref. 132)

Scheme 122

shown in Scheme 94, a.[87] 1-Lithio-1-seleno-3-methylpropene[45,87] reacts exclusively from its α-site with benzaldehyde (Scheme 18, c and d), whereas a mixture of the two regioisomers is produced when the potassium analog is used instead (Scheme 18, e).[45] Presumably the metal coordinates to the less-hindered carbon and the reaction takes place through a six-membered cycle, implying a precoordination of the oxygen of the carbonyl compound with the metal counterion.

Although carbonyl compounds add at the γ-site of phenylselenoallyllithium, the reversed regioselectivity is realized[81,82] by using a triethylaluminum ate complex (Scheme 16, compare b with c and d with e; Scheme 128). Unfortunately, with benzaldehyde and acetophenone the degree of regiochemical convergence is not so high,[82] presumably owing to the steric effect of the carbonyl group. The *threo*

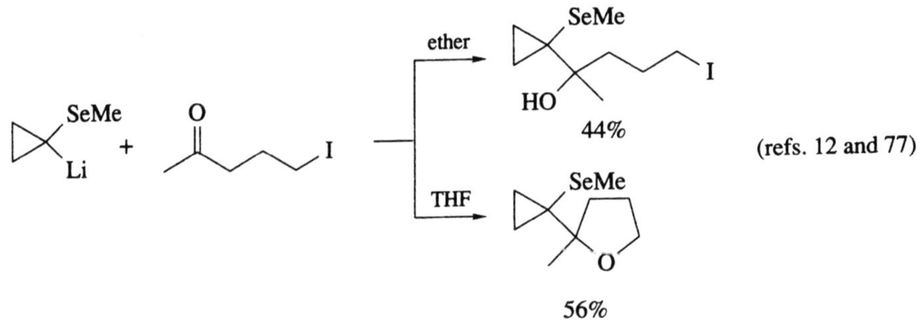

(refs. 12 and 77)

ether: 44%; THF: 56%

Scheme 123

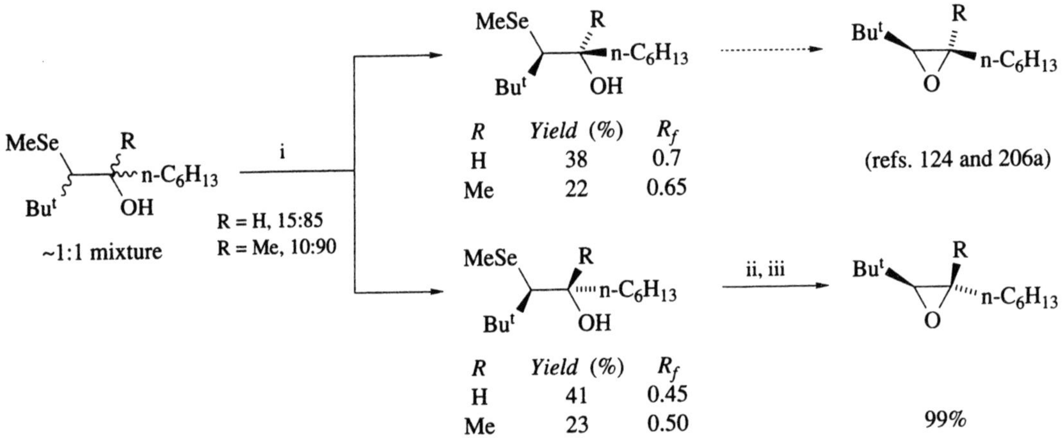

(refs. 124 and 206a)

R	*Yield (%)*	R_f
H	38	0.7
Me	22	0.65

~1:1 mixture; R = H, 15:85; R = Me, 10:90

R	*Yield (%)*	R_f
H	41	0.45
Me	23	0.50

99%

i, TLC, SiO_2, ether/pentane; ii, $MeSO_3F$, ether; iii, 10% KOH, ether

Scheme 124

isomer is predominantly produced and results[82] probably from the *trans* geometry of the intermediate ate complex. This reaction has been successfully applied[82] to the synthesis of 9,11-dodecadien-1-yl acetate, a pheromone of *Disparosis castanea* (Scheme 128). The reaction is highly chemoselective, since ordinary organic halides, trimethylsilyl chloride and trimethylsilyl acetates, which usually react[80] with the lithium derivative, are unreactive towards the aluminum ate complex.[82]

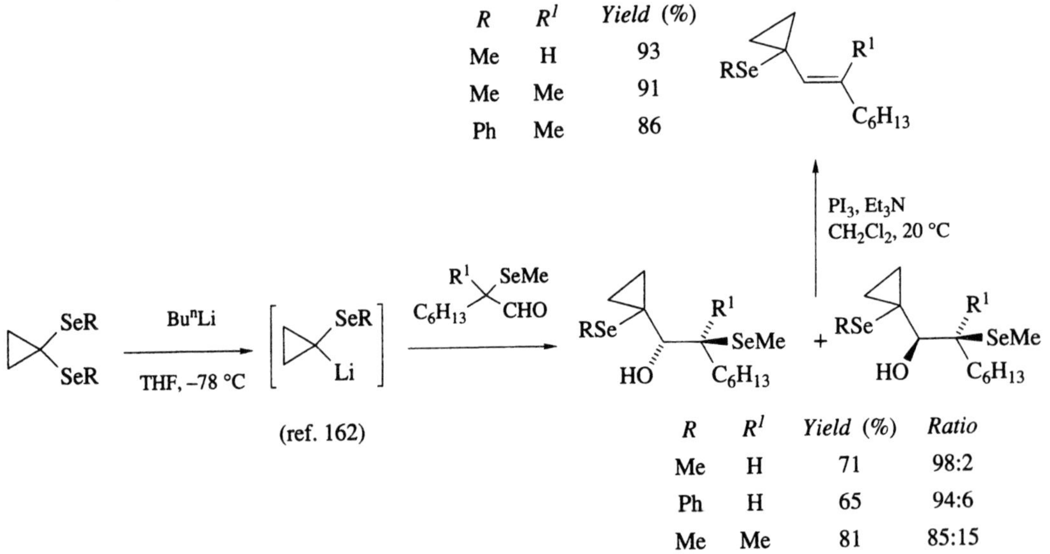

R	R^1	*Yield (%)*
Me	H	93
Me	Me	91
Ph	Me	86

PI_3, Et_3N, CH_2Cl_2, 20 °C

Bu^nLi, THF, –78 °C

(ref. 162)

R	R^1	*Yield (%)*	*Ratio*
Me	H	71	98:2
Ph	H	65	94:6
Me	Me	81	85:15

Scheme 125

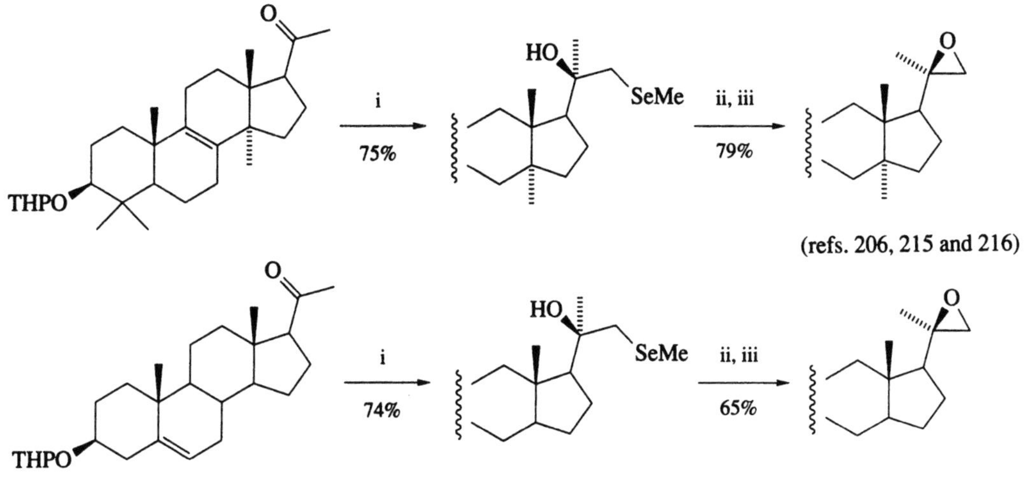

(refs. 206, 215 and 216)

i, $MeSeCH_2Li$, ether–hexane; ii, $MeSO_3F$, 0 to 20 °C, 1 h; iii, 10% aq. KOH, ether, 20 °C, 16 h

Scheme 126

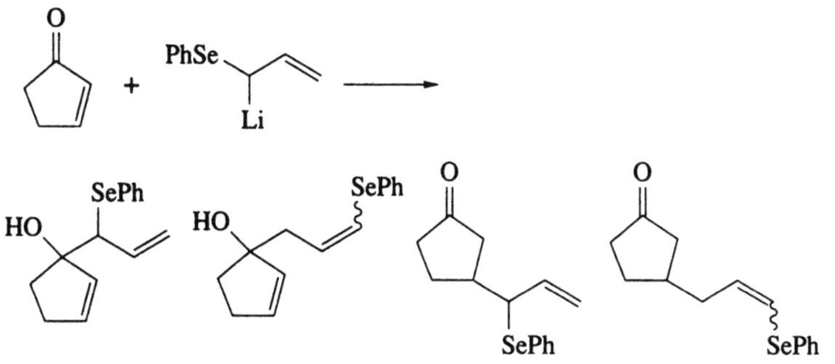

	Conditions	*Yield (%) (stereoisomeric ratio)*			(refs. 83 and 90)
(a)	i, THF, –78 °C; ii, H^+	46 (48:52)	16 (48:52)	10	8 (41:59)
(b)	i, THF–HMPA, –78 °C; ii, H^+	—	—	81 (54:46)	5

Scheme 127

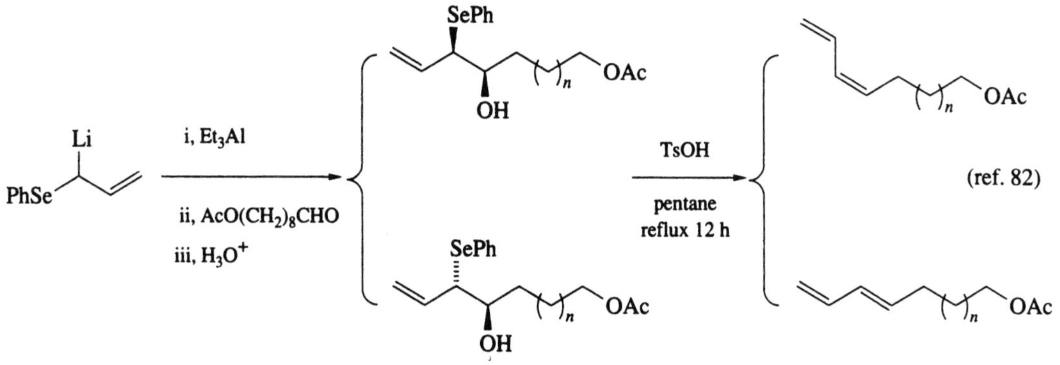

i, Et_3Al; ii, $AcO(CH_2)_8CHO$; iii, H_3O^+

TsOH, pentane, reflux 12 h

70% overall
86:14 *(syn:anti)*

77% *(E):(Z)* = 86:14

(ref. 82)

pheromone of *Diparopsis castanea* (*E:Z* ~ 90:10)

Scheme 128

α-Heterosubstituted allyl metal–boronate complexes[82] usually behave like their aluminum analogs. This is not the case of allyl metals bearing a seleno moiety.[217] These in fact undergo a facile migration from boron to the α-carbon. The allylic rearrangement of the resulting boron–selenium complex (**33** to **34**; Scheme 129) is slow[217] in comparison with the usual allylic boranes rearrangement. Thus the trialkylborane–α-phenylselenoallylithium complex produced at –78 °C rapidly leads to the ate complex (**33**) after 1 h at 0 °C and to the complex (**34**) after 12 h at 20 °C. Each one reacts regioselectively with carbonyl compounds at the position where the boron was attached (Scheme 129). Otherwise, 1-lithio-1,3-bis(phenylseleno)-2-propene bearing an α-trimethylsilyl group regioselectively reacts with acetone at the γ-carbon (Scheme 33, a),[85] whereas a mixture of regioisomers is produced from benzaldehyde and 1,1,3-tris(methylseleno)propene (Scheme 130).[87]

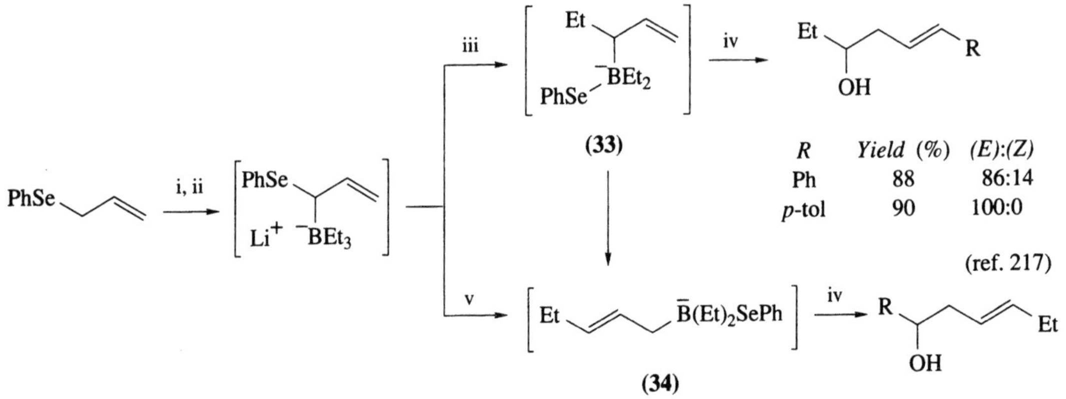

R	*Yield (%)*	*(E):(Z)*
Ph	88	86:14
p-tol	90	100:0

(ref. 217)

(33) **(34)**

R	*Yield (%)*	*Erthyro:threo*
Ph	89	24:76
p-tol	92	36:64
p-$NO_2C_6H_4$	93	18:82
Pr^n	85	25:75

i, LDA, –78 °C; ii, Et_3B, –78 °C; iii, 0 °C, 1 h; iv, RCHO; v,excess BEt_3, 20 °C, 12 h

Scheme 129

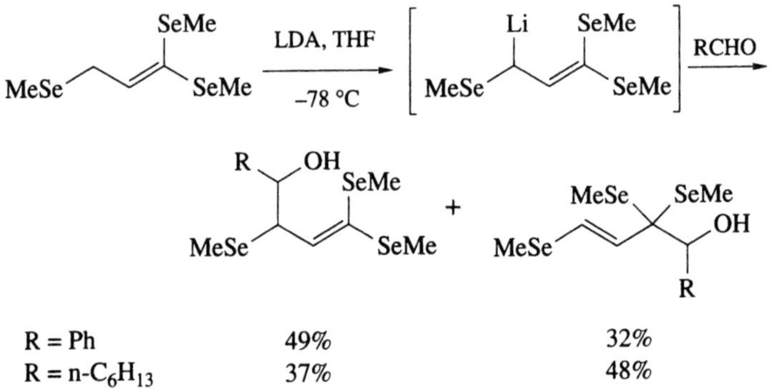

(ref. 87)

R = Ph	49%	32%
R = n-C_6H_{13}	37%	48%

Scheme 130

2.6.3.5 Control of the Regiochemistry of Addition of α-Selenoalkyl Metals to Enones, Enals and Enoates

2.6.3.5.1 Generalities

The control of the regiochemistry of addition of organometallic α-enones, α-enals and α-ene esters has attracted the attention of many chemists from the early age of organic chemistry. Not only have α-selenoalkyllithiums not escaped the rule, but fundamental discoveries made at the occasion of this work have been successfully extended to other organometallics.[7]

The structure of the organometallic and of the α,β-unsaturated carbonyl compound, as well as the experimental conditions, play a crucial role in controlling the regiochemistry of the reaction. In general, enals possess a higher tendency to react at their C-1 site than enones and ene esters and increasing the substitution around the C-1 (or the C-3) site favors the reaction at the C-3 (or at the C-1) site of the enones. Furthermore, the compounds which possess either a privileged cisoid conformation, or the lowest half-wave potential of electrolytic reduction,[218] or the lower level of their LUMO orbitals,[219,220] react predominantly at the C-3 site (PhCH═CHC(═O)Ph >> PhCH═CHC(═O)Me ≈ MeCH═CHC(═O)Ph >> MeCH═CHC(═O)Me).

Performing the reaction at a higher temperature and in a more polar solvent usually increases[7] the amount of the C-3 over the C-1 adduct. The C-1:C-3 ratio is also highly dependent upon the nature of the α-selenoalkyllithium. In this respect, organometallics where the carbanion is directly linked to a selenium atom can be classified into three different categories, each of which behaves differently towards α,β-unsaturated carbonyl compounds: (i) α-selenoalkyllithiums bearing hydrogens or alkyl groups on the carbanionic center which possess, whatever the solvent and the conditions used, the highest propensity to add at the C-1 site of enones; (ii) 1,1-bis(seleno)-1-alkyllithiums where the C-1:C-3 reactivity can be adjusted under kinetically controlled conditions by the proper choice of the solvent; and (iii) enolates derived from α-seleno esters which possess a tendency to add at the C-1 site of α,β-unsaturated carbonyl compounds under kinetically controlled conditions and at the C-3 site under thermodynamic control.

2.6.3.5.2 Reactivity of α-selenoalkyl metals, α-selenoxy-γ-alkyl metals and selenium ylides with enals and enones

α-Selenoalkyllithiums bearing hydrogen or alkyl groups on the carbanionic center possess the highest tendency to react at the C-1 site of enals (Schemes 117, 131 and 132)[12,221–225] and enones (Schemes 117, 133 and 134).[56,77,221,225–227] They usually provide stereoisomeric mixtures of β-hydroxy-γ-alkenyl selenides in good yields.

The reaction takes another course[221] with chalcone since a mixture of β-hydroxyalkyl selenides and γ-selenoalkyl phenyl ketones resulting from C-1 and C-3 attack, respectively, is produced (Scheme 135, compare b and c with a). This enone is the only one among those tested to react in such a way.[221] Interes-

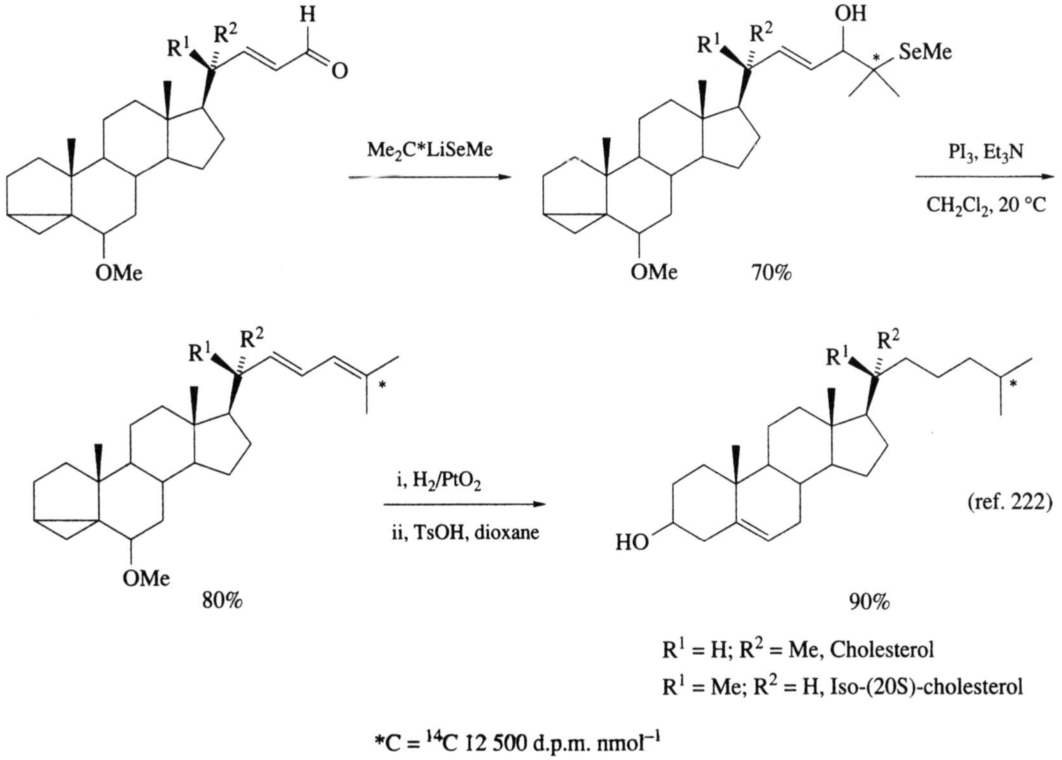

$Me_2C^*LiSeMe$ → 70%; PI_3, Et_3N, CH_2Cl_2, 20 °C → 80%; i, H_2/PtO_2; ii, TsOH, dioxane → 90% (ref. 222)

R^1 = H; R^2 = Me, Cholesterol

R^1 = Me; R^2 = H, Iso-(20S)-cholesterol

*C = ^{14}C 12 500 d.p.m. nmol^{-1}

Scheme 131

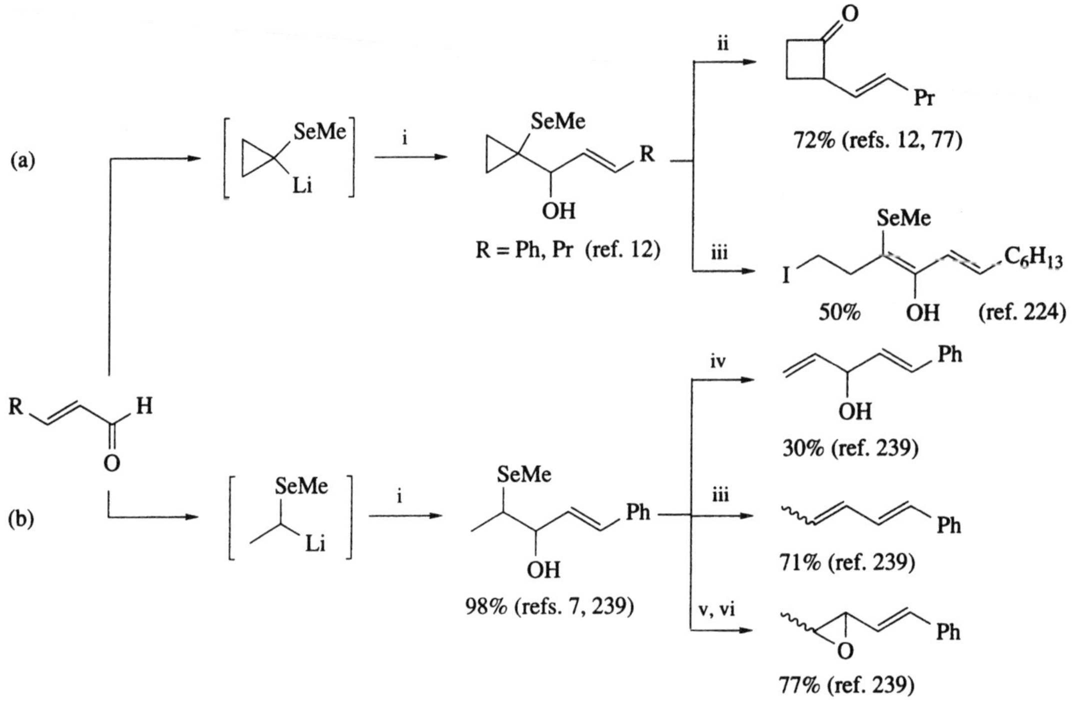
(a)
SeMe
Li
i
SeMe
R
OH
R = Ph, Pr (ref. 12)
ii
O
Pr
72% (refs. 12, 77)
iii
SeMe
I
C6H13
50%
OH
(ref. 224)
R
H
O
(b)
SeMe
Li
i
SeMe
Ph
OH
98% (refs. 7, 239)
iv
Ph
OH
30% (ref. 239)
iii
Ph
71% (ref. 239)
v, vi
Ph
O
77% (ref. 239)

i, THF, –78 °C; ii, TsOH, wet benzene, 80 °C; iii, PI_3, Et_3N, CH_2Cl_2, 20 °C; iv, Bu^{t}OOH, Al_2O_3, THF, 55 °C; v, MeI (neat), 20 °C; vi, aq. KOH, ether, 20 °C

Scheme 132

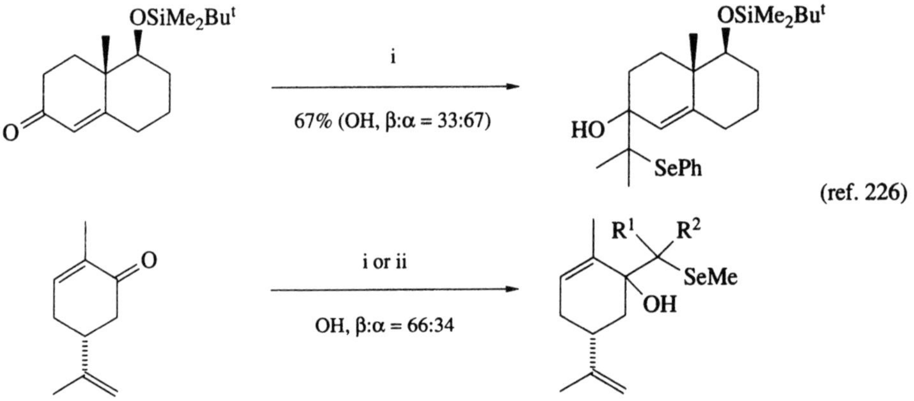
OSiMe2But
O
i
67% (OH, β:α = 33:67)
OSiMe2But
HO
SePh
(ref. 226)
O
i or ii
OH, β:α = 66:34
R1
R2
SeMe
OH

i, $R^1 = R^2$ = Me; 82%: ii, R^1 = H, $R^2 = C_5H_{11}$; 90%

i, $PhSeCMe_2Li$, THF, –78 °C; ii, $PhSeCH(C_5H_{11})Li$

Scheme 133

tingly, the higher percentage of C-3 attack is observed[221] with the α-selenoalkyllithium which possesses the more alkyl substituted carbanionic center. Furthermore, in all these cases the best yields have been observed when the reactions are performed at low temperature (–78 °C) in ether, but it has been noticed that the C-1:C-3 ratio of attack is not affected by the type of solvent (ether, THF or THF–HMPA), the temperature at which the reaction is performed or the reaction time.

α-Selenoxyalkyllithiums in THF (Scheme 136)[97] and methylenedimethylselenurane (generated from trimethylselenonium iodide and potassium *t*-butoxide) in DMSO (Scheme 59, c)[132] also react at the C-1 site of α,β-unsaturated carbonyl compounds. It is interesting to note that whereas methylenedimethylselenurane adds[132] at the C-1 site of chalcone and produces the corresponding α,β-unsaturated epoxide (Scheme 59, c), the more stabilized phenacylmethylenedimethylselenurane exclusively adds[142] at the C-3 site of chalcone and leads to the corresponding cyclopropane (Scheme 68b).

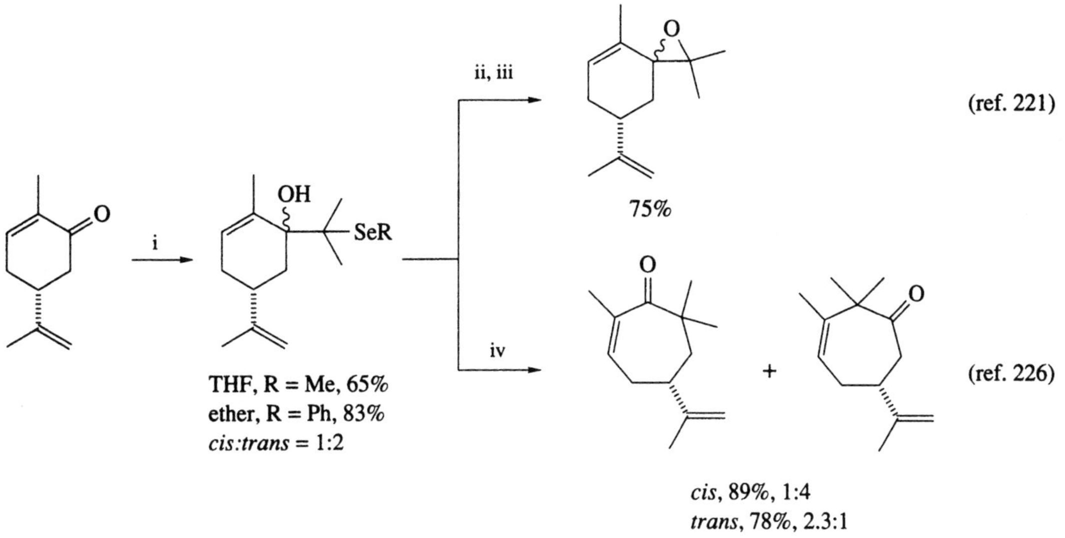

O i → OH SeR

ii, iii → O, 75% (ref. 221)

iv → O + O (ref. 226)

THF, R = Me, 65%
ether, R = Ph, 83%
cis:trans = 1:2

cis, 89%, 1:4
trans, 78%, 2.3:1

i, $Me_2CLi(SeR)$, solvent (THF or ether), –78 °C; ii, MeI; iii, $KOBu^t$, DMSO; iv, TlOEt (5.5 equiv.), $CHCl_3$

Scheme 134

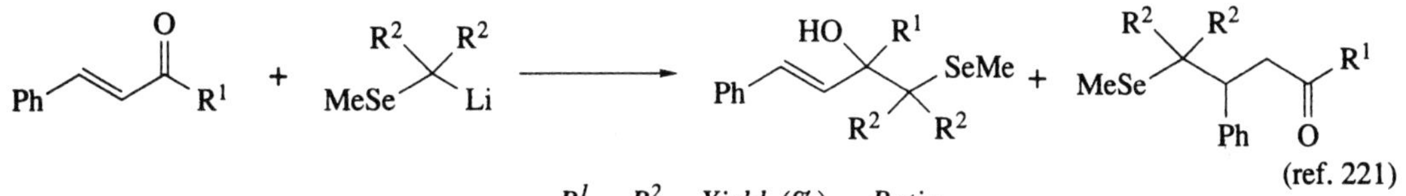

Ph–CH=CH–C(O)–R^1 + R^2 R^2 MeSe Li → Ph HO R^1 SeMe R^2 R^2 + R^2 R^2 MeSe Ph O R^1

(ref. 221)

	R^1	R^2	*Yield* (%)	*Ratio*
(a)	Me	H	87	100:0
(b)	Ph	H	70	50:50
(c)	Ph	Me	68	30:70

Conditions: THF, –78 °C, or THF, –78 to 20 °C, or ether –78 °C

Scheme 135

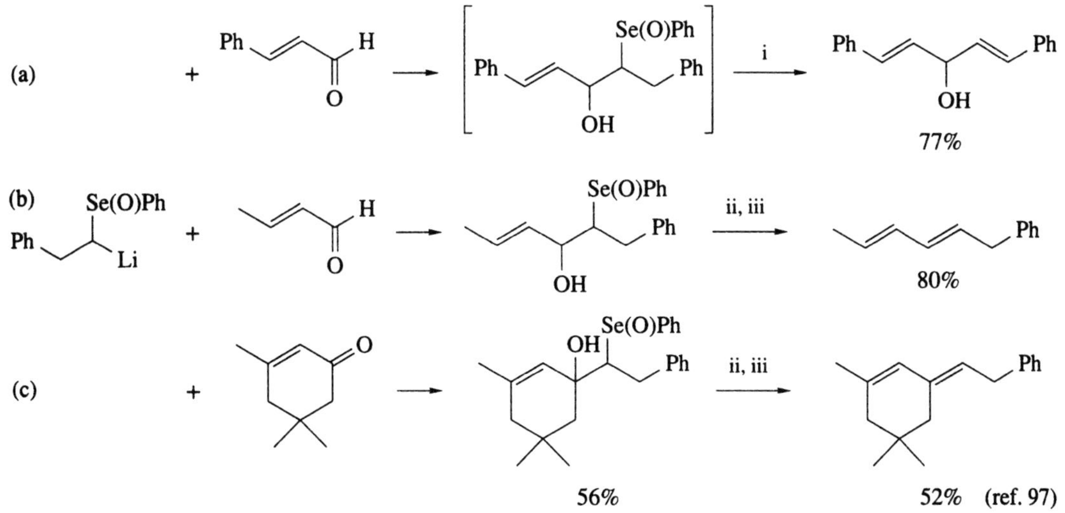

(a) + Ph H O → [Ph Se(O)Ph Ph OH] i → Ph Ph OH 77%

(b) Se(O)Ph Ph Li + H O → Se(O)Ph Ph OH ii, iii → Ph 80%

(c) + O → OH Se(O)Ph Ph ii, iii → Ph

56% 52% (ref. 97)

i, Δ; ii, $EtCO_2H$, $P(OMe)_3$; iii, MsCl, Et_3N

Scheme 136

A great deal of attention has been paid to the addition of α-selenoalkyl metals at the C-3 site of enones. Performing the reaction in the presence of the copper iodide–dimethyl sulfide complex does not lead to the expected results, although a novel species is produced[135] which exhibits different behavior from α-selenoalkyllithiums towards, for example, allyl halides.[135] This novel species, which has even been observed[135] by ^{77}Se NMR, does not react at –78 °C with enones and decomposes in THF or in ether at around –45 °C with no addition to enones. In fact, α-selenoalkyllithiums react at low temperature with the copper iodide–dimethyl sulfide complex in THF or ether to provide,[16,135,228] around –45 °C, alkenes resulting formally from the reductive coupling of the organometallic followed by elimination of the two seleno moieties present on the adjacent carbons (Scheme 137). This reaction is reasonably general since it allows the synthesis of α,β-disubstituted and tetrasubstituted alkenes as mixtures of stereoisomers from α-selenoalkyllithiums bearing one or two alkyl substituents on the carbanionic center. The reaction takes another course when the putative α-selenoalkylcuprate is reacted[229] with the enone in the presence of trimethylsilyl chloride (Scheme 138, compare c with a, b and d), since after hydrolysis the C-3 adducts are obtained from α-selenoalkyllithiums with carbanionic centers that are monoalkyl substituted. Unfortunately, this reaction is not general and those organometallics which bear two alkyl groups on the carbanionic center do not react with enones under such conditions.[229]

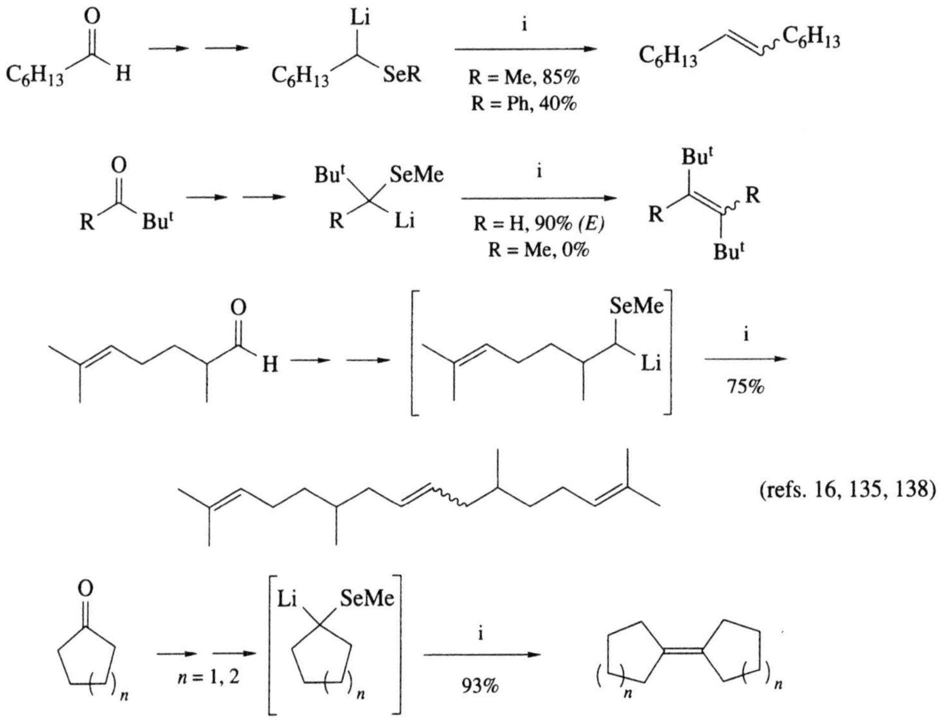

(refs. 16, 135, 138)

i, CuI•SMe$_2$ (0.5 equiv.), ether, –78 to 20 °C

Scheme 137

2.6.3.5.3 Reactivity of 1,1-bis(seleno)-1-alkyl metals and α-selenocarbonyl compounds with enals and enones

(i) Generalities

α-Selenoalkyl metals with carbanionic centers that are substituted with another seleno moiety[7,73,230–234] or an electron-withdrawing group such as an ester[112,113] or a nitrile[117] behave differently. In these cases the nature of the solvent, the seleno moiety attached to the carbanionic center and the metal have an influence on the regiochemical outcome of the reaction. However, whereas the temperature does not affect the regioisomeric ratio with 1-metallo-1,1-bis(seleno)alkanes,[230–232] it plays an especially crucial role with ester enolates derived from α-seleno esters.[112,113,157]

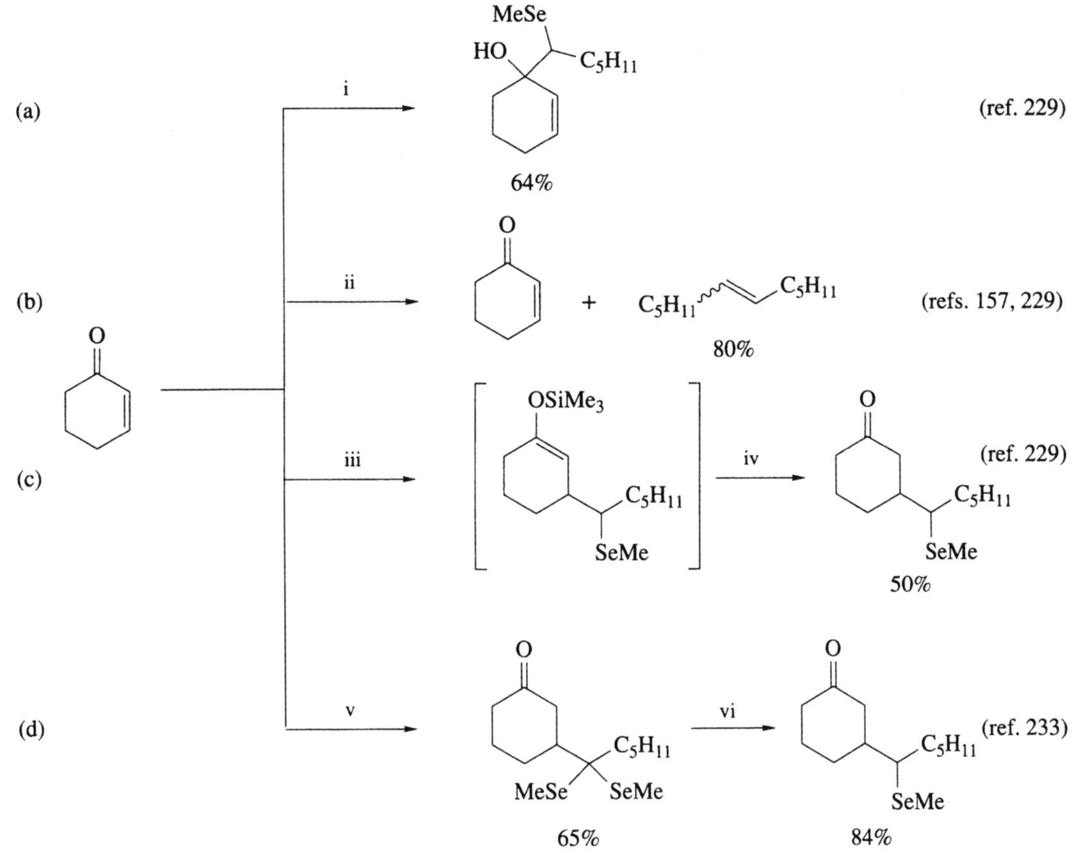

(a) i → 64% (ref. 229)

(b) ii → + 80% (refs. 157, 229)

(c) iii → [ ] iv → 50% (ref. 229)

(d) v → 65% vi → 84% (ref. 233)

i, $C_5H_{11}CH(SeMe)Li$, THF, –78 °C; ii, $[C_5H_{11}CH(SeMe)]_2CuLi$, Me_2S, –78 to 30 °C;
iii, $[C_5H_{11}CH(SeMe)]_2CuLi$, Me_2S, Me_3SiCl, –78 °C; iv, H_3O^+; v, $C_5H_{11}C(SeMe)_2Li$, THF–HMPA;
vi, Bu_3SnH, toluene, 100 °C

Scheme 138

Furthermore, although similar behavior will be often observed with these two classes of compounds, some fundamental differences exist. In fact with 1,1-bis(seleno)alkyllithiums regiochemical control both at C-1 and C-3 has been achieved under kinetically controlled conditions,[7,16,157,230,231] whereas the ester enolates tend to add at C-1 under kinetically controlled conditions[112,113,157] and at C-3 under thermodynamic control.[7,16,112,113,157] The case of 1-lithio-1-(phenylseleno)cyanoalkanes is somewhat different, since they already react at –78 °C in THF with, for example, cyclohexenone to give the C-3 adduct exclusively (Scheme 40, c), and C-1 attack has not yet been achieved.

(ii) Reactivity of 1,1-bis(seleno)-1-alkyl metals with enals and enones

(a) Generalities. 1,1-Bis(seleno)alkyl metals are valuable synthetic intermediates which *inter alia* play the role of acyl anion equivalent. Some synthetic uses of these intermediates are displayed in Schemes 139 and 140, whereas various results concerning the reactivity of such species with enals and enones are gathered in Schemes 141–147.[157]

(b) Control of the C-1 attack. The formation of the C-1 over the C-3 adduct is usually favored if the reactions are carried out: (i) in the less polar and basic solvent compatible with the formation of the organometallic (compare Scheme 140, a to b in Scheme 142, a to b; Scheme 143, a to b; Scheme 144, a to b, c and d; Scheme 145, a with Scheme 146, c and d) THF–hexane is often reasonably good.[230,231] but ether–hexane is even better (Schemes 140, a; Scheme 141; Scheme 144, a; Scheme 145).[233] Generally the better overall yields are obtained when the reactions are carried out in the less polar solvent and at the

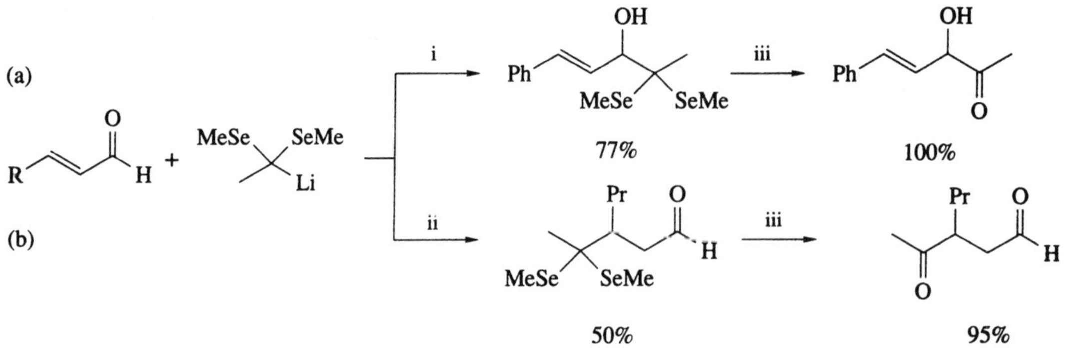

i, THF, –78 °C, 0.2 h; ii, THF, HMPA (2 equiv.), –78 °C, 0.2 h; iii, $CuCl_2$, CuO (4 equiv.), aq. acetone, 20 °C, 0.1 h (ref. 157)

Scheme 139

lowest temperature; (ii) with 1,1-bis(methylseleno)alkyllithiums rather than their phenylseleno analogs (compare Scheme 147, a with Scheme 146, c); and (iii) when potassium[73] rather than lithium salts are involved (compare Scheme 147, a and c).

Enals[232] have the highest propensity to react (Scheme 142, a; Scheme 143, a; compare to Scheme 144, b) at C-1 than straight chain ketones,[157] and this is also the case for cyclohexenones (Scheme 146)[7,230,231] as compared to cyclopentenones[157] and cycloheptenones.[157] Chalcone, however, is a very special compound. It apparently shows a different behavior since the C-1 adduct is undoubtedly favored in the more basic solvents (Scheme 148).[234] In all the cases cited above it has been proven[231] that the reactions proceed under kinetic control, so that once the C-1 adduct is formed it is stable and does not produce the C-3 adduct if the temperature of the medium is raised or if a more basic solvent is added before hydrolysis of the alcoholate.

(c) Control of the C-3 attack. In general, the C-3 adducts can be produced when the reactions are carried out in a more polar or basic solvent, such as THF containing at least 1.1 equiv. of HMPA per mol of organometallic (Scheme 139, b; Scheme 140, b–e; Scheme 141; Scheme 142, b; Scheme 143, b; Scheme 144, c and d; Scheme 146, b, d and f; Scheme 147, b).[230,231] Use of more HMPA, which usually increases the C-3:C-1 ratio, often leads,[231] due to competing enolization, to a much lower yield of γ-oxo selenoacetal (Scheme 141).

HMPA is not the only additive which leads to the formation of the C-3 adduct. Performing the reaction in ether but in the presence of an 18-crown-6 ether (Kryptofix, trade name (Merck) for hexaoxa-4,7,13,16,21,24-diaza-1,10-bicyclo[8.8.8]hexaiosane) or in DME also favors the formation of the C-3 adduct. However, HMPA is by far the best and the more reliable alternative.

The conditions mentioned above are applicable to various enals (Scheme 139, compare b to a; Scheme 141, compare e to d; Scheme 142, compare b to a; Scheme 143, compare b to a)[157,232,233] and straight chain ketones (Scheme 144, compare c and d to a and b),[157,233] although the regiochemical control is less spectacular than with cyclic enones, especially cyclopentenones and cyclohexenones (Scheme 140, compare b–e with a; Scheme 146, compare b, d and f with a, c and e; Scheme 147, compare b to c; Scheme 149).[157,230,231,233] Surprisingly, 1-metallo-1,1-bis(methylseleno)alkanes are more prone than their phenylseleno analogs to react under these conditions at the C-3 site of enals[232] when the reactions are carried out in THF–HMPA, although the reverse was found with cyclohexenone in THF (compare Scheme 147, b with Scheme 146, c; and Scheme 142, c and d with b). Finally, and surprisingly, the addition of 1-lithio-1,1-bis(seleno)alkanes at the C-3 site of chalcone can be achieved[234] if the reaction is performed in a less basic solvent (Scheme 148, compare a to b).

The addition of 1,1-bis(seleno)alkyllithiums at the C-3 site of enals and enones produces enolates which can be trapped with various electrophiles (Scheme 140, b–d; Schemes 149 and 150).[157,230,233] Silylation of the lithium enolates with trimethylsilyl chloride leads to the corresponding silyl enol ethers (Scheme 140, c),[157,230] which can then be subjected to further reactions.

Methylation of the lithium enolates derived from cyclohexen-2-one and 1,1-bis(methylseleno)ethyllithium takes place[230] regioselectively but requires the use of more than one equivalent of HMPA (5 equiv., 0 °C, 1 h). The product consists[230] quite exclusively of the *trans* stereoisomer (99%) with a trace (1%) of the *cis* (Scheme 140, d and e). The latter can be obtained as a single stereoisomer on hydrolysis of the adduct resulting from the reaction of the same organometallics with 2-methylcyclohexen-2-one

i, ether, –78 °C, 0.2 h; ii, $CuCl_2$–CuO (1:4), aq. acetone, 20 °C; iii, $NaIO_4$, EtOH, 20 °C; iv, THF, HMPA (1.1 equiv.), –78 °C, 0.2 h; v, H_3O^+; vi, $CuCl_2$–CuO (1:4), aq. acetone, 20 °C; vii, THF, HMPA (1.1 equiv.), –78 °C, 0.2 h; viii, Me_3SiCl; ix, Bu^nLi, THF, –78 °C; x, $C_5H_{11}CHO$, then H_3O^+; xi, MeLi, 20 °C, then H_3O^+; xii, P_2I_4, Et_3N, CH_2Cl_2, 20 °C, xiii, THF–HMPA (1.1 equiv.), –78 °C, 0.2 h; xiv, MeI (5 equiv.), HMPA (5.5 equiv.), 0 °C, 1 h; xv, H_3O^+; xvi, Bu_3SnH (3 equiv.), AIBN (0.1 equiv.), toluene, 80 °C, 0.3 h; xvii, $CuCl_2$–CuO (1:4), aq. acetone, 20 °C, 0.1 h; xviii, $HgCl_2$, MeCN, 85 °C, 0.3 h

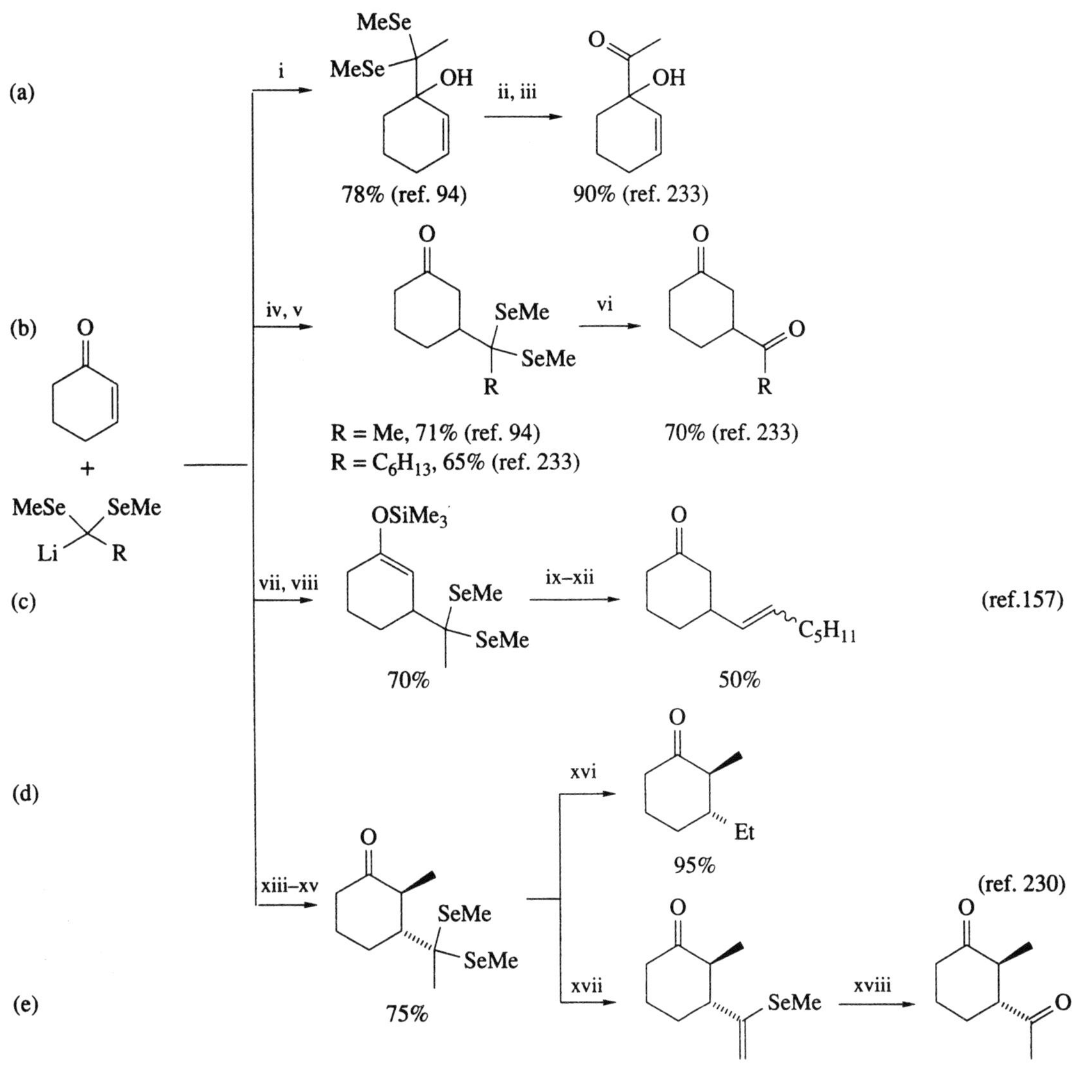

Scheme 140

(Scheme 149).[230] Unfortunately, the methylation of the lithium enolate derived from cyclopentenone is less regioselective (Scheme 150).[157]

The reaction between 1,1-bis(methylseleno)-1-propyllithium and other methylcyclohexenones has also been investigated (Scheme 151).[230] No adduct was obtained[230] from 3-methylcyclohexen-2-ones whatever the conditions used (THF–HMPA or DME), whereas γ-keto selenoacetals are produced from 4-, 5- and 6-methyl-2-cyclohexen-2-ones in yields ranging from 58 to 76% when the reactions are carried out in THF–HMPA (Scheme 151). Yields are lowered by 10% when DME is used instead.[230]

Most of the above mentioned derivatives have been reduced[230] by tributyltin hydride to methyl-3-ethylcyclohexanones. These have been produced with a stereochemical control often superior to that observed from the copper chloride catalyzed addition of ethylmagnesium bromide in ether to the same ketones (Scheme 149, b; Scheme 151, a). Reaction of the C-1 and C-3 adducts with copper chloride (1

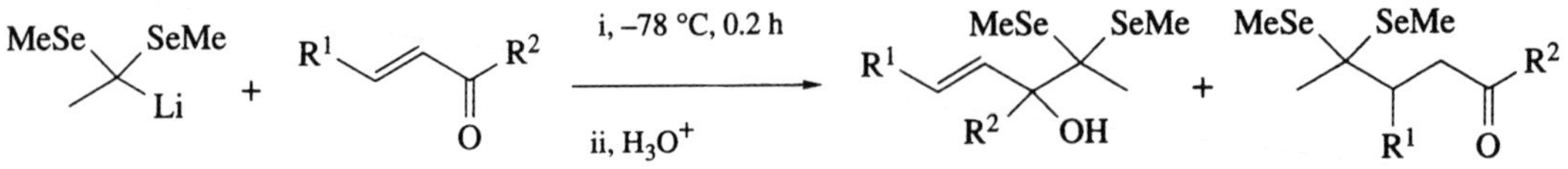

i, –78 °C, 0.2 h; ii, H_3O^+

	Conditions[a]	SM is cyclohexanone Yield (%)	Ratio	SM is hexenal Yield (%)	Ratio
(a)	Et_2O	80	98:2		
(b)	Et_2O, HMPA (0.5 equiv.)[b]	74	75:25		
(c)	Et_2O, Kryptofix 2.2.2	60	28:72		
(d)	THF	63	73:27	74	100:0
(e)	THF–HMPA (1:1.1)	78	1:99	68	28:72
(f)	THF–HMPA (1:4.4)			27	18:82
(g)	THF–Kryptofix 2.2.2				86:14
(h)	THF–Kryptofix 1.1.1				68:32
(i)	THF–Ph_3PO	68	23:77		
(j)	THF–TPPT[c]			58	24:76
(k)	DME	61	13:87		83:17
(l)	DME–HMPA (1:1.1)				24:76

[a] Hexane present in each case. [b] Refers to the additive:organometallic ratio.
[c] TPPT = tripyrrolidinophosphoroimide.

Scheme 141[157, 231]

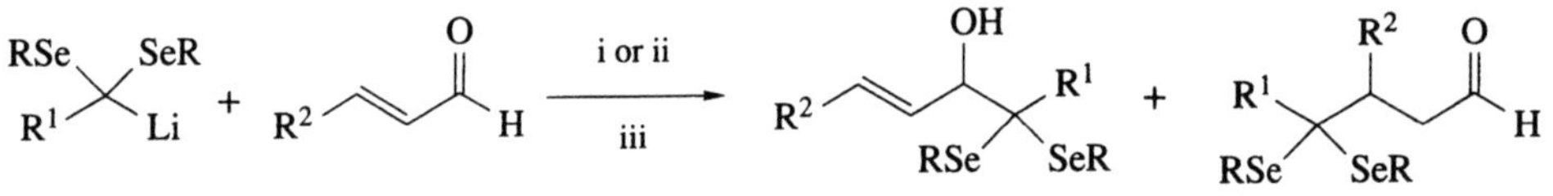

	R	R[1]	R[2]	Conditions	Yield (%)	Ratio
(a)	Me	Me	Pr	i	74	100:0
	Me	Me	Ph	i	74	100:0
(b)	Me	Me	Pr	ii	63	28:72
	Me	Me	Ph	ii	67	57:43
(c)	Ph	H	Pr	ii	55	100:0
(d)	Ph	Me	Pr	ii	60	100:0

i, THF, –78 °C, 0.2 h; ii, THF, HMPA (1.1 equiv.), –78 °C, 0.2 h; iii, NH_4Cl, –78 °C

Scheme 142[118, 157]

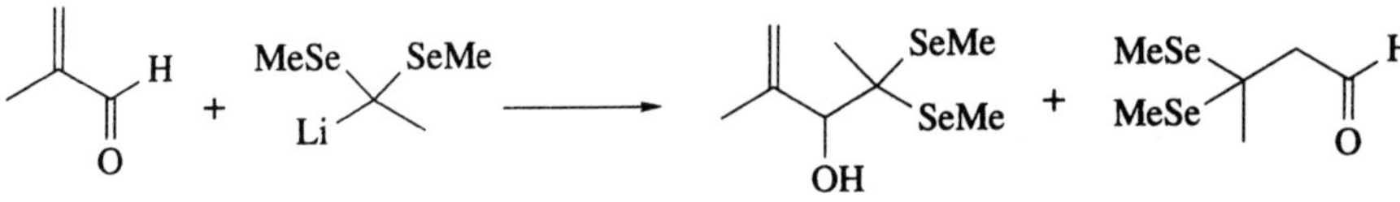

(ref. 157)

	Conditions	Yield (%)	Ratio
(a)	THF, –78 °C, 0.2 h	80	100:0
(b)	THF, HMPA (1.1 equiv.), –78 °C, 0.2 h	56	25:75

Scheme 143

equiv.) and copper oxide (4 equiv.) in aqueous acetone (1%) allows the synthesis of the corresponding α-hydroxy-β,γ-unsaturated carbonyl compounds[73,233] or γ-ketocarbonyl compounds.[233] Therefore,

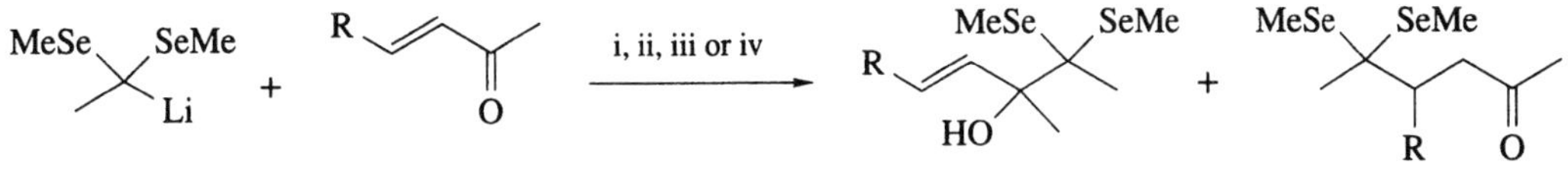

MeSe SeMe / Li + R-CH=CH-C(O)-Me —(i, ii, iii or iv)→ 1,2-adduct + 1,4-adduct (ref. 157)

	R	Conditions	Yield (%)	Ratio
(a)	Me	i	73	100:0
	Pr	i	81	100:0
(b)	Me	ii	51	75:25
	Pr	ii	54	95:5
(c)	Me	iii	70	15:85
	Pr	iii	60	40:60
(d)	Pr	iv	38	35:65

i, Et_2O, –78 °C, 0.2 h; ii, THF, –78 °C, 0.2 h; iii, THF, HMPA (1.1 equiv.), –78 °C, 0.2 h;
iv, THF, HMPA (2.2 equiv.), –78 °C, 0.2 h

Scheme 144

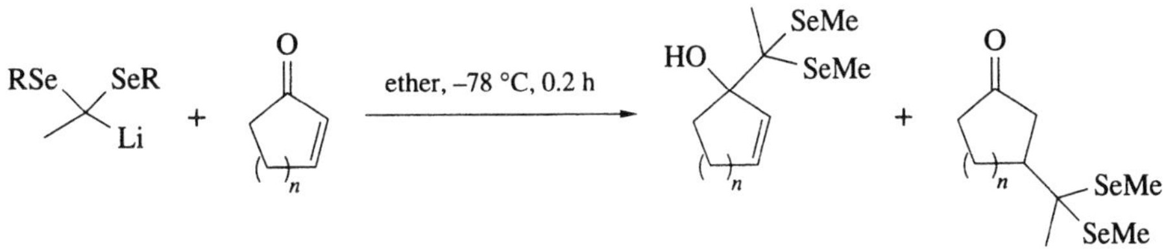

RSe SeR / Li + cycloalkenone —(ether, –78 °C, 0.2 h)→ 1,2-adduct + 1,4-adduct (ref. 94)

	R	n	Yield (%)	Ratio
(a)	Me	1	68	100:0
(b)	Me	2	78	97:3
(c)	Ph	2	71	93:7

Scheme 145

1,1-bis(seleno)alkyl metals play the role of valuable acyl anion equivalents (Scheme 139; Scheme 140, a, b and e; Scheme 149, b).

Regiochemical control of the addition of 1-phenylselenoallyllithium to cyclopentenone has been achieved[83,90] in a manner similar to that mentioned above. In THF alone the C-1 adducts resulting from the attack of the ambident nucleophile at its α- and γ-sites prevail (Scheme 127, a),[83,90] whereas the C-3 adducts resulting from the almost exclusive reaction of the α-selenoallyllithium at its α-site are produced[83,90] if the reaction is performed instead in the presence of HMPA (Scheme 127, b). Therefore, this solvent mixture allows the simultaneous control of the sites of attack of these two ambident species.

(iii) Reactivity of enolates derived from α-selenocarbonyl compounds

Lithium enolates derived from α-seleno esters have a marked tendency to produce the C-3 adduct, and in fact the C-1 adducts are the most difficult to obtain. Representative results are collected in Schemes 152–154.

The best control of addition of these organometallics at the C-1 site of enones is achieved[157,231] by performing the reaction in a less basic solvent, such as ether or THF, at a lower temperature (–110 or –78 °C), and within a shorter time, so that not only the formation of the C-1 adducts is favored, but also there is a lower chance of them isomerizing to the more stable C-3 adducts (Scheme 152, compare f to g; Scheme 153, compare a with b–d; Scheme 154, compare c to a). This control is particularly efficient with cyclohexenone (Scheme 153, a and b)[112,157] and straight chain enones (Scheme 152, e)[112] and particularly poor with cyclopentenone (Scheme 154, a and b).[157] Such control cannot be achieved with chalcone (Scheme 152, a–d).[113]

The formation of the C-3 adduct relies on the fact that the α,β-unsaturated alcoholate, produced, as mentioned above, by attack of the organometallic at the α-site of the α,β-unsaturated carbonyl compound, can rearrange to the enolate resulting from C-3 attack. This isomerization has been more conveniently achieved by raising the temperature to 20 °C (Scheme 153, compare d to b; Scheme 154 compare c to b).[112] It has also been performed by adding HMPA[112] to the medium (Scheme 152, g; Schemes 153 and 154).[157] Otherwise, the C-3 adduct can be obtained directly, and even at –78 °C, by using potassium

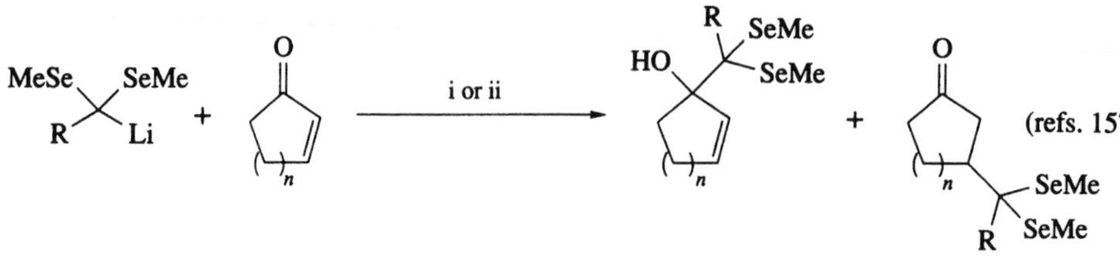

	R	Conditions	n	Yield (%)	C-1:C-3
(a)	H	i	1	55	75:25
	H	i	2	75	92:8
	H	i	3	67	87:13
(b)	H	ii	1	75	0:100
	H	ii	2	78	0:100
	H	ii	3	29	20:80
(c)	Me	i	1	31	70:30
	Me	i	2	63	73:27
	Me	i	3	50	80:20
(d)	Me	ii	1	78	0:100
	Me	ii	2	78	0:100
	Me	ii	3	72	11:89
(e)	C_6H_{13}	i	1	22	40:60
	C_6H_{13}	i	2	63	100:0
(f)	C_6H_{13}	ii	1	68	0:100
	C_6H_{13}	ii	2	59	0:100

i, THF, –78 °C, 0.2 h; ii, THF–HMPA (1:1.1), –78 °C, 0.2 h

(refs. 157 and 231)

Scheme 146

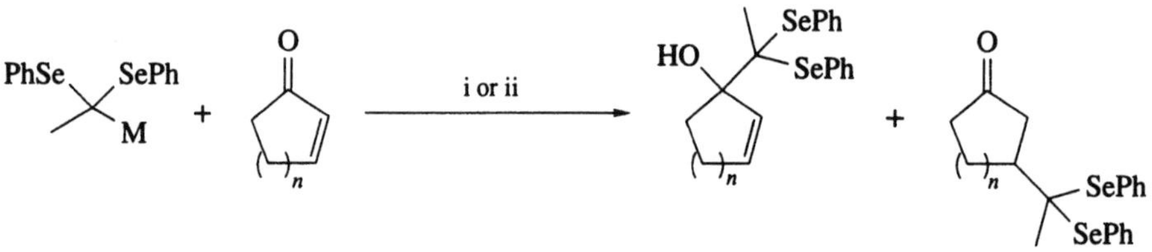

	M	Conditions	n	Yield (%)	Ratio	Ref.
(a)	Li	i	2	56	0:100	94
	Li	i	3	80	0:100	94
(b)	Li	ii	2	94	0:100	94
	Li	ii	3	80	0:100	94
(c)	K	ii	2	76	100:0	73

i, THF, –78 °C, 0.2 h; ii, THF, HMPA (1.1 equiv.), –78 °C, 0.2 h

Scheme 147

instead of lithium enolates derived from α-seleno esters (Schemes 152, d; Scheme 153, e and f; Scheme 154, d).

The mechanism of this C(1)–C(3) isomerization has been investigated,[113] starting from the alcohols resulting from the hydrolysis of the C-1 adducts of cyclohexenone. Reaction of these alcohols with bases leads to the alcoholates which then rearrange to the C-3 adducts. This rearrangement takes place[113] almost instantaneously with potassium alcoholates and is substantially less efficient with the lithium analogs (Scheme 155). In the latter case it is faster[113] with phenylseleno than with methylseleno derivatives and with adducts bearing an alkyl group on the carbon next to the ester group (Scheme 155, compare a to b).

Each of the two stereoisomeric C-1 adducts shown in Scheme 156, when subjected to the reactions mentioned above, leads to a different ratio of the C-3 stereoisomers (Scheme 156, a).[113] It was also found

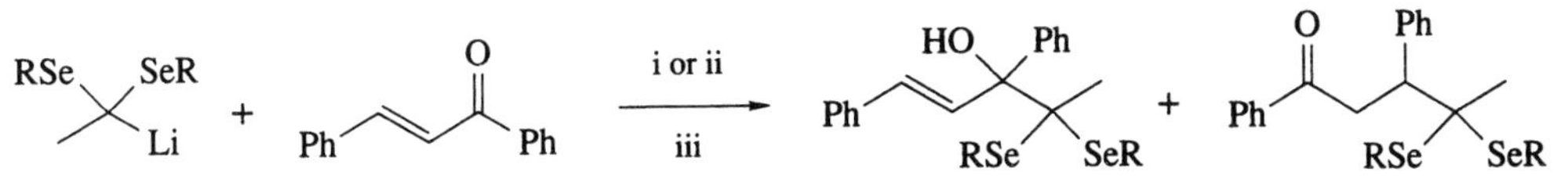

	R	*Conditions*	*Yield* (%)	*Ratio*
(a)	Me	i	91	25:75
	Ph	i	60	55:45
(b)	Me	ii	77	95:5
	Ph	ii	63	98:2

(ref. 234)

i, THF, –78 °C, 0.2 h, ii, THF, HMPA (1.1 equiv.), –78 °C, 0.2 h, iii, H_3O^+, –78 °C

Scheme 148

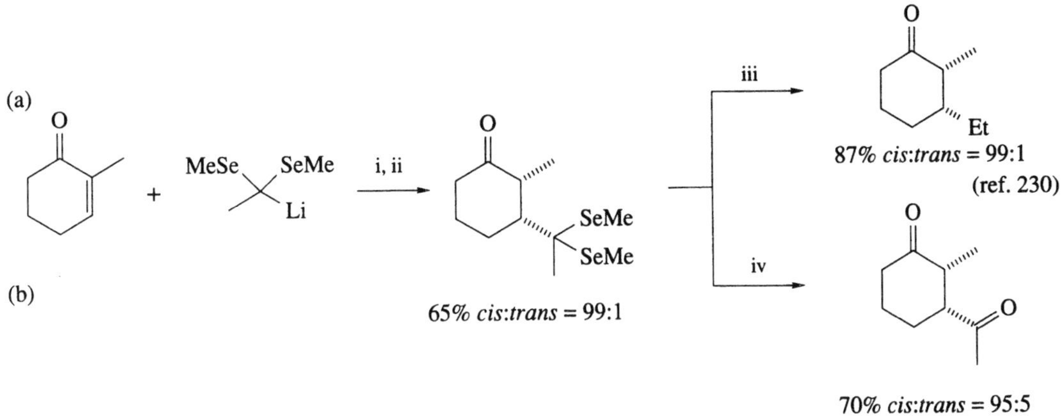

(a) 87% *cis*:*trans* = 99:1 (ref. 230)

(b) 70% *cis*:*trans* = 95:5

65% *cis*:*trans* = 99:1

i, THF–HMPA, –78 °C, 0.2 h, ii, H_3O^+; iii, Bu_3SnH (3 equiv.), AIBN (0.1 equiv.), toluene, 80 °C; iv, CuCl (1 equiv.), CuO (4 equiv.), aq. acetone, 20 °C, 0.1 h

Scheme 149

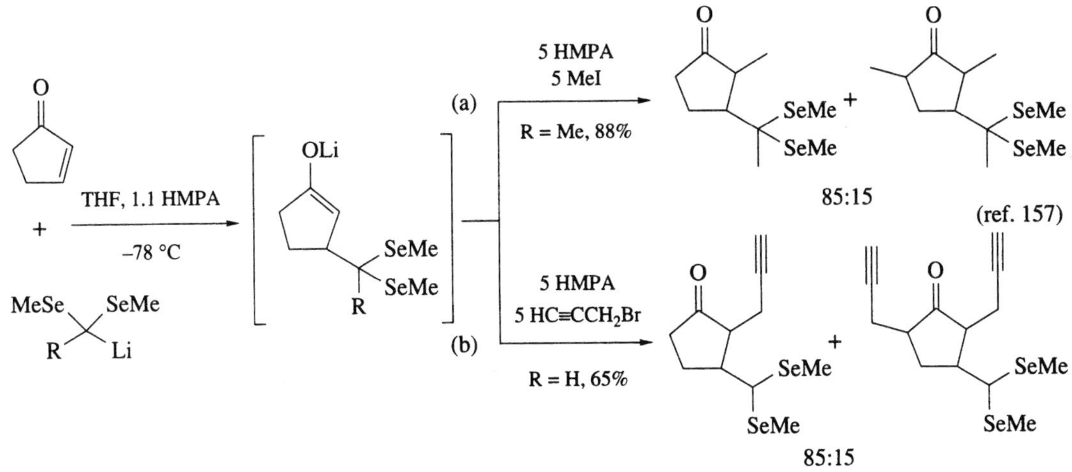

THF, 1.1 HMPA, –78 °C

(a) 5 HMPA, 5 MeI, R = Me, 88%; 85:15 (ref. 157)

(b) 5 HMPA, 5 HC≡CCH$_2$Br, R = H, 65%; 85:15

Scheme 150

that the C(3)–C(1) isomerization involves the dissociation of the C-1 adduct to cyclohexenone and α-metallo-α-seleno esters.[113] Since the isomerization is performed in the presence of chalcone the ester enolate moiety is transferred to the latter enone (Scheme 156, b).[113]

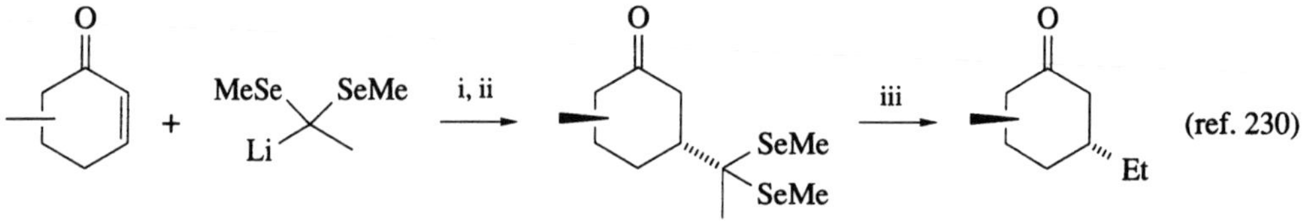

(ref. 230)

4-Me	58%, 95:5[a]	74%
5-Me	76%, 13:87	90%
6-Me	74%, 32:68	85%

[a] The first number in the ratio refers to the compound shown.

i, THF–HMPA (1:1.1), –78 °C; ii, aq. NH_4Cl; iii, Bu_3SnH (3 equiv.), AIBN (0.1 equiv.), toluene, 0.3 h

Scheme 151

2.6.3.5.4 Reaction of organometallics bearing an α-seleno carbanion with carboxylic acid derivatives and related compounds

(i) Reactions involving α-selenoalkyllithiums and α-selenoxyalkyllithiums

α-Selenoalkyllithiums have also been reacted with other carbonyl compounds such as esters,[235] acid chlorides,[235,236] carboxylic anhydrides,[235] dimethylformamide[12,162,235,237] and nitriles,[42] as well as chlorocarbonates,[235] carbonic anhydride (Scheme 157, a–d; Scheme 158)[12,77,235] and α,β-unsaturated lactones.[238] In the last case the C-3 adduct, produced when the reaction is performed in THF–HMPA, has been successfully used in an efficient synthesis of pederin (Scheme 159).[238] Finally, it is possible to acylate[98] α-selenoxyalkyllithiums but, except for special cases, yields are unsatisfactory (Scheme 157, e).

(ii) Reactions involving 1,1-bis(seleno)alkyllithiums

1,1-Bis(seleno)alkyllithiums exclusively add[232] at the C-1 site of α,β-unsaturated esters in THF (Scheme 160, a).[157] The reaction does not in fact stop at that stage since the organometallic reacts further on the α′,α′-bis(seleno)-α-enone intermediate to give an α-seleno-α-enone and 1,1,1-tris(methylseleno)-

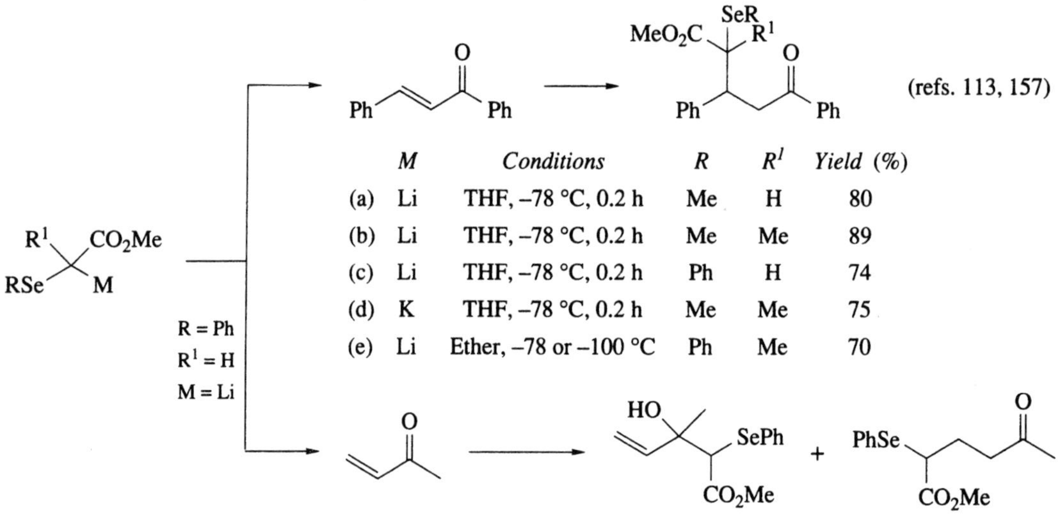

(refs. 113, 157)

	M	Conditions	R	R[1]	Yield (%)
(a)	Li	THF, –78 °C, 0.2 h	Me	H	80
(b)	Li	THF, –78 °C, 0.2 h	Me	Me	89
(c)	Li	THF, –78 °C, 0.2 h	Ph	H	74
(d)	K	THF, –78 °C, 0.2 h	Me	Me	75
(e)	Li	Ether, –78 or –100 °C	Ph	Me	70

R = Ph
R^1 = H
M = Li

(ref. 112)

	Conditions	Yield (%)	Ratio
(f)	THF, –78 °C, 0.5 h	45	81:19
(g)	THF–HMPA, –78 °C, 0.5 h	41	15:85

Scheme 152

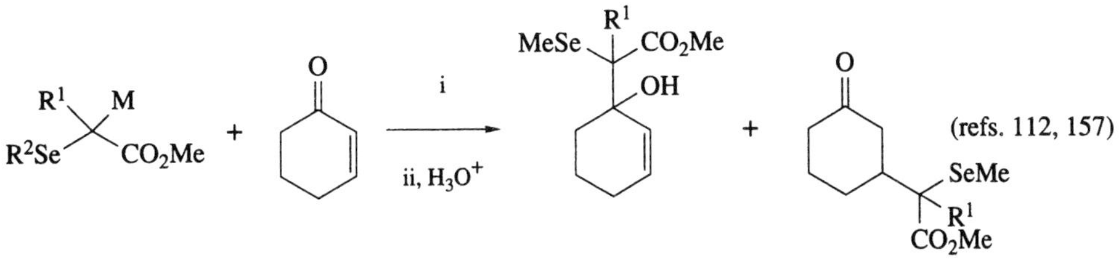

(refs. 112, 157)

	M	*Condition i*[a]	*R*	*R*[1]	*Yield (%)*	*Ratio*
(a)	Li	Ether–hexane (80:20)	Me	Me	71	100:0
	Li	Ether–hexane (80:20)	Ph	Me	66	100:0
(b)	Li	THF–hexane (80:20)	Me	Me	75	85:15
	Li	THF–hexane (80:20)	Ph	Me	85	85:15
(c)	Li	THF–HMPA–hexane (70:7:23)	Me	H	54	60:40
(d)	Li	THF–hexane (80:20), –78 to 20 °C, 1 h	Me	Me	72	0:100
	Li	THF–hexane (80:20), –78 to 20 °C, 1 h	Ph	Me	70	0:100
(e)	Na	THF–pentane (80:20)	Me	Me	65	0:100
	K	THF–pentane (80:20)	Me	Me	69	0:100
(f)	K	Ether–pentane (80:20), –100 °C	Me	Me	74	0:100

[a] At –78 °C, 0.2 h unless otherwise stated.

Scheme 153

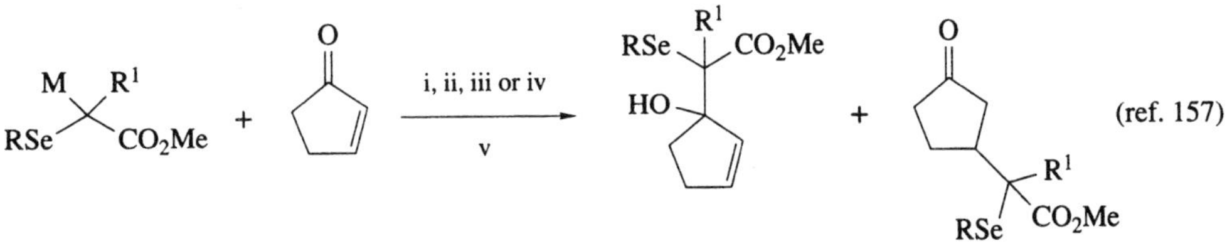

(ref. 157)

	Conditions	*R*	*R*[1]	*Yield (%)*	*Ratio*
(a)	i	Me	H	40	75:25
	i	Ph	H	46	63:37
(b)	ii	Me	H	69	54:46
	ii	Ph	H	68	41:59
	ii	Me	Me	73	35:65
	ii	Ph	Me	66	0:100
(c)	iii	Me	H	61	0:100
(d)	iv	Ph	H	68	0:100

i, M = Li, ether, –78 °C, 0.2 h; ii, M = Li, THF, –78 °C, 0.2 h; iii, M = Li, THF, –78 °C, 0.2 h, then 20 °C, 0.2 h; iv, M = K, THF, –78 °C, 0.2 h; v, H_3O^+

Scheme 154

alkane after hydrolysis.[157] If the reaction is performed instead in THF–HMPA or in DME, an alkyl γ,γ-(dimethylseleno)alkylcarboxylate, resulting from the addition of the organometallic at the C-3 site of the α,β-unsaturated ester, is produced exclusively (Scheme 160, b).[157]

Otherwise, α-selenoxyalkyllithiums have been reacted with aromatic esters,[97,98] and 1-metallo-1-selenoalkenes efficiently react with acyl bromides,[173] dimethylformamide (Schemes 43 and 101),[122,123,176] chlorocarbonates[176] and carbonic anhydride (Schemes 101 and 103).[65,123,176]

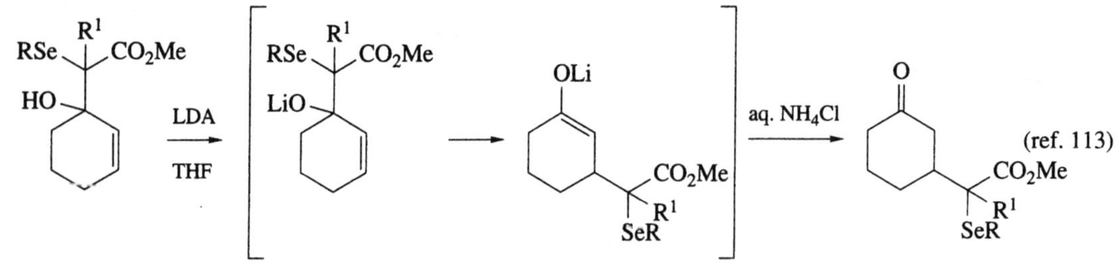

(ref. 113)

	R	R^1	*Temperature* (°C)	*Time* (h)[a]
(a)	Ph	H	–25	1
(b)	Ph	Me	–60	0.2
	Ph	Me	–40	0.1
	Ph	Me	–25	Instantaneous

[a] Minimum required for complete C(1)–C(3) isomerization.

Scheme 155

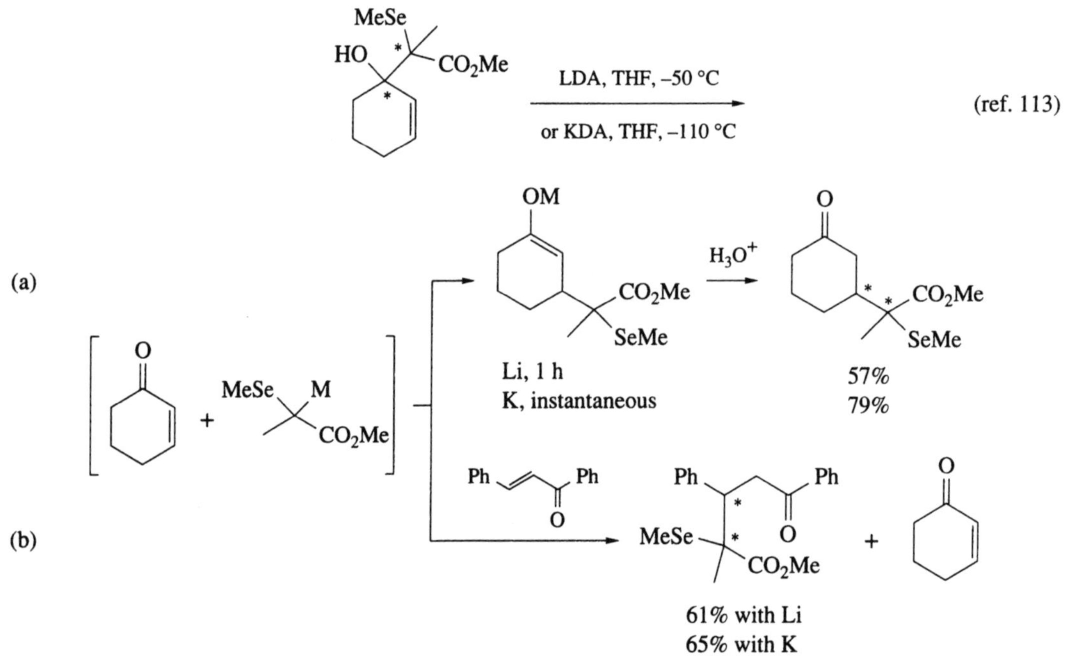

(ref. 113)

Scheme 156

2.6.4 β-HYDROXYALKYL SELENIDES AS VALUABLE SYNTHETIC INTERMEDIATES

2.6.4.1 Generalities

β-Hydroxyalkyl selenides are valuable intermediates for the synthesis of various selenium-free compounds. Although typical reactions of alcohols and selenides can often be achieved on such functionalized derivatives, some other reactions involve at the same time both functionalities. In fact, β-hydroxyalkyl selenides can be viewed as 'super' pinacols, since they possess two very well differentiated functionalities. The oxygen atom is hard and is linked to a hydrogen, whereas the selenium atom is softer and is attached to an alkyl or an aryl group. Typical reactions of β-hydroxyalkyl selenides, including those that are functionalized, are displayed in Schemes 96 and 161–167. β-Hydroxyalkyl selenides have been, amongst other things: (i) oxidized to α-selenocarbonyl compounds, β-hydroxyalkyl selenoxides (precursors of allyl alcohols) and β-hydroxyalkyl selenones (precursors of epoxides or rearranged

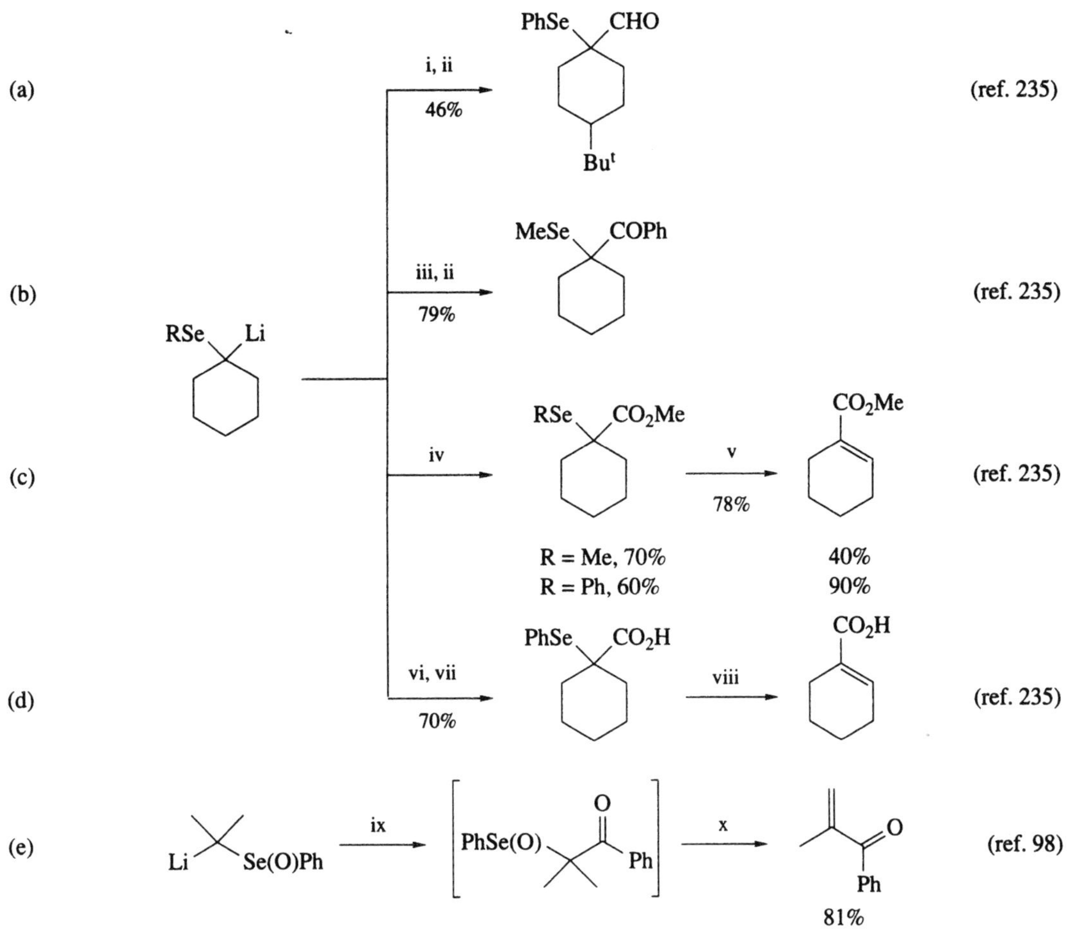

(a) RSe, Li cyclohexyllithium —i, ii (46%)→ 1-PhSe-1-CHO-4-Bu^t-cyclohexane (ref. 235)

(b) —iii, ii (79%)→ 1-MeSe-1-COPh-cyclohexane (ref. 235)

(c) —iv→ 1-RSe-1-CO_2Me-cyclohexane —v (78%)→ 1-CO_2Me-cyclohexene (ref. 235)

R = Me, 70% 40%
R = Ph, 60% 90%

(d) —vi, vii (70%)→ 1-PhSe-1-CO_2H-cyclohexane —viii→ 1-CO_2H-cyclohexene (ref. 235)

(e) Li–C(Me)₂–Se(O)Ph —ix→ [PhSe(O)–C(Me)₂–CO–Ph] —x→ CH_2=C(Me)–CO–Ph 81% (ref. 98)

i, Me_2NCHO; ii, H_3O^+; iii, PhCOCl; iv, MeOCOCl; v, H_2O_2, THF; vi, CO_2; vii, H_3O^+; viii, H_2O_2/THF; ix, $PhCO_2Me$; x, Δ

Scheme 157

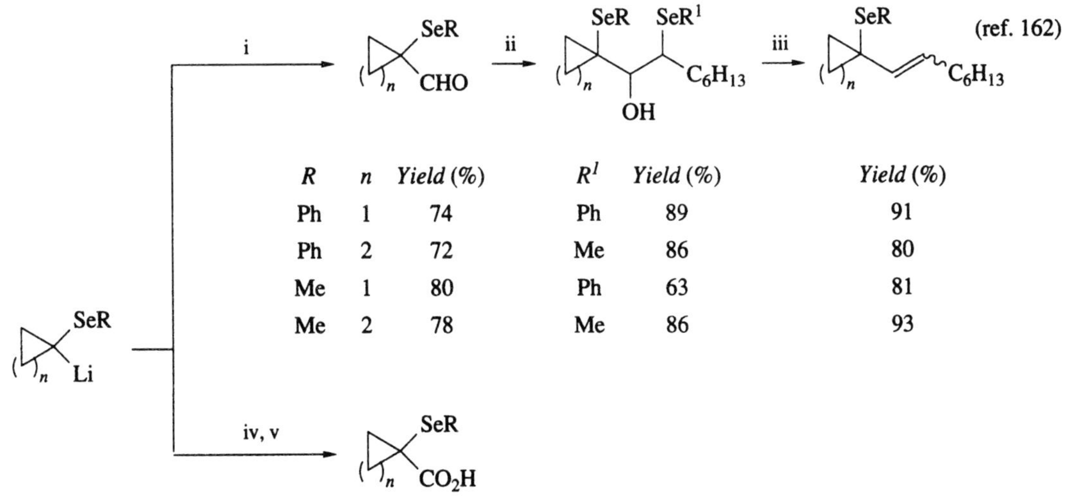

1-SeR-1-Li-cycloalkane ($(CH_2)_n$) —i→ 1-SeR-1-CHO —ii→ 1-SeR-1-[CH(OH)CH(SeR^1)C_6H_{13}] —iii→ 1-SeR-1-(CH=CHC_6H_{13}) (ref. 162)

R	*n*	*Yield* (%)	*R*[1]	*Yield* (%)	*Yield* (%)
Ph	1	74	Ph	89	91
Ph	2	72	Me	86	80
Me	1	80	Ph	63	81
Me	2	78	Me	86	93

—iv, v→ 1-SeR-1-CO_2H

n = 1, R = Ph, 42% (ref. 97); *n* = 2, R = Ph, 88%; R = Me, 79% (ref. 94)

i, Me_2NCHO; ii, $C_6H_{13}CH(SeR^1)Li$; iii, PI_3/Et_3N, CH_2Cl_2, 20 °C; iv, CO_2; v, H_3O^+

Scheme 158 [12,136,162]

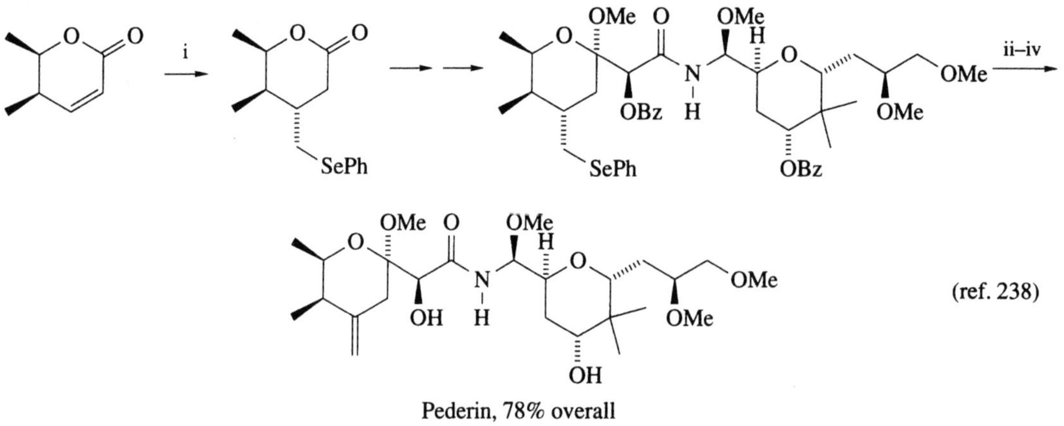

Pederin, 78% overall

(ref. 238)

i, $PhSeCH_2Li$, THF–HMPA, –85 °C, 1 h; ii, $NaIO_4$, $MeOH–H_2O$, 20 °C; iii, benzene, Et_3N, reflux, 0.04 h; iv, LiOH, $MeOH–H_2O$, 50 °C, 5 h

Scheme 159

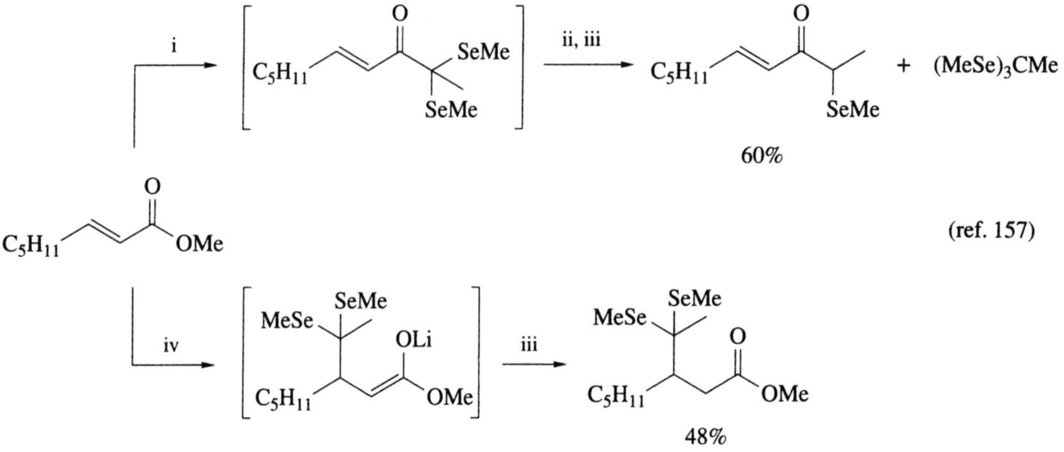

(ref. 157)

i, $(MeSe)_2C(Me)Li$, THF, –78 °C; ii, $(MeSe)_2C(Me)Li$; iii, H_3O^+; iv, $(MeSe)_2C(Me)Li$, THF–HMPA (1:1.1)

Scheme 160

ketones); (ii) reduced to alcohols; (iii) transformed to halohydrins, β-haloalkyl selenides, vinyl selenides, alkenes, epoxides and rearranged ketones; and (iv) transformed to β-hydroxyalkylselenonium salts (precursors, *inter alia*, of epoxides).

Although some reactions, such as the transformation of β-hydroxyalkyl selenides to β-haloalkyl selenides (Scheme 161, b)[7] or to vinyl selenides,[7] enones (Scheme 161, f),[7,240] α,α-dihalocyclopropanes (Scheme 162, f)[198] or β-hydroxyalkyl halides (Scheme 161, h; Scheme 162, g),[223,241] have been occasionally described or found only with specific types of β-hydroxyalkyl selenides, especially those having a strained ring [*e.g.* their transformation to allyl selenides (Scheme 163, b),[12,161] 1-selenocyclobutenes (Scheme 163, c)[12,161] and cyclobutanones (Scheme 163, f),[12,159,160,163]] others are far more general. This is particularly the case of their reductions to alcohols (Scheme 161, a; Scheme 162, a; Scheme 163, a; Scheme 164, a; Scheme 165, a)[7,188,189,246] or alkenes (Scheme 161, c; Scheme 162, c; Scheme 163, d; Scheme 164, c; Scheme 165, a; Scheme 166, c),[7,188,189,194,239] their transformation to allyl alcohols (Scheme 161, e; Scheme 162, b; Scheme 164, b; Scheme 166, b),[188,195] epoxides (Scheme 161, g; Scheme 162, d; Scheme 163, e; Scheme 164, d; Scheme 165, b; Scheme 166, d)[7,188,189,206] and rearranged carbonyl compounds (Scheme 162, e; Scheme 164, e; Scheme 165, a, c; Scheme 166, e),[188,198,244] as well as oxidation to α-selenocarbonyl compounds (Scheme 161, d).[240,242,248–251]

2.6.4.2 Reduction of β-Hydroxyalkyl Selenides to Alcohols

The reduction of β-hydroxyalkyl selenides to alcohols has been achieved[7,12,188,189,247,254] by lithium in ethylamine (Scheme 161, a; Scheme 162, a; Scheme 163, a; Scheme 167)[246] or triphenyl- or tributyl-tin hydride in toluene,[188,247,254] with or without AIBN. Most of these reactions proceed through radicals. The reactions involving tin hydrides can be carried out thermally around 120 °C[247] or photochemically at much lower temperature (0–20 °C).[148,252] The cleavage of the C—SePh bond is faster[255] than that of the

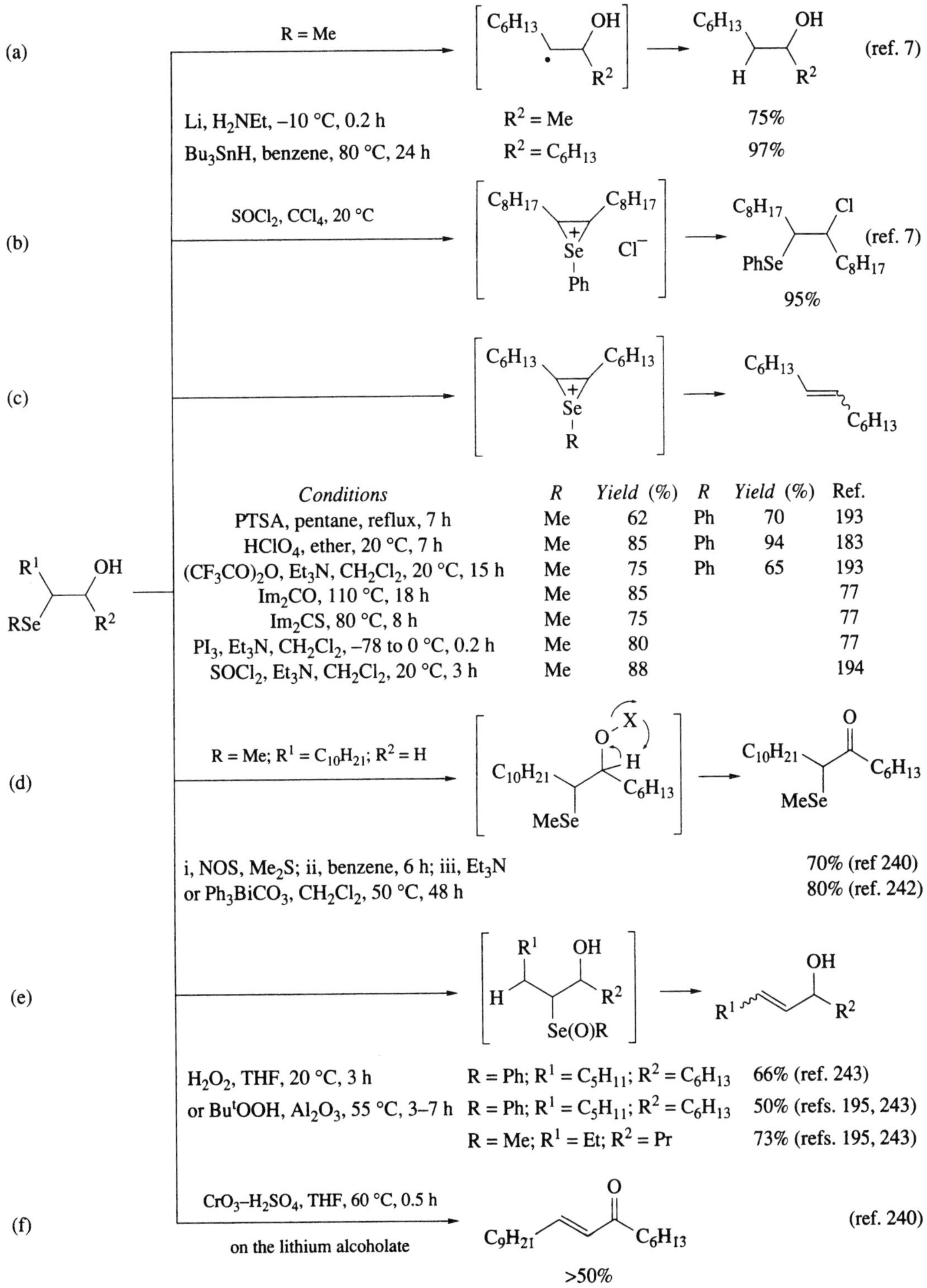

(a) R = Me → [C_6H_{13} radical, OH, R^2] → C_6H_{13}, OH, H, R^2 (ref. 7)

Li, H_2NEt, –10 °C, 0.2 h — R^2 = Me — 75%
Bu_3SnH, benzene, 80 °C, 24 h — $R^2 = C_6H_{13}$ — 97%

(b) $SOCl_2$, CCl_4, 20 °C → [C_8H_{17}, C_8H_{17}, Se+, Ph, Cl^-] → C_8H_{17}, Cl, PhSe, C_8H_{17} (ref. 7) 95%

(c) → [C_6H_{13}, C_6H_{13}, Se+, R] → C_6H_{13}, C_6H_{13}

R^1, OH, RSe, R^2

Conditions	R	Yield (%)	R	Yield (%)	Ref.
PTSA, pentane, reflux, 7 h	Me	62	Ph	70	193
$HClO_4$, ether, 20 °C, 7 h	Me	85	Ph	94	183
$(CF_3CO)_2O$, Et_3N, CH_2Cl_2, 20 °C, 15 h	Me	75	Ph	65	193
Im_2CO, 110 °C, 18 h	Me	85			77
Im_2CS, 80 °C, 8 h	Me	75			77
PI_3, Et_3N, CH_2Cl_2, –78 to 0 °C, 0.2 h	Me	80			77
$SOCl_2$, Et_3N, CH_2Cl_2, 20 °C, 3 h	Me	88			194

(d) R = Me; $R^1 = C_{10}H_{21}$; R^2 = H → [$C_{10}H_{21}$, MeSe, O, X, H, C_6H_{13}] → $C_{10}H_{21}$, MeSe, O, C_6H_{13}

i, NOS, Me_2S; ii, benzene, 6 h; iii, Et_3N — 70% (ref 240)
or Ph_3BiCO_3, CH_2Cl_2, 50 °C, 48 h — 80% (ref. 242)

(e) → [R^1, OH, H, R^2, Se(O)R] → R^1, OH, R^2

H_2O_2, THF, 20 °C, 3 h — R = Ph; $R^1 = C_5H_{11}$; $R^2 = C_6H_{13}$ — 66% (ref. 243)
or Bu^tOOH, Al_2O_3, 55 °C, 3–7 h — R = Ph; $R^1 = C_5H_{11}$; $R^2 = C_6H_{13}$ — 50% (refs. 195, 243)
R = Me; R^1 = Et; R^2 = Pr — 73% (refs. 195, 243)

(f) CrO_3–H_2SO_4, THF, 60 °C, 0.5 h on the lithium alcoholate → C_9H_{21}, O, C_6H_{13} >50% (ref. 240)

Scheme 161

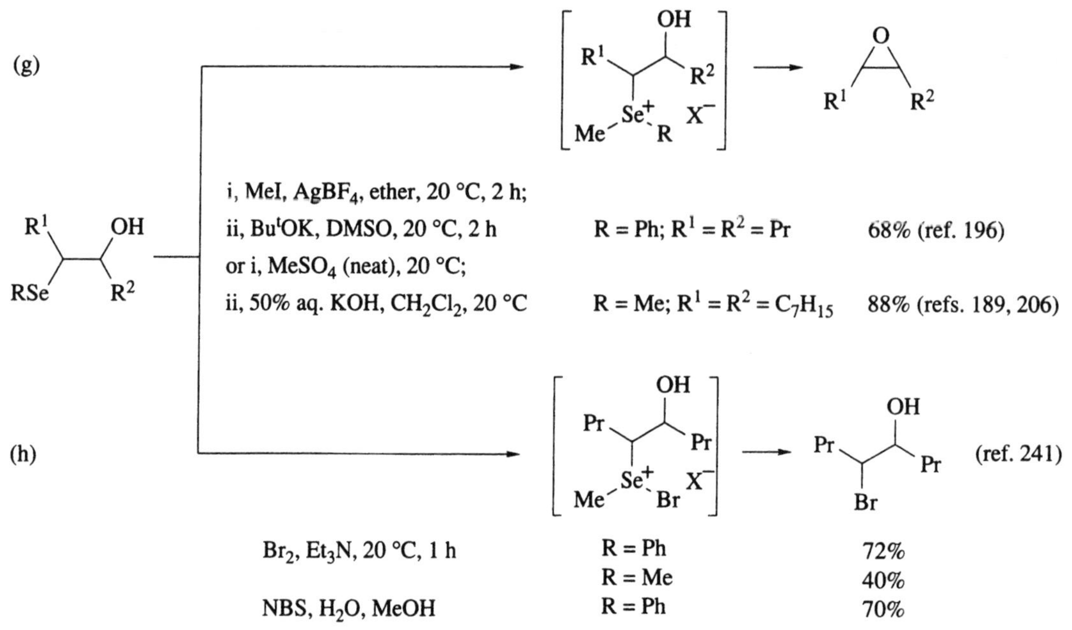

(g) [R^1CH(Se$^+$(Me)R)CH(OH)R^2 X^-] → epoxide of R^1, R^2

R^1CH(SeR)CH(OH)R^2

i, MeI, $AgBF_4$, ether, 20 °C, 2 h;
ii, Bu^tOK, DMSO, 20 °C, 2 h
or i, $MeSO_4$ (neat), 20 °C;
ii, 50% aq. KOH, CH_2Cl_2, 20 °C

R = Ph; $R^1 = R^2$ = Pr — 68% (ref. 196)

R = Me; $R^1 = R^2 = C_7H_{15}$ — 88% (refs. 189, 206)

(h) [PrCH(Se$^+$(Me)Br)CH(OH)Pr X^-] → PrCH(Br)CH(OH)Pr (ref. 241)

Br_2, Et_3N, 20 °C, 1 h	R = Ph	72%
	R = Me	40%
NBS, H_2O, MeOH	R = Ph	70%

Scheme 161 *(continued)*

C—SeMe and the C—Cl bonds and often faster than that of the C—Br bond. The reduction is highly chemoselective and leads to alcohols usually in almost quantitative yield (Scheme 161, a; Scheme 164, a; Scheme 168, a and b). In rare cases, however, such as when a carbon–carbon double or triple bond is present in a suitable position, the formation of a five- or six-membered ring takes place by trapping of the radical intermediate (Scheme 118).[203] Tin hydride reduction has been advantageously extended[176] to β-hydroxy-γ-alkenyl and β-hydroxy-α-alkenyl selenides displayed in Scheme 166 (a) and Scheme 168 (a and b) and derived from α-selenoalkyllithiums and enenones, and from 1-seleno-1-alkenyl metals and carbonyl compounds, respectively.

2.6.4.3 Reductive Elimination of β-Hydroxyalkyl Selenides to Alkenes

β-Hydroxyalkyl selenides are valuable precursors of alkenes. This reduction reaction, the mechanism of which is tentatively presented in Scheme 169, is observed[194] when the hydroxy group is transformed to a better leaving group and occurs by formal *anti* elimination of the hydroxy and selenyl moieties. It takes advantage of: (i) the difference in reactivity between the hard oxygen and the soft selenium atom which allows the selective activation of the hydroxy group when a hard electrophile is used; (ii) the high propensity of the selenium atom to favor the substitution of the oxygen-containing moiety by participation, thus allowing the intermediate formation of a seleniranium ion; and (iii) the attack of an exogeneous nucleophile on the charged selenium atom of the seleniranium ion which finally leads to the alkene.

This reaction has been performed on various β-hydroxyalkyl selenides bearing a phenylseleno (Scheme 21, b; Scheme 33, a; Schemes 38, 39, 53 and 128; Scheme 136, b and c; Scheme 161, c; Scheme 170),[16,56,74,82,85,97,136,194,236,256] *p*-chlorophenylseleno (Scheme 170),[74] (*m*-trifluoromethyl)phenylseleno (Scheme 21, a and c; Scheme 128)[56,65] or a methylseleno moiety (Scheme 31, c; Scheme 96, c; Scheme 98, b–d; Schemes 125 and 131; Scheme 132, b; Scheme 140, c; Scheme 161, c; Scheme 162, c; Scheme 163, d; Scheme 164, c; Scheme 165, a; Scheme 166, c; Scheme 168, c; Schemes 169 and 170),[7,12,77,108,136,159,162,189,193,194,221,222,236] with different reagents able to react selectively on the hydroxy group, such as perchloric or *p*-toluenesulfonic acids[193] in pentane or ether, as well as mesyl chloride,[56,65,97,166,192,202,257] thionyl chloride,[108,194,207,236,258] trifluoroacetic anhydride,[193] phosphorus oxychloride,[74,112,171,239] or diphosphorus tetraiodide[108,188,245] or phosphorus triiodide[108,136,159,162,188,189,245,259] in the presence of an amine or trimethylsilyl chloride and sodium iodide in acetonitrile.[260] The reactions usually proceed at room temperature with the above reagents (Scheme 158, a and b; Scheme 161, c; Scheme 162, c; Scheme 163, d; Scheme 164, c), but require heating to 50 °C or to 110 °C when *p*-toluenesulfonic acid (Scheme 128; Scheme 161, c; Scheme 165, a)[82,193,236] or carbonyl or thiocarbonyl diimidazole (Scheme 161, c; Scheme 163, d),[7,77,159] respectively, is used.

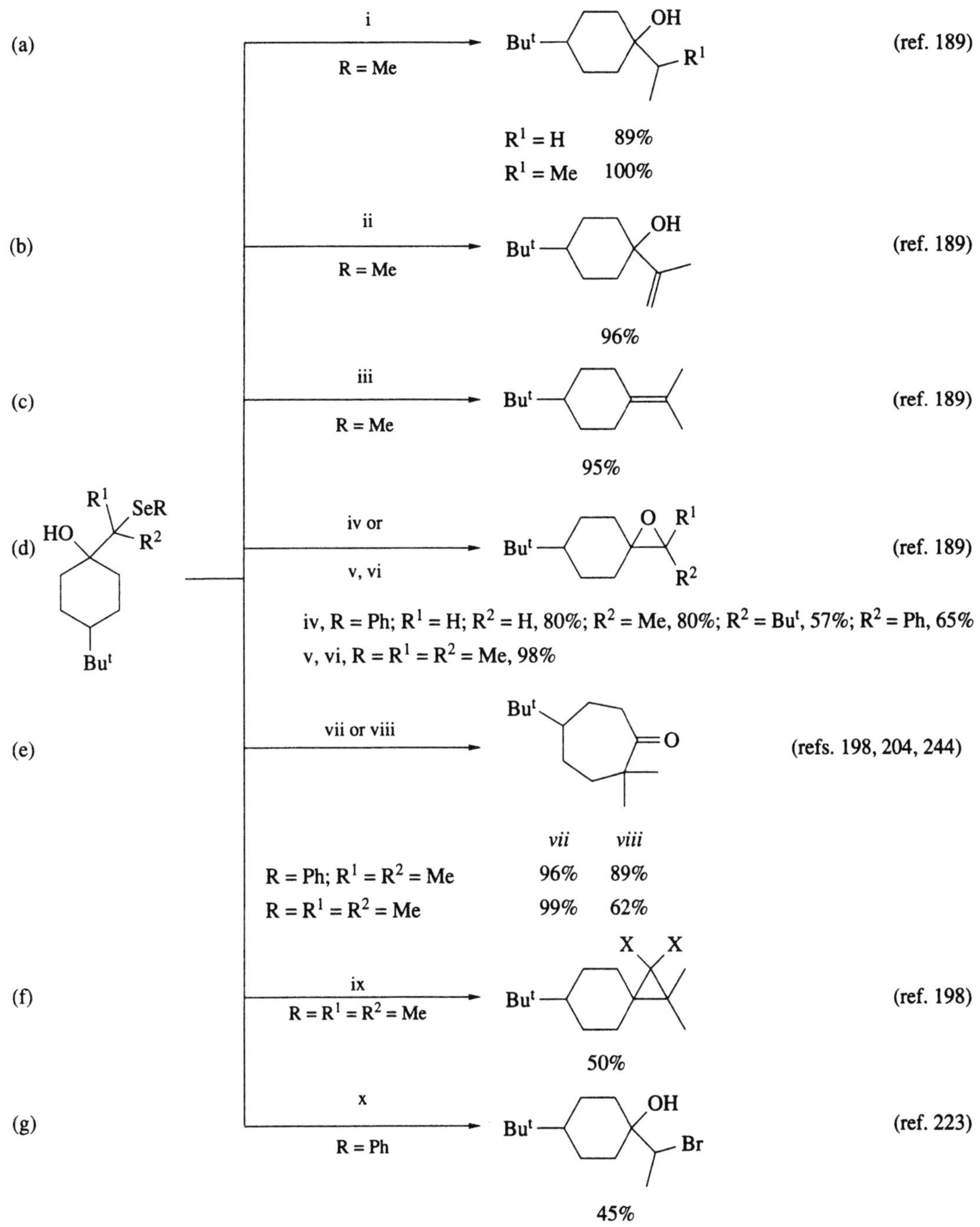

(a) i, R = Me (ref. 189)

R^1 = H 89%
R^1 = Me 100%

(b) ii, R = Me (ref. 189)

96%

(c) iii, R = Me (ref. 189)

95%

(d) iv or v, vi (ref. 189)

iv, R = Ph; R^1 = H; R^2 = H, 80%; R^2 = Me, 80%; R^2 = Bu^t, 57%; R^2 = Ph, 65%
v, vi, R = R^1 = R^2 = Me, 98%

(e) vii or viii (refs. 198, 204, 244)

	vii	*viii*
R = Ph; R^1 = R^2 = Me	96%	89%
R = R^1 = R^2 = Me	99%	62%

(f) ix, R = R^1 = R^2 = Me (ref. 198)

50%

(g) x, R = Ph (ref. 223)

45%

i, Li, $EtNH_2$, –10 °C; ii, Bu^{t}OOH, Al_2O_3, THF, 55 °C; iii, PI_3, Et_3N, CH_2Cl_2, 20 °C; iv, TlOEt (5.5 equiv.), $CHCl_3$, 20 °C, 20 h; v, $MeSO_3F$, ether; vi, 10% aq. KOH, ether; vii, TlOEt (5.5 equiv.), $CHCl_3$, 20 °C, 8 h; viii, $AgBF_4$ (1.3 equiv.), $CHCl_3$; ix, Bu^{t}OK (10 equiv.), $CHCl_3$; x, TlOEt (5.5 equiv.), $CHBr_3$, 20 °C

Scheme 162

Except in rare cases which will be discussed below, all these methods work equally well and allow the synthesis of almost all of the different types of alkene such as terminal, α,α- and α,β-disubstituted, tri- and tetra-substituted, including alkylidenecyclobutanes[77,136] and alkylidenecyclohexanes (Scheme 162, c),[39,77,159,189,192,202] especially those that are particularly hindered, as depicted in Scheme 164.[159,188] Specific examples are gathered in: Scheme 53; Scheme 125; Scheme 140, c; Scheme 158, a and b; Scheme 161, c; Scheme 162, c; Scheme 163, d; Scheme 164, c; Scheme 165, a; Scheme 166, c; Scheme 170 and Scheme 171.

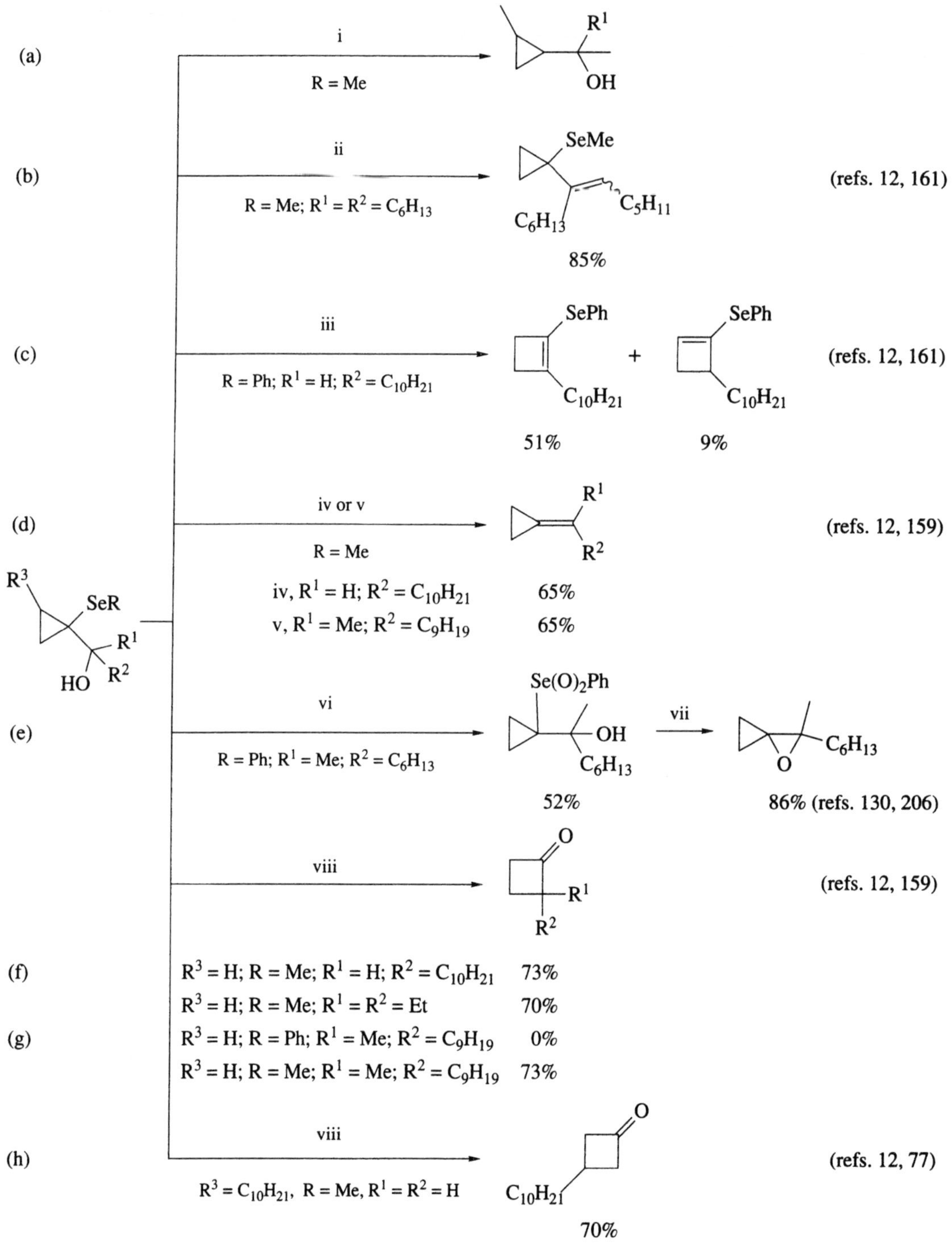

(a) i; R = Me; OH; R^1

(b) ii; R = Me; $R^1 = R^2 = C_6H_{13}$; SeMe; C_6H_{13}; C_5H_{11}; 85% (refs. 12, 161)

(c) iii; R = Ph; R^1 = H; $R^2 = C_{10}H_{21}$; SePh; $C_{10}H_{21}$; 51% + SePh; $C_{10}H_{21}$; 9% (refs. 12, 161)

(d) iv or v; R = Me; R^1; R^2 (refs. 12, 159)

iv, R^1 = H; $R^2 = C_{10}H_{21}$ 65%

v, R^1 = Me; $R^2 = C_9H_{19}$ 65%

R^3; SeR; R^1; HO; R^2

(e) vi; R = Ph; R^1 = Me; $R^2 = C_6H_{13}$; $Se(O)_2Ph$; OH; C_6H_{13}; 52%; vii; O; C_6H_{13}; 86% (refs. 130, 206)

viii; O; R^1; R^2 (refs. 12, 159)

(f) R^3 = H; R = Me; R^1 = H; $R^2 = C_{10}H_{21}$ 73%

R^3 = H; R = Me; $R^1 = R^2$ = Et 70%

(g) R^3 = H; R = Ph; R^1 = Me; $R^2 = C_9H_{19}$ 0%

R^3 = H; R = Me; R^1 = Me; $R^2 = C_9H_{19}$ 73%

(h) viii; $R^3 = C_{10}H_{21}$, R = Me, $R^1 = R^2$ = H; O; $C_{10}H_{21}$; 70% (refs. 12, 77)

i, Li, NH_3; ii, MeO_2CNSO_2, Et_3N, toluene, 110 °C; iii, MeO_2CNSO_2, Et_3N; iv, Im_2CO, toluene, 110 °C; v, PI_3, Et_3N, CH_2Cl_2, 20 °C; vi, MCPBA (22 equiv.), CH_2Cl_2, 20 °C, 36 h; vii, Bu^tOK, THF, 20 °C, 2 h; viii, TsOH, C_6H_6, H_2O, 80 °C

Scheme 163

The reactions are usually easier with methylseleno[199] than with phenylseleno,[39,199] *p*-chlorophenylseleno[74] or (*p*-trifluoromethyl)phenylseleno[39,56] derivatives, and allenes[176] and alkylidenecyclopropanes[159] are much more difficult to synthesize than other alkenic compounds. Thus, although allyl selenides resulting from selective elimination of the hydroxy and the methylseleno moiety are usually produced from β-(methylseleno)-β′-(phenylseleno) alcohols,[77,162,199] in the case of 1-seleno-1-(1′-hydroxy-2′-seleno)-

cyclopropanes the elimination takes place away from the cyclopropane ring and exclusively leads to 1-seleno-1-vinylcyclopropanes, and not to alkylidenecyclopropanes, whatever the nature of the seleno moiety (Scheme 125; Scheme 158, a and b).

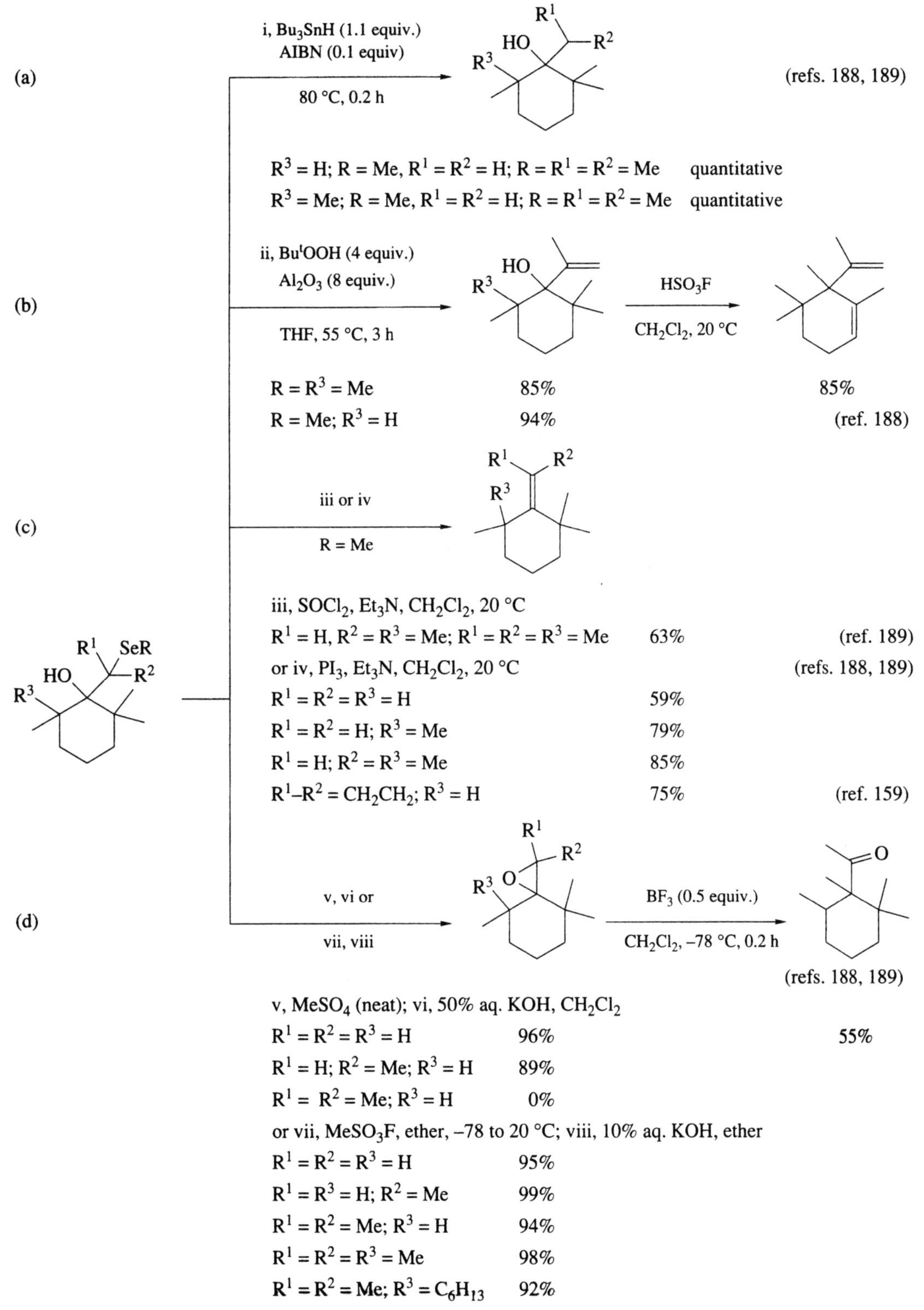

(a) i, Bu_3SnH (1.1 equiv.) AIBN (0.1 equiv) 80 °C, 0.2 h (refs. 188, 189)

R^3 = H; R = Me, $R^1 = R^2$ = H; R = $R^1 = R^2$ = Me quantitative
R^3 = Me; R = Me, $R^1 = R^2$ = H; R = $R^1 = R^2$ = Me quantitative

(b) ii, Bu^tOOH (4 equiv.) Al_2O_3 (8 equiv.) THF, 55 °C, 3 h; HSO_3F, CH_2Cl_2, 20 °C

R = R^3 = Me 85% 85%
R = Me; R^3 = H 94% (ref. 188)

(c) iii or iv, R = Me

iii, $SOCl_2$, Et_3N, CH_2Cl_2, 20 °C
R^1 = H, $R^2 = R^3$ = Me; $R^1 = R^2 = R^3$ = Me 63% (ref. 189)
or iv, PI_3, Et_3N, CH_2Cl_2, 20 °C (refs. 188, 189)
$R^1 = R^2 = R^3$ = H 59%
$R^1 = R^2$ = H; R^3 = Me 79%
R^1 = H; $R^2 = R^3$ = Me 85%
R^1–R^2 = CH_2CH_2; R^3 = H 75% (ref. 159)

(d) v, vi or vii, viii; BF_3 (0.5 equiv.), CH_2Cl_2, –78 °C, 0.2 h (refs. 188, 189)

v, $MeSO_4$ (neat); vi, 50% aq. KOH, CH_2Cl_2
$R^1 = R^2 = R^3$ = H 96% 55%
R^1 = H; R^2 = Me; R^3 = H 89%
R^1 = R^2 = Me; R^3 = H 0%
or vii, $MeSO_3F$, ether, –78 to 20 °C; viii, 10% aq. KOH, ether
$R^1 = R^2 = R^3$ = H 95%
$R^1 = R^3$ = H; R^2 = Me 99%
$R^1 = R^2$ = Me; R^3 = H 94%
$R^1 = R^2 = R^3$ = Me 98%
$R^1 = R^2$ = Me; R^3 = C_6H_{13} 92%

Scheme 164

(e) 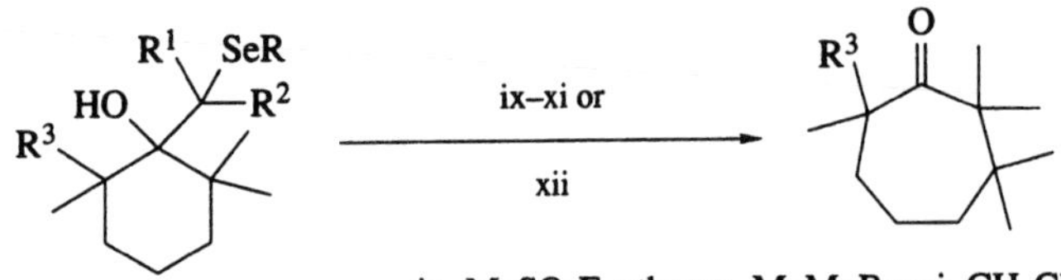 ix–xi or xii (ref. 188)

ix, $MeSO_3F$, ether; x, MeMgBr; xi, CH_2Cl_2, 50 °C R^3 = Me, 95%
or xii, $MeSO_3F$, CH_2Cl_2, 20 °C R^3 = H, 95%

Scheme 164 *(continued)*

Acidic conditions are,[193,194] in most cases, as good as others but they have,[166,194] at least in the case of β-hydroxyalkyl selenides bearing two alkyl substituents at the selenium-bearing carbon, a tendency to produce rearranged compounds besides the expected alkenes (Scheme 165, a). These become the exclusive compounds obtained[12,159,160] when 1-seleno-1-(1′-hydroxyalkyl)cyclopropanes are involved (Schemes 132, a; Scheme 163, f–h). This behavior is general for almost all cyclopropylcarbinols bearing a heterosubstituent (*e.g.* OR, NR_2, SR, SeR) at that position on the cyclopropane ring (see Section 2.6.4.5.3).[12]

In general, the best combination, at least in our hands, involves phosphorus triiodide or diphosphorus tetraiodide and β-hydroxyalkyl methyl selenides. These conditions allow the high yield synthesis of: (i) highly hindered alkenes[188,189] derived from, for example, bulky selenoalkyllithiums such as those bearing two alkyl groups on the carbanionic center, and particularly hindered 2,2,6-trimethylcyclohexanone, 2,2,6,6-tetramethylcyclohexanone (Scheme 164) and di-*t*-butyl ketone; (ii) highly strained and hindered alkylidenecyclopropanes[12,159] derived from 1-methylselenocyclopropyllithium and ketones including 2,2,6-trimethylcyclohexanone (Scheme 163, d); and (iii) allenes from β-hydroxyalkenyl methyl selenides derived from 1-seleno-1-alkenyllithiums (Scheme 167, c).[176]

These successful conditions take advantage of (i) the high affinity of trivalent phosphorus for oxygen; (ii) the ease with which a methylseleno moiety (compared to phenylseleno) is able to participate in order

(a) 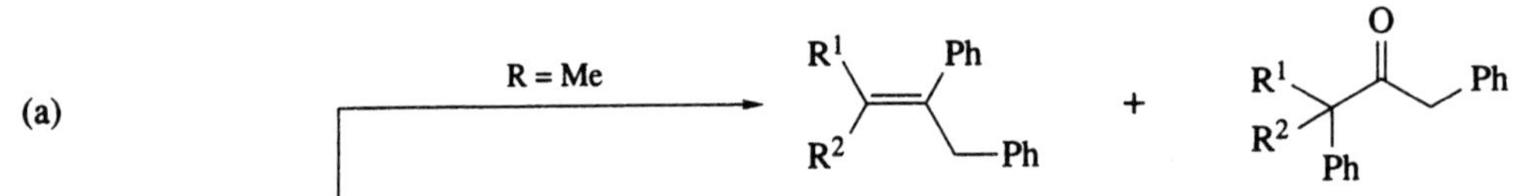 R = Me

R^1	R^2	*Conditions*	*Yield (%)*	*Ratio*	*Ref.*
Me	Me	TsOH, pentane, 50 °C	96	50:50	194
Me	Me	$(CF_3CO)_2O$, Et_3N, 20 °C, 1 h	93	100:0	194
Me	Me	$SOCl_2$, Et_3N, 20 °C	85	100:0	194
CH_2CH_2		PI_3, Et_3N, 20 °C	75	100:0	159

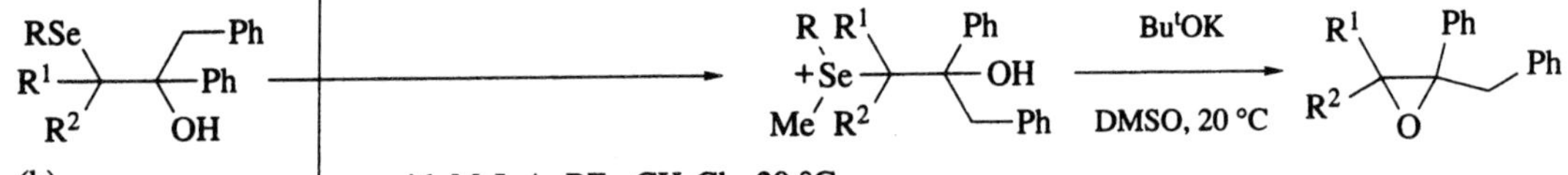

(b) Bu^tOK, DMSO, 20 °C

with MeI, $AgBF_4$, CH_2Cl_2, 20 °C

R = Ph; $R^1 = R^2$ = H	93%	46%	(ref. 196)
R = Ph; R^1 = H; R^2 = Me	73%	51%	

with MeI or Me_2SO_4 (neat), 20 °C (ref. 55)

R = Me; R^1 = H; R^2 = Me	73%	98%
R = R^1 = R^2 = Me	88%	90%

(c) 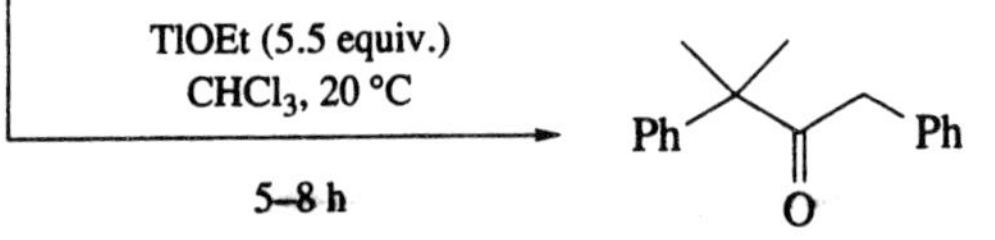 TlOEt (5.5 equiv.) $CHCl_3$, 20 °C 5–8 h (ref. 223)

R = Ph; $R^1 = R^2$ = Me 85%

Scheme 165

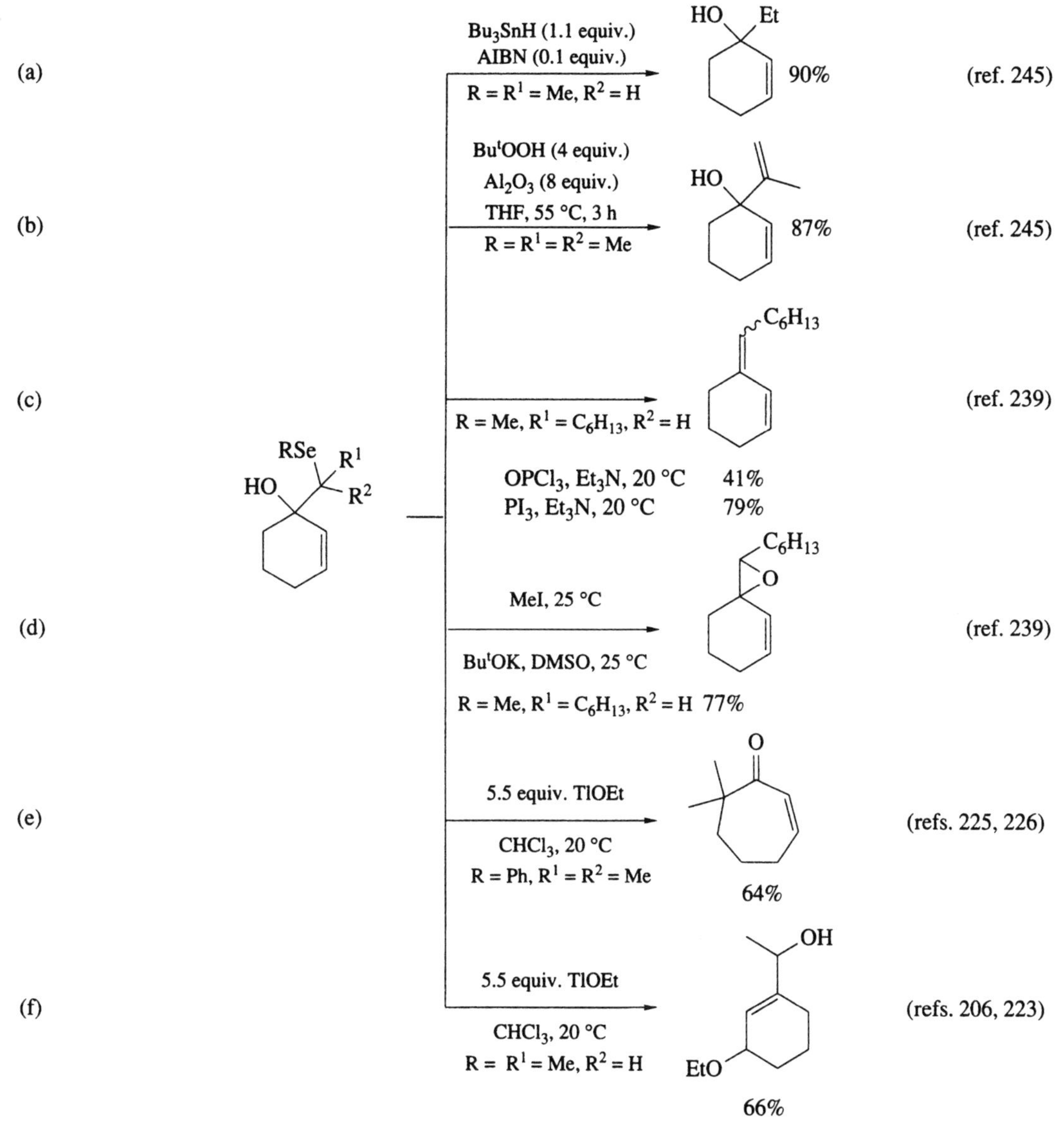

(a) Bu_3SnH (1.1 equiv.), AIBN (0.1 equiv.); $R = R^1 = Me$, $R^2 = H$ — 90% (ref. 245)

(b) Bu^tOOH (4 equiv.), Al_2O_3 (8 equiv.), THF, 55 °C, 3 h; $R = R^1 = R^2 = Me$ — 87% (ref. 245)

(c) $R = Me$, $R^1 = C_6H_{13}$, $R^2 = H$ — $OPCl_3$, Et_3N, 20 °C 41%; PI_3, Et_3N, 20 °C 79% (ref. 239)

(d) MeI, 25 °C; Bu^tOK, DMSO, 25 °C; $R = Me$, $R^1 = C_6H_{13}$, $R^2 = H$ 77% (ref. 239)

(e) 5.5 equiv. TlOEt, $CHCl_3$, 20 °C; $R = Ph$, $R^1 = R^2 = Me$ — 64% (refs. 225, 226)

(f) 5.5 equiv. TlOEt, $CHCl_3$, 20 °C; $R = R^1 = Me$, $R^2 = H$ — 66% (refs. 206, 223)

Scheme 166

to expel the activated oxygen-containing group; and (iii) the softness of the iodide counterion which favors its reaction on the selenium atom rather than on the carbon atom of the intermediate seleniranium ion (see, for example, Scheme 169).

The reductive elimination reported above allows[56,74,171,193,194,222,236,259] the stereoselective synthesis of each of the two stereoisomers of several alkenes, including those that are disubstituted, trisubstituted and functionalized, from each of the two stereoisomers of β-hydroxyalkyl selenides (Schemes 170–172).

The reductive elimination has been successfully applied to the synthesis of various functionalized alkenes such as: (i) dienes, including α,ω-dienes (Scheme 11)[60,245] and 1,3-dienes (Schemes 128 and 131; Scheme 132, d; Scheme 136, b and c; Scheme 166)[7,56,82,97,194,221,222,239] at the exclusion of allylidenecyclopropanes (Scheme 132, a);[224] (ii) vinyl ethers (Scheme 21, a; Scheme 97);[56] (iii) vinylsilanes (Scheme 96);[171] (iv) vinyl selenides (Scheme 21, b and c; Scheme 98, b–d)[56,65,108] and 1,3-dienyl selenides (Scheme 33, a);[85] (v) ketene selenoacetals (Scheme 31, c);[108] (vi) allyl sulfides;[77,199] (vi) allylsilanes;[12,199,202] (vii) allyl selenides (Scheme 125; Scheme 158, a and b);[12,77,162,199] and (viii) α,β-unsaturated esters (Schemes 39 and 170)[56,74] and lactones (Scheme 38).[74]

Particularly interesting results have been observed from β-hydroxy-α-trimethylsilylalkyl selenides since vinylsilanes are exclusively produced[171] under conditions where both the selenyl and the silyl moieties are removed with the hydroxy moiety. Apparently the selenyl moiety seems to stabilize a β-carbenium ion to a better extent than a silyl group.

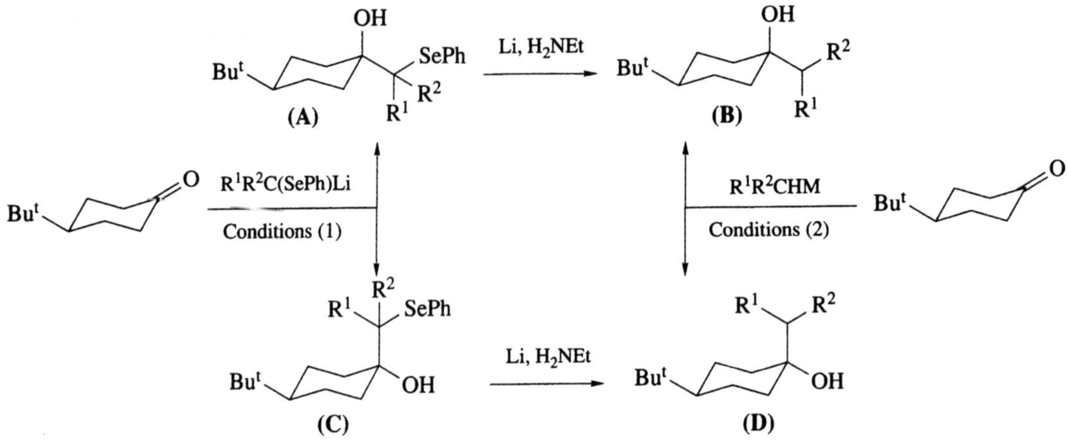

Bu^t (cyclohexanone) O — $R^1R^2C(SePh)Li$, Conditions (1) → **(A)** Bu^t, OH, SePh, R^1, R^2 + **(C)** Bu^t, OH, R^1, R^2, SePh

(A) — Li, H_2NEt → **(B)** Bu^t, OH, R^1, R^2

(C) — Li, H_2NEt → **(D)** Bu^t, OH, R^1, R^2

Bu^t (cyclohexanone) O — R^1R^2CHM, Conditions (2) → **(B)** + **(D)**

R^1	R^2	*Conditions (1)*	*Yield (%)* [a]	*Yield (%)* [b]	*Conditions (2)*	*Yield (%)* [c]
H	H	THF, –78 °C, 1 h	68, 85:15	93, 85:15	MeMgI, ether	95, 56:44
					MeLi, ether	—, 68:12
H	Me	Ether, –78 °C, 2 h	78, 100:0	89, 100:0	EtMgBr, ether	80, 68:12
Me	Me	Ether, –78 °C, 2 h	75, 100:0	92, 100:0	Pr^iMgBr, ether	47, 83:17
					Bu^tMgCl	22, 100:0

[a] Yield is **(A)** + **(C)**, ratio is **(A)**:**(C)**. [b] Yield is **(B)** + **(D)**, ratio is **(B)**:**(D)**, from reactions **(A)** to **(B)** and **(C)** to **(D)**. [c] Yield is **(B)** + **(D)**, ratio **(B)**:**(D)**.

Scheme 167 [210]

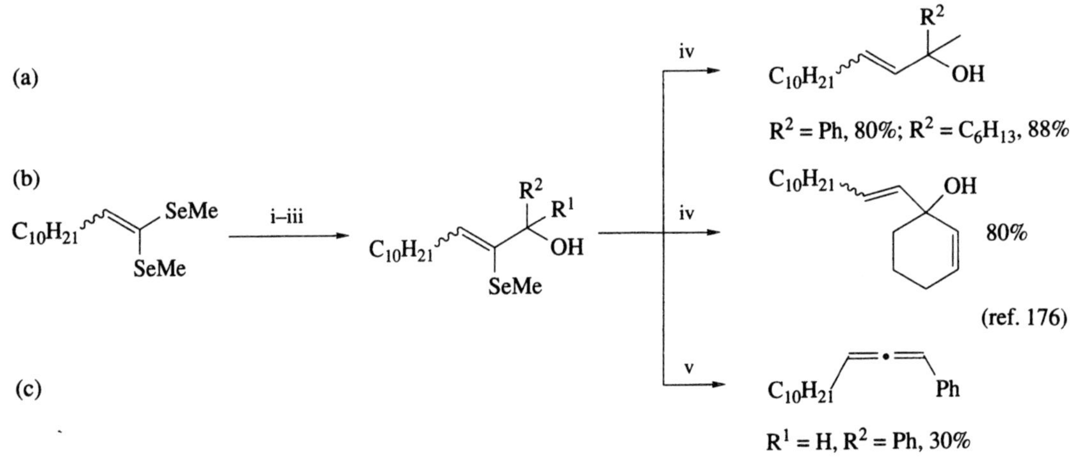

$C_{10}H_{21}$, SeMe, SeMe — i–iii → $C_{10}H_{21}$, R^2, R^1, OH, SeMe

(a) — iv → $C_{10}H_{21}$, R^2, OH

R^2 = Ph, 80%; R^2 = C_6H_{13}, 88%

(b) — iv → $C_{10}H_{21}$, OH, 80%

(ref. 176)

(c) — v → $C_{10}H_{21}$, Ph

R^1 = H, R^2 = Ph, 30%

i, Bu^nLi, THF; ii, R^1R^2CO; iii, H_3O^+; iv, Bu_3SnH, AIBN, toluene, 110 °C, 3 h; v, P_2I_4, Et_3N, 20 °C, 1.5 h

Scheme 168

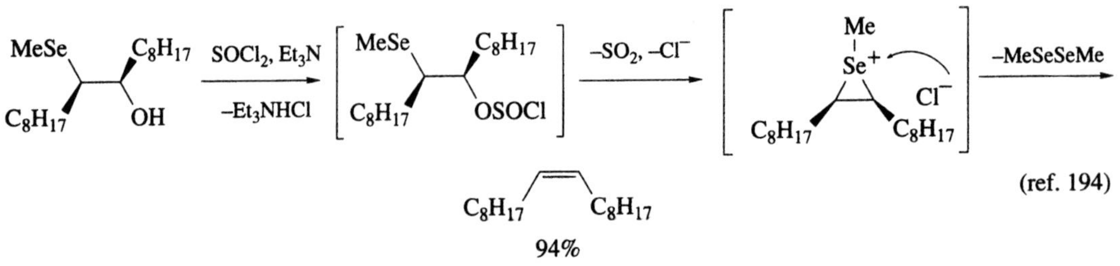

MeSe, C_8H_{17}, C_8H_{17}, OH — $SOCl_2$, Et_3N, $-Et_3NHCl$ → [MeSe, C_8H_{17}, C_8H_{17}, OSOCl] — $-SO_2$, $-Cl^-$ → [Me, Se^+, Cl^-, C_8H_{17}, C_8H_{17}] — –MeSeSeMe → C_8H_{17}, C_8H_{17}

(ref. 194)

94%

Scheme 169

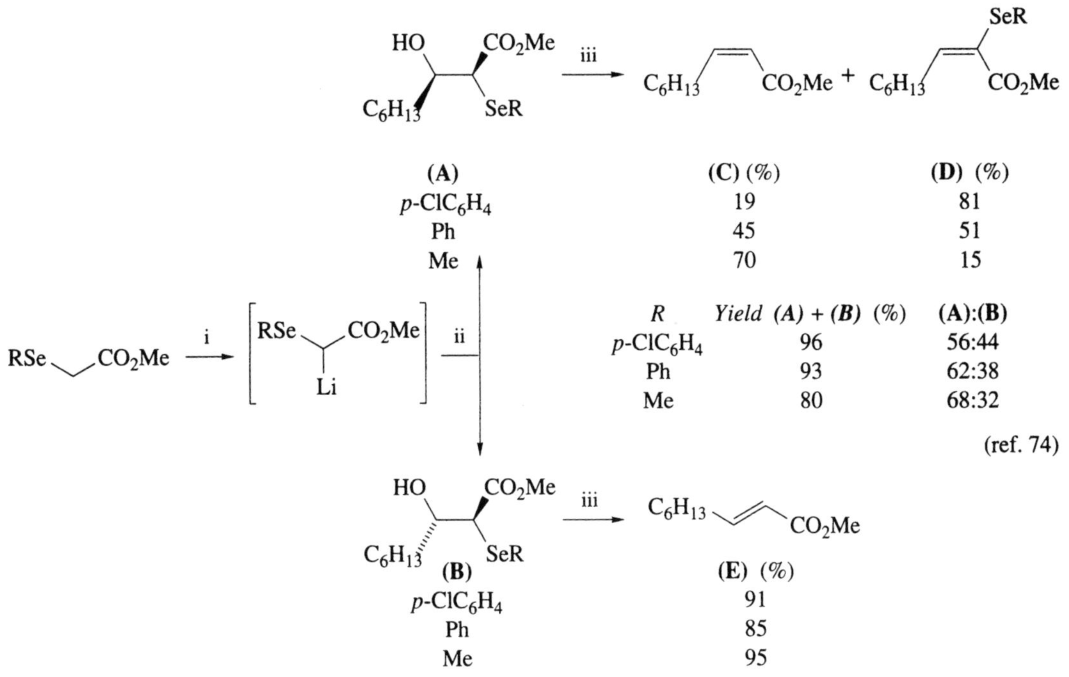

(A)	(C) (%)	(D) (%)
p-ClC_6H_4	19	81
Ph	45	51
Me	70	15

R	Yield (A) + (B) (%)	(A):(B)
p-ClC_6H_4	96	56:44
Ph	93	62:38
Me	80	68:32

(ref. 74)

(B)	(E) (%)
p-ClC_6H_4	91
Ph	85
Me	95

$R_{fB} - R_{fA}$ = 0.1–0.2; SiO_2, ether–pentane (2:8)

i, LDA, THF, –78 °C; ii, $C_6H_{13}CHO$; iii, $OPCl_3$, Et_3N, 20 °C, 2 h

Scheme 170

The elimination is completely chemoselective with 1-(1-hydroxyalkyl)-δ-lactones[74] and with methyl β-hydroxy-α-selenoalkylcarboxylates (precursors of (*E*)-α,β-unsaturated methyl carboxylates; Scheme 39, b and Scheme 170),[56,74] but not with the other stereoisomer which produces a mixture of (*Z*)-α,β-unsaturated methyl carboxylates (by formal selenenic elimination) and α-seleno-α,β-unsaturated methyl carboxylates (by formal elimination of water).

Elimination of water is particularly favored from *p*-chlorophenyl[74] and (*p*-trifluoromethyl)phenyl[56] selenides, whereas elimination of seleninic acid is mainly observed from methylseleno derivatives. Here, two different effects work in the same direction. The arylseleno group, more than the methylseleno group, makes the α-hydrogen acidic, and has at the same time a lower tendency to stabilize a β-carbenium ion *via* the formation of a seleniranium ion.

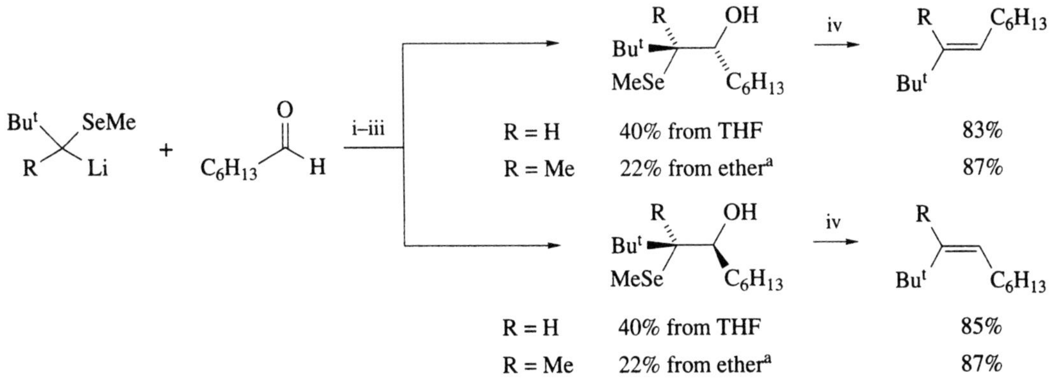

	Adduct	Alkene
R = H	40% from THF	83%
R = Me	22% from ether[a]	87%
R = H	40% from THF	85%
R = Me	22% from ether[a]	87%

[a] Refers to the yield after separation of the 1:1 mixture of stereoisomers

i, –78 °C, 1–1.5 h; ii, H_3O^+; iii, separation by TLC; iv, PI_3, Et_3N

Scheme 171 [189]

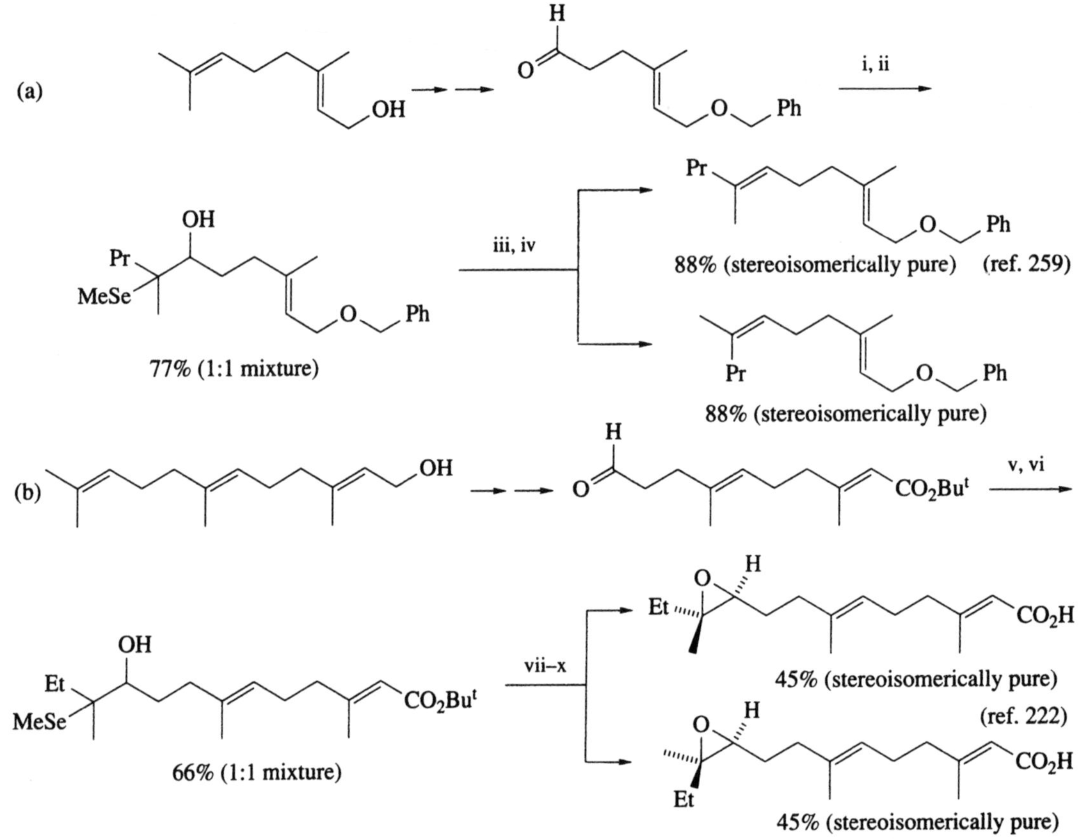

i, PrC(SeMe)(Me)Li, THF, –78 °C; ii, H_3O^+; iii, separation (SiO_2, hexane–ethyl acetate); iv, PI_3, Et_3N, CH_2Cl_2 on each stereoisomer; v, EtC(SeMe)(Me)Li, THF, –78 °C; vi, H_3O^+; vii, separation; viii, $MeSO_3F$; ix, aq. KOH, ether; x, H_3O^+

Scheme 172

2.6.4.4 Synthesis of Allyl Alcohols

β-Hydroxyalkyl selenoxides directly available from α-lithioalkyl selenoxides or by oxidation of β-hydroxyalkyl selenides are valuable precursors of allyl alcohols (Schemes 51 and 52; Scheme 54, a; Scheme 105; Scheme 161, e; Scheme 162, b; Scheme 164, b; Scheme 166, b; Schemes 173 and 174).[7,40,42,48,49,97,98,120,127,128,136,188,189,195,253,261] In the former case the reaction is usually performed on thermolysis at around 70 °C in carbon tetrachloride[98] and in the presence of an amine[97,243] able to trap the selenenic acid concomitantly produced.

In general, the elimination rection is easier: (i) with arylseleno derivatives bearing an electron-withdrawing group at the *ortho* or *para* position of the phenyl ring than with their phenylseleno or methylseleno analogs; and (ii) when it provides a relatively more-substituted carbon–carbon double bond.

On β-hydroxyalkyl phenyl selenides, the oxidation–elimination is best achieved with an excess of 30% aqueous hydrogen peroxide according to the original description of Sharpless (Scheme 174, c and d).[261] Although 1-(1′-hydroxyalkyl)cyclohexyl phenyl selenide produces[49] under these conditions a hydroperoxide (Scheme 174, a), the desired elimination reaction has been successfully achieved[49] with sodium periodate in ethanol (Scheme 174, b).

The above conditions are not suitable for β-hydroxyalkyl methyl selenides (Scheme 175, a)[195,243] or for β-hydroxyalkyl phenyl selenides bearing a primary hydroxy group (Scheme 176, a).[195] Several other conditions including ozone and *t*-butyl hydroperoxide are not efficient for that purpose (Schemes 175 and 176).[195,243] The best reagent is *t*-butyl hydroperoxide supported on basic alumina. The reaction takes place at reflux in THF, and allows the synthesis from β-hydroxyalkyl methyl selenides of a large variety

(a) $R^1 = R^2 = Me$, 81%
$R^1 = Ph$; $R^2 = H$, 72%

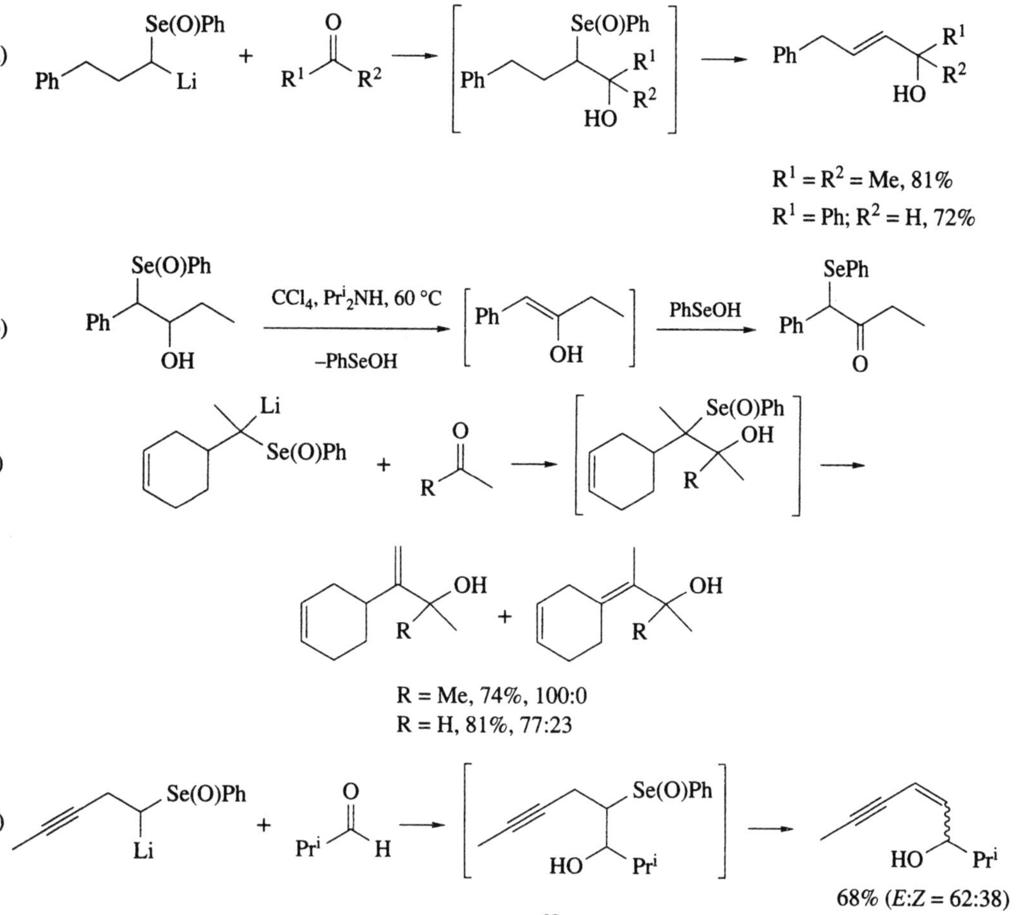

(b) CCl_4, Pr^i_2NH, 60 °C; –PhSeOH; PhSeOH

(c) R = Me, 74%, 100:0
R = H, 81%, 77:23

(d) 68% (*E*:*Z* = 62:38)

Scheme 173 [97]

of allyl alcohols, even those bearing a primary hydroxy group (Scheme 161, e; Scheme 162, b; Scheme 164, b; Scheme 174, c; Scheme 175, c; Scheme 176, c).[195]

This reaction, which is related to the well-known amine oxide pyrolysis,[262] involves a *syn* elimination of the selenenic acid, but proceeds under milder conditions (20 °C to 60 °C instead of 400 °C). It is more regioselective, takes place away from the oxygen and leads to allyl alcohols instead of enols, except when the only available hydrogen is the one α to the hydroxy group (Scheme 173, b).[97]

Selenoxides normally show[97] a preference for elimination towards the least-substituted carbon (Scheme 173, c), and for the formation of conjugated over unconjugated alkenes.

In the case presented in Scheme 51 the differences in behavior of the β-hydroxyalkyl selenoxides of different origin may be due to different diastereoisomeric ratios the equilibration of which is hampered by steric crowding. Therefore, there is a pronounced stereochemical effect on the regiochemistry of selenoxide *syn* elimination reactions. This elimination reaction leads to allyl alcohols where the α,β-disubstituted double bond possesses exclusively the (*E*)-stereochemistry when reasonably bulky groups are present (Scheme 52; Scheme 105, c; Scheme 174, d).[7,40,42,97,195,243] Those substituted with smaller groups such as a primary hydroxy group (Scheme 176, c)[195] or an ethynyl moiety (Scheme 173, d)[97] lead, however, to a mixture of stereoisomers in which the (*E*)-isomer prevails.

This reaction allows the synthesis of: (i) β,β′-dienols by oxidation of β-hydroxy-γ-alkenyl selenides[239,245] or more conveniently from α-lithioalkyl selenoxides and enones (Scheme 136 and 166);[97] (ii) β,δ-dienols from 1-lithio-3-alkenyl phenyl selenoxides and carbonyl compounds (Scheme 177);[97] and (iii) 2-(1′-hydroxyalkyl)-1,3-butadienes from 1-methylselenocyclobutyllithium and carbonyl compounds (Scheme 178).[136] α,β-Unsaturated alcohols bearing a methylselenoxy or a phenylselenoxy group at the α-position do not lead on thermolysis to propargyl alcohols; however, those bearing a (trifluoromethylphenyl)selenoxy moiety at the α-position are valuable precursors of such compounds (Scheme 179).[103]

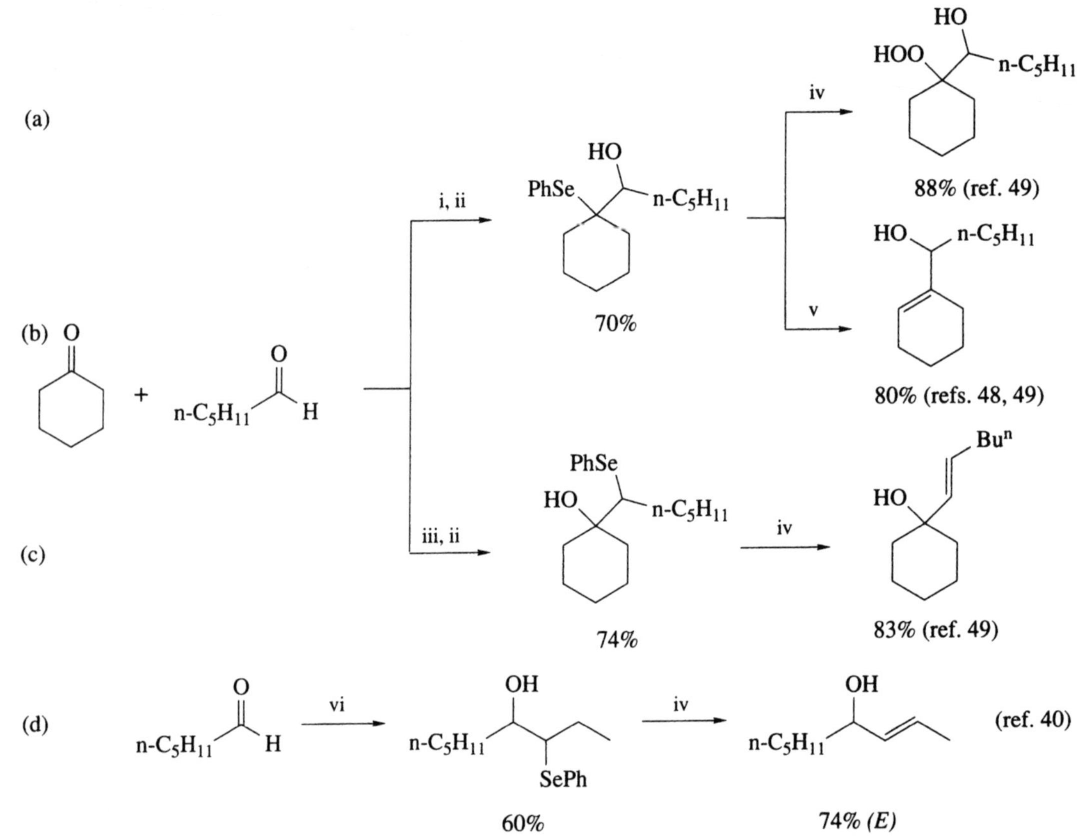

i, cyclohexanone, PhSeH, HCl; ii, Bu^nLi, THF, –78 °C; iii, hexanal, PhSeH, $ZnCl_2$; iv, 30% H_2O_2, THF, 20 °C, 0.2 h; v, $NaIO_4$, MeOH–H_2O, 0 °C, 1 h, then 20 °C, 1 h; vi, EtC(Li)SePh

Scheme 174

Finally, 1-(1-selenoalkyl)cyclobutanols provide the expected 1-(1′-alkenyl)cyclobutanols (Scheme 54, a),[127,128] whereas a ring enlargement reaction leading to a mixture of cyclopentanones and α-phenylselenocyclopentanones (Scheme 54, b) occurs with the corresponding alcoholates.[127,128] The last reaction is strictly limited to strained compounds such as cyclobutanols and cyclopropanols, and does not take place with higher homologs.

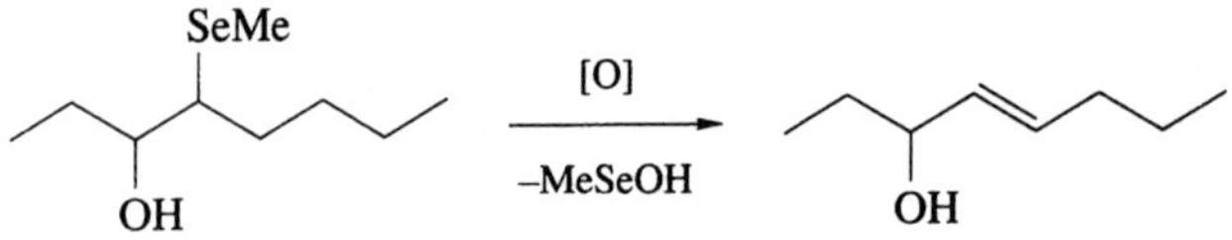

	Conditions	Yield (%) at 20 °C	Yield (%) at 55 °C
(a)	H_2O_2, THF, 5 h	14	27
(b)	H_2O_2–Al_2O_3, THF, 4 h	43	63
(c)	Bu^tOOH, Al_2O_3, THF, 3 h	39	43
(d)	O_3, $CDCl_3$, 2 h	—	50 (50% SM)
(e)	i, O_3, $CDCl_3$, –78 °C; ii, 3 Et_3N, CH_2Cl_2	—	62 (40 °C)

Scheme 175 [188, 195]

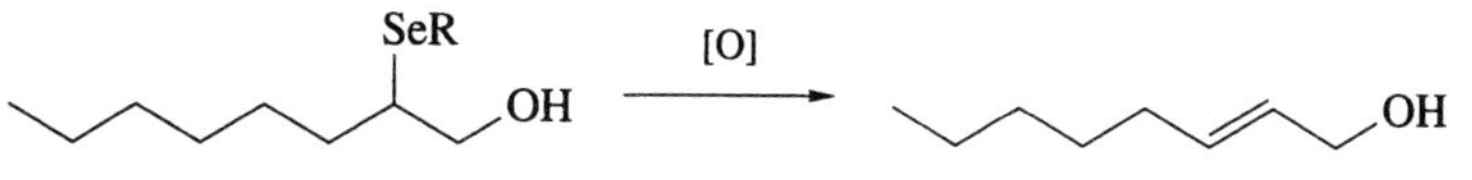

SeR, OH → [O] → OH (ref. 195)

	Conditions	Yield (%) R = Ph	R = Me
(a)	30% H_2O_2, THF, 16 h	35	—
(b)	30% H_2O_2, Al_2O_3, THF, 55 °C, 3–6 h	74	47
(c)	$Bu^tOOH–Al_2O_3$, THF, 55 °C, 5 h	78	80 (E:Z = 75:25)

Scheme 176

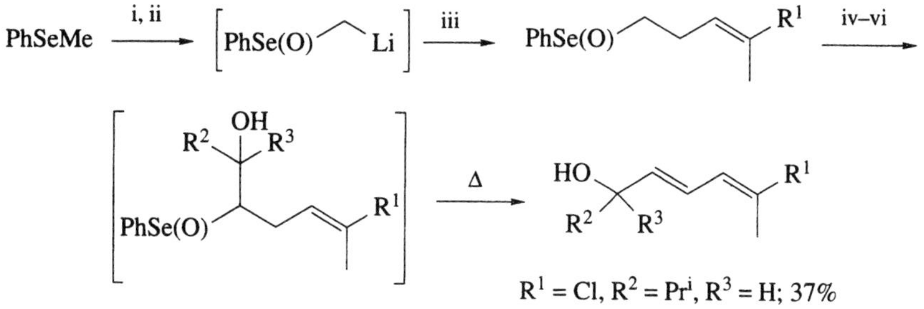

PhSeMe →(i, ii) [PhSe(O)CH₂Li] →(iii) PhSe(O)…R^1 →(iv–vi) [PhSe(O)…C(OH)R^2R^3…R^1] →(Δ) HOCR^2R^3…R^1 (ref. 97)

R^1 = Cl, R^2 = Pr^i, R^3 = H; 37%

R^1 = Me, R^2 = Et, R^3 = Me; 69%

i, MCPBA, THF, –78 °C; ii, LDA, –78 °C; iii, (E)-R^1(Me)C=$CHCH_2X$; iv, LDA; v, R^2R^3CO; vi, H_3O^+

Scheme 177

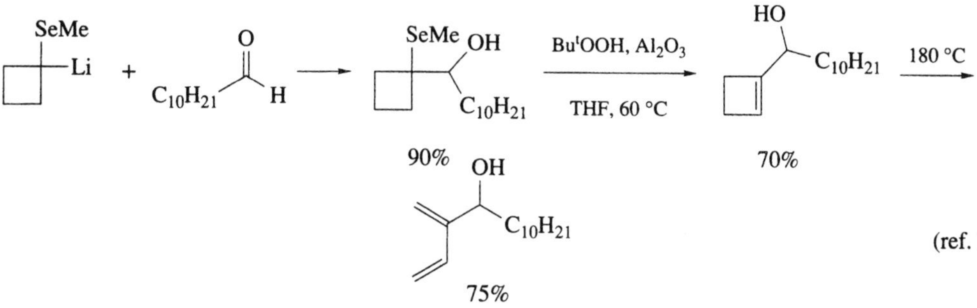

SeMe, Li + $C_{10}H_{21}$CHO → SeMe, OH, $C_{10}H_{21}$ (90%) →(Bu^tOOH, Al_2O_3, THF, 60 °C) HO, $C_{10}H_{21}$ (70%) →(180 °C) OH, $C_{10}H_{21}$ (75%) (ref. 136)

Scheme 178

Se(O)Ar →(i, LDA, THF, –78 °C; ii, O, Ph, 4) Se(O)Ar, (CH_2)$_4$Ph, OH (67%) →(DABCO, 95 °C) (CH_2)$_4$Ph, OH (83%)

Ar = 4-$CF_3C_6H_4$

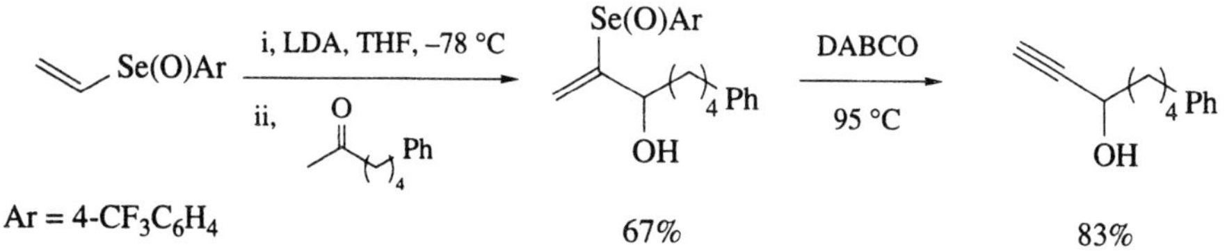

Scheme 179 [103]

2.6.4.5 β-Hydroxyalkyl Selenides as Precursors of Epoxides and Carbonyl Compounds

2.6.4.5.1 Generalities

The presence of two potential leaving groups β to each other makes β-hydroxyalkyl selenides valuable candidates for epoxide synthesis,[206a] as well as for pinacol-type rearrangement.[206b] These concurrent transformations, which usually require the selective activation of the selenyl moiety to a better leaving group, have been successfully achieved by three distinct ways: (i) by alkylation to β-hydroxyalkyl selenonium salts; (ii) by oxidation to β-hydroxyalkyl selenones; and (iii) by complexation with metal salts where the metal is a soft acceptor. The first two methods can provide either the epoxide or the rearranged carbonyl compounds, whereas the last one exclusively leads to rearranged carbonyl compounds only when the carbon bearing the seleno moiety is alkyl disubstituted.

2.6.4.5.2 Synthesis of epoxides[7,12,16,199,206]

The transformation of β-hydroxyalkyl selenides to epoxides (Schemes 124, 132b, 134a, 161g, 162d, 163e, 164d, 165b, 166d and 180–185) has been achieved under three different sets of conditions: (i) by alkylation of the selenium atom, leading to stable β-hydroxyalkylselenonium salts, with MeI,[55,136,163,206,221,236,239,263] MeI–$AgBF_4$[55,163,196,206,221] Me_2SO_4,[55,189,196,206,265] Et_3OBF_4[206,258,264] or $MeFSO_3$.[163,188,206,215,216,222] The selenonium salts are then cyclized on reaction with a base such as Bu^{t}OK/DMSO,[55,136,163,196,206,221,236,239,265] 50% aqueous KOH/CH_2Cl_2[189,206] 10% aqueous KOH/Et_2O,[136,163,206,215,216,222,239,263] KH/DME[258] or NaH/THF;[264] (ii) by reaction of dichlorocarbene, generated[206,266] from chloroform and thallous ethoxide (5.5 equiv. T1OEt, $CHCl_3$, 20 °C, 8–24 h), or more conveniently from 50% aqueous potassium hydroxide solution in the presence of a phase transfer catalyst (50% aq. KOH, $CHCl_3$, 20 °C, 1–2 h);[206,266] and (iii) by oxidation to β-hydroxyalkyl selenones (2.5 equiv. MCPBA, $CHCl_3$),[130,201,206] which are cyclized in the presence of a base (Bu^{t}OK/DMSO or $K_2CO_3/CHCl_3$).[130] All these reactions proceed stereoselectively by complete inversion of the configuration at the substituted carbon.[206,236,263,264,266]

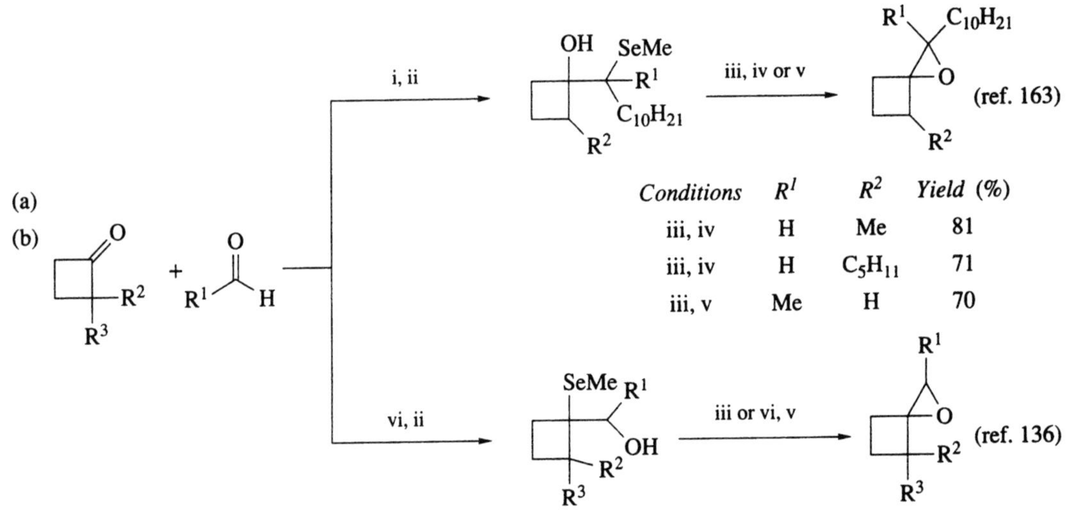

(a) i, ii → OH SeMe R¹ C₁₀H₂₁ R² → iii, iv or v → R¹ C₁₀H₂₁ O R² (ref. 163)

(b) O R² R³ + O R¹ H

Conditions	R^1	R^2	Yield (%)
iii, iv	H	Me	81
iii, iv	H	C_5H_{11}	71
iii, v	Me	H	70

(c) vi, ii → SeMe R¹ OH R² R³ → iii or vi, v → R¹ O R² R³ (ref. 136)

(d)

Conditions	R^1	R^2	R^3	Yield (%)
iii, iv	$C_{10}H_{21}$	H	H	65
vi, v	H	Me	C_9H_{19}	67

i, R^1CHO, MeSeH, $ZnCl_2$; ii, Bu^{n}Li, THF, –78 °C; iii, MeI (neat), 20 °C; iv, Bu^{t}OK, DMSO, 20 °C; v, 10% KOH, ether, 20 °C; vi, cyclobutanone, MeSeH, $ZnCl_2$; vi, MeI•$AgBF_4$

Scheme 180

The reaction involving the alkylation of β-hydroxyalkyl selenides to give β-hydroxyalkylselenonium salts which are then cyclized with a base is by far the most general. It allows the synthesis of a large variety of epoxides such as terminal, α,α- and α,β-disubstituted, tri-[55,196] and tetra-substituted,[55,188] as well as oxaspiro[2.0.*n*]-hexanes,[136,163] -heptanes[189] and -octanes (Scheme 161, g; Scheme 162, d; Scheme 164, d; Scheme 165, b)[55,188,196] and vinyl oxiranes (Schemes 166 and 181)[221,239] from both β-hydroxyalkyl methyl[55,136,163,215,221,222,236,263] and phenyl selenides.[163,196,258,264]

The most successful combination involves methylseleno derivatives as starting materials, methyl iodide (neat) as the alkylating agent and 10% aqueous potassium hydroxide solution in ether as the base.[188] In some difficult cases, such as those involving hindered β-hydroxyalkyl selenides,[188] where the

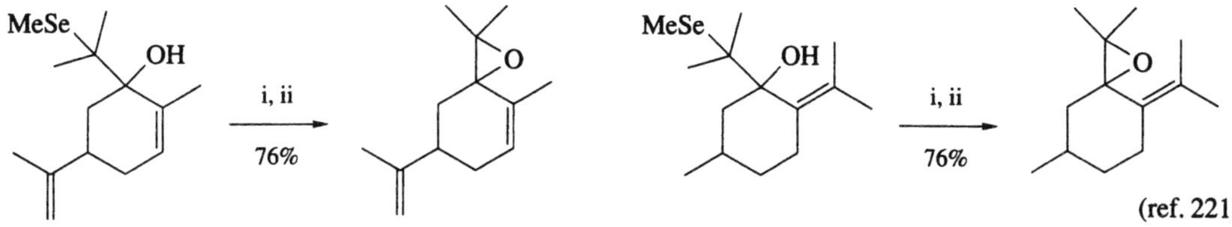

MeSe OH → i, ii 76% → O; MeSe OH → i, ii 76% → O

(ref. 221)

i, neat MeI, 20 °C; ii, Bu^{t}OK, DMSO, 20 °C

Scheme 181

seleno moiety is linked to a cyclohexane ring[55,189] or to a primary alcohol[189] (Scheme 164, d; Schemes 182 and 183),[189] the consecutive use of methyl fluorosulfonate and aqueous potassium hydroxide in ether or pentane is required in order to avoid the concomitant formation of carbonyl compounds, allyl alcohols[188] or alkenes[189] which are observed when other conditions are used.

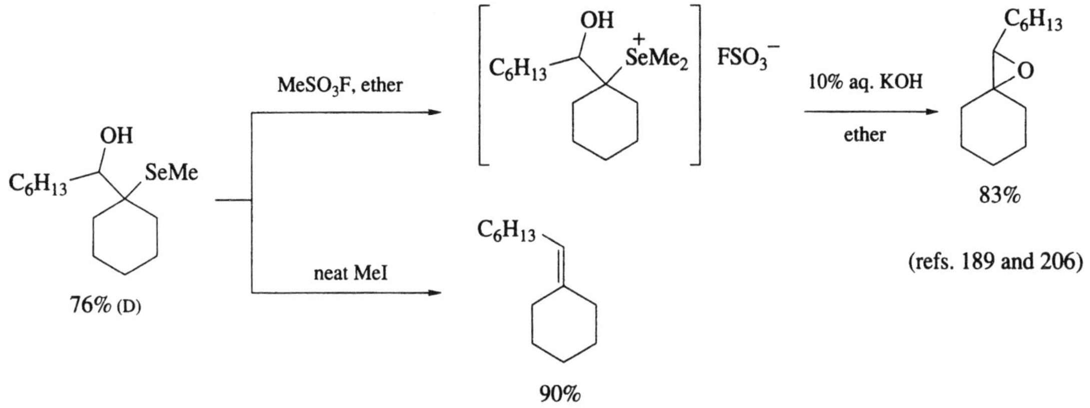

C_6H_{13}, OH, SeMe, 76% (D) → $MeSO_3F$, ether → [C_6H_{13}, OH, $\overset{+}{Se}Me_2$] FSO_3^- → 10% aq. KOH, ether → C_6H_{13}, O, 83%

neat MeI → C_6H_{13}, 90%

(refs. 189 and 206)

Scheme 182

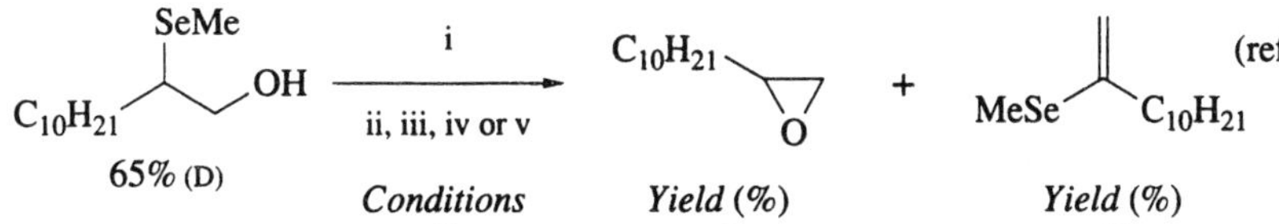

$C_{10}H_{21}$, SeMe, OH, 65% (D) → i; ii, iii, iv or v → $C_{10}H_{21}$ epoxide (O) + MeSe, $C_{10}H_{21}$ (refs. 189 and 206)

	Conditions	*Yield (%)*	*Yield (%)*
(a)	ii	5	45
(b)	iii	65	30
(c)	iv	85	15
(d)	v	78	3

i, $MeSO_3F$, ether; ii, 50% aq. KOH/CH_2Cl_2; iii, aq. K_2CO_3/ether; iv, 10% aq. KOH/ether; v, 10% aq. KOH/pentane

Scheme 183

Methylseleno derivatives offer the advantage of being more easily methylated than their phenylseleno analogs (MeI instead of MeI–$AgBF_4$) and allowing the synthesis of a wide range of β-hydroxyalkylselenonium salts, including those which have two alkyl groups on the carbon bearing the selenium atom. These are unavailable from phenylseleno derivatives;[196] therefore, tetrasubstituted epoxides can only be produced from methylseleno compounds.[55]

The reaction involving dichlorocarbene also provides a wide range of epoxides,[206,266] including terminal, α,α- and α,β-dialkyl-substituted and trialkyl-substituted compounds, from a large variety of β-hydroxyalkyl methyl and phenyl selenides (Scheme 162, d). The thallous ethoxide reaction, although it takes place more slowly (especially with phenylseleno derivatives) than under phase transfer catalysis conditions,[206,266] has to be in several instances preferred since it avoids the concomitant formation of alkenic or dihalocyclopropane derivatives often observed under the latter conditions.[206,223,266] This reaction is less general than that involving the discrete formation of β-hydroxyalkylselenonium salts since it takes another course with β-hydroxyalkyl selenides where the carbon bearing the seleno moiety is dialkyl substituted, and instead delivers rearranged ketones (see Section 2.6.4.5.3). This method, therefore, does not allow the synthesis of tetraalkyl-substituted epoxides (Scheme 162, compare d and e). However it offers the definite advantage of taking place in one step and avoiding the use of expensive silver tetrafluoroborate[196] or tricthyloxonium tetrafluoroborate.[264]

The reaction involving β-hydroxyalkyl selenones[130,201,206] is by far the less attractive route to epoxides among those cited above due to the difficulties usually encountered in the oxidation step and the easy rearrangement of the β-hydroxyalkyl selenones to carbonyl compounds (Scheme 184, b and c).[130] This method is, however, particularly useful for the synthesis of oxaspiropentanes[130,206] (especially for those bearing a fully alkyl-substituted epoxide ring) from 1-seleno-1-(1′-hydroxyalkyl)cyclopropanes (Scheme 163, e; Scheme 185), which cannot be obtained under other conditions.[12]

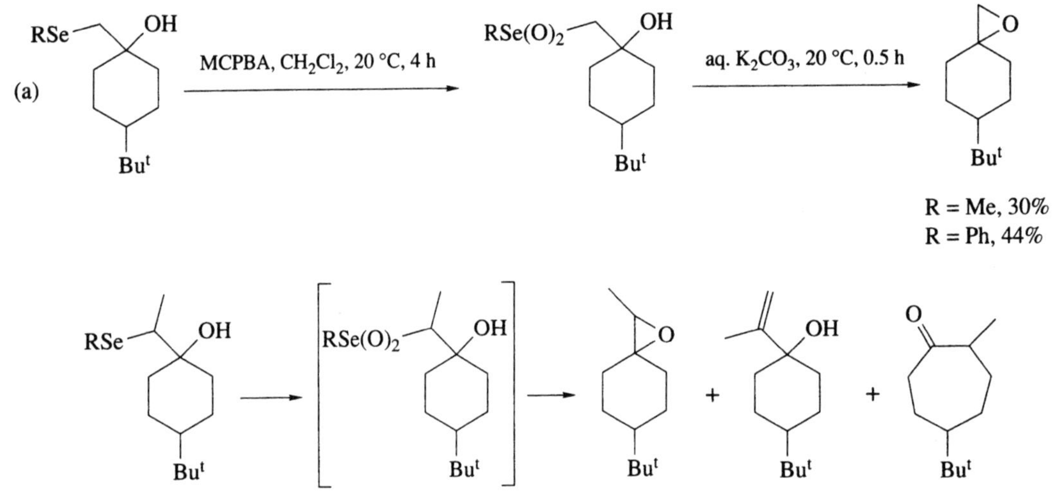

(a) MCPBA, CH_2Cl_2, 20 °C, 4 h; aq. K_2CO_3, 20 °C, 0.5 h

R = Me, 30%
R = Ph, 44%

	Conditions	Yield (%)		
(b)	MCPBA (3 equiv.), CH_2Cl_2, 20 °C, 3 h	4	21	56
(c)	excess K_2CO_3, MCPBA (3 equiv.), CH_2Cl_2, 20 °C, 3 h	44	—	9

Scheme 184

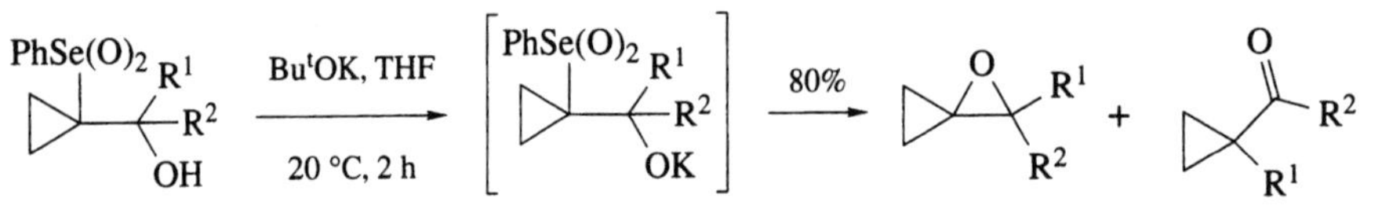

Bu^tOK, THF, 20 °C, 2 h; 80%

R^1 = H, R^2 = Ph 33:67

Scheme 185 [130]

2.6.4.5.3 *Rearrangement of β-hydroxyalkyl selenides to carbonyl compounds*[16,206b]

The rearrangement of β-hydroxyalkyl selenides to carbonyl compounds has been carried out under different conditions, which have to be chosen with regard to the structure of the starting material and especially the substitution pattern around the carbon bearing the seleno moiety. The rearrangement of β-hydroxyalkyl selenides possessing two hydrogen atoms on the carbon bearing the seleno moiety has not yet been achieved. However, the rearrangement of monoalkyl-substituted derivatives takes place on oxidation with *m*-chloroperoxybenzoic acid (2.5 equiv. of MCPBA in $CHCl_3$[130] or in MeOH;[201] Scheme 184), and the rearrangement of dialkyl-substituted compounds occurs on reaction with silver tetrafluoroborate[190,191,204,227] in dichloromethane and chloroform or silver tetrafluoroborate supported on basic alumina[197,204] in the same solvents (Scheme 162, e; Scheme 186, a; Scheme 187), or silver nitrate,[267] or with dichlorocarbene generated from chloroform and thallous ethoxide[197,225–227,244] or a 50% aqueous solution of potassium hydroxide in the presence of a phase transfer catalyst ($PhNEt_3Cl$, TEBAC)

(Scheme 162, e; Scheme 165, c; Scheme 166, e; Scheme 186, b; Scheme 188, a; Scheme 189, a and b).[197,198,227]

The above conditions allow the synthesis of a large variety of α,α-dialkyl substituted ketones including spiro derivatives and cyclopentanones, cyclohexanones, cycloheptanones and cyclododecanones bearing two alkyl groups at the α-position (Scheme 162, e; Scheme 165, c; Scheme 166, e; Scheme 186, a and b; Scheme 187; Scheme 188, a; Scheme 189, a and b).[160,198,204,244]

The reactions are less selective when the dichlorocarbene is generated from bromodichloromethane[223] and lead to *gem*-dichlorocyclopropanes if the dichlorocarbene is produced from chloroform and potassium *t*-butoxide (Scheme 162, f; Scheme 188, b; Scheme 189, c).[198]

Particularly crowded or strained derivatives lead to carbonyl compounds under special conditions. Thus, β-hydroxyalkylselenonium salts, obtained on alkylation of particularly hindered β-hydroxyalkyl selenides and the selenium-bearing carbon of which is dialkyl substituted, have been rearranged[160,188] to ketones by heating or by simple dissolution in dichloromethane (Scheme 164, e; Scheme 186, c).

Strained β-oxidoalkyl phenyl selenoxides, such as 1-oxido-1-(1′-phenylselenoxyalkyl)cyclopropanes, derived from oxaspiropentanes with tetraalkyl-substituted oxirane rings,[268] and 1-(1′-hydroxyalkyl)-1-selenoxycyclobutanes,[127,128] obtained on oxidation of the corresponding selenides or on reaction of α-lithioalkyl selenoxides with cyclobutenones, possess a high propensity to rearrange to cyclobutanones

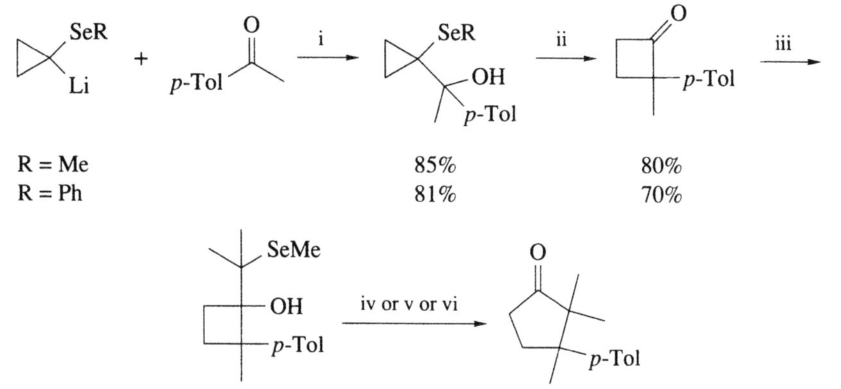

SeR, Li + p-Tol–CO–Me —i→ (SeR, OH, p-Tol) —ii→ (cyclobutanone, p-Tol) —iii→

R = Me: 85%, 80%
R = Ph: 81%, 70%

(SeMe, OH, p-Tol) —iv or v or vi→ (cyclopentanone, p-Tol)

66% from ether
0% from THF

	Conditions	Yield (%)
(a)	iv	69
(b)	v	57
(c)	vi	82

i, ether, –78 °C; ii, *p*-TsOH, benzene/H_2O, 80 °C, 12 h; iii, $Me_2C(SeMeLi)$, solvent, –78 °C; iv, $AgBF_4$–Al_2O_3, CH_2Cl_2; v, TlOEt,$CHCl_3$; vi, $MeSO_3F$, ether

Scheme 186 [160]

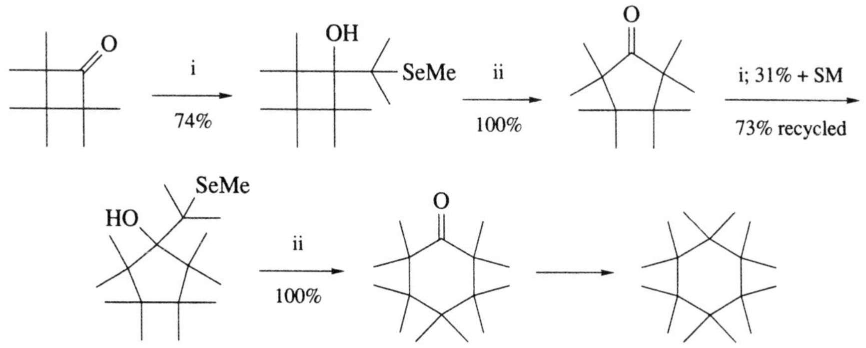

i, 74%; SeMe, OH; ii, 100%; i; 31% + SM, 73% recycled; HO, SeMe; ii, 100%

i, $Me_2C(SeMe)Li$, ether, –78 °C; ii, $AgBF_4$, $CHCl_3$

Scheme 187 [190]

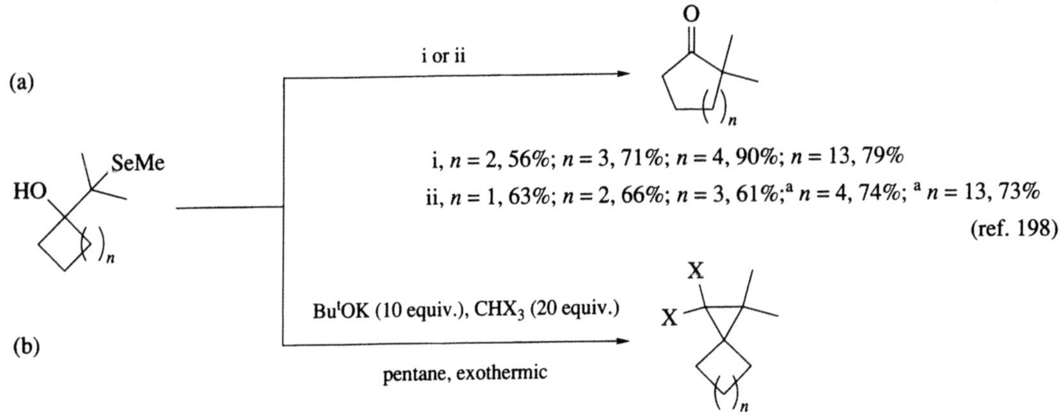

(a) i or ii

i, $n = 2$, 56%; $n = 3$, 71%; $n = 4$, 90%; $n = 13$, 79%
ii, $n = 1$, 63%; $n = 2$, 66%; $n = 3$, 61%;[a] $n = 4$, 74%; [a] $n = 13$, 73%
(ref. 198)

(b) Bu^tOK (10 equiv.), CHX_3 (20 equiv.), pentane, exothermic

X = Cl; $n = 3$, 39%; $n = 4$, 50%; $n = 12$, 40%
X = Br, $n = 3$, 46%

i, TlOEt (5.5 equiv.), $CHCl_3$, 20 °C, 8 h; ii, 50% aq. KOH, $CHCl_3$, TEBAC, 20 °C, 1 h
[a] Allyl alcohols also produced (4–13%)

Scheme 188

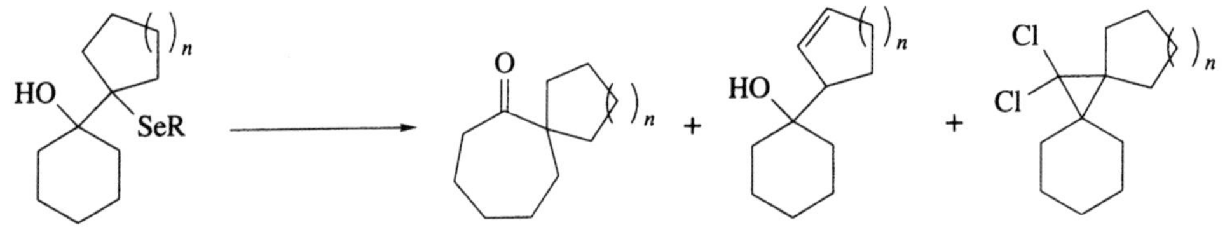

	Conditions	*R*	*n*	*Yield* (%)						*Ref.*
(a)	i	Ph	1	87	100	:	0	:	0	198, 244
	i	Ph	2	92	100	:	0	:	0	198, 244
(b)	ii	Ph	1	89	68	:	32	:	0	198
	ii	Me	1	91	78	:	22	:	0	198
(c)	iii	Me	1	47	40	:	17	:	43	198

i, TlOEt (5.5 equiv.), $CHCl_3$, 20 °C, 8 h; ii, 10% aq. KOH, $CHCl_3$ (3 equiv.), TEBAC, CH_2Cl_2, 20 °C;
iii, Bu^tOK (10 equiv.), $CHCl_3$ (10 equiv.), pentane, 20 °C

Scheme 189

(Scheme 190)[268] and cyclopentanones (Schemes 54b and 121),[127,128] respectively, rather than produce the corresponding allyl alcoholates *via* the well-established seleninic acid elimination (Section 2.6.4.4). The latter reaction, however, becomes important with oxaspiropentanes where the oxirane ring is trialkyl substituted (Scheme 191).[268]

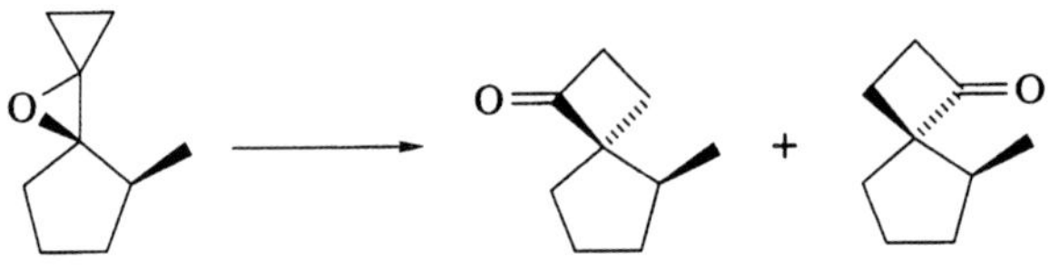

(ref. 268)

	Conditions	*Yield* (%)	*Ratio*
(a)	PhSeNa, EtOH, then MCPBA	48	4:96
(b)	$LiBF_4$	61	100:0

Scheme 190

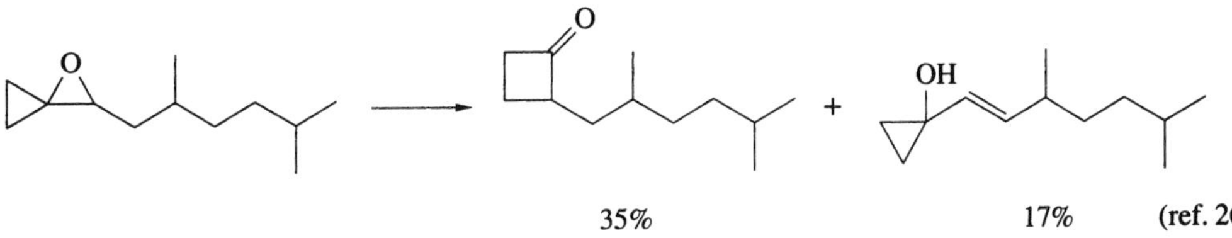

35% 17% (ref. 268)

i, PhSeNa, EtOH; ii, MCPBA, CH_2Cl_2, –25 °C, 4.5 h, then 0 °C, 1 h

Scheme 191

The stereochemistry of the spirocyclobutanones shown in Scheme 190 seems to be determined by the rate of the ring expansion compared with that of bond rotation.[268] Indeed, in several cases high stereoselectivity is observed (Scheme 190, a),[268] which is quite opposite to that found[268] with lithium tetrafluoroborate (Scheme 190, b).

Finally, 1-(1′-hydroxyalkyl)-1-(methylseleno)cyclopropanes and some rare phenylseleno analogs belonging to the phenylcarbinol series produce cyclobutanones on reaction with *p*-toluenesulfonic acid in a similar manner to the phenylthio analogs.[12] This reaction has allowed the synthesis of a large variety of cyclobutanones, including those that are polyalkylated (Scheme 163; f–h),[12,77,159,160,199] aryl substituted or vinyl substituted (Scheme 132, a; Scheme 186),[12,77,160] by selective migration of the more-substituted carbon (Scheme 163, compare h with f).[12,77,160]

The conditions involving dichlorocarbene generated from chloroform and thallium ethoxide give the best chemoselectivity, yields and reproducibility (Scheme 162, e; Scheme 165, c; Scheme 166, e; Scheme 186, b; Scheme 188, a; Scheme 189, a).[160,197,225–227,244] The reactions are usually carried out at room temperature and require at least 8 h to go to completion.[244] They are faster with methylseleno than with phenylseleno derivatives when performed at a higher temperature.[244] The reactions performed under phase transfer catalysis conditions are valuable[198] owing to the simplicity of the reagents used and the particularly rapid reaction rate (20 °C, 1–2 h). They lead,[198] however, to appreciable amounts of allyl alcohol (5–20%) besides the rearranged ketones (Scheme 188, a; Scheme 189, b). They are valuable for the synthesis of the most water soluble and volatile ketones, such as 2,2-dimethylcyclopentanone, since they avoid the presence of ethanol and chloroform, difficult to remove from the desired compounds. The same transformation can also be performed[204] with silver tetrafluoroborate or silver tetrafluoroborate supported on basic alumina. The latter conditions are usually better than the former since they avoid the presence of fluoroboric acid, responsible for the concomitant formation of alkenes (by formal elimination of RSeOH).

When two different groups are susceptible to migrate on reaction of β-hydroxyalkyl selenides with dichlorocarbene or silver salts, the tendency to migrate follows the order CH_2SiMe_3 < H < Ph < alkyl (Scheme 192; Scheme 165, c).[197,223,244,267] In general, the more alkyl substituted carbon possesses the highest propensity to migrate whatever the conditions used (Scheme 192). In the case of 2-methylcyclohexanols, however, low selectivity has been observed[197] with thallium(I) ethoxide (Scheme 192, a and b) and, surprisingly, when the reaction is instead performed with silver tetrafluoroborate, the less-substituted carbon has the highest tendency to migrate (Scheme 192, c).[197]

The case of β-hydroxy-γ-alkenyl selenides merits further comments.[225–227] The rearrangement efficiently takes place using the thallium(I) ethoxide method and the presence of an additional double bond in the reactant does not introduce a serious problem associated with unwanted reaction with the dichlorocarbene intermediate.[225,226] This is not the case when silver tetrafluoroborate is used.

The thallium ethoxide method almost selectively provides α,α-dialkyl-β-enones by vinyl migration in sterically unconstrained straight chain derivatives (Scheme 193),[225,226] whereas alkyl migration leading to α-enones takes place almost exclusively with the 2-cyclohexen-1-ol derivatives[225,226] shown in Scheme 194. This is probably due to interrelationships of steric and conformational influences associated with proper antiperiplanar alignment of the $RSeCCl_2H$ leaving group. However, such high selectivity in favor of the alkyl over the vinyl migration is not observed[225] with 2-cycloalken-1-ols of lower (2-cyclopenten-1-ols) and of higher (2-cycloocten-1-ols) ring size (Scheme 194), and although the alkyl migration is favored[225] with 6-alkyl-substituted 2-cyclohexen-1-ols, the vinyl migration is observed[225,226] if the cyclohexene ring is substituted at the 2- or 3-position (Schemes 134 and 195).

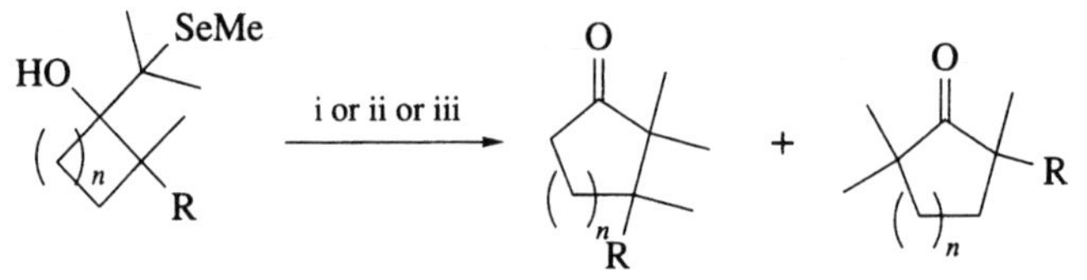

	R	Conditions	n	Yield (%)	Ratio	Ref.
(a)	H	i	2	76	90:10	197, 244
	H	i	3	77	60:40	197, 244
	H	i	4	65	78:22	197, 244
(b)	H	ii	2	69	92:8	197
	H	ii	3	58	62:38	197
	H	ii	4	56	84:16	197
(c)	H	iii	2	74	90:10	204
	H	iii	3	54	22:78	204
	H	iii	4	49	72:28	204
(d)	Me	i	3	81	100:0	197, 244
	Me	i	4	81	99:1	197, 244
(e)	Me	ii	3	54	100:0	197
	Me	ii	4	52	99:1	197
(f)	Me	iii	3	65	100:0	204
	Me	iii	4	55	70:30	204

i, TlOEt (5.5 equiv.), $CHCl_3$, 20 °C; ii, 10% aq. KOH, $CHCl_3$, TBEAC, 20 °C; iii, $AgBF_4$–Al_2O_3, CH_2Cl_2, 20 °C

Scheme 192

2.6.4.5.4 Mechanism of epoxide and ketone formation from β-hydroxyalkyl selenides

Striking differences exist between silver tetrafluoroborate and dichlorocarbene mediated reactions since they proceed differently in several instances.[200,206] Although the dichlorocarbene reaction takes place with almost all kinds of β-hydroxyalkyl selenides,[160,197,225–227,244] the silver tetrafluoroborate mediated reaction only works with those β-hydroxyalkyl selenides which possess[190,191,204] two alkyl groups on the carbon bearing the seleno moiety and on the condition that no unsaturation is present around the active site. Furthermore, detailed investigations carried out on the β-hydroxyalkyl selenides (**35**) and (**36**) shown in Scheme 196 shed some light on the intimate mechanism of these reactions and on the subtle differences which exist between dichlorocarbene and silver tetrafluoroborate mediated reactions, although both often lead to identical products from the same starting materials.[20,200,206]

Thus, the β-hydroxyalkyl selenides (**35a**) and (**35b**) react[266] with dichlorocarbene and stereoselectively lead to *trans*- and *cis*-9,10-epoxyoctadecene (**37a** and **37b**; Scheme 196). Under similar conditions the β-hydroxyalkyl selenide (**36a**) regio- and stereoselectively leads to the cycloheptanone (**38a**) resulting from the migration of the most-substituted carbon, and its diastereoisomer (**36b**) produces a mixture of the regioisomeric cycloheptanones (**38b**) and (**39a**) in which the latter, arising from the migration of the least-substituted carbon, prevails (Scheme 196, c and e).[200] In all cases the reactions take place apparently in a concerted manner with inversion of the configuration at the substituted carbon,[200,206,266] and in the

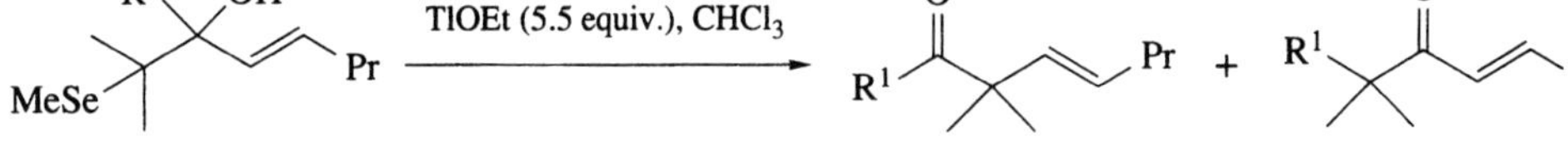

TlOEt (5.5 equiv.), $CHCl_3$ (ref. 225)

R^1 = Me, 43%, 100:0
R^1 = Et, 71%, 78:22

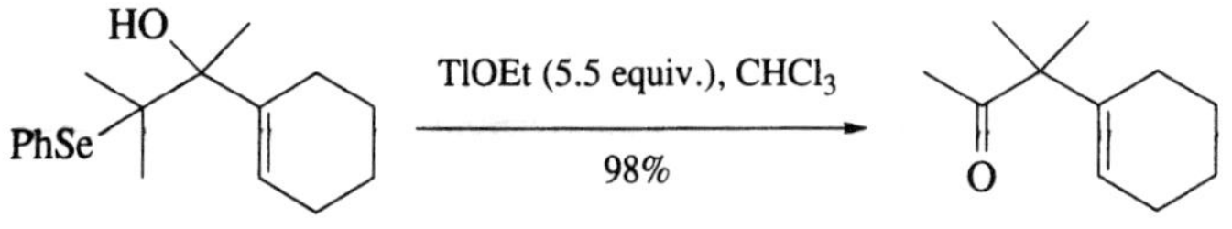

TlOEt (5.5 equiv.), $CHCl_3$, 98% (ref. 226)

Scheme 193

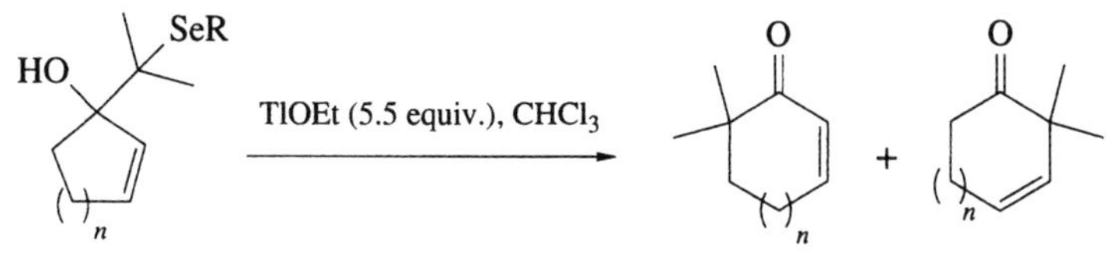

TlOEt (5.5 equiv.), $CHCl_3$

	n	*R*	*Yield (%)*	*Ratio*
(a)	1	Me	—	43:57
(a)	2	Me	68	91:9
(b)	2	Ph	84	95:5
(a)	3	Me	69	51:49
(b)	3	Ph	88	56:44
(a)	8	Me	87	42:58

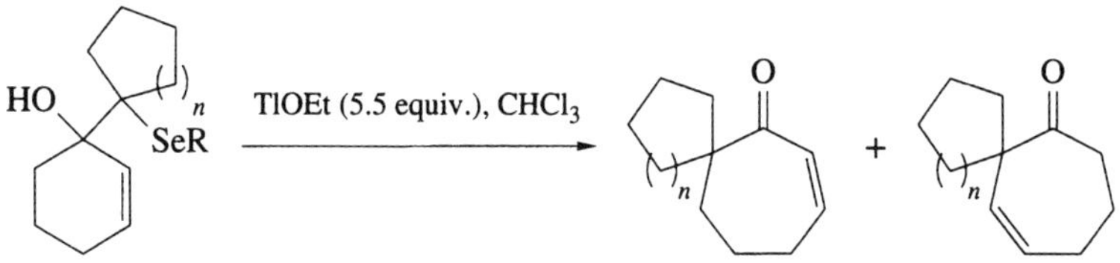

TlOEt (5.5 equiv.), $CHCl_3$

	n	*R*	*Yield (%)*	*Ratio*
(c)	1	Ph	95	80:20
(d)	2	Me	80	90:10

(refs. 225 and 226)

Scheme 194

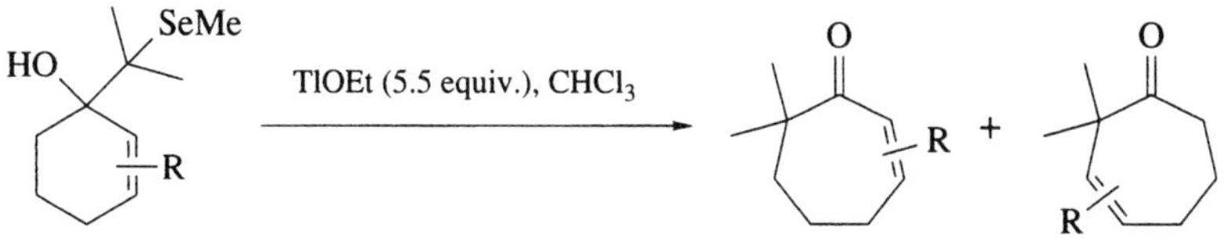

TlOEt (5.5 equiv.), $CHCl_3$ (ref. 225)

		Yield (%)	*Ratio*
(a)	2-Me	64	78:22
	3-Me	67	73:27
	4-Me	77	91:9
	6-Me	71	100:0
(b)	6-Ph	70	100:0

Scheme 195

case of the ring enlargement reaction with complete retention of the configuration at the migrating carbon.[200]

The reactions of the same β-hydroxyalkyl selenides (**35**) and (**36**) with silver tetrafluoroborate take a somewhat different course. Although cycloheptanones are produced from (**36**), (**35**) is unreactive and gives only the starting materials on heating (Scheme 196). Furthermore, with the β-hydroxyalkyl selenides (**36**) the structure and stereochemistry of the cycloheptanones obtained are different from those disclosed when the reaction is carried out with dichlorocarbene. Not only the least-substituted carbon now migrates, regardless of whether (**36a**) or (**36b**) is involved, but also the ring enlargement proceeds with almost complete racemization from (**36b**) (leading to a mixture of **39a** and **39b**; Scheme 196) and with apparent complete retention of configuration from (**36a**) (leading to **39a**; Scheme 196). The latter might be the result of two consecutive inversions involving the intermediate formation of an epoxide, but this remains to be confirmed.

The reaction of dichlorocarbene with β-hydroxyalkyl selenides is believed[200,206,226,266] to take place selectively on the selenium atom, thus producing the corresponding β-hydroxyalkyldichloromethyleneselenonium ylide (**40**; Scheme 197). This, after an intramolecular proton transfer leading to (**41**; Scheme 197), which probably occurs in a six-membered transition state, collapses selectively to the epoxide[206,266] or to the carbonyl compound depending upon the substitution pattern around the substituted carbon.[200] The regiochemistry of the latter reaction depends upon the stereochemistry of the starting material, and the regioisomeric ratio without doubt reflects the ratio of the two conformers (**43** and **44**; Scheme 197)

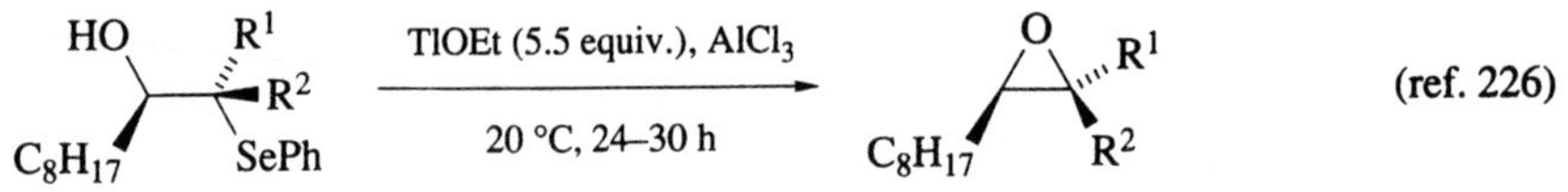

HO, R^1, R^2, C_8H_{17}, SePh → TlOEt (5.5 equiv.), $AlCl_3$, 20 °C, 24–30 h → O, R^1, C_8H_{17}, R^2 (ref. 226)

(37)

(a)	**(35a)**	$R^1 = C_8H_{17}$, H	50%
(b)	**(35b)**	$R^2 = H, C_8H_{17}$	55%

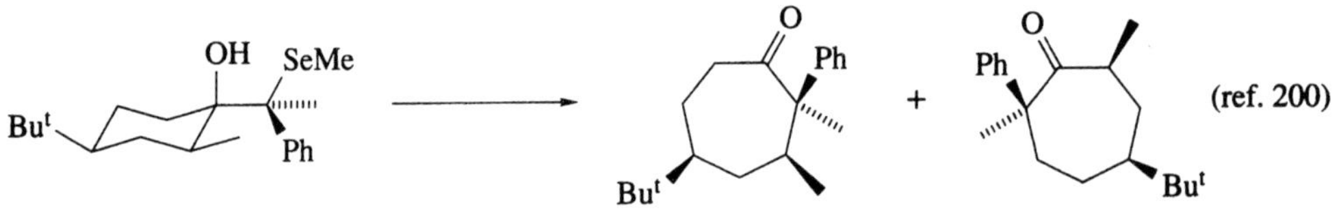

Bu^t, OH, SeMe, Ph → O, Ph, Bu^t + O, Ph, Bu^t (ref. 200)

			(38a)	**(39a)**
(c)	**(36a)**	TlOEt (5.6 equiv.), $CHCl_3$	25%	—
(d)	**(36a)**	$AgBF_4$ (1.3 equiv.)/Al_2O_3	—	48%

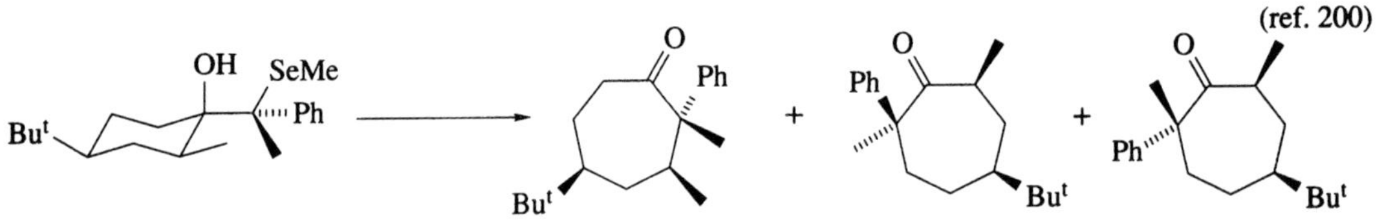

Bu^t, OH, SeMe, Ph → O, Ph, Bu^t + O, Ph, Bu^t + O, Ph, Bu^t (ref. 200)

			(38b)	**(39a)**	**(39b)**
(e)	**(36b)**	TlOEt (5.6 equiv.), $CHCl_3$	14%	26%	—
(f)	**(36b)**	$AgBF_4$ (1.3 equiv.)/Al_2O_3	—	32%	27%

Scheme 196

which have the different groups involved in the rearrangement properly aligned for antiparallel migration.[200,226] These two conformations are not isoenergetic and will not be attained with equal probability.[200,226]

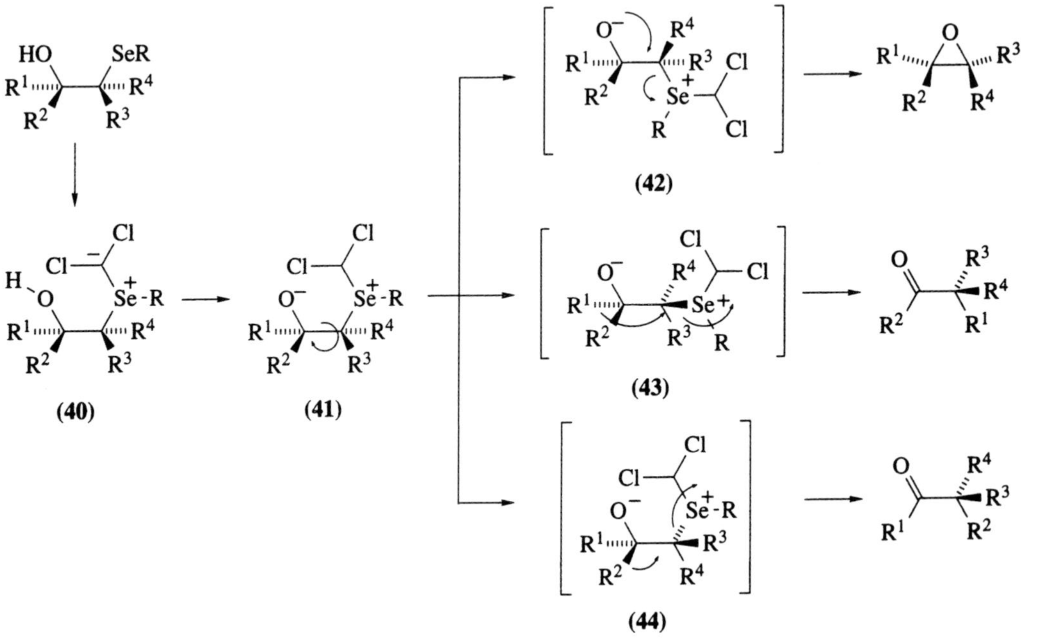

HO, SeR, R^1, R^4, R^2, R^3

↓

H, O, Cl, Cl, $\overset{+}{Se}$-R, R^1, R^4, R^2, R^3 **(40)** → Cl, Cl, O^-, $\overset{+}{Se}$-R, R^1, R^4, R^2, R^3 **(41)**

→ [O^-, R^4, R^1, R^3, Cl, R^2, $\overset{+}{Se}$, R, Cl] **(42)** → R^1, O, R^3, R^2, R^4

→ [Cl, O^-, R^4, Cl, R^1, Se^+, R^2, R^3, R] **(43)** → O, R^3, R^4, R^2, R^1

→ [Cl, Cl, $\overset{+}{Se}$-R, O^-, R^1, R^3, R^2, R^4] **(44)** → O, R^4, R^3, R^1, R^2

Scheme 197

2.6.4.6 Synthetic Uses of β-Hydroxyalkyl Selenides: Comparison with Well-established Related Reactions

2.6.4.6.1 Generalities

The combination of reactions described above (Sections 2.6.4.2 to 2.6.4.5) allows the selective synthesis of a large variety of alcohols, allyl alcohols, alkenes, epoxides and carbonyl compounds from β-hydroxyalkyl selenides. These products often can be obtained from two carbonyl compounds by activation of one of them as an α-selenoalkyllithium (Schemes 161–196).

2.6.4.6.2 Synthesis of alcohols

The whole process involving the reaction of an α-selenoalkyllithium with a carbonyl compound and further reduction of the C—Se bond in the resulting β-hydroxyalkyl selenide leads to an alcohol which can be directly produced from an alkyl metal and the same carbonyl compound (Section 2.6.4.2). This two-step procedure offers, in some cases, interesting advantages due to the larger size of α-selenoalkyllithiums, which favors a better stereochemical control (see, for example, Scheme 167 for comparison between the two approaches).[210]

2.6.4.6.3 Synthesis of allyl alcohols

Allyl alcohols have been prepared from α-selenoalkyllithiums or from α-selenoxyalkyllithiums and carbonyl compounds (Section 2.6.4.4). Although the former route is less connective than the latter, it usually provides allyl alcohols in similar yields (Scheme 198) and offers the advantage of a higher nucleophilicity of the organometallic, which becomes valuable for the more-hindered carbonyl compounds (Scheme 164, b). Both α-selenoalkyllithiums and α-selenoxyalkyllithiums in these processes are synthetically equivalent to vinyl anions. The principal advantage over the classical alkenyllithium route is in the availability of the starting materials (selenoacetals and alkyl selenides), in contrast to the difficulty of obtaining the simplest vinyl halides. Furthermore, since allyl alcohols can be obtained[7,49] *via* the selenoacetal route from two carbonyl compounds, the activation of each of the two derivatives allows the regioselective synthesis of one of the two isomeric allyl alcohols (Scheme 174, b and c).[7,49]

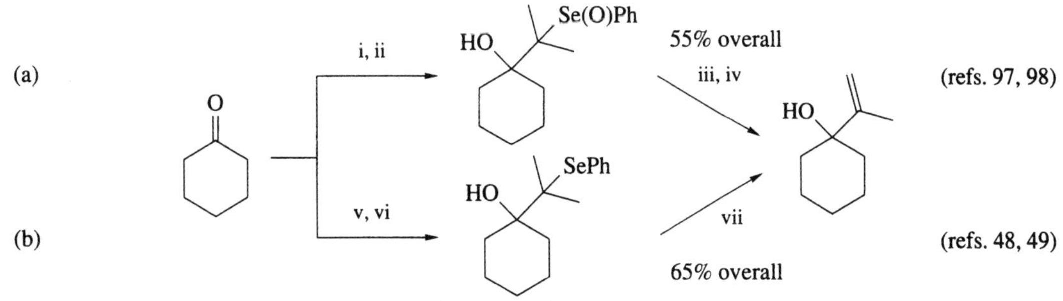

(a) i, ii → HO, Se(O)Ph; 55% overall, iii, iv → HO (refs. 97, 98)

(b) v, vi → HO, SePh; vii, 65% overall (refs. 48, 49)

i, $LiC(Me)_2Se(O)Ph$, THF; ii, 10% AcOH, THF; iii, CCl_4, 70 °C; iv, 20 °C; v, $LiC(Me)_2SePh$, THF; vi, H_3O^+; vii, H_2O_2, THF, 20 °C

Scheme 198

2.6.4.6.4 Synthesis of epoxides, alkenes and carbonyl compounds

Alkenes, epoxides and carbonyl compounds have been conveniently prepared from β-hydroxyalkyl selenides (Sections 2.6.4.3 and 2.6.4.5). These often can, in turn, be obtained from two carbonyl compounds by activation of one of them as an α-selenoalkyllithium. Although each of the two carbonyl compounds can be formally transformed to an α-selenoalkyllithium, in practice a judicious selection is often of some value and even in some cases crucial for the success of the transformations. The following rules might help for the choice. It is better to transform to an α-selenoalkyllithium the carbonyl compound which is: (i) the most readily available; (ii) the least substituted; (iii) the least hindered; and (iv) not aryl or vinyl substituted. Since α-selenoalkyllithiums belong to the well-known family of α-hetero-substituted organometallics,[7] some of them, such as phosphorus ylides[7,25,269–275] or α-silylalkyllithiums,[7,16,31,276] sulfur ylides[7,25,32,33] or α-thioalkyllithiums,[7,16,33] and diazoalkanes[277] have been intensively and successfully used for the one-pot synthesis from carbonyl compounds of alkenes and epoxides, as well as for the ring enlargement of cyclic ketones. The selective synthesis of the same compounds can also be achieved from α-selenoalkyllithiums by the multistep sequences already discussed, which involve the formation of the same bonds (shown in bold in Scheme 199).

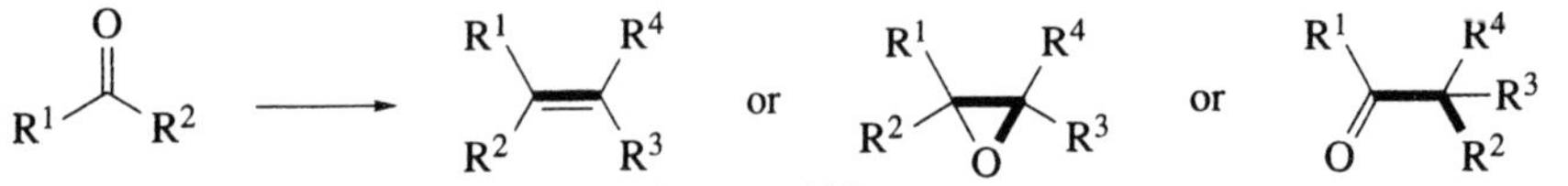

Scheme 199

Although in many instances the well-established, one-step procedures presented above are very efficient and should be preferred to the α-selenoalkyllithium route, there remain several cases for which the latter approach offers definite advantages.

The synthesis of the starting α-selenoalkyllithiums is, as already mentioned, easy and straightforward. Differently substituted reagents, including those bearing two alkyl groups on the carbanionic center, are available from carbonyl compounds and *via* a sequence of reactions which regioselectively allow the replacement of the carbonyl oxygen atom by the seleno moiety and the lithium atom. This compares very well to the synthesis of: (i) phosphorus and sulphur ylides, which require the use of lengthy procedures when *s*-alkyl-phosphonium[7,25,269–275] or -sulphonium salts[7,25,32,33] (with the exclusion of 2-propyl and cyclopropyl derivatives) are needed; and (ii) α-silyl-[7,16,31,276] or α-thio-alkyllithiums,[7,16,33] which is limited to the unsubstituted and monoalkyl-substituted derivatives.

The high nucleophilicity of α-selenoalkyllithiums towards carbonyl compounds, even those that are the most hindered or enolizable, such as 2,2,6-trimethyl- and 2,2,6,6-tetramethyl-cyclohexanone (Schemes 113 and 164),[188] di-*t*-butyl ketone,[189] permethylcyclobutanone,[190,191] permethylcyclopentanone (Schemes 113 and 187)[190,191] and deoxybenzoin (Schemes 115, 116 and 165),[159,194,196,223] allows the synthesis of related alkenes, epoxides and rearranged ketones which are not available from the same carbonyl compounds on reaction with phosphorus or sulfur ylides[278,279] or diazoalkanes.[206a]

The high tendency of α-selenoalkyllithiums, even those that bear alkyl substituents on the carbanionic center, to add at the C-1 site of enones allows the high yield synthesis of vinyl oxiranes (Schemes 132, 134 and 166).[221,239] These cannot be prepared from sulfur ylides since they add across the carbon–carbon double bond of enones to produce cyclopropyl ketones (Scheme 200).[280]

(ref. 280) i 74% ii–v 53% (ref. 221)

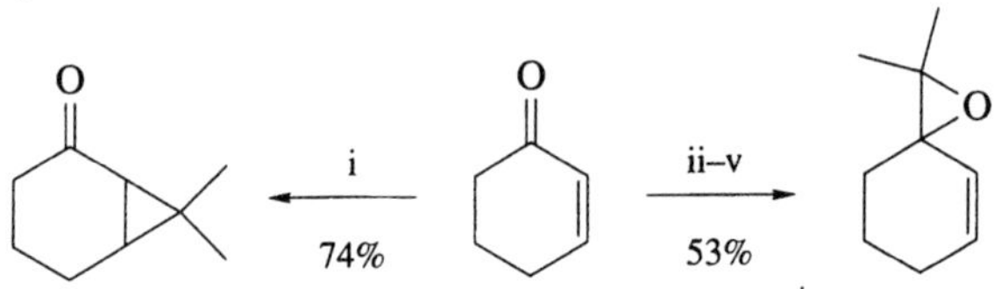

i, $Me_2C{=}SPh_2$, $LiBF_4$, DME; ii, $Me_2C(SeMe)Li$, THF; iii, H_3O^+; iv, MeI; v, Bu^tOK, DMSO

Scheme 200

In the rearrangement reaction, the multistep sequence offers the definite advantage of avoiding the polyhomologation of cyclic ketones, often encountered with diazoalkanes[277] due to the presence of the reagent with both the starting and rearranged ketone.

The stereoisomeric mixture of β-hydroxyalkyl selenides resulting from the reaction of the α-selenoalkyllithium and the carbonyl compound has been often cleanly and easily separated into its constituents by liquid chromatography on silica gel (Schemes 124, 133, 134, and 170–172).[200,206,222,226,229,258,259] This has, therefore, allowed the synthesis of each of the two stereoisomers of various di- and tri-substituted alkenes (Schemes 124, 170 and 171; Scheme 172, a) and epoxides (Scheme 124; Scheme 172, b),[206] which are otherwise obtained as intractable mixtures of stereoisomers through the conventional phosphorus or sulfur ylide methods. Last but not least, 2-lithio-2-methylselenopropane can be used as the precursor of various compounds bearing *gem* dimethyl substituted carbons, such as squalene,[222] oxidosqualene,[206,222,229] lanosterol[222] and cholesterol.[222] Use of commercially available perdeuterated or $^{14}C_1$ or $^{14}C_1$—$^{14}C_2$ acetone allows the straightforward synthesis of the corresponding labelled compounds (Schemes 131 and 201), which otherwise would have required a more lengthy synthetic route.

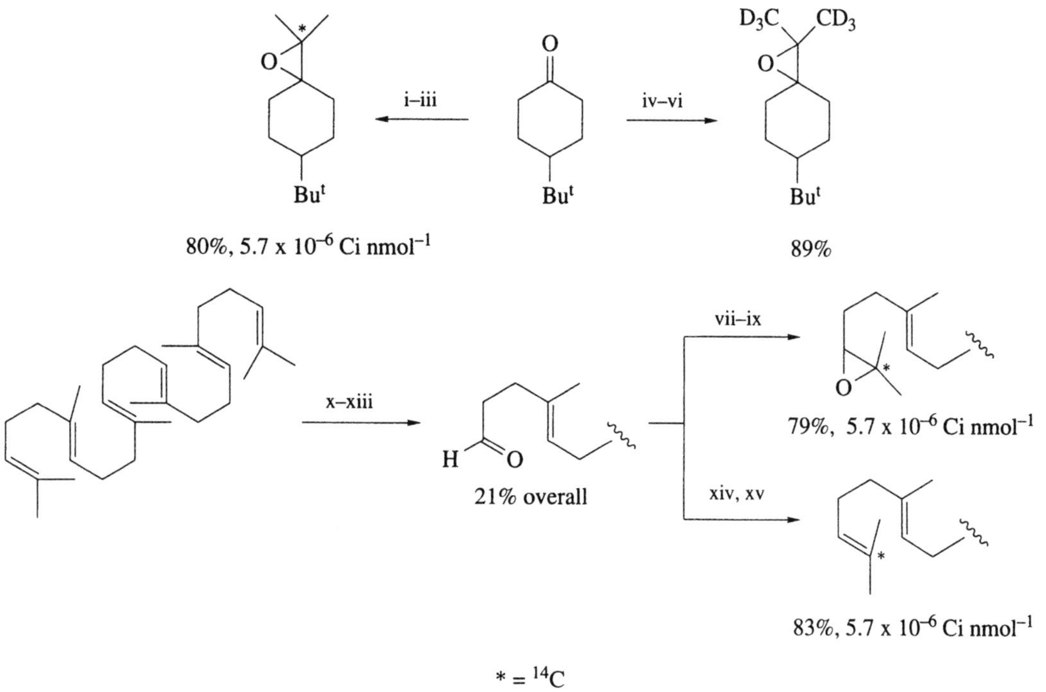

80%, 5.7 x 10^{-6} Ci nmol^{-1}

89%

21% overall

79%, 5.7 x 10^{-6} Ci nmol^{-1}

83%, 5.7 x 10^{-6} Ci nmol^{-1}

* = ^{14}C

i, $MeSe^{14}CMe_2Li$; ii, MeI, 20 °C; iii, 10% aq. KOH, ether; iv, $MeSeC(CD_3)_2Li$, ether; v, MeI, 20 °C; vi, 10% aq. KOH, ether; vii, $MeSe^{14}CMe_2Li$; viii, $MeSO_3F$, ether; ix, 10% aq. KOH, ether; x, NBS (1.1 equiv.), DME, H_2O; xi, 10% aq. KOH; xii, $HClO_4$; xiii, $NaIO_4$; xiv, $MeSe^{14}CMe_2Li$; xv, PI_3, Et_3N, CH_2Cl_2

Scheme 201 [222]

2.6.5 MISCELLANEOUS REACTIONS INVOLVING FUNCTIONALIZED α-SELENOALKYL METALS AND CARBONYL COMPOUNDS

2.6.5.1 α-Selenoalkyl and α-Selenoxyalkyl Metals as Vinyl Anion Equivalents

α-Selenoalkyllithiums and α-selenoxyalkyllithiums have at several occasions played the role of vinyllithium equivalents. Thus, various β-seleno-[235] and β-selenoxy-carbonyl[98] compounds have been transformed to α,β-unsaturated carbonyl compounds by taking advantage of the well-established elimination of the corresponding selenoxide (Scheme 157, c, d and e).

2.6.5.2 1,1-Bis(seleno)alkyl Metals and 1-Silyl-1-selenoalkyl Metals as Acyl Anion Equivalents

2.6.5.2.1 1,1-Bis(seleno)alkyl metals as acyl anion equivalents

1,1-Bis(seleno)alkyl metals are valuable acyl anion equivalents. Thus β-hydroxybis(seleno)alkanes,[281] as well as β-hydroxy-γ,δ-unsaturated bis(seleno)alkanes or γ-oxobis(seleno)alkanes, selectively obtained on reaction of enals, or enones and 1-metallo-1,1-bis(seleno)alkanes in ether or THF–HMPA, immediately react[73,94,233,281] with copper chloride–copper oxide in acetone and lead, respectively, to β-hydroxy-β,γ-unsaturated carbonyl[73,94,233] compounds or γ-diketones (Scheme 139; Scheme 140, a, b and e; Scheme 149, b).[94,233] In some rare cases (see Scheme 140, e), the reaction proceeds[233] differently and instead leads to a vinyl selenide, which can be further transformed to a carbonyl compound when reacted with mercury(II) chloride.

2.6.5.2.2 1,1-Silyl-1-selenoalkyl metals as acyl anion equivalents

α-Hydroxycarbonyl compounds have also been synthesized[171] from β-hydroxy-α-silylalkyl selenides and hydrogen peroxide (Scheme 96, d). The reaction proceeds smoothly under mild conditions and might involve a Pummerer-type rearrangement.

2.6.5.3 Homologation of Carbonyl Compounds from α-Heterosubstituted-α-selenoalkyl Metals

α-Selenoalkyl metals bearing on the carbanionic center a phosphono,[106] phosphonato (Scheme 28),[102] silyl (Scheme 96, e)[171] or seleno moiety (Scheme 98, b) allow the homologation of carbonyl compounds. The three former reagents immediately lead to vinyl selenides on reaction with carbonyl compounds, whereas the latter requires the reductive elimination of the β-hydroxybis(seleno)alkane intermediate to provide the same vinyl selenides (Scheme 98, b).[108] These have been later transformed to carbonyl compounds on reaction with mercury(II) chloride.[11,102,106,108,171]

2.6.5.4 Conclusion

Most of the reactions described in this section have not been carried out on a large array of compounds. It is therefore difficult to compare their scope to the related reactions which use other α-heterosubstituted alkyl metals.[7]

2.6.6 GENERAL CONCLUSION

In conclusion, α-selenoalkyllithiums and α-selenoxyalkyl metals are versatile building blocks, readily available, and particularly prone to produce β-hydroxyalkyl selenides and β-hydroxyalkyl selenoxides which, after proper activation, have allowed the selective synthesis of various selenium-free compounds.

2.6.7 REFERENCES

1. K. C. Nicolaou and N. A. Petasis, in 'Selenium in Natural Product Synthesis', CIS, Philadelphia, 1984.
2. H. Rheinboldt, *Methoden Org. Chem. (Houben–Weyl)*, 1967, **IX**, 917.
3. D. L. Klayman and W. H. H. Günther (eds.), 'Organic Selenium Compounds: Their Chemistry and Biology', Wiley, New York, 1973.
4. D. L. J. Clive, *Tetrahedron*, 1978, **34**, 1049.
5. D. Liotta (ed.), 'Organoselenium Chemistry', Wiley-Interscience, New York, 1987.
6. S. Patai and Z. Rappoport (eds.), 'The Chemistry of Selenium and Tellurium Compounds', Wiley, Chichester, 1987, vol. 2.
7. A. Krief, *Tetrahedron*, 1980, **36**, 2531.
8. H. J. Reich, *Acc. Chem. Res.*, 1979, **12**, 22.
9. H. J. Reich, in 'Oxidation in Organic Chemistry', ed. W. S. Trahanovsky, Academic Press, New York, 1978, part C, p. 1.
10. P. Magnus, in 'Comprehensive Organic Chemistry', ed. D. H. R. Barton and W. D. Ollis, Pergamon Press, Oxford, 1979, vol. 3, p. 491.
11. J. V. Comasseto, *J. Organomet. Chem.*, 1983, **253**, 131.
12. A. Krief, *Top. Curr. Chem.*, 1987, **135**, 1.
13. A. Krief and L. Hevesi, *Janssen Chimica Acta*, 1984, **2**, 3.
14. S. Patai and Z. Rappoport (eds.), in 'The Chemistry of Selenium and Tellurium Compounds', Wiley, Chichester, 1986, vol. 1, p. 675.
15. C. Paulmier, in 'Selenium Reagents and Intermediates in Organic Synthesis', Pergamon Press, Oxford, 1986.
16. A. Krief, in 'The Chemistry of Selenium and Tellurium Compounds', ed. S. Patai and Z. Rappoport, Wiley, Chichester, 1987, vol. 2.
17. H. J. Reich, in 'Organoselenium Chemistry', ed. D. Liotta, Wiley-Interscience, New York, 1987, p. 243.
18. D. Seebach and K.-H. Geiss, in 'New Applications of Organometallic Reagents in Organic Synthesis', ed. D. Seyferth, Elsevier, Amsterdam, 1976.
19. D. J. Peterson, *Organomet. Chem. Rev., Sect. A*, 1972, **7**, 295.
20. T. Kauffmann, *Top. Curr. Chem.*, 1980, **92**, 109.
21. T. Kauffmann, *Angew. Chem., Int. Ed. Engl.*, 1982, **21**, 410.
22. U. Schöllkopf, *Angew. Chem., Int. Ed. Engl.*, 1970, **9**, 763.
23. D. Seebach, *Angew. Chem., Int. Ed. Engl.*, 1969, **8**, 639.
24. O. W. Lever, Jr., *Tetrahedron*, 1976, **32**, 1943.
25. A. W. Johnson, in 'Ylid Chemistry', Academic Press, New York, 1966.
26. P. Beak and D. B. Reitz, *Chem. Rev.*, 1978, **78**, 275.

27. D. Seebach and D. Enders, *Angew. Chem., Int. Ed. Engl.*, 1975, **14**, 15.
28. D. Hoppe, *Angew. Chem., Int. Ed. Engl.*, 1974, **13**, 789.
29. U. Schöllkopf, *Angew. Chem., Int. Ed. Engl.*, 1977, **16**, 339.
30. J. F. Biellmann and J. B. Ducep, *Org. React. (N.Y.)*, 1982, **27**, 1.
31. T. H. Chan, *Acc. Chem. Res.*, 1977, **10**, 442.
32. B. M. Trost and L. S. Melvin, Jr., 'Sulfur Ylides: Emerging Synthetic Intermediates', ed. A. T. Blomquist and H. H. Wasserman, Academic Press, New York, 1975, vol. 31.
33. E. Block, in 'Reactions of Organosulfur Compounds', Academic Press, New York, 1978, vol. 37.
34. D. Liotta, *Acc. Chem. Res.*, 1984, **17**, 28.
35. D. Seebach and N. Peleties, *Angew. Chem., Int. Ed. Engl.*, 1969, **8**, 450.
36. D. Seebach and N. Peleties, *Chem. Ber.*, 1972, **105**, 511.
37. F. J. Webb, *Iowa State Coll. J. Sci.*, 1942, **17**, 152.
38. H. Gilman and F. J. Webb, *J. Am. Chem. Soc.*, 1949, **71**, 4062.
39. H. Gilman and R. L. Bebb, *J. Am. Chem. Soc.*, 1939, **61**, 109.
40. W. Dumont, P. Bayet and A. Krief, *Angew. Chem., Int. Ed. Engl.*, 1974, **13**, 804.
41. M. Sevrin, J.-N. Denis and A. Krief, *Angew. Chem., Int. Ed. Engl.*, 1978, **17**, 526.
42. S. Raucher and G. A. Koolpe, *J. Org. Chem.*, 1978, **43**, 4252.
43. A. Krief and P. Barbeaux, *J. Chem. Soc., Chem. Commun.*, 1987, 1214.
44. M. Clarembeau, Ph.D. Thesis, Facultés Universitaire Notre Dame, Namur, 1988.
45. M. Clarembeau and A. Krief, *Tetrahedron Lett.*, 1984, **25**, 3629.
46. M. Clarembeau and A. Krief, *Tetrahedron Lett.*, 1985, **26**, 1093.
47. A. Anciaux, A. Eman, W. Dumont and A. Krief, *Tetrahedron Lett.*, 1975, 1617.
48. D. Seebach and A. K. Beck, *Angew. Chem., Int. Ed. Engl.*, 1974, **13**, 806.
49. D. Seebach, N. Meyer and A. K. Beck, *Justus Liebigs Ann. Chem.*, 1977, 846.
50. W. Dumont and A. Krief, *Angew. Chem., Int. Ed. Engl.*, 1976, **15**, 161.
51. B. J. Wakefield, 'The Chemistry of Organolithium Compounds', Pergamon Press, Oxford, 1974.
52. D. Seebach and H. Heumann, *Chem. Ber.*, 1974, **107**, 847.
53. M. Clarembeau, A. Cravador, W. Dumont, L. Hevesi, A. Krief, J. Lucchetti and D. Van Ende, *Tetrahedron*, 1985, **41**, 4793.
54. A. Krief, W. Dumont, M. Clarembeau, G. Bernard and E. Badaoui, *Tetrahedron*, 1989, **45**, 2005.
55. D. Van Ende, W. Dumont and A. Krief, *Angew. Chem., Int. Ed. Engl.*, 1975, **14**, 700.
56. H. J. Reich, F. Chow and S. K. Shah, *J. Am. Chem. Soc.*, 1979, **101**, 6638.
57. D. Seebach, J. Gabriel and R. Hassig, *Helv. Chim. Acta*, 1984, **67**, 1083.
58. M. Clarembeau and A. Krief, *Tetrahedron Lett.*, 1986, **27**, 4917.
59. J.-M. Lehn, G. Wipff and J. Demuynck, *Helv. Chim. Acta*, 1977, **60**, 1239.
60. A. Krief, W. Dumont, M. Clarembeau and E. Badaoui, *Tetrahedron*, 1989, **45**, 2023.
61. A. I. Shatenshtein and H. A. Gvozdeva, *Tetrahedron*, 1969, **25**, 2749.
62. G. Pourcelot and C. Georgoulis, *Bull. Soc. Chim. Fr.*, 1964, 866.
63. G. Pourcelot and J.-M. Cense, *Bull. Soc. Chim. Fr.*, 1976, **9–10**, 1578.
64. G. D. Meakins, R. K. Percy, E. E. Richards and R. N. Youg, *J. Chem. Soc.*, 1968, 1106.
65. H. J. Reich, W. W. Willis, Jr. and P. D. Clark, *J. Org. Chem.*, 1981, **46**, 2775.
66. F. G. Bordwell, J. E. Bares, J. E. Bartmess, G. E. Drucker, J. Gerhold, G. J. McCollum, M. Van Der Puy, N. R. Vanier and W. S. Matthews, *J. Org. Chem.*, 1977, **42**, 326.
67. W. E. Doering and A. K. Hoffmann, *J. Am. Chem. Soc.*, 1955, **77**, 521.
68. A. I. Shatenshtein, N. N. Magdesiava, Y. I. Ranneva, I. O. Shapiro and A. I. Serebryanskaya, *Teor. Eksp. Khim.*, 1967, **3**, 343.
69. H. J. Reich and W. W. Willis, Jr., *J. Org. Chem.*, 1980, **45**, 5227.
70. H. J. Reich and S. K. Shah, *J. Org. Chem.*, 1977, **42**, 1773.
71. D. Van Ende, A. Cravador and A. Krief, *J. Organomet. Chem.*, 1979, **177**, 1.
72. M. Clarembeau and A. Krief, *Tetrahedron Lett.*, 1986, **27**, 1723.
73. S. Raucher and G. A. Koolpe, *J. Org. Chem.*, 1978, **43**, 3794.
74. J. Lucchetti and A. Krief, *Tetrahedron Lett.*, 1978, 2693.
75. K. B. Sharpless, R. F. Lauer and A. Y. Teranishi, *J. Am. Chem. Soc.*, 1973, **95**, 6137.
76. F. S. Guziec, Jr., in 'Organoselenium Chemistry', ed. D. Liotta, Wiley-Interscience, New York, 1987, p. 277.
77. S. Halazy, Ph.D. Thesis, Facultés Universitaire Notre Dame, Namur, 1982.
78. R. H. Mitchell, *J. Chem. Soc., Chem. Commun.*, 1974, 990.
79. R. H. Mitchell, *Can. J. Chem.*, 1980, **58**, 1398.
80. H. J. Reich, *J. Org. Chem.*, 1975, **40**, 2570.
81. Y. Yamamoto, Y. Saito and K. Maruyama, *Tetrahedron Lett.*, 1982, **23**, 4597.
82. Y. Yamamoto, H. Yatagai, Y. Saito and K. Maruyama, *J. Org. Chem*, 1984, **49**, 1096.
83. R. N. Binns, R. K. Haynes, T. L. Houston and W. R. Jackson, *Tetrahedron Lett.*, 1980, **21**, 573.
84. H. J. Reich, M. C. Schroeder, and I. L. Reich, *Isr. J. Chem.*, 1984, **24**, 157.
85. H. J. Reich, M. C. Clark and W. W. Willis, Jr., *J. Org. Chem.*, 1982, **47**, 1618.
86. H. Wetter, *Helv. Chim. Acta*, 1978, **61**, 3072.
87. L. Hevesi, K. M. Nsunda and M. Renard, *Bull. Soc. Chim. Belg.*, 1985, **94**, 1039.
88. R. G. Shea, J. N. Fitzner, J. E. Fankhauser, A. Spaltenstein, P. A. Carpino, R. M. Peevey, D. V. Pratt, B. J. Tenge and P. B. Hopkins, *J. Org. Chem.*, 1986, **51**, 5243.
89. A. Spaltenstein, P. A. Carpino, F. Miyake and P. B. Hopkins, *J. Org. Chem.*, 1987, **52**, 3759.
90. M. R. Binns and R. K. Haynes, *J. Org. Chem.*, 1981, **46**, 3790.
91. T. Kauffmann and K. R. Gaydoul, *Tetrahedron Lett.*, 1985, **26**, 4071.
92. H. J. Reich and S. K. Shah, *J. Am. Chem. Soc.*, 1977, **99**, 263.
93. H. J. Reich, S. K. Shah, P. M. Gold and R. E. Olson, *J. Am. Chem. Soc.*, 1981, **103**, 3112.
94. P. Bouhy, Memoire de Licence, Facultés Universitaire Notre Dame, Namur, 1980.

95. B. T. Gröbel and D. Seebach, *Chem. Ber.*, 1977, **110**, 852.
96. H. J. Reich and M. L. Cohen, *J. Am. Chem. Soc.*, 1979, **101**, 1307.
97. H. J. Reich, S. K. Shah and F. Chow, *J. Am. Chem. Soc.*, 1979, **101**, 6648.
98. H. J. Reich and S. K. Shah, *J. Am. Chem. Soc.*, 1975, **97**, 3250.
99. K. Sachdev and H. S. Sachdev, *Tetrahedron Lett.*, 1976, 4223; 1977, 814.
100. D. Van Ende, W. Dumont and A. Krief, *J. Organomet. Chem.*, 1978, **149**, C10.
101. C. A. Brandt, J. V. Comasseto, W. Nakamura and N. Petragnani, *J. Chem. Res. (S)*, 1983, 156.
102. J. V. Comasseto and N. Petragnani, *J. Organomet. Chem.*, 1978, **152**, 295.
103. H. J. Reich and W. W. Willis, Jr., *J. Am. Chem. Soc.*, 1980, **102**, 5967.
104. S. Halazy, J. Lucchetti and A. Krief, *Tetrahedron Lett.*, 1978, 3971.
105. U. Schöllkopf and H. Küppers, *Tetrahedron Lett.*, 1963, 105.
106. N. Petragnani, R. Rodrigues and J. V. Comasseto, *J. Organomet. Chem.*, 1976, **114**, 281.
107. N. Petragnani, J. V. Comasseto, R. Rodrigues and T. J. Brocksom, *J. Organomet. Chem.*, 1977, **124**, 1.
108. J.-N. Denis, S. Desauvage, L. Hevesi and A. Krief, *Tetrahedron Lett.*, 1981, **22**, 4009.
109. T. Takahashi, H. Nagashima and J. Tsuji, *Tetrahedron Lett.*, 1978, 799.
110. P. A. Grieco, M. Nishizawa, T. Oguri, S. D. Burke and N. Marinovic, *J. Am. Chem. Soc.*, 1977, **99**, 5773.
111. D. Liotta, C. S. Barnum and M. Saindane, *J. Org. Chem.*, 1981, **46**, 4301.
112. J. Lucchetti and A. Krief, *Tetrahedron Lett.*, 1978, 2697.
113. J. Lucchetti and A. Krief, *J. Chem. Soc., Chem. Commun.*, 1982, 127.
114. P. A. Grieco and M. Miyashita, *J. Org. Chem.*, 1974, **39**, 120.
115. N. Petragnani and H. M. C. Ferraz, *Synthesis*, 1978, **6**, 476.
116. Y. Masuyama, Y. Ueno and M. Okawara, *Chem. Lett.*, 1977, **7**, 835.
117. P. A. Grieco and Y. Yokoyama, *J. Am. Chem. Soc.*, 1977, **99**, 5210.
118. M. W. Rathke and A. Lindert, *Tetrahedron Lett.*, 1971, 3995.
119. M. W. Rathke and A. Lindert, *J. Am. Chem. Soc.*, 1971, **93**, 2318.
120. T. Sakakibara, S. Ikuta and R. Sudoh, *Synthesis*, 1982, 261.
121. K. Hirai, Y. Iwano and K. Fujimoto, *Tetrahedron Lett.*, 1982, **23**, 4021.
122. J. Morel, C. Paulmier, D. Semard and P. Pastour, *C. R. Hebd. Seances Acad. Sci., Ser. C*, 1970, **270**, 825.
123. L. Christiaens, R. Dufour and M. Renson, *Bull. Soc. Chim. Belg.*, 1970, **79**, 143.
124. T. Kauffmann, H. Ahlers, H.-J. Tilhard and A. Woltermann, *Angew. Chem., Int. Ed. Engl.*, 1977, **16**, 710.
125. J.-N. Denis, Ph.D. Thesis, Facultés Universitaire Notre Dame, Namur, 1983.
126. A. Toshimitsu, H. Owada, K. Terao, S. Uemura and M. Okano, *J. Chem. Soc., Perkin Trans. 1*, 1985, 373.
127. R. C. Gadwood, *J. Org. Chem.*, 1983, **48**, 2098.
128. R. C. Gadwood, I. M. Mallick and A. J. DeWinter, *J. Org. Chem.*, 1987, **52**, 774.
129. A. Krief, W. Dumont and J.-N. Denis, *J. Chem. Soc., Chem. Commun.*, 1985, 571.
130. A. Krief, W. Dumont and J.-L. Laboureur, *Tetrahedron Lett.*, 1988, **29**, 3265.
131. A. Krief, W. Dumont and A. F. De Mahieu, *Tetrahedron Lett.*, 1988, **29**, 3269.
132. W. Dumont, P. Bayet and A. Krief, *Angew. Chem., Int. Ed. Engl.*, 1974, **13**, 274.
133. S. Halazy and A. Krief, *Tetrahedron Lett.*, 1979, 4233.
134. F. Zutterman and A. Krief, *J. Org. Chem.*, 1983, **48**, 1135.
135. M. Clarembeau, J. L. Bertrand and A. Krief, *Isr. J. Chem.*, 1984, **24**, 125.
136. S. Halazy and A. Krief, *Tetrahedron Lett.*, 1980, **21**, 1997.
137. P. G. Gassman, T. Miura and A. Mossman, *J. Org. Chem.*, 1982, **47**, 954.
138. T. Q. Lê, Memoire de Licence, Facultés Universitaire Notre Dame, Namur, 1988.
139. A. Krief, W. Dumont, J.-N. Denis, G. Evrard and B. Norberg, *J. Chem. Soc., Chem. Commun.*, 1985, 569.
140. K. B. Sharpless, M. W. Young and R. F. Lauer, *Tetrahedron Lett.*, 1973, 1979.
141. S. Halazy and A. Krief, *Tetrahedron Lett.*, 1981, **22**, 2135.
142. W. W. Lotz and J. Gosselck, *Tetrahedron*, 1973, **29**, 917.
143. A. Cravador, A. Krief and L. Hevesi, *J. Chem. Soc., Chem. Commun.*, 1980, 451.
144. A. Cravador and A. Krief, *C. R. Hebd. Seances Acad. Sci., Ser. C*, 1979, **289**, 267.
145. W. Dumont and A. Krief, *Angew. Chem., Int. Ed. Engl.*, 1977, **16**, 540.
146. D. L. J. Clive and S. M. Menchen, *J. Chem. Soc., Chem. Commun.*, 1978, 356.
147. D. L. J. Clive and S. M. Menchen, *J. Org. Chem.*, 1979, **44**, 4279.
148. M. J. Calverley, *Tetrahedron Lett.*, 1987, **28**, 1337.
149. D. L. J. Clive and S. M. Menchen, *J. Org. Chem.*, 1979, **44**, 1883.
150. N. Petragnani and G. Schill, *Chem. Ber.*, 1970, **103**, 2271.
151. R. Pellicciari, M. Curini, P. Ceccherelli and R. Fringuelli, *J. Chem. Soc., Chem. Commun.*, 1979, 440.
152. P. J. Giddings, D. I. John and E. J. Thomas, *Tetrahedron Lett.*, 1980, **21**, 399.
153. K. Fujimoto, Y. Iwano and K. Hirai, *Bull. Chem. Soc. Jpn.*, 1986, **59**, 1887.
154. A. Anciaux, A. Eman, W. Dumont, D. Van Ende and A. Krief, *Tetrahedron Lett.*, 1975, 1613.
155. P. A. Zoretic and P. Soja, *J. Org. Chem.*, 1976, **41**, 3587.
156. C. Paulmier and P. Lerouge, *Tetrahedron Lett.*, 1982, **23**, 1557.
157. J. Lucchetti, Ph.D. Thesis, Facultés Universitaire Notre Dame, Namur, 1983.
158. I. Kuwajima, S. Hoshino, T. Tanaka and M. Shimizu, *Tetrahedron Lett.*, 1980, **21**, 3209.
159. S. Halazy and A. Krief, *J. Chem. Soc., Chem. Commun.*, 1979, 1136.
160. S. Halazy, F. Zutterman and A. Krief, *Tetrahedron Lett.*, 1982, **23**, 4385.
161. S. Halazy and A. Krief, *Tetrahedron Lett.*, 1981, **22**, 1829.
162. S. Halazy and A. Krief, *Tetrahedron Lett.*, 1981, **22**, 1833.
163. S. Halazy and A. Krief, *J. Chem. Soc., Chem. Commun.*, 1982, 1200.
164. M. Clarembeau and A. Krief, *Tetrahedron Lett.*, 1986, **27**, 1719.
165. H. M. J. Gillissen, P. Schipper, P. J. J. M. Van Ool and H. M. Buck, *J. Org. Chem.*, 1980, **45**, 319.
166. H. M. J. Gillissen, P. Schipper and H. M. Buck, *Recl. Trav. Chim. Pays-Bas*, 1980, **99**, 346.

167. (a) A. Krief, G. Evrard, E. Badaoui, V. De Beys and R. Dieden, *Tetrahedron Lett.* 1989, **30**, 5635; (b) H. J. Reich and M. D. Bowe, *J. Am. Chem. Soc.* 1990, **112**, 8994; (c) A. Krief, E. Badaoui, W. Dumont, L. Hevesi, B. Herman and R. Dieden, *Tetrahedron Lett.* 1991, **32**, 3231; (d) W. R. Hoffmann and M. Bewersdorf, *Tetrahedron Lett.* 1990, **31**, 67; (e) R. W. Hoffmann, M. Julius and K. Oltmann, *Tetrahedron Lett.* 1990, **31**, 7419.
168. M. Hobe, unpublished results from our laboratory, 1989.
169. A. Krief, D. Surleraux, W. Dumont, P. Pasau, P. Lecomte and P. Barbeaux, in 'Proceedings of the NATO Conference on ''Strain and its Implications in Organic Chemistry', August 1988', ed. A. de Meijere and S. Blechert, Klumer, Dordrecht, 1989, 333; A. Krief and P. Barbeaux, *J. Chem. Soc., Chem. Commun.*, 1987, 1214; A. Krief and P. Barbeaux, *Synlett*, 1991, 511; A. Krief and P. Barbeaux, *Tetrahedron Lett.*, 1991, **32**, 417.
170. P. Barbeaux, unpublished results from our laboratory, 1989; A. Krief, P. Barbeaux and E. Guittet, *Synlett*, 1991, 509.
171. W. Dumont, D. Van Ende and A. Krief, *Tetrahedron Lett.*, 1979, 485.
172. P. J. Stang, K. A. Roberts and L. E. Lynch, *J. Org. Chem.*, 1984, **49**, 1653.
173. B. T. Grübel and D. Seebach, *Chem. Ber.*, 1977, **110**, 867.
174. J.-N. Denis and A. Krief, *Tetrahedron Lett.*, 1982, **23**, 3407.
175. K. M. Nsunda and L. Hevesi, *J. Chem. Soc., Chem. Commun.*, 1985, 1000.
176. J.-N. Denis and A. Krief, *Tetrahedron Lett.*, 1982, **23**, 3411.
177. W. Dumont, M. Sevrin and A. Krief, *Angew. Chem., Int. Ed. Engl.*, 1977, **16**, 541.
178. M. Sevrin, W. Dumont and A. Krief, *Tetrahedron Lett.*, 1977, 3835.
179. S. Raucher, *J. Org. Chem.*, 1977, **42**, 2950.
180. M. Sevrin, Ph.D. Thesis, Facultés Universitaire Notre Dame, Namur, 1980.
181. T. Kauffmann, H. Ahlers, A. Hamsen, H. Schulz, H.-J. Tilhard and A. Vahrenhorst, *Angew. Chem., Int. Ed. Engl.*, 1977, **16**, 119.
182. T. Kauffmann, H. Ahlers, K.-J. Echsler, H. Schulz and H.-J. Tilhard, *Chem. Ber.*, 1985, **118**, 4496.
183. G. Zima, C. S. Barnum and D. Liotta, *J. Org. Chem.*, 1980, **45**, 2736.
184. M. Shimizu and I. Kuwajima, *J. Org. Chem.*, 1980, **45**, 2921.
185. M. Shimizu and I. Kuwajima, *J. Org. Chem.*, 1980, **45**, 4063.
186. R. Ando, T. Sugawara and I. Kuwajima, *J. Chem. Soc., Chem. Commun.*, 1983, 1514.
187. M. Tiecco, D. Chianelli, M. Tingoli, L. Testaferri and D. Bartoli, *Tetrahedron*, 1986, **42**, 4897.
188. D. Labar and A. Krief, *J. Chem. Soc., Chem. Commun.*, 1982, 564.
189. D. Labar, Ph.D. Thesis, Facultés Universitaire Notre Dame, Namur, 1985.
190. L. Fitjer, H.-J. Scheuermann and D. Wehle, *Tetrahedron Lett.*, 1984, **25**, 2329.
191. L. Fitjer, D. Wehle and H.-J. Scheuermann, *Chem. Ber.*, 1986, **119**, 1162.
192. H. J. Reich and F. Chow, *J. Chem. Soc., Chem. Commun.*, 1975, 790.
193. J. Remion, W. Dumont and A. Krief, *Tetrahedron Lett.*, 1976, 1385.
194. J. Remion and A. Krief, *Tetrahedron Lett.*, 1976, 3743.
195. D. Labar, W. Dumont, L. Hevesi and A. Krief, *Tetrahedron Lett.*, 1978, 1145.
196. W. Dumont and A. Krief, *Angew. Chem., Int. Ed. Engl.*, 1975, **14**, 350.
197. A. Krief and J.-L. Laboureur, *Tetrahedron Lett.*, 1987, **28**, 1545.
198. A. Krief, J.-L. Laboureur and W. Dumont, *Tetrahedron Lett.*, 1987, **28**, 1549.
199. A. Krief, W. Dumont, A. Cravador, J.-N. Denis, S. Halazy, L. Hevesi, D. Labar, J. Lucchetti, J. Rémion, M. Sevrin and D. Van Ende, *Bull. Soc. Chim. Fr., Part 2*, 1980, 519.
200. A. Krief, J.-L. Laboureur, G. Evrard, B. Norberg and E. Guittet, *Tetrahedron Lett.*, 1989, **30**, 575.
201. S. Uemura, K. Ohe and N. Sugita, *J. Chem. Soc., Chem. Commun.*, 1988, 111.
202. L. A. Paquette, T.-H. Yan and G. J. Wells, *J. Org. Chem.*, 1984, **49**, 3610.
203. L. Set, D. R. Cheshire and D. L. J. Clive, *J. Chem. Soc., Chem. Commun.*, 1985, 1205.
204. D. Labar, J.-L. Laboureur and A. Krief, *Tetrahedron Lett.*, 1982, **23**, 983.
205. S. Halazy, W. Dumont and A. Krief, *Tetrahedron Lett.*, 1981, **22**, 4737.
206. (a) A. Krief, W. Dumont, D. Van Ende, S. Halazy, D. Labar, J.-L. Laboureur and T. Q. Lê, *Heterocycles*, 1989, **28**, 1203; (b) A. Krief, J.-L. Laboureur, W. Dumont and D. Labar, *Bull Soc. Chim. Fr.*, 1990, **127**, 681.
207. A. I. Meyers, T. Sohda and M. F. Loewe, *J. Org. Chem.*, 1986, **51**, 3108.
208. F. Zutterman and A. Krief, Facultés Universitaire Notre Dame, unpublished results, 1980.
209. J.-N. Denis, W. Dumont and A. Krief, *Tetrahedron Lett.*, 1979, 4111, 4336.
210. D. Labar, A. Krief, B. Norberg, G. Evrard and F. Durant, *Bull. Soc. Chim. Belg.*, 1985, **94**, 1083.
211. D. J. Cram and F. Ahmed Abd Elhafez, *J. Am. Chem. Soc.*, 1952, **74**, 5828.
212. T. J. Leitereg and D. J. Cram, *J. Am. Chem. Soc.*, 1968, **90**, 4019.
213. M. Cherest, H. Felkin and N. Prudent, *Tetrahedron Lett.*, 1968, 2199.
214. M. Cherest and H. Felkin, *Tetrahedron Lett.*, 1968, 2205.
215. J. R. Schauder and A. Krief, *Tetrahedron Lett.*, 1982, **23**, 4389.
216. J. R. Schauder, Ph.D. Thesis, Facultés Universitaire Notre Dame, Namur, 1983.
217. Y. Yamamoto, Y. Saito and K. Maruyama, *J. Org. Chem.*, 1983, **48**, 5408.
218. H. O. House, L. E. Huber and M. J. Umen, *J. Am. Chem. Soc.*, 1972, **94**, 8471.
219. M. Cossentini, B. Deschamps, N. Trong Anh and J. Seyden-Penne, *Tetrahedron*, 1977, **33**, 409.
220. B. Deschamps and J. Seyden-Penne, *Tetrahedron*, 1977, **33**, 413.
221. D. Van Ende and A. Krief, *Tetrahedron Lett.*, 1976, 457.
222. A. Krief, W. Dumont, D. Van Ende, D. Labar, J. R. Schauder, J.-L. Laboureur and G. Chaboteaux, in 'Proceedings of the Third International Conference on Chemistry and Biotechnology, Sofia', Belgrade, 1985.
223. J.-L. Laboureur, unpublished results from our laboratory, 1989.
224. J.-N. Denis and A. Krief, *J. Chem. Soc., Chem. Commun.*, 1983, 229.
225. A. Krief and J.-L. Laboureur, *J. Chem. Soc., Chem. Commun.*, 1986, 702.
226. L. A. Paquette, J. R. Peterson and R. J. Ross, *J. Org. Chem.*, 1985, **50**, 5200.

227. C. Schmit, S. Sahraoui-Taleb, E. Differding, C. G. Dehasse-De Lombaert and L. Ghosez, *Tetrahedron Lett.*, 1984, **25**, 5043.
228. J. Lucchetti, J. Remion and A. Krief, *C. R. Hebd. Seances Acad. Sci., Ser. C*, 1979, **288**, 553.
229. W. Dumont, unpublished results from our laboratory, 1989.
230. J. Lucchetti and A. Krief, *Tetrahedron Lett.*, 1981, **22**, 1623.
231. J. Lucchetti, W. Dumont and A. Krief, *Tetrahedron Lett.*, 1979, 2695.
232. J. Lucchetti and A. Krief, *J. Organomet. Chem.*, 1980, **194**, C49.
233. J. Lucchetti and A. Krief, *Synth. Commun.*, 1983, **13**, 1153.
234. W. Dumont, J. Lucchetti and A. Krief, *J. Chem. Soc., Chem. Commun.*, 1983, 66.
235. J.-N. Denis, W. Dumont and A. Krief, *Tetrahedron Lett.*, 1976, 453.
236. A. M. Leonard-Coppens and A. Krief, *Tetrahedron Lett.*, 1976, 3227.
237. T. Di Giamberardino, S. Halazy, W. Dumont and A. Krief, *Tetrahedron Lett.*, 1983, **24**, 3413.
238. T. M. Willson, P. J. Kocieński, A. Faller and S. Campbell, *J. Chem. Soc., Chem. Commun.*, 1987, 106.
239. S. Spreutels, Memoire de Licence, Facultés Universitaire Notre Dame, Namur, 1979.
240. J. Lucchetti and A. Krief, *C. R. Hebd. Seances Acad. Sci., Ser. C*, 1979, **288**, 537.
241. M. Sevrin, W. Dumont, L. Hevesi and A. Krief, *Tetrahedron Lett.*, 1976, 2647.
242. D. H. R. Barton, D. J. Lester, W. B. Motherwell and M. T. Barros Papoula, *J. Chem. Soc., Chem. Commun.*, 1980, 246.
243. D. Labar, L. Hevesi, W. Dumont and A. Krief, *Tetrahedron Lett.*, 1978, 1141.
244. J.-L. Laboureur and A. Krief, *Tetrahedron Lett.*, 1984, **25**, 2713.
245. M. Beaujean, unpublished results from our laboratory, 1989.
246. M. Sevrin, D. Van Ende and A. Krief, *Tetrahedron Lett.*, 1976, 2643.
247. D. L. J. Clive, G. J. Chittattu, V. Farina, W. A. Kiel, S. M. Menchen, C. G. Russell, A. Singh, C. K. Wong and N. J. Curtis, *J. Am. Chem. Soc.*, 1980, **102**, 4438.
248. K. Katsuura and V. Snieckus, *Tetrahedron Lett.*, 1985, **26**, 9.
249. R. Baudat and M. Petrzilka, *Helv. Chim. Acta*, 1979, **62**, 1406.
250. G. H. Posner and M. J. Chapdelaine, *Tetrahedron Lett.*, 1977, 3227.
251. J. P. Konopelski, C. Djerassi and J. P. Raynaud, *J. Med. Chem.*, 1980, **23**, 722.
252. J. C. Scaiano, P. Schmid and K. U. Ingold, *J. Organomet. Chem.*, 1976, **121**, C4.
253. E. J. Corey, H. L. Pearce, I. Szekely and M. Ishiguro, *Tetrahedron Lett.*, 1978, 1023, 1524.
254. D. L. J. Clive, G. J. Chittattu and C. K. Wong, *J. Chem. Soc., Chem. Commun.*, 1978, 41.
255. A. L. J. Beckwith and P. E. Pigou, *Aust. J. Chem.*, 1986, **39**, 77.
256. D. L. J. Clive and C. G. Russell, *J. Chem. Soc., Chem. Commun.*, 1981, 434.
257. D. L. J. Clive, C. G. Russell and S. C. Suri, *J. Org. Chem.*, 1982, **47**, 1632.
258. G. D. Crouse and L. A. Paquette, *J. Org. Chem.*, 1981, **46**, 4272.
259. P. Pasau, unpublished results from our laboratory, 1989.
260. D. L. J. Clive and V. N. Kale, *J. Org. Chem.*, 1981, **46**, 231.
261. K. B. Sharpless and R. F. Lauer, *J. Am. Chem. Soc.*, 1973, **95**, 2697.
262. A. C. Cope, E. Ciganek and J. Lazar, *J. Am. Chem. Soc.*, 1962, **84**, 2591.
263. J.-N. Denis, J. J. Vicens and A. Krief, *Tetrahedron Lett.*, 1979, 2697.
264. K. B. Sharpless, K. M. Gordon, R. F. Lauer, S. P. Singer and M. W. Young, *Chem. Scr.*, 1975, **8A**, 9.
265. D. Labar, A. Krief and L. Hevesi, *Tetrahedron Lett.*, 1978, 3967.
266. J.-L. Laboureur, W. Dumont and A. Krief, *Tetrahedron Lett.*, 1984, **25**, 4569.
267. K. Nishiyama, T. Kitajima, A. Yamamoto and K. Itoh, *J. Chem. Soc., Chem. Commun.*, 1982, 1232.
268. B. M. Trost and P. H. Scudder, *J. Am. Chem. Soc.*, 1977, **99**, 7601; 1978, **100**, 1327.
269. A. Maercker, *Org. React. (N.Y.)*, 1965, **14**, 270.
270. W. S. Wadsworth, Jr., *Org. React. (N.Y.)*, 1977, **25**, 73.
271. H. Pommer, *Angew. Chem., Int. Ed. Engl.*, 1977, **16**, 423.
272. H. J. Bestmann, *Angew. Chem., Int. Ed. Engl.*, 1965, **4**, 583.
273. H. J. Bestmann, *Angew. Chem., Int. Ed. Engl.*, 1965, **4**, 645.
274. H. J. Bestmann, *Angew. Chem., Int. Ed. Engl.*, 1965, **4**, 830.
275. U. Schollkopf, *Angew. Chem.*, 1959, **71**, 260.
276. E. W. Colvin, *Chem. Soc. Rev.*, 1978, **7**, 15.
277. C. D. Gutsche, *Org. React. (N.Y.)*, 1954, **8**, 364.
278. C. R. Johnson, C. W. Schroek and J. R. Shanklin, *J. Am. Chem. Soc.*, 1973, **95**, 7424.
279. E. J. Corey and M. Chaykovsky, *J. Am. Chem. Soc.*, 1965, **87**, 1353.
280. E. J. Corey and M. Jautelat, *J. Am. Chem. Soc.*, 1967, **89**, 3912.
281. A. Burton, L. Hevesi, W. Dumont, A. Cravador and A. Krief, *Synthesis*, 1979, 877.

3.1
Alkene Synthesis

SARAH E. KELLY
Pfizer Central Research, Groton, CT, USA

3.1.1 INTRODUCTION

In this chapter, the various methods of converting a carbonyl derivative to an alkene are discussed. Particular focus has been given to recently developed methods and their use in the context of natural product synthesis. Formally this volume in the series is devoted to nonstabilized carbanion additions, but this mechanistic definition has not been strictly adhered to in this chapter. The methods are categorized according to the type of alkene synthesized, and the stabilizing group attached to the anion or carbene. In addition, all of the methods discussed involve the addition of a carbon unit to the carbonyl, followed by some type of elimination or cleavage to the alkene. To further narrow the scope of this chapter, the added carbon unit and the original carbonyl carbon must form the units of the alkene, as shown in Scheme 1.

Aldehydes, ketones, esters and lactones, and amides and lactams are all featured in these sections, depending on the synthetic method being discussed.

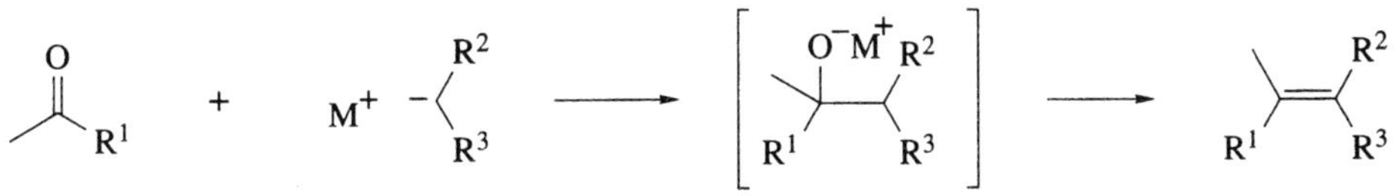

Scheme 1

3.1.2 METHYLENATION REACTIONS: INTRODUCTION

The methylenation of ketones and aldehydes by the Wittig reaction is a well-established and selective methodology. Unlike addition–elimination methods of alkene formation, the Wittig proceeds in a defined sense, producing an alkene at the original site of the carbonyl. The Wittig reaction is not considered here, but is used as the standard by which the methods discussed are measured.[1] The topics covered in the methylenation sections include the Peterson alkenation, the Johnson sulfoximine approach, the Tebbe reaction and the Oshima–Takai titanium–dihalomethane method.

3.1.3 SILICON-STABILIZED METHYLENATION: THE PETERSON ALKENATION

The conversion of a carbonyl to an alkene can be effected by the addition of a α-silyl-substituted alkyl anion to an aldehyde or ketone, followed by either alkoxide attack on the silicon and elimination of silanolate, or acid-catalyzed elimination (Scheme 2).[2]

Scheme 2

In the case of base-induced elimination, the Peterson alkenation relies on the strong bond formed between silicon and oxygen, and the ready propensity for silicon to be attacked by alkoxide, to drive the reaction.[3] In the original study by Peterson, the β-silylcarbinols were prepared by the addition of (trimethylsilyl)methylmagnesium chloride to the carbonyl. The carbinols were subsequently eliminated by treatment with sodium or potassium hydride or with sulfuric acid to form the methylene derivatives in excellent yield. The Peterson reaction has proven to be of general utility in the synthesis of alkenes.[4]

3.1.3.1 Methylenation in Comparison with the Wittig Reaction

The Peterson reagent is more basic and is sterically less hindered than the phosphorus ylide.[4] Therefore, the reagent has the advantage of being more reactive than the Wittig reagent and the side product, hexamethyldisiloxane, is simpler to remove than the various phosphorus by-products of the Wittig reaction. One of the most frequently cited comparisons of the Wittig reaction with the Peterson is the synthesis of β-gorgonene (**2**) by Boeckman and coworkers, where the Wittig failed and the Peterson methodology was successful (equation 1).[5]

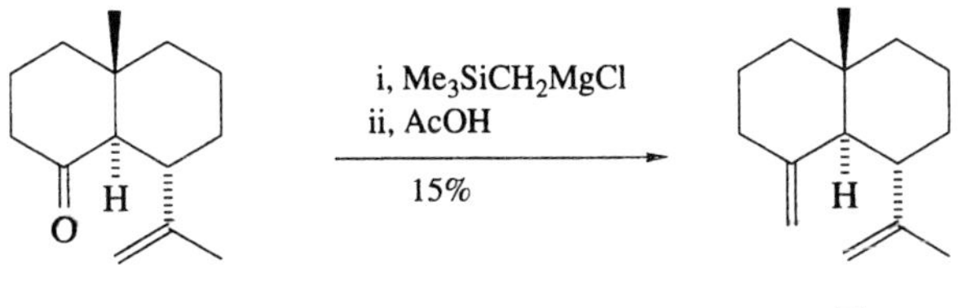

i, Me_3SiCH_2MgCl
ii, AcOH
15% (1)

(1) **(2)**

In an interesting comparison of the propensity for alkoxide attack on silicon, given the option of elimination in the direction of phosphorus or silicon, it is the silicon elimination that prevails, to give the vinylphosphonium salt (Scheme 3).[2] For use with enolizable carbonyls, the basicity of the reagent can be greatly reduced by the use of cerium chloride, as discussed in Section 3.1.3.4.2.

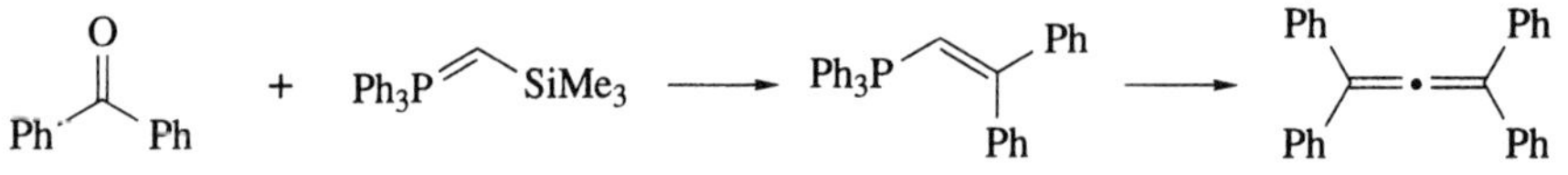

Scheme 3

3.1.3.2 Elimination Conditions

As Peterson outlined in his preliminary communication of the method, either basic (KH, KOBu^t or NaH) or acidic conditions (acetic acid, sulfuric acid or boron trifluoride etherate) may be utilized to effect the elimination of the silylcarbinol.[2] Alternatively the initial adduct may be treated *in situ* with thionyl or acetyl chloride.[6] This procedure may be advantageous in cases where isomerization of the alkene is problematic, and is particularly useful in the synthesis of terminal alkenes. As discussed in Section 3.1.3.4.2, the Johnson group has successfully employed aqueous HF to effect the elimination and this method may also have advantages in situations complicated by base-catalyzed isomerization.[7]

3.1.3.3 Variation of the Metal Anion

The most prevalent applications of the Peterson methodology are with the readily available lithium or magnesium anions, which have similar reactivity with both aldehydes and ketones.[8] Recent examples of the Peterson methylenation in organic synthesis are presented in Table 1.

In addition to the examples in Table 1, the Peterson methylenation has been used in several interesting natural product syntheses, as the examples in equation (2)–(6) indicate. Danishefsky and coworkers used the Peterson reaction in an approach to mitomycins (**4**; equation 2).[16] This application demonstrated the use of unique elimination conditions. The hydroxysilane intermediate was stable to direct Peterson elimination. Therefore, the removal of the silyl protecting group and the elimination of the silyloxy group were carried out with DDQ in quantitative yield.

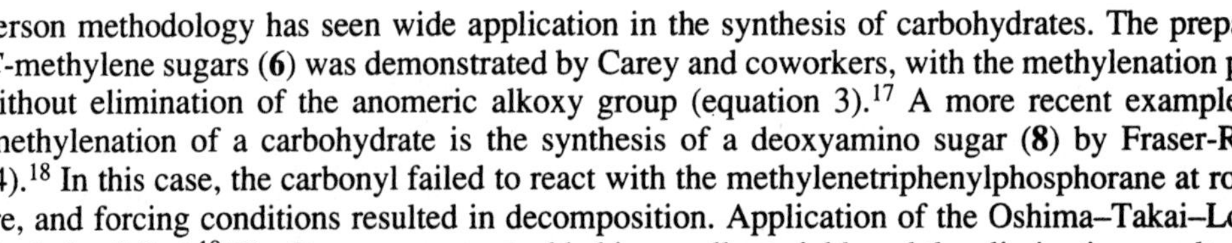

i, $LiCH_2SiMe_3$ high yield; ii, DDQ, THF quantitative (2)

(**3**) R = H, Me (**4**)

The Peterson methodology has seen wide application in the synthesis of carbohydrates. The preparation of 3-*C*-methylene sugars (**6**) was demonstrated by Carey and coworkers, with the methylenation proceeding without elimination of the anomeric alkoxy group (equation 3).[17] A more recent example of Peterson methylenation of a carbohydrate is the synthesis of a deoxyamino sugar (**8**) by Fraser-Reid (equation 4).[18] In this case, the carbonyl failed to react with the methylenetriphenylphosphorane at room temperature, and forcing conditions resulted in decomposition. Application of the Oshima–Takai–Lombardo method also failed.[19] The Peterson reagent added in excellent yield, and the elimination resulted in the formation of the desired methylene (**8a**) as well as the vinylsilane (**8b**).

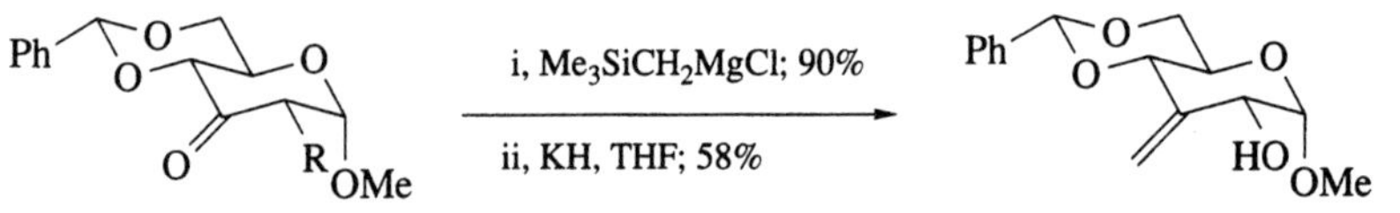

i, Me_3SiCH_2MgCl; 90%; ii, KH, THF; 58% (3)

(**5**) R = OCOPh (**6**)

Further examples of the Peterson methylenation successfully being applied when other techniques failed are shown in equations (5) and (6). In the synthesis of bicyclomycin, a diazabicyclodecane dione, both the Wittig and the Tebbe procedures failed.[20] (Trimethylsilyl)methylmagnesium chloride was added to ketone (**9**) in 52% yield. The usual elimination procedures failed. The trifluoroacetate was formed, and this derivative was treated with Bu^n_4NF to effect the elimination. Following deprotection, the desired methylene derivative (**10**) was isolated in 81% yield (equation 5). In studies directed toward the synthesis of senoxepin (**12**), $CH_2Br_2/Zn/TiCl_4$ methylenation destroyed the oxepin.[21] The Peterson reagent was successfully added, and further elaboration of the ring skeleton was carried out to complete the synthesis (equation 6).

Table 1 Methylenation by the Addition of Lithium or Magnesium Trimethylsilylmethyl Anion to Carbonyls

Substrate	*Conditions*	*Yield (%)*	*Ref.*
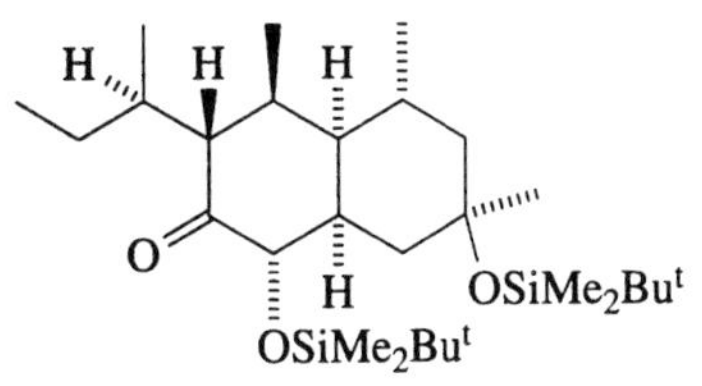 H, H, H, O, H, $OSiMe_2Bu^t$, $OSiMe_2Bu^t$	i, $LiCH_2SiMe_3$; ii, KH, THF; iii, Bu^n_4NF no epimerization	35 (3 steps)	9
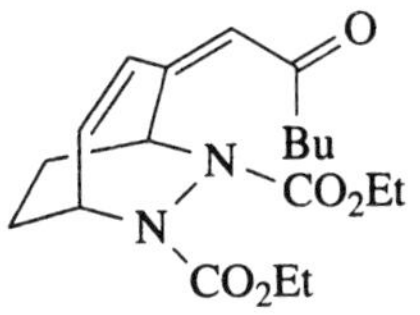 O, N, Bu, CO_2Et, N, CO_2Et	i, $ClMgCH_2SiMe_3$; ii, $SOCl_2$	50–60	10
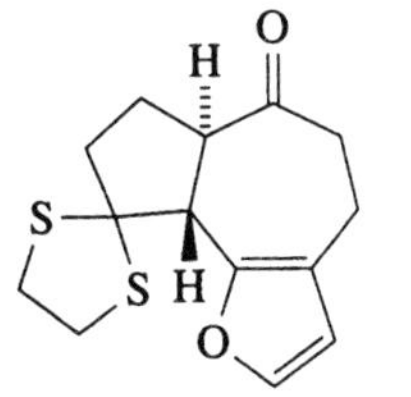 H, O, S, S, H, O	i, $LiCH_2SiMe_3$; ii, NCS, $AgNO_3$	80	11
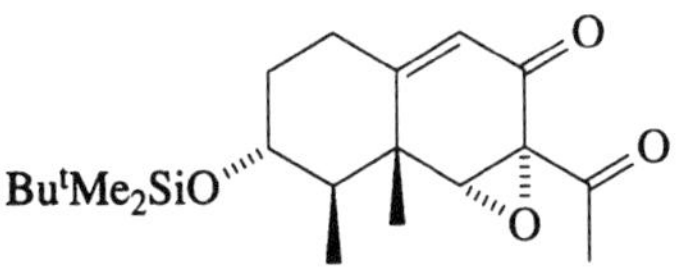 O, Bu^tMe_2SiO, O, O	i, $ClMgCH_2SiMe_3$; ii, H_2SO_4, THF; iii, HF, MeCN	88[a] (3 steps)	12
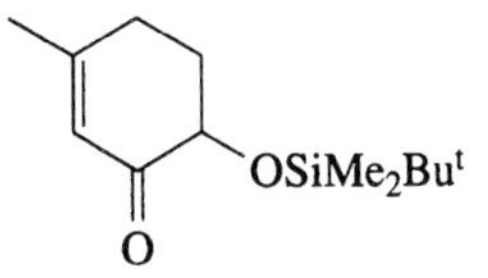 $OSiMe_2Bu^t$, O	i, $ClMgCH_2SiMe_3$; ii, cat. oxalic acid, MeOH base not effective	72	13
 CHO	i, $ClMgCH_2SiMe_3$; ii, MeCOCl, $MeCO_2H$	60	14
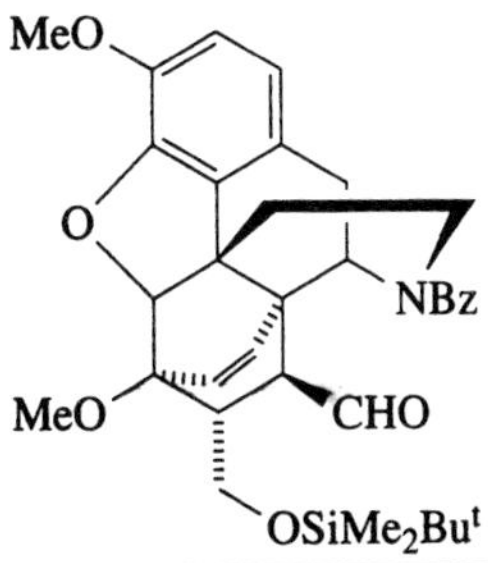 MeO, O, NBz, MeO, CHO, $OSiMe_2Bu^t$	i, $ClMgCH_2SiMe_3$, 84%; ii, KH, THF, 89%	75	15

[a] Yield of product methylenated at side chain carbonyl only.

i, Me_3SiCH_2MgCl; 90%
ii, $SOCl_2$, pyridine; 86%

(4)

(**7**) Cbz = benzyloxycarbonyl

(**8a**) R = H; 75%
(**8b**) R = Me_3Si; 25%

i, Me_3SiCH_2MgCl; 52%
ii, $(CF_3CO)_2O$, then
Bu^n_4NF, KF, MeOH; 81%

(5)

(**9**) (**10**)

i, Me_3SiCH_2MgCl; 48%
ii, MCPBA; 82%
iii, NBS, AIBN, then
NaI, acetone; 22%

(6)

(**11**) (**12**)

3.1.3.4 Chemoselectivity

3.1.3.4.1 Titanium

Recently, the chemoselective addition of the α-silylmethyl anion to aldehydes has been accomplished with titanium (equation 7). In studies by Kauffmann, the Grignard derivative was treated with $TiCl_4$ to prepare the titanium species *in situ*.[22] As indicated in Table 2,[22] this reagent added in good yield to aldehydes to produce the desired methylene compounds but was ineffective for the conversion of ketones to the corresponding methylene compound.[23]

$TiCl_4$, Me_3SiCH_2MgCl
Et_2O

(7)

Table 2 Titanium Peterson Methylenation[22]

Substrate	*Yield (%)*	*Substrate*	*Yield (%)*
Hexanal	60	Benzaldehyde	59
Heptanal	65	3,3-Dimethyl-3-butanone	0
Octanal	59	4-*t*-Butylcyclohexanone	8
Nonanal	61	Acetophenone	3

3.1.3.4.2 Cerium

Johnson has studied the cerium modification of the Peterson methylenation.[24] (Trimethylsilyl)methyllithium was added to $CeCl_3$ and the addition carried out in the presence of TMEDA (equation 8). The cerium lithium reagent proved to be superior to the lithium, magnesium or cerium magnesium mixed species on reaction with base sensitive aldehydes and ketones.[25] The addition is chemoselective for aldehydes and ketones in the presence of amides, esters and halides. The elimination was carried out with HF or KH. Aqueous hydrogen fluoride (with or without pyridine) gave superior yields of the desired products without isomerization. Several examples are listed in Table 3.[24,26]

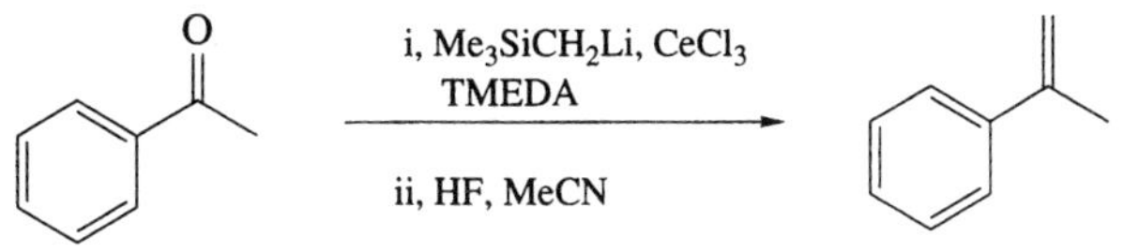

i, Me_3SiCH_2Li, $CeCl_3$ TMEDA; ii, HF, MeCN

(13) → **(14)** (8)

Table 3 Cerium Modification of the Peterson Methylenation[24]

Carbonyl	*Me_3SiCH_2Li*	*$Me_3SiCH_2Li/CeCl_3$*	*HF*
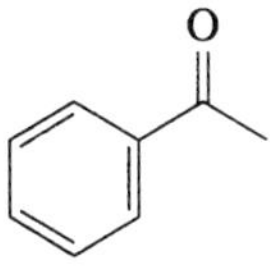 (acetophenone)	78%	93%	84%
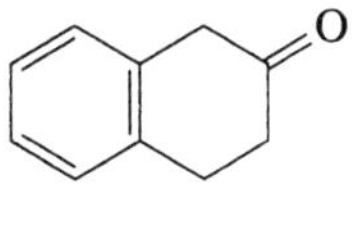 (2-tetralone)	0%	82%	80%
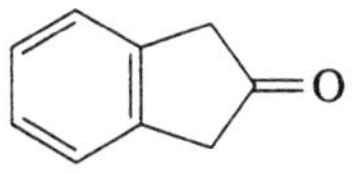 (2-indanone)	6%	83%	87%
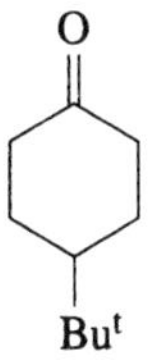 (4-Bu^t-cyclohexanone)	78%	91%	94%
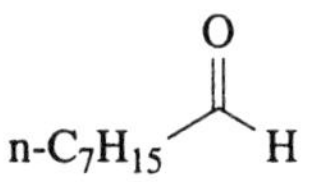 n-C_7H_{15}CHO	56%	86%	93%

In addition, this approach has been utilized in the synthesis of allylsilanes, as shown in Scheme 4. An ester was reacted with an excess of the reagent prepared from $TMSCH_2MgCl$ and $CeCl_3$ to give the bis(silylmethyl)carbinol (**16**).[27] Treatment with silica gel resulted in the formation of the allylsilane (**17**). Examples are listed in Table 4.

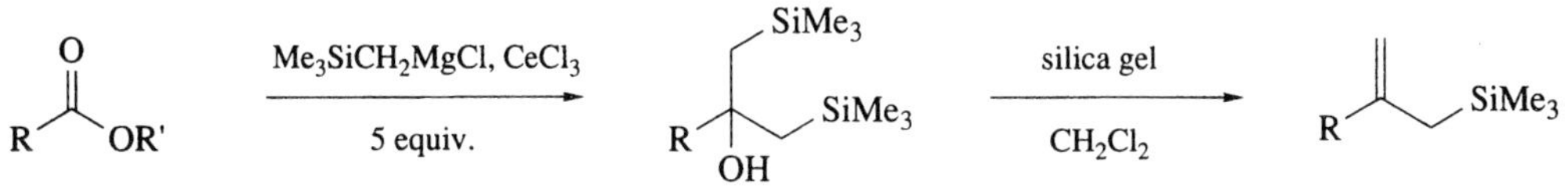

RCO-OR' **(15)** —(Me_3SiCH_2MgCl, $CeCl_3$, 5 equiv.)→ R-C(OH)(CH_2SiMe_3)$_2$ **(16)** —(silica gel, CH_2Cl_2)→ R-C(=CH_2)-CH_2SiMe_3 **(17)**

Scheme 4

As in the work by Johnson, a distinct difference was noted in the magnesium- and lithium-derived cerium species. The lithium/cerium complex has been used by Fuchs and coworkers to synthesize allylsilane derivatives (**20**) from acyl chlorides (**18**; Scheme 5).[28] This work is summarized in Table 5.

Table 4 Reaction with Esters[27]

Ester	*Yield of allylsilane (%)*	*Ester*	*Yield of allylsilane (%)*
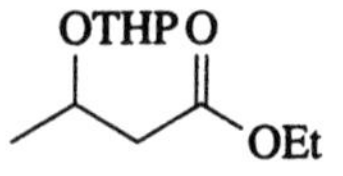 OTHP O, OEt	90	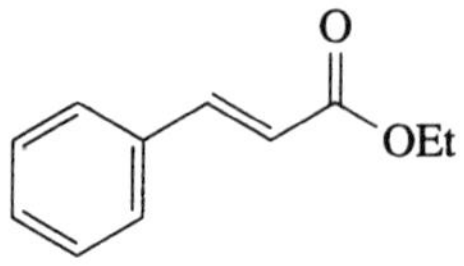 O, OEt	93
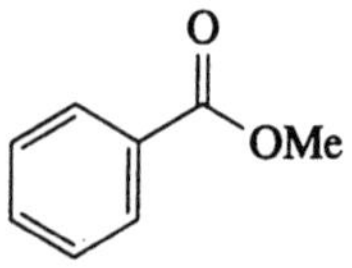 O, OMe	95	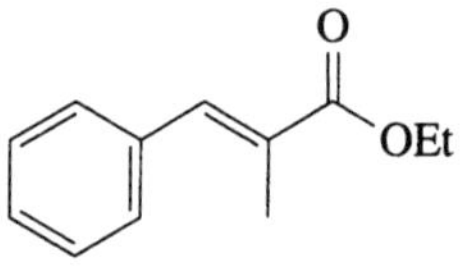 O, OEt	92

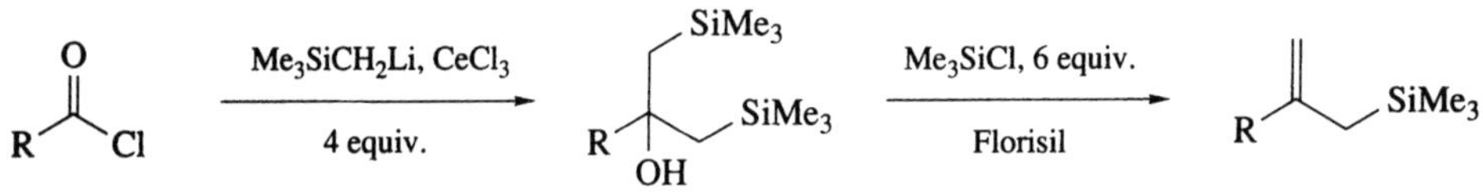

O, R, Cl (18) — Me_3SiCH_2Li, $CeCl_3$, 4 equiv. → $SiMe_3$, R, $SiMe_3$, OH (19) — Me_3SiCl, 6 equiv., Florisil → R, $SiMe_3$ (20)

Scheme 5

Table 5 The Addition of the Cerium Peterson Reagent to Acid Chlorides to Form Allylsilanes[28]

Substrate	Me_3SiCH_2MgCl	Me_3SiCH_2Li	$Me_3SiCH_2Li/CeCl_3$
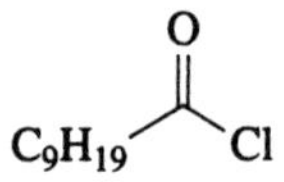 C_9H_{19}, O, Cl	20%	61%	87%
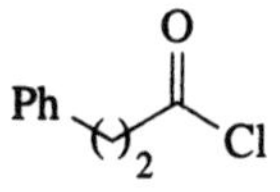 Ph, 2, O, Cl	0%	40%	72%
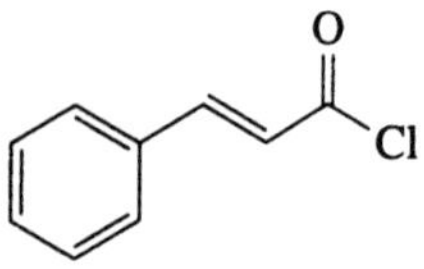 O, Cl	22%	40%	90%
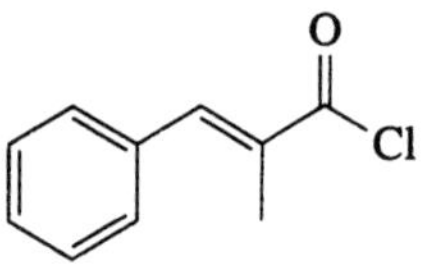 O, Cl	7%	64%	80%

Interestingly, the lithium/cerium reagent was not effective on reaction with esters. Besides the differences in counterion, the Fuchs and Bunnelle groups utilized different methods for forming the cerium complex. Johnson and Bunnelle formed the cerium Peterson reagent by heating $CeCl_3 \cdot 7H_2O$ to 140–150 °C under high vacuum for 2 h.[24,27] The flask is cooled and THF is added. The $CeCl_3$ in THF is stirred at room temperature for 2 h, then cooled to –78 °C, and the lithium or magnesium Peterson reagent is added. After this solution has been stirred for 30 min to 1 h the carbonyl derivative is added. The Fuchs anion is derived from commercially available $CeCl_3$.[28] It is stirred at room temperature in THF for 12–24 h, and then the solution is cooled and the Peterson anion added. It is unknown what effect the different preparations has on the additions to the various carbonyl derivatives. The cerium reagent has proven to be highly advantageous for the synthesis of allylsilanes from esters or acid chlorides that are sterically hindered, without competitive enolization.

3.1.3.5 Substitution on Silicon[29]

(Phenyldimethylsilyl)methylmagnesium chloride has recently been demonstrated to be an effective alternative to the trimethylsilyl derivative.[30] This reagent has been used in the methylenation of pyranosides (**21**). It forms a stable α-hydroxysilane that can be purified by silica gel chromatography. The intermediate may be oxidized to the diol or quantitatively cleaved to the methylene derivative (**22**; equation 9).

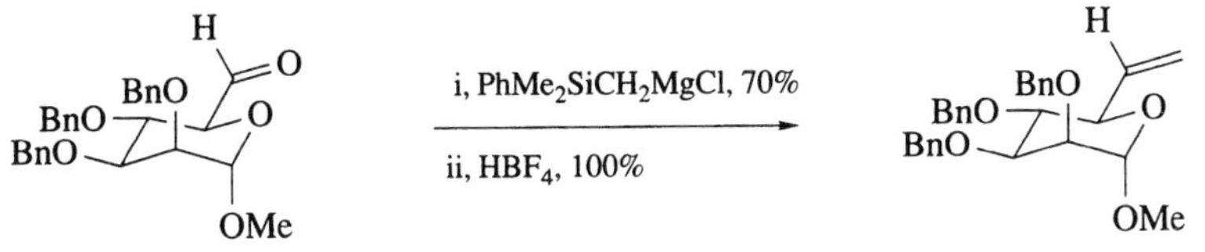

i, $PhMe_2SiCH_2MgCl$, 70%; ii, HBF_4, 100%

(**21**) → (**22**) (9)

3.1.3.6 Other Reactions

α,β-Epoxysilanes can be synthesized directly from an aldehyde or a ketone by the addition of chloromethyl(trimethylsilyl)lithium (Scheme 6).[31] This functional group can be used to form the homologated aldehyde or ketone (**24**), or, as discussed in Section 3.1.10, additions may be carried out on the epoxide, and the β-hydroxysilane eliminated to form the alkene (**25**).

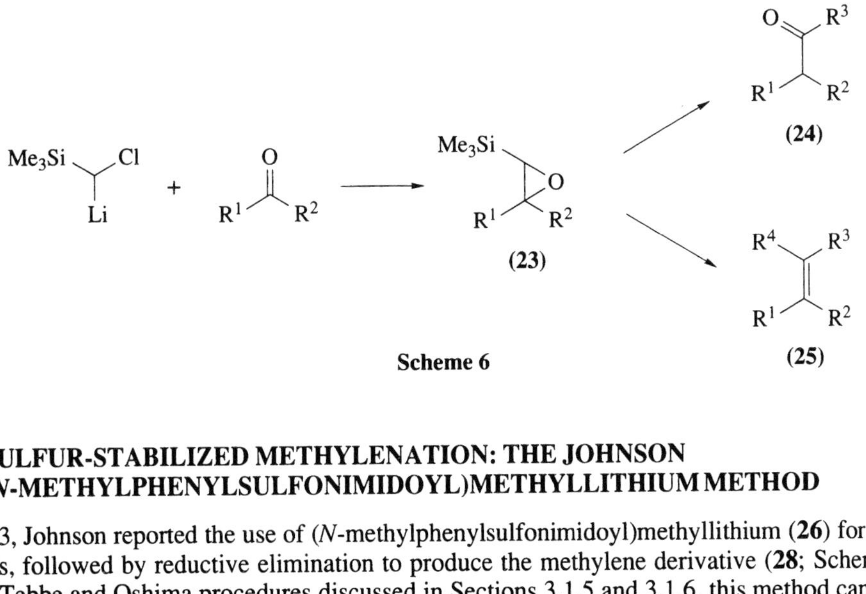

Scheme 6

3.1.4 SULFUR-STABILIZED METHYLENATION: THE JOHNSON (*N*-METHYLPHENYLSULFONIMIDOYL)METHYLLITHIUM METHOD

In 1973, Johnson reported the use of (*N*-methylphenylsulfonimidoyl)methyllithium (**26**) for addition to carbonyls, followed by reductive elimination to produce the methylene derivative (**28**; Scheme 7).[32] As with the Tebbe and Oshima procedures discussed in Sections 3.1.5 and 3.1.6, this method can be applied to enones, ketones and, with comparatively diminished efficiency, aldehydes.[33] The anion appears to be more nucleophillic than methylenetriphenylphosphorane, and there are several examples, detailed below, in which the Wittig reaction failed but the Johnson procedure succeeded. The addition and reductive cleavage can be combined into a single operation without isolation of the β-hydroxysulfoximine.[33]

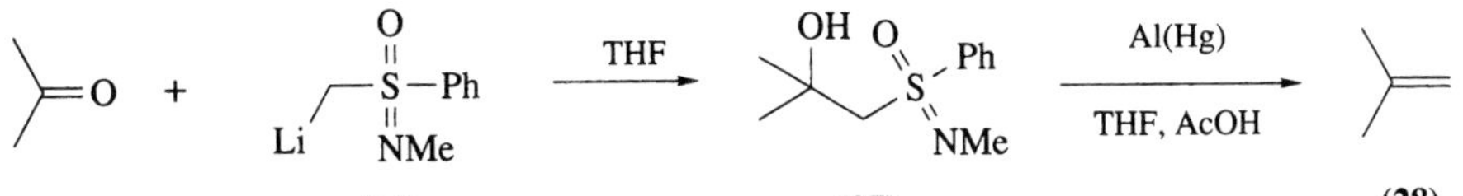

THF; Al(Hg), THF, AcOH

(**26**) (**27**) (**28**)

Scheme 7

3.1.4.1 Reductive Elimination

In addition to aluminum amalgam, several other methods were explored for the elimination of the β-hydroxysulfoximine.[33] Acetylation, followed by attempted elimination under a variety of conditions, was not effective, and therefore reducing agents were explored. Of the various metals investigated, magnesium amalgam, zinc and chromium did not have high enough oxidation potentials to give good ratios of the desired alkene. While sodium, magnesium, aluminum and aluminum amalgam were all effective, only the latter reagent, in the presence of acetic acid, was high yielding and gave acceptable ratios of alkene to alcohol. The proposed mechanism of reductive cleavage entails a two-electron transfer from the aluminum to the sulfoximine (**29**). Collapse of this intermediate results in a carbanion (**31**) and the sulfinamide (**32**). The sulfur derivative may be further reduced to thiophenol and the carbanion can either be protonated to yield the alcohol or eliminated to yield the alkene (**33**; Scheme 8).[33] It was found that the methylenation of aryl aldehydes is complicated by the reaction of the styrene (**34**) with the thiophenol produced in the reduction to give the sulfide (**35**; equation 10). While this side reaction can be abated by the use of less aluminum amalgam, this gave diminished yields, so the best choice would be an alternative method of methylenation.

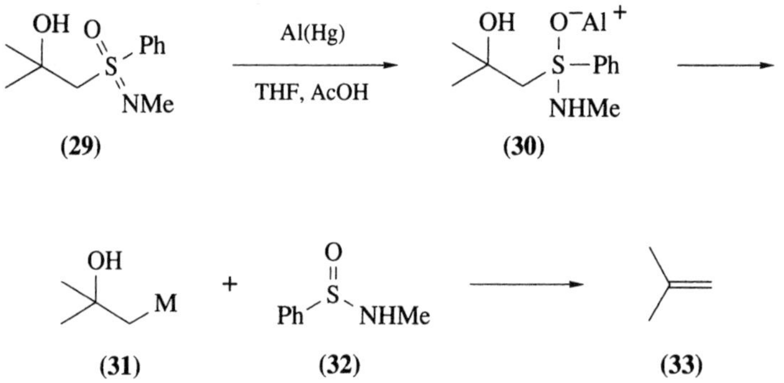

(29) —Al(Hg), THF, AcOH→ (30) → (31) + (32) → (33)

Scheme 8

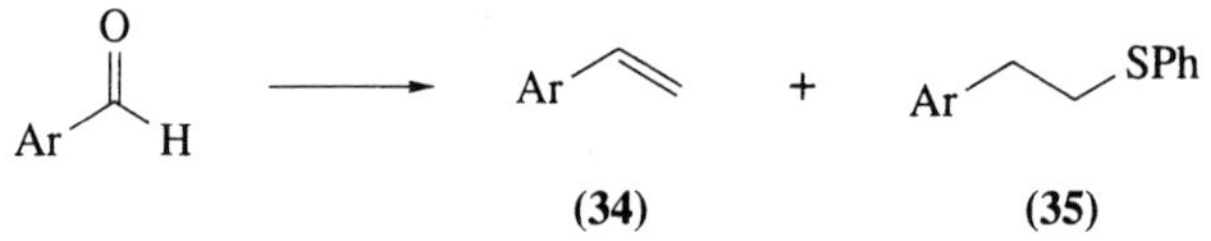

ArCHO → (34) + (35) (10)

3.1.4.2 Examples of the Johnson Methylenation Procedure

Examples of the Johnson methylenation procedure are outlined in Table 6. As the examples indicate, the methodology is compatible with a wide variety of ethers, as well as amides and esters.

3.1.4.3 Comparison with the Wittig Procedure

Of particular note in comparison with the Wittig procedure is Boeckman's total synthesis of gascardic acid (equation 11). It was found that direct Wittig methylenation was low yielding in the reaction with (**36**).[39] It was hypothesized that this was because of interference with the two-carbon side chain. The problem was solved by hydrolyzing the esters, followed by addition and elimination of the sulfoximine and reesterification to obtain the desired adduct (**37**) in 70% overall yield. The Johnson method was also utilized to selectively methylenate the less-hindered ketone (**38**) in the synthesis of (–)-picrotoxinin (equation 12).[40]

3.1.4.4 Resolution

The tremendous advantage of this method of methylene formation over the various other procedures is that it is capable of producing optically pure material through the addition of an enantiomer of (methyl-

Table 6 Reaction of *N*-Methylphenylsulfonimidoylmethyl Anion with Ketones

Entry	*Substrate*	*Reagent*	*Yield of RR'CCH$_2$ (%)*	*Ref.*
1	[structure]	$BrMgCH_2SOPhNMe$	75	34
2	[structure]	$LiCH_2SOPhNMe$	R = COPh 59; R = CH_2Ph 50	35
3	[structure]	$LiCH_2SOPhNMe$	72	36
4	[structure]	$LiCH_2SOPhNMe$	66	37
5	[structure]	$LiCH_2SOPhNMe$	67	38

(36) → (37) i–iii, 70% (11)

i, Na_2CO_3, aq. MeOH; ii, $PhSO(NMe)CH_2Li$, THF, –78 °C, then *in situ* Al(Hg), AcOH; iii, CH_2N_2, Et_2O

phenylsulfonimidoyl)methyllithium to the ketone.[41] As demonstrated in the synthesis of (–)-β-panasinsene (**43**), the diastereomers (**41**) and (**42**) can be separated (frequently by simple column chromatography) and taken on to the enantiomeric methylenes (**43**) and (**44**).[42] Johnson observed excellent facial diastereoselectivity in the addition of the sulfoximine to the racemic ketone (**40**; Scheme 9). In another interesting comparison to the Wittig reaction, the same ketone was reported to be unreactive to the Wittig reagent in DMSO.[42]

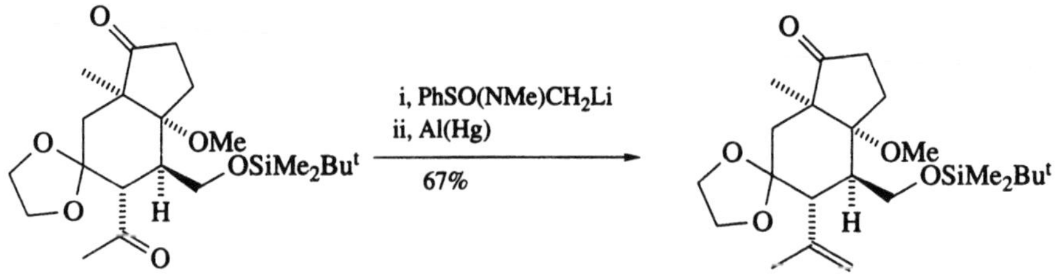

i, PhSO(NMe)CH$_2$Li
ii, Al(Hg)
67%

(38) (39) (12)

i, (S)-PhSO(NMe)CH$_2$Li
ii, diastereomers separated by flash chromatography

(±)-(40) (+)-(41) 42% + (+)-(42) 33%

Al(Hg) Al(Hg)

(43) 96% (44) 92%

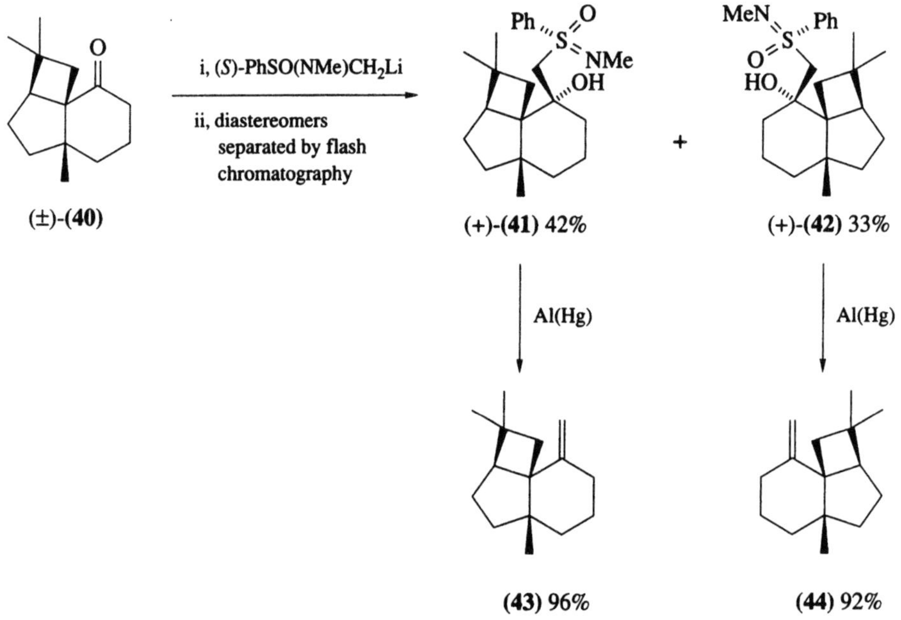

Scheme 9

In addition to synthesizing optically pure alkenes, this method can be utilized to produce chiral ketones and alcohols as shown in Scheme 10.[43] The intermediate sulfoximine (**45**) can either be reduced with Raney nickel to the chiral alcohol (**46**) or, because it is not stable to thermolysis, heated to revert back to the ketone (**47**).

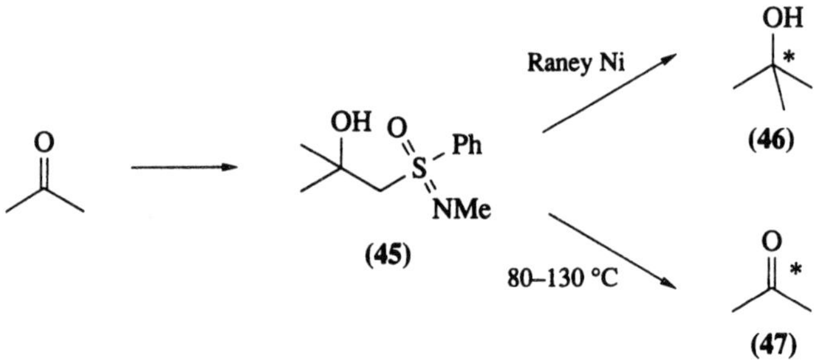

Raney Ni
80–130 °C

(45) (46) (47)

Scheme 10

As detailed in Scheme 11, Paquette has taken advantage of both the methylenation and the thermolysis in studies concerning the nucleus of cerorubenic acid.[44] Since the absolute configuration of the natural product was unknown, the route made use of the symmetry of the diketone (**48**) to produce either enantiomer of the ketoalkenes (**52**) and (**53**). Paquette observed the same high facial selectivity that Johnson had observed on the addition of the sulfoximine to the ketone.[43b] The diastereomers produced, (**49**) and (**50**), could be readily separated by silica gel chromatography, reduced with Raney nickel and dehydrated with the Burgess reagent. Although the reductive cleavage with Al(Hg) was attempted, in this example it proved unsuccessful.[44] The overall efficiency of the process was improved by first carrying out the Wittig reaction to produce methylene (**51**), followed by the addition of the sulfoximine. Once again the diastereomers could be separated and, having served its purpose of resolution, the sulfoximine was cleaved thermally back to the ketoalkenes (**52**) and (**53**).

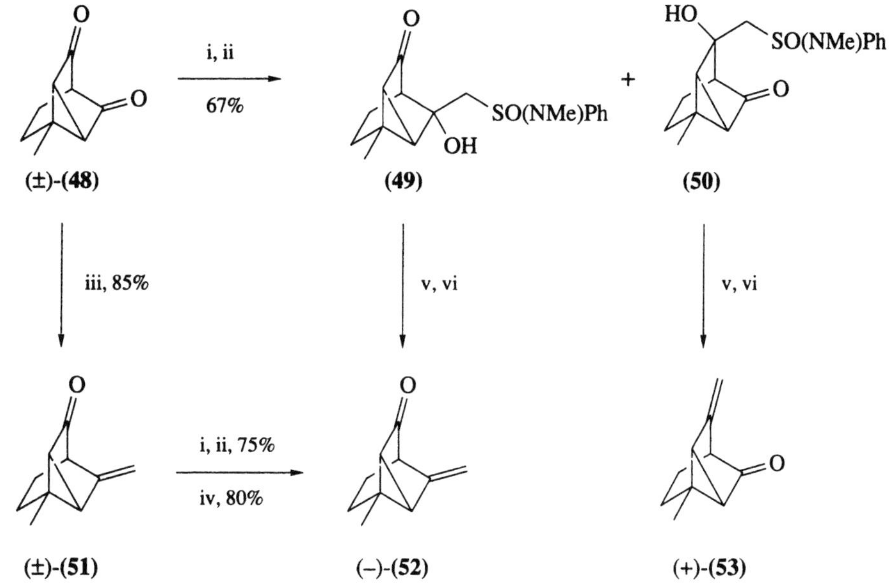

(±)-(**48**) —i, ii, 67%→ (**49**) + (**50**)

(±)-(**48**) —iii, 85%→ (±)-(**51**); (**49**) —v, vi→ (–)-(**52**); (**50**) —v, vi→ (+)-(**53**)

(±)-(**51**) —i, ii, 75%; iv, 80%→ (–)-(**52**)

i, (+)-PhSO(NMe)CH_2Li; ii, separate diastereomers; iii, Ph_3PCH_2;
iv, 130 °C, PhH; v, Raney Ni, EtOH; vi, $MeO_2C\bar{N}SO_2\overset{+}{N}Et_3$

Scheme 11

Besides serving as a useful tool for resolution, the sulfoximine addition adduct can be utilized to direct such reactions as the Simmons–Smith cyclopropanation and OsO_4 addition to an alkene.[43]

3.1.4.5 Halogen Incorporation

Finch has demonstrated that the sulfoximine approach is a viable alternative for fluoromethylenation (Scheme 12). The fluorosulfoximine (**54**) is deprotonated with LDA in THF and the aldehyde or ketone added to the anion. Conversion to the alkene is carried out with the standard aluminum amalgam procedure to yield a 1:1 mixture of (*E*)- and (*Z*)-alkenes (**56**). The reaction is very effective for aromatic and aliphatic aldehydes and aliphatic and alicyclic ketones, but, while aromatic and α,β-unsaturated ketones give good yields of the addition adduct, the reductive elimination results in a variable amount of product formation. This method was applied to the synthesis of prostaglandin 9-fluoromethylene (**58**; equation 13).[45]

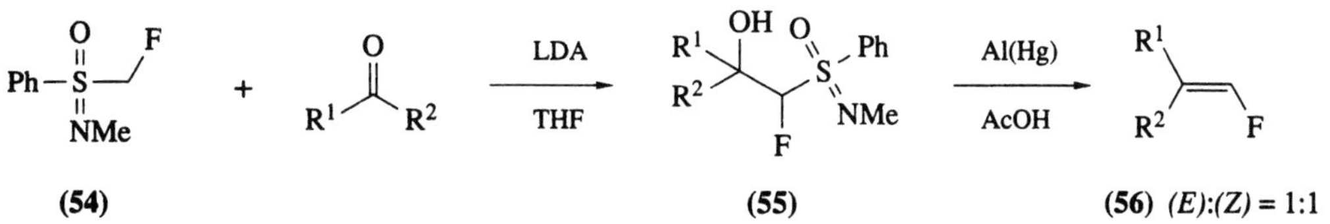

(**54**) + R^1COR^2 —LDA, THF→ (**55**) —Al(Hg), AcOH→ (**56**) (*E*):(*Z*) = 1:1

Scheme 12

i, $PhSO(NMe)CH_2F$, LDA; ii, CH_2N_2, 100%; Al/Hg, THF, AcOH, 57%

(57) → (58) (E):(Z) = 1:1 (13)

3.1.4.6 Di-, Tri- and Tetra-substituted Alkenes

The Johnson procedure may not be applied to tetrasubstituted alkenes. The explanation is that the reduced nucleophilicity of the sulfoximine causes it to function as a strong base, rather than add to the ketone.[33] The reaction of higher analogs of the methyl sulfoximine may be applied to the synthesis of di- and tri-substituted alkenes but, while the reactions proceed to give *trans*-alkene as the major product, the *trans*/*cis* selectivity is not high.[33] An excellent comparison between the sulfoximine and Wittig reactions is found in the synthesis of 6a-carbaprostaglandin I_2 (**60**; equation 14).[46] Initial attempts to incorporate the top portion by a Wittig reaction were complicated by apparent enolization of the carbonyl (**59**). While this was effectively solved by employing the more nucleophilic Johnson sulfoximine procedure, the *trans*/*cis* ratio was approximately 1:1. It was subsequently discovered that the difficulties encountered with the Wittig could be solved if an excess of the ylide was utilized.[46]

(59) → i, ii, 48% → iii, iv, v, vi, 62% → (60) (14)

i, $PhSO(NMe)CH_2(CH_2)_4OTHP$, MeMgBr; ii, Al/Hg, THF, AcOH; iii, $Bu^n{}_4NF$; iv, Ac_2O, Py; v, AcOH; vi, Jones reagent

It has been demonstrated that the mechanism (see Section 3.1.4.1) for the reductive cleavage of the sulfoximine explains the mix of alkenes.[33] Other effects, such as interconversion of the hydroxysulfoximines, alcohol elimination or equilibration of the alkenes, were experimentally eliminated as the cause of the *trans*/*cis* ratios.[33]

3.1.4.7 Phenylphosphinothioic Amide

In addition to the sulfoximines, Johnson has studied phosphinothioic amides as alkene precursors.[47] This reagent has not achieved the popularity of the sulfoximines. It can be utilized for ketone methylenation with resolution, as well as for the synthesis of more highly substituted alkenes.[48] Rigby has found that in the synthesis of guaianolides this reagent was effective where Peterson and Wittig reactions gave only β-elimination.[49]

3.1.5 TITANIUM-STABILIZED METHYLENATION: THE TEBBE REACTION

The application of Wittig technology to higher oxidation state carbonyls, such as esters and amides, is complicated by the undesired cleavage of the ester or amide bond. In addition, any basic method of alkene formation has the inherent possibility of enolizing a sensitive carbonyl derivative. In the course of studies directed toward reagents for the alkene metathesis reaction, a secondary observation was made by Schrock that lead to the discovery of higher oxidation state metal analogs of the Wittig reagent. These alkylidene complexes have filled an important gap in synthetic methodology.

Schrock discovered that the *t*-butylalkylidene complex of tantalum (and in lower yield, niobium) was a structural analog to the phosphorus ylide.[50] The complex proved to be a reagent for *t*-butylalkene formation. Of the Wittig-type reactions tried, most notable was the ability of the complex to react with esters and amides to form the corresponding *t*-butylalkenes in good yields.

3.1.5.1 Methylenation

Stimulated by these findings, the titanium methylenation reagent, known as the Tebbe reagent **(61)**, was first introduced in 1978 and has been widely used for the one-carbon homologation of carbonyls.[51] As discussed in a footnote in the Tebbe paper, the reagent reacts with aldehydes, ketones and esters to produce the methylene derivatives. This observation was expanded by Evans and Grubbs to include a wide range of both esters and lactones.[52] It was observed that the rate of reaction was increased by the presence of donor ligands such as THF and pyridine. This effect has been noted in the reaction of the Tebbe reagent with alkenes.[53] The explaination provided is that the Lewis acid removes Me_2AlCl from the metallocycle, allowing the reactive $[Cp_2TiCH_2]$ fragment to be trapped by the carbonyl or alkene. In addition, Evans and Grubbs discovered that the Tebbe reagent tolerated ketal and alkene functionality. The stereochemical integrity of unsaturated carbonyls is maintained. Pine and coworkers found that the reactivity of the reagent towards ketones was higher than that towards esters,[52] allowing for selective methylenation of dicarbonyl compounds.[54] In addition to ester functionality, the reaction can be performed on amides to produce the corresponding enamines. The methylenation reaction is summarized in Scheme 13. Inspection of Tables 7–10 reveals the wide variety of substrates and functional groups with which the reaction is compatible.

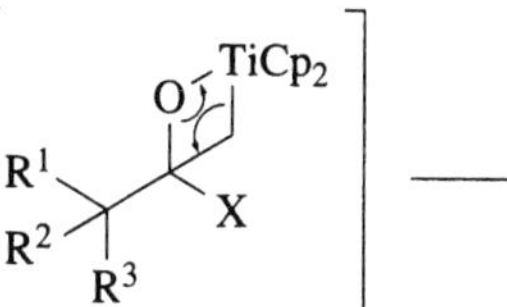

Cp_2Ti AlMe$_2$ Cl

(61)

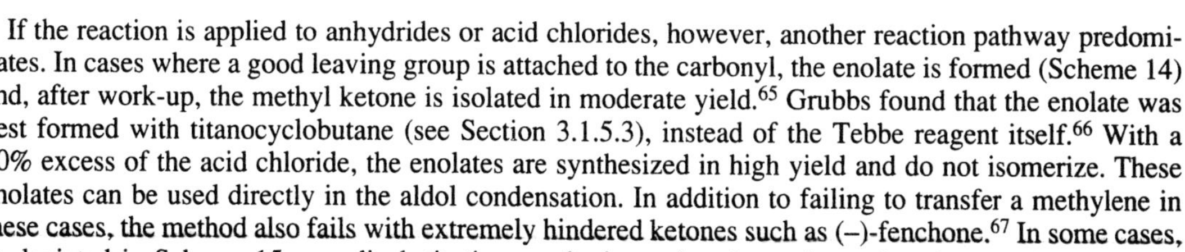

$Cp_2TiCH_2AlClMe_2$

X = H, OR, NR_2, CR_3

$+ [Cp_2Ti{=}O]_n$

Scheme 13

If the reaction is applied to anhydrides or acid chlorides, however, another reaction pathway predominates. In cases where a good leaving group is attached to the carbonyl, the enolate is formed (Scheme 14) and, after work-up, the methyl ketone is isolated in moderate yield.[65] Grubbs found that the enolate was best formed with titanocyclobutane (see Section 3.1.5.3), instead of the Tebbe reagent itself.[66] With a 50% excess of the acid chloride, the enolates are synthesized in high yield and do not isomerize. These enolates can be used directly in the aldol condensation. In addition to failing to transfer a methylene in these cases, the method also fails with extremely hindered ketones such as (–)-fenchone.[67] In some cases, as depicted in Scheme 15, α,α-disubstitution results in enolate formation rather than methylene transfer.[67] Anhydrides form enolates as well, but, unlike the products from acid chlorides, they are not efficiently utilized in subsequent reactions.[68] The anhydride enolates can react with starting materials, resulting in mixtures of products.

Table 7 Tebbe Reaction on Lactones

Entry	*Substrate*	*Solvent*		*Yield of RR'CCH$_2$ (%)*	*Ref.*
1		Toluene/THF	$n = 1$	85	52
			$n = 2$	85	
2		THF/DMAP		Not isolated	55
3	Bu^{t}Me$_2$SiO	Toluene/THF		76	56
4	Bu^n	Toluene/THF/pyridine		Not isolated	57
5	OSiMe$_2$Bu^t	Toluene/THF		85	58
6	RO, RO, OR, OR	Toluene/THF/pyridine	R = CH$_2$Ph	82	59
			R = SiMe$_3$	54	
			R = SiEt$_3$	86	
7	PhMe$_2$SiO, Me$_3$SiO, OSiMe$_3$	Toluene/THF/pyridine		70	59
8		Toluene/THF/pyridine		92	60
9[a]	R, R'	Toluene/THF/pyridine		Not isolated	60

Table 7 *(continued)*

Entry	*Substrate*	*Solvent*	*Yield of RR'CCH$_2$ (%)*	*Ref.*
10[b]	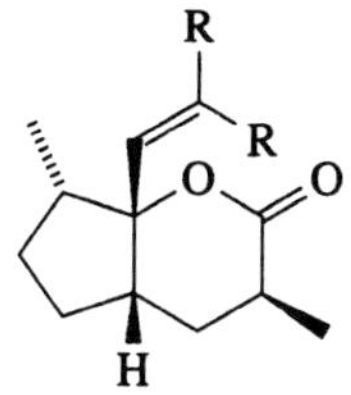 R, R, O, O, H	Toluene/THF/pyridine	Not isolated	60

[a] R, R' = Me, H. [b] R = H, Me.

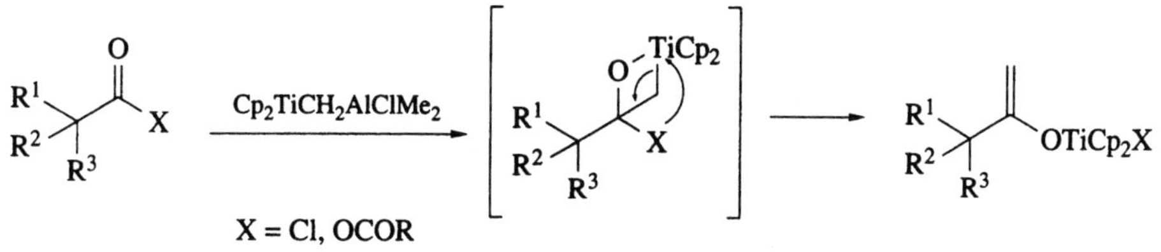

O, R^1, R^2, R^3, X

$Cp_2TiCH_2AlClMe_2$

X = Cl, OCOR

[O, $TiCp_2$, R^1, R^2, R^3, X] → R^1, R^2, R^3, $OTiCp_2X$

Scheme 14

O, H, R^2, R^3, X

$Cp_2TiCH_2AlClMe_2$

X = bulky alkyl group

[$TiCp_2$, O, H, R^2, R^3, X] → R^2, R^3, X, $OTiCp_2Me$

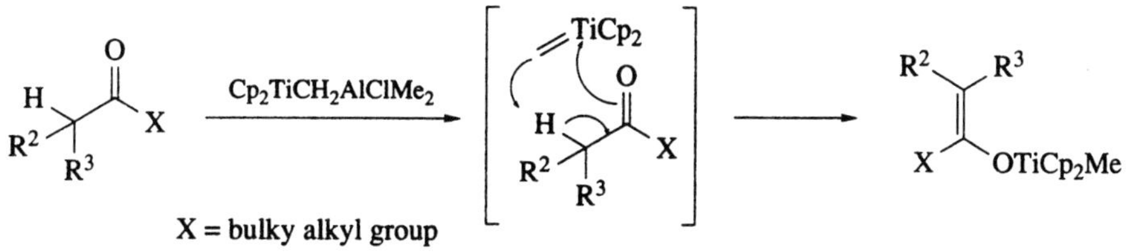

Scheme 15

Succinimides (**62**) react to form the corresponding mono- (**63**) or di-methylenated products (**64**). If two alkyl substituents are present in the α-position, a high degree of regioselectivity is observed for reaction at the less-hindered carbon. All of the diaddition compounds have the potential of isomerization to the pyrrole (**65**; Scheme 16). Piperidinediones (**66**), however, react predominantly by the enolization pathway to give (**67**; equation 15).[68]

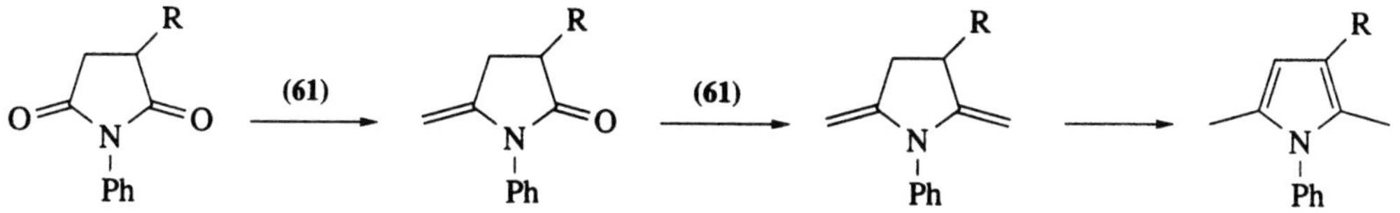

R, O, N, O, Ph (**62**) —(**61**)→ R, N, O, Ph (**63**) —(**61**)→ R, N, Ph (**64**) → R, N, Ph (**65**)

Scheme 16

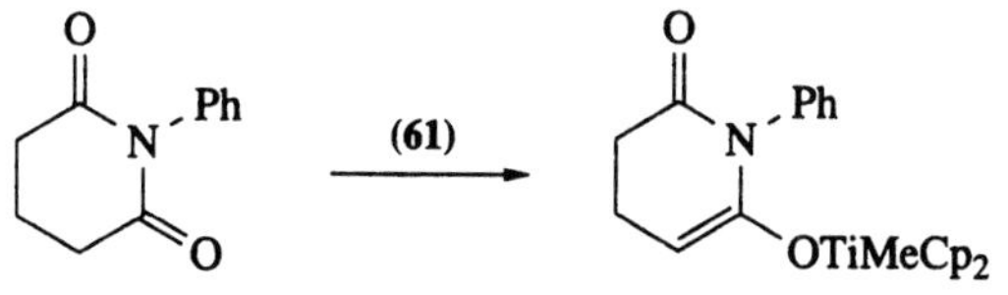

O, N, Ph, O (**66**) —(**61**)→ O, N, Ph, $OTiMeCp_2$ (**67**) (15)

Table 8 Tebbe Reactions on Ketones

Entry	*Substrate*	*Solvent*		*Yield of RR'CCH$_2$ (%)*	*Ref.*
1	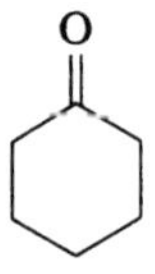	Toluene		65	51
2	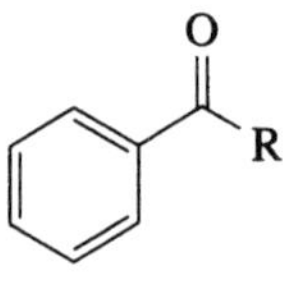 R	Toluene/THF	R = Me	88	54
			R = Ph	97	
			R = Bu^t	96	
			R = CF$_3$	50	
3	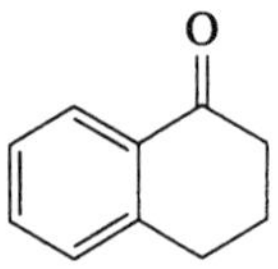	Toluene/THF		73	54
4	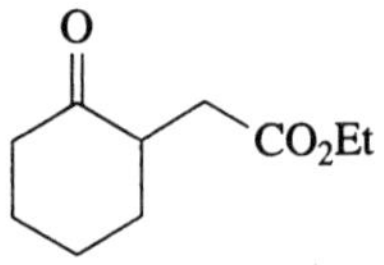 CO$_2$Et	Toluene/THF		67	54
5	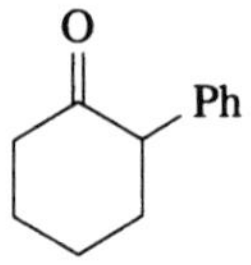 Ph	THF		93	67
6	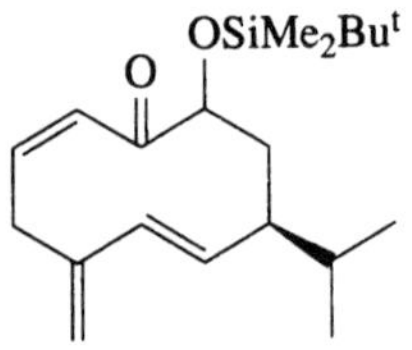 OSiMe$_2$Bu^t	THF/pyridine		45	61
7	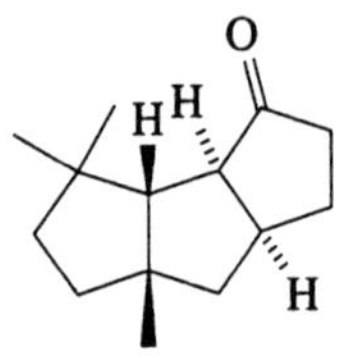	Et$_2$O/DMAP		93	62

3.1.5.2 Metals Other Than Aluminum

The $AlMe_3$ used in the production of the Tebbe reagent (**61**) is thought to inhibit decomposition of Cp_2TiMe_2. Further, the use of Al dictates the abstraction of the hydrogen from the methyl group rather than the cyclopentadienyl.[51] Other metals have been utilized in conjunction with [Cp_2TiCH_2], including Me_2Zn,[51] $CH_2(ZnI)_2$[69] and $MgBr_2$.[70] In addition Grubbs has studied a variety of metal complexes with the Cp_2TiCH_2 system.[71] These complexes provide interesting mechanistic information on the Tebbe reaction. Some of the reagents exhibit modified reactivity compared to the original reagent. For example, Eisch's Zn compound does not form the methyl ketone as the major product when reacted with acid

Table 9 Tebbe Reactions on Esters[a]

Entry	*Substrate*		*Yield (%)*	*Ref.*
1	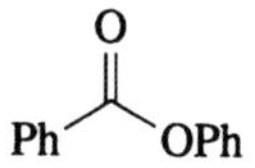 O, Ph, OPh		94	52
2	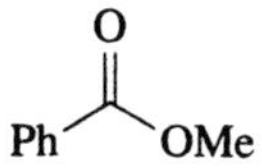 O, Ph, OMe		81	52
3	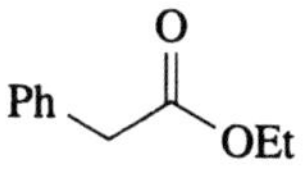 Ph, O, OEt		90	52
4	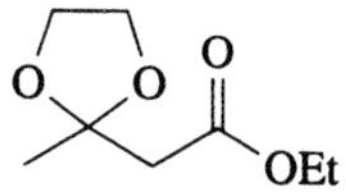 O, O, O, OEt		87	52
5	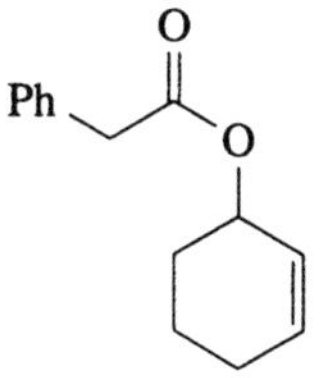 Ph, O, O		96	52
6	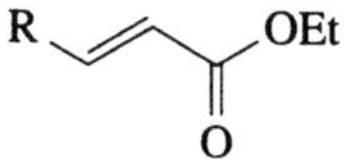 R, OEt, O	R = Me	82	52
		R = Ph	96	
7	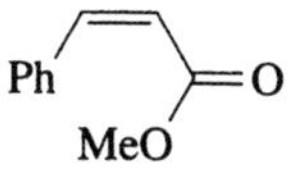 Ph, O, MeO		79	52
8	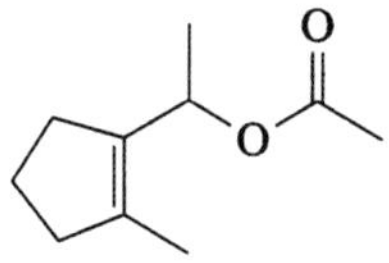 O, O		85	80a
9	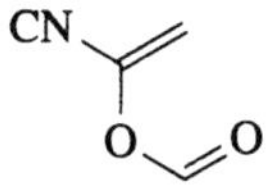 CN, O, O		No yield	63
10	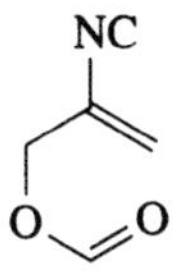 NC, O, O		No yield	63
11	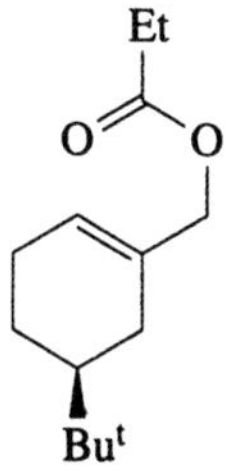 Et, O, O, Bu^t		86	64

[a] All reactions were carried out in toluene/THF

Table 10 Tebbe Reactions with Amides

Entry	*Substrate*	*Solvent*		*Yield (%)*	*Ref.*
1	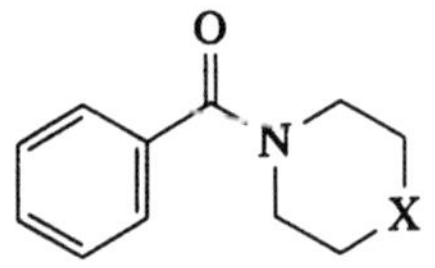 N-benzoyl (X = CH₂ / O) ring amide	Benzene/toluene	X = CH_2	76	54
			X = O	67	
2	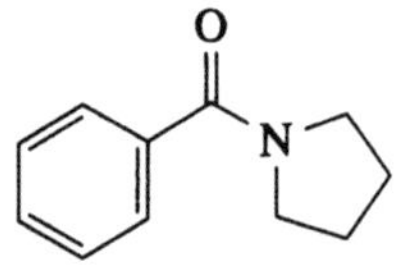 N-benzoylpyrrolidine	Benzene/toluene		80	54
3	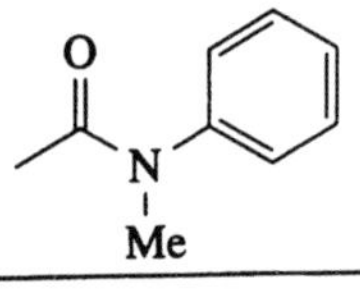 N-methyl-N-phenylacetamide	Benzene/toluene		97	54

chlorides.[69] By and large, these modified reagents do not appear to offer any significant synthetic advantage over the Tebbe reagent, and have not found application in organic synthesis.

3.1.5.3 Metallacyclobutane Complexes

Grubbs discovered that the reaction of the Tebbe reagent (**61**) with alkenes to form metallacycles that can then be used for methylene transfer, has distinct advantages over the Tebbe reagent itself.[72] These complexes are stable to air and are more easily handled that the Tebbe reagent. In addition, the aqueous work-up required in the Tebbe reaction (it is typical to quench with a 15% aqueous NaOH solution) can result in isomerization.[73] A comparative study on the two methods found that similar yields were obtained with both.[67,68] Grubbs applied both types of reactivity to a synthesis of capnellene.[74] As depicted in Scheme 17, the Tebbe reagent (**61**) first reacts with the alkene to form the metallacycle (**69**). Heating the reaction results in the ring opening to the alkene alkylidene (**70**). The *t*-butyl ester is then trapped intramolecularly to form the cyclobutene enol ether (**71**).

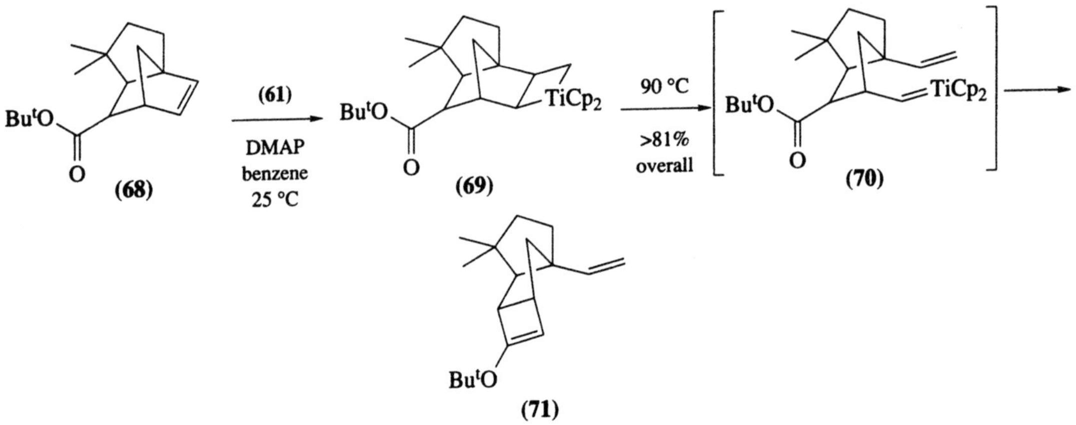

Bu^{t}O (**68**) —(**61**), DMAP, benzene, 25 °C→ Bu^{t}O TiCp$_2$ (**69**) —90 °C, >81% overall→ [Bu^{t}O TiCp$_2$ (**70**)] → Bu^{t}O (**71**)

Scheme 17

3.1.5.4 Methylene *versus* Alkylidene Transfer

Unfortunately the Tebbe reagent (**61**) cannot be extended to higher alkyl analogs. When the complex is formed with β-hydrogens present on the aluminum reagent, alternative products are formed.[75] Tebbe found that the use of Et_3Al resulted in a bridged cyclopentadienyl group in a dimeric complex.[76] Although the problems of higher order complex synthesis can be circumvented,[78] these reagents do not react with the high conversion of the Tebbe reagent, and have not been utilized in total synthesis.[77] As discussed in Section 3.1.12, there are many metals that form alkylidene complexes.[78] The Tebbe reagent has proven to be of widespread generality and use in organic synthesis. Schwartz has developed a Zr alkylidene reagent that will add to carbonyls in analogy to the Tebbe reaction in high yield.[317] Many of these complexes require special techniques for synthesis. Takai has discovered that alkylidenation occurs with a dibromoalkyl in the presence of $TiCl_4$, Zn and TMEDA *in situ*, resulting in the preparation of (*Z*)-alkenyl ethers.[79] This reagent is discussed in Section 3.1.12.2

3.1.5.5 Miscellaneous Reactivity

As originally pointed out by Evans and Grubbs, one of the advantages of the Tebbe reagent is the facile preparation of the intermediate allyl vinyl ethers for the Claisen rearrangement.[52] It has also been observed that the Tebbe reagent itself can behave as a Lewis acid catalyst for the rearrangement.[80] Negishi and Grubbs have studied the ability of the Tebbe reagent to form allene derivatives.[81]

3.1.6 TITANIUM–ZINC METHYLENATION[82]

A reaction that appears to be mechanistically similar to the Tebbe reaction was developed by Oshima in 1978.[83] Diiodomethane or dibromomethane in the presence of zinc is treated with a Lewis acid to form, presumably, a divalent complex (**72**), which reacts with aldehydes and ketones to produce the corresponding methylene derivative (**73**; Scheme 18).[84] This reagent complements the reactivity of the Tebbe reagent, in that the zinc methylenation is not reactive towards esters or lactones.[85] Because it is an electrophilic reagent, it is suitable for the methylenation of enolizable ketones and aldehydes.

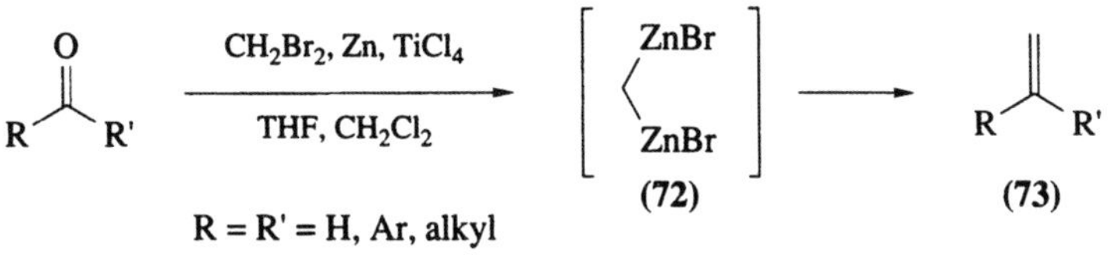

CH_2Br_2, Zn, $TiCl_4$ / THF, CH_2Cl_2 → [ZnBr / ZnBr] (**72**) → (**73**)

R = R' = H, Ar, alkyl

Scheme 18

3.1.6.1 Other Lewis Acids

In addition to titanium tetrachloride, a variety of Lewis acids have been utilized by Oshima.[86] The choice centers around the type of reactivity desired, and this is further discussed below in Section 3.1.6.2. Dibromomethane and zinc combined with $TiCl_4$ is the most commonly used reagent for the methylenation of ketones.[87] Lombardo discovered, during work on the gibberellins (**74**; equation 16), a method of synthesizing the $CH_2Br_2/Zn/TiCl_4$ reagent that results in an extremely active catalyst that can be stored in the freezer.[87] Lombardo's catalyst may have advantages over the original procedure for

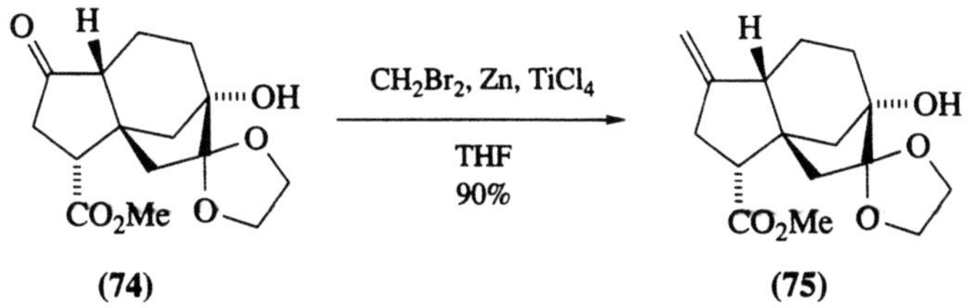

(**74**) CH_2Br_2, Zn, $TiCl_4$ / THF / 90% → (**75**) (16)

compounds that are sensitive to titanium tetrachloride, but no comparative study of the catalyst formed by the two methods has been done.[87]

3.1.6.2 Chemoselectivity

It is possible to exclusively methylenate a ketone in the presence of an aldehyde by precomplexing the aldehyde (*e.g.* **76**) with $Ti(NEt_2)_4$, followed by treatment with the usual methylene zinc/$TiCl_4$ reagent (equation 17).[88] Takai also studied the chemoselective methylenation of aldehydes (**78**) in the presence of ketones, and found the use of diiodomethane, zinc and titanium isopropoxide or trimethylaluminum to be effective (equation 18).[88]

(**76**) —i, ii, 76%→ (**77**) (17)

i, $Ti(NEt_2)_4$, CH_2Cl_2; ii, CH_2I_2, Zn, $TiCl_4$, THF

(**78**) → (**79**) (18)

CH_2I_2, Zn, $Ti(OPr^i)_4$, 83%; CH_2I_2, Zn, Me_3Al, 96%

3.1.6.3 Examples of the Reaction with Aldehydes

It is advantageous to utilize either titanium isopropoxide or trimethylaluminum complexes with aldehydes in general, because pinacol-coupled diols form with the $Zn/CH_2Br_2/TiCl_4$ systems as minor side products.[86] No evidence of Simmons–Smith-type side products was observed with any of the methylenation reagents.[83] Additional examples of the reaction with aldehydes are presented in Table 11.

3.1.6.4 Examples of the $TiCl_4/CH_2X_2/Zn$ Reaction Compared with the Wittig

Some of the more intriguing examples of the use of the $TiCl_4/CH_2X_2/Zn$ reagent are in cases where the Wittig or other methods of methylenation have failed, and the Oshima method has proved a successful solution. Specific examples are outlined in Table 12.

3.1.6.5 Examples of the $TiCl_4/CH_2Br_2/Zn$ Reaction with Ketones Where Stereochemistry is Preserved

As the examples indicate, the reaction is compatible with a wide range of functional groups, including alcohols, esters, acetates, carboxylic acids, ethers, halogens, silyl groups, acetinides, lactones, alkenes, ketals and amines. The methylenation proceeds with allylic carbonyls, without loss of stereochemistry. One of the tremendous advantages of the $TiCl_4/CH_2X_2/Zn$ reaction is that sensitive ketones can be methylenated without loss of stereochemistry. Additional examples of the reactions of ketones with the reagent are summarized in Table 13. The $CH_2Br_2/Zn/TiCl_4$ reagent has proven to be of broad utility in natural products synthesis.

3.1.6.6 Isotopic Labeling

When the methylenation reagent is generated using zinc, CD_2Br_2 or CD_2Cl_2, and titanium tetrachloride, it can be utilized to synthesize the deuterated analogs (**82**). In addition, Trost has demonstrated that a ^{13}C label may be incorporated in a ketone (**81**) using $H_2{}^{13}CI_2$ (Scheme 19).[108]

Table 11 Zinc, CH_2X_2 Reaction with Aldehydes

Entry	*Substrate*	*Reagent*	*Yield of RR'CCH₂ (%)*	*Ref.*
1	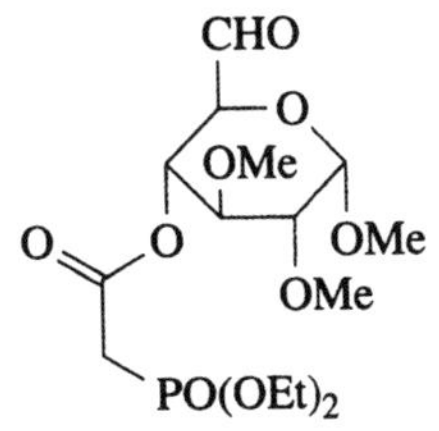 CHO; O; OMe; O; O; OMe; OMe; $PO(OEt)_2$	$TiCl_4–CH_2Br_2–Zn$	60–75	89
2	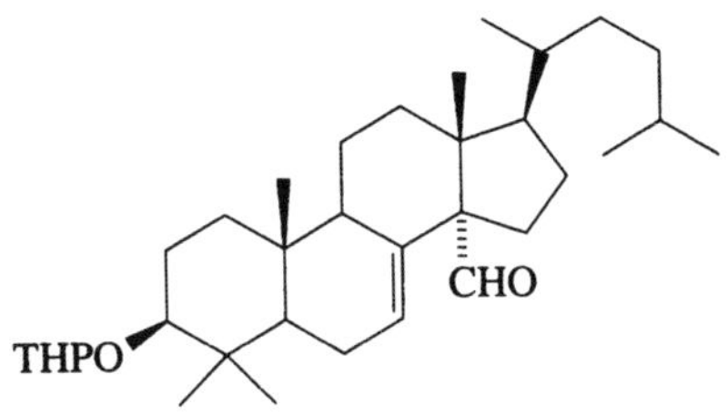 CHO; THPO	$TiCl_4–CH_2Br_2–Zn$	—	90
3	 MeO_2C; CHO	$TiCl_4–CH_2Br_2–Zn$	70	91
4	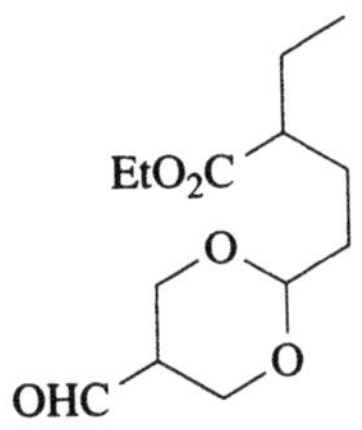 EtO_2C; O; O; OHC	$TiCl_4–CH_2Br_2–Zn$	35, 2 steps	92
5	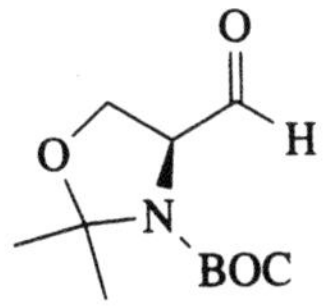 O; H; O; N; BOC	$AlMe_3–CH_2I_2–Zn$	75, >99% *ee*	93

3.1.6.7 Halogen Incorporation

The zinc procedure has been extended to CF_3CCl_3 to form 2-chloro-1,1,1,-trifluoro-2-alkenes (**84**; equation 19). and α-fluoro-α,β-unsaturated carboxylic acid methyl esters (**87**) with methyl dichlorofluoroacetate (**85**; equation 20).[109] Both of these transformations are carried out with zinc and acetic anhydride. Several variations of this reaction have appeared in the literature.[109] The mechanism appears to be dramatically different from the Oshima methylenation. In this case, the reaction proceeds through the alcohol intermediate, which is converted to the acetate and reductively cleaved.

3.1.7 OTHER METHODS OF METHYLENATION

3.1.7.1 Samarium-induced Methylenation

Inanaga has studied the use of SmI_2 deoxygenation of aldehydes and ketones.[110] The reaction proceeds by a two-step mechanism, where SmI_2 and CH_2I_2 react with the carbonyl to form an O—SmIX species (**88**), which is converted to a better leaving group and reductively eliminated with SmI_2.[111] As the examples depicted in Scheme 20 indicate, no studies of chemoselectivity have been undertaken.[112]

Table 12 Examples of $TiCl_4$–CH_2Br_2–Zn Reaction Compared with Wittig

Entry	*Substrate*	*Solvent*	*Product*	*Yield (%)*	*Ref.*
1	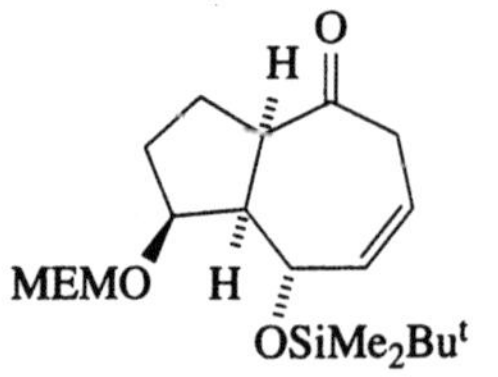 H, O, MEMO, H, $OSiMe_2Bu^t$	THF, CH_2Cl_2	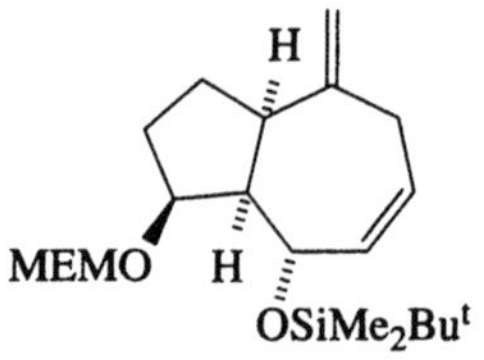 H, MEMO, H, $OSiMe_2Bu^t$	99[a]	94
2	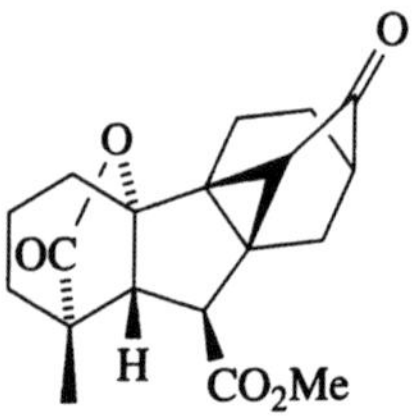 O, OC, H, CO_2Me	CH_2Cl_2	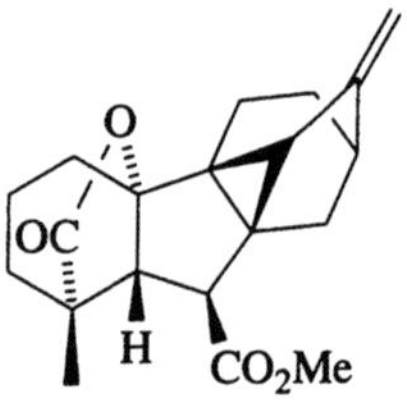 O, OC, H, CO_2Me	87	95
			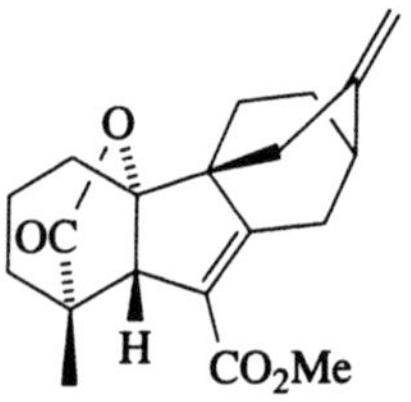 O, OC, H, CO_2Me	b	95
3	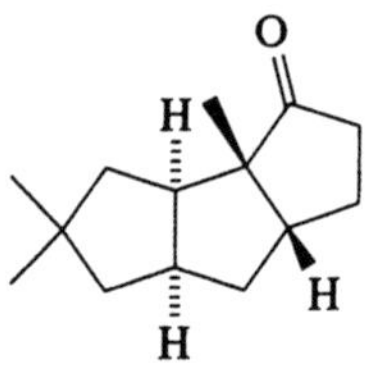 O, H, H, H	CH_2Cl_2	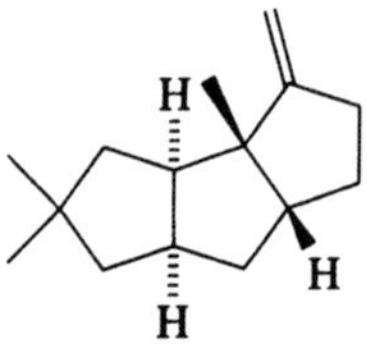 H, H, H	98[c]	96
4	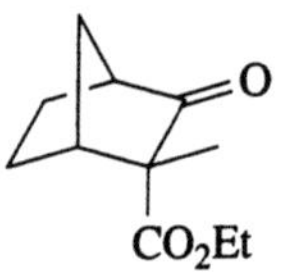 O, CO_2Et	THF	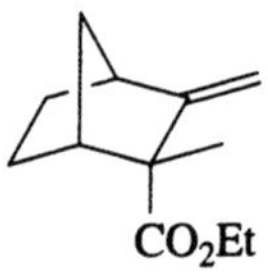 CO_2Et	40[d]	97
5	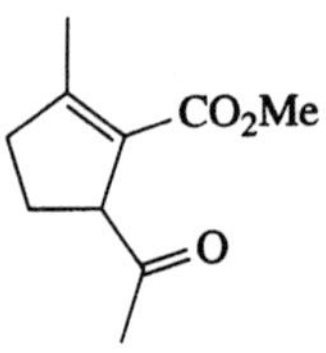 CO_2Me, O	THF, CH_2Cl_2	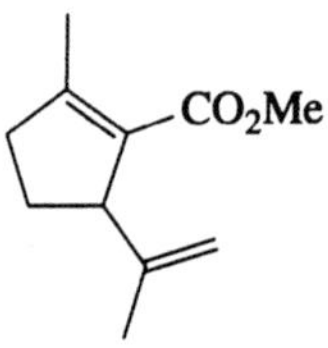 CO_2Me	95[e]	98
6	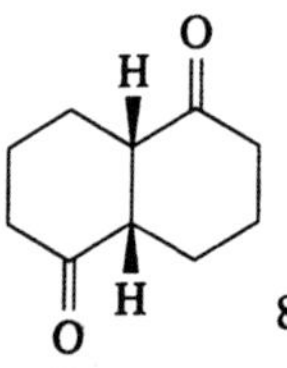 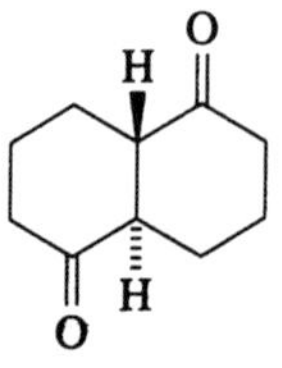 H, O, H, O + H, O, H, O (85:15)		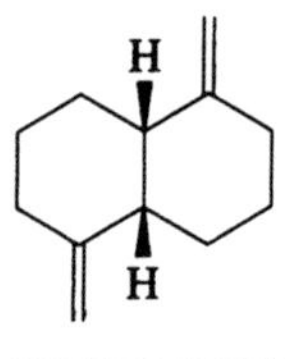 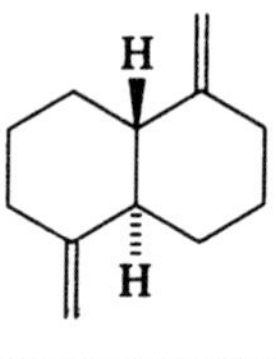 H, H + H, H (65:35)	75[f]	99

[a] Wittig and Johnson sulfoxime fail. [b] Wittig results in retro Michael (poor recovery). [c] Wittig 70%, capricous. [d] Wittig and acetalization conditions fail. [e] Wittig yield only 22%. [f] Wittig fails.

Table 13 Examples of $TiCl_4$–CH_2Br_2–Zn Reaction with Ketones Where Stereochemistry is Preserved

Entry	*Substrate*		*Yield (%)*	*Ref.*
1	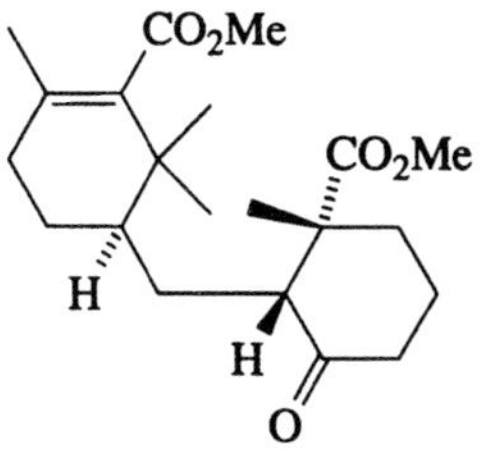 CO_2Me; CO_2Me; H; H; O		60	100
2	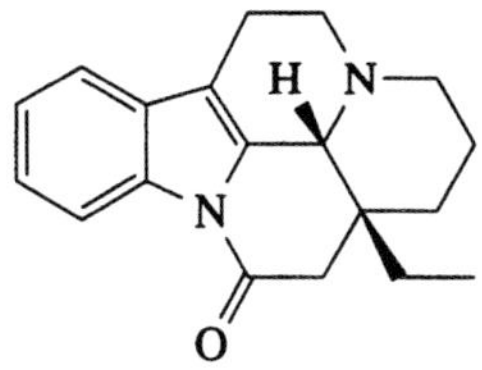 H; N; N; O		85	101
3	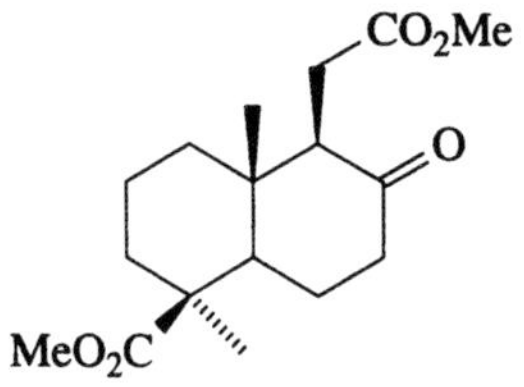 CO_2Me; O; MeO_2C		86	102
4	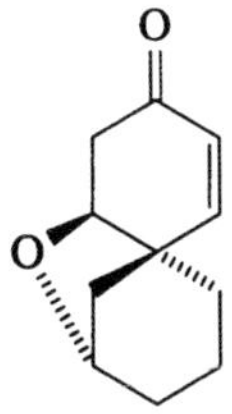 O; O	used CH_2I_2	62	103
5	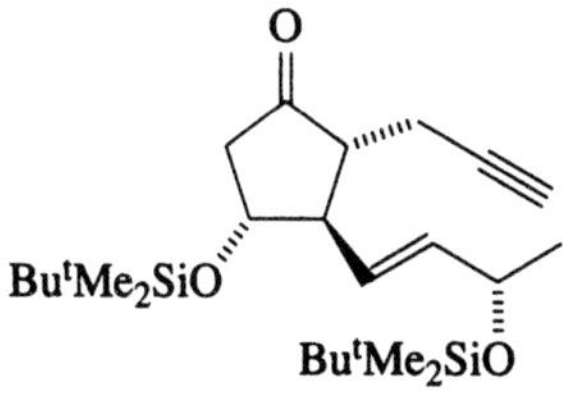 O; Bu^tMe_2SiO; Bu^tMe_2SiO		78	104
6	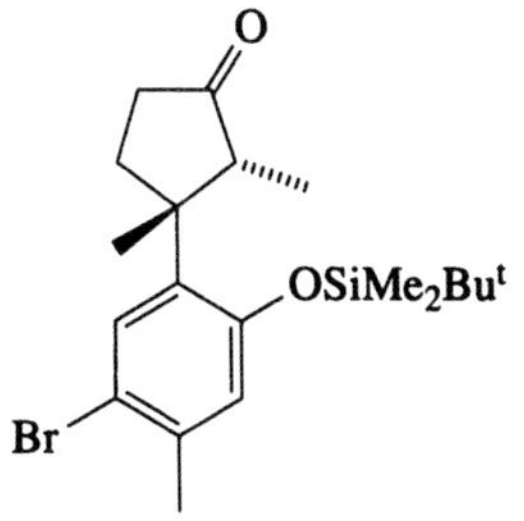 O; $OSiMe_2Bu^t$; Br		84	105

Table 13 *(continued)*

Entry	*Substrate*	*Yield (%)*	*Ref.*
7	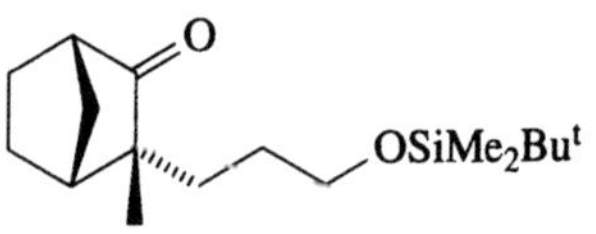 $OSiMe_2Bu^t$	75	106
8	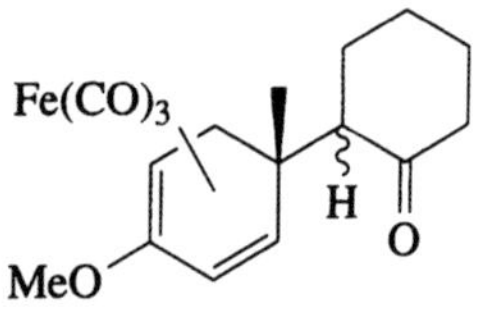 $Fe(CO)_3$, H, O, MeO	80	107

MeO_2C, MeO_2C, $^{13}CH_2$ $\xleftarrow[\text{Zn, TiCl}_4]{H_2{}^{13}CI_2}$ MeO_2C, MeO_2C, O $\xrightarrow[\text{Zn, TiCl}_4]{D_2CI_2}$ MeO_2C, MeO_2C, CD_2

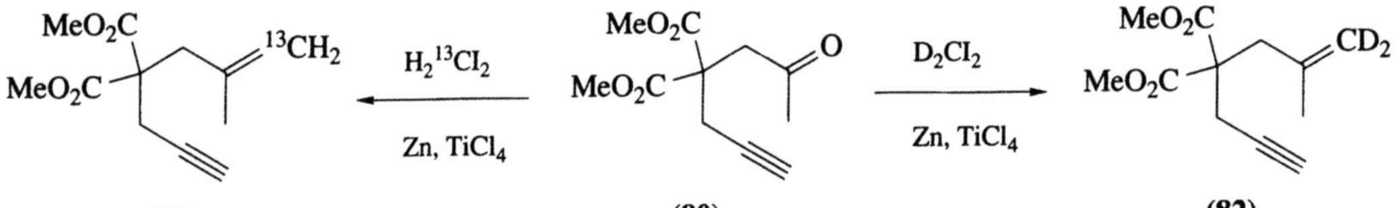

(81) **(80)** **(82)**

Scheme 19

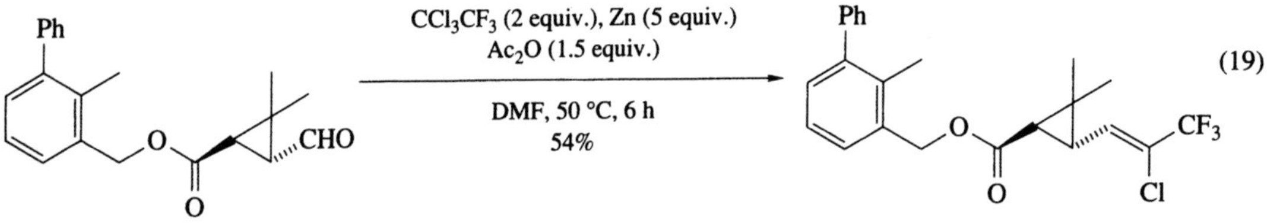

Ph, O, O, CHO **(83)** $\xrightarrow[\text{DMF, 50 °C, 6 h; 54\%}]{CCl_3CF_3 \text{ (2 equiv.), Zn (5 equiv.), } Ac_2O \text{ (1.5 equiv.)}}$ Ph, O, O, CF_3, Cl **(84)** (19)

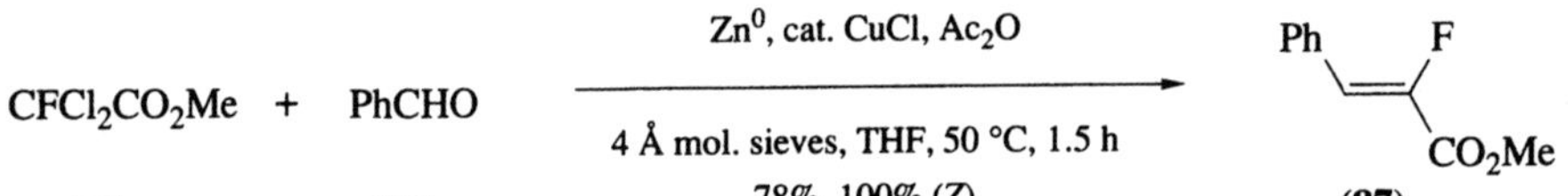

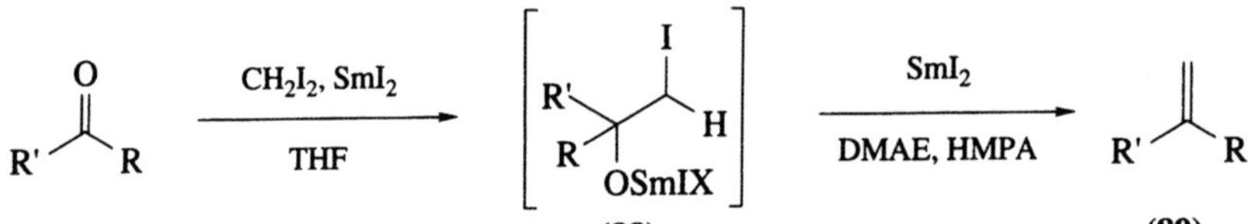

$CFCl_2CO_2Me$ **(85)** + PhCHO **(86)** $\xrightarrow[\text{4 Å mol. sieves, THF, 50 °C, 1.5 h; 78\%, 100\% (Z)}]{Zn^0\text{, cat. CuCl, } Ac_2O}$ Ph, F, CO_2Me **(87)** (20)

O, R', R $\xrightarrow[\text{THF}]{CH_2I_2\text{, }SmI_2}$ [I, R', H, R, OSmIX] **(88)** $\xrightarrow[\text{DMAE, HMPA}]{SmI_2}$ R', R **(89)**

R' = $Me(CH_2)_{10}$, R = H; 73%
R, R' = —$(CH_2)_{10}$— ; 80%

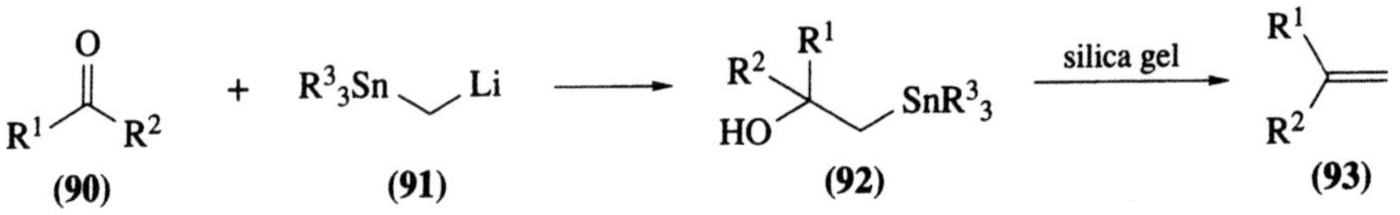

O, R^1, R^2 **(90)** + $R^3{}_3Sn$ Li **(91)** ⟶ R^2, R^1, HO, $SnR^3{}_3$ **(92)** $\xrightarrow{\text{silica gel}}$ R^1, R^2 **(93)**

R^1 = Ph, R^2 = Et; 78%
R^1, R^2 = 1-tetralone; 91%
R^1 = *p*-$MeOC_6H_4$, R^2 = H; 94%
R^1, R^2 = *cis*-1-decalone; 91%
R^1, R^2 = *trans*-1-decalone; 96%

Scheme 20

3.1.7.2 Tin-induced Methylenation

In direct analogy to the Peterson methylenation, the triaryl- and trialkyl-stannylmethyllithium reagent (**91**) can be added to aldehydes and ketones (**90**), followed by elimination of the hydroxystannane (**92**) to obtain the methylene derivative (**93**; Scheme 20).[113] This reaction, like the titanium and cerium modifications of the Peterson methylenation of Kaufmann and Johnson, may prove advantageous, in comparison to the Wittig reaction, for enolizible substrates.

3.1.8 ALKENE FORMATION: INTRODUCTION

Many of the methods discussed in the previous sections may be extended to more complex substitution patterns. The sections covering alkene formation are divided according to the stabilizing substituent on the anion component. For each stabilizing element, its application to the synthesis of alkenes with precise geometries is discussed.

3.1.9 PHOSPHORUS-STABILIZED ALKENATION

The Wittig reaction is one of the most effective and general methods of alkene formation from carbonyl derivatives.[114] Prior to the development of phosphorus-stabilized anion addition and elimination, the synthesis of an alkene from a carbonyl entailed anion addition and subsequent elimination with nonspecific alkene position and configuration. The Wittig reaction proceeds with defined positional selectivity, in addition to chemo- and stereo-selectivity. It has become the standard by which all subsequent methodology is judged. This section is organized according to the type of Wittig reagent used. Section 3.1.9.1 discusses phosphonium ylides and is further subdivided depending on the presence of stabilizing or conjugating functionality in the Wittig reagent. Phosphoryl-stabilized carbanions are covered in Sections 3.1.9.2, 3.1.9.3 and 3.1.9.4; this discussion includes the phosphonate and phosphine oxide carbanions. Since the Wittig reaction has been the subject of many excellent reviews, this section will briefly discuss expected stereochemical trends and will emphasize new methods, with particular focus on attaining (*E*)- and (*Z*)-selectivity.[115]

3.1.9.1 Phosphonium Ylides

The general representation of the classic Wittig reaction is presented in equation (21). The (*E*)- and (*Z*)-selectivity may be controlled by the choice of the type of ylide (**95**), the carbonyl derivative (**94**), the solvent and the counterion for ylide formation. As a general rule, the use of a nonstabilized ylide (**95**; X and Y are H or alkyl substituents and R^3 is phenyl) and salt-free conditions in a nonprotic, polar solvent favors the formation of the (*Z*)-alkene isomer (**96**) in reactions with an aldehyde. A stabilized ylide with strongly conjugating substituents such as an ester, nitrile or sulfone forms predominantly the (*E*)-alkene.

$$R^1R^2C{=}O \;(\mathbf{94}) + R^3{}_3P{=}CXY \;(\mathbf{95}) \longrightarrow R^1R^2C{=}CXY \;(\mathbf{96}) + R^3{}_3P{=}O \;(\mathbf{97}) \qquad (21)$$

3.1.9.1.1 Mechanism

Mechanistic studies have been the subject of a great deal of recent work.[116] Although at one time the Wittig reaction was thought to occur through the formation of zwitterionic betaine intermediates (**100**) and (**101**), the reaction of a nonstabilized triphenylphosphorus ylide (**99**) with an aldehyde forms observable (by NMR) 1,2-oxaphosphetanes (**104**) and (**105**), which eliminate to produce the alkene (**102**) and phosphine oxide (**103**) (Scheme 21).[117]

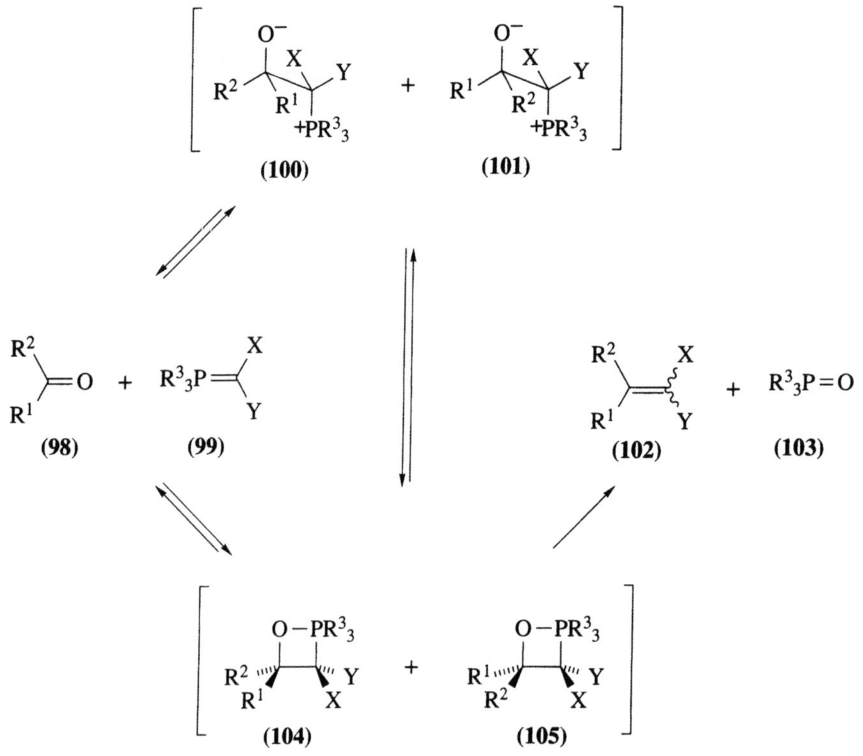

Scheme 21

There are two steps to the reaction that define the stereochemical outcome. The first is the intial addition of ylide and carbonyl, with inherent preferences for the formation of *cis*- and *trans*-oxaphosphetane intermediates (**104**) and (**105**), and the second is the ability of the intermediates to equilibrate. Maryanoff has studied numerous examples in which the final (*E*)/(*Z*) ratio of the alkene (**102**) produced does not correspond to the initial ratios of oxaphosphetanes (**104**) and (**105**) and has termed this phenomenon 'stereochemical drift'.[116b] The intermediate oxaphosphetanes are thought to interconvert by reversal to reactants (**98**) and (**99**), followed by recombination. In this case the final ratio of alkene can be substantially different from the initial addition ratio.

Interconversion and the intermediacy of betaine structures (**100**) and (**101**) are still matters of debate and ongoing research. There are distinct differences in the reactions of aliphatic and aromatic aldehydes, and also of aromatic and aliphatic phosphonium ylides, with regard to reversibility of the initial addition adducts.[116] Vedejs and coworkers have evidence that salt-free Wittig reactions with unbranched aliphatic aldehydes occur with less than 2% equilibration and are therefore under kinetic control.[116d–f] These studies suggest the (*Z*):(*E*) alkene ratios correspond to the kinetic selectivity of the initial addition step. While *cis*-oxaphosphetanes can equilibrate, the more stabile *trans* isomer does not. Elimination of phosphine oxide occurs stereospecifically *syn*. Factors that are known to enhance equilbration of the *cis*-oxaphosphetane are the presence of an alkyl or donor ligand on phosphorous, lithium salts, and steric hindrance in either the aldehyde or ylide.[118] Mechanistic studies have been done largely on nonstabilized ylides, due to the fact that intermediates can be followed by NMR. Vedejs has developed a unified theory to explain the stereoselectivity of stabilized and nonstabilized Wittig reactions.[116f] Nonstabilized ylides add in an early transition state to give a high ratio of *cis*- to *trans*-oxaphosphetanes, while stabilized ylides have a later transition state and a larger portion of *trans* isomer. Equilibration is thought to occur only under special circumstances, and both reactions are thought to be under kinetic control.

3.1.9.1.2 Nonstabilized ylides

(i) Nonstabilized ylides giving (Z)-stereoselectivity

Typically, nonstabilized ylides are utilized for the synthesis of (*Z*)-alkenes. In 1986, Schlosser published a paper summarizing the factors that enhance (*Z*)-selectivity.[119] Salt effects have historically been defined as the response to the presence of soluble lithium salts.[114] Any soluble salt will compromise the (*Z*)-selectivity of the reaction, and typically this issue has been resolved by the use of sodium amide or sodium or potassium hexamethyldisilazane (NaHMDS or KHMDS) as the base. Solvent effects are also vital to the stereoselectivity. In general, ethereal solvents such as THF, diethyl ether, DME and *t*-butyl methyl ether are the solvents of choice.[119] In cases where competitive enolate formation is problematic, toluene may be utilized. Protic solvents, such as alcohols, as well as DMSO, should be avoided in attempts to maximize (*Z*)-selectivity. Finally, the dropwise addition of the carbonyl to the ylide should be carried out at low temperature (–78 °C). Recent applications of phosphonium ylides in natural product synthesis have been extensively reviewed by Maryanoff and Reitz.[114]

(ii) α-Oxygenated substrates

As discussed in the following sections, there are notable exceptions to the general rules of selectivity in the Wittig reaction. Carbonyl derivatives with α-oxygenation reverse normal selectivity with stabilized ylides. This effect does not predominate in the reactions of nonstabilized ylides. A systematic study of Wittig reactions of ethylidenetriphenylphosphorane and derivatives of hydroxyacetone (**106**) was undertaken by Still.[120] Optimal (*Z*)-selectivity was obtained with KHMDS as the base with 10% HMPA in THF (equation 22). This procedure was applied to the synthesis of α-santalol (**109**), with alkene formation proceeding in 85% yield and greater than 99% stereochemical purity (equation 23).[121] Other recent examples of successful (*Z*)-alkenations in the presence of oxygenated substrates are the syntheses of isoquinuclidines by Trost, in which alkenation of an epoxy ketone occurred to produce only the (*Z*)-isomer, and of the C(7)–C(13) fragment of erythronolide by Burke, in which a pyran derivative was homologated to the ethylene with high (*Z*)-selectivity.[122]

O; R; OR'; Ph_3PCHMe; KHDMS, HMPA; THF; R; OR' (22)

(**106**) (**107**)

R = H, R' = CH_2Ph; *(Z):(E)* = 12:1
R = H, R' = $SiMe_2Bu^t$; *(Z):(E)* = 14:1
R = H, R' = THP; *(Z):(E)* = 41:1, 83%
R = Me, R' = THP; *(Z):(E)* = 200:1, 95%

PPh_3; i, O; O; OMe; ii, H_3O^+; 85%; OH (23)

(**108**) (**109**) >99% *(Z)*

(iii) (E)-Selective alkenation

Application of the Wittig reaction of a nonstabilized ylide to the synthesis of an (*E*)-alkene is practically and effectively carried out by the Schlosser modification.[123] Alternatively, the use of a trialkylphosphonium ylide can produce high ratios of (*E*)-alkene.[116] Recently, Vedejs has developed a reagent using dibenzophosphole ylides (**110**) to synthesize (*E*)-disubstituted alkenes (**111**) from aldehydes (equation 24).[124] The initial addition of ylide occurs at –78 °C, but the intermediate oxaphosphetane must be heated to induce alkene formation. The stereoselectivity in the process is excellent, particularly for aldehydes with branched substitution α to the reacting center. Both the ethyl and butyl ylides have been utilized.

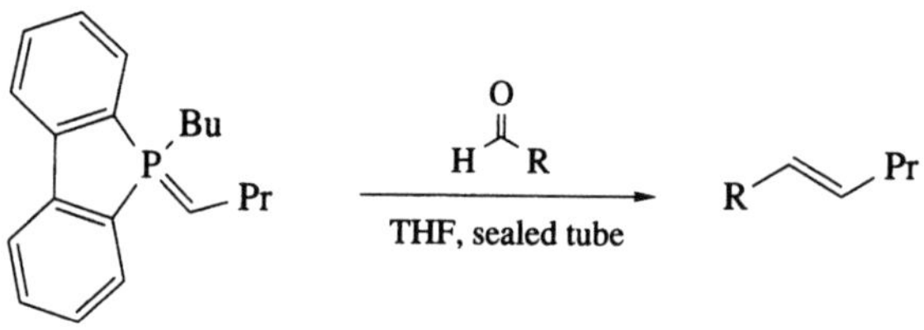

(24)

(**110**) (**111**)

R = CH_2CH_2Ph; *(E):(Z)* = 20:1, 66%
R = CMe_2CH_2Ph; *(E):(Z)* = 12:1, 82%
R = $CHMeC_9H_{19}$; *(E):(Z)* = 124:1, 91%
R = c-C_6H_{11}; *(E):(Z)* = 84:1, 97%

The presence of an oxido, carboxy or amide group in the ylide (**112**) can shift selectivity to produce (*E*)-alkenes. The effect of various nucleophilic groups at different distances away from the phosphorus has been systematically studied by Maryanoff.[116a] There is an optimal chain length for the nucleophile in order to maximize (*E*)-alkene formation. In the case of oxido ylides, short chain lengths and the use of lithium as the counterion maximized (*E*)-selectivity. Oxido ylides form significant amounts of (*E*)-alkene with both benzaldehyde (*E*:*Z* = 94:6) and hexanal (*E*:*Z* = 52:48) relative to an ylide that does not possess an nucleophilic group. Carboxy ylides, on the other hand, only showed dramatic increases in (*E*)-selectivity with benzaldehyde (*E*:*Z* = 93:7). In addition, the effect of short chain length was critical, but interestingly the selectivity was not strongly dependent on the type of counterion. Significant effects with amino substitution on the ylide were observed if 2 equiv. of base were added to generate the amido species (*E*:*Z* = 87:13 with benzaldehyde).

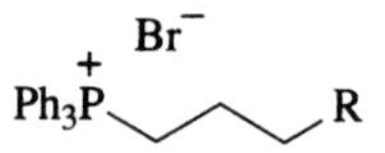

(**112**) R = OH, CO_2H, $CONH_2$

3.1.9.1.3 Semistabilized ylides

As a general rule, ylides with allylic or benzylic functionality do not proceed with a high degree of stereoselectivity.[125] There have been recent examples of arachidonic acid derivatives in which the coupling of an allylic phosphonium salt with an unsaturated aldehyde resulted in (*Z*)-selective alkenation.[126] As in the case of nonstabilized ylides, replacing aromatic phosphorus substituents with allylic (**113**; equation 25) or alkyl (**115**; equation 26) groups dramatically increases the production of the (*E*)-alkene.[127]

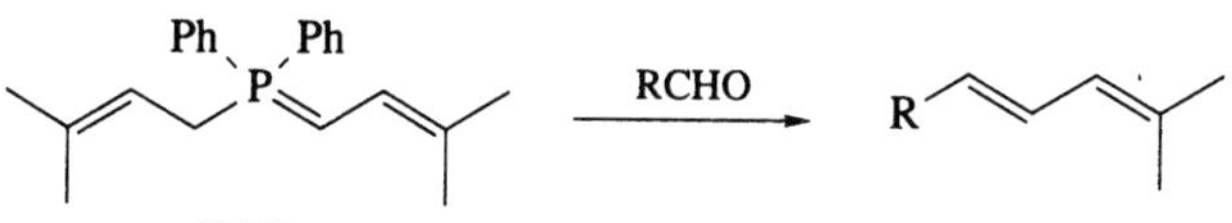

(25)

(**113**) (**114**)

R = Ph, $NaNH_2$ as base; *(E):(Z)* = 15:1, 93%
R = c-C_6H_{11}, $NaNH_2$ as base; *(E):(Z)* = 40:1, 99%

(115) $\xrightarrow{RCHO}$ (116) (26)

R = Ph, $KOBu^t$ as base; *(E):(Z)* = 2:1, 96%
R = c-C_6H_{11}, $NaNH_2$ as base; *(E):(Z)* = 26:1, 73%

In cases were the allylic fragment is inexpensive, the phosphonium salt is formed by reacting lithiated diphenylphosphine with 2 equiv. of the allylic bromide. Alternatively, diphenylmethylphosphonium salts can be selectively deprotonated in the allylic position. The highest ratios were obtained with sodium/ammonia as the base, but $KOBu^t$ in THF provides a practical and selective route to (*E*)-1,3-dienes. Tamura and coworkers have demonstrated that allylic nitro or acetate compounds (**117**) can be converted to allylic tributylphosphonium ylides by palladium(0) catalysis.[128] This procedure may be carried out in one pot by forming the ylide in methanol and THF, followed by addition of $KOBu^t$ and the aldehyde (equation 27). These reactions proceed in good overall yield, and stereoselectivity is high for bulky phosphonium ylides and a range of aldehydes. A comparison of the selectivity of tributyl- and triphenyl-phosphonium ylides has been carried out.[129] Le Corre has demonstrated the use of diphenylphosphinopropanoic acid in the semistabilized Wittig reaction.[130] Enhanced ratios of (*E*)-isomer were observed and because of the increased water solubility of the phosphine oxide produced, work-up is simplified.

(117) $\xrightarrow[\text{ii, KOBu}^t\text{, RCHO}]{\text{i, PBu}_3\text{, Pd(PPh}_3)_4\text{, MeOH, THF}}$ (118) (27)

R = Ph; *(E):(Z)* = 95:1, 83%
R = $Me(CH_2)_5CHO$; *(E):(Z)* = 93:7, 82%

3.1.9.1.4 Stabilized ylides

Ylides which have conjugating functionality present tend to produce (*E*)-alkenes stereoselectively.[114] A significant departure from this anticipated selectivity is the reaction of α-alkoxy substrates with stabilized phosphonium ylides to produce (*Z*)-conjugated esters.[114] In an application of the Wittig reaction to aldehydosugars (**119**), two important factors were noted to obtain high selectivity.[131] The sugars which had polar groups, such as ether substituents with lone pairs, *cis* to the aldehyde in the β-position formed (*Z*)-alkenes (**120**; equation 28). This stereoselectivity is solvent dependent and (*Z*)-alkene formation can be minimized by using DMF and maximized in methanol. Several comparative studies have been carried out with sugar derivatives.[132] It is important that the α-oxgenated functionality be protected as an ether. The reaction of stabilized ylides with α-hydroxy ketones forms the normal (*E*)-alkene.[133] Wilcox and coworkers have studied the reaction of lactols and found that in dichloromethane the lactol (**121**) selectively produced the (*Z*)-isomer (**122**), while the methyl ether (**123**) formed the corresponding *trans*-alkene (**124**) (Scheme 22).[134]

(119) $\xrightarrow{Ph_3PCHCO_2Et}$ (120) (28)

DMF; *(Z):(E)* =14:86
$CHCl_3$; *(Z):(E)* = 60:40
MeOH; *(Z):(E)* = 92:8

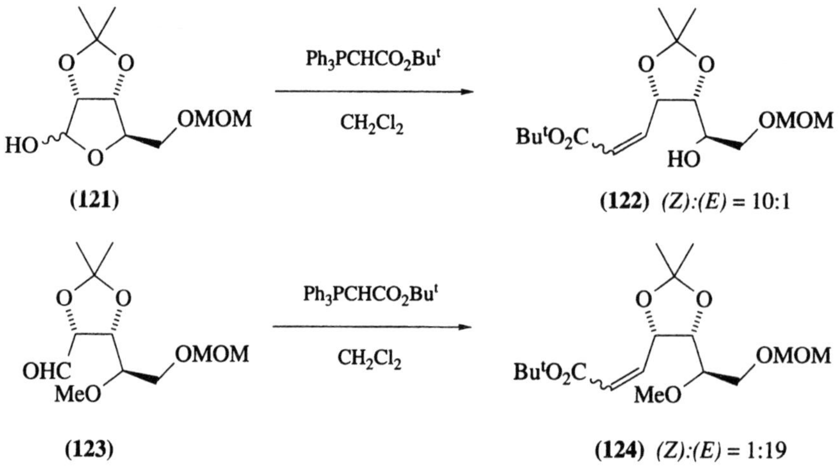

$Ph_3PCHCO_2Bu^t$, CH_2Cl_2

(121) → **(122)** *(Z):(E)* = 10:1

$Ph_3PCHCO_2Bu^t$, CH_2Cl_2

(123) → **(124)** *(Z):(E)* = 1:19

Scheme 22

Although this reaction is general for ester-stabilized Wittig reagents, (formylmethylene)triphenylphosphorane was reacted with a pyranose (**125**) to form the expected (*E*)-alkene (**126**; Scheme 23).[135] Reduction of (**126**) and (**127**) supplied (*E*)- and (*Z*)-allylic alcohols selectively. Precise rules for the prediction of (*E*)- and (*Z*)-selectivity with α-oxygenated substrates are difficult to formulate.[114]

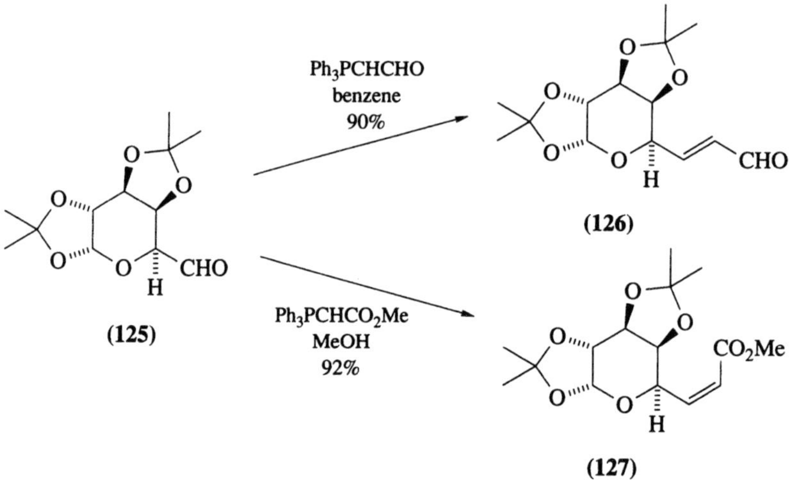

(125) → $Ph_3PCHCHO$, benzene, 90% → **(126)**

(125) → Ph_3PCHCO_2Me, MeOH, 92% → **(127)**

Scheme 23

The synthesis of conjugated (*Z*)-enones (**129**) can be undertaken with a stabilized ylide (**128**) and an aliphatic aldehyde.[136] The reactivity of the ylide is altered by *in situ* deprotonation with 2 equiv. of NaH in THF with a small amount of water (equation 29). Stereoselectivity was approximately 9:1 in favor of the (*Z*)-isomer. This ratio could be improved with aldehydes with α-branched substitution. As discussed

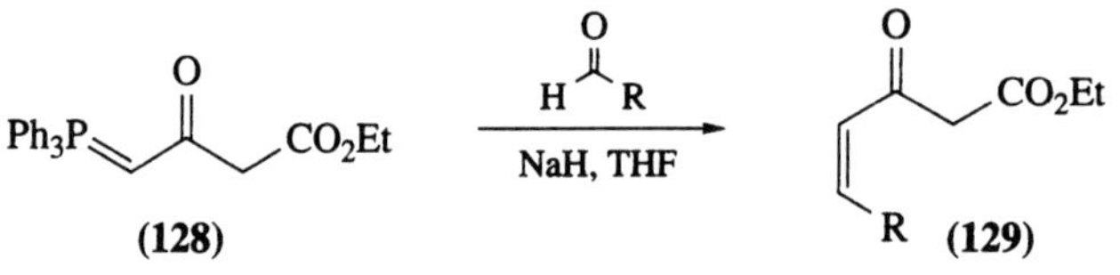

(128) → RCHO, NaH, THF → **(129)** (29)

R = Et; *(Z):(E)* = 85:15, 65%

R = $(CH_2)_9Me$; *(Z):(E)* = 86:14, 91%

R = Pr^i; *(Z):(E)* = 84:16, 90%

with the nonstabilized and semistabilized ylides, substitution of phenylphosphonium ylides with alkyl functionality has been studied on stabilized ylides.[137] Both the ylide and the dianion species demonstrate enhanced reactivity with ketones to produce (*E*)-alkenes (**131**; equation 30). Replacement of the phenyl ligands on the phosphorus by alkyl groups increases the proportion of (*E*)-alkene, even in the absence of stabilizing functionality, by increased reversibility of the *cis*-oxaphosphetane for these substrates.[116]

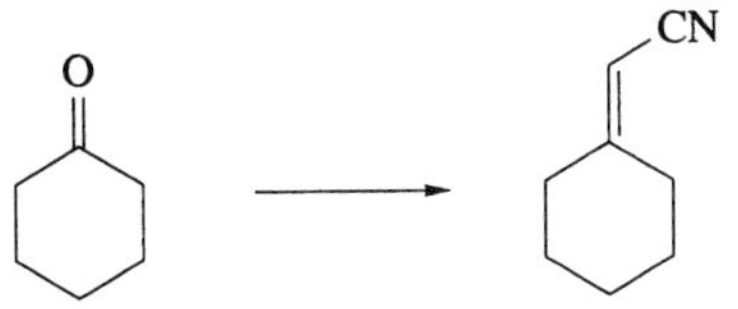

(30)

(130) **(131)**

Ph_3PCH_2CN, Bu^nLi, THF; 6%
$Ph_2P(CH_2CN)_2$, Bu^nLi, THF; 81%

3.1.9.2 Phosphoryl-stabilized Carbanions

The use of anions derived from a phosphine oxide (**132**) or a diethyl phosphonate (**133**) to form alkenes was originally described by Horner.[138] Although these papers laid the foundations for the use of phosphoryl-stabilized carbanions for alkene synthesis, it was not until Wadsworth and Emmons published a more detailed account of the general applicability of the reaction that phosphonates became widely used.[139] Since the work of Wadsworth and Emmons was significant and crucial to the acceptance of this methodology, the reaction of a phosphonate carbanion with a carbonyl derivative to form an alkene is referred to as a 'Horner–Wadsworth–Emmons' reaction (abbreviated HWE).[114] The phosphine oxide variation of the Wittig alkenation is called the 'Horner' reaction.

(132) **(133)**

3.1.9.3 Phosphonates: the Horner–Wadsworth–Emmons Reaction

Phosphonates are the most commonly used phosphoryl-stabilized carbanions.[140] These reagents are more nucleophilic than the corresponding phosphonium ylides. Additional advantages of the HWE reagents are that the by-products of the alkenation are water soluble and reaction conditions can be altered to yield either the (*E*)- or the (*Z*)-isomer. The disadvantage of the phosphonate reagents is that a stabilizing group must be present in the α-position, unless a two-step addition and elimination strategy is employed. The stabilizing functionality is frequently a carboxyl derivative; however, aryl, vinyl, sulfide, amine and ether functionalities have also proven to stabilize the anion sufficiently for alkene formation to take place.

3.1.9.3.1 Mechanism

The mechanism of phosphonate anion (**135**) addition to carbonyl derivatives is similar to the phosphonium ylide addition; however, there are several notable features to these anion additions that distinguish the reactions from those of the classical Wittig. The addition of the anion gives a mixture of the *erythro* (**136** and **137**) and *threo* (**139** and **140**) isomeric β-hydroxyphosphonates (Scheme 24). In the case of phosphine oxides, the initial oxyanion intermediates may be trapped. The anion intermediates decompose by a *syn* elimination of phosphate or phosphinate to give the alkene. The elimination is stereospecific, with the *erythro* isomer producing the *cis*-alkene (**138**), and the *threo* addition adduct producing the *trans*-alkene (**141**). The ratio of (*E*)- to (*Z*)-alkene is dependent on the initial ratio of the *erythro* and *threo* adducts formed, as well as the ability of these intermediates to equilibrate.

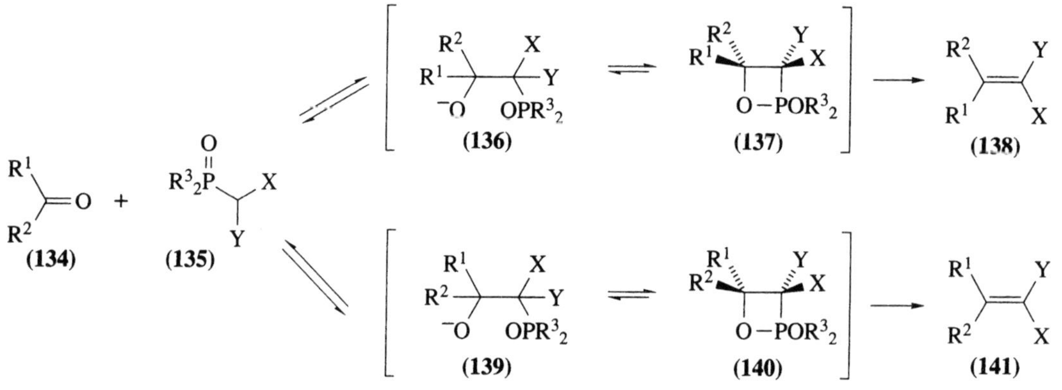

Scheme 24

3.1.9.3.2 *Formation of (*E*)-alkenes*

The most common applications of the HWE reaction are in the synthesis of disubstituted (*E*)-alkenes. The stereoselectivity of the reaction can be maximized by increasing the size of the substituents on the phosphoryl portion. In cases where the stabilizing functionality is a carboxy group, the size of the ester may be 'tuned' to enhance (*E*)-alkene formation.

The HWE reaction can be carried out on a ketone, but often the stereoselectivity is not as good as the reaction of a substituted phosphonate carbanion with the corresponding aldehyde. Because of the greater reactivity of the phosphonate reagent relative to the phosphonium carbanion, the HWE reaction has proven to be effective with hindered ketones that were unreactive toward classical Wittig ylides.

3.1.9.3.3 *(*E*)-Selectivity, effect of phosphonate size*

Several studies have described the successful enhancement of the (*E*)-selectivity in alkene formation by increasing the steric requirements of the phosphonate. This selectivity may be derived by increasing the selective formation of the *threo*-β-hydroxyphosphonate.

Kishi and coworkers noted the importance of the phosphonate structure in the synthesis of monensin, where a methyl phosphonate was used to maximize (*Z*)-alkene formation.[141] These observations were expanded upon in the synthesis of rifamycin S.[142] Several interesting Wittig reactions from this synthesis are presented in Table 14.

Table 14 Rifamycin S Studies by Kishi and Coworkers

Carbonyl	*Reagent*	*(Z):(E) ratio*
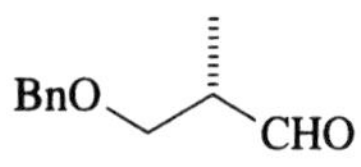 BnO, CHO	$Ph_3PCH_2CO_2Et$, CH_2Cl_2, 0 °C	1:7
	$(Pr^iO)_2POCH_2CO_2Et$, $KOBu^t$, THF, –78 °C	5:95
	$(MeO)_2POCH_2CO_2Me$, $KOBu^t$, THF, –78 °C	3:1
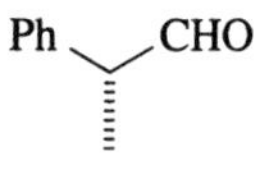 Ph, CHO	$Ph_3PCH(Me)CO_2Et$, CH_2Cl_2, r.t.	5:95
	$Ph_3PCH(Me)CO_2Et$, MeOH, r.t.	15:85
	$(MeO)_2POCH(Me)CO_2Me$, $KOBu^t$, THF, –78 °C	95:5
	$(MeO)_2POCH(Me)CO_2Et$, $KOBu^t$, THF, –78 °C	90:10
	$(EtO)_2POCH(Me)CO_2Et$, $KOBu^t$, THF, –78 °C	60:40
	$(Pr^iO)_2POCH(Me)CO_2Et$, $KOBu^t$, THF, –78 °C	10:90
	$(Pr^iO)_2POCH(Me)CO_2Pr^i$, $KOBu^t$, THF, –78 °C	5:95

In order to maximize the (*E*)-selectivity, it was found best to use the bulky isopropyl substitutent on the phosphonate as well as on the stabilizing ester. Alternatively, in the case of an aromatic aldehyde, the stabilized phosphonium ylide provided high (*E*)-alkene formation. When the (*Z*)-isomer was desired, it was best to use a methyl substituent on the phosphonate and ester functionalities. Strongly dissociating basic conditions and a hindered aldehyde can result in reversal of the normal (*E*)-selectivity of a phosphonate. This effect is strongly dependent on the reactants and is further discussed in Section 3.1.9.3.4. The importance of steric effects were exemplified in the synthesis of the C–D fragment of amphotericin B by Masmune and coworkers.[143] The methyl and ethyl phosphonates, respectively, formed 1:1.2 and 1.75:1 ratios of (*E*)- to (*Z*)-alkenes. By using isopropyl or 3-pentyl phosphonates, the (*E*)-isomer (**144**) was formed exclusively. The yield of the reaction was improved by using lithium tetramethylpiperidide (**143**) in THF (equation 31). The diisopropyl phosphonate was recently applied to the synthesis of brefeldin C.[144] In this example, the (*E*)-alkene (**146**) was selectively synthesised without epimerization (equation 32).

Et_3SiO $OSiMe_2Bu^t$ CHO — $(RO)_2P(O)(CH_2)_5CO_2Et$, (**143**) → Et_3SiO $OSiMe_2Bu^t$ $(CH=CH)(CH_2)_6CO_2Et$ (31)

(**142**) R = Pr^i, $CHEt_2$ Li (**143**) (**144**) 68%

OMPM, H, CHO, $OSiMe_2Bu^t$, H — $(Pr^iO)_2P(O)CH_2C(O)OMe$, $KOBu^t$, THF, 72% → OMPM, H, CO_2Me $OSiMe_2Bu^t$, H (32)

(**145**) (**146**)

3.1.9.3.4 *Formation of (Z)-alkenes*

As noted in the discussion of (*E*)-selective alkene formation, Kishi has found that α-substituted aldehydes reacted with trimethylphosphonopropionate and $KOBu^t$ to produce the (*Z*)-alkene selectively. A strongly dissociating base is critical to this approach. In addition to the examples already presented in the discussion of (*E*)-alkene formation, the (*Z*)-selective reaction has recently been applied to the synthesis of macrolide antibiotics.[145] In this example, a trisubstituted alkene was formed and closed to the lactone (**148**; equation 33). In an application to diterpenoids, Piers encountered an example of how substrate-specific the alkene formation can be.[146] With α-dimethoxyphosphonyl-γ-butyrolactone (**150**), the reactions with simple aldehydes proceeded with very high selectivity [(*Z*):(*E*) = 99:1]. On application of the reaction to the more complex aldehyde (**149**) the (*Z*):(*E*) stereoselectivity dropped to 3:1 in 58% yield (equation 34). No selectivity was observed on reaction with benzaldehyde. Although for hindered substrates, strongly basic conditions with a dimethyl phosphonate can be a simple and effective method for the synthesis of (*Z*)-isomers, the reaction is not general. In 1983, Still and coworkers introduced methodology that used bis(trifluoroethyl)phosphonoesters (**153**) to provide a facile approach to (*Z*)-alkenes (**154**) when reacted with aldehydes (equation 35).[147]

MPMO, HC(O)O, CHO, OBn — i, $(MeO)_2P(O)CH(CH_3)CO_2Me$, NaH, THF; ii, K_2CO_3, MeOH, 68% → MPMO, lactone, OBn (33)

(**147**) (**148**)

The (*Z*)-selectivity is presumed to occur because of rapid elimination of the β-hydroxyphosphonate before equilibration can take place. Several base and solvent combinations were explored, including triton B, K_2CO_3 and $KOBu^t$, but potassium hexamethyldisilazide with 18-crown-6 in THF gave the most

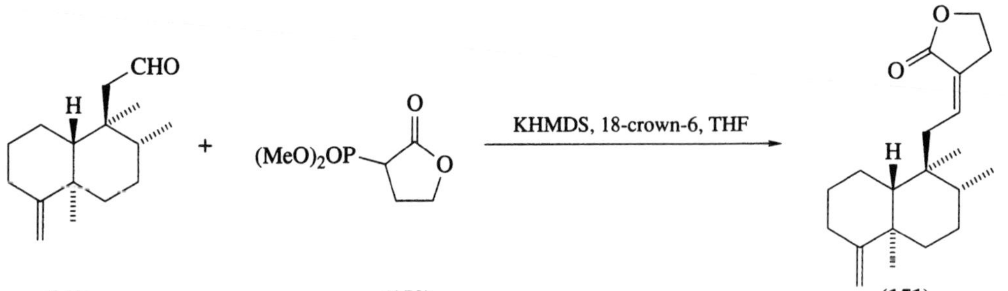

(149) + (150) $\xrightarrow{\text{KHMDS, 18-crown-6, THF}}$ (151) (34)

(150) + PhCHO; *(Z):(E)* = 1:1, 91%
(150) + Bu^iCHO; *(Z):(E)* = 99:1, 86%
(150) + n-$C_6H_{13}CHO$; *(Z):(E)* = 99:1, 94%

(Z):(E) = 3:1, 58%

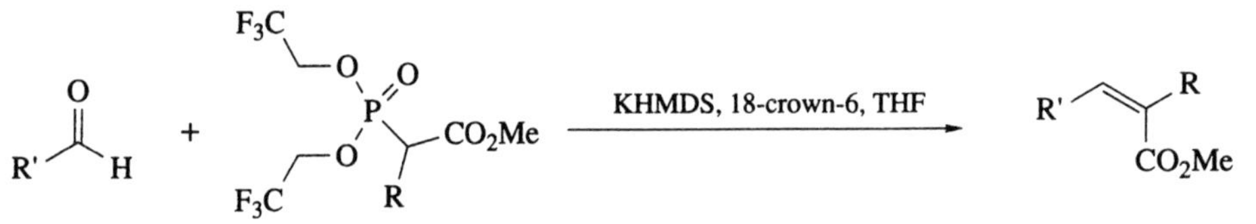

(152) + (153) R = H, Me $\xrightarrow{\text{KHMDS, 18-crown-6, THF}}$ (154) (35)

consistent and highest (*Z*)-selectivities. A comparison was carried out with trimethyl phosphonoacetate under the same conditions and (*Z*)-alkene formation occurred to a high degree [(*Z*):(*E*) = 12:1] in the reaction with cyclohexanal, which was consistent with the observations of Kishi and coworkers.[147] This was the only instance in which the trimethyl phosphonate formed the (*Z*)-alkene with higher selectivity than the trifluoroethyl analog. The diminished selectivity for α-alkyl-substituted aldehydes does not appear to be general, as two recent syntheses by Baker and Roush demonstrate. Baker synthesized macbecin 1 using a Still phosphonate to form the (*Z*)-alkene (**156**) and a standard stabilized ylide to form the (*E*)-linkage (**157**; Scheme 25).[148] Roush used the trifluoroethyl phosphonate in a synthesis of the C(1)–C(15) segment of streptovaricin, and in this case (**159**) the (*Z*):(*E*) selectivity was at least 10:1 (equation 36).[149]

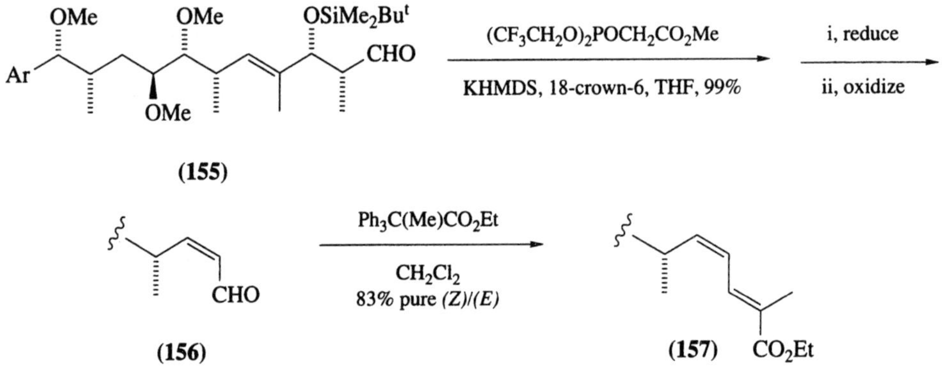

(155) $\xrightarrow[\text{KHMDS, 18-crown-6, THF, 99\%}]{(CF_3CH_2O)_2POCH_2CO_2Me}$ $\xrightarrow[\text{ii, oxidize}]{\text{i, reduce}}$ (156) $\xrightarrow[\text{CH}_2\text{Cl}_2\text{, 83\% pure (Z)/(E)}]{Ph_3C(Me)CO_2Et}$ (157)

Scheme 25

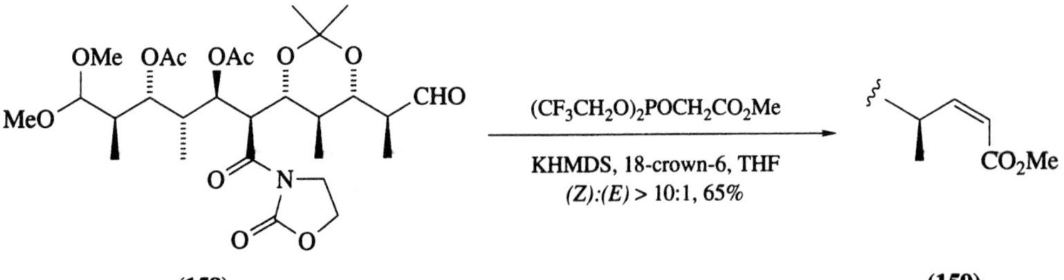

(158) $\xrightarrow[\text{KHMDS, 18-crown-6, THF; (Z):(E) > 10:1, 65\%}]{(CF_3CH_2O)_2POCH_2CO_2Me}$ (159) (36)

An aldehyde with α-ether functionality was used in the synthesis of *N*-acetylneuraminic acid by Danishefsky (equation 37).[150] A series of *cis*- and *trans*-alkenes with α-oxygen functionality were synthesized by Cinquinin and coworkers.[151] Of the examples generated, the HWE reaction proceeded in the expected (*E*)-selective manner (**164**), while the trifluoroethyl phosphonate was used to form the (*Z*)-alkenes (**163**) selectively (Scheme 26). Interestingly, 18-crown-6 was not used for (*Z*)-alkene formation.

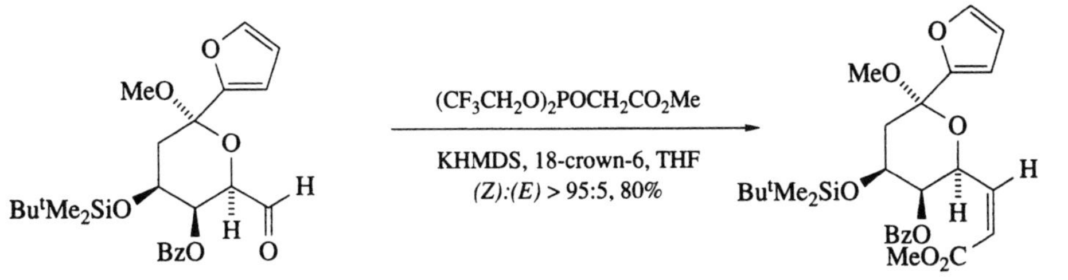

MeO, Bu^tMe_2SiO, BzO, H, O, H

$(CF_3CH_2O)_2POCH_2CO_2Me$

KHMDS, 18-crown-6, THF

(*Z*):(*E*) > 95:5, 80%

MeO, Bu^tMe_2SiO, BzO, H, H, MeO_2C

(37)

(160) **(161)**

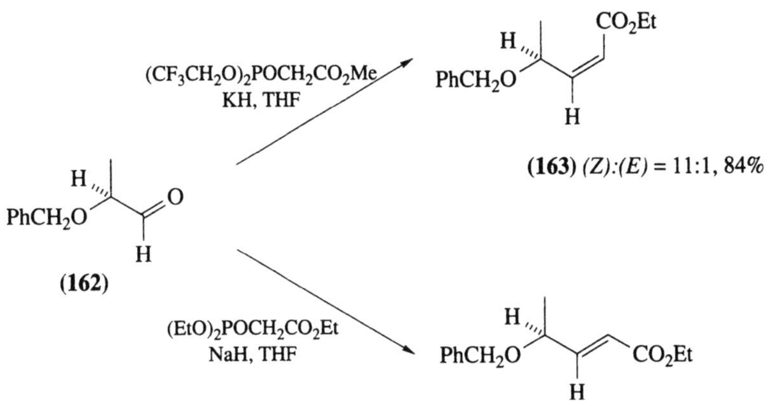

$(CF_3CH_2O)_2POCH_2CO_2Me$

KH, THF

H, $PhCH_2O$, CO_2Et, H

(163) (*Z*):(*E*) = 11:1, 84%

H, $PhCH_2O$, O, H

(162)

$(EtO)_2POCH_2CO_2Et$

NaH, THF

H, $PhCH_2O$, CO_2Et, H

(164) (*E*):(*Z*) = 12:1, 83%

Scheme 26

Several examples are presented in Table 15 which indicate that α-heteroatom substitution is compatible with this reagent. The tremendous advantage of the trifluoroethyl phosphonate reagent is the selective formation of (*Z*)-alkene with aromatic aldehydes, while the trimethyl phosphonate gives the normal selectivity.

The trifluoroethyl phosphonate reagent is effective with a number of carbanion-stabilizing groups, including carboalkoxy, cyano and vinylogous cyano. Liu and coworkers have utilized a conjugated nitrile phosphonate (**166**) to synthesize all-*cis*-retinal (**167**; equation 38).[165] In this example, the Wittig reaction gives a mixture of products that were separated by HPLC. The crude yield was 45%, with 72% of the product having the *cis* configuration at the alkene formed. The type of ester stabilizing the phosphonate does not appear to have an large effect on the selectivity. Both the methyl and ethyl esters are routinely used. Recently, Boeckman used the allyl ester in the synthesis of (+)-ikarugamicin to produce the desired (*Z*)-alkene (**169**), the (*Z*):(*E*) ratio being 19:1 (equation 39).[166]

Still's original communication demonstrated that this approach was effective for the synthesis of trisubstituted alkenes by the reaction of a methyl-substituted Wittig reagent with an aldehyde. Examples of the synthesis of methyl-trisubstituted alkenes are presented in Table 16.

Fuchs has published a comparative study of the methyl, ethyl, isopropyl and trifluoroethyl phosphonates, stabilized with either an ester or nitrile, in reaction with an α-amino aldehyde (**170**; equation 40).[174] Although the use of strongly basic conditions improved the ratio of (*Z*)-alkene produced with methyl phosphonate, the (*Z*):(*E*) ratios were 2:3 with either potassium or sodium anions. Interestingly, the HWE reagent stabilized with a nitrile demonstrated far less sensitivity to the size of the phosphonate functionality. With sodium hydride as the base, the diisopropyl phosphonate gave higher (*Z*)-selectivity (7:1) than that of the corresponding ethyl derivative (3.3:1). Fuchs hypothesized that the sterically less demanding nitrile produces a higher ratio of the *erythro* intermediate than the ester-stabilized phosphonate. The trisubstituted alkene was best prepared by using the trifluoroethyl phosphonate, (*Z*):(*E*) = 70:1,

Table 15 Synthesis of Disubstituted Alkenes by Trifluoroethyl Phosphonate Esters in THF with KHMDS and 18-Crown-6[a]

Carbonyl	*Yield (%)*	*(Z):(E) ratio*	*Notes*	*Ref.*
MeO; NHBOC; CHO	>80	All *(Z)*		152
CHO	78	7:1		153
$OSiEt_3$; CHO	70	All *(Z)*	In Et_2O	143
$OSiMe_2Bu^t$; CHO	56	All *(Z)*	*(E)* with Ph_3PCHCO_2Me, $CHCl_3$, 90%, *(Z):(E)* = 1:12	154
$SiMe_3$; CHO	>61	9.4:1		155
I; O; O; O; OHC	75	All *(Z)*	*(E)* with Ph_3PCHCO_2Et, toluene, 95%	156
$CH(OMe)_2$; CHO	94	5:1		157
OTHP; CHO	75	15:1	No 18-Crown-6, 1:1	158
CHO	75	7:1		159
O; O; CHO; $n-C_{15}H_{31}$	77	9% *(E)*		160
OHC; BOC - N; O	>87	*(Z)* >65%		161
O; O; CHO	89	5:1	In Et_2O with NaH	162

Table 15 *(continued)*

Carbonyl	*Yield (%)*	*(Z):(E) ratio*	*Notes*	*Ref.*
Me_3Si, CHO	>62	25:1		163
CHO, BnO, O	80			164

[a] Exceptions to these reaction conditions are noted.

(165) CHO → $(CF_3CH_2O)_2P(O)$, CN (166); KHMDS, 18-crown-6, THF → (167) CN (38)

(168) CHO → $(CF_3CH_2O)_2POCH_2CO_2CH{=}CH_2$; K_2CO_3, 18-crown-6, toluene; *(Z):(E)* = 19:1, >78% → (169) (39)

even in the absence of a crown ether. Marshall and coworkers investigated the extension of the trifluoroethyl phosphonate to trisubstituted alkenes with more complex substitution patterns.[175] The reaction of aldehydes lacking α-substitution with methyl α-(dimethylphosphono)propionate and KHMDS/18-crown-6 gave an increased percentage of (*Z*)-isomer, but still gave predominantly the (*E*)-alkene. In examining Kishi's conditions, Marshall found it critical to have a large excess of phosphonate for optimum (*Z*)-selectivity. Alkyl substituents larger than methyl appear to slow the rate of phosphonate oxide elimination. This in turn increases the amount of equilibration that occurs and compromises the (*Z*)-selectivity. The best results observed were for the trifluoroethyl phosphonate (**172**), which formed the (*Z*)-alkene (**173**) on reaction with nonanal in 94% yield and (*Z*):(*E*) = 87:13 (equation 41).

Several other examples of highly substituted applications of the Still phosphonate have appeared. In work very similar to Marshall's, a substituted methyl phosphonate directed toward furancembraolides demonstrated a (*Z*):(*E*) selectivity of 4:5, while the trifluoroethyl phosphonate gave 4:1.[176] Recently an application to squalinoids was demonstrated with a (*Z*)-selectivity similar to Marshall's examples [(*Z*):(*E*) = 6.5:1].[177] Marshall has also investigated an intramolecular ring closure for formation cembraolides and in this application the Still phosphonate did not have any inherent advantage over an alkyl phosphonate in terms of yield or (*E*)-selectivity.[178] Recently, two other examples of substituted phosphonate applications have appeared. Oppolzer has used the methodology to generate an allylic acetate (**175**) with excellent yield and high (*Z*)-selectivity (equation 42).[179]

In an application of (*Z*)-selective alkene formation to enolizable aldehydes, it was noted that the combination of LiCl and DBU was effective for deprotonation by lithium complexation of the Still phosphonate.[180] In this example, the cyclopropyl aldehyde (**176**) reacted chemoselectively in the presence of the ketone (equation 43). In addition, the (*E*)-alkene could be synthesized by lithium coordination with a standard HWE methyl phosphonate. As this example illustrates, the trifluoroethyl phosphonate can fill an important void by providing trisubstituted alkenes with sensitive substrates in good selectivity. From the examples of Marshall and Oppolzer it appears that the application of the reaction to higher order trisubstituted alkenes is selective for the (*Z*)-isomer. The magnitude of the selectivity is substrate specific and dependent on the rapid rate of *erythro*-α-oxyphosphonate decomposition.

Table 16 Synthesis of Trisubstituted Alkenes by $(CF_3CH_2O)_2POCH(Me)CO_2R$ in THF with KHMDS and 18-Crown-6[a]

Carbonyl	*Yield (%)*	*(Z):(E) ratio*	*Notes*	*Ref.*
 BnO CHO	60	58% *(Z)*		167
 H, CHO, NHBOC	68	99:1		168
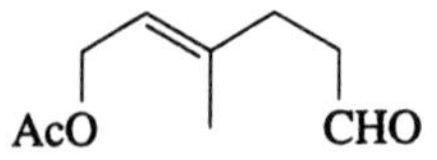 AcO, CHO	71	>99% *(Z)*		169
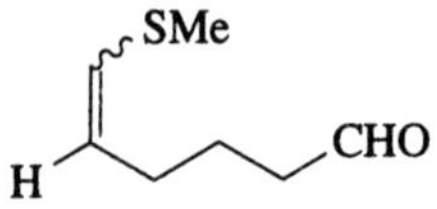 SMe, H, CHO	93	98:2	*(E)* with $Ph_3PC(Me)CO_2Me$, 86%, *(E)* only	170
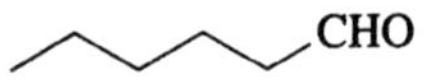 CHO	63	91:9	*In situ* DIBAL and Wittig *(E)* with diethyl phosphonate, 77%, 18:82 *(Z)*:*(E)*	171
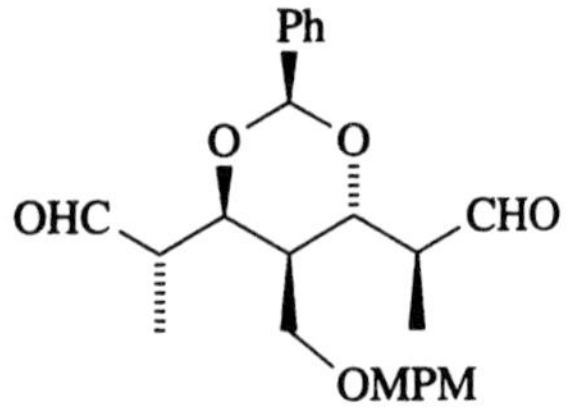 Ph, O, O, OHC, CHO, OMPM	78 (4 steps)			172
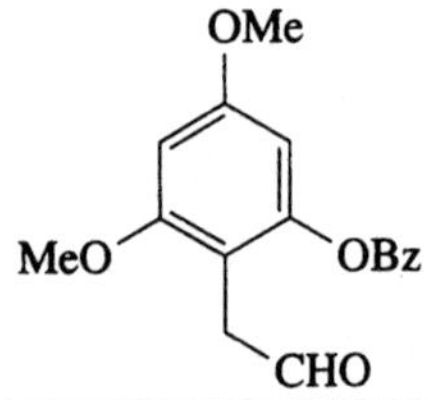 OMe, MeO, OBz, CHO	77	All *(Z)*		173

[a] Exceptions to these conditions are noted.

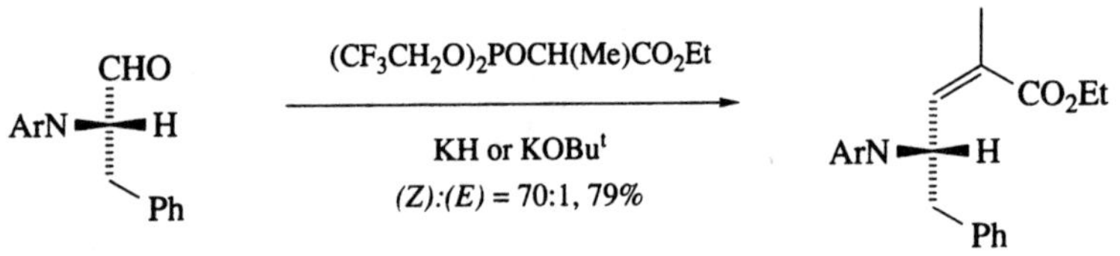

CHO, ArN, H, Ph → $(CF_3CH_2O)_2POCH(Me)CO_2Et$; KH or $KOBu^t$; *(Z)*:*(E)* = 70:1, 79% → CO_2Et, ArN, H, Ph (40)

(170) Ar = Ts, Bn → **(171)**

CO_2Me, $)_9$, $P(OCH_2CF_3)_2$, O **(172)** → KHMDS, 18-crown-6; O, H, n-C_8H_{17}; *(Z)*:*(E)* = 87:13, 94% → CO_2Me, $)_9$, n-C_8H_{17} **(173)** (41)

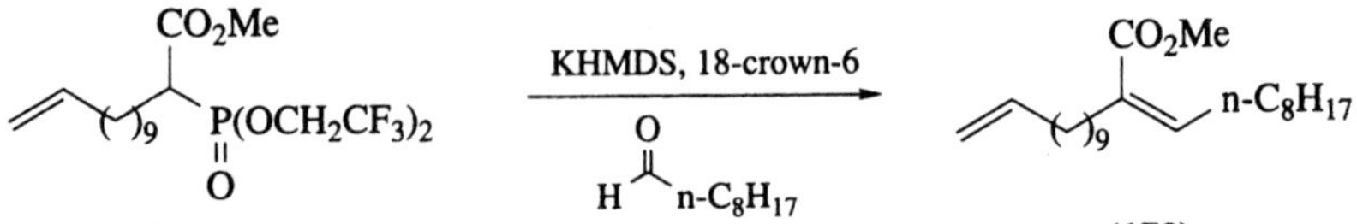

$(CF_3CH_2O)_2OP$ CO_2Me **(174)** —i–iii, (Z):(E) = 94:6, 86%→ OAc, Ph **(175)** (42)

i, KHMDS, 18-crown-6, THF, $Ph(CH_2)_2CHO$; ii, DIBAL-H; iii, Ac_2O, Et_3N

CHO, O **(176)** + MeO_2C, $(RO)_2OP$ **(177)** → MeO_2C **(178)** (43)

R = Me; LiHMDS, DME; (Z):(E) = 14:86, 100%
R = CF_3CH_2; DBU, LiCl, MeCN; (Z):(E) = 75:25, 100%

3.1.9.3.5 Effect of bases

As noted in the work of Seyden-Penne, and demonstrated in numerous examples such as those by Still and Kishi, higher (*Z*)-selectivity is observed by using base systems that have minimally complexing counterions in order to increase the rate of elimination relative to equilibration.[181] Coordination has been used to advantage by Massamune and Roush, who found that the addition of lithium chloride to a phosphonate formed a tight complex (**180**; equation 44) that could be deprotonated with a weak base such as DBU or diisopropylethylamine.[182] This method is a mild and extremely effective modification for substrates in which racemization or β-elimination are processes competitive with alkene formation. Rathke and coworkers have demonstrated that TEA and LiBr or $MgBr_2$ may also be used to form a reactive HWE reagent.[183] Acetonitrile and THF are the most frequently used solvents. These conditions do not alter the normal course of stereoselectivity for alkene formation. As noted above, the method was utilized with the Still phosphonate to produce (*Z*)-alkenes. There is some indication that this technique may enhance inherent (*E*)-stereoselectivity in alkenations with a standard HWE reagent.[182] Other stabilizing functionalities are compatible with this method. As outlined in Table 17, ester, ketone, sulfone, amide and allylic ketone groups have been demonstrated to be effective as stabilizing functionalities.

$(EtO)_2P(O)CH_2C(O)OEt$ **(179)** —LiCl→ Li, O, O, $(EtO)_2P$, OEt **(180)** (44)

This reaction has been utilized in the context of natural product synthesis. A recent example is the synthesis of colletodiol by Keck, shown in equation (45).[194] In this example, no problems were encountered with epimerization or ester cleavage. The desired (*E*)-ester (**182**) was synthesized in 80% yield. Two examples are outlined (in equations 46 and 47) in which epimerization was a substantial problem with sodium or potassium salts, while the LiCl/amine method effectively suppressed this side reaction. When the phosphonate was allowed to react with the cyclohexanal derivative (**183**), the sodium salt gave epimerized material. Use of LiCl and diisopropylethylamine gave an 88% yield of alkene (**184**), free of epimer (equation 46).[195] In the synthesis of norsecurinine, Heathcock found that the phosphonate anion

Table 17 Synthesis of Alkenes Utilizing LiX and Amine Base

Carbonyl	*Phosphonate*	*Comments*	*Ref.*
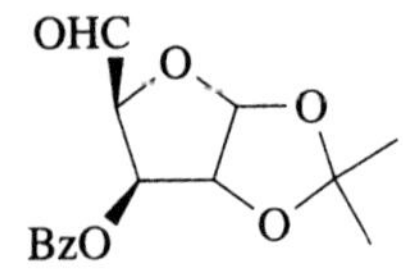 OHC, O, O, O, BzO	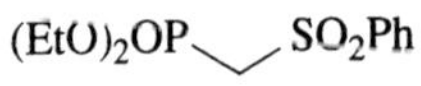 $(EtO)_2OP$ SO_2Ph	LiCl, $Pr^i{}_2NEt$, 63%, *(E)*	184
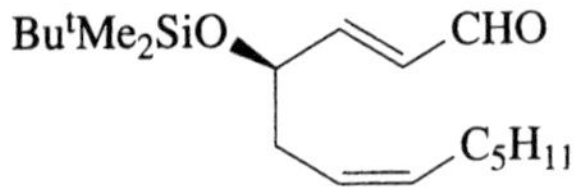 Bu^tMe_2SiO CHO C_5H_{11}	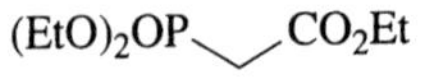 $(EtO)_2OP$ CO_2Et	LiCl, DBU, 60%, *(E)*	185
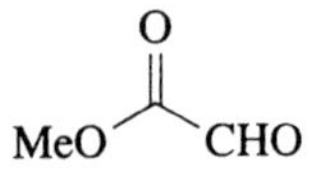 O, MeO CHO	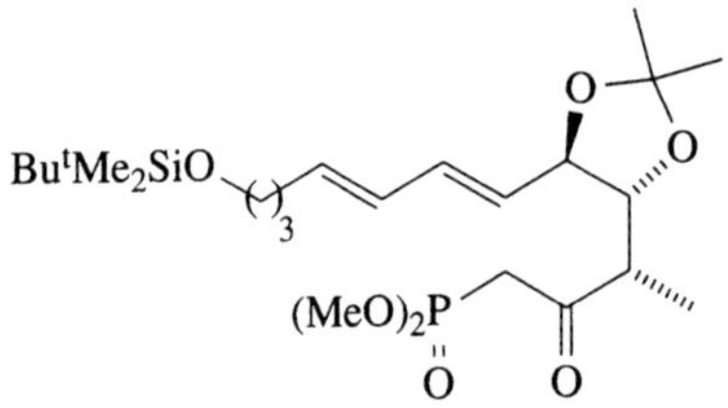 Bu^tMe_2SiO 3, O, O, $(MeO)_2P$, O, O	LiCl, $Pr^i{}_2NEt$, 99%, *(E)*	186
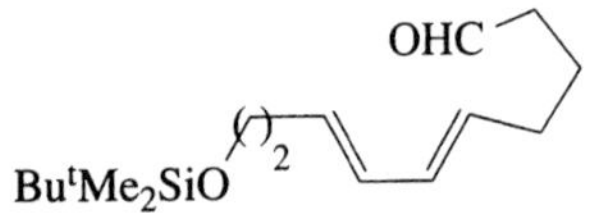 OHC, 2, Bu^tMe_2SiO	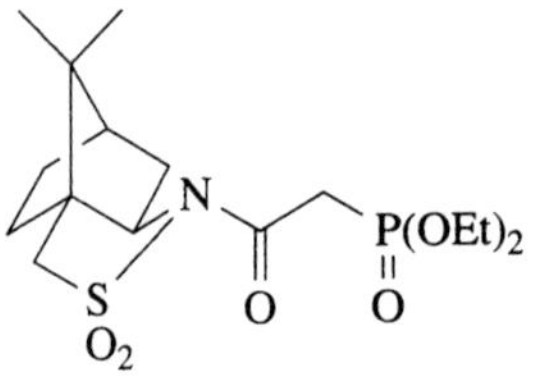 N, S, O_2, O, $P(OEt)_2$, O	LiCl, DBU, 89%, *(E)*	187
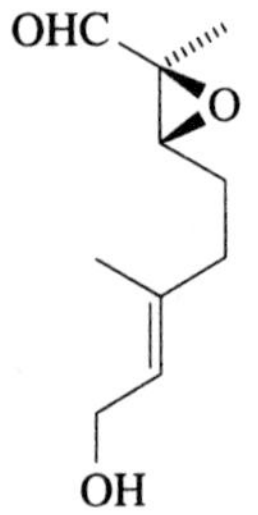 OHC, O, OH	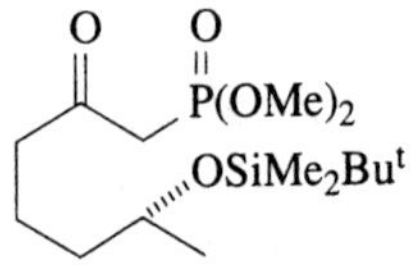 O, O, $P(OMe)_2$, $OSiMe_2Bu^t$	LiCl, $Pr^i{}_2NEt$, 88%, *(E)*	188
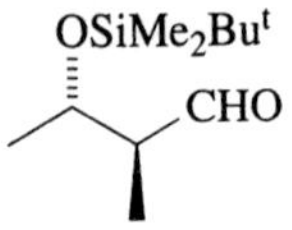 $OSiMe_2Bu^t$, CHO	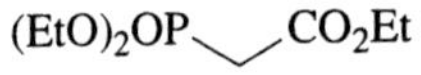 $(EtO)_2OP$ CO_2Et	LiCl, DBU, 90%, *(E)*	189
PhCHO	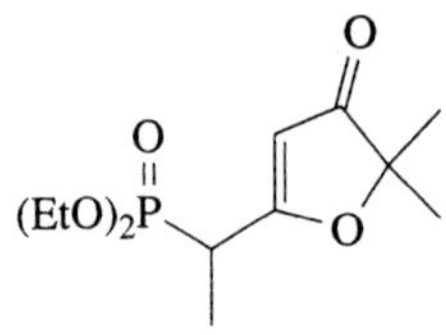 O, $(EtO)_2P$, O, O	LiBr, Et_3N, 85%, 83:7 *(E)*	190
Pr^iCHO	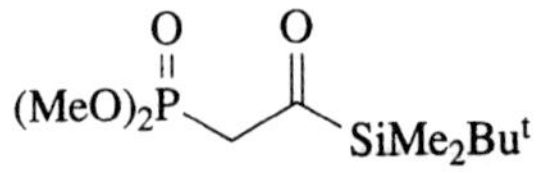 O, O, $(MeO)_2P$ $SiMe_2Bu^t$	LiCl, DBU, 87%, >95% *(E)*	191

Table 17 *(continued)*

Carbonyl	*Phosphonate*	*Comments*	*Ref.*
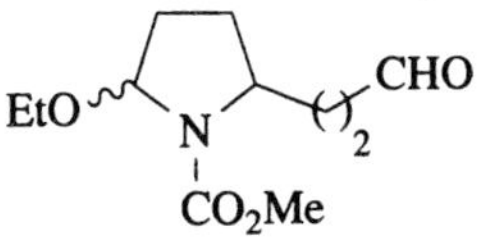	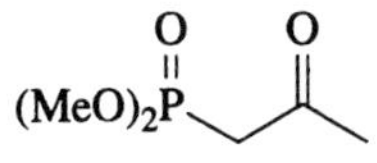	LiCl, Pr^i_2NEt, 83% *(E)*	192
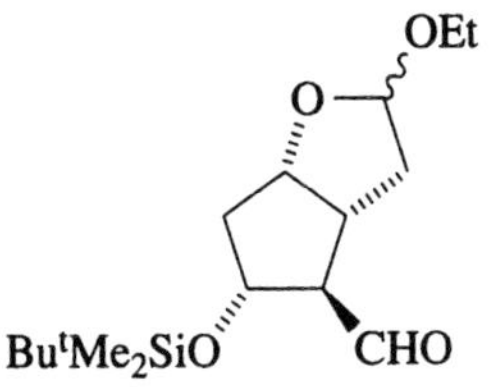	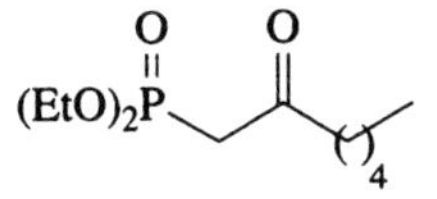	LiCl, Pr^i_2NEt, 93% *(E)*	193

generated with $KOBu^t$ gave an excellent (96%) yield of the desired alkene (**186**), but the material was racemic.[196] Use of DBU and LiCl produced (**186**) in 84% yield and 93% *ee* (equation 47).

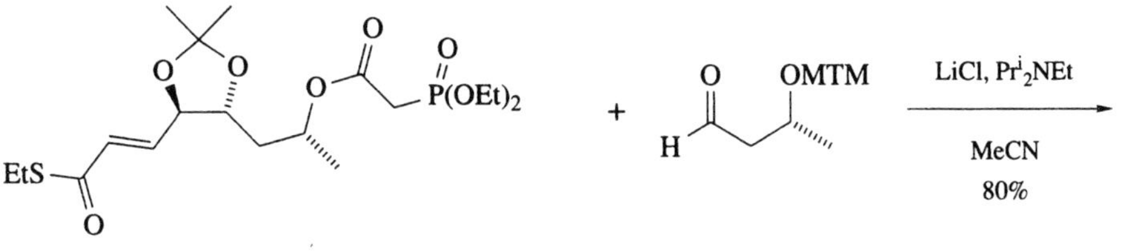

LiCl, Pr^i_2NEt, MeCN, 80%

(**181**) 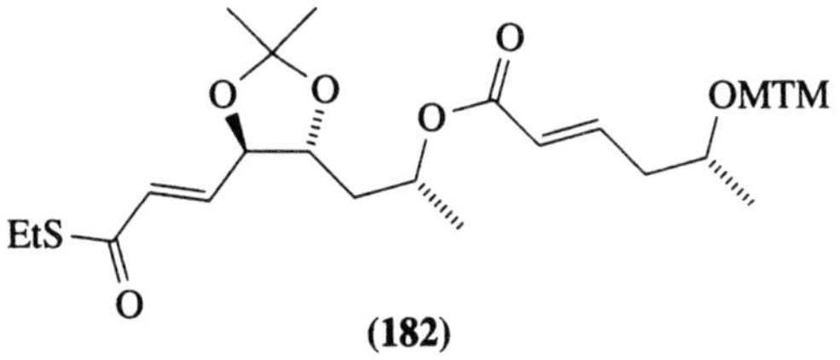 → (**182**) (45)

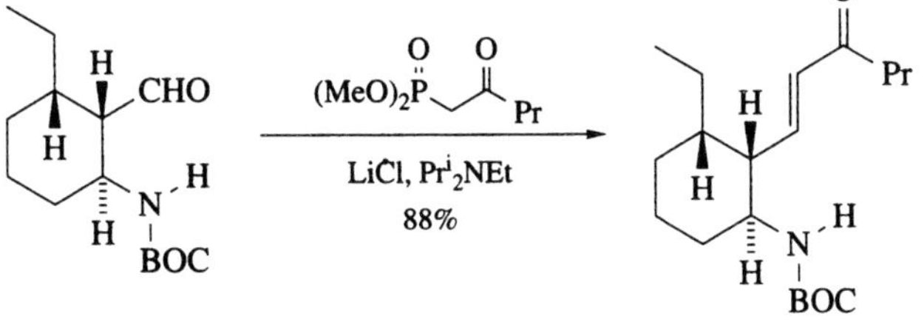

$(MeO)_2P(O)CH_2COPr$, LiCl, Pr^i_2NEt, 88%

(**183**) → (**184**) (46)

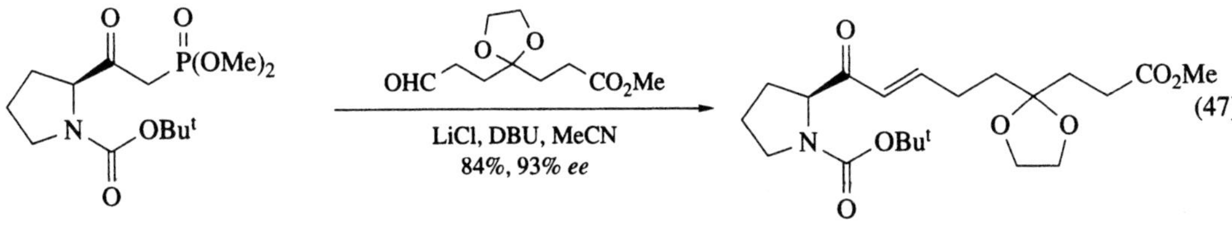

OHC–(dioxolane)–CO_2Me, LiCl, DBU, MeCN, 84%, 93% *ee*

(**185**) → (**186**) (47)

As with all reactions with phosphonates, these conditions are sensitive to the steric environment of the carbonyl and phosphonate. In a reaction directed at intermediates for synthesis of the erythronolides, Paterson and coworkers found that the unsubstituted phosphonate (**188**) added in 82% yield in the presence of molecular sieves (equation 48).[197] When R was methyl the reaction failed with DBU because of competitive elimination of $OSiMe_2Bu^t$. Model studies with the phosphonate and isopropyl aldehyde were successful, providing the alkene in 78% yield in an (*E*):(*Z*) ratio of 8:1. The Roush–Masamune modification of the HWE reaction was utilized by Heathcock and coworkers in the synthesis of mevinic acids.[198] The coupling was very clean and resulted in 35–60% yields of alkene (**192**), along with 35–50% of recovered aldehyde (**191**; equation 49). This methodology is also effective for macrolide synthesis by intramolecular ring closure. An example of the utility of this approach is the synthesis of amphotericin B by Nicolaou.[199] In this example, the macrolide could be formed from (**193**) either with potassium carbonate and 18-crown-6, or with LiCl and DBU at 0.01 M in 70% yield to form the (*E*)-alkene. Other applications of intramolecular cyclization have appeared for the cembranolides[200] and rubradirins.[201]

$OSiMe_2Bu^t$, CHO, OH (**187**) + R, $PO(OMe)_2$, O, PhS, $OSiMe_2Bu^t$ (**188**) —LiCl, $Pr^i{}_2NEt$, MeCN; 4 Å sieves; R = H, (*E*) only, 82%→ $OSiMe_2Bu^t$, OH, O, PhS, $OSiMe_2Bu^t$ (**189**) (48)

Bu^tMe_2SiO, O, MeO_2C, $PO(OMe)_2$ (**190**) + O, O, H, CHO (**191**) —LiCl; DBU, MeCN→ MeO_2C, $OSiMe_2Bu^t$, O, O, O, H (**192**) 35–60% (49)

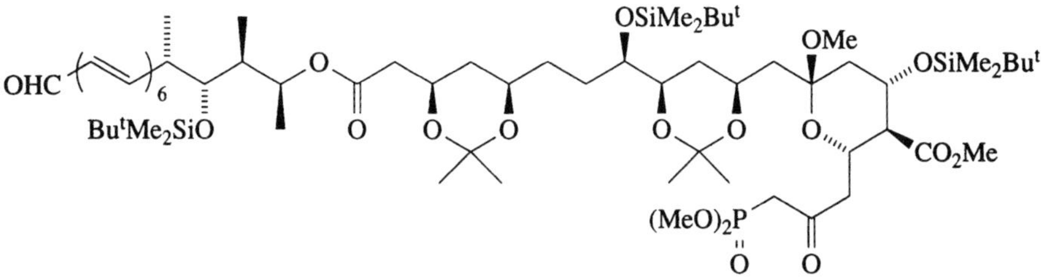

(193)

3.1.9.3.6 Asymmetric Horner–Wadsworth–Emmons

Hanessian and coworkers have prepared a homochiral bicyclic phosphonamide (**194**) that reacted with cyclohexanone derivatives to form (*E*)-ethylenes with good optical purity.[202] The phosphonamide (**194**) was deprotonated with KDA in THF and reacted with 4-*t*-butylcyclohexanone (**195**) to give a 82% yield of 90% optically pure alkene (**196**; Scheme 27). Either the (*R*,*R*)- or the (*S*,*S*)-phosphonamide (**194**) can be synthesized, and on reaction with (+)-3-methylcyclohexanone (**197**) it is possible to obtain (*E*),(3*R*)-alkene (**198**) with (*R*,*R*)-reagent and (*Z*),(3*R*)-alkene with the (*S*,*S*)-phosphonamide. Gais and Rehwinkel have demonstrated the use of a chiral phosphonate in which the stabilizing carbonyl is an 8-phenylmenthyl ester.[203] This methology was applied to the synthesis of carbacylins (**200** and **201**; equation 50). Whereas a methyl phosphonate gives a 1:1 mixture of (*E*)- and (*Z*)-alkenes, the presence of a chiral ester gives good selectivity for either the (*E*)- or (*Z*)-alkene, depending on the enantiomer of the phenylmenthyl chosen. This reaction is temperature sensitive and, if heated, the (*E*)-alkene predominates.[203b] The choice of solvent is not as critical a factor as temperature, and better ratios were observed with potassium as the counterion. Other esters were examined, including (+)-menthyl and (–)-*trans*-2-phenylcyclohexyl, but the phenylmenthyl ester proceeded with higher selectivity.[203a] This reaction has been applied to achiral ketones to produce diastereomeric esters.[203]

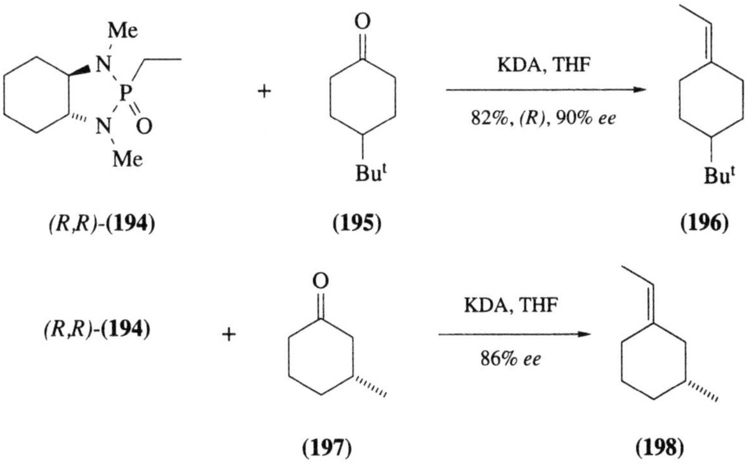

(*R*,*R*)-(**194**) + (**195**) → KDA, THF; 82%, (*R*), 90% *ee* → (**196**)

(*R*,*R*)-(**194**) + (**197**) → KDA, THF; 86% *ee* → (**198**)

Scheme 27

3.1.9.4 Phosphine Oxides: the Horner Reaction

In Horner's original work, phosphine oxides (**202**) were treated with potassium *t*-butoxide or sodamide and allowed to react with an aldehyde or ketone to form the alkene (**203**) directly (Scheme 28). Horner observed that the use of a lithium anion resulted in the isolation of the β-hydroxyphosphine oxide (**204**).[204] In addition, he found that the intermediate hydroxyphosphine oxide could be obtained by LAH reduction of the ketophosphine oxide. Warren and coworkers have utilized and expanded upon these techniques by isolating and separating the diastereomeric, frequently crystalline, β-hydroxyphosphine

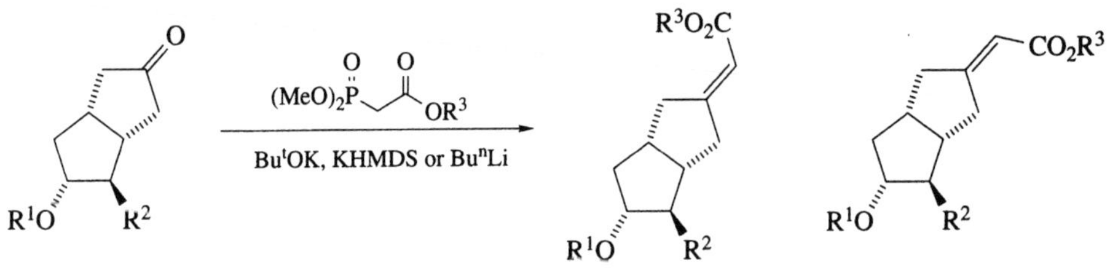

(50)

(199) (E)-(200) (Z)-(201)

$R^1 = SiMe_2Bu^t$, $R^2 = OSiMe_2Bu^t$, R^3 = (+)-8-phenylmenthyl; (E):(Z) = 86:14, 95%

$R^1 = SiMe_2Bu^t$, $R^2 = OSiMe_2Bu^t$, R^3 = (–)-8-phenylmenthyl; (E):(Z) = 23:77, 89%

R^1 = H, R^2 = 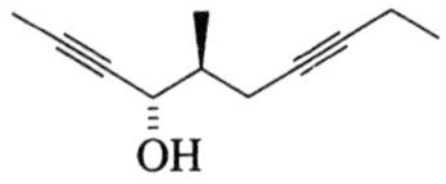 , R^3 = (+)-8-phenylmenthyl; (E):(Z) = 86:14, 92%

R^1 = H, R^2 = 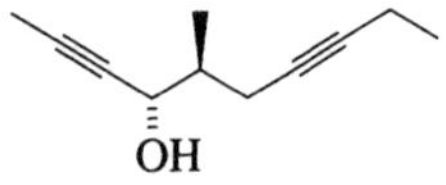 , R^3 = (+)-8-phenylmenthyl; (E):(Z) = 15:85, 95%

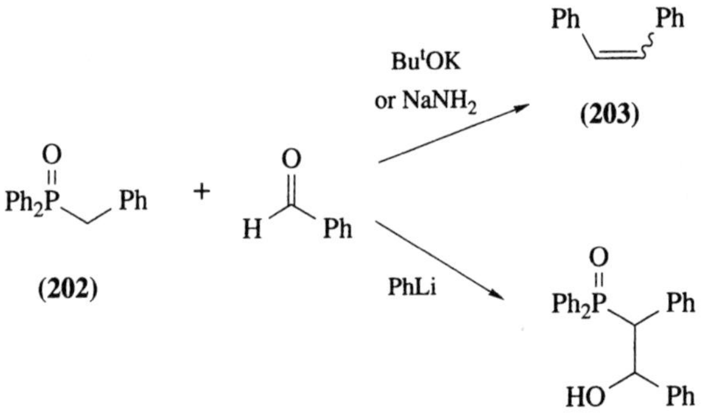

(202) (203) (204)

Scheme 28

oxides (**206**) and (**207**).[205] The *erythro* and *threo* adducts are then subjected to *syn* elimination, with the *erythro* isomer producing the (*Z*)-alkene (**208**), and the *threo* the (*E*)-isomer (**209**; Scheme 29).[206]

Unlike the Peterson alkenation, which is in principle similar, the phosphine oxide anion addition can be controlled to produce predominantly the *erthyro* isomer (**206**). The *threo* isomer can be obtained by selective reduction of the α-ketophosphine oxide (**210**), allowing highly stereoselective alkene formation. Since a two-step sequence is employed, this reaction does not require a stabilizing functionality to be conjugated to the phosphine oxide in order to produce the alkene. In fact, unlike the phosphonate HWE reagents, the reaction of a ketophosphine oxide (**211**) with a carbonyl derivative does not occur to produce the unsaturated carbonyl (**213**; Scheme 30).[207] The addition step is presumably too rapidly reversible and the elimination of phosphine oxide too slow.

The use of diphenylcyanomethylphosphine oxide is effective for the synthesis of (*E*)-α,β-unsaturated nitriles.[208] Phosphine oxides can be used to synthesize a variety of functionalized alkenes, including vinyl ethers (**215**; equation 51),[209] vinyl sulfides (**217**; equation 52),[210] allylic amines (**219**) and amides (equation 53),[211] ketene acetals (**221**; equation 54)[212] and ketene thioketals (**223**; equation 55).[213] In the examples of α-thio substitution, the alkenes are formed directly.

3.1.9.4.1 Elimination

The intermediate β-hydroxyphosphine oxide is isolated only if lithium is used to deprotonate the phosphine oxide. Sodium or potassium anions eliminate *in situ* to form the alkene directly. Eliminations of

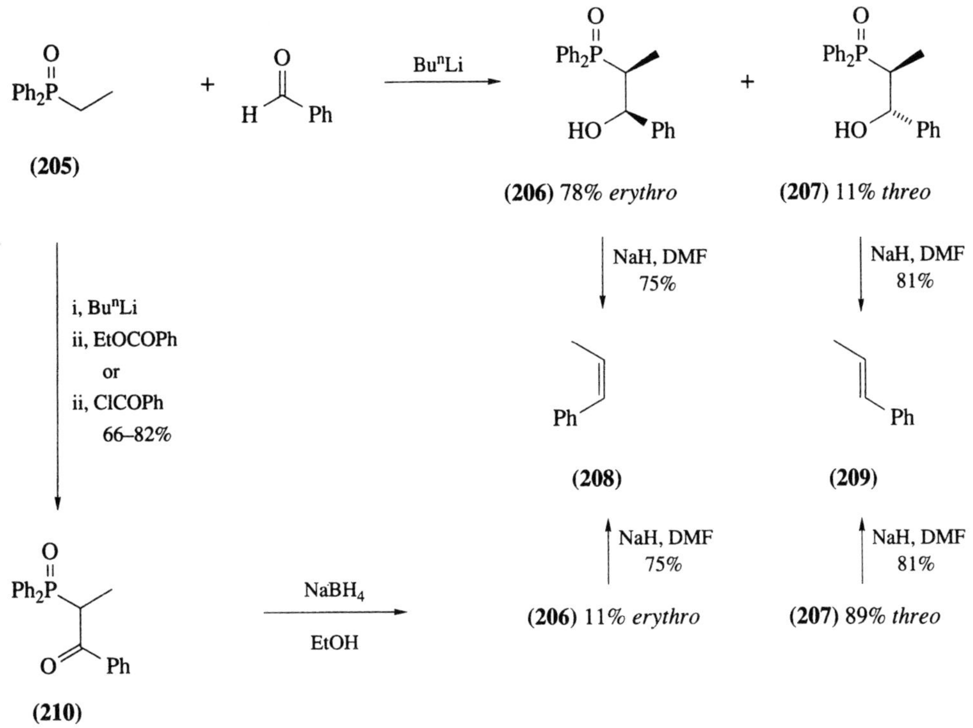

$Ph_2P(O)CH_2CH_3$ + PhCHO

(205)

Bu^nLi

(206) 78% *erythro* + (207) 11% *threo*

NaH, DMF 75% → (208)

NaH, DMF 81% → (209)

i, Bu^nLi
ii, EtOCOPh
or
ii, ClCOPh
66–82%

(210)

$NaBH_4$, EtOH

(206) 11% *erythro*; NaH, DMF 75% → (208)

(207) 89% *threo*; NaH, DMF 81% → (209)

Scheme 29

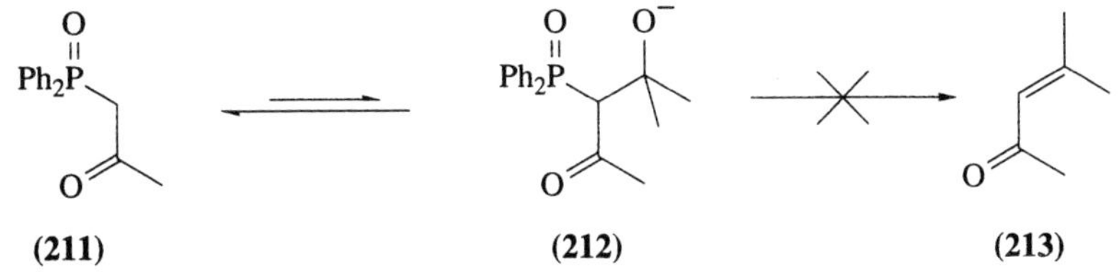

(211) ⇌ (212) ✕→ (213)

Scheme 30

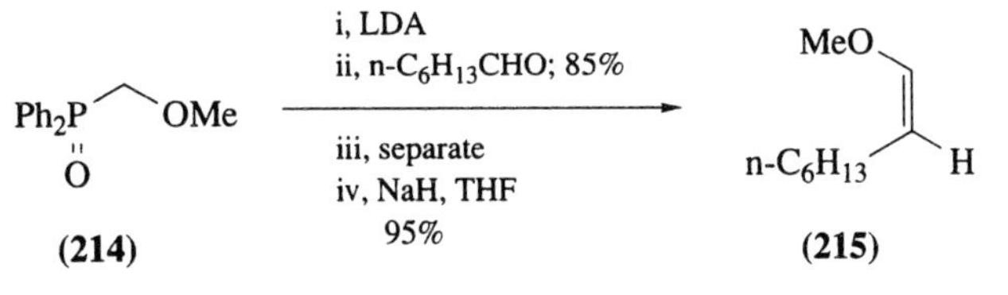

(214) → i, LDA; ii, n-C_6H_{13}CHO; 85%; iii, separate; iv, NaH, THF 95% → (215) (51)

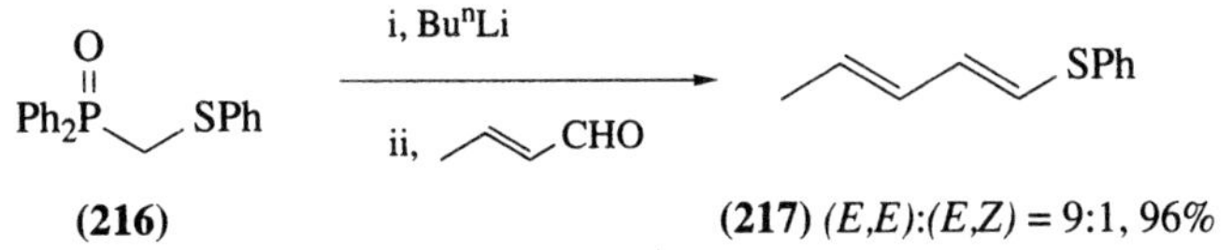

MeO, n-C_6H_{13}, H

(216) → i, Bu^nLi; ii, CH₃CH=CHCHO → (217) *(E,E)*:*(E,Z)* = 9:1, 96% (52)

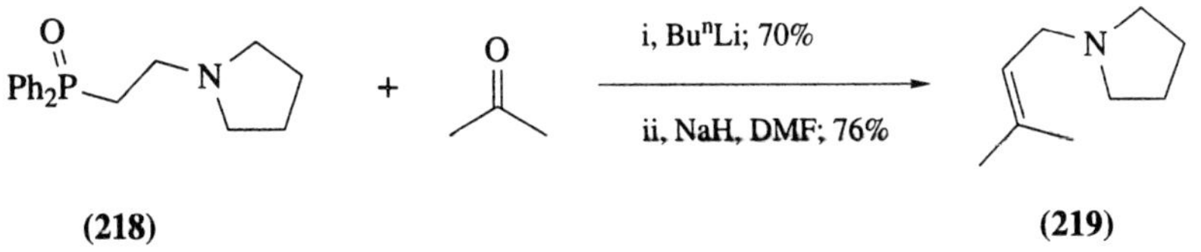

(218) + acetone → i, Bu^nLi; 70%; ii, NaH, DMF; 76% → (219) (53)

(220) i, LDA; ii, cyclohexanone; ii, Bu^{t}OK; 90% → (221) (54)

(222) i, Bu^{n}Li; ii, PhCHO; 80% → (223) (55)

the *threo*-hydroxyphosphine oxide intermediates are stereospecifically *syn* and produce the (*E*)-alkene, but the reaction of *erythro* adducts can be complicated by equilibration with starting materials, particularly in the case of aromatic aldehydes.[214] Typically, bases such as NaH, KOH and KOBu^t are used in DMF, DMSO or THF to effect elimination.

3.1.9.4.2 **Erythro** ***selectivity,*** **cis*-alkene***

The addition of the phosphine oxide anion to the carbonyl is dramatically affected by solvent, base and temperature.[214] These conditions can be modified in order to maximize the *erythro* isomer formation. In nonpolar solvents the addition proceeds with virtually no selectivity. Substantial improvements are seen by the use of ethers, and the highest ratios of *erythro* adduct (**225**) are obtained in THF with the lithium complexing reagent TMEDA present at –78 °C or lower temperatures (equation 56).

(224) + PhCHO —Bu^{n}Li→ (225) + (226) (56)

Solvent	(**225**) *erythro* (%)	(**226**) *threo* (%)
Pentane	55	45
THF	85	15
THF, TMEDA, –78 °C	88	12

Substituent effects are also important to the selectivity. Branching α to the phosphine oxide can significantly erode the ratios of *erythro* to *threo* intermediates. For example, changing the α-substituent from methyl to isopropyl gives a 64:36 ratio of *erythro* to *threo* isomers on reaction with benzaldehyde.[214] Increasing the branching of the aldehyde component also diminishes the selectivity, but the effect is smaller. Cyclohexanal combines with the lithium anion of ethylphosphine oxide to give a 79:21 mixture of *erythro* to *threo* adducts in 79% yield. Although the elimination of phosphine oxide to form the alkene is *syn*, a certain amount of (*E*)-alkene is formed from *erythro* intermediates with a conjugating functionality α to the phosphine oxide.[206] This loss of selectivity is thought to occur by cleavage back to the starting anion and the carbonyl, resulting in equilibration to higher ratios of *threo* adduct and therefore (*E*)-alkene. The equilibration can be minimized by using a polar solvent such as DMF or DMSO and higher temperatures to ensure rapid elimination of phosphinate. An additional solution is demonstrated in the selective synthesis of (*Z*)-stilbene (**229**) using dibenzophosphole oxide in the Horner reaction (Scheme 31).[215] The cyclic phosphine oxide shifts the equilibrium so that phosphinate elimination is favored relative to the reverse aldol process. This is because of the marked acceleration of elimination when phosphorus is incorporated in a five-membered ring. The *erythro* adduct was generated by ring-opening the epoxide. Elimination with NaH in DMSO gives a 91% yield with a (*Z*):(*E*) ratio of 89:11. This ratio can be dramatically improved to >99:1 by using DBU.

Warren has also studied dibenzophosphole oxides.[216] The ketophosphine oxide (**230**) substrate can be formed and selective reduction to either the *erythro* (**233**) or the *threo* (**231**) adducts carried out. The normal $NaBH_4$ conditions were used for reduction to the *threo* isomer and $CeCl_3$ was added to obtain the *erythro* adduct. This methodology was applied to the synthesis of (*E*)- and (*Z*)-isosafroles (**232**) and

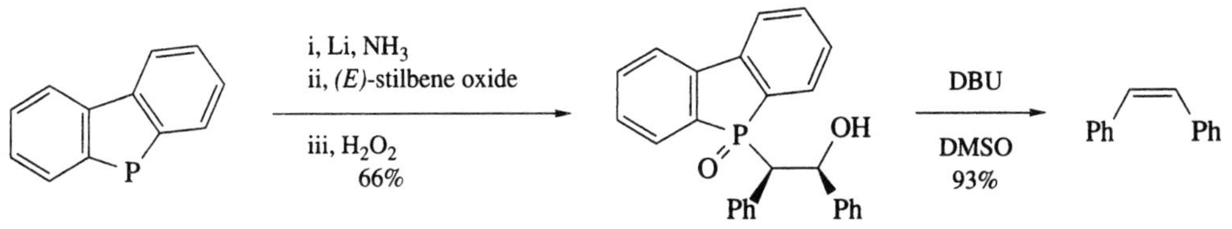

(227) (228) (229) (Z):(E) > 99:1

i, Li, NH_3
ii, (E)-stilbene oxide
iii, H_2O_2
66%

DBU
DMSO
93%

Scheme 31

(**234**), respectively, from a single ketophosphine oxide intermediate (**230**; Scheme 32). The application of this approach to diphenylphosphine oxides would provide a method by which either alkene isomer could be readily obtained from a single intermediate. Unfortunately, the Luche reduction, which is vital to this strategy, is highly substrate specific.[216b]

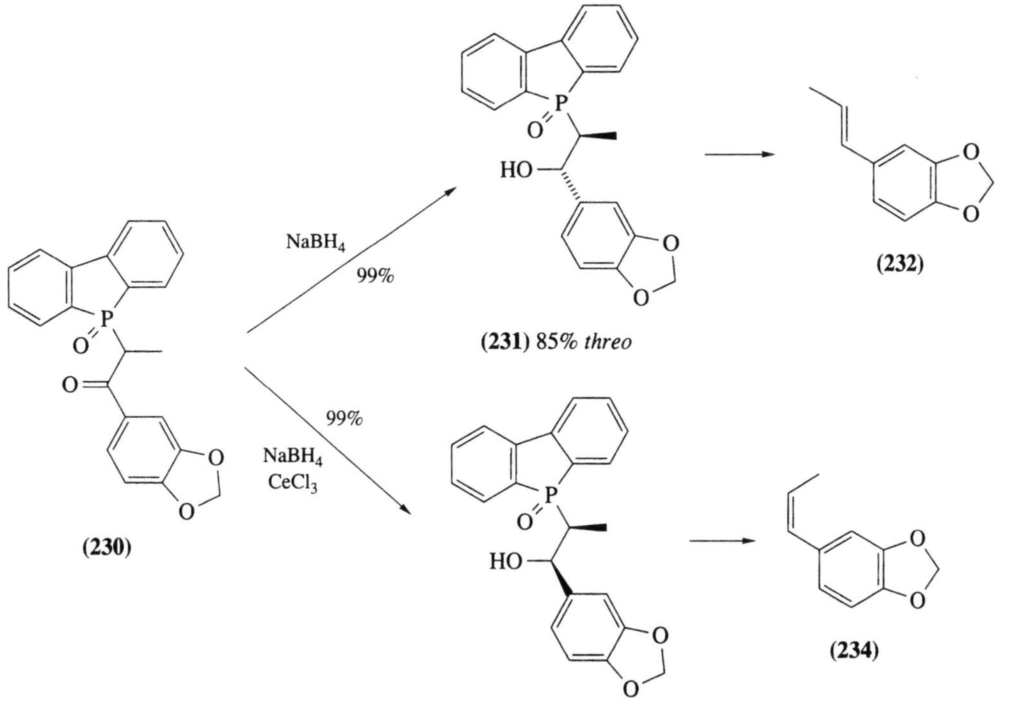

$NaBH_4$ 99%

$NaBH_4$ $CeCl_3$ 99%

(**230**)

(**231**) 85% *threo*

(**232**)

(**233**) 85% *erythro*

(**234**)

Scheme 32

3.1.9.4.3 Threo *selectivity,* trans-*alkene*

Trans-alkenes can be synthesized by the Horner reaction from the ketophosphine oxide, which is reduced selectively to the *threo* adduct.[217] This intermediate is typically formed by reaction of the phosphine oxide with an ester or acid chloride. Alternatively, the keto intermediate may be obtained by oxidation of β-hydroxyphosphine oxides. This sequence was applied to the synthesis of the pure (*E*)-triene (**237**) by Warren (equation 57).[217]

Unlike the elimination of *erythro* adducts, the elimination of the *threo* intermediate is stereospecific for aromatic and aliphatic substituents. A study of various methods of reduction found that $NaBH_4$ in EtOH gave the best combination of *threo* selectivity and yield.[218]

This methodology has been applied to the synthesis of oudemansins by Kallmerten.[219] In this example, the Wittig alkenation gave mixtures of (*E*)- and (*Z*)-products, as well as epimerizing the ether position. These problems were solved by acylating the phosphine oxide (**239**) and carrying out reduction and

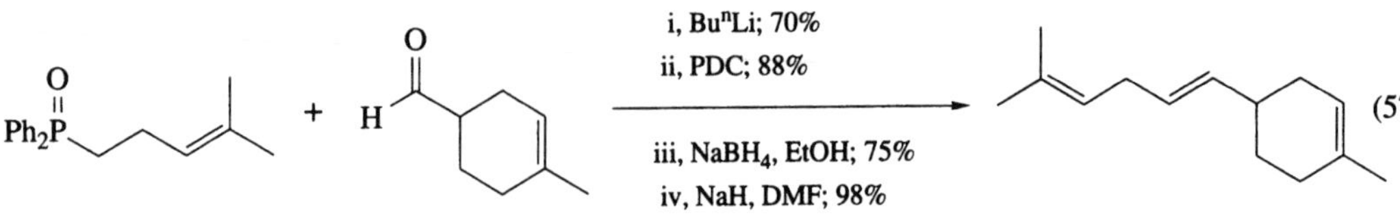

i, Bu^nLi; 70%
ii, PDC; 88%
iii, $NaBH_4$, EtOH; 75%
iv, NaH, DMF; 98%

(57)

(235) **(236)** **(237)**

elimination to (**240**) with $LiBH_4$ (Scheme 33). It was not anticipated that the reduction conditions would also cause elimination. It is unknown whether this effect is general for conjugated alkenes. Recently, Warren has studied the effect of additional chiral centers on the reduction of the ketophosphine oxide intermediate.[220] In simple alkyl-substituted cases (**241**) it is possible to reduce the intermediate with high diastereoselectivity to produce (*E*)-alkenes (**242**). Alternatively an anion addition can be carried out on the ketophosphine oxide (**243**) to synthesize trisubstituted alkenes (**245**). Yields of these processes are good and alkene ratios are high (Scheme 34).

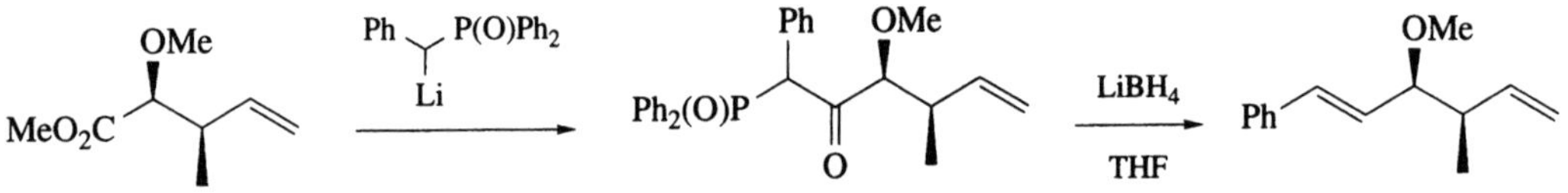

(238) **(239)** **(240)** (*E*) only, 32%

LiBH$_4$, THF

Scheme 33

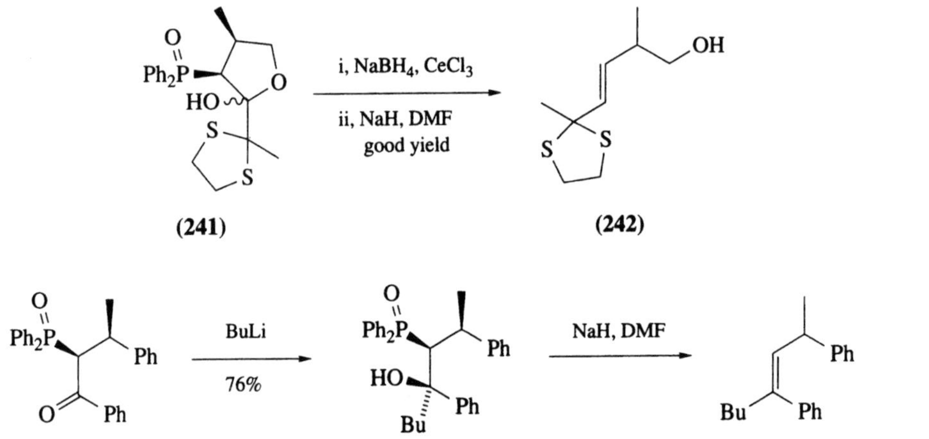

i, $NaBH_4$, $CeCl_3$
ii, NaH, DMF
good yield

(241) **(242)**

BuLi, 76%

NaH, DMF

(243) **(244)** **(245)**

Scheme 34

3.1.9.4.4 Disubstituted alkenes

For the synthesis of disubstituted alkenes by the Horner reaction, there are general guidelines that can be followed to maximize selectivity, as shown in structure (**246**).[221] It is best to have the larger of the two substituents of the double bond to be formed derived from the carbonyl moiety. This is particularly important if the substituent is anion-stabilizing since this will erode the selectivity at the elimination. These rules can be followed, regardless as to the stereochemistry of the alkene desired.

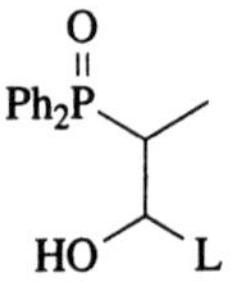

(246) L = large, conjugating group

The Wittig and Horner processes can be combined in a cyclic process to form (*E,Z*)-dienes (**249**; Scheme 35).[222] The initial Wittig process proceeded in 63% yield with 68:32 to (*Z*)-selectivity. Deprotonation of the phosphine oxide (**248**) and reaction with benzaldehyde gave exclusively the *erythro* isomer in 82% yield. The phosphine oxide was eliminated with NaH in DMF to give the (*E,Z*)-1,6-diene in an isomer ratio of 92:8 with the (*E,E*)-diene.

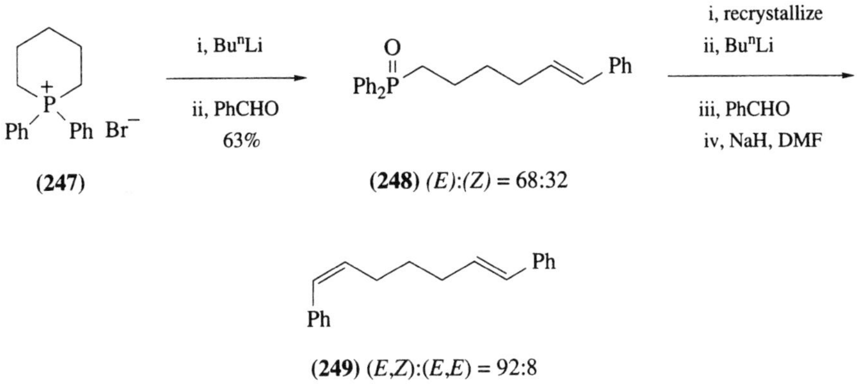

(**247**) → i, Bu^nLi; ii, PhCHO; 63% → (**248**) (*E*):(*Z*) = 68:32 → i, recrystallize; ii, Bu^nLi; iii, PhCHO; iv, NaH, DMF → (**249**) (*E,Z*):(*E,E*) = 92:8

Scheme 35

The Horner–Wittig process has been utilized in the synthesis of vitamin D and its metabolites. Recently, a process was developed for the synthesis of hydrindanols by the 1,4-addition of the phosphine oxide to cyclopentenone.[223] After further elaboration, the phosphine oxide formed (**250**) can be utilized to incorporate side chains (**251**; equation 58).

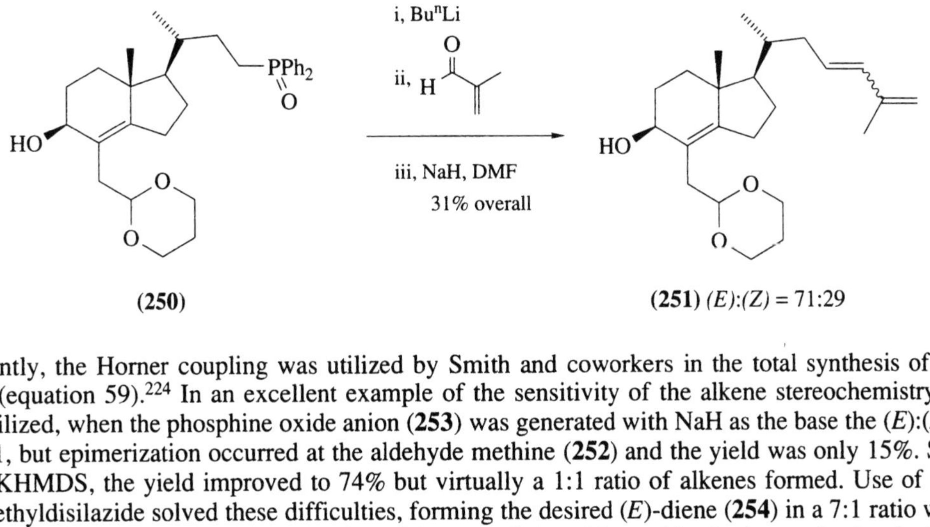

(**250**) → i, Bu^nLi; ii, (aldehyde); iii, NaH, DMF; 31% overall → (**251**) (*E*):(*Z*) = 71:29 (58)

Recently, the Horner coupling was utilized by Smith and coworkers in the total synthesis of milbemycin (equation 59).[224] In an excellent example of the sensitivity of the alkene stereochemistry to the base utilized, when the phosphine oxide anion (**253**) was generated with NaH as the base the (*E*):(*Z*) ratio was 7:1, but epimerization occurred at the aldehyde methine (**252**) and the yield was only 15%. Switching to KHMDS, the yield improved to 74% but virtually a 1:1 ratio of alkenes formed. Use of sodium hexamethyldisilazide solved these difficulties, forming the desired (*E*)-diene (**254**) in a 7:1 ratio with the (*Z*)- in 85–95% yield. Additional examples of the use of phosphine oxides in the synthesis of milbemycins and FK-506 are presented in Section 3.1.11.4.

Warren has applied the Horner reaction to the synthesis of isoxazoles (equation 60).[225] Either the 3-alkyl or the 5-alkyl substitution is effective, and in either case good yields of (*E*)-alkenes are obtained. The isoxazoles can be cleaved with $Mo(CO)_6$. The Horner reaction can be utilized in the synthesis of polyenes. For example, Nicolaou utilized this methodology in the synthesis of a pentaene (equation 61).[226] The addition was carried out with LDA and the elimination effected with $KOBu^t$ to give the (*E*)-isomer (**258**).

3.1.9.4.5 Trisubstituted alkenes

The Horner reaction can be applied to the synthesis of trisubstituted alkenes. As in the case of HWE reactions, the yield obtained by adding a disubstituted phosphine oxide to an aldehyde is frequently higher than that obtained by adding an anion to a ketone. This methodology was applied to the synthesis

(252)

+

(253)

$NaN(SiMe_3)_2$

THF

85–95%

(254) (*E*):(*Z*) = 7:1

(59)

(255)

i, BuLi

ii, PhCHO

78%

(256)

(60)

(257)

i, $Ph_2(O)P$, LDA

ii, Bu^tOK

55%

(258)

(61)

of (*Z*)-α-bisabolene (**261**; equation 62).[227] As stated above, heteroatom substitution is compatable with the formation of alkenes by phosphine oxide. This technique has been used by Ley and coworkers to synthesize intermediates for spiroketal formation (**263**; equation 63).[228] The Horner coupling has been utilized by many groups for the synthesis of the diene portion of vitamin D and its metabolites (**266**; equation 64).[229] These reactions occur with excellent stereoselectivity for the diene formation.

(259) + (260)

i, Bu^nLi

ii, separate

53% *erythro*, 4% *threo*

iii, NaH, DMF; 91%

(261)

(62)

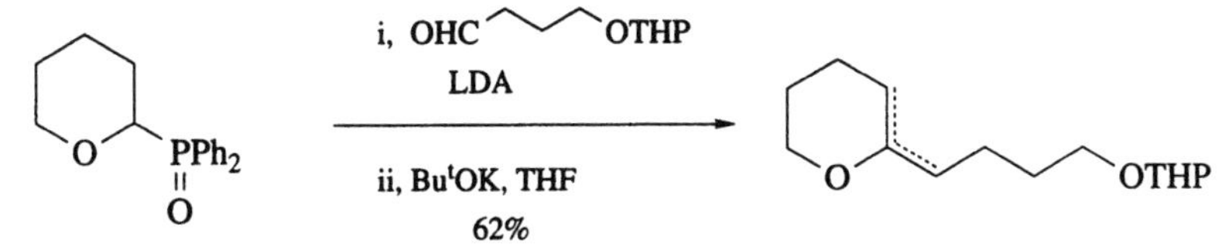

(63)

(262) (263)

(64)

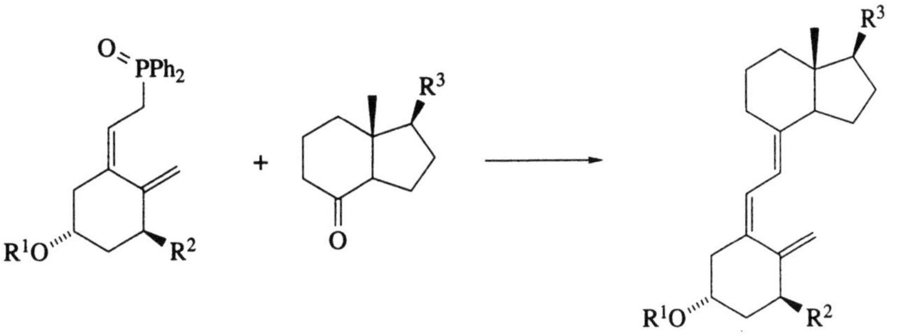

(264) (265) (266)

R^1 = CHMeOEt, R^2 = H, R^3 = C_8H_{17}; 60%

R^1 = $SiMe_2Bu^t$, R^2 = H, R^3 =  ; 61%

R^1 = $SiMe_2Bu^t$, R^2 = $OSiMe_2Bu^t$, R^3 = 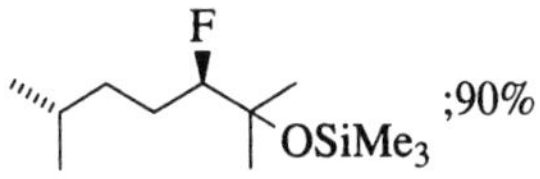 ;90%

3.1.9.4.6 *Alternative approaches*

In analogy to the Peterson alkenation, the intermediate hydroxyphosphine oxides (**269**) can be prepared by addition to epoxide derivatives (**268**; Scheme 36).[230] Overall yields are high for this process, and this sequence can be applied to the synthesis of phosphonate intermediates as well. Warren has studied hydroxy-directed epoxidation.[230b] Provided the allylic phosphine oxide is trisubstituted, as is (**270**) in equation (65), these oxidations proceed with good selectivity. Ring opening can then be undertaken to generate the hydroxyphosphine oxide.

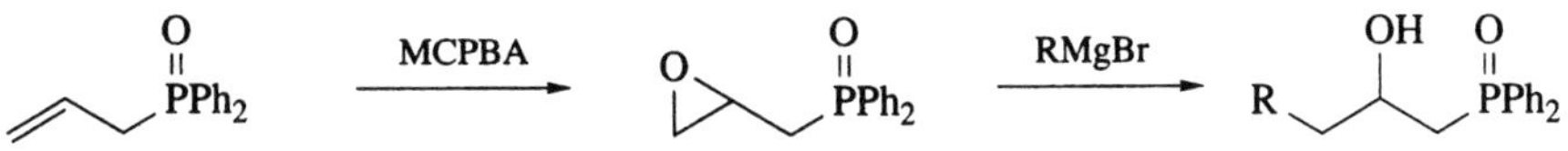

(267) (268) (269)

Scheme 36

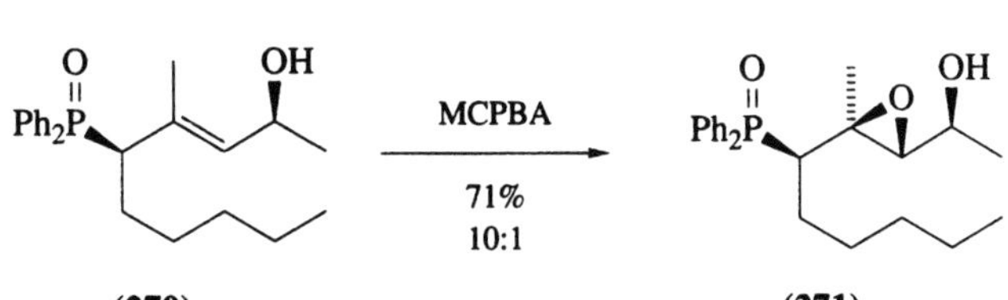

(65)

(270) (271)

Acylated phosphine oxides have been used as intermediates to unsaturated acids such as (**276**; Scheme 37).[231] These compounds cannot be formed by the direct addition of an acid derivative of the phosphine oxide to a carbonyl (**273**). Instead, the ketophosphine oxide (**274**) is reduced and the lactones (**275**) are separated. The sequence is completed by treatment with KOH in aqueous THF followed by elimination in DMSO. The application of the Horner reaction to the synthesis of *homo*-allylic alcohols (**279**) was studied by Warren and coworkers.[232] These derivatives were formed by an intramolecular acyl transfer to the ketophosphine oxide (**278**; Scheme 38). The acyl group may be alkyl or aromatic and the reduction of the ketone, followed by elimination to the alkene, takes place under the standard protocol. This reaction was also applied to systems with chiral centers present in the phosphine oxide chain.[233] The intermediate phosphine oxide can also be converted to a hydroxy ketone and a cyclopropyl ketone by treatment with base. An alternative approach to *erythro*-phosphine oxides is to add the phosphine imide anion (**280**) in a selective manner to the aldehyde and hydrolyze to the hydroxyphosphine oxide (**282**; Scheme 39).[234] In comparing the imide to the phosphine oxide, the *erythro* selectivity of the phosphine imide addition was higher (98:2 *versus* 88:12 for the methyl- and phenyl-substituted alkene), and it remains to be demonstrated whether this alternative will prove to have practical applications in organic synthesis.

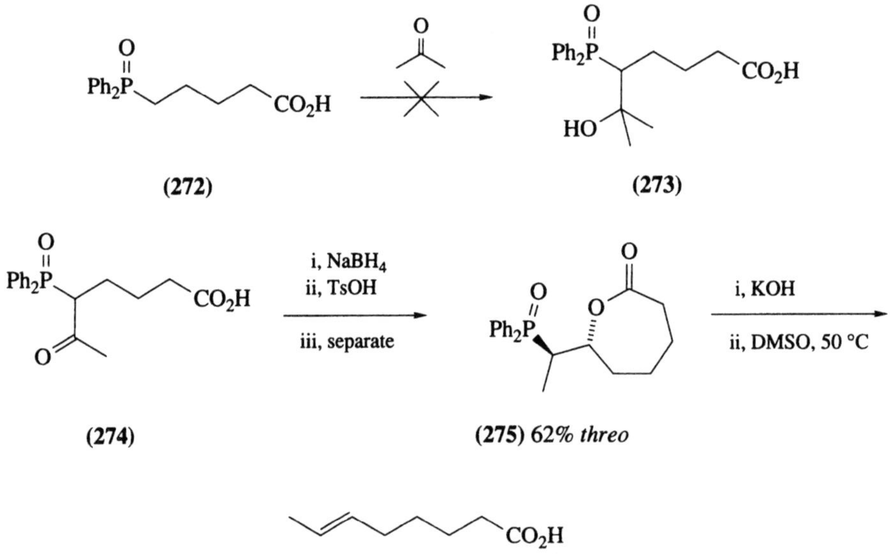

(**272**) (**273**)

i, $NaBH_4$; ii, TsOH; iii, separate — i, KOH; ii, DMSO, 50 °C

(**274**) (**275**) 62% *threo*

(**276**) 93%

Scheme 37

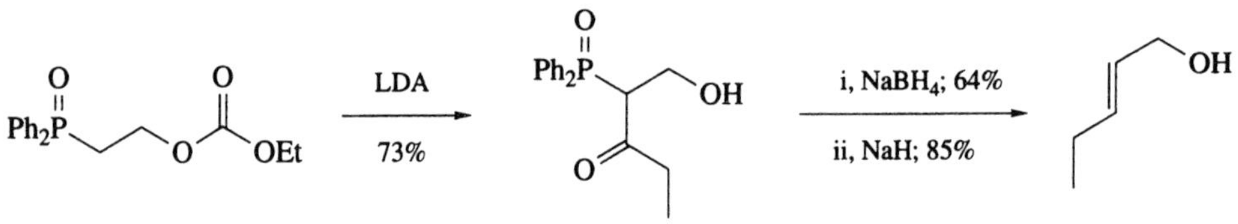

LDA, 73% — i, $NaBH_4$; 64%; ii, NaH; 85%

(**277**) (**278**) (**279**)

Scheme 38

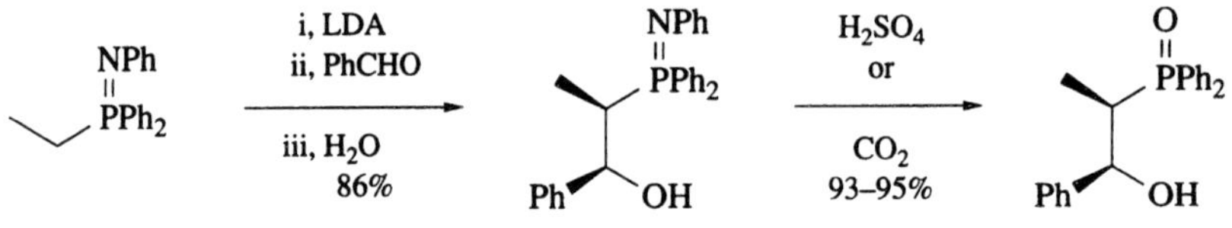

i, LDA; ii, PhCHO; iii, H_2O, 86% — H_2SO_4 or CO_2, 93–95%

(**280**) (**281**) (**282**)

Scheme 39

3.1.10 SILICON-MEDIATED ALKENE FORMATION

In Section 3.1.3.1, the advantages of the Peterson alkenation in comparison to the Wittig reaction were detailed. The by-product (hexamethyldisiloxane) is volatile and is easier to remove than the phosphine oxides, and it is a more basic and more reactive anion. On the other hand, the anion can be more arduous to obtain than the Wittig anion for more complex applications, and the stereochemistry of the elimination is such that control of alkene isomers requires separation of the β-silylcarbinols.

3.1.10.1 Anion Formation

The α-silyl carbanions necessary to apply the Peterson reaction to higher substituted examples are limited by the ability to efficiently produce the anion. A clever example of an alternative to the Wittig reaction for ethylidene formation to give (**285**) with α-(trimethylsilyl)vinyllithium was utilized by Jung in the synthesis of coronafacic acid (Scheme 40).[235]

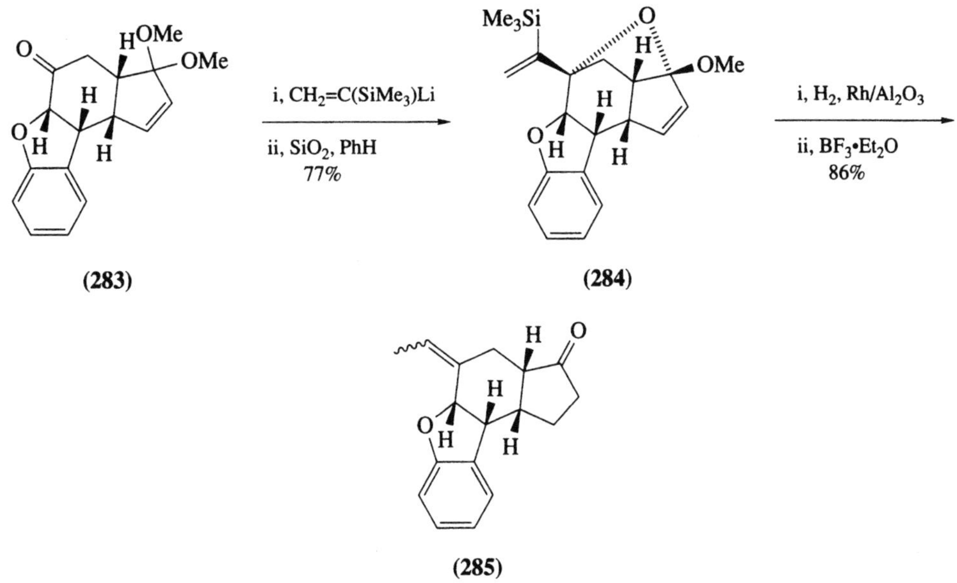
HOMe
OMe
O
H
H
H
i, $CH_2{=}C(SiMe_3)Li$
ii, SiO_2, PhH
77%
Me_3Si
O
H
OMe
i, H_2, Rh/Al_2O_3
ii, $BF_3{\cdot}Et_2O$
86%
(283)
(284)
(285)

Scheme 40

As a general rule, unless an anion-stabilizing group, such as phenyl, or a heteroatom such as sulfur is present, the alkylsilane is not readily deprotonated.[236] The α-halosilane can be deprotonated but, unlike the readily available chloromethyltrimethylsilane, there are few general methods to this approach. Alkyllithium reagents add to vinylsilanes (**286**) to produce the carbanion (**287**).[237] Silyl derivatives with heteroatoms, such as sulfur, selenium, silicon or tin, in the α-position (**288**) may be transmetallated (Scheme 41).[4d] Besides the difficulty in synthesizing the anion, alkene formation lacks specificity for simple di- and tri-alkyl-substituted alkenes. As a result, the Peterson reaction of an α-silyl carbanion with a carbonyl has found the greatest utility in the synthesis of methylene derivatives, (as discussed in Section 3.1.3), heterosubstituted alkenes and α,β-unsaturated esters, aldehydes and nitriles.

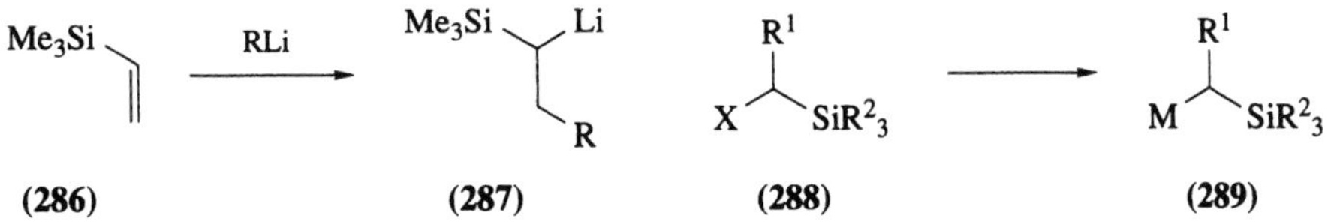
Me_3Si
RLi
Me_3Si
Li
R
R^1
X
SiR^2_3
R^1
M
SiR^2_3
(286)
(287)
(288)
(289)

$X = SR^3, SeR^3, SnR^3_3, SiR^2_3$

Scheme 41

3.1.10.2 Elimination

In a study by Hudrlik, the elimination was demonstrated to be stereospecific.[238] Diastereomeric β-hydroxysilanes were synthesized by the reduction of the corresponding ketones. In this way, each distinct diastereomer (**290**) and (**293**) was monitored in the elimination reaction. The elimination of the silane was stereospecific, with the acid-promoted eliminations being *anti* and the base-induced reaction following the *syn* pathway (Scheme 42).

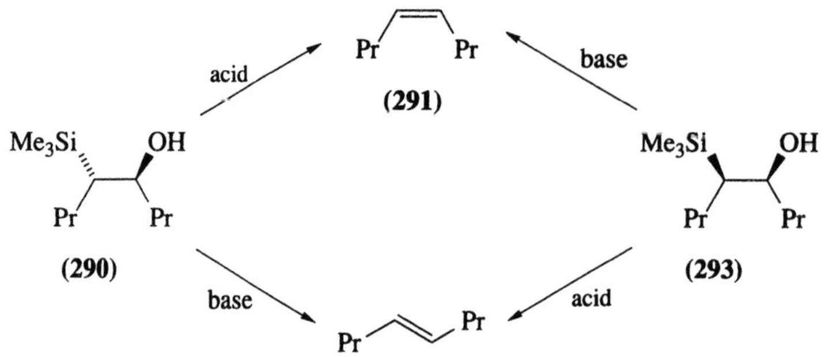

(290) (291) (292) (293)

Scheme 42

The implication of this observation is that if one can form the silylcarbinol selectively, the ratios of *cis*- and *trans*-alkenes should be controllable and high. Unfortunately, the addition of the silyl anion to a carbonyl does not result in formation of a single carbinol. Unlike the Wittig reaction, the initial addition step is not reversible (the reaction is under kinetic control), therefore the inherent ratios in the addition step define the ratio of *cis*- to *trans*-alkenes produced.[239] The reaction is not influenced significantly by solvent, counterion effects, added salts or temperature. Because of these combined factors, the Peterson alkenation generally yields equal amounts of *cis*- and *trans*-alkenes.[240] To take advantage of the stereospecific elimination, either the substrate must be synthesized by an approach other than anion addition to the carbonyl, or the diasteromeric β-silylcarbinols must be separated.[241] Examples of successful formation of the stereodefined carbinol are the selective reduction of the α-silyl carbonyl (**294**; equation 66),[242] the addition of an alkyl anion to an α-silyl carbonyl (**296**; equation 67),[243] the addition of a silyl anion to an epoxide (**298**; equation 68)[4d] and the ring opening of a silyl epoxide (**300**; equation 69).[244]

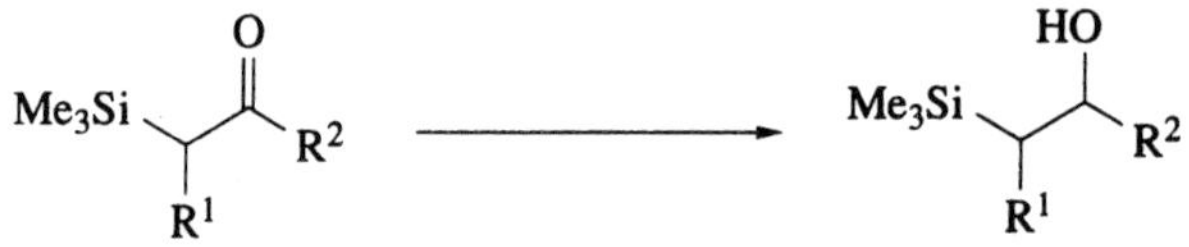

(294) → (295) (66)

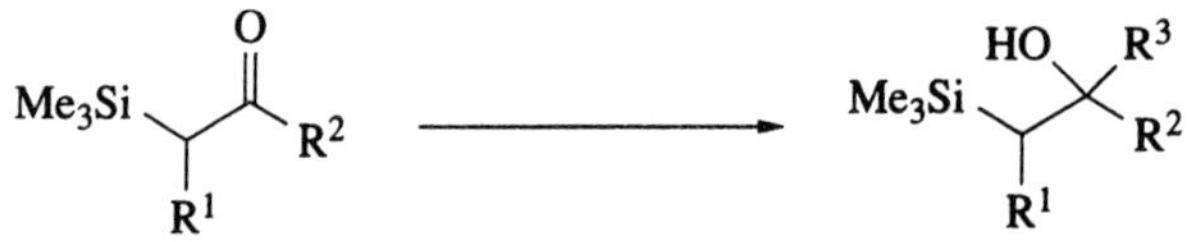

(296) → (297) (67)

(298) → (299) (68)

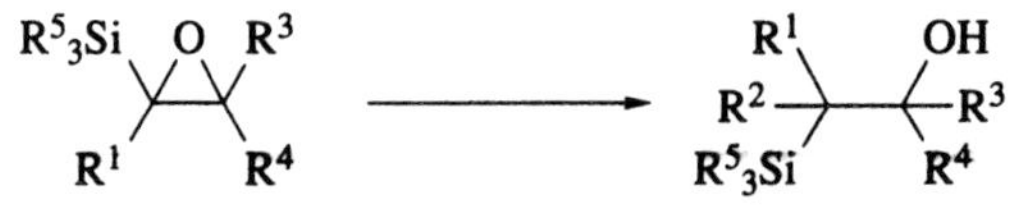

(300) → (301) (69)

The term 'Peterson alkenation' has been used to describe the elimination of a functionalized organosilicon compound with alkene formation for substrates synthesized by these methods. In accordance with the mechanistic definition outlined in the introduction to this chapter, such topics are not considered in detail in this review.[245] For the purpose of this discussion, the Peterson alkenation will be considered as the addition of an anion derivative to a carbonyl compound, followed by elimination to the alkene.

3.1.10.3 Mechanism

As mentioned in the previous section, the Peterson reaction proceeds by an irreversible addition of the silyl-substituted carbanion to a carbonyl. It has generally been assumed that an intermediate β-oxidosilane is formed and then eliminated. In support of this mechanistic hypothesis, if an anion-stabilizing group is not present in the silyl anion, the β-hydroxysilanes can be isolated from the reaction, and elimination to the alkene carried out in a separate step. Recent studies by Hudrlik indicate that, in analogy to the Wittig reaction, an oxasiletane (**304**) may be formed directly by simultaneous C—C and Si—O bond formation (Scheme 43).[246] The β-hydroxysilanes were synthesized by addition to the silyl epoxide. When the base-induced elimination was carried out, dramatically different ratios of *cis*- to *trans*-alkenes were obtained than from the direct Peterson alkenation. While conclusions of the mechanism in general await further study, the Peterson alkenation may prove to be more closely allied with the Wittig reaction than with β-elimination reactions.

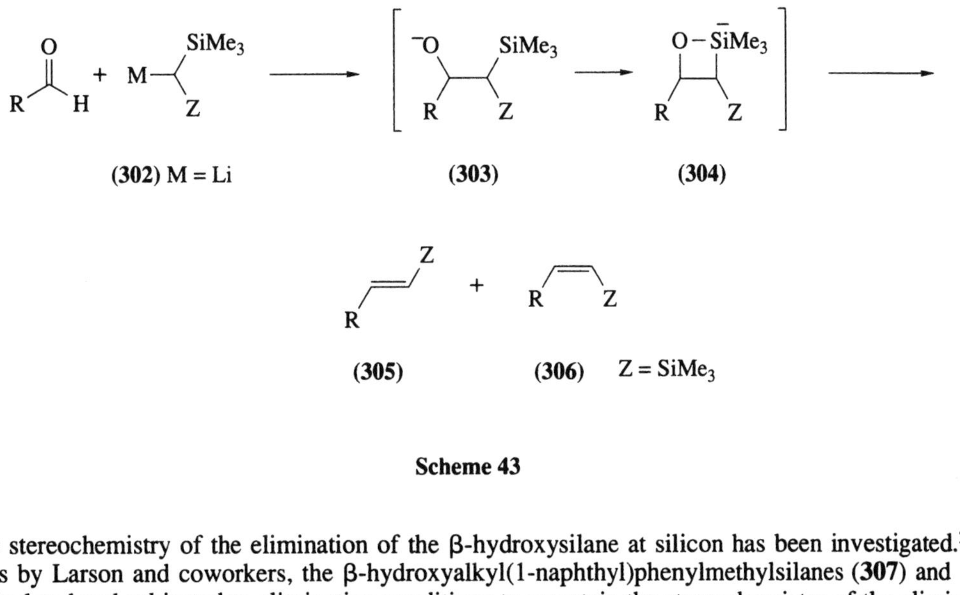

(**302**) M = Li (**303**) (**304**)

(**305**) (**306**) Z = $SiMe_3$

Scheme 43

The stereochemistry of the elimination of the β-hydroxysilane at silicon has been investigated.[247] In studies by Larson and coworkers, the β-hydroxyalkyl(1-naphthyl)phenylmethylsilanes (**307**) and (**309**) were isolated and subjected to elimination conditions to ascertain the stereochemistry of the elimination on the silyl group (Scheme 44). The acid-catalyzed eliminations proceed with inversion of stereochemistry at silicon, while the base-catalyzed elimination occurred with retention. These results are in agreement with the mechanism proposed of *anti* elimination under acidic conditions and *syn* elimination under basic. While the optically pure silicon was useful for determining the course of the elimination, it could not be utilized in asymmetric synthesis. Addition of the anion to various carbonyls afforded virtually no diastereoselectivity, and it was not possible to separate the diastereomers formed either by crystallization or by chromatography.

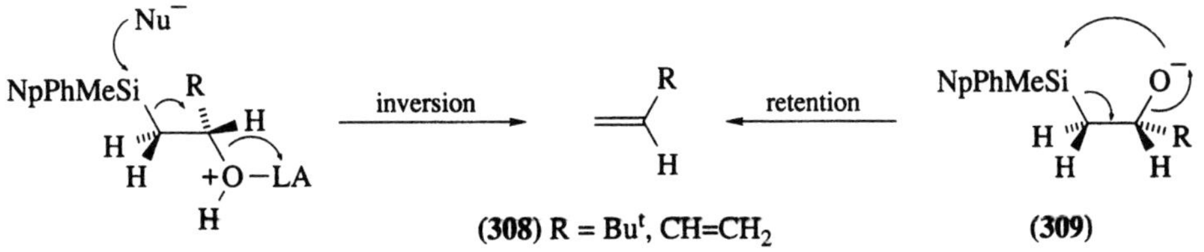

(**307**) inversion (**308**) R = Bu^t, CH=CH_2 retention (**309**)

Scheme 44

3.1.10.4 Alkene Formation

The Peterson reaction, as shown in equation (70), has been applied to the synthesis of alkenes that are hindered and difficult to form by the Wittig reaction. In the case of trisubstituted alkenes (**311**) in which R^1 and R^2 are components of a ring or are identical, the reaction may prove to be the method of choice.

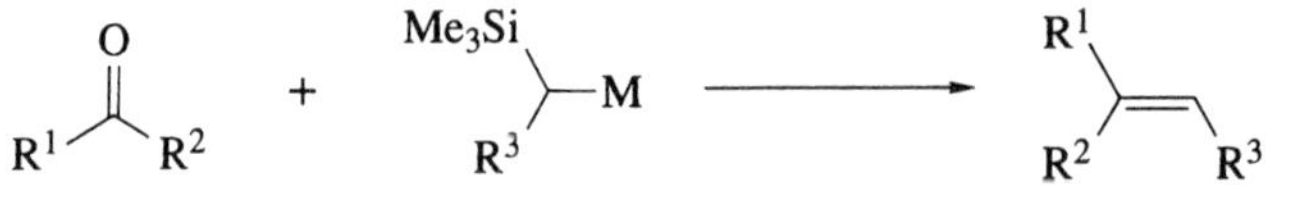

$$R^1C(O)R^2 + Me_3SiCH(R^3)M \longrightarrow R^1R^2C{=}CHR^3 \quad (70)$$

(**310**) (**311**)

An example is the preparation of allylidenecyclopropanes (Scheme 45).[248] The 1-(trimethylsilyl)cyclopropane (**312**) is reductively lithiated with lithium 1-(dimethylamino)naphthalenide (LDMAN) followed by addition of an aldehyde to form the β-silylcarbinol (**314**).[249] This method of anion formation is general[17a] but has seen greatest application in the synthesis of cyclopropyl compounds. The intermediate can be eliminated *in situ* with KOBu^t to form the alkene (**315**). Yields are good but, as discussed in the mechanistic section, in unsymmetrical cases a mixture of products results. This reaction has been extended by Halton and Stang to the synthesis of cycloproparenes.[250]

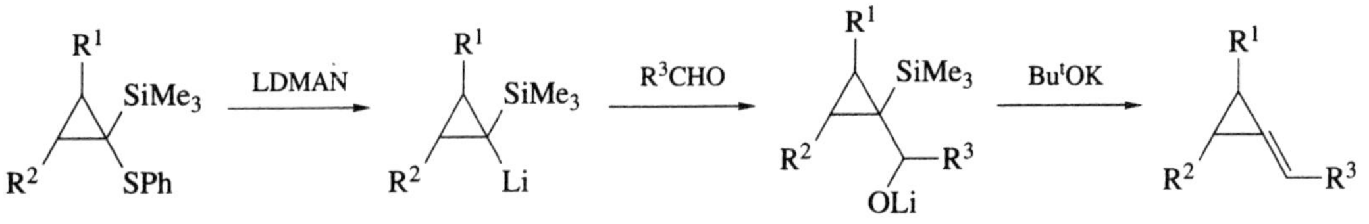

(**312**) $\xrightarrow{\text{LDMAN}}$ (**313**) $\xrightarrow{R^3CHO}$ (**314**) $\xrightarrow{Bu^tOK}$ (**315**)

Scheme 45

3.1.10.5 Heteroatom Substitution

3.1.10.5.1 Silicon

The Peterson reaction can be used to synthesize a number of heterosubstituted alkenes. Methoxydimethylsilyl(trimethylsilyl)methyllithium (**316**) can be added to aldehydes and ketones, including enolizable substrates, to form the vinylsilane (**317**; equation 71). Modest (*E*):(*Z*)-selectivity was observed in unsymmetrical cases. This reagent may represent an improvement over the bis(trimethylsilyl)methyl anion, which is ineffective for enolizable substrates.

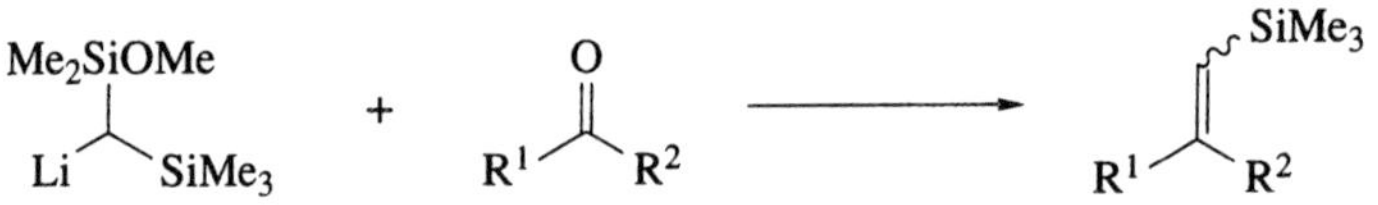

$$Me_2SiOMe(Li)CHSiMe_3 + R^1C(O)R^2 \longrightarrow R^1R^2C{=}CHSiMe_3 \quad (71)$$

(**316**) (**317**)

3.1.10.5.2 Sulfur

In the original study by Peterson, the alkenation procedure was found to be compatible with sulfur and phosphorus substitution.[251] The alkenation reaction has been applied successfully to a variety of substituted alkenes.[252] Because of the anion-stabilizing nature of the thiophenyl, the β-hydroxysilane is not isolated and the elimination to the alkene takes place directly to form a 1:1 mixture of (*E*)- and (*Z*)-isomers. Ager studied the reaction of the lithio anions of phenyl (trimethylsilyl)methyl sulfides (**318**) with a variety of carbonyl compounds (equation 72).[14] Yields of this process were good, and addition occurred even with enolizable substrates. This reaction was extended to vinyl sulfones. In contrast to the sulfide case, the substituted sulfone silyl anion behaves as a base, leading to undesired enolization. The best yields were observed for the case where R^1 is a hydrogen or phenyl.

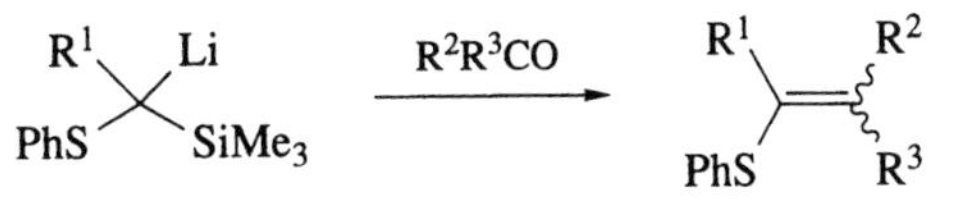

(72)

(318) (319)

R^1, R^2, R^3 = H, Me, Et, Bu, C_5H_{11}, Ph, *etc.*

Ley and coworkers have done studies with phenyl (trimethylsilyl)methyl sulfones (**320**; equation 73).[253] The lithio anion was generated with Bu^nLi in DME to form vinyl sulfones (**321**) in good to excellent yields as isomeric mixtures. There is some indication that the reaction should best be carried out in DME at –78 °C, rather than in THF as in the initial Ager work.[16] Trapping the intermediate alkoxide as the acetate, followed by attempts at stereospecific elimination did not prove to be successful in forming a single alkene isomer.

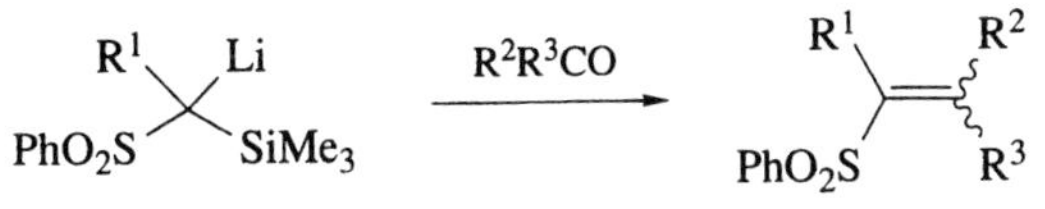

(73)

(320) (321)

R^1, R^2, R^3 = H, Me, C_5H_{11}, Ph, *etc.*

Addition of sulfides and sulfones to acid derivatives has been investigated by Agawa.[254] Phenyl and methyl (trimethylsilyl)methyl sulfides and sulfones added to amides (**323**) to produce the aminovinyl sulfides (**324**; equation 74) and sulfone (**326**; equation 75). The reactions proceeded in good yield with some examples of stereocontrolled synthesis of the (*E*)-isomer. Other acid derivatives such as esters, carbonates and ureas were investigated, but gave inconsistent results.

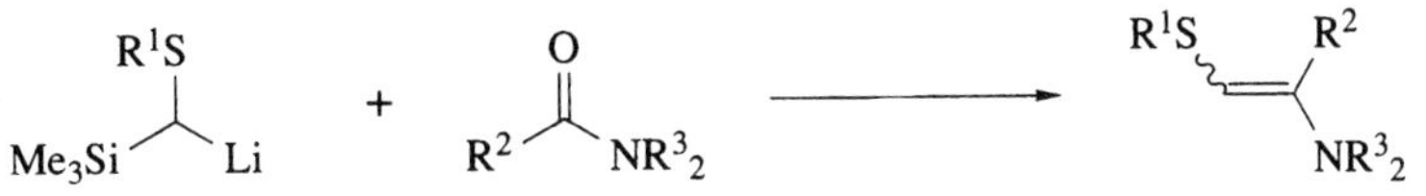

(74)

(322) (323) (324)

(75)

(325) (326)

In studies directed toward intermediates for the synthesis of quadrone, Livinghouse demonstrated the utility of lithiated methoxy(phenylthio)(trimethylsilyl)methane (**327**) for the conversion of aldehydes and ketones to ketene *O,S*-acetals (**328**) in good to excellent yields (Scheme 46).[255] These Peterson alkenations gave predominantly the (*E*)-double bond isomer. As the example depicted in the scheme demonstrates, this procedure may be used to homologate a carbonyl to the phenyl thioester (**329**) in excellent yields.

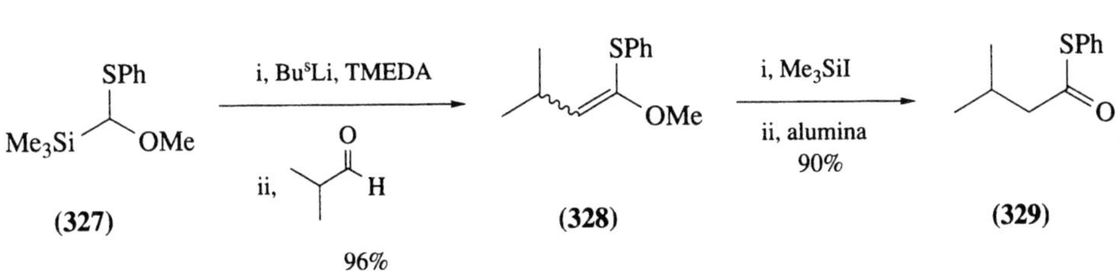

i, Bu^sLi, TMEDA; ii, (aldehyde), 96%

i, Me_3SiI; ii, alumina, 90%

(327) (328) (329)

Scheme 46

The use of sulfur in the Peterson reaction can be extended to the optically pure lithio anion of *S*-phenyl-*S*-(trimethylsilyl)methyl-*N*-tosylsulfoximine (**330**; equation 76).[256] Unlike most Peterson alkenations this reaction is selective for the formation of the (*E*)-alkene isomer (**331**) with aldehydes. In addition, the stereochemistry of the sulfoximine is maintained.

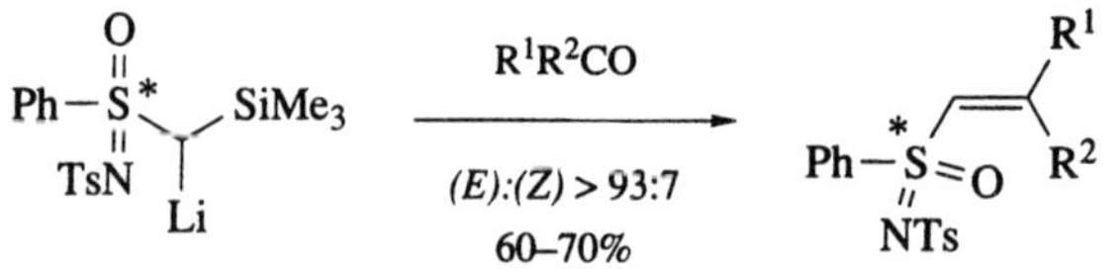

R^1R^2CO; (*E*):(*Z*) > 93:7; 60–70% (76)

(**330**) R^1, R^2 = H, Me, Ph, Pr^i, Bu^t, *etc.* (**331**)

3.1.10.5.3 Phosphorus

As in the case of sulfur-substituted analogs to the Peterson alkenation, the compatibility of the reaction with a phosphorus substituent was demonstrated in the original work of Peterson. An *in situ* procedure for the preparation and reaction of the lithio anion of (trimethylsilyl)alkylphosphonates has been developed (**333**; Scheme 47).[257] The phosphonate (**332**) was treated with 2 equiv. of LDA followed by TMS-Cl and after warming to –20 °C, the aldehyde. The overall yields were good, but (*E*):(*Z*) selectivities for this reaction were low. Zbiral and coworkers developed a method for the homologation of a carbonyl compound to an α-hydroxy ester with a silylphosphonate derivative (**335**; Scheme 48).[258] The reactions produced a 1:1 mixture of alkene isomers (**336**). The intermediate vinylphosphonate was oxidized with OsO_4 to form the α-hydroxy ester (**337**). Recently, both sulfur and phosphorus functionalities were combined in a Peterson reagent.[259] The lithium anion of the (methylthio)phosphonate (**338**) reacted with high (*E*)-selectivity with aldehydes (equation 77). The reaction was ineffective with ketones.

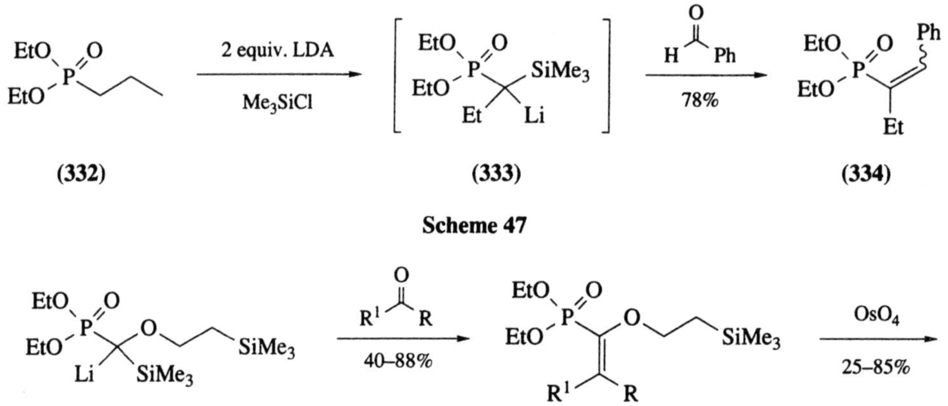

2 equiv. LDA, Me_3SiCl; PhCHO, 78%

(**332**) (**333**) (**334**)

Scheme 47

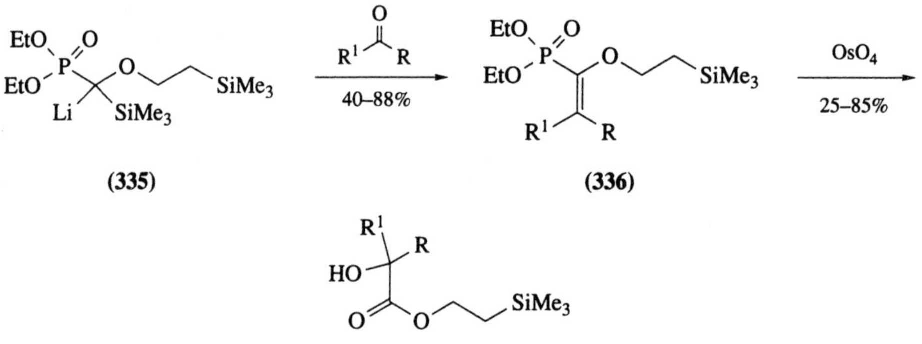

R^1COR, 40–88%; OsO_4, 25–85%

(**335**) (**336**) (**337**)

Scheme 48

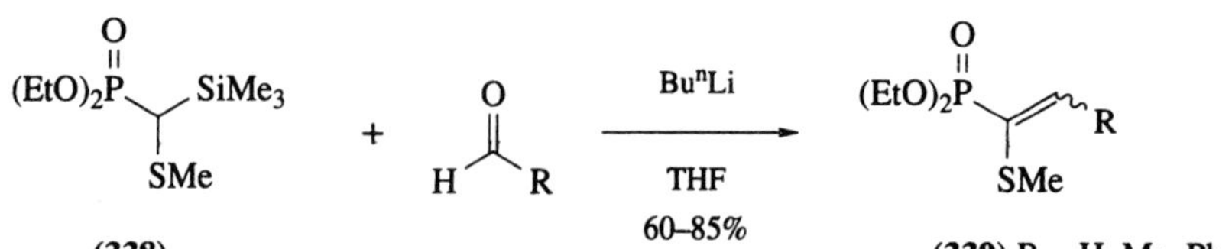

(**338**) + RCHO; Bu^nLi, THF, 60–85% (77)

(**339**) R = H, Me, Ph

3.1.10.6 Synthesis of Conjugated Alkenes

In the application of the Peterson alkenation to the synthesis of α,β-unsaturated esters, the full advantage of the greater reactivity of the silyl-stabilized anion as well as the ease of by-product removal can be realized. In addition, the Peterson reagent can be directly formed by deprotonation if an activating group, in these cases a carbonyl or similar conjugating functionality, is present, making the anion readily accessible. Selectivity for (*E*)- and (*Z*)-alkene isomers is highly substrate specific and no general predictive rule can be stated, but selectivity is enhanced by increasing the steric bulk of the silyl substituent, and in some instances by altering the counterion, as the following examples illustrate.

The reaction of a silylacetate derivative with an aldehyde or ketone was initially studied by Rathke and Yamamoto.[260] Rathke and coworkers studied the addition of the lithium anion of *t*-butyl (trimethylsilyl)acetate (**340**) with a variety of aldehydes and ketones (equation 78). The anion can be formed directly from the silyl compound on treatment with LDA. The reaction proceeded to give the conjugated alkenes in excellent yields. Unsaturated compounds reacted *via* 1,2-addition. No discussion of alkene geometry was presented. In the Yamamoto work, the ethyl (trimethylsilyl)acetate derivative (**342**) was used in a variety of reactions with aldehydes and ketones (equation 79). The anion was formed with dicyclohexylamide in THF. It was stated in the experimental section that the (*E*):(*Z*) ratios of alkenes were dependent on the reaction conditions. In all the examples presented in this work, the (*E*)-isomer was predominantly formed.

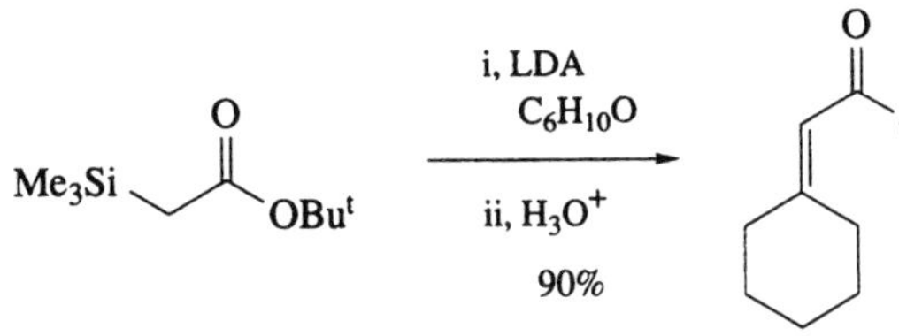

Me_3Si O OBu^t → i, LDA, $C_6H_{10}O$; ii, H_3O^+; 90% → O OBu^t (78)

(340) **(341)**

Me_3Si O OEt → 84% → O OEt (79)

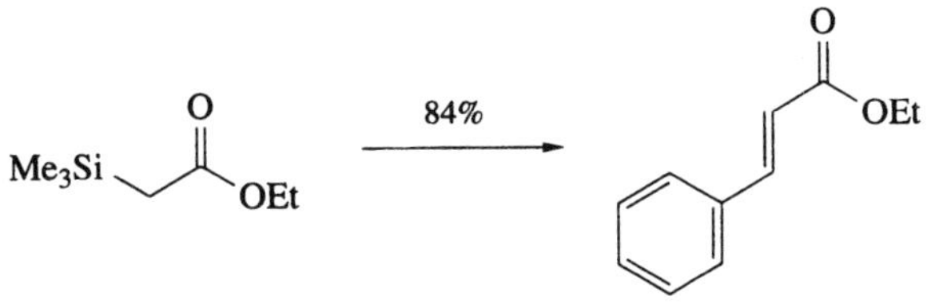

(342) **(343)** *(E)*:*(Z)* = 3:1 to 9:1

The ability to define the alkene geometry by modification of the anion was investigated by Debal.[261] In this work, the intermediate β-hydroxysilanes were either isolated and the elimination carried out with $BF_3 \cdot Et_2O$ to synthesize the (*E*)-isomers (**347**), or the addition adduct (formed in the presence of $MgBr_2$) was treated with HMPT and eliminated *in situ* to give predominantly the (*Z*)-isomer (**346**), as shown in Scheme 49.[262b]

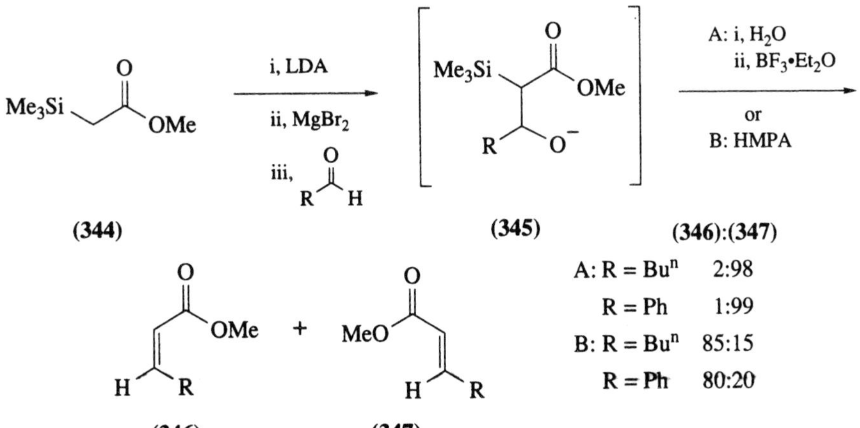

Me_3Si O OMe → i, LDA; ii, $MgBr_2$; iii, RCHO → [Me_3Si O OMe, R O$^-$] → A: i, H_2O; ii, $BF_3 \cdot Et_2O$ or B: HMPA

(344) **(345)**

O OMe, H R + MeO O, H R

(346) **(347)**

		(346):(347)
A:	R = Bu^n	2:98
	R = Ph	1:99
B:	R = Bu^n	85:15
	R = Ph	80:20

Scheme 49

Comparative examples of the Wittig reaction and the Peterson alkenation with ketones (**353**; equation 82) and (**355**; equation 83), epoxy ketones (**351**; equation 81), or protected α-hydroxy ketones (**348**; equation 80) have appeared.[262] The reactions can proceed with high kinetic control for the (*Z*)-isomer and, as a result, the Peterson technology may form complementary isomers to the Wittig reaction.[114]

(348) →[i, PPh_3CHCO_2Et or ii, $Me_3SiCH(Li)CO_2Et$] (349) + (350) (80)

i,	$n = 1$, 52%	97:3
	$n = 2$, 80%	96:4
	$n = 3$, 27%	94:6
ii,	$n = 1$, 64%	33:67
	$n = 2$, 68%	14:86
	$n = 3$, 50%	8:92

(351) →[i, $(EtO)_2P(O)CHNaCO_2Et$, (*Z*):(*E*) = 10:90, 85% or ii, $Me_3SiCH(Li)CO_2Et$, (*Z*):(*E*) = 90:10, 90%] (352) (81)

(353) →[$Me_3SiCH(Li)CO_2Et$, (*Z*):(*E*) = 89:11, 86%] (354) (82)

(355) →[i, $Me_3SiCH(Li)CO_2Et$, (*Z*):(*E*) = 67:33, 82% or ii, $Me_3SiCH(Li)CO_2Bu^t$, (*Z*):(*E*) = 82:18, 56%] (356) (83)

The reaction has been extended to α-silyl lactones and lactams.[263] Other stabilizing groups have been demonstrated to be effective. For example, the Peterson reagent formed from bis(trimethylsilyl)propyne (**357**) and that formed from α-silyl acetonitrile derivatives (**358**) both give the (*Z*)-alkene as the predominant product (equation 84).[264]

$R^2R^3R^4Si$–CH(M)R^1 + cyclohexyl–CHO → (*Z*)-cyclohexyl–CH=CH–R^1 (84)

(357) $R^1 = C{\equiv}CSiMe_3$, $R^2 = Bu^t$, $R^3 = R^4 = Me$, M = MgBr; (*Z*):(*E*) = 30:1, 75%

(358) $R^1 = CN$, $R^2 = R^3 = R^4 = Ph$, M = MgI; (*Z*):(*E*) = 9:1, 80%

Increasing the steric bulk of the silyl group enhances the ratios of (Z)- to (E)-alkene isomers. The size of the ester group can also affect the (*E*):(*Z*) ratios of alkenes.[265] An example is the synthesis of butenolides (**360**) by Peterson alkylation of α-keto acetals (**359**; equation 85). Increasing the ester from methyl to isopropyl and *t*-butyl resulted in a corresponding increase in (*Z*)-alkene formation.

(85)

(359) → **(360)** ($Me_3SiCH_2CO_2R$)

R = Me; *(Z):(E)* = 43:57, 70%
R = Pr^i; *(Z):(E)* = 80:20, 73%
R = Bu^t; *(Z):(E)* = 88:12, 77%

Boeckman and coworkers studied the reaction of bis(trimethylsilyl) ester (**361**) with aldehydes to form the silyl-substituted unsaturated ester (**362**; equation 86).[266] The anion was formed with potassium or lithium diisopropylamide. Other metals, such as magnesium or aluminum, were introduced by treating the lithium anion with Lewis acids. The addition step produced a single diastereomer, enabling the effects of counterion and steric bulk on the elimination to be ascertained. Excellent selectivity for the (*E*)-isomer (**362**) may be obtained by using K or Li cations and a sterically hindered aldehyde. In studies directed toward the synthesis of substituted pseudomonic acid esters, the Peterson alkenation was utilized to form a mixture of (*Z*)- and (*E*)-alkene isomers, one example of which (**365**) is depicted in equation (87).[267] In this example the conditions were optimized to form the highest degree of selectivity for the (*Z*)-alkene.

$Bu^tO_2CCH(SiMe_3)_2$ + RCHO —M→ (362) + (363) (86)

(361) R = Pr^i, Bu^t; M = K, Li **(362)** >100:1 **(363)**

(364) —$MeCH(SiMe_3)CO_2Bu^t$, LDA, THF, 70%→ **(365)** *(Z):(E)* = 7.2:1 (87)

A comparison of the Wittig and Peterson alkenation procedures for the conversion of an aldehyde to its α,β-unsaturated vinylog was carried out by Schlessinger (Scheme 50).[268] The Wittig methodology was more effective for the base-sensitive lactone substrate (**366**), while the Peterson reaction was the preferred method for the hindered aldehyde (**368**). An additional comparative example of the Peterson and Wittig methods to synthesize an unsaturated aldehyde is found in studies on the streptogramin antibiotics by Meyers (Scheme 51).[269] Higher yields and (*E*)-alkene isomer selectivity (**371**) were obtained with the Wittig reagent. The Schlessinger method was recently used in the synthesis of FK-506 by Mills and coworkers (equation 88).[296a] Addition to the hindered aldehyde (**372**) produced the α,β-unsaturated aldehyde (**373**) in 78% yield and an (*E*):(*Z*) ratio of >100:1.

While these examples point to the possible use of the Wittig and the Peterson reactions as complementary methods for (*Z*)- and (*E*)-alkene formation in cases with conjugating functionality, it must be emphasized that no systematic predictive rule can be applied to the possible selectivity, and this area remains one of active research.

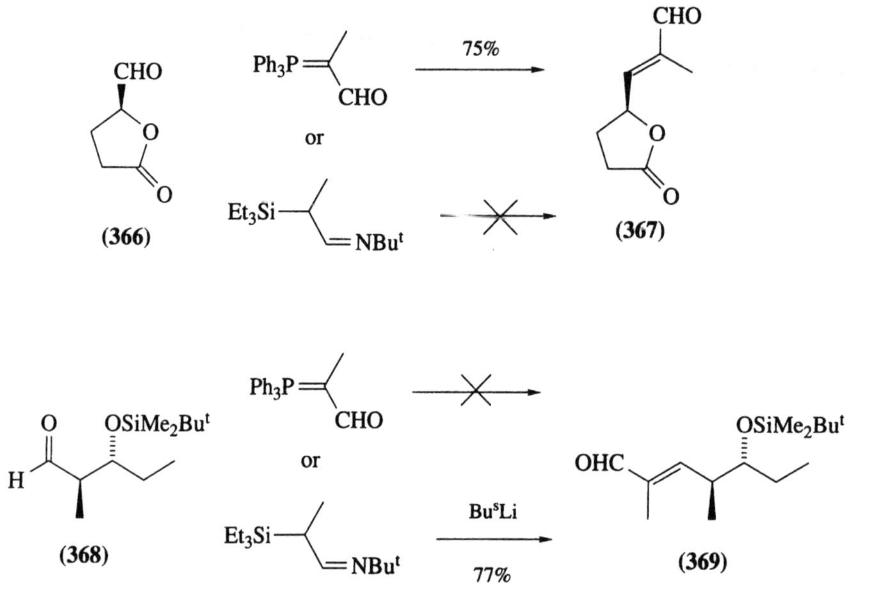

CHO
(366)
Ph_3P= CHO
or
Et_3Si =NBu^t
75%
(367)

$OSiMe_2Bu^t$
(368)
Ph_3P= CHO
or
Et_3Si =NBu^t
Bu^sLi
77%
OHC
(369)

Scheme 50

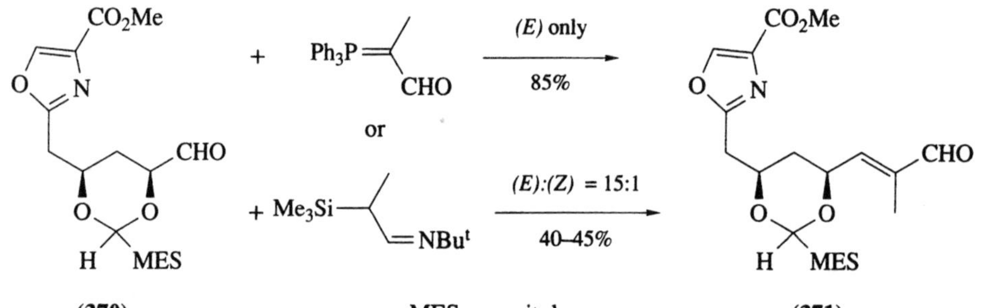

CO_2Me
(370)
+ Ph_3P= CHO
or
+ Me_3Si =NBu^t
(*E*) only
85%
(*E*):(*Z*) = 15:1
40–45%
(371)
MES = mesityl

Scheme 51

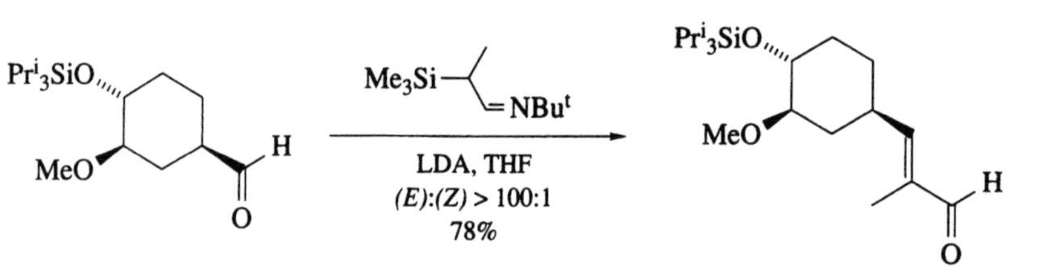

$Pr^i{}_3SiO$
MeO
(372)
Me_3Si =NBu^t
LDA, THF
(*E*):(*Z*) > 100:1
78%
(373)
(88)

3.1.10.7 Lewis Acid Catalysis

The use of cerium trichloride for the Peterson alkenation has been applied by Ueda and coworkers for nucleoside synthesis (Scheme 52).[270] The anion was generated with LDA/cerium trichloride and condensed with the aldehyde (**375**). The product was treated with KH in THF and the pyrimidopyridine (**377**) isolated in 50% yield.

3.1.11 SULFUR-STABILIZED ALKENATIONS: THE JULIA COUPLING

In 1973, Julia introduced the reaction depicted in Scheme 53, that bears his name.[271] The sulfone derivative is metallated (**378**) and added to the carbonyl, followed by functionalization (**380**), and reductive elimination, to produce the alkene (**381**). The yield for the transformation from carbonyl to alkene is ex-

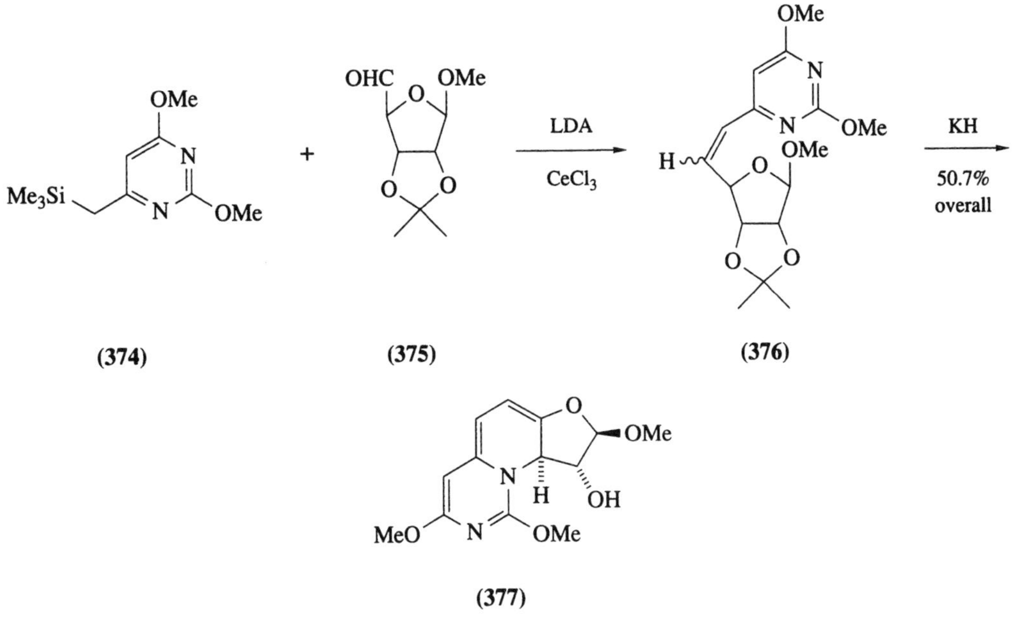

LDA, $CeCl_3$ → KH, 50.7% overall

(374) **(375)** **(376)**

(377)

Scheme 52

tremely high; usually greater than 80% overall. In the original communication, the Julia coupling was applied to the synthesis of mono-, di- and tetra-substituted alkenes, and since then has been employed to solve many challenging synthetic problems.[272] The selectivity can be extremely high for the production of disubstituted (*E*)-alkenes and it is this particular aspect of the Julia coupling in the synthesis of complex molecules that will be considered below.

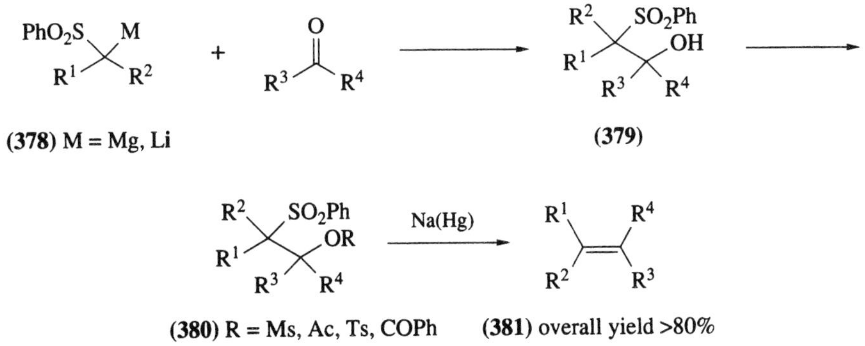

(378) M = Mg, Li **(379)**

Na(Hg)

(380) R = Ms, Ac, Ts, COPh **(381)** overall yield >80%

Scheme 53

3.1.11.1 (*E*)-/(*Z*)-Selectivity

A study carried out by Kocienski and Lythgoe first demonstrated the *trans* selectivity of the Julia coupling process.[273] The authors found the reductive elimination could best be carried out with the acetoxy or benzoyloxy sulfones. If the lithio sulfone derivative is used for addition to the carbonyl, the reaction can be worked up with acetic anhydride or benzoyl chloride to obtain the alkene precursor. In cases where enolization of the carbonyl is a complication, the magnesium derivative can frequently be used successfully.[274] A modification of the reductive elimination was found to be most effective. Methanol, ethyl acetate/methanol or THF/methanol were the solvents of choice and a temperature of –20 °C was effective at suppressing the undesired elimination of the acetoxy group to produce the vinyl sulfone. With these modifications of the original procedure, the ability of the reaction to produce dienes as well as *trans*-disubstituted alkenes was demonstrated. The diastereoisomeric *erythro*- and *threo*-acetoxy sulfones could be separated and it was demonstrated that both isomers were converted to the *trans*-alkene. It was hypothesized that the (*E*)-selectivity is derived from the reductive removal of the phenylsulfonyl group, generating an anion (**383**) that assumes the low energy *trans* configuration before loss of the acylate anion (Scheme 54). As demonstrated by numerous examples, the mechanism for reductive elimination is consistent with the finding that the alkenes obtained are the thermodynamic mixture and that increased branching at the site of elimination should, for steric reasons, increase the *trans* selectivity.[273c,d]

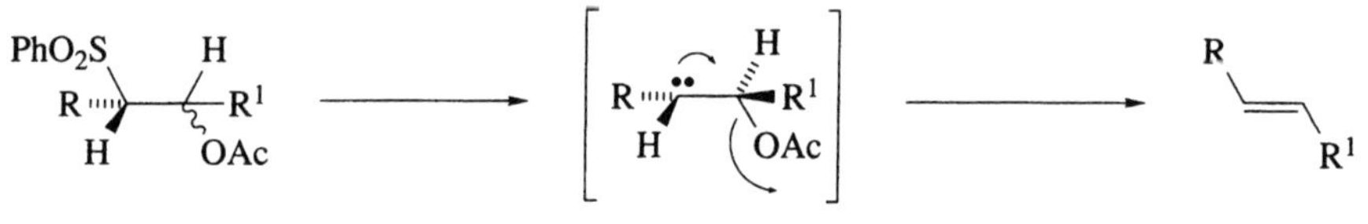

(**382**) (**383**) (**384**)

Scheme 54

3.1.11.2 Reductive Cleavage

The reduction of the β-acyloxy sulfone is most often carried out with sodium amalgam, as the examples below indicate. The reductive elimination can be buffered with disodium hydrogenphosphate for sensitive substrates.[275] In certain applications it has proven advantageous to utilize lithium or sodium in ammonia. For example, Keck's synthesis of pseudomonic acid C made use of the lithium/ammonia reductive elimination to simultaneously form an alkene and deprotect a benzyl ether.[281] In studies directed toward the same target, Williams made use of a reductive elimination procedure developed by Lythgoe, involving the formation of the xanthate ester followed by reduction with tri-*n*-butyltin hydride.[276]

3.1.11.3 The Synthesis of (*E*)-Disubstituted Alkenes: Comparison with the Wittig Reaction

3.1.11.3.1 The synthesis of pseudomonic acid C

The Julia coupling can be utilized as an alternative to the Schlosser–Wittig reaction to form (*E*)-alkenes.[277] Several reported syntheses of pseudomonic acid C (**385**) have provided interesting clues as to variability of applications of the Julia coupling in the context of natural product synthesis.

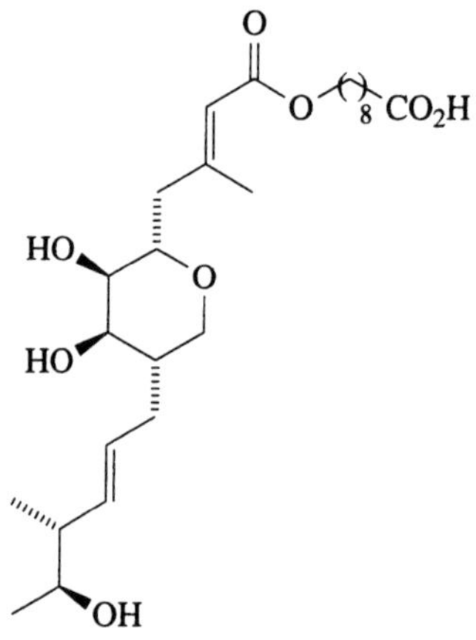

(**385**)

The (*E*)-disubstituted alkene has been extensively studied with several applications of the Julia coupling attempted.[278] The first synthesis of the natural product was accomplished by Kozikowski.[279] In this approach, the aldehyde (**386**) was reacted with 2 equiv. of the Wittig reagent (**387**) to produce (**388**), with an (*E*):(*Z*) ratio of 60:40 (equation 89; no yield given). This first synthesis establishes the baseline selectivity achievable with the Wittig reaction and is useful for subsequent comparison to the Julia coupling. Further studies by Kozikowski on methyl deoxypseudomonate B utilized the Julia coupling to form a similar alkene (equation 90).[280] The aldehyde (**389**) was coupled with the anion of the sulfone (**390**) and the hydroxyl group isolated as the benzoate. The alkene (**391**) was produced by treatment with Na(Hg) to give the desired (*E*)-alkene (no yield specified).

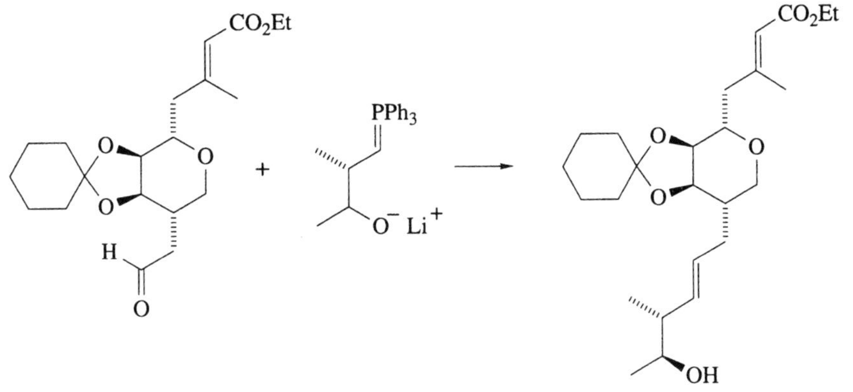

(89)

(**386**) (**387**) (**388**) (*E*):(*Z*) = 60:30

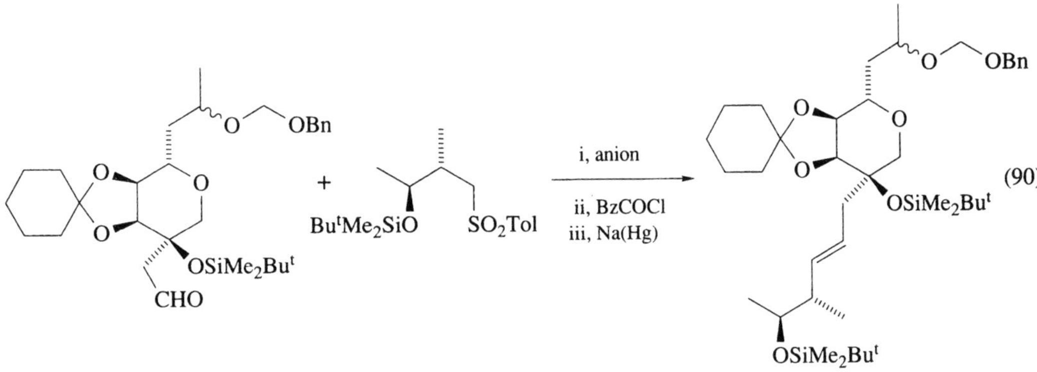

i, anion; ii, BzCOCl; iii, Na(Hg) (90)

(**389**) (**390**) (**391**)

Keck attempted to apply the Julia coupling to the synthesis of pseudomonic acid C.[281] Despite the success of the sulfone (**393**) in reactions with simple aldehydes, only modest yields of the desired coupling were observed. This problem was solved by reversing the aldehyde (**395**) and sulfone components (**394**), as shown in Scheme 55. The anion was formed with LDA in THF and condensed with the aldehyde. The β-hydroxysulfone was converted to the mesylate, and the reduction and simultaneous deprotection of the benzylglycoside was carried out with lithium and ammonia to produce the (*E*)-alkene (**396**), in 37% overall yield with excellent selectivity.

Williams carried out a Julia coupling similar to the Keck example. With the removal of the acetal functionality, the coupling step of the Julia reaction was efficient, but the usual reductive elimination procedure failed.[282] As an alternative to the acetylation and reductive elimination procedure, the β-sulfonyl xanthate was formed by quenching the addition reaction with carbon disulfide and methyl iodide. Reductive elimination was then carried out with tri-*n*-butyltin hydride to yield the desired (*E*)-alkene (**399**) in an 85:15 ratio with the (*Z*)-alkene in 83% overall yield (equation 91).

Finally, White has utilized a substrate (**400**) with which the Wittig reagent (**401**) gave a 37% yield of a 57:43 mixture of (*E*)- and (*Z*)-isomers (Scheme 56). Using the preparation of the sulfone (**393**) developed by Keck, White formed the anion with *n*-butyllithium in THF and condensed it with the aldehyde (**400**). The β-hydroxy sulfone was acetylated and reduced with sodium amalgam to produce a 17:3 ratio of (*E*)- to (*Z*)-alkenes (**402**) in 62% overall yield.

These four examples of the successful application of the Julia coupling in natural product synthesis indicate the sensitivity of various substrates to the anionic conditions. The solutions, interchanging the aldehyde and sulfone portions, modification of the substrate or altering the reductive elimination conditions, are all techniques that can enable the successful use of the Julia coupling for (*E*)-alkene synthesis.

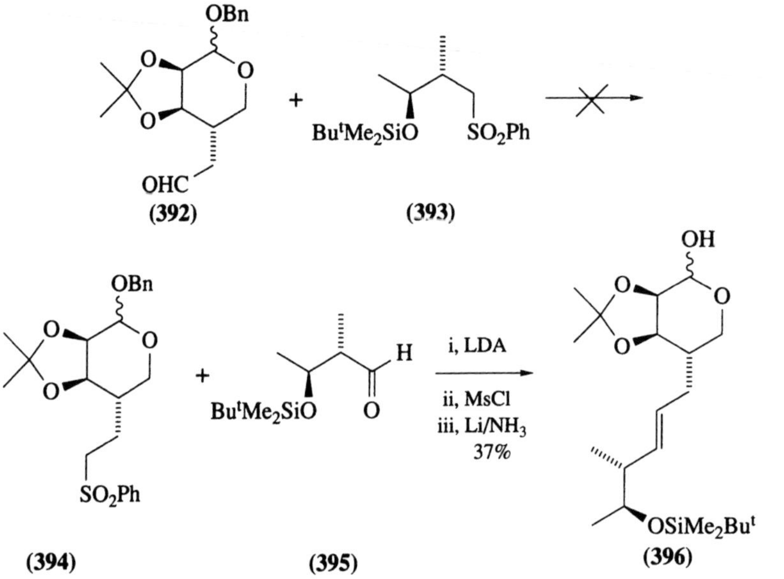

(392) + (393) —✕→

(394) + (395) → (396)

i, LDA
ii, MsCl
iii, Li/NH_3
37%

Scheme 55

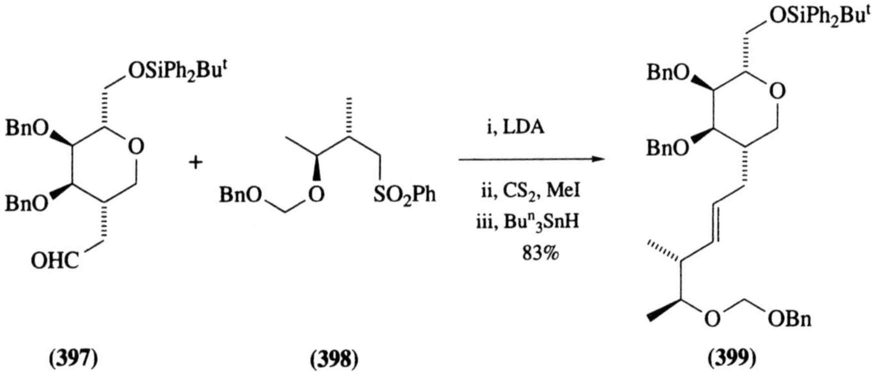

(397) + (398) → (399) (91)

i, LDA
ii, CS_2, MeI
iii, $Bu^n{}_3SnH$
83%

(400) + (393) → (402) (E):(Z) = 17:3

i, Bu^nLi; 84%
ii, Ac_2O; 93%
iii, Na(Hg), Na_2HPO_4; 79%

(400) + (401) → (E):(Z) = 57:43, 37%

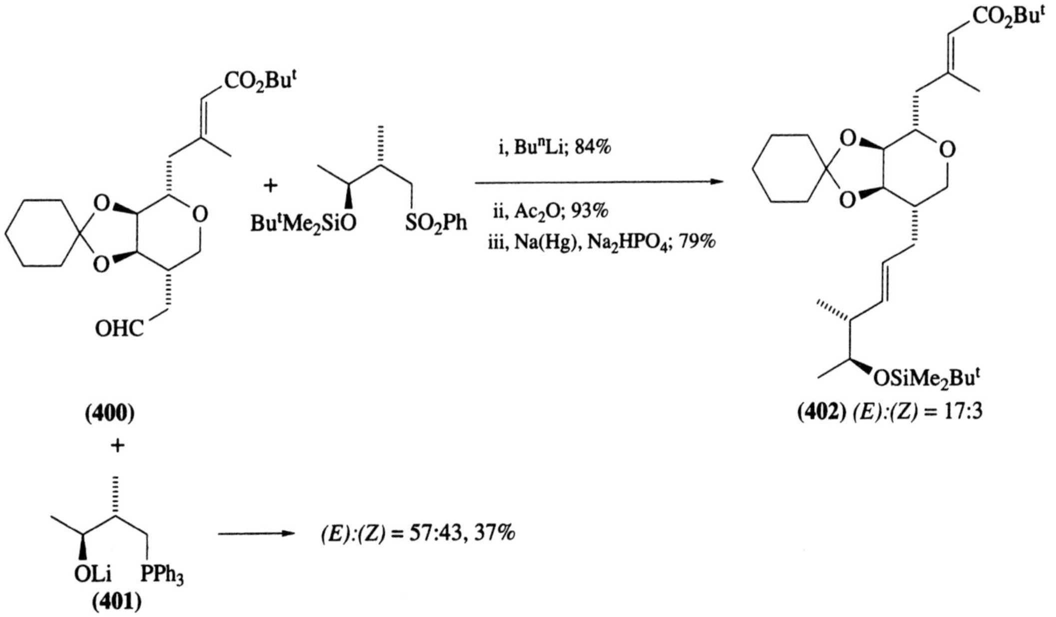

Scheme 56

3.1.11.3.2 Other examples

An interesting comparative example of the unstabilized Wittig reaction and the Julia coupling is found in the synthesis of the capsaicinoids by Gannett.[283] A selective route to both the (*E*)- and (*Z*)-isomers (**405**) and (**406**), respectively, is depicted in equation (92). The Wittig was carried out with potassium *t*-butoxide in DMF as the base, to form the (*Z*)-alkene (**406**) as the major product in a 91:9 ratio to the (*E*)-isomer. The Julia coupling and subsequent elimination produced the isomer (**405**) as a 9:1 mixture with the (*Z*)-isomer in 70–80% yields. Other examples of the Julia coupling for the synthesis of (*E*)-disubstituted alkenes are outlined in Table 18. From the examples cited, is apparent that the sulfone anion adds chemoselectively to aldehydes in the presence of esters and amides. The reagent is also compatible with a wide range of protecting groups, including ethers, acetals and amines.

(**403**) + (**404**) →

R = SO_2Ph, PPh_3

(**405**) + (**406**) (92)

3.1.11.4 (*E*)-Trisubstituted Alkene Synthesis

The Julia coupling has also been successfully utilized for the synthesis of more complex alkenes.[277] There are limitations to the application of the method to tri- and tetra-substituted alkenes, since the addition of the sulfone anion to a highly substituted ketone forms a β-alkoxy sulfone that is difficult to trap and isolate.[273d] There is a tendency for highly substituted β-alkoxy sulfones to revert back to the ketone sulfone. There have been several recent examples of the synthesis of trisubstituted (*E*)-alkenes worthy of note.

3.1.11.4.1 The milbemycins and avermectins

The milbemycins and the structurally related avermectins contain an (*E*)-trisubstituted alkene linkage that has been formed by the Julia coupling. The first application of the Julia coupling to avermectin was undertaken by Hanessian.[292] In this synthesis, the spiroketal portion was added as the sulfone (**407**) to the ketone (**408**; equation 93). The yield for the anion addition was 40%, but based on recovered sulfone (**407**) was 95%. The β-hydroxy sulfone was directly reduced to the (*E*)-trisubstituted alkene with excellent selectivity. In the synthesis of milbemycin, Barrett carried out a very similar transformation, but with the two components reversed (equation 94).[293] The sulfone (**410**) was metallated and condensed with the aldehyde (**411**). The adduct was isolated as the acetate in 86% yield and the mixture of isomers reduced with sodium amalgam in 86% overall yield. Although the conversion to the alkene (**412**) was efficient, the (*E*)- to (*Z*)-selectivity was 5:3. Hirama and coworkers have also applied the Julia coupling to the trisubstituted alkene portion of the avermectins (equation 95).[294] In this case, the addition of the sulfone anion (**414**) to the ketone (**413**; R = Me) failed and the starting materials were recovered. It was hypothesized that the lack of reactivity was due to the presence of oxygen substituents in the ketone which coordinate with the metal cation. In support of this theory, 3-methyl-2-butanone did undergo coupling with the sulfone and subsequent reductive elimination in satisfactory yield; however, the geometric selectivity in forming the trisubstituted alkene was a disappointing 2:1. In this case, the problem was solved by addition of the sulfone (**414**) to the aldehyde (**413**; R = H). The β-hydroxy sulfone (**415**) was converted to the enol triflate and displaced with dimethylcuprate.[295]

Table 18 Julia Coupling to Produce *(E)*-Disubstituted Alkenes[284]

Entry	*Carbonyl*	*Sulfone*	*Yield (%)*	*(E):(Z) ratio*	*Ref.*
1	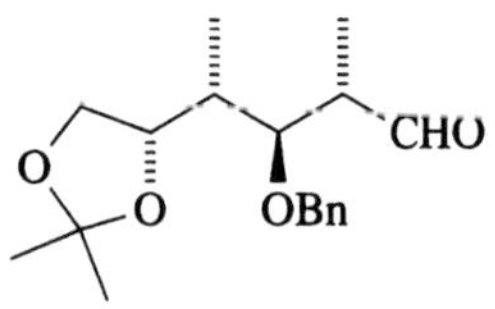 O, O, OBn, CHO	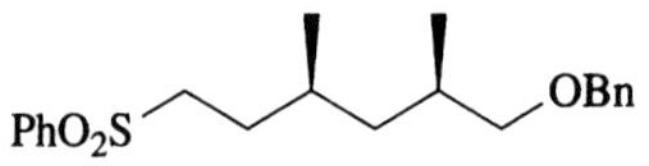 PhO_2S, OBn	60	9:1	285
2	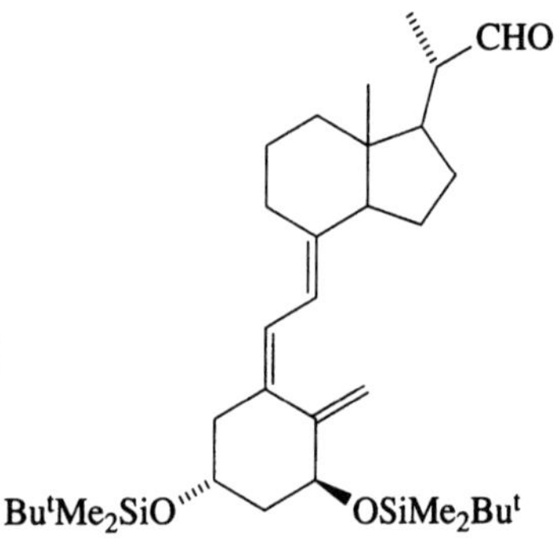 CHO, Bu^tMe_2SiO, $OSiMe_2Bu^t$	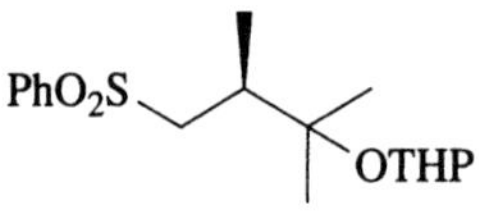 PhO_2S, OTHP	36	*(E)*	286
3	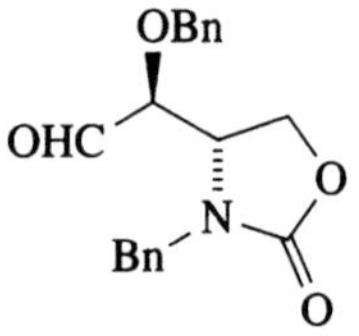 OBn, OHC, O, N, Bn, O	$PhO_2S-C_{14}H_{29}$	28	*(E)*	287
4	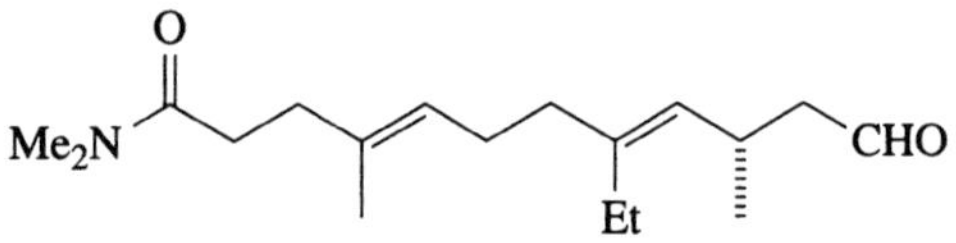 O, Me_2N, Et, CHO	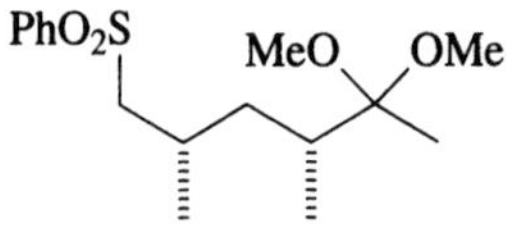 PhO_2S, MeO, OMe	68	4:1	288
5	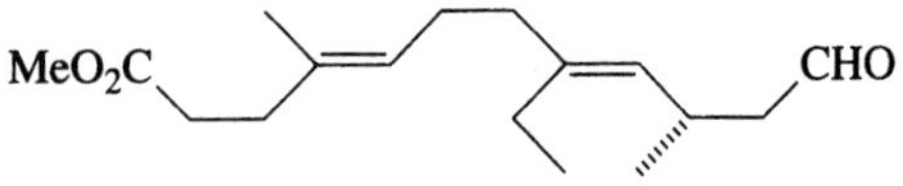 MeO_2C, CHO	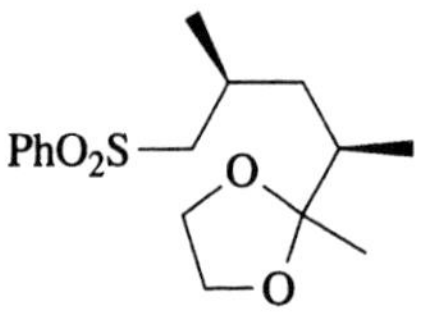 PhO_2S, O, O	35	*(E)*	289
6	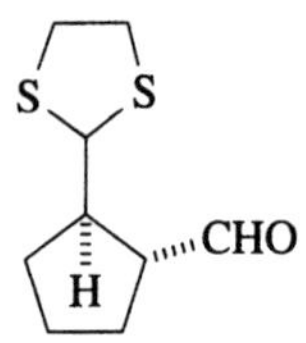 S, S, H, CHO	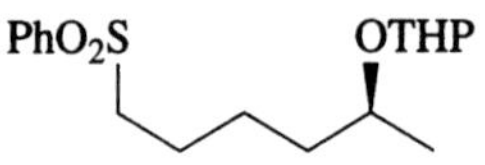 PhO_2S, OTHP	46	*(E)*	290
7	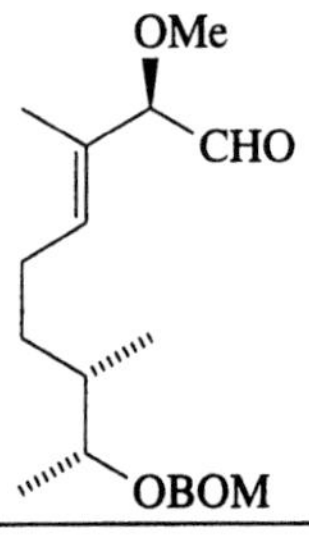 OMe, CHO, OBOM	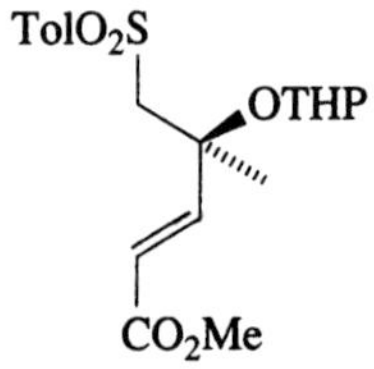 $TolO_2S$, OTHP, CO_2Me	30	*(E)*	291

(407) + (408) → (409) — i, Bu^nLi, 40%; ii, Na(Hg), MeOH, THF, KH_2PO_4, 40% (93)

(410) + (411) → (412) (E):(Z) = 5:3 — 86% (94)

(413) + (414) → (415) — R = Me (no reaction); R = H (95)

3.1.11.4.2 FK-506

Most recently, the immunosuppressive agent FK-506 (**416**) has been the target of total synthesis. To date several approaches to the trisubstituted alkene region at C-19 and C-20 have appeared. These preliminary studies allow the comparison between the Warren phosphine oxide approach and the Julia coupling. In the first total synthesis of FK-506, Jones and coworkers at Merck formed the the alkene by deprotonation of the phosphine oxide (**418**) and condensation with the aldehyde (**417**).[296] The hydroxyphosphine oxides were formed in a ratio of 1:1 in 77% yield. The less polar diastereomer was treated with base to obtain the (*E*)-alkene (**419**) in 32% overall yield from the aldehyde (equation 96). Danishefsky utilized the Julia coupling for the formation of the trisubstituted alkene region.[297] The sulfone anion (**420**) was treated with isobutyraldehyde as a model, followed by acetylation and reductive elimination to

(**416**)

give a 2–2.5:1 (*E*)- to (*Z*)-alkene mixture (**421**) in 83% yield for the addition and 33% yield for the acetylation and reductive elimination (equation 97).

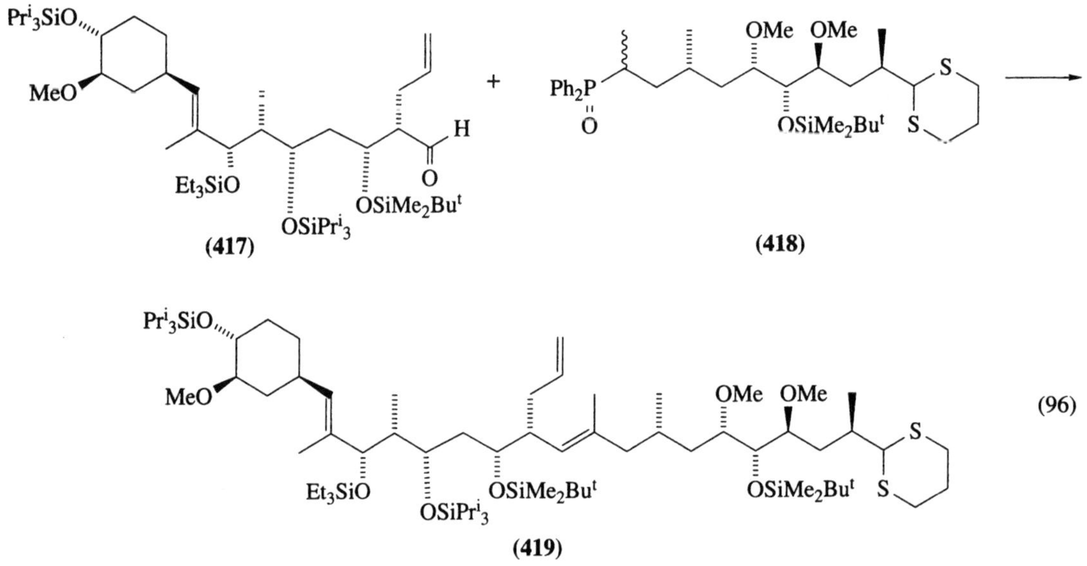

(**417**) + (**418**) → (**419**) (96)

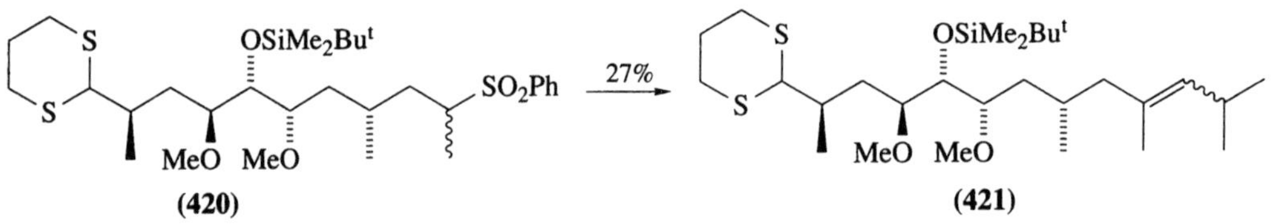

(**420**) —27%→ (**421**) (97)

In a more direct comparison of the phosphine oxide elimination with the sulfone, Schreiber employed a identical system to Danishefsky, but used the phosphine oxide (**422**).[298] Reaction with isobutyraldehyde and subsequent elimination resulted in a 1:1 mixture of the (*E*)- and (*Z*)-alkenes (**421**; equation 98). It appears from the more complex example of the Merck synthesis and from this example, that the Julia coupling proceeds with higher (*E*)-selectivity, in similar yield.

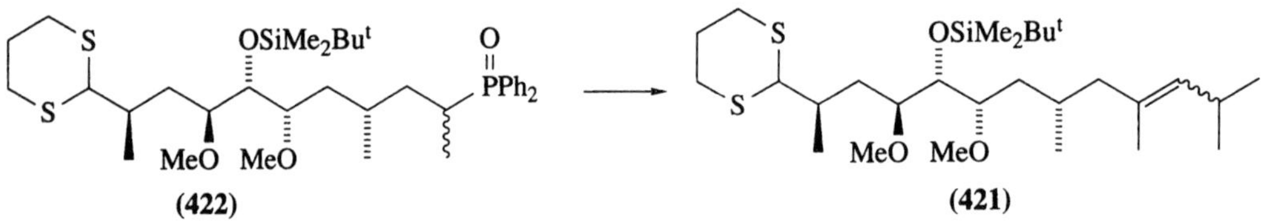

(**422**) → (**421**) (98)

3.1.11.5 Diene Synthesis

Use of the Julia coupling for complex natural product synthesis has provided some of the most significant examples of the broad utility of the reagent. The coupling procedure can be a very selective method of (*E*,*E*)-diene synthesis, as the examples below indicate.

3.1.11.5.1 X-14547 A

Ley synthesized the antibiotic X-14547 A using Julia coupling to form the (*E*,*E*)-diene portion (**424**; equation 99).[299] The sulfone anion was generated with Bu^nLi in THF/HMPA and the addition adduct trapped with benzoyl chloride. The alkene generated from the mixture of sulfones was found to be exclusively the (*E*,*E*)-isomer (**424**). In studies directed to the same target by Roush, an interesting comparative study of the Julia coupling, Horner–Wadsworth–Emmons, Wittig and phosphine oxide methods of alkene formation was discussed (see Scheme 57).[300] The synthesis of the key polyene intermediate was carried out by the addition of the phosphonate (**426**) to the aldehyde (**425**) to produce a 95% yield of a 95:5 mixture of the desired (*E*)-alkene (**427**). The corresponding phosphorane and phosphine oxide proceed with significantly lower selectivity (3.5:1) and lower yield (37%) respectively. As an alternative route, the Julia coupling of (**429**) and (**428**) formed the alkene in 49% overall yield and an (*E*):(*Z*) selectivity of 10:1.

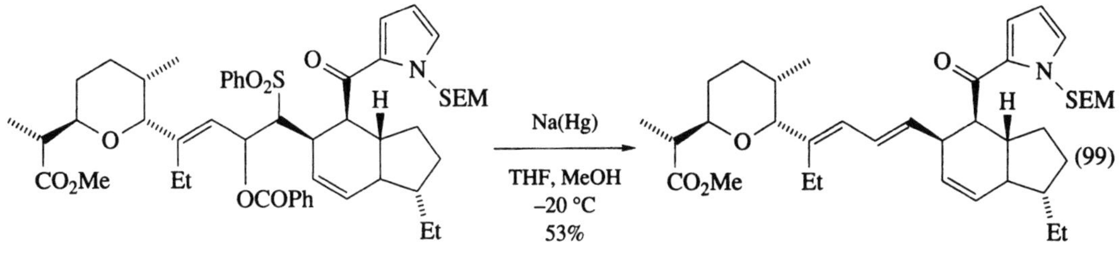

(**423**) → Na(Hg), THF, MeOH, −20 °C, 53% → (**424**) (99)

(**425**) + (**426**) → KOBu^t, DME; (*E*):(*Z*) = 95:5, 95% → (**427**)

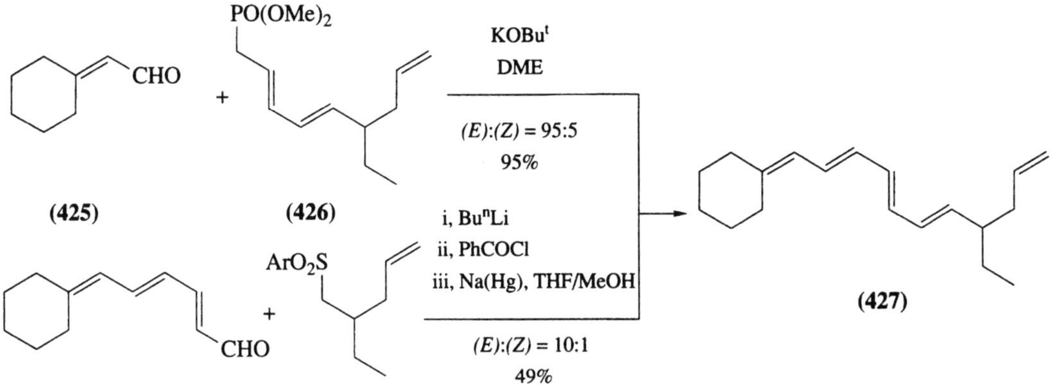

(**428**) + (**429**) → i, Bu^{n}Li; ii, PhCOCl; iii, Na(Hg), THF/MeOH; (*E*):(*Z*) = 10:1, 49% → (**427**)

Scheme 57

3.1.11.5.2 The diene portions of avermectin and milbemycin

The diene portions of avermectin and milbemycin have been synthesized by application of the Julia coupling. For the total synthesis of milbemycin β3 by Baker and coworkers, the aromatic ring was incorporated as the aldehyde (**431**) and the spiroketal portion added as the sulfone (**430**; equation 100).[301] The overall yield was 70–80% of the (*E,E*)-alkene (**432**), exclusively. The identical bond disconnection was studied by Kocienski, but with the aldehyde (**433**) and sulfone (**434**) components reversed (equation 101).[302] The anion was formed with LDA and, following functionalization and reductive elimination, the alkene was isolated in 39% yield in a 5:1 ratio of the (*E*)- and (*Z*)-isomers (**435**).

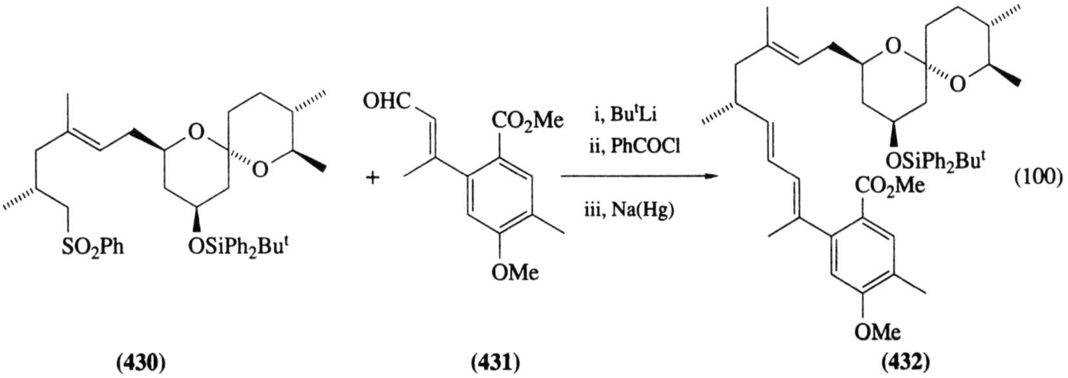

(**430**) + (**431**) → i, Bu^{t}Li; ii, PhCOCl; iii, Na(Hg) → (**432**) (100)

An interesting variation upon these approaches was published by Ley and coworkers.[303] In studies directed toward the synthesis of milbemycin β1, a vinyl sulfone (**437**) was deprotonated to form the diene

(433) + (434) → (435)

i, LDA, PhCOCl
ii, Na(Hg), THF–MeOH
(*E*):(*Z*) = 5:1
39%

(101)

unit (**438**; equation 102). The anion couples at the α-carbon with the aldehyde (**436**), and the β-hydroxy sulfone is derivatized with benzoyl chloride and subjected to reductive cleavage to form the desired diene (**438**) without contamination from the other isomers, in 25% overall yield.

(436) + (437) → (438)

i, Bu^tLi
ii, PhCOCl
iii, Na(Hg)
25%

(102)

In Hanessian's approach to avermectin discussed in Section 3.1.11.4.1, the Julia coupling was used for the trisubstituted alkene and the diene portion of the molecule.[292] The sulfone (**439**) was deprotonated with Bu^nLi and the aldehyde (**440**) added to it to obtain a 47% yield (77% based on recovered sulfone) of β-hydroxy sulfones (equation 103). The alcohol was converted to the chloride and the reductive cleavage carried out with sodium amalgam in 35% yield. The desired diene was the only detectable isomer (**441**). From the examples cited, it is apparent that the synthesis of (*E*,*E*)-dienes by the Julia coupling is an extremely successful process, in terms of both yield and selectivity.

3.1.11.5.3 The double elimination of the sulfone and the hydroxy component

An alternative approach to diene synthesis using the basic methodology of the Julia coupling has been studied by Otera. Polyenes (**445**) and alkynes (**443**) can be formed by double elimination of the sulfone and the hydroxy component (Scheme 58).[304]

The addition portion of the Julia coupling is run as usual. The β-hydroxy sulfone is functionalized and treated with $KOBu^t$ to effect the elimination. If an allylic hydrogen is present in the substrate, the

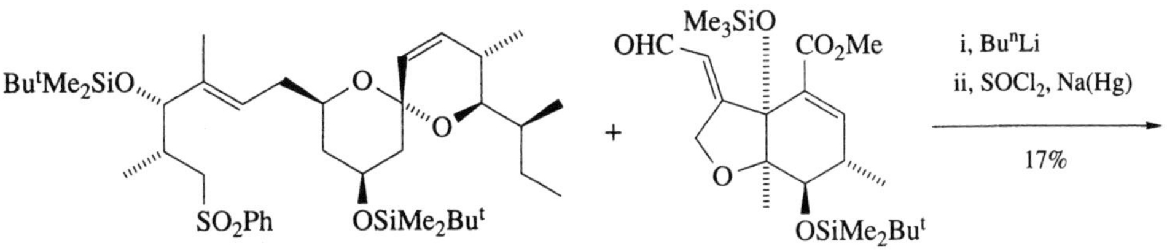

Bu^{t}Me$_2$SiO, SO$_2$Ph, OSiMe$_2$Bu^t **(439)** + OHC, Me$_3$SiO, CO$_2$Me, OSiMe$_2$Bu^t **(440)** → i, Bu^{n}Li; ii, SOCl$_2$, Na(Hg); 17%

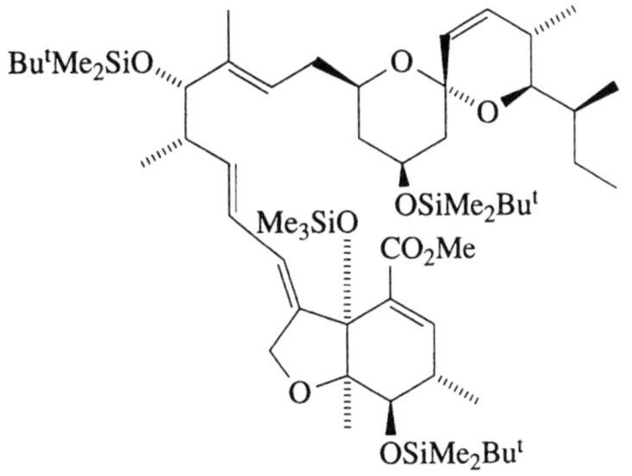

Bu^{t}Me$_2$SiO, Me$_3$SiO, OSiMe$_2$Bu^t, CO$_2$Me, OSiMe$_2$Bu^t **(441)** (103)

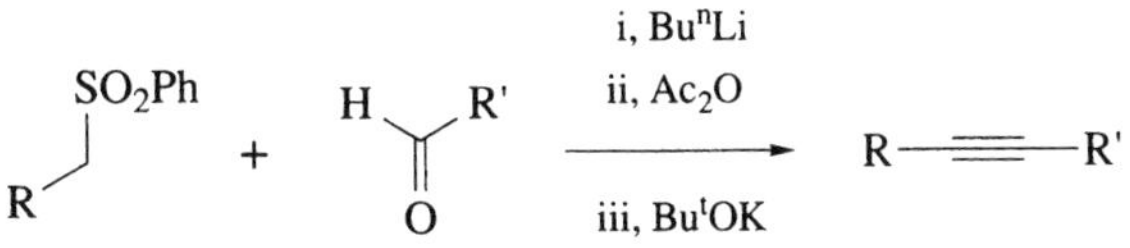

R–CH$_2$SO$_2$Ph **(442)** + HC(O)R' → i, Bu^{n}Li; ii, Ac$_2$O; iii, Bu^{t}OK → R—≡—R' **(443)**

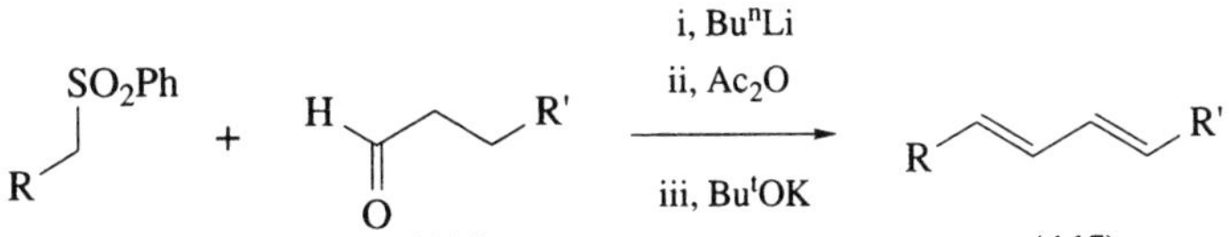

R–CH$_2$SO$_2$Ph + HC(O)CH$_2$CH$_2$R' **(444)** → i, Bu^{n}Li; ii, Ac$_2$O; iii, Bu^{t}OK → R–CH=CH–CH=CH–R' **(445)**

Scheme 58

polyene is formed, otherwise the alkyne is the product. Mechanistically, the alkyne formation proceeds by elimination of the acetoxy or alkoxy group to the vinyl sulfone, followed by elimination of the sulfone.[305] In the case of the polyene, the vinyl sulfone is formed, followed by isomerization to produce allyl sulfone. This substrate then undergoes 1,4-elimination to form the polyene. This sequence has been applied to the synthesis of muscone, methyl retinoate and vitamin A.[306]

In cases where the alkene is conjugated to a carbonyl, the polyene is formed with virtually exclusively the (*E*)-geometry.[307] In the case of the dienamides (**447**) the selectivity was 88–95% for the production of the (*E*,*E*)-isomer (equation 104).

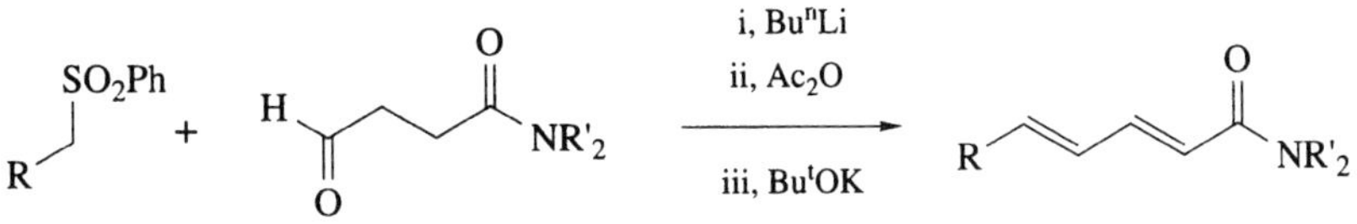

R–CH$_2$SO$_2$Ph + HC(O)CH$_2$CH$_2$C(O)NR'$_2$ **(446)** → i, Bu^{n}Li; ii, Ac$_2$O; iii, Bu^{t}OK → R–CH=CH–CH=CH–C(O)NR'$_2$ **(447)** (104)

3.1.11.6 Reaction with Esters

As discussed in the sections above, the sulfonylmethane anion reacts with ketones and aldehydes chemoselectively in the presence of an ester or amide. It is possible to obtain the keto sulfone on addition to an ester (**448**) when the reaction is carried out without competing carbonyl groups (equation 105). The keto sulfone (**450**) can either be reduced to the β-hydroxy sulfone (**451**) and carried through to the alkene (**452**), or the ketone can be trapped as the enol phosphinate (**453**), and either reduced to the alkene (**452**), eliminated to the vinyl sulfone (**454**)[308] or cleaved to the alkyne (**455**),[313] as shown in Scheme 59.

(**448**) X = OR^3, Cl + (**449**) → (**450**) (105)

Scheme 59

3.1.11.6.1 *Alkene formation*

As discussed in Section 3.1.11.1, which covers the reductive cleavage of the β-hydroxy sulfone derivatives to alkenes, the Julia reaction proceeds by the formation of an anion that is able to equilibrate to the thermodynamic mixture prior to elimination. Therefore, there is no inherent advantage in producing the *erthyro-* or *threo*-β-hydroxy sulfone selectively from the keto sulfone. The (*E*)/(*Z*)-mixture of alkenes should be the same.[273] This method is used to produce alkenes in cases where the acid derivative is more readily available or more reactive. The reaction of the sulfone anion with esters to form the keto sulfone, followed by reduction with metal hydrides has been studied.[308] The steric effects in the reduction do become important for the reaction to produce vinyl sulfones, which are formed from the *anti* elimination of the β-hydroxy sulfone adduct, as mentioned in Section 3.1.11.6.2. Some examples of the use of esters are presented below.

An interesting intramolecular application of the Julia coupling through an ester was carried out by Kang and coworkers in the total synthesis of trinoranastreptene (**458**; Scheme 60).[309] The reduction produced isomeric alcohols, which were separated by chromatography. The major isomer was carried through the sequence shown in Scheme 60, while the minor isomer was resistant to functionalization.

(**456**) →(i, 88%) (**457**) →(ii–iv, 77%) (**458**)

i, LDA, THF, 0 °C; ii, $NaBH_4$, Et_2O, MeOH; iii, Ac_2O; iv, Na(Hg), EtOAc, MeOH

Scheme 60

The reaction of lactone (**459**) with *p*-tolylsulfonylmethylmagnesium iodide, followed by reduction of the keto sulfone to the β-hydroxy sulfone has been studied.[310] In this paper, the β-hydroxy sulfone was reduced to the alkene (**460**) electrochemically (equation 106). The method was applied to a wide variety of esters and lactones. The yield of the reduction to the methylene derivative was 70–87%.

i, $ArSO_2CH_2MgI$
ii, $NaBH_4$
iii, e^-
73% overall

(459) (460) (106)

Both the enol phosphate method, and the reduction of the keto sulfone to the β-hydroxy sulfone followed by reductive cleavage to produce (*E*)-alkenes, have been applied to the synthesis of brefeldin A. Gais and coworkers added the sulfone (**462**) to the lactone (**461**), formed the enol phosphate and reduced with sodium and ammonia to the (*E*)-alkene (**463**), in a 65% overall yield (equation 107).[311]

i, LiHMDS
ii, PhCOCl, 90%
iii, $ClPO(PPh)_2$, 94%
iv, Na, NH_3, 76%

(461) (462) (463) (107)

Trost added the sulfone (**465**) to the ester (**464**), and reduced the ketone to the alcohol.[312] In this example the formation of the enol derivative of the keto sulfone failed. The hydroxy group was functionalized with acetic anhydride, and the reduction to the alkene (**466**) carried out with the standard sodium amalgam conditions in 54% overall yield, as shown in equation (108).

i, LDA, THF, 89%
ii, $NaBH_4$
iii, Ac_2O, 94%
iv, Na(Hg), 64%

(464) (465) (466) (108)

3.1.11.6.2 Vinyl sulfone formation

Julia has undertaken studies on the selective reduction of keto sulfones, followed by the *trans* elimination to the vinyl sulfone.[308] These intermediates can then be reacted with Grignard reagents to produce trisubstituted alkenes.[295]

3.1.11.6.3 Alkyne formation

In 1978, Bartlett published a study of the reaction of esters and acid chlorides to produce β-keto sulfones (**467**), which could be converted to the enol phosphinates (**468**) and reduced with sodium in ammonia to the alkyne (**469**; Scheme 61).[313] The overall yields for the process are good. As discussed in

Section 3.1.11.5.3, the double elimination developed by Otera can also be employed in the synthesis of alkynes.

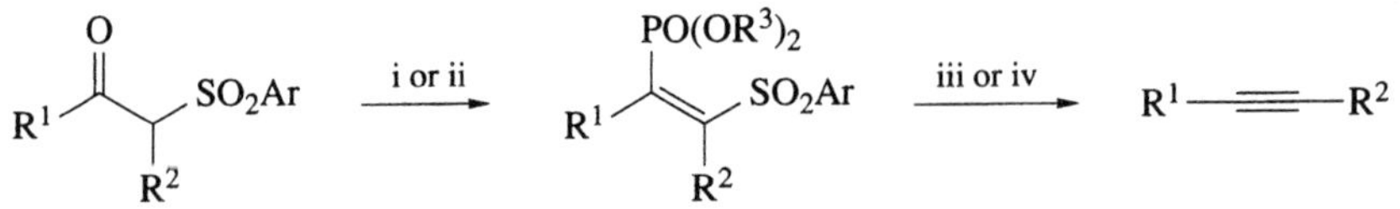

i, NaH or KH, $(R^3O)_2POCl$; ii, DMAP, Et_3N, $(R^3O)_2POCl$; iii, Na/NH_3; iv, Na(Hg)/THF

Scheme 61

3.1.11.7 Lewis Acid Catalyzed Additions of Sulfone Anions to Carbonyls

In a study by Wicha directed to the synthesis of prostaglandins from the Corey lactone, the use of $BF_3{\cdot}Et_2O$ to catalyze the addition of the lithium sulfone anion (**470**) to aldehydes was demonstrated (equation 109).[314] The use of Lewis acid catalysis results in significantly improved yields for the addition component of the Julia coupling. In this example, the addition of either the lithium or the magnesium sulfone anion proceeded in low yield. With the addition of $BF_3{\cdot}Et_2O$, the β-hydroxy sulfone can either be isolated, or directly converted to an alkene in one pot. This sequence was originally developed to deal with the specific problem of α-hydroxy aldehydes, and the difficulty of sulfone anion addition to these adducts. Other problems with addition of the sulfone adduct may be amenable to this solution as well.

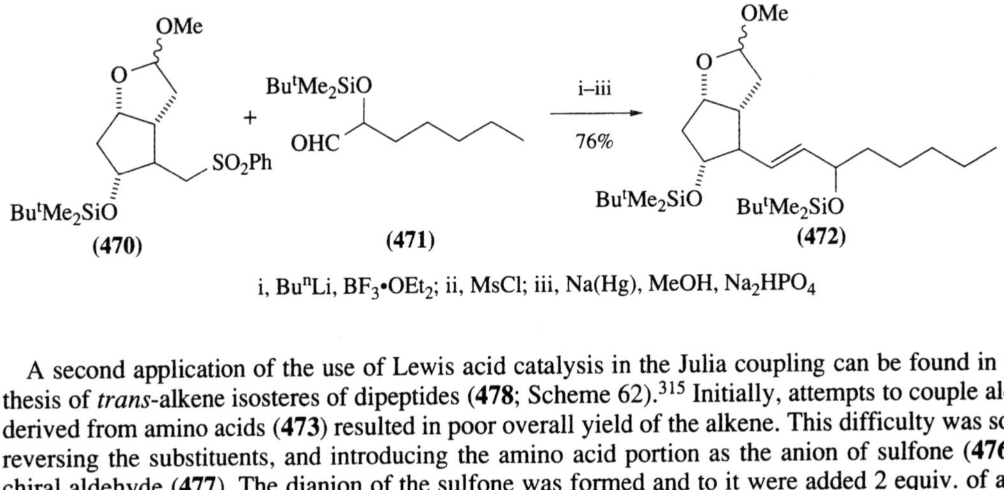

(109)

i, Bu^nLi, $BF_3{\bullet}OEt_2$; ii, MsCl; iii, Na(Hg), MeOH, Na_2HPO_4

A second application of the use of Lewis acid catalysis in the Julia coupling can be found in the synthesis of *trans*-alkene isosteres of dipeptides (**478**; Scheme 62).[315] Initially, attempts to couple aldehydes derived from amino acids (**473**) resulted in poor overall yield of the alkene. This difficulty was solved by reversing the substituents, and introducing the amino acid portion as the anion of sulfone (**476**) to the chiral aldehyde (**477**). The dianion of the sulfone was formed and to it were added 2 equiv. of aldehyde and 1 equiv. of diisobutylaluminum methoxide. The resulting β-hydroxy sulfone was taken on to the reductive elimination step to produce the desired (*E*)-alkene (**478**), in 74% overall yield.

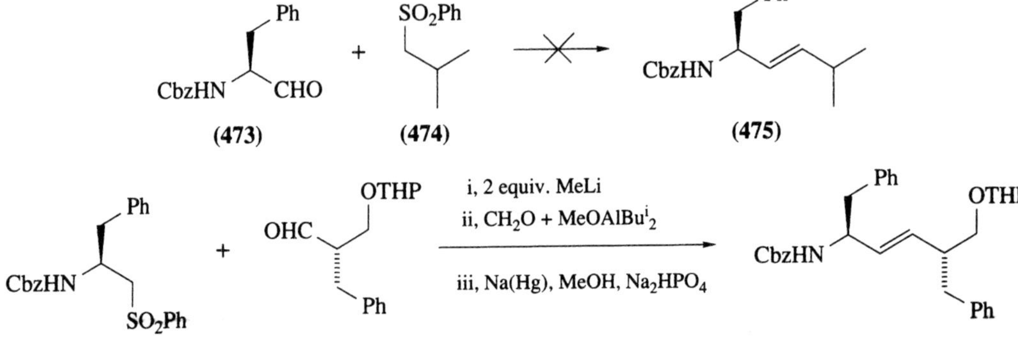

i, 2 equiv. MeLi; ii, CH_2O + $MeOAlBu^i_2$; iii, Na(Hg), MeOH, Na_2HPO_4

(**478**) 74% overall

Scheme 62

3.1.12 METHODS OF ALKENE FORMATION BY METAL CARBENE COMPLEXES

There are many metals that form alkylidene complexes.[316] Among these, the Tebbe reagent has proven to be of widespread generality and use in organic synthesis for the transfer of a methylene unit. As discussed in Section 3.1.6, a method using titanium and zinc has recently been applied for the methylenation of aldehydes and ketones. In the quest to extend this technology to longer chain alkyl groups, other metals have been investigated. Unfortunately the dimetallo precursors to metal carbene complexes are prone to β-hydride elimination, and this has limited the application of this methodology to alkylidenation. In addition, the methods of forming the metal carbene complexes can be arduous.[316] Consequently, these methods have not found widespread application in the context of natural product synthesis. Recently some mild and practical solutions to this problem have been discovered. These results are discussed in the following sections.

3.1.12.1 Chromium

A chromium carbene generated *in situ* from $CrCl_2$ and a dihaloalkyl derivative, has been utilized to form a variety of (*E*)-disubstituted alkene compounds (**480**). The general reaction is summarized in equation (110). Alkenyl halides, sulfides and silanes, as well as dialkyl-substituted alkenes, have been synthesized by this method.

$$\underset{\textbf{(479)}}{RCHO} \xrightarrow{CrCl_2,\ CHYX_2} \underset{\textbf{(480)}}{(E)\text{-}RCH{=}CHY} \qquad (110)$$

3.1.12.1.1 Alkenyl halides

Aldehydes may be converted to (*E*)-alkenyl halides[317] by the reaction of $CrCl_2$ with a haloform in THF.[318] The highest overall yields for the conversion were with iodoform, but somewhat higher (*E*):(*Z*) ratios were observed with bromoform or chloroform. Other low-valent metals, such as tin, zinc, manganese and vanadium, were ineffective. As the examples in Table 19 indicate, the reaction is selective for the (*E*)-isomer, except in the case of an α,β-unsaturated aldehyde. In addition, the reaction with ketones is sufficiently slow for chemoselectivity to be observed for mixed substrates.

Table 19 The Reaction of $CrCl_2$ with Dihalo Derivatives to Produce Alkenes

Substrate	*Reagent*	*Yield (%)*	*(E):(Z)*	*Ref.*
PhCHO	CHI_3	87	94:6	318
	$CHCl_3$	43	95:5	318
	$TMSCHBr_2$	82	(*E*) only	322
	$PhSCHCl_2$, LiI	83	82:18	322
	$PrCHI_2$	87	88:12	324
	Pr^iCHI_2	79	88:12	324
	Bu^tCHI_2	80	96:4	324
	CH_2I_2	70	—	324
$Me(CH_2)_7CHO$	CHI_3	82	83:17	318
	$TMSCHBr_2$	82	(*E*) only	322
	$PhSCHCl_2$, LiI	68	71:29	322
cyclohexyl-CHO	CHI_3	78	89:11	318
	$TMSCHBr_2$	81	(*E*) only	322
$MeCO(CH_2)_5CHO$	CHI_3	75	81:19	318
	$TMSCHBr_2$	76	(*E*) only	322
$Ph(CH_2)_2CHO$	$MeCHI_2$	85	97:3	324
Et_2CHCHO	$MeCHI_2$	99	98:2	324

This procedure has been utilized by Nicolaou to prepare the vinyl iodide (**482**) in the synthesis of lipotoxins A_5 and B_5 (equation 111).[319] The synthesis of vinyl iodides was also utilized by Bestmann in a stereospecific synthesis of (4*E*,6*E*,11*Z*)-hexadecatrienal. A Wittig reaction was used to create the (*E*)-alkene (**484**), and the (*E*)-diene portion was synthesized *via* vinyl iodide (**485**) formation and coupling (Scheme 63).[320] This method has been applied to the synthesis of ^{13}C-labeled vinyl iodide (**488**; equation 112).[321] Labeled iodoform (**486**) was made and the reaction conditions altered from those employed by Takai and coworkers in order to use smaller quantities of the valuable iodoform, relative to $CrCl_2$.

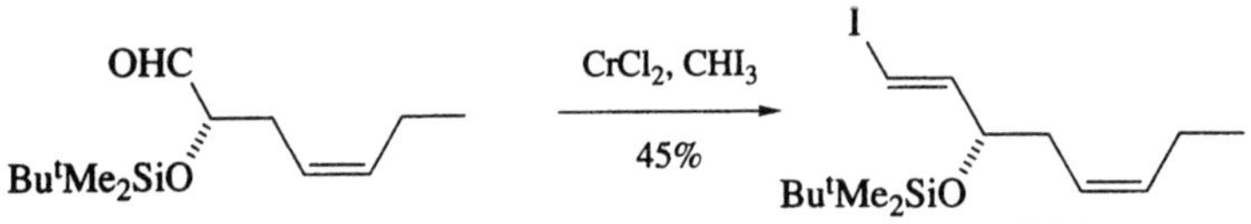

OHC, Bu^{t}Me$_2$SiO (**481**) —$CrCl_2$, CHI_3, 45%→ I, Bu^{t}Me$_2$SiO (**482**) (111)

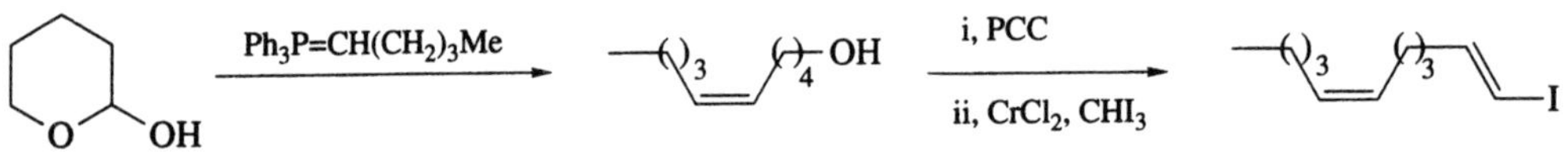

(**483**) O, OH —$Ph_3P=CH(CH_2)_3Me$→ (**484**) OH —i, PCC; ii, $CrCl_2$, CHI_3→ (**485**) I

Scheme 63

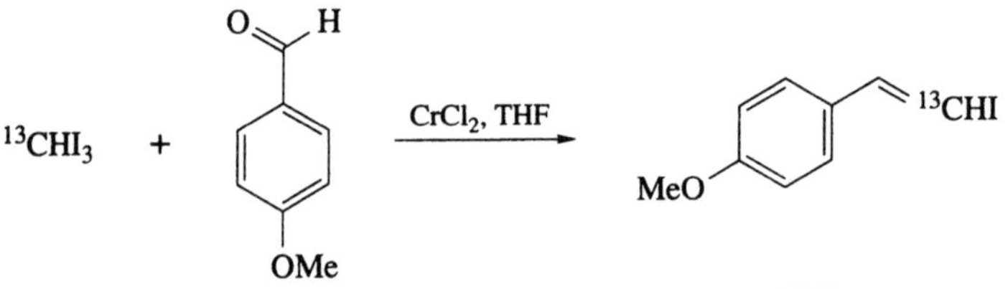

$^{13}CHI_3$ (**486**) + O, H, OMe (**487**) —$CrCl_2$, THF→ MeO, ^{13}CHI (**488**) (112)

3.1.12.1.2 *Substituted examples*

This reaction has been extended to alkenylsilanes and alkenyl sulfides, as summarized in Table 19.[322] In addition to various heteroatom-substituted examples, the reagent formed from $CrCl_2$ and a *gem*-diiodoalkane can be reacted with an aldehyde to selectively form the (*E*)-alkene. For the transfer of groups larger than ethylene with an aldehyde, it was found advantageous to add DMF. The conditions are compatible with nitriles, esters, acetals, alkynes and alkyl halides.[323] In the case of ethylidenation, the chromium carbene reacts with ketones in good yields.[324] It has not proven to be of general utility for the synthesis of trisubstituted alkenes either by the addition of a Cr derivative to a ketone or by the addition of a disubstituted Cr species to an aldehyde. An interesting comparative study of the Wittig reaction and the alkylidenation with Cr was carried out by Maxwell in the synthesis of several long chain ketones.[325] The all-(*Z*)-isomers (**490**) were synthesized by the salt-free Wittig reaction, while the all-(*E*)-isomers (**491**) were obtained by the $CrCl_2$ coupling of the diiodide (Scheme 64). The reaction has been applied to the synthesis of a segment of macbecins.[326] The aldehyde was treated with $CrCl_2$ and $MeCHI_2$ to produce the desired alkene (**493**) in 91% yield with 99% (*E*)-selectivity (equation 113).

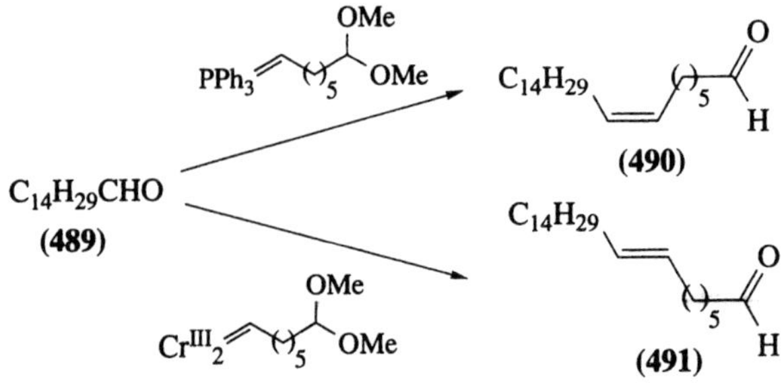

$C_{14}H_{29}CHO$ (**489**) —PPh_3, OMe, OMe→ $C_{14}H_{29}$, O, H (**490**)

$C_{14}H_{29}CHO$ (**489**) —$Cr^{III}{}_2$, OMe, OMe→ $C_{14}H_{29}$, O, H (**491**)

Scheme 64

(492) → (493): i, DIBAL-H; ii, $CrCl_2$, $MeCHI_2$; 91% (113)

As pointed out in this work, the mild conditions of the Cr reaction resulted in no observed epimerization, and offered an excellent alternative to the Schlosser modification of the Wittig reaction. Both the $CrCl_2$ chemistry and the $TiCl_4/Zn$ method discussed in Section 3.1.12.1.2 have the advantage of preparing the metallocarbene complex *in situ*.

3.1.12.2 Titanium–Zinc

The chemistry of Oshima and Takai, previously discussed in Section 3.1.6, has been extended to the incorporation of alkylidene (**495**) units when reacted with esters (**494**; equation 114).[327] The essential additive is TMEDA. The reaction is highly selective, producing the disubstituted (*Z*)-enol ether in good yield. The ratio of (*Z*)- to (*E*)-isomer is enhanced by increasing the steric repulsion of substituents R^1 and R^2 relative to the size of the ester functionality. In addition, the rate of reaction was increased with aromatic or unsaturated esters.[330] Although diiodoalkanes can be used in the reaction, better yields were obtained with the dibromo compounds. This reaction is complementary to the Tebbe reagent discussed in Section 3.1.5. The $CH_2Br_2/Zn/TiCl_4$ system does not react with esters even in the presence of TMEDA.[328] Therefore this method can be utilized to selectively form more highly substituted enol ethers, while the Tebbe reaction may be used to synthesize methylene derivatives. Examples are presented in Table 20.

(494) + (495) → (496): $TiCl_4$, Zn; THF, TMEDA (114)

The reaction is effective with electron-rich carbonyls such as trimethylsilyl esters and thioesters, as Table 20 indicates.[329] Lactones are substrates for alkylidenation; however, hydroxy ketones are formed as side products, and yields are lower than with alkyl esters.[327,330] Amides are also effective, but form the (*E*)-isomer predominantly. This method has been applied to the synthesis of precursors to spiroacetals (**499**) by Kocienski (equation 115).[330] The reaction was found to be compatible with THP-protected hydroxy groups, aromatic and branched substituents, and alkene functionality, although complex substitution leads to varying rates of reaction for alkylidenation. Kocienski and coworkers found the intramolecular reaction to be problematic.[330] As with the $CrCl_2$ chemistry, this reaction cannot be used with a disubstituted dibromoalkane to form the tetrasubstituted enol ether. Attempts were made to apply this reaction to alkene formation by reaction with aldehydes and ketones, but unfortunately the (*Z*):(*E*)-ratio of the alkenes formed is virtually 1:1.[327]

(497) + (498) → (499): $TiCl_4$, Zn; THF, TMEDA (115)

R = H, 88%; R = Et, 89%

(499)

Table 20 Synthesis of Enol Ether Derivatives by Zn, $TiCl_4$ and TMEDA

Carbonyl compound	*Alkyl halide*	*Yield (%)*	*(Z):(E)*	*Ref.*
$PhCO_2Me$	$MeCHBr_2$	86	92:8	327
$PhCO_2Me$	$n\text{-}C_5H_{11}CHBr_2$	89	95:5	327
$PhCO_2Me$	$c\text{-}C_6H_{11}CHBr_2$	61	90:10	327
$PhCO_2Bu^t$	$MeCHBr_2$	81	71:29	327
Pr^iCO_2Me	$n\text{-}C_5H_{11}CHBr_2$	89	100:0	327
$MeCHCHCO_2Et$	$n\text{-}C_5H_{11}CHBr_2$	90	94:6	327
$BuCO_2CH_2CHCH_2$	$n\text{-}C_5H_{11}CHBr_2$	52	92:8	327
$PhCO_2TMS$	$MeCHBr_2$	90	73:27	329a
	$BuCHBr_2$	84	73:27	329a
	$c\text{-}C_6H_{11}CHBr_2$	74	80:20	329a
$n\text{-}C_9H_{19}CO_2TMS$	$MeCHBr_2$	77	86:14	329a
$c\text{-}C_6H_{11}CO_2TMS$	$MeCHBr_2$	80	100:0	329a
$PhCHCHCO_2TMS$	$MeCHBr_2$	65	96:4	329a
$PhCOSMe$	$MeCHBr_2$	77	80:20	329b
	Bu^nCHBr_2	75	84:16	329b
$n\text{-}C_8H_{17}COSMe$	$MeCHBr_2$	94	73:27	329b
$c\text{-}C_6H_{11}COSMe$	$MeCHBr_2$	88	94:6	329b
	$PhCH_2CHBr_2$	95	100:0	329b
$PhC(O)N(CH_2)_5$ (N-acylpiperidine structure)	$MeCHBr_2$	70	2:98	329b
	$PhCH_2CHBr_2$	87	1:99	329b
$n\text{-}C_8H_{17}C(O)N(CH_2)_5$ (N-acylpiperidine structure)	$MeCHBr_2$	79	1:99	329b
$c\text{-}C_6H_{11}C(O)N(CH_2)_5$ (N-acylpiperidine structure)	$MeCHBr_2$	82	47:53	329b

3.1.13 REFERENCES

1. For a recent discussion comparing the Wittig methylenation to the methods discussed below see L. Fitjer and U. Quabeck, *Synth. Commun.*, 1985, **15**, 855.
2. D. J. Peterson, *J. Org. Chem.*, 1968, **33**, 780.
3. For a recent mechanistic study see P. F. Hudrlik, E. L. O. Agwaramgbo and A. M. Hudrlik, *J. Org. Chem.*, 1989, **54**, 5613; see also Section 3.1.10 for a more complete discussion of the mechanism.
4. For reviews see (a) T. H. Chan, *Acc. Chem. Res.*, 1977, **10**, 442; (b) P. Magnus, *Aldrichimica Acta*, 1980, **13**, 43; (c) W. P. Weber, 'Silicon Reagents for Organic Synthesis', Springer-Verlag, Berlin, 1983, vol. 14, p. 58; (d) D. J. Ager, *Synthesis*, 1984, 384; (e) R. Anderson, *Synthesis*, 1985, 717; (f) E. W. Colvin, 'Silicon Reagents in Organic Synthesis', Academic Press, New York, 1988, p. 63; (g) G. L. Larson, *J. Organomet. Chem.*, 1989, **360**, 39.
5. R. K. Boeckman, Jr. and S. M. Silver, *Tetrahedron Lett.*, 1973, 3497.
6. T. H. Chan and E. Chang, *J. Org. Chem.*, 1974, **39**, 3264.
7. For a discussion of the stereochemistry of the elimination see Section 3.1.10.
8. For a discussion of the preparation of silicon anions see refs. 4b, c and f; for a discussion of experimental procedures see refs. 4d and f.
9. D. V. Pratt and P. B. Hopkins, *J. Org. Chem.*, 1988, **53**, 5885.
10. R. J. Bushby and C. Jarecki, *Tetrahedron Lett.*, 1988, **29**, 2715.
11. S. P. Tanis, G. M. Johnson and M. C. McMills, *Tetrahedron Lett.*, 1988, **29**, 4521.
12. T. Kitahara, H. Kurata and K. Mori, *Tetrahedron*, 1988, **44**, 4339.
13. J. Lin, M. M. Nikaido and G. Clark, *J. Org. Chem.*, 1987, **52**, 3745.
14. E. Dunach, R. L. Halterman and K. P. C. Vollhardt, *J. Am. Chem. Soc.*, 1985, **107**, 1664.
15. P. J. Maurer and H. Rapoport, *J. Med. Chem.*, 1987, **30**, 2016.
16. G. B. Feigelson and S. J. Danishefsky, *J. Org. Chem.*, 1988, **53**, 3391.
17. F. A. Carey and W. Frank, *J. Org. Chem.*, 1982, **47**, 3548.
18. U. E. Udodong and B. Fraser-Reid, *J. Org. Chem.*, 1989, **54**, 2103.
19. See Section 3.1.6.
20. M. Yamaura, T. Suzuki, H. Hashimoto, J. Yoshimura and C. Shin, *Bull. Chem. Soc. Jpn.*, 1985, **58**, 2812; see also Section 3.1.5.

21. A. Cleve and F. Bohlmann, *Tetrahedron Lett.*, 1989, **30**, 1241; the titanium reagent is discussed in Section 3.1.6.
22. T. Kauffmann, R. Konig, C. Pahde and A. Tannert, *Tetrahedron Lett.*, 1981, **22**, 5031; the $CrCl_2$ derivative was also studied, but yields were superior with titanium.
23. For a complementary approach of protecting the aldehyde as the aldimine and forming the methylene derivative of the ketone see G. P. Zecchini, M. P. Paradisi and I. Torrini, *Tetrahedron*, 1983, **39**, 2709.
24. C. R. Johnson and B. D. Tait, *J. Org. Chem.*, 1987, **52**, 281.
25. For a study of Grignard additions catalyzed by cerium see (a) T. Imamoto, N. Takiyama and K. Nakamura, *Tetrahedron Lett.*, 1985, **26**, 4763; (b) T. Imamoto, N. Takiyama, K. Nakamura, T. Hatajima and Y. Kamiya, *J. Am. Chem. Soc.*, 1989, **111**, 4392.
26. See Section 3.1.10 for substituted examples of the cerium Peterson alkenation.
27. B. A. Narayanan and W. H. Bunnelle, *Tetrahedron Lett.*, 1987, **28**, 6261.
28. M. B. Anderson and P. L. Fuchs, *Synth. Commun.*, 1987, **17**, 621.
29. See Section 3.1.10.3 for further discussion.
30. G. J. P. H. Boons, G. A. van der Marel and J. H. Boom, *Tetrahedron Lett.*, 1989, **30**, 229.
31. C. Burford, F. Cooke, G. Roy and P. Magnus, *Tetrahedron*, 1983, **39**, 867.
32. C. R. Johnson, J. R. Shanklin and R. A. Kirchhoff, *J. Am. Chem. Soc.*, 1973, **95**, 6462.
33. For full experimental details and numerous examples see C. R. Johnson and R. A. Kirchhoff, *J. Am. Chem. Soc.*, 1979, **101**, 3602.
34. G. L. Bundy, *Tetrahedron Lett.*, 1975, 1957.
35. T. K. Schaaf, D. L. Bussolotti, M. J. Parry and E. J. Corey, *J. Am. Chem. Soc.*, 1981, **103**, 6502.
36. P. A. Aristoff and A. W. Harrison, *Tetrahedron Lett.*, 1982, **23**, 2067.
37. M. F. Ansell, J. S. Mason, and, M. P. L. Caton, *J. Chem. Soc., Perkin Trans. 1*, 1984, 1061.
38. H. Niwa, T. Hasagawa, N. Ban and K. Yamada, *Tetrahedron*, 1987, **43**, 825.
39. R. K. Boeckman, Jr., D. M. Blum and S. D. Arthur, *J. Am. Chem. Soc.*, 1979, **101**, 5060.
40. H. Niwa, K. Wakamatsu, T. Hida, K. Niiyama, H. Kigoshi, M. Yamada, H. Nagase, M. Suzuki and K. Yamada, *J. Am. Chem. Soc.*, 1984, **106**, 4547.
41. For the preparation of enantiomers of *N,S*-dimethyl-*S*-phenylsulfoxime see C. S. Shiner and A. H. Berks, *J. Org. Chem.*, 1988, **53**, 5542, and references cited therein.
42. C. R. Johnson and N. A. Meanwell, *J. Am. Chem. Soc.*, 1981, **103**, 7667.
43. For excellent reviews of all the types of application of this reagent see (a) C. R. Johnson, *Aldrichimica Acta*, 1985, **18**, 3; (b) C. R. Johnson, M. R. Barbachyn, N. A. Meanwell, C. J. Stark, Jr. and J. R. Zeller, *Phosphorus Sulfur*, 1985, **24**, 151; (c) C. R. Johnson, *Pure Appl. Chem.*, 1987, **59**, 969.
44. M.-A. Poupart and L. A. Paquette, *Tetrahedron Lett.*, 1988, **29**, 269, 273.
45. M. L. Boys, E. W. Collington, H. Finch, S. Swanson and J. F. Whitehead, *Tetrahedron Lett.*, 1988, **29**, 3365, and references cited therein.
46. D. R. Morton, Jr. and F. C. Brokaw, *J. Org. Chem.*, 1979, **44**, 2880.
47. C. R. Johnson and R. C. Elliott, *J. Am. Chem. Soc.*, 1982, **104**, 7041.
48. C. R. Johnson, R. C. Elliott and N. A. Meanwell, *Tetrahedron Lett.*, 1982, **23**, 5005.
49. J. H. Rigby and J. Z. Wilson, *J. Am. Chem. Soc.*, 1984, **106**, 8217.
50. R. R. Schrock, *J. Am. Chem. Soc.*, 1976, **98**, 5399.
51. F. N. Tebbe, G. W. Parshall and G. S. Reddy, *J. Am. Chem. Soc.*, 1978, **100**, 3611; for preparation of the reagent, see also ref. 52, and L. Cannizzo and R. H. Grubbs, *J. Org. Chem.*, 1985, **50**, 2386.
52. S. H. Pine, R. Zahler, D. A. Evans and R. H. Grubbs, *J. Am. Chem. Soc.*, 1980, **102**, 3270.
53. See refs. 51 and 52, and T. R. Howard, J. B. Lee and R. H. Grubbs, *J. Am. Chem. Soc.*, 1980, **102**, 6876.
54. S. H. Pine, R. J. Pettit, G. D. Geib, S. G. Cruz, C. H. Gallego, T. Tijerina and R. D. Pine, *J. Org. Chem.*, 1985, **50**, 1212.
55. J. S. Clark and A. B. Holmes, *Tetrahedron Lett.*, 1988, **29**, 4333.
56. R. E. Ireland, S. Thaisrivongs and P. H. Dussault, *J. Am. Chem. Soc.*, 1988, **110**, 5768.
57. F. E. Ziegler, A. Kneisley, J. K. Thottathil and R. T. Wester, *J. Am. Chem. Soc.*, 1988, **110**, 5434.
58. C. S. Wilcox, G. W. Long and H. Suh, *Tetrahedron Lett.*, 1984, **25**, 395.
59. T. V. RajanBabu and G. S. Reddy, *J. Org. Chem.*, 1986, **51**, 5458.
60. W. A. Kinney, M. J. Coghlan and L. A. Paquette, *J. Am. Chem. Soc.*, 1985, **107**, 7352.
61. H. Hauptmann, G. Muhlbauer and H. Suss, *Tetrahedron Lett.*, 1986, **27**, 6189.
62. J. R. Stille and R. H. Grubbs, *J. Am. Chem. Soc.*, 1986, **108**, 855.
63. C. J. Burrows and B. K. Carpenter, *J. Am. Chem. Soc.*, 1981, **103**, 6983.
64. R. E. Ireland and M. D. Varney, *J. Org. Chem.*, 1983, **48**, 1829.
65. T.-S. Chou and S.-B. Huang, *Tetrahedron Lett.*, 1983, **24**, 2169.
66. J. R. Stille and R. H. Grubbs, *J. Am. Chem. Soc.*, 1983, **105**, 1664.
67. L. Clawson, S. L. Buchwald and R. H. Grubbs, *Tetrahedron Lett.*, 1984, **25**, 5733.
68. L. Cannizzo and R. H. Grubbs, *J. Org. Chem.*, 1985, **50**, 2316.
69. J. J. Eisch and A. Piotrowski, *Tetrahedron Lett.*, 1983, **24**, 2043.
70. B. J. J. Van De Heisteeg, G. Schat, O. S. Akkerman and F. Bickelhaupt, *Tetrahedron Lett.*, 1987, **28**, 6493.
71. P. B. Mackenzie, K. C. Ott and R. H. Grubbs, *Pure Appl. Chem.*, 1984, **56**, 59; for a recent study of methylenation with dimethyltitanocene, see N. A. Petasis and E. I. Bzowej, *J. Am. Chem. Soc.*, 1990, **112**, 6392.
72. K. A. Brown-Wensley, S. L. Buchwald, L. Cannizzo, L. Clawson, S. Ho, D. Meinhardt, J. R. Stille, D. Straus and R. H. Grubbs, *Pure Appl. Chem.*, 1983, **55**, 1733.
73. This problem can be circumvented by not isolating before continuing with subsequent steps; see entries 2 and 4, Table 7; for a case in which isomerization was used to advantage see E. Wenkert, A. J. Marsaioli and P. D. R. Moeller, *J. Chromatogr.*, 1988, **440**, 449.
74. J. R. Stille and R. H. Grubbs, *J. Am. Chem. Soc.*, 1986, **108**, 855.
75. F. W. Hartner and J. Schwartz, *J. Am. Chem. Soc.*, 1981, **103**, 4979.

76. F. N. Tebbe and L. J. Guggenberger, *J. Chem. Soc., Chem. Commun.*, 1973, 227.
77. See also T. Yoshida, *Chem. Lett.*, 1982, 429. The reaction proceeds in moderate yield to give substituted alkenes, but with low stereoselectivity.
78. See Section 3.1.12 and (a) K. A. Brown-Wensley, S. L. Buchwald, L. Cannizzo, L. Clawson, S. Ho, D. Meinhardt, J. R. Stille, D. Straus and R. H. Grubbs, *Pure Appl. Chem.*, 1983, **55**, 1733; (b) J. H. Freudenberger and R. R. Schrock, *J. Organomet. Chem.*, 1986, **5**, 398, and ref. 50; (c) T. Kauffmann, M. Enk, W. Kaschube, E. Toliopoulos and D. Wingbermühle, *Angew. Chem., Int. Ed. Engl.*, 1986, **25**, 910; (d) T. Kauffmann, R. Abeln, S. Welke and D. Wingbermühle, *Angew. Chem., Int. Ed. Engl.*, 1986, **25**, 909; (e) A. Dormond, A. El Bouadili and C. Moïse, *J. Org. Chem.*, 1987, **52**, 688.
79. T. Okazoe, K. Takai, K. Oshima and K. Utimoto, *J. Org. Chem.*, 1987, **52**, 4410.
80. See ref. 73 and (a) J. W. S. Stevenson and T. A. Bryson, *Tetrahedron Lett.*, 1982, **23**, 3143; (b) P. A. Bartlett, Y. Nakagawa, C. R. Johnson, S. H. Reich and A. Luis, *J. Org. Chem.*, 1988, **53**, 3195.
81. (a) E. Negishi and T. Takahashi, *Synthesis*, 1988, 1; (b) S. L. Buchwald and R. H. Grubbs, *J. Am. Chem. Soc.*, 1983, **105**, 5490.
82. For an excellent review on the use of low-valent transition metal complexes see J.-M. Pons and M. Santelli, *Tetrahedron*, 1988, **44**, 4295.
83. See K. Takai, Y. Hotta, K. Oshima and H. Nozaki, *Tetrahedron Lett.*, 1978, 2417 for preparation of diiodomethane and trimethylaluminum as well as dibromomethane and titanium tetrachloride. For additional procedures, see ref. 87.
84. For an excellent discussion of this reaction and a hypothesis regarding the organometallic species involved see M. T. Reetz, 'Organotitanium Reagents in Organic Synthesis', Springer-Verlag, Berlin, 1986, p. 223.
85. See Section 3.1.12.2 for examples of successful application of this method to substituted alkenyl ethers.
86. K. Takai, Y. Hotta, K. Oshima and H. Nozaki, *Bull. Chem. Soc. Jpn.*, 1980, **53**, 1698.
87. See ref. 83; for dibromomethane use and preparation see (a) L. Lombardo, *Tetrahedron Lett.*, 1982, **23**, 4293; (b) L. Lombardo, *Org. Synth*, 1987, **65**, 81; for diiodomethane use and preparation see (c) J.-i. Hibino, T. Okazoe, K. Takai and H. Nozaki, *Tetrahedron Lett.*, 1985, **26**, 5579.
88. See ref. 83, and T. Okazoe, J.-i. Hibino, K. Takai and H. Nozaki, *Tetrahedron Lett.*, 1985, **26**, 5581; in addition, the use of diiodomethane and zinc irradiated with ultrasound results in an active catalyst for methylenation of aldehydes, see J. Yamashita, Y. Inoue, T. Kondo and H. Hasukichi *Bull. Chem. Soc. Jpn.*, 1984, **57**, 2335.
89. D. Nicoll-Griffith and L. Weiler, *J. Chem. Soc., Chem. Commun.*, 1984, 659.
90. L. L. Frye and C. H. Robinson, *J. Chem. Soc., Chem. Commun.*, 1988, 129.
91. K. P. Shelly and L. Weiler, *Can. J. Chem.*, 1988, **66**, 1359.
92. I. T. Kay and D. Bartholomew, *Tetrahedron Lett.*, 1984, **25**, 2035.
93. T. Moriwake, S. Hamano, S. Saito and S. Torii, *Chem. Lett.*, 1987, 2085.
94. J. H. Rigby and J. A. Zbur Wilson, *J. Org. Chem.*, 1987, **52**, 34.
95. M. Furber and L. N. Mander, *J. Am. Chem. Soc.*, 1988, **110**, 4084.
96. S. V. Ley, P. J. Murray and B. D. Palmer, *Tetrahedron*, 1985, **41**, 4765.
97. Y. Arai, M. Yamamoto and T. Koizumi, *Bull. Chem. Soc. Jpn.*, 1988, **61**, 467.
98. T. Imagawa, T. Sonobe, H. Ishiwari, T. Akiyama and M. Kawanisi, *J. Org. Chem.*, 1980, **45**, 2005.
99. B. H. Buding, B. Fuchs and H. Musso, *Chem. Ber.*, 1985, **118**, 4613.
100. A. S. Kende, S. Johnson, P. Sanfilippo, J. C. Hodges and L. N. Jungheim, *J. Am. Chem. Soc.*, 1986, **108**, 3513.
101. K. Hakam, M. Thielmann, T. Thielmann and E. Winterfeldt, *Tetrahedron*, 1987, **43**, 2035.
102. R. C. Cambie, H. M. McNally, J. D. Robertson, P. S. Rutledge and P. D. Woodgate, *Aust. J. Chem.*, 1984, **37**, 409.
103. C. Iwata, T. Akiyama and K. Miyashita, *Chem. Pharm. Bull.*, 1988, **36**, 2879.
104. M. Suzuki, H. Koyano and R. Noyori, *J. Org. Chem.*, 1987, **52**, 5583.
105. M. B. Gewali and R. C. Ronald, *J. Org. Chem.*, 1982, **47**, 2792.
106. R. L. Snowden, P. Sonnay and G. Ohloff, *Helv. Chim. Acta*, 1981, **64**, 25.
107. E. Mincione, S. Corsano and P. Bovicelli, *Gazz. Chim. Ital.*, 1984, **114**, 85.
108. See ref. 87a, and B. M. Trost and G. J. Tanoury, *J. Am. Chem. Soc.*, 1988, **110**, 1636.
109. (a) M. Fujita and T. Hiyama, *Tetrahedron Lett.*, 1986, **27**, 3655; (b) M. Fujita and T. Hiyama, *Bull. Chem. Soc. Jpn.*, 1987, **60**, 4377; (c) M. Fujita, K. Kondo and T. Hiyama, *Bull. Chem. Soc. Jpn.*, 1987, **60**, 4385; (d) T. Ishihara and M. Kuroboshi, *Chem. Lett.*, 1987, 1145.
110. M. Matsukawa, T. Tabuchi, J. Inanaga and M. Yamaguchi, *Chem. Lett.*, 1987, 2101.
111. The iodohydrin can be isolated, see T. Tabuchi, J. Inanaga and M. Yamaguchi, *Tetrahedron Lett.*, 1986, **27**, 3891.
112. For other reactions with Sm see P. Girard, J.-L. Namy and H. B. Kagan, *J. Am. Chem. Soc.*, 1980, **102**, 2693.
113. (a) T. Kauffmann, R. Kriegesmann and A. Woltermann, *Angew. Chem., Int. Ed. Engl.*, 1977, **16**, 862; (b) E. Muriyama, T. Kikuchi, K. Saski, N. Sootome and T. Sato, *Chem. Lett.*, 1984, 1897.
114. The Wittig reaction has been extensively reviewed. This section relied heavily on an excellent recent review that includes a discussion of the mechanism, see B. E. Maryanoff and A. B. Reitz, *Chem. Rev.*, 1989, **89**, 863. For additional reviews see ref. 115.
115. See ref. 114; for a review of the classic Wittig reaction, see (a) A. Maercker, *Org. React. (N. Y.)*, 1965, **14**, 270; for a review of the application of the Wittig reaction to natural product synthesis see (b) H. J. Bestmann and O. Vostrowsky, *Top. Curr. Chem.*, 1983, **109**, 85; for a review of industrial applications see (c) H. Pommer and P. C. Thieme, *Top. Curr. Chem.*, 1983, **109**, 165; for general reviews of alkene synthesis see (d) J. Reucroft and P. G. Sammes, *Q. Rev., Chem. Soc.*, 1971, **25**, 135; and (e) D. J. Faulkner, *Synthesis*, 1971, 175; for a review of the application of the Wittig reaction to cycloalkenes see (f) K. B. Becker, *Tetrahedron Report*, 1980, **36**, 1717; for a review of phosphonate carbanions see (g) J. Boutagy and R. Thomas, *Chem. Rev.*, 1974, **74**, 87; and (h) B. J. Walker, in 'Organophosphorus Reagents in Organic Synthesis', ed. J. I. G. Cadogan, Academic Press, New York, 1979, 155; for general reviews of all aspects of the Wittig reaction see (i) M. Schlosser, *Top. Stereochem.*, 1970, **5**, 1; (j) I. Gosney, and A. G. Rowley, in 'Organophosphorus

Reagents in Organic Synthesis', ed. J. I. G. Cadogan, Academic Press, New York, 1979, p. 17; (k) H. J. Bestmann, *Pure Appl. Chem.*, 1980, **52**, 771; and (l) M. Schlosser, R. Oi and B. Schaub, *Phosphorus Sulfur*, 1983, **18**, 171; for a review of the Wittig reaction with carbonyl derivatives other than aldehydes and ketones see (m) P. J. Murphy and J. Brennan, *Chem. Soc. Rev.*, 1988, **17**, 1.

116. (a) B. E. Maryanoff, A. B. Reitz and B. A. Duhl-Emswiler, *J. Am. Chem. Soc.*, 1985, **107**, 217; (b) B. E. Maryanoff, A. B. Reitz, M. S. Mutter, R. R. Inners and H. R. Almond, Jr., *J. Am. Chem. Soc.*, 1985, **107**, 1068; (c) B. E. Maryanoff, A. B. Reitz, M. S. Mutter, R. R. Inners, H. R. Almond, Jr., R. R. Whittle and R. A. Olofson, *J. Am. Chem. Soc.*, 1986, **108**, 7664; (d) E. Vedejs, C. F. Marth, and R. Ruggeri, *J. Am. Chem. Soc.*, 1988, **110**, 3940; (e) E. Vedejs and C. F. Marth, *J. Am. Chem. Soc.*, 1988, **110**, 3948; (f) E. Vedejs and T. J. Fleck, *J. Am. Chem. Soc.*, 1989, **111**, 5861.
117. See ref. 114 for an excellent summary of current thought with regards to the mechanism of the Wittig reaction.
118. B. E. Maryanoff, A. B. Reitz, D. W. Graden and H. R. Almond, Jr., *Tetrahedron Lett.*, 1989, **30**, 1361.
119. M. Schlosser, B. Schaub, J. de Oliveira-Neto and S. Jeganathan, *Chimia*, 1986, **40**, 244.
120. C. Sreekumar, K. P. Darst and W. C. Still, *J. Org. Chem.*, 1980, **45**, 4260.
121. For an alternative synthesis from the aldehyde by electrophilic addition to the Wittig intermediate see E. J. Corey and H. Yamamoto, *J. Am. Chem. Soc.*, 1970, **92**, 226, 3523.
122. (a) B. M. Trost and A. G. Romero, *J. Org. Chem.*, 1986, **51**, 2332; (b) S. D. Burke, F. J. Schoenen and M. S. Nair, *Tetrahedron Lett.*, 1987, **28**, 4143.
123. M. Schlosser and K. F. Christmann, *Angew. Chem., Int. Ed. Engl.*, 1966, **5**, 126; see also ref. 115.
124. E. Vedejs and C. F. Marth, *Tetrahedron Lett.*, 1987, **28**, 3445.
125. For a recent comparative study of a nonstabilized ylide reacted with an unsaturated aldehyde and a semistabilized ylide with a saturated aldehyde in the synthesis of dienes, see R. Ideses and A. Shani, *Tetrahedron*, 1989, **45**, 3523.
126. (a) E. J. Corey and R. Nagata, *Tetrahedron Lett.*, 1987, **28**, 5391; (b) S. R. Baker and D. W. Clissold, *Tetrahedron Lett.*, 1988, **29**, 991.
127. (a) E. Vedejs and H. W. Fang, *J. Org. Chem.*, 1984, **49**, 210; see also (b) E. G. McKenna and B. J. Walker, *Tetrahedron Lett.*, 1988, **29**, 485.
128. R. Tamura, M. Kato, K. Saegusa, M. Kakihana and D. Oda, *J. Org. Chem.*, 1987, **52**, 4121.
129. R. Tamura, K. Saegusa, M. Kakihana and D. Oda, *J. Org. Chem.*, 1988, **53**, 2723.
130. H. Daniel and M. Le Corre, *Tetrahedron Lett.*, 1987, **28**, 1165.
131. J. M. J. Tronchet and B. Gentile, *Helv. Chim. Acta*, 1979, **62**, 2091.
132. (a) S. Valverde, M. Martin-Lomas, B. Herradon and S. Garcia-Ochoa, *Tetrahedron*, 1987, **43**, 1895; (b) F. J. Lopez Herrera, M. S. Pino Gonzalez, M. Nieto Sampedro and R. M. Dominguez Aciego, *Tetrahedron*, 1989, **45**, 269.
133. P. Garner and S. Ramakanth, *J. Org. Chem.*, 1987, **52**, 2631.
134. T. H. Webb, L. M. Thomasco, S. T. Schlachter, J. J. Gaudino and C. S. Wilcox, *Tetrahedron Lett.*, 1988, **29**, 6823.
135. J. S. Brimacombe, R. Hanna, A. K. M. S. Kabir, F. Bennett and I. D. Taylor, *J. Chem. Soc., Perkin Trans. 1*, 1986, 815.
136. (a) K. M. Pietrusiewicz, and J. Monkiewicz, *Tetrahedron Lett.*, 1986, **27**, 739; (b) C. M. Moorhoff and D. F. Schneider, *Tetrahedron Lett.*, 1987, **28**, 4721.
137. E. G. McKenna and B. J. Walker, *J. Chem. Soc., Chem. Commun.*, 1989, 568.
138. (a) L. Horner, H. M. R. Hoffmann and H. G. Wippel, *Chem. Ber.*, 1958, **91**, 61; (b) L. Horner, H. M. R. Hoffmann, H. G. Wippel and G. Klahre, *Chem. Ber.*, 1959, **92**, 2499.
139. W. S. Wadsworth, Jr. and W. D. Emmons, *J. Am. Chem. Soc.*, 1961, **83**, 1733.
140. For a summary of current work on the mechanism of phosphoryl-stabilized Wittig reactions as well as the preparation of these reagents, see ref. 114.
141. G. Schmidt, T. Fukuyama, K. Akasaka and Y. Kishi, *J. Am. Chem. Soc.*, 1979, **101**, 260.
142. H. Nagaoka and Y. Kishi, *Tetrahedron*, 1981, **37**, 3873.
143. D. Boschelli, T. Takemasa, Y. Nishitani and S. Masamune, *Tetrahedron Lett.*, 1985, **26**, 5239.
144. S. Hatakeyama, K. Osanai, H. Numata and S. Takano, *Tetrahedron Lett.*, 1989, **30**, 4845.
145. T. Tanaka, Y. Oikawa, T. Hamada and O. Yonemitsu, *Chem. Lett.*, 1987, **35**, 2209.
146. E. Piers and J. S. M. Wai, *J. Chem. Soc., Chem. Commun.*, 1988, 1245.
147. W. C. Still and C. Gennari, *Tetrahedron Lett.*, 1983, **24**, 4405.
148. R. Baker and J. L. Castro, *J. Chem. Soc., Chem. Commun.*, 1989, 378.
149. W. R. Roush and A. D. Palkowitz, *J. Org. Chem.*, 1989, **54**, 3009.
150. S. J. Danishefsky, M. P. DeNinno and S. Chen, *J. Am. Chem. Soc.*, 1988, **110**, 3929.
151. R. Annunziata, M. Cinquini, F. Cozzi, G. Dondio and L. Raimondi, *Tetrahedron*, 1987, **43**, 2369.
152. S. Jedham and B. C. Das, *Tetrahedron Lett.*, 1988, **29**, 4419; see also H. Kogen and T. Nishi, *J. Chem. Soc., Chem. Commun.*, 1987, 311.
153. T. A. Engler and W. Falter, *Synth. Commun.*, 1988, **18**, 783.
154. P. A. Bartlett and P. C. Ting, *J. Org. Chem.*, 1986, **51**, 2230.
155. S. Hatekeyama, H. Namata, K. Osanai and S. Takamo, *J. Chem. Soc., Chem. Commun.*, 1989, 1893.
156. B.-W. A. Yeung, J. L. M. Contelles and B. Fraser-Reid, *J. Chem. Soc., Chem. Commun.*, 1989, 1163.
157. K. Baettig, A. Marinier, R. Pitteloud and P. Deslongchamps, *Tetrahedron Lett.*, 1987, **28**, 5253.
158. G. O'Kay, *Synth. Commun.*, 1989, **19**, 2125.
159. C.-Q. Sun and D. H. Rich, *Tetrahedron Lett.*, 1988, **29**, 5205.
160. T. Nakata, T. Suenaga, K. Nakashima and T. Oishi, *Tetrahedron Lett.*, 1989, **30**, 6529.
161. M. Yanagida, K. Hashimoto, M. Ishida, H. Shinozaki and H. Shirahama, *Tetrahedron Lett.*, 1989, **30**, 3799.
162. H. Kotsuki, Y. Ushio, I. Kadota and M. Ochi, *J. Org. Chem.*, 1989, **54**, 5153.
163. L. E. Overman and A. S. Thompson, *J. Am. Chem. Soc.*, 1988, **110**, 2248.
164. J. A. Marshall, J. D. Trometer, B. E. Blough and T. D. Crute, *J. Org. Chem.*, 1988, **53**, 4274.

165. A. Trehan and R. S. H. Liu, *Tetrahedron Lett.*, 1988, **29**, 419, and references sited therein.
166. R. K. Boeckman, Jr., C. H. Weidner, R. B. Perni and J. J. Napier, *J. Am. Chem. Soc.*, 1989, **111**, 8036.
167. J. A. Marshall, J. D. Trometer, B. E. Blough and T. D. Crute, *J. Org. Chem.*, 1988, **53**, 4274.
168. T. Fujii, M. Ohba and M. Sakari, *Heterocycles*, 1988, **27**, 2077.
169. J. A. Marshall, J. Lebreton, B. S. DeHoff and T. M. Jenson, *J. Org. Chem.*, 1987, **52**, 3883.
170. S. E. Denmark and J. A. Sternberg, *J. Am. Chem. Soc.*, 1986, **108**, 8277.
171. J. M. Takacs, M. A. Helle, and F. L. Seely, *Tetrahedron Lett.*, 1986, **27**, 1257.
172. S. L. Schreiber and Z. Wang, *Tetrahedron Lett.*, 1988, **29**, 4085.
173. K. S. Reddy, O.-H. Ko, D. Ho, P. E. Persons and J. M. Cassady, *Tetrahedron Lett.*, 1987, **28**, 3075.
174. M. J. Hensel and P. L. Fuchs, *Synth. Commun.*, 1986, **16**, 1285.
175. J. A. Marshall, B. S. DeHoff and D. G. Cleary, *J. Org. Chem.*, 1986, **51**, 1735.
176. A. Kondo, T. Ochi, H. Iio, T. Tokoroyama and M. Siro, *Chem. Lett.*, 1987, 1491.
177. J. C. Medina, R. Guajardo and K. S. Kyler, *J. Am. Chem. Soc.*, 1989, **111**, 2310.
178. J. A. Marshall and B. S. DeHoff, *Tetrahedron Lett.*, 1986, **27**, 4873.
179. W. Oppolzer, R. E. Swenson and J.-M. Gaudin, *Tetrahedron Lett.*, 1988, **29**, 5529.
180. G. Hammond, M. Blagg Cox and D. F. Weimer, *J. Org. Chem.*, 1990, **55**, 128.
181. For a review see J. Seyden-Penne, *Bull. Soc. Chim. Fr.*, 1988, 238.
182. M. A. Blanchette, W. Choy, J. T. Davis, A. P. Essenfeld, S. Masamune, W. R. Roush and T. Sakai, *Tetrahedron Lett.*, 1984, **25**, 2183.
183. M. W. Rathke and M. Nowak, *J. Org. Chem.*, 1985, **50**, 2624.
184. B. M. Trost, P. Seoane, S. Mignani and M. Accmoglu, *J. Am. Chem. Soc.*, 1989, **111**, 7487.
185. Y. Le Merrer, C. Gravier-Pelletier, D. Micas-Languin, F. Mestre, A. Dureault and J.-C. Depezay, *J. Org. Chem.*, 1989, **54**, 2409.
186. J. W. Coe and W. R. Roush, *J. Org. Chem.*, 1989, **54**, 915.
187. W. Oppolzer, D. Dupuis, G. Poli, T. M. Raynham and G. Bernadinelli, *Tetrahedron Lett.*, 1988, **29**, 5885.
188. I. Paterson and I. Boddy, *Tetrahedron Lett.*, 1988, **29**, 5301.
189. B. M. Trost, P. Metz and J. T. Hane, *Tetrahedron Lett.*, 1986, **27**, 5691.
190. O. Tsuge, S. Kanemase and H. Suga, *Chem. Lett.*, 1987, 323.
191. J. S. Nowick and R. L. Danheiser, *J. Org. Chem.*, 1989, **54**, 2798.
192. K. H. Melching, H. Hiemstra, W. J. Klaver and W. N. Speckamp, *Tetrahedron Lett.*, 1986, **27**, 4799.
193. G. Stork, P. M. Sher and H.-L. Chen, *J. Am. Chem. Soc.*, 1986, **108**, 6384.
194. G. E. Keck, E. P. Boden and M. R. Wiley, *J. Org. Chem.*, 1989, **54**, 896.
195. P. J. McCloskey and A. G. Schultz, *J. Org. Chem.*, 1988, **53**, 1380.
196. C. H. Heathcock and T. W. VonGeldern, *Heterocycles*, 1987, **25**, 75.
197. I. Paterson, D. O. P. Laffan and D. J. Rawson, *Tetrahedron Lett.*, 1988, **29**, 1461.
198. C. H. Heathcock, C. R. Hadley, T. Rosen, P. D. Theisen and S. J. Hecker, *J. Med. Chem.*, 1987, **30**, 1859.
199. K. C. Nicolaou, R. A. Daines, T. K. Chakraborty and Y. Ogawa, *J. Am. Chem. Soc.*, 1988, **110**, 4685.
200. (a) J. A. Marshall and B. S. DeHoff, *Tetrahedron*, 1987, **43**, 4849; (b) M. A. Tius and A. Fauq, *J. Am. Chem. Soc.*, 1986, **108**, 6389.
201. A. P. Kozikowski and Y. Xia, *J. Org. Chem.*, 1987, **52**, 1375.
202. S. Hanessian, D. Delorme, S. Beaudoin and Y. Leblanc, *J. Am. Chem. Soc.*, 1984, **106**, 5754.
203. (a) H. Rehwinkel, J. Skupsch and H. Vorbruggen, *Tetrahedron Lett.*, 1988, **29**, 1775; (b) H.-J. Gais, G. Schiedl, W. A. Ball, J. Bund, G. Hellmann and I. Erdelmeier, *Tetrahedron Lett.*, 1988, **29**, 1773.
204. L. Horner, H. M. R. Hoffmann, H. G. Wippel and G. Klahre, *Chem. Ber.*, 1959, **92**, 2499.
205. (a) A. D. Buss and S. Warren, *J. Chem. Soc., Chem. Commun.*, 1981, 100; (b) A. D. Buss and S. Warren, *J. Chem. Soc., Perkin Trans. 1*, 1985, 2307.
206. A. D. Buss, W. B. Cruse, O. Kennard and S. Warren, *J. Chem. Soc., Perkin Trans. 1*, 1984, 243.
207. R. S. Torr, and S. Warren, *J. Chem. Soc., Perkin Trans. 1*, 1983, 1173.
208. For an excellent comparative study of phosphonate and phosphine oxide approaches to unsaturated nitriles see G. Etemad-Moghadam and J. Seyden-Penne, *Synth. Commun.*, 1984, **14**, 565.
209. (a) C. Earnshaw, C. J. Wallis and S. Warren, *J. Chem. Soc., Chem. Commun.*, 1977, 314; (b) C. Earnshaw, C. J. Wallis and S. Warren, *J. Chem. Soc., Perkin Trans. 1*, 1979, 3099; (c) E. F. Birse, A. McKenzie and A. W. Murray, *J. Chem. Soc., Perkin Trans. 1*, 1988, 1039.
210. (a) J. I. Grayson and S. Warren, *J. Chem. Soc., Perkin Trans. 1*, 1977, 2263; (b) S. Warren and A. T. Zaslona, *Tetrahedron Lett.*, 1982, **23**, 4167; (c) J. I. Grayson, S. Warren and A. T. Zaslona, *J. Chem. Soc., Perkin Trans. 1*, 1987, 967.
211. (a) D. Cavalla and S. Warren, *Tetrahedron Lett.*, 1982, **23**, 4505; (b) D. Cavalla and S. Warren, *Tetrahedron Lett.*, 1983, **24**, 295; (c) D. Cavalla, W. B. Cruse and S. Warren, *J. Chem. Soc., Perkin Trans. 1*, 1987, 1883.
212. T. A. M. van Schaik, A. V. Henzen and A. van der Gen, *Tetrahedron Lett.*, 1983, **24**, 1303.
213. E. Juaristi, B. Gordillo and L. Valle, *Tetrahedron*, 1986, **42**, 1963.
214. A. D. Buss and S. Warren, *Tetrahedron Lett.*, 1983, **24**, 3931.
215. T. G. Roberts and G. H. Whitham, *J. Chem. Soc., Perkin Trans. 1*, 1985, 1953.
216. (a) J. Elliott and S. Warren, *Tetrahedron Lett.*, 1986, **27**, 645; (b) J. Elliott, D. Hall and S. Warren, *Tetrahedron Lett.*, 1989, **30**, 601.
217. A. D. Buss, R. Mason and S. Warren, *Tetrahedron Lett.*, 1983, **24**, 5293.
218. See refs. 205b, 216 and 217.
219. (a) J. Kallmerten and M. D. Wittman, *Tetrahedron Lett.*, 1986, **27**, 2443; (b) M. D. Wittman and J. Kallmerten, *J. Org. Chem.*, 1987, **52**, 4303.
220. P. M. Ayrey and S. Warren, *Tetrahedron Lett.*, 1989, **30**, 4581.
221. (a) A. D. Buss and S. Warren, *J. Chem. Soc., Perkin Trans. 1*, 1985, 2307; (b) A. D. Buss, N. Greeves, R. Mason and S. Warren, *J. Chem. Soc., Perkin Trans. 1*, 1987, 2569.
222. (a) I. Yamamoto, T. Fujimoto, K. Ohta and K. Matsuzaki, *J. Chem. Soc., Perkin Trans. 1*, 1987, 1537; (b) I. Yamamoto, S. Tanaka, T. Fujimoto and K. Ohta, *J. Org. Chem.*, 1989, **54**, 747.

223. R. K. Haynes, S. C. Vonwiller and T. W. Hambley, *J. Org. Chem.*, 1989, **54**, 5162.
224. S. R. Schow, J. D. Bloom, A. S. Thompson, K. N. Winzenberg and A. B. Smith, III, *J. Am. Chem. Soc.*, 1986, **108**, 2662.
225. E. W. Collington, J. G. Knight, C. J. Wallis and S. Warren, *Tetrahedron Lett.*, 1989, **30**, 877.
226. (a) K. C. Nicolaou, R. Zipkin and D. Tanner, *J. Chem. Soc., Chem. Commun.*, 1984, 349; (b) D. Caine, B. Stanhop and S. Fiddler, *J. Org. Chem.*, 1988, **53**, 4124.
227. A. D. Buss and S. Warren, *Tetrahedron Lett.*, 1983, **24**, 111.
228. (a) S. V. Ley and B. Lygo, *Tetrahedron Lett.*, 1984, **25**, 113; (b) S. V. Ley, B. Lygo, H. M. Organ and A. Wonnacott, *Tetrahedron*, 1985, **41**, 3825.
229. (a) B. Lythgoe, T. A. Moran, M. E. N. Nambudiry, J. Tideswell and P. W. Wright, *J. Chem. Soc., Perkin Trans. 1*, 1978, 590; (b) H. T. Toh, and W. H. Okamura, *J. Org. Chem.*, 1983, **48**, 1414; (c) S.-J. Shiuey, J. J. Partridge and M. R. Uskokovic, *J. Org. Chem.*, 1988, **53**, 1040.
230. (a) T. Nagase, T. Kawashima and N. Inamoto, *Chem. Lett.*, 1984, 1997; (b) A. B. McElroy and S. Warren, *Tetrahedron Lett.*, 1985, **26**, 2119, 5709.
231. (a) D. Levin and S. Warren, *Tetrahedron Lett.*, 1985, **26**, 505; (b) D. Levin and S. Warren, *Tetrahedron Lett.*, 1986, **27**, 2265; (c) D. Levin and S. Warren, *J. Chem. Soc., Perkin Trans. 1*, 1988, 1799.
232. P. Wallace and S. Warren, *J. Chem. Soc., Perkin Trans. 1*, 1988, 2971.
233. P. M. Ayrey and S. Warren, *Tetrahedron Lett.*, 1989, **30**, 4581.
234. J. Barluenga, F. Lopez and F. Palacios, *Synthesis*, 1988, 562.
235. M. E. Jung and J. P. Hudspeth, *J. Am. Chem. Soc.*, 1980, **102**, 2463.
236. For summaries of conditions of anion formation with stabilizing groups present see refs. 4a and 4b.
237. For an excellent summary of the methods of silyl anion formation see ref. 4c.
238. P. F. Hudrlik and D. J. Peterson, *J. Am. Chem. Soc.*, 1975, **97**, 1464.
239. A. R. Bassindale, R. J. Ellis, J. C. Y. Lau and P. G. Taylor, *J. Chem. Soc., Perkin Trans. 2*, 1986, 593.
240. The reaction is sensitive to steric effects and large substituents on the silicon increase ratios of *cis*-alkenes, see A. R. Bassindale, R. J. Ellis and P. G. Taylor, *Tetrahedron Lett.*, 1984, **25**, 2705, and ref. 239.
241. For an excellent review and experimental procedures for a variety of silyl anions as well as alkene formation see ref. 4f.
242. P. F. Hudrlik and D. J. Peterson, *Tetrahedron Lett.*, 1974, 1133.
243. (a) P. F. Hudrlik and D. J. Peterson, *Tetrahedron Lett.*, 1972, 1785; (b) Y. Takeda, T. Matsumoto and F. Sato, *J. Org. Chem.*, 1986, **51**, 4728; (c) G. L. Larson, I. Montes de Lopez-Cepero and L. Rodriguez Mieles, *Org. Synth*, 1989, **67**, 125.
244. (a) P. F. Hudrlik, D. J. Peterson and R. J. Rona, *J. Org. Chem.*, 1975, **40**, 2264; (b) Y. Takeda, T. Matsumoto and F. Sato, *J. Org. Chem.*, 1986, **51**, 4731; (c) S. Okamoto, T. Shimazaki, Y. Kobayashi and F. Sato, *Tetrahedron Lett.*, 1987, **28**, 2033.
245. Consult ref. 4 for further details.
246. P. F. Hudrlik, E. L. O. Agwaramgbo and A. M. Hudrlik, *J. Org. Chem.*, 1989, **54**, 5613.
247. G. L. Larson, J. A. Prieto and E. Ortiz, *Tetrahedron*, 1988, **44**, 3781.
248. (a) T. Cohen and M. Bhupathy, *Acc. Chem. Res.*, 1989, **22**, 152; (b) T. Cohen, S.-H. Jung, M. L. Romberger and D. W. McCullough, *Tetrahedron Lett.*, 1988, **29**, 25; (c) T. Cohen, J. P. Sherbine, S. A. Mendelson and M. Myers, *Tetrahedron Lett.*, 1985, **26**, 2965.
249. (a) T. Cohen, J. P. Serbine, J. R. Matz, R. R. Hutchins, B. M. McHenry and P. R. Willey, *J. Am. Chem. Soc.*, 1984, **106**, 3245; for other examples of transmetallation of thiosilanes as well as the synthesis of vinyl sulfides and sulfones see (b) D. J. Ager, *J. Chem. Soc., Perkin Trans. 1*, 1986, 183; for additional information on sulfones see (c) D. J. Ager, *J. Chem. Soc., Chem. Commun.*, 1984, 486.
250. (a) B. Halton and P. J. Stang, *Acc. Chem. Res.*, 1987, **20**, 443; (b) B. Halton, C. J. Randall, G. J. Gainsford and P. J. Stang, *J. Am. Chem. Soc.*, 1986, **108**, 5949.
251. See ref. 2 as well as the synthesis of diethyl phosphonate and phenylthio derivatives, F. A. Carey and A. S. Scott, *J. Org. Chem.*, 1972, **37**, 939.
252. For a review of the chemistry of organosulfur–silicon compounds see E. Block and M. Aslam, *Tetrahedron*, 1988, **44**, 281.
253. (a) D. Craig, S. V. Ley, N. S. Simpkins, G. H. Whitham and M. J. Prior, *J. Chem. Soc., Perkin Trans. 1*, 1985, 1949; (b) S. V. Ley and N. S. Simpkins, *J. Chem. Soc., Chem. Commun.*, 1983, 1281.
254. T. Agawa, M. Ishikawa, M. Komatsu and Y. Ohshiro, *Bull. Chem. Soc. Jpn.*, 1982, **55**, 1205.
255. S. Hackett and T. Livinghouse, *J. Org. Chem.*, 1986, **51**, 879; see also W. Chamchaang, V. Prankprakma, B. Tarnchompoo, C. Thebtaranonth and Y. Thebtaranonth, *J. Chem. Soc., Chem. Commun.*, 1982, 579 for the use of lithiotrimethylsilyl-1,3-dithiane.
256. I. Erdelmeier and H.-J. Gais, *Tetrahedron Lett.*, 1985, **26**, 4359.
257. E. E. Aboujaoude, S. Lietje and N. Collignon, *Synthesis*, 1986, 934.
258. J. Binder and E. Zbiral, *Tetrahedron Lett.*, 1986, **27**, 5829.
259. M. Mikolajczyk and P. Balczewski, *Synthesis*, 1989, 101.
260. (a) S. L. Hartzell, D. F. Sullivan, and M. W. Rathke, *Tetrahedron Lett.*, 1974, 1403; (b) H. Taguchi, K. Shimoji, H. Yamamoto and H. Nozaki, *Bull. Chem. Soc. Jpn.*, 1974, **47**, 2529; (c) K. Shimoji, H. Taguchi, K. Oshima, H. Yamamoto and H. Nozaki, *J. Am. Chem. Soc.*, 1974, **96**, 1620.
261. M. Larcheveque and A. Debal, *J. Chem. Soc., Chem. Commun.*, 1981, 877.
262. (a) G. L. Larson, J. A. Prieto and A. Hernandez, *Tetrahedron Lett.*, 1981, **22**, 1575; (b) L. Strekowski, M. Visnick and M. A. Battiste, *Tetrahedron Lett.*, 1984, **25**, 5603.
263. See, for example, (a) G. L. Larson and R. M. Betancourt de Perez, *J. Org. Chem.*, 1985, **50**, 5257; (b) S. Kano, T. Ebata, K. Funaki and S. Shibuya, *Synthesis*, 1978, 746.
264. Y. Yamakado, M. Ishiguro, N. Ikeda and H. Yamamoto, *J. Am. Chem. Soc.*, 1981, **103**, 5568.
265. See, for example, M. Larcheveque, C. Legueut, A. Debal and J. Y. Lallemand, *Tetrahedron Lett.*, 1981, **22**, 1595.
266. R. K. Boeckman, Jr. and R. L. Chinn, *Tetrahedron Lett.*, 1985, **26**, 5005.

267. M. T. Crimmins, P. J. O'Hanion and N. H. Rogers, *J. Chem. Soc., Perkin Trans. 1*, 1985, 541.
268. R. H. Schlessinger, M. A. Poss, S. Richardson and P. Lin, *Tetrahedron Lett.*, 1985, **26**, 2391.
269. A. I. Meyers, J. P. Lawson, D. G. Walker and R. J. Linderman, *J. Org. Chem.*, 1986, **51**, 5111.
270. Y. Yoshimura, A. Matsuda and T. Ueda, *Chem. Pharm. Bull.*, 1989, **37**, 660.
271. M. Julia, and J.-M. Paris, *Tetrahedron Lett.*, 1973, 4833.
272. For reviews of the Julia coupling see (a) P. J. Kocienski, *Phosphorus Sulfur*, 1985, **24**, 97; (b) M. Julia, *Pure Appl. Chem.*, 1985, **57**, 763; (c) B. M. Trost, *Bull. Chem. Soc. Jpn.*, 1988, **61**, 107.
273. (a) P. J. Kocienski, B. Lythgoe and S. Ruston, *J. Chem. Soc., Perkin Trans. 1*, 1978, 829; (b) P. J. Kocienski, B. Lythgoe and D. A. Roberts, *J. Chem. Soc., Perkin Trans. 1*, 1978, 834; (c) P. J. Kocienski, B. Lythgoe and I. Waterhouse, *J. Chem. Soc., Perkin Trans. 1*, 1980, 1045; (d) P. J. Kocienski, *Chem. Ind. (London)*, 1981, 548.
274. See also Section 3.1.11.7
275. B. M. Trost, H. C. Arndt, P. E. Strege and T. R. Verhoeven, *Tetrahedron Lett.*, 1976, 3477.
276. B. Lythgoe and I. Waterhouse, *Tetrahedron Lett.*, 1977, 4223.
277. See ref. 272a for an excellent discussion of all aspects of the Julia coupling, as well as additional examples of (*E*)-di- and -tri-substituted alkenes, and (*E,E*)-dienes.
278. Unpublished results of P. J. Kocienski are discussed in ref. 272a.
279. A. P. Kozikowski, R. J. Schmiesing and K. L. Sorgi, *J. Am. Chem. Soc.*, 1980, **102**, 6577.
280. A. P. Kozikowski and K. L. Sorgi, *Tetrahedron Lett.*, 1984, **25**, 2085.
281. (a) G. E. Keck, D. F. Kachensky and E. J. Enholm, *J. Org. Chem.*, 1984, **49**, 1464; (b) G. E. Keck, D. F. Kachensky and E. J. Enholm, *J. Org. Chem.*, 1985, **50**, 4317.
282. D. R. Williams, J. L. Moore and M. Yamada, *J. Org. Chem.*, 1986, **51**, 3918.
283. P. Gannett, D. L. Nagel, P. J. Reilly, T. Lawson, J. Sharpe and B. Toth, *J. Org. Chem.*, 1988, **53**, 1064.
284. The yields in this table are the overall yields of addition, functionalization and reductive elimination of alkenes. In cases where a specific (*E*):(*Z*) ratio was not reported, or where it was determined to be entirely the (*E*)-isomer, the ratio is reported as '(*E*) only'.
285. S. Hanessian and P. J. Murray, *Can. J. Chem.*, 1986, **64**, 2231.
286. S. H. Kang, W. J. Kim and Y. B. Chae, *Tetrahedron Lett.*, 1988, **29**, 5169.
287. M. E. Findeis and G. M. Whitesides, *J. Org. Chem.*, 1987, **52**, 2838.
288. D. A. Evans and M. DiMare, *J. Am. Chem. Soc.*, 1986, **108**, 2476.
289. D. V. Patel, F. VanMiddlesworth, J. Donaubauer, P. Gannett and C. J. Sih, *J. Am. Chem. Soc.*, 1986, **108**, 4603.
290. S. L. Schrieber and H. V. Meyers, *J. Am. Chem. Soc.*, 1988, **110**, 5198.
291. D. Tanner and P. Somfai, *Tetrahedron*, 1987, **43**, 4395.
292. S. Hanessian, A. Ugolini, D. Dube, P. J. Hodges and C. Andre, *J. Am. Chem. Soc.*, 1986, **108**, 2776.
293. A. G. M. Barrett, R. A. E. Carr, S. V. Attwood, G. Richardson and N. D. A. Walshe, *J. Org. Chem.*, 1986, **51**, 4840.
294. M. Hirama, T. Nakamine and S. Ito, *Tetrahedron Lett.*, 1988, **29**, 1197.
295. Formation of the vinyl sulfone from the β-hydroxy sulfone followed by displacement with a Grignard reagent or cuprate is an alternative to the Julia coupling sequence for trisubstituted alkenes; see ref. 272a.
296. (a) S. Mills, R. Desmond, R. A. Reamer, R. P. Volante and I. Shinkai, *Tetrahedron Lett.*, 1988, **29**, 281; (b) T. K. Jones, S. Mills, R. A. Reamer, D. Askin, R. Desmond, R. P. Volante and I. Shinkai, *J. Am. Chem. Soc.*, 1989, **111**, 1157.
297. A. Villalobos and S. J. Danishefsky, *J. Org. Chem.*, 1989, **54**, 12.
298. S. L. Schreiber, T. Sammakia and D. E. Uehling, *J. Org. Chem.*, 1989, **54**, 15.
299. M. P. Edwards, S. V. Ley, S. G. Lister and B. D. Palmer, *J. Chem. Soc., Chem. Commun.*, 1983, 630.
300. W. R. Roush, and S. M. Peseckis, *Tetrahedron Lett.*, 1982, **23**, 4879.
301. (a) R. Baker, M. J. O'Mahony and C. J. Swain, *J. Chem. Soc., Chem. Commun.*, 1985, 1326; (b) R. Baker, M. J. O'Mahony and C. J. Swain, *J. Chem. Soc., Perkin Trans. 1*, 1987, 1623.
302. P. J. Kocienski, S. D. A. Street, C. Yeates and S. Campbell, *J. Chem. Soc., Perkin Trans. 1*, 1987, 2171.
303. N. J. Anthony, P. Grice and S. V. Ley, *Tetrahedron Lett.*, 1987, **28**, 5763.
304. T. Mandai, T. Yanagi, K. Araki, Y. Morisaki, M. Kawada and J. Otera, *J. Am. Chem. Soc.*, 1984, **106**, 3670.
305. J. Otera, H. Misawa and K. Sugimoto, *J. Org. Chem.*, 1986, **51**, 3830.
306. J. Otera, H. Misawa, T. Onishi, S. Suzuki and Y. Fujita, *J. Org. Chem.*, 1986, **51**, 3834.
307. T. Mandai, T. Moriyama, K. Tsujimoto, M. Kawada and J. Otera, *Tetrahedron Lett.*, 1986, **27**, 603.
308. (a) M. Julia, M. Launay, J.-P. Stancino and J.-N. Verpeaux, *Tetrahedron Lett.*, 1982, **23**, 2465; (b) J. S. Grossert, H. R. W. Dhararatne, T. S. Cameron and B. R. Vincent, *Can. J. Chem.*, 1988, **66**, 2860.
309. S. H. Kang, W. J. Kim and Y. B. Chae, *Tetrahedron Lett.*, 1988, **29**, 5169.
310. T. Shono, Y. Matsumura and S. Kashimura, *Chem. Lett.*, 1978, 69.
311. H.-J. Gais and T. Lied, *Angew. Chem., Int. Ed. Engl.*, 1984, **23**, 145.
312. B. M. Trost, J. Lynch, P. Renaut and D. H. Steinman, *J. Am. Chem. Soc.*, 1986, **108**, 284.
313. P. A. Bartlett, F. R. Green, III and E. H. Rose, *J. Am. Chem. Soc.*, 1978, **100**, 4852. See also ref. 272a for discussion of examples.
314. (a) B. Achmatowicz, E. Baranowska, A. R. Daniewski, J. Pankowski and J. Wicha, *Tetrahedron Lett.*, 1985, **26**, 5597; (b) B. Achmatowicz, E. Baranowska, A. R. Daniewski, J. Pankowski and J. Wicha, *Tetrahedron*, 1988, **44**, 4989.
315. A. Spaltenstein, P. A. Carpino, F. Miyake and P. B. Hopkins, *Tetrahedron Lett.*, 1986, **27**, 2095; (b) A. Spaltenstein, P. A. Carpino, F. Miyake and P. B. Hopkins, *J. Org. Chem.*, 1987, **52**, 3759.
316. See ref. 78, and (a) F. W. Hartner, J. Schwartz and S. M. Clift, *J. Am. Chem. Soc.*, 1983, **105**, 640; (b) S. M. Clift and J. Schwartz, *J. Am. Chem. Soc.*, 1984, **106**, 8300; (c) J. Schwartz, G. M. Arvanitis, J. A. Smegel, I. K. Meier, S. M. Clift and D. Van Engen, *Pure Appl. Chem.*, 1988, **60**, 65.
317. Stork has recently published a Wittig route to (*Z*)-iodoalkenes, see G. Stork and K. Zang, *Tetrahedron Lett.*, 1989, **30**, 2173.

318. K. Takai, K. Nitta and K. Utimoto, *J. Am. Chem. Soc.*, 1986, **108**, 7408.
319. K. C. Nicolaou, S. E. Webber, J. Ramphal and Y. Abe, *Angew. Chem., Int. Ed. Engl.*, 1987, **26**, 1019.
320. H. J. Bestmann, A. B. Attygalle, J. Schwartz, W. Garbe, O. Vostrowsky and I. Tomida, *Tetrahedron Lett.*, 1989, **30**, 2911.
321. K. V. Baker, J. M. Brown and N. A. Cooley, *J. Labelled Compd. Radiopharm.*, 1988, **25**, 1229.
322. K. Takai, Y. Kataoka, T. Okazoe and K. Utimoto, *Tetrahedron Lett.*, 1987, **28**, 1443.
323. Allylic acetals can be treated with $CrCl_2$ to obtain the allylic chromium reagent, see K. Takai, K. Nitta and K. Utimoto, *Tetrahedron Lett.*, 1988, **29**, 5263.
324. T. Okazoe, K. Takai and K. Utimoto, *J. Am. Chem. Soc.*, 1987, **109**, 951.
325. J. A. Rechka and J. R. Maxwell, *Tetrahedron Lett.*, 1988, **29**, 2599.
326. R. Baker and J. L. Castro, *J. Chem. Soc., Perkin Trans. 1*, 1989, 190.
327. T. Okazoe, K. Takai, K. Oshima and K. Utimoto, *J. Org. Chem.*, 1987, **52**, 4410.
328. For a recent review of the chemistry of titanium see C. Betschart and D. Seebach, *Chimia*, 1989, **43**, 39.
329. (a) K. Takai, Y. Kataoka, T. Okazoe and K. Utimoto, *Tetrahedron Lett.*, 1988, **29**, 1065; (b) K. Takai, O. Fujimura, Y. Kataoka and K. Utimoto, *Tetrahedron Lett.*, 1989, **30**, 211.
330. M. Mortimore and P. J. Kocienski, *Tetrahedron Lett.*, 1988, **29**, 3357.

3.2
Epoxidation and Related Processes

JEFFREY AUBÉ
University of Kansas, Lawrence, KS, USA

3.2.1 INTRODUCTION

This chapter concerns the net conversion of a C=X π-bond (X = O and N) to various three-membered ring heterocycles. The most common such process is the conversion of an aldehyde or ketone to a homologous epoxide. These reactions will be discussed along with the analogous process which takes place on imines and related compounds. Additionally, methods to effect the addition of elements other than carbon across either unsaturated system will be considered. The synthesis of thiiranes by the addition of carbon across the C=S π-bond is the subject of a recent comprehensive review[1] and will not be covered here.

3.2.2 ADDITIONS TO C=O π-BONDS

3.2.2.1 General Considerations

The particular attributes of carbonyl epoxidation reactions, in addition to the importance of forming any carbon–carbon bond, result from the fact that the newly introduced carbon substituent is functionalized and, due to its presence in a three-membered ring, activated. The lability of epoxides toward nucleophiles is well known and accounts for the bulk of their utility in synthesis.[2,3] In addition, epoxides undergo a variety of rearrangement reactions, such as chain elongation or ring expansion processes. Emphasis is placed here on techniques which allow the conversion of a carbonyl to a homologous isolable epoxide in a one-pot procedure, although pertinent multistep methods will be briefly noted.

All of these methods require the formation of an alkoxide with a leaving group X^n (the formal charge n is usually 0 or +1) in the β-position which collapses with the expulsion of an X^{n-1} species to afford the epoxide, as shown in Scheme 1. Rearrangement processes often compete with epoxidation and sometimes constitute the major reaction pathway. The bulk of these approaches entail the formation of this intermediate by the addition of an anionic species $R^3R^4XC^-$ to the electrophilic carbonyl group (path a). Very often, the leaving group serves the additional function of stabilizing the carbanionic species as well. Some mention will be made of alternative routes in which a nucleophile attacks a carbonyl compound which contains an adjacent leaving group (path b).

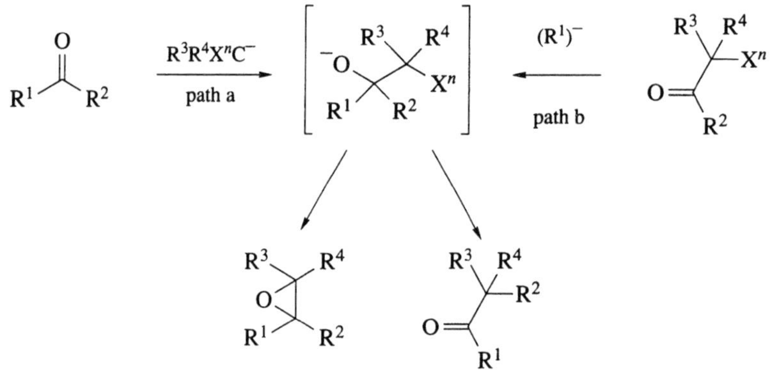

$R^3R^4X^nC^-$ path a; $(R^1)^-$ path b

Scheme 1

3.2.2.2 Sulfur Ylides

The most commonly used reagents to effect the addition of a methylene group to an aldehyde or ketone are sulfur ylides such as dimethylsulfonium methylide (**1**) or dimethyloxosulfonium methylide (**2**)[4] (Corey–Chaykovsky reaction).[5] This reaction is well reviewed in standard treatises of organic synthesis[6,7] and several useful monographs.[8,9] This update will concentrate on progress attained from 1975. The reader is also encouraged to consult reviews on the chemistry of the related sulfoximine-derived ylides such as (**3**).[10,11]

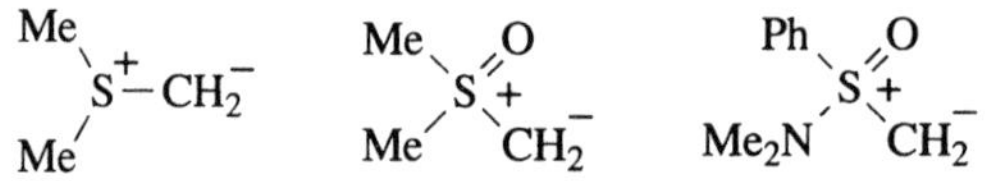

(**1**) (**2**) (**3**)

Sulfur ylides are primarily synthesized by deprotonation of a sulfonium salt, which in turn is usually prepared by alkylation of a sulfide.[12] Branched chain sulfonium salts are not generally preparable by simple sulfide displacement, although some can be made by other routes.[9,12] While ylide solutions are most commonly prepared and used immediately, frozen stock solutions of the more stable (**2**) can be stored for several months in DMSO at –20 °C.[13]

The reaction between sulfur ylides and carbonyl compounds entails attack of the ylide on the carbonyl to form a betaine, which then collapses with expulsion of the neutral sulfide or sulfoxide (Scheme 1; $X^n = R_2S^+$).[6,7,9] A theoretical study of this mechanism has appeared.[14] Ylides belonging to the general classes of (**1**) and (**2**) differ in stability and in the relative rates of the two mechanistic steps. Specifically, the more stable (**2**) reacts reversibly with carbonyl groups, whereas (**1**) undergoes a kinetic addition to the substrate followed by a rapid collapse of the betaine to an epoxide. Differences in chemoselectivity and stereoselectivity between the ylides are attributed to this key difference.[6,7,9]

The structural variety of epoxides available through the above technology is limited by the availability of the starting sulfonium salt, the reactivity of the desired ylide, and the propensity of either species to undergo a variety of side reactions. For simple methylene addition, (**1**) and (**2**) are the most general and reliable reagents available. The rearrangement reactions observed with other reagents (most notoriously, diazoalkanes) are much less common with sulfur ylides. Even so, some electron-rich epoxides are prone to undergo rearrangement prior to isolation, yielding aldehydes or ketones.[15] In addition, (**2**) or (**3**) reacts with epoxides under rather forcing conditions (50 °C, 3 d) giving rise to oxetanes.[16,17]

A particularly interesting and useful group of ylides are those derived from diphenylcyclopropyl sulfonium halides, such as (**4**) and (**5**).[18] These reagents react with ketones and aldehydes to afford oxaspiropentanes, which are versatile synthetic intermediates.[19] Ylide (**4**) exhibits stereochemical and regiochemical tendencies similar to (**1**).

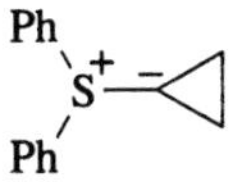
Ph
S
Ph

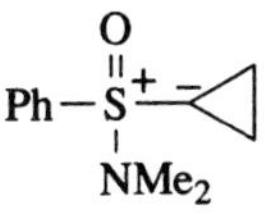
O
Ph—S
NMe2

(**4**) (**5**)

New methods for the generation of ylide (**1**) in the presence of a substrate have been introduced. Solid potassium hydroxide in nonpolar solvents containing trace amounts of water has proven a highly effective medium for the promotion of methylene[20] and benzylidene[21] transfer reactions (equation 1). The counterion of the sulfonium salt and the base used has a substantial effect on the success of the reaction.[22]

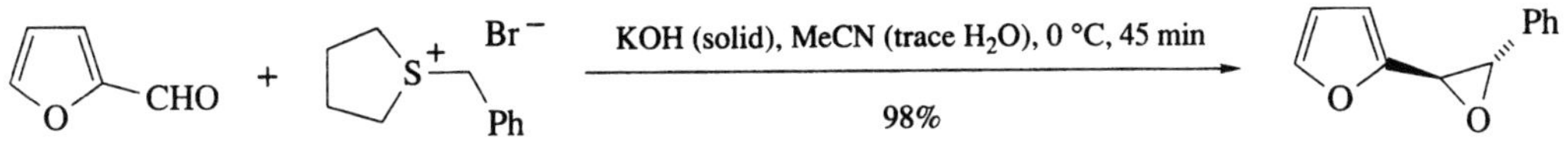
CHO
Br⁻
Ph
KOH (solid), MeCN (trace H2O), 0 °C, 45 min
98%
Ph

(1)

Two-phase systems, usually consisting of a strongly alkaline aqueous phase and methylene chloride, are also effective and convenient. Trialkylsulfonium salts with methylsulfate counterions can be used alone under these conditions,[23] whereas those with halide counterions generally benefit from added phase transfer catalysts (equation 2).[24] Dodecyldimethylsulfonium chloride was found to be an effective phase transfer reagent.[24] A sulfonium salt covalently attached to a polymeric resin was found most effective when a phase transfer agent was also employed.[25]

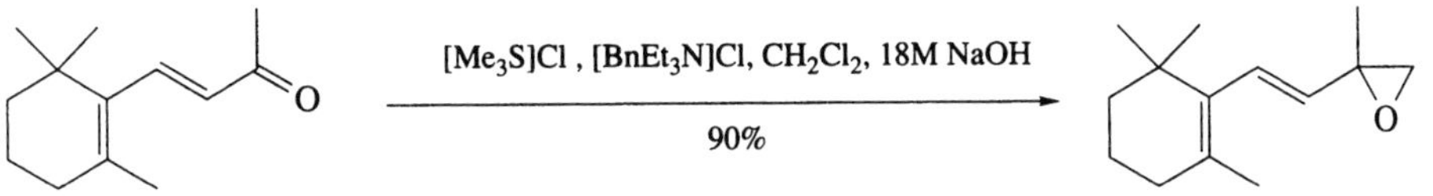
[Me3S]Cl , [BnEt3N]Cl, CH2Cl2, 18M NaOH
90%

(2)

Epoxidations can be effected by adsorption of sulfonium salts and substrate on potassium fluoride impregnated alumina.[26] Soluble fluoride ion allowed the generation of a methylide from a TMS-substituted precursor, which underwent equilibration to the more stable benzylide prior to addition (equation 3).[27] The reaction failed with ketones or imines.

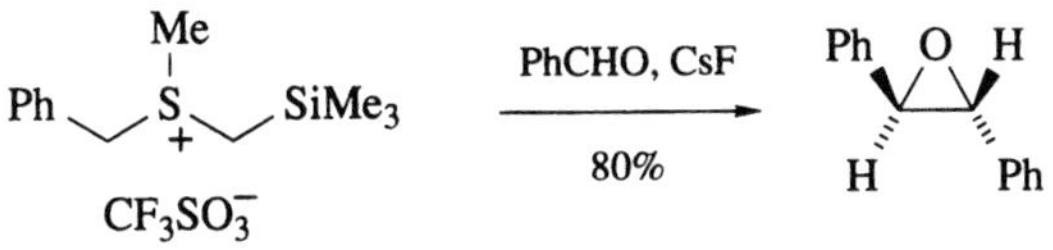
Me
Ph
SiMe3
CF3SO3⁻
PhCHO, CsF
80%
Ph O H
H Ph

(3)

Several groups have investigated the use of complex sulfur ylide reagents for a key carbon–carbon bond formation step in leukotriene synthesis (equation 4).[28,29] For example, tetraene (**6**) reacted with methyl 5-oxopentanoate to give an 85% yield of the desired epoxides as a mixture of *cis* and *trans* isomers. Significantly, only small amounts of by-products resulting from elimination or rearrangement of the sulfonium salt were observed.

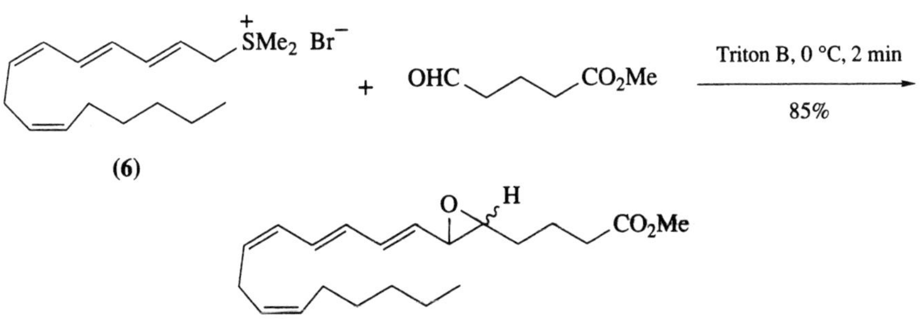
SMe2 Br⁻
OHC
CO2Me
Triton B, 0 °C, 2 min
85%
(6)
O H
CO2Me

(4)

Several examples of intramolecular epoxide formation have been reported.[30,31] In general, a keto sulfide is reacted with a trialkyloxonium tetrafluoroborate to yield a sulfonium salt, which cyclizes upon treatment with base. A particularly clever variation involves the *in situ* generation of an allylic ylide (Scheme 2).[32]

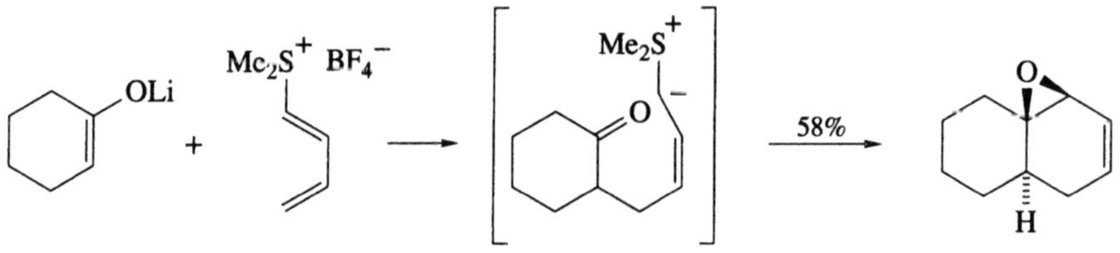

Me_2S^+ BF_4^- OLi + 58%

Scheme 2

Compared to unstabilized ylides, ylides bearing an electron-withdrawing group on the adjacent carbon can be generated under milder conditions and are considerably more stable. However, most reactive aldehydes or ketones are not epoxidized by these reagents and classical Darzens-type reactions are usually preferred for the synthesis of epoxides bearing an electron-stabilizing group (see Volume 2, Chapter 1.13). However, ylide (**7**)[33] does react with ketones to afford α,β-epoxy carboxylic acids (equation 5).[34] Notably, ketone (**8**) did not react with standard Darzens reagents.

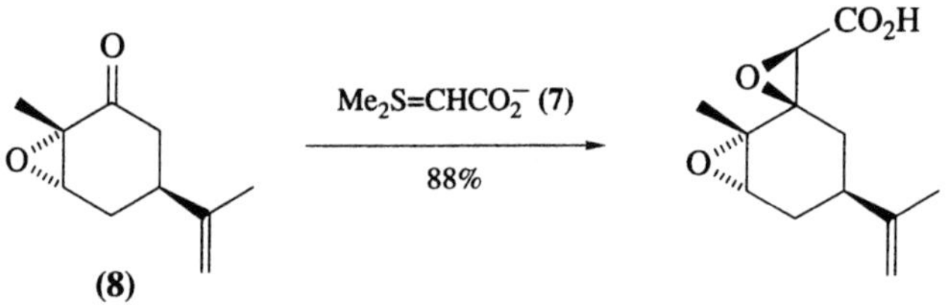

$Me_2S{=}CHCO_2^-$ (**7**), 88% (5)

(**8**)

Early efforts to probe the stereochemistry of sulfur ylide additions have been reviewed.[35] Ylide (**1**) reacts with unhindered cyclohexanone derivatives to give products resulting from axial attack, whereas (**2**) gives rise to net equatorial addition. This is again due to the tendency of (**1**) to undergo kinetically controlled reactions and (**2**) to give products resulting from thermodynamic control. However, equatorial attack may prevail even in kinetically controlled reactions if axial attack is rendered sufficiently hindered, *e.g.* by 1,3-diaxial interactions or by additional steric bulk adjacent to the reactive carbonyl.[6,7,9] For example, the addition of (**1**) to ketone (**9**) affords the β-epoxide shown in equation (6) exclusively.[36] Other cyclic systems react with fairly predictable stereoselectivities in accord with the above observations.[9,19]

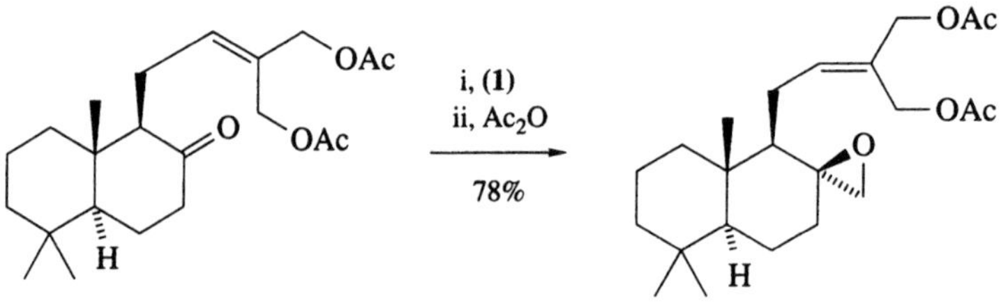

i, (**1**); ii, Ac_2O; 78% (6)

(**9**)

Ylide (**1**) adds selectively to the least hindered side of a spirocyclic cyclobutanone to afford epoxide (**10**) as the sole product in ≥65% yield (Scheme 3).[37] Where only steric factors are likely to be important, complementary stereochemical results are often possible by methylene addition to a carbonyl group or alternatively, by carrying out an epoxidation reaction on the derived alkene. In each case, the reagent approaches a double bond from the same stereochemical face. This point is nicely illustrated by the synthesis of epoxide (**11**).

Several workers have observed differences in the stereoselectivity of the reactions of (**1**) and (**2**) with ketones related to the trichothecene family of natural products (Scheme 4).[38,39] Reaction of ketone (**12**) with (**1**) stereoselectively led to epoxide (**13**), whereas ylide (**2**) gave (**14**) instead. Very high diastereoselection was obtained in the late stages of a synthesis of (+)-phyllanthocin: treatment of ketone

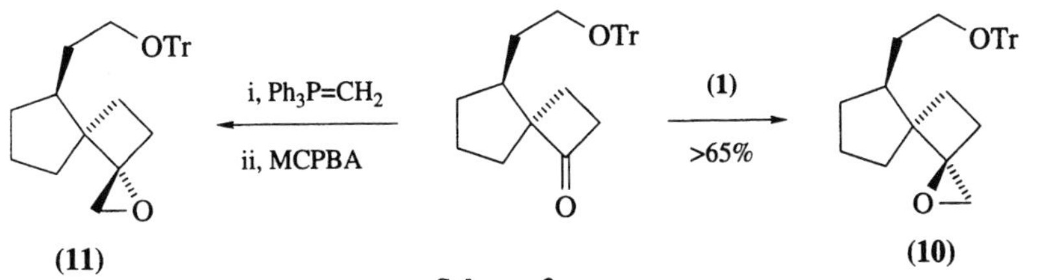

Scheme 3

(15) with a large excess of (2) afforded a 98% yield of epoxide (16) in 97% diastereomeric purity (equation 7).[13]

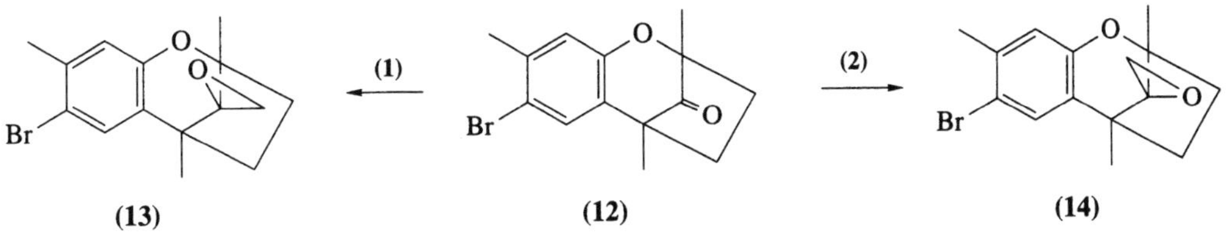

Scheme 4

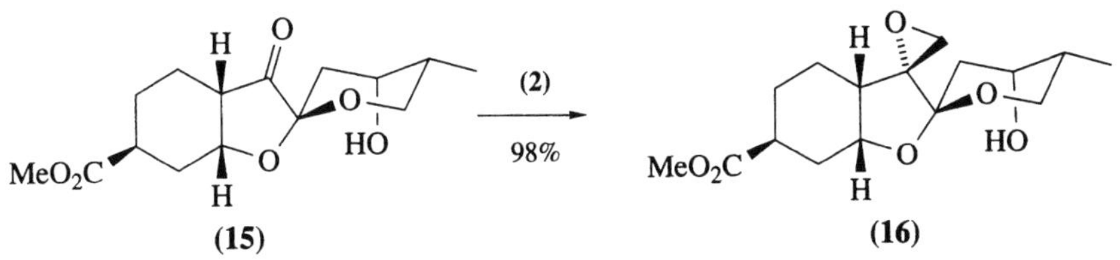

(7)

In contrast with the considerable effort which has gone into delineating the stereochemistry of sulfur ylide additions to cyclic ketones, studies in acyclic stereoselection have been sparse, and generally anecdotal. An interesting example is shown in Scheme 5.[40] Here treatment of ketone (17) with (2) gave an approximately 78:22 preference for (18); addition of zinc chloride or magnesium chloride reversed the sense of the diastereoselection (with ratios of (18):(19) ranging from 23:77 to 10:90). The authors propose cyclic intermediates with significant bond formation between nitrogen and sulfur in the uncatalyzed reaction. The selectivity was thus explained by the preferential decomposition of the betaine containing an *exo*-methyl group. The Lewis acids were postulated to disrupt the cyclic structures.

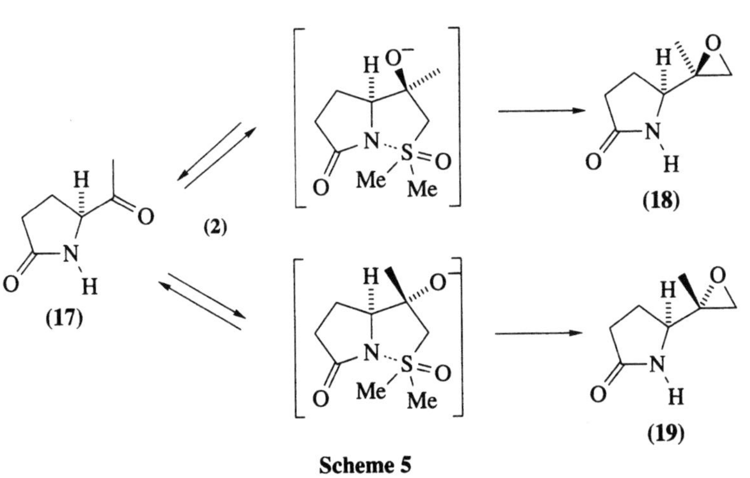

Scheme 5

Other examples of stereoselective additions of sulfoxonium methylide (2) to chiral aldehydes are shown in equations (8),[41,42] (9)[43] and (10).[44a] At present, no one model is able to account for all of these results. Equations (10a)[44b] and (10b)[44c] show the effect of a nitrogen protecting group on the stereoselectivity of addition of unstabilized ylide (1) to α-amino acid derived aldehydes. The major products of epoxidation of *N,N*-dibenzylamino aldehydes result from nonchelation-controlled attack of the nucleophile.

(20) —(2), 39%→ (21) 68% + 32% (8)

—(2), 66%→ (9)

C_6H_{11} ... $^+SEt_2$ —CHO, 51%→ 49% + 8% + 43% (10)

Ph ... CHO, NHC(O)OBu^t —(1), 46%→ NHC(O)OBu^t 50% + NHC(O)OBu^t 50% (10a)

Ph ... CHO, Ph N Ph —(1), 75%→ 86% + 14% (10b)

In a synthetic effort directed toward a segment of erythronolide A, the addition of (**2**) to aldehyde (**22**) gave, after treatment with MeMgBr/CuI, an approximately 80:20 mixture of ring-opened products (**23**) and (**24**; equation 11).[45] Interestingly, direct alkylation of this aldehyde (as a mixture of double bond isomers) with ethyllithium gave an 18:82 mixture of adducts. The factors responsible for the complementary face selectivity shown by (**2**) *versus* ethyllithium are unclear. Comparisons are particularly difficult due to the fact that most organolithium additions to carbonyl compounds are irreversible, kinetically controlled processes, whereas reactions of (**2**) can be reversible.

(**22**) BOM = CH_2OBn —i, (2); ii, MeMgBr, CuI→ (**23**) + (**24**) (11)

Reactions with substituted ylides give rise to questions of relative stereochemistry in the product epoxides. (*E*)/(*Z*)-Selectivity is usually low in the absence of other stereochemical control elements.[35] An exception is diphenylsulfonium benzylide, which reacts stereoselectively with aldehydes to give *trans*-stilbene oxides (*e.g.* see equations 1 and 3). Another example of modest stereocontrol in a cyclic example is shown in equation (12).[46] Arsenic ylides are generally more stereoselective.

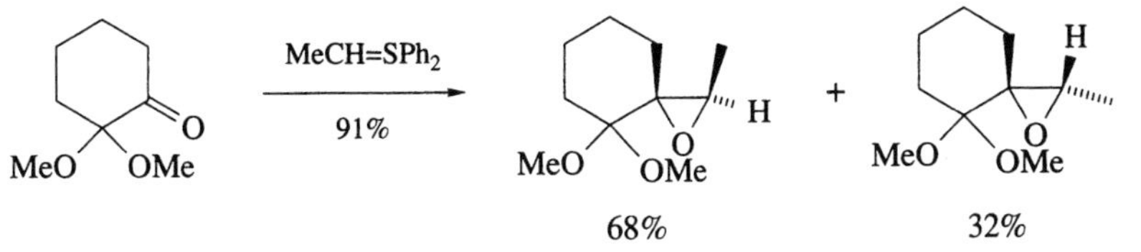

MeO OMe O —($MeCH{=}SPh_2$, 91%)→ MeO OMe O H (68%) + MeO OMe O H (32%) (12)

Another stereochemical feature of potential utility arises from the basic nature of many ylide generation conditions, in which enolizable ketones may undergo epimerization faster than epoxidation.[19] This means that a stereochemically heterogeneous ketone can undergo equilibration prior to reaction. A recent example is shown in equation (13).[47]

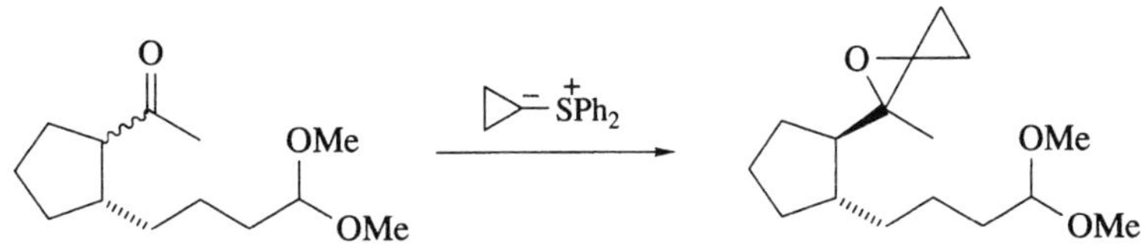

O OMe OMe —($\overset{-}{\triangleright}\overset{+}{S}Ph_2$)→ O OMe OMe (13)

The few published attempts at the asymmetric epoxidation of carbonyl compounds with chiral sulfur ylides have been reviewed.[48] Thus far, such processes have not been very useful synthetically. For example, reaction of benzaldehyde with an optically pure sulfoximine ylide only afforded an epoxide in 20% enantiomeric excess.[49] More recently, chiral sulfur methylides have provided *trans*-stilbene oxides in up to 83% *ee*.[50] An example of optical induction observed in reactions taking place with a chiral phase transfer reagent was reported,[51] but later disputed.[24]

3.2.2.3 Other Ylides

A number of elements form ylides which are analogous to the sulfur-based reagents discussed above, and these reagents display chemistry which is very similar to that obtained with the sulfur compounds. The following discussion will emphasize processes which appear to offer real advantages to better-known techniques.

Arsonium ylides were discovered near the turn of the century, but their reactions with carbonyl compounds did not become elucidated until the 1960s.[8,52] In a broad sense, arsonium ylides are midway in chemical behavior between ylides of phosphorus and those of sulfur. Stabilized arsonium ylides react with carbonyl compounds to afford alkenes, whereas the unstabilized analogs give rise to epoxides. More subtly, the nature of the substituents on either the ylide arsenic or carbon atom can alter the course of the reaction; the choice of solvent can exert a similar effect.[53]

The synthesis of the arsonium ylides most commonly involves deprotonation of arsonium salts utilizing a variety of bases. Simple arsonium salts are available *via* direct alkylation of triphenylarsine, but more highly substituted compounds require specialized methods. These include the use of highly electrophilic triflate salts, alkylation of triphenylarsonium methylide, and double alkylation of lithiodiphenylarsine. For some purposes, formation of the tetrafluoroborate salts is desirable and can be effected in good yield by cation exchange.[54]

The reactions of substituted arsonium ylides with carbonyl compounds can be carried out with high stereoselectivity in favor of *trans*-disubstituted epoxides (equation 14).[54] Equatorial attack is observed for addition to 4-*t*-butylcyclohexanone. Good stereoselectivity (≥9:1) was observed for the addition of triphenylarsonium methylide to some (*N,N*-dibenzyl)amino aldehydes at –78 °C in THF.[44c] Interestingly, the initial hydroxy tetraalkylarsonium adducts were isolated under these conditions, and had to be cyclized under the action of sodium hydride in a separate step.

Another arsonium ylide reaction involves a notable 'transylidation' reaction between a phosphorus and an arsenic ylide (Scheme 6).[55] A useful arsenic ylide which provides a hydroxymethyl epoxide has been reported (equation 15); note the use of biphasic reaction conditions for ylide generation.

Selenium and tellurium ylides take part in chemistry which is analogous to the reactions discussed thus far, and the subject has been well reviewed.[56,57] Both alkenes and epoxides are formed in their reactions,

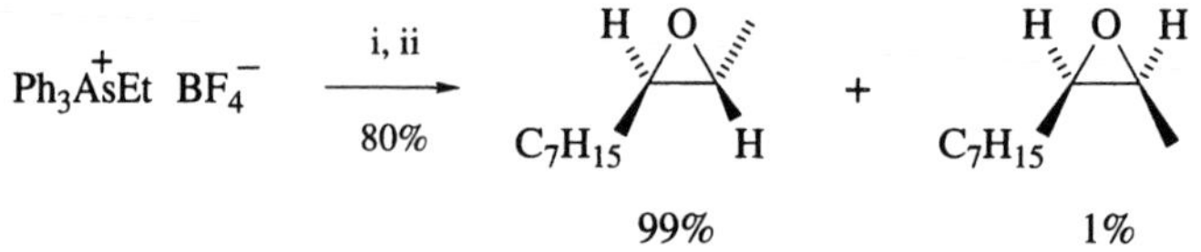

$Ph_3\overset{+}{A}sEt\ BF_4^-$ $\xrightarrow[80\%]{i,\ ii}$ (99%) + (1%) (14)

i, $KN(SiMe_3)_2$, THF/HMPA, 40 °C; ii, $Me(CH_2)_6CHO$

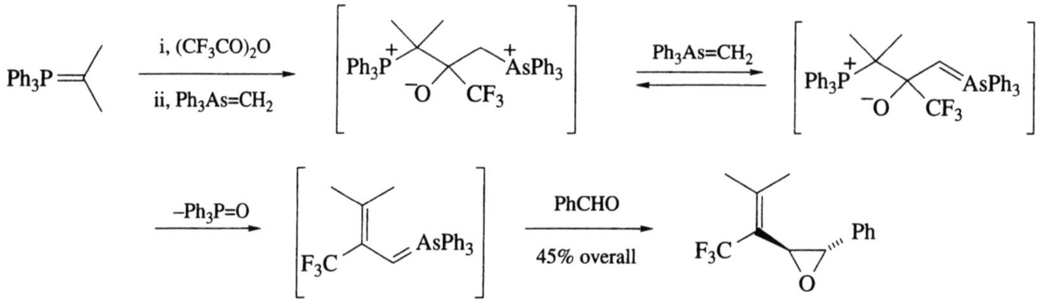

Ph_3P i, $(CF_3CO)_2O$; ii, $Ph_3As{=}CH_2$; $Ph_3As{=}CH_2$; $-Ph_3P{=}O$; PhCHO, 45% overall

Scheme 6

$Ph_3\overset{+}{A}s$ OH Br^- $\xrightarrow[83\%]{i}$ (89%) + (11%) (15)

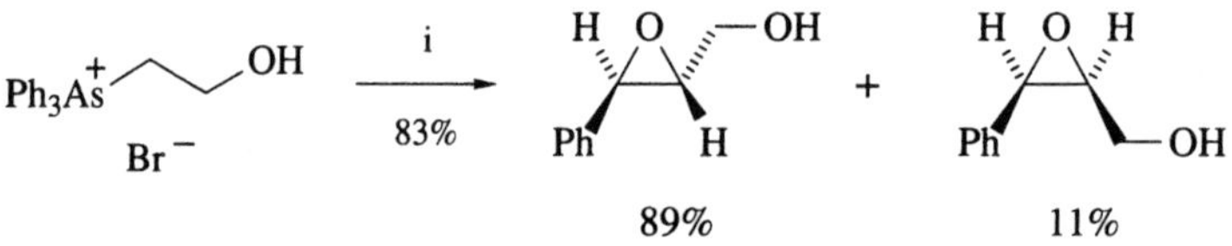

i, PhCHO, KOH (solid), THF (trace water)

as with the arsonium ylides above. The stereoselectivities obtained using these ylides are generally less satisfactory than those observed in the arsenic counterparts.

3.2.2.4 Addition of Anionic Species

Strategies employing α-heterosubstituted organometallic reagents often provide useful alternatives to ylide chemistry for methylene transfer.[58] The carbonyl compound is treated with a heteroatom-stabilized organometallic reagent, which gives rise to an isolable β-hydroxy adduct. These are transformed into the desired epoxide either: (a) spontaneously; (b) under the action of base; or (c) following an intermediate activation step in which X is converted into a better leaving group.

A powerful class of epoxidation reagents in this category are the anions derived from *N*-*p*-tolylsulfonylsulfoximines[10,11] (equation 16).[59] These reagents form epoxides directly upon addition to carbonyl compounds and are available in a number of alkyl substitution patterns. Their chemistry is similar to that of dimethyloxosulfonium methylide, to which they often provide a practical alternative due to their greater nucleophilicity.

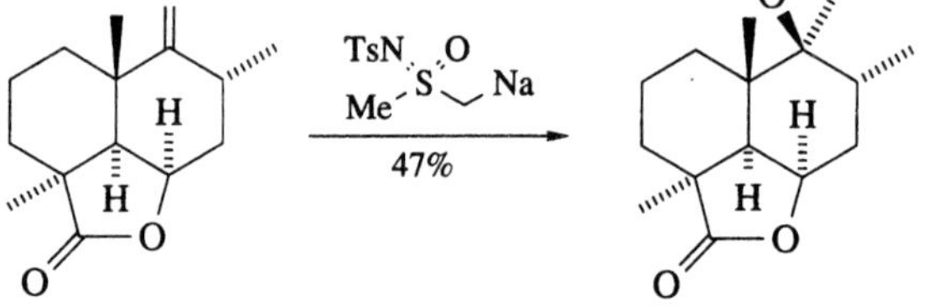

TsN(O)S(Me)CH₂Na, 47% (16)

Deprotonation of thioanisole and addition of a carbonyl compound affords an isolable hydroxy sulfide which can then be alkylated (Scheme 7).[60] Treatment with base generates the same betaine that would have been formed using the sulfur ylide approach, and effects the intramolecular displacement reaction. The addition of methylthiomethyllithium to 2-cyclohexenone exclusively provides the 1,2-addition product;[61] the dianions derived from phenylmethanethiol[62] or allylthiol[63] react in a similar manner. In one case a 6-keto steroidal substrate was found to undergo smooth methylenation using this procedure, whereas attempted epoxidation using either ylide (**1**) or (**2**) exclusively gave rearrangement products such as (**25**; equation 17).

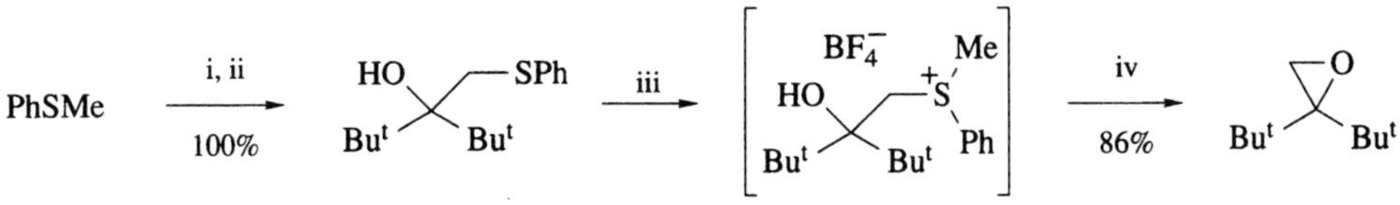

i, Bu^nLi, DABCO; ii, $Bu^t_2C{=}O$; iii, $[Me_3O]BF_4$; iv, CH_2Cl_2, 0.5M NaOH

Scheme 7

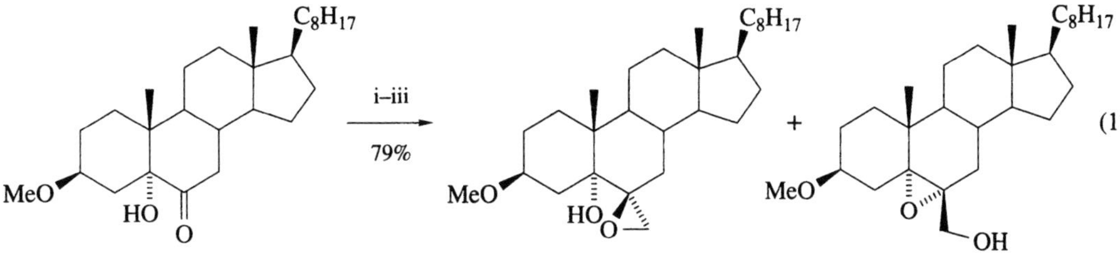

(17)

89% (**25**) 11%

i, $PhSCH_2Li$; ii, $[Me_3O]BF_4$; iii, aq. NaOH

The lithiation of ethyl allyl sulfide followed by transmetallation with titanium isopropoxide engenders an allyltitanium reagent formulated as (**26**; Scheme 8).[64a] This and related reagents add to aldehydes or ketones to afford hydroxy sulfides, which are converted to epoxides as shown. The power of this method for the stereoselective generation of even trisubstituted epoxides is evident from Scheme 8 and equation (18). Reagent (**26a**), prepared as shown in Scheme 8a, undergoes addition to ketone (**26b**) to afford product exclusively resulting from chelation-controlled diastereofacial addition (as a mixture of epimers at the position shown).[64b]

Early efforts to synthesize optically active epoxides using chiral α-lithio sulfoxide anions or chiral complexes of α-lithio sulfides have been reviewed.[65] These approaches usually yield epoxides of low optical purity. More successful, if indirect, approaches are outlined in Section 3.2.2.7.

Chloromethyl phenyl sulfone and related compounds can undergo deprotonation and addition to ketones under a variety of conditions.[66,67] These procedures appear most prominently as part of one-

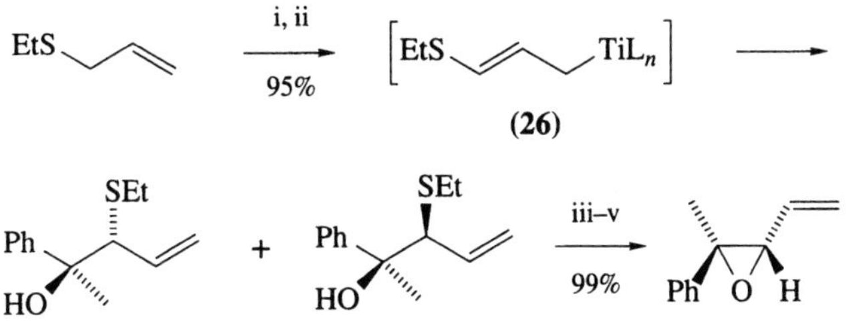

(**26**)

>97% <3%

i, Bu^tLi, $Ti(OPr^i)_4$; ii, PhC(O)Me; iii, separate; iv, $[Me_3O]BF_4$; v, NaOH

Scheme 8

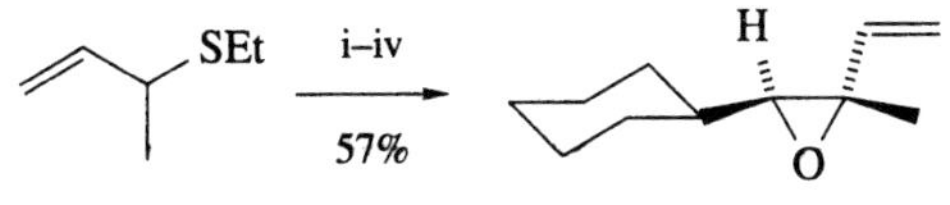

(18)

i, Bu^tLi, $Ti(OPr^i)_4$; ii, $C_6H_{11}CHO$; iii, $[Me_3O]BF_4$; iv, Bu^nLi, –78 °C

i, Bu^tLi; ii, $Ti(OPr^i)_4$ → $(OPr^i)_4Ti^-$ ← SEt **(26a)**

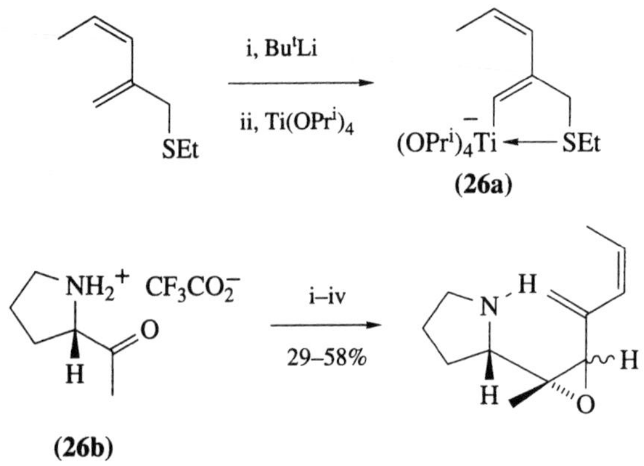

(26b) → i–iv, 29–58%

i, **(26a)**; ii, $(CF_3CO)_2O$; iii, $[Me_3O]BF_4$; iv, NaOH

Scheme 8a

carbon homologation procedures; alternatively, the sulfur substituent can be removed by treatment with *n*-butyllithium at low temperature (Scheme 9).[68] This reaction has recently been featured in a stereoselective approach to a functionalized steroid side chain.[69] Attempts to extend this reaction to asymmetric epoxide synthesis using chiral phase transfer catalysis[70] or a chirally substituted sulfone[71] have been reported.

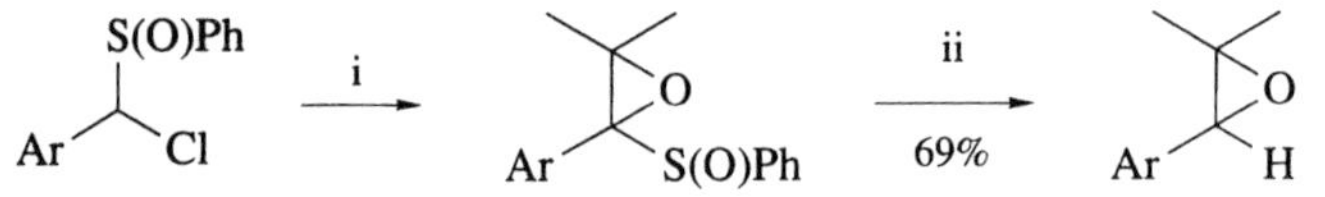

Ar = *p*-$MeOC_6H_4$

i, NaOH, MeCN, acetone; ii, Bu^nLi, –100 °C

Scheme 9

Procedures which utilize selenides are similar, but α-lithio selenides are not generally preparable *via* simple deprotonation chemistry, due to facile selenium–lithium exchange.[56–58,72,73] Selenium-stabilized anions are available, however, by transmetallation reactions of selenium acetals and add readily to carbonyl compounds. The use of branched selenium-stabilized anions has been shown to result exclusively in 1,2-addition to unhindered cyclohexenones, in contrast to the analogous sulfur ylides. The resulting β-hydroxy selenides undergo elimination by treatment with base after activation by alkylation[58,73] or oxidation[74,75] (Scheme 10). An alternative method of activating either β-hydroxy selenides or sulfides toward elimination involves treatment of a chloroform solution of the adduct with thallium ethoxide (Scheme 11).[76] A mechanism involving the intermediacy of a selenium ylide is proposed.

Although much less is known about the chemistry of selenones, these compounds can be converted into epoxides upon reaction with carbonyl compounds in a single-step procedure, which succeeds due to the greater leaving group ability of the benzeneselenonyl anion.[77] A series of cyclopropylselenonyl anions have been prepared and they afford oxaspiropentanes in excellent yields.[78]

α-Chloro-substituted silicon[79] and germanium[80] anions also afford substituted epoxides upon treatment with base and carbonyl compounds. The reactions give disubstituted epoxides with modest levels of *trans* selectivity (equation 19). These reagents undergo addition to chiral aldehydes and ketones with face selectivities analogous to the sulfur ylides described earlier.[79] The resulting silicon-substituted epoxides are useful precursors for chain elongation and ring expansion processes, and can be stereospecifically converted to silyl enol ethers.[81]

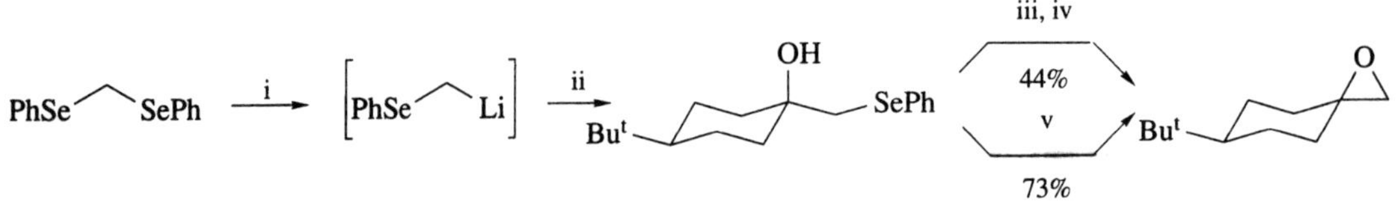

i, Bu^nLi; ii, 4-*t*-butylcyclohexanone; iii, MeI; iv, base; v, MCPBA

Scheme 10

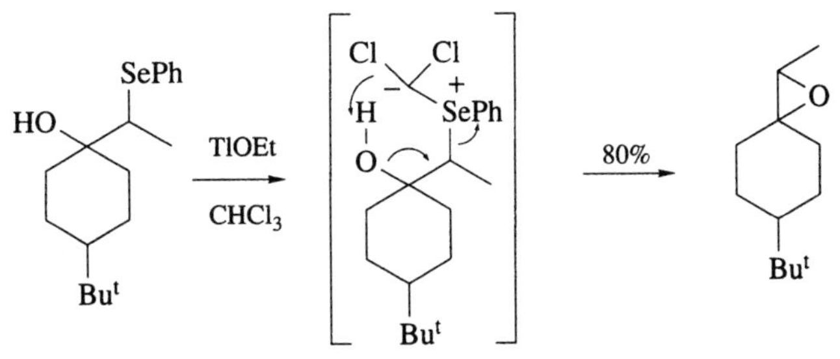

Scheme 11

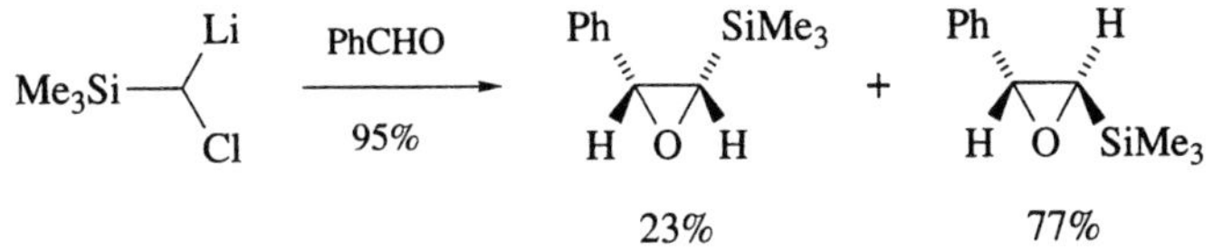

(19)

The additions of metallated 2-allyloxybenzimidazoles to carbonyl compounds are tunable according to the nature of the transition metal counterion. The parent allyl ethers can be deprotonated by *n*-butyllithium and converted to the corresponding allyl-cadmium[82] or -aluminate[83] species by transmetallation (Scheme 12). The authors speculate that the allylcadmium reagent exists primarily in the (*Z*)-configuration, whereas the allylaluminum species is mostly (*E*). Each is presumed to react through a boat-like transition state to yield *erythro* or *threo* adducts, respectively, along with varying amounts of γ-attack. Ring closure with sodium hydride completes the sequence. Addition of the allylcadmium reagent to a chiral aldehyde predominantly afforded the Cram–Felkin isomer.[84]

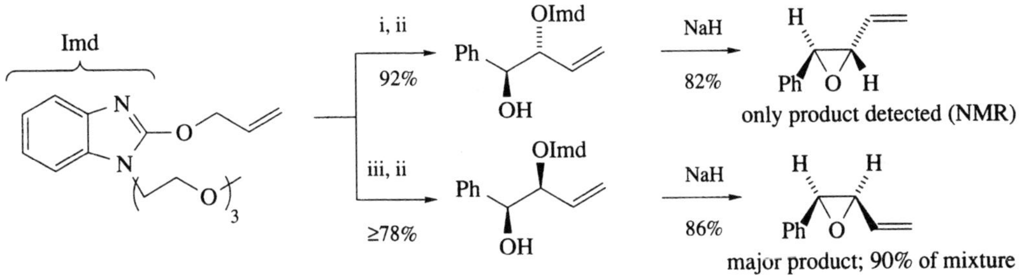

i, Bu^nLi, CdI_2; ii, PhCHO; iii, Bu^nLi, Et_3Al

Scheme 12

Other α-alkoxylithium carbanions have been generated by tin–lithium exchange at low temperature,[85] by metallation of (alkoxymethyl)trimethylsilane,[86] or by using samarium iodide.[87] These reagents do not directly form epoxides due to the poor nucleofugicity of alkoxide anion. However, the resultant protected diols can be considered as particularly robust precursors to epoxides when deprotected and manipulated in any number of standard ways. A notable example of this strategy is shown in Scheme 13.[88]

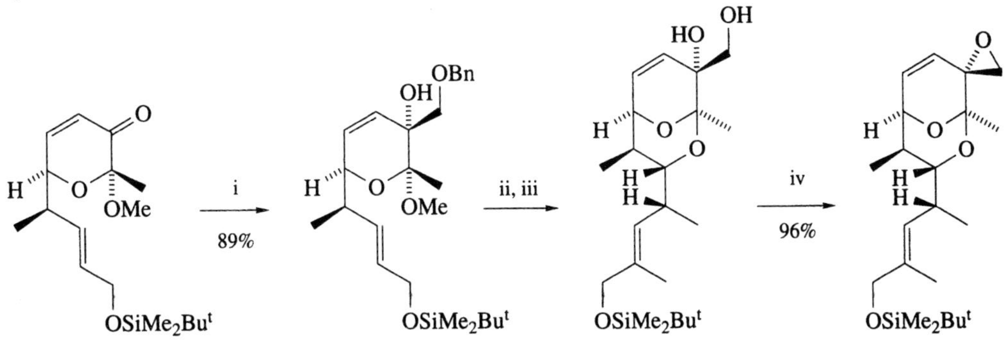

i, $BnOCH_2Li$; ii, 9 steps; iii, debenzylation; iv, (tolylsulfonyl)imidazole, NaH

Scheme 13

3.2.2.5 Halogen-stabilized Carbenoids

Another important group of stabilized anions which have a built-in leaving group suitable for epoxide formation are carbenoids (Köbrich reagents). These reagents are both highly reactive and thermally unstable. Several reviews on the varied chemistry of these reagents are available.[89,90]

The mechanistic outline of carbenoid/carbonyl reactivity follows the paradigm illustrated at the outset of this chapter (Scheme 1; X = halogen). The nucleophilic lithium species adds to the carbonyl compound and suffers elimination to provide the epoxide. Competition from molecular rearrangements emanating from the intermediate halohydrin or the product epoxides is sometimes a problem, particularly with cyclic ketones. Also, the initial adduct frequently fails to cyclize when the reaction is quenched at low temperature, but it is usually a simple matter to effect ring closure by treatment of the halohydrin with mild base in a separate step.

Lithium halocarbenoids are generated by either direct metallation (deprotonation), or lithium–halide exchange. Early efforts to use the former route were limited to vinyl halides or carbenoids bearing at least two chlorine atoms. Lithium–halide exchange has the advantages of offering a greater variety of structural types, and more importantly, often occurring even at ≤ -100 °C. In the case of especially unstable carbenoids, such as those bearing additional alkyl substitution, those having a single chlorine atom, or when X = Br or I, this method of generation is crucial.

Solvent plays an important role in these reactions. A highly basic solvent, usually THF, is used to provide stabilization of the lithium halocarbenoid, due to coordination which disrupts the internal Li–Br interaction which would otherwise lead to α-elimination (equation 20).[91] The extreme low temperature of many of the reactions requires special solvent systems, commonly a mixture of THF, diethyl ether, and petroleum ether or pentane in a ratio of 4:1:1 (Trapp mixture). Solvent has also been noted to alter the mechanistic course of the reaction; for example, reaction of ketones with lithiobromomethane in hexane was observed to give rise to alkenes rather than epoxides.[92]

$$R_2C(Li)Br \longrightarrow R_2C\colon + LiBr \qquad (20)$$

Contemporary work on the development of *in situ* methods for the generation of the classical carbenoid species or providing additional stabilization by other transition metals has permitted the more convenient use of these reagents. A chief advance involved the generation of the lithiohalocarbenoid by exchange with *n*-butyllithium, lithium metal, or lithium amalgam in the presence of the carbonyl partner at –78 °C.[90,92] Simple methylene transfer or epoxidations involving more highly substituted epoxides can be effected under these conditions. For example, the use of chloroiodomethane as the carbenoid precursor affords epoxides in very high yield (equation 21).[93,94] Epoxidations involving more highly substituted epoxides can also be effected under these conditions.[90]

$$PhCOMe + ICH_2Cl \xrightarrow[99\%]{MeLi/LiBr} \text{2-methyl-2-phenyloxirane} \qquad (21)$$

A later modification of the original Köbrich procedure entails the generation of carbenoids in the presence of an added equivalent of lithium bromide salt (equation 22).[91] Apparently, the Lewis acid character of the added salt helps to stabilize the carbenoid by coordination with the halogen atom. The lithium salt is also implicated in Lewis acid complexation to the oxygen atom of the C=O bond, and seems to assist in the cyclization step as well.

$$n\text{-}C_5H_{11}CH(Li)Br \xrightarrow[77\%]{\text{i, LiBr, }-110\ ^\circ\text{C; ii, cyclohexanone}} \text{spiro epoxide} \qquad (22)$$

Lithium–bromine exchange of bromochloromethane can be accomplished using sonication at temperatures as high as –15 °C.[95] The apparent generation of a dichloromethyl anion was effected at room temperature using phase transfer catalysis (equation 23).[96] A practical route involving the selective deprotonation of several dihaloalkane solutions containing a ketone or aldehyde substrate using hindered lithium amide bases has been reported (equation 24).[97] Fluoride catalysis liberates a 'naked' carbenoid anion from a TMS-containing precursor, which adds in good yields to a variety of aldehydes at 0 °C to room temperature.[98] Addition to 2-phenylpropanal predominantly afforded the isomer predicted by Cram's rule (equation 25). Similar results were obtained by generating (presumably) the same anion under electrochemical conditions.[99]

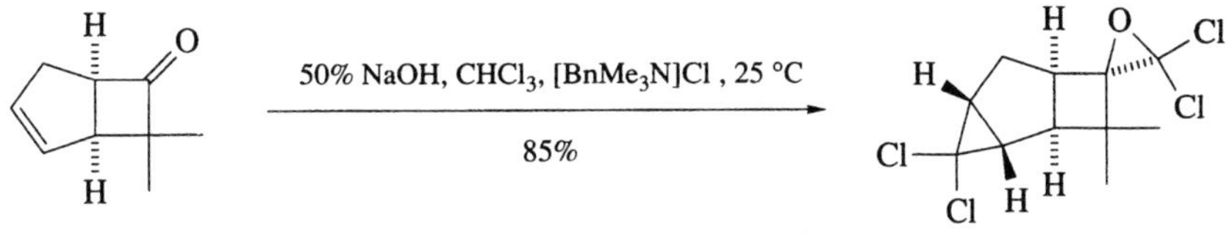

(23)

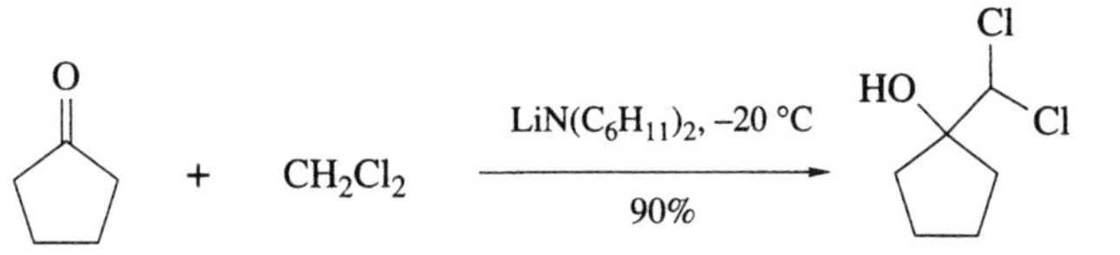

(24)

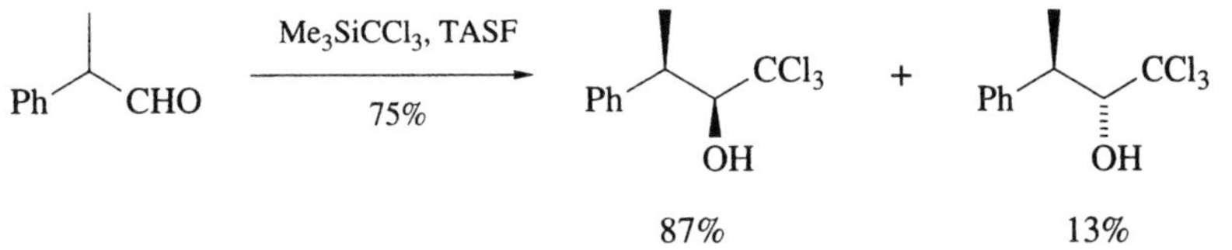

(25)

A number of methods to increase the thermal stability of Köbrich reagents have centered around replacement of the lithium counterion normally present with a variety of transition metals, chiefly titanium, hafnium and copper.[100] For example, the reagent derived from transmetallation of lithium dichloromethane with titanium isopropoxide could be reacted with various ketones and aldehydes to afford halohydrin adducts in good yields at temperatures as high as 0 °C. Competition experiments established the chemoselectivity of this reagent (Scheme 14). The direct activation of several allylic halides has been accomplished with tin[101] and chromium[102] reagents (equation 26). Transmetallation of an alkyllead precursor has also been noted.[103]

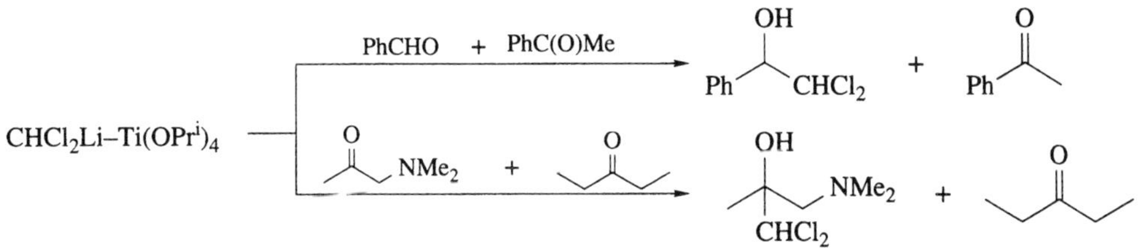

Scheme 14

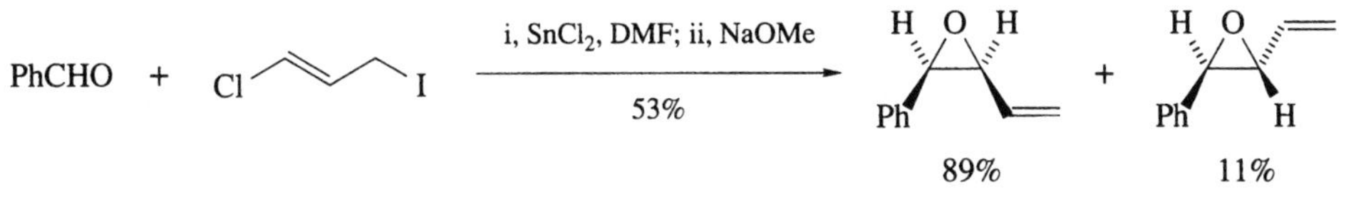

(26)

Activation of dibromo- and diiodo-methanes can also be effected using samarium powder, which generates SmI_2 *in situ*. This procedure allows the isolation of the formal adduct of iodomethane to ketones and aldehydes at room temperature with very short reaction times.[104,105] The iodohydrin so isolated can, of course, be readily converted to the corresponding epoxide. The reaction is thought to occur by a radical chain process. The stereoselectivity of the reaction was briefly investigated (equations 27 and 28).

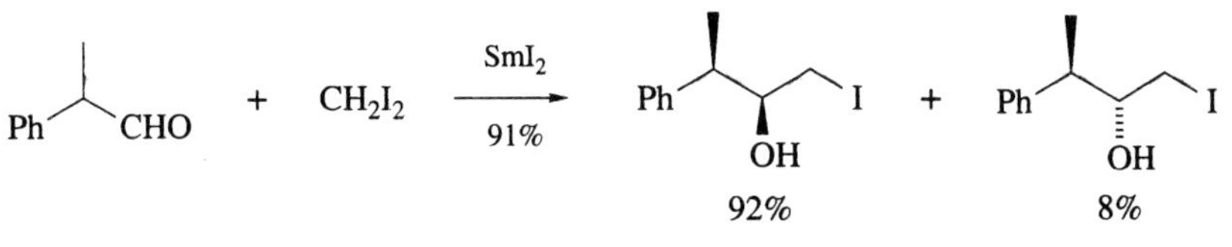

(27)

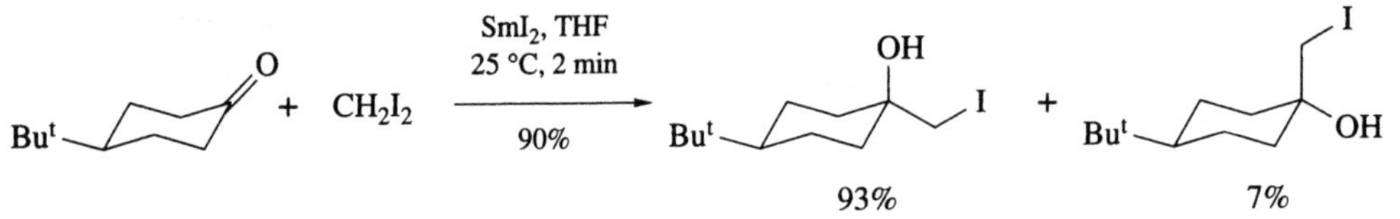

(28)

There are few reports which deal with questions of chemo- and stereo-selectivity of Köbrich reagents. Lithiobromomethane reacted preferentially with a saturated ketone in the presence of an α,β-unsaturated ketone in a steroidal substrate, although the reaction was not stereoselective.[92] However, addition of a related reagent to an acyclic chiral ketone was reported to occur with complete diastereoselectivity (equation 29).[106]

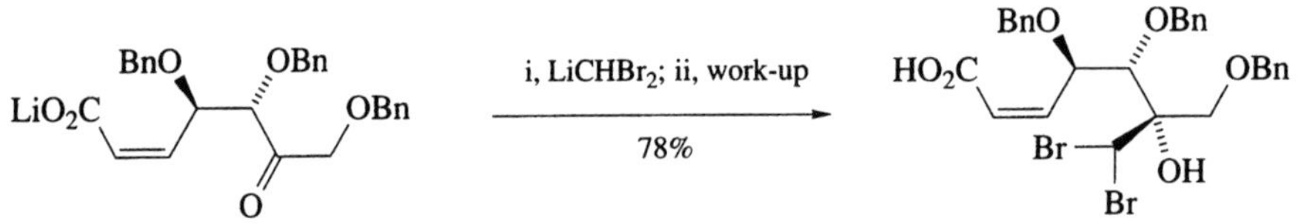

(29)

Chemically distinct halogen atoms can undergo selective exchange in substrates containing a Lewis base site capable of internal coordination.[107] A provocative example involves an exchange reaction which occurs with diastereotopic group selectivity (Scheme 15).[108]

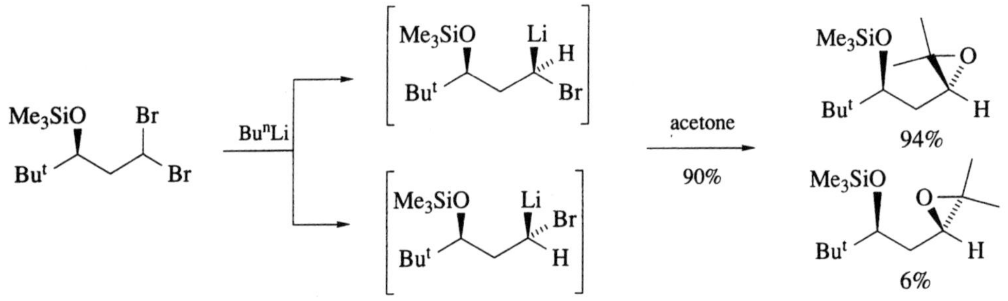

Scheme 15

3.2.2.6 Diazoalkane Reactions

Although historically important, the epoxidation of ketones and aldehydes using diazoalkanes is of less contemporary interest as a synthetic tool. This is due to the much greater propensity of the adducts resulting from addition of the diazo compound to the carbonyl group to react along other pathways. Products arising from chain elongation and ring expansion tend to predominate in these reactions, particularly with cyclic ketones. The interested reader is directed to several good reviews on the subject,[109,110] an older but still useful *Organic Reactions* chapter,[111] and a review dedicated to diazomethane chemistry.[112]

The mechanism for the addition of diazoalkanes to a C=O double bond is generally written along lines similar to those discussed so far (Scheme 1; $X^n = N_2^+$). The initial adduct is also the progenitor of the various rearrangement pathways. However, the subject of mechanism is by no means settled, with 1,3-dipolar cycloadditions[109] and carbonyl ylide formation[113] considered to be prominent alternatives. In general, successful epoxidation of carbonyl compounds improves with increasing electron-poor character of the C=O bond. When the diazoalkane is electron poor, yields of epoxide diminish.

An epoxidation reaction utilizing a complex diazo compound is shown in equation (30).[114] The addition of diazomethane to highly electron-deficient esters has been reported to yield 2-alkoxy-2-substituted epoxides (equation 31).[115] The stereoselectivity of the addition of diazomethane to pentulose derivative (**20**) was shown to be superior to that obtained using sulfur ylides, giving ratios of 95:5 in favor of (**21**; equation 8). However, the yields were unstated, and the products were accompanied in most cases by significant amounts of homologous ketone.[41]

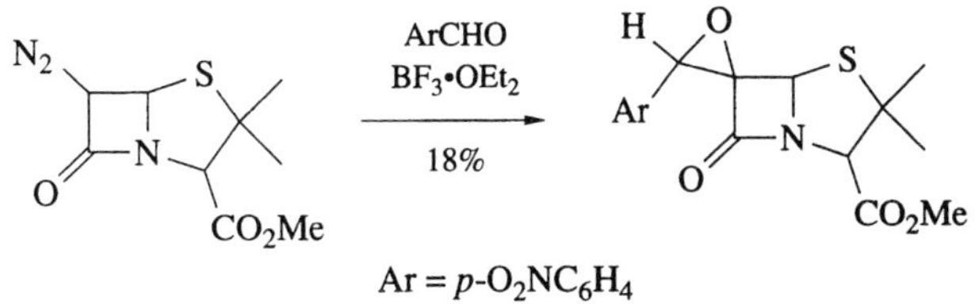

$$\xrightarrow[18\%]{ArCHO,\ BF_3\bullet OEt_2} \quad (30)$$

Ar = *p*-$O_2NC_6H_4$

$$\xrightarrow[56\%]{CH_2N_2} \quad (31)$$

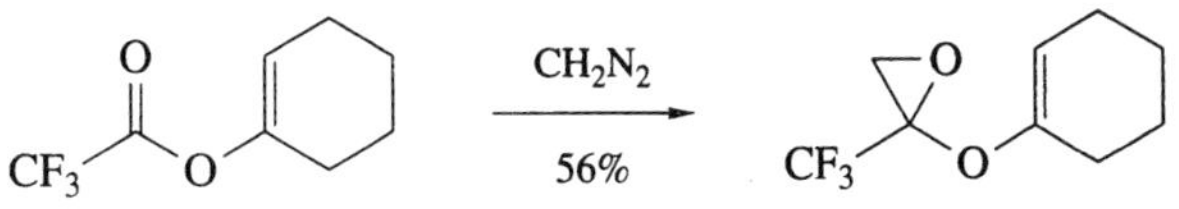

3.2.2.7 Additions to C═O π-Compounds Bearing Adjacent Leaving Groups

Although the bulk of this section is concerned with the formal addition of a single atom across both ends of a C═O double bond, halohydrins also result from the addition of a nucleophile (H^- or R^-) to carbonyl compounds bearing an adjacent leaving group or its equivalent (path b, Scheme 1). The use of this approach for epoxide synthesis was pioneered by Cornforth and coworkers as early as 1959.[116] The scope of this process is very wide, so only a few highlights will be covered here.

The most simple example of this reaction type is the reduction of a ketone $RC(O)CH_2X$ to afford a monosubstituted epoxide. Recently, several strategies for performing this reaction to give epoxides in optically active form have appeared. The corresponding β-keto sulfoxides are available in optically pure form by the deprotonation of resolved *p*-tolylmethyl sulfoxide and addition to esters (Scheme 16).[117] The adducts can be stereospecifically reduced[118] to afford either diastereomer, depending upon the reaction conditions: DIBAL-H in THF gives a 95:5 mixture of (**27**) and (**28**), whereas carrying out the reaction in the presence of zinc chloride[118,119] reversed the ratio to 5:95. The stereochemistry of the former reduction is rationalized by an *anti* conformation resulting from dipole–dipole repulsions between the S—O and C═O bonds, whereas a chelation-controlled transition state best explains the latter result. Reduction to the corresponding sulfide, alkylation and elimination afforded optically pure styrene oxide in high yield. An even more direct approach utilizes a chiral reducing agent (**29**) (equation 32).[120]

The preparation of disubstituted epoxides is similarly accomplished by the addition of nucleophiles to more highly substituted ketones. Selective reduction of keto sulfides with either L-Selectride or zinc

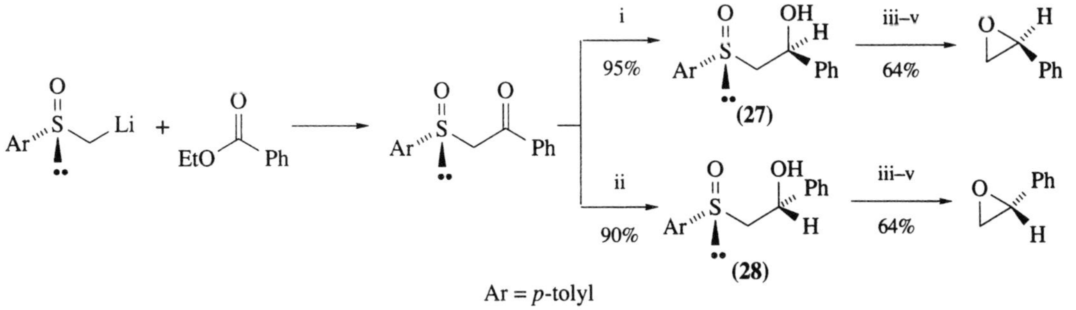

i, 95%: (**27**); iii–v, 64%

ii, 90%: (**28**); iii–v, 64%

Ar = *p*-tolyl

i, DIBAL-H; ii, DIBAL-H, $ZnCl_2$; iii, $LiAlH_4$; iv, $[Me_3O]BF_4$; v, NaOH

Scheme 16

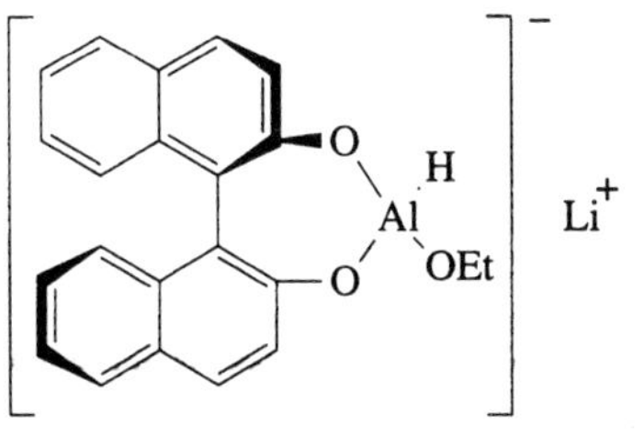

Li^+

(**29**)

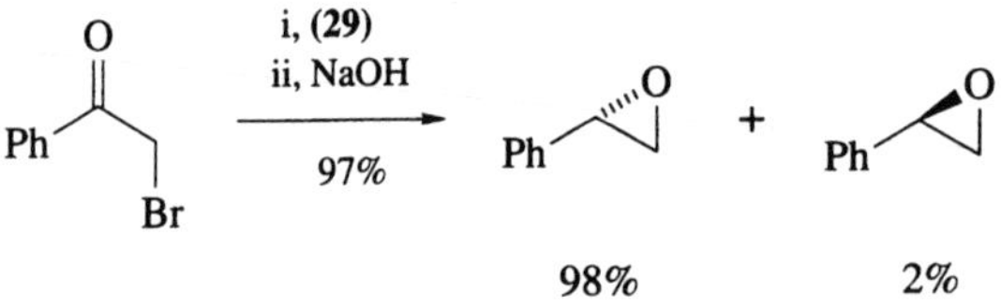

i, **(29)**
ii, NaOH
97%

98% 2%

(32)

borohydride ultimately affords *cis-* or *trans*-epoxides, respectively, as the major products (Scheme 17).[121a] This route was extended to the preparation of optically active epoxides.[121b] The optically active keto sulfides were prepared in 50–85% enantiomeric excess by carrying out an asymmetric sulfenylation reaction on the parent ketones. β-Ketosulfonium salts can also be stereoselectively reduced to give *trans*-epoxides after cyclization.[122]

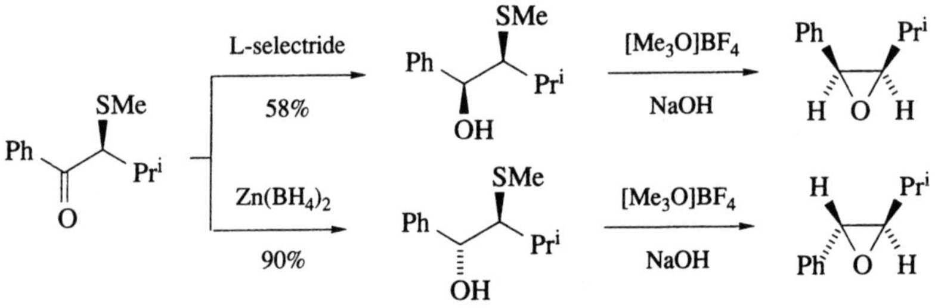

L-selectride 58%
$[Me_3O]BF_4$ NaOH
$Zn(BH_4)_2$ 90%
$[Me_3O]BF_4$ NaOH

Scheme 17

A three-step procedure leads to trisubstituted epoxides from an acid chloride (Scheme 18).[123] A palladium-catalyzed addition of an allyltin reagent to α-chloro ketones was reported to yield trisubstituted epoxides in moderate yield.[124]

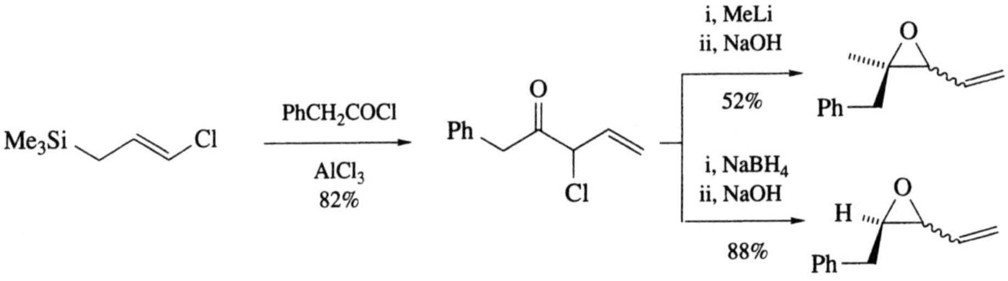

$PhCH_2COCl$ $AlCl_3$ 82%
i, MeLi ii, NaOH 52%
i, $NaBH_4$ ii, NaOH 88%

Scheme 18

3.2.2.8 Other Heterocyclic Syntheses

The formal addition of an oxygen atom across the carbonyl group gives rise to dioxiranes (equation 33). In practice, this reaction is effected with Oxone, and dimethyldioxirane (**30**) and other dioxiranes have been generated in solutions of their parent ketones.[125] Dioxirane (**30**) has been implicated in oxidations of alkenes,[126] sulfides[127] and imines.[128] The formal addition of nitrogen across a carbon–oxygen double bond to afford oxaziridines has been reviewed (equation 34).[129,130] There are also many methods available for the indirect conversion of carbonyl compounds to aziridines[131] and thiiranes[1] using multi-step conversions.

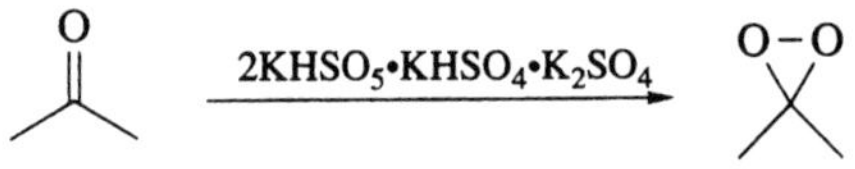

$2KHSO_5{\cdot}KHSO_4{\cdot}K_2SO_4$

(30)

(33)

NH_2Cl
85%

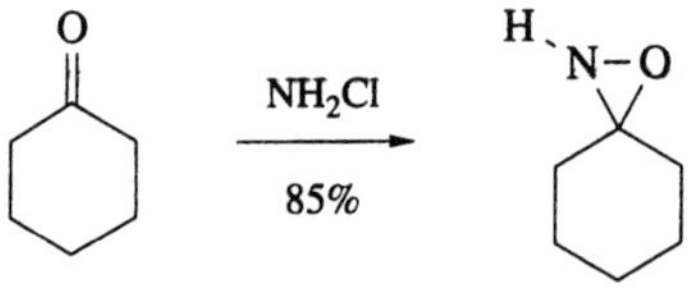

(34)

A number of techniques are available for the one-pot conversions of carbonyl substrates to thiiranes.[1] These entail the addition of a lithiated thiol derivative to the substrate, transfer of either an acyl group or heterocycle to the oxygen atom, and cyclization. A recent example is depicted in Scheme 19.[132]

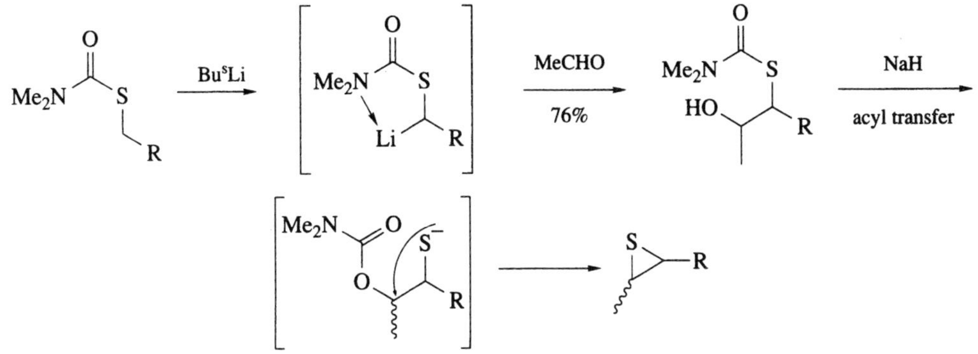

Me2N
S
R
BusLi
Li
MeCHO
76%
HO
NaH
acyl transfer
O

Scheme 19

3.2.3 ADDITIONS TO C═N π-BONDS

3.2.3.1 Addition of Carbon

The addition of a carbon atom to the C═N double bond has been reported using sulfur (equation 35)[135] and arsenic ylides, halogen-stabilized carbenes, and diazo compounds.[131,133] The best substrates are imines in which the nitrogen is substituted with an aromatic ring, substituted oximes, and hydrazones.

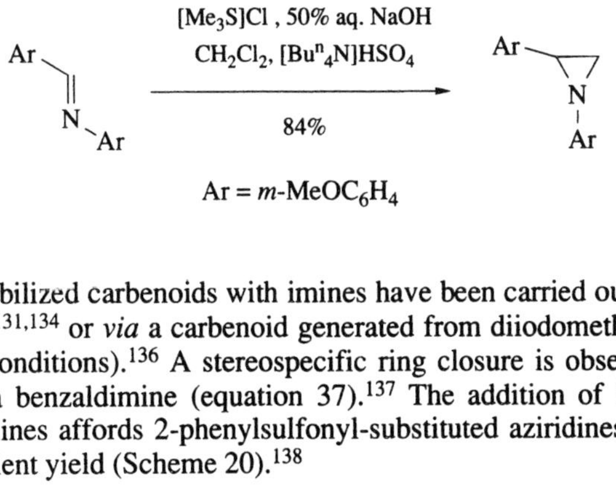

[Me3S]Cl , 50% aq. NaOH
CH2Cl2, [Bun4N]HSO4
84%
Ar
N

(35)

Ar = m-MeOC$_6$H$_4$

Reactions of halogen-stabilized carbenoids with imines have been carried out using preformed lithium species (*e.g.* equation 36),[131,134] or *via* a carbenoid generated from diiodomethane utilizing zinc–copper couple (Simmons–Smith conditions).[136] A stereospecific ring closure is observed after the addition of lithiodichloromethane to a benzaldimine (equation 37).[137] The addition of lithiochloro(phenylsulfonyl)methane to aromatic imines affords 2-phenylsulfonyl-substituted aziridines, which can be deprotonated and alkylated in excellent yield (Scheme 20).[138]

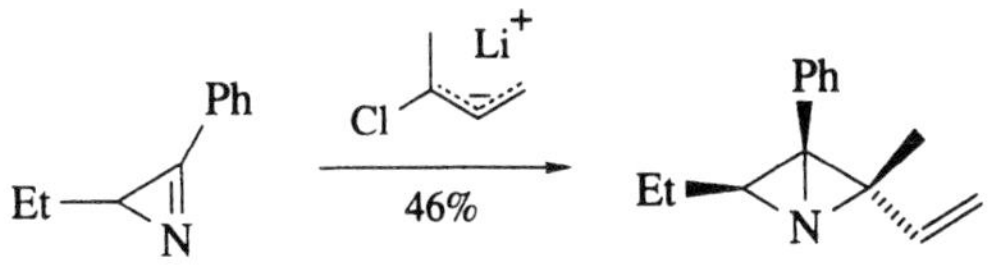

Ph
Et
N
Cl
Li+
46%

(36)

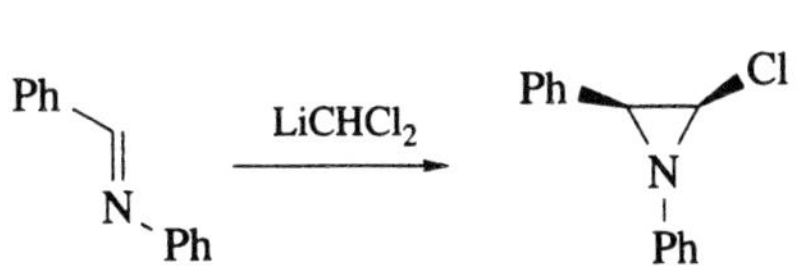

Ph
N
LiCHCl2
Cl

(37)

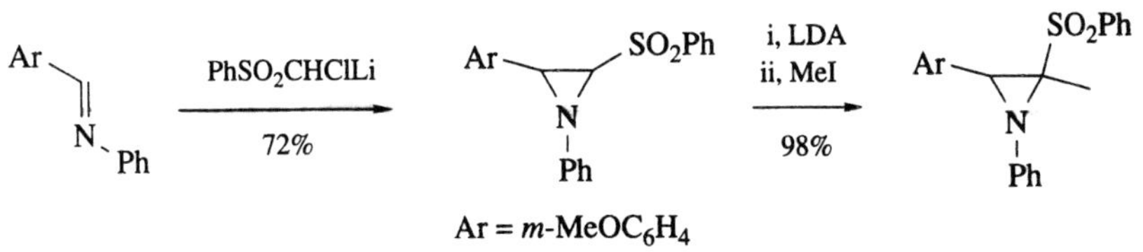

Ar
N
Ph
PhSO2CHClLi
72%
SO2Ph
i, LDA
ii, MeI
98%

Ar = m-MeOC$_6$H$_4$

Scheme 20

The anion of (**31**) reacts with *O*-methyl oximes[139] or nitrones[140] to afford *trans* *N*-unsubstituted aziridines in modest yields (*e.g.* equation 38). Nitrones also react with stabilized phosphonate esters (equation 39).[131,141]

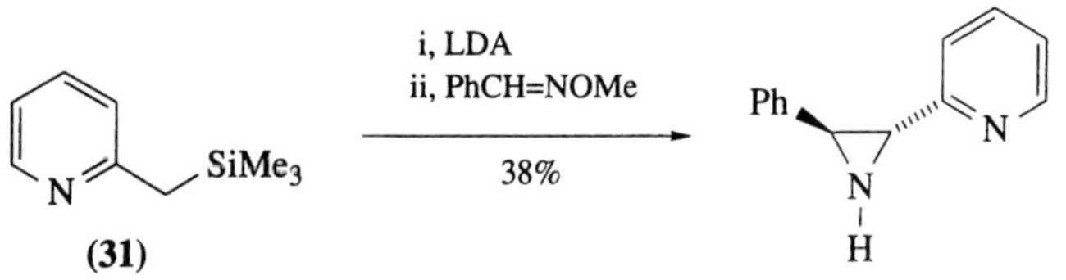

(**31**) — i, LDA; ii, PhCH=NOMe; 38% (38)

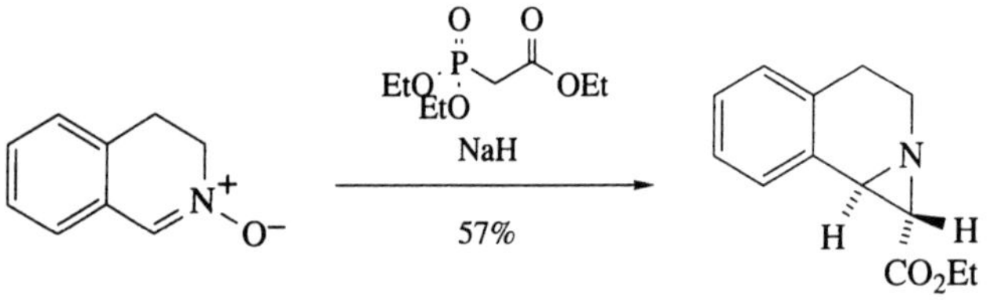

NaH; 57% (39)

The reactions of diazoalkanes and imines occur in generally low yields, but Lewis acid catalysis is of some help (equation 40).[142] The reaction involves a 1,3-dipolar mechanism which proceeds through an isolable triazoline intermediate (Scheme 21). Treatment of the latter with acid affords aziridine in modest yield.

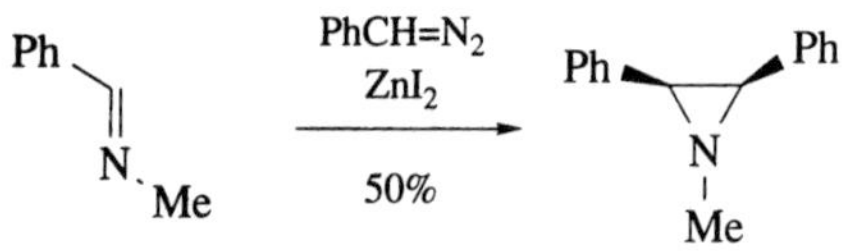

$PhCH{=}N_2$, ZnI_2; 50% (40)

CH_2N_2; 92% — H^+; 42%

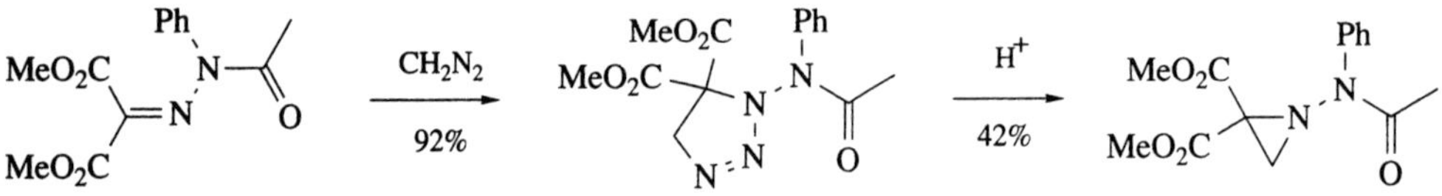

Scheme 21

In contrast, the addition of various diazoalkanes to preformed iminium ions can be a highly efficient process, leading to aziridinium salts, which may undergo further chemistry prior to isolation.[131,143] The intermediacy of a diazonium ion which may react with nearby nucleophilic centers prior to cyclization has been postulated (Scheme 22).[144] This and similar stereoselective reactions have been investigated as potential approaches to morphinoid analgesics.[145,146]

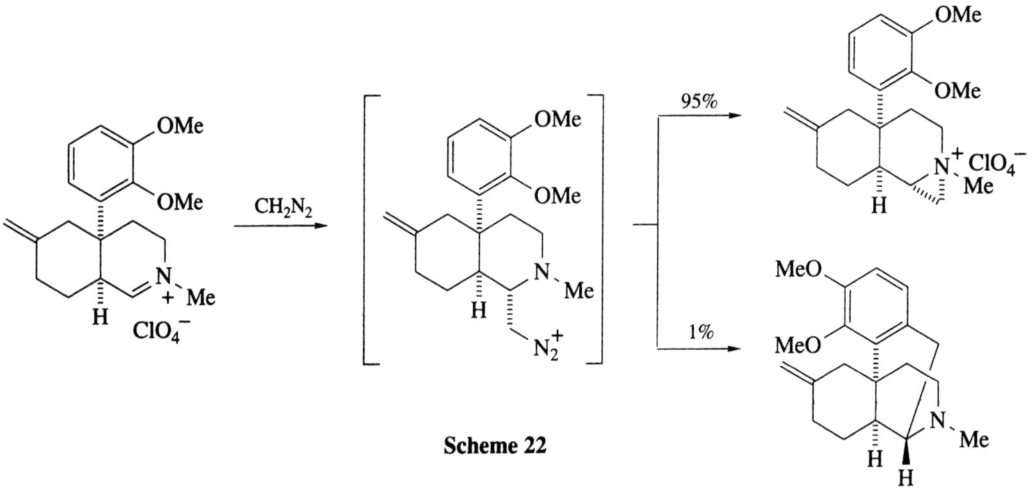

CH_2N_2; 95%; 1%

Scheme 22

The asymmetric synthesis of aziridines *via* these routes has been sparingly investigated. Aziridines bearing an *N*-alkoxy substituent can be resolved into nitrogen stereoisomers. Modest asymmetric induction was observed in the reaction of a chiral *O*-alkyl oxime with dimethylsulfonium methylide (equation 41).[147] The analogous reaction with diazomethane was not stereospecific. Earlier work in this area has been reviewed.[148]

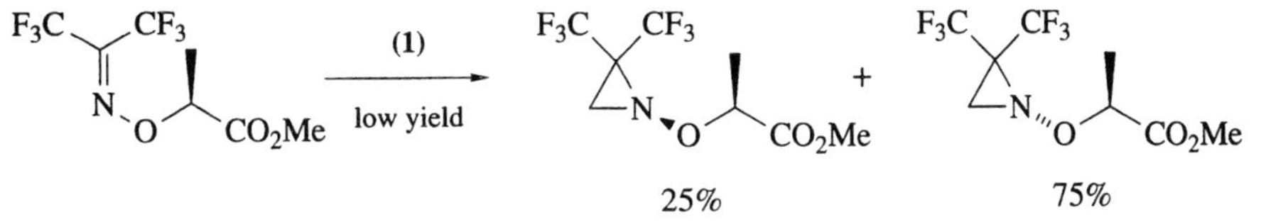

(1), low yield; 25%, 75% (41)

Synthetic methods which utilize the addition of nucleophiles to α-halo oximes and similar compounds (the Hoch–Campbell reaction) have been reviewed.[131]

3.2.3.2 Addition of Oxygen and Nitrogen

The addition of oxygen across the C=N double bond is a common and synthetically useful process.[149] The product oxaziridines are of interest for theoretical reasons (largely due to the high configurational stability of the chiral nitrogen), as reagents (such as oxygen transfer moieties), and as synthetic intermediates.[150] The stereochemical aspects of oxaziridine synthesis and reactivity have been reviewed.[148]

Oxaziridines can be conveniently classified by the substituent on nitrogen. *N*-Alkyloxaziridines and *N*-aryloxaziridines often differ in behavior from those compounds bearing an electron-poor nitrogen substituent, usually an *N*-sulfonyl or *N*-sulfamyl group.

The nitrogen atom in oxaziridines usually constitutes a stable stereogenic center. The nitrogen atoms in *N*-alkyloxaziridines and *N*-aryloxaziridines have barriers to inversion of the order of *ca.* 30–33 kcal mol^{-1} (1 cal = 4.18 J),[148] rendering these compounds configurationally stable at ambient temperature. Alternatively, the various *N*-sulfonyloxaziridines appear to be more readily epimerized, with inversion barriers of *ca.* 19–21 kcal mol^{-1}.[151,152] 3-Alkoxyoxaziridines readily epimerize at room temperature, possibly through zwitterionic intermediates.[153]

The most common method for the synthesis of *N*-alkyloxaziridines and *N*-aryloxaziridines from imines is through the oxidation of imines with an organic peroxy acid.[129,154] Two mechanisms for this reaction have been advanced (Scheme 23). The concerted mechanism is exactly analogous to that usually accepted for the oxidation of alkenes with electron-deficient peracids. Unlike the latter conversion, however, the oxidation reaction of imines is not stereospecific. To account for this observation, it is necessary to presume that the imine isomers equilibrate more rapidly than the oxidation occurs, and that the *cis* isomer is more reactive.[155] Neither of these possibilities can be convincingly ruled out. However, an alternative, Baeyer–Villiger type of mechanism seems more likely. Addition of the active oxygen atom to the carbon of the C=N double bond followed by expulsion of the acid by-product gives rise to oxaziridine. Intermediate (**32**) must be sufficiently long lived to allow nitrogen inversion to occur prior to ring closure. The two-step mechanism has been supported by an *ab initio* molecular orbital study.[156]

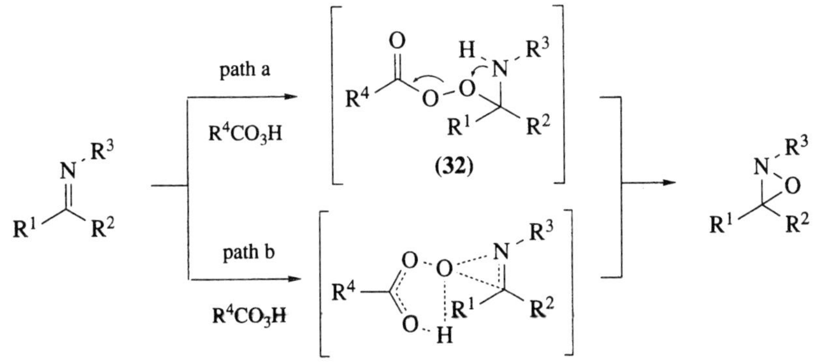

path a, R^4CO_3H; (**32**); path b, R^4CO_3H

Scheme 23

Methods utilizing MCPBA[157,158] or Oxone[159] in buffered biphasic systems usually suffice for the stereoselective synthesis of *trans*-sulfonyl- and sulfamyl-substituted oxaziridines. The stereoselectivity is rationalized by the greater ability of the *N*-sulfonyl moiety to stabilize a negative charge in the intermediate resulting from addition to the imine carbon atom, allowing more time for bond rotation or nitrogen inversion before the ring closure step.[160]

The oxidation of imines derived from substituted cyclohexanones occurs predominantly from the equatorial direction. However, the product oxaziridines can undergo subsequent equilibration to favor a more stable conformation which places the bulkier nitrogen substituent in an equatorial conformation (Scheme 24).[161] A particularly useful oxaziridine derived from camphor was formed exclusively *via exo* approach of the reagent (equation 42).[162]

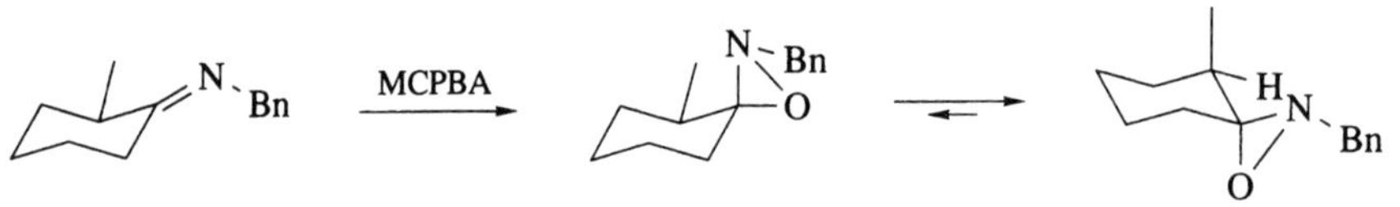

N Bn MCPBA N Bn O H N Bn O

Scheme 24

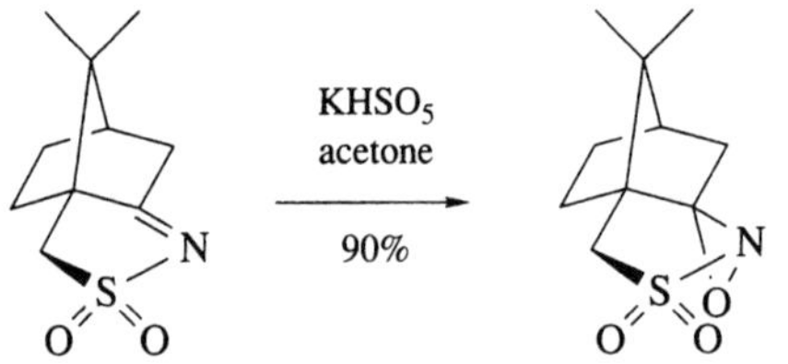

$KHSO_5$ acetone 90% (42)

A chiral substituent on nitrogen can direct the predominant attack of an oxidizing agent to one diastereotopic face of an imine. Ratios as high as 97:3 were observed in a series of imines derived from α-methylbenzylamine (equation 43).[163] The oxidation of achiral imines with optically active peroxy acids, most notably monoperoxycamphoric acid [(+)-MPCA], do afford optically active oxaziridines, but not generally in synthetically useful optical ratios,[148] but ratios as high as 80:20 have been reported.[164]

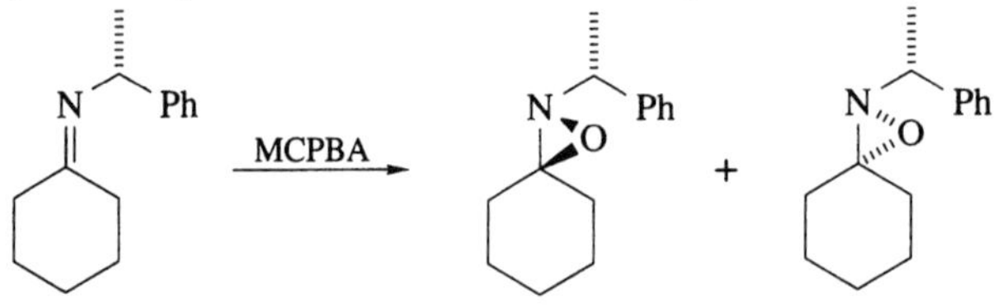

N Ph MCPBA N Ph O + N Ph O (43)

Reactions involving the use of chiral substrates in conjunction with chiral reagents have been reported. A seminal experiment was the oxidation of both enantiomers of a chiral imine with (+)-MPCA.[148,165] In another example, the cumulative effect of equatorial attack in prochiral cyclohexanones with diastereoselectivity induced by a chiral nitrogen substituent allowed the synthesis of spirocyclic oxaziridines with a high induction of axial disymmetry (Scheme 25).[166,167] The major oxaziridine isomer **(33)** results from both the favored equatorial attack and oxidation *anti* to the chiral nitrogen substituent (as drawn in Scheme 25). The isomers **(34)** and **(35)** result from a combination of one favored and one disfavored process and are minor products, whereas **(36)**, which would result from a doubly disfavored trajectory, is not observed. The oxidation of a cyclic imine which involved a synergism between control by the chiral nitrogen substituent and attack from the convex face of the ring system was reported (equation 44).[168] Oxaziridines of the type **(37)** have been utilized in the synthesis of racemic[169,170] and optically active[167,168] indole alkaloids.

Imines and substituted oximes react with several nitrogen-containing nucleophiles to provide a general synthesis of diaziridines or diazirines.[171]

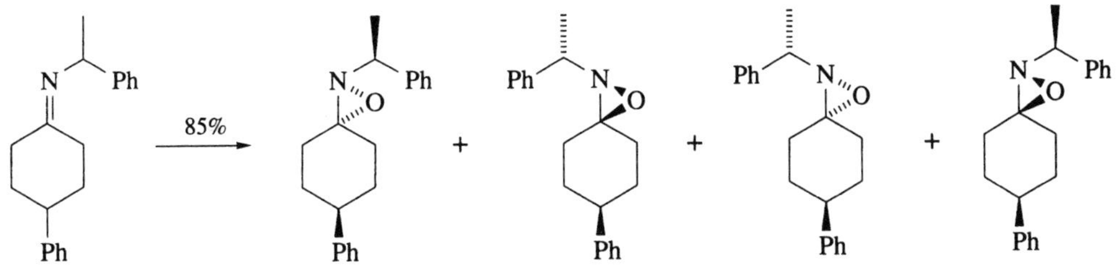

85%

(33) 70% + (34) 22% + (35) 8% + (36) 0%

Scheme 25

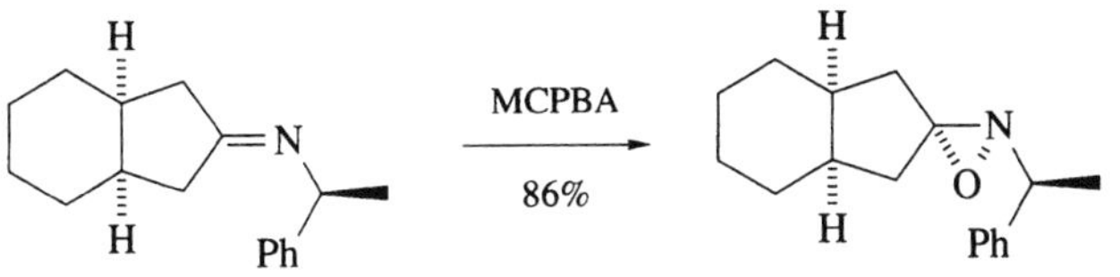

MCPBA 86%

mixture of C=N isomers → (37) 66% + 3 other isomers 34% (44)

3.2.4 REFERENCES

1. U. Zoller, in 'Small Ring Heterocycles', ed. A. Hassner, Wiley, New York, 1983, part 1, p. 333.
2. A. S. Rao, S. K. Paknikar and J. G. Kirtane, *Tetrahedron*, 1983, **39**, 2323.
3. M. Bartók and K. L. Láng, in 'Small Ring Heterocycles', ed. A. Hassner, Wiley, New York, 1985, part 3, p. 1.
4. Y. G. Gololobov, A. N. Nesmeyanov, V. P. Lysenko and I. E. Boldeskul, *Tetrahedron*, 1987, **43**, 2609.
5. E. J. Corey and M. Chaykovsky, *J. Am. Chem. Soc.*, 1965, **87**, 1353.
6. H. O. House, 'Modern Synthetic Reactions', 2nd edn., Benjamin, Menlo Park, CA, 1972, p. 709.
7. C. R. Johnson, in 'Comprehensive Organic Chemistry', ed. D. H. R. Barton and W. D. Ollis, Pergamon Press, Oxford, 1979, vol. 3, p. 247.
8. A. W. Johnson, 'Ylide Chemistry', Academic Press, New York, 1966.
9. B. M. Trost and L. S. Melvin, Jr., 'Sulfur Ylides: Emerging Synthetic Intermediates', Academic Press, New York, 1975.
10. C. R. Johnson, *Acc. Chem. Res.*, 1973, **6**, 341.
11. C. R. Johnson, *Aldrichimica Acta*, 1985, **18**, 3.
12. G. C. Barrett, in 'Comprehensive Organic Chemistry', ed. D. H. R. Barton and W. D. Ollis, Pergamon Press, Oxford, 1979, vol. 3, p. 105.
13. D. R. Williams and S.-Y. Sit, *J. Am. Chem. Soc.*, 1984, **106**, 2949.
14. F. Volatron and O. Eisenstein, *J. Am. Chem. Soc.*, 1987, **109**, 1.
15. A. Kumar, R. Singh and A. K. Mandal, *Synth. Commun.*, 1982, **12**, 613.
16. K. Okuma, Y. Tanaka, S. Kaji and H. Ohta, *J. Org. Chem.*, 1983, **48**, 5133.
17. K. Okuma, K. Nishimura and H. Ohta, *Chem. Lett.*, 1984, 93.
18. B. M. Trost, *Acc. Chem. Res.*, 1974, **7**, 85.
19. B. M. Trost, *Top. Curr. Chem.*, 1986, **133**, 3.
20. H. Bouda, M. E. Borredon, M. Delmas and A. Gaset, *Synth. Commun.*, 1987, **17**, 503.
21. M. E. Borredon, M. Delmas and A. Gaset, *Tetrahedron*, 1987, **43**, 3945.
22. M. E. Borredon, J. V. Sinisterra, M. Delmas and A. Gaset, *Tetrahedron*, 1987, **43**, 4141.
23. P. Mosset and R. Grée, *Synth. Commun.*, 1985, **15**, 749.
24. M. Rosenberger, W. Jackson and G. Saucy, *Helv. Chim. Acta*, 1980, **63**, 1665.
25. M. J. Farrall, T. Durst and J. M. J. Fréchet, *Tetrahedron Lett.*, 1979, 203.
26. F. Texier-Boullet, D. Villemin, M. Ricard, H. Moison and A. Foucaud, *Tetrahedron*, 1985, **41**, 1259.
27. A. Padwa and J. R. Gasdaska, *Tetrahedron*, 1988, **44**, 4147.
28. (a) E. J. Corey, Y. Arai and C. Mioskowski, *J. Am. Chem. Soc.*, 1979, **101**, 6748; (b) J. Rokach, Y. Girard, Y. Guindon, J. G. Atkinson, M. Larue, R. N. Young, P. Masson and G. Holme, *Tetrahedron Lett.*, 1980, **21**, 1485.
29. M. Rosenberger, C. Newkom and E. R. Aig, *J. Am. Chem. Soc.*, 1983, **105**, 3656.
30. J. K. Crandall, H. S. Magaha, R. K. Widener and G. A. Tharp, *Tetrahedron Lett.*, 1980, **21**, 4807.
31. M. E. Garst, B. J. McBride and A. T. Johnson, *J. Org. Chem.*, 1983, **48**, 8.
32. M. E. Garst and P. Arrhenius, *J. Org. Chem.*, 1983, **48**, 16.
33. J. Adams, L. Hoffman, Jr. and B. M. Trost, *J. Org. Chem.*, 1970, **35**, 1600.
34. E. G. Baggiolini, B. M. Hennessy, J. A. Iacobelli and M. R. Uskokovic, *Tetrahedron Lett.*, 1987, **28**, 2095.
35. G. Berti, *Top. Stereochem*, 1973, **7**, 93.
36. T. H. Kim and S. Isoe, *J. Chem. Soc., Chem. Commun.*, 1983, 730.
37. B. M. Trost and L. H. Latimer, *J. Org. Chem.*, 1978, **43**, 1031.
38. E. W. Colvin and S. Cameron, *J. Chem. Soc., Chem. Commun.*, 1986, 1642.
39. D. J. Goldsmith, T. K. John, C. D. Kwong and G. R. Painter, III, *J. Org. Chem.*, 1980, **45**, 3989.

40. M. J. Fray, E. J. Thomas and J. D. Wallis, *J. Chem. Soc., Perkin Trans. 1*, 1983, 395.
41. S. Hagen, T. Anthonsen and L. Kilaas, *Tetrahedron*, 1979, **35**, 2583.
42. S. Hagen, W. Lwande, L. Kilaas and T. Anthonsen, *Tetrahedron*, 1980, **36**, 3101.
43. T. Masamune, H. Murase, H. Matsue and A. Murai, *Bull. Chem. Soc. Jpn.*, 1979, **52**, 135.
44. (a) M. Tanaka, K. Tomioka and K. Koga, *Tetrahedron Lett.*, 1985, **26**, 6109; (b) B. E. Evans, K. E. Rittle, C. F. Homnick, J. P. Springer, J. Hirshfield and D. F. Veber, *J. Org. Chem.*, 1985, **50**, 4615; (c) M. T. Reetz and J. Binder, *Tetrahedron Lett.*, 1989, **30**, 5425.
45. S. Hoagland, Y. Morita, D. L. Bai, H.-P. Märki, K. Kees, L. Brown and C. H. Heathcock, *J. Org. Chem.*, 1988, **53**, 4730.
46. P. Lescure and F. Huet, *Synthesis*, 1987, 404.
47. B. M. Trost and D. C. Lee, *J. Am. Chem. Soc.*, 1988, **110**, 6556.
48. M. R. Barbachyn and C. R. Johnson, in 'Asymmetric Synthesis', ed. J. D. Morrison and J. W. Scott, Academic Press, New York, 1984, vol. 4, p. 227.
49. C. R. Johnson, C. W. Schroeck and J. R. Shanklin *J. Am. Chem. Soc.*, 1973, **95**, 7424.
50. (a) N. Furukawa, Y. Sugihara and H. Fujihara, *J. Org. Chem.*, 1989, **54**, 4222; (b) L. Breau, W. W. Ogilvie and T. Durst, *Tetrahedron Lett.*, 1990, **31**, 35.
51. T. Hiyama, T. Mishima, H. Sawada and H. Nozaki, *J. Am. Chem. Soc.*, 1975, **97**, 1626.
52. D. Lloyd, I. Gosney and R. A. Ormiston, *Chem. Soc. Rev.*, 1987, **16**, 45.
53. J. B. Ousset, C. Mioskowski and G. Solladié, *Synth. Commun.*, 1983, **13**, 1193.
54. W. C. Still and V. J. Novack, *J. Am. Chem. Soc.*, 1981, **103**, 1283.
55. Y. Shen, Q. Liao and W. Qiu, *J. Chem. Soc., Chem Commun.*, 1988, 1309.
56. (a) A. Krief, in 'The Chemistry of Organic Selenium and Tellurium Compounds', ed. S. Patai, Wiley, New York, 1987, vol. 5, p. 675; (b) A. Krief, W. Dumont, D. Van Ende, S. Halazy, D. Labar, J. L. Laboureur and T. Q. Lê, *Heterocycles*, 1989, **28**, 1203.
57. D. L. J. Clive, *Tetrahedron*, 1978, **34**, 1049.
58. A. Krief, *Tetrahedron*, 1980, **36**, 2531.
59. S. C. Welch, A. S. C. P. Rao, J. T. Lyon and J.-M. Assercq, *J. Am. Chem. Soc.*, 1983, **105**, 252.
60. J. R. Shanklin, C. R. Johnson, J. Ollinger and R. M. Coates, *J. Am. Chem. Soc.*, 1973, **95**, 3429.
61. S. P. Tanis, M. C. McMills and P. M. Herrinton, *J. Org. Chem.*, 1985, **50**, 5887.
62. K.-H. Geiss, D. Seebach and B. Seuring, *Chem. Ber.*, 1977, **110**, 1833.
63. M. Pohmakotr, K.-H. Geiss and D. Seebach, *Chem. Ber.*, 1979, **112**, 1420.
64. (a) Y. Ikeda, K. Furuta, N. Meguriya, N. Ikeda and H. Yamamoto, *J. Am. Chem. Soc.*, 1982, **104**, 7663; (b) B. M. Trost and T. S. Scanlon, *J. Am. Chem. Soc.*, 1989, **111**, 4988.
65. G. Solladié, in 'Asymmetric Synthesis', ed. J. D. Morrison, Academic Press, New York, 1983, vol. 2, p. 157.
66. D. R. White, *Tetrahedron Lett.*, 1976, 1753.
67. M. Adamczyk, E. K. Dolence, D. S. Watt, M. R. Christy, J. Reibenspies and O. P. Anderson, *J. Org. Chem.*, 1984, **49**, 1378.
68. T. Satoh, Y. Kaneko and K. Yamakawa, *Tetrahedron Lett.*, 1986, **27**, 2379.
69. E. K. Dolence, M. Adamczyk, D. S. Watt, G. B. Russell and D. H. S. Horn, *Tetrahedron Lett.*, 1985, **26**, 1189.
70. S. Colonna, R. Fornasier and U. Pfeiffer, *J. Chem. Soc., Perkin Trans. 1*, 1978, 8.
71. M. H. H. Nkunya and B. Zwanenburg, *Recl. Trav. Chim. Pays-Bas*, 1985, **104**, 253.
72. H. J. Reich, in 'Organoselenium Chemistry', ed. D. Liotta, Wiley, New York, 1987, p. 243.
73. A. Krief, *Top. Curr. Chem.*, 1987, **135**, 1.
74. A. Krief, W. Dumont and J.-L. Laboureur, *Tetrahedron Lett.*, 1988, **29**, 3265.
75. S. Uemura, K. Ohe and N. Sugita, *J. Chem. Soc., Chem. Commun.*, 1988, 111.
76. J.-L. Laboureur, W. Dumont and A. Krief, *Tetrahedron Lett.*, 1984, **25**, 4569.
77. A. Krief, W. Dumont and J.-N. Denis, *J. Chem. Soc., Chem. Commun.*, 1985, 571.
78. A. Krief, W. Dumont and J. L. Laboureur, *Tetrahedron Lett.*, 1988, **29**, 3265.
79. C. Burford, F. Cooke, G. Roy and P. Magnus, *Tetrahedron*, 1983, **39**, 867.
80. T. Kauffmann, R. König and M. Wensing, *Tetrahedron Lett.*, 1984, **25**, 637.
81. P. F. Hudrlik, A. M. Hudrlik and A. K. Kulkarni, *J. Am. Chem. Soc.*, 1985, **107**, 4260.
82. M. Yamaguchi and T. Mukaiyama, *Chem. Lett.*, 1979, 1279.
83. M. Yamaguchi and T. Mukaiyama, *Chem. Lett.*, 1982, 237.
84. M. Yamaguchi and T. Mukaiyama, *Chem. Lett.*, 1981, 1005.
85. W. C. Still, *J. Am. Chem. Soc.*, 1978, **100**, 1481.
86. P. Magnus and G. Roy, *J. Chem. Soc., Chem. Commun.*, 1979, 822.
87. T. Imamoto, T. Takeyama and M. Yokoyama, *Tetrahedron Lett.*, 1984, **25**, 3225.
88. R. E. Ireland and M. G. Smith, *J. Am. Chem. Soc.*, 1988, **110**, 854.
89. G. Köbrich, *Angew. Chem., Int. Ed. Engl.*, 1972, **11**, 473.
90. H. Siegel, *Top. Curr. Chem.*, 1982, **106**, 55.
91. J. Villieras, B. Kirschleger, R. Tarhouni and M. Rambaud, *Bull. Soc. Chim. Fr.*, 1986, 470.
92. G. Cainelli, A. U. Ronchi, F. Bertini, P. Grasselli and G. Zubiani, *Tetrahedron*, 1971, **27**, 6109.
93. K. M. Sadhu and D. S. Matteson, *Tetrahedron Lett.*, 1986, **27**, 795.
94. J. Barluenga, J. L. Fernández-Simón, J. M. Concellón and M. Yus, *J. Chem. Soc., Chem. Commun.*, 1987, 915.
95. C. Einhorn, C. Allavena and J. L. Luche, *J. Chem. Soc., Chem. Commun.*, 1988, 333.
96. H. Greuter, T. Winkler and D. Bellus, *Helv. Chim. Acta*, 1979, **62**, 1275.
97. H. Taguchi, H. Yamamoto and H. Nozaki, *Bull. Chem. Soc. Jpn.*, 1977, **50**, 1588.
98. M. Fujita and T. Hiyama, *J. Am. Chem. Soc.*, 1985, **107**, 4085.
99. T. Shono, N. Kise and T. Suzumoto, *J. Am. Chem. Soc.*, 1984, **106**, 259.
100. T. Kauffmann, R. Fobker and M. Wensing, *Angew. Chem., Int. Ed. Engl.*, 1988, **27**, 943.
101. J. Augé and S. David, *Tetrahedron Lett.*, 1983, **24**, 4009.
102. J. Augé, *Tetrahedron Lett.*, 1988, **29**, 6107.
103. A. Doucoure, B. Mauzé and L. Miginiac, *J. Organomet. Chem.*, 1982, **236**, 139.

104. T. Imamoto, T. Takeyama and H. Koto, *Tetrahedron Lett.*, 1986, **27**, 3243.
105. T. Tabuchi, J. Inanaga and M. Yamaguchi, *Tetrahedron Lett.*, 1986, **27**, 3891.
106. C. S. Wilcox and J. J. Gaudino, *J. Am. Chem. Soc.*, 1986, **108**, 3102.
107. K. G. Taylor, *Tetrahedron*, 1982, **38**, 2751.
108. R. W. Hoffmann, M. Bewersdorf, K. Ditrich, M. Krüger and R. Stürmer, *Angew. Chem., Int. Ed. Engl.*, 1988, **27**, 1176.
109. G. W. Cawell and A. Ledwith, *Q. Rev., Chem. Soc.*, 1970, **24**, 119.
110. D. S. Wulfman, G. Linstrumelle and C. F. Cooper, in 'The Chemistry of Diazonium and Diazo Groups', ed. S. Patai, Wiley, New York, 1978, p. 821.
111. C. D. Gutsche, *Org. React. (N. Y.)*, 1954, **8**, 364.
112. J. S. Pizey, 'Synthetic Reagents', Horwood, Chichester, 1974, vol. 2, p. 65.
113. R. Huisgen and P. de March, *J. Am. Chem. Soc.*, 1982, **104**, 4953.
114. V. J. Jephcote, D. I. John, P. D. Edwards, K. Luk and D. J. Williams, *Tetrahedron Lett.*, 1984, **25**, 2915.
115. P. Strazzolini, G. Verardo and A. G. Giumanini, *J. Org. Chem.*, 1988, **53**, 3321.
116. J. D. Morrison and H. S. Mosher, 'Asymmetric Organic Reactions', American Chemical Society, Washington, D.C., 1971, p. 98.
117. T. Durst, in 'Comprehensive Organic Chemistry', ed. D. H. R. Barton and W. D. Ollis, Pergamon Press, Oxford, 1979, vol. 3, p. 121.
118. G. Solladié, G. Demailly and C. Greck, *Tetrahedron Lett.*, 1985, **26**, 435.
119. M. Kosugi, H. Konta and H. Uda, *J. Chem. Soc., Chem. Commun.*, 1985, 211.
120. R. Noyori, I. Tomino, M. Yamada and M. Nishizawa, *J. Am. Chem. Soc.*, 1984, **106**, 6717.
121. (a) M. Shimagaki, T. Maeda, Y. Matsuzaki, I. Hori, T. Nakata and T. Oishi, *Tetrahedron Lett.*, 1984, **25**, 4775; (b) T. Yura, N. Iwasawa, R. Clark and T. Mukaiyama, *Chem. Lett.*, 1986, 1809.
122. M. Shimagaki, Y. Matsuzaki, I. Hori, T. Nakata and T. Oishi, *Tetrahedron Lett.*, 1984, **25**, 4779.
123. M. Ochiai and E. Fujita, *Tetrahedron Lett.*, 1980, **21**, 4369.
124. M. Kosugi, H. Arai, A. Yoshino and T. Migita, *Chem. Lett.*, 1978, 795.
125. R. W. Murray and R. Jeyaraman, *J. Org. Chem.*, 1985, **50**, 2847.
126. G. Cicala, R. Curci, M. Fiorentino and O. Laricchiuta, *J. Org. Chem.*, 1982, **47**, 2670.
127. R. W. Murray, R. Jeyaraman and M. K. Pillay, *J. Org. Chem.*, 1987, **52**, 746.
128. F. A. Davis, S. Chattopadhyay, J. C. Towson, S. Lal and T. Reddy, *J. Org. Chem.*, 1988, **53**, 2087.
129. M. J. Haddadin and J. P. Freeman, in 'Small Ring Heterocycles', ed. A. Hassner, Wiley, New York, 1985, part 3, p. 283.
130. L. L. Müller and J. Hamer, '1,2-Cycloaddition Reactions', Wiley, New York, 1967, p. 103.
131. J. A. Deyrup, in 'Small Ring Heterocycles', ed. A. Hassner, Wiley, New York, 1983, part 1, p. 1.
132. P. Beak and P. D. Becker, *J. Org. Chem.*, 1982, **47**, 3855.
133. S. Trippett and M. A. Walker, *J. Chem. Soc. C*, 1971, 1114.
134. B. Mauzé, *Tetrahedron Lett.*, 1984, **25**, 843.
135. R. S. Tewari, A. K. Awasthi and A. Awasthi, *Synthesis*, 1983, 330.
136. P. Baret, H. Buffet and J. L. Pierre, *Bull. Soc. Chim. Fr.*, 1972, 825.
137. J. A. Deyrup and R. B. Greenwald, *Tetrahedron Lett.*, 1965, 321.
138. V. Reutrakul, V. Prapansiri and C. Panyachotipun, *Tetrahedron Lett.*, 1984, **25**, 1949.
139. T. Konakahara, M. Matsuki and K. Sato, *Heterocycles*, 1984, **22**, 1319.
140. O. Tsuge, K. Sone, S. Urano and K. Matsuda, *J. Org. Chem.*, 1982, **47**, 5171.
141. S. Zbaida and E. Breuer, *J. Org. Chem.*, 1982, **47**, 1073.
142. R. Bartnik and G. Mlostón, *Synthesis*, 1983, 924.
143. D. R. Crist and N. J. Leonard, *Angew. Chem., Int. Ed. Engl.*, 1969, **8**, 962.
144. D. A. Evans and C. H. Mitch, *Tetrahedron Lett.*, 1982, **23**, 285.
145. J. E. McMurry, V. Farina, W. J. Scott, A. H. Davidson, D. R. Sumners and A. Shenvi, *J. Org. Chem.*, 1984, **49**, 3803.
146. W. H. Moos, R. D. Gless and H. Rapoport, *J. Org. Chem.*, 1983, **48**, 227.
147. A. V. Prosyanik, A. I. Mishchenko, N. L. Zaichenko, G. V. Shustov, P. N. Belov and R. G. Kostyanovskii, *Izv. Akad. Nauk SSSR, Ser. Khim.*, 1984, 596.
148. F. A. Davis and R. H. Jenkins, Jr., in 'Asymmetric Synthesis', ed. J. D. Morrison and J. W. Scott, Academic Press, Orlando, 1984, vol. 4, p. 313.
149. M. J. Haddadin and J. P. Freeman, in 'Small Ring Heterocycles', ed. A. Hassner, Wiley, New York, 1983, part 3, p. 283.
150. F. A. Davis and A. C. Sheppard, *Tetrahedron*, 1989, **45**, 5703.
151. W. B. Jennings, S. P. Watson and M. S. Tolley, *J. Am. Chem. Soc.*, 1987, **109**, 8099.
152. W. B. Jennings, S. P. Watson and D. R. Boyd, *J. Chem. Soc., Chem. Commun.*, 1988, 931.
153. O. Gonzalez C., D. E. Gallis and D. R. Crist, *J. Org. Chem.*, 1986, **51**, 3266.
154. R. Paredes, H. Bastos, R. Montoya, A. L. Chavez, W. R. Dolbier, Jr. and C. R. Burkholder, *Tetrahedron*, 1988, **44**, 6821.
155. M. Bucciarelli, A. Forni, I. Moretti and G. Torre, *J. Chem. Soc., Perkin Trans. 2*, 1977, 1339.
156. A. Ažman, J. Koller and B. Plesničar, *J. Am. Chem. Soc.*, 1979, **101**, 1107.
157. L. C. Vishwakarma, O. D. Stringer and F. A. Davis, *Org. Synth.*, 1988, **66**, 203.
158. F. A. Davis, J. Lamendola, Jr., U. Nadir, E. W. Kluger, T. C. Sedergran, T. W. Panunto, R. Billmers, R. Jenkins, Jr., I. J. Turchi, W. H. Watson, J. S. Chen and M. Kimura, *J. Am. Chem. Soc.*, 1980, **102**, 2000.
159. F. A. Davis, S. Chattopadhyay, J. C. Towson, S. Lal and T. Reddy, *J. Org. Chem.*, 1988, **53**, 2087.
160. M. Bucciarelli, A. Forni, I. Moretti and G. Torre, *J. Chem. Soc., Perkin Trans. 2*, 1983, 923.
161. (a) E. Oliveros, M. Rivière and A. Lattes, *J. Heterocycl. Chem.*, 1980, **17**, 107; (b) J. Aubé, M. Hammond, E. Gherardini and F. Takusagawa, *J. Org. Chem.*, 1991, **56**, 499.
162. F. A. Davis, J. C. Towson, M. C. Weismiller, S. Lal and P. J. Carroll, *J. Am. Chem. Soc.*, 1988, **110**, 8477.
163. C. Belzecki and D. Mostowicz, *J. Org. Chem.*, 1975, **40**, 3878.

164. W. H. Pirkle and P. L. Rinaldi, *J. Org. Chem.*, 1977, **42**, 2080.
165. D. Mostowicz and C. Belzecki, *J. Org. Chem.*, 1977, **42**, 3917.
166. A. Lattes, E. Oliveros, M. Rivière, C. Belzecki, D. Mostowicz, W. Abramskj, C. Piccinni-Leopardi, G. Germain and M. van Meerssche, *J. Am. Chem. Soc.*, 1982, **104**, 3929.
167. J. Aubé, Y. Wang, M. Hammond, M. Tanol, F. Takusagawa and D. Van der Velde, *J. Am. Chem. Soc.*, 1990, **112**, 4879.
168. J. Aubé, *Tetrahedron Lett.*, 1988, **29**, 4509.
169. Y. Langlois, A. Pouilhès, D. Génin, R. Z. Andriamialisoa and N. Langlois, *Tetrahedron*, 1983, **39**, 3755.
170. M. E. Kuehne and W. H. Parsons, *Tetrahedron*, 1983, **39**, 3763.
171. H. W. Heine, in 'Small Ring Heterocycles', ed. A. Hassner, Wiley, New York, 1983, part 3, p. 547.

3.3
Skeletal Reorganizations: Chain Extension and Ring Expansion

PETER M. WOVKULICH
Hoffmann-La Roche, Nutley, NJ, USA

3.3.1 INTRODUCTION

The utilization of homologation reactions as a general synthesis strategy may be considered for various reasons, such as a greater accessibility of the lower homolog, the opportunity to introduce additional functionality, or simply the need for a regular series of homologs. The merit of an expansion/extension reaction which may be considered for incorporation into a synthetic plan will, to a large extent, be measured by the efficiency, technical simplicity, and regio- and/or stereo-selectivity of the overall operation. The persistent growth of new methodologies is testimony to the continuing demand for strategies which offer greater selectivity or versatility.

An exhaustive review[1] on the general topic of carbocyclic ring expansions was recorded over twenty years ago, while a number of more specialized reviews,[2a–g] each not necessarily devoted to but incorporating some aspects of homologation reactions, have appeared during the interim. The intention of this chapter is to present a variety of methodologies with illustrative examples, focusing on the processes which achieve an overall homologation with preservation of the original carbonyl group, as illustrated in Scheme 1. This criterium emphasizes rearrangement processes over simple homologation protocols such as the methoxymethylenation/hydrolysis of carbonyl groups, alkylation with acyl anion equivalents, carbonyl group transpositions and so on. Since the orientation is primarily towards synthetic transformations, detailed mechanistic discussions will be held to a minimum, although in-depth analyses may usually be found in the accompanying references. The intention is to cover a variety of methods with sufficient examples to display both the advantages and disadvantages of the process, in the hope that the advantages will be exploited and the disadvantages will be taken as opportunities for improvement.

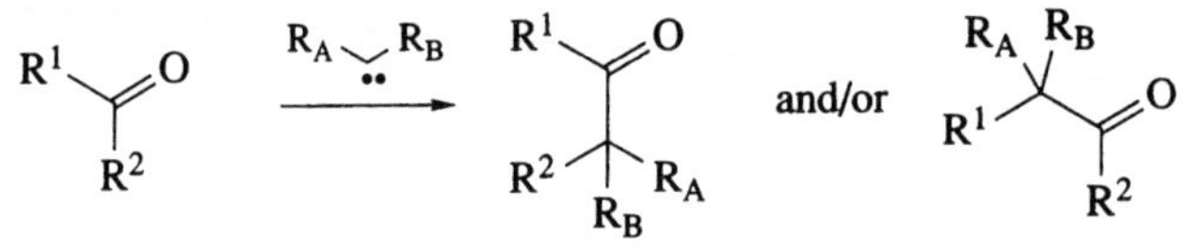

Scheme 1

3.3.2 DIAZOALKANES AND RELATED REACTIONS

3.3.2.1 Arndt–Eistert Homologation

One of the more commonly applied chain extension reactions for carboxylic acids utilizes the unique reactivity of diazoalkanes. This sequence, generally referred to as the Arndt–Eistert synthesis, is a two-step process which, in the first step, involves the formation of an α-diazo ketone by reaction of the corresponding acyl chloride with an excess of diazoalkane (Scheme 2).[3a–c] In the second stage of the sequence, the α-diazo ketone is induced to undergo rearrangement by photochemical or thermal means or by exposure to metal salts (*e.g.* Ag, Cu), with concomitant loss of nitrogen to produce an intermediate ketene, which reacts *in situ* with water or alcohol to form the homologated acid or ester. This rearrangement of α-diazo ketones, known as the Wolff rearrangement, has been reviewed elsewhere in sufficient depth with regard to mechanism and synthetic utility that no comprehensive treatment will be given here.[4,5a–d]

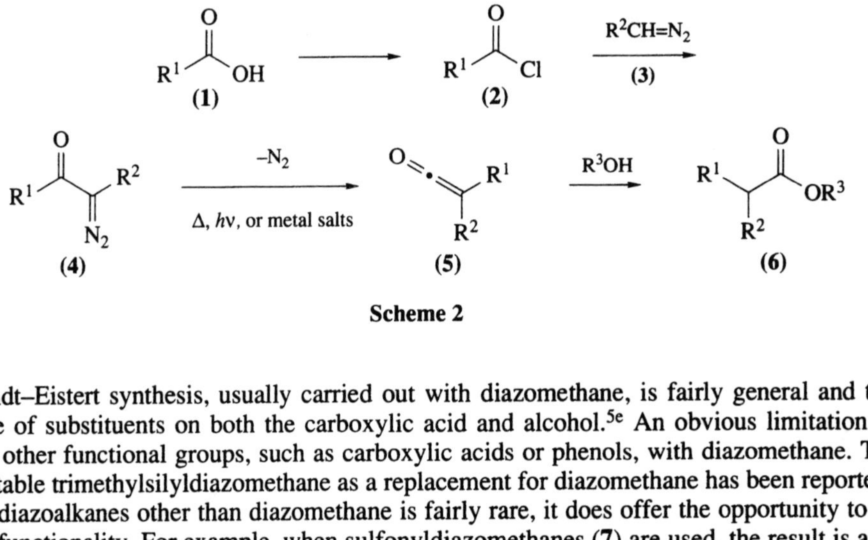

Scheme 2

The Arndt–Eistert synthesis, usually carried out with diazomethane, is fairly general and tolerates a wide range of substituents on both the carboxylic acid and alcohol.[5e] An obvious limitation is the reactivity of other functional groups, such as carboxylic acids or phenols, with diazomethane. The use of the more stable trimethylsilyldiazomethane as a replacement for diazomethane has been reported.[6] While the use of diazoalkanes other than diazomethane is fairly rare, it does offer the opportunity to introduce additional functionality. For example, when sulfonyldiazomethanes (**7**) are used, the result is an α-sulfonyl carboxylic ester (Scheme 3).[7]

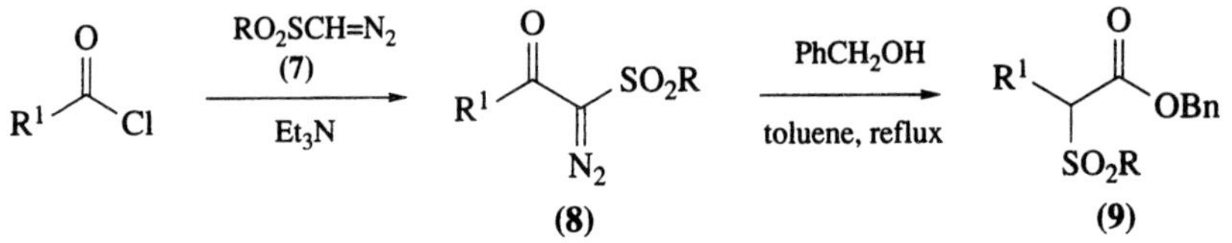

R, R^1 = alkyl, aryl

Scheme 3

3.3.2.2 Reactions of Aldehydes and Ketones

The reaction of diazomethane derivatives with aldehydes and ketones, though a fairly well-studied field now, continues to be a source of practical extension/homologation technology.[8a–f,9] The general reaction is outlined in Scheme 4. The diazoalkane, shown in resonance forms (**11a**) and (**11b**), reacts with the carbonyl carbon to form an intermediary betaine (**12**). Depending on the circumstances, the intermediate may be isolated in the form of the β-hydroxy diazo compound (**13**) or may lose nitrogen to produce an intermediate (**14**), which may then undergo rearrangement with migration of a substituent R^2 or R^1 to give the homologated compounds (**15**) and (**16**) respectively. Alternatively, simple collapse of (**14**) without migration of a substituent leads to epoxide (**17**). In general, the course of the reaction is influenced by the substituents and by the reaction conditions.

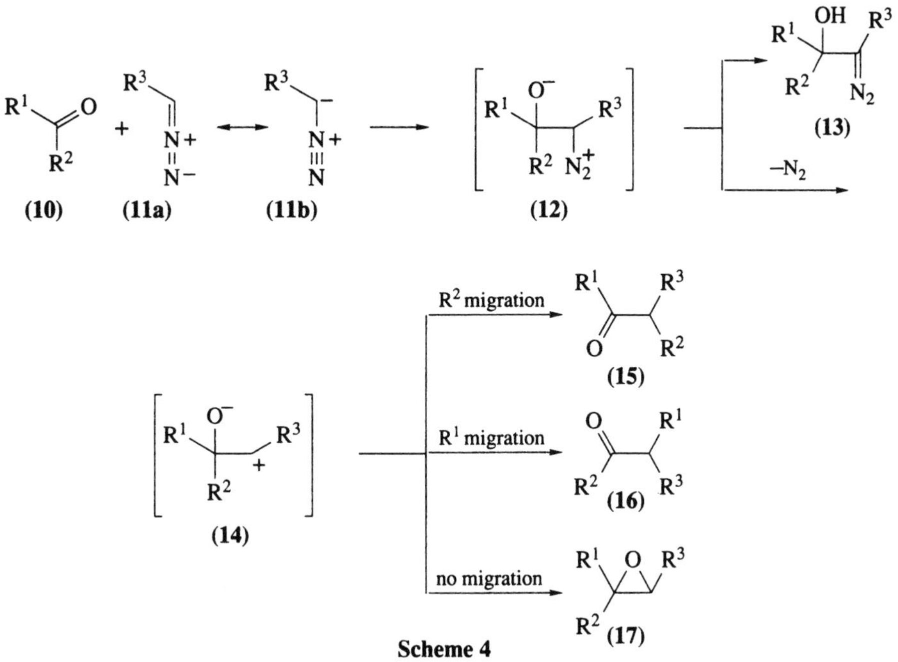

Scheme 4

While the reaction of aldehydes with diazomethane is, in general, not a practical method for homologating aldehydes, it illustrates the influence of the relative migratory aptitudes of substituents on the outcome of the reaction. With aliphatic aldehydes, the preferential migration of hydrogen *versus* an alkyl group produces fair to good yields of the methyl ketones (**18**), along with varying amounts of epoxides (Scheme 5). Aromatic aldehydes are more variable, being influenced by substituents and solvent. For example, the addition of methanol or water favors aryl group migration; however, the resulting aldehyde reacts further to produce the homologated methyl ketone (**21**).[10] A compilation of these reactions may be found in refs 8b and 9.

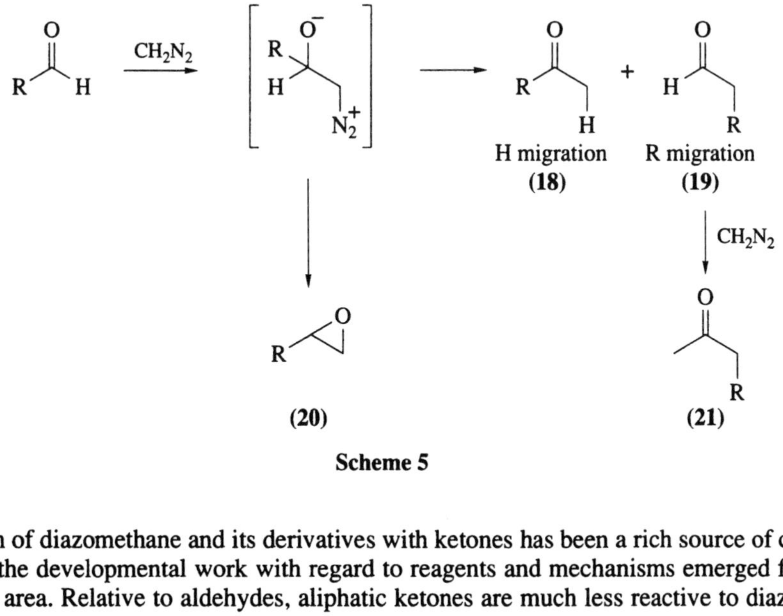

Scheme 5

The reaction of diazomethane and its derivatives with ketones has been a rich source of chemistry. The major part of the developmental work with regard to reagents and mechanisms emerged from extensive studies in this area. Relative to aldehydes, aliphatic ketones are much less reactive to diazomethane and yield, in addition to homologation products, substantial amounts of epoxides.[9] The use of Lewis acids to activate the ketone carbonyl has greatly improved the situation, by increasing the rate of reaction and by reducing the degree of epoxide formation. The general trend for the migration of substituents, aryl > methyl ≥ substituted alkyl, may be seen in Scheme 6.[11,12]

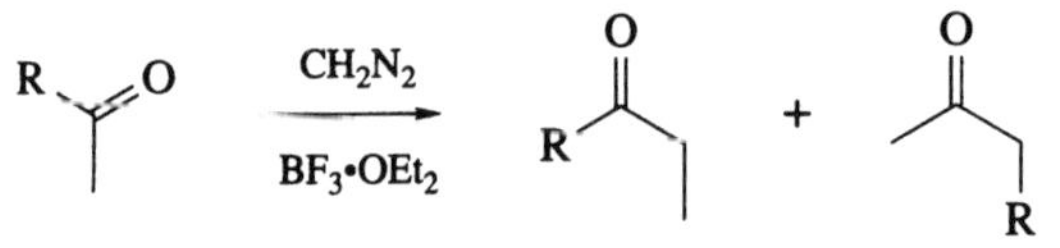

CH_2N_2, $BF_3{\cdot}OEt_2$

methyl *versus* R migration

R	
Pr^n	50:50
Pr^i	67:33
Bu^t	74:26
$PhCH_2$	78:22
Ph	22:78
$CH{=}CMe_2$	29:71

Scheme 6

Similar results were obtained on application of the related Tiffeneau–Demjanov reaction. This semipinacol-type reaction,[13,14] an extension of the Demjanov rearrangement, involves the rearrangement of a diazonium ion (**25**; Scheme 7), which is generated by the diazotization of the corresponding amino alcohol (**24**).[15] The amino alcohol is obtained from the ketone by reduction of a nitromethane adduct (**23a**),[16] cyanohydrin (**23b**) or trimethylsilyl cyanohydrin (**23c**).[17] This procedure allows for a controlled addition–rearrangement sequence in cases where the use of diazomethane is complicated by the further reaction of the product ketone.

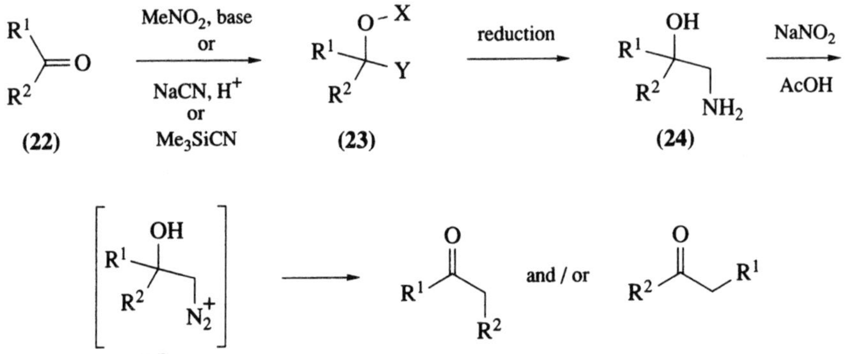

(**22**) $MeNO_2$, base or NaCN, H^+ or Me_3SiCN → (**23**) reduction → (**24**) $NaNO_2$, AcOH → (**25**) → and / or

(a) X = H, Y = CH_2NO_2
(b) X = H, Y = CN
(c) X = $SiMe_3$, Y = CN

Scheme 7

The hydroxide-catalyzed addition of α-acyldiazomethane derivatives, while quite facile with aldehydes to produce stable aldol adducts, is poor with aliphatic ketones.[18a–c] However, the lithium or magnesium anion (**26**) adds smoothly to produce the corresponding aldol adducts (**27**) in good yield (Scheme 8). On treatment with acid,[19a,b] thermolysis[18a] or metal catalysts,[20] these adducts lose nitrogen and rearrange to the β-keto esters (**28**) and (**29**). The analogous reaction with α-diazo ketones has also been carried out; however, with bulkier substituents, the retro-aldol reaction competes with rearrangement.

In a mechanistic study on the Lewis acid catalyzed addition of ethyl diazoacetate to ketones a similar profile of rearrangement to the β-keto esters was observed (Scheme 9).[21] In the same reaction with acetophenones, substitution on the benzene ring was found to only slightly affect the otherwise 90:10 preference for migration of aryl *versus* methyl.[22]

Application of the above series of homologation reactions to cyclic ketones provides a route to ring-expanded products. Due to the constraints of ring systems, the preference for the migration of one bond over the other during these reactions is influenced by an additional set of factors, among which include

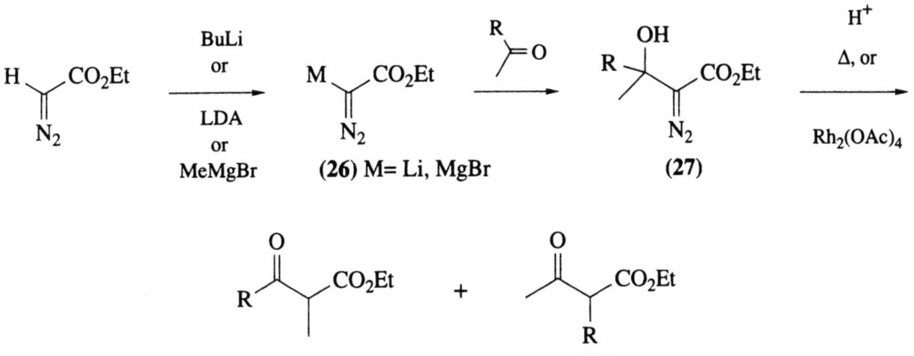

R	*Adduct yield (%)*	*Overall yield (%)*	**(28)**:**(29)**	*Ref.*
Me	93	74	—	19b
Ph	94	72	0:100	18
Et	65	87	55:45	20
Bu	80	68	55:45	20
Bu^t	76	58	100:0	20
$PhCH_2$	–	73	38:62	20
Ph	47	80	8:92	20

Scheme 8

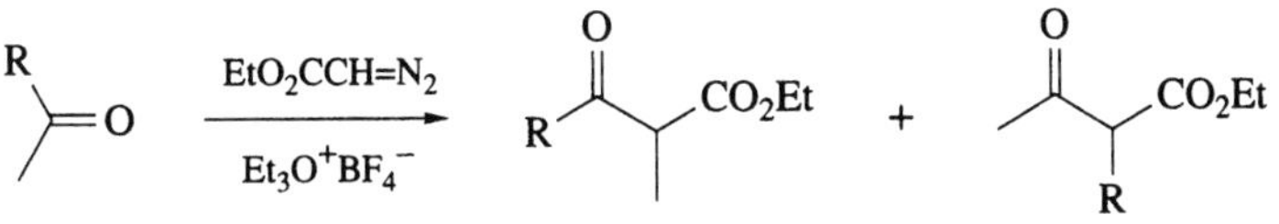

R	*Yield (%)*	*Product ratio*
Me	78	–
Et	89	50:50
Pr^i	54	66:33
Bu^t	10	95:5
$PhCH_2$	96	62:38
Ph	78	10:90
$CH=CMe_2$	10	2:98

Scheme 9

ring strain, steric effects, stereochemical relationships in the approach of the diazoalkane, and conformational effects in the intermediate betaine (Scheme 10). Earlier work in this area has been reviewed.[9,15,23]

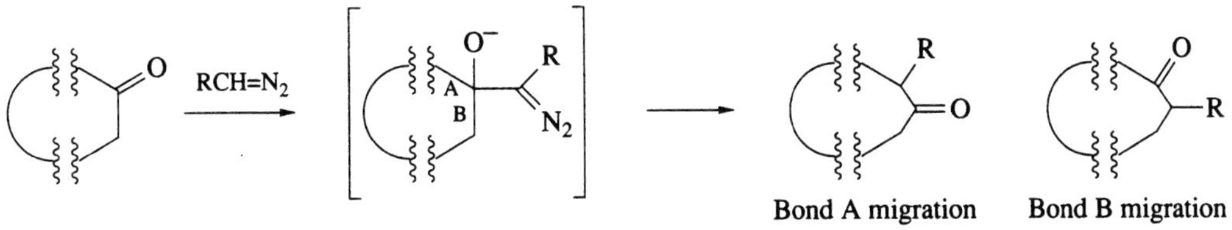

Bond A migration Bond B migration

Scheme 10

For monocyclic and many fused bicyclic ketones the general order of reactivity of $3 > 4 > 6 > 7 \geq 5$ has been observed in the reaction with diazomethane and derivatives.[1,9] Cyclopropanones react smoothly with diazomethane and diazoethane in the absence of catalysts to form cyclobutanones (equations 1 and 2).[24]

$RCHN_2$, CH_2Cl_2 (1)

R = H	43	:	57, 85%
R = Me	43	:	57, 85%

$RCHN_2$, CH_2Cl_2 (2)

R = H	33	:	67, 80%
R = Me	70	:	30, 80%

The ready availability of cyclobutanones from the [2 + 2] cycloaddition of ketenes with alkenes makes cyclobutanones attractive precursors to cyclopentanones. These react with diazomethane in ether/methanol to give the ring-expanded products (equations 3[25] and 4[26]). The presence of an α-halo substituent both accelerates the reaction of ketones with diazomethane and strongly shifts the migratory preference to the nonhalogenated carbon (equation 5a).[27] It is noteworthy that the Tiffeneau–Demjanov expansion produces the opposite regiochemistry for expansion in the nonhalogenated case (equation 5b).[15] Additional examples in mono-, bi- and tri-cyclic cyclobutanones are given in equations (6)–(10).

CO_2R, OH, CH_2N_2 — 66:33, 75% (3)

MeO_2C, H, CH_2N_2 — 50:50 (4)

CH_2N_2, $MeOH/Et_2O$ (5a)

X = Y = H	55:45
X = H, Y = Cl	90:10
X = Y = Cl	95:5

H OH NH_2, $NaNO_2$, AcOH — X = Y = H 16:84 (5b)

i, $R^2CH{=}N_2$; ii, Zn, AcOH (ref. 27) (6)

R^1 = Ph, R^2 = H	82% overall
R^1 = Ph, R^2 = Me	74%
R^1 = C_8H_{17}, R^2 = H	75%

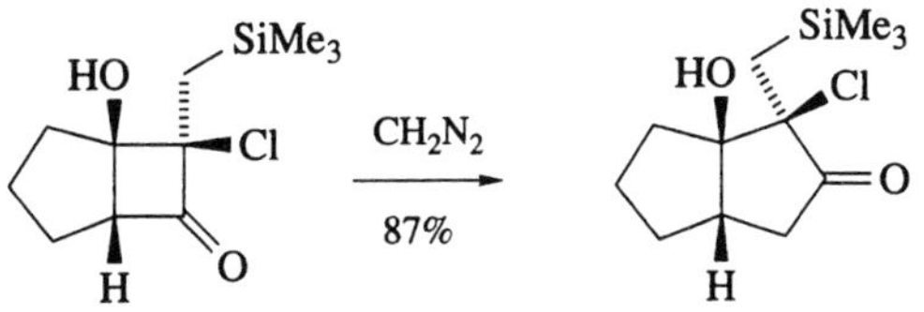

(ref. 28) (7)

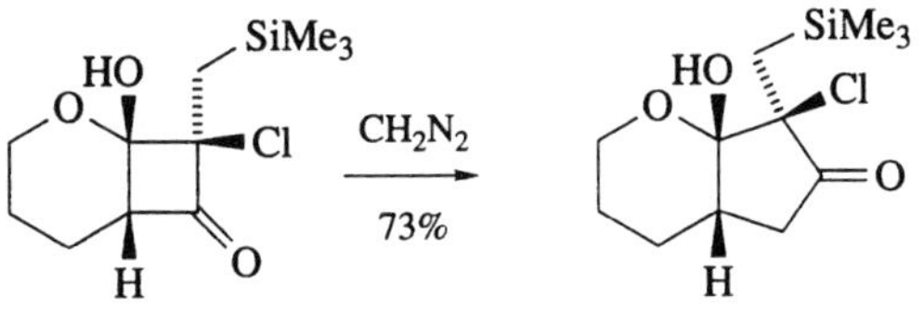

(ref. 28) (8)

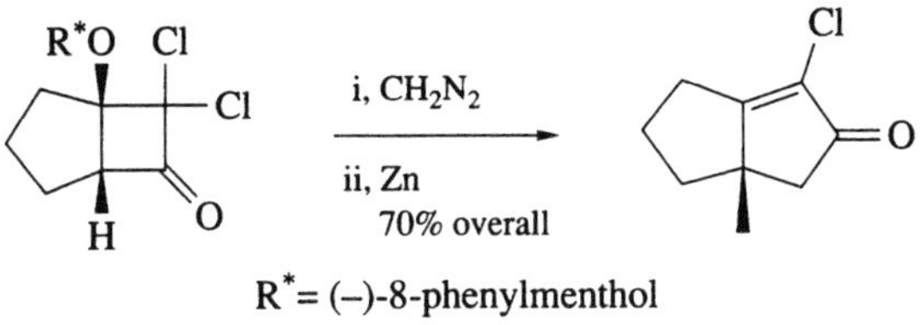

(ref. 29) (9)

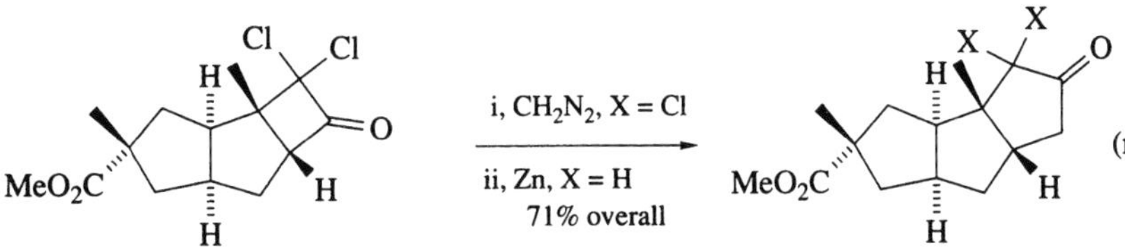

(ref. 26) (10)

The Lewis acid catalyzed expansion with ethyl diazoacetate also gives good regioselectivity, with a preference for migration of the non-ring-fusion carbon (equations 11–16). The use of $SbCl_5$ as the Lewis acid at –78 °C was reported to improve the regioselectivity in one case (equation 14).[26]

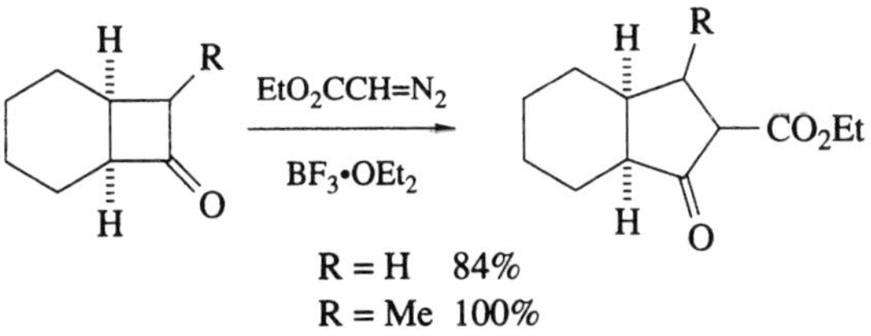

(ref. 30) (11)

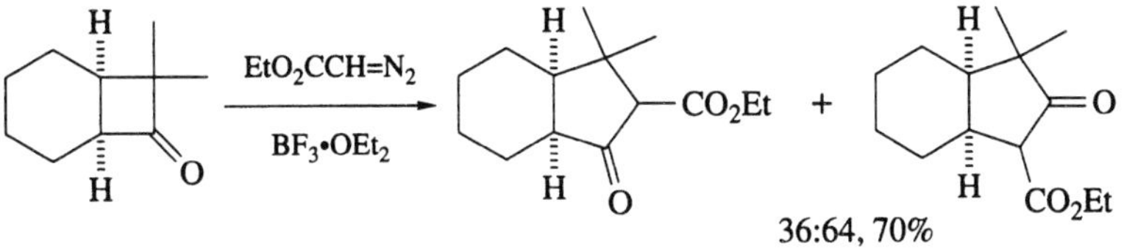

(ref. 30) (12)

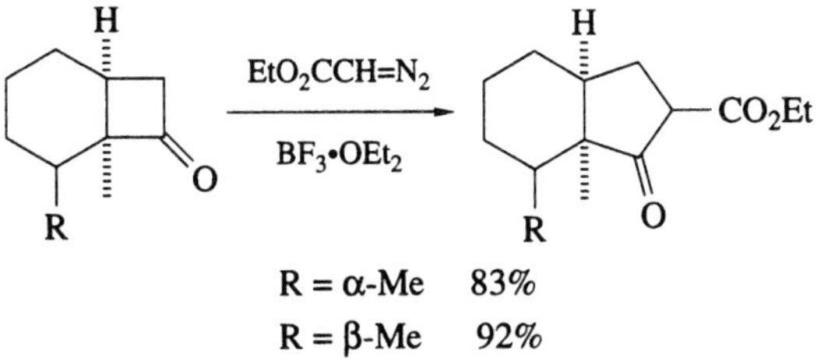

(ref. 30) (13)

The diminished reactivity of cyclopentanones toward diazomethane, while an asset in the ring expansion of cyclobutanones, has also proven to be a disadvantage. Some examples in the older literature demonstrate the problem of further reaction to higher homologs. The successful expansion reactions with

$EtO_2CCH{=}N_2$, Lewis acid (ref. 26) (14)

$BF_3{\bullet}OEt_2$ 68:32
$SbCl_5$, −78 °C 98:2

$EtO_2CCH{=}N_2$, $BF_3{\bullet}OEt_2$ (ref. 31) (15)

83:17, 73%

$EtO_2CCH{=}N_2$, $BF_3{\bullet}OEt_2$ (ref. 32) (16)

89:11, 91%

ethyl diazoacetate are a likely consequence of the relatively unreactive β-keto ester products (equations 17 and 18). The sequential addition of the lithium anion of ethyl diazoacetate (**26**) to form the aldol product followed by rearrangement has also been found to be a viable tactic for the ring expansion of cyclopentanones (Scheme 11).[19b]

$EtO_2CCH{=}N_2$, $BF_3{\bullet}OEt_2$ (ref. 33) (17)

R = H 70:30, 95%
R = Me 100:0, 69%

$EtO_2CCH{=}N_2$, $BF_3{\bullet}OEt_2$ (ref. 34) (18)

79:21
(66% after decarboethoxylation)

There is an abundance of literature dealing with the diazomethane, diazomethane derivatives and Tiffeneau–Demjanov expansions of six-membered and larger ring ketones, with most of the simpler examples reported in the older literature.[9,15] For example, the diazomethane expansion of cyclohexanone

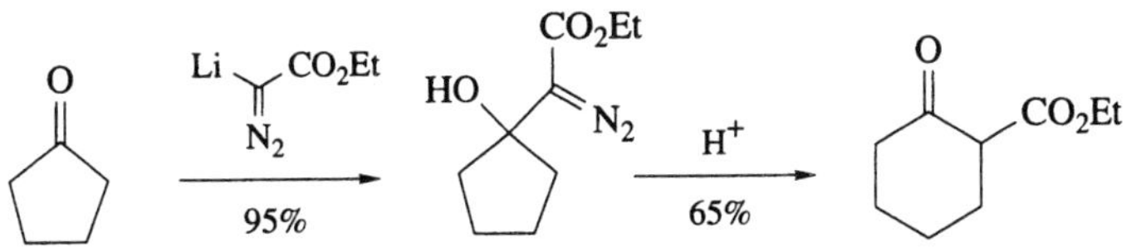

Li–C(N_2)–CO_2Et, 95%; HO, $C(N_2)CO_2Et$; H^+, 65%; CO_2Et

Scheme 11

produces cycloheptanone in 33–36% yield on a preparative scale,[35] and the corresponding Tiffeneau–Demjanov expansion provides a 41% overall yield.[16] The preparative scale expansion of cyclohexanone with phenyldiazomethane to 2-phenylcycloheptanone has also been recorded and provides a route to 2-arylcycloheptanones.[36] A series of 4-substituted cyclohexanones have been expanded with diazoethane to yield the 2-methylcycloheptanones in good yield as *ca.* 1:1 *cis/trans* mixtures (equation 19).[37] As in the case with diazomethane, the addition of alcohol to the reaction mixture with diazoethane had a beneficial effect.

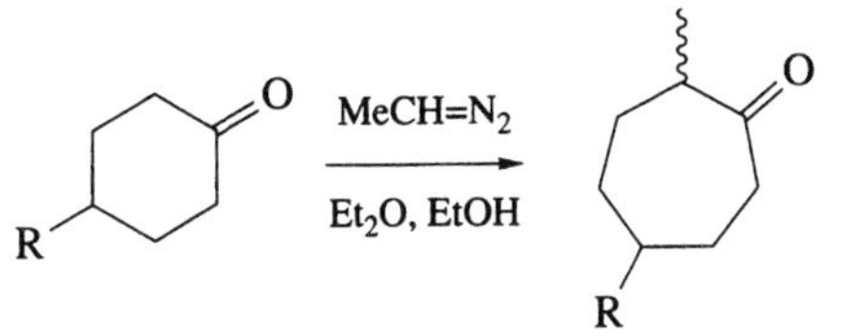

$$\text{4-R-cyclohexanone} \xrightarrow[\text{Et}_2\text{O, EtOH}]{\text{MeCH=N}_2} \text{2-methyl-5-R-cycloheptanone} \quad (19)$$

R	*Yield (%)*	*R*	*Yield (%)*
H	84%	Me_2COH	81%
Me	81%	CO_2Et	89%
Pr^i	75%		

The use of Lewis acids for the diazomethane and diazoethane homologations of a series of C_6–C_{14} ketones has been reported to reduce the amount of epoxide formation, although as the ring size increases the amount of unreacted starting material increases (Scheme 12).[12] The expansion of 2-allylcyclohexanone with diazomethane in the presence of $AlCl_3$ proceeds with migration of the less-substituted bond,[12] in contrast to the Tiffeneau–Demjanov expansion of 2-methyl- and 2-isopropyl-cyclohexanone wherein the alkyl-substituted bond migrates preferentially (Scheme 13).[38,39]

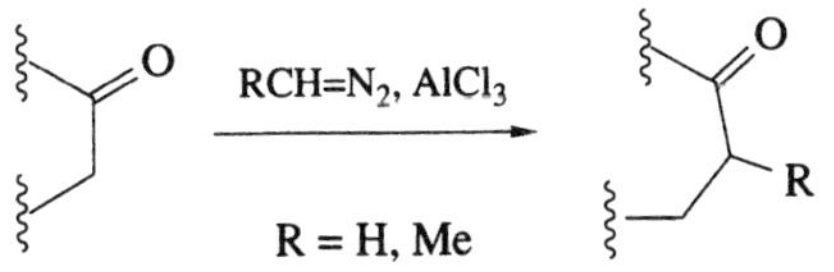

RCH=N_2, $AlCl_3$; R = H, Me

	R = H	*Me*
$C_6 \rightarrow C_7$	58%	-
$C_7 \rightarrow C_8$	50%	82%
$C_8 \rightarrow C_9$	46%	66%
$C_9 \rightarrow C_{10}$	45%	72%
$C_{10} \rightarrow C_{11}$	25%	57%
$C_{12} \rightarrow C_{13}$	26%	66%

Scheme 12

The trimethylsilyldiazomethane–$BF_3 \cdot Et_2O$ homologation of 2-methylcyclohexanone also proceeds with high regioselectivity to give 2-methylcycloheptanone in 69% yield. A particular advantage for this reagent over diazomethane appears to be for the homologation of larger rings where the yields are good (equations 20[40] and 21[41]). The Tiffeneau–Demjanov expansion also displays good regioselectivity with larger ketones containing a fused benzene ring (equations 22[42] and 23[43]), although a loss of selectivity was noted in a 12-membered ring ketone (equation 24).[44]

The one-carbon expansion of cyclohexanones with introduction of a carbonyl group, trifluromethyl, cyano, phosphonate or benzenesulfonate has also been reported to proceed in the presence of Lewis acids (Scheme 14). The analogous reactions of larger ring ketones [expressed as ring size (yield)] with ethyl diazoacetate[21] [7 (81%), 8 (85%)] and phenylsulfonyldiazomethane[47] [7 (86%), 8 (27%), 12 (43%), 14 (48%), 15 (54%) and 16 (58%)] have also been reported. In the reaction of ethyl diazoacetate and

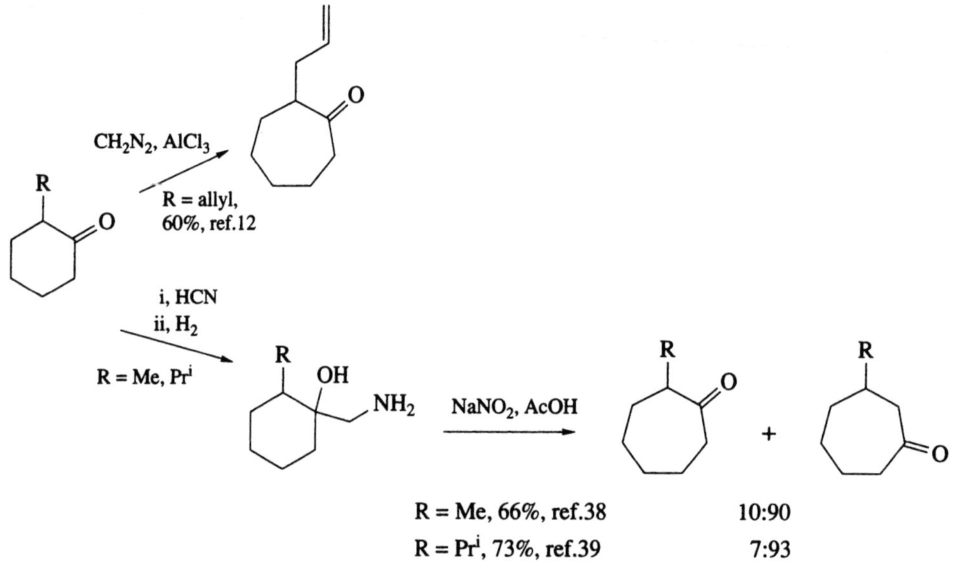

CH_2N_2, $AlCl_3$

R = allyl, 60%, ref.12

i, HCN
ii, H_2

R = Me, Pr^i

OH
NH_2

$NaNO_2$, AcOH

R = Me, 66%, ref.38 10:90

R = Pr^i, 73%, ref.39 7:93

Scheme 13

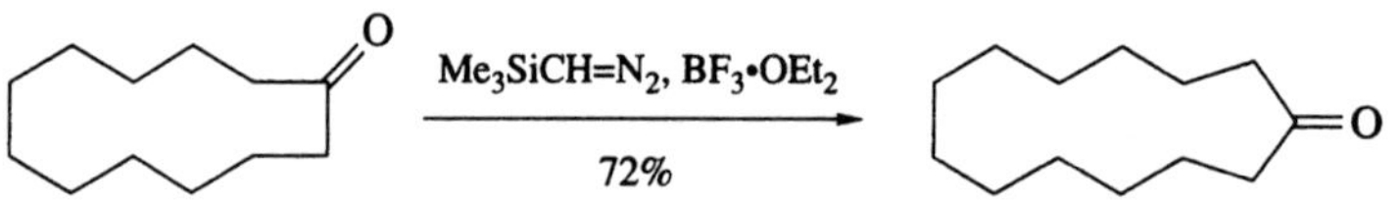

$Me_3SiCH{=}N_2$, $BF_3{\bullet}OEt_2$

72%

(ref. 40) (20)

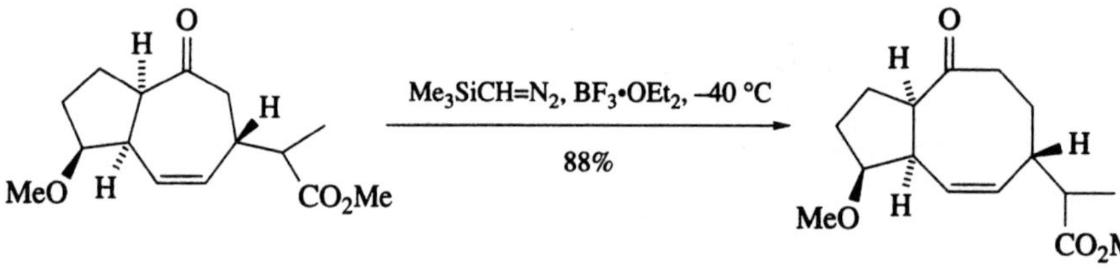

$Me_3SiCH{=}N_2$, $BF_3{\bullet}OEt_2$, –40 °C

88%

(ref. 41) (21)

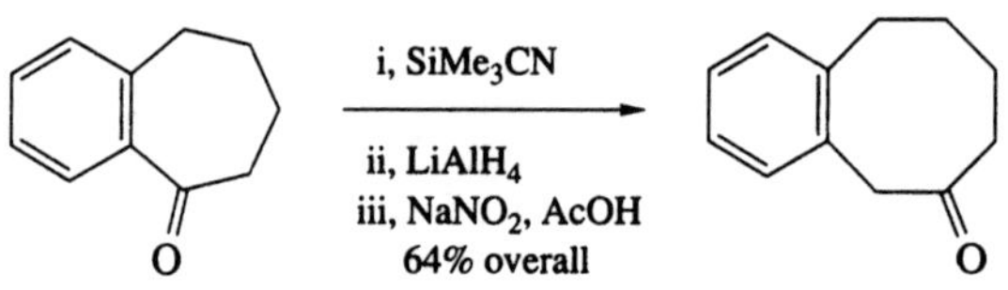

i, $SiMe_3CN$
ii, $LiAlH_4$
iii, $NaNO_2$, AcOH
64% overall

(22)

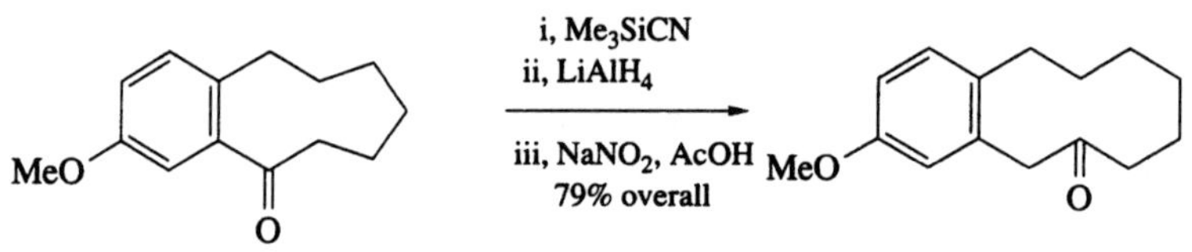

i, Me_3SiCN
ii, $LiAlH_4$
iii, $NaNO_2$, AcOH
79% overall

(23)

i, Me_3SiCN
ii, $LiAlH_4$
iii, $NaNO_2$, AcOH

52:48

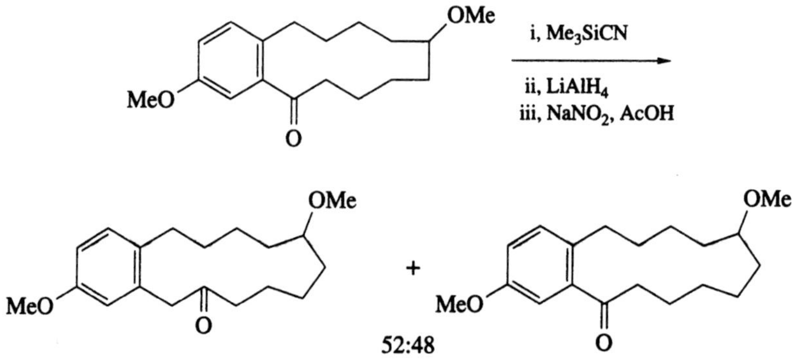

(24)

2-substituted cyclohexanones, a preference for the migration of the unsubstituted carbon was observed (Scheme 15). The regioselectivity of this rearrangement promoted by $BF_3 \cdot OEt_2$ was improved by the use of $SbCl_5$ at –78 °C.

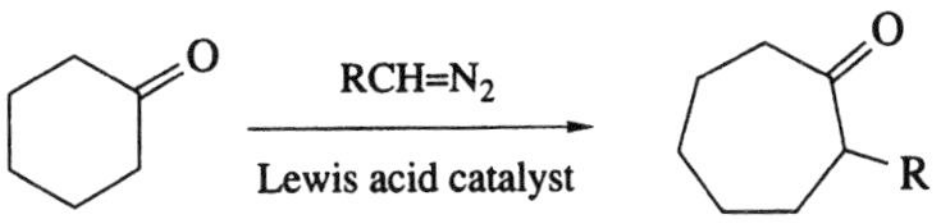

$RCH=N_2$, Lewis acid catalyst

R	*Catalyst*	*Yield (%)*	*Ref.*
CO_2Et	$Et_3O^+BF_4^-$	90	21
CO_2Bn	$BF_3 \cdot Et_2O$	66	45
CO_2Bu^t	$Et_3O^+BF_4^-$	46	21
CN	$Et_3O^+BF_4^-$	58	21
CF_3	$Et_3O^+BF_4^-$	85	21
$PO(OMe)_2$	$Et_3O^+BF_4^-$	65	21
COC_6H_4OMe	$BF_3 \cdot Et_2O$	92	46
SO_2Ph	$TiCl_4$	71	47

Scheme 14

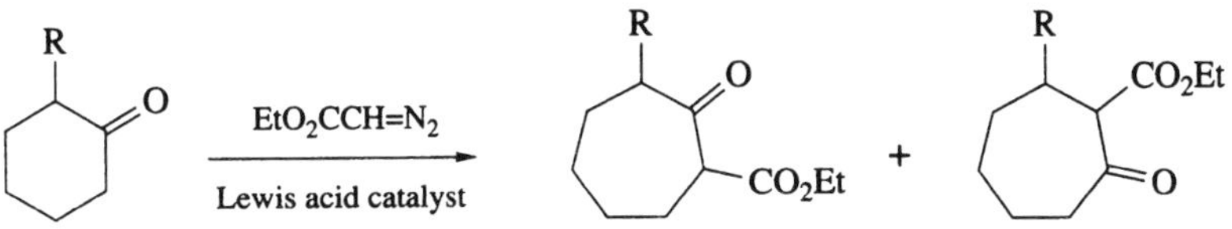

$EtO_2CCH=N_2$, Lewis acid catalyst

R	*Catalyst*	*Ratio*	*Yield (%)*	*Ref.*
Cl	$Et_3O^+BF_4^-$	100:0	90	21
Me	$Et_2O \cdot BF_3$	83:17	66	33
Me	$Et_3O^+BF_4^-$	85:15	46	21
Me	$SbCl_5$	94:6	58	21
Et	$Et_3O^+BF_4^-$, $Et_2O \cdot BF_3$	80:20	85	21, 33
Pr^i	$Et_3O^+BF_4^-$	90:10	65	21
Bu^t	$Et_3O^+BF_4^-$	73:27	92	21
Ph	$Et_3O^+BF_4^-$	>95:5	71	21

Scheme 15

The powerful directing effect of halogen substitution has been studied in the reaction of a series of cholestane derivativers with ethyl diazoacetate. In both the 5α and 5β series, essentially no selectivity was observed for the nonhalogenated series, while with halogen substitution only the products arising from migration of the unhalogenated carbon were observed (Scheme 16).[48] A similar effect was observed for an acetoxy substituent.

The sequential lithioethyl diazoacetate addition–Rh catalyzed rearrangement proved to be advantageous in a pseudoguanolide synthesis, where the corresponding one-step ethyl diazoacetate–$BF_3 \cdot OEt_2$ method produced an 80:20 mixture of regioisomers (Scheme 17).[49] An investigation for a guaiane synthesis utilized a similar sequence, again with only one regioisomer being observed (Scheme 18).[50]

The reaction of diazomethane and its derivatives with bridged polycyclic ketones continues to provide a convenient source of the next higher homologs, and also serves as a reactivity probe with which the effects of ring strain and steric and electronic interactions may be unveiled. In cases where further ring expansion occurs due to a lower reactivity of the starting ketone relative to the product, the Tiffeneau–Demjanov ring expansion is especially useful. Moreover, this sequence allows for the study of bond migration preferences in relation to the stereochemistry of leaving groups when the intermediate amino alcohols can be isolated as separate stereoisomers. The reader is referred to an excellent review on the ring expansion of bridged bicyclic compounds for a detailed analysis of the factors which influence rearrangements in general for this class of compounds.[2c] Illustrative examples are provided in Table 1.

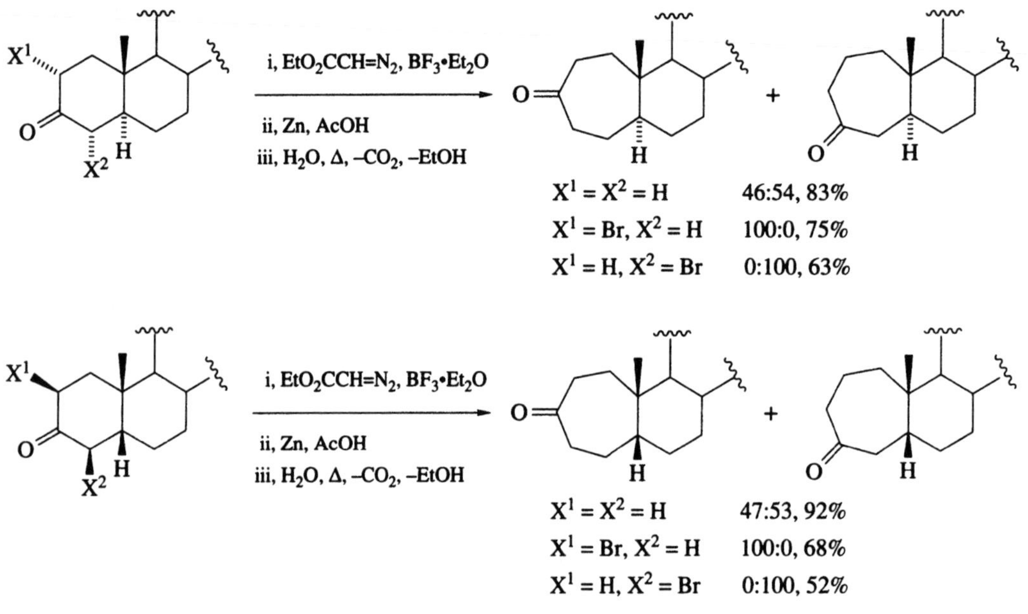

X^1 O X^2 H

i, $EtO_2CCH{=}N_2$, $BF_3{\bullet}Et_2O$
ii, Zn, AcOH
iii, H_2O, Δ, $-CO_2$, $-EtOH$

O H + O H

$X^1 = X^2 = H$ 46:54, 83%
$X^1 = Br, X^2 = H$ 100:0, 75%
$X^1 = H, X^2 = Br$ 0:100, 63%

X^1 O X^2 H

i, $EtO_2CCH{=}N_2$, $BF_3{\bullet}Et_2O$
ii, Zn, AcOH
iii, H_2O, Δ, $-CO_2$, $-EtOH$

O H + O H

$X^1 = X^2 = H$ 47:53, 92%
$X^1 = Br, X^2 = H$ 100:0, 68%
$X^1 = H, X^2 = Br$ 0:100, 52%

Scheme 16

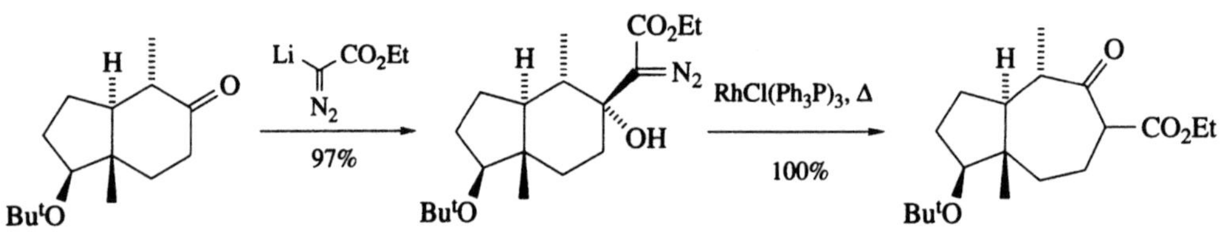

H O Bu^tO

Li CO_2Et N_2
97%

H CO_2Et N_2 OH Bu^tO

$RhCl(Ph_3P)_3$, Δ
100%

H O CO_2Et Bu^tO

Scheme 17

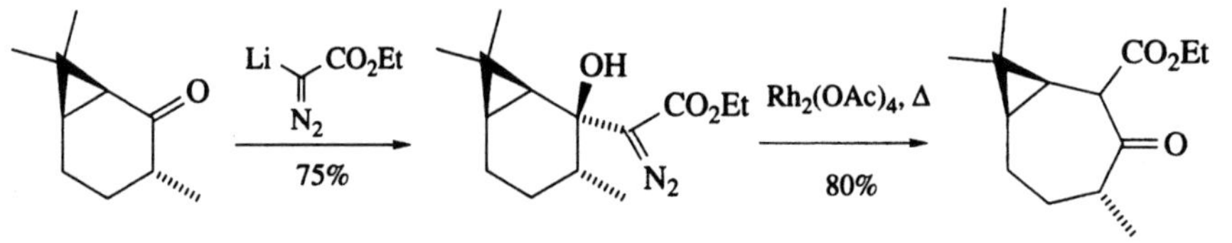

O

Li CO_2Et N_2
75%

OH CO_2Et N_2

$Rh_2(OAc)_4$, Δ
80%

CO_2Et O

Scheme 18

O N_2 $(\ )_n$ → $(\ )_n$ O

$n = 1$, 65%; $n = 2$, 81%

N_2 O R → O H R + O R

R = H 66:34
R = CO_2Et 70:30, 60%

N_2 O R → O R + O R

R = H 0:100
R = CO_2Et 91:9, 70%

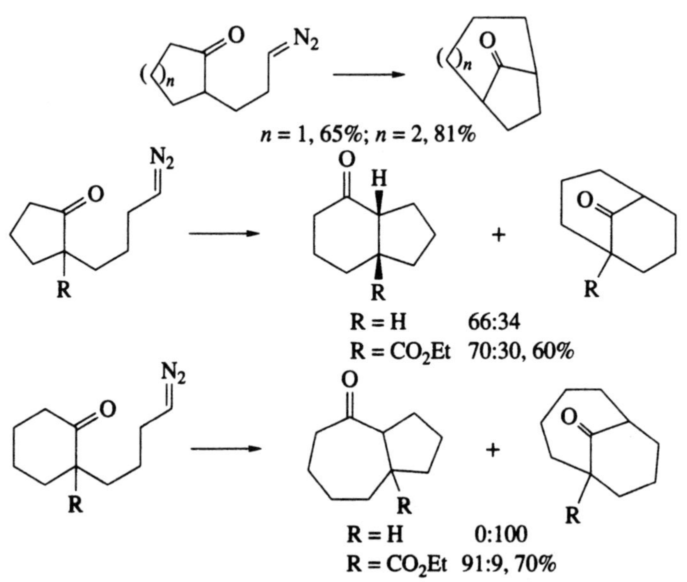

Scheme 19

Table 1 Ring Expnasion of Bridged Bicyclic Ketones with Diazoalkanes or Tiffeneau–Demjanov Rearrangement

Ketone	*Conditions*	*Intermediates if isolated*	*Conditions*	*Products*		*Yield (%)*	*Ref.*
	a		b	+			51
				R = H	81:19	85	
				R = Me	93:7	80	
				R = F	100:0	65	
				R = Cl	100:0	66	
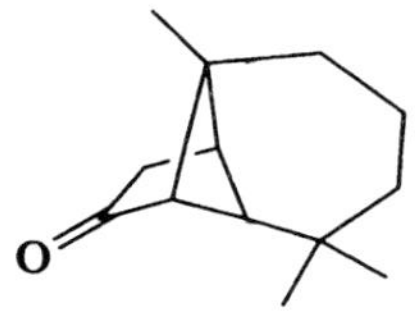	c	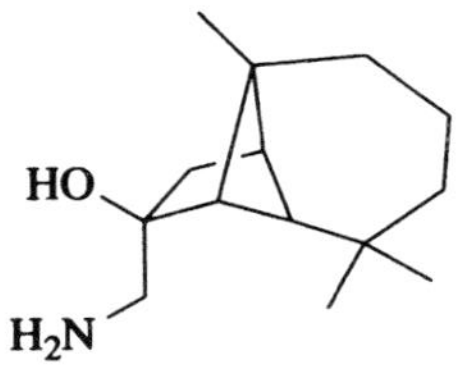	b	+	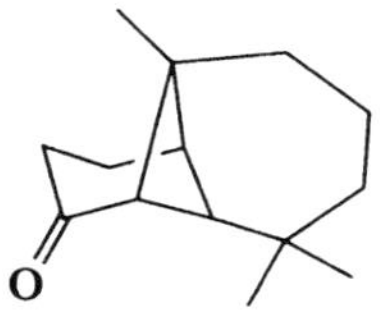 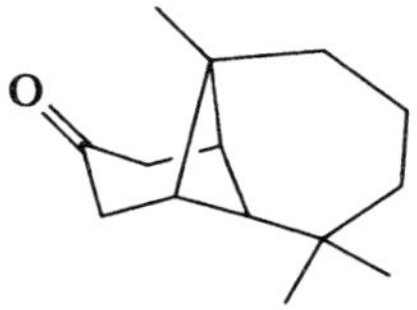 94:6		52
	d	—	—	+			53
				single bond	60:40		
				double bond	42:58		
	e		b	+			53
				single bond	62:38		
				double bond	50:50		

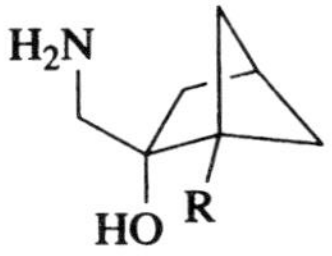

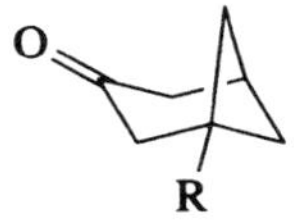

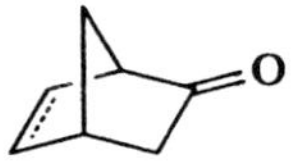

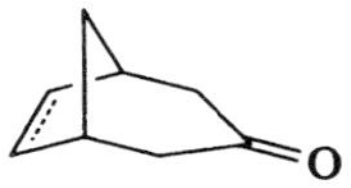

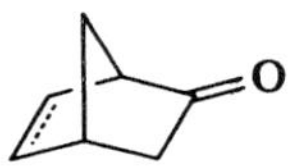

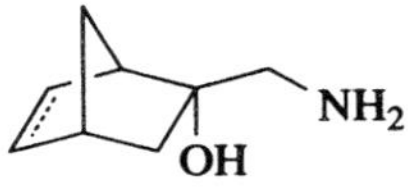

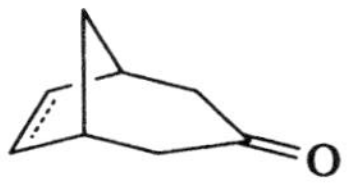

Table 1 *(continued)*

Ketone	*Conditions*	*Intermediates if isolated*	*Conditions*	*Products*	*Yield (%)*	*Ref.*
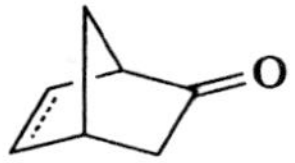	e	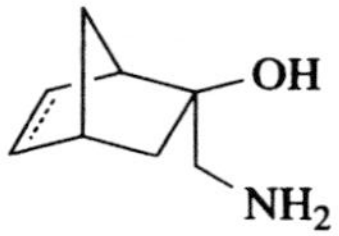 OH, NH_2	b	 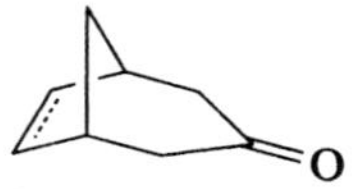 + ; single bond 91:9; double bond 77:23		53
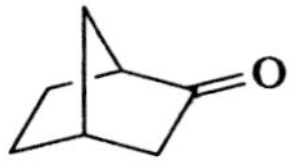	f	—	—	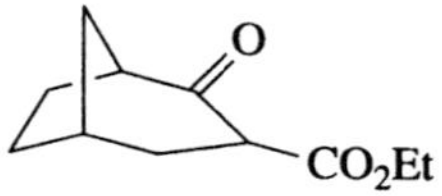 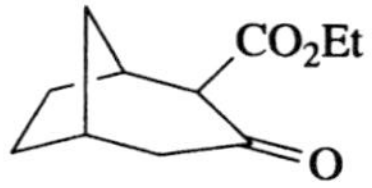 CO_2Et + CO_2Et; 51:49	52	22
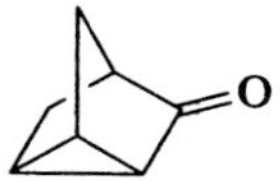	g	—	—	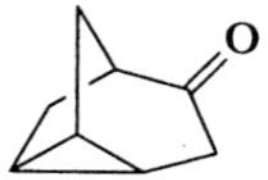 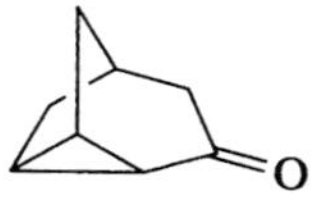 + ; 90:10	42	54
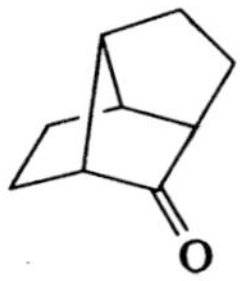	a	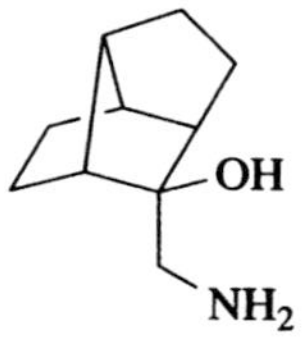 OH, NH_2; 95%	b	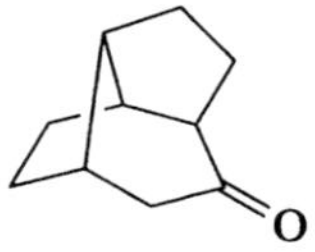	99	55
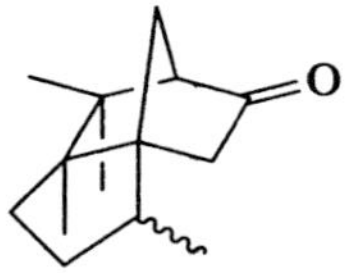	a	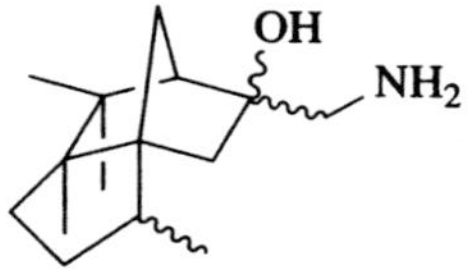 OH, NH_2	b	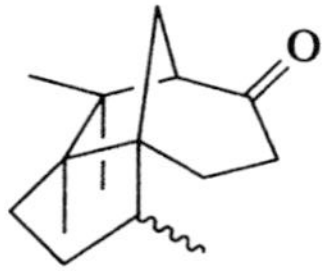  + ; 78:22	93	56

Table 1 *(continued)*

Ketone	*Conditions*	*Intermediates if isolated*	*Conditions*	*Products*	*Yield (%)*	*Ref.*
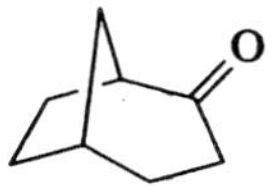 O	e	—	—	 O + O 68:32		57
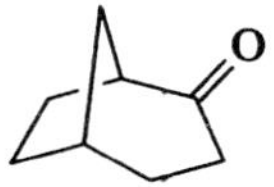 O	e	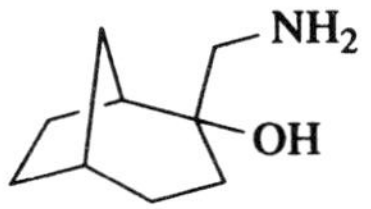 NH_2 OH and 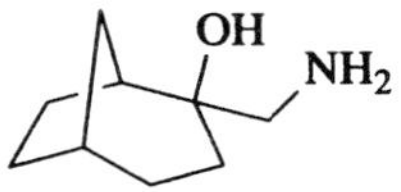	b	 O + O 88:12		57
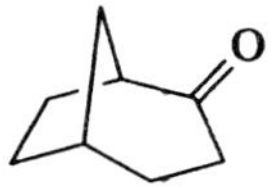 O	e	OH NH_2	b	 O + O 50:50		57
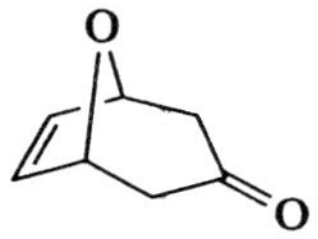 O O	a	—	b	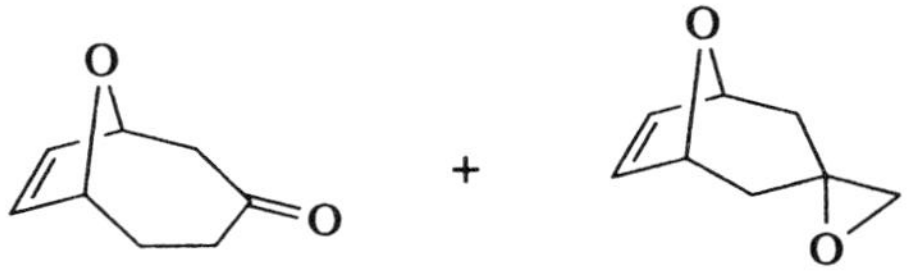 O O + O O 75:25	71	58
 Ar R O	i	—	—	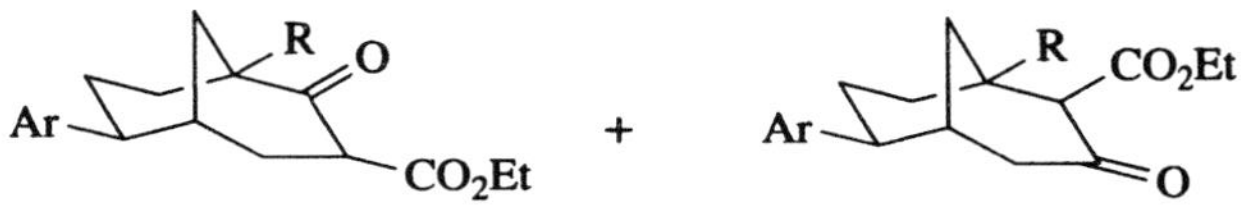 Ar R O CO_2Et + Ar R CO_2Et O ~75:25 (R = H, Me)		59

Table 1 (*continued*)

Ketone	*Conditions*	*Intermediates if isolated*	*Conditions*	*Products*	*Yield (%)*	*Ref.*
O	j	—	—	O	70	60
O, Br, R	g	—	—	O, Br, R R = CO_2Me R = Br R = Me	 50 45 45	61
O	g	—	—	O	75	62
 O	k	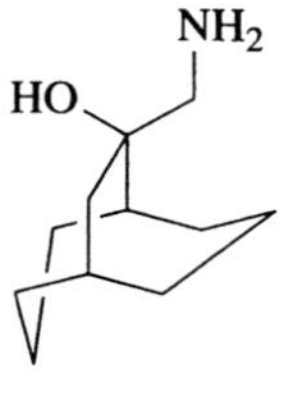 HO, NH_2	b	 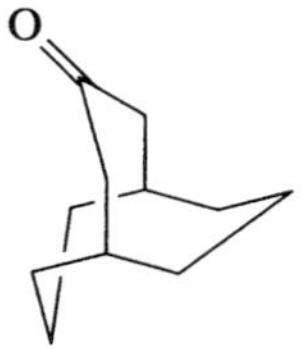 O + O 66:34		62

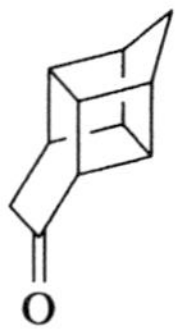

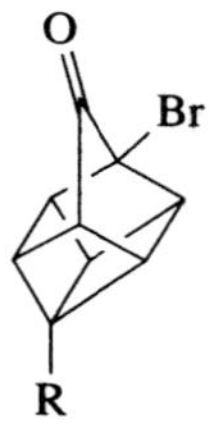

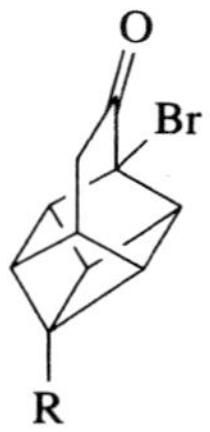

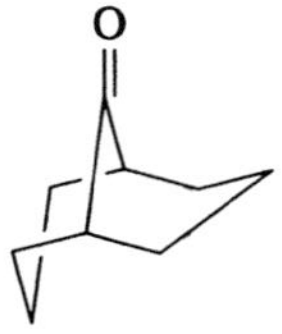

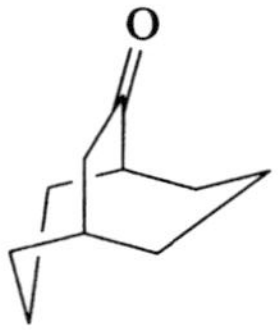

Table 1 *(continued)*

Ketone	*Conditions*	*Intermediates if isolated*	*Conditions*	*Products*	*Yield (%)*	*Ref.*
O, O	g	—	—	O, O	80	63
O	l	—	—	O, R; R = H	85	64
				R = CO_2Et	63	21
O	a	HO NH_2	b	O	70	65
O	g	—	—	O	90	66
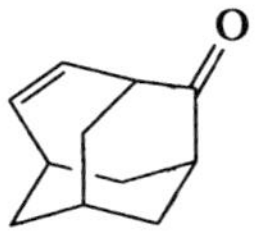 O	a	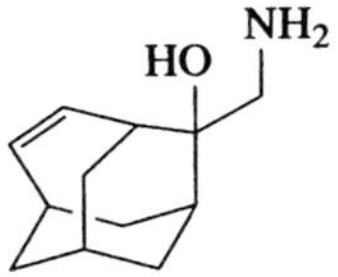 HO NH_2	b	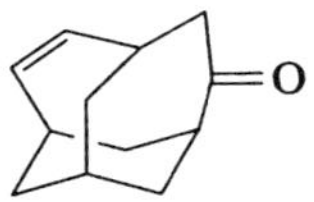 O + O (90:10) 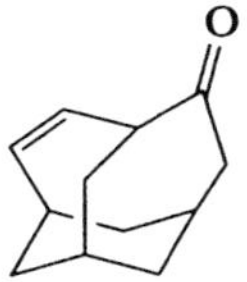	75	67

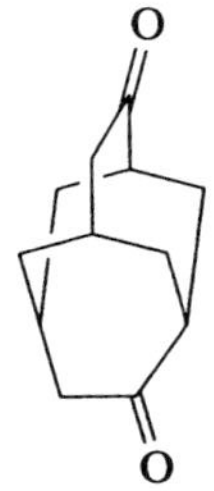

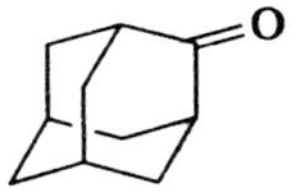

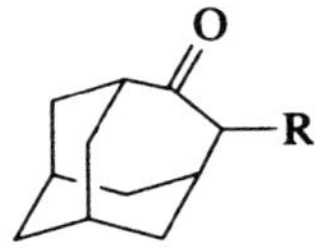

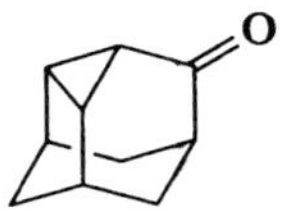

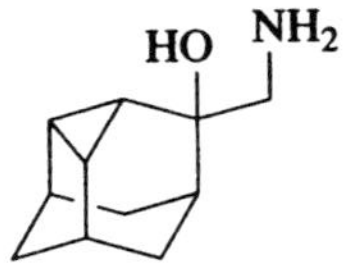

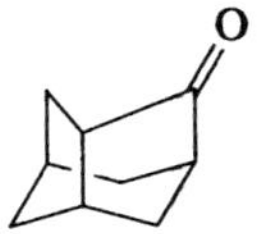

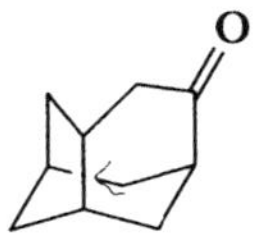

Table 1 *(continued)*

Ketone	*Conditions*	*Intermediates if isolated*	*Conditions*	*Products*	*Yield (%)*	*Ref.*
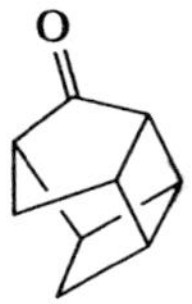	m	—	—	 + next higher homolog 72:28	64	68
	g	—	—	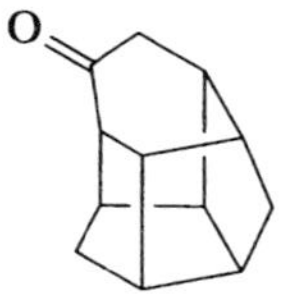	56	69
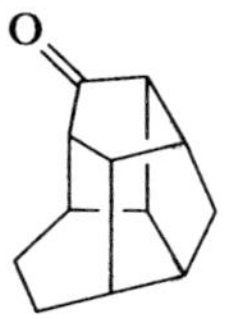	m	—	—	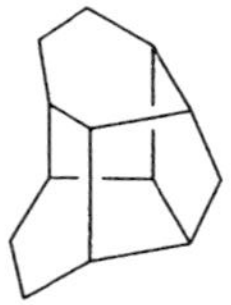	71 overall	70

[a] i, Me_3SiCN; ii, $LiAlH_4$. [b] $NaNO_2$. [c] i, $Me_2C{=}CH_2$; ii, NaN_3; iii, H_2, PtO_2. [d] CH_2N_2, 17–20% completion. [e] i, HCN; ii, H_2, PtO_2. [f] Ethyl diazoacetate/$BF_3{\bullet}OEt_2$. [g] Excess CH_2N_2. [h] 0.06 equiv. CH_2N_2. [i] Ethyl diazoacetate, $Et_3O^+ BF_4^-$. [j] CH_2N_2, Et_2O, MeOH. [k] i, $Ph_3P{=}CH_2$, ii, MCPBA; iii, NaN_3; iv, H_2, PtO_2. [l] $RCH{=}N_2$ (R = H or CO_2Et). [m] i, CH_2N_2; ii, Wolff–Kishner reduction.

The intramolecular reaction of diazo groups with ketones has also been examined in several systems. Early work studied the variation in ring size, substitution and chain length on the reaction outcome (Scheme 19).[71] Despite the interesting synthetic potential for an intramolecular ring expansion reaction, relatively few examples have been reported (equations 25,[21] 26,[72] 27[73] and 28[73]).

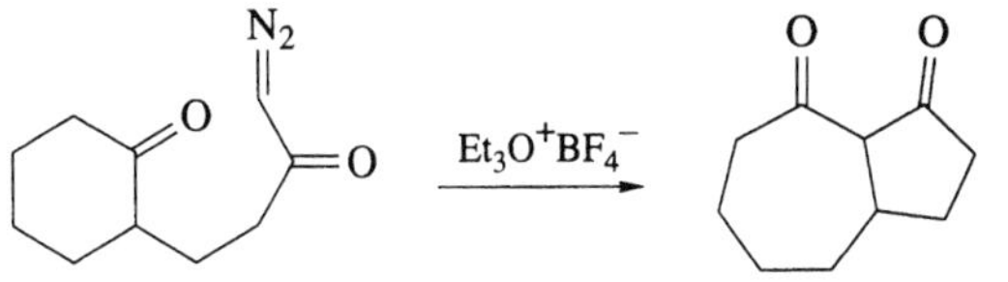

N_2 O O $Et_3O^+BF_4^-$ (25)

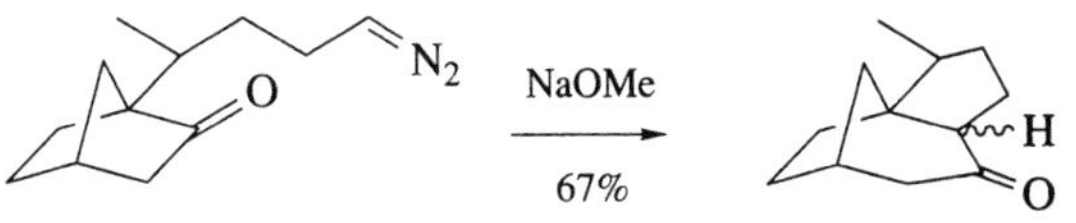

N_2 O NaOMe 67% H O (26)

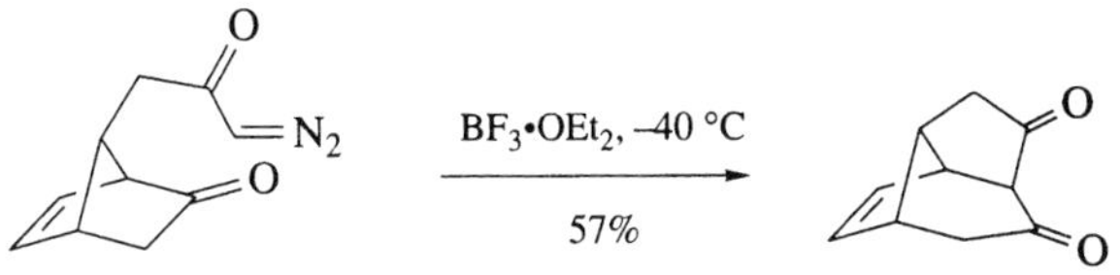

O N_2 O BF_3•OEt_2, –40 °C 57% O O (27)

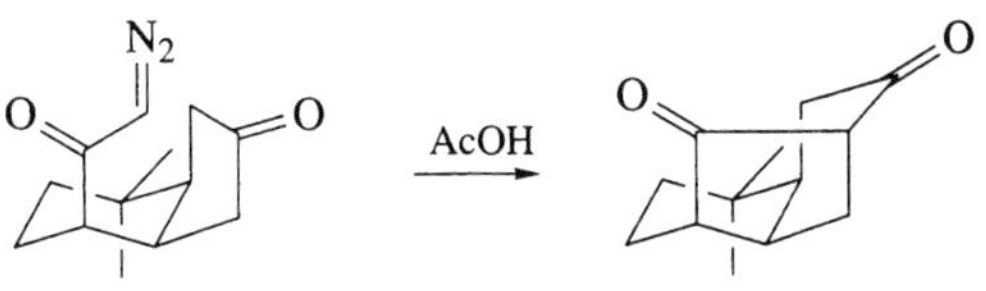

N_2 O O AcOH O O (28)

3.3.3 PINACOL-TYPE REACTIONS OF β-HYDROXY SULFIDES AND SELENIDES

In a related addition–elimination with rearrangement sequence, the chain extension–ring expansion reactions of ketones with sulfur- and selenium-stabilized anions has, relative to diazoalkanes, only been examined fairly recently. The general reaction involves the addition of a sulfur- or selenium-stabilized carbanion (**30**) to the ketone carbonyl to form adduct (**31**; Scheme 20). The expulsion of the sulfur or selenium may be induced by alkylation, complexation with metal salts, reaction with carbenes or oxidation, which leads to the rearranged products and/or epoxides. While the mechanistic details of the rearrangement aspect may differ substantially, depending on the reaction conditions, this set of expansion reactions is organized according to a common first step, the addition of a sulfur- or selenium-stabilized anion to a carbonyl group and will be presented in the order of the number of heteroatoms on the nucleophile (**30**).

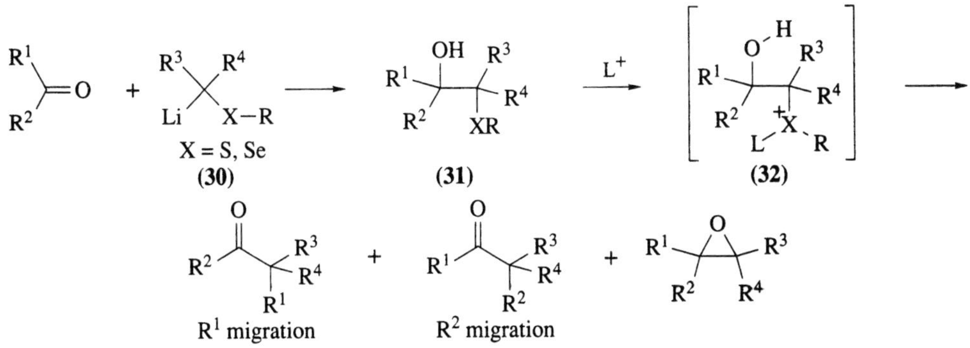

R^1 R^2 =O + R^3 R^4 Li X–R X = S, Se (**30**) → OH R^3 R^1 R^4 R^2 XR (**31**) L^+ → [O H R^3 R^1 R^4 R^2 X^+ L R] (**32**) →

O R^2 R^3 R^4 R^1 R^1 migration + O R^1 R^3 R^4 R^2 R^2 migration + O R^1 R^3 R^2 R^4

Scheme 20

When one heteroatom is present on the nucleophilic unit, the overall result is similar to the carbon insertion with diazomethane. While the addition of $PhSeCH_2Li$ or $MeSeCH_2Li$[74] to ketones is an efficient process, the elimination step using methyl iodide and base (*e.g.* KOH), dichlorocarbene generated from thallium(I) ethoxide and chloroform, or oxidation with peroxy acids leads, in general, to high yields of epoxides from aliphatic ketones. For cyclobutanones the resulting epoxides have been rearranged to the cyclopentanones with preferential migration of the more-substituted carbon (Schemes 21[75a,b] and 22[76]).

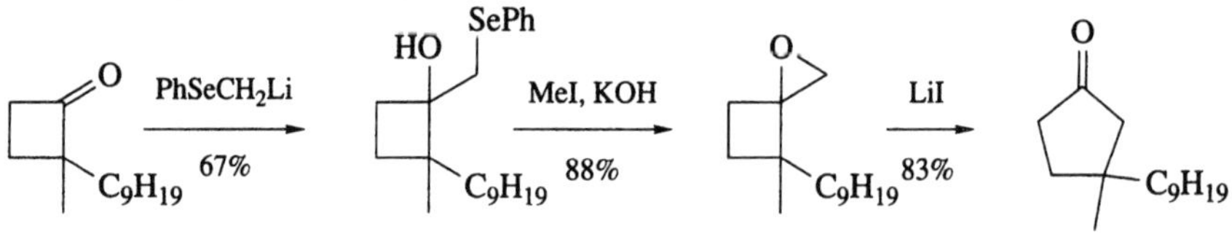

PhSeCH$_2$Li, 67%; MeI, KOH, 88%; LiI, 83%; C_9H_{19}; HO; SePh; O

Scheme 21

PhSeCH$_2$Li, 75%; TlOEt, CHCl$_3$, 88%; LiI, 12-crown-4; Tol; HO; SePh; O

94:6, 95%

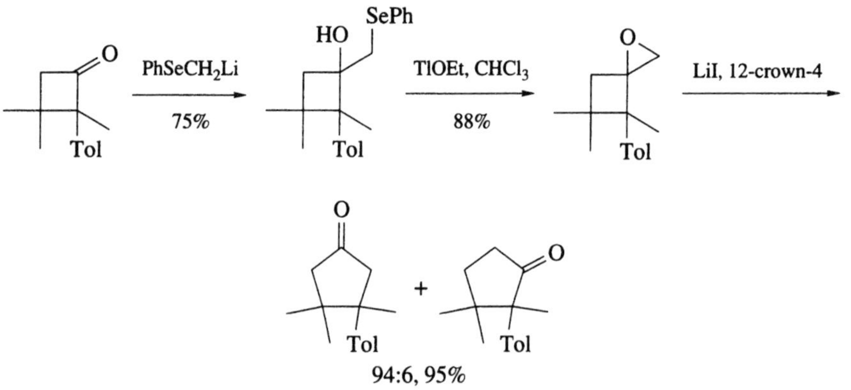

Scheme 22

The β-hydroxy phenyl selenide products from the reaction of aryl-substituted ketones and $PhSeCH_2Li$ eliminate the phenylselenyl group on treatment with peroxy acid, with migration of the aromatic group to give the one-carbon extended ketones or, in the case of cyclic ketones, the ring-expanded products (Schemes 23 and 24).[77] Cyclobutanones have been ring-expanded in a two-step procedure by reaction of the lithium anion of methyl 2-chlorophenyl sulfoxide followed by heating the β-hydroxy sulfoxide with potassium hydride in tetrahydrofuran. Rearrangement with elimination of sulfur proceeds with the exclusive migration of the more-substituted carbon atom (Scheme 25). This method has also been applied to cyclobutanones with substituted sulfoxides to provide 2-substituted cyclopentanones (Table 2).[78] The precise mechanistic details have yet to be elucidated, as well as the extension of the reaction to other ketones.

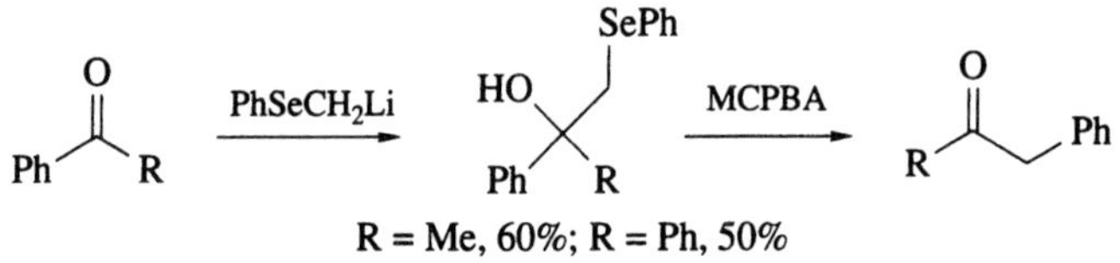

PhSeCH$_2$Li; MCPBA; Ph; R; HO; SePh; O

R = Me, 60%; R = Ph, 50%

Scheme 23

PhSeCH$_2$Li; MCPBA; HO; SePh; O; $)_n$

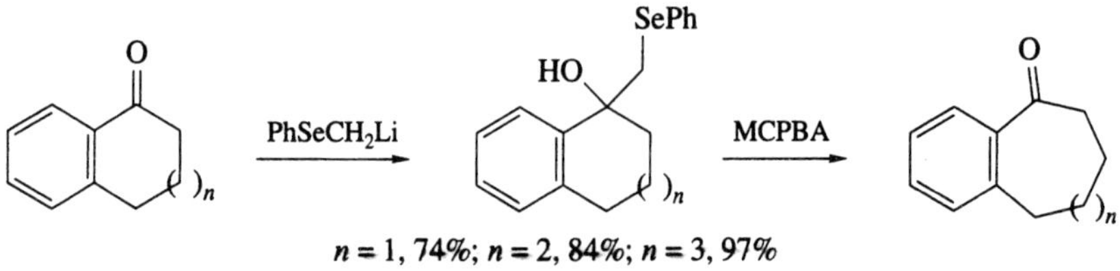

n = 1, 74%; n = 2, 84%; n = 3, 97%

Scheme 24

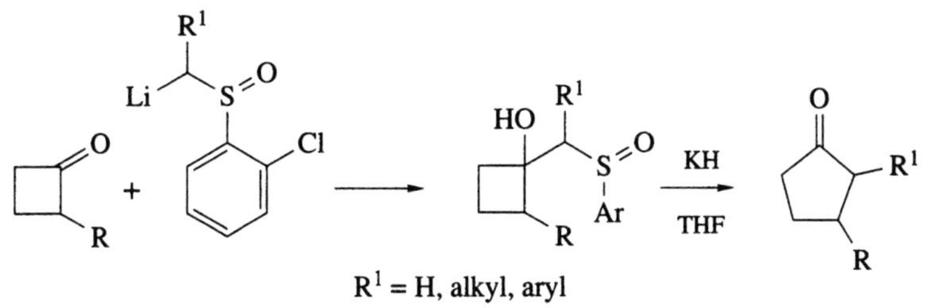

R^1 = H, alkyl, aryl

Scheme 25

Table 2 Ring Expansion of Cyclobutanones with Lithium Alkyl Aryl Sulfoxides

Sulfoxide RCH_2SOAr [a]	*Cyclobutanone*	*Product*	*Yield (%)*
R = H	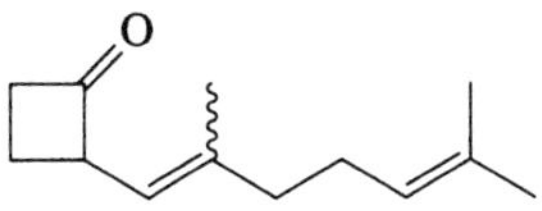	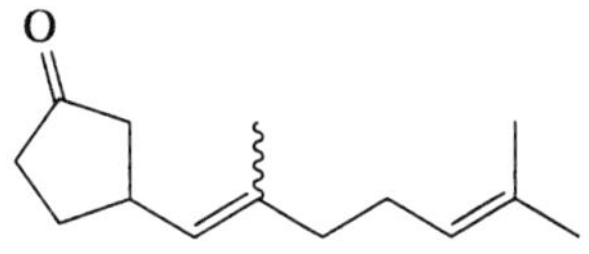	58
R = H	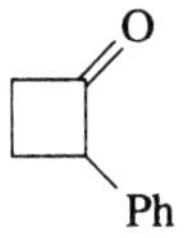	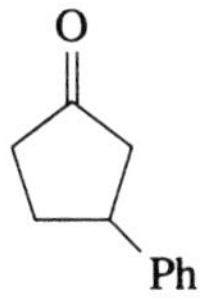	52
R = H	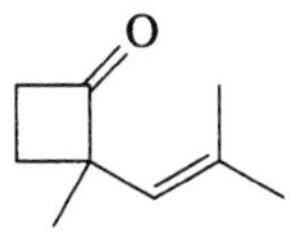	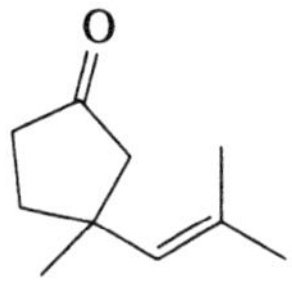	72
R = H	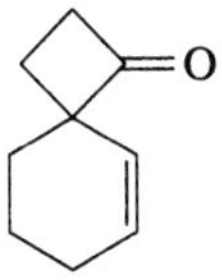	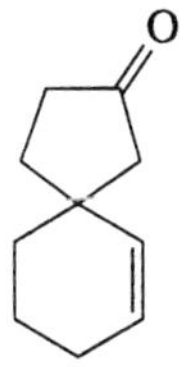	94
R = H	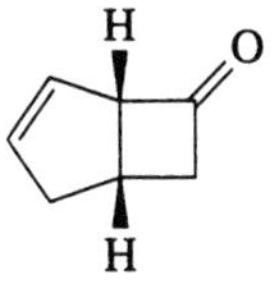	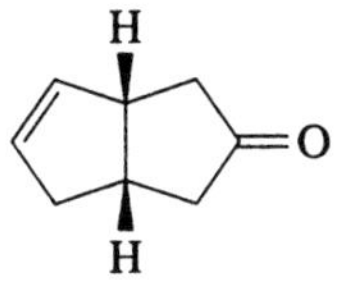	63
R = Me	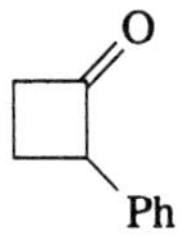	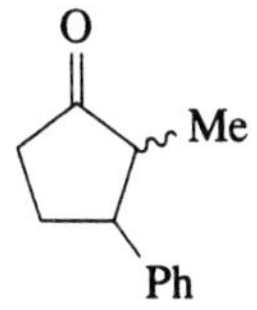	52
R = Me	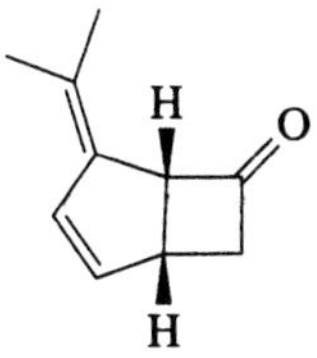	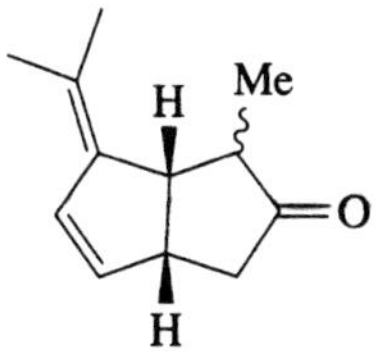	52

Table 2 *(continued)*

Sulfoxide RCH_2SOAr [a]	*Cyclobutanone*	*Product*	*Yield (%)*
$R = CH_2CH_2CH{=}CH_2$	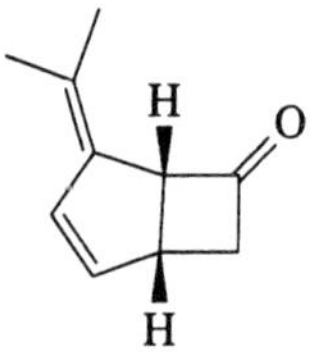	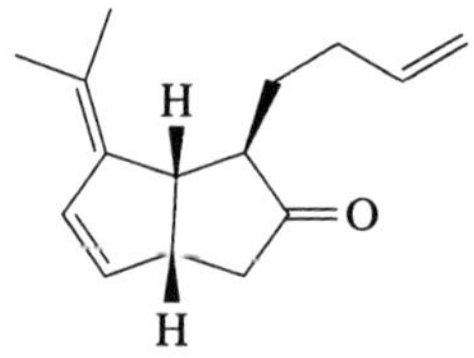	69
$R = (CH_2)_7OSiMe_2Bu^t$	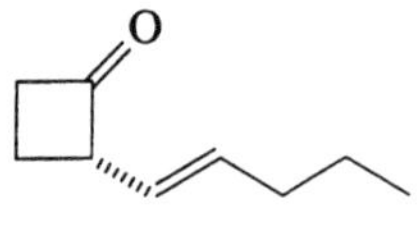	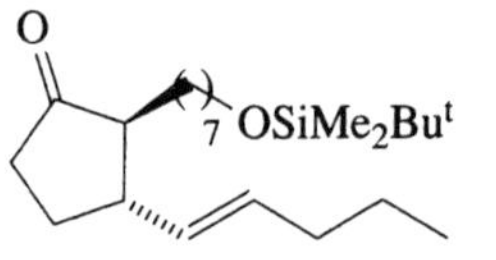	53
$R = CH_2CH_2Ph$	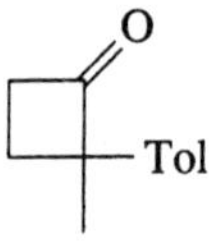	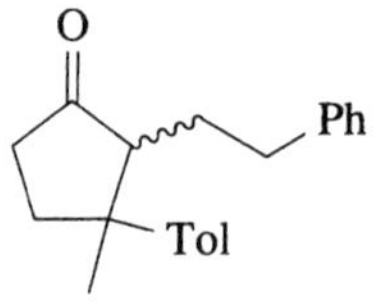	53

[a] Ar = 2-chlorophenyl.

The selenium-based methodology is particularly well suited for the homologation of ketones with the incorporation of a geminally disubstituted carbon because, in this instance, the formation of epoxides is minimal. Representative examples are given in Table 3. The regioselectivity of the expansion reaction is sensitive to steric effects and stereochemical relationships, as well as to the method used for activation of the selenium for elimination. For cyclic enones, ring sizes larger than six begin to show a decline in regioselectivity. It is noteworthy that the thallium(I) ethoxide/chloroform generated dichlorocarbene can be utilized in the presence of double bonds. The last entry of Table 3 is of particular interest with regard to the mechanistic aspects of the overall reaction. This example indicates that the configuration of the migrating carbon is preserved regardless of the elimination method used, while the center from which the selenium departs undergoes net inversion with $TlOEt/CHCl_3$ and varying amounts of net retention or inversion with the silver tetrafluoroborate induced elimination.

When the lithium reagent added to the carbonyl group bears two heteroatoms, the net result following rearrangement is the incorporation of a carbon bearing a heteroatom. The hydroxy diphenyl dithioacetals, readily formed from the addition of the anion of the formaldehyde diphenyl dithioacetal to ketones, have been shown to rearrange in the presence of copper(I) triflate to produce the α-benzenesulfenyl homologated ketones. The preferred migration of the more-substituted carbon atom is observed for aliphatic and cyclic ketones, while hydrogen migration is still preferred in the case of aldehyde adducts (Table 4).[85] The presence of an epoxide intermediate has been observed and may bear some relevance to the failure of the rearrangement process to homologate cycloheptanone. Complementary to the copper(I) triflate induced elimination is the base-induced loss of thiophenol. Treatment of the hydroxy diphenyl dithioacetal adduct with two equivalents of *s*-butyllithium at –78 °C produces a dianion, which on warming undergoes a carbenoid-type reaction (*cf.* the next section for a discussion) to the ring-expanded ketone (Scheme 26).[86] In general the yields and regioselectivity for the bond migrations appear to be comparable to the copper(I) triflate reaction, with the notable improvement that ring size does not appear to be a restriction.

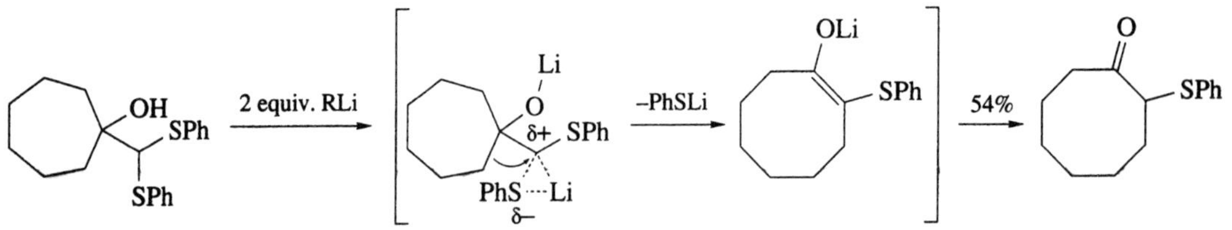

2 equiv. RLi → –PhSLi → 54%

Scheme 26

Another variation on the Lewis acid catalyzed homologation process is to raise the oxidation level on one of the sulfur groups to the level of sulfone.[87] The addition of the less nucleophilic lithium anion of (phenylthio)methyl phenyl sulfone to ketones requires the presence of a Lewis acid (diethylaluminum chloride). The rearrangement step, which did not occur in ether solvents, is conducted at low temperature in dichloromethane in the presence of diethylaluminum chloride and provides good yields of the ring-enlarged ketones. The general trend of the more-substituted bond migration also follows here, except for a bridged bicyclic system, the last entry of Table 5.

A nitro group has been found to perform the same leaving group function as the phenyl sulfone in an analogous procedure using the addition of the dianion of $PhSCH_2NO_2$ to ketones.[88] The rearrangement step is catalyzed by the action of aluminum chloride, is apparently general for four- through seven-membered ring ketones and also displays a high regioselectivity for migration of the more-substituted carbon. An illustrative example is shown in Scheme 27.

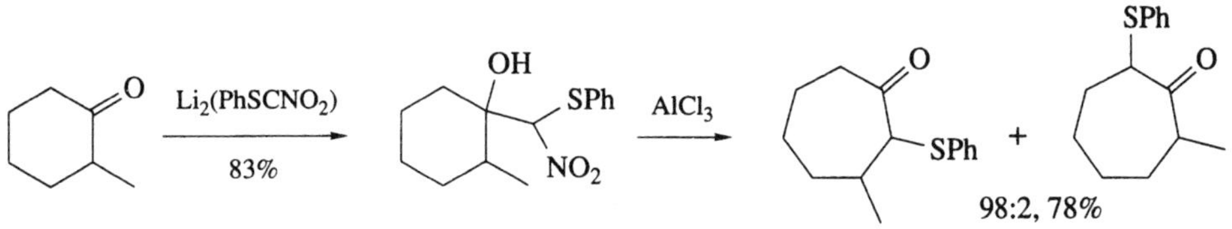

Scheme 27

This general procedure has also been extended to allow for the insertion of a methoxy-containing carbon. The uncatalyzed addition of the lithium anion of methoxymethyl phenyl sulfone to ketones proceeds readily at low temperature in dimethoxyethane to form the intermediate adduct. Addition of a Lewis acid (ethylaluminum dichloride or diisobutylaluminum diisopropylamide) directly to the reaction mixture effects the rearrangement reaction to produce the ring-expanded α-methoxy ketone. This sequence, illustrated by the example in Scheme 28, is limited to the expansion of four- and five-membered ring ketones.[87]

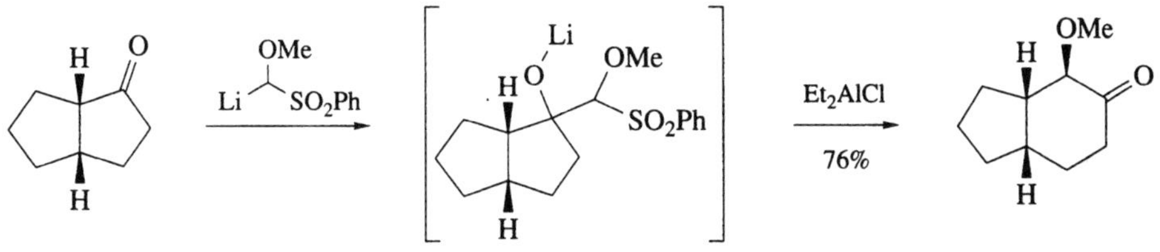

Scheme 28

Insertion of a latent carbonyl group, in the form of a dithioketal, may be achieved through the same sulfur-based strategy by employing a nucleophile containing three heteroatoms such as the lithium anion of tris(methylthio)methane. The rearrangement, with loss of one thiomethyl group, has been carried out in a separate step on the anion of the adduct in the presence of $CuClO_4{\cdot}4MeCN$, $CuBF_4{\cdot}4MeCN$ in toluene or in a strained case simply by treatment with acid (Scheme 29). The reaction proceeds well for the expansion of four- and five-membered ring ketones, but is diverted to nonexpanded products for cyclohexanone and other ketones, a result rationalized by the isomerization through epoxides. As may be seen from the examples in Table 6, the regioselectivities in general are high.

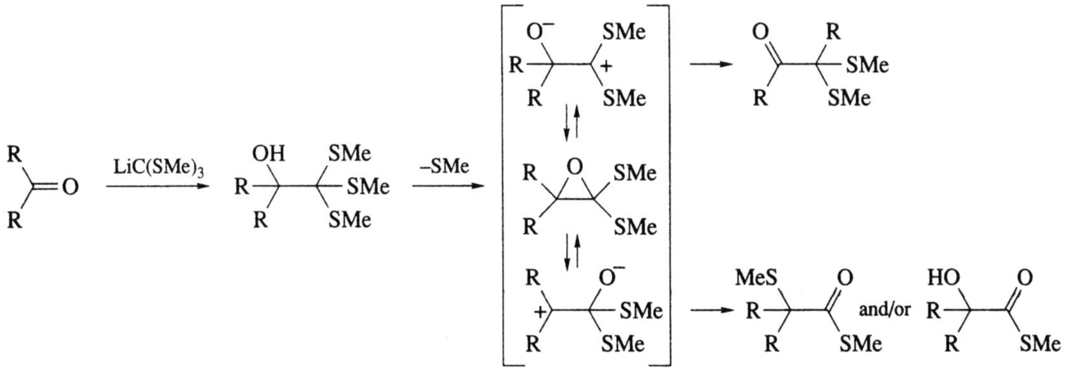

Scheme 29

Table 3 Ring Expansion of Ketones by Se-based Addition–Rearrangement Reactions

Lithioselenide	*Ketone*	*Adducts*	*Yield (%)*	*Rearrangement method*	*Product(s)*	*Yield (%)*	*Ref.*
 Li SeMe	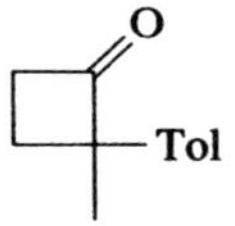 O, Tol	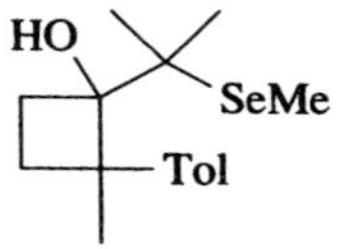 HO, SeMe, Tol	66	a c	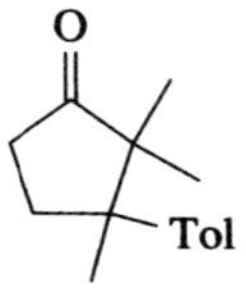 O, Tol	57 83	79
 Li SeMe	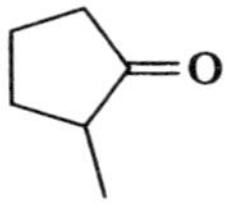 O	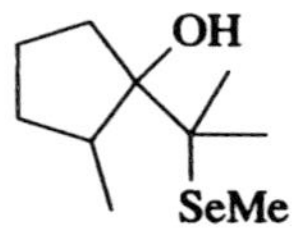 OH, SeMe	20	a	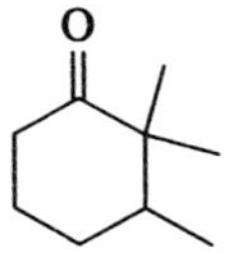 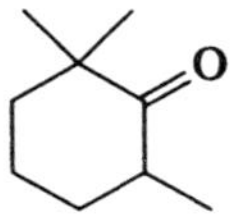 O + O 90:10	76	79
 Li SeMe	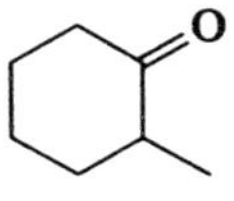 O	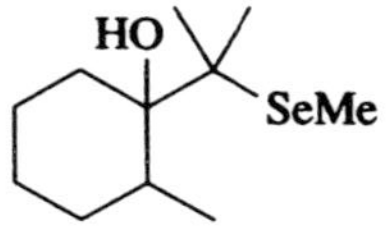 HO, SeMe		a b	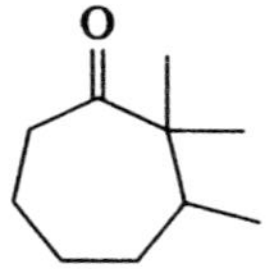 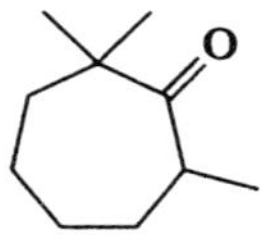 O + O 60:40 22:78	77 54	79
 Li SeMe	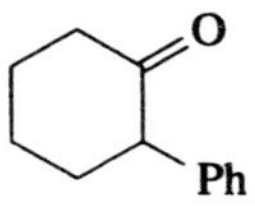 O, Ph	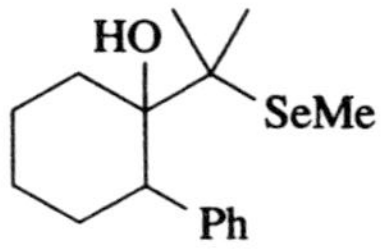 HO, SeMe, Ph	89	a	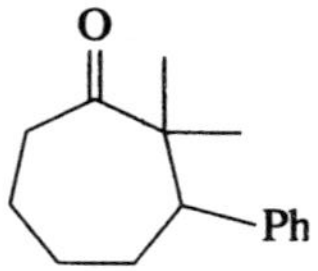 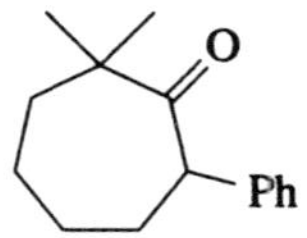 O, Ph + O, Ph 81:19	79	79
 Li SeMe	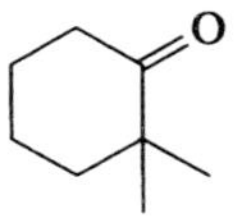 O	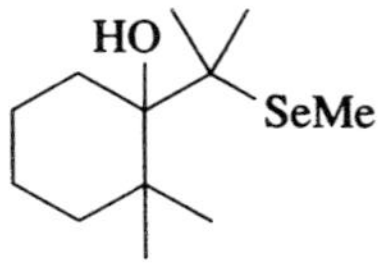 HO, SeMe	68	a	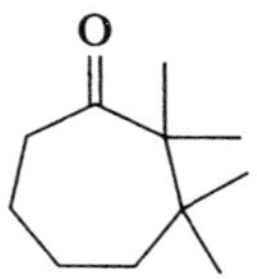 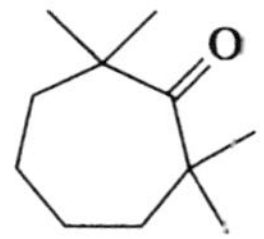 O + O 99:1	81	79

Table 3 *(continued)*

Lithioselenide	*Ketone*	*Adducts*	*Yield (%)*	*Rearrangement method*	*Product(s)*	*Yield (%)*	*Ref.*
 Li SeMe	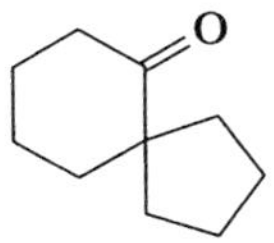 O	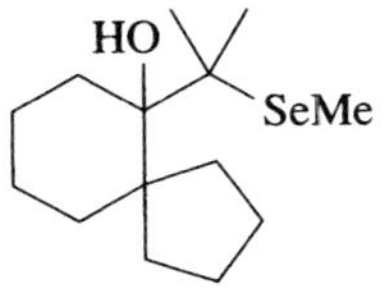 HO SeMe	70		 O + O		79
				a	99:1	77	
				b	34:66	45	
 Li SeMe	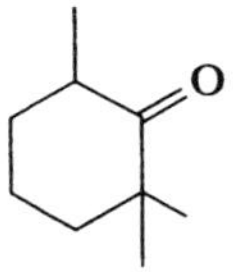 O	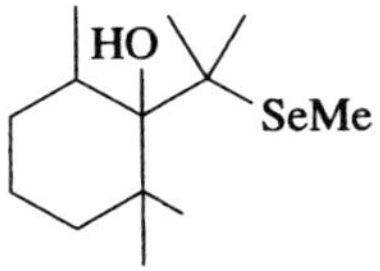 HO SeMe		c	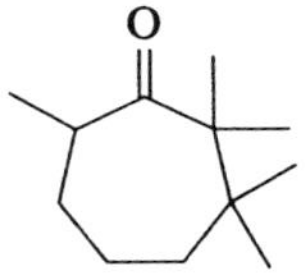 O	88	79
 Li SeMe	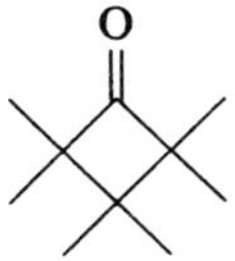 O	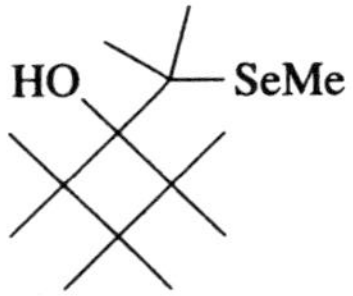 HO SeMe		a	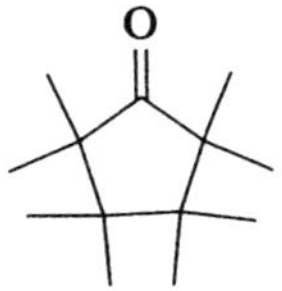 O	71	80
 Li SeMe	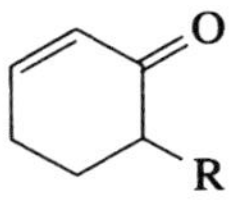 O R	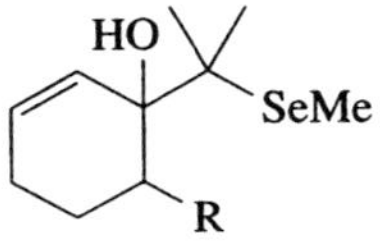 HO SeMe R		a	 O R + O R		81
	R = H		58		91:9	68	
	R = Me		70		100:0	71	
	R = Ph		74		100:0	67	

Table 3 *(continued)*

Lithioselenide	*Ketone*	*Adducts*	*Yield (%)*	*Rearrangement method*	*Product(s)*	*Yield (%)*	*Ref.*
Li SeMe	O	HO SeMe	39	a	 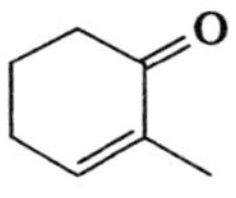 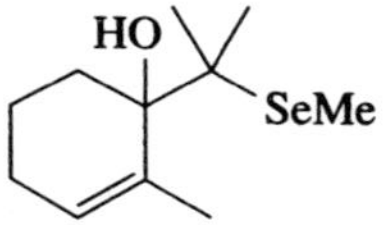 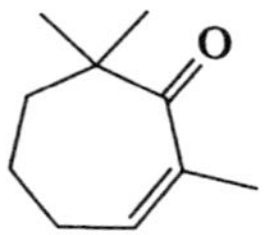 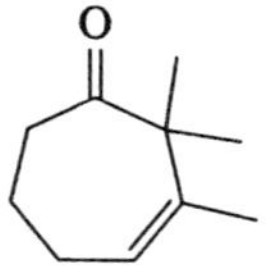 O + O 78:22	64	81
Li SeMe	O	HO SeMe	51	a	 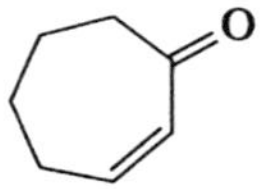 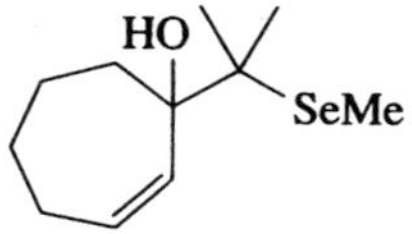 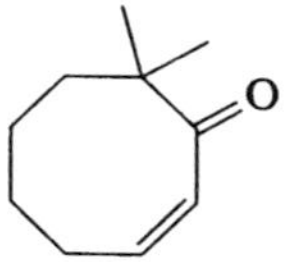 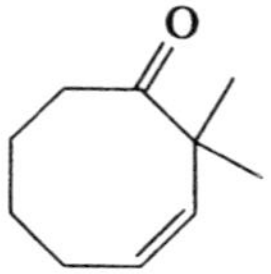 O + O 56:44	88	81
Li SePh	O	OH SePh		a	O	98	82
Li SePh	OR O	OR PhSe OH (a) (a):(b) = 34:66 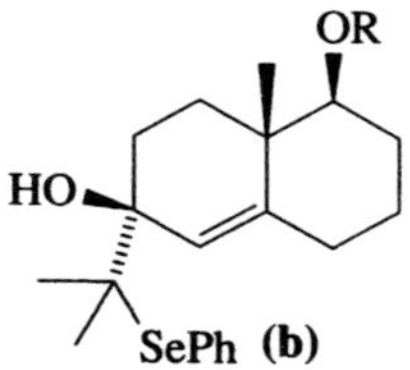 OR HO SePh (b)	67	a a	94:6 OR O + OR O 77:23	93 72	82

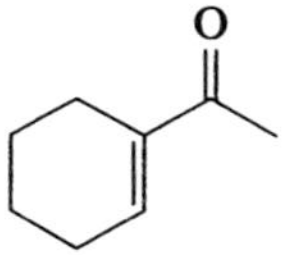

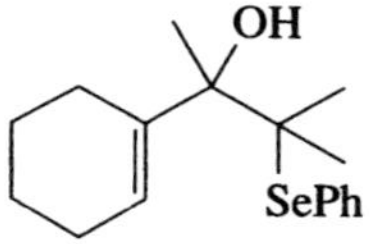

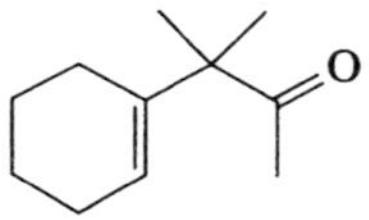

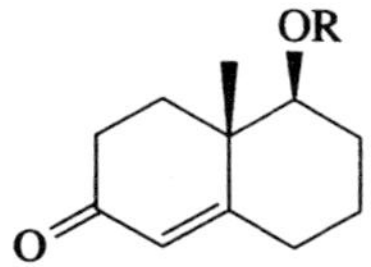

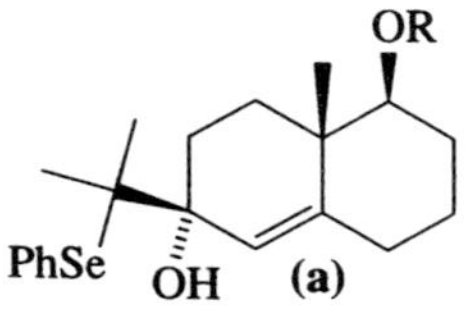

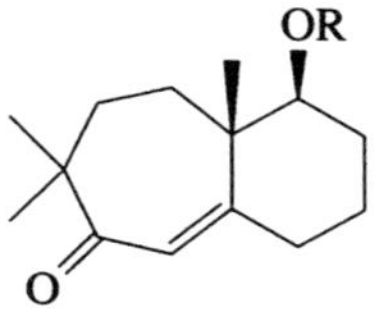

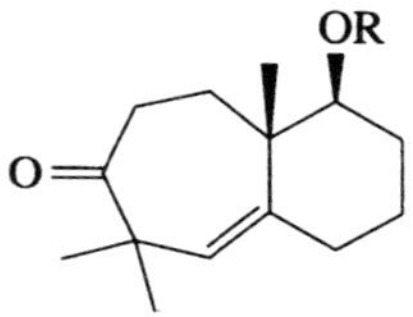

Table 3 *(continued)*

Lithioselenide	*Ketone*	*Adducts*	*Yield (%)*	*Rearrangement method*	*Product(s)*	*Yield (%)*	*Ref.*
 Li, SePh	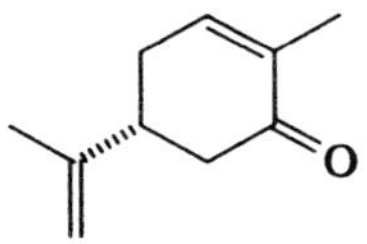 O	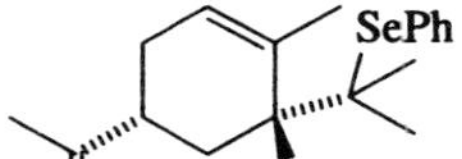 SePh, HO (a)		a	20:80	89	
		(a):(b) = 34:66	83		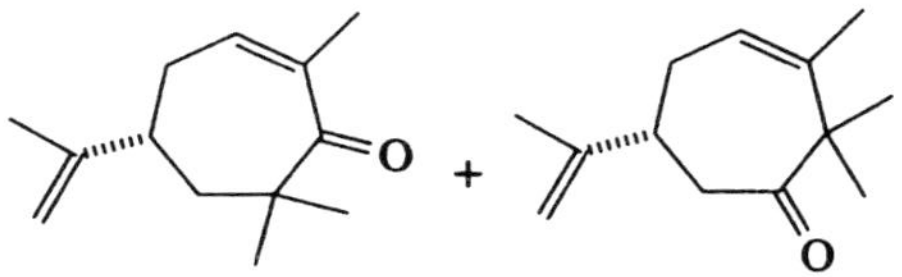 O + O		82
		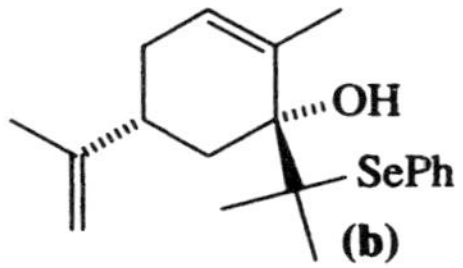 OH, SePh (b)		a	70:30	78	
	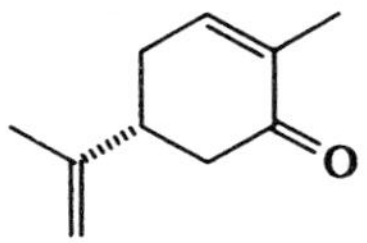	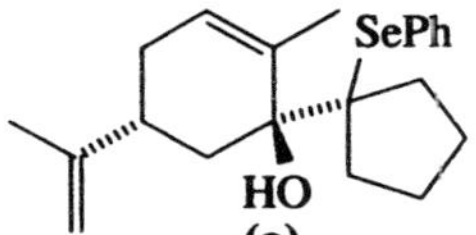 SePh, HO (a)		a	4:96	65	
Li, SePh	O	(a):(b) = 34:66			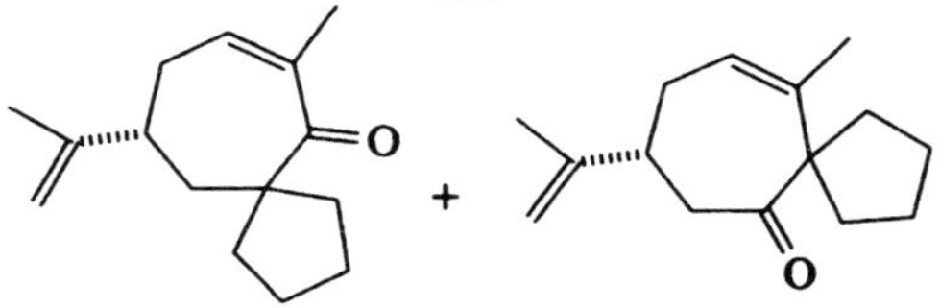 O + O		82
		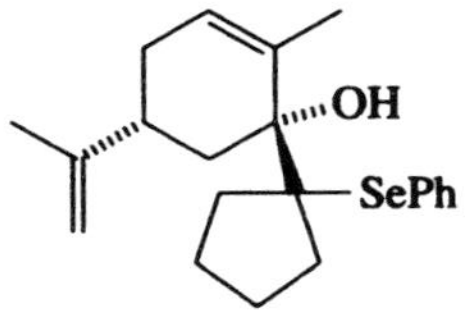 OH, SePh (b)		a	32:68	20	
 Li, SePh	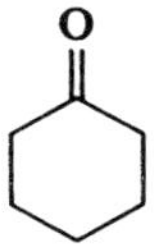 O	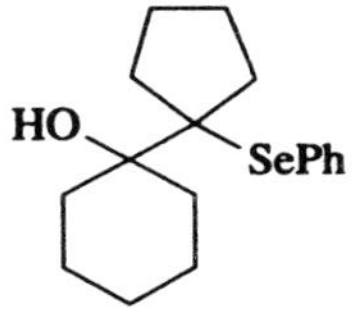 HO, SePh		a	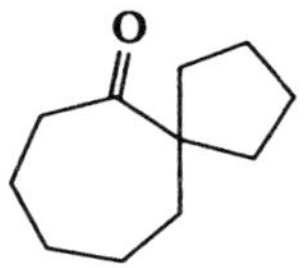 O	87	83

Table 3 *(continued)*

Lithioselenide	*Ketone*	*Adducts*	*Yield (%)*	*Rearrangement method*	*Product(s)*	*Yield (%)*	*Ref.*
MeSe, Ph, Li	O, Bu^t	HO, SeMe, Bu^t, Ph **(a)**; **(a)**:**(b)** = 50:50; HO, SeMe, Bu^t, Ph **(b)**	84	a b	Ph, O, Bu^t + Ph, O, Bu^t 100:0 0:100	25 48	84
				a b	O, Ph, Bu^t + Ph, O, Bu^t + Ph, O, Bu^t 35:65:0 0:54:46	40 89	

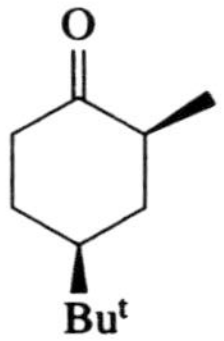

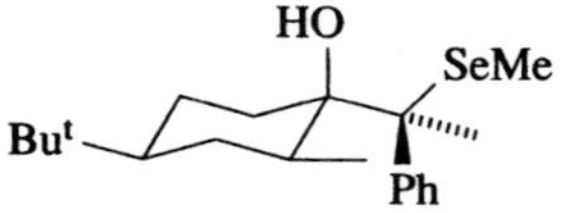

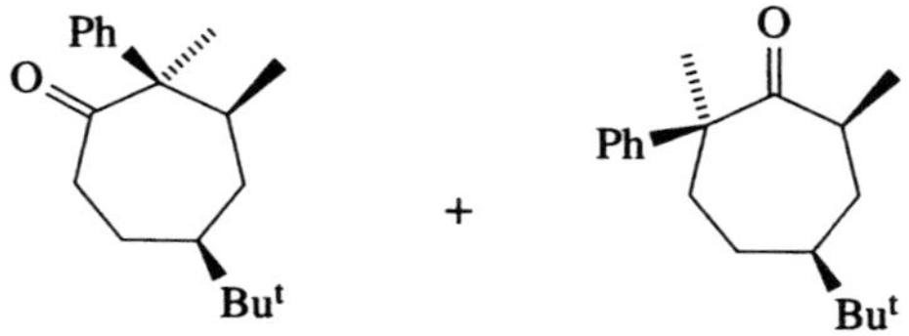

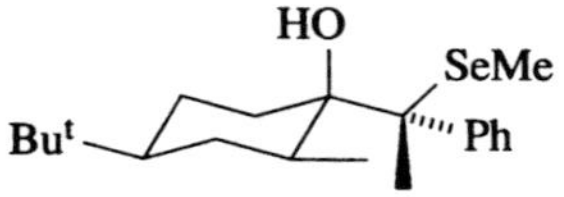

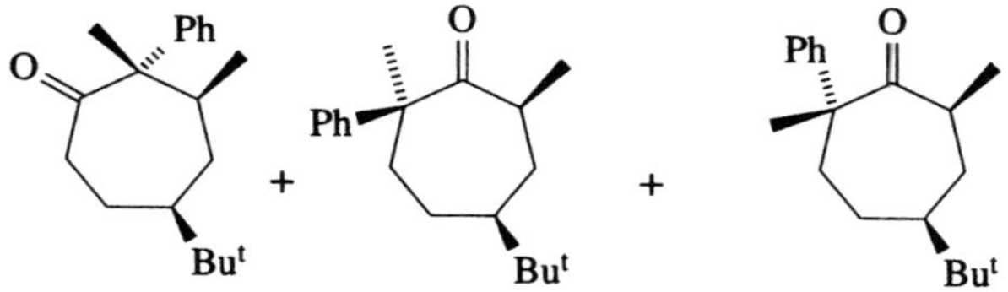

[a] $TlOEt/CHCl_3$. [b] $AgBF_4/CH_2Cl_2$. [c] $MeSO_3F$.

Table 4 Homologation of Carbonyls by Reaction with $(PhS)_2CHLi$ and Rearrangement with Cu^+

Carbonyl	*Adduct yield (%)*	*Homologated ketone*	*Yield (%)*
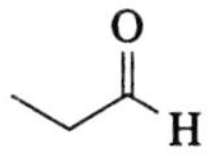	80 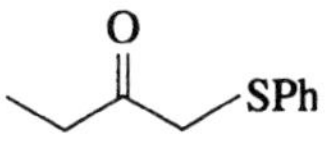	SPh	66
	90	SPh	88
	85	SPh	81
	86	SPh	72
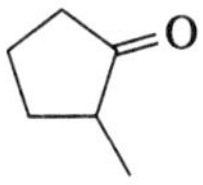	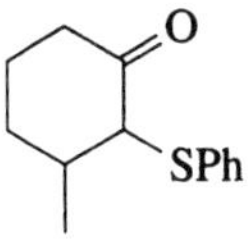 84	SPh + SPh (89:11)	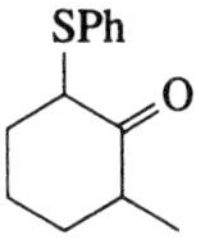 83
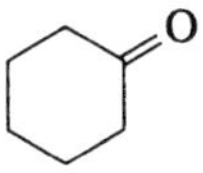	74	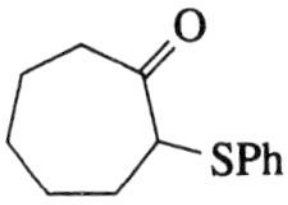 SPh + SPh, CHO (82:18)	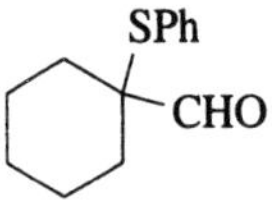 71

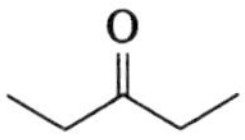

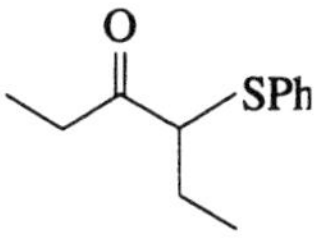

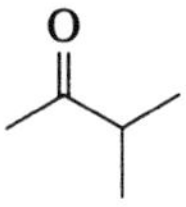

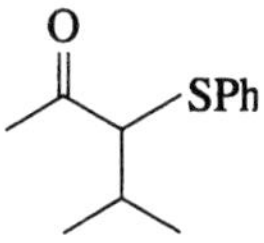

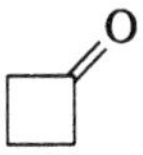

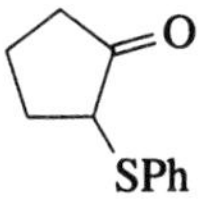

Table 5 Ring Expansion of Ketones with $PhS(PhSO_2)CHLi$

Ketone	*Adduct*	*Yield (%)*	*Expansion product*	*Overall yield (%)*
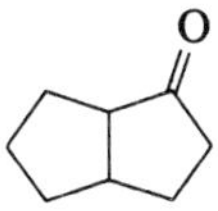	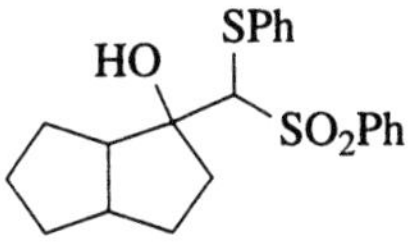 HO, SPh, SO_2Ph	92	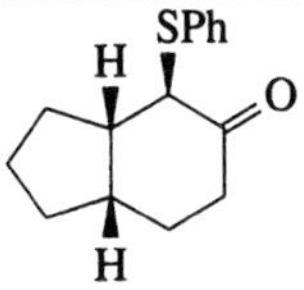 H, SPh, H	79
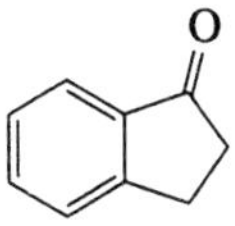	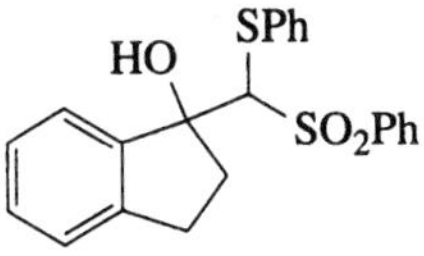 HO, SPh, SO_2Ph	94	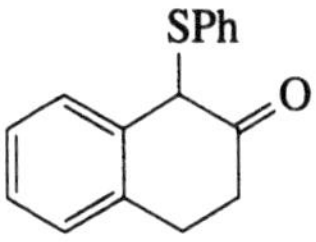 SPh	77
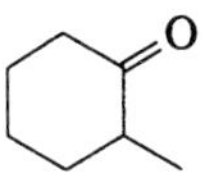	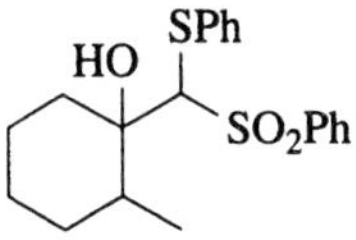 HO, SPh, SO_2Ph	92	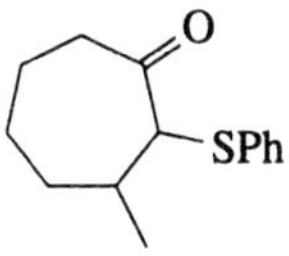 SPh	69
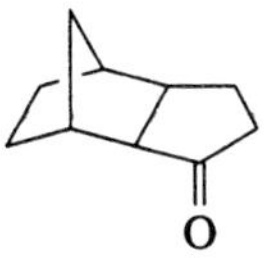	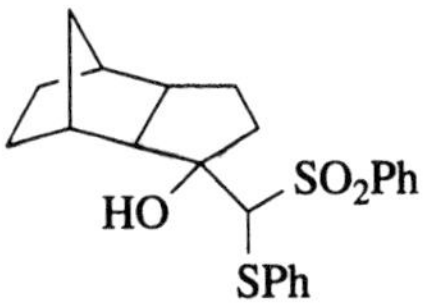 HO, SO_2Ph, SPh	93	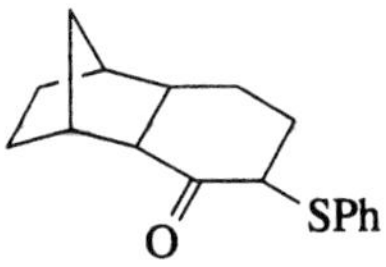 SPh	74

Table 6 Ketone Ring Expansions *via* $(MeS)_3CLi$

Ketone adduct	*Yield (%)*	*Expansion product*	*Yield (%)*	*Ref.*
OH, $C(SMe)_3$	73	O, SMe, SMe	73	89
OH, $C(SMe)_3$	85	O, SMe, SMe	58	89
$C(SMe)_3$, OH	95	O, MeS, SMe	61	89
SEMO, H, $C(SMe)_3$, H, OH	95	SEMO, H, O, SMe, H, SMe	84	89
SEMO, H, $C(SMe)_3$, H, OH **(a)**; **(a)**:**(b)** = 48:52; SEMO, H, OH, H, $C(SMe)_3$ **(b)**	87	SEMO, H, O, SMe, H, SMe + SEMO, H, SMe, SMe, H, O; 21:79 (a); 69:31 (b)	92 (a); 83 (b)	89
HO, $C(SMe)_3$	74	MeS, SMe, O + MeS, O, SMe; 73:27	99	90
OH, $C(SMe)_3$	75	SMe, SMe, O	71	90
$C(SMe)_3$, OH		SMe, SMe, O	66	91

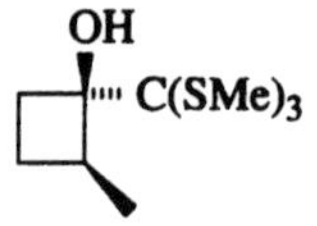

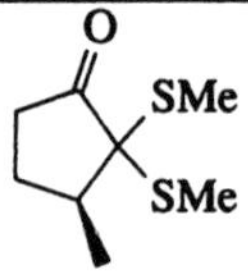

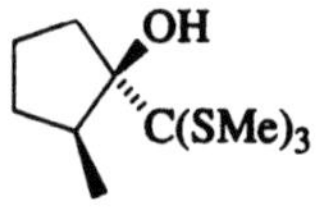

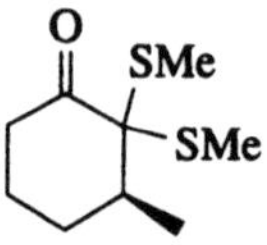

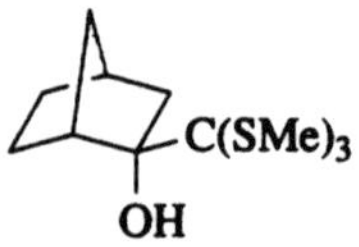

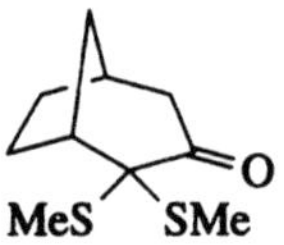

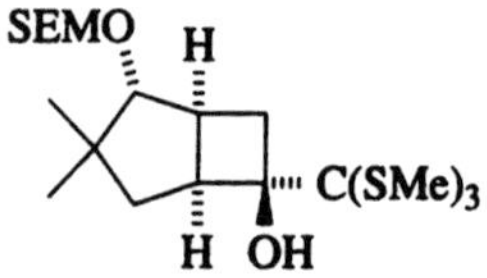

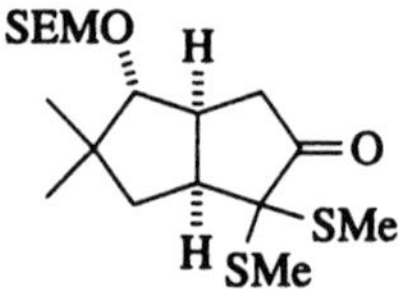

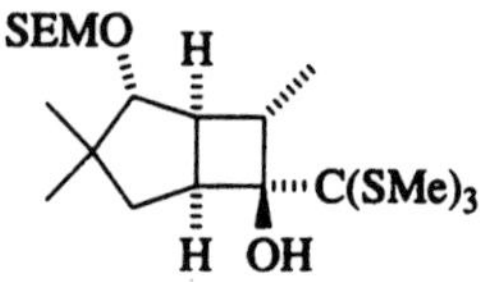

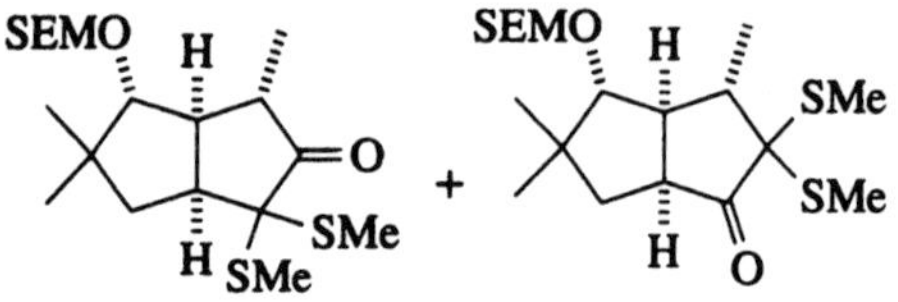

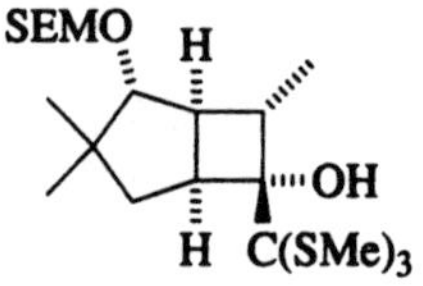

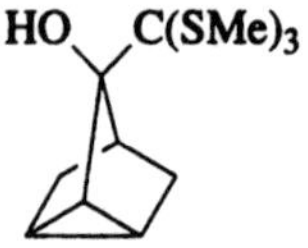

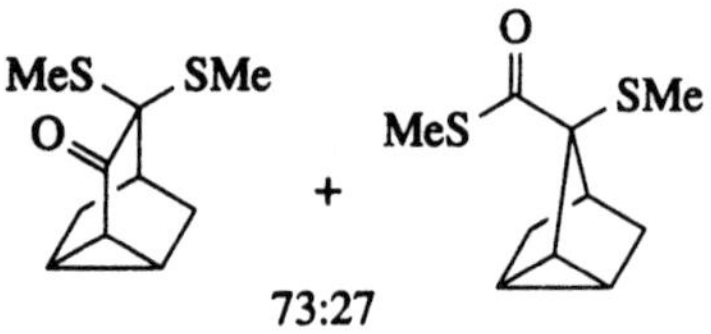

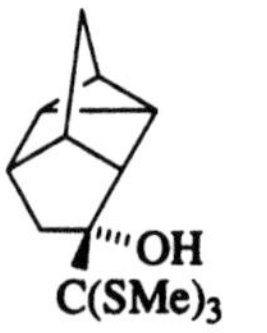

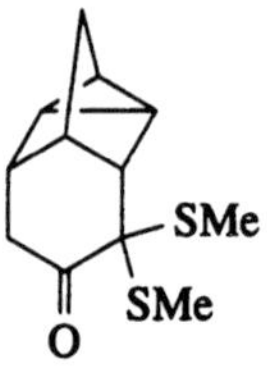

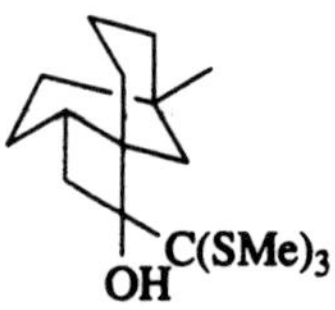

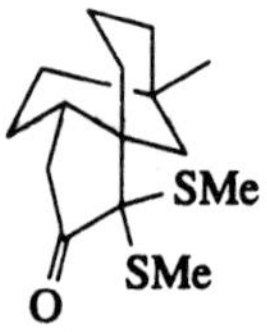

3.3.4 REARRANGEMENT OF β-OXIDO CARBENOIDS

A novel homologation method arises from the base-induced rearrangement of dihalohydrins.[92] On an historical note, this reaction is mechanistically distinct from an earlier ring expansion protocol that employed the known tendency of the magnesium salts of monohalohydrins to undergo rearrangement. That procedure involved the addition of benzylmagnesium chloride to ketones and bromination of the resulting alcohols at the benzylic position with *N*-bromosuccinimide to form the bromohydrin.[93] On heating with one equivalent of isopropylmagnesium bromide, the bromohydrin underwent rearrangement to the homologated ketone, either through a pinacol-like reaction, or possibly *via* an epoxide. Examples are shown in Schemes 30 and 31. While the regioselectivity for bond migration is moderate for the 2-methylcyclopentanone expansion, the corresponding expansion for the 2-methylcyclohexanone is near unity.

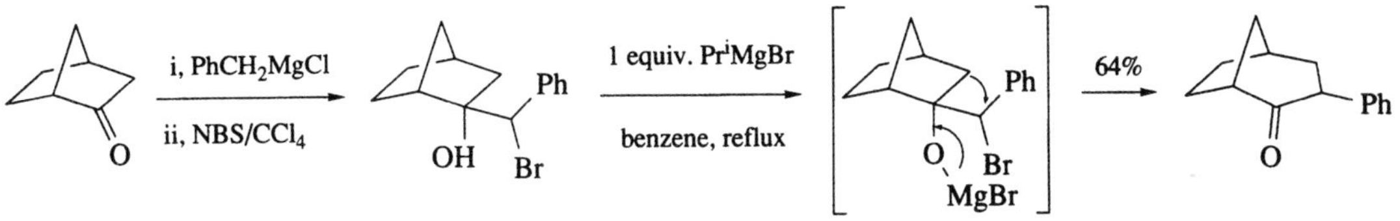

i, $PhCH_2MgCl$; ii, NBS/CCl_4; 1 equiv. Pr^iMgBr, benzene, reflux; 64%

Scheme 30

i, $PhCH_2MgCl$; ii, NBS/CCl_4; 1 equiv. Pr^iMgBr, benzene, reflux; 17:83, 58%

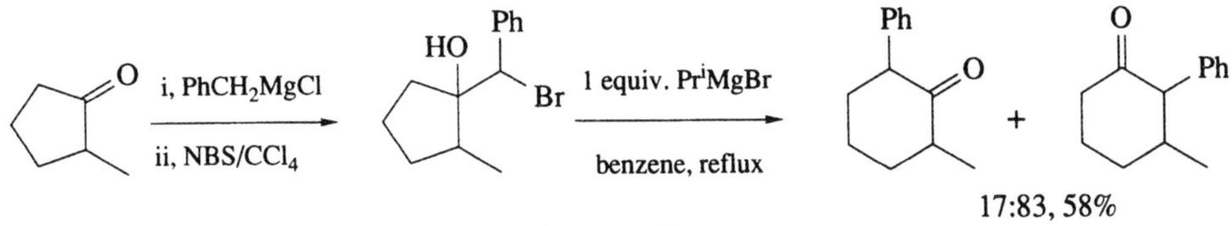

Scheme 31

On the other hand, dichlorohydrins, produced by the addition of dichloromethyllithium to ketones, may be deprotonated at low temperature by two equivalents of alkyllithiums or lithium dialkylamides to form the corresponding dianion. This dianion loses LiCl on warming, and rearranges either directly or through a carbenoid intermediate to the enolate of the homologated α-chloro ketone (Scheme 32).[92,94] As a homologation, the reaction proceeds well for aliphatic (Scheme 33)[92,95a] and cyclic ketones (Scheme 34),[92,95b] but suffers with aldehydes due to the preferred migration of hydrogen ~ phenyl >> alkyl.

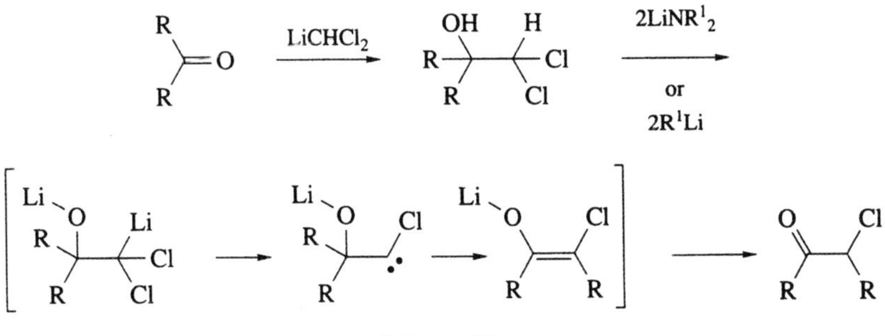

$LiCHCl_2$; $2LiNR^1_2$ or $2R^1Li$

Scheme 32

$LiCHCl_2$; LiTMP, −70 °C to r.t.

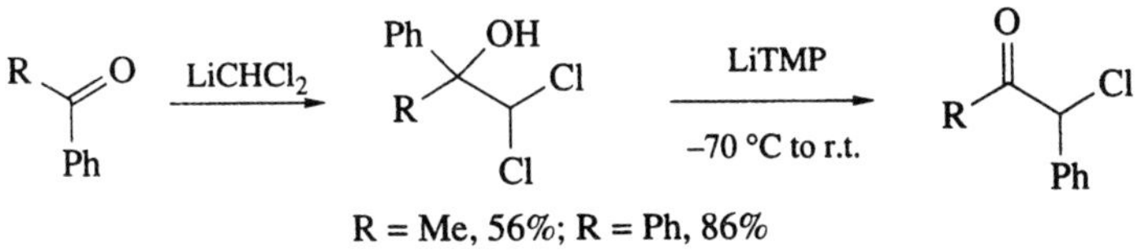

R = Me, 56%; R = Ph, 86%

Scheme 33

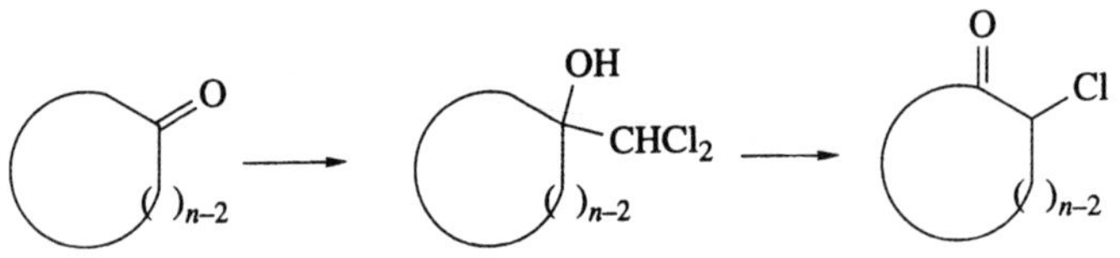

$n = 5$, 90%; $n = 6$, 70%; $n = 7$, 68%; $n = 8$, 48%

Scheme 34

In contrast, when the analogous dibromohydrins are exposed to two equivalents of butyllithium at low temperature, lithium–halogen exchange occurs to produce the monobromo dianion. On warming, this intermediate loses LiBr to give a carbene, which rearranges with migration of a substituent to form the enolate of the expanded ketone (Scheme 35).[96] This procedure appears to be fairly general for the one-carbon expansion of cyclic ketones (Table 7). It is noteworthy that the intermediate enolate anion has been trapped as the trimethylsilyl enol ether. The preferred regioselectivity for the migration of the more-substituted carbon atom has been rationalized in terms of a preferred orientation of the dianion in which the departing bromine atom is situated opposite the more sterically hindered center (Scheme 36).[96b] For the few cases examined, it has been determined that the configuration at the migrating carbon atom is preserved: examples are given in Table 7.

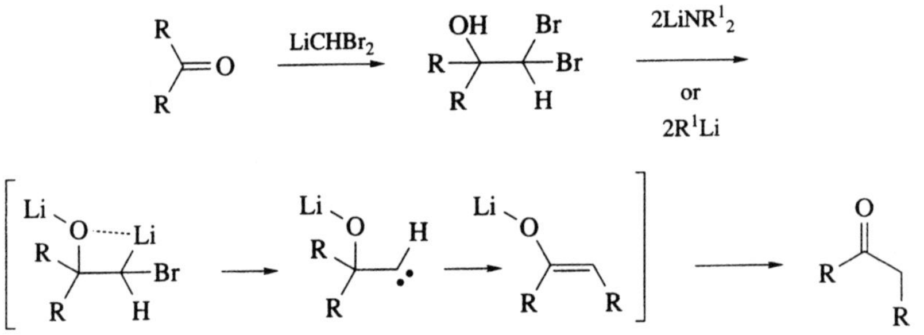

Scheme 35

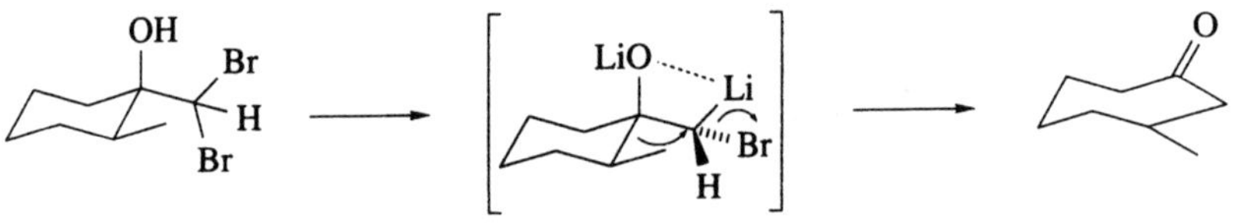

Scheme 36

This rearrangement process has also been developed into a homologation procedure for esters. The low temperature addition of dibromomethyllithium to esters produces the primary adduct, which, depending on substitution may collapse to the dibromo enolate (Scheme 37). A rapid metal–halogen exchange occurs on addition of butyllithium to produce the dilithio anion, which then undergoes a rearrangement on warming. The resulting lithium ynolate is quenched with acidic ethanol to give the homologated ester. This one-flask process appears to be general for esters and lactones (Table 8).[103] The rearrangement aspect of the reaction has been verified by carbon-13 labeling, and proceeds with retention of configuration at the migrating carbon. An interesting adjunct to this homologation procedure is the observation that the intermediate ynolate may be reduced to the enolate by lithium hydride generated *in situ* from butyllithium and 1,3-cyclohexadiene (Scheme 38).[104] The esters shown in Table 8 have been converted to the corresponding aldehydes or enol acetates in 50–60% overall yield by this procedure.

Table 7 β-Oxido Carbenoid Ring Expansion

Ketone	*Adduct*	*Product(s)*	*Yield (%)*	*Ref.*
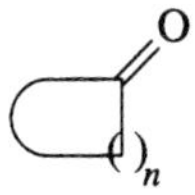 O, ( )$_n$	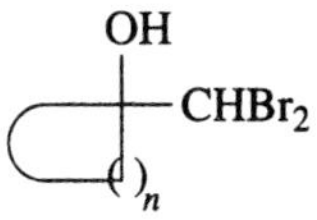 OH, $CHBr_2$, ( )$_n$	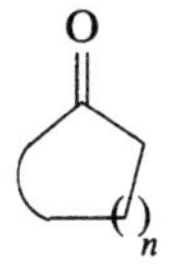 O, ( )$_n$		96b
		$n = 5$	92	
		$n = 6$	70	
		$n = 7$	80	
		$n = 8$	87	
		$n = 12$	89	
	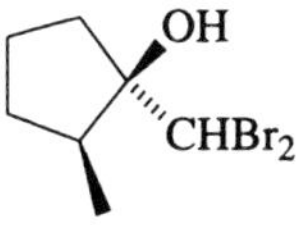 OH, $CHBr_2$ **(a)**	69:31	90	
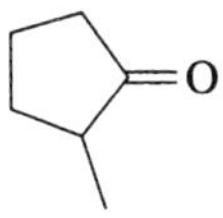 O	**(a)**:**(b)** = 70:30	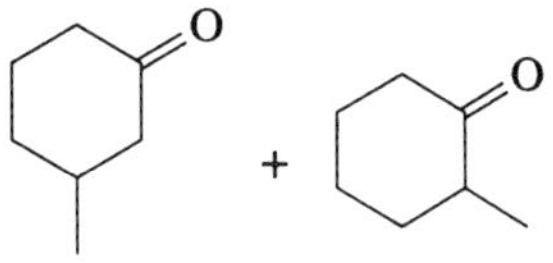 O + O		96b
	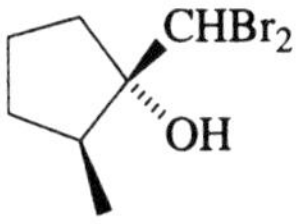 $CHBr_2$, OH **(b)**	49:51	73	
O	OH, $CHBr_2$	O + O	86	96b
		97:3		
O	OH, $CHBr_2$	O + O		
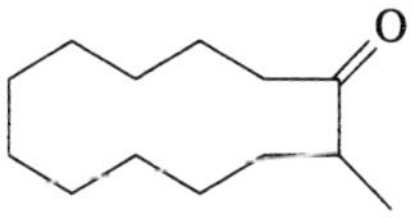	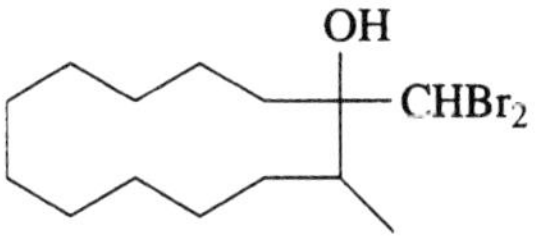	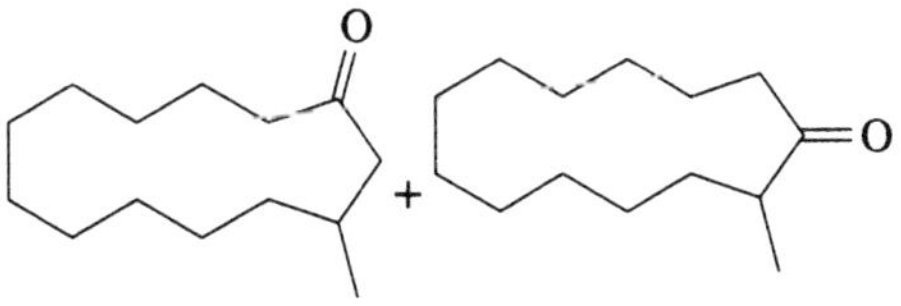 99:1	96	96b
O	OH, $CHBr_2$	O + O	76	96b
		95:5		
O, Me_3SiO	OH, $CHBr_2$, Me_3SiO	$OSiMe_3$, Me_3SiO	85	97

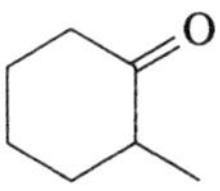

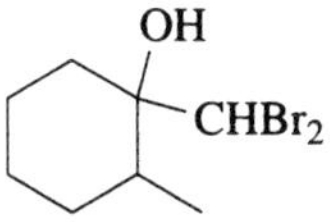

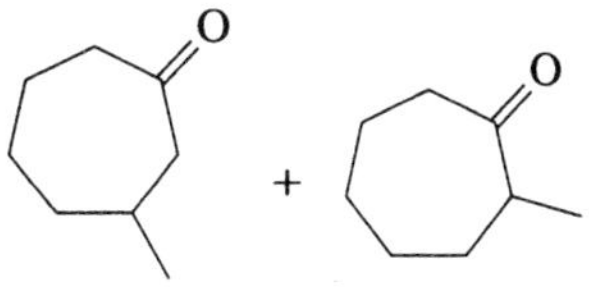

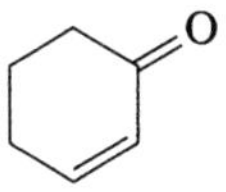

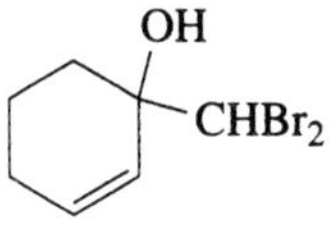

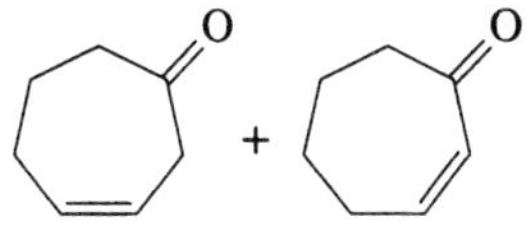

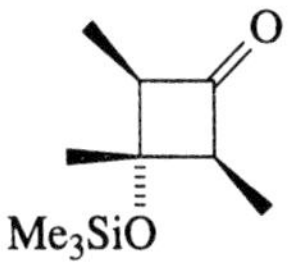

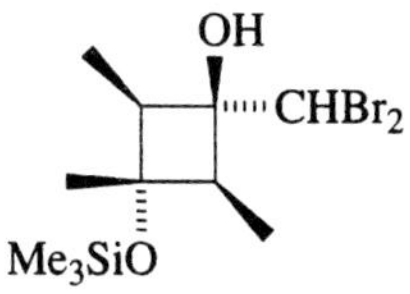

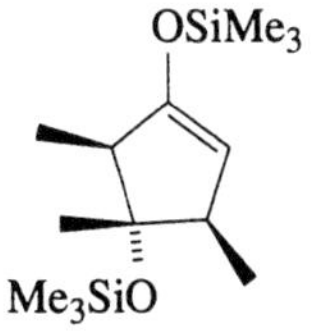

Table 7 *(continued)*

Ketone	*Adduct*	*Product(s)*	*Yield (%)*	*Ref.*
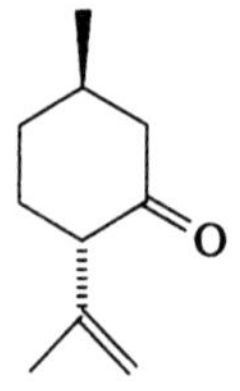 O	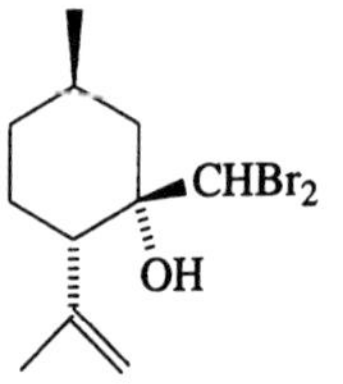 $CHBr_2$ OH	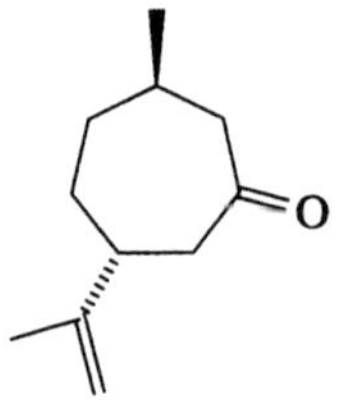 O	60	50
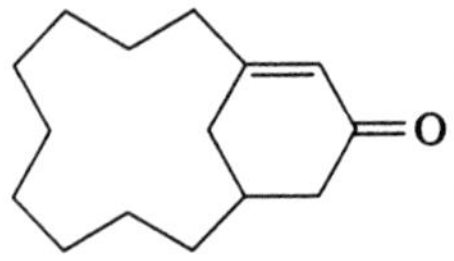 O	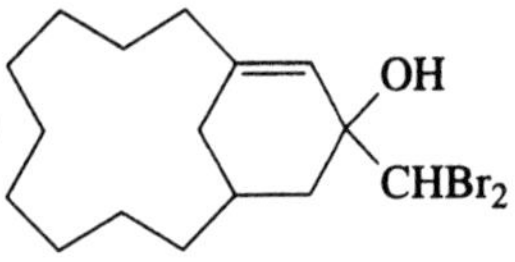 OH $CHBr_2$	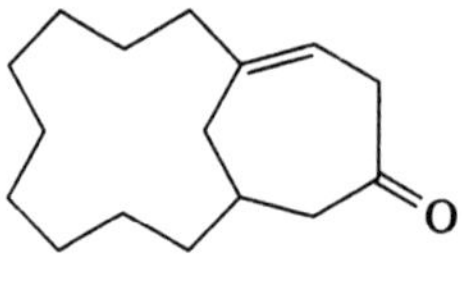 O	50	98
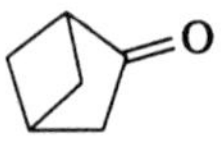 O	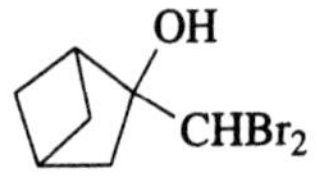 OH $CHBr_2$ 98%	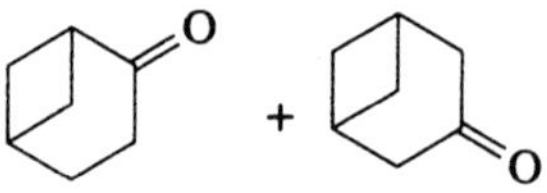 O + O 85:15	68	99
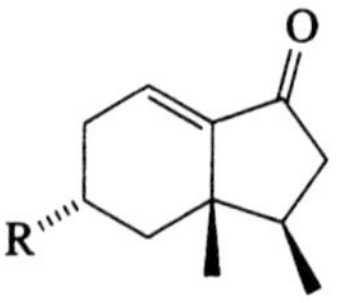 O R	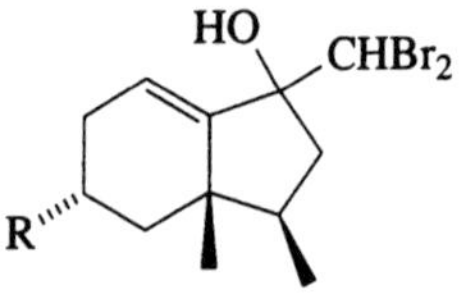 HO $CHBr_2$ R	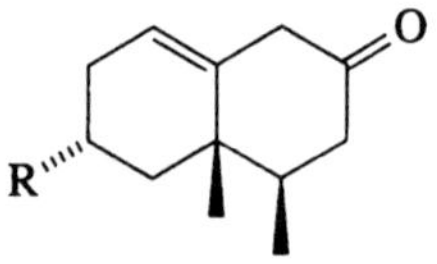 O R		100
R = H	86%		41	
R = MeC(=CH_2)–	60%		83	
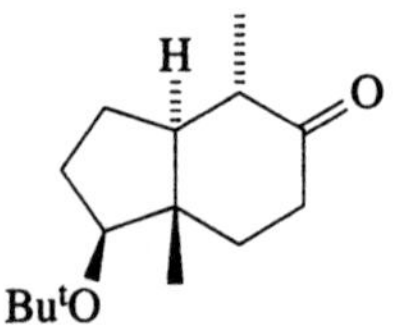 H O Bu^{t}O	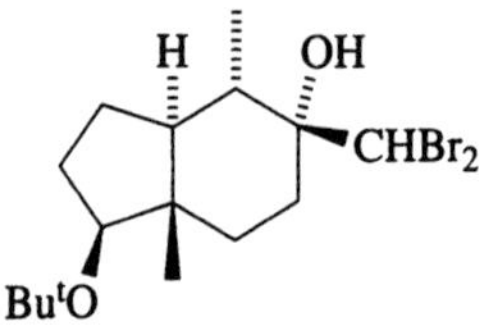 H OH $CHBr_2$ Bu^{t}O	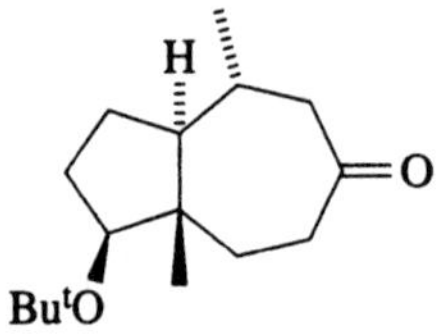 H O Bu^{t}O	58	49
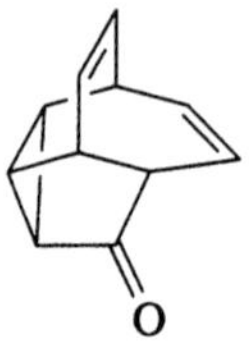 O	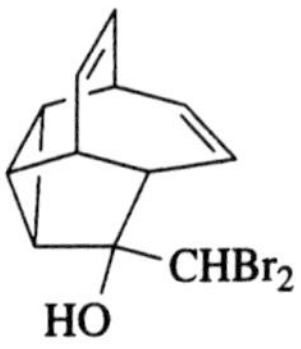 $CHBr_2$ HO	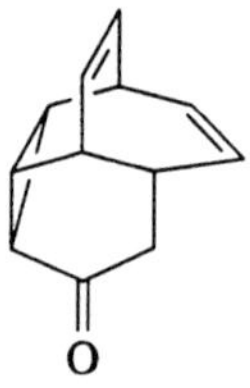 O	53	101
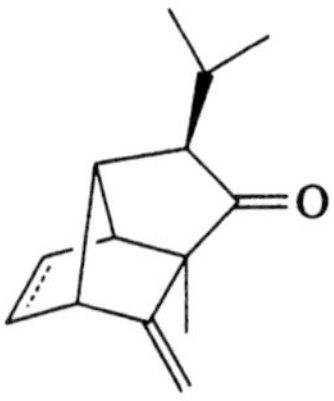 O	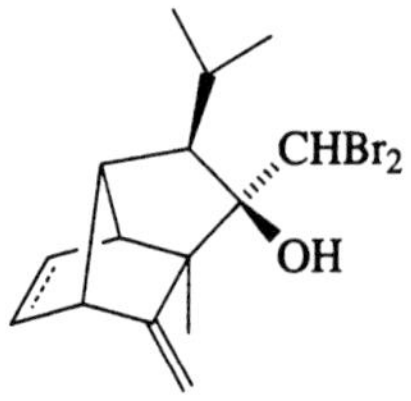 $CHBr_2$ OH	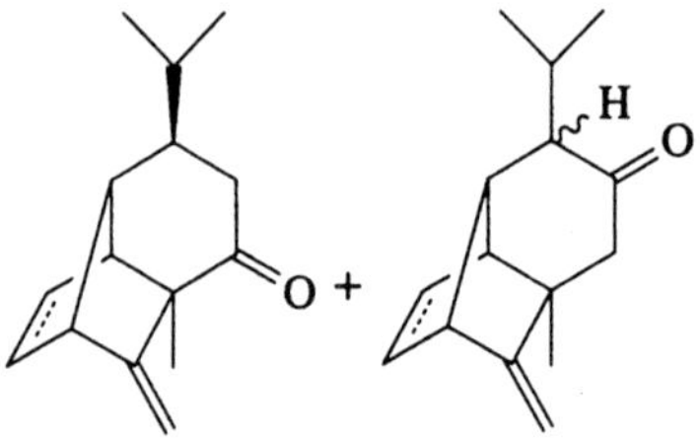 O + H O		102
double bond	63%	73:27	56	
single bond	62%	80:20	50	

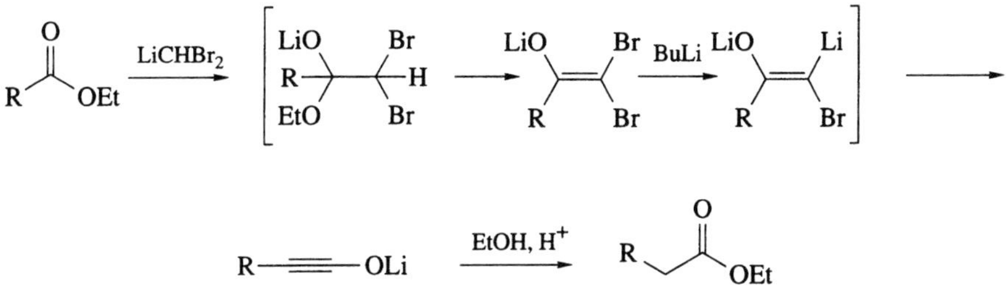

Scheme 37

Table 8 Ester Homologation

Starting ester	*Product*	*Yield (%)*
$PhCO_2Et$	$PhCH_2CO_2Et$	65
$PhCH_2CO_2Et$	$PhCH_2CH_2CO_2Et$	74
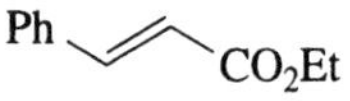	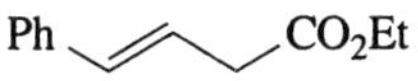	53
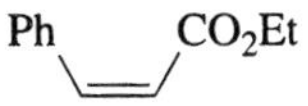	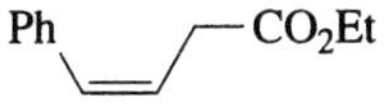	55
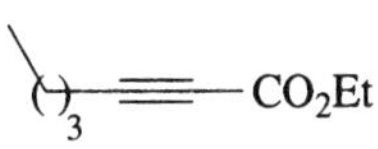		60
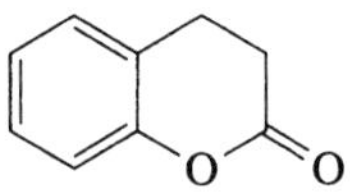	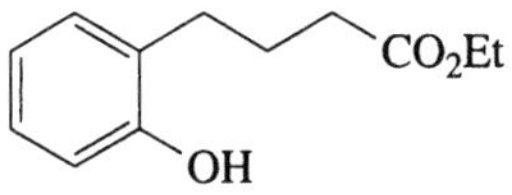	72
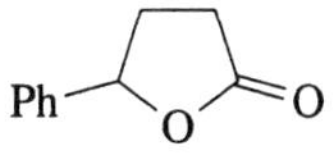	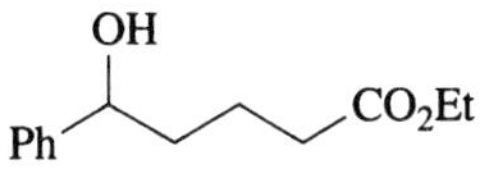	75
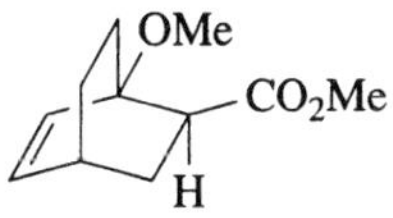	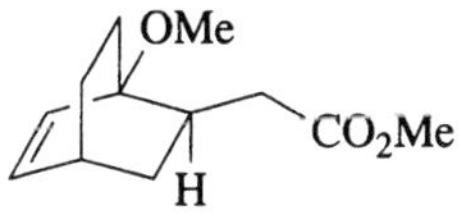	57
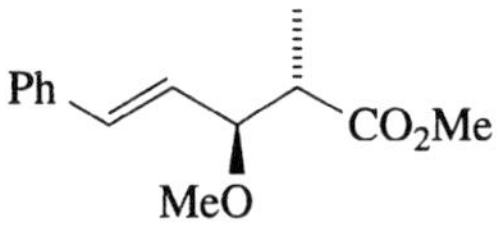	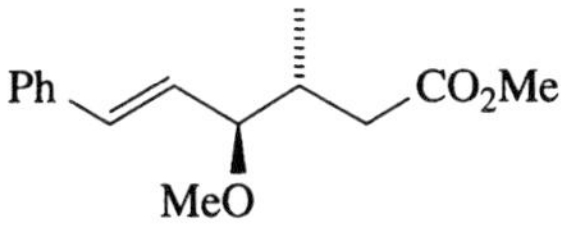	60
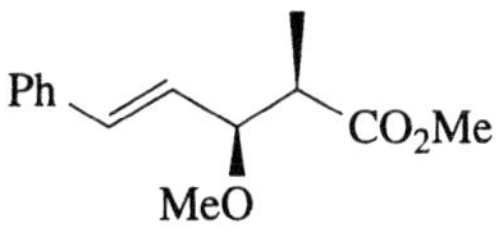	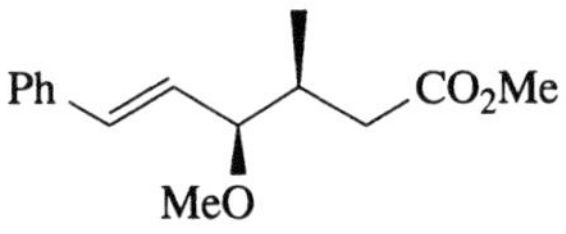	65

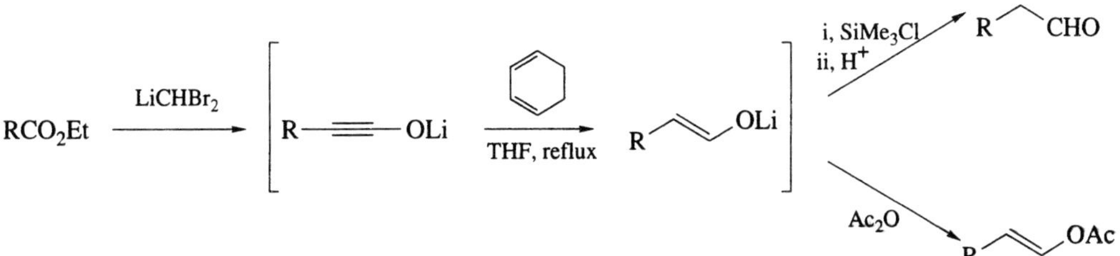

Scheme 38

3.3.5 MISCELLANEOUS HOMOLOGATION METHODS

In this section are homologation/expansion procedures which do not fit a strict addition–elimination with rearrangement type of process such as those described above. Included among these are known carbon–carbon bond-forming processes which, when applied to cyclic ketones, achieve ring expansions by forming, either transiently or in a stepwise manner, bicyclic systems that suffer fragmentations to the homologated ketone (Scheme 39). Several of these protocols provide the opportunity to introduce more than one carbon atom at a time.

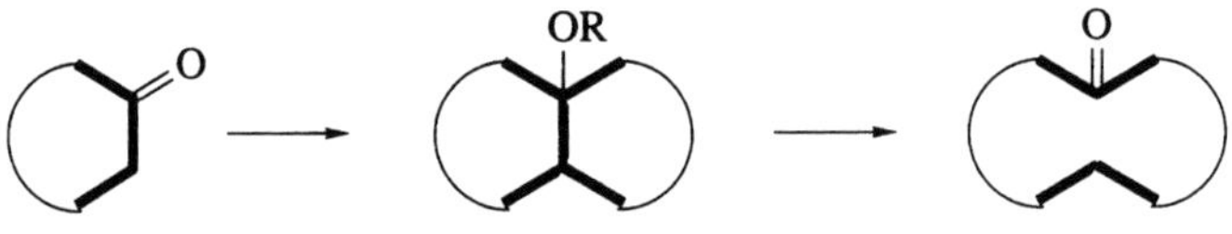

OR O

Scheme 39

3.3.5.1 Rearrangement of Cyclopropanol Derivatives from Ketone Enols

This method is based on the known tendency of dihalocyclopropanol ethers to undergo solvolytic rearrangements.[105] Addition of dihalocarbenes[106] to enol ethers and acetates of cyclic ketones produces the corresponding dihalocyclopropane derivatives (Scheme 40). The cyclopropyl derivative may undergo an ionic ring-opening, either thermally or in the presence of silver salts, to provide either the α-halo-α,β-unsaturated ketone or the corresponding enol ether. Early studies with ethyl enol ethers indicated that, while homologation for the six-membered ring worked well, expansions of seven- and eight-membered ring ketones were not successful.[107] When the ring-opening step was carried out on the corresponding acetate derivatives in the presence of $LiAlH_4$, conditions which may alter the mechanistic aspects of the reaction, the expanded α-halo-α,β-unsaturated ketones could be obtained as the corresponding alcohols in good yields for five- through eight-membered ring ketones.[108]

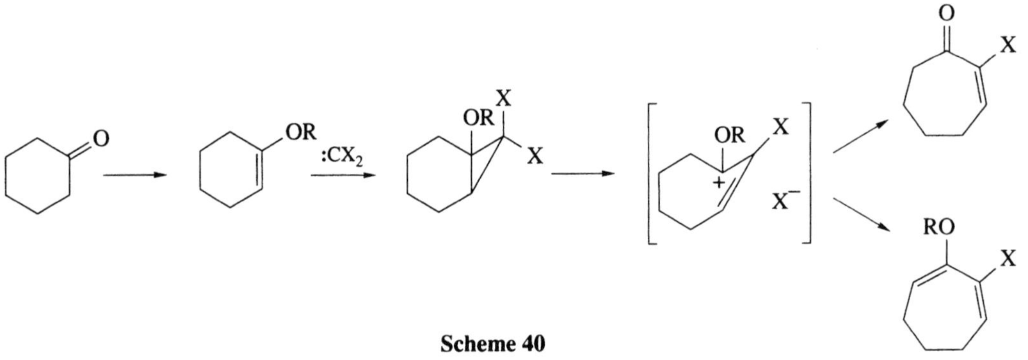

O OR $:CX_2$ OR X X OR X + X^- O X RO X

Scheme 40

Along these lines, cyclic enol esters, after reaction with dichlorocarbene, were found to undergo smooth expansion to the 2-chlorocycloheptenones on room temperature treatment with ethanolic potassium carbonate (Scheme 41).[109]

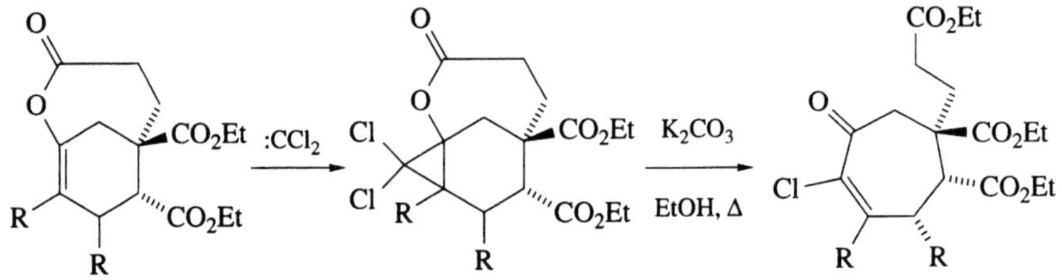

O O CO_2Et $:CCl_2$ Cl Cl O O CO_2Et K_2CO_3 EtOH, Δ CO_2Et O CO_2Et CO_2Et Cl R R R R R CO_2Et R R

R	*Yield* (%)	*Yield* (%)
Me	80	62
$(CH_2)_3$	66	65
$(CH_2)_4$	80	55

Scheme 41

Similarly, the rearrangement of trialkylsilyloxycyclopropanes, obtained from the reaction of the silyl enol ethers of cyclic ketones with halocarbenes, may be carried out under either acidic (alcohol/HCl) or basic (alcohol/Et_3N) conditions. Since the regiochemistry of the expansion is fixed by the position of the starting enol ether, this variation has the further advantage that methods for the convenient regioselective generation of the silyl enols are well established.[110] Examples showing rearrangements from mono- and dihalo-cyclopropyl derivatives are given in Table 9.[111a–e]

Table 9 Ring Expansion of Ketones *via* Halo Carbene Addition to Silyl Enol Ethers

Silyl enol ether	*Carbene adduct*	*Product*	*Yield (%)*	*Ref.*
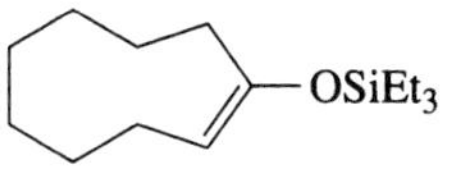 $OSiEt_3$	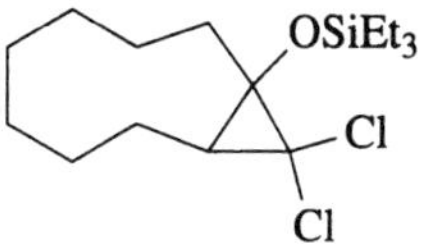 $OSiEt_3$, Cl, Cl	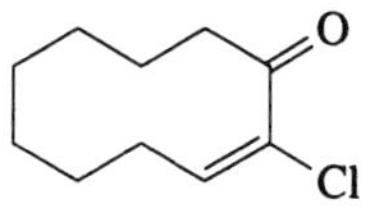 O, Cl	66	111a
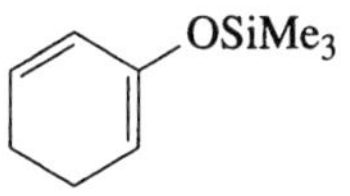 $OSiMe_3$	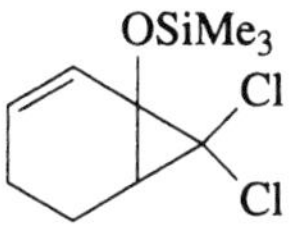 $OSiMe_3$, Cl, Cl	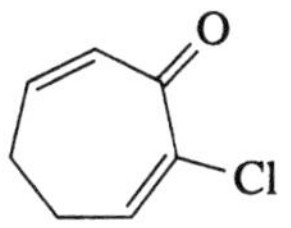 O, Cl	73	111b
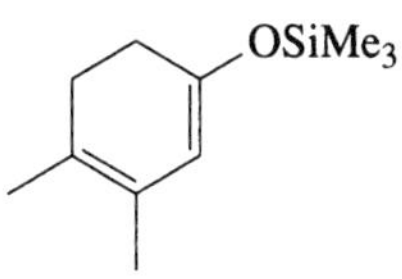 $OSiMe_3$	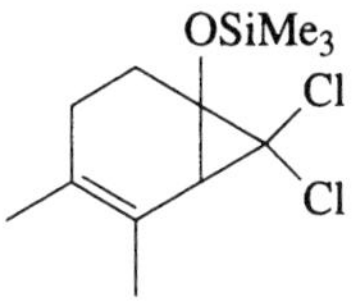 $OSiMe_3$, Cl, Cl	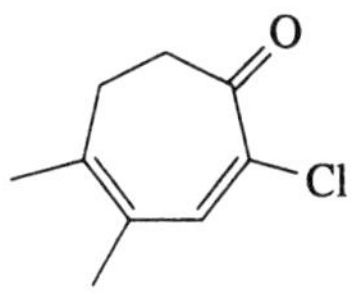 O, Cl	72	111b
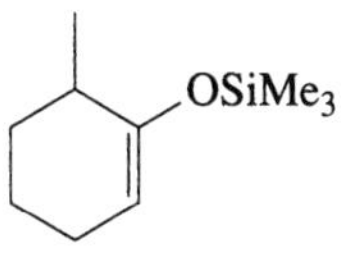 $OSiMe_3$	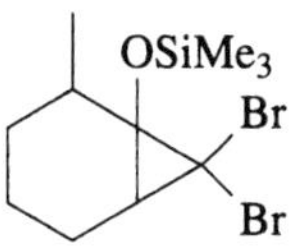 $OSiMe_3$, Br, Br	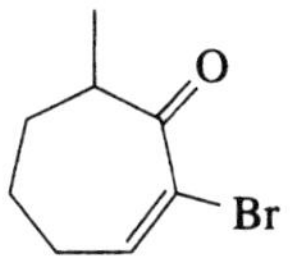 O, Br	40	111c
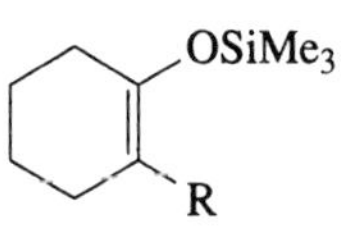 $OSiMe_3$, R	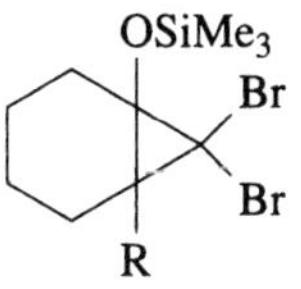 $OSiMe_3$, Br, Br, R	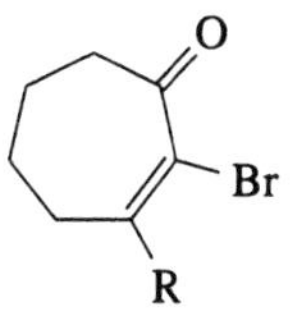 O, Br, R		111c
R = H			90	
R = Me			40	
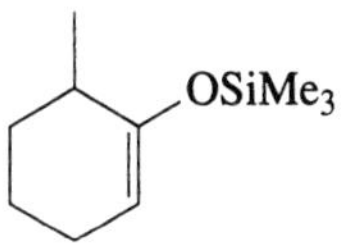 $OSiMe_3$	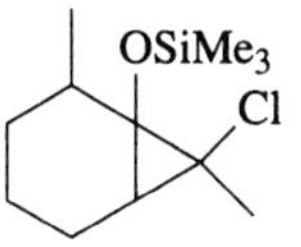 $OSiMe_3$, Cl	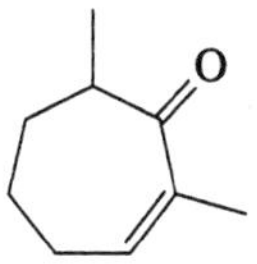 O	56	111d
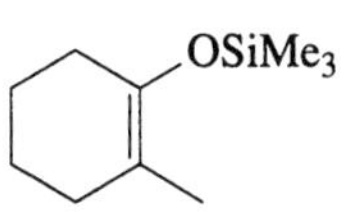 $OSiMe_3$	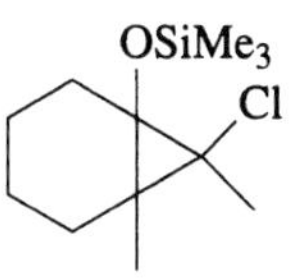 $OSiMe_3$, Cl	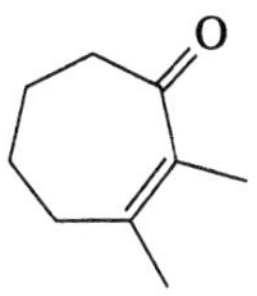 O	56	111d
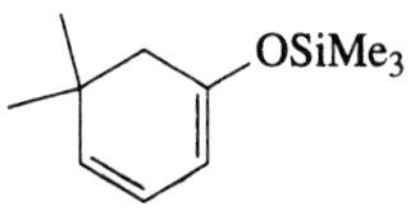 $OSiMe_3$	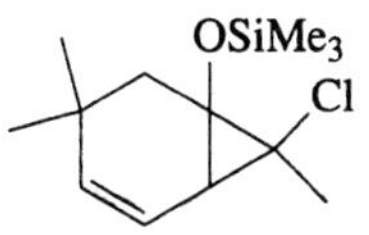 $OSiMe_3$, Cl	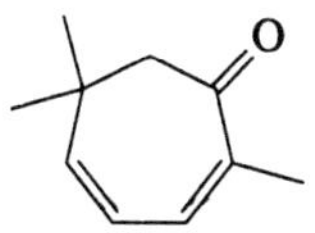 O	52	111e

A similar, but mechanistically different, ring expansion sequence has been described using the iron(III) chloride induced opening of nonhalogenated cyclopropyl derivatives. In this case, the silyl enol ethers are cyclopropanated under Simmons–Smith conditions[112] and then exposed to $FeCl_3$ in dimethylformamide. The ring fission, rationalized in terms of a radical mechanism, produces the β-chloro-expanded ketone which, either spontaneously or with sodium acetate–methanol, eliminates chloride to yield the corresponding expanded enone (Scheme 42).[113] This protocol has been demonstrated for the homologation of five-, six-, seven- and twelve-membered ring ketones. An example which demonstrates a regioselective expansion by taking advantage of the *in situ* trapping of a regiospecifically generated enolate is shown in Scheme 43.[114]

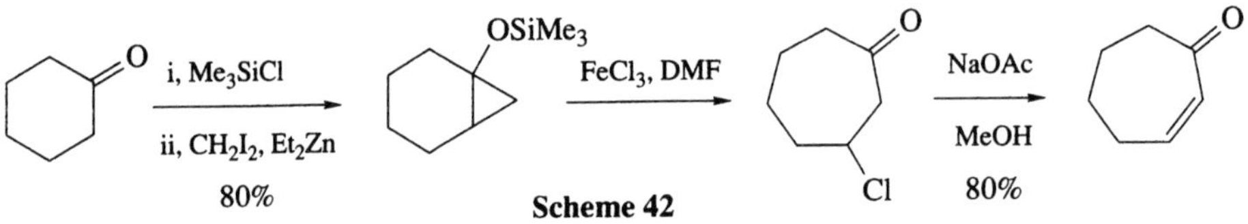

i, Me_3SiCl; ii, CH_2I_2, Et_2Zn 80% → $OSiMe_3$ → $FeCl_3$, DMF → Cl → NaOAc, MeOH 80%

Scheme 42

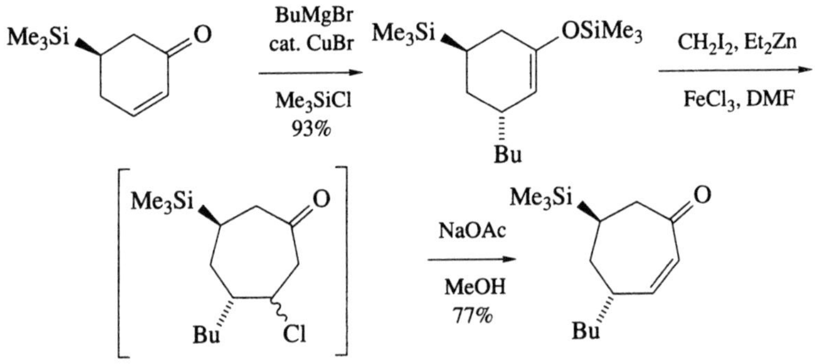

Me_3Si → BuMgBr, cat. CuBr, Me_3SiCl 93% → Me_3Si, $OSiMe_3$, Bu → CH_2I_2, Et_2Zn; $FeCl_3$, DMF → [Me_3Si, O, Bu, Cl] → NaOAc, MeOH 77% → Me_3Si, O, Bu

Scheme 43

3.3.5.2 Homologations *via* 3,3-Rearrangements

The 3,3-rearrangement of allylvinylcarbinols and their derivatives, also referred to as the oxy-Cope rearrangement, is a well-studied bond reorganization process that has found substantial use in synthesis. Since this topic has already been reviewed in depth in this series with regard to mechanistic detail and general synthetic application, the coverage here will be limited to those particular instances where simple ring homologations have been achieved.[115] The overall two-step process, illustrated in Scheme 44, first involves the addition of a vinyl organometallic to a ketone bearing a vinyl substituent. The thermal 3,3-rearrangement step is carried out on the carbinol (R = H), the ether derivatives (R = alkyl or Me_3Si) or, with a substantial rate increase,[116] the alkoxide (R = Na, K) to produce the enol derivative, which is subsequently transformed to the ketone. The net result of this operation is the insertion of a four-carbon unit, which for cyclic ketones (R^1 and R^2 are joined) translates to a four-carbon ring expansion.

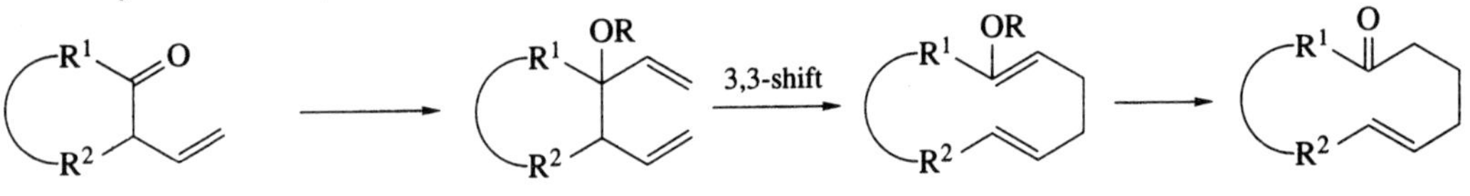

R^1, R^2, O → OR, R^1, R^2 → 3,3-shift → OR, R^1, R^2 → O, R^1, R^2

Scheme 44

The placement of the substituents on the ring establishes the regiochemistry of the expansion process, while the stereochemistry of the vinylcarbinol adduct influences the double bond geometry in the products and is illustrated in the following example. Addition of a vinyl organometallic to 2-vinylcyclohexanone produces two vinylcarbinols (Scheme 45).[117] The major isomer on thermolysis (220 °C, 3 h) produces the 10-membered ring ketone as a single isomer in 90% yield, while, on the other hand, the minor carbinol isomer produces a mixture of *cis*- and *trans*-enones. These results have been interpreted in terms of the preferred conformations of the the carbinols during rearrangement assuming a chair-like transition state.

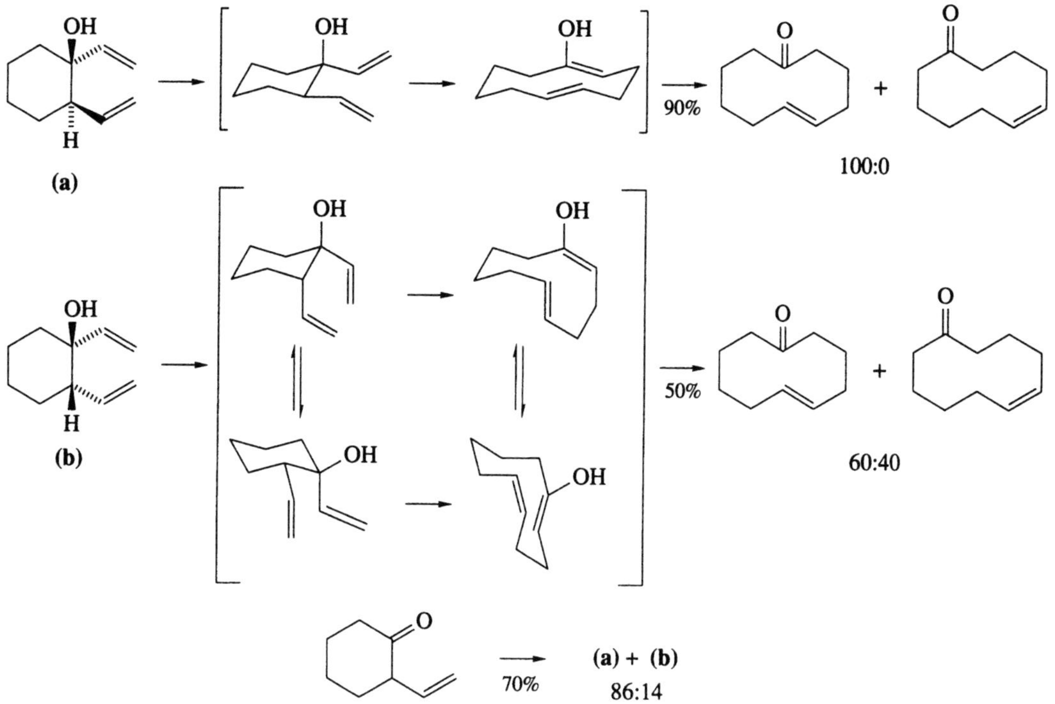

(a)

90%

100:0

(b)

50%

60:40

70%

(a) + (b)
86:14

Scheme 45

The ring size also influences the course of the reaction. For example, both the *cis*- and *trans*-divinylcyclopentanols lead to a single (*E*)-nonenone, while for the divinyldodecanols the *trans* isomer produces only the (*E*)-enone, whereas the *cis* isomer produces mostly the (*Z*)-enone (Scheme 46).[118] Relief of ring strain can provide a considerable driving force for the rearrangement, such as in the example shown in Scheme 47, where rearrangement of the carbinol occurs at only 50 °C.[119]

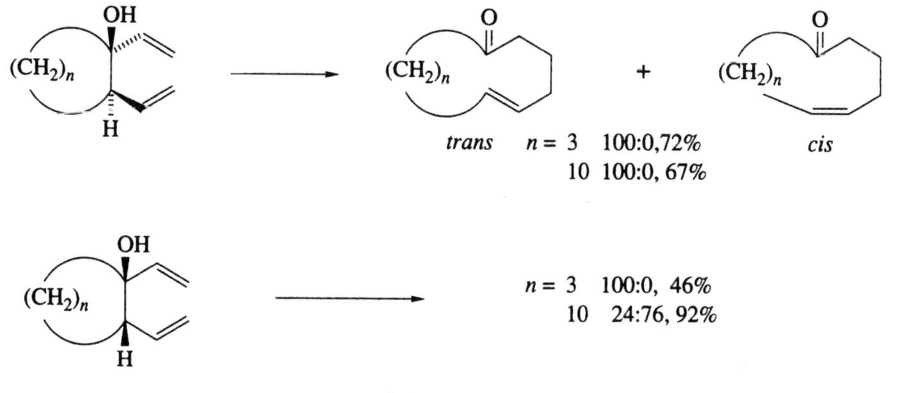

trans $n = 3$ 100:0,72% *cis*
10 100:0, 67%

$n = 3$ 100:0, 46%
10 24:76, 92%

Scheme 46

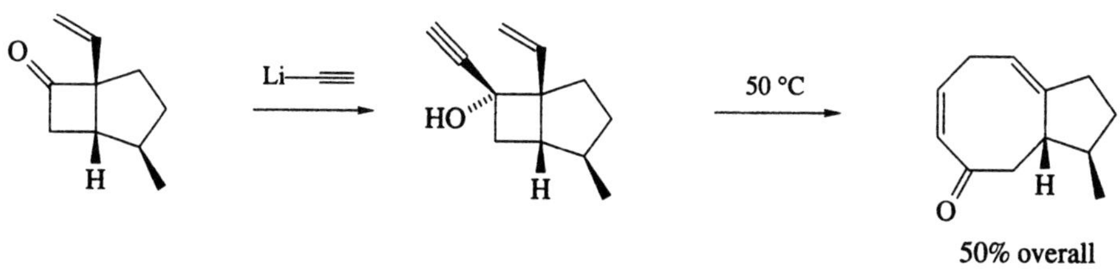

50 °C

50% overall

Scheme 47

This basic strategy has found application in the synthesis of germacranes (Schemes 48,[120] 49[121] and 50[122]), where the initial addition of the vinyl organometallic to the ketone proceeded with a high level of stereoselectivity, which in turn led to the selective 3,3-rearrangement of the potassium alkoxides. The sequence depicted in Scheme 50 demonstrates the trapping of the regiospecifically generated enolate as a trimethylsilyl ether, and its subsequent reaction to form an α-hydroxy ketone as a single isomer. The regiospecific generation of the enol has also allowed for the regiospecific repetitive ring expansion process shown in Scheme 51.[123]

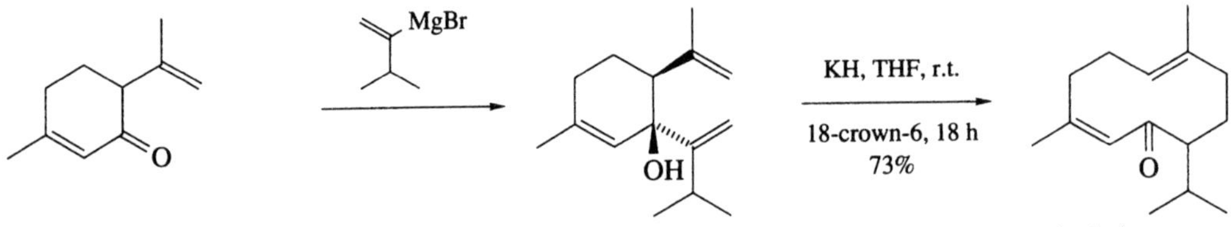
MgBr
KH, THF, r.t.
18-crown-6, 18 h
73%
O
OH
O
single isomer

Scheme 48

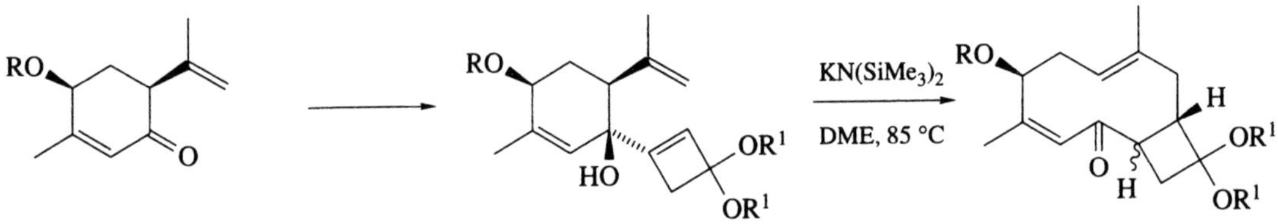
RO
O
RO
HO
OR1
OR1
KN(SiMe3)2
DME, 85 °C
RO
H
OR1
O
H
OR1

Scheme 49

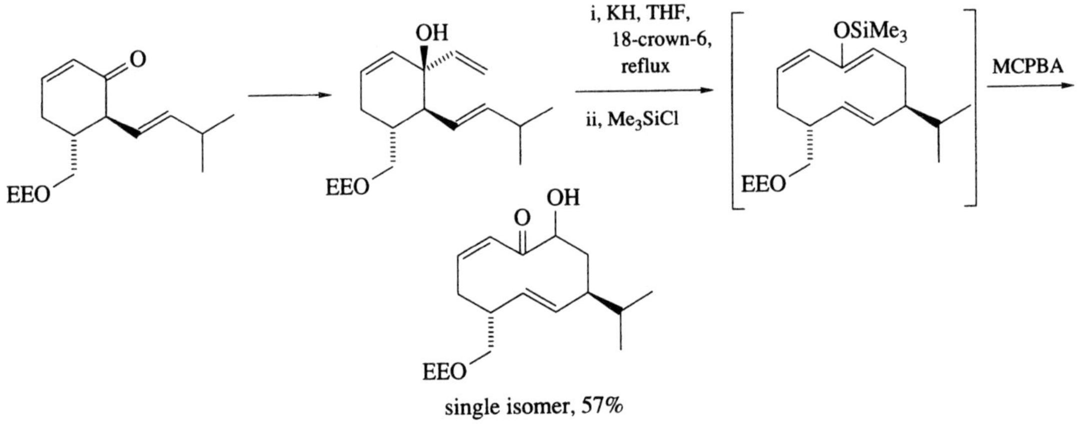
O
EEO
OH
EEO
i, KH, THF,
18-crown-6,
reflux
ii, Me3SiCl
OSiMe3
EEO
MCPBA
O
OH
EEO
single isomer, 57%

Scheme 50

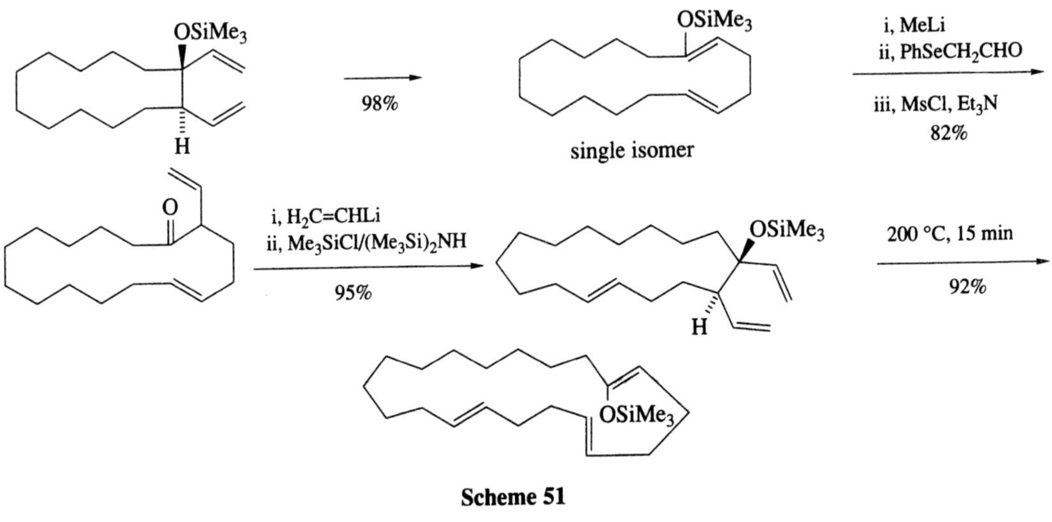
OSiMe3
H
98%
OSiMe3
single isomer
i, MeLi
ii, PhSeCH2CHO
iii, MsCl, Et3N
82%
O
i, H2C=CHLi
ii, Me3SiCl/(Me3Si)2NH
95%
OSiMe3
H
200 °C, 15 min
92%
OSiMe3

Scheme 51

A highly efficient route to the ophiobolin nucleus is a noteworthy application of the *in situ* enolate-trapping strategy (Scheme 52).[124] The 3,3-rearrangement of the intermediate divinyl alkoxide, produced from the addition of a cyclopentenyl anion to the bicyclic ketone, proceeds at low temperature *via* a boat-like transition state to generate the enolate. Subsequent trapping with methyl iodide provides the expanded, alkylated ketones in good overall yield. Transannular alkylations are feasible when an alkylating moiety resides on the reacting molecule, as illustrated in Scheme 53.[125] A stereochemical dependence for the transannular alkylation step has been observed in an unusual use of allylic ethers as alkylating agents. While the stereospecific 3,3-rearrangement proceeds for both the (*E*)- and (*Z*)-isomers shown in Scheme 54,[126] only the intermediate from the (*Z*)-enol ether undergoes the internal alkylation.

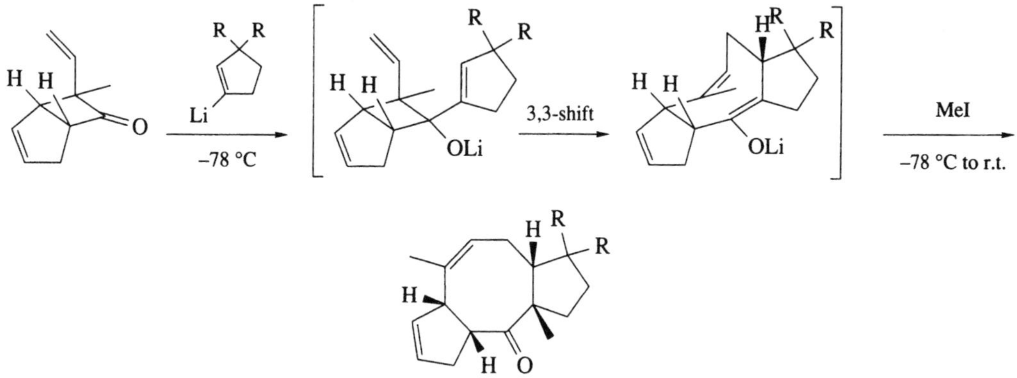

R = H, 96%; R = SCH_2CH_2S, 71%

Scheme 52

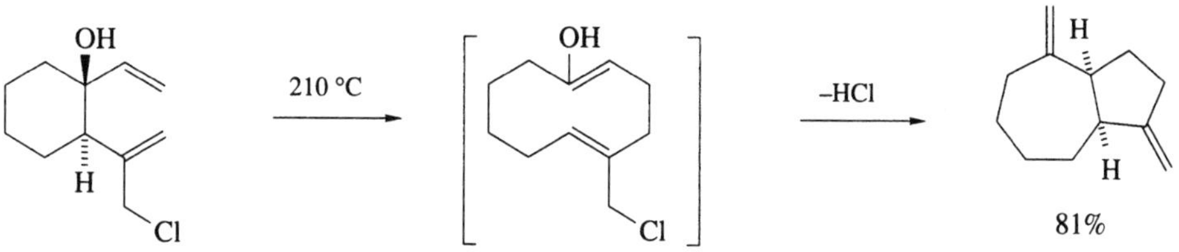

Scheme 53

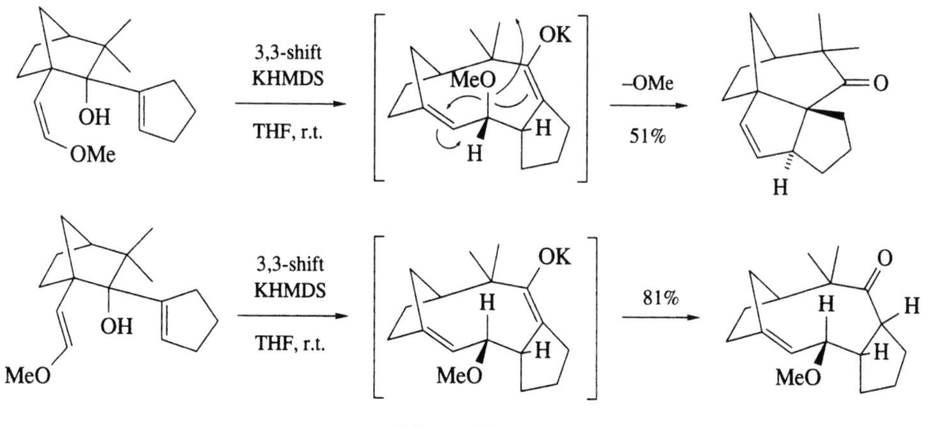

Scheme 54

Other variations of the 3,3-rearrangement have been reported where the vinyl unit is incorporated into a ring. Addition of vinyllithium to a cyclohexanespirocyclobutanone produces a mixture of two stereoisomeric alkoxides, one of which on warming to room temperature undergoes a smooth anionic oxy-Cope rearrangement to the octahydrobenzocyclooctene, while the other suffers ring fission (Scheme 55).[127] Interestingly, if the carbinols are isolated first by a low temperature quench and then transformed to the potassium alkoxides, the isomer with the *trans*-disposed vinyl units undergoes rearrangement to the hexahydronaphthalene ketone. This alternate reaction course, rationalized in terms of either a fragmentation–recombination process or a 1,3-rearrangement, has been utilized in other systems as a ring expansion process (see Section 3.3.5.3). The formation of bicyclo[5.3.1]undecenes has been made feasible *via* the anionic oxy-Cope rearrangement arising from either spirocyclobutanone systems or bicyclo[2.2.2]octane systems (Table 10). Significantly, a base-induced stereoisomerization has been suggested to explain the fact that, in the case of the spirocyclobutanones, either stereoisomer of the vinylcarbinol undergoes the rearrangement to the same product.[127]

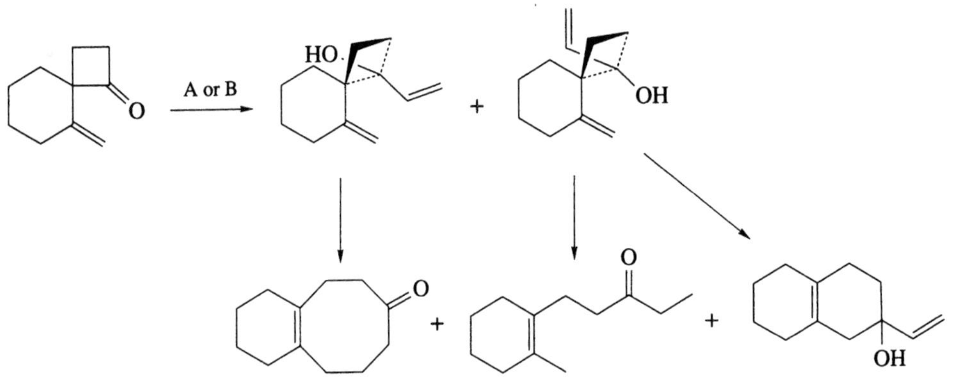

(A) i, H_2C=CHLi, −78 °C to r.t. 44:56:0 78%
(B) i, H_2C=CHLi, ii, quench, iii, KH, THF 44:0:56 43%

Scheme 55

Table 10 Ring Expansion of Ketones *via* 1,3-Rearrangements

Ketone adduct	*Rearrangement product*		*Yield (%)*	*Ref.*
		R = H	80	128
		R = Me	72	127
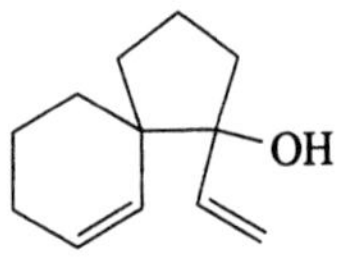	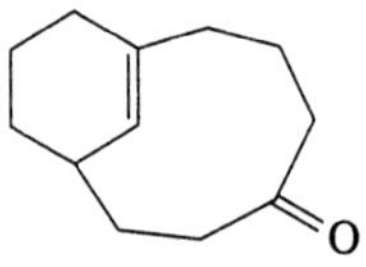		83	129
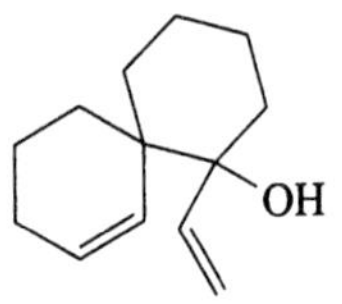	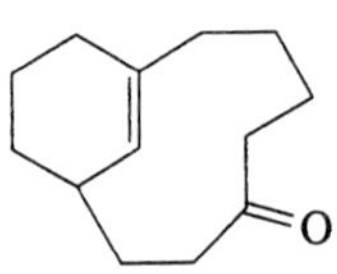		79	129
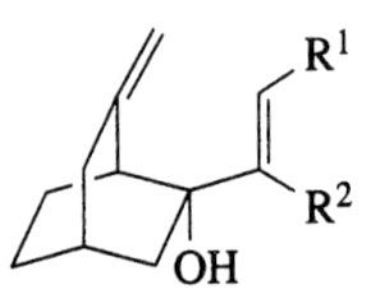 $R^1 = R^2$ = H; R^1 = H, R^2 = Me; $R^1 + R^2 = -(CH_2)_4-$	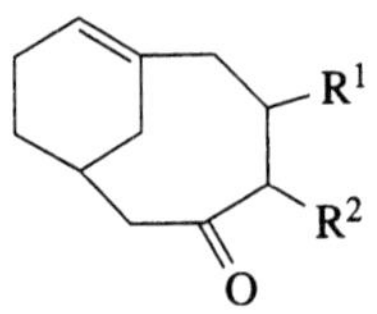		70–85	130

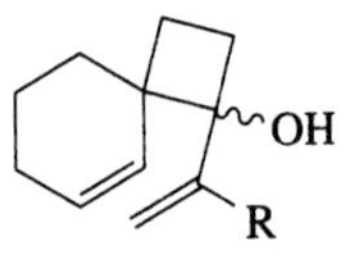

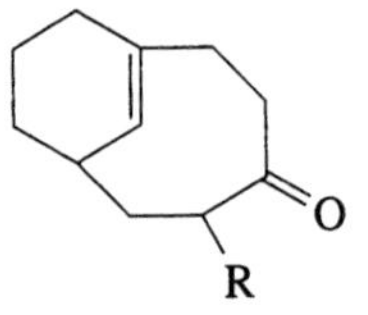

Table 10 *(continued)*

Ketone adduct	*Rearrangement product*		*Yield (%)*	*Ref.*
HO	H, H, O	R = H	80	128
		R = Me	72	127
			90	131
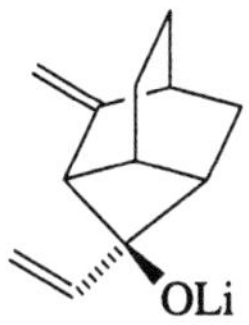 OLi	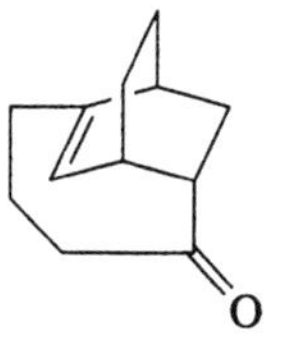 O		62	132

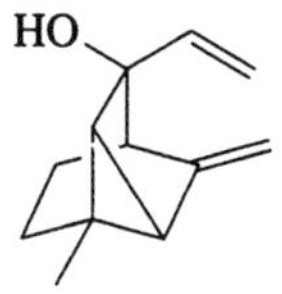

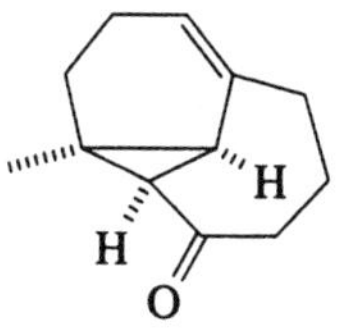

Even larger ring expansions are possible when bis(butadienyl) systems are employed. As exemplified in Schemes 56,[133a] 57[133b] and 58[133c] the process involves the addition of a butadienyl organometallic to a ketone bearing a butadienyl side chain. The 5,5-bond reorganization process of the resulting carbinol, which may in fact be the outcome of two consecutive 3,3-rearrangements, is conducted thermally as for the 3,3-rearrangements above.[133a–c]

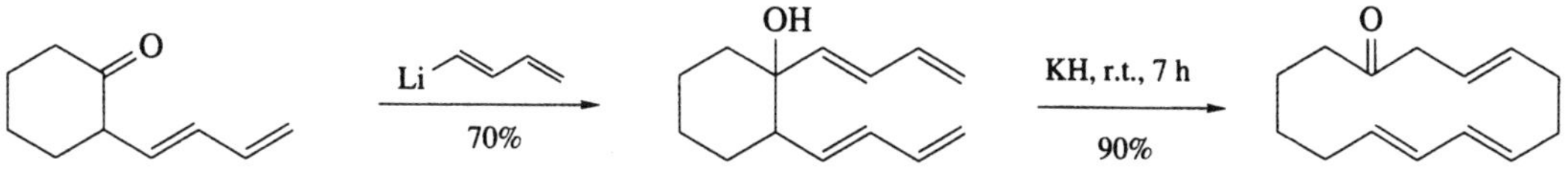

O; Li; 70%; OH; KH, r.t., 7 h; 90%; O

Scheme 56

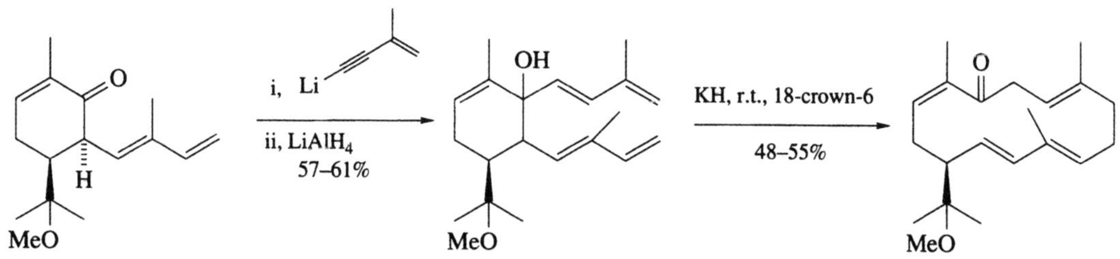

O; H; MeO; i, Li; ii, $LiAlH_4$; 57–61%; OH; MeO; KH, r.t., 18-crown-6; 48–55%; O; MeO

Scheme 57

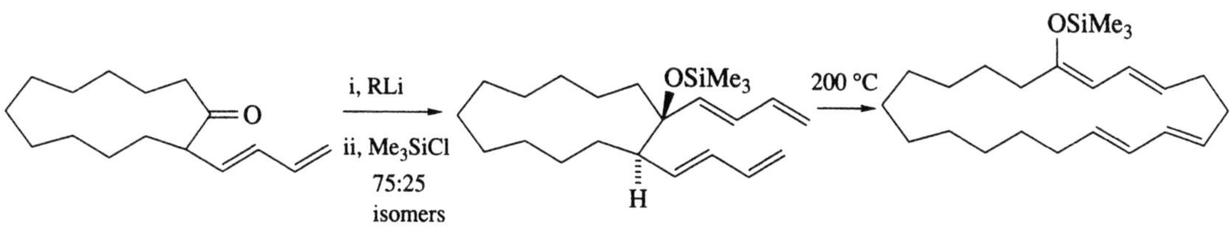

O; i, RLi; ii, Me_3SiCl; 75:25 isomers; $OSiMe_3$; H; major isomer; 200 °C; $OSiMe_3$; 95% single isomer

Scheme 58

3.3.5.3 Homologations *via* 1,3-Rearrangements

During the course of investigations on the 3,3-rearrangement of allylvinylcarbinols, some cases were found where 1,3-reaction pathways predominate (Scheme 59).[134] This alternative reaction sequence has been used to advantage for the two-carbon ring expansion in a number of ring systems. As in the 3,3-rearrangements, a substantial rate acceleration has been observed in the 1,3-rearrangements for the corresponding alkoxides, although this modification may well alter the reaction mechanism. Shown in Scheme 60 are examples of systems where the ring expansion *via* 1,3-rearrangement predominates in a synthetically useful manner over the alternative nonexpanding 3,3-rearrangement.[135]

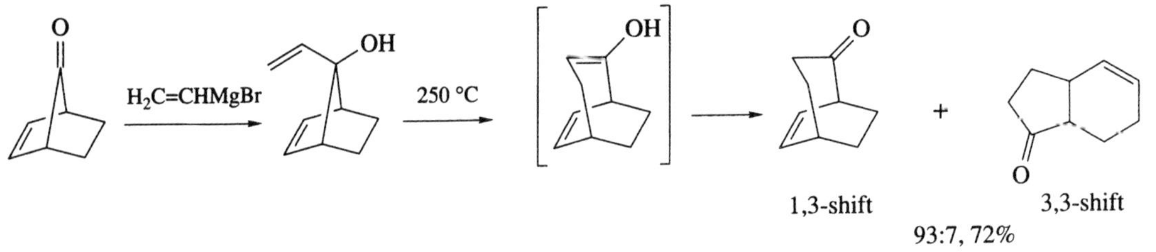

$H_2C{=}CHMgBr$; 250 °C

1,3-shift : 3,3-shift = 93:7, 72%

Scheme 59

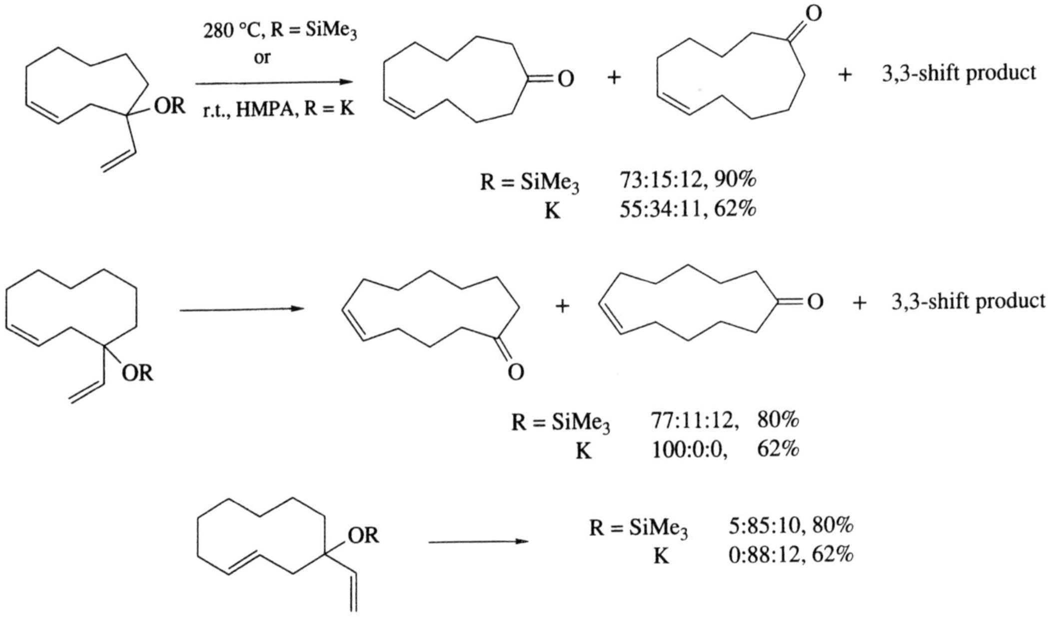

280 °C, R = $SiMe_3$ or r.t., HMPA, R = K → + 3,3-shift product

R = $SiMe_3$ 73:15:12, 90%
K 55:34:11, 62%

→ + 3,3-shift product

R = $SiMe_3$ 77:11:12, 80%
K 100:0:0, 62%

R = $SiMe_3$ 5:85:10, 80%
K 0:88:12, 62%

Scheme 60

This process for large rings is somewhat substrate sensitive. For example, the corresponding saturated ring systems do not undergo the rearrangement; however, the presence of a phenyl group on the carbon adjacent to the carbinol restores the facility of the 1,3-rearrangement.[136] In similar fashion, incorporation of an aromatic moiety into the ring system, also adjacent to the original ketone, allows for the execution of a two-carbon ring expansion (Schemes 61 and 62).[42,43] In the 13- to 15-membered ring ketone expansion, the competition between 1,3- and 3,3-rearrangement is dependent on both the rearrangement conditions and substitution patterns on the vinyl moiety (Scheme 63).[137]

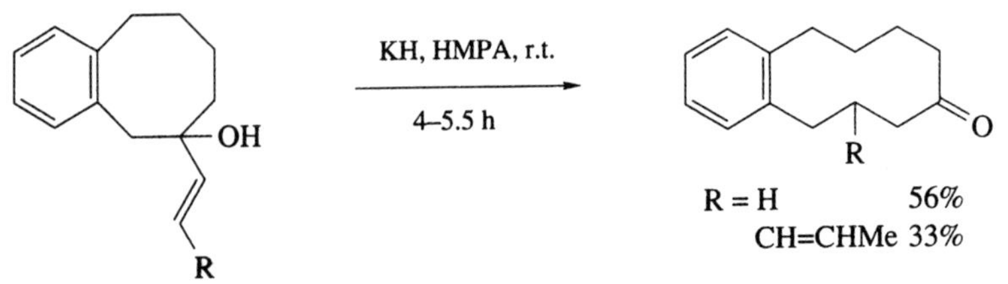

KH, HMPA, r.t., 4–5.5 h

R = H 56%
CH=CHMe 33%

Scheme 61

KH, HMPA, r.t.
7–8 h

$R^1 = R^2 = H$ 70%
$R^1 = H, R^2 = OMe$ 90%
$R^1 = OMe, R^2 = H$ 54%

Scheme 62

1,3-shift
+ 3,3-shift product

R^1	R^2	R^3	*via TMS ether*	*via KH, HMPA*
H	H	H	46:54, 79%	13:87, 70%
Me	H	H	15:85, 47%	18:82, 62%
H	Me	H	98:2 , 47%	55:45, 39%
H	H	Me	98:2, 55%	46:54, 62%

Scheme 63

The relief of ring strain in cyclobutanols also provides the driving force for alkoxide-induced 1,3-rearrangements (Scheme 64).[138] The reduction of α-vinyl-substituted cyclobutanones with borohydride reagents followed by methyllithium generates the intermediate alkoxides which, on warming, undergo the 1,3-rearrangement to the cyclohexanol derivatives. Examples are provided in Table 11. A reversal of the stereochemical outcome of the 1,3-rearrangement has been achieved by altering the reaction conditions. Further investigation of the intermediate isomeric alcohols has revealed that, for the example shown in Scheme 65, the *cis*-alcohol isomerizes to the *trans*-alcohol, which then undergoes the 1,3-rearrangement.[139] Moreover, this isomerization is accelerated by the addition of 18-crown-6, a modification which also changes the product distribution. This effect has been rationalized in terms of a fragmentation-recombination process where the geometry of the recombination step is changed according to whether the potassium ion is intramolecularly coordinated by the carbonyl oxygen or externally coordinated by the crown ether.

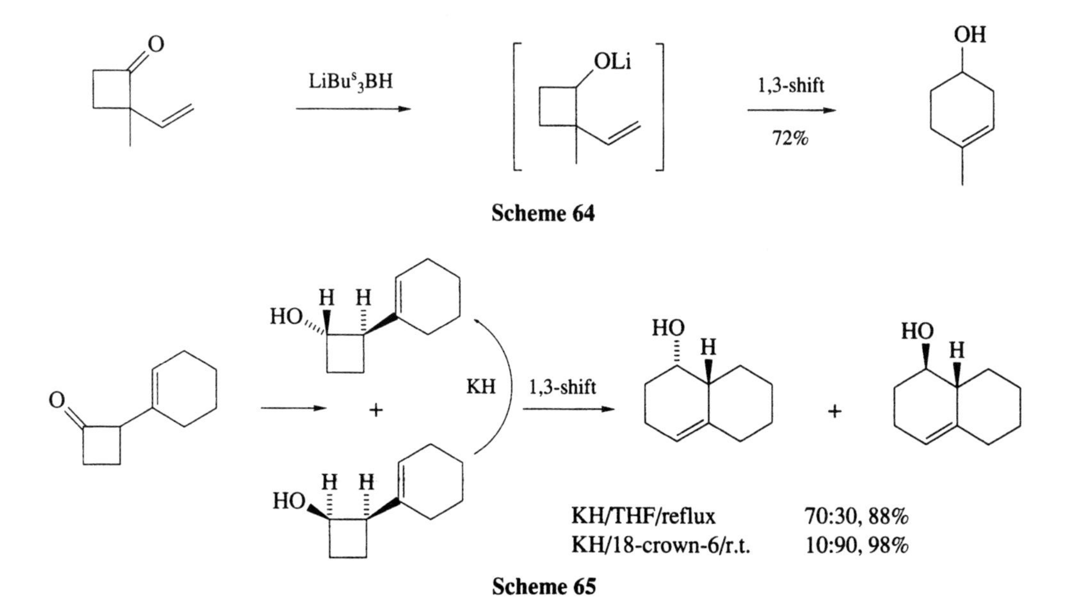

Scheme 64

Scheme 65

Table 11 Ring Expansion of Cyclobutanones *via* 1,3-Rearrangements

Ketone	*Products*	*Yield* (%)	*Ref.*
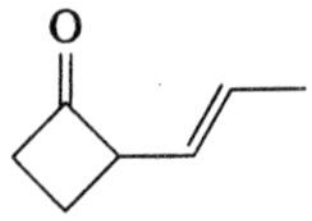	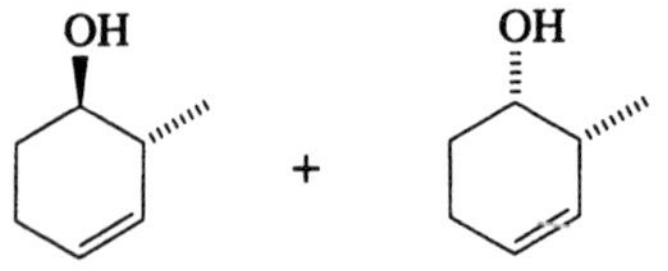 84:16	65	138
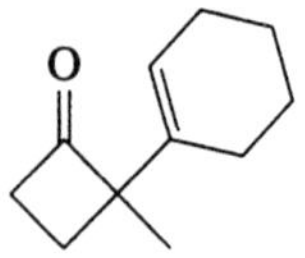	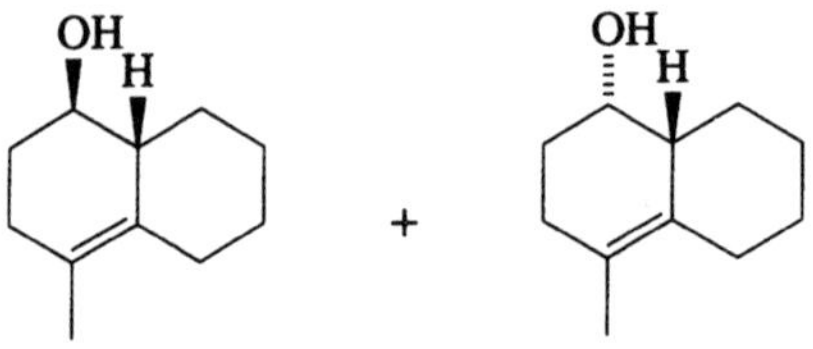 94:6	75	138
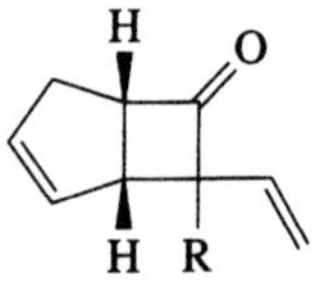	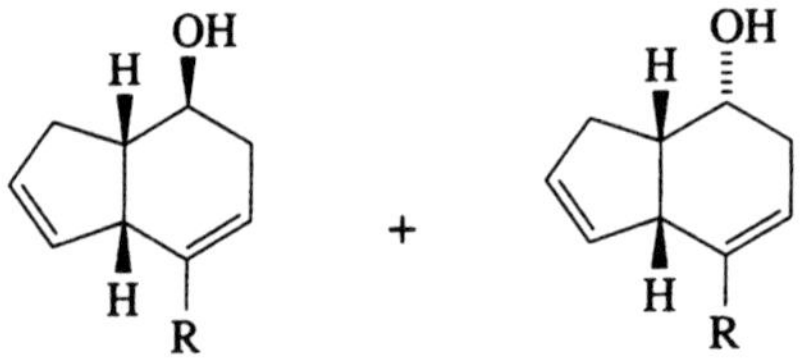		138
R = H	17:83	65	
R = Me	83:17	70	
R = Bu	90:10	89	
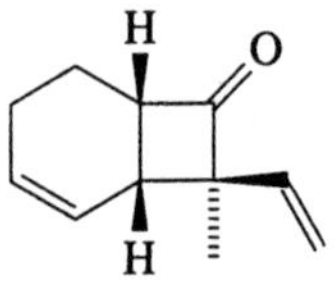	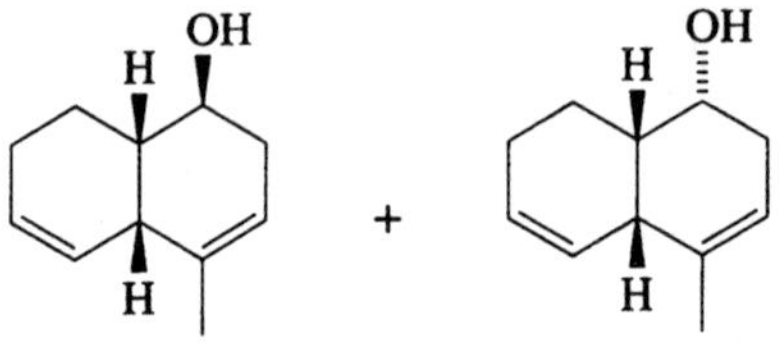 97:3		138
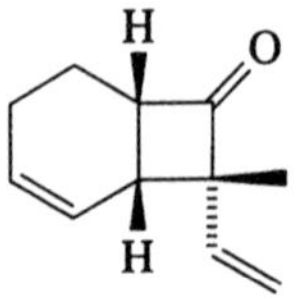	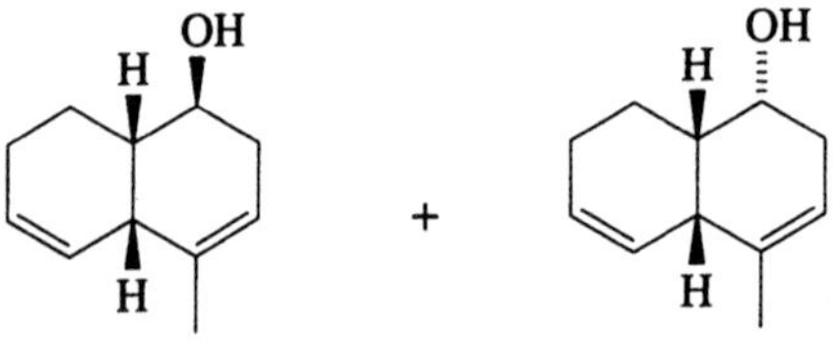 22:78		138
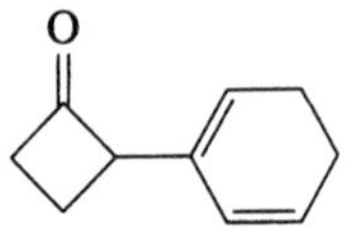	 *trans* [a] 92:8; *cis* [a] 72:28		139

[a] Intermediate alcohols were isolated and treated separately with KH/THF.

3.3.5.4 Cationic Variations of the 3,3-Rearrangement

Ring expansions of appropriately α-substituted ketones *via* their vinyl carbinol derivatives has also been carried out under cationic conditions. The basic sequence, outlined in Scheme 66, begins with the addition of a vinyl organometallic to an α-substituted ketone, where X is carbon or a heteroatom. Departure of a leaving group Y then produces a cationic species, which may react further to form products. Two main pathways appear to be operative: one is a cationic alkene cyclization, followed by a pinacol-like rearrangement; while the other is a 3,3 or 3,3-like rearrangement, followed by an intramolecular alkylation of the intermediate enol.

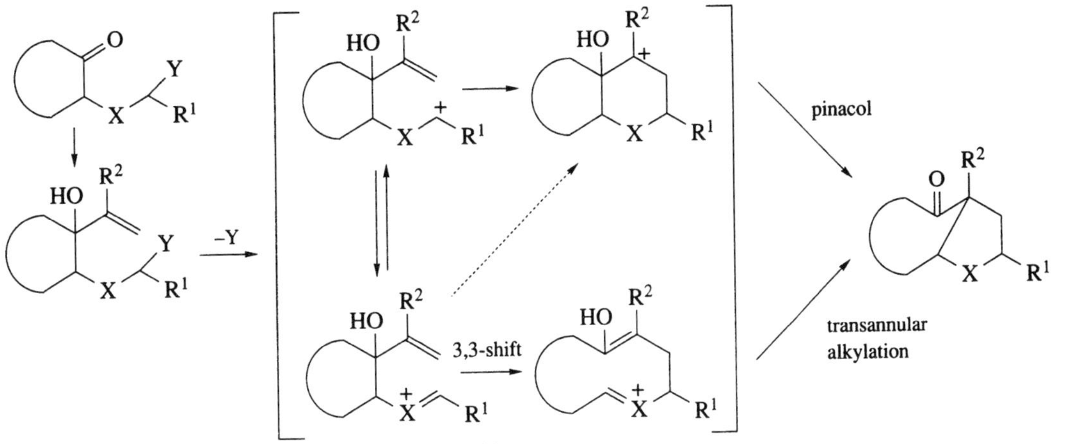

Scheme 66

When X is nitrogen, the overall process achieves ring expansion with a pyrrolidine annulation. Representative of this protocol is the addition of aryl-substituted vinyllithium reagents to the α-amino-substituted ketones, as shown in Scheme 67.[140] The vinylcarbinols, produced as *cis/trans* mixtures, are reduced to the secondary amines with sodium cyanoborohydride. On exposure of either isomer to paraformaldehyde and acid in dimethyl sulfoxide, intermediate iminium ions are formed that rearrange preferentially to the *cis*-fused bicyclic systems. Alternatively, the iminium ions may be generated through the acid or silver ion induced loss of cyanide from the related *N*-cyanomethyl derivatives. The process, shown in Scheme 68, involves the addition of a vinyllithium reagent in a *trans* fashion to the α-amino-substituted ketone.[141] Treatment with camphorsulfonic acid in refluxing benzene or silver nitrate at room temperature yields the *cis*-fused ring-expanded product. Thc procedure appears to be general for four- to five-, five- to six- and six- to seven-membered ring expansion–annulations, and has found utility in alkaloid synthesis, examples of which are depicted in Schemes 69[142] and 70.[143]

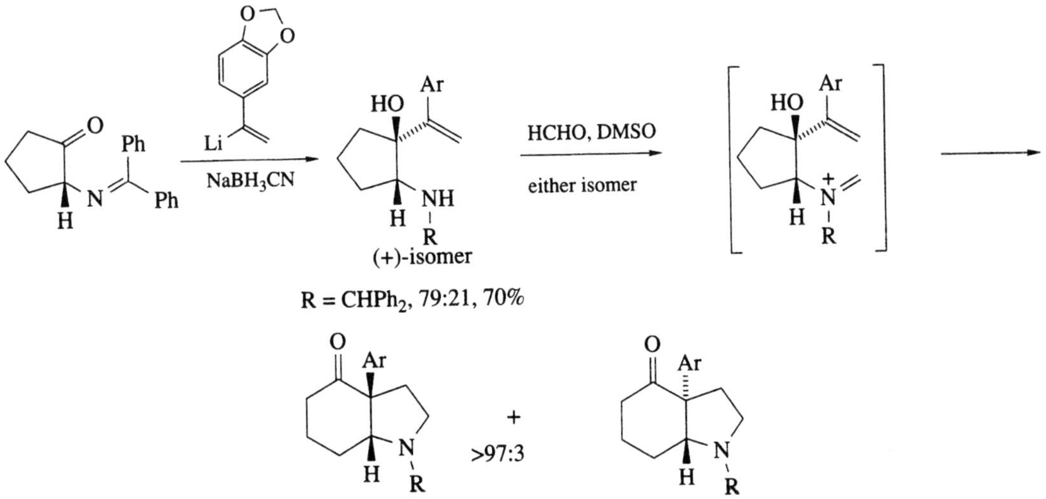

Scheme 67

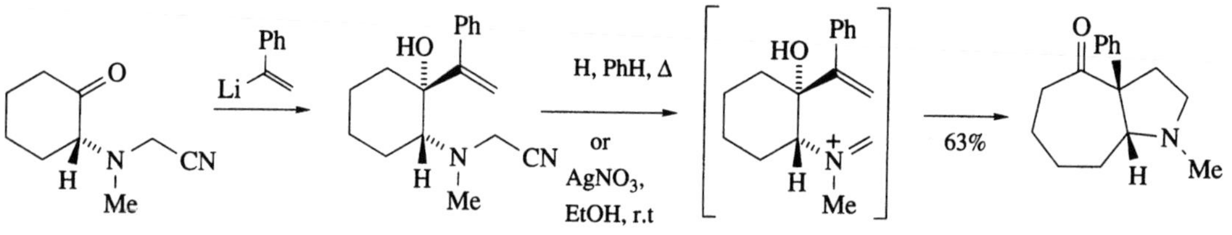

Scheme 68

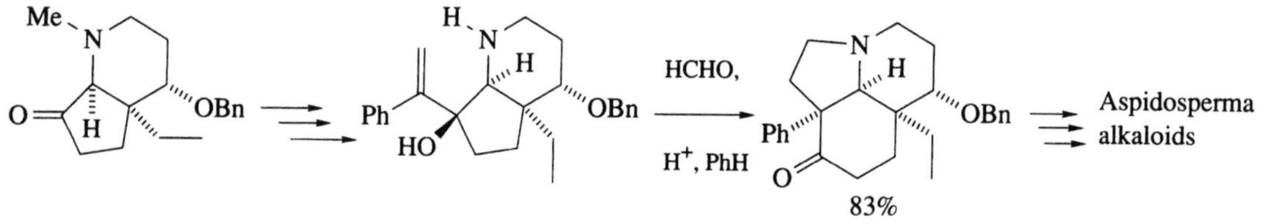

Scheme 69

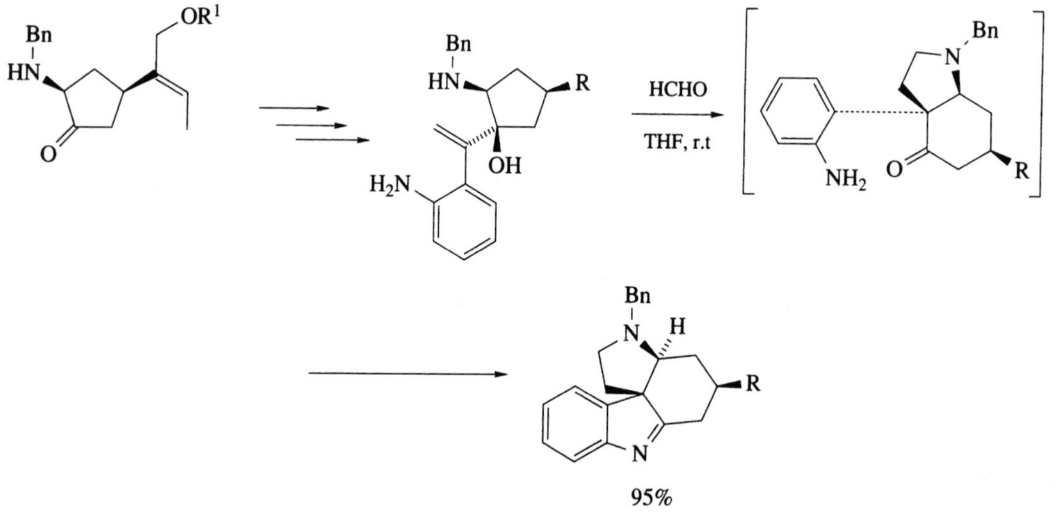

Scheme 70

The mechanistic interpretation of these results has favored the 3,3-like rearrangement–transannular alkylation pathway. Detailed conformational analyses on this basis have been advanced for the stereochemical outcome of the reaction.[140–144] Evidence in favor of the 3,3-rearrangement comes in part from the example in Scheme 71.[144] If a cationic alkene cyclization–pinacol-like rearrangement mechanism were operative and the substrate and product did not racemize under the reaction conditions, the optically active amino alcohol derivative would be expected to produce an optically active product. That a racemic product is formed suggests that the intermediate enol–iminium ion, which if formed in a chiral conformation by a 3,3-rearrangement, undergoes racemization by rotation around the single bond prior to ring closure. Supporting evidence comes from the example presented in Scheme 72,[145] where the rearrangement of optically active starting material produced optically pure product. The preservation of chirality is attributed to the restricted bond rotation imposed by the cyclic structure of the asymmetric enol conformation prior to the alkylation step.

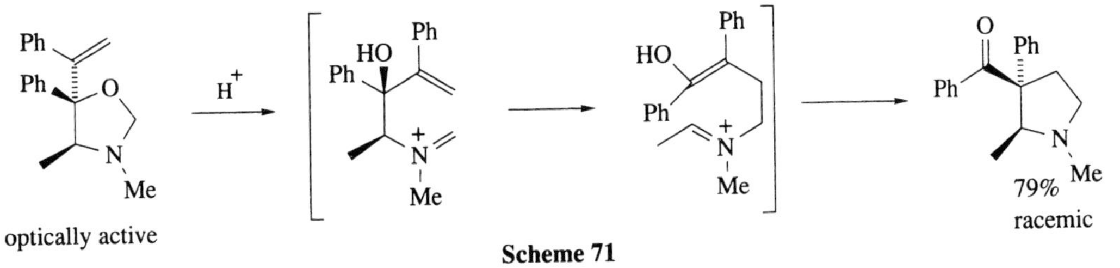

optically active

79% racemic

Scheme 71

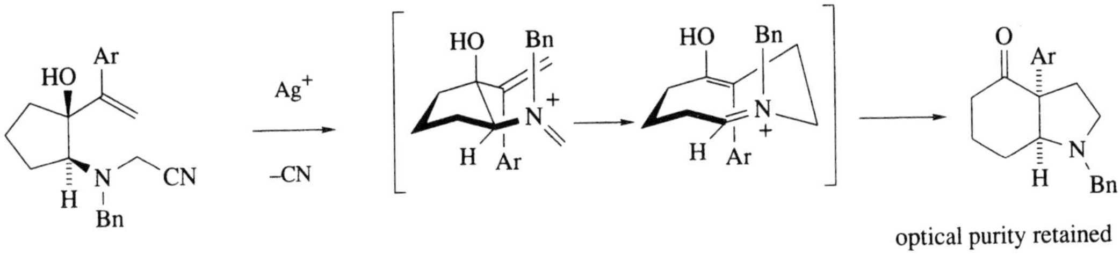

optical purity retained

Scheme 72

In contrast, when the X group is oxygen, the reaction appears to follow the cationic alkene cylization–pinacol-like rearrangement pathway. In this case, reaction of an optically pure acyclic oxygen-containing analog of the system in Scheme 71 leads to a product with preservation of optical activity.[146] This reaction protocol, which accomplishes an overall ring expansion with a tetrahydrofuran annulation, has been examined for α-hydroxy-cyclopentanones and -cyclohexanones (Scheme 73 and equations 29–31).[147]

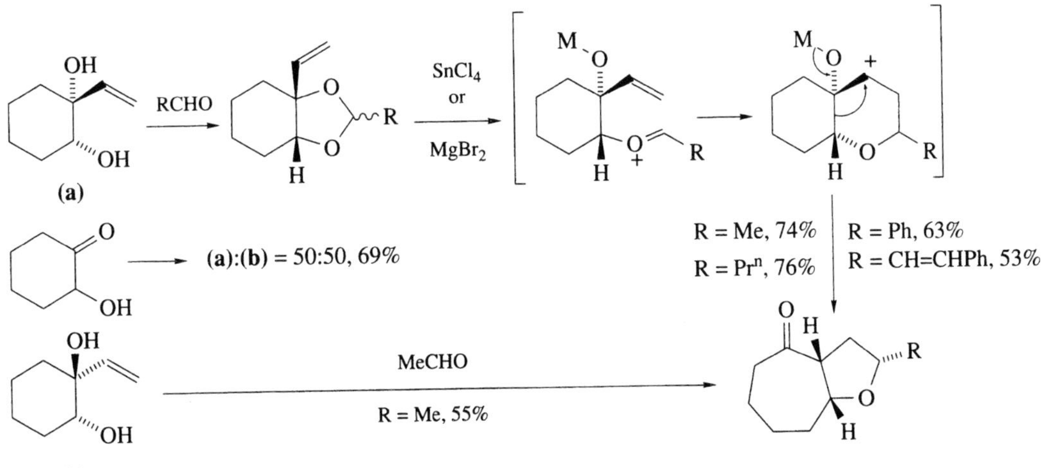

(a):(b) = 50:50, 69%

R = Me, 74%; R = Pr^n, 76%; R = Ph, 63%; R = CH=CHPh, 53%

MeCHO, R = Me, 55%

Scheme 73

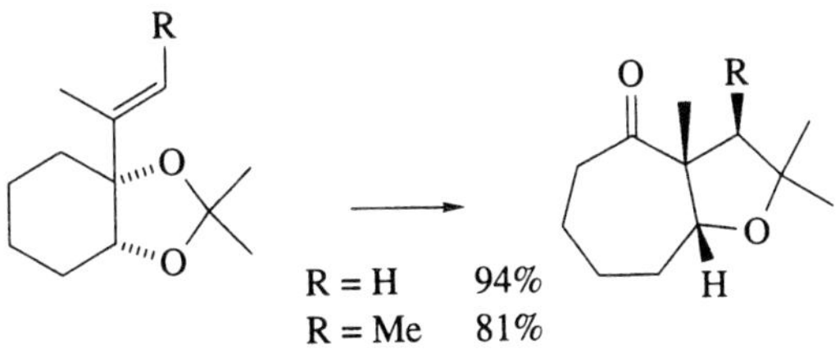

R = H 94%
R = Me 81%

(29)

80%

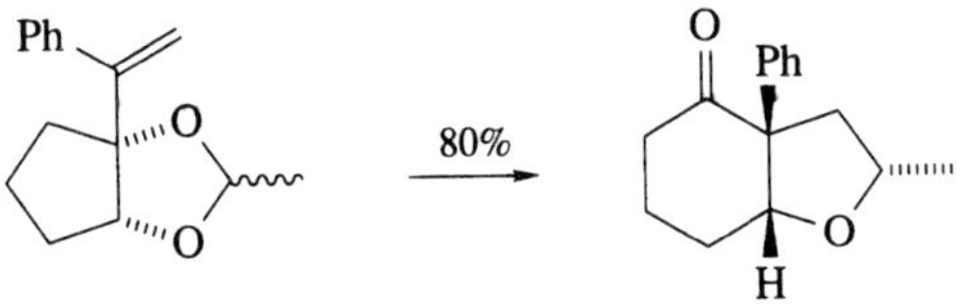

(30)

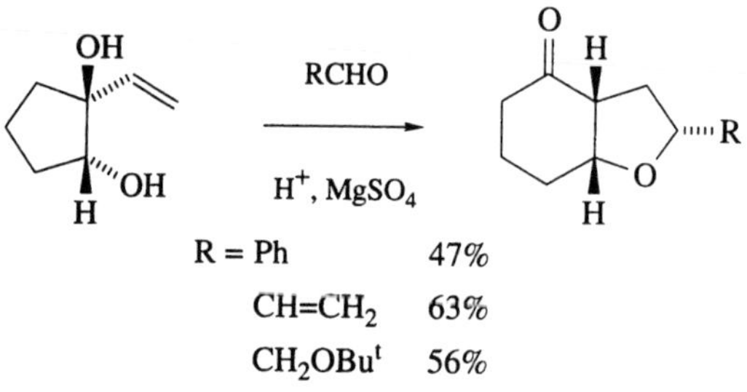

OH, OH, H — RCHO, H^+, $MgSO_4$ → O, H, R, O, H (31)

R = Ph 47%
CH=CH$_2$ 63%
CH$_2$OBu^t 56%

The process has also been extended to the X = C series. The ring expansion–cyclopentane annulation provides good yields for the five- to six- (82%), six- to seven- (90%) and seven- to eight-membered ring (75%) ketone series studied (Scheme 74 and equations 32 and 33).[148,149]

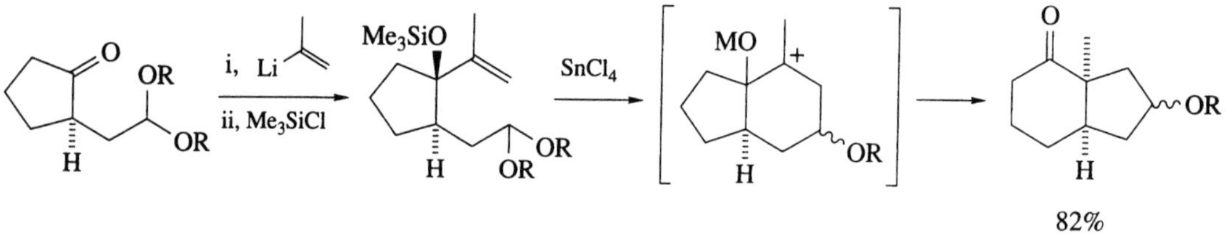

O, OR, OR, H — i, Li; ii, Me_3SiCl → Me_3SiO, OR, OR, H — $SnCl_4$ → [MO, +, OR, H] → O, OR, H

82%

Scheme 74

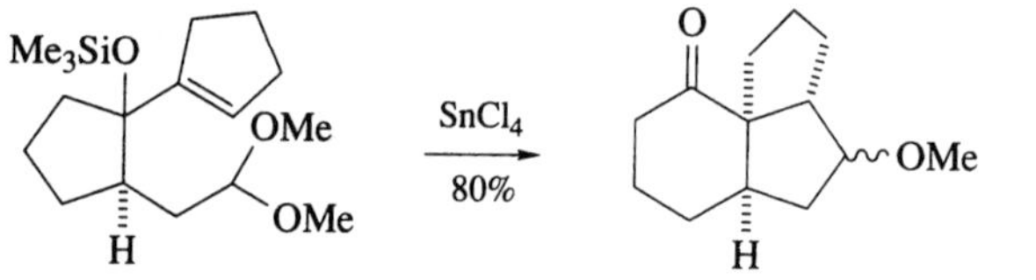

Me_3SiO, OMe, OMe, H — $SnCl_4$, 80% → O, OMe, H (32)

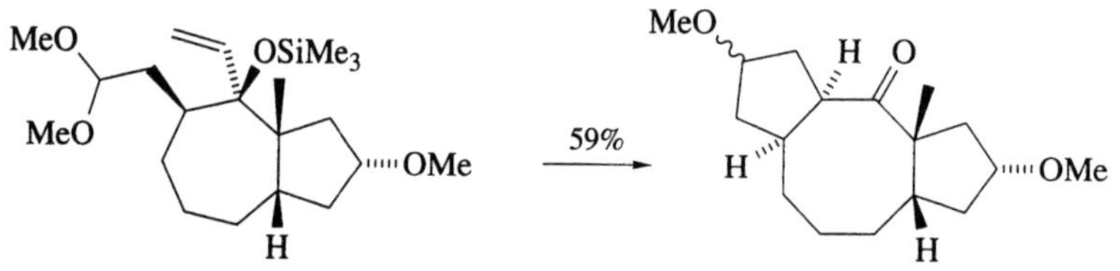

MeO, MeO, $OSiMe_3$, OMe, H — 59% → MeO, H, O, H, OMe, H (33)

3.3.5.5 Expansion by Intramolecular Addition–Fragmentation

Another homologation strategy is the intramolecular addition of nucleophiles to the ketone carbonyl followed by a fragmentation reaction, which may be facilitated by a charge-stabilizing group X (Scheme 75).[150] The reaction, with X = SO_2Ph, NO_2 or CN, has been examined mostly for the expansion of cyclic ketones, although a few acyclic examples exist. The sequence does have the virtue that the presence of the electron-withdrawing group X allows for the convenient attachment of the requisite side chain. The sulfone-based methodology utilizes the fluoride-induced intramolecular addition of an allyl silane to the ketone catalytically in the first stage and stoichiometrically for the fragmentation step (Scheme 76). The reaction works well for the three-carbon expansion of five-, eight- and twelve-membered ring ketones, as well as an acyclic case (equation 34), but fails at the fragmentation step for six- and seven-membered ring ketones.[151]

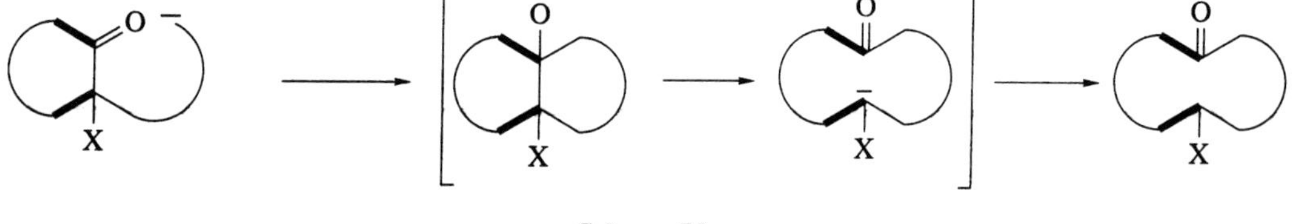

O, –, X → [O^-, X → O, –, X] → O, X

Scheme 75

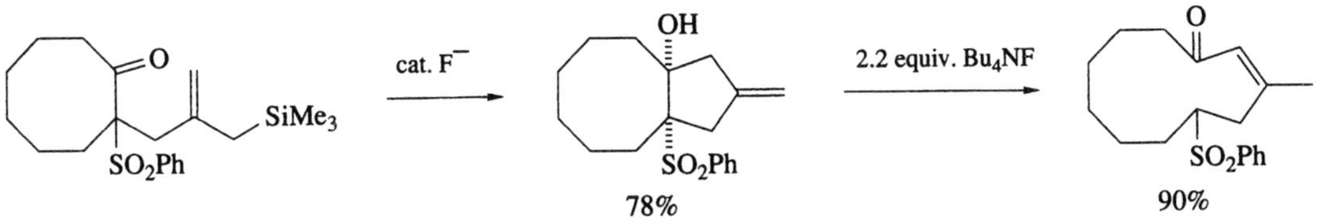

cat. F^- ; 78% ; 2.2 equiv. Bu_4NF ; 90%

Scheme 76

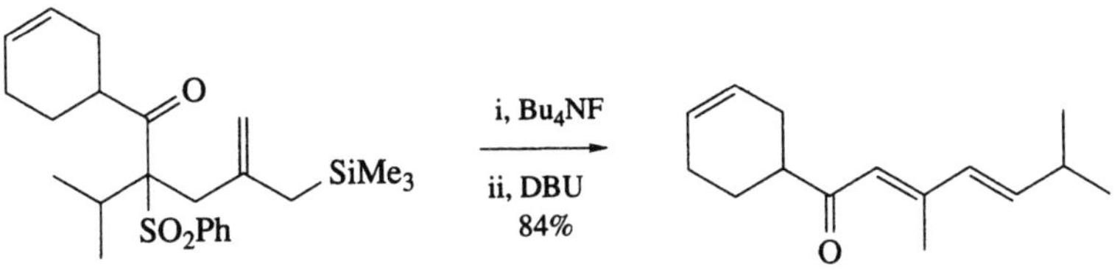

i, Bu_4NF ; ii, DBU ; 84% (34)

A four-carbon ring expansion, where X is nitro, has been reported to proceed well for seven- and eight-membered ring ketones, poorly for the 12-membered ring ketone, and fails at the fragmentation step for five- and six-membered ring ketones (equation 35).[152] A successful three-carbon expansion has been reported for the octanone where X is cyano (equation 36).[153] For these anionic type addition–fragmentation reactions, the success of the sequence is dependent to a large extent on the fragmentation of the intermediate carbinol derivative, which in turn has been correlated in part with the release of ring strain.

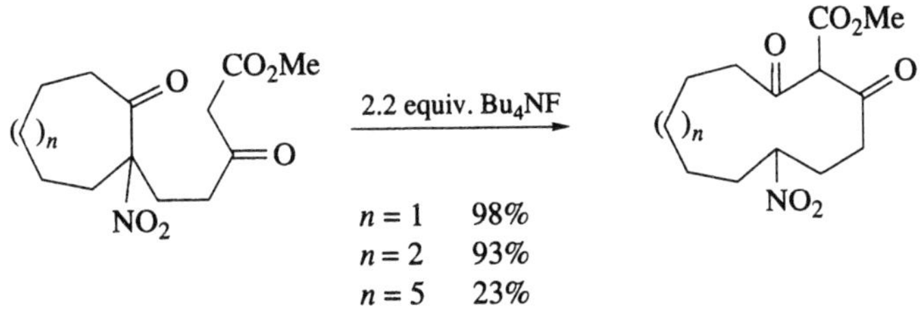

2.2 equiv. Bu_4NF (35)

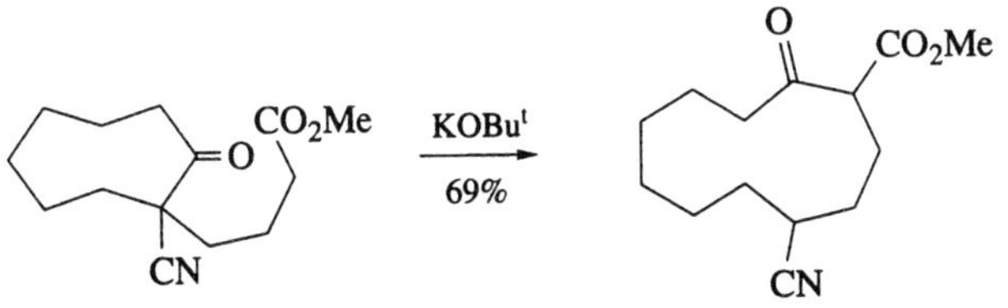

$n = 1$	98%
$n = 2$	93%
$n = 5$	23%

KOBu^t ; 69% (36)

More recently, a radical-mediated variation of this addition–fragmentation has been explored. The reaction, summarized in Scheme 77 for a one-carbon expansion, involves the generation of a radical at the terminus of a chain by homolytic cleavage of a carbon–heteroatom bond. Addition of the radical to the carbonyl produces a bicyclic intermediate, which on cleavage of the alternate bond regenerates the ketone carbonyl group with formation of a new radical. The sequence is terminated by the reduction of the radical with the tributyltin hydride reagent. The near neutral conditions of the reaction avoid the reclo-

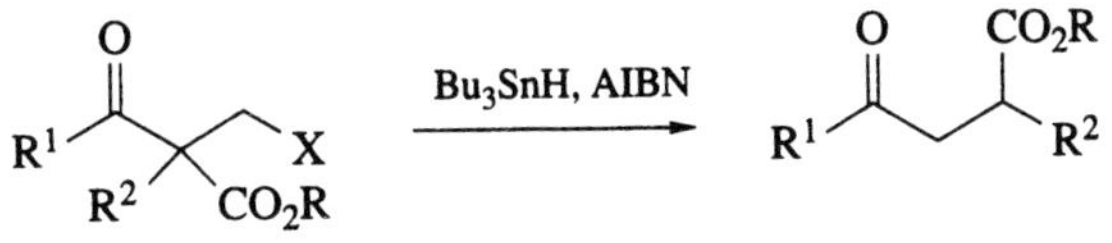

Bu_3SnH, AIBN (37)

R[1]	*R[2]*	*X*	*Yield (%)*
CO_2Et	H	Br	76
Me	H	Br	64
SEt	Me	I	91

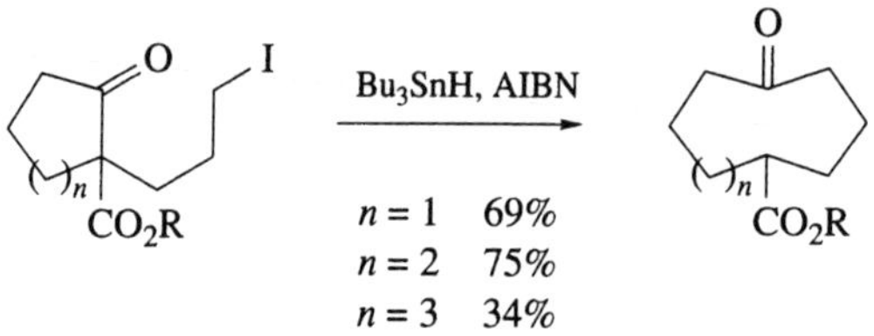

Bu$_3$SnH, AIBN

$n = 1$ 69%
$n = 2$ 75%
$n = 3$ 34%

(38)

Bu$_3$SnH, AIBN

$n = 1$ 36%
$n = 2$ 71%
$n = 3$ 45%

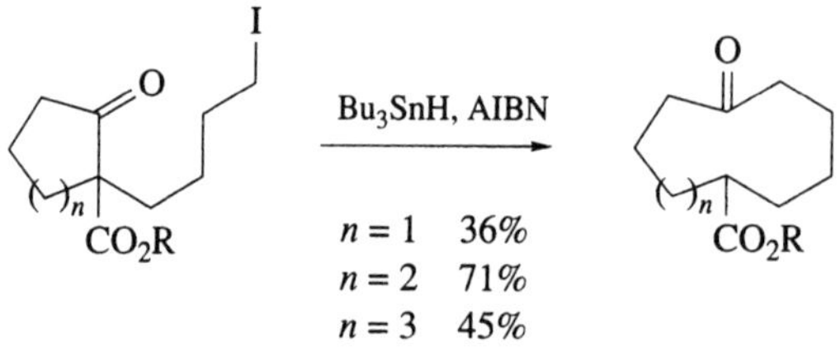

(39)

sure or nonfragmentation complication encountered in the anionic series above. The sequence proceeds in good yield for the one-carbon expansion of the five-, six-, seven- and eight-membered ring ketones studied (Scheme 77) as well as acyclic ketones (equation 37).[154,155] While the corresponding two-carbon homologation has not been successful, three- and four-carbon expansions are viable, although in some cases reduction at the chain terminus competes with the expansion reaction (equations 38 and 39).[154–156]

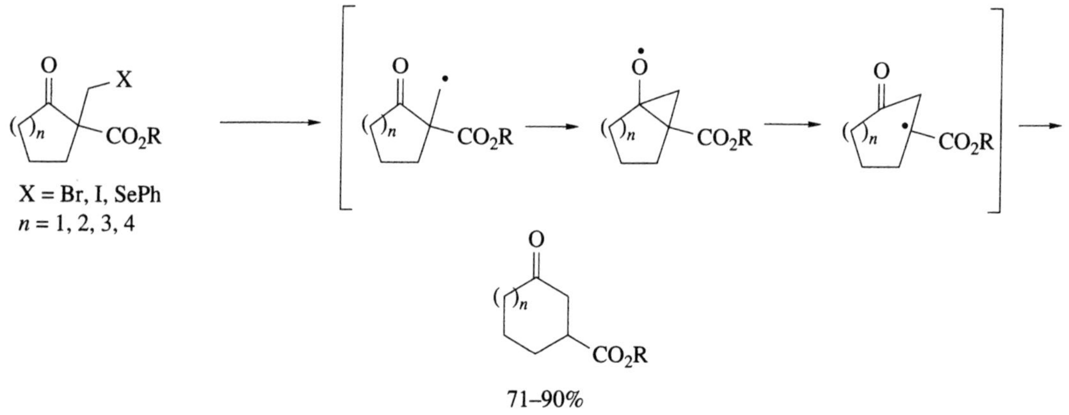

X = Br, I, SePh
$n = 1, 2, 3, 4$

71–90%

Scheme 77

A slightly different approach has been utilized for the three-carbon expansion in a system which does not contain a carboalkoxy group. The initial intramolecular addition step is carried out separately with butyllithium or samarium diiodide to give a bicyclic alcohol (Scheme 78).[157] The radical-induced fragmentation step is carried out photochemically in the presence of mercury(II) oxide and iodine to provide the expanded iodo ketone.

An alternate radical-induced addition–fragmentation sequence relies on the 1,4-elimination of a leaving group to drive the reaction.[158] In this sequence, because the fragmentation is accompanied by the expulsion of a tributylstannyl radical, the requirement for the initial tributyltin hydride is only catalytic (Scheme 79).[159] The stereospecificity of the reaction course has been rationalized in terms of a conformational preference for the elimination step, as depicted in Scheme 79 where the *trans*-substituted cyclohexanone forms only one double bond geometry in the product.[160] As shown in equations (40) and (41), the *cis*-substituted cyclohexanones undergo the four- and three-carbon expansions stereospecifically to give the (*Z*)-alkenes.

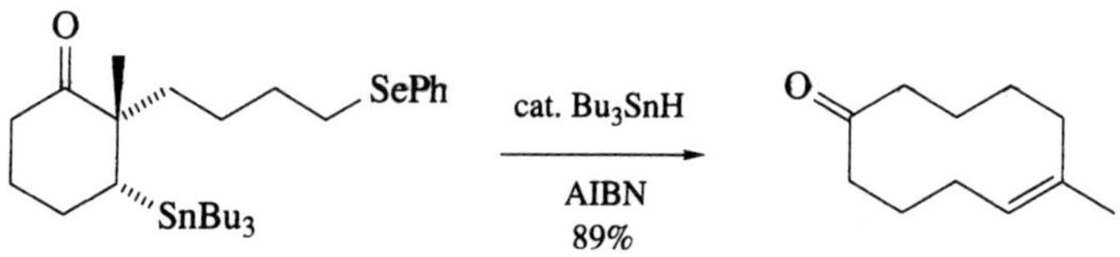

cat. Bu$_3$SnH

AIBN
89%

(40)

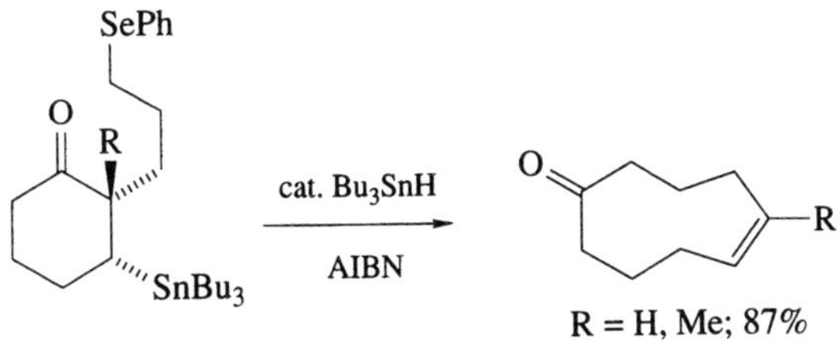

cat. Bu_3SnH, AIBN

R = H, Me; 87%

(41)

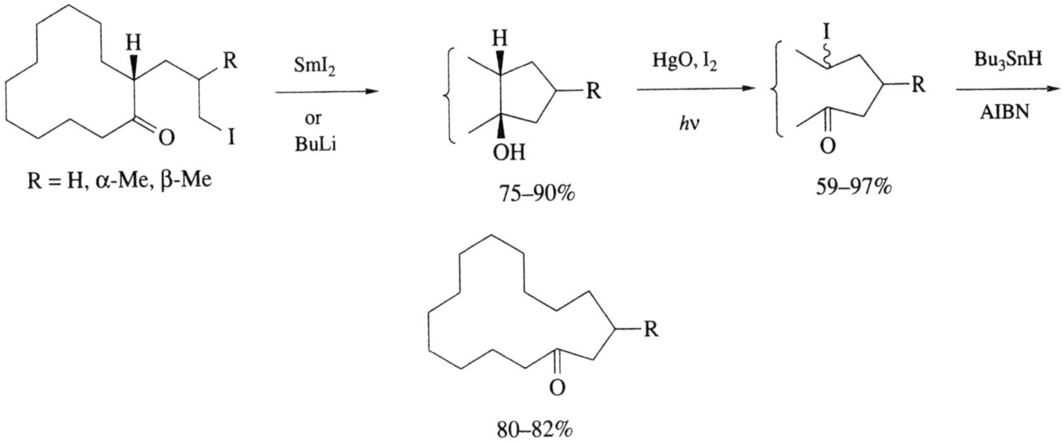

R = H, α-Me, β-Me

SmI_2 or BuLi; 75–90%

HgO, I_2, *hν*; 59–97%

Bu_3SnH, AIBN

80–82%

Scheme 78

Bu_3SnH, AIBN, Δ, 2–72 h

$Bu_3Sn•$

R = H, 47%; D, 75%; Me, 85%; CO_2Me, 72%

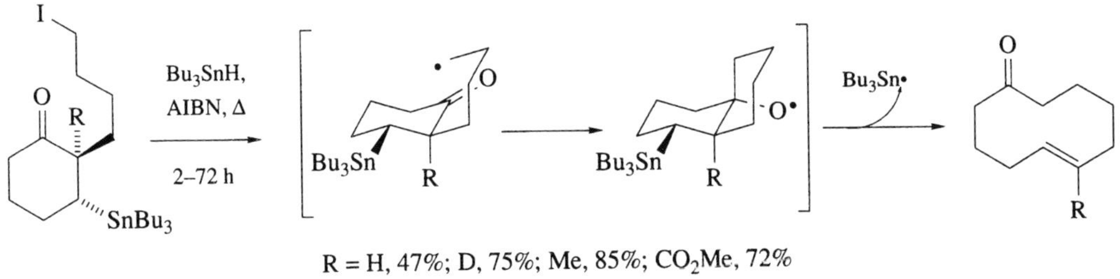

Scheme 79

3.3.6 REFERENCES

1. C. D. Gutsche and D. Redmore, *Adv. Alicyclic Chem.*, 1968, **2** (suppl. 1), 1.
2. (a) H. Stach and M. Hesse, *Tetrahedron*, 1988, **44**, 1573; (b) D. Bellus and B. Ernst, *Angew. Chem., Int. Ed. Engl.*, 1988, **27**, 797; (c) G. R. Krow, *Tetrahedron*, 1987, **43**, 3; (d) M. Ramaiah, *Synthesis*, 1984, 529; (e) H. Heimgartner, *Chimia*, 1980, **34**, 333; (f) A. P. Krapcho, *Synthesis*, 1976, 425; (g) M. Hesse, 'Ring Enlargement in Organic Chemistry', VCH, Weinheim, 1991.
3. (a) W. E. Bachmann and W. S. Struve, *Org. React. (N.Y.)*, 1942, **1**, 38; (b) B. Eistert, 'Newer Methods of Preparative Organic Chemistry', Interscience, New York, 1948, vol. 1, p. 513; (c) F. Weygand and H. J. Bestmann, 'Newer Methods of Preparative Organic Chemistry', Academic Press, New York, 1964, vol. 3, p. 451.
4. G. B. Gill, in 'Comprehensive Organic Synthesis', ed. B. M. Trost, Pergamon Press, Oxford, vol. 3, chap. 39.
5. (a) M. Regitz and G. Maas, 'Diazo Compounds', Academic Press, New York, 1986, p. 185; (b) H. Meier and K. Zeller, *Angew. Chem., Int. Ed. Engl.*, 1975, **14**, 32; (c) W. Kirmse, 'Carbene Chemistry', 2nd ed., Academic Press, New York, 1971, p. 475; (d) P. de Mayo, 'Molecular Rearrangements', Interscience, New York, 1963, vol. 1, p. 528; (e) for an example of a preparative scale Arndt–Eistert homologation (*i.e.* naphthoic acid to ethyl naphthylacetate) see V. Lee and M. S. Newman, *Org. Synth., Coll. Vol.*, 1988, **6**, 613.
6. T. Aoyama and T. Shioiri, *Chem. Pharm. Bull.*, 1981, **29**, 3249.
7. Y. Kuo, T. Aoyama and T. Shioiri, *Chem. Pharm. Bull.*, 1982, **30**, 526, 899, 2787.
8. For earlier reviews on diazoalkane chemistry which cover this topic in part see (a) B. Eistert and M. Regitz, *Methoden Org. Chem. (Houben-Weyl)*, 1976, **7/2b**, 1852; (b) B. Eistert and M. Regitz, *Methoden Org. Chem. (Houben-Weyl)*, 1973, **7/2a**, 672; (c) M. Regitz and G. Maas, 'Diazo Compounds', Academic Press, New York, 1986, p. 156; (d) J. S. Pizey, 'Synthetic Reagents', Wiley, New York, 1974, vol. 2, p. 65; (e) G. W.

Cowell and A. Ledwith, *Q. Rev., Chem. Soc.*, 1970, **24**, 119; (f) B. Eistert, M. Regitz, G. Heck and H. Schwall, *Methoden Org. Chem. (Houben-Weyl)*, 1968, **10/4**, 473.
9. C. D. Gutsche, *Org. React. (N.Y.)*, 1954, **8**, 364.
10. For the analogous reaction of aldehydes and phenyldiazomethane to produce benzyl ketones, see C. A. Loeschorn, M. Nakajima, P. J. McCloskey and J. Anselme, *J. Org. Chem.*, 1983, **48**, 4407.
11. H. O. House, E. J. Grubbs and W. F. Gannon, *J. Am. Chem. Soc.*, 1960, **82**, 4099.
12. The use of $AlCl_3$, in a similar series, has been reported to show a preference of substituted alkyl over methyl, see E. Müller and M. Bauer, *Justus Liebigs Ann. Chem.*, 1962, **654**, 92.
13. D. J. Coveney in 'Comprehensive Organic Synthesis', ed. B. M. Trost, Pergamon Press, Oxford, 1991, vol. 3, chap. 3.4.
14. B. Rickborn, in 'Comprehensive Organic Synthesis', ed. B. M. Trost, Pergamon Press, Oxford, 1991, vol. 3, chap. 3.2.
15. For a review on the application of the Tiffeneau–Demjanov reaction to cyclic ketones, see P. A. S. Smith, D. R. Baer, *Org. React. (N.Y.)*, 1960, **11**, 157.
16. For a representative example, see H. J. Dauben, H. J. Ringold, R. H. Wade, D. L. Pearson and A. G. Anderson, *Org. Synth., Coll. Vol.*, 1963, **4**, 221.
17. D. A. Evans, G. L. Carroll and L. K. Truesdale, *J. Org. Chem.*, 1974, **39**, 914; see also P. G. Gassman and J. J. Talley, *Org. Synth.*, 1981, **60**, 14; T. Livinghouse, *Org. Synth.*, 1981, **60**, 126.
18. (a) E. Wenkert and C. A. McPherson, *J. Am. Chem. Soc.*, 1972, **94**, 8084; (b) R. Pellicciari, R. Fringuelli, P. Ceccherelli and E. Sisani, *J. Chem. Soc., Chem. Commun.*, 1979, 959; (c) R. Pellicciari, R. Fringuelli, E. Sisani and M. Curini, *J. Chem. Soc., Perkin Trans. 1*, 1981, 2566.
19. (a) U. Schöllkopf and H. Frasnelli, *Angew. Chem., Int. Ed., Engl.*, 1970, **9**, 301; (b) U. Schöllkopf, B. Banhidai, H. Frasnelli, R. Meyer and H. Beckhaus, *Justus Liebigs Ann. Chem.*, 1974, 1767.
20. K. Nagao, M. Chiba and S. Kim, *Synthesis*, 1983, 197.
21. W. L. Mock and M. E. Hartman, *J. Org. Chem.*, 1977, **42**, 459.
22. W. L. Mock and M. E. Hartman, *J. Org. Chem.*, 1977, **42**, 466.
23. See refs. 1,2c,7b and 8f.
24. N. J. Turro and R. B. Gagosian, *J. Am. Chem. Soc.*, 1970, **92**, 2036 and references therein for a discussion on mechanism.
25. A. E. Greene, J. Deprès, H. Nagano and P. Crabbé, *Tetrahedron Lett.*, 1977, 2365.
26. A. E. Greene, M. Luche and A. A. Serra, *J. Org. Chem.*, 1985, **50**, 3957.
27. A. E. Greene and J. Deprès, *J. Am. Chem. Soc.*, 1979, **101**, 4003.
28. L. A. Paquette, R. S. Valpey and G. D. Annis, *J. Org. Chem.*, 1984, **49**, 1317.
29. A. E. Greene and F. Charbonnier, *Tetrahedron Lett.*, 1985, **26**, 5525.
30. H. J. Liu and T. Ogino, *Tetrahedron Lett.*, 1973, 4937.
31. J. R. Stille and R. H. Grubbs, *J. Am. Chem. Soc.*, 1986, **108**, 855; see also A. De Mesmaeker, S. J. Veenstra and B. Ernst, *Tetrahedron Lett.*, 1988, **29**, 459.
32. S. J. Veenstra, A. De Mesmaeker and B. Ernst, *Tetrahedron Lett.*, 1988, **29**, 2303.
33. H. J. Liu and S. P. Majumdar, *Synth. Commun.*, 1975, **5**, 125; see also ref. 21.
34. S. Ohuchida, N. Hamanaka and M. Hayashi, *Tetrahedron*, 1983, **39**, 4269.
35. T. J. de Boer and H. J. Backer, *Org. Synth., Coll. Vol.*, 1963, **4**, 225.
36. C. D. Gutsche and H. E. Johnson, *Org. Synth., Coll. Vol.*, 1963, **4**, 780 and references cited therein.
37. J. A. Marshall and J. J. Partridge, *J. Org. Chem.*, 1968, **33**, 4090.
38. B. Tchoubar, *Bull. Soc. Chim. Fr.*, 1949, 164.
39. M. Miyashita, S. Hara and A. Yoshikoshi, *J. Org. Chem.*, 1987, **52**, 2602.
40. N. Hashimoto, T. Aoyama and T. Shioiri, *Chem. Pharm. Bull.*, 1982, **30**, 119.
41. J. H. Rigby and C. Senanayake, *J. Org. Chem.*, 1987, **52**, 4634.
42. R. W. Thies and E. P. Seitz, *J. Org. Chem.*, 1978, **43**, 1050.
43. R. W. Thies and J. R. Pierce, *J. Org. Chem.*, 1982, **47**, 798.
44. R. W. Thies and S. T. Yue, *J. Org. Chem.*, 1982, **47**, 2685.
45. S. W. Baldwin and N. G. Landmesser, *Synth. Commun.*, 1978, **8**, 413.
46. T. Ibata, K. Miyauchi and S. Nakata, *Bull. Chem. Soc. Jpn.*, 1979, **52**, 3467.
47. S. Toyama, T. Aoyama and T. Shioiri, *Chem. Pharm. Bull.*, 1982, **30**, 3032.
48. V. Dave and E. W. Warnhoff, *J. Org. Chem.*, 1983, **48**, 2590.
49. K. Nagoa, I. Yoshimura, M. Chiba and S. Kim, *Chem. Pharm. Bull.*, 1983, **31**, 114.
50. R. Baudouy, J. Gore and L. Ruest, *Tetrahedron*, 1987, **43**, 1099.
51. E. W. Della and P. E. Pigou, *Aust. J. Chem.*, 1983, **36**, 2261.
52. M. Miyashita and A. Yoshikoshi, *J. Am. Chem. Soc.*, 1974, **96**, 1917.
53. M. A. McKinney and P. P. Patel, *J. Org. Chem.*, 1973, **38**, 4059 and references therein.
54. R. R. Sauers, J. A. Beisler and H. Feilich, *J. Org. Chem.*, 1967, **32**, 569.
55. A. Nickon and A. G. Stern, *Tetrahedron Lett.*, 1985, **26**, 5915.
56. E. G. Breitholle and A. G. Fallis, *J. Org. Chem.*, 1978, **43**, 1964.
57. E. Volpi, F. Del Cima and F. Pietra, *J. Chem. Soc., Perkin Trans. 1*, 1974, 703 and references therein.
58. P. Buchs and C. Ganter, *Helv. Chim. Acta*, 1980, **63**, 1420.
59. R. Chakraborti, B. C. Ranu and U. R. Ghatak, *Synth. Commun.*, 1987, **17**, 1539.
60. K. Hirao, Y. Kajikawa, E. Abe and O. Yonemitsu, *J. Chem. Soc., Perkin Trans. 1*, 1983, 1791 and references therein.
61. A. J. C. van Seters, M. Buza, A. J. H. Klunder and B. Zwanenburg, *Tetrahedron*, 1981, **37**, 1027.
62. J. C. Coll, D. R. Crist, M. G. Barrio and N. J. Leonard, *J. Am. Chem. Soc.*, 1972, **94**, 7092 and references therein.
63. D. Škare and Z. Majerski, *Tetrahedron Lett.*, 1972, 4887.
64. P. von R. Schleyer, E. Funke and S. H. Liggero, *J. Am. Chem. Soc.*, 1969, **91**, 3965.
65. R. K. Murray and T. M. Ford, *J. Org. Chem.*, 1979, **44**, 3504.

66. M. Farcasiu, D. Farcasiu, J. Slutsky and P. von R. Schleyer, *Tetrahedron Lett.*, 1974, 4059.
67. J. S. Polley and R. K. Murray, *J. Org. Chem.*, 1976, **41**, 3294.
68. K. Naemura, T. Katoh, R. Fukunaga, H. Chikamatsu and M. Nakazaki, *Bull. Chem. Soc. Jpn.*, 1985, **58**, 1407.
69. M. Nakazaki, K. Naemura, N. Arashiba and M. Iwasaki, *J. Org. Chem.*, 2433, **44**, 1979; see also M. Nakazaki, K. Naemura, Y. Kondo, S. Nakahara and M. Hashimoto, *J. Org. Chem.*, 1980, **45**, 4440.
70. M. Nakazaki, K. Naemura, H. Chikamatsu, M. Iwasaki and M. Hashimoto, *J. Org. Chem.*, 1981, **46**, 2300.
71. C. D. Gutsche and H. R. Zandstra, *J. Org. Chem.*, 1974, **39**, 324 and references therein.
72. R. M. Coates and R. L. Sowerby, *J. Am. Chem. Soc.*, 1972, **94**, 5386.
73. L. N. Mander and C. Wilshire, *Aust. J. Chem.*, 1979, **32**, 1975 and references therein.
74. For a review on the preparation and use of α-S- or Se-substituted organometallics, see A. Krief, *Tetrahedron*, 1980, **36**, 2531. See also A. Krief, J.-L. Labouruer, W. Dumont and D. Labar, *Bull. Soc. Chim. Fr.*, 1990, **127**, 681.
75. (a) S. Halazy and A. Krief, *J. Chem. Soc., Chem. Commun.*, 1982, 1200; (b) for rearrangement of small ring spiro epoxides, see T. W. Hart and M. Comte, *Tetrahedron Lett.*, 1985, **26**, 2713; B. M. Trost and L. H. Latimer, *J. Org. Chem.*, 1978, **43**, 1031; B. M. Trost, *Acc. Chem. Res.*, 1974, **7**, 85 and references therein.
76. S. Halazy, F. Zutterman and A. Krief, *Tetrahedron Lett.*, 1982, **23**, 4385.
77. S. Uemura, K. Ohe and N. Sugita, *J. Chem. Soc., Chem. Commun.*, 1988, 111.
78. R. C. Gadwood, I. M. Mallick and A. J. DeWinter, *J. Org. Chem.*, 1987, **52**, 774.
79. A. Krief and J. L. Laboureur, *Tetrahedron Lett.*, 1987, **28**, 1545.
80. L. Fitjer, D. Wehle and H. Scheuermann, *Chem. Ber.*, 1986, **119**, 1162.
81. A. Krief and J. L. Laboureur, *J. Chem. Soc., Chem. Commun.*, 1986, 702.
82. L. A. Paquette, J. R. Peterson and R. J. Ross, *J. Org. Chem.*, 1985, **50**, 5200.
83. J. L. Laboureur and A. Krief, *Tetrahedron Lett.*, 1984, **25**, 2713; A. Krief, J. L. Laboureur and W. Dumont, *Tetrahedron Lett.*, 1987, 28, 1549.
84. A. Krief, J. L. Laboureur, G. Evrard, B. Norberg and E. Guittet, *Tetrahedron Lett.*, 1989, **30**, 575.
85. T. Cohen, D. Kuhn and J. R. Falck, *J. Am. Chem. Soc.*, 1975, **97**, 4749.
86. W. D. Abraham, M. Bhupathy and T. Cohen, *Tetrahedron Lett.*, 1987, **28**, 2203.
87. B. M. Trost and G. K. Mikhail, *J. Am. Chem. Soc.*, 1987, **109**, 4124; an earlier example at the sulfoxide level allowed for the ring expansion of 2-methylcyclobutanone adducts with aqueous acid, K. Ogura, M. Yamashita, M. Suzuki and G. Tsuchihashi, *Chem. Lett.*, 1982, 93. See also H. Finch, A. M. M. Mjalli, J. G. Montana, S. M. Roberts and R. J. K. Taylor, *Tetrahedron*, 1990, **46**, 4925.
88. S. Kim and J. H. Park, *Chem. Lett.*, 1988, 1323.
89. S. Knapp, A. F. Trope, M. S. Theodore, N. Hirata and J. J. Barchi, *J. Org. Chem.*, 1984, **49**, 608 and references therein.
90. R. J. Israel and R. K. Murray, *J. Org. Chem.*, 1985, **50**, 1573, 4703.
91. R. L. Funk, P. M. Novak and M. M. Abelman, *Tetrahedron Lett.*, 1988, **29**, 1493.
92. J. Villieras, C. Bacquet and J. F. Normant, *J. Organomet. Chem.*, 1972, **40**, C1; G. Köbrich and J. Grosser, *Tetrahedron Lett.*, 1972, 4117; H. Taguchi, H. Yamamoto and H. Nozaki, *Tetrahedron Lett.*, 1972, 4661.
93. A. J. Sisti, *Tetrahedron Lett.*, 1967, 5327; A. J. Sisti and G. M. Rusch, *J. Org. Chem.*, 1974, **39**, 1182 and references therein.
94. For a review on carbenes and carbenoids with neighboring heteroatoms, see K. G. Taylor, *Tetrahedron*, 1982, **38**, 2751.
95. (a) G. Köbrich and J. Grosser, *Chem. Ber.*, 1973, **106**, 2626; (b) J. Villieras, C. Bacquet and J. F. Normant, *C. R. Hebd. Seances, Acad. Sci., Ser. C*, 1973, **276**, 433; J. Villieras, C. Bacquet and J. F. Normant, *Bull. Soc. Chim. Fr.*, 1975, 1797.
96. H. Taguchi, H. Yamamoto and H. Nozaki, *J. Am. Chem. Soc.*, 1974, **96**, 6510; (b) H. Taguchi, H. Yamamoto and H. Nozaki, *Bull. Chem. Soc. Jpn*, 1977, **50**, 1592; H. Taguchi, H. Yamamoto and H. Nozaki, *Tetrahedron Lett.*, 1976, 2617.
97. E. Vedejs and S. D. Larsen, *J. Am. Chem. Soc.*, 1984, **106**, 3030.
98. H. Yamaga, Y. Fujise and S. Itô, *Tetrahedron Lett.*, 1987, **28**, 585.
99. K. C. Nicolaou, R. L. Magolda and D. A. Claremon, *J. Am. Chem. Soc.*, 1980, **102**, 1404.
100. T. Hiyama, M. Shinoda, H. Saimoto and H. Nozaki, *Bull. Chem. Soc. Jpn*, 1981, **54**, 2747.
101. E. Vedejs and R. A. Shepherd, *J. Org. Chem.*, 1976, **41**, 742.
102. R. Sigrist, M. Rey and A. S. Dreiding, *Helv. Chim. Acta*, 1988, **71**, 788.
103. C. J. Kowalski and K. W. Fields, *J. Am. Chem. Soc.*, 1982, **104**, 321; C. J. Kowalski, M. S. Haque and K. W. Fields, *J. Am. Chem. Soc.*, 1985, **107**, 1429.
104. C. J. Kowalski and G. S. Lal, *J. Am. Chem. Soc.*, 1986, **108**, 5356; C. J. Kowalski and M. S. Haque, *J. Am. Chem. Soc.*, 1986, **108**, 1325.
105. See, for example, W. E. Parham and E. E. Schweizer, *Org. React. (N.Y.)*, 1963, **13**, 55; for a review on the chemistry of cyclopropanols, see D. H. Gibson and C. H. DePuy, *Chem. Rev.*, 1974, **74**, 605.
106. W. Kirmse, 'Carbene Chemistry', 2nd edn., Academic Press, New York, 1971.
107. W. E. Parham, R. W. Soeder, J. R. Throckmorton, K. Kuncl and R. M. Dodson, *J. Am. Chem. Soc.*, 1965, **87**, 321 and references therein.
108. G. Stork, M. Nussim and B. August, *Tetrahedron Suppl.*, 1966, **8** part 1, 105; R. C. DeSelms, *Tetrahedron Lett.*, 1966, 1965.
109. K. J. Shea, W. M. Fruscella and W. P. England, *Tetrahedron Lett.*, 1987, **28**, 5623.
110. For an extensive review on silyl enol ethers, see P. Brownbridge, *Synthesis*, 1983, 1, 85.
111. (a) G. Stork and T. L. Macdonald, *J. Am. Chem. Soc.*, 1975, **97**, 1264; (b) T. L. Macdonald, *J. Org. Chem.*, 1978, **43**, 4241; (c) P. Amice, L. Blanco and J. M. Conia, *Synthesis*, 1976, 196; (d) L. Blanco, P. Amice and J. M. Conia, *Synthesis*, 1981, 289; (e) L. Blanco, N. Slougui, G. Rousseau and J. M. Conia, *Tetrahedron Lett.*, 1981, **22**, 645.
112. For a review on the Simmons–Smith reaction, see H. E. Simmons, T. L. Cairns, S. A. Vladuchick and C. M. Hoiness, *Org. React. (N.Y.)*, 1973, **20**, 1.

113. Y. Ito, S. Fujii, M. Nakatsuka, F. Kawamoto and T. Saegusa, *Org. Synth.*, 1979, **59**, 113 and references therein.
114. M. Asoaka, K. Takenouchi and H. Takei, *Chem. Lett.*, 1988, 921.
115. R. K. Hill, in 'Comprehensive Organic Synthesis', ed. B. M. Trost, Pergamon Press, Oxford, 1991, vol. 5, chap. 7.1; see also S. J. Rhodes and N. R. Raulins, *Org. React. (N.Y.)*, 1975, **22**, 1.
116. D. A. Evans and A. M. Golob, *J. Am. Chem. Soc.*, 1975, **97**, 4765; D. A. Evans, D. J. Baillargeon and J. V. Nelson, *J. Am. Chem. Soc.*, 1978, **100**, 2242; D. A. Evans, A. M. Golob, N. S. Mandell and G. S. Mandell, *J. Am. Chem. Soc.*, 1978, **100**, 8170; D. A. Evans and D. J. Baillargeon, *Tetrahedron Lett.*, 1978, 3315, 3319; M. L. Steigerwald, W. A. Goddard and D. A. Evans, *J. Am. Chem. Soc.*, 1979, **101**, 1994; D. A. Evans and J. V. Nelson, *J. Am. Chem. Soc.*, 1980, **102**, 774; see also ref. 115.
117. E. N. Marvell and W. Whalley, *Tetrahedron Lett.*, 1970, 509.
118. T. Kato, H. Kondo, M. Nishino, M. Tanaka, G. Hata and A. Miyake, *Bull. Chem. Soc. Jpn.*, 1980, **53**, 2958.
119. R. C. Gadwood, R. Lett and J. E. Wissinger, *J. Am. Chem. Soc.*, 1986, **108**, 6343 and references therein.
120. W. C. Still, *J. Am. Chem. Soc.*, 1977, **99**, 4186.
121. W. C. Still, S. Murata, G. Revial and K. Yoshihara, *J. Am. Chem. Soc.*, 1983, **105**, 625.
122. W. C. Still, *J. Am. Chem. Soc.*, 1979, **101**, 2493.
123. D. L. J. Clive, A. G. Angoh, S. C. Suri, S. N. Rao and C. G. Russell, *J. Chem. Soc., Chem. Commun.*, 1982, 828.
124. L. A. Paquette, J. A. Colapret and D. R. Andrews, *J. Org. Chem.*, 1985, **50**, 201.
125. M. Sworin and K. Lin, *J. Am. Chem. Soc.*, 1989, **111**, 1815; M. Sworin and K. Lin, *J. Org. Chem.*, 1987, **52**, 5640.
126. L. A. Paquette, J. Reagan, S. L. Schreiber and C. A. Teleha, *J. Am. Chem. Soc.*, 1989, **111**, 2331.
127. R. C. Gadwood, R. Lett, *J. Org. Chem.*, 1982, **47**, 2268.
128. Refs. 126, and 128 and M. Kahn, *Tetrahedron Lett.*, 1980, **21**, 4547.
129. S. G. Levine and R. L. McDaniel, *J. Org. Chem.*, 1981, **46**, 2199.
130. S. F. Martin, J. B. White and R. Wagner, *J. Org. Chem.*, 1982, **47**, 3190.
131. L. A. Paquette and M. Poupart, *Tetrahedron Lett.*, 1988, **29**, 273.
132. B. B. Snider and R. B. Beal, *J. Org. Chem.*, 1988, **53**, 4508.
133. (a) P. A. Wender and S. M. Sieburth, *Tetrahedron Lett.*, 1981, **22**, 2471; (b) P. A. Wender and D. A. Holt, *J. Am. Chem. Soc.*, 1985, **107**, 7771; (c) A. G. Angoh and D. L. J. Clive, *J. Chem. Soc., Chem. Commun.*, 1984, 534.
134. J. A. Berson and M. Jones, *J. Am. Chem. Soc.*, 1964, **86**, 5017, 5019; J. A. Berson and E. J. Walsh, *J. Am. Chem. Soc.*, 1968, **90**, 4729; see also L. A. Paquette, F. Pierre and C. E. Cottrell, *J. Am. Chem. Soc.*, 1987, **109**, 5731.
135. R. W. Thies and E. P. Seitz, *J. Chem. Soc., Chem. Commun.*, 1976, 846; R. W. Thies and J. E. Billigmeier, *J. Am. Chem. Soc.*, 1974, **96**, 200; R. W. Thies, *J. Am. Chem. Soc.*, 1972, **94**, 7074; see also ref. 42.
136. R. W. Thies and Y. B. Choi, *J. Org. Chem.*, 1973, **38**, 4067.
137. R. W. Thies and K. P. Daruwala, *J. Org. Chem.*, 1987, **52**, 3798 and references therein.
138. R. L. Danheiser, C. Martinez-Davila and H. Sard, *Tetrahedron*, 1981, **37**, 3943.
139. M. Bhupathy and T. Cohen, *J. Am. Chem. Soc.*, 1983, **105**, 6978; T. Cohen, M. Bhupathy and J. R. Matz, *J. Am. Chem. Soc.*, 1983, **105**, 520.
140. (a) L. E. Overman and L. T. Mendelson, *J. Am. Chem. Soc.*, 1981, **103**, 5579; (b) L. E. Overman, L. T. Mendelson and E. J. Jacobsen, *J. Am. Chem. Soc.*, 1983, **105**, 6629; (c) L. E. Overman, L. T. Mendelson and L. A. Flippin, *Tetrahedron Lett.*, 1982, **23**, 2733.
141. (a) L. E. Overman, M. E. Okazaki and E. J. Jacobsen, *J. Org. Chem.*, 1985, **50**, 2403; (b) L. E. Overman, E. J. Jacobsen and R. J. Doedens, *J. Org. Chem.*, 1983, **48**, 3393; (c) L. E. Overman and E. J. Jacobsen, *Tetrahedron Lett.*, 1982, **23**, 2741; see also ref. 141b.
142. L. E. Overman, M. Sworin and R. M. Burk, *J. Org. Chem.*, 1983, **48**, 2685; L. E. Overman, M. Sworin, L. S. Bass and J. Clardy, *Tetrahedron*, 1981, **37**, 4041.
143. L. E. Overman and S. R. Angle, *J. Org. Chem.*, 1985, **50**, 4021 and references therein.
144. E. J. Jacobsen, J. Levin and L. E. Overman, *J. Am. Chem. Soc.*, 1988, **110**, 4329.
145. Footnote 33 in ref. 144; see also L. E. Overman and S. Sugai, *Helv. Chim. Acta*, 1985, **68**, 745.
146. M. H. Hopkins and L. E. Overman, *J. Am. Chem. Soc.*, 1987, **109**, 4748.
147. P. M. Herrinton, M. H. Hopkins, P. Mishra, M. J. Brown and L. E. Overman, *J. Org. Chem.*, 1987, **52**, 3711.
148. G. C. Hirst, P. N. Howard and L. E. Overman, *J. Am. Chem. Soc.*, 1989, **111**, 1514.
149. These reactions are distinct from the Pd^{2+}- and Hg^{2+}-catalyzed 3,3-rearrangements where internal alkylation does not occur to an appreciable extent; N. Bluthe, M. Malacria and J. Gore, *Tetrahedron Lett.*, 1983, **24**, 1157; N. Bluthe, M. Malacria and J. Gore, *Tetrahedron*, 1984, **40**, 3277. Pd^{2+} has also been utilized for the ring expansion of vinylcyclobutanols; G. R. Clark and S. Thiensathit, *Tetrahedron Lett.*, 1985, **26**, 2503; M. Demuth, B. Pandey, B. Wietfeld, H. Said and J. Viader, *Helv. Chim. Acta*, 1988, **71**, 1392.
150. This strategy has been successfully employed in expansions to macrocyclic lactones and lactams; see ref. 2a.
151. B. M. Trost and J. E. Vincent, *J. Am. Chem. Soc.*, 1980, **102**, 5680.
152. Y. Nakashita and M. Hesse, *Helv. Chim. Acta*, 1983, **66**, 845.
153. M. Süsse, J. Hájíček and M. Hesse, *Helv. Chim. Acta*, 1985, **68**, 1986. See also Z.-F. Xie and K. Sakai, *J. Org. Chem.*, 1990, **55**, 820.
154. P. Dowd and S. Choi, *J. Am. Chem. Soc.*, 1987, **109**, 3493; P. Dowd and S. Choi, *Tetrahedron*, 1989, **45**, 77; P. Dowd and S.-C. Choi, *Tetrahedron Lett.*, 1989, **30**, 6129.
155. A. L. J. Beckwith, D. M. O'Shea, S. Gerba and S. W. Westwood, *J. Chem. Soc., Chem. Commun.*, 1987, 666; A. L. J. Beckwith, D. M. O'Shea and S. W. Westwood, *J. Am. Chem. Soc.*, 1988, **110**, 2565.
156. P. Dowd and S. Choi, *J. Am. Chem. Soc.*, 1987, **109**, 6548; A. Nishida, H. Takahashi, H. Takeda, N. Takeda and O. Yonemitsu, *J. Am. Chem. Soc.*, 1990, **112**, 902.
157. H. Suginome and S. Yamada, *Tetrahedron Lett.*, 1987, **28**, 3963.

158. For discussions on 1,4-elimination reactions with scission of the 2,3-bond, see C. A. Grob and P. W. Schiess, *Angew. Chem., Int. Ed. Engl.*, 1967, **6**, 1; J. B. Hendrickson, *J. Am. Chem. Soc.*, 1986, **108**, 6748; P. Deslongchamps, 'Stereoelectronic Effects in Organic Chemistry', Pergamon Press, Oxford, 1983, p. 257.
159. J. E. Baldwin, R. M. Adlington and J. Robertson, *J. Chem. Soc., Chem. Commun.*, 1988, 1404; J. E. Baldwin, R. M. Adlington and J. Robertson, *Tetrahedron*, 1989, **45**, 909.
160. The functionally related stereospecific fragmentation of β-stannyl and β-silyl cyclic ketones to acyclic alkenes under oxidative conditions has been observed previously, see K. Nakatani and S. Isoe, *Tetrahedron Lett.*, 1984, **25**, 5335; M. Ochiai, T. Ukita, Y. Nagao and E. Fujita, *J. Chem. Soc., Chem. Commun.*, 1985, 637; S. R. Wilson, P. A. Zucker, C. Kim and C. A. Villa, *Tetrahedron Lett.*, 1985, **26**, 1969.

Author Index

This Author Index comprises an alphabetical listing of the names of over 7000 authors cited in the references listed in the bibliographies which appear at the end of each chapter in this volume.

Each entry consists of the author's name, followed by a list of numbers, each of which is associated with a superscript number. For example

Abbott, D. E., 6[12,12c], 10[40], 573[53,54]

The numbers indicate the text pages on which references by the author in question are cited; the superscript numbers refer to the reference number in the chapter bibliography. Citations occurring in the text, tables and chemical schemes and equations have all been included.

Although much effort has gone into eliminating inaccuracies resulting from the use of different combinations of initials by the same author, the use by some journals of only one initial, and different spellings of the same name as a result of transliteration processes, the accuracy of some entries may have been affected by these factors.

Subject Index